Safety Symbols

Safety symbols in the following table are used in the lab activities to indicate possible hazards. Learn the meaning of each symbol. **It is recommended that you wear safety goggles and apron at all times in the lab. This might be required in your school district.**

Safety Symbols	Hazard	Examples	Precaution	Remedy
Disposal	Special disposal procedures need to be followed.	certain chemicals, living organisms	Do not dispose of these materials in the sink or trash can.	Dispose of wastes as directed by your teacher.
Biological	Organisms or other biological materials that might be harmful to humans	bacteria, fungi, blood, unpreserved tissues, plant materials	Avoid skin contact with these materials. Wear mask or gloves.	Notify your teacher if you suspect contact with material. Wash hands thoroughly.
Extreme Temperature	Objects that can burn skin by being too cold or too hot	boiling liquids, hot plates, dry ice, liquid nitrogen	Use proper protection when handling.	Go to your teacher for first aid.
Sharp Object	Use of tools or glassware that can easily puncture or slice skin	razor blades, pins, scalpels, pointed tools, dissecting probes, broken glass	Practice common-sense behavior and follow guidelines for use of the tool.	Go to your teacher for first aid.
Fume	Possible danger to respiratory tract from fumes	ammonia, acetone, nail polish remover, heated sulfur, moth balls	Be sure there is good ventilation. Never smell fumes directly. Wear a mask.	Leave foul area and notify your teacher immediately.
Electrical	Possible danger from electrical shock or burn	improper grounding, liquid spills, short circuits, exposed wires	Double-check setup with teacher. Check condition of wires and apparatus. Use GFI-protected outlets.	Do not attempt to fix electrical problems. Notify your teacher immediately.
Irritant	Substances that can irritate the skin or mucous membranes of the respiratory tract	pollen, moth balls, steel wool, fiberglass, potassium permanganate	Wear dust mask and gloves. Practice extra care when handling these materials.	Go to your teacher for first aid.
Chemical	Chemicals that can react with and destroy tissue and other materials	bleaches such as hydrogen peroxide; acids such as sulfuric acid, hydrochloric acid; bases such as ammonia, sodium hydroxide	Wear goggles, gloves, and an apron.	Immediately flush the affected area with water and notify your teacher.
Toxic	Substance may be poisonous if touched, inhaled, or swallowed.	mercury, many metal compounds, iodine, poinsettia plant parts	Follow your teacher's instructions.	Always wash hands thoroughly after use. Go to your teacher for first aid.
Flammable	Flammable chemicals may be ignited by open flame, spark, or exposed heat.	alcohol, kerosene, potassium permanganate	Avoid open flames and heat when using flammable chemicals.	Notify your teacher immediately. Use fire safety equipment if applicable.
Open Flame	Open flame in use, may cause fire.	hair, clothing, paper, synthetic materials	Tie back hair and loose clothing. Follow teacher's instruction on lighting and extinguishing flames.	Notify your teacher immediately. Use fire safety equipment if applicable.

Eye Safety
Proper eye protection should be worn at all times by anyone performing or observing science activities.

Clothing Protection
This symbol appears when substances could stain or burn clothing.

Animal Safety
This symbol appears when safety of animals and students must be ensured.

Radioactivity
This symbol appears when radioactive materials are used.

Handwashing
After the lab, wash hands with soap and water before removing goggles.

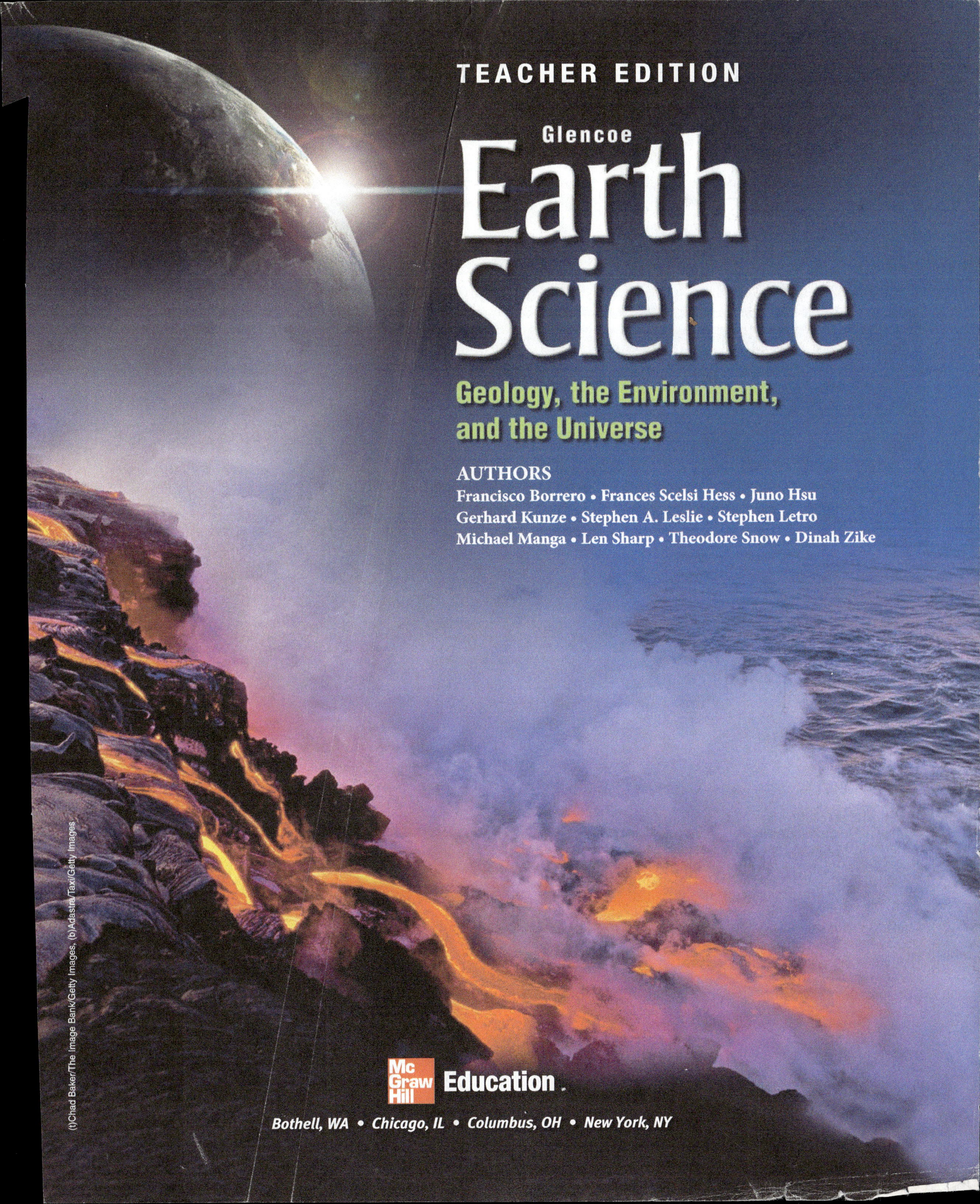

TEACHER EDITION

Glencoe

Earth Science

Geology, the Environment, and the Universe

AUTHORS
Francisco Borrero • Frances Scelsi Hess • Juno Hsu
Gerhard Kunze • Stephen A. Leslie • Stephen Letro
Michael Manga • Len Sharp • Theodore Snow • Dinah Zike

McGraw Hill Education

Bothell, WA • Chicago, IL • Columbus, OH • New York, NY

About the Photo

The lava photo on the cover was taken in Hawaii Volcanoes National Park on the Big Island of Hawaii. The lava in the photo is flowing from active vents on the flank of the Kilauea Volcano. When lava flows into the sea, sulfuric acid in the lava mixes with chlorides in the saltwater to form a mist of water vapor and hydrochloric acid.

Cover (earth) Chad Baker/The Image Bank/Getty Images,
(lava) Adastra/Taxi/Getty Images,
(ferns) Paul Chesley/Photographer's Choice/Getty Images

The McGraw-Hill Companies

connectED.mcgraw-hill.com

McGraw-Hill **Education**

Send all inquiries to:
McGraw-Hill Education
STEM Learning Solutions Center
8787 Orion Place
Columbus, OH 43240-4027

Student Edition
ISBN: 978-0-07-658713-1
MHID: 0-07-658713-4

Teacher Edition
ISBN: 978-0-07-658714-8
MHID: 0-07-658714-2

Printed in the United States of America.

6 7 8 9 10 11 QVS 20 19 18 17 16

STEM

McGraw-Hill is committed to providing instructional materials in Science, Technology, Engineering, and Mathematics (STEM) that give all students a solid foundation, one that prepares them for college and careers in the 21st century.

TEACHER HANDBOOK

Table of Contents

Teacher Edition Preview 2T–7T
Classroom Solutions 8T–27T
Teacher Classroom Resources: Print Ancillaries 28T–29T
Teacher Classroom Resources: Technology 30T–31T
Professional Development 32T
National Science Content Standards 33T
Course Planning Guide 34T
Suppliers 35T
Laboratory Safety 36T–37T
Laboratory Materials 38T–42T

BananaStock/PictureQuest

Program Framework

Welcome to the Teacher Edition of *Earth Science: Geology, the Environment, and the Universe*. We have created this teacher edition based on input from experienced science teachers and educational consultants. Our goal is to provide you with research-based teaching strategies and activities, which are labeled for you at point-of-use.

Point-of-Use
- Strategies and activities apply directly to content.
- Concepts in Motion and Video icons indicate where your students can interact with tables and figures or use interactive online resources.

Review and Reinforcement
- Scaffolding—the introduction and reinforcement of skills and content—is incorporated throughout the lessons.
- Assessments check student understanding of key concepts and provide opportunities for reteaching at the end of each section.

Standards-Based Instruction

Differentiated Instruction
- Leveled activities and options for differentiated instruction help meet the needs of all your students, including English learners.

Hierarchical Structure
- **Level 1: Themes** The book is organized around five themes—change, structures, geologic time, systems, and scientific inquiry.
- **Level 2: BIGIDEA** Each chapter has a Big Idea, which summarizes the chapter content in an overarching statement.
- **Level 3: MAINIDEA** Each section of the chapter has a Main Idea that describes the focus of the section.

Assessment and Intervention
- Lessons provide standards practice.
- Assessments gauge student mastery of standards.
- Additional resources provide intervention options.

Differentiated Instruction

Activity Leveling

Teaching strategies and activities have been coded for ability-level appropriateness. A competency level is given for each activity using the following code:

AL Activities for students working above grade level

OL Activities for students working on grade level

BL Activities for students working below grade level

EL Activities for English learners

COOP LEARN Activities or strategies that promote cooperative learning

Activity

Recognize Minerals Have students work with partners to examine a pile of crushed granite. Instruct students to use tweezers and a microscope or magnifying lens to separate the crushed material into different mineral groups. Students should be able to distinguish three or four types of distinctly different minerals. Have students describe in their Earth science journals what criteria they used to classify the minerals. Answers will vary. Students will likely say color, luster, or texture, or some combination of all three. Accept all reasonable answers. OL COOP LEARN

CAREERS IN EARTH SCIENCE WebQuest

Teacher Content Support

Useful Graphite Some motor oil is dark gray. The color is caused by the mineral graphite. Graphite is added to the motor oil because its softness makes it a good lubricant for the moving parts in motors. Graphite also can be used to lubricate moving parts in door locks. A powdered form of the mineral is squeezed into the door mechanism.

READING CHECK Luster describes the way light reflects off the surface of the mineral.

Caption Question Fig. 7 According to Mohs scale of hardness, glass is harder than a fingernail so the mineral that scratches glass is harder.

CAREERS IN EARTH SCIENCE

Lapidary A lapidary is someone who cuts, polishes, and engraves precious stones. He or she studies minerals and their properties in order to know which minerals are the best for certain projects. WebQuest

Identifying Minerals

Geologists rely on several simple tests to identify minerals. These tests are based on a mineral's physical and chemical properties: crystal form, luster, hardness, cleavage, fracture, streak, color, texture, density, specific gravity, and special properties. As you will learn in the GeoLab at the end of this chapter, it is usually best to use a combination of tests instead of just one to identify minerals.

Crystal form Some minerals form such distinct crystal shapes that they are immediately recognizable. Halite—common table salt—always forms perfect cubes. Quartz crystals, with their double-pointed ends and six-sided crystals, are also readily recognized. However, as you learned earlier in this section, perfect crystals are not always formed, so identification based only on crystal form is rare.

Luster The way that a mineral reflects light from its surface is called **luster.** There are two types of luster—metallic luster and nonmetallic luster. Silver, gold, copper, and galena have shiny surfaces that reflect light, like the chrome trim on cars. Thus, they are said to have a metallic luster. Not all metallic minerals are metals. If their surfaces have shiny appearances like metals, they are considered to have a metallic luster. Sphalerite, for example, is a mineral with a metallic luster that is not a metal.

Minerals with nonmetallic lusters, such as calcite, gypsum, sulfur, and quartz, do not shine like metals. Nonmetallic lusters might be described as dull, pearly, waxy, silky, or earthy. Differences in luster, shown in **Figure 6,** are caused by differences in the chemical compositions of minerals. Describing the luster of nonmetallic minerals is a subjective process. For example, a mineral that appears waxy to one person might not appear waxy to another. Using luster to identify a mineral should usually be used in combination with other physical characteristics.

READING CHECK **Define** the term *luster.*

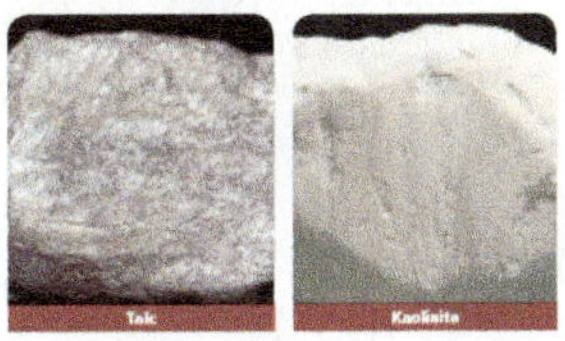

Figure 6 The flaky and shiny nature of talc gives it a pearly luster. Another white mineral, kaolinite, contrasts sharply with its dull, earthy luster.

Demonstration

Recognize Metallic Luster Show students various objects made of metal and note their color, shine, etc. Then show students a variety of metallic minerals and those nonmetallic minerals that are often misidentified as metallic—sphalerite, biotite and muscovite mica, etc. Help students to see and understand the difference in these last [illegible] BL OL

Explore Mohs scale with an interactive table. Concepts in Motion

Table 2 Mohs Scale of Hardness

Mineral	Hardness	Hardness of Common Objects
Diamond	10	
Corundum	9	
Topaz	8	
Quartz	7	streak plate = 7
Feldspar	6	steel file = 6.5
Apatite	5	glass = 5.5
Fluorite	4	iron nail = 4.5
Calcite	3	piece of copper = 3.5
Gypsum	2	fingernail = 2.5
Talc	1	

Hardness One of the most useful and reliable tests for identifying minerals is hardness. **Hardness** is a measure of how easily a mineral can be scratched. German geologist Friedrich Mohs developed a scale by which an unknown mineral's hardness can be compared to the known hardness of ten minerals. The minerals in the Mohs scale of mineral hardness were selected because they are easily recognized and, with the exception of diamond, readily found in nature.

READING CHECK **Explain** what hardness measures.

Talc is one of the softest minerals and can be scratched by a fingernail; therefore, talc represents 1 on the Mohs scale of hardness. In contrast, diamond is so hard that it can be used as a sharpener and cutting tool, so diamond represents 10 on the Mohs scale of hardness. The scale, shown in **Table 2,** is used in the following way: a mineral that can be scratched by your fingernail has a hardness less than 2.5. A mineral that cannot be scratched by your fingernail and cannot scratch glass has a hardness value between 5.5 and 2.5. Finally, a mineral that scratches glass has a hardness greater than 5.5. Using other common objects, such as those listed in the table, can help you determine a more precise hardness and provide you with more information with which to identify an unknown mineral. Sometimes more than one mineral is present in a sample. If this is the case, it is a good idea to test more than one area of the sample. This way, you can be sure that you are testing the hardness of the mineral you are studying. **Figure 7** shows two minerals that have different hardness values.

READING CHECK Hardness measures a mineral's resistance to scratching.

Figure 7 The mineral on top can be scratched with a fingernail. The mineral on the bottom easily scratches glass.
Determine *Which mineral has greater hardness?*

DIFFERENTIATED INSTRUCTION

Struggling Learners Provide students with index cards on which have been written common terms used to describe luster: *earthy, shiny, greasy, pearly, glassy, waxy,* silky, and hand samples of minerals that exhibit these lusters. Have students create a display that matches up the mineral samples with the correct terms. EL

Identify Misconceptions

Many people mistakenly believe that diamond is the only mineral hard enough to scratch glass, and that this ability to scratch glass is a foolproof way to identify diamond.

Uncover the Misconception Ask students whether they believe that only diamond can scratch glass. A number of students will likely say yes. Tell the class that glass has a hardness of 5.5 on the Mohs scale of hardness. Therefore, any mineral with a hardness greater than 5.5 will scratch a glass surface.

Demonstrate the Concept Rake a sample of crystalline quartz across a piece of glass. Have students observe the scratch left on glass by the quartz.

Assess New Knowledge Tell students to use a field guide to minerals or the *Reference Handbook* to create a list of minerals that will scratch glass. Possible minerals include diamond, feldspar, topaz, corundum, quartz, and amethyst.

Reinforcement

Use Hardness to Identify Minerals Have some students demonstrate how the Mohs scale of hardness can be used to help identify an unknown mineral. Give the students a glass slide, a copper penny, and unlabeled samples of fluorite, olivine, and gypsum. Point out that their fingernails are also testing instruments. By using these objects, the students should be able to identify the minerals. The hardness of olivine is 6.5, so it will scratch glass. Fluorite will scratch the penny but not the glass. Gypsum will not scratch the penny and can be scratched by a fingernail. Remind students that other tests might also need to be done in order to accurately identify a mineral.

90 Chapter 4 • Minerals

Section 1 • What is a mineral? 91

Answers and Additional Support

- Along the side and bottom of the Teacher Edition, you will find answers to questions in the student edition,
- Demonstrations and activities that help you quickly and easily address key concepts,
- Teacher Content Support and elements that provide you with additional background information, and
- Differentiated Instruction strategies that help you meet the needs of all students.

Planning the Chapter

Planning pages appear at the beginning of each chapter.

Chapter Organizers
detail the Big Idea, the Essential Questions from each section, standards covered, and resources and materials needed to teach the chapter.

Suggested Pacing
provides lesson suggestions for the chapter. When used in conjunction with the Course Planning Guide starting on page 34T, you can tailor your instruction to the individual needs of your classes.

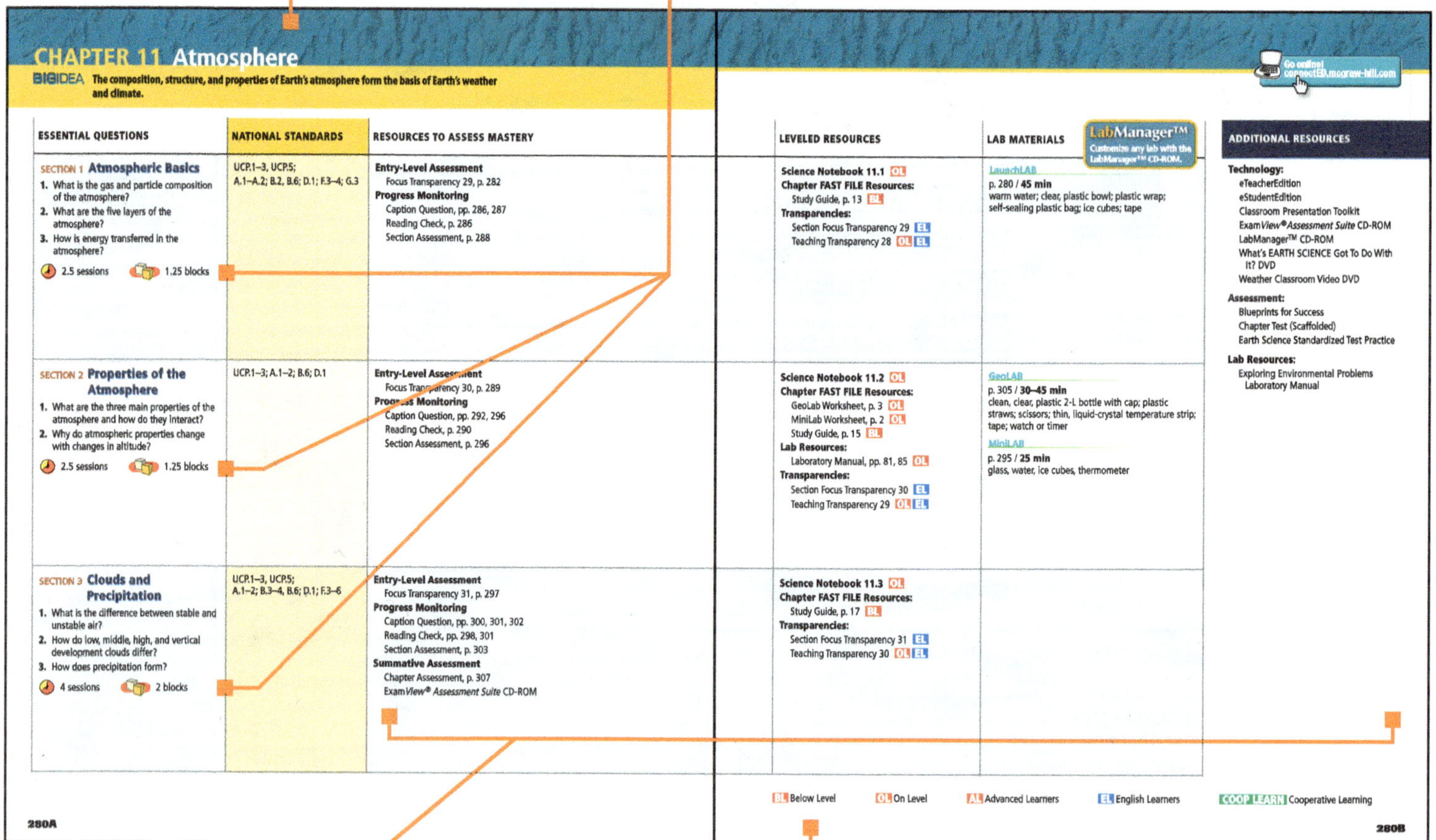

CHAPTER 11 Atmosphere

BIGIDEA The composition, structure, and properties of Earth's atmosphere form the basis of Earth's weather and climate.

ESSENTIAL QUESTIONS	NATIONAL STANDARDS	RESOURCES TO ASSESS MASTERY
SECTION 1 Atmospheric Basics 1. What is the gas and particle composition of the atmosphere? 2. What are the five layers of the atmosphere? 3. How is energy transferred in the atmosphere? 2.5 sessions 1.25 blocks	UCP.1–3, UCP.5; A.1–A.2; B.2, B.6; D.1; F.3–4; G.3	Entry-Level Assessment Focus Transparency 29, p. 282 Progress Monitoring Caption Question, pp. 286, 287 Reading Check, p. 286 Section Assessment, p. 288
SECTION 2 Properties of the Atmosphere 1. What are the three main properties of the atmosphere and how do they interact? 2. Why do atmospheric properties change with changes in altitude? 2.5 sessions 1.25 blocks	UCP.1–3; A.1–2; B.6; D.1	Entry-Level Assessment Focus Transparency 30, p. 289 Progress Monitoring Caption Question, pp. 292, 296 Reading Check, p. 290 Section Assessment, p. 296
SECTION 3 Clouds and Precipitation 1. What is the difference between stable and unstable air? 2. How do low, middle, high, and vertical development clouds differ? 3. How does precipitation form? 4 sessions 2 blocks	UCP.1–3, UCP.5; A.1–2; B.3–4, B.6; D.1; F.3–6	Entry-Level Assessment Focus Transparency 31, p. 297 Progress Monitoring Caption Question, pp. 300, 301, 302 Reading Check, pp. 298, 301 Section Assessment, p. 303 Summative Assessment Chapter Assessment, p. 307 ExamView® Assessment Suite CD-ROM

280A

Go online! connectED.mcgraw-hill.com

LEVELED RESOURCES	LAB MATERIALS
Science Notebook 11.1 OL Chapter FAST FILE Resources: Study Guide, p. 13 BL Transparencies: Section Focus Transparency 29 EL Teaching Transparency 28 OL EL	LaunchLAB p. 280 / 45 min warm water; clear, plastic bowl; plastic wrap; self-sealing plastic bag; ice cubes; tape
Science Notebook 11.2 OL Chapter FAST FILE Resources: GeoLab Worksheet, p. 3 OL MiniLab Worksheet, p. 2 OL Study Guide, p. 15 BL Lab Resources: Laboratory Manual, pp. 81, 85 OL Transparencies: Section Focus Transparency 30 EL Teaching Transparency 29 OL EL	GeoLAB p. 305 / 30–45 min clean, clear, plastic 2-L bottle with cap; plastic straws; scissors; thin, liquid-crystal temperature strip; tape; watch or timer MiniLAB p. 295 / 25 min glass, water, ice cubes, thermometer
Science Notebook 11.3 OL Chapter FAST FILE Resources: Study Guide, p. 17 BL Transparencies: Section Focus Transparency 31 EL Teaching Transparency 30 OL EL	

LabManager™ Customize any lab with the LabManager™ CD-ROM.

ADDITIONAL RESOURCES

Technology:
eTeacherEdition
eStudentEdition
Classroom Presentation Toolkit
ExamView® Assessment Suite CD-ROM
LabManager™ CD-ROM
What's EARTH SCIENCE Got To Do With It? DVD
Weather Classroom Video DVD

Assessment:
Blueprints for Success
Chapter Test (Scaffolded)
Earth Science Standardized Test Practice

Lab Resources:
Exploring Environmental Problems Laboratory Manual

BL Below Level OL On Level AL Advanced Learners EL English Learners COOP LEARN Cooperative Learning

280B

Resource Lists
identify at a glance the lab materials, ancillaries, and technology resources needed to teach the chapter.

Leveling Key
describes the differentiated instruction used throughout the Teacher Edition.

Teaching the Chapter

The Big Idea and Teacher Content Support at the beginning of the chapter help you teach the standards.

Introduce the BIGIDEA

Weather Ask the students why Earth has weather. Earth's surface receives uneven heating by the Sun due to its axial tilt, which, in turn, changes the solar angle and length of the daylight period from day to day. Uneven heating leads to regions of warm rising air and regions of cold sinking air. Additionally, the warmer air carries with it evaporated moisture which contributes to clouds and precipitation in those regions. Rising and sinking air contribute to pressure differences (highs and lows), and the air moving between pressure regions creates winds. If heat was distributed evenly, none of this would occur.

Teacher Content Support

Water Vapor Water vapor transports large quantities of heat from Earth's surface to the atmosphere. Heat is absorbed by water at the surface, causing water to change its state to a gas. The gas is water vapor, and the process is evaporation. The energy contained in the vapor molecule is called latent (hidden) heat. When water vapor cools at altitude, the water vapor reverts back to visible droplets of water or crystals of ice, forming clouds by the process of condensation. The latent heat in the water vapor is released during condensation to the surrounding air, warming it. This process is the first leg of the water cycle and is responsible for the production of Earth's freshwater and for partially cooling Earth. The recycling occurs when precipitation returns to Earth.

BIGIDEA activities help students understand the conceptual structure of the chapter—starting with the Big Idea overarching the chapter to the Main Ideas that are the focus of each section.

Teacher Content Support is a more in-depth look at an important concept in the chapter.

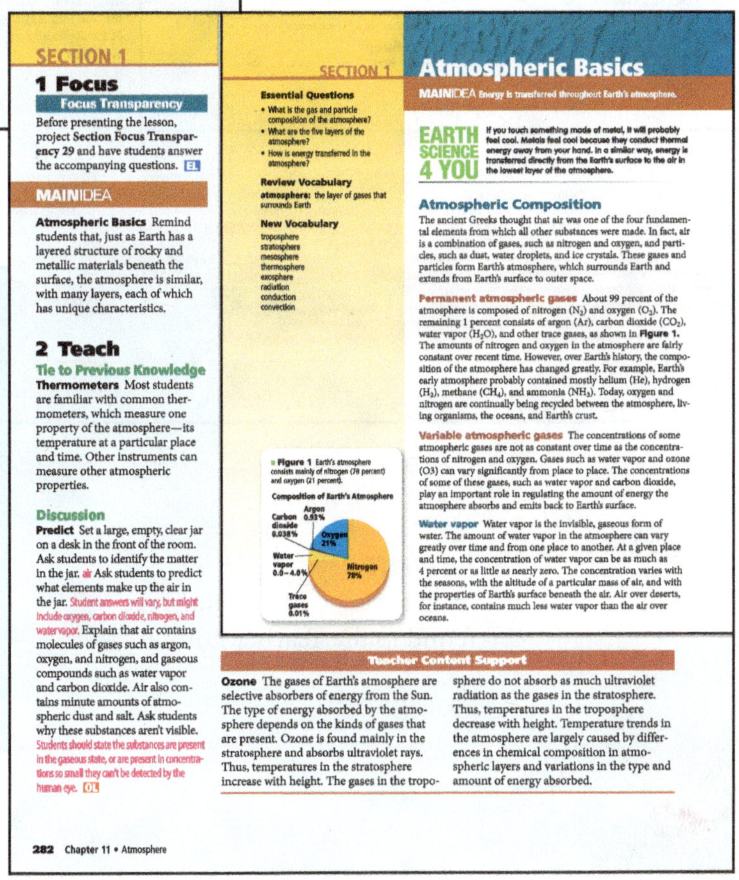

SECTION 1

1 Focus

Focus Transparency

Before presenting the lesson, project **Section Focus Transparency 29** and have students answer the accompanying questions. EL

MAINIDEA

Atmospheric Basics Remind students that, just as Earth has a layered structure of rocky and metallic materials beneath the surface, the atmosphere is similar, with many layers, each of which has unique characteristics.

2 Teach

Tie to Previous Knowledge

Thermometers Most students are familiar with common thermometers, which measure one property of the atmosphere—its temperature at a particular place and time. Other instruments can measure other atmospheric properties.

Discussion

Predict Set a large, empty, clear jar on a desk in the front of the room. Ask students to identify the matter in the jar. Ask students to predict what elements make up the air in the jar. Student answers will vary, but might include oxygen, carbon dioxide, nitrogen, and water vapor. Explain that air contains molecules of gases such as argon, oxygen, and nitrogen, and gaseous compounds such as water vapor and carbon dioxide. Air also contains minute amounts of atmospheric dust and salt. Ask students why these substances aren't visible. Students should state the substances are present in the gaseous state, or are present in concentrations so small they can't be detected by the human eye. OL

Atmospheric Basics

MAINIDEA Energy is transferred throughout Earth's atmosphere.

EARTH SCIENCE 4 YOU If you touch something made of metal, it will probably feel cool. Metals feel cool because they conduct thermal energy away from your hand. In a similar way, energy is transferred directly from the Earth's surface to the air in the lowest layer of the atmosphere.

Atmospheric Composition

The ancient Greeks thought that air was one of the four fundamental elements from which all other substances were made. In fact, air is a combination of gases, such as nitrogen and oxygen, and particles, such as dust, water droplets, and ice crystals. These gases and particles form Earth's atmosphere, which surrounds Earth and extends from Earth's surface to outer space.

Permanent atmospheric gases About 99 percent of the atmosphere is composed of nitrogen (N_2) and oxygen (O_2). The remaining 1 percent consists of argon (Ar), carbon dioxide (CO_2), water vapor (H_2O), and other trace gases, as shown in **Figure 1**. The amounts of nitrogen and oxygen in the atmosphere are fairly constant over recent time. However, over Earth's history, the composition of the atmosphere has changed greatly. For example, Earth's early atmosphere probably contained mostly helium (He), hydrogen (H_2), methane (CH_4), and ammonia (NH_3). Today, oxygen and nitrogen are continually being recycled between the atmosphere, living organisms, the oceans, and Earth's crust.

Variable atmospheric gases The concentrations of some atmospheric gases are not as constant over time as the concentrations of nitrogen and oxygen. Gases such as water vapor and ozone (O3) can vary significantly from place to place. The concentrations of some of these gases, such as water vapor and carbon dioxide, play an important role in regulating the amount of energy the atmosphere absorbs and emits back to Earth's surface.

Water vapor Water vapor is the invisible, gaseous form of water. The amount of water vapor in the atmosphere can vary greatly over time and from one place to another. At a given place and time, the concentration of water vapor can be as much as 4 percent or as little as nearly zero. The concentration varies with the seasons, with the altitude of a particular mass of air, and with the properties of Earth's surface beneath the air. Air over deserts, for instance, contains much less water vapor than the air over oceans.

Teacher Content Support

Ozone The gases of Earth's atmosphere are selective absorbers of energy from the Sun. The type of energy absorbed by the atmosphere depends on the kinds of gases that are present. Ozone is found mainly in the stratosphere and absorbs ultraviolet rays. Thus, temperatures in the stratosphere increase with height. The gases in the troposphere do not absorb as much ultraviolet radiation as the gases in the stratosphere. Thus, temperatures in the troposphere decrease with height. Temperature trends in the atmosphere are largely caused by differences in chemical composition in atmospheric layers and variations in the type and amount of energy absorbed.

282 Chapter 11 • Atmosphere

The three-step teaching model in each section gives you direction when you want it and options to enrich the section or to meet the diverse needs of your students.

1 Focus The first step when teaching a set of concepts or skills is to focus the students' attention.

2 Teach Instruct students through a variety of active-learning strategies.

3 Assess Skill, Knowledge, and Performance assessment strategies provide a variety of assessment for all your students.

Assessment: Sections

Student activities and questions throughout the book provide opportunities for ongoing assessment and remediation.

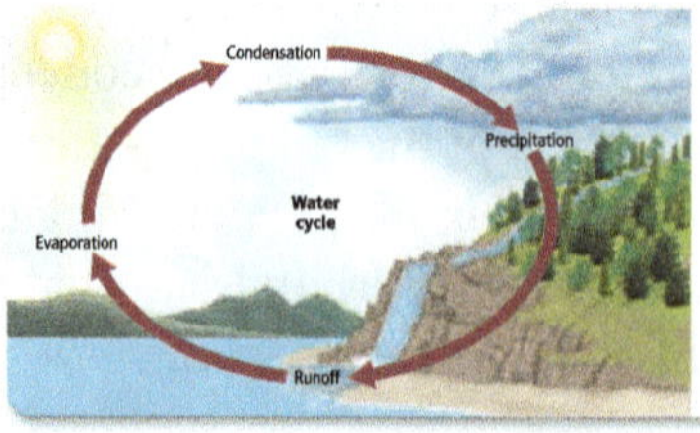

Figure 23 Water moves from Earth to the atmosphere and back to Earth in the water cycle.

The water cycle More than 97 percent of Earth's water is in the oceans. At any one time, only a small percentage of water is present in the atmosphere. Still, this water is vitally important because, as it continually moves between the atmosphere and Earth's surface, it nourishes living things. The constant movement of water between the atmosphere and Earth's surface is known as the water cycle.

The water cycle is summarized in **Figure 23.** Radiation from the Sun causes liquid water to evaporate. Water evaporates from lakes, streams, and oceans and rises into Earth's atmosphere. As water vapor rises, it cools and condenses to form clouds. Water droplets combine to form larger drops that fall to Earth as precipitation. This water soaks into the ground and enters lakes, streams, and oceans, or it falls directly into bodies of water and eventually evaporates, continuing the water cycle.

SECTION 3 REVIEW

Section Self-Check

Section Summary

- Clouds are formed as warm, moist air is forced upward, expands, and cools.
- An air mass is stable if it tends to return to its original height after it starts rising.
- Cloud droplets form when water vapor is cooled to the dew point and condenses on condensation nuclei.
- Clouds are classified by their shapes and the altitudes at which they form.
- Cloud droplets collide and coalesce into larger droplets that can fall to Earth as rain, snow, sleet, or hail.

Understand Main Ideas

1. MAINIDEA **Summarize** the differences between low clouds, middle clouds, and high clouds.
2. **Describe** how precipitation forms.
3. **Determine** the reason precipitation will fall as snow rather than rain.
4. **Compare** stable and unstable air.

Think Critically

5. **Evaluate** how a reduction in the number of condensation nuclei in the troposphere would affect precipitation. Explain your reasoning.

WRITING IN ▶ Earth Science

6. Describe the path a drop of rain might follow throughout the water cycle.

3 Assess

Check for Understanding

Discussion Ask students to identify the major cloud groups and at least two subcategories of each major group. major groups—stratiform and cumuliform Identify the probable heights at which the cloud bases form. <2000 m, 2000–6000 m and >6000 m Ask: Are the clouds composed of liquid water or ice crystals? cirriform, made of ice; all others liquid Are they likely to produce precipitation? high and middle level, less likely; lower forms, more likely

Reteach

Concept Map Construct concept maps showing the life cycle of a water droplet. Concept maps should include evaporation of water from the ocean surface; lifting of the warm, less-dense, moist air; cooling of the air to its LCL; saturation; condensation; coalescence; precipitation; and finally, the return of the water droplet to the ocean through a waterway.

Assessment

Knowledge Ask students the following question: Of the following, which describes a layered cloud that is dense enough to block out the Sun and is composed entirely of ice crystals?

a. stratus
b. cirrostratus
c. altostratus

Answer b: The prescence of ice crystals is the deciding factor.

SECTION 3 REVIEW

1. Low and middle clouds are below the freezing level, while high clouds are above it. Low clouds are made of water droplets, middle clouds are made of water droplets and ice, and high clouds are always made of ice.
2. Answers will vary and should include the idea that water vapor is condensing to a liquid form or forming ice crystals of some form.
3. If the precipitation is always in an air layer that is at or below freezing, snow falls.
4. The stability of air depends on how the temperature of the air mass changes relative to its surroundings. If the air mass is always cooler than its surroundings and tends to sink, then it is stable. If the air mass is always warmer and less dense than its surroundings, it is unstable.
5. Because precipitation forms around condensation nuclei, a reduction in nuclei would result in a reduction in precipitation. Air without particles would have difficulty in producing much precipitation.
6. Answers will vary, but should include evaporation, condensation, and precipitation, in that order. Answers can also reference runoff as a return mechanism for precipitation to a water body.

Rubric

Section 3 • Clouds and Precipitation 303

3 Assess

provides an evaluation of a key concept and an activity to reteach for students struggling to meet that learning objective.

SECTION REVIEW

provides students with summary statements and questions that tie to the MAINIDEA and Essential Questions for that section.

Answers

to all assessment questions are found in the Teacher Edition.

Assessment: Chapters

CONSTRUCTED RESPONSE and **THINK CRITICALLY** require students to demonstrate higher-order thinking and use their and writing skills.

UNDERSTAND KEY CONCEPTS and **VOCABULARY REVIEW** assess comprehension of the terms and key concepts in each section.

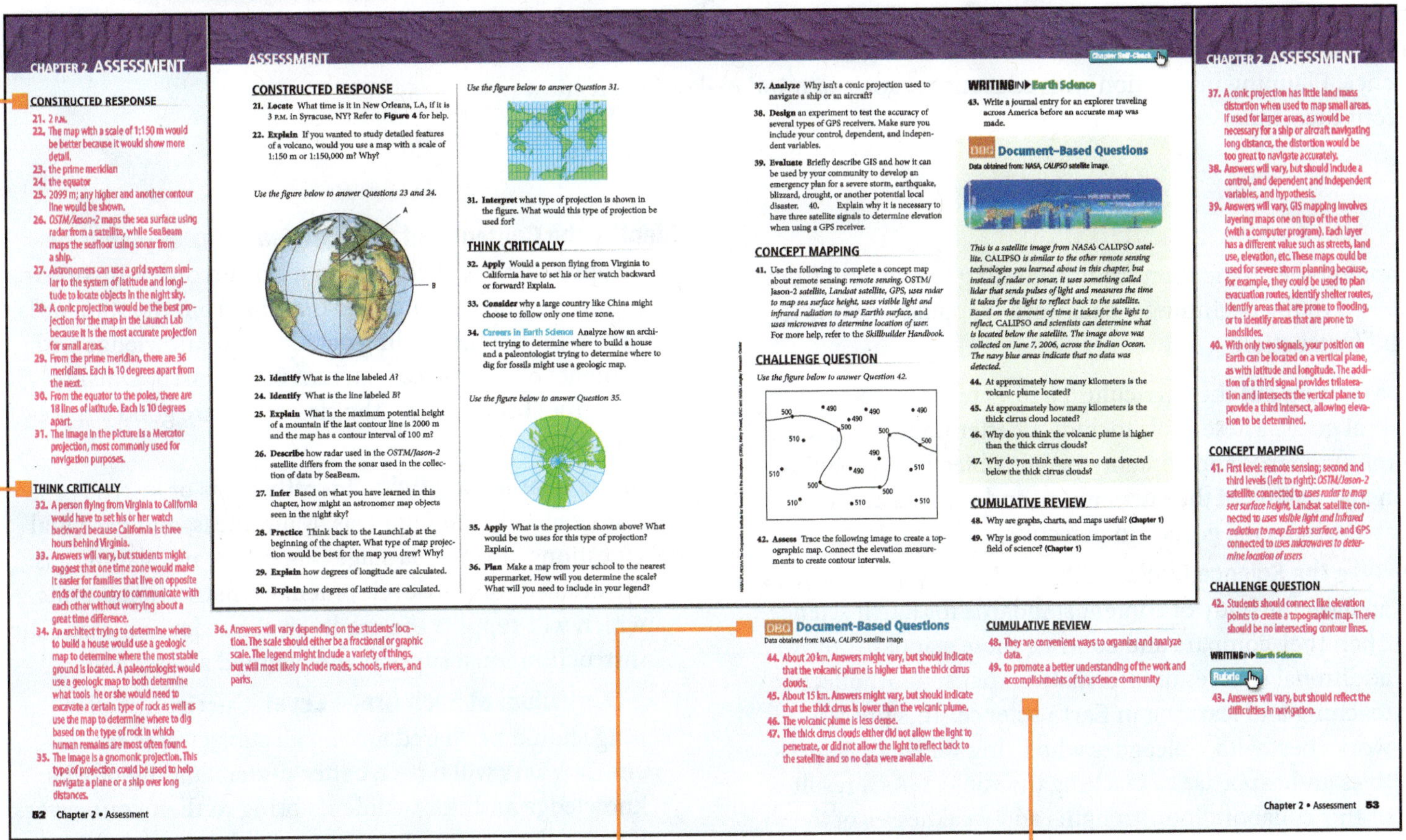

CHAPTER 2 ASSESSMENT

CONSTRUCTED RESPONSE

21. 2 P.M.
22. The map with a scale of 1:150 m would be better because it would show more detail.
23. the prime meridian
24. the equator
25. 2099 m; any higher and another contour line would be shown.
26. *OSTM/Jason-2* maps the sea surface using radar from a satellite, while SeaBeam maps the seafloor using sonar from a ship.
27. Astronomers can use a grid system similar to the system of latitude and longitude to locate objects in the night sky.
28. A conic projection would be the best projection for the map in the Launch Lab because it is the most accurate projection for small areas.
29. From the prime meridian, there are 36 meridians. Each is 10 degrees apart from the next.
30. From the equator to the poles, there are 18 lines of latitude. Each is 10 degrees apart.
31. The image in the picture is a Mercator projection, most commonly used for navigation purposes.

THINK CRITICALLY

32. A person flying from Virginia to California would have to set his or her watch backward because California is three hours behind Virginia.
33. Answers will vary, but students might suggest that one time zone would make it easier for families that live on opposite ends of the country to communicate with each other without worrying about a great time difference.
34. An architect trying to determine where to build a house would use a geologic map to determine where the most stable ground is located. A paleontologist would use a geologic map to both determine what tools he or she would need to excavate a certain type of rock as well as use the map to determine where to dig based on the type of rock in which human remains are most often found.
35. The image is a gnomonic projection. This type of projection could be used to help navigate a plane or a ship over long distances.

52 Chapter 2 • Assessment

ASSESSMENT

CONSTRUCTED RESPONSE

21. **Locate** What time is it in New Orleans, LA, if it is 3 P.M. in Syracuse, NY? Refer to **Figure 4** for help.

22. **Explain** If you wanted to study detailed features of a volcano, would you use a map with a scale of 1:150 m or 1:150,000 m? Why?

Use the figure below to answer Questions 23 and 24.

23. **Identify** What is the line labeled *A*?

24. **Identify** What is the line labeled *B*?

25. **Explain** What is the maximum potential height of a mountain if the last contour line is 2000 m and the map has a contour interval of 100 m?

26. **Describe** how radar used in the *OSTM/Jason-2* satellite differs from the sonar used in the collection of data by SeaBeam.

27. **Infer** Based on what you have learned in this chapter, how might an astronomer map objects seen in the night sky?

28. **Practice** Think back to the LaunchLab at the beginning of the chapter. What type of map projection would be best for the map you drew? Why?

29. **Explain** how degrees of longitude are calculated.

30. **Explain** how degrees of latitude are calculated.

Use the figure below to answer Question 31.

31. **Interpret** what type of projection is shown in the figure. What would this type of projection be used for?

THINK CRITICALLY

32. **Apply** Would a person flying from Virginia to California have to set his or her watch backward or forward? Explain.

33. **Consider** why a large country like China might choose to follow only one time zone.

34. **Careers in Earth Science** Analyze how an architect trying to determine where to build a house and a paleontologist trying to determine where to dig for fossils might use a geologic map.

Use the figure below to answer Question 35.

35. **Apply** What is the projection shown above? What would be two uses for this type of projection? Explain.

36. **Plan** Make a map from your school to the nearest supermarket. How will you determine the scale? What will you need to include in your legend?

37. **Analyze** Why isn't a conic projection used to navigate a ship or an aircraft?

38. **Design** an experiment to test the accuracy of several types of GPS receivers. Make sure you include your control, dependent, and independent variables.

39. **Evaluate** Briefly describe GIS and how it can be used by your community to develop an emergency plan for a severe storm, earthquake, blizzard, drought, or another potential local disaster. 40. Explain why it is necessary to have three satellite signals to determine elevation when using a GPS receiver.

CONCEPT MAPPING

41. Use the following to complete a concept map about remote sensing: *remote sensing*, OSTM/Jason-2 *satellite, Landsat satellite, GPS, uses radar to map sea surface height, uses visible light and infrared radiation to map Earth's surface,* and *uses microwaves to determine location of user.* For more help, refer to the *Skillbuilder Handbook.*

CHALLENGE QUESTION

Use the figure below to answer Question 42.

42. **Assess** Trace the following image to create a topographic map. Connect the elevation measurements to create contour intervals.

Chapter Self-Check

WRITING IN ▸ Earth Science

43. Write a journal entry for an explorer traveling across America before an accurate map was made.

DBQ Document-Based Questions

Data obtained from: NASA, *CALIPSO* satellite image.

This is a satellite image from NASA's CALIPSO *satellite.* CALIPSO *is similar to the other remote sensing technologies you learned about in this chapter, but instead of radar or sonar, it uses something called lidar that sends pulses of light and measures the time it takes for the light to reflect back to the satellite. Based on the amount of time it takes for the light to reflect,* CALIPSO *and scientists can determine what is located below the satellite. The image above was collected on June 7, 2006, across the Indian Ocean. The navy blue areas indicate that no data was detected.*

44. At approximately how many kilometers is the volcanic plume located?

45. At approximately how many kilometers is the thick cirrus cloud located?

46. Why do you think the volcanic plume is higher than the thick cirrus clouds?

47. Why do you think there was no data detected below the thick cirrus clouds?

CUMULATIVE REVIEW

48. Why are graphs, charts, and maps useful? **(Chapter 1)**

49. Why is good communication important in the field of science? **(Chapter 1)**

36. Answers will vary depending on the students' location. The scale should either be a fractional or graphic scale. The legend might include a variety of things, but will most likely include roads, schools, rivers, and parks.

DBQ Document-Based Questions

Data obtained from: NASA, *CALIPSO* satellite image

44. About 20 km. Answers might vary, but should indicate that the volcanic plume is higher than the thick cirrus clouds.
45. About 15 km. Answers might vary, but should indicate that the thick cirrus is lower than the volcanic plume.
46. The volcanic plume is less dense.
47. The thick cirrus clouds either did not allow the light to penetrate, or did not allow the light to reflect back to the satellite and so no data were available.

CUMULATIVE REVIEW

48. They are convenient ways to organize and analyze data.
49. to promote a better understanding of the work and accomplishments of the science community

CHAPTER 2 ASSESSMENT

37. A conic projection has little land mass distortion when used to map small areas. If used for larger areas, as would be necessary for a ship or aircraft navigating long distance, the distortion would be too great to navigate accurately.
38. Answers will vary, but should include a control, and dependent and independent variables, and hypothesis.
39. Answers will vary. GIS mapping involves layering maps one on top of the other (with a computer program). Each layer has a different value such as streets, land use, elevation, etc. These maps could be used for severe storm planning because, for example, they could be used to plan evacuation routes, identify shelter routes, identify areas that are prone to flooding, or to identify areas that are prone to landslides.
40. With only two signals, your position on Earth can be located on a vertical plane, as with latitude and longitude. The addition of a third signal provides trilateration and intersects the vertical plane to provide a third intersect, allowing elevation to be determined.

CONCEPT MAPPING

41. First level: remote sensing; second and third levels (left to right): *OSTM/Jason-2* satellite connected to *uses radar to map sea surface height,* Landsat satellite connected to *uses visible light and infrared radiation to map Earth's surface,* and GPS connected to *uses microwaves to determine location of users*

CHALLENGE QUESTION

42. Students should connect like elevation points to create a topographic map. There should be no intersecting contour lines.

WRITING IN ▸ Earth Science

Rubric

43. Answers will vary, but should reflect the difficulties in navigation.

Chapter 2 • Assessment 53

DBQ **Document-Based Questions** connect students to real-world applications as they evaluate real data from current research. Students analyze graphs, charts, and other displays of data from recognized scientific journals and classic historic documents.

CUMULATIVE REVIEW questions assess student retention of material from earlier chapters.

Backward Mapping

How can my instruction help students succeed in a standards-based system?

by Emily M. Schell, Ed.D.

Content standards articulate what students should know and be able to do in every science classroom. Effective instructional planning based in the standards and maximizing available resources is essential for meaningful teaching and learning of Earth science. Planning instruction with educational goals in mind makes for the most effective teaching.

How do I map my curriculum?

Mapping the curriculum from beginning to end, and from the end to the beginning—backward mapping—makes for solid instruction.

Mapping out the curriculum allows teachers to achieve several goals. These goals include a better understanding of the standards and content-specific objectives, organization and pacing of the curriculum, and focused assessment related to specific goals and objectives.

Analyze the Science Content Standards To begin, teachers analyze the body of content standards for Earth science. Then they compare and contrast these standards to additional sources of information that support effective teaching and learning in Earth science. This process works best with colleagues who bring varying perspectives and expertise to teaching the subject. As a result of this collaboration, strengths and weaknesses of the standards become apparent. Teachers will have a better understanding of the standards and identify concerns and questions for follow-up while mapping.

Analyze the Organization of the Standards-Based Content Most science teachers agree with researchers that science is best taught in order from simple topics to complex topics. However, some state standards either do not or cannot present the content in order from simple to complex. Rich discussions about themes and concepts tend to emerge, and teachers identify meaningful methods for presenting complicated and overlapping information. In this way, students will see the connections that transcend the "simple-to-complex" order.

Identify the Content and Order of Teaching A plan is developed to present content in a certain order. Incorporating content that is either missing from the standards or is essential in building background knowledge with students enters the curriculum map as well. Outside resources brought into the classroom are good supplements.

Separate Overlapping Units Identify areas of instruction related to the Big Ideas, Main Ideas, and Essential Questions. It is at this stage that backward planning is introduced for the development of instructional units, which will support the grade-level curriculum map. The instruction must support the planned assessment.

Map Curriculum at Each Grade Level Curriculum planning should be shared among all subject-area teachers. Teachers will have a better understanding of what knowledge and skills students bring to their coursework if they take into consideration what has been learned previously.

How do I use backward mapping?

After a year-long course of study is mapped out, further develop each unit through backward mapping. Start with the end in mind—know your curricular goals and objectives at the outset, which are often found in the content standards and articulated in the curriculum maps.

Once goals have been determined, teachers develop assessments that will show progress toward those goals and objectives. In the final step of this backward mapping process, teachers determine meaningful teaching and learning strategies and identify useful resources that support the assessment.

To use backward mapping in developing your units of instruction, consider the following steps:

Step One:

Know Your Targets

First, identify exactly what students must know and do in this unit. Analyze content standards and any other resources that support curricular goals and objectives for this unit. As you plan, ask yourself:

- What do I want my students to know as a result of this unit?
- What skills will students develop during the course of this unit?
- How do I describe these goals clearly and concisely to my students so they understand where we should be at the end of this unit?
- What essential knowledge will students need to access to make sense of this information?
- Do my instructional goals align with strategies identified in the curriculum map?
- Have I introduced any Big Ideas that are pertinent to this content?

Step Two:

Identify and Develop Assessments

Second, consider the multiple forms of assessments that will help you determine to what degree each student has achieved the goals and objectives seen in Step One. Some assessments are embedded throughout the instructional unit, while others come at the end of the unit. Some assessments are performance-based, while others are not. Some are authentic applications of information and skills, while others require the formal recall of information. Ask yourself:

- What do I want to know and see from each student?
- What are the best methods for students to demonstrate what they know and can do based on the goals and objectives?
- How many assessments do I need to determine what students know and can do?
- How will I balance informal and formal assessments?
- How will I assess students with diverse learning styles, skills, and abilities?
- How will these assessments promote student understanding of Earth science?
- At what time(s) during the unit will I administer these assessments?

Step Three:

Develop Meaningful Instruction

After the assessments for the unit have been determined, consider the meaningful and effective teaching strategies that will support learning and student achievement on assessments. While developing lesson plans for instruction, ask yourself:

- How will students learn what they are expected to know?
- How will I engage students in the concepts of this unit?
- In what ways might students relate or connect to this information?
- What research-based strategies will be most effective with my students and in these studies?
- How will I differentiate my instruction to meet the diverse needs of my students?
- How will I scaffold or provide access to the curriculum for my English learners?
- What vocabulary requires attention in this unit?
- How much time will I have to effectively teach this unit?

Step Four:

Locate and Manage Resources

Effective teaching and learning of Earth science requires the use of multiple forms of text and varied resources. Consider what you have available in your classroom, including your textbook, and identify resources you will add in order to teach this unit successfully. Ask yourself:

- What parts of the textbook are required for the lessons determined in Step Three?
- What ancillary materials are needed for the lessons in this unit?
- What Web sites will I recommend to students to support these lessons?
- Do I need to contact guest speakers or obtain outside resources?
- What literature resources are available to support this unit?

Emily Schell is a visiting professor at San Diego State University and social studies education director at SDSU City Heights Educational Collaborative, San Diego, CA

Meeting the Diverse Needs Of Your Students

by Douglas Fisher, Ph.D.

Today's classroom contains students from a variety of backgrounds with a variety of learning styles, strengths, and challenges. As teachers, we face the challenge of helping students reach their educational potential. With careful planning, you can address the needs of all students in the Earth science classroom. The basis for this planning is universal access. When classrooms are planned with universal access in mind, fewer students require specific accommodations.

What is Universal Design?

Universal design was first conceived in architectural studies when businesspeople, engineers, and architects began making considerations for physical access to buildings. The idea was to plan the environment in advance to ensure that everyone had access. As a result, the environment would not have to be changed later for people with physical disabilities, people pushing strollers, workers who had injuries, or others for whom the environment would be difficult to negotiate. The Center for Universal Design, www.design.ncsu.edu/cud, defines universal design as "the design of products and environments to be usable by all people, to the greatest extent possible, without the need for adaptation or specialized design."

Universal Design and Access in Education

Researchers, teachers, and parents have expanded the development of built-in adaptations and inclusive accommodations from architectural space to the educational experience, especially in the area of curriculum development.

In 1998, the National Center to Improve the Tools of Educators (NCITE), in partnership with the Center for Applied Special Technology (CAST), proposed an expanded definition of universal design focused on education: *In terms of learning, universal design means the design of instructional materials and activities that allow the learning goals to be achievable by individuals with wide differences in their abilities to see, hear, speak, move, read, write, understand English, attend, organize, engage, and remember.*

How Does Universal Design Work in Education?

Universal design and access, as they apply to education and schooling, suggest the following:

Inclusive Classroom Participation Curricula should be designed with all students and their needs in mind. *Earth Science: Geology, the Environment, and the Universe* was designed for a wide range of students. For example, because English learners and students who struggle with reading will use this textbook, vocabulary is specifically taught and reinforced. Similarly, the teacher-support materials provide multiple instructional points to be used depending on the needs of the students in each class. Further, the Main Ideas are identified for all learners. Throughout the text, there are multiple opportunities to activate students' prior knowledge. Connections between what students know and think about are made throughout the text.

(t)courtesy Douglas Fisher, (b)BananaStock/Jupiterimages

Maximum Text Readability In universally designed classrooms that provide access for all students, texts use direct language, clear noun-verb agreement, and clear construct-based wording. In addition to these factors, the *Earth Science: Geology, the Environment, and the Universe* text uses embedded definitions for difficult terms, provides for specific instruction in reading skills, uses a number of visual representations, and includes note-taking strategies.

Adaptability and Accommodation The content in this textbook can be easily translated, read aloud, or otherwise changed to meet the needs of students in the classroom. The section and end-of-chapter assessments provide students with multiple ways of demonstrating their content knowledge while also ensuring that they have practice with thinking in terms of multiple-choice questions. Critical thinking and analysis skills also are practiced.

How is Differentiated Instruction the Key to Universal Access?

To differentiate instruction, teachers must acknowledge student differences in background knowledge and current reading, writing, and English language skills. They also must consider student learning styles and preferences, interests, and needs, and react accordingly. There are a number of general guidelines for differentiating instruction in the classroom to reach all students, including:

Link Assessment with Instruction Assessments should occur before, during, and after instruction to ensure that the curriculum is aligned with what students do and do not know. Using assessments in this way allows you to plan instruction for whole groups, small groups, and individual students. Backward mapping, in which you establish the assessment before you begin instruction, is also important.

Clarify Key Concepts and Generalizations Students need to know what is essential and how this information can be used in their future learning. In addition, students need to develop a sense of the Big Ideas—ideas that transcend time and place.

Emphasize Critical and Creative Thinking The content, process, and products used or assigned in the classroom should require students to think about what they are learning. While some students may require support, additional motivation, varied tasks, materials, or equipment, the overall focus on critical and creative thinking allows for students to participate in the lesson.

Include Teacher- and Student-Selected Tasks A differentiated classroom includes both teacher- and student-selected activities and tasks. At some points in the lesson or day, the teacher must provide instruction and assign learning activities. In other parts of the lesson, students should have choices in how they engage with the content. This balance increases motivation, engagement, and learning.

Below is an example of a classroom activity for teaching about the periodic table. It is followed by an example of the methods this text provides for differentiating instruction to meet all students' needs.

Classroom Activity

Display a map of the world as it now exists and several versions from different geologic time periods. Discuss the history of the development of the theory of continental drift and describe the reasons scientists realized that the continents must have been linked or connected sometime in the geologic past.

Strategies for Differentiating This Activity

- Have students pretend to be Wegener and write a letter to a relative explaining in simple terms what he had developed with his theory of continental drift.
- Have students investigate the historical context of Wegener's accomplishment by researching the original reactions of scientists to the theory of continental drift. What were the competing theories of the time? How long did it take for Wegener's theory to be adopted?
- Group students into small groups of three or four and have each group choose a continent. Have half of the group write about what the chosen continent has in common with the other continents today, including life-forms, climate, and topography; have the other student in each pair write about what the continents had in common with the other continents 135 mya.
- Ask students to draw a chart about Wegener's continental drift theory, addressing what was correct and incorrect in his original theory. Some questions to consider include: Did Wegener's theory have the ability to correctly estimate the movement of continents in the future? Was he correct to say that the continents plowed through the ocean crust as they moved? What evidence is available today that was not available at the time he came up with this theory?

How do I support individual students?

The majority of students will thrive in a classroom based on universal access and differentiated instruction. However, wise teachers recognize that no single option will work for all students and that there might be students who require unique systems of support to be successful.

Tips for Instruction

The following tips for instruction can support your efforts to help all students reach their maximum potential.

- Survey students to discover their individual differences. Use interest inventories of their unique talents so you can encourage contributions in the classroom.
- Be a model for respecting others. Adolescents crave social acceptance. The student with learning differences is especially sensitive to correction and criticism, particularly when it comes from a teacher. Your behavior will set the tone for how students treat one another.
- Expand opportunities for success. Provide a variety of instructional activities that reinforce skills and concepts.
- Establish measurable objectives and decide how you can best help students meet them.
- Celebrate successes and praise "work in progress."
- Keep it simple. Point out problem areas if doing so can help a student effect change. Avoid overwhelming students with too many goals at one time.
- Assign cooperative group projects that challenge all students to contribute to solving a problem or creating a product.

How do I reach students who have learning disabilities?

- Provide support and structure. Clearly specify rules, assignments, and responsibilities.
- Practice skills frequently. Use games and drills to help maintain student interest.
- Incorporate many modalities into the learning process. Provide opportunities to say, hear, write, read, and act out important concepts and information. Link new skills and concepts to those already mastered.
- If possible, allow students to record answers using an audio-recording device.
- Allow extra time to complete assessments and assignments.
- Let students demonstrate proficiency with alternative presentations, including oral reports, role plays, and art or musical projects.
- Provide outlines, notes, or audio recordings of lecture material.
- Pair students with peer helpers, and provide class time for pair interaction.

How do I reach students who have behavioral challenges?

- Provide a structured environment with simple and clearly defined schedules, rules, seat assignments, and safety procedures.
- Reinforce appropriate behavior and model it for students.
- Cue distracted students back to the task through verbal and nonverbal signals and teacher proximity.
- Set small goals that can be achieved in the short term. Work for long-term improvement in the big areas.

How do I reach students who have physical challenges?

- Openly discuss with the student any uncertainties you have about when to offer aid.
- Ask parents or therapists and students what special devices or procedures are needed and whether any special safety precautions need to be taken.
- Welcome students with physical challenges into all class activities, including field trips, special events, and classroom and community projects.
- Provide information to assist class members and parents in their understanding of support needed.

How do I reach students who have visual impairments?

- Facilitate independence. Modify assignments as needed.
- Teach classmates how and when to serve as visual guides.
- Limit unnecessary noise in the classroom if it distracts the student with visual impairments.
- Provide tactile models whenever possible.
- Foster a spirit of inclusion. Describe people and events as they occur in the classroom. Remind classmates that the student with visual impairments cannot interpret gestures and other forms of nonverbal communication.

How do I reach students who have hearing impairments?

- Seat students where they can see your lip movements easily and where they can avoid any visual distractions.
- Avoid standing with your back to the window or other light source.
- Use an overhead projector so you can maintain eye contact while writing information for students.
- Make sure students sit where they can see all speakers.
- Post all assignments on the board, or hand out written instructions.
- If the student has a manual interpreter, allow both student and interpreter to select the most favorable seating arrangements.
- Teach students to look directly at each other when they speak.

How do I reach English learners?

- Remember, students' abilities to speak English do not reflect their academic abilities.
- Try to incorporate students' cultural experience into your instruction. The help of a bilingual aide may be effective.
- Avoid any references in your instruction that could be construed as cultural stereotypes.
- Preteach important vocabulary and concepts.
- Encourage students to preview text before they begin reading, noting headings.
- Remind students not to ignore graphic organizers, photographs, charts, graphs, and tables, since there is much information in these visuals.
- Use demonstrations and specimens whenever possible to build background knowledge and understanding. For example, you can use ball-and-stick models when describing the structures of molecules.

How do I reach students who are working above grade level?

- Make arrangements for students to take selected subjects early and to work on independent projects.
- Ask "what if" questions to develop high-level thinking skills. Establish an environment safe for risk-taking in your classroom.
- Emphasize concepts, theories, ideas, relationships, and generalizations about the content.
- Promote interest in Earth science by inviting students to make connections to other disciplines that interest them.
- Let students express themselves in alternative ways, such as creative writing, acting, debates, simulations, drawing, or music.
- Provide students with a catalogue of helpful resources, including agencies that provide free and inexpensive materials, appropriate community services and programs, and community experts who might be called upon to speak to your students.
- Assign extension projects that allow students to solve real-life problems related to their communities.

Douglas Fisher is a professor at San Diego State University, San Diego, CA.

Tim Fuller

Culturally Responsive Teaching

By Jacqueline Jordan Irvine and Beverly Jeanne Armento

The growing number of culturally diverse students, particularly African American, Hispanic, and Native American students, in United States schools is well documented. However, on most indicators and measures of academic achievement, African American and Hispanic American students' performance lags behind their white and Asian peers. Although African American students have shown some increased performance on standardized test scores, the gains have been relatively small and inconsistent over time. The data from the National Assessment of Educational Progress (NAEP) reveal that only 19 percent of white twelfth graders scored below basic proficiency level in reading as compared to 48 percent of African Americans and 42 percent of Hispanic students. Similar achievement discrepancies were found in mathematics, writing, history, and science (Nettles 1997).

In addition to the disproportionate school failure of culturally diverse students, there are other compelling data that speak to the significance and urgency of implementing a culturally responsive pedagogy. Demographic data confirm that by the year 2020 about 40 percent of the nation's school-age population will be students of color, and students of color already represent 70 percent of the student population in the 20 largest school districts. However, as the number of culturally diverse students increases, the number of teachers of color is decreasing. The American Association of Colleges of Teacher Education reports (1995) that 80 percent of preservice teachers are white females who are unfamiliar with the cultural experiences of their diverse students. Consequently, these teachers often feel unprepared to deliver effective instruction for the multicultural student populations they instruct.

What is culturally responsive pedagogy?

The term *culturally responsive* is used interchangeably with several terms, such as *culturally responsible, culturally appropriate, culturally congruent, culturally compatible, culturally relevant,* and *multicultural,* to describe a variety of effective teaching strategies in culturally diverse classrooms. These terms all imply that teachers should be responsive to their students by incorporating elements of the students' culture in their teaching. All teachers should be responsive to their students. *Responsive* simply means reacting appropriately in the instructional context. The teaching effectiveness research literature informs us that a responsive teacher is sensitive to the needs, interests, learning preferences, and abilities of his or her students. Responsive teachers do not blindly follow one teaching method or use the same teaching methods and materials for all students. Instead, these teachers modify their knowledge and training, paying attention to classroom contexts and to individual student needs and experiences. For example, responsive teachers provide the correct amount of structure for those students who need structure. They use appropriate classroom materials and methods such as peer tutoring and cooperative learning for students who prefer to work with others. Students who have difficulty with abstractions are provided with models, manipulatives, and other concrete representations.

Culturally responsive teachers spend considerable classroom and nonclassroom time developing a personal relationship with their students (Irvine 1990). These relationship-building exchanges are recurrent and spontaneous daily events. These teachers understand that teaching is a social interaction that involves the development and maintenance of relationships as well as more widely accepted activities such as the planning, delivery, and evaluation of instruction. Culturally responsive teachers listen patiently to their students and allow them to share personal stories and anecdotes during classroom time. Teachers ask students to share stories about their families, weekend activities, hobbies, and so forth. Students are encouraged to express themselves openly, and storytelling delivered with high affect and positive emotions are not prohibited. Responsive teachers also share appropriate stories about their personal lives and connect their own lives and their students' lives to the content of the instruction.

Principles of a Culturally Responsive Curriculum

Just what does a culturally responsive classroom look like? Imagine a classroom where all students and educators feel a sense of caring, security, trust, and genuine value; where each person is treated with a sense of dignity and an expectation for outstanding performance; where each person truly wants to be present and is excited about learning and teaching; where all persons feel proud of their growth, are curious about the world, and are adventuresome and aggressive about their own learning.

The demographic changes in today's schools alone (Shinagawa and Jung 1998) demand new ways of organizing and implementing instruction. Diverse students bring a range of learning preferences, a range of life experiences and typical ways of relating to other people and their teachers, and basic beliefs about learning and school that may be different from the teacher's knowledge, beliefs, and views and certainly different from the dominant norms of the school.

Basic Beliefs of a Culturally Responsive Educator

There are at least ten basic beliefs that underlie the development of a culturally responsive learning environment. Since these beliefs form the foundation of any curriculum development or implementation, it is important to reflect on these basic ideas and think about ways we might nurture these attitudes in ourselves.

You'll find Cultural Diversity features throughout your Teacher Edition, providing opportunities for students to appreciate cultural diversity and scientific achievement.

CULTURAL DIVERSITY

Ayers Rock—Uluru In Kata Tjuta National Park, southwest of Alice Springs in the Northern Territory of Australia, a huge sandstone monolith rises 345 m above the desert. It is 2 km wide and semi-oval in shape. It is called Ayers Rock, or Uluru, as the Aboriginals (native Australians) named it. Uluru is a sacred site for the Aboriginal people. They believe that there is an energy source emanating from the rock and that the area around it is inhabited by ancestral beings. The rock seems to change color from blue to red, depending on the time of day and the quality of light. Uluru bears many carvings and paintings made by the Aboriginals who conduct sacred rites there. The Australian government restored ownership of Uluru and the Kata Tjuta National Park to the Aboriginals in 1985.

Culturally Responsive Educators

- Hold high academic and personal expectations for each child.
- Provide for each child equitable access to the necessary learning resources and sufficient opportunities to learn.
- Ensure that learning outcomes are meaningful, relevant, useful, and important to each child.
- Nurture learning-support communities for each child (families, peers, homework hotlines, community centers).
- Facilitate the maximum growth of each learner by making informed academic adaptations that match and build upon the learner's prior knowledge, experiences, skills, and beliefs.
- Build positive and supportive school and classroom learning environments that are grounded in mutual and genuine respect for cultural diversity.
- Promote classroom climates built on social justice, democracy, and equity.
- Promote individual empowerment, self-efficacy, positive self-regard, and a belief in societal reform.
- Value diversity as well as human commonalities.
- Believe that it is their role and responsibility to provide effective and empowering instruction for each child.

These ten basic beliefs of culturally responsive educators form the foundation for effective teaching and learning for all students and are the basis of equity and excellence in the classroom. Educators who hold these basic beliefs are well on their way to establishing wholesome classrooms where each child is able to flourish and where students will learn the essence of democratic ways of behaving and thinking.

Jacqueline Jordan Irvine is a professor of multicultural education and urban teacher education in the Division of Educational Studies at Emory University, Atlanta, GA

Beverly Jeanne Armento is a research professor and department chair at the Georgia State University College of Education, Atlanta, GA

Academic Vocabulary

How can I help my students learn academic vocabulary?

What is academic English?

Academic English is the language used by the educated and by leaders in business, academic, and other professional disciplines. It is the language used in courts of law and in professional books, including textbooks. This type of English contains specific linguistic features that are associated with academic disciplines like Earth science. Proficiency in reading and using academic English is especially related to long-term success in all parts of life.

Academic vocabulary is the basis for academic English. By reinforcing academic vocabulary and academic English, teachers can help learners to access authentic, academic texts—not simplified texts that "dummy down" the content. In this way, they can provide information that will help build their students' background knowledge rapidly.

What is academic vocabulary?

By the time children have completed elementary school, they must have acquired the knowledge needed to understand academic vocabulary. How many words should they acquire to be able to access their textbooks? A basic 2,000-word vocabulary of high-frequency words makes up 87 percent of the vocabulary of academic texts. Eight hundred other academic words comprise an additional 8 percent of the words. Three percent of the remaining words are technical words. The remaining 2 percent are low-frequency words. There might be as many as 123,000 low-frequency words in academic texts.

Why should students learn academic vocabulary?

English learners who have mastered a basic 2,000-word vocabulary are ready to acquire the majority of general words found in their academic texts.

Knowledge of academic words combined with continued acquisition of general words can significantly boost an English learner's comprehension level of academic texts. English learners who learn and practice these words before they graduate from high school are likely to be able to master academic material with more confidence and speed. They waste less time and effort in guessing words or consulting dictionaries than those who only know the basic 2,000 words that characterize general conversation. Also, consider academic success in terms of measurement and assessment—state standards-based assessments, the SAT, and the ACT—with regard to word mastery. All demand an understanding of academic vocabulary.

How do I include academic vocabulary and academic English in my teaching?

- Teachers can provide their students with rich samples of academic vocabulary and help students understand and attend to the academic English in their texts.
- To develop academic English, learners must have already acquired a basic proficiency in the grammar of everyday English.
- Academic English should be taught within contexts that make sense. Teaching academic English includes providing students with access to core curriculum—in this case, Earth science.
- Academic English arises not only from knowledge of a linguistic code and cognition, but also from social practices in which academic English is used to accomplish communicative goals. The acquisition of academic vocabulary and grammar is necessary to advance the development of academic English.

Digital Vision/Getty Images

Tips for Teaching Academic Vocabulary

Expose students to academic vocabulary. Provide students with sufficient exposure to academic words.

Do not ignore English learners in this process. They can learn academic vocabulary before they are completely fluent in everyday English.

Encourage broader learning by helping students build academic vocabulary. Students who have mastered the basic academic vocabulary are ready to continue acquiring words from the rest of the groups. To help determine which words are in the 2,000-word basic group, refer to Coxhead's Academic Word List.

Guidelines for Teaching Academic Vocabulary:

There are a number of guidelines that teachers can use when teaching academic English and vocabulary.

- direct and planned instruction
- models that have increasingly difficult language
- attention to form—pointing out linguistic features of words
- practice
- motivation
- instructional feedback
- assessment—on a regular basis

Classroom Activity: Writing About Energy

As an example of teaching academic vocabulary, when the class studies the topic of energy, you could give students an impromptu writing assignment. Ask them to write a short essay about one of the topics listed below in the left column. Have students use as many of the academic vocabulary words in the right column as they can in their essays. Give students a time limit for their writing. When students have completed the assignment, ask student volunteers to share their writing. Help them use academic words correctly.

Topic	Academic Vocabulary
Erosion	mechanical weathering
	rock cycle
	frost wedging
	chemical weathering
	temperature
Identifying minerals	pressure
	acid precipitation
	water
	luster
	compound
	element
	texture
	density
	reflect
	bonds

Matt Meadows

Test–Taking Strategies

How can I help my students excel on tests?

It's not enough for students to learn Earth science facts and concepts—they must be able to show what they know in a variety of test-taking situations.

How can I help my students do well on objective tests?

Objective tests might include multiple choice, true/false, and matching questions. The following strategies can help students do their best on objective tests.

Multiple-Choice Questions

Students should read the directions carefully to learn what answer the test requires—the best answer or the right answer. This is especially important when answer choices include "all of the above" or "none of the above."

- Advise students to watch for negative words in the questions, such as *not, except, unless,* and *never.*
- Students should try to mentally answer the questions before reading the answer choices.
- Students should read all the answer choices and cross out those that are obviously wrong. Then they should choose an answer from those that remain.

True/False Questions

- It is important that students read the entire question before answering. For an answer to be true, the entire statement must be true. If one part of a statement is false, the answer should be marked false.
- Remind students to watch for words like *all, never, every,* and *always.* Statements containing absolute words such as these are often false.

Matching Questions

- Students should read through both lists before they mark any answers.
- Unless an answer can be used more than once, students should cross out each choice as they use it.
- Student should use what they know about grammar to find the right answer. When matching a word with its definition, the definition is often the same part of speech (noun or verb, for example) as the word.

image100 Ltd

How can I help my students do well on essay tests?

Essay tests require students to write a thorough and well-organized answer to a question or questions. Help students use the following strategies on essay tests.

Read the Question

The key to writing successful essays lies in reading and interpreting questions correctly. Teach students to identify and underline key words in the questions, and to use these words to guide them in understanding what the question asks. Help students understand the meaning of some of the most common key words listed in the chart below.

Analyze	To **analyze** means to systematically and critically examine all parts of an issue or event.
Classify or categorize	To **classify** or **categorize** means to put people, things, or ideas into groups, based on a common set of characteristics.
Compare and contrast	To **compare** is to show how things are similar or alike. To **contrast** is to show how things are different.
Describe	To **describe** means to present a sketch or impression. Rich details help to flesh out a description.
Discuss	To **discuss** means to systematically write about all sides of an issue or event.
Evaluate	To **evaluate** means to make a judgment and support it with evidence.
Explain	To **explain** means to clarify or make plain.
Illustrate	To **illustrate** means to provide examples or to show with a picture or other graphic.
Infer	To **infer** means to read between the lines or to use knowledge and experience to draw conclusions or make a generalization.
Justify	To **justify** means to prove or to support a position with specific facts and reasons.
Predict	To **predict** means to tell what will happen in the future, based on an understanding of prior events and behaviors.
State	To **state** means to briefly and concisely present information.
Summarize	To **summarize** means to give a brief overview of the main points of an issue or event.

Plan and Write the Essay

After students understand the question, they should follow the steps below to develop and write their essays.

Step 1	Step 2	Step 3	Step 4	Step 5
Map out an answer. Make lists, webs, or an outline to plan the response.	Write an opening statement that directly responds to the essay question.	Write the essay. Expand on the opening statement. Support key points with specific facts, details, and reasons.	Write a closing statement that brings the main points together.	Proofread to check for spelling, grammar, and punctuation.

How can I help my students prepare for the SAT, the ACT, and other standardized tests?

- **Read About the Test** Students can familiarize themselves with the format of the test, the types of questions that will be asked, and the amount of time they will have to complete the test.
- **Review the Content** Consistent study throughout the school year will help students build science content knowledge and understanding. If there are specific objectives or standards that are tested on the exam, help students review these facts or skills to be sure they are proficient.
- **Practice** Provide practice, ideally with released tests, to build students' familiarity with the content, format, and timing of the actual exam. Students should practice all the types of questions they will encounter on the test—multiple choice, short answer, and extended response.
- **Analyze Practice Results** Help students improve test-taking performance by analyzing their test-taking strengths and weaknesses. Spend time discussing students' completed practice tests, explaining why particular answers are right or wrong. Look for patterns in errors and then tailor your instruction to the appropriate skills or science content.

Alternative Assessment Strategies

How can I go beyond tests to assess students' understanding of Earth science facts and concepts?

In response to the growing demand for accountability in the classroom, educators must use multiple assessment measures to accurately gauge student performance. In addition to quizzes, tests, essay exams, and standardized tests, assessment today uses a variety of performance-based measures and portfolio opportunities.

What are some typical performance-based assessments?

There are many kinds of performance-based assessments. They all share one common characteristic: they challenge students to create written or oral reports that demonstrate what they know. One good way to present a performance assessment is in the form of an open-ended question.

- **Creative Stories** Students integrate scientific events into a piece of fiction.
- **Poems and Songs** Students follow the conventions of a particular type of song or poem as they tell about a scientist or scientific event.
- **Research Reports** Students synthesize information from a variety of sources into a well-developed report.

Writing

Performance-based writing assessments challenge students to apply their knowledge of Earth science concepts and information in various ways. Writing activities are most often completed by one student, rather than by a group.

- **Journals** Students write from the perspective of a scientist, either current or historical.
- **Letters** Students compose a letter from one scientist to another or from a scientist to a family member or other audience.
- **Position Paper or Editorial** Students explain a controversial issue and present their own opinion and recommendations, supported with strong evidence and convincing reasons.
- **Newspaper** Students write a variety of stories from the perspective of a reporter.
- **Biographies and Autobiographies** Students write about scientists either from the third-person point of view (biography) or from the first person (autobiography).

Oral Presentations

Oral presentations allow students to demonstrate their scientific literacy before an audience. Oral presentations are often group efforts, although this does not need to be the case.

- **Simulations** Students hold simulations, or reenactments, of actual events, such as famous experiments or discoveries.
- **Debates** Students debate two or more sides to a scientific policy or issue. Students can debate from a contemporary perspective or through role-playing, from the viewpoint of a historical character.
- **Interviews** Students conduct a mock interview of a scientist.
- **Oral Reports** Students present the results of research efforts in an oral report.
- **Skits and Plays** Students use scientific events as the basis for a play or skit.

Visual Presentations

Visual presentations allow students to demonstrate their scientific understanding in a variety of visual formats. Visual presentations can be either group or individual projects.

- **Model** Students make a model to demonstrate or represent a process or structure.
- **Museum Exhibit** Students create a rich display of materials around a topic. Typical displays might include models, illustrations, photographs, videos, writings, and audio-recorded presentations.
- **Graph or Chart** Students analyze and represent scientific data in a line graph, bar graph, table, or other chart format.
- **Drawing** Students represent or interpret a scientific event or period through illustration, including political cartoons.

- **Posters and Murals** Posters and murals might include graphs, charts, tables, maps, time lines, diagrams, illustrations, photographs, and text that reflect students' understanding of scientific information.
- **Quilt** Students sew or draw a design for a patchwork quilt that shows a variety of perspectives, events, or issues related to a key topic.
- **Videotapes or DVDs** Students film a video or DVD to preserve a simulation of a scientific event. Students can also film plays they have written that incorporate Earth science in some way.
- **Multimedia Presentation or Slideshow** Students create a computer-generated multimedia presentation containing scientific information and analysis.

How are performance assessments scored?

There are a variety of means available to evaluate performance tasks. Some or all of the following methods can be used.

- **Scoring Rubrics** A scoring rubric is a set of guidelines for assessing the quality of a process or product. It sets out criteria used to distinguish acceptable responses from unacceptable ones, generally along a scale from excellent to poor.
- **Rubric** A variety of modifiable rubrics are available on your eTeacherEdition.
- **Models of Excellent Work** Teacher-selected models of excellent work give a concrete illustration of what is expected and help students set goals for their own projects.
- **Student Self-Assessment** Common methods of self-assessment include having students rank their work in relation to the model, use a scoring rubric, and write their own goals and then evaluate how well they have met these goals. Regardless of the method or methods students use, they should be encouraged to evaluate their behaviors, processes, and the finished product.
- **Peer or Audience Assessment** Many of the performance tasks target an audience other than the classroom teacher. If possible, an audience of peers should give the students feedback. Have the class work together to create rubrics for specific projects.
- **Observation** As students carry out their performance tasks, you might want to formally observe students at work. Start by developing a checklist, identifying the specific behaviors and knowledge you expect students to demonstrate. Then observe students as they carry out performance tasks and check off these items on your checklist as you observe them.
- **Interviews** As a form of ongoing assessment, you might want to conduct interviews with students, asking them to analyze, explain, and assess their participation in performance tasks. When projects take place over an extended period of time, you can hold periodic interviews as well as exit interviews. In this way, you can gauge the status of the project and guide students' efforts along the way.

Noel Hendrickson/Getty Images

Web Strategies

How can I use the Internet to teach Earth science?

From the Internet to round-the-clock live newscasts, teachers and students have never before had so much information at their fingertips. Yet never before has it been so confusing to determine where to turn for reliable content and what to do with it once you have found it. In today's world, Earth science teachers must not only use the Internet as a source of up-to-the-minute information for students; they must also teach students how to find and evaluate sources on their own.

What's available on the Internet?

- **Teacher-Focused Web Sites** These Web sites provide teaching tips, detailed lesson plans, and links to other sites of interest to teachers and students.
- **Scientific Data** Research data has been catalogued and placed on the Web. These sites are rich depositories of statistics for all scientific endeavors.
- **Geographical Information** The Web holds a variety of geographical resources, from historical and physical maps to interactive mapping programs to information about plants, animals, people, and places in the United States and around the world.
- **Statistics** Government (.gov) Web sites are rich depositories for statistics of all kinds, including information about populations, habitats, bioinformatics, diseases, and other topics relating to Earth science.
- **Reference Sources** Students can access full-text versions of encyclopedias, dictionaries, atlases, and other reference books, as well as databases containing millions of journal and newspaper articles.
- **News** Traditional media sources—including television, radio, newspapers, and newsmagazines—sponsor Internet sites that provide almost-instantaneous news updates, as well as in-depth news coverage and analysis. Extensive archives facilitate research on past news stories.
- **Topical Information** Among the most numerous Web sites are those organized around a particular topic or issue, such as the periodic table. These Internet pages may contain essays, analyses, and other commentaries, as well as primary source documents, maps, photographs, video and audio clips, bibliographies, and links to related online resources.
- **Organizations** Many organizations, such as research institutes, post Web pages that provide online exhibits, archives, and other information.

Glencoe Earth Science: Geology, the Environment, and the Universe

Visit connectED.mcgraw-hill.com for interactive, Web-based Earth science curriculum.

- Online Learning Centers challenge students to apply newly acquired skills.
- Concepts in Motion and Earth science videos provide students with interactive learning tools to reiterate concepts.
- Students can assess their knowledge and review for tests using self-check quizzes and practice tests.
- Students can participate in inquiry-oriented Web research using safe, prescreened Internet sites.

For more information about ConnectED, see page 30T.

Finding Things on the Internet

The greatest asset of the Internet—its vast array of materials —is also its greatest deterrent. Many excellent science-specific sites provide links to relevant content. Using Internet search engines can also help you find what you need.

- A search engine is an Internet search tool. You type in a keyword, name, or phrase, and the search engine lists the URLs for Web sites that match your search. However, a search engine could find things that are not at all related or may miss sites that you would consider of interest. The key is to find ways to define your search.
- Not all search engines are the same. Each seeks out information a little bit differently. Different search engines use different criteria to determine what constitutes a "match" for your search topic. The Internet contains numerous articles that compare search engines and offer guidelines for choosing those that best meet your needs.

How do I teach students to evaluate Web sites?

Anyone can develop a Web site. Web content is easy to change, so webmasters constantly update their Web sites by adding, modifying, and removing content. These characteristics make evaluating Web sites more challenging than evaluating traditional print resources. Teach students to evaluate Web resources critically, using the questions and criteria below.

- **Authority:** Who wrote the material or created the Web site? What qualifications does this person or group have? Who has ultimate responsibility for the site? If the site is sponsored by an organization, are the organization's goals clearly stated?
- **Accuracy:** How reliable is the information? Are sources listed so that they can be verified? Is the Web page free from surface errors in spelling and grammar? How does it compare with other sources you've found on the Web and in print?

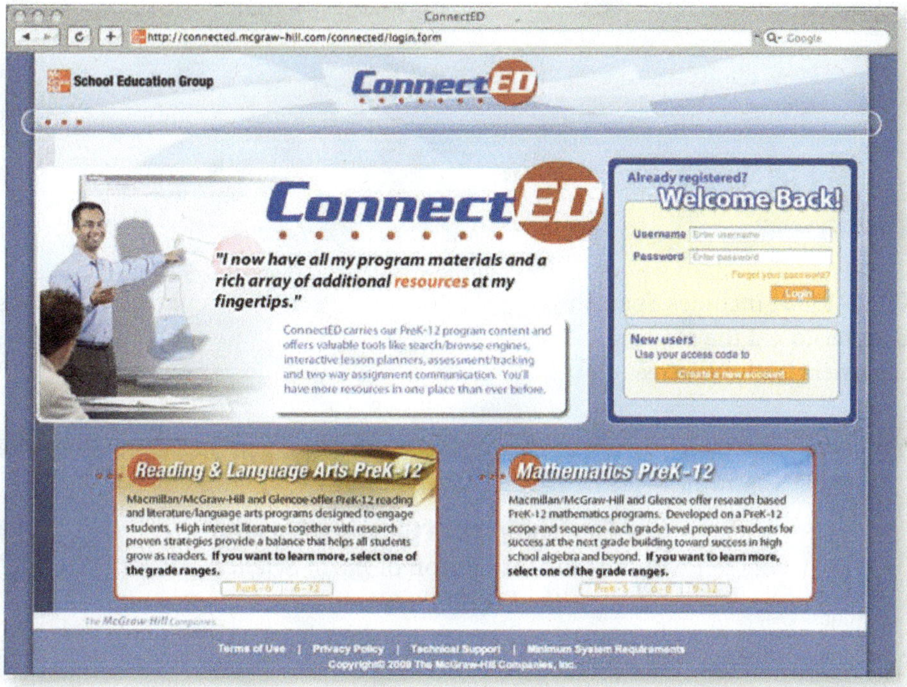

- **Purpose:** What is the purpose of the Web site or Web page? Is it an informational Web page, a news site, a business site, an advocacy site, or a personal Web page? Many sites serve more than one purpose. For instance, a news site may provide current events accompanied by banner ads that market the products advertisers think readers might want.
- **URL:** What is the URL or Web address? Where does the site originate? That can sometimes tell you about the group or business behind the Web page. For example, URLs whose domain names end in *.edu* and *.gov* indicate that the site is connected to an educational institution or a government agency, respectively. A *.com* suffix usually means that a commercial or business interest hosts the Web site, but it might also indicate a personal Web page. A nonprofit organization's Web address might end with *.org*.
- **Objectivity** If the site presents itself as an informational site, is the material free from bias? If there is advertising, is it easy to tell the difference between the ads and other features? If the site mixes factual information with opinion, can you spot the difference between the two? If the site advocates an opinion or viewpoint, is the opinion clearly stated and logically defended?
- **Currency:** When was the information first placed online? Is the site updated on a regular basis? When was the last revision? If the information is time-sensitive, are the updates frequent enough?
- **Coverage:** What topics are covered on the Web site? What is the depth of coverage? Are all sides of an issue presented? How does the coverage compare with other online and print sources?

English Learners (EL)

How can I reach English learners in the Earth science classroom?

American classrooms reflect the rich and diverse cultural heritage of the American people. Students come from different ethnic backgrounds and different cultural experiences into a common classroom that must assist all of them in learning. Multicultural and/or bilingual students often speak English as a second language or not at all. In providing for EL students, the focus needs to be on overcoming the language barrier.

It is important not to confuse ability in speaking and reading English with academic ability or intelligence. In general, the best method to assist EL students is to provide them with a variety of ways to learn, apply, and be assessed on the concepts. *Earth Science: Geology, the Environment, and the Universe* has risen to this challenge with a full complement of ancillaries and technology for English learners.

eStudent Edition

Textbook · Audio · Workbooks · and More!

The **StudentWorks**™ Plus DVD-ROM includes Spanish audio summaries of each chapter to aid the reading comprehension of English learners in the science classroom.

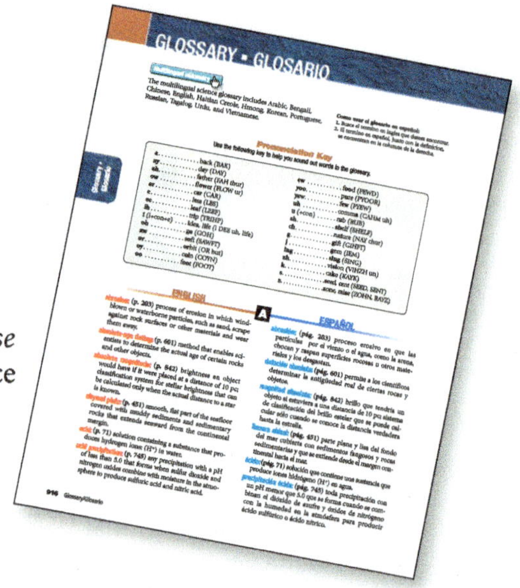

The **English/Spanish Glossary/Glosario** in the Student Edition of *Earth Science: Geology, the Environment, and the Universe* helps native Spanish speakers learn science vocabulary.

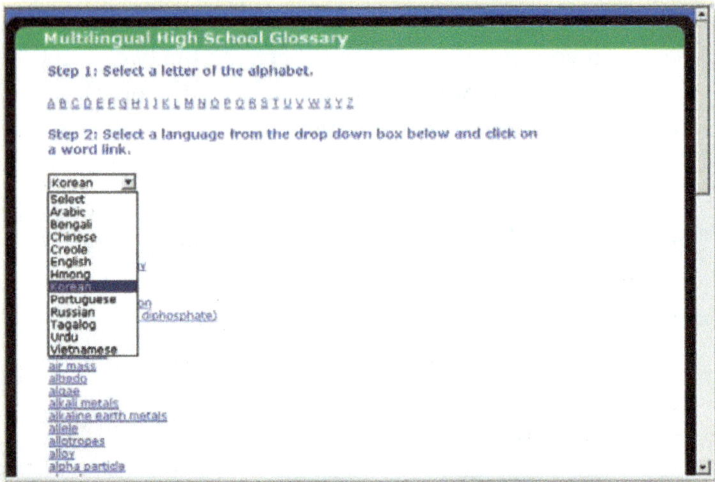

The online **Multilingual Science Glossary,** which helps speakers of many languages learn science vocabulary, is available at connectED.mcgraw-hill.com. The glossary includes the following languages: Arabic, Bengali, Chinese, English, Haitian Creole, Hmong, Korean, Portuguese, Russian, Spanish, Tagalog, Urdu, and Vietnamese.

Science Notebook is a note-taking guide designed to help students succeed in learning science content. It contains note-taking tools based on the Cornell Note-Taking System.

Glencoe Science *Blueprints for Success: Science Classrooms that Work* provides practical tools and suggests modifications that can help students master science concepts while developing skills for reading and writing in science. This all-in-one professional development resource will help you discover alternative ways to teach science concepts to English-language learners, connect classroom content to language arts and literacy, and provide all students with activities for building science vocabulary.

EL strategies in the Teacher Edition provide teachers with additional support for EL students in the science classroom. Blue EL boxes on the B page before each chapter indicate which ancillary products would be useful to EL students.

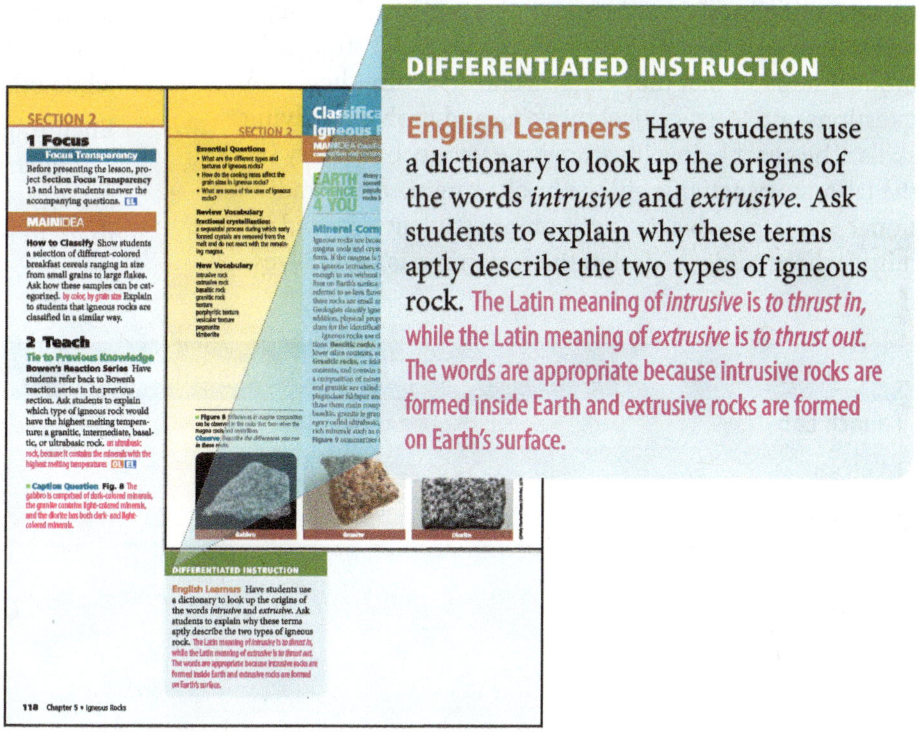

Inquiry-Based Instruction

***How is inquiry used in* Earth Science: Geology, the Environment, and the Universe?**

Earth Science: Geology, the Environment, and the Universe is based on six specific, research-based strategies, all of which are particularly important in a science inquiry teaching approach. These strategies are:

- using prior knowledge to learn new information;
- practicing important tasks;
- using visuals to communicate, organize, and reinforce science learning;
- getting students motivated to achieve;
- developing decoding strategies and reading comprehension proficiencies;
- learning by using study strategies.

The process of inquiry models scientific practice, encouraging problem-solving strategies and developing critical thinking skills. Inquiry gets students actively involved in the learning process by allowing them to determine materials, procedures, or the topics and questions they want to investigate.

Inquiry Continuum

Inquiry-based instruction can be understood as a continuum with several basic levels, ranging from a very structured activity in which students are guided through the process of a scientific investigation to a student-directed inquiry investigation. It is important to gradually build students' inquiry skills so that they can learn to structure their own questions and learn critical thinking and problem-solving skills. The chart below illustrates the variety of inquiry in the laboratory activities in the book, ranging from traditional hands-on lab activities to more student-directed inquiry, where students design the experiments themselves.

Teaching Tips for Incorporating Inquiry

Inquiry in science takes time, just as it would in a research lab. However, it is time well spent as students learn the content and build problem-solving skills in a way this is engaging and meaningful. Use the following techniques to help efficiently implement inquiry in the science classroom.

- In the beginning of the school year, some students may need more guidance if they are unfamiliar with the practice of scientific inquiry. You may need to teach them how to ask testable questions, make observations, collect reliable data and use evidence to support their conclusions.
- Early on, plan and set up long-term experiments and projects with students that they can observe and manipulate throughout the year.
- Give your students a more guided activity that introduces hard-to-understand concepts and skills. Then allow them to explore on their own with a wider variety of materials.
- Have students brainstorm questions they would like to explore. As a class, choose one or two reasonable questions that each group will explore on its own. (This is very helpful if you are trying to cover a specific topic or standard.)
- Divide classes into groups and have them explore different phases of an experiment. One group may require more supervision or may work more closely with you, while others work more independently.
- Encourage students to rely on the data they collect and evidence they gather rather than on what they think the answer should be. If their data are unexpected, help them to problem-solve.

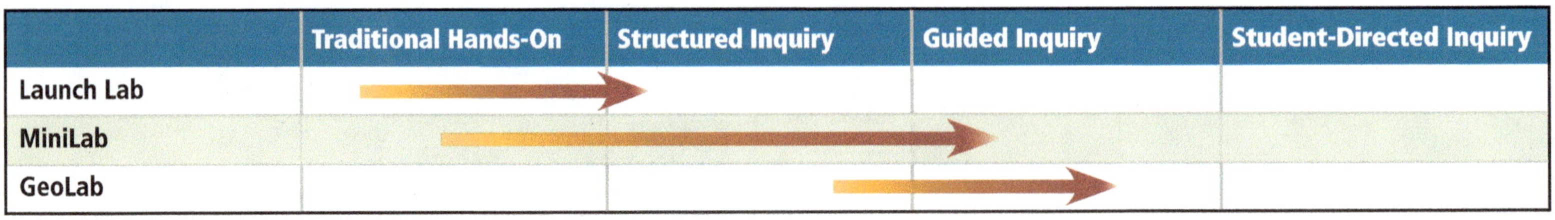

Foldables

Improving Active Reading and Study Skills

Foldables are easy-to-make, three-dimensional, interactive graphic organizers that students create out of simple sheets of paper. These unique hands-on tools for studying and reviewing were created exclusively for Glencoe by education specialist Dinah Zike.

The FOLDABLES® icon that appears in the margin at point-of-use in the body of a chapter tells students when to add information to their Foldable.

FOLDABLES® Build Prereading Skills

- enable students to organize and process information
- encourage students to observe and read
- create an interactive snapshot of the chapter

FOLDABLES® Encourage Reading and Writing

- provide practice in basic reading and writing skills
- develop skills in finding and reporting main ideas
- organize information
- provide review of key vocabulary terms

FOLDABLES® Summarize Content for Review

- create a comprehensive, interactive snapshot of the chapter
- provide preparation support for chapter, unit, and end-of-course exams, as well as standardized tests

FOLDABLES® Help with Assessment

- probe the depth of your students' understanding of chapter concepts
- pinpoint what your students understand
- identify misconceptions

Additional Resources

Dinah Zike's Teaching with Foldables: Science offers

- presentations of Foldables instructions
- ideas on how to incorporate Foldables into your lessons
- easy-to-read folding instruction pages

Dinah Zike's Big Book of Science, which contains even more presentations, ideas, and color instruction pages, can be ordered from:

Dinah-Might Adventures, LP
P.O. Box 690328
San Antonio, Texas 78269
1-800-99-DINAH
www.dinah.com

FOLDABLES® Study Organizer

Mineral Identification

Make a layered-look book and label the tabs with the names of the tests used to identify minerals. Use it to organize your notes on mineral identification.

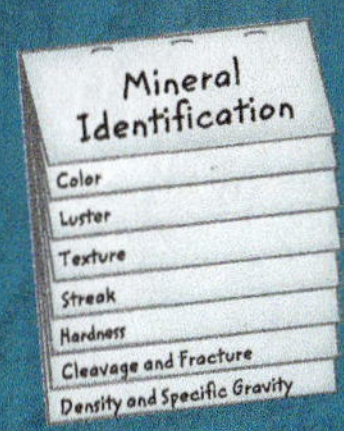

Hands-On Learning

***Glencoe's Earth Science: Geology, the Environment, and the Universe* offers a variety of print products to enhance your students' science experience.**

Resources For All Your Needs

In addition to the wide array of instructional options provided in the student and teacher editions, *Earth Science: Geology, the Environment, and the Universe* offers an extensive list of support materials and program resources. Some of these materials offer alternative ways of presenting your program, others provide tools for reinforcing core concepts and evaluating student learning, and still others will help you extend and enrich your course. You won't have time to use them all, but the ones you use will help you make the best use of the time you have.

Laboratory Manual, Student Edition and Teacher Edition

If you want more hands-on options, the Laboratory Manual offers you two more labs for each chapter, which provide students with a variety of laboratory experiences that reinforce the principles in the text.

Additional Lab Manuals

The *Exploring Environmental Problems* Laboratory Manual offers still more laboratory experiences for your students.

Performance Assessment

Challenge students to solve the more advanced problems in *Performance Assessment in Earth Science.* They are especially suited to students performing at or above grade level.

Science Notebook

Science Notebook is a note-taking guide designed to help students succeed in learning science content. It contains note-taking tools based on the Cornell Note-taking System.

Section Focus Transparency Package

The full-color transparencies provide a way for you to begin each lesson by capturing the attention of your students with simple activities. Each numbered section begins with a Section Focus strategy that lists the number of the transparency for that section. The package also contains teaching strategies for all transparencies.

Teaching Transparency Package

These full-color teaching transparencies reinforce concepts taught in the student text. A blackline master and worksheet for each transparency can be found in the FAST FILE Unit Resources.

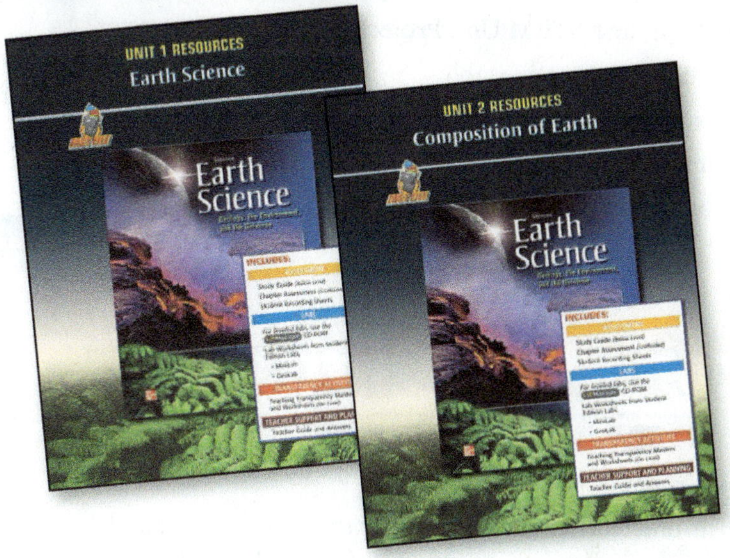

Earth Science Standardized Test Practice

This workbook is designed to cover the National Science Content Standards (NSCS) that relate to Earth science, and provide additional chapter content review of the Glencoe textbook, *Earth Science: Geology, the Environment, and the Universe.* The Teacher Edition is comprised of an answer sheet master, review questions, annotations, and cross-referenced NSCS.

Earth Science: Geology, the Environment, and the Universe Unit Resources

The unit resource books contain effective content, lab, and assessment activity worksheets that will strengthen every student's understanding of *Earth Science: Geology, the Environment, and the Universe.*

Each FAST FILE includes:

- MiniLab Worksheets
- GeoLab Worksheets
- Study Guides
- Teaching Transparency Masters and Worksheets
- Chapter Assessments
- STP Recording Sheets

Technology Resources

Earth Science: Geology, the Environment, and the Universe incorporates the latest educational technology resources into your classroom, providing one-click access to teaching presentations, additional problems, student assessments, laboratory investigations, video assets and more!

Go online!
connectED.mcgraw-hill.com

Visit connectED.mcgraw-hill.com to connect to a host of online resources.

- interactive activities that review chapter concepts
- online, interactive student edition
- self-check quizzes and vocabulary games
- *WebQuests* and STEM Unit Projects
- animations, interactive tables
- videos
- multilingual science glossary
- editable student worksheets

eTeacherEdition
access to complete teacher resources

eTeacherEdition
An all-in-one teacher resource center to access student and teacher resources, connect to the Internet, and more!

In the iLab Station at connectED.mcgraw-hill.com, you will find laboratory investigations, procedures, and worksheets:

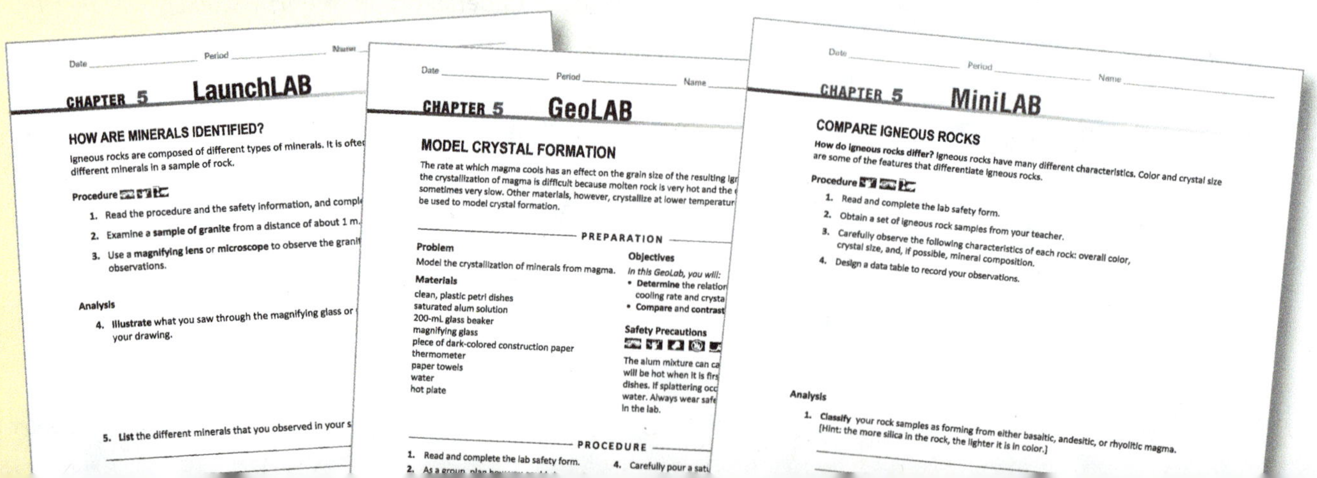

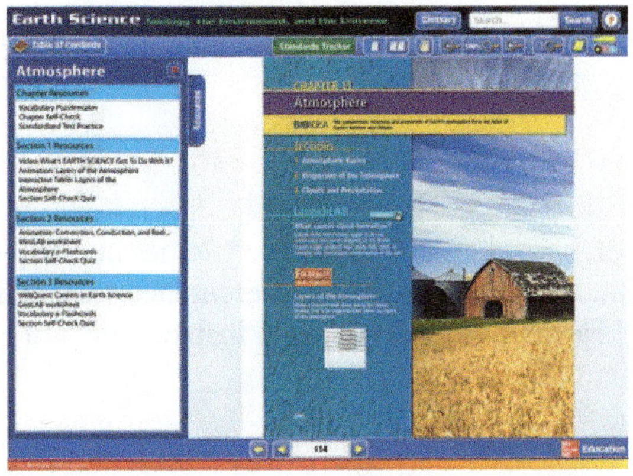

eStudent Edition

Provide students with a complete interactive Student Edition that integrates audio, links to online resources, and access to worksheets.

Classroom Presentation Toolkit

Lesson planning is simple with editable Microsoft® PowerPoint® presentations. Use this tool to teach key concepts from each section or customize your own presentations.

Exam*View*

Create and customize tests with this convenient assessment platform. Track student performance, correlate assessment to state standards, and even translate tests into Spanish with one click!

What's EARTH SCIENCE Got To Do With It?

Engage your students with these videos that connect Earth science concepts with real-world applications.

LabManager

Customize any lab with this easy-to-use activity database. Search by topic, time requires, materials, concepts, and key words. Choose from thousands of labs and edit them to suite your needs.

Glencoe Professional Development

Science, YES! (Your Education Solution)

Educational Strategies DVD and Workshop Facilitation Guide

This set of five videos emphasizes key issues in science education: Literacy Strategies, Differentiated Instruction, English Learners, Standards-Based Instruction, and Assessment Strategies.

Educational Strategies DVD

Each video begins with a discussion on the topic, providing a clear overview of the key issue and its importance in science education. Next, viewers see lesson excerpts from various science classrooms that demonstrate instructional strategies and techniques used to address the video topic. In place of the classroom demonstrations, the *Standards-Based Instruction and Assessment Strategies* videos present excerpts of science department meetings. In these segments, viewers watch department meeting collaboration to determine local curricula and assessment.

Workshop Facilitation Guide

The *Facilitation Guide* contains workshop activities and information for conducting the workshops. Workshop activities foster participants' discussion and collaboration. Each workshop activity provides opportunities for science teachers to evaluate, discuss, and extend the strategies demonstrated in the video.

Educational Strategies Online Professional Development

Science, YES! Educational Strategies Online is comprised of five individual modules. These online modules, including Literacy Strategies, Differentiated Instruction, English Learners, Standards-Based Instruction, and Assessment Strategies, emphasize techniques used to teach science. Each online module contains video, interactive content, and a variety of online learning tools.

Online learners watch video clips of educational commentary and science classroom video delivered over the Web. Tools, such as a glossary, standards links, and lists of Web resources, are included throughout. Module content encourages online learners to discuss ideas with others in their online community through the discussion board.

The modules may be taken individually or they may be taken together as a university course. An electronic portfolio can be used to verify module completion for university credit. They can be combined with other modules to develop custom courses. A module takes 6–8 hours to complete and contains six content segments—Introduce, Model, Practice, Assess, Classroom Connection and Textbook Tie. Segments divide the content into shorter manageable chunks of learning, making it easy for a busy teacher to complete the module.

Science Content Standards

Unifying Concepts and Processes

UCP.1 Systems, order, and organization
UCP.2 Evidence, models, and explanation
UCP.3 Change, constancy, and measurement
UCP.4 Evolution and equilibrium
UCP.5 Form and function

Science as Inquiry

A.1 Abilities necessary to do scientific inquiry
A.2 Understandings about scientific inquiry

Physical Science

B.1 Structure of atoms
B.2 Structure and properties of matter
B.3 Chemical reactions
B.4 Motions and forces
B.5 Conservation of energy and increase in disorder
B.6 Interactions of energy and matter

Life Science

C.1 The cell
C.2 Molecular basis of heredity
C.3 Biological evolution
C.4 Interdependence of organisms
C.5 Matter, energy, and organization in living systems
C.6 Behavior of organisms

Earth and Space Science

D.1 Energy in the Earth system
D.2 Geochemical cycles
D.3 Origin and evolution of the Earth system
D.4 Origin and evolution of the universe

Science and Technology

E.1 Abilities of technological design
E.2 Understandings about science and technology

Science in Personal and Social Perspectives

F.1 Personal and community health
F.2 Population growth
F.3 Natural resources
F.4 Environmental quality
F.5 Natural and human-induced hazards
F.6 Science and technology in local, national, and global challenges

History and Nature of Science

G.1 Science as a human endeavor
G.2 Nature of scientific knowledge
G.3 Historical perspectives

COURSE PLANNING GUIDE

As the teacher, you are in the best position to set the pace at which the content is covered and determine what material should be given emphasis. To assist you, Glencoe has provided two Course Planning Guides—one for a Core Course that covers core areas (noted with a *) in depth, and one for a Comprehensive Course that provides instruction on a broad range of topics. Either course can be used in a full-year, two-semester program comprised of 160 single periods, or a half-year, one-semester program comprised of 80 block periods.

Core Earth Science Course

This option provides teachers with the ability to spend more time on the core areas, as indicated by the (*).

Core Course

Chapter	Single Periods (45 minutes)	Block Periods (90 minutes)
1	8	4
2*	7	3.5
3	8	4
4*	6	3
5*	7	3.5
6*	7	3.5
7*	8	4
8*	8	4
9*	8	4
10*	7	3.5
11*	9	4.5
12*	7	3.5
13*	8	4
14*	10	5
15*	7	3.5
16	8	4
17*	8	4
18*	7	3.5
19*	7	3.5
20*	8	4
21*	9	4.5
22	8	4
23	8	4
24*	6	3
25	8	4
26	11	5.5
27*	7	3.5
28*	8	4
29	7	3.5
30	12	6

* These core chapters can be taught in 160 single periods or 80 block periods.

Comprehensive Earth Science Course

This option offers a more accelerated pace and covers more material than the basic course.

Comprehensive Course

Chapter	Single Periods (45 minutes)	Block Periods (90 minutes)
1	5	2.5
2	5	2.5
3	5	2.5
4	4	2
5	5	2.5
6	3	1.5
7	6	3
8	5	2.5
9	5	2.5
10	5	2.5
11	6	3
12	5	2.5
13	6	3
14	6	3
15	5	2.5
16	6	3
17	5	2.5
18	5	2.5
19	4	2
20	6	3
21	7	3.5
22	6	3
23	6	3
24	4	2
25	5	2.5
26	7	3.5
27	4	2
28	6	3
29	5	2.5
30	8	4

SUPPLIERS

Equipment Suppliers

American Science & Surplus
P.O. Box 1030
Skokie, IL 60076
(847) 647-0011
www.sciplus.com

Bio-Rad Laboratories
Life Science Group
2000 Alfred Nobel Dr.
Hercules, CA 94547
(800) 424-6723
www.bio-rad.com

Carolina Biological Supply Co.
2700 York Rd.
Burlington, NC 27215
(800) 334-5551
www.carolina.com

Edmund Scientific Company
60 Pearce Ave.
Tonawanda, NY 14150-6711
(800) 728-6999
www.scientificsonline.com

Fisher Science Education
Educational Materials Division
4500 Turnberry Dr.
Hanover Park, IL 60133
(800) 955-1177
www.fisheredu.com

Nasco Science
901 Janesville Avenue,
P.O. Box 901
Fort Atkinson, WI 53538-0901
(800) 558-9595
www.enasco.com

Nebraska Scientific
3823 Leavenworth St.
Omaha, NE 68105-1180
(800) 228-7117
www.nebraskascientific.com

PASCO Scientific
10101 Foothills Blvd.
P.O. Box 619011
Roseville, CA 95747
(800) 772-8700
www.pasco.com

Sargent-Welch/VWR Scientific Products
P.O. Box 4130
Buffalo, NY 14217
(800) SAR-GENT
www.SargentWelch.com

Science Kit and Boreal Laboratories
777 East Park Dr.
Tonawanda, NY 14150
(800) 828-7777
www.sciencekit.com

Vernier Software & Technology
13979 SW Millikan Way
Beaverton, OR 97005-2886
(800) 837-6437
www.vernier.com

WARD's Natural Science Est.
5100 W. Henrietta Rd.
P.O. Box 92912
Rochester, NY 14692-9012
(800) 962-2660
www.wardsci.com

Audiovisual Distributors

Bullfrog Films
P.O. Box 149
Oley, PA 19547
(800) 543-FROG
www.bullfrogfilms.com

Phoenix Learning Group Coronet/MTI Film & Video
2349 Chaffee Dr.
St Louis, MO 63146
(800) 221-1274
www.phoenixlearninggroup.com

Discovery Channel School
8516 Georgia Ave.
Silver Springs, MD 20910
(240) 662-2000
www.discoveryschool.com

Films for the Humanities and Sciences
P.O. Box 2053
Princeton, NJ 08543
(800) 257-5126
www.filmsmediagroup.com

Flinn Scientific
P.O. Box 219
770 N. Raddant Rd.
Batavia, IL 60510
(800) 452-1261
www.flinnsci.com

Frey Scientific
P.O. Box 8101
100 Paragon Parkway
Mansfield, OH 44903
(800) 225-FREY
www.freyscientific.com

National Geographic Society Educational Services
1145 17th Street, N.W.
Washington, DC 20036
(800) 368-2728
www.nationalgeographic.com

Optical Data/SRA/MCGraw-Hill
12600 Deerfield Pkwy.
Alpharetta, GA 30004
(800) 201-7103
www.opticaldata.com

Scholastic, Inc.
555 Broadway
New York, NY 10012-3999
(800) 724-6527
www.scholastic.com

Videodiscovery, Inc.
920 N. 34th St.
Suite 300
Seattle, WA 98103
(800) 548-3472

Software Distributors

Boreal Laboratories, Ltd.
399 Vansickle Rd.
St. Catherines, Ontario
L25 3T4 CANADA
(800) 387-9393
www.boreal.com

Educational Activities, Inc.
1937 Grand Ave.
Baldwin, NY 11510
(800) 797-3223
www.edact.com

IBM Education
www.solutions.ibm.com/K12

J. Weston Walch, Publisher
321 Valley St.
P.O. Box 658
Portland, ME 04104-0658
(800) 341-6094
www.walch.com

Scholastic, Inc.
555 Broadway
New York, NY 10012-3999
(800) 724-6527
www.scholastic.com

SciCity
8100 Shaffer Pkwy.
Suite 130
Littleton, CO 80127
(303) 792-5615

Sunburst Technology
1550 Executive Dr.
Eglin, IL 60123
(888) 492-8817
www.SUNBURST.com

Safety in the Laboratory

The activities in *Earth Science: Geology, the Environment, and the Universe* are designed to minimize dangers in the laboratory. Careful planning and preparation as well as awareness of hazards can keep accidents to a minimum. Practice good laboratory housekeeping and management by observing these guidelines.

Personal Protection

The use of personal protection equipment is required when potentially hazardous material is present. Personal protection equipment includes eyewear, protective gloves, and laboratory aprons and coats.

Eyewear

Safety (Chemical Splash) goggles are required for science laboratory and field activities involving any hazardous chemical, which could cause damage if splashed or rubbed into the eye. Chemical Splash goggles provide eye protection from fine dusts, liquids, splashes, mists, and sprays. They also prevent splashes and sprays from body fluids or dangerous chemicals.

Safety (Chemical Splash) goggles should be large enough to protect and form a seal around the eyes. If not able to seal, goggles should contain side shields to prevent contamination to the eyes.

Eyewear should meet the ANSI Standard Z87.1-*Practice for Occupational and Educational Eye and Face Protection.* Eyewear meeting this standard will bear markings such as "Z87.1" on the frames, and the lens will be marked with the manufacturer's trademark.

Eye protection may also be provided with safety glasses. Safety glasses with side shields will not provide adequate protection from chemical splashes. They are designed primarily to protect the eyes from flying objects. Safety glasses should also have the "Z87.1 ANSI standard markings on the frames.

Protective Gloves

Gloves protect hands from heat, absorb perspiration, provide a shield from corrosive chemicals and body fluids, and prevent the transmission of microorganisms from person to person. Always check gloves to be sure that there are no tears, punctures, or holes. When removing gloves, peel the gloves off your hand, starting at the wrists and working toward the fingers. Keep the working surface of the gloves from contact with the skin during removal. (Some students and teachers are allergic to rubber or latex gloves. Alternatives must be provided in this case.)

Protective Gloves and Their Function

Glove Type	Function
Plastic	protects against light corrosives and irritants
Latex	provides protection against biological materials; should be changed as soon as they are soiled NOTE: *Some people may have an allergic reaction to latex, which can lead to serious medical problems.*
Natural Rubber	protects against electric shock and light corrosive material
Neoprene	use when working with solvents, oils, or light corrosive material
Cotton	Absorbs perspiration; wear under latex gloves.
Asbestos	insulates against heat NOTE: *Asbestos gloves are labeled with a warning about the danger of cancer. Asbestos is a known carcinogen.*

Laboratory Aprons and Coats

Laboratory aprons and coats are designed to protect clothing and skin from splashed and spilled chemicals and biological materials. They should fit the wearer properly to provide maximum protection. A laboratory coat or apron should be worn at all times in the laboratory.

Aprons are usually listed as "bib type," which are suitable for laboratory use. Aprons should be worn over clothing that covers the arms and body. Laboratory coats are usually fire retardant and made of cotton or paper. They are good for protection against flying objects, sharp or rough edges, splashes and spills, and fire. (Make sure aprons are the correct length in that they can present a trip fall hazard if too long.)

Fire Protection

Fire is one of the most frequent mishaps in the science laboratory. The first line of defense from a fire is fire prevention. Effective fire prevention centers on thorough understanding of combustion and the required ingredients. As long as air is present, oxygen will be available for combustion to take place. The areas where prevention measures are best exercised are the fuel and ignition sources.

Fires are classified by the chemical properties of the fuel. The basic classifications are grouped as follows.

- Class A — ordinary combustible (i.e., paper, wood)
- Class B — organic solvents or Flammables (i.e., acetone, alcohols, ethers)
- Class C — electrical wiring or static charges
- Class D — active metals (i.e., sodium, potassium, magnesium)

These symbols are accepted for the different classifications of fire. They are applied to fire extinguishers and extinguisher locations to indicate their suitability in extinguishing the different types of fires.

The following precautions should be taken to prevent fires from occurring in the science classroom, laboratory, storage, and preparation area.

- Be aware of ignition sources in your laboratory area (open flames, heat, and electrical equipment).
- Purchase and store flammable reagents in the smallest quantities possible.
- Do not store flammable liquids in standard refrigerators (an explosion-proof refrigerator should be used). (Use appropriate signage on the door of the refrigerator —e.g.: For laboratory chemicals only, no edible foods allowed!)
- Store flammable liquids in appropriate safety cabinets and/or safety cans.
- Do not store incompatible reagents together (e.g., acids with flammables).
- Do not store ethers for extended periods of time (no more than one year) as explosive peroxides can form.
- Make sure that all electrical cords are in good condition. All electrical outlets should be grounded and should accommodate a 3-pronged plug. (All circuits should be GFCI protected. Do not use frayed or exposed wires missing insulation.)

Each science classroom, laboratory, storage room and preparation area should have a fire blanket and an appropriate fire extinguisher.

Fire Extinguishers

In most school environments, hand-held, portable fire extinguishers are the first fire-extinguishing agent used. Therefore, a multipurpose ABC fire extinguisher must be located in each science classroom, laboratory, storage room, and preparation area. Extinguishers must be

- located in an open and highly visible area;
- inspected on a regular basis;
- used by well-trained teachers and students.

Fire extinguishers are labeled in accordance with NFPA standards. Have appropriate signage posted at levels easy to observe.

Fire Blankets

Actual fire control requires proper types of control devices such as a fire blanket. Fire blankets are made of specially treated fabric and should be located at strategic areas for all science laboratories where hazardous chemicals are stored and used. Students can use fire blankets if they are unable to reach the safety shower. Have appropriate signage posted at levels easy to observe.

Electrical Protection

Electrical safety needs must be considered for all new, old, and renovated science classrooms, laboratories, storage rooms, and preparation areas. Prevention should be the emphasis of electrical safety. Minimum considerations for electrical safety include the following.

- Ground-fault interrupters (GFI) should be installed to protect against major shock and electrical fires by preventing short circuits.
- All outlets must be grounded to prevent electrical accidents. Sufficient outlets should be provided to eliminate the need for extension cords. If floor boxes are used, they should not be located near water sources or areas where water is used.
- Surge protectors should be used to protect computers and other electronic devices from power surges.
- Emergency shut-off controls (electricity, gas, and water) should be located in an area that is easily accessible for laboratory occupants.
- Circuits should not be overloaded.
- Use only spark-free refrigeration in laboratories, storage rooms, and preparation areas for storage of flammable chemicals.
- Avoid the use of extension cords.
- Wiring should not have frayed or bare areas.

In developing the school's safety program, include provisions for handling electrical emergencies. All teachers of science should know where the master electrical cutoff switch and the control box are located, and how to operate both of these. Before an activity is conducted that requires the use of an electrical device, the teacher and students must be familiar with its operation and safety features.

Equipment and Materials List

This table of equipment and inexpensive, easily accessible materials can help you prepare your Earth science classes for the year. The numbers on the table represent the page numbers in the book. The number in parentheses is the quantity of the material used in the lab per lab group. Refer to the Chapter Organizer at the beginning of each chapter for a list of equipment and materials used for each lab in the chapter.

Non-Consumables

Item	Launch Lab	MiniLab	GeoLab
baking pan		p. 210	p. 77
balance			p. 20, 185
beaker (1000-mL)		p. 564	
beaker (100-mL)		p. 712, 740	p. 153, 429 (4)
beaker (150-mL)		p. 631	
beaker (200-mL)			p. 125
beaker (250-mL)	p. 58, 162 (2), 250 (2), 618, 706	p. 453	p. 77 (2)
beaker (500-mL)	p. 588		p. 429
beaker (600-mL)	p. 498	p. 416	
beam-balance			p. 153
box (small)		p. 505	
brick	p. 646		
calculator			p. 305, 429, 456, 490, 553, 699, 821
cardboard box		p. 394	p. 725
catch basin		p. 172	
clamp		p. 505	p. 243
clipboard	p. 828		
colored pencils	p. 28		p. 490, 519, 611, 639
containers	p. 192 (3), 560	p. 12 (3)	p. 243
cylinder (large)			p. 20
drafting compass			p. 456, 553
dropper	p. 646		p. 102, 429
flashlight		p. 315	p. 725
funnel		p. 740	
glass		p. 295, 631	p. 725
glass plate			p. 102
globe	p. 404	p. 32, 481	
granite	p. 110		
grease pencil			p. 243
hose			p. 243

Non-Consumables

Item	Launch Lab	MiniLab	GeoLab
hot plate	p. 706		p. 77, 125
ice cube tray		p. 362	
jar (200-mL, with lid)		p. 136	
jar (baby food)		p. 695 (6)	
jar (glass)		p. 394	
jar (plastic, with lid)			p. 185
magnet	p. 58		p. 102
magnifying glass	p. 58, 84, 110		p. 102, 125, 153
map of U.S.			p. 553
marker (low VOC)	p. 404		p. 505, 821
marker (felt-tip)		p. 873	
meter stick		p. 843, 873	p. 214, 243, 397, 821
microscope	p. 84, 436		
microscope slide	p. 84		
mirror			p. 725
mortar and pestle	p. 436		
newspaper		p. 505	
paper punch	p. 342		p. 243
pencil	p. 58, 312		p. 578
petri dish		p. 631	p. 125
plastic bowl	p. 280		
plastic container	p. 222 (2)	p. 240 (3), 265	
plumb bob			p. 243
protractor		p. 92, 776, 843	p. 243
psychrometer			p. 397
quartz crystal	p. 84		
ring stand			p. 243
rock	p. 374 (2)		p. 20 (5), 639, 725
rolling pin	p. 58		
ruler	p. 404, 466, 560	p. 597, 801	p. 20, 48, 270, 335, 429, 456, 490, 553, 578, 639, 752, 786, 853
scale			p. 77, 429
scale (spring)			p. 20
scissors	p. 762	p. 505, 597	p. 305, 490, 725
sieve		p. 453 (5)	
sponge	p. 588		
steel file			p. 102

Non-Consumables

Item	Launch Lab	MiniLab	GeoLab
stirring rod	p. 162, 588		p. 77
streak plate			p. 102
tape measure			p. 821
telescope	p. 828		
temperature strip	p. 312		p. 305
thermometer		p. 12 (2), 295, 394 (2)	p. 125, 397, 429, 725
thumbtacks	p. 526 (10)		
timer	p. 162	p. 453	p. 185, 243, 305
tray	p. 192		
umbrella	p. 374		
wind sock			p. 397
window screen	p. 222 (2)		

Consumables

Item	Launch Lab	MiniLab	GeoLab
aluminum foil			p. 725 (30 x 60 cm)
aquarium gravel		p. 712 (30 mL)	
balloon	p. 342	p. 505, 873	
bleach		p. 631 (15 mL)	
calcium chloride		p. 416 (1.1 g)	
cardboard		p. 801 (30 cm^2)	
cereal	p. 58 (250 g)		
chalk (natural)	p. 436		
clay	p. 250 (250 mL)	p. 240 (4500 mL), 265 (1500 mL)	
coffee filter		p. 740	
construction paper	p. 374 (2)		p. 125
cooking oil		p. 453 (250 mL), 712 (20 mL)	
copper			p. 102
corn syrup		p. 564 (1000 mL)	
dishwasher detergent bottle		p. 362	
distilled water		p. 416 (965.57 g), 695 (20 mL)	p. 77 (150 mL)
fabric	p. 342 (30 x 60 cm)		p. 725 (30 x 60 cm)
flour		p. 505 (6 cups)	
food coloring			p. 429 (2 mL)
glue			p. 725 (10 mL)
graph paper	p. 28		p. 20, 270, 429, 456, 578
gravel		p. 240 (4500 mL)	
halite			p. 77 (54 g)
halite chip			p. 185 (100 g)

Consumables

Item	Launch Lab	MiniLab	GeoLab
hydrochloric acid (dilute)			p. 102 (1 mL)
ice cubes	p. 280, 312	p. 295 (10)	
ice with sediment		p. 210 (4)	
igneous rocks		p. 115	
limestone			p. 153
liquid soap		p. 695 (10 mL)	
magnesium chloride		p. 416 (4.98 g)	
marble			p. 153
masking tape		p. 12 (10 cm)	p. 821 (50 cm)
milk	p. 618 (50 mL)		
milk carton	p. 588		
minerals		p. 92 (5)	p. 102
modeling clay		p. 653 (250 g)	
paint			p. 725 (60 mL per color)
paper	p. 762, 828	p. 315, 541, 801	p. 48, 153, 243, 490, 639, 725, 786
paper (large)		p. 481	
paper clip			p. 102
paper towels			p. 125 (10), 185 (10)
pins		p. 801 (2)	
plastic bag	p. 58, 280		
plastic cup		p. 564	
plastic straw		p. 265 (3)	p. 305 (4)
plastic wrap	p. 280 (30 cm²)		p. 77 (30 cm²)
polystyrene			p. 725
popsicle stick		p. 210	
poster board			p. 611
potassium bromide		p. 416 (0.1 g)	
potassium chloride		p. 416 (0.66 g)	
quartzite			p. 153
rubber tubing		p. 505 (10 cm)	
salt	p. 84 (10 g), 498 (10 g), 588 (100 mL)		p. 429 (7.5 g)
sand	p. 192 (225 mL), 250 (500 mL), 588 (500 mL)	p. 240 (4500 mL), 265 (3 kg), 712 (3 kg), 740 (3000 mL)	
sandpaper	p. 526 (2)		
sandstone	p. 646		p. 153
saturated alum			p. 125 (150 mL)
sediment		p. 136 (100 mL), 453 (5)	
shale			p. 153
slate			p. 153
soap (bar)		p. 172	
sod	p. 222 (8 cm x 16 cm)		

Consumables			
Item	**Launch Lab**	**MiniLab**	**GeoLab**
sodium bicarbonate		p. 416 (0.19 g)	
sodium chloride		p. 416 (23.48 g)	
sodium sulfate		p. 416 (3.92 g)	
soil	p. 222 (8 cm x 16 cm)		
steel wool		p. 631 (2 g)	
string	p. 404 (1 m)	p. 801 (1 m), 843 (1 m), 873 (1 m)	p. 20 (1 m), 48 (1 m)
sugar (granulated)	p. 162 (5 mL)		
sugar cube	p. 162 (5 mL)		
tape		p. 505 (50 cm)	p. 305 (10 cm), 725 (10 cm)
toothpick		p. 172	
transparent paper			p. 270
vegetable oil	p. 498 (80 mL), 618 (50 mL), 646 (300 mL)		
vinyl gutter pipe			p. 243 (1m)
water	p. 58 (150 mL), 162 (500 mL), 192 (220 mL), 222 (1000 mL), 250 (750 mL), 280 (125 mL), 498 (300 mL), 560, 588 (250 mL), 618 (50 mL), 646 (100 mL), 706 (200 mL)	p. 12, 136 (150 mL), 240 (1500 mL), 265, 295 (350 mL), 362 (1000 mL), 631 (80 mL), 695 (20 mL), 712 (30 mL), 740 (50 mL)	p. 21 (1000 mL), 125 (150 mL), 153, 185 (300 mL), 429 (400 mL)
white sand		p. 631 (40 mL)	
wood blocks	p. 526 (2), 560 (3)		

CONTENTS IN BRIEF

Online Guide iv

Earth Science

1 The Nature of Science 4
2 Mapping Our World 28

Composition of Earth

3 Matter and Change 58
4 Minerals 84
5 Igneous Rocks 110
6 Sedimentary and Metamorphic Rocks 132

Surface Processes on Earth

7 Weathering, Erosion, and Soil 162
8 Mass Movements, Wind, and Glaciers 192
9 Surface Water 222
10 Groundwater 250

The Atmosphere and the Oceans

11 Atmosphere 280
12 Meteorology 312
13 The Nature of Storms 342
14 Climate 374
15 Earth's Oceans 404
16 The Marine Environment 436

The Dynamic Earth

17 Plate Tectonics 466
18 Volcanism 498
19 Earthquakes 526
20 Mountain Building 560

Geologic Time

21 Fossils and the Rock Record 588
22 The Precambrian Earth 618
23 The Paleozoic, Mesozoic, and Cenozoic Eras 646

Resources and the Environment

24 Earth Resources 676
25 Energy Resources 706
26 Human Impact on Resources 732

Beyond Earth

27 The Sun-Earth-Moon System 762
28 Our Solar System 794
29 Stars 828
30 Galaxies and the Universe 860

Student Resources 890

Stocktrek Images, Inc./Alamy

A² DEVELOPMENT PROCESS

Why A²?

Accuracy Assurance is central to McGraw-Hill's commitment to high-quality, learner-oriented, real-world, and error-free products. Also at the heart of our A² Development Process is a commitment to make the text **Accessible** and **Approachable** for both students and teachers. A collaboration among authors, content editors, academic advisors, and classroom teachers, the A² Development Process provides opportunities for continual improvement through customer feedback and thorough content review.

The A² Development Process begins with a review of the previous edition and a look forward to state and national standards. The authors for *Earth Science: Geology, the Environment, and the Universe* combine expertise in teacher training and education with a mastery of science content knowledge. As manuscript is created and edited, consultants review the accuracy of the content while our Teacher Advisory Board members examine the program from the points of view of both teacher and student. Student labs and teacher demonstrations are reviewed for both accuracy of content and safety. As design elements are applied, chapter content is again reviewed, as are photos and diagrams.

Throughout the life of the program, Glencoe/McGraw-Hill continues to troubleshoot and incorporate improvements. Our goal is to deliver to you a program that has been created, refined, tested, and validated as a successful tool for your continued **Academic Achievement**.

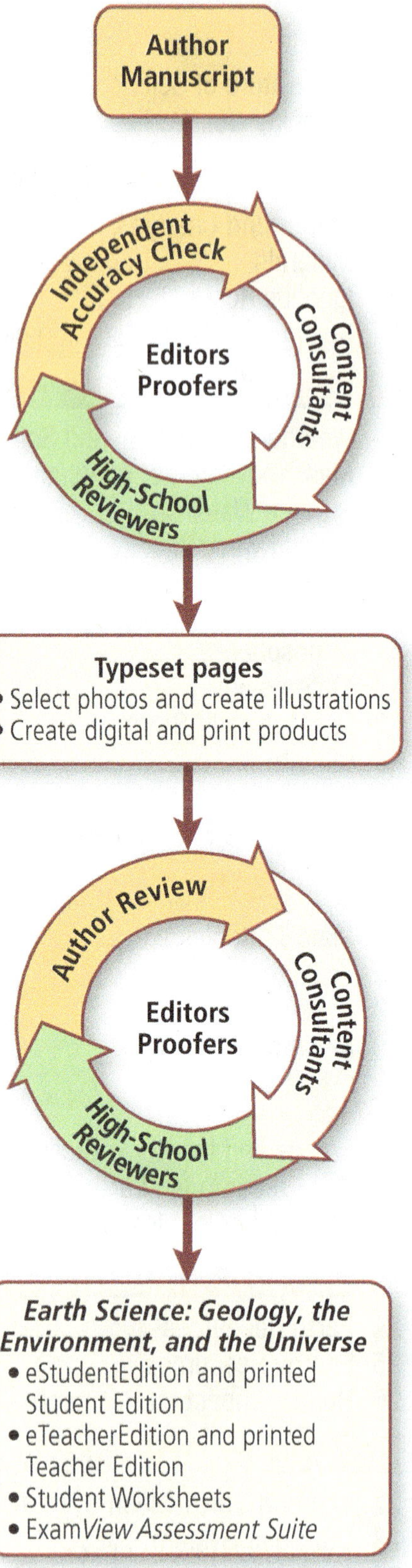

Authors

The authors of *Earth Science: Geology, the Environment, and the Universe* used their content knowledge and teaching expertise to craft manuscript that is accessible and accurate, geared toward student achievement.

Dr. Francisco Borrero

is a research associate and Adjunct Curator of Mollusks at Cincinnati Museum Center in Cincinnati, Ohio. He has taught Earth science and Spanish for over 20 years, most recently at the Cincinnati Country Day School. Dr. Borrero holds a B.S. in zoology from Universidad del Valle, Colombia, and M.S. and Ph.D. degrees in biological sciences from the University of South Carolina at Columbia. Dr. Borrero's research examines the relationship between physical habitat characteristics and the diversity and distribution of natural populations of mollusks.

Dr. Frances Scelsi Hess

is an educational consultant following many years of teaching Earth science and advance placement environmental science at Cooperstown High School in New York. She received her B.S. and M.S. in science education from the State University at Oneonta, and her Ed.D. from Columbia University. Dr. Hess is a Fellow of the Science Teachers Association of New York State, and has received numerous teaching awards, including the Phi Delta Kappa Reed Travel Scholarship to Australia and New Zealand, the Presidential Award for Excellence in Science Teaching, and National Board Certification.

Dr. Chia Hui (Juno) Hsu

currently works as a project scientist at University of California, Irvine. She holds a B.S. in physics and Earth science from National Taiwan Normal University, an M.S. in atmospheric sciences from National Taiwan University, and a PhD in atmospheric sciences from Massachusetts Institute of Technology. Before beginning her graduate work, Dr. Hsu taught ninth-grade Earth science. Her research interests include the dynamics of monsoons, climate regime shifts, and modeling global-scale atmospheric chemistry.

Dr. Gerhard Kunze

is professor emeritus of geology at the University of Akron in Ohio. He has a B.S. in science and a Ph.D. in geophysics from Penn State University. He was an NRC research associate at Johnson Space Center, Houston, Texas from 1973 to 1974. In 1990, Dr. Kunze was awarded a senior Fulbright scholarship to teach geophysics at the Institute of Geophysics, a department of the University of Kiel in Germany. His current research interests include engineering geophysical surveys and digital modeling of geophysical anomalies.

Dr. Stephen A. Leslie

is the professor and department head of geology and environmental science at James Madison University in Harrisburg, Virginia. He was formerly an associate professor of geology at the University of Arkansas in Little Rock. His areas of research include paleontology, stratigraphy, and the evolution of early life on Earth. He has a B.S. in geology from Bowling Green State University, an M.S. in geology from the University of Idaho, and a Ph.D. in geology from The Ohio State University.

(t to b)courtesy of Francisco Borrero, courtesy of Frances Scelsi Hess, courtesy of Juno Hsu, courtesy of Gerhard Kunze, courtesy of Stephen Leslie

A² DEVELOPMENT PROCESS

Stephen Letro

has been a meteorologist for the National Weather Service, the media, and private industry since 1971. He currently serves as the Meteorologist-in-Charge of the National Weather Service office in Jacksonville, Florida. He received his B.S. in meteorology from Florida State University with an emphasis on tropical meteorology. He is a member of the National Hurricane Center's Hurricane Liaison Team, and has received numerous awards, including an award for his role in restructuring the National Weather Service.

Dr. Michael Manga

is a professor of Earth and planetary science at U.C. Berkeley. He has a B.S. in geophysics from McGill University and a Ph.D. in Earth science from Harvard University. His areas of research include volcanology, the internal evolution and dynamics of planets, and hydrogeology. He is a MacArthur Fellow, and has received the Donath medal from the Geological Society of America and the Macelwane medal from the American Geophysical Union.

Len Sharp

taught Earth science at Liverpool High School, New York, for 30 years. He has a B.S. in secondary education and an M.S. in science education from Syracuse University. Mr. Sharp was president of the Science Teachers Association of New York from 1991 to 1992, and president of the National Earth Science Teachers Association from 1992 to 1994. He was a Presidential Awardee in 1995, and received the 2005 Distinguished Teacher Award from NSTA and the 2006 NAGT—Eastern Section, Outstanding Earth Science Teacher award.

Dr. Theodore Snow

is a professor of astronomy at the University of Colorado. He has a B.A. from Yale University, and an M.S. and Ph.D. in astrophysics from the University of Washington. Dr. Snow is a founder and former director of the Center for Astrophysics and Space Astronomy at the University of Colorado. He has led instrument-development programs for space-based telescopes, and was a member of the Science Team for an ultraviolet spectrograph that was installed aboard the *Hubble Space Telescope* in 2008. Dr. Snow's research examines the gas and dust between the stars, called diffuse interstellar bands (DIBs).

Dinah Zike

is an international curriculum consultant and inventor who has developed educational products and three-dimensional, interactive graphic organizers for over 30 years. As president and founder of Dinah-Might Adventures, L.P., Dinah is the author of more than 100 award-winning educational publications, including *The Big Book of Science.* Dinah has a BS and an MS in educational curriculum and instruction from Texas A&M University. Dinah Zike's *Foldables* are an exclusive feature of McGraw-Hill textbooks.

(t to b)courtesy of Stephen Letro, courtesy of Michael Manga, courtesy of Len Sharp, courtesy of Theodore Snow, Dinah Zike

Teacher Advisory Board

The Teacher Advisory Board gave the editorial staff and design team feedback on the content and design of both the Student Edition and Teacher Edition. We thank these teachers for their hard work and creative suggestions.

Francisco Borrero
Cincinnati Country Day High School
Cincinnati, OH

Bill Brown
Grandview Heights High School
Columbus, OH

Carmen S. Dixon
East Knox High School
Howard, OH

Joel Heuberger
Waite High School
Toledo, OH

Jane Karabaic
Steubenville City Schools
Steubenville, OH

Terry Stephens
Edgewood High School
Trenton, OH

Teacher Reviewers

Each teacher reviewed selected chapters of *Earth Science: Geology, the Environment, and the Universe* and provided feedback and suggestions regarding the effectiveness of the instruction.

Mark Brazo
Lincoln High School
Portland, OR

Gayle R. Dawson
Blackman High School
Murfreesboro, TN

William Dicks
Northville High School
Northville, MI

Alvin Echeverria
Del Sol High School
Las Vegas, NV

Wendy Elkins
Blue Valley Northwest High School
Overland Park, KS

Carolyn C. Elliot
South Iredell High School
Statesville, NC

Sandra Forster-Terrell
Atherton High School
Louisville, KY

Carol L. Jarocha
Northville High School
Northville, MI

Steve Kluge
Fox Lane High School
Bedford, NY

Sussan Nwabunachi Oladipo
Wells Academy High School
Chicago, IL

Michael J. Passow
White Plains Middle School
White Plains, NY

Jeremy Richardson
Lewis and Clark High School
Spokane, WA

Angela Jones Rizzo
AC Flora High School
Columbia, SC

Terry A. Stephens
Edgewood High School
Trenton, OH

DEVELOPMENT PROCESS

Content Consultants

Content consultants each reviewed selected chapters of *Earth Science: Geology, the Environment, and the Universe* for content accuracy and clarity.

Anastasia Chopelas, PhD
Research Professor of Earth and Space Sciences
University of Washington
Seattle, WA

Diane Clayton, PhD
University of California at Santa Barbara
Santa Barbara, CA

Sarah Gille, PhD
Associate Professor
Scripps Institution of Oceanography and Department of Mechanical and Aerospace Engineering
University of California San Diego
San Diego, CA

Alan Gishlick, PhD
National Center for Science Education
Oakland, CA

Janet Herman, PhD
Professor and Director of Program of Interdisciplinary Research in Contaminant Hydrogeology
University of Virginia
Charlottesville, VA

David Ho, PhD
Storke-Doherty Lecturer & Doherty Associate Research Scientist
Lamont-Doherty Earth Observatory
Columbia University
New York, NY

Jose Miguel Hurtado, PhD
Associate Professor of Geology
University of Texas at El Paso
El Paso, TX

Monika Kress, PhD
Assistant Professor of Physics and Astronomy
San Jose State University
San Jose, CA

Amy Leventer, PhD
Associate Professor of Geology
Colgate University
Hamilton, NY

Amala Mahadevan, PhD
Associate Research Professor
Department of Earth Sciences
Boston University
Boston, MA

Nathan Niemi, PhD
Assistant Professor of Geological Sciences
University of Michigan
Ann Arbor, MI

Anne Raymond, PhD
Professor of Geology and Geophysics
Texas A&M University
College Station, TX

Safety Consultant

The safety consultant reviewed labs and lab materials for safety and implementation.

Kenneth Russell Roy, Ph.D.
Director of Environmental Health and Safety
Glastonbury Public Schools
Glastonbury, CT

Digital Light Source, Richard Hutchings/McGraw-Hill Education

Introducing iLab Station

At connectED.mcgraw-hill.com, you will find this one-stop online resource for laboratory investigations, procedures, and worksheets. Look for the iLab Station button throughout your eStudentEdition for links to these laboratory experiences:

- LaunchLABs
- MiniLABs
- GeoLABs

Online, editable lab worksheets from the iLab Station will let your teacher tailor the content to your specific needs.

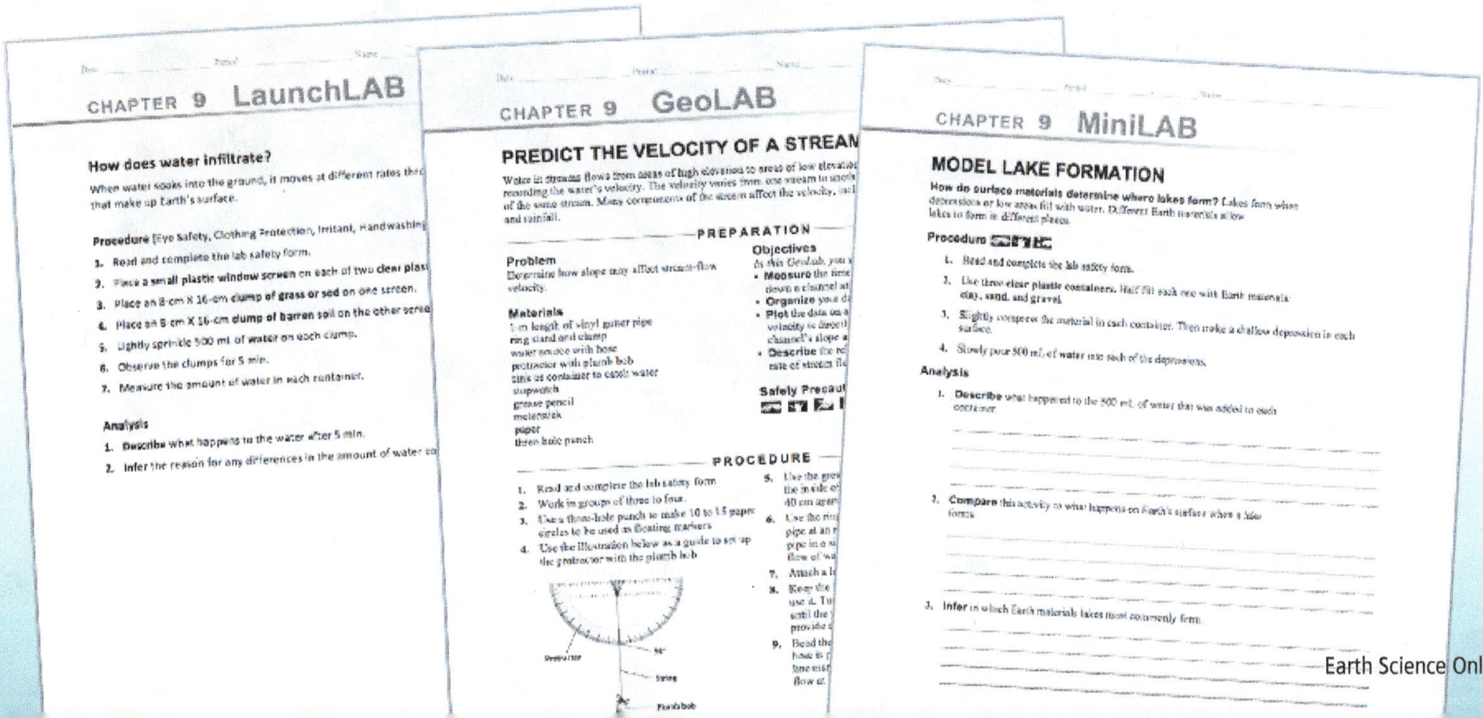

ANIMATIONS & STUDY TOOLS

Animations, assessments, and study tools provide opportunities for self-assessment, review, and additional practice.

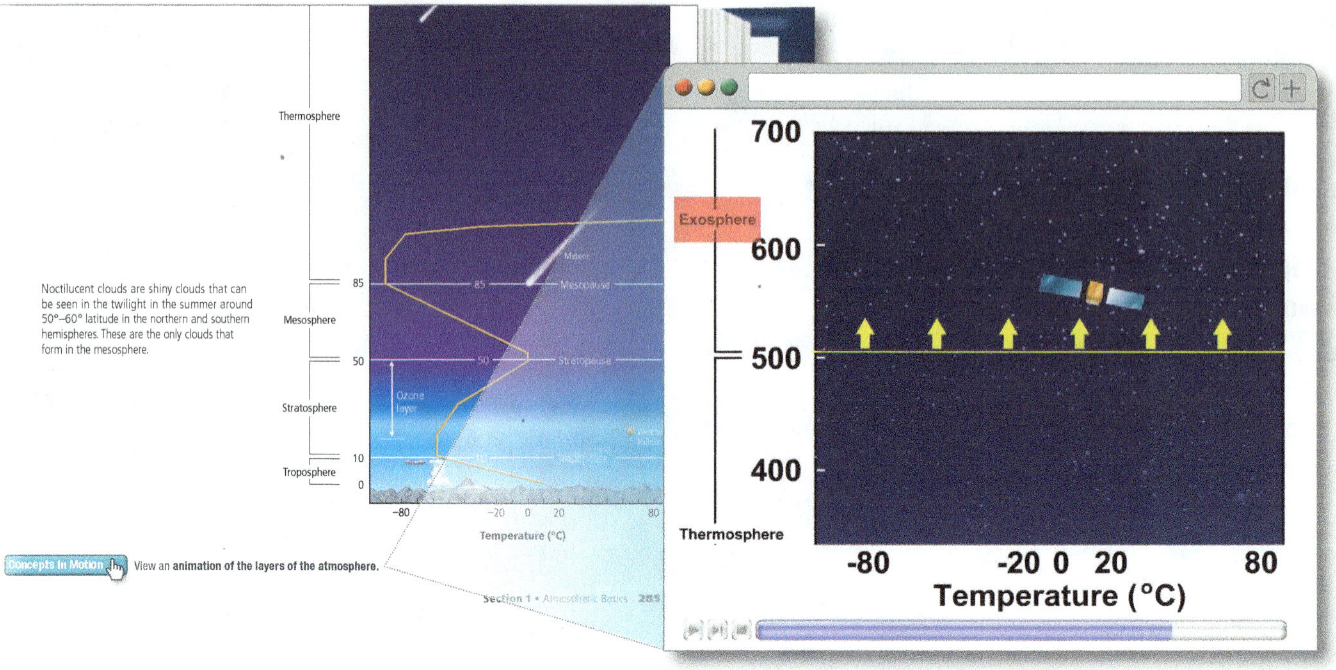

Concepts In Motion

See earth science content come to life in **animated figures** and moving diagrams.

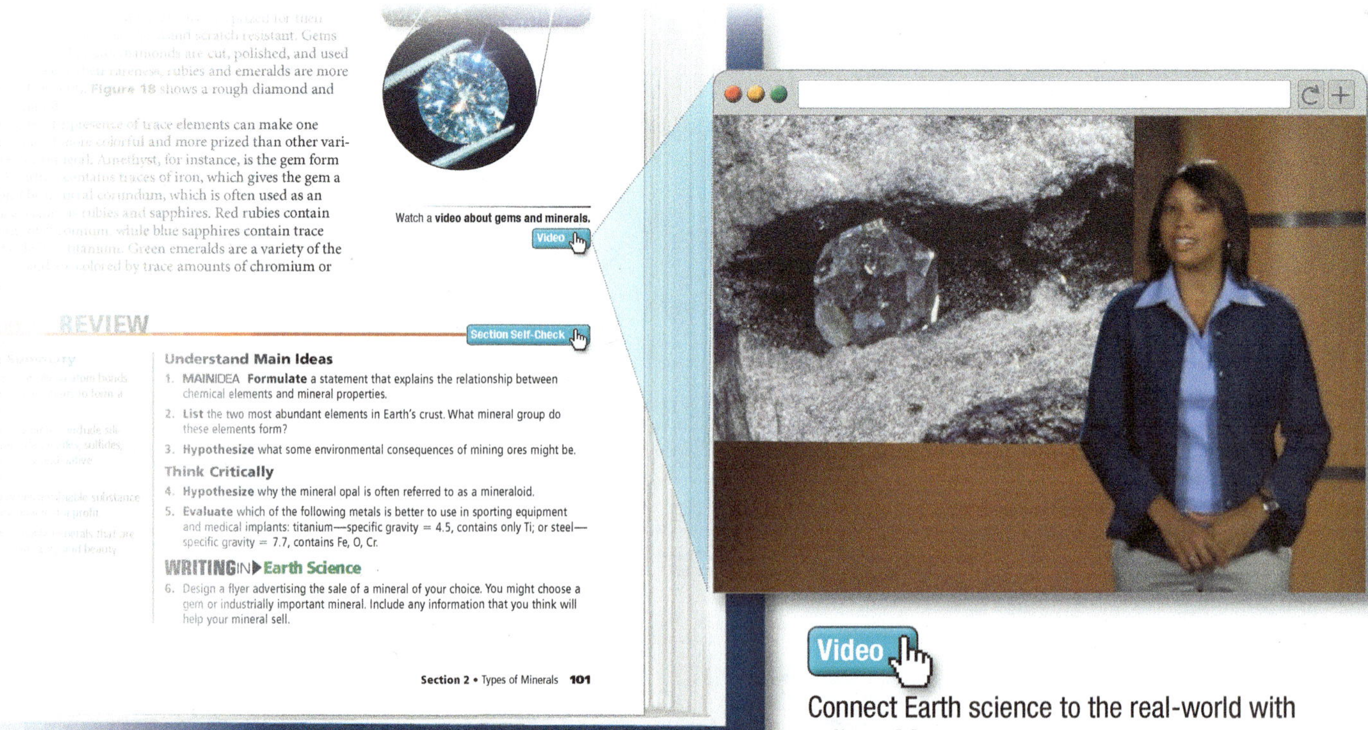

Video

Connect Earth science to the real-world with online **videos**.

(l)Image State/Alamy Images, (r)©Doug Sherman/Geofile

Go online!
connectED.mcgraw-hill.com

"The e-book includes **VIDEOS**, **ANIMATIONS**, and tools to help me **LEARN**."

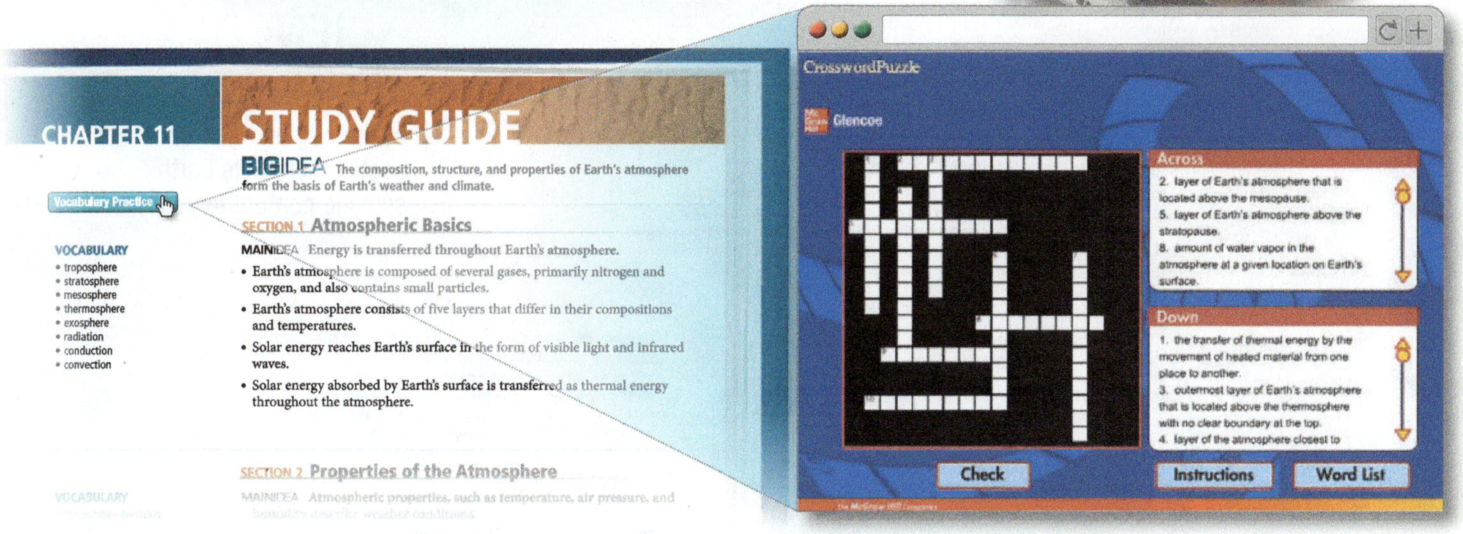

Vocabulary Practice

The **multilingual e-Glossary** and **vocabulary study tools** drive home important concepts.

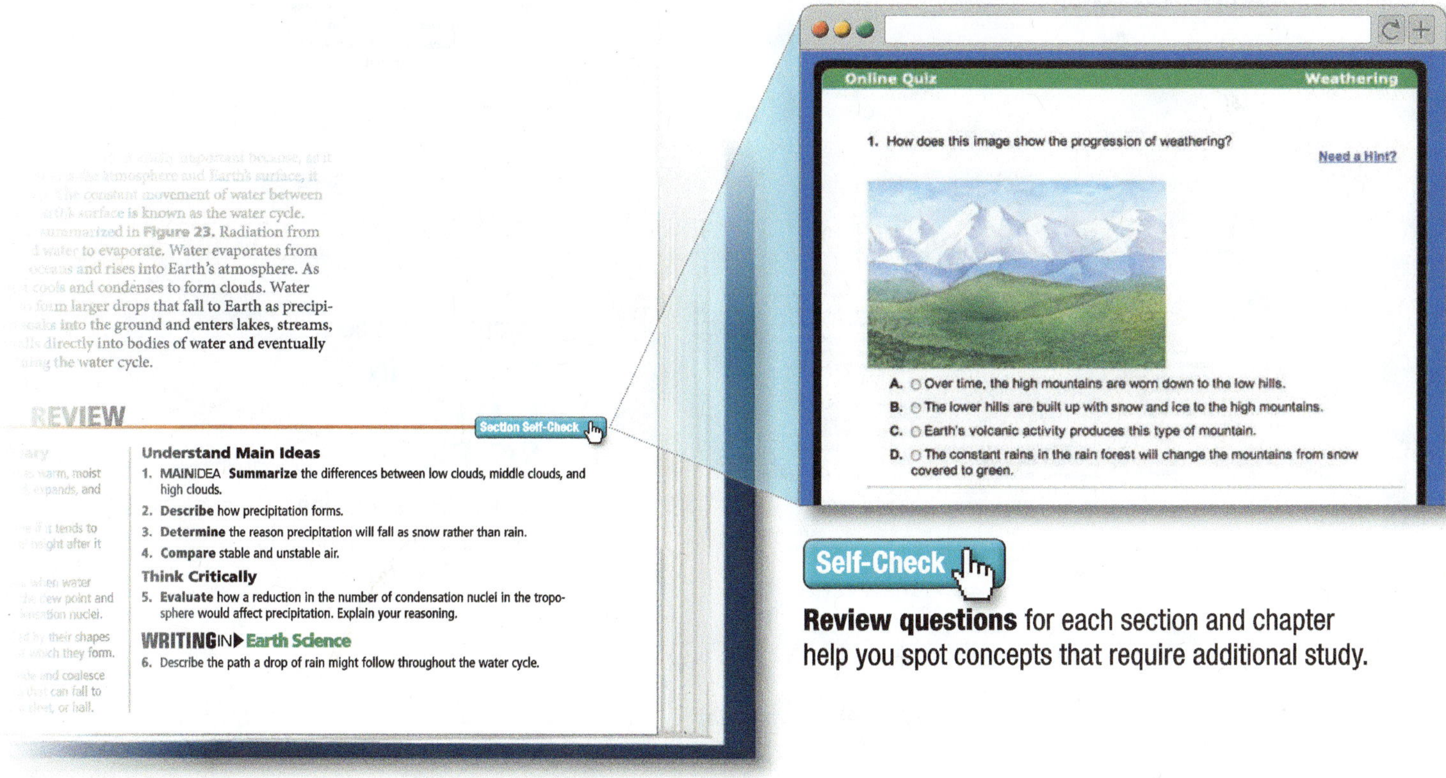

Self-Check

Review questions for each section and chapter help you spot concepts that require additional study.

Ocean/CORBIS

REAL-WORLD SCIENCE

Earth Science: Geology, the Environment, and the Universe **connects Earth science to your world. Throughout the text, find personal science connections, surprising examples of Earth science in careers, and connections to the environment.**

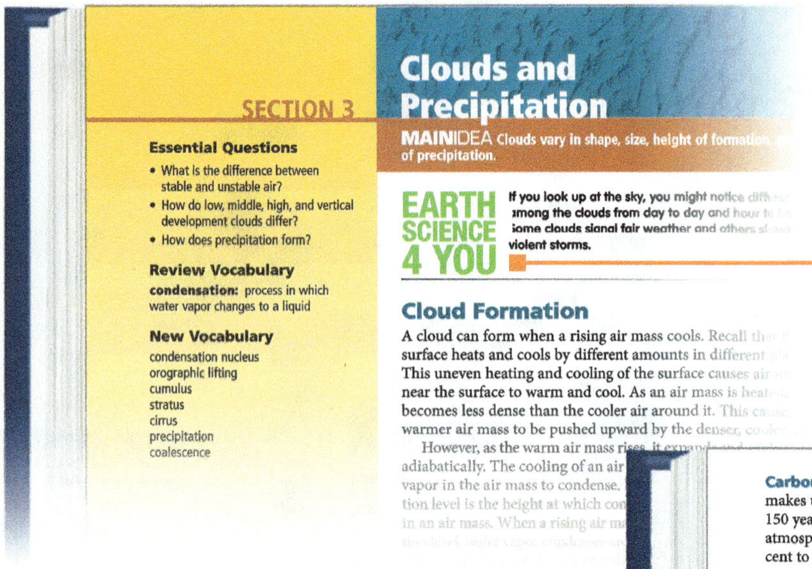

SECTION 3

Clouds and Precipitation

MAINIDEA Clouds vary in shape, size, height of formation, and type of precipitation.

Essential Questions

- What is the difference between stable and unstable air?
- How do low, middle, high, and vertical development clouds differ?
- How does precipitation form?

Review Vocabulary

condensation: process in which water vapor changes to a liquid

New Vocabulary

condensation nucleus
orographic lifting
cumulus
stratus
cirrus
precipitation
coalescence

EARTH SCIENCE 4 YOU If you look up at the sky, you might notice differences among the clouds from day to day and hour to hour. Some clouds signal fair weather and others signal violent storms.

Cloud Formation

A cloud can form when a rising air mass cools. Recall that Earth's surface heats and cools by different amounts in different places. This uneven heating and cooling of the surface causes air near the surface to warm and cool. As an air mass is heated, it becomes less dense than the cooler air around it. This causes the warmer air mass to be pushed upward by the denser, cooler air.

However, as the warm air mass rises, it expands and cools adiabatically. The cooling of an air mass causes water vapor in the air mass to condense. The condensation level is the height at which condensation occurs in an air mass. When a rising air mass ...

EARTH SCIENCE 4 YOU

at the beginning of each section connects Earth science content to your life.

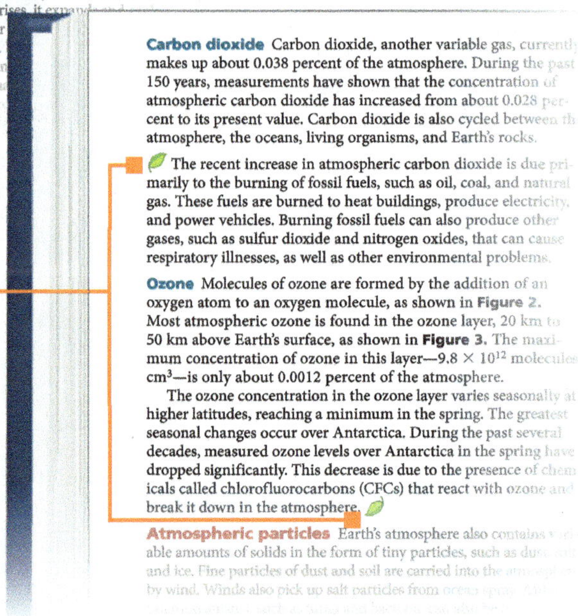

Carbon dioxide Carbon dioxide, another variable gas, currently makes up about 0.038 percent of the atmosphere. During the past 150 years, measurements have shown that the concentration of atmospheric carbon dioxide has increased from about 0.028 percent to its present value. Carbon dioxide is also cycled between the atmosphere, the oceans, living organisms, and Earth's rocks.

The recent increase in atmospheric carbon dioxide is due primarily to the burning of fossil fuels, such as oil, coal, and natural gas. These fuels are burned to heat buildings, produce electricity, and power vehicles. Burning fossil fuels can also produce other gases, such as sulfur dioxide and nitrogen oxides, that can cause respiratory illnesses, as well as other environmental problems.

Ozone Molecules of ozone are formed by the addition of an oxygen atom to an oxygen molecule, as shown in **Figure 2.** Most atmospheric ozone is found in the ozone layer, 20 km to 50 km above Earth's surface, as shown in **Figure 3.** The maximum concentration of ozone in this layer—9.8×10^{12} molecules/cm^3—is only about 0.0012 percent of the atmosphere.

The ozone concentration in the ozone layer varies seasonally at higher latitudes, reaching a minimum in the spring. The greatest seasonal changes occur over Antarctica. During the past several decades, measured ozone levels over Antarctica in the spring have dropped significantly. This decrease is due to the presence of chemicals called chlorofluorocarbons (CFCs) that react with ozone and break it down in the atmosphere.

Atmospheric particles Earth's atmosphere also contains variable amounts of solids in the form of tiny particles, such as dust and ice. Fine particles of dust and soil are carried into the atmosphere by wind. Winds also pick up salt particles from ocean spray.

Environmental Connections

point out paragraphs that emphasize real-world environmental applications of Earth science.

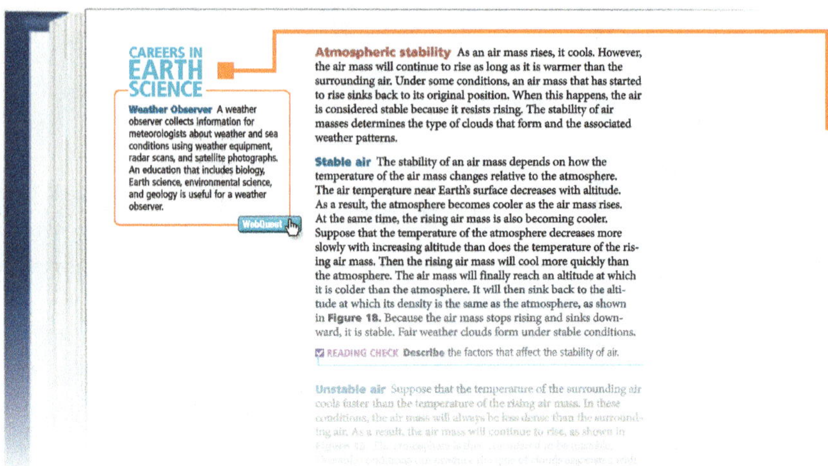

CAREERS IN EARTH SCIENCE

Weather Observer A weather observer collects information for meteorologists about weather and sea conditions using weather equipment, radar scans, and satellite photographs. An education that includes biology, Earth science, environmental science, and geology is useful for a weather observer.

WebQuest

Atmospheric stability As an air mass rises, it cools. However, the air mass will continue to rise as long as it is warmer than the surrounding air. Under some conditions, an air mass that has started to rise sinks back to its original position. When this happens, the air is considered stable because it resists rising. The stability of air masses determines the type of clouds that form and the associated weather patterns.

Stable air The stability of an air mass depends on how the temperature of the air mass changes relative to the atmosphere. The air temperature near Earth's surface decreases with altitude. As a result, the atmosphere becomes cooler as the air mass rises. At the same time, the rising air mass is also becoming cooler. Suppose that the temperature of the atmosphere decreases more slowly with increasing altitude than does the temperature of the rising air mass. Then the rising air mass will cool more quickly than the atmosphere. The air mass will finally reach an altitude at which it is colder than the atmosphere. It will then sink back to the altitude at which its density is the same as the atmosphere, as shown in **Figure 18.** Because the air mass stops rising and sinks downward, it is stable. Fair weather clouds form under stable conditions.

READING CHECK **Describe** the factors that affect the stability of air.

Unstable air Suppose that the temperature of the surrounding air cools faster than the temperature of the rising air mass. In these conditions, the air mass will always be less dense than the surrounding air. As a result, the air mass will continue to rise, as shown in ...

Throughout the book,

CAREERS IN EARTH SCIENCE

demonstrates how the chapter content applies to real-world careers.

End-of-chapter features highlight Earth science as it applies to careers, how it connects to the real world, and what today's scientists are doing to learn more about the planet.

eXpeditions!

Get an inside look at exciting places and scientists doing real-world Earth science investigations.

Earth Science & TECHNOLOGY

Discover recent technological advancements that have influenced the field of Earth science.

Earth Science & ENVIRONMENT

Explore the environmental issues that Earth scientists are working to understand and address.

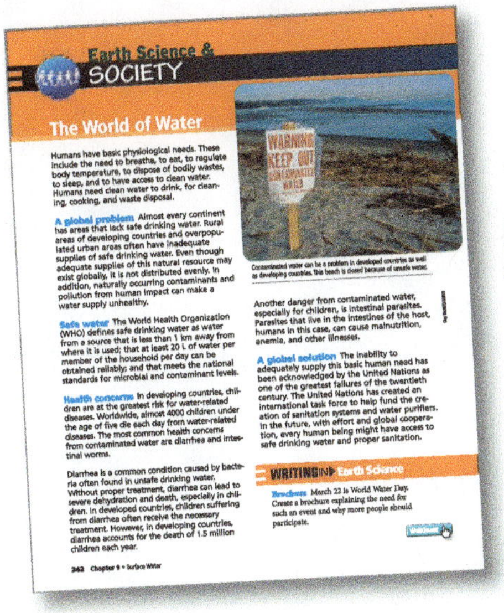

Earth Science & SOCIETY

Learn about Earth science in the news and sharpen your debating skills on complex issues in Earth science.

UNDERSTANDING EARTH SCIENCE

At the start of each chapter, you will see the BIGIDEA that will help you understand how what you are about to investigate fits into the big picture of science.

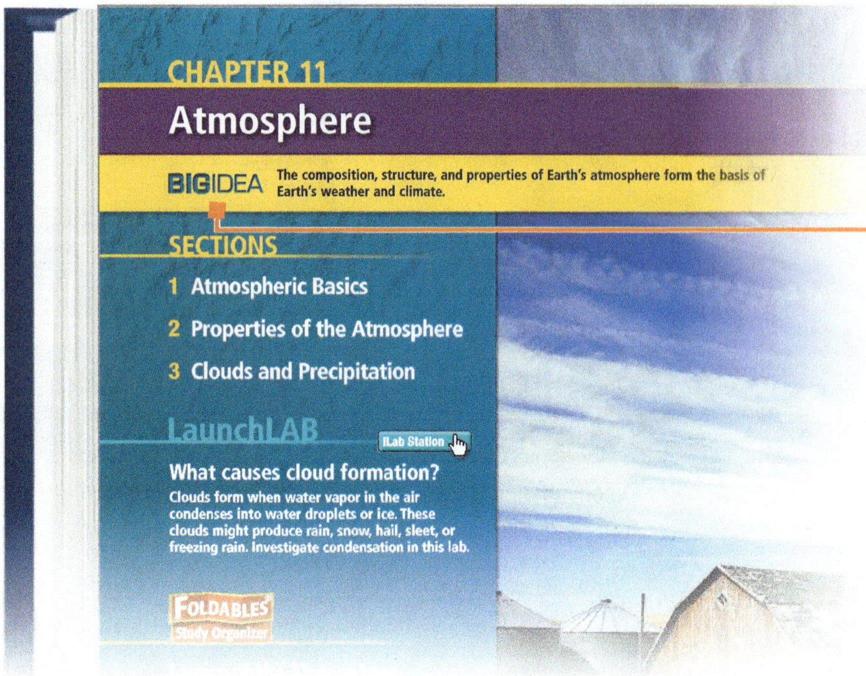

CHAPTER 11

Atmosphere

BIGIDEA The composition, structure, and properties of Earth's atmosphere form the basis of Earth's weather and climate.

SECTIONS

1 Atmospheric Basics

2 Properties of the Atmosphere

3 Clouds and Precipitation

LaunchLAB

iLab Station

What causes cloud formation?

Clouds form when water vapor in the air condenses into water droplets or ice. These clouds might produce rain, snow, hail, sleet, or freezing rain. Investigate condensation in this lab.

FOLDABLES Study Organizer

The BIGIDEA is the focus of the chapter. The labs, text, and other chapter content will build an in-depth understanding of these major concepts.

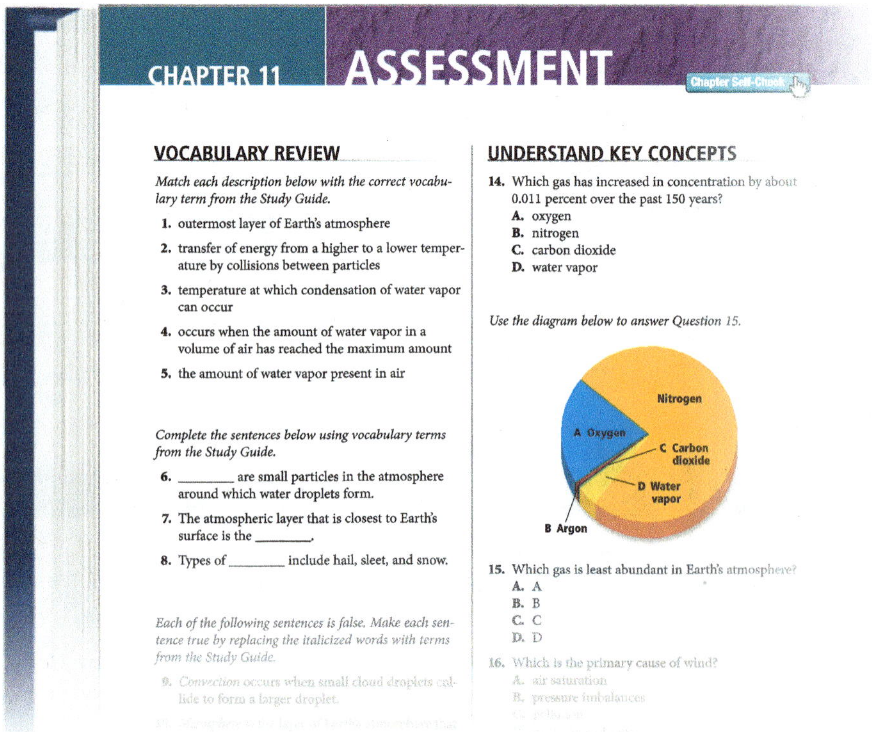

CHAPTER 11 ASSESSMENT

Chapter Self-Check

VOCABULARY REVIEW

Match each description below with the correct vocabulary term from the Study Guide.

1. outermost layer of Earth's atmosphere
2. transfer of energy from a higher to a lower temperature by collisions between particles
3. temperature at which condensation of water vapor can occur
4. occurs when the amount of water vapor in a volume of air has reached the maximum amount
5. the amount of water vapor present in air

Complete the sentences below using vocabulary terms from the Study Guide.

6. ________ are small particles in the atmosphere around which water droplets form.
7. The atmospheric layer that is closest to Earth's surface is the ________.
8. Types of ________ include hail, sleet, and snow.

Each of the following sentences is false. Make each sentence true by replacing the italicized words with terms from the Study Guide.

9. *Convection* occurs when small cloud droplets collide to form a larger droplet.

UNDERSTAND KEY CONCEPTS

14. Which gas has increased in concentration by about 0.011 percent over the past 150 years?
    A. oxygen
    B. nitrogen
    C. carbon dioxide
    D. water vapor

Use the diagram below to answer Question 15.

15. Which gas is least abundant in Earth's atmosphere?
    A. A
    B. B
    C. C
    D. D

16. Which is the primary cause of wind?
    A. air saturation
    B. pressure imbalances

The Chapter Assessment will help you evaluate your understanding of the BIGIDEA.

At the start of each section, you will find a reading preview that summarizes what you will learn while exploring the section.

The **MAIN**IDEA is the core concept covered in the section. Together, the Main Ideas from all the sections in the chapter support the chapter's Big Idea.

Essential Questions reflect the important goals of the section. Together, an understanding of these questions will lead toward understanding of the section's Main Idea.

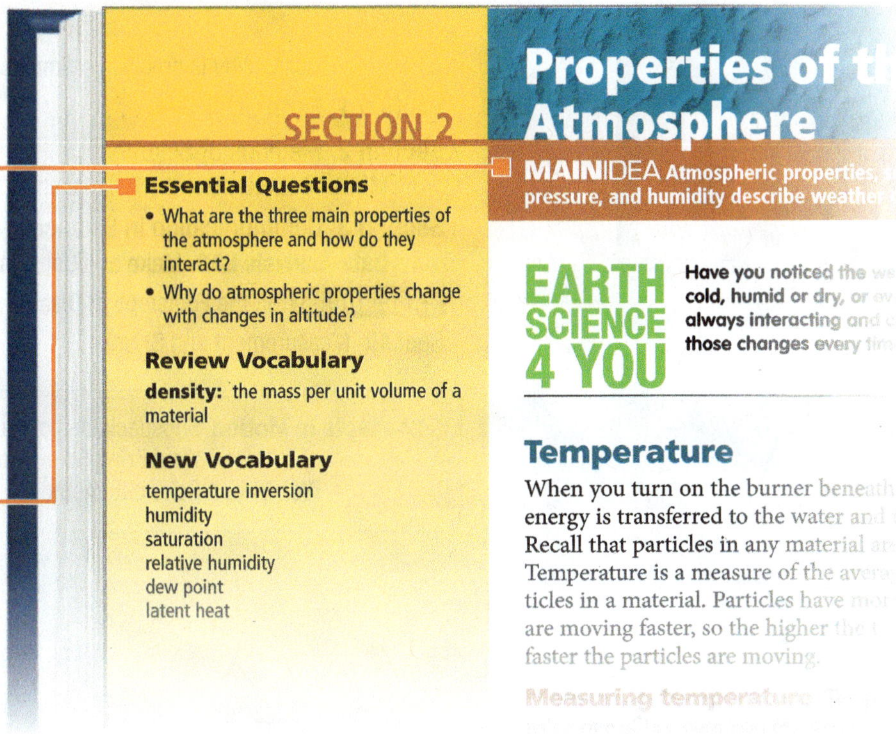
SECTION 2

Properties of th
Atmosphere

MAINIDEA Atmospheric properties, s
pressure, and humidity describe weathe

Essential Questions

- What are the three main properties of the atmosphere and how do they interact?
- Why do atmospheric properties change with changes in altitude?

Review Vocabulary

density: the mass per unit volume of a material

New Vocabulary

temperature inversion
humidity
saturation
relative humidity
dew point
latent heat

EARTH SCIENCE 4 YOU

Have you noticed the w
cold, humid or dry, or ev
always interacting and c
those changes every tim

Temperature

When you turn on the burner beneath
energy is transferred to the water and
Recall that particles in any material ar
Temperature is a measure of the avera
ticles in a material. Particles have mor
are moving faster, so the higher the t
faster the particles are moving.

In the Section Review, you will find a question that will help you to assess your understanding of the section's **MAIN**IDEA.

Essential Questions are assessed by the remaining review questions.

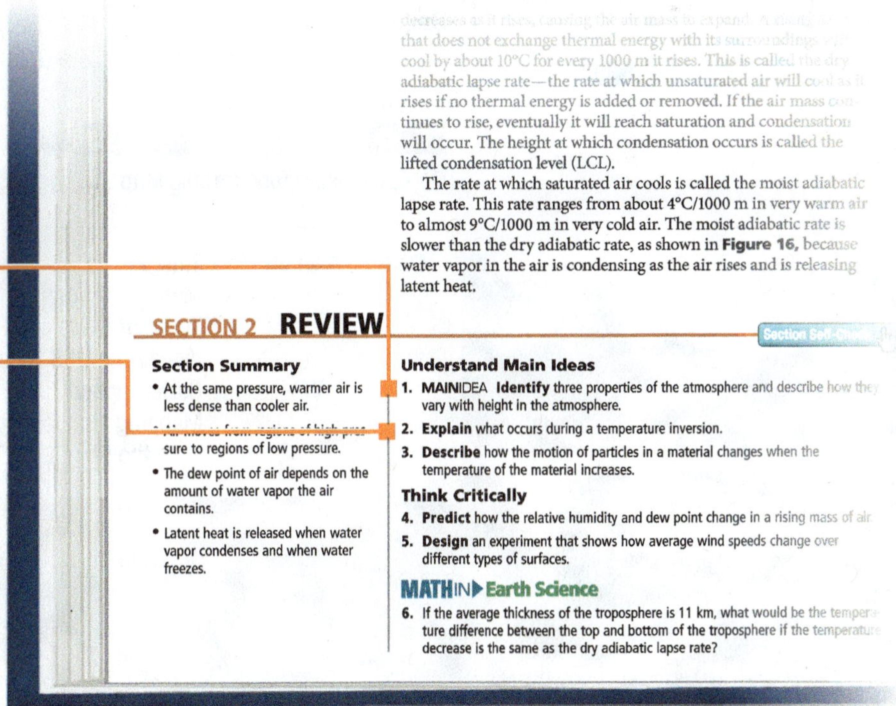
that does not exchange thermal energy with its surroundings w
cool by about 10°C for every 1000 m it rises. This is called the dry
adiabatic lapse rate—the rate at which unsaturated air will cool as it
rises if no thermal energy is added or removed. If the air mass con-
tinues to rise, eventually it will reach saturation and condensation
will occur. The height at which condensation occurs is called the
lifted condensation level (LCL).

The rate at which saturated air cools is called the moist adiabatic lapse rate. This rate ranges from about 4°C/1000 m in very warm air to almost 9°C/1000 m in very cold air. The moist adiabatic rate is slower than the dry adiabatic rate, as shown in **Figure 16**, because water vapor in the air is condensing as the air rises and is releasing latent heat.

SECTION 2 **REVIEW**

Section Summary

- At the same pressure, warmer air is less dense than cooler air.
- Air moves from regions of high pressure to regions of low pressure.
- The dew point of air depends on the amount of water vapor the air contains.
- Latent heat is released when water vapor condenses and when water freezes.

Understand Main Ideas

1. **MAIN**IDEA **Identify** three properties of the atmosphere and describe how they vary with height in the atmosphere.
2. **Explain** what occurs during a temperature inversion.
3. **Describe** how the motion of particles in a material changes when the temperature of the material increases.

Think Critically

4. **Predict** how the relative humidity and dew point change in a rising mass of air
5. **Design** an experiment that shows how average wind speeds change over different types of surfaces.

MATH IN Earth Science

6. If the average thickness of the troposphere is 11 km, what would be the tempera-ture difference between the top and bottom of the troposphere if the temperature decrease is the same as the dry adiabatic lapse rate?

TABLE OF CONTENTS

Earth Science

Unit 1 **STEM Project** The Nature of Science — Go online!

CHAPTER 1

The Nature of Science **4**

LaunchLAB Why is precise communication important? 4

Section 1 Earth Science 6

Section 2 Methods of Scientists 10

MiniLAB Determine the Relationship Between Variables 12

Section 3 Communication in Science 17

Data Analysis LAB Make and Use Graphs 18

eXpeditions! In the Footsteps of Disaster 20

GeoLAB Measurement and SI Units 21

Go online!

Concepts in Motion Subspecialties of Earth Science
Visualizing Scientific Methods

WebQuest Scientific method

CHAPTER 2

Mapping Our World **28**

LaunchLAB Can you make an accurate map? 28

Section 1 Latitude and Longitude 30

MiniLAB Locate Places on Earth 32

Section 2 Types of Maps 34

Problem-Solving LAB Calculate Gradients 37

Section 3 Remote Sensing 41

Earth Science & TECHNOLOGY Mapping Disaster Zones 47

GeoLAB Use a Topographic Map 48

Go online!

Concepts in Motion Time Zones
Map Projections
Types of Maps and Projections
Visualizing GPS Satellites

WebQuest Cartographer
Mapping technologies

Video Take the Road Less Traveled

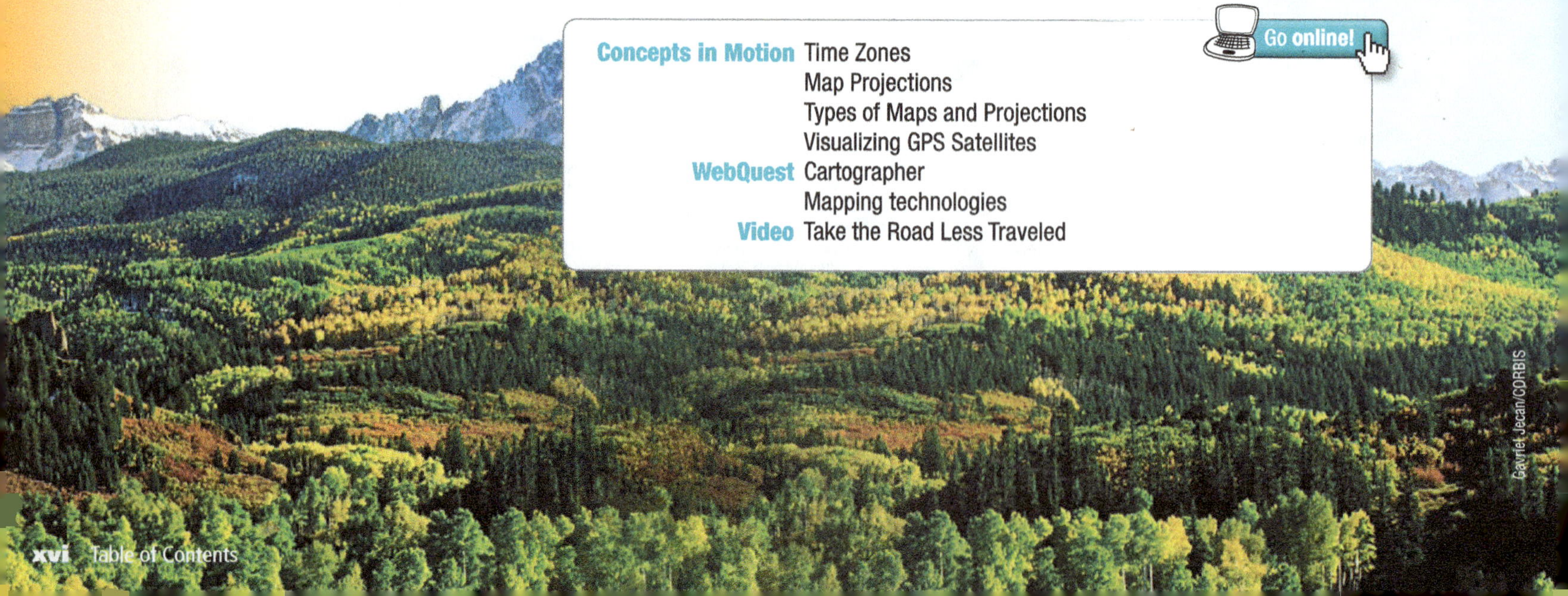

Garret Jecan/CORBIS

Go online! connectED.mcgraw-hill.com

Composition of Earth

Unit 2 **STEM Project** Composition of Earth — Go online!

CHAPTER 3

Matter and Change **58**

LaunchLAB What do fortified cereals contain? 58

Section 1 Matter 60

MiniLAB Identify Elements 62

Section 2 Combining Matter 66

Problem-Solving LAB Interpret Scientific Illustrations 70

Section 3 States of Matter 73

Earth Science & TECHNOLOGY Liquid Crystal Displays 76

GeoLAB Precipitate Salts 77

Go online!

Concepts in Motions Electron Cloud
Periodic Table of Elements
Ionic Bonding
Electron Flow
Visualizing Bonds
pH Scale

WebQuest Electronic visual displays

CHAPTER 4

Minerals **84**

LaunchLAB What shapes do minerals form? 84

Section 1 What is a mineral? 86

MiniLAB Recognize Cleavage and Fracture 92

Data Analysis LAB Make and Use a Table 94

Section 2 Types of Minerals 96

eXpeditions! Crystals at Large in Mexico 102

GeoLAB Make a Field Guide For Minerals 103

Go online!

Concepts in Motion Mohs Scale of Hardness
Special Properties of Minerals
Visualizing the Silicon-Oxygen Tetrahedron

WebQuest Lapidary
Crystal caves

Video A Rainbow of Gems

Doug Wilson/CORBIS

Composition of Earth

CHAPTER 5

Igneous Rocks **110**
LaunchLAB How are minerals identified? 110
Section 1 What are igneous rocks? 112
MiniLAB Compare Igneous Rocks 115
Section 2 Classification of Igneous Rocks 118
Problem-Solving LAB Interpret Scientific Illustrations 122
Earth Science & ENVIRONMENT Moon Rocks 124
GeoLAB Model Crystal Formation 125

Go online!
Concepts in Motion Types of Magma
Visualizing Fractional Crystallization and Crystal Settling
WebQuest Lunar Rocks

CHAPTER 6

Sedimentary and Metamorphic Rocks **132**
LaunchLAB What happened here? 132
Section 1 Formation of Sedimentary Rocks 134
MiniLAB Model Sediment Layering 136
Section 2 Types of Sedimentary Rocks 141
Section 3 Metamorphic Rocks 145
Problem-Solving LAB Interpret Scientific Illustrations 148
eXpeditions! Geology in Central Park 152
GeoLAB Interpret Changes in Rocks 153

Go online!
Concepts in Motion Visualizing Cross-Bedding and Ripple Marks
Classification of Sedimentary Rocks
WebQuest Sedimentologist
Rocks: Everyday uses

Doug Sherman/Geofile

Go online! connectED.mcgraw-hill.com

Surface Processes on Earth

Unit 3 STEM Project Surface Processes on Earth — Go online!

CHAPTER 7

Weathering, Erosion, and Soil **162**

LaunchLAB How does change relate to surface area? 162

Section 1 Weathering. 164

Section 2 Erosion and Deposition. 171

MiniLAB Model Erosion 172

Section 3 Soil 176

Data Analysis LAB Interpret the Data 182

Earth Science & TECHNOLOGY Space-Age Technology Shapes Modern Farming. 184

GeoLAB Model Mineral Weathering 185

Go online!

Concepts in Motions Frost Wedging
Visualizing Soil Orders

WebQuest Landscaper
GPS/GIS applications

CHAPTER 8

Mass Movements, Wind, and Glaciers **192**

LaunchLAB How does water affect sediments on slopes? 192

Section 1 Mass Movements. 194

Section 2 Wind 201

Section 3 Glaciers 207

Data Analysis LAB Interpret the Data 208

MiniLAB Model Glacial Deposition. 210

Earth Science & SOCIETY Slipping Away. 213

GeoLAB Map a Landslide. 214

Go online!

Concepts in Motion Rockslides
Dune Migration
Type of Dunes
Glacier Formation
Visualizing Continental Glacial Features

WebQuest Mass movements

William Manning/CORBIS

Surface Processes on Earth

CHAPTER 9

Surface Water **222**
LaunchLAB How does water infiltrate? 222
Section 1 Surface Water Movement 224
Problem-Solving LAB Interpret the Graph 227
Section 2 Stream Development 232
Section 3 Lakes and Freshwater Wetlands 238
MiniLAB Model Lake Formation 240
Earth Science & SOCIETY The World of Water 242
GeoLAB Predict the Velocity of a Stream 243

Go online!

Concepts in Motion Water Cycle
Meander Formation
Visualizing Erosion and Deposition in a Meander
WebQuest Geochemist Technician
Water quality

CHAPTER 10

Groundwater **250**
LaunchLAB How is water stored underground? 250
Section 1 Movement and Storage of Groundwater 252
Section 2 Groundwater Weathering and Deposition 259
Section 3 Groundwater Supply 263
Problem-Solving LAB Make a Topographic Profile 264
MiniLAB Model an Artesian Well 265
Earth Science & ENVIRONMENT Watcher of the Water 269
GeoLAB Track Groundwater Pollution 270

Go online!

Concepts in Motion World's Water Supply
Visualizing Springs
Saltwater Contamination
WebQuest Hydrogeologist
Aquifers
Video What are you drinking?

Eric Bean/The Image Bank/Getty Images

Go online! connectED.mcgraw-hill.com

The Atmosphere and the Oceans

Unit 4 **STEM Project** The Atmosphere and the Oceans — Go online!

CHAPTER 11

Atmosphere **280**

LaunchLAB What causes cloud formation? 280

Section 1 Atmospheric Basics 282

Section 2 Properties of the Atmosphere 289

Problem-Solving LAB Interpret the Graph 294

MiniLAB Investigate Dew Formation 295

Section 3 Clouds and Precipitation 297

Earth Science & ENVIRONMENT Ozone Variation 304

GeoLAB Interpret Pressure-Temperature Relationships 305

Go online!

Concepts in Motions Components of the Atmosphere
Visualizing the Layers of the Atmosphere
Conduction, Convection, and Radiation

WebQuest Weather Observer
Ozone

CHAPTER 12

Meteorology **312**

LaunchLAB How does a cold air mass form? 312

Section 1 The Causes of Weather 314

MiniLAB Compare the Angles of Sunlight to Earth 315

Section 2 Weather Systems 318

Section 3 Gathering Weather Data 324

Section 4 Weather Analysis and Prediction 329

Problem-Solving LAB Interpret a Scientific Illustration 330

Earth Science & TECHNOLOGY Weather Forecasting—Precision from Chaos 333

GeoLAB Interpret a Weather Map 334

Go online!

Concepts in Motion Air Mass Characteristics
Visualizing the Coriolis Effect
Fronts

WebQuest Meteorologist
Weather forecasts

The Atmosphere and the Oceans

TABLE OF CONTENTS

CHAPTER 13

The Nature of Storms 342

LaunchLAB Why does lightning form? 342

Section 1 Thunderstorms 344

Section 2 Severe Weather 350

Section 3 Tropical Storms 355

Section 4 Recurrent Weather 361

MiniLAB Model Flood Conditions 362

Data Analysis LAB Interpret the Table 364

eXpeditions! Storm Spotters 366

GeoLAB Track a Tropical Cyclone 367

Go online!

Concepts in Motion Thunderstorm Development
Tornado Formation
Enhanced Fujita Tornado Damage Scale
Tropical Cyclones
Visualizing Cyclone Formation
The Heat Index

WebQuest Hurricane Hunter
Severe weather

Video Storm Chasers

CHAPTER 14

Climate 374

LaunchLAB How can you model could cover? 374

Section 1 Defining Climate 376

Data Analysis LAB Interpret the Data 377

Section 2 Climate Classification 381

Section 3 Climatic Changes 387

Section 4 Impact of Human Activities 393

MiniLAB Model the Greenhouse Effect 394

Earth Science & SOCIETY Effects of Global Warming on the Arctic 396

GeoLAB Identify a Microclimate 397

Go online!

Concepts in Motion Visualizing Worldwide Climates
Seasons
Greenhouse Effect

WebQuest Climatologist
Arctic warming

Video Climate's History

Go online! connectED.mcgraw-hill.com

The Atmosphere and the Oceans

CHAPTER 15

Earth's Oceans **404**

LaunchLAB How much of Earth's surface is covered by water? 404

Section 1 An Overview of Oceans 406

Section 2 Seawater 413

MiniLAB Model Seawater 416

Section 3 Ocean Movements 421

Data Analysis LAB Graph Data 423

Earth Science & ENVIRONMENT Bacterial Counts and Full Moons 428

GeoLAB Model Water Mass 429

Go online!

Concepts in Motions Visualizing the Salt Cycle
Removal of Sea Salts
Waves

WebQuest Oceanographer
Tidal patterns

CHAPTER 16

The Marine Environment **436**

LaunchLAB Where does chalk form? 436

Section 1 Shoreline Features 438

Section 2 Seafloor Features 447

Problem-Solving LAB Interpret Graphs 449

MiniLAB Measure Sediment Settling Rates 453

eXpeditions! Surveying the Deep Ocean Floor 455

GeoLAB Identify Coastal Landforms 456

Go online!

Concepts in Motion Longshore Currents
Visualizing the Ocean Floor

WebQuest Deep sea exploring

Tan Yilmaz/Moment/Getty Images

The Dynamic Earth

Unit 5 **STEM Project** The Dynamic Earth — Go online!

CHAPTER 17

Plate Tectonics **466**
LaunchLAB Is California moving? 466
Section 1 Drifting Continents 468
Section 2 Seafloor Spreading 473
Section 3 Plate Boundaries 480
MiniLAB Model Ocean-Basin Formation 481
Problem-Solving LAB Interpret Scientific Illustrations 484
Section 4 Causes of Plate Motions 486
Earth Science & ENVIRONMENT Vailulu'u Seamount 489
GeoLAB Model Plate Boundaries and Isochrons 490

Go online!
Concepts in Motion Continental Drift
Visualizing Seafloor Spreading
Summary of Convergent Boundaries
WebQuest Marine Geologist
Island formation
Video Being Stressed Out

CHAPTER 18

Volcanism **498**
LaunchLAB What makes magma rise? 498
Section 1 Volcanoes 500
Data Analysis LAB Interpret the Graph 501
MiniLAB Model a Caldera 505
Section 2 Eruptions 508
Section 3 Intrusive Activity 514
eXpeditions! Hawaiian Volcano Observatory 518
GeoLAB Predict the Safety of a Volcano 519

Go online!
Concepts in Motion Subduction
Divergent Plate Boundaries
Caldera Formation
Types of Volcanoes
Visualizing Eruptions
WebQuest Monitoring volcanoes

Dr. Morley Read/Photo Researchers

Go online! connectED.mcgraw-hill.com

The Dynamic Earth

CHAPTER 19

Earthquakes 526

LaunchLAB What can cause an earthquake? 526

Section 1 Forces Within Earth 528

Section 2 Seismic Waves and Earth's Interior 534

Section 3 Measuring and Locating Earthquakes 539

MiniLAB Make a Map 541

Data Analysis LAB Interpret the Data 543

Section 4 Earthquakes and Society 545

Earth Science & SOCIETY Learning from the Past 552

GeoLAB Relate Epicenters and Plate Tectonics 553

Go online!

Concepts in Motions Faults
Types of Faults
Seismic Waves
Seismometers
P-Waves and S-Waves
Visualizing Seismic Waves
Modified Mercalli Scale
Tsunami

WebQuest Earthquake prediction

CHAPTER 20

Mountain Building 560

LaunchLAB How does crust displace the mantle? 560

Section 1 Crust-Mantle Relationships 562

MiniLAB Model Isostatic Rebound 564

Problem-Solving LAB Make and Use a Graph 565

Section 2 Orogeny 567

Section 3 Other Types of Mountain Building 574

eXpeditions! Hiking the Appalachian Trail 577

GeoLAB Make a Map Profile 578

Go online!

Concepts in Motion Isostacy
Island Formation
Convergence
Folding Rocks
Visualizing the Rise and Fall of the Appalachians

WebQuest Petrologist
Orogenies

R.E. Wallace/USGS

Geologic Time

Unit 6 STEM Project Geologic Time — Go online!

CHAPTER 21

Fossils and the Rock Record 588

LaunchLAB How are fossils made? 588
Section 1 The Rock Record 590
Section 2 Relative-Age Dating 595
MiniLAB Determine Relative Age 597
Problem-Solving LAB Interpret the Diagram 599
Section 3 Absolute-Age Dating 601
Section 4 Fossil Remains 606
Earth Science & TECHNOLOGY Drilling into the Past 610
GeoLAB Interpret History-Shaping Events 611

Go online!
Concepts in Motions Visualizing the Geologic Time Scale
Angular Unconformity
Alpha Decay
Half-Lives
Half-Lives of Selected Radioactive Isotopes
WebQuest Petroleum Geologist
Fossil discoveries
Video Clues to the Past

CHAPTER 22

The Precambrian Earth 618

LaunchLAB How do liquids of different densities model early Earth? 618
Section 1 Early Earth 620
Section 2 Formation of the Crust and Continents 623
Section 3 Formation of the Atmosphere and Oceans 628
Problem-Solving LAB Calculate Profits 630
MiniLAB Model Red Bed Formation 631
Section 4 Early Life on Earth 633
Earth Science & TECHNOLOGY Martian Microenvironments 638
GeoLAB Map Continental Growth 639

Go online!
Concepts in Motions Visualizing Continent Formation
Miller-Urey Experiment
How Life Might Have Begun on Earth: Three Hypotheses
WebQuest Planetary Geologist
Early Earth

CHAPTER 23

The Paleozoic, Mesozoic, and Cenozoic Eras 646

LaunchLAB How is oil stored in rocks? 646
Section 1 The Paleozoic Era 648
Data Analysis LAB Interpret the Table 652
MiniLAB Model Continental Shelf Area 653
Section 2 The Mesozoic Era 655
Section 3 The Cenozoic Era 660
eXpeditions! Digging for Dinosaurs 666
GeoLAB Solve Dinosaur Fossil Puzzles 667

Go online!
Concepts in Motions Major Extinctions in the Phanerozoic
Visualizing the Basin and Range Province
WebQuest Paleoecologist
Phanerozoic life

Go online! connectED.mcgraw-hill.com

Resources and the Environment

Unit 7 **STEM Project** Resources and the Environment — Go online!

CHAPTER 24 Earth Resources 676

LaunchLAB What natural resources do you use in your classroom? 676
Section 1 Natural Resources 678
Section 2 Resources from Earth's Crust 682
Section 3 Air Resources 687
Data Analysis LAB Interpret Graphs 688
Section 4 Water Resources 693
MiniLAB Determine the Hardness of Water 695
Earth Science & SOCIETY The Price of Water 698
GeoLAB Monitor Daily Water Usage 699

Go online!
Concepts in Motions Visualizing Carbon and Nitrogen Cycles
Distillation
WebQuest Materials Engineer
Water supply

CHAPTER 25 Energy Resources 706

LaunchLAB Can you identify sources of energy? 706
Section 1 Conventional Energy Resources 708
MiniLAB Model Oil Migration 712
Section 2 Alternative Energy Resources 714
Section 3 Conservation of Energy Resources 720
Data Analysis LAB Make and Use Graphs 724
Earth Science & ENVIRONMENT Bacteria Power! 724
GeoLAB Design an Energy-Efficient Building 725

Go online!
Concepts in Motions Visualizing Coal
Geothermal Power
Fission Reactor
Advantages of Public Transportation
WebQuest Environmental Consultant
Alternative energy

CHAPTER 26 Human Impact on Resources 732

LaunchLAB What resources are used in classroom items? 732
Section 1 Populations and the Use of Natural Resources 734
Section 2 Human Impact on Land Resources 737
MiniLAB Model Nutrient Loss 740
Section 3 Human Impact on Air Resources 743
Data Analysis LAB Interpret the Data 746
Section 4 Human Impact on Water Resources 748
Earth Science & TECHNOLOGY Measuring and Modeling Climate Change 751
GeoLAB Pinpoint a Source of Pollution 752

Go online!
Concepts in Motions Carrying Capacity
Visualizing Agricultural Practices
How Smog Forms
Ozone Depletion
WebQuest Hydrologist
Climate change
Video Living Green
Habitat Restoration

Beyond Earth

Unit 8 **STEM Project** Beyond Earth — Go online!

CHAPTER 27

The Sun-Earth-Moon System **762**

LaunchLAB How can the Sun-Earth-Moon system be modeled? 762

Section 1 Tools of Astronomy 764

Section 2 The Moon 770

Section 3 The Sun-Earth-Moon System 775

MiniLAB Predict the Sun's Summer Solstice Position 776

Problem-Solving LAB Interpret Scientific Illustrations 782

eXpeditions! Living in Space 785

GeoLAB Determine Relative Ages of Lunar Features 786

Go online!

Concepts in Motion The Moon and Earth
Moon Impact Theory
Visualizing the Phases of the Moon
Eclipse

WebQuest Space Engineer
Space exploration

CHAPTER 28

Our Solar System **794**

LaunchLAB What can be learned from space missions? 794

Section 1 Formation of the Solar System 796

MiniLAB Explore Eccentricity 801

Section 2 The Inner Planets 804

Problem-Solving LAB Apply Kepler's Third Law 807

Section 3 The Outer Planets 811

Section 4 Other Solar System Objects 816

Earth Science & TECHNOLOGY Water in the Solar System 820

GeoLAB Model the Solar System 821

Go online!

Concepts in Motion Physical Data of the Planets
Gravitational Attraction
Visualizing Other Objects in the Solar System

WebQuest Planetologist
Water in the solar system

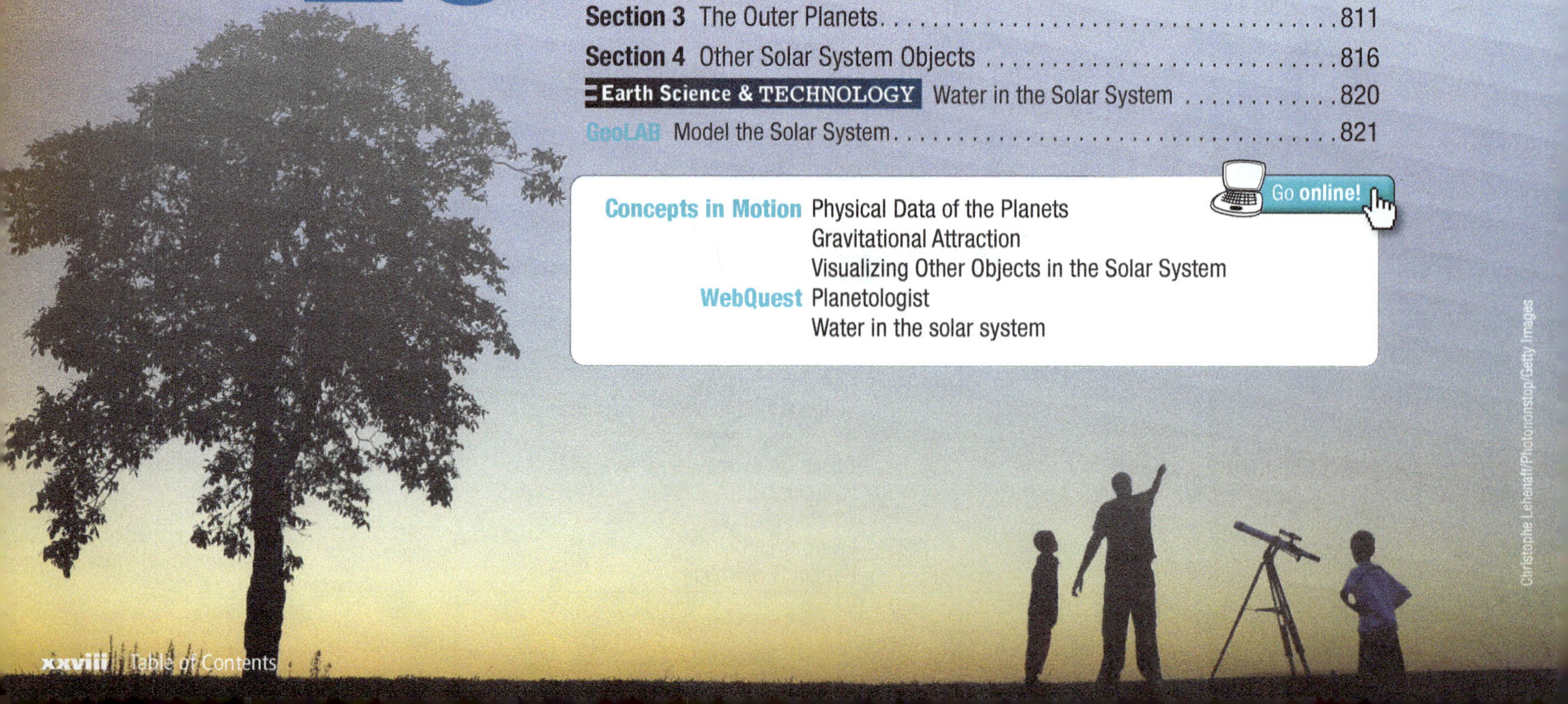

Christophe Lehenaff/Photononstop/Getty Images

Go online! connectED.mcgraw-hill.com

Beyond Earth

CHAPTER 29

Stars **828**
LaunchLAB How can you observe sunspots? 828
Section 1 The Sun 830
Data Analysis LAB Interpret Data 835
Section 2 Measuring the Stars 837
MiniLAB Model Parallax 843
Section 3 Stellar Evolution 847
Earth Science & TECHNOLOGY Space Weather and Earth Systems 852
GeoLAB Identify Stellar Spectral Lines 853

Go online!

Concepts in Motions Relative Properties of the Sun
Visualizing Star Groupings
Doppler Effect
Parallax
Relationships of Spectral Types of Stars
Properties of Main-Sequence Stars
Star Formation
Helium Core

WebQuest Spectroscopist
Space weather

Video In the Dark

CHAPTER 30

Galaxies and the Universe **860**
LaunchLAB How big is the Milky Way? 860
Section 1 The Milky Way Galaxy 862
Section 2 Other Galaxies in the Universe 869
MiniLAB Model Expansion 873
Problem-Solving LAB Make and Use Graphs 874
Section 3 Cosmology 878
Earth Science & TECHNOLOGY Black Holes Are Green? 882
GeoLAB Classify Galaxies 883

Go online!

Concepts in Motion Population I and II
Stars of the Milky Way
Visualizing the Local Group

WebQuest Computer Programmer
Black holes

NOAO/SPL/Photo Researchers

STUDENT RESOURCES

Skillbuilder Handbook 891

Problem-Solving Skills 891

Make Comparisons.... 891
Analyze Information 892
Synthesize Information.... 893
Take Notes and Outline.... 894
Understand Cause and Effect 895
Read a Time Line 896
Analyze Media Sources 897
Use Graphic Organizers 898
Debate Skills 899

Math Skills 900

Measure in SI.... 900
Convert Temperature 900
Make and Use Tables 901
Make and Use Graphs 901

Reference Handbook 904

Safety in the Laboratory.... 904
Physiographic Map of Earth 906
Topographic Symbols.... 908
Weather Map Symbols.... 909
Periodic Table of the Elements 910
Relative Humidity.... 911
Minerals.... 912
Rocks.... 914
Solar System Charts.... 915

Glossary ▪ Glosario 916

Index 954

Rubberball/Getty Images

FOLDABLES® by Dinah Zike

Folding Instructions

The following pages offer step-by-step instructions to make the Foldables study guides.

Layered-Look Book

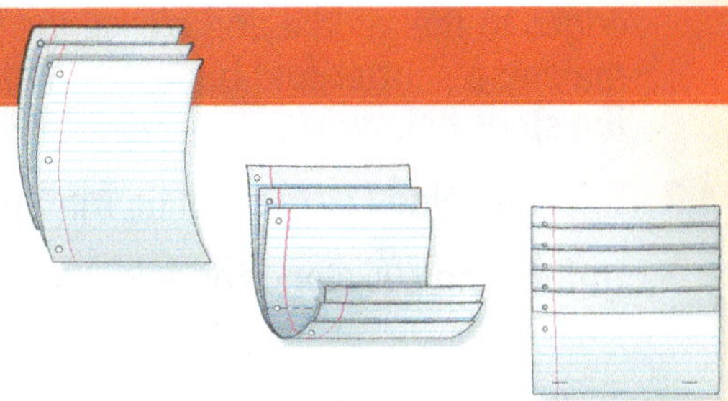

1. Collect three sheets of paper and layer them about 1 cm apart vertically. Keep the edges level.
2. Fold up the bottom edges of the paper to form 6 equal tabs.
3. Fold the papers and crease well to hold the tabs in place. Staple along the fold. Label each tab.

Trifold Book

1. Fold a vertical sheet of paper into thirds.
2. Unfold and label each row.

Three-Tab Book

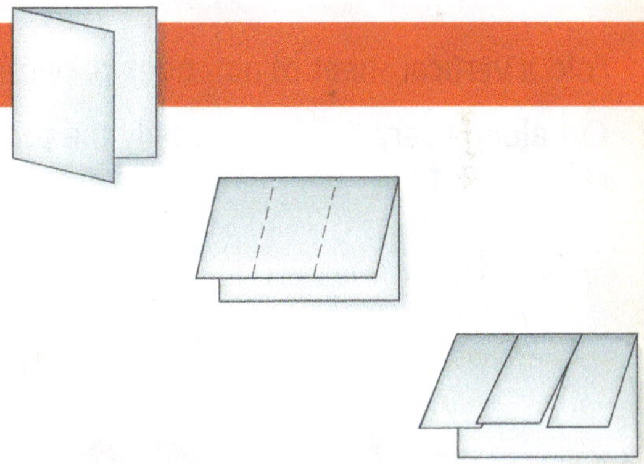

1. Fold a vertical sheet of paper from side to side. Make the front edge about 2 cm shorter than the back edge.
2. Turn length-wise and fold into thirds.
3. Unfold and cut only the top layer along both folds to make three tabs. Label each tab.

Two- and Four-Tab Books

1. Fold a sheet of paper in half.
2. Fold in half again. If making a four-tab book, then fold in half again to make three folds.
3. Unfold and cut only the top layer along the folds to make two or four tabs. Label each tab.

FOLDABLES®

by Dinah Zike

Shutter-Fold and Four-Door Books

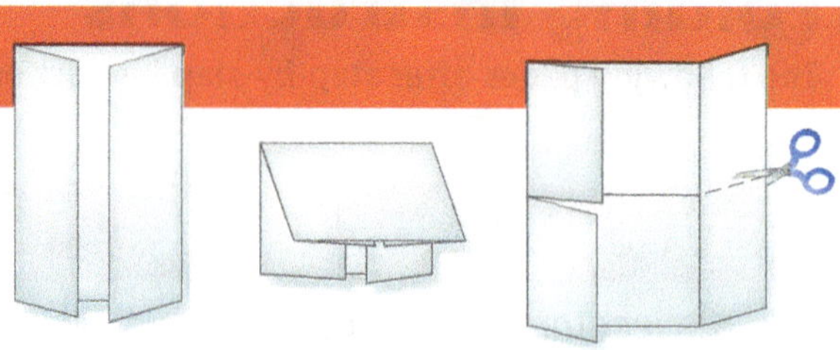

1. Find the middle of a horizontal sheet of paper. Fold both edges to the middle and crease the folds. Stop here if making a shutter-fold book. For a four-door book, complete the steps below.
2. Fold the folded paper in half, from top to bottom.
3. Unfold and cut along the fold lines to make four tabs. Label each tab.

Concept-Map Book

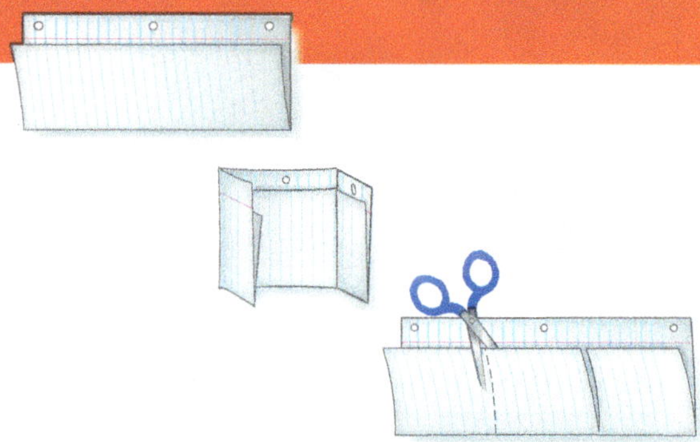

1. Fold a vertical sheet of paper from top to bottom. Make the top edge about 2 cm shorter than the bottom edge.
2. Turn length-wise and fold into thirds.
3. Unfold and cut only the top layer along both folds to make three tabs. Label the top and each tab.

Vocabulary Book

1. Fold a vertical sheet of notebook paper in half.
2. Cut along every third line of only the top layer to form tabs. Label each tab.

Folded Chart

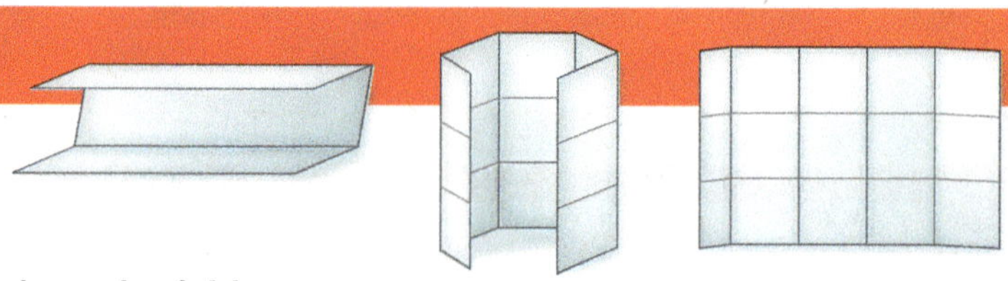

1. Fold a sheet of paper length-wise into thirds.
2. Fold the paper width-wise into fifths.
3. Unfold, lay the paper length-wise, and draw lines along the folds. Label the table.

Pocket Book

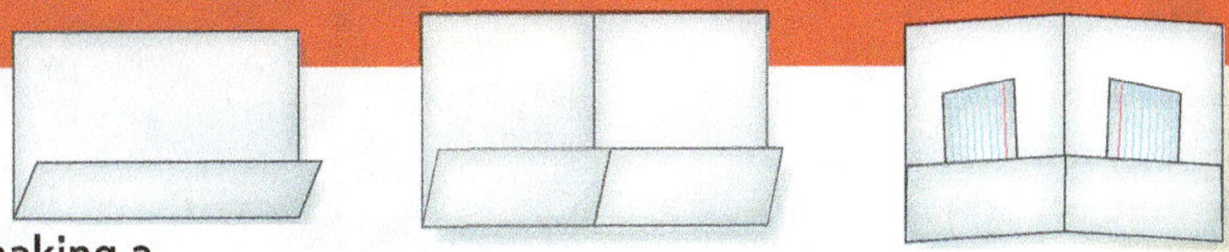

1. Fold the bottom of a horizontal sheet of paper up about 3 cm.
2. If making a two-pocket book, fold in half. If making a three-pocket book, fold in thirds.
3. Unfold once and dot with glue or staple to make two pockets. Label each pocket.

Bound Book

1. Fold several sheets of paper in half to find the middle. Hold all but one sheet together and make a 3-cm cut at the fold line on each side of the paper.
2. On the final page, cut along the fold line to within 3-cm of each edge.
3. Slip the first few sheets through the cut in the final sheet to make a multi-page book.

Top-Tab Book

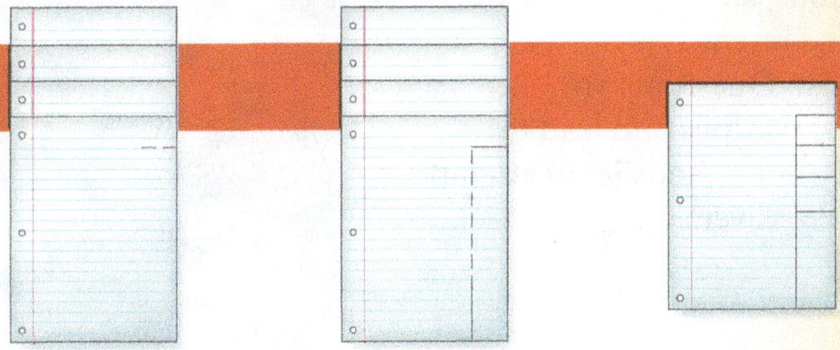

1. Layer multiple sheets of paper so that about 2–3 cm of each can be seen.
2. Make a 2–3-cm horizontal cut through all pages a short distance (3 cm) from the top edge of the top sheet.
3. Make a vertical cut up from the bottom to meet the horizontal cut.
4. Place the sheets on top of an uncut sheet and align the tops and sides of all sheets. Label each tab.

Accordion Book

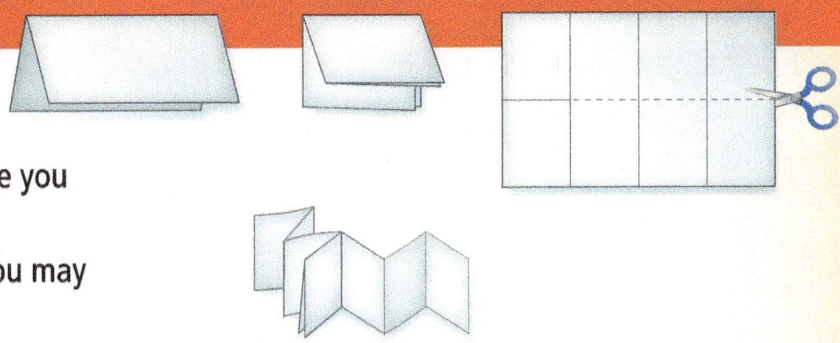

1. Fold a sheet of paper in half. Fold in half and in half again to form eight sections.
2. Cut along the long fold line, stopping before you reach the last two sections.
3. Refold the paper into an accordion book. You may want to glue to double pages together.

Earth Science

Themes

Change Improvements in technology allow scientists to make more accurate measurements and communicate their data worldwide.

Structures Satellites and improved imagery in many wavelengths provide scientists detail about Earth's surface.

Geologic Time Scientists are able to measure geologic time more accurately with new methods and technology.

Systems Consistent methods of study and new combinations of technology have developed into a system of learning about Earth and its features.

Scientific Inquiry Asking questions, such as how Earth's systems interact, is the job of a scientist. Questions are stimulated by observations and lead to a variety of methods in an attempt to find answers.

UNIT 1

Earth Science

CHAPTERS

1 The Nature of Science

2 Mapping Our World

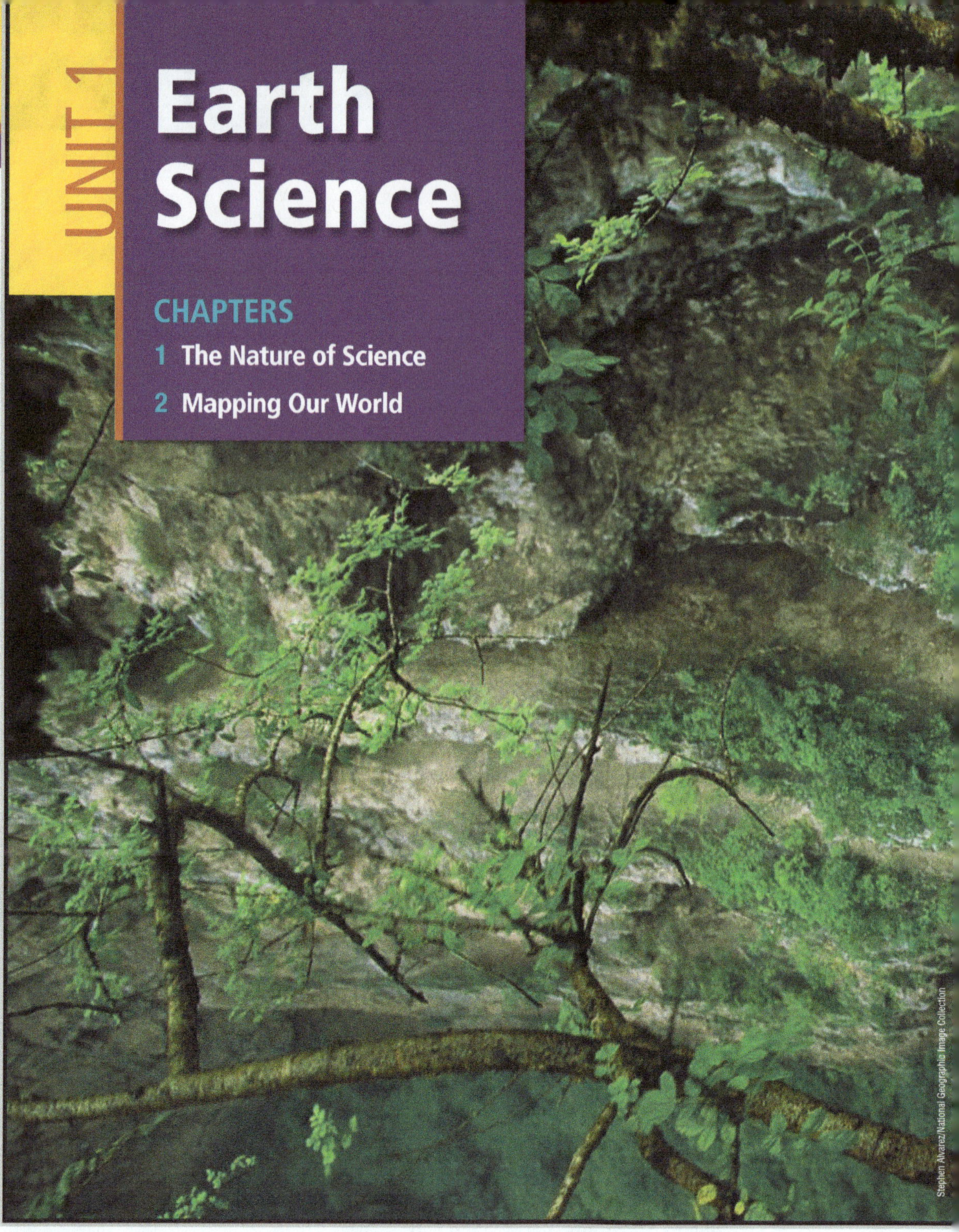

Stephen Alvarez/National Geographic Image Collection

STEM Project

CAREERS IN EARTH SCIENCE **Speleologist**

This **speleologist,** a scientist who studies caves, descends into a 200-m-deep sinkhole. Speleologists use scientific methods to make maps, collect samples, and make observations of incredible landforms resulting from geologic processes.

Introduce the Unit

Our Dynamic Planet The enormous sinkhole in this photograph is Tawi Attair, the Well of the Birds, in Oman. It formed long ago when the roof of a cave collapsed. This sinkhole is deep enough hold a 50-floor office building. The geologist pictured, Louise Hose, is part of a research team that is using GPS and lasers to generate a detailed three-dimensional map of the cave system. The limestone rocks that make up the walls of the sinkhole are part of one of Earth's major systems: the geosphere. Earth's geosphere continuously changes as it interacts with the atmosphere, the hydrosphere, and organisms of the biosphere. Have students speculate about how interactions among Earth's major systems change the planet.

Make a Model Ask students to describe how the features shown in the photo could be represented on a piece of paper. Discuss how a two-dimensional representation would differ from the actual three-dimensional scene. Have a variety of maps available for students to study.

CHAPTER 1 The Nature of Science

BIGIDEA Earth scientists use specific methods to investigate Earth and beyond.

ESSENTIAL QUESTIONS	NATIONAL STANDARDS	RESOURCES TO ASSESS MASTERY
SECTION 1 **Earth Science** 1. How do the areas of study within Earth science compare? 2. What are Earth's systems? 3. What are the relationships among Earth's systems? 4. Why is technology important? 1 session 0.5 block	UCP.1; A.1–2; C.4–5; D1–3; E1–2; F.6; G1–3	**Entry-Level Assessment** Focus Transparency 1, p. 6 **Progress Monitoring** Caption Question, p. 7 Section Review, p. 9
SECTION 2 **Methods of Scientists** 1. What are independent and dependent variables? 2. How does experimentation and investigation differ? 3. What are the differences between mass and weight? 4. What is scientific notation and how is it used? 4 sessions 2 blocks	A.1–2; B.2, B.6; G.2	**Entry-Level Assessment** Focus Transparency 2, p. 10 **Progress Monitoring** Reading Check, pp. 12, 14 Section Review, p. 16
SECTION 3 **Communication in Science** 1. Why is precise communication crucial in science? 2. What are the differences between scientific theories and scientific laws? 3. When is it appropriate to use a graph or a model? 3 sessions 1.5 blocks	UCP.2–3; A.1–2; B.2; D.1, D.3; E.1–2; G.1–3	**Entry-Level Assessment** Focus Transparency 3, p. 17 **Progress Monitoring** Caption Question, pp. 17, 18, 19 Section Review, p. 19 **Summative Assessment** Chapter Assessment, p. 23 Exam*View*® *Assessment Suite* CD-ROM

Go online!
connectED.mcgraw-hill.com

LEVELED RESOURCES	LAB MATERIALS
Science Notebook 1.1 OL **Chapter FAST FILE Resources:** Study Guide, p. 13 BL **Transparencies:** Section Focus Transparency 1 EL Teaching Transparency 1 OL EL	**LaunchLAB** p. 4 / **10 min** object
Science Notebook 1.2 OL **Chapter FAST FILE Resources:** MiniLab Worksheet, p. 2 OL GeoLab Worksheet, p. 3 OL Study Guide, p. 15 BL **Lab Resources:** Laboratory Manual, pp. 1, 5 OL **Transparencies:** Section Focus Transparency 2 EL Teaching Transparency 2 OL EL	**MiniLAB** p. 12 / **75 min** soil, containers, water, thermometers, masking tape, sunny windowsill **GeoLAB** p. 21 / **45 min** water, large graduated cylinder or beaker, graph paper, balance, pieces of string, spring scale, rock samples, ruler
Science Notebook 1.3 OL **Chapter FAST FILE Resources:** Study Guide, p. 17 BL **Transparencies:** Section Focus Transparency 3 EL Teaching Transparency 3 OL EL	

LabManager™
Customize any lab with the LabManager™ CD-ROM.

ADDITIONAL RESOURCES

Technology:
- eTeacherEdition
- eStudentEdition
- Classroom Presentation Toolkit
- Exam*View®Assessment Suite* CD-ROM
- LabManager™ CD-ROM
- What's EARTH SCIENCE Got To Do With It? DVD
- Weather Classroom Video DVD

Assessment:
- Blueprints for Success
- Chapter Test (Scaffolded)
- Earth Science Standardized Test Practice

Lab Resources:
- Exploring Environmental Problems Laboratory Manual

BL Below Level OL On Level AL Advanced Learners EL English Learners COOP LEARN Cooperative Learning

Chapter 1

LaunchLAB

iLab Station | Rubric

Why is precise communication important?

Process Skills describe, communicate

Safety Precaution Approve lab safety forms before work begins.

Teaching Strategies
- Collect a number of small objects that students can describe, such as rocks, minerals, buttons, and coins.
- Make sure that each student has an Earth science journal in which to record the results of this lab as well as other labs and activities that will be conducted throughout this course.

Procedure
1. Have students read and complete the lab safety form and follow the procedure below.
2. Obtain an **object** from your teacher. Do not show it to your partner.
3. Write one sentence that accurately describes the object in detail without identifying or naming the object.
4. Give your partner the description and allow him or her a few minutes to identify your object.
5. Now use your partner's description to identify his or her object.

Analysis
1. **Identify** Were you and your partner able to identify each others' objects? Why or why not? Answers will vary. Students should note which terms were descriptive and which terms were ambiguous.
2. **Error Analysis** Work together to rewrite each description in your science journals to make them as accurate as possible. Answers will vary. Students should pay more attention to detail in the second description.
3. **Compare** Trade the new descriptions with another pair of students. Did this pair of students have an easier time determining the objects than you and your partner did? Why or why not? Answers will vary. Students should recognize that accurate descriptions are important in communication.

CHAPTER 1
The Nature of Science

BIGIDEA Earth scientists use specific methods to investigate Earth and beyond.

SECTIONS

1 Earth Science

2 Methods of Scientists

3 Communication in Science

LaunchLAB

iLab Station

Why is precise communication important?

Have you ever explained something to someone only later to find out that what you thought was a clear explanation was confusing, misleading, or even incorrect? Precise communication is an important skill. Practice your communication skills with this activity.

FOLDABLES Study Organizer

Earth's Systems

Make a four-tab book. Label the tabs *Geosphere, Hydrosphere, Atmosphere,* and *Biosphere.* Use it to organize your notes on Earth's main systems.

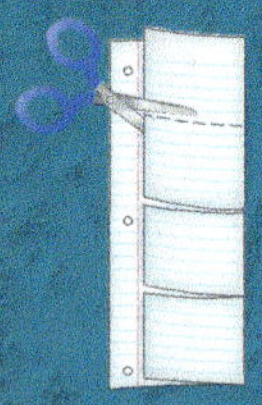

Assessment

Knowledge Have students use the results of this lab to discuss the importance of logical and objective communication in science.

LabManager™
Customize any lab with the LabManager™ CD-ROM.

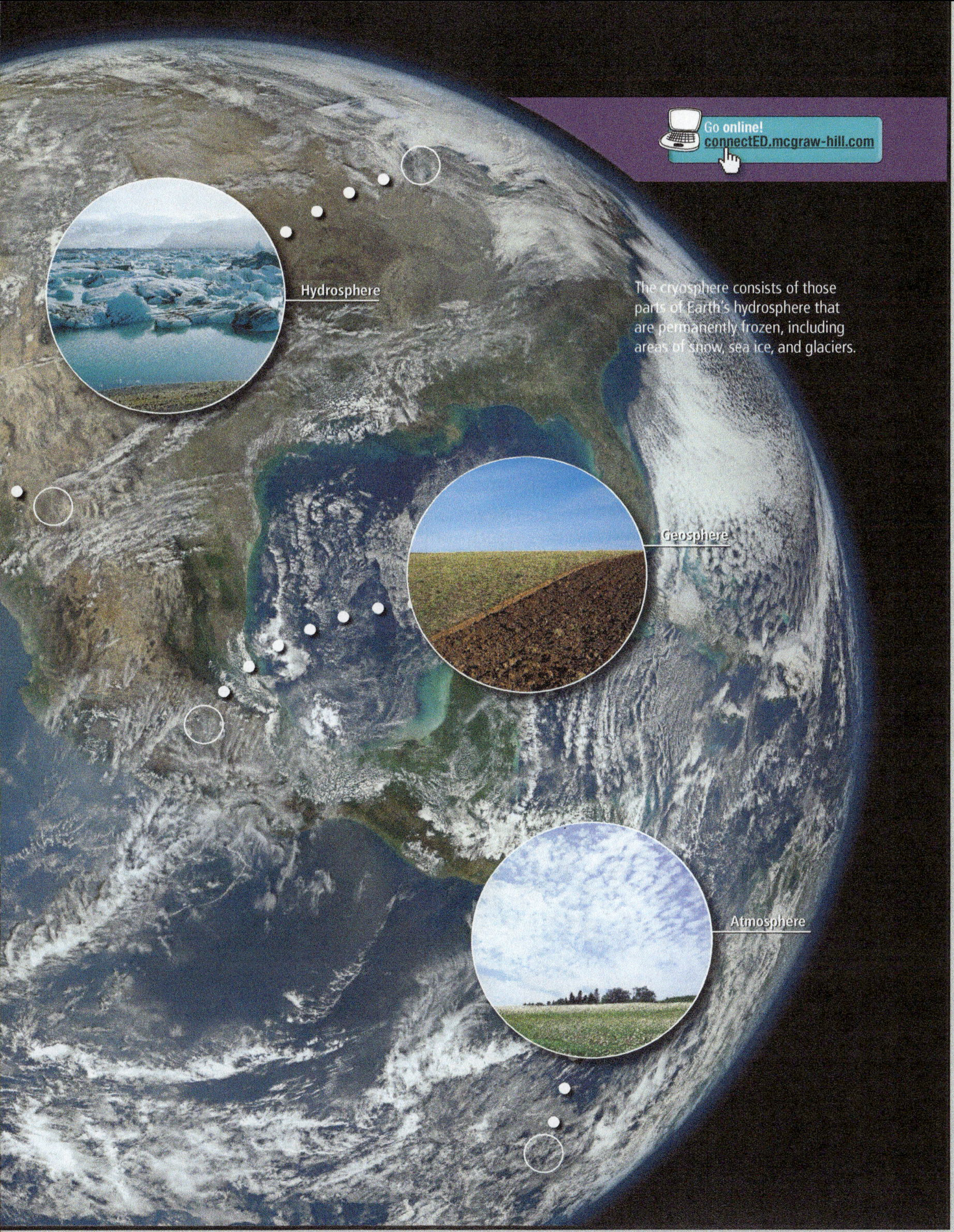

Introduce the BIG IDEA

Scientific Methods Ask students to describe a question or problem they faced recently. How did they answer the question or solve the problem? As a class, analyze whether the process they used was logical, efficient, and successful. Then, list the steps of a scientific method on the board. Using the steps as a guide, ask students to suggest more efficient ways of addressing the problem under discussion. Afterward, point out that scientific methods can be used to solve everyday problems. They are not limited to experiments conducted by scientists in laboratories.

Teacher Content Support

Earth from Space Tell students that many photos of Earth from space are taken by satellites or by astronauts in orbit around Earth. These photos can be used to study different aspects of our planet. Ask students to examine the photo, then brainstorm a list of things scientists might study using a view of Earth from space. Sample answer: Scientists might study cloud patterns, storm movement, drainage patterns, and volcanic eruptions.

SECTION 1

1 Focus

Focus Transparency

Before presenting the lesson, project **Section Focus Transparency 1** and have students answer the accompanying questions. EL

MAINIDEA

Earth Science Ask students how they think Earth science affects their everyday lives. Answers might include observing clouds and relating these observations to changes in weather; seeing loose materials slide down a hillside; observing that soils in different places vary in color and texture; and gazing at objects in the night sky. From their responses, students should realize that the study of Earth science is broad. Tell them this is one reason why Earth science is divided into different areas of study, as they will learn in this section.

2 Teach

Teacher Content Support

Extraterrestrial Life
Discoveries of life-forms in increasingly diverse places on and in Earth have led to speculation that life, at least at the microscopic level, might be able to form and survive in extraterrestrial environments. Within the solar system, at least two places have been identified as either having or once having had conditions less hostile than those in some of the places where life exists on Earth. Mars, for example, is thought to have had a warmer climate than it does now, with liquid water on its surface. Jupiter's moon Europa has a subsurface ocean of liquid water.

SECTION 1

Earth Science

MAINIDEA Earth science encompasses five areas of study: astronomy, meteorology, geology, oceanography, and environmental science.

Essential Questions
- How do the areas of study within Earth science compare?
- What are Earth's systems?
- What are the relationships among Earth's systems?
- Why is technology important?

Review Vocabulary

technology: the application of knowledge gained from scientific research to solve society's needs and problems

New Vocabulary

astronomy
meteorology
geology
oceanography
environmental science
geosphere
atmosphere
hydrosphere
cryosphere
biosphere

EARTH SCIENCE 4 YOU From the maps you use when traveling, to the weather report you use when deciding whether or not to carry an umbrella, Earth science is part of your everyday life.

The Scope of Earth Science

The scope of Earth science is vast. This broad field can be broken into five major areas of specialization: astronomy, meteorology, geology, oceanography, and environmental science.

Astronomy The study of objects beyond Earth's atmosphere is called **astronomy.** Prior to the invention of sophisticated instruments, such as the telescope shown in **Figure 1,** many astronomers merely described the locations of objects in space in relation to each other. Today, Earth scientists study the universe and everything in it, including galaxies, stars, planets, and other bodies they have identified.

Meteorology The study of the forces and processes that cause the atmosphere to change and produce weather is **meteorology.** Meteorologists also try to forecast the weather and learn how changes in weather over time might affect Earth's climate.

Figure 1 The Keck I and Keck II telescopes are part of the Mauna Kea Observatories in Hawaii. One of the Keck telescopes is visible here in its protective dome.

ACROSS THE CURRICULUM

History The idea that Earth is a dynamic planet gained support in the 1700s. Detailed observations of the landscape led a Scottish amateur geologist named James Hutton to propose that processes such as erosion could alter Earth over time. Hutton concluded that Earth is much different today than it was in the geologic past. He hypothesized that our planet changes gradually as the result of Earth processes. Hutton was correct in his hypothesis that Earth changes, but incorrect in his assumption that only gradual changes occur. Most Earth scientists today agree that Earth evolves sporadically, with some changes taking place more rapidly than others.

Geology The study of the materials that make up Earth, the processes that form and change these materials, and the history of the planet and its life-forms since its origin is the branch of Earth science known as **geology.** Geologists identify rocks and fossils, study glacial movements, interpret clues to Earth's 4.6-billion-year history, and determine how forces change our planet.

Oceanography The study of Earth's oceans, which cover nearly three-fourths of the planet, is called **oceanography.** Oceanographers study the creatures that inhabit salt water, measure different physical and chemical properties of the oceans, and observe various processes in these bodies of water. When oceanographers are conducting field research, they often have to dive into the ocean to gather data, as shown in **Figure 2.**

Figure 2 Oceanographers study the life and properties of the ocean.
Investigate *What kind of training would this Earth scientist need?*

Environmental science The study of the interactions of organisms and their surroundings is called **environmental science.** Environmental scientists study how organisms impact the environment both positively and negatively. The topics an environmental scientist might study include the use of natural resources, the effects of pollution, alternative energy sources, and the impact of humans on the atmosphere.

Subspecialties The study of our planet is a broad endeavor, and as such, each of the five major areas of Earth science consists of a variety of subspecialties, such as climatology, paleontology, and environmental chemistry. The descriptions of several subspecialties of Earth science are listed in **Table 1.**

Explore the **scope of Earth science with an interactive table.** Concepts In Motion

Table 1 Subspecialties of Earth Science

Major Area of Study	Subspecialty	Subjects Studied
Astronomy	astrophysics	physics of the universe, including the physical properties of objects found in space
	planetary science	planets of the solar system and the processes that form them
Meteorology	climatology	patterns of weather over a long period of time
	atmospheric chemistry	chemistry of Earth's atmosphere, and the atmospheres of other planets
Geology	paleontology	remains of organisms that once lived on Earth; ancient environments
	geochemistry	Earth's composition and the processes that change it
Oceanography	physical oceanography	physical characteristics of oceans, such as salinity, waves, and currents
	marine geology	geologic features of the ocean floor, including plate tectonics of the ocean
Environmental science	environmental soil science	interactions between humans and the soil, such as the impact of farming practices; effects of pollution on soil, plants, and groundwater
	environmental chemistry	chemical alterations to the environment through pollution and natural means

Alexis Rosenfeld/Photo Researchers

Use Science Terms

Word Parts Have students use their dictionaries to find out that the suffix *-ology* has a Greek origin and means *the study of.* Challenge students to determine the meanings of the following word parts used on these two pages: *meteor-*, *geo-*, *climat-*, *paleo-*, *hydro,* and *eco-*. Inform students that all of these terms come from the Greek language and reflect the pioneering exploration of many sciences during the time of the Greek empire, more than 2000 years ago. OL EL

Enrichment

Careers in Science Ask each student to choose one of the subspecialties of Earth science and write a brief summary of what people in that field do. In addition to describing the type of work done in the selected discipline, each student summary should include a description of the educational background needed in the field and the kinds of institutions where most people in that field are employed. This activity will make students aware of the possibilities for careers in Earth science and what is necessary to prepare for such careers. OL

Caption Question Fig. 2 Answers will vary. Students should indicate that some sort of higher degree is necessary.

EARTH SCIENCE JOURNAL

Apply Science Ask each student to find at least three examples of Earth science applications in the news or in their daily lives and to write a description of each application in their Earth science journals. Examples might include weather reports, construction projects, obtaining resources, global warming, pollution, and astronomical discoveries in the news. BL

DIFFERENTIATED INSTRUCTION

Struggling Learners Earth processes that are violent and abrupt typically hold most students' attention. Have students with learning disabilities find pictures or reports of cataclysmic events related to Earth science, such as volcanic eruptions, storms, floods, and earthquakes. Suggest that students paste the data in their Earth science journals and highlight any facts that they find interesting.

Teacher Content Support

Technology Studies in the subdisciplines of Earth science would not be possible without sophisticated and expensive technology, such as deep-sea probes, scanning microscopes, remote sensing, and spacecraft. Only a limited view of Earth would be possible without these technologies.

Collaborative Learning

Earth's Systems Divide the class into four groups and have each group prepare a report, which should include illustrations and photographs, about one of the four major Earth systems. OL COOP LEARN

FOLDABLES® Rubric

FOLDABLES® Incorporate information from this section into your Foldable.

VOCABULARY
SCIENCE USAGE V. COMMON USAGE
Crust
Science usage: the thin, rocky, outer layer of Earth

Common usage: the hardened exterior or surface part of bread

Earth's Systems

Scientists who study Earth have identified four main Earth systems: the geosphere, atmosphere, hydrosphere, and biosphere. Each system is unique, yet each interacts with the others.

Geosphere The area from the surface of Earth down to its center is called the **geosphere.** The geosphere is divided into three main parts: the crust, mantle, and core. These three parts are illustrated in **Figure 3.**

The rigid outer shell of Earth is called the crust. There are two kinds of crust—continental crust and oceanic crust. Just below the crust is Earth's mantle. The mantle differs from the crust both in composition and behavior. The mantle ranges in temperature from 100°C to 4000°C—much warmer than the temperatures found in Earth's crust. Below the mantle is Earth's core. Temperatures in the core may be as high as 7000°C.

Atmosphere The blanket of gases that surrounds our planet is called the **atmosphere.** Earth's atmosphere contains about 78 percent nitrogen and 21 percent oxygen. The remaining 1 percent of gases in the atmosphere include water vapor, argon, carbon dioxide, and other trace gases. Earth's atmosphere provides oxygen for living things, protects Earth's inhabitants from harmful radiation from the Sun, and helps to keep the planet at a temperature suitable for life.

Hydrosphere All the water on Earth, including the water in the atmosphere, makes up the **hydrosphere.** About 97 percent of Earth's water exists as salt water, while the remaining 3 percent is freshwater contained in lakes and rivers, beneath Earth's surface as groundwater, and in glaciers. The region of permanently frozen water on Earth is called the **cryosphere.** Only a fraction of Earth's total amount of freshwater is in lakes and rivers.

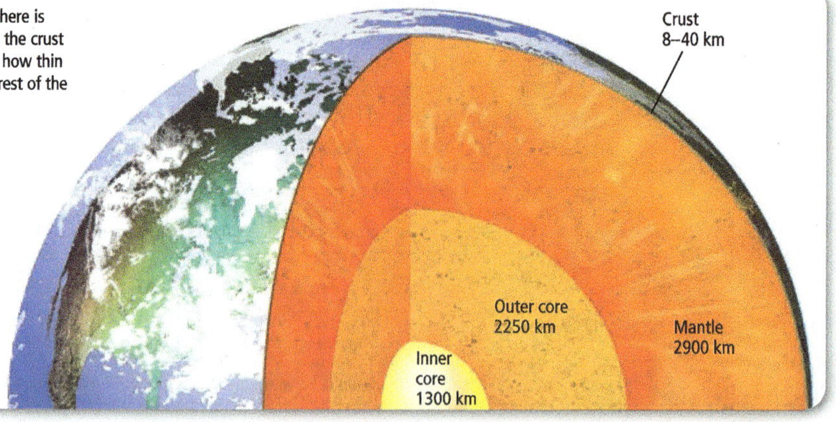

Figure 3 Earth's geosphere is composed of everything from the crust to the center of Earth. Notice how thin the crust is in relation to the rest of the geosphere's components.

IN THE FIELD

Written in the Stars Some Egyptologists hypothesize that the positions of various pyramids coincide with certain constellations. The relative positions of the Giza pyramids, for example, align with the positions of the stars that form the belt of the hunter, Orion. Also, the chamber of Pharaoh Khufu points directly to Polaris. Have interested students research the importance of astronomy in the Mayan, Anasazi, and Aztec civilizations.

Biosphere The **biosphere** includes all organisms on Earth as well as the environments in which they live. Most organisms live within a few meters of Earth's surface, but some exist deep beneath the ocean's surface, and others live high atop Earth's mountains. All of Earth's life-forms require interaction with at least one of the other systems for their survival.

As illustrated in **Figure 4,** Earth's biosphere, geosphere, hydrosphere, and atmosphere are interconnected and interdependent systems. For example, Earth's present atmosphere formed millions of years ago through interactions with the geosphere, hydrosphere, and biosphere. Organisms in the biosphere, including humans, continue to change the atmosphere through their activities and natural processes.

Figure 4 All of Earth's systems are interdependent. Notice how water from the hydrosphere enters the atmosphere, falls on the biosphere, and soaks into the geosphere.

Technology

The study of science, including Earth science, has led to many discoveries that have been applied to solve society's needs and problems. The application of scientific discoveries is called technology. Technology is transferable, which means that it can be applied to new situations. Freeze-dried foods, ski goggles, laptops, and the ultralight materials used to make many pieces of sports equipment were created from technologies used in our space program. Technology is not used only to make life easier. It can also make life safer. Smoke detectors are placed in houses and buildings to help warn people if there is a fire. Smoke detectors were also invented as part of the space program and were adapted for use in everyday life.

SECTION 1 REVIEW

Section Self-Check

Section Summary

- Earth is divided into four main systems: the geosphere, hydrosphere, atmosphere, and biosphere.
- Earth systems are all interdependent.
- Identifying the interrelationships between Earth systems leads to specialties and subspecialties.
- Technology is important, not only in science, but in everyday life.
- Earth science has contributed to the development of many items used in everyday life.

Understand Main Ideas

1. MAINIDEA **Explain** why it is helpful to identify specialties and subspecialties of Earth science.
2. **Apply** What are three items you use on a daily basis that have come from research in Earth science?
3. **Compare and contrast** Earth's geology and geosphere.
4. **Hypothesize** about human impact on each of Earth's systems.
5. **Compare and contrast** the hydrosphere and biosphere.

Think Critically

6. **Predict** what would happen if the makeup of the hydrosphere changed. What would happen if the atmosphere changed?

WRITING IN ▶ Earth Science

7. Research a subspecialty of Earth science. Make a brochure about a career in this field.

Interpret the Illustration

Interactions Among Earth's Systems Ask students to write down all of the interactions among the four main Earth systems that they are able to observe or infer from **Figure 4.** Answers will vary. Students should note that each system interacts with all the others. BL

3 Assess

Check for Understanding

Reinforcement Ask pairs of students to summarize the components and compositions of each of Earth's four major systems in a data table. Major points should include: 1. geosphere—surface of Earth to its center; 2. atmosphere—clouds; gases and liquids; 3. hydrosphere—lakes, rivers, oceans, seas, glaciers; water and water vapor; 4. biosphere—living things and their environments.

Reteach

Activity Show students a photograph from a nature magazine that shows components of Earth's major systems. Have students identify the components of each system and explain how the systems interact.

Assessment

Performance Ask small groups of students to debate whether astronomy should be included as a field of Earth science.

SECTION 1 REVIEW

1. Sample answer: The different specialties and subspecialties of Earth science add structure to the larger field of study. They add focus, allowing for more detailed information and organization of this information.
2. Answers will vary, but could include cell phones, MP3 players, computers, GPS units, the Internet, and so on.
3. The geosphere includes Earth and everything in it. Geology is the study of the materials that make up Earth, the processes that form these materials, and the history of the planet and its life-forms.
4. Answers will vary, but might include human's contribution to air, water, and soil pollution, which in turn affects Earth's organisms.
5. The hydrosphere includes all the water found on Earth, as well as water molecules in the atmosphere. The biosphere is made of all living things and their environments, including those found in the hydrosphere. Without the hydrosphere, living things could not survive.
6. Answers will vary. Students should indicate that all Earth's systems are interrelated, so a change in one would affect all other systems.
7. Answers will vary. Encourage students to research the subspecialty that interests them the most.

Rubric

SECTION 2

1 Focus

Focus Transparency

Before presenting the lesson, project **Section Focus Transparency 2** and have students answer the accompanying questions. EL

MAINIDEA

Solve Problems Propose a question that a scientist could have, such as: Would it be more energy-efficient to paint a house a dark color or a light color? Ask students how they think the scientist should go about finding the answer to this question.

2 Teach

Tie to Previous Knowledge

Assess TV Ads Discuss television commercial claims for various products such as soaps, detergents, and paper towels. Have students propose how they might test an advertiser's claims about the effectiveness of the same product made by different manufacturers. AL

SECTION 2

Methods of Scientists

MAINIDEA Scientists use scientific methods to structure their experiments and investigations.

Essential Questions

- What are independent and dependent variables?
- How does experimentation and investigation differ?
- What are the differences between mass and weight?
- What is scientific notation and how is it used?

Review Vocabulary

experiment: procedure performed in a controlled setting to test a hypothesis and collect precise data

New Vocabulary

scientific methods
hypothesis
independent variable
dependent variable
control
Le Système International d'Unités (SI)
scientific notation

EARTH SCIENCE 4 YOU **Have you ever seen a distinct rock formation and wondered how it formed? Have you ever wondered why the soil near your home might be different from the soil in your schoolyard? If so, you have already begun to think like a scientist. Scientists often ask questions and make observations to begin their investigations.**

The Nature of Scientific Investigations

Scientists work in many different places to gather data. Some work in the field, and some work in a lab, as shown in **Figure 5.** No matter where they work, they all use similar methods to gather data and communicate information. These methods are referred to as scientific methods. As illustrated in **Figure 6,** **scientific methods** are a series of problem-solving procedures that help scientists conduct experiments.

Whatever problem a scientist chooses to pursue, he or she must gather background information on the topic. Once the problem is defined and the background research is complete, a hypothesis is made. A **hypothesis** is a testable explanation of a situation that can be supported or disproved by careful procedures.

It is important to note that scientific methods are not rigid, step-by-step outlines to solve problems. Scientists can take many different approaches to performing a scientific investigation. In many scientific investigations, for example, scientists form a new hypothesis after observing unexpected results. A researcher might modify a procedure, or change the control mechanism. And a natural phenomenon might change the direction of the investigation.

Figure 5 Whether a meteorologist gathers storm data in the field or an environmental scientist analyzes microbial growth in a lab, scientific methods provide an approach to problem-solving and investigation.

Meteorologist

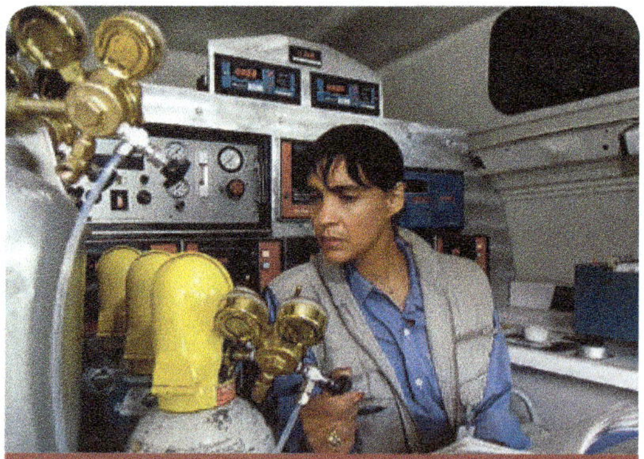

Environmental scientist

(l)David Hay Jones/Photo Researchers, (r)Dwayne Newton/PhotoEdit

ACROSS THE CURRICULUM

Language Arts Ask students that are English language learners to share words in their native language from Earth science. For example, what are their words for the Earth, Sun, Moon, planets and stars? Is star-gazing something they did as kids? What planets and constellations could they see from their windows? Students might recognize similarities between Earth science terms from different languages. EL

VISUALIZING Scientific Methods

Figure 6 Scientific methods are used by scientists to help organize and plan their experiments and investigations. The flow chart below outlines some of the methods commonly used by scientists.

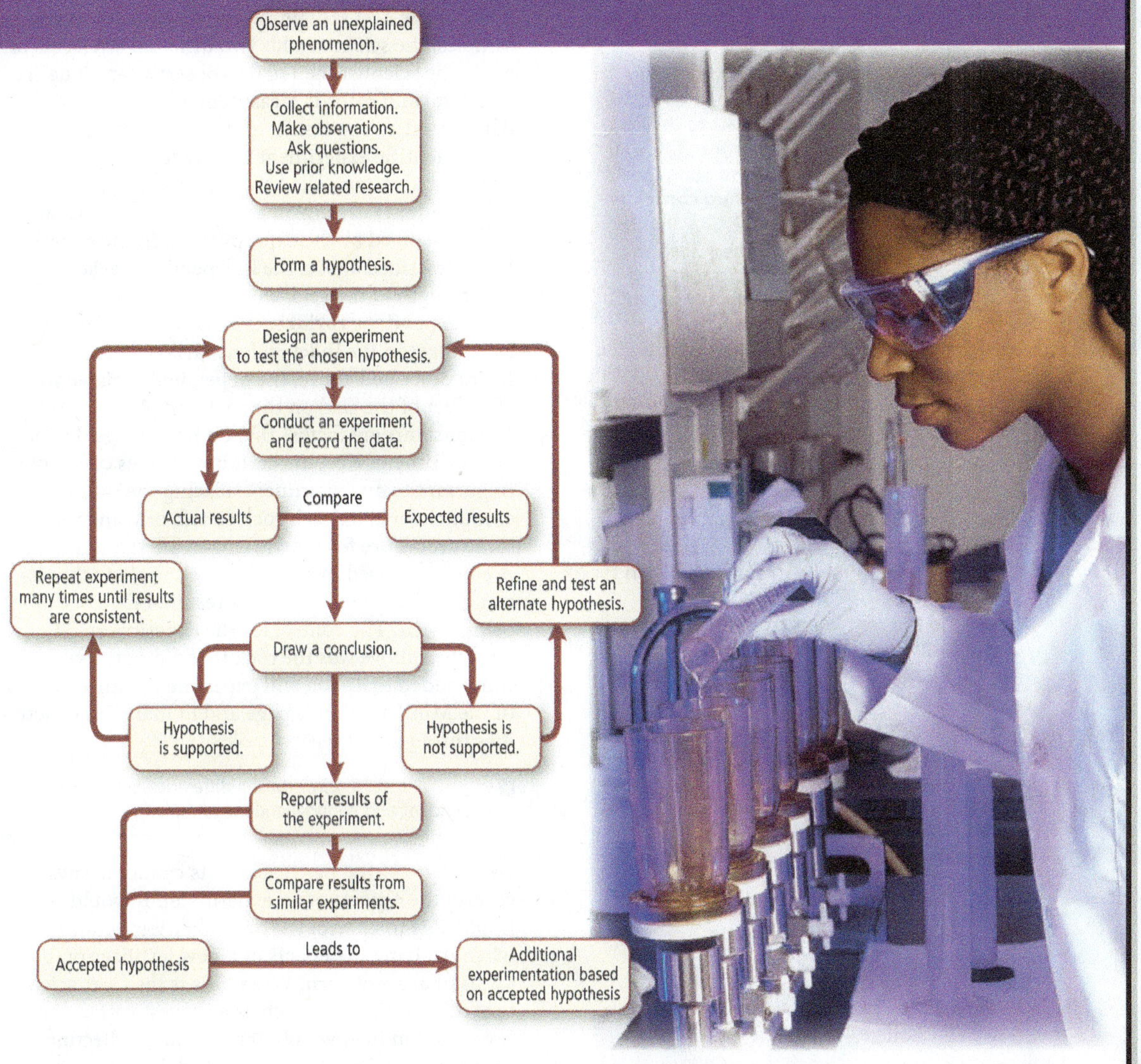

David Wasserman/Brand X/CORBIS

Concepts In Motion View an **animation of scientific methods.**

VISUALIZING

Purpose

Students will explore scientific methods.

Project

Design Experiments Have pairs of students design their own experiments to test some aspect of Earth science. Have students describe in writing how they would carry out their experiments. Each description should clearly indicate the purpose, the hypothesis, the independent and dependent variables, and the expected results. Some students might wish to perform the experiments. Make sure to preapprove any experiment before it is conducted. **AL** **COOP LEARN**

Use an Analogy

Solve Crimes Explain that a scientific experiment can be likened to the work of a detective solving a crime. A detective's purpose is to determine who committed the crime, and at some point, the detective forms a hypothesis, which can be tested by logical analysis. Sometimes, a detective will even make predictions about where evidence could be hidden or what the suspect in a crime might do next. Finally, a detective, like a scientist, must be prepared to reject a hypothesis that is found to be inconsistent with known facts or that is not supported by further investigation.

DIFFERENTIATED INSTRUCTION

Struggling Learners Some students might be more familiar with a linear process for scientific methods or a more simplified version. For those students, provide them with a simplified version and ask them to determine similarities and differences when compared to the flow chart above. Ask students to identify the benefits of using a nonlinear flow chart.

READING CHECK The independent variable is the factor that changes in a experiment. The dependent variable is the factor that is affected by changes in the independent variable.

MiniLAB

iLab Station | Rubric

Purpose Students will manipulate variables to measure heat transfer and retention in soil and water.

Process Skills observe; recognize cause and effect; use variables, constants, and controls

Safety Precautions Approve lab safety forms before work begins. Tell students to alert you immediately if the thermometer breaks. Stress that they should NOT touch the thermometer. Find out where to properly dispose of broken glass.

Teaching Strategies

- If direct sunlight is not available, use a 150-W floodlight as the light source.
- Have students make multiple line graphs of their results.

Expected Results The soil will heat up and cool down faster than water.

Analysis

1. The soil absorbed heat more quickly than the water did. The soil cooled more quickly than the water did.
2. The independent variable was the medium—soil or water. The dependent variable was temperature.
3. The control was the empty container.

Assessment

Knowledge Ask students to hypothesize what would have happened if both containers had been left in sunlight so long that the temperature in both the soil and the water reached a steady-state value. The water would have had a higher steady-state temperature because water has a higher heat capacity, which means that it can store more heat per cubic meter than soil can.

MiniLAB

iLab Station

Determine the Relationship Between Variables

How do the rates of heat absorption and release vary between soil and water? Different substances absorb and release heat at different rates.

Procedure

1. Read and complete the lab safety form.
2. Read the procedure and create a data table to record your temperature results.
3. Pour **soil** into **one container** until it is half full. Pour **water** into a **second container** until it is half full. Leave a **third container** empty.
4. Place **one thermometer** in the soil so that the bulb is barely covered. Use **masking tape** to secure **another thermometer** about 1 cm above the top of the soil.
5. Repeat Step 4 for the container with water.
6. In the empty container, place the bulb of one thermometer halfway into the cup and secure it with masking tape. Use the tape to secure another thermometer bulb about 2 cm higher than the first thermometer bulb.
7. Put the containers on a **sunny windowsill.** Record the temperature shown on each thermometer. Write these values in a table. Record temperature readings every 5 min for 30 min.
8. Remove the containers from the windowsill and continue to record the temperature on each thermometer every 5 min for 30 min.

Analysis

1. **Determine** Which substance absorbed heat more quickly? Which substance lost heat more quickly?
2. **Specify** What was your independent variable? What was your dependent variable?
3. **Identify** your control.

Experimentation An experiment is classified as an organized procedure that involves making observations and measurements to test a hypothesis. Collecting good qualitative and quantitative data is vital to the success of an experiment.

Imagine a scientist is conducting an experiment on the effects of acid on the weathering of rocks. In this experiment, there are three different samples of identical rock pieces. The scientist does not add anything to the first sample. To the second and third samples, the scientist adds two different strengths of acid. The scientist then makes observations (qualitative data) and records measurements (quantitative data) based on the results of the experiment.

A scientific experiment usually tests only one changeable factor, called a variable, at a time. The **independent variable** in an experiment is the factor that is changed by the experimenter. In the experiment described above, the independent variable was the strength of the acid.

A **dependent variable** is a factor that is affected by changes in the independent variable. In the experiment described above, the dependent variable was the effect of the acid on the rock samples.

Constants are factors that do not change during an experiment. Keeping certain variables constant is important to an experiment. Placing the same amount of acid on each rock tested, or using the same procedure for measurement, are two examples. A **control** is used in an experiment to show that the results of an experiment are a result of the condition being tested. The control for the experiment described above was the rock that did not have anything added to it. You will experiment with variables in the MiniLab on this page and in many other activities throughout this textbook.

READING CHECK **Explain** the difference between a dependent and an independent variable.

Investigation Earth scientists cannot always control the aspects of an experiment. It would be impossible to control the rainfall or temperature when studying the effects of a new fertilizer on thousands of acres of corn. When this is the case, scientists refer to their research as an investigation. An investigation involves observation and collecting data but does not include a control. Investigations can often lead scientists to design future experiments based on the observations they have made.

LabManager™ Customize any lab with the LabManager™ CD-ROM.

Safety Many of the experiments and investigations in this book will require that you handle various materials and equipment. When conducting any scientific investigation, it is important to use all materials and equipment only as instructed. Refer to the *Reference Handbook* for additional safety information and a table of safety symbols.

Analysis and conclusions New ideas in science are carefully examined by the scientist who made the initial discovery and by other scientists in the same field. Processes, data, and conclusions must be examined to eliminate influence by expectations or beliefs, which is called bias. During a scientific experiment, all data are carefully recorded. Once an experiment is complete, graphs, tables, and charts are commonly used to display data. These data are then analyzed so that a conclusion can be drawn. Many times, a conclusion does not support the original hypothesis. In such a case, the hypothesis must be reevaluated and further research must be conducted.

VOCABULARY

ACADEMIC VOCABULARY

Bias

to influence in a particular, typically unfair, direction; prejudice

Their choice of teammates showed a bias toward their friends.

Measurement

Scientific investigations often involve making measurements. A measurement includes both a number and a unit of measure. Scientific investigations use a standard system of units called **Le Système International d'Unités** (SI), which is a modern version of the metric system. SI is based on a decimal system that uses the number 10 as the base unit. See **Table 2** for information on SI and metric units of measure commonly used in science.

Length The standard SI unit to measure length is the meter (m). The distance from a doorknob to the floor is about 1 m. The meter is divided into 100 equal parts called centimeters (cm). Thus, 1 cm is 1/100 of 1 m. One millimeter (mm) is smaller than 1 cm. There are 10 mm in 1 cm. Longer distances are measured in kilometers (km). There are 1000 m in 1 km.

Table 2 Measurement and Units

Measurement	SI and Metric Units Commonly Used in Science
Length	millimeter (mm), centimeter (cm), meter (m), kilometer (km)
Mass	gram (g), kilogram (kg), metric ton
Area	square meter (m^2), square centimeter (cm^2)*
Volume	cubic meter (m^3)*, milliliter (mL), liter (L)#
Density	grams per cubic centimeter (g/cm^3)*, grams per milliliter (g/mL)#, kilograms per cubic meter (kg/m^3)*
Time	second (s), hour (h)#
Temperature	kelvin (K)

* units derived from SI units   # commonly used metric units

Teacher Content Support

Measurement To evaluate critically the merit and validity of scientific research and experiments, scientists must use the same system of measurement. Most scientists use le Système International d'Unités, or SI. The basic units in this system are the kilogram, the meter, and the second. Originally, the kilogram and meter were defined by prototype samples—a kilogram mass and a platinum meterstick, both of which were kept in a vault in France. Copies of these standards were made and distributed so that scientists everywhere could use identical measurements. However, because this proved to be impractical, in 1983, the meter was redefined as the distance that light travels in a vacuum during the time period of one 1/c, where c is the speed of light, 299,792,458 m/s. Nothing could be done about the kilogram, however, and this unit is still defined by the prototype block in the vault in France.

DIFFERENTIATED INSTRUCTION

Advanced Learners Ask students to each find a description of an actual scientific experiment in an Earth science technical journal. Have students identify the purpose, the hypothesis, the procedure, the independent and dependent variables, and the conclusions of the experiment.

Identify Misconceptions

Students often use the terms *mass* and *weight* synonymously.

Uncover the Misconception Ask students how the weight of an astronaut changes when he or she travels to the Moon.

Demonstrate the Concept Remind students that weight is the force exerted on an object by the local gravitational field. An astronaut on the Moon weighs one-sixth of his or her sea-level weight on Earth, but has the same mass. Mass is the quantity of matter contained in an object and does not depend on the position of the object.

Assess New Knowledge Ask students to explain why the standard conversion of 1 kg to 2.2 lb can be used everywhere on Earth and why this conversion would be invalid on the Moon.

READING CHECK Weight is the measure of a gravitational force on an object. Mass is the amount of matter in an object.

GeoLAB iLab Station

The GeoLab at the end of the chapter can be used at this point in the lesson.

Mass The amount of matter in an object is called mass. Mass depends on the number and types of atoms that make up the object. The mass of an object is the same no matter where the object is located in the universe. The SI unit of mass is the kilogram (kg).

Weight Weight is a measure of the gravitational force on an object. Weight is typically measured with some type of scale. Unlike mass, weight varies with location. For example, the weight of an astronaut while on the Moon is about one-sixth the astronaut's weight on Earth. This is because the gravitational force exerted by the Moon on the astronaut is one-sixth the force exerted by Earth on the astronaut. Weight is a force, and the SI unit for force is the newton (N). A 2-L bottle of soft drink with a mass of 2 kg weighs about 20 N on Earth.

READING CHECK **Compare** mass and weight.

Area and volume Some measurements, such as area, require a combination of SI units. Area is the amount of surface included within a set of boundaries and is expressed in square units of length, such as square meters (m^2).

The amount of space occupied by an object is the object's volume. The SI units for volume, like those for area, are derived from the SI units used to measure length. The basic SI unit of volume for a solid object is the cubic meter (m^3). Measurements for fluid volumes are usually made in milliliters (mL) or liters (L). Liters and milliliters are metric units that are commonly used to measure liquid volumes. Volume can also be expressed in cubic centimeters (cm^3)—1 cm^3 equals 1 mL.

Figure 7
Major Events in Earth Science
Many discoveries during the twentieth and early twenty-first centuries revolutionized our understanding of Earth and its systems.

1907 Scientists begin using radioactive decay to determine that Earth is billions of years old. This method will be used to develop the first accurate geological time scale.

1913 French physicists discover the ozone layer in Earth's upper atmosphere and propose that it protects Earth from the Sun's ultraviolet radiation.

1925 Cecilia Payne's analysis of the spectra of stars reveals that hydrogen and helium are the most abundant elements in the universe.

1936 Inge Lehmann proposes that Earth's center consists of a solid inner core and a liquid outer core based on her studies of seismic waves.

1955 Louis Essen invents a highly accurate atomic clock that tracks radiation emitted and absorbed by cesium atoms.

1900 1920 1940 1960

ACROSS THE CURRICULUM

History Have students research the history of the metric system and report on significant steps in its development and adoption around the world. Students will find that the United States played a significant role in the development of the metric system. The United States was the first nation to adopt a decimal currency, and a federal law was passed in 1866 urging the use of the metric system for weights and measures. Challenge students to find out why the United States is the only major nation resisting full-scale adoption of the metric system. OL

Density The measure of the amount of matter that occupies a given space is density. Density is calculated by dividing the mass of the matter by its volume. Density is often expressed in grams per cubic centimeter (g/cm^3), grams per milliliter (g/mL), or kilograms per cubic meter (kg/m^3).

Time The interval between two events is time. The SI unit of time is the second. In the activities in this book, you will generally measure time in seconds or minutes. Time is usually measured with a watch or clock. The atomic clock provides the most precise measure of time currently known. Known as UTC, Coordinated Universal Time is based on the atomic clock element cesium-133 and is adapted to the astronomical demarcation of day and night. See **Figure 7** for more information on the invention of the atomic clock and other advances in Earth science.

Vocabulary

Academic vocabulary

Interval
space of time between two events or states
The interval for pendulum swings was three seconds.

Temperature A measure of the average kinetic energy of the particles that make up a material is called temperature. A mass made up of particles that vibrate quickly generally has a higher temperature than a mass whose particles vibrate more slowly. Temperature is measured in degrees with a thermometer. Scientists often measure temperature using the Celsius (°C) scale. On the Celsius scale, a comfortable room temperature is about 21°C, and the normal temperature of the human body is about 37°C.

The SI unit for temperature is the kelvin (K). The coldest possible temperature, absolute zero, was established as 0 K or −273 °C. Since both temperature units are the same size, the difference between the two scales (273) is used to convert from one scale to another. For example, the temperature of the human body is 37°C, to which you would add 273 to get 310 K.

NASA/epa/CORBIS

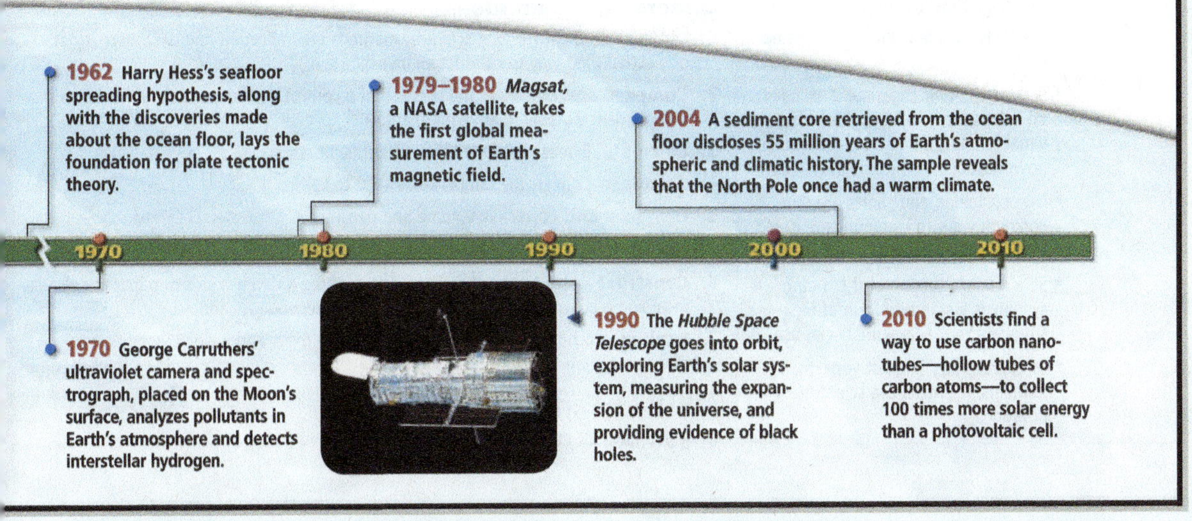

Teacher Content Support

Experimental Errors Uncertainty in measurement is unavoidable and must be accounted for when drawing conclusions. Experimental error can be reduced by making repeated measurements of the same quantity and averaging the results to cancel random errors. Make students aware of this as they do experiments and activities throughout this course.

Activity

Convert Temperatures Have students make the following conversions between the Celsius and absolute temperature scales: 50°C to K; −49°C to K; 77 K to °C; 500 K to °C. 50°C = 323 K; −49°C = 224 K; 77 K = −196°C; 500 K = 227°C OL

Teacher Content Support

Temperature Scales Despite the adoption of the Celsius temperature scale by most countries, in the United States, the Fahrenheit scale is more commonly used. To convert from Celsius to Fahrenheit, multiply by 9/5 and then add 32°. To convert from Fahrenheit to Celsius, subtract 32°, then multiply by 5/9.

Demonstration

Experimental Error To illustrate that experimental error is always present in measurements, have students take turns measuring the same quantity, such as the width of a desk or tabletop, with a meterstick. Have each student record his or her value. Compile the results on the chalkboard so that students can see that there is some variation in measured values. BL COOP LEARN

3 Assess

Check for Understanding

Activity Ask students to go through an entire day using SI units. Have students convert every quantitative value that they encounter into SI units, including weather report temperatures, distances traveled, speeds of vehicles, distances and masses encountered in a physical education class or on an athletic field, and so on. Each student should record all of these quantities in his or her Earth science journal.

Reteach

Activity Have students name the SI unit for each of the following quantities: mass, temperature, weight, length, area, volume, time, and density.

Assessment

Knowledge Have students write the following values in scientific notation: 299,792,458 m/s (the speed of light) and 0.000 000 000 000 000 000 000 000 000 000 911 kg (the mass of an electron). 2.99792458×10^8 and 9.11×10^{-31} kg

■ **Figure 8** On a 5-km-long beach, such as the one shown above, there might be 8×10^{15} grains of sand. The average size of a grain of sand is 0.5 mm.

Scientific Notation

In many branches of science, some commonly used numbers are very small, while others are very large. To express these numbers conveniently, scientists use a type of shorthand called **scientific notation,** in which a number is expressed as a value between 1 and 10 multiplied by a power of 10. The power of 10 is the number of places the decimal point must be shifted so that only a single digit remains to the left of the decimal point.

If the decimal point must be shifted to the left, the exponent of 10 is positive. **Figure 8** shows a beach covered in sand. The number of grains of sand on Earth has been estimated to be approximately 4,000,000,000,000,000,000,000. In scientific notation, this number is written as 4×10^{21}. In this case, 21 refers to the number of places the decimal point must be moved to the left, or in other words, the number of decimal places to the right of the 4.

In astronomy, masses and distances are usually so large that writing out the numbers would be cumbersome. For example, the mass of Earth at 5,973,600,000,000,000,000,000,000 kg would be written as 5.9736×10^{24} kg in scientific notation.

If the decimal point in a number must be shifted to the right, the exponent of 10 is negative. The diameter of an atom in meters, for example, which is approximately 0.0000000001 m, is written as 1×10^{-10} m.

SECTION 2 REVIEW

Section Self-Check

Section Summary

- Scientists work in many ways to gather data.
- A good scientific experiment includes an independent variable, dependent variable, and control. An investigation, however, does not include a control.
- Graphs, tables, and charts are three common ways to communicate data from an experiment.
- SI, a modern version of the metric system, is a standard form of measurement that all scientists can use.
- To express very large or very small numbers, scientists use scientific notation.

Understand Main Ideas

1. MAINIDEA **Explain** why scientific methods are important and why there is not one established way to conduct an investigation.
2. **Compare and contrast** the purpose of a control, an independent variable, and a dependent variable in an experiment.
3. **Calculate** Express 0.00049386 in scientific notation.
4. **Calculate** Convert the temperature 49°C to kelvin.
5. **Compare and contrast** volume and density.

Think Critically

6. **Construct** a plan to test the absorption of three different kinds of paper towels, including a control, dependent variable, and independent variable.
7. **Explain** which is more useful when comparing mass and weight on different planets.

MATH IN ▶ Earth Science

8. If you have 20 mL of water, how many cubic centimeters of water do you have?

(inset)David Scharf/Photo Researchers, (bkgd)Bloomimage/CORBIS

SECTION 2 REVIEW

1. Scientific methods are important because they guide scientists as they conduct experiments and investigations. Scientists sometimes form new hypotheses after obtaining unexpected results or pursue new ideas that arise in the midst of an investigation. Thus, there is no one correct way to design an investigation.
2. A control is a standard for comparison. A dependent variable is a quantity whose value is determined by other quantities. An independent variable is a quantity whose value might vary without depending on other quantities, but which might determine the values of other quantities.
3. 4.9386×10^{-4}
4. 322 K
5. Volume is the amount of space an object occupies. Density is the measure of the amount of matter that occupies a given space.
6. Answers will vary but should identify a control, dependent variable, and dependent variable.
7. Weight is useful when comparing gravities. Mass is useful when comparing objects.
8. 20 cm^3

Communication in Science

SECTION 3

Essential Questions

- Why is precise communication crucial in science?
- What are the differences between scientific theories and scientific laws?
- When is it appropriate to use a graph or a model?

Review Vocabulary

hypothesis: testable explanation of a situation

New Vocabulary

scientific model
scientific theory
scientific law

MAINIDEA Precise communication is crucial for scientists to share their results effectively with each other and with society.

EARTH SCIENCE 4 YOU If you read an advertisement for a product called "Glag" without any description, would you know whether to eat it or wear it? When a scientist does an investigation, he or she has to describe every part of it precisely so that everyone can understand his or her conclusions.

Communicating Results

There are many ways to communicate information, such as newspapers, magazines, TV, the Internet, and scientific journals. Think back to the Launch Lab from the beginning of the chapter. Although you and your lab partner both used the same form of communication, were your descriptions identical? Scientists have the responsibility to truthfully and accurately report their methods and results. To keep them ethical, a system of peer review is used in which scientists in the same field verify each other's results and examine procedures and conclusions for bias. Communicating scientific data and results, as the scientists are shown doing in **Figure 9,** also allows others to learn of new discoveries and conduct new investigations that build on previous investigations.

Lab reports Throughout this book, you will conduct many Earth science investigations and experiments. During and after each, you will be asked to record and analyze the information that you collected and to draw conclusions based on your data. Your written account of each lab is your lab report. This will be used by your teacher to assess your understanding. You might also be asked to compare your results with those of other students to help you find both similarities and differences among the results.

■ **Figure 9** Scientists, like those shown in the photo, communicate data and discoveries with each other to maintain accuracy in methods and reporting.
Infer *what could happen if scientists did not compare results.*

EARTH SCIENCE JOURNAL

Scientific Publications Have students research the steps involved in publishing a scientific paper. Students should share their results with the class. When a scientist develops a theory or makes a discovery, the results are scrutinized by the scientist's peers as part of the processes of testing and validation. Normally, the scientist who formulates the new model or makes the discovery writes a paper describing the work, which can be published in a professional journal. Before the paper is published, however, it is reviewed by other scientists in the same field who determine whether the work is new and whether there are any flaws in the logic or errors in the analysis of the data. Once the paper passes this often-stringent review process, it is published in a scientific journal, where other scientists can read it and either verify the work or challenge it. Thus, scientific research is a self-regulating process aimed at ensuring that only viable models and theories are put forward. AL

SECTION 3

1 Focus

Focus Transparency

Before presenting the lesson, project **Section Focus Transparency 3** and have students answer the accompanying questions. EL

MAINIDEA

Communication Ask students to think of a time when they had to explain something such as a video game to another person. Help them realize that understanding something is different from communicating the idea to someone else.

2 Teach

Identify Misconceptions

People often use the terms *hypothesis* and *theory* synonymously, but they are different concepts.

Uncover the Misconception Ask students to define *hypothesis* and *theory.* Write some responses on the board.

Demonstrate the Concept Use the following scenario to demonstrate the concept: A scientist tests a hypothesis. The results of the experiment support her hypothesis. If the experiment is repeated many times and the same results occur, the hypothesis becomes a theory.

Assess New Knowledge Ask students to revise their definitions of *hypothesis* and *theory.*

■ **Caption Question Fig. 9** Scientists would have a difficult time learning of new discoveries, verifying results, and examining conclusions for bias.

■ **Caption Question Fig. 10** As gas volume increases, temperature increases.

Data Analysis LAB

About the Lab

- Explain that the increase in Earth's surface temperature is called global warming and is thought to be mainly the result of burning fossil fuels.
- To find out more about climate change, visit NOAA's Web site.

Think Critically

1. Students should make a line graph.
2. 13.76°C, 13.77°C, 13.97°C, 14.02°C, 13.98°C, 14.33°C, 14.59°C
3. approximately 14.45°C
4. approximately 14.9°C

Activity

Extrapolate Data Have students extrapolate the data in the Data Analysis Lab to determine what Earth's surface temperature might be every 100 years for the next 1000 years. AL

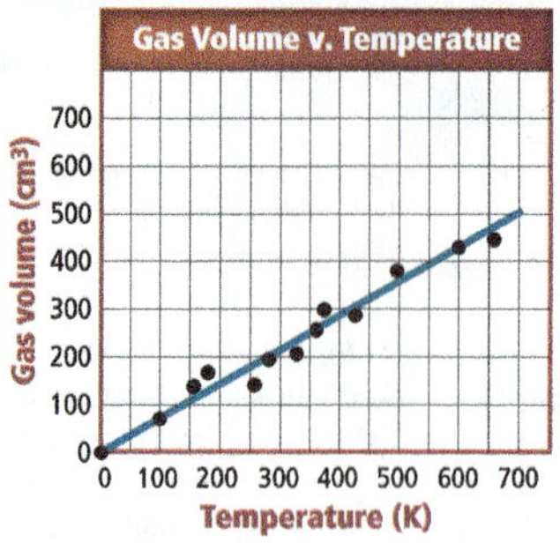

■ **Figure 10** A line graph shows the relationship between two variables.
Determine *Based on this graph, what is the relationship between gas volume and temperature?*

Graphs By graphing data in a variety of ways, scientists can more easily show the relationships among data sets. Graphs also allow scientists to represent trends in their data. You will be asked to graph the results of many experiments and activities in this book. There are three types of graphs you will use in this book.

Line graphs A visual display that shows how two variables are related is called a line graph. As shown in **Figure 10,** on a line graph, the independent variable is plotted on the horizontal (x) axis, and the dependent variable is plotted on the vertical (y) axis.

Circle graphs To show a fixed quantity, scientists often use a circle graph, also called a pie graph. The circle represents the total and the slices represent the different parts of the whole. The slices are usually presented as percentages.

Bar graphs To represent quantitative data, bar graphs use rectangular blocks called bars. The length of the bar is determined by the amount of the variable you are measuring as well as the scale of the bar graph. See the *Skillbuilder Handbook* for examples of all the types of graphs described above.

Models In some of the investigations, you will be making and using models. A **scientific model** is an idea picture, a system, or a mathematical expression that represents the concept being explained. While a model might not have all of the components of a given idea, it should be a fairly accurate representation.

Data Analysis LAB

Based on Real Data*

Make and Use Graphs

How can graphs help interpret data? The table shows the average surface temperature of Earth over the past 125 years. The data in the table are global, average surface temperatures, in kelvin, starting in the year 1880.

Think Critically

1. **Construct** a line graph from the average surface temperatures in the data table.
2. **Convert** each temperature from kelvin to degrees Celsius by subtracting 273 from each value. Place both on your graph.
3. **Determine** from your graph the average surface temperature for 1988 in degrees Celsius.
4. **Extrapolate,** in Celsius, what the average surface temperature will be in the year 2100 if this trend continues.

Data and Observations

Average Global Surface Temperatures

Years	Average surface temperature (K)
1880–1899	286.76
1900–1919	286.77
1920–1939	286.97
1940–1959	287.02
1960–1979	286.98
1980–1999	287.33
2000–2004	287.59

*Data obtained from Goddard Institute for Space Studies, NASA Goddard Space Flight Center

DIFFERENTIATED INSTRUCTION

Visually Impaired Provide students who are visually impaired with enlarged copies of the completed graph from the Data Analysis Lab to assist them in interpreting the data.

ACROSS THE CURRICULUM

Language Arts Have students define and use the terms *theory* and *law* in everyday vernacular. Have students develop working definitions of these terms as they apply to Earth science. Correct any errors in the definitions as you teach this section. EL OL

Models can change when more data are gathered. As shown in **Figure 11,** early astronomers thought that Earth was the center of the solar system. This model was changed as the result of observations of the motions of the Sun and the planets in the night sky. The observations showed that the planets in our solar system orbit the Sun.

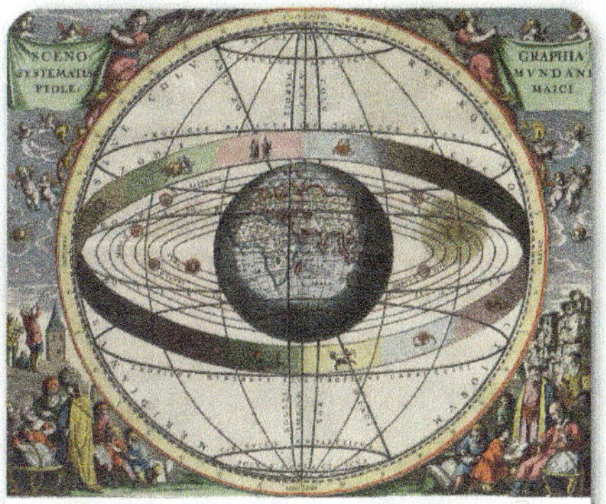

■ **Figure 11** Scientific models, like this ancient one of the solar system, are used to represent a larger idea or system. As scientists gather new information, models can change or be revised.
Explain *what is wrong with this model.*

■ **Caption Question Fig. 11** Earth is shown as the center of the universe.

Theories and Laws

A **scientific theory** is an explanation based on many observations during repeated investigations. A scientific theory is valid only if it meets the following criteria: is consistent with observations and is supported by experimental or factual evidence, makes predictions that can be tested, and is the simplest explanation of observations. Like a scientific model, a theory can be changed or modified with the discovery of new data.

A **scientific law** is a principle that describes the behavior of a natural phenomenon. A scientific law can be thought of as a rule of nature, even though the cause of the law might not be known. The events described by a law are observed to be the same every time. An example of a scientific law is Newton's first law of motion, which states that an object at rest or in motion stays at rest or in motion unless it is acted on by an outside force. This law explains why Earth and other planets in our solar system remain in orbit around the Sun. Theories are often used to explain scientific laws.

In this book, you will communicate your observations and draw conclusions based on scientific data. You will also read that many of the models, theories, and laws used by Earth scientists to explain various processes and phenomena grow from the work of other scientists and sometimes develop from unexpected discoveries.

3 Assess

Check for Understanding

Activity Have each student write a short paragraph that compares and contrasts the following scientific terms: *model, theory,* and *law.* Allow a few volunteers to read their paragraphs aloud. Have other students constructively critique the paragraphs.

Reteach

Activity Provide students with an incomplete concept map that summarizes the information presented in this section. Allow students to work in pairs to complete their maps.

Assessment

Knowledge Ask students to compare the format of their lab reports with the elements of a scientific method. Students will discover that the formats are similar. Most student lab-report guidelines call for a sequence of steps, starting with a statement of the problem, a hypothesis, an experiment to test the hypothesis, and a conclusion. These steps closely parallel the main steps of a scientific method.

SECTION 3 REVIEW

Section Self-Check

Section Summary

- Scientists communicate data so others can learn the results, verify the results, examine conclusions for bias, and conduct new experiments.
- There are three main types of graphs scientists use to represent data: line graphs, circle graphs, and bar graphs.
- A scientific model is an accurate representation of an idea or theory.
- Scientific theories and scientific laws are sometimes discovered accidentally.

Understand Main Ideas

1. MAINIDEA **Explain** what might happen if a scientist inaccurately reported data from his or her experiment.
2. **Describe** the difference between scientific theory and scientific law.
3. **Apply** Why is it important to compare your data from a lab with that of your classmates?

Think Critically

4. **Interpret** Why would a model be important when studying the solar system?
5. **Explain** when to use a line graph, a circle graph, and a bar graph.

WRITING IN ▶ Earth Science

6. Research scientific laws and theories, and write a concise example of each.

SECTION 3 REVIEW

1. Answers will vary. Students might suggest that the scientist's reputation would suffer, he or she could lose funding, future experiments based on his or her work would be flawed, and conclusions drawn from the study could put people at risk.
2. A scientific theory is the best available explanation of a phenomenon. A scientific law is a rule of nature whose cause is not always understood but which is accepted based on observations or experiments.
3. It helps to verify results and identify possible sources of error.
4. The solar system is vast, and scientists cannot easily travel to faraway planets to study them directly.
5. Line graphs are generally used to show trends in data, such as a change in temperature over time. Circle graphs are used to show parts of a whole, or percentages. Circle graphs might be used extensively in population studies. Bar graphs are used to compare groups, such as comparing the number of squirrels in different neighborhoods.
5. Answers will vary, but might include the law of gravity, and the theory of plate tectonics.

Rubric

eXpeditions!

Purpose

Students will learn about the efforts of an environmental engineer to investigate the causes and effects of tsunamis.

Teacher Content Support

Tsunamis A tsunami usually occurs when an earthquake displaces water. The energy from the quake is transferred to the water, which is pushed above normal sea level. Once the wave or tsunami forms, gravity forces the energy out in concentric circles. The waves are barely noticeable above the waterline, but they travel much faster than ordinary waves. As they approach the shore, they slow down, transferring their speed into wave height. As the height increases, the distance between waves shortens. Often, observers of the tsunami will first notice the tide receding more than usual.

Teaching Strategy

Place students into four groups. Assign each group a natural disaster, such as a tsunami, landslide, flood, and tornado. Ask each group to develop a plan showing how they could use the disaster to gather data. What data would they collect? How would they use the data to create a rescue plan? Afterward, critique the plans as a class.

eXpeditions!

ON SITE: IN THE FOOTSTEPS OF DISASTER

The tsunami destroyed many homes and buildings, leaving few of the structures standing.

On December 26, 2004, a massive earthquake rattled the seafloor of the Indian Ocean. A tsunami was generated by the earthquake, which devastated the landscape and killed almost 300,000 people in 12 countries. After humanitarian efforts were underway, many Earth scientists mobilized to collect data before the area was changed by cleanup efforts.

Planning the investigation Jose Borrero, an environmental engineer at University of Southern California, wanted to determine the height of the waves associated with the tsunami, how far inland they traveled, the number of waves, and the distance between them. This information would determine where to rebuild towns and assist in the development of a warning system and a hazard plan.

Taking measurements To measure heights of the waves and the following rush of water, Borrero looked for mud or watermarks on the buildings that were left standing. He then placed a 5-m pole next to the watermark to measure the height the water reached. The closer he got to the coast, however, the less he was able to measure accurately. The water had surged up over 5 m deep, so he relied on visual estimates and photos for documentation. With each measurement, he recorded the location on the satellite-based Global Positioning System (GPS).

After a six-day study of the devastation, Borrero had more than 150 data points. Upon returning to the United States, scientists used these data to determine that the waves reached 15–30 m high in Banda Aceh, and almost 3.2 km inland.

Using models It is impossible and unethical to simulate natural disasters on an actual scale, so scientists use the data collected from real incidents to create models of those events to learn more about how nature behaves. Using scientific methods and data gathered, scientists are able to provide information for model building or computer simulation. Back at the lab, Borrero applies the data to study other possible tsunami scenarios. He uses data to predict wave height and the area of inundation along the coast, should a tsunami hit the United States.

He hopes that the data collected will enable better detection and prevent widespread devastation from a natural tsunami disaster.

WRITING IN ▸ Earth Science

Journal Imagine you are a geologist who is accompanying a team of scientists to the site of a natural disaster. Describe the way you will use the scientific method to gather data for your report.

WebQuest

WRITING IN ▸ Earth Science

Rubric

Journal Answers will vary, but should include some of the steps listed in **Figure 6.** The order of the steps can vary, depending on the investigation.

WebQuest

GeoLAB

iLab Station

Measurement and SI Units

Background: **Suppose someone asked you to measure the area of your classroom in square cubits. What would you use? A cubit is an ancient unit of length equal to the distance from the elbow to the tip of the middle finger. Today, SI is used as a standard system of measurement.**

Question: *Why are standard units of measure important?*

Materials

water
large graduated cylinder or beaker
graph paper
balance
pieces of string
spring scale
rock samples
ruler

Safety Precautions

Procedure

1. Read and complete the lab safety form.
2. Obtain a set of rock samples from your teacher.
3. Measure the weight and length of two rock samples using a nonstandard unit of measure. You might use your pinky, a paper clip, or anything you choose.
4. Record your measurements.
5. Working with a partner, explain your units of measure and which samples you measured. Ask your partner to measure the rocks using your units.
6. Record your partner's measurements.
7. Use the information in the *Skillbuilder Handbook* to design a data table in which to record the following measurements for each rock sample: area, volume, mass, weight, and density.
8. Carefully trace the outline of each rock onto a piece of graph paper. Determine the area of each sample and record the values in your data table.
9. Secure each rock with a piece of dry string. Place the string loop over the hook of the spring scale to determine the weight of each rock sample. Record the values in your data table.
10. Pour water into a large graduated cylinder until it is half full. Record this volume in the table. Slowly lower the sample by its string into the cylinder. Record the volume of the water. Subtract the two values to determine the volume of the rock sample.
11. Repeat Steps 9 and 10 for each rock. Make sure the original volume of water for each rock is the same as when you measured your first sample.
12. Follow your teacher's instructions about how to use the balance to determine the mass of each rock. Record the measurements in your table.

Analyze and Conclude

1. **Interpret** How did the results of your initial measurements (Step 4) compare with your lab partner's (Step 6)? If they were different, why were they?
2. **Propose** What does this tell you about the importance of standard units of measure?
3. **Compare** the area of each of your samples with the volumes determined for the same rock. Which method of measurement was more accurate? Explain.
4. **Calculate** the density of each sample using this formula: density = mass/volume. Record these values in your data table.
5. **Explain** Does mass depend on the size or shape of a rock? Explain.
6. **Identify** the variables you used to determine the volume of each sample.
7. **List** the standard units you used in this investigation and explain the standard unit advantages over your measurement units.

INQUIRY EXTENSION

Inquiry How could you find the volume of a rock, such as pumice, that floats in water? Design an investigation to test your prediction.

GeoLAB

iLab Station Rubric

Preparation

Time Allotment 45 min

Process Skills measure in SI; use numbers; collect, organize, and interpret data

Safety Precautions Approve lab safety forms before work begins. Remind students to handle glassware carefully. Stress to students that the rocks must be gently lowered into glass beakers to prevent the beakers from breaking.

Preparation of Materials

- Review the basic SI units that will be used in this lab.
- Demonstrate how to use the balance and the spring scale properly.

Procedure

- This is an appropriate activity for small groups, but it can also be done individually if sufficient materials are available.
- Explain that measuring the areas of the outlines of the rock samples requires counting squares because the shapes are irregular. You might need to review this procedure with students.
- **Troubleshooting** Tell students to make multiple measurements and average their results as a means of checking for errors and improving accuracy.

Analyze and Conclude

1. Answers will vary. Without a standard unit, variance in measurement will increase.
2. A standard unit of measure is crucial in science so that scientists can compare results from experiments, results are easily understood, and experiments can be replicated.
3. Measuring volume is more accurate than measuring area. Area measurements will depend on the orientation of the sample when its outline is traced and on the ability of students to measure the area of an irregular solid.
4. Allow students to use calculators if necessary.
5. Mass depends only on the size of the samples because it is a measure of the quantity of matter in a sample.
6. The variables were the size of the samples and the change in the level of the water in the beaker.
7. Standard units are metric and SI units. They make it easier to compare and replicate results.

INQUIRY EXTENSION

Inquiry Answers will vary. It is less important that students identify the correct way to test for the density of the pumice, and more important that they accurately indicate the process they would follow to test their method. To find the volume of an object that floats in water, one could use a lower-density liquid, such as alcohol, or use a narrow stick to hold the sample below the water level and record the volume change. In the second method, one must assume that the volume of the stick is negligible compared to the volume of the sample.

LabManager™
Customize any lab with the LabManager™ CD-ROM.

MAINIDEAS **Summary statements can be used by students to review the major concepts of the chapter.**

Students can review with these online resources.

Vocabulary Practice

Vocabulary eGames
Vocabulary eFlashcards
Vocabulary PuzzleMaker

Section Self-Check

Chapter Self-Check

Online Test Practice

Use the Exam*View*® *Assessment Suite* CD-ROM to:

- create multiple versions of tests
- create modified tests with one mouse click
- edit existing questions and add your own questions
- build tests aligned with state standards using built-in state curriculum tags
- change English tests to Spanish with one mouse click
- track students' progress using the Teacher Management System

CHAPTER 1 STUDY GUIDE

BIGIDEA **Earth scientists use specific methods to investigate Earth and beyond.**

Vocabulary Practice

SECTION 1 Earth Science

VOCABULARY
- astronomy
- meteorology
- geology
- oceanography
- environmental science
- geosphere
- atmosphere
- hydrosphere
- cryosphere
- biosphere

MAINIDEA **Earth science encompasses five areas of study: astronomy, meteorology, geology, oceanography, and environmental science.**

- Earth is divided into four main systems: the geosphere, hydrosphere, atmosphere, and biosphere.
- Earth systems are all interdependent.
- Identifying the interrelationships between Earth systems leads to specialties and subspecialties.
- Technology is important, not only in science, but in everyday life.
- Earth science has contributed to the development of many items used in everyday life.

SECTION 2 Methods of Scientists

VOCABULARY
- scientific methods
- hypothesis
- independent variable
- dependent variable
- control
- Le Système International d'Unités (SI)
- scientific notation

MAINIDEA **Scientists use scientific methods to structure their experiments and investigations.**

- Scientists work in many ways to gather data.
- A good scientific experiment includes an independent variable, dependent variable, and control. An investigation, however, does not include a control.
- Graphs, tables, and charts are three common ways to communicate data from an experiment.
- SI, a modern version of the metric system, is a standard form of measurement that all scientists can use.
- To express very large or very small numbers, scientists use scientific notation.

SECTION 3 Communication in Science

VOCABULARY
- scientific model
- scientific theory
- scientific law

MAINIDEA **Precise communication is crucial for scientists to share their results effectively with each other and with society.**

- Scientists communicate data so others can learn the results, verify the results, examine conclusions for bias, and conduct new experiments.
- There are three main types of graphs scientists use to represent data: line graphs, circle graphs, and bar graphs.
- A scientific model is an accurate representation of an idea or theory.
- Scientific theories and scientific laws are sometimes discovered accidentally.

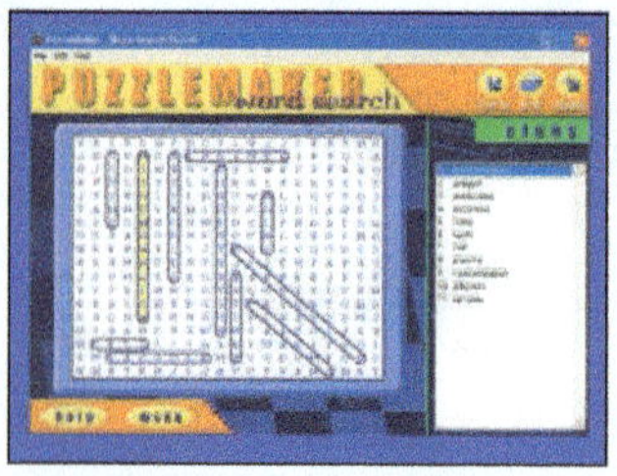

Vocabulary Practice

For additional practice with vocabulary, have students access the Vocabulary PuzzleMaker.

Chapter Self-Check

VOCABULARY REVIEW

Explain the relationship between the vocabulary terms below.

1. geosphere, mantle
2. hydrosphere, atmosphere
3. oceanography, cryosphere
4. meteorology, atmosphere
5. geology, biosphere

For Questions 6 to 9, fill in the blanks with the correct vocabulary terms from the Study Guide.

6. When conducting experiments, scientists use ________ to help guide their processes.
7. The ________ is the one factor that can be manipulated by the experimenter.
8. Scientists use a form of shorthand called ________ to express very large or very small numbers.
9. Most scientific studies and experiments use a standard system of units called ________.

Write a sentence using the following vocabulary terms.

10. scientific theory
11. scientific law
12. scientific model

Fill in the blanks with a vocabulary term from the Study Guide.

13. In the field of ________, scientists measure temperature, pressure, and humidity.
14. Their measurements come from features of the ________ and hydrosphere, and they look at how weather affects the ________ and geosphere.
15. The units of their measurements come from ________ and the metric system.
16. The numbers generally are not large, so ________ is not used.

Roger Ressmeyer/CORBIS

UNDERSTAND KEY CONCEPTS

17. Which one of these is NOT a specialized area of Earth science?
    A. astronomy
    B. environmental science
    C. technology
    D. oceanography

Use the figure below to answer Questions 18 and 19.

18. Which type of scientist is shown above?
    A. oceanographer
    B. geologist
    C. astronomer
    D. meteorologist
19. Which type of research is this scientist conducting?
    A. field research
    B. lab research
    C. library research
    D. biological research
20. Which is a sequence of steps a scientist might use to conduct an investigation?
    A. analysis, test, question, conclude
    B. test, question, conclude, analysis
    C. question, test, analysis, conclude
    D. conclude, test, question, analysis

CHAPTER 1 ASSESSMENT

VOCABULARY REVIEW

1. The mantle is part of the geosphere.
2. The hydrosphere is all the water on Earth, including that in the atmosphere. The atmosphere includes the gases that surround Earth.
3. The cryosphere consists of the permanently frozen water on Earth, including sea ice. Oceanography is the study of Earth's oceans.
4. Meteorology is the study of how weather patterns affect both everyday life and long-term climate changes on Earth. The atmosphere is an integral part of this study.
5. Geology is the study of the materials that make up Earth. The biosphere, which includes all living things and their environments, changes as the geology of an area changes.
6. scientific methods
7. independent variable
8. scientific notation
9. SI
10. Students' sentences should demonstrate an understanding of scientific theories.
11. Students' sentences should demonstrate an understanding of scientific laws.
12. Students' sentences should demonstrate an understanding of scientific models.
13. meteorology
14. atmosphere, biosphere
15. SI
16. scientific notation

UNDERSTAND KEY CONCEPTS

17. C
18. B
19. A
20. C

21. A
22. B
23. B
24. C
25. C

CONSTRUCTED RESPONSE

26. Technological advances help scientists make new discoveries and learn more about the world. Technology can also be transferred from science to everyday life.
27. length = cm, m; area = cm^2; volume = cm^3
28. Earth's main systems are geosphere, hydrosphere, atmosphere, and biosphere. The geosphere includes Earth from the crust to the center of the inner core. The hydrosphere is all the water on Earth, including that in the atmosphere. The atmosphere is the gases that surround our planet. The biosphere is all the living things on the planet and their environments. All the systems are interconnected and relate to one another.
29. An investigation involves many of the same components as an experiment but does not utilize a control. An experiment always contains a control.
30. Sample answer: A graph would be helpful to show the relationship among data sets. It is sometimes easier to show this relationship visually than to write it in words. Graphed data can also better reveal trends in the data.
31. The melting of ice is a law, or "rule" of nature that has been observed numerous times. The change of state is observed to be the same every time ice melts.

THINK CRITICALLY

32. A meteorologist would have to understand how Earth's water moves through the water cycle, where it is found, and how it ultimately affects weather and climate.
33. Answers will vary, but should list variables and a control.

Use the figure below to answer Questions 21 and 22.

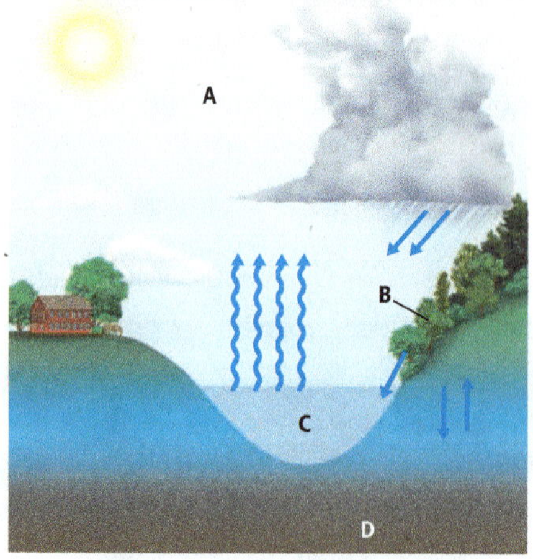

21. Identify the Earth system that is labeled *A*.
A. atmosphere
B. biosphere
C. hydrosphere
D. geosphere

22. Identify the Earth system that is labeled *B*.
A. atmosphere
B. biosphere
C. hydrosphere
D. geosphere

23. Which type makes up 97 percent of Earth's water?
A. groundwater
B. salt water
C. freshwater
D. spring water

24. Which is true of scientific models?
A. They never change.
B. They must be true for at least ten years.
C. They will be modified with new observations and data.
D. They are generally the work of one scientist.

25. Select the correct scientific notation for 150,000,000 km.
A. 150×10^6 km
B. 15×10^7 km
C. 1.5×10^8 km
D. 0.15×10^9 km

CONSTRUCTED RESPONSE

26. Explain how technology relates to science.

Use the photo below to answer Question 27.

27. Identify the SI units that would be used to measure each of the above items.

28. Summarize each of Earth's systems and explain their relationships to each other.

29. Compare and contrast an investigation and an experiment.

30. Apply Why might a graph be more helpful in explaining data than just writing the results in words?

31. Apply When ice is heated above 0°C, it melts. Is this a theory or a law? Explain.

THINK CRITICALLY

32. CAREERS IN EARTH SCIENCE Why would a meteorologist need an understanding of Earth's hydrosphere?

33. Design an Experiment Suppose you want to find the effect of sunlight on the temperature of a room with the shade up and the shade down. Describe how you would test this hypothesis. What would be your variables? What would you use as a control?

34. Propose An ecologist wants to study the effects of pollution on plant growth. The scientist uses two groups of plants. To the first group, a type of pollutant is added. To the second group, nothing is added. The scientist records plant growth for each plant for two weeks. What is the purpose of the second group in the scientist's study?

Use the table below to answer Question 35.

Some SI Conversions				
1 m	= ______	mm	= ______	km
1 g	= ______	mg	= ______	kg
1 cm^3	= ______	m^3	= ______	mL
3.5 km	= ______	m	= ______	cm

35. Calculate Copy the table into your notebook. Complete the table. Once you have made your conversions, express each answer in scientific notation.

CONCEPT MAPPING

36. Use the following terms to make a concept map summarizing the units used to measure each quantity discussed in the chapter: *time, density, temperature, volume, mass, weight, length, area, °C, g/mL, km, s, cm³, m², kg,* and *N*. For help, refer to the *Skillbuilder Handbook*.

CHALLENGE QUESTION

37. Evaluate A scientist is researching a new cancer drug. Fifty patients have been diagnosed with the type of cancer the drug is designed to treat. If a control is used, the patients might not receive any medication. The patients do not know if they are receiving the placebo or the new medication. For this reason, the patients are allowed to also receive traditional treatment if they choose. How will this impact the research? How should the scientist account for this information in the results? Should the scientists be allowed to discourage patients from receiving additional treatment?

WRITING IN ▶ Earth Science

38. Imagine you are writing an explanation of the scientific methods for someone who has never done a scientific investigation before. Explain what the scientific methods are and why they are so important.

DBQ Document–Based Questions

Data obtained from: Annual mean sunspot numbers 1700–2002. *National Geophysical Data Center.*

Use the graphs below to answer Questions 39–41.

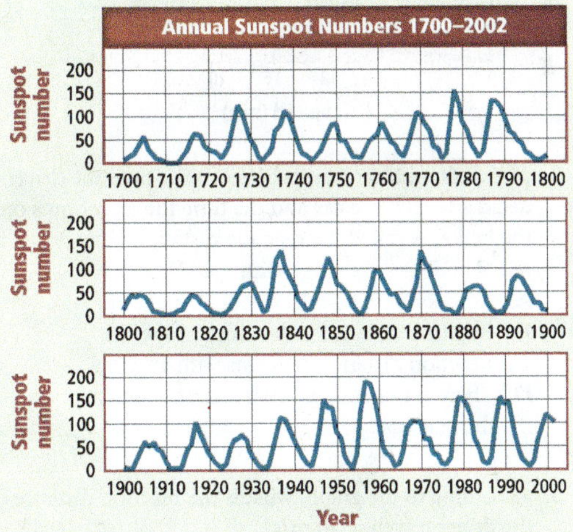

39. Is there a consistent pattern in the graphs? If so, what is the pattern showing?

40. What do the graphs express regarding the number of sunspots that have been seen and recorded since the 1700s?

41. What would you predict would be the pattern for the years 2000 to 2100?

CUMULATIVE REVIEW

In the following Chapters, Cumulative Review questions will help you review and check your understanding of concepts discussed in previous chapters.

34. The second group is the control.

35. 1 m = 1000 mm (1.0×10^3) = .001 km (1.0×10^{-3})
1 g = 1000 mg (1.0×10^3) = .001 kg (1.0×10^{-3})
1 cm^3 = .000001 m^3 (1.0×10^{-6}) = 1 mL
3.5 km = 3500m (3.5×10^3) = 350,000 cm (3.5×10^5)

CONCEPT MAPPING

36. time: s, density: g/mL, temperature: °C, volume: cm^3, mass: kg, weight: N, length: km, area: m^2

CHALLENGE QUESTION

37. Answers will vary, but students should indicate that the results could be skewed because patients could be receiving drugs other than the test drugs. Scientists would have to acknowledge this in their published findings. Students might say that patients either should or should not receive additional treatment. Either response is acceptable if the reasoning is explained.

WRITING IN ▶ Earth Science

Rubric

38. Answers will vary, but should include the following terms: *hypothesis, experiment, analysis,* and *conclusion.*

DBQ Document–Based Questions

Data obtained from: Annual mean sunspot numbers 1700–2002. *National Geophysical Data Center.*

39. Yes, there is a consistent pattern. The number of sunspots rises and falls about every 10 years.

40. The number of observable sunspots repeatedly increases and then decreases, and the height of the peaks vary.

41. a pattern similar to that in the graphs

CHAPTER 1 STANDARDIZED TEST PRACTICE

MULTIPLE CHOICE

1. C
2. C
3. B
4. A
5. D
6. A
7. B
8. D
9. C
10. A

CHAPTER 1 STANDARDIZED TEST PRACTICE

MULTIPLE CHOICE

1. Identify the type of Earth science that involves the study of the materials that make up Earth.
   A. astronomy
   B. meteorology
   C. geology
   D. oceanography

Use the graph below to answer Questions 2 and 3.

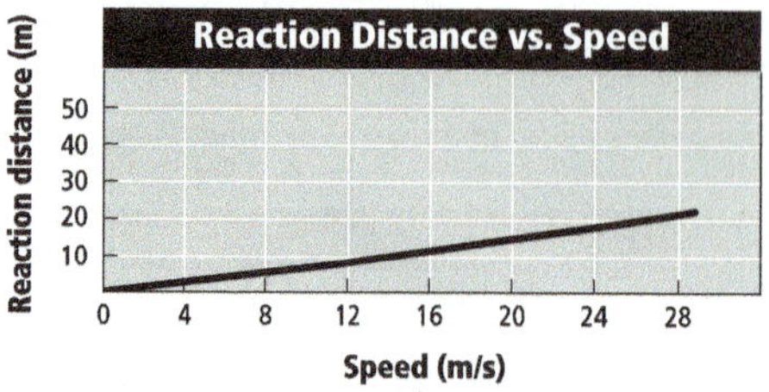

2. The distance a car travels between the time the driver decides to stop the car and the time the driver puts on the brakes is called the reaction distance. How does the reaction distance change with speed?
   A. Reaction distance decreases with speed.
   B. Reaction distance is the same as speed.
   C. Reaction distance increases with speed.
   D. There is not enough information to answer the question.

3. According to the graph, what is the reaction distance of the driver traveling 20 m/s?
   A. 3 m   C. 20 m
   B. 15 m   D. 28 m

4. Which lists Earth's layers from the inside out?
   A. inner core, outer core, mantle, crust
   B. crust, mantle, outer core, inner core
   C. crust, inner core, outer core, mantle
   D. mantle, outer core, inner core, crust

5. A block is 2 cm wide, 5.4 cm deep, and 3.1 cm long. The density of the block is 8.5 g/cm^3. What is the mass of the block?
   A. 33.48 g   C. 399.3 g
   B. 85.10 g   D. 284.58 g

6. If a conclusion is supported by data, but does not support an original hypothesis, what should a scientist do?
   A. The scientist should reevaluate the original hypothesis.
   B. The scientist should redesign the experiment.
   C. The scientist should not change anything.
   D. The scientist should modify the conclusion.

Use the illustration below to answer Questions 7 and 8.

7. This sign was found at the entrance to a chemistry laboratory. Why is this an important sign?
   A. Goggles help chemists see better.
   B. Chemicals can seriously damage eyes and skin.
   C. Accidents rarely happen in laboratories.
   D. Chemists will be fined if they do not obey the rules.

8. Why are safety rules posted, like this sign, or stated when conducting experiments?
   A. Safety rules are used to scare students.
   B. The goal of safety rules is to make an experiment exciting.
   C. Safety rules are just suggestions as to how to behave during an experiment.
   D. The safety rules are given for scientists' protection.

9. What should you always do when conducting an experiment?
   A. You should clean up broken glass yourself.
   B. You should unplug cords by pulling on the cord, not the plug.
   C. You should report spills immediately.
   D. You should flush your eyes at the eyewash station.

10. Which of the following are Sir Isaac Newton's ideas on motion considered to be?
    A. scientific law   C. scientific model
    B. scientific theory   D. hypothesis

Online Test Practice

SHORT ANSWER

Use the graph below to answer Questions 11–13.

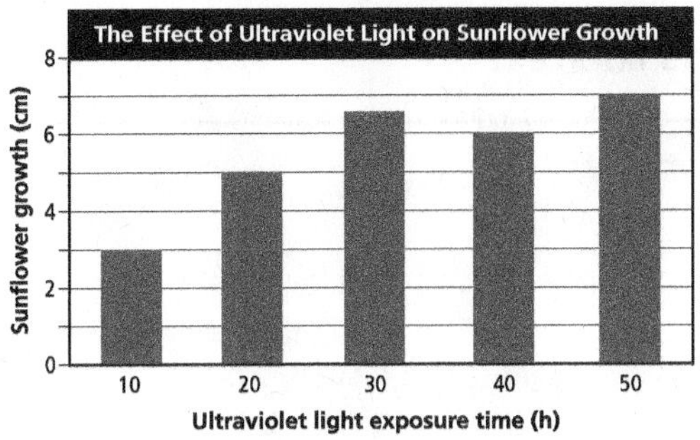

11. According to the graph, what was the greatest growth observed?

12. What type of graph is this? Why is this the best way to represent the data?

13. What are some variables that might affect the outcome of the experiment?

14. Describe the difference between the terms *astronomy* and *meteorology.*

15. Analyze the idea that technology is transferable. How is this beneficial?

16. Explain the importance of making a hypothesis before conducting an experiment.

17. Justine wants to measure how far an ant moves across a table in 1-min intervals. What would be the independent variable in this example?

READING FOR COMPREHENSION

Investigation Steps

Michael conducted an experiment to test if matter is conserved after a phase change. He filled an empty bottle with 50 mL of water and placed it in a sunny window until the liquid water changed to water vapor. The steps of the activity are listed in the table below but might not be in the correct order.

Investigation Steps	
1	Find the mass of the bottle, lid, and water vapor.
2	Pour 50 mL of water into an empty bottle.
3	Find the mass of the bottle, lid, and 50 mL of water.
4	Place a lid on the opening of the bottle to tightly seal it.

18. Which shows the investigation steps in the correct order?
A. 1, 2, 3, 4
B. 2, 4, 3, 1
C. 4, 2, 1, 3
D. 2, 4, 1, 3

19. According to the text, what scientific idea is Michael testing?
A. the rate of evaporation
B. the conservation of energy
C. the conservation of matter
D. the equilibrium of a system

20. Why does Michael put a lid on the bottle?
A. to prevent water vapor from leaving the system
B. to make the water evaporate faster
C. to prevent bacterial growth in the system
D. to establish equilibrium

NEED EXTRA HELP?																	
If You Missed Question . . .	1	2	3	4	5	6	7	8	9	10	11	12	13	14	15	16	17
Review Section . . .	1.1	1.3	1.3	1.1	1.2	1.3	1.2	1.2	1.2	1.3	1.3	1.3	1.3	1.1	1.1	1.2	1.2

SHORT ANSWER

11. The greatest growth observed was 7 cm after 50 hours of ultraviolet light exposure.

12. Bar graph; the data is quantitative so a bar graph shows the individual data for each amount of growth as compared to time.

13. Sample answers: amount of water given each plant, the location of each plant, the health of the original sunflower seed, etc.

14. Astronomy is the study of objects beyond Earth's atmosphere, whereas meteorology is the study of the atmosphere.

15. It means that technology can be applied to new situations. Employing technology in a new way or a new situation allows results more quickly and the user doesn't have to invent new technology for the experiment.

16. A hypothesis is a statement based on background information relating to the question posed. It can help to guide how an experiment is conducted, proved or disproved, and help find possible answers to the original question.

17. Time continues to change regardless of the other variables, so it is independent.

READING FOR COMPREHENSION

18. B
19. C
20. A

CHAPTER 2 Mapping Our World

BIGIDEA **Earth scientists use mapping technologies to investigate and describe the world.**

ESSENTIAL QUESTIONS	NATIONAL STANDARDS	RESOURCES TO ASSESS MASTERY
SECTION 1 **Latitude and Longitude** **1.** What is the difference between latitude and longitude? **2.** Why is it important to give a city's complete coordinates when describing its location? **3.** Why are there different time zones from one geographical area to the next? 2 sessions 1 block	UCP.1–3, UCP.5; A.1–2; E.1, E.2; F.6; G.3	**Entry-Level Assessment** Focus Transparency 4, p. 30 **Progress Monitoring** Caption Question, p. 33 Section Review, p. 33
SECTION 2 **Types of Maps** **1.** What are the similarities and differences between different types of maps? **2.** Why are different maps used for different purposes? **3.** How are gradients calculated on a topographic map? 2 sessions 1 block	UCP. 3, UCP.5; A.1; E.1: F.6	**Entry-Level Assessment** Focus Transparency 5, p. 34 **Progress Monitoring** Caption Question, p. 36 Reading Check, pp. 37, 39, 40 Section Review, p. 40
SECTION 3 **Remote Sensing** **1.** What are some of the different types of remote sensing? **2.** How are satellites and sonar used to map Earth's surface and its oceans? **3.** What is the Global Positioning System and how does it work? 3 sessions 1.5 blocks	UCP.1, UCP.3, UCP.5; A.1; B.4; C.6; E.1, E.2; F.6; G.1, G.3	**Entry-Level Assessment** Focus Transparency 6, p. 41 **Progress Monitoring** Caption Question, p. 41 Reading Check, pp. 43, 44 Section Review, p. 46 **Summative Assessment** Chapter Assessment, p. 51 Exam*View®Assessment Suite* CD-ROM

Go online! connectED.mcgraw-hill.com

LEVELED RESOURCES	LAB MATERIALS
Science Notebook 2.1 OL **Chapter FAST FILE Resources:** MiniLab Worksheet, p. 28 OL Study Guide, p. 39 BL **Transparencies:** Section Focus Transparency 4 EL Teaching Transparency 4 OL EL	LaunchLAB p. 28 / **15 min** graph paper, colored pencils MiniLAB p. 32 / **15 min** world map or globe
Science Notebook 2.2 OL **Chapter FAST FILE Resources:** GeoLab Worksheet, p. 29 OL Study Guide, p. 41 BL **Lab Resources:** Laboratory Manual, pp. 9, 13 OL **Transparencies:** Section Focus Transparency 5 EL Teaching Transparency 5 OL EL	GeoLAB p. 48 / **45 min** ruler, string, piece of paper
Science Notebook 2.3 OL **Chapter FAST FILE Resources:** Study Guide, p. 44 BL **Transparencies:** Section Focus Transparency 6 EL Teaching Transparency 6 OL EL	

LabManager™
Customize any lab with the LabManager™ CD-ROM.

ADDITIONAL RESOURCES

Technology:
- eTeacherEdition
- eStudentEdition
- Classroom Presentation Toolkit
- Exam*View® Assessment Suite* CD-ROM
- LabManager™ CD-ROM
- What's EARTH SCIENCE Got To Do With It? DVD
- Weather Classroom Video DVD

Assessment:
- Blueprints for Success
- Chapter Test (Scaffolded)
- Earth Science Standardized Test Practice

Lab Resources:
- Exploring Environmental Problems Laboratory Manual

BL Below Level   OL On Level   AL Advanced Learners   EL English Learners   COOP LEARN Cooperative Learning

CHAPTER 2

LaunchLAB

iLab Station Rubric

Can you make an accurate map?

Teaching Strategies

- You might find most students already know where their classmates live. In this case, have students map local destinations, such as a park or the county courthouse, relative to the school.
- Select several good examples of student maps. Make transparencies of these maps or enlarge them for display on the board.

Procedure

1. Have students read and complete the lab safety form and follow the procedure below.
2. With a classmate, choose a location in your school or schoolyard.
3. Use a sheet of **graph paper** and **colored pencils** to draw a map from your classroom to the location you chose. Include landmarks such as drinking fountains and restrooms.
4. Share your map with a classmate. Compare the landmarks you chose and the path each of you chose to get to your locations. If they were different, explain why.
5. Follow your map to the location you and your partner chose. Was your map correct? Were there details you left out that might have been helpful?

Analysis

1. **Discuss** with your classmate how you could improve your maps. Students might say they could improve their maps by developing a system for illustrating elevation or making a legend to illustrate map symbols.
2. **Examine** What details could you add? Students could include scale and direction and use color to represent various features.

Assessment

Skill Have students compare and contrast their maps of particular areas and actual maps of the areas. Students should write brief reports in their Earth science journals describing how the maps differ.

Rubric

LabManager™
Customize any lab with the LabManager™ CD-ROM.

CHAPTER 2

Mapping Our World

BIGIDEA Earth scientists use mapping technologies to investigate and describe the world.

SECTIONS

1 Latitude and Longitude

2 Types of Maps

3 Remote Sensing

LaunchLAB

iLab Station

Can you make an accurate map?

If you have ever been asked to give someone directions, you know that it is important to include as many details as possible so that the person asking for directions will not get lost. Perhaps you drew a detailed map of the destination in question. Test your map-making skills in this lab.

FOLDABLES®
Study Organizer

Types of Mapping Technologies

Make a four-door book using the labels *Landsat, GPS/GIS,* OSTM/Jason–2, and *SeaBeam.* Use it to organize your notes on mapping technologies.

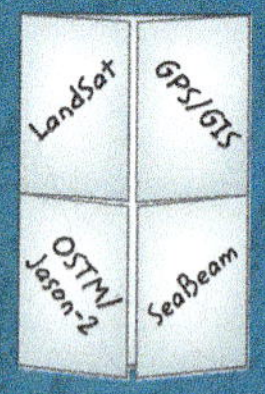

Digital Vision/Getty Images

Go online!
connectED.mcgraw-hill.com

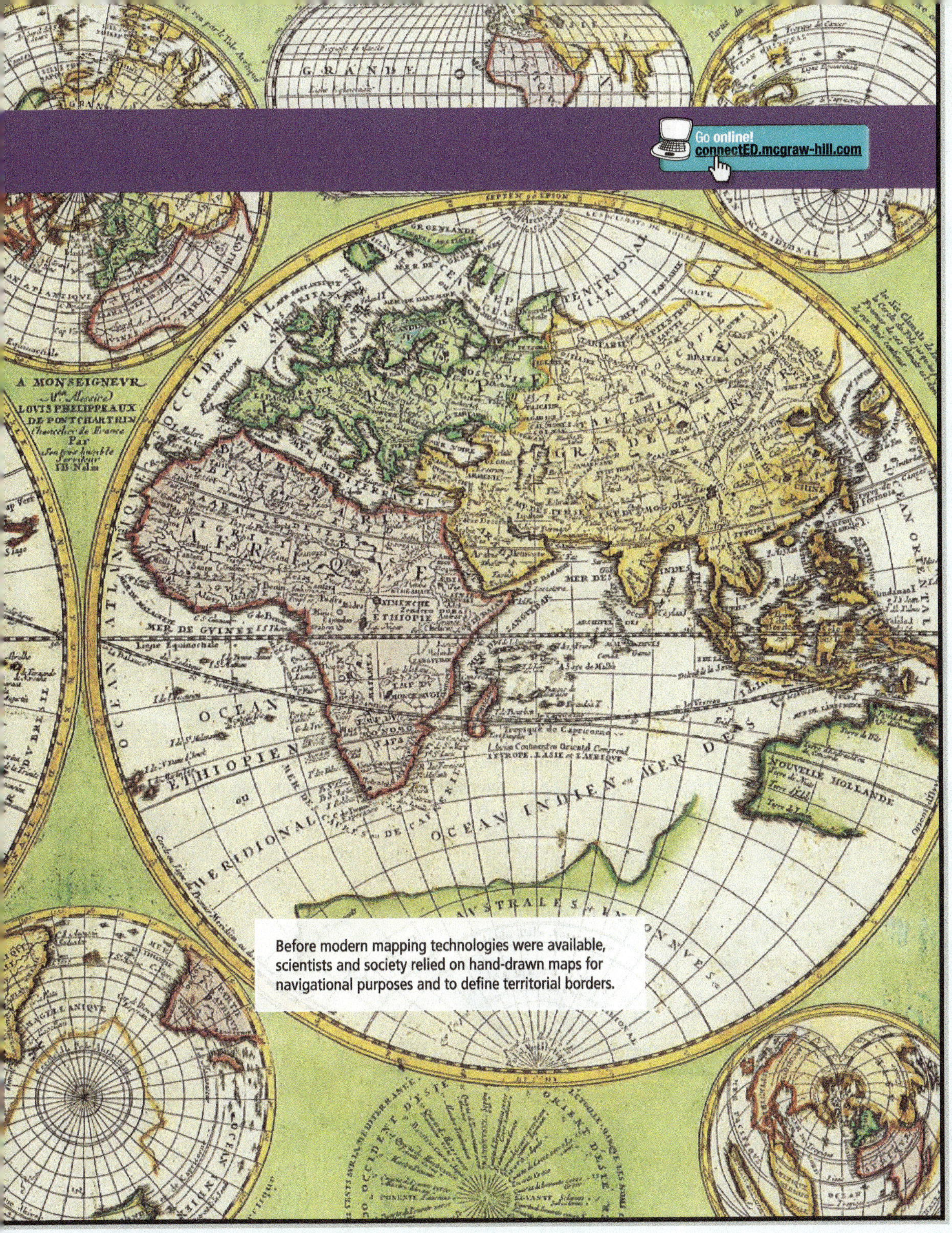

Before modern mapping technologies were available, scientists and society relied on hand-drawn maps for navigational purposes and to define territorial borders.

Introduce the BIG IDEA

Maps in Science Write the following branches of science on the board: geology, astronomy, environmental studies, land-use planning, climatology, and meteorology. Ask students to name one piece of equipment each of the branches uses. Help students understand that maps have broad applications, and they are used in these branches of science.

Teacher Content Support

Theater of the World The world map shown here was made in 1570 by Abraham Ortels. He was a geographer and mapmaker and is the author of what is considered to be the first modern world atlas, *Theatrum Orbis Terrarum* (Theater of the World). It consisted of 70 maps on 53 sheets with accompanying text. From its publication until 1612, the Theatrum was continuously revised to reflect new geographic discoveries. For example, the world map shown here shows an incorrect interpretation of the coastlines of the continents. More than 7300 copies of the Theatrum were printed in 31 editions and 7 languages. The first four imprints of the 1570 edition are held in the collections at the Library of Congress. The entire third imprint, which was taken apart for conservation and subsequently scanned, can be viewed online.

SECTION 1

1 Focus

Focus Transparency

Before presenting the lesson, project **Section Focus Transparency 4** and have students answer the accompanying questions. EL

MAINIDEA

Describe Locations Ask students to name as many ways as possible to describe the location of their school. Keep a tally on the board. Students might list street address, intersection of two streets, state, hemisphere, continent, etc. Ask students which suggestion is the most accurate. If longitude and latitude are not listed, add this method to the tally at the end of the discussion.

2 Teach

Interpret the Illustration

Latitude and Longitude Have students look at **Figures 1** and **2** and compare and contrast latitude and longitude. Students should mention that latitude is distance in degrees north and south of the equator, and longitude is distance in degrees east and west of the prime meridian. Students also might mention that lines of latitude are parallel, while lines of longitude are semicircles.

Tie to Previous Knowledge

Grids in Cities and on Earth Show students a map of their city or any city that has a well-defined grid system of streets and avenues. Tell students lines of latitude and longitude make up a similar grid system on a much larger scale. BL EL

SECTION 1

Latitude and Longitude

MAINIDEA Lines of latitude and longitude are used to locate places on Earth.

Essential Questions

- What is the difference between latitude and longitude?
- Why is it important to give a city's complete coordinates when describing its location?
- Why are there different time zones from one geographic area to the next?

Review Vocabulary

time zone: a geographic region within which the same standard time is used

New Vocabulary

cartography
equator
latitude
longitude
prime meridian
International Date Line

EARTH SCIENCE 4 YOU Imagine you were traveling from New York City, New York, to Los Angeles, California. How would you know where to go? Many people use maps to help them plan the quickest route.

Latitude

Maps are flat models of three-dimensional objects. For thousands of years people have used maps to define borders and to find places. The map at the beginning of this chapter was made in 1570. What do you notice about the size and shape of the continents? Today, more information is available to create more accurate maps. The science of mapmaking is called **cartography.**

Cartographers use an imaginary grid of parallel lines to locate exact points on Earth. In this grid, the **equator** horizontally circles Earth halfway between the North and South poles. The equator separates Earth into two equal halves called the northern hemisphere and the southern hemisphere.

Lines on a map running parallel to the equator are called lines of **latitude.** Latitude is the distance in degrees north or south of the equator as shown in **Figure 1.** The equator, which serves as the reference point for latitude, is numbered 0° latitude. The poles are each numbered 90° latitude. Latitude is thus measured from 0° at the equator to 90° at the poles.

Locations north of the equator are referred to by degrees north latitude (N). Locations south of the equator are referred to by degrees south latitude (S). For example, Syracuse, New York, is located at 43° N, and Christchurch, New Zealand, is located at 43° S.

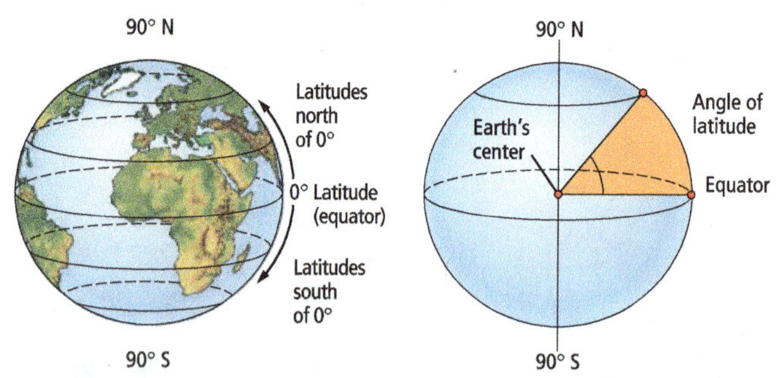

Figure 1 Lines of latitude are parallel to the equator. The value in degrees of each line of latitude is determined by measuring the imaginary angle created between the equator, the center of Earth, and the line of latitude as seen in the globe on the right.

DIFFERENTIATED INSTRUCTION

Visually Impaired Have students carefully tape precut pieces of string over the lines of latitude and longitude on a world map. Then have students with visual impairments trace these lines with their fingers. This should help both groups of students acquire a deeper understanding of the grid system used to locate places on Earth. COOP LEARN

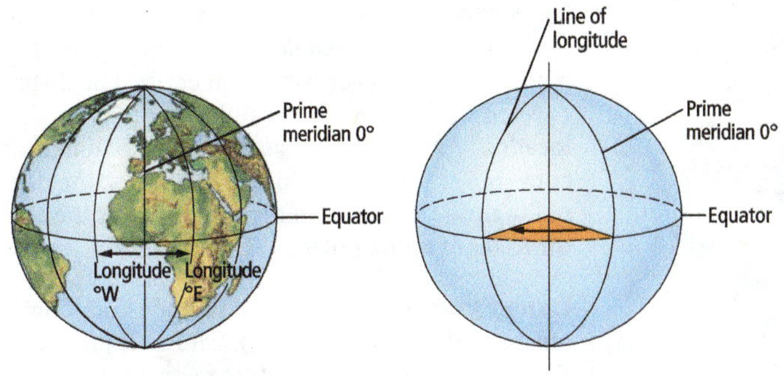

Figure 2 The reference line for longitude is the prime meridian. The degree value of each line of longitude is determined by measuring the imaginary angle created between the prime meridian, the center of Earth, and the line of longitude as seen on the globe on the right.

Degrees of latitude Each degree of latitude is equivalent to about 111 km on Earth's surface. How did cartographers determine this distance? Earth is a sphere and can be divided into 360°. The circumference of Earth is about 40,000 km. To find the distance of each degree of latitude, cartographers divided 40,000 km by 360°.

To locate positions on Earth more precisely, cartographers break down degrees of latitude into 60 smaller units, called minutes. The symbol for a minute is ′. The actual distance on Earth's surface of each minute of latitude is 1.85 km, which is obtained by dividing 111 km by 60′.

A minute of latitude can be further divided into seconds, which are represented by the symbol ″. Longitude is also divided into degrees, minutes, and seconds.

VOCABULARY

SCIENCE USAGE V. COMMON USAGE

Minute

Science usage: a unit used to indicate a portion of a degree of latitude

Common usage: a unit of time comprised of 60 seconds

Longitude

To locate positions in east and west directions, cartographers use lines of longitude, also known as meridians. As shown in **Figure 2,** **longitude** is the distance in degrees east or west of the prime meridian, which is the reference point for longitude.

The **prime meridian** represents 0° longitude. In 1884, astronomers decided that the prime meridian should go through Greenwich, England, home of the Royal Naval Observatory. Points west of the prime meridian are numbered from 0° to 180° west longitude (W); points east of the prime meridian are numbered from 0° to 180° east longitude (E).

Semicircles Unlike lines of latitude, lines of longitude are not parallel. Instead, they are large semicircles that extend vertically from pole to pole. For instance, the prime meridian runs from the North Pole through Greenwich, England, to the South Pole.

The line of longitude on the opposite side of Earth from the prime meridian is the 180° meridian. There, east lines of longitude meet west lines of longitude. This meridian is also known as the International Date Line, and will be discussed later in this section.

ACROSS THE CURRICULUM

History Have students research and discuss how the ancient Polynesians of the Marshall Islands of the South Pacific navigated from one island to another without the aid of navigational instruments. The early Polynesians used their knowledge of winds, waves, and currents to navigate. They also developed maps of star positions and ocean currents.

Model

Use a Sextant Have students work in pairs to research how latitude can be determined using the North Star and a sextant. Show students photographs of a sextant or have one available in class for students to examine. Then have student pairs construct a sextant using a pin, a protractor, string, and a straw, and explain how it works. Students should take care when handling sharp pins. To make and use the sextant, students will pin the straw to the protractor so that the straw pivots at the center point of the protractor. Then, they hold the protractor up to their eyes so that its base is parallel to the ground. Students can then look along the straw to find the North Star. The angle made by the straw is their latitude. OL

COOP LEARN

Teacher Content Support

Equator and Prime Meridian The use of the equator as the reference line for latitude was established at least as early as A.D. 150, when it appeared in a world atlas by Ptolemy, an Egyptian astronomer. The use of the prime meridian as the reference point for longitude came into practice in the late 1700s after the publication of the Nautical Almanac by English royal astronomer Nevil Maskelyne.

Enrichment

Distance and Latitude Give students the following problem: Imagine you are on an airplane. Your plane has flown from 30° N to 42° N. How far have you flown? The difference between 30° N and 42° N is 12°. Each degree of latitude is roughly 111 km. Thus, 12° multiplied by 111 km equals 1332 km. AL

MiniLAB

iLab Station Rubric

Purpose Students will use lines of latitude and longitude to locate specific places on Earth's surface.

Process Skills observe and infer, interpret scientific illustrations, communicate

Safety Precaution Approve lab safety forms before work begins.

Teaching Strategies
- Have students work in groups of three or four.
- Obtain an atlas that lists places by latitude and longitude.

Expected Results Students will identify places on Earth's surface using latitude and longitude coordinates.

Analysis

1. Mount St. Helens: 46°12′ N, 122°11′ W; Niagara Falls: 43°05′ N, 79°03′ W; Mt. Everest: 27°59′ N, 86°56′ E; Great Barrier Reef: 18°00′ S, 145°50′ E
2. 0°03′ S, 90°30′ W: Galápagos Islands; 27°07′ S, 109°22′ W: Easter Island; 41°10′ N, 112°30′ W: Great Salt Lake; 35°02′ N, 111°02′ W: Meteor Crater; 3°04′ S, 37°22′ E: Mt. Kilimanjaro
3. Students might suggest travel time, distance measured from place to place, and landmarks such as large trees or cliffs.

Assessment

Performance Give students reduced copies of blank world maps that show only lines of latitude and longitude and unlabeled continents. Have students locate their answers from questions 1 and 2 on their maps.

LabManager™
Customize any lab with the LabManager™ CD-ROM.

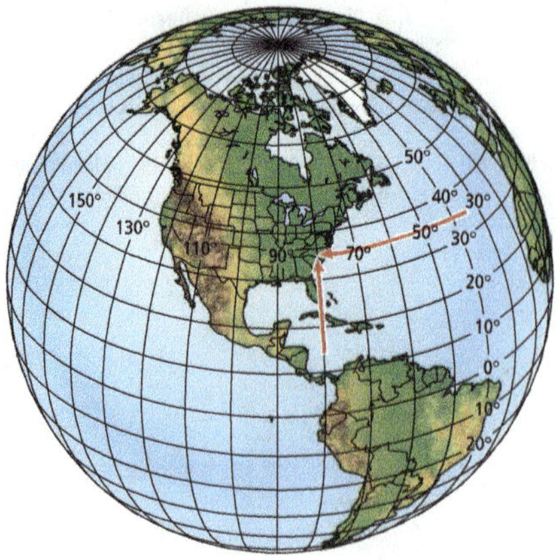

Figure 3 The precise location of Charlotte, North Carolina is 35°14′N, 80°50′W. Note that latitude comes first in reference to the coordinates of a particular location.

Degrees of longitude Degrees of latitude cover relatively consistent distances. Thus, the distance between any two latitude lines is approximately the same everywhere on earth. The distances covered by degrees of longitude, however, vary with location. As shown in **Figure 2,** lines of longitude converge at the poles into a point. Thus, one degree of longitude varies from about 111 km at the equator to 0 km at the poles.

Using coordinates Both latitude and longitude are needed to locate positions on Earth precisely. For example, it is not sufficient to say that Charlotte, North Carolina, is located at 35°14′ N because that measurement includes any place on Earth located along the 35°14′ line of north latitude.

The same is true of the longitude of Charlotte; 80°50′ W could be any point along that longitude from pole to pole. To locate Charlotte, use its complete coordinates—latitude and longitude—as shown in **Figure 3.**

Time zones Earth is divided into 24 time zones. Why 24? Earth takes about 24 hours to rotate once (360°) on its axis. Thus, there are 24 times zones, each representing a different hour. Every hour the Earth spins approximately 15°, so each time zone is 15° wide, corresponding roughly to lines of longitude. To avoid confusion, however, time zone boundaries have been adjusted in local areas so that cities and towns are not split into different time zones.

MiniLAB

iLab Station

Locate Places on Earth

How can you locate specific places on Earth with latitude and longitude?

Procedure

1. Read and complete the lab safety form.
2. Use a **world map** or **globe** to locate the prime meridian and the equator.
3. Take a few moments to become familiar with the grid system. Examine lines of latitude and longitude on the map or globe.

Analysis

1. **Locate** the following places:
   - Mount St. Helens, Washington; Niagara Falls, New York; Mount Everest, Nepal; Great Barrier Reef, Australia
2. **Locate** the following coordinates, and record the names of the places there:
   - 0°03′S, 90°30′W; 27°07′S, 109°22′W; 41°10′N, 112°30′W; 35°02′N, 111°02′W; 3°04′S, 37°22′E
3. **Analyze** How might early cartographers have located cities, mountains, or rivers without latitude and longitude lines?

IN THE FIELD

Polar Explorer Mathew Hensen, an African-American explorer, was the first American to reach the North Pole and officially map its location. He went to sea at age 12 and learned the skills of an explorer: navigation, mathematics, and cartography. At age 21, Hensen met a fellow explorer, Robert Peary, who hired him as an assistant. In 1907, they set off to achieve a lifelong dream of charting the North Pole. To survive in the rugged climate, they relied on Inuit guides. On April 9, 1909, Hensen became the first American to reach the North Pole, followed by Peary and the Inuit guides. At the time, Peary, the leader of the expedition, was given credit for this accomplishment. Hensen's role in the expedition was officially recognized in 1988.

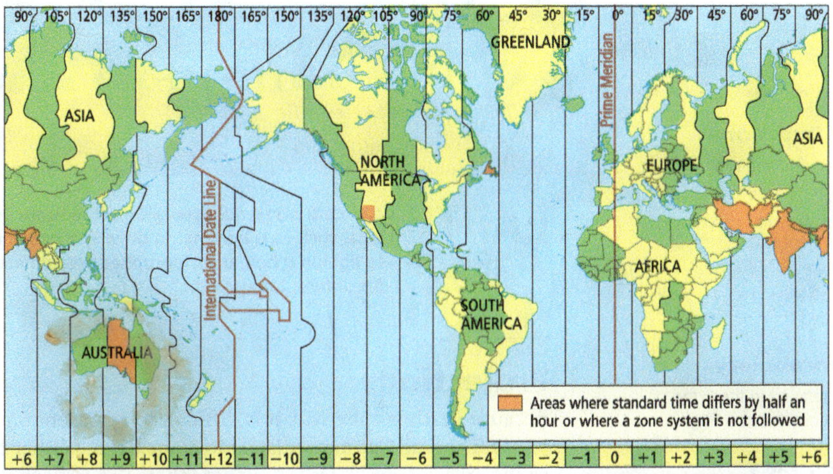

Figure 4 In most cases, each time zone represents a different hour. However, there are some exceptions.
Identify *two areas where the time zone is not standard.*

View an **animation of time zones.**

Concepts In Motion

For example all of Morton County, North Dakota, operates within the central time zone, even though the western part of the county is within the mountain-time-zone boundary. As shown in **Figure 4,** there are six time zones in the United States.

International Date Line Each time you travel through a time zone, you gain or lose time until, at some point, you gain or lose an entire day. The **International Date Line,** which is 180° meridian, serves as the transition line for calendar days. This imaginary line runs through the Pacific Ocean, as shown in **Figure 4.** If you were traveling west across the International Date Line, you would advance your calendar one day. If you were traveling east, you would move your calendar back one day.

SECTION 1 REVIEW

Section Self-Check

Section Summary

- Latitude lines run parallel to the equator.
- Longitude lines run east and west of the prime meridian.
- Both latitude and longitude lines are necessary to locate exact places on Earth.
- Earth is divided into 24 time zones, each 15° wide, that help regulate daylight hours across the world.

Understand Main Ideas

1. MAINIDEA **Explain** why it is important to give both latitude and longitude when giving coordinates.
2. **Describe** how the distance of a degree of longitude varies from the equator to the poles.
3. **Estimate** the time difference between your home and places that are 60° east and west longitude of your home.

Think Critically

4. **Evaluate** If you were flying directly south from the North Pole and reached 70° N, how many degrees of latitude would be between you and the South Pole?

WRITING IN ▶ Earth Science

5. Imagine what it would be like to fly from where you live to Paris, France. Describe what it would be like to adjust to the time difference.

SECTION 1 REVIEW

1. Without giving both latitude and longitude, the location could be anywhere on the entire latitude or longitude line.
2. At the equator, lines of longitude are 111 km apart. At the poles, the lines converge and are only a point apart.
3. 4 hours earlier to the west, 4 hours later to the east.
4. 160°
5. Answers will vary, but should accurately state the change in time zone from your area.

Rubric

Caption Question Fig. 4 part of Australia, all of India, Iran, Afghanistan, Myanmar (formerly Burma), Nepal

Project

Time Zones Have students work in pairs to draw and label the 24 time zones on a world map. Students should locate their town or city on the map. Have students use their location as a reference point and calculate the time in various places around the world.

3 Assess

Check for Understanding

Reinforcement Ask students the following question: If it is 2 P.M. in New York City, New York, what time is it in Quito, Peru? It is 2 P.M. because Quito is in the same time zone as New York.

Reteach

Explain Have students explain why time zone boundaries do not strictly follow lines of longitude. The boundaries of time zones are sometimes adjusted for geographic and political reasons.

Assessment

Performance Have pairs of students model lines of latitude. Each pair will need a letter-sized sheet of paper, a compass, a ruler, and scissors. Have students draw a horizontal line halfway down the paper and label it *equator.* Next, students should draw nine parallel lines both above and below the equator at intervals of 6 mm. Each line should be labeled in 10° increments, starting with 0° at the equator and going up to 90° N above the equator and down to 90° S below the equator. Tell students to place the compass point in the center of the equator and open the compass so that the tip of the pencil touches 90° N. Have students draw a circle and cut it out. Students can then add continents and oceans to their models.

SECTION 2

1 Focus

Focus Transparency

Before presenting the lesson, project **Section Focus Transparency 5** and have students answer the accompanying questions. EL

MAINIDEA

Map Design Ask students: What makes a map, a map? What is required on it, and what is optional? Show students several examples of maps ranging from simple to complex.

2 Teach

Collaborative Learning

Moon Exploration Show students a globe of the Moon. Choose two locations and provide the following scenario: The moon rover needs to travel from Point A to Point B. What type of map would be best to use? Students' answers will vary depending on what factors they choose to emphasize. Accept all reasonable responses. COOP LEARN

SECTION 2

Essential Questions

- What are the similarities and differences between different types of maps?
- Why are different maps used for different purposes?
- How are gradients on a topographic map calculated?

Review Vocabulary

parallel: extending in the same direction and never intersecting

New Vocabulary

Mercator projection
conic projection
gnomonic projection
topographic map
contour line
contour interval
geologic map
map legend
map scale

Figure 5 In a Mercator projection, points and lines on a globe are transferred onto cylinder-shaped paper. Mercator projections show true direction but distort areas near the poles.

View an **animation of map projections.**

Concepts In Motion

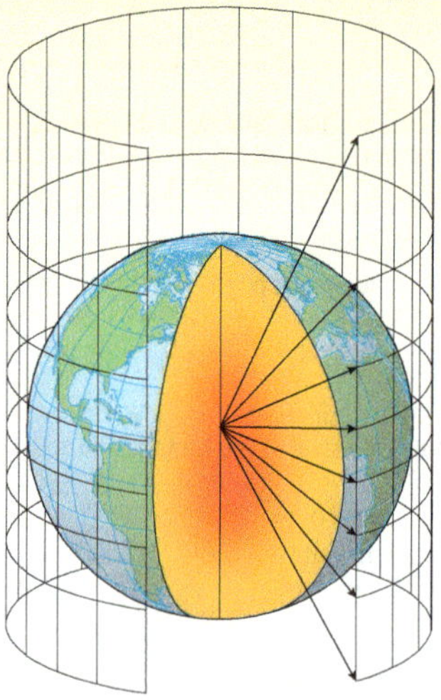

Types of Maps

MAINIDEA Maps are flat projections that come in many different forms.

EARTH SCIENCE 4 YOU Just as a carpenter uses different tools for different jobs, such as a hammer to drive in a nail and a wrench to tighten a bolt, a cartographer uses different maps for different purposes.

Projections

Because Earth is spherical, it is difficult to represent on a piece of paper. Thus, all flat maps distort to some degree either the shapes or the areas of landmasses. Cartographers use projections to make maps. A map projection is made by transferring points and lines on a globe's surface onto a sheet of paper.

Mercator projections A **Mercator projection** is a map that has parallel lines of latitude and longitude. Recall that lines of longitude meet at the poles. When lines of longitude are projected as being parallel on a map, landmasses near the poles are exaggerated. Thus, in a Mercator projection, the shapes of the landmasses are correct, but their areas are distorted.

As shown in **Figure 5,** Greenland appears much larger than Australia. In reality, Greenland is much smaller than Australia. Because Mercator projections show the correct shapes of landmasses and also clearly indicate direction in straight lines, they are most commonly used for the navigation of ships.

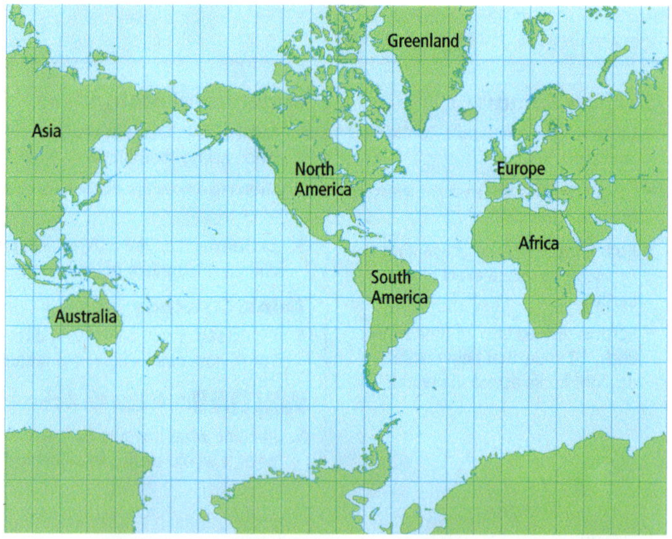

Demonstration

Compare and Contrast Maps Obtain several examples of different types of map projections, such as Mercator, conic, gnomonic, Robinson, and azimuthal. Display the maps on the board. Discuss the advantages and disadvantages of each map projection. Have students outline the main ideas of the discussion in their Earth science journals.

DIFFERENTIATED INSTRUCTION

Advanced Learners Assign each student a different type of map projection to research, such as gnomonic, Mercator, conic, azimuthal, or Robinson. Ask students to each write a report describing how the projection is made, the best uses of the projection, and problems associated with its use. Have students construct models of their projections and share their results with the class.

Conic projections A **conic projection** is made by projecting points and lines from a globe onto a paper cone, as shown in **Figure 6.** The cone touches the globe at a particular line of latitude. There is little distortion in the areas or shapes of landmasses that fall along this line of latitude. Distortion is evident, however, near the top and bottom of the projection. As shown in **Figure 6,** the landmass at the top of the map is distorted. Because conic projections have a high degree of accuracy for limited areas, they are excellent for mapping small areas. Hence, they are used to make road maps and weather maps.

Gnomonic projections A **gnomonic** (noh MAHN ihk) **projection** is made by projecting points and lines from a globe onto a piece of paper that touches the globe at a single point. At the single point where the map is projected, there is no distortion, but outside of this single point, great amounts of distortion are visible both in direction and landmass, as shown in **Figure 7.**

Because Earth is a sphere, it is difficult to plan long travel routes on a flat projection with great distortion, such as a conic projection. To plan such a trip, a gnomonic projection is most useful. Although the direction and landmasses on the projection are distorted, it is useful for navigation. A straight line on a gnomonic projection is the straightest route from one point to another when traveled on Earth.

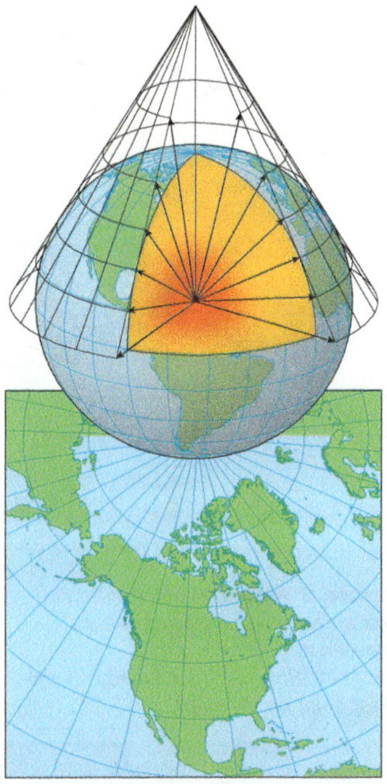

Figure 6 In a conic projection, points and lines on a globe are projected onto cone-shaped paper. There is little distortion along the line of latitude touched by the paper.

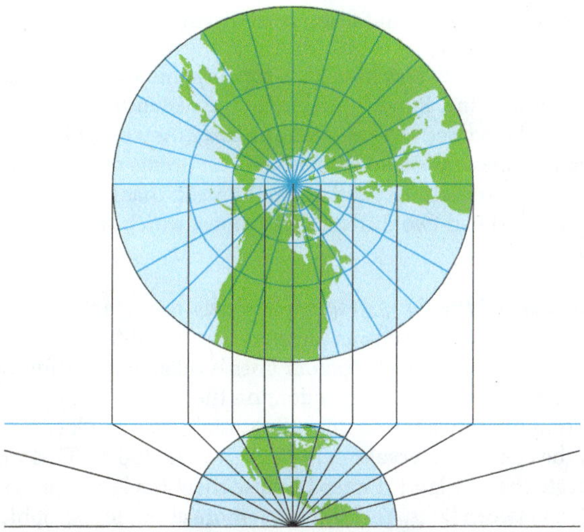

Figure 7 In a gnomonic projection, points and lines from a globe are projected onto paper that touches the globe at a single point.

Interpret the Illustration

Use Conic Projections Have students study **Figure 6** and answer the following questions: How is a conic map projection made? Points and lines from a globe are projected onto a cone that touches the globe at one line of latitude. For what are conic projections used, and why? Because conic projections have a high degree of accuracy for limited areas, they are used to map small areas, and thus make excellent road maps and weather maps. OL AL

Teacher Content Support

3-D Views of Earth Maps are two-dimensional models of Earth. Geologists, however, often need to know what Earth looks like in three dimensions. With computers, topographic maps are digitized to make the top or surface layer of three-dimensional maps of Earth. Digitizing is a process by which points are located on a coordinate grid. Geologists use data obtained from drilling into Earth's crust to generate the other layers of three dimensional maps.

ACROSS THE CURRICULUM

History Mapmaking can be considered an art, although the style has changed dramatically throughout history. Use atlases from the library and the Internet to assemble a collection of maps. Compare some early maps of the western hemisphere with those of today. Compare how the information was presented then and now, and what information was included then and now.

Apply Earth Science

Orienteering Invite a local orienteering club member to visit your class. Ask the person to explain how orienteering involves the use of a compass and map to find specific locations. If possible, arrange a field trip on a local trail so students can apply their new knowledge.

Caption Question Fig. 8 slightly higher than 2500 m

Teacher Content Support

Topographic Maps In the United States, topographic maps are published by the United States Geological Survey (USGS), the Department of Defense, the Department of Agriculture, and the Forest Service. Standard maps produced by the USGS are called *quadrangles.* The boundaries of quadrangle maps are formed by lines of latitude and longitude. The maps are named for obvious features within the area shown on the map. No two quadrangles of the same series in the same state can have the same name.

The USGS uses standard colors to represent natural or cultural features in topographic maps. Black is used for human-made features such as houses and other buildings, and for the names of places. Green represents wooded or heavily vegetated areas. Red is used to mark important highways. Blue is used for all bodies of water, including rivers, ponds, reservoirs, swamps, lakes, and oceans. The contour lines that indicate the elevations of hills and valleys are brown.

Topographic maps are periodically updated. Purple is used on updated maps to reflect any natural or human-made changes in the landscape.

Figure 8 Points of elevation on Earth's surface are projected onto paper to make a topographic map.
Interpret *How many meters high is the highest point on the map?*

Topographic Maps

Detailed maps showing the hills and valleys of an area are called topographic maps. **Topographic maps** show changes in elevation of Earth's surface, as shown in **Figure 8.** They also show mountains, rivers, forests, and bridges, among other features. Topographic maps use lines, symbols, and colors to represent changes in elevation and features on Earth's surface.

Contour lines Elevation on a topographic map is represented by a contour line. Elevation refers to the distance of a location above or below sea level. A **contour line** connects points of equal elevation. Because contour lines connect points of equal elevation, they never cross. If they did, it would mean that the point where they crossed had two different elevations, which would be impossible.

Contour intervals As **Figure 8** shows, topographic maps use contour lines to show changes in elevation. The difference in elevation between two side-by-side contour lines is called the **contour interval.** The contour interval is dependent on the terrain.

For mountains, the contour lines might be very close together, and the contour interval might be as great as 100 m. This would indicate that the land is steep because there is a large change in elevation between lines. You will learn more about topographic maps in the Mapping GeoLab at the end of this chapter.

DIFFERENTIATED INSTRUCTION

Visually Impaired Have topographic relief maps available for students with visual impairments. These students can touch the maps to discern changes in elevation. Be sure to tell students the scale of the map they are touching.

EARTH SCIENCE JOURNAL

Topographic Map Symbols Have students examine a topographic map that shows quarries and gravel pits. Have them compare and contrast how these features are represented on topographic maps. A quarry is symbolized by a pair of crossed picks, and a gravel pit is represented by a pair of crossed shovels. Both quarries and gravel pits are shown by depression contour lines. BL

Rubric

Index contours To aid in the interpretation of topographic maps, some contour lines are marked by numbers representing their elevations. These contour lines are called index contours, and they are used hand-in-hand with contour intervals to help determine elevation.

If you look at a map with a contour interval of 5 m, you can determine the elevations represented by other lines around the index contour by adding or subtracting 5 m from the elevation indicated on the index contour. Learn more about contour maps and index contours in the Problem-Solving Lab on this page.

☑ READING CHECK **Analyze** If you were looking at a topographic map with a contour interval of 50 m and the contour lines were far apart, would this indicate a rapid increase or slow increase in elevation?

Depression contour lines The elevations of some features such as volcanic craters and mines are lower than that of the surrounding landscape. Depression contour lines are used to represent such features.

On a map, depression contour lines look like regular contour lines, but have hachures, or short lines at right angles to the contour line, to indicate depressions. As shown in **Figure 9,** the hachures point toward lower elevations. When going uphill, note that the first depression contour line is the same elevation as the contour line before it.

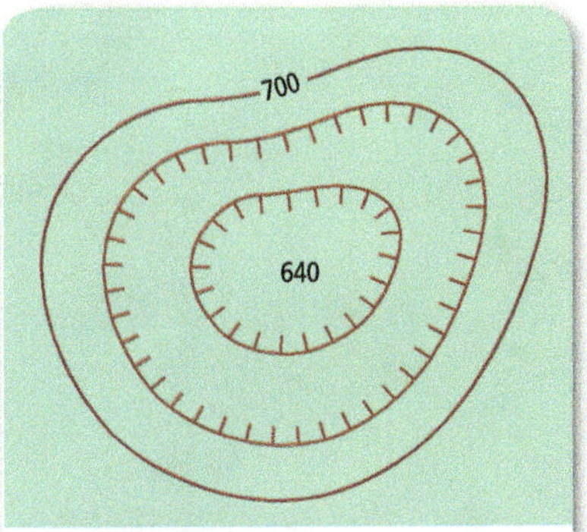

■ **Figure 9** The depression contour lines shown here indicate that the center of the area has a lower elevation than the outer portion of the area. The short lines pointing inward are called hachures and indicate the direction of the elevation change.

Problem-Solving LAB

Calculate Gradients

How can you analyze changes in elevation? Gradient refers to the steepness of a slope. To measure gradient, divide the change in elevation between two points on a map by the distance between the two points. Use the map to answer the following questions, and convert your answers to SI units.

Analysis

1. **Determine** the distance from Point A to Point B using the map scale.
2. **Record** the change in elevation.
3. **Calculate** If you were to hike the distance from Point A to Point B, what would be the gradient of your climb?

Think Critically

4. **Explain** Would it be more difficult to hike from Point A to Point B, or from Point B to Point C?

Topographic Map of Burr Hill

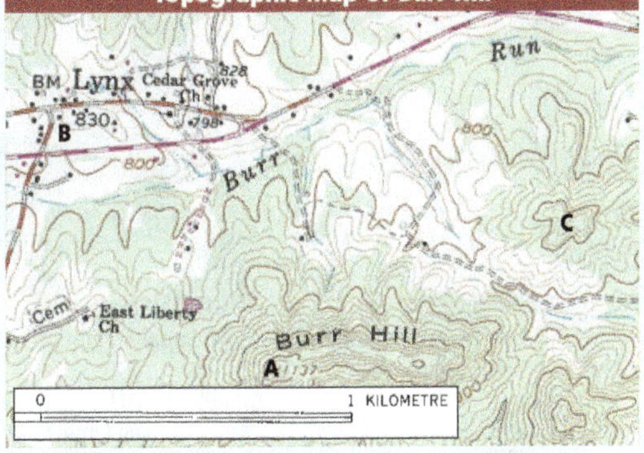

5. **Calculate** Between Point A and Point C, where is the steepest part of the hike? How do you know?

USGS

Problem-Solving LAB

Purpose Students will use a topographic map to analyze changes in elevation in terms of gradient.

Process Skills measure, use numbers, explain, draw conclusions

Teaching Strategy If possible, take students outside and show them an actual slope. Have one student stand at the top of the slope and another student stand at the bottom. Tell students the gradient of the slope can be determined by dividing the change in elevation between the two students by the distance between the two students.

Analysis

1. The distance between Point A and Point B is approximately 0.9 km.
2. The change in elevation is 1137 ft − 830 ft = 307 ft; 307 ft = 93.6 m.
3. 307 ft ÷ 0.5 mi = 614 ft/mi; or 93.6m ÷ 0.93 km = 100.6 m/km

Think Critically

4. Answers will vary. The distance between Point B and Point C is approximately 1 mi. The change in elevation is 960 ft − 830 ft = 130 ft. The gradient is 130 ft ÷ 1 mi = 130 ft./mi. In SI units: 1 mi = 1.61 km. The change in elevation is 292.6 m − 253.0 m = 39.6 m. The gradient is 39.6 m ÷ 1.61 km = 26.6 m/km. The hike from Point B to Point C would be less steep, but longer than the hike from Point A to Point B.
5. The steepest part of the hike would be from the 900-ft contour line to Point A. The close contour lines indicate a sharp rise in elevation.

☑ READING CHECK Contour lines that are far apart would represent a gentle increase in elevation.

GeoLAB

iLab Station

The GeoLab located at the end of the chapter can be used at this point in the lesson.

Identify Misconceptions

Students might see an illustration of a compass on a map and think that real compasses point to true geographic north, as shown on the map.

Uncover the Misconception

Ask students whether a compass points to geographic north or magnetic north. Many students might say geographic north. Make sure students understand geographic north and magnetic north differ by 11.5°.

Demonstrate the Concept

Show students the magnetic declination on several topographic maps. Tell them magnetic declination is the difference between the true, geographic North Pole and the magnetic north pole, and it varies with longitude.

Assess New Knowledge

Ask students to explain why compasses point to magnetic north. The needles of the compasses react to Earth's magnetic field. Also, have students research how they can determine the direction of true north. Answers will vary. Students might say they could locate the North Star in the night sky. The North Star is located about 0.5° away from Earth's geographic North Pole.

Geologic Maps

A useful tool for a geologist is a geologic map. A **geologic map** is used to show the distribution, arrangement, and type of rocks located below the soil. A geologic map can also show features such as fault lines, bedrock, and geologic formations.

Using the information contained on a geologic map, combined with data from visible rock formations, geologists can infer how rocks might look below Earth's surface. They can also gather information about geologic trends, based on the type and distribution of rock shown on the map.

Geologic maps are most often superimposed over topographic maps and color coded by the type of rock formation, as shown in **Figure 10.** Each color corresponds to the type of rock present in a given area. There are also symbols that represent mineral deposits and other structural features. Refer to **Table 1** on the following page to compare geologic maps to the other maps you have learned about in this chapter.

Figure 10 Geologic maps show the distribution of surface geologic features. Notice the abundance of Older Precambrian rock formations. Extensive uplift and erosion have resulted in the exposure of these ancient rocks.

Geologic Map of Grand Canyon

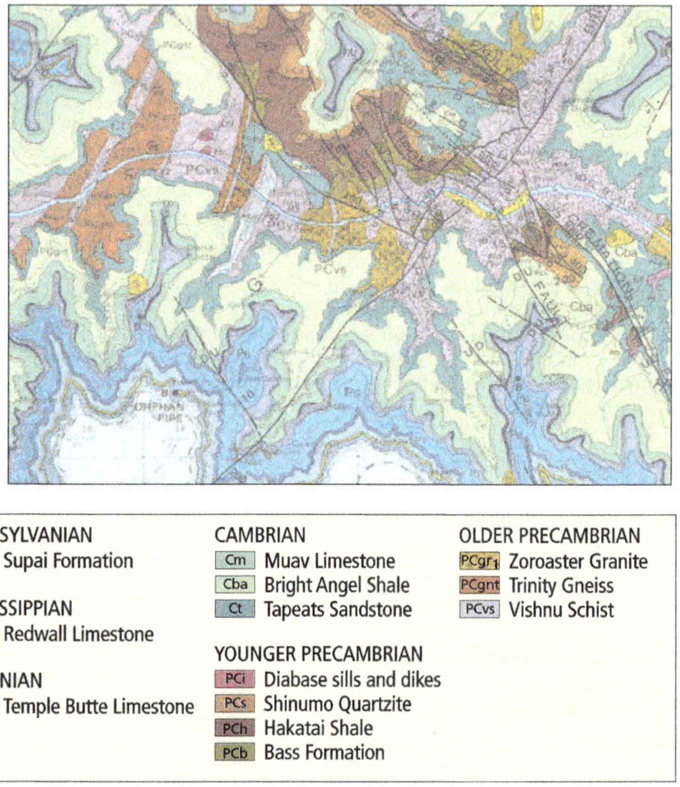

Demonstration

Interpret the Geology Show students a topographic map of the Grand Canyon and the geologic map in **Figure 10.** Point out the contour lines that are superimposed on the geologic map. Ask how these contour lines help geologists to interpret local geology. The spacing of contour lines provides information on the steepness of an area. Knowing the rock type from the geologic map and the topography from the topographic map allows geologists to make a more complete interpretation of the area. The Redwall Limestone and the Hermit Shale can be used to illustrate this point.

Explore **maps and projections with an interactive table.** Concepts In Motion

Table 1 Types of Maps and Projections

Map or Projection	Common Uses	Distortions
Mercator projection	navigation of planes and ships	The land near the poles is distorted.
Conic projection	road and weather maps	The areas at the top and bottom of the map are distorted.
Gnomonic projection	great circle routes	The direction and distance between landmasses is distorted.
Topographic map	to show elevation changes on a flat projection	It depends on the type of projection used.
Geologic map	to show the types of rocks present in a given area	It depends on the type of projection used.
Three-dimensional map	to aid in conceptualizing geologic structures and processes	It depends on the type of projection used.

Three-dimensional maps Topographic and geologic maps are two-dimensional models of Earth's surface. Sometimes, scientists need to visualize Earth three-dimensionally. To do this, scientists often rely on computers to digitize features such as rivers, mountains, valleys, and hills.

Map Legends

Most maps include both human-made and natural features located on Earth's surface. These features are represented by symbols, such as black dotted lines for trails, solid red lines for highways, and small black squares and rectangles for buildings. A **map legend,** such as the one shown in **Figure 11,** explains what the symbols represent. For more information about the symbols in map legends, see the *Reference Handbook.*

☑ READING CHECK **Apply** If you made a legend for a map of your neighborhood, what symbols would you include?

Map Scales

When using a map, you need to know how to measure distances. This is accomplished by using a map scale. A **map scale** is the ratio between distances on a map and actual distances on the surface of Earth. Normally, map scales are measured in SI, but as you will see on the map in the GeoLab, sometimes they are measured in different units such as miles and inches. There are three types of map scales: verbal scales, graphic scales, and fractional scales.

■ **Figure 11** Map legends explain what the symbols on maps represent.

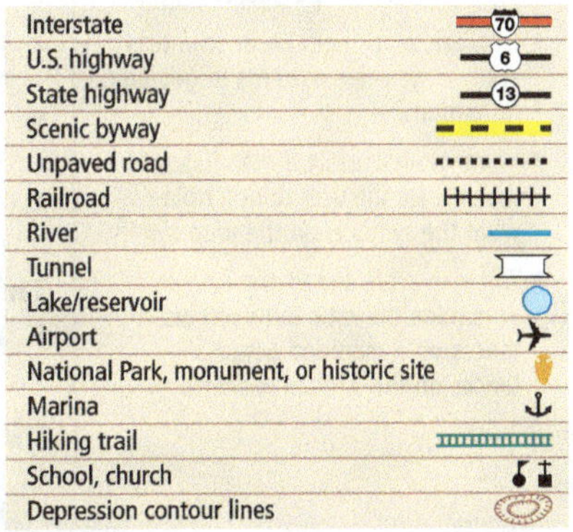

Interpret the Illustration

Subsurface Geology Have students review the geologic map shown in **Figure 10** and discuss the information it provides. Then ask students to write what they think a map of the subsurface would look like. Students should recognize that a subsurface map would not look the same as the one shown, but would instead represent a cross-section. Students might recognize that the rock formations would form layers in a subsurface map, with older units on the bottom and younger units on top in an undisturbed sequence. AL

Reinforcement

Use Map Scales Draw graphic map scales and fractional map scales on the board. They should represent different distances. Have students create verbal scales for each of the examples given. Ask students to state which map scale would show the most detail on a map and which would show the least detail. Students should be able to use a fractional scale to relate distance on the map to distance on the ground. They should be able to relate the size of the graphic scale to the distance represented on the map. Larger fractional scales show less detail than smaller fractional scales.

☑ READING CHECK Answers will vary depending on students' locations.

EARTH SCIENCE JOURNAL

Design a Map Imagine that you are working on a new map series. Choose 8 to 10 items that will be included on your map. Create a brochure describing these new symbols and how they are to be interpreted.

Rubric

READING CHECK The scale used depends on how much detail is desired. If very fine detail is desired, then the scale will be small. If a large area is to be shown, then the scale should be large.

3 Assess

Check for Understanding

Discussion Have students describe the differences among verbal, graphic, and fractional map scales. Students should give examples of each type of scale.

Reteach

Outline Have students outline the main ideas of this section in their Earth science journals. Ask students to provide specific examples in support of each main idea.

Assessment

Performance Place students in groups of three. Obtain enough topographic maps so that each group has one. Have groups answer the following questions about their maps: What symbols are included on the map? What do the symbols represent? What is the fractional scale of the map? What is the contour interval? What are the highest and lowest elevations on the map? What is the gradient between the highest and lowest places? Answers will vary depending on maps used. **COOP LEARN**

VOCABULARY

ACADEMIC VOCABULARY

Ratio
the relationship in quantity, amount, or size between two or more things
The ratio of girls to boys in the class was one to one.

Verbal scales To express distance as a statement, such as "one centimeter is equal to one kilometer," cartographers and Earth scientists use verbal scales. The verbal scale, in this example, means that one centimeter on the map represents one kilometer on Earth's surface.

Graphic scales Instead of writing the map scale out in words, graphic scales consist of a line that represents a certain distance, such as 5 km or 5 miles. The line is labeled, and then broken down into sections with hash marks, and each section represents a distance on Earth's surface. For instance, a graphic scale of 5 km might be broken down into five sections, with each section representing 1 km. Graphic scales are the most common type of map scale.

READING CHECK **Infer** why an Earth scientist might use different types of scales on different types of maps.

Fractional scales Fractional scales express distance as a ratio, such as 1:63,500. This means that one unit of distance on the map represents 63,500 units of distance on Earth's surface. One centimeter on a map, for instance, would be equivalent to 63,500 centimeters on Earth's surface. Any unit of distance can be used in fractional scales, but the units on each side of the ratio must always be the same.

A large ratio indicates that the map represents a large area, while a small ratio indicates that the map represents a small area. A map with a large fractional scale such as 1:100,000 km would therefore show less detail than a map with a small fractional scale such as 1:1000 km.

SECTION 2 REVIEW

Section Self-Check

Section Summary

- Different types of projections are used for different purposes.
- Geologic maps help Earth scientists study large-scale patterns in geologic formations.
- Maps often contain a map legend that allows the user to determine what the symbols on the map signify.
- The map scale allows the user to determine the ratio between distances on a map and actual distances on the surface of Earth.

Understand Main Ideas

1. **MAIN IDEA** **Explain** why distortion occurs at different places on different types of projections.
2. **Describe** how a conic projection is made. Why is this type of projection best suited for mapping small areas?
3. **Determine** On a Mercator projection, where does most of the distortion occur? Why?
4. **Compare and contrast** Mercator and gnomonic projections. What are these projections commonly used for?

Think Critically

5. **Predict** how a geologic map could help a city planner decide where to build a city park.

MATH IN Earth Science

6. Determine the gradient of a slope that starts at an elevation of 55 m and ends 20 km away at an elevation of 15 m.

SECTION 2 REVIEW

1. Distortion naturally occurs when a curved surface is projected onto a flat piece of paper. Accuracy can be maintained in only one area, for instance, land mass.
2. A conic projection is made by projecting points and lines from a globe onto a cone that touches the globe at a particular line of latitude. There is very little distortion along this line of latitude. Thus, conic projections are excellent for mapping limited areas.
3. On a Mercator projection, the majority of the distortion occurs at the poles. This is because lines of longitude are shown parallel rather than meeting at a single point. This causes the land masses at the poles to appear to be spread out.
4. A Mercator projection is made by projecting points and lines from a globe onto a cylinder. A gnomonic projection is made by projecting points and lines from a globe onto a piece of paper that touches the globe at a single point. Both are used for navigation.
5. A city planner could look at a geologic map to determine the location of a fault line or other geologic feature that would make it difficult to build a park in that location.
6. The change in elevation is 55 m − 15 m = 40m ÷ 20 km = 2 m/1 km

SECTION 3

Essential Questions

- What are some of the different types of remote sensing?
- How are satellites and sonar used to map Earth's surface and its oceans?
- What is the Global Positioning System and how does it work?

Review Vocabulary

satellite: natural or human-made object that orbits Earth, the Moon, or another celestial body

New Vocabulary

remote sensing
Landsat satellite
sonar
Global Positioning System
Geographic Information System

Remote Sensing

MAINIDEA New technologies have changed the appearances and uses of maps.

EARTH SCIENCE 4 YOU Many years ago, if you wanted a family portrait, it would be painted by an artist over many hours. Today, cameras can create a photo in seconds. Cartography has also changed. Cartographers use digital images to create maps with many more details that can be updated instantly.

Landsat Satellite

Advanced technology has changed the way maps are made. The process of gathering data about Earth using instruments mounted on satellites, airplanes, or ships is called **remote sensing.**

One form of remote sensing is detected with satellites. Features on Earth's surface, such as rivers and forests, radiate warmth at slightly different frequencies. **Landsat satellites** record reflected wavelengths of energy from Earth's surface. These include wavelengths of visible light and infrared radiation. One example of a Landsat image is shown in **Figure 12.**

To obtain such images, each Landsat satellite is equipped with a moving mirror that scans Earth's surface. This mirror has rows of detectors that measure the intensity of energy received from Earth. This information is then converted by computers into digital images that show landforms in great detail.

Landsat satellites are often used to aid in natural disaster relief planning. Landsat data are also used to study pollution, the movements of Earth's plates, and the melting of glaciers and ice caps.

Figure 12 Notice the differences between the two Landsat photos of New Orleans.
Interpret *Which image was taken after Hurricane Katrina in 2005? Explain.*

USGS

SECTION 3

1 Focus

Focus Transparency

Before presenting the lesson, project **Section Focus Transparency 6** and have students answer the accompanying questions. EL

MAINIDEA

Compare Old and New Maps Use your library or the Internet to obtain several maps of the same area that were made at least 30 years apart. Compare points of interest using information such as elevations and distances. Ask students to explain why the numbers have changed. Advances in technology have made measurements more precise.

2 Teach

Concept Development

Satellites and Nature Ask students to research how the Landsat satellites have been used in environmental studies. Students should find images from Landsat have been used to study the causes and effects of El Niño, to locate mineral resources, to plot changes in vegetation, to study ecological conditions following natural and human-made hazards, and to study the effects of overpopulation. Accept all reasonable answers. OL

Caption Question Fig. 12 The image on the bottom was taken after Hurricane Katrina occurred. Students should note the water's encroachment onto the the land in this image, especially in the upper left portion of the photo.

EARTH SCIENCE JOURNAL

Interpret Landsat Images To encourage students' understanding of the data on a Landsat image, have them write a script for a television documentary starring themselves describing the changes that New Orleans experienced as evidenced by the Landsat images in **Figure 12.**

Rubric

Collaborative Learning

Remote Sensing Devices Place students in groups of four. Assign each group a remote-sensing device to research, such as the *OSTM/Jason-2* satellite, *MOLA*, the *GOES* satellite, the *Hubble Space Telescope*, Landsat, or Sea-Beam. Have each group write a report that explains the system's design, how it uses electromagnetic frequencies, and its applications. Groups should share their reports with the class.
COOP LEARN

Concept Development

Compare Remote Sensing Data Use the Internet to download images and photographs from remote-sensing technology. Show students the type of information that is being sent back to Earth for analysis. Ask students to compare and contrast the images and photographs of various remote-sensing devices. For instance, some remote-sensing satellites use infrared imagery, while others use visible-light imagery. Students should observe an appreciable difference in the images.

FOLDABLES® Rubric

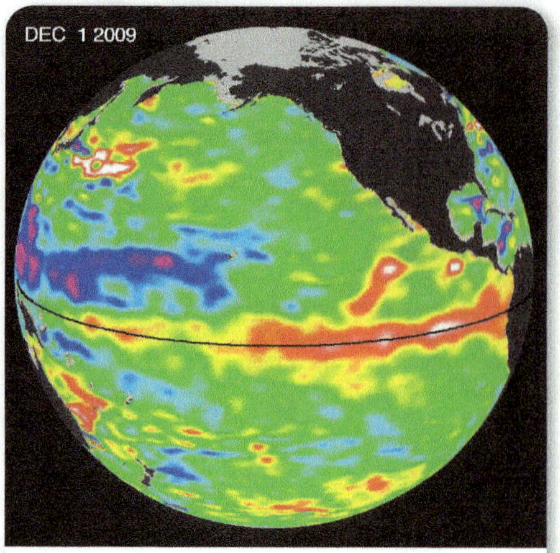

■ **Figure 13** This image, which focuses on the Pacific Ocean, was created with data from *OSTM/Jason-2*. The red color along the equator shows the rise in ocean depth and temperature during an El Niño event relative to normal.

FOLDABLES® Incorporate information from this section into your Foldable.

OSTM/Jason-2 Satellite

One satellite that uses radar to measure and map sea surface height is the *OSTM/Jason-2* satellite. **OSTM** stands for **O**cean **S**urface **T**opography **M**ission and is a follow-on to the *TOPEX/Poseidon* and *Jason-1* satellites. Radar uses high-frequency signals that are transmitted from the satellite to the surface of the ocean. A receiving device then picks up the returning echo as it is reflected off the water.

The distance to the water's surface is calculated using the known speed of light and the time it takes for the signal to be reflected. This data is important for measuring variations in sea level and monitoring changes in global ocean currents and heat transfer. These changes are reflected in satellite-to-sea measurements and result in images such as the one shown in **Figure 13.**

Using *OSTM/Jason-2* data, scientists are able to estimate global sea levels with an accuracy of just a few millimeters. Scientists can use this data combined with other existing data to create maps of ocean-floor features. For instance, ocean water bulges over seafloor mountains and forms depressions over seafloor valleys.

The *OSTM/Jason-2*, *Jason-1*, and *TOPEX/Poseidon* satellites have also been used for forecasting El Niño events, an unusual warming in the eastern Pacific Ocean. **Figure 14** below shows additional technological advances in cartography.

■ **Figure 14**
Mapping Technology
Advances in mapping have relied on technological developments.

1500 B.C. — A.D. 1 — 1500

1300 B.C. An ancient Egyptian scribe draws the oldest surviving topographical map.

150 B.C. The ancient Greek scientist Ptolemy creates the first map using a coordinate grid. It depicted Earth as a sphere and included Africa, Asia, and Europe.

A.D. 1154 Arab scholar Al-Idrisi creates a world map used by European explorers for several centuries. Earlier medieval maps showed Jerusalem as the center of a flat world.

1569 Flemish geographer Gerhardus Mercator devises a way to project the globe onto a flat map using lines of longitude and latitude.

DIFFERENTIATED INSTRUCTION

Struggling Learners Rather than writing reports about remote-sensing technology as suggested in the Collaborative Learning activity on this page, students can make posters that illustrate remote-sensing devices that have been used or are being used to collect data about the solar system. Examples might include the *Voyager* space probes, the *Hubble Space Telescope*, and the Mars Rover. Encourage students to choose topics that pique their interests.

SeaBeam

SeaBeam technology is similar to the *OSTM/Jason-2* satellite in that it is also used to map the ocean. However, SeaBeam is located on a ship rather than on a satellite, and maps the ocean floor rather than the surface. **Figure 15** shows an example of a map created with information gathered with SeaBeam technology. To map ocean-floor features, SeaBeam relies on **sonar,** which is the use of sound waves to detect and measure objects underwater.

You might have heard of sonar before. It is often used to detect objects like ships or submarines under water. This same technology allows scientists to detect changes in water depth or calculate distances between features on the ocean floor.

First, to gather the information needed to map the seafloor, a sound wave is sent from a ship toward the ocean floor. A receiving device then picks up the returning echo when it bounces off the seafloor.

Computers on the ship calculate the distance from the ship to the ocean floor using the speed of sound in water and the time it takes for the sound to be reflected. SeaBeam technology is used by fishing fleets, deep-sea drilling operations, and scientists such as oceanographers, volcanologists, and archaeologists.

☑ READING CHECK **Compare and contrast** SeaBeam images with *OSTM/Jason-2* images and how each might be used.

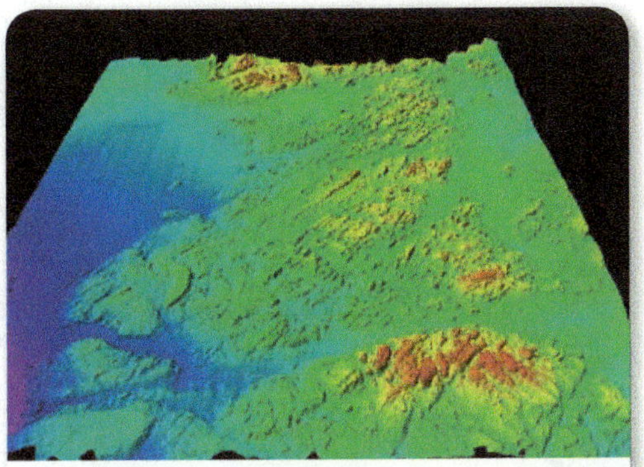

■ **Figure 15** This offshore image near Plymouth, MA was created with data from SeaBeam. The change in color indicates a change in elevation. The red-orange colors are the peaks, and the blue colors are the lowest elevations.

1752 A French cartographer first uses contour lines to represent elevation and marine depth for sailors exploring the New World.

2000 Space shuttle *Endeavour* collects the most complete topographical data of Earth, mapping almost 80 percent of Earth's land surface.

1700 — 1800 — 1900 — 2000

1875 Ella Eliza Fitz invents a method to mount a globe that shows the position of the Sun and the length of nights and days.

1966 Harvard University researchers develop the first computerized grid-based mapping system, the forerunner of GIS.

2008 *OSTM/Jason-2* launches. Used to help scientists study climate change, the satellite will scan 95% of Earth's ice-free oceans.

Teacher Content Support

What is SONAR? Sonar (sound navigation and ranging) was first used in World War I to detect submarines using echoes that bounced off their hulls. Today, it is widely used to calculate ocean depth. In addition, a type of sonar called side-scan sonar can be used to map ocean-floor features. Side-scan sonar directs sound waves to the seafloor at an angle, so that the sides of underwater hills and other topographic features can be mapped.

Interpret the Photo

Remote Imaging Divide the class into groups of three or four. Have students examine **Figure 15** and discuss how the image of the volcano was created. Have them compare the acquisition of this image to a sonogram of a baby in its mother's womb. Students might point out or speculate that sonograms on tissues are also acquired with sound waves. As long as the waves have something to bounce off of, an image can be created.

☑ READING CHECK *OSTM/Jason-2* uses radar on a satellite to map the ocean surface and SeaBeam uses sonar on a ship to map the ocean floor.

EARTH SCIENCE JOURNAL

Historical Events Have students research and add a data point to the time line in **Figure 14.** Topics might include people, events, or technology. Students should summarize their findings in their Earth science journals.

Rubric

ACROSS THE CURRICULUM

Math Sonar uses the speed of sound in water to calculate the depth of the ocean floor or to map ocean-floor features. A sound wave travels much faster in water than in air because the water molecules are closer together. The velocity of sound at 0°C in dry air is about 332 m/s. In freshwater, it is 1454 m/s.

Activity

Use GPS Have students research and write reports about the Global Positioning System (GPS). Students' reports should include information about the system's operation, its accuracy, the number of satellites in the system, its orbiting distance above Earth, and uses of the system. You might want to direct students to *The Global Positioning System and ArcGIS, Third Edition,* by Michael Kennedy, CRC Press, 2009. OL

CAREERS IN EARTH SCIENCE

WebQuest

READING CHECK GPS receivers must have signals from at least three satellites to determine position, while *OSTM/Jason-2* can gather data on its own. *OSTM/Jason-2* does not maintain communication with receivers on Earth at all times. *OSTM/Jason-2* gathers data to measure sea surface height. GPS satellites transmit signals to allow users with GPS receivers to locate their position on Earth.

Watch a **video about GPS.** Video

The Global Positioning System

The **Global Positioning System (GPS)** is a satellite navigation system that allows users to locate their approximate position on Earth. There are 24 satellites orbiting Earth, as shown in **Figure 16,** for use with GPS units. The satellites are positioned around Earth, and are constantly orbiting so that signals from at least three or four satellites can be picked up at any given moment by a GPS receiver.

To use GPS to find your location on Earth, you need a GPS receiver. The receiver calculates your approximate latitude and longitude—usually within 10 m—by processing the signals emitted by the satellites. If enough information is present, these satellites can also relay information about elevation, direction of movement, and speed. With signals from three satellites, a GPS receiver can calculate location on Earth without elevation, while four satellite signals will allow a GPS receiver to calculate elevation also. For more information on how the satellites are used to determine location, see **Figure 16.**

CAREERS IN EARTH SCIENCE

Cartographer An Earth scientist who works primarily with maps is called a cartographer. A cartographer might make maps, interpret maps, or research mapping techniques and procedures.

WebQuest

Uses for GPS technology GPS technology is used extensively for navigation by airplanes and ships. However, as you will read later, it is also used to help detect earthquakes, create maps, and track wildlife.

GPS technology also has many applications for everyday life. GPS receivers are often placed in cars to help navigate to preprogrammed destinations such as restaurants, hotels, and homes. Portable handheld GPS systems are also used in hiking, biking, and other travels. They allow for finding destinations more quickly and can help in determining specific locations on a map. Cell phones may also contain GPS systems that aid in finding locations.

READING CHECK **Compare** GPS satellites with *OSTM/Jason-2.*

The Geographic Information System

The **Geographic Information System (GIS)** combines many of the traditional types and styles of mapping described in this chapter. GIS mapping uses a database of information gathered by scientists, professionals, and students like you from around the world to create layers, or "themes," of information that can be placed one on top of the other to create a comprehensive map. These "themes" are often maps that were created with information gathered by remote sensing.

VOCABULARY

ACADEMIC VOCABULARY

Comprehensive
covering completely or broadly
The teacher gave the students a comprehensive study guide for the final exam.

Scientists from many disciplines use GIS technologies. A geologist might use GIS mapping when studying a volcano to help track historical eruptions. An ecologist might use GIS mapping to track pollution or to follow animal or plant population trends of a given area.

What's EARTH SCIENCE Got To Do With It?

Video *Take the Road Less Traveled*

Demonstration

GIS Maps Use the Internet to access the GIS home Web site. Choose an example to illustrate how this information is used. For example: a developer uses information to determine what properties might be at risk for flooding. Ask students to suggest information that should be included on this type of map. Students might suggest proximity of rivers, flood data, soil and rock analyses, rainfall data, etc.

VISUALIZING GPS Satellites

Figure 16 GPS receivers detect signals from the 24 GPS satellites orbiting Earth. Using signals from at least three satellites, the receiver can calculate location within 10 m.

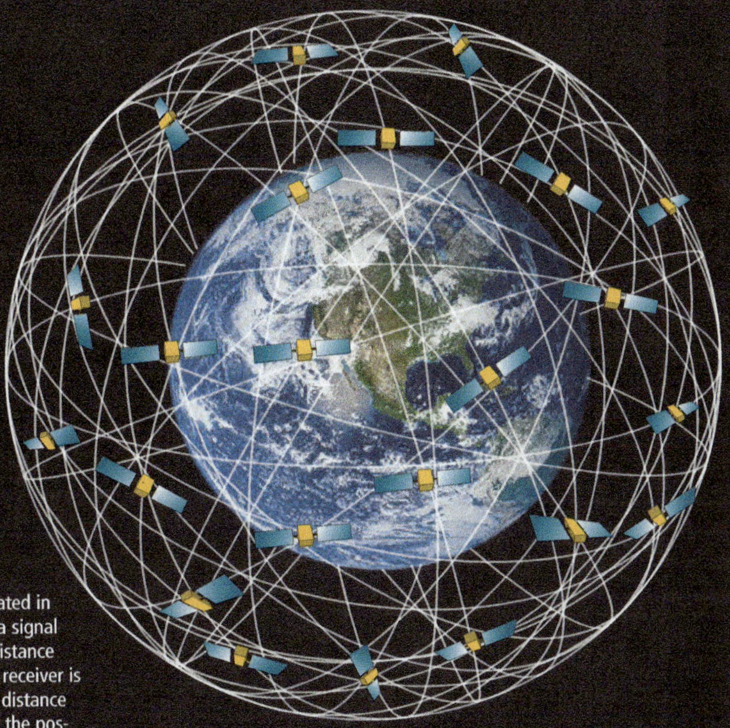

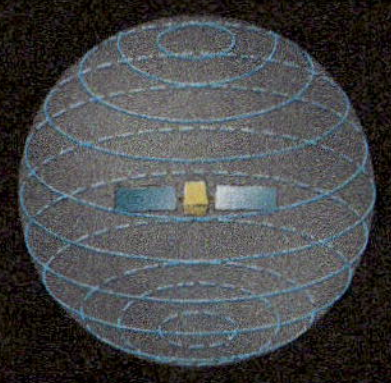

First, a GPS receiver, located in New York City, receives a signal from one satellite. The distance from the satellite to the receiver is calculated. Suppose the distance is 20,000 km. This limits the possible location of the receiver to anywhere on a sphere 20,000 km from the satellite.

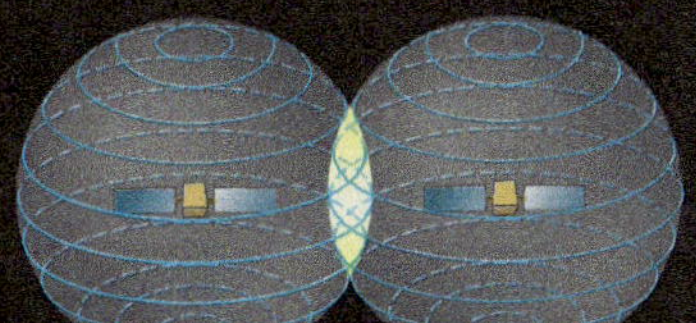

Next, the receiver measures the distance to a second satellite. Suppose this distance is calculated to be 21,000 km away. The location of the receiver has to be somewhere on the area where the two spheres intersect, shown here in yellow.

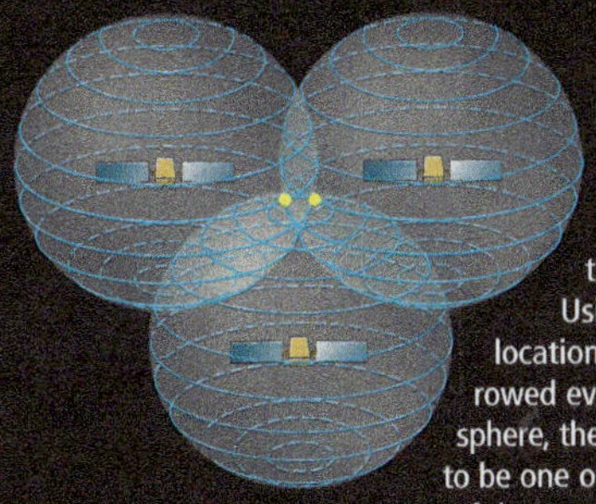

Finally, the distance to a third satellite is calculated. Using this information, the location of the receiver can be narrowed even further. By adding a third sphere, the location can be calculated to be one of two points as shown. Often one of these points can be rejected as an improbable or impossible location.

Concepts In Motion View an **animation of GPS satellites.**

VISUALIZING

Purpose

Students will learn how GPS satellites and systems work.

Enrichment

Use GPS Devices Bring in a handheld GPS device. Divide the class into small groups. Have them choose a nearby location and write directions from the school to the location. Then compare their directions with that given by the GPS device. You might use a car in the school parking lot if a handheld device is not available. **OL**

Teacher Content Support

GPS Satellites There are currently 24 satellites in six circular orbits 20,200 km above Earth. The satellites are positioned at an angle of 55 degrees and have a 12-hour period. They are spaced out so that at any given moment of the day, six satellites are visible to users anywhere in the world. GPS satellites broadcast position and time continuously.

3 Assess

Check for Understanding

Discussion Ask students to define remote sensing and explain how satellites use the reflected energy from Earth's surface to collect data about Earth.

Reteach

Demonstrate Obtain a handheld GPS unit. Take students outside, turn on the GPS unit, and walk around the school grounds. Pause every 50 m or so to show students that the latitude and longitude values change as the position of the GPS user changes. Have students explain why the coordinates constantly change. The handheld GPS unit continually receives information from an array of satellites in space. This information is used to continually update the user's position.

Assessment

Performance Have students draw a diagram to illustrate how satellite mirrors and computers collect and process data from Earth's surface.

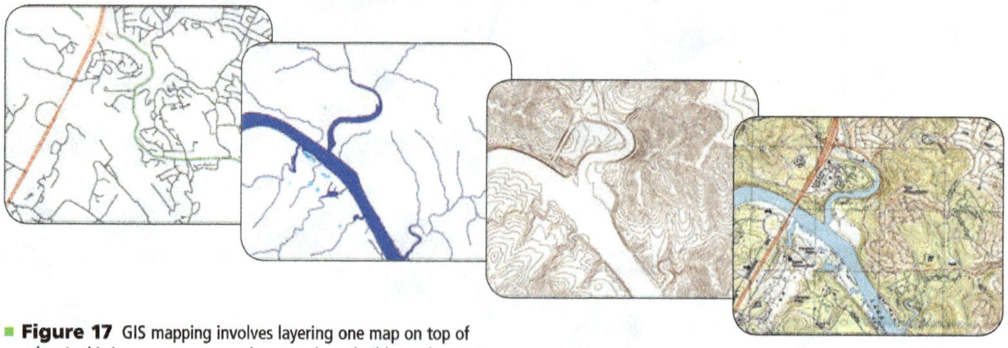

■ **Figure 17** GIS mapping involves layering one map on top of another. In this image, you can see how one layer builds on the next.

GIS maps might contain many layers of information compiled from several different types of maps, such as a geologic map and a topographic map. As shown in **Figure 17,** maps of rivers, topography, roads, cities, and other notable landforms from the same geographic area can be layered on top of each other to create one comprehensive map.

One major difference between GIS mapping and traditional mapping is that a GIS map can be updated as new information is loaded into the database. Once a map is created, the layers are still linked to the original information. If this information changes, the GIS layers also change. The result is a map that is always up-to-date—a valuable resource for people who rely on current information.

SECTION 3 REVIEW

Section Self-Check

Section Summary

- Remote sensing is an important part of modern cartography.
- Satellites are used to gather data about features of Earth's surface.
- Sonar is also used to gather data about features of Earth's surface.
- GPS is a navigational tool that is now used in many everyday items.

Understand Main Ideas

1. MAINIDEA **Describe** how remote sensing works and why it is important in cartography.
2. **Apply** Why is GPS navigation important to Earth scientists?
3. **Explain** the different types of information that can be gathered with satellites.
4. **Predict** why it might be important to be able to add and subtract map layers as with GIS mapping.

Think Critically

5. **Infer** How could GIS mapping be helpful in determining where to build a housing development?
6. **Explain** why it is important to have maps of the ocean floor, such as those gathered with SeaBeam technology.

WRITING IN ▶ Earth Science

7. Write an article describing how GPS satellites help you locate your position on Earth.

USGS

SECTION 3 REVIEW

1. Remote sensing is the practice of gathering data about Earth from instruments aboard ships, satellites, and aircraft. Different technologies such as radar and sonar are used to gather the data. This data is important in cartography because it lends new information that is both current and more complete than data gathered directly.
2. GPS navigation is important to record specific locations of geologic interest, such as soil samples, rock formations, or water samples. GPS navigation can also be used to verify the accuracy of maps or other scientific data.
3. Possible answers: ocean surface topography, vegetation, water, geologic features
4. Different people need to obtain different kinds of information from maps. By adding and subtracting layers, GIS mapping makes it easier to customize maps for individual purposes.
5. With GIS mapping, you would be able to include all the features you needed to consider for a housing development, including current land use, roads, rivers and lakes, geologic features, etc.
6. Seafloor maps can be compared to track changes on the ocean floor; they can also be used to track changes in ocean depth.
7. Answers will vary, but should include information about trilateration, GPS receivers, and error range of about 10 m.

Rubric

Earth Science & TECHNOLOGY

Mapping Disaster Zones

On August 29, 2005, Hurricane Katrina hit the New Orleans area, causing an estimated $96 billion in damage and resulting in the deaths of nearly 2000 people. With such widespread devastation, how did relief workers reach the damaged areas? Mapping technologies helped workers to identify priority areas and create a plan to aid those affected.

GPS and disaster relief Global Positioning System (GPS) satellites send signals back to Earth telling the receiver the exact location of the user. The satellites travel at approximately 14,000 km/h, and are powered by solar energy. During Katrina, GPS signals provided up-to-the-minute information regarding destruction detail and locations of survivors and aid workers.

Using GIS Another important mapping tool used during disasters is the Geographic Information System Technology (GIS). This technology captures, stores, records, and analyzes data dependent on geography and location. As a result, many important decisions about environmental issues or relief efforts can be made using GIS data. After Katrina, GIS data provided relief workers with images of area hospitals within a small geographic area. This enabled emergency workers to get injured individuals to medical facilities quickly.

Other imaging systems Other mapping software packages provide actual pictorial images of Earth. These images show the damaged areas as well as buildings that can be appropriate for setting up relief sites.

Synthetic Aperture Radar (SAR) polarimetry is an imaging technology that is able to rapidly detect disaster zones. With other satellite images, views of the affected landscape can be blocked by clouds, darkness, smoke, or dust.

USGS

This aerial image shows some of the flooding and destruction caused by Hurricane Katrina. Images like this help workers navigate through the altered landscape.

By using radar, SAR mapping is not affected by these things, thus making the images readily available to relief workers.

Mapping areas affected by natural disasters with satellite and aerial images makes these areas accessible by relief workers. They are better able to prepare for the changes in local geography, destruction of buildings, and other physical challenges in the disaster zone. Continued improvements in mapping technologies and increased accessibility are important for continued improvement of disaster relief programs.

WRITING IN ▶ Earth Science

Mapping Applications Research a recent natural disaster. Write a news article that describes the disaster based on the images of the disaster you find. Include several images in your news article. WebQuest

WRITING IN ▶ Earth Science

Rubric

Mapping Applications Ask students to explain to the class how officials could have mitigated damages done by the disaster they chose.

WebQuest

Earth Science & TECHNOLOGY

Purpose

Students will learn how various remote sensing technology is used during disaster relief operations.

Teacher Content Support

GPS and Weather Prediction

The GPS system not only was involved in the relief effort during the aftermath of Hurricane Katrina but also provided up-to-the-minute meteorological information, maritime and weather data for people working in the area. Better weather predictions and more informed decision making regarding water vapor was made possible due to GPS information. The data was accurate within centimeters and precise in terms of positioning. The system consisted of 85 transmitting sites located throughout the United States, including Puerto Rico, Alaska, and Hawaii.

Teaching Strategies

- Use a world map to show students where hurricanes occur. Include the occurrence of typhoons and tropical cyclones to illustrate the United States is not the only country that experiences such severe storms.
- In order to ensure students understand how dangerous hurricanes can be, hold a discussion to point out hurricane occurrence, characteristics, and potential damage. Have students add to the discussion by asking questions.

GeoLAB

iLab Station | Rubric

Preparation

Time Allotment 45 min

Process Skills interpret scientific illustrations, use numbers, communicate, analyze data, draw conclusions

Safety Precaution Approve lab safety forms before work begins.

Procedure

- To increase student proficiency in SI, have them give their answers in both English units and SI units.
- Team students with visual impairments with students who can easily discern small print.
- If possible, make a transparency of the map to enhance the class discussion following the activity.
- Remind students that gradient is calculated by dividing the change in elevation by the change in distance.

LabManager™
Customize any lab with the LabManager™ CD-ROM.

GeoLAB

iLab Station

Mapping: Use a Topographic Map

Background: Topographic maps show two-dimensional representations of Earth's surface. With these maps, you can determine the slope of a hill, what direction streams flow, and where mines and other features are located. In this lab, you will use the topographic map on the following page to determine elevation for several routes and to create a profile showing elevation.

Question: *How can you use a topographic map to interpret information about an area?*

Materials

ruler
string
piece of paper

Procedure

1. Read and complete the lab safety form.
2. Take a piece of paper and lay it on the map so that it intersects Point A and Point B.
3. On this piece of paper, draw a small line at each place where a contour line intersects the line from Point A to Point B. Also note the elevation at each hash mark and any rivers crossed.
4. Copy the table shown on this page into your science journal.
5. Now take your paper where you marked your lines and place it along the base of the table.
6. Mark a corresponding dot on the table for each elevation.
7. Connect the dots to create a topographic profile.
8. Use the map to answer the following questions. Be sure to check the map's scale.
9. Use the string to measure distances between two points that are not in a straight line. Lay the string along curves, and then measure the distance by laying the string along the ruler. Remember that elevations on United States Geological Survey (USGS) maps are given in feet.

Analyze and Conclude

1. **Determine** What is the contour interval?
2. **Identify** what type of map scale the map utilizes.
3. **Calculate** the stream gradient of Big Wildhorse Creek from the Gravel Pit in Section 21 to where the creek crosses the road in Section 34.
4. **Calculate** What is the highest elevation of the jeep trail? If you followed the jeep trail from the highest point to where it intersects an unimproved road, what would be your change in elevation?
5. **Apply** If you started at the bench mark (BM) on the jeep trail and hiked along the trail and the road to the Gravel Pit in section 21, how far would you hike?
6. **Analyze** What is the straight line distance between the two points in Question 4? What is the change in elevation?
7. **Predict** Does Big Wildhorse Creek flow throughout the year? Explain your answer.
8. **Calculate** What is the shortest distance along roads from the Gravel Pit in Section 21 to the secondary highway?

	820
	810
	800
	790
	780
	770
	760
	750
	740
	730
	720
	710
	700

INQUIRY EXTENSION

Make a Map Using what you have learned in this lab, create a topographic map of your hometown. Compare your map to your classmates' maps.

Analyze and Conclude

1. The contour interval is 10 ft.
2. This map utilizes both a fractional and a graphic scale.
3. The change in elevation is 750 ft − 710 ft = 40 ft. The change in distance is 4 mi. The stream gradient is 40 ft ÷ 4 mi = 10 ft/mi.
4. The highest elevation = 1071 ft. The change in elevation = 1071 ft − 750 ft = 321 ft.
5. You would have hiked approximately 5.25 mi.
6. The straight line distance = 2 mi. The change in elevation = 1071 ft − 750 ft = 321 ft.
7. No; the topographic map symbol for the stream indicates it is an intermittent stream.
8. The shortest distance is slightly more than 3 mi.

INQUIRY EXTENSION

Make a Map Answers will vary depending on the students' locations.

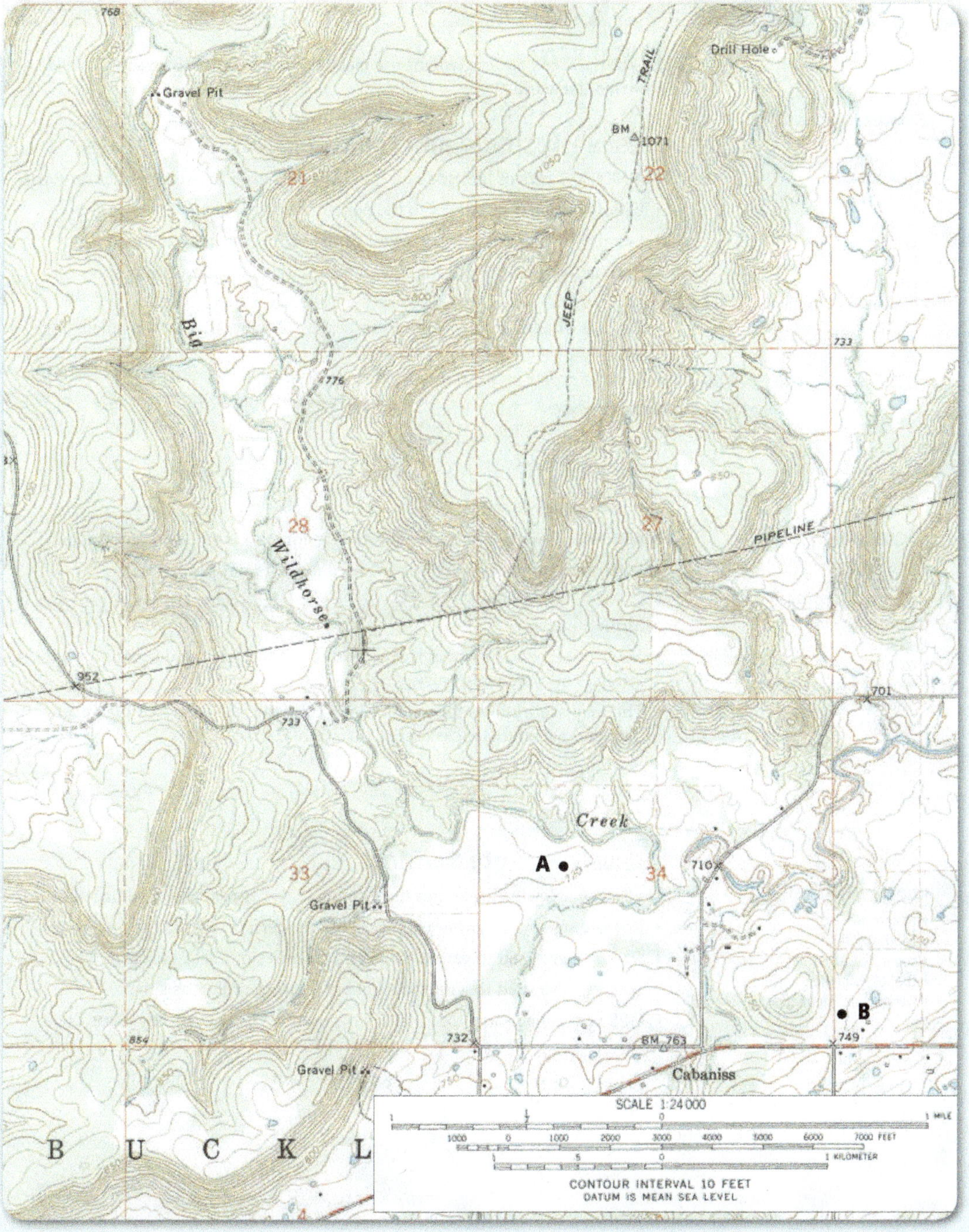
Gravel Pit
Drill Hole
BM 1071
JEEP TRAIL
Big
Wildhorse
Creek
PIPELINE
A
B
710
701
733
952
732
BM 763
749
Cabaniss
Gravel Pit
Gravel Pit
B U C K L
SCALE 1:24 000
CONTOUR INTERVAL 10 FEET
DATUM IS MEAN SEA LEVEL
USGS

MAINIDEAS Summary statements can be used by students to review the major concepts of the chapter.

Students can review with these online resources.

Vocabulary Practice

Vocabulary eGames
Vocabulary eFlashcards
Vocabulary PuzzleMaker

Section Self-Check

Chapter Self-Check

Online Test Practice

Use the ExamView® *Assessment Suite* CD-ROM to:

- create multiple versions of tests
- create modified tests with one mouse click
- edit existing questions and add your own questions
- build tests aligned with state standards using built-in state curriculum tags
- change English tests to Spanish with one mouse click
- track students' progress using the Teacher Management System

CHAPTER 2 STUDY GUIDE

BIGIDEA **Earth scientists use mapping technologies to investigate and describe the world.**

Vocabulary Practice

SECTION 1 Latitude and Longitude

VOCABULARY

- cartography
- equator
- latitude
- longitude
- prime meridian
- International Date Line

MAINIDEA Lines of latitude and longitude are used to locate places on Earth.

- Latitude lines run parallel to the equator.
- Longitude lines run east and west of the prime meridian.
- Both latitude and longitude lines are necessary to locate exact places on Earth.
- Earth is divided into 24 time zones, each 15° wide, that help regulate daylight hours across the world.

SECTION 2 Types of Maps

VOCABULARY

- Mercator projection
- conic projection
- gnomonic projection
- topographic map
- contour line
- contour interval
- geologic map
- map legend
- map scale

MAINIDEA Maps are flat projections that come in many different forms.

- Different types of projections are used for different purposes.
- Geologic maps help Earth scientists study large-scale patterns in geologic formations.
- Maps often contain a map legend that allows the user to determine what the symbols on the map signify.
- The map scale allows the user to determine the ratio between distances on a map and actual distances on the surface of Earth.

SECTION 3 Remote Sensing

VOCABULARY

- remote sensing
- Landsat satellite
- sonar
- Global Positioning System
- Geographic Information System

MAINIDEA New technologies have changed the appearance and use of maps.

- Remote sensing is an important part of modern cartography.
- Satellites are used to gather data about features of Earth's surface.
- Sonar is also used to gather data about features of Earth's surface.
- GPS is a navigational tool that is now used in many everyday items.

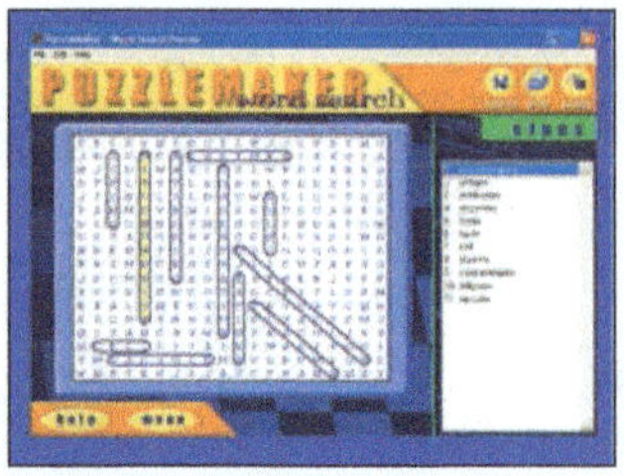

Vocabulary Practice

For additional practice with vocabulary, have students access the Vocabulary PuzzleMaker.

Chapter Self-Check

VOCABULARY REVIEW

Each of the following sentences is false. Make each sentence true by replacing the italicized word with a vocabulary term from the Study Guide.

1. The study of mapmaking is called *topology*.
2. A *gnomonic projection* is a map that has parallel lines of latitude and longitude.
3. The process of collecting data about Earth from far above Earth's surface is called *planetology*.
4. *Landsat satellite* uses sonar waves emitted from a ship to map the ocean floor.
5. A *map scale* explains what the symbols on the map represent.

Replace the underlined words with the correct vocabulary term from the Study Guide.

6. Latitude lines run north to south and are measured from the prime meridan.
7. A map legend shows the ratio between distances on a map.
8. GPS mapping combines many traditional types of maps into one.
9. GIS technology helps determine a user's exact location.

Choose the correct vocabulary term from the Study Guide to complete the following sentences.

10. Zero longitude is known as the ________.
11. The difference in elevation between two side-by-side contour lines on a topographic map is called the ________.
12. ________ is the use of sound waves to detect and measure objects underwater.
13. The ________ serves as the transition line for calendar days.
14. A(n) ________ is used on a topographic map to indicate elevation.

UNDERSTAND KEY CONCEPTS

Use the figure below to answer Questions 15 and 16.

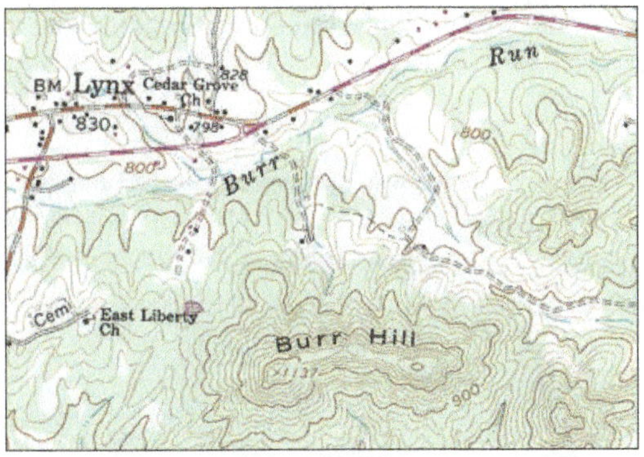

USGS

15. What is shown in this image?
    A. a Landsat image
    B. a topographic map
    C. a gnomonic projection
    D. a GIS map
16. What are the lines in the figure called?
    A. hachures
    B. contour lines
    C. latitude lines
    D. longitude lines
17. Refer to **Figure 4.** How many time zones are there in Australia?
    A. 5
    B. 1
    C. 3
    D. 10
18. Which is a use of SeaBeam?
    A. to map continents
    B. to map the ocean floor
    C. to map Antarctica
    D. to map mountains and valleys
19. On a topographic map, which do hachures point toward?
    A. higher elevations
    B. lakes
    C. no change in elevation
    D. lower elevations
20. Which is not usually included in map legends?
    A. interstates
    B. people
    C. rivers
    D. railroads

CHAPTER 2 ASSESSMENT

VOCABULARY REVIEW

1. cartography
2. Mercator projection
3. remote sensing
4. SeaBeam
5. map legend
6. longitude
7. map scale
8. GIS
9. GPS
10. prime meridian
11. contour interval
12. Sonar
13. International Date Line
14. contour line

UNDERSTAND KEY CONCEPTS

15. B
16. B
17. C
18. B
19. D
20. B

CONSTRUCTED RESPONSE

21. 2 P.M.
22. The map with a scale of 1:150 m would be better because it would show more detail.
23. the prime meridian
24. the equator
25. 2099 m; any higher and another contour line would be shown.
26. *OSTM/Jason-2* maps the sea surface using radar from a satellite, while SeaBeam maps the seafloor using sonar from a ship.
27. Astronomers can use a grid system similar to the system of latitude and longitude to locate objects in the night sky.
28. A conic projection would be the best projection for the map in the Launch Lab because it is the most accurate projection for small areas.
29. From the prime meridian, there are 36 meridians. Each is 10 degrees apart from the next.
30. From the equator to the poles, there are 18 lines of latitude. Each is 10 degrees apart.
31. The image in the picture is a Mercator projection, most commonly used for navigation purposes.

THINK CRITICALLY

32. A person flying from Virginia to California would have to set his or her watch backward because California is three hours behind Virginia.
33. Answers will vary, but students might suggest that one time zone would make it easier for families that live on opposite ends of the country to communicate with each other without worrying about a great time difference.
34. An architect trying to determine where to build a house would use a geologic map to determine where the most stable ground is located. A paleontologist would use a geologic map to both determine what tools he or she would need to excavate a certain type of rock as well as use the map to determine where to dig based on the type of rock in which human remains are most often found.
35. The image is a gnomonic projection. This type of projection could be used to help navigate a plane or a ship over long distances.

CONSTRUCTED RESPONSE

21. Locate What time is it in New Orleans, LA, if it is 3 P.M. in Syracuse, NY? Refer to **Figure 4** for help.

22. Explain If you wanted to study detailed features of a volcano, would you use a map with a scale of 1:150 m or 1:150,000 m? Why?

Use the figure below to answer Questions 23 and 24.

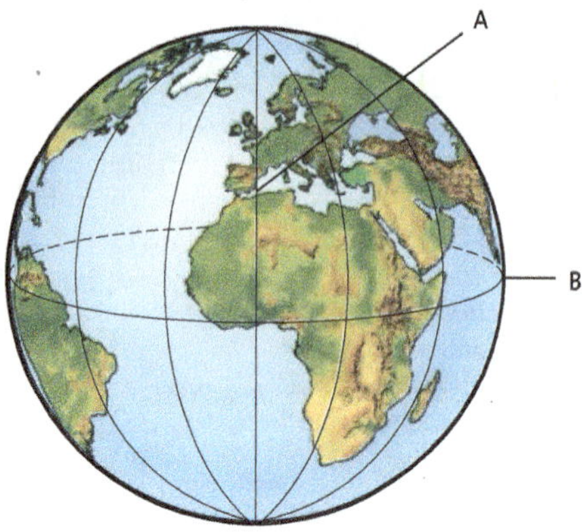

23. Identify What is the line labeled *A*?

24. Identify What is the line labeled *B*?

25. Explain What is the maximum potential height of a mountain if the last contour line is 2000 m and the map has a contour interval of 100 m?

26. Describe how radar used in the *OSTM/Jason-2* satellite differs from the sonar used in the collection of data by SeaBeam.

27. Infer Based on what you have learned in this chapter, how might an astronomer map objects seen in the night sky?

28. Practice Think back to the LaunchLab at the beginning of the chapter. What type of map projection would be best for the map you drew? Why?

29. Explain how degrees of longitude are calculated.

30. Explain how degrees of latitude are calculated.

Use the figure below to answer Question 31.

31. Interpret what type of projection is shown in the figure. What would this type of projection be used for?

THINK CRITICALLY

32. Apply Would a person flying from Virginia to California have to set his or her watch backward or forward? Explain.

33. Consider why a large country like China might choose to follow only one time zone.

34. Careers in Earth Science Analyze how an architect trying to determine where to build a house and a paleontologist trying to determine where to dig for fossils might use a geologic map.

Use the figure below to answer Question 35.

35. Apply What is the projection shown above? What would be two uses for this type of projection? Explain.

36. Plan Make a map from your school to the nearest supermarket. How will you determine the scale? What will you need to include in your legend?

36. Answers will vary depending on the students' location. The scale should either be a fractional or graphic scale. The legend might include a variety of things, but will most likely include roads, schools, rivers, and parks.

Chapter Self-Check

37. **Analyze** Why isn't a conic projection used to navigate a ship or an aircraft?

38. **Design** an experiment to test the accuracy of several types of GPS receivers. Make sure you include your control, dependent, and independent variables.

39. **Evaluate** Briefly describe GIS and how it can be used by your community to develop an emergency plan for a severe storm, earthquake, blizzard, drought, or another potential local disaster.

40. **Explain** why it is necessary to have three satellite signals to determine elevation when using a GPS receiver.

CONCEPT MAPPING

41. Use the following to complete a concept map about remote sensing: *remote sensing,* OSTM/Jason-2 *satellite, Landsat satellite, GPS, uses radar to map sea surface height, uses visible light and infrared radiation to map Earth's surface,* and *uses microwaves to determine location of user.* For more help, refer to the *Skillbuilder Handbook.*

CHALLENGE QUESTION

Use the figure below to answer Question 42.

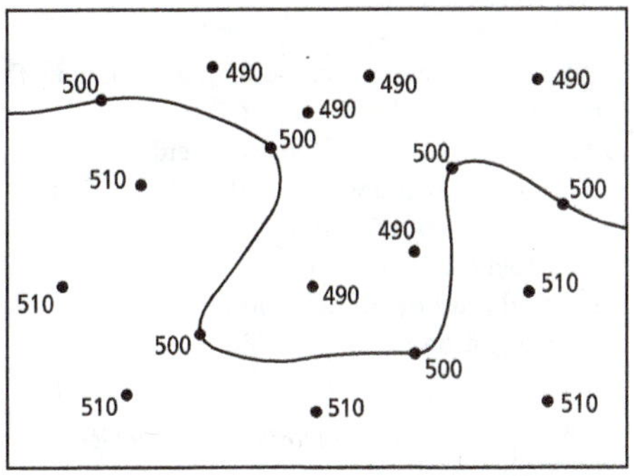

42. **Assess** Trace the following image to create a topographic map. Connect the elevation measurements to create contour intervals.

WRITING IN ▶ Earth Science

43. Write a journal entry for an explorer traveling across America before an accurate map was made.

DBQ Document–Based Questions

Data obtained from: NASA, *CALIPSO* satellite image.

This is a satellite image from NASA's CALIPSO *satellite.* CALIPSO *is similar to the other remote sensing technologies you learned about in this chapter, but instead of radar or sonar, it uses something called lidar that sends pulses of light and measures the time it takes for the light to reflect back to the satellite. Based on the amount of time it takes for the light to reflect,* CALIPSO *and scientists can determine what is located below the satellite. The image above was collected on June 7, 2006, across the Indian Ocean. The navy blue areas indicate that no data was detected.*

44. At approximately how many kilometers is the volcanic plume located?

45. At approximately how many kilometers is the thick cirrus cloud located?

46. Why do you think the volcanic plume is higher than the thick cirrus clouds?

47. Why do you think there was no data detected below the thick cirrus clouds?

CUMULATIVE REVIEW

48. Why are graphs, charts, and maps useful? **(Chapter 1)**

49. Why is good communication important in the field of science? **(Chapter 1)**

CHAPTER 2 ASSESSMENT

37. A conic projection has little land mass distortion when used to map small areas. If used for larger areas, as would be necessary for a ship or aircraft navigating long distance, the distortion would be too great to navigate accurately.
38. Answers will vary, but should include a control, and dependent and independent variables, and hypothesis.
39. Answers will vary. GIS mapping involves layering maps one on top of the other (with a computer program). Each layer has a different value such as streets, land use, elevation, etc. These maps could be used for severe storm planning because, for example, they could be used to plan evacuation routes, identify shelter routes, identify areas that are prone to flooding, or to identify areas that are prone to landslides.
40. With only two signals, your position on Earth can be located on a vertical plane, as with latitude and longitude. The addition of a third signal provides trilateration and intersects the vertical plane to provide a third intersect, allowing elevation to be determined.

CONCEPT MAPPING

41. First level: remote sensing; second and third levels (left to right): *OSTM/Jason-2* satellite connected to *uses radar to map sea surface height,* Landsat satellite connected to *uses visible light and infrared radiation to map Earth's surface,* and GPS connected to *uses microwaves to determine location of users*

CHALLENGE QUESTION

42. Students should connect like elevation points to create a topographic map. There should be no intersecting contour lines.

WRITING IN ▶ Earth Science

Rubric

43. Answers will vary, but should reflect the difficulties in navigation.

DBQ Document–Based Questions

Data obtained from: NASA, *CALIPSO* satellite image

44. About 20 km. Answers might vary, but should indicate that the volcanic plume is higher than the thick cirrus clouds.
45. About 15 km. Answers might vary, but should indicate that the thick cirrus is lower than the volcanic plume.
46. The volcanic plume is less dense.
47. The thick cirrus clouds either did not allow the light to penetrate, or did not allow the light to reflect back to the satellite and so no data were available.

CUMULATIVE REVIEW

48. They are convenient ways to organize and analyze data.
49. to promote a better understanding of the work and accomplishments of the science community

MULTIPLE CHOICE

1. C
2. D
3. C
4. B
5. C
6. B
7. D
8. A

CUMULATIVE STANDARDIZED TEST PRACTICE

MULTIPLE CHOICE

Use the map to answer Questions 1 and 2.

1. What is the latitude and longitude of the location pointed out by the arrows?
A. 30° N, 100° W
B. 45° N, 105° W
C. 30° N, 90° W
D. 10° N, 90° W

2. Roughly how many degrees of latitude does the United States cover?
A. 10°
B. 15°
C. 20°
D. 25°

3. Which would be most useful if you were lost in the Sahara desert?
A. Landsat satellite
B. *OSTM/Jason-2* satellite
C. Global Positioning System
D. topographic map of Africa

4. What is the reference point for lines of longitude?
A. the equator
B. the prime meridian
C. the International Date Line
D. the 360th meridian

5. Why do cartographers break down degrees of longitude and latitude into minutes and seconds
A. to get a better time frame of how long it takes to get from one place to the next
B. to help travelers with planning trips
C. to locate positions on Earth more precisely
D. to make cartography easier to understand

Use the map below to answer Questions 6 and 7.

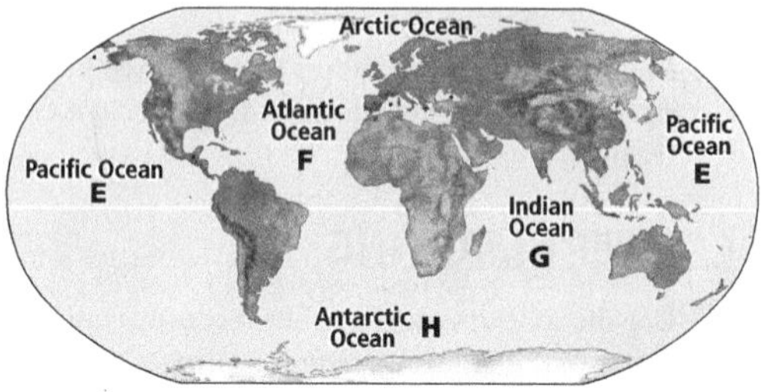

6. What problem do cartographers encounter when creating maps such as the one shown above?
A. placing all of the continents in the correct position
B. transferring a three-dimensional Earth onto a flat piece of paper
C. naming all of the important locations on the map
D. placing lines of latitude and longitude at the correct locations

7. What improvements could be made to make this map more helpful to sailors?
A. Distort the size of the continents.
B. Show only the water locations and not the locations of the land.
C. Label the various continents.
D. Add lines of latitude and longitude for navigation.

8. For what purpose are conic projection maps typically used?
A. road and weather maps
B. showing changes in elevation
C. plotting long distance trips
D. showing one specific point on Earth

Online Test Practice

SHORT ANSWER

Use the map below to answer Questions 9–11.

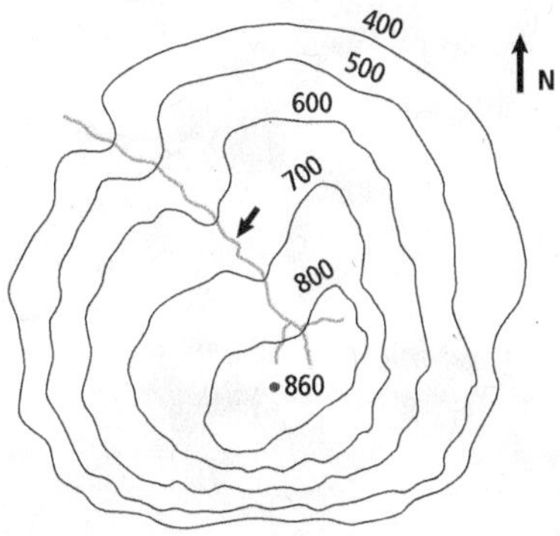

9. What is the map above showing?

10. What do the numbers on the map represent?

11. How might a hiker use this map in creating a route to get to the top?

12. Why would a ship find SeaBeam technology beneficial?

13. The distance from Earth to the Sun is 149,500,000 km. Rewrite this number using scientific notation.

14. Why is it important to include legends on a map?

15. Jenna measured the temperature of solutions before, during, and after an exothermic reaction. Which type of display would show the changes in temperature throughout the reaction most clearly and why?

READING FOR COMPREHENSION

Map Likely Fake, Experts Say

Recently, a Chinese map, including North America, Antarctica, and Australia, was unveiled. This map purported to show that a Chinese explorer discovered America in 1418, but has been met with skepticism from cartographers and historians alike. Antiquities collector Liu Gang, who unveiled the map in Beijing, says it proves that Chinese seafarer Zheng discovered America more than 70 years before Christopher Columbus set foot in the New World. But experts have dismissed the map as a fake. They say the map resembles a French seventeenth-century world map with its depiction of California as an island. That China is not shown in the center also suggests the Chinese did not make the map, one expert says.

Article obatined from: Lovgren, S. "Chinese Columbus" map likely fake, experts say. *National Geographic News.* January 23, 2006.

16. Why might a seventeenth-century map show California as an island?
    - **A.** California really was an island back then.
    - **B.** America had not been explored well enough to know that California was actually connected.
    - **C.** California was so different from the rest of America that they assumed it was an island.
    - **D.** A river was mistakenly drawn to look like part of the ocean.

17. What can be inferred from this passage?
    - **A.** China should be put in the center of every map drawn.
    - **B.** The map is an exact copy of the seventeenth-century world map.
    - **C.** Liu Gang wants people to believe that the Chinese first discovered America.
    - **D.** Liu Gang drew the map himself.

NEED EXTRA HELP?

If You Missed Question . . .	1	2	3	4	5	6	7	8	9	10	11	12	13	14	15
Review Section . . .	2.1	2.1	2.1	2.3	2.1	2.2	2.2	2.2	2.2	2.2	2.2	2.3	1.2	2.2	1.3

SHORT ANSWER

9. The map is showing the contour lines of a location that rises in elevation.
10. The numbers show the rise in elevation from one location to the next; 860 represents the highest point on the map.
11. By reading the contour lines, the hiker would know that the south side is the steepest side while the north side is more gradual and would be easier to climb.
12. Sample answer: SeaBeam technology shows the depth of the water and detects the location of objects underwater. By having this technology, the ship would be better able to navigate around any possible hazards.
13. 1.495×10^8
14. Legends show what each symbol on a map represents so that the reader of the map can more easily interpret it.
15. line graph, because it can show changes between the measurements

READING FOR COMPREHENSION

16. B
17. C

UNIT 2

Composition of Earth

Themes

Change Elements become minerals; minerals become rocks; rocks change from one kind to another due to Earth's processes.

Structures The crust of Earth is composed of rock in various states and conditions.

Geologic Time Igneous rock was the first to form as Earth cooled from a molten mass to a rocky planet.

Systems The production and recycling of Earth's crust and materials is never ending.

Scientific Inquiry Scientists discovered information about Earth's age, origin, and composition by questioning and investigating the origin of soil. They also discovered how soil is created and how it fits into the rock cycle.

UNIT 2

Composition of Earth

CHAPTERS

3 Matter and Change

4 Minerals

5 Igneous Rocks

6 Sedimentary and Metamorphic Rocks

STEM Project

David Boyer/National Geographic Image Collection

CAREERS IN EARTH SCIENCE **Geologist**

This **geologist** is exploring the internal structures of this giant cave. Geologists like this one might collect samples of the rocks and minerals to help describe the origins of the geologic features within the cave.

Introduce the Unit

Camuy Cave Formation This photograph shows a view of Camuy (KAH mwi) Caves in Puerto Rico. The Camuy cave system is up to 60 m tall and more than a kilometer long and is known as the largest in the western hemisphere. The view in this photo shows the opening of a sinkhole that was carved out by the action of the Camuy River, one of the world's largest underground rivers. Ask students what types of environmental or geologic events could have resulted in the present condition of Camuy Caves. Ask whether they see any evidence in this photo of volcanic activity or erosion by wind or water. What kind of material do students think this rock is made of?

Geology of Camuy Cave The Camuy Caves are part of a large area of karst topography in Northern Puerto Rico. This part of the Puerto Rican island has a layer of limestone that is as much as 1.7 km thick in some areas. The limestone overlies intensely folded basaltic rock that dates to the Cretaceous Period. The Camuy River once traveled above ground and joined the Tanama River, but over time it sank underground and now runs through the cave system and flows directly to the Atlantic Ocean.

CHAPTER 3 Matter and Change

BIGIDEA **The variety of substances on Earth results from the way that atoms are arranged and combined.**

ESSENTIAL QUESTIONS	NATIONAL STANDARDS	RESOURCES TO ASSESS MASTERY
SECTION 1 Matter 1. How are atoms and their components described? 2. How are the energy levels of atoms related to the chemical properties of elements? 3. What are isotopes? 2 sessions 1 block	A.1; B.1–2	**Entry-Level Assessment** Focus Transparency 7, p. 60 **Progress Monitoring** Caption Question, pp. 62, 65 Reading Check, p. 64 Section Review, p. 65
SECTION 2 Combining Matter 1. What are the different types of chemical bonds that unite atoms to form compounds? 2. How is the nature of chemical bonds that hold compounds together related to the physical structures of compounds? 3. What are the different types of mixtures and solutions? 3 sessions 1.5 blocks	UCP.2–3; B.1–3	**Entry-Level Assessment** Focus Transparency 8, p. 66 **Progress Monitoring** Reading Check, pp. 68, 71 Section Review, p. 72
SECTION 3 States of Matter 1. What are the different states of matter on Earth? 2. Why does matter exist in these states? 3. How is thermal energy related to changes in state of matter? 3 sessions 1.5 blocks	UCP.1, UCP.3, UCP.5; A.1–2; B.1–6; E.1, E.2	**Entry-Level Assessment** Focus Transparency 9, p. 73 **Progress Monitoring** Caption Question, pp. 74, 75 Reading Check, p. 74 Section Review, p. 75 **Summative Assessment** Chapter Assessment, p. 79 Exam*View®Assessment Suite* CD-ROM

Go online!
connectED.mcgraw-hill.com

LEVELED RESOURCES	LAB MATERIALS
Science Notebook 3.1 OL **Chapter FAST FILE Resources:** MiniLab Worksheet, p. 2 OL Study Guide, p. 11 BL **Transparencies:** Section Focus Transparency 7 EL Teaching Transparency 7 OL EL	LaunchLAB p. 58 / **20 min** small, strong magnet; pencil; dry, fortified cereal; small, plastic bag; rolling pin; 250-mL glass beaker; tap water; magnifying lens MiniLAB p. 62 / **10 min** iron or steel tools, file cabinets, keys, paper clips, electrical wires, coins, earrings, pencils, soft-drink cans, fluorescent lights
Science Notebook 3.2 OL **Chapter FAST FILE Resources:** GeoLab Worksheet, p. 3 OL Study Guide, p. 13 BL **Lab Resources:** Laboratory Manual, p. 21 OL **Transparencies:** Section Focus Transparency 8 EL Teaching Transparency 8 OL EL	GeoLAB p. 77 / **30–40 min** halite; 250-mL glass beaker (2); distilled water; plastic wrap; laboratory scale; hot plate; shallow glass baking dish; refrigerator; glass stirring rod
Science Notebook 3.3 OL **Chapter FAST FILE Resources:** Study Guide, p. 15 BL **Lab Resources:** Laboratory Manual, p. 17 OL **Transparencies:** Section Focus Transparency 9 EL	

LabManager™
Customize any lab with the LabManager™ CD-ROM.

ADDITIONAL RESOURCES

Technology:
- eTeacherEdition
- eStudentEdition
- Classroom Presentation Toolkit
- Exam*View® Assessment Suite* CD-ROM
- LabManager™ CD-ROM
- What's EARTH SCIENCE Got To Do With It? DVD
- Weather Classroom Video DVD

Assessment:
- Blueprints for Success
- Chapter Test (Scaffolded)
- Earth Science Standardized Test Practice

Lab Resources:
- Exploring Environmental Problems Laboratory Manual

BL Below Level OL On Level AL Advanced Learners EL English Learners COOP LEARN Cooperative Learning

CHAPTER 3

LaunchLAB

iLab Station | Rubric

What do fortified cereals contain?

Teaching Strategies

- Make sure students stir the cereal-water mixture very slowly for the last minute to allow the iron particles to adhere to the magnet.
- The amount of iron collected on the magnet will depend upon the type of cereal used. Have students test several different brands of cereal and share their data with the class.

Procedure

1. Have students read and complete the lab safety form and follow the procedure below.
2. Tape a **small, strong magnet** to the eraser end of a **pencil.**
3. Pour 250 g of **dry, fortified cereal** into a **small, plastic bag.** Smooth the bag as you close it to release excess air.
4. Using a **rolling pin,** thoroughly crush the cereal in the bag.
5. Pour the crushed cereal into a **250-mL glass beaker.** Add 150 mL of **tap water** to the beaker.
6. Using the pencil-magnet as a stirrer, stir the cereal/water mixture for 10 min, stirring slowly for the last minute.
7. Remove the stirrer from the mixture and examine the magnet end of the stirrer with a **magnifying lens.**

Analysis

1. **Describe** what you see on the magnet. Small black pieces of iron should be visible on the end of the magnet.
2. **Determine** Study the cereal box to determine what the substance on the magnet might be. Students will read that iron is an added ingredient in fortified cereals.

CHAPTER 3

Matter and Change

BIGIDEA The variety of substances on Earth results from the way that atoms are arranged and combined.

SECTIONS

1 Matter

2 Combining Matter

3 States of Matter

LaunchLAB

iLab Station

What do fortified cereals contain?

Everything is made up of matter; different types of matter have different properties. Some metals, such as iron, cobalt, and nickel, are attracted to magnets. In this lab, investigate dry cereal to learn what metals they contain.

FOLDABLES® Study Organizer

States of Matter

Make a four-tab book using the labels shown. Use it to organize your notes on states of matter.

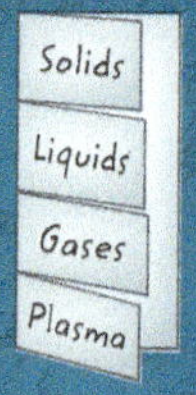

Assessment

Performance Ask students to make comprehensive lists of vitamin and mineral additives, carbohydrates, fats, and calories contained in the various brands of cereals tested. Have students post the lists so teams of students can study them and reach a conclusion about the most healthful brand. BL COOP LEARN

LabManager™
Customize any lab with the LabManager™ CD-ROM.

Introduce the BIG IDEA

How small can you go? Show the class several centimeter-sized crystals of halite. Describe the major properties of the crystals. Smash one of the crystals to a powder, place some of the powder under a microscope, and let students observe it. Ask students whether the pieces still look like salt. Then ask whether the pieces could be split into smaller and smaller fragments indefinitely.

Teacher Content Support

Changes in Matter All matter undergoes change. Have students look at the big photo. Ask them how long they think the mailbox has been outside. How long do they think it took to become rusty? How long will it take before the mailbox is completely rusted? Would the same result occur if the mailbox were in a different part of the country? Explain that rate, type and degree of changes in matter are dependent on the specific conditions to which matter is exposed.

SECTION 1

1 Focus

Focus Transparency

Before presenting the lesson, project **Section Focus Transparency** 7 and have students answer the accompanying questions. EL

MAINIDEA

Components of Matter Display a piece of rock salt on the table. Using a rock hammer, gently tap the rock until it breaks apart. Continue hitting until the sample consists of small grains. Place a few grains in a beaker of warm water, and stir until it dissolves. Ask the students to explain where the rock salt went. Ask them if the composition of the rock salt changed during this process. The rock salt remained the same until it dissolved in the water. At that point, it broke apart into the elements that make it, sodium and chloride.

2 Teach

Teacher Content Support

Element Names Tell students they can figure out the names of many elements from their chemical symbols, for example O for oxygen and Al for aluminum. Explain the Latin origin of the chemical symbols Na, K, Au, Ag, and Fe, for sodium, potassium, gold, silver, and iron, respectively: natrium, kalium, aurum, argentum, and ferrum.

SECTION 1

Essential Questions

- How are atoms and their components described?
- How are the energy levels of atoms related to the chemical properties of elements?
- What are isotopes?

Review Vocabulary

atom: the smallest particle of an element that retains all the properties of that element

New Vocabulary

matter
element
nucleus
proton
neutron
electron
atomic number
mass number
isotope
ion

Matter

MAINIDEA Atoms are the basic building blocks of all matter.

EARTH SCIENCE 4 YOU Gold, which is often used in jewelry, is so soft that it can be molded, hammered, sculpted, or drawn into wire. Whatever its size or shape, the gold is still gold. Gold is a type of matter.

Atoms

Matter is anything that has volume and mass. Everything in the physical world that surrounds you is composed of matter. On Earth, matter usually occurs as a solid, a liquid, or a gas. All matter is made of substances called elements. An **element** is a substance that cannot be broken down into simpler substances by physical or chemical means. For example, gold is still gold whether it is a gold brick, coins, or a statue.

Each element has unique physical and chemical properties. Although aluminum has different properties than gold, both aluminum and gold are elements that are made up of atoms. All atoms consist of even smaller particles—protons, neutrons, and electrons. **Figure 1** shows one method of representing an atom. The center of an atom is called the nucleus (NEW klee us) (plural, nuclei). The **nucleus** of an atom is made up of protons and neutrons. A **proton** (p) is a tiny particle that has mass and a positive electric charge. A **neutron** (n) is a particle with approximately the same mass as a proton, but it is electrically neutral; that is, it has no electric charge. All atomic nuclei have a positive charge because they are composed of protons with positive electric charges and neutrons with no electric charges.

Figure 1 In this representation of an atom, the fuzzy area surrounding the nucleus is referred to as an electron cloud.

View an **animation of the electron cloud.**

Concepts In Motion

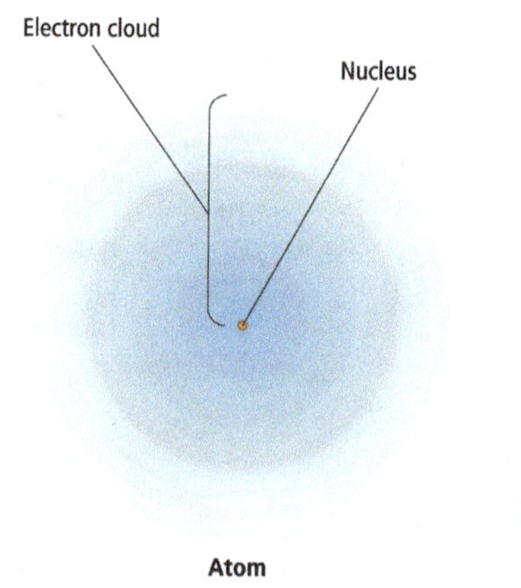

Demonstration

Loose Electrons Charge a glass rod with static electricity by rubbing it with a silk cloth. The silk cloth will remove electrons. Produce electric sparks by holding the rod close to a metallic object. Explain that these sparks are caused by electrons returning to the positively charged rod. Also show that the rod can pick up small bits of paper by electrostatic attraction. OL

EARTH SCIENCE JOURNAL

What is an atom? To get students thinking about what they know about atoms, have each student write a list describing what they think makes up atoms, why matter looks the way it does, and how the presence of atoms affects our lives. Students might or might not be able to list the three parts of atoms, describe how they bond, and state the fact that atoms are the basic building blocks of all matter.

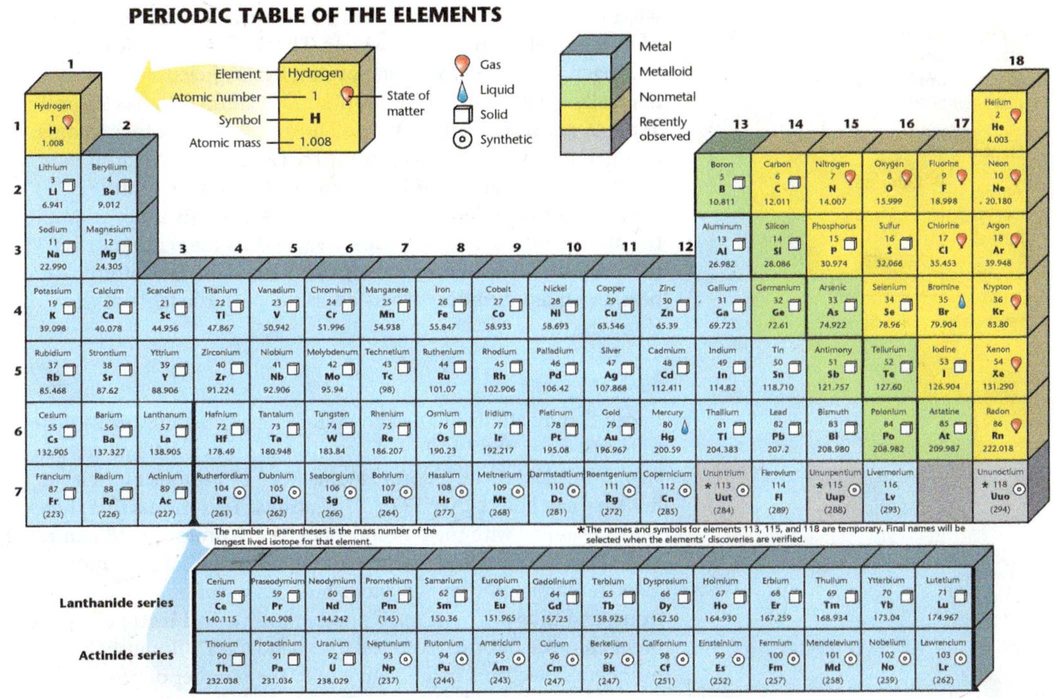

Figure 2 The periodic table of the elements is arranged so that a great deal of information about all of the known elements is provided in a small space.

View an **animation of the periodic table of elements.**

Concepts In Motion

Explore **updates to the periodic table.**

Periodic Table

Surrounding the nucleus of an atom are smaller particles called electrons. An **electron** (e^-) has little mass, but it has a negative electric charge that is exactly the same magnitude as the positive charge of a proton. When an atom has an equal number of protons and electrons, the electric charge of an electron cancels the positive charge of a proton to produce an atom that has no overall charge. Notice that the electrons in **Figure 1** are shown as a cloudlike region surrounding the nucleus. This is because electrons are in constant motion around an atom's nucleus, and their exact positions at any given moment cannot be determined.

Symbols for elements There are 92 elements that occur naturally on Earth and in the stars. Other elements have been produced in laboratory experiments. Generally, each element is identified by a one-, two-, or three-letter abbreviation known as a chemical symbol. For example, the symbol H represents the element hydrogen, C represents carbon, and O represents oxygen. Elements identified in ancient times, such as gold and mercury, have symbols of Latin origin. For example, gold is identified by the symbol Au for its Latin name, *aurum.* All elements are classified and arranged according to their chemical properties in the periodic table of the elements, shown in **Figure 2.**

Interpret the Illustration

Atomic Model **Figure 1** shows a conceptual model of an atom. Real atoms do not look like this. The relative sizes of the nucleus and electron cloud are greatly exaggerated. For example, the actual size of a proton or neutron is about 10^{-15} m, and the diameter of a hydrogen atom is about 10^{-10} m. If the protons in these atoms are represented by 1-mm spheres, then the model hydrogen atom should have a diameter of about 100 m, roughly the length of a football field. However, scientists aren't sure whether protons are in fact spherical. OL AL

Apply Earth Science

Elements in Everyday Items Separate the class into six groups and, using rows 1–6, assign each group one row from the periodic table. Have each group research and develop a list of at least one item that is made from or contains each element in their row. Students researching elements from the higher-numbered rows will find fewer items that contain them than those researching elements from the lower-numbered rows. OL

ACROSS THE CURRICULUM

History During the Middle Ages, alchemists attempted to make gold from other substances through various chemical reactions. All these attempts failed. Have students discuss why obtaining gold in this way is impossible. Chemical reactions don't affect the nuclei of elements. As long as the number of protons stays the same, the element will not change.

Caption Question **Fig. 3** The balloon indicates the element is in a gaseous state at room temperature.

MiniLAB

iLab Station | Rubric

Purpose Students will identify common elements in their surroundings.

Process Skills collect and interpret data, observe and infer, compare and contrast, classify

Safety Precaution Approve lab safety forms before work begins.

Additional Materials iron or steel tools, file cabinets, keys, paper clips, electrical wires, coins, earrings, pencils, soft-drink cans, fluorescent lights

Teaching Strategy Have students work in groups of three. Have the groups exchange their information.

Expected Results Students will identify the atomic number of several elements as well as objects that contain those elements. Possible answers: iron, atomic number 26: file cabinets, keys, paper clips; copper (no. 29): wires, pennies; aluminum (no. 13): soft-drink cans; carbon (no. 6): pencil lead; neon (no. 10): fluorescent lights.

Analysis

1. Some examples include: solids—most metals; liquids—mercury, water, and gasoline; and gases—neon, helium, and air.
2. Solids have definite shapes and resist deformation, liquids deform readily and flow, and gases expand and fill all available space.

Assessment

Knowledge Have students describe the physical properties of some elements identified in this lab.

LabManager™ Customize any lab with the LabManager™ CD-ROM.

Figure 3 The element chlorine is atomic number 17.
Infer *In what state is chlorine at room temperature?*

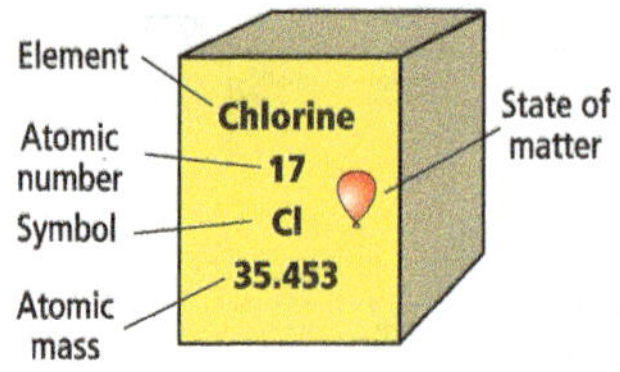

Mass number The number of protons and neutrons in atoms of different elements varies widely. The lightest of all atoms is hydrogen, which has only one proton in its nucleus. The heaviest naturally occurring atom is uranium. Uranium-238 has 92 protons and 146 neutrons in its nucleus. The number of protons in an atom's nucleus is its **atomic number.** The sum of the protons and neutrons is its **mass number.** Because electrons have little mass, they are not included in determining mass number. For example, the atomic number of uranium is 92, and its mass number is 238 (92 protons + 146 neutrons). **Figure 3** explains how atomic numbers and mass numbers are listed in the periodic table of the elements.

Isotopes

Recall that all atoms of an element have the same number of protons. However, the number of neutrons of an element's atoms can vary. For example, all chlorine atoms have 17 protons in their nuclei, but they can have either 18 or 20 neutrons. This means that there are chlorine atoms with mass numbers of 35 (17 protons + 18 neutrons) and 37 (17 protons + 20 neutrons). Atoms of the same element that have the same number of protons but a different number of neutrons, and thus different mass numbers, are called **isotopes.** The element chlorine has two isotopes: Cl-35 and Cl-37. Because the number of electrons in an atom equals the number of protons, isotopes of an element have the same chemical properties.

Look again at the periodic table in **Figure 2.** Scientists have measured the mass of atoms of elements. The atomic mass of an element is the average of the mass numbers of the isotopes of an element. Most elements are mixtures of isotopes. For example, notice in **Figure 2** that the atomic mass of chlorine is 35.453. This number is the average of the mass numbers of the naturally occurring isotopes of chlorine-35 and chlorine-37.

MiniLAB

iLab Station

Identify Elements

What elements are in your classroom? Most substances on Earth occur in the form of elements or chemical compounds. Around your classroom, there are numerous objects or substances that consist mostly of a single element.

Procedure

1. Read and complete the lab safety form.
2. Create a data table with the following column headings: Article, Element, Atomic Number, Properties.
3. Name three objects in your classroom and the three different elements of which they are made.
4. List the atomic numbers of these elements and describe some of their properties.

Analysis

1. **Categorize** List two examples of a solid, a liquid, and a gaseous object or substance.
2. **Compare and contrast** solids, liquids, and gases.

DIFFERENTIATED INSTRUCTION

English Learners Ask students to compare the English names of some of the elements in **Figure 2** with the names of those elements in their native languages. Have students list the elements whose English and non-English names are similar or identical.

Radioactive isotopes The nuclei of some isotopes are unstable and tend to break down. When this happens, the isotope also emits energy in the form of radiation. Radioactive decay is the spontaneous process through which unstable nuclei emit radiation. In the process of radioactive decay, a nucleus will either lose protons and neutrons, change a proton to a neutron, or change a neutron to a proton. Because the number of protons in a nucleus identifies an element, decay changes the identity of an element. For example, the isotope polonium-218 decays at a steady rate over time into bismuth-214. The polonium originally present in a rock is gradually replaced by bismuth. The process of radioactive decay is often used to calculate the ages of rocks.

Electrons in Energy Levels

Although the exact position of an electron cannot be determined, scientists have discovered that electrons occupy areas called energy levels. Look again at **Figure 1.** The volume of an atom is mostly empty space. However, the size of an atom depends on the number and arrangement of its electrons.

Filling energy levels **Figure 4** presents a model to help you visualize atomic particles. Note that electrons are distributed over one or more energy levels in a predictable pattern. Keep in mind that the electrons are not sitting still in one place. Each energy level can hold only a limited number of electrons. For example, the smallest, innermost energy level can hold only two electrons, as illustrated by the oxygen atom in **Figure 4**. The second energy level is larger, and it can hold up to eight electrons. The third energy level can hold up to 18 electrons and the fourth energy level can hold up to 32 electrons. Depending on the element, an atom might have electrons in as many as seven energy levels surrounding its nucleus.

Figure 4 Electrons occupy one energy level in hydrogen, two energy levels in oxygen, and three energy levels in aluminum.

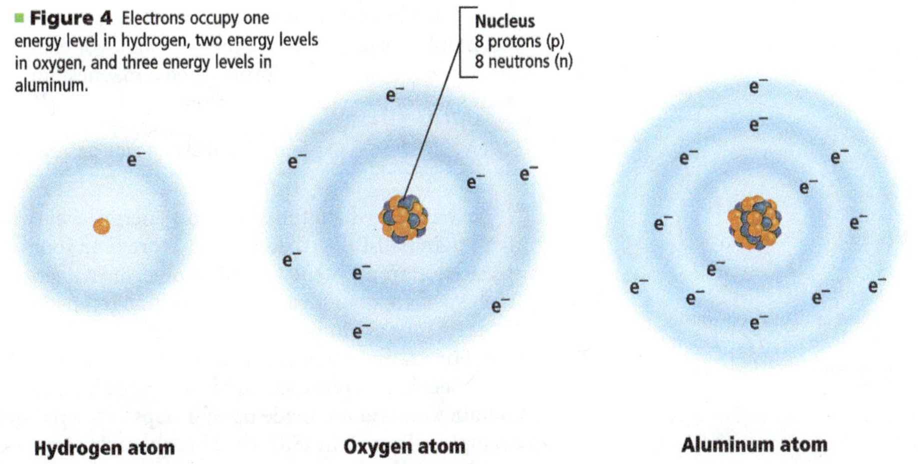

Teacher Content Support

Dangerous Gas The noble gas radon is radioactive. It is produced by the radioactive decay of uranium in soil and bedrock. Radon gas is heavier than air, and in some areas, it seeps into and accumulates in the basements of buildings. High concentrations of radioactive radon have been linked to cancer.

Identify Misconceptions

A common misconception is that isotopes are radioactive substances.

Uncover the Misconception Ask students what they think isotopes are. Also ask whether all isotopes are radioactive.

Demonstrate the Concept Explain that many elements are mixtures of isotopes, and that most isotopes are not radioactive. Examples of radioactive isotopes are the isotopes of uranium. Examples of nonradioactive isotopes are carbon-12 and carbon-13.

Assess New Knowledge Ask students to define *isotope* and to give examples of nonradioactive isotopes.

Teacher Content Support

Electron Energy Levels An electron can occupy only specific energy levels. Numbered one through seven, each energy level number, n, is called the principal quantum number and corresponds to the energy levels 1, 2, 3, and so on. The greatest number of electrons possible in any one level is $2n^2$. Thus, the number of electrons possible in energy level 1 is $2(1)^2 = 2$. The number of electrons possible in energy level 2 is $2(2)^2 = 8$, and so on, to energy level 7, where $2(7)^2 = 98$.

ACROSS THE CURRICULUM

Biology Our bodies consist mostly of the following elements: oxygen, 65%; carbon, 18.5%; hydrogen, 10.5%; nitrogen, 3.3%; calcium, 1.5%; phosphorus, 1.0%; potassium, 0.4%; sulfur, 0.3%; sodium, 0.2%; chlorine, 0.2%; and magnesium, 0.1%. Note that four of these elements—oxygen, carbon, hydrogen, and nitrogen, make up over 96 percent of the human body. Most of the oxygen and hydrogen in our bodies is in the form of water. Our bodies contain traces of iron, zinc, copper, iodine, fluorine and many other elements.

Teacher Content Support

Noble Gases For many years after their discovery, inert gases were believed to be chemically unreactive. However, in 1962, the first compound of an inert gas was synthesized. Because inert compounds can now be made fairly easily, inert gases are now known as noble gases.

Concept Development

A Complex Ion A silicate ion consists of a small silicon (Si^{4+}) ion surrounded by four large oxygen (O^{2-}) ions in the form of a tetrahedron. The net charge of a silicate ion is, therefore, –4 (+4 –2 –2 –2 –2). Silicate ions are the building blocks of most of Earth's minerals. They are discussed further in the next chapter.

Model

Ion Shape Have students build models of a silicate ion and a sodium chloride crystal (with 64 units) out of large and small plastic foam spheres. The large NaCl crystal could be assembled from eight subcrystals, each containing eight units built by several (up to eight) groups of students. OL EL

Discussion

Small Ions Have students draw model diagrams of the following ions: Na^+, Li^+, and H^+. Have them discuss their diagrams, specifically their H^+ models. Ask students what the differences are between the H^+ ion and the other ions. The H^+ ion has no electrons, and no neutrons. Ask students exactly what an H^+ ion is. a proton Then ask what they think makes it so extremely reactive. Because of its small size, it can penetrate and disrupt any structure.

READING CHECK The loss of electrons makes ions positive.

Caption Question Fig. 6 Most of the hydrogen and helium in the universe is found in the stars.

Figure 5 The inert nature of argon makes it an ideal gas to use inside an incandescent light bulb because it does not react with the extremely hot filament.

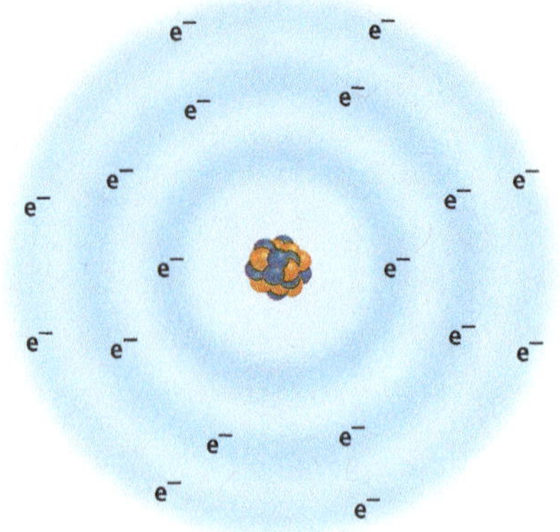

Argon atom

Valence electrons The electrons in the outermost energy level determine the chemical behavior of the different elements. These outermost electrons are called valence electrons. Elements with the same number of valence electrons have similar chemical properties. For example, on the left side of the periodic table, both a sodium atom and a potassium atom, with different atomic numbers, have one valence electron. Thus both sodium and potassium exhibit similar chemical behavior. These elements are highly reactive metals, which means that they combine easily with many other elements.

On the right side of the periodic table elements such as helium and argon have full outermost energy levels. For example, an argon atom, shown in **Figure 5,** has 18 electrons, with two electrons in the first energy level and eight electrons in the second and third (outermost) energy levels. Elements that have full outermost energy levels are highly unreactive.

Ions

Sometimes atoms gain or lose electrons from their outermost energy levels. Recall that atoms are electrically neutral when the number of electrons, which have negative charges, balances the number of protons, which have positive charges. An atom that gains or loses an electron has a net electric charge and is called an **ion.** In general, an atom in which the outermost energy level is less than half-full—that is, it has fewer than four valence electrons—tends to lose its valence electrons. When an atom loses valence electrons, it becomes positively charged. In chemistry, a positive ion is indicated by a superscript plus sign. For example, a sodium ion is represented by Na^+. If more than one electron is lost, that number is placed before the plus sign. For example, a magnesium ion, which forms when a magnesium atom has lost two electrons, is represented by Mg^{2+}.

READING CHECK **Explain** what makes an ion positive.

VOCABULARY
ACADEMIC VOCABULARY
Exhibit
to show or display outwardly
The dog exhibited aggression by baring its teeth and growling.

An atom in which the outermost energy level is more than half-full—that is, it has more than four valence electrons—tends to fill its outermost energy level. Such an atom forms a negatively charged ion. Negative ions are indicated by a superscript minus sign. For example, a nitrogen atom that has gained three electrons is represented by N^{3-}. Some substances contain ions that are made up of groups of atoms—for example, silicate ions $(SiO_4)^{4-}$. These complex ions are important constituents of most rocks and minerals.

IN THE FIELD

St. Elmo Brady St. Elmo Brady was the first African-American to earn a PhD in chemistry. He received his bachelor's degree from Fisk University in Tennessee in 1904 and his PhD from the University of Illinois in 1916. His research on the diatomic oxygen molecule contributed greatly to his field. Brady had a long and distinguished academic career. He was responsible for creating or greatly improving the science departments of several historically African-American colleges such as Howard and Fisk, where he became head of the chemistry departments and taught both general and organic chemistry. At Tougaloo College in Mississippi, he designed a completely new chemistry program and attracted an outstanding faculty.

Abundance of Elements

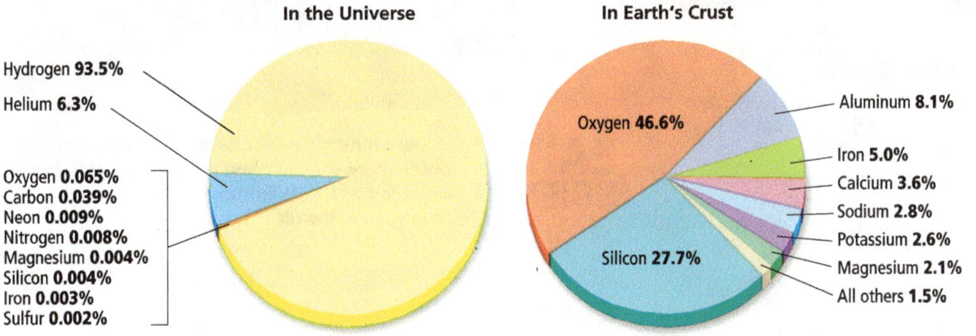

Figure 6 The most abundant elements in the universe are greatly different from the most abundant elements on Earth.

Hypothesize ***Where might most of the hydrogen and helium in the universe be found?***

What elements are most abundant?

Astronomers have identified the two most abundant elements in the universe as hydrogen and helium. All other elements account for less than 1 percent of all atoms in the universe, as shown in **Figure 6.** Analyses of the composition of rocks and minerals on Earth indicate that the percentages of elements in Earth's crust differ from the percentages in the universe. As shown in **Figure 6,** 98.5 percent of Earth's crust is made up of only eight elements. Two of these elements, oxygen and silicon, account for almost 75 percent of the crust's composition. This means that most of the rocks and minerals on Earth's crust contain oxygen and silicon. Minerals that contain oxygen and silicon are called silicates. The next two most abundant elements in Earth's crust are aluminum and iron.

SECTION 1 REVIEW

Section Self-Check

Section Summary

- Atoms consist of protons, neutrons, and electrons.
- An element consists of atoms that have a specific number of protons in their nuclei.
- Isotopes of an element differ by the number of neutrons in their nuclei.
- Elements with full outermost energy levels are highly unreactive.
- Ions are electrically charged atoms or groups of atoms.

Understand Main Ideas

1. **MAINIDEA Differentiate** among the three particles of an atom in terms of their location, charge, and mass.
2. **Infer** why the elements magnesium and calcium have similar properties.
3. **Illustrate** how a neutral atom becomes an ion.
4. **Compare and contrast** these isotopes: uranium-239, uranium-238, and uranium-235.

Think Critically

5. **Illustrate** a model of a calcium atom, including the number and position of protons, neutrons, and electrons in the atom.
6. **Interpret** the representation of magnesium in the periodic table. Explain why the atomic mass of magnesium is not a whole number.

MATH IN ▶ Earth Science

7. As the radioactive isotope radium-226 decays, it emits two protons and two neutrons. How many protons and neutrons are now left in the nucleus? What is the atom's new atomic number? What is the name of this element?

Enrichment

Uranium Decay As the radioactive isotope uranium-235 (atomic number 92) decays, it emits two protons and two neutrons. How many protons and neutrons are left in the nucleus after these neutrons and protons have been ejected? Subtracting 2 protons and 2 neutrons leaves 90 protons and 141 neutrons. What is the new atomic number? 90 What is the name of this element? thorium **AL**

3 Assess

Check for Understanding

Reinforcement Ask students the following questions: How many protons, neutrons, and electrons are in a sodium-23 atom? 11 protons, 12 neutrons, 11 electrons In which energy levels are the electrons? 2 in the innermost energy level, 8 in the next energy level, 1 in the outermost energy level

Reteach

Explain Ask students to explain how isotopes, for example He-3 and He-4, differ from each other, and why they have the same chemical behavior.

Assessment

Skill Have students draw models of the electron configurations of the first 12 elements in the periodic table. Ask them to compare and contrast the models, and to record their work in their Earth science journals.

SECTION 1 REVIEW

1. proton: positive charge, located in the nucleus, affects the atomic mass; neutron: neutral charge, located in the nucleus, affects the atomic mass; electron: negative charge, located in the cloud area surrounding the nucleus, has negligible mass
2. Both magnesium and calcium have two valence electrons, so the way they bond with other elements is the same.
3. The illustration should show an atom that has gained or lost an electron, giving it either a positive or negative charge, respectively.
4. All are the element uranium. They differ in the number of neutrons in their nuclei: 147, 146, and 143 respectively.
5. The model should be similar in design to those shown in Figure 3.4. Calcium has 20 protons and 20 neutrons in its nucleus and 20 electrons, 2 of which are valence electrons.
6. Magnesium is a solid. Its atoms contain 22 protons. It consists of different isotopes. The average mass of all the isotopes equals 24.305.
7. There are 86 protons and 136 neutrons in the nucleus. The new atomic number is 86 and the element is radon.

SECTION 2

1 Focus

Focus Transparency

Before presenting the lesson, project **Section Focus Transparency 8** and have students answer the accompanying questions. EL

MAINIDEA

Mixtures Use clear glass beakers to mix up the following: lemonade from a powdered mix, oil and water, clay powder and water. Ask students to observe and predict which mixtures will stay mixed and which will separate. Ask: Why do some mixtures stay together while others don't? It depends on the types of bonds that form between the components and if the individual components can separate into the liquid.

2 Teach

Teacher Content Support

Write Chemical Formulas
Chemical formulas of compounds follow the convention of writing the symbol(s) of the positive (metal) ion(s) first, and the symbol(s) of the negative ion(s) last, for example, NaCl, not ClNa, and Mg_2SiO_4, not SiO_4Mg_2.

Activity

Analyze Chemical Formulas
Have students analyze the following chemical formulas: NaCl, $MgCl_2$, CaO, K_2O, Al_2O_3, Fe_2O_3, $CaCO_3$, and Mg_2SiO_4. Have students list the chemical symbols and the charges of the positive and negative ions involved. This activity could also be a collaborative learning exercise or a homework assignment. OL

SECTION 2

Combining Matter

MAINIDEA Atoms combine through electric forces, forming molecules and compounds.

Essential Questions

- What are the different types of chemical bonds that unite atoms to form compounds?
- How is the nature of chemical bonds that hold compounds together related to the physical structures of compounds?
- What are the different types of mixtures and solutions?

Review Vocabulary

ion: an electrically charged atom

New Vocabulary

compound
chemical bond
covalent bond
molecule
ionic bond
metallic bond
chemical reaction
solution
acid
base

EARTH SCIENCE 4 YOU Is there a rusty mailbox or bicycle on your street? Nearly everywhere you look, you can see iron objects that have become rusty. Rust forms when iron is exposed to water or oxygen in the air.

Compounds

Can you identify the substances in **Figure 7?** The greenish gas in the flask is the element chlorine, which is poisonous. The solid, silvery metal is the element sodium, which is highly reactive. These two elements combine chemically to form the third material in the photograph—table salt. How can two dangerous elements combine to form a material that you sprinkle on your popcorn?

Table salt is a compound, not an element. A **compound** is a substance that is composed of atoms of two or more different elements that are chemically combined. Water is another example of a compound because it is composed of two elements—hydrogen and oxygen. Most compounds have different properties from the elements of which they are composed. For example, both oxygen and hydrogen are highly flammable gases at room temperature, but in combination they form water—a liquid.

Chemical formulas Compounds are represented by chemical formulas. These formulas include the symbol for each element followed by a subscript number that stands for the number of atoms of that element in the compound. If there is only one atom of an element, no subscript number follows the symbol. Thus, the chemical formula for table salt is NaCl. The chemical formula for water is H_2O.

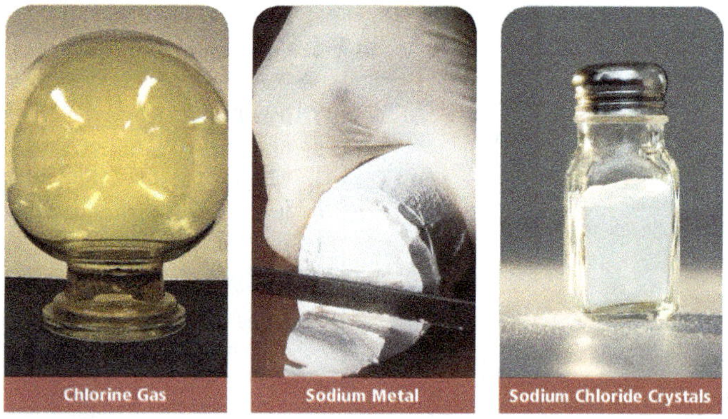

Figure 7 Sodium is a silvery metal that is soft enough to cut with a knife. Chlorine is a green, poisonous gas. When they react, they produce sodium chloride, a white solid.

Demonstration

Combine Atoms Use a safety shield for this demo. Fill an inverted 100-mL beaker with one part hydrogen and three parts air. Carefully ignite the mixture—it will explode! Let the beaker cool. Ask students what the liquid condensate is. Explain that the oxygen and hydrogen atoms combined to form water. EL

Covalent Bonds

Recall that an atom is chemically stable when its outermost energy level is full. A state of stability is achieved by some elements by forming chemical bonds. A **chemical bond** is the force that holds together the elements in a compound. One way in which atoms fill their outermost energy levels is by sharing electrons. For example, individual atoms of hydrogen each have just one electron. Each atom becomes more stable when it shares its electron with another hydrogen atom so that each atom has two electrons in its outermost energy level. **Figure 8** shows an example of this bond. How do these two atoms stay together? The nucleus of each atom has one proton with a positive charge, and the two positively charged protons attract the two negatively charged electrons. This attraction of two atoms for a shared pair of electrons that holds the atoms together is called a **covalent bond.**

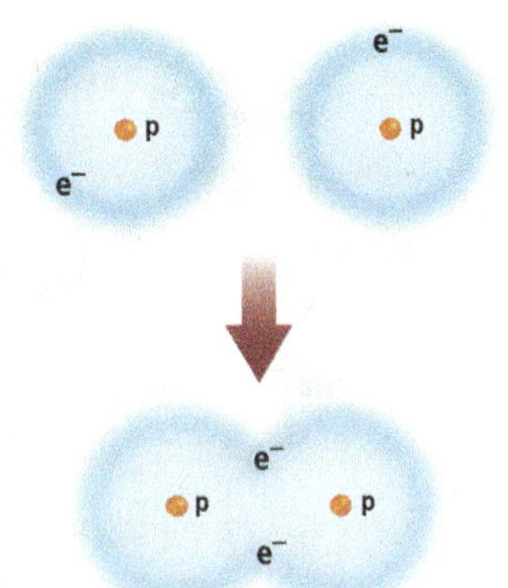

Figure 8 In this covalent bond example, notice the positions of the electrons in the outermost energy levels. They can now be considered as part of each atom.

Molecules A **molecule** is composed of two or more atoms held together by covalent bonds. Molecules have no overall electric charge because the total number of electrons equals the total number of protons. Water is an example of a compound whose atoms are held together by covalent bonds, as illustrated in **Figure 9.** The chemical formula for a water molecule is H_2O because, in this molecule, two atoms of hydrogen, each of which need to gain an electron to become stable, are combined with one atom of oxygen, which needs to gain two electrons to become stable. A compound comprised of molecules is called a molecular compound.

Polar molecules Although water molecules are held together by covalent bonds, the atoms do not share the electrons equally. As shown in **Figure 9,** the shared electrons in a water molecule are attracted more strongly by the oxygen atom than by the hydrogen atoms. As a result, the electrons spend more time near the oxygen atom than they do near the hydrogen atoms. This unequal sharing of electrons results in polar molecules. A polar molecule has a slightly positive end and a slightly negative end.

VOCABULARY

SCIENCE USAGE V. COMMON USAGE

Polar

Science usage: the unequal sharing of electrons

Common usage: locations of or near the north or south pole, or the ends of a magnet

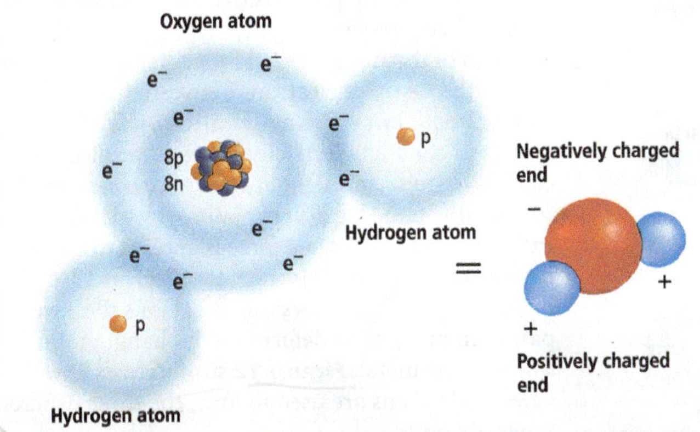

Figure 9 Polar molecules are similar to bar magnets. At one end of a water molecule, the hydrogen atoms have a positive charge, while at the opposite end, the oxygen atom has a negative charge.

Teacher Content Support

Nonreactive Gases Noble gases have either full outer energy levels or eight electrons (filled *s* and *p* subshells) in their outer energy levels. Other elements are stable if they have the same electron configuration as the nearest inert gas in the periodic table. Such a configuration can be achieved by the gain or loss of electrons (the formation of ions) or by the sharing of electrons (covalent bonding). For instance, the ions O^{2-}, F^-, Na^+, Mg^{2+}, and Al^{3+} and the noble gas Ne all have the same electron configuration: two electrons in the first energy level and eight electrons in the second (outer) energy level.

Concept Development

Molecule Defined The smallest unit of a compound with the properties of that compound is called a molecule. However, ionic compounds don't consist of molecules. More precisely, molecules are basic units consisting of two or more atoms bound together by covalent bonds, such as H_2O. These molecules can exist in the solid, liquid, or gaseous state. Solids consisting of such molecules are sometimes called molecular compounds. This is not the same as a covalent compound, in which all the atoms are held together by covalent bonds, as in a diamond. In molecular compounds, the molecules are held together by other, weaker bonds, such as hydrogen bonds in ice.

ACROSS THE CURRICULUM

Chemistry Water, as a polar solvent, can not be used to clean stains that do not dissolve in polar liquids. Dry cleaning has traditionally been done with the nonpolar solvent perchloroethylene, a harmful and potentially carcinogenic liquid. Two possible replacements have been developed. Wet cleaning immerses garments normally cleaned in nonpolar solvents. Care is taken to avoid fading and shrinkage. Liquid carbon dioxide is a nonpolar solvent that is most effective in removing oil and grease.

Use Science Terms

Origin of *Ion* Have students research the use and origin of the term *ionic* and provide examples of when the word is used. *Ionic* is a variant of the word *ion* derived from the Greek word meaning *to go.* Ions go toward an electrode of the opposite charge. Ionic columns were designed by Ionian Greeks during the mid-6th century BCE. The two terms are related only in their Greek origin.

☑ READING CHECK Ionic bonds form when electrons from the outer level of one atom fill the outer level of another atom.

Identify Misconceptions

Chemistry has the reputation of being difficult and of dealing with strange, smelly, and possibly poisonous substances.

Uncover the Misconception

Ask students what they think chemistry is.

Demonstrate the Concept

Demonstrate some common activities that involve chemistry, and ask students if they think chemistry is involved. Possible demos include mixing lemonade, toasting bread, lighting a candle.

Assess New Knowledge

Have groups of students create lists of common and obscure activities that involve chemistry. Lists may include cooking, cannon fire, fireworks, makeup, video games, cleaning grease, etc.

GeoLAB

iLab Station

The GeoLab located at the end of the chapter can be used at this point in the lesson.

Ionic Bonds

As you might expect, positive and negative ions attract each other. An **ionic bond** is the attractive force between two ions of opposite charge. **Figure 10** illustrates an ionic bond between a positive ion of sodium and a negative ion of chlorine called chloride. The chemical formula for common table salt is NaCl, which consists of equal numbers of sodium ions (Na^+) and chloride ions (Cl^-) . Note that positive ions are always written first in chemical formulas.

Within the compound NaCl, there are as many positive ions as negative ions; therefore, the positive charge on the sodium ion equals the negative charge on the chloride ion, and the net electric charge of the compound NaCl is zero. Magnesium and oxygen ions combine in a similar manner to form the compound magnesium oxide (MgO)—one of the most common compounds on Earth. Compounds formed by ionic bonding are called ionic compounds. Other ionic compounds have different proportions of ions. For example, oxygen and sodium ions combine in the ratio shown by the chemical formula for sodium oxide (Na_2O), in which there are two sodium ions to each oxygen ion.

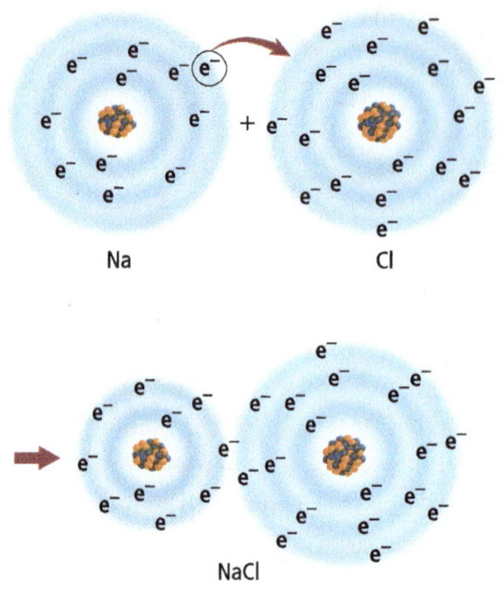

Figure 10 The single valence electron in a sodium atom is used to form an ionic bond with a chlorine atom. Once an ionic bond is formed, the negatively charged ion is slightly larger than the positively charged ion.

View an **animation of ionic bonds.**

Concepts In Motion

☑ READING CHECK **Describe** how ionic bonds form.

Metallic Bonding

Most compounds on Earth are held together by ionic or covalent bonds, or by a combination of these bonds. Another type of bond is shown in **Figure 11.** In metals, the valence electrons are shared by all the atoms, not just by adjacent atoms as they are in covalent compounds. You could think of a metal as a group of positive ions surrounded by a sea of freely moving negative electrons. The positive ions of the metal are held together by the attraction to the negative electrons between them. This type of bond, known as a **metallic bond,** allows metals to conduct electricity because the electrons can move freely throughout the entire solid metal.

Metallic bonding also explains why metals are so easily deformed. When a force is applied to a metal, such as the blow of a hammer, the electrons are pushed aside. This allows the metal ions to move past each other, thus deforming or changing the shape of the metal. **Figure 12** summarizes how valence electrons are used to form the three different types of bonds.

Figure 11 Metallic bonds are formed when valence electrons are shared equally among all the positively charged atoms. Because the electrons flow freely among the positively charged ions, you can visualize electricity flowing through electrical wires.

View an **animation of electron flow.**

Concepts In Motion

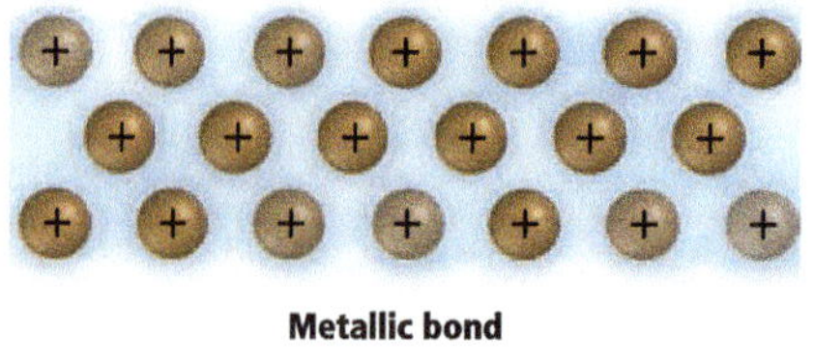

DIFFERENTIATED INSTRUCTION

Visually Impaired Have students with visual impairments handle the plastic foam models constructed in the Modeling activity in Section 1. Explain the role of the different-sized spheres to them. BL EL

VISUALIZING Bonds

Figure 12 Atoms gain stability by sharing, gaining, or losing electrons to form ions and molecules. The properties of metals can be explained by metallic bonds.

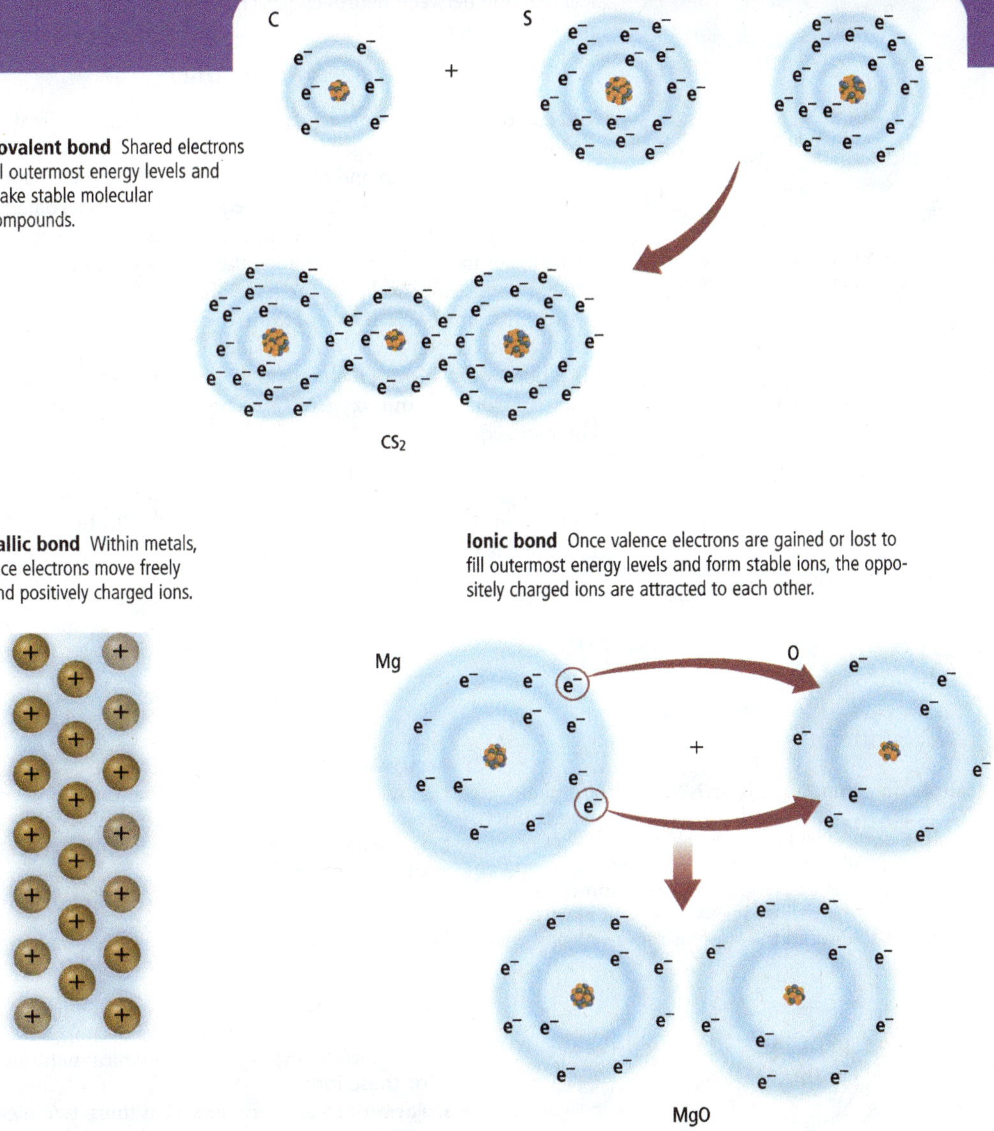

Covalent bond Shared electrons fill outermost energy levels and make stable molecular compounds.

Metallic bond Within metals, valence electrons move freely around positively charged ions.

Ionic bond Once valence electrons are gained or lost to fill outermost energy levels and form stable ions, the oppositely charged ions are attracted to each other.

Concepts In Motion View an **animation of chemical bonding.**

Demonstration

Bond Strength Ionic compounds are characterized by high melting points, whereas covalent compounds are characterized by low melting points. Demonstrate the difference between these two types of compounds by melting salt, an ionic compound, and solid shortening, a covalent compound. Use a safety shield for this demonstration. Place a small amount of each substance on the end of a metal spatula and place it in the flame of a laboratory burner. Salt does not melt at this temperature, but solid shortening does. Point out that ionic bonds give substances the property of high melting points.

VISUALIZING

Purpose

Students will learn about the three types of chemical bonds.

Concept Development

Bond Differences Ask students to explain the difference in bonding between ionic, covalent and metallic bonds. Have students list properties that are unique to metals, such as malleability and electrical conductivity.

Discussion

Metals and Metallic Bonds Using **Figure 12,** review the free-flowing nature of electrons in a metallic bond. Have students make inferences about how the electron arrangement results in the special properties of metals. The freely flowing electrons result in the ductility and malleability of metals.

Activity

Demonstrate Metallic Bonds Give each student a metal paper clip. Ask: How can you demonstrate the presence of metallic bonds with this paper clip? Students might choose to deform the paper clip, thus demonstrating the ability of the particles to flow past each other. They might also choose to heat one end and demonstrate that the entire clip will get hot; again demonstrating the freely moving nature of the electrons in transferring heat from one point to another.

Problem-Solving LAB

Purpose Students will identify the formation of positive and negative ions, and of ionic compounds.

Process Skills compare and contrast, predict, communicate

Teaching Strategies
- Discuss the formation of ions before assigning this lab.
- Have students use the periodic table to get started.

Analysis
1. A: 12 electrons; B: 16 electrons
2. A: 12 protons; B: 16 protons
3. element A: magnesium, Mg; element B: sulfur, S

Think Critically

4. Yes. A can lose 2 electrons and thus become a positive ion; Mg^{2+}. B can gain 2 electrons and thus become a negative ion; S^{2-}.
5. MgS, magnesium sulfide

Enrichment

Illustrate Ions Ask students to draw model diagrams of magnesium and sulfur ions, and of argon atoms. Have them do the same with sodium, fluorine, and neon for homework.

Teacher Content Support

Chemical Products and Reactants Double arrows in chemical equations signify that chemical reactions can go in either direction. For instance, even as CO_2 and H_2O combine to form carbonic acid (H_2CO_3), some of the H_2CO_3 breaks apart into CO_2 and H_2O molecules.

Figure 13 When a copper wire is placed in the solution of silver nitrate in the beaker, a chemical reaction occurs in which silver replaces copper in the wire and an aqua-colored copper nitrate solution forms.

Chemical Reactions

You have learned that atoms gain, lose, or share electrons to become more stable and that these atoms form compounds. Sometimes, compounds break down into simpler substances. The change of one or more substances into other substances, such as those in **Figure 13,** is called a **chemical reaction.** Chemical reactions are described by chemical equations. For example, water (H_2O) is formed by the chemical reaction between hydrogen gas (H_2) and oxygen gas (O_2). The formation of water can be described by the following chemical equation.

$$2H_2 + O_2 \rightarrow 2H_2O$$

You can read this chemical equation as "two molecules of hydrogen and one molecule of oxygen react to yield two molecules of water." In this reaction, hydrogen and oxygen are the reactants and water is the product. When you write a chemical equation, you must balance the equation by showing an equal number of atoms for each element on each side of the equation. Therefore, the same amount of matter is present both before and after the reaction. Note that there are four hydrogen atoms on each side of the above equation ($2 \times 2 = 4$). There are also two oxygen atoms on each side of the equation.

Another example of a chemical reaction, one that takes place between iron (Fe) and oxygen (O), is represented by the following chemical equation.

$$4Fe + 3O_2 \rightarrow 2Fe_2O_3$$

You will examine how compounds form in the Problem-Solving Lab on this page.

Problem-Solving LAB

Interpret Scientific Illustrations

How do compounds form? Many atoms gain or lose electrons in order to have eight electrons in the outermost energy level. In the diagram, energy levels are indicated by the circles around the nucleus of each element. The colored spheres in the energy levels represent electrons, and the spheres in the nucleus represent protons and neutrons.

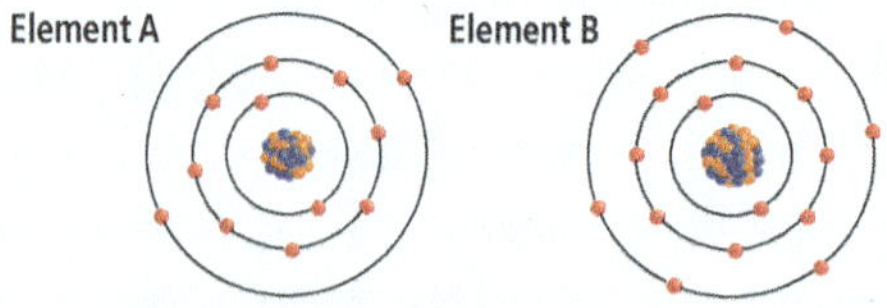

Analysis
1. How many electrons are present in atoms of Element A? Element B?
2. How many protons are present in the nuclei of these atoms?
3. Use the periodic table on page 61 to determine the name and symbol of Element A and Element B.

Think Critically

4. **Decide** if these elements can form ions. If so, what would be the electric charges (magnitude and sign) and chemical symbols of these ions?
5. **Formulate** a compound from these two elements. What is the chemical formula of the compound?

Demonstration

Neutralize Acids Assemble different brands of antacid tablets and liquids, plus water, baking soda, vinegar (to simulate stomach acid), and litmus paper. Ask students to predict which over-the-counter antacid will work the best in neutralizing stomach acid and thus an upset stomach. Add the recommended dose to the same volume of vinegar and test for pH. Record the results on the board and discuss.

Mixtures and Solutions

Unlike a compound, in which the atoms combine and lose their identities, a mixture is a combination of two or more components that retain their identities. When a mixture's components are easily recognizable, it is called a heterogeneous mixture. For example, beach sand, shown in **Figure 14,** is a heterogeneous mixture because its components are still recognizable—shells, small pieces of broken shells, grains of minerals, and so on. In a homogeneous mixture, which is also called a **solution,** the component particles cannot be distinguished from each other, even though they still retain their original properties.

A solution can be liquid, gaseous, or solid. Seawater is a solution consisting of water molecules and ions of many elements that exist on Earth. Molten rock is also a liquid solution; it is composed of ions representing all atoms that were present in the crystals of the rock before it melted. Air is a solution of gases, mostly nitrogen and oxygen molecules together with other atoms and molecules. Metal alloys, such as bronze and brass, are also solutions. Bronze is a homogeneous mixture of copper and tin atoms; brass is a similar mixture of copper and zinc atoms. Such solid homogeneous mixtures are called solid solutions.

Figure 14 Not all mixtures of beach sand and shells are alike. Mixtures from the Atlantic Ocean will contain components that are different from mixtures that form in the Pacific Ocean.

READING CHECK **Describe** three examples of solutions.

Acids Many chemical reactions that occur on Earth involve solutions called acids and bases. An **acid** is a solution containing a substance that produces hydrogen ions (H^+) in water. Recall that a hydrogen atom consists of one proton and one electron. When a hydrogen atom loses its electron, it becomes a hydrogen ion (H^+). The pH scale, shown in **Figure 15,** is based on the amount of hydrogen ions in a solution, referred to as the concentration. A value of 7 is considered neutral; distilled water usually has a pH of 7. A solution with a pH reading below 7 is considered to be acidic. The lower the number, the more acidic the solution.

View an **animation of the pH scale.**

Concepts In Motion

Figure 15 The pH scale is not only reserved for science class. All substances have a pH value, as you can see by the common household substances shown here.

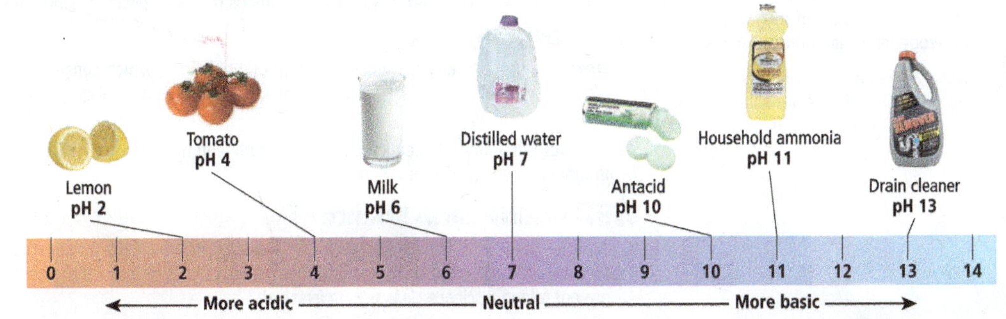

Environmental Connection

Cave Formation Caves form as a result of chemical reactions in the environment. The dissolution of carbon dioxide in water forms a weak acid—carbonic acid—which aids the formation of caves. Have students brainstorm other examples of how chemical reactions in the environment modify the landscape. Answers will vary, but might include acid rain dissolving limestone and marble structures.

READING CHECK Possible answers include air, metal alloys, powdered drink mix, and water.

Identify Misconceptions

A common misconception is that all compounds containing carbon are organic compounds.

Uncover the Misconception Ask students whether all compounds containing carbon are organic compounds.

Demonstrate the Concept Explain that all organic compounds contain carbon, but many carbon-containing compounds are inorganic, for example, $CaCO_3$ in some limestone, and CO_2 in volcanic gases. Food, fuel, medicines, cosmetics, and clothing consist of organic compounds.

Assess New Knowledge Ask students to compare and contrast the compounds Na_2CO_3, bicarbonate of soda, and $C_6H_{12}O_6$, a sugar. Which is organic and which is inorganic? Bicarbonate of soda is inorganic, and sugar is organic.

EARTH SCIENCE JOURNAL

Salty Water Seawater is a solution that contains many different salts. Have students research and list the six most abundant ions in seawater (Cl^-, Na^+, SO_4^{2-}, Mg^{2+}, Ca^{2+}, and K^+) and their relative abundances. Also ask students to name several possible ionic compounds that could form when seawater evaporates, and to write their formulas. OL AL

Demonstration

Everyday pH Bring in a variety of household and food items that have varying pH values. Enlarge and display **Figure 15.** Use litmus paper to determine the pH of each item, and mark its position on the pH scale. Explain that all solutions have a pH value and awareness of this is important for safety and health reasons.

CAREERS IN EARTH SCIENCE

WebQuest

3 Assess

Check for Understanding

Reinforcement Ask students the following questions: What does the superscript in Ca^{2+} mean? The superscript 2+ means this calcium ion has lost two valence electrons. What do the superscript and subscript in CO_3^{2-} mean? The subscript means there are three oxygen atoms in this complex ion. The superscript means the complex ion has gained two electrons. What is the chemical formula of calcium carbonate? $CaCO_3$

Reteach

Review Starting with the atomic number of calcium, explain the electron structure, number of valence electrons of calcium, and formation of a calcium ion and its charge. Repeat for chlorine. Explain the chemical equation that shows the formation of calcium chloride: $Ca^{2+} + 2Cl^- \rightarrow CaCl_2$.

Assessment

Knowledge Have students determine what ions are formed by oxygen, sodium, and magnesium. O^{2-}, Na^+, and Mg^{2+} Also have them determine the formulas of sodium oxide and magnesium oxide. Na_2O and MgO Have students explain and record in their Earth science journals the step-by-step procedure used in these determinations.

CAREERS IN EARTH SCIENCE

Geochemistry Some geochemists study the interaction of rocks, minerals and the environment. They can help mining companies reduce the amount of contamination from waste piles by understanding how the rocks and minerals break down and how toxic the byproducts might be.

WebQuest

The most common acid in Earth's environment is carbonic acid (H_2CO_3), which is produced when carbon dioxide (CO_2) is dissolved in water (H_2O) by the following reaction.

$$H_2O + CO_2 \rightarrow H_2CO_3$$

Some of the carbonic acid (H_2CO_3) in the water ionizes, or breaks apart, into hydrogen ions (H^+) and bicarbonate ions (HCO_3^-), as represented by the following equation.

$$H_2CO_3 \rightarrow H^+ + HCO_3^-$$

These two equations play a major role in the dissolution and precipitation of limestone and the formation of caves. In addition, while distilled water usually has a pH of 7, rainwater is slightly acidic, with a pH of 5.0 to 5.6. Many of the reaction rates involved in geological processes are very slow. For example, it might take thousands of years for enough carbonic acid mixed with groundwater to dissolve limestone and form a cave.

Bases A **base** is a substance that produces hydroxide ions (OH^-) in water. A base can neutralize an acid because hydrogen ions (H^+) from the acid react with the hydroxide ions (OH^-) from the base to form water through the following reaction.

$$H^+ + OH^- \rightarrow H_2O$$

Refer again to **Figure 15.** A solution with a reading above 7 is considered to be basic. The higher the number, the more basic the solution. Many household cleaning products have pH values from 11 to 13, placing them on the more basic end of the pH scale. The pH values of some basic substances are shown in **Figure 15.**

SECTION 2 REVIEW

Section Self-Check

Section Summary

- Atoms of different elements combine to form compounds.
- Covalent bonds form from shared electrons between atoms.
- Ionic compounds form from the attraction of positive and negative ions.
- There are two types of mixtures—heterogeneous and homogeneous.
- Acids are solutions containing hydrogen ions. Bases are solutions containing hydroxide ions.

Understand Main Ideas

1. MAINIDEA **Explain** why molecules do not have electric charges.
2. **Differentiate** between molecules and compounds.
3. **Calculate** the number of atoms needed to balance the following equation: $CaCO_3 + HCl \rightarrow CO_2 + H_2O + CaCl$
4. **Diagram** how an acid can be neutralized.
5. **Compare and contrast** mixtures and solutions by using specific examples of each.

Think Critically

6. **Design** a procedure to demonstrate whether whole milk, which consists of microscopic fat globules suspended in a solution of nutrients, is a homogeneous or heterogeneous mixture.
7. **Predict** what kind of chemical bond forms between nitrogen and hydrogen atoms in ammonia (NH_3). Sketch this molecule.

WRITING IN ▶ Earth Science

8. Antacids are used to relieve indigestion and upset stomachs. Write an advertisement for a new antacid product. Explain how the product works in terms that people who are not taking a science class will understand.

SECTION 2 REVIEW

1. Individual molecules contain the same number of electrons and protons. They cancel out each other.
2. Molecules form when atoms bond together with covalent bonds. Compounds are two or more different elements bonded together with ionic or covalent bonds.
3. The balanced equation reads: $CaCO_3 + 2HCl \rightarrow CO_2 + H_2O + CaCl_2$
4. Possible answer: acid with H^+ ions added to base with OH^- ions produces water and by-products.
5. Mixtures and solutions both contain more than one substance. The components of mixtures, such as granola cereal, can be easily separated and are visible. Components of solutions, such as sugar water, cannot be separated.
6. Possible answer: put milk in container, stir milk to mix thoroughly, draw one dropper full and put on microscope slide. Examine and describe what is seen. Students will see that the fat globules are still distinguishable, indicating that milk is a heterogeneous mixture.
7. NH_3 molecules have covalent bonding. Diagram should show a nitrogen atom with 5 valence electrons encircled by 3 hydrogen atoms with 1 valence electron. The valence electrons total 8, which makes a covalent molecule.
8. Answers should demonstrate an understanding of pH, acids, and bases.

SECTION 3

States of Matter

MAINIDEA All matter on Earth and in the universe occurs in the form of a solid, a liquid, a gas, or plasma.

Essential Questions

- What are the different states of matter on Earth?
- Why does matter exist in these states?
- How is thermal energy related to changes in states of matter?

Review Vocabulary

chemical reaction: the change of one or more substances into another substance

New Vocabulary

crystalline structure
glass
evaporation
plasma
condensation
sublimation

EARTH SCIENCE 4 YOU When your skin is wet, even on a hot day, it usually feels cool—especially if there is a wind blowing. How can warm air feel cold? When the water evaporates, it absorbs heat from your skin. The harder the wind blows, the more water evaporates and the colder your skin becomes.

Solids

Solids are substances with densely packed particles, which can be ions, atoms, or molecules. Most solids are **crystalline structures** because the particles of a solid are arranged in regular geometric patterns. Examples of crystals are shown in **Figure 16.** Because of their crystalline structures, solids have both a definite shape and volume.

Perfectly formed crystals are rare. When many crystals form in the same space at the same time, crowding prevents the formation of smooth, well-defined crystal faces. The result is a mass of intergrown crystals called a polycrystalline solid. Most solid substances on Earth, including rocks, are polycrystalline solids. **Figure 16** shows the polycrystalline nature of the rock granite.

Some solid materials have no regular internal patterns. **Glass** is a solid that consists of densely packed atoms arranged randomly. Glasses form when molten material is chilled so rapidly that atoms do not have enough time to arrange themselves in a regular pattern. These solids do not form crystals, or their crystals are so small that they cannot be seen. Window glass consists mostly of disordered silicon and oxygen (SiO_2).

■ **Figure 16** This granite is composed of mineral crystals that fit together like interlocking puzzle pieces. The minerals that make up the rock are composed of individual atoms and molecules that are aligned in a repeating pattern.

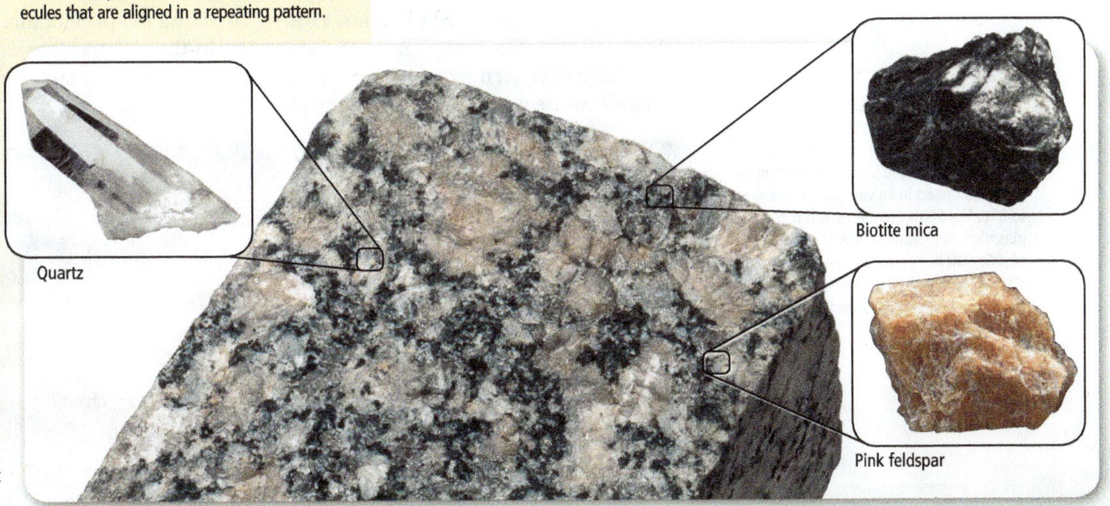

SECTION 3

1 Focus

Focus Transparency

Before presenting the lesson, project **Section Focus Transparency 9** and have students answer the accompanying questions. EL

MAINIDEA

Changes in State Place an ice cube in a clear, glass beaker on a hot plate. Turn on the hot plate and have the class describe the changes that they see. Ask the students to explain why the changes occur. The heat energy added to the system causes the bonds that are holding the molecules together to break apart, allowing the solid to become liquid, then gas.

2 Teach

Project

Properties of Metals Have student groups research the properties of metals, including melting and boiling points, crystal type, and uses. Have each group display a poster of its results. OL COOP LEARN

DIFFERENTIATED INSTRUCTION

Struggling Learners Have students arrange marbles in a neat, hexagonal pattern in a clear, plastic container. Tell them that this is the way atoms are arranged in some solids, such as magnesium. Next, have students shake the container so that the marbles roll around randomly, and explain that the marbles represent atoms in a liquid. Then have students stop shaking the container so that the marbles come to rest in a disorderly pattern. Explain to students that this pattern represents atoms in a glass.

Activity

Temperature and Phase Change Give pairs of students a small beaker containing a small amount of water and an ice cube. At one-minute intervals until two minutes after the ice melts, have students measure the temperature of the water in the beaker and record their data in a table with these headings: *Group, Time, Temperature,* and *Ice? (Y or N)*. Have the class discuss the results. Water temperature should stay near 0°C until the ice melts. BL COOP LEARN

Enrichment

Temperature Scales The Fahrenheit scale of temperature measurement was invented in Germany by Gabriel D. Fahrenheit (1686–1736). Have interested students research the origins of the Fahrenheit and Celsius scales and the relationship between the scales. Have students present their findings in class.

READING CHECK Increasing temperature causes the atoms to vibrate faster and possibly break apart, forming a liquid.

Caption Question Fig. 17 The levels differ because liquids do not have a definite shape. They are able to spread out into the containers they are placed in.

Caption Question Fig. 18 Students may note that the Sun's corona is 500 times hotter than lightning.

FOLDABLES® Rubric

Figure 17 Each of these containers has the same volume of liquid in it.
Explain *why the liquids are not all at the same level in the containers.*

Liquids

At any temperature above absolute zero (−273°C), the atoms in a solid vibrate. Because these vibrations increase with increasing temperature, they are called thermal vibrations. At the melting point of the material, these vibrations become vigorous enough to break the forces holding the solid together. The particles can then slide past each other, and the substance becomes liquid. Liquids take the shape of the container they are placed in, as you can see in **Figure 17.** However, liquids do have definite volume.

READING CHECK **Explain** the effect that increasing temperature has on the atoms in solids.

Gases

The particles in liquids vibrate vigorously. As a result, some particles can gain sufficient energy to escape the liquid. This process of change from a liquid to a gas at temperatures below the boiling point is called **evaporation**. When any liquid reaches its boiling point, it vaporizes quickly as a gas.

In gases, the particles are separated by relatively large distances and they travel at high speeds in one direction until they bump into another gas particle or the walls of a container. Gases, like liquids, have no definite shape. Gases also have no definite volume unless they are restrained by a container or a force such as gravity. For example, Earth's gravity keeps gases in the atmosphere from escaping into space.

FOLDABLES® Incorporate information from this section into your Foldable.

Plasma

When matter is heated to a temperature greater than 5000°C, the collisions between particles are so violent that electrons are knocked away from atoms. Such extremely high temperatures exist in stars and, as a result, the gases of stars consist entirely of positive ions and free electrons. These hot, highly ionized, electrically conducting gases are called **plasmas. Figure 18** shows the plasma that forms the Sun's corona. You have seen matter in the plasma state if you have ever seen lightning or a neon sign. Both lightning and the matter inside a neon tube are in the plasma state.

Figure 18 The Sun's temperature is often expressed in kelvins; 273 K is equal to 0°C. The Sun's corona, which is a plasma, has a temperature of about 15,000,000 K.
Compare *the temperature of the corona to lightning, which is 30,000 K.*

(t)John Evans, (b)ESA/NASA/SOHO

ACROSS THE CURRICULUM

Astronomy Auroras, the waves of colored lights that appear in the night sky at high latitudes, are caused by the interaction of plasma and Earth's magnetic field. The Sun emits streams of plasma, also called the solar wind. When the streams meet the edge of the magnetic field, some particles become trapped in the field and are forced down into the ionosphere. Here, they interact with gases which begin to glow, producing the colors that are seen as the auroras. The constant movement of Earth's magnetic field and the solar wind causes the wavy character of the auroras. Up to 1 trillion (1×10^{12}) watts of electricity can be generated during one auroral display. This is what causes the interference in radio, television and satellite transmissions.

Changes of State

Solids melt when they absorb enough thermal energy to cause their orderly internal crystalline arrangements to break down. This happens at the melting point. When liquids are cooled, they solidify at that same temperature and release thermal energy. The temperature at which liquids solidify is called the freezing point. Freezing is the reverse of melting.

When a liquid is heated to the boiling point and absorbs enough thermal energy, vaporization occurs and it becomes a gas. When a gas is cooled to the boiling point, it becomes a liquid in a process called **condensation,** shown in **Figure 19.** Condensation is the reverse of vaporization. Energy that was absorbed during vaporization is released upon condensation.

Vaporization can occur below the boiling point when thermal vibrations enable individual atoms or molecules to escape from a solid. This process is called evaporation. You might have noticed that even on winter days with temperatures below freezing, snow gradually disappears. This slow change of state from a solid (ice crystals) to a gas (water vapor) without an intermediate liquid state is called **sublimation.**

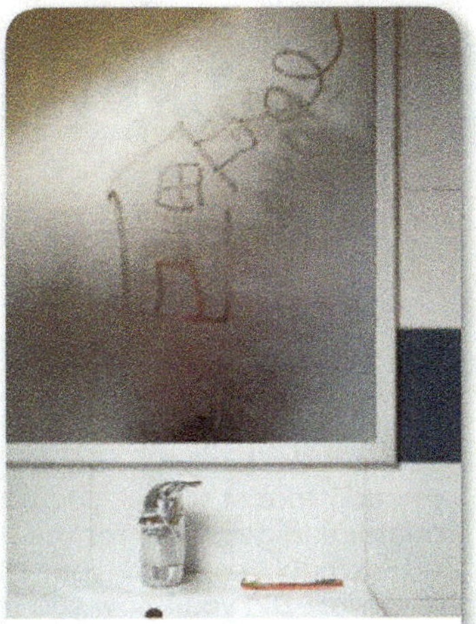

Figure 19 As the hot, moist air from the shower encounters the cool glass of the mirror, the water vapor in the air condenses on the glass.
Predict ***What would happen if the glass were the same temperature as the air?***

PhotoAlto/SuperStock

Conservation of Energy

The identity of matter can be changed through chemical reactions and nuclear processes, and its state can be changed under different thermal conditions. You have learned that a chemical equation must be balanced because matter cannot be created or destroyed. This fundamental fact is called the law of conservation of matter. Like matter, energy cannot be created or destroyed, but it can be changed from one form to another. For example, electric energy might be converted into light energy. This law, called the conservation of energy, is also known as the first law of thermodynamics.

SECTION 3 REVIEW

Section Self-Check

Section Summary

- Changes of state involve thermal energy.
- The law of conservation of matter states that matter cannot be created or destroyed.
- The law of conservation of energy states that energy is neither created nor destroyed.

Understand Main Ideas

1. MAINIDEA **Explain** how thermal energy is involved in changes of state.
2. **Evaluate** the nature of the thermal vibrations in each of the four states of matter.
3. **Apply** what you know about thermal energy to compare evaporation and condensation.

Think Critically

4. **Infer** how the boiling point of water (100°C) would change if water molecules were not polar molecules.
5. **Consider** glass and diamond—two clear, colorless solids. Why does glass shatter more easily than diamond?

MATH IN ▸ Earth Science

6. Refer to **Figure 18.** Calculate the corona's temperature in degrees. Remember that 273 K is equal to 0°C.

Caption Question Fig. 19 Because the air would not cool when it touches the glass, condensation would not occur.

3 Assess

Check for Understanding

Discussion Ask the class what happens to the water in a wet beach towel that is hung out to dry. The water evaporates. Also ask where the water goes. into the atmosphere as water vapor Then ask why water can vaporize at temperatures below the boiling point. because some molecules vibrate vigorously enough to break loose from others

Reteach

Summarize In solid ice, all molecules are bound into an orderly, 3-D framework by hydrogen bonds. In liquid water, enough of these bonds are broken so the molecules can slide around. In water vapor, all bonds between molecules are broken. Bonds break because of thermal vibrations. At the boiling point, the vibrations are vigorous enough to break all bonds.

Assessment

Performance Have students each assemble a collage of pictures showing examples of the states of water—glaciers, icebergs, snowflakes, hoar frost, dew, fog, clouds, streams, the sea, and any other examples.

SECTION 3 REVIEW

1. Changes of state happen as a result of a change in thermal energy, which is measured by changes in temperature. When the temperature of ice reaches the melting point, the ice becomes liquid water. When the temperature of water reaches the boiling point, it changes from a liquid to a gas.
2. solid—vibrations very slow or not at all; liquid—active vibrations, strong enough to break bonds; gas—strong vibrations that allow particles to escape from the liquid; plasma—vibrations violent enough to knock electrons from the nucleus' hold
3. Thermal energy is absorbed during evaporation. Thermal energy is released during condensation.
4. The boiling point would be lower because nonpolar molecules are less attracted to each other and it would be easier for the molecules to leave the liquid phase.
5. The internal arrangement of atoms in diamonds is orderly and the bonds are strong. Glass does not have an internal arrangement of atoms; the bonds holding it together are weak.
6. The temperature of the corona is 15,000,000 K, which converts to 14,999,727°C.

Earth Science &
TECHNOLOGY

Purpose

Students will learn about liquid crystal technology, the structure of liquid crystals and their uses.

Teacher Content Support

Liquid Crystal Technology

Liquid crystals were first discovered in 1904, but it wasn't until 1968 that a functional LCD device was created. The fact that liquid crystal displays can be small, are lightweight, and consume little power makes them highly useful in display technology. Computer screens, televisions, projectors, computer monitors, and wristwatches are just a few examples of technologies using LCDs. While wristwatches and other numeric displays have been around for a while, the color screen applications, such as computer monitors and television screens, are growing as the technology advances. Since they are lightweight, there are distinct advantages over typical cathode-ray technologies (most television and computer monitors are of this variety.) They do not require as much space, and the picture quality is as good, or better. However, there are some quirks that need to be worked out before LCD screens replace standard cathode-ray displays. Limited viewing range is only one of the drawbacks.

Teaching Strategy

Have students work in pairs to generate lists of current and future uses of LCD technology. Use the lists to start a classroom discussion about future LCD technology.

Earth Science &
TECHNOLOGY

Liquid Crystal Displays

You wake up in the morning, and get ready for school. You grab your music player and dash out the door, glancing at your watch as you go. Once at school, you pull out your calculator and get ready for the big math exam. After school, you check your email and watch television. Did you know you may have used liquid crystal display (LCD) technology five times already? You have more than likely heard that term before, but do you know what it means?

What is a liquid crystal? You know that liquids and crystals are two states of matter; but how is it possible to be both a liquid and a crystal? Recall that particles in a liquid can slide past each other in a container, while particles in a solid are packed together and cannot move separately. Liquid crystals are long molecules that keep their orientation—if they were oriented side-to-side in a thin layer on a glass plate, they would keep that side-to-side orientation. Because of their liquid property, the crystals can move around almost like a school of fish. Therefore, they share characteristics with both solids and liquids. This unique property makes them useful for a variety of electronic applications.

How do LCDs work? Consider a digital watch, for example, like the one shown in the photo above right. If you look closely at it, you can see the numbers, even when they are not darkened. These are the tracks that are engraved in the middle layer of a display "sandwich." Two plates of glass make up the outer portion of this sandwich. The inner portion of the sandwich—the tracks—contains liquid crystals that are in their natural, "relaxed" state. In the relaxed state, light from within the watch passes through the plates of glass, and is reflected out.

Digital watch displays are made possible through LCD technology. The inset photograph shows a polarized light micrograph of a LCD.

If an electric current is applied across a track of liquid crystals, the liquid crystals become charged and lose their original orientation. As long as a small current passes through them, light entering the plates of glass will not be reflected. In other words, that track will appear black.

Seems simple enough, right? That is why LCD technology is so popular. LCD is common display technology, used often because it is thin, lightweight, and energy efficient. The next generation of liquid crystal displays might be even more thin and flexible. LCDs that you can roll up like a piece of paper may be commonplace in the next several years.

WRITING IN ▶ Earth Science

Diagram Research the different layers of an LCD. Create a drawing showing all the different layers and how they fit together.

WebQuest

WRITING IN ▶ Earth Science

Rubric

Diagram Students' diagrams should show an understanding of how liquid crystal technology works.

WebQuest

GeoLAB

iLab Station

Precipitate Salts

Background: Many rocks on Earth form from salts precipitated from seawater. Salts precipitate when a salt solution becomes saturated. Solubility is the ability of a substance to dissolve in a solution. When a solution is saturated, no more of that substance can be dissolved.

Question: *Under what conditions do salt solutions become saturated, and under what conditions does salt precipitate out of solution?*

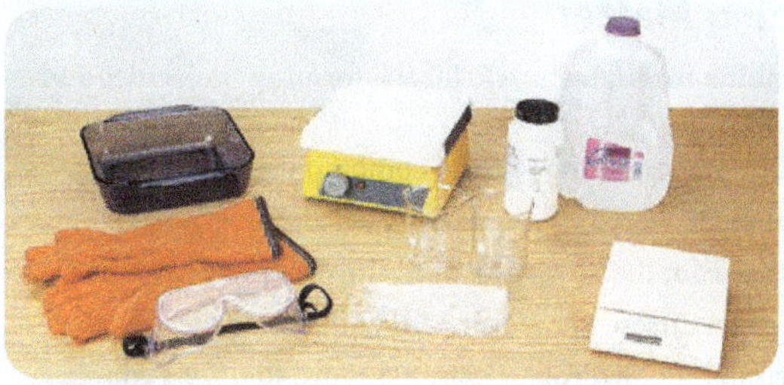

Matt Meadows

Suggested materials

Materials

halite (sodium chloride)
250-mL glass beakers (2)
distilled water
plastic wrap
laboratory scale
hot plate
shallow glass baking dish
refrigerator
glass stirring rod

Safety Precautions

Procedure

1. Read and complete the lab safety form.
2. Make a data table to record your observations.
3. Pour 150 mL of distilled water into a 250-mL glass beaker. Add 54 g of sodium chloride and stir until only a few grains remain on the bottom of the beaker.
4. Place the beaker on the hot plate, and turn on the hot plate. Stir the solution until the last few grains of sodium chloride dissolve. The salt solution will then be saturated.
5. Pour 50 mL of the warm, saturated solution into the second 250-mL glass beaker, and cover it with plastic wrap so that it forms a seal. Put this beaker in the refrigerator.
6. Pour 50 mL of the saturated solution into the glass baking dish. Place the dish on the hot plate and heat the salt solution until all the liquid evaporates. **WARNING:** ***The baking dish will be hot. Handle with care.***
7. Place the original beaker with 50 mL of the remaining solution on a shelf or windowsill. Do not cover the beaker.
8. Observe both beakers one day later. If crystals have not formed, wait another day.
9. Once crystals have formed in all three containers, observe the size and shape of the crystals. Write your observations in your data table.

Analyze and Conclude

1. **Describe** the shape of the precipitated crystals in the three containers. Does the shape of the crystals alone identify them as sodium chloride?
2. **Infer** how heating the salt solution affected the solubility of the sodium chloride.
3. **Interpret** what effect cooling has on the solubility of salt. What effect does evaporation have on the solubility of salt?
4. **Evaluate** the relationship between rate of cooling and crystal size.

INQUIRY EXTENSION

Use Other Substances Design an experiment to investigate other soluble substances. Test to see how much of the substance can be dissolved in a given amount of water, how long it takes for the solution to evaporate, and what crystal shapes form. Prepare a short report to share with your class.

Analyze and Conclude

1. The crystals are cubic, and some might be in the form of flattened squares. No; compounds other than NaCl might have similar crystals.
2. Heating increased the solubility, because increased thermal energy keeps salt ions from precipitating.
3. Cooling decreases the solubility of salt. Evaporation increases the concentration of salt ions in the remaining liquid. The solution eventually becomes supersaturated and salt crystals precipitate.
4. Faster cooling rates result in smaller crystal sizes.

INQUIRY EXTENSION

Use Other Substances Encourage students to try a variety of things that dissolve in water. Remind them to wear eye protection and gloves if using potentially irritating or caustic substances.

GeoLAB

iLab Station Rubric

Preparation

Time Allotment 30–40 min (check results after one or two days)

Process Skills observe and infer, draw conclusions, recognize cause and effect, compare and contrast, communicate

Safety Precautions Approve lab safety forms before work begins. Advise students to be careful with hot plates. Remind them to wear safety goggles.

Preparation of Materials Avoid using regular table salt, which contains iodine.

Procedure

- Have students work in groups of three.
- Students will observe that at room temperature, 150 mL of water can dissolve only about 52 g of salt.
- Students will note that large salt crystals, several millimeters in size, form in the beakers, while small salt crystals, only a fraction of a millimeter, form in the baking dish.
- **Troubleshooting** Salt crystals might not be able to nucleate in the beakers even if the salt solution becomes supersaturated. If this happens, add some small seed crystals and explain the problem to students.
- Remind students that sugar dissolves much more readily in hot tea than in iced tea.

MAINIDEAS Summary statements can be used by students to review the major concepts of the chapter.

Students can review with these online resources.

Vocabulary Practice

Vocabulary eGames
Vocabulary eFlashcards
Vocabulary PuzzleMaker

Section Self-Check

Chapter Self-Check

Online Test Practice

Use the ExamView® *Assessment Suite* CD-ROM to:

- create multiple versions of tests
- create modified tests with one mouse click
- edit existing questions and add your own questions
- build tests aligned with state standards using built-in state curriculum tags
- change English tests to Spanish with one mouse click
- track students' progress using the Teacher Management System

CHAPTER 3 STUDY GUIDE

BIGIDEA The variety of substances on Earth results from the way that atoms are arranged and combined.

Vocabulary Practice

SECTION 1 Matter

VOCABULARY
- matter
- element
- nucleus
- proton
- neutron
- electron
- atomic number
- mass number
- isotope
- ion

MAINIDEA Atoms are the basic building blocks of all matter.
- Atoms consist of protons, neutrons, and electrons.
- An element consists of atoms that have a specific number of protons in their nuclei.
- Isotopes of an element differ by the number of neutrons in their nuclei.
- Elements with full outermost energy levels are highly unreactive.
- Ions are electrically charged atoms or groups of atoms.

SECTION 2 Combining Matter

VOCABULARY
- compound
- chemical bond
- covalent bond
- molecule
- ionic bond
- metallic bond
- chemical reaction
- solution
- acid
- base

MAINIDEA Atoms combine through electric forces, forming molecules and compounds.
- Atoms of different elements combine to form compounds.
- Covalent bonds form from shared electrons between atoms.
- Ionic compounds form from the attraction of positive and negative ions.
- There are two types of mixtures—heterogeneous and homogeneous.
- Acids are solutions containing hydrogen ions. Bases are solutions containing hydroxide ions.

SECTION 3 State of Matter

VOCABULARY
- crystalline structure
- glass
- evaporation
- plasma
- condensation
- sublimation

MAINIDEA All matter on Earth and in the universe occurs in the form of a solid, a liquid, a gas, or plasma.
- Changes of state involve thermal energy.
- The law of conservation of matter states that matter cannot be created or destroyed.
- The law of conservation of energy states that energy is neither created nor destroyed.

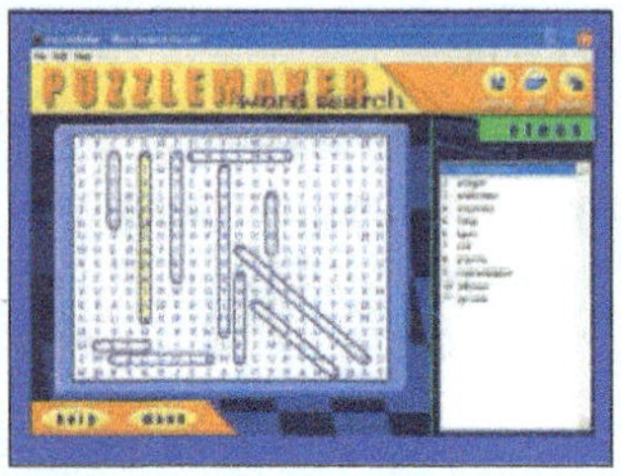

Vocabulary Practice

For additional practice with vocabulary, have students access the Vocabulary PuzzleMaker.

CHAPTER 3 ASSESSMENT

Chapter Self-Check

VOCABULARY REVIEW

Fill in the blank with the correct vocabulary term from the Study Guide.

1. The electrically neutral particles in the nucleus of an atom are called ____________
2. The ____________ of an element is equal to the number of ____________ in the nucleus of its atoms.
3. Atoms of an element that differ by their mass numbers are called ____________

Explain how both terms in each set below are related.

4. ionic, covalent
5. homogeneous mixture, solution
6. acid, base

Arrange each set of vocabulary terms into a meaningful and true sentence.

7. solid, glass
8. molecules, ions, plasma, gas
9. evaporation, condensation
10. electrons, metallic bond

UNDERSTAND KEY CONCEPTS

Use the figure below to answer Questions 11 to 13.

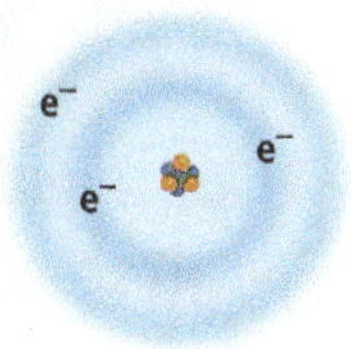

11. What is the atomic number of this atom?
    A. 3 C. 5
    B. 4 D. 6
12. How many valence electrons does this atom have?
    A. 1 C. 3
    B. 2 D. 4
13. Which element does this atom represent? (Refer to the periodic table of the elements in **Figure 2.**)
    A. helium
    B. beryllium
    C. lithium
    D. nitrogen
14. What ionic compound is formed by the ions Al^{3+} and O^{2-}?
    A. Al_3O_2 C. Al_2O_3
    B. Al_2O D. AlO

Use the figure below to answer Question 15.

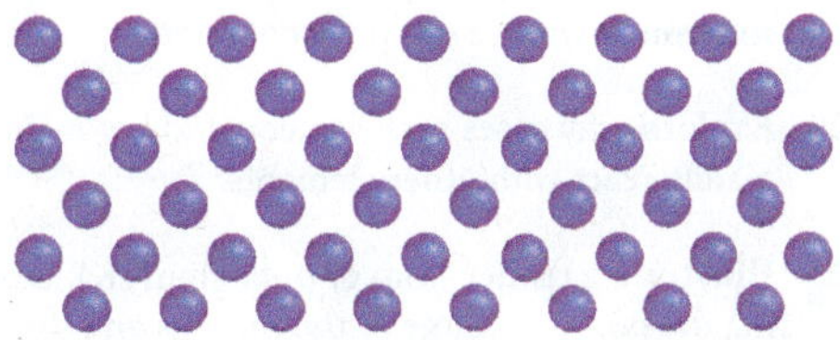

15. The figure shows the arrangement of atoms in a substance. What is this substance?
    A. gas
    B. glass
    C. liquid
    D. solid
16. Which is an example of a heterogeneous mixture?
    A. coffee
    B. soil
    C. gelatin
    D. air
17. What is ice converted into during sublimation?
    A. hydrogen ions and hydroxide ions
    B. hydrogen
    C. water
    D. water vapor
18. Many musical instruments are made of brass, which is a mixture of copper and zinc atoms. What is brass an example of?
    A solid solution
    B. ionic compound
    C. chemical reaction
    D. base

CHAPTER 3 ASSESSMENT

VOCABULARY REVIEW

1. neutrons
2. atomic number, protons
3. isotopes
4. Both are types of chemical bonds.
5. The terms mean the same thing. The components in the mixtures cannot be distinguished.
6. In a solution of water, acids donate H^+ ions, bases donate OH^- ions.
7. Solids have ordered internal structures while glasses do not.
8. A gas contains molecules, and plasma contains ions.
9. During changes of state, evaporation absorbs thermal energy and condensation releases it.
10. In metallic bonds, electrons flow freely around and between the positive ions.

UNDERSTAND KEY CONCEPTS

11. A
12. A
13. C
14. C
15. D
16. B
17. D
18. A

19. A
20. D

CONSTRUCTED RESPONSE

21. NaCl is an ionic compound. All of its electrons are bonded to its ions and can't move about as they do in metallic bonds.
22. The outermost energy level is full.
23. Potassium is a metal; diagrams should show 19 protons in the nucleus, 4 electron energy levels with electrons as follows: 2, 8, 8, 1.
24. All have 1 valence electron and form ions that have a +1 charge.
25. The atomic mass value is the average mass of all known isotopes of each element.
26. a metallic bond
27. Both are noble gases. Neon atoms are larger and heavier.
28. The bond is covalent. Both atoms have filled outer energy levels because the valence electrons are being shared.
29. In a metallic bond, some of the valence electrons would be disassociated from the atoms and moving freely around the solid.
30. Water molecules containing deuterium are heavier.
31. The statement is true. A plasma is usually hotter because it takes more thermal energy to ionize molecules into a plasma phase.

THINK CRITICALLY

32. By converting 1 neutron to a proton, the new atom has 7 protons and 7 neutrons in its nucleus. The new atom is nitrogen-14.

19. What happens to a gas when it condenses and forms a liquid?
A. It releases thermal energy.
B. It absorbs thermal energy.
C. It increases in temperature.
D. It decreases in temperature.

20. What kind of ion is present in an acid?
A. oxygen ion
B. negative ion
C. hydroxide ion
D. hydrogen ion

CONSTRUCTED RESPONSE

21. Explain why table salt does not conduct electricity.

22. Explain why gases such as neon and argon do not readily react with other elements.

23. Illustrate a model atom of potassium (K), indicating the positive charge of the nucleus and the idealized positions of the electrons in the various energy levels. Is potassium a metal or nonmetal? Refer to the periodic table of the elements in **Figure 2.**

Use the figure below to answer Questions 24 and 25.

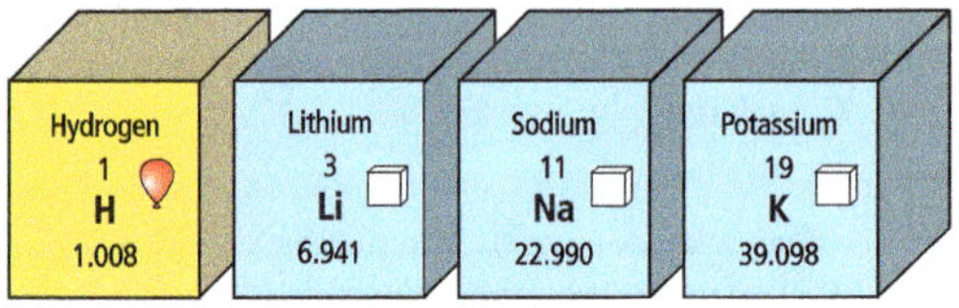

24. Detect What do these elements have in common?

25. Explain why the atomic masses of these elements are not whole numbers.

26. Distinguish which kind of chemical bond produces a solid that readily conducts heat and electricity.

27. Compare and contrast the physical properties of the elements helium and neon.

Use the figure below to answer Questions 28 and 29.

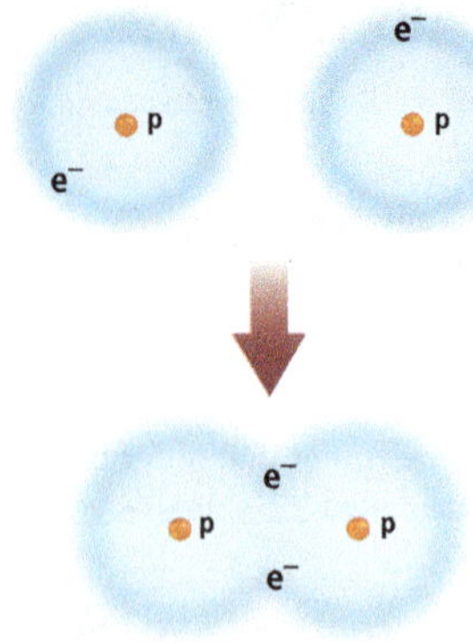

28. Identify the type of bond shown in the figure. Explain your reasoning.

29. Compare this bond to a metallic bond. Use an illustration to clarify your answer.

30. Deduce what the difference would be between water molecules containing deuterium and those containing ordinary hydrogen atoms. *(Hint: Deuterium is an isotope of hydrogen with mass number two. It forms the same chemical compounds as other hydrogen atoms, including water.)*

31. Evaluate the statement: Plasma is usually hotter than gas.

THINK CRITICALLY

Use the figure below to answer Question 32.

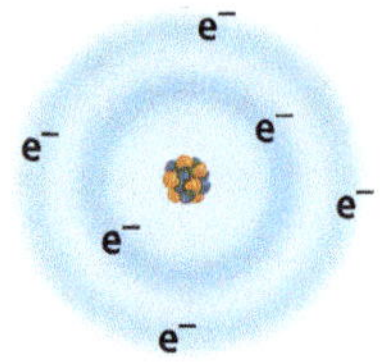

32. Deduce The figure shows an atom of carbon-14. This radioactive isotope decays by converting one of its neutrons to a proton. What element and isotope is produced by the radioactive decay of carbon-14?

33. **Illustrate** Use an illustration to show why water is effective in dissolving ionic solids such as table salt.

34. **Group** and list some of the properties that all metals have in common.

35. **Assess** the correctness of the following statement: Snow that covers the ground can disappear on cold days even when the temperature remains below 0°C.

36. **Arrange** When hydrochloric acid (HCl) is added to the sedimentary rock limestone ($CaCO_3$), carbon dioxide (CO_2), calcium chloride ($CaCl_2$) and water (H_2O) are given off. Arrange the chemical compounds listed above into a balanced equation that shows this chemical reaction.

37. **Infer** Earth's upper atmosphere—the ionosphere—conducts electricity. Infer about the state of matter in the ionosphere.

38. Careers in Earth Science Assess the importance of understanding chemical reactions in order to interpret the conditions of rocks and minerals that are present on other planets.

39. **Estimate** Air at sea level has a density of 0.13 g/L. Estimate how much air, in kilograms, fills your classroom. *(Hint: to start, multiply the length, width, and height of your classroom to calculate the room's volume.)*

CONCEPT MAPPING

40. Create a concept map using the following terms or phrases: *ionic bond, covalent bond, metallic bond, shared electrons, gain or lose electrons, a sea of electrons, molecule,* and *compound.*

CHALLENGE QUESTION

41. An atom is mostly empty space. A typical atom has a diameter of 10^{-10} m with a nucleus of diameter 10^{-15} m. To visualize this, enlarge this atom by a factor of 10^{13} (10 trillion) so that its nucleus has the size of a marble (1 cm). What would be the diameter of this enlarged atom? Would this atom fit into a football field?

WRITING IN ▸ Earth Science

42. Prepare a news release reporting on the discovery of a new chemical element. The element has 121 protons in its nucleus. Be sure to include the characteristics of this element and its location in the periodic table.

DBQ Document–Based Questions

Data obtained from: Kramer, D. A. 2005. Mineral resource of the month: Magnesium. *Geotimes* 50 (November): 57.

Magnesium is lightweight and has a high strength-to-weight ratio. It constitutes about 2 percent of Earth's crust and its concentration in seawater is 0.13 percent. Magnesium is present in more than 60 minerals and is produced from magnesium-bearing ores, seawater, and brines.

Magnesium is made into an alloy with aluminum to increase strength and corrosion resistance, especially in beverage cans. Its light weight makes it useful in aircrafts, cars, chain saws, lawn mowers, and other machine parts. Annual world magnesium production is 584,000 metric tons. China produces the most at 426,000 metric tons. Yearly production in the United States is 43,000 metric tons and U.S. consumption is 140,000 metric tons per year. Canada, China, Israel, and Russia supply 92 percent of U.S. magnesium imports. Recycling covers about 15 percent of U.S. magnesium consumption.

43. Determine the amount of magnesium in 1 m^3 of seawater. Express your answer in kilograms. (Hint: The density of seawater is 1025 kg/m^3.)

44. Analyze and explain the role of magnesium in the manufacture of cars and beverage cans.

45. Compare U.S. production of magnesium to that of other countries. How does U.S. dependence on imports affect these other countries?

CUMULATIVE REVIEW

46. Why is the concept of time and scale in the study of Earth science difficult to understand? **(Chapter 1)**

47. In the collection of data, measurements must follow what general guidelines? **(Chapter 1)**

33. Water is a polar molecule. Its negative end attracts positive ions, and its positive end attracts negative ions. The illustration should show the uneven distribution of electrons across the molecule.
34. Metals are shiny, dense, conduct electricity and heat, and can be molded into various shapes.
35. The statement is true. Sublimation of snow occurs at temperatures below the freezing point.
36. The correct equation is: $2HCl + CaCO_3 \longrightarrow CO_2 + H_2O + CaCl_2$
37. The ionosphere is a mixture of gas and plasma because it is partly ionized. The ions and electrons present conduct electricity.
38. The chemical reactions that occur tell us about the materials that are present. For example, rust is produced from the reaction of water and oxygen. If rust is present on another planet, it can be concluded that water and oxygen are, or once were, present on the planet.
39. Possible answer: a room 10 m long, 10 m wide, and 3 m high has a volume of 300 m^3, or 300,000 L. 300,000 L $\times$ 0.13 g/L = 39,000 g = 39 kg of air.

CONCEPT MAPPING

40. Students' maps should have lines drawn from ionic bond to gain or lose electrons and compound; covalent bond to shared electrons, molecule and compound; metallic bond to sea of electrons and compound.

CHALLENGE QUESTION

41. If the nucleus were the size of a marble, then the diameter of the atom would be 1000 m. No, this would not fit in a football field.

WRITING IN ▸ Earth Science

Rubric

42. Answers should show an understanding of atomic structure, atomic number and basic arrangement of the periodic table, including the position of the element in the column beneath actinium, that it would have three valence electrons, and eight energy levels.

DBQ Document-Based Questions

Data obtained from: Kramer, D.A. 2005. Mineral resource of the month: Magnesium. *Geotimes* 50 (November): 57.

43. The mass of 1 m^3 of seawater is 1025 kg. So the mass of Mg = 1025 kg $\times$ 0.0013 = 1.33 kg.
44. Magnesium is used because it adds strength to aluminum, which although lightweight, is not strong by itself.
45. U.S. production is well below other countries, yet our consumption is high, making us rely on other countries for our needs. This helps to boost the economies of the magnesium-producing countries.

CUMULATIVE REVIEW

46. Humans live such a relatively short period of time and are of relatively small size in comparison to Earth's and the universe's length of time of existence and size.
47. Measurements must be accurate, precise, and truthful.

MULTIPLE CHOICE

1. A
2. B
3. A
4. C
5. C
6. D
7. B
8. D
9. C

CUMULATIVE STANDARDIZED TEST PRACTICE

MULTIPLE CHOICE

Use the table below to answer Questions 1–3.

Atomic Structure		
Element	**Atomic Number**	**Atomic Mass**
Beryllium	4	9.01
Calcium	20	40.08
Silicon	14	28.09
Scandium	21	44.96
Titanium	22	47.88
Zirconium	40	91.22

1. If titanium has 22 protons in its nucleus, how many neutrons are present in the nucleus of its most common isotope?
   A. 26
   B. 28
   C. 48
   D. 60

2. If the most common isotope of scandium has 24 neutrons in its nucleus, how many protons does scandium have?
   A. 13
   B. 21
   C. 45
   D. 66

3. If calcium's most common isotope has 20 neutrons in its nucleus, how many neutrons can be found in another naturally occurring isotope of calcium?
   A. 21
   B. 30
   C. 41
   D. 60

4. Determine the number of valence electrons that oxygen has.
   A. 2
   B. 4
   C. 6
   D. 9

5. How should a city that is located between two time zones establish a time?
   A. Have two different times within the same city.
   B. Allow the people to choose what time zone they want to go by.
   C. Move the time zone split outside of the city.
   D. Divide the time in half to split the difference of the two times.

Use the illustration below to answer Questions 6–8.

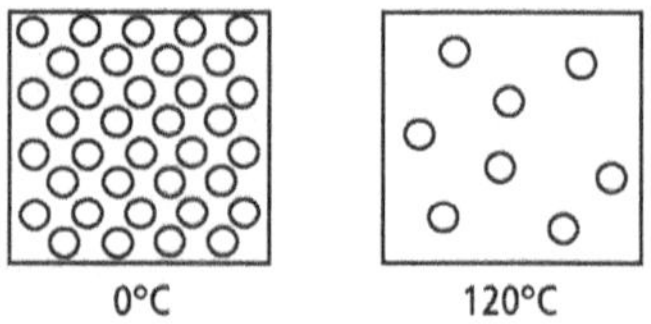

6. In a cup, an ice cube melts in liquid water. Which is true at the moment the ice melts?
   A. The water has less thermal energy than the ice.
   B. The water has more thermal energy than the ice.
   C. The water is at a higher temperature than the ice.
   D. The water is at the same temperature as the ice.

7. According to the illustration, what happens to water molecules when water is heated?
   A. Their energy levels decrease.
   B. They move farther apart.
   C. They move more slowly.
   D. They stop moving.

8. In order for a liquid to change to a gaseous state, what must it reach?
   A. its freezing point
   B. its condensation point
   C. its melting point
   D. its boiling point

9. Which is the most acidic?
   A. banana (pH 4.7)
   B. celery (pH 5.9)
   C. grape (pH 3.0)
   D. lettuce (pH 6.9)

SHORT ANSWER

Use the graph below to answer Questions 10–12.

Heating Curve

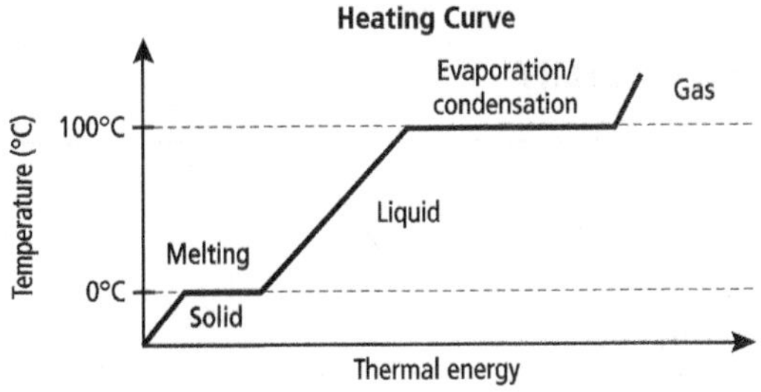

10. The diagram represents a sample of water. What does it demonstrate?

11. At which point does the water have the least amount of thermal energy?

12. What do the level lines in the diagram represent?

13. Differentiate between a theory and a hypothesis.

14. Why would a geologic map be important to a scientist studying earthquakes?

15. Does silicon have any isotopes? Explain your answer.

16. Ethical scientific researchers accurately report the data on which they base their conclusions. Why is this important?

READING FOR COMPREHENSION

Anthocyanin Pigments

Red cabbage contains a pigment molecule called flavin (an anthocyanin). This water-soluble pigment is also found in apple skin, plums, poppies, cornflowers, and grapes. Acidic solutions will turn anthocyanin a red color. Neutral solutions result in a purplish color. Basic solutions appear in greenish-yellow. Therefore, it is possible to determine the pH of a solution based on the color it turns the anthocyanin pigments in red cabbage juice.

How to make red cabbage pH indicator. *About: Chemistry.* (Online resource accessed February 12, 2007.)

17. How does red cabbage act as an acid/base indicator?
- **A.** Its pigment changes color based on the acid or base with which it comes in contact.
- **B.** Its pigment will not change when it comes into contact with a neutral solution.
- **C.** It always stays red.
- **D.** Its pigment releases water when it comes in contact with an acid or base.

18. What can be inferred from this passage?
- **A.** Red cabbage is the only food that can act as an indicator of acids and bases.
- **B.** Anthocyanin pigments in red cabbage juice change color when exposed to acids or bases.
- **C.** It is safe to eat the cabbage after using it for an acid/base experiment.
- **D.** The change in color does not indicate an acid or base.

19. What color does the cabbage become when exposed to a base?
- **A.** purplish color
- **B.** red color
- **C.** blue color
- **D.** greenish-yellow color

NEED EXTRA HELP?

If You Missed Question . . .	1	2	3	4	5	6	7	8	9	10	11	12	13	14	15	16
Review Section . . .	3.1	3.1	3.1	3.1	2.1	3.3	3.3	3.3	3.2	3.3	3.3	3.3	1.3	2.2	3.1	1.3

SHORT ANSWER

10. The diagram demonstrates how water changes states as thermal energy is added to it.

11. The water has the least amount of thermal energy in its solid stage as ice.

12. The level lines represent the point at which the water changes state from solid to liquid and then liquid to gas.

13. A theory is an explanation of a process or event that is based on observations and collected data. A hypothesis is a possible explanation for a problem that comes from what you know and observe; it is tested by gathering data.

14. Geologic maps can show the location of fault lines along which earthquakes might occur.

15. Silicon has an atomic mass of 28.09. Because the answer is not a whole number, it indicates that the atomic mass was calculated using the average masses of all forms, or isotopes, of silicon.

16. It allows other scientists to evaluate their conclusions.

READING FOR COMPREHENSION

17. A
18. B
19. D

CHAPTER 4 Minerals

BIGIDEA **Minerals are an integral part of daily life.**

ESSENTIAL QUESTIONS	NATIONAL STANDARDS	RESOURCES TO ASSESS MASTERY
SECTION 1 **What is a mineral?** **1.** How are minerals defined? **2.** How do minerals form? **3.** How are minerals classified? 2 sessions 1 block	UCP.1, UCP.3, UCP.5; A.1, A.2; B.2; D.2–3; F.3; G.3	**Entry-Level Assessment** Focus Transparency 10, p. 86 **Progress Monitoring** Caption Question, pp. 89, 91 Reading Check, pp. 87, 89, 90, 91, 93 Section Review, p. 95
SECTION 2 **Types of Minerals** **1.** What are the major groups of minerals? **2.** How is the silicon-oxygen tetrahedron illustrated? **3.** How are minerals used? 4 sessions 2 blocks	UCP.1–3, UCP5; A.1, A.2; B2–3; D.3; E.2; F.3–4, F.6; G.3	**Entry-Level Assessment** Focus Transparency 11, p. 96 **Progress Monitoring** Caption Question, p. 96 Section Review, p. 101 **Progress Monitoring** Chapter Assessment, p. 105 Exam*View*® *Assessment Suite* CD-ROM

Go online!
connectED.mcgraw-hill.com

LEVELED RESOURCES	LAB MATERIALS	ADDITIONAL RESOURCES
Science Notebook 4.1 OL **Chapter FAST FILE Resources:** MiniLab Worksheet, p. 26 OL GeoLab Worksheet, p. 2 OL Study Guide, p. 35 BL **Lab Resources:** Laboratory Manual, p. 25 OL **Transparencies:** Section Focus Transparency 10 EL Teaching Transparency 9 OL EL	**LabManager™** Customize any lab with the LabManager™ CD-ROM. LaunchLAB p. 84 / **15 min** table salt, microscope slide, microscope, magnifying lens, quartz crystal MiniLAB p. 92 / **20 min** mineral samples (7), protractor GeoLAB p. 103 / **60 min** mineral samples, magnifying lens, glass plate, streak plate, the Mohs scale of mineral hardness, steel file or nail, piece of copper, paper clip, magnet, dilute HCl, dropper, *Reference Handbook*	**Technology:** eTeacherEdition eStudent Edition Classroom Presentation Toolkit Exam*View®Assessment Suite* CD-ROM LabManager™ CD-ROM What's EARTH SCIENCE Got To Do With It? DVD Weather Classroom Video DVD **Assessment:** Blueprints for Success Chapter Test (Scaffolded) Earth Science Standardized Test Practice **Lab Resources:** Exploring Environmental Problems Laboratory Manual
Science Notebook 4.2 OL **Chapter FAST FILE Resources:** Study Guide, p. 15 BL **Lab Resources:** Laboratory Manual, p. 29 OL **Transparencies:** Section Focus Transparency 11 EL Teaching Transparency 10 OL EL		

BL Below Level OL On Level AL Advanced Learners EL English Learners COOP LEARN Cooperative Learning

CHAPTER 4

Minerals

BIGIDEA Minerals are an integral part of daily life.

SECTIONS

1 What is a mineral?

2 Types of Minerals

LaunchLAB

iLab Station

What shapes do minerals form?

Although there are thousands of minerals in Earth's crust, each type of mineral has unique characteristics. These characteristics are clues to a mineral's composition and to the way it formed. Physical properties can also be used to distinguish one type of mineral from another. Examine different crystal shapes in this lab.

FOLDABLES® Study Organizer

Mineral Identification

Make a layered-look book and label the tabs with the names of the tests used to identify minerals. Use it to organize your notes on mineral identification.

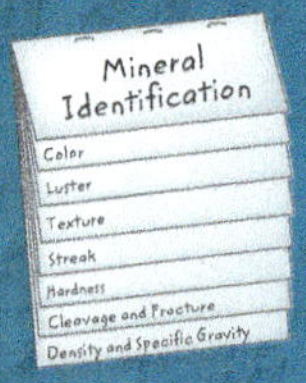

CHAPTER 4

LaunchLAB

iLab Station | Rubric

What shapes do minerals form?

Process Skills observe and infer, communicate, compare and contrast

Safety Precaution Approve lab safety forms before work begins.

Teaching Strategy

- Do not use sugar as an alternate material. Sugar is organic and thus not a mineral.
- Premeasure the salt. Place a container on a table for students to use for salt disposal. The salt can be stored and reused in other activities.

Procedure

1. Have students read and complete the lab safety form and follow the procedure below.
2. Place a few grains of **table salt** (the mineral halite) on a **microscope slide.** Place the slide on the **microscope** stage. Or, observe the grains with a **magnifying lens.**
3. Focus on one grain at a time. Count the number of sides of each grain. Make sketches of the grains.
4. Next, examine a **quartz crystal** with the microscope or magnifying lens. Count the number of sides of the quartz crystal. Sketch the shape of the quartz crystal.

Analysis

1. **Compare and contrast** the shapes of the samples of halite and quartz. Both the halite and quartz have distinctive crystal shapes. Halite crystals are cubic with four sides. Quartz crystals have six sides and can be double-pointed.
2. **Describe** some other properties of your mineral samples. Both are usually colorless, although the halite might be slightly translucent. Students might recognize other properties such as luster and texture.
3. **Infer** what might account for the differences you observed. Answers will vary. Students might know that the environments in which these minerals form are different. Halite forms by evaporation of chlorine and sodium-rich water; quartz forms by crystallization of molten material, or by precipitation from water.

Assessment

Knowledge Have groups compare and contrast their lists of the physical properties of the samples. Write a comprehensive list of the mineral properties on the board. Have students discuss which properties are more easily recognizable than others.

LabManager™
Customize any lab with the LabManager™ CD-ROM.

Stalactites and other cave formations are usually composed of the mineral calcite, and take thousands of years to form. One estimate is that a stalactite will grow only 10 cm in 1000 years. That is equal to 0.1 mm each year!

Introduce the BIGIDEA

Mineral Use Tell students that relatively few of the 3000 minerals present in Earth's crust have economic value. However, these few minerals are widely used for a variety of purposes. Have students write descriptions of seven objects around the classroom that contain minerals or that were obtained from minerals. Tell them to refer to the *Reference Handbook* for guidance.

Teacher Content Support

Caves and Tourism This cave is located in Carlsbad Caverns, New Mexico. The formations are delicate. Ask: Should this cave be open to the public? What are the advantages and disadvantages to such a venture? Advantages include increased tourism, revenue, and opportunities for people to see something beautiful; disadvantages include damage to the cave and surrounding ecosystems.

SECTION 1

1 Focus

Focus Transparency

Before presenting the lesson, project **Section Focus Transparency 10** and have students answer the accompanying questions. EL

MAINIDEA

Understand the Definition of Organic Students may question why sugar is not considered a mineral. Explain to them that although sugar has never been alive, it is an organic compound. An organic compound is a member of a group of compounds based on carbon bonded to itself and hydrogen, as well as some other elements.

2 Teach

Enrichment

Internal Atomic Structure
Encourage interested students to research the atomic structures and crystal shapes of minerals other than those shown in the Demo below. Have them present their findings to the class. AL

Concept Development

Mineral or Not? Show students samples of quartz, mica, salt, amber, pearl, coal, and sugar. Ask students to classify the samples as minerals or non-minerals. Record their answers on the board. Then tell students the definition of a mineral: naturally occurring, inorganic solids with specific crystal structure and definite chemical compositions. Refer again to the samples and have students discuss whether the samples fit the definition of minerals. Reclassify any errors. Tell students that many substances, such as pearls, coal, ivory, coral, and synthetic gems are often erroneously classified as minerals.

SECTION 1

Essential Questions
- How are minerals defined?
- How do minerals form?
- How are minerals classified?

Review Vocabulary
element: a pure substance that cannot be broken down into simpler substances by chemical or physical means

New Vocabulary
mineral
crystal
luster
hardness
cleavage
fracture
streak
specific gravity

What is a mineral?

MAINIDEA Minerals are naturally occurring, solid, inorganic compounds or elements.

EARTH SCIENCE 4 YOU Look around your classroom. The metal in your desk, the graphite in your pencil, and the glass in the windows are just three examples of how modern humans use products made from minerals.

Mineral Characteristics

Earth's crust is composed of about 3000 minerals. Minerals play important roles in forming rocks and in shaping Earth's surface. A select few have helped shape civilization. For example, great progress in prehistory was made when early humans began making tools from iron.

A **mineral** is a naturally occurring, inorganic solid, with a specific chemical composition and a definite crystalline structure. This crystalline structure is often exhibited by the crystal shape itself. Examples of mineral crystal shapes are shown in **Figure 1.**

Naturally occurring and inorganic Minerals are naturally occurring, meaning that they are formed by natural processes. Such processes will be discussed later in this section. Thus, synthetic diamonds and other substances developed in labs are not minerals. All minerals are inorganic. They are not alive and never were alive. Based on these criteria, salt is a mineral, but sugar, which is harvested from plants, is not. What about coal? According to the scientific definition of minerals, coal is not a mineral because millions of years ago, it formed from organic materials.

Figure 1 The shapes of these mineral crystals reflect the internal arrangements of their atoms.

Pyrite | Calcite

Demonstration

Atomic Arrangement and Crystal Form Show students samples of cubic minerals such as galena, pyrite, and halite and corresponding models of their atomic structures. Use these to demonstrate how the internal structure is reflected outward by these crystal forms.

Definite crystalline structure The atoms in minerals are arranged in regular geometric patterns that are repeated. This regular pattern results in the formation of a crystal. A **crystal** is a solid in which the atoms are arranged in repeating patterns. Sometimes, a mineral will form in an open space and grow into one large crystal. The well-defined crystal shapes shown in **Figure 1** are rare. More commonly, the internal atomic arrangement of a mineral is not apparent because the mineral formed in a restricted space. **Figure 2** shows a sample of quartz that formed in a restricted space.

Figure 2 This piece of quartz most likely formed in a restricted space, such as within a crack in a rock.

☑ READING CHECK **Describe** the atomic arrangement of a crystal.

Solids with specific compositions The fourth characteristic of minerals is that they are solids. Recall that solids have definite shapes and volumes, while liquids and gases do not. Because of this, no gas or liquid can be considered a mineral.

Each type of mineral has a chemical composition unique to that mineral. This composition might be specific, or it might vary within a set range of compositions. A few minerals, such as copper, silver, and sulfur, are composed of single elements. The vast majority, however, are made from compounds. The mineral quartz (SiO_2), for example, is a combination of two atoms of oxygen and one atom of silicon. Although other minerals might contain silicon and oxygen, the arrangement and proportion of these elements in quartz are unique to quartz.

VOCABULARY
ACADEMIC VOCABULARY
Restricted
small space; to have limits
The room was so small that it felt very restricted.

DIFFERENTIATED INSTRUCTION

Advanced Learners Show students pieces of lead crystal glass. Tell them that this glass is called lead crystal. Have them research the nature of lead crystal, how it is made, if it truly is a crystal and why it is called crystal.

☑ READING CHECK Atoms are arranged in regular, orderly and repeating patterns in crystals.

Tie to Previous Knowledge

Room for Crystal Growth Ask students whether they have ever blown large bubbles with bubble gum. Tell them to imagine trying to blow such bubbles if their hands were cupped over their mouths. Relate this to the formation of minerals in unrestricted and restricted spaces. Tell students minerals form either from magma or from solution. However, the size and shape of a mineral crystal is determined by the type of environmental conditions in which the mineral forms. Make sure students understand that most minerals do not have visible, well-defined geometric shapes, even though the atoms are arranged in a regular, orderly, repeating pattern that defines the specific mineral. The wide variety of mineral crystal sizes and shapes that we see exists because many minerals form within restricted spaces.

Collaborative Learning

Grow Crystals Have students work in groups of three to make supersaturated solutions of water and table salt. Students should allow the solution to evaporate over the course of several days. Then, have students observe the remaining residue under a microscope or with a magnifying lens. What shapes are the salt crystals? Are the crystals large or small? Did they grow in a restricted or unrestricted space? How do the students know? Ask students to record their observations in their Earth science journals. Table salt (sodium chloride) forms cubic crystals. If the solution evaporates quickly, the crystals will be small. If the solution evaporates slowly, the crystals will be large. By observing the shape of the crystals, it can be determined whether the crystals grew in a restricted or unrestricted environment. Crystals that grow in restricted environments should not look like perfect cubes. Crystals that grow in unrestricted environments form perfect cubes because of abundant space for crystal growth. COOP LEARN

Activity

Earth's Elements Have students use a computer-graphics program to make graphs that show the relative percentages of the eight most common elements in Earth's crust. OL AL

Interpret the Table

Variations in Composition Have students review **Table 1.** Ask: What are the possible results of the variations in composition among individual minerals? The most obvious is color. Students might also suggest density, hardness, luster, and stability. OL

Project

Hot Springs and Minerals Have students research the history of Mammoth Hot Springs in Yellowstone National Park, shown in **Figure 5.** They should include its size, its age, how fast it is growing, and any changes that have occurred since its discovery. If possible, photos should be used to illustrate key points. Students can present their findings to the class. OL AL

Figure 3 The mineral albite is a sodium-rich feldspar, while anorthite is calcium-rich. Oligoclase and labradorite contain both sodium and calcium in varying compositions.

Variations in Composition In some minerals, chemical composition can vary slightly depending on the temperature at which the mineral crystallizes. The plagioclase feldspar, shown in **Figure 3,** ranges from sodium-rich albite (AHL bite) at low temperatures to calcium-rich anorthite (uh NOR thite) at high temperatures. The difference in the mineral's appearance is due to a slight change in the chemical composition and a difference in growth pattern as the temperature changes. At intermediate temperatures, both calcium and sodium are incorporated into the crystal structure. This builds up alternating layers that allow light to refract or scatter, producing a range of colors, as shown in the labradorite in **Figure 3.**

Rock-Forming Minerals

Although about 3000 minerals occur in Earth's crust, only about 30 of these are common. Eight to ten of these minerals are referred to as rock-forming minerals because they make up most of the rocks in Earth's crust. They are primarily composed of the eight most common elements in Earth's crust. This is illustrated in **Table 1.**

Table 1 Most Common Rock-Forming Minerals

Quartz	Feldspar	Mica	Pyroxene*
SiO_2	$NaAlSi_3O_8 - CaAl_2Si_2O_8$ $KAlSi_3O_8$	$K(Mg,Fe)_3(AlSi_3O_{10})(OH)_2$ $KAl_2(AlSi_3O_{10})(OH)_2$	$MgSiO_3$ $Ca(Mg,Fe)Si_2O_6$ $NaAlSi_2O_6$
Amphibole*	**Olivine**	**Garnet***	**Calcite**
$Ca_2(Mg,Fe)_5Si_8O_{22}(OH)_2$ $Fe_7Si_8O_{22}(OH)_2$	$(Mg,Fe)_2SiO_4$	$Mg_3Al_2Si_3O_{12}$ $Fe_3Al_2Si_3O_{12}$ $Ca_3Al_2Si_3O_{12}$	$CaCO_3$

*representative mineral compositions

IN THE FIELD

Larimar In 1974, a new mineral was discovered in the Dominican Republic and is now adding to the country's economy. Called larimar, this turquoiose-colored mineral is found in only one locality in a fairly inaccesible region. Larimar is volcanic in nature, meaning that it formed as molten material cooled and crystallized. Chemical analysis has revealed it to be similar to the mineral pectolite, although enough variation occurs to suggest that it might be something else. Larimar is currently used in jewelry of all kinds.

Minerals from magma Molten material that forms and accumulates below Earth's surface is called magma. Magma is less dense than the surrounding solid rock, so it can rise upward into cooler layers of Earth's interior. Here, the magma cools and crystallizes. The type and number of elements present in the magma determine which minerals will form. The rate at which the magma cools determines the size of the mineral crystals. If the magma cools slowly within Earth's heated interior, the atoms have time to arrange themselves into large crystals. If the magma reaches Earth's surface, comes in contact with air or water, and cools quickly, the atoms do not have time to arrange themselves into large crystals. Thus, small crystals form from rapidly cooling magma, and large crystals form from slowly cooling magma. The mineral crystals in the granite shown in **Figure 4** are the result of slowly cooling magma.

☑ READING CHECK **Explain** how contact with water affects crystal size.

Minerals from solutions Minerals are often dissolved in water. For example, the salts that are dissolved in ocean water make it salty. When a liquid becomes full of a dissolved substance and it can dissolve no more of that substance, the liquid is saturated. If more solute is added, the solution is called supersaturated and conditions are right for minerals to form. At this point, individual atoms bond together and mineral crystals precipitate, which means that they form into solids from the solution.

Minerals also crystallize when the solution in which they are dissolved evaporates. You might have experienced this if you have ever gone swimming in the ocean. As the water evaporated off your skin, the salts were left behind as mineral crystals. Minerals that form from the evaporation of liquid are called evaporites. The rock salt in **Figure 4** was formed from evaporation. **Figure 5** shows Mammoth Hot Springs, a large evaporite complex in Yellowstone National Park.

Granite

Rock salt

■ **Figure 4** The crystals in these two samples formed in different ways.
Describe *the differences you see in these rock samples.*

■ **Figure 5** This large complex of evaporite minerals is in Yellowstone National Park. The variation in color is the result of the variety of elements that are dissolved in the water.

Model

Precipitation and Evaporation

Give groups of students the following supplies: 2 clear 100-mL beakers, sugar, 50 mL warm water, spoons, and stirring rods. Ask them to think about the different ways that minerals form from solutions. Then have students use the supplies to model precipitation from a solution and precipitation as a result of evaporation. Have them record their procedure and results in their Earth science journals. Students should model the precipitation from a solution by dissolving as much sugar as they can in the water, thus creating a supersaturated solution. Precipitation from evaporation can be modeled by dissolving a lesser amount of sugar in the water and then allowing it to evaporate. OL AL

☑ READING CHECK It causes faster cooling which results in small crystals.

■ **Caption Question Fig. 4**
Students might suggest differences in composition, method of formation (igneous v. evaporation) and rate of crystal formation.

DIFFERENTIATED INSTRUCTION

Advanced Learners Ask students to research and prepare presentations on one of the following topics: the environmental conditions under which the vast salt beds present in the rocks of New York and Ohio were formed, or the environmental conditions under which gold is formed.

Identifying Minerals

Geologists rely on several simple tests to identify minerals. These tests are based on a mineral's physical and chemical properties: crystal form, luster, hardness, cleavage, fracture, streak, color, texture, density, specific gravity, and special properties. As you will learn in the GeoLab at the end of this chapter, it is usually best to use a combination of tests instead of just one to identify minerals.

Crystal form Some minerals form such distinct crystal shapes that they are immediately recognizable. Halite—common table salt—always forms perfect cubes. Quartz crystals, with their double-pointed ends and six-sided crystals, are also readily recognized. However, as you learned earlier in this section, perfect crystals are not always formed, so identification based only on crystal form is rare.

Luster The way that a mineral reflects light from its surface is called **luster.** There are two types of luster—metallic luster and nonmetallic luster. Silver, gold, copper, and galena have shiny surfaces that reflect light, like the chrome trim on cars. Thus, they are said to have a metallic luster. Not all metallic minerals are metals. If their surfaces have shiny appearances like metals, they are considered to have a metallic luster. Sphalerite, for example, is a mineral with a metallic luster that is not a metal.

Minerals with nonmetallic lusters, such as calcite, gypsum, sulfur, and quartz, do not shine like metals. Nonmetallic lusters might be described as dull, pearly, waxy, silky, or earthy. Differences in luster, shown in **Figure 6,** are caused by differences in the chemical compositions of minerals. Describing the luster of nonmetallic minerals is a subjective process. For example, a mineral that appears waxy to one person might not appear waxy to another. Using luster to identify a mineral should usually be used in combination with other physical characteristics.

☑ READING CHECK **Define** the term *luster.*

CAREERS IN EARTH SCIENCE

Lapidary A lapidary is someone who cuts, polishes, and engraves precious stones. He or she studies minerals and their properties in order to know which minerals are the best for certain projects.

WebQuest

■ **Figure 6** The flaky and shiny nature of talc gives it a pearly luster. Another white mineral, kaolinite, contrasts sharply with its dull, earthy luster.

Activity

Recognize Minerals Have students work with partners to examine a pile of crushed granite. Instruct students to use tweezers and a microscope or magnifying lens to separate the crushed material into different mineral groups. Students should be able to distinguish three or four types of distinctly different minerals. Have students describe in their Earth science journals what criteria they used to classify the minerals. Answers will vary. Students will likely say color, luster, or texture, or some combination of all three. Accept all reasonable answers. OL COOP LEARN

CAREERS IN EARTH SCIENCE

WebQuest

Teacher Content Support

Useful Graphite Some motor oil is dark gray. The color is caused by the mineral graphite. Graphite is added to the motor oil because its softness makes it a good lubricant for the moving parts in motors. Graphite also can be used to lubricate moving parts in door locks. A powdered form of the mineral is squeezed into the door mechanism.

☑ READING CHECK Luster describes the way light reflects off the surface of the mineral.

■ **Caption Question Fig. 7** According to Mohs scale of hardness, glass is harder than a fingernail so the mineral that scratches glass is harder.

Demonstration

Recognize Metallic Luster Show students various objects made of metal and note their color, shine, etc. Then show students a variety of metallic minerals and those nonmetallic minerals that are often misidentified as metallic—sphalerite, biotite and muscovite mica, etc. Help students to see and understand the difference in these lusters. BL OL

Explore **Mohs scale with an interactive table.** Concepts In Motion

Table 2 Mohs Scale of Hardness

Mineral	Hardness	Hardness of Common Objects
Diamond	10	
Corundum	9	
Topaz	8	
Quartz	7	streak plate = 7
Feldspar	6	steel file = 6.5
Apatite	5	glass = 5.5
Fluorite	4	iron nail = 4.5
Calcite	3	piece of copper = 3.5
Gypsum	2	fingernail = 2.5
Talc	1	

Hardness One of the most useful and reliable tests for identifying minerals is hardness. **Hardness** is a measure of how easily a mineral can be scratched. German geologist Friedrich Mohs developed a scale by which an unknown mineral's hardness can be compared to the known hardness of ten minerals. The minerals in the Mohs scale of mineral hardness were selected because they are easily recognized and, with the exception of diamond, readily found in nature.

READING CHECK **Explain** what hardness measures.

Talc is one of the softest minerals and can be scratched by a fingernail; therefore, talc represents 1 on the Mohs scale of hardness. In contrast, diamond is so hard that it can be used as a sharpener and cutting tool, so diamond represents 10 on the Mohs scale of hardness. The scale, shown in **Table 2,** is used in the following way: a mineral that can be scratched by your fingernail has a hardness less than 2.5. A mineral that cannot be scratched by your fingernail and cannot scratch glass has a hardness value between 5.5 and 2.5. Finally, a mineral that scratches glass has a hardness greater than 5.5. Using other common objects, such as those listed in the table, can help you determine a more precise hardness and provide you with more information with which to identify an unknown mineral. Sometimes more than one mineral is present in a sample. If this is the case, it is a good idea to test more than one area of the sample. This way, you can be sure that you are testing the hardness of the mineral you are studying. **Figure 7** shows two minerals that have different hardness values.

Figure 7 The mineral on top can be scratched with a fingernail. The mineral on the bottom easily scratches glass.
Determine *Which mineral has greater hardness?*

Matt Meadows

Identify Misconceptions

Many people mistakenly believe that diamond is the only mineral hard enough to scratch glass, and that this ability to scratch glass is a foolproof way to identify diamond.

Uncover the Misconception Ask students whether they believe that only diamond can scratch glass. A number of students will likely say yes. Tell the class that glass has a hardness of 5.5 on the Mohs scale of hardness. Therefore, any mineral with a hardness greater than 5.5 will scratch a glass surface.

Demonstrate the Concept Rake a sample of crystalline quartz across a piece of glass. Have students observe the scratch left on glass by the quartz.

Assess New Knowledge Tell students to use a field guide to minerals or the *Reference Handbook* to create a list of minerals that will scratch glass. Possible minerals include diamond, feldspar, topaz, corundum, quartz, and amethyst.

Reinforcement

Use Hardness to Identify Minerals Have some students demonstrate how the Mohs scale of hardness can be used to help identify an unknown mineral. Give the students a glass slide, a copper penny, and unlabeled samples of fluorite, olivine, and gypsum. Point out that their fingernails are also testing instruments. By using these objects, the students should be able to identify the minerals. The hardness of olivine is 6.5, so it will scratch glass. Fluorite will scratch the penny but not the glass. Gypsum will not scratch the penny and can be scratched by a fingernail. Remind students that other tests might also need to be done in order to accurately identify a mineral.

READING CHECK Hardness measures a mineral's resistance to scratching.

DIFFERENTIATED INSTRUCTION

Struggling Learners Provide students with index cards on which have been written common terms used to describe luster: *earthy, shiny, greasy, pearly, glassy, waxy,* silky, and hand samples of minerals that exhibit these lusters. Have students create a display that matches up the mineral samples with the correct terms. EL

MiniLAB

iLab Station | Rubric

Purpose Students will learn to recognize cleavage in minerals.

Process Skills observe, measure, interpret, predict

Safety Precaution Approve lab safety forms before work begins.

Teaching Strategies

- Use the idea of parallel planes of atoms to help students understand that parallel cleavage planes count as one cleavage plane.
- For Step 2, use the following: Sample 1, biotite (1 cleavage plane); Sample 2, feldspar (2 cleavage planes); Sample 3, galena (3 cleavage planes); Sample 4, massive quartz; Sample 5, chalcedony, magnetite, chalcopryte, or hematite (specular or red).
- For Step 5, use the following: Sample 6, calcite (cleavage planes at 60° and 120°); Sample 7, halite (perfect cubic cleavage).

Expected Results Students should be able to recognize two groups of minerals: those with cleavage and those without. They should also recognize that the clear, nearly square minerals are not the same type of minerals.

Analysis

1. The following properties are present among the seven samples: 1, 2, and 3 cleavage planes, fracture, and conchoidal fracture.
2. The angles are not the same, thus the minerals are not the same.
3. The mica most likely will just bend, and the galena might not form cubes. Other minerals with cleavage will break along their weakest bonds, forming new cleavage planes in the smaller pieces mirroring the larger pieces. Minerals with no cleavage will fracture.

Figure 8 Halite has perfect cleavage in three directions; it breaks apart into pieces that have 90° angles. The strong bonds in quartz prevent cleavage from forming. Conchoidal fractures are characteristic of microcrystalline minerals such as flint.

Cleavage and fracture Atomic arrangement also determines how a mineral will break. Minerals break along planes where atomic bonding is weak. A mineral that splits relatively easily and evenly along one or more flat planes is said to have **cleavage.** To identify a mineral according to its cleavage, geologists count the number of cleaved planes and study the angle or angles between them. For example, mica has perfect cleavage in one direction. It breaks in sheets because of weak atomic bonds. Halite, shown in **Figure 8,** has cubic cleavage, which means that it breaks in three directions along planes of weak atomic attraction. Quartz and flint do not have natural planes of separation. They fracture instead of cleave.

MiniLAB

iLab Station

Recognize Cleavage and Fracture

How is cleavage used? Cleavage forms when a mineral breaks along a plane of weakly bonded atoms. If a mineral has no cleavage, it exhibits fracture. Recognizing the presence or absence of cleavage and determining the number of cleavage planes is a reliable method of identifying minerals.

Procedure

Part 1

1. Read and complete the lab safety form.
2. Obtain five **mineral samples** from your teacher. Separate them into two sets—those with cleavage and those without cleavage.
3. Arrange the minerals that have cleavage in order from fewest to most cleavage planes. How many cleavage planes does each sample have? Identify these minerals if you can.
4. Examine the samples that have no cleavage. Describe their surfaces. Identify these minerals if you can.

Part 2

5. Obtain two more samples from your teacher. Are these the same mineral? How can you tell?
6. Use a **protractor** to measure the cleavage plane angles of both minerals. Record your measurements.

Analysis

1. **Record** the number of cleavage planes or presence of fracture for all seven samples.
2. **Compare** the cleavage plane angles for Samples 6 and 7. What do they tell you about the mineral samples?
3. **Predict** the shape each mineral would exhibit if you were to hit each one with a hammer.

Assessment

Knowledge Ask students to explain why parallel cleavage planes are only counted once. Each cleavage plane is an external representation of the atomic structure. Parallel planes of atoms result in parallel cleavage planes. Therefore, there is only one direction in which the minerals break.

LabManager™
Customize any lab with the LabManager™ CD-ROM.

EARTH SCIENCE JOURNAL

Books of Mica Show students relatively large samples of mica and demonstrate how the samples cleave. Tell students that such samples are often called *books.* Have students write an explanation of the term in their Earth science journals. Sketches of cleaved samples should be incuded. The term refers to the fact that individual sheets of mica can be separated, similar to pages in a book.

Quartz, shown in **Figure 8,** breaks unevenly along jagged edges because of its tightly bonded atoms. Minerals that break with rough or jagged edges are said to have **fracture.** Flint, jasper, and chalcedony (kal SEH duh nee) (microcrystalline forms of quartz) exhibit a unique fracture with arclike patterns resembling clamshells, also shown in **Figure 8.** This fracture is called conchoidal (kahn KOY duhl) fracture and is diagnostic in identifying the rocks and minerals that exhibit it.

Streak A mineral rubbed across an unglazed porcelain plate will sometimes leave a colored powdered streak on the surface of the plate. **Streak** is the color of a mineral when it is broken up and powdered. The streak of a nonmetallic mineral is usually white. Streak is most useful in identifying metallic minerals.

Sometimes, a metallic mineral's streak does not match its external color, as shown in **Figure 9.** For example, the mineral hematite occurs in two different forms, resulting in two distinctly different appearances. Hematite that forms from weathering and exposure to air and water is a rusty red color and has an earthy feel. Hematite that forms from crystallization of magma can be silver and metallic in appearance. However, both forms make a reddish-brown streak when tested. The streak test can be used only on minerals that are softer than a porcelain plate. This is another reason why streak cannot be used to identify all minerals.

Figure 9 Despite the fact that these pieces of hematite appear remarkably different, their chemical compositions are the same. Thus, the streak that each makes is the same color.

☑ READING CHECK **Explain** which type of mineral can be identified using streak.

Color One of the most noticeable characteristics of a mineral is its color. Color is sometimes caused by the presence of trace elements or compounds within a mineral. For example, quartz occurs in a variety of colors, as shown in **Figure 10.** These different colors are the result of different trace elements in the quartz samples. Red jasper, purple amethyst, and orange citrine contain different amounts and forms of iron. Rose quartz contains manganese or titanium. However, the appearance of milky quartz is caused by the numerous bubbles of gas and liquid trapped within the crystal. In general, color is one of the least reliable clues of a mineral's identity.

FOLDABLES®
Incorporate information from this section into your Foldable.

Figure 10 These varieties of quartz all contain silicon and oxygen. Trace elements determine their colors.

Collaborative Learning

Variation in Quartz Have students work with partners to examine the following samples of quartz: amethyst, citrine, agate, massive quartz, and smoky quartz. Ask students to list the physical properties of each sample in a data table. At a minimum, data tables should include crystal shape, color, texture, and luster. Have pairs of students share their data with other pairs. OL COOP LEARN

Apply Earth Science

Mineralogy and Art Have students research the Smithsonian Museum of Natural History—specifically, the Hall of Minerals. Have students explore the different ways that gems and minerals have been used by artists. Students should write brief reports describing their research.

Activity

Tricky Colors To illustrate that color is not the most reliable way to identify a mineral, obtain samples of colorless and different-colored calcite, such as orange, blue, green, and reddish-brown. Do not identify the samples for students yet. Instead, ask students to compare and contrast the physical properties of the samples. Then tell students the minerals are all calcite. The color of each sample differs because of the presence of trace elements in the minerals. OL

ACROSS THE CURRICULUM

Biology Human eyes see color because our eyes have special receptors within the retina that detect light and color. Rods detect light, and cones detect color. Because the cones and rods vary in form within individual eyes, each person sees light and color differently. Some people have difficulty differentiating between certain pairs of colors such as red and green.

☑ READING CHECK Generally, metallic minerals can be identified using streak. Some nonmetallic minerals can also be identified in this way.

FOLDABLES® Rubric

Activity

Special Properties Give groups of students samples of calcite, Iceland spar, and ulexite. First, have students place their samples over the words *special properties* above. Students should be able to clearly observe the property of double refraction when using Iceland Spar as they see a double image of the words *special properties*. Next, have students place the ulexite over a page in their textbooks and describe what they see. This property is called telescoping and is a result of the fibrous nature of the mineral. Last, have one student from each group put on the gloves, goggles, and apron. Have them place a drop of dilute hydrochloric acid on the sample of calcite. Have students describe the chemical reaction. Tell them that calcite is calcium carbonate ($CaCO_3$). The acid reacts with the calcium carbonate and produces carbon dioxide in the form of fizzing bubbles. AL

Explore **minerals** with an interactive table. Concepts In Motion

Table 3 Special Properties of Minerals

Property	**Double refraction** occurs when a ray of light passes through the mineral and is split into two rays.	**Effervescence** occurs when reaction with hydrochloric acid causes calcite in limestone to fizz.	**Magnetism** occurs between minerals that contain iron; only magnetite and pyrrhotite are strongly magnetic.	**Iridescence**—a play of colors, caused by the bending of light rays.	**Fluorescence** occurs when some minerals are exposed to ultraviolet light, which causes them to glow in the dark.
Mineral	Calcite—Variety Iceland Spar	Limestone	Magnetite Pyrrhotite	Labradorite	Calcite
Example	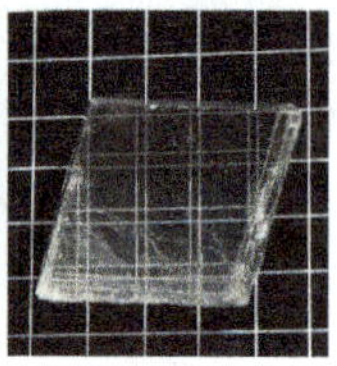				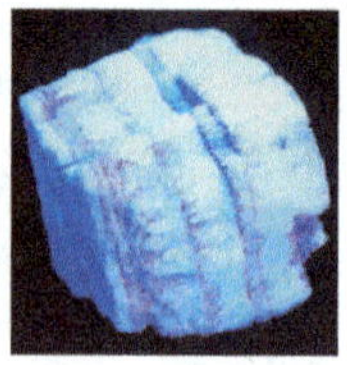

Special properties Several special properties of minerals can also be used for identification purposes. Some of these properties are magnetism, iridescence, double refraction, effervescence with hydrochloric acid, and fluorescence, shown in **Table 3**. For example, Iceland spar is a form of calcite that exhibits double refraction. The arrangement of atoms in this type of calcite causes light to be bent in two directions when it passes through the mineral. The refraction of the single ray of light into two rays creates the appearance of two images.

Data Analysis LAB

Based on Real Data*

Make and Use a Table

What information should you include in a mineral identification chart?

Mineral Identification Chart

Mineral Color	Streak	Hardness	Breakage Pattern
copper red		3	hackly, fracture
	red or reddish brown	6	irregular fracture
pale to golden yellow	yellow		
	colorless	7.5	conchoidal fracture
gray, green or white			two cleavage planes

Analysis

1. Copy the data table and use the *Reference Handbook* to complete the table.
2. Expand the table to include the names of the minerals, other properties, and uses.

Think Critically

3. **Determine** which of these minerals will scratch glass? Explain.
4. **Identify** which of these minerals might be present in both a painting and your desk.
5. **Identify** any other information you could include in the table.

*Data obtained from: Klein, C. 2002. *The Manual of Mineral Science.*

Data Analysis LAB

About the Lab

- Discuss the various physical properties that are used to identify a mineral. Review which properties are more helpful than others. This should help students decide what column heads to include in their data tables.
- See also Pough, F. ed. 1998. *Peterson first guide to rocks and minerals*. Boston: Houghton Mifflin.

Think Critically

1. The minerals, described in order, are copper (copper red streak), hematite, gold (2.5–3 hardness, hackly fracture), garnet, and plagioclase feldspar (colorless streak, 6 hardness).
2. The names of minerals are given in Question 1; properties and uses are listed in the *Reference Handbook*.
3. Hematite, garnet, and plagioclase feldspar will scratch glass. They all have a hardness greater than the hardness of glass (5.5).
4. Hematite and copper could be used as a pigment in paint colors. Iron or steel in desks would be made from hematite, the major iron ore.
5. Students might have included mineral uses, crystal shape, luster, special properties, and mineral name. The minerals' chemical formulas also could have been included in the table. Refer to the *Reference Handbook* for the answers to complete the table.

DIFFERENTIATED INSTRUCTION

Visually Impaired Tell students with visual impairments that the texture of minerals can be described using a variety of terms, such as *soapy, glassy, greasy, smooth,* or *ragged*. Have samples of minerals with distinct textures available for students to examine. Have students come up with their own terms to describe mineral texture based on the feel of the samples. OL

Texture Texture describes how a mineral feels to the touch. This, like luster, is subjective. Therefore, texture is often used in combination with other tests to identify a mineral. The texture of a mineral might be described as smooth, rough, ragged, greasy, or soapy. For example, fluorite, shown in **Figure 11,** has a smooth texture, while the texture of talc, shown in **Figure 6,** is greasy.

Density and specific gravity Sometimes, two minerals of the same size have different weights. Differences in weight are the result of differences in density, which is defined as mass per unit of volume. Density is expressed as follows.

$$D = \frac{M}{V}$$

In this equation, D = density, M = mass and V = volume. For example, pyrite has a density of 5.2 g/cm^3, and gold has a density of 19.3 g/cm^3. If you had a sample of gold and a sample of pyrite of the same size, the gold would have greater weight because it is denser.

Density reflects the atomic mass and structure of a mineral. Because density is not dependent on the size or shape of a mineral, it is a useful identification tool. Often, however, differences in density are too small to be distinguished by lifting different minerals. Thus, for accurate mineral identification, density must be measured. The most common measure of density used by geologists is **specific gravity,** which is the ratio of the mass of a substance to the mass of an equal volume of water at 4°C. For example, the specific gravity of pyrite is 5.2. The specific gravity of pure gold is 19.3.

Figure 11 Textures are interpreted differently by different people. The texture of this fluorite is usually described as smooth.

SECTION 1 REVIEW

Section Self-Check

Section Summary

- A mineral is a naturally occurring, inorganic solid with a specific chemical composition and a definite crystalline structure.
- A crystal is a solid in which the atoms are arranged in repeating patterns.
- Minerals form from magma, supersaturated solutions, or evaporation of solutions in which they are dissolved.
- Minerals can be identified based on physical and chemical properties.
- The most reliable way to identify a mineral is by using a combination of several tests.

Understand Main Ideas

1. MAINIDEA **List** two reasons why petroleum is not a mineral.
2. **Define** *naturally occurring* in terms of mineral formation.
3. **Contrast** the formation of minerals from magma and their formation from solution.
4. **Differentiate** between subjective and objective mineral properties.

Think Critically

5. **Develop** a plan to test the hardness of a sample of feldspar using the following items: glass plate, copper penny, and streak plate.
6. **Predict** the success of a lab test in which students plan to compare the streak colors of fluorite, quartz, and feldspar.

MATH IN Earth Science

7. Calculate the volume of a 5-g sample of pure gold.

Teacher Content Support

The mineral labradorite in **Table 3** displays labradorescence, a specific type of iridescence, which occurs only in this mineral.

Enrichment

Math Have students work with partners to determine the density of several minerals. Students should show their work and record their data in a table format. Also, tell students the density of a mineral, and have them calculate the mass of the sample, then solve for volume.

3 Assess

Check for Understanding

Discussion Have students explain why a synthetic diamond is not a mineral. A synthetic diamond is not naturally occurring.

Reteach

Review Write the definition of a mineral on the board. Provide a visual illustration of each characteristic. For instance, use several different-sized crystals of the same mineral to show that the crystalline stucture is consistent and definite.

Assessment

Knowledge Ask: What physical properties are the most reliable for identifying minerals? hardness, specific gravity What physical properties are least reliable? color, luster

SECTION 1 REVIEW

1. Petroleum is a liquid, not a solid. It was formed from once-living things, so it is not inorganic.
2. The object cannot have been artificially made in a laboratory. It must have been formed in nature.
3. Minerals form when magma cools beneath, at, or near Earth's surface. Minerals also form as a result of the precipitation from a supersaturated solution, or when the solution in which they are dissolved evaporates. All result in the formation of crystals.
4. Subjective properties are open to interpretation; objective properties are factual. Subjective properties are texture, color, and luster. Objective properties are hardness and cleavage.
5. Start with the softest testing material and work your way up. The first one to scratch the mineral will give an idea of the hardness of the mineral. The order would be: copper penny, glass plate, streak plate.
6. This will not work. None of these minerals are metallic; therefore, their streak colors will be colorless.
7. The answer is 0.26 cm^3.

SECTION 2

1 Focus

Focus Transparency

Before presenting the lesson, project **Section Focus Transparency 11** and have students answer the accompanying questions. EL

MAINIDEA

Classifications Ask students to name several types of groups. Ask them to list how each of these is defined. Students might list dog breeds, sports teams, cities, students, etc. Explain that minerals are also classified into groups, based on certain characteristics and chemical composition.

2 Teach

Teacher Content Support

Silicate Minerals The largest mineral group on Earth is the silicates, which makes up almost 96 percent of the known minerals on Earth. Feldspar and quartz, both silicates, are the first- and second-most abundant minerals present in Earth's crust, respectively. Other common silicates are muscovite, biotite, talc, serpentine, zircon, and topaz.

Caption Question Fig. 12 There are 5 atoms in a tetrahedron—4 oxygen atoms and 1 silicon atom.

SECTION 2

Essential Questions

- What are the major mineral groups?
- How is the silicon-oxygen tetrahedron illustrated?
- How are minerals used?

Review Vocabulary

chemical bond: the force that holds two atoms together

New Vocabulary

silicate
tetrahedron
ore
gem

Types of Minerals

MAINIDEA Minerals are classified based on their chemical properties and characteristics.

EARTH SCIENCE 4 YOU Everything on Earth is classified into various categories. Food, animals, and music are all classified according to certain properties or features. Minerals are no different; they, too, are classified into groups.

Mineral Groups

You have learned that elements combine in many different ways and proportions. One result is the thousands of different minerals present on Earth. In order to study these minerals and understand their properties, geologists have classified them into groups. Each group has a distinct chemical nature and specific characteristics.

Silicates Oxygen is the most abundant element in Earth's crust, followed by silicon. Minerals that contain silicon and oxygen, and usually one or more additional elements, are known as **silicates.** Silicates make up approximately 96 percent of the minerals present in Earth's crust. The two most common minerals, feldspar and quartz, are silicates. The basic building block of the silicates is the silicon-oxygen tetrahedron, shown in **Figure 12.** A **tetrahedron** (plural, tetrahedra) is a geometric solid having four sides that are equilateral triangles, resembling a pyramid. Recall that the electrons in the outermost energy level of an atom are called valence electrons. The number of valence electrons determines the type and number of chemical bonds an atom will form. Because silicon atoms have four valence electrons, silicon has the ability to bond with four oxygen atoms. As shown in **Figure 13,** silicon-oxygen tetrahedra can share oxygen atoms. This structure allows tetrahedra to combine in a number of ways, which accounts for the large diversity of structures and properties of silicate minerals.

Figure 12 The silicate polyatomic ion SiO_4^{4-} forms a tetrahedron in which a central silicon atom is covalently bonded to oxygen atoms.
Specify *How many atoms are in one tetrahedron?*

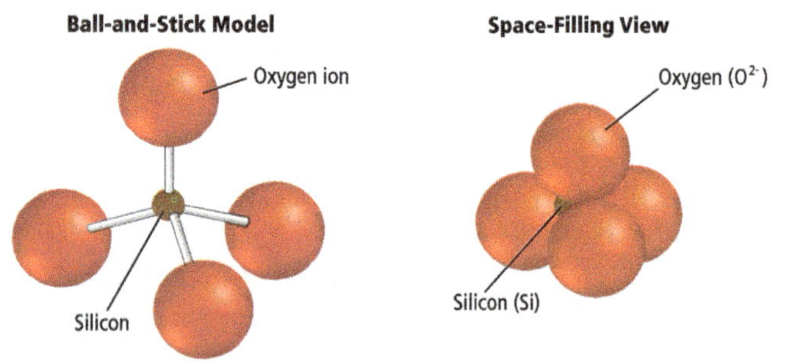

DIFFERENTIATED INSTRUCTION

Struggling Learners Supply students with a box of straws and a ball of string. Have students work with partners to construct models of basic silicon-oxygen tetrahedra. Students should use **Figure 12** as a guide. Foam balls of different sizes and pipe cleaners for connectors can also be used to construct tetrahedra models. For an added challenge, students can refer to **Figure 13** and assemble models of other silicon-oxygen tetrahedron arrangements, such as sheets or networks. EL

VISUALIZING the Silicon-Oxygen Tetrahedron

Figure 13 A silicon-oxygen tetrahedron contains four oxygen atoms bonded to a central silicon atom. Chains, sheets, and complex structures form when tetrahedra share oxygen atoms. These structures and the types of metal ions bonded to them determine the numerous silicate minerals that are present on Earth.

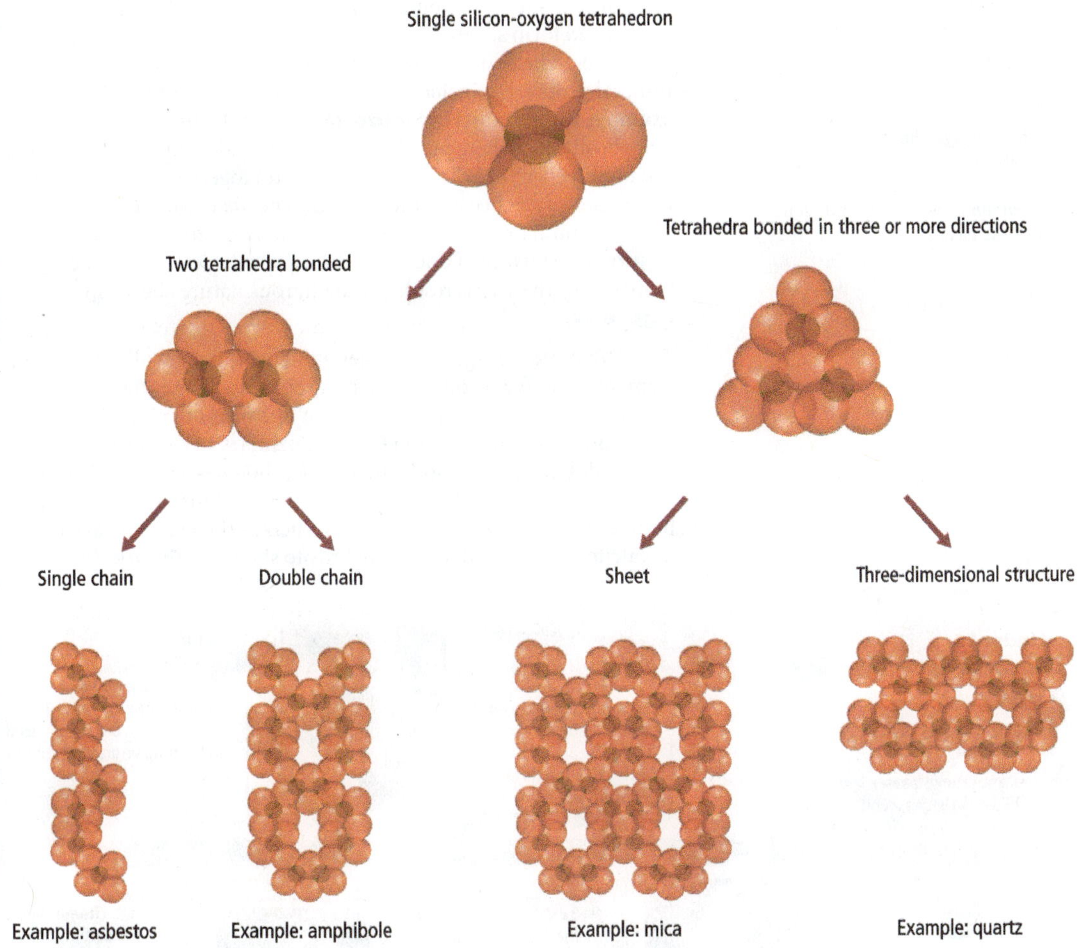

Concepts In Motion View an **animation of bonding of the silicon-oxygen tetrahedron.**

VISUALIZING

Purpose

Students will compare and contrast the different ways that the silica tetrahedron bonds together and the resulting mineral groups that form.

Enrichment

Other Tetrahedra Remind students that the silicate group is composed of silicon and oxygen atoms bonded together. Ask students if silicon is the only element that can bond to oxygen. no Tell students that phosphorous (P) and sulfur (S) also bond to oxygen atoms. Have students draw molecules that are composed of P and S atoms and determine the shape of the structures that form. Students will find that tetrahedra are formed with these elements. Have students determine the names of these molecules and the name of the minerals that they form. These will be minerals from the phosphate and sulfate groups.

Concept Development

Complex Tetrahedron Arrangements Have students study **Figure 13** and note the different tetrahedra arrangements shown. Have students assemble models of each type of tetrahedral arrangment. OL

Enrichment

Gold Rush History Tell students that there have been several gold rushes in American history. Assign half the class to research the 1849 gold rush at Sutter's Mill, California. The other half of the class should research the Alaskan gold rush of the 1890s. Students should write group reports that include the following information: dates of discovery, location of gold fields, methods of extraction, and human-interest stories about life in the gold camps. Students should illustrate their reports with photographs, if possible, and share their research with the class.

Project

Asbestos Dangers Have students research the connection between the minerals asbestos and talc, and why the use of talc in dusting powders and crayons has been questioned. Students can present their findings in the media of their choice. OL

Interpret the Illustration

Tetrahedral Bonds and Properties Have students study **Figure 14** and describe in their own words how the bonds between silicon-oxygen tetrahedra affect mineral properties, such as the way a mineral breaks.

Figure 14 The differences in silicate minerals are due to the differences in the arrangement of their silicon-oxygen tetrahedra. Certain types of asbestos consist of weakly bonded double chains of tetrahedra, while mica consists of weakly bonded sheets of tetrahedra.

Asbestos

Mica

Vocabulary
Science usage v. Common usage
Phyllo
Science usage: the sheets of silica tetrahedra

Common usage: sheets of dough used to make pastries and pies

Individual tetrahedron ions are strongly bonded. They can bond together to form sheets, chains, and complex three-dimensional structures. The bonds between the atoms help determine several mineral properties, including a mineral's cleavage or fracture. For example, mica, shown in **Figure 14,** is a sheet silicate, also called a phyllosilicate, where positive potassium or aluminum ions bond the negatively charged sheets of tetrahedra together. Mica separates easily into sheets because the attraction between the tetrahedra and the aluminum or potassium ions is weak. Asbestos, also shown in **Figure 14,** consists of double chains of tetrahedra that are weakly bonded together. This results in the fibrous nature shown in **Figure 14.**

Carbonates Oxygen combines easily with almost all other elements, and forms other mineral groups, such as carbonates. Carbonates are minerals composed of one or more metallic elements and the carbonate ion CO_3^{2-}. Examples of carbonates are calcite, dolomite, and rhodochrosite. Carbonates are the primary minerals found in rocks such as limestone and marble. Some carbonates have distinctive colorations, such as the colorful varieties of calcite and the pink of rhodochrosite shown in **Figure 16.**

Figure 15
Mineral Use Through Time
The value and uses of minerals have changed over time.

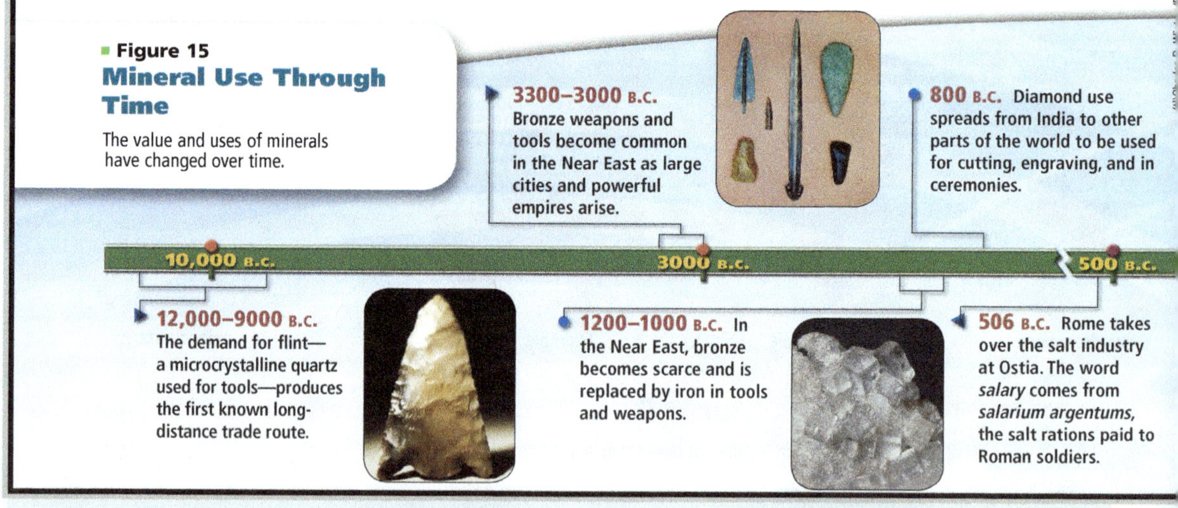

CULTURAL DIVERSITY

Ancient Peruvian Gold Before the 1500s, gold was mined mainly in Greece and Italy. When these gold supplies became limited, European explorers sought gold in the Americas. In 1532, Francisco Pizarro invaded Peru and encountered cultures such as the Chimu, the Chavin, the Nazca, and the Inca, who were experienced miners and goldsmiths. As early as 1200 B.C., these cultures were collecting gold from the rivers of the Andes Mountains. By 500 B.C., gold was being hammered into fine sheets and engraved. Sophisticated casting and filigree techniques were also developed. Unfortunately, Pizarro had many of the gold pieces melted down to ship them back to Spain. Only about 10,000 pieces of these ancient artifacts are preserved in the Museo de Oro in Bogotá, Colombia.

Figure 16 Carbonates such as calcite and rhodochrosite occur in distinct colors due to trace elements found in them.

Oxides Oxides are compounds of oxygen and a metal. Hematite (Fe_2O_3) and magnetite (Fe_3O_4) are common iron oxides and good sources of iron. The mineral uraninite (UO_2) is valuable because it is the major source of uranium, which is used to generate nuclear power.

Other Groups Other major mineral groups are sulfides, sulfates, halides, and native elements. Sulfides, such as pyrite (FeS_2), are compounds of sulfur and one or more elements. Sulfates, such as anhydrite ($CaSO_4$), are composed of elements and the sulfate ion SO_4^{2-}. Halides, such as halite (NaCl), are made up of chloride or fluoride along with calcium, sodium, or potassium. A native element such as silver (Ag) or copper (Cu), is made up of one element only.

Economic Minerals

Minerals are virtually everywhere. They are used to make computers, cars, televisions, desks, roads, buildings, jewelry, beds, paints, sports equipment, and medicines, in addition to many other things. You can learn about the uses of minerals throughout history by examining **Figure 15.**

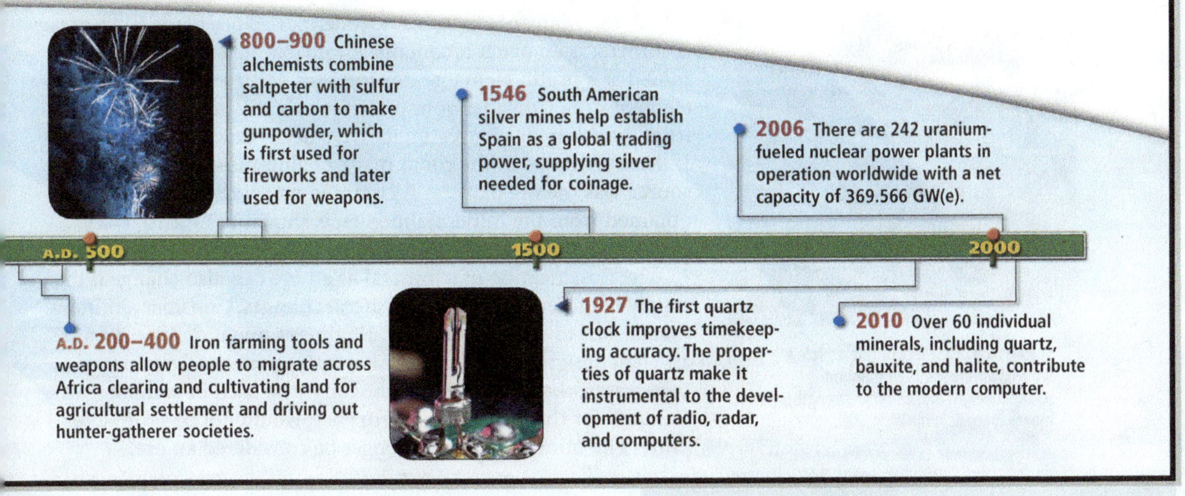

EARTH SCIENCE JOURNAL

Mineral Products Have students work in groups to compile lists of objects and materials that contain or are obtained from the following mineral groups: native elements, silicates, carbonates, and oxides. Have a field guide to minerals available for students to use. Students should record their findings in their Earth science journals. OL COOP LEARN

Use Scientific Terms

Mineral Groups Have students use chemical compositions to distinguish the following groups of minerals: silicates, carbonates, sulfates, native elements, and oxides. Students should list mineral examples for each group. Have students write their responses in their Earth science journals.

Enrichment

What are quarries and mines like? If possible, arrange a guided field trip to a local quarry or mine. Have students prepare questions beforehand to ask the person conducting the tour. If no mine or quarry is within driving distance, have students contact and interview a mining expert via the telephone or other source. Encourage students to ask a wide variety of questions. For instance, in addition to questions about extraction methods, students can ask about safety and reclamation issues. BL OL

Concept Development

Colorful Fireworks Have students list the colors of fireworks that they have seen. Then list on the board, some of the chemicals that are used to create these colors. strontium carbonate—$SrCO_3$, lithium carbonate—$LiCO_3$; calcium chloride—$CaCl_2$, calcium sulfate—$CaSO_4 \times H_2O$; copper chloride—CuCl; sodium nitrate—$NaNO_3$; barium oxide—BaO, barium chloride—BaCl. Ask student volunteers to match the chemical with the color it produces. strontium carbonate—red, lithium carbonate—red; calcium chloride—orange, calcium sulfate—orange; copper chloride—blue; sodium nitrate—yellow; barium oxide—white, barium chloride—green

Concept Development

State Mineral Reserves Assign each student a specific state to research. Tell students that a good place to start is to find the address for their state's Department of Natural Resources office. Ask students to list the major minerals found in their states and their uses. Also, ask students to request assistance in obtaining mineral samples from their state's Department of Natural Resources. Students should share their results with the class. Keep an ongoing record of the major minerals for each state. After the reports have been completed, have students construct a U.S. map that shows the major minerals in each state. Then, have student groups review the map for patterns. For instance, does a particular mineral occur exclusively in one state or one region? Do some minerals occur throughout the country? OL

Environmental Connection

Occurrence and Use of Ores Divide the class into groups of four. Assign each group an ore to research. Possible mineral ores include gold, zinc, iron, silver, aluminum, sulfur, tin, and uranium. Student reports should include the uses of the ore, where the ore occurs, and methods of extracting the ore. Encourage students to develop multimedia presentations to share their results. COOP LEARN

Explore **major mineral groups with an interactive table.** Concepts In Motion

Table 4 Major Mineral Groups

Group	Examples	Economic Use
Silicates	mica (biotite) olivine (Mg_2SiO_4) quartz (SiO_2) vermiculite	furnace windows gem (as peridot) timepieces potting soil additive; swells when wet
Sulfides	pyrite (FeS_2) marcasite (FeS_2) galena (PbS) sphalerite (ZnS)	used to make sulfuric acid; often mistaken for gold (fool's gold) jewelry lead ore zinc ore
Oxides	hematite (Fe_2O_3) corundum (Al_2O_3) uraninite (UO_2) ilmenite ($FeTiO_3$) chromite ($FeCr_2O_4$)	iron ore; red pigment abrasive, gem (as in ruby or sapphire) uranium source titanium source; pigment; replaced lead in paint chromium source, plumbing fixtures, auto accessories
Sulfates	gypsum ($CaSO_4{\cdot}2H_2O$) anhydrite ($CaSO_4$)	plaster, drywall; slows drying in cement plaster; name indicates absence of water
Halides	halite (NaCl) fluorite (CaF_2) sylvite (KCl)	table salt, stock feed, weed killer, food preparation and preservative steel manufacturing, enameling cookware fertilizer
Carbonates	calcite ($CaCO_3$) dolomite ($CaMg(CO_3)_2$)	Portland cement, lime, chalk Portland cement, lime; source of calcium and magnesium in vitamin supplements
Native elements	gold (Au) copper (Cu) silver (Ag) sulfur (S) graphite (C)	monetary standard, jewelry coinage, electrical wiring, jewelry coinage, jewelry, photography sulfa drugs and chemicals; match heads; fireworks pencil lead, dry lubricant

Figure 17 Parts of this athlete's wheelchair are made of titanium. Its light weight and extreme strength makes it an ideal metal to use.

Ores Many of the items just mentioned are made from ores. A mineral is an **ore** if it contains a valuable substance that can be mined at a profit. Hematite, for instance, is an ore that contains the element iron. Consider your classroom. If any items are made of iron, their original source might have been the mineral hematite. If there are items in the room made of aluminum, their original source was the ore bauxite. A common use of the metal titanium, obtained from the mineral ilmenite, is shown in **Figure 17.** **Table 4** summarizes the mineral groups and their major uses.

The classification of a mineral as an ore can also change if the supply of or demand for that mineral changes. Consider a mineral that is used to make computers. Engineers might develop a more efficient design or a less costly alternative material. In either of these cases, the mineral would no longer be used in computers. Demand for the mineral would drop. It would not be profitable to mine. The mineral would no longer be considered an ore.

ACROSS THE CURRICULUM

Engineering Diamonds have a hardness of 10 on the Mohs scale of hardness. Because diamonds are so hard, they are often used to coat the tips of drill bits and other cutting instruments. These diamond-tipped instruments can cut through steel and rock. Engineers use methane gas and microwaves to make the synthetic diamonds used on the cutting instruments. Each methane molecule contains a carbon atom and four hydrogen atoms. Microwaves are used to strip the hydrogen atoms from the molecules. The carbon atoms then bond on the surface of the cutting instrument and form tiny rows of diamonds. This process is used to make diamond-tipped scalpels, dental drills, and computer parts.

Mines Ores that are located deep within Earth's crust are removed by underground mining. Ores that are near Earth's surface are obtained from large, open-pit mines. When a mine is excavated, unwanted rock and minerals, known as gangue, are dug up along with the valuable ore. The overburden must also be removed before the ore can be used. Removing the overburden can be expensive and, in some cases, harmful to the environment. If the cost of removing the overburden or separating the gangue becomes higher than the value of the ore itself, the mineral will no longer be classified as an ore. It would no longer be economical to mine.

Gems What makes a ruby more valuable than mica? Rubies are rarer and more visually pleasing than mica. Rubies are thus considered gems. **Gems** are valuable minerals that are prized for their rarity and beauty. They are very hard and scratch resistant. Gems such as rubies, emeralds, and diamonds are cut, polished, and used for jewelry. Because of their rareness, rubies and emeralds are more valuable than diamonds. **Figure 18** shows a rough diamond and a polished diamond.

In some cases, the presence of trace elements can make one variety of a mineral more colorful and more prized than other varieties of the same mineral. Amethyst, for instance, is the gem form of quartz. Amethyst contains traces of iron, which gives the gem a purple color. The mineral corundum, which is often used as an abrasive, also occurs as rubies and sapphires. Red rubies contain trace amounts of chromium, while blue sapphires contain trace amounts of cobalt or titanium. Green emeralds are a variety of the mineral beryl, and are colored by trace amounts of chromium or vanadium.

Figure 18 The real beauty of gemstones is revealed once they are cut and polished.

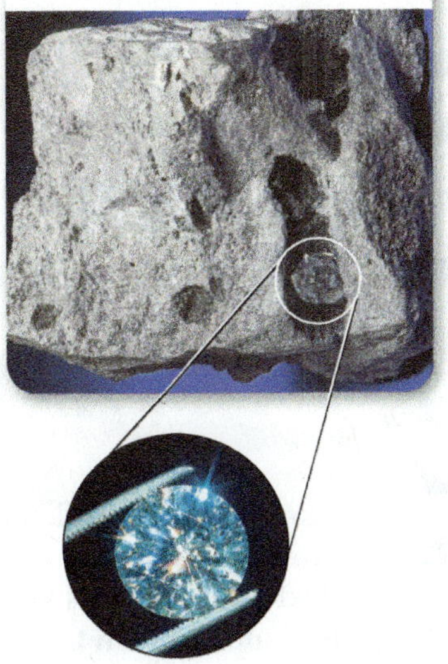

Watch a **video about gems and minerals.**

Video

SECTION 2 REVIEW

Section Self-Check

Section Summary

- In silicates, one silicon atom bonds with four oxygen atoms to form a tetrahedron.
- Major mineral groups include silicates, carbonates, oxides, sulfides, sulfates, halides, and native elements.
- An ore contains a valuable substance that can be mined at a profit.
- Gems are valuable minerals that are prized for their rarity and beauty.

Understand Main Ideas

1. **MAIN**IDEA **Formulate** a statement that explains the relationship between chemical elements and mineral properties.
2. **List** the two most abundant elements in Earth's crust. What mineral group do these elements form?
3. **Hypothesize** what some environmental consequences of mining ores might be.

Think Critically

4. **Hypothesize** why the mineral opal is often referred to as a mineraloid.
5. **Evaluate** which of the following metals is better to use in sporting equipment and medical implants: titanium—specific gravity = 4.5, contains only Ti; or steel—specific gravity = 7.7, contains Fe, O, Cr.

WRITING IN ▸ Earth Science

6. Design a flyer advertising the sale of a mineral of your choice. You might choose a gem or industrially important mineral. Include any information that you think will help your mineral sell.

3 Assess

Check for Understanding

Demo Place samples of minerals on a table at the front of the classroom. Have student volunteers identify and describe the mineral samples. Students should explain what criteria they used to identify the samples.

Reteach

Outline Have students outline the main ideas of this chapter in their Earth science journals. Ask students to provide specific examples in support of each main idea.

Assessment

Performance Have students conduct research and write reports about mining, mining operations, and reclamation policies. Students can also contact the U.S. Environmental Protection Agency, U.S. Geological Survey Mineral Information, and their state's Department of Natural Resources. Student reports should include photographs, diagrams, ore samples, and references.

What's EARTH SCIENCE Got To Do With It?

Video *A Rainbow of Gems*

SECTION 2 REVIEW

1. Possible answer: Elements bond together in many ways. The way in which they bond determines mineral characteristics, such as hardness, color, and cleavage.
2. silicon and oxygen; the group is the silicates
3. Possible answers include destruction of wildlife and pollution from runoff and waste.
4. Possible answer: The word *mineraloid* implies that it is something other than a mineral. Opal possesses some qualities of minerals (naturally occurring, solid, specific chemical composition) but not others (it has no set internal structure).
5. The specific gravity value indicates that titanium is lighter in weight. The chemical components indicates that steel, with iron and oxygen, will rust over time. The better metal to use is titanium.
6. Students' answers will vary, but should be scientifically accurate.

Rubric

eXpeditions!

Purpose

Students will learn about a remarkable cave environment in which unusually large crystals grow.

Teacher Content Support

Wet Caves Cave formations such as stalactites and stalagmites can grow only when a cave is a wet cave. That is, even though the water table may be below the level of the caverns, there is enough water seepage from groundwater to keep the caverns humid and wet. The constant dripping of mineral-saturated water and subsequent precipitation of calcium carbonate—both at the end of the droplets of water and on the ground where they land—ultimately results in the formation of stalactites and stalagmites. However, the Cave of Crystals provided an ideal environment for the continuous and uninterrupted growth of the crystals because the cave was completely submerged in mineral-rich water with a stable temperature. Today, humans can get into the cave only because a mining company's pumping operations in the area keep the water out of the cave. After the mining operation is closed, the cave will fill up with water and the minerals will again begin to grow.

Teaching Strategy

Help students to understand that not all caves contain fantastic formations. Students might find that the cave they choose to research could be a dry cave, or might have become dry after formations were formed. Have students compare their findings to help illustrate this point.

eXpeditions!

ON SITE: CRYSTALS AT LARGE IN MEXICO

Eloy and Javier Delgado walk slowly into the Naica Cave in Chihuahua, Mexico. The cave is very hot, making it difficult for them to breathe. They enter a room in the cave and before them are huge 4.5-m crystals that are clear and brilliant. How did these crystals grow this large? What kinds of conditions make these crystals possible?

The climate inside the cave Large crystals of gypsum are present in the Cave of Crystals, a room in Naica Cave, located 300 m below Earth's surface. Temperatures in the cave hover around 50°C. The air here has a relative humidity of 90 to 100 percent. These extreme conditions mean that anyone entering the cave can remain only for a few minutes at a time.

Crystal formations in the cave The crystals in the Naica Cave are a crystalline form of gypsum called selenite. The crystals in this cave grow into three distinct shapes. Crystals that grow from the floor of the cave are plantlike in appearance. They are grayish in color from the mud that seeps into them as they grow. Swordlike crystals cover the walls of the cave. These crystals grow to lengths of 0.5 m to 1 m and are opaque white in color. Within the main room of the cave, there are crystals with masses of up to 50,000 kg and up to 11 m long and 1 m wide.

Cave of Crystals, part of Naica Cave in Chihuahua, Mexico is known for its spectacular crystals.

How did these crystals form? Crystals need several things in order to form. First, they need a space—in this case, a cave. Caves form as a result of water circulating along weak planes in a rock. Over time, the rock dissolves and a cave is formed. Second, crystals need a source of water that is rich in dissolved minerals. Crystal formation also depends on factors such as pressure, temperature, the water level in the cave, and the chemistry of the mineral-rich water.

Geologists have determined that the crystals' massive sizes resulted from the steady temperature of about 55°C while the cave was full of mineral-rich water. As long as the crystals remained in this environment, they continued to grow. In 1985, miners lowered the water table and unknowingly drained the cave, halting the growth of the crystals. Scientists hypothesize that the largest crystals are about 600,000 years old.

WRITING IN ▶ Earth Science

Illustrate Research the processes that form crystals in a cave. Choose a cave and design a brochure describing and illustrating the types of crystals found there.

WebQuest

WRITING IN ▶ Earth Science

Rubric

Research Students' reports should include the cave's location, its size, types of formations, and what minerals make up the formations or crystals found in the cave.

WebQuest

GeoLAB

iLab Station

Design Your Own: Make a Field Guide for Minerals

Background: **Have you ever used a field guide to identify a bird, flower, rock, or insect? If so, you know that field guides include more than photographs. A typical field guide for minerals might include background information about minerals in general and specific information about the formation, properties, and uses of each mineral.**

Question: *Which mineral properties should be included in a field guide to help identify unknown minerals?*

Materials

Choose materials that would be appropriate for this lab.

mineral samples
magnifying lens
glass plate
streak plate
the Mohs scale of mineral hardness
steel file or nail
piece of copper
paper clip
magnet
dilute hydrochloric acid
dropper
Reference Handbook

Safety Precautions

Procedure

1. Read and complete the lab safety form.
2. As a group, list the steps that you will take to create your field guide. Keep the available materials in mind as you plan your procedure.
3. Should you test any of the properties more than once for any of the minerals? How will you determine whether certain properties indicate a specific mineral?
4. Design a data table to summarize your results. Be sure to include a column to record whether or not a particular test will be included in the guide. You can use this table as the basis for your field guide.
5. Read over your entire plan to make sure that all steps are in a logical order.
6. Have you included a step for additional research? You might have to use the library or the Internet to gather all the necessary information for your field guide.
7. What additional information will be included in the field guide? Possible data include how each mineral formed, its uses, its chemical formula, and a labeled photograph or drawing of the mineral.
8. Make sure your teacher approves your plan before you proceed.

Analyze and Conclude

1. **Interpret** Which properties were most reliable for identifying minerals? Which properties were least reliable? Discuss reasons that one property is more useful than others.
2. **Observe and Infer** What mineral reacted with the hydrochloric acid? Why did the mineral bubble? Write the balanced equation that describes the chemical reaction that took place between the mineral and the acid.
3. **Summarize** What information did you include in the field guide? What resources did you use to gather your data? Describe the layout of your field guide.
4. **Evaluate** the advantages and disadvantages of field guides.
5. **Conclude** Based on your results, is there any one definitive test that can always be used to identify a mineral? Explain your answer.

WRITING IN ▸ Earth Science

Peer Review Trade field guides with another group and test them out by using them to identify a new mineral. Provide feedback to the authors of the guide that you use.

GeoLAB

iLab Station | Rubric

Preparation

Time Allotment 60 min

Process Skills observe and infer, classify, analyze data, compare and contrast, communicate, interpret data, interpret scientific illustrations

Safety Precautions Approve lab safety forms before work begins. Hardness should be tested using a glass plate that is flat on the lab table. Tell students to handle the steel nail or file carefully. It might be sharp.

Preparation of Materials Some of the data for this lab can be found in the *Reference Handbook*.

Procedure

- Students might hypothesize that a combination of properties can be used to identify minerals.
- Students might note that these properties, along with other special properties, should be included in their field guide.
- Possible order of procedures: determine metallic or nonmetallic luster, hardness (a penny may be used instead of a piece of copper), streak, cleavage or fracture, special properties (reaction to HCl, magnetism, etc.)
- In addition to testing the samples, encourage students to access additional data from the library, museums, mineral collectors and any other sources.

Analyze and Conclude

1. Answers will vary. Special properties are most reliable because usually only one or two minerals share specific special properties. The least reliable properties are color, luster, and texture, because many minerals share these same properties.
2. Calcite reacts with HCl. The HCl and calcium carbonate present in the calcite react to release carbon dioxide gas in the form of bubbles. The equation is $CaCO_3 + 2HCl = CaCl_2 + H_2O + CO_2$.
3. Students should list each mineral's name, its properties, uses, chemical formula, and a photograph or sketch of the mineral. Resources might include data from tests, the library, the Internet, magazines, field guides, and mineral collections. Layouts should be clear and easy to follow.
4. Advantage: it can be used to identify and classify objects using both physical and chemical properties. Disadvantage: photos of perfect specimens do not resemble field samples.
5. No. Students will likely find that a combination of tests worked better than any one particular test.

WRITING IN ▸ Earth Science

Rubric

Peer Review Students should be able to identify a mineral using another group's field guide.

MAINIDEAS Summary statements can be used by students to review the major concepts of the chapter.

Students can review with these online resources.

Vocabulary Practice

- Vocabulary eGames
- Vocabulary eFlashcards
- Vocabulary PuzzleMaker

Section Self-Check

Chapter Self-Check

Online Test Practice

Use the Exam*View*® *Assessment Suite* CD-ROM to:

- create multiple versions of tests
- create modified tests with one mouse click
- edit existing questions and add your own questions
- build tests aligned with state standards using built-in state curriculum tags
- change English tests to Spanish with one mouse click
- track students' progress using the Teacher Management System

CHAPTER 4 STUDY GUIDE

BIGIDEA **Minerals are an integral part of daily life.**

Vocabulary Practice

SECTION 1 What is a mineral?

VOCABULARY

- mineral
- crystal
- luster
- hardness
- cleavage
- fracture
- streak
- specific gravity

MAINIDEA **Minerals are naturally occurring, solid, inorganic compounds or elements.**

- A mineral is a naturally occurring, inorganic solid with a specific chemical composition and a definite crystalline structure.
- A crystal is a solid in which the atoms are arranged in repeating patterns.
- Minerals form from magma, supersaturated solutions, or from evaporation of solutions in which they are dissolved.
- Minerals can be identified based on physical and chemical properties.
- The most reliable way to identify a mineral is by using a combination of several tests.

SECTION 2 Types of Minerals

VOCABULARY

- silicate
- tetrahedron
- ore
- gem

MAINIDEA **Minerals are classified based on their chemical properties and characteristics.**

- In silicates, one silicon atom bonds with four oxygen atoms to form a tetrahedron.
- Major mineral groups include silicates, carbonates, oxides, sulfides, sulfates, halides, and native elements.
- An ore contains a valuable substance that can be mined at a profit.
- Gems are valuable minerals that are prized for their rarity and beauty.

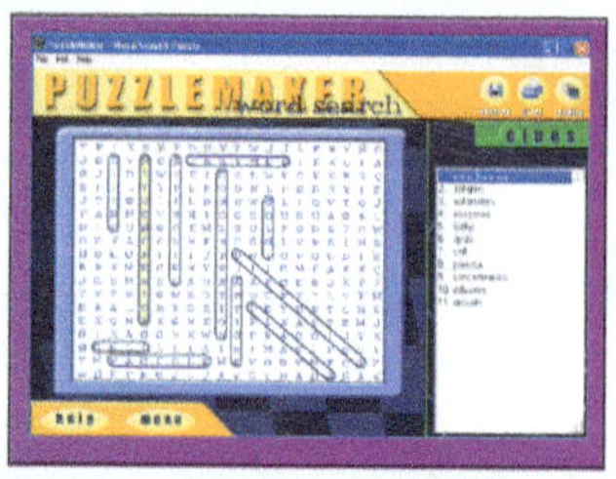

Vocabulary Practice

For additional practice with vocabulary, have students access the Vocabulary PuzzleMaker.

CHAPTER 4 ASSESSMENT

Chapter Self-Check

VOCABULARY REVIEW

Use what you know about the vocabulary terms listed on the Study Guide to answer the following questions.

1. What is a naturally occurring, solid, inorganic compound or element?
2. What term refers to the regular, geometric shapes that occur in many minerals?
3. What is the term for minerals containing silicon and oxygen?

Explain the relationship between the vocabulary terms in each pair.

4. ore, gem
5. silicate, tetrahedron

Complete the sentences below using vocabulary terms from the Study Guide.

6. Minerals that break randomly exhibit ________.
7. The ________ test determines what materials a mineral will scratch.

UNDERSTAND KEY CONCEPTS

Use the photo below to answer Question 8.

Matt Meadows

8. Which mineral property is being tested?
   - A. texture
   - B. hardness
   - C. cleavage
   - D. fluorescence
9. Which property causes the mineral galena to break into tiny cubes?
   - A. density
   - B. crystal structure
   - C. hardness
   - D. luster
10. What characteristic is used for classifying minerals into individual groups?
    - A. internal atomic structure
    - B. presence or absence of silica tetrahedrons
    - C. chemical composition
    - D. density and hardness
11. A mineral has a mass of 100 g and a volume of 50 cm^3. What is its density?
    - A. 5000 g/cm^3
    - B. 2 g/cm^3
    - C. 5 g/cm^3
    - D. 150 g/cm^3
12. What is the correct chemical formula for a silicon-oxygen tetrahedron?
    - A. SiO_2
    - B. $Si_2O_2^{+4}$
    - C. SiO_4^{-4}
    - D. Si_2O_2

Use the diagram below to answer Questions 13 and 14.

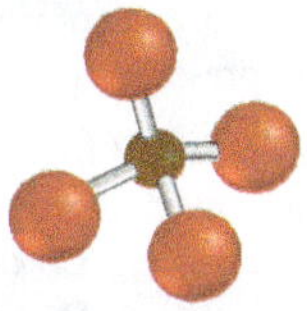

13. Where do the tetrahedra bond to each other?
    - A. the center of the silicon atom
    - B. at any oxygen atom
    - C. only the top oxygen atom
    - D. only the bottom oxygen atoms
14. What group of minerals is composed mainly of these tetrahedra?
    - A. silicates
    - B. oxides
    - C. carbonates
    - D. sulfates
15. Which is an example of a mineral whose streak cannot be determined with a porcelain streak plate?
    - A. hematite
    - B. gold
    - C. feldspar
    - D. magnetite

CHAPTER 4 ASSESSMENT

VOCABULARY REVIEW

1. mineral
2. crystal
3. silicate
4. Ores and gems are both economic minerals; their value depends on their cost of production and on the supply and demand.
5. Silicates are minerals that are formed by the bonding of silicon-oxygen tetrahedrons.
6. fracture
7. hardness

UNDERSTAND KEY CONCEPTS

8. B
9. B
10. C
11. B
12. C
13. B
14. A
15. C

16. B
17. B
18. C
19. C
20. A
21. D

CONSTRUCTED RESPONSE

22. They have small amounts of different trace elements that cause the color differences.
23. The word *geology* would appear double. The light rays are bent in two different directions when they pass through the Iceland spar. This double refraction of light is caused by the internal arrangement of atoms.
24. As the tea evaporates, the concentration of sugar increases. Eventually the tea will become saturated with sugar and then supersaturated. When that happens, sugar crystals will precipitate out of the tea.
25. crystal form—reflects the internal pattern; hardness—it depends on the strength of the bonds between the atoms; cleavage—splits the crystal along places of weak bonds; density—depends on how tightly packed the atoms are
26. They have different properties which makes them different minerals. Graphite is soft and not pretty. Diamond is hard, pretty, and can be cut into many desirable shapes. It is also much rarer in occurrence.

THINK CRITICALLY

27. They would differ in color.
28. An illustration should show atoms lined up in a repeating geometric pattern of a cube.
29. Corundum has a hardness of 9 and thus would make excellent sandpaper. Topaz and quartz, with hardnesses of 8 and 7 respectively, might also make good sandpaper.

ASSESSMENT

16. Which is one of the three most common elements in Earth's crust?
A. sodium
B. silicon
C. iron
D. carbon

Use the table below to answer Question 17.

Mineral Formulas	
Name	**Formula**
Quartz	SiO_2
Feldspar	$NaAlSi_3O_8$— $CaAl_2Si_2O_8$ & $KalSi_3O_8$
Amphibole	$Ca_2(Mg,Fe)_5Si_8O_{22}(OH)_2$ $Fe_7Si_8O_{22}(OH)_2$
Olivine	$(Mg,Fe)_2SiO_4$

17. What is the main factor that determines the formation of the minerals listed in the table?
A. rate of magma cooling
B. temperature of the magma
C. presence or absence of water
D. changes in pressure

18. Calcite is the dominant mineral in the rock limestone. In which mineral group does it belong?
A. silicates
B. oxides
C. carbonates
D. sulfates

19. What mineral fizzes when it comes in contact with hydrochloric acid?
A. quartz
B. gypsum
C. calcite
D. fluorite

20. *Dull, silky, waxy, pearly,* and *earthy* are terms that best describe which property of minerals?
A. luster
B. color
C. streak
D. cleavage

21. For a mineral to be considered an ore, which requirement must it meet?
A. It must be a common mineral.
B. Its production must not generate pollution.
C. It must be naturally occurring.
D. Its production must generate a profit.

CONSTRUCTED RESPONSE

22. Explain why rubies and sapphires, which are both forms of the mineral corundum, are different colors.

23. Describe the visual effect of placing a piece of clear, Iceland spar on top of the word *geology* in a textbook.

24. Summarize the process of sugar crystals forming in a glass of sugar-sweetened hot tea.

25. Hypothesize which mineral properties are the direct result of the arrangement of atoms or ions in a crystal. Explain your answer.

26. Compare and Contrast Diamond and graphite have the same chemical composition. Compare and contrast these two to explain why diamond is a gem and graphite is not.

THINK CRITICALLY

27. Describe the differences that might be exhibited by the garnets listed in **Table 1.**

Use the figure below to answer Question 28.

28. Illustrate what the atomic structure might be if the crystal shape is an external reflection of it.

29. Recommend which minerals, other than diamond, would be best for making sandpaper. Explain your answer. Refer to **Table 2.**

Chapter Self-Check

30. **Decide** which of the following materials are not minerals, and explain why: petroleum, wood, coal, steel, concrete, and glass.

31. **Infer** how early prospectors used density to determine whether they had found gold or pyrite in a mine.

32. **Assess** Imagine that a new gem is discovered that is more beautiful than the most stunning diamond or ruby. Assess the factors that will determine its cost compared to other known gems.

Use the figure below to answer Questions 33–34.

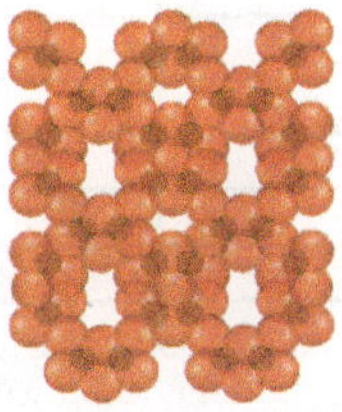

33. **Infer** Mica is a mineral with a sheet silicate structure. The atomic arrangement is shown above. Infer what is holding these sheets, which consist of negatively charged silicon-oxygen tetrahedra, together.

34. **Describe** the type of cleavage that occurs in minerals with the atomic arrangement shown.

CONCEPT MAPPING

35. Create a concept map using the following terms: *silicates, oxides, halides, sulfates, sulfides, native elements,* and *carbonates.* Add any other terms that are helpful. For more help, refer to the *Skillbuilder Handbook.*

CHALLENGE QUESTION

36. **Arrange** In addition to sheet silicates, there are chain silicates, tectosilicates, and cyclosilicates. Arrange six silicon-oxygen tetrahedra in a cyclosilicate form. Be sure to bond the oxygen atoms correctly.

WRITING IN ▶ Earth Science

37. Imagine that you are planning a camping trip. What tools should you pack if you want to identify interesting minerals? How would you use these tools?

DBQ Document–Based Questions

Data obtained from: Plunkert, P.A. 2005. Mineral resource of the month: Aluminum. *Geotimes* 50:57.

Aluminum is an abundant metallic element in Earth's crust. It is lightweight, ductile [bendable], corrosion resistant, and a good conductor of electricity. It is used most often in the manufacture of cars, buses, trailers, ships, aircraft, railway and subway cars. Other uses include beverage cans, aluminum foil, machinery, and electrical equipment.

Aluminum is produced from bauxite (hydrated aluminum-oxide) deposits, located mostly in Guinea, Australia, and South America. The United States does not have bauxite deposits; it imports it from Brazil, Guinea, and Jamaica. Total world aluminum production is approximately 30 million metric tons per year. U.S. aluminum production is less than U.S. aluminum consumption. Leading aluminum producers are China and Russia. A major part (3 million metric tons per year) of the U.S. aluminum supply comes from recycling.

38. Interpret the relationship between aluminum's resistance to corrosion and its use in transportation vehicles.

39. Propose a plan for how the United States can increase aluminum production without increasing the amount it imports.

40. Predict the possible effects an increase in U.S. production would have on Guinea, Jamaica, and China.

CUMULATIVE REVIEW

41. How do different isotopes of an element differ from each other? **(Chapter 3)**

42. Why is an understanding of the study of Earth science important to us as residents of Earth? **(Chapter 1)**

30. Petroleum is neither solid nor inorganic; wood and coal are organic, steel and concrete don't occur naturally; and glass (whether natural or synthetic) does not have a crystalline structure. Window glass and glass utensils do not occur naturally.

31. Gold has a higher density. By weighing the samples and/or floating the samples in water, the prospectors would be able to distinguish gold from pyrite.

32. Abundance or rarity, cost of production, and supply and demand will determine the cost of this new gem.

33. The positive metal ions such as K or Al bond the sheets together.

34. It has a sheetlike cleavage in one direction.

CONCEPT MAPPING

35. The map should have each of the choices as a separate branch with properties of each group added to the branch.

CHALLENGE QUESTION

36. The oxygen atoms of one side should be bonded to other oxygen atoms in a ringlike structure.

WRITING IN ▶ Earth Science

Rubric

37. Answers will vary. Students' lists could include a pocket-knife, plastic bottle of HCl, magnifying lens, penny (or piece of copper) pencil and paper, and a small streak plate.

DBQ Document-Based Questions

Data obtained from: Plunkert, P.A. 2005. Mineral resource of the month: aluminum. *Geotimes* 50:57.

38. Because many vehicles are outside and exposed to water and air, resistance to corrosion ensures that rust is kept to a minimum and the vehicles last a long time.

39. The United States could increase the amount of aluminum it recycles.

40. It would affect jobs in these countries because less Al would need to be produced. The smaller countries would be affected more. The price could also be driven up because as less Al is needed and possibly less is produced, a price increase would help keep their profits the same.

CUMULATIVE REVIEW

41. by the number of neutrons in their nuclei and mass number

42. It helps us understand how to protect the world in which we live to ensure species survival and a good quality of life.

MULTIPLE CHOICE

1. C
2. D
3. C
4. A
5. A
6. C
7. A
8. D
9. A
10. B

CUMULATIVE STANDARDIZED TEST PRACTICE

MULTIPLE CHOICE

1. What is the second most abundant element in Earth's crust?
   A. nitrogen
   B. oxygen
   C. silicon
   D. carbon

Use the table below to answer Questions 2 and 3.

Mineral Characteristics			
Mineral	Hardness	Specific Gravity	Luster/Color
Feldspar	6–6.5	2.5–2.8	nonmetallic/colorless or white
Fluorite	4	3–3.3	nonmetallic/yellow, blue, purple, rose, green, or brown
Galena	2.5–2.75	7.4–7.6	metallic/grayish black
Quartz	7	2.65	nonmetallic/colorless in pure form

2. What is the hardest mineral in the table?
   A. feldspar
   B. fluorite
   C. galena
   D. quartz

3. Which mineral has a metallic luster?
   A. feldspar
   B. fluorite
   C. galena
   D. quartz

4. What can be inferred about an isotope that releases radiation?
   A. It has unstable nuclei.
   B. It has stable nuclei.
   C. It has the same mass number as another element.
   D. It is not undergoing decay.

5. How do electrons typically fill energy levels?
   A. from lowest to highest
   B. from highest to lowest
   C. in no predictable pattern
   D. all in one energy level

6. Which is the most reliable clue to a mineral's identity?
   A. color
   B. streak
   C. hardness
   D. luster

Use the table below to answer Questions 7 and 8.

Mineral	Hardness
Talc	1
Gypsum	2
Calcite	3
Fluorite	4
Apatite	5
Feldspar	6
Quartz	7
Topaz	8
Corundum	9
Diamond	10

7. Which mineral will scratch feldspar but not topaz?
   A. quartz
   B. calcite
   C. apatite
   D. diamond

8. What can be implied about diamond based on the table?
   A. It is the heaviest mineral.
   B. It is the slowest mineral to form.
   C. It has the most defined crystalline structure.
   D. It cannot be scratched by any other mineral.

9. A well-planned experiment must have all of the following EXCEPT
   A. technology
   B. a control
   C. a hypothesis
   D. collectible data

10. What name is given to the imaginary line circling Earth halfway between the North and South poles?
    A. prime meridian
    B. equator
    C. latitude
    D. longitude

Online Test Practice

SHORT ANSWER

Use the conversion factor and table below to answer Questions 11–13.

1.0 carat = 0.2 grams

Diamond	Carats	Grams
Uncle Sam: largest diamond found in United States	40.4	?
Punch Jones: second largest; named after boy who discovered it	?	6.89
Theresa: discovered in Wisconsin in 1888	21.5	4.3
2001 diamond production from western Australia	21,679,930	?

11. List the three diamonds from least to greatest according to carats, and list the carats.

12. How many kilograms of diamonds were produced in western Australia in 2001?

13. Why would a diamond excavator want to convert the diamond measurement from carats to grams?

14. Why are map scales important parts of a map?

15. Discuss how a scientist might use a Landsat satellite image to determine the amount of pollution being produced by a city.

16. Why might a mineral no longer be classified as an ore?

READING FOR COMPREHENSION

Silicon Valley

Silicon (Si) is the second most abundant element in Earth's crust, but we didn't hear much about it until Silicon Valley. It is present in measurable amounts in nearly every rock, in all natural waters, as dust in the air, in the skeletons of many plants and some animals, and even in the stars. Silicon is never found in the free state like gold or silver, but is always with oxygen (O), aluminum (Al), magnesium (Mg), calcium (Ca), sodium (Na), potassium (K), iron (Fe), or other elements in combinations called the silicates. Silicates are the largest and most complicated group of minerals. Silicon is dull gray in appearance and has a specific gravity of 2.42. It has valence electrons like carbon (C) and can form a vast array of chemical compounds like silicon carbide abrasive, silicon rubber and caulking, oils and paints. Pure silicon is used in semiconductors, as solar panels to generate electricity from light, and in microchips for transistors.

Information obtained from: Ellison, B. Si and SiO_2...or what a difference a little O makes. (online resource accessed October 2006.)

17. According to the text, what is the most challenging aspect of silicon?
   A. It has valence electrons.
   B. It is dull gray in appearance.
   C. It is never found in its free state.
   D. It is present in many places.

18. Which is NOT a use of silicon as a chemical compound given in this passage?
   A. silicon rubber and caulking
   B. silicon carbide abrasive
   C. microchips for transistors
   D. oils and paints

19. Why can it be said that the occurrence of silicon is universal?

NEED EXTRA HELP?

If You Missed Question . . .	1	2	3	4	5	6	7	8	9	10	11	12	13	14	15	16
Review Section . . .	4.2	4.1	4.1	3.1	3.6	4.1	4.1	4.1	1.2	2.1	4.2	4.2	1.2	2.2	2.3	3.2

SHORT ANSWER

11. Theresa (21.5), Punch Jones ($6.89 \div 0.2 = 34.45$), Uncle Sam (40.4)
12. 21,679,930 carats $\times$ 0.2 = 4,336 kg
13. He would want to convert the measurement because grams are a more commonly known and understood unit of measure and would help people more easily understand the mass of the diamond.
14. Because a map is a scaled-down representation of part of Earth, a map scale gives the ratio between distances on the map and actual distances on Earth's surface. Without a map scale, a person would not be able to determine actual distance.
15. The Landsat satellite is able to show areas of warmth in different colors. By taking an image of the city with the Landsat satellite and comparing it to surrounding areas, a scientist would be able to see how much more warmth the pollution is causing than in the surrounding, less polluted areas.
16. If the cost of separating waste material from the ore becomes more expensive than the ore itself, or if the supply or demand for a particular ore changes, then that mineral might no longer be classified as an ore.

READING FOR COMPREHENSION

17. C
18. C
19. Silicon is found both on Earth and in the stars.

CHAPTER 5 Igneous Rocks

BIGIDEA **Igneous rocks were the first rocks to form as Earth cooled from a molten mass to the crystalline rocks of the early crust.**

ESSENTIAL QUESTIONS	NATIONAL STANDARDS	RESOURCES TO ASSESS MASTERY
SECTION 1 **What are igneous rocks?** **1.** How do igneous rocks form? **2.** How can the composition of magma be described? **3.** What are the factors that affect how rocks melt and crystallize? 2 sessions 1 block	UCP.1–5; A.1–2; B.2; D.1–3; F.3; G.3	**Entry-Level Assessment** Focus Transparency 12, p. 112 **Progress Monitoring** Caption Question, p. 114 Reading Check, pp. 113, 114 Section Review, p. 117
SECTION 2 **Classification of Igneous Rocks** **1.** What are the different types and textures of igneous rocks? **2.** How do the cooling rates affect the grain sizes in igneous rocks? **3.** What are some of the uses of igneous rocks? 5 sessions 2.5 blocks	UCP.1–3, UCP.5; A.1–2; B.2; D.1–2; E.1; F.3	**Entry-Level Assessment** Focus Transparency 13, p. 118 **Progress Monitoring** Caption Question, pp. 118, 121 Reading Check, pp. 120, 121 Section Review, p. 123 **Summative Assessment** Chapter Assessment, p. 127 Exam*View® Assessment Suite* CD-ROM

Go online!
connectED.mcgraw-hill.com

LEVELED RESOURCES	LAB MATERIALS
Science Notebook 5.1 OL **Chapter FAST FILE Resources:** MiniLab Worksheet, p. 50 OL Study Guide, p. 59 BL **Lab Resources:** Laboratory Manual, p. 33 OL **Transparencies:** Section Focus Transparency 12 EL Teaching Transparency 11 OL EL	LaunchLAB p. 110 / **10 min** samples of granite, magnifying lens or microscope MiniLAB p. 115 / **15 min** igneous rock samples, paper, pencil
Science Notebook 5.2 OL **Chapter FAST FILE Resources:** GeoLab Worksheet, p. 51 OL Study Guide, p. 62 BL **Lab Resources:** Laboratory Manual, p. 37 OL **Transparencies:** Section Focus Transparency 13 EL Teaching Transparency 12 OL EL	GeoLAB p. 125 / **90 min** clean, plastic petri dishes; saturated alum solution; 200-mL glass beaker; magnifying lens; dark-colored construction paper; thermometer; paper towels; water; hot plate

LabManager™
Customize any lab with the LabManager™ CD-ROM.

ADDITIONAL RESOURCES

Technology:
- eTeacherEdition
- eStudent Edition
- Classroom Presentation Toolkit
- Exam*View*® *Assessment Suite* CD-ROM
- LabManager™ CD-ROM
- What's EARTH SCIENCE Got To Do With It? DVD
- Weather Classroom Video DVD

Assessment:
- Blueprints for Success
- Chapter Test (Scaffolded)
- Earth Science Standardized Test Practice

Lab Resources:
- Exploring Environmental Problems Laboratory Manual

BL Below Level | OL On Level | AL Advanced Learners | EL English Learners | COOP LEARN Cooperative Learning

CHAPTER 5

LaunchLAB

iLab Station | Rubric

How are minerals identified?

Teaching Strategies

- A local company that sells granite countertops or flooring might be willing to give you broken samples or scraps that have polished surfaces.
- Student sketches will vary. Typically, granite contains four to five different minerals. Quartz is clear to gray, potassium feldspar is pinkish, biotite is black, and plagioclase feldspar is usually white. Another black mineral, hornblende, is often present. It looks similar to biotite. Biotite, however, can be scratched by a metal pin; hornblende cannot. Most of the mineral crystals will be about the same size and blocky in shape.

Procedure

1. Have students read and complete the lab safety form and follow the procedure below.
2. Examine a **sample of granite** from a distance of about 1 m. Record your observations.
3. Use a **magnifying lens** or **microscope** to observe the granite sample. Record your observations.

Analysis

1. **Illustrate** what you saw through the magnifying glass or microscope. Include a scale for your drawing. Students answers will vary. Drawings might show 3 to 5 different minerals and interlocking crystals.
2. **List** the different minerals that you observed in your sample. Possible minerals include quartz, potassium and plagioclase feldspar, biotite, and hornblende.
3. **Describe** the sizes and shapes of the mineral crystals. The sizes will vary depending on the sample provided. Students should notice that the crystals fit together like puzzle pieces and might be irregular in outline. Flat crystal faces or cleavage surfaces can also be observed.
4. **Describe** any evidence that suggests that these crystals formed from molten rock. Students' answers will vary. Students might note the interlocking crystals as evidence of formation from cooling magma.

CHAPTER 5

Igneous Rocks

BIGIDEA Igneous rocks were the first rocks to form as Earth cooled from a molten mass to the crystalline rocks of the early crust.

SECTIONS

1 What are igneous rocks?

2 Classification of Igneous Rocks

LaunchLAB

iLab Station

How are minerals identified?

Igneous rocks are composed of different types of minerals. In this lab, you will learn that it is often possible to identify the different minerals in a sample of rock.

FOLDABLES® Study Organizer

Types of Igneous Rocks

Make a pocket book using the labels shown. Use it to organize your notes on the types of igneous rocks and how they form.

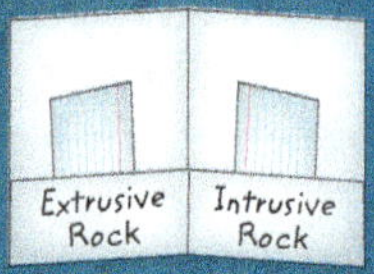

When completed, this monument of Chief Crazy Horse will be more than 170 m tall and 195 m long. Nearly 10,000,000 metric tons of rock have already been blasted away.

Assessment

Knowledge Ask students the following questions: Which mineral in granite has the greatest hardness? quartz Based on the minerals you identified, what are the most abundant elements in granite? silicon, oxygen, and aluminum

LabManager™ Customize any lab with the LabManager™ CD-ROM.

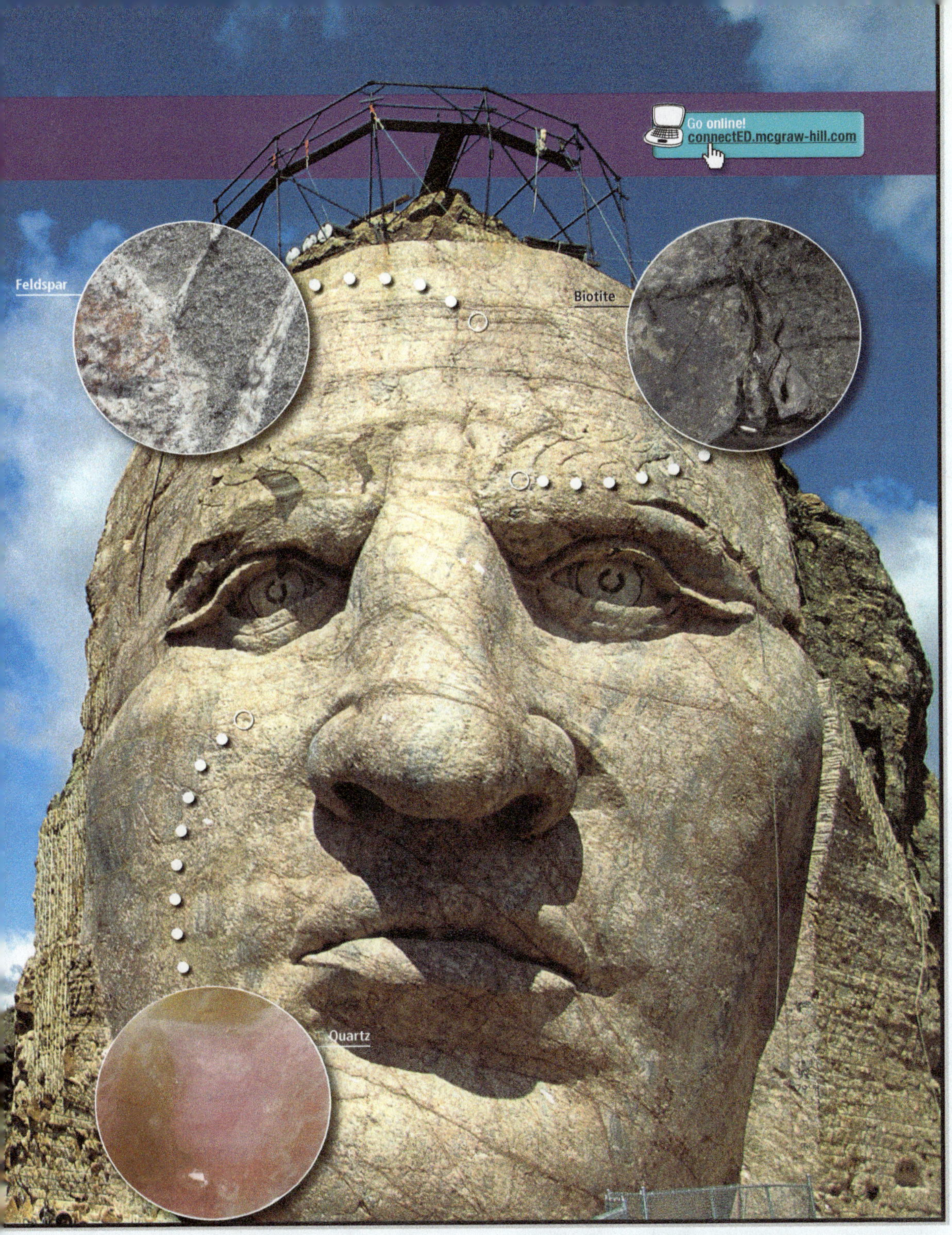

Introduce the BIG IDEA

Rock Formation Show students a container of gravel or sand. Ask them to write what they think would happen if the material was heated to its melting temperature and then allowed to slowly cool. After students have completed this chapter, have them review their predictions.

Teacher Content Support

Chief Crazy Horse Memorial
The pegmatite granite that makes up this mountain exhibits several igneous rock characteristics, such as crystal size variation and mineral content. Ask students what minerals are identified in the photos. quartz, feldspar, and biotite What characteristics or features do they observe in the photos? Possible answer: The minerals vary in color and appearance. The rock can be sculpted. Different mineral veins run across the face of Chief Crazy Horse. Tell students the sculptors use detonator cord to blast out the general shape of the sculpture. Why might explosives need to be used instead of a hammer and a chisel? The minerals quartz and feldspar are hard minerals and require much energy to sculpt. Tell students that once the general shape is apparent, the sculptors use a torch to smooth the rock surface. Along with smoothing the surface, the heat reacts with the minerals in the granite enhancing their natural colors. Ask students to look at the photos of the minerals and determine what main color they think is enhanced in the sculpture. A pinkish rose color is enhanced in the sculpture due to the pink color of the rose quartz and potassium feldspar.

SECTION 1

1 Focus

Focus Transparency

Before presenting the lesson, project **Section Focus Transparency 12** and have students answer the accompanying questions. EL

MAINIDEA

Molten Earth Show students an artist's rendering of what early Earth might have looked like. Discuss with students that when Earth first formed, it was a mass of molten material. Ask: What happened to this material as Earth cooled? The material cooled and crystallized to form the first rocks, igneous rocks. OL

2 Teach

Enrichment

Glass Production Have students research how glass is made and answer the following questions: What minerals or materials are used? Glass is often made from quartz sand and sodium. How is the process of glass formation similar to the formation of an igneous rock? During glass production, quartz is melted, poured into molds, and cooled quickly. The similarity is that both glass and igneous rocks form from molten material. OL

SECTION 1

Essential Questions

- How do igneous rocks form?
- How can the composition of magma be described?
- What are the factors that affect how rocks melt and crystallize?

Review Vocabulary

silicate: mineral that contains silicon and oxygen, and usually one or more other elements

New Vocabulary

lava
igneous rock
partial melting
Bowen's reaction series
fractional crystallization

What are igneous rocks?

MAINIDEA Igneous rocks are the rocks that form when molten material cools and crystallizes.

EARTH SCIENCE 4 YOU At any given point in time, igneous rocks are forming somewhere on Earth. The location and the conditions that are present determine the types of igneous rocks that form.

Igneous Rock Formation

If you live near an active volcano, you can literally watch igneous rocks form. A hot, molten mass of rock can solidify into solid rock overnight. As you have learned, magma is molten rock below Earth's surface. **Lava** is magma that flows out onto Earth's surface. **Igneous rocks** form when lava or magma cools and crystallizes.

In the laboratory, most rocks must be heated to temperatures of 800°C to 1200°C before they melt. In nature, these temperatures are present in the upper mantle and lower crust. Where does this heat come from? Scientists theorize that the remaining energy from Earth's molten formation and the heat generated from the decay of radioactive elements are the sources of Earth's thermal energy.

Composition of magma The type of igneous rock that forms depends on the composition of the magma. Magma is often a slushy mix of molten rock, dissolved gases, and mineral crystals. The common elements present in magma are the same major elements that are in Earth's crust: oxygen (O), silicon (Si), aluminum (Al), iron (Fe), magnesium (Mg), calcium (Ca), potassium (K), and sodium (Na). Of all the compounds present in magma, silica is the most abundant and has the greatest effect on magma characteristics. As summarized in **Table 1,** magma is classified as basaltic, andesitic, or rhyolitic, based on the amount of silica it contains. Silica content affects melting temperature and impacts a magma's viscosity, or resistance to flow. Rhyolitic magma has a higher viscosity than basaltic magma.

Explore **magma composition with an interactive table.** Concepts In Motion

Table 1 Types of Magma

Group	Silica Content	Example Location
Basaltic	45–52%	Hawaiian Islands
Andesitic	52–66%	Cascade Mountains, Andes Mountains
Rhyolitic	more than 66%	Yellowstone National Park

DIFFERENTIATED INSTRUCTION

Visually Impaired Help students to understand the interlocking nature of crystalline rocks by providing them with puzzle pieces of various sizes. Have another student help to fit the pieces together and to visualize that there are large crystals connected together, small crystals connected together, and mixtures of large and small crystals connected together.

Once magma is free of the overlying pressure of the rock layers around it, dissolved gases are able to escape into the atmosphere. Thus, the chemical composition of lava is slightly different from the chemical composition of the magma from which it developed.

Magma formation Magma can be formed either by melting of Earth's crust or by melting within the mantle. The four main factors involved in the formation of magma are temperature, pressure, water content, and the mineral content of the crust or mantle. Temperature generally increases with depth in Earth's crust. This temperature increase, known as the geothermal gradient, is plotted in **Figure 1.** Oil-well drillers and miners have firsthand experience with the geothermal gradient. Drill bits, such as the one shown in **Figure 2,** can encounter temperatures in excess of 200°C when drilling deep oil wells.

Pressure also increases with depth. This is a result of the weight of overlying rock. Laboratory experiments show that as pressure on a rock increases, its melting point also increases. Thus, a rock that melts at 1100°C at Earth's surface will melt at 1400°C at a depth of 100 km.

The third factor that affects the formation of magma is water content. Rocks and minerals often contain small percentages of water, which changes the melting point of the rocks. As water content increases, the melting point decreases.

☑ READING CHECK **List** the main factors involved in magma formation.

Mineral content In order to better understand how the types of elements and compounds present give magma its overall character, it is helpful to discuss this fourth factor in more detail. Different minerals have different melting points. For example, rocks such as basalt, which are formed of olivine, calcium feldspar, and pyroxene (pi RAHK seen), melt at higher temperatures than rocks such as granite, which contain quartz and potassium feldspar. Granite has a melting point that is lower than basalt's melting point because granite contains more water and minerals that melt at lower temperatures. In general, rocks that are rich in iron and magnesium melt at higher temperatures than rocks that contain higher levels of silicon.

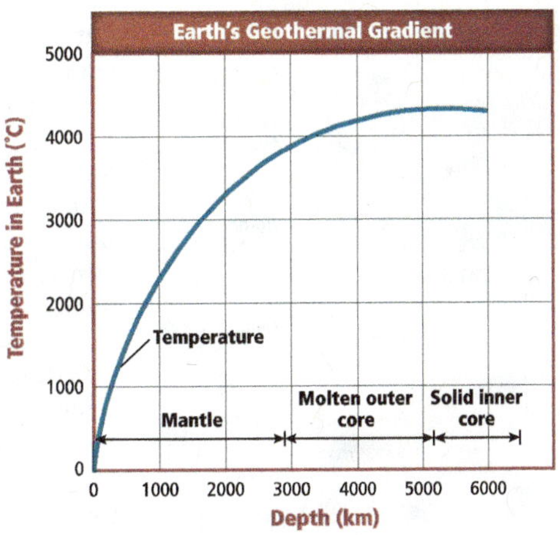

Figure 1 The average geothermal gradient in the crust is about 25°C/km, but scientists think that it drops sharply in the mantle to as low as 1°C/km.

Figure 2 The temperature of Earth's upper crust increases with depth by about 30°C for each 1 km. At a depth of 3 km, this drill bit will encounter rock that is close to the temperature of boiling water.

Teacher Content Support

Dangers of High Pressure

Deep-mining operations are designed with pressure and temperature conditions in mind. Both pressure and temperature increase with depth. While the high temperatures make working conditions uncomfortable, the high pressures caused by the rocks above are far more dangerous. Lack of roof and wall support can cause rock to shatter and blow out into the mine tunnel from the roof and walls. Steel plates are sometimes bolted to the roof and walls to control this problem.

Use an Analogy

Pressure and Deep Water

Relate increasing pressure with depth inside Earth to increasing pressure with depth in the ocean. Tell students that when someone dives deep into the ocean, the weight of the overlying water presses in on the diver's eardrums and sinus membranes, causing uncomfortable pressure. In a similar way, the weight of overlying rocks inside Earth presses down on the rocks below. BL

Discussion

Wet Rocks Discuss the factors that affect how rocks melt, particularly water content. Ask students to infer how water can exist in rocks at the high temperatures required for melting. The more water contained in the minerals that make up the rocks, the lower the temperature needed to melt the rocks. Water can be locked inside the crystal structure of the minerals. OL AL

☑ READING CHECK The four factors are temperature, pressure, water content, and mineral content.

ACROSS THE CURRICULUM

Chemistry Review what happens to atoms and molecules when matter changes state from a solid to a liquid. Ask students to explain why increasing pressure causes the melting temperatures of rocks to increase. When matter changes from a solid to a liquid, atoms and molecules loosen their bonds and move farther apart. High pressure makes it more difficult for this to occur, and thus more thermal energy—or a higher melting temperature—is required for the matter to change state. OL AL

Identify Misconceptions

Students might think that a rock, like an ice cube, melts all at once.

Uncover the Misconception Ask students to describe what might happen if the granite sample from the Launch Lab was slowly heated to 1200°C.

Demonstrate the Concept Explain that minerals have different melting temperatures. Minerals with low melting temperatures will melt before minerals with higher melting temperatures. Because rock is composed of a variety of minerals, which melt at different temperatures, rocks can be partially melted, as shown in **Figure 3.**

Assess New Knowledge After you complete this section, have students list the melting order of the minerals in granite.

Collaborative Learning

Different Melting Points Have groups of students brainstorm other mixtures besides wax and ice that might model the partial melting behavior of rocks. One example is ice cream with chocolate chips. BL COOP LEARN

Model

Partial Melting Have students perform the analogy about wax and ice described in the student text. Students should freeze bits of colored wax and water in an ice cube tray, then melt the resulting ice cubes in a beaker placed on a hot plate. The wax and water will melt at different rates, thus modeling partial melting. OL

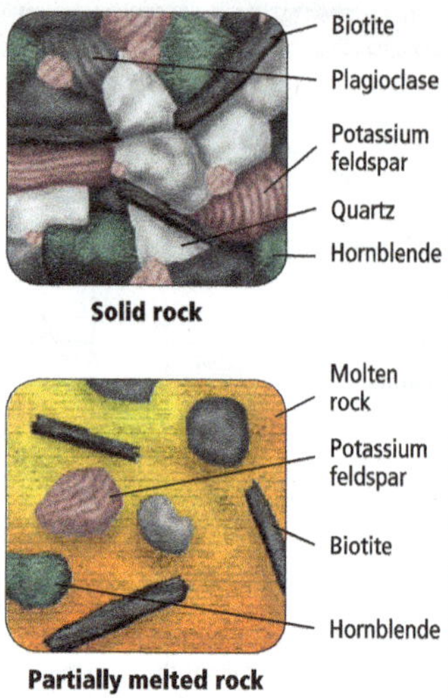

Figure 3 As the temperature increases in an area, minerals begin to melt.
Determine *What can you suggest about the melting temperature of quartz based on this diagram?*

Partial melting Suppose you froze melted candle wax and water in an ice cube tray. If you took the tray out of the freezer and left it at room temperature, the ice would melt, but the candle wax would not. This is because the two substances have different melting points. Rocks melt in a similar way because the minerals they contain have different melting points. Not all parts of a rock melt at the same temperature. This explains why magma is often a slushy mix of crystals and molten rock. The process whereby some minerals melt at relatively low temperatures while other minerals remain solid is called **partial melting.** Partial melting is illustrated in **Figure 3.** As each group of minerals melts, different elements are added to the magma mixture thereby changing its composition. If temperatures are not high enough to melt the entire rock, the resulting magma will have a different composition than that of the original rock. This is one way in which different types of igneous rocks form.

READING CHECK **Summarize** the formation of magma that has a different chemical composition from the original rock.

Bowen's Reaction Series

In the early 1900s, Canadian geologist N. L. Bowen demonstrated that as magma cools and crystallizes, minerals form in predictable patterns in a process now known as the **Bowen's reaction series. Figure 4** illustrates the relationship between cooling magma and the formation of minerals that make up igneous rock. Bowen discovered two main patterns, or branches, of crystallization. The right-hand branch is characterized by a continuous, gradual change of mineral compositions in the feldspar group. An abrupt change of mineral type in the iron-magnesium groups characterizes the left-hand branch.

Figure 4 On the left side of Bowen's reaction series, minerals rich in iron and magnesium change abruptly as the temperature of the magma decreases.
Compare *How does this compare to the feldspars on the right side of the diagram?*

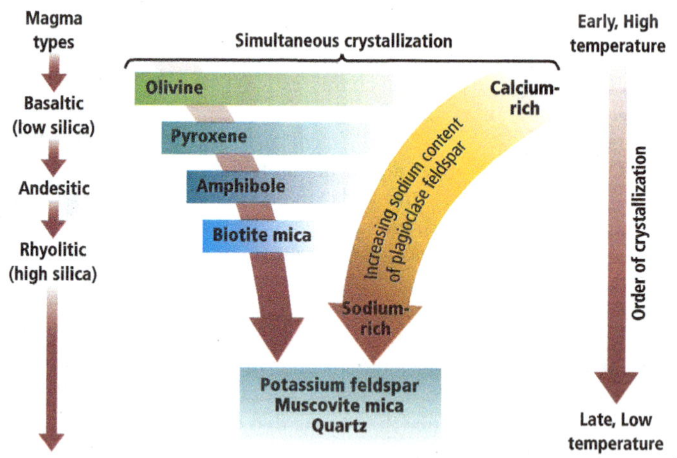

EARTH SCIENCE JOURNAL

Melting and Crystallization Have students each write an explanation of why the first minerals to crystallize are the last to melt when a rock is heated. The minerals that crystallize first have high melting temperatures. It takes a great deal of thermal energy to keep these minerals in a molten state. Thus, when temperatures decrease, these minerals are the first to crystallize. Conversely, temperatures must be high for the minerals to melt. OL

Caption Question Fig. 3 Quartz melts at a lower temperature than the other minerals present.

READING CHECK This magma will form if the temperature is not hot enough to melt all of the rock. In this case, the magma will not contain the same elements—and thus minerals—as the rock.

Caption Question Fig. 4 Feldspars change gradually.

Iron-rich minerals The left branch of Bowen's reaction series represents the iron-rich minerals. These minerals undergo abrupt changes as magma cools and crystallizes. For example, olivine is the first mineral to crystallize when magma that is rich in iron and magnesium begins to cool. When the temperature decreases enough for a completely new mineral, pyroxene, to form, the olivine that previously formed reacts with the magma and is converted to pyroxene. As the temperature decreases further, similar reactions produce the minerals amphibole and biotite mica.

Feldspars In Bowen's reaction series, the right branch represents the plagioclase feldspars, shown in **Figure 5.** These undergo a continuous change of composition. As magma cools, the first feldspars to form are rich in calcium. As cooling continues, these feldspars react with magma, and their calcium-rich compositions change to sodium-rich compositions. In some instances, such as when magma cools rapidly, the calcium-rich crystals are unable to react completely with the magma. The result is a zoned crystal with a calcium-rich core and sodium-rich outer layers.

Figure 5 Plagioclase feldspars undergo a continuous change of composition in Bowen's reaction series.

Fractional Crystallization

When magma cools, it crystallizes in the reverse order of partial melting. That is, the first minerals that crystallize from magma are the last minerals that melt. This process, called **fractional crystallization,** is similar to partial melting in that the composition of magma can change. In this case, however, early formed crystals are removed from the magma and cannot react with it. As minerals form and their elements are removed from the remaining magma, it becomes concentrated in silica as shown in **Figure 4.**

FOLDABLES®
Incorporate information from this section into your Foldable.

MiniLAB

iLab Station

Compare Igneous Rocks

How do igneous rocks differ? Igneous rocks have many different characteristics. Color and crystal size are some of the features that differentiate igneous rocks.

Procedure

1. Read and complete the lab safety form.
2. Obtain a set of **igneous rock samples** from your teacher.
3. Carefully observe the following characteristics of each rock: overall color, crystal size, and, if possible, mineral composition.
4. Design a **data table** to record your observations.

Analysis

1. **Classify** your samples as forming from either basaltic, andesitic, or rhyolitic magma. [Hint: The more silica in the rock, the lighter it is in color.]
2. **Compare and contrast** your samples using the data from the data table. How do they differ? What characteristics do each of the groups share?
3. **Speculate** in which order the samples crystallized. [Hint: Use Bowen's reaction series as a guide.]

FOLDABLES® Rubric

MiniLAB

iLab Station Rubric

Purpose Students will compare and contrast the different characteristics of igneous rocks.

Process Skills observe and infer, compare and contrast, make and use tables, classify

Safety Precaution Approve lab safety forms before work begins.

Teaching Strategies

- Samples should be labeled. If materials are limited, have students work in small groups.
- Stress the need for students to examine the samples closely to observe differences.

Expected Results Data tables should have four to five columns that list the name of each sample, its color, grain size, texture, and mineral composition.

Analysis

1. Students' answers will vary depending on the samples provided. Basaltic: dark-colored rocks—basalt and gabbro; rhyolitic: light-colored rocks—granite and rhyolite; andesitic: the rocks that fall in between.
2. Differences include color, mineral content and crystal size. Similarities include the fact that they are all crystalline.
3. Basaltic rocks contain iron and magnesium, and according to Bowen's reaction series, these crystallize first. The order would have been basaltic, andesitic, then rhyolitic.

Assessment

Skill Have students arrange and classify their samples using a different characteristic, such as crystal size. Ask them which classification seems to work better.

LabManager™
Customize any lab with the LabManager™ CD-ROM.

VISUALIZING

Purpose

Students will explore a real-world example of fractional crystallization and crystal settling.

Enrichment

Demo Partial Melting Ask students to design ways to demonstrate how the concept of partial melting might be used to separate the components of a mixture. Designs will vary. Students might say they could carefully heat the mixture until one component melts, and then filter out the remaining solid portion. OL

Project

Order of Crystallization Have groups of students each make a poster of Bowen's reaction series. Then, have students choose one of the three types of magma and illustrate the crystallization sequence for an intrusive igneous rock formed from that magma. BL

Activity

Composition Changes Write the chemical formulas of the minerals olivine and augite on the board. Then, have students determine the compositional changes that would need to occur for olivine to be converted to augite. The chemical formula of olivine is $(Mg,Fe)_2SiO_4$. The chemical formula of augite is $(Ca,Na)(Mg,Fe,Al)(Si,Al)_2O_6$. For olivine to be converted to augite, additional Ca, Na, and Al would have to be added. AL COOP LEARN

VISUALIZING Fractional Crystallization and Crystal Settling

Figure 6 The Palisades Sill in the Hudson River valley of New York and New Jersey is a classic example of fractional crystallization and crystal settling. In this basaltic intrusion, small crystals formed in the chill zone as the outer areas of the intrusion cooled more quickly than the interior.

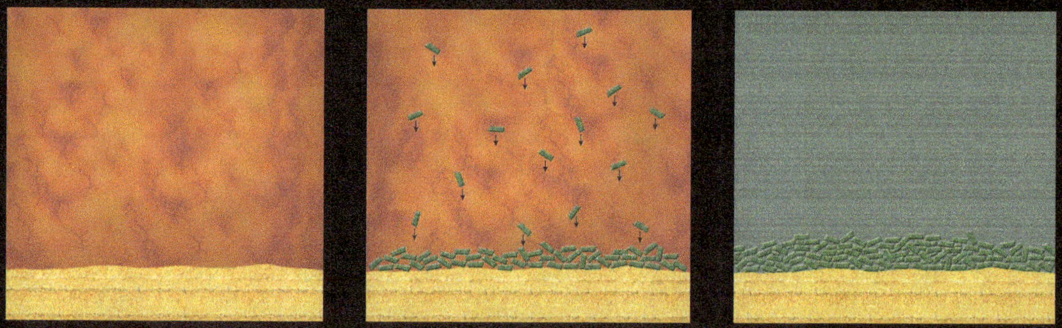

As magma in an intrusion begins to cool, crystals form and settle to the bottom. This layering of crystals is fractional crystallization.

View an **animation of the Palisades Sill.** Concepts In Motion

Demonstration

Separating Substances Choose one or two methods described in the Enrichment activity above and demonstrate it for the class. Ask students to predict the success of the method prior to starting. OL

EARTH SCIENCE JOURNAL

Original Composition Have pairs of students study **Figure 6** and note the composition of the different layers. Have them speculate what the original composition of the magma might have been and record their thoughts in their Earth science journals. The composition of the chill zone represents the original magma composition. Students might suggest minerals that are high in Fe and Mg, such as olivine and pyroxene.

As is often the case with scientific inquiry, the discovery of Bowen's reaction series led to more questions. For example, if olivine converts to pyroxene during cooling, why is olivine present in some rocks? Geologists hypothesize that, under certain conditions, newly formed crystals are separated from the cooling magma, and the chemical reactions between the magma and the minerals stop. This can occur when crystals settle to the bottom of the magma body, and when liquid magma is squeezed from the crystal mush. This results in the formation of two distinct igneous bodies with different compositions. **Figure 6** illustrates this process and the concept of fractional crystallization with an example from the Hudson River valley in New York and New Jersey. This is one way in which the magmas listed in **Table 1** are formed.

As fractional crystallization continues, and more and more crystals are separated from the magma, the magma becomes more concentrated in silica, aluminum, and potassium. For this reason, the last two minerals to crystallize out of a cooling body of magma are potassium feldspar and quartz. Potassium feldspar is one of the most common feldspars in Earth's crust. Quartz often occurs in veins, as shown in **Figure 7,** because it crystallizes while the last liquid portion of magma is squeezed into rock fractures.

■ **Figure 7** These quartz veins represent the last remnants of a magma body that cooled and crystallized.

SECTION 1 REVIEW

Section Self-Check

Section Summary

- Magma consists of molten rock, dissolved gases, and mineral crystals.
- Magma is classified as basaltic, andesitic, or rhyolitic, based on the amount of silica it contains.
- Different minerals melt and crystallize at different temperatures.
- Bowen's reaction series defines the order in which minerals crystallize from magma.

Understand Main Ideas

1. MAINIDEA **Predict** the appearance of an igneous rock that formed as magma cooled quickly and then more slowly.
2. **List** the eight major elements present in most magmas. Include the chemical symbol of each element.
3. **Summarize** the factors that affect the formation of magma.
4. **Compare and contrast** magma and lava.

Think Critically

5. **Predict** If the temperature increases toward the center of Earth, why is the inner core solid?
6. **Infer** the silica content of magma derived from partial melting of an igneous rock. Would it be higher, lower, or about the same as the rock itself? Explain.

WRITING IN ▶ Earth Science

7. A local rock collector claims that she has found the first example of pyroxene and sodium-rich feldspar in the same rock. Write a commentary about her claim for publication in a rock collector society newsletter.

Reinforcement

Order of Crystallization Ask students the following questions: Which elements are removed first from a solidifying magma? the elements present in the iron-magnesium minerals and the calcium-rich plagioclase In the wax-ice analogy, which material represents the quartz-rich portion of crystallizing magma? Why? the ice, because it crystallizes last BL

3 Assess

Check for Understanding

Reinforcement Ask students to describe how temperature and pressure would change as a mass of magma moved upward through Earth's crust. Temperature and pressure would both decrease. What effect would this have on crystallization? The decrease in pressure would lower the melting temperature of the molten material. The decrease in temperature would cause the magma to crystallize.

Reteach

Outline Have students use the headings in the textbook to outline the major points in this section.

Assessment

Skill Assess students' critical-thinking skills by having students answer the following question: In a rhyolitic magma with low levels of iron and magnesium, which minerals in Bowen's reaction series would you not expect to form? olivine, pyroxene, and amphibole

SECTION 1 REVIEW

1. The rock will have equal-sized, small crystals. Because the magma cooled quickly, there was no time for larger crystals to form. By the time it began to cool more slowly, there would have been no room for large crystals to form.
2. oxygen (O), silicon (Si), aluminum (Al), iron (Fe), magnesium (Mg), calcium (Ca), potassium (K), and sodium (Na)
3. temperature, pressure, water content, and mineral content; different combinations of these factors result in the formation of different magmas.
4. Magma is molten, forms belowground and is under pressure. Lava is molten, accumulates aboveground, is not under pressure, and differs in chemical composition from the magma from which it formed because the gases that were dissolved under pressure have escaped.
5. Pressure is too great here and temperature is not high enough to melt or keep the core molten.
6. Because quartz melts first, the silica content of the magma would be higher than the rock itself. Thus the magma would have more silica at the outset.
7. Possible comments might include that although coexistence is possible, it is unlikely based on Bowen's reaction series and their temperatures of crystallization. Ca-rich feldspar and pyroxene is the more likely combination.

SECTION 2

1 Focus

Focus Transparency

Before presenting the lesson, project **Section Focus Transparency 13** and have students answer the accompanying questions. EL

MAINIDEA

How to Classify Show students a selection of different-colored breakfast cereals ranging in size from small grains to large flakes. Ask how these samples can be categorized. by color, by grain size Explain to students that igneous rocks are classified in a similar way.

2 Teach

Tie to Previous Knowledge

Bowen's Reaction Series Have students refer back to Bowen's reaction series in the previous section. Ask students to explain which type of igneous rock would have the highest melting temperature: a granitic, intermediate, basaltic, or ultrabasic rock. an ultrabasic rock, because it contains the minerals with the highest melting temperatures OL EL

Caption Question Fig. 8 The gabbro is comprised of dark-colored minerals, the granite contains light-colored minerals, and the diorite has both dark- and light-colored minerals.

SECTION 2

Classification of Igneous Rocks

MAINIDEA Classification of igneous rocks is based on mineral composition and texture.

Essential Questions

- What are the different types and textures of igneous rocks?
- How do the cooling rates affect the grain sizes in igneous rocks?
- What are some of the uses of igneous rocks?

Review Vocabulary

fractional crystallization: a sequential process during which early formed crystals are removed from the melt and do not react with the remaining magma.

New Vocabulary

intrusive rock
extrusive rock
basaltic rock
granitic rock
texture
porphyritic texture
vesicular texture
pegmatite
kimberlite

EARTH SCIENCE 4 YOU **Many statues, floors, buildings, and countertops have something in common. Many of them are made of the popular rock type granite—one of the most abundant rocks in Earth's crust.**

Mineral Composition of Igneous Rocks

Igneous rocks are broadly classified as intrusive or extrusive. When magma cools and crystallizes below Earth's surface, **intrusive rocks** form. If the magma is injected into the surrounding rock, it is called an igneous intrusion. Crystals of intrusive rocks are generally large enough to see without magnification. Magma that cools and crystallizes on Earth's surface forms **extrusive rocks.** These are sometimes referred to as lava flows or flood basalts. The crystals that form in these rocks are small and difficult to see without magnification. Geologists classify igneous rocks by their mineral compositions. In addition, physical properties such as grain size and texture serve as clues for the identification of various igneous rocks.

Igneous rocks are classified according to their mineral compositions. **Basaltic rocks,** also called mafic rocks, are dark-colored, have lower silica contents, and contain mostly plagioclase and pyroxene. **Granitic rocks,** or felsic rocks, are light-colored, have high silica contents, and contain mostly quartz and feldspar. Rocks that have a composition of minerals that is somewhere in between basaltic and granitic are called intermediate rocks. They consist mostly of plagioclase feldspar and hornblende. **Figure 8** shows examples from these three main compositional groups of igneous rocks: gabbro is basaltic, granite is granitic, and diorite is intermediate. A fourth category called ultrabasic, or ultramafic, contains rocks with only iron-rich minerals such as olivine and pyroxene and are always dark. **Figure 9** summarizes igneous rock identification.

Figure 8 Differences in magma composition can be observed in the rocks that form when the magma cools and crystallizes.
Observe *Describe the differences you see in these rocks.*

DIFFERENTIATED INSTRUCTION

English Learners Have students use a dictionary to look up the origins of the words *intrusive* and *extrusive.* Ask students to explain why these terms aptly describe the two types of igneous rock. The Latin meaning of *intrusive* is *to thrust in,* while the Latin meaning of *extrusive* is *to thrust out.* The words are appropriate because intrusive rocks are formed inside Earth and extrusive rocks are formed on Earth's surface.

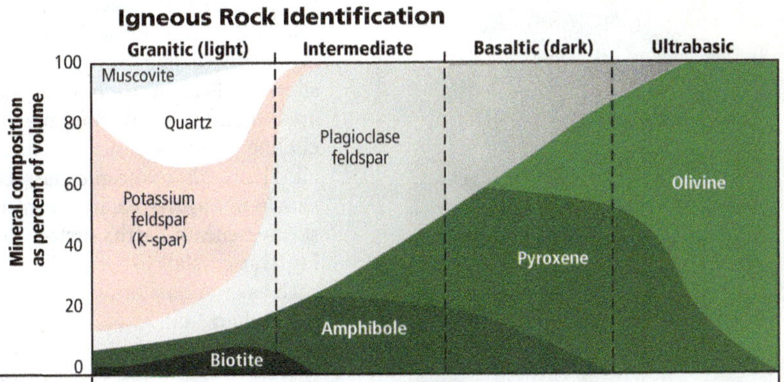

Origin	Texture	Rock Names			
Intrusive	coarse-grained	granite	diorite	gabbro	peridotite
	porphyritic	porphyritic rhyolite or granite	porphyritic andesite or diorite	porphyritic basalt or gabbro	
Extrusive	fine-grained	rhyolite	andesite	basalt	
	glassy	obsidian			
	vesicular	pumice		scoria (vesicular basalt)	

■ **Figure 9** Rock type can be determined by estimating the relative percentages of minerals in the rocks.

Texture

In addition to differences in their mineral compositions, igneous rocks differ in the sizes of their grains or crystals. **Texture** refers to the size, shape, and distribution of the crystals or grains that make up a rock. For example, as shown in **Figure 10,** the texture of rhyolite can be described as fine-grained, while granite can be described as coarse-grained. The difference in crystal size can be explained by the fact that one rock is extrusive and the other is intrusive.

■ **Figure 10** Rhyolite, granite, and obsidian have different textures because they formed in different ways. Obsidian's glassy texture is a result of rapid cooling.

Interpret the Illustration

Mineral Composition Have students study **Figure 9** and describe the mineral compositions of several different types of igneous rocks. For instance, granite, an intrusive igneous rock contains potassium feldspar, quartz, plagioclase feldspar, biotite, and amphibole. **OL**

Discussion

Compositional Differences Ask students whether every diorite has the same composition. no Have groups of students use **Figure 9** to list the mineral compositions of two rocks that represent the extreme range of the diorite group. Rock 1 (at the dashed line between granitic and intermediate): 8% quartz, 75% total feldspar, 5% biotite, 12% amphibole, 0% pyroxene; Rock 2 (at the dashed line between intermediate and basaltic): 0% quartz, 52% total feldspar, 0% biotite, 18% amphibole, 30% pyroxene. **OL**

Use Science Terms

Alternate Terminology The terms *felsic* and *mafic* are also used to describe rocks of granitic and basaltic compositions, respectively. Have students research these words and explain how they are applied. *Felsic* is a combination of *Fe* for feldspar and *Si* for silica. It is applied to rocks having abundant light-colored minerals. *Mafic* is used to describe rocks that contain abundant ferromagnesium minerals. These minerals contain magnesium (Mg) and iron (Fe) and are dark-colored. **AL**

IN THE FIELD

Early Tools Archaeologists have learned much about the cultures of ancient Native Americans through the study of tools made of obsidian. The early Native Americans for instance, used this volcanic glass to butcher game and to make arrowheads, spear points, and large ceremonial objects. Archaeological records show that obsidian tools were used more than 10,000 years ago. Scientists have found evidence of vast trade networks that ranged across the United States.

Teacher Content Support

Types of Volcanic Glass Volcanic glass contains no mineral grains and therefore cannot be classified based on mineral percentage. Chemical analysis is used to classify volcanic glasses. Some typical chemical compositions are as follows: Rhyolite glass usually has around 76 percent SiO_2, 12 percent Al_2O_3, 1 percent FeO, and 0.1 percent MgO. Andesite glass is about 58 percent SiO_2, 17 percent Al_2O_3, 4 percent FeO, and 3 percent MgO. Basaltic glass has around 46 percent SiO_2, 9 percent Al_2O_3, 10 percent FeO, and 15 percent MgO.

Enrichment

Porphyritic Composition Show students a porphyritic basalt with large crystals of plagioclase feldspar surrounded by small crystals of feldspar. Have students infer how the compositions of the two different-sized feldspars might differ. The larger crystals might be more rich in calcium if they did not have time to react with the magma while the finer-grained crystals were forming. **AL**

Discussion

Porphyry Formation Ask students which mineral would more likely form porphyritic textures and why: plagioclase feldspar or potassium feldspar. Plagioclase feldspar would be more likely to form porphyritic textures; it crystallizes earlier. The large crystals present in porphyritic textures must form early in the crystallization sequence. **AL**

GeoLAB

iLab Station

The GeoLab located at the end of the chapter can be used at this point in the lesson.

☑ READING CHECK Holes are formed when gas bubbles burst in lava or when gas is trapped in lava.

■ **Figure 11** Rock textures provide information about a rock's formation. Evidence of the rate of cooling and the presence or absence of dissolved gases is preserved in the rocks shown here.

Crystal size and cooling rates When lava flows on Earth's surface, it cools quickly and there is not enough time for large crystals to form. The resulting extrusive igneous rocks, such as rhyolite, which is shown in **Figure 10,** have crystals so small that they are difficult to see without magnification. Sometimes, cooling occurs so quickly that crystals do not form at all. The result is volcanic glass, called obsidian, also shown in **Figure 10.** In contrast, when magma cools slowly beneath Earth's surface, there is sufficient time for large crystals to form. Thus, intrusive igneous rocks, such as granite, diorite, and gabbro, can have crystals larger than 1 cm.

Porphyritic rocks Look at the textures of the rocks shown in **Figure 11.** The top photo shows a rock with different crystal sizes. This rock has a **porphyritic** (por fuh RIH tihk) **texture,** which is characterized by large, well-formed crystals surrounded by finer-grained crystals of the same mineral or different minerals.

What causes minerals to form both large and small crystals in the same rock? Porphyritic textures indicate a complex cooling history during which a slowly cooling magma suddenly began cooling rapidly. Imagine a magma body cooling slowly, deep in Earth's crust. As it cools, the first crystals to form grow large. If this magma were to be suddenly moved higher in the crust, or if it erupted onto Earth's surface, the remaining magma would cool quickly and form smaller crystals.

Vesicular rocks Magma contains dissolved gases that escape when the pressure on the magma lessens. If the lava is thick enough to prevent the gas bubbles from escaping, holes called vesicles are left behind. The rock that forms looks spongy. This spongy appearance is called **vesicular texture.** Pumice and vesicular basalt are examples shown in **Figure 11.**

☑ READING CHECK **Explain** what causes holes to form in igneous rocks.

Thin Sections

It is usually easier to observe the sizes of mineral grains than it is to identify the mineral. To identify minerals, geologists examine samples that are called thin sections. A thin section is a slice of rock, generally 2 cm × 4 cm and only 0.03 mm thick. Because it is so thin, light is able to pass through it.

DIFFERENTIATED INSTRUCTION

Advanced Learners Have students research and explain the use of polarized light and crossed-polarizing filters to identify minerals in thin sections. Light travels at different speeds through crystals depending upon the orientation of individual crystals. This causes certain wavelengths to interfere with and cancel each other. The remaining wavelengths show up as different colors in thin sections.

ACROSS THE CURRICULUM

History The Sierra Nevada were formed by large granitic intrusions that, in turn, formed large numbers of gold-rich quartz veins. Ask students to research and write reports about how these geologic features affected the westward migration and settlement of California. **BL** **OL**

■ **Figure 12** The minerals that make up this piece of granite can be identified in a thin section.

When viewed through a special microscope, called a petrographic microscope, mineral grains exhibit distinct properties. These properties allow geologists to identify the minerals present in the rock. For example, feldspar grains often show a distinct banding called twinning. Quartz grains might appear wavy as the microscope stage is rotated. Calcite crystals become dark, or extinguish, as the stage is rotated. **Figure 12** shows the appearance of a thin section of granite under a petrographic microscope.

Igneous Rocks as Resources

The cooling and crystallization history of igneous rocks sometimes results in the formation of unusual but useful minerals. These minerals can be used in many fields, including construction, energy production, and jewelry making. Some of these uses are described in the following paragraphs.

Veins As you have learned, ores are minerals that contain a useful material that can be mined for a profit. Valuable ore deposits often occur within igneous intrusions. At other times, ore minerals are found in the rocks surrounding intrusions. These types of deposits sometimes occur as veins. Recall from Bowen's reaction series that the fluid left during magma crystallization contains high levels of silica and water. This fluid also contains any leftover elements that were not incorporated into the common igneous minerals. Some important metallic elements that are not included in common minerals are gold, silver, lead, and copper. These elements, along with the dissolved silica, are released at the end of magma crystallization in a hot, mineral-rich fluid that fills cracks and voids in the surrounding rock. This fluid solidifies to form metal-rich quartz veins, such as the gold-bearing veins in the Sierra Nevada. An example of gold formed in a quartz vein is shown in **Figure 13.**

☑ READING CHECK **Explain** why veins have high amounts of quartz.

■ **Figure 13** Gold and quartz are extracted from mines together. The two are later separated.
Infer ***What can you determine from this photo about the melting temperature of gold?***

Apply Earth Science

Igneous Rock Use Have students investigate their homes and yards to see whether any igneous rocks have been used for decorative or construction purposes. Students should describe the rocks and how they are being used in their Earth science journals. BL EL

Environmental Connection

Ceramic Burners Toxic chemicals can be destroyed by incineration at high temperatures. Incinerators often contain ceramic burners, which are made from fired clay minerals. Have interested students research the types of minerals used to make ceramic burners.

Teacher Content Support

Minerals in Thin Section Under crossed polarized light, minerals can appear to change colors, as in **Figure 12.** Another feature that minerals may exhibit in thin section is twinning. Plagioclase feldspar crystals might appear to be mirror images of each other, and as the microscope stage is rotated, one half becomes extinct, or goes dark. As the stage is rotated further, the other half becomes extinct, and the first half becomes light. Twinning can also appear as parallel lines which also exhibit alternating extinction patterns. A further pattern is called tartan twinning. The crystals appear to have a plaid pattern to them.

☑ READING CHECK Veins have high amounts of quartz because silicon and oxygen are left as a magma completely crystallizes. This liquid gets squeezed into the cracks in the rocks.

■ **Caption Question** **Fig. 13** It is approximately equal to quartz.

Demonstration

Thin Sections Demonstrate the use of thin sections in mineral identification. Show students a series of photos showing a thin section being rotated on a petrographic microscope stage. Point out the changes that the mineral crystals exhibit as the slide is turned. Ask students to describe the characteristics of and changes that quartz, feldspar, and calcite exhibit in thin section.

EARTH SCIENCE JOURNAL

Arteries or Veins Have students research the definitions of the words *artery* and *vein.* Then have them write and illustrate an explanation supporting the use of the word *vein* in geology.

Rubric

Problem-Solving LAB

Purpose Students will use a photo of a thin section of igneous rock to estimate the mineral composition of the rock.

Process Skills make and use tables, apply concepts, think critically, interpret scientific photos

Teaching Strategies

- Remind students that their estimated percentages should add up to 100 percent.
- Emphasize that mineral composition is being estimated, not the number of mineral grains. The areas of the mineral grains should be taken into consideration when estimating mineral composition.

Analysis

1. Make a copy of **Figure 12.** Mark off a section of the rock about one-fourth the size of the whole rock. Estimate the amount of gray quartz and muscovite, pink feldspar, and black amphibole and biotite that is in the marked-off section.
2. The minerals in the data table will include quartz and muscovite, feldspar, and amphibole and biotite. The percentages will vary depending on which section of the rock was used for the estimate, but the percentages should fall within the following ranges: quartz and muscovite (18–35%), feldspar (58–80%), amphibole and biotite (2–17%). The quartz and muscovite will be together in the table. They are both gray in color and can not be differentiated from a photograph. The same is true for amphibole and biotite. They are both black in color and can not be differentiated from a photograph.

Think Critically

3. The abundance of quartz and potassium feldspar indicates that this is a granitic rock.
4. Answers will vary depending on student estimations. Possible sources of error include too few data points, difficulty in estimating the area of different mineral shapes, and errors in calculation.
5. Usually, the more data points used, the more accurate the estimate.

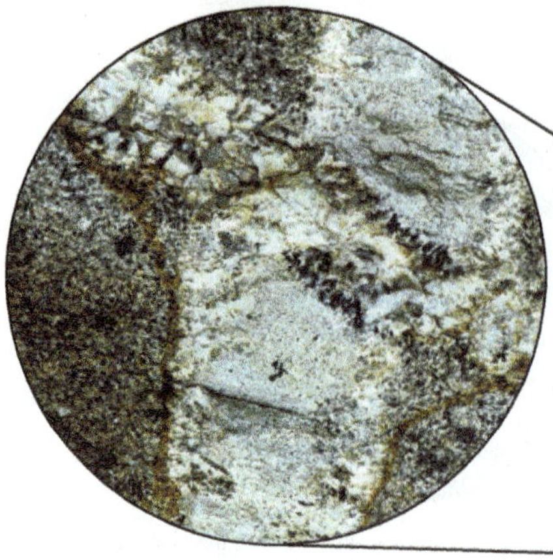

Figure 14 Pegmatite veins cut through much of the rock from which Mount Rushmore National Memorial is carved. You can see the veins running across Thomas Jefferson's face.

Pegmatites Vein deposits can contain other valuable resources in addition to metals. Igneous rocks that are made of extremely large-grained minerals are called **pegmatites,** and are usually found as igneous intrusions or veins. Ores of rare elements, such as lithium (Li) and beryllium (Be), often form in pegmatites. In addition to ores, pegmatites can produce beautiful crystals. Because veins fill cavities and fractures in rock, minerals grow into voids and retain their shapes. Some of the world's most beautiful minerals have been found in pegmatite veins. A famous pegmatite is the rock source for the Mount Rushmore National Memorial located near Keystone, South Dakota. A close-up view of President Thomas Jefferson, shown in **Figure 14,** reveals the huge mineral veins that run through the rock.

Problem-Solving LAB

Interpret Scientific Illustrations

How do you estimate mineral composition? Igneous rocks are classified by their mineral compositions. In this activity, you will use the thin section in **Figure 12** to estimate the different percentages of minerals in the sample.

Analysis

1. Design a method to estimate the percentages of the minerals in the rock sample shown in **Figure 12.**
2. Make a data table that lists the minerals and their estimated percentages.

Think Critically

3. **Interpret Figure 9** to determine where in the chart this rock sample fits.
4. **Compare** your estimates of the percentages of minerals in the rock with those of your classmates. Why do the estimates vary? What are some possible sources of error?
5. **Propose** a method to improve the accuracy of your estimate.

ACROSS THE CURRICULUM

History The pegmatite granite into which Mount Rushmore is carved is highly fractured in some places. Photos of the dedication of the Washington head in 1930 show the beginnings of the Jefferson head to the left of George Washington. All four heads are farther back into the mountain than originally planned. Discuss what might have necessitated this change. Have students research and present information on the changes that were made to the design while the monument was being made. BL OL

Kimberlites Diamond is a valuable mineral found in rare, ultrabasic rocks known as **kimberlites,** named after Kimberly, South Africa, where the intrusions were first identified. These unusual rocks are a variety of peridotite. They most likely form in the mantle at depths of 150 to 300 km. This is because diamonds and other minerals present in kimberlites can form only under very high pressure.

Geologists hypothesize that kimberlite magma is intruded rapidly upward toward Earth's surface, forming long, narrow, pipelike structures. These structures extend many kilometers into the crust, but they are only 100 to 300 m in diameter. For this reason, they are often called kimberlite pipes. Most of the world's diamonds come from South African mines, such as the one shown in **Figure 15.** Many kimberlites have been discovered in the United States, but diamonds have been found only in Arkansas and Colorado. The diamond mine in Colorado is the only diamond mine currently in operation in the United States.

Figure 15 Diamonds are mined from kimberlite in mines like this one in Richtersveld, Northern Cape, South Africa.

Igneous rocks in construction Igneous rocks have several characteristics that make them especially useful as building materials. The interlocking grain textures of igneous rocks make them strong. In addition, many of the minerals present in igneous rocks are resistant to weathering. Granite is among the most durable of igneous rocks. You have probably seen many items, such as countertops, floors, and statues, made from the wide variety of granite that has formed on Earth.

SECTION 2 REVIEW

Section Self-Check

Section Summary

- Igneous rocks are either ultrabasic, basaltic, intermediate, or granitic.
- The rate of cooling determines crystal size.
- Ores often occur in pegmatites. Diamonds occur in kimberlites.
- Some igneous rocks are used as building materials because of their strength, durability, and beauty.

Understand Main Ideas

1. **MAINIDEA Infer** why obsidian, which is black or red in color, usually has a granitic composition.
2. **Describe** the three major compositional groups of igneous rocks.
3. **Apply** what you know about cooling rates to explain differences in crystal sizes.
4. **Distinguish** between andesite and diorite using two physical properties of igneous rocks.

Think Critically

5. **Speculate** why there are almost no extrusive ultrabasic rocks in Earth's crust.
6. **Determine** whether quartz or plagioclase feldspar is more likely to form a well-shaped crystal in an igneous rock. Explain.

MATH IN Earth Science

7. A granite slab has a density of 2.7 g/cm^3. What is the mass of a 2-cm-thick countertop that is 0.6 m × 2.5 m? How many grams is this?

3 Assess

Check for Understanding

Discussion Ask students the following question: Is it possible for an igneous rock to demonstrate both intrusive and extrusive textures? Explain. Yes, a rock with a porphyritic texture can have large crystals like an intrusive rock and fine-grained crystals like an extrusive rock.

Reteach

Summarize Have students each make a data table summarizing the classification system for igneous rocks.

Assessment

Performance Have groups of students develop an igneous rock collection based on the rocks listed in **Figure 9.** Students can obtain actual rock samples or use photographs and sketches in their collections. Each rock in the collections should be labeled. Also, the collections should be accompanied by brief reports on the mineral compositions of each rock.

SECTION 2 REVIEW

1. The dark color of obsidian is the result of small amounts of mafic minerals, such as iron and magnesium. However, obsidian is primarily silica, hence the term *volcanic glass.*
2. granitic—rich in quartz and other light-colored minerals; basaltic—rich in iron- and magnesium-bearing dark-colored minerals; intermediate—a mixture of both light- and dark-colored minerals
3. Slow cooling rates result in large, often well-formed crystals; fast cooling rates result in small crystals that might or might not be well formed.
4. Mineral composition: andesite and diorite have similar mineral compositions. Crystal size: andesite has small crystals, diorite has larger crystals.
5. Ultrabasic rocks consist of minerals with high melting points. An ultrabasic magma would cool and crystallize long before reaching the surface.
6. Plagioclase; it forms early and there is room in the magma for the crystals to grow. Quartz crystallizes later and fills in the spaces between previously formed mineral grains.
7. 0.6 m = 60 cm; 2.5 m = 250 cm; 2 cm × 60 cm × 250 cm = 30,000 cm^3; 30,000 cm^3 × 2.7 g/cm^3 = 81,000 g, or 81 kg

Earth Science &
ENVIRONMENT

Purpose

Students will learn about the various rocks that are present on the Moon. They will learn where the rocks are stored and under what type of conditions.

Teacher Content Support

Moon Rocks The samples collected during the various Apollo missions contain rocks with varying chemistries.

The *Apollo 11* mission landed on the Mare Tranquillitatus. This area of the Moon is quite flat. The samples contain basalts with high titanium concentrations—9 to 12% TiO_2. There are no basalts on Earth that are similar to these basalts.

The *Apollo 12* site is located on the Oceanus Procellarum south of Copernicus. It is also flat, and the high iron content of the basalts in the area suggests that the area sustained an impact from an iron meteorite.

The *Apollo 15* landing site was located in the hills east of the Mare Imbrium lava lake and was chosen because it might be a collapsed lava tube. It was here that the oldest moon rock, about 4.5 billion years old, was collected. The rock is anorthosite—a plagioclase-feldspar-rich intrusive igneous rock. The *Apollo 17* landing site was near Shorty Crater. Tiny spheres of orange and black volcanic glass were found around the rim of the crater. The orange color is due to the presence of TiO_2. The age of the glass is 3.64 billion years. A lava flow buried the volcanic glass until the impact that formed the crater exposed it.

Earth Science &
ENVIRONMENT

Moon Rocks

During each of the six Apollo missions, lunar rocks were collected with the hope of providing information about the Moon's origin, history, and environment. How do moon rocks compare with rocks on Earth?

Moon rock types Between 1969 and 1972, astronauts collected approximately 380 kg of lunar rocks. The 2415 individual pieces range in size from a grain of sand to a basketball.

Generally, moon rocks vary in color from gray to black to white to green. Some rocks are glassy, some are hard, and others are fragile. Analysis of the rocks has revealed at least three different rock types on the Moon. Basaltic rocks formed from lava flows and volcanic ash that reached the surface through cracks and fissures caused by meteorite impacts. Breccias formed when meteorites shattered rocks and then fused the pieces together with the heat generated by the impact. Pristine rock is rock that has not been hit by meteorites. Pristine rock is commonly composed of calcium-rich plagioclase feldspar and is gray in color.

Moon rock composition Moon rocks are unique in two ways. First, they are not oxidized. Considering how much iron is contained in the rocks, this is a sharp contrast to weathered and rusty iron-bearing rocks on Earth. Second, the surfaces of some moon rocks are covered with tiny pockmarks called zap pits. These are caused by micrometeoroids that impact the rocks on the Moon's surface. Zap pits do not occur on Earth rocks because friction from Earth's atmosphere causes tiny meteoroids to burn up long before they reach Earth's surface.

Moon rock classification Scientists use the same categories for classifying lunar rocks as they use for igneous rocks on Earth.

Roger Ressmeyer/CORBIS

This scientist is studying a piece of basalt that was collected from the lunar surface during the *Apollo 15* mission.

Based on mineral composition, scientists named a new class of moon rocks called KREEP rocks. These contain high amounts of potassium (K), rare Earth elements (REE), and phosphorus (P).

Water in moon rocks For over 40 years, scientists thought that the moon was dry. However, in 2010, researchers identified water in lunar rock samples collected during NASA's *Apollo* missions in the 1970s. This water likely came from comets. Scientists continue to pose questions about these rocks as they study the Moon's origin and history.

WRITING IN ▶ Earth Science

Lunar Rock Game Use resources to design a game that involves the collection and analysis of lunar rocks by scientists. Trade games with classmates to increase your understanding of lunar rocks.

WebQuest

Teaching Strategy

Have students play one another's games in class. Be sure they evaluate the game for accuracy, thoroughness of background research, and detailed scientific illustrations. Once the games have been played, process with the class what was learned about lunar rocks by playing the games.

WRITING IN ▶ Earth Science

Rubric

Lunar Rock Game Students should acquire new knowledge of the Moon and its composition by playing the games.

WebQuest

GeoLAB

iLab Station

Design Your Own: Model Crystal Formation

Background: The rate at which magma cools affects the grain size of the resulting igneous rock. Observing the crystallization of magma is difficult because molten rock is very hot and the crystallization process is sometimes very slow. Other materials, however, crystallize at lower temperatures. These materials can be used to model crystal formation.

Question: *How do minerals crystallize from magma?*

Matt Meadows

Materials

clean, plastic petri dishes
saturated alum solution
200-mL glass beaker
magnifying lens
dark-colored construction paper
thermometer
paper towels
water
hot plate

Safety Precautions

WARNING: *The alum solution can cause skin irritation and will be hot when it is first poured into the petri dishes. If splattering occurs, wash skin with cold water.*

Procedure

1. Read and complete the lab safety form.
2. As a group, plan how you will change the cooling rate of a hot solution poured into a petri dish. Each group member should choose a petri dish in a predetermined location to observe during the investigation. Make sure your teacher approves your plan before you begin.
3. Place a piece of dark-colored construction paper on a level surface where it will not be disturbed. Be sure to put the paper in all of the predetermined locations. Place the petri dishes on top of the paper.
4. Using the glass beaker, obtain about 150 mL of saturated alum solution from your teacher. The temperature should be about 95°C to 98°C, just below boiling temperature.
5. Carefully pour some of the solution into each petri dish so that it is half full. Use caution when pouring the hot liquid to avoid splatters and burns.
6. Every 5 min for 30 min, record your observations of your petri dish. Make drawings of any crystals that begin to form.

Analyze and Conclude

1. **Compare** your methods of cooling with those of other groups. Did some methods appear to work better than others? Explain.
2. **Examine** your alum crystals. What do the crystals look like? Are they all the same size? Do all the crystals have the same shape?
3. **Draw** the most common crystal shape in your science journal. Compare your drawings with those of other groups. Describe any patterns that you see.
4. **Deduce** what factors affected the size of the crystals in the different petri dishes. How do you know?
5. **Infer** why the crystals changed shape as they grew.
6. **Compare and contrast** this experiment with magma crystallization.
7. **Evaluate** the relationship between cooling rate and crystal formation.

SHARE YOUR DATA

Peer Review Post a summary of your data. Compare and contrast your results with those of other students who have completed this lab.

Analyze and Conclude

1. Answers will vary depending on the cooling method used. Possible methods include: ice water bath, refrigeration, and fanning the dish.
2. Alum crystals are tabular and have six sides. They look like triangles with the corners cut off. The crystals might be different sizes.
3. The majority of the crystals will have the same shape until they begin to grow together.
4. The cooling rate affected the crystal size. The crystals in the petri dishes where the solution cooled slowly were larger than the crystals in the petri dishes where cooling occurred quickly.
5. The crystals grew larger but maintained a similar shape until they began to interfere with each other. Their shapes became distorted as they grew together.
6. The experiment is different from magma crystallization in that magma crystallization involves the cooling of melted minerals, whereas this experiment involves the cooling of a hot solution containing dissolved minerals. The experiment is similar to magma crystallization in that the cooling rate affected crystal size in both cases and the crystals grew by adding atoms to their surfaces.
7. A fast cooling rate results in small crystals; a slow cooling rate results in large crystals.

GeoLAB

iLab Station Rubric

Preparation

Time Allotment 90 min

Process Skills analyze data, design an experiment, observe and infer, compare and contrast

Safety Precautions Approve lab safety forms before work begins. The alum solution will be hot. Students should not try to move the petri dishes after the solution has been poured. Students should wear safety goggles and aprons.

Preparation of Materials Prepare the saturated solution by dissolving as much alum as possible in a large beaker of nearly boiling water. Use approximately 400–500 mL of alum for each 1000 mL of water. The solution could become cloudy if it boils. This will not affect the crystallization process.

Procedure

- Discuss the concept of saturated solutions. Hot water can dissolve more material than cold water can. As a hot, saturated solution cools, the water molecules lose energy and cannot keep the alum atoms in solution. The atoms bind together in a set pattern, forming solid crystals with a characteristic shape.
- **Troubleshooting** Drop a few grains of alum into a dish to trigger crystallization.

SHARE YOUR DATA

Peer Review Students should discover that even with different cooling methods, the crystals of the alum solution grow into predictable shapes.

MAINIDEAS **Summary statements can be used by students to review the major concepts of the chapter.**

Students can review with these online resources.

Vocabulary Practice

- Vocabulary eGames
- Vocabulary eFlashcards
- Vocabulary PuzzleMaker

Section Self-Check

Chapter Self-Check

Online Test Practice

Use the ExamView® *Assessment Suite* CD-ROM to:

- create multiple versions of tests
- create modified tests with one mouse click
- edit existing questions and add your own questions
- build tests aligned with state standards using built-in state curriculum tags
- change English tests to Spanish with one mouse click
- track students' progress using the Teacher Management System

CHAPTER 5 STUDY GUIDE

BIGIDEA **Igneous rocks were the first rocks to form as Earth cooled from a molten mass to the crystalline rocks of the early crust.**

Vocabulary Practice

SECTION 1 What are igneous rocks?

VOCABULARY
- lava
- igneous rock
- partial melting
- Bowen's reaction series
- fractional crystallization

MAINIDEA **Igneous rocks are the rocks that form when molten material cools and crystallizes.**

- Magma consists of molten rock, dissolved gases, and mineral crystals.
- Magma is classified as basaltic, andesitic, or rhyolitic, based on the amount of silica it contains.
- Different minerals melt and crystallize at different temperatures.
- Bowen's reaction series defines the order in which minerals crystallize from magma.

SECTION 2 Classification of Igneous Rocks

VOCABULARY
- intrusive rock
- extrusive rock
- basaltic rock
- granitic rock
- texture
- porphyritic texture
- vesicular texture
- pegmatite
- kimberlite

MAINIDEA **Classification of igneous rocks is based on mineral composition and texture.**

- Igneous rocks are either ultrabasic, basaltic, intermediate, or granitic.
- The rate of cooling determines crystal size.
- Ores often occur in pegmatites. Diamonds occur in kimberlites.
- Some igneous rocks are used as building materials because of their strength, durability, and beauty.

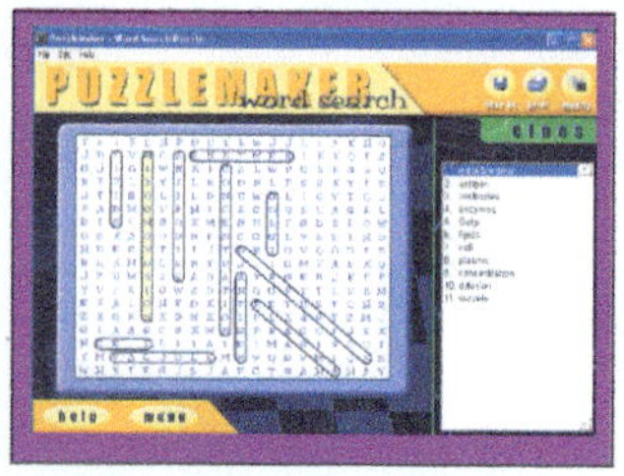

Vocabulary Practice

For additional practice with vocabulary, have students access the Vocabulary PuzzleMaker.

CHAPTER 5 ASSESSMENT

Chapter Self-Check

VOCABULARY REVIEW

The sentences below are incorrect. Make each sentence correct by replacing the italicized word or phrase with a vocabulary term from the Study Guide.

1. Gases escape from *magma* as it flows out onto Earth's surface.
2. *Mohs scale of hardness* describes the order in which minerals crystallize.
3. *Lava* forms deep beneath Earth's crust.

Complete the sentences by filling in the blank with the correct vocabulary term from the Study Guide.

4. An igneous texture characterized by large crystals embedded in a fine-grained background is called a ________.
5. Igneous rocks that form under conditions of fast cooling are said to be ________.
6. Light-colored rocks high in silica are said to be ________.

UNDERSTAND KEY CONCEPTS

7. Which is the first mineral to form in cooling magma?
   A. quartz **C.** potassium feldspar
   B. mica **D.** olivine

Use the diagram below to answer Question 8.

Bob Coyle/McGraw-Hill Education

8. Which process is occurring in the diagram?
   A. fractional separation
   B. crystal separation
   C. fractional crystallization
   D. partial melting

9. Which minerals are associated with the right-hand branch of Bowen's reaction series?
   A. olivine and pyroxene
   B. feldspars
   C. mica and feldspars
   D. quartz and biotite

10. Which magma type contains the greatest amount of silica?
    A. basaltic **C.** rhyolitic
    B. andesitic **D.** peridotic

11. Which does not affect the formation of magma?
    A. volume **C.** pressure
    B. temperature **D.** mineral composition

12. Which intrusive rock has the same composition as andesite?
    A. granite **C.** obsidian
    B. basalt **D.** diorite

Use the figure below to answer Question 13.

13. Which process formed this rock?
    A. slow cooling
    B. fast cooling
    C. very fast cooling
    D. slow, then fast cooling

14. Which type of ultrabasic rock sometimes contains diamonds?
    A. pegmatite
    B. kimberlite
    C. granite
    D. rhyolite

CHAPTER 5 ASSESSMENT

VOCABULARY REVIEW

1. lava
2. Bowen's reaction series
3. magma
4. porphyritic texture
5. extrusive
6. granitic

UNDERSTAND KEY CONCEPTS

7. D
8. D
9. B
10. C
11. A
12. D
13. C
14. B

15. A
16. B
17. A

CONSTRUCTED RESPONSE

18. Answers will vary. Students might list kitchen countertops, floors, grave markers, monuments, etc.
19. Plagioclase in basaltic rocks formed at higher temperatures and contains more calcium than sodium. Plagioclase in granitic rocks contains more sodium than calcium.
20. Possible answer: gases dissolved in magma; lava flows out onto surface, pressure decreases; gases escape and form bubbles in lava; lava hardens, preserving the bubbles.
21. Students might suggest that the large amount of air space in the rock helps to keep it afloat.
22. The diagram should reflect the following information: When iron-rich olivine forms, the growing olivine crystals remove iron from the magma, and the iron content of the magma decreases.
23. Minerals melt and crystallize at different temperatures. Thus, some minerals might remain solid, while others melt.
24. Rock 2 is most likely granite. The percentages of quartz, feldspar, biotite, and amphibole place it in the granite section of the identification chart.
25. The percentages of minerals in Rock 4 indicate that it is a medium-colored rock, making it intermediate. The fine-grained nature suggests that the rock is andesite.

THINK CRITICALLY

26. Obsidian is volcanic glass and has no internal crystalline structure. Therefore, it shatters when hit with a hammer. Granite is composed of minerals that all have internal crystalline structures. Although these minerals might break, they simply form smaller versions of the mineral and do not shatter.
27. This is possible during partial melting. The first minerals to melt might have a high silica content and will thus generate magma with more silica than the parent rock.
28. The dominant minerals that make up granite, quartz, and feldspar have a greater hardness than stainless steel.

ASSESSMENT

15. What effect does a fast cooling rate have on grain size in igneous rocks?
A. It forms fine-grained crystals.
B. It forms large-grained crystals.
C. It forms light crystals.
D. It forms dark crystals.

16. What term describes igneous rocks that crystallize inside Earth?
A. magma
B. intrusive
C. lava
D. extrusive

17. Which minerals are most common in granite?
A. quartz and feldspar
B. plagioclase feldspar and amphibole
C. olivine and pyroxene
D. quartz and olivine

CONSTRUCTED RESPONSE

18. List some uses of igneous rocks in the construction industry.

19. Explain how and why the plagioclase feldspar in basaltic rocks differs from that in granitic rocks.

Use the photos below to answer Questions 20 and 21.

20. Draw a flowchart documenting the formation of the holes in this sample of vesicular basalt.

21. Speculate on the reasons that samples of pumice are able to float in water.

22. Illustrate how fractional crystallization changes the composition of magma, using the formation of iron-rich olivine to illustrate the point.

23. Apply the concepts of temperature and crystallization to explain why magma is often described as a slushy mixture of crystals and molten rock.

Use the table below to answer Questions 24 and 25.

Rock Composition

Mineral	Mineral Percentage			
	Rock 1	Rock 2	Rock 3	Rock 4
Quartz	5	35	0	0
Potassium feldspar	0	15	0	0
Plagioclase feldspar	55	25	0	55
Biotite	15	15	0	10
Amphibole	25	10	0	30
Pyroxene	0	0	40	5
Olivine	0	0	60	0

24. Analyze the data in the table, and explain which rock is most likely granite.

25. Incorporate Use the data for Rock 4 and the fact that it is fine-grained to determine the name of Rock 4.

THINK CRITICALLY

26. Compare obsidian and granite to explain why granite is more easily carved into statues and monuments.

27. Evaluate this statement: It is possible for magma to have a higher silica content than the melted rock from which it forms.

28. Apply what you know about mineral hardness to explain why stainless steel knives do not harm granite cutting boards.

Chapter Self-Check

29. Infer Kimberlites are the source of most diamonds. Infer why scientists study kimberlites to learn more about Earth's mantle.

30. Assess Rocks generally consist of minerals. When molten rock is chilled rapidly, it becomes a glass. Volcanic glass is an extrusive igneous rock. Assess whether this rock contains minerals. Explain your answer. *[Hint: Recall the definition of a mineral.]*

31. Infer why rocks that are composed of minerals that crystallize first according to Bowen's reaction series are unstable and break down quickly at Earth's surface.

32. Hypothesize what the Palisades Sill would look like if the magma that formed it was granitic in composition.

CONCEPT MAPPING

33. Use the following terms to create a concept map showing the relationship among position in Earth's crust and mantle, crystal size, and rock type: *fast, slow, slowest, intrusive, extrusive, magma, lava, granite, rhyolite, basalt, gabbro, obsidian,* and *pumice.*

CHALLENGE QUESTION

Use the diagram below to answer Question 34.

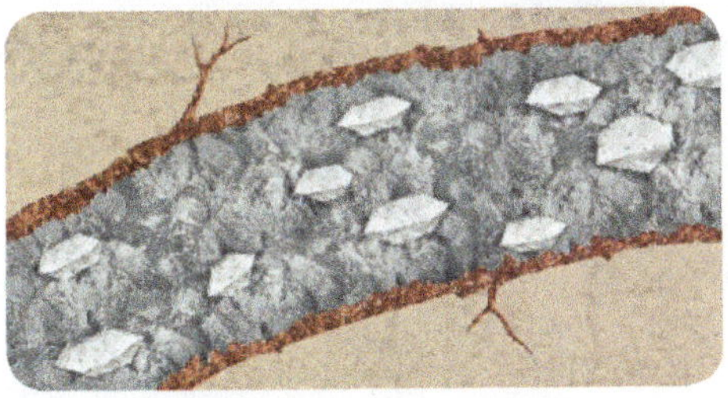

34. Determine The diagram shows a cross section of the Leopard Lode, an igneous rock unit in Wyoming. Determine the formation history of this rock unit.

WRITING IN ▶ Earth Science

35. Building stone is expensive. Suppose you are selling kitchen countertops that look like granite, but consist of a less-expensive synthetic material. List the specific characteristics of granite that your customers would look for in the imitation granite.

DBQ Document–Based Questions

Data obtained from: Gerya, T.V., et al. 2003. Cold fingers in a hot magma: numerical modeling of country-rock diapirs in the Bushveld Complex, South Africa. *Geology* 31 (9): 753.

The Bushveld Complex is the world's largest layered intrusion. It was injected as a hot, dense basaltic magma between overlying volcanic and underlying sedimentary rocks. Modeling of this event indicates that finger-shaped bodies of heated, metamorphosed sedimentary rocks subsequently intruded the overlying igneous layers. The model assumed the igneous rock properties shown in the table.

Igneous Rock Properties

Rock Type	Density (kg/m^3)	T Solidus (°C)	T Liquidus (°C)
Granitic	2700 (solid) 2400 (molten)	675	925
Basaltic/ ultrabasic	3000 (solid) 2900 (molten)	950	1100

36. Compare and contrast the density of solid and molten rocks in this model.

37. Speculate about why the overlying rhyolitic rocks could not penetrate, or sink into, the basaltic magma.

38. Infer the meaning of the terms *liquidus* and *solidus.* At what temperature do the first crystals in granitic rocks melt?

CUMULATIVE REVIEW

39. What is a molecule? **(Chapter 3)**

40. Name a gemstone that consists of corundum. **(Chapter 4)**

29. Diamonds form in the mantle. Thus, their host rock composition might represent the composition of the mantle.

30. The rock does not contain minerals. Glass is not considered a mineral because it lacks a crystalline structure.

31. Minerals that form at the highest temperatures and pressures are unstable at Earth's surface where conditions are substantially different.

32. Students might note that the minerals in the sill would be the lighter-colored minerals such as quartz, potassium feldspar, muscovite mica, and sodium-rich plagioclase feldspar. They might suggest that the chill margin would be composed of the original composition, the layer of crystals that settled out first would be composed of the sodium feldspar, and the center part of the sill would be composed of the quartz, mica, and remaining feldspar.

CONCEPT MAPPING

33. The terms should be linked together as follows: Slowest, intrusive, magma, granite, gabbro; slow, intrusive or extrusive, lava, rhyolite, basalt; fast, obsidian, pumice.

CHALLENGE QUESTION

34. This is a porphyritic rock with three distinct sections. The edges are referred to as chill margins. The crystals are small and are of a different color than the fine-grained background of the rest of the rock. They formed when the molten rock came in contact with the surrounding, colder rock. The magma cooled quickly and formed small crystals of a specific composition. As the remaining magma cooled, first the large, well-formed crystals formed. Finally, cooling sped up and the remaining magma cooled quickly, forming the fine-grained background.

WRITING IN ▶ Earth Science

Rubric

35. Students' answers will vary. Possible answers include a realistic mix of colors, interlocking crystals, realistic crystal sizes and shapes, hardness, and durability.

DBQ Document–Based Questions

Data obtained from: Gerya, T. V., et al. 2003. Cold fingers in a hot magma: Numerical modeling of country-rock diapirs in the Bushveld Complex, South Africa. *Geology* 31 (9): 753.

36. Molten rocks have lower densities than solid rocks.

37. The density of the granitic rocks is less than that of the basaltic magma.

38. The temperature at which the first crystals form in a cooling magma is called the liquidus. The rocks begin to melt at 675°C. This is called the solidus.

CUMULATIVE REVIEW

39. A molecule is two or more atoms held together by covalent bonds.

40. ruby or sapphire

CUMULATIVE STANDARDIZED TEST PRACTICE

MULTIPLE CHOICE

1. A
2. C
3. D
4. C
5. D
6. B
7. A
8. D
9. C
10. D

CUMULATIVE STANDARDIZED TEST PRACTICE

MULTIPLE CHOICE

Use the table below to answer Questions 1 and 2.

Characteristics of Rocks			
	Color	Silica Content	Composition
Rock A	light	high	quartz and feldspars
Rock B	dark	low	iron and magnesium

1. Rock A is most likely what kind of rock?
   A. granitic
   B. basaltic
   C. ultrabasic
   D. intermediate

2. Which of the following is Rock B?
   A. granite
   B. diorite
   C. gabbro
   D. rhyolite

3. Which is most abundant in magma and has the greatest effect on its characteristics?
   A. O
   B. Ca
   C. Al
   D. SiO_2

4. Which process describes how minerals in igneous rocks form in predictable sequences?
   A. partial melting
   B. fractional crystallization
   C. Bowen's reaction series
   D. geothermal gradient

5. Which is NOT a feature used for identifying minerals?
   A. hardness
   B. color
   C. density
   D. volume

6. Which is distorted on a Mercator projection map?
   A. shapes of the landmasses
   B. areas of the landmasses
   C. latitude lines
   D. longitude lines

Use the graph below to answer Questions 7 and 8.

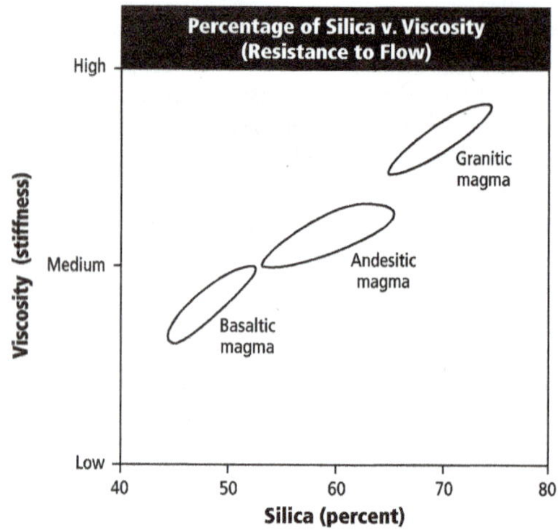

7. What relationship can be inferred from the graph?
   A. Magmas that have more silica are more viscous.
   B. Magmas that have less silica are more viscous.
   C. Magmas always have low viscosity.
   D. There is no relationship between silica content and viscosity (resistance to flow).

8. Which is a true statement about rhyolitic magma?
   A. Rhyolitic magma is heavier than the other two types of magma.
   B. Rhyolitic magma is lighter than the other two types of magma.
   C. Rhyolitic magma flows more quickly than the other two types of magma.
   D. Rhyolitic magma flows more slowly than the other two types of magma.

9. Which is a combination of two or more components that retain their identities?
   A. chemical
   B. solution
   C. mixture
   D. element

10. Which is the lightest of all atoms?
    A. uranium atom
    B. oxygen atom
    C. carbon atom
    D. hydrogen atom

SHORT ANSWER

Use the picture below to answer Questions 11–13.

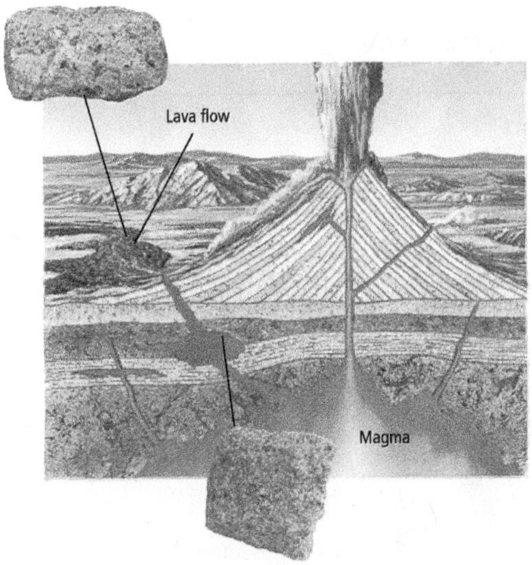

11. What type of igneous rock is located at the bottom of the picture? State a common example of that type of rock and explain how that rock is formed.

12. What type of igneous rock is located at the top of the picture? State a common example of that type of rock and explain how that type of rock is formed.

13. Contrast the formation of the two types of igneous rock.

14. What does it mean to say that minerals are naturally occurring and inorganic?

15. Why are some minerals classified as gems?

16. Why are both latitude and longitude lines necessary when identifying a location?

READING FOR COMPREHENSION

Mariana Island Research

Billowing ash plumes, molten sulfur droplets, feisty shrimp feasting on fish killed by noxious gases, and red lava jetting from a vent are all part of the action recently filmed at an underwater volcano in the western Pacific Ocean. The images are the first ever direct observations of an active, submarine-arc volcano. Unlike volcanic activity at mid-ocean ridges, island-arc volcanoes can remain fixed over their magma sources for thousands of years, allowing them to sometimes grow above water level and become islands. The new studies at the Mariana Islands are giving scientists a firsthand look into this formation process. The volcano has been going through nearly constant low-level eruptions since at least 2004, when it was first observed, Embley says. It could potentially keep erupting for decades, giving scientists the opportunity to monitor its growth.

Article obtained from: Roach, J. "Deep-Sea Volcano Erupts on Film—A First" *National Geographic News*. 24 May 2006.

17. What are the benefits of the new studies at the Mariana Islands?
    - A. The studies give scientists a firsthand look into the formation process.
    - B. The studies reveal that the volcano could potentially keep erupting for decades.
    - C. The studies show life near the vent.
    - D. The studies are the first ever direct observations of an active submarine arc-volcano.

18. What can you infer from this passage?
    - A. Volcanoes constantly erupt at some level of intensity.
    - B. Volcanic activity occurs only at mid-ocean ridges.
    - C. Fish and shrimp can live near underwater volcanoes.
    - D. There are many active submarine volcanoes.

NEED EXTRA HELP?

If You Missed Question . . .	1	2	3	4	5	6	7	8	9	10	11	12	13	14	15	16
Review Section . . .	5.2	5.2	5.1	5.1	4.1	2.2	5.1	5.1	3.2	3.1	5.2	5.2	5.2	4.1	4.2	2.1

SHORT ANSWER

11. Intrusive igneous rock; sample answer: granite. Magma flows into cracks and spaces in the crust and slowly solidifies into rock.
12. Extrusive igneous rock; sample answer: rhyolite. Lava flows across the crust and quickly solidifies.
13. Intrusive igneous rock forms when magma cools and crystallizes slowly under Earth's surface. Extrusive igneous rock forms when lava cools and crystallizes quickly on Earth's surface.
14. It means that minerals are formed by natural processes and that they are not alive and never were alive in any part of their existence.
15. Minerals that are rarer and have more beauty than other minerals are classified as gems.
16. Two intersecting lines are needed to locate a point in space. With only a latitude (or longitude) line, the location could be anywhere on Earth along that line. But by adding the longitude (or latitude) line, the location is pinpointed to that one place where the two cross.

READING FOR COMPREHENSION

17. A
18. C

CHAPTER 6 Sedimentary and Metamorphic Rocks

BIGIDEA **Most rocks are formed from preexisting rocks through external and internal geologic processes.**

ESSENTIAL QUESTIONS	NATIONAL STANDARDS	RESOURCES TO ASSESS MASTERY
SECTION 1 **Formation of Sedimentary Rocks** **1.** How are sedimentary rocks formed? **2.** What is the process of lithification? **3.** What are the main features of sedimentary rocks? 1.5 sessions 0.75 block	UCP.2, UCP.3; A.1–2; B.2–4; C.6; D.1, D.3; F.3	**Entry-Level Assessment** Focus Transparency 14, p. 134 **Progress Monitoring** Caption Question, p. 134 Reading Check, p. 135 Section Review, p. 140
SECTION 2 **Types of Sedimentary Rocks** **1.** How can the different types of clastic sedimentary rocks be described? **2.** How do chemical sedimentary rocks form? **3.** What are biochemical sedimentary rocks? 1 session 0.5 block	UCP.1, UCP.3; A.1–2; B.2–4; C.5; F.3	**Entry-Level Assessment** Focus Transparency 15, p. 141 **Progress Monitoring** Caption Question, p. 141 Reading Check, p. 143 Section Review, p. 144
SECTION 3 **Metamorphic Rocks** **1.** What are the different types and causes of metamorphism? **2.** How are metamorphic textures described? **3.** How do mineral and compositional changes occur during metamorphism? **4.** How are rocks classifed using the rock cycle? 4 sessions 2 blocks	UCP.1–3, UCP.5; A.1–2; B.2, B.4; D.2; F.1, F.3, F.5	**Entry-Level Assessment** Focus Transparency 16, p. 145 **Progress Monitoring** Caption Question, pp. 145, 149 Reading Check, pp. 146, 150 Section Review, p. 151 **Summative Assessment** Chapter Assessment, p. 155 Exam*View®* *Assessment Suite* CD-ROM

Go online!
connectED.mcgraw-hill.com

LEVELED RESOURCES	LAB MATERIALS
Science Notebook 6.1 OL **Chapter FAST FILE Resources:** MiniLab Worksheet, p. 74 OL Study Guide, p. 85 OL **Lab Resources:** Laboratory Manual, p. 41 OL **Transparencies:** Section Focus Transparency 14 EL Teaching Transparency 13 OL EL	LaunchLAB p. 132 / **10 min** photograph of footprints preserved in sedimentary rock MiniLAB p. 136 / **15 min** sediment, 200-mL jar with lid, water
Science Notebook 6.2 OL **Chapter FAST FILE Resources:** Study Guide, p. 87 BL **Lab Resources:** Laboratory Manual, p. 45 OL **Transparencies:** Section Focus Transparency 15 EL Teaching Transparency 14 OL EL	
Science Notebook 6.3 OL **Chapter FAST FILE Resources:** GeoLab Worksheet, p. 75 OL Study Guide, p. 88 BL **Transparencies:** Section Focus Transparency 16 EL Teaching Transparency 15 OL EL	GeoLAB p. 153 / **60–120 min** samples of granite, sandstone, shale, limestone, quartzite, gneiss, slate and marble; magnifying lens; paper; beam balance; 100-mL graduated cylinder or beaker; water

LabManager™
Customize any lab with the LabManager™ CD-ROM.

ADDITIONAL RESOURCES

Technology:
- eTeacherEdition
- eStudentEdition
- Classroom Presentation Toolkit
- Exam*View*® *Assessment Suite* CD-ROM
- LabManager™ CD-ROM
- What's EARTH SCIENCE Got To Do With It? DVD
- Weather Classroom Video DVD

Assessment:
- Blueprints for Success
- Chapter Test (Scaffolded)
- Earth Science Standardized Test Practice

Lab Resources:
- Exploring Environmental Problems Laboratory Manual

BL Below Level OL On Level AL Advanced Learners EL English Learners COOP LEARN Cooperative Learning

CHAPTER 6

LaunchLAB

iLab Station | Rubric

What happened here?

Teaching Strategy
Have students focus on the obvious first. For example, they should determine the number of tracks and the direction of travel before attempting to interpret more complex interactions.

Procedure
1. Have students read and complete the lab safety form and follow the procedure below.
2. Obtain a **photograph of a set of footprints that have been preserved in sedimentary rock.**
3. Write a description of how these tracks might have been made.
4. Draw your own diagram of a set of fossilized footprints that records the interactions of organisms in the environment.
5. Give your diagram to another student and have him or her interpret what happened.

Analysis
1. **Determine** the number of animals that made these tracks. at least two
2. **Infer** types of information that can be obtained by studying fossil footprints. types of animals, stride lengths and weights of the animals, number of toes, direction of movement, which tracks were made first, etc.
3. **Interpret** another group's diagram. Is your answer the same as theirs? What might have caused any differences? Interpretation of diagrams will vary. Possible reasons include: different animal interactions might leave similar types of tracks, students might interpret the appearance differently.

CHAPTER 6

Sedimentary and Metamorphic Rocks

BIGIDEA Most rocks are formed from preexisting rocks through external and internal geologic processes.

SECTIONS

1 Formation of Sedimentary Rocks

2 Types of Sedimentary Rocks

3 Metamorphic Rocks

LaunchLAB

iLab Station

What happened here?

Fossils are the remains and traces of once-living plants and animals. In this activity, you will interpret animal activity from the pattern of fossil footprints.

FOLDABLES® Study Organizer

The Rock Cycle

Make a six-door book using the labels shown. Use it to show possible paths of rock formation.

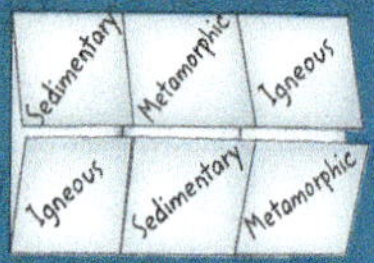

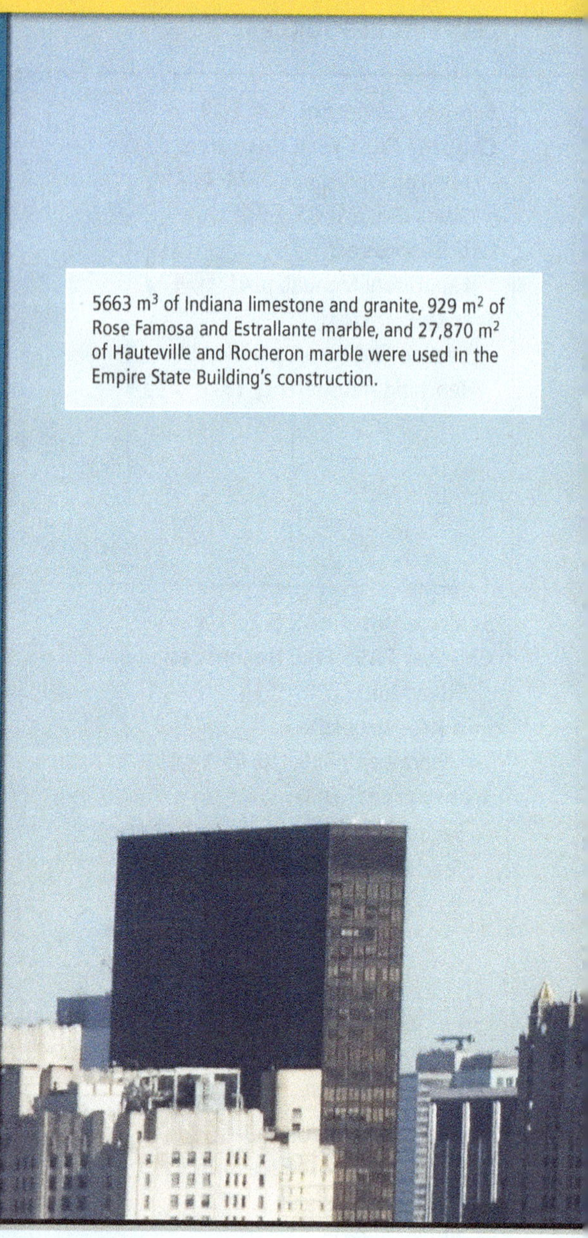

5663 m^3 of Indiana limestone and granite, 929 m^2 of Rose Famosa and Estrallante marble, and 27,870 m^2 of Hauteville and Rocheron marble were used in the Empire State Building's construction.

Assessment

Skill Have students list the types of information that cannot be determined from fossilized footprints. Fossilized footprints do not contain any information about the animal's appearance: they contain no bones, no skeletal structures, and no evidence of any part of the animal that did not touch the ground.

LabManager™
Customize any lab with the LabManager™ CD-ROM.

Introduce the BIGIDEA

Recognize Characteristics Hold up a rice cake and a sample of coarse sandstone. Ask students how they are the same and how they are different. They both appear to be made of individual pieces. The rice cake pieces are all approximately the same size; the sandstone might contain pieces of differing sizes. Neither one exhibits visible cement. Point out that the most common type of sedimentary rocks, clastic sedimentary rocks, are composed of individual particles cemented together.

Teacher Content Support

Building Stone The natural building stones used to build the Empire State Building are limestone, marble, and granite. Ask students why certain types of rock are used in buildings and others are not. strength, cost, durability, appearance, and availability Limestone is made of the mineral calcite. Ask students why it might be important to control acid rain in the city. Students have learned that calcite dissolves in hydrochloric acid. They should make the connection that acid rain can dissolve building stone as well.

SECTION 1

1 Focus

Focus Transparency

Before presenting the lesson, pro-ject **Section Focus Transparency 14** and have students answer the accompanying questions. EL

MAINIDEA

What makes a rock? Prepare the following samples: Pack wet sand into a small paper cup and carefully remove from the mold. Repeat this procedure using the same amount of sand mixed with diluted white glue, which has then been allowed to dry. Have students visually examine the samples. Ask from where the sand might have come. Students might suggest eroded mountains, eroded rocks, etc. Ask if the samples are both rocks. What makes them rocks? Gently tap each sample with a hammer to show that one will easily break apart. Explain that although the samples are compacted and solid, cement is necessary to make the sediment a rock.

2 Teach

Tie to Previous Knowledge

Mineral Stability Refer students to Bowen's reaction series. Tell students minerals that crystallize early are more susceptible to chemical weathering than those that crystallize later. Ask students which type of mineral grain they would expect to be most common in a clastic sedimentary rock. the most resistant minerals, such as quartz and feldspar

▪ **Caption Question Fig. 1** Quartz will be the most resistant because it is the last mineral to crystallize and it forms closest to conditions that are present on Earth's surface.

SECTION 1

Essential Questions

- How are sedimentary rocks formed?
- What is the process of lithification?
- What are the main features of sedimentary rocks?

Review Vocabulary

texture: the physical appearance or feel of a rock

New Vocabulary

sediment
lithification
cementation
bedding
graded bedding
cross-bedding

Formation of Sedimentary Rocks

MAINIDEA Sediments produced by weathering and erosion form sedimentary rocks through the process of lithification.

EARTH SCIENCE 4 YOU **Whenever you are outside, you might see pieces of broken rock, sand, and soil on the ground. What happens to this material? With one heavy rain, these pieces of broken rock, sand, and soil could be on their way to becoming part of a sedimentary rock.**

Weathering and Erosion

Wherever rock is exposed at Earth's surface, it is continuously being broken down by weathering—a set of physical and chemical processes that breaks rock into smaller pieces. **Sediments** are small pieces of rock that are moved and deposited by water, wind, glaciers, and gravity. When sediments become glued together, they form sedimentary rocks. The formation of sedimentary rocks begins when weathering and erosion produce sediments.

Weathering Weathering produces rock and mineral fragments known as sediments. These sediments range in size from huge boulders to microscopic particles. Chemical weathering occurs when the minerals in a rock are dissolved or otherwise chemically changed. What happens to more-resistant minerals during weathering? While the less-stable minerals are chemically broken down, the more-resistant grains are broken off of the rock as smaller grains. During physical weathering, however, minerals remain chemically unchanged. Rock fragments break off of the solid rock along fractures or grain boundaries. The rock in **Figure 1** has been chemically and physically weathered.

▪ **Figure 1** When exposed to both chemical and physical weathering, granite eventually breaks apart and might look like the decomposed granite shown here.
Infer *which of the three common minerals —quartz, feldspar and mica—will be most resistant to chemical weathering.*

Demonstration

Physical v. Chemical Weathering

Place a piece of granite on the table alongside a metal file, hammer, container of water, dilute hydrochloric acid, and other items you might choose. Use each item on the granite and ask students which type of weathering was modeled. physical—hammer, file, water; chemical—water, hydrochloric acid
BL EL

Erosion The removal and transport of sediment is called erosion. **Figure 2** shows the four main agents of erosion: wind, moving water, gravity, and glaciers. Glaciers are large masses of ice that move across land. Visible signs of erosion are all around you. For example, water in streams becomes muddy after a storm because eroded silt and clay-sized particles have been mixed in it. You can observe erosion in action when a gust of wind blows soil across the infield at a baseball park. The force of the wind removes the soil and carries it away.

After rock fragments and sediments have been weathered out of the rock, they often are transported to new locations through the process of erosion. Eroded material is almost always carried downhill. Although wind can sometimes carry fine sand and dust to higher elevations, particles transported by water are almost always moved downhill. Eventually, even windblown dust and fine sand are pulled downhill by gravity.

☑ READING CHECK **Summarize** what occurs during erosion.

■ **Figure 2** Rocks and sediment are weathered and transported by the main agents of erosion—wind, moving water, gravity, and glaciers.

Tie to Previous Knowledge

Hardness and Cleavage Revisited Remind students about mineral characteristics. Ask students which mineral characteristics have the greatest effect on the shape of grains after erosion and transport. Hardness and cleavage; the grains of hard minerals take a long time to become smooth and round, and minerals with strong cleavage have a tendency to be angular.

Discussion

Weathering v. Erosion Ask students how weathering differs from erosion. Weathering loosens and breaks down particles; erosion transports particles.

Interpret the Photo

Agents of Erosion Have students study **Figure 2.** Ask students which of the four agents of erosion move the largest particles. Also ask students in which environment shown they would expect to find well-sorted sediments. Landslides and glaciers move the largest particles. The wind-blown deposits contain well-sorted sediments.

Use Science Terms

Sediment Ask students what words they can think of that are similar to *sediment* and how they might be related. sedimentation, sedate, and sedative; *Sedimentation* is another form of the noun *sediment; sedative* and *sedate* are not related.

☑ READING CHECK Rocks are broken apart into pieces, and the pieces are removed from the immediate area.

ACROSS THE CURRICULUM

Geography—Beneficial Flooding Not all floods are bad. When a river floods, quiet water can spread horizontally a great distance from the riverbanks. These floodwaters deposit a layer of fine silt, clay, and organic material that improves the fertility of the soil. The dams on the Nile River in Egypt have interrupted this cycle, and soil fertility along the river is declining.

EARTH SCIENCE JOURNAL

Creeping Sand Dunes Migration of sand dunes across irrigated fields and construction of housing developments is a big problem in parts of the world. Have students write about their ideas on how such sand encroachment might be stopped. Encourage students to write about the ways in which sand-sized grains are transported and deposited.

MiniLAB

iLab Station Rubric

Purpose Students will observe how sediments are deposited in layers.

Process Skills observe and infer, draw a conclusion

Safety Precaution Approve lab safety forms before work begins.

Teaching Strategies

- Best results will be obtained using a mixture of gravel, sand, and fine sand or clay. Use of soil will result in the formation of mud and slow settling rates.
- The best layering occurs when the sediment settles through as much water as possible. Have students hesitate momentarily with the jar upside down before tipping it upright for the last time.
- Explain to students that smaller, lighter grains fall more slowly as a result of the friction with water and viscosity of water.
- Prepare a demonstration settlement column ahead of time for students to observe if they fail to obtain the expected results.

Expected Results Sediments that have been deposited by water can be sorted into layers of similar grain size. The coarsest sediment will settle out first. The topmost layers will be composed of clay and silt.

Analysis

1. Diagrams should show the coarsest material on the bottom, with successive layers consisting of finer material.
2. coarse sediment—they are the heaviest grains present
3. Clay and silt. Answers should explain that smaller, lighter grains fall more slowly as a result of friction and the viscosity of water.

MiniLAB

iLab Station

Model Sediment Layering

How do layers form in sedimentary rocks? Sedimentary rocks usually form in layers. In this activity, you will investigate how layers form from particles that settle in water.

Procedure

1. Read and complete the lab safety form.
2. Obtain 50 mL of **sediment** from a location specified by your teacher.
3. Place the sediment in a **200 mL jar with a lid.**
4. Add **water** to the jar until it is three-fourths full.
5. Place the lid on the jar securely.
6. Pick up the jar with both hands and turn it upside down several times to mix the water and sediment. Hesitate briefly with the jar upside down before tipping it up for the last time. Place the jar on a flat surface.
7. Let the jar sit for about 5 min.
8. Observe the settling process.

Analysis

1. **Illustrate** what you observed in a diagram.
2. **Describe** what types of particles settle out first. Explain.
3. **Describe** what types of particles form the topmost layers. Explain.

Deposition When transported sediments are deposited on the ground or sink to the bottom of a body of water, deposition occurs. During the MiniLab, what happened when you stopped turning the jar full of sediment and water? The sediment sank to the bottom and was deposited in layers with the largest grains at the bottom and the smallest grains at the top. Similarly, sediments in nature are deposited when transport stops. Perhaps the wind stops blowing or a river enters a quiet lake or an ocean. In each case, the particles being carried will settle out, forming layers of sediment with the largest grains at the bottom.

Energy of transporting agents Fast-moving water can transport larger particles better than slow-moving water. As water slows down, the largest particles settle out first, then the next largest, and so on, so that different-sized particles are sorted into layers. Such deposits are characteristic of sediment transported by water and wind. Wind, however, can move only small grains. For this reason, sand dunes are commonly made of fine, well-sorted sand, as shown in **Figure 3.** Not all sediment deposits are sorted. Glaciers, for example, move all materials with equal ease. Large boulders, sand, and mud are all carried along by the ice and dumped in an unsorted pile as the glacier melts. Landslides create similar deposits when sediment moves downhill in a jumbled mass.

Lithification

Most sediments are ultimately deposited on Earth in low areas such as valleys and ocean basins. As more sediment is deposited in an area, the bottom layers are subjected to increasing pressure and temperature. These conditions cause **lithification,** the physical and chemical processes that transform sediments into sedimentary rocks. *Lithify* comes from the Greek word *lithos*, which means *stone.*

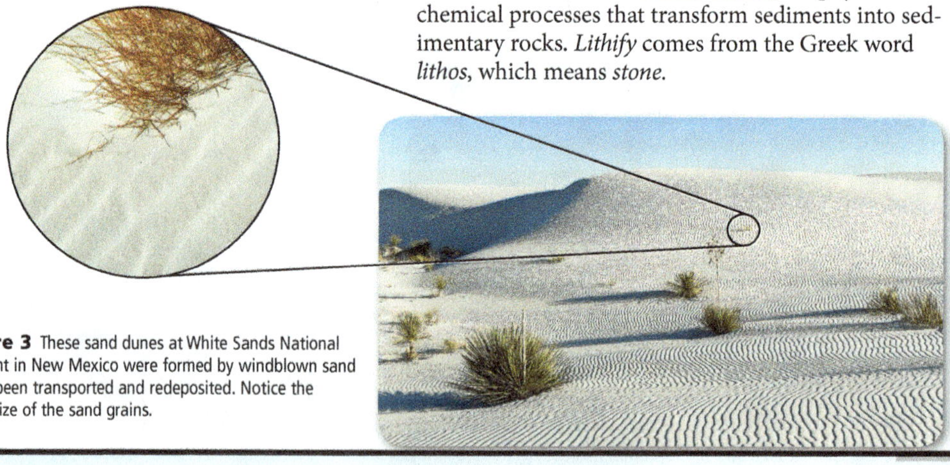

Figure 3 These sand dunes at White Sands National Monument in New Mexico were formed by windblown sand that has been transported and redeposited. Notice the uniform size of the sand grains.

Assessment

Performance Most sedimentary rock layers do not contain all the different grain sizes. Ask students to diagram and describe how layers of only coarse sand and gravel might be formed. The water might be carrying only sand and gravel, or the sediments were deposited in moving water with enough energy to prevent the finer particles from settling out.

LabManager™
Customize any lab with the LabManager™ CD-ROM.

Compaction Lithification begins with compaction. The weight of overlying sediments forces the sediment grains closer together, causing the physical changes shown in **Figure 4.** Layers of mud can contain up to 60 percent water, and these shrink as excess water is squeezed out. Sand does not compact as much as mud during burial. One reason is that individual sand grains, usually composed of quartz, do not deform under normal burial conditions. Grain-to-grain contacts in sand form a supporting framework that helps maintain open spaces between the grains. Groundwater, oil, and natural gas are commonly present in these spaces in sedimentary rocks.

Cementation Compaction is not the only force that binds the grains together. **Cementation** occurs when mineral growth glues sediment grains together into solid rock. This occurs when a new mineral, such as calcite ($CaCO_3$) or iron oxide (Fe_2O_3), grows between sediment grains as dissolved minerals precipitate out of groundwater. This process is illustrated in **Figure 5.**

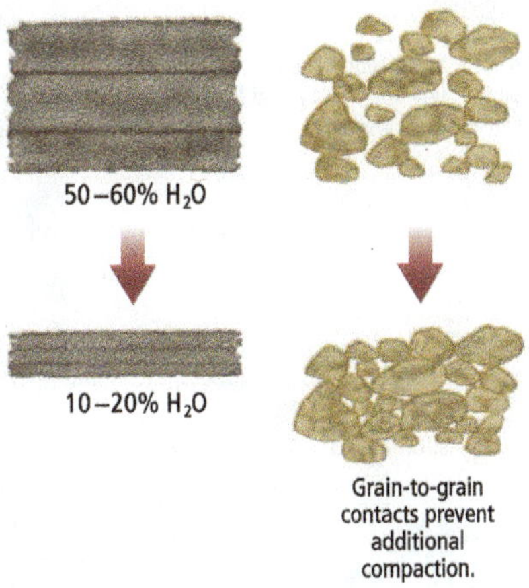

Figure 4 The high water content and flat shape of particles in mud cause it to compact greatly when subjected to the weight of overlying sediments.

Sedimentary Features

Just as igneous rocks contain information about the history of their formation, sedimentary rocks also have features and characteristics that help geologists interpret how they formed and the history of the area in which they formed.

Bedding The primary feature of sedimentary rocks is horizontal layering called **bedding.** This feature results from the way sediment settles out of water or wind. Individual beds can range in thickness from a few millimeters to several meters. There are two different types of bedding, each dependent upon the method of transport. However, the size of the grains and the material within the bedding depend upon many other factors.

FOLDABLES®
Incorporate information from this section into your Foldable.

Figure 5 Minerals precipitate out of water as it flows through pore spaces in the sediment. These minerals form the cement that glues the sediments together.

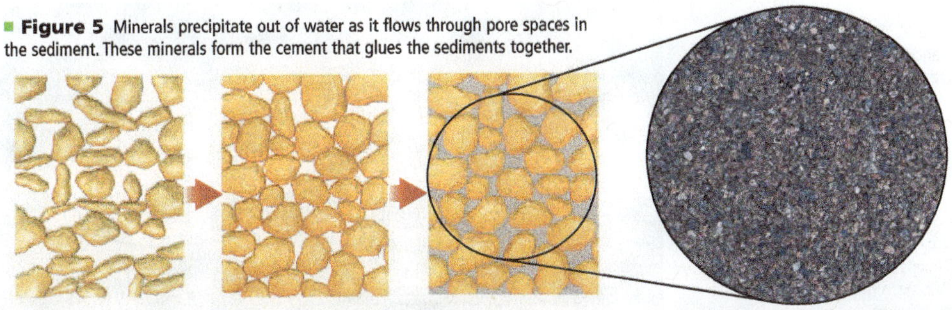

Albert J Copley/Getty Images

Enrichment

Seasonal Deposition Winter storm waves commonly have higher levels of energy than summer storm waves do. Ask students to predict how beach sediments deposited during the winter might be different from those deposited during the summer. Compared to summer waves, the higher-energy winter waves deposit coarser particles farther inland from the ocean water.

Use Science Terms

Lithos Have students find other words that have the Greek prefix *lithos*. lithograph, lithography, lithosphere, lithology, and lithophyte

Use an Analogy

Grains and Cement Ask students to explain how marshmallow-and-cereal treats are analogous to cementation. The marshmallow cement is equivalent to cement forming or precipitating between existing grains.

Project

Investigate Porosity Have groups of students design an investigation to measure the amount of pore space in a sample of coarse sand. Students might choose to use water and graduated cylinders to measure the volume of the samples. OL COOP LEARN

Tie to Previous Knowledge

What makes cement? Remind students that different minerals have different characteristics. Ask students how they might distinguish between calcite-cemented sandstone and quartz-cemented sandstone. Dilute HCl would fizz when applied to calcite-cemented sandstone.

FOLDABLES® Rubric

DIFFERENTIATED INSTRUCTION

Advanced Learners Ask students to research how quartz and calcite become common cements for sedimentary rocks, while other minerals, such as pyroxene, do not. Weathered quartz grains dissolve under the high pressure of accumulating sediments. The dissolved quartz then precipitates between sediment grains in areas of lower pressure and cements the grains together. Calcite is dissolved by carbonic acid and reprecipitates out of solution.

ACROSS THE CURRICULUM

Biology Ask students how their respiratory systems deal with airborne dust. Nasal hairs filter out dust; mucus on the lining of lungs traps particles.

Identify Misconceptions

Students might think that deposition of sediments occurs only when transport stops.

Uncover the Misconception

Have students make a list of different environments where they would expect sedimentation to occur. Have students write explanations of why the sediments are deposited in each environment.

Demonstrate the Concept

Explain to students that deposition occurs during sediment transport, as transport slows down, and when transport stops. For example, as water velocity slows down, the larger particles can be deposited even while the finer particles are still being carried away.

Assess New Knowledge

Have students explain how a well-sorted gravel bar could form in the middle of a river transporting sand and gravel. Students should be able to explain that if water velocity slows down, the gravel can be deposited in a bar while the sand-sized materials are carried farther downstream. Therefore, a decrease in velocity causes a decrease in the size of particles and the amount of sediment transported.

CAREERS IN EARTH SCIENCE

WebQuest

Figure 6 The graded bedding shown in this close-up of the Furnace Creek Formation in Death Valley, California, records an episode of deposition during which the water that carried these sediments slowed and lost energy.

CAREERS IN EARTH SCIENCE

Sedimentologist Studying the origin and deposition of sediments and their conversion to sedimentary rocks is the job of a sedimentologist. Sedimentologists are often involved in searching for and finding oil, natural gas, and economically important minerals.

WebQuest

Graded bedding Bedding in which the particle sizes become progressively finer and lighter toward the top layers is called **graded bedding.** Graded bedding is often observed in marine sedimentary rocks that were deposited by underwater landslides. As the sliding material slowly came to rest underwater, the largest and heaviest material settled out first and was followed by progressively finer material. An example of graded bedding is shown in **Figure 6.**

Cross-bedding Another characteristic feature of sedimentary rocks is cross-bedding. **Cross-bedding,** such as that shown in **Figure 7,** is formed as inclined layers of sediment are deposited across a horizontal surface. When these deposits become lithified, the cross-beds are preserved in the rock. This process is illustrated in **Figure 8.** Small-scale cross-bedding forms on sandy beaches and along sandbars in streams and rivers. Most large-scale cross-bedding is formed by migrating sand dunes.

Ripple marks When sediment is moved into small ridges by wind or wave action or by a river current, ripple marks form. The back-and-forth movement of waves forms ripples that are symmetrical, while a current flowing in one direction, such as in a river or stream, produces asymmetrical ripples. If a rippled surface is buried gently by more sediment without being disturbed, it might later be preserved in solid rock. The formation of ripple marks is illustrated in **Figure 8.**

Figure 7 The large-scale cross-beds in these ancient dunes at Zion National Park were deposited by wind.

Teacher Content Support

Turbidites and Graded Bedding Turbidites are thick sedimentary sequences containing multiple layers of graded bedding. These cyclic deposits of sandstone and shale can be thousands of meters thick, with each layer ranging from 1 cm to 1 m. Each sandstone layer has a sharp base and fines upward to be capped by a shale layer. Turbidites are formed by underwater landslides. As the current wanes, an upward-fining sequence of sediment is deposited. Subsequent turbidite flows cover preceding flows. Because each layer is deposited in deep water, there is little disturbance between events. Turbidites have been identified in rocks of all ages and are forming today along the continental shelves. These deposits could have great importance to oil and gas exploration in offshore regions.

VISUALIZING Cross-Bedding and Ripple Marks

Figure 8 Moving water and loose sediment result in the formation of sedimentary structures such as cross-bedding and ripple marks.

Cross-Bedding

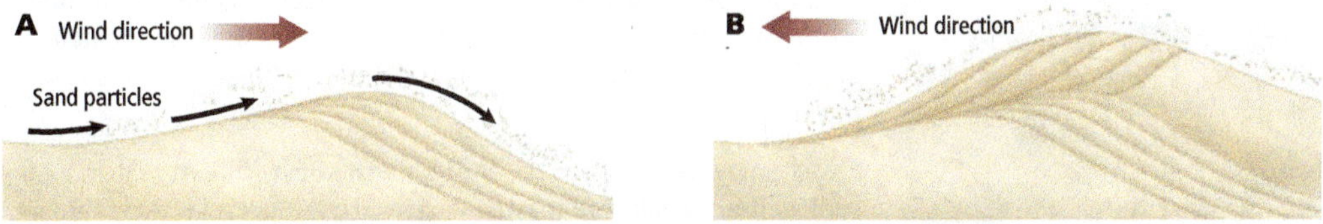

Sand carried by wind gets deposited on the downwind side of a dune. As the wind changes direction, cross-bedding is formed that records this change in direction.

Sediment on the river bottom gets pushed into small hills and ripples by the current. Additional sediment gets deposited at an angle on the downcurrent side of these hills forming cross-beds. Eventually, it levels out or new hills form and the process begins again.

Symmetrical Ripple Marks

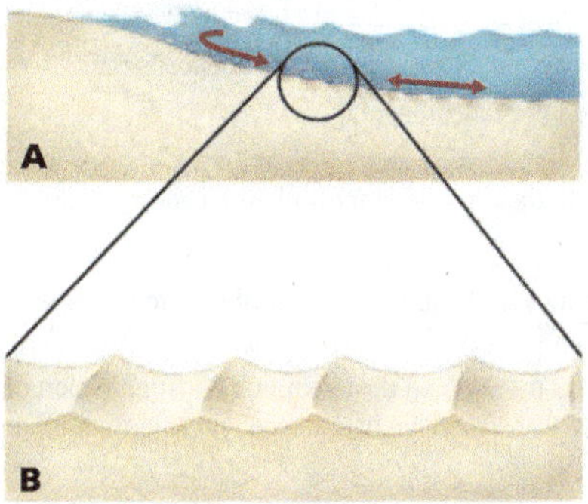

The back-and-forth wave action on a shore pushes the sand on the bottom into symmetrical ripple marks. Grain size is evenly distributed.

Asymmetrical Ripple Marks

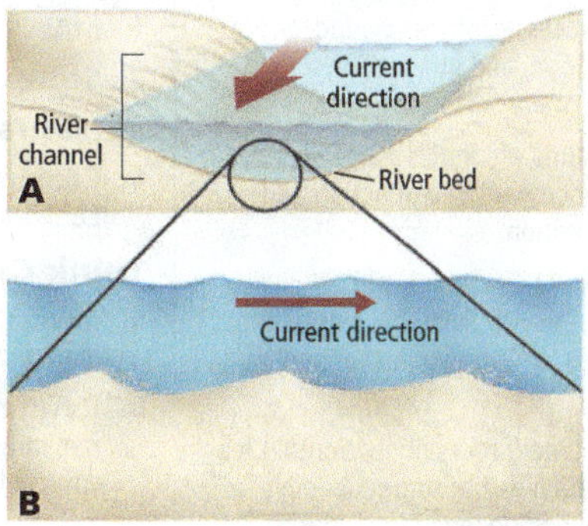

Current that flows in one direction, such as that of a river, pushes sediment on the bottom into asymmetrical ripple marks. They are steeper upstream and contain coarser sediment on the upstream side.

Concepts In Motion View an **animation of cross-bedding and ripple marks.**

VISUALIZING

Purpose

Students will learn how cross-bedding and ripple marks are formed by wind and water respectively.

Activity

Make Ripple Marks Divide the class into four groups and assign them either wind- or water-generated ripple marks. For the water ripple marks: Give each group a long, clear plastic storage box that has been filled ¼ with medium-grained sand and then half-filled with water. Place the box on a large plastic sheet on the floor. Have students smooth the sand, and then use the lid of the box to generate wave currents in the water. Encourage them to try to make both symmetrical and asymmetrical ripple marks.

For the wind-generated ripple marks: Give each group a clear plastic storage box that has had a 4–6 cm hole drilled into one end about halfway up from the bottom. The box should be filled about ¼ with fine dry sand and the lid should be firmly attached. Have students use a hair dryer to create ripple marks inside the box. Encourage them to use different speeds and directions to create ripples. Have the groups present their results and findings to the rest of the class.

EARTH SCIENCE JOURNAL

Ripples on Land and Sea After studying the Concepts in Motion on this page, have students imagine they are standing on the shore of a lake or ocean. Ask them to describe—using words and illustrations—the ripple marks they see in the water in front of them and the ripple marks they can see in the sand behind them. OL

Rubric

3 Assess

Check for Understanding

Discussion Ask students how the type of transport and deposition affects how sediment is sorted. Sediment deposited by water is usually well- sorted, sediment deposited by wind is well- sorted, and sediment deposited by ice or by landslides is not sorted.

Reteach

Outline Have students use the headings and vocabulary terms to outline the major concepts of this section.

Assessment

Skill Have students review the processes of erosion and deposition. Ask students to explain why not all organisms become fossils when they die. Erosion and transport can destroy fragile organic material; deposition and burial must occur quickly before the remains are disturbed.

Figure 9 The carbonate sand has sharp, jagged pieces and is not as rounded and smooth as the quartz sand.

Angular vs. rounded Close examination of individual sediment grains reveals that some have jagged, angular edges and some are rounded. When a rock breaks apart, the pieces are initially angular in shape. As the sediment is transported away from its source, individual pieces knock into each other. The edges are broken off and, over time, the pieces become rounded. Thus the amount of rounding is influenced by how long the sediment has been in transport, and consequently, how far the sediment has traveled. Additionally, harder minerals with little to no cleavage have a better chance of becoming rounded before they break apart. As shown in **Figure 9,** quartz sand on beaches is nearly round while carbonate sand, which is made up of seashells and calcite, is usually more angular because it is deposited closer to the source of the sediment.

Evidence of past life Probably the best-known features of sedimentary rocks are fossils. Fossils are the preserved remains, impressions, or any other evidence of once-living organisms. When an organism dies, if its remains are buried without being disturbed, it might be preserved as a fossil. During lithification, parts of the organism can be replaced by minerals and turned into rock, such as shells that have been mineralized. Fossils are of great interest to Earth scientists because fossils provide evidence of the types of organisms that lived in the distant past, the environments that existed in the past, and how organisms have changed over time. You learned first-hand how fossils can be used to interpret past events when you completed the Launch Lab at the beginning of this chapter.

SECTION 1 REVIEW

Section Self-Check

Section Summary

- The processes of weathering, erosion, deposition, and lithification form sedimentary rocks.
- Sediments are lithified into rock by the processes of compaction and cementation.
- Fossils are the remains or other evidence of once-living organisms that are often preserved in sedimentary rocks.
- Sedimentary rocks might contain features such as horizontal bedding, cross-bedding, and ripple marks.

Understand Main Ideas

1. **MAINIDEA Describe** how sediments are produced by weathering and erosion.
2. **Sequence** Use a flowchart to show why sediment deposits tend to form layers.
3. **Illustrate** the formation of graded bedding.
4. **Compare** temperature and pressure conditions at Earth's surface and below Earth's surface, and relate them to the process of lithification.

Think Critically

5. **Evaluate** this statement: It is possible for a layer of rock to show both cross-bedding and graded bedding.
6. **Determine** whether you are walking upstream or downstream along a dry mountain stream if you notice that the shape of the sediment is getting more angular as you continue walking. Explain.

WRITING IN ▸ Earth Science

7. Imagine you are designing a display for a museum based on a sedimentary rock that contains fossils of corals and other ocean-dwelling animals. Draw a picture of what this environment might have looked like, and write the accompanying description that will be posted next to the display.

SECTION 1 REVIEW

1. Physical and chemical weathering break rocks apart. The pieces that break off become sediment and are eroded or transported away.
2. The flowchart should show transportation of sediment, deposition under the influence of gravity creating horizontal layers and continued depositon.
3. Diagrams should contain the following information: grain sizes decreasing toward the top of a deposit and water energy decreasing toward the top of the deposit.
4. Pressure and temperature increase below the surface. These increases cause compaction and the start of lithification of the grains.
5. It is possible that each layer will show an upward-fining sequence. Also, if cross-bedding is being formed while water velocity is decreasing, the grain size can decrease from one cross-bed to another. However, usually cross-beds are well sorted.
6. Because sediment becomes more rounded as it travels away from its source, you can determine that you are walking upstream, toward the source of the sediments.
7. The picture should contain a variety of corals and other sea animals with which students are familiar. Any reasonable description is acceptable.

Rubric

SECTION 2

Essential Questions

- How can the different types of clastic sedimentary rocks be described?
- How do chemical sedimentary rocks form?
- What are biochemical sedimentary rocks?

Review Vocabulary

saturated: the maximum possible content of dissolved minerals in solution

New Vocabulary

clastic sedimentary rock
clastic
porosity
evaporite

Types of Sedimentary Rocks

MAINIDEA Sedimentary rocks are classified by their mode of formation.

EARTH SCIENCE 4 YOU **If you have ever walked along the beach or along a riverbank, you might have noticed different sizes of sediments. The grain size of the sediment determines what type of sedimentary rock it can become.**

Clastic Sedimentary Rocks

The most common sedimentary rocks, **clastic sedimentary rocks,** are formed from the abundant deposits of loose sediments that accumulate on Earth's surface. The word **clastic** comes from the Greek word *klastos,* meaning *broken.* These rocks are further classified according to the sizes of their particles. As you read about each rock type, refer to **Table 1** on the next page, which summarizes the classification of sedimentary rocks based on grain size, mode of formation, and mineral content.

Coarse-grained rocks Sedimentary rocks consisting of gravel-sized rock and mineral fragments are classified as coarse-grained rocks, samples of which are shown in **Figure 10.** Conglomerates have rounded, gravel-sized particles. Because of its relatively large mass, gravel is transported by high-energy flows of water, such as those generated by mountain streams, flooding rivers, some ocean waves, and glacial meltwater. During transport, gravel becomes abraded and rounded as the particles scrape against one another. This is why beach and river gravels are often well rounded. Lithification turns these sediments into conglomerates.

In contrast, breccias are composed of angular, gravel-sized particles. The angularity indicates that the sediments from which they formed did not have time to become rounded. This suggests that the particles were transported only a short distance and deposited close to their source. Refer to **Table 1** to see how these rocks are named.

Conglomerate | Breccia

Figure 10 Conglomerates and breccias are made of coarse sediments that have been transported by high-energy water.
Infer *the circumstances that might cause the types of transport necessary for each to form.*

CULTURAL DIVERSITY

Ayers Rock—Uluru In Kata Tjuta National Park, southwest of Alice Springs in the Northern Territory of Australia, a huge sandstone monolith rises 345 m above the desert. It is 2 km wide and semi-oval in shape. It is called Ayers Rock, or Uluru, as the Aboriginals (native Australians) named it. Uluru is a sacred site for the Aboriginal people. They believe that there is an energy source emanating from the rock and that the area around it is inhabited by ancestral beings. The rock seems to change color from blue to red, depending on the time of day and the quality of light. Uluru bears many carvings and paintings made by the Aboriginals who conduct sacred rites there. The Australian government restored ownership of Uluru and the Kata Tjuta National Park to the Aboriginals in 1985.

SECTION 2

1 Focus

Focus Transparency

Before presenting the lesson, project **Section Focus Transparency 15** and have students answer the accompanying questions. EL

MAINIDEA

Different Origins Ask students to restate where sediment comes from and where it ends up. Ask if they think that all sedimentary rocks form from sediments. Show students a sample of rock salt and tell them it is composed of the mineral halite. Ask those students who said that all sedimentary rocks form from sediments to defend or revise their answers.

2 Teach

Enrichment

Pan for Gold Have students research placer (PLA sur) gold deposits. Ask students how these deposits are related to the transport of sediments. Placer deposits are formed by sorting during sediment transport. Have students discuss the characteristics of gold that cause it to collect in placer deposits. Gold is highly resistant to weathering and has a high density. These characteristics cause gold to be left in pockets during transport and deposition.

Concept Development

Placer Deposits Have interested students research the origin of the word *placer.*

Caption Question Fig. 10 high-energy water, fast-flowing water, flash flooding, etc.

Activity

Windblown Particles Have students use magnifying lenses and microscopes to observe dust samples from windowsills or elsewhere. Have students list the types of particles they can identify. Ask students to predict what percentage of the particles are silt and clay grains.

Teacher Content Support

Permeability and Oil Exploration Groundwater geologists and petroleum geologists are usually more interested in the permeability of a sandstone than its porosity. Permeability is a measure of how well liquids flow through a rock and is determined by how interconnected the pore spaces are. Oil might not flow from high-porosity sandstone if it has low permeability. In some cases, explosives are used to shatter a rock formation in an oil well to improve permeability and production rates.

Explore **sedimentary rock formation with an interactive table.** Concepts In Motion

Table 1 Classification of Sedimentary Rocks

Classification	Texture/Grain Size	Composition	Rock Name
Clastic	coarse (> 2 mm)	Fragments of any rock type—quartz, chert and quartzite common } rounded angular	conglomerate breccia
	medium (1/16 mm to 2 mm)	quartz and rock fragments quartz, potassium feldspar and rock fragments	sandstone arkose
	fine (1/256 mm–1/16 mm)	quartz and clay	siltstone
	very fine (< 1/256 mm)	quartz and clay	shale
Biochemical	microcrystalline with conchoidal fracture	calcite ($CaCO_3$) quartz (SiO_2)	micrite chert
	abundant fossils in micrite matrix	calcite ($CaCO_3$)	fossiliferous limestone
	shells and shell fragments loosely cemented	calcite ($CaCO_3$)	coquina
	microscopic shells and clay	calcite ($CaCO_3$)	chalk
	variously sized fragments	highly altered plant remains, some plant fossils	coal
Chemical	ooids (small spheres of calcium carbonate)	calcite ($CaCO_3$)	oolitic limestone
	fine to coarsely crystalline	calcite ($CaCO_3$)	crystalline limestone
	fine to coarsely crystalline	dolomite $(Ca,Mg)CO_3$ (will effervesce if powdered)	dolostone
	very finely crystalline	quartz (SiO_2)—light colored; dark colored calcite ($CaCO_3$)	chert; flint micrite
	fine to coarsely crystalline	gypsum ($CaSO_4 \bullet 2H_2O$)	rock gypsum
	fine to coarsely crystalline	halite (NaCl)	rock salt

Medium-grained rocks Stream and river channels, beaches, and deserts often contain abundant sand-sized sediments. Sedimentary rocks that contain sand-sized rock and mineral fragments are classified as medium-grained clastic rocks. Refer to **Table1** for a listing of rocks with sand-sized particles. Sandstone usually contains several features of interest to scientists. For example, because ripple marks and cross-bedding indicate the direction of current flow, geologists use sandstone layers to map ancient stream and river channels.

VOCABULARY

ACADEMIC VOCABULARY

Reservoir

a subsurface area of rock that has enough porosity to allow for the accumulation of oil, natural gas, or water

The newly discovered reservoir contained large amounts of natural gas and oil.

Another important feature of sandstone is its relatively high porosity. **Porosity** is the percentage of open spaces between grains in a rock. Loose sand can have a porosity of up to 40 percent. Some of these open spaces are maintained during the formation of sandstone, often resulting in porosities as high as 30 percent. When pore spaces are connected to one another, fluids can move through sandstone. This feature makes sandstone layers valuable as underground reservoirs of oil, natural gas, and groundwater.

Demonstration

Porosity and Permeability Put some coarse gravel in a clear container to demonstrate grain-to-grain contacts and pore spaces. Have students use magnifying lenses to examine the sediment in more detail. Tell students magnified sand grains would look the same. BL

DIFFERENTIATED INSTRUCTION

Visually Impaired Have students with visual impairments feel sediment of different sizes, including gravel, sand, silt, and clay, to differentiate them based on grain size. BL OL

Fine-grained rocks Sedimentary rocks consisting of silt- and clay-sized particles, such as siltstone and shale, are called fine-grained rocks. These rocks represent environments like swamps, ponds, and deep oceans which have still or slow-moving waters. In the absence of strong currents and wave action, these sediments settle to the bottom where they accumulate in thin horizontal layers. Shale often breaks along thin layers, as shown in **Figure 11.** Unlike sandstone, fine-grained sedimentary rock has low porosity and often forms barriers that hinder the movement of groundwater and oil. **Table 1** shows how these rocks are named.

☑ READING CHECK **Identify** the types of environments in which fine-grained rocks form.

■ **Figure 11** The very fine-grained sediment that formed this shale was deposited in thin layers in still waters.

Chemical and Biochemical Sedimentary Rocks

The formation of chemical and biochemical rocks involves the processes of evaporation and precipitation of minerals. During weathering, minerals can be dissolved and carried into lakes and oceans. As water evaporates from the lakes and oceans, the dissolved minerals are left behind. In arid regions, high evaporation rates can increase the concentration of dissolved minerals in bodies of water. The Great Salt Lake, shown in **Figure 12,** is an example of a lake that has high concentrations of dissolved minerals.

Chemical sedimentary rocks When the concentration of dissolved minerals in a body of water reaches saturation, crystals can precipitate out of solution and settle to the bottom. As a result, layers of chemical sedimentary rocks form, most of which are called **evaporites.** Evaporites primarily form in arid regions, drainage basins on continents that have low water flow, and in coastal settings. Because these areas usually have minimal freshwater input and high rates of evaporation, the concentration of dissolved minerals remains high. Over time, thick layers of evaporite minerals can accumulate on basin floors, as illustrated in **Figure 12.**

■ **Figure 12** The constant evaporation from a body of salt water results in precipitation of large amounts of salt. This process has been occurring in the Great Salt Lake in Utah for approximately 18,000 years.

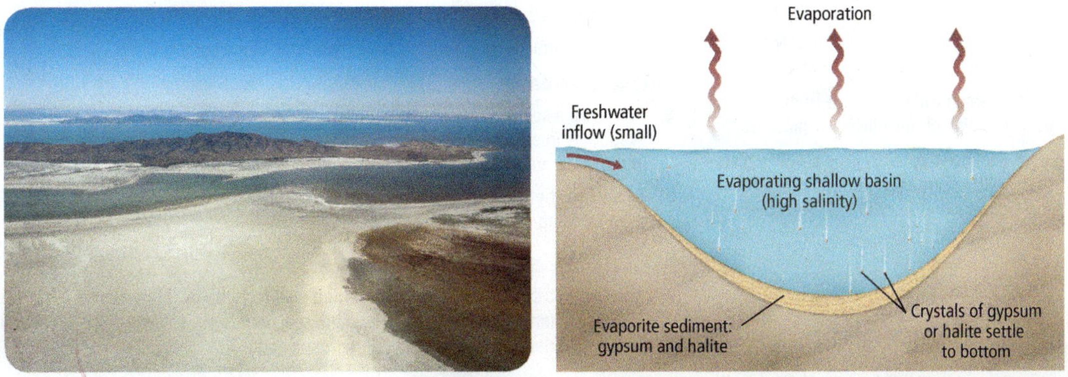

Discussion

Oil Reservoirs Ask students why it might be important to have both fine-grained layers and sandstone layers in an oil field. The sandstone layers have a high porosity that allows fluid to move easily. The fine-grained layers have low porosity and thus trap the oil and gas in the high porosity layer. Without the fine-grained layers, the resources would migrate away.

Discussion

Solubility Ask students what common characteristic all evaporite minerals share. They are all water soluble over time.

Model

Evaporite Formation Have groups of students design an investigation to model an evaporite basin. Have them pay specific attention to how they might create thick layers of evaporites. AL COOP LEARN

Activity

Salt Precipitation Have students dissolve a measured amount of salt in water. Ask: After the water evaporates, is all the original salt recovered? How could you be sure? All of the original salt should be recovered. If students determine the mass of the container beforehand, they will be able to determine the mass of the salt left behind after evaporation. OL

Concept Development

Sedimentary Rock Types Display clear containers of materials that represent the three main groups of sedimentary rocks: a sand-gravel mix (clastic), a shell-plant mix (organic), and rock salt (evaporite).

☑ READING CHECK Fine-grained rocks form in environments with slow-moving water, such as ponds, swamps, and the deep ocean.

ACROSS THE CURRICULUM

History Thick limestone beds underlie the Yucatán Peninsula. In many locations, cave roofs have collapsed, forming water-filled sinkholes called cenotes (sih NOH teez). Ask students to research what role these cenotes played in the Mayan civilization. They were involved in some religious ceremonies, and they provided a source of freshwater in a region with few above-ground rivers.

DIFFERENTIATED INSTRUCTION

Advanced Learners Have students design experiments to compare the porosity of gravel to that of medium-grained sand. Students might also be interested in finding out the effect of sorting on porosity.

Interpret the Photo

Microfossils Have students study **Figure 13.** Ask: How can there be different-sized fossils in the same rock? Organisms of all sizes can live and die in the same area. When marine organisms die, their shells accumulate on the ocean floor and can form limestone with various sized fossils.

3 Assess

Check for Understanding

Analogy Ask students the following question: If a sedimentary rock provides a "snapshot" of past surface conditions, how might a scientist obtain a "video" of past surface conditions? A thick sequence of sediments represents changing depositional conditions over time.

Reteach

Understand Vocabulary Have students use the vocabulary terms to make an outline of the major concepts in this section.

Assessment

Performance Have students make a data table that lists the three types of sedimentary rocks, their mode of formation, and their composition.

■ **Figure 13** Limestone can contain many different fossil organisms. Geologists can interpret where and when the limestone formed by studying the fossils within the rock.

Biochemical sedimentary rocks Biochemical sedimentary rocks are formed from the remains of once-living organisms. The most abundant of these rocks is limestone, which is composed primarily of calcite. Some organisms that live in the ocean use the calcium carbonate that is dissolved in seawater to make their shells. When these organisms die, their shells settle to the bottom of the ocean and can form thick layers of carbonate sediment. During burial and lithification, calcium carbonate precipitates out of the water, crystallizes between the grains of carbonate sediment, and forms limestone.

Limestone is common in shallow water environments, such as those in the Bahamas, where coral reefs thrive in 15 to 20 m of water just offshore. The skeletal and shell materials that are currently accumulating there will someday become limestone as well. Many types of limestone contain evidence of their biological origin in the form of abundant fossils. As shown in **Figure 13,** these fossils can range from large-shelled organisms to microscopic, unicellular organisms. However, not all limestone contains fossils or is biochemical in origin. Some limestone has a crystalline texture, or consists of tiny spheres of carbonate sand called ooids. These are listed in **Table 1.**

Other organisms make their shells out of silica, or microcrystalline quartz. After these organisms die and settle to the bottom of the ocean, their shells can form sediment that is often referred to as siliceous ooze because it is rich in silica. Siliceous ooze becomes lithified into the sedimentary rock chert, which is also listed in **Table 1.**

SECTION 2 REVIEW

Section Self-Check

Section Summary

- Sedimentary rocks can be clastic, chemical, or biochemical.
- Clastic rocks form from sediments and are classified by particle size and shape.
- Chemical rocks form primarily from minerals precipitated from water.
- Biochemical rocks form from the remains of once-living organisms.
- Sedimentary rocks provide geologists with information about surface conditions that existed in Earth's past.

Understand Main Ideas

1. **MAINIDEA State** the type of sedimentary rock that is formed from the erosion and transport of rocks and sediments.
2. **Infer** why coal is a biochemical sedimentary rock.
3. **Calculate** the factor by which grain size increases with each texture category.
4. **Analyze** the environmental conditions to explain why most chemical sedimentary rocks form mainly in areas that have high rates of evaporation.

Think Critically

5. **Propose** a scenario to explain how it is possible to form additional layers of evaporites in a body of seawater when the original amount of dissolved minerals in the water was enough to form only a thin evaporite.
6. **Examine** the layers of shale in **Figure 11** and explain why shale contains no cross-bedding or ripple marks.

MATH IN ▶ Earth Science

7. Assume that the volume of a layer of mud will decrease by 35 percent during deposition and compaction. If the original sediment layer is 30 cm thick, what will be the thickness of the shale layer after compaction?

SECTION 2 REVIEW

1. clastic sedimentary rock
2. Coal is formed from once-living plant material.
3. According to **Table 1,** each texture category increases by a factor of 16.
4. Evaporation continuously removes freshwater from the body of water. This helps to maintain high concentrations of minerals in the remaining water. These minerals precipitate out to form chemical sedimentary rock.
5. As evaporation continues, additional seawater flows in to replace the evaporating water. This water also contains dissolved minerals. Evaporation of this additional seawater causes precipitation of additional minerals. This process keeps repeating.
6. Shale is composed of flat grains of mud and clay. These grains stick together in flat layers. It is not possible for them to accumulate on an angle to form either cross-beds or ripple marks.
7. The final thickness will be 65 percent of the original thickness. $30 \text{ cm} \times 0.65 = 19.5 \text{ cm}$.

SECTION 3

Metamorphic Rocks

MAINIDEA Metamorphic rocks form when preexisting rocks are exposed to increases in temperature and pressure and to hydrothermal solutions.

Essential Questions

- What are the different types and causes of metamorphism?
- How are metamorphic textures described?
- How do mineral and compositional changes occur during metamorphism?
- How are rocks classified using the rock cycle?

Review Vocabulary

intrusive: rocks that form from magma that cooled and crystallized slowly beneath Earth's surface

New Vocabulary

foliated
nonfoliated
regional metamorphism
contact metamorphism
hydrothermal metamorphism
rock cycle

EARTH SCIENCE 4 YOU When you make a cake, all of the individual ingredients that you put into the pan change into something new. When rocks are exposed to high temperatures, their individual characteristics also change into something new and form a completely different rock.

Recognizing Metamorphic Rock

The rock layers shown in **Figure 14** have been metamorphosed (meh tuh MOR fohzd)—this means that they have been changed. How do geologists know that this has happened? Pressure and temperature increase with depth. When temperature and pressure becomes high enough, rocks melt and form magma. But what happens if the rocks do not reach the melting point? When temperature and pressure combine and change the texture, mineral composition, or chemical composition of a rock without melting it, a metamorphic rock forms. The word *metamorphism* is derived from the Greek words *meta,* meaning *change,* and *morphé,* meaning *form.* During metamorphism, a rock changes form while remaining solid.

The high temperatures required for metamorphism are ultimately derived from Earth's internal heat, either through deep burial or from nearby igneous intrusions. The high pressures required for metamorphism come from deep burial or from compression during mountain building.

Tony Waltham/Robert Harding World Imagery/CORBIS

Figure 14 Strong forces were required to bend these rock layers into the shape they are today.
Hypothesize *the changes that occurred to the sediments after they were deposited.*

SECTION 3

1 Focus

Focus Transparency

Before presenting the lesson, project **Section Focus Transparency 16** and have students answer the accompanying questions. EL

MAINIDEA

Metamorphism Ask students to define the word *metamorphose.* to change form Ask students to list ways in which things change. Write the list on the board. grow, change color, bake, weather, die, change shape, etc. Be sure the list includes some changes that involve temperature and pressure. Tell students that under the right conditions, increases in temperature and/or pressure might cause rocks to metamorphose.

2 Teach

Tie to Previous Knowledge

Recall Bowen's Reaction Series Have students review how rocks melt. Ask students the following question: As rocks reach high-grade metamorphism and are close to melting, which minerals will be most affected? the minerals that crystallize last, such as quartz, potassium feldspar, and sodium-rich plagioclase feldspar

Caption Question Fig. 14 Sediment was deposited, lithified, then the layers were uplifted and folded during metamorphism.

Demonstration

Metamorphosis and Baking Remind students that baking is similar to metamorphism. Place the ingredients for brownies on the front table. Have volunteers help to mix the brownies, while explaining that although the items are no longer individually recognizable, the matter from which they are made is still present. If possible, bake the brownies at school so you can show your students the immediate results of adding temperature to the mixture. If you have more than one class, you can use the baked brownies from the class before as your final result for the next class. Stress again the relationship of increased temperature and changing a mixture. Point out that rocks undergo similar changes when they are exposed to increases in temperature. BL EL

Model

Grain Orientation Have students model the effect of pressure on grain orientation. First, mix a batch of salt dough using equal amounts of the following ingredients: flour, salt, and water. Mix salt and flour thoroughly, then add water a little at a time until you get the proper consistency. Food coloring can be added to give contrast with the rice. Next, have students mix uncooked rice grains into a ball of dough, then cut open the ball with a plastic knife and draw the orientation of the grains. Have students re-form the ball and then flatten it down on a desk. When they cut open the dough again, students should observe that most of the rice grains are oriented with their long axes perpendicular to the direction of pressure. OL

Tie to Previous Knowledge

Solid Solution in Metamorphic Rocks Remind students about the plagioclase feldspar continuous reaction series. Have them explain how this is an example of a solid-state alteration. The solid plagioclase crystals react with the surrounding liquid and continuously change their chemical composition without melting.

READING CHECK Metamorphic minerals are minerals that form during metamorphism. They are stable under different conditions than other minerals.

■ **Figure 15** Metamorphic minerals, such as mica, staurolite, garnet, and talc (shown above, clockwise from top left), occur in many colors, shapes, and crystal sizes. Colors can be dark or light and crystal form can be unique.

Metamorphic minerals How do minerals change without melting? Think back to the concept of fractional crystallization. Bowen's reaction series shows that all minerals are stable at certain temperatures and they crystallize from magma along a range of different temperatures. Scientists have discovered that these stability ranges also apply to minerals in solid rock. During metamorphism, the minerals in a rock change into new minerals that are stable under the new temperature and pressure conditions. Minerals that change in this way are said to undergo solid-state alterations. Scientists have conducted experiments to identify the metamorphic conditions that create specific minerals. When the same minerals are identified in rocks, scientists are able to interpret the conditions inside the crust during the rocks' metamorphism. **Figure 15** shows some common metamorphic minerals.

READING CHECK **Explain** what metamorphic minerals are.

Metamorphic textures Metamorphic rocks are classified into two textural groups: foliated and nonfoliated. Geologists use metamorphic textures and mineral composition to identify metamorphic rocks. **Figure 16** shows how these two characteristics are used in the classification of metamorphic rocks.

Foliated rocks Layers and bands of minerals characterize **foliated** metamorphic rocks. High pressure during metamorphism causes minerals with flat or needlelike crystals to form with their long axes perpendicular to the pressure, as shown in **Figure 17.** This parallel alignment of minerals creates the layers observed in foliated metamorphic rocks.

■ **Figure 16** Increasing grain size parallels changes in composition and development of foliation. Grain size is not a factor in nonfoliated rocks.

Metamorphic Rock Identification Chart

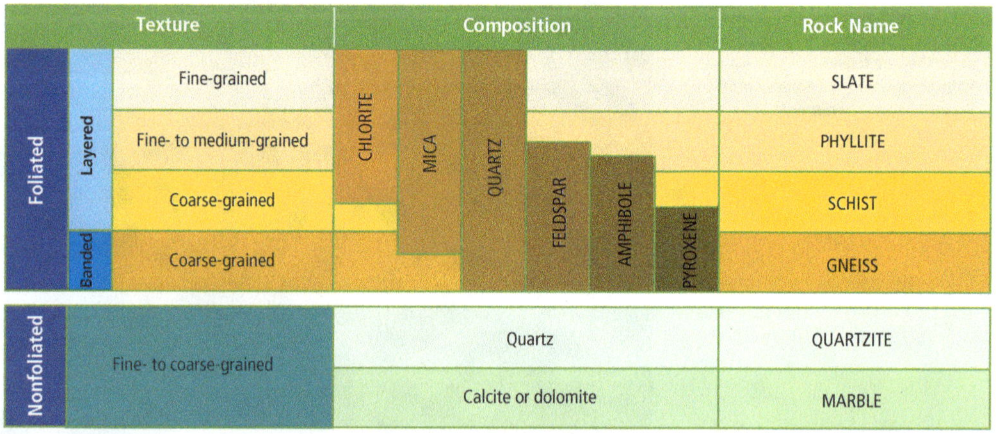

ACROSS THE CURRICULUM

Literature and Mythology Staurolite is a unique metamorphic mineral that consists of two intersecting crystals in the shape of a cross. Known as fairy stone, it is named for the Greek words *stauros* and *lithos,* meaning *cross stone.* The crystals intersect at two different angles. The 90° crystals are called Greek crosses, and the 60° crystals are called St. Andrew's crosses. Staurolite has been considered a good-luck charm for centuries and is often worn as an amulet. One legend of its origin centers around Cherokee Native Americans. Staurolite crystals are said to be the tears of the Cherokee people as they walked the "Trail of Tears" when they were forced to leave their ancestral homeland in the East. Cherokee County, North Carolina, is a major source of staurolite crystals.

■ **Figure 17** Foliation develops when pressure is applied from opposite directions. The foliation develops perpendicular to the pressure direction.

Nonfoliated rocks Unlike foliated rocks, **nonfoliated** metamorphic rocks are composed mainly of minerals that form with blocky crystal shapes. Two common examples of nonfoliated rocks, shown in **Figure 18,** are quartzite and marble. Quartzite is a hard, often light-colored rock formed by the metamorphism of quartz-rich sandstone. Marble is formed by the metamorphism of limestone or dolomite. Some marbles have smooth textures that are formed by interlocking grains of calcite. These marbles are often used in sculptures. Fossils are rarely preserved in metamorphic rocks.

Under certain conditions, new metamorphic minerals can grow large while the surrounding minerals remain small. The large crystals, which can range in size from a few millimeters to a few centimeters, are called porphyroblasts. Although these crystals resemble the very large crystals that form in pegmatite granite, they are not the same. Instead of forming from magma, they form in solid rock through the reorganization of atoms during metamorphism. Garnet, shown in **Figure 18,** is, a mineral that commonly forms porphyroblasts.

■ **Figure 18** As a result of the extreme heat and pressure during metamorphism, marble rarely contains fossils. Metamorphism does not, however, always destroy cross-bedding and ripple marks, which can be seen in some quartzites. Garnet porphyroblasts can grow to be quite large in some rocks.

Concept Development

Pavement Materials In the past, roads were sometimes paved with rock. Ask students which type of metamorphic rock would make the best street paver: quartzite, gneiss, schist, or marble? Why? Quartzite would make the best street paver. It is composed of quartz, a hard and stable mineral at surface temperatures. Marble is composed of calcite, and thus it is soft. The foliation in gneiss and schist create planes of weakness along the foliations.

Use an Analogy

Visualize Foliation Tell students mineral grains in a nonfoliated rock make a pattern like the pieces of a jigsaw puzzle. The mineral grains in a foliated rock are more like a stack of playing cards. BL

Teacher Content Support

Ancient Roads The streets of the Roman Forum and ancient cities such as Pompeii and Herculaneum are paved with blocks of basalt. Despite basalt's hardness and durability, chariot tracks were worn into the stones. They are still visible on some of the roadways 2000 years after the last chariot passed by.

Interpret the Photo

Growth of Porphyroblasts

Have students carefully observe the crystals around the garnet porphyroblast shown in **Figure 18.** Ask: What evidence is there to suggest that the porphyroblast grew into the rock? The overlapping white crystals show that the growing garnet formed between the matrix crystals and pushed them aside instead of replacing them. This is evidence of the migration of atoms and fluids during metamorphism.

GeoLAB

iLab Station

The GeoLab located at the end of the chapter can be used at this point in the lesson.

Teacher Content Support

Garnets Most people think of garnet as being a gemstone. However, few garnets are of gem quality. The vast majority of metamorphic garnets are used as abrasives. Garnet is a hard mineral that breaks with an irregular fracture. This makes it especially useful in sandpaper. The light-orange sand-paper sold in hardware stores is often called garnet paper because it is made from crushed garnets.

DIFFERENTIATED INSTRUCTION

Advanced Learners Foliation is not the same as layered bedding or cross-stratification. Challenge students to demonstrate that foliation can form perpendicular to layered bedding. Students might use layers of modeling clay and apply pressure from different sides to simulate the formation of foliation. They will observe that pressure that is perpendicular to the bedding will form foliation that also is perpendicular to the bedding.

Problem-Solving LAB

Purpose Students will compare metamorphic mineral assemblages formed under different conditions from different parent rocks.

Process Skills interpret scientific illustrations, compare and contrast, think critically

Teaching Strategy Explain to students how to read the chart. Each horizontal band represents the stability range of a single mineral. A vertical line from each metamorphic grade will intersect the minerals that will form the characteristic mineral assemblage for those conditions.

Analysis

1. chlorite and sodium-rich plagioclase
2. sillimanite

Think Critically

3. shale—muscovite, biotite, garnet, staurolite, kyanite, albite; basalt—chlorite, epidote, amphibole, garnet, plagioclase
4. Shale has a higher Al and Si content, while basalt has more Fe and Mg. Aluminum-silicate minerals such as staurolite, kyanite, and sillimanite form from shale, while more Fe- and Mg-rich minerals such as amphiboles and pyroxenes form from basalt.
5. Limestone contains few elements other than Ca, C, and O, so minerals other than calcite cannot form. Also, calcite is stable over a wide range of temperatures and pressures.

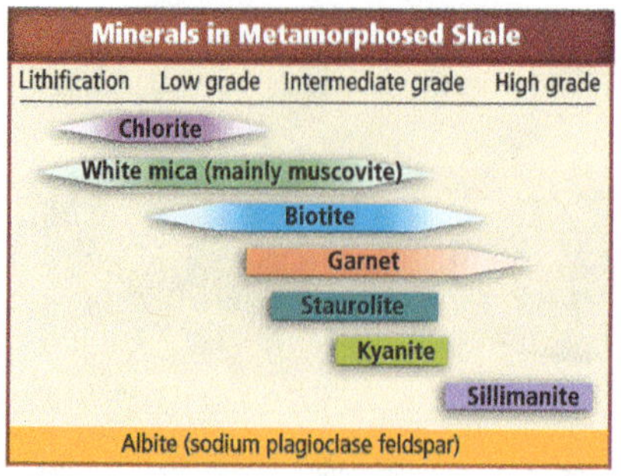

Figure 19 Metamorphism of shale results in the formation of minerals that provide the wide variety of color observed in slate.

Grades of Metamorphism

Different combinations of temperature and pressure result in different grades of metamorphism. Low-grade metamorphism is associated with low temperatures and pressures and a particular suite of minerals and textures. High-grade metamorphism is associated with high temperatures and pressures and a different suite of minerals and textures. Intermediate-grade metamorphism is in between low- and high-grade metamorphism.

Figure 19 shows the minerals present in metamorphosed shale. Note the change in composition as conditions change from low-grade to high-grade metamorphism. Geologists can create metamorphic maps by plotting the location of metamorphic minerals. Knowing the temperatures that certain areas experienced when rocks were forming helps geologists locate valuable metamorphic minerals such as garnet and talc. Studying the distribution of metamorphic minerals helps geologists to interpret the metamorphic history of an area.

Types of Metamorphism

The effects of metamorphism can be the result of contact metamorphism, regional metamorphism, or hydrothermal metamorphism. The minerals that form and the degree of change in the rocks provide information as to the type and grade of metamorphism that occurred.

Problem-Solving LAB

Interpret Scientific Illustrations

Which metamorphic minerals will form? The minerals that form in metamorphic rocks depend on the metamorphic grade and composition of the original rock. The figure below and **Figure 19** show the mineral groups that form under different metamorphic conditions.

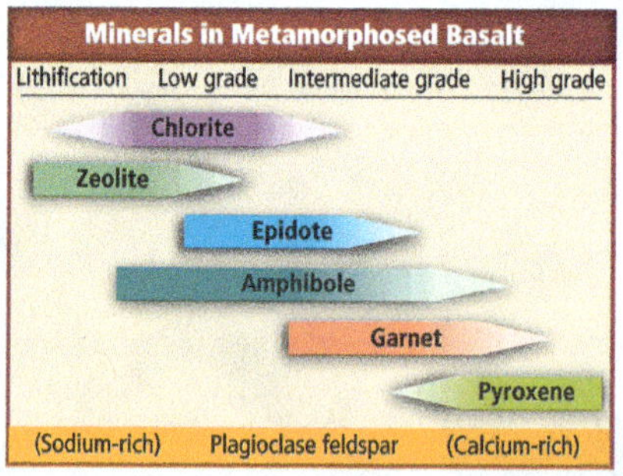

Analysis

1. What minerals are formed when shale and basalt are exposed to low-grade metamorphism?
2. Under high-grade metamorphism, what mineral is formed in shale but not in basalt?

Think Critically

3. **Compare** the mineral groups that you would expect to form from intermediate-grade metamorphism of shale and basalt.
4. **Describe** the major compositional differences between shale and basalt. How are these differences reflected in the minerals formed during metamorphism?
5. **Explain** When limestone is metamorphosed, there is little change in mineral composition. Calcite is still the dominant mineral. Explain why this happens.

DIFFERENTIATED INSTRUCTION

Advanced Learners Have students research contact metamorphism in limestone and dolomite. What are these deposits called? What types of valuable minerals are found in these deposits? These are known as skarn deposits and sometimes contain tungsten, copper, zinc, and gold.

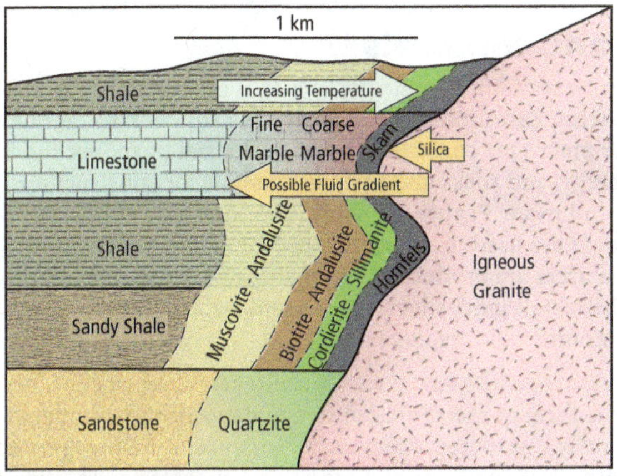

■ **Figure 20** Contact metamorphism from the intrusion of this granite batholith has caused zones of metamorphic minerals to form.

Apply ***what you know about contact metamorphism to determine the type of rock that would form if the granite intrusion was metamorphosed.***

Regional metamorphism When temperature and pressure affect large regions of Earth's crust, they produce large belts of **regional metamorphism.** The metamorphism can range in grade from low to high grade. Results of regional metamorphism include changes in minerals and rock types, foliation, and folding and deforming of the rock layers that make up the area. The folded rock layers shown in **Figure 14** experienced regional metamorphism.

VOCABULARY
SCIENCE USAGE V. COMMON USAGE
Intrusion
Science usage: the placement of a body of magma into preexisting rock

Common usage: joining or coming into without being invited

Contact metamorphism When molten material, such as that in an igneous intrusion, comes in contact with solid rock, a local effect called **contact metamorphism** occurs. High temperature and moderate-to-low pressure form mineral assemblages that are characteristic of contact metamorphism. **Figure 20** shows zones of different minerals surrounding an intrusion. Because temperature decreases with distance from an intrusion, metamorphic effects also decrease with distance. Recall that minerals crystallize at specific temperatures. Metamorphic minerals that form at high temperatures occur closest to the intrusion, where it is hottest. Because lava cools too quickly for the heat to penetrate far into surface rocks, contact metamorphism from extrusive igneous rocks is limited to thin zones.

Hydrothermal metamorphism When very hot water reacts with rock and alters its chemical and mineral composition, **hydrothermal metamorphism** occurs. The word *hydrothermal* is derived from the Greek words *hydro,* meaning *water,* and *thermal,* meaning *heat.* As hot fluids migrate in and out of the rock during metamorphism, the original mineral composition and texture of the rock can change. Chemical changes are common during contact metamorphism near igneous intrusions and active volcanoes. Valuable ore deposits of gold, copper, zinc, tungsten, and lead are formed in this manner. The gold deposited in the quartz shown in **Figure 21** is the result of hydrothermal metamorphism.

■ **Figure 21** When the hydrothermal solution in the quartz cooled, gold veins formed.

Enrichment

Brickmaking and Metamorphism Have students research how bricks are made. To what type of metamorphism is this process most similar? This is a high-temperature, low-pressure process, so it is most similar to contact metamorphism. What determines the characteristics of the final brick? The temperature, the duration of firing, and the type of clay used are factors that affect the bricks' characteristics.

■ **Caption Question Fig. 20** gneiss

Teacher Content Support

Skarn Deposits Contact metamorphism occurs whenever rocks are intruded by magma. The type of minerals that form in the area around the intrusion depends on the type of parent rocks. A special type of contact metamorphism, called a skarn deposit, occurs when a siliceous magma intrudes dirty carbonate rocks, such as silty limestone or calcareous shale. In this situation, there are a great variety of minerals and elements present, and significant changes can occur between the magma and the carbonate rocks. Skarn deposits are usually rich in calcium and silica, forming primarily garnet and pyroxene. Some skarns contain valuable concentrations of metals. Important sources of iron, gold, copper, zinc, and wollastonite have been found in skarn deposits.

Collaborative Learning

Interpret Field Observations Have groups of students discuss the following scenario: Some geologists find small intrusions that are surrounded by rings of erosion-resistant rock. What might be an explanation for these features? Contact metamorphism closest to the intrusion formed rings of harder rock. OL

Discussion

Rock Cycle Conduct a class discussion to clarify the following concepts: Some students might think that rocks have to melt to become metamorphic rocks. Others might think that only sedimentary rocks can become metamorphic rocks. If a rock crystallizes from molten material, it is an igneous rock. Any type of rock—igneous, metamorphic, or sedimentary—can be recrystallized in the solid state under high temperatures and pressures to become metamorphic.

Discussion

Deep Metamorphism Ask students what grade of regional metamorphism they would expect to see in the eroded roots of mountain ranges. Because these regions were buried at great depth, they will show evidence of high- grade metamorphism.

Enrichment

Talc and Asbestos Have students research the occurrence of asbestos, its uses, and its dangers. Students should also include information on what the government has done to remove asbestos from buildings and its use in general. OL

Teacher Content Support

Slate Roofs Some old, historic buildings and some new buildings have roofs covered with pieces of slate. Blocks of slate are quarried from outcrops, then the tiles are separated along the slaty cleavage. Although these tiles are heavy and brittle, they are nearly indestructible to weathering and are completely fireproof. Slate tile quarries are still in operation to meet the demand stemming from restoration work on historic buildings and also on other buildings where slate roofs are desired.

Economic Importance of Metamorphic Rocks and Minerals

The modern way of life is made possible by a great number of naturally occurring Earth materials. We need salt for cooking, gold for trade, other metals for construction and industrial purposes, fossil fuels for energy, and rocks and various minerals for construction, cosmetics, and more. **Figure 22** shows two examples of how metamorphic rocks are used in construction. Many of these economic mineral resources are produced by metamorphic processes. Among these are the metals gold, silver, copper, and lead, as well as many significant nonmetallic resources.

Metallic mineral resources Metallic resources occur mostly in the form of metal ores, although deposits of pure metals are occasionally discovered. Many metallic deposits are precipitated from hydrothermal solutions and are either concentrated in veins or spread throughout the rock mass. Native gold, silver, and copper deposits tend to occur in hydrothermal quartz veins near igneous intrusions or in contact metamorphic zones. However, most hydrothermal metal deposits are in the form of metal sulfides such as galena (PbS) or pyrite (FeS_2). The iron ores magnetite and hematite are oxide minerals often formed by precipitation from iron-bearing hydrothermal solutions.

☑ READING CHECK **State** what resources hydrothermal metamorphism produces.

Nonmetallic mineral resources Metamorphism of ultrabasic igneous rocks produces the minerals talc and asbestos. Talc, with a hardness of 1, is used as a dusting powder, as a lubricant, and to provide texture in paints. Because it is not combustible and has low thermal and electric conductivity, asbestos has been used in fireproof and insulating materials. Prior to the recognition of its cancer-causing properties, it was also widely utilized in the construction industry. Many older buildings still have asbestos-containing materials. Graphite, the main ingredient of the lead in pencils, may be formed by the metamorphism of coal.

■ **Figure 22** Marble and slate are metamorphic rocks that have been used in construction for centuries.

(l)Altrendo Travel/Altrendo/Getty Images, (r)Pixtal/SuperStock

☑ READING CHECK Metallic ores such as gold, silver, and copper, plus metal sulfides such as galena and pyrite form from hydrothermal metamorphism.

EARTH SCIENCE JOURNAL

Rock Recycling Have each student write an analogy comparing the rock cycle to a loop for recycling materials used in their everyday lives. For example, recycling glass by remelting it is similar to the formation of igneous rocks; using pressure to melt and form recycled plastic beads into new items is similar to metamorphism, and shredding newspaper to use as a material in drywall is similar to the formation of sedimentary rocks.

The Rock Cycle

Metamorphic rocks form when other rocks change. The three types of rock—igneous, sedimentary, and metamorphic—are grouped according to how they form. Igneous rocks crystallize from magma underneath Earth's surface, or from lava on Earth's surface; sedimentary rocks form from cemented or precipitated minerals and sediments; and metamorphic rocks form from changes in temperature and pressure.

Once a rock forms, does it remain the same type of rock always? Possibly, but it most likely will not. Heat and pressure can change an igneous rock into a metamorphic rock. A metamorphic rock can be changed into another metamorphic rock or melted to form an igneous rock. Alternately, the metamorphic rock can be weathered and eroded into sediments that might become cemented into a sedimentary rock. In fact, any rock can be changed into any other type of rock. The continuous changing and remaking of rocks is called the **rock cycle.** The rock cycle is summarized in **Figure 23.** The arrows represent the different processes that change rocks into different types.

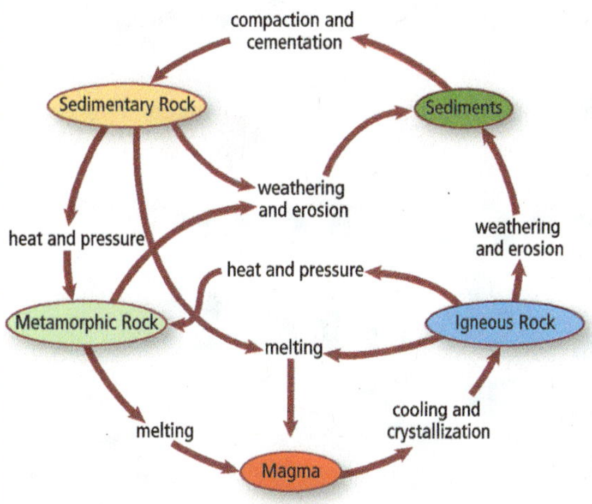

Figure 23 Rocks are continually being changed above and beneath Earth's surface. The rock cycle shows some of the series of changes rocks undergo.

SECTION 3 REVIEW

Section Self-Check

Section Summary

- The three main types of metamorphism are regional, contact, and hydrothermal.
- The texture of metamorphic rocks can be foliated or nonfoliated.
- During metamorphism, new minerals form that are stable under the increased temperature and pressure conditions.
- The rock cycle is the set of processes through which rocks continuously change into other types of rocks.

Understand Main Ideas

1. MAINIDEA **Summarize** how temperature increases can cause metamorphism.
2. **Summarize** what causes foliated metamorphic textures to form.
3. **Apply** the concept of the rock cycle to explain how the three main types of rocks are classified.
4. **Compare and contrast** the factors that cause the three main types of metamorphism.

Think Critically

5. **Infer** which steps in the rock cycle are skipped when granite metamorphoses to gneiss.
6. **Predict** the location of an igneous intrusion based on the following mineral data. Muscovite and chlorite were collected in the northern portion of the area of study; garnet and staurolite were collected in the southern portion of the area.

MATH IN Earth Science

7. Gemstones often form as porphyroblasts. Gemstones are described in terms of carat weight. A carat is equal to 0.2 g or 200 mg. A large garnet discovered in New York in 1885 weighs 4.4 kg and is 15 cm in diameter. What is the carat weight of this gemstone?

3 Assess

Check for Understanding

Draw Have each student draw an example of a foliated and nonfoliated rock and list three characteristics of each. foliated: long, platy crystals, layered or banded, more than one mineral; nonfoliated: blocky, square crystals, no layering, may form porphyroblasts

Reteach

Summarize Have each student make a data table listing the characteristics of the three types of metamorphism.

Assessment

Skill Have each student make a list in which they compare and contrast the rock-cycle processes. They should divide their lists into two categories: those processes that occur at Earth's surface and those processes that occur at depth. Ask students to identify those processes that can occur in both places. Some examples include the following: chemical weathering can occur at Earth's surface but also at depth from hydrothermal alteration; igneous rocks form both at depth and at Earth's surface; and sediments can be cemented at Earth's surface—they do not have to be buried.

SECTION 3 REVIEW

1. Increases in temperature cause reorganization of the atoms in the elements that make up the minerals. This results in the formation of new minerals or causes mineral crystals to grow larger.
2. High pressure causes minerals with flat or linear crystals to grow with a preferred or unidirectional orientation.
3. The rock cycle illustrates that rocks form in specific environments and through specific processes. The three main types of rocks—igneous, sedimentary, and metamorphic—are classified by their mode of formation.
4. regional metamorphism—temperature and pressure affect large areas of Earth's crust; contact metamorphism—local effects caused by the heat from nearby igneous intrusions; hydrothermal metamorphism—alteration of rocks as a result of reactions with very hot water
5. uplift, weathering and erosion, deposition and lithification
6. The igneous intrusion is most likely located to the south of the garnet and staurolite minerals. They form at higher temperatures than muscovite and chlorite, indicating that the temperature was hotter to the south.
7. The carat weight of the garnet is 22,000 carats.

eXpeditions!

Purpose

Students will learn about urban examples of the different types of rock—igneous, metamorphic, and sedimentary.

Teacher Content Support

Famous Features The Obelisk in New York's Central Park is also known as Cleopatra's Needle. The Obelisk was moved from Egypt to the United States when the then viceroy of Egypt gave it to the United States as a gesture to increase economic interest in Egypt. The base of the Obelisk is supported by four enormous bronze crabs, each weighing 400 kg.

The Gapstow Bridge was originally built from wood in 1874. The bridge was replaced with local Manhattan schist just 22 years later because the wood decayed.

Teaching Strategies

- Have students review the rock cycle and list examples of the different rock types—igneous, metamorphic, and sedimentary.
- Discuss with students why certain kinds of rock, such as granite, hold up to weathering better than other kinds, such as limestone.

eXpeditions!

ON SITE: GEOLOGY IN CENTRAL PARK

Some people travel to remote locations of the world to see different types of rock. However, examples of rocks often can easily be found in urban areas. Central Park in New York City is an excellent place to find examples of igneous, sedimentary, and metamorphic rock, both naturally occurring and used for sculptures, monuments, and bridges.

Maine Monument

The Maine Monument Located at the main entrance to Central Park, the Maine Monument is an immense structure made of marble, limestone, and bronze. The massive bow of a ship that makes up the base of the monument was sculpted out of marble, a type of metamorphic rock. A bronze statue sits atop a 15-m limestone pylon.

Schist and gneiss These two types of metamorphic rock occur naturally in Central Park. Outcroppings of these rocks, formed from sedimentary or igneous rock under intense heat and pressure, can be found throughout the park. The Gapstow Bridge was constructed using the local bedrock.

Gapstow Bridge

The Obelisk Weighing 221 tons and standing 21 m high, Cleopatra's Needle is the oldest human-made object in Central Park. The granite was quarried in Egypt more than 3000 years ago in 1475 B.C. The sculpture remained in Egypt until 1879, when it was moved to the United States. Granite is more resistant to weathering than other types of rock, and engravings made in granite can be read for hundreds of years, making it an excellent rock for the construction of monuments.

Cleopatra's Needle

WRITING IN ▶ Earth Science

Promotional Brochure Research more information about the type of rock used to build structures and that occur naturally in your area. Create a promotional brochure that describes a tour focused on local geology.

WebQuest

WRITING IN ▶ Earth Science

Rubric

Promotional Brochure Brochures should describe a geology tour that focuses on examples of different types of rock that can be found naturally or used in building structures in the local area.

WebQuest

GeoLAB

iLab Station

Interpret Changes In Rocks

Background: As the rock cycle continues and rocks change from one type to another, more changes occur than meet the eye. Color, grain size, texture, and mineral composition are easily observed and described visually. Yet, with mineral changes come changes in crystal structure and density. How can these be accounted for and described? Studying pairs of rocks can show you how.

Question: *How do the characteristics of sedimentary or igneous rocks compare to metamorphic rocks?*

Materials

samples of sandstone, shale, limestone, granite, quartzite, slate, marble, and gneiss
magnifying lens
paper
beam balance
100-mL graduated cylinder or beaker that is large enough to hold the rock samples
water

Safety Precautions

Procedure

1. Read and complete the lab safety form.
2. Prepare a data table similar to the one at the right. Adjust the width of the columns as needed.
3. Observe each rock sample. Record your observations in the data table.
4. Recall that density = mass/volume. Make a plan that will allow you to measure the mass and volume of a rock sample.
5. Determine the density of each rock sample, and record this information in the data table.

Sample Data Table

Sample Number	1	2	3	4
Rock type				
Specific characteristics				
Mass				
Volume				
Density				

Analyze and Conclude

1. **Compare and contrast** sandstone and quartzite.
2. **Describe** how the grain size of sandstone changes during metamorphism.
3. **Describe** the textural differences you observe between shale and slate.
4. **Infer** Compare your calculated densities to those calculated by other students. Infer why yours might differ.
5. **Explain** why the color of a sedimentary rock may change during metamorphism.
6. **Evaluate** the changes in density between shale and slate, sandstone and quartzite, limestone and marble, and granite and gneiss. Does density always change? Explain your results.

SHARE YOUR DATA

Peer Review Discuss your results with other groups in your class. Speculate on the reasons for variations in mass, volume, and density.

GeoLAB

iLab Station | Rubric

Preparation

Time Allotment 60–120 min, depending on the number of rock samples

Process Skills compare and contrast, measure and use numbers, observe and infer

Safety Precaution Approve lab safety forms before work begins.

Procedure

- If supplies are short, organize the samples in sets, with each set being a sedimentary-metamorphic series, such as sandstone and quartzite; shale and slate; or limestone and marble, or an igneous-metamorphic series, such as granite and gneiss. Have student groups work on one set at a time instead of all at once.
- After students have made their plans for determining volume, remind them of the proper techniques for water displacement to avoid splashing or breakage.
- Ask students to think about how immersing sandstone samples might affect their density.

Analyze and Conclude

1. sandstone—lighter color, medium grained, thicker layers or massive; quartzite—possibly similar color to sandstone, grains fused together, thick layers or massive
2. The grains become larger as they grow together and the individual sand grains are no longer felt.
3. Slate has thinner foliated layers and might have a smoother feel and shinier luster because of the presence of metamorphic mica minerals.
4. The calculated densities will vary. Possible sources of error are mathematical mistakes, mass differences between wet and dry samples, lack of precision in volume measurements, and slight differences between samples.
5. The formation of new metamorphic minerals may cause the color to change.
6. Each pair of rocks increased in density. If the new metamorphic minerals have greater density than the original mineral makeup, the density of the metamorphic rocks will be higher in density too.

SHARE YOUR DATA

Peer Review Data from each lab group should be similar. Variations can be accounted for by mineral composition, porosity, and permeability and by human error.

LabManager™
Customize any lab with the LabManager™ CD-ROM.

MAINIDEAS Summary statements can be used by students to review the major concepts of the chapter.

Students can review with these online resources.

Vocabulary Practice

Vocabulary eGames
Vocabulary eFlashcards
Vocabulary PuzzleMaker

Section Self-Check

Chapter Self-Check

Online Test Practice

Use the ExamView® *Assessment Suite* CD-ROM to:

- create multiple versions of tests
- create modified tests with one mouse click
- edit existing questions and add your own questions
- build tests aligned with state standards using built-in state curriculum tags
- change English tests to Spanish with one mouse click
- track students' progress using the Teacher Management System

CHAPTER 6 STUDY GUIDE

BIGIDEA **Most rocks are formed from preexisting rocks through external and internal geologic processes.**

Vocabulary Practice

SECTION 1 Formation of Sedimentary Rocks

VOCABULARY
- sediment
- lithification
- cementation
- bedding
- graded bedding
- cross-bedding

MAINIDEA **Sediments produced by weathering and erosion form sedimentary rocks through the process of lithification.**

- The processes of weathering, erosion, deposition, and lithification form sedimentary rocks.
- Sediments are lithified into rock by the processes of compaction and cementation.
- Fossils are the remains or other evidence of once-living organisms that are often preserved in sedimentary rocks.
- Sedimentary rocks might contain features such as horizontal bedding, cross-bedding, and ripple marks.

SECTION 2 Types of Sedimentary Rocks

VOCABULARY
- clastic sedimentary rock
- clastic
- porosity
- evaporite

MAINIDEA **Sedimentary rocks are classified by their mode of formation.**

- Sedimentary rocks can be clastic, chemical, or biochemical.
- Clastic rocks form from sediments and are classified by particle size and shape.
- Chemical rocks form primarily from minerals precipitated from water.
- Biochemical rocks form from the remains of once-living organisms.
- Sedimentary rocks provide geologists with information about surface conditions that existed in Earth's past.

SECTION 3 Metamorphic Rocks

VOCABULARY
- foliated
- nonfoliated
- regional metamorphism
- contact metamorphism
- hydrothermal metamorphism
- rock cycle

MAINIDEA **Metamorphic rocks form when preexisting rocks are exposed to increases in temperature and pressure and to hydrothermal solutions.**

- The three main types of metamorphism are regional, contact, and hydrothermal.
- The texture of metamorphic rocks can be foliated or nonfoliated.
- During metamorphism, new minerals form that are stable under the increased temperature and pressure conditions.
- The rock cycle is the set of processes through which rocks continuously change into other types of rocks.

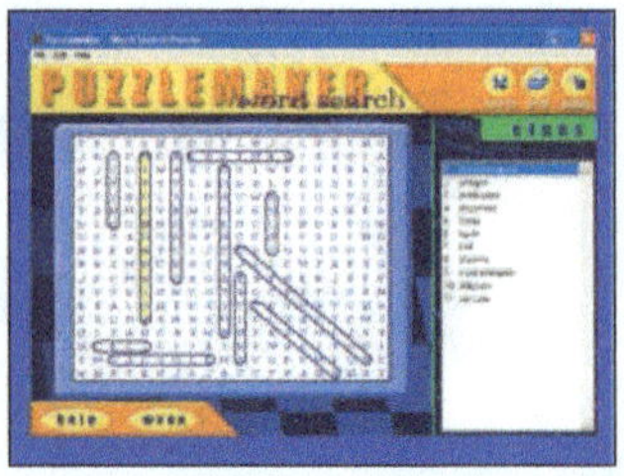

Vocabulary Practice

For additional practice with vocabulary, have students access the Vocabulary PuzzleMaker.

Chapter Self-Check

VOCABULARY REVIEW

Complete the sentences below using vocabulary terms from the Study Guide.

1. Compaction and cementation of clastic sediments result in ________.
2. Sedimentary layers that are deposited on an angle are called ________.
3. Cooling and crystallization, igneous rocks, uplift, and weathering and erosion describe a path along the ________.
4. Hot fluids that come in contact with solid rock result in ________.

Replace the italicized word with the correct vocabulary term from the Study Guide.

5. *Cementation* occurs when sediment gets deposited as the energy of the water decreases.
6. *Foliated* rocks have square, blocky crystals.

Write a sentence using each pair of words.

7. contact metamorphism, regional metamorphism
8. porosity, clastic sedimentary rock
9. sediment, bedding
10. clastic, evaporite

UNDERSTAND KEY CONCEPTS

11. Which clastic sediment has the smallest grain size?
 A. sand
 B. clay
 C. pebbles
 D. silt

12. Which is a coarse-grained clastic rock that contains angular fragments?
 A. limestone
 B. conglomerate
 C. sandstone
 D. breccia

13. Which is often a biochemical rock containing fossils?
 A. chert
 B. limestone
 C. sandstone
 D. breccia

14. Which process forms salt beds?
 A. deposition
 B. cementation
 C. evaporation
 D. lithification

15. Which does not cause metamorphism?
 A. lithification
 B. hydrothermal solutions
 C. heat
 D. pressure

Use the diagram below to answers Questions 16 and 17.

16. Which term best describes this rock's texture?
 A. crystalline
 B. nonfoliated
 C. foliated
 D. clastic

17. From what igneous rock does this sample usually form?
 A. obsidian
 B. basalt
 C. granite
 D. gabbro

CHAPTER 6 ASSESSMENT

VOCABULARY REVIEW

1. lithification
2. cross-bedding
3. rock cycle
4. hydrothermal metamorphism
5. graded bedding
6. nonfoliated rocks
7. Contact metamorphism is confined to a small area, while regional metamorphism covers a large area.
8. Some clastic sedimentary rocks have high porosity.
9. When sediment is deposited in horizontal layers, bedding forms.
10. A sedimentary rock that is an evaporite is not clastic.

UNDERSTAND KEY CONCEPTS

11. B
12. D
13. B
14. C
15. A
16. C
17. C

18. D
19. A
20. B

CONSTRUCTED RESPONSE

21. Dissolved minerals in groundwater precipitate out of solution as it moves through the sediment. They precipitate onto the grains and cement them together.
22. Coquina is composed of shells loosely cemented together; fossiliferous limestone is composed of fossil shells and lime mud cemented together.
23. Recall that 1 liter = 1000 cm^3. The answer is 300 liters.
24. The illustration should show a rock with elongated or platy minerals and pressure pushing on the rock perpendicular to the minerals.
25. Sand becomes lithified mostly by cementation; mud becomes lithified mostly by compaction.
26. poorly sorted: glacial deposits, landslide material; well-sorted: dune sand, beach sand
27. Precipitation of these minerals causes cementation of the sediment and lithification into rock. Iron oxide cement would cause a red color and calcite cement would fizz in acid.
28. Conglomerate: rounded fragments, transported far distances; Breccia: angular fragments, forms closer to source of sediments. Both contain abundant quartz and quartzite.
29. The water could have a higher salt concentration due to the addition of sea water, and perhaps the evaporite layers would become thicker. Or an opening to the ocean could cause the lake to drain.

THINK CRITICALLY

30. Marble consists of equidimensional, blocky calcite crystals. They are not elongated or platy, so they cannot form foliation.
31. Coal consists of organic plant material. Minerals are composed of inorganic elements or compounds.
32. Sedimentologists recommend sand and gravel based on its intended use. If a highly porous sand was needed, they could not recommend the mixture described in the question.

ASSESSMENT

18. Which agent of erosion can usually move only sand-sized or smaller particles?
A. landslides
B. glaciers
C. water
D. wind

19. Which would you expect to have the greatest porosity?
A. sandstone
B. gneiss
C. shale
D. quartzite

20. By what process are surface materials removed and transported from one location to another?
A. weathering
B. erosion
C. deposition
D. cementation

CONSTRUCTED RESPONSE

Use the diagram to answer Question 21.

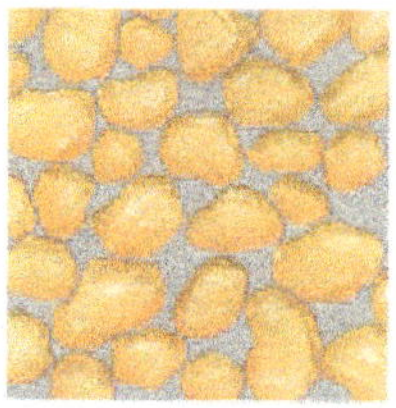

21. Describe how the grains in the diagram become glued together.

22. Summarize the main difference between coquina and fossiliferous limestone. Use **Table 1** for help.

23. Calculate A sandstone block has a volume of 1 m^3 and a porosity of 30 percent. How many liters of water can this block hold?

24. Illustrate the conditions necessary to form a foliated metamorphic rock.

25. Compare and contrast the modes of lithification for sand and mud.

26. Classify the following types of sediments as either poorly sorted or well sorted: dune sand, landslide material, glacial deposits, and beach sand.

27. Analyze the effect that precipitation of calcite or iron oxide minerals has on clastic sediments.

28. Compare and contrast the character and formation of breccia and conglomerate.

Use the diagram below to answer Question 29.

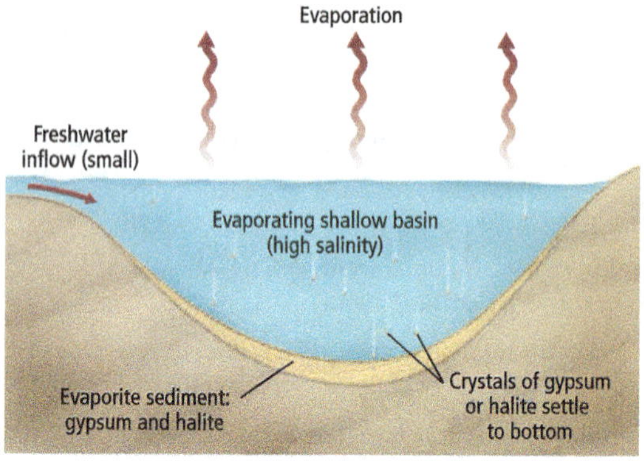

29. Evaluate the effect that an opening to the ocean would have on this environment.

THINK CRITICALLY

30. Incorporate what you know about crystal form to explain why marble, even if formed under high pressure, does not show foliation.

31. Compose a statement to explain why the sedimentary rock coal does not meet the standard definition of a rock—an aggregate of minerals.

32. Careers in Earth Science Some sedimentologists work in sand and gravel pits where they analyze the material to best decide where and how it should be used. Infer why it is important for the sedimentologists to understand what would happen to the porosity of sand if finer-grained sediment were mixed in with the sand.

33. Illustrate an oil reservoir made up of layers of sandstone and shale. Indicate the position of the oil within the rocks.

33. The diagram should show sandstone layers below shale layers. The oil should be contained within the sandstone, not as a chamber or pool within the layers.

Chapter Self-Check

34. Assess whether ripple marks and animal footprints preserved in sandstone are fossils. Explain your reasoning.

Use the figure below to answer Questions 35 and 36.

Doug Sherman/Geofile

35. Evaluate the sediment in the layers in the figure. What type of bedding is this, and how well is it sorted? Explain.

36. Infer Look at **Figure 2** and explain which agents of erosion can produce the layers shown.

37. Deduce why glass on a quartz sand beach becomes rounded and frosted, while glass on a carbonate sand beach stays sharp and glassy.

CONCEPT MAPPING

38. Use the following terms to create a concept map that organizes sedimentary features: *ripple marks, graded bedding, horizontal bedding, asymmetrical, symmetrical, river current, wave action, wind deposited,* and *water deposited.* Some terms can be used more than once.

CHALLENGE QUESTION

39. Hypothesize At an approximate ocean depth of 4000 m, the carbonate compensation depth occurs. Below this depth, no calcium carbonate precipitates and no shells accumulate on the ocean floor. Hypothesize why this condition exists.

WRITING IN ▸ Earth Science

40. Imagine that you are planning a geologic walking tour of your community. Create a brochure highlighting the various natural building stones that are used in homes and buildings in your town or neighborhood.

DBQ Document–Based Questions

Data obtained from: Mineral Commodity Summaries. January 2006. *United States Geological Survey.*

Dimension stone is natural rock material used in construction, for monuments, and home interiors, such as kitchen countertops and floors. The principal rock types used are granite, limestone, marble, sandstone, and slate. Global resources of dimension stone are virtually limitless. Production of dimension stone in the United States and elsewhere has been steadily increasing.

Dimension Stone Production	U.S. Sold or Used (tonnage)	U.S. Sales or Uses (by value)
Limestone	39%	34%
Granite	29%	39%
Sandstone	14%	9%
Misc. stone	10%	7%
Marble	7%	6%
Slate	1%	5%

41. Construct a graph comparing the amount of dimension stone used by the value of the types of dimension stone.

42. Propose an explanation for why the value of granite is the highest of the dimension stones listed.

CUMULATIVE REVIEW

43. Compare and contrast the terms *science* and *technology.* **(Chapter 1)**

44. What is the formula of the ionic compound magnesium chloride? **(Chapter 3)**

45. Explain the concepts of partial melting and fractional crystallization. **(Chapter 5)**

34. Fossils are evidence of past life-forms. Animal footprints are evidence of past life; ripple marks are not such evidence.

35. This is graded bedding; it is well sorted with particle sizes becoming smaller toward the top.

36. All of the agents of weathering can carry different grain sizes, but only wind and water are able to sort sediments as they are transported. As wind and water slow down and lose energy, the heavier particles settle out and create graded bedding.

37. Glass has a hardness of 5.5. Quartz sand, with a hardness of 7, can easily wear down and polish the glass. Carbonate sand, being composed of calcite, has a hardness of 3. It cannot polish the glass, so the glass remains sharp.

CONCEPT MAPPING

38. The terms might be linked as follows: horizontal bedding, water deposited, wind deposited; graded bedding, water deposited; ripple marks, water deposited or wind deposited, river current then asymmetrical, wave action then symmetrical.

CHALLENGE QUESTION

39. Water is colder at depth. Calcium carbonate is soluble in cold water, so it can't precipitate at this depth and temperature.

WRITING IN ▸ Earth Science

Rubric

40. Students' answers will vary. Encourage them to document as many rock types as possible.

DBQ Document-Based Questions

Data obtained from: Mineral Commodity Summaries. January 2006. *United States Geological Survey.*

41. Student graphs will vary depending on the scale they choose and which axis represents which variable.

42. Possible answers might include: it is the most expensive to quarry, the demand is high so the price can be high.

CUMULATIVE REVIEW

43. *Science* refers to knowledge derived from objective investigating processes using the scientific method. *Technology* is the application of what we learn in science into the design of instruments and tools which serve to promote further study as well as to improve our quality of life.

44. $MgCl_2$

45. Partial melting refers to the melting of only those minerals of a rock that have a low melting point. Fractional crystallization refers to the crystallization of high temperature minerals while the rest of the magma remains molten.

MULTIPLE CHOICE

1. D
2. C
3. A
4. B
5. D
6. B
7. B
8. D
9. C
10. A

CUMULATIVE STANDARDIZED TEST PRACTICE

MULTIPLE CHOICE

Use the illustration below to answer Questions 1 and 2.

1. Which rocks are most likely to metamorphose from the lava flow?
A. only the rocks in the crater of the volcano, where the lava is hottest
B. rocks in the crater and rocks along the top half of the mountain
C. all the rocks on the mountain
D. all the rocks reached by the lava flow

2. As the lava cools and crystallizes, what type of rock will form?
A. sedimentary
B. metamorphic
C. extrusive igneous
D. intrusive igneous

3. What is NaCl commonly known as?
A. table salt **C.** water
B. sugar **D.** natural chlorine

4. What initiates the process that changes sediments into sedimentary rocks?
A. bedding **C.** cementation
B. burial **D.** compaction

5. Identify the unit that is NOT accepted for use with Le Système International D'Unités (SI).
A. metric ton **C.** ampere
B. kilogram **D.** Fahrenheit

6. Which rocks are composed of minerals that form with blocky crystal shapes?
A. foliated **C.** porphyroblasts
B. nonfoliated **D.** gneiss

Use the diagram below to answer Questions 7 and 8.

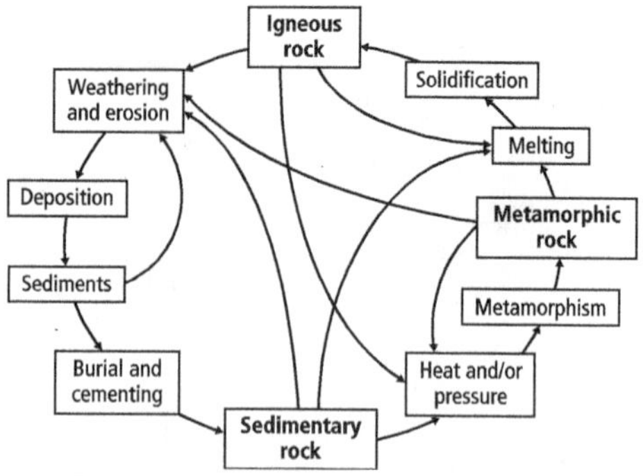

7. Based on the diagram, which is the most reasonable hypothesis?
A. Igneous rocks have layers caused by deposition.
B. Sedimentary rocks contain grains of other rocks.
C. Metamorphic rocks never have layers.
D. Sedimentary rocks are always the same color.

8. According to the rock cycle shown above, what most likely happens after the deposition of sediment?
A. Weathering forms more sediment.
B. Magma cools and forms igneous rock.
C. Heat and pressure cause the sediment to melt.
D. Cementation occurs and forms sedimentary rock.

9. Where are valence electrons located?
A. every energy level
B. middle energy levels
C. the outermost energy level
D. the innermost energy level

10. Which of the following is a fine-grained, clastic sedimentary rock?
A. shale
B. sandstone
C. phosphate
D. limestone

SHORT ANSWER

Use the illustration below to answer Questions 11 and 12.

11. What do you notice about the formation of sedimentary rock above?

12. Does this process represent compaction or cementation? Describe the difference between the two.

13. The results of an experiment show that as temperature increases, enzyme activity decreases. Describe what a line graph made from this data would look like.

14. Define *luster.* Why is it difficult to use luster to identify minerals?

15. What process does Bowen's reaction series illustrate?

16. Boron has an atomic number of 5. Describe an atom of boron with a mass number of 10 and an atom of boron with a mass number of 11 in terms of their atomic particles. What is unique about these two atoms of boron?

17. Briefly describe the process by which magma becomes igneous rock.

18. How does studying sedimentary rock layers and understanding how they form help paleontologists learn about Earth's history?

READING FOR COMPREHENSION

Sedimentary Rock Layers

Paleontologists wanted to study the sedimentary rock layers and their contents of a particular area. The diagram shows a cross section of the rock layers they studied. The table shows the data the scientists were able to collect.

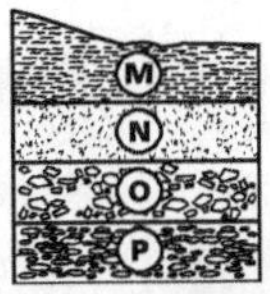

Age of Sedimentary Rock Layers

Layer	Composition	Estimated Age (years)	Depth (meters)
M	sedimentary rock	100,000	0–4
N	sedimentary rock	Unknown	5–7
O	sedimentary rock	6 million	8–9
P	sedimentary rock	6.1 million	9–10

19. What could the paleontologists have recorded to improve their study?
    A. time of year
    B. age of layer N
    C. location of the work site
    D. mass of the sedimentary rocks

20. If fossils of a species were found in Layers O and P, but not M and N, which could you conclude?
    A. The species does not exist anywhere on Earth today.
    B. The species evolved into a completely different species.
    C. The species became extinct less than 100,000 years ago.
    D. The species disappeared from the area around 6 mya.

NEED EXTRA HELP?

If You Missed Question . . .	1	2	3	4	5	6	7	8	9	10	11	12	13	14	15	16	17	18
Review Section . . .	6.3	5.1	3.2	6.1	1.2	6.3	6.3	6.3	3.1	6.2	6.1	6.1	1.3	4.1	5.1	3.1	5.1	6.1

SHORT ANSWER

11. The space between the sediments decreases, causing them to become more tightly compacted.

12. This is the process of compaction. In compaction, the space gets smaller between sediments, while in cementation, sediments are held together when minerals crystallize between the grains of sediment.

13. With temperature as the independent variable and enzyme activity as the dependent variable, the line graph would go from the upper left corner of the graph to the bottom right corner of the graph.

14. Luster is the way a mineral reflects light from its surface. While luster is based upon differences in chemical compositions, it is subjective and should be used along with other methods of identifying minerals.

15. Bowen's reaction series illustrates the relationship between cooling magma and mineral formation.

16. Boron with a mass number of 10 has 5 protons and 5 neutrons. Boron with a mass number of 11 has 5 protons and 6 neutrons. These two atoms are isotopes of each other, meaning they have the same number of protons but a different number of neutrons. Therefore, they have the same chemical properties.

17. Magma is the molten rock below Earth's surface. As it moves closer to Earth's surface, it cools and hardens to form intrusive igneous rock. When it flows out of an erupting volcano, it is called lava. As this lava cools and hardens, minerals crystallize to become extrusive igneous rock.

18. Sample answer: Sedimentary rock forms when sediments—pieces of solid material deposited on Earth's surface—are compacted together. The rocks build on top of each other, thus causing the oldest material to be at the bottom and the newer material at the top. Paleontologists study fossils that occur in sedimentary rock layers. By understanding the ancient environments in which the sediments were deposited, a paleontologist can infer the environment in which the organism lived. Hence, the geologic history of Earth is being discovered one sedimentary rock layer at a time.

READING FOR COMPREHENSION

19. B
20. D

UNIT 3

Surface Processes on Earth

Themes

Change The continuous effects of weather on Earth's crust create conditions for a variety of erosion forms and features.

Structures Evidence of the effect of weather on Earth's surface can be observed in the form of dunes, arches, caves, soil slumps, and landslides.

Geologic Time Crustal rock is constantly exposed to the changing conditions of the weather. Over time, mountains can erode into flat plains, creating a supply of sand for beaches around the world.

Systems Water is the key substance involved in eroding Earth's crust. Ice, sediment-carrying rivers, and acidic pools break, scour, and eat away the rock all around us.

Scientific Inquiry Observing the processes acting on Earth's surface led scientists to theories of weathering and erosion by wind, water, and gravity. In turn, they were then able to speculate on how water acts within Earth's surface to form caves and aquifers.

UNIT 3

Surface Processes on Earth

CHAPTERS

7 Weathering, Erosion, and Soil

8 Mass Movements, Wind, and Glaciers

9 Surface Water

10 Groundwater

STEM Project

CAREERS IN EARTH SCIENCE **Glaciologist**

This glaciologist is studying the Antarctic ice sheet by recording its vibrations. Glaciologists study the movement, formation, and effects of glaciers on landscapes. Information gathered by glaciologists provides insight into Earth's geologic history as well as its future.

Galen Rowell/CORBIS

Introduce the Unit

Preconceptions In the background, this picture shows part of Antarctica's Royal Society Range, which reaches an elevation of 4025 m. Glaciers on either side of the Royal Society mountains lead to the Ross Ice Shelf located in the area of Antarctica south of New Zealand. The ice in Antarctica has accumulated in glaciers such as these for at least the past 40 million years. What reactions and images do glaciers bring up in the students' minds? The foreground shows a landscape that has been shaped by past glaciation. Ask for specific examples of how glaciers affected this landscape. Ask students who have lived in an area that was glaciated, such as portions of the Midwest, to describe how the environment was affected by past glaciers.

Historical Examples During December of 2006, the Philippines experienced heavy rains from a typhoon that caused an unprecedented amount of destruction to many communities, homes, businesses, and natural ecosystems. Over 1800 people perished from the mudflows caused by the rain combined with volcanic ash from Mayon. In this unit, students will explore how Earth's landscape features are changed by water, wind, and glaciers and how these processes affect human populations.

A System of External Processes Collect and display a variety of photographs from periodicals and newspapers that show landslides, avalanches, caves, rivers, and streams. Stress how external forces such as mass movement, surface water, groundwater, wind, and glaciers are constructive as well as destructive. How are some lakes formed by glacial activity? Why are humans still vulnerable to dangerous floods, mudslides, and avalanches? All of these processes are connected.

CHAPTER 7 Weathering, Erosion, and Soil

BIGIDEA **Weathering and erosion are agents of change on Earth's surface.**

ESSENTIAL QUESTIONS	NATIONAL STANDARDS	RESOURCES TO ASSESS MASTERY
SECTION 1 **Weathering** **1.** How do mechanical and chemical weathering differ? **2.** What are the different factors that affect mechanical and chemical weathering? **3.** What variables affect the rate of weathering? 3.5 sessions 1.75 blocks	UCP.1–5; A.1, A.2; B.4; D.1–3; F.3–4; G.3	**Entry-Level Assessment** Focus Transparency 17, p. 164 **Progress Monitoring** Caption Question, pp. 164, 168 Reading Check, pp. 166, 168 Section Review, p. 170
SECTION 2 **Erosion and Deposition** **1.** What is the relationship of gravity to all agents of erosion? **2.** What features are characteristic of the different types of erosion? **3.** How do living and nonliving things impact the processes of weathering and erosion? 1 session 0.5 block	UCP.1–4; A.1, A.2; B.4; D.1–3; F.6	**Entry-Level Assessment** Focus Transparency 18, p. 171 **Progress Monitoring** Caption Question, pp. 171, 172, 174, 175 Reading Check, p. 172 Section Review, p. 175
SECTION 3 **Soil** **1.** How does soil form? **2.** What are the different soil horizons in a soil profile? **3.** What factors affect soil formation? 4 sessions 2 blocks	UCP.1–5; A.1, A.2; B.4; C.4; D.1–3; E.2; F.3–4; G.3	**Entry-Level Assessment** Focus Transparency 19, p. 176 **Progress Monitoring** Caption Question, pp. 176, 179, 180 Reading Check, pp. 177, 178, 180 Section Review, p. 183 **Summative Assessment** Chapter Assessment, p. 187 Exam*View*® *Assessment Suite* CD-ROM

Go online!
connectED.mcgraw-hill.com

LEVELED RESOURCES	LAB MATERIALS
Science Notebook 7.1 OL **Chapter FAST FILE Resources:** GeoLab Worksheet, p. 3 OL Study Guide, p. 13 BL **Lab Resources:** Laboratory Manual, p. 49 OL **Transparencies:** Section Focus Transparency 17 EL Teaching Transparency 16 OL EL	LaunchLAB p. 162 / **20 min** 250-mL beakers (2), water, sugar cube, granulated sugar (5 mL), stirring rod, stopwatch GeoLAB p. 185 / **60 min** plastic jar with lid, water (300 mL), halite chips (100 g), balance, timer, paper towels
Science Notebook 7.2 OL **Chapter FAST FILE Resources:** MiniLab Worksheet, p. 2 OL Study Guide, p. 15 BL **Transparencies:** Section Focus Transparency 18 EL Teaching Transparency 17 OL EL	MiniLAB p. 172 / **20 min** bar of soap, toothpick, catch basin, water, metric ruler, balance
Science Notebook 7.3 OL **Chapter FAST FILE Resources:** Study Guide, p. 17 BL **Lab Resources:** Laboratory Manual, p. 53 OL **Transparencies:** Section Focus Transparency 19 EL Teaching Transparency 18 OL EL	

LabManager™
Customize any lab with the LabManager™ CD-ROM.

ADDITIONAL RESOURCES

Technology:
- eTeacherEdition
- eStudentEdition
- Classroom Presentation Toolkit
- Exam*View® Assessment Suite* CD-ROM
- LabManager™ CD-ROM
- What's EARTH SCIENCE Got To Do With It? DVD
- Weather Classroom Video DVD

Assessment:
- Blueprints for Success
- Chapter Test (Scaffolded)
- Earth Science Standardized Test Practice

Lab Resources:
- Exploring Environmental Problems Laboratory Manual

BL Below Level   OL On Level   AL Advanced Learners   EL English Learners   COOP LEARN Cooperative Learning

CHAPTER 7

LaunchLAB

iLab Station | Rubric

How does change relate to surface area?

Process Skills observe and infer, recognize cause and effect, communicate, model, predict, measure

Safety Precautions Approve lab safety forms before work begins. Remind students not to put any material used in the lab in their mouths, including the sugar.

Teaching Strategy Have students record time to the nearest second.

Procedure

1. Have students read and complete the lab safety form and follow the procedure below.
2. Fill two **250-mL beakers** with **water** at room temperature.
3. Drop a **sugar cube** in one beaker and 5 mL of **granulated sugar** in the other beaker at the same time. Record the time.
4. Slowly and continuously use a **stirring rod** to stir the solution in each beaker.
5. Observe the sugar in both beakers. Using a **stopwatch,** record the amount of time it takes for the sugar to completely dissolve in each beaker of water.

Analysis

1. **Describe** what happened to the sugar cube and the granulated sugar. The granulated sugar dissolved more quickly than the sugar cube.
2. **Explain** why one form of sugar dissolved faster than the other. Granulated sugar has more surface area than a sugar cube, so it dissolves more quickly.
3. **Infer** how you could decrease the time required for the slower-dissolving form of sugar to dissolve. To decrease the amount of time to dissolve the sugar cube, you could break it up into smaller pieces by hand, which would increase the surface area.

CHAPTER 7

Weathering, Erosion, and Soil

BIGIDEA Weathering and erosion are agents of change on Earth's surface.

SECTIONS

1 Weathering

2 Erosion and Deposition

3 Soil

LaunchLAB

iLab Station

How does change relate to surface area?

Surface area is a measure of the interface between an object and its environment. An object having more surface area can be affected more rapidly by its surroundings. Observe the relationship between weathering and surface area in this lab.

FOLDABLES® Study Organizer

Types of Weathering

Make a half-book using the labels shown. Use it to organize your notes on the types of weathering.

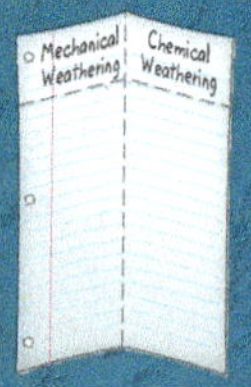

Changes in temperature and pressure can cause rocks on Earth's surface to crack and split.

Assessment

Knowledge Have students answer the following question: Which size of hard candy will dissolve most quickly?

a. 5 cm^3 **b.** 15 cm^3
c. 25 cm^3 **d.** 10.3 cm^3

The answer is a.

LabManager™
Customize any lab with the LabManager™ CD-ROM.

Introduce the BIGIDEA

Identify Processes Show students photos of local interest including any of the following: sand dune, glacier, landslide, canyon, peeling paint from a house, a pothole in the road. Ask: How would you describe the material in the photo? What sort of process was responsible for shaping the material in the photo? Answers will vary. What sort of climate might be responsible for shaping the material in the photo? Answers will vary.

Teacher Content Support

Weathering Weathering occurs in a number of different ways, a few of which are pictured to the right. Pose the following questions: What do you see happening in the photos? Rocks are cracking because of ice and plants in the top and bottom photos. Rocks are peeling away in sheets in the middle photo. What might cause the rocks to look like this? rain, ice, snow, increase or decrease in pressure.

SECTION 1

1 Focus

Focus Transparency

Before presenting the lesson, project **Section Focus Transparency 17** and have students answer the accompanying questions. **EL**

MAINIDEA

Model Weathering Take a piece of chalk and smash it on a paper plate with a hammer. Ask students the following questions: What happened to the chalk? It was broken by the hammer. What might this model in the real world? An active force disintegrating a material.

2 Teach

Identify Misconceptions

Students might confuse the term *weathering* with weather.

Uncover the Misconception Ask students to give the definitions of *weather* and *weathering.*

Demonstrate the Concept Explain to students that weather is the condition of the atmosphere at a given time and includes such variables as temperature, humidity, wind direction, etc. Then explain weathering as the breakdown of Earth's solid crust.

Assess New Knowledge Ask students which is true: Weather can affect weathering. true Weathering can affect weather. false

Caption Question Fig. 1 Frost wedging will continue to break the boulder into smaller and smaller pieces.

SECTION 1

Essential Questions

- How do mechanical and chemical weathering differ?
- What are the different factors that affect mechanical and chemical weathering?
- What variables affect the rate of weathering?

Review Vocabulary

acid: solution that contains hydrogen ions

New Vocabulary

weathering
mechanical weathering
frost wedging
exfoliation
chemical weathering
oxidation

Weathering

MAINIDEA Weathering breaks down materials on or near Earth's surface.

EARTH SCIENCE 4 YOU While walking on a sidewalk, you might notice that it has been pushed upward in places. In areas where trees are close to sidewalks, tree roots can cause the sidewalk to rise, buckle, and break.

Mechanical Weathering

Weathering is the process in which materials on or near Earth's surface break down and change. **Mechanical weathering** is a type of weathering in which rocks and minerals break down into smaller pieces. This process is also called physical weathering. Mechanical weathering does not involve any change in a rock's composition, only changes in the size and shape of the rock. A variety of factors are involved in mechanical weathering, including changes in temperature and pressure.

Effect of temperature Temperature plays a role in mechanical weathering. When water freezes, it expands and increases in volume by 9 percent. You have observed this increase in volume if you have ever frozen water in an ice cube tray. In many places on Earth's surface, water collects in the cracks of rocks and rock layers. If the temperature drops to the freezing point, water freezes, expands, exerts pressure on the rocks, and can cause the cracks to widen, as shown in **Figure 1.** When the temperature increases, the ice melts in the cracks of rocks and rock layers. The freeze-thaw cycles of water in the cracks of rocks is called **frost wedging.** Frost wedging is responsible for the formation of potholes in many roads in the northern United States where winter temperatures vary frequently between freezing and thawing.

Figure 1 Frost wedging begins in hairline fractures of a rock. Repeated cycles of freeze and thaw cause the crack to expand over time.
Predict *the results of additional frost wedging on this boulder.*

View an **animation of frost wedging.** Concepts In Motion

DIFFERENTIATED INSTRUCTION

Struggling Learners To demonstrate frost wedging, drill a series of holes two centimeters deep across the middle of a brick. Keep the brick in the freezer. Each day for a week, remove the brick and cover the top with water, making sure the holes are filled; then return the brick to the freezer until the next class session. Have students observe the size of the holes and the condition and strength of the brick each day. Have them relate this demonstration to how potholes form.

Effect of pressure Another factor involved in mechanical weathering is pressure. Roots of trees and other plants can exert pressure on rocks when they wedge themselves into the cracks in rocks. As the roots grow and expand, they exert increasing amounts of pressure which often causes the rocks to split, as shown in **Figure 2.**

On a much larger scale, pressure also functions within Earth. Bedrock at great depths is under tremendous pressure from the overlying rock layers. A large mass of rock, such as a batholith, may originally form under great pressure from the weight of several kilometers of rock above it. When the overlying rock layers are removed by processes such as erosion or even mining, the pressure on the bedrock is reduced. The bedrock surface that was buried expands, and long, curved cracks can form. These cracks, also known as joints, occur parallel to the surface of the rocks. Reduction of pressure also allows existing cracks in the bedrock to widen. For example, when several layers of overlying rocks are removed from a deep mine, the sudden decrease of pressure can cause large pieces of rocks to explode off the walls of the mine tunnels.

After a rock body is uplifted as a result of geological processes, fine cracks may develop in the rock due to a decrease in pressure. Over time, the outer layers of rock can be stripped away in succession, similar to the way an onion's layers can be peeled. The process by which outer rock layers are stripped away is called **exfoliation.** Exfoliation often results in dome-shaped formations, such as Moxham Mountain in New York and Half Dome in Yosemite National Park in California, shown in **Figure 3.**

Figure 2 Tree roots can grow within the cracks and joints in rocks and eventually cause the rocks to split.

FOLDABLES® Incorporate information from this section into your Foldable.

Figure 3 The rock that makes up Half Dome in Yosemite National Park fractures along its outer surface in a process called exfoliation. Over time this has resulted in the dome shape of the outcrop.

Teacher Content Support

Weathering, Erosion, and Soil The processes of weathering, erosion, and formation of soil are closely interrelated, as are many processes on Earth. Weathering includes both the disintegration and decomposition of surface or near-surface rock material. Disintegration is the physical or mechanical breakdown of Earth materials, and decomposition is the chemical altering of the composition of Earth materials. A change in location or exposure to changing atmospheric conditions can affect the rate of weathering of Earth materials. Weathering is an unending process that plays an important role in the rock cycle. The results of weathering change Earth's landscapes and reveal resources that help to meet humans' energy demands.

FOLDABLES® Rubric

Teacher Content Support

Exfoliation In the photo to the left, the top and right sides of Half Dome in Yosemite National Park are formed by exfoliation, but the flat portion to the left was formed by glaciers.

Demonstration

Model Pressure Unloading Cut a piece of polyurethane foam into large, irregular pieces. Keep the pieces in place with the pressure of your hands, using a desk or table as a base. Have students observe that the model stays together in one piece. Then, release the pressure of your hands so that the pieces separate and fall away from one another. This is similar to the pressure unloading that occurs when pressure on joints is reduced above and around bedrock.

Enrichment

Polarity of Water A water molecule is composed of one oxygen atom and two hydrogen atoms. They are arranged so that the two hydrogen atoms are attached on one end of the molecule and the oxygen atom is on the opposite end of the molecule. The end with the hydrogen atoms has a positive charge and the opposite end has a negative charge. This polarity of the water molecule is responsible for water's ability to dissolve substances. AL

Apply Earth Science

Rust Pose the following questions to students: What items have you seen rusted? cars, bicycles, fences, and so on. What could you do to such objects to prevent or reduce rusting? paint the objects, keep the objects out of the rain and snow Students should recognize that painting will prevent oxygen from coming in contact with the metal. BL

Use Science Terms

Hydrolysis and Oxidation *Hydrolysis* and *oxidation* are terms that students will encounter in chemistry. It might be a good idea to talk with the chemistry teacher at your school to make sure that his or her descriptions of these processes are consistent with yours. Students who readily learn the chemical aspects of this chapter should be encouraged to continue in the sciences and study chemistry. OL

READING CHECK Chemical weathering changes the chemical composition of rocks.

Figure 4 This statue has been chemically weathered by acidic water and atmospheric pollutants.

Chemical Weathering

Chemical weathering is the process by which rocks and minerals undergo changes in their composition. Agents of chemical weathering include water, oxygen, carbon dioxide, and acid precipitation. The interaction of these agents with rock can cause some substances to dissolve, and some new minerals to form. The new minerals have properties different than those that were in the original rock. For example, iron often combines with oxygen to form iron oxide, such as in hematite.

READING CHECK **Express** in your own words the effect that chemical weathering has on rocks.

The composition of a rock determines the effects that chemical weathering will have on it. Some minerals, such as calcite, which is composed of calcium carbonate, can decompose completely in acidic water. Limestone and marble are made almost entirely from calcite, and are therefore greatly affected by chemical weathering. Buildings and monuments made of these rocks usually show signs of wear as a result of chemical weathering. The statue in **Figure 4** is an example of chemical weathering from acid precipitation.

Temperature is another significant factor in chemical weathering because it influences the rate at which chemical interactions occur. Chemical reaction rates increase as temperature increases. With all other factors being equal, the rate of chemical weathering reactions doubles with each 10°C increase in temperature.

Effect of water Water is an important agent in chemical weathering because it can dissolve many kinds of minerals and rocks. Water also plays an active role in many reactions by serving as a medium in which the reactions can occur. Water can also react directly with minerals in a chemical reaction. In one common reaction with water, large molecules of the mineral break down into smaller molecules. This reaction decomposes and transforms many silicate minerals. For example, potassium feldspar decomposes into kaolinite, a fine-grained clay mineral common in soils.

Effect of oxygen An important element in chemical weathering is oxygen. The chemical reaction of oxygen with other substances is called **oxidation.** Approximately 21 percent of Earth's atmosphere is oxygen gas. Iron in rocks and minerals combines with oxygen in the water and air to form minerals with the oxidized form of iron. A common mineral that contains the oxidized form of iron is hematite.

ACROSS THE CURRICULUM

Math Ask students to calculate the increase in the rate of chemical weathering for a substance whose temperature increases to 30°C from 10°C. four times faster

DIFFERENTIATED INSTRUCTION

Struggling Learners Ask students to make a word search for a partner using the vocabulary of weathering, including *mechanical weathering, frost wedging, exfoliation, chemical weathering, hydrolysis,* and *oxidation.* Ask them to provide definitions for each term so their partners will know for what words to search.

Effect of carbon dioxide Another atmospheric gas that contributes to the chemical weathering process is carbon dioxide. Carbon dioxide is a gas that occurs naturally in the atmosphere as a product of living organisms. When carbon dioxide combines with water in the atmosphere, it forms a very weak acid called carbonic acid that falls to Earth's surface as precipitation.

Precipitation includes rain, snow, sleet, and fog. Natural precipitation has a pH of 5.6. The slight acidity of precipitation causes it to dissolve certain rocks, such as limestone.

Decaying organic matter and respiration produce high levels of carbon dioxide. When slightly acidic water from precipitation seeps into the ground and combines with carbon dioxide in the soil, carbonic acid becomes a stronger agent in the chemical weathering process. Carbonic acid slowly reacts with minerals such as calcite in limestone and marble to dissolve rocks. After many years, limestone caverns can form where the carbonic acid flowed through cracks in limestone rocks and reacted with calcite.

Effect of acid precipitation Another agent of chemical weathering is acid precipitation, which is caused by sulfur dioxide, carbon dioxide, and nitrogen oxides. These compounds are released into the atmosphere, often by human activities. Sulfur dioxide and carbon dioxide are primarily the product of burning fossil fuels. Motor vehicle exhaust contributes to the emissions of nitrogen oxides. These three gases combine with oxygen and water in the atmosphere and form strong sulfuric, nitric, and carbonic acids.

Recall that the acidity of a solution is described using the pH scale. Acid precipitation is precipitation that has a pH value below 5.6—the pH of normal rainfall. Because strong acids can be harmful to many organisms and destructive to human-made structures, acid precipitation often creates problems. Many plant and animal populations, such as the forest shown in **Figure 5,** cannot survive even slight changes in acidity. Acid precipitation is a serious issue in New York, West Virginia, and much of Pennsylvania.

VOCABULARY

ACADEMIC VOCABULARY

Process
a natural phenomenon marked by gradual changes that lead toward a particular result
The process of growth changes a seedling into a tree.

Piotr Zawisza/E+/Getty Images

Figure 5 Forests around the world have been damaged by the effects of acid precipitation. Acid precipitation can make forests more vulnerable to disease and damage by insects.

Project

Identify pH Have students work in groups of three or four to determine the pH levels of common materials found at home. Ask each student to bring in one common household item, such as vinegar, milk, orange juice, apple juice, or shampoo to test. Make sure students do not bring in hazardous chemicals such as bleach, ammonia, or lye. Have students use litmus paper to determine the pH levels of the items. Then, compile the data on the board for students to record in their Earth science journals. BL COOP LEARN

Environmental Connection

Acid Precipitation The effects of acid precipitation on people are different from its effects on aquatic species. Swimming in an acidic lake might not harm a human, but it will harm a frog or a fish. However, the gases in the atmosphere that cause acid precipitation are harmful to humans. These gases, sulfur dioxide and nitrogen oxides, irritate and damage people's lungs.

ACROSS THE CURRICULUM

Chemistry Acids and bases are standard topics in the study of chemistry. Acids have pH values less than 7, and bases have pH values greater than 7. A pH of 7 is neutral. Some plants and animals can thrive only within specific pH ranges, which is why acid precipitation can be damaging to both terrestrial and aquatic organisms.

Project

Weathering Take the class on a field trip to a cemetery. Have students work in groups of three to four. Each group will need to bring a notebook, pencil, paper with which to make tombstone rubbings, metric ruler, camera, and compass. Each group will collect information and make a report that includes the following:

1. the average rate of weathering of at least ten tombstones (based on tombstone dates)
2. a map of the cemetery with the locations of the tombstones noted and a legend with a directional component
3. photographs and accompanying rubbings of the tombstones studied
4. a group report that includes the purpose of the investigation, the procedures followed, data, and conclusions OL

Concept Development

Rates of Weathering In addition to the effects of climate, the rates of weathering are affected by the type and structure of rock. For example, rocks composed of quartz are more resistant to weathering than rocks made of less resistant minerals, such as feldspars. Rocks of high porosity (void spaces) and high permeability (ease with which fluid moves through) are readily weathered. Fractures and bedding planes in rocks act as natural passageways for water, which accelerates weathering processes.

☑ READING CHECK Variables of climate including precipitation, temperature, and evaporation contribute to different rates of weathering.

▪ **Caption Question Fig. 6** Colder, drier climates with less vegetation experience less chemical weathering.

Rate of Weathering

The natural weathering of earth materials occurs slowly. For example, it can take 2000 years to weather 1 cm of limestone, and most rocks weather at even slower rates. Certain conditions and interactions can accelerate or slow the weathering process, as demonstrated in the GeoLab at the end of this chapter.

Effects of climate on weathering Climate is the major influence on the rate of weathering of earth materials. Precipitation, temperature, and evaporation are factors of climate. The interaction between temperature and precipitation in a given climate determines the rate of weathering in a region.

☑ READING CHECK **Explain** why different climates have different rates of weathering.

Rates of chemical weathering Chemical weathering is rapid in climates with warm temperatures, abundant rainfall, and lush vegetation. These climatic conditions can produce soils that are rich in organic matter. Water from heavy rainfalls combines with the carbon dioxide in soil organic matter and produces high levels of carbonic acid. The resulting carbonic acid accelerates the weathering process. Chemical weathering has the greatest effects along the equator, where rainfall is plentiful and the temperature tends to be high, as shown in **Figure 6.**

▪ **Figure 6** The impact of chemical weathering is related to a region's climate. Warm, lush areas such as the tropics experience the fastest chemical weathering.
Infer *what parts of the world experience less chemical weathering.*

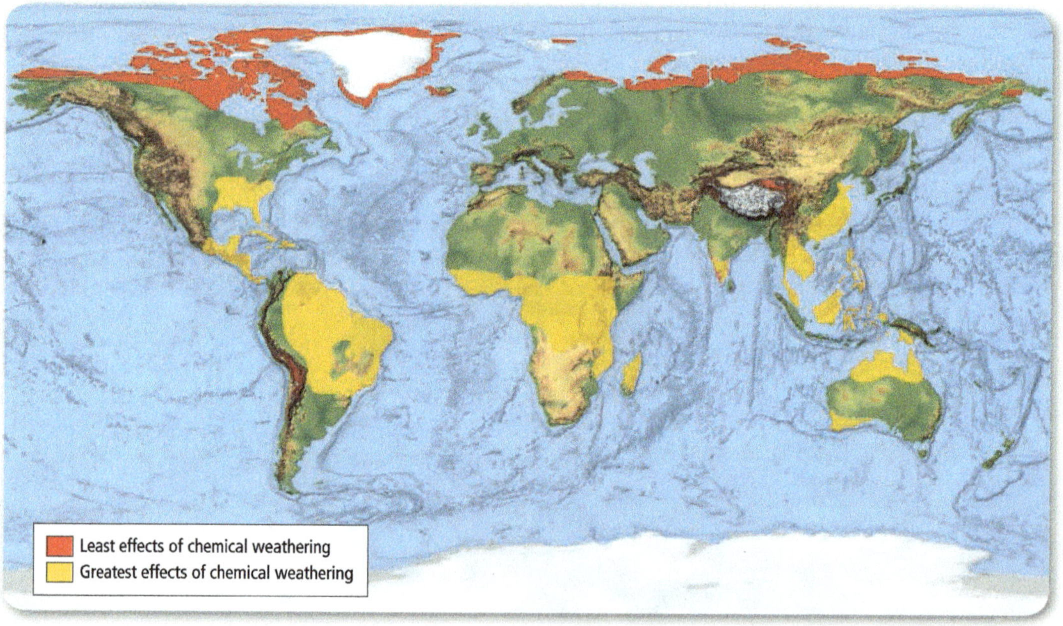

EARTH SCIENCE JOURNAL

Weathering and Materials Ask students to research how a variety of materials would weather in their climate, including a granite obelisk, a limestone pillar at the front of a building, a marble statue outdoors, and a wooden deck chair. They should describe in their Earth science journals how each of these objects would be affected by the climate in your area and make predictions about which object would weather the most and least if left outside for one year. OL

Rubric

Rates of physical weathering Conversely, physical weathering can break down rocks more rapidly in cool climates. Physical weathering rates are highest in areas where water in cracks within the rocks undergoes repeated freezing and thawing. Conditions in such climates do not favor chemical weathering because cool temperatures slow or inhibit chemical reactions. Little or no chemical weathering occurs in areas that are frigid year-round.

The different rates of weathering caused by different climatic conditions can be emphasized by a comparison of Asheville, North Carolina, and Phoenix, Arizona. Phoenix has dry, warm, conditions; temperatures do not drop below the freezing point of water, and humidity is low. In Asheville, temperatures can drop below freezing during the winter months, and Asheville has more monthly rainfall and higher levels of humidity than Phoenix. Because of these differences in their climates, rocks and man-made structures in Asheville experience higher rates of mechanical and chemical weathering than those in Phoenix.

Figure 7 shows how rates of weathering are dependent on climate. Both Egyptian obelisks were carved from granite more than one thousand years ago. For more than a thousand years, they stood in Egypt's dry climate, showing few effects of weathering. In 1881, Cleopatra's Needle was transported from Egypt to New York City. In the time that has passed since then, the acid precipitation and the repeated cycles of freezing and thawing in New York City accelerated the processes of chemical and physical weathering. In comparison, the obelisk that remains in Egypt appears unchanged.

Rock type and composition. Not all the rocks in the same climate weather at the same rate. The effects of climate on the weathering of rock also depends on the rock type and composition. For example, rocks containing mostly calcite, such as limestone and marble, are more easily weathered than rocks containing mostly quartz, such as granite and quartzite.

Figure 7 The climate of New York City caused the obelisk on the left to weather rapidly. The obelisk on the right has been preserved by Egypt's dry, warm climate.

Interpret the Photo

Rates of Weathering Ask students to carefully study these two photos in **Figure 7** as you ask the following questions: Which type of weathering would be involved in weathering each obelisk? chemical weathering Which obelisk would you expect to have weathered the most in another 125 years? Justify your answer. The obelisk in New York City will weather more than the obelisk in Egypt because acid precipitation and the climate in New York City accelerate the weathering process. OL

GeoLAB

iLab Station

The GeoLab located at the end of the chapter can be used at this point in the lesson.

ACROSS THE CURRICULUM

Biology Many children are told to chew their food thoroughly before swallowing. The purpose of chewing food is to break it down into smaller pieces. Smaller pieces have more total surface area, which provides more contact area for chemical digestive processes to take place. The same concept applies to the rate of weathering. OL

3 Assess

Check for Understanding

Observe and Infer Select photographs of different regions of the world depicting areas that show extreme or familiar sites. Place one large photograph on each lab table and have groups rotate from one table to another. In each case, the group is to determine from the photograph whether the area is likely to experience low, moderate, or high levels of both physical and chemical weathering. Groups should include reasons for their answers.

Reteach

Communicate Have small groups of students each select one topic within the section to present to the class for a review. Each group will be responsible for developing five quiz questions for classmates to answer. **COOP LEARN**

Assessment

Performance Have students decide on their best individual and group work for this section, collect it, and arrange it in a display case or on a bulletin board. The assemblage can then be used for individual and class review activities for upcoming assessments.

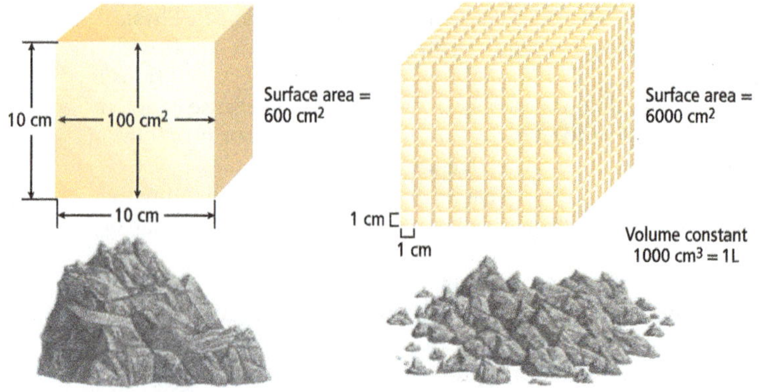

Figure 8 When the same object is broken into two or more pieces, the surface area increases. The large cube has a volume of 1000 cm^3. When it is broken into 1000 pieces, the volume is unchanged, but the surface area is increased one thousand times.

Surface area The rate of weathering also depends on the surface area that is exposed. Mechanical weathering breaks rocks into smaller pieces. As the pieces get smaller, their surface area increases, as illustrated in **Figure 8.** When this happens, there is more total surface area available for chemical weathering. The result is that weathering has more of an effect on multiple, smaller particles, as opposed to a single, large rock as you learned in the Launch Lab.

Topography The slope of a landscape also determines the rate of weathering. Rocks on level areas are likely to remain in place over time, whereas the same rocks on slopes tend to move downslope as a result of gravity. Steep slopes therefore promote erosion and continually expose more rocks to mechanical and chemical weathering.

SECTION 1 REVIEW

Section Self-Check

Section Summary

- Mechanical weathering changes a rock's size and shape.
- Frost wedging and exfoliation are forms of mechanical weathering.
- Chemical weathering changes the composition of a rock.
- The rate of chemical weathering depends on the climate, rock type, surface area, and topography.

Understand Main Ideas

1. MAINIDEA **Distinguish** between the characteristics of an unweathered rock and those of a highly weathered rock.
2. **Describe** the factors that control the rate of chemical weathering and those that control the rate of physical weathering.
3. **Compare** chemical weathering to mechanical weathering.
4. **Analyze** the relationship between surface area and weathering.

Think Critically

5. **Infer** which would last longer, the engraving in a headstone made of marble, or an identical engraving in a headstone made of granite.

MATH IN Earth Science

6. Infer the relationship between weathering and surface area by graphing the relationship between the rate of weathering and the surface area of a material.

SECTION 1 REVIEW

1. An unweathered rock has fewer cracks, and its minerals have not reacted with the air or been affected by water compared to a highly weathered rock.
2. Chemical weathering is rapid in climates with warm temperatures, abundant rainfall, and lush vegetation. Physical weathering is rapid in cool climates where water in cracks undergoes repeated freezing and thawing.
3. Chemical weathering changes the composition of a rock, either by dissolving part of it, or by chemical reactions that change the composition of the minerals. Mechanical weathering changes only the shape of the rock.
4. As surface area that is available for weathering increases, the amount of weathering can increase.
5. An engraving on a granite headstone will last longer, because granite is composed of minerals that do not weather rapidly. Marble is made of the mineral calcite, which reacts with acid in rainwater, so a marble headstone decomposes much more quickly.
6. Graphs should reflect an increase in the rate of weathering with an increase in surface area. The x- and y-axes should be labeled *surface area* and *weathering rate*, respectively.

SECTION 2

Erosion and Deposition

MAINIDEA Erosion transports weathered materials across Earth's surface until they are deposited.

Essential Questions

- What is the relationship of gravity to all agents of erosion?
- What features are characteristic of the different types of erosion?
- How do living and nonliving things impact the processes of weathering and erosion?

Review Vocabulary

gravity: a force of attraction between objects due to their masses

New Vocabulary

erosion
deposition
rill erosion
gully erosion

EARTH SCIENCE 4 YOU Have you ever noticed the mud that collects on sidewalks and streets after a heavy rainfall? Water carries sediment to the sidewalks and streets and deposits it as mud.

Gravity's Role

Recall that the process of weathering breaks rock and soil into smaller pieces, but never moves it. The removal of weathered rock and soil from its original location is a process called **erosion.** Erosion can remove material through a number of different agents, including running water, glaciers, wind, ocean currents, and waves. These agents of erosion can carry rock and soil thousands of kilometers away from their source. After the materials are transported, they are dropped in another location in a process known as **deposition.**

Gravity is associated with many erosional agents because the force of gravity tends to pull all materials downslope. Without gravity, neither streams nor glaciers would flow. In the process of erosion, gravity pulls loose rock downslope. **Figure 9** shows the effects of gravity on the landscape of Fall Creek Falls in Tennessee. The effects of gravity on erosion by running water can often produce dramatic landscapes with steep valleys.

■ **Figure 9** Within about 2,600 m, the stream descends 78 m at Fall Creek Falls in Tennessee.
Calculate *the average descent of the stream per meter along the river.*

SECTION 2

1 Focus

Focus Transparency

Before presenting the lesson, project **Section Focus Transparency 18** and have students answer the accompanying questions. EL

MAINIDEA

Model Erosion Crush a piece of chalk with a hammer on a paper plate. This models weathering. Then sweep the crushed material with your hand. Ask: What might this movement represent? erosion Then drop the chalk particles into the trash can. Ask: What might this process represent? deposition

2 Teach

Teacher Content Support

Erosion Erosion can occur only after other Earth processes such as weathering have broken down solid Earth material. Energy from the movement of agents of erosion dislodges, carries, and transports the particles. Hence, the sizes and masses of the particles being moved are limited by the amount of energy and force associated with each erosional agent. The erosional-depositional process is an energy system in which energy cycles between potential and kinetic energy.

■ **Caption Question Fig. 9** For every 33 meters of horizontal distance, the river descends 1 meter.

EARTH SCIENCE JOURNAL

Weathering and Erosion To familiarize students with the way water weathers rock and erodes it, ask students to describe in their Earth science journals how gravity and water have carved the landscape of Watkins Glen State Park in **Figure 9.** Ask students to create a sketch of the weathering before and after in their Earth science journals. OL

DIFFERENTIATED INSTRUCTION

English Learners Students from other areas of the world might have rich experiences to share with the class. Students might have observed different types of erosion in different places in the world. Encourage students from other cultures to share their own observations of erosion and deposition with the class.

MiniLAB

iLab Station Rubric

Purpose Students will demonstrate how a model of a rock erodes when exposed to environmental conditions.

Process Skills model, recognize cause and effect, communicate, interpret data, observe and infer

Additional Materials metric ruler, balance

Safety Precautions Approve lab safety forms before work begins. Care should be taken with toothpicks to avoid puncturing skin.

Teaching Strategy Some soaps are softer than others and will "weather" more quickly. Pretest the brand of soap you use.

Expected Results Results should indicate a decrease in the depth of the letters with the flow of water.

Analysis

1. Their depth decreased.
2. The shape of the bar of soap became more rounded, the size of the bar became smaller, and the mass of the bar of soap decreased.
3. Answers will vary, but should include various ways to detect the presence of dissolved soap in the catch basin.

Assessment

Performance Have students design and conduct an experiment to compare weathering rates of different brands of soap and present their data in a chart and graph format.

LabManager™
Customize any lab with the LabManager™ CD-ROM.

Figure 10 Rill erosion can occur in an agricultural field. Gully erosion often develops from rills.
Suggest *land management practices that can slow or prevent the development of gully erosion.*

Erosion by Water

Moving water is perhaps the most powerful agent of erosion. Stream erosion can reshape entire landscapes. Stream erosion is greatest when a large volume of water is moving rapidly, such as during spring thaws and torrential downpours. Water flowing down steep slopes has additional erosive potential resulting from gravity, causing it to cut downward into the slopes, carving steep valleys and carrying away rock and soil. Water that flows swiftly or in large volumes can independently carry more material. The Mississippi River carries over 400,000 metric tons of sediment each day from thousands of kilometers away due to the volume of water in the river.

READING CHECK **Predict** what time of year water has the most potential for erosion.

Erosion by water can have destructive results. For example, water flowing downslope can carry away fertile soil, thus affecting agricultural areas. **Rill erosion** develops when running water cuts small channels into the side of a slope, as shown in **Figure 10.** When a channel becomes deep and wide as a result of further erosion, rill erosion evolves into **gully erosion,** also shown in **Figure 10.** The channels formed in gully erosion can transport much more water, and consequently more soil, than rills. Gullies can be more than 3 m deep and can cause major problems in farming and grazing areas.

MiniLAB

iLab Station

Model Erosion

How do rocks erode? When rocks are weathered by their surrounding environment, particles can be carried away by erosion.

Procedure

1. Read and complete the lab safety form.
2. Carve your name deeply into a **bar of soap** with a **toothpick.** Measure the mass of the soap.
3. Measure and record the depth of the letters carved into the soap.
4. Place the bar of soap on its edge in a **catch basin.**
5. Slowly pour **water** over the bar of soap until a change occurs in the depth of the carved letters.
6. Measure and record the depth of the carved letters.

Analysis

1. **Describe** how the depth of the letters carved into the bar of soap changed.
2. **Infer** whether the shape, size, or mass of the bar of soap changed.
3. **Consider** what additional procedure you could follow to determine whether any soap wore away.

READING CHECK Erosion by water is more likely in the spring, when there is an increase in the amount of water available.

Caption Question **Fig. 10** Answers will vary, but should include ways of slowing or preventing the flow of water downslope.

DIFFERENTIATED INSTRUCTION

Struggling Learners Have students create a daily diary for a small pebble or a grain of sand as it is moved from one location to another by water. Have students note the speed of movement and points of deposition. If students have difficulty writing a diary, let them record their diary using a tape recorder.

Rivers and streams Each year, streams carry billions of metric tons of sediments and weathered material to coastal areas. Once a river enters the ocean, the current slows down, which reduces the potential of the stream to carry sediment. As a result, streams deposit large amounts of sediments in the region where they enter the ocean. The buildup of sediments over time forms deltas, such as the Colorado River Delta, shown in **Figure 11.** The volume of river flow and the action of tides determines the shapes of deltas, most of which contain fertile soil. The Colorado River Delta shows the classic fan shape associated with many deltas.

Figure 11 Streams slow down when they meet the ocean. In these regions, sediments are deposited by the river, resulting in the development of a delta.

Wave action Erosion of materials also occurs along the ocean floor and at continental and island shorelines. The work of ocean currents, waves, and tides carves out cliffs, arches, and other features along the continents' edges. In addition, sand particles accumulate on shorelines and form dunes and beaches. The constant movement of water and the availability of accumulated weathered material result in a continuous erosional process, especially along ocean shorelines. Sand along a shoreline is repeatedly picked up, moved, and deposited by ocean currents. As a result, sandbars form from offshore sand deposits. If the sandbars continue to be built up with sediments, they can develop into barrier islands. Many barrier islands, such as the Outer Banks of North Carolina shown in **Figure 12,** have formed along both the Gulf and Atlantic Coasts of the United States.

Just as shorelines are built by the process of deposition in some areas, they are reduced by the process of coastal erosion in other areas. Changing tides and conditions associated with coastal storms can also have a great impact on coastal erosion. Human development and population growth along shorelines have led to attempts to control the erosion of sand. However, efforts to keep the sand on one beachfront disrupt the natural migration of sand along the shore, depleting sand from another area.

Figure 12 The Outer Banks of North Carolina have been built over time by deposition of sand and sediments.

Teacher Content Support

Erosion and Deposition In Earth science, perhaps no other two words are more closely related than erosion and deposition. The processes are dependent on one another. Distinguishing one from the other is a means to enhance understanding and to emphasize the importance of all aspects of this Earth cycle. Erosion and deposition are primarily responsible for the changing of Earth's surface features. Various erosional agents pick up Earth materials, carry them to other locations, and deposit the relocated materials in a variety of settings or environments.

The depositional environment of a sedimentary rock provides geologists with valuable clues to the geologic history of an area. Determining the depositional environment is also a crucial step in the exploration for petroleum and other geological commodities.

Erosion is a destructive process that wears down Earth's surface, while deposition is a constructive process that builds up Earth's surface. Together, they help maintain the dynamic equilibrium of Earth's surface features.

Project

Changes to Coastlines Have students work in small groups to research an east coast area of coastline. Have students find dated documentation of changes to the coastline over time. They can make a visual time line by using illustrations or copied photos of the coastal area over time. AL

EARTH SCIENCE JOURNAL

Erosional and Depositional Features

To familiarize students with erosional and depositional features, have them research to find at least two erosional features and two depositional features. Students should create a written description of each feature in their Earth science journals that includes how the feature formed and a small sketch. Cliffs and arches are erosional features and sandbars and sand dunes are depositional features. AL

Rubric

Concept Development

Glacial Activity The last glacial period occurred so recently in geologic time in North America, South America, and Eurasia that other erosional processes such as weathering and erosion have not had time to significantly change the landscape formed by glacial activity. As a result, what is observable on Earth's surface today, aside from the cover of vegetation, remains much as it was formed by the most recent glacial activity. Thus, scientists are able to reconstruct the shapes of glaciers and infer their movements.

Enrichment

Little Ice Age Evidence obtained through the study of old documents indicates that during the sixteenth and seventeenth centuries, there was a so-called little ice age. The expansion of glaciers over small villages in the Alps, and generally cooler climates have also been documented from the mid-thirteenth century through the mid-nineteenth century. Small-scale ice ages between major ice ages appear to be a natural phenomenon. AL

Environmental Connection

Wind Erosion Wind barriers are an effective tool in decreasing the amount of airborne dust on a construction site. Wind barriers on construction sites are constructed out of potted trees, fences, berms, and even large construction equipment. To reduce dust and wind erosion on a construction site, wind barriers must be used along with the application of water.

Caption Question Fig. 14 These trees can serve as a wind break for 300 m.

Figure 13 Iceberg Lake in Glacier National Park, Montana, was formed by glaciers.

Glacial Erosion

Although glaciers currently cover about 10 percent of Earth's surface, they have covered over 30 percent of Earth's surface in the past. Glaciers left their mark on much of the landscape, and their erosional effects are large-scale and dramatic. Glaciers scrape and gouge out large sections of Earth's landscape. Because they can move as dense, enormous rivers of slowly flowing ice, glaciers have the capacity to carry huge rocks and piles of debris over great distances and grind the rocks beneath them into flour-sized particles. Glacial movements scratch and grind surfaces. The features left in the wake of glacial movements include steep U-shaped valleys and lakes, such as the one shown in **Figure 13.**

The effects of glaciers on the landscape also include deposition. For example, soils in the northern parts of the United States are formed from material that was transported and deposited by glaciers. Although the most recent ice age ended about 12,000 years ago, glaciers continue to affect erosional processes on Earth.

Wind Erosion

Wind can be a major erosional agent, especially in arid and coastal regions. Such regions tend to have little vegetation to hold soil in place. Wind can easily pick up and move fine, dry particles. The effects of wind erosion can be both dramatic and devastating. The abrasive action of windblown particles can damage both natural features and human-made structures. Winds can blow against the force of gravity and easily move fine-grained sediments and sand uphill.

Wind barriers One farming method that can reduce the effects of wind erosion is the planting of wind barriers, also called windbreaks, shown in **Figure 14.** Windbreaks are trees or other vegetation planted perpendicular to the direction of the wind. A wind barrier might be a row of trees along the edge of a field. In addition to reducing erosion, wind barriers can trap blowing snow, conserve moisture, and protect crops from the effects of the wind.

Figure 14 A windbreak can reduce the speed of the wind for distances up to 30 times the height of the tree.
Calculate *If these trees are 10 m tall, what is the distance over which they can serve as a windbreak?*

ACROSS THE CURRICULUM

Astronomy Wind erosion is a major influence on the formation of surface features on Mars. Mars has been known to have dust storms that have lasted for months at a time. Photographs from the *Phoenix Mars Lander* and *Mars Reconnaissance Orbiter* space missions have provided evidence of Mars's hostile, desert environment.

EARTH SCIENCE JOURNAL

Glacial Activity Assign students to write essays in their Earth science journals about the positive effects of past glacial activity on civilization today. Student essays might discuss landscape beauty, enjoyable and useful lakes, outstanding skiing and mountain-climbing opportunities, and transport of sediments that become fertile soils. BL

Figure 15 In this construction project, the landscape was considerably altered. **Analyze** *the results of this alteration of the landscape.*

Erosion by Living Things

Plants and animals also play a role in erosion. As plants and animals carry out their life processes, they frequently move Earth's surface materials from one place to another. For example, rocks and sediments are moved by the roots of plants, or when animals burrow into soil. Humans also play a role in erosion when excavating large areas and moving soil from one location to another, as shown in **Figure 15.** Planting a garden, developing a new athletic field, and building a highway are all examples of human activities that result in the moving of earth materials from one place to another.

SECTION 2 REVIEW

Section Self-Check

Section Summary

- The processes of erosion and deposition have shaped Earth's landscape in many ways.
- Gravity is the driving force behind major agents of erosion.
- Agents of erosion include running water, waves, glaciers, wind, and living things.

Understand Main Ideas

1. MAINIDEA **Discuss** how weathering and erosion are related.
2. **Describe** how gravity is associated with many erosional agents.
3. **Classify** the type of erosion that could move sand along a shoreline.
4. **Compare and contrast** rill erosion and gully erosion.

Think Critically

5. **Generalize** about which type of erosion is most significant in your area.
6. **Diagram** a design for a wind barrier to prevent wind erosion.

WRITING IN Earth Science

7. Research how a development in your area has alleviated or contributed to erosion. Present your results to the class, including which type of erosion occurred, and where the eroded materials will eventually be deposited.

Caption Question Fig. 15 Humans have moved soil and altered the landscape, which might increase erosion.

3 Assess

Check for Understanding

Compare and Contrast Lead a class discussion in which erosion on Earth and on the Moon are compared. As you proceed with the discussion, develop a data table on the board entitled "A Comparison of Erosion." Subdivide the table into one column for Earth and another for the Moon. Use the following row headings: Gravity, Running Water, Glaciers, Wind, and Plants and Animals. Lead students to see that the lack of erosion on the Moon is a result of its relatively weak gravitation force and its lack of an atmosphere, running water, and life.

Reteach

Outline Have students outline the main points of this section in their Earth science journals. Ask students to find pictures in magazines to paste into their Earth science journals as examples.

Assessment

Knowledge Ask students to describe the similarities and differences among erosion caused by water, wind, and glaciers. How are these agents of erosion similar to and different from erosion caused by plants, animals, and humans?

SECTION 2 REVIEW

1. Weathering breaks down rocks and soil into smaller fragments. Weathered material is more susceptible to the forces of erosion. Erosion transports loose material via agents such as water, wind, and glaciers.
2. Gravity moves all materials downslope.
3. wave action
4. Rill erosion and gully erosion are both caused by water. Rill erosion has small channels, while gully erosion has deep channels.
5. Answers will vary depending on your location.
6. Answers will vary, but designs should include a method of slowing wind.
7. Answers should include a description of the type of erosion and the location of the eroded materials as well as a description of the development.

Rubric

SECTION 3

1 Focus

Focus Transparency

Before presenting the lesson, project **Section Focus Transparency 19** and have students answer the accompanying questions. EL

MAINIDEA

Describe Soil Bring a bag of soil into class. Ask the students to describe the characteristics of the soil sample in terms of color and texture.

2 Teach

Activity

Compare and Contrast Samples Have students each bring in a plastic bag full of soil. Each bag should have a label identifying the student's name, the date, and a description of the location from which it was taken (for example, along a stream bank or on a hillside). Set up a display in the classroom. Have students note similarities and differences among the soil samples. EL

Caption Question Fig. 16 Animals can bring organic material such as leaves and grass into their burrows; animals also leave behind fecal matter that decomposes in the soil.

SECTION 3

Essential Questions

- How does soil form?
- What are the different soil horizons in a soil profile?
- What factors affect soil formation?

Review Vocabulary

organism: anything that has or once had all the characteristics of life

New Vocabulary

soil
residual soil
transported soil
soil profile
soil horizon

Soil

MAINIDEA Soil forms slowly as a result of mechanical and chemical processes.

EARTH SCIENCE 4 YOU What color is soil? Soils can be many different colors—dark brown, light brown, red, or almost white. Soils develop through the interaction of a number of factors, which determine the color of soil.

Soil Formation

What is soil? It is found almost everywhere on Earth's surface. Weathered rock alone is not soil. **Soil** is the loose covering of weathered rock particles and decaying organic matter, called humus, overlying the bedrock of Earth's surface, and serves as a medium for the growth of plants. Soil is the product of thousands of years of chemical and mechanical weathering and biological activity.

Soil development The soil-development process often begins when weathering breaks solid bedrock into smaller pieces. These pieces of rock continue to weather and break down into smaller pieces. Worms and other organisms help break down organic matter and add nutrients to the soil as well as create passages for air and water, as shown in **Figure 16.**

As nutrients are added to the soil, its texture changes, and the soil's capacity to hold water increases. While all soil contains some organic matter in various states of decay, the amount varies widely among different types of soil. For example, as much as 5 percent of the volume of prairie soils is organic matter, while most desert soils have almost no organic matter.

Figure 16 Organisms in the soil change the soil's structure over time by adding nutrients and passages for air.
Infer *how animals also alter the soil by adding organic material.*

EARTH SCIENCE JOURNAL

Soil Formation Ask students to describe in their Earth science journals how soils form in your area. Students should research parent material in the area so they make the connection between the type of rock and the type of soil. Suggest that they begin with a boulder and end with soil in a farmer's field so that they describe the complete process. AL

Soil Layers

During the process of soil formation, layers develop in the soil. Most of the volume of soil is formed from the weathered products of a source rock, called the parent material. The parent material of a soil is often the bedrock. As the parent material weathers, the weathering products rest on top of the parent material. Over time, a layer of the smallest pieces of weathered rock develops above the parent material. Eventually, living organisms such as plants and animals become established, and use nutrients and shelter available in the material. Rainwater seeps through this top layer of materials and dissolves soluble minerals, carrying them into the lower layers of the soil.

A soil whose parent material is the local bedrock is called **residual soil.** Kentucky's bluegrass soil is an example of residual soil, as are the red soils in Georgia. Not all soil develops from local bedrock. **Transported soil,** shown in the valley in **Figure 17,** is soil that develops from parent material that has been moved far from its original location. Agents of erosion transport parent material from its place of origin to new locations. For example, glaciers have transported sediments from Canada to many parts of the United States. Streams and rivers, especially during times of flooding, also transport sediments downstream to floodplains. Winds also carry sediment to new locations. Over time, processes of soil formation transform these deposits into mature soil layers.

READING CHECK **Explain** how residual soils are different from transported soils.

CAREERS IN EARTH SCIENCE

Landscaper A landscaper uses his or her knowledge of soils and performs tests to evaluate soils at different sites. Landscapers use the information they gather to choose plants that are appropriate to the soil conditions.

WebQuest

Figure 17 In a stream valley, transported soils are often found in the flood plain. Residual soils are often found in the higher, mountainous regions.

Pixtal/AGE Fotostock

Use an Analogy

Parent Bedrock Parent bedrock or parent material weathers into rock pieces that eventually form soil. The characteristics of a soil are dependent on the characteristics of the parent bedrock. An analogy can be drawn between parent bedrock and soil and children and their parents. Characteristics that children have are often similar to those of their parents. A child with black hair is likely to have at least one parent with black hair. A tall child usually has a tall parent. As is the case with children, soils have characteristics similar to those of their parent bedrock. BL

READING CHECK The primary difference between residual and transported soils is in the current location of the parent material. Residual soils form from bedrock underneath, while transported soils are moved away from parent bedrock.

CAREERS IN EARTH SCIENCE

WebQuest

IN THE FIELD

Pedro Sanchez Pedro Sanchez is a Cuban-born soil scientist who earned his PhD at Cornell University. He is the Director of the Tropical Agriculture and the Rural Environment Program, as well as the Co-Chair of the Hunger Task Force of the Millennium Project, an advisory body to the United Nations. He is best known for finding ways to restore soil fertility to some of the world's most degraded soils. Sanchez has researched tropical soils in the Philippines, Peru, Columbia, and Kenya.

Model

Soil Profile Have students draw, label, and color a model of the vertical layers of a soil profile. Take students out to the schoolyard so they can compare their models with an actual soil profile. Use a garden trowel to dig a narrow but deep pit (about 10 cm). Pull the soil away from the pit so students can see the different layers. While they are outdoors, have students sketch a real soil profile based on the pit. They should record observations about organic material, texture, size of particles, and color. Take a small sample back to the classroom to classify the soil type. **AL**

☑ **READING CHECK** A mature soil has well-developed soil horizons, while an immature soil does not have well-developed soil horizons.

■ **Figure 18** An undeveloped soil has few, if any, distinct layers, while mature soils are characterized by several soil horizons that have developed over time.

Soil profiles Digging a deep hole in the ground will reveal a soil profile. A **soil profile** is a vertical sequence of soil layers. Some soil profiles have more distinct layers than others. Relatively new soils that have not yet developed distinct layers are called undeveloped soils, shown in **Figure 18.** It can take tens of thousands of years for distinct layers to form in a soil. Those soils are called mature. An example is shown in **Figure 18.**

☑ **READING CHECK** **Explain** the difference between a mature and an undeveloped soil.

Soil horizons A distinct layer within a soil profile is called a **soil horizon.** There are typically four major soil horizons in mature soils, O, A, B, and C. The O-horizon is the top layer of organic material, which is made of humus and leaf litter. Below that, the A-horizon is a layer of weathered rock combined with a rich concentration of dark brown organic material. The B-horizon, also called the zone of accumulation, is a red or brown layer that has been enriched over time by clay and minerals deposited by water flowing from the layers above, or percolating upward from layers below. Usually the clay gives a blocky structure to the B-horizon. Accumulations of certain minerals can result in a hard layer called hardpan. Hardpan can be so dense that it allows little or no water to pass through it. The C-horizon contains little or no organic matter, and is often made of broken-down bedrock. The development of each horizon depends on the factors of soil formation.

DIFFERENTIATED INSTRUCTION

Visually Impaired Have students who are visually impaired construct soil profiles in plastic shoeboxes or other containers using cardboard and paper. Make sure they note the textures and moisture content of each horizon in the profiles. Have students explain their soil profile to a partner so that they communicate the differences in horizons. **OL**

Factors of Soil Formation

Five factors influence soil formation: climate, topography, parent material, biological activity, and time. These factors combine to produce different types of soil, called soil orders, that differ by region. Soil taxonomy (tak SAH nuh mee) is the system that scientists use to classify soils into orders and other categories. The five factors of soil formation result in 12 different soil orders.

Climate Climate is the most significant factor controlling the development of soils. Temperature, wind, and the amount of rainfall determines the type of soil that can develop.

Recall from Section 1 that rocks tend to weather rapidly under humid, temperate conditions, such as those found in climates along the eastern United States. Weathering results in soils that are rich in aluminum and iron oxides. Water from abundant rainfall moves downward, carrying dissolved minerals into the B-horizon. In contrast, the soils of arid regions are so dry that water from below ground moves up through evaporation, and leaves an accumulation of white calcium carbonate in the B-horizon. Tropical areas experience high temperatures and heavy rainfall. These conditions lead to the development of intensely weathered soils where all but the most insoluble minerals have been flushed out.

Topography Topography, which includes the slope and orientation of the land, affects the type of soil that forms. On steep slopes, weathered rock is carried downhill by agents of erosion. As a result, hillsides tend to have shallow soils, while valleys and flat areas develop thicker soils with more organic material. The orientation of slopes also affects soil formation. In the northern hemisphere, slopes that face south receive more sunlight than other slopes. The extra sunlight allows more vegetation to grow. Slopes without vegetation tend to lose more soil to erosion. **Figure 19** shows how the orientation and slope of a landscape can affect the formation of soil.

■ **Figure 19** The slope on the right side faces south, and the slope on the left side faces north.
Interpret *why one slope has more vegetation than the other.*

Identify Misconceptions

Some students are likely to think that the only importance of soil to humans is as a substance in which to grow plants.

Uncover the Misconception Ask students leading questions to help them identify common benefits of soil and materials gained either directly or indirectly from soil.

Demonstrate the Concept Describe how soil is used to grow trees, which are harvested for wood, which is shaped into furniture.

Assess New Knowledge Ask students to describe the role of soil in producing a cotton shirt.

■ **Caption Question Fig. 19** The south-facing slope receives more sunlight, and, therefore has more vegetation.

EARTH SCIENCE JOURNAL

Effects of Soil Movement Ask students to each write a letter in their Earth science journals from the point of view of a farmer who lives on a hill. The letter is to be addressed to another farmer (one who lives in the valley) and should discuss the problems that the farmer who lives on the hill is having with soil, including soil thickness, fertility, and erosional problems. BL

Teacher Content Support

Soil Soil is the product of a series of processes that begin with the weathering of Earth materials. The rate of soil formation is controlled by interdependent environmental conditions. The weathering and erosion of rock and mineral fragments that are found on Earth's surface, called regolith, begin the process. Of all the major factors that influence the rate of soil formation, including parent material and climatic conditions, time is one of the most crucial. Other factors that influence soil formation include the topography and the types of organisms present—including plants, animals, and microorganisms. Mature soil consists of identifiable horizons, which become further defined with time. These horizons can be further subdivided based on their characteristics, such as texture and fertility. Soil often includes material from both the A- and B-horizons, which together constitute what is called the solum.

☑ READING CHECK Microorganisms decompose dead plants and animals, which add organic material to soil.

■ **Caption Question** **Fig. 20** North Carolina has ultisols, which have been heavily weathered because of the high moisture content in the low-lying coastal plain areas.

Parent material Recall that a soil can be either residual or transported. If the soil is residual, it will have the same chemical composition as the local bedrock. For example, in regions near volcanoes, the soils form from weathered products of lava and ash. Volcanic soils tend to be rich in the minerals that were present in the lava. If the soil is transported, the minerals in the soil are likely to be different from those in the local bedrock.

Biological activity Organisms including fungi and bacteria, as well as plants and animals, interact with soil. Microorganisms decompose dead plants and animals. Plant roots can open channels, and when they decompose, they add organic material to the soil. Different types of living organisms in a soil can result in different soil orders. Mollisols (MAH lih sawlz), which are called prairie soils, and alfisols (AL fuh sawlz), also called woodland soils, both develop from the same climate, topography, and parent material. The different sets of organisms result in two soils with entirely different characteristics. For example, the activity of prairie organisms in mollisols produces a thick A-horizon, rich in organic matter. Some of the most fertile agricultural lands in the Great Plains region are mollisols.

☑ READING CHECK **Describe** how microorganisms affect soil formation.

Time The effects of time alone can determine the characteristics of a soil. New soils, such as entisols (EN tih sawlz), are often found along rivers, where sediment is deposited by periodic flooding. This type of soil is shown as a light blue color in **Figure 20.** These soils have had little time to weather and develop soil horizons. The effects of time on soil can be easy to recognize. After tens of thousands of years of weathering, most of the original minerals in a soil are changed or washed away. Minerals containing aluminum and iron remain, which can give older soils, such as ultisols (UL tih sawlz), a red color. **Figure 21** shows the locations of the 12 soil orders in the United States.

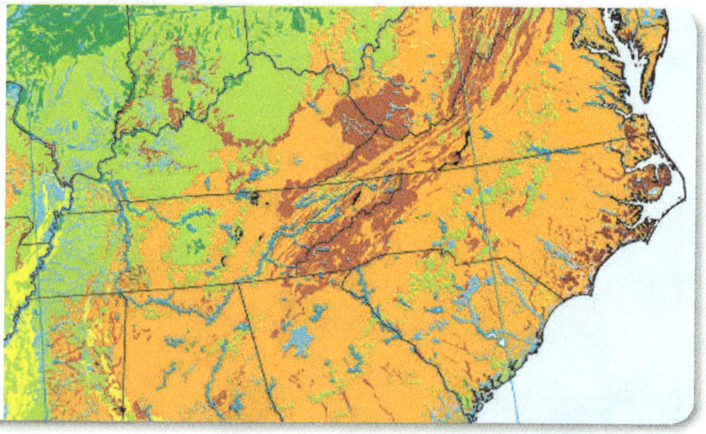

■ **Figure 20** Soil types vary widely from one area to the next, depending on the local climate, topography, parent material, organisms, and age of the soil. Entisols are shown in light blue and ultisols are shown in orange on this map.
Infer *how differences in topography have affected the types of soils in North Carolina.*

DIFFERENTIATED INSTRUCTION

Struggling Learners Ask students to make a concept map of the five factors of soil formation: climate, topography, parent material, biological activity, and time, to show how the factors are related. Students should show that time interacts with climate, topography, parent material, and biological activity, but that the other four factors are not necessarily connected.

VISUALIZING Soil Orders

Figure 21 The five factors of soil formation determine how the soil orders are distributed across the United States. Soil profiles of three soil orders from different parts of the country are shown. Each soil profile has soil horizons expressed differently.

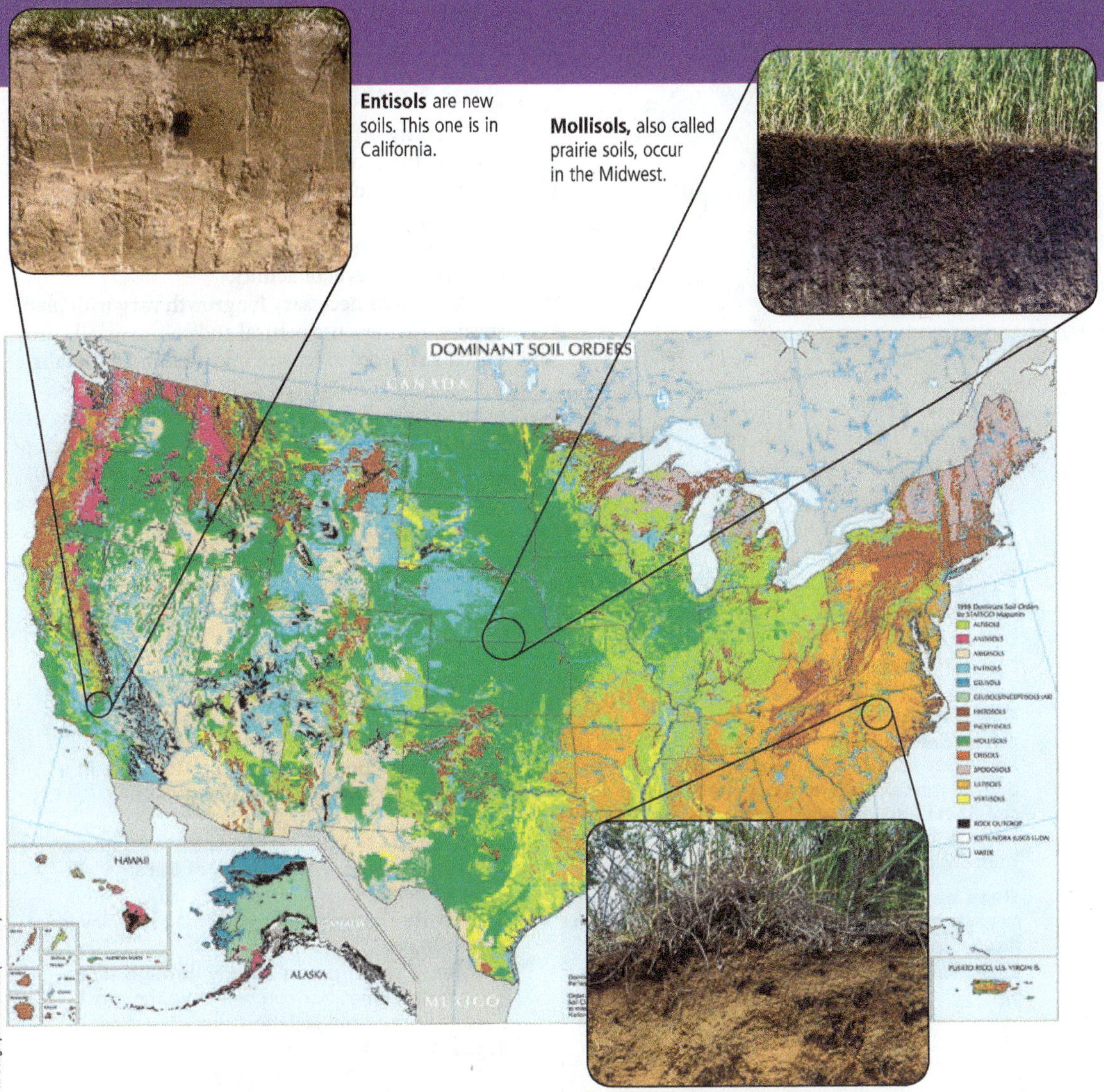

Entisols are new soils. This one is in California.

Mollisols, also called prairie soils, occur in the Midwest.

Ultisols are highly weathered soils. This one is in North Carolina.

(bkgd)State Soil Geographic Database (STATSGO)/NRCS/USDA

Concepts In Motion View an **animation of soil orders.**

VISUALIZING

Purpose

Students will understand differences between soils and where different orders are found in the U.S.

Teacher Content Support

Soil Classification One older method of classifying soils is based on the soils' compositions. With this method, most soils belong in one of three broad categories: pedalfers, pedocals, or laterites. The term *pedalfer* is a combination of *ped,* meaning *soil, Al,* meaning *aluminum,* and *Fe,* meaning *iron.* Pedalfers are common in humid regions. Most soluble materials leach out of pedalfers, leaving accumulations of aluminum-rich clays and iron oxides. Pedalfers are common in the eastern portion of the United States.

The term *pedocal* is a combination of *ped,* meaning *soil,* and *cal,* meaning *calcite* (calcium carbonate). Pedocals are found in the dry regions in the western portion of the United States. Pedocals have accumulations of calcium carbonate.

Laterites are wet, tropical soils resulting from intense weathering which leaves the soil with little humus. The calcite and silica has been leached, leaving this soil type with high concentrations of iron and aluminum.

Demonstration

Model a Soil Profile Use an old aquarium to show a model of a soil profile for a soil type in your area. Each side of the aquarium can be labeled as a different soil order and exhibit the particular horizons of that soil type. Once the model is set up, it can be left in the classroom for student reference and review. You can either make the model yourself and display it, or have students construct it as a class demonstration as the topic is being presented.

Data Analysis LAB

About the Lab

- Go through examples with the class as a whole and in small groups to be sure that students can use the soil-textural triangle.
- Ask students what particle sizes fertile soil contains and why. Fertile soil has a combination of all particle sizes to provide good water retention and drainage.
- See also Gee, W., J. W. Bauder. 1986. Particle Size Analysis. *Methods of soil analysis, part I, physical and mineralogical methods.* Madison, WI: American Society of Agronomy.

Think Critically

1. A, 41% sand; B, 28% silt, clay; C, 9% clay, silty loam.
2. Sample B has the greatest percent (67%) of smallest-size particles (clay).
3. 40 percent clay.
4. The B horizon will have the highest moisture capacity since it has the highest amount of clay and silt.
5. Answers will vary, but should include permeability, ability to compact, and vulnerability to erosion by wind.

USDA Soil Classification

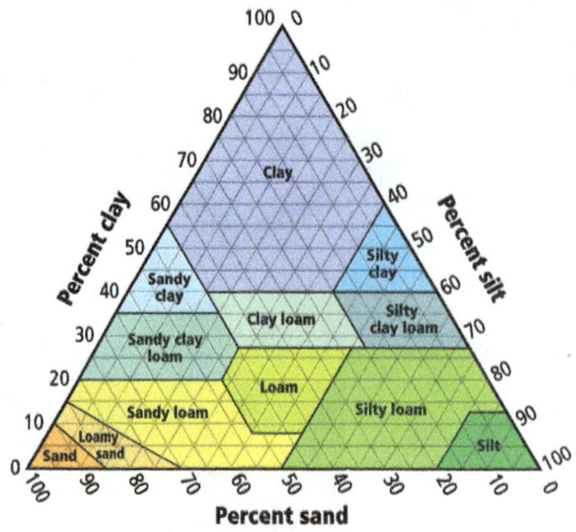

■ **Figure 22** A soil textural triangle is used to determine a soil's texture.

Soil Texture

Particles of soil are classified according to size as clay, silt, or sand, with clay being the smallest and sand being the largest. The relative proportions of particle sizes determine a soil's texture, as shown in **Figure 22.** Soil texture affects a soil's capacity to accept and retain moisture and therefore its ability to support plant growth. Soil texture also varies with depth.

Soil Fertility

Soil fertility is the measure of how well a soil can support the growth of plants. Factors that affect soil fertility include the topography, availability of minerals and nutrients, the number of microorganisms present, the amount of precipitation available, and the level of acidity.

Conditions necessary for growth vary with plant species. Farmers use natural and commercially produced fertilizers to replace minerals and maintain soil fertility. Commercial fertilizers add nitrates, potassium, and phosphorus to soil. The planting of legumes, such as beans and clover, allows bacteria to grow on plant roots and replace nitrates in the soil. Pulverized limestone is often added to soil to reduce acidity and enhance crop growth.

Data Analysis LAB

Based on Real Data*

Interpret the Data

How can you determine a soil's texture? Soils can be classified with the use of a soil textural triangle. Soil texture is determined by the percentages of the sand, silt, and clay that make up the soil. These also vary with depth, from one soil horizon to another. Below are data from three horizons of a soil in North Carolina.

Data and Observations

Soil Sample	Percent Clay	Percent Silt	Percent Sand	Texture
A	11	48		Loam
B	67		5	
C		53	38	

Data obtained from: Soil Survey Staff. 2006. National Soil Survey Characterization Data. Soil Survey Laboratory. National Soil Survey Center. (November 9) USDA-NRCS-Lincoln, NE

Think Critically

1. **Examine** the soil texture triangle shown in **Figure 22** to complete the data table. Record the percentages of particle sizes in the soil samples and the names of their textures.
2. **Infer** from the data table which soil sample has the greatest percentage of the smallest-sized particles.
3. **Identify** the maximum percentage of clay in clay loam.
4. **Infer,** if water passes quickly through sand particles, what horizon will have the most capacity to hold soil moisture.
5. **Identify** one characteristic of soil, other than water-holding capacity, that is determined by the soil's particle size.

ACROSS THE CURRICULUM

Math Millimeters can be converted to centimeters by moving the decimal point one place to the left. Meters can be converted to millimeters by moving the decimal point three places to the right. Meters can be converted to kilometers by moving the decimal point three places to the left. When a number is being converted from small units to larger units, the decimal moves to the left. When a number is being converted from larger units to smaller units, the decimal moves to the right. Have students convert the following numbers to units of meters, centimeters, and millimeters: 1 m, 2 cm, and 3 mm.

= 1 m, 0.02 m, and 0.003 m
= 100 cm, 2 cm, and 0.3 cm
= 1000 mm, 20 mm, and 3 mm BL

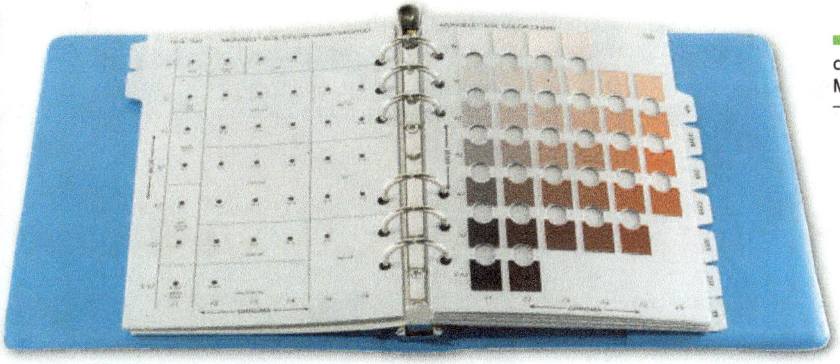

■ **Figure 23** Hue, value, and chroma can be found using the Munsell System of Color Notation.

Soil Color

The minerals, organic matter, and moisture in each soil horizon determine its color. An examination of the color of a soil can reveal many of its properties. For example, the layers that compose the O-horizon and A-horizon are usually dark-colored because they are rich in organic material. Red and yellow soils might be the result of oxidation of iron minerals. Yellow soils are usually poorly drained and are often associated with environmental problems. Grayish or bluish soils are common in poorly drained regions where soils are constantly wet, might be leached of minerals, and lack oxygen.

Scientists use the Munsell System of Color Notation, shown in **Figure 23,** to describe soil color. This system consists of three parts: hue (color), value (lightness or darkness), and chroma (intensity). Each color is shown on a chip from a soil book. Using the components of hue, value, and chroma, a soil's color can be precisely described.

SECTION 3 REVIEW

Section Self-Check

Section Summary

- Soil consists of weathered rock and humus.
- Soil is either residual or transported.
- A typical soil profile has O-horizon, A-horizon, B-horizon, and C-horizon.
- Five factors influence soil formation: climate, topography, parent material, biological activity, and time.
- Characteristics of soil include texture, fertility, and color.

Understand Main Ideas

1. MAINIDEA **Describe** how soil forms.
2. **Summarize** the features of each horizon of soil.
3. **Classify** a soil profile based on whether it is mature or immature.
4. **Generalize** the effect that topography has on soil formation.

Think Critically

5. **Infer** Soil scientists discover that a soil in a valley has a C-horizon of sand that is 1 km deep. Is this a transported soil or a residual soil? Justify your answer.
6. **Hypothesize** what type of soil exists in your area, and describe how you would determine whether your hypothesis is correct.

WRITING IN ▸ Earth Science

7. Soil in a portion of a garden is found to be claylike and acidic. Design a plan for improving the fertility of this soil.

3 Assess

Check for Understanding

Observe and Infer Using the soil-textural triangle as a guide, make three soil samples to test for water-holding and drainage capacity. The water-holding capacity can be determined by using the same volume of each soil sample and adding a measured amount of water to each sample until the water begins to drain or drip out of the bottom of an open tube or funnel containing the sample. The drainage rate of the samples can be determined by timing how long it takes a given amount of water to drain through saturated samples.

Reteach

Communicate Divide the class into groups, and have each group select a topic from this section to develop into a news report. Videotape student reports and play them back on the classroom television. COOP LEARN

Assessment

Skill Have students look through the samples of soil that they brought in at the beginning of this section and categorize the samples into general types of soil such as loam, sand, and clay.

SECTION 3 REVIEW

1. Soil forms from mechanical and chemical weathering as well as biological activity over time.
2. O-horizon contains humus and leaf litter. A-horizon contains concentrations of organic material. B-horizon contains subsoils enriched with clay minerals. C-horizon contains broken down bedrock.
3. A mature soil profile is one in which the horizons are well-developed, especially the B-horizon. An undeveloped soil profile is one with profiles that are not yet formed, or not easily seen. Undeveloped soil profiles are usually characterized by lack of a B-horizon.
4. Topography is one of the five factors of soil formation. It includes the slope and orientation of the land. A steep slope leads to more erosion on the slope and more deposition below.
5. The soil is a transported soil—it must have been formed by the deposition of sediment that slowly filled this valley.
6. Answers will vary depending on your location.
7. Answers will vary. Students may suggest adding fertilizers, soil that is a different texture, or basic materials to neutralize the acid.

Rubric

Earth Science &
TECHNOLOGY

Purpose

Students will understand the importance and use of Global Positioning Systems.

Teacher Content Support

Global Positioning System

The Global Positioning System, called GPS, uses satellites that orbit Earth at 20,200 km (12,600 miles). There are a total of 27 satellites—24 operational and 3 in reserve—that have a 12-hour orbit around Earth. They constantly send signals to Earth that are monitored by ground stations located all around the globe. Each satellite is equipped with a precise clock that keeps time to the nanosecond. The accuracy is important because GPS measures the length of time it takes for a satellite beam to reach a receiver and then calculate the position. It calculates the signal from three or four satellites in order to determine location. The receiver calculates how long the signal took to reach it, and multiplies that by the speed of light to determine the distance of the receiver from the satellite. Using the data from each of the satellites, which are always in known locations, the receiver can determine where it is located. Whether it is in a car, on a tractor, or in a handheld device, GPS technology has revolutionized many fields, including farming and disaster rescue.

Earth Science &
TECHNOLOGY

Space-Age Technology Shapes Modern Farming

Many years ago, farmers planted and plowed with their hands, a few tools, and sometimes large animals, such as horses. Since then, new technology has revolutionized the work of farmers. In the United States, agriculture is a multi-billion dollar industry, in part because of something called precision farming.

Precision farming Precision farming, which is also called site-specific farming, is a method of farming that involves giving special attention to certain areas of a field.

The fields across a farm can vary greatly. Soil fertility might differ from one area to the next, some areas might retain water more easily than others, and the topography might vary. In the past, a farmer would have made decisions about planting, fertilizing, irrigation, and pesticide applications based on the average characteristics of a field. So some areas of the field would then receive too much fertilizer, while other areas of the field would not receive enough. Precision farming allows farmers to account for the differences across the field, which can increase crop yields, reduce waste, and protect natural resources. Precision farming relies on tools called Geographic Information Systems (GIS) and the Global Positioning System (GPS).

GIS mapping GIS helps farmers plot many types of information onto a computerized map of their fields. Farmers can record areas on a field that are prone to pest infestations, or areas where there is a change in elevation. Images of the field taken from satellites can be combined with observations made by the farmer. A computer program incorporates all of the information that is added, and creates GIS map layers.

Satellites give information to farmers about their exact locations.

These layers are used to create detailed maps of the farm which can be used to plan for future crops, and to help plan where fertilizer or herbicides should be applied.

GPS navigation A system of satellites in orbit around Earth constantly relay their signals to Earth's surface. Specialized devices called GPS receivers can pick up the signals from these satellites, and use them to instantly calculate their exact location on a GIS map. This technology is used in many ways, including helping farmers find their location within a few centimeters' accuracy. Using GPS, farmers can program their tractors to plow rows that are perfectly straight, and know exactly how much fertilizer to apply to the soil.

WRITING IN ▸ Earth Science

Write a journal entry about what it would be like to run a farm where all the tractors were operated remotely. Describe the systems that you would use and how they work.

WebQuest

Teaching Strategy

Ask the class if any students know what GPS is. Then ask students to explain what it is. Ask students to identify uses for technology and possibly for farming.

WebQuest

WRITING IN ▸ Earth Science

Rubric

Journal Correct answers will demonstrate an understanding that remote operation would be possible using GPS receivers to monitor the position of farm equipment, and GIS maps to determine techniques such as where to apply fertilizers.

GeoLAB

iLab Station

Model Mineral Weathering

Background: **Many factors affect the rate of weathering of Earth materials. Two major factors that affect the rate at which a rock weathers include the length of time it is exposed to a weathering agent and the composition of the rock.**

Question: *What is the relationship between exposure time and weathering?*

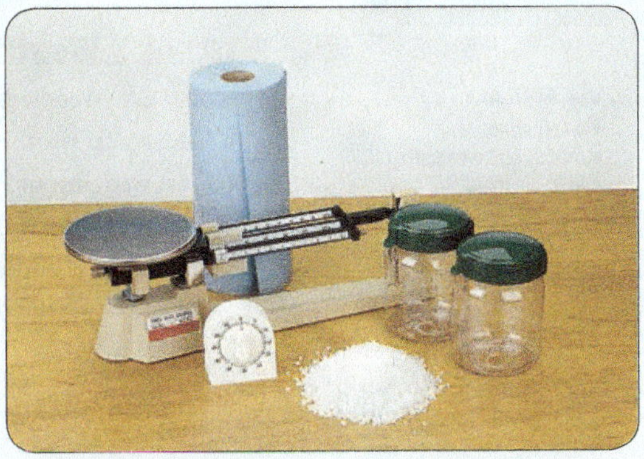

Materials

plastic jar with lid
water (300 mL)
halite chips (100 g)
balance
timer
paper towels

Safety Precautions

Procedure

1. Read and complete the lab safety form.
2. Soak 100 g of halite chips in water overnight.
3. As a class, decide on a uniform method of shaking the jars.
4. Pour off the water, and use paper towels to gently dry the halite chips. Divide them into four piles on the paper towel.
5. Use a balance to find the starting mass of one pile of the chips.
6. Place the halite chips in the plastic jar.
7. Add 300 mL of water to the jar.
8. Secure the lid on the jar, and shake the jar for the assigned period of time.
9. Pour the water from the jar.
10. Use paper towels to gently dry the halite chips.
11. Use a balance to find the final mass of the chips. Record your measurement in a data table similar to the one provided.
12. Subtract the final mass from the starting mass to calculate the change in mass of the halite chips.
13. Repeat Steps 4 to 12 using a fresh pile of halite chips for each period of time.

Weathering Data

Shaking Time (min)	Starting Mass of Chips (g)	Final Mass of Chips (g)	Change in Mass of Chips (g)
2			
4			
6			
8			

Analyze and Conclude

1. **State** What real-world process did you model in this investigation?
2. **Infer** Why did you need to soak the halite chips before conducting the experiment?
3. **Compare** the lab procedure with actual weathering processes. What did the halite represent? What process did shaking the jar represent?
4. **Deduce** How would acid precipitation affect this process in the real world?
5. **Conclude** How would the results of your investigation be affected if you used pieces of quartz instead of halite?

INQUIRY EXTENSION

Design an Experiment This lab demonstrated the relationship between exposure time and weathering. Consider other factors that affect weathering. Design an experiment to measure the effects of those factors.

GeoLAB

iLab Station Rubric

Preparation

Time Allotment 60 min

Process Skills acquire and analyze information, make and use graphs, make and use tables, measure and use numbers, observe and infer, predict, recognize cause and effect, think critically

Safety Precautions Approve lab safety forms before work begins. Remind students to always wear safety goggles when conducting laboratory investigations.

Preparation of Materials Soak the rock chips before using them to reduce water absorption during the activity. Different rock chips can be used if their dissolving rate can be accomplished within the time frame of a class period.

Procedure

- Having students work in small groups will allow them to develop an appreciation of the importance of teamwork in scientific investigations.
- Remind students to make sure the lid is tightly sealed before shaking.
- Drying rocks thoroughly before weighing them is important to avoid sources of error.
- Shaking in a uniform manner helps to provide some level of control for the investigation.
- **Troubleshooting** Use halite chips made for water softeners so that the chips do not dissolve completely.

LabManager™
Customize any lab with the LabManager™ CD-ROM.

Analyze and Conclude

1. mechanical and physical weathering of rocks in moving water
2. Soaking in water is important so that the chips have already absorbed all the water they can. Absorption of water during the lab would lead to inconsistencies in data.
3. The halite represented minerals carried by water. Shaking the jar represented movement of water, such as stream movement, or waves.
4. In the real world, acid precipitation would speed up the rate of weathering of some types of rocks by making the water more acidic. Note that halite would not react with acid precipitation. This can be a topic for discussion in which students can consider how the type of parent material can affect the impact that acid precipitation has on a region. (Higher concentrations of calcite in the bedrock can help neutralize acid rain.)
5. The pieces of quartz would not wear away in the time allotted for this investigation.

INQUIRY EXTENSION

Design an Experiment Answers will vary. Students may demonstrate the effects of temperature, pressure, oxygen, or acid on various types of rocks.

MAINIDEAS Summary statements can be used by students to review the major concepts of the chapter.

Students can review with these online resources.

Vocabulary Practice

Vocabulary eGames
Vocabulary eFlashcards
Vocabulary PuzzleMaker

Section Self-Check

Chapter Self-Check

Online Test Practice

Use the Exam*View*® *Assessment Suite* CD-ROM to:

- create multiple versions of tests
- create modified tests with one mouse click
- edit existing questions and add your own questions
- build tests aligned with state standards using built-in state curriculum tags
- change English tests to Spanish with one mouse click
- track students' progress using the Teacher Management System

CHAPTER 7 STUDY GUIDE

BIGIDEA **Weathering and erosion are agents of change on Earth's surface.**

Vocabulary Practice

SECTION 1 Weathering

VOCABULARY
- weathering
- mechanical weathering
- frost wedging
- exfoliation
- chemical weathering
- oxidation

MAINIDEA **Weathering breaks down materials on or near Earth's surface.**

- Mechanical weathering changes a rock's size and shape.
- Frost wedging and exfoliation are forms of mechanical weathering.
- Chemical weathering changes the composition of a rock.
- The rate of chemical weathering depends on the climate, rock type, surface area, and topography.

SECTION 2 Erosion and Deposition

VOCABULARY
- erosion
- deposition
- rill erosion
- gully erosion

MAINIDEA **Erosion transports weathered materials across Earth's surface until they are deposited.**

- The processes of erosion and deposition have shaped Earth's landscape in many ways.
- Gravity is the driving force behind major agents of erosion.
- Agents of erosion include running water, waves, glaciers, wind, and living things.

SECTION 3 Soil

VOCABULARY
- soil
- residual soil
- transported soil
- soil profile
- soil horizon

MAINIDEA **Soil forms slowly as a result of mechanical and chemical processes.**

- Soil consists of weathered rock and humus.
- Soil is either residual or transported.
- A typical soil profile has O-horizon, A-horizon, B-horizon, and C-horizon.
- Five factors influence soil formation: climate, topography, parent material, biological activity, and time.
- Characteristics of soil include texture, fertility, and color.

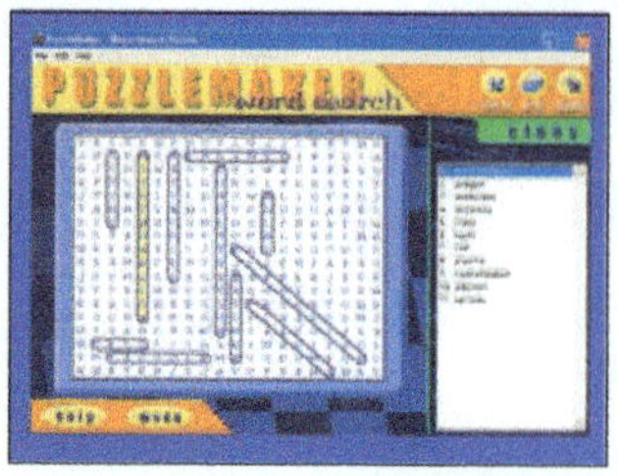

Vocabulary Practice

For additional practice with vocabulary, have students access the Vocabulary PuzzleMaker.

Chapter Self-Check

VOCABULARY REVIEW

Match the correct vocabulary term from the Study Guide to the following definitions.

1. the process of breaking down and changing rocks on or near Earth's surface
2. the removal of weathered materials from a location by running water, wind, ice, or waves
3. the fracturing of rock along curved lines that results when pressure is removed from bedrock

Each of the following sentences is false. Make each sentence true by replacing the italicized words with the correct vocabulary term from the Study Guide.

4. *Weathering* is caused by water flowing down the side of a slope.
5. The process by which eroded materials are left at a new location is called *physical weathering*.
6. *Mechanical weathering* is the process during which smaller eroded channels become deep and wide.
7. *Soil horizon* is formed from parent material that was moved away from its original source by water, wind, or a glacier.
8. A *soil profile* is a distinct layer or zone within a cross section of Earth's surface.
9. *Humus* is the loose covering of broken rock particles and decaying organic matter overlying the bedrock of Earth's surface.

Distinguish between the vocabulary terms in each pair.

10. weathering, erosion
11. chemical weathering, mechanical weathering
12. gully erosion, rill erosion
13. soil horizon, soil profile
14. erosion, deposition
15. residual soil, transported soil

UNDERSTAND KEY CONCEPTS

16. Approximately what percent of Earth's surface is presently covered by glaciers?
    A. 5 percent
    B. 10 percent
    C. 20 percent
    D. 50 percent
17. In which horizon is humus most concentrated?
    A. A-horizon
    B. B-horizon
    C. C-horizon
    D. O-horizon
18. Which is usually the primary factor that affects the rate of weathering?
    A. topography
    B. volume
    C. climate
    D. biological activity

Use the figure below to answer Questions 19 and 20.

19. Which process most likely produced the present appearance of this feature found in Arches National Park, Utah?
    A. chemical weathering
    B. mechanical weathering
    C. earthquake activity
    D. acid precipitation
20. Which erosional agent was most likely responsible for the appearance of this feature?
    A. water
    B. wind
    C. glaciers
    D. living organisms

CHAPTER 7 ASSESSMENT

VOCABULARY REVIEW

1. weathering
2. erosion
3. exfoliation
4. Rill erosion
5. deposition
6. Gully erosion
7. Transported soil
8. soil horizon
9. Soil
10. Weathering is the decomposition of material; erosion is the process that carries it away from a place.
11. Chemical weathering involves a change in the mineral composition of a rock; mechanical weathering involves a change in the size of the rock.
12. Gully erosion is the advanced stage after rill erosion has progressed.
13. A soil horizon is a layer in a soil profile.
14. Erosion is the process involving removal of material; deposition is the process of that material being left in a new location.
15. Residual soil develops from parent material that is native to that location; transported soil develops from parent material that was transported from another geologic area.

UNDERSTAND KEY CONCEPTS

16. B
17. D
18. C
19. B
20. B

21. A
22. D
23. D
24. C
25. B
26. D
27. B
28. B

CONSTRUCTED RESPONSE

29. The greater the total surface area, the more rapidly weathering occurs.
30. The rate of chemical or physical weathering is greatly influenced by precipitation, temperature, and evaporation.
31. Answers will vary, but should include building wind barriers and increasing the surface roughness, or covering the crops with straw or manure.

21. Frost wedging primarily relies on which process(es)?
A. freezing and thawing
B. gravity
C. oxidation
D. depth

22. Which is not considered a factor of soil formation?
A. topography
B. parent material
C. time
D. chemistry

23. Which does not contribute to the rate of weathering?
A. rock type
B. rock composition
C. climate
D. fossils

Use the photo below to answer Questions 24 and 25.

24. Which agent of erosion is shown in the picture of Letchworth State Park, known as the Grand Canyon of the East, in central New York?
A. glaciers
B. wind
C. running water
D. earthquakes

25. Which is the underlying force behind the agent of erosion in Letchworth State Park?
A. pressure
B. gravity
C. temperature
D. light

26. Which describes a residual soil?
A. soil from sediment deposited by glaciers
B. sand that has collected in a floodplain
C. fine-grained sediment that was deposited by wind
D. layers of material that weathered from bedrock below

27. Which soil horizon is a zone of accumulation consisting of soluble minerals that have been carried by water from above?
A. A-horizon
B. B-horizon
C. C-horizon
D. O-horizon

28. A mature soil most likely possesses which characteristic?
A. thin B-horizons
B. thick B-horizons
C. fertility
D. dark color

CONSTRUCTED RESPONSE

29. Analyze the relationship between surface area and rate of mechanical weathering of a rock.

30. Classify how different climates affect the way rocks weather.

Use the figure below to answer Question 31.

31. Design a method that would have prevented the erosion occurring at this location.

Chapter Self-Check

32. **List** the factors that control the formation of soil, and give an example of the effects of each.

33. CAREERS IN EARTH SCIENCE A soil scientist stated that the soil in your area is acidic. Suggest a solution for the local gardeners.

THINK CRITICALLY

34. **Examine** how the processes of erosion and deposition cause barrier islands to migrate.

Use the figure below to answer Question 35.

35. **Summarize** how the processes of erosion and deposition have resulted in this landscape feature.

36. **Create** a poster that illustrates the effects of erosion and deposition in your community.

37. **Draw and label** a soil profile of a mature soil containing an O-horizon, A-horizon, B-horizon, C-horizon, and bedrock. Describe how each layer of the soil was developed.

CONCEPT MAPPING

38. Create a concept map using the following terms: *weathering, erosion, deposition, chemical weathering, mechanical weathering, gully erosion,* and *rill erosion*. Refer to the *Skillbuilder Handbook* for more information.

CHALLENGE QUESTION

39. **Critique** this statement: Weathering, erosion and deposition are all parts of the same process.

WRITING IN ▸ Earth Science

40. Imagine that you are a soil scientist studying a sample in the lab. Write a journal entry describing the soil sample. Include information about what you can infer from the soil sample.

DBQ Document–Based Questions

Data obtained from: United States Department of Agriculture, Natural Resources Conservation Service. Honeoye–New York State Soil, 2006.

Honeoye [HON ee yah] soils are exceptionally fertile soils that occur in New York. The word Honeoye *is from the Iroquois* Hay-e-a-yeah.

41. Using the photograph, create an illustration of the Honeoye soil and label the following layers: *A-horizon, B-horizon,* and *C-horizon*.

42. Describe the soil profile.

43. Is the soil pictured above undeveloped or mature? How can you tell?

CUMULATIVE REVIEW

44. What is the difference between latitude and longitude? **(Chapter 2)**

45. What is a mineral? **(Chapter 4)**

46. Which common chemical sedimentary rock consists of calcite? **(Chapter 5)**

---

32. climate, topography, parent material, organisms, and time; Examples will vary.
33. Add pulverized limestone to the soil.

THINK CRITICALLY

34. Wind and ocean currents carry the sand grains away and deposit them downwind, or down current. Over time, this affects the bulk of barrier islands, causing the islands to move.
35. Water flowed along the surface of the land, eroded material, and transported and deposited it as a delta.
36. Posters will vary.
37. O-horizon is a layer of dead leaves and humus. A-horizon has high concentrations of organic matter and humus mixed with weathered soil. B-horizon contains subsoils that are enriched with clay minerals which have drained from A-horizon, or leached up from C-horizon. C-horizon is directly above bedrock and contains weathered parent material.

CONCEPT MAPPING

38. Answers will vary. Check students' maps for accuracy.

CHALLENGE QUESTION

39. Weathering, erosion and deposition are closely related.

WRITING IN ▸ Earth Science

Rubric

40. Answers will vary, but might include water-holding capacity, fertility, color, and vulnerability to erosion.

DBQ Document-Based Questions

Data obtained from: United States Department of Agriculture, Natural Resources Conservation Service. Honeoye–New York State Soil, 2006.

41. Answers will vary, but illustration should include A-, B-, and C-horizons.
42. High concentrations of organic matter and humus are found in A-horizon, which is dark-colored. B-horizon contains soil that is enriched with clay minerals. C-horizon is generally composed of weathered parent material from the bedrock:
43. The Honeoye soil profile in the photograph above is a mature soil, because the horizons are distinct.

CUMULATIVE REVIEW

44. Latitude lines are circles that run east to west around the world and are measured north and south of the equator. Longitude lines are semicircles that run north and south from the north to south poles and are measured east and west of Prime Meridian.
45. A mineral is a naturally occurring, inorganic, solid element or compound, with a specific crystal structure.
46. limestone

MULTIPLE CHOICE

1. B
2. C
3. D
4. D
5. B
6. A
7. D
8. D
9. A

MULTIPLE CHOICE

1. Which farming method is used to reduce wind erosion?
   A. planting different crops
   B. planting wind barriers
   C. building earth mounds
   D. building stone walls

Use the figure below to answer Questions 2–4.

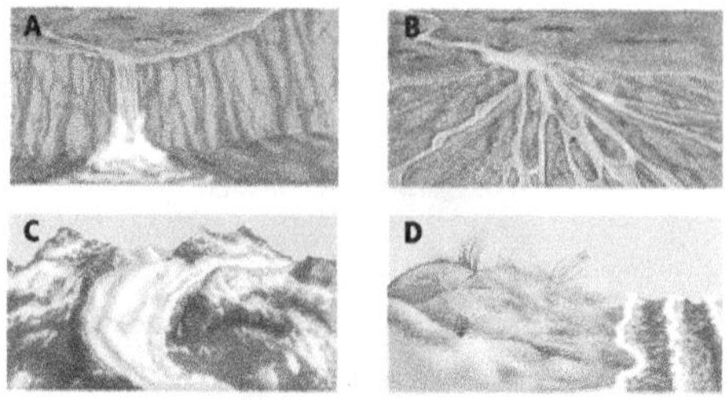

2. Which image shows the erosional agent that was responsible for leaving behind U-shaped valleys, hanging valleys, lakes, and deposits of sediment in New England and New York State?
   A. A C. C
   B. B D. D

3. Which image shows the erosional agent responsible for dunes formed along the Gulf and Atlantic coasts of the United States?
   A. A C. C
   B. B D. D

4. What common factor is responsible for three of the four erosional processes pictured?
   A. wind
   B. heat
   C. human intervention
   D. gravity

5. What is the best-known feature of sedimentary rocks?
   A. ripple marks
   B. fossils
   C. graded bedding
   D. cross-bedding

6. How does granite differ from gabbro in coloring and silica content?
   A. Granite is lighter colored with higher silica content.
   B. Granite is darker colored with lower silica content.
   C. Granite is darker colored with higher silica content.
   D. Granite is lighter colored with lower silica content.

7. Which is NOT an agent of chemical weathering?
   A. water C. carbon dioxide
   B. oxygen D. wind

Use the map below to answer Questions 8 and 9.

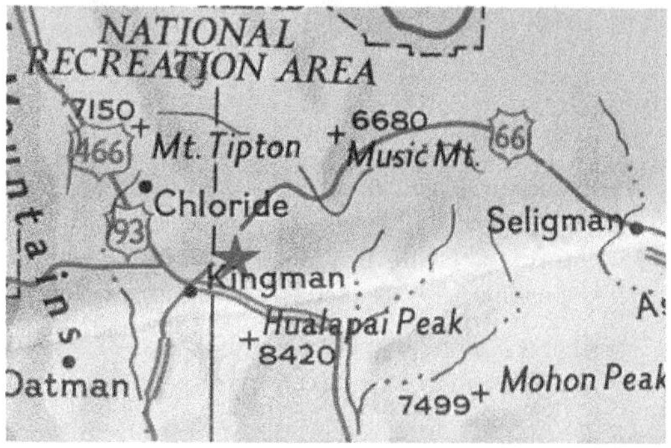

8. What can you infer about the location of the area shown in the road map?
   A. It is largely uninhabited.
   B. It is in a major city.
   C. It is mostly impassable terrain.
   D. It is a mountainous area.

9. Pikes Peak in Colorado is about 14,100 ft high. If a hiker wanted to climb an equivalent height in the area located on this map, what two mountains should he climb?
   A. Music Mountain and Mohon Peak
   B. Music Mountain and Mount Tipton
   C. Hualapai Peak and Mohon Peak
   D. Mount Tipton and Mohon Peak

Online Test Practice

SHORT ANSWER

Use the graph below to answer Questions 10–12.

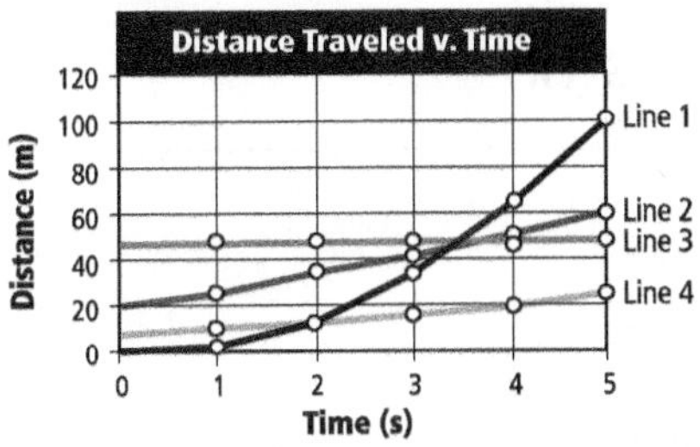

10. What was the average speed of the object represented by Line 1 during the 5 s its time was recorded?

11. Explain why Line 3 is horizontal.

12. What are the independent and dependent variables in this graph?

13. Describe how limestone forms.

14. Explain whether or not coal is a mineral.

15. When does regional metamorphism occur and what are its results?

READING FOR COMPREHENSION

Agricultural Land Use

Food production takes up almost half of Earth's land surface and threatens to consume the fertile land that still remains. The global impact of farming on the environment is revealed in new maps, which show that 40 percent of Earth's land is used for agriculture. Navin Ramankutty, a land-use researcher with Wisconsin-Madison's Center for Sustainability and the Global Environment (SAGE), posed the following question: "How can we continue to produce food from the land while preventing negative environmental consequences, such as deforestation, water pollution, and soil erosion?" One potential solution could be "precision farming." This model uses new technology to improve productivity while reducing the use of water and the application of fertilizer and other potentially harmful chemicals. The precision system, currently being developed by NASA geoscientists, would use satellite data to help farmers decide how to use their resources with pinpoint accuracy based on the requirements of different areas of each field.

Article obtained from: Owen, J. Farming claims almost half of Earth's land, new maps show. *National Geographic News.* December 9, 2005.

16. According to this passage, which statement is false?
A. Farming can harm Earth.
B. Satellite data can improve farming.
C. Farming does not cause pollution.
D. "Precision farming" is a solution.

17. Which one is not a negative environmental consequence of farming listed in the passage?
A. deforestation
B. air pollution
C. water pollution
D. soil erosion

18. What can be inferred from this text?
A. There are solutions to improve farming and its effects on the land.
B. Wisconsin is the only state with farming problems.
C. People need to eat less so that less land is needed for food.
D. There is no fertile land left to cultivate.

NEED EXTRA HELP?

If You Missed Question . . .	1	2	3	4	5	6	7	8	9	10	11	12	13	14	15
Review Section . . .	7.2	7.1	7.1	7.2	6.1	5.2	7.2	2.2	2.2	1.2	1.3	1.2	6.2	4.1	6.3

SHORT ANSWER

10. Speed = distance ÷ time
100 m ÷ 5 s = 20 m/s

11. Line 3 is horizontal because during the 5 s, the object did not move.

12. The independent variable is time, and the dependent variable is distance.

13. It forms three different ways. The rarest way it forms is from the shells of some ocean organisms that are made out of calcium carbonate. When the organisms die, their shells sink to the bottom and form carbonate sediment, which is cemented by calcium carbonate from groundwater or ocean water. Another type of formation occurs during burial and lithification as calcium carbonate separates out of the water, crystallizes between the grains, and replaces the carbonate sediment. A third type of formation occurs when calcium carbonate mud solidifies, becoming limestone.

14. Coal is not a mineral because, millions of years ago, it formed from an organic process. To be considered a mineral, it would have to be both naturally occurring and inorganic.

15. Regional metamorphism occurs when high temperature and pressure affect large regions of Earth's crust, such as in a mountain-building episode. This results in changes in rock types and minerals and causes the rock layers in the area to form and deform.

READING FOR COMPREHENSION

16. C
17. B
18. A

CHAPTER 8 Mass Movements, Wind, and Glaciers

BIGIDEA **Movements due to gravity, winds, and glaciers shape and change Earth's surface.**

ESSENTIAL QUESTIONS	NATIONAL STANDARDS	RESOURCES TO ASSESS MASTERY
SECTION 1 Mass Movements **1.** What is the relationship between gravity and mass movements? **2.** What factors affect mass movements? **3.** What are the different types of mass movements and how are they described? **4.** How do mass movements affect people? 3 sessions 1.5 blocks	UCP.1–4; A.1–2; B.4; D.1, D.3; F.5–6	**Entry-Level Assessment** Focus Transparency 20, p. 194 **Progress Monitoring** Caption Question, pp. 195, 196, 198, 200 Reading Check, pp. 194, 196, 198 Section Review, p. 200
SECTION 2 Wind **1.** What are the conditions that contribute to the likelihood that an area will experience wind erosion? **2.** What features are characteristic of wind erosion and deposition? **3.** How do dunes form and migrate? 1 session 0.5 block	UCP.1–5; A.1–2; B.4; D.1, D.3; F.5; G.3	**Entry-Level Assessment** Focus Transparency 21, p. 201 **Progress Monitoring** Caption Question, pp. 201, 204 Reading Check, pp. 202, 203 Section Review, p. 206
SECTION 3 Glaciers **1.** How do glaciers form? **2.** What are the similarities and differences between valley glaciers and continental glaciers? **3.** How do glaciers modify landscapes? **4.** What features are characteristic of glacial erosion and deposition? 4 sessions 2 blocks	UCP.1–5; A.1–2; B.4; D.1, D.3; E.1–2; F.6; G.3	**Entry-Level Assessment** Focus Transparency 22, p. 207 **Progress Monitoring** Caption Question, pp. 207, 210, 212 Reading Check, p. 208 Section Review, p. 212 **Summative Assessment** Chapter Assessment, p. 217 Exam*View® Assessment Suite* CD-ROM

LEVELED RESOURCES	LAB MATERIALS
Science Notebook 8.1 OL **Chapter FAST FILE Resources:** GeoLab Worksheet, p. 29 OL Study Guide, p. 39 BL **Transparencies:** Section Focus Transparency 20 EL Teaching Transparency 19 OL EL	LaunchLAB p. 192 / **25 min** sand, containers (3), water GeoLAB p. 214 / **45 min** metric ruler
Science Notebook 8.2 OL **Chapter FAST FILE Resources:** Study Guide, p. 41 BL **Lab Resources:** Laboratory Manual, p. 57 OL **Transparencies:** Section Focus Transparency 21 EL Teaching Transparency 20 OL EL	
Science Notebook 8.3 OL **Chapter FAST FILE Resources:** MiniLab Worksheet, p. 28 OL Study Guide, p. 43 BL **Lab Resources:** Laboratory Manual, p. 61 OL **Transparencies:** Section Focus Transparency 22 EL Teaching Transparency 21 OL EL	MiniLAB p. 210 / **30 min** glaciers (4), baking pan, popsicle stick, textbook, Earth science journal

LabManager™
Customize any lab with the LabManager™ CD-ROM.

ADDITIONAL RESOURCES

Technology:
- eTeacherEdition
- eStudentEdition
- Classroom Presentation Toolkit
- Exam*View*® *Assessment Suite* CD-ROM
- LabManager™ CD-ROM
- What's EARTH SCIENCE Got To Do With It? DVD
- Weather Classroom Video DVD

Assessment:
- Blueprints for Success
- Chapter Test (Scaffolded)
- Earth Science Standardized Test Practice

Lab Resources:
- Exploring Environmental Problems Laboratory Manual

BL Below Level OL On Level AL Advanced Learners EL English Learners COOP LEARN Cooperative Learning

CHAPTER 8

LaunchLAB

iLab Station Rubric

How does water affect sediments on slopes?

Process Skills measure, observe and infer, recognize cause and effect, communicate, model, predict

Teaching Strategies

- Very fine or very coarse sand could require a modification of the amount of water added. To ensure student success, do the lab first yourself with the sand that will be used.
- Some students might try to make shapes that are too complex. It is best for them to stick with simple models such as pyramid or mountain shapes.

Procedure

1. Have students read and complete the lab safety form and follow the procedure below.
2. Place 225 mL of **sand** in each of three separate **containers,** such as aluminum pie plates.
3. Add 20 mL of **water** to the first container of sand, and mix well. Add 100 mL of water to the second container of sand, and mix well. Add 200 mL of water to the third container of sand, and mix well.
4. Tilt each pan to test the effect of slopes. Start with a slight tilt and increase until the sand begins to move.
5. Test each mixture for its ability to be molded and retain its shape. Compare your results for the three samples.

WARNING: *Wipe up any spilled water.*

Analysis

1. **Describe** how the addition of water affected the sand's ability to be molded in the three samples. Too much water and too little water did not allow the sand to be molded.
2. **Explain** why one mixture was better able to maintain its shape than the others. The second container was able to maintain its shape because it had enough water to allow the sand to bind together, but not so much that the sand liquefied.
3. **Explain** how water affects sediments on slopes. Too much water can make the slope liquefy, while too little will not allow the slope to hold shape for long.

CHAPTER 8

Mass Movements, Wind, and Glaciers

BIGIDEA Movements due to gravity, winds, and glaciers shape and change Earth's surface.

SECTIONS

1 Mass Movements

2 Wind

3 Glaciers

LaunchLAB

iLab Station

How does water affect sediments on slopes?

Water has a significant effect on sediments on slopes. In this activity, you will demonstrate how the addition of water affects how sediments are held together.

FOLDABLES® Study Organizer

External Processes that Shape Earth

Make a pocket book using the labels shown. Use it to organize your notes on processes that shape Earth's surface.

Assessment

Performance Ask students which of the sand and water mixtures held its molded shape best. Ask them to explain the results that they obtained.

LabManager™
Customize any lab with the LabManager™ CD-ROM.

Introduce the BIGIDEA

Compare Features Show students photos of a rock perched precariously, a sand dune, and a glacier. Ask students to identify how the features were formed. Explain that every feature was formed by movement due to gravity, winds, and glaciers on Earth's surface.

Teacher Content Support

Glacial Erosion Glaciers shape the landscape by eroding, transporting, and depositing rocks and sediment, often as glacial till that has been sorted by glacial meltwater.

SECTION 1

1 Focus

Focus Transparency

Before presenting the lesson, project **Section Focus Transparency 20** and have students answer the accompanying questions. EL

MAINIDEA

Landslides Show the students photos of landslides, including a local photo if available. Divide the class into groups of two or three. Ask them to discuss what factors might trigger a landslide, and how a landslide might be prevented.

2 Teach

Tie to Previous Knowledge

Maps Have a volunteer give a definition of a map and an example of how it might be useful. Students have learned that maps are models of Earth's surface. Maps can be used to highlight areas of high risk. For example, the USGS and the Colorado Geological Survey have recently worked together to make detailed digital maps of 36 areas in western Colorado. These maps are important because of the number of active landslides in this rapidly developing area.

READING CHECK Gravity moves weathered materials downslope.

SECTION 1

Essential Questions

- What is the relationship between gravity and mass movements?
- What factors affect mass movements?
- What are the different types of mass movements and how are they described?
- How do mass movements affect people?

Review Vocabulary

gravity: the force every object exerts on every other object due to their masses

New Vocabulary

mass movement
creep
mudflow
landslide
slump
avalanche

Mass Movements

MAINIDEA Mass movements alter Earth's surface over time due to gravity moving sediment and rocks downslope.

EARTH SCIENCE 4 YOU How fast can you travel on a waterslide? A number of factors might come into play, including the angle of the slide, the amount of water on the slide, the material of the slide, friction, and your own mass. These factors also affect mass movements on Earth's surface.

Mass Movements

How do landforms, such as mountains, hills, and plateaus, wear down and change? Landforms can change through processes involving wind, ice, and water, and sometimes through the force of gravity alone. The downslope movement of soil and weathered rock resulting from the force of gravity is called **mass movement.** Recall that weathering processes weaken and break rock into smaller pieces. Mass movements often carry the weathered debris downslope. Because climate has a major effect on the weathering activities that occur in a particular area, climatic conditions determine the extent of mass movement.

All mass movements, such as the one shown in **Figure 1,** occur on slopes. Because few places on Earth are completely flat, almost all of Earth's surface undergoes mass movement. Mass movements range from motions that are barely detectable to sudden slides, falls, and flows. The Earth materials that are moved range in size from fine-grained mud to large boulders.

READING CHECK **Describe** how gravity causes a mass movement.

Figure 1 Mass movements can cause tree trunks to curve in order to continue growing opposite the pull of gravity, which is toward the center of Earth.

DIFFERENTIATED INSTRUCTION

Struggling Learners Have students make ice cream sundaes to demonstrate the variables that influence mass movement. Have them cover ice cream with sprinkles and crushed peanuts and note that the peanuts, which weigh more, fall first. Shake the sundae dish to observe how "tremors" affect the movement of the toppings. Conclude the activity by allowing students to consume their mass movements. **WARNING:** *Check for allergies to peanuts and make sure you have sugar-free ice cream or frozen yogurt available for students with diabetes.*

Factors that Influence Mass Movements

Several factors influence the mass movements of Earth's material. One factor is the material's weight, which works to pull the material downslope. A second factor is the material's resistance to sliding or flowing, which depends on the amount of friction, how cohesive the material is, and whether it is anchored to the bedrock. A third factor is a trigger, such as an earthquake, that shakes material loose. Mass movement occurs when the forces pulling material downslope are stronger than the material's resistance to sliding, flowing, or falling.

Water is a fourth variable that influences mass movements. The landslide shown in **Figure 2** occurred after days of heavy rains. Saturation by water greatly increases the weight of soils and sediments. In addition, as the water fills the tiny open spaces between grains, it acts as a lubricant between the grains, reducing the friction between them.

■ **Figure 2** Mass movements like the one shown here can significantly alter landscapes.
Summarize *the factors that might have been involved in this mass movement.*

Types of Mass Movements

Mass movements are classified as creep, flows, slides, and rockfalls. Mass movements move different types of materials in various ways.

Creep The slow, steady, downhill flow of loose, weathered Earth materials, especially soils, is called **creep.** Because movement might be as little as a few centimeters per year, the effects of creep are usually noticeable only over long periods of time. One way to tell whether creep has occurred is to observe the positions of structures and objects. As illustrated in **Figure 3,** creep can cause once-vertical utility poles and fences to tilt, and trees and walls to break. Loose materials on almost all slopes undergo creep.

Creep that usually occurs in regions of permafrost, or permanently frozen soil, is called solifluction (SOH luh fluk shun). The material moved in solifluction is a mud-like liquid that is produced when water is released from melting permafrost during the warm season. The water saturates the surface layer of soil and is unable to move downward. As a result, the surface layer can slide slowly downslope.

■ **Figure 3** All slopes undergo creep to some extent. Tilting of vertical objects is often the result.

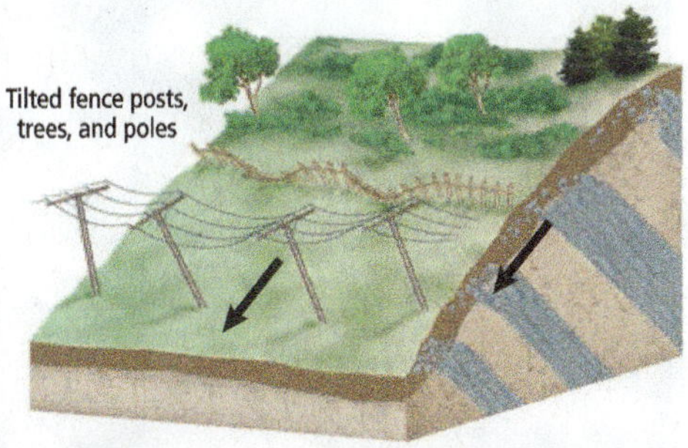

Teacher Content Support

The Force of Gravity The internal strength of earth materials is usually able to maintain the materials in their relative positions on Earth's surface, even though the materials are always experiencing the downward pull of the force of gravity. Particles of earth material move downslope, even along imperceptible slopes, when the force of gravity exceeds the opposing forces of cohesion and friction between the surfaces of the particles. The presence of water in and around the particles reduces the contact between the particles and therefore reduces the amount of friction that must be overcome for movement to occur.

Use an Analogy

Creep Compare the movement of creep with what would happen to a person standing on a rug if the rug were pulled out from under his or her feet. Although the rug would be pulled at a much faster rate than creep, have students imagine taking a video of the action and playing it back frame by frame. Just as a person's body would begin to fall backward as the rug was pulled, fences and gravestones tilt backward as soil at the base of the structure moves downslope. OL

■ **Caption Question Fig. 2** The material's weight, resistance to flow, a trigger, and the addition of water might have been factors.

Demonstration

Observe Water and Friction Place a wooden board on an incline by propping up one side with a few textbooks. On the board, place a small object such as a cork. Have students note that the cork stays in its position. Then, slowly run some water down the board. The cork should move downslope. This demonstration will show students how water reduces the friction that keeps particles in one place. BL EL

Tie to Previous Knowledge

Run and Slide The movement of Earth particles in wet conditions is similar to a student's ability to run and slide more easily on a wet surface than on a dry one. It also is easier to slide on ice than on a dry surface because there is less friction.

Model

Mass Movements Have small groups of students each make a clay model that demonstrates a form of mass movement. Once the models have been completed, have the groups rotate to the different models and identify the form of mass movement exhibited by each model. Models can then be left on display for review and reinforcement. OL EL

Caption Question Fig. 4 The lahar appears to have buried the town.

READING CHECK A lahar can be triggered by volcanic activity.

Figure 4 The city of Armero, in Colombia, was covered in mud and debris by a lahar that contained snowmelt and volcanic material.
Describe *the effect of the lahar on the city shown above.*

Figure 5 Mudflows can be extremely destructive and can result in severe property damage, road closures, and power outages.

Flows In some mass movements, Earth materials flow as if they were a thick liquid. The materials might move as slowly as a few centimeters per year or as rapidly as one hundred kilometers per hour. Earth flows are moderately slow movements of soils, whereas **mudflows** are swiftly moving mixtures of mud and water. Mudflows can be triggered by earthquakes or similar vibrations and are common in volcanic regions where the heat from a volcano melts snow on nearby slopes that have fine sediment and little vegetation. The meltwater fills the spaces between the small particles of sediment and allows them to slide readily over one another and move downslope.

A lahar (LAH har) is a type of mudflow that occurs after a volcanic eruption. Often a lahar results when a snow-topped volcanic mountain erupts and melts the snow on top of a mountain. The melted snow mixes with ash and flows downslope. **Figure 4** shows how a lahar that originated from Nevado del Ruiz, one of the volcanic mountains in the Andes, devastated a town. The Nevado del Ruiz is 5389 m high and covered with 25 km^2 of snow and ice, which melted when it erupted. Four hours after Nevado del Ruiz erupted, lahars had traveled more than 100 km downslope. As a result of these lahars, which occurred in 1985, approximately 23,000 people were killed, 5000 were injured, and 5000 homes were destroyed.

READING CHECK **Determine** what triggers a lahar.

Mudflows are also common in sloped, semi-arid regions that experience intense, short-lived rainstorms. The Los Angeles Basin in Southern California is an example of an area where mudflows are common. In such areas, periods of drought and forest fires leave the slopes with little protective vegetation. When heavy rains eventually fall in these areas, they can cause massive, destructive mudflows because there is little vegetation to anchor the soil. Mudflows are especially destructive in areas where urban development has spread to the bases of mountainous areas. These mudflows can bury homes, as shown in **Figure 5.**

ACROSS THE CURRICULUM

Math Have students determine how long it will take a mudflow traveling at 30 km/h to reach a location 195 km away. 195/30 = 6.5 hours. Then have students determine how far this mudflow would migrate in just one minute. 30/60 = 0.5 km AL

■ **Figure 6** Typical of landslides, this soil moved in a large block.

Slides A rapid, downslope movement of Earth materials that occurs when a relatively thin block of soil, rock, and debris separates from the underlying bedrock is called a **landslide,** shown in **Figure 6.** The material rapidly slides downslope as one block, with little internal mixing. A landslide mass eventually stops and becomes a pile of debris at the bottom of a slope, sometimes damming rivers and causing flooding. Landslides are common on steep slopes, especially when soils and weathered bedrock are fully saturated by water. This destructive form of mass movement causes damage costing almost 2 billion dollars and 25 to 50 associated deaths per year in the United States alone. You will explore the movement of a landslide in the GeoLab at the end of this chapter.

A rockslide is a type of landslide that occurs when a sheet of rock moves downhill on a sliding surface. During a rockslide, some blocks of rock are broken into smaller blocks as they move downslope, as shown in **Figure 7.** Often triggered by earthquakes, rockslides can move large amounts of material.

■ **Figure 7** During this rockslide, blocks of rock were broken into smaller blocks as they moved downslope.

View an **animation of a rockslide.**
Concepts In Motion

ACROSS THE CURRICULUM

Math Ask students if they think they could outrun a landslide. Have them determine how fast they would have to run to outrun a landslide moving at 200 km/h. Answers will vary. The landslide is moving too fast for a human to escape. Then ask students whether it would be possible to outdistance this landslide in an automobile. If students have difficulty with metric units, it might be helpful to convert kilometers to miles. 200 km/h is equal to approximately 124 mph. OL

Teacher Content Support

Change on Earth's Surface

Earth's surface continually experiences forces that are both constructive (building up) and destructive (wearing down). This is because earth material is continuously being transported from one location to another. Movement of earth material is influenced by environmental factors such as wind and water. The force of gravity pulls earth material downward. The downward movement of earth material is referred to as mass movement and can take many forms.

The slope of a surface and the amount of water present greatly affect the stability of earth material. Steep slopes and great amounts of water allow the force of gravity to easily move material downslope. The presence of vegetation can hinder downslope movement and reduce mass movements.

Most mass movements are natural, spontaneous occurrences. However, human activities can instigate and accelerate mass movements in isolated situations.

Collaborative Learning

Triggers of Mass Movements

Have students work in groups to plan a demonstration to show how triggers suddenly start mass movements. Each group can demonstrate for the class how its trigger model works. Students can experiment with sound, tremors, or objects moving over Earth's surface. EL COOP LEARN

GeoLAB

iLab Station

The GeoLab at the end of the chapter can be used at this point in the lesson.

Identify Misconceptions

Students are likely to think of snow as being light and fluffy and to believe that being buried under snow is not dangerous.

Uncover the Misconception

Ask students if they think being buried in snow is dangerous and if it is difficult to escape from being buried in snow. Have students write their responses in their Earth science journals.

Demonstrate the Concept

The amount of snow that can become part of an avalanche can be enormous in mass and can hit objects with the power of a freight train. The force can destroy buildings and bury large trees. In western states, avalanches cause more human deaths than tornadoes or floods. Most people caught in an avalanche die of suffocation or trauma. It is important to heed warnings during periods when conditions favor avalanches and to pay attention to unusual changes in the weather. Most avalanches occur during or after a storm. Others occur in the early spring, many during the months of February or March.

Assess New Knowledge

Once again, ask students to write in their Earth science journals how dangerous it can be to be buried under snow. In addition, ask students to include some precautions to take in a place that might be prone to an avalanche.

☑ READING CHECK Slumps can be triggered by an earthquake or heavy rain.

Caption Question Fig. 9 Avalanches are more likely to occur on snow-covered slopes with angles between 30° and 45°.

Figure 8 Slumps leave distinct crescent-shaped scars on hillsides as the soil rotates downward.

Slumps When the mass of material in a landslide moves along a curved surface, a **slump** results. Material at the top of the slump moves downhill, and slightly inward, while the material at the bottom of the slump moves outward. Slumps can occur in areas that have thick soils on moderate-to-steep slopes. Sometimes, slumps occur along highways where the slopes of soils are extremely steep. Slumps are common after rains, when water reduces the frictional contact between grains of soil and acts as a lubricant between surface materials and underlying layers. The weight of the additional water pulls material downhill. As with other types of mass movement, slumps can be triggered by earthquakes. Slumps leave crescent-shaped scars on slopes, as shown in **Figure 8.**

☑ READING CHECK **Describe** what conditions can cause a slump.

Avalanches Landslides that occur in mountainous areas with thick accumulations of snow are called **avalanches.** About 10,000 avalanches occur each year in the mountains of the western United States. Radiation from the Sun can melt surface snow, which then refreezes at night into an icy crust. Snow that falls on top of this crust can eventually build up, become heavy, slip off, and slide downslope as an avalanche. Avalanches can happen in early winter when snow accumulates on the warm ground. The snow in contact with the warm ground melts, then refreezes into a layer of jagged, slippery snow crystals.

Avalanches of dangerous size, like the one shown in **Figure 9,** occur on slope angles between 30° and 45°. When the angle of a slope is greater than 45°, enough snow cannot accumulate to create a large avalanche. At angles less than 30°, the slope is not steep enough for snow to begin sliding. A vibrating trigger, even from a single skier, can send this unstable layer sliding down a mountainside. Avalanches pose significant risks in places such as Switzerland, where more than 50 percent of the population lives in avalanche terrain.

Figure 9 Vibrations from a single skier can trigger an avalanche.
Identify *the conditions that make a landscape more vulnerable to avalanches.*

EARTH SCIENCE JOURNAL

Avalanche Have students complete the following creative writing assignment in their Earth science journals. Ask students to imagine that they witnessed an avalanche. Ask them to describe the avalanche, its results, and feelings that survivors could be experiencing as they see the destruction left in the wake of the event. OL

Rubric

Figure 10 This rockfall in Topanga Canyon, California, was unusual in that it involved mainly one large rock.

Rockfalls On high cliffs, rocks are loosened by physical weathering processes, such as freezing and thawing, and by plant growth. As rocks break up and fall directly downward, they can bounce and roll, ultimately producing a cone-shaped pile of coarse debris, called talus, at the base of the slope. Rockfalls, such as the one shown in **Figure 10,** commonly occur at high elevations, in steep road cuts, and on rocky shorelines. Rockfalls are less likely to occur in humid regions where the rock is typically covered by a thick layer of soil, vegetation, and loose materials. On human-made rock walls, such as road cuts, rockfalls are particularly common.

Mass Movements Affect People

While mass movements are natural processes, human activities often contribute to the factors that cause mass movements. Activities such as the construction of buildings, roads, and other structures can make slopes unstable. In addition, poor maintenance of septic systems, which often leak, can trigger slides. In 2006, mudslides in the Philippines, shown in **Figure 11,** were triggered after ten days of torrential rains delivered 5 m of precipitation. Four years later, a massive storm once again caused mudslides, affecting thousands of people.

Figure 11 The mudflow on the island of Luzon, in the Philippines, occurred after days of rain.

Enrichment

Development Ask students just how steep a slope they think is too steep on which to live. Many planning regulations discourage development on slopes greater than 10 percent. Have students engage in a study of slopes and the difficulty of climbing slopes with a bicycle. This activity will help students develop a physical understanding of the steepness of slopes. AL

Apply Earth Science

Development Ask students what they think about allowing people to build homes in areas that could be potentially dangerous. Allow students to share their varying opinions. Then, ask students how they think the range of their opinions compares with the opinions of people who live in such areas. OL

Environmental Connection

Mudflows and Vegetation Mudflows disrupt surface vegetation and change the landscape dramatically. It takes many years for a surface devastated by a mudflow to become covered with protective vegetation again. In the meantime, erosional forces carry surface materials to new locations and further change the landscape of the area.

Demonstration

Observe Stability Make three piles of rocks on a desk or table. Make one pile out of rocks that have different sizes and shapes. In another pile, place rocks that are all round. In the third pile, use flat rocks. Then, bump the table. Have students observe which pile of rocks is most stable and which is least stable. Rock piles composed of round rocks are least stable, and those composed of flat rocks are most stable. Ask students what measures can be taken to stabilize hillsides covered with loose rocks. Hillsides can be stabilized with human-made abutments and vegetation. BL EL

Caption Question Fig. 12
Rockfalls might be prevented by steel nets.

3 Assess

Check for Understanding

Draw Conclusions Ask students what they think will have happened to Earth's surface in their community when their grandchildren are in high school, keeping in mind the types of changes discussed in this section. Student answers should all reflect that some changes will occur, some will be more noticeable than others, and some will be bigger threats to humans.

Reteach

Think Critically Students can listen to an audiotape of the section to review the material covered. After students listen to the tape, ask them if they heard anything on the tape that they do not remember being covered in class. If so, check their level of understanding of the particular concept.

Assessment

Performance Have students make models of their community with locations of possible mass movements and escape routes labeled. COOP LEARN

Figure 12 Covering hillsides with steel nets can reduce risks of mass movements and harm to humans.
Identify *the type of mass movement that these steel nets help prevent.*

Reducing the risks Catastrophic mass movements are most common on slopes greater than 25° that experience annual rainfall of over 90 cm. Risk increases if that rainfall tends to occur in a short period of time. Humans can minimize the destruction caused by mass movements by not building structures on or near the base of steep and unstable slopes.

Although preventing mass-movement disasters is not easy, some actions can help reduce the risks. For example, a series of trenches can be dug to divert running water around a slope and control its drainage. Landslides and rockfalls can be controlled by covering steep slopes with materials such as steel nets, shown in **Figure 12**, and constructing fences along highways in areas where mass movements are common. Other approaches involve the installation of retaining walls to support the bases of weakened slopes. Most of these efforts at slope stabilization and mass-movement prevention are only temporarily successful.

The best way to reduce the number of disasters related to mass movements is to continue to monitor mass movements, and educate people about the problems of building on steep slopes. For example, The United States Geological Survey (USGS) collects data about landslides in an effort to learn more about where and when landslides will occur. This information helps people decide where they can safely build homes or businesses.

SECTION 1 REVIEW

Section Self-Check

Section Summary

- Mass movements are classified in part by how rapidly they occur.
- Factors involved in the mass movement of Earth materials include the material's weight, its resistance to sliding, the trigger, and the presence of water.
- Mass movements are natural processes that can affect human life and activities.
- Human activities can increase the potential for the occurrence of mass movements.

Understand Main Ideas

1. **MAINIDEA Organize** the following types of mass movements in order of increasing speed: slides, creep, flows, and rockfalls.
2. **Identify** the underlying force behind all forms of mass movement.
3. **Analyze** how water affects mass movements by using two examples of mass movement.
4. **Appraise** the effects of one type of mass movement on humans.

Think Critically

5. **Generalize** in which regions of the world mudflows are more common.
6. **Evaluate** how one particular human activity can increase the risk of mass movement and suggest a solution to the problem.

WRITING IN ▸ Earth Science

7. Make a poster that compares and contrasts solifluction and a slump. Consider the way soil moves and the role of water.

SECTION 1 REVIEW

1. creep, flow, slide, rockfall
2. Gravity is the force behind all mass movements.
3. Answers will vary. In solifluction, water saturates the upper layer of soil above the permafrost, causing this layer to slide down hill. In flows, water fills the spaces between particles of soil, which lubricates them so they can slide across one another.
4. Answers will vary. Flows can quickly cover and destroy towns and cause many deaths. Landslides can destroy property and cause deaths in a more localized area. Rock falls cause problems by blocking roads. Over long periods of time, creep can break walls and pipelines.
5. In areas where there are short and intense rainfalls, mudflows are more common because built-up debris becomes saturated.
6. Humans venture into snow-covered slopes with angles between 30° and 45°, where avalanches are more likely, and which could be avoided by humans.
7. Posters should indicate that solifluction is a slow and steady movement in regions of permafrost, while a slump is a fast movement along a curved surface. Water is involved in both types of mass movements.

Rubric

SECTION 2

Wind

MAINIDEA Wind modifies landscapes in all areas of the world by transporting sediment.

Essential Questions

- What are the conditions that contribute to the likelihood that an area will experience wind erosion?
- What features are characteristic of wind erosion and deposition?
- How do dunes form and migrate?

Review Vocabulary

velocity: the speed of an object and its direction of motion

New Vocabulary

deflation
abrasion
ventifact
dune
loess

EARTH SCIENCE 4 YOU **If you have ever been on a beach on a windy day, you might have felt the stinging of sand on your face. Sand travels in the wind if the wind is fast enough.**

Wind Erosion and Transport

A current of rapidly moving air can pick up and carry sediment in the same way that water does. However, except for the extreme winds of hurricanes, tornadoes, and other strong storms, winds cannot generally carry particles as large as those transported by moving water. Regardless, wind is a powerful agent of erosion.

Winds transport materials by causing their particles to move in different ways. For example, wind can move sand on the ground in a rolling motion. A method of transport by which strong winds cause small particles to stay airborne for long distances is called suspension. Another method of wind transport, called saltation, causes a bouncing motion of larger particles. Saltation accounts for most sand transport by wind. Limited precipitation leads to an increase in the amount of wind erosion because precipitation holds down sediments and allows plants to grow. Thus, wind transport and erosion primarily occur in areas with little vegetative cover, such as deserts, semiarid areas, seashores, and some lakeshores. Wind erosion is a problem in many parts of the United States, as shown in **Figure 13.**

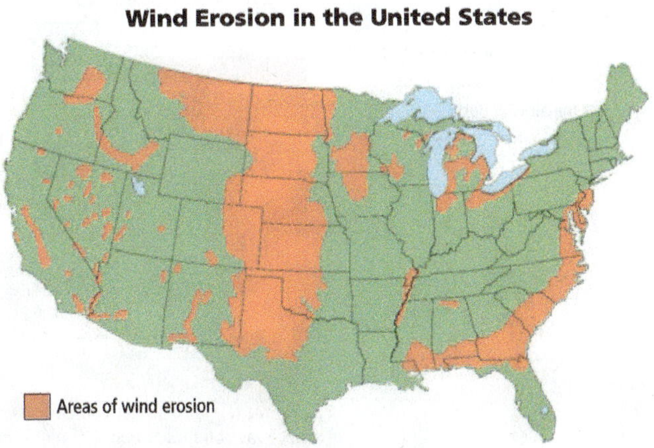

Figure 13 Wind erosion does not affect all areas of the United States equally. **Observe** *which areas are subject to wind erosion.*

SECTION 2

1 Focus

Focus Transparency

Before presenting the lesson, project **Section Focus Transparency 21** and have students answer the accompanying questions. EL

MAINIDEA

The Dust Bowl Ask students: What was the Dust Bowl? Where did it occur? Why did it happen? Students' answers will vary. The Dust Bowl was a drying up of rich farmlands in the interior of the United States due to a severe drought in the 1930s. The drought, combined with persistent winds and poor conservation practices, created vast areas of windblown silt. This illustrates how wind modifies landscapes.

2 Teach

Teacher Content Support

Wind The uneven heating of Earth's surface results in the movement of air from one place to another, which is called wind. Winds can be local or global. Local winds are affected by terrain and vegetation, and global winds are influenced by major air masses and Earth's rotation. Over short time periods, large continental areas experience greater temperature variations than expansive oceanic areas. As a result, although winds can be more sustained over the ocean, they are more changeable over continental areas.

Caption Question Fig. 13 The Southeast, Midwest around the Great Lakes, the Great Plains, as well as areas of the West such as the San Joaquin Valley (California) are all subject to erosion.

ACROSS THE CURRICULUM

History Have students write a brief research report on the importance of wind to the economy of a region. In particular, direct their attention to the economic importance of loess soils in the United States and China. Ask students to include in their reports their opinions about how the agricultural history of the region might have been different if wind erosion and deposition had been absent. OL

Teacher Content Support

Describe Wind Wind is described in terms of both speed and direction. The wind vane, an old yet still commonly used weather tool, points in the direction from which the wind is coming. Winds are named according to the direction from which they are blowing. For example, winds blowing from the west are known as westerlies. Forces that affect wind include gravity, the Coriolis effect, centrifugal force, and friction. In addition, the landscape over which wind moves can influence both its speed and direction.

Collaborative Learning

Debate the Dust Bowl Divide the class into three groups: farmers, economists, and scientists. Have each group role-play a panel discussion held in 1938 on the topic "Planning for Our Future: The Dust Bowl." COOP LEARN

☑ READING CHECK Deflation removes surface particles by wind.

Interpret the Photograph

Deflation Have students examine **Figure 14.** Ask: How does wind deflate the land in the photo? Wind removes the surface particles over a period of time, which lowers the surface of the land.

■ **Figure 14** Through deflation, the wind can create a bowl-shaped blowout.

Deflation The lowering of the land surface that results from the wind's removal of surface particles is called **deflation.** During the 1930s, portions of the Great Plains region, which stretches from Montana to Texas, experienced severe drought. The area was already suffering from the effects of poor agricultural practices, in which large areas of natural vegetation were removed to clear the land for farming. Strong winds readily picked up the dry surface particles, which lacked any protective vegetation. Severe dust storms resulted in daytime skies that were often darkened, and the region became known as the Dust Bowl.

Today, the Great Plains are characterized by thousands of shallow depressions known as deflation blowouts. Many are the result of the removal of surface sediment by wind erosion during the 1930s. The depressions range in size from a few meters to hundreds of meters in diameter. Deflation blowouts are also found in other areas that have sandy soil, as shown in **Figure 14.** Wind erosion continues today throughout the world, as shown by the duststorm in **Figure 15.**

☑ READING CHECK **Explain** how deflation removes surface particles.

■ **Figure 15** A duststorm in a desert region fills the air with dust.

Demonstration

Compare Erosion In a tray, place a layer of soil. Place a piece of sod on top of the soil in half of the tray. On the other half, put a layer of dry sand over the layer of soil. Sprinkle some dry sand on the sod. Turn a fan on low speed, making sure to face the fan away from all students before turning it on. Allow the fan to blow air evenly across the tray. Ask students to observe which conditions cause more wind erosion. Students should notice that the sand in the sod does not move as readily with the wind as does the sand in the other half of the tray. Students should recognize from the demonstration the importance of vegetation in reducing wind erosion. BL

Deflation is a major problem in many agricultural areas of the world as well as in deserts, where wind has been consistently strong for thousands of years. In areas of intense wind erosion, coarse gravel and pebbles are usually left behind as the finer surface material is removed by winds. The coarse surface left behind is called desert pavement.

Abrasion Another process of erosion, called **abrasion,** occurs when particles such as sand rub against the surface of rocks or other materials. Abrasion occurs as part of the erosional activities of winds, streams, and glaciers. In wind abrasion, wind picks up materials such as sand particles and blows them against anything in their path. Because sand is often made of quartz, a hard mineral, wind abrasion can be an effective agent of erosion—windblown sand particles eventually wear away rocks. Over long periods of time, wind erosion can produce unique structures such as those shown in **Figure 16.** Structures, such as telephone poles, can also be worn away or undermined by wind abrasion, and paint and glass on homes and vehicles can be damaged by windblown sand.

Materials that are exposed to wind abrasion show unique characteristics. For example, windblown sand causes rocks to become pitted and grooved. With continued abrasion, rocks become polished on the windward side and develop smooth surfaces with sharp edges. In areas of shifting winds, abrasion patterns correspond to wind shifts, and different sides of rocks become polished and smooth. Rocks that have been shaped by windblown sediments are called **ventifacts.** They range in size from pebbles to boulders.

☑ READING CHECK **Identify** the unique characteristics of materials shaped by abrasion.

Figure 16 Arches and pillars with caprock form in different types of environments, but most commonly in arid climates where wind is the dominant erosional force.

Use an Analogy

Abrasion Ask students how cleaning a cooking pan with scouring powder is similar to how Earth materials undergo abrasion. Students should recognize that the grease is worn off the pan in the same way that rocks are worn away: by abrasion. BL

Apply Earth Science

Abrasion Ask students to recall the last time they had their teeth cleaned. The hygienist used an abrasive mixture to wear away the plaque from their teeth. Regular toothpaste has the same type—but a lower amount—of abrasive material. Using the same level of abrasive material daily as is used for dental cleaning would wear away a person's teeth just as wind continuously blowing abrasive material over rocks wears them away. BL

☑ READING CHECK Pits, grooves, and polishing are all characteristics of materials shaped by abrasion.

DIFFERENTIATED INSTRUCTION

Visually Impaired Have students with visual impairments feel and handle rock samples before and after they have been smoothed and polished by abrasion. Ask students to describe in writing how the rock samples have changed with polishing, and ask them to envision how this might affect a landscape. BL

■ **Caption Question Fig. 17** The wind blows from the right side to the left side in the photograph.

Teacher Content Support

Dune Migration Dunes can migrate great distances. With less sand to be moved, small dunes can migrate farther and faster than large dunes. Coastal dunes can move out to sea and extend the coastline. A German freighter that became stranded off Africa's southwestern coast in the early 1900s is now landlocked by migrating coastal dunes more than a kilometer from the shoreline.

Discussion

Protect the Shorelines Ask students whether they have ever been to a beach along an ocean and noticed any attempts to protect sand dunes. Were people staying off the dunes? Have students share any problems that they think might be involved in protecting shoreline dunes. Enforcement, importance of tourism, the local economy, and costs might be some issues brought up by students. OL

■ **Figure 17** Great Sand Dunes National Monument, in southern Colorado, contains North America's highest sand dunes of more than 228.6 m.
Identify *the dominant direction of wind in the figure.*

Wind Deposition

Wind deposition occurs in areas where wind velocity decreases. As the wind velocity slows down, some of the windblown sand and other materials cannot stay airborne, and they drop out of the air stream to form a deposit on the ground.

VOCABULARY
ACADEMIC VOCABULARY
migrate
to move from one location to another
Dunes migrate as wind blows over sand.

Dunes In windblown environments, sand particles tend to accumulate where an object, such as a rock, landform, or piece of vegetation, blocks the forward movement of the particles. Sand continues to be deposited as long as winds blow in one general direction. Over time, the pile of windblown sand develops into a **dune,** as shown in **Figure 17.** All dunes have a characteristic profile. The gentler slope of a dune, located on the side from which the wind blows, is called the windward side. The steeper slope, on the side protected from the wind, is called the leeward side. The conditions under which a dune forms determine its shape. These conditions include the availability of sand, wind velocity, wind direction, and the amount of vegetation present. The different types of dunes are shown in **Table 1.**

Dune migration As long as winds continue to blow, dunes will migrate. As shown in **Figure 18,** dune migration is caused when prevailing winds continue to move sand from the windward side of a dune to its leeward side, causing the dune to move slowly over time.

■ **Figure 18** Dune migration is caused by wind.

View an **animation of dune migration.**
Concepts In Motion

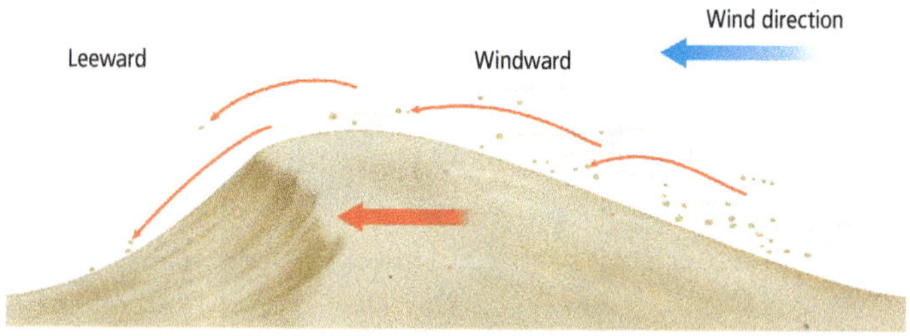

EARTH SCIENCE JOURNAL

Sand Dunes In their Earth science journals, have students write a scenario about the development of a sand dune, beginning with the solid granite in Earth's crust through dune migration. Ask students to include in their writing where each event occurs. OL

Rubric

Explore **sand dunes with an interactive table.** Concepts In Motion

Table 1 Types of Dunes

Example of Dune	Description
	**Barchan Dunes** • form solitary, crescent shapes • form from a small amount of sand • covered by minimal or no vegetation • form in flat areas of constant wind direction • crests point downwind • reach maximum size of 30 m
	**Transverse Dunes** • form series of ridge shapes • form from a large amount of sand • covered by minimal or no vegetation • form in ridges that are perpendicular to the direction of the strong wind • reach maximum size of 25 m
	**Parabolic Dunes** • form U-shapes • form from a large amount of sand • covered by minimal vegetation • form in humid areas with moderate winds • crests point upwind • reach maximum size of 30 m
	**Longitudinal Dunes** • form series of ridge shapes • form from small or large amounts of sand • covered by minimal or no vegetation • form parallel to variable wind direction • reach maximum height of 300 m

Interpret the Illustration

Dunes and Wind Ask students to study **Table 1.** Then ask students to draw illustrations of each type of dune, including the direction of wind movement.

Teacher Content Support

Parabolic Dunes Parabolic dunes are also referred to as U-shaped, hairpin, or blowout dunes. Parabolic dunes have elongated arms that follow the dune as a result of the vegetation holding some of the sand in place. The longest trailing arms recorded are 12 km long.

Enrichment

Gypsum Dunes In New Mexico, at the northern end of the Chihuahuan Desert, the brilliant white gypsum dunes of the Tularosa Basin formed from a shallow sea that covered the area 250 mya. The Tularosa Basin is surrounded by mountain ranges, including the San Andres and Sacramento Mountains. Because there was no river to carry dissolved gypsum to the ocean, gypsum became trapped in the basin and settled there. AL

ACROSS THE CURRICULUM

English Provide students the opportunity to use the topic of wind for a poem or other creative writing piece. Read a poem about wind to provide an example for students. BL

3 Assess

Check for Understanding

Observe and Infer Use a video camera to record how particles of sand move when they are exposed to wind. Shoot the video against a black background so that the movement of the particles is readily observable. Use the camera's zoom capabilities, and play back the movement of the particles in slow motion or stop-frame fashion. Have students identify the skip-hop motion, the rolling, and the suspension of the sand particles. You might want to expand this project to show dune development. You might also print and display still photographs of key frames and use them for testing purposes.

Reteach

Think Critically Have students debate the issue of restrictions on the use of and development on sand dunes.

Assessment

Performance Have students work in small groups to develop a bulletin board highlighting dune types, formation of dunes, locations of particular interest, and environmental issues associated with this section. COOP LEARN

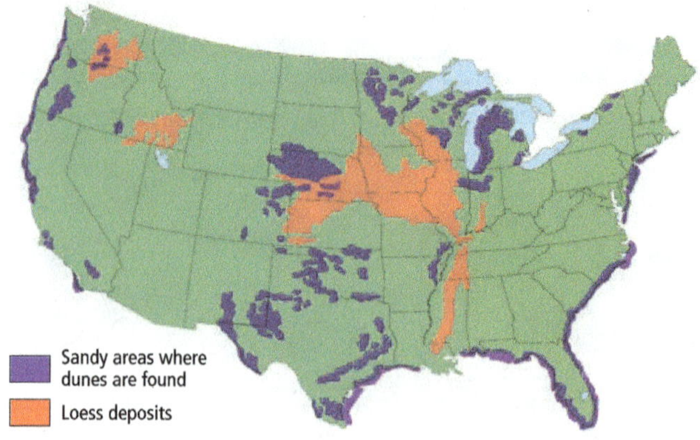

■ **Figure 19** This map shows the location of loess deposits in the continental United States.

Loess Wind can carry fine, lightweight particles such as silt and clay in great quantities and for long distances. Many parts of Earth's surface are covered by thick layers of yellow-brown windblown silt, which are thought to have accumulated as a result of thousands of years of dust storms. The source of these silt deposits might have been the fine sediments that were exposed when glaciers melted after the last ice age, more than 10,000 years ago. These thick, windblown silt deposits are known as **loess** (LUSS). Loess soils are some of the most fertile soils because they contain abundant minerals and nutrients. **Figure 19** shows the states where agriculture has benefited from the loess deposits.

SECTION 2 REVIEW

Section Self-Check

Section Summary

- Wind is a powerful agent of erosion.
- Wind can transport sediment in several ways, including suspension and saltation.
- Dunes form when wind velocity slows down and windblown sand is deposited.
- Dunes migrate as long as winds continue to blow.

Understand Main Ideas

1. **MAIN**IDEA **Distinguish** the various types of landforms formed by wind and how these landforms are created.
2. **Identify** conditions that can contribute to an increase in wind erosion.
3. **Examine** why loess can travel much greater distances than sand.
4. **Classify** the four types of dunes as they are related to wind, vegetation, and amount of sand available.

Think Critically

5. **Infer** how the movement of sand grains by saltation affects the overall movement of dunes.
6. **Evaluate** why wind erosion is an effective agent of erosion.

WRITING IN ▶ Earth Science

7. Explain in a paragraph how human activities directly affect wind erosion on coastlines.

SECTION 2 REVIEW

1. Desert pavement is made as a result of deflation. Ventifacts are formed by the abrasion of windblown sediment. Dunes develop from wind-transported sand, which is then stopped by vegetation and obstacles.
2. An increase in wind velocity and loss of vegetation increases wind erosion.
3. Loess is composed of much smaller particle sizes than sand, so it can remain airborne longer and with weaker winds.
4. Barchan dunes form where there is little sand, little vegetation, and a constant wind direction. Transverse dunes are found where there is no vegetation, lots of sand, and strong winds from the same direction. Parabolic dunes form around clumps of vegetation with moderate winds. Longitudinal dunes are found in areas with variable winds and little sand.
5. Sand grains bounce up the windward side of a dune through saltation and then fall over to the leeward side and come to rest. This continued movement of sand from one side of a dune to the other allows the dune to migrate.
6. Wind is an effective agent of erosion because wind can transport sediments uphill as well as downhill, and can modify arid and coastal landscapes.
7. Paragraphs may include that property development on a coast can change wind patterns, remove dunes and plants, affecting wind erosion.

Glaciers

MAINIDEA Glaciers modify landscapes by eroding and depositing rocks.

Essential Questions
- How do glaciers form?
- What are the similarities and differences between valley glaciers and continental glaciers?
- How do glaciers modify landscapes?
- What features are characteristic of glacial erosion and deposition?

Review Vocabulary

latitude: distance in degrees north and south of the equator

New Vocabulary

glacier
valley glacier
continental glacier
cirque
moraine
outwash plain
drumlin
esker
kame
kettle

EARTH SCIENCE 4 YOU Have you ever wondered what formed the landscape around you? Glaciers might have left deposits of sediment as well as carved features in rock that you see every day.

Moving Masses of Ice

A large mass of moving ice is called a **glacier.** Glaciers form near Earth's poles and in mountainous areas at high elevations. They currently cover about 10 percent of Earth's surface, as shown in **Figure 20.** In the past, glaciers were more widespread than they are today. During the last ice age, which began about 1.6 mya and ended more than 10,000 years ago, ice covered about 30 percent of Earth.

Areas at extreme northern and southern latitude, such as Greenland and Antarctica, and areas of high elevations, such as the Alps, have temperatures near 0°C year-round. Cold temperatures keep fallen snow from completely melting, and each year the snow that has not melted accumulates in an area called a snowfield. Thus, the total thickness of the snow layer increases as the years pass. The accumulated snow develops into a glacier. The weight of the top layers of snow eventually exerts enough downward pressure to force the accumulated snow below to recrystallize into ice. A glacier can develop in any location that provides the necessary conditions. Glaciers can be classified as one of two types—valley glaciers or continental glaciers.

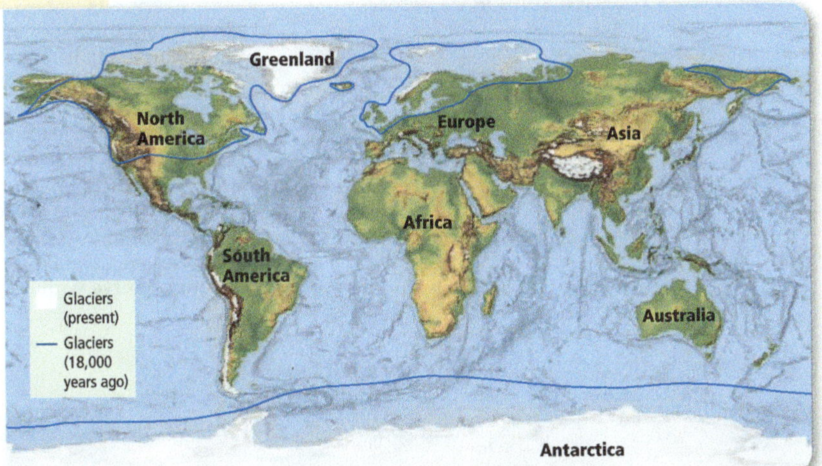

■ **Figure 20** Glaciers around the world have changed in distribution throughout geologic time.
Infer *what changes have occurred in the distribution of glaciers around the world.*

View an **animation of glacier formation.**

Concepts In Motion

IN THE FIELD

Preserved in the Glacial Ice He might have been a traveling salesman, a sheepherder, or a hunter; what is fairly certain is that he lived around 3000 B.C. This Copper Age "Ice Man," discovered in 1991 by a group of hikers in the Alps on the border of Italy and Austria, is one of the oldest, most complete, and best-preserved mummified bodies ever found. His tool kit and fragments of his clothes, including a leather boot stuffed with grass, were found to be amazingly intact. Scientists theorize that the body was naturally preserved in an airtight pocket of glacial ice. The remains are kept in a bulletproof refrigerated case in a northern Italian town. Hundreds of tourists visit every year to wonder about the identity, the cultural origins, and the tragic fate of the mysterious Ice Man.

SECTION 3

1 Focus

Focus Transparency

Before presenting the lesson, project **Section Focus Transparency 22** and have students answer the accompanying questions. **EL**

MAINIDEA

Glacial Features Show students an older photo of a valley glacier in a valley and a more recent photo after a valley glacier receded. Ask them to identify the features left by the glacier. U-shaped valley, scratches and grooves, outwash plain, hanging valleys

2 Teach

Teacher Content Support

Glaciers Most of Earth's freshwater supply is locked up in the form of glaciers. Glaciers are found in areas where climatic conditions can sustain their formation and preservation year-round. Today, most of the glaciers on Earth exist in Greenland and Antarctica, but they can also be found at high elevations on most continents.

Earth has experienced ice ages for approximately 20–30 percent of its history. The most recent ice age occurred during the Pleistocene Epoch and ended approximately 10,000 years ago. Since then, Earth has been experiencing an interglacial period with temperatures approximately 15°C warmer, resulting in the melting of most of the glaciers formed during that time.

■ **Caption Question Fig. 20** There are fewer glaciers now than in the past.

Teacher Content Support

Glacial Budget Glaciers form when more snow falls annually than melts or evaporates. Freshly fallen snow has a low density; generally, its volume consists of as much as 90 percent air. However, the snow quickly settles and compacts to the point at which air might comprise 50 percent or less of its volume. As the snow becomes packed, along with some melting and refreezing, the snow becomes even more compact. The accumulation of snow over the years leads to the formation of large snowfields. The pressure of overlying snow results in the recrystallization of the snow into ice and thus the formation of a glacier. Close inspection of glacial ice has revealed a crystal pattern similar to the patterns of coarse-grained rocks such as granite.

Data Analysis LAB

About the Lab

- Ask students whether they know of any method used to determine what Earth's climates were like hundreds of years ago.
- See also Etheridge, D. M., G. I. Pearman, and P. J. Fraser. 1992. Changes in tropospheric methane between 1841 and 1978 from a high accumulation rate Antarctic ice core. *Tellus B* 44 (4): 282–294.

Think Critically

1. The highest amount of radioactivity was between 410–425 cm. The lowest amount was between 115–150 cm.
2. Radioactivity levels dropped in the ice cores.
3. The amount of radioactivity in the ice cores probably decreased unless there were other nuclear accidents.
4. Ice cores contain whatever material was in the atmosphere and indicate human activities such as nuclear testing and natural activities such as volcanic eruptions.

Valley glaciers Glaciers that form in valleys in high, mountainous areas are called **valley glaciers.** The movement of a valley glacier occurs when the growing ice mass becomes so heavy that the ice maintains its rigid shape and begins to flow. For most valley glaciers, flow begins when the accumulation of snow and ice exceeds 40 m in thickness. As a valley glacier moves, deep cracks in the surface of the ice, called crevasses, can form.

The speed of a valley glacier's movement is affected by the slope of the valley floor, the temperature and thickness of the ice, and the shape of the valley walls. The sides and bottom of a valley glacier move more slowly than the middle because friction slows down the sides and bottom where the glacier comes in contact with the ground. Movement downslope is usually slow—less than a few millimeters per day. Over time, as valley glaciers flow downslope, their powerful carving action transitions V-shaped stream valleys into U-shaped glacial valleys.

FOLDABLES® Incorporate information from this section into your Foldable.

☑ READING CHECK **Describe** how V-shaped valleys become U-shaped.

Continental glaciers Glaciers that cover broad, continent-sized areas are called **continental glaciers.** These glaciers form in cold climates where snow accumulates over many years. A continental glacier is thickest at its center. The weight of the center forces the rest of the glacier to flatten in all directions. In the past, when Earth experienced colder average temperatures than it does today, continental glaciers covered huge portions of Earth's surface. Today, they are confined to Greenland and Antarctica.

Data Analysis LAB

Based on Real Data*

Interpret the Data

How much radioactivity is in ice cores? Glaciologists have found that ice cores taken from the arctic region contain preserved radioactive fallout. Data collected from the study of these ice cores have been plotted on the graph.

Data and Observations

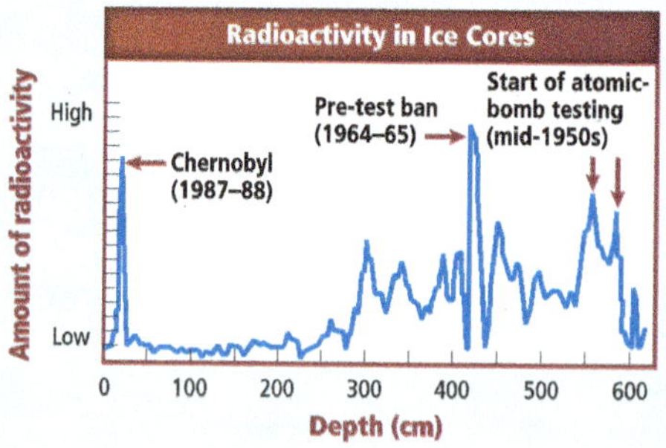

Think Critically

1. **Determine** the depth in the ice cores where the highest and lowest amounts of radioactivity were found.
2. **Describe** what happened to the amount of radioactivity in the ice cores between the pretest ban and Chernobyl.
3. **Infer** what happened to the amount of radioactivity in the ice cores after Chernobyl.
4. **Explain** what information or material other than radioactive fallout you think ice cores might preserve within them.

*Data obtained from: Mayewski, et al. 1990. Beta radiation from snow. *Nature* 345:25.

☑ READING CHECK Glacial erosion transforms a V-shaped valley to a U-shaped valley.

FOLDABLES® Rubric

EARTH SCIENCE JOURNAL

Valley and Continental Glaciers Ask students to use their textbooks and other resources to make a table in their Earth science journals to compare valley and continental glaciers. Students should include the characteristics of the glaciers, including thickness and coverage. OL

Rubric

■ **Figure 21** Glacial erosion by valley glaciers creates features such as cirques, horns, and hanging valleys.

Glacial movement Both valley glaciers and continental glaciers move outward when snow gathers at the zone of accumulation, a location in which more snow falls than melts, evaporates, or sublimates. For valley glaciers, the zone of accumulation is at the top of mountains, while for continental glaciers, the zone of accumulation is the center of the ice sheet. Both types of glaciers recede when the ends melt faster than the zone of accumulation builds up snow and ice.

Glacial Erosion

Of all the erosional agents, glaciers are the most powerful because of their great size, weight, and density. When a valley glacier moves, it breaks off pieces of rock through a process called plucking. When glaciers with embedded rocks move over bedrock, they act like the grains on a piece of sandpaper, grinding parallel scratches into the bedrock. Small scratches are called striations, and larger ones are called grooves. Striations and grooves provide evidence of a glacier's history and indicate its direction of movement.

Glacial erosion by valley glaciers can create features like those shown in **Figure 21.** At the high elevations where snow accumulates, valley glaciers also scoop out deep, bowl-shaped depressions, called **cirques.** Where two cirques on opposite sides of a valley meet, they form a sharp, steep ridge called an arête. When there are glaciers on three or more sides of a mountaintop, the carving action creates a steep, pyramid-shaped peak called a horn. The most famous example of this feature is Switzerland's Matterhorn.

Valley glaciers can also leave hanging valleys in the glaciated landscape. Hanging valleys are formed when higher tributary glaciers converge with the lower primary glaciers and later retreat. The primary glacier is so thick that it meets the height of the smaller tributary glacier. When the glaciers melt, the valley is left hanging high above what is now a river in the primary valley floor. Hanging valleys today are often characterized by waterfalls where the tributary glacier used to be.

Identify Misconceptions

Students might think that Earth's surface generally stays the same in most places.

Uncover the Misconception
Ask students to describe how the local landscape has changed over 100, 1000, 10,000, and 100,000 years.

Demonstrate the Concept
Show the class some evidence of change in examples such as the following: an old photograph of a familiar place where change is evident, a rock with striations on it, and samples of soil and rocks. Explain that all these examples provide evidence that change is a constant occurrence on Earth's crust.

Assess New Knowledge
Ask students to describe in their Earth science journals how Earth's surface changes.

ACROSS THE CURRICULUM

Art Have students each draw with pencil or charcoal, or paint with tempera or watercolor paints, a landscape that includes glacial features. The glacial features should be labeled.

Demonstration

Model Downslope Movement
Mix flour with a little water to make a thick, dough-like mixture. Place the dough on an inclined board and have students check the position of the dough several times during the class period. Place toothpicks in different locations on the dough so students can observe changes in the positions of the toothpicks. OL

■ **Caption Question Fig. 22** On a topography map, a drumlin will resemble a hill with one side steeper than the other.

MiniLAB

iLab Station | Rubric

Purpose Students will model glacial deposition by comparing four glaciers of different materials.

Process Skills observe and infer, describe, think critically

Additional Material Earth science journal

Safety Precaution Approve lab safety forms before work begins.

Teaching Strategy

- Save time by preparing the ice cubes before class. Prepare the glaciers by filling slots of an ice cube tray 1/2 full with either sand, soil, or aquarium gravel. Fill the rest with water and freeze overnight.
- Depending on the temperature of the classroom, the ice cubes will take 15–35 minutes to melt.
- Have students set up their trays at the beginning of the class period and then make their observations at the end of the period.
- Placing the ice cubes under a lamp will accelerate the rate of melting. Have students work in groups to compare glacial melting.

Expected Results Materials will be transported and deposited differently according to material type.

Analysis

1. Answers will vary depending on the materials used.
2. Meltwater formed from the melting ice cube just as meltwater forms from the melting glacier.
3. The materials embedded in the ice cube represented glacial till.
4. The ice cubes containing various sediments represented glaciers that had picked up rocks and other debris. The materials deposited by the ice cubes created small ridges that modeled moraines.

■ **Figure 22** Elongated landforms called drumlins can be grouped together as a drumlin field in areas once covered by continental glaciers.
Describe *how you could identify a drumlin on a topographic map.*

Glacial Deposition

Glacial till is the unsorted rock, gravel, sand, and clay that glaciers carry embedded in their ice and on their tops, sides, and front edges. Glacial till is formed from the grinding action of the glacier on underlying rock. Glaciers deposit unsorted ridges of till called **moraines** when the glacier melts. Terminal moraines are found along the edge where the retreating glacier melts, and lateral moraines are located parallel to the direction of a valley glacier flow.

Outwash When the farthest ends of a glacier melt and the glacier begins to recede, meltwater floods the valley below. Meltwater contains gravel, sand, and fine silt. When this sediment is deposited by meltwater carried away from the glacier, it is called outwash. Because of the way water transports sediment, outwash is always sorted by particle size. The area at the leading edge of the glacier where the meltwater flows and deposits outwash is called an **outwash plain.**

Drumlins, eskers, and kames Continental glaciers that move over older moraines form the material into elongated landforms called **drumlins,** shown in **Figure 22.** A drumlin's steeper slope faces the direction from which the glacier came. Streams flowing under melting glaciers leave long, winding ridges of layered sediments called **eskers,** shown in **Figure 23.** A **kame** is a mound of layered sediment that forms when till gets washed into depressions or openings in the melting ice. When the ice finally melts, a cone-shaped hill or mound is left. Kames are also shown in **Figure 23.**

MiniLAB

iLab Station

Model Glacial Deposition

How do glaciers deposit different types of rocks and sediments? Glaciers are powerful forces of erosion. As they move across the land, they pick up rocks and sediments, and carry them to new locations. When a glacier melts, these materials are left behind and deposits form in different shapes.

Procedure

1. Read and complete the lab safety form.
2. Work with a group of 2 to 3 other students. One student should obtain four **glaciers** from your teacher.
3. Place the glaciers on a **baking pan.** In front of each glacier, place a **popsicle stick** (to prevent the glacier from sliding down the pan).
4. Place a **textbook** under one end of the baking pan (your glaciers should be toward the elevated end of the pan).
5. Observe what happens as the glaciers melt. Record your observations in your science journal.
6. Dispose of your materials as your teacher instructs.

Analysis

1. **Discuss** Did the materials differ in the way they were deposited by the melting ice cubes? Were your results similar to those of your classmates? Explain.
2. **Explain** how this activity modeled the formation of meltwater.
3. **Apply** Which materials in this activity modeled glacial till?
4. **Apply** How did this activity model glacial deposition and the formation of a moraine?

Assessment

Performance Have students use ice cubes with frozen sand to model the formation of lateral moraines, terminal moraines, and medial moraine.

LabManager™ Customize any lab with the LabManager™ CD-ROM.

Demonstration

Model Moraines Have students create their own moraines by pushing a block of ice through a mixture of sand, soil, and pebbles on a slightly slanted board. They should identify the terminal and lateral moraines. As the ice melts, discuss outwash and have students identify where an outwash plain might occur. BL OL

VISUALIZING Continental Glacial Features

Figure 23 Continental glaciers carve out vast regions of landscape, leaving behind distinctive features such as kames, eskers, drumlins, and moraines.

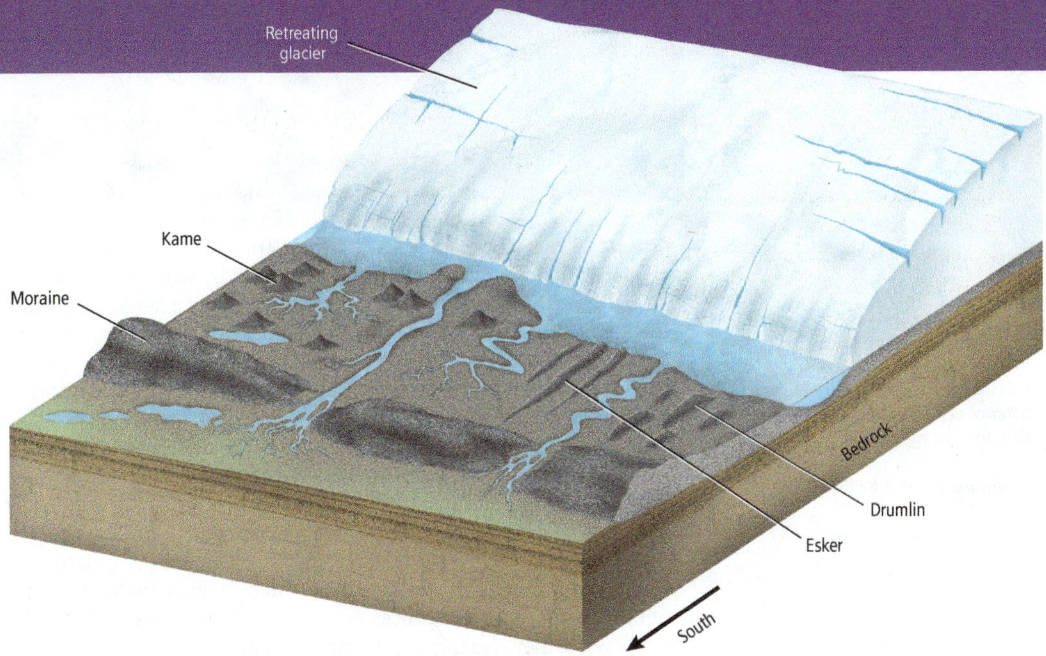

Kames are cone-shaped mounds of sand and gravel. They form when material is washed into a depression or hole in the ice. The material settles into the distinct shape as the ice beneath it melts.

Eskers are long ridges of sorted deposits. They are shaped from outwash deposited by water flowing through tunnels in the glacier.

Drumlins are shaped as the glacier moves over old moraines. They are made of unsorted till.

Concepts In Motion View an **animation of glacial features.**

VISUALIZING

Purpose

Students will compare and contrast continental glacial features.

Discussion

Glacial Melting Share with students the fact that investigators at NASA have used satellite imagery, airborne laser altimeter flights, and the Global Positioning System to determine that Greenland's ice sheets are experiencing an increase in the rate of melting. Ask students to share their thoughts about the reason for the apparent increase in melting rate. Responses might include global warming and El Niño. Students might wish to know what information is available about coastline changes resulting from the increase in melting rates. Effects on shorelines have been negligible, but if the melting rate continues to increase, changes in coastlines will occur. AL

Model

Glacial Landscapes Have students use clay to model glacial landscapes. List the features that students should include in their models, such as moraines, drumlins, eskers, and kames. BL EL

Demonstration

Model Outwash Place a plastic bag full of water in a freezer. Once it is frozen, remove the ice from the plastic bag and place it on a large tray that has soil evenly spread over it. Slightly incline the tray by placing a book under one end. Sprinkle dirt randomly on the block of ice. Have students monitor the changes in the block of ice. Allow the block of ice to melt, and then have students compare the patterns of runoff and deposition to those of a glacial landscape. BL

■ **Caption Question Fig. 24** Kettles would appear as irregular-shaped depressions on a topographic map, with or without water.

3 Assess

Check for Understanding

Summarize Have each student develop five review questions. Collect all questions and randomly select questions for members of two opposing teams to answer.

Reteach

Observe and Infer Show a video about glaciers that will review the features covered in the section. Provide students with a list of features to check off as they view the video.

Assessment

Skill Have students each develop a concept map using the following terms and phrases: *valley glacier, continental glacier, V-shaped valley, U-shaped valley, cirque, more snow falls than melts each year, high elevations, large broad areas,* and *till.* More snow falls than melts each year → 1) high elevations → valley glacier → cirque → V-shaped valley → U-shaped valley → till; 2) large broad areas → continental glaciers → till.

■ **Figure 24** These kettle lakes in South Dakota are a result of glacial retreat. **Describe** *how you might be able to locate kettles on a topographic map.*

VOCABULARY
SCIENCE USAGE V. COMMON USAGE
Kettle
Science usage: a steep-sided depression formed by a glacier

Common usage: a metallic pot used for cooking

Glacial lakes Sometimes, a large block of ice breaks off a continental glacier and the surrounding area is covered by sediment. When the ice block melts, it leaves behind a depression called a kettle hole. After the ice block melts, the kettle hole fills with water from precipitation and runoff to form a kettle lake. **Kettles** or kettle lakes, such as those shown in **Figure 24,** are common in New England, New York, and Wisconsin. With valley glaciers, cirques can also fill with water, and they become cirque lakes. Because of their altitude, cirque lakes are usually frozen much of the year, and thaw in the brief alpine summer. When a terminal moraine blocks off a valley, the valley fills with water to form a lake. Moraine-dammed lakes include the Great Lakes and the Finger Lakes of northern New York, which are long and narrow.

Mass movements, wind, and glaciers all contribute to the changing of Earth's surface. These processes erode landforms constantly, and in many ways, they also impact human populations and activities.

SECTION 3 REVIEW

Section Self-Check

Section Summary

- Glaciers are large moving masses of ice that form near Earth's poles and in mountain areas.
- Glaciers can be classified as valley glaciers or continental glaciers.
- Glaciers modify the landscape by erosion and deposition.
- Features formed by glaciers include U-shaped valleys, hanging valleys, moraines, drumlins, and kettles.

Understand Main Ideas

1. MAINIDEA **Describe** two examples of how glaciers modify landscapes.
2. **Explain** how glaciers form.
3. **Compare and contrast** the characteristics of valley glaciers and continental glaciers.
4. **Differentiate** among different glacial depositional features.

Think Critically

5. **Evaluate** the evidence of past glaciers that can be found on Earth today.
6. **Infer** whether valley glaciers or continental glaciers have shaped more of the landscape of the United States.

WRITING IN ▶ Earth Science

7. Draw a picture of each glacial feature, and write a description to accompany each feature.

SECTION 3 REVIEW

1. Through erosion, glaciers can create cirques, horns, or hanging valleys. Through deposition, glaciers can create drumlins, eskers, and kames.
2. Glaciers form when cold temperatures prevent fallen snow from completely melting, so that the accumulated snow develops into a glacier. The weight of the top layers of snow exert pressure to force the snow to recrystallize.
3. Valley glaciers are much smaller than continental glaciers. They form in the valleys of high mountains and move down the valley when they become too heavy to contain their ridged shape. Continental glaciers cover continents. They are thickest in the center and thin out around the edges.
4. The outwash plain is the surface over which meltwater flows. Drumlins are elongated mounds, eskers are winding ridges, and kames are mounds.
5. Evidence of past glacial activity includes depositional features, such as drumlins, eskers, and kames, and erosional features, such as cirques and horns.
6. Continental glaciers have the power to shape great parts of the landscape, and have affected much of the continents of the northern hemisphere. Valley glaciers can carve dramatic features, but affect less in terms of area.
7. Drawings and answers should include eskers, kames, dumlins, and moraines.

Rubric

Earth Science &
SOCIETY

Slipping Away

On the morning of January 10, 2005, the residents of La Conchita, California, awoke to find the highway out of town closed in both directions, due to landslides. Around 12:30 P.M. many residents heard an ominous roar as the bluff above the town unleashed 600,000 metric tons of dirt and mud, covering four blocks in 10 m of debris. Scientists went to the scene to discover exactly what had caused this enormous landslide and whether one could happen again.

The setting La Conchita is built on a narrow swatch of land between the highway and a huge bluff. The bluff is held together weakly, so it is susceptible to being loosened by heavy water content, such as a prolonged, heavy rain. The slope is further weakened from the effects of regular landslides, as well as being on a fault line.

In the two weeks prior to the landslide, the area had received a record amount of rain—about 35 cm—the amount it normally receives in a year! The excess water caused the earth to literally slide off the face of the mountain.

A history of landslides This event was not, however, the first landslide to hit the area. In fact, the mountain bluff is scarred with the evidence of many landslides. Ten years earlier, in March of 1995, two devastating landslides hit the area in the span of a week. These landslides were also caused by a large amount of rain, but the movement of the earth was relatively slow, so residents were able to get away. The 2005 landslide was a continuation of the 1995 slide—the soil that was deposited by the earlier slide was loosened by the rainwater and slipped down the slope. After the 1995 slide, the state government erected a retaining wall to keep the landslides at bay. However, soil, mud, and debris from the 2005 disaster passed right over parts of the wall.

Mark Reid/USGS

The mass movement at La Conchita, California, in 2005 killed ten people.

The debate Could the 2005 landslide have been detected and the people warned in time to prevent loss of life? Most likely, yes. In fact, some of the residents of the town are suing the government for failure to protect their citizens, as well as failure to adequately notify them of the impending danger.

Are governments responsible for providing warnings and protection to citizens who move into areas that are prone to natural disasters? Or, does the responsibility lie with the citizens that might not have understood the dangers of living in the area? These questions and more are sure to be considered by the residents and government of La Conchita, as well as cities and local governments of disaster- prone areas throughout the United States for years to come.w

WRITING IN ▶ Earth Science

Debate Research information about a natural disaster that has occurred near your location. Hold a classroom debate on the topic of why people should, or should not, live in an area where natural disasters have occurred.

WebQuest

Earth Science &
SOCIETY

Purpose

Students will learn about the causes and devastation of a landslide in La Conchita, California.

Teacher Content Support

Landslides Landslides occur when a segment of a hill or mountain collapses. The landslide can consist of earth, mud, or other debris. A mudslide occurs when the soil is heavily saturated with water. The deluge of water and earth can crush anything in its path: houses, trees, cars, and people. In La Conchita, the 2005 landslide was a continuation of the 1995 landslide; there was little or no new soil involved. The cliff above the town is especially well-suited for landslides because of poor rock formation and poor cementation of layers. On top of this, a fault line runs through the mountain. All these factors together leave the community of La Conchita extremely vulnerable to landslides, especially after heavy rainfall.

Teaching Strategies

- Ask students where they think landslides occur. Then ask if they can identify what causes landslides.
- Ask students to build models with household materials of an area where a landslide is likely to occur.

WRITING IN ▶ Earth Science

Rubric

Debate Have students form groups to research both sides of the issue of whether or not people should live in areas where natural disasters have occurred. Then have groups list their top three points for debate.

WebQuest

GeoLAB

iLab Station Rubric

Preparation

Time Allotment 45 min

Process Skills communicate, observe and infer, interpret scientific illustrations, measure and use numbers, recognize cause and effect, sequence, draw a conclusion

Safety Precaution Approve lab safety forms before work begins.

Procedure

- Review the use of compass directions on maps.
- To increase student proficiency in using SI, have students give their answers in both English units and SI units.
- Measurements should be to the nearest tenth.
- Team students with visual impairments with students who can easily discern small print.
- Encourage student partners to discuss their work as they proceed to help each other with comprehension and understanding of content and procedures.

LabManager™
Customize any lab with the LabManager™ CD-ROM.

GeoLAB

iLab Station

Mapping: Map a Landslide

This image shows the Tully Valley landslide three days after it occurred. The Tully Farms Road is covered up to 5 m deep with clay.

Background: Around midday on April 27, 1993, in a normally quiet, rural area of New York, the landscape dramatically changed. Unexpectedly, almost 1 million m^3 of earth debris slid 300 m down the lower slope of Bare Mountain and into Tully Valley. The debris flowed over the road and buried nearby homes. The people who lived there had no knowledge of any prior landslides occurring in the area. This landslide was the largest to occur in New York in more than 75 years.

Question: *How can you use a drawing based on a topographic map to infer how the Tully Valley Landslide occurred?*

Materials

metric ruler

Procedure

Imagine that you work for the United States Geological Survey (USGS) specializing in mass movements. You have just been asked to evaluate the Tully Valley Landslide.

1. Read and complete the lab safety form.
2. Check the map's scale.
3. Measure the length and width of the Tully Valley in kilometers. Double-check your results.

Analyze and Conclude

1. **Interpret Data** What does the shape of the valley tell you about how it formed?
2. **Determine** In what direction did the landslide flow?
3. **Determine** In what direction does the Onondaga Creek flow?
4. **Infer** from the map which side of Tully Valley has the steepest valley walls.
5. **Deduce** What conditions must have been present for the landslide to occur?
6. **Infer** At the time of the Tully Valley Landslide, the trees were bare. How could this have affected the conditions that caused the landslide?

WRITING IN ▶ Earth Science

Explain why the mass movement event you examined in this GeoLab is classified as a landslide. Differentiate a landslide from a creep, slump, flow, avalanche, and rockfall.

Analyze and Conclude

1. The shape of the valley indicates that it was formed by glacial activity.
2. west to east
3. north
4. Students should infer from the map that the western side has experienced landslides in the past and is steepest.
5. The ground had to be saturated with water.
6. Trees with leaves would have soaked up more of the water in the ground, and the landslide might not have occurred.

WRITING IN ▶ Earth Science

Rubric

Explain Answers should include definition of a landslide and differentiation from other mass movements.

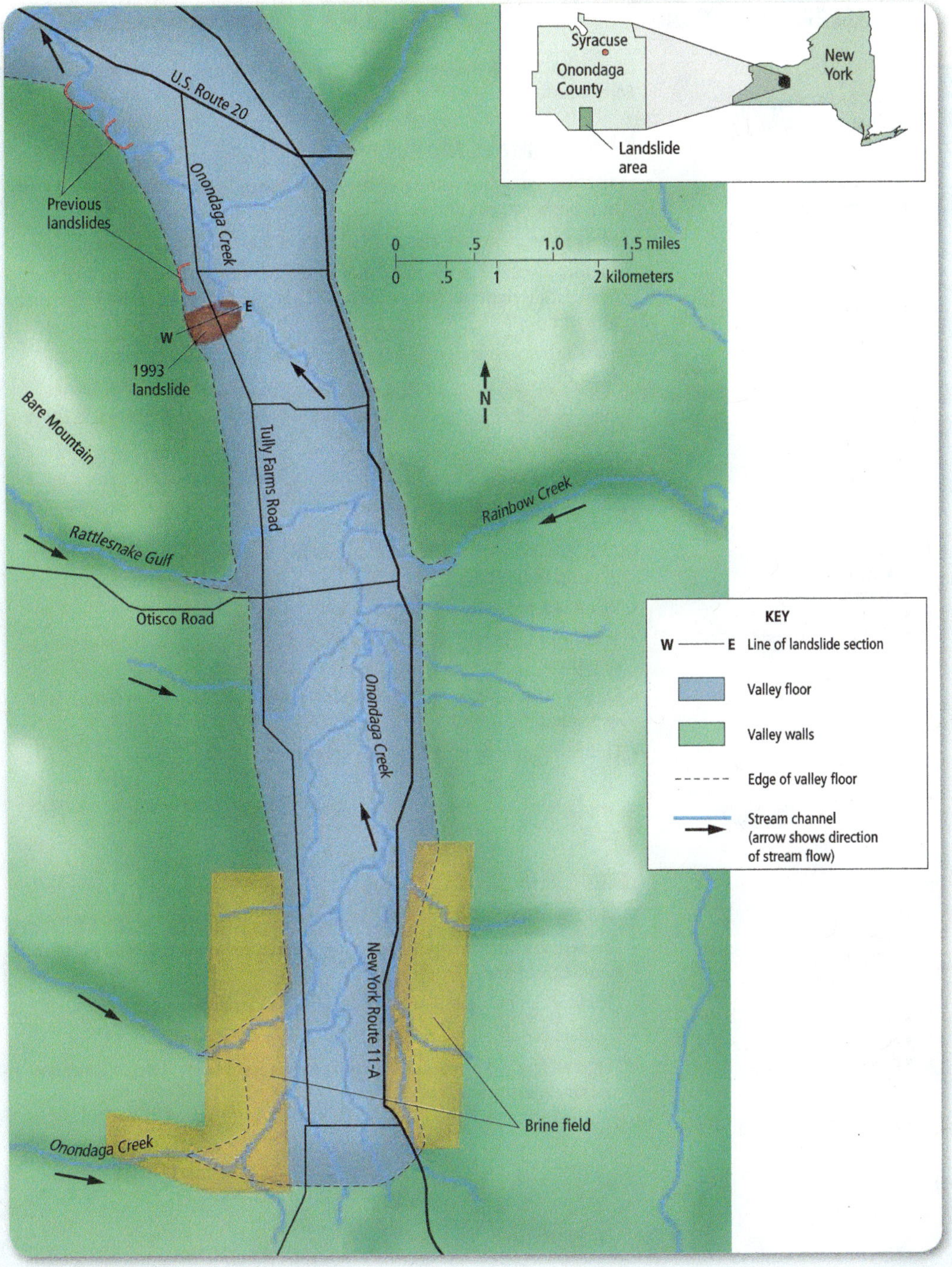
Syracuse
Onondaga County
New York
Landslide area
U.S. Route 20
Previous landslides
Onondaga Creek
0 .5 1.0 1.5 miles
0 .5 1 2 kilometers
E
W
1993 landslide
Bare Mountain
N
Tully Farms Road
Rainbow Creek
Rattlesnake Gulf
Otisco Road
KEY
W —— E Line of landslide section
Valley floor
Valley walls
Edge of valley floor
Stream channel (arrow shows direction of stream flow)
Onondaga Creek
New York Route 11-A
Brine field
Onondaga Creek

MAINIDEAS

Summary statements can be used by students to review the major concepts of the chapter.

Students can review with these online resources.

Vocabulary Practice

- Vocabulary eGames
- Vocabulary eFlashcards
- Vocabulary PuzzleMaker

Section Self-Check

Chapter Self-Check

Online Test Practice

Use the Exam*View*® *Assessment Suite* CD-ROM to:

- create multiple versions of tests
- create modified tests with one mouse click
- edit existing questions and add your own questions
- build tests aligned with state standards using built-in state curriculum tags
- change English tests to Spanish with one mouse click
- track students' progress using the Teacher Management System

CHAPTER 8 STUDY GUIDE

BIGIDEA **Movements due to gravity, winds, and glaciers shape and change Earth's surface.**

Vocabulary Practice

SECTION 1 Mass Movements

VOCABULARY

- mass movement
- creep
- mudflow
- landslide
- slump
- avalanche

MAINIDEA **Mass movements alter Earth's surface over time due to gravity moving sediment and rocks downslope.**

- Mass movements are classified in part by how rapidly they occur.
- Factors involved in the mass movement of Earth materials include the material's weight, its resistance to sliding, the trigger, and the presence of water.
- Mass movements are natural processes that can affect human life and activities.
- Human activities can increase the potential for the occurrence of mass movements.

SECTION 2 Wind

VOCABULARY

- deflation
- abrasion
- ventifact
- dune
- loess

MAINIDEA **Wind modifies landscapes in all areas of the world by transporting sediment.**

- Wind is a powerful agent of erosion.
- Wind can transport sediment in several ways, including suspension and saltation.
- Dunes form when wind velocity slows down and windblown sand is deposited.
- Dunes migrate as long as winds continue to blow.

SECTION 3 Glaciers

VOCABULARY

- glacier
- valley glacier
- continental glacier
- cirque
- moraine
- outwash plain
- drumlin
- esker
- kame
- kettle

MAINIDEA **Glaciers modify landscapes by eroding and depositing rocks.**

- Glaciers are large moving masses of ice that form near Earth's poles and in mountain areas.
- Glaciers can be classified as valley glaciers or continental glaciers.
- Glaciers modify the landscape by erosion and deposition.
- Features formed by glaciers include U-shaped valleys, hanging valleys, moraines, drumlins, and kettles.

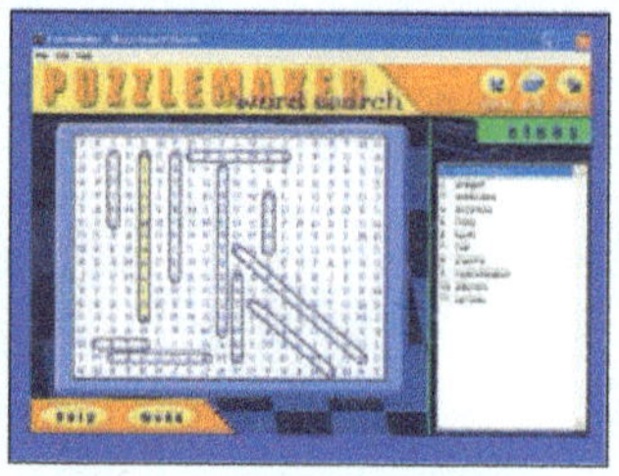

Vocabulary Practice

For additional practice with vocabulary, have students access the Vocabulary PuzzleMaker.

Chapter Self-Check

VOCABULARY REVIEW

Match the correct vocabulary term from the Study Guide to the following definitions.

1. rapid downslope movement of a mass of loose sediment
2. rapidly flowing, often destructive mixtures of mud and water
3. slow, steady downhill movement of loose, weathered Earth materials

Replace each underlined word with the correct vocabulary term from the Study Guide.

4. Barchans are rocks shaped by windblown sediments.
5. Thick, windblown, fertile deposits of silt that contain high levels of nutrients and minerals are known as desert pavement.
6. Deflation occurs when particles such as sand rub against the surface of rocks.

Explain the differences between the vocabulary terms in the following sets.

7. valley glacier, continental glacier
8. esker, kame
9. moraine, outwash plain

UNDERSTAND KEY CONCEPTS

10. What are elongated landforms made of older moraines over which glaciers have moved?
    A drumlins
    B. kettle lakes
    C. eskers
    D. outwash plains
11. Which particles can wind move most easily?
    A. sand
    B. pebbles
    C. silt
    D. gravel
12. Which is the underlying force that causes all forms of mass movement?
    A. friction
    B. gravity
    C. magnetism
    D. the Coriolis Effect
13. The last continental ice age covered approximately what percent of Earth's surface?
    A. 10 percent
    B. 20 percent
    C. 30 percent
    D. 50 percent
14. Where are large deposits of glacial loess primarily found?
    A. eastern United States
    B. southeastern United States
    C. southwestern United States
    D. midwestern United States
15. Which has the fastest movement?
    A. solifluction
    B. creep
    C. mudflow
    D. avalanche

Use the photo below to answer Questions 16 and 17.

16. Which formed the structure in the photo?
    A. ice **C.** wind
    B. water **D.** organisms
17. Which process formed the structure in the photo?
    A. abrasion **C.** deposition
    B. deflation **D.** migration

CHAPTER 8 ASSESSMENT

VOCABULARY REVIEW

1. landslide
2. mudflow
3. creep
4. Ventifacts
5. loess
6. Abrasion
7. Valley glaciers form in mountainous areas, while continental glaciers cover entire continents.
8. Eskers are ridges formed by glaciers, while kames are conical hills of till that form in the depressions of a glacier.
9. A moraine is a ridge of till deposited by a glacier, while an outwash plain is the leading edge of the glacier where the meltwater flows.

UNDERSTAND KEY CONCEPTS

10. A
11. C
12. B
13. C
14. D
15. D
16. C
17. A

18. C
19. C
20. B
21. D
22. C
23. D

CONSTRUCTED RESPONSE

24. If particles are airborne for long periods of time, they are carried in suspension. If they bounce from place to place, they are moved by saltation.
25. Water acts as a lubricant between the grains of sand to reduce friction between them.
26. cirque, horn, hanging valley, arête
27. Answers will vary, but should include movement from windward to leeward side.
28. In a flow, Earth materials move as if they were liquid. In a slide, a relatively thin block of soil, rock, or debris separates from the underlying bedrock. In a rockfall, rocks break up and fall downward.
29. Solifluction is a type of creep that occurs mostly in regions of permafrost.
30. earthflows and mudflows
31. Wind-eroded particles are pitted, with angular, faceted surfaces, while water-eroded particles are smooth and round.
32. Building in coastal-dune areas and removing dune vegetation disrupt dune growth and damage dunes.
33. A dune is produced when sand and smaller particles are deposited by wind. The shape of the dune depends on the amount of sand and the wind direction and speed. Drumlins are stream-lined hills produced from glacial till of all sizes. Their shape is determined by the movement of the glacier and the flow of meltwater.

18. Which range of slope angles is associated with producing an avalanche?
A. 10 to 20 degrees
B. 20 to 35 degrees
C. 30 to 45 degrees
D. 45 to 60 degrees

19. Which statement best describes sediments deposited by glaciers and rivers?
A. Both glacial and river deposits are sorted.
B. Glacial deposits are sorted, and river deposits are unsorted.
C. Glacial deposits are unsorted, and river deposits are sorted.
D. Both glacial and river deposits are unsorted deposits.

Use the photo below to answer Question 20.

20. Which most likely created the valley?
A. running water
B. glacial ice
C. landslide
D. strong prevailing winds

21. Which is a way to reduce the risk of mass movements?
A. Develop hillsides with roads so they become stable.
B. Allow septic systems to run unmaintained so that they provide a source of nutrients for the soil.
C. Build homes in steep terrain in order to stabilize the slope.
D. Avoid construction and structures on vulnerable slopes.

22. Which property is used to classify dunes?
A. age
B. composition
C. shape
D. density

23. Which does NOT affect the speed of a valley glacier's movement?
A. slope of the valley floor
B. shape of the valley wall
C. temperature and thickness of the ice
D. internal chemistry of the glacier

CONSTRUCTED RESPONSE

24. Compare and contrast suspension and saltation as they relate to transport of materials by wind.

25. Infer What happens to sand particles as the sand becomes saturated with water?

Use the figure below to answer Question 26.

26. Identify the features of the glacial landscape.

27. Diagram and label a migrating sand dune. Indicate the prevailing wind direction.

28. Contrast a slide, flow, and rockfall.

29. Describe the relationship between permafrost and solifluction.

30. Identify which mass movements are dependent on the addition of water.

31. Describe how particles eroded by wind differ from particles eroded by water.

32. Infer how human activities could affect the formation and migration of sand dunes in coastal areas.

33. Compare and contrast the formation, shape, and size of particles of a sand dune and a drumlin. How are these features used to indicate the direction of wind and glacial movement?

Chapter Self-Check

THINK CRITICALLY

Use the figure below to answer Question 34.

34. **Hypothesize** what kind of mass movements might have occurred after the eruption of Mount St. Helens.

35. **Analyze** the conditions that contribute to the likelihood that an area will experience wind erosion, and identify at least three areas in the United States that are prone to wind erosion.

36. **Infer** why wind abrasion is such an effective agent of erosion.

37. **Predict** the shape of a lake formed by a valley glacier.

CONCEPT MAPPING

38. Create a concept map to compare the terms *drumlin, esker,* and *kame.* For more help, refer to the *Skillbuilder Handbook.*

CHALLENGE QUESTION

39. **Hypothesize** how the Dust Bowl of the 1930s might have been avoided.

WRITING IN ▶ Earth Science

40. Write an editorial for a newspaper explaining why laws are needed to prevent developers from building homes on relatively steep and loosely consolidated hillside areas.

DBQ Document–Based Questions

Data obtained from: Natural Hazards-Landslides Information Sheet. 2006. *USGS.*

The photo below shows the potential for landslides across the continental United States.

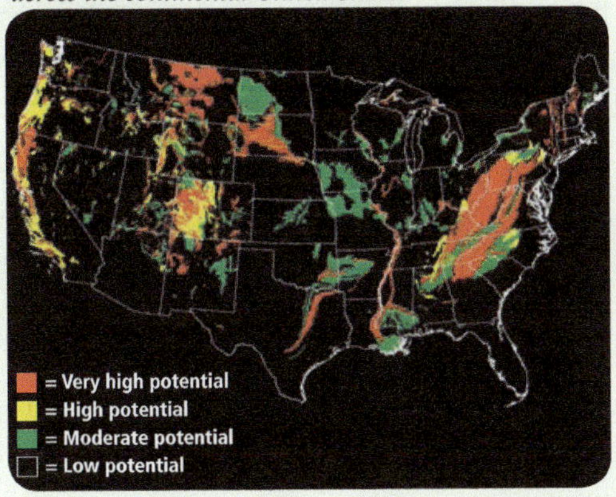

41. Identify landscapes or possible triggers for two areas that have very high potential for landslides.

42. Infer why the potential for landslides occurring in Florida is low.

43. What can be done to reduce the number of deaths due to landslides?

CUMULATIVE REVIEW

44. How many valence electrons does oxygen (atomic number 8) have? **(Chapter 3)**

45. Which compositional type of igneous rock has the lowest silica content? **(Chapter 5)**

46. What are fossils? **(Chapter 6)**

THINK CRITICALLY

34. After a volcanic eruption, meltwater might cause a mudflow or lahar.
35. Areas of little vegetative cover, such as deserts, semi-arid areas, seashores, and lakeshores are particularly vulnerable. Areas that are vulnerable to wind erosion include the Midwest, the Southeast, and around the Great Lakes.
36. Wind abrasion, which can polish and smooth different kinds of rocks and can wear away the base of telephone poles and remove paint on homes, constantly bombards structures with wind-borne particles of hard substances such as quartz.
37. round

CONCEPT MAPPING

38. Concept map should show that drumlin, esker, and kame are all sorted glacial depositional features, but that a drumlin is an elongate ridge, an esker is a sinuous ridge, and a kame is a mound or hill.

CHALLENGE QUESTION

39. The Dust Bowl of the 1930s might have been avoided if natural vegetation had not been removed to clear land for farming.

WRITING IN ▶ Earth Science

Rubric

40. Answers should include a discussion about the high probability of property damage and injury due to a landslide.

DBQ Document–Based Questions

Data obtained from: Natural Hazards-Landslides Information Sheet. 2006. *USGS.*

41. Possible answers: steep slopes in the Appalachian Mountains, earthquakes in northern California, high amounts of rainfall in northern California
42. The conditions of the landscape in Florida and the occurrence of triggers such as earthquakes make the possibility of landslides low.
43. Increasing the public awareness of landslides as threats should decrease the number of deaths.

CUMULATIVE REVIEW

44. six
45. Ultrabasic (also known as ultramafic) igneous rocks have the lowest silica content.
46. any trace or remnant of past life

MULTIPLE CHOICE

1. C
2. A
3. A
4. B
5. B
6. C
7. C
8. D
9. A
10. B

CUMULATIVE STANDARDIZED TEST PRACTICE

MULTIPLE CHOICE

1. What is the strongest factor that controls the development of soils?
   A. parent material
   B. topography
   C. climate
   D. time

Use the table below to answer Questions 2 and 3.

Region	Characteristics
A	semiarid; experiences intense but brief rainstorms
B	permafrost; much loose, waterlogged material
C	mountainous; thick accumulations of snow
D	thick soils on semi-steep and steep slopes; occasional earthquake activity
E	arid; high cliffs and rocky shorelines

2. Which mass movement is most likely to occur in Region A?
   A. mudflow
   B. avalanche
   C. slump
   D. rockfall

3. Which is most likely to occur in Region B?
   A. solifluction
   B. mudflow
   C. avalanche
   D. slump

4. Which branch of science studies humans' interactions with the environment?
   A. planetary science
   B. environmental science
   C. oceanography
   D. geology

5. When do minerals precipitate out of a solution?
   A. when the solution is saturated
   B. when the solution is supersaturated
   C. when the solution is unsaturated
   D. when the solution is ultrasaturated

6. Why are the 24 time zones located approximately 15° apart?
   A. to line up with the equator
   B. to roughly match lines of latitude
   C. to roughly match lines of longitude
   D. to line up with the prime meridian

7. Identify the term used to describe wind transportation of materials by a bouncing motion of particles.
   A. suspension
   B. deflation
   C. saltation
   D. abrasion

Use the geologic cross section below to answer Questions 8 and 9.

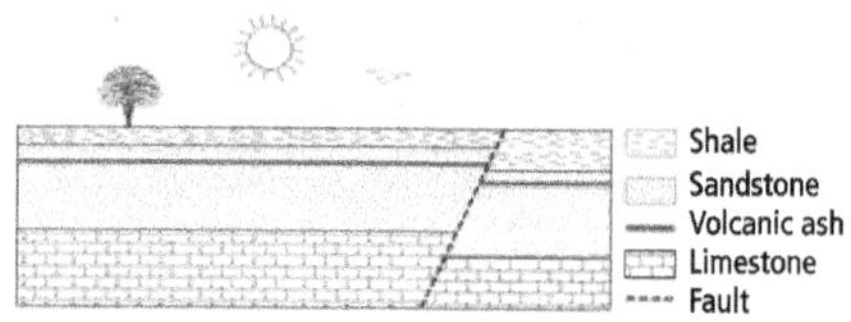

8. Assuming the rock layers shown are in the same orientation that they were deposited, which layer is the oldest?
   A. shale
   B. sandstone
   C. volcanic ash
   D. limestone

9. Which layer was probably created by sediments deposited by slow-moving water?
   A. shale
   B. sandstone
   C. volcanic ash
   D. limestone

10. Which is NOT a feature of valley glaciers?
    A. cirque
    B. loess
    C. moraine
    D. arête

Online Test Practice

SHORT ANSWER

Use the table below to answer Questions 11 and 12.

Liquid	Final Color of Litmus Paper
ammonia	blue
lemon juice	red
tea	red
vinegar	red

11. What conclusions can you draw from the results of the litmus paper tests on the liquids as shown in the table?

12. If an unknown liquid did not change the color of litmus paper, what could you infer?

13. Describe the formation of soil.

14. Evaluate the negative impact of building in coastal-dune areas.

15. Distinguish between weathering and erosion.

16. What is one reason granite is commonly used in construction?

17. What are some benefits of communicating scientific results?

18. What are isotopes of an element?

READING FOR COMPREHENSION

Arctic Ice Levels

The amount of sea ice in the Arctic shrank dramatically this summer and is now smaller than it has been in a century of record-keeping, new research reveals. Scientists say rising temperatures brought on by human-made global warming is probably to blame for the melting trend. Most scientists attribute this warming to human activities such as burning fossil fuels. The shift could lead to increased coastal erosion and shrinking of habitat for animals like polar bears. Melting sea ice may lead to greater coastal erosion, because Arctic storms could produce much larger waves on the open ocean. As the sea ice continues to melt, polar habitat also continues to shrink. If the decline in sea ice continued, summers in the Arctic could become completely ice-free before the end of this century, scientists warn.

Article obtained from Lovgren, S. Arctic ice levels at record low, may keep melting, study warns. *National Geographic News.* October 3, 2005.

19. How could melting sea ice possibly lead to greater coastal erosion?
- **A.** Polar bears would use more of the land.
- **B.** Humans would use the exposed land for fossil fuels.
- **C.** It would increase global temperatures, ruining the land.
- **D.** Arctic storms could produce larger waves to erode the shoreline.

20. What is causing the sea ice to melt?
- **A.** increased temperatures
- **B.** arctic storms
- **C.** polar bears
- **D.** time

21. What can be inferred from this text?

NEED EXTRA HELP?

If You Missed Question . . .	1	2	3	4	5	6	7	8	9	10	11	12	13	14	15	16	17	18
Review Section . . .	7.3	8.1	8.1	1.1	4.1	2.1	8.2	6.1	6.1	8.3	3.2	3.2	7.3	8.2	7.1	5.2	1.3	3.1

SHORT ANSWER

11. Lemon juice, tea, and vinegar are all acids, while ammonia is the only base.

12. You could infer that the liquid you tested was a neutral substance.

13. Soil forms when rock is broken into smaller pieces by weathering. The rock continues to break down and organisms inhabit these materials. As they die and decay, they add nutrients to form soil.

14. Building in coastal-dune areas removes dune vegetation and increases beach erosion and flooding.

15. Weathering is the process of breaking down materials on or near Earth's surface, while erosion is the process that transports and deposits these materials in different locations.

16. Granite is an igneous rock that has interlocking grains, which give it strength, and certain minerals that make it resistant to weathering.

17. The communication of scientific results helps scientists identify any possible error or bias by verifying the results, introduces new theories or discoveries to others, and could lead to further experimentation.

18. Isotopes of an element occur when atoms of the same element have the same number of protons but different mass numbers due to the different amounts of neutrons.

READING FOR COMPREHENSION

19. D

20. A

21. Sample answer: Human activities such as burning fossil fuels could be contributing to rising temperatures and melting the arctic ice. If this trend continues, much of the land could be lost through coastal erosion, and some species of animals, such as polar bears, could lose their habitats and become extinct.

CHAPTER 9 Surface Water

BIGIDEA **Surface water moves materials produced by weathering and shapes the surface of Earth.**

ESSENTIAL QUESTIONS	NATIONAL STANDARDS	RESOURCES TO ASSESS MASTERY
SECTION 1 **Surface Water Movement** **1.** How can surface water move weathered materials? **2.** How does a stream carry its load? **3.** How does a floodplain develop? 2 sessions 1 block	UCP.1–3; A.1–2; B.2, B.4; D.1–2; F.3, F.5–6; G.3	**Entry-Level Assessment** Focus Transparency 23, p. 224 **Progress Monitoring** Caption Question, pp. 224, 227, 230 Reading Check, pp. 224, 227, 230 Section Review, p. 231
SECTION 2 **Stream Development** **1.** What physical features are characteristic of stream development? **2.** What is the relationship between meanders and stream flow? **3.** How is the process of rejuvenation in stream development explained? 2 sessions 1 block	UCP.2–5; B.4; D.1, D.3; G.3	**Entry-Level Assessment** Focus Transparency 24, p. 232 **Progress Monitoring** Caption Question, p. 233 Reading Check, p. 236 Section Review, p. 237
SECTION 3 **Lakes and Freshwater Wetlands** **1.** How do freshwater lakes and wetlands form? **2.** How is the process of eutrophication described? **3.** What are the effects of human activity on lake development? 4 sessions 2 blocks	UCP.1–5; A.1–2; B.4; C.4; D.1, D.3; F.4–5; G.3	**Entry-Level Assessment** Focus Transparency 25, p. 238 **Progress Monitoring** Reading Check, pp. 239, 240 Section Review, p. 241 **Summative Assessment** Chapter Assessment, p. 245 Exam*View*® *Assessment Suite* CD-ROM

Go online! connectED.mcgraw-hill.com

LEVELED RESOURCES	LAB MATERIALS	ADDITIONAL RESOURCES
Science Notebook 9.1 OL **Chapter FAST FILE Resources:** Study Guide, p. 65 BL **Lab Resources:** Laboratory Manual, p. 65 OL **Transparencies:** Section Focus Transparency 23 EL Teaching Transparency 22 OL EL	LaunchLAB p. 222 / **15 min** small, plastic window screen; clear plastic containers (2); clump of grass or sod; clump of barren soil; water	**Technology:** eTeacherEdition eStudentEdition Classroom Presentation Toolkit Exam*View® Assessment Suite* CD-ROM LabManager™ CD-ROM What's EARTH SCIENCE Got To Do With It? DVD Weather Classroom Video DVD **Assessment:** Blueprints for Success Chapter Test (Scaffolded) Earth Science Standardized Test Practice **Lab Resources:** Exploring Environmental Problems Laboratory Manual
Science Notebook 9.2 OL **Chapter FAST FILE Resources:** GeoLab Worksheet, p. 55 OL Study Guide, p. 67 BL **Lab Resources:** Laboratory Manual, p. 69 OL **Transparencies:** Section Focus Transparency 24 EL Teaching Transparency 23 OL EL	GeoLAB p. 243 / **45 min** 1-m length of vinyl gutter pipe, ring stand and clamp, water source with hose, protractor with plumb bob, sink or container to catch water, stopwatch, grease pencil, meterstick, paper, three-hole punch	
Science Notebook 9.3 OL **Chapter FAST FILE Resources:** MiniLab Worksheet, p. 54 OL Study Guide, p. 69 BL **Transparencies:** Section Focus Transparency 25 EL Teaching Transparency 24 OL EL	MiniLAB p. 240 / **15 min** clear plastic containers (3), clay, sand, gravel, water	

LabManager™
Customize any lab with the LabManager™ CD-ROM.

BL Below Level   OL On Level   AL Advanced Learners   EL English Learners   COOP LEARN Cooperative Learning

CHAPTER 9

LaunchLAB

iLab Station | Rubric

How does water infiltrate?

Process Skills observe and infer, recognize cause and effect, communicate, model, measure, use numbers

Safety Precautions Approve lab safety forms before work begins. Make sure students wear safety goggles.

Teaching Strategies

- Make sure the plastic containers do not leak.
- Provide a bucket for muddy water to be collected and discarded outside the school building onto a grassy area.

Procedure

1. Have students read and complete the lab safety form and follow the procedure below.
2. Place a **small plastic window screen** on each of two **clear plastic containers.**
3. Place an 8-cm × 16-cm **clump of grass or sod** on one screen.
4. Place an 8-cm × 16-cm **clump of barren soil** on the other screen.
5. Lightly sprinkle 500 mL of **water** on each clump.
6. Observe the clumps for 5 min.
7. Measure the amount of water in each container.

Analysis

1. **Describe** what happens to the water after 5 min. Water runs through the barren soil more easily than the clump of sod.
2. **Infer** the reason for any differences in the amount of water collected in each container. Plants slow water flow and increase contact with soil particles, thus aiding infiltration.

CHAPTER 9

Surface Water

BIGIDEA Surface water moves materials produced by weathering and shapes the surface of Earth.

SECTIONS

1 Surface Water Movement

2 Stream Development

3 Lakes and Freshwater Wetlands

LaunchLAB

iLab Station

How does water infiltrate?

When water soaks into the ground, it moves at different rates through the different materials that make up Earth's surface. Explore how in this lab.

FOLDABLES® Study Organizer

Stream Development

Make a six-page book using the label shown. Use it to describe and illustrate the steps in stream development.

Assessment

Performance Ask students which will hold more water from precipitation: a surface with or without vegetation. The surface with vegetation holds more water than the surface without vegetation.

LabManager™
Customize any lab with the LabManager™ CD-ROM.

The United States has approximately 5,600,000 km of rivers. The Missouri River is about 4087 km long, making it the longest river in North America.

Introduce the BIG IDEA

Surface Water Movement Ask students: What types of bodies of water constitute surface water? rivers, lakes, streams, ponds, seas Write their answers on the board. Ask students: From where does surface water come? rain, ice, snow, under the ground Write their answers on the board. Ask students: Where does surface water go? under the ground, into the ocean, evaporates into the sky Write their answers on the board.

Teacher Content Support

Wild and Scenic Rivers In 1968, Congress created the Wild and Scenic Rivers Act, which protects a number of rivers in the United States. Only one-quarter of 1 percent of the rivers in the United States are protected with this designation. Ask students: why might rivers need to be protected? Rivers might need to be protected from being diverted, being dammed, or being polluted.

SECTION 1

1 Focus

Focus Transparency

Before presenting the lesson, project **Section Focus Transparency 23** and have students answer the accompanying questions. EL

MAINIDEA

Running Water Write the following terms on the board and ask students to organize them in a concept map: *rain, slope, ocean, flat, silt, river, valley, sediments, downstream, pond, flow.* Answers will vary.

2 Teach

Identify Misconceptions

Most students think clouds are made of water vapor.

Uncover the Misconception

Ask students what makes up clouds.

Demonstrate the Concept

Tell students cloud droplets form when air becomes saturated with water vapor and the water vapor condenses on small, solid particles called condensation nuclei, such as dust, pollen, or salt. Tell students that these liquid drops are suspended in the atmosphere, and when they occur in concentrations large enough to block out varying degrees of sunlight, they can be recognized as a mass in themselves. The aggregate of droplets is referred to as a cloud. Hence, clouds are composed of liquid drops of water, not water vapor.

Assess New Knowledge

Ask students to answer the following question: What are clouds made of? Clouds are made up of liquid drops of water.

SECTION 1

Essential Questions

- How can surface water move weathered materials?
- How does a stream carry its load?
- How does a floodplain develop?

Review Vocabulary

solution: a homogeneous mixture in which the component particles cannot be distinguished

New Vocabulary

runoff
watershed
divide
suspension
bed load
discharge
flood
floodplain

Surface Water Movement

MAINIDEA Running water is an agent of erosion, carrying sediments in streams and rivers and depositing them downstream.

EARTH SCIENCE 4 YOU Have you ever noticed that sometimes a river is muddy but other times it is clear? In floods, rivers can carry greater amounts of materials, which makes them muddy. Under normal conditions, they often carry less sediment, which makes them clearer.

The Water Cycle

Earth's water supply is recycled in a continuous process called the water cycle, shown in **Figure 1.** Water molecules move continuously through the water cycle following many pathways: they evaporate from a body of water or the surface of Earth, condense into cloud droplets, fall as precipitation back to Earth's surface, and infiltrate the ground. As part of a continuous cycle, the water molecules eventually evaporate back to the atmosphere, form clouds, fall as precipitation, and the cycle repeats. Understanding the mechanics of the water cycle will help you understand the reasons for variations in the amount of freshwater that is available throughout the world.

Often, a water molecule's pathway involves time spent within a living organism or as part of a snowfield, glacier, lake, or ocean. Although water molecules might follow a number of different pathways, the overall process is one of repeated evaporation and condensation powered by the Sun's energy.

☑ READING CHECK **Explain** What happens once water reaches Earth's surface?

Figure 1 The water cycle, also referred to as the hydrologic cycle, is a never-ending, natural circulation of water through Earth's systems.
Identify *the driving force for the water cycle.*

View an **animation of the water cycle.**

Concepts In Motion

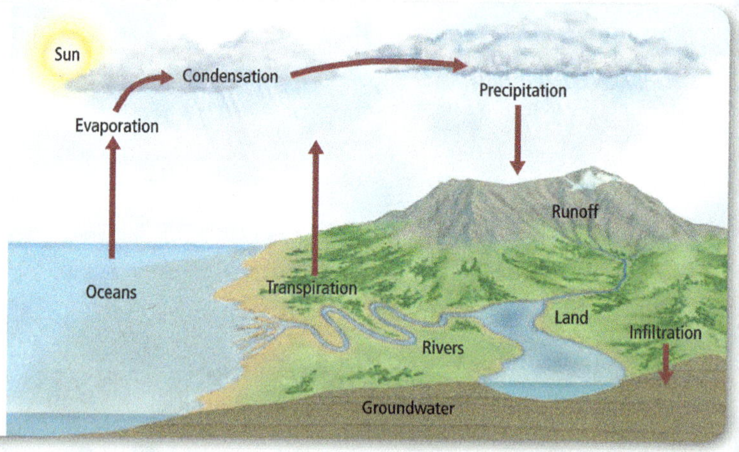

☑ READING CHECK Water can infiltrate or run off when it reaches the surface.

Caption Question Fig. 1 The driving force of the water cycle is the Sun.

EARTH SCIENCE JOURNAL

Local Brooks and Streams Ask students to locate local brooks or streams near their homes and to describe them in their Earth science journals, including where students think they start and where they end. Then encourage students to consult maps of their area to discover the routes of the local brooks and describe the actual routes in their Earth science journals. OL

Runoff

Water flowing downslope along Earth's surface is called **runoff.** Runoff might reach a stream, river, or lake, it might evaporate, or it might accumulate as puddles in small depressions and infiltrate the ground. During and after heavy rains, you can observe these processes in your yard or local park. Water that infiltrates Earth's surface becomes groundwater.

A number of conditions determine whether water on Earth's surface will infiltrate the ground or become runoff. For water to enter the ground, there must be large enough pores or spaces in the soil and rock to accommodate the water's volume, as in the loose soil illustrated in **Figure 2.** If the pores already contain water, the newly fallen precipitation will either remain in puddles on top of the ground or, if the area has a slope, run downhill. Water standing on the surface of Earth eventually evaporates, flows away, or slowly enters the groundwater.

Soil composition The physical and chemical composition of soil affects its water-holding capacity. Soil consists of decayed organic matter, called humus, and minerals. Humus creates pores in the soil, thereby increasing a soil's ability to retain water. The minerals in soil have different particle sizes, which are classified as sand, silt, or clay. Recall that the percentages of particles of each size vary from soil to soil. Soil with a high percentage of coarse particles, such as sand, has relatively large pores between its particles that allow water to enter and pass through the soil quickly. In contrast, soil with a high percentage of fine particles, such as clay, clumps together and has few or no spaces between the particles. Small pores restrict both the amount of water that can enter the ground and the ease of movement of water through the soil.

Rate of precipitation Light, gentle precipitation can infiltrate dry ground. However, the rate of precipitation might temporarily exceed the rate of infiltration. For example, during heavy precipitation, water falls too quickly to infiltrate the ground and becomes runoff. Thus, a gentle, long-lasting rainfall is more beneficial to plants and causes less erosion by runoff than a torrential downpour. If you have a garden, remember that more water will enter the ground if you water your plants slowly and gently.

VOCABULARY

ACADEMIC VOCABULARY

Accommodate
to hold without crowding or inconvenience
The teacher said she could accommodate three more students in her classroom.

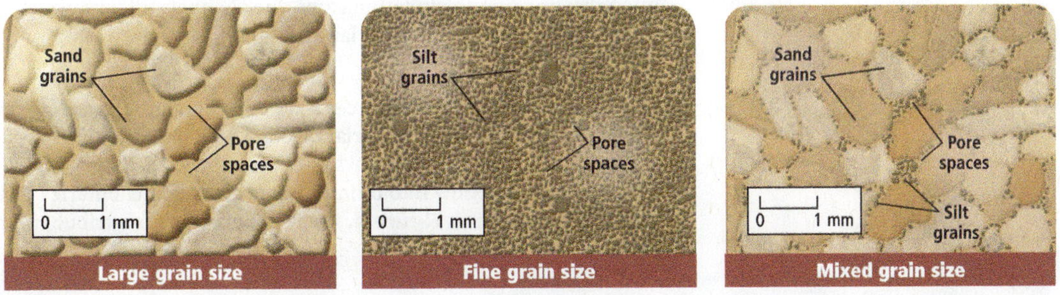

Figure 2 Soil that has open surface pores allows water to infiltrate. The particle size that makes up a soil helps determine the pore space of the soil.

Teacher Content Support

Runoff Runoff that consolidates in long, narrow depressions produces channel or stream flows. Runoff that flows downslope in generally broad sheets is called overland flow. There are different types of overland flows. When runoff flows over smooth surfaces in continuous, thin sheets, it is called sheet flow. Runoff that flows over steep slopes of bare soil or less resistant bedrock flows in long, slender channels called shoestring rills.

Tie to Previous Knowledge

Water and Temperature Ask students what they notice when a glass of ice water sits on a table during the summer. The glass seems to sweat and water appears around the glass. Explain to students what happens to the glass is part of the water cycle. The cooled container of water cools the air around it and causes water vapor in the air to condense on the outside of the glass of ice water. OL

Interpret the Photo

Infiltration Tell students to study **Figure 2.** Ask students: Which soil type will allow more water to infiltrate? large grains Which soil type will allow the least amount of water to infiltrate? fine grains

Demonstration

Model Infiltration Pack soil into a large, clear container. Fill another clear container loosely with soil. Sprinkle equal amounts of water into both containers of soil. Ask students to note which container of soil allows the most water to penetrate. The loosely packed soil allows more water to penetrate because it has spaces available for the water. The packed soil has few or no spaces for the water to enter.

Use Science Terms

Stream and River The words *stream* and *river* can be used interchangeably in Earth science for the most part. Although the word *river* generally indicates a large stream, water moves in the same manner in both small and large bodies of moving water. As a result, the word *stream* is commonly used to refer to moving bodies of water of various sizes.

Collaborative Learning

Stream Systems Have groups of students develop and make clay models of landscapes that show a stream system, the system's watershed, and at least one divide. Ask each group to develop questions about its watershed model at specific areas noted on the model. EL COOP LEARN

Identify Misconceptions

Students often underestimate the power of moving water.

Uncover the Misconception Ask students if moving water poses a threat to human lives.

Demonstrate the Concept Students who have experienced the force of the moving water associated with waves along an ocean shore might need to be reminded of the strength of the force they felt at the time. Students might not relate these experiences to the danger of driving through a flooded street or to the possibility of being swept away by floodwater, which can rise quickly. Point out that waves breaking against the shores of a lake during a storm not only poses a threat to humans, but also can damage or destroy infrastructure.

Assess New Knowledge Ask students to describe the threats posed by moving water.

Figure 3 Vegetation can slow the rate of runoff of surface water. Raindrops are slowed when they strike the leaves of trees or blades of grass, and they trickle down slowly.

Vegetation Soils that contain grasses or other vegetation allow more water to enter the ground than do soils with no vegetation. Precipitation falling on vegetation slowly flows down leaves and branches and eventually drops gently to the ground, where the plants' root systems help maintain the pore space needed to hold water, as shown in **Figure 3.** In contrast, precipitation strikes with far more force on barren land. In such areas, soil particles clump together and form dense aggregates with little space between them. The force of falling rain can then push the soil clumps together, thereby closing pores and allowing less water to enter.

Slope The slope of a land area plays a significant role in determining the ability of water to enter the ground. Water from precipitation falling on slopes flows to areas of lower elevation. The steeper the slope, the faster the water flows. There is also greater potential for erosion on steep slopes. In areas with steep slopes, much of the precipitation is carried away as runoff.

Stream Systems

Precipitation that does not enter the ground usually runs off the surface quickly. Some surface water flows in thin sheets and eventually collects in small channels, which are the physical areas where streams flow. As the amount of runoff increases, the channels widen, deepen, and become longer. Although these small channels often dry up after precipitation stops, the channels fill with water each time it rains and become larger and longer.

Tributaries All streams flow downslope to lower elevations. However, the path of a stream can vary considerably, depending on the slope and the type of material through which the stream flows. Some streams flow into lakes, while others flow directly into the ocean. Rivers that flow into other streams are called tributaries. For example, as shown in **Figure 4,** the Missouri River is a tributary of the Mississippi River.

DIFFERENTIATED INSTRUCTION

Struggling Learners Prepare several cardboard boxes with approximately 15 cm of various types of soil in each. You could use packed clay, sand, topsoil, and gravel. Have students slowly pour 40 mL of water into each box and place the boxes on separate cookie sheets. Have students check the bottoms of the boxes at one-minute intervals for signs that moisture has begun to leak through. Have students discuss their observations.

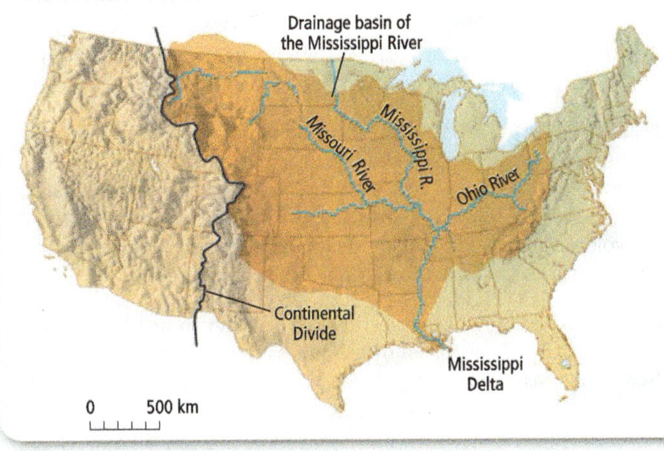

■ **Figure 4** The watershed of the Mississippi River includes many stream systems, including the Mississippi, Missouri and Ohio Rivers. The Continental Divide marks the western boundary of the watershed.
Identify ***what portion of the continental United States eventually drains into the Mississippi River.***

Watersheds and divides All of the land area whose water drains into a stream system is called the system's **watershed.** Watersheds can be relatively small or extremely large in area. A **divide** is an elevated land area that separates one watershed from another. In a watershed, the water flows away from the divide, as this is the high point of the watershed.

Each tributary in a stream system has its own watershed and divides, but they are all part of the larger stream system to which the tributary belongs. The watershed of the Mississippi River, shown in **Figure 4,** is the largest in North America.

☑ READING CHECK **Describe** what a divide is and what role it plays in a watershed.

Problem-Solving LAB

Interpret the Graph

How do sediments move in a stream? The critical velocity of water determines the size of particles that can be moved. The higher the stream velocity, the larger the particles that can be transported.

Think Critically

1. **Identify** at what velocity flowing water would pick up a pebble.
2. **Identify** at what range of velocities flowing water would carry a pebble.
3. **Infer** which object would not fall into the same size range as a pebble: an egg, a baseball, a golf ball, a table tennis ball, a volleyball, and a pea. How would you test your conclusions?

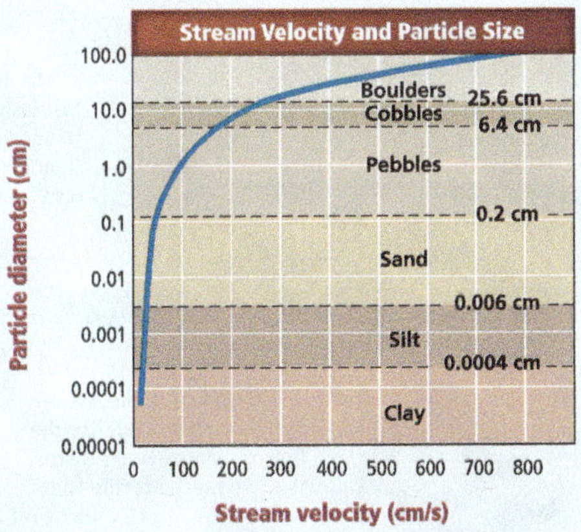

☑ **READING CHECK** A divide is a high area of land that separates one watershed from another.

■ **Caption Question Fig. 4** Roughly a third of the continental United States eventually drains into the Mississippi.

Problem-Solving LAB

Purpose Students will determine how different-sized sediments move in a stream.

Process Skills Make and use graphs, observe and infer

Teaching Strategies

- Ask students whether they have ever dropped something into a stream and had it carried away by the stream. Have students share their experiences.
- Review with students the function of the *x*-axis and the *y*-axis on the graph.

Think Critically

1. at 50 cm/s
2. from 50 to 160 cm/s (note: flowing water with a velocity greater than 160 cm/s would continue to carry a pebble)
3. Baseball and volleyball; measure the objects to see if they fall between 0.2 and 6.4 cm.

Model

Watersheds Have students work in groups of four to make a model of a watershed. This can be done using a stream table or a large tray with a mound of dirt placed at one end. Have students sprinkle water from a hose or watering can over the mass of dirt and allow the water to form rills and gullies. Larger streams will naturally form with tributaries and meanders. The erosional and depositional features of stream action will also be evident in the student models. BL EL

Concept Development

Stream Load What is carried as stream load depends on many factors. One factor is the land over which the water is moving. If the water is moving over soluble material, it will pick up minerals more easily than it would if it were flowing over less-soluble material. The temperature of the water also affects how much material can be dissolved in it. Warmer water usually can dissolve more material from Earth's surface than colder water can. The temperature of the water can also affect the amount of gases that can remain in solution. In general, colder water can hold more gases than warmer water. The types and numbers of plants and animals living in the water can also affect what materials, both gases and particulates, are carried in the stream's load.

Discussion

Stream Load Ask students why groundwater adds most of a stream's dissolved load, while runoff adds only a small amount. Groundwater, in general, moves over and through rocks for a longer period of time than runoff moves over Earth's surface. As a result, groundwater has more opportunity to come in contact with the minerals in rocks and dissolve them. Runoff, on the other hand, not only spends less time moving over Earth's surface, but it also might not flow directly on the rocks themselves but might instead flow over vegetation. OL

Figure 5 Particles rub, scrape, and grind against one another in a streambed, which can create potholes and other erosional features.

Stream Load

The material that a stream carries is known as stream load. Stream load is carried in three ways.

Materials in suspension **Suspension** is the method of transport for all particles small enough to be held up by the turbulence of a stream's moving water. Particles such as silt, clay, and sand are part of a stream's suspended load. The stream's suspended load varies based on the volume and velocity of the stream. Rapidly moving water carries larger particles in suspension than slowly moving water.

Bed load Sediment that is too large or heavy to be held up by turbulent water is transported by streams as the bed load. A stream's **bed load** consists of sand, pebbles, and cobbles that the stream's water can roll or push along the bed of the stream. The faster the water moves, the larger the particles it can carry. As the particles move, they rub against one another or the solid rock of the streambed, which can erode the surface of the streambed, as shown in **Figure 5.**

Materials in solution Minerals that are dissolved in a stream's water are called materials in solution. When water runs through or over rocks with soluble minerals, it dissolves small amounts of the minerals and carries them in solution. Groundwater adds the majority of the dissolved load to streams. The amount of dissolved material that water carries is often expressed in parts per million (ppm). For example, a measurement of 10 ppm means that there are 10 parts of dissolved material for every 1 million parts of water. The total concentration of materials in solution in streams averages 115–120 ppm, although some streams carry as little dissolved material as 10 ppm. Values greater than 10,000 ppm have been observed for streams draining desert basins.

Figure 6

Floods in Focus

Floods have shaped the landscape and affected human lives.

1927 Heavy rains flood the Mississippi River from Illinois to Louisiana leaving more than 600,000 people homeless.

1931 China's Yellow River floods when heavy rain causes the river's large silt deposits to shift and block the channel.

1900 — 1925 — 1950

1902 In Egypt, the Aswan Dam is built to stabilize the flow of annual flood waters that create the fertile Nile Delta.

1958 Following a flood that claimed almost 2000 lives, Holland begins creating a vast network of dams, dikes, and barriers, shortening its coastline by 700 km.

Demonstration

Bed Load Run water over a mound of soil and have students observe the bed load with a magnifying lens. Ask them to describe how the sediment particles are moving. They should observe that the rounder sediments roll and the flatter sediments slide in the water. The smaller sediments might be suspended in the water, while the heavier sediments will roll along the bottom. BL

Stream Carrying Capacity

The ability of a stream to transport material, referred to as its carrying capacity, depends on both the velocity and the amount of water moving in the stream. The channel's slope, depth, and width all affect the speed and direction the water moves within it. A stream's water moves more quickly where there is less friction; consequently, deep and smooth-sided channels with steep slopes allow water to move the most rapidly. The total volume of moving water also affects a stream's carrying capacity. **Discharge,** shown in **Figure 7,** is the measure of the volume of stream water that flows past a particular location within a given period of time. Discharge is commonly expressed in cubic meters per second (m^3/s). The following formula is used to calculate the discharge of a stream.

$$\underset{(m^3/s)}{\text{discharge}} = \underset{(m)}{\text{average width}} \times \underset{(m)}{\text{average depth}} \times \underset{(m/s)}{\text{average velocity}}$$

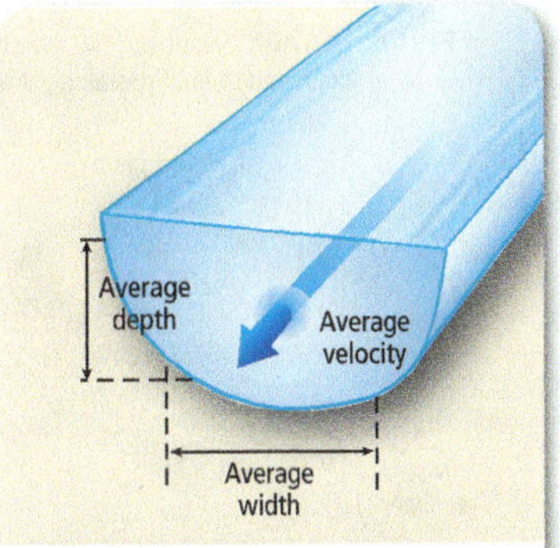

Figure 7 Stream discharge is the product of a stream's average width, average depth, and the velocity of the water.

The largest river in North America, the Mississippi River, has a huge average discharge of about 17,000 m^3/s. The Amazon River, the largest river in the world, has a discharge of about ten times that amount. The discharge from the Amazon River over a two-hour period would supply New York City's water needs for an entire year!

As a stream's discharge increases, its carrying capacity also increases. Both water velocity and volume increase during times of heavy precipitation, rapid melting of snow, and flooding. In addition to increasing a stream's carrying capacity, these conditions heighten a stream's ability to erode the land over which it passes. As a result of an increase in erosional power, a streambed can widen and deepen, adding to the stream's carrying capacity. Streams shape the landscape both during periods of normal flow and during floods, as highlighted in **Figure 6.**

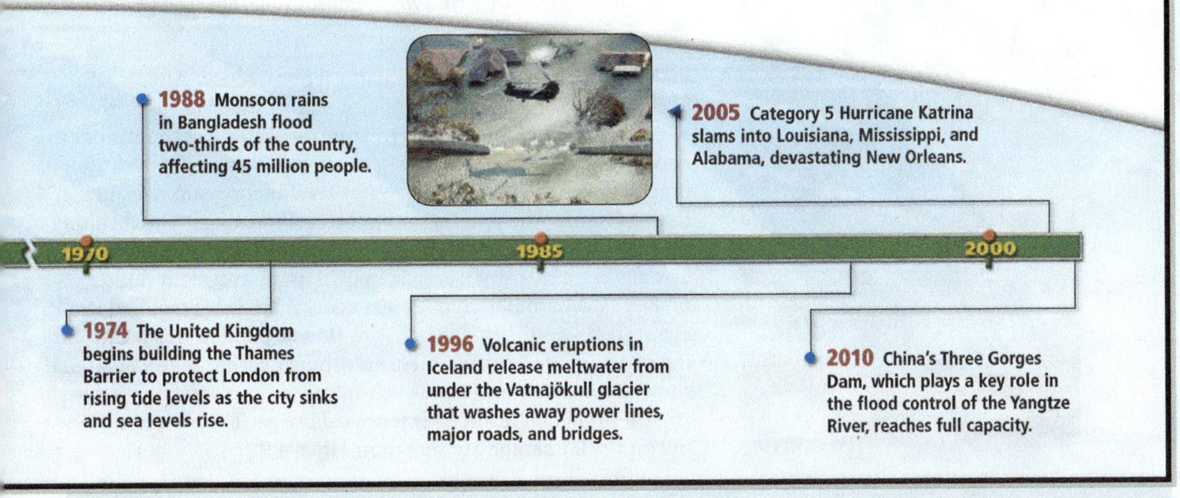

Teacher Content Support

Competence and Capacity

The competence and capacity of a stream determine how much load the stream can carry. The maximum size of the particles that a stream can carry is referred to as its competence. The larger the particles a stream can carry, the greater the stream's competence.

The competence of a stream is determined by the stream's velocity. The faster the water is moving, the more power it has to carry larger particles. In general, the competence of a stream increases as the square of the stream's velocity. For example, if a stream's velocity doubles, the competence will increase fourfold, and if the velocity triples, the competence will increase nine times.

The total load that can be carried by a stream is its capacity, which is related to the stream's discharge. The greater the volume of water that a stream is moving, the greater the stream's capacity. During times of flooding, an increase in both the velocity and volume of water occurs, and as a result, both the competence and capacity of a stream increase. Hence, floodwaters gain a tremendous amount of destructive and erosive power.

ACROSS THE CURRICULUM

Chemistry The study of water is closely connected to chemistry. With the use of chemical testing kits, the amount of dissolved oxygen and other components of water, including pH level, can be determined. Students can apply chemistry concepts and skills by using such common chemistry techniques as titrations, using litmus paper, or reading pH meters. Understanding how materials can be in solution is another connection between the section content and chemistry. This is a good time to check with the chemistry teacher to determine common themes and terminology, as well as to discover a possible source of chemicals, supplies, and equipment that might be accessible to Earth science classes.

■ **Caption Question** **Fig. 9** River banks prevent floodwater from re-entering the stream and allow it to saturate the soil. Sediments in the water can improve the soil's fertility by adding nutrients.

☑ READING CHECK As floodwaters recede, the volume and speed in the river decreases and the water drops its sediment load.

Activity

Flood Model Students will need water, a spray bottle, 2 dry sponges, and a dishpan for this activity. Have students place one dry sponge in a tilted dishpan. They should spray the sponge for exactly one minute on the spray nozzle setting, and then repeat with the other dry sponge and the mist setting. Ask students to record their observations. How do the two different nozzle settings relate to the natural environment and floods? EL

Apply Earth Science

FEMA Maps Show students a FEMA flood insurance rate map (FIRM) or the equivalent digital Q3 map. Ask students to make observations about the flood zones and compare the FIRM map with a topographic map of the same area. Explain what the flood lines mean. FEMA FIRMs can be obtained by phone, fax, or mail for your area: Map Service Center, P.O. Box 1038, Jessup, MD 20794–1038, Tel: (800) 358–9616, Fax: (800) 358–9620. OL

■ **Figure 8** When rivers overflow their banks, the floodwater deposits sediment. Over time, sediment accumulates along the edges of a river, resulting in natural levees.

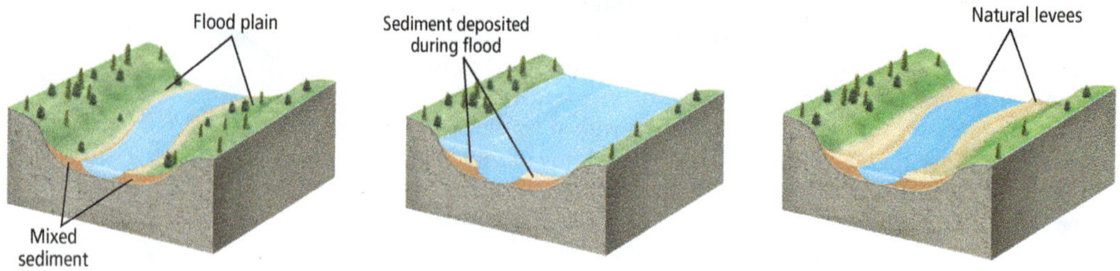

Floods

The amount of water being transported in a particular stream at any given time varies with weather conditions. Sometimes, more water pours into a stream than the banks of the stream channel can hold. A **flood** occurs when water spills over the sides of a stream's banks onto the adjacent land. The broad, flat area that extends out from a stream's bank and is covered by excess water during times of flooding is known as the stream's **floodplain.**

Floodwater carries along with it a great amount of sediment eroded from Earth's surface and the sides of the stream channel. As floodwater recedes and its volume and speed decrease, the water drops its sediment load onto the stream's floodplain. After repeated floods over time, sediments deposited by floods tend to accumulate along the banks of the stream. These develop into continuous ridges along the sides of a river, called natural levees, as shown in **Figure 8.** Floodplains develop highly fertile soils as more sediment is deposited with each subsequent flood. The fertile soils of floodplains make some of the best croplands in the world.

☑ READING CHECK **Describe** what happens when floodwaters recede.

■ **Figure 9** This flood was caused by heavy rainfall upstream. Notice the farm fields that have been covered in floodwater.
Analyze ***What long-term effects might this flood have on the crops grown in this area?***

Flood stages Floods are a natural occurrence. After a rain event or snowmelt, it takes time for runoff water to reach the streams. As water enters the streams, the water level continues to rise and might reach its highest point, called its crest, days after precipitation ends. When the water level in a stream rises higher than its banks, the river is said to be at flood stage. The resulting flooding might occur over localized areas or across large regions. The flooding of a small area is known as an upstream flood.

Heavy accumulation of excess water from large regional drainage systems results in downstream floods. Such floods occur during or after long-lasting, intense storms or spring thaws of large snowpacks. The tremendous volume of water involved in a downstream flood can result in extensive damage. The effects of flooding on the landscape are shown in **Figure 9.**

ACROSS THE CURRICULUM

History Have students each write a report about how streams have affected the history of a country. Students could write about the role that a stream played in a battle, a fight for water rights, the use of a stream as a boundary, or the problems associated with the need to cross streams. Encourage participants to share their reports in small groups so they hear of the importance of water throughout history. AL

■ **Figure 10** Gaging stations, like this one, can send data to meteorologic stations. There, scientists can process the information and alert the public to potential floods.

Flood Monitoring and Warning Systems

In order to provide warnings for people at risk, government agencies, such as the National Weather Service, monitor potential flood conditions. Earth-orbiting weather satellites photograph Earth and collect and transmit information about weather conditions, storms, and streams. In addition, the U.S. Geological Survey (USGS) has established approximately 9200 gaging stations in the United States, like the one shown in **Figure 10.** These gaging stations provide a continuous record of the water level in each monitored stream. Gaging systems often transmit data to satellites and telephone lines where the information is then sent to the local monitoring office.

In areas that are prone to severe flooding, warning systems are the first step in implementing emergency management plans. Flood warnings and emergency plans often allow people to safely evacuate an area in advance of a flood.

SECTION 1 REVIEW

Section Self-Check

Section Summary

- Infiltration of water into the ground depends on the number of open pores.
- All the land area that drains into a stream system is the system's watershed.
- Elevated land areas called divides separate one watershed from another.
- A stream's load is the material the stream carries.
- Flooding occurs in small, localized areas as upstream floods or in large downstream floods.

Understand Main Ideas

1. MAINIDEA **Analyze** ways in which moving water can carve a landscape.
2. **Describe** the three ways in which a stream carries its load.
3. **Analyze** the relationship between the carrying capacity of a stream and its discharge and velocity.
4. **Determine** why little water from runoff infiltrates the ground in areas with steep slopes.

Think Critically

5. **Determine** how a floodplain forms and why people live on floodplains.
6. **Analyze** how levees form.

MATH IN Earth Science

7. Design a data table that compares how silt, clay, sand, and large pebbles settle to the bottom of a stream as the velocity of water decreases.

USGS

Environmental Connection

Water Monitoring Monitoring concentrations of various substances at various stream locations can indicate whether harmful material is being added to the water from various sources.

3 Assess

Check for Understanding

Recognize Cause and Effect Have students list the factors that affect runoff. They include slope, vegetation, rate of precipitation, and soil composition.

Reteach

Communication Have students each make a glossary of terms from this section in their Earth science journals. Students may choose to include diagrams to accompany written definitions. Each term should also be used in a sentence.

Assessment

Performance Have students each select a specific stream and use resources to research significant facts about the stream, its characteristics, and its importance to people locally or regionally. Each student report should include maps and photographs of the stream. Reports should indicate possible sources of pollution.

Rubric

SECTION 1 REVIEW

1. Streams carve downward toward their base level. The bed load abrades the channel, making it deeper and wider, and potholes can be formed.
2. A stream's load is composed of particles in solution, particles in suspension, and the bed load. Material in solution is dissolved while the suspended load is held up by the flowing water. The bed load is rolled and pushed along the stream's bottom.
3. When velocity and discharge increase, so does a stream's carrying capacity.
4. In areas of steep slope, the water tends to run off because it does not have time to seep into the ground.
5. A floodplain forms from excess water during times of flood, which leaves sediment behind. Floodplains often contain fertile soils, so they are often used for croplands and settled.
6. As water overflows the stream's banks, it slows down. Large sediments are dropped closer to the stream's edge, building the levee. Smaller sediments create the fertile cropland of the floodplain.
7.

Stream velocity	Sediments that settle
high velocity	large pebbles
	sand
	silt
low velocity	clay

SECTION 2

1 Focus

Focus Transparency

Before presenting the lesson, project **Section Focus Transparency 24** and have students answer the accompanying questions. EL

MAINIDEA

Stream Movement Ask students these questions: How do streams dislodge material from the ground? chemically by dissolution, or, if current is strong enough, by pushing materials and moving them along What happens to the materials dislodged? Dislodged materials are deposited when the current decreases. What happens to the ground from which materials were dislodged? The surface becomes lower than before materials were dislodged.

2 Teach

Teacher Content Support

Stream Development A stream is part of a larger system called a drainage basin. A drainage basin consists of all the interconnecting streams that drain runoff into a particular river. The high areas between drainage basins are called divides. Divides can occur on a small or large scale. The Rocky Mountains are an example of a large divide. On the western side of the Rocky Mountains, runoff drains into the Pacific Ocean, while on the eastern side, runoff flows into the Gulf of Mexico. Drainage systems are often a network of streams. The patterns that streams make as they flow into one another depend on the bedrock over which the water is flowing. If the bedrock is generally uniform, the stream pattern is similar to tree branches. This pattern, called a dendritic pattern, is the most common.

SECTION 2

Stream Development

MAINIDEA Streams erode paths through sediment and rock, forming V-shaped stream valleys.

Essential Questions
- What physical features are characteristic of stream development?
- What is the relationship between meanders and stream flow?
- How is the process of rejuvenation in stream development explained?

Review Vocabulary

abrasion: process of erosion in which windblown or waterborne particles, such as sand, scrape against rock surfaces or other materials and wear them away

New Vocabulary

stream channel
stream bank
base level
meander
delta
rejuvenation

EARTH SCIENCE 4 YOU When was the last time you saw water flow uphill? Water in all rivers travels downslope to the lowest point. This allows geologists to predict the path of the river based on the features of an area.

Supply of Water

Stream formation relies on an adequate water supply. Precipitation provides water for the beginnings of stream formation. Streams can also be fed by underground reservoirs of water. As a stream develops, it changes width and size, and shapes the land over which it flows.

Stream channels The region where water first accumulates to supply a stream is called the headwaters. It is common for a stream's headwaters to be high in the mountains. Falling precipitation accumulates in small gullies at these higher elevations and forms briskly moving streams. As surface water begins its flow, its path might not be well defined. In time, the moving water carves a narrow pathway into the sediment or rock called the **stream channel.** The channel widens and deepens as more water accumulates and cuts into Earth's surface. **Stream banks** hold the moving water within them.

When small streams erode away the rock or soil at the head of a stream, it is known as headward erosion. These streams move swiftly over rough terrain and often form waterfalls and rapids as they flow over steep inclines. Sometimes a stream erodes the high area separating two drainage basins, joining another stream. It then draws water away from the other stream in a process called stream capture, as shown in **Figure 11.**

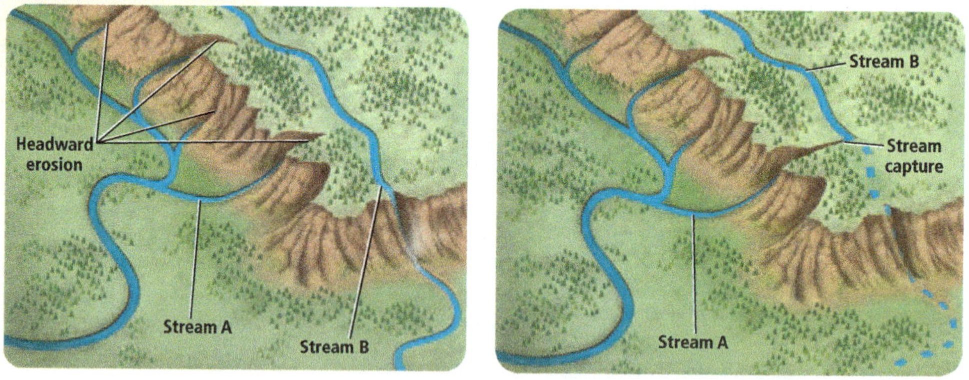

Figure 11 The headward erosion of Stream A cuts into Stream B and draws its water away into one stream.

Demonstration

Parts of a Stream Spread soil evenly over the bottom of a large baking pan. Prop up one side of the pan. Pour water from a cup or beaker onto the higher side of the pan. Have students note what happens to the water. After the water has moved toward the lower end of the pan, ask students to examine the source, channel, and banks.

ACROSS THE CURRICULUM

Math Give students topographic maps of two types of streams, and ask them to compare the stream gradients of each. The gradient is determined by the change in elevation, in meters, divided by the change in distance, in kilometers [(elevation A—elevation B) / (total distance between point A and B)]. This is easily remembered by the phrase "rise over run." AL

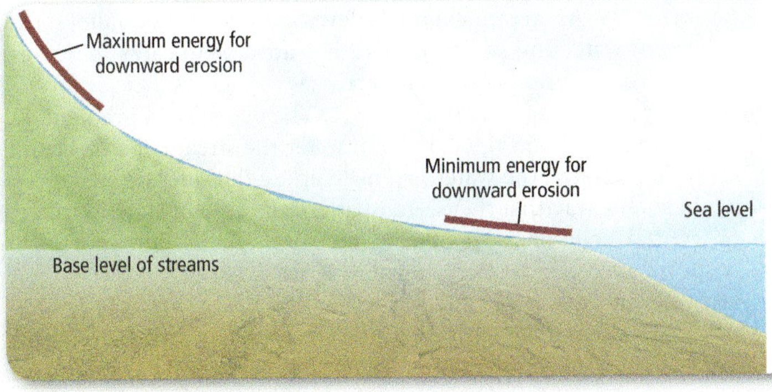

■ **Figure 12** The height of a stream above its base level determines how much downcutting energy the stream will have.

Formation of Stream Valleys

The driving force of a stream is the force of gravity on water. This means that the energy of a stream comes from the movement of water down a slope called a stream gradient. When the gradient of a stream is steep, water in the stream moves downhill rapidly, cutting steep valleys. The gradient of the stream depends on its **base level,** which is the elevation at which it enters another stream or body of water. The lowest base level possible for any stream is sea level, the point at which the stream enters the ocean, as shown in **Figure 12.**

Far from its base level, a stream actively erodes a path through the sediment or rock, and a V-shaped channel develops. V-shaped channels have steep sides and sometimes form canyons or gorges. The Yellowstone River in Wyoming flows through an impressive example of this type of narrow, deep gorge carved by a stream. **Figure 13** shows the classic V-shaped valley. As a stream approaches its base level, it has less energy for downward erosion. Instead, streams that are near their base level tend to erode at the sides of the stream channel, and over time result in broader valleys with gentle slopes, as shown in **Figure 13.**

FOLDABLES®
Incorporate information from this section into your Foldable.

■ **Figure 13** A V-shaped valley is formed by the downcutting of a stream. A wide, broad valley is a result of stream erosion over a long period of time.
Identify *which river is closer to its base level.*

Teacher Content Support

Aggradation When sediments are carried in greater amounts than a stream can hold, the sediments are deposited along the bottom of the stream's channel. This process is called aggradation. The contrasting process, in which the stream is capable of carrying more load than it currently is carrying and results in downcutting, is called degradation. Aggradated streams are likely to have shallow, broad stream channels that often are characterized by the deposition of long bars of gravel or sand that develop into braided streams.

Project

Grand Canyon Have students collect information about the Grand Canyon. Much information can be collected by writing to the National Park Service. As information is collected, put groups of students in charge of posting the information on a bulletin board. Add to the bulletin board a topographic map that shows the Grand Canyon area. COOP LEARN

■ **Caption Question Fig. 13** The river on the right is closer to its base level because it is close to the level at which it can enter another body of water.

FOLDABLES® Rubric

EARTH SCIENCE JOURNAL

The Grand Canyon Ask students to each write a story in the form of a camper's log about a camping trip to the Grand Canyon. Each story should be written in a notebook with additional material taped or pasted next to Earth science journal entries.

Rubric

DIFFERENTIATED INSTRUCTION

Struggling Learners Pack 2-m sections of aluminum gutter with soil, gravel, sand, clay, and other materials to simulate areas where a stream might form. Tip the gutter into a sink containing a large bucket. Use a hose to send a stream of water down the gutter. Have students note the shape of the stream channel that forms and areas where the stream changes direction.

Model

Oxbow Lake Development Have students use clay to make models showing the stages in the development of an oxbow lake. Stages should include the beginning meander, the more decisive meander, the cutoff, and the drying up of the oxbow lake. Use stages of different student models to assess student understanding. Letter each of the different stages and randomly place them on a demonstration table. Ask students to each list the stages of development in order. In a follow-up question, ask students why oxbow lakes are not always permanent. Students should recognize that once the oxbow lake is cut off, it loses its source of water. OL EL

Use an Analogy

Racecar Driving The way in which a racecar moves around the curves of a racetrack can be compared to the movement of water in a meander. The racecar cannot go as fast on the inside of a curve as it can on the outside of a curve. In addition, when a racecar passes another car on the outside of a curve, it has to go faster than it would if it were passing on the inside of a curve. Like a racecar on a track, water also moves with the least energy on the inside of the curves of a stream.

VOCABULARY
SCIENCE USAGE V. COMMON USAGE

Meander

Science usage: a bend or curve in a stream channel caused by moving water

Common usage: a winding path or course

Meanders As stream channels develop into broader valleys, the volume of water and sediment that they are able to carry increases. In addition, a stream's gradient decreases as it nears its base level, and the channel gets wider as a result. The decrease in gradient causes an increase in the volume of water the stream channel can carry. Sometimes, the water begins to erode the sides of the channel in such a way that the overall path of the stream starts to bend or wind. A bend or curve in a stream channel caused by moving water is called a **meander,** shown in **Figure 14.**

Water in the straight parts of a stream flows at different velocities, depending on its location in the channel. In a straight length of a stream, water in the center of the channel flows at the maximum velocity. Water along the bottom and sides of the channel flows more slowly because it experiences friction as it moves against the land.

In contrast, the water moving along the outside of a meander curve experiences the greatest velocity within the meander. The water that flows along this outside part of the curve continues to erode away the sides of the streambed, thus making the meander larger. Along the inside of the meander, the water moves more slowly and deposition is dominant forming point bars, as shown in **Figure 15.** These differences in the velocity within meanders cause the meanders to become more accentuated over time.

Oxbow lakes Meanders continue to develop and become larger and wider over time. After enough winding, however, it is common for a stream to cut off a meander and once again flow along a straighter path. The stream then deposits material along the adjoining meander and eventually blocks off its water supply, as shown in **Figure 14.** The blocked-off meander becomes an oxbow lake, which eventually dries up.

As a stream approaches a larger body of water or its endpoint, the ocean, the streambed's gradient flattens out and its channel becomes very wide. The area of the stream that leads into the ocean or another large body of water is called the mouth.

Figure 14 As the path of the stream bends and winds, it creates meanders and eventually oxbow lakes.

View an **animation of meander formation.**

Concepts In Motion

IN THE FIELD

Ancient Riverworks Artificial rivers have been discovered beneath the Sahara and in Turkey, Iran, Afghanistan, parts of Russia, and sections of China. These human-made rivers are called foggara in Arabic, qanat in Turkish, and quarez in Persian. No one knows exactly who made these ancient riverworks and aqueducts, but Iranian people still use similar structures today. Two or three "rivers" supply all the water necessary to support a village. Modern surveyors dig shafts and trenches through which groundwater runs. Scientists theorize the ancient aqueducts were built similarly. Some qanats are several thousand years old.

VISUALIZING Erosion and Deposition in a Meander

Figure 15 As the water travels down a meander, the area of maximum velocity changes. As shown in cross-section A, when the meander is straight, the maximum velocity is located near the center. When the meander curves, the maximum velocity shifts to the outside of the curve, as shown in cross-section B. As the meander travels around to cross-section C, the maximum velocity shifts again to the outside of the curve. Erosion occurs around curves in the meander in areas of high velocity. The high velocity of the water carries the sediment downstream and deposits it where the velocity decreases, on the inside of a curve. The area where the erosion occurs is called a cutbank and the area where the deposition occurs is called a point bar.

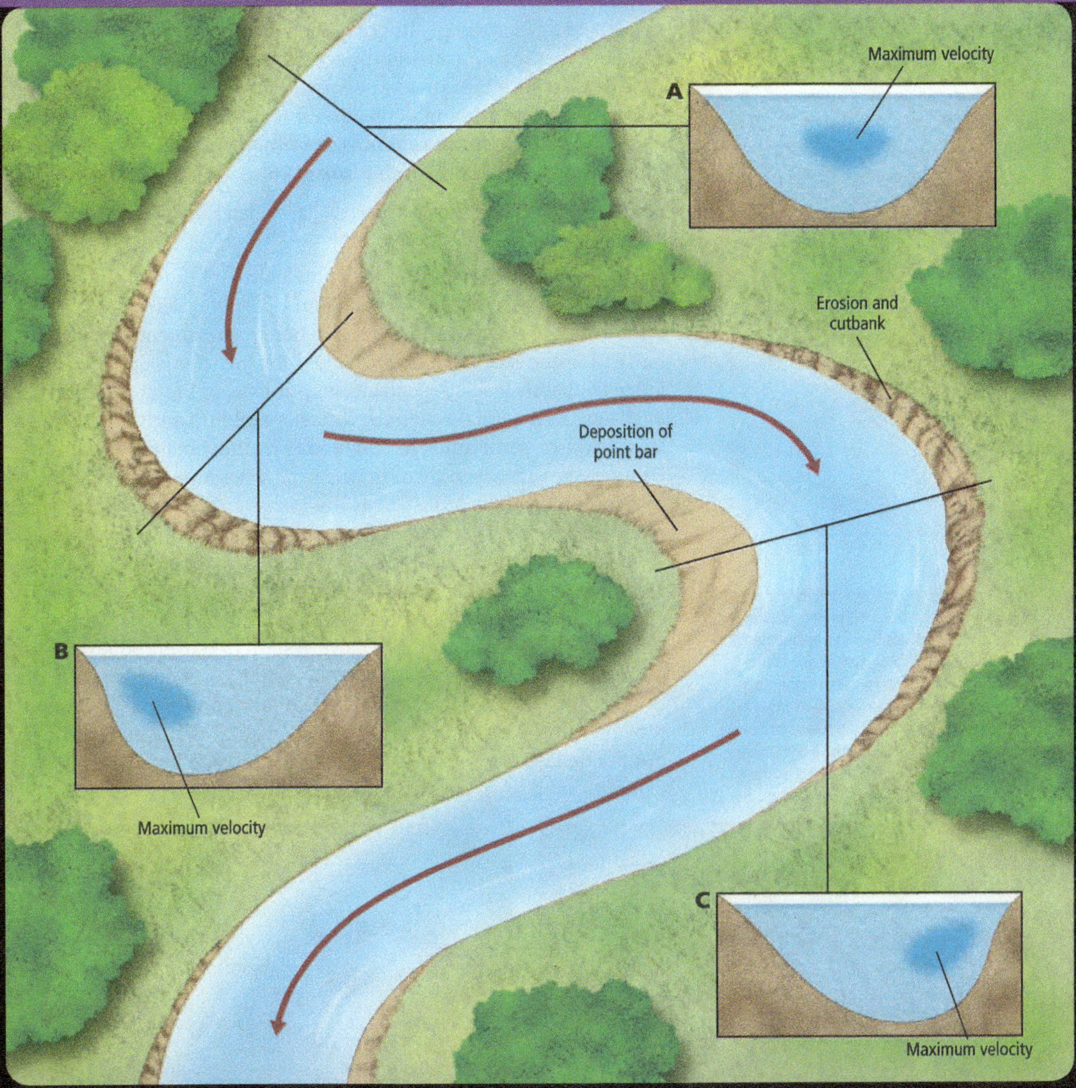

Concepts In Motion View an **animation of erosion and deposition.**

VISUALIZING

Purpose

Students will compare erosion and deposition along a meander.

Collaborative Learning

Meander It might be difficult for students to understand that water within a meander moves at different velocities. To help students understand the rate of water movement along a stream's curve, take the class outside to a large open field. Have students stand in a line and link elbows with the person next to them on both sides. Have a student at one end act as a pivot point and turn slowly. The farther away students are from the pivot point, the faster they will need to move to stay in a straight line.

This activity simulates the varying speeds at which water moves along a curve. Students near the pivot point will move most slowly and represent the water molecules moving on the inside of a stream's curve. Students farthest from the pivot point represent water molecules along the outside of a stream's curve.

Care must be taken with this activity so that enthusiastic students do not try to move the line too quickly. An alternative to having students form one line is to have shorter lines of five students each. This will reduce the overall speed and lessen the chance of students falling and possibly injuring themselves. **COOP LEARN**

ACROSS THE CURRICULUM

Language Arts Have students each write a poem centered on the theme of streams. The poems can concentrate on a stream's aesthetic, economic, or recreational aspects. Display students' poems around the classroom on "Stream Appreciation Day" and celebrate by having each student recite his or her poem. **OL**

Project

Stream Features Have students work in small groups to make diagrams of hypothetical islands with varied topographies. Have students indicate on their diagrams where they think streams, deltas, and alluvial fans would be found. If students wish to test their ideas, they can make models of their hypothetical islands and test them accordingly. COOP LEARN

Tie to Previous Knowledge

Topographic Maps Ask students to think about how an alluvial fan would appear on a topographic map and to describe the morphology in writing. Hand out topographic maps of an area in the American Southwest with long sloping alluvial fans. Ask students to draw a cross section of a predetermined line of section and to note whether the two cross sections are similar. OL

READING CHECK An alluvial fan is created when streams lose their sediment on a broad flat valley floor.

GeoLAB

iLab Station

The GeoLab at the end of the chapter can be used at this point in the lesson.

Figure 16 An alluvial fan is a fan-shaped depositional feature.

Deposition of Sediment

The velocity of a stream determines how much sediment it can transport. Rapidly flowing streams have the energy to transport sediment as large as gravel. When streams lose velocity, they lose some of the energy needed to transport sediment, and deposition of sediment occurs.

Alluvial fans A stream's velocity lessens and its sediment load is deposited when its gradient abruptly decreases. In dry regions such as the North American Southwest, mountain streams flow intermittently down steep, rocky slopes and then flatten out onto expansive dry lake beds. In areas such as these, a stream's gradient suddenly decreases, causing the stream to drop its sediment at the base of the mountain in a fan-shaped deposit called an alluvial fan. Alluvial fans are sloping depositional features formed at the bases of slopes and are composed mostly of sand and gravel. An example of an alluvial fan is shown in **Figure 16.**

READING CHECK **Describe** how an alluvial fan is formed.

Deltas Streams also lose velocity and some of their capacity to carry sediment when they join larger bodies of quiet water. The often triangular deposit that forms where a stream enters a large body of water is called a **delta,** named for the triangle-shaped Greek letter delta (Δ). Delta deposits usually consist of layers of silt and clay particles. As a delta develops, sediments build up and slow the stream water, sometimes even blocking its movement. Smaller distributary streams then form to carry the stream water through the developing delta. Deltas, such as the Mississippi River Delta, are normally areas where the stream flow changes direction frequently.

Over the course of thousands of years, the Mississippi River Delta has frequently changed. Today, any small change in the drainage channels can result in catastrophic flooding for local communities. To prevent floods, an extensive system of dams and levees is in place to protect people and economic activities. A consequence of flood control is the decrease in the regular deposition of sediment throughout the delta. In the absence of regular deposition throughout the delta, normal processes of coastal erosion have caused the delta to shrink over time, as shown in **Figure 17.**

Figure 17 The Mississippi River Delta formed from the deposition of river sediments. The area in the top left of both images is a marshland used for both recreation and business. Since 1973, waters upstream of the Mississippi River have been dammed, reducing the sediment flow. Over the course of 30 years, the area of the marshland decreased as the sediment input from upstream decreased.

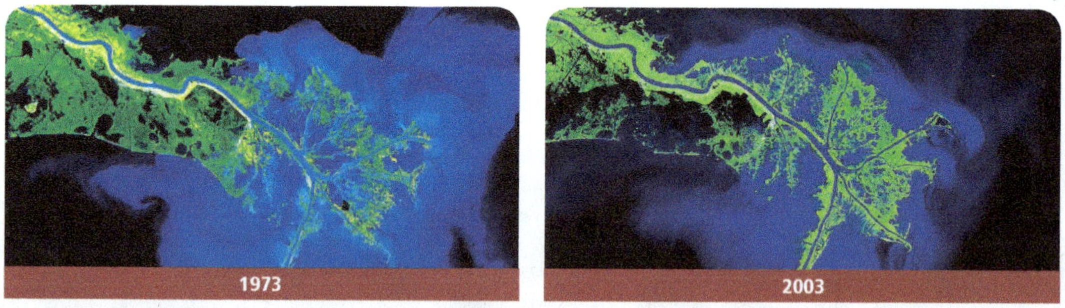

Demonstration

Alluvial Fan Formation In a leakproof, shallow container, make a mound of soil to represent a mountain. Slowly pour water on the top of the mound and let the water flow down one side of it. As the water flows down the mound, it will pick up sediment and carry it down the slope. As the water loses its velocity at the bottom of the hill, it will drop the sediment it has picked up along its journey in the shape of a fan. This is a model of the formation of an alluvial fan. You can also model the formation of a delta by allowing water-carrying sediment to enter a larger body of water and slow down. Both alluvial fans and deltas form readily on artificial models.

Rejuvenation

During the process of stream formation, downcutting can occur. Downcutting is the wearing away of the streambed and is a major erosional process that influences the stream until it reaches its base level. If the base level drops as a result of geologic processes, the stream undergoes rejuvenation.

Rejuvenation means *to make young again.* During **rejuvenation**, a stream actively resumes the process of downcutting toward its base level. This causes an increase in the stream's velocity and the stream's channel once again cuts downward into the existing meanders. Rejuvenation can cause deep-sided canyons to form. A well-known example of rejuvenation is the Grand Canyon, shown in **Figure 18.**

Millions of years ago, the Colorado River was near its base level, like much of the Mississippi River today. Then the land was uplifted compared to the level of the ocean, which caused the base level of the Colorado River to drop. This caused the process of rejuvenation, in which renewed erosion occurred as the river began cutting downward into the existing meanders. The result is the 1.6-km-deep canyons, which attract millions of visitors each year from all over the world. Rejuvenation is still occurring today in the Grand Canyon, as the Colorado River continues downcutting toward its base level.

Figure 18 Rejuvenation shaped the Grand Canyon when the base level of the Colorado River changed and the river began downcutting into existing meanders.

SECTION 2 REVIEW

Section Self-Check

Section Summary

- Water from precipitation gathers in gullies at a stream's headwaters.
- Stream water flows in channels confined by the stream's banks.
- Alluvial fans and deltas form when stream velocity decreases and sediment is deposited.
- Alluvial fans are fan-shaped and form where water flows down steep slopes onto flat plains.
- Deltas are often triangular and form when streams enter wide, relatively quiet bodies of water.

Understand Main Ideas

1. MAINIDEA **Describe** how a V-shaped valley is formed.
2. **Identify** four changes that a stream undergoes before it reaches the ocean.
3. **Compare** the velocity on the inside of a meander curve with that on the outside of the curve.

Think Scientifically

4. **Analyze** how the type of bedrock over which a stream flows affects the time it takes for the stream to reach its base level.
5. **Infer** how you can tell that rejuvenation has modified the landscape.

MATH IN Earth Science

6. Create a line graph that plots the direction of change in a hypothetical stream's rate of flow at the stream's headwaters, at midstream, and at its mouth.

3 Assess

Check for Understanding

Measure Ask students in small groups to develop methods to measure the speed at which water in a stream is flowing. Take the groups to a nearby stream to test their plans. Upon returning to the classroom, have the various groups critique their own setups and make suggestions for improvements. COOP LEARN

Reteach

Communication Assign small groups of students the task of creating a game with a game board that represents the water flowing on Earth's surface. Students must use facts from the chapter that will help them review terms and concepts. Once the games are completed, they can be rotated from one group to another for students to play and test the games.

Assessment

Skill Ask each student to develop a question about the material in this section and to write the question on an index card with the answer on the reverse side. Collect the cards and use them in a whole-class review.

SECTION 2 REVIEW

1. A V-shaped valley forms on a steep slope where it is downcut by a river over a long period of time.
2. As a stream travels towards the ocean, its slope decreases, it becomes wider, its volume increases, and it becomes less turbulent.
3. On the inside of a meander bend, the velocity is at its minimum, while on the outside, the velocity is at its maximum.
4. If the bedrock is hard and does not dissolve easily, it will take longer to reach base level.
5. Rejuvenation can be seen in a meandering stream that has carved into a canyon.
6. Answers will vary, but the rate will be fastest at the headwaters and slowest at the mouth.

SECTION 3

1 Focus

Focus Transparency

Before presenting the lesson, project **Section Focus Transparency 25** and have students answer the accompanying questions. EL

MAINIDEA

Lakes Divide students into groups of four. Ask them to discuss what might change the level of water in a lake. evaporation, addition of water from nearby streams, addition of sediment, growth of vegetation

2 Teach

Teacher Content Support

Lakes All lakes consist of water and the materials contained within the water. If a lake bottom is porous, water can leach out of the depression. The depression will contain water only if the water table remains stable, during times of heavy rain, or when there is excessive runoff from spring thaws. A depression that receives more water than it loses from leaching, evaporation, or use by people will remain as a lake for a long period of time. From a geological perspective, lakes are temporary water-holding areas.

Activity

Uses of Lakes Divide the class into teams of four or five people. Ask each team to write down as many uses of lakes as it can. After a given amount of time, ask each team to report one use on its list. Give a point for each use that was not listed by the other teams. This will encourage students to come up with as many different uses as possible and get them ready to learn about lakes. BL OL

SECTION 3

Lakes and Freshwater Wetlands

MAINIDEA As the amount of water changes and the amount of sediments increases, lakes can be transformed into wetlands and eventually into dry land.

Essential Questions

- How do freshwater lakes and wetlands form?
- How is the process of eutrophication described?
- What are the effects of human activity on lake development?

Review Vocabulary

kettle: a depression resulting from the melting of an ice block left behind by a glacier

New Vocabulary

lake
eutrophication
wetland

EARTH SCIENCE 4 YOU Have you ever felt the bottom of a lake with your feet? It was probably soft and squishy from deposits of fine sediments. Lakes and ponds receive materials that are carried by rivers from upland areas. Over time, accumulation of these sediments changes the characteristics of the lake.

Origins of Lakes

Natural **lakes**, bodies of water surrounded by land, form in different ways in surface depressions and in low areas. As you learned in Section 2, oxbow lakes form when streams cut off meanders and leave isolated channels of water. Lakes also form when stream flow becomes blocked by sediment from landslides or other sources. Still other lakes have glacial origins. The basins of these lakes formed when glaciers gouged out the land during the ice ages. Many of the lakes in the northernmost parts of Europe and North America are in recently glaciated areas. Glacial moraines originally dammed some of these depressions and restricted the outward flow of water. The lakes that formed as a result are known as moraine-dammed lakes. In another process, cirques carved high in the mountains by valley glaciers filled with water to form cirque lakes. Other lakes formed as blocks of ice left on the outwash plain ahead of melting glaciers eventually melted, leaving depressions called kettles. When these depressions filled with water, they formed kettle lakes such as those shown in **Figure 19.**

Figure 19 Lakes such as these in Minnesota were formed from blocks of ice that melted after glaciers retreated.

Phil Schermeister/CORBIS

ACROSS THE CURRICULUM

Biology Water on Earth's surface consists of living and nonliving components as well as water molecules. A lake is an ecosystem that contains organic and inorganic components, such as dissolved oxygen. Living organisms in the water use oxygen from and add waste products to the water. In addition, the decay of dead organisms depletes oxygen supplies in the water. The amount of dissolved oxygen (DO) and the biochemical oxygen demand (BOD) are two of the major factors that determine the quality of water for living things.

Lakes Undergo Change

Water from precipitation, runoff, and underground sources can maintain a lake's water supply. Some lakes contain water only during times of heavy rain or excessive runoff from spring thaws. A depression that receives more water than it loses to evaporation or use by humans will exist as a lake for a long period of time. However, most lakes are temporary water-holding areas; over hundreds of thousands of years, lakes usually fill in with sediment and become part of a new landscape.

Eutrophication Through the process of photosynthesis, plants such as green algae add oxygen to lake water. Animals that live in a lake need oxygen in the water. Oxygen is also consumed during the decay process of animal waste, and during the decay process that occurs after plants and animals living in the body of water die. Scientists use the amount of dissolved oxygen present in a body of water to assess the overall quality of the water. Dissolved oxygen is one quality a body of water must have to support life.

The process by which the surrounding watershed enriches bodies of water with nutrients that stimulate excessive plant growth is called **eutrophication.** Although eutrophication is a natural process, it can be sped up with the addition of nutrients, such as fertilizers, that contain nitrogen and phosphorus. Other major sources of nutrients that concentrate in lakes are animal wastes and phosphate detergents.

When eutrophication occurs, the animal and plant communities in the lake can change rapidly. Algae growing at the surface of the water can suddenly multiply very quickly. The excessive algae growth in a lake or pond appears as a green blanket, as shown in **Figure 20.** Other organisms that eat the algae can multiply in numbers as well. In addition, the population of algae on the surface can block sunlight from penetrating to the bottom of the lake, causing sunlight-dependent plants and other organisms below the surface to die. The resulting overpopulation and, later, the decay of a large number of plants and animals depletes the water's dissolved oxygen supply. Fish and other sensitive organisms might die as a result of the lack of dissolved oxygen in the water. In some cases, the algae can also release toxins into the water that are harmful to the other organisms.

☑ READING CHECK **Identify** the effects of eutrophication on the aquatic animals in an affected lake system.

Figure 20 Eutrophication is a natural process that can be accelerated with the addition of nitrogen and phosphorus to a body of water. Once the process begins, it can cause rapid changes in the plant and animal communities in the affected body of water.

Teacher Content Support

Desert Lakes Lakes can form in desert regions following rare but heavy rainstorms. Such lakes are usually temporary, and when they dry up, they leave a dry lake bed on the desert called a playa. *Playa* is a Spanish word meaning *beach.* Repeated filling and drying of these lakes lead to the significant buildup of evaporite minerals that can be used for industrial purposes.

Environmental Connection

Water Pollution While eutrophication sometimes occurs naturally, it can be caused or increased by the addition of nutrients to a water supply via water pollution. The main nutrients involved in eutrophication are nitrogen and phosphorus. Both of these nutrients are the main ingredients in many chemical fertilizers. Ask students: What are some ways that humans can reduce the amount of these nutrients that reach bodies of water? Answers will vary, but could include: reduce the amount of fertilizers used, use different types of fertilizers, divert runoff that contains these pollutants.

Enrichment

Eutrophication Allow students who wish to work in small groups to research a nearby lake area for signs of eutrophication. Encourage students to visit the lake, under the supervision of an adult, and to make a video or take photographs and a sampling of the vegetation, if permissible. AL COOP LEARN

☑ READING CHECK Eutrophication increases the amount of algae. This results in overpopulation. Fish could die because of lack of oxygen.

EARTH SCIENCE JOURNAL

Local Lakes Ask students to write the name of a lake they know about or have visited in their Earth science journals. Ask students to research how the lake formed and give some details, such as location, size, use by humans, and lifeforms that it contains. Tell students that during this section of the chapter, they should be collecting information about this lake for a future assignment. OL

CAREERS IN EARTH SCIENCE

WebQuest

☑ READING CHECK A bog receives water from precipitation.

MiniLAB

iLab Station | Rubric

Purpose Students will demonstrate how different types of material found on Earth's surface determine where a lake can form.

Process Skills model, recognize cause and effect, communicate, interpret data, observe and infer, measure in SI

Safety Precaution Approve lab safety forms before work begins.

Teaching Strategy Have students work in pairs to develop ideas, note observations, and clean up.

Expected Results Students will observe that gravel allows water to penetrate through it most easily and clay least easily.

Analysis

1. The water should quickly flow through the gravel. The water will flow through the sand less quickly. The water will remain on top of the clay the longest.
2. When water collects in an area where the spaces between the particles of surface material are small, the water is more likely to form a lake.
3. surface materials that usually do not allow water to easily pass through them, such as clay

Assessment

Knowledge Ask students to describe the characteristics of a material that does not easily allow water to pass through it. The material is likely composed of small particles with few spaces between them.

LabManager™
Customize any lab with the LabManager™ CD-ROM.

CAREERS IN EARTH SCIENCE

Geochemist Technician Some geochemist technicians take core samples from lakes to analyze the pollutants in lake sediments.

WebQuest

Freshwater wetlands A **wetland** is any land area that is covered with water for a part of the year. Wetlands include environments commonly known as bogs, marshes, and swamps. They have certain soil types and support specific plant species. Their soil types depend on the degree of water saturation.

Bogs Bogs are not stream-fed but instead receive their water from precipitation. The waterlogged soil tends to be rich in *Sphagnum,* also called peat moss. The breakdown of peat moss produces acids, thereby contributing to the soil's acidity. The waterlogged, acidic soil supports unusual plant species, including insect-eating pitcher plants such as sundew and Venus flytrap.

☑ READING CHECK **Identify** how a bog receives water.

Marshes Freshwater marshes frequently form along the mouths of streams and in areas with extensive deltas. The constant supply of water and nutrients allows for the lush growth of marsh grasses. The shallow roots of the grasses anchor deposits of silt and mud on the delta, thereby slowing the water and expanding the marsh area. Grasses, reeds, sedges, and rushes, along with abundant wildlife, are common in marsh areas.

Swamps Swamps are low-lying areas often located near streams. Swamps can develop from marshes that have filled in sufficiently to support the growth of shrubs and trees. As these larger plants grow and begin to shade the marsh plants, the marsh plants die. Swamps that existed about 300 mya during the Carboniferous Period developed into present-day coal reserves that are common in Pennsylvania and many other locations in the United States and around the world.

MiniLAB

iLab Station

Model Lake Formation

How do surface materials determine where lakes form? Lakes form when depressions or low areas fill with water. Different Earth materials allow lakes to form in different places.

Procedure

1. Read and complete the lab safety form.
2. Use three **clear plastic containers.** Half fill each one with Earth materials: **clay, sand,** and **gravel.**
3. Slightly compress the material in each container. Make a shallow depression in each surface.
4. Slowly pour 500 mL of **water** into each of the depressions.

Analysis

1. **Describe** what happened to the 500 mL of water that was added to each container.
2. **Compare** this activity to what happens on Earth's surface when a lake forms.
3. **Infer** in which Earth materials lakes most commonly form.

DIFFERENTIATED INSTRUCTION

Advanced Learners Ask students to each make a list of the detergents used both at home and in the school. Along with the name of each detergent, have students list the detergent's ingredients to determine which, if any, contain phosphate or other materials that may be harmful to the environment. Have students decide on an approved course of action if they do indeed find that cleaning materials that are harmful to the environment are being used.

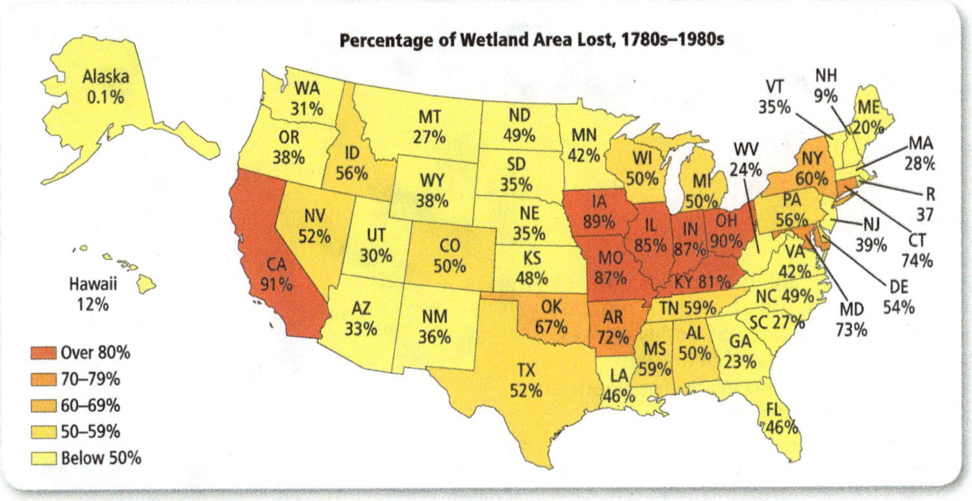

Figure 21 The area of wetlands in the United States was drastically reduced until the 1980s. Since then, efforts have been made to preserve wetlands.

Wetlands and water quality Wetlands play a valuable role in improving water quality. They serve as a filtering system that traps pollutants, sediments, and pathogenic bacteria contained in water sources. Wetlands also provide vital habitats for migratory waterbirds and homes for an abundance of other wildlife. In the past, it was common for wetland areas to be filled in to create more land on which to build. Government data reveal that from the late 1700s to the mid-1980s, the continental United States lost 50 percent of its wetlands, as shown in **Figure 21.** By 1985, it was estimated that 50 percent of the wetlands in Europe were drained. Now, however, the restoration and preservation of existing wetland areas, as well as the creation of new wetlands as replacements for old wetlands that were drained or filled, has become a global concern.

SECTION 3 REVIEW

Section Self-Check

Section Summary

- Lakes form in a variety of ways when depressions on land fill with water.
- Eutrophication is a natural nutrient-enrichment process that can be accelerated when nutrients from fertilizers, detergents, or sewage are added.
- Wetlands are low-lying areas that are periodically saturated with water and support specific plant species.

Understand Main Ideas

1. MAINIDEA **Explain** the transformation process that a lake might undergo as it changes to dry land.
2. **Describe** the conditions necessary for the formation of a natural lake.
3. **Identify** human activities that might affect the process of eutrophication in a lake near you.

Think Critically

4. **Organize** a data table to compare various types of lakes and their origins.
5. **Analyze** a situation in which protection of wetlands might conflict with human plans for land use.

WRITING IN ▶ Earth Science

6. Write an essay explaining the role wetlands play in improving water quality.

Environmental Connection

Wetlands In 2006, a study by the Fish and Wildlife Service found that wetlands have been restored and formed at a rate of 12,900 ha (32,000 acres) per year.

3 Assess

Check for Understanding

Model Have students work in small groups to develop dioramas that depict the stages of change that a lake might undergo. Each diorama will require the development of an explanatory script that is recorded on audiotape. Once the dioramas are set up around the room, they can serve as learning and review for students. COOP LEARN

Reteach

Communication Have students each develop a list of vocabulary words to define in their Earth science journals. Review the terms as a class and have students add other students' words to their lists.

Assessment

Knowledge Show the class slides or photographs that represent particular stages of development or eutrophication of lakes. Ask students to relate each photograph to concepts covered in this section. Have students develop one question and answer for each photograph.

SECTION 3 REVIEW

1. Answers will vary. A lake might hold less and less water as it fills with sediment. Eventually, it might not hold any water at all and would be dry land.
2. In order for a lake to form, there must be a depression where water can gather, a source of water, and impermeable surface material.
3. Humans cause eutrophication when septic tanks leak untreated sewage, industries release toxins, and farms use fertilizers which get into the water supply.
4. Oxbow lakes are formed from cutoff meanders. Landslide lakes are formed from landslides, which block streams. Glacial lakes are gouged by glaciers, dammed by moraines, or left as cirques or kettles.
5. Developers might want to use a plot of land to build houses, but in doing so would lose the wetland.
6. Answers will vary, but should include information about how wetlands filter the water.

Rubric

Earth Science & SOCIETY

Purpose

Students will learn about the importance of clean drinking water on Earth.

Teacher Content Support

Clean Drinking Water Lack of sanitary conditions and lack of clean drinking water are two of the many issues Earth is facing as our population continues to increase. Poor hygiene and lack of proper sanitation confine nearly half of the world's population (nearly 3 billion people) to a life that is comparable to life in the Dark Ages. They do not have access to even the most basic of latrines and cannot practice basic hygiene. In some developing countries, people in rural areas may walk up to an average of 6 km to get fresh water, and there is no guarantee that the water will be safe for drinking. On average, an African person uses 20 L of water per day. A European uses 150 L per day, while someone in the United States uses 400 L.

Teaching Strategy

Safe Drinking Water Ask: How many people do you think do not have access to clean drinking water? about 1 billion people worldwide Then ask: In what ways do you think this can be a problem? This is a problem for sanitation and waterborne illness, particularly in young children. Finally ask: What are some problems that can occur if you do not have access to clean drinking water? Answers will vary, but could include sickness and disputes over existing water supplies.

Earth Science & SOCIETY

The World of Water

Humans have basic physiological needs. These include the need to breathe, to eat, to regulate body temperature, to dispose of bodily wastes, to sleep, and to have access to clean water. Humans need clean water to drink, for cleaning, cooking, and waste disposal.

A global problem Almost every continent has areas that lack safe drinking water. Rural areas of developing countries and overpopulated urban areas often have inadequate supplies of safe drinking water. Even though adequate supplies of this natural resource may exist globally, it is not distributed evenly. In addition, naturally occurring contaminants and pollution from human impact can make a water supply unhealthy.

Safe water The World Health Organization (WHO) defines safe drinking water as water from a source that is less than 1 km away from where it is used; that at least 20 L of water per member of the household per day can be obtained reliably; and that meets the national standards for microbial and contaminant levels.

Health concerns In developing countries, children are at the greatest risk for water-related diseases. Worldwide, almost 4000 children under the age of five die each day from water-related diseases. The most common health concerns from contaminated water are diarrhea and intestinal worms.

Diarrhea is a common condition caused by bacteria often found in unsafe drinking water. Without proper treatment, diarrhea can lead to severe dehydration and death, especially in children. In developed countries, children suffering from diarrhea often receive the necessary treatment. However, in developing countries, diarrhea accounts for the death of 1.5 million children each year.

Contaminated water can be a problem in developed countries as well as developing countries. This beach is closed because of unsafe water.

Another danger from contaminated water, especially for children, is intestinal parasites. Parasites that live in the intestines of the host, humans in this case, can cause malnutrition, anemia, and other illnesses.

A global solution The inability to adequately supply this basic human need has been acknowledged by the United Nations as one of the greatest failures of the twentieth century. The United Nations has created an international task force to help fund the creation of sanitation systems and water purifiers. In the future, with effort and global cooperation, every human being might have access to safe drinking water and proper sanitation.

WRITING IN ▸ Earth Science

Brochure March 22 is World Water Day. Create a brochure explaining the need for such an event and why more people should participate.

WebQuest

WRITING IN ▸ Earth Science

Rubric

Brochure Many people in industrialized nations don't realize the seriousness of the limited water availability on Earth. Having such a day will increase awareness and influence people to conserve their water.

WebQuest

GeoLAB

iLab Station

Predict the Velocity of a Stream

Background: Water in streams flows from areas of high elevation to areas of low elevation. Stream flow is measured by recording the water's velocity. The velocity varies from one stream to another and also in different areas of the same stream. Many components of the stream affect the velocity, including sediment load, slope, and rainfall.

Question *How does slope affect velocity?*

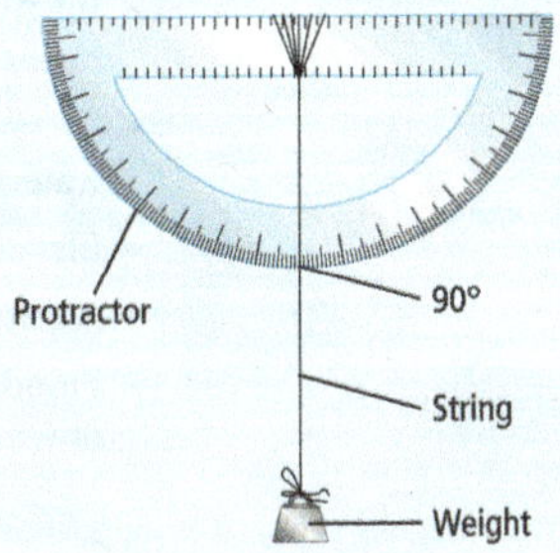

Materials

1-m length of vinyl gutter pipe
ring stand and clamp
water source with hose
protractor with plumb bob
sink or container to catch water
stopwatch
grease pencil
meterstick
paper
three-hole punch

Safety Precautions

Procedure

1. Read and complete the lab safety form.
2. Work in groups of three to four.
3. Use a three-hole punch to make 10 to 15 paper circles to be used as floating markers.
4. Use the illustration as a guide to set up the protractor with the plumb bob.
5. Use the grease pencil to mark two lines across the inside of the gutter pipe at a distance of 40 cm apart.
6. Use the ring stand and clamp to hold the gutter pipe at an angle of 10°. Place the end of the pipe in a sink or basin to collect the discharged flow of water.
7. Attach a long hose to a water faucet in the sink.
8. Keep the hose in the sink until you are ready to use it. Turn on the water and adjust the flow until the water moves quickly enough to provide a steady flow.
9. Bend the hose to block the water flow until the hose is positioned at least 5 cm above the top line marked on the pipe. Allow the water to flow at the same rate for all slope angles.
10. Drop a floating marker approximately 4 cm above the top line on the pipe into the flowing water.
11. Measure the time it takes for the floating marker to move from the top line to the bottom line. Record the time in your science journal.
12. Repeat Step 9 two more times.
13. Repeat Steps 9 and 10, but change the slope to 20°, 30°, and then 40°.
14. Make a line graph of the average velocity.

Analyze and Conclude

1. **Interpret Data** What is the relationship between the velocity and the angle of the slope?
2. **Apply** Describe one reason that a stream's slope might change.
3. **Infer** Where would you expect to find streams with the highest velocity?
4. **Predict** Using your graph, predict the velocity for a 35° slope.

INQUIRY EXTENSION

Design Your Own As discussed in the chapter, the texture of the streambed can affect the rate of stream flow. Design an experiment to test this variable.

GeoLAB

iLab Station Rubric

Preparation

Time Allotment 45 min

Process Skills observe and infer, interpret data, make a model, analyze data, interpret scientific illustrations, communicate, recognize cause and effect, compare and contrast

Safety Precautions Approve lab safety forms before work begins. Make sure students wear safety goggles and aprons during this lab.

Preparation of Materials This investigation works best with students working in groups of four. One student can be in charge of maintaining water flow. One student can drop the floating marker. A third student can be in charge of water drainage, and the fourth can time the rate of flow. It might be helpful to have student groups make their protractors with plumb bobs ahead of time.

Procedure

- Caution students to make sure the clamps are securely attached to the gutter pipe and the hose is securely attached to the faucet.
- Instruct students to slowly turn on the water to avoid spurts and excessive water spills.
- Caution students to watch for twists in the water hose that could temporarily impede water flow.

Analyze and Conclude

1. There is a direct relationship between the rate of flow and the angle of the slope. Velocity increases as the angle of the slope increases.
2. As a stream erodes its base, the stream's slope is reduced.
3. at higher elevations
4. The rate of water flow should be between velocities for 30° and 40°.

INQUIRY EXTENSION

Design Your Own Student experiments should show adequate assessment of texture as a variable.

MAINIDEAS **Summary statements can be used by students to review the major concepts of the chapter.**

Students can review with these online resources.

Vocabulary Practice

- Vocabulary eGames
- Vocabulary eFlashcards
- Vocabulary PuzzleMaker

Section Self-Check

Chapter Self-Check

Online Test Practice

Use the Exam*View*® *Assessment Suite* CD-ROM to:

- create multiple versions of tests
- create modified tests with one mouse click
- edit existing questions and add your own questions
- build tests aligned with state standards using built-in state curriculum tags
- change English tests to Spanish with one mouse click
- track students' progress using the Teacher Management System

CHAPTER 9 STUDY GUIDE

BIGIDEA **Surface water moves materials produced by weathering and shapes the surface of Earth.**

Vocabulary Practice

SECTION 1 Surface Water Movement

VOCABULARY
- runoff
- watershed
- divide
- suspension
- bed load
- discharge
- flood
- floodplain

MAINIDEA **Running water is an agent of erosion, carrying sediments in streams and rivers and depositing them downstream.**

- Infiltration of water into the ground depends on the number of open pores.
- All the land area that drains into a stream system is the system's watershed.
- Elevated land areas called divides separate one watershed from another.
- A stream's load is the material the stream carries.
- Flooding occurs in small, localized areas as upstream floods or in large downstream floods.

SECTION 2 Stream Development

VOCABULARY
- stream channel
- stream bank
- base level
- meander
- delta
- rejuvenation

MAINIDEA **Streams erode paths through sediment and rock, forming V-shaped stream valleys.**

- Water from precipitation gathers in gullies at a stream's headwaters.
- Stream water flows in channels confined by the stream's banks.
- Alluvial fans and deltas form when stream velocity decreases and sediment is deposited.
- Alluvial fans are fan-shaped and form where water flows down steep slopes onto flat plains.
- Deltas are triangular and form when streams enter wide, relatively quiet bodies of water.

SECTION 3 Lakes and Freshwater Wetlands

VOCABULARY
- lake
- eutrophication
- wetland

MAINIDEA **As the amount of water changes and the amount of sediments increases, lakes can be transformed into wetlands and eventually into dry land.**

- Lakes form in a variety of ways when depressions on land fill with water.
- Eutrophication is a natural nutrient-enrichment process that can be accelerated when nutrients from fertilizers, detergents, or sewage are added.
- Wetlands are low-lying areas that are periodically saturated with water and support specific plant species.

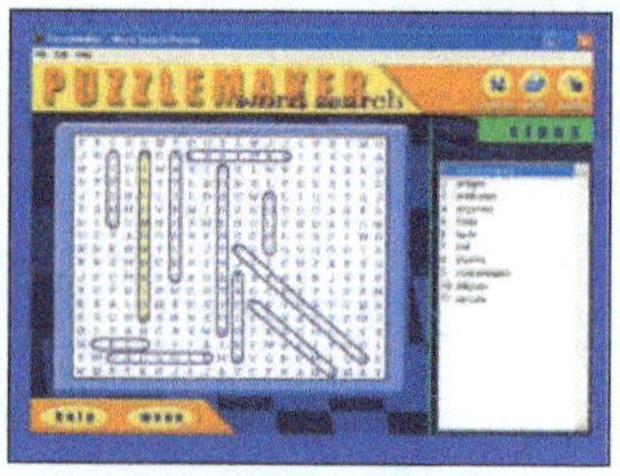

Vocabulary Practice

For additional practice with vocabulary, have students access the Vocabulary PuzzleMaker.

CHAPTER 9 ASSESSMENT

Chapter Self-Check

VOCABULARY REVIEW

Choose the vocabulary term from the Study Guide that best describes each phrase.

1. influenced by vegetation, precipitation, soil composition, and slope
2. the land area whose water drains into a stream
3. sediments that are transported by streams but are too large to be held up in suspension
4. the measure of the volume of stream water that flows over a particular location within a given period of time
5. the triangular deposit of sediment that forms where a stream enters a large body of water such as a lake or ocean
6. the narrow channel carved over time by the water of a stream into sediment or rock layers
7. the process by which a stream resumes downcutting toward its base level

The sentences below include terms that have been used incorrectly. Make the sentences true by replacing each italicized word with a vocabulary term from the Study Guide.

8. A depression that receives more water than is removed will exist as a *stream* for a long period of time.
9. *Enrichment* is the process by which lakes become rich in nutrients, resulting in a change in the kinds of organisms in the lake.
10. Marshes, swamps, and bogs are all types of *inundated* areas.
11. *Solution* is the method of transport for all particles small enough to be held up by the turbulence of a stream's moving water.
12. The area that extends from a stream's bank and is covered by excess water during times of flooding is known as a *wetland.*

Yann Arthus-Bertrand/CORBIS

UNDERSTAND KEY CONCEPTS

13. Which scenario most likely formed most large lakes in North America and Europe?
    - **A.** Lakes formed from stream meanders that were cut off.
    - **B.** Landslides blocked the flow of streams, creating lakes.
    - **C.** Glacial activity on continental masses scoured the landscape and left depressions behind.
    - **D.** Reservoirs were created for the purpose of storing water for communities.
14. What is the driving force of a stream?
    - **A.** velocity
    - **B.** gravity
    - **C.** discharge
    - **D.** downcutting

Use the figure below to answer Question 15.

15. Which part of a river is shown?
    - **A.** the headwater
    - **B.** the main channel
    - **C.** the streambed
    - **D.** the mouth
16. When does downcutting of the streambed stop during stream formation?
    - **A.** when the water reaches a porous layer
    - **B.** when the water reaches a base level
    - **C.** when the water reaches a new level
    - **D.** when the water reaches a impermeable layer

CHAPTER 9 ASSESSMENT

VOCABULARY REVIEW

1. runoff
2. watershed
3. bed load
4. discharge
5. delta
6. stream channel
7. rejuvenation
8. lake
9. eutrophication
10. wetland
11. suspension
12. floodplain

UNDERSTAND KEY CONCEPTS

13. C
14. B
15. D
16. B

17. A
18. B
19. B
20. B
21. C
22. A
23. D
24. A

CONSTRUCTED RESPONSE

25. A river delta is a depositional feature that forms when a river ends in a large lake or the ocean. As it reaches the lake or ocean, the water slows down and loses its carrying capacity, thus depositing fine sediments. Alluvial fans are deposited in a similar manner. They form on land at the base of mountains when a river flows out of a narrow canyon and into a valley. They consist mostly of sand and gravel.
26. on the outside of the meander curve
27. on the inside of the meander curve where the point bar forms
28. on the outside of the meander curve where the velocity is highest
29. 22,500 m^3/s

ASSESSMENT

Use the figure below to answer Question 17.

17. How did this terrain feature form?
 A. It formed by a meander of a stream that has been cut off.
 B. It formed from a glacier that scoured out the land.
 C. It is the result of a flood.
 D. It is the result of eutrophication.

18. What type of streams form V-shaped valleys?
 A. streams that carry a lot of sediment
 B. streams that are far from ultimate base level
 C. streams that meander
 D. streams that carry no bed load

19. Which is not a way in which lakes are typically formed?
 A. from cutoff meanders of streams
 B. from asteroid craters
 C. from landslides that block rivers
 D. from glacial carving

20. Which substance plays a major role in the eutrophication process?
 A. iron
 B. phosphorus
 C. ozone
 D. salt

21. Which factors determine the discharge of a stream?
 A. width, length, depth
 B. width, length, velocity
 C. width, depth, velocity
 D. length, depth, runoff

22. Which process would result in rejuvenation of a stream?
 A. lifting of existing base level
 B. lowering of existing base level
 C. lowering of the land
 D. deposition on the stream banks

23. Which characteristic of the soil in a depression aids in the formation of a lake?
 A. high content of organic material
 B. high content of mostly of inorganic material
 C. gravelly soil
 D. relatively nonporous, clay-rich layer of soil

24. If a stream is carrying sand, large boulders, clay, and small pebbles, which type of particle is deposited last as the stream begins to slow down?
 A. clay
 B. sand
 C. large boulders
 D. small pebbles

CONSTRUCTED RESPONSE

25. **Compare and contrast** the formation of a river delta to an alluvial fan.

Use the following aerial view of a stream to answer Questions 26 to 28.

26. **Identify** the location at which the stream has the greatest velocity.
27. **Identify** the location at which deposition most actively occurs.
28. **Identify** the location at which erosion most actively occurs.
29. **Calculate** the discharge of a stream that has a velocity of 300 m/s and is 25 m wide and 3 m deep.

THINK CRITICALLY

30. The water in a stream and in the atmosphere are both part of the water cycle. Water vapor contained in clouds can condense and precipitate onto the ground. Some of the water from precipitation can make its way to streams via channels on sloping ground. The water in streams can evaporate back into the atmosphere and form clouds.

Chapter Self-Check

THINK CRITICALLY

30. **Analyze** how water in a stream is related to water in the atmosphere.

Use the figure below to answer Question 31.

31. **Interpret** the features in this pond in terms of the processes that might have formed them.
32. **Hypothesize** how an increase in a river's turbulence could decrease its bed load.
33. **Analyze** how specific soil characteristics determine how much water from precipitation infiltrates or runs off.
34. **Infer** why upstream tributaries often have relatively small yet turbulent flow, whereas downstream portions usually have larger but smooth-flowing discharges.
35. **Discuss** which areas of a stream are most likely to contain fertile soil.

CONCEPT MAPPING

36. Construct a concept map of the parts of the water cycle and illustrate the relationships among them.

CHALLENGE QUESTION

37. Recommend measures that a town whose wastewater runs into a large lake should take to ensure the long-term quality of its lake water.

WRITING IN ▶ Earth Science

38. Write a newspaper article to explain recent flood destruction in your town. In the article, explain how erosion and transportation of sediment might be different during a flood.

DBQ Document–Based Questions

Data obtained from: Varis, O. 2005. Are floods growing? *Ambio* 34 (August): 478–480.

Dramatic changes in river flow have taken place with water flow to the sea in China's Huang He River, spanning a period of almost 30 years.

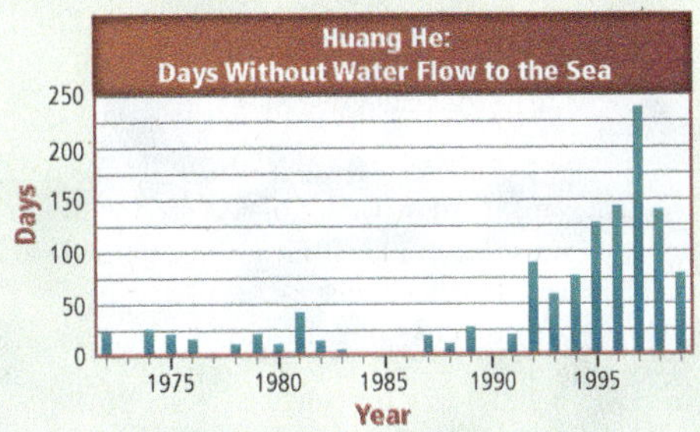

39. Based on the graph, beginning in approximately what year were the first signs evident of a pattern of progressively decreasing river flow?
40. In the worst year of records presented, approximately how many days was there a water flow from the Huang He River to the sea?
41. What kinds of conditions might cause such dramatic reductions in water flow?

CUMULATIVE REVIEW

42. Describe a topographic map and its uses. **(Chapter 2)**
43. Which of the following is NOT a mineral: quartz, ice, coal, or native gold? **(Chapter 4)**
44. Explain how water affects the process of mass movement. **(Chapter 8)**

DBQ Document-Based Questions

Data obtained from: Varis, O. 2005. Are floods growing? *Ambio* 34 (August): 478–480.

39. from 1992 on (prior to that river flow was variable but stable)
40. The year is 1997. There was water flow during approximately 122 days.
41. Answers will vary but should include: 1) increased uptake of water for irrigation (agriculture), municipal water use, and others (industrial and hydroelectric); 2) extended drought (1992–on); 3) deforestation and degradation in headwaters.

CUMULATIVE REVIEW

42. Topographic maps are detailed maps showing the hills and valleys of an area. Topographic maps show changes in elevation of Earth's surface. They also show mountains, rivers, forests, and human-made features such as bridges, buildings, highways, and railroads.
43. coal
44. Water contributes to the process of mass movement by adding weight and filling the spaces between the sediment particles, thereby allowing the particles to more easily slide over one another as they move downslope.

THINK CRITICALLY

31. The algal bloom is a result of eutrophication, which can occur when too much phosphorus or nitrogen are added.
32. A turbulent flow might carry so much energy that large particles could be sustained in suspension.
33. Vegetation, porous soils, and a slow rate of precipitation lead to infiltration. A steep slope, compacted soil, and rapid rate of precipitation leads to runoff.
34. Upstream tributaries carry a smaller volume of water through steep and narrow channels which leads to turbulent flow. Downstream portions of rivers have a greater amount of water, but they are wider and the slope of the stream is flatter which leads to smooth-flowing water.
35. The floodplain of the stream most likely has the most fertile soil.

CONCEPT MAPPING

36. Concept map should show the connection of the following terms: precipitation → infiltration (leads to groundwater) and runoff (which flows through stream systems to the ocean) → transpiration and evaporation → condensation → precipitation

CHALLENGE QUESTION

37. Answers will vary. The wastewater should be tested and treated before being released into the lake. Levels of dissolved materials such as phosphates and silicates should be monitored and controlled to a healthful level. The presence of bacteria should be monitored.

WRITING IN ▶ Earth Science

Rubric

38. Answers will vary. The recent flood has caused vast destruction, for with the increased width, depth, and velocity of the stream, the discharge has greatly increased. With its high velocity the carrying capacity of the stream has increased, so large boulders and trees were swept downstream. The stream has eroded its banks, increasing the size of the stream bed. As the flood subsided, large amounts of sand and debris were left all over the town.

MULTIPLE CHOICE

1. B
2. D
3. C
4. A
5. C
6. D
7. A
8. B
9. B
10. C

MULTIPLE CHOICE

1. Which condition would create the most runoff?
- **A.** land covered with vegetation
- **B.** plants in densely packed soil
- **C.** light precipitation
- **D.** soil with a high percentage of sand

Use the photo below to answer Questions 2 and 3.

2. What most likely caused the odd shape of the boulder?
- **A.** a rock slide
- **B.** glacier erosion
- **C.** wind deflation
- **D.** wind abrasion

3. What clue would help scientists determine the method of erosion for this boulder?
- **A.** The boulder has smooth surfaces with smooth edges.
- **B.** The boulder has a coarse surface.
- **C.** The boulder is polished on the windward side.
- **D.** The boulder has a rough surface and rough edges.

4. If you were creating a model of rock formation, you would represent the different layers of rock. In this model, which type of rock would represent particles that have been compressed and hardened?
- **A.** sedimentary
- **B.** volcanic
- **C.** igneous
- **D.** intrusive

5. As the velocity of a stream decreases, which transported particle size would settle to the stream's bottom first?
- **A.** clay
- **B.** silt
- **C.** pebble
- **D.** sand

6. Which condition helps determine the quality of lake water?
- **A.** the amount of nitrogen
- **B.** the amount of dissolved calcium carbonate
- **C.** the amount of potassium
- **D.** the amount of dissolved oxygen

7. Which state of matter are all minerals?
- **A.** solids
- **B.** liquids
- **C.** gases
- **D.** plasma

Use the table below to answer Questions 8–10.

Texture Data for a Soil Profile

Horizon	Percent		
	Sand	Silt	Clay
A	16.2	54.4	29.4
B	10.5	50.2	39.3
C	31.4	48.4	20.2
R (bedrock)	31.7	50.1	18.2

8. What inferences can scientists make from this soil profile?
- **A.** The soil is newly layered.
- **B.** The soil is well developed and mature.
- **C.** The soil is poorly developed.
- **D.** The soil profile came from the West.

9. Which horizon most likely contains the hard material known as hardpan?
- **A.** A-horizon
- **B.** B-horizon
- **C.** C-horizon
- **D.** R-horizon

10. O-horizon was not listed on this table. What could be the reason?
- **A.** It was too deep to be studied.
- **B.** It only contained sand and not silt or clay.
- **C.** It was insignificant to the study containing only humus and leaf litter.
- **D.** It contained only clay and not sand or silt.

Online Test Practice

SHORT ANSWER

Use the table below to answer Questions 11 and 12.

Mineral Characteristics				
Mineral	Color	Streak	Hardness	Specific Gravity
Sulfur	yellow	yellow	2	2.1
Schorl	black	white	7	3.2
Topaz	blue	colorless	9	3.6
Zinc	white	light gray	2	6.9

11. How is this table organized?

12. From this table, what can you infer about hardness and specific gravity?

13. According to the water cycle, what happens after water molecules evaporate and condense into cloud droplets?

14. Compare and contrast conglomerates and breccias.

15. What causes dune migration?

16. What are some possible strategies a road construction crew could use to protect highways located at the bottom of steep slopes susceptible to landslides?

READING FOR COMPREHENSION

Inland Flooding

According to the National Weather Service, inland flooding is one of the deadliest effects of hurricanes. Below is a list of steps to help reduce your risk of being caught in inland flooding when you hear about a potential hurricane and live in a potential flood zone.

- If advised to evacuate, do so immediately. Move to a safe area before access is cut off by flood water.
- Keep abreast of road conditions through the news media.
- Do not attempt to cross flowing water. As little as six inches of water might cause you to lose control of your vehicle—two feet of water will carry most cars away.
- Develop a flood emergency action plan with your community leaders.

Article obtained from: Hurricane flooding: a deadly danger. *NOAA's National Weather Service.* March 2001. (Online resource accessed October 2006)

17. What is important to know about water?
    A. It is safe to drive on a road with flowing water.
    B. Flowing water is safer than standing water.
    C. Six inches of water will do no harm.
    D. Two feet of water can carry most cars away.

18. According to the text, which is not a step to take to ensure your safety from inland flooding?
    A. Do not attempt to cross flowing water.
    B. Move to the highest level of your house.
    C. Evacuate when advised to do so.
    D. Develop a flood emergency action plan.

19. What is the goal of the National Weather Service in distributing this list?
    A. to discuss hurricanes
    B. to offer advice for people to read and use if they want
    C. to inform people of the hazards of inland flooding and offer steps to protect themselves
    D. to scare people

20. Suppose you live in a flood zone that could possibly be affected by inland flooding. Develop a strategy that you would follow to stay safe.

NEED EXTRA HELP?																
If You Missed Question . . .	1	2	3	4	5	6	7	8	9	10	11	12	13	14	15	16
Review Section . . .	9.1	2.2	2.2	4.1	9.2	9.3	4.1	7.3	7.3	7.3	4.1	4.1	9.1	8.2	8.2	8.1

SHORT ANSWER

11. The table is organized according to increasing specific gravity.
12. From this table, you can infer that, as specific gravity increases, hardness does not have to increase as well.
13. The next part of the water cycle is for the water droplets to fall to the Earth as precipitation and soak into the ground, where the process begins again.
14. Both conglomerates and breccias are coarse-grained sedimentary rocks. Conglomerates are primarily composed of rounded grains, suggesting that the grains were transported by water and traveled far from their source. Breccias are composed of mostly angular grains, suggesting that the particles did not travel far before deposition.
15. Dune migration results when prevailing winds continue to move sand from the windward side of a dune to its leeward side, thereby causing the dune to move slowly over time.
16. Sample Answer: Construction crews could cover the steep slopes with materials such as steel nets or construct protective fences at the bottom of the slopes to hold back the material from the highway in the event of a landslide.

READING FOR COMPREHENSION

17. D
18. B
19. C
20. Answers will vary. Possible answers: having a battery powered radio to keep informed of what is going on, a cell phone in case emergency calls need to be made, finding a high spot nearby, in case they have to move out quickly, and having an action plan that the entire family can follow.

CHAPTER 10 Groundwater

BIGIDEA **Precipitation and infiltration contribute to groundwater, which is stored in underground reservoirs until it surfaces as a spring or is drawn from a well.**

ESSENTIAL QUESTIONS	NATIONAL STANDARDS	RESOURCES TO ASSESS MASTERY
SECTION 1 **Movement and Storage of Groundwater** 1. How do groundwater storage and underground movement relate to the water cycle? 2. How are aquifers and aquicludes related? 3. How are the components of aquifers related to the presence of springs? 1.5 sessions 0.75 block	UCP. 1–4; A.1–2; B.4; D.1; F.3	**Entry-Level Assessment** Focus Transparency 26, p. 252 **Progress Monitoring** Caption Question, pp. 253, 254, 257, 258 Reading Check, pp. 253, 256 Section Review, p. 258
SECTION 2 **Groundwater Weathering and Deposition** 1. How does groundwater dissolve and deposit rocks and minerals? 2. How do caves form? 3. What features are characteristic of karst topography? 2 sessions 1 block	UCP. 1–3; D.1–2	**Entry-Level Assessment** Focus Transparency 27, p. 259 **Progress Monitoring** Caption Question, pp. 260, 261, 262 Reading Check, p. 260 Section Review, p. 262
SECTION 3 **Groundwater Supply** 1. How is groundwater withdrawn from aquifers? 2. What are the major problems that threaten groundwater supplies? 4 sessions 2 blocks	UCP.1–3, UCP.5; A.1–2; D.1–2; E.1; F.3–5	**Entry-Level Assessment** Focus Transparency 28, p. 263 **Progress Monitoring** Caption Question, pp. 264, 265, 267 Reading Check, pp. 265, 266, 267 Section Review, p. 268 **Summative Assessment** Chapter Assessment, p. 273 Exam*View*® *Assessment Suite* CD-ROM

Go online! connectED.mcgraw-hill.com

LEVELED RESOURCES	LAB MATERIALS	ADDITIONAL RESOURCES
Science Notebook 10.1 OL **Chapter FAST FILE Resources:** Study Guide, p. 91 BL **Lab Resources:** Laboratory Manual, p. 73 OL **Transparencies:** Section Focus Transparency 26 EL Teaching Transparency 25 OL EL	LaunchLAB p. 250 / **20 min** 250-mL graduated cylinders (2), fine sand, water, coarse sand, clay	**Technology:** eTeacherEdition eStudentEdition Classroom Presentation Toolkit Exam*View® Assessment Suite* CD-ROM LabManager™ CD-ROM What's EARTH SCIENCE Got To Do With It? DVD Weather Classroom Video DVD **Assessment:** Blueprints for Success Chapter Test (Scaffolded) Earth Science Standardized Test Practice **Lab Resources:** Exploring Environmental Problems Laboratory Manual
Science Notebook 10.2 OL **Chapter FAST FILE Resources:** Study Guide, p. 93 BL **Transparencies:** Section Focus Transparency 27 EL Teaching Transparency 26 OL EL		
Science Notebook 10.3 OL **Chapter FAST FILE Resources:** MiniLab Worksheet, p. 80 OL GeoLab Worksheet, p. 81 OL Study Guide, p. 95 BL **Lab Resources:** Laboratory Manual, p. 77 OL **Transparencies:** Section Focus Transparency 28 EL Teaching Transparency 27 OL EL	MiniLAB p. 265 / **20 min** plastic container, sand, water, clay, book, straw, clear straw GeoLAB p. 270 / **45 min** U. S. Geological Survey topographic map of Forest City, FL; transparent paper; ruler; graph paper; calculator	

LabManager™
Customize any lab with the LabManager™ CD-ROM.

BL Below Level   OL On Level   AL Advanced Learners   EL English Learners   COOP LEARN Cooperative Learning

CHAPTER 10

LaunchLAB

iLab Station Rubric

How is water stored underground?

Safety Precautions Make sure students wear safety goggles.

Teaching Strategies

- The volume of water in the sand cylinder equals the original volume in the second cylinder minus the volume remaining in the second cylinder. To determine the volume of water in 1 m^3 of sand, it is best for students to use ratios, based on the fact that 1 cm^3 = 1 mL.

Procedure

1. Have students read and complete the lab safety form and follow the procedure below.
2. Fill a **250-mL graduated cylinder** with **fine, dry sand.**
3. Fill **another 250-mL graduated cylinder** with **water.**
4. Pour water from the second cylinder into the sand-filled cylinder until the water level is flush with the surface of the sand. Measure and record the volume of saturated sand in the cylinder.
5. Measure and record how much water is left in the second cylinder.
6. Repeat the experiment twice using **coarse sand** and **clay.**

Analysis

1. **Describe** how much water is present in the saturated fine sand, coarse sand, and clay. Answers will vary, but clay should hold the most water, coarse sand the second most, and fine sand holds the least. Possible answers: clay, 133 mL; coarse sand, 96 mL; fine sand, 86 mL.
2. **Calculate** the ratio of water volume to the volume of fine sand, coarse sand, and clay, and express the value as a percentage. The three samples will have approximately the following percentages: fine sand: 35%; coarse sand: 38%; clay: 53%
3. **Infer** how many liters of water could be stored in a cubic meter of each sediment. (a cubic meter = 1000 L) Fine sand: 350 L; coarse sand: 380 L; clay: 530 L water

CHAPTER 10

Groundwater

BIGIDEA Precipitation and infiltration contribute to groundwater, which is stored in underground reservoirs until it surfaces as a spring or is drawn from a well.

SECTIONS

1 Movement and Storage of Groundwater

2 Groundwater Weathering and Deposition

3 Groundwater Supply

LaunchLAB

iLab Station

How is water stored underground?

Beneath your feet, there are vast amounts of water. This water fills in the pore spaces and fractures in rock and unconsolidated sediment. In this activity, you will model groundwater storage.

FOLDABLES® Study Organizer

Threats to the Water Supply

Make a six-tab book and label it as you read. Use it to organize your notes on the major problems that threaten groundwater supplies.

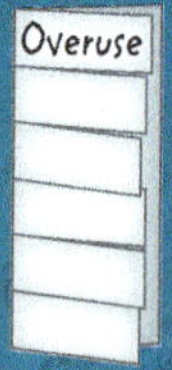

Assessment

Performance Ask students to calculate which contains more water: a large swimming pool with a volume of 1 million L or a saturated cube of sand 10 m on each side. the swimming pool; the sand cube is not 100 percent water.

LabManager™
Customize any lab with the LabManager™ CD-ROM.

The Strokkur geyser in Iceland erupts every 5 to 10 minutes. The eruptions can reach heights of more than 30 m.

Introduce the BIGIDEA

Groundwater Have students work in pairs to describe the series of events that must take place so water from rain can be stored in groundwater deposits. Then, ask students to organize the series of events on the board. Although students might not know all of the terminology, students' answers might include: Rain/snow falling (precipitation) → wetting of the ground → percolating water through the surface soil (infiltration) → infiltrating water traveling downward through permeable layers of soil → water reaching an impermeable layer that will stop water from continuing downward → saturation of upper layers with water.

Teacher Content Support

Geysers This is a photograph of the Strokkur geyser in the Haukadalur Valley in Iceland. The Strokkur geyser is one of the most predictable geysers in the world. There are two types of geysers: cone geysers and fountain geysers. Strokkur geyser is a cone geyser, which means it erupts from a cone of siliceous material in the center of the geyser. A fountain geyser erupts from a pool of water.

SECTION 1

1 Focus

Focus Transparency

Before presenting the lesson, project **Section Focus Transparency 26** and have students answer the accompanying questions. EL

MAINIDEA

Groundwater Movement Ask students: How does water circulate on Earth? The water cycle describes how water precipitates, infiltrates, runs off, evaporates, condenses, and then precipitates again.

2 Teach

Teacher Content Support

Location of Groundwater Groundwater is present only in the upper part of Earth's crust, because the pressure of the overlying rocks gradually compresses the pore spaces. Below a depth of about 10 km, all pore spaces are essentially closed, and rocks contain only traces of water. The deeper groundwater (below the top 1–2 km) is always salty, because most of it is ancient seawater trapped in the marine sediments that later became continental sedimentary rocks through tectonic processes.

SECTION 1

Movement and Storage of Groundwater

MAINIDEA Groundwater reservoirs provide water to streams and wetlands wherever the water table intersects the surface of the ground.

Essential Questions

- How do groundwater storage and underground movement relate to the water cycle?
- How are aquifers and aquicludes related?
- How are the components of aquifers related to the presence of springs?

Review Vocabulary

hydrologic cycle: a never-ending natural circulation of water through Earth's systems

New Vocabulary

infiltration
zone of saturation
water table
zone of aeration
permeability
aquifer
aquiclude
spring
hot spring
geyser

EARTH SCIENCE 4 YOU Have you ever noticed that a stream flows even when it has not rained in a long time? Rainfall contributes to the flow in a stream, but much of the water comes from beneath the ground.

The Hydrosphere

The water on and in Earth's crust makes up the hydrosphere, named after *hydros*, the Greek word for *water*. You have learned about the hydrosphere in the context of Earth's systems, including the geosphere, hydrosphere, atmosphere, and biosphere. About 97 percent of the hydrosphere is contained in the oceans. The water contained by landmasses—nearly all of it freshwater—makes up only about 3 percent of the hydrosphere.

Freshwater is one of Earth's most abundant and important renewable resources. However, of all the freshwater, about 70 percent is held in polar ice caps and glaciers. All the rivers, streams, and lakes on Earth represent only a small fraction of Earth's liquid freshwater. The distribution of the world's water is shown in **Table 1.** Recall that water in the hydrosphere moves through the water cycle.

Explore **Earth's water supply with an interactive table.** Concepts In Motion

Table 1 Estimated World's Water Supply

Location	Percentage of Total Water	Water Volume (km^3)	Estimated Average Residence Time of Water
Oceans	97	1,316,700,000	thousands of years
Ice caps and glaciers	2.1	28,700,000	tens of thousands of years and longer
Groundwater	0.31	4,200,000	hundreds to many thousands of years
Lakes	0.15	2,100,000	tens of years
Atmosphere	0.1	1,140,000	nine days
Rivers and streams	0.003	40,000	two weeks

DIFFERENTIATED INSTRUCTION

Struggling Learners Ask students to create an illustration that represents the information in **Table 1.** Illustrations should include the oceans, rivers, streams, groundwater, lakes, and glaciers—each labeled with the appropriate amount. Draw students' attention to the water volume for each element so that their proportions are roughly correct.

Demonstration

Water Circulation Add water to a one-gallon clear container that contains mostly potting soil with a layer of modeling clay underneath, which is sealed to sides of the container. Make sure students observe that water flows through the soil, but not through the clay.

Groundwater and Precipitation

The ultimate source of all water on land is the oceans. Evaporation of seawater cycles water into the atmosphere in the form of invisible water vapor and visible clouds. Winds and weather systems move this atmospheric moisture all over Earth, with much of it concentrated over the continents. Precipitation brings atmospheric moisture back to Earth's surface. Some of this precipitation falls directly into the oceans and some falls on land.

Infiltration is the process by which precipitation that has fallen on land trickles into the ground and becomes groundwater. Only a small portion of precipitation becomes runoff and is returned directly to the oceans through streams and rivers. Groundwater slowly moves through the ground, eventually returns to the surface through springs and seepage into wetlands and streams, and then flows back to the oceans.

☑ READING CHECK **Identify** the ultimate source of all water on land.

Groundwater Storage

Puddles of water that are left after it rains quickly disappear, partly by infiltrating the ground. On sandy soils, rain soaks into the ground almost immediately. Where does that water go? The water seeps into small openings within the ground. Although Earth's crust appears solid, it is composed of soil, sediment, and rock that contain countless small openings, called pore spaces.

Pore spaces make up large portions of some of these materials. The amount of pore space in a material is its porosity. The greater the porosity, the more water can be stored in the material. Subsurface materials have porosities ranging from 2 percent to more than 50 percent. For example, the porosity of well-sorted sand is 30 percent; however, in poorly sorted sediment, smaller particles occupy some of the pore spaces and reduce the overall porosity of the sediment, as shown in **Figure 1**. Similarly, the cement that binds the grains of sedimentary rocks together reduces the rocks' porosity. Because of the enormous volume of sediment and rock beneath Earth's surface, enormous quantities of groundwater are stored in the pore spaces.

Figure 1 Porosity depends on the size and variety of particles in a material.
Compare *the porosities shown in each sample.*

Activity

Porosity Porosity is the fraction of a material occupied by void spaces. It can be expressed as a percentage or as a decimal, for example, 30% or 0.30. Poorly sorted sediments have a lower porosity than well-sorted sediments. Have students work in groups to compare and contrast the porosity of well-sorted and poorly sorted materials by completing the following exercise: Assume that the porosity of well-sorted sand and well-sorted gravel is the same: 30 percent (0.30). Determine the porosity of a mixture of sand and gravel, assuming that fine sand with a porosity of 30 percent fills all the pore spaces of the gravel (30 percent of the gravel). In other words, what is 30 percent of 30 percent? 9 percent

This exercise can be done mathematically ($0.3 \times 0.3 = 0.09$), or graphically, using columns to represent 100 percent. The empty spaces in the gravel are represented by a column divided vertically into a solid section 70 units long and a void section 30 units long. The sand-filled gravel column should have a solid gravel section 70 units long, a solid sand section 21 units long (70 percent of the original void section), and a void section 9 units long. Demonstrate this concept on the board using 50 percent of 50 percent = 25 percent.

AL COOP LEARN

ACROSS THE CURRICULUM

Math Residence time is the average length of time that a substance spends in a reservoir. For instance, if water is exchanged (added and removed) in a bathtub holding 600 L at a rate of 0.5 L/s, the residence time (or exchange time) of that water is 600 L/0.5 L/s, or 1200 s (20 min). Have students calculate the residence time in years of Earth's groundwater, given that the total amount of groundwater is 4,000,000 km^3, and the infiltration (replacement) rate is 2000 km^3/y. 2000 y AL

Caption Question Fig. 1 The unsorted sample in the middle contains the lowest porosity. The large sand grains have slightly higher porosity than the fine sand grains.

☑ READING CHECK The ultimate source of all water on land is the oceans.

Model

The Zone of Saturation Have students draw a hypothetical cross-sectional view of the ground that includes the following features: soil moisture zone, zone of aeration, water table, and the zone of saturation. Ask students to draw a second cross section of the same hypothetical location shortly after it has rained.

Ask students to address the following questions by writing in their Earth science journals: What happens to the thickness of the zone of aeration and the zone of saturation? The zone of aeration decreases and the zone of saturation increases after it rains. What happens to the position of the water table? The water table rises after it rains. What might happen in the second cross-sectional view of the ground if it rained again? If it rains again, the zone of aeration will decrease further, the zone of saturation will increase further, and the water table will rise more.

Caption Question **Fig. 2** The zone of aeration is above the zone of saturation. The water table is the boundary between the two zones.

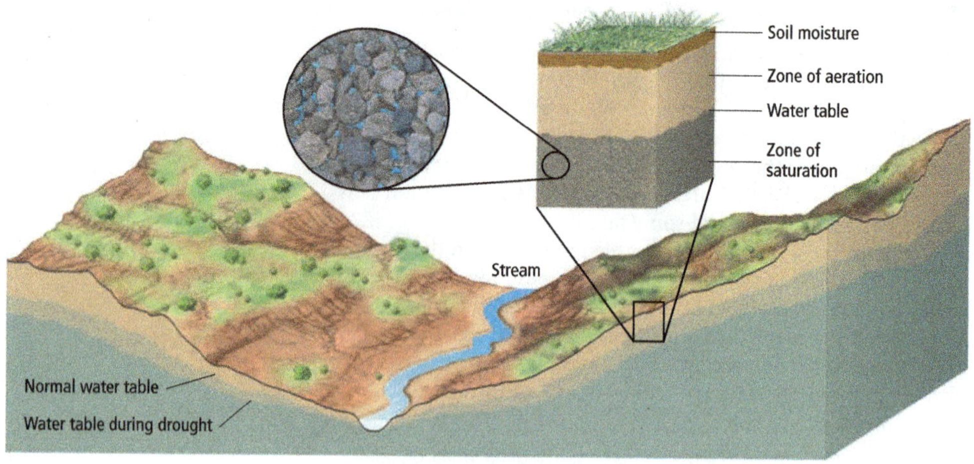

Figure 2 The zone of saturation is where groundwater completely fills all the pores of a material below Earth's surface. **Describe** *what is above the zone of saturation.*

The Zone of Saturation

The region below Earth's surface in which groundwater completely fills all the pores of a material is called the **zone of saturation.** The upper boundary of the zone of saturation is the **water table,** shown in **Figure 2.** Strictly speaking, only the water in the zone of saturation is called groundwater. In the **zone of aeration,** which is above the water table, materials are moist, but because they are not saturated with water, air occupies much of the pores.

Water movement Water in the zone of saturation and zone of aeration can be classified as either gravitational water or capillary water. Gravitational water is water that trickles downward as a result of gravity. Capillary water is water that is drawn upward through capillary action above the water table and is held in the pore spaces of rocks and sediment because of surface tension. Capillary action can be seen when the tip of a paper towel is dipped into water and the water seems to climb up through the fibers of the paper towel.

The water table The depth of the water table often varies depending on local conditions. For example, in stream valleys, groundwater is relatively close to Earth's surface, and thus the water table can be only a few meters deep. In swampy areas, the water table is at Earth's surface, whereas on hilltops or in arid regions, the water table can be tens to hundreds of meters or more beneath the surface. As shown in **Figure 2,** the topography of the water table generally follows the topography of the land above it. For example, the slope of the water table corresponds to the shape of valleys and hills on the surface above.

Because of its dependence on precipitation, the water table fluctuates with seasonal and other weather conditions. It rises during wet seasons, usually in spring, and drops during dry seasons, often in late summer.

DIFFERENTIATED INSTRUCTION

Advanced Learners Tell students there is an oceanful of groundwater beneath our feet. Ask interested students to estimate how deep that ocean would be if all groundwater were at Earth's surface. Specifically, have students calculate the thickness of a water layer containing all the groundwater in the upper part of Earth's continental crust, which is 10 km thick, assuming that the crust has an average porosity of 5 percent. $10 \text{ km} \times 0.05 = 0.5 \text{ km} = 500 \text{ m}$

Groundwater Movement

Groundwater flows downhill in the direction of the slope of the water table. Usually, this downhill movement is slow because the water has to flow through numerous tiny pores in the subsurface material. The ability of a material to let water pass through it is its **permeability.** Materials with large, connected pores, such as sand and gravel, have high permeability and permit relatively high flow velocities up to hundreds of meters per hour. Other permeable subsurface materials include highly fractured bedrock, sandstone, and limestone.

Permeability Groundwater flows through permeable sediment and rock, called **aquifers,** such as the one shown in **Figure 3.** In aquifers, the pore spaces are large and connected. Fine-grained materials have low permeabilities because their pores are small. These materials are said to be impermeable. Groundwater flows so slowly through impermeable materials that the flow is often measured in millimeters per day. Some examples of impermeable materials include silt, clay, and shale. Clay is so impermeable that a clay-lined depression will hold water. For this reason, clay is often used to line artificial ponds and landfills. Impermeable layers, called **aquicludes,** are barriers to groundwater flow.

Flow velocity The flow velocity of groundwater depends on the slope of the water table and the permeability of the material through which the groundwater is moving. The force of gravity pulling the water downward is greater when the slope of the water table surface is steeper. Water also flows faster through a large opening than through a small opening. The flow velocity of groundwater is proportional to both the slope of the water table and the permeability of the material through which the water flows.

Figure 3 An aquifer is a layer of permeable subsurface material that is saturated with water. This aquifer is located between two impermeable layers called aquicludes.

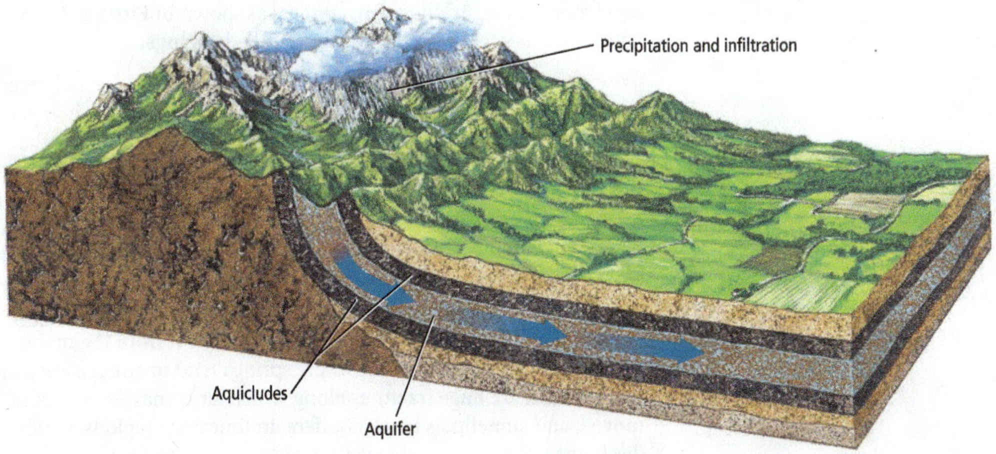

Teacher Content Support

Groundwater Flow Groundwater is the subsurface water below the water table. Nowhere is the water table completely horizontal, so groundwater moves gradually downslope under the pull of gravity. The flow paths of the moving water are not parallel to the water table, but are concave upward between the points of recharge and discharge.

The speed of groundwater motion is given by Darcy's law: $V = k(h/L)$, where V is the hypothetical flow velocity, k is the hydraulic conductivity, and h/L is the average slope of the water table (h is the rise; L is the run) between the two points connecting the flow path. The flow velocity, therefore, is proportional to the slope of the water table; high slopes produce higher flow velocities than low slopes, as might be expected.

The hydraulic conductivity, k, is related to the permeability of the subsurface material and ranges over many orders of magnitude for different geological materials. Representative values for k are 400 m/day for gravel, 12 m/day for sand, 3 m/day for sandstone, 0.1 m/day for silt, and 0.0002 m/day for clay. This means that coarse-grained, permeable materials such as sand and gravel permit higher flow velocities than do impermeable materials such as silt and clay, no matter what their slope.

Demonstration

Confined Aquifer Have a student hold a rubber tube in a U-shape with one end lower than the other. Pour water into the upper end of the tube until it flows out the lower end. Ask students to describe the movement of water in the tube. It flowed with gravity.

Draw a U-shaped cross section of a confined aquifer on the board and explain that similar natural discharges of groundwater by springs can take place in areas of folded sedimentary rocks.

Identify Misconceptions

Many people are under the erroneous impression that groundwater is confined to underground rivers or natural conduits, and wells must be drilled into specific conduits.

Uncover the Misconception Ask students to define a spring. Many students will say it is where an underground river surfaces.

Demonstrate the Concept While this might be partially true in areas where there are substantial deposits of limestone or bodies of crystalline rock, explain to students groundwater is present everywhere beneath the water table, and all materials below the water table are completely saturated with water so a spring can originate from many types of groundwater deposits. As long as a well is drilled into a permeable material below the water table, it will produce water.

Assess New Knowledge Ask students to draw a cross section of the ground, including groundwater and a spring.

READING CHECK Groundwater tends to emerge wherever the water table intersects Earth's surface. These intersections tend to occur in areas where the ground slopes.

Figure 4 Springs occur at points where the water table intersects Earth's surface.

Springs

Groundwater moves slowly but continuously through aquifers and eventually returns to Earth's surface. In most cases, groundwater emerges wherever the water table intersects Earth's surface. Such intersections commonly occur in areas that have sloping surface topography. The exact places where groundwater emerges depend on the arrangement of aquifers and aquicludes in an area.

READING CHECK **Explain** the relationship between the slope of the land and where groundwater emerges.

As you learned on the previous page, aquifers are permeable underground layers through which groundwater flows easily, and aquicludes are impermeable layers. Aquifers are commonly composed of layers of sand and gravel, sandstone, and limestone. In contrast, aquicludes, such as layers of clay or shale, block groundwater movement. As a result, groundwater tends to discharge at Earth's surface where an aquifer and an aquiclude are in contact, as shown in **Figure 4.** These natural discharges of groundwater are called **springs.**

Emergence of springs The volume of water that is discharged by a spring might be a mere trickle or it might form a stream. In some regions called karst regions, an entire river might emerge from the ground. Such a superspring is called a karst spring. Karst springs occur in limestone regions where springs discharge water from underground pathways. In regions of nearly horizontal sedimentary rocks, springs often emerge on the sides of valleys at about the same elevation, at the bases of aquifers, as shown in **Figure 5.** Springs might also emerge at the edges of perched water tables. In a perched water table, a zone of saturation that overlies an aquiclude separates it from the main water table below. Other areas where springs tend to emerge are along faults, which are huge fractures along which large masses of rock have moved, and sometimes block aquifers. In limestone regions, springs discharge water from underground pathways as karst springs.

IN THE FIELD

Springs For tens of thousands of years, human beings have chosen to build their settlements along sources of water. Archaeological sites are often found adjacent to springs. Not only do springs provide a source of water for human beings, but they provide water for other animals, who might serve as food for human beings.

VISUALIZING Springs

Figure 5 A spring is the result of groundwater that emerges at Earth's surface. Springs can be caused by a variety of conditions.
Compare and contrast *the origin of the four types of springs.*

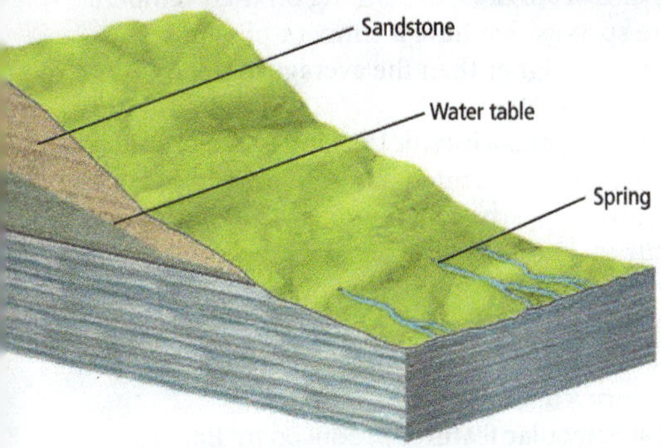

A spring forms where a permeable layer and an impermeable layer come together.

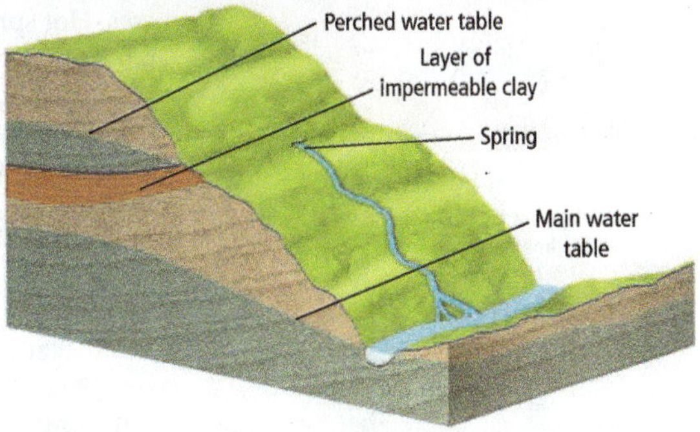

A layer of impermeable rock or clay can create a perched water table. Springs can result where groundwater emerges from a perched water table.

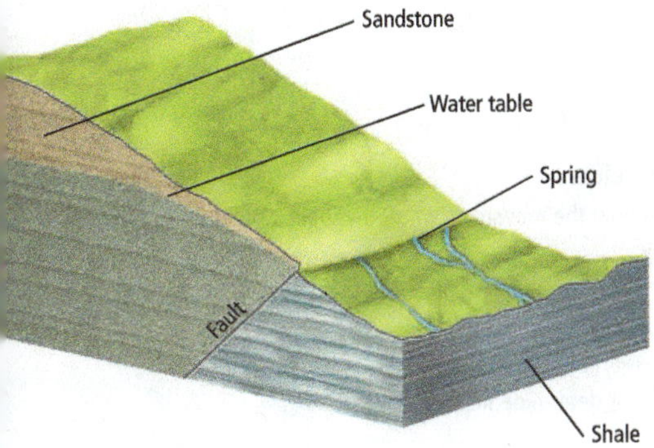

Some springs form where a fault has brought together two different types of bedrock, such as a porous rock and a nonporous rock.

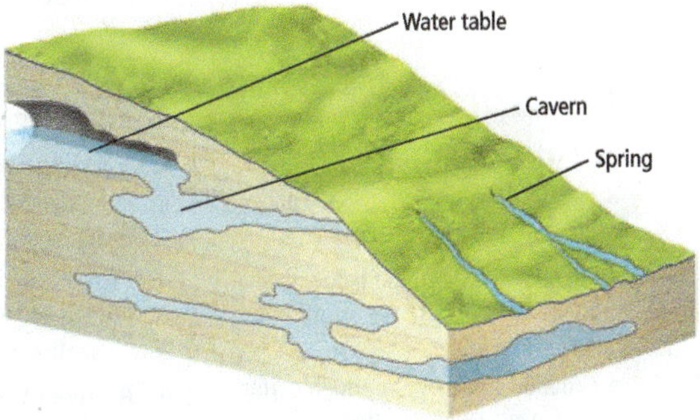

Karst springs form where groundwater weathers through limestone bedrock, and water in the underground caverns emerges at Earth's surface.

Concepts In Motion View an **animation of springs.**

VISUALIZING

Purpose

Students will compare and contrast four types of springs.

Teacher Content Support

Springs Springs can have a constant discharge or a varying discharge. Some springs are permanent, while others are seasonal. Many different gases and minerals can be dissolved within a spring. Large springs are often found in limestone. Central Florida has many springs that cover several acres. Many of these springs are interconnected through underground flow and through sinkholes.

Caption Question **Fig. 5** The contact spring and the perched water table contact spring originate when a spring emerges at the same level as the ground. A fault spring forms when a fault brings an aquifer to the surface, and a karst spring forms in a limestone area when springs discharge water from underground pathways.

Tie to Previous Knowledge

Sedimentary Rocks Have students recall the properties of sedimentary rocks. Ask students to compare the textures, porosities, and mineral composition of sandstone, shale, and limestone, and to infer how these properties affect the development of springs.

EARTH SCIENCE JOURNAL

Types of Springs Ask students to create a table in their Earth science journals that compares the characteristics of different types of springs: contact springs, perched water table contact springs, fault springs, and karst springs. They might wish to include an illustration based on **Figure 5.**

Rubric

■ **Caption Question Fig. 6** A geyser is created when water is heated and the aquifer vaporizes, which produces pressure.

3 Assess

Check for Understanding

Explain Ask students what happens to rainwater as it infiltrates. Infiltrating water trickles through the zone of aeration to the water table, enters the zone of saturation, and slowly moves downslope through aquifers until it is discharged through springs to the surface and flows to the ocean.

Reteach

Define Operationally Have students propose an operational definition for each of the following terms: *infiltration, porosity, zone of saturation, water table, zone of aeration, permeability, aquifer, aquicludes, spring, hot spring,* and *geyser.*

Assessment

Performance Have students compare and contrast porosity and permeability, and how grain size, grain shape, cementation, and sorting influence these two properties. Porosity is independent of grain size, decreases with grain angularity, decreases with cementation, and increases with sorting. Permeability increases with grain size, decreases with grain angularity, decreases with cementation, and increases with sorting.

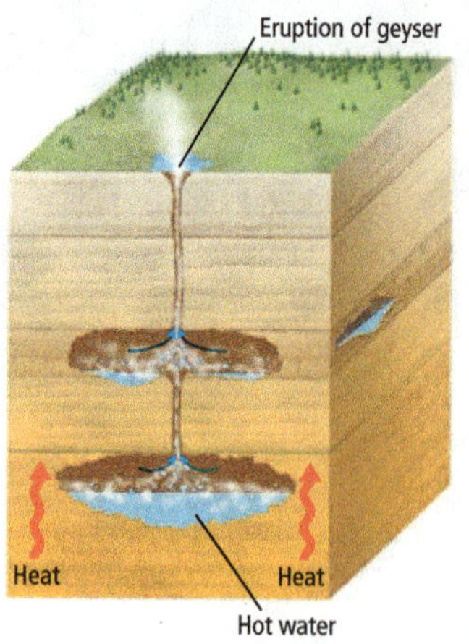

■ **Figure 6** A geyser is a type of hot spring from which very hot water and vapor erupt at the surface.
Identify *the origin of a geyser.*

Temperature of springs People usually think of spring water as being cool and refreshing. But the temperature of groundwater that is discharged through a spring is generally the average annual temperature of the region in which it is located. Thus, springs in New England have temperatures of about 10°C, while springs in the Gulf states have temperatures of about 20°C.

Compared to air temperatures, groundwater is generally colder in the summer and warmer in the winter. However, in some regions around the world, springs discharge water that is much warmer than the average annual temperature. These springs are called warm springs or **hot springs,** depending on their temperatures. Hot springs are springs that have a temperature above 36.6°C, though are usually higher than the average temperature of the human body, which is 37°C.

There are thousands of hot springs in the United States. Most of them are located in the western United States in areas where the subsurface is still hot from nearby igneous activity. A number of hot springs also occur in some eastern states. These hot springs emerge from aquifers that descend to tremendous depths in Earth's crust and through which deep, hot water rises. The water is hot because temperatures in Earth's upper crust increase by an average of 25°C for every km of depth.

Among the most spectacular features produced by Earth's underground thermal energy in volcanic regions are geysers, shown in **Figure 6**. **Geysers** are explosive hot springs. In a geyser, water is heated past its boiling point, causing it to vaporize. The resulting water vapor builds up tremendous pressure. This pressure is what fuels the eruptions. One of the world's most famous geysers, Old Faithful, is located in Yellowstone National Park, Wyoming. Old Faithful erupts every 44 to 125 minutes, and can reach heights of 27 to 56 m.

SECTION 1 REVIEW

Section Self-Check

Section Summary

- Some precipitation infiltrates the ground to become groundwater.
- Groundwater is stored below the water table in pore spaces of rocks and sediment.
- Groundwater moves through permeable layers called aquifers and is trapped by impermeable layers called aquicludes.
- Groundwater emerges from the ground where the water table intersects Earth's surface.

Understand Main Ideas

1. MAINIDEA **Explain** how the movement of groundwater is related to the water cycle.
2. **Illustrate** how the relative positions of an aquifer and aquiclude can result in the presence of a spring.
3. **Describe** how the water in hot springs gets hot.
4. **Analyze** the factors that determine flow velocity.

Think Critically

5. **Differentiate** between porosity and permeability in subsurface materials.
6. **Infer** why it is beneficial for a community to have an aquiclude located beneath the aquifer from which it draw its water supply.

WRITING IN ▶ Earth Science

7. Develop a set of guidelines in which you describe where you would be most likely to find groundwater.

SECTION 1 REVIEW

1. When water precipitates and infiltrates, it joins groundwater. When groundwater emerges from springs and wells, it joins surface runoff and can eventually return to the ocean.
2. Illustrations will vary. Answers should include a spring emerging where an aquifer overlies an aquiclude and both intersect the surface.
3. Groundwater can be heated where recent igneous activity has heated subsurface material. The heated groundwater tends to rise because it is less dense compared to the cooler groundwater.
4. Flow velocity is determined by the slope of the water table and the permeability of the soil.
5. The porosity of a material depends on the amount of pore space in the material, while the permeability is the ability of water to flow though the pore spaces of a material. To be permeable, the pore spaces of a material must be interconnected.
6. The aquiclude could help trap infiltrating water, and prevent it from traveling too deep for wells to reach easily.
7. Answers will vary. Answer should include that groundwater is most likely to be found above an aquiclude and in a well-sorted sediment with large grain sizes.

Rubric

SECTION 2

Essential Questions

- How does groundwater dissolve and deposit rocks and minerals?
- How do caves form?
- What features are characteristic of karst topography?

Review Vocabulary

hydrolysis: chemical reaction of water with other substances

New Vocabulary

cave
sinkhole
karst topography
stalactite
stalagmite

Groundwater Weathering and Deposition

MAINIDEA Chemical weathering of limestone by water causes the characteristic topography of karst areas.

EARTH SCIENCE 4 YOU You might have seen an old gravestone, statue, or sculpture that has been weathered by acidic water. Similar processes form limestone caves underground.

Carbonic Acid

Acids are aqueous solutions that contain hydrogen ions. Most groundwater is slightly acidic due to carbonic acid. Carbonic acid forms when carbon dioxide gas dissolves in water and combines with water molecules. This happens when precipitation falls through the atmosphere and interacts with carbon dioxide gas or when groundwater infiltrates the products of decaying organic matter in soil. As a result of these processes, groundwater is usually slightly acidic and attacks carbonate rocks, especially limestone. Limestone mostly consists of calcite, also called calcium carbonate, which reacts with any kind of acid. The results of this reaction over time are shown in **Figure 7.** This process occurs above ground and below ground.

Figure 7 Carbonic acid has dissolved large portions of this limestone. This resulting formation is the Stone Forest in China.

Dissolution by Groundwater

The process by which carbonic acid forms and dissolves calcite, can be described by three simple chemical reactions.

In the first reaction, carbon dioxide (CO_2) and water (H_2O) combine to form carbonic acid (H_2CO_3), as represented by the following equation.

$$CO_2 + H_2O \rightarrow H_2CO_3$$

In the second reaction, carbonic acid splits into hydrogen ions (H^+) and bicarbonate ions (HCO_3^-). This process is represented by the following equation.

$$H_2CO_3 \rightarrow H^+ + HCO_3^-$$

In the third reaction, the hydrogen ions (H^+) react with calcite ($CaCO_3$) and form calcium ions (Ca^{2+}) and bicarbonate ions (HCO_3^-).

$$CaCO_3 + H^+ \rightarrow Ca^{2+} + HCO_3^-$$

Demonstration

Dissolution This demonstration requires several pieces of sandstone (quartz-cemented), limestone, and granite; a small bottle of dilute acid (HCl); and a dropper. Drop some of the acid on each rock sample. Point out only the acid dropped on limestone fizzes, that limestone is mostly calcium carbonate, and the fizzing gas bubbles are carbon dioxide released during the dissolution of calcium carbonate. Explain again that groundwater is slightly acidic and slowly dissolves limestone.

SECTION 2

1 Focus

Focus Transparency

Before presenting the lesson, project **Section Focus Transparency 27** and have students answer the accompanying questions. EL

MAINIDEA

Erosion and Deposition Show students a photo of a cave. Ask students: How was the cave formed? How could you describe the process? erosion and deposition

2 Teach

Teacher Content Support

Calcium Carbonate Most of the geologic activity of groundwater involves the dissolution and precipitation of calcium carbonate. The dissolution of calcium carbonate produces caves and karst topography in limestone regions. The precipitation of calcium carbonate forms stalactites and stalagmites in caves, as well as the cements of many sedimentary rocks.

Groundwater that is acidic dissolves calcium carbonate. The most common acid in groundwater is carbonic acid, which forms from carbon dioxide dissolved in groundwater through the reaction $CO_2 + H_2O \leftrightarrow H_2CO_3$. The double arrow indicates that all three substances (CO_2, H_2O, and H_2CO_3) are present in groundwater, and that, under equilibrium conditions, the rate of formation of H_2CO_3 substances by the forward reaction (arrow to the right) equals the rate of dissociation into CO_2 and H_2O substances by the reverse reaction (arrow to the left).

Identify Misconceptions

Most people believe that dripstone deposits in caves, such as stalactites, form when drops of water coating the cave surfaces evaporate, leaving behind minerals.

Uncover the Misconception

Ask students to explain the formation of dripstone deposits in caves.

Demonstrate the Concept

Inform students that the air in caves is saturated with moisture, so humidity is 100 percent, and therefore water can't evaporate. Explain that carbon dioxide in the water dripping from the cave ceiling is lost to the cave atmosphere; this reduces the concentration of carbonic acid in the water, which allows the calcium carbonate previously kept dissolved in the water by the carbonic acid content to precipitate out of solution and form dripstones.

Assess New Knowledge

Ask students what would happen if a cave was filled with carbon-dioxide gas. Carbon dioxide would diffuse into the water and make it more acidic, the dripstone formations would dissolve, and the cave would gradually grow in size.

Use Science Terms

Cave and Cavern *Cave* is a more general term than *cavern*, which describes a cave that is very deep and whose full extent might not be known.

READING CHECK Most caves form when the slight acidity in groundwater dissolves limestone and, over time, produces hollow chambers in the passages where the water flowed.

Caption Question Fig. 8
$CO_2 + H_2O \rightarrow H_2CO_3$, $H_2CO_3 \rightarrow H^+ + HCO_3^-$, $CaCO_3 + H^+ \rightarrow Ca^{2+} + HCO_3^-$

The resulting calcium ions (Ca^{2+}) and bicarbonate ions (HCO_3^-) are then carried away in the groundwater. Eventually, they precipitate, which means they crystallize out of the solution, somewhere else. Precipitation occurs when the groundwater evaporates or when the carbon dioxide gas leaves the water. The processes of dissolving, called dissolution, and precipitation of calcite both play a major role in the formation of limestone caves, such as those shown in **Figure 8.**

Caves A natural underground opening with a connection to Earth's surface is called a **cave** or a cavern. Some caves form three-dimensional mazes of passages, shafts, and chambers that stretch for many kilometers. Some caves are dry, while some contain underground streams or lakes. Others are totally flooded and can be explored only by cave divers. Mammoth Cave in Kentucky, shown in **Figure 8,** is composed of a series of connected underground passages.

Most caves are formed when groundwater dissolves limestone. The development of most caves begins in the zone of saturation just below the water table. As groundwater infiltrates the cracks and joints of limestone formations, it gradually dissolves the adjacent rock and enlarges these passages to form an interconnected network of openings. As the water table is lowered, the cave system becomes filled with air. New caves then form beneath the lowered water table. If the water table continues to drop, the thick limestone formations eventually become honeycombed with caves. This is a common occurrence in limestone regions that have been uplifted by tectonic forces.

READING CHECK **Explain** how most caves form.

Figure 8 Groundwater dissolution and precipitation result in a variety of features in caves.
Identify *which chemical reactions might be at work.*

Carlsbad Caverns, New Mexico

Mammoth Cave, Kentucky

ACROSS THE CURRICULUM

Biology Ask students to investigate and write a one-page report on an animal that lives in underground streams or caves. Scientists have found over 200 species of animals in the caves of Mammoth Cave National Park. These unique habitats are not only for bats.

DIFFERENTIATED INSTRUCTION

Struggling Learners Have students find photos of caves showing prominent dripstone deposits, including stalactites and stalagmites, and identify them. Students can select their photos from National Geographic magazines or other resources. They should enter brief reports about the photos in their Earth science journals. **EL**

■ **Figure 9** Karst topography is characterized by a landscape of sinkholes formed by dissolution of limestone.
Identify *what controls the rate of dissolution of bedrock in karst topography.*

Karst topography **Figure 9** shows some of the surface features produced by the dissolution of limestone bedrock. One of the main features is a **sinkhole**—a depression in the ground caused by the collapse of a cave or by the direct dissolution of limestone by acidic water. Another type of feature, called a disappearing stream, forms when a surface stream drains into a cave system and continues flowing underground, leaving a dry valley above. Disappearing streams sometimes reemerge on Earth's surface as karst springs.

Limestone regions that have sinkholes and disappearing streams are said to have **karst topography.** The word *karst* comes from the name of a region in Croatia where these features are especially well developed. Prominent karst regions in the United States are located in Kentucky, Indiana, Florida, and Missouri. The Mammoth Cave region in Kentucky has karst topography that contains tens of thousands of sinkholes.

In karst areas, sinkholes proliferate, grow, and eventually join to form wide valleys. Most of the original surface has been dissolved, with the exception of scattered mesas and small buttes. The rate of the dissolution process varies greatly among locations, depending on factors such as humidity and soil composition. In humid areas, where there is more precipitation, more water infiltrates areas of porous soil, and dissolves the limestone in the subsurface.

Groundwater Deposits

Calcium ions eventually precipitate from groundwater and form new calcite minerals. These minerals create spectacular natural features.

Dripstones The most remarkable features produced by groundwater are the rock formations called dripstone that decorate many caves above the water table, as shown in **Figure 10.** These formations are built over time as water drips through caves. Each drop of water hanging on the ceiling of a cave loses some carbon dioxide and precipitates some calcite. A form of dripstone, called a **stalactite,** hangs from the cave's ceiling like icicles and forms gradually. As the water drips to the floor of the cave, it may also slowly build mound-shaped dripstone called **stalagmites.**

■ **Figure 10** Stalactites are dripstones produced by a buildup of minerals precipitated from groundwater.

Teacher Content Support

Karst Topography Water and carbon dioxide can be considered agents of change in an area of limestone bedrock. Areas of the country with limestone bedrock exhibit morphologies associated with the processes of $CaCO_3$ dissolution and precipitation. Caves and sinkholes are a result of the process of dissolution. Often found inside caves, large pillars made of $CaCO_3$ are a result of the deposition of $CaCO_3$.

Collaborative Learning

Chemical Reactions Have students act out the chemical reactions by which carbonic acid forms and dissolves calcite. You will need 11 volunteers to represent two hydrogen atoms, six oxygen atoms, two carbon atoms, and one calcium atom, identified by appropriate name tags such as H (hydrogen), O (oxygen), C (carbon), and Ca (calcium). Have students form the following three substances by linking arms: carbon dioxide (CO_2), water (H_2O), and calcium carbonate ($CaCO_3$). Next, have students simulate the three reactions: the CO_2 molecule holds hands with the H_2O molecule to form H_2CO_3 substances; an H ion breaks away from the H_2CO_3, leaving a bicarbonate ion (HCO_3^-); and the single H atom attacks the $CaCO_3$ molecule, pulls away the Ca atom, and takes its place, forming another bicarbonate ion (HCO_3^-). EL

■ **Caption Question** **Fig. 9** Soil composition and humidity control the rate of dissolution.

EARTH SCIENCE JOURNAL

Karst Topography Have students research and report on the areas in the United States that have karst topography. Students can research karst springs as water supplies, karst caves as unique habitats, and problems associated with city planning and karst topography. OL

Rubric

■ **Caption Question Fig. 11** A likely precipitate in the pipe is calcite. Other answers can include other minerals containing calcium, magnesium, or iron.

3 Assess

Check for Understanding

Explain Ask students to explain the formation of caves. Caves form through the dissolution of limestone by groundwater.

Reteach

Summarize Explain the formation of caves in general terms, and have students read an article of their choice about caves in a nature or science magazine. Have students write brief summaries of the articles they read in their Earth science journals.

Assessment

Explain Have students explain how a geologist is able to identify a limestone region, such as Central Florida, from the air or from a satellite image. Have them list and describe at least two possible indicators. Karst regions are identified by sinkholes, which are closed depressions formed by bedrock dissolution or the collapse of a cave; disappearing streams, which are drained off into a cave system; dry valleys left by disappearing streams; karst springs, which are large streams emerging from a cave; and generally irregular topography with disrupted drainage. Caves themselves are generally not visible from the air.

■ **Figure 11** Hard water contains high concentrations of minerals., which leave precipitates in household water pipes such as this one.
Identify *one of the likely precipitates in this pipe.*

Over time, stalactites and stalagmites can meet and grow into one another to form dripstone columns. Increasingly, researchers are finding abundant and varied microorganisms associated with dripstone formations in caves. It is possible that these organisms play an important role in the deposition of at least some of the materials found in caves.

Hard water You are probably aware that tap water contains various dissolved solids. While some of these materials are added by water treatment facilities, others come from the dissolution of minerals in soils and subsurface rock and sediment. Water that contains high concentrations of calcium, magnesium, or iron is called hard water. Hard water is common in areas where the subsurface rock is limestone. Because limestone is made of mostly calcite, the groundwater in these areas contain significant amounts of dissolved calcite. Hard water used in households can sometimes cause problems. Just as calcite precipitates in caves, it can also precipitate in water pipes, as shown in **Figure 11,** and on the heating elements of appliances. Over time, deposits of calcite can clog water pipes and render some electrical appliances useless.

SECTION 2 REVIEW

Section Self-Check

Section Summary

- Groundwater dissolves limestone and forms underground caves.
- Sinkholes form at Earth's surface when bedrock is dissolved or when caves collapse.
- Irregular topography caused by groundwater dissolution is called karst topography.
- The precipitation of dissolved calcite forms stalactites and stalagmites in caves.

Understand Main Ideas

1. MAINIDEA **Analyze** how limestone is weathered, and identify the features that are formed as a result of this dissolution.
2. **Identify** the acid that is most common in groundwater.
3. **Illustrate** in a series of pictures how caves are formed.
4. **Examine** Why is hard water more common in some areas than others?

Think Critically

5. **Compare and contrast** the formation of stalactites and stalagmites.
6. **Analyze** how you might be able to tell an area of karst topography on a topographic map.

WRITING IN ▸ Earth Science

7. Explain how subsurface limestone is related to karst topography.

SECTION 2 REVIEW

1. Caves are formed in limestone areas when H+ ions dissolve the calcium carbonate. Features such as stalactites and stalagmites are formed as the calcium carbonate is precipitated in the caves.
2. Carbonic acid is most common.
3. Illustrations will vary. Answers should include groundwater dissolving limestone in the zone of saturation below the water table.
4. Hard water is common in limestone areas where the groundwater is saturated with calcium carbonate.
5. Stalactites are formed on the ceiling of a cave as drops of water lose some carbon dioxide and deposit calcium carbonate. Stalagmites are formed when drops of water splash to the floor of the cave.
6. Karst topography could be seen by a series of depressions pockmarking an area on a topographic map. Some might be filled with water and some might not be.
7. The dissolution of limestone creates sinkholes, sinks, and sinking streams, which become interconnected and create karst topography.

Rubric

SECTION 3

Groundwater Supply

MAINIDEA Water is not always available in the quantities and in the locations where it is needed and might be compromised by pollution.

Essential Questions

- How is groundwater withdrawn from aquifers?
- What are the major problems that threaten groundwater supplies?

Review Vocabulary

runoff: water flowing downslope along Earth's surface

New Vocabulary

well
drawdown
recharge
artesian well

EARTH SCIENCE 4 YOU If you have a bank account, can you withdraw as much money as you want? Of course not. Like a bank account, groundwater can be withdrawn, but only in the amount that has been deposited there.

Wells

Wells are holes dug or drilled into the ground to reach an aquifer. There are two main types of wells: ordinary wells and artesian wells.

Ordinary wells The simplest wells are those that are dug or drilled below the water table, into what is called a water-table aquifer, as shown in **Figure 12.** In a water-table aquifer, the level of the water in the well is the same as the level of the surrounding water-table. As water is drawn out of a well, it is replaced by surrounding water in the aquifer.

Overpumping occurs when water is drawn out of the well at a rate that is faster than that at which it is replaced. Overpumping of the well lowers the local water level and results in a cone of depression around the well, as shown in **Figure 12.** The difference between the original water-table level and the water level in the pumped well is called the **drawdown.** If many wells withdraw water from a water-table aquifer, the cones of depression can overlap and cause an overall lowering of the water table, causing shallow wells to become dry. Water from precipitation replenishes the water content of an aquifer in the process of **recharge.** Groundwater recharge from precipitation and runoff sometimes replaces the water withdrawn from wells. However, if withdrawal of groundwater exceeds the aquifer's recharge rate, the drawdown increases until all wells in the area become dry.

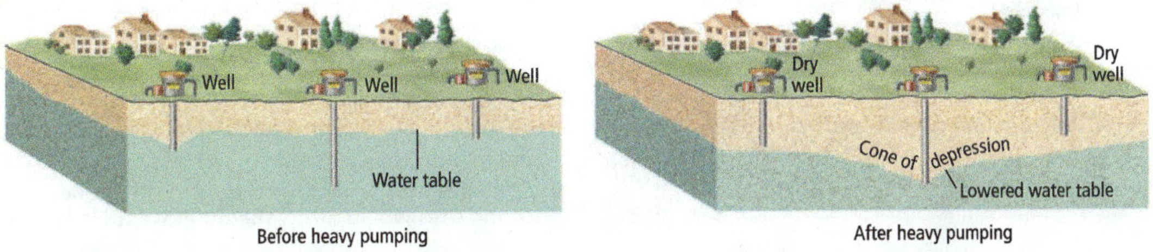

Figure 12 Overpumping from one well or multiple wells can result in a cone of depression and a general lowering of the water table.

Demonstration

Cone of Depression To demonstrate a cone of depression, obtain a Plexiglas box or a large aquarium tank. Fill the box or aquarium with very fine sand and water that has been colored with food coloring such that the water table is a few centimeters below the level sand surface. Rapidly withdraw water from a simulated well at the center of the box or near the center of one of the long aquarium walls, using a water-filled tube with one end stuck in the sand below the water table and the other end at a lower elevation than the box or aquarium. Point out to students the resulting drawdown and cone of depression in the water table. Then, return or recharge the water and have students observe the recovery of the water table.

SECTION 3

1 Focus

Focus Transparency

Before presenting the lesson, project **Section Focus Transparency 28** and have students answer the accompanying questions. EL

MAINIDEA

Groundwater Withdrawal Ask students to describe the various ways in which humans use water. As they say them aloud, list the uses of water on the board, separating them into categories such as crop production, health and hygiene, production of energy, animal husbandry, and others. Stress the fact that many human activities are limited to certain areas where water is available.

2 Teach

Interpret the Illustration

Cone of Depression Ask students to study **Figure 12,** and then ask: What is the difference between the illustration on the left and right? What has changed? The illustration on the right shows a drawdown cone, the result of overpumping.

Teacher Content Support

Saltwater Intrusion In coastal areas and on islands, fresh groundwater floats in hydrostatic equilibrium on top of denser salt water within the zone of saturation. Seawater has a density of about 1.025 g/cm^3, while freshwater has a density of 1.000 g/cm^3 or less. This means that a 40-m column of seawater weighs as much as a 41-m column of lower-density freshwater. Therefore, if the freshwater thickness in a coastal water-table aquifer is 41 m, the water table must be 1 m above sea level.

Problem-Solving LAB

Purpose Students will enhance their understanding of artesian aquifers.

Process Skills make and use graphs, measure and use numbers, apply concepts, compare and contrast, predict, think critically

Teaching Strategies

- Review artesian springs and wells with the class.
- Show students how to plot a similar cross section on the board using fictitious numbers.

Analysis

1. Cross sections should reflect the levels in the data table.
2. The topographic profile can be created by plotting the elevations against the site numbers. Use different lines (dotted, dashed, solid) to connect the elevation points for each column.

Think Critically

3. They will rise to the pressure surface elevation; Site 1, 2 m; Site 2, 1 m; Site 3, overflow 1 m.
4. The water would rise above the surface by 2 m at Site 3.
5. Because Site 3 is the only site where an artesian well could occur, the drawdown of the artesian well could cause the other two wells to become dry, depending on their depths.

Caption Question Fig. 13 An artesian well is pressurized, while an ordinary well is not.

Figure 13 An artesian aquifer contains water under pressure.
Identify *the features that cause the primary difference between an ordinary well and an artesian well.*

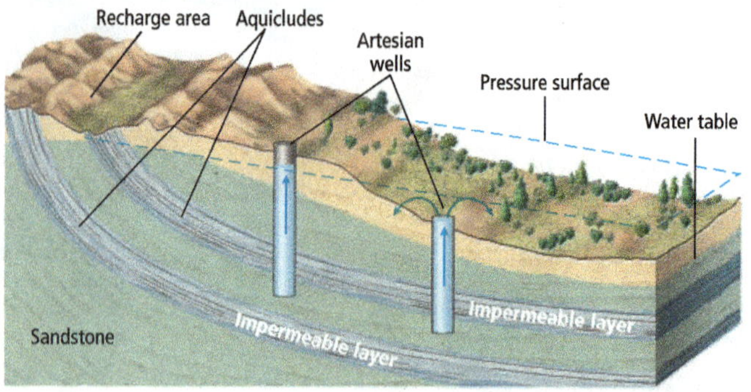

Artesian wells An aquifer's area of recharge is often at a higher elevation than the rest of the aquifer. An aquifer located between aquicludes, called a confined aquifer, can contain water that is under pressure. This is because the water at the top of the slope exerts gravitational force on the water downslope, as you will learn in the Problem-Solving Lab on this page. An aquifer that contains water under pressure is called an artesian aquifer. When the rate of recharge is high enough, the pressurized water in a well drilled into an artesian aquifer can spurt above the land surface in the form of a fountain known as an **artesian well.** The level to which water in an open well can rise is called its pressure surface, as shown in **Figure 13.** Similarly, a spring that discharges pressurized water is called an artesian spring. The name *artesian* is derived from the French province of Artois, where such wells were first drilled almost 900 years ago.

Problem-Solving LAB

Make a Topographic Profile

How does water level vary in an artesian well? Artesian aquifers contain water under pressure. The table provides data about an artesian aquifer for three sites that are spaced 100 m apart along a survey line. It shows the following: elevations of the land surface, the water table, the upper surface of the aquiclude on top of the artesian aquifer, and artesian pressure surface.

Analysis

1. Plot the elevation data on a graph with the sites on the *x*-axis and the elevations on the *y*-axis.
2. Make a topographic profile of the survey line from Site 1 to Site 3. Use a heavy line to indicate land surface.

Aquifer Data

Site	Surface Elevation (m)	Water Table Elevation (m)	Aquiclude Elevation (m)	Pressure Surface (m)
1	396	392	388	394
2	394	390	386	393
3	390	388	381	392

Think Critically

3. **Analyze** How close to the surface will water rise in a well drilled at each site?
4. **Evaluate** what would happen if a well were drilled into the confined aquifer at Site 3.
5. **Consider** how drilling an artesian well at one of the sites would affect the other wells.

EARTH SCIENCE JOURNAL

Apply an Analogy Ask students to describe in writing what would happen if they drilled a small hole into the bottom of a boat. Ask them to apply this analogy to the water level in a recharge area. The water would spurt about as high as the water level outside the boat; this level is the invisible pressure surface inside the boat. The level of the lake is analogous to the water table in a recharge area, the bottom of the boat to the upper aquiclude, the water under the boat to an artesian aquifer, and the hole to an artesian spring. AL

Rubric

Threats to Our Water Supply

Freshwater is Earth's most precious natural resource. Human demands for freshwater are enormous, because it is essential for life. Water is also used extensively in agriculture and industry. **Figure 14** shows freshwater usage in the United States. Groundwater supplies much of this water.

☑ READING CHECK **Summarize** why freshwater is Earth's most precious natural resource.

Estimates of water supplies are the result of a dynamic equilibrium between various factors. These factors include the amounts of precipitation and infiltration, the surface drainage, the porosity and permeability of subsurface rock or sediment, and the volume of groundwater naturally discharged back to the surface. Several of these factors vary naturally over time, and several can be affected by human activities. Changes to groundwater supplies can lead to environmental issues such as a lowered water table, subsidence, and pollution.

An important aquifer in the United States is the Ogallala Aquifer, which underlies the Great Plains. This aquifer delivers water to a huge area stretching from South Dakota to Texas. The recharge areas of the Ogallala Aquifer are located in the Black Hills and the Rocky Mountains.

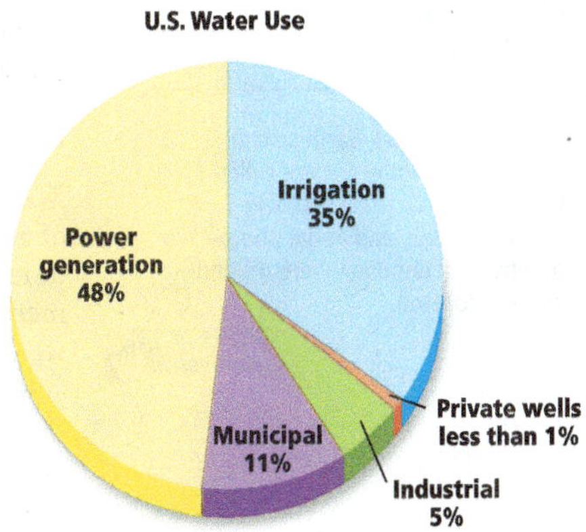

■ **Figure 14** Municipal water and private wells supply your daily water needs.
Identify *how you are involved with water use in each of the other areas.*

MiniLAB

iLab Station

Model an Artesian Well

How does an artesian well form? What causes the water to rise above the ground surface?

Procedure

1. Read and complete the lab safety form.
2. Half fill a **plastic container** with **sand.** Add enough **water** to saturate the sand. Cover the sand completely with a 1- or 2-cm layer of **clay** or a similar impermeable material.
3. Tilt the container at an angle of about 10°. Use a **book** for a prop.
4. Using a **straw**, punch three holes through the clay, one near the low end, one near the middle, and one near the high end of the container. Insert a **clear straw** through each hole into the sand below. Seal the holes around the straws.

Analysis

1. **Observe** the water levels in the straws. Where is the water level the highest? The lowest?
2. **Identify** the water table in the container.
3. **Analyze** Where is the water under greatest pressure? Explain.
4. **Predict** what will happen to the water table and the surface if the water flows from one of the straws.

MiniLAB

iLab Station · Rubric

Purpose Students will demonstrate the nature of artesian wells.

Process Skills compare and contrast, observe and infer, recognize cause and effect, hypothesize

Safety Precaution Approve lab safety forms before work begins.

Teaching Strategies

- Review artesian aquifers and piezometric (pressure) surfaces prior to the lab.
- Use a latex-based putty to avoid volatile organic compounds.
- If water does not rise in any straw, add some water under the clay by lifting a small portion of the clay layer at the elevated end of the shoe box.

Expected Results The water levels in the straws should be highest at the low end of the container.

Analysis

1. The water levels are the highest in the straws toward the bottom of the slope, and lowest toward the top of the slope.
2. The water table is the level of the water in the sand.
3. The water is under the greatest pressure at the bottom of the slope, because that is where it experiences the most weight of overlying water.
4. Both the water table in the sand and the pressure surface will be lowered.

Assessment

Knowledge Ask students where the water pressure is greater: at the bottom of a lake 10 m below the water surface, or in an artesian limestone aquifer 10 m below the water table in the recharge area. The pressures should be equal.

LabManager™
Customize any lab with the LabManager™ CD-ROM.

ACROSS THE CURRICULUM

Chemistry Pollutants found in groundwater include various organic and inorganic chemicals. Some inorganic chemicals dissolved in groundwater are acids, salts, nitrates, phosphates, and compounds of toxic metals. Organic chemicals include hydrocarbons, such as oil and gasoline, and synthetic chemicals, pesticides, and solvents, among others.

■ **Caption Question Fig. 14** Water for industry provides us with many products for personal use, irrigation provides foodstuffs for people and animals, and power generation provides heating, cooling, cooking, lighting, entertainment, and recreation.

☑ READING CHECK Freshwater is essential for many aspects of survival.

CAREERS IN EARTH SCIENCE

WebQuest

Environmental Connection

Arsenic One of the most popular poisons in murder mysteries is arsenic. Surprise your students by informing them they drink arsenic every day: drinking water contains arsenic. Minute concentrations of arsenic are present in groundwater everywhere. Some of it is introduced by industrial waste and mining activities, but most of it occurs naturally, having been leached from arsenic-bearing rocks and minerals underground.

Concentrations of arsenic in groundwater typically are several parts per billion (ppb). Arsenic concentrations tend to be higher in the western states than in the eastern and southern states, but in all states, arsenic concentrations are much too small to cause poisoning. However, recent studies have linked high arsenic concentrations in drinking water to lung and bladder cancer. In 2002, the Arsenic Rule was established by the EPA, which limits concentrations in drinking water to 10 ppb, a value exceeded by at least some groundwater systems.

☑ READING CHECK Water-table aquifers are most vulnerable to pollution. Pollutants reach the groundwater by infiltrating the ground along with precipitation.

GeoLAB

iLab Station

The GeoLab located at the end of the chapter can be used at this point in the lesson.

CAREERS IN EARTH SCIENCE

Hydrogeologist Earth scientists who map groundwater are called hydrogeologists. They use field methods, maps, and aerial photographs to determine where groundwater is located.

WebQuest

Overuse Groundwater supplies can be depleted. If groundwater is pumped out at a rate greater than the recharge rate, the groundwater supply will decrease and the water table will drop. This is happening to the Ogallala Aquifer. Its water, used mostly for irrigation, is being withdrawn at a rate much higher than the recharge rate.

Subsidence Another problem caused by the excessive withdrawal of groundwater is ground subsidence—the sinking of land. The volume of water underground helps support the weight of the soil, sediment, and rock above. When the height of the water table drops, the weight of the overlying material is increasingly transferred to the aquifer's mineral grains, which then squeeze together more tightly. As a result, the land surface above the aquifer sinks.

A dramatic example of subsidence can be seen along parts of the Gulf coast of Texas, where heavy usage of groundwater over many decades resulted in a wide-scale drop in the ground level. In a region of 12,000 km^2, the average subsidence was 15 cm, while some areas dropped by as much as 3 m. This has presented flooding hazards for much of the coastal region.

Watch a **video about water contamination.**

Video

Pollution in groundwater In general, the most easily polluted groundwater reservoirs are water-table aquifers, which lack a confining layer above them. Confined aquifers are affected less frequently by local pollution because they are protected by impermeable barriers. When the recharge areas of confined aquifers are polluted, however, those aquifers can also become contaminated.

☑ READING CHECK **Identify** which kind of aquifer is more vulnerable to pollution.

Sources of groundwater pollution include sewage from faulty septic tanks and farms, landfills, and other waste disposal sites. Pollutants usually enter the ground above the water table, but they eventually infiltrate to the water table. In highly permeable aquifers, pollutants can spread quickly in a specific direction, such as toward the wells shown in **Figure 15.**

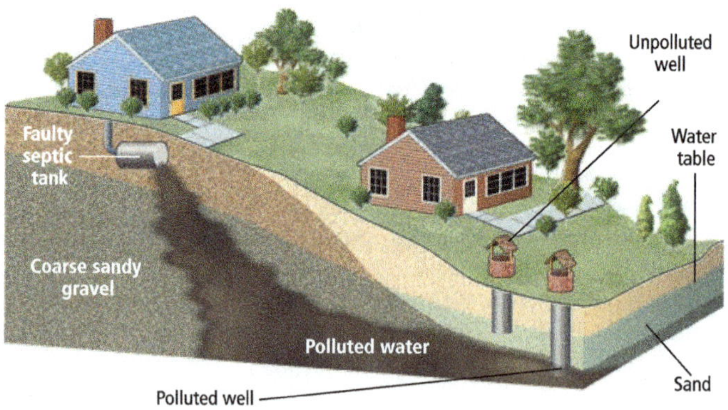

Figure 15 Pollutants can spread rapidly through a highly permeable aquifer. Note how the polluted well has drawn the pollution toward it as it has withdrawn water from the water table.

DIFFERENTIATED INSTRUCTION

Struggling Learners Lead a discussion with students about how people can conserve groundwater. Ask students to list at least five ways they can reduce usage of water in their homes and why it is important.

What's EARTH SCIENCE Got To Do With It?

Video *What Are You Drinking?*

Chemicals Because chemicals dissolved and transported with groundwater are submicroscopic in size, they can travel through the smallest pores of fine-grained sediment. For this reason, chemicals such as arsenic can contaminate any type of aquifer. The chemicals generally move downslope from a source in the form of a pollution plume, which is a mass of contaminants that spreads through the aquifer. Once chemical contaminants have entered groundwater, they cannot be easily removed. In the GeoLab at the end of the chapter, you will learn more about how geologists predict the risks of chemical contamination of groundwater based on a region's topography.

READING CHECK **Explain** why chemicals such as arsenic can contaminate any kind of aquifer.

Sewage, landfills, and other waste disposal sites can include a variety of contaminants. Chemical contaminants can be leached, meaning dissolved by infiltrating groundwater. When chemical and biological contaminants enter the groundwater, they flow through the aquifer at the same rate as the rest of the groundwater. Over time, an entire aquifer can become contaminated and toxic to humans. Aquifers are particularly vulnerable to pollution in humid areas where the water table is shallow and can more easily come in contact with waste.

Salt Not all pollutants are toxic or unhealthful in and of themselves. For example, ordinary table salt is used to season food, but water is undrinkable when its salt content is too high. In like manner, groundwater is unusable after the intrusion of salt water. Salt pollution is one of the major threats to groundwater supplies, especially in coastal areas, where the intrusion of salt water into groundwater is a major problem. In coastal areas, salty seawater, which is denser than freshwater, underlies the groundwater near Earth's surface, as shown in **Figure 16.** The overpumping of wells can cause the underlying salt water to rise into the wells and contaminate the freshwater aquifer.

VOCABULARY

ACADEMIC VOCABULARY

Transport
to move from one place to another
Airplanes transport packages across the country.

Figure 16 Freshwater aquifers can become contaminated with salt water.
Identify *how overpumping can cause the underlying salt water to rise in wells.*

View an **animation of salt water contamination.**

Concepts In Motion

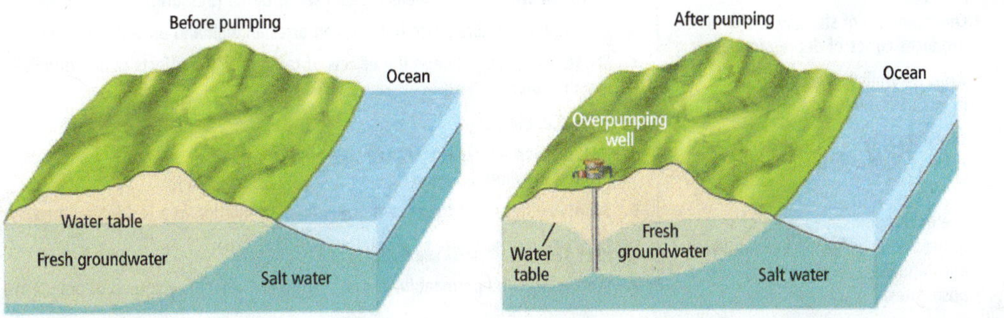

Teacher Content Support

Saltwater Infiltration It doesn't take much contamination by seawater to make well water undrinkable. According to World Health Organization standards, the maximum allowable concentration of total dissolved solids (TDS) in drinking water is 1500 ppm. Water exceeding this limit is considered to be brackish. The average salinity of seawater is 35,000 ppm.

If nine parts of distilled water are mixed with one part seawater, the resulting mixture will have a salinity of 3500 ppm, well above the allowable limit. A mixture containing 4 percent seawater would have a salinity of 1400 ppm. However, most fresh groundwater already contains several hundred ppm of dissolved ions. If, for instance, groundwater containing 500 ppm of dissolved ions is mixed with seawater, then a mixture containing less than 3 percent seawater would exceed the allowable limit of 1500 ppm. As a result, many wells in coastal areas contain brackish water.

READING CHECK Chemicals can contaminate any kind of aquifer because they are dissolved in the water and travel wherever the water goes.

Caption Question Fig. 16 Overpumping removes the lighter fresh groundwater so that the underlying salt water rises into wells.

EARTH SCIENCE JOURNAL

Groundwater Supply Have each student compile a newspaper article about problems that affect groundwater supplies using resources available. Ask them to present their reports to the class. OL

Rubric

3 Assess

Check for Understanding

Explain Ask students to explain why artesian wells are less prone to pollution than other wells. Artesian wells are protected from surface pollution by aquicludes.

Reteach

Use Models Have the class plan a lab-sized model of an artesian aquifer overlain by a water-table aquifer. Have students develop the plan on the board and include the various permeable and impermeable geological materials they would use to construct the aquifer. Have the class discuss the geometry of the model, the necessary layering, the location of the recharge area, and the placement of model wells to show flowing and nonflowing artesian wells and a water-table well. COOP LEARN

Assessment

Performance Have students compare and contrast artesian wells with wells drilled into a water-table aquifer. Have each student include a drawing showing the basic parameters: recharge area, pressure surface, ground surface, confined aquifer, local water table, flowing artesian well, water-table well, and cone of depression.

Table 2 Groundwater Pollution Sources
Infiltration from fertilizers
Leaks from storage tanks
Drainage of acid from mines
Seepage from faulty septic tanks
Saltwater intrusion into aquifers near shorelines
Leaks from waste disposal sites
Radon

Radon Another source of natural groundwater pollution is radon gas, which is one of the leading causes of lung cancer in the United States. Radon found in groundwater is one of the products of the radioactive decay of uranium in rocks and sediment, and it usually occurs in very low concentrations in all groundwater. However, some rocks, especially granite and shale, contain more uranium than others. Therefore the groundwater in areas where these rocks are present contains higher levels of radon.

Some radon can seep into houses, and, because it is heavier than air, it can accumulate in poorly ventilated basements. The United States Environmental Protection Agency (EPA) advises homeowners in radon-prone regions to have their homes tested regularly for radon gas.

Protecting Our Water Supply

There are a number of ways by which groundwater resources can be protected and restored. First, major pollution sources, many of which are listed in **Table 2,** need to be located, identified and eliminated. Pollution can enter groundwater resources via runoff and infiltration, or directly from underground. Pollution plumes that already exist can be monitored with observation wells and other techniques. Most pollution plumes spread slowly providing adequate time for alternate water supplies to be found. In some cases, pollution plumes can be stopped by building impermeable underground barriers around the polluted area. Sometimes, polluted groundwater can be pumped out for chemical treatment on the surface.

While these measures can have limited success, they alone cannot save Earth's water supply. Humans must be aware of how their activities impact the groundwater system so that they can protect the water supply.

SECTION 3 REVIEW

Section Self-Check

Section Summary

- Wells are drilled into the zone of saturation to provide water.
- Overpumping of shallow wells produces cones of depression.
- Artesian wells tap confined aquifers in which water is under pressure.
- When groundwater withdrawal exceeds recharge, it lowers the water table
- The most common sources of groundwater pollution include sewage, landfills, and other waste disposal sites.

Understand Main Ideas

1. MAINIDEA **Evaluate** the problems associated with overpumping wells.
2. **Explain** why artesian wells contain water under pressure.
3. **Illustrate** the difference between an artesian well and an ordinary well.
4. **Differentiate** between the effects of radon and the effects of salt dissolved in groundwater.

Think Critically

5. **Formulate** an experiment which would test if there were impermeable barriers around a polluted area.
6. **Analyze** how best to prevent groundwater pollution in a residential area.

WRITING IN Earth Science

7. Predict how the permeability of an aquifer can affect the spread of pollutants.

SECTION 3 REVIEW

1. If wells are overpumped, then groundwater supplies decrease and the water table drops. If wells are overpumped near the ocean, salt water can infiltrate.
2. An artesian well is under pressure because the confined aquifer is under pressure.
3. Illustrations will vary. Answers should show that an artesian well draws from a confined aquifer, while an ordinary well draws from a water-table aquifer.
4. Radon is a leading cause of cancer in the United States. Radon occurs in low concentrations in all groundwater. On the other hand, salt intrusion makes groundwater undrinkable, but salt is not toxic or unhealthful in and of itself.
5. Experiments will vary, but should include knowledge that the aquiclude is an impermeable barrier.
6. Groundwater pollution could be reduced in a residential area by decreasing use of road salt, monitoring for radon, and decreasing use of pesticides and other chemicals.
7. The higher the permeability of an aquifer, the more easily pollutants move into (and out of) it.

Rubric

Earth Science &

ENVIRONMENT

Watcher of the Water

Safe drinking water is something that many people take for granted. Most of the water that is used for human consumption comes from groundwater. Who ensures that groundwater sources remain safe?

Hydrogeologists A groundwater scientist, called a hydrogeologist, is responsible for finding and monitoring groundwater sources to ensure the water supply is free of contaminants and is not used faster than it is replaced. What does a typical day in the life of a hydrogeologist look like? One day might be spent in the field conducting tests on the water levels. The next day might be spent evaluating the data in the office. The day after that might involve looking for trouble in the water-supply line of a house.

Aquifer case study Suppose a farmer wants to install an irrigation system, which involves digging a new well. First, the water level in the area's aquifer must be checked to ensure that a new well will not cause shortages for other users. The hydrogeologist finds an active well nearby and hooks it up to a pump that continuously draws water for 24 hours. Periodic checks of other wells in the area determine the changes in the water level and quality. From the data gathered, he or she computes how much water the aquifer contains and determines the amount of water available for a new well.

Suppose that, after the farm starts using the irrigation system, a house down the road loses its water supply. The hydrogeologist goes to the house and checks for technical problems, such as a hole in the well casing. If the cause is not technical, he or she will reassess the irrigation system by rechecking the water supply in the aquifer.

These hydrogeologists collect water from a well to determine whether or not it has been contaminated.

Kevin Fleming/CORBIS

Quality assurance Hydrogeologists are also responsible for checking water quality. If the water from a particular aquifer develops a strange taste and odor, the residents would want to ensure the water is safe to drink. The hydrogeologist gathers samples and sends them to a lab to test for various contaminants, such as sewage, pesticides, dissolved metals, or organic material. If a contaminant is found, the hydrogeologist will advise the residents not to drink the water until the source is discovered and the problem is resolved. The hydrogeologist will then begin investigating the problem and searching for clues to find and stop the contamination.

WRITING IN ▶ Earth Science

Journal Research more about hydrogeologists. Then, imagine you are accompanying a groundwater scientist on a day on the job. Describe what you saw, what you did, and what you learned about aquifers.

WebQuest

Earth Science &

ENVIRONMENT

Purpose

Students will learn how a hydrogeologist studies water use and water quality to ensure there is enough clean water to support life.

Teacher Content Support

Aquifers An aquifer is a source of underground water. It occurs when water takes up the empty spaces in the earth, and can be tapped to provide a water source. The water is drawn from the aquifer through capillary action. Hydrogeology is a multidisciplinary science, which incorporates the chemical, physical, biological, and legal aspects of groundwater to make sure there is a safe amount of water for everyone.

Hydrogeologists use complex mathematical equations and computer programs to calculate the amount of water in an aquifer and conduct sophisticated studies to ensure the water is not depleted faster than it can be restored. Groundwater is vulnerable to seepage from waste management sites, chemical pollutants, and human waste pollution. Part of the job of the hydrogeologist is to ensure water is safe.

Teaching Strategy

Have the students read about aquifers before class. A general background is helpful when building their own. Have each group construct their own aquifer and well system. Give the students enough time to discuss how to build it and how to get the well to work. After they have built a working model, have them present their model to the class. Provide examples for students of work done by hydrogeologists.

WRITING IN ▶ Earth Science

Rubric

Journal Answers will vary. Students should demonstrate they have fully researched all the aspects of a hydrogeologist's occupation. Give the students an opportunity to share what they have learned with the class.

WebQuest

EARTH SCIENCE JOURNAL

Hydrogeology Ask students to describe in writing three ways that hydrogeologists ensure water supply. Ask students to make predictions about what might happen if an irrigation system for a farm fails. You might find that suggesting an illustration or drawing helps students illustrate their predictions about water supply. BL

Rubric

GeoLAB

iLab Station Rubric

Preparation

Time Allotment 45 min

Process Skills collect and interpret data, measure, use numbers, make and use graphs, apply concepts, draw a conclusion, predict

Safety Precaution Approve lab safety forms before work begins.

Procedure

- Review map scales and contour lines with the class.
- Have students work in groups of three or four. If possible, each group should include a student who is an advanced learner.
- Make sure students follow the instructions step by step.
- **Troubleshooting** Some students might have trouble locating all the lakes in the data table. Be prepared for questions. Some lake elevations are not given. These can be estimated from the topographic map contours. The water-table contours are not well defined everywhere by data points. Be sure students produce reasonable contour maps.
- Jim's Gas Station is represented on the map by a magenta rectangle.

LabManager™
Customize any lab with the LabManager™ CD-ROM.

GeoLAB

iLab Station

Mapping: Track Groundwater Pollution

Background: You can use a topographic map to estimate the direction of groundwater flow. Groundwater pollution spreads out from its source and follows the flow of groundwater. The spread and movement of the pollution resembles a plume that stems from its source.

Question: *How can you determine the movement of a pollution plume?*

Materials

U. S. Geological Survey topographic map of Forest City, Florida
transparent paper
ruler
graph paper
calculator

Procedure

Imagine that Jim's Gas Station has discovered a major gasoline leak from one of its underground tanks. As the local hydrogeologist, you are asked to determine the path that the gasoline will take through the groundwater, and to notify the residents of the areas that might be affected by the contamination.

1. Read and complete the lab safety form.
2. Identify the lakes and swamps in the southwest corner of the map and list their names and elevations in a data table. (**Note:** *The elevations are given or can be estimated from the contour lines. The elevation of the water table in each area can be estimated from the elevations of nearby bodies of water*).
3. Note the location of Jim's Gas Station on Forest City Rd., about 1400 feet north of the Seminole County line (at the 96-foot elevation mark).
4. Take out a piece of paper to construct a cross section of the surface topography and the water table. Lay the paper on the map from Lake Lotus to Lake Lucien (through Jim's Gas Station).
5. On this piece of paper, mark the location of Jim's Gas Station.
6. Draw a small line at each place where a contour line intersects the line from Lake Lotus to Lake Lucien. Also note the elevation at each hash mark and any rivers crossed.
7. Draw a table to use for your topographic profile, using the width representing the distance between Lake Lotus to Lake Lucien. For the y-axis, use the elevations 60, 70, 80, 90, and 100 ft.
8. Now take your paper where you marked your lines and place it along the base of the table.
9. Mark a corresponding dot on the table for each elevation, and mark the position of Jim's Gas Station.
10. Connect the dots to create a topographic profile.
11. Note the elevations of the nearby bodies of water to approximate the distance from the ground surface to the water table. Use dots to indicate those distances on the topographic profile. Connect the dots to draw the water table on the topographic profile.

Analyze and Conclude

1. **Calculate** the slope of the ground surface on either side of Jim's Gas Station.
2. **Estimate** the slope of the water table at Jim's Gas Station.
3. **Infer** the direction toward which the pollution plume will move.
4. **Identify** the houses and bodies of water that are threatened by this pollution plume.
5. **Conclude** Prepare a written statement to present to the local community. Explain the path the plume is predicted to take, and how this was determined.

APPLY YOUR SKILL

Design Using what you have learned in this lab and in the chapter, develop a plan for stopping the pollution plume. Make a map showing where your plan will be implemented. Indicate the sites where water quality will be monitored regularly.

Analyze and Conclude

1. 44 ft/mi on either side
2. at most, 1/100 (10'/1000')
3. northwest
4. none
5. Answers will vary. Students should demonstrate an understanding of the downslope direction that groundwater, and consequently pollution plumes, will flow in areas with varying elevations.

APPLY YOUR SKILL

Design Answers will vary, but should include a map with sites where water quality will be monitored.

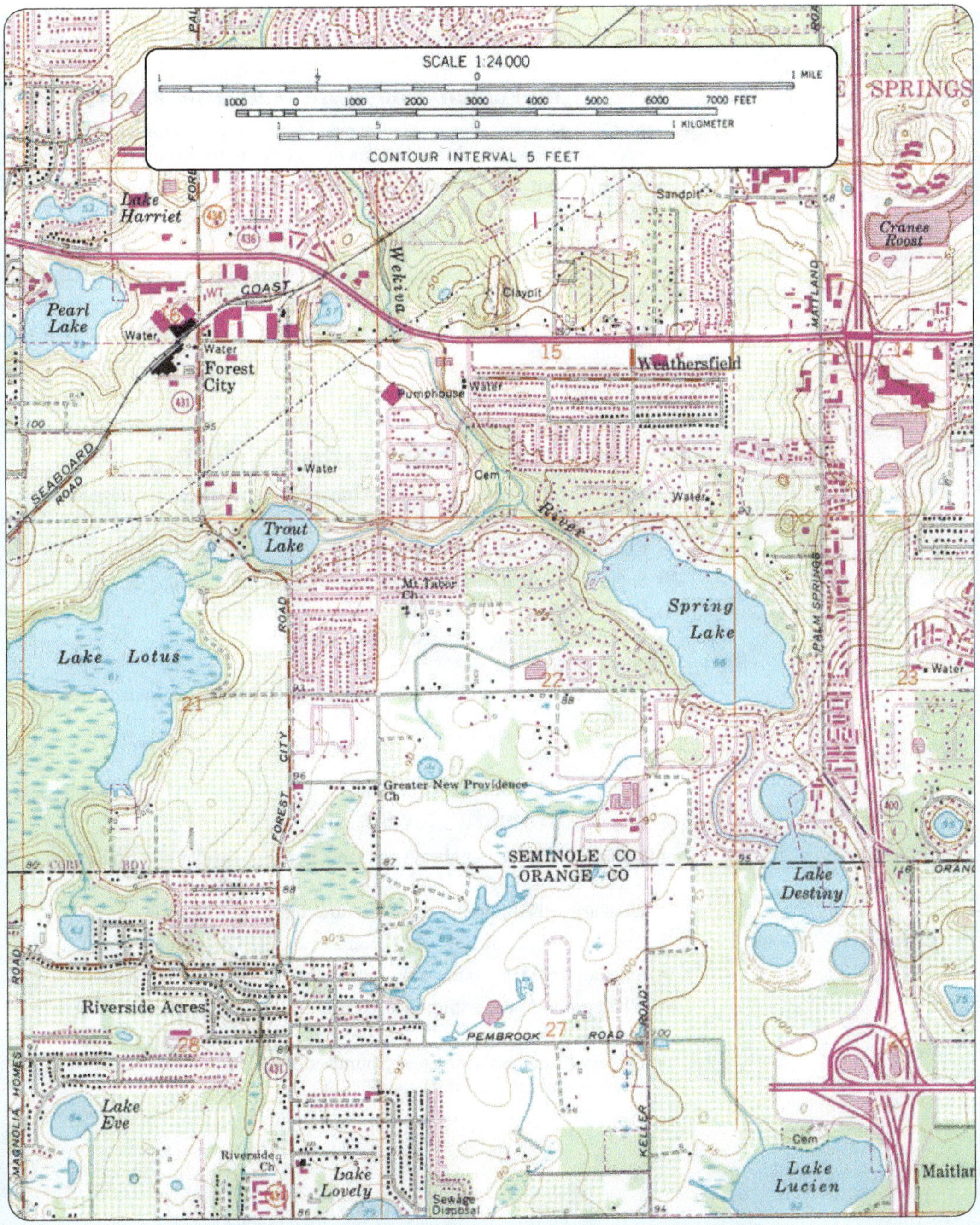

SCALE 1:24 000
1 MILE
7000 FEET
1 KILOMETER
CONTOUR INTERVAL 5 FEET
Lake Harriet
Pearl Lake
Forest City
Weathersfield
Trout Lake
Spring Lake
Lake Lotus
Greater New Providence Ch
SEMINOLE CO
ORANGE CO
Lake Destiny
Riverside Acres
PEMBROOK ROAD
Lake Eve
Lake Lovely
Lake Lucien
Cranes Roost
SPRINGS
Sewage Disposal

USGS

CHAPTER 10 STUDY GUIDE

MAINIDEAS Summary statements can be used by students to review the major concepts of the chapter.

Students can review with these online resources.

Vocabulary Practice

Vocabulary eGames
Vocabulary eFlashcards
Vocabulary PuzzleMaker

Section Self-Check

Chapter Self-Check

Online Test Practice

Use the Exam*View*® *Assessment Suite* CD-ROM to:

- create multiple versions of tests
- create modified tests with one mouse click
- edit existing questions and add your own questions
- build tests aligned with state standards using built-in state curriculum tags
- change English tests to Spanish with one mouse click
- track students' progress using the Teacher Management System

CHAPTER 10 STUDY GUIDE

BIGIDEA **Precipitation and infiltration contribute to groundwater, which is stored in underground reservoirs until it surfaces as a spring or is drawn from a well.**

Vocabulary Practice

SECTION 1 Movement and Storage of Groundwater

VOCABULARY
- infiltration
- zone of saturation
- water table
- zone of aeration
- permeability
- aquifer
- aquiclude
- spring
- hot spring
- geyser

MAINIDEA **Groundwater reservoirs provide water to streams and wetlands wherever the water table intersects the surface of the ground.**

- Some precipitation infiltrates the ground to become groundwater.
- Groundwater is stored below the water table in pore spaces of rocks and sediment.
- Groundwater moves through permeable layers called aquifers and is trapped by impermeable layers called aquicludes.
- Groundwater emerges from the ground where the water table intersects Earth's surface.

SECTION 2 Groundwater Weathering and Deposition

VOCABULARY
- cave
- sinkhole
- karst topography
- stalactite
- stalagmite

MAINIDEA **Chemical weathering of limestone by water causes the characteristic topography of karst areas.**

- Groundwater dissolves limestone and forms underground caves.
- Sinkholes form at Earth's surface when bedrock is dissolved or when caves collapse.
- Irregular topography caused by groundwater dissolution is called karst topography.
- The precipitation of dissolved calcite forms stalactites and stalagmites in caves.

SECTION 3 Groundwater Supply

VOCABULARY
- well
- drawdown
- recharge
- artesian well

MAINIDEA **Water is not always available in the quantities and in the locations where it is needed and might be compromised by pollution.**

- Wells are drilled into the zone of saturation to provide water.
- Overpumping of shallow wells produces cones of depression.
- Artesian wells tap confined aquifers in which water is under pressure.
- When groundwater withdrawal exceeds recharge, it lowers the water table.
- The most common sources of groundwater pollution include sewage, landfills, and other waste disposal sites.

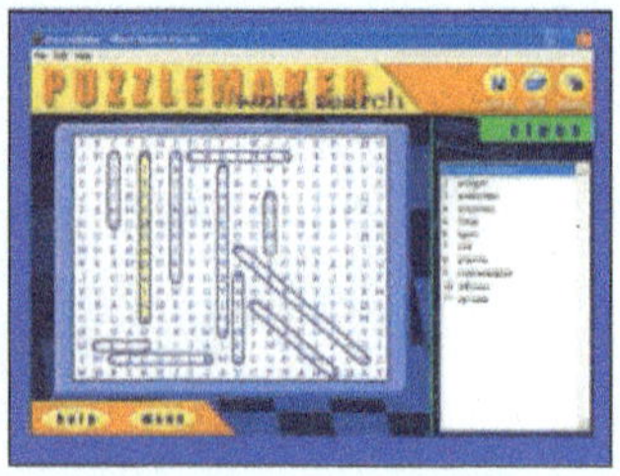

Vocabulary Practice

For additional practice with vocabulary, have students access the Vocabulary PuzzleMaker.

Chapter Self-Check

VOCABULARY REVIEW

Match each phrase with a vocabulary term from the Study Guide.

1. the depth below Earth's surface at which all pores in layers of soil are filled with water
2. the vertical movement of water through ground layers
3. all of the permeable layers at a location
4. the replacement of water content in an aquifer

Each of the following sentences is false. Make each sentence true by replacing the italicized words with a vocabulary term from the Study Guide.

5. *Drawdown* is produced in limestone regions that have sinkholes and sinking streams.
6. *Stalagmites* are icicle-shaped deposits hanging from the ceiling of caves.
7. Collapsing caves or dissolution of bedrock at the surface produce *systems of caves.*

Use what you know about the vocabulary terms found on the Study Guide to answer the following questions.

8. What two features are most often associated with the formation of springs?
9. What is the main difference between regular springs and artesian springs?
10. What are explosive hot springs that develop in volcanic areas?

UNDERSTAND KEY CONCEPTS

11. Which single source of freshwater represents the largest volume of freshwater worldwide readily available for use by humans?
    A. ice caps and glaciers
    B. freshwater lakes
    C. rivers and streams
    D. groundwater deposits

12. Sinkholes may eventually join to form
    A. wide valleys
    B. zone of aeration
    C. dripstones
    D. aquifer

13. What is the name of a layer of sediment or rock that does not allow water to pass through it?
    A. a permeable layer
    B. an aquiclude
    C. an aquifer
    D. a nonaqueous layer

Use the diagram below to answer Questions 14 and 15.

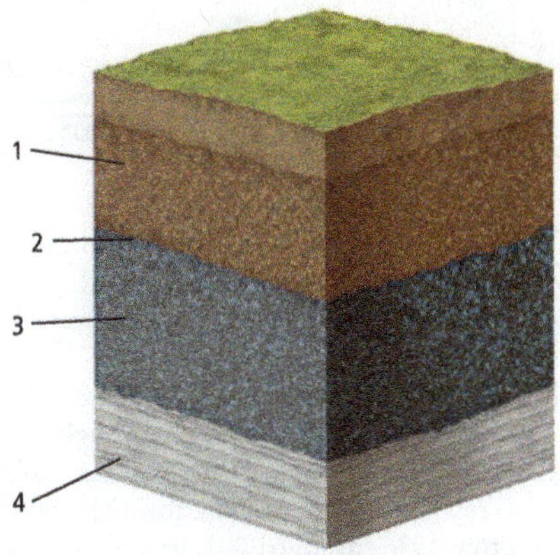

14. Which sequence of terms correctly labels the features shown in the diagram?
    A. 2: water table, 3: impermeable layer
    B. 3: surface zone, 4: impermeable layers
    C. 1: zone of aeration, 3: zone of saturation
    D. 1: zone of saturation, 3: zone of aeration

15. In which layer do the pores contain mostly air, although the materials are moist?
    A. layer 1
    B. layer 2
    C. layer 3
    D. layer 4

CHAPTER 10 ASSESSMENT

VOCABULARY REVIEW

1. zone of saturation
2. infiltration
3. aquifer
4. recharge
5. Karst topography
6. Stalactites
7. sinkholes
8. aquifer and aquiclude
9. In artesian springs, water surfaces due to pressure, while regular springs are not under pressure.
10. Geysers are explosive hot springs in volcanic areas.

UNDERSTAND KEY CONCEPTS

11. D
12. A
13. B
14. C
15. A

16. C
17. D
18. D
19. C
20. C
21. C

CONSTRUCTED RESPONSE

22. The water table is at the surface in an area with lakes and wetlands, whereas it is under the surface in an area with no standing water.
23. An aquifer must be confined and under pressure.
24. In humid areas, the water table is more likely to be close to the surface. In arid areas, the water table is more likely to be far below the surface.
25. The cement that binds grains in sedimentary rock affects the amount of pore space between grains as well as the degree to which water can flow through the pore space.
26. A small aquifer would not experience any recharge during a drought and the water table might drop causing the aquifer to go dry.
27. Caves are more likely to form in an area containing limestone bedrock because groundwater is slightly acidic and can attack carbonate rocks, such as limestone.
28. Disposal of toxic waste into a sinkhole poses threats to the water supply because once in the sinkhole, the pollution can enter the groundwater and travel through the water supply.

16. Which characteristics do most areas with karst topography share?
A. they are dry areas; limestone bedrock
B. they are humid areas; granite bedrock
C. they are humid areas; limestone bedrock
D. they are dry areas; granite bedrock

Use the graph below from a single well in North Carolina to answer Questions 17 and 18.

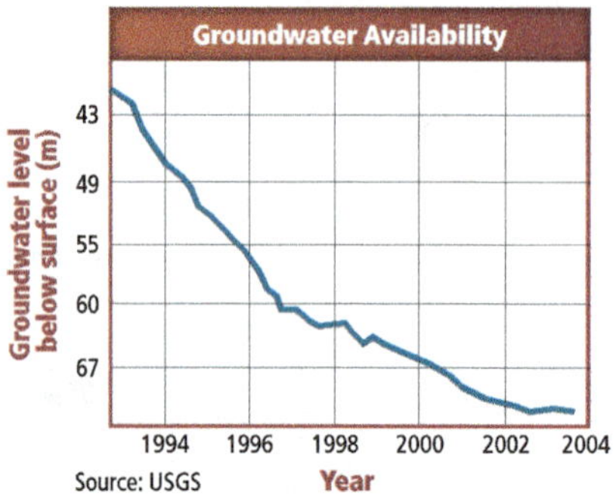

17. Which statement is a logical conclusion that can be drawn from information in the graph?
A. From 1993 through 2003, groundwater availability at this well has increased.
B. From 2002 through 2003, the water table has fallen faster than from 1993 through 1994.
C. From 1993 through 1994, the water table has fallen less than from 2002 through 2003.
D. From 1993 through 2003, groundwater availability at this well has declined.

18. What year was the groundwater level the highest?
A. 2004 **C.** 1996
B. 2003 **D.** 1993

19. What forms when carbon dioxide dissolves in water?
A. calcite **C.** carbonic acid
B. acid rain **D.** hydrogen ions

20. What characteristic must porous rocks have for them to be permeable?
A. They must be above the water table.
B. Their pores must be large.
C. Their pores must be interconnected.
D. They must be below the water table.

Use the diagram below to answer Question 21.

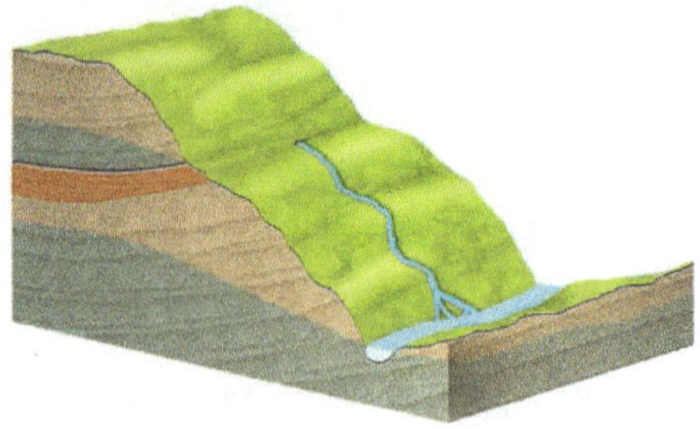

21. What conditions are required for the formation of the spring?
A. defined areas of aeration, saturation, and an impermeable layer
B. an aquiclude holding water above defined areas of aeration and saturation
C. an aquiclude holding water above the main water table, and recharged from above
D. an aquiclude defining a main water table, and recharged from above

CONSTRUCTED RESPONSE

22. **Classify** the locations of the water table in a lake or wetland and the water table in a region with no surface water.

23. **Identify** the two features an aquifer must have to be a source of artesian water.

24. **Compare and contrast** how the water table differs between humid and arid regions.

25. **Examine** how the cement that binds the grains of sedimentary rocks affects the porosity and permeability of the rock.

26. **Predict** how a small aquifer will be affected by a multiyear drought.

27. **Generalize** whether caves are more likely to develop in a region containing limestone bedrock or sandstone bedrock. Justify your answer.

28. **Explain** why disposal of toxic waste into a sinkhole can pose serious hazards for local drinking water.

Chapter Self-Check

THINK CRITICALLY

29. **Formulate** an explanation for why stalactites have a tapering shape whereas stalagmites usually have less regular shapes and broader bases.

30. **Hypothesize** the effect that a severely lowered water table would have on the emergence of springs on a hillside.

31. **Infer** why caves often include dry chambers although most caves develop in the zone of saturation just below the water table.

Use the photo below to answer Question 32.

32. **Consider** the stream emerging from Earth's surface. Diagram a scenario that would explain the role of groundwater in the photo.

33. **Assess** what would be an important consequence of sea level rise on groundwater supplies in coastal areas.

CONCEPT MAPPING

34. Make a concept map using the following terms: *ordinary well, artesian well, aquiclude, confined, unconfined,* and *water-table aquifer.*

CHALLENGE QUESTION

35. **Infer** the effect that increased atmospheric CO_2 concentration might have on structures made of calcite and the development of karst topography.

WRITING IN ▶ Earth Science

36. Write a short story to demonstrate how shared groundwater resources could cause conflict between neighboring states or countries.

DBQ Document–Based Questions

Data obtained from: Lerch, R.N., C.M Wicks, and P.L. Moss. 2006. Hydrological characterization of two karst recharge areas in Boone County, Missouri. *Journal of Cave and Karst Studies* 67 (3): 158–173.

In the graphs, monthly precipitation (bar graph) and monthly discharge (line graph) are shown for Devil's Icebox cave streams in Missouri.

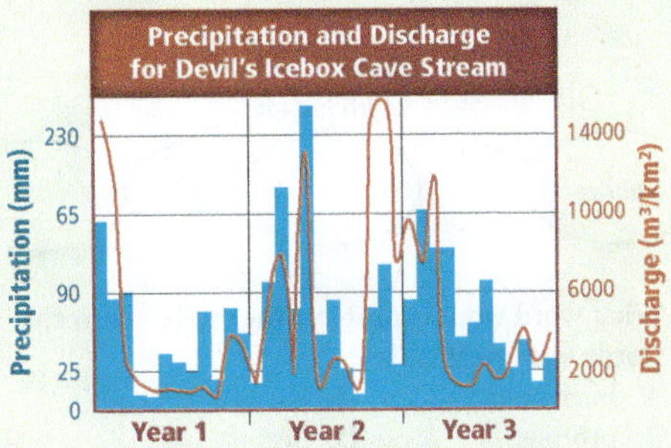

37. Which year had the most precipitation?

38. What is the general relationship between precipitation and discharge according to the graph?

39. Identify an exception to the general relationship described in Question 38. Suggest a possible explanation for this exception.

CUMULATIVE REVIEW

40. Compare and contrast the ionic bond and the covalent bond. **(Chapter 3)**

41. What is the significance of the rock cycle? **(Chapter 6)**

42. List factors that affect the rate of weathering. Underline the factor you think is the most important in chemical weathering and explain why. **(Chapter 8)**

THINK CRITICALLY

29. Stalactites taper because gravity pulls the water downward, but leaves calcium carbonate to precipitate; stalagmites form on many kinds of uneven surfaces.
30. Springs would not emerge on a hillside with a severely lowered water table.
31. Caves often include dry chambers because partially dry chambers can be left above the downward flow of water, tectonic uplifting might have affected the cave, and the water table might have become lowered due to drought or overuse.
32. Diagrams will vary, but should include the location of a spring where the groundwater intersects the ground.
33. Seal level rise might decrease the water supply in coastal areas with the infiltration of salt water into the groundwater.

CONCEPT MAPPING

34. Answers will vary. Concept maps should show that both types of wells tap into aquifers, but that artesian wells tap into confined aquifers.

CHALLENGE QUESTION

35. An increase in atmospheric carbon dioxide might lead to an increase in carbonic acid, which forms when carbon dioxide gas dissolves in water and combines with water molecules. Carbonic acid would erode the limestone in caves and karst topography.

WRITING IN ▶ Earth Science

Rubric

36. Answers will vary, but should include the fact that groundwater is a water supply below the surface of Earth where it might not be clearly delineated, as well as the fact that there are limited amounts of freshwater on Earth.

DBQ Document-Based Questions

Data obtained from: Lerch, R.N., C.M. Wicks, and P.L. Moss. 2006. Hydrological characterization of two karst recharge areas in Boone County, Missouri. *Journal of Cave and Karst Studies* 67 (3): 158–173.

37. Year 2
38. Precipitation and discharge are closely related and have the same general trend in the graph over time.
39. At the end of Year 2, discharge is much higher than precipitation. This might have been caused by a flood or other exceptional event upstream from where discharge is measured.

CUMULATIVE REVIEW

40. The ionic bond is formed by the electrostatic attraction of oppositely charged ions; the covalent bond is formed by electron-sharing.
41. It illustrates how different rock types can form new rock types.
42. Climate, rock type and composition, surface area, topography, organisms; Climate is the major influence on the rate of chemical weathering of Earth materials. Areas of high precipitation and precipitation with lush vegetation, such as those found in the tropics, exhibit the greatest amounts of chemical weathering.

MULTIPLE CHOICE

1. C
2. A
3. B
4. A
5. C
6. A
7. D
8. A
9. D
10. C

CUMULATIVE STANDARDIZED TEST PRACTICE

MULTIPLE CHOICE

1. Which materials would be best suited for lining a pond?
A. gravel
B. limestone
C. clay
D. sand

Use the concept map to answer Questions 2 and 3.

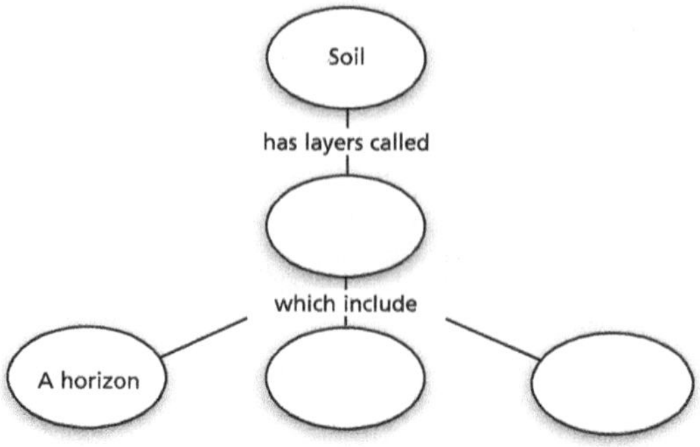

2. What word would complete the circle below the words *has layers called?*
A. horizons
B. profiles
C. levels
D. humus

3. The bottom two circles should be filled in with ________ and ________.
A. O horizon; R horizon
B. B horizon; C horizon
C. A level; A profile
D. humus; litter

4. Which is NOT a value of wetlands?
A. feeding lakes and deltas with nutrients and oxygen-rich water
B. filtering water by trapping pollutants, sediments, and pathogenic bacteria
C. providing habitats for migratory birds and other wildlife
D. preserving fossils due to the anaerobic and acidic conditions

5. What is the reaction of water with other substances known as?
A. aquification
B. oxidation
C. hydrolysis
D. carbonation

6. Which water sources are the most easily polluted?
A. water-table aquifers
B. confined aquifers
C. artesian wells
D. hot springs

Use the table below to answer Questions 7 and 8.

Year	Erosion (meters)
2001	2
2002	18
2003	7
2004	5

7. Which could have caused the unusual level of sand loss in 2002 as shown in the table?
A. lower than usual temperatures
B. higher than usual temperatures
C. lower than usual storm activity
D. higher than usual storm activity

8. What human intervention could have caused the drop in erosion from 2002 to 2003?
A. dune building
B. removing dune vegetation
C. constructing buildings along the coast
D. building fences along the coast

9. What are natural structures hanging from a cave's ceiling?
A. geyserites
B. travertines
C. stalagmites
D. stalactites

10. In which part of a meander does the water travel the fastest?
A. along the inside curve of the meander
B. along the bottom of the meander
C. along the outside curve of a meander
D. all parts of the meander are equal

SHORT ANSWER

Use the illustration below to answer Questions 11–13.

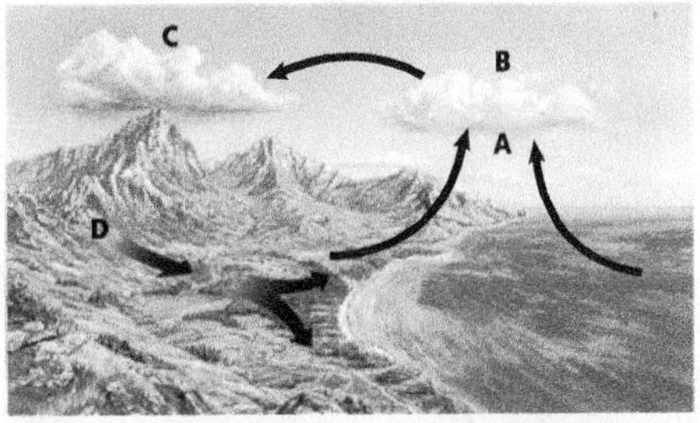

11. Explain the process being illustrated. Be sure to include the name of the process.

12. Why are there two arrows rising during the stage labeled with the letter *A*?

13. What process is occurring from Step C to Step D?

14. How does the silicon-oxygen tetrahedron affect silicate minerals?

15. What is the risk with overpumping a well?

16. How can human activities trigger mass movements?

READING FOR COMPREHENSION

Living Caves

Cool air billowed from a crack in Arizona's desert and lured cave hunters underground. There, formations of rock sprouted from the ground and hung from the ceiling. The explorers discovered a so-called living cave. Tufts and Tenen were the first known to set foot in the Kartchner Caverns, which are among the world's top show caves. The caverns are *living,* a term used to describe active caves. "The cave formations still have water on them; they're still continuing to grow," said Rick Toomey, a staff scientist at the Kartchner Caverns in Benson. Rainwater from the surface seeps through the ground, absorbing calcium carbonate along the way. Inside the cave, the mixture drips from the ceiling. As it hardens, it forms the icicle-like stalactites on the ceiling and sproutlike stalagmites on the floor. With the exception of small cracks, the caverns are closed off to the outside world. The isolation allows the caverns to maintain an average temperature of 20°Celsius and near 99 percent humidity.

Article obtained from: Roach, J. Arizona tried tourism to save 'living cave.' *National Geographic News.* April 19, 2005.

17. According to the passage, what makes a cavern living?
    A. There are animals in the cave.
    B. People can go into the cave.
    C. There are plants in the cave.
    D. Cave formations continue to grow.

18. What is the first ingredient in the process that allows the stalactites and stalagmites to grow as listed in the text?
    A. soil
    B. rainwater
    C. calcium carbonate
    D. rocks

19. What can be inferred from this passage?
    A. Arizona is the only state that has living caves.
    B. The temperature in the caves is quite cold.
    C. Being isolated from the world has protected the cave.
    D. Tufts and Tenen were the first people to discover a cave.

NEED EXTRA HELP?

If You Missed Question . . .	1	2	3	4	5	6	7	8	9	10	11	12	13	14	15	16
Review Section . . .	10.1	7.3	7.3	9.3	7.1	10.3	7.2	8.2	10.2	9.2	9.1	9.1	9.1	4.2	10.2	8.1

SHORT ANSWER

11. The illustration represents the water cycle. Water passes through the steps of evaporation, lifting, condensation, precipitation, runoff, infiltration, and transpiration during this process.

12. This stage involves evaporation. Water is not just evaporated from the ocean and other bodies of water, but from the surface of Earth as well, so two arrows are needed.

13. Precipitation is falling from the clouds in Step C and is being absorbed by the earth or flowing into bodies of water in Step D.

14. The tetrahedron structure allows molecules to combine chemically and structurally in many different ways, thus producing a wide variety of silicates.

15. Overpumping a well can cause the water level in that well to drop and lower the water table in the surrounding area as well. This drop could cause the well and other shallow wells around it to dry up.

16. When humans build structures such as heavy buildings or roads on or along slopes, it can cause the slope materials to become unstable, which can trigger slides.

READING FOR COMPREHENSION

17. D
18. B
19. C

UNIT 4

The Atmosphere and the Oceans

Themes

Change Heat stored in the oceans creates currents that affect the hydrosphere and the atmosphere. Storms on Earth's surface vary in strength and intensity due to these and other factors.

Structures Density differences of the atmosphere and ocean create layers with differing characteristics. Determining and predicting the processes within these layers is the job of meteorologists and oceanographers.

Geologic Time As scientists study Earth's past, one key factor that helps to identify divisions on the geologic time scale is the effect climate change has on life and on land formations.

Systems The water cycle traces the path of water on the surface of Earth into the atmosphere and its return to the oceans.

Scientific Inquiry Accurate weather prediction is the goal of most meteorologists and climatologists. The combined efforts of storm chasers, scientists, and even damage claims estimators have led to more precise information about causes of storms, storm features, and storm safety.

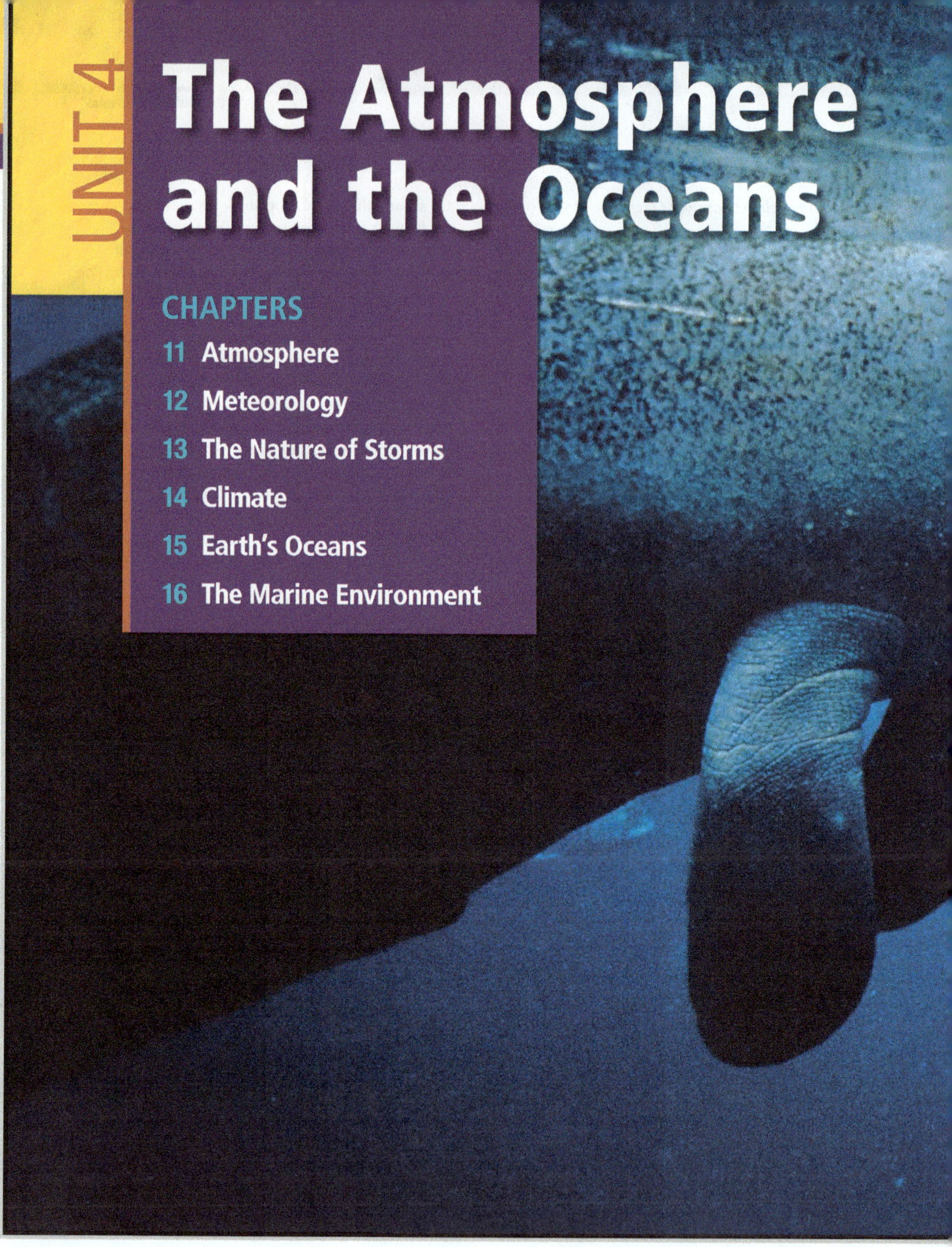

UNIT 4

The Atmosphere and the Oceans

CHAPTERS

11 Atmosphere

12 Meteorology

13 The Nature of Storms

14 Climate

15 Earth's Oceans

16 The Marine Environment

STEM Project

CAREERS IN EARTH SCIENCE **Marine Scientist**

This marine scientist is studying a young manatee to learn more about its interaction with the environment. Marine scientists study the ocean to classify and conserve underwater life.

Douglas Faulkner/Photo Researchers

Introduce the Unit

Preconceptions Tsunamis are large, destructive ocean waves generated by earthquakes in ocean basins or by underwater landslides. They are often mistakenly called "tidal waves" in the media and popular literature. Tidal waves, however, usually are relatively small, calm bulges of water generated by the gravitational attraction of the Sun and the Moon. These bulges of water, commonly known as tides, break against the shores as waves. Tell students that tsunamis and tidal waves have some similarities—both travel through the ocean as shallow-water waves. In a tsunami, however, the wavelength, or horizontal distance between the highest points of two successive waves, can be hundreds of kilometers long. Tidal waves have wavelengths that are thousands of kilometers long.

Ocean-Air Interactions Ask students whether they have ever been to the ocean. Have those students who answer in the affirmative share their experiences. Encourage students to describe the movement of the breaking waves on the shore. Also, have them describe the weather conditions. Was there a cool ocean breeze? Generally, the weather along coastal areas is cooler than that of areas farther inland. Tell students that the ocean often affects weather and vice versa. In this unit, students will learn about the factors that affect Earth's atmosphere and oceans.

CHAPTER 11 Atmosphere

BIGIDEA **The composition, structure, and properties of Earth's atmosphere form the basis of Earth's weather and climate.**

ESSENTIAL QUESTIONS	NATIONAL STANDARDS	RESOURCES TO ASSESS MASTERY
SECTION 1 **Atmospheric Basics** 1. What is the gas and particle composition of the atmosphere? 2. What are the five layers of the atmosphere? 3. How is energy transferred in the atmosphere? 2.5 sessions 1.25 blocks	UCP.1–3, UCP.5; A.1–A.2; B.2, B.6; D.1; F.3–4; G.3	**Entry-Level Assessment** Focus Transparency 29, p. 282 **Progress Monitoring** Caption Question, pp. 286, 287 Reading Check, p. 286 Section Review, p. 288
SECTION 2 **Properties of the Atmosphere** 1. What are the three main properties of the atmosphere and how do they interact? 2. Why do atmospheric properties change with changes in altitude? 2.5 sessions 1.25 blocks	UCP.1–3; A.1–2; B.6; D.1	**Entry-Level Assessment** Focus Transparency 30, p. 289 **Progress Monitoring** Caption Question, pp. 292, 296 Reading Check, p. 290 Section Review, p. 296
SECTION 3 **Clouds and Precipitation** 1. What is the difference between stable and unstable air? 2. How do low, middle, high, and vertical development clouds differ? 3. How does precipitation form? 4 sessions 2 blocks	UCP.1–3, UCP.5; A.1–2; B.3–4, B.6; D.1; F.3–6	**Entry-Level Assessment** Focus Transparency 31, p. 297 **Progress Monitoring** Caption Question, pp. 300, 301, 302 Reading Check, pp. 298, 301 Section Review, p. 303 **Summative Assessment** Chapter Assessment, p. 307 Exam*View*® *Assessment Suite* CD-ROM

Go online!
connectED.mcgraw-hill.com

LEVELED RESOURCES	LAB MATERIALS
Science Notebook 11.1 OL **Chapter FAST FILE Resources:** Study Guide, p. 13 BL **Transparencies:** Section Focus Transparency 29 EL Teaching Transparency 28 OL EL	**LaunchLAB** p. 280 / **45 min** warm water; clear, plastic bowl; plastic wrap; self-sealing plastic bag; ice cubes; tape
Science Notebook 11.2 OL **Chapter FAST FILE Resources:** GeoLab Worksheet, p. 3 OL MiniLab Worksheet, p. 2 OL Study Guide, p. 15 BL **Lab Resources:** Laboratory Manual, pp. 81, 85 OL **Transparencies:** Section Focus Transparency 30 EL Teaching Transparency 29 OL EL	**GeoLAB** p. 305 / **30–45 min** clean, clear, plastic 2-L bottle with cap; plastic straws; scissors; thin, liquid-crystal temperature strip; tape; watch or timer **MiniLAB** p. 295 / **25 min** glass, water, ice cubes, thermometer
Science Notebook 11.3 OL **Chapter FAST FILE Resources:** Study Guide, p. 17 BL **Transparencies:** Section Focus Transparency 31 EL Teaching Transparency 30 OL EL	

LabManager™
Customize any lab with the LabManager™ CD-ROM.

ADDITIONAL RESOURCES

Technology:
- eTeacherEdition
- eStudentEdition
- Classroom Presentation Toolkit
- Exam*View® Assessment Suite* CD-ROM
- LabManager™ CD-ROM
- What's EARTH SCIENCE Got To Do With It? DVD
- Weather Classroom Video DVD

Assessment:
- Blueprints for Success
- Chapter Test (Scaffolded)
- Earth Science Standardized Test Practice

Lab Resources:
- Exploring Environmental Problems Laboratory Manual

BL Below Level OL On Level AL Advanced Learners EL English Learners COOP LEARN Cooperative Learning

CHAPTER 11

LaunchLAB

iLab Station Rubric

What causes cloud formation?

Teaching Strategy

- Do not overfill the bag with ice, as it will be difficult to control when placed on the wrap.
- The ice in the bag will drop the temperature of the wrap below the dew point of the air below the wrap. As the air beneath the wrap is cooled, it will condense droplets of moisture on the underside of the wrap. In a similar way, dew forms when moist air near the ground cools.

Procedure

1. Have students read and complete the lab safety form and follow the procedure below.
2. Pour about 125 mL of **warm water** into a **clear, plastic bowl.**
3. Loosely cover the top of the bowl with **plastic wrap.** Overlap the edges of the bowl by about 5 cm.
4. Fill a **self-sealing plastic bag** with **ice cubes,** seal it, and place it in the center of the plastic wrap on top of the bowl. Push the bag of ice down so that the plastic wrap sags in the center but does not touch the surface of the water.
5. Use **tape** to seal the plastic wrap around the bowl.
6. Observe the surface of the plastic wrap directly under the ice cubes every 10 min for 30 min, or until the ice melts.

Analysis

1. **Infer** What formed on the underside of the wrap? Why did this happen? Water droplets formed as water vapor in the rising, warm air condensed on the underside of the wrap near the ice.
2. **Relate** your observations to processes in the atmosphere. Water molecules gain enough thermal energy to separate from other water molecules and enter the atmosphere as water vapor during evaporation. Water vapor condenses into droplets of liquid water and eventually falls as precipitation.
3. **Predict** what would happen if you repeated this activity with hot water in the bowl. The warmed air would have risen more quickly, which in turn would have caused a faster rate of condensation. More precipitation would have formed more rapidly.

CHAPTER 11

Atmosphere

BIGIDEA The composition, structure, and properties of Earth's atmosphere form the basis of Earth's weather and climate.

SECTIONS

1 Atmospheric Basics

2 Properties of the Atmosphere

3 Clouds and Precipitation

LaunchLAB

iLab Station

What causes cloud formation?

Clouds form when water vapor in the air condenses into water droplets or ice. These clouds might produce rain, snow, hail, sleet, or freezing rain. Investigate condensation in this lab.

FOLDABLES® Study Organizer

Layers of the Atmosphere

Make a layered-look book using the labels shown. Use it to organize your notes on layers of the atmosphere.

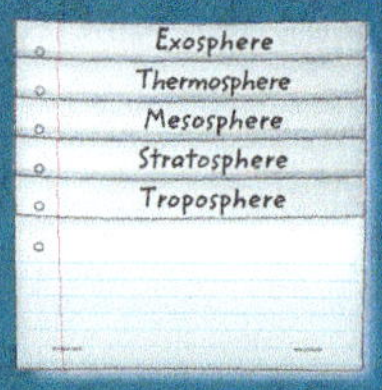

Assessment

Performance Have students test their inferences. Did moisture appear on the underside of the wrap at different locations? Students should compare their inferences with their results.

LabManager™
Customize any lab with the LabManager™ CD-ROM.

Introduce the BIG IDEA

Weather Ask the students why Earth has weather. Earth's surface receives uneven heating by the Sun due to its axial tilt, which, in turn, changes the solar angle and length of the daylight period from day to day. Uneven heating leads to regions of warm rising air and regions of cold sinking air. Additionally, the warmer air carries with it evaporated moisture which contributes to clouds and precipitation in those regions. Rising and sinking air contribute to pressure differences (highs and lows), and the air moving between pressure regions creates winds. If heat was distributed evenly, none of this would occur.

Teacher Content Support

Water Vapor Water vapor transports large quantities of heat from Earth's surface to the atmosphere. Heat is absorbed by water at the surface, causing water to change its state to a gas. The gas is water vapor, and the process is evaporation. The energy contained in the vapor molecule is called latent (hidden) heat. When water vapor cools at altitude, the water vapor reverts back to visible droplets of water or crystals of ice, forming clouds by the process of condensation. The latent heat in the water vapor is released during condensation to the surrounding air, warming it. This process is the first leg of the water cycle and is responsible for the production of Earth's freshwater and for partially cooling Earth. The recycling occurs when precipitation returns to Earth.

SECTION 1

1 Focus

Focus Transparency

Before presenting the lesson, project **Section Focus Transparency 29** and have students answer the accompanying questions. EL

MAINIDEA

Atmospheric Basics Remind students that, just as Earth has a layered structure of rocky and metallic materials beneath the surface, the atmosphere is similar, with many layers, each of which has unique characteristics.

2 Teach

Tie to Previous Knowledge

Thermometers Most students are familiar with common thermometers, which measure one property of the atmosphere—its temperature at a particular place and time. Other instruments can measure other atmospheric properties.

Discussion

Predict Set a large, empty, clear jar on a desk in the front of the room. Ask students to identify the matter in the jar. air Ask students to predict what elements make up the air in the jar. Student answers will vary, but might include oxygen, carbon dioxide, nitrogen, and water vapor. Explain that air contains molecules of gases such as argon, oxygen, and nitrogen, and gaseous compounds such as water vapor and carbon dioxide. Air also contains minute amounts of atmospheric dust and salt. Ask students why these substances aren't visible. Students should state the substances are present in the gaseous state, or are present in concentrations so small they can't be detected by the human eye. OL

SECTION 1

Atmospheric Basics

MAINIDEA Energy is transferred throughout Earth's atmosphere.

Essential Questions

- What is the gas and particle composition of the atmosphere?
- What are the five layers of the atmosphere?
- How is energy transferred in the atmosphere?

Review Vocabulary

atmosphere: the layer of gases that surrounds Earth

New Vocabulary

troposphere
stratosphere
mesosphere
thermosphere
exosphere
radiation
conduction
convection

EARTH SCIENCE 4 YOU **If you touch something made of metal, it will probably feel cool. Metals feel cool because they conduct thermal energy away from your hand. In a similar way, energy is transferred directly from the Earth's surface to the air in the lowest layer of the atmosphere.**

Atmospheric Composition

The ancient Greeks thought that air was one of the four fundamental elements from which all other substances were made. In fact, air is a combination of gases, such as nitrogen and oxygen, and particles, such as dust, water droplets, and ice crystals. These gases and particles form Earth's atmosphere, which surrounds Earth and extends from Earth's surface to outer space.

Permanent atmospheric gases About 99 percent of the atmosphere is composed of nitrogen (N_2) and oxygen (O_2). The remaining 1 percent consists of argon (Ar), carbon dioxide (CO_2), water vapor (H_2O), and other trace gases, as shown in **Figure 1.** The amounts of nitrogen and oxygen in the atmosphere are fairly constant over recent time. However, over Earth's history, the composition of the atmosphere has changed greatly. For example, Earth's early atmosphere probably contained mostly helium (He), hydrogen (H_2), methane (CH_4), and ammonia (NH_3). Today, oxygen and nitrogen are continually being recycled between the atmosphere, living organisms, the oceans, and Earth's crust.

Figure 1 Earth's atmosphere consists mainly of nitrogen (78 percent) and oxygen (21 percent).

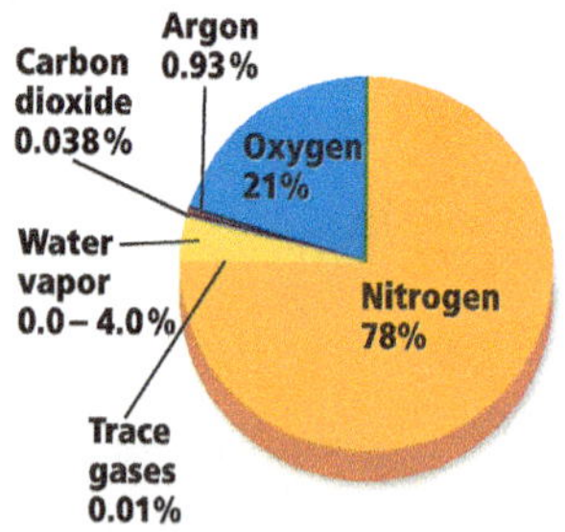

Variable atmospheric gases The concentrations of some atmospheric gases are not as constant over time as the concentrations of nitrogen and oxygen. Gases such as water vapor and ozone (O3) can vary significantly from place to place. The concentrations of some of these gases, such as water vapor and carbon dioxide, play an important role in regulating the amount of energy the atmosphere absorbs and emits back to Earth's surface.

Water vapor Water vapor is the invisible, gaseous form of water. The amount of water vapor in the atmosphere can vary greatly over time and from one place to another. At a given place and time, the concentration of water vapor can be as much as 4 percent or as little as nearly zero. The concentration varies with the seasons, with the altitude of a particular mass of air, and with the properties of Earth's surface beneath the air. Air over deserts, for instance, contains much less water vapor than the air over oceans.

Teacher Content Support

Ozone The gases of Earth's atmosphere are selective absorbers of energy from the Sun. The type of energy absorbed by the atmosphere depends on the kinds of gases that are present. Ozone is found mainly in the stratosphere and absorbs ultraviolet rays. Thus, temperatures in the stratosphere increase with height. The gases in the troposphere do not absorb as much ultraviolet radiation as the gases in the stratosphere. Thus, temperatures in the troposphere decrease with height. Temperature trends in the atmosphere are largely caused by differences in chemical composition in atmospheric layers and variations in the type and amount of energy absorbed.

Carbon dioxide Carbon dioxide, another variable gas, currently makes up about 0.038 percent of the atmosphere. During the past 150 years, measurements have shown that the concentration of atmospheric carbon dioxide has increased from about 0.028 percent to its present value. Carbon dioxide is also cycled between the atmosphere, the oceans, living organisms, and Earth's rocks.

The recent increase in atmospheric carbon dioxide is due primarily to the burning of fossil fuels, such as oil, coal, and natural gas. These fuels are burned to heat buildings, produce electricity, and power vehicles. Burning fossil fuels can also produce other gases, such as sulfur dioxide and nitrogen oxides, that can cause respiratory illnesses, as well as other environmental problems.

Ozone Molecules of ozone are formed by the addition of an oxygen atom to an oxygen molecule, as shown in **Figure 2.** Most atmospheric ozone is found in the ozone layer, 20 km to 50 km above Earth's surface, as shown in **Figure 3.** The maximum concentration of ozone in this layer—9.8×10^{12} molecules/cm^3—is only about 0.0012 percent of the atmosphere.

The ozone concentration in the ozone layer varies seasonally at higher latitudes, reaching a minimum in the spring. The greatest seasonal changes occur over Antarctica. During the past several decades, measured ozone levels over Antarctica in the spring have dropped significantly. This decrease is due to the presence of chemicals called chlorofluorocarbons (CFCs) that react with ozone and break it down in the atmosphere.

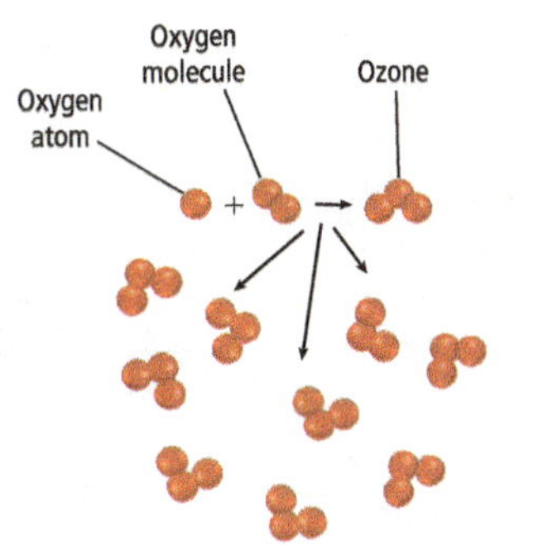

Figure 2 Molecules of ozone are formed by the addition of an oxygen atom to an oxygen molecule.

Atmospheric particles Earth's atmosphere also contains variable amounts of solids in the form of tiny particles, such as dust, salt, and ice. Fine particles of dust and soil are carried into the atmosphere by wind. Winds also pick up salt particles from ocean spray. Airborne microorganisms, such as fungi and bacteria, can also be found attached to microscopic dust particles in the atmosphere.

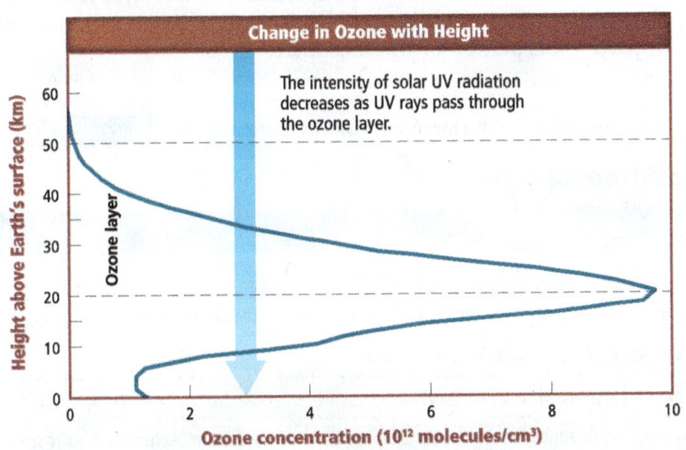

Figure 3 The ozone layer blocks harmful ultraviolet rays from reaching Earth's surface. Ozone concentration is highest at about 20 km above Earth's surface, in the ozone layer.

Environmental Connection

Ozone Layer Damage Evidence suggests that compounds called cholorfluorocarbons (CFCs), which were used in hair sprays, spray paints, and refrigerants, have damaged the ozone layer, especially above Antarctica. CFCs are destructive to atmospheric ozone. In the upper atmosphere, CFCs are broken down by ultraviolet light into their components, one of which is chlorine. A chlorine atom alone, or in combination with an oxygen atom as chlorine monoxide, breaks down ozone molecules. It is also estimated that CFCs remain in the atmosphere, capable of destroying ozone, for as long as 75 years.

Although the damage caused by CFCs is often referred to as a "hole" by the media, it is more properly called a thinning of the ozone layer. Thinning of stratospheric ozone could potentially cause widespread health problems. Have students research ways to stop or reduce the damage to the ozone layer.

Teacher Content Support

Exosphere The exosphere, which begins at a height of around 600 km, marks the transition of Earth's atmosphere to sparse interplanetary gases. In the exosphere, molecules of hydrogen and atoms of helium absorb solar radiation. This radiant energy increases the movement of these particles, and they escape from Earth's gravitational field into space.

DIFFERENTIATED INSTRUCTION

Struggling Learners Use a hot plate to heat a cup of water until it boils. **WARNING:** ***Boiling water can burn exposed skin. Be careful when handling the cup.*** Also, pour water into a glass, and put a few ice cubes on a plate. Place the cup of steaming water, the water in the glass, and the plate with ice cubes on a table. Have students observe water as a solid, a liquid, and a gas. Encourage them to describe liquid water, ice, and water vapor. Be sure students do not put their hands near the steam. Relate student observations to water in the atmosphere. Tell them water occurs as solid in snow, it occurs as liquid in clouds/fog or as rain/drizzle, and it occurs as a gas known as water vapor.

Model

Insolation Have groups model what happens to incoming solar radiation using a flashlight to model solar radiation, a pan of water (reflected back to space), a sheet of dark construction paper (absorption of energy by Earth's surface), and a white sheet of paper held between the dark sheet of paper and the flashlight (both the absorptive and reflective characteristics of clouds). Describe each model part and the effects that could not be modeled, such as reradiation by clouds and absorption of energy by the atmosphere itself. All of these processes are critical to radiation distribution. BL EL

Discussion

Solar Radiation Have students describe the three methods by which solar energy that reaches Earth is transferred. It is transferred through space by radiation, then absorbed by Earth's surface. The energy is then transferred to the first few molecules of the atmosphere through conduction, and it is moved throughout the troposphere through convection. OL

Apply Earth Science

Temperature Change Have you ever been to the top of a mountain? If yes, describe what you noticed about the temperature. It is much cooler at the top of the mountain than it is at the bottom, in any season.

FOLDABLES® Rubric

FOLDABLES® Incorporate information from this section into your Foldable.

Atmospheric Layers

The atmosphere is classified into five different layers, as shown in **Table 1** and **Figure 4.** These layers are the troposphere, stratosphere, mesosphere, thermosphere, and exosphere. Each layer differs in composition and temperature profile.

Troposphere The layer closest to Earth's surface, the **troposphere,** contains most of the mass of the atmosphere. Weather occurs in the troposphere. In the troposphere, air temperature decreases as altitude increases. The altitude at which the temperature stops decreasing is called the tropopause. The height of the tropopause varies from about 16 km above Earth's surface in the tropics to about 9 km above it at the poles. Temperatures at the tropopause can be as low as –60°C.

Stratosphere Above the tropopause is the **stratosphere,** a layer in which the air temperature mainly increases with altitude and contains the ozone layer. In the lower stratosphere below the ozone layer, the temperature stays constant with altitude. However, starting at the bottom of the ozone layer, the temperature in the stratosphere increases as altitude increases. This heating is caused by ozone molecules, which absorb ultraviolet radiation from the Sun. At the stratopause, air temperature stops increasing with altitude. The stratopause is about 50 km above Earth's surface. About 99.9 percent of the mass of Earth's atmosphere is below the stratopause.

Mesosphere Above the stratopause is the **mesosphere,** which is about 50 km to 85 km above Earth's surface. In the mesosphere, air temperature decreases with altitude, as shown in **Figure 4.** This temperature decrease occurs because very little solar radiation is absorbed in this layer. The top of the mesosphere, where temperatures stop decreasing with altitude, is called the mesopause.

Thermosphere The **thermosphere** is the layer between about 85 km and 600 km above Earth's surface. In this layer, the extremely low density of air causes the temperature to rise. This will be discussed further in Section 2. Temperatures in this layer can be up to 2000°C. The ionosphere, which is made of electrically charged particles, is part of the thermosphere.

Explore **the layers of the atmosphere with an interactive table.** Concepts In Motion

Table 1 Components of the Atmosphere

Atmospheric Layer	Components
Troposphere	layer closest to Earth's surface, ends at the tropopause
Stratosphere	layer above the troposphere, contains the ozone layer, and ends at the stratopause
Mesosphere	layer above the stratosphere, ends at the mesopause
Thermosphere	layer above the mesosphere, absorbs solar radiation
Exosphere	outermost layer of Earth's atmosphere, transitional space between Earth's atmosphere and outer space

Teacher Content Support

Earth's Primitive Atmosphere Earth's early atmosphere was probably composed of methane and ammonia. In the first billion years after Earth formed, its surface was more volcanically active than it is today. This primitive atmosphere changed over geologic time as erupting volcanoes emitted gases: water vapor, chlorine, carbon dioxide, hydrogen, and nitrogen. Over millions of years, as the planet cooled, the water vapor condensed and absorbed most of the carbon dioxide. Clouds formed and torrential rains began to fall. It is hypothesized that this water filled basins on Earth's surface and formed the planet's oceans. Oxygen was probably formed from the dissociation of water molecules and by photosynthesis of primitive cyanobacteria.

VISUALIZING the Layers of the Atmosphere

Figure 4 Earth's atmosphere is made up of five layers. Each layer is unique in composition and temperature. As shown, air temperature changes with altitude. When you fly in a plane, you might be flying at the top of the troposphere, or you might enter into the stratosphere.

In the exosphere, gas molecules can be exchanged between the atmosphere and space.

Noctilucent clouds are shiny clouds that can be seen in the twilight in the summer around 50°–60° latitude in the northern and southern hemispheres. These are the only clouds that form in the mesosphere.

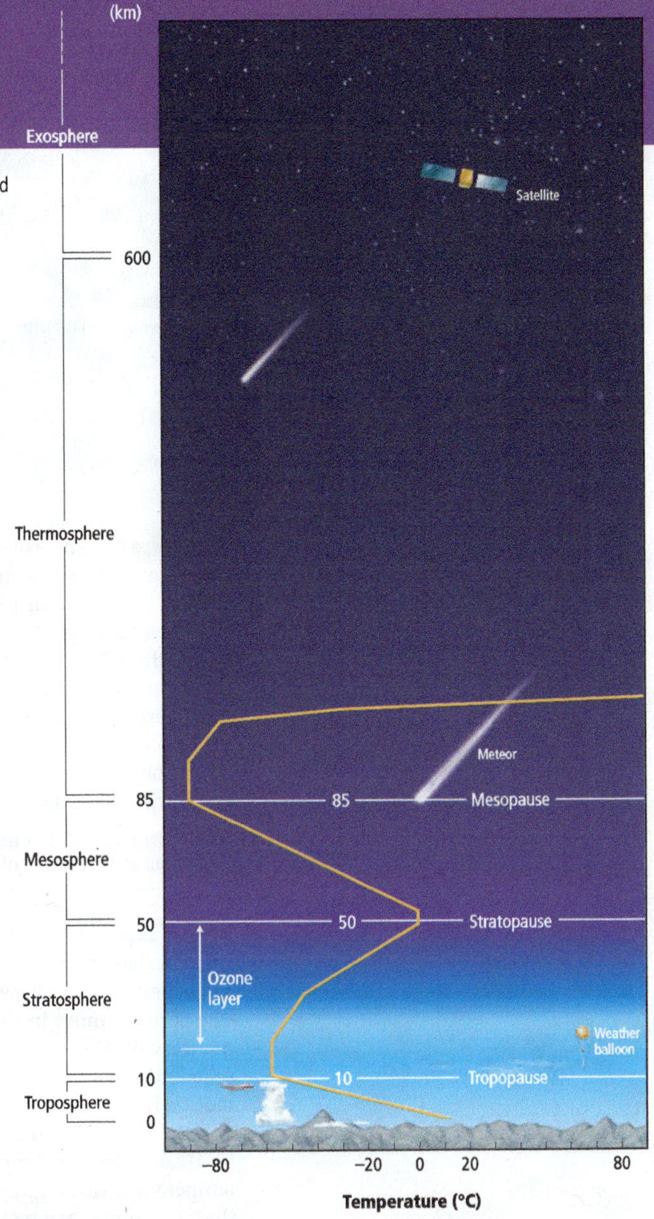

Concepts in Motion View an **animation of the layers of the atmosphere.**

VISUALIZING

Purpose

Students will learn how layers of the atmosphere are differentiated based on vertical temperature profile. Students will understand what objects may appear in each layer.

Enrichment

Atmospheric Layers Tell students that temperature is only one way to classify layers in the atmosphere. Ask students what other variables could be used to identify atmospheric layers. composition, ionic activity

Activity

Absorption Help students understand that different surfaces absorb the Sun's energy at different rates. On a warm, sunny afternoon, have students record air temperatures over an asphalt parking lot at heights ranging from a couple of centimeters to several meters. Have students graph the temperature changes with height. Repeat the procedure over a grassy area. Have students explain any differences in temperature trends between the two locations. OL

DIFFERENTIATED INSTRUCTION

Advanced Learners Have students study the temperature profile of the atmosphere. Ask students to infer why the profile in **Figure 4** ends at the base of the thermosphere. Have students recall the relationship between temperature and density. Air in the thermosphere and exosphere becomes increasingly less dense. Therefore, incoming radiation from the Sun causes a large temperature increase. Tell students that even though the temperature in these layers can be greater than 2000°C, the air would still feel cold to our skin because of the low density of air molecules. Beyond the exosphere, the vacuum of space has no temperature, although objects traveling through space do.

EARTH SCIENCE JOURNAL

Atmospheric Basics Have students write a description of how temperature the inside of their home can vary using the principles of radiation and convection. From this, explain which parts of the house are likely to be warmer or cooler in summer or winter. OL

Rubric

Teacher Content Support

Transfer of Energy There are three means of transfer of energy: radiation, convection, and conduction. Transfer by radiation does not involve matter, but rather electromagnetic waves. In the case of Earth, there is radiation emitted or reflected from Earth itself and incoming radiation from the Sun. It is the radiation from Earth that is largely responsible for heating Earth's atmosphere. Direct radiation from the Sun is secondary. The transfer of energy by convection distributes energy through the movement of a fluid until equilibrium is reached within that fluid. Earth's atmosphere is such a large volume of fluid that equilibrium is never attained. Many weather events are directly related to convection. The final means of transfer is conduction, which requires molecular contact. The atmosphere gains some of its energy through direct contact between Earth's surface and the atmosphere.

Use an Analogy

Greenhouse Effect Why does the inside of a closed parked car heat up, while the windows stay cool? The windows are much like the atmosphere—they allow incoming radiation to pass through without much absorption. The inside of the car, though, is like Earth's surface—it absorbs the incoming energy and turns it into thermal energy. This thermal energy, however, cannot pass back through the windows, and, thus, the inside of the car warms up.

READING CHECK With increase in altitude, the troposphere cools, the stratosphere warms, the mesosphere cools, and the thermosphere warms.

Caption Question Fig. 5 Helicopter–1 layer, 747–2 layers, Space Ship–1–3 layers, Space Shuttle–3 layers, possibly 4 layers

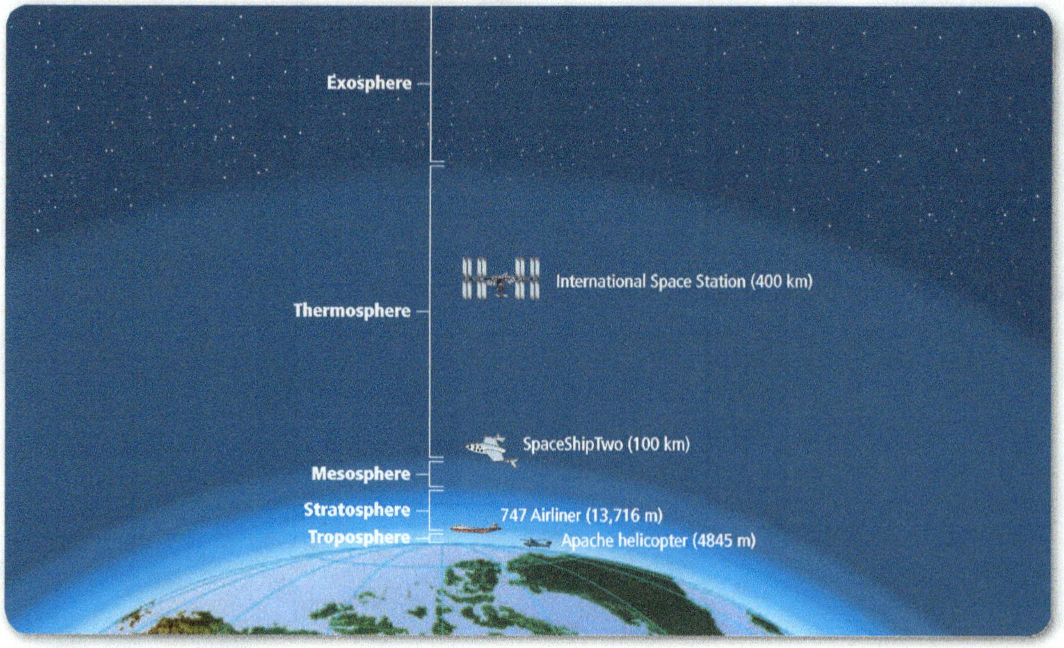

Figure 5 Different spacecraft can traverse the various layers of the atmosphere. **Compare** *the number of atmospheric layers each spacecraft can reach in its flight path.*

Exosphere The **exosphere** is the outermost layer of Earth's atmosphere, as shown in **Figure 5.** The exosphere extends from about 600 km to more than 10,000 km above Earth's surface. There is no clear boundary at the top of the exosphere. Instead, the exosphere can be thought of as the transitional region between Earth's atmosphere and outer space. The number of atoms and molecules in the exosphere becomes very small as altitude increases.

In the exosphere, atoms and molecules are so far apart that they rarely collide with each other. In this layer, some atoms and molecules are moving fast enough that they are able to escape into outer space.

READING CHECK **Summarize** how temperature varies with altitude in the four lowest layers of the atmosphere.

Energy Transfer in the Atmosphere

All materials are made of particles, such as atoms and molecules. These particles are always moving, even if the object is not moving. The particles move in all directions with various speeds—a type of motion called random motion. A moving object has a form of energy called kinetic energy. As a result, the particles moving in random motion have kinetic energy. The total energy of the particles in an object due to their random motion is called thermal energy.

Heat is the transfer of thermal energy from a region of higher temperature to a region of lower temperature. In the atmosphere, thermal energy can be transferred by radiation, conduction, and convection.

ACROSS THE CURRICULUM

Biology Have students research and discuss the role some of the key atmospheric gases such as nitrogen, oxygen, and carbon dioxide play in the growth of plant and animal life. Ask how plant and animal life might change if the percentages of those gases were significantly different.

DIFFERENTIATED INSTRUCTION

Struggling Learners Assuming a dry adiabatic lapse rate of 10°C per thousand meters, have students estimate the temperature at 5000 m, 10,000 m, and 15,000 m if the surface temperature is 30°C. Also have them estimate the temperature outside an airplane flying at 11,000 m.

Radiation Light from the Sun heats some portions of Earth's surface at all times, just as the heat lamp in **Figure 6** uses the process of radiation to warm food. **Radiation** is the transfer of thermal energy by electromagnetic waves. The heat lamp emits visible light and infrared waves that travel from the lamp and are absorbed by the food. The thermal energy carried by these waves causes the temperature of the food to increase. In the same way, thermal energy is transferred from the Sun to Earth by radiation. The solar energy that reaches Earth is absorbed and reflected by Earth's atmosphere and Earth's surface.

Figure 6 A heat lamp transfers thermal energy by radiation. Here, the thermal energy helps to keep the french fries hot.

Absorption and reflection Most of the solar energy that reaches Earth is in the form of visible light waves and infrared waves. Almost all of the visible light waves pass through the atmosphere and strike Earth's surface. Most of these waves are absorbed by Earth's surface. As the surface absorbs these visible light waves, it also emits infrared waves. The atmosphere absorbs some infrared waves from the Sun and emits infrared waves with different wavelengths, as shown in **Figure 7.**

About 30 percent of solar radiation is reflected into space by Earth's surface, the atmosphere, or clouds. Another 20 percent is absorbed by the atmosphere and clouds. About 50 percent of solar radiation is absorbed directly or indirectly by Earth's surface and keeps Earth's surface warm.

Rate of absorption The rate of absorption for any particular area varies depending on the physical characteristics of the area and the amount of solar radiation it receives. Different areas absorb energy and heat at different rates. For example, water heats and cools more slowly than land. Also, as a general rule, darker objects absorb energy faster than light-colored objects. For instance, a black asphalt driveway heats faster on a sunny day than a light-colored concrete driveway.

Figure 7 Incoming solar radiation is either reflected back into space or absorbed by Earth's atmosphere or its surface.
Trace *the pathways by which solar radiation is absorbed and reflected.*

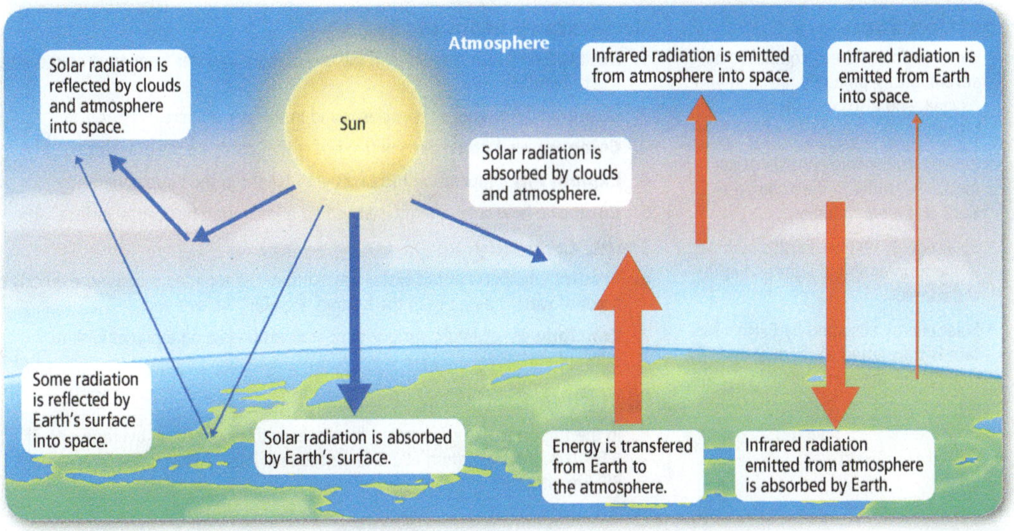

Apply Earth Science

Cloud Cover Increased cloud cover during the day will prevent solar radiation from reaching Earth's surface and keep temperatures low. Increased cloud cover overnight will prevent terrestrial radiation from escaping Earth's surface and keep temperatures high. The result is mild temperatures during cloudy periods. Clouds have a complicated effect on Earth's temperatures. This is an area of current research.

Caption Question Fig. 7
Radiation is absorbed by clouds, the atmosphere, and the Earth's surface; radiation is reflected by clouds and the Earth's surface.

Demonstration

Measure Humidity Use cobalt chloride test paper to show moisture in the air. Place pieces of the test paper in different areas inside and outside the school. Have students record the color of each piece of paper. Pink indicates high relative humidity; blue indicates low relative humidity.

IN THE FIELD

Biology The primary danger resulting from the thinning of the ozone layer is the increase in ultraviolet radiation that reaches Earth's surface. Ask students to research which parts of their bodies would most likely be directly affected by this increased radiation. Answers should include damage to the skin, possibly causing skin cancer, and damage to the eyes, possibly causing cataracts or blindness.

3 Assess

Check for Understanding

Discussion How is conduction related to cold air temperatures at the poles, which are covered with ice and snow? As light-colored materials, ice and snow reflect most solar radiation and therefore absorb and retain little thermal energy. They cannot conduct much thermal energy into the atmosphere, so the air in those areas does not warm up as much as in areas that are thermal energy absorbers.

Reteach

Linguistics Have students outline the main points of this section in their Earth science journals. Ask them to include specific examples in support of each point.

Assessment

Performance Have students develop posters showing the different layers of the atmosphere and the different characteristics of each layer. Posters should also show how solar energy is transferred throughout the atmosphere.
COOP LEARN

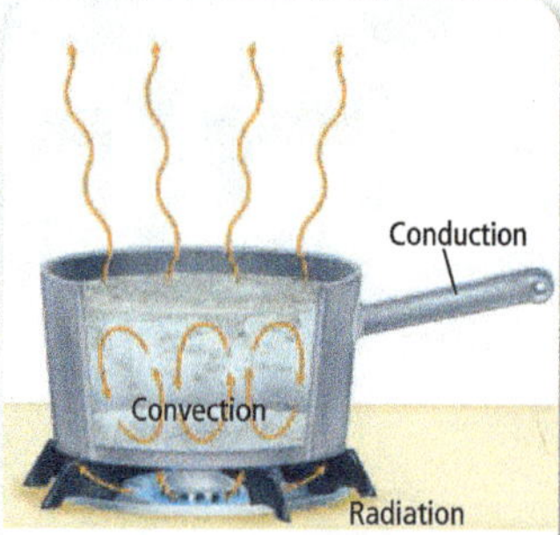

■ **Figure 8** Thermal energy is transferred to the burner from the heat source by radiation. The burner transfers the energy to the atoms in the bottom of the pot, which collide with neighboring atoms. As these collisions occur, thermal energy is transferred by conduction to other parts of the pot, including the handle.

View an **animation of energy transfer.**

Concepts In Motion

Conduction Another process of energy transfer can occur when two objects at different temperatures are in contact. **Conduction** is the transfer of thermal energy between objects when their atoms or molecules collide, as shown in **Figure 8.** Conduction can occur more easily in solids and liquids, where particles are close together, than in gases, where particles are farther apart. Because air is a mixture of gases, it is a poor conductor of thermal energy. In the atmosphere, conduction occurs between Earth's surface and the lowest part of the atmosphere.

Convection Throughout much of the atmosphere, thermal energy is transferred by a process called convection. The process of convection occurs mainly in liquids and gases. **Convection** is the transfer of thermal energy by the movement of heated material from one place to another. **Figure 8** illustrates the process of convection in a pot of water. As water at the bottom of the pot is heated, it expands and becomes less dense than the water around it. Because it is less dense, the warmer water is forced upward. As it rises, it transfers thermal energy to the cooler water around it, and cools. It then becomes denser than the water around it and sinks to the bottom of the pot, where it is reheated.

A similar process occurs in the atmosphere. Parcels of air near Earth's surface are heated, become less dense than the surrounding air, and rise. As the warm air rises, it cools and its density increases. When it cools below the temperature of the surrounding air, the air parcel becomes denser than the air around it and sinks. As it sinks, it warms again, and the process repeats. Convection currents, as these movements of air are called, are the main mechanism for energy transfer in the atmosphere.

SECTION 1 REVIEW

Section Self-Check

Section Summary

- Earth's atmosphere is composed of several gases, primarily nitrogen and oxygen, and also contains small particles.
- Earth's atmosphere consists of five layers that differ in their compositions and temperatures.
- Solar energy reaches Earth's surface in the form of visible light and infrared waves.
- Solar energy absorbed by Earth's surface is transferred as thermal energy throughout the atmosphere.

Understand Main Ideas

1. MAINIDEA **Rank** the gases in the atmosphere in order from most abundant to least abundant.
2. **Name** the four types of particles found in the atmosphere.
3. **Compare and contrast** the five layers that make up the atmosphere.
4. **Explain** why temperature increases with height in the stratosphere.
5. **Compare** how solar energy is absorbed and emitted by Earth's surface.

Think Critically

6. **Predict** whether a pot of water heated from the top would boil more quickly than a pot of water heated from the bottom. Explain your answer.
7. **Conclude** What might surface temperatures be like on a planet with no atmosphere?

MATH IN Earth Science

8. In the troposphere, temperature decreases with height at an average rate of 6.5°C/km. If temperature at 2.5 km altitude is 7.0°C, what is the temperature at 5.5 km altitude?

SECTION 1 REVIEW

1. nitrogen, oxygen, argon, carbon dioxide, water vapor [Note: The amount of water vapor varies; students may correctly rank it after oxygen.]
2. dust, salt, ice, and airborne microorganisms
3. Each layer is marked by a cessation of temperature change called a "pause". The troposphere and mesosphere cool with increased altitude, while the stratosphere and thermosphere warm with increased altitude. Each layer is different in composition.
4. Temperature increases with height in the stratosphere because ozone molecules absorb UV radiation from the Sun.
5. Earth's surface absorbs solar radiation in the form of visible light, and emits radiation in the form of infrared waves.
6. Convection wouldn't occur in a pot heated from the top because warm, less dense water would be on top and would not sink. Because convection doesn't occur, the thermal energy would be transferred more slowly. As a result, the pot heated from above would take longer to boil.
7. A planet with no atmosphere would most likely be very cold.
8. The temperature is −12.5°C.

SECTION 2

Properties of the Atmosphere

MAINIDEA Atmospheric properties, such as temperature, air pressure, and humidity describe weather conditions.

Essential Questions
- What are the three main properties of the atmosphere and how do they interact?
- Why do atmospheric properties change with changes in altitude?

Review Vocabulary

density: the mass per unit volume of a material

New Vocabulary

temperature inversion
humidity
saturation
relative humidity
dew point
latent heat

EARTH SCIENCE 4 YOU Have you noticed the weather today? Maybe it is hot or cold, humid or dry, or even windy. These properties are always interacting and changing, and you can observe those changes every time you step outside.

Temperature

When you turn on the burner beneath a pot of water, thermal energy is transferred to the water and the temperature increases. Recall that particles in any material are in random motion. Temperature is a measure of the average kinetic energy of the particles in a material. Particles have more kinetic energy when they are moving faster, so the higher the temperature of a material, the faster the particles are moving.

Measuring temperature Temperature is usually measured using one of two common temperature scales. These scales are the Fahrenheit (°F) scale, used mainly in the United States, and the Celsius (°C) scale. The SI temperature scale used in science is the Kelvin (K) scale. **Figure 9** shows the differences among these temperature scales. The Fahrenheit and Celsius scales are based on the freezing point and boiling point of water. The zero point of the Kelvin scale is absolute zero—the lowest temperature that any substance can have.

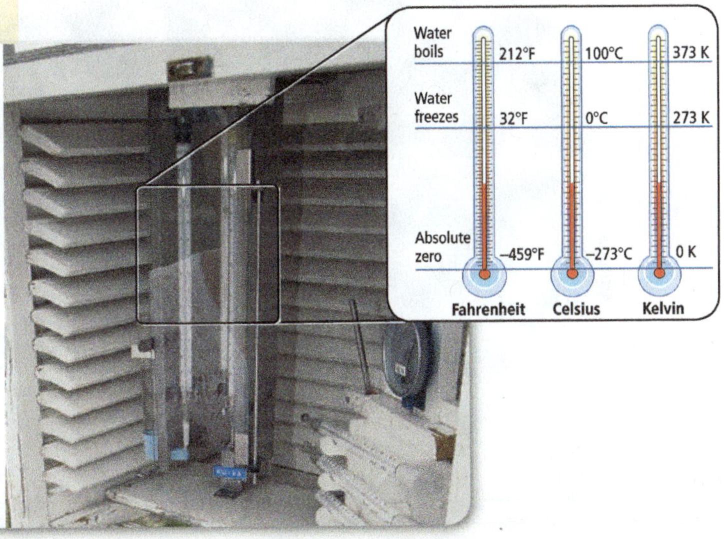

Figure 9 Temperature can be measured in degrees Fahrenheit, degrees Celsius, or in kelvin. The Kelvin scale starts at 0 K, which corresponds to –273°C and –459°F.

David Hays Jones/Photo Researchers

SECTION 2

1 Focus

Focus Transparency

Before presenting the lesson, project **Section Focus Transparency 30** and have students answer the accompanying questions. EL

MAINIDEA

Properties of the Atmosphere Ask students to use media sources, such as the evening television news, to record temperature, relative humidity, and wind at a specific time each day for a week. Point out that just as any of these properties change from day to day, they will learn that changes in one property can cause changes in the others as well.

2 Teach

Tie to Previous Knowledge

Thermal Energy Absorption Students know that different surfaces absorb thermal energy at different rates. Make sure students understand that different surfaces lose thermal energy at different rates as well. For instance, materials such as grass lose thermal energy quickly at night. The dewpoint temperature, in turn, is reached quickly, and dew forms on the grass.

ACROSS THE CURRICULUM

Biology Many organisms that live in the desert depend on dew as their only source of water. The dew comes from the small amount of moisture in the air and in plants. Plant transpiration releases water into the air and onto plant surfaces. Nocturnal organisms collect the water at night by licking the wet plant surfaces.

EARTH SCIENCE JOURNAL

Atmospheric Properties Have students describe how temperature and wind would be likely to change going from sea level to the top of a 3000-m mountain. Be sure to explain why those changes probably took place. OL

Rubric

Reinforcement

Lapse Rate Ask students to explain what happens to temperature with increased altitude through the atmosphere. Then ask the same question as it relates to pressure. Make sure students understand that as long as density remains constant, as temperature increases or decreases, pressure does too, and vice versa. OL

Discussion

Pressure and Temperature Ask students the following question: Which of the following statements about the pressure-temperature-density relationship is true?

a. Temperature is directly proportional to pressure.
b. Temperature is directly proportional to density.
c. Temperature, pressure, and density decrease with height throughout the atmosphere.
d. Decreasing density with height is called an inversion.

Statement "a" is true; temperature is directly proportional to pressure.

READING CHECK Your body pressure pushes back with a force equal to air pressure; also, pressure in fluids comes from all directions, while that of solids comes from only one direction due to the pull of gravity or a force applied from one direction.

VOCABULARY
SCIENCE USAGE V. COMMON USAGE
Force
Science usage: an influence that might cause a body to accelerate

Common usage: violence, compulsion, or strength exerted upon or against a person or thing

Air Pressure

If you hold your hand out in front of you, Earth's atmosphere exerts a downward force on your hand due to the weight of the atmosphere above it. The force exerted on your hand divided by its area is the pressure exerted on your hand. Air pressure is the pressure exerted on a surface by the weight of the atmosphere above the surface.

Because pressure is equal to force divided by area, the units for pressure are Newtons per square meter (N/m^2). Air pressure is often measured in units of millibars (mb), where 1 mb equals 100 N/m^2. At sea level, the atmosphere exerts a pressure of about 100,000 N/m^2, or 1000 mb. As you go higher in the atmosphere, air pressure decreases as the mass of the air above you decreases. **Figure 10** shows how pressure in the atmosphere changes with altitude.

READING CHECK **Deduce** why air pressure does not crush a human.

Density of air The density of a material is the mass of material in a unit volume, such as 1 m^3. Atoms and molecules become farther apart in the atmosphere as altitude increases. This means that the density of air decreases with increasing altitude, as shown in **Figure 10.** Near sea level, the density of air is about 1.2 kg/m^3. At the average altitude of the tropopause, or about 12 km above Earth's surface, the density of air is about 25 percent of its sea-level value. At the stratopause, or about 48 km above Earth's surface, air density has decreased to only about 0.2 percent of the air density at sea level.

Figure 10 The density and pressure of the atmosphere decrease as altitude increases.

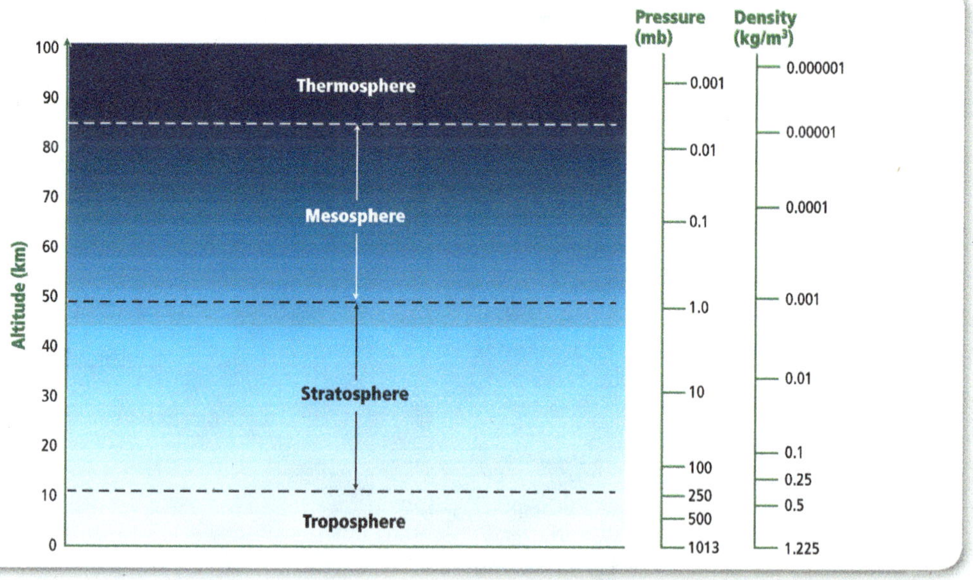

Demonstration

Absorption Explain that Earth's surface absorbs much more solar energy than air does. You can demonstrate this by placing a dark cotton shirt in a brightly lit window, then allowing students to feel the warmth of the shirt, particularly in comparison to the air around it. Tell students that the shirt is a model of Earth's surface.

DIFFERENTIATED INSTRUCTION

Advanced Learners Oceanic layers show temperature changes similar to those in the atmosphere. Have students compare and contrast the oceanic equivalent of other atmospheric properties such as wind, pressure, and density. Ask them if any of these might be related to those in the atmosphere. For example, the driving forces behind some oceanic currents are large–scale wind patterns.

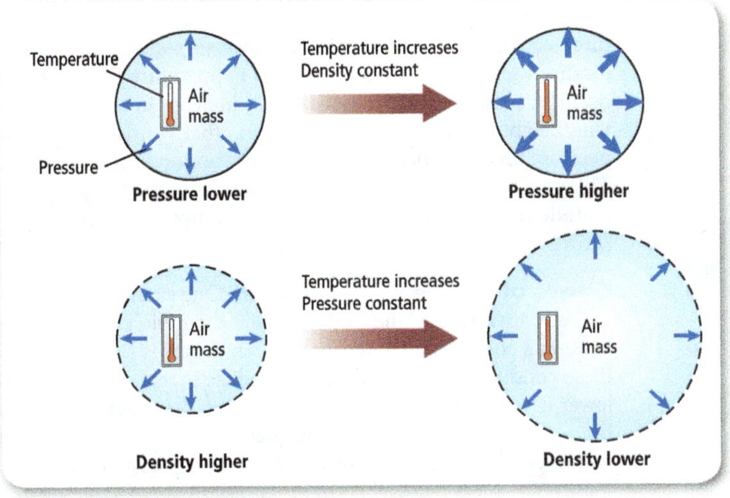

■ **Figure 11** Temperature, pressure, and density are all related to one another. If temperature increases, but density is constant, the pressure increases. If the temperature increases and the pressure is constant, the density decreases.

Pressure-temperature-density relationship In the atmosphere, the temperature, pressure, and density of air are related to each other, as shown in **Figure 11.** Imagine a sealed container containing only air. The pressure exerted by the air inside the container is related to the air temperature inside the container and the air density. How does the pressure change if the air temperature or density changes?

Air pressure and temperature The pressure exerted by the air in the container is due to the collisions of the gas particles in the air with the sides of the container. When these particles move faster due to an increase in temperature, they exert a greater force when they collide with the sides of the container. The air pressure inside the container increases. This means that for air with the same density, warmer air is at a higher pressure than cooler air.

Air pressure and density Imagine that the temperature of the air does not change, but that more air is pumped into the container. Now there are more gas particles in the container, and therefore, the mass of the air in the container has increased. Because the volume has not changed, the density of the air has increased. Now there are more gas particles colliding with the walls of the container, and so more force is being exerted by the particles on the walls. This means that at the same temperature, air with a higher density exerts more pressure than air with a lower density.

Temperature and density Heating a balloon causes the air inside to move faster, causing the balloon to expand and increase in volume. As a result, the air density inside the balloon decreases. The same is true for air masses in the atmosphere. At the same pressure, warmer air is less dense than cooler air.

VOCABULARY
ACADEMIC VOCABULARY
Exert
to put forth (as strength)
Susan exerted a lot of energy playing basketball.

Identify Misconceptions

Students often misunderstand vertical temperature change.

Uncover the Misconception
Students might think that if a thermometer reads 20°C, it is 20°C throughout the atmosphere. They might even believe that the atmosphere gets warmer with height because of increased nearness to the Sun.

Demonstrate the Concept
Mountaintops are often snow-capped. If you travel up into the mountains, temperatures become cooler. Clouds usually form at altitudes where the air cools enough to change water vapor back to tiny cloud droplets.

Assess New Knowledge
Have students describe an experiment or activity to illustrate this concept.

Reinforcement

Temperature Change Tell students that the temperature today somewhere in your area is –30°C. Ask them to think where this might be. Based on the understanding that temperature decreases with height, students should realize that this is true well above Earth's surface.

Teacher Content Support

Properties of Air Air is composed of matter. The substances in air can absorb and retain thermal energy, move, hold moisture, and exert pressure. The main source of atmospheric thermal energy is radiation from Earth's surface. Usually, air near the surface is warmest and it decreases with height through the troposphere. The moisture air holds can vary in amount and exist in all three states of matter. The density of air varies, depending on how many components of air are present in a certain space. The density of a parcel of air depends on its temperature and moisture content; thus, changes in these properties affect how much pressure air exerts.

Use Science Terms

Winds A wind blowing down the slope of a mountain is known as a downslope wind. An example of a downslope wind that warms during its journey is the Santa Ana winds of Southern California. They often cause grass fires and forest fires to spread rapidly as they descend the western slope of the Sierra Nevadas. Another type of warm downslope wind is the chinook, which blows down the eastern slopes of the Rocky Mountains. Chinooks can raise air temperature as much as 30°C in as little as 15 minutes.

Interpret the Photo

Figure 12 Ask students to infer why a temperature inversion can worsen air pollution. They should realize that air rises only when it is less dense than the air around it. In a temperature inversion, a layer of cooler air exists under a layer of warmer air. This cool air is denser than the warm air above, and therefore it cannot rise through the less-dense air to be dispersed throughout the atmosphere. The cool air and the pollutants contained in the air are trapped in the area below the less-dense air.

Caption Question Fig. 13 Pollutants, such as smoke, rise easily through air that is more dense than they are. However, pollutants are held below inversion because the inversion is warm, and therefore less dense or equal in density to the pollutant.

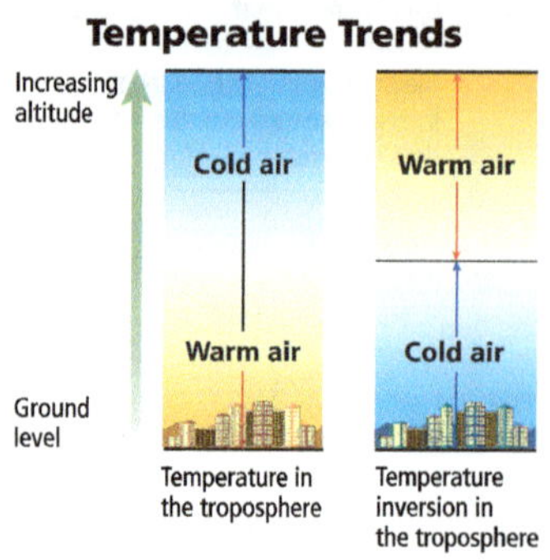

Figure 12 In a temperature inversion, the warm air is located on top of the cooler air.

Temperature inversion In the troposphere, air temperature decreases as height increases. However, sometimes over a localized region in the troposphere, a temperature inversion can occur. A **temperature inversion** is an increase in temperature with height in an atmospheric layer. In other words, when a temperature inversion occurs, warmer air is on top of cooler air. This is called a temperature inversion because the temperature-altitude relationship is inverted, or turned upside down, as shown in **Figure 12.**

Causes of temperature inversion One example of a temperature inversion in the troposphere is the rapid cooling of land on a cold, clear, winter night when the air is calm. Under these conditions, the land does not radiate thermal energy to the lower layers of the atmosphere. As a result, the lower layers of air become cooler than the air above them, so that temperature increases with height and forms a temperature inversion.

Effects of temperature inversion If the sky is very hazy, there is probably an inversion somewhere in the lower atmosphere. A temperature inversion can lead to fog or low-level clouds. Fog is a significant factor in lowering visibility in many coastal cities, such as San Francisco. In some cities, such as the one shown in **Figure 13,** a temperature inversion can worsen air-pollution problems. Because air rises as long as it is warmer than the air above it, the cool pollution-filled air becomes trapped under the warm layer. Pollutants are consequently unable to be lifted from Earth's surface. Temperature inversions that remain over an industrial area for a long time usually result in episodes of severe smog—a type of air pollution—that can cause respiratory problems.

Figure 13 A temperature inversion in New York City traps air pollution above the city.
Describe *the effect of temperature inversion on air quality in metropolitan areas.*

DIFFERENTIATED INSTRUCTION

Advanced Learners Have students research places in the world where air pollution is a major problem, including their own area. Have them determine whether temperature inversions are contributing to the pollution problem in these areas. Students should share their research with the class.

ACROSS THE CURRICULUM

Physics Wind is sometimes described as "air moving from one temperature or pressure to another." Ask students to discuss the validity of this statement using the pressure and temperature relationship. Have them consider how this relationship might affect winds blowing up or down the sides of mountains.

Wind Imagine you are entering a large, air-conditioned building on a hot summer day. As you open the door, you feel cool air rushing past you out of the building. This sudden rush of cool air occurs because the warm air outside the building is less dense and at a lower pressure than the cooler air inside the building. When the door opens, the difference in pressure causes the cool, dense air to rush out of the building. The movement of air is commonly known as wind.

Wind and pressure differences In the example above, the air in the building moves from a region of higher density to a region of lower density. In the lower atmosphere, air also generally moves from regions of higher density to regions of lower density. These density differences are produced by the unequal heating and cooling of different regions of Earth's surface. In the atmosphere, air pressure generally increases as density increases, so regions of high and low density are also regions of high and low air pressure respectively. As a result, air moves from a region of high pressure to a region of low pressure, resulting in wind.

Wind speed and altitude Wind speed and direction change with height in the atmosphere. Near Earth's surface, wind is constantly slowed by the friction that results from contact with surfaces including trees, buildings, and hills, as shown **Figure 14.** Even the surface of water affects air motion. Higher up from Earth's surface, air encounters less friction and wind speeds increase. Wind speed is usually measured in miles per hour (mph) or kilometers per hour (km/h). Ships at sea usually measure wind in knots. One knot is equal to 1.85 km/h.

Figure 14 When wind blows over a forested area by a coast, it encounters more friction than when it blows over flatter terrain. This occurs because the wind encounters friction from the mountains, trees, and then the water, slowing the wind's speed.

Activity

Heating Earth Because solar energy feels warm on our skin, students might suppose that it heats the atmosphere in a similar fashion. In reality, air is primarily heated indirectly by energy emitted by Earth's surface. Have students draw diagrams showing the path of energy from the Sun. Diagrams should show how much energy is absorbed and reflected by clouds, how much is absorbed directly by the atmosphere, and how much reaches Earth's surface. OL

GeoLAB

iLab Station

The GeoLab located at the end of the chapter can be used at this point in the lesson.

ACROSS THE CURRICULUM

Physics An object in motion will not immediately stop when the force that set it in motion is removed, because the object has momentum. Air that rises through the atmosphere has momentum. Thus, it won't stop rising immediately after cooling to the surrounding air temperature. Rather, it will rise a short distance farther. In the atmosphere, that short distance is critical because it allows condensation to continue to the point that enough latent heat is released to keep the air rising on its own.

Apply Earth Science

Medical Conditions Explain that many people have medical conditions that require them to live in dry climates or climates with low rates of relative humidity. Have students research which U.S. cities have low rates of relative humidity and compare them to rates for their own area. **OL**

Concept Development

Latent Heat Make sure students understand the release of latent heat allows a cumulus cloud to continue to grow until it reaches the cumulonimbus stage. Also, reinforce this growth is largely dependent upon the amount of moisture the cloud is able to draw into itself to continue the condensation process necessary for the release of latent heat.

Problem-Solving LAB

Purpose Use the graphical method for relative humidity determination.

Process Skills read a graph, analyze data, infer

Teaching Strategy Stress that relative humidity is a ratio of actual water vapor content to possible vapor content. The graph shows the possible water vapor content.

Think Critically

1. approximately 13 g/m^3; approximately 23 g/m^3
2. 10 g/m^3 ÷ 18 g/m^3 = approximately 55.5%
3. No. Because relative humidity is the ratio of moisture in the air to the maximum amount of moisture the air could hold, the value will never exceed 100%.

Humidity

The distribution and movement of water vapor in the atmosphere play an important role in determining the weather of any region. **Humidity** is the amount of water vapor in the atmosphere at a given location on Earth's surface. Two ways of expressing the water vapor content of the atmosphere are relative humidity and dew point.

Relative humidity Consider a flask containing water. Some water molecules evaporate, leaving the liquid and becoming part of the water vapor in the flask. At the same time, other water molecules condense, returning from the vapor to become part of the liquid. Just as the amount of water vapor in the flask might vary, so does the amount of water vapor in the atmosphere. Water on Earth's surface evaporates and enters the atmosphere and condenses to form clouds and precipitation.

In the example of the flask, if the rate of evaporation is greater than the rate of condensation, the amount of water vapor in the flask increases. **Saturation** occurs when the amount of water vapor in a volume of air has reached the maximum amount. A saturated solution cannot hold any more of the substance that is being added to it. When a volume of air is saturated, it cannot hold any more water.

The amount of water vapor in a volume of air relative to the amount of water vapor needed for that volume of air to reach saturation is called **relative humidity.** Relative humidity is expressed as a percentage. When a certain volume of air is saturated, its relative humidity is 100 percent. If you hear a weather forecaster say that the relative humidity is 50 percent, it means that the air contains 50 percent of the water vapor needed for the air to be saturated.

Problem-Solving LAB

Interpret the Graph

How do you calculate relative humidity? Relative humidity is the ratio of the actual amount of water vapor in a volume of air relative to the maximum amount of water vapor needed for that volume of air to reach saturation. Use the graph at the right to answer the following questions.

Think Critically

1. **Compare** the maximum amount of water vapor 1 m^3 of air could hold at 15°C and 25°C.
2. **Calculate** the relative humidity of 1 m^3 of air containing 10 g/m^3 at 20°C.
3. **Analyze** Can relative humidity be more than 100 percent? Explain your answer.

Data and Observations

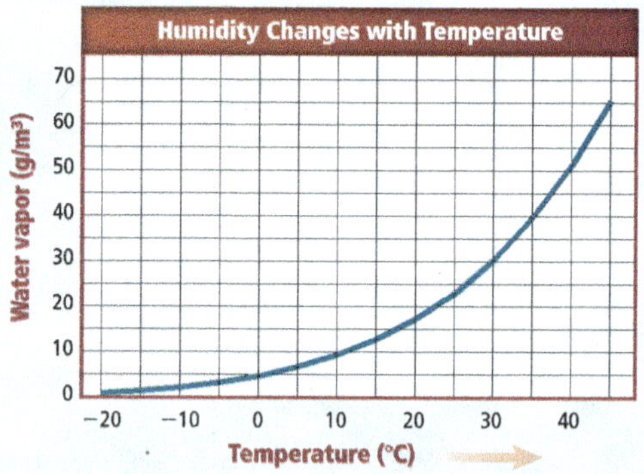

Teacher Content Support

The Water Cycle Worldwide each year, about 500,000 km^3 of water evaporates. About 110,000 km^3 of this water eventually falls on land as precipitation; the rest falls as rain, snow, or sleet into the oceans. Some of the water that falls on land is absorbed by plants, filters into the soil, fills lakes, or becomes runoff that eventually flows to the oceans. During evaporation, thermal energy causes the water molecules to move apart. The water then changes from a liquid to a gas. During condensation, water molecules lose energy and move closer together. The water changes from a gas to a liquid.

Dew point Another common way of describing the moisture content of air is the dew point. The **dew point** is the temperature to which air must be cooled at constant pressure to reach saturation. The name *dew point* comes from the fact that when the temperature falls to this level, water vapor condenses and dew begins to form. If the dew point is nearly the same as the air temperature, then the relative humidity is high.

Latent heat As water vapor in the air condenses, thermal energy is released. Where does this energy come from? To change liquid water to water vapor, thermal energy is added to the water by heating it. The water vapor then contains more thermal energy than the liquid water. This is the energy that is released when condensation occurs. The extra thermal energy contained in water vapor compared to liquid water is called **latent heat.**

When condensation occurs, as in **Figure 15,** latent heat is released and warms the air. At any given time, the amount of water vapor present in the atmosphere is a significant source of energy because it contains latent heat. When water vapor condenses, the latent heat released can provide energy to a weather system, such as a hurricane, increasing its intensity.

Condensation level An air mass can change temperature without being heated or cooled. A process in which temperature changes without the addition or removal of thermal energy from a system is called an adiabatic process. An example of an adiabatic process is the heating of air in a bicycle pump as the air is compressed. In a similar way, an air mass heats up as it sinks and cools off as it rises. Adiabatic heating occurs when air is compressed, and adiabatic cooling occurs when air expands.

MiniLAB

iLab Station

Investigate Dew Formation

How does dew form? Dew forms when moist air near the ground cools and the water vapor in the air condenses into water droplets.

Procedure

1. Read and complete the lab safety form.
2. Fill a **glass** about two-thirds full of **water**. Record the temperature of the room and the water.
3. Add **ice cubes** until the glass is full. Record the temperature of the water at 10-s intervals.
4. Observe the outside of the glass. Note the time and the temperature at which changes occur on the outside of the glass.
5. Repeat the investigation outside. Record the temperature of the water and the air outside.

Analysis

1. **Compare and contrast** what happened to the outside of the glass when the investigation was performed in your classroom and when it was performed outside. If there was a difference, explain.
2. **Relate** your observations to the formation of dew.

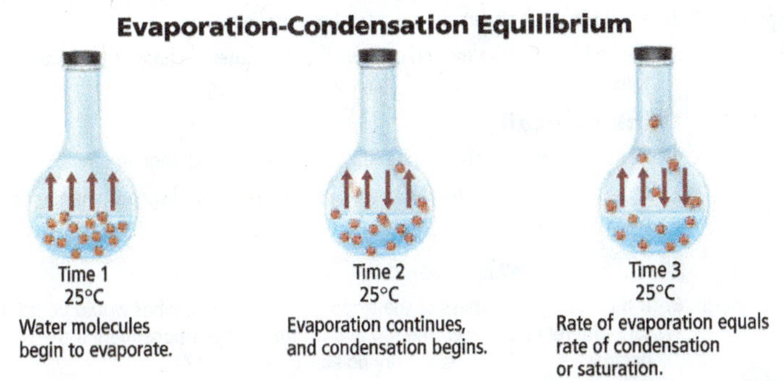

Figure 15 During evaporation, water molecules escape from the surface of the liquid and enter the air as water vapor. During condensation, water molecules return to the liquid state. At equilibrium, evaporation and condensation continue, but the amount of water in the air and amount of water in the liquid form remain constant.

MiniLAB

iLab Station Rubric

Purpose Students will model the formation of condensation.

Additional Material
thermometer

Teaching Strategies

- Prepare the data table in advance.
- Use different sizes of glasses and/or glasses of different materials (glass, metal, plastic, polystyrene) to introduce variables.
- Have students watch the outside of the glass for any kind of change.

Expected Results A fine mist of condensation will begin to form when the temperature of the outside surface of the glass drops to the dew point. The temperature of the ice-water mix at this time is the dew point. There should be no difference in the dew point temperatures, but some materials are insulators and therefore reach the dew point in longer periods of time or not at all. The outside of an insulated glass (polystyrene cup) might never reach the dew point.

Analysis

1. In both cases, condensation will form. The glasses may differ if the locations differed in relative humidities. The more humid air produces the greater amount of condensation.
2. In a similar way, dew forms when moist air near the ground cools and condenses.

LabManager™
Customize any lab with the LabManager™ CD-ROM.

Assessment

Knowledge Ask students why any surface, such as eyeglasses, the windshield of a car, or an iced-tea glass, sometimes fogs up. The surface's temperature equals the dew point. Why do some surfaces frost up? The dew-point temperature is below freezing. How can one get rid of the condensation/frost? Warm the surface to a temperature above the dew point; for example, use the defroster setting in a car's heating/cooling system or switch on the tiny wires in the rear window of a car.

■ **Caption Question Fig. 16**
Latent heat is released during condensation to offset some of the cooling of the dry adiabatic lapse rate.

3 Assess

Check for Understanding

Discussion What is the relationship between temperature and pressure? Temperature and density? Temperature is directly proportional to pressure and inversely proportional to density.

Reteach

Use Data Tables Have students use the relative humidity table in the *Reference Handbook* to determine whether two areas with the same relative humidity also have the same temperature. Have students explain their answers. No, they do not. Relative humidity in the table is based on the difference in temperature between the dry-bulb thermometer and the wet-bulb thermometer.

Assessment

Knowledge What is the relationship between temperature and humidity and the probable air pressure in a region? Warm, humid air has less density and less air pressure; cool, dry air is more dense and has a higher pressure.

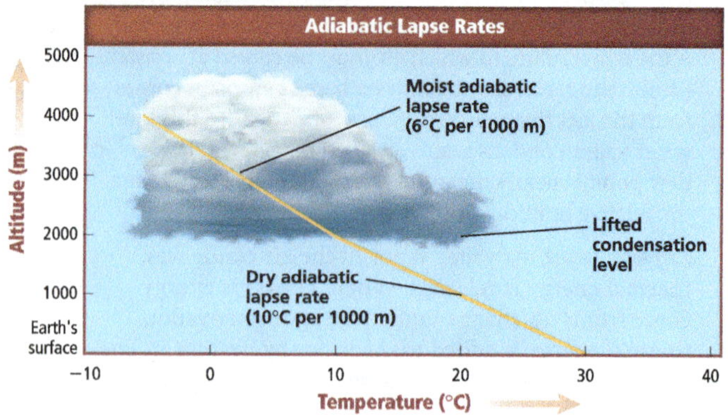

■ **Figure 16** Condensation occurs at the lifted condensation level (LCL). Air above the LCL is saturated and thus cools more slowly than air below the LCL.
Explain *why air above the LCL cools more slowly than air below the LCL.*

A rising air mass cools because the air pressure around it decreases as it rises, causing the air mass to expand. A rising air mass that does not exchange thermal energy with its surroundings will cool by about 10°C for every 1000 m it rises. This is called the dry adiabatic lapse rate—the rate at which unsaturated air will cool as it rises if no thermal energy is added or removed. If the air mass continues to rise, eventually it will reach saturation and condensation will occur. The height at which condensation occurs is called the lifted condensation level (LCL).

The rate at which saturated air cools is called the moist adiabatic lapse rate. This rate ranges from about 4°C/1000 m in very warm air to almost 9°C/1000 m in very cold air. The moist adiabatic rate is slower than the dry adiabatic rate, as shown in **Figure 16,** because water vapor in the air is condensing as the air rises and is releasing latent heat.

SECTION 2 REVIEW

Section Self-Check

Section Summary

- At the same pressure, warmer air is less dense than cooler air.
- Air moves from regions of high pressure to regions of low pressure.
- The dew point of air depends on the amount of water vapor the air contains.
- Latent heat is released when water vapor condenses and when water freezes.

Understand Main Ideas

1. MAINIDEA **Identify** three properties of the atmosphere and describe how they vary with height in the atmosphere.
2. **Explain** what occurs during a temperature inversion.
3. **Describe** how the motion of particles in a material changes when the temperature of the material increases.

Think Critically

4. **Predict** how the relative humidity and dew point change in a rising mass of air.
5. **Design** an experiment that shows how average wind speeds change over different types of surfaces.

MATH IN ▶ Earth Science

6. If the average thickness of the troposphere is 11 km, what would be the temperature difference between the top and bottom of the troposphere if the temperature decrease is the same as the dry adiabatic lapse rate?

SECTION 2 REVIEW

1. Air pressure and density decrease with height in the atmosphere. Temperature decreases in the troposphere and mesosphere. Relative humidity decreases with increased altitude.
2. In a temperature inversion, temperature increases with height.
3. The particles—atoms and molecules—move faster when the temperature increases.
4. Relative humidity increases with altitude because the air is cooling and its capacity for humidity is decreasing. The dew point is decreasing, but only slightly. The decreasing temperature of rising, cooling air approaches the dew point. When both are the same, condensation begins.
5. Experiments will differ. Students can use a fan to generate wind over different types of surfaces. The surfaces should differ in roughness.
6. Dry adiabatic lapse rate is 10°C per km. Over a height of 11 km, the temperature would drop 110°C.

SECTION 3

Clouds and Precipitation

MAINIDEA Clouds vary in shape, size, height of formation, and type of precipitation.

Essential Questions

- What is the difference between stable and unstable air?
- How do low, middle, high, and vertical development clouds differ?
- How does precipitation form?

Review Vocabulary

condensation: process in which water vapor changes to a liquid

New Vocabulary

condensation nucleus
orographic lifting
cumulus
stratus
cirrus
precipitation
coalescence

EARTH SCIENCE 4 YOU If you look up at the sky, you might notice differences among the clouds from day to day and hour to hour. Some clouds signal fair weather and others signal violent storms.

Cloud Formation

A cloud can form when a rising air mass cools. Recall that Earth's surface heats and cools by different amounts in different places. This uneven heating and cooling of the surface causes air masses near the surface to warm and cool. As an air mass is heated, it becomes less dense than the cooler air around it. This causes the warmer air mass to be pushed upward by the denser, cooler air.

However, as the warm air mass rises, it expands and cools adiabatically. The cooling of an air mass as it rises can cause water vapor in the air mass to condense. Recall that the lifted condensation level is the height at which condensation of water vapor occurs in an air mass. When a rising air mass reaches the lifted condensation level, water vapor condenses around condensation nuclei, as shown in **Figure 17.** A **condensation nucleus** is a small particle in the atmosphere around which water droplets can form. These particles are usually less than about 0.001 mm in diameter and can be made of ice, salt, dust, and other materials. The droplets that form can be liquid water or ice, depending on the surrounding temperature. When the number of these droplets is large enough, a cloud is visible.

Figure 17 Clouds form when a mass of rising air becomes saturated and water collects on condensation nuclei.

Jason Weingart Photography

SECTION 3

1 Focus

Focus Transparency

Before presenting the lesson, project **Section Focus Transparency 31** and have students answer the accompanying questions. EL

MAINIDEA

Clouds and Precipitation Have students consider why some clouds look different than others. Explain that though they are all clouds, they can appear different and indicate varied types of weather because of differences in the processes that created them.

2 Teach

Teacher Content Support

Dynamic Lifting In addition to the processes of thermal and orographic lifting, there is another way for air to be lifted. In fact, it is one of the more common methods. Called dynamic lifting, it occurs because of the large-scale rising and sinking of air associated with large-scale weather systems, such as cyclones, anticyclones, and fronts.

Activity

Cloud Formation To prepare students for the discussion of clouds, have them describe what happens to air that is warmer than the surrounding air. Tell students that once the rising air is saturated, the water vapor in the air condenses into water droplets. These droplets coalesce, and a cloud forms. COOP LEARN

Discussion

Latent Heat and Clouds By now, students should understand that a parcel of air will rise if it is warmer than the surrounding air. Conversely, a parcel of air will sink if it is cooler than its surrounding air. Students should also realize that air cools as it rises—eventually, it will reach the same temperature as the surrounding air and stop rising. It's common, however, for air to rise well above the point where it should have cooled to the temperature of the surrounding air. Challenge students to infer why. Help them to understand that other processes, such as the release of latent heat, cause the parcel of air to continue rising.

READING CHECK The stability of an air mass depends on how the temperature and moisture of the air mass changes relative to its surroundings.

CAREERS IN EARTH SCIENCE

WebQuest

CAREERS IN EARTH SCIENCE

Weather Observer A weather observer collects information for meteorologists about weather and sea conditions using weather equipment, radar scans, and satellite photographs. An education that includes biology, Earth science, environmental science, and geology is useful for a weather observer.

WebQuest

Atmospheric stability As an air mass rises, it cools. However, the air mass will continue to rise as long as it is warmer than the surrounding air. Under some conditions, an air mass that has started to rise sinks back to its original position. When this happens, the air is considered stable because it resists rising. The stability of air masses determines the type of clouds that form and the associated weather patterns.

Stable air The stability of an air mass depends on how the temperature of the air mass changes relative to the atmosphere. The air temperature near Earth's surface decreases with altitude. As a result, the atmosphere becomes cooler as the air mass rises. At the same time, the rising air mass is also becoming cooler. Suppose that the temperature of the atmosphere decreases more slowly with increasing altitude than does the temperature of the rising air mass. Then the rising air mass will cool more quickly than the atmosphere. The air mass will finally reach an altitude at which it is colder than the atmosphere. It will then sink back to the altitude at which its density is the same as the atmosphere, as shown in **Figure 18.** Because the air mass stops rising and sinks downward, it is stable. Fair weather clouds form under stable conditions.

READING CHECK **Describe** the factors that affect the stability of air.

Unstable air Suppose that the temperature of the surrounding air cools faster than the temperature of the rising air mass. In these conditions, the air mass will always be less dense than the surrounding air. As a result, the air mass will continue to rise, as shown in **Figure 18.** The atmosphere is then considered to be unstable. Unstable conditions can produce the type of clouds associated with thunderstorms.

Figure 18 Stable air has a tendency to resist movement. Unstable air does not resist vertical displacement. When the temperature of a mass of air is greater than the temperature of the surrounding air, the air mass rises. When the temperature of the surrounding air is greater than that of the air mass, it sinks.

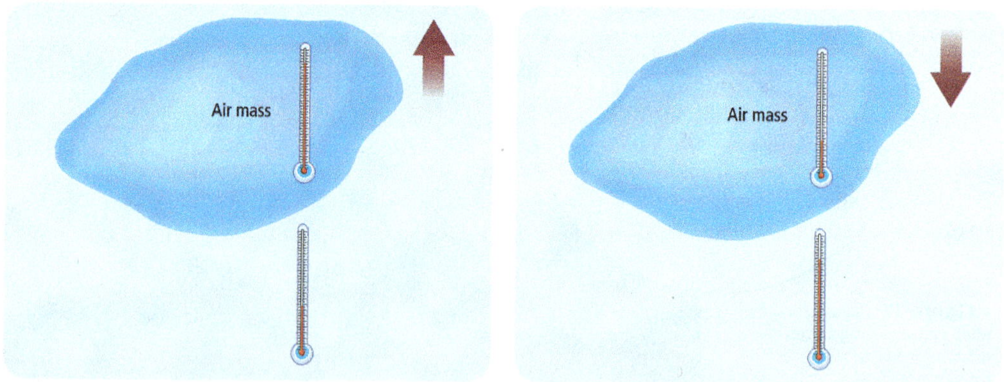

IN THE FIELD

Classify Clouds English naturalist Luke Howard is credited with providing the framework for the cloud classification system used today. However, a French naturalist, Jean Baptiste Lamarck, produced the first published classification of clouds in 1802, a year earlier than Howard. Lamarck did not classify all cloud types, however, and his work was discredited because it appeared in a publication that also printed forecasts based on astrological data. Further, he chose to publish in French; Howard published in Latin, which, at the time, was a more universal language. Howard's system was expanded in 1840 by German meteorologist Kaemtz, then again by French scientist Renou in 1855. These expansions resulted in the basic system used today.

■ **Figure 19** Orographic lifting occurs when warm, moist air is cooled because it is forced to rise over a topographic barrier, such as a mountain or cliff as shown here.

Atmospheric lifting Clouds can form when moist air rises, expands, and cools. Air rises when it is heated and becomes warmer than the surrounding air. This process is known as convective lifting. Clouds can also form when air is forced upward or lifted by mechanical processes. Two of these processes are orographic lifting and convergence.

Orographic lifting Clouds can form when air is forced to rise over elevated land or other topographic barriers. This can happen, for example, when an air mass approaches a mountain range. **Orographic lifting** occurs when an air mass is forced to rise over a topographic barrier, as shown in **Figure 19.** The rising air mass expands and cools, with water droplets condensing when the temperature reaches the dew point. Many of the rainiest places on Earth are located on the windward sides of mountain slopes, such as the coastal side of the Sierra Nevadas. The formation of clouds and the resulting heavy precipitation along the west coast of Canada are also primarily due to orographic lifting.

Convergence Air can be lifted by convergence, which occurs when air masses move into the same area from different directions. Then some of the air is forced upward. This process is even more pronounced when air masses at different temperatures collide. When a warm air mass and a cooler air mass collide, the warmer, less-dense air is forced upward over the denser, cooler air. As the warm air rises, it cools adiabatically. If the rising air cools to the dew-point temperature, then water vapor can condense on condensation nuclei and form a cloud. This cloud formation mechanism is common at middle latitudes where severe storm systems form as cold polar air collides with warmer air. Convergence also occurs near the equator where the trade winds meet at the intertropical convergence zone.

Use an Analogy

Hot-Air Balloons To keep a hot-air balloon aloft, a pilot must periodically supply more heat to the air in the balloon. This is because the air inside the balloon interacts with the air outside the balloon and cools through conduction. A rising parcel of warm air follows the same principle. It cools through conduction and needs additional input of energy to continue rising. The atmospheric equivalent of the pilot firing a burner is the release of latent heat when condensation occurs.

Concept Development

High Clouds Students have read that, as a rising parcel of air is heated near Earth's surface and reaches its LCL, a visible cloud can form if sufficient moisture is present. How, though, does this explain the development of clouds that are usually well above the LCL, such as middle altocumulus clouds and cirriform clouds made of ice crystals? Large-scale weather disturbances cause air to rise at many levels of the atmosphere, and over large areas. Regardless of the level, if enough lift exists to cool the air to its saturation temperature, condensation can occur and clouds can form.

Demonstration

Coalescence Place a piece of waxed paper on a board. Next, sprinkle some water on the waxed paper. **WARNING:** ***Wipe up spills immediately to avoid accidents.*** Incline the board slightly. At this point, the water droplets should not move. Gently manipulate the waxed paper so that the droplets start to move around and touch each other. Eventually, the drops will coalesce and become large enough to roll down the incline and fall off the paper. Have students relate this process to cloud formation and precipitation. Small water droplets coalesce to form large drops. These drops can eventually become too large to be held aloft by rising air, and they will fall to the ground as precipitation.

Interpret the Illustration

Figure 20 Show students photographs of different types of clouds, and have students classify the clouds according to the information provided in **Figure 20.**

Apply Earth Science

Cloud Seeding Cloud seeding can cause a developing or ongoing storm to produce more rain. This is done by seeding clouds with microscopic particles that act as condensation nuclei. Cloud seeding allows more raindrops to form, and the storm releases all the moisture it can. Ask students to identify the main drawback of cloud seeding. Cloud seeding enhances the chance of rainfall in an area but it can create drought conditions downwind. **OL**

Caption Question Fig. 20 Cirrus clouds are high clouds that tend to be thin and wispy and are composed of ice crystals; stratus clouds are low, layered clouds that can take up most of the sky.

Types of Clouds

You have probably noticed that clouds have different shapes. Some clouds look like puffy cotton balls, while others have a thin, feathery appearance. These differences in cloud shape are due to differences in the processes that cause clouds to form. Cloud formation can also take place at different altitudes—sometimes even right at Earth's surface, in which case the cloud is known as fog.

Clouds are generally classified according to a system developed in 1803, and only minor changes have been made since it was first introduced. **Figure 20** shows the most common types of clouds. This system classifies clouds by the altitudes at which they form and by their shapes. There are three classes of clouds based on the altitudes at which they form: low, middle, and high. Low clouds typically form below 2000 m. Middle clouds form mainly between 2000 m and 6000 m. High clouds form above 6000 m. One type of low cloud—cumulonimbus—can develop upward vertically through the middle and high cloud levels. Such clouds are known as clouds with vertical development.

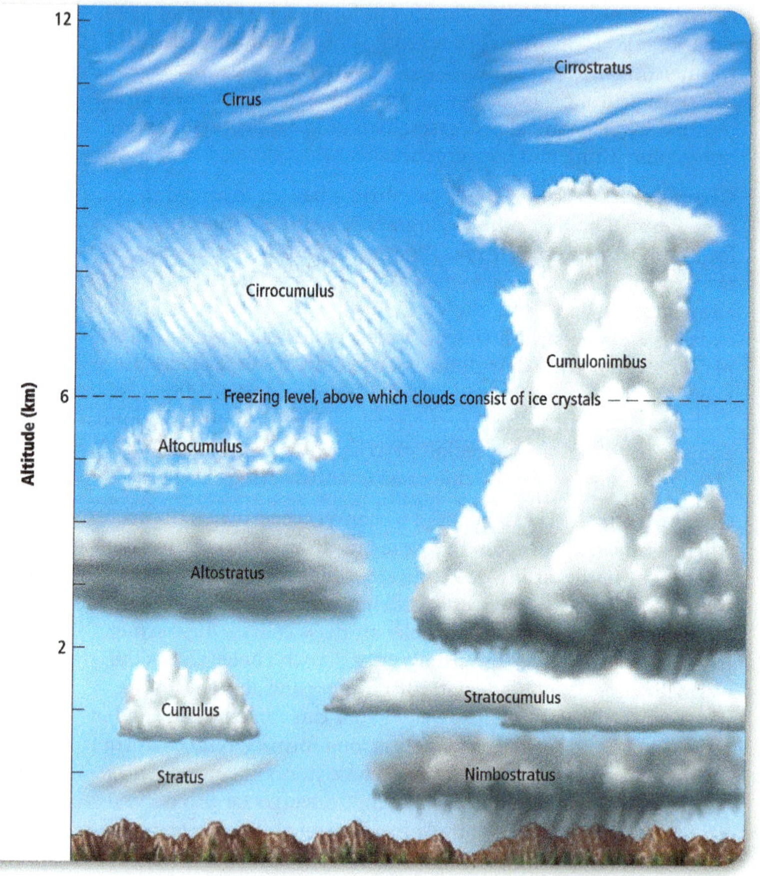

Figure 20 Clouds form at different altitudes and in different shapes.
Compare and contrast *cirrus and stratus clouds.*

ACROSS THE CURRICULUM

Geology The movement of water through the water cycle can have great effects on shaping geologic features. Have students identify ways that water helps shape the landscape. Some examples might include the existence and paths of current and former rivers, the formation of valleys, the erosion of hills or mountains, and the effects of the advance or retreat of glaciers.

EARTH SCIENCE JOURNAL

Clouds and Precipitation Have students write an Earth science journal entry describing a droplet of water vapor from the time it starts to rise near Earth's surface to its return as a hailstone falling from a large cumulonimbus cloud. Have them describe at least one reason it started to rise. **OL**

Rubric

Low clouds Clouds can form when warm, moist air rises, expands, and cools. If conditions are stable, the air mass stops rising at the altitude where its temperature is the same as that of the surrounding air. If a cloud has formed, it will flatten out and winds will spread it horizontally into stratocumulus or layered cumulus clouds, as shown in **Figure 20.** **Cumulus** (KYEW myuh lus) clouds are puffy, lumpy-looking clouds that usually occur below 2000 m. Also below 2000 m, layered sheet-like **stratus** (STRAY tus) clouds can cover much or all of the sky in a given area. Stratus clouds often form when fog lifts away from Earth's surface. A nimbostratus cloud produces precipitation.

Middle clouds Altocumulus and altostratus clouds form at altitudes between 2000 m and 6000 m. They are made up of ice crystals and water droplets due to the colder temperatures generally present at these altitudes. Middle clouds are usually layered. Altocumulus clouds are white or gray in color and form large, round masses or wavy rows. Altostratus clouds have a gray appearance, and they form thin sheets of clouds. Middle clouds sometimes produce mild precipitation.

High clouds High clouds, made up of ice crystals, form at heights above 6000 m where temperatures are below freezing. Some, such as **cirrus** (SIHR us) clouds, often have a wispy, indistinct appearance. A second type of high cloud, cirrostratus, forms as a continuous layer in the sky, varying from almost transparent to dense enough to block out the Sun or the Moon. A third type of high cloud, cirrocumulus, has a rippled appearance.

☑ READING CHECK **Identify** types of low, middle, and high clouds.

Vertical development clouds If the air that makes up a cumulus cloud is unstable, the cloud will be warmer than the surrounding air and will continue to grow upward. As it rises, water vapor condenses, and the air continues to increase in temperature due to the release of latent heat. The cloud can grow through middle altitudes as a towering cumulonimbus, as shown in **Figure 21,** and, if conditions are right, it can reach the tropopause. Its top is then composed entirely of ice crystals. Strong winds can spread the top of the cloud into an anvil shape. What began as a small mass of unstable moist air is now an atmospheric giant, capable of producing the torrential rains, strong winds, and hail characteristic of some thunderstorms.

■ **Figure 21** Cumulonimbus clouds, such as the large, puffy cloud here, are associated with thunderstorms.
Describe *how a cumulonimbus cloud can form.*

Use Science Terms

Virga Sometimes, rain or snow falls through a layer of air so dry that the precipitation evaporates. The precipitation can be seen falling from the base of the cloud, but it disappears into filmy wisps as it evaporates. This is known as virga, a type of precipitation that falls but doesn't reach the ground. Virga is most often observed in the United States in the Great Plains, the Rocky Mountain region, and the deserts of the Southwest. In these places, layers of dry air are more common than in coastal regions.

■ **Caption Question Fig. 21** Cumulonimus clouds form in very unstable air where there is a large amount of water vapor available.

☑ READING CHECK Low clouds are those below 2000 m; middle clouds are between 2000–6000 m and have the (alto–) prefix; high clouds are above 6000 m and have the prefix (cirro–).

EARTH SCIENCE JOURNAL

Measure Rainfall Technology for estimating the amount of precipitation that falls ranges from simple rain gauges to sophisticated satellite systems. Information obtained from these sources is used in part to forecast floods. Have students research and write brief reports about the different types of equipment used to measure rainfall. OL

DIFFERENTIATED INSTRUCTION

Advanced Learners There are three main cloud classifications, grouped according to the heights at which their bases form. However, each of these three groups is further subdivided into nine subgroups. Have students research these subgroups and then create data tables detailing the main cloud classifications and their subgroups along with descriptions and illustrations or photos.

Collaborative Learning

What happens to precipitation? Have groups of students research all the things that might happen to water when it falls to Earth as precipitation. One group might examine how surface runoff makes its way through various waterways to the oceans. Another group could research what happens to water absorbed by Earth's surface. Yet another group could focus on annual variations in the amount of water stored in ice fields near the poles. Groups should share their results with the class. OL COOP LEARN

■ **Caption Question** **Fig. 22** The hailstone must make multiple trips above and below the freezing level in a thunderstorm; below, it takes on a new layer of water vapor which freezes above the freezing level.

Precipitation

All forms of water that fall from clouds to the ground are **precipitation.** Rain, snow, sleet, and hail are the four main types of precipitation. Clouds contain water droplets that are so small that the upward movement of air in the cloud can keep the droplets from falling. In order for these droplets to become heavy enough to fall, their size must increase by 50 to 100 times.

Coalescence One way that cloud droplets can increase in size is by coalescence. In a warm cloud, coalescence is the primary process responsible for the formation of precipitation. **Coalescence** koh uh LEH sunts)occurs when cloud droplets collide and join together to form a larger droplet. These collisions occur as larger droplets fall and collide with smaller droplets. As the process continues, the droplets eventually become too heavy to remain suspended in the cloud and fall to Earth as precipitation. Rain is precipitation that reaches Earth's surface as a liquid. Raindrops typically have diameters between 0.5 mm and 5 mm.

Snow, sleet, and hail The type of precipitation that reaches Earth depends on the vertical variation of temperature in the atmosphere. In cold clouds where the air temperature is far below freezing, ice crystals can form that finally fall to the ground as snow. Sometimes, even if ice crystals form in a cloud, they can reach the ground as rain if they fall through air warmer than 0°C and melt.

In some cases, air currents in a cloud can cause cloud droplets to move up and down through freezing and nonfreezing air, forming ice pellets that fall to the ground as sleet. Sleet can also occur when raindrops freeze as they fall through freezing air near the surface.

If the up-and-down motion in a cloud is especially strong and occurs over large stretches of the atmosphere, large ice pellets known as hail can form. **Figure 22** shows a sample of hail. Most hailstones are smaller in diameter than a dime, but some stones have been found to weigh more than 0.5 kg. Larger stones are often produced during severe thunderstorms.

■ **Figure 22** Hail is precipitation in the form of balls or lumps of ice that is produced by intense thunderstorms.
Infer *How might the layers in the cross section of the hailstone form?*

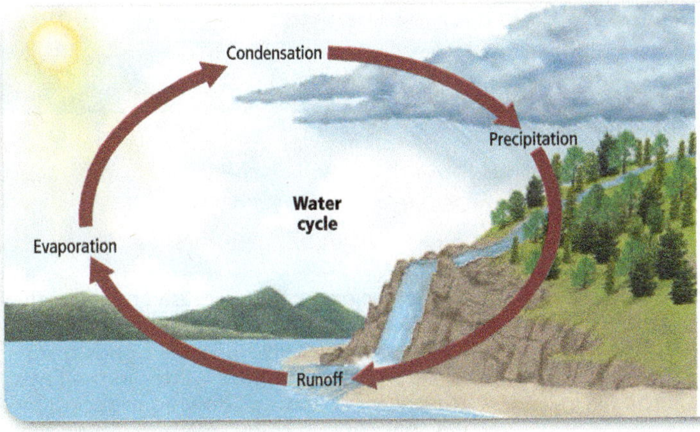

■ **Figure 23** Water moves from Earth to the atmosphere and back to Earth in the water cycle.

The water cycle About 97 percent of Earth's water is in the oceans. At any one time, only a small percentage of water is present in the atmosphere. Still, this water is vitally important because, as it continually moves between the atmosphere and Earth's surface, it nourishes living things. The constant movement of water between the atmosphere and Earth's surface is known as the water cycle.

The water cycle is summarized in **Figure 23.** Radiation from the Sun causes liquid water to evaporate. Water evaporates from lakes, streams, and oceans and rises into Earth's atmosphere. As water vapor rises, it cools and condenses to form clouds. Water droplets combine to form larger drops that fall to Earth as precipitation. This water soaks into the ground and enters lakes, streams, and oceans, or it falls directly into bodies of water and eventually evaporates, continuing the water cycle.

SECTION 3 REVIEW

Section Self-Check

Section Summary

- Clouds are formed as warm, moist air is forced upward, expands, and cools.
- An air mass is stable if it tends to return to its original height after it starts rising.
- Cloud droplets form when water vapor is cooled to the dew point and condenses on condensation nuclei.
- Clouds are classified by their shapes and the altitudes at which they form.
- Cloud droplets collide and coalesce into larger droplets that can fall to Earth as rain, snow, sleet, or hail.

Understand Main Ideas

1. MAINIDEA **Summarize** the differences between low clouds, middle clouds, and high clouds.
2. **Describe** how precipitation forms.
3. **Determine** the reason precipitation will fall as snow rather than rain.
4. **Compare** stable and unstable air.

Think Critically

5. **Evaluate** how a reduction in the number of condensation nuclei in the troposphere would affect precipitation. Explain your reasoning.

WRITING IN ▶ Earth Science

6. Describe the path a drop of rain might follow throughout the water cycle.

3 Assess

Check for Understanding

Discussion Ask students to identify the major cloud groups and at least two subcategories of each major group. major groups—stratiform and cumuliform Identify the probable heights at which the cloud bases form. <2000 m, 2000–6000 m and >6000 m Ask: Are the clouds composed of liquid water or ice crystals? cirriform, made of ice; all others liquid Are they likely to produce precipitation? high and middle level, less likely; lower forms, more likely

Reteach

Concept Map Construct concept maps showing the life cycle of a water droplet. Concept maps should include evaporation of water from the ocean surface; lifting of the warm, less-dense, moist air; cooling of the air to its LCL; saturation; condensation; coalescence; precipitation; and finally, the return of the water droplet to the ocean through a waterway.

Assessment

Knowledge Ask students the following question: Of the following, which describes a layered cloud that is dense enough to block out the Sun and is composed entirely of ice crystals?

a. stratus
b. cirrostratus
c. altostratus

Answer b: The prescence of ice crystals is the deciding factor.

SECTION 3 REVIEW

1. Low and middle clouds are below the freezing level, while high clouds are above it. Low clouds are made of water droplets, middle clouds are made of water droplets and ice, and high clouds are always made of ice.
2. Answers will vary and should include the idea that water vapor is condensing to a liquid form or forming ice crystals of some form.
3. If the precipitation is always in an air layer that is at or below freezing, snow falls.
4. The stability of air depends on how the temperature of the air mass changes relative to its surroundings. If the air mass is always cooler than its surroundings and tends to sink, then it is stable. If the air mass is always warmer and less dense than its surroundings, it is unstable.
5. Because precipitation forms around condensation nuclei, a reduction in nuclei would result in a reduction in precipitation. Air without particles would have difficulty in producing much precipitation.
6. Answers will vary, but should include evaporation, condensation, and precipitation, in that order. Answers can also reference runoff as a return mechanism for precipitation to a water body.

Rubric

Earth Science & ENVIRONMENT

Purpose

Students will learn about the history of ozone depletion and the results of studies on the recovery of the ozone layer due to the reduction of CFCs.

Teaching Strategy

Have students discuss the importance of a worldwide effort to reduce and end the use of CFCs in order to protect the ozone layer. Ninety-five percent of the world's countries signed the Montreal Protocol. Why is a worldwide effort needed to successfully solve environmental issues that concern the atmosphere?

Teacher Content Support

Ozone Ozone depletion varies with latitude. The most severe depletion has occurred in the highest latitudes in the southern hemisphere, where stratospheric temperature is low enough. Regions near the equator have experienced very little depletion. Measurements of ozone depletion also vary with seasons. The greatest depletion occurs in the winter and spring.

Earth Science & ENVIRONMENT

Ozone Variation

Atoms, such as chlorine and bromine, when located in the stratosphere, can destroy ozone molecules. The decline in stratospheric ozone measured since the early 1980s is showing signs of recovery perhaps due to a decrease in stratospheric chlorine.

Variations in ozone amounts The total amount of ozone in the atmosphere over Earth's surface varies with location and also changes with time. Total ozone increases with latitude, being low at the equator and highest in the polar regions. Ozone amounts also vary seasonally, usually decreasing from winter to summer. The largest seasonal changes occur at high latitudes, particularly in the polar regions.

The Antarctic ozone hole Over Antarctica, the lowest ozone amounts occur in early spring. This decrease in ozone over Antarctica is called the Antarctic ozone hole.

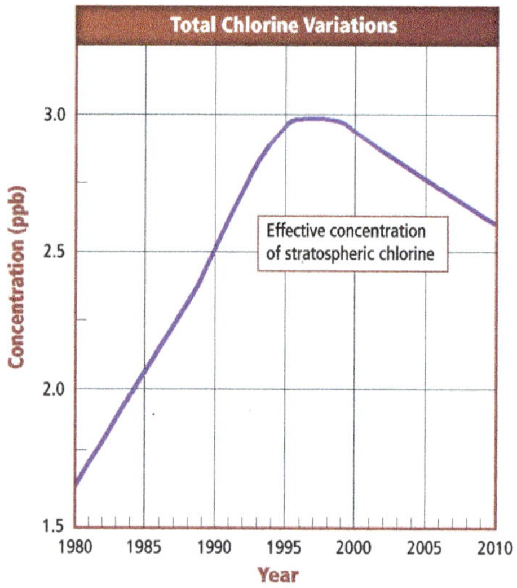

The Antarctic ozone hole is caused by chlorofluorocarbons (CFCs) and the presence of polar stratospheric clouds (PSCs). These clouds form over Antarctica during the winter in the lower stratosphere. CFCs break down, producing molecules that contain chlorine atoms. These molecules undergo chemical reactions on ice crystals in the PSCs, producing chlorine and other compounds that destroy ozone.

The Montreal Protocol Satellite measurements beginning in the late 1970s also showed a decrease in global ozone. Concerns over decreasing ozone led to the adoption of the Montreal Protocol in 1987. This international agreement requires countries to phase out the production and use of CFCs and similar chemicals. As a result, levels of chlorine and other ozone-destroying chemicals in the stratosphere have been declining since the late 1990s, as shown in the graph.

Signs of recovery Since 1996, the decrease in total ozone has leveled off in most regions, and in some regions the ozone layer is showing signs of early recovery. This recovery might be due to the Montreal Protocol, as well as natural causes such as changes in atmospheric wind patterns and solar variability. Careful monitoring of the ozone over the next several years will be needed to assure the ozone layer continues to recover.

WRITING IN ▶ Earth Science

Magazine Article Research how natural processes, such as volcanic eruptions, solar activity, and air movements affect ozone levels in the stratosphere. Write a magazine article that reports what you found.

WebQuest

WRITING IN ▶ Earth Science

Rubric

Magazine Article Volcanic eruptions usually lead to a decrease in ozone levels. The sulfur particles released in eruptions remain in the upper atmosphere for years and increase the effectiveness of halogen gases at breaking down ozone molecules. Sunspots lead to increased levels of ozone. Wind moves air in the tropics with high levels of ozone toward the poles.

WebQuest

GeoLAB

iLab Station

Interpret Pressure-Temperature Relationships

Background: **As you go up a mountain, both temperature and air pressure decrease. Temperature decreases as you get farther away from the atmosphere's heat source—Earth's surface. Pressure decreases as you ascend the mountain because there are fewer particles in the air above you. Pressure and temperature are also related through the expansion and compression of air, regardless of height.**

Question: *How does the expansion and compression of air affect temperature?*

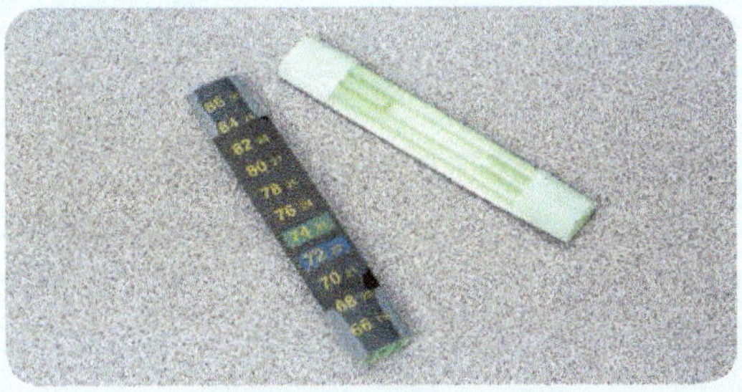

Materials

clean, clear, plastic 2-L bottle with cap
plastic straws
scissors
thin, liquid-crystal temperature strip
tape
watch or timer

Safety Precautions

Procedure

1. Read and complete the lab safety form.
2. Working with a partner, cut two pieces of straw, each the length of the temperature strip. Then cut two 2-cm pieces of straw. **WARNING:** ***Scissors pose a skin puncture or cut hazard.***
3. Lay the two long pieces on a table. Place the two shorter pieces within the space created by the longer pieces so that the four pieces form a support for the temperature strip as shown in the figure.
4. Tape the four pieces of straw together. Place the temperature strip lengthwise on the straws. Tape the strip to the straws.
5. Slide the temperature-strip-straw assembly into the clean, dry bottle. Screw the cap on tightly.
6. Place the sealed bottle on the table so that the temperature strip faces you and is easy to read. Do not handle the bottle any more than is necessary so that the temperature will not be affected by your hands.
7. Record the temperature of the air inside the bottle as indicated by the temperature strip.
8. Position the bottle so that half its length extends beyond the edge of the table. Placing one hand on each end of the bottle, push down on both ends so that the bottle bends in the middle. Hold the bottle this way for 2 min while your partner records the temperature every 15 s.
9. Release the pressure on the bottle. Observe and record the temperature every 15 s for the next 2 min.

Analyze and Conclude

1. **Interpret Data** What was the average temperature of the air inside the bottle as you applied pressure? How did this differ from the average temperature of the bottled air when you released the pressure?
2. **Graph** the temperature changes.
3. **Explain** how these temperature changes are related to changes in pressure.
4. **Predict** how the experiment would change if you took the cap off the bottle.
5. **Infer** Given your observations and what you know about the behavior of warm air, would you expect the air over an equatorial desert at midday to be characterized by high or low pressure?

WRITING IN ▶ Earth Science

Research more about how pressure changes can affect the daily weather. Share your findings with your classmates.

GeoLAB

iLab Station Rubric

Preparation

Time Allotment 30–45 min

Process Skills observe and infer, interpret and analyze data, recognize cause and effect, interpret scientific illustrations, compare and contrast, model

Safety Precautions Advise students to be careful when using scissors. Approve lab safety forms before work begins.

Procedure

- Inform students their activities here will simulate the changes in temperature that take place when atmospheric pressure increases or decreases.
- Bottles and temperature strips should not be handled any more than necessary to avoid transfer of body heat. Once the strips are placed inside the bottles and the bottles are capped, students should leave the bottles undisturbed for about five minutes to allow the temperature inside to stabilize.
- Students might want to take turns pressing down on the bottle.
- **Troubleshooting** Are the bottles clean and dry?

LabManager™
Customize any lab with the LabManager™ CD-ROM.

Analyze and Conclude

1. Average temperatures will differ. However, in all cases, temperature should have decreased when pressure was released.
2. Graphs will vary, but all should show that temperatures increased when pressure was applied and decreased when pressure was released.
3. As pressure increases, the molecules are packed more tightly together. This creates more collisions and produces more heat.
4. Air would escape when pressure was applied on the bottle. There would be no change in pressure and thus no change in temperature.
5. Low pressure; in the atmosphere, the heated air would be less dense than the air around it and would, therefore, rise. When density decreases, there are fewer particles pushing down over an area, resulting in lower pressure.

WRITING IN ▶ Earth Science

Rubric

Research Students should discuss how low pressure creates uplift, causing more clouds and precipitation. High pressure creates clear skies, low wind speeds, and dry conditions.

MAINIDEAS Summary statements can be used by students to review the major concepts of the chapter.

Students can review with these online resources.

Vocabulary Practice

- Vocabulary eGames
- Vocabulary eFlashcards
- Vocabulary PuzzleMaker

Section Self-Check

Chapter Self-Check

Online Test Practice

Use the Exam*View*® *Assessment Suite* CD-ROM to:

- create multiple versions of tests
- create modified tests with one mouse click
- edit existing questions and add your own questions
- build tests aligned with state standards using built-in state curriculum tags
- change English tests to Spanish with one mouse click
- track students' progress using the Teacher Management System

CHAPTER 11 STUDY GUIDE

BIGIDEA The composition, structure, and properties of Earth's atmosphere form the basis of Earth's weather and climate.

Vocabulary Practice

SECTION 1 Atmospheric Basics

VOCABULARY
- troposphere
- stratosphere
- mesosphere
- thermosphere
- exosphere
- radiation
- conduction
- convection

MAINIDEA Energy is transferred throughout Earth's atmosphere.

- Earth's atmosphere is composed of several gases, primarily nitrogen and oxygen, and also contains small particles.
- Earth's atmosphere consists of five layers that differ in their compositions and temperatures.
- Solar energy reaches Earth's surface in the form of visible light and infrared waves.
- Solar energy absorbed by Earth's surface is transferred as thermal energy throughout the atmosphere.

SECTION 2 Properties of the Atmosphere

VOCABULARY
- temperature inversion
- humidity
- saturation
- relative humidity
- dew point
- latent heat

MAINIDEA Atmospheric properties, such as temperature, air pressure, and humidity describe weather conditions.

- At the same pressure, warmer air is less dense than cooler air.
- Air moves from regions of high pressure to regions of low pressure.
- The dew point of air depends on the amount of water vapor the air contains.
- Latent heat is released when water vapor condenses and when water freezes.

SECTION 3 Clouds and Precipitation

VOCABULARY
- condensation nucleus
- orographic lifting
- cumulus
- stratus
- cirrus
- precipitation
- coalescence

MAINIDEA Clouds vary in shape, size, height of formation, and type of precipitation.

- Clouds are formed as warm, moist air is forced upward, expands, and cools.
- An air mass is stable if it tends to return to its original height after it starts rising.
- Cloud droplets form when water vapor is cooled to the dew point and condenses on condensation nuclei.
- Clouds are classified by their shapes and the altitudes at which they form.
- Cloud droplets collide and coalesce into larger droplets that can fall to Earth as rain, snow, sleet, or hail.

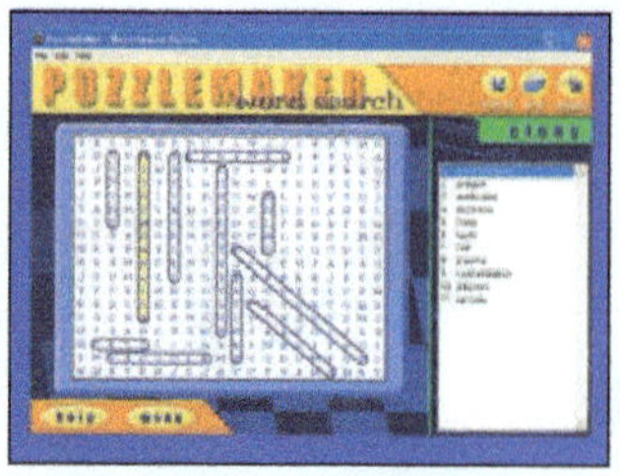

Vocabulary Practice

For additional practice with vocabulary, have students access the Vocabulary PuzzleMaker.

Chapter Self-Check

VOCABULARY REVIEW

Match each description below with the correct vocabulary term from the Study Guide.

1. outermost layer of Earth's atmosphere
2. transfer of energy from a higher to a lower temperature by collisions between particles
3. temperature at which condensation of water vapor can occur
4. occurs when the amount of water vapor in a volume of air has reached the maximum amount
5. the amount of water vapor present in air

Complete the sentences below using vocabulary terms from the Study Guide.

6. ________ are small particles in the atmosphere around which water droplets form.
7. The atmospheric layer that is closest to Earth's surface is the ________.
8. Types of ________ include hail, sleet, and snow.

Each of the following sentences is false. Make each sentence true by replacing the italicized words with terms from the Study Guide.

9. *Convection* occurs when small cloud droplets collide to form a larger droplet.
10. *Mesosphere* is the layer of Earth's atmosphere that contains the ozone layer.
11. The transfer of energy in matter or space by electromagnetic waves is called *latent heat.*
12. When the bottom of a pan of water is heated and the water expands, becoming less dense than the surrounding water, it is forced upward. As it rises, the water cools and sinks back to the bottom of the pan. This process is called *precipitation.*
13. When *saturation* occurs, an air mass is forced to rise over a topographic barrier.

UNDERSTAND KEY CONCEPTS

14. Which gas has increased in concentration by about 0.011 percent over the past 150 years?
    - **A.** oxygen
    - **B.** nitrogen
    - **C.** carbon dioxide
    - **D.** water vapor

Use the diagram below to answer Question 15.

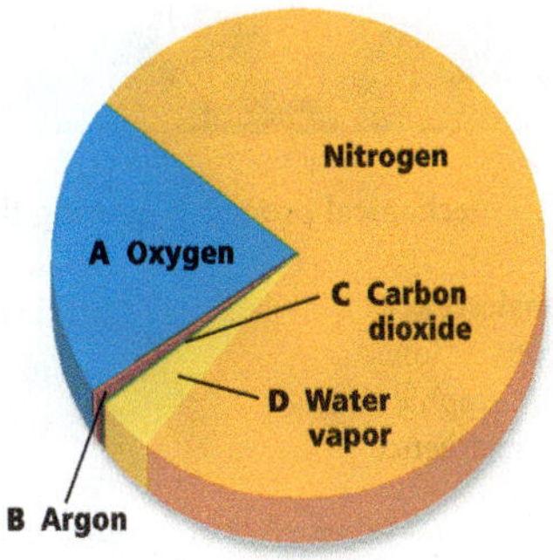

15. Which gas is least abundant in Earth's atmosphere?
    - **A.** A
    - **B.** B
    - **C.** C
    - **D.** D
16. Which is the primary cause of wind?
    - **A.** air saturation
    - **B.** pressure imbalances
    - **C.** pollution
    - **D.** movement of water
17. Which process takes up latent heat?
    - **A.** condensation of water vapor
    - **B.** evaporation of water vapor
    - **C.** adiabatic heating
    - **D.** pressure increase
18. Wind speed on Earth is reduced by which?
    - **A.** temperature
    - **B.** friction
    - **C.** weather
    - **D.** convergence

VOCABULARY REVIEW

1. exosphere
2. conduction
3. dew point
4. saturation
5. humidity
6. Condensation nuclei
7. troposphere
8. precipitation
9. Coalescence
10. Stratosphere
11. radiation
12. convection
13. orographic lifting

UNDERSTANDING KEY CONCEPTS

14. C
15. C
16. B
17. B
18. B

VOCABULARY REVIEW

1. exosphere
2. conduction
3. dew point
4. saturation
5. humidity
6. Condensation nuclei
7. troposphere
8. precipitation
9. Coalescence
10. Stratosphere
11. radiation
12. convection
13. orographic lifting

19. C
20. A
21. D
22. D
23. A

CONSTRUCTED RESPONSE

24. Precipitation from cumulonimbus clouds often begins as lumps of ice that melt as they fall. Such lumps contain more water than a simple raindrop formed by the collision-coalescense process in a nimbostratus cloud.
25. Evaporation converts water to water vapor, and condensation converts water vapor back to water, usually at altitudes where clouds form, causing precipitation.
26. Latent heat is stored during evaporation and released as sensible heat during condensation.
27. Unstable air rises rapidly, taking large quantities of water vapor with it. Latent heat release allows the cloud to be warmer than its surroundings, thus enabling further growth.
28. The island would have a higher relative humidity because of its proximity to an abundant supply of water vapor.
29. Earth's atmosphere is heated by Earth's surface through conduction and distributed vertically by convection.
30. Conduction transfers energy through collisions of molecules; convection transfers energy through the circulation of a fluid.
31. Troposphere cools with increased altitude, while the stratosphere warms. The troposphere has a consistent percentage composition throughout, while the stratosphere has higher amounts of ozone.
32. The temperature of the air nearest Earth; if the air is below freezing, the precipitation will be in a frozen form. If the air is above freezing, the precipitation will be in a liquid form.
33. When the dew point is equal to air temperature, then the air is saturated.
34. Water vapor is the main ingredient in cloud formation. It also releases latent heat during condensation.

Use the diagram below to answer Question 19.

19. Which mechanical process is causing the air to rise?
A. coalescence
B. convection
C. orographic lifting
D. convergence

20. Which is a cloud of vertical development?
A. cumulonimbus
B. cirrus
C. stratus
D. altocumulus

21. Almost all weather, clouds, and storms occur in which layer of the atmosphere?
A. thermosphere
B. mesosphere
C. stratosphere
D. troposphere

22. What color would be best for a home designed to absorb energy?
A. red
B. white
C. gray
D. black

23. Which temperature is coldest?
A. 32°F
B. 10°C
C. 280 K
D. 5°C

CONSTRUCTED RESPONSE

24. Explain why precipitation from a cumulonimbus cloud is generally heavier than that from a nimbostratus cloud.

25. Identify the role that evaporation and condensation play in Earth's water cycle.

26. Compare what happens to latent heat in the atmosphere during evaporation to what occurs during condensation.

Use the figure below to answer Question 27.

27. Describe the process that causes the cloud type shown to reach heights of over 6000 m.

28. Determine whether the average relative humidity on a small island in the ocean would likely be higher or lower than 100 km inland on a continent.

29. Explain If clouds absorb only a small amount of solar radiation, how is Earth's atmosphere heated?

30. Distinguish between convection and conduction as methods of transferring energy in the atmosphere.

31. Compare the temperature and composition of the troposphere and the stratosphere.

32. Determine what causes precipitation to fall as rain or snow.

33. Relate dew point and saturation.

34. Describe the importance of water vapor in the atmosphere.

Chapter Self-Check

THINK CRITICALLY

35. Careers in Earth Science Research information about the workday of a weather observer.

36. Predict how the concentration of ozone molecules would change if the concentration of oxygen molecules decreased.

37. Infer Using the idea that almost all weather occurs in the troposphere, infer why many airliners usually fly at altitudes of 11 km or higher.

38. Predict whether afternoon summertime temperatures near the beach would be warmer or cooler than temperatures farther inland. Explain.

39. Predict why spring is often the windiest time of the year based on your knowledge of temperature and wind.

40. Predict how the energy absorbed by the Arctic Ocean would change if the amount of the sea ice covering the ocean is reduced. Keep in mind that sea ice reflects more incoming solar energy than water does.

41. Assess which cloud type would be of most interest to a hydrologist who is concerned with possible heavy rain and flooding over large regions. Why?

42. Analyze why relative humidity usually decreases after the Sun rises and increases after the Sun sets.

CONCEPT MAPPING

43. Use the following terms and phrases to construct a concept map that describes the process of the water cycle: *water cycle; evaporation; condensation; precipitation; water changes from liquid to gas; water changes from gas to liquid; water falls as rain, snow, sleet, or hail.* For more help, refer to the *Skillbuilder Handbook.*

CHALLENGE QUESTION

44. Based on what you know about radiation and conduction, what conclusion might you make about summer temperatures in a large city compared with those in the surrounding countryside?

WRITING IN ▶ Earth Science

45. Write and illustrate a short story for elementary students that describes cumulonimbus cloud formation and the kinds of weather patterns they produce.

DBQ Document–Based Questions

Data obtained from: Climatological normals 1971–2000. *National Oceanographic and Atmospheric Administration, National Climatic Data Center.*

The graphs show the monthly variations in temperature and precipitation at three locations in the United States. Use the data to answer the questions below.

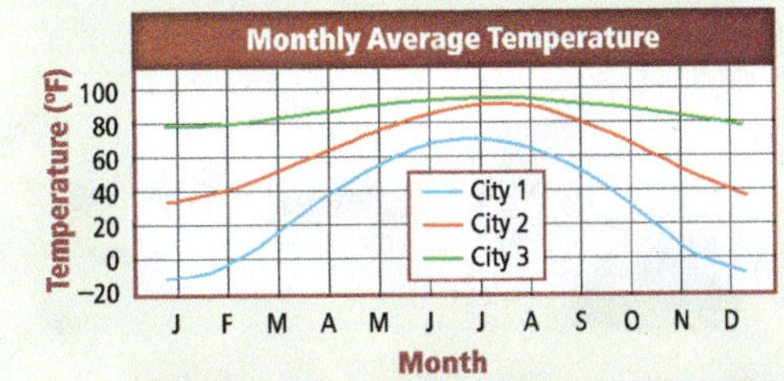

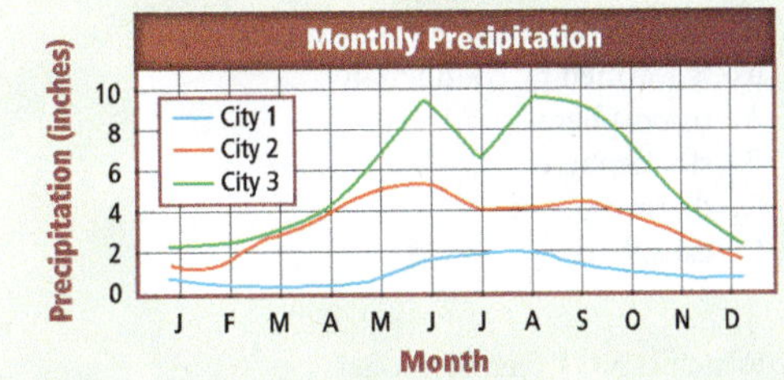

46. Estimate from the data which location probably receives the least annual solar radiation.

47. In which location would you expect heavy precipitation?

48. Deduce from the graphs which station probably receives the most annual snowfall.

CUMULATIVE REVIEW

49. Describe the properties of a contour line. **(Chapter 2)**

50. What process is explained by Bowen's reaction series? **(Chapter 5)**

WRITING IN ▶ Earth Science

Rubric

45. Answers will vary, but should include ideas connecting cumulonimbus clouds with cold fronts, tornadoes, and thunderstorms.

DBQ Document-Based Questions

Data obtained from: Climatological normals 1971—2000. *National Oceanographic and Atmospheric Administration, National Climatic Data Centers.*

46. City 1
47. City 3
48. City 2

CHAPTER 11 ASSESSMENT

THINK CRITICALLY

35. Answers will vary, but should include the technology used by a weather observer.

36. If there were less oxygen available, ozone would most likely decrease.

37. At this altitude, the flight is smoother and not plagued with weather concerns.

38. It would be cooler at the beach. Land heats faster than water, allowing for rising air over the land. Sea breezes then form which bring cooler air off the water.

39. There are greater temperature and pressure gradients formed due to the differences in the prevailing springtime air masses.

40. Energy absorbed would increase due to the reduced albedo (sea ice) in the region. This would result in increased temperatures.

41. cumulonimbus, because they produce the greatest amounts of precipitation over the shortest periods of time

42. Sunrise brings warmer temperatures. Warmer air can hold more humidity and assuming no inputs of new moisture, relative humidity will drop. Sunset is followed by cooler temperatures and reduced capacity for humidity.

CONCEPT MAPPING

43. Concept maps should be constructed in the following order. The first level is water cycle. In the next levels, in any order, evaporation linked to water changes from liquid to gas; condensation linked to water changes from gas to liquid; precipitation linked to falling water.

CHALLENGE QUESTION

44. The temperature in the large city would most likely be higher for reasons including concrete and pavement absorbing radiation and releasing it slowly.

CUMULATIVE REVIEW

49. Contour lines connect points of equal value in a field of data points.

50. The order of formation of the principle minerals in an igneous rock.

MULTIPLE CHOICE

1. B
2. A
3. C
4. D
5. A
6. B
7. C
8. A
9. C
10. D

CUMULATIVE STANDARDIZED TEST PRACTICE

MULTIPLE CHOICE

1. What is the composition of dripstone formations?
   A. gravel
   B. limestone
   C. clay
   D. sand

Use the diagram to answer Questions 2 and 3.

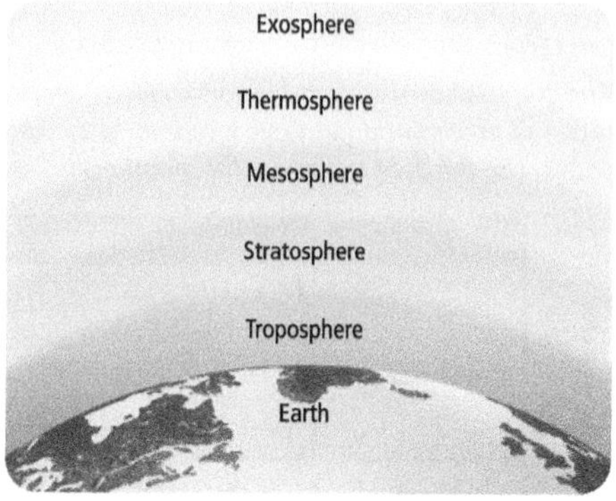

2. In which layer of Earth's atmosphere is air most likely warmed by conduction?
   A. troposphere
   B. stratosphere
   C. thermosphere
   D. exosphere

3. Which is NOT true of ozone?
   A. It absorbs ultraviolet radiation.
   B. Its concentration is decreasing.
   C. It is concentrated in the atmospheric layer called the mesosphere.
   D. It is a gas formed by the addition of one oxygen atom to an oxygen molecule.

4. Which describes the temperature of groundwater flowing through a natural spring?
   A. hotter than the region's average temperature
   B. cooler than the region's average temperature
   C. the same temperature no matter where the spring is located
   D. the same temperature as the region's average temperature

5. Why do deserts experience wind erosion?
   A. There is limited rain to allow plants to grow and hold down sediment.
   B. Saltation does not occur readily in desert areas.
   C. The increased amount of heat increases wind patterns.
   D. Wind can carry larger particles than water.

6. Which is NOT a significant agent of chemical weathering?
   A. oxygen
   B. nitrogen
   C. carbon dioxide
   D. water

Use the table below to answer Questions 7 and 8.

Population of Unknown Organisms				
	Spring	Summer	Autumn	Winter
1995	564	14,598	25,762	127
1996	750	16,422	42,511	102
1997	365	14,106	36,562	136

7. What inference can be made based upon the data?
   A. Scientists have a hard time consistently tracking the organism.
   B. The organism migrates yearly.
   C. The organism is most abundant during summer and fall.
   D. The organism should be placed on the endangered species list.

8. What would be the best graphical representation of this data?
   A. bar graph
   B. line graph
   C. circle graph
   D. model

9. Which is most likely to cause orographic lifting?
   A. a sandy beach
   B. a flowing river
   C. a rocky mountain
   D. a sand dune

10. Why are many of the lakes in Central Florida considered to be part of the karst topography?
    A. They are depressions in the ground near caves.
    B. They are part of a sinking stream.
    C. They are layered with limestone.
    D. They are sinkholes.

Online Test Practice

SHORT ANSWER

Use the illustration below to answer Questions 11–13.

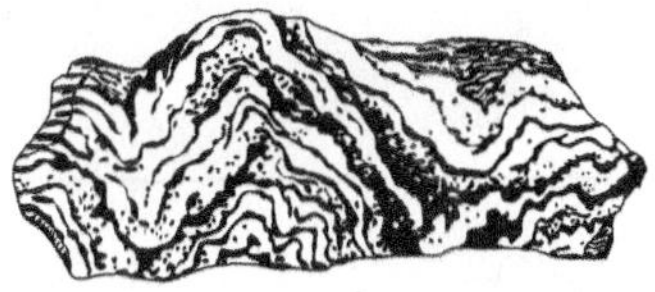

11. What type of rock is shown above? What features indicate this?

12. Hypothesize how this sample of rock formed.

13. According to the rock cycle, what changes could occur in this rock? What new type of rock would be produced?

14. Describe temperature and heat.

15. Define ion and explain how an ion is formed.

16. Describe how a flood might cause residual soil to become transported soil.

READING FOR COMPREHENSION

Ozone Layer Recovery

Damage to the ozone layer, caused by chlorofluorocarbon (CFC) chemicals and other pollutants, is starting to reverse itself, according to data collected by NASA satellites. Ozone degradation continues despite global bans on ozone-depleting pollutants. The rate has slowed enough in the upper stratosphere that scientists think ozone is starting to be replenished there.

Evidence suggests that international efforts to reduce chlorofluorocarbon (CFC) pollution are working. Some predictions suggest that the ozone layer will have recovered to preindustrial levels by the late twenty-first century, though total recovery could happen within 50 years.

17. According to the passage, what is the major cause of the replenishing of the ozone layer?
    A. the ban of chlorofluorocarbons
    B. preindustrial pollution
    C. the upper stratosphere
    D. NASA satellites

18. What can be inferred from this passage?
    A. The ozone layer is recovering, but will never be fully restored.
    B. CFC pollution is no longer occurring.
    C. The upper stratosphere is the only layer with ozone depletion.
    D. Ozone depletion in the upper stratosphere has slowed down.

19. According to the text, how long could it take for a full recovery of the ozone layer?
    A. a decade
    B. until the late twenty-first century
    C. 50 years
    D. several years

20. Why is it important that the ozone layer in the upper stratosphere is replenished?

NEED EXTRA HELP?

If You Missed Question . . .	1	2	3	4	5	6	7	8	9	10	11	12	13	14	15	16
Review Section . . .	10.2	11.1	11.1	10.1	8.2	7.1	1.2	1.3	11.3	10.2	6.3	6.3	6.3	11.2	3.1	7.3

SHORT ANSWER

11. The rock is metamorphic because it is foliated with distorted banding.

12. This rock formed when high temperature and pressure combined to change the texture or chemical composition of the rock without melting it. The minerals in the rock changed into new minerals that are stable under the new temperature and pressure conditions.

13. If enough heat and pressure were continually applied to this rock, the rock would melt turning into magma. Once this magma cooled it would become igneous rock. It can also change into another metamorphic rock. The rock could be weathered and eroded into sediment that may become cemented into sedimentary rock.

14. Temperature is a measurement of the kinetic energy of a substance. Faster moving particles have more thermal energy, leading to a higher temperature. Heat is the transfer of energy from a substance at a higher temperature to a substance at lower temperature.

15. An ion is an atom with a negative or positive charge. An ion forms when an atom gains or loses one or more electrons.

16. Residual soil is located above its parent material. During a flood, the water rises over the soil, dissolving and transporting it away from its location, then later depositing it in a new location. The soil is now considered transported soil because it has been moved away from its parent material.

READING FOR COMPREHENSION

17. A

18. D

19. C

20. Sample answer: Ozone absorbs ultraviolet radiation from the Sun. Without the ozone layer, humans and other organisms would not be able to survive the increased amount of exposure of the Sun. Because findings show that the ozone layer is being replenished, this means that some of the damage caused by humans is being reversed.

CHAPTER 12 Meteorology

BIGIDEA **Weather patterns can be observed, analyzed, and predicted.**

ESSENTIAL QUESTIONS	NATIONAL STANDARDS	RESOURCES TO ASSESS MASTERY
SECTION 1 **The Causes of Weather** **1.** What is the difference between weather and climate? **2.** How do imbalances in the heating of Earth's surface create weather? **3.** How do air masses form? **4.** What are the five types of air masses? 2 sessions 1 block	UCP.1–3; A.1–2; D.1; F.4	**Entry-Level Assessment** Focus Transparency 32, p. 314 **Progress Monitoring** Caption Question, p. 315 Reading Check, pp. 314, 316 Section Review, p. 317
SECTION 2 **Weather Systems** **1.** What are the similarities and differences between the three major wind systems? **2.** What are the four types of fronts? **3.** How do high- and low-pressure systems differ? 1 session 0.5 block	UCP. 1–3; D.1	**Entry-Level Assessment** Focus Transparency 33, p. 318 **Progress Monitoring** Caption Question, p. 322 Reading Check, p. 320 Section Review, p. 323
SECTION 3 **Gathering Weather Data** **1.** Why is accurate weather data important? **2.** What are some of the instruments used to collect weather data from Earth's surface? **3.** What are the strengths and weaknesses of weather radar and weather satellites? 1 session 0.5 block	UCP. 1–3; B.6; D.1; E.1–2; F.6	**Entry-Level Assessment** Focus Transparency 34, p. 324 **Progress Monitoring** Caption Question, p. 327 Reading Check, pp. 325, 326 Section Review, p. 328
SECTION 4 **Weather Analysis and Prediction** **1.** What information is on a basic surface weather chart? **2.** How do digital and analog forecasting differ? **3.** What are problems associated with long-term forecasts? 3 sessions 1.5 blocks	UCP. 1–3; D.1; E.1–2; F.5–6	**Entry-Level Assessment** Focus Transparency 35, p. 329 **Progress Monitoring** Caption Question, pp. 329, 330 Reading Check, p. 331 Section Review, p. 332 **Summative Assessment** Chapter Assessment, p. 337 Exam*View®Assessment Suite* CD-ROM

Go online!
connectED.mcgraw-hill.com

LEVELED RESOURCES	LAB MATERIALS	ADDITIONAL RESOURCES
Science Notebook 12.1 OL **Chapter FAST FILE Resources:** MiniLab Worksheet, p. 28 OL Study Guide, p. 39 BL **Transparencies:** Section Focus Transparency 32 EL Teaching Transparency 31 OL EL	LaunchLAB p. 312 / **20 min** full tray of ice cubes, 3 pencils, 2 liquid-crystal temperature strips MiniLAB p. 315 / **15 min** flashlight, paper	**Technology:** eTeacherEdition eStudentEdition Classroom Presentation Toolkit Exam*View®Assessment Suite* CD-ROM LabManager™ CD-ROM What's EARTH SCIENCE Got To Do With It? DVD Weather Classroom Video DVD **Assessment:** Blueprints for Success Chapter Test (Scaffolded) Earth Science Standardized Test Practice **Lab Resources:** Exploring Environmental Problems Laboratory Manual
Science Notebook 12.2 OL **Chapter FAST FILE Resources:** Study Guide, p. 40 BL **Lab Resources:** Laboratory Manual, p. 89 OL **Transparencies:** Section Focus Transparency 33 EL Teaching Transparency 32 OL EL		
Science Notebook 12.3 OL **Chapter FAST FILE Resources:** Study Guide, p. 42 BL **Transparencies:** Section Focus Transparency 34 EL Teaching Transparency 33 OL EL		
Science Notebook 12.4 OL **Chapter FAST FILE Resources:** GeoLab Worksheet, p. 29 OL Study Guide, p. 43 BL **Lab Resources:** Laboratory Manual, p. 93 OL **Transparencies:** Section Focus Transparency 35 EL	GeoLAB p. 334 / **45 min** ruler, *Reference Handbook,* Weather Map Symbols	

LabManager™
Customize any lab with the LabManager™ CD-ROM.

BL Below Level OL On Level AL Advanced Learners EL English Learners COOP LEARN Cooperative Learning

CHAPTER 12

LaunchLAB

iLab Station | Rubric

How does a cold air mass form?

Safety Precautions Approve lab safety forms before work begins. Tell students to wipe up spills immediately to prevent accidents.

Teaching Strategies

- Have students use plastic or wooden pencils, not metal or mechanical types.
- Ice cubes should not touch either of the temperature strips.
- Ice cubes, by conduction, cooled the air immediately around them, causing the temperature of the air above and beneath the tray to drop. The temperature of the strip beneath the tray dropped more because the cooled air, being dense, settled to the surface of the table. Sinking cold air is key to the formation of cold air masses.

Procedure

1. Have students read and complete the lab safety form and follow the procedure below.
2. Place a **full tray of ice cubes** on a table. Place a **pencil** under each end of the tray to raise it off the table.
3. Slide a **liquid-crystal temperature strip** under the ice-cube tray.
4. Place **two pencils** across the top of the tray, and **another temperature strip** across them.
5. Record the temperature of each strip at 1-min intervals for about 5 min.
6. Make a graph of the temperature changes over time for each temperature strip.

CHAPTER 12
Meteorology

BIGIDEA Weather patterns can be observed, analyzed, and predicted.

SECTIONS

1 The Causes of Weather

2 Weather Systems

3 Gathering Weather Data

4 Weather Analysis and Prediction

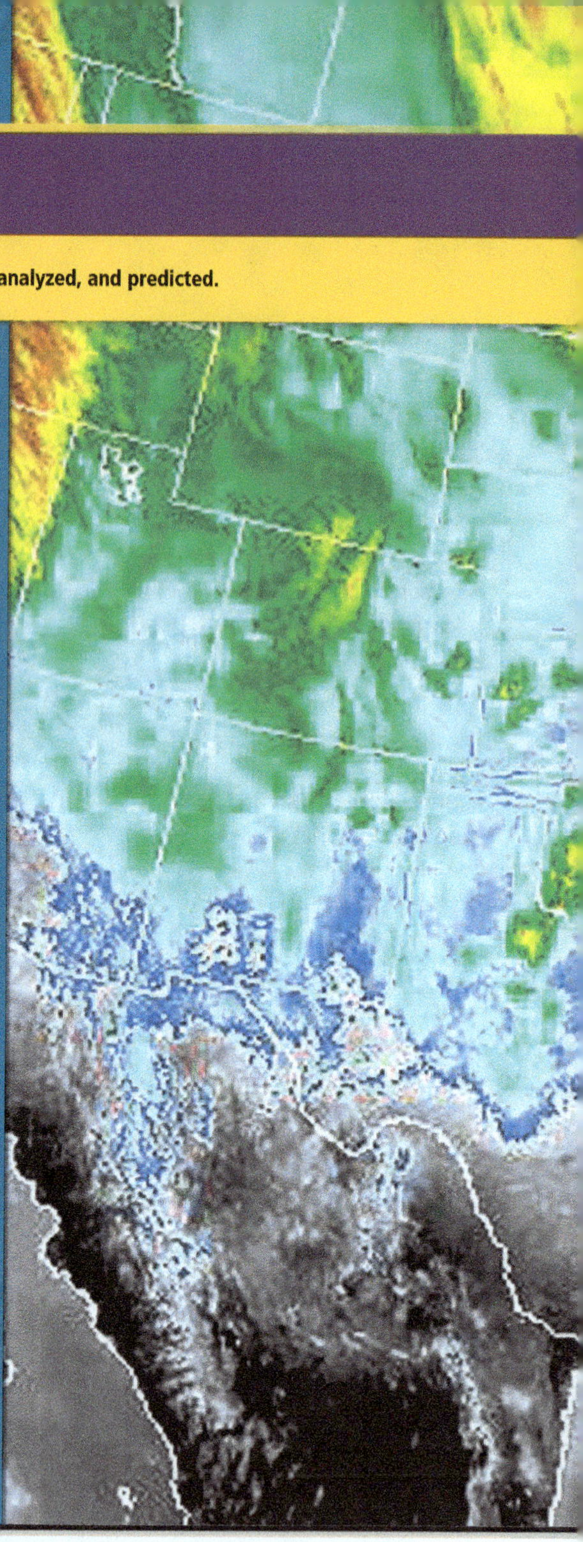

LaunchLAB

iLab Station

How does a cold air mass form?

An air mass is a large volume of air that has the characteristics of the area over which it formed. Model the formation of a cold air mass in this activity.

FOLDABLES® Study Organizer

Types of Fronts

Make a side-tab book using the labels shown. Use it to organize your notes on the four types of fronts.

Analysis

1. **Describe** what happened to the temperatures above and below the tray. Through conduction, the ice cubes cooled the air immediately around them as seen by lower temperatures on both thermometers. The temperature of the strip beneath the tray was lower because the dense, cooled air settled to the surface of the table.
2. **Explain** how this models a mass of cold air. The sinking cold air is analogous to the key component in the formation of cold air masses. Cold air settles to and hugs the ground due to its greater density.

Assessment

Performance Infer what happened to the temperature of the air 2 cm above and below the ice-cube tray. Students can repeat the experiment to determine if their inferences were supported.

LabManager™
Customize any lab with the LabManager™ CD-ROM.

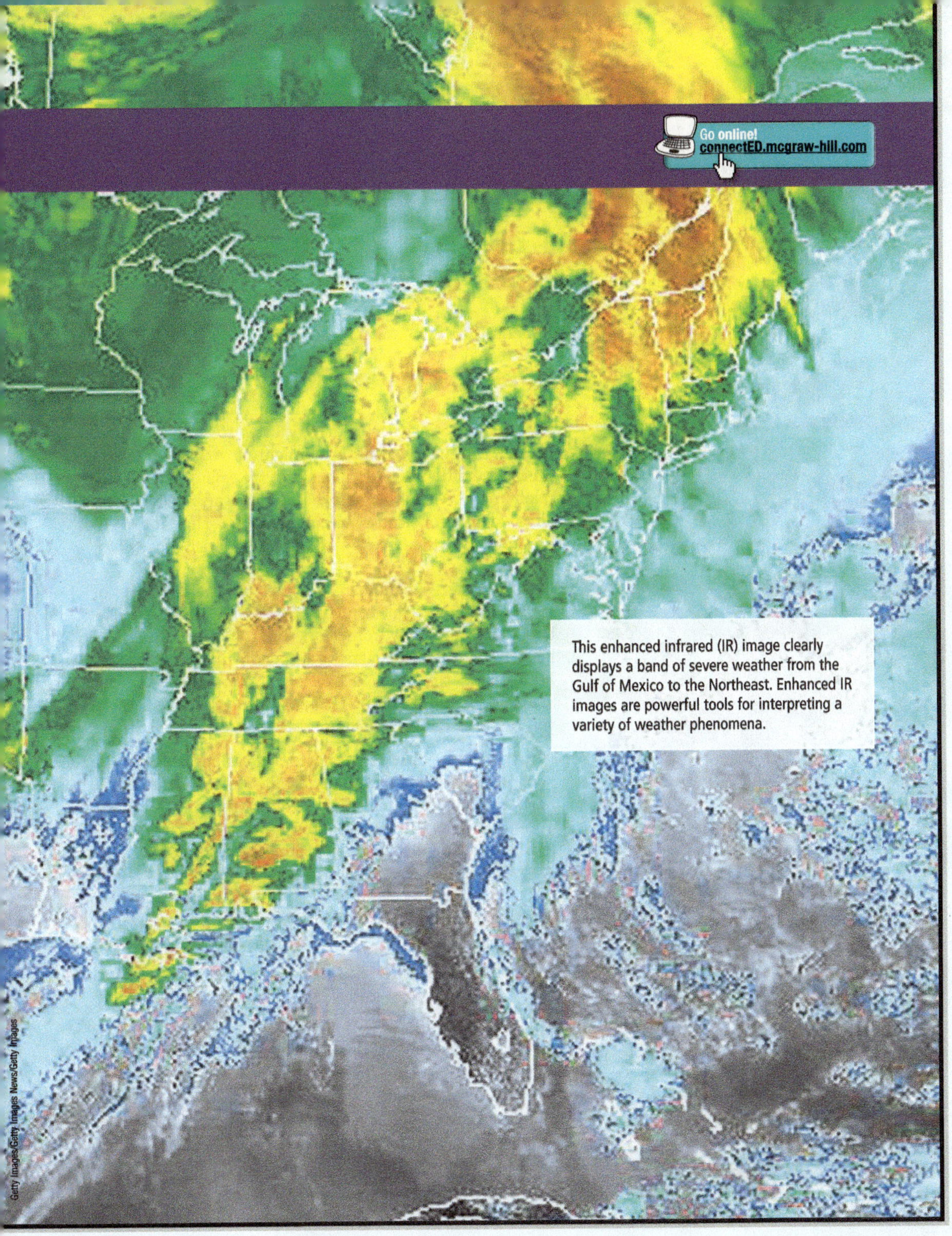

This enhanced infrared (IR) image clearly displays a band of severe weather from the Gulf of Mexico to the Northeast. Enhanced IR images are powerful tools for interpreting a variety of weather phenomena.

Introduce the BIGIDEA

Weather Factors How do meteorologists predict the weather? Note students' answers; they should center on making observations. Have the class monitor local daily weather forecasts on television, the newspaper, or the Internet for one week. Ask them also to track the weather in four other cities from across the country. Students should record high and low temperatures, and levels of precipitation for all cities. Ask students to compare the weather in the different cities. Ask if they notice any patterns in the weather across the locations. At the end of the week, have students present their findings to the class.

Teacher Content Support

Images Meteorologists use satellite and radar images to identify weather systems and their structures, strength, and movement. Images can be gathered in different wavelengths of radiation. The image shown here is an enhanced infrared image. Combined with other data sources and images, this type of information forms the basis of the weather analysis and forecasting process.

SECTION 1

1 Focus

Focus Transparency

Before presenting the lesson, project **Section Focus Transparency 32** and have students answer the accompanying questions. EL

MAINIDEA

Causes of Weather Ask students how spending a summer afternoon in the desert would be different than on a tropical island. Explain the different weather in these locations is related to how heat energy is absorbed by different surfaces.

2 Teach

Concept Development

Absorption Different surfaces absorb solar energy at different rates. Dark, moist soil absorbs solar energy efficiently, while ice and snow reflect a great deal of solar radiation. Have students discuss what different surfaces are found in their area and how the surfaces affect energy absorption.

Teacher Content Support

Motions Meteorological processes entail atmospheric motions, such as convection, on different time and spatial scales. A warm air current rising from the ground on a summer afternoon and the growth of a powerful thunderstorm are both the result of the rising and sinking of air. Atmospheric motions also are found in the formation of air masses and wind systems. These common processes demonstrate how seemingly unrelated events share the same underlying processes.

SECTION 1

The Causes of Weather

MAINIDEA Air masses have different temperatures and amounts of moisture because of the uneven heating of Earth's surface.

Essential Questions

- What is the difference between weather and climate?
- How do imbalances in the heating of Earth's surface create weather?
- How do air masses form?
- What are the five types of air masses?

Review Vocabulary

heat: transfer of thermal energy from a warmer material to a cooler material

New Vocabulary

weather
climate
air mass
source region

EARTH SCIENCE 4 YOU Have you ever walked barefoot on cool grass and then stepped onto hot pavement on a sunny summer day? Around the world, the Sun heats the different surfaces on Earth to different extents. This uneven heating causes weather.

What is meteorology?

What do you enjoy doing on a summer afternoon? Do you like to watch clouds move across the sky, listen to leaves rustling in a breeze, or feel the warmth of sunlight on your skin? Clouds, breezes, and the warmth of sunlight are examples of atmospheric phenomena. Meteorology is the study of the physics, chemistry, and dynamics of atmospheric phenomena. The root word of *meteorology* is the Greek word *meteoros,* which means *high in the air.*

Atmospheric phenomena are often classified as types of meteors. Cloud droplets and precipitation—rain, snow, sleet, and hail—are types of hydrometeors (hi droh MEE tee urz). Smoke, haze, dust, and other particles suspended in the atmosphere are lithometeors (lih thuh MEE tee urz). Examples of electrometeors are thunder and lightning—signs of atmospheric electricity that you can hear or see. Meteorologists study these various meteors.

Weather vs. climate Short-term variations in atmospheric phenomena that interact and affect the environment and life on Earth are called **weather.** These variations can take place over minutes, hours, days, weeks, months, or years. **Climate** is the long-term average of variations in weather for a particular area. Meteorologists use weather-data averages over 30 years to define an area's climate, such as that of the desert shown in **Figure 1.** Weather is what determines the clothes we wear outside on a given day, whereas climate determines what crops we plant in a given region.

☑ READING CHECK **Differentiate** between weather and climate.

Figure 1 A desert climate is dry with extreme variations in day and night temperatures. Only organisms adapted to these conditions, such as this ocotillo, can survive there.

Heating Earth's Surface

As you have learned, sunlight, which is a part of solar radiation, is always heating some portion of Earth's surface. Over the course of a year, the amount of thermal energy that Earth receives is the same as the amount that Earth radiates back to space. In meteorology, a crucial question is how solar radiation is distributed around Earth.

☑ READING CHECK Weather is the short-term variations in the atmospheric phenomena that interact and affect the environment and life on Earth. Climate is the long-term average variations in weather for a particular area.

EARTH SCIENCE JOURNAL

Causes of Weather Describe the types of air masses that affect your area; each should include at least two weather factors such as temperature and humidity. Infer where the air masses that affect you most often probably originated.

Rubric

Imbalanced heating Why are average January temperatures warmer in Miami, Florida, than in Detroit, Michigan? Part of the explanation is that Earth's axis of rotation is tilted relative to the plane of Earth's orbit. Therefore, the number of hours of daylight and amount of solar radiation is greater in Miami during January than in Detroit.

Another factor is that Earth is a sphere and different places on Earth are at different angles to the Sun, as shown in **Figure 2.** For most of the year, the amount of solar radiation that reaches a given area at the equator covers a larger area at latitudes nearer the poles. The greater the area covered, the smaller amount of heat per unit of area. Because Detroit is farther from the equator than Miami is, the same amount of solar radiation that heats Miami will heat Detroit less. Investigate this relationship in the MiniLab on this page.

Thermal energy redistribution Areas around Earth maintain about the same average temperatures over time due to the constant movement of air and water among Earth's surfaces, oceans, and atmosphere. The constant movement of air on Earth's surface, and currents within Earth's oceans, redistributes thermal energy around the world. Weather—from thunderstorms to large-scale weather systems—is part of the constant redistribution of Earth's thermal energy.

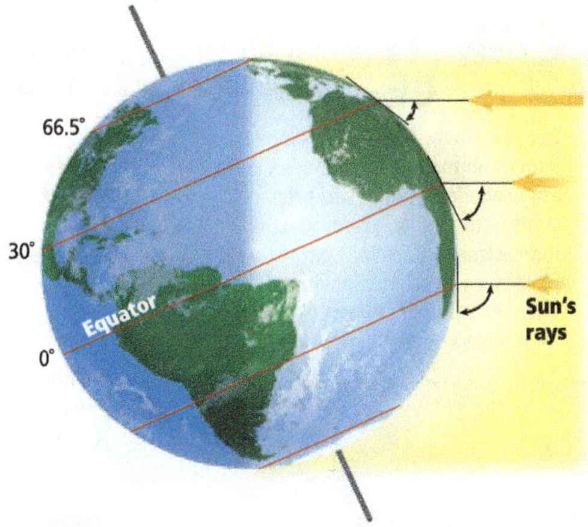

Figure 2 Solar radiation is unequal partly due to the changing angle of incidence of the sunlight. In this example it is perpendicular to Earth's surface south of the equator, 60° at the equator, 40° north of the equator.

Explain *why average temperatures decline from the equator to the poles.*

MiniLAB

iLab Station

Compare the Angles of Sunlight to Earth

What is the relationship between the angle of sunlight and amount of heating? The angle at which sunlight reaches Earth's surface varies with latitude. This results in uneven heating of Earth.

Procedure

1. Read and complete the lab safety form.
2. Turn on a **flashlight,** and hold it 20 cm above a piece of **paper.** Point the flashlight straight down.
3. Use a pencil to trace the outer edge of the light on the paper. This models the angle of sunlight to Earth at the equator.
4. Keep the flashlight the same distance above the paper, but rotate it about 30°.
5. Trace the outer edge of the light. This is similar to the angle of sunlight to Earth at latitudes nearer the poles.

Analysis

1. **Describe** how the outline of the light differed between Step 3 and Step 5. Explain why it differed.
2. **Compare** the amount of energy per unit of area received near the equator to the amount at latitudes nearer the poles.

MiniLAB

iLab Station Rubric

Purpose Students will demonstrate how changes in the angle of the Sun affect the amount of solar energy received in the tropics and the polar regions.

Process Skills model, recognize cause and effect, observe and infer

Safety Precaution Approve lab safety forms before work begins.

Teaching Strategy Students should work in groups of three: one to hold the flashlight, one to verify its angle and distance from the paper, and one to trace the outline of the light on the paper.

Expected Results As the angle of the flashlight changes, the light is spread over a wider area. Solar energy also is spread over a wider area as the angle of latitude increases.

Analysis

1. The outline of light was larger in Step 5 because the light covered a larger area.
2. The amount of energy received at any one place decreases when the energy covers a larger area due to the lower solar angle.

Assessment

Performance Challenge students to model the angle at which the Sun's rays strike the poles during winter. Students should increase the angle of the flashlight until the light is almost parallel to the paper.

LabManager™
Customize any lab with the LabManager™ CD-ROM.

ACROSS THE CURRICULUM

Biology and Geology Balances are common in nature. In biology, balances develop between predators and prey. In geology, the shifting of tectonic plates and the relieving of stress created by earthquakes are ways in which the forces acting on Earth's crust are balanced. This pattern can also be seen in the balance of heat throughout Earth's atmosphere and oceans.

Caption Question Fig. 2 As the angle at which the Sun's rays hits Earth's surface decreases, the same amount of heat is dispersed over a greater area.

Identify Misconceptions

Students often have the impression that the weather they are experiencing developed locally. Warm and cool spells in an area can be the result of air masses traveling into the region from far away.

Uncover the Misconception Why can the temperature change dramatically from one day to the next? The air mass that determined the weather might have moved from one day to the next, changing the weather pattern.

Demonstrate the Concept Remind students that anomalously warm days can occur during winter. The main cause of weather on any given day is the presence of an air mass that might have originated in the warm tropics or cold polar region.

Assess New Knowledge Ask students what type of air mass would be needed to change the current weather in their area. For example, if it is currently hot, an air mass of polar origin would be required to change the weather.

Reinforcement

Air Mass Weather What type of air mass brings cloudy, rainy weather to the Pacific Northwest?
a. cT **b.** mP **c.** cP **d.** mT
Answer: b BL EL

Model

Air Masses Have students describe what happens when they take a hot shower in a closed room. This represents a model of a maritime tropical (mT) air mass, where warm air combines with abundant moisture. The combination of warm temperatures and moisture produces air masses with high relative humidity. Brainstorm other scenarios that represent different types of air masses.

CAREERS IN EARTH SCIENCE

Meteorologist A meteorologist studies air masses and other atmospheric conditions to prepare short-term and long-term weather forecasts. An education that includes physics, Earth science, environmental science, and mathematics is useful for a meteorologist.

WebQuest

Air Masses

You have learned that air over a warm surface can be heated by conduction. This heated air rises because it is less dense than the surrounding air. On Earth, this process can take place over thousands of square kilometers for days or weeks. The result is the formation of an air mass. An **air mass** is a large volume of air that has the same characteristics, such as humidity and temperature, as its **source region**—the area over which the air mass forms. Most air masses form over tropical regions or polar regions.

Types of air masses The five types of air masses, listed in **Table 1**, influence weather in the United States. These air masses are all common in North America because there is a source region nearby.

Tropical air masses The origins of maritime tropical air are tropical bodies of water, listed in **Table 1**. In the summer, they bring hot, humid weather to the eastern two-thirds of North America. The southwestern United States and Mexico are a source region of continental tropical air, which is hot and dry, especially in summer.

Polar air masses Maritime polar air masses form over the cold waters of the North Atlantic and North Pacific. The one that forms over the North Pacific primarily affects the West Coast of the United States, occasionally bringing heavy rains in winter. Continental polar air masses form over the interior of Canada and Alaska. In winter, these air masses can carry frigid air southward. In the summer, however, cool, relatively dry, continental polar air masses bring relief from hot, humid weather.

☑ READING CHECK **Compare and contrast** tropical and polar air masses.

Explore **air masses with an interactive table.** Concepts In Motion

Table 1 Air Mass Characteristics

Air Mass Type	Weather Map Symbol	Source Region	Characteristics	
			Winter	Summer
Arctic	A	Siberia, Arctic Basin	bitter cold, dry	cold, dry
Continental polar	cP	interiors of Canada and Alaska	very cold, dry	cool, dry
Continental tropical	cT	southwest United States, Mexico	warm, dry	hot, dry
Maritime polar	mP	North Pacific Ocean	mild, humid	mild, humid
		North Atlantic Ocean	cold, humid	cool, humid
Maritime tropical	mT	Gulf of Mexico, Caribbean Sea, tropical and subtropical Atlantic Ocean and Pacific Ocean	warm, humid	hot, humid

☑ READING CHECK Tropical air masses form over warm land or water. Polar air masses form over cold land or water.

CAREERS IN EARTH SCIENCE

WebQuest

DIFFERENTIATED INSTRUCTION

English Learners Construct a dictionary of weather terms and words that are weather-related in languages other than English.

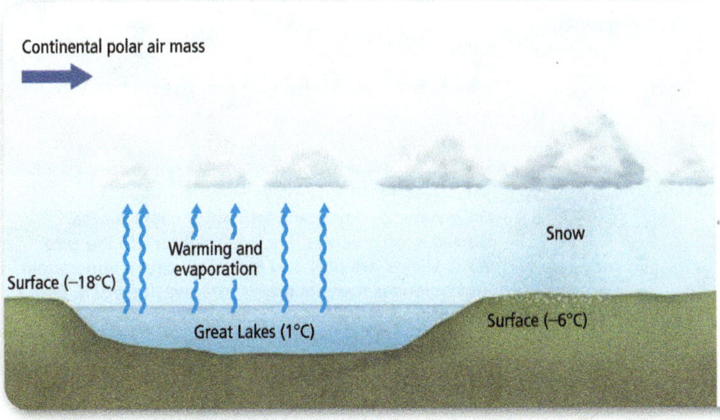

■ **Figure 3** As the cold, continental polar air moves over the warmer Great Lakes, the air gains thermal energy and moisture. This modified air cools as it is uplifted because of convection and topographic features, and produces lake-effect snows.

Arctic air masses Earth's ice- and snow-covered surfaces above 60° N latitude in Siberia and the Arctic Basin are the source regions of arctic air masses. During part of the winter, these areas receive almost no solar radiation but continue to radiate thermal energy. As a result, they become extremely cold and can bring the most frigid temperatures during winter.

Air mass modification Air masses do not stay in one place indefinitely. Eventually, they move, transferring thermal energy from one area to another. When an air mass travels over land or water that has characteristics different from those of its source region, the air mass can acquire some of the characteristics of that land or water. For example, when a polar air mass travels over a relatively warm body of water, as shown in **Figure 3,** the modified air mass might produce lake-effect snows. When an air mass undergoes modification, it exchanges thermal energy, moisture, or both with the surface over which it travels.

SECTION 1 REVIEW

Section Self-Check

Section Summary

- Meteorology is the study of atmospheric phenomena.
- Solar radiation is unequally distributed between Earth's equator and its poles.
- An air mass is a large body of air that takes on the moisture and temperature characteristics of the area over which it forms.
- Each type of air mass is classified by its source region.

Understand Main Ideas

1. MAINIDEA **Summarize** how an air mass forms.
2. **Explain** the process that prevents the poles from steadily cooling off and the tropics from heating up over time.
3. **Distinguish** between the causes of weather and climate.
4. **Differentiate** among the five types of air masses.

Think Critically

5. **Predict** which type of air mass you would expect to become modified more quickly: an arctic air mass moving over the Gulf of Mexico in winter or a maritime tropical air mass moving into the southeastern United States in summer.

WRITING IN ▶ Earth Science

6. Describe in a paragraph how a maritime polar air mass formed over the North Pacific is modified as it moves west over North America.

Collaborative Learning

Air Mass Modification Have students form four groups, and assign each group a different air mass to research from among the following: mT, cT, cP, and mP. Groups should describe how their air masses are modified as they move over areas with different characteristics. COOP LEARN

3 Assess

Check for Understanding

Discussion What makes one air mass different from another? temperature and humidity content Where do air masses form? over regions that have long-term stable conditions

Reteach

Make and Use Tables Have students make data tables showing the different types of air masses in the northern hemisphere, their origins, their characteristics, and the areas that they affect.

Assessment

Knowledge Have students match the main air mass types with places in the United States that are influenced by those air masses. Example: arctic—Alaska; cT—Arizona.

SECTION 1 REVIEW

1. Due to the processes of conduction and evaporation, an air mass takes on the characteristics, such as moisture and temperature, of the surface below.
2. Thermal energy is continually transferred from the equatorial regions to polar regions, which redistributes energy among Earth's surface, oceans, and atmosphere.
3. The causes of weather include transient features such as the movement of air masses, while the causes of climate are longer term and more stable, such as the amount of solar energy received.
4. See **Table 1.**
5. The arctic air mass; the waters of the Gulf of Mexico would be warm compared to the arctic air. The warm water would rapidly modify the cold air. The mT air would move over warm land, and would be modified less quickly.
6. The air mass would mix with dry continental air and gradually take on the characteristics of the continental air mass.

Rubric

SECTION 2

1 Focus

Focus Transparency

Before presenting the lesson, project **Section Focus Transparency 33** and have students answer the accompanying questions. EL

MAINIDEA

Weather Systems Have students plot daily high temperatures for the last month. Explain that the changes they see are largely the result of the passage of different air masses.

2 Teach

Tie to Previous Knowledge

Convection The three convection cells in each hemisphere are large versions of the convection cells that form cumulus clouds. These large cells have both horizontal and vertical movements of wind.

Concept Development

Meteorological Equator

Students might think of the equator only as a place halfway between the poles. The intertropical convergence zone serves as a "meteorological equator" or boundary zone where the two hemispheric wind systems meet. Students will learn that this large mass of rising air contributes to cloudy conditions and thunderstorms, and it can contain tropical disturbances, which can grow into tropical cyclones.

SECTION 2

Weather Systems

MAINIDEA Weather results when air masses with different pressures and temperatures move, change, and collide.

Essential Questions

- What are the similarities and differences between the three major wind systems?
- What are the four types of fronts?
- How do high- and low-pressure systems differ?

Review Vocabulary

convection: the transfer of thermal energy by the flow of a heated substance

New Vocabulary

Coriolis effect
polar easterlies
prevailing westerlies
trade winds
jet stream
front

EARTH SCIENCE 4 YOU On a summer day, you might enjoy cool breezes. However, on a winter day, you might avoid the cold wind. Winds are part of a global air circulation system that balances thermal energy around the world.

Global Wind Systems

If Earth did not rotate on its axis, two large air convection currents would cover Earth, as shown in **Figure 4.** The colder and more dense air at the poles would sink to the surface and flow toward the tropics. There, the cold air would force warm, equatorial air to rise. This air would cool as it gained altitude and flowed back toward the poles. However, Earth rotates from west to east, which prevents this situation.

The directions of Earth's winds are influenced by Earth's rotation. This **Coriolis effect** results in fluids and objects moving in an apparent curved path rather than a straight line. Thus, as illustrated in **Figure 5,** moving air curves to the right in the northern hemisphere and curves to the left in the southern hemisphere. Together, the Coriolis effect and the heat imbalance on Earth create distinct global wind systems. They transport colder air to warmer areas near the equator and warmer air to colder areas near the poles. Global wind systems help to equalize the thermal energy on Earth.

There are three basic zones, or wind systems, at Earth's surface in each hemisphere. They are polar easterlies, prevailing westerlies, and trade winds.

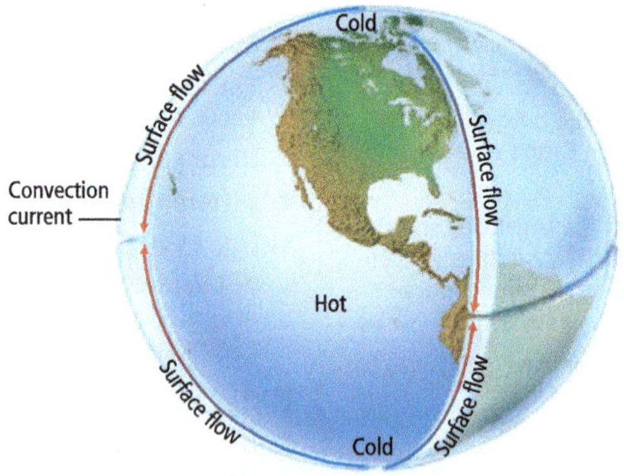

Figure 4 If Earth did not rotate, two large convection currents would form as denser polar air moved toward the equator. These currents would warm and rise as they approached the equator, and cool as they moved toward each pole.

DIFFERENTIATED INSTRUCTION

Struggling Learners On a warm day, have students place pans of water, soil, rock, and/or ice in a sunny window sill and a piece of dark construction paper next to a pan. After the Sun has been shining on the pans and paper for about 30 minutes, have students touch each pan and paper to observe which is warmer. The experiment models different types of air masses forming over different surfaces. Students can present reports on the types of air masses that they modeled.

VISUALIZING the Coriolis Effect

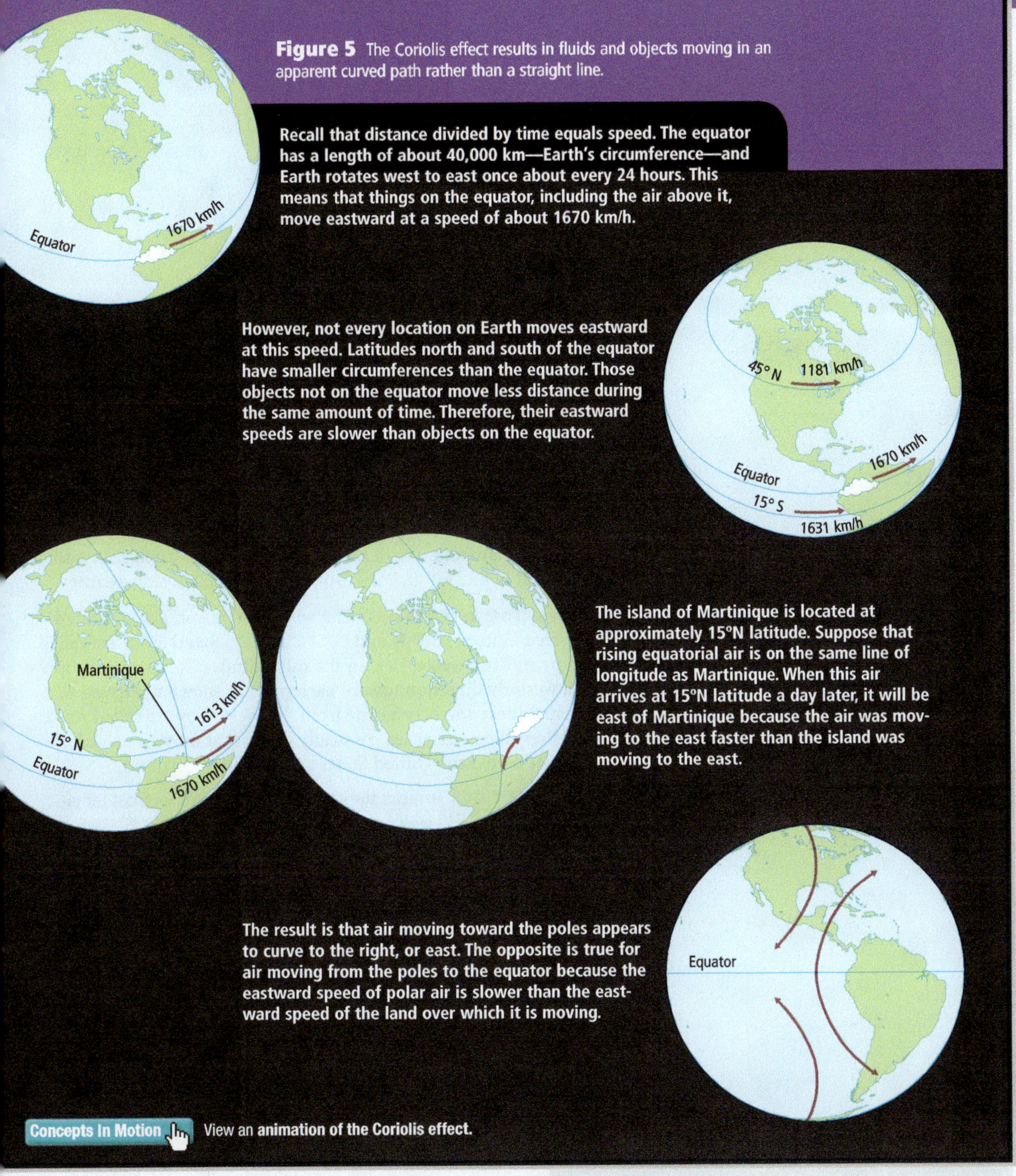

Figure 5 The Coriolis effect results in fluids and objects moving in an apparent curved path rather than a straight line.

Concepts In Motion View an animation of the Coriolis effect.

ACROSS THE CURRICULUM

History The tropical easterlies are also known as the trade winds. Sailing ships used the constant winds in this belt to travel to the New World. Because these ships frequently brought vast quantities of goods for trade, the winds became known as the trade winds. Many sailors learned that these same trade winds also traveled through the Atlantic and Pacific hurricane belts.

VISUALIZING

Purpose

Students will understand the concept of the Coriolis effect through the application of a specific example.

Teacher Content Support

Coriolis Effect All masses moving freely over Earth's surface are affected by the Coriolis effect. The effect is not equal for all regions. The Coriolis effect is more pronounced at higher latitudes. For this reason, the rate of curvature an object has is greater near the poles than at the equator. The effect is also greater the faster the mass is moving. A missile fired from Cape Canaveral will exhibit more curvature than a hurricane moving through the Atlantic Ocean.

The consequences of the Coriolis effect must be dealt with when firing missiles or cannon shells and tracking weather systems such as hurricanes. With regard to hurricanes and the Coriolis effect, when the storms are moving slowly westward near the equator, the effect is slight, hence the hurricane track often is straight, or due west. The storm over time may curve to the right, or clockwise, carrying it into higher latitudes with a more pronounced Coriolis effect. Couple this with the effects of global wind belts, subtropical jet streams, variations in water temperature, high altitude wind shear, and the effects of mid-latitude cyclones on the hurricane and it is no wonder that predicting a hurricane's path is more art than science.

Reinforcement

Fronts and Air Masses Review the way in which air masses are replaced as cold and warm fronts move in. For instance, when a warm front advances, it does so by pushing the cool air it encounters out of its way and replacing the cool air with warmer air. Clouds form because the warm air runs over the cool air.

☑ READING CHECK from the southwest to the northeast

Teacher Content Support

Monsoons Monsoons are seasonal winds that occur in tropical areas because of temperature and pressure differences between land and oceans. During the summer, when the air above the heated land is warm and less dense than the air over the cooler oceans, winds blow from the ocean toward the land. When these winds are strong, they transport moisture inland and can trigger torrential rains and floods. During the winter, when the air over the land is cooler and drier than the air over the oceans, the winds reverse. Winter is the dry season for these tropical areas. Much of southeast Asia experiences monsoons, which are critical to the region's food supply and economy. Every few weeks of delay in the onset of the summer monsoon can spell disaster to the food supply. Rice is the only variety of grain that can tolerate the extreme rainfall of a monsoon, thus the weather determined the grain of choice for these people. During the winter monsoon, dry season crops such as wheat are grown. The U.S. southwest also experiences monsoon. During the summer months, the Gulf of California supplies moisture for a rainy season in Arizona and parts of New Mexico, Colorado, and Utah.

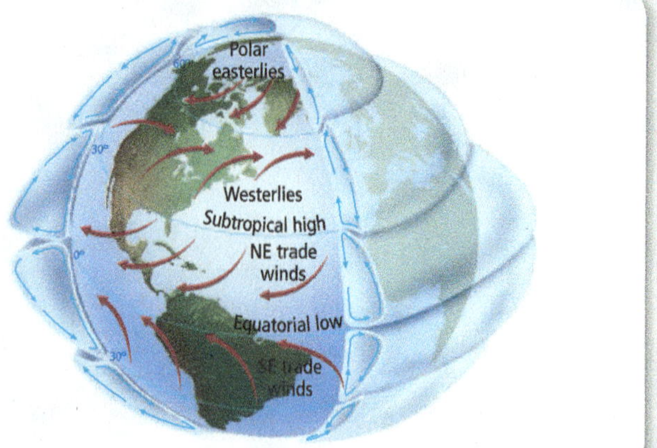

Figure 6 The directions of Earth's wind systems, such as the polar easterlies and the trade winds, vary with the latitudes in which they occur. Note that a wind is named for the direction from which it blows. A north wind blows from the north.

Polar easterlies The wind zones between 60° N latitude and the North Pole, and 60° S latitude and the South Pole are called the **polar easterlies,** as shown in **Figure 6.** Polar easterlies begin as dense polar air that sinks. As Earth spins, this cold, descending air is deflected in an easterly direction away from each pole. The polar easterlies are typically cold winds. Unlike the prevailing westerlies, these polar easterlies are often weak and sporadic.

Between polar easterlies and prevailing westerlies is an area called a polar front. Earth has two polar fronts located near latitudes 60° N and 60° S. Polar fronts are areas of stormy weather.

Vocabulary
Science usage v. Common usage
Circulation
Science usage: movement in a circle or circuit

Common usage: condition of being passed about and widely known; distribution

Prevailing westerlies The wind systems on Earth located between latitudes 30° N and 60° N, and 30° S and 60° S are called the **prevailing westerlies.** In the midlatitudes, surface winds move in a westerly direction toward each pole, as shown in **Figure 6.** Because these winds originate from the West, they are called westerlies. Prevailing westerlies are steady winds that move much of the weather across the United States and Canada.

☑ READING CHECK **Predict** the direction of movement for most tornadoes in the United States.

Trade winds Between latitudes 30° N and 30° S are two circulation belts of wind known as the **trade winds,** which are shown in **Figure 6.** Air in these regions sinks and moves toward the equator in an easterly direction. When the air reaches the equator, it warms, rises and moves back toward latitudes 30° N and 30° S, where it sinks and the process repeats.

Horse latitudes Near latitudes 30° N and 30° S, the sinking air associated with the trade winds creates an area of high pressure. This results in a belt of weak surface winds called the horse latitudes. Earth's major deserts, such as the Sahara, are under these high-pressure areas.

EARTH SCIENCE JOURNAL

Monsoons Many countries in tropical regions have agricultural economies that depend on monsoon rainfall. Recent discoveries have led to more accurate predictions of the arrival and strength of monsoons. Have students research how variations in the strength of monsoons in any given year have affected the economies of tropical nations. Ask students to research modern methods of predicting the strength of monsoons. Students should explain how these predictions can benefit those economies.

Rubric

Intertropical convergence zone Near the equator, trade winds from the North and the South meet and join, as shown in **Figure 6.** The air is forced upward, which creates an area of low pressure. This process, called convergence, can occur on a small or large scale. Near the equator, it occurs over a large area called the intertropical convergence zone (ITCZ). The ITCZ drifts south and north of the equator as seasons change. In general, it follows the positions of the Sun in relation to the equator. In March and September it is directly over the equator. Because the ITCZ is a region of rising air, it has bands of cloudiness and thunderstorms, which deliver moisture to many of the world's tropical rain forests.

VOCABULARY
ACADEMIC VOCABULARY
Generate (JE nuh rayt)
to bring into existence
Wind is generated as air moves from an area of high pressure to an area of low pressure.

Jet Streams

Atmospheric conditions and events that occur at the boundaries between wind zones strongly influence Earth's weather. On either side of these boundaries, both surface air and upper-level air differ greatly in temperature and pressure. Recall that warmer air has higher pressure than cooler air, and that the difference in air pressure causes wind. Wind is the movement of air from an area of high pressure to an area of low pressure.

A large temperature gradient in upper-level air combined with the Coriolis effect results in strong westerly winds called jet streams. A **jet stream,** shown in **Figure 7,** is a narrow band of fast wind. Its speed varies with the temperature differences between the air masses at the wind zone boundaries. A jet stream can have a speed up to 400 km/h at altitudes of 10.7 km to 12.2 km.

The position of a jet stream varies with the season. It generally is located in the region of strongest temperature differences on a line from the equator to a pole. The jet stream can move almost due south or north, instead of following its normal westerly direction. It can also split into branches and re-form later. Whatever form or position it takes, the jet stream represents the strongest core of winds.

Types of jet streams The major jet streams, called the polar jet streams, separate the polar easterlies from the prevailing westerlies in the northern and southern hemispheres. The polar jet streams occur at about latitudes 40° N to 60° N and 40° S to 60° S, and move west to east. The minor jet streams are the subtropical jet streams. They occur where the trade winds meet the prevailing westerlies, at about latitudes 20° N to 30° N and 20° S to 30° S.

Jet streams and weather systems Storms form along jet streams and generate large-scale weather systems. These systems transport cold surface air toward the tropics and warm surface air toward the poles. Weather systems generally follow the path of jet streams. Jet streams also affect the intensity of weather systems by moving air of different temperatures from one region of Earth to another.

Figure 7 Weather in the middle latitudes is strongly influenced by fast-moving, high-altitude jet streams.

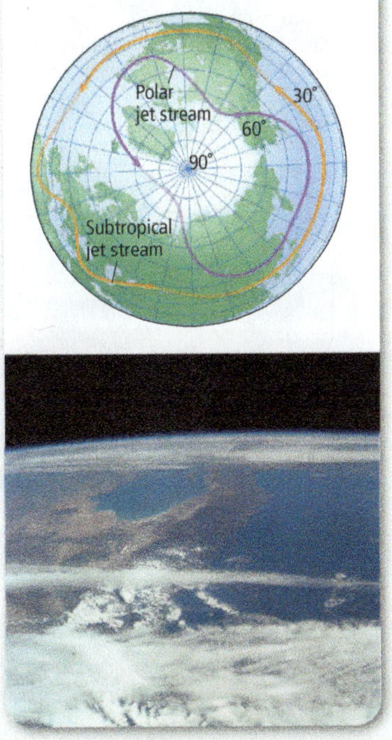

Apply Earth Science

Jet Streams The strength and position of jet streams have significant effects on the aviation industry. Because jet streams often attain speeds in excess of 320 km/h, they can provide either favorable tailwinds or extremely strong headwinds. Tailwinds result in the savings of hundreds of thousands of dollars in fuel costs. Conversely, headwinds increase fuel costs. Have students make a list of airline routes across the country and identify how travel along these routes might be impacted by the presence of a strong jet stream. **AL**

Teacher Content Support

Steering Currents Students should not lose sight of the fact that traveling weather systems and the jet streams associated with them are interrelated. The motion and intensity of a weather system are directly connected to changes that are taking place in the jet stream itself. In other words, unlike a real railroad track, this atmospheric "track" can speed up or slow down, change positions, and sometimes disappear altogether.

Project

Accuracy Have students research the number of surface observation sites and upper-air observation sites in the United States. surface observation sites: roughly 1000; upper-air observation sites: roughly 100 Lead students to understand that without accurate and thorough measurements of upper-atmospheric conditions, weather forecasts are likely to contain inaccuracies.

DIFFERENTIATED INSTRUCTION

Visually Impaired To help students understand the concepts of fronts and air masses, place a bowl of ice on a table. Place a fan near the bowl so that the fan blows cold air from the ice on the students. Explain that the rush of cold air is similar to how a cold front moves through an area, and that the cool air that remains behind is the equivalent of a cool air mass that moves in behind a front.

Collaborative Learning

Fronts Have students form four teams. Assign each team a different kind of front to research including the kinds of hazardous weather that the front might be expected to bring to an area and the times of year the front is strongest and most frequent. Teams should then combine their results to produce a chart detailing the relative frequency of the different kinds of fronts in your area and the kinds of hazardous weather associated with each front. OL COOP LEARN

Activity

Low Pressure Have students research several different types of low-pressure systems that occur around the world, including blizzards, hurricanes, tornadoes, waterspouts, and dust devils. Students should provide a brief description of each system, listing its average size, intensity, and potential for damage or loss of life. Students should then identify which of these cyclonic storms affect their area. OL

Caption Question **Fig. 8** warm front

Figure 8 The type of front formed depends on the types of air masses that collide.
Identify *the front associated with high cirrus clouds.*

View an **animation of fronts.**
Concepts In Motion

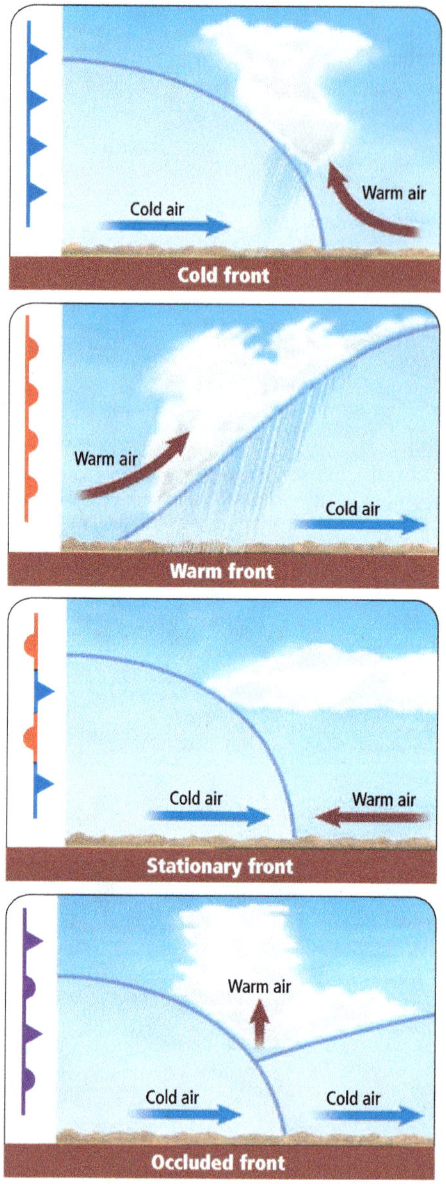

Fronts

Air masses with different characteristics can collide and result in dramatic weather changes. A collision of two air masses forms a **front**—a narrow region between two air masses of different densities. Recall that the density of an air mass results from its temperature, pressure, and humidity. Fronts can form across thousands of kilometers of Earth's surface.

Cold front When cold, dense air displaces warm air, it forces the warm air, which is less dense, up along a steep slope, as shown in **Figure 8.** This type of collision is called a cold front. As the warm air rises, it cools and water vapor condenses. Intense precipitation and sometimes thunderstorms are common with cold fronts. A blue line with evenly spaced blue triangles represents a cold front on a weather map. The triangles point in the direction of the front's movement.

Warm front Advancing warm air displaces cold air along a warm front. A warm front develops a gradual boundary slope, as illustrated in **Figure 8.** A warm front can cause widespread light precipitation. On a weather map, a red line with evenly spaced, red semicircles pointing in the direction of the front's movement indicates a warm front.

Stationary front When two air masses meet but neither advances, the boundary between them stalls. This front—a stationary front, as shown in **Figure 8**—frequently occurs between two modified air masses that have small temperature and pressure gradients between them. The air masses can continue moving parallel to the front. Stationary fronts sometimes have light winds and precipitation. A line of evenly spaced, alternating cold- and warm-front symbols pointing in opposite directions represents a stationary front on a weather map.

Occluded front Sometimes, a cold air mass moves so rapidly that it overtakes a warm front and forces the warm air upward, as shown in **Figure 8.** As the warm air is lifted, the advancing cold air mass collides with another cold air mass that was in front of the warm air. This is called an occluded front. Strong winds and heavy precipitation are common along an occluded front. An occluded front is shown on a weather map as a line of evenly spaced, alternating purple triangles and semicircles pointing in the direction of the occluded front's movement.

SECTION 2 REVIEW

1. Answers should summarize and explain cold, warm, stationary, and occluded fronts.
2. Trade winds occur in the tropics and move to the west. Prevailing westerlies occur in the mid-latitudes and move to the east. Polar easterlies occur at latitudes near the poles and move east to west.
3. Because Earth is a sphere, latitudes have different circumferences, and therefore, different eastward speeds. The greatest eastward speed is at the equator. Air rising at the equator and moving toward the poles appears to curve east because it is moving faster than the part of Earth over which it is moving. The smallest circumferences are at the poles. Air moving from the poles toward the equator appears to curve west because it is moving slower than the part of Earth over which it is moving.
4. The ITCZ is near the equator and is characterized by rising air over a large area, which creates clouds and abundant precipitation for many of the world's tropical forests.

Pressure Systems

You have learned previously that at Earth's surface, sinking air is associated with high pressure and rising air is associated with low pressure. Air always flows from an area of high pressure to an area of low pressure. Sinking or rising air, combined with the Coriolis effect, results in the formation of rotating high- and low-pressure systems in the atmosphere. Air in these systems moves in a circular motion around either a high- or low-pressure center.

Low-pressure systems In surface low-pressure systems, air rises. When air from outside the system replaces the rising air, this air spirals inward toward the center and then upward. Air in a low-pressure system in the northern hemisphere moves in a counterclockwise direction, as shown in **Figure 9.** The opposite occurs in the southern hemisphere. Air in a low-pressure system in the southern hemisphere moves in a clockwise direction. As air rises, it cools and often condenses into clouds and precipitation. Therefore, a low-pressure system, whether in the northern or southern hemisphere, is often associated with cloudy weather and precipitation.

High-pressure systems In a surface high-pressure system, sinking air moves away from the system's center when it reaches Earth's surface. The Coriolis effect causes the sinking air to move to the right, making the air circulate in a clockwise direction in the northern hemisphere and in a counter clockwise direction in the southern hemisphere. High-pressure systems are usually associated with fair weather. They dominate most of Earth's subtropical oceans and provide generally pleasant weather.

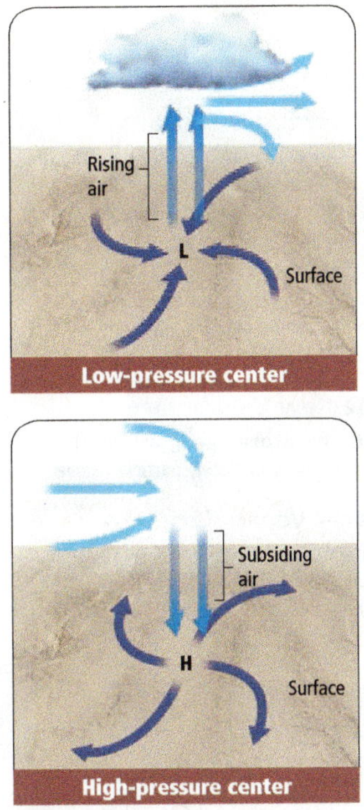

■ **Figure 9** In the northern hemisphere, winds move counterclockwise around a low-pressure center, and clockwise around a high-pressure center.

SECTION 2 REVIEW

Section Self-Check

Section Summary

- The three major wind systems are the polar easterlies, the prevailing westerlies, and the trade winds.
- Fast-moving, high-altitude jet streams greatly influence weather in the middle latitudes.
- The four types of fronts are cold fronts, warm fronts, occluded fronts, and stationary fronts.
- Air moves in a generally circular motion around either a high- or low-pressure center.

Understand Main Ideas

1. MAINIDEA **Summarize** information about the four types of fronts. Explain how they form and lead to changes in weather.
2. **Distinguish** among the three main wind systems.
3. **Describe** the Coriolis effect.
4. **Explain** why most tropical rain forests are located near the equator.
5. **Describe** how a jet stream affects the movement of air masses.
6. **Compare and contrast** high-pressure and low-pressure systems.

Think Critically

7. **Analyze** why most of the world's deserts are located between latitudes 10°N to 30°N and 10°S to 30°S.

WRITING IN ▶ Earth Science

8. Write a summary about how the major wind systems form.

3 Assess

Check for Understanding

Discussion Tell students that clashes between different air masses cause tornadoes. Ask students to use this information to explain why tornadoes are more common in the U.S. Central Plains than in Puerto Rico. Puerto Rico is located in a tropical ocean dominated by only one type of air mass, mT. The Central Plains frequently experience different types of air masses, and the clashes between them spawn tornadoes.

Reteach

Make and Use Tables Have students make data tables listing the various types of fronts, and the types of air masses and weather that are associated with each.

Assessment

Skill Have students diagram why precipitation occurs along different types of fronts. Diagrams should show air rising over frontal boundaries, air cooling as it rises, air saturating, water vapor condensing into clouds, and precipitation.

5. Surface weather systems that move air masses travel along the jet stream, and the intensity of these systems depends on the strength of the jet stream.
6. A low-pressure system consists of air moving inward and upward in a counterclockwise direction in the northern hemisphere. It is characterized by rising air, cloudy weather, and precipitation. A high-pressure system consists of air moving downward and outward, a clockwise direction, in the northern hemisphere. It is characterized by sinking air and fair weather.
7. Air sinks, dries, and warms by compression in these areas, creating hot and dry conditions over large portions of landmasses.
8. Answers should include information about the Coriolis effect, polar easterlies, prevailing westerlies, and tradewinds.

Rubric

1 Focus

Focus Transparency

Before presenting the lesson, project **Section Focus Transparency 34** and have students answer the accompanying questions. **EL**

MAINIDEA

Gather Weather Data Ask students to predict the outcome of an upcoming sporting event. Then ask them if their prediction would be the same if all the star players couldn't play because of illness. Use this example to demonstrate that all predictions, including weather, are based on having accurate data.

2 Teach

Teacher Content Support

Rawinsonde Observation The ground-based instrument that tracks a radiosonde also computes the radiosonde's range and angle constantly. This process, known as rawinsonde observation, can be used to determine wind speed and direction at different atmospheric levels because the position of the radiosonde relative to the ground station is known millions of times a second. This allows for the development of a triangle, with the three vertices representing the ground station, the location of the radiosonde at the previous minute, and the location of the radiosonde at the current microsecond. The use of trigonometry allows the computation of the direction of movement and the distance instantaneously.

Gathering Weather Data

MAINIDEA Accurate measurements of atmospheric properties are a critical part of weather analysis and prediction.

Essential Questions

- Why is accurate weather data important?
- What are some of the instruments used to collect weather data from Earth's surface?
- What are the strengths and weaknesses of weather radar and weather satellites?

Review Vocabulary

temperature: the measurement of how rapidly or slowly particles move

New Vocabulary

thermometer
barometer
anemometer
hygrometer
radiosonde
Doppler effect

EARTH SCIENCE 4 YOU **Before a doctor can make a diagnosis, he or she must accurately assess the patient's state of health. This usually includes measuring body temperature and blood pressure. Similarly, in order to forecast the weather, meteorologists must have accurate measurements of the atmosphere.**

Data from Earth's Surface

Meteorologists measure atmospheric conditions, such as temperature, air pressure, wind speed, and relative humidity. The quality of the data is critical for complete weather analysis and precise predictions. Two important factors in weather forecasting are the accuracy of the data and the amount of available data.

Temperature and air pressure A **thermometer,** shown in **Figure 10,** measures temperature using either the Fahrenheit or Celsius scale. Thermometers in most homes are liquid-in-glass or bimetallic-strip thermometers. Liquid-in-glass thermometers usually contain a column of alcohol sealed in a glass tube. The liquid expands when heated, causing the column to rise, and contracts when it cools, causing the column to fall. A bimetallic-strip thermometer has a dial with a pointer. It contains a strip of metal made from two different metals that expand at different rates when heated. The strip is long and coiled into a spiral, making it more sensitive to temperature changes.

A **barometer** measures air pressure. Some barometers have a column of mercury in a glass tube. One end of the tube is submerged in an open container of mercury. Changes in air pressure change the height of the column. Another type of barometer is an aneroid barometer, shown in **Figure 10.** It has a sealed, metal chamber with flexible sides. Most of the air is removed, so the chamber contracts or expands with changes in air pressure. A system of levers connects the chamber to a pointer on a dial.

Liquid-in-glass thermometer

Bimetallic-strip thermometer

Aneroid barometer

Figure 10 Thermometers and barometers are common weather instruments.

IN THE FIELD

Henry Diaz Born in Cuba, where hurricanes are a frequent danger, Henry Diaz became interested in the study of weather phenomena. When his family moved to the United States for political reasons, Diaz went to the University of Colorado to study geography and climatology. He later worked for the National Oceanic and Atmospheric Administration, becoming an expert in the study of El Niño, the warming system that originates in the South Pacific Ocean and affects weather systems around the world. Diaz and a team of other scientists are working to help people prepare for the long-term effects of climate-changing conditions such as El Niño.

Wind speed and relative humidity

An **anemometer** (a nuh MAH muh tur), shown in **Figure 11,** measures wind speed. The simplest type of anemometer has three or four cupped arms, positioned at equal angles from each other, that rotate as the wind blows. The wind's speed can be calculated using the number of revolutions of the cups over a given time. Some anemometers also have a wind vane that shows the direction of the wind.

A **hygrometer** (hi GRAH muh tur), such as the one in **Figure 11,** measures humidity. This type of hygrometer has wet-bulb and dry-bulb thermometers and requires a conversion table to determine relative humidity. When water evaporates from the wet bulb, the bulb cools. The temperatures of the two thermometers are read at the same time, and the difference between them is calculated. The relative humidity table lists the specific relative humidity for the difference between the thermometers.

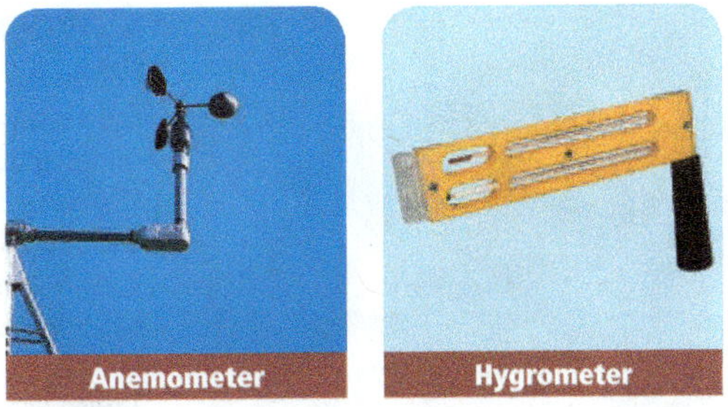

Figure 11 Anemometers are used to measure wind speed based on the rotation of the cups as the wind blows. Hygrometers measure humidity using techniques such as finding the temperature difference between the wet bulb and the dry bulb.

☑ READING CHECK **Analyze** the relationship between the amount of moisture in air and the temperature of the wet bulb in a hygrometer.

Automated Surface Observing System

Meteorologists need a true "snapshot" of the atmosphere at one particular moment to develop an accurate forecast. To obtain this, meteorologists analyze and interpret data gathered at the same time from weather instruments at many different locations. Coordinating the collection of this data was a complicated process until late in the twentieth century. With the development of reliable automated sensors and computer technology, instantaneously collecting and broadcasting accurate weather-related data became possible.

In the United States, the National Weather Service (NWS), the Federal Aviation Administration, and the Department of Defense jointly established a surface-weather observation network known as the Automated Surface Observing System (ASOS). It gathers data in a consistent manner, 24 hours a day, every day. It began operating in the 1990s and more than doubled the number of full-time observation sites, such as the one shown in **Figure 12.** ASOS provides essential weather data for aviation, weather forecasting, and weather-related research.

Figure 12 This weather station in the United Kingdom consists of several instruments that measure atmospheric conditions.

Teacher Content Support

Relative Humidity One of the most difficult weather variables to accurately measure is relative humidity. This is because the best way to measure relative humidity directly is dependent on the absorptive properties of some material, often hair. Unfortunately, the absorptive properties of most materials change with age, atmospheric conditions, and other factors. For this reason, relative humidity is not usually measured directly. Rather, it is calculated using properties that can be measured accurately, such as air temperature, wet-bulb temperature, and dew-point temperature.

Tie to Previous Knowledge

Hair and Humidity Hair that turns frizzy in moist weather is a direct demonstration of the absorption properties of hair. Hair reacts to changes in relative humidity, which is the basic principle behind the hair-type hygrometer.

Discussion

Weather Prediction Have students infer why meteorologists must first know the current state of weather variables not only at Earth's surface, but also throughout the atmosphere before changes in any weather variables can be predicted. Only then can meteorologists predict how air will move and what changes are likely to occur as a result of those movements.

☑ READING CHECK Evaporation requires an input of heat; therefore, the drier the air, the greater the rate of evaporation and the greater the temperature drop in the wet bulb.

DIFFERENTIATED INSTRUCTION

Visually Impaired Allow students with visual impairments to carefully handle different types of weather instruments such as thermometers and barometers. Take care that mercury thermometers and barometers do not break. BL EL

EARTH SCIENCE JOURNAL

Gather Weather Data Have students research and document how methods of observing the weather have changed over the last 100 years, including those made possible by new technologies. Ask them to explain how the changes that have taken place have improved the quality and quantity of available weather data.

Rubric

Project

Satellites Have students research the history of weather satellites and the different types used today. Students can present their research in posters that include the different types of satellites and the different types of imagery used by the satellites. OL

Activity

Weather Instruments There are many other instruments used to record weather data than those discussed here. Have students research other instruments, then give a short presentation to the class. Make sure they include what each instrument is used for, advantages and disadvantages of each instrument, and how the instrument works. Examples of possible instruments are Profilers, Lidars, RASS, and GPS systems.

READING CHECK to gather data about upper-atmospheric conditions

Figure 13 Radiosondes gather upper-level weather data such as air temperature, pressure, and humidity.

Data from the Upper Atmosphere

While surface-weather data are important, the weather is largely the result of changes that take place high in the troposphere. To make accurate forecasts, meteorologists must gather data at high altitudes, up to 30,000 m. This task is more difficult than gathering surface data, and it requires sophisticated technology.

An instrument used for gathering upper-atmospheric data is a **radiosonde** (RAY dee oh sahnd), shown in **Figure 13.** It consists of a package of sensors and a battery-powered radio transmitter. These are suspended from a balloon that is about 2 m in diameter and filled with helium or hydrogen. A radiosonde's sensors measure the air's temperature, pressure, and humidity. Radio signals constantly transmit these data to a ground station that tracks the radiosonde's movement. If a radiosonde also measures wind direction and speed, it is called a rawinsonde (RAY wuhn sahnd), **ra**dar + **win**d + radio**sonde**.

Tracking is a crucial component of upper-level observations. The system used since the 1980s has been replaced with one that uses Global Positioning System (GPS) and the latest computer technology. Meteorologists can determine wind speed and direction by tracking how fast and in what direction a rawinsonde moves. The various data are plotted on a chart that gives meteorologists a profile of the temperature, pressure, humidity, wind speed, and wind direction of a particular part of the atmosphere. Such charts are used to forecast atmospheric changes that affect surface weather.

READING CHECK **Describe** the function of a radiosonde.

Vocabulary
Academic vocabulary
Compute (kuhm PYEWT)
to perform mathematical operations
Jane used a calculator to compute the answers for her math homework.

Weather Observation Systems

There are many surface and upper-level observation sites across the United States. However, data from these sites cannot be used to locate exactly where precipitation falls without the additional help of data from weather radars and weather satellites.

Weather radar A weather radar system detects specific locations of precipitation. The term *radar* stands for **ra**dio **d**etection **a**nd **r**anging. How does radar work? A radar system generates radio waves and transmits them through an antenna at the speed of light. Recall that radio waves are electromagnetic waves with wavelengths greater than 10^{-3} m. The transmitter is programmed to generate waves that only reflect from particles larger than a specific size. For example, when the radio waves encounter raindrops, some of the waves scatter. Because an antenna cannot send and receive signals at the same time, radars send a pulse and wait for the return before another pulse is sent. An amplifier increases the received wave signals, and then a computer processes and displays them on a monitor. From these data, the distance to precipitation and its location relative to the receiving antenna.

ACROSS THE CURRICULUM

Physics The Doppler effect is used in astronomy as evidence that the universe is expanding. In 1924, Edwin Hubble used a spectrograph to study light from other galaxies. He noticed a redshift in the light from galaxies beyond the Local Group, which is the cluster that includes the Milky Way. The Doppler effect in light causes redshifts, or the lengthening of the wavelengths of light that are moving away from Earth. Galaxies that are approaching Earth demonstrate blueshifts, or the shortening of their wavelengths of light. Because all galaxies outside of the Local Group show redshifts in their spectra, the galaxies are moving away from Earth—evidence that the universe is expanding.

Doppler weather radar You have probably noticed that the pitch produced by the horn of an approaching car gets higher as it comes closer to you and lower as it passes and moves away from you. This sound phenomenon is called the Doppler effect. The **Doppler effect** is the change in pitch or frequency that occurs due to the relative motion of a wave, such as sound or light, as it comes toward or goes away from an observer.

The NWS uses Weather Surveillance Radar-1988 Doppler (WSR-88D), shown in **Figure 14,** based on the Doppler effect of moving waves. Analysis of Doppler radar data can be used to determine the speed at which precipitation moves toward or away from a radar station. Because the movement of precipitation is caused by wind, Doppler radar can also provide a good estimation of the wind speeds associated with precipitation areas, including those with severe weather, such as thunderstorms and tornadoes. The ability to measure wind speeds gives Doppler radar a distinct advantage over conventional weather radar systems.

Weather satellites In addition to communications, one of the main uses of satellites orbiting Earth is to observe weather. Cameras mounted aboard a weather satellite take images of Earth at regular intervals. A weather satellite can use infrared, visible-light, or water-vapor imagery to observe the atmosphere.

Infrared imagery Some weather satellites use infrared imagery to make observations at night. Objects radiate thermal energy at slightly different frequencies. Infrared imagery detects these different frequencies, which enables meteorologists to map either cloud cover or surface temperatures. Different frequencies are distinguishable in an infrared image, as shown in **Figure 15.**

As you have learned, clouds form at different altitudes and have different temperatures. Using infrared imagery, meteorologists can determine the cloud's temperature, its type, and its altitude. Infrared imagery is especially useful in detecting strong thunderstorms that develop and reach high altitudes. Since the troposphere cools with increasing altitude, they appear as very cold areas on an infrared image. Because the strength of a thunderstorm is related to the altitude that it reaches, infrared imagery can be used to establish a storm's potential to produce severe weather.

■ **Figure 14** Norman, Oklahoma, was the site of the first Doppler radar installation.
Relate *the importance of this location to severe weather conditions.*

■ **Figure 15** This infrared image shows cloud cover across most of the midwestern United States.

Collaborative Learning

Doppler Radar Doppler radar measures the motion of atmospheric particles toward and away from a radar receiver. The velocity and direction of motion is shown by the color of the pixels on the radar screen. Have students explain how Doppler radar can be used to predict a tornado within a developing thunderstorm. If atmospheric particles moving both toward and away from the radar show up on the radar as pixels immediately next to each other, then they indicate the presence of the tight wind circulation that can produce a tornado.

Use an Analogy

Satellites Have students imagine trying to take a photograph in a dark closet to demonstrate the need for infrared satellite imagery. The photo would not turn out because of the lack of light. Meteorologists face the same problem when they try to photograph the dark side of Earth. The technology of infrared imagery was developed in part to obtain images of all parts of Earth's surface during day and night.

■ **Caption Question** **Fig. 14** Severe thunderstorms, and often tornadoes, form in this area. Doppler radar enables more precise detection and prediction of such storms and quicker notification of the public.

Demonstration

Weather Radar Show students a weather radar image and ask them to describe what they see. Some will realize that they are seeing areas of precipitation. Explain that weather radar sends out pulses of energy designed to bounce off droplets that are large enough to fall as precipitation. The radar ignores smaller cloud droplets. This is accomplished by controlling the wavelength of the radar beam, which is usually set at 10 cm for the WSR-88D.

3 Assess

Check for Understanding

Discussion Have students use what they have learned in this section to describe in their Earth science journals the equipment they would need to set up a fully functioning weather observatory. The observatory should be capable of monitoring surface and upper-air data, as well as large-scale cloud and precipitation patterns.

Reteach

Instrumentation Ask students to each make a data table listing the various weather instruments about which they learned in this section and what the instruments are used to measure.

Assessment

Performance On a partly cloudy day, have students take photographs or draw sketches of the sky once every 5 minutes for about 30 minutes. Students should take each photo or draw each sketch from the same spot and the same angle. Have students create a time-lapse view of how the sky changes over the course of 30 minutes. Students should relate this project to the use of satellite imagery to detect the motion of large-scale weather systems.

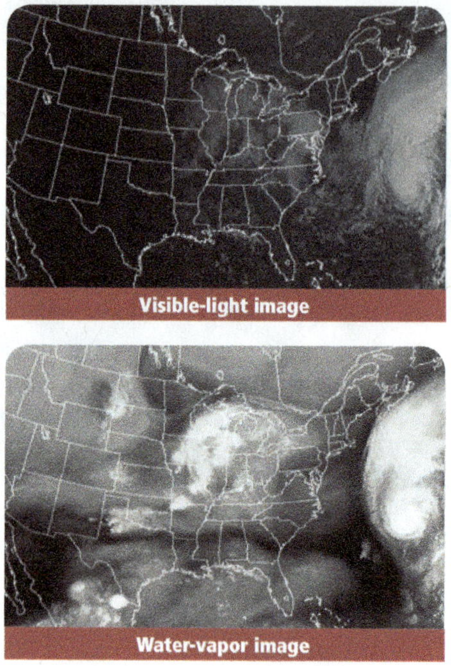

■ **Figure 16** These images were taken at the same time as the one in **Figure 15.** Each type of image shows different atmospheric characteristics. Together, they help meteorologists accurately analyze and predict weather.

Visible-light imagery Some satellites use cameras that require visible light to photograph Earth. These digital photos, like the one in **Figure 16,** are sent back to ground stations, and their data are plotted on maps. Unlike weather radar, which tracks precipitation but not clouds, these satellites track clouds but not necessarily precipitation. They also show differences in the shading of clouds, which can relate to cloud thickness. By combining radar and visible imagery data, meteorologists can determine where both clouds and precipitation are occurring.

Water-vapor imagery Another type of satellite imagery that is useful in weather analysis and forecasting is called water-vapor imagery, also shown in **Figure 16.** Water vapor is an invisible gas and cannot be photographed directly, but it absorbs and emits infrared radiation at certain wavelengths. Many weather satellites have sensors that are able to provide a measure of the amount of water vapor present in the atmosphere.

Water-vapor imagery is a valuable tool for weather analysis and prediction because it shows moisture in the atmosphere, not just cloud patterns. Because air currents that guide weather systems are often well defined by trails of water vapor, meteorologists can closely monitor the development and change in storm systems even when clouds are not present.

SECTION 3 REVIEW

Section Self-Check

Section Summary

- To make accurate weather forecasts, meteorologists analyze and interpret data gathered from Earth's surface by weather instruments.
- A radiosonde collects upper-atmospheric data.
- Doppler radar locates where precipitation occurs.
- Weather satellites use infrared, visible-light, or water-vapor imagery to observe and monitor changing weather conditions on Earth.

Understand Main Ideas

1. MAINIDEA **Identify** two important factors in collecting and analyzing weather data in the United States.
2. **Compare and contrast** methods for obtaining data from Earth's surface and Earth's upper atmosphere.
3. **State** the main advantage of Doppler radar over conventional weather radar.
4. **Summarize** the three kinds of weather satellite imagery using a graphic organizer.

Think Critically

5. **Predict** whether you would expect weather forecasts to be more accurate for the state of Kansas or a remote Caribbean island, based on what you know about weather observation systems. Explain.

WRITING IN ▶ Earth Science

6. Write a newspaper article about the use of water-vapor imagery to detect water on the planet Mars.

SECTION 3 REVIEW

1. Two of the most important factors in weather forecasting are the accuracy and the density, or amount, of the data.
2. Surface weather instruments include devices such as thermometers (temperature), barometers (air pressure), anemometers (wind speed), and hygrometers (relative humidity). Radiosondes measure upper-atmospheric data such as temperature and humidity. Weather radars and weather satellites can collect more precise upper-atmospheric data such as those about rainfall location and storm development.
3. Doppler radar can detect wind speeds.
4. Answers should summarize the information on pages 327–328 about the three types of satellite imagery: visible-light, infrared, and water.
5. Kansas; it is in the middle of the United States with a well-established weather-observation network.
6. Answers should mention how water-vapor imagery could be used to detect clouds in the Martian atmosphere, which would suggest evaporation of water on Mars's surface.

Rubric

Weather Analysis and Prediction

SECTION 4

Essential Questions
- What information is on a basic surface weather chart?
- How do digital and analog forecasting differ?
- What are problems associated with long-term forecasts?

Review Vocabulary

model: an idea, system, or mathematical expression that represents an idea

New Vocabulary

station model
isobar
isotherm
digital forecast
analog forecast

MAINIDEA Several methods are used to develop short-term and long-term weather forecasts.

EARTH SCIENCE 4 YOU It is usually easier to predict what you will be doing later today than what you will be doing a week from now. Weather predictions also are easier for shorter time spans than for longer time spans.

Surface Weather Analysis

Newspapers, radio and television stations, and Web sites often give weather reports. These data are plotted on weather charts and maps and are often accompanied by radar and satellite imagery.

Station models After weather data are gathered, meteorologists plot the data on a map using station models for individual cities or towns. A **station model** is a record of weather data for a particular site at a particular time. Meteorological symbols, such as the ones shown in **Figure 17,** are used to represent weather data in a station model. A station model allows meteorologists to fit a large amount of data into a small space. It also gives meteorologists a uniform way of communicating weather data.

Plotting station model data Station models provide information for individual sites. To plot data nationwide and globally, meteorologists use lines called isopleths that connect points of equal or constant values. The values represent different weather variables, such as pressure or temperature. Lines of equal pressure, for example, are called **isobars,** while lines of equal temperature are called **isotherms.** The lines themselves are similar to the contour lines—lines of equal elevation—on a topographic map.

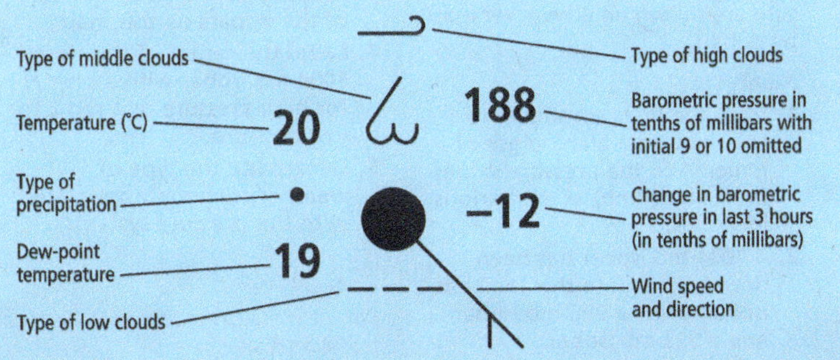

Figure 17 A station model shows temperature, wind direction and speed, and other weather data for a particular location at a particular time.
Explain *the advantage of using meteorological symbols.*

SECTION 4

1 Focus

Focus Transparency

Before presenting the lesson, project **Section Focus Transparency 35** and have students answer the accompanying questions. EL

MAINIDEA

Weather Analysis and Prediction Ask students to predict what the weather will be like in 10 minutes, then in 10 hours, and lastly in 10 days; ask them how confident they are in each prediction. Predictions become more uncertain as time spans increase.

Environmental Connection

Variation in Forecasts Ask students to speculate on the different ways that human activity might influence local weather and therefore also could impact weather forecasts on different time scales. Some examples of this would include changes in absorption of heat energy in large urban areas, the greenhouse effect, and atmospheric pollution.

Caption Question Fig. 17 Symbols allow meteorologists to fit much more weather information around each location on a map.

Teacher Content Support

Weather Forecasts The weather forecasts on television use computer imaging to project images of weather maps onto televisions. Using a process called chroma key, a weather forecaster stands in front of a special, single-color backdrop. This backdrop is blank. Weather information is shown on TV monitors located on either side of the backdrop. The monitors show the same weather maps that audiences at home see on their television. The forecaster looks at these TV monitors to see where to point on the backdrop. Skilled forecasters make it appear as though they are pointing to changing information on the backdrop itself.

Tie to Previous Knowledge

Station Models Have students review the concept of map legends. Station models are map legends that show weather variables in symbolic forms.

Activity

Isobars Have students draw a set of isobars to show a region with low pressure, using arrows to show wind direction around the center of low pressure; repeat the drawing for a region with high pressure. Tell students that the diagrams should represent pressure regions in the northern hemisphere. Wind flows counterclockwise and inward around a low-pressure center, and clockwise and outward around a high-pressure center. OL

Teacher Content Support

Weather Analysis Once the current state of the atmosphere has been measured, the physical laws that govern atmospheric motion can be applied to predict how the atmosphere will change. This process, weather forecasting, is rife with inaccuracies; meteorologists cannot possibly measure the state of the atmosphere over the entire world at all levels of the atmosphere.

Caption Question **Fig. 18** where isobars are closer together

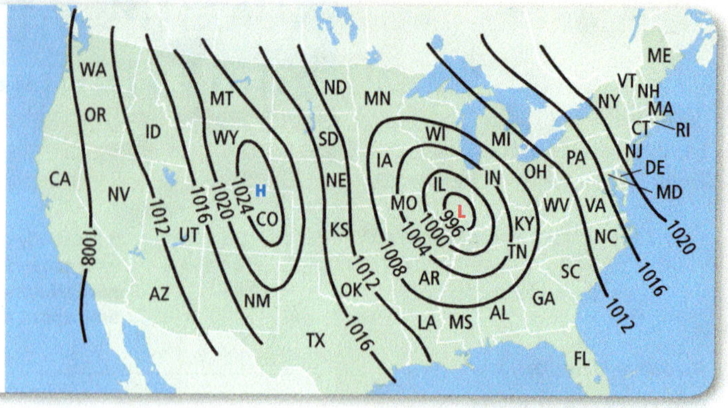

Figure 18 The weather map shows isobars and air pressure data for the continental United States. Air pressure is measured in millibars (mb).
Determine *where on the weather map you would expect the strongest winds.*

Interpreting station model data Recall that inferences about elevation can be made by studying contour intervals on a map. Inferences about weather, such as wind speed, can be made by studying isobars and isotherms on a map. Isobars that are close together indicate a large pressure difference over a small area, which means strong winds. Isobars that are far apart indicate a small difference in pressure and light winds. As shown in **Figure 18,** isobars also indicate the locations of high- and low-pressure systems. Combining this information with that of isotherms helps meteorologists to identify fronts.

Using isobars, isotherms, and station-model data, meteorologists can detect and analyze current weather conditions for a particular location. This is important because meteorologists must understand current weather conditions before they can forecast the weather.

Problem-Solving LAB

Interpret a Scientific Illustration

How do you analyze a weather map? Areas of high and low pressure are shown on a weather map by isobars.

Analysis

1. Trace the diagram shown to the right on a blank piece of paper. Add the pressure values in millibars (mb) at the various locations.
2. A 1004-mb isobar has been drawn. Complete the 1000-mb isobar. Draw a 996-mb isobar and a 992-mb isobar.

Think Critically

3. **Identify** the contour interval of the isobars on this map.
4. **Label** the center of the closed 1004-mb isobar with a blue *H* for high pressure or a red *L* for low pressure.
5. **Determine** the type of weather commonly associated with this pressure system.

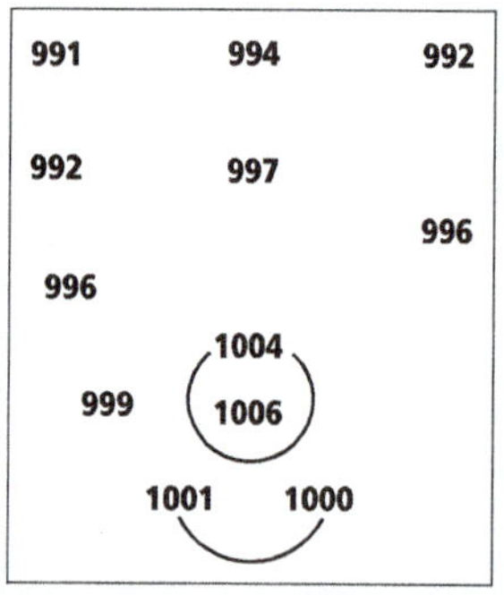

Problem-Solving LAB

Purpose Students will analyze pressure data on a surface weather map.

Process Skills interpret scientific illustrations, analyze data, predict

Teaching Strategies

- Review topographic maps and contour lines.
- Tell students that isobars are similar to contour lines in that they connect points of equal values.

Analysis

1 and 2. Isobars on diagrams should connect points of equal air pressure.

Think Critically

3. 4 mb
4. A blue H should be drawn inside the 1004 isobar.
5. A high-pressure system is normally associated with clear skies and fair weather.

Types of Forecasts

A meteorologist, shown in **Figure 19,** must analyze data from different levels in the atmosphere, based on current and past weather conditions, to produce a reliable forecast. Two types of forecasts are digital forecasts and analog forecasts.

Digital forecasts The atmosphere behaves like a fluid. Physical principles that apply to a fluid, such as temperature, pressure, and density, can be applied to the atmosphere and its variables. In addition, they can be expressed as mathematical equations to determine how atmospheric variables change over time.

A **digital forecast** is created by applying physical principles and mathematics to atmospheric variables and then making a prediction about how these variables will change over time. Digital forecasting relies on numerical data. Its accuracy is related directly to the amount of available data. It would take a long time for meteorologists to solve atmospheric equations on a global or national scale. Fortunately, computers can do the job quickly. Digital forecasting is the main method used by present-day meteorologists.

■ **Figure 19** This meteorologist is analyzing data from various sources to prepare a weather forecast.

☑ READING CHECK **State** the relationship between the accuracy of a digital forecast and the data on which it is based.

Analog forecasts Another type of forecast, an **analog forecast,** is based on a comparison of current weather patterns to similar weather patterns from the past. Meteorologists coined the term *analog forecasting* because they look for a pattern from the past that is similar, or analogous, to a current pattern. To ensure the accuracy of an analog forecast, meteorologists must find a past event that had similar atmosphere, at all levels and over a large area, to a current event.

The main disadvantage of analog forecasting is the difficulty in finding the same weather pattern in the past. Still, analog forecasting is useful for conducting monthly or seasonal forecasts, which are based mainly on the past behavior of cyclic weather patterns.

Short-Term Forecasts

The most accurate and detailed forecasts are short term because weather systems change directions, speeds, and intensities over time. For hourly forecasts, extrapolation is a reliable forecasting method because small-scale weather features that are readily observable by radar and satellites dominate current weather.

One- to three-day forecasts are no longer based on the movement of observed clouds and precipitation, which change by the hour. Instead, these forecasts are based on the behavior of larger surface and upper-level features, such as low-pressure systems. A one- to three-day forecast is usually accurate for expected temperatures, and for when and how much precipitation will occur. For this time span, however, the forecast will not be able to pinpoint an exact temperature or sky condition at a specific time.

VOCABULARY

ACADEMIC VOCABULARY

Extrapolation
(ihk stra puh LAY shun)
the act of inferring a probable value from an existing set of values
Short-term weather forecasts can be extrapolated from data collected by radar and satellites.

Use an Analogy

Prediction Dilemma Have students imagine heating alphabet soup in a pot until it boils, then trying to predict where each of the letters will be a minute later. This is similar to weather forecasting in that meteorologists must attempt to predict how all the variables in the atmosphere at various levels will change with time.

Activity

Computers Challenge students to count to 1 million. After they have reached 100 or 200 and realized how tedious this task is, tell them that weather forecasting on a national or global level involves numerous weather variables for thousands of places. The interactions of these weather variables can be expressed as mathematical equations, but the task of mentally calculating the equations would take an extremely long time. That's why high-speed computers are used in digital forecasting, a type of weather forecasting that students will learn about in this section.

Teacher Content Support

Digital Models Digital methods have been applied to different models of Earth's climate to analyze the effects of global warming, the greenhouse effect, volcanic eruptions, and changes in climate caused by variations in Earth's orbit and rotation.

☑ READING CHECK The denser the data, the more accurate a digital forecast will be.

DIFFERENTIATED INSTRUCTION

Advanced Learners Gifted students might want to research upper-level weather-analysis methods. Students can get information on upper-air analyses from the local National Weather Service office.

EARTH SCIENCE JOURNAL

Weather Analysis and Prediction Using internet or other media, have students record the forecasted temperatures for a date seven days from now. Do this each day for that same date. When that day arrives, have them calculate how far off each day's forecast was from the recorded temperatures on the chosen date. How did the errors change as the date got closer?

GeoLAB

iLab Station

The GeoLab located at the end of the chapter can be used at this point in the lesson.

3 Assess

Check for Understanding

Discussion New satellite technology enables better moisture-level measurements to be made throughout the atmosphere. Ask students how this might help meteorologists produce digital weather forecasts. Additional moisture data would result in more accurate digital forecasts, because the density of the data would be improved.

Reteach

Forecasts Have students each make a data table that lists the two main types of weather forecasts, the strengths and weaknesses of each forecast method, and which method would be most accurate for short-term and long-term forecasts.

Assessment

Skill For three consecutive days, obtain a weather map from a newspaper or the Internet. On the fourth day, students should use the extrapolation method to predict the location of high- and low-pressure systems and fronts. Also, have students predict the weather for their area and compare their predictions to actual weather conditions.

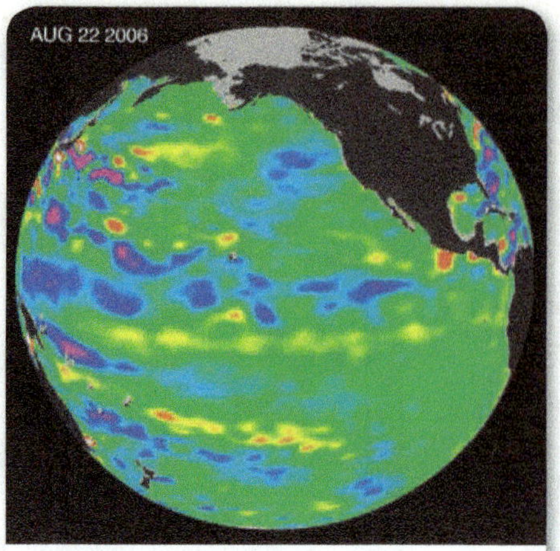

Figure 20 La Niña occurs when stronger-than-normal trade winds carry the colder water (blue) from the coast of South America to the equatorial Pacific Ocean. This happens about every three to five years and can affect global weather patterns.

Long-Term Forecasts

Because it is impossible for computers to model all of the different variables that affect the weather at a given time and place, all long-term forecasts are less reliable than short-term forecasts. Recall that features on Earth's surface, such as lakes or snow cover, affect the amount of thermal energy absorbed at any location. This affects the pressure at that location, which, in turn, affects the wind. Wind influences cloud formation and virtually all other aspects of the weather in that location. Over time, these factors interact and create more complicated weather scenarios.

Meteorologists use changes in surface weather systems based on circulation patterns throughout the troposphere and lower stratosphere for four- to seven-day forecasts. They can estimate each day's weather but cannot pinpoint when or what specific weather conditions will occur. One- to two-week forecasts are based on changes in large-scale circulation patterns. Thus, these forecasts are vague and are based mainly on similar conditions that have occurred in the past.

Forecasts for months and seasons are based mostly on weather cycles or patterns. These cycles, such as the one shown in **Figure 20,** can involve changes in the atmosphere, ocean currents, and solar activity that might occur at the same time. Improvements in weather forecasts depend on identifying the influences of the cycles involved, understanding how they interact, and determining their ultimate effect on weather over longer time periods.

SECTION 4 REVIEW

Section Self-Check

Section Summary

- A station model is used to plot different weather variables.
- Meteorologists plot lines on a map that connect variables of equal value to represent nationwide and global trends.
- Two kinds of forecasts are digital and analog.
- The longer the prediction period, the less reliable the weather forecast.

Understand Main Ideas

1. MAINIDEA **Describe** the methods used for illustrating weather forecasts.
2. **Identify** some of the symbols used in a station model.
3. **Model** how temperature and pressure are shown on a weather map.
4. **Compare and contrast** analog and digital forecasts.
5. **Explain** why long-term forecasts are not as accurate as short-term forecasts.

Think Critically

6. **Assess** which forecast type—digital or analog—would be more accurate for three days or less.

MATH IN Earth Science

7. Using a newspaper or other media sources, find and record the high and low temperatures in your area for five days. Calculate the average high and low temperatures for the five-day period.

SECTION 4 REVIEW

1. Surface data is plotted on weather charts and weather maps using station models, isobars, and isotherms. These plots are often accompanied by radar and satellite imagery for making forecasts.
2. See **Figure 17** for symbols including temperature, type of precipitation, dew point, type of clouds, air pressure, and wind speed and wind direction.
3. Temperature and pressure are shown on weather maps by drawing lines connecting points with equal temperature or pressure. These lines are called isotherms and isobars, respectively.
4. Analog forecasting compares the current positions and strengths of weather systems with similar, or analogous, past weather events. Digital forecasting relies on principles of physics and mathematics to predict how atmospheric variables will change with time.
5. Small errors in the observations can grow with the length of the forecast. In short-term forecasts, these errors do not create large problems; in long-term forecasts, these errors can lead to large forecast errors.
6. The analog method might work well during the first day. However, it could not accurately predict further changes in the size, intensity, and movement of a weather system. The digital method would be most accurate over a three-day period.
7. Answers should accurately reflect the weather in your area.

Earth Science &
SOCIETY

Weather Forecasting—Precision from Chaos

On a rainy evening in New Jersey, four teens went out to play soccer. They began to play, expecting the rain to clear before the game got into full swing. However, as the game progressed, the clouds darkened to a charcoal grey and thickened. When the thunder and lightning began, the teens decided to leave the field. As they walked from the field, they were struck by lightning. Two of the teens died in the hospital a few hours later. The deaths rocked the community. The storm had not been predicted in the weather forecast. Why isn't weather forecasting more predictable?

Chaos and weather systems In 1963, a meteorologist named Edward Lorenz first presented chaos theory, which states that formulated systems are dependent on initial conditions and that the precision of initial measurements has an exponential impact on the expected outcome.

Years after Lorenz published his findings in meteorology journals, other scientists recognized the importance of his work. The simplified equations Lorenz created through his studies helped form the basis of modern weather forecasting.

The beginning of a forecast Weather forecasting begins with observations. Data are collected from various sources and fed into supercomputers, which create mathematical models of the atmosphere. In the United States, the National Weather Service operates these computers and releases their data to local and regional forecasters.

Meteorologists generally agree that useful day-to-day forecasts are limited to only five days. Most meteorologists also agree that reliable forecasts of day-to-day weather for up to six or seven days ahead are not now possible.

Weather forecasts are created from data collected from the atmosphere.

Meteorologists hope that improved measurements, computer technology, and weather models might someday predict day-to-day weather up to three weeks in advance.

Limitations of long-range forecasting Meteorologists generally find that day-to-day forecasts for more than a week in the future are unreliable. Their approach to long-range forecasting is based instead on comparisons of current and past weather patterns, as well as global ocean temperatures, to determine the probability that temperature and precipitation values will be above or below normal ranges. The National Weather Service's Climate Prediction Center, as well as other organizations, offers monthly and seasonal predictions for these values.

WRITING IN ▶ Earth Science

Evaluate Use a newspaper or other local news source to obtain a weather report for the next seven days. Record the temperature and weather conditions for your city during the next week and compare the forecasted weather with the observed weather. Write a summary to share your observations with your class.

WebQuest

Earth Science &
SOCIETY

Purpose

Students will describe how weather forecasting is done, the limitations of long-range predictions, and how chaos theory is involved in the process of making forecasts.

Teacher Content Support

Weather Forecasting Weather forecasting dates back to the Babylonians in 650 B.C. Aristotle wrote a book about weather in 340 B.C. in which he described the formation of rain, clouds, hail, wind, and hurricanes. He made many accurate observations about weather in this book which stood as the authoritative guide on weather for almost 2000 years. Weather-observing and recording stations did not surface until the 1860s. In 1906, the linking of mathematical equations with weather predicting began. By the 1940s, computers were used for the first time to calculate numerical weather forecasts, and by 1950, accurate 24-hour forecasts were being made.

Teaching Strategies

- Have students create their own forecasts for a city and present them to the class. Encourage students in the audience to pose questions regarding the way the forecast was composed after each presentation.
- Ask why 1–3 day forecasts are more accurate than 10-day forecasts. Will we ever have accurate 30-day forecasts? Why or why not?

WRITING IN ▶ Earth Science

Rubric

Evaluate Have students check different numerical models when they record the forecast. Do the models tend to agree or disagree? How do they think model agreement influences forecasts and their accuracy?

WebQuest

GeoLAB

iLab Station | Rubric

Preparation

Time Allotment 45 min

Process Skills analyze data, communicate, interpret data, predict, interpret scientific illustrations

Safety Precaution Approve lab safety forms before work begins.

Preparation of Materials Review the weather symbols for high- and low-pressure systems and fronts. Be sure that students have completed the Problem-Solving Lab from Section 4.

Procedure

- To increase student proficiency in using SI, have students give their answers in both English units and SI units when appropriate.
- Students with visual impairments can be teamed with students who can easily discern small print.

GeoLAB

iLab Station

Mapping: Interpret a Weather Map

Background: The surface weather map on the following page shows actual weather data for the United States. In this activity, you will use the station models, isobars, and pressure systems on the map to forecast the weather.

Question: *How can you use a surface weather map to interpret information about current weather and to forecast future weather?*

Materials

ruler
Reference Handbook, Weather Map Symbols, p. 959

Procedure

1. Read and complete the lab safety form.
2. The map scale is given in nautical miles. Refer to the scale when calculating distances.
3. The unit for isobars is millibars (mb). In station models, pressure readings are abbreviated. For example, 1021.9 mb is plotted on a station model as 219 but read as 1021.9.
4. Wind shafts point in the direction from which the wind is blowing. Refer to Weather Map Symbols in the table on the right and the *Reference Handbook* to learn about the symbols that indicate wind speed.
5. Each number around a city represents a different atmospheric measure. By convention, the same atmospheric measure is always in the same relative location in a station model. Refer to **Figure 17** and Weather Map Symbols in the *Reference Handbook* to learn what numbers represent in a station model.

Analyze and Conclude

1. **Identify** the contour interval of the isobars.
2. **Find** the highest and lowest isobars and where they are located.
3. **Describe** the winds across Texas and Louisiana.
4. **Determine** and **record** with their locations the coldest and warmest temperatures on the map.
5. **Infer** whether the weather in Georgia and Florida is clear or rainy. Explain.
6. **Predict** Low-pressure systems in eastern Canada and off the Oregon coast are moving east at about 24 km/h. Predict short-term weather forecasts for northern New York and Oregon.

Symbols Used in Plotting Report	
Fronts and Pressure Systems	
(H) or High **(L) or Low**	Center of high- or low-pressure systems
	Cold front
	Warm front
	Occluded front
	Stationary front

APPLY YOUR SKILL

Forecasting Find your area on the map. Based on the data shown in the map, use the extrapolation method to forecast the next day's weather for your location.

Analyze and Conclude

1. 4 mb
2. The highest isobar is 1040 mb and is around the high-pressure center in south-central Canada. The lowest isobar is 988 mb and is around the low-pressure center off the coast of Oregon.
3. They are gently blowing mainly from the south or southeast.
4. The coldest is −31°F at Hudson Bay Saskatchewan. The warmest is 72°F in the Gulf of Mexico.
5. It would probably be clear, because the high-pressure system along the coast would cause air to sink and dry out.
6. The low-pressure systems would cause air to rise and produce clouds and rain.

APPLY YOUR SKILL

Forecasting Using today's weather map, have students use the extrapolation method to predict tomorrow's weather. How does their forecast compare to the official forecast for your area? Have a discussion tomorrow about how accurate the forecast was.

LabManager™
Customize any lab with the LabManager™ CD-ROM.

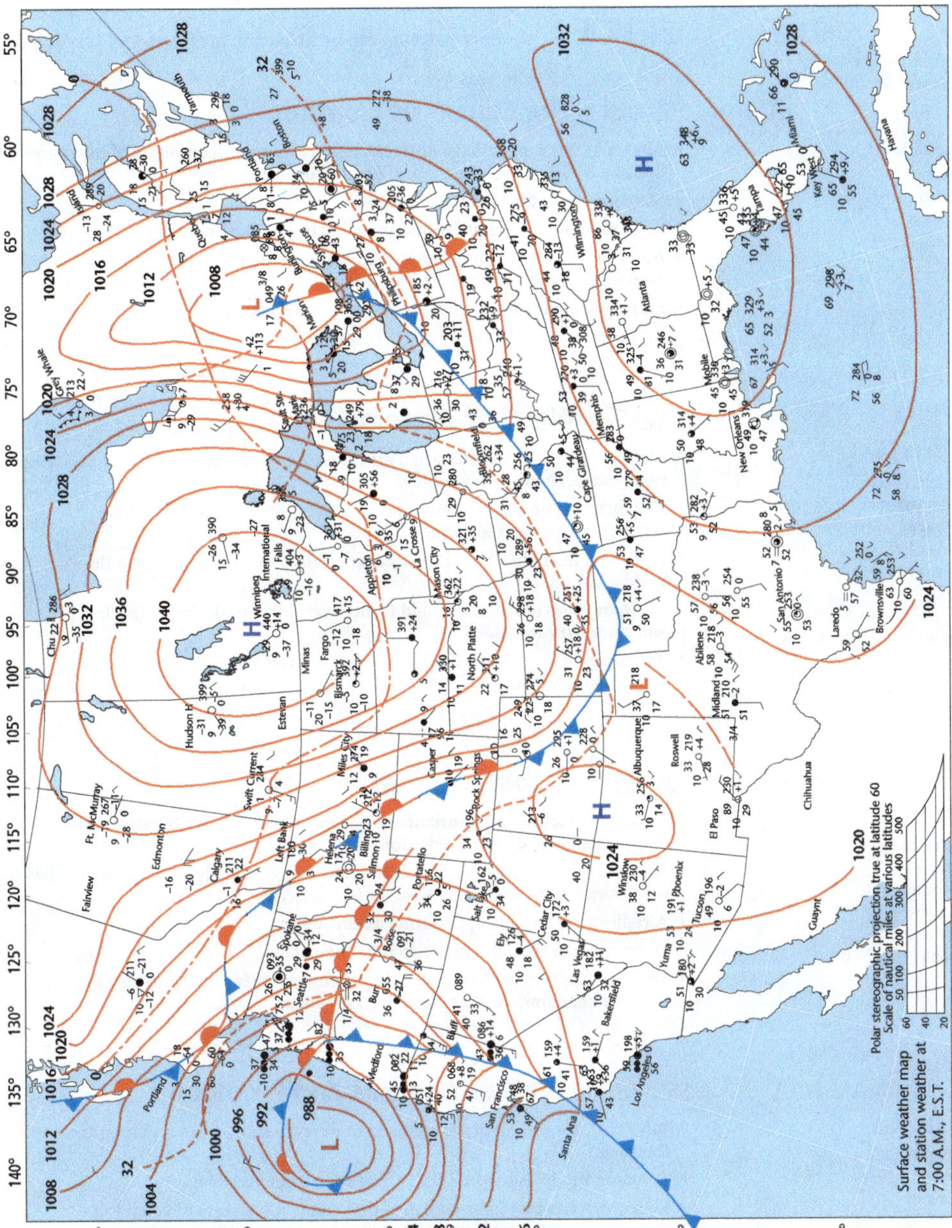
Surface weather map
and station weather at
7:00 A.M., E.S.T.
Polar stereographic projection true at latitude 60
Scale of nautical miles at various latitudes

MAINIDEAS Summary statements can be used by students to review the major concepts of the chapter.

Students can review with these online resources.

Vocabulary Practice

Vocabulary eGames
Vocabulary eFlashcards
Vocabulary PuzzleMaker

Section Self-Check

Chapter Self-Check

Online Test Practice

Use the ExamView® *Assessment Suite* CD-ROM to:

- create multiple versions of tests
- create modified tests with one mouse click
- edit existing questions and add your own questions
- build tests aligned with state standards using built-in state curriculum tags
- change English tests to Spanish with one mouse click
- track students' progress using the Teacher Management System

CHAPTER 12 STUDY GUIDE

BIGIDEA **Weather patterns can be observed, analyzed, and predicted.**

Vocabulary Practice

SECTION 1 The Causes of Weather

VOCABULARY
- weather
- climate
- air mass
- source region

MAINIDEA **Air masses have different temperatures and amounts of moisture because of the uneven heating of Earth's surface.**

- Meteorology is the study of atmospheric phenomena.
- Solar radiation is unequally distributed between Earth's equator and its poles.
- An air mass is a large body of air that takes on the moisture and temperature characteristics of the area over which it forms.
- Each type of air mass is classified by its source region.

SECTION 2 Weather Systems

VOCABULARY
- Coriolis effect
- polar easterlies
- prevailing westerlies
- trade winds
- jet stream
- front

MAINIDEA **Weather results when air masses with different pressures and temperatures move, change, and collide.**

- The three major wind systems are the polar easterlies, the prevailing westerlies, and the trade winds.
- Fast-moving, high-altitude jet streams greatly influence weather in the middle latitudes.
- The four types of fronts are cold fronts, warm fronts, occluded fronts, and stationary fronts.
- Air moves in a generally circular motion around either a high- or low-pressure center.

SECTION 3 Gathering Weather Data

VOCABULARY
- thermometer
- barometer
- anemometer
- hygrometer
- radiosonde
- Doppler effect

MAINIDEA **Accurate measurements of atmospheric properties are a critical part of weather analysis and prediction.**

- To make accurate weather forecasts, meteorologists analyze and interpret data gathered from Earth's surface by weather instruments.
- A radiosonde collects upper-atmospheric data.
- Doppler radar locates where precipitation occurs.
- Weather satellites use infrared, visible-light, or water-vapor imagery to observe and monitor changing weather conditions on Earth.

SECTION 4 Weather Analysis and Prediction

VOCABULARY
- station model
- isobar
- isotherm
- digital forecast
- analog forecast

MAINIDEA **Several methods are used to develop short-term and long-term weather forecasts.**

- A station model is used to plot different weather variables.
- Meteorologists plot lines on a map that connect variables of equal value to represent nationwide and global trends.
- Two kinds of forecasts are digital and analog.
- The longer the prediction period, the less reliable the weather forecast.

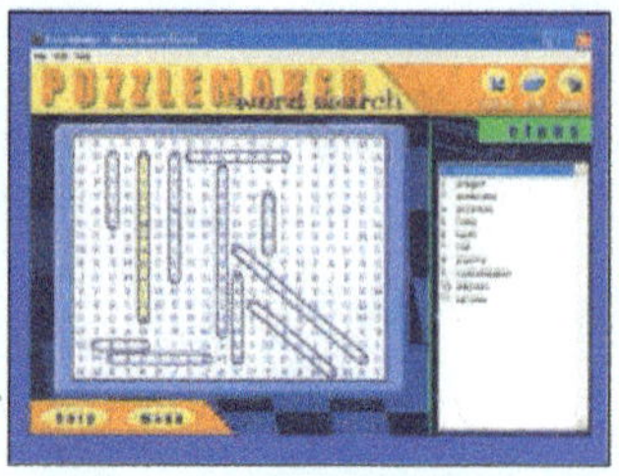

Vocabulary Practice

For additional practice with vocabulary, have students access the Vocabulary PuzzleMaker.

Chapter Self-Check

VOCABULARY REVIEW

Match each description below with the correct vocabulary term from the Study Guide.

1. lines of equal pressure on a weather map
2. current state of the atmosphere
3. a forecast that relies on numerical data
4. long-term variations in weather conditions over a particular area
5. large volume of air that takes on the characteristics of the area over which it forms
6. depiction of weather data for a particular location at a particular time

Complete the sentences below using vocabulary terms from the Study Guide.

7. A ________ is used to measure relative humidity.
8. ________ describes the narrow region separating two air masses of different densities.
9. The deflection of air due to the rotation of Earth is called the ________.
10. Lines of equal temperature on a weather map are called ________.

Each of the following sentences is false. Make each sentence true by replacing the italicized words with vocabulary terms from the Study Guide.

11. The *horse latitudes* are two belts of surface winds that occur between latitudes 30° N and 60° N, and 30° S and 60° S.
12. Meteorologists use a special kind of radar, which is based on the *polar easterlies,* to plot the movement of precipitation.
13. Narrow bands of fast winds are called *trade winds.*
14. A balloon-transported package of sensors is called a *source region.*
15. An instrument that measures wind speed is called a *barometer.*

UNDERSTAND KEY CONCEPTS

16. What does a large temperature gradient at high altitudes of the atmosphere cause?
    A. trade winds
    B. Coriolis effect
    C. ITCZ
    D. polar jet streams

Use the diagram below to answer Questions 17 and 18.

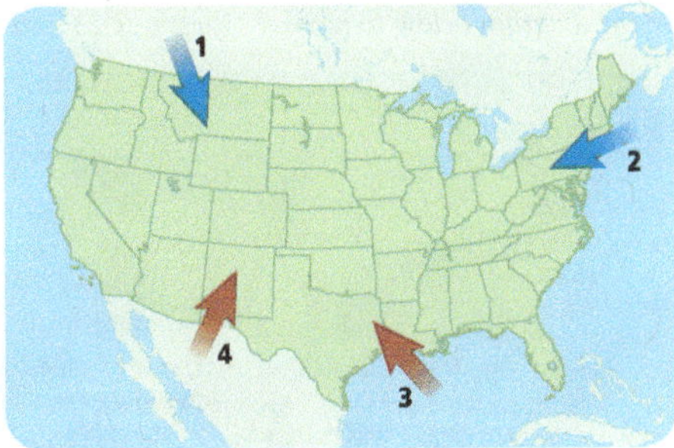

17. Which is probably the coldest air mass?
    A. 1
    B. 2
    C. 3
    D. 4
18. Which air mass is hot and dry?
    A. 1
    B. 2
    C. 3
    D. 4
19. Which is not one of Earth's three basic wind systems or zones?
    A. polar easterlies
    B. polar jet streams
    C. trade winds
    D. prevailing westerlies
20. Which location on Earth receives the most solar radiation in any given year?
    A. the poles
    B. the oceans
    C. the tropics
    D. the continents

CHAPTER 12 ASSESSMENT

VOCABULARY REVIEW

1. isobars
2. weather
3. digital forecast
4. climate
5. air mass
6. station model
7. hygrometer
8. Front
9. Coriolis effect
10. isotherms
11. prevailing westerlies
12. Doppler effect
13. jet steams
14. radiosonde
15. anenometer

UNDERSTAND KEY CONCEPTS

16. D
17. A
18. D
19. B
20. C

UNDERSTAND KEY CONCEPTS

21. D
22. C
23. B
24. D
25. B
26. D

CONSTRUCTED RESPONSE

27. The slope of a warm front is gradual, so clouds and precipitation occur over a larger area. Also, warm fronts generally move more slowly than cold fronts, so it takes longer for the bad weather to move through.
28. since the change was drastic, probably a cold front
29. It would be impossible to tell if weather differences between the stations were based on actual weather changes or simply the difference in observation times.
30. The horse latitudes lie inside a high-pressure belt where air sinks and winds are light.
31. A continental polar air mass is cold and dry, while a maritime tropical air mass is warm and moist.
32. It allows for the assimilation and processing of vast amounts of data at many locations and heights throughout the atmosphere.
33. One could check the surface ASOS observation to see if it indicated rain, or check the Doppler radar image to see if rain was falling in that area.
34. Isobars are lines of equal pressure, and isotherms are lines of equal temperature.
35. The best choice is maritime tropical because it is warm and carries more moisture.

21. Which is an example of climate?
A. today's high temperature
B. yesterday's rainfall
C. tomorrow's highest wind speed
D. average rainfall over 30 years

22. Which instrument is not used to measure surface weather?
A. barometer
B. hygrometer
C. radiosonde
D. thermometer

Use the diagram below to answer Questions 23 and 24.

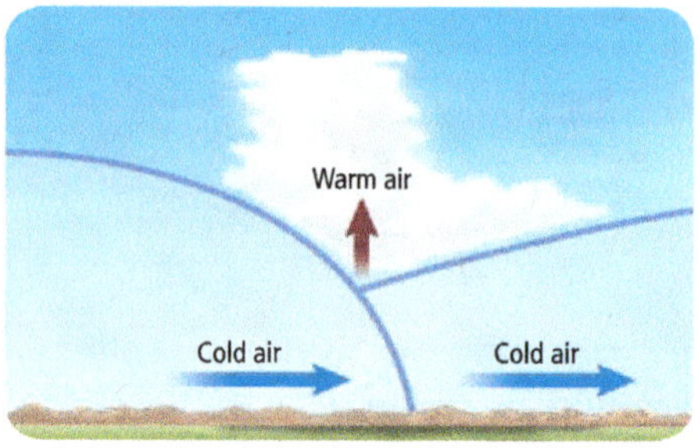

23. Which type of front is illustrated above?
A. cold front
B. occluded front
C. precipitation front
D. stationary front

24. Which weather conditions occur as a result of this type of front?
A. warm temperatures and precipitation
B. cool temperatures and thunderstorms
C. light winds and precipitation
D. strong winds and precipitation

25. Which is the most accurate forecast?
A. long-term digital forecast
B. short-term digital forecast
C. long-term analog forecast
D. short-term analog forecast

26. What is a station model used to create?
A. digital forecast
B. depiction of a jet stream
C. long-term forecast
D. surface weather map

CONSTRUCTED RESPONSE

27. **Infer** why weather ahead of a warm front might be cloudier and rainier for a longer period of time than weather ahead of a cold front.

28. **Identify** the weather feature that might be indicated by a drastic temperature change over a short distance on a surface analysis.

29. **Generalize** the problems that could result from making a weather analysis based on observations at several locations made at different times.

30. **Discuss** why light to no winds characterize the horse latitudes and why they occur at those latitudes.

31. **Compare and contrast** the temperature and moisture properties of a continental polar air mass and a maritime tropical air mass.

32. **State** the main benefit of the digital forecast method.

33. **Describe** two different weather-data methods you could use to determine if it is a rainy day at a given location.

34. **Distinguish** between isobars and isotherms.

Use the table below to answer Question 35.

Air Mass Descriptions		
Air Mass	**Source Region**	**Summer**
Arctic	Siberia, Arctic Basin	cold, dry
Continental polar	interiors of Canada and Alaska	cool, dry
Continental tropical	southwest United States, Mexico	hot, dry
Maritime polar	North Pacific Ocean	mild, humid
	North Atlantic Ocean	cool, humid
Maritime tropical	Gulf of Mexico, Caribbean Sea, tropical and subtropical Atlantic Ocean and Pacific Ocean	hot, humid

35. **Choose** the summertime air mass that would most likely be associated with significant precipitation. Explain your choice.

Chapter Self-Check

THINK CRITICALLY

36. Propose an ideal weather-data collection system for your school.

Use the diagram below to answer Question 37.

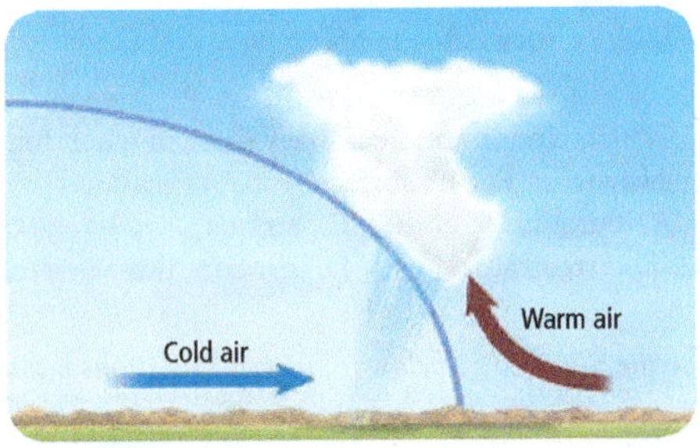

37. Sequence the weather changes that a person on the ground will observe for the front shown above.

38. Compare the challenges of forecasting weather for Seattle, Washington, with those of forecasting weather for New York City.

39. CAREERS IN EARTH SCIENCE Develop a one-day forecast for your city using the weather map from the GeoLab. Role-play a television meteorologist, and give your weather report.

40. Determine the type of modification to an arctic air mass moving southward over the north Atlantic during the summer.

41. Evaluate whether temperature readings taken near an asphalt parking lot on a summer day would represent those for the entire city.

CONCEPT MAPPING

42. Create a concept map showing the relationships among the types of weather data collection.

CHALLENGE QUESTION

43. Explain why cold fronts become stationary and break down as they move through Florida.

WRITING IN ▶ Earth Science

44. Research a local weather-related organization, and write a short essay about the kind of analyses that it performs.

DBQ Document–Based Questions

Data obtained from: National Weather Service, National Center for Environmental Prediction. January 2006. *NOAA.*

The graphs below show the accuracy of three numerical forecast models (A, B, and C) in predicting maximum daily temperatures during January 2006 for up to a period of eight days.

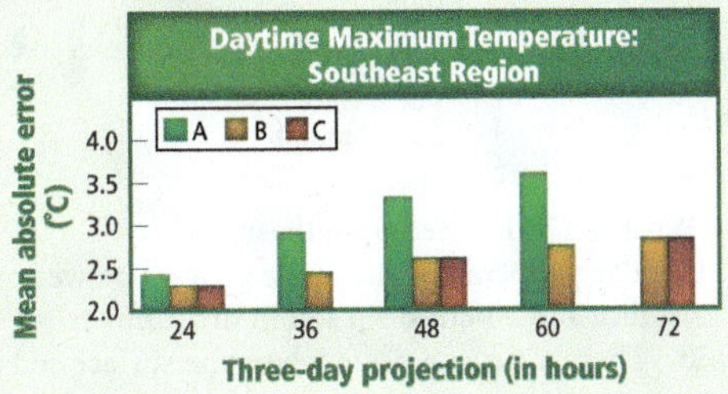

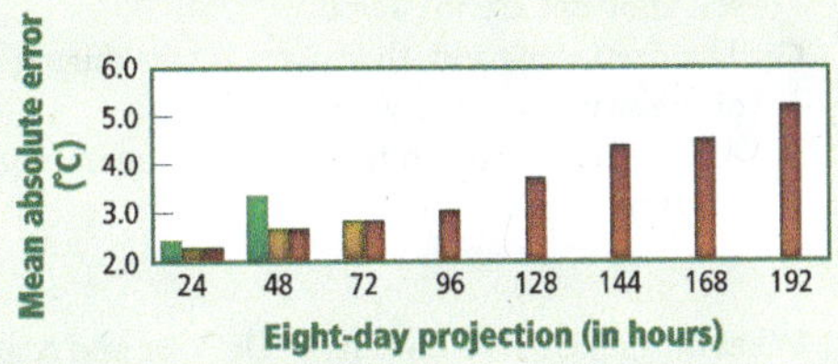

45. Which model had the greatest mean absolute error over the first 60 hours?

46. Which model could be used to find a maximum-temperature forecast for a week from now?

47. Which of the three models shown is the most valuable overall? Explain your answer.

CUMULATIVE REVIEW

48. How is a hand-held GPS receiver used to locate a position? **(Chapter 2)**

49. What determines the maximum height to which water from an artesian well will rise? **(Chapter 10)**

THINK CRITICALLY

36. Possible answer: an ASOS station

37. The weather becomes cloudier and slightly warmer as the cold front approaches. As the front moves through, it is accompanied by showers or thunderstorms and shifting winds. Behind the front, the sky clears and it turns drastically colder and less humid.

38. Seattle is trickier. Since most weather moves from west to east, the weather affecting Seattle comes from the Pacific Ocean, which is a more data-sparse area than the area west of New York.

39. Answers will vary but should accurately reflect information available for your city.

40. It would warm and gain moisture.

41. The thermometer would be measuring the extra-heated parking lot air but not the air of an entire area.

CONCEPT MAPPING

42. Check students' concept maps for accuracy.

CHALLENGE QUESTION

43. By the time a cold front reaches Florida, it is so far from its cold, dry source region and has undergone so much modification that there are no longer significant density differences.

WRITING IN ▶ Earth Science

Rubric

44. Answers will vary, but should accurately report information on a local weather-related organization.

DBQ Document-Based Questions

Data obtained from: National Weather Service, National Center for Environmental Prediction. January 2006. *NOAA.*

45. A

46. C; since it shows verification information to 192 hours.

47. C; it has lower mean absolute errors than A and can be used up to 192 hours in the future.

CUMULATIVE REVIEW

48. The handheld GPS receiver continually receives information from an array of satellites in space. This information is used continually to update the user's position.

49. The maximum height is determined by the artesian pressure surface, which is a straight downsloping line extending from the elevated recharge area of the aquifer to the location of the artesian well.

MULTIPLE CHOICE

1. A
2. B
3. A
4. B
5. C
6. C
7. B
8. A

CUMULATIVE STANDARDIZED TEST PRACTICE

MULTIPLE CHOICE

Use the graph to answer Questions 1–3.

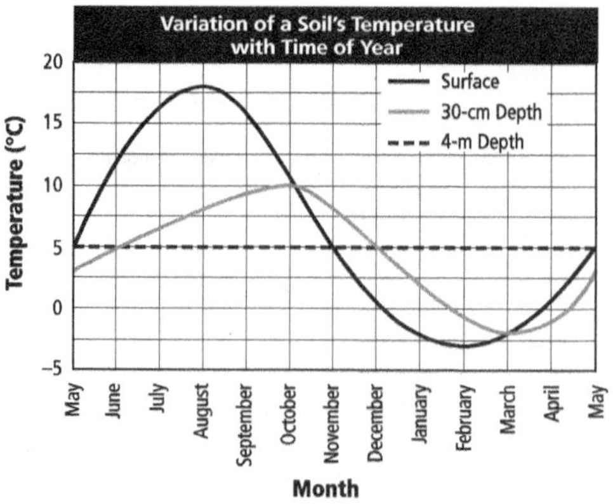

1. What can be inferred from the graph?
   A. The temperature varies the greatest between surface and 30-cm-deep soil in the summer.
   B. There is never a point where the surface soil and soil at 30 cm are the same.
   C. The deeper the soil, the cooler it gets during all of the seasons of the year.
   D. All soil layers vary in temperature depending on the time of year.

2. What is unique to the months of October and March?
   A. The surface and 30-cm-deep soils are warmer than the 4-m-deep soil.
   B. The surface and 30-cm-deep soils have the same temperature.
   C. The surface and 30-cm-deep soils are colder than the 4-m-deep soil.
   D. The soil is the same temperature at all depths.

3. Why is a soil depth of 4 m an important spot to record?
   A. The temperature of the soil never changes there.
   B. No other layer of soil ever reaches that temperature.
   C. It is the only soil that is always frozen.
   D. It is the deepest that soil goes below Earth's surface.

4. Which observation about a rock could lead you to identify it as igneous?
   A. The rock has well defined layers.
   B. The rock has a glassy texture.
   C. The rock contains pebbles.
   D. The rock is made of calcite.

5. Which clouds are most likely to form when fog lifts away from Earth's surface?
   A. cumulus
   B. cirrostratus
   C. stratus
   D. altocumulus

Use the illustration below to answer Questions 6 and 7.

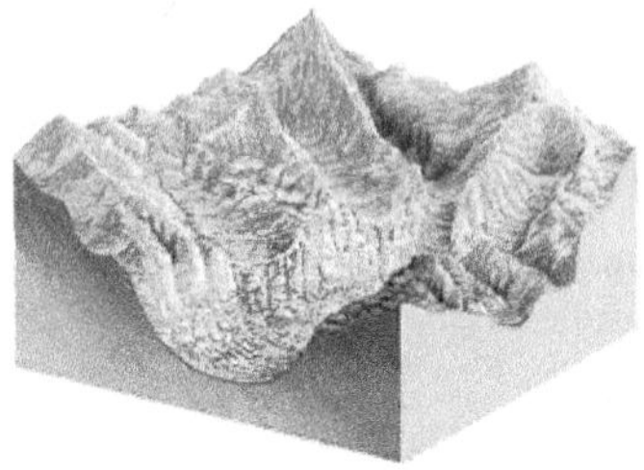

6. What event created the features shown above?
   A. water erosion
   B. wind erosion
   C. glacial erosion
   D. asteroid impact

7. What can scientists learn by studying areas similar to the illustration?
   A. how rivers create U-shaped valleys
   B. glacier history and its direction of movement
   C. why glaciers moved
   D. the impact of wind on mountainous features

8. Which statement is true about fossils found in previously undisturbed strata of sedimentary rock?
   A. Fossils in the upper strata are younger than those in the lower strata.
   B. Fossils in the upper strata are older than those in the lower strata.
   C. Fossils in the upper strata generally are less complex than those in the lower strata.
   D. There are no fossils in the upper strata that resemble those in the lower strata.

Online Test Practice

SHORT ANSWER

Use the image below to answer Questions 9 and 10.

9. What weather instrument is shown in the image above? How does it work?

10. Why is it important for meteorologists to use tools like the one shown for gathering weather data?

11. Suppose surface water that is not absorbed forms a channel and quickly dries up. What can happen over time?

12. Discuss ground subsidence as it relates to our water supply.

13. Suppose a stream is 6 m wide and 3 m deep with a velocity of 12 m/s. How could you determine the stream's discharge?

14. Differentiate between the mass of a brick and the weight of the same brick.

READING FOR COMPREHENSION

Another Use for Radar

Doppler radar tracks moving objects, such as raindrops, through the atmosphere, by bouncing electromagnetic energy off them and measuring the amplitude as well as the change in frequency. Radar doesn't distinguish between bats and hailstones. To the radar, the millions of bats emerging from their caves look like a huge storm that starts at a point on the ground and spreads rapidly up and over the landscape. "It didn't take long for word to get around among bat researchers that we could view bat colonies on the new radar," recalls Jim Ward, science and operations officer at New Braunfels. "We saw bats flying as high as 10,000 feet." Since bats in other locales pursue insects close to the ground, we wondered why the free-tails were flying as high as 10,000 feet. Again, Doppler radar offered clues by detecting the billions of insects that swarm high above Texas. Since the 1980s, researchers have used radar to map the flight patterns of some of North America's most destructive agricultural pests—fall armyworms, beet armyworms, tobacco budworms, and corn earworms.

Article obtained from: McCracken G. F. and J. K Westbrook. 2002. Bat patrol. *National Geographic Magazine* (April): 1.

15. What can be inferred from this passage?
    A. Doppler radar is not useful for monitoring weather.
    B. Doppler radar can track both a major storm and a swarm of bats.
    C. Doppler radar should be used only for studying weather.
    D. Doppler radar should be used only for studying bats.

16. What important insight about technology can you gain by reading this article?

NEED EXTRA HELP?														
If You Missed Question . . .	1	2	3	4	5	6	7	8	9	10	11	12	13	14
Review Section . . .	7.3	7.3	7.3	5.1	11.3	8.3	8.3	6.1	12.3	12.3	9.1	10.3	9.1	1.2

SHORT ANSWER

9. The instrument is a radiosonde. It is a balloon-type instrument which holds sensors that collect upper-level weather data. These sensors measure temperature, air pressure, and humidity.

10. The weather we experience on Earth is mostly the result of changes that occur high in the troposphere. By using instruments such as the radiosonde, meteorologists are able to see what changes are occurring and make more precise weather predictions.

11. Over time, as water continually collects in the channel, the channel will become wider, deeper, and longer. If the amount of water is sufficient, it will begin to flow more permanently, developing into a stream.

12. Ground subsidence—the sinking of land—can be caused by excessive withdrawal of groundwater. Groundwater pressure helps to keep land in place. If too much ground water is removed, the pressure drops, causing the land to sink, blocking the groundwater and lowering the water table, making it harder to get to the water we need.

13. To determine the discharge of the stream multiply the width, depth, and velocity. $6m \times 3m \times 12m/s = 216\ m^3/s$

14. Mass is a measure of the amount of matter in the brick. Weight is the gravitational force on the brick.

READING FOR COMPREHENSION

15. B

16. Sample answer: It is important to see that technology, which has been developed for one field of science, does not need to be confined only to that field. By applying technology in different ways, many areas of science benefit.

CHAPTER 13 The Nature of Storms

BIGIDEA **The exchange of thermal energy in the atmosphere sometimes occurs with great violence that varies in form, size, and duration.**

ESSENTIAL QUESTIONS	NATIONAL STANDARDS	RESOURCES TO ASSESS MASTERY
SECTION 1 **Thunderstorms** **1.** How do thunderstorms form? **2.** What are the different types of thunderstorms? **3.** What is the life cycle of a thunderstorm? 1.5 sessions 0.75 block	UCP.1–3; A.1–2; D.1; F.5	**Entry-Level Assessment** Focus Transparency 36, p. 344 **Progress Monitoring** Caption Question, pp. 344, 346, 348 Reading Check, p. 345 Section Review, p. 349
SECTION 2 **Severe Weather** **1.** Why are some thunderstorms more severe than others? **2.** What are the dangers of severe weather? **3.** How do tornadoes form? 1 session 0.5 block	UCP.1, UCP.3; B.6; D.1; F.5–6	**Entry-Level Assessment** Focus Transparency 37, p. 350 **Progress Monitoring** Caption Question, p. 352 Reading Check, p. 352 Section Review, p. 354
SECTION 3 **Tropical Storms** **1.** How do tropical cyclones form? **2.** What is the life cycle of a tropical cyclone? **3.** What are the dangers associated with hurricanes? 2 sessions 1 block	UCP.1, UCP.3; D.1; F.5–6; G.3	**Entry-Level Assessment** Focus Transparency 38, p. 355 **Progress Monitoring** Reading Check, p. 356 Section Review, p. 360
SECTION 4 **Recurrent Weather** **1.** What are the problems associated with recurring weather patterns? **2.** What atmospheric events cause recurring weather patterns? **3.** How do heat waves and cold waves differ? 4 sessions 2 blocks	UCP.1–3; A.1–2; D.1; E.1–2; F.5–6	**Entry-Level Assessment** Focus Transparency 39, p. 361 **Progress Monitoring** Caption Question, pp. 361, 365 Reading Check, pp. 363, 364 Section Review, p. 365 **Summative Assessment** Chapter Assessment, p. 369 Exam*View*® *Assessment Suite* CD-ROM

Go online!
connectED.mcgraw-hill.com

LEVELED RESOURCES	LAB MATERIALS
Science Notebook 13.1 OL **Chapter FAST FILE Resources:** Study Guide, p. 67 BL **Transparencies:** Section Focus Transparency 36 EL Teaching Transparency 34 OL EL	**LaunchLAB** p. 342 / **10 min** paper, paper punch, balloon, permanent marker, fabric
Science Notebook 13.2 OL **Chapter FAST FILE Resources:** Study Guide, p. 69 BL **Transparencies:** Section Focus Transparency 37 EL Teaching Transparency 35 OL EL	
Science Notebook 13.3 OL **Chapter FAST FILE Resources:** GeoLab Worksheet, p. 55 OL Study Guide, p. 71 BL **Lab Resources:** Laboratory Manual, p. 101 OL **Transparencies:** Section Focus Transparency 38 EL Teaching Transparency 36 OL EL	**GeoLAB** p. 367 / **90 min** internet access, calculator
Science Notebook 13.4 OL **Chapter FAST FILE Resources:** MiniLab Worksheet, p. 54 OL Study Guide, p. 72 BL **Lab Resources:** Laboratory Manual, p. 97 OL **Transparencies:** Section Focus Transparency 39 EL Teaching Transparency 37 OL EL	**MiniLAB** p. 362 / **25 min** ice cube tray, sink or tub, water, clean plastic dishwashing-detergent bottle

LabManager™
Customize any lab with the LabManager™ CD-ROM.

ADDITIONAL RESOURCES

Technology:
- eTeacherEdition
- eStudentEdition
- Classroom Presentation Toolkit
- Exam*View*® *Assessment Suite* CD-ROM
- LabManager™ CD-ROM
- What's EARTH SCIENCE Got To Do With It? DVD
- Weather Classroom Video DVD

Assessment:
- Blueprints for Success
- Chapter Test (Scaffolded)
- Earth Science Standardized Test Practice

Lab Resources:
- Exploring Environmental Problems Laboratory Manual

BL Below Level OL On Level AL Advanced Learners EL English Learners COOP LEARN Cooperative Learning

CHAPTER 13

The Nature of Storms

BIGIDEA The exchange of thermal energy in the atmosphere sometimes occurs with great violence that varies in form, size, and duration.

SECTIONS

1 Thunderstorms

2 Severe Weather

3 Tropical Storms

4 Recurrent Weather

LaunchLAB

iLab Station

Why does lightning form?

You have probably felt the shock of static electricity when you scuff your feet on a rug and then touch a doorknob. Your feet pick up additional electrons, which are negatively charged. These electrons are attracted to the positively charged protons of the doorknob metal, causing a small electrical current to form. The current causes you to feel a small shock. In this lab, explore why lightning forms.

FOLDABLES® Study Organizer

Thunderstorm Development

Make a trifold book using the labels shown. Use it to summarize the stages of thunderstorm development.

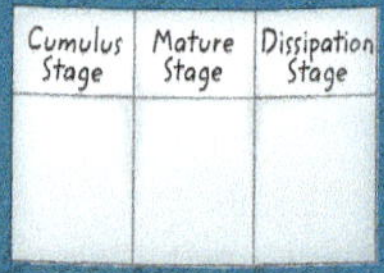

Hurricanes are given names to provide ease in communication between forecasters and the general public. Names for hurricanes, such as Hurricane Katrina, are selected from a list maintained by the World Meteorological Organization.

CHAPTER 13

LaunchLAB

iLab Station Rubric

Why does lightning form?

Process Skills model, infer

Additional Material permanent marker

Safety Precautions Approve lab safety forms before work begins. Students who are allergic to latex should not touch latex balloons.

Teaching Strategy Have students identify the independent variable in the lab. The independent variable is the treatment of the balloon; one side is rubbed against the fabric, while the other side is not.

Procedure

1. Have students read and complete the safety lab form, and follow the procedure below.
2. With a **paper punch,** create 10 **paper circles.**
3. Place the circles in two piles of 5 on your desk.
4. Blow up a small **balloon** and mark one side with an *X*.
5. Rub the *X* side of the balloon on some **fabric.**
6. Hold the *X* side of the balloon 2 cm above one pile of paper circles.
7. Turn the balloon over, opposite the *X*, and hold it 2 cm above the other pile of paper circles.

Analysis

1. **Describe** what happened to the paper circles. The circles were attracted to the side of the balloon with the *X*.
2. **Explain** what happened when you rubbed the balloon on the fabric. Electrons were stripped from the fabric, setting up charge differences between the balloon and the paper. The charge attractions caused the paper circles to cling to the balloon.
3. **Infer** how the static attracting the paper is similar to the static electricity you produced on a rug. The balloon becomes negatively charged because it has picked up electrons from the fabric. These electrons are attracted to the protons in the paper.
4. **Infer** what causes lightning to jump from spot to spot. Regions of air with opposite charges. Lightning moves to areas of electrical imbalance.

Assessment

Performance Have students design an experiment to test whether the amount of time the balloon is rubbed against the fabric affects the amount of electrical imbalance created. Have students carry out the experiment using the same materials they used in this lab.

LabManager™ Customize any lab with the LabManager™ CD-ROM.

Introduce the BIGIDEA

Severe Weather On any given day, a portion of the United States is likely experiencing some sort of inclement weather. Have students monitor weather reports for one week. Students should list different types of severe weather occurring around the country, as well as any weather-related injuries and property damage.

Teacher Content Support

Storms Hurricanes, like the one in the Gulf of Mexico shown in this photo, are one example of the potentially extreme nature of storms. They are capable of producing extreme winds, rainfall, and coastal flooding. They can produce as much energy in one day as the U.S. uses in almost six months.

SECTION 1

1 Focus

Focus Transparency

Before presenting the lesson, project **Section Focus Transparency 36** and have students answer the accompanying questions. **EL**

MAINIDEA

Thunderstorms Ask students to think about how a heat engine works. Fuel produces heat, which creates energy to do work. Thunderstorms are similar. The "fuel" is warm air and moisture, which creates energy that produces weather events such as rain, wind, and lightning. Like heat engines, the "fuel" that creates thunderstorms can come from different sources.

2 Teach

Teacher Content Support

The Study of Thunderstorms
Much of what we know about thunderstorms comes from information obtained during the U.S. Thunderstorm Project, a joint experiment that took place among the former U.S. Weather Bureau and several other agencies between 1946 and 1949. This project first defined a thunderstorm as a collection of convection cells characterized by vigorous columns of rising and descending air. The project also was responsible for identifying the different stages in the life cycle of a thunderstorm: the cumulus stage, the mature stage, and the dissipation stage.

SECTION 1

Thunderstorms

MAINIDEA The intensity and duration of thunderstorms depend on the local conditions that create them.

Essential Questions

- How do thunderstorms form?
- What are the different types of thunderstorms?
- What is the life cycle of a thunderstorm?

Review Vocabulary

latent heat: stored energy in water vapor that is not released to warm the atmosphere until condensation occurs

New Vocabulary

air-mass thunderstorm
mountain thunderstorm
sea-breeze thunderstorm
frontal thunderstorm
stepped leader
return stroke

EARTH SCIENCE 4 YOU Think about how an engine processes fuel to produce energy that powers an automobile. A thunderstorm is an atmospheric engine that uses heat and moisture as fuel and expends its energy in the form of clouds, rain, lightning, and wind.

Overview of Thunderstorms

At any given moment, nearly 2000 thunderstorms are in progress around the world. Most do little more than provide welcome relief on a muggy summer afternoon, or provide a spectacle of lightning. Some, however, grow into atmospheric monsters capable of producing hail the size of baseballs, swirling tornadoes, and surface winds of more than 160 km/h. These severe thunderstorms can also provide the energy for nature's most destructive storms—hurricanes. These severe thunderstorms, regardless of intensity, have certain characteristics in common. **Figure 1** shows which areas of the United States experience the most thunderstorms annually.

How thunderstorms form Recall that the stability of the air is determined by whether or not an air mass can lift. Cooling air masses are stable and those that are warmed from the land or water below them are not. Under the right conditions, convection can cause a cumulus cloud to grow into a cumulonimbus cloud. The conditions that produce cumulonimbus clouds are the same conditions that produce thunderstorms. For a thunderstorm to form, three conditions must exist: a source of moisture, lifting of the air mass, and an unstable atmosphere.

Average Number of Thunderstorm Days Annually

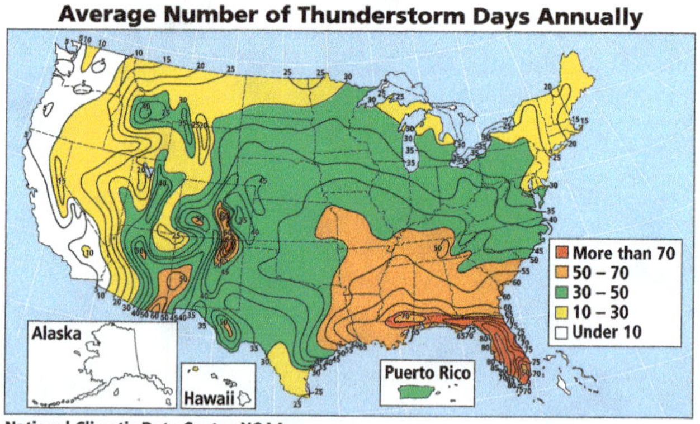

National Climatic Data Center, NOAA

■ **Figure 1** Both geography and air mass movements make thunderstorms most common in the southeastern United States.
Predict *why the Pacific Coast has so few thunderstorms and Florida has so many.*

■ **Caption Question Fig. 1** Air over the Pacific coast is cooler, influenced by the cold ocean current that runs up the coast. Air over the Floridian coast is warmer, influenced by the warm underlying land and Gulf Stream current.

■ **Figure 2** This cumulus cloud is growing as a result of unstable conditions. As the cloud continues to develop into a cumulonimbus cloud, a thunderstorm might develop.

Moisture First, for a thunderstorm to form, there must be an abundant source of moisture in the lower levels of the atmosphere. Air masses that form over tropical oceans or large lakes become more humid from water evaporating from the surface below. This humid air is less dense than the surrounding dry air and is lifted. The water vapor it contains condenses into droplets forming clouds. Latent heat, which is released from the water vapor during the process of condensation, warms the air causing it to rise further, cool further, and condense more of its water vapor.

Lifting Second, there must be some mechanism for condensing moisture to release its latent heat. This occurs when a warm air mass is lifted into a cooler region of the atmosphere. Dense, cold air along a cold front can push warmer air upward, just like an air mass does when moving up a mountainside. Warm land areas, heat islands such as cities, and bodies of water can also provide heat for lifting an air mass. Only when the water vapor condenses can it release latent heat and keep the cloud rising.

Stability Third, if the surrounding air remains cooler than the rising air mass, the unstable conditions can produce clouds that grow upward. This releases more latent heat and allows continued lifting. However, when the density of the rising air mass and the surrounding air are nearly the same, the cloud stops growing. **Figure 2** shows a cumulus cloud that is on its way to becoming a cumulonimbus cloud that can produce thunderstorms.

☑ READING CHECK **Describe** the three conditions for thunderstorm growth.

Limits to thunderstorm growth The conditions that limit thunderstorm growth are the same ones that form the storm. Conditions that create lift, condense water vapor, and release latent heat keep the air mass warmer than the surrounding air. The air mass will continue to rise until it reaches a layer of equal density that it cannot overcome. Because atmospheric stability increases with height, most cumulonimbus clouds are limited to about 12,000 m. Thunderstorms are also limited by duration and size.

Interpret the Illustration

Location of Thunderstorms

Have students study **Figure 1.** Have them identify the areas on the map where thunderstorms are most likely to occur. Most storms occur near the South Atlantic and Gulf coasts and the southern states.

Reinforcement

Rising Air Remind students that air will continue to rise only if it is warmer than the surrounding air. The atmosphere cools with height, which severely limits how high air can rise before it cools to a temperature below that of the surrounding air. This is why the release of latent heat and an abundant supply of moisture are so crucial to rising air. As more moisture is condensed, more latent heat is released, which allows the air to remain warm and continue to rise for a longer period of time.

☑ READING CHECK abundant moisture in the lower atmosphere, a mechanism for lifting the air, and the rising air mass must remain warmer than the surrounding air

DIFFERENTIATED INSTRUCTION

Advanced Learners Thunderstorms occur most often near moist air masses. Ask students to explain why more thunderstorms don't occur along the Pacific Coast. The California Current along the Pacific Coast brings cold water from British Columbia down the coastline. This cold water does not contain enough thermal energy to induce thunderstorm genesis.

Tie to Previous Knowledge

Cold Fronts Students have learned that cold fronts move forward and spread cold air under warmer, less-dense air, thereby forcing the warm air to rise. This process, which produces thunderstorms, is similar to the formation of storms by sea breezes. As the cooler air moves inland, it displaces the warmer air, thereby producing clouds and usually precipitation.

■ **Caption Question Fig. 3** Land heats up quickly during the day, while the water heats up slowly. At night, the land gives up its heat quickly and the water gives up heat slowly.

Concept Development

Convection Cells A key factor in the development of clouds and thunderstorms is the environment surrounding the convection cell. If the environment is unstable, the cell will likely continue to grow. If the environment is stable, the cell will have a short lifetime. In unstable conditions, rising air continues to encounter cooler air, and the rising air therefore remains buoyant. In stable conditions, rising air quickly encounters warm air, and the rising air is unable to rise any farther.

■ **Figure 3** Temperature differences exist over land and water and vary with the time of day.
Infer *why water is warmer than the land at night.*

During the day, the temperature of land increases faster than the temperature of water. The warm air over land expands and rises, and the colder air over the sea moves inland and replaces the warm air. These conditions can produce strong updrafts that result in thunderstorms.

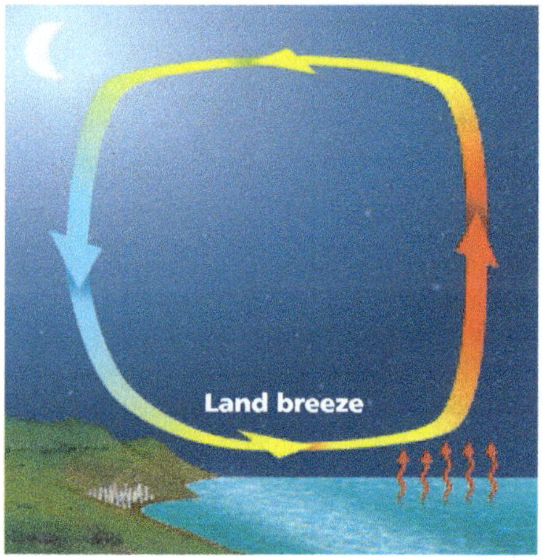

At night, conditions are reversed. The land cools faster than water, so the warmer sea air rises, and cooler air from above land moves over the water and replaces it. Nighttime conditions are considered stable.

Types of Thunderstorms

Thunderstorms are often classified according to the mechanism that causes the air mass that formed them to rise. There are two main types of thunderstorms: air-mass and frontal.

Air-mass thunderstorms When air rises because of unequal heating of Earth's surface within one air mass, the thunderstorm is called an **air-mass thunderstorm.** The unequal heating of Earth's surface reaches its maximum during mid-afternoon, so it is common for air-mass thunderstorms, also called pop-up storms, to occur.

There are two kinds of air-mass thunderstorms. **Mountain thunderstorms** occur when an air mass rises by orographic lifting, which involves air moving up the side of a mountain. **Sea-breeze thunderstorms** are local air-mass thunderstorms that occur because land and water store and release thermal energy differently. Sea-breeze thunderstorms are common along coastal areas during the summer, especially in the tropics and subtropics. Because land heats and cools faster than water, temperature differences can develop between the air over coastal land and the air over water, as shown in **Figure 3.**

Frontal thunderstorms The second main type is **frontal thunderstorms,** which are produced by advancing cold fronts and, more rarely, warm fronts. In a cold front, dense, cold air pushes under warm air, which is less dense, rapidly lifting it up a steep cold-front boundary. This rapid upward motion can produce a thin line of thunderstorms, sometimes hundreds of kilometers long, along the leading edge of the cold front. Cold-front thunderstorms get their initial lift from the push of the cold air. Because they are not dependent on daytime heating for their initial lift, cold-front thunderstorms can persist long into the night. Flooding from soil saturation is common with these storms. Floods are the main cause of thunderstorm-related deaths in the United States each year.

Less frequently, thunderstorms can develop along the advancing edge of a warm front. In a warm-front storm, a warm air mass slides up and over a gently sloping cold air mass. If the warm air behind the warm front is unstable and moisture levels are sufficiently high, a relatively mild thunderstorm can develop.

EARTH SCIENCE JOURNAL

Dangers of Floods Explain to students that flood-related deaths often occur when people drive their automobiles into moving water. Have students research and explain why driving a car through swiftly moving water is dangerous. Have students include safety tips explaining what to do when floodwaters are encountered.

Rubric

Thunderstorm Development

A thunderstorm usually has three stages: the cumulus stage, the mature stage, and the dissipation stage. The stages are classified according to the direction the air is moving.

Cumulus stage In the cumulus stage, air starts to rise vertically, as shown in **Figure 4.** The updrafts are relatively localized and cover an area of about 5–8 km. This creates updrafts, which transport water vapor to the cooler, upper regions of the cloud. The water vapor condenses into visible cloud droplets and releases latent heat. As the cloud droplets coalesce, they become larger and heavier until the updrafts can no longer sustain them and they fall to Earth as precipitation. This begins the mature stage of a thunderstorm.

Mature stage In the mature stage, updrafts and downdrafts exist side by side in the cumulonimbus cloud. Precipitation, composed of water and ice droplets that formed at high, cool levels of the atmosphere, cools the air as it falls. The newly cooled air is more dense than the surrounding air, so it sinks rapidly to the ground along with the precipitation. This creates downdrafts. As **Figure 4** shows, the updrafts and downdrafts form a convection cell which produces the surface winds associated with thunderstorms. The average area covered by a thunderstorm in its mature stage is 24 km.

Dissipation stage The convection cell can exist only if there is a steady supply of warm, moist air at Earth's surface. Once that supply is depleted, the updrafts slow down and eventually stop. In a thunderstorm, the cool downdrafts spread in all directions when they reach Earth's surface. This cools the areas from which the storm draws its energy, the updrafts cease, and clouds can no longer form. The storm is then in the dissipation stage shown in **Figure 4.** This stage will last until all of the previously formed precipitation has fallen.

FOLDABLES® Incorporate information from this section into your Foldable.

View an **animation of thunderstorm development.**

Concepts In Motion

Figure 4 The cumulus stage of a thunderstorm is characterized mainly by updrafts. The mature stage is characterized by strong updrafts and downdrafts. The storm loses energy in the dissipation stage.

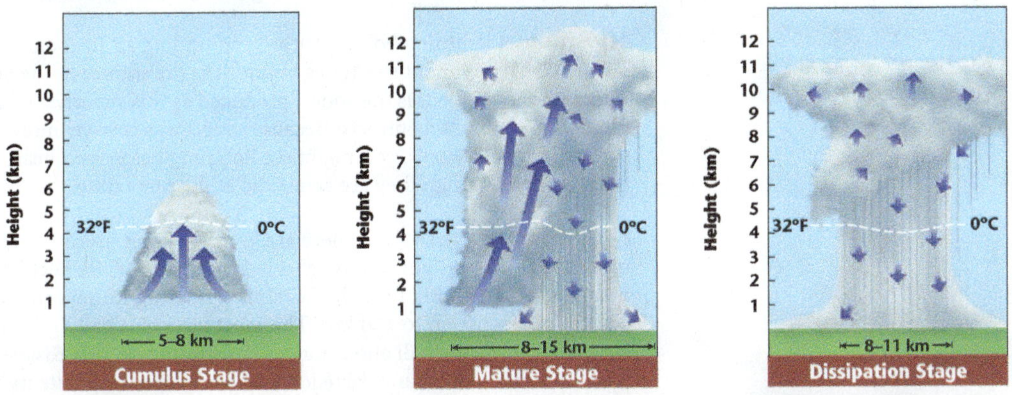

Discussion

Updrafts Ask students what happens when the updrafts in a thunderstorm cease. Make sure students understand that the storm has entered the dissipation stage, but it is not yet dead and can continue to produce rain for several hours. Tell students it takes time for the upward momentum within the storm cell to fade completely, and thus the coalescence of already-formed water droplets can take some time to cease. Even when the lower levels of the storm no longer produce updrafts, the remaining middle levels can still produce light rainfall.

FOLDABLES® Rubric

ACROSS THE CURRICULUM

Chemistry In many ways, the life cycle of a thunderstorm is similar to a chemical reaction. When chemicals mix, a reaction begins, much like the cumulus stage of a thunderstorm when heat and moisture form a convection cell. Just as a chemical reaction has a point of peak efficiency, a thunderstorm intensifies at the point at which it most efficiently utilizes heat and moisture. Just as a chemical reaction eventually slows, a thunderstorm dissipates when it runs out of heat or moisture.

Project

Types of Lightning There are many types of lightning, including cloud-to-ground, cloud-to-cloud, sheet, and ball lightning. Have students research the different types of lightning, and then make presentations based on their research. Students should include labeled photos of different types of lightning. OL

Collaborative Learning

Origins of Lightning Have students debate whether lightning starts from the cloud and moves down or from the ground and moves up. Technically, both answers are correct because a stepped leader and a return stroke are both required to discharge the lightning bolt. The return stroke from the ground, however, causes the illumination and discharge called lightning. COOP LEARN

Caption Question Fig. 5 Outlines will vary but should include the following basic steps: clouds become charged, a stepped leader forms, and a return stroke surges from the ground to meet the stepped leader.

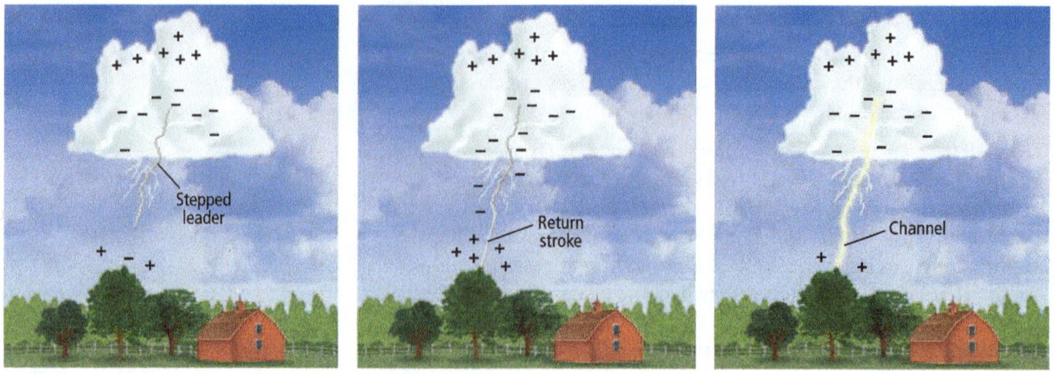

Figure 5 When a stepped leader nears an object on the ground, a powerful surge of electricity from the ground moves upward to the cloud and lightning is produced.
Sequence ***Make an outline sequencing the steps of lightning formation.***

Lightning

Have you ever touched a metal object on a dry winter day and been zapped by a spark from static electricity? The static electricity was generated from friction, and the spark is similar to lightning. Lightning is the transfer of electricity generated by the rapid rushes of air in a cumulonimbus cloud. Clouds become charged when friction between the updrafts and downdrafts removes electrons from some of the atoms in the cloud. The atoms that lose electrons become positively charged ions. Other atoms receive the extra electrons and become negatively charged ions. As **Figure 5** shows, this creates regions of air with opposite charges. Eventually, the differences in charges break down, and a branched channel of partially charged air is formed between the positive and negative regions. The channel of partially charged air is called a **stepped leader,** and it generally moves from the center of the cloud toward the ground. When the stepped leader nears the ground, a branched channel of positively charged particles, called the **return stroke,** rushes upward to meet it. The return stroke surges from the ground to the cloud, illuminating the connecting channel with about 100 million volts of electricity. That illumination is the brightest part of lightning.

Thunder A lightning bolt heats the surrounding air to about 30,000°C. That is about five times hotter than the surface of the Sun. The thunder you hear is the sound produced as this superheated air rapidly expands and contracts. Because sound waves travel more slowly than light waves, you might see lightning before you hear thunder, even though they are generated at the same time.

Lightning variations There are several names given to lightning effects. Sheet lightning is reflected by clouds, while heat lightning is sheet lightning near the horizon. Spider lightning can crawl across the sky for up to 150 km. The most bizarre is ball lightning which is a hovering ball about the size of a pumpkin that disappears in a fizzle or a bang. Blue jets and red sprites originate in clouds and rise rapidly toward the stratosphere as cones or bursts.

DIFFERENTIATED INSTRUCTION

Advanced Learners The National Weather Service issues severe thunderstorm warnings for high winds and hail but not for lightning, which kills more people. Ask students to infer why. Warnings for lightning would have to be issued for every thunderstorm that develops because a thunderstorm, by definition, exhibits lightning.

■ **Figure 6** Five times hotter than the surface of the Sun, a lightning bolt can be spectacular. But when an object such as this pine tree is struck, it can be explosive.

Thunderstorm and lightning safety Each year in the United States, lightning causes about 12,000 wildfires, which result in the burning of thousands of square kilometers of land. In addition, lightning strikes in the United States cause a yearly average of 300 injuries and 58 deaths to humans. **Figure 6** indicates how destructive a lightning strike might be.

Avoid putting yourself in danger of being struck by lightning. If you are outdoors and feel your hair stand on end, squat low on the balls of your feet. Duck your head and make yourself the smallest target possible. Small sheds, isolated trees, and convertible automobiles are hazardous as shelters. Be aware that using electrical appliances and telephones during a lightning storm can lead to electric shock. Stay out of boats and away from water during a thunderstorm.

SECTION 1 REVIEW

Section Self-Check

Section Summary

- The cumulus stage, the mature stage, and the dissipation stage comprise the life cycle of a thunderstorm.
- Clouds form as water is condensed and latent heat is released.
- Thunderstorms can be produced either within air masses or along fronts.
- From formation to dissipation, all thunderstorms go through the same stages.
- Lightning is a natural result of thunderstorm development.

Understand Main Ideas

1. **MAIN**IDEA **List** the conditions needed for a thunderstorm's cumulus stage.
2. **Explain** how a thunderstorm is formed along a front.
3. **Differentiate** between a sea-breeze thunderstorm and a mountain thunderstorm.
4. **Identify** what causes a thunderstorm to dissipate.
5. **Compare and contrast** how a cold front and a warm front can create thunderstorms.
6. **Describe** two different types of lightning.

Think Critically

7. **Infer** Lightning occurs mainly during which stage of thunderstorm formation?
8. **Determine** the conditions in thunderstorm formation that creates lightning.

WRITING IN ▶ Earth Science

9. Write a setting for a movie using a storm as part of the opening scene.

3 Assess

Check for Understanding

Compare and Contrast Ask students to compare and contrast different types of thunderstorms.

Reteach

Think Critically Have students divide a map of the United States into several regions. Then, have them develop a thunderstorm profile for each region. The profiles should include the relative frequency of thunderstorms, the probable season of occurrence, and the probable types of thunderstorms that occur in each particular region.

Assessment

Knowledge Have students outline the three stages in the development of a thunderstorm.

SECTION 1 REVIEW

1. source of moisture, lifting of the air mass, unstable atmosphere
2. Cold air pushes under warm air, rapidly lifting it up the steep cold-front boundary. This rapid upward motion can produce a line of thunderstorms hundreds of kilometers long.
3. Mountain thunderstorms occur as moist air is lifted up the side of a mountain, while sea-breeze storms are caused by large temperature differences between land and sea air masses.
4. The convection cell can exist only if there is a steady, updraft supply of warm, moist air at Earth's surface. Once the production of downdrafts starts, the updrafts slow and eventually stop.
5. A cold air mass pushes a warm air mass up rapidly, setting up thunderstorms along a cold front. In a warm front, the warm air is lifted slowly and a milder form of thunderstorm can be triggered.
6. Heat lightning is distant sheet lightning. Spider lightning can crawl across the sky for up to 150 km. Ball lightning is a ball that hovers over the ground until it pops out of sight.
7. mature stage
8. Anything that causes air to gain or lose electrons, such as updraft friction or wind friction, sets up the conditions for lightning to occur.
9. Answers will vary, but should be creative and scientifically accurate.

Rubric

1 Focus

Focus Transparency

Before presenting the lesson, project **Section Focus Transparency 37** and have students answer the accompanying questions. **EL**

MAINIDEA

Storm Damage Have students consider the effects of a common thunderstorm. Then show them pictures of damage caused by flooding, high winds, or tornadoes. Remind them these can also be caused by thunderstorms, pointing out how different the effects can be.

2 Teach

Teacher Content Support

Supercells Most thunderstorms are characterized by heavy rain, thunder, lightning, and gusty winds. Under the right conditions, these storms can grow into raging tempests known as supercells. The transformation takes place as a result of various mechanisms that act to enhance the convective currents that power a storm. These currents can be strengthened by the addition of heat to the lower portions of the storm or by the cooling of the upper layers—any mechanism, in fact, that increases the vertical temperature differences in the storm's environment and therefore decreases the stability of the air that fuels the storm.

Severe Weather

MAINIDEA All thunderstorms produce wind, rain, and lightning, which can have dangerous and damaging effects under certain circumstances.

Essential Questions
- Why are some thunderstorms more severe than others?
- What are the dangers of severe weather?
- How do tornadoes form?

Review Vocabulary

air mass: large body of air that takes on the characteristics of the area over which it forms

New Vocabulary

supercell
downburst
tornado
Enhanced Fujita Tornado Damage scale

EARTH SCIENCE 4 YOU **Sliding down a park slide might seem mild and safe compared to a roller coaster's wild and chaotic ride. Similarly, while a gentle rain is appreciated by many, the same weather processes can create thunderstorms on a massive atmospheric scale, resulting in disaster.**

Severe Thunderstorms

All thunderstorms are not created equal. Some die out within minutes, while others flash and thunder throughout the night. What makes one thunderstorm more severe than another? The increasing instability of the air intensifies the strength of a storm's updrafts and downdrafts, which makes the storm severe.

Supercells Severe thunderstorms can produce some of the most violent weather conditions on Earth. They can develop into self-sustaining, extremely powerful storms called **supercells.** Supercells are characterized by intense, rotating updrafts taking 10 to 20 minutes to reach the top of the cloud. These furious storms can last for several hours and can have updrafts as strong as 240 km/h. It is not uncommon for a supercell to spawn long-lived tornadoes. **Figure 7** shows an illustration of a supercell. Notice the anvil-shaped cumulonimbus clouds associated with these severe storms. The tops of the supercells are chopped off by wind shear. Of the estimated 100,000 thunderstorms that occur each year in the United States, only about 10 percent are considered to be severe, and fewer still reach supercell proportions.

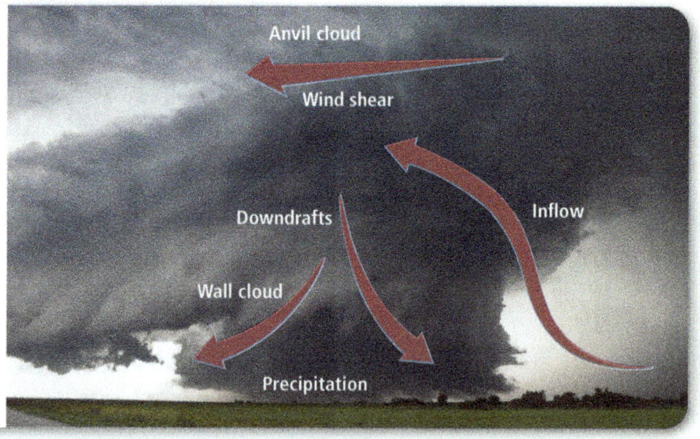

Figure 7 An anvil-shaped cumulonimbus cloud is characteristic of many severe thunderstorms. The most severe thunderstorms are called supercells.

Demonstration

Downbursts Ask a volunteer to hold his or her arms straight out at chest level. Carefully load books onto the student's arms. **WARNING:** ***The student should not be wearing open-toed sandals.*** Eventually, the load of books will become so heavy that the student will drop them. Relate this to the development of a downburst. Updrafts continually pump air upward into a storm, until eventually, the mass of air becomes too great for the updrafts to keep aloft. At that point, the entire mass of air is released at once, and it plummets to Earth as a downdraft.

Strong Winds

Recall that rain-cooled downdrafts descend to Earth's surface during a thunderstorm and spread out as they reach the ground. Sometimes, instead of dispersing that downward energy over a large area underneath the storm, the energy becomes concentrated in a local area. The resulting winds are exceptionally strong, with speeds of more than 160 km/h. Violent downdrafts that are concentrated in a local area are called **downbursts.**

Based on the size of the area they affect, downbursts are classified as either macrobursts or microbursts. Macrobursts can cause a path of destruction up to 5 km wide. They have wind speeds of more than 200 km/h and can last up to 30 minutes. Smaller in size, though deadlier in force, microbursts affect areas of less than 4 km but can have winds exceeding 250 km/h. Despite lasting fewer than 10 minutes on average, a microburst is especially deadly because its small size makes it extremely difficult to predict and detect. **Figure 8** shows a microburst.

Figure 8 A microburst, such as this one in Colorado, can be as destructive as a tornado.

Hail

Each year in the United States, almost one billion dollars in damage is caused by hail—precipitation in the form of balls or lumps of ice. Hail can do tremendous damage to crops, vehicles, and rooftops, particularly in the central United States where hail occurs most frequently. Hail is most common during the spring growing season. **Figure 9** shows some conditions associated with hail.

Hail forms because of two characteristics common to thunderstorms. First, water droplets enter the parts of a cumulonimbus cloud where the temperature is below freezing. When these supercooled water droplets encounter ice pellets, the water droplets freeze on contact and cause the ice pellets to grow larger. The second characteristic that allows hail to form is an abundance of strong updrafts and downdrafts existing side by side within a cloud. The growing ice pellets are caught alternately in the updrafts and downdrafts, so that they constantly encounter more supercooled water droplets. The ice pellets keep growing until they are too heavy for even the strongest updrafts to keep aloft, and they finally fall to Earth as hail.

Figure 9 This hail storm in Sydney, Australia, caused slippery conditions for the traffic as well as damage to property.

Enrichment

Wind Damage While wind damage is frequently attributed to tornadoes, downbursts are often the real cause of damage. Remind students that downbursts produce straight-line winds at ground level, and that tornadoes rotate. Then, ask students to propose ways to distinguish downburst damage from tornado damage. Wind damage caused by a downburst is usually in a straight line from a center point, while wind damage caused by a tornado comes from several directions over a small area.

Teacher Content Support

Development of a Supercell The supercell concept was proposed by the late Dr. Theodore Fujita. His studies revealed that, in extreme convection cells, updrafts lift enough air to lower surface pressure and create a so-called meso-low at Earth's surface. This low, being cyclonic, acquires a counterclockwise rotation in the northern hemisphere; thus, the thunderstorm itself exhibits rotation in the updrafts. The resulting violent updrafts can produce severe weather because their ability to hold large ice pellets aloft facilitates the formation of hail, and also because they are a crucial component in the development of tornadoes.

IN THE FIELD

Weather Expert Sulochana Gadgil is an atmospheric scientist from India best known for her studies on monsoons, seasonal winds that bring severe storms to Asia. She has investigated the climatic changes, the effects on Asia's agriculture, and the rainfall patterns associated with the monsoon winds. Gadgil earned her PhD in mathematics from Harvard University, but later changed her area of study to meteorology. In 1971, she returned to her birthplace to teach at the Indian Institute of Science in Bangalore. Gadgil has also served as chair of the Center for Atmospheric Sciences at the Indian Institute since 1989. She is the author of more than 40 scientific articles on climatology, atmospheric and oceanic circulation, and monsoon research.

Interpret the Illustration

Wind Shear Have students examine **Figure 10.** Explain that wind shear is a change in wind direction or speed that can occur either vertically or horizontally. In regard to the formation of a tornado, the vertical change in direction is most important because it leads to the development of the horizontal tube, or vortex. As this vortex is incorporated into the parent thunderstorm's strong updrafts, it tilts in a vertical direction and creates a funnel-shaped rotation that can develop into a tornado.

Use Science Terms

Funnel Clouds Sometimes, the developing stages of a tornado take place well above the ground, and the funnel never reaches the ground. When this happens, the swirling vortex is called a funnel cloud. A funnel cloud becomes a tornado if it reaches the ground. Tell students a tornado can usually be identified by the presence of swirling dust or debris on the ground at the end of the funnel. If such a swirl is not visible, the vortex is most likely a funnel cloud rather than a tornado.

☑ READING CHECK A tornado is a violent, whirling column of air in contact with the ground.

▪ **Caption Question Fig. 10** Possible answer: Updrafts are created by warm air. As a cold front approaches a warm air mass, the warm air is pushed up as an updraft. The greater the temperature difference, the faster the air moves.

Watch a **video about tornadoes.**

Video

VOCABULARY

ACADEMIC VOCABULARY

Phenomenon
an object or aspect known through the senses rather than by thought or intuition
Students observing the phenomenon realized later that the powerful wind was a microburst.

Tornadoes

In some parts of the world, the most dangerous form of severe weather is the tornado. A **tornado** is a violent, whirling column of air in contact with the ground. When a tornado does not reach the ground, it is called a funnel cloud. Tornadoes are often associated with supercells—the most severe thunderstorms. The air in a tornado is made visible by dust and debris drawn into the swirling column, sometimes called the vortex, or by the condensation of water vapor into a visible cloud.

☑ READING CHECK **Define** the term *tornado.*

Development of tornadoes A tornado forms when wind speed and direction change suddenly with height, a phenomenon called wind shear. Current thinking suggests that tornadoes form when pockets of rising air are rotated horizontally like a rolling-pin between layers of air with varying wind speeds or directions, as shown in **Figure 10.** If this rotation occurs close enough to the thunderstorm's updrafts, the twisting column of wind can be tilted from a horizontal to a vertical position. As updrafts stretch the column, the rotation is accelerated. Air is removed from the center of the column, which in turn lowers the air pressure in the center. The extreme pressure gradient between the center and the outer portion of the tornado produces the violent winds associated with tornadoes. Although tornadoes rarely exceed 200 m in diameter and usually last only a few minutes, they can be extremely destructive. A tornado is classified according to its destructive force.

View an **animation of tornado formation.**

Concepts In Motion

▪ **Figure 10** Tornado formation is associated with changes in wind speed and direction.
Infer *what would cause the updrafts.*

A change in wind direction and speed creates a horizontal rotation in the lower atmosphere.

Strong updrafts tilt the rotating air from a horizontal to a vertical position.

A tornado forms within the rotating winds.

What's EARTH SCIENCE Got To Do With It?

Video *Storm Chasers*

ACROSS THE CURRICULUM

Geography While tornadoes are common in the United States, some parts of the country experience few of these storms. For example, tornadoes virtually never strike Alaska or Hawaii, and rarely strike the Pacific Northwest. Ask students to infer why. Conditions are not right to generate the updrafts needed for tornadoes to develop from thunderstorms.

Explore the **Enhanced Fujita Tornado Damage scale with an interactive table.** Concepts In Motion

Table 1 Enhanced Fujita Tornado Damage Scale

Enhanced Fujita scale tornadoes	Weak (EF0 and EF1) 69 percent of all tornadoes Wind speed: 105–177 km/h	Strong (EF2 and EF3) 29 percent of all tornadoes Wind speed: 178–266 km/h	Violent (EF4 and EF5) 2 percent of all tornadoes Wind speed: 267–322+ km/h
Photo of tornado			

Tornado classification Tornadoes vary greatly in size and intensity. The **Enhanced Fujita Tornado Damage scale,** which ranks tornadoes according to their destruction and estimated wind speed, is used to classify tornadoes. The Enhanced Fujita scale is an update to the original Fujita scale that was named for Japanese tornado researcher Dr. Theodore Fujita in the 1970s. The Enhanced Fujita scale ranges from EF0, which is characterized by winds of up to 137 km/h, to the incredibly violent EF5, which can pack winds of more than 322 km/h. Most tornadoes do not exceed the EF1 category. In fact, only about 2 percent reach EF4 or EF5. Those that do, however, can lift entire buildings from their foundations and toss automobiles and trucks around like toys. The Enhanced Fujita Tornado Damage scale is shown in **Table 1.**

Tornado distribution While tornadoes can occur at any time and at any place, there are some times and locations where they are more likely to form. Most tornadoes—especially violent ones—form in the spring during the late afternoon and evening, when the temperature contrasts between polar air and tropical air are the greatest. Large temperature contrasts occur most frequently in the central United States, where cold continental polar air collides with maritime tropical air moving northward from the Gulf of Mexico. These large temperature contrasts often spark the development of supercells, which are each capable of producing several strong tornadoes. More than 1000 tornadoes touch down each year in the United States. Many of these occur in a region called "Tornado Alley," which extends from northern Texas through Oklahoma, Kansas, and Missouri.

Discussion

Tornado Strength Tornadoes have been the subject of several movies. One such movie depicted a couple surviving an EF5 tornado by tying themselves to large irrigation pipes in the middle of a field, while the buildings around them were destroyed. Ask students to discuss whether this is a realistic depiction of what might happen during an EF5 tornado. This is a far-fetched scenario. Violent winds would toss debris from the buildings and surrounding area that could easily strike and kill the couple. Also, the couple could have been sandblasted to death by dirt driven at them at speeds in excess of 322 km/h.

Teacher Content Support

Tornado Warnings While EF4 and EF5 tornadoes account for only a small percentage of the total number of tornadoes, they cause most deaths and tornado-related property damage. More than 90 percent of all tornadoes are classified as EF0 or EF1, but they account for only a small percentage of total tornado damage. It is hard to issue advance warnings for weak tornadoes because they are small and brief, and therefore are not easily detected by weather radar. They might not be detected in time for advance warnings to be of use. In contrast, the conditions that are associated with violent tornadoes are usually easily detected by weather radar, and warnings can usually be issued in time for people to take protective actions.

DIFFERENTIATED INSTRUCTION

Struggling Learners To help students become comfortable with the Enhanced Fujita Tornado Damage Scale, have them use the information on the scale in **Table 1** to classify the following examples of two tornadoes: In 2004, a tornado occurred in Victoria, Australia, with wind speeds between 115–150 km/h. What class of tornado was this? EF1 In 2011, a tornado occurred in Joplin, Missouri, with wind speeds estimated at 320 km/h. What class of tornado was this? EF5

3 Assess

Check for Understanding

Describe Have students describe the development of a tornado, beginning with the formation of the parent thunderstorm.

Reteach

Summarize Have students each make a data table showing the hazards associated with thunderstorms. Student tables should include the types of hazards, their relative severity, and the basic processes that cause them to develop.

Assessment

Performance Have students each research a well-known tornado, such as the one that occurred in Xenia, Ohio, in 1974. Students should develop a multimedia presentation that includes data on the damage path of the tornado; statistics on deaths, injuries, and property damage; and suggestions about what could have been done to mitigate the damage.

■ **Figure 11** In some areas, tornado shelters are common. If you are caught in a tornado, take shelter in the southwest corner of a basement, a small downstairs room or closet, or a tornado shelter like this one.

Tornado safety In the United States, an average of 80 deaths and 1500 injuries result from tornadoes each year. In an ongoing effort to reduce tornado-related fatalities, the National Weather Service issues tornado watches and warnings before a tornado strikes. These advisories are broadcast on radio and television stations when conditions are conducive to the development of tornadoes, or when tornadoes are indicated on weather radar or spotted in the region. During a severe thunderstorm, the presence of dark, greenish skies, a towering wall of clouds, large hailstones, and a loud, roaring noise similar to that of a freight train are signs of an approaching or developing tornado.

The National Weather Service stresses that despite advanced tracking systems, some tornadoes develop very quickly. In these cases, advance warnings might not be possible. However, the threat of tornado-related injury can be substantially decreased when people seek shelter, such as the one shown in **Figure 11,** at the first sign of threatening skies.

SECTION 2 REVIEW

Section Self-Check

Section Summary

- Intense rotating updrafts are associated with supercells.
- Downbursts are strong winds that result in damage associated with thunderstorms.
- Hail is precipitation in the form of balls or lumps of ice that accompany severe storms.
- The worst storm damage comes from a vortex of high winds that moves along the ground as a tornado.

Understand Main Ideas

1. MAINIDEA **Identify** the characteristics of a severe storm.
2. **Describe** two characteristics of thunderstorms that lead to hail formation.
3. **Explain** how some hail can become baseball-sized.
4. **Compare and contrast** a macroburst and a microburst.
5. **Identify** the steps that change wind shear into a tornado.
6. **Identify** the conditions that lead to high winds, hail, and lightning.

Think Critically

7. **Explain** Why are there more tornado-producing storms in flat plains than in mountainous areas?
8. **Analyze** the data of the Enhanced Fujita Tornado Damage scale, and determine why EF5 tornadoes often have a longer path than EF1 tornadoes.

WRITING IN ▶ Earth Science

9. Design a pamphlet about tornado safety.

SECTION 2 REVIEW

1. Severe storms occur when there is a continuous supply of surface moisture, upper- level low-pressure systems causing unstable air, and large temperature differences between air masses.
2. Water droplets exist in the liquid state in the parts of a cumulonimbus cloud where the temperature is below freezing and an abundance of strong updrafts and downdrafts exist side by side.
3. Water droplets in a downdraft can freeze as they fall through cold air. They then can be lifted by entering an updraft. When they start to fall again, more water can condense and freeze on the outside. Repeating this process many times can create baseball-sized hail.
4. Macrobursts can cause a path of destruction up to 5 km wide, have wind speeds of more than 200 km/h, and can last up to 30 min. Microbursts affect areas of less than 4 km but can have winds exceeding 250 km/h.
5. Wind shear causes the air to roll over the surface of the ground. If one end happens to pass over a warm place, that end can be lifted and turn the rolling air mass into a tornado.
6. Increasing instability of the air intensifies the strength of a storm's updrafts and downdrafts.
7. Although the big differences in temperatures between air masses exist in both areas, there also must be open areas for tornados to form. Mountainous areas can break up tornado formation.
8. Tornado paths are determined by several factors including terrain, temperature of the surface, and obstructions. EF5 tornados have a lot of energy to release before they dissipate, whereas EF1 tornadoes break up quickly.
9. Answers will vary, but should mention taking shelter in a basement, small room, closet, or tornado shelter.

Rubric

SECTION 3

Essential Questions

- How do tropical cyclones form?
- What is the life cycle of a tropical cyclone?
- What are the dangers associated with hurricanes?

Review Vocabulary

Coriolis effect: caused by Earth's rotation, moving particles, such as air, are deflected to the right north of the equator, and to the left south of the equator

New Vocabulary

tropical cyclone
eye
eyewall
Saffir-Simpson Hurricane Wind scale
storm surge

View an **animation of tropical cyclones.**

Concepts In Motion

Tropical Storms

MAINIDEA Normally peaceful, tropical oceans are capable of producing one of Earth's most violent weather systems—the tropical cyclone.

EARTH SCIENCE 4 YOU **If you try mixing cake batter in a shallow bowl, you might find that a low speed works well, but a high speed creates a big mess. Tropical cyclones form from processes similar to other storm systems, but their high winds can bring devastation to locations in their path.**

Overview of Tropical Cyclones

During the summer and fall, the tropics experience conditions ideal for the formation of large, rotating, low-pressure tropical storms called **tropical cyclones.** In different parts of the world, the largest of these storms are known as hurricanes, typhoons, and cyclones.

Cyclone location Favorable conditions for tropical cyclone formation exist in all tropical oceans except the South Atlantic Ocean and the Pacific Ocean off the west coast of South America. The water in these areas is somewhat cooler and these areas contain regions of nearly permanently stable air. As a consequence, tropical cyclones do not normally occur in these areas. They do occur in the large expanse of warm waters in the western Pacific Ocean where they are known as typhoons. To people living near the Indian Ocean, they are known as cyclones. In the North Atlantic Ocean, the Caribbean Sea, the Gulf of Mexico, and along the western coast of Mexico, the strongest of these storms are called hurricanes. **Figure 12** shows where tropical cyclones generally form.

Figure 12 Tropical cyclones are common in all of Earth's tropical oceans except in the relatively cool waters of both the South Pacific and South Atlantic oceans.

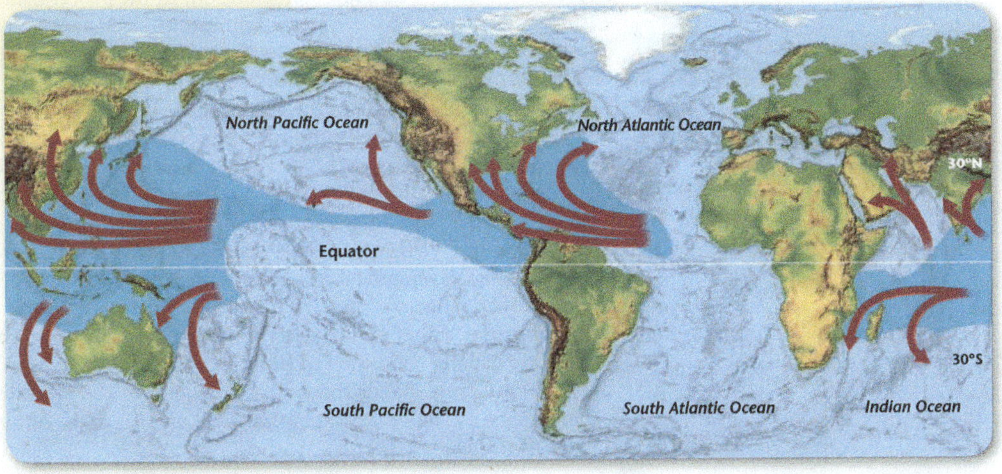

Demonstration

Model a Tropical Cyclone Using a pencil and a piece of paper, trace the edge of an oval platter in a clockwise direction, starting with a long, curved edge closest to you, to model how a tropical cyclone moves in the northern hemisphere. Tell students that the platter represents the Bermuda High, and the traced line represents the path of a tropical cyclone. The Bermuda High steers storms around it, much like the platter steered the pencil.

SECTION 3

1 Focus

Focus Transparency

Before presenting the lesson, project **Section Focus Transparency 38** and have students answer the accompanying questions. EL

MAINIDEA

Tropical Cyclones Ask students to picture themselves on the beach on a tropical island. Explain that the same warm ocean and air temperatures they would enjoy also represent a tremendous source of heat energy. Point out that under the right conditions, this energy can be harnessed and transformed into one of nature's largest and deadliest natural phenomena—the tropical cyclone, or hurricane.

2 Teach

Teacher Content Support

Hurricanes Why are hurricanes more common in some years than others? One reason is the amount of wind shear present in the tropics. When wind shear in the tropics is large—that is, there is a large change in wind direction or speed with height—the small tropical disturbances that can potentially grow into hurricanes become tilted with height and cannot develop into a vertical column through which condensation—and the heat it releases—can be sustained. This prohibits further growth of the disturbance.

Interpret the Photo

Eye of the Storm Have students study **Figure 13.** Seen from above, a tropical cyclone is easily identified by its distinct features. Explain that a mature tropical cyclone has an eye, which is the calm—and at times, nearly clear—circular center of the storm. Also, a tropical cyclone does not exhibit a long front of clouds, as do other storms. This is because a tropical cyclone derives its energy from warm ocean waters, not from the energy generated by clashing air masses.

Teacher Content Support

Storm Safety People often seek to escape the harsh winters of northern climates by settling along the coasts in warmer latitudes. Unfortunately, these areas are frequently hit by violent tropical cyclones. The increasing population along coastal areas presents a problem for emergency planners. The evacuation of large groups of people can be a logistical nightmare. Safety measures have become increasingly dependent on a greater understanding of the nature of tropical cyclones and the dangers associated with them.

Project

El Niño Studies have shown that when the phenomenon known as El Niño occurs in the Pacific Ocean, there are fewer hurricanes than normal in the Atlantic, Caribbean, and Gulf of Mexico basins. Have students research the effects of El Niño and report on why El Niño causes fewer tropical cyclones to occur in these basins. The presence of El Niño increases wind shear across these basins, and thus prevents small tropical disturbances from developing into hurricanes. OL

Figure 13 The characteristic rotating nature of cyclonic storms is evident in this hurricane that formed over the Atlantic Ocean.

Cyclone formation Tropical cyclones require two basic conditions to form: an abundant supply of warm ocean water and some sort of mechanism to lift warm air and keep it rising. Tropical cyclones thrive on the tremendous amount of energy in warm, tropical oceans. As water evaporates from the ocean surface, latent heat is stored in water vapor. This latent heat is later released when the air rises and water vapor condenses.

The air usually rises because of some sort of existing weather disturbance moving across the tropics. Many disturbances originate along the equator. Others are the result of weak, low-pressure systems called tropical waves. Tropical disturbances are common during the summer and early fall. Regardless of their origin, only a small percentage of tropical disturbances develop into cyclones. There are three stages in the development of a full tropical cyclone.

READING CHECK **Infer** what is produced when water vapor condenses.

Formative stage The first indications of a building tropical cyclone is a moving tropical disturbance. Less-dense, moist air is lifted, triggering rainfall and air circulation. As these disturbances produce more precipitation, more latent heat is released. In addition, the rising air creates an area of low pressure at the ocean surface. As more warm, dense air moves toward the low-pressure center to replace the air that has risen, the Coriolis effect causes the moving air to turn counterclockwise in the northern hemisphere. This produces the cyclonic (counterclockwise) rotation of a tropical cyclone, as shown in **Figure 13.** When a disturbance over a tropical ocean acquires a cyclonic circulation around a center of low pressure, it has reached the developmental stage and is known as a tropical depression, as illustrated in **Figure 14.**

VOCABULARY

SCIENCE USAGE V. COMMON USAGE

Depression

Science usage: a pressing down or lowering, the low spot on a curved line

Common usage: a state of feeling sad

Mature stage As the moving air approaches the center of the growing storm, it rises, rotates, and increases in speed as more energy is released through condensation. In the process, air pressure in the center of the system continues to decrease. As long as warm air is fed into the system at the surface and removed in the upper atmosphere, the storm will continue to build and the winds of rotation will increase as the air pressure drops.

When wind speeds around the low-pressure center of a tropical depression exceed 62 km/h, the system is called a tropical storm. If air pressure continues to fall and winds around the center reach at least 119 km/h, the storm is officially classified as a tropical cyclone. Once winds reach these speeds, another phenomenon occurs—the development of a calm center of the storm called the **eye,** shown in **Figure 14.** The eye of the cyclone is often 30 to 60 km of calm weather and blue sky. The strongest winds in a hurricane are usually concentrated in the **eyewall**—a tall band of strong winds and dense clouds that surrounds the eye. The eyewall is visible because of the clouds that form there and mark the outward edge of the eye.

READING CHECK When water condenses, it becomes the water droplets that make up clouds, and it releases latent heat to the atmosphere.

DIFFERENTIATED INSTRUCTION

Advanced Learners Some tropical cyclones have altered the course of history, usually by interfering with naval fleets. Have interested students each research one such occurrence and provide details about the location of the event, the estimated strength of the tropical cyclone, the damage it caused, and how it affected historical events.

VISUALIZING Cyclone Formation

Figure 14 Like most storms, cyclones begin with warm, moist air rising.

Moving air starts to spin as a result of the Coriolis effect.

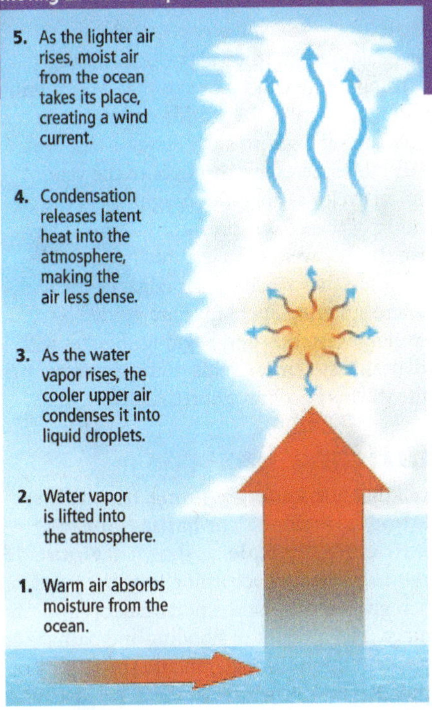

Tropical Depression The first indications of a building storm are a tropical depression with notable circulation, thunderstorms, and sustained winds of up to 62 km/h.

Tropical Storm As winds increase to speeds of 63–118 km/h, strong thunderstorms develop and become well defined. They are now tropical storms.

Tropical Cyclone With sustained winds of more than 119 km/h, an intense tropical weather system with well-defined circulation becomes a cyclone, also called a typhoon or hurricane.

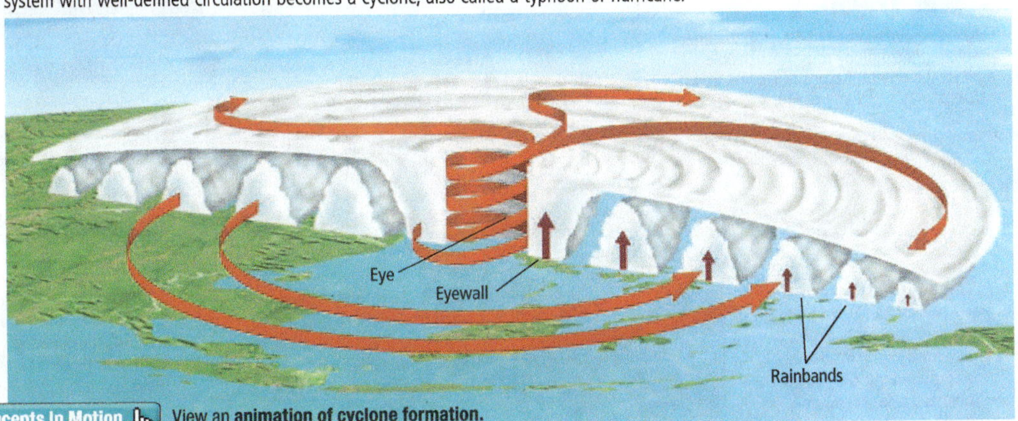

Concepts In Motion View an **animation of cyclone formation.**

VISUALIZING

Purpose

Students will learn the steps involved in the formation of cyclones.

Tie to Previous Knowledge

Heat Balance Students have been introduced to the concept of heat balance on Earth, which requires the transfer of energy from the tropics to the poles. In this section, students are introduced to one of the more dramatic methods by which this transfer takes place: the tropical cyclone. Fueled almost entirely by energy drawn from warm tropical waters, these massive storms play an important part in transporting energy from the warm tropics to the colder polar regions.

Concept Development

Coriolis Effect Like any other moving body on Earth's surface, a tropical cyclone is subject to the Coriolis effect. Thus, in the northern hemisphere, it will always try to move to the right of the direction in which it is traveling. This is why a storm moving westward across the tropical waters of the northern hemisphere eventually turns northward. Initially, its path is usually blocked by the large high-pressure systems that dominate these ocean areas. Once the storm reaches the edge of those systems, it is able to turn right and follow a recurving path, providing that no stronger weather systems interfere.

Teacher Content Support

Project Stormfury In the 1960s and early 1970s, the U.S. government initiated a program to weaken hurricanes. Called Project Stormfury, the experiment involved seeding a tropical cyclone with silver iodide. Massive amounts of condensation nuclei—in this case, silver iodide—were infused into the storm so that it would expend its energy through rainfall instead of wind. The results of the experiment were inconclusive.

Identify Misconceptions

Some students might wonder why hurricanes cannot be stopped while they are still weak.

Uncover the Misconception

Ask students to consider whether it is a good idea to try to destroy hurricanes before they grow strong enough to be a problem.

Demonstrate the Concept

Point out that even a weak tropical cyclone is so huge that not even an atomic device could destroy it. Next, point out that hurricanes play an important role in maintaining Earth's heat balance. The consequences could be severe if hurricanes were not allowed to fulfill this role.

Assess New Knowledge

Help students realize that better preparation, rather than destruction, is the key to dealing effectively with hurricanes. Encourage students to suggest safety measures that can be taken before and during a hurricane. Also, have them discuss problems related to overdevelopment in hurricane-prone regions.

Saffir-Simpson Hurricane Wind Scale

Category	Winds (km/h)	Damage
5	>250	catastrophic
4	210–249	extreme
3	178–209	extensive
2	154–177	moderate
1	119–153	minimal

Figure 15 The Saffir-Simpson Hurricane Wind scale classifies hurricanes according to wind speed, which implies potential for property damage.

Dissipation stage A cyclone will last until it can no longer produce enough energy to sustain itself. This usually happens when the storm has moved either over land or over colder water. During its life cycle, a cyclone can undergo several fluctuations in intensity as it interacts with other atmospheric systems.

Tropical cyclone movement Like all large-scale storms, tropical cyclones move according to the wind currents that steer them. Recall that many of the world's oceans are home to subtropical high-pressure systems that are present to some extent throughout the year. Tropical cyclones are often caught up in the circulation of these high-pressure systems. They move steadily west, then eventually turn poleward when they reach the far edges of the high-pressure systems. There they are guided by prevailing westerlies and begin to interact with midlatitude systems. At this point, the movement of the storms becomes unpredictable.

Hurricane Hazards

Recall that hurricanes are strong, regionally specific tropical cyclones. The **Saffir-Simpson Hurricane Wind scale,** as shown in **Figure 15,** classifies hurricanes according to wind speed, which gives an idea of the potential for property damage. The amount of flooding and damage are dependent upon shore depth and the density of population and structures in the affected area.

Figure 16
Storm Tracking
Scientists have worked to develop weather prediction technology to protect people against the different types of storms that cause damage and loss of life.

1888 A three-day blizzard dumps 125 cm of snow on the northeast United States, creating 17-m snowdrifts, burying houses and trains, killing 400 people, and sinking 200 ships.

1850 — 1900 — 1925

1861 An English newspaper publishes the first daily weather forecasts based on countrywide data that is compiled via the recently invented telegraph.

1900 A Category 4 hurricane hits Texas. Five-m waves sweep over Galveston Island, killing more than 8000 people and washing away half the homes on the island.

1925 An EF5 tornado rips through Missouri, Illinois, and Indiana, covering 352 km in three hours.

ACROSS THE CURRICULUM

Social Studies Have students investigate the use of the words *hurricane, typhoon,* and *cyclone* in different parts of the world. Students should determine where in the world each of the words is used. Have students make a simple world map and mark the regions in which each word is used.

Damage Hurricanes can cause extensive damage, particularly along coastal areas, which tend to be where human populations are the most dense. Evidence of storm damage is documented in **Figure 16** by a photo from a hurricane that hit Galveston, Texas, in 1900.

Winds Much of the damage caused by hurricanes is associated with violent winds. The strongest winds in a hurricane are usually located at the eyewall. Outside of the eyewall, winds taper off as distance from the center increases, although winds of more than 60 km/h can extend as far as 400 km from the center of a hurricane.

Storm surge Strong winds moving onshore in coastal areas are partly responsible for the largest hurricane threat—storm surges. A **storm surge** occurs when hurricane-force winds drive a mound of ocean water toward coastal areas where it washes over the land. Storm surges can sometimes reach 6 m above normal sea level, as shown in **Figure 17.** When this occurs during high tide, the surge can cause enormous damage. In the northern hemisphere, a storm surge occurs primarily on the right side of a storm relative to the direction of its forward motion. That is where the strongest onshore winds occur. This is due to the counterclockwise rotation of the storm.

Hurricanes produce great amounts of rain because of their continuous uptake of warm, moist ocean water. Thus, floods from intense rainfall are an additional hurricane hazard, particularly if the storm moves over mountainous areas, where orographic lifting enhances the upward motion of air and the resulting condensation of water vapor.

Figure 17 Storm surges can sometimes reach 6 m above normal sea level and cause enormous damage.

1949 Lightning sparks a wildfire in Helena National Forest, Montana, that claims the lives of 13 firefighters and destroys 20 km² of land in five days.

1970 Large hailstones, more than 13 cm in diameter, fell on Coffeyville, Kansas.

2005 The Atlantic hurricane season unleashes the most hurricanes and Category 5 storms in history.

1950 — 1975 — 2000

1960 The U.S. government launches TIROS, the first weather satellite.

1990 The United States deploys its first operational Doppler radar system after more than 30 years of research.

Model

Storm Surge A storm surge occurs mainly because extreme winds blowing onshore in coastal areas pile up water, although the low pressure at the center of the storm also makes a contribution by allowing a small rise in sea level. Have students model a storm surge by filling a small pan of water nearly to the top and then blowing hard on the water. **WARNING:** ***Wipe up spills immediately to avoid accidents.*** Make sure students blow outward from the center of the pan toward the edge. The water should pile up along the edge of the pan and overflow, simulating a storm surge. BL EL

Use an Analogy

Hurricanes Tell students that a weakening hurricane is similar to a car that has run out of fuel. Just as the car's forward speed slows when it runs out of gas, a tropical cyclone's wind speed weakens when it runs out of the warm water from which it derives its energy.

GeoLAB

iLab Station

The GeoLab at the end of the chapter can be used at this point in the lesson.

EARTH SCIENCE JOURNAL

Effect of Tropical Cyclones Have students each choose a location in the hurricane belt that is as near as possible to their location. Students should research how past tropical cyclones have affected that location. Have students record in their Earth science journals the names of the storms, when they struck land, their strength, and the damage they inflicted. OL

3 Assess

Check for Understanding

Think Critically Tell students that some of the strongest hurricanes on record have moved through the warm waters of the western Caribbean Sea. Have students explain why this area produces such strong storms. This warm water contains a tremendous amount of energy for hurricanes to utilize.

Reteach

Sequence Have students work in groups to develop a chart that shows the path of an imaginary hurricane that strikes the Atlantic Coast of the United States. Students should show the developmental stages of the storm as it moves from the ocean to land and finally weakens. Students' hurricane charts should reflect knowledge of the content in this chapter.

Assessment

Performance Have students work in groups to prepare evacuation plans for residents of a coastal area that will soon be struck by a hypothetical hurricane. Tell students they have two days to set the plans in action. Have them use maps to show the areas that will be evacuated and the evacuation routes. Students can present their plans in booklets or pamphlets.

CAREERS IN EARTH SCIENCE

WebQuest

Figure 18 This residential area has been engulfed in debris left behind from the flood waters of Hurricane Katrina. Most of the deaths associated with a hurricane come from flooding, not high winds.

CAREERS IN EARTH SCIENCE

Hurricane Hunter A hurricane hunter flies an instrument-laden airplane into a hurricane to measure wind speed and gather weather data on the features of a hurricane.

WebQuest

Hurricane advisories and safety The National Hurricane Center, located in Miami, Florida, is responsible for tracking and forecasting the intensity and motion of tropical cyclones in the western hemisphere. The center issues a hurricane warning at least 36 hours before a hurricane is predicted to strike. The center also issues regular advisories that indicate a storm's position, strength, and movement. Using this information, people can then track a storm on a hurricane-tracking chart, such as the one you will use in the GeoLab at the end of this chapter. Awareness, combined with proper safety precautions, has greatly reduced death tolls associated with hurricanes in recent years. **Figure 18** shows debris and destruction left by hurricane flooding; loss of life can be prevented by evacuating residents before the storm hits.

SECTION 3 REVIEW

Section Self-Check

Section Summary

- Tropical cyclones rotate counterclockwise in the northern hemisphere.
- Tropical cyclones are also known as hurricanes and typhoons.
- Tropical cyclones go through the same stages of formation and dissipation as other storms.
- Tropical cyclones are moved by various wind systems after they form.
- The most dangerous part of a tropical cyclone is the storm surge.
- Hurricane alerts are given at least 36 hours before the hurricane arrives.

Understand Main Ideas

1. MAINIDEA **Identify** the three main stages of a tropical cyclone.
2. **Describe** the changing wind systems that guide a tropical cyclone as it moves from the tropics to the midlatitudes.
3. **Identify** two conditions that must exist for a tropical cyclone to form.
4. **Explain** what causes a cyclone to dissipate.

Think Critically

5. **Analyze** Imagine that you live on the eastern coast of the United States and are advised that the center of a hurricane is moving inland 70 km north of your location. Would a storm surge be a major problem in your area? Why or why not?
6. **Compare** the Saffir-Simpson scale with the Enhanced Fujita scale. How are they different? Why?

MATH IN Earth Science

7. Determine the average wind speed for each hurricane category shown in **Figure 15**.

SECTION 3 REVIEW

1. formative, mature, dissipation
2. Tropical cyclones move according to the wind currents that steer them. Cyclones circulate in high-pressure systems, moving steadily toward the west, then eventually turn poleward. They are guided by prevailing westerlies and interact with midlatitude systems.
3. an abundant supply of warm ocean water, and some sort of disturbance to lift warm air and keep it rising
4. Heat is the mechanism of lift for any storm. Once a cyclone moves over land, much of that heat is taken away, and the storm falls apart gradually.
5. The potential for storm surge and flood damage would depend more on the location of the eye making landfall than distance. If your location is north of the eye's landfall, there can be a danger up to 70 km away, while south of that landfall there is little surge. The hurricane eye is north of your position, so there is less danger.
6. The Enhanced Fujita Tornado Damage scale rates high winds and damage over a small land area. The Saffir-Simpson Hurricane Wind scale rates high winds over a tremendously large water area.
7. Category 1 = 136 kph, Category 2 = 165.5 kph, Category 3 = 193.5 kph, Category 4 = 229.5 kph, Category 5 = 250+ kph

Essential Questions

- What are the problems associated with recurring weather patterns?
- What atmospheric events cause recurring weather patterns?
- How do heat waves and cold waves differ?

Review Vocabulary

Fahrenheit scale: a temperature scale in which water freezes at 32° and boils at 212°

New Vocabulary

drought
heat wave
cold wave
windchill index

Recurrent Weather

MAINIDEA Even a relatively mild weather system can become destructive and dangerous if it persists for long periods of time.

EARTH SCIENCE 4 YOU Have you ever eaten so much candy you made yourself sick? Too much of any specific type of weather—cold, wet, warm, or dry—can also be unwelcome because of the serious consequences that can result from it.

Floods

An individual thunderstorm can unleash enough rain to produce floods, and hurricanes can also cause torrential downpours, which result in extensive flooding. Floods can also occur, however, when weather patterns cause even mild storms to persist over the same area for an extended period of time. For example, a storm with a rainfall rate of 1.5 cm/h is not much of a problem if it lasts only an hour or two. If this same storm were to remain over one area for 18 hours, however, the total rainfall would be 27 cm, which is enough to create flooding in most areas. In the spring of 2010, a two-day storm caused flooding throughout much of Tennessee, Kentucky, and Mississippi. Over 10,000 properties were damaged in Nashville, TN, which experienced extensive flooding, as shown in **Figure 19.**

Low-lying areas are most susceptible to flooding, making coastlines particularly vulnerable to storm surges during hurricanes. Rivers in narrow-walled valleys can rise rapidly, creating high-powered and destructive walls of water. Building in the floodplain of a river or stream can be inconvenient and potentially dangerous during a flood.

Figure 19 Streets, cars, and buildings were flooded after heavy rains in Nashville, Tennessee.
Infer *What areas are most affected by flooding?*

Demonstration

Floods Demonstrate how torrential rainfall can cause flooding. Fill a small pan with dirt and a glass salt shaker with water. Sprinkle the water slowly onto the dirt. It will be absorbed, with little change to the surface of the dirt. Next, repeat the process, only this time take off the top of the shaker and pour the water onto the dirt. The dirt will be washed away in one area—the equivalent of a flood in a torrential downpour.

SECTION 4

1 Focus

Focus Transparency

Before presenting the lesson, project **Section Focus Transparency 39** and have students answer the accompanying questions. EL

MAINIDEA

Precipitation Provide students with the normal annual total precipitation for your area. Have them discuss how the impacts on your area might differ if all that rain fell in one month, or one week, or one day. In each scenario, be sure to also discuss the consequences of having no more rain the rest of the year.

2 Teach

Tie to Previous Knowledge

The Jet Stream Students have learned how the jet stream changes positions and steers weather systems. Sometimes, however, the position of the jet stream does not change appreciably with time. This can result in recurring weather patterns.

Use Science Terms

Flash Flood The term *flash flood* is frequently used to describe floods, but it applies only to hilly or mountainous areas. A flash flood occurs when a normally dry or nearly dry riverbed receives torrential rainfall in the area of its headwaters. As the water travels downstream, it becomes a raging torrent that is capable of washing away people, cars, roads, and even buildings. The flooding that occurs in urban areas as a result of poor drainage is not flash flooding.

Caption Question **Fig. 19** Coastal and low-lying areas along rivers and streams are more likely to flood.

MiniLAB

iLab Station | Rubric

Purpose Students will demonstrate how slow-moving weather systems can cause floods.

Process Skills model, observe and infer, analyze

Safety Precautions Approve lab safety forms before work begins. Wipe up spills immediately to avoid accidents.

Teaching Strategy Let students practice maintaining a steady flow of water prior to conducting the experiment.

Expected Results Moving the bottle slowly will produce a greater depth of water in the tray than moving the bottle quickly does.

Analysis

1. The average depth decreased from Step 5 to Step 7 because the water fell over each compartment in a shorter amount of time.
2. Storms that move slowly can release more rain over any one place than can storms that move quickly.
3. speed up or slow down the motion of the water bottle; open or close the twist top slightly to simulate heavier or lighter rainfall

Assessment

Knowledge Tell students that the ability of the ground to absorb water has a major effect on flooding. Ask them to infer which would cause more flooding: saturated ground or dry ground. Saturated ground would cause more flooding than dry ground, which would absorb some of the rain.

LabManager™
Customize any lab with the LabManager™ CD-ROM.

■ **Figure 20** Cotton plants struggle to survive in dried, cracked mud during a drought.

MiniLAB

iLab Station

Model Flood Conditions

How can mild rains cause floods? Flooding can result from repeated, slow-moving storms that drop rain over the same area for a long period of time.

Procedure

1. Read and complete the lab safety form.
2. Place an **ice cube tray** on the bottom of a **large sink or tub.**
3. Pour **water** into a clean, **plastic dishwashing-detergent bottle** until it is two-thirds full. Replace the cap on the bottle.
4. Hold the bottle upside down with the cap open about 8 cm above one end of the ice cube tray. Gently squeeze the bottle to maintain a constant flow of water into the tray.
5. Slowly move the bottle from one end of the tray to the other over the course of 30 s. Try to put approximately equal amounts of water in each ice cube compartment.
6. Measure the depth of water in each compartment. Calculate the average depth.
7. Repeat Steps 2 to 4, but move the bottle across the ice cube tray in 15 s.

Analysis

1. **Compare** How did the average depth of the water differ in Steps 5 and 7? How might you account for the difference?
2. **Infer** Based on these results, infer how the speed of a moving storm affects the amount of rain received in any one area.
3. **Deduce** How could you alter the experiment to simulate different rates of rainfall?

Droughts

Too much dry weather can cause nearly as much damage as too much rainfall. **Droughts** are extended periods of well-below-average rainfall. One of the most extreme droughts in American history occurred during the 1930s in the central United States. This extended drought put countless farmers out of business, as rainfall was inadequate to grow crops.

Droughts are usually the result of shifts in global wind patterns that allow large, high-pressure systems to persist for weeks or months over continental areas. Under a dome of high pressure, air sinks on a large scale. Because the sinking air blocks moisture from rising through it, condensation cannot occur, and drought sets in until global patterns shift enough to move the high-pressure system. **Figure 20** shows one of the impacts of long-term drought.

Heat waves An unpleasant side effect of droughts often comes in the form of **heat waves,** which are extended periods of above-average temperatures. Heat waves can be formed by the same high-pressure systems that cause droughts. As the air under a large high-pressure system sinks, it warms by compression and causes above-average temperatures. The high-pressure system also blocks cooler air masses from moving into the area, so there is little relief from the heat. Because it is difficult for condensation to occur under the sinking air of the high-pressure system, there are few, if any, clouds to block the blazing sunshine. The jet stream, or "atmospheric railway," that weather systems normally follow is farther poleward and weaker during the summer. Thus, any upper-air currents that might guide the high-pressure system are so weak that the system barely moves.

ACROSS THE CURRICULUM

History Floods and droughts affect food availability and prices. When these weather events occur in agricultural areas, they can wipe out entire crops over large areas. This creates a shortage of the affected crops, which in turn drives up food prices. Have students research an example from history in which a flood or drought affected food availability and prices. Have students share the results of their research with the rest of the class in an oral presentation.

Heat index Increasing humidity can add to the discomfort and potential danger of a heat wave. Human bodies cool by evaporating moisture from the surface of the skin. In the process, thermal energy is removed from the body. If air is humid, the rate of evaporation is reduced, which diminishes the body's ability to regulate internal temperature. During heat waves, this can lead to serious health problems such as heatstroke, sunstroke, and even death.

Because of the dangers posed by the combination of heat and humidity, the National Weather Service (NWS) routinely reports the heat index, shown in **Table 2.** Note that the NWS uses the Fahrenheit scale in the heat index, as well as several other indices it produces, because most United States citizens are more familiar with this scale.

The heat index assesses the effect of the body's increasing difficulty in regulating its internal temperature as relative humidity rises. This index estimates how warm the air feels to the human body based on the actual air temperature and relative humidity. For example, an air temperature of 85°F (29°C) combined with relative humidity of 80 percent would require the body to cool itself at the same rate as if the air temperature were 97°F (36°C).

READING CHECK **Identify** the cause of serious health problems associated with heat waves.

Explore the **heat index with an interactive table.** Concepts In Motion

Table 2 The Heat Index

Relative Humidity (%)	Air Temperature (°F)										
	70	75	80	85	90	95	100	105	110	115	120
	Apparent Temperature (°F)										
0	64	69	73	78	83	87	91	95	99	103	107
10	65	70	75	80	85	90	95	100	105	111	116
20	66	72	77	82	87	93	99	105	112	120	130
30	67	73	78	84	90	96	104	113	123	135	148
40	68	74	79	86	93	101	110	123	137	151	
50	69	75	81	88	96	107	120	135	150		
60	70	76	82	90	100	114	132	149			
70	70	77	85	93	106	124	144				
80	71	78	86	97	113	136					
90	71	79	88	102	122						
100	72	80	91	108							

Source: National Weather Service, NOAA

Environmental Connection

Heat Waves Droughts and heat waves often occur together because the same sinking air that blocks precipitation also causes warming by air compression. This combination significantly impacts ecosystems. As streams, creeks, and ponds dry up, oxygen levels in the water decrease, and aquatic life suffers.

Project

Heat Index Heat-index values are classified according to their threat to health and safety. Values between 80° and 90°F are in the "Caution" range; those between 90° and 105°F are in the "Extreme Caution" range, those between 105° and 120°F are in the "Danger" range, and those above 120°F fall into the "Extreme Danger" range. Have students research average summer temperatures and humidity levels for their location. Students can use the heat index to determine the classification of summer temperatures and humidity levels in their area. OL AL

READING CHECK The rate of evaporation of sweat from the skin is reduced when humidity is high, which can result in the body having difficulty regulating its internal temperature. If body temperature gets too high, heatstroke can occur.

DIFFERENTIATED INSTRUCTION

Advanced Learners Tell students that most heat-wave related deaths typically take place in cities among the elderly and those with respiratory problems. Ask students to research why this is so. As humidity increases, it is more difficult for the body to cool itself through the evaporation of perspiration. Elderly people can have less tolerance for increased body temperature. The sinking air that causes the heat also traps pollutants near the ground. This can worsen respiratory problems.

Data Analysis LAB

About the Lab

- Ask students whether they have ever experienced a severe heat wave. Those students who answer in the affirmative can share their experiences with the class.

Analysis

1. Day 1 = 27.5°C; Day 2 = 30.5°C; Day 3 = 34°C; Day 4 = 34°C; Day 5 = 31°C; Day 6 = 29°C; Day 7 = 27.5°C
2. Graphs should show that the maximum temperature increased, peaking on Day 3, then decreased.

Think Critically

3. The heat wave began on Day 2. The heat wave lasted 4 days (Day 2–Day 5).
4. The average temperature during the heat wave was 32.3°C. This is from 3 to 5° higher than the average temperatures on the other days.

Interpret the Photo

Cold Waves Have students study **Figure 21.** Have them describe the impacts of cold waves on human activities. Possible answer: A severe cold wave can pose a hazard to health, temporarily halt all forms of transportation and recreation, and close businesses and schools.

☑ READING CHECK The snow-covering of the Arctic region reflects the sunlight back into space so that the air cannot be heated convectively.

■ **Figure 21** Prolonged cold or recurrent cold waves can create conditions such as these that fell on Denver in 2006.

Cold Waves

The opposite of a heat wave is a **cold wave,** which is an extended period of below-average temperatures. Interestingly, like heat waves, cold waves are also brought on by large, high-pressure systems. However, cold waves are caused by systems of continental polar or arctic origin. During the winter, little sunlight is available to provide warmth. At the same time, the snow-covered surface is constantly reflecting the sunlight back to space. The combined effect of these two factors is the development of large pools of extremely cold air over polar continental areas. Because cold air sinks, the pressure near the surface increases, creating a strong high-pressure system.

Because of the location and the time of year in which they occur, winter high-pressure systems are much more influenced by the jet stream than are summer high-pressure systems. Moved along by the fast-moving jet stream, these high-pressure systems rarely linger in any area. However, the winter location of the jet stream can remain essentially unchanged for days or even weeks. This means that several polar high-pressure systems can follow the same path and subject the same areas to continuous numbing cold until the position of the jet stream changes. One effect of prolonged periods of cold weather is shown in **Figure 21.**

☑ READING CHECK **Explain** why the Sun's energy has little effect on air temperature in the arctic.

Data Analysis LAB

Based on Real Data*

Interpret the Table

How can you calculate a heat wave? The following data represent the daily maximum and minimum temperatures for seven consecutive summer days in Chicago. A heat wave in this area is defined as two or more days with an average temperature of 29.4°C or higher.

Analysis

1. Calculate the average temperature for each day in your table.
2. Plot the daily maximum and minimum temperatures on a graph with the days on the *x*-axis and the maximum temperatures on the *y*-axis. Using the data points, draw a curve to show how the temperatures changed over the seven-day period. Add the average temperatures.

Think Critically

3. **Determine** What day did the city heat wave begin? How long did it last?
4. **Compare** the average temperature for the days of the heat wave to the average temperature of the remaining days.

Data and Observations

Daily Temperatures

Day	Maximum (°C)	Minimum (°C)	Average (°C)
1	32	23	
2	37	24	
3	41	27	
4	39	29	
5	37	25	
6	34	24	
7	32	23	

* Data obtained from: Klinenberg, E. 2002. *Heat Wave: A social autopsy of disaster in Chicago, IL.* Chicago: University of Chicago Press.

EARTH SCIENCE JOURNAL

Environmental Effects Explain to students that in recent years, there have been increasing environmental problems caused by recurring weather, which includes floods, droughts, cold waves, and heat waves. Have students research and report on the environmental impacts of these events in farming areas and how they impact natural resource use, such as fossil fuels for heating and cooling.

Rubric

Windchill Chart

Temperature (°F) \ Wind (mph): Calm	5	10	15	20	25	30	35	40	45	50	55	60
40	36	34	32	30	29	28	28	27	26	26	25	25
35	31	27	25	24	23	22	21	20	19	19	18	17
30	25	21	19	17	16	15	14	13	12	12	11	10
25	19	15	13	11	9	8	7	6	5	4	4	3
20	13	9	6	4	3	1	0	-1	-2	-3	-3	-4
15	7	3	0	-2	-4	-5	-7	-8	-9	-10	-11	-11
10	1	-4	-7	-9	-11	-12	-14	-15	-16	-17	-18	-19
5	-5	-10	-13	-15	-17	-19	-21	-22	-23	-24	-25	-26
0	-11	-16	-19	-22	-24	-26	-27	-29	-30	-31	-32	-33
-5	-16	-22	-26	-29	-31	-33	-34	-36	-37	-38	-39	-40
-10	-22	-28	-32	-35	-37	-39	-41	-43	-44	-45	-46	-48
-15	-28	-35	-39	-42	-44	-46	-48	-50	-51	-52	-54	-55
-20	-34	-41	-45	-48	-51	-53	-55	-57	-58	-60	-61	-62
-25	-40	-47	-51	-55	-58	-60	-62	-64	-65	-67	-68	-69

Frostbite times ■ 30 min ■ 10 min ■ 5 min

■ **Figure 22** The windchill chart was designed to show the dangers of cold and wind.
What *wind speed and temperature is the same as 10°F on a calm day?*

Windchill index The effects of cold air on the human body are magnified by wind. Known as the windchill factor, this phenomenon is measured by the **windchill index** in **Figure 22.** The index estimates how cold the air feels to the human body based on the actual air temperature and wind speed. While the windchill index is helpful, it does not account for individual variations in sensitivity to cold, the effects of physical activity, or relative humidity. In 2001, the National Weather Service revised the calculations to utilize advances in science, technology, and computer modeling. These revisions provide a more accurate, understandable, and useful index for estimating the dangers caused by winter winds and freezing temperatures.

SECTION 4 REVIEW

Section Self-Check

Section Summary

- Too much heat and too little precipitation causes droughts.
- Extended periods of above-average temperatures often occur with droughts.
- Heat index estimates the effect on the human body when the air is hot and the humidity is high.
- Too little heat and a stalled jet stream can cause a cold wave, or an extended period of cold weather.
- Windchill index tells you how wind and temperature affect your body in winter.

Understand Main Ideas

1. MAINIDEA **Explain** how everyday weather can become recurrent and dangerous.
2. **Describe** how relatively light rain could cause flooding.
3. **Compare and contrast** a cold wave and a heat wave.
4. **Explain** why one type of front would be more closely associated with flooding than another.

Think Critically

5. **Explain** why air in a winter high-pressure system is very cold despite compressional warming.
6. **Compare** the data of the heat-index scale and the windchill scale. What variables influence each scale?

MATH IN Earth Science

7. A storm stalls over Virginia, dropping 0.75 cm of rain per hour. If the storm lingers for 17 hours, how much rain will accumulate?

3 Assess

Check for Understanding

Explain Ask students to identify various recurring weather events and to explain some dangers associated with these events.

Reteach

Explain Ask several students to each state a safety rule for floods, heat waves, or cold waves. Students should defend their safety rules with scientific reasoning.

Assessment

Performance Have students research and report on a heat wave or cold wave that occurred in their area. For information, they can contact the local branch of the National Weather Service.

Caption Question Fig. 22 30°F with 60-mph winds

SECTION 4 REVIEW

1. When weather patterns stall due to jet streams or slow-moving systems, or similar air masses pass over the same area, recurrent patterns can be established. Flooding and extended periods of heat or cold can occur.
2. Floods can also occur when weather patterns cause mild storms to persist over the same area.
3. A cold wave is an extended period of below-normal temperatures. The winter location of the jet stream can remain essentially unchanged for days or even weeks, meaning that several polar high-pressure systems can follow the same path and subject the same areas to bouts of cold. Heat waves are extended periods of above-normal temperatures formed as air sinks in a high pressure system that is warmed by compression. The high pressure blocks cooler air masses from moving into the area, making it difficult for condensation to occur. The jet stream is farther north and weaker during the summer, so high-pressure systems scarcely move.
4. A cold front can cause tremendous storms, but they are usually of short duration over large areas. A warm front is more likely to stall and remain in one spot for a longer time and can produce lots of rain in that time period. The exception is a hurricane.
5. When a high-pressure system forms in the winter, little sunlight is available to provide warmth. At the same time, the snow-covered surface is constantly reflecting and radiating its limited heat back to space. The combined effect of these factors is the development of large pools of extremely cold air over continental areas.
6. The windchill scale is the combined effect of temperature and wind while the heat index is the combined effect of temperature and humidity.
7. 12.75 cm

eXpeditions!

Purpose

Students will learn about storm spotters who volunteer for the National Weather Service.

Teacher Content Support

Storm Spotters The invention of the radio and the telephone made it possible to warn people in advance when they are located in the path of a storm. Since the 1950s, the United States has relied on volunteer storm spotters as part of a severe weather early warning system. Storm spotters perform an essential service, giving the National Weather Service (NWS) the best picture of what is happening on the ground. Spotters travel to key lookout points on the edge of town and use amateur radio to report their observations into their local network through a Net Control Station (NCS). The NCS summarizes these reports and notifies the NWS, which uses this information and the radar data to issue a warning. Over the past 60 years, this cooperation between the NWS, local emergency personnel, and spotter networks has saved hundreds of thousands of lives.

Teaching Strategy

Have students discuss the risks involved in being a storm spotter. What safety precautions can be taken by spotters to avoid getting injured by a storm? Possible answers: Educate themselves on what to do when lightning strikes or when a tornado is coming while they are outside.

eXpeditions!

ON SITE: STORM SPOTTERS

Storm chasers in a specially built armored vehicle study a severe storm in Kearney, Nebraska.

When storm spotters hear that severe weather is approaching the area, do they seek the safety of their house or basement like most people do? No, they head out to the edge of town or to a high point to check on the exact wind and weather conditions.

Volunteers for the NWS Storm spotters work as volunteers for the National Weather Service (NWS) to help give NWS forecasters a clear picture of what is really happening on the ground. Although Doppler radar and other systems are sophisticated data collectors, these devices can only detect weather conditions that might produce a severe thunderstorm or a tornado. The NWS typically uses this information to issue a severe storm or tornado watch. When a watch is issued, spotters travel to key lookout points and report their observations. The observations made on the ground by storm spotters are essential to the NWS in upgrading watches to warnings.

Making Reports The NWS trains spotters to assess certain weather conditions such as wind speed, hail size, and cloud formation. For example, if large tree branches begin to sway, umbrellas are difficult to use, and the wind creates a whistling noise along telephone wires, spotters know that wind speed is between 40 and 50 km/h. If trees are uprooted, wind speed is estimated to be between 85 and 115 km/h.

Spotters study the clouds to determine where hail is falling, where a tornado might develop, and in what direction the storm is headed. When they call in, they report the event, its location, its direction, and whether there is need for emergency assistance.

High Risk Mobile spotters risk their own safety in order to protect their community. The major risks they face stem from driving in bad weather and standing on a high spot where lightning might strike. Spotters always travel with a partner, so that one person can drive and the other can watch the sky. To stay safe, spotters keep watch in all directions, keep the car engine running, and have an escape plan.

The combination of technology and the work of spotters has saved many lives since the volunteer system was started by the NWS in the 1970s. The number of deaths as a result of tornadoes and other severe weather has decreased significantly since the program began.

WRITING IN ▶ Earth Science

Make a Pamphlet Research more information about how to become a storm spotter and the training involved. Write and illustrate a pamphlet about storm spotting that includes this information.

WebQuest

WRITING IN ▶ Earth Science

Rubric

Make a Pamphlet Training to become a storm spotter is provided by the National Weather Service and completed through local agencies. A person must successfully complete the training before he or she can be certified as a storm spotter.

WebQuest

fStop/Getty Images

GeoLAB

iLab Station

Track a Tropical Cyclone

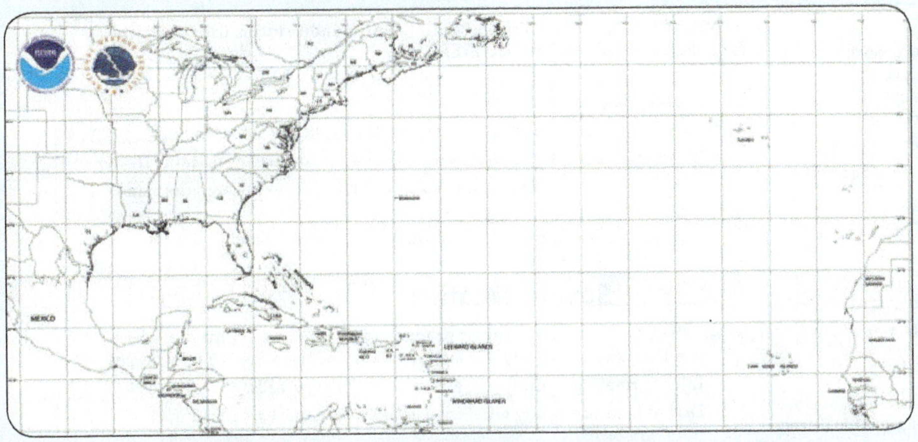

Atlantic Basin Hurricane Tracking Chart This chart is used by the National Hurricane Center to track active tropical cyclones in the Atlantic basin.

NOAA

Background: **Tropical cyclones form very violent storms, so it is important to have advanced warning before they hit land. By tracking the changing position of a storm on a chart and connecting these positions with a line, you can model or predict a cyclone's path.**

Question: *What information can you obtain by studying the path of a tropical cyclone?*

Procedure

1. Read and complete the lab safety form.
2. Form a hypothesis about how a tropical cyclone's path can be used to predict the strength of the storm and where the most damage might be inflicted.
3. Review the tropical cyclone data that your teacher has provided.
4. Choose the track of a tropical cyclone that has occurred during the past five years.
5. Plot the position, air pressure, wind speed, and stage of the tropical cyclone at 6-h intervals throughout its existence.
6. Plot the changing position of the tropical cyclone on your hurricane-tracking chart.
7. Incorporate your research into a data table. Add any additional information that you think is important.

Analyze and Conclude

1. **Identify** What was the maximum wind speed in knots that the tropical cyclone reached?
2. **Calculate** Multiply the value from Question 1 by 1.85 to find the wind speed in kilometers per hour. Based on this value, how would the tropical cyclone be classified on the Saffir-Simpson Hurricane Wind scale shown in **Figure 15?**
3. **List** the landmasses over which the tropical cyclone passed.
4. **Identify** What was the life span of your tropical cyclone? What was the name of your cyclone?
5. **Infer** Where would you expect the storm surge to have been the greatest? Explain.
6. **Examine** How was the tropical cyclone's strength affected when its center passed over land?

SHARE YOUR DATA

Peer Review Post a summary of your data. Compare your data to other data collected for this investigation.

GeoLAB

iLab Station Rubric

Preparation

Time Allotment 90 min

Process Skills make and use tables, communicate, collect data, interpret data, observe and infer, interpret scientific illustrations, predict

Additional Materials internet access, calculator

Safety Precaution Approve lab safety forms before work begins.

Procedure

- If you do not have access to the Internet, use historical data available from the library or from the local National Weather Service office.
- This lab can be done in groups of three or as an individual project. If students are working in groups, one student can obtain data, another can develop the data table, and a third can plot the hurricane's track. Students can trade roles for each hurricane they research.
- Review the concepts of latitude and longitude before proceeding with this lab.
- Data tables should include six hourly latitude and longitude positions, maximum wind speeds, and minimum central pressures. Tracking charts should show the six hourly positions connected by a straight line to form a track.

Analyze and Conclude

1–4. Answers will vary depending on the tropical cyclone chosen.

5. The storm surge would be strongest on the side of the storm where onshore winds developed as the storm center moved inland.

6. Wind speeds decrease rapidly when a storm center moves inland, because the storm becomes cut off from its energy source and weakens.

SHARE YOUR DATA

Peer Review Help students post their data. Compare your students' findings with those of other classrooms around the country.

LabManager™
Customize any lab with the LabManager™ CD-ROM.

MAINIDEAS Summary statements can be used by students to review the major concepts of the chapter.

Students can review with these online resources.

Vocabulary Practice

- Vocabulary eGames
- Vocabulary eFlashcards
- Vocabulary PuzzleMaker

Section Self-Check

Chapter Self-Check

Online Test Practice

Use the Exam*View*® *Assessment Suite* CD-ROM to:

- create multiple versions of tests
- create modified tests with one mouse click
- edit existing questions and add your own questions
- build tests aligned with state standards using built-in state curriculum tags
- change English tests to Spanish with one mouse click
- track students' progress using the Teacher Management System

CHAPTER 13 STUDY GUIDE

BIGIDEA **The exchange of thermal energy in the atmosphere sometimes occurs with great violence that varies in form, size, and duration.**

Vocabulary Practice

SECTION 1 Thunderstorms

VOCABULARY
- air-mass thunderstorm
- mountain thunderstorm
- sea-breeze thunderstorm
- frontal thunderstorm
- stepped leader
- return stroke

MAINIDEA The intensity and duration of thunderstorms depend on the local conditions that create them.

- The cumulus stage, the mature stage, and the dissipation stage comprise the life cycle of a thunderstorm.
- Clouds form as water is condensed and latent heat is released.
- Thunderstorms can be produced either within air masses or along fronts.
- From formation to dissipation, all thunderstorms go through the same stages.
- Lightning is a natural result of thunderstorm development.

SECTION 2 Severe Weather

VOCABULARY
- supercell
- downburst
- tornado
- Enhanced Fujita Tornado Damage scale

MAINIDEA All thunderstorms produce wind, rain, and lightning, which can have dangerous and damaging effects under certain circumstances.

- Intense rotating updrafts are associated with supercells.
- Downbursts are strong winds that result in damage associated with thunderstorms.
- Hail is precipitation in the form of balls or lumps of ice that accompany severe storms.
- The worst storm damage comes from a vortex of high winds that moves along the ground as a tornado.

SECTION 3 Tropical Storms

VOCABULARY
- tropical cyclone
- eye
- eyewall
- Saffir-Simpson Hurricane Wind scale
- storm surge

MAINIDEA Normally peaceful, tropical oceans are capable of producing one of Earth's most violent weather systems—the tropical cyclone.

- Tropical cyclones rotate counterclockwise in the northern hemisphere.
- Tropical cyclones are also known as hurricanes and typhoons.
- Tropical cyclones go through the same stages of formation and dissipation as other storms.
- Tropical cyclones are moved by various wind systems after they form.
- The most dangerous part of a tropical cyclone is the storm surge.
- Hurricane alerts are given at least 36 hours before the hurricane arrives.

SECTION 4 Recurring Weather

VOCABULARY
- drought
- heat wave
- cold wave
- windchill index

MAINIDEA Even a relatively mild weather system can become destructive and dangerous if it persists for long periods of time.

- Too much heat and too little precipitation causes droughts.
- Extended periods of above-average temperatures often occur with droughts.
- Heat index estimates the effect on the human body when the air is hot and the humidity is high.
- Too little heat and a stalled jet stream can cause a cold wave, or an extended period of cold weather.
- Windchill index tells you how wind and temperature affect your body in winter.

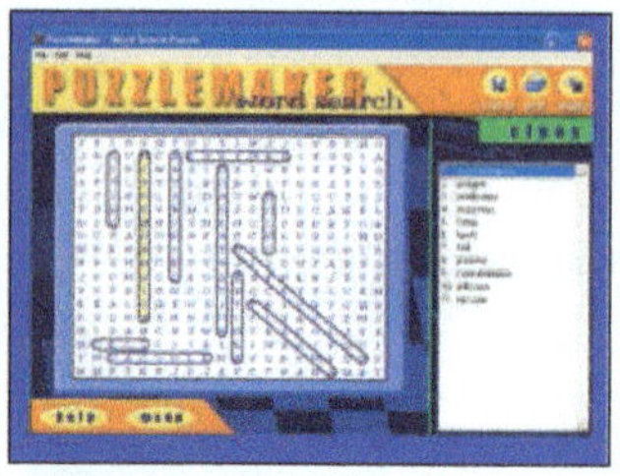

Vocabulary Practice

For additional practice with vocabulary, have students access the Vocabulary PuzzleMaker.

Chapter Self-Check

VOCABULARY REVIEW

Choose the correct italicized vocabulary term to complete each sentence.

1. A(n) ________ thunderstorm is characterized by temperature differences within a mass of air. *frontal, severe, air-mass*

2. Intense, self-sustaining thunderstorms are known as ________. *downbursts, tornadoes, supercells*

Compare and contrast the following pairs of vocabulary terms.

3. cold wave, windchill factor

4. eye, eyewall

5. air-mass thunderstorm, frontal thunderstorm

6. stepped leader, return stroke

Replace the underlined term with the correct one from the vocabulary list on the Study Guide.

7. More lives are lost during a hurricane's <u>heat wave</u> than from its winds.

8. A microburst is a form of <u>weather</u> that is difficult to predict.

9. Another name for a typhoon is a <u>forecast</u>.

10. A <u>weather map</u> is an extended period of extreme cold in one area.

11. The <u>Enhanced Fujita scale</u> indicates the wind speeds associated with hurricanes.

UNDERSTAND KEY CONCEPTS

12. Which would work against the development of a thunderstorm?
A. rising air
B. stable air
C. moisture
D. unstable air

13. Which does not describe a type of damaging thunderstorm wind?
A. downburst
B. microburst
C. land breeze
D. macroburst

Use the diagram below to answer Question 14.

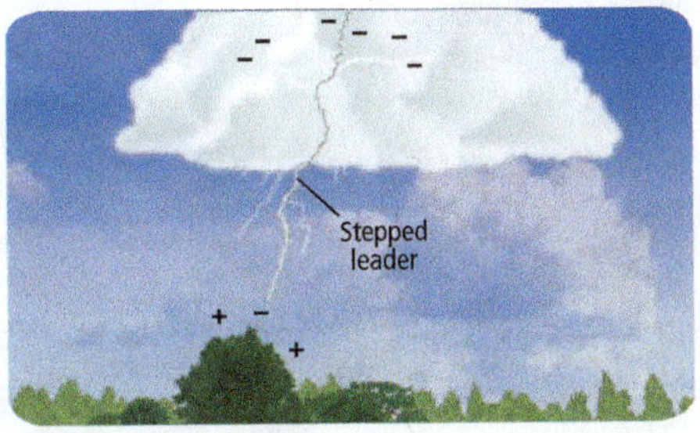

14. What phrase describes a stepped leader?
A. return stroke
B. partially charged air
C. positive charge
D. downdraft

15. Which does not play a key role in the development of hail?
A. supercooled water
B. freezing temperatures
C. warm ocean water
D. strong updrafts

16. Heat waves involve high-pressure systems that cause air to sink and warm by which process?
A. compression
B. conduction
C. evaporation
D. condensation

17. Which weather hazard involves a lack of moisture?
A. hail
B. drought
C. storm surge
D. flood

CHAPTER 13 ASSESSMENT

VOCABULARY REVIEW

1. air-mass
2. supercells
3. A cold wave is an extended period of below-average temperatures. The wind chill tells you what the temperature feels like to your body.
4. The hurricane eye is the low-pressure, calm wind center of a hurricane. The eyewall is the outer edge of the eye where the highest wind speeds are recorded in these storms.
5. Air-mass thunderstorms are produced by large changes in temperature within a single air mass, while frontal thunderstorms are produced by advancing cold fronts and sometimes advancing warm fronts.
6. The stepped leader of a lightning stroke is invisible and moves downward. The return stroke is the visible strike and moves generally upward.
7. storm surge
8. downburst
9. hurricane or tropical cyclone
10. cold wave
11. Saffir-Simpson Hurricane Wind scale

UNDERSTAND KEY CONCEPTS

12. B
13. C
14. B
15. C
16. A
17. B

18. A
19. C
20. A
21. B
22. D
23. B

CONSTRUCTED RESPONSE

24. When humidity reaches 50 percent or higher, the temperature seems higher than it is because of the reduction in evaporation of moisture from the skin.
25. Wind speed of a Category 4 hurricane can be double that of a Category 1 hurricane.
26. The storm surge will be to the right of the hurricane eye regardless of the approach angle due to the rotation. The evacuation will be to higher ground, but will also be in directions away from landfall; remember, the Coriolis effect will drag the storm to the right from its current direction. Area of landfall will vary with the storm.
27. Hurricanes are severe storms that form over water, last for days, can be several hundred km in diameter, and produce winds in excess of 250 km/h. Tornadoes are severe storms that form over land, usually last for an hour or less, can be several km in diameter, and produce winds in excess of 300 km/h. Hurricanes can spawn tornadoes.
28. Macrobursts cause a path of destruction up to 5 km wide, have wind speeds of more than 200 km/h, and can last up to 30 min. Microbursts affect areas of less than 4 km, can have winds exceeding 250 km/h, and last less than 10 min.
29. Plans should include safety tips such as seeking shelter during a tornado or moving to higher ground during a flood.

ASSESSMENT

18. Flooding is most likely to take place because of rains associated with which type of storm?
A. Category 5 hurricane moving at 25 m/s
B. EF2 tornado moving at 10 m/s
C. A stationary tropical storm
D. thunderstorm moving at 2 m/s

Use this photo to answer Question 19.

19. Which part of the tropical cyclone is not visible in the photo above?
A. eye
B. eyewall
C. storm surge
D. cirrus overcast

20. What percentage of tornadoes are classified as EF4 or EF5 on the Enhanced Fujita scale?
A. 1 percent
B. 10 percent
C. 50 percent
D. 75 percent

21. Which factor, if increased, would increase the chance that a severe thunderstorm would occur?
A. upper-level temperature
B. surface moisture
C. strength of the jet stream
D. conduction

22. In which ocean would you not expect to experience a tropical cyclone?
A. West Pacific
B. Indian
C. North Atlantic
D. South Atlantic

23. What weather events are cold waves most often associated with?
A. floods
B. polar high-pressure systems
C. tropical high-pressure systems
D. droughts

CONSTRUCTED RESPONSE

24. Determine the effect humidity has on the heat index.

25. Compare and contrast wind speeds of Category 1 and Category 4 hurricanes using **Figure 15.**

Use the photo below to answer Question 26.

26. Predict the area of landfall, the areas affected by storm surge, and evacuation routes for this hurricane.

27. Compare and contrast tornadoes and hurricanes.

28. Point out the differences between a microburst and a macroburst.

29. Develop a plan of safety for a family that might encounter flooding, lightning, and a tornado in the area where they live.

THINK CRITICALLY

30. Explain how temperature and condensation are limiting factors to the growth of a thunderstorm.

31. Distinguish how an investigator would differentiate between a microburst and a tornado.

32. Point out the features of the South Atlantic that might deter the formation of hurricanes.

THINK CRITICALLY

30. A thunderstorm will continue to grow as condensation occurs and latent heat is released. Cool atmospheric temperatures cause condensation, but the release of latent heat creates instability as it produces temperature differences between the cloud and atmosphere.
31. Having the wind speed and strength of an EF2 or EF3 tornado, a microburst only affects the area similar to an EF0 or EF1 tornado. Microbursts do not exhibit a twisting pattern of the debris they create, and their duration is generally half that of a tornado.
32. The main features of the South Atlantic are cooler surface waters and air temperatures that are not conducive to storm building as well as higher levels of wind shear.

Chapter Self-Check

33. **Explain** why supercells that produce tornadoes also often produce large hailstones.

34. **Predict** why extreme cold waves are more common in the northern hemisphere than in the southern hemisphere.

35. **Describe** how cold fronts are more likely to produce severe thunderstorms than warm fronts.

Use the diagram below to answer Questions 36 and 37.

36. **Explain** how the wind shears that produce a tornado go from horizontal to vertical.

37. **Identify** what causes the lift for the vertical motion.

38. **Careers in Earth Science** Imagine you work for the National Weather Service and it is your job to write public service announcements. Write a safety plan for people who live in places where hurricanes are frequent.

CONCEPT MAPPING

39. Make a concept map to differentiate among thunderstorms, tornadoes, and tropical cyclones.

CHALLENGE QUESTION

40. **Appraise** the use of the windchill factor to determine school delays and work closings versus temperature, daylight, or icy conditions alone.

WRITING IN ▶ Earth Science

41. Imagine that you are a weather research scientist researching a way to stop hurricanes. Tell about your research that includes ways to cool the ocean's surface.

DBQ Document–Based Questions

Data obtained from: National Oceanic and Atmospheric Administration, 2004. *Hurricane Research Division.*

When a hurricane passes over the surface of the ocean, the sea surface temperature (SST) can become several degrees cooler. This is the result of cooler water being churned up to the surface from lower levels. The radial distance is the distance from the eye (TC center) measured in degrees of latitude.

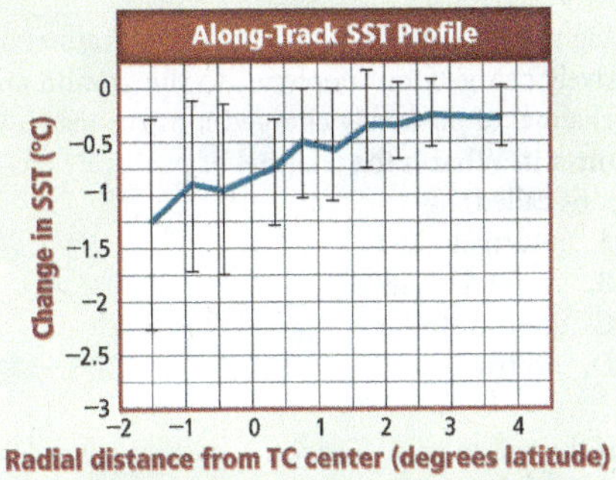

42. What is the estimated range of temperature change for each radial distance indicated?

43. A difference of 0.5°C can be the difference between a storm that intensifies and one that stops developing. At what distance from the TC center would that range be most critical?

44. What factors in a hurricane might cause the increased temperature changes to happen closer to the eye than at the edges?

CUMULATIVE REVIEW

45. What are the two most abundant elements in Earth's crust? **(Chapter 4)**

46. Describe why radiosonde data is important. **(Chapter 12)**

CHAPTER 13 ASSESSMENT

33. Strong updrafts are required for both.
34. Polar continental high-pressure areas form over land near the poles. There is a larger percent of land mass in the northern hemisphere and much of it is near the poles, so there is more opportunity for cold waves to form.
35. In a cold front, warm air is moved upward rapidly. The larger the difference in temperature and the more rapidly the warm air is lifted and condensed, the greater the incidence of severe storm formation. In a warm front, warm air masses are lifted more gradually over longer periods of time, and there is less chance of severe weather forming.
36. As the wind shears cause an air mass to roll, it can be lifted into a vertical column to form a tornado.
37. Possible answer: convection currents
38. Answers will vary, but should include evacuation to higher ground and away from the coast.

CONCEPT MAPPING

39. Maps should have features of each.

CHALLENGE QUESTION

40. Answers will vary. Look for information about surface temperature cooling.

WRITING IN ▶ Earth Science

Rubric

41. Writing answers will vary, but should include plausible scenarios.

DBQ Document-Based Questions

Data obtained from: National Oceanic and Atmospheric Administration, 2004. *Hurricane Research Division.*

42.

Dist.	Range
−2	−0.25 to −2.25
−1	−0.1 to −1.75
0	−0.2 to −1.8
1	0 to −1.1
2	0.1 to −0.75
3	0.1 to −0.5
4	0 to −0.5

43. −1.5° latitude
44. The regions closest to the eye are where the storm is most intense. The area from the eye (−2) to about 1° is where the temperature changes are greatest (steepest slope). Higher wind speeds near the eye would cause more churning, cooling, and lifting than at the edges.

CUMULATIVE REVIEW

45. oxygen and silicon
46. Radiosondes gather temperature, pressure, moisture and wind data from upper portions of the atmosphere, including the jet stream where important weather changes take place.

MULTIPLE CHOICE

1. D
2. A
3. A
4. D
5. B
6. A
7. B
8. B
9. A

CUMULATIVE STANDARDIZED TEST PRACTICE

MULTIPLE CHOICE

Use the illustration to answer Questions 1 and 2.

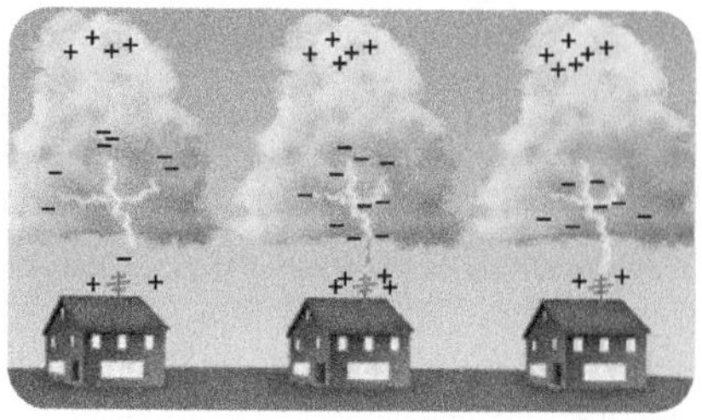

1. With what type of cloud is lightning associated?
A. altocumulus
B. stratocumulus
C. cirrus
D. cumulonimbus

2. Lightning occurs when an invisible channel of negatively charged ions descends to the ground and a channel of positively charged ions rushes upward to meet it. What is the channel of positively charged ions called?
A. return stroke
B. stepped leader
C. ground stroke
D. electronic leader

3. If the soil in an A-horizon is dark in color, what does this imply?
A. The layer of soil is rich in humus.
B. The layer of soil is extremely fertile.
C. The layer of soil is from a poorly drained area.
D. The layer of soil is rich with iron materials.

4. Where is the majority of freshwater found?
A. in the atmosphere
B. underground
C. in rivers, streams, and lakes
D. in polar ice caps and glaciers

5. Why is the Geographic Information System (GIS) beneficial to science?
A. It is very similar to traditional mapping.
B. It can be used by scientists in many different disciplines.
C. It limits maps to just one layer of information.
D. It does not change with new information.

6. What does Doppler radar monitor?
A. the motion of moving raindrops
B. atmospheric pressure
C. temperature, air pressure, and humidity
D. the height of cloud layers

7. The data gathered by Doppler radar can be used to make a type of forecast that relies on numerical data. What is this type of forecast called?
A. an analog forecast
B. a digital forecast
C. an isopleth
D. ASOS

Use the illustration below to answer Questions 8 and 9.

8. What process is being demonstrated by the circulating arrows?
A. radiation
B. a convection current
C. a wind current
D. a conduction current

9. Describe the process of radiation to warm Earth's surface.
A. It is one of the methods that transfer energy from the Sun through different forms of electromagnetic waves to warm Earth.
B. It transfers energy through the collision of molecules to warm Earth.
C. It transfers energy through the flow of a heated substance to warm Earth.
D. It allows the atmosphere to absorb the Sun's rays or reflect them back into space.

SHORT ANSWER

Use the diagram below to answer Questions 11–13.

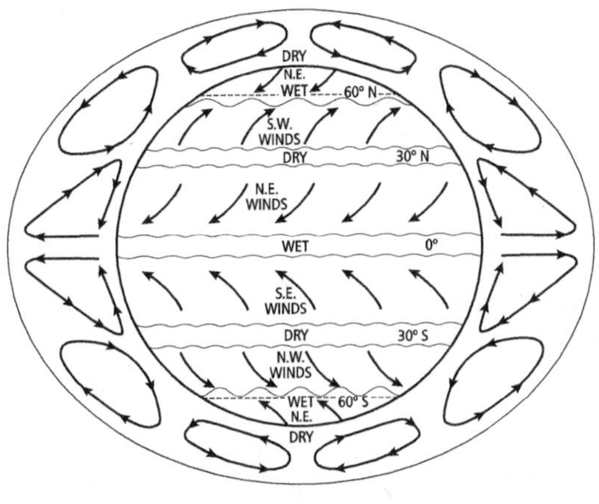

10. How does the rotation of Earth affect wind systems?

11. Analyze the wind pattern occurring between the equator and 30° north and south latitudes.

12. Why would it benefit sailors to know what type of wind system they are traveling in?

13. Why do most tornadoes form in the spring during the late afternoon and evening?

14. Describe a step farmers can take to improve soil fertility.

15. Explain the properties of a geyser.

16. How does the amount of water on Earth change as a result of the water cycle?

READING FOR COMPREHENSION

Hurricane Preparedness

The Saffir-Simpson scale, used by the National Weather Service since the 1970s to classify hurricane strength, ranks storm strength from Level 1 to 5, and is used to give an estimate of the possible property damage and flooding expected when the hurricane makes landfall. A Category 5 storm maintains winds in excess of 155 mph and is capable of causing widespread damages such as destroying or causing extensive structural damage to homes. It can also trigger a storm surge more than 18 ft above normal.

Two of the most fundamental steps people can take are assembling a disaster supplies kit and making an emergency plan. They are simple measures, but can make all the difference when disasters strike.

Article obtained from: Booth, M. Hurricane Isabel strengthens to Category 5. *In the News.* September 12, 2006. (Online resource accessed October 2006.)

17. For what purpose is the Saffir-Simpson scale used?
    - **A.** estimating property damage and flooding when a hurricane makes landfall
    - **B.** estimating the height of a storm surge above normal
    - **C.** determining who should be told to evacuate before a hurricane hits
    - **D.** measuring the strength of an approaching hurricane

18. What can be inferred from this passage?
    - **A.** Only Category 5 hurricanes will cause major damage.
    - **B.** Only people living along the shoreline are in danger during hurricanes.
    - **C.** Category 5 hurricanes have the potential to cause major damage.
    - **D.** Category 5 hurricanes always cause flooding.

NEED EXTRA HELP?

If You Missed Question . . .	1	2	3	4	5	6	7	8	9	10	11	12	13	14	15	16
Review Section . . .	13.1	13.2	7.3	10.1	2.3	12.3	12.3	11.1	11.1	12.2	12.2	12.2	13.2	7.3	10.1	11.3

SHORT ANSWER

10. Earth's rotation creates the Coriolis effect, causing air in the northern hemisphere to move right and move left in the southern hemisphere. This movement, combined with uneven heating of Earth, leads to wind systems.

11. The winds located between the equator and 30° north and south are known as the trade winds. Here the air sinks, warms, and moves in an easterly direction toward the equator. At the equator, it rises again and moves back toward latitude 30°.

12. In the different wind systems, the air is moving in different directions and with varying levels of strength. By knowing the wind system in which they were sailing, sailors could best determine what route to travel in order to move the most quickly.

13. During this time of year, the temperature contrasts between polar air and tropical air are the greatest. These contrasts can develop supercells, which are capable of producing several tornadoes.

14. Sample answer: Pulverized limestone can be added to the soil. This reduces the acidity in the soil, enhancing crop growth.

15. A geyser is an explosive hot spring that erupts at regular intervals. It is produced by Earth's underground heat located near a volcanic region.

16. It remains constant.

READING FOR COMPREHENSION

17. A

18. C

CHAPTER 14 Climate

BIGIDEA **The different climates on Earth are influenced by natural factors as well as human activities.**

ESSENTIAL QUESTIONS	NATIONAL STANDARDS	RESOURCES TO ASSESS MASTERY
SECTION 1 Defining Climate 1. What are limits associated with the use of normals? 2. Why do climates vary? 3. How do temperatures in different regions on Earth differ? 2 sessions — 1 block	UCP.1–3; A.1, A.2; D.1	**Entry-Level Assessment** Focus Transparency 40, p. 376 **Progress Monitoring** Caption Question, p. 378 Reading Check, p. 377, p. 379 Section Review, p. 380
SECTION 2 Climate Classification 1. What are the criteria used to classify climates? 2. How are different climates described? 3. What are microclimates? 2 sessions — 1 block	UCP.1–3; D.1; F.5	**Entry-Level Assessment** Focus Transparency 41, p. 381 **Progress Monitoring** Caption Question, pp. 381, 382, 383, 385 Reading Check, pp. 383, 384 Section Review, p. 386
SECTION 3 Climatic Changes 1. What is the difference between long-term and short-term climatic changes? 2. What are natural causes of climate change? 3. Why do climatic changes occur? 2 sessions — 1 block	UCP.1–3; D.1–3; F.5–6; G.3	**Entry-Level Assessment** Focus Transparency 42, p. 387 **Progress Monitoring** Caption Question, pp. 387, 390, 392 Reading Check, p. 391 Section Review, p. 392
SECTION 4 Impact of Human Activities 1. What is the greenhouse effect? 2. What is global warming? 3. How do humans impact climate? 4 sessions — 2 blocks	UCP.1–3; A.1, A.2; D.1, D.3; F.3–6; G.3	**Entry-Level Assessment** Focus Transparency 43, p. 393 **Progress Monitoring** Caption Question, p. 395 Reading Check, p. 394 Section Review, p. 395 **Summative Assessment** Chapter Assessment, p. 399 Exam*View®Assessment Suite* CD-ROM

Go online!
connectED.mcgraw-hill.com

LEVELED RESOURCES	LAB MATERIALS
Science Notebook 14.1 OL **Chapter FAST FILE Resources:** Study Guide, p. 95 BL **Lab Resources:** Laboratory Manual, pp. 105 OL **Transparencies:** Section Focus Transparency 40 BL EL Teaching Transparency 38 OL EL	LaunchLAB p. 374 / **15 min lab set up, 5 min for 4 days, 15 min last day** dark construction paper (2), rock, umbrella
Science Notebook 14. 2 OL **Chapter FAST FILE Resources:** GeoLab Worksheet, p. 83 OL Study Guide, p. 96 BL **Lab Resources:** Laboratory Manual, pp. 109 OL **Transparencies:** Section Focus Transparency 41 BL EL Teaching Transparency 39 OL EL	GeoLAB p. 397 / **90 min** thermometer, psychrometer, paper strip or wind sock, meterstick, relative humidity chart
Science Notebook 14. 3 OL **Chapter FAST FILE Resources:** Study Guide, p. 97 BL **Transparencies:** Section Focus Transparency 42 BL EL Teaching Transparency 40 OL EL	
Science Notebook 14. 4 OL **Chapter FAST FILE Resources:** MiniLab Worksheet, p. 82 OL Study Guide, p. 99 BL **Transparencies:** Section Focus Transparency 43 BL EL Teaching Transparency 41 OL EL	MiniLAB p. 394 / **45 min** cardboard box, thermometers (2), clean glass jar

LabManager™
Customize any lab with the LabManager™ CD-ROM.

ADDITIONAL RESOURCES

Technology:
- eTeacherEdition
- eStudentEdition
- Classroom Presentation Toolkit
- Exam*View® Assessment Suite* CD-ROM
- LabManager™ CD-ROM
- What's EARTH SCIENCE Got To Do With It? DVD
- Weather Classroom Video DVD

Assessment:
- Blueprints for Success
- Chapter Test (Scaffolded)
- Earth Science Standardized Test Practice

Lab Resources:
- Exploring Environmental Problems Laboratory Manual

BL Below Level | OL On Level | AL Advanced Learners | EL English Learners | COOP LEARN Cooperative Learning

CHAPTER 14

LaunchLAB

iLab Station | Rubric

How can you model cloud cover?

Process Skills observe and infer, recognize cause and effect, model

Safety Precautions Approve lab safety forms before work begins. Caution students to be careful when handling the umbrella; its pointed tip can cause injuries.

Teaching Strategies

- Monitor weather forecasts to identify the best nights to conduct this lab. One of the four days should be a calm, cool night. A variety of weather patterns will give the students data to compare and contrast.
- Tell students to be sure the umbrella completely covers the paper.

Procedure

1. Have students read and complete the lab safety form and follow the procedure below.
2. Lay two sheets of **dark construction paper** on the grass in an open area. Place a **rock** on each sheet of paper to prevent it from blowing away.
3. Open an **umbrella** and anchor it in the ground over one of the sheets of paper.
4. During each of the next four days, observe what happens to the sheets of paper.

Analysis

1. **Describe** any differences in dew formation that you observed each day. No dew or frost forms on the paper under the umbrella, while the paper exposed to the open sky collects dew or frost.
2. **Explain** How is the umbrella in this activity similar to clouds in the atmosphere? The umbrella models cloud cover by trapping surface heat and reflecting solar radiation.
3. **Infer** how temperatures during the night might differ between climates with extensive cloud cover and climates with few clouds. On cloudy nights, clouds trap heat from Earth's surface. Temperatures in the cloudy region would be warmer at night than temperatures in the less cloudy region.

CHAPTER 14

Climate

BIGIDEA The different climates on Earth are influenced by natural factors as well as human activities.

SECTIONS

1 Defining Climate
2 Climate Classification
3 Climatic Changes
4 Impact of Human Activities

Temperate rain forest

The temperate rain forests of Olympic National Park in Washington state, receive up to 500 cm of precipitation each year. In contrast, desert areas receive less than 25 cm of precipitation annually.

LaunchLAB

iLab Station

How can you model cloud cover?

Some areas are generally cloudier than others. This affects both the temperature and the amount of precipitation that these areas receive. Discover how cloud cover forms in this lab.

FOLDABLES® Study Organizer

Climate Classification

Make a layered-look book using the labels shown. Use it to organize your notes on the five main types of climates.

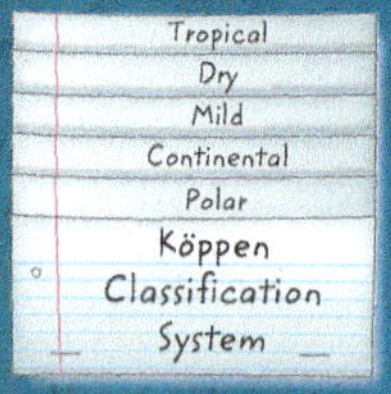

Assessment

Knowledge Ask students to infer what might be the best place to pitch a tent on a camping trip to avoid having a wet tent in the morning.

LabManager™
Customize any lab with the LabManager™ CD-ROM.

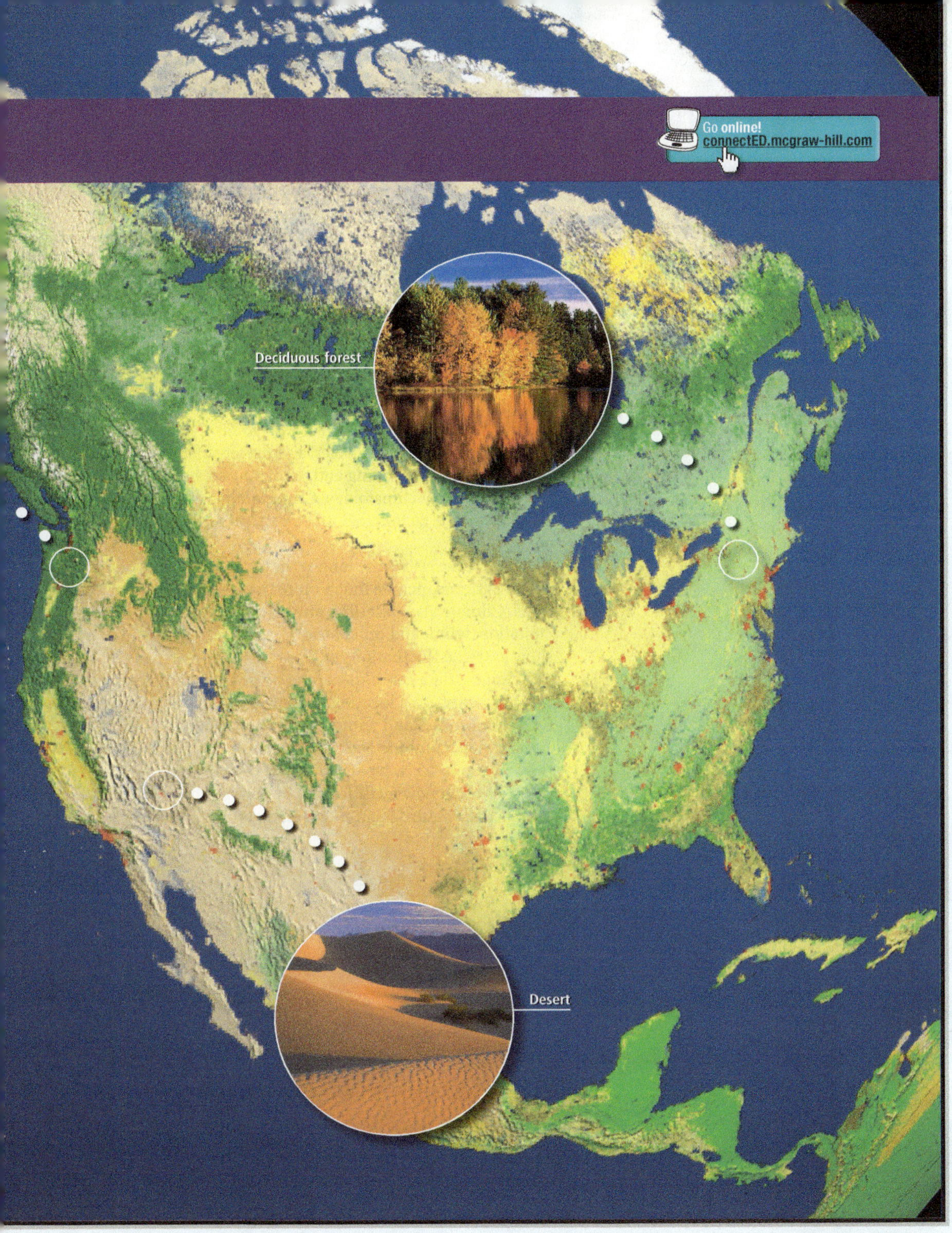

Introduce the BIGIDEA

Differences in Climates No one sitting in the middle of a raging blizzard would be likely to mistake his or her location for a beach in sunny Florida or Southern California. This is because we know that different parts of the world have distinctive types of weather. Have students give examples of and describe different types of climate. Descriptions should include details about temperature and the amount of precipitation a region receives.

Teacher Content Support

Characteristics of Climates The photos show examples of climate regions in North America. Explain that temperate rain forests occur in the Pacific Northwest in a marine west-coast climate characterized by mild winters, cool summers, and high amounts of precipitation. Remind students that deciduous trees are trees that lose their leaves during winter months. What characteristics can be inferred about the climate in which deciduous trees live? There are temperature and precipitation differences between each season. Have students describe the characteristics of a desert. low precipitation, high temperatures during the day

SECTION 1

1 Focus

Focus Transparency

Before presenting the lesson, project **Section Focus Transparency 40** and have students answer the accompanying questions. EL

MAINIDEA

Define Climate Ask students if they would enjoy a trip to sunbathe on the beach—in January. Discuss how in Florida, weather conditions would be appropriate for this activity; however, they would not be in Maine. Point out this knowledge comes from understanding the climate differences between the two locations.

2 Teach

Teacher Content Support

Climate Change Climate is the average weather of an area over a long period of time. Average weather, however, is based on values that can vary widely. For example, a climate with an average annual temperature of 18°C could easily experience temperatures as high as 38°C or as low as –5°C during a given year. A change in long-term average annual temperature of just 2°C over a period of 30 years, however, would be considered a massive climatic change. Climate, then, involves the time scale over which changes take place, while weather reflects the impacts of short-term changes.

SECTION 1

Defining Climate

MAINIDEA Climate is affected by several factors including latitude and elevation.

Essential Questions

- What are limits associated with the use of normals?
- Why do climates vary?
- How do temperatures in different regions on Earth differ?

Review Vocabulary

jet stream: a high-altitude, narrow, westerly wind band that occurs above large temperature changes

New Vocabulary

climatology
normal
tropics
temperate zones
polar zones

EARTH SCIENCE 4 YOU Just because you observed someone eating a steak dinner, you probably would not assume that they ate steak for every meal. In nature, taking a one-day "snapshot" of the weather does not necessarily describe what that location experiences over the course of many days.

Annual Averages and Variations

Fifty thousand years ago, the United States had much different weather patterns than those that exist today. The average temperature was several degrees cooler, and the jet stream was probably farther south. Understanding and predicting such climatic changes are the basic goals of climatology. **Climatology** is the study of Earth's climate and the factors that cause past, present, and future climatic changes.

Climate describes the long-term weather patterns of an area. These patterns include much more than average weather conditions. Climate also describes annual variations of temperature, precipitation, wind, and other weather variables. Studies of climate show extreme fluctuations of these variables over time. For example, climatic data can indicate the warmest and coldest temperatures recorded for a location. **Figure 1** shows weather differences between summer and winter in Chicago, Illinois. This type of information, combined with comparisons between recent conditions and long-term averages, can be used by businesses to decide where to build new facilities and by people who have medical conditions that require them to live in certain climates.

Figure 1 Climate data include the warmest and coldest temperatures recorded for a location. The highest temperature on record for Chicago, IL, is 41°C, which occurred in July 1934. The lowest temperature on record for Chicago, IL, is –33°C, which occurred in January 1985.

Chicago, IL, in the summer

Chicago, IL, in the winter

EARTH SCIENCE JOURNAL

Degree Days Tell students that a degree day is defined as the difference between the actual daily average temperature and an arbitrary daily average value—usually 18°C. Ask students to research and write reports in their Earth science journals that describe cooling- and heating-degree days, and their uses. A cooling-degree day occurs when the actual average daily temperature is warmer than 18°C; the day is classified as a cooling-degree day because temperatures must cool off to reach 18°C. A heating-degree day occurs when the actual average daily temperature is colder than 18°C. Degree days are particularly useful for energy and utility companies that must plan for increases in consumer energy demands.

Rubric

Normals The data used to describe an area's climate are compiled from meteorological records, which are continuously gathered at thousands of locations around the world. These data include daily high and low temperatures, amounts of rainfall, wind speed and direction, humidity, and air pressure. The data are averaged on a monthly or annual basis for a period of at least 30 years to determine the **normals,** which are the standard values for a location.

☑ READING CHECK **Identify** data that can be used to calculate normals.

Limitations of normals While normals offer valuable information, they must be used with caution. Weather conditions on any given day might differ widely from normals. For instance, the normal high temperature in January for a city might be 0°C. However, it is possible that no single day in January had a high of exactly 0°C. Normals are not intended to describe usual weather conditions; they are the average values over a long period of time.

While climate describes the average weather conditions for a region, normals apply only to the specific place where the meteorological data were collected. Most meteorological data are gathered at airports, which cannot operate without up-to-date, accurate weather information. However, many airports are located outside city limits. When climatic normals are based on airport data, they might differ from actual weather conditions in nearby cities. Changes in elevation and other factors, such as proximity to large bodies of water, can cause climates to vary.

CAREERS IN EARTH SCIENCE

Climatologist Scientists who study long-term trends in climate are called climatologists. Climatologists might collect data by sampling ice cores or taking ocean water temperatures.

WebQuest

Data Analysis LAB

Based on Real Data*

Interpret the Data

What is the temperature in Phoenix, Arizona? The table contains temperature data for Phoenix, Arizona, based on data collected from July 1, 1948, through April 30, 2010.

Analysis

1. Plot the monthly values for average maximum temperatures. Place the month on the *x*-axis and temperature on the *y*-axis.
2. Repeat Step 1 using the monthly values for the average minimum temperatures.

Think Critically

3. **Identify** the months that were warmer than the average maximum temperature.
4. **Identify** the months that were colder than the average minimum temperature.
5. **Infer** What is the climate of Phoenix, Arizona, based on average temperatures?

Data and Observations

Monthly Temperature Summary for Phoenix, AZ

Temperature (°C)	Jan	Feb	Mar	Apr	May	Jun	Jul	Aug	Sep	Oct	Nov	Dec	Average
Average maximum	19	21	24	29	34	39	41	40	37	31	24	19	29.8
Average minimum	5	7	9	13	18	23	27	26	23	16	9	5	15.1

*Data obtained from: Western Regional Climate Center. 2010.

☑ READING CHECK daily high and low temperatures, amounts of rainfall, wind speed and direction, humidity, air pressure

CAREERS IN EARTH SCIENCE

WebQuest

Data Analysis LAB

About the Lab

- Students will become familiar with the use of climatological data.
- Students will need graph paper, a pencil, and a calculator to complete the lab.
- Ask students whether they have ever heard weather forecasters say that temperature or precipitation is above or below normal. Explain that the forecasters are comparing current data to a set of established long-term averages, or normals, and that changes in these values can be used to determine climatic trends.
- More information about average temperatures can be found on NOAA's Web site.

Think Critically

3. May, June, July, August, September, and October
4. January, February, March, April, November, and December
5. Temperatures are warm during the winter months and hot in the summer during the day. Temperatures at night during the winter are cooler.

Collaborative Learning

Calculate Normals Temperature and precipitation normals are calculated over a period of 30 years. Ask students to discuss why such a long period is used. Annual temperatures can fluctuate greatly from year to year. A period of 30 years is necessary to filter out short-term changes and to identify only those changes that truly represent long-term trends. COOP LEARN

Interpret the Illustration

The Effect of Latitude on Climate Have students study **Figure 2.** Students have learned that the tropics are the source of warm air masses and that the poles are the source of cold air masses. Tell students climates follow a similar pattern. Average temperatures in polar regions are generally quite cold, while those in the tropics are quite warm. Based on what students know about air masses in the middle latitudes and about prevailing westerlies, ask students to infer the probable climate of those latitudes. Midlatitudes exhibit characteristics of both the tropics and the poles because warm, tropical air masses and cold, polar air masses often invade the region. For this same reason, midlatitudes also have the greatest contrast in temperatures.

Teacher Content Support

Early Climatology The first nucleus of a climatological network in the United States began on May 2, 1814, when then Army Surgeon General, Dr. J. Tilton, ordered hospital surgeons to "keep a diary of the Weather." The resulting reports were first published in 1826 by Surgeon General Dr. J. Lovell. In 1842, another young surgeon, Dr. S. Forry, produced the first scientific climatological study of the United States, titled "The Climate of the United States and Its Endemic Influence."

Caption Question **Fig. 2** The angle decreases.

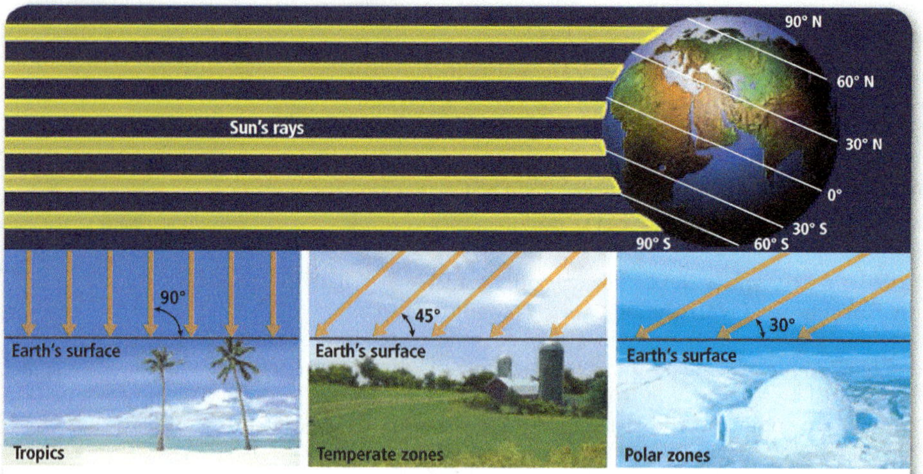

Figure 2 Latitude has a great effect on climate. The amount of solar radiation received on Earth decreases from the equator to the poles.
Describe *what happens to the angle at which the Sun's rays hit Earth's surface as one moves from the equator to the poles.*

Causes of Climate

You probably know from watching the weather reports that climates around the country vary greatly. For example, on average, daily temperatures are much warmer in Dallas, Texas, than in Minneapolis, Minnesota. There are several reasons for such climatic variations, including differences in latitude, topography, closeness of lakes and oceans, availability of moisture, global wind patterns, ocean currents, and air masses.

Latitude Recall that different parts of Earth receive different amounts of solar radiation. The amount of solar radiation received by any one place varies because Earth is tilted on its axis, and this affects how the Sun's rays strike Earth's surface. The area between 23.5° S and 23.5° N of the equator is known as the **tropics.** As **Figure 2** shows, tropical areas receive the most solar radiation because the Sun's rays are nearly perpendicular to Earth's surface. As you might expect, temperatures in the tropics are generally warm year-round. For example, Caracas, Venezuela, located at about 10° N, enjoys average maximum temperatures between 24°C and 27°C year-round. The **temperate zones** lie between 23.5° and 66.5° north and south of the equator. As their name implies, temperatures in these regions are moderate. The **polar zones** are located from 66.5° north and south of the equator to the poles. Solar radiation strikes the polar zones at a low angle. Thus, polar temperatures tend to be cold. Thule, Greenland, located at 77° N, has average maximum temperatures between −20°C and 8°C year-round.

VOCABULARY
ACADEMIC VOCABULARY
Imply
to indicate by association rather than by direct statement
The title of the movie implied that it was a love story.

DIFFERENTIATED INSTRUCTION

Struggling Learners Place a thermometer in a pan of warm water and put the pan in a freezer. Place another thermometer in the freezer; it should not touch the water. Record the temperatures of the thermometers five minutes later. Tell students that the wet thermometer, which represents the oceans, recorded a relatively high temperature because water gains and loses heat slowly. In contrast, the dry thermometer cooled off quickly; it represents a landmass in this model.

Topographic effects Water heats up and cools down more slowly than land. Thus, large bodies of water affect the climates of coastal areas. Many coastal regions are warmer in the winter and cooler in the summer than inland areas at similar latitudes.

Also, temperatures in the lower atmosphere generally decrease with altitude. Thus, mountain climates are usually cooler than those at sea level. In addition, climates often differ on either side of a mountain. Air rises up one side of a mountain as a result of orographic lifting. The rising air cools, condenses, and drops its moisture, as shown in **Figure 3.** The climate on this side of the mountain—the windward side—is usually wet and cool. On the opposite side of the mountain—the leeward side—the air is drier, and it warms as it descends. For this reason, deserts are common on the leeward side of mountains.

☑ READING CHECK **Explain** how large bodies of water affect the climate of coastal areas.

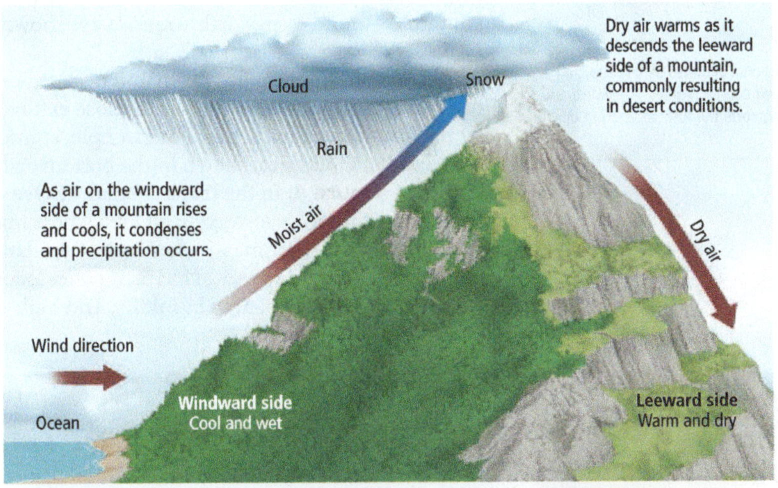

■ **Figure 3** Orographic lifting leads to rain on the windward side of a mountain. The leeward side is usually dry and warm.

Windward side of mountains on Maui, Hawaii

Leeward side of mountains on Maui, Hawaii

Project

Climatological Data For most large cities, the National Climatic Data Center records climatological data. Have students each select a large city and research that city's climatological data. Students should record maximum and minimum temperatures, monthly average temperatures, and average annual rainfall and snowfall. Have students compare their data with those of other students. OL

Apply Earth Science

Orographic Lifting Have students research the monthly temperature and precipitation averages for two cities on opposite sides of the Cascade Mountain range: Seattle, WA and Spokane, WA. Which city is on the windward side of the mountain? Explain your answer. Seattle is on the windward side. It has a higher amount of precipitation per year and lower temperatures in the summer compared to Spokane.

☑ READING CHECK Water heats up and cools down more slowly than land. Coastal areas are warmer in the winter and cooler in the summer than inland areas at similar latitudes.

ACROSS THE CURRICULUM

Biology Organisms that live in the desert have physiological and behavioral adaptations that help them survive in extreme conditions. For example, kangaroo rats live in burrows and are nocturnal, moving around at night when it is cooler. Other animals are active and search for food during dawn and dusk, conserving energy and staying cool during the hottest hours of the day.

3 Assess

Check for Understanding

Identify Ask students to identify three different factors that can influence climate. variations in latitude, proximity to large bodies of water, and mountains

Reteach

Collect Data Divide students into three groups. Have each group research the climate of a different location in the tropics, the temperate zones, and the polar regions, respectively. Ask each group of students to develop a sample set of monthly average temperatures for their location. Student groups should share their data with other groups. COOP LEARN

Assessment

Knowledge Ask students to explain why climatological normals do not necessarily describe normal weather conditions. Normals represent standard averages rather than any one actual value.

Major Air Masses Over North America

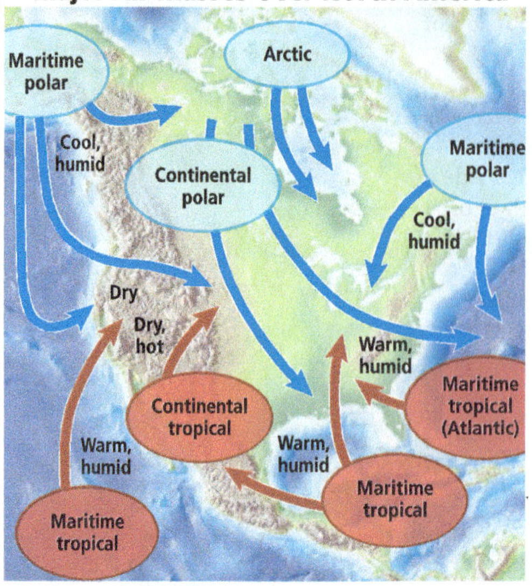

Figure 4 Air masses affect regional climates by transporting the temperature and humidity of their source regions. The warm and humid maritime tropical air mass supports the lush vegetation on the island of Dominica.

Lush vegetation on the Caribbean island of Dominica

Air masses Two of the main causes of weather are the movement and interaction of air masses. Air masses also affect climate. Recall that air masses have distinct regions of origin, caused primarily by differences in the amount of solar radiation. The properties of air masses also depend on whether they formed over land or water. The air masses commonly found over North America are shown in **Figure 4.**

Average weather conditions in and near regions of air-mass formation are similar to those exhibited by the air masses themselves. For example, consider the island of Dominica, shown in the bottom right corner of **Figure 4,** in the tropical Atlantic Ocean. Because this island is located in an area where maritime tropical (mT) air masses dominate, the island's climate has maritime tropical characteristics, such as warm temperatures, high humidity, and high amounts of precipitation.

SECTION 1 REVIEW

Section Self-Check

Section Summary

- Climate describes the long-term weather patterns of an area.
- Normals are the standard climatic values for a location.
- Temperatures vary among tropical, temperate, and polar zones.
- Climate is influenced by several different factors.
- Air masses have distinct regions of origin.

Understand Main Ideas

1. MAINIDEA **Describe** two factors that cause variations in climate.
2. **Identify** What are some limits associated with the use of normals?
3. **Compare and contrast** temperatures in the tropics, temperate zones, and polar zones.
4. **Infer** how climate data can be used by farmers.

Think Critically

5. **Assess** Average daily temperatures for City A, located at 15° S, are 5°C cooler than average daily temperatures for City B, located at 30° S. What might account for the cooler temperatures in City A, even though it is closer to the equator?

WRITING IN ▶ Earth Science

6. Write a hypothesis that explains why meteorological data gathered at an airport would differ from data gathered near a large lake. Assume all other factors are constant.

SECTION 1 REVIEW

1. Large bodies of water and mountains can both cause variations in climate. Because water heats up and cools down more slowly than land, many coastal areas are warmer in the winter and cooler in the summer than inland areas at similar latitudes. As a result of orographic lifting, the windward side of a mountain is usually cool and wet. The leeward side is drier and warmer.
2. Actual weather conditions can vary widely from normals. Also, normals apply only to the specific place where the data is collected.
3. Temperatures in the tropics are generally warm. In temperate zones, temperatures are moderate. The poles are nearly always cold.
4. Data can be used to determine average temperatures and rates of precipitation, as well as the length of the growing season.
5. Possible answer: City A is cooler because it is at a higher elevation than City B.
6. A large body of water heats up and cools down more slowly than land. Daily high temperatures might be slightly lower at the lake; overnight temperatures might be slightly higher at the lake.

Rubric

Essential Questions
- What are the criteria used to classify climates?
- How are different climates described?
- What are microclimates?

Review Vocabulary

precipitation: all solid and liquid forms of water—including rain, snow, sleet, and hail—that fall from clouds

New Vocabulary

Köppen classification system
microclimate
heat island

Climate Classification

MAINIDEA Climates are categorized according to the average temperatures and precipitation amounts.

EARTH SCIENCE 4 YOU What sort of place comes to mind when you think of a vacation in a tropical climate? A place with hot weather and a lot of rain? If so, you already know something about a tropical climate, even if you have never visited one.

Köppen Classification System

The graph on the left in **Figure 5** shows climate data for a desert in Reno, Nevada. The graph on the right shows climate data for a tropical rain forest in New Guinea. What criteria are used to classify the climates described in the graphs? Temperature is an obvious choice, as is amount of precipitation. The **Köppen classification system** is a classification system for climates that is based on the average monthly values of temperature and precipitation. Developed by German climatologist Wladimir Köppen, the system also takes into account the distinct vegetation found in different climates.

Köppen decided that a good way to distinguish among different climatic zones was by natural vegetation. Palm trees, for instance, are not located in polar regions, but instead are largely limited to tropical and subtropical regions. Köppen later realized that quantitative values would make his system more objective and therefore more scientific. Thus, he revised his system to include the numerical values of temperature and precipitation. A map of global climates according to a modified version of Köppen's classification system is shown in **Figure 6**.

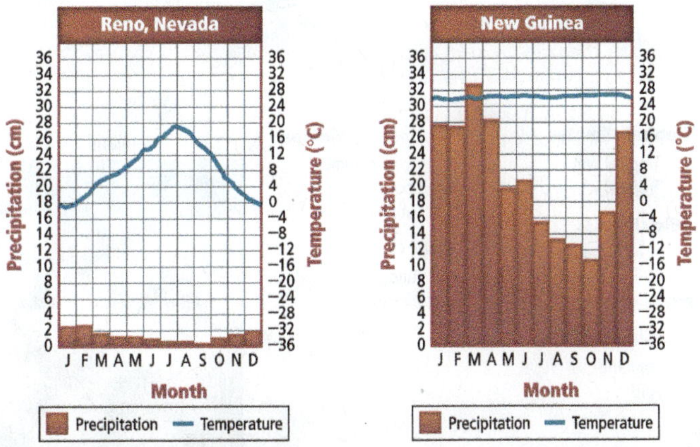

Figure 5 These graphs show temperature and precipitation for two different climates—a desert in Reno, Nevada, and a tropical rain forest in New Guinea. **Describe** *the difference in temperature between these two climates.*

SECTION 2

1 Focus

Focus Transparency

Before presenting the lesson, project **Section Focus Transparency 41** and have students answer the accompanying questions. EL

MAINIDEA

Climate Classification Have students pick five cities around the world at random and research each city's main weather features. Ask students to group the cities according to the characteristics they have in common. Explain they have just developed a basic climate classification system that is probably similar to the one used by scientists.

2 Teach

Project

Climates Around the World Divide the class into seven groups, and assign each group a continent. As students study this section, have each group keep an ongoing record of the various climates of its continent. After the section is completed, student groups should present reports about the climates of their continents. OL COOP LEARN

Caption Question Fig. 5 The temperatures in New Guinea are nearly constant year-round, at about 27°C. The temperatures in Reno fluctuate with the seasons; it is cooler in the winter and warmer in the summer.

DIFFERENTIATED INSTRUCTION

Struggling Learners Give students about a dozen beads of different sizes and colors, and have them develop a classification system for the beads. This strategy will help students realize that subgroups are part of a common whole, and in this way will further their understanding of climate classification. After the activity, tell students that climates are classified, too, but according to average weather conditions over long periods of time.

VISUALIZING

Purpose

Students will learn about the five main climate divisions based on the Köppen classification.

Discussion

Climate Descriptions The five main climate divisions include tropical climates, dry climates, mild climates, continental climates, and polar climates. Have students provide a description of each climate type, including information about temperatures, precipitation, changes in season, and the type of vegetation that can be found in each. Have students compare and revise their descriptions based on the information about each climate type on the following pages in this section.

Interpret the Illustration

The Köppen Classification System **Figure 6** shows the five main climate types according to a modified version of the Köppen classification system. Ask students to infer why the greatest variation in climate type occurs in the midlatitudes. The combined influence of the polar, tropical, moist, and dry air masses that move through midlatitudes causes climates to vary widely.

■ **Caption Question** **Fig. 6** The correct answer is 6 percent. However, student answers will be based on estimates and thus will likely vary. Answers between 5 to 10 percent are acceptable.

VISUALIZING Worldwide Climates

Highland climate, Canada

Arid climate, Australia

Figure 6 Köppen's classification system, shown here in a modified version, is made up of five main divisions based on temperature and precipitation. **Estimate** ***Use the map to determine the approximate percentage of land covered by tropical wet climates.***

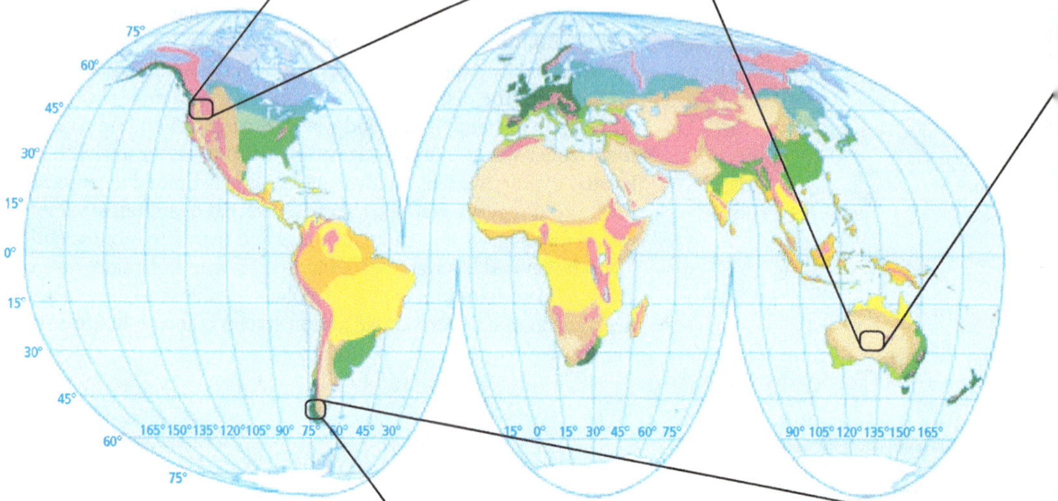

Semiarid climate, Argentina

View an **animation of climate classification.** Concepts In Motion

ACROSS THE CURRICULUM

History Ancient Greeks developed one of the first known climate classification systems. They divided each hemisphere into three zones: torrid, temperate, and frigid. This simple scheme was based on Earth-Sun relationships. The boundaries were the four astronomically important parallels: the Tropic of Cancer, the Tropic of Capricorn, the Arctic circle, and the Antarctic circle.

Tropical climates Year-round high temperatures characterize tropical climates. In tropical wet climates, the locations of which are shown in **Figure 6,** high temperatures are accompanied by up to 600 cm of rain each year. The combination of warmth and rain produces tropical rain forests, which contain some of the most dramatic vegetation on Earth. Tropical regions are almost continually under the influence of maritime tropical air.

The areas that border the rainy tropics to the north and south of the equator are transition zones, known as the tropical wet and dry zones. Tropical wet and dry zones include savannas. These tropical grasslands are found in Africa, among other places. These areas have distinct dry winter seasons as a result of the seasonal influx of dry continental air masses. **Figure 7** shows the average monthly temperature and precipitation readings for Normanton, Australia—a savanna in northeast Australia.

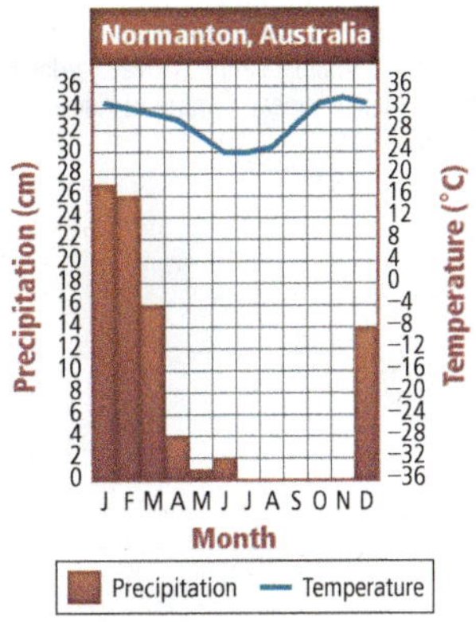

Figure 7 The graph shows the temperature and precipitation readings for a tropical savanna in Australia.
Analyze *How does the rainfall in this area differ from that of a tropical rain forest?*

☑ READING CHECK **Explain** the difference between tropical wet and tropical wet and dry climate zones.

Dry climates Dry climates, which cover about 30 percent of Earth's land area, make up the largest climatic zone. Most of the world's deserts, such as the Sahara, the Gobi, and the Australian, are classified as dry climates. In these climates, continental tropical (cT) air dominates, precipitation is low, and vegetation is scarce. Many of these areas are located near the tropics. Thus, intense solar radiation results in high rates of evaporation and few clouds. Overall, evaporation rates exceed precipitation rates. The resulting moisture deficit gives this zone its name. Within this classification, there are two subtypes: arid regions, called deserts, and semiarid regions, called semideserts. Semideserts, like the one shown in **Figure 8,** are usually more humid than deserts. They generally separate arid regions from bordering wet climates.

Figure 8 This semidesert in Kazakhstan is another example of a transition zone. It separates deserts from bordering climates that are more humid.

Teacher Content Support

Tropical Climates One of the most distinguishing characteristics of tropical climates is the lack of variability in temperatures from month to month. Cold air masses rarely penetrate the deep tropics. In addition, the Sun's rays strike the tropics from almost directly overhead throughout the entire year. Additionally, the lengths of days and nights remain nearly the same throughout the year, a factor that helps to maintain constant temperatures.

Collaborative Learning

Deserts A common fallacy about deserts is that they are lifeless. Have students work in groups to research desert plants and the adaptations these plants have that help them survive the harsh conditions of the desert. Desert plants are highly resistant to drought. Many have waxy leaves, stems, or branches, or thickened outer layers to reduce water loss. Others have extensive root systems that tap moisture deep in the ground. Often, plant stems are thickened by a spongy tissue that stores water.
COOP LEARN

☑ READING CHECK Tropical wet climates receive up to 600 cm of rain each year and have high temperatures. Tropical wet and dry climates are transition zones that have a dry season in the winter.

Caption Question Fig. 7 It is much drier during the winter months.

EARTH SCIENCE JOURNAL

Dry Climates v. Tropical Climates

Have students summarize the differences between tropical climates and dry climates. Encourage them to research and include information about the difference in plant and animal life between the two climates and to find an example of a city in each climate zone.

Rubric

Identify Misconceptions

Some students might think that a continental climate does not experience marine influences.

Uncover the Misconception Ask students to speculate why some of the most severe storms take place in continental climates.

Demonstrate the Concept Explain that even though moist air masses do not originate over continents, weather systems frequently transport moist air into areas dominated by continental climates. The collision of these moisture-laden air masses from maritime climates with the often cooler and drier air of the continental climate often results in the formation of severe thunderstorms and tornadoes.

Assess New Knowledge Have students research the term "Tornado Alley" where some of the most frequent and damaging tornadoes occur. Have them assess the likely sources of the dry and moist air masses that collide in this area, and identify their probable climates of origin.

Tie To Previous Knowledge

Hurricanes Using the concepts learned about hurricanes, have students discuss the likelihood of hurricane formation in continental climates and explain their reasoning.

Figure 9 Olive trees thrive in the warm, dry summers and cool, rainy winters of the Mediterranean climate of Huesca, Spain.

Mild climates Mild climates can be classified into three subtypes: humid subtropical climates, marine west-coast climates, and Mediterranean climates. Humid subtropical climates are influenced by the subtropical high-pressure systems that are normally found over oceans in the summer. The southeastern United States has this type of climate. There, warm, muggy weather prevails during the warmer months and dry, cool conditions predominate during the winter. The marine west-coast climates are dominated by the constant inland flow of air off the ocean, which creates mild winters and cool summers, with abundant precipitation throughout the year. Mediterranean climates, named for the climate that characterizes much of the land around the Mediterranean Sea, are also found in California and parts of South America. An example of this type of climate is shown in **Figure 9.** Summers in Mediterranean climates are generally warm and dry because of their nearness to the dry midlatitude climates from the south. Winters are cool and rainy as a result of the midlatitude weather systems that bring storm systems from the north.

☑ READING CHECK **Compare and contrast** humid subtropical and marine west-coast climates.

Figure 10 Tornadoes, such as this one in Kansas, occur in continental climates.

Continental climates Continental climates are also classified into three subtypes: warm summer climates, cool summer climates, and subarctic climates. Tropical and polar air masses often form fronts as they meet in continental climates. Thus, these zones experience rapid and sometimes violent changes in weather, including severe thunderstorms or tornadoes like the one shown in **Figure 10.** Both summer and winter temperatures can be extreme because the influence of polar air masses is strong in winter, while warm tropical air dominates in summer. The presence of warm, moist air causes summers to be generally more wet than winters, especially in latitudes that are relatively close to the tropics.

ACROSS THE CURRICULUM

Biology Have students research and then compare and contrast the types of vegetation, including trees, that are most likely to be found in continental and Mediterranean climates. Where might you find vegetation that is common to both climates? on the boundary, between climate regions

☑ READING CHECK Humid subtropical climates are warm and muggy in the summer and cool in the winter. Marine west-coast climates have cool summers and mild winters.

Polar climates To the north of subarctic climate lies one of the polar climates—the tundra. Just as the tropics are known for their year-round warmth, tundra is known for its low temperatures—the mean temperature of the warmest month is usually less than 10°C. There are no trees in the tundra and precipitation is generally low because cold air contains less moisture than warm air. Also, the amount of heat radiated by Earth's surface is too low to produce the strong convection currents needed to release heavy precipitation. The ice-cap polar climate, found at the highest latitudes in both hemispheres, does not have a single month in which average temperatures rise above 0°C. No vegetation grows in an ice-cap climate and the land is permanently covered by ice and snow. **Figure 11** shows an ice-cap polar climate.

A variation of the polar climate, called a highland climate, is found at high elevations. This type of climate includes parts of the Andes Mountains in South America, which lie near the equator. The intense solar radiation found near such equatorial regions is offset by the decrease in temperature that occurs with altitude.

■ **Figure 11** Icebergs float in the sea in the ice-cap polar climate of Greenland.

Microclimates

Sometimes the climate of a small area can be much different than that of the larger area surrounding it. A localized climate that differs from the main regional climate is called a **microclimate.** If you climb to the top of a mountain, you can experience a type of microclimate; the climate becomes cooler with increasing elevation. **Figure 12** shows a microclimate created by the buildings and concrete in a city.

FOLDABLES® Incorporate information from this section into your Foldable.

Heat islands Sometimes the presence of buildings can create a microclimate in the area immediately surrounding it. Many concrete buildings and large expanses of asphalt can create a **heat island,** where the climate is warmer than in surrounding rural areas, as shown in **Figure 12.** This effect was first recognized in the early nineteenth century when Londoners noted that the temperature in the city was noticeably warmer than in the surrounding countryside.

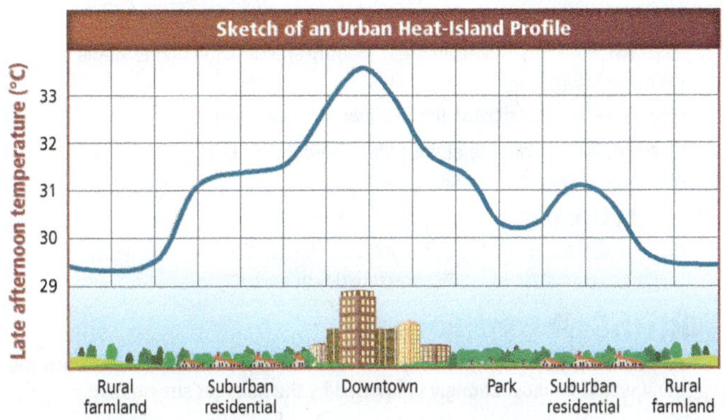

■ **Figure 12** This diagram shows the difference in temperature between the downtown area of a city and the surrounding suburban and rural areas.
Analyze *How much warmer is it in the city compared to the rural areas?*

Discussion

Polar Climates Ask students to explain why the polar climates are cold. Very little solar radiation is received during polar winters. Even during the long days of summer, the angle of the Sun's rays is so low that solar radiation is dispersed over a wide area and thus does not warm these regions.

Concept Development

Ice Caps and Tundra Polar climates can be divided into two subtypes: ice caps and tundra. The boundary of the tundra marks the limit of tree growth; vegetation is mainly grasses, sedges, mosses, and lichens. Temperatures in the ice-cap climate are below freezing year-round. The surface is covered by permanent ice and snow.

Teacher Content Support

Heat Islands The development of heat islands is an excellent example of the effects that humans have on climate. Heat islands impact more than temperature and precipitation levels. The heat combines with industrial pollution to worsen smog problems. Ask students to brainstorm other effects that heat islands have on climate.

■ **Caption Question Fig. 12**
about 4°C

GeoLAB

iLab Station

The GeoLab at the end of the chapter can be used at this point in the lesson.

FOLDABLES® Rubric

Demonstration

Heat-Island Effect The heat-island effect can be demonstrated using a lamp, a small chunk of concrete, and some leaves or grass. **WARNING:** ***The lamp will get hot.*** Position the light just above both the concrete and the vegetation for at least 30 minutes, and then have students touch both surfaces. The concrete will be noticeably warmer, showing that areas covered by asphalt and concrete absorb heat more efficiently than the surrounding countryside does.

3 Assess

Check for Understanding

Compare and Contrast Have students compare and contrast the main climate types in the Köppen classification system.

Reteach

Illustrate Have students make posters that illustrate the main characteristics of the climatic zone of their state. If the state is made up of several different climatic zones, divide students into groups and have each group make a poster illustrating the different climatic zones.

Assessment

Performance Have students research a major urban area to determine whether a heat-island effect is apparent in the city's climatological data. Students can do this by comparing the city's data with those of surrounding rural areas. Tell students to gather information about the cities of their choice. Have students present their results to the class. Encourage students to use a variety of media in their presentations.

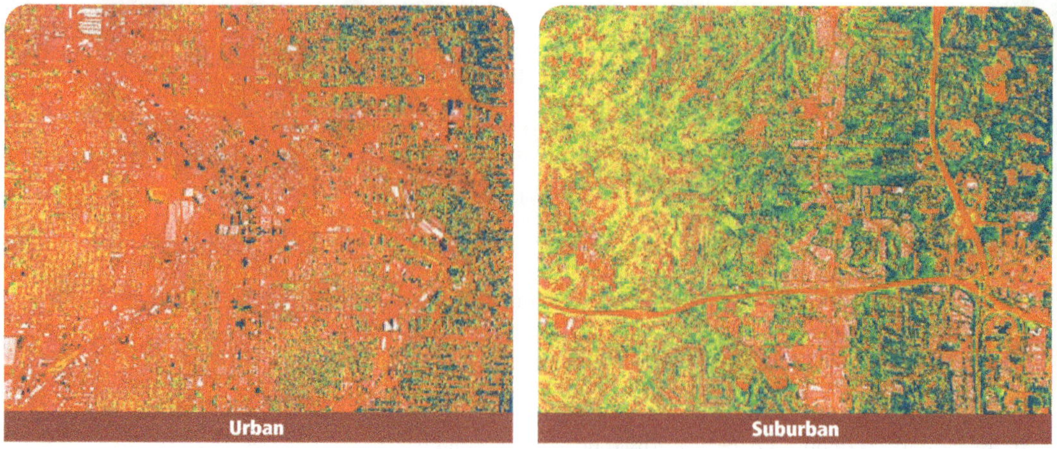

Figure 13 These thermal images show differences in daytime temperatures between an urban area and a suburban area. The coolest temperatures are represented by blue; the warmest temperatures are represented by red.

Pavement, buildings, and roofs that are made of dark materials, such as asphalt, absorb more energy from the Sun than the surrounding vegetation does. This causes the temperature of these objects to increase, heating the air around them. This also causes mean temperatures in large cities to be significantly warmer than in surrounding areas, as shown in **Figure 13.** The heat-island effect also causes greater changes in temperature with altitude, which sparks strong convection currents. This, in turn, produces increased cloudiness and up to 15 percent more total precipitation in cities.

Heat islands are examples of climatic change on a small scale. In Sections 3 and 4, you will examine large-scale climatic changes caused by both natural events and human activities.

SECTION 2 REVIEW

Section Self-Check

Section Summary

- German scientist Wladimir Köppen developed a climate classification system.
- There are five main climate types: tropical, dry, mild, continental, and polar.
- Microclimates can occur within cities.

Understand Main Ideas

1. MAINIDEA **Describe** On what criteria is the Köppen climate classification system based?
2. **Explain** What are microclimates? Identify and describe one example of a microclimate.
3. **Compare and contrast** the five main climate types.
4. **Categorize** the climate of your area. In which zone do you live? Which air masses generally affect your climate?

Think Critically

5. **Construct** Make a table of the Köppen climate classification system. Include major zones, subzones, and characteristics of each.

WRITING IN ▶ Earth Science

6. Write a short paragraph that explains which of the different climate types you think would be most strongly influenced by the polar jet stream.

SECTION 2 REVIEW

1. vegetation, temperature, and precipitation
2. Microclimates are localized climates that differ from the main regional climate. Heat islands cause more heat to be absorbed in cities.
3. Tropical climates have high temperatures; some are characterized by abundant rainfall. Dry climates have low precipitation and scarce vegetation. Mild climates have generally warm winters and cool summers. Continental climates have extreme temperatures in summer and winter. Polar climates are cold year-round.
4. Answers will vary depending on students' locations.
5. Tables should include the main characteristics of each climate type and subtype discussed in this section.
6. The mild or continental climates; they are found well north of the tropics and south of the poles—regions where the polar jet stream is most active.

Rubric

SECTION 3

Climatic Changes

MAINIDEA Earth's climate is constantly changing on many different timescales.

Essential Questions

- What is the difference between long-term and short-term climatic changes?
- What are natural causes of climate change?
- Why do climatic changes occur?

Review Vocabulary

glacier: large, moving mass of ice that forms near Earth's poles and in mountainous regions at high elevations

New Vocabulary

ice age
season
El Niño
Maunder minimum

EARTH SCIENCE 4 YOU You might not notice changes in a friend's physical appearance from day to day; however, if you only see someone once a year, he or she might appear to have changed a lot. Climate changes on long timescales with differences that might not be noticed day to day.

Long-Term Climatic Changes

Some years might be warmer, cooler, wetter, or drier than others, but during the average human lifetime, climates do not appear to change significantly. However, a study of Earth's history over hundreds of thousands of years shows that climates have always been, and currently are, in a constant state of change. These changes usually take place over long time periods.

Ice ages A good example of climatic change involves glaciers, which have alternately advanced and retreated over the past 2 million years. At times, much of Earth's surface was covered by vast sheets of ice. During these periods of extensive glacial coverage, called **ice ages,** average global temperatures decreased by an estimated 5°C. Global climates became generally colder and snowfall increased, which sparked the advance of existing ice sheets. Ice ages alternate with warm periods—called interglacial intervals—and Earth is currently experiencing such an interval. The most recent ice age, shown in **Figure 14,** ended only about 10,000 years ago. In North America, glaciers spread from the east coast to the west coast and as far south as Indiana and Ohio.

Figure 14 The last ice age covered large portions of North America, Europe, and Asia. Average global temperatures were roughly 5°C lower than they are today.
Explain *how decreased global temperatures can lead to an ice age.*

Watch a **video about climate history.** Video

SECTION 3

1 Focus

Focus Transparency

Before presenting the lesson, project **Section Focus Transparency 42** and have students answer the accompanying questions. EL

MAINIDEA

Climatic Changes Ask students to think about how different climates might be if Earth were closer to or farther away from the Sun.

2 Teach

Identify Misconceptions

Some students might think that individual climates do not vary.

Uncover the Misconception Ask students if the climate regions on Earth are the same as they were 15,000 years ago.

Demonstrate the Concept Explain that Earth's climates have undergone major changes throughout time as a result of many different processes. Changes continue to occur.

Assess New Knowledge Have students make analogies to compare long-term climate change to other changes on Earth that take place on geologic time scales. Possible answers: mountain building, weathering, erosion

Caption Question Fig. 14 As global temperatures decrease, global climates become colder and snowfall increases, producing widespread glaciation.

What's EARTH SCIENCE Got To Do With It?

Video *Climate's History*

Interpret the Illustration

Ice Ages Have students study **Figure 14.** Explain that the reflection of solar radiation by ice and snow, as well as the inability of ice to effectively absorb and retain solar radiation, caused lowered temperatures around the world.

Enrichment

Solstices The solstices, which are the times of year when solar radiation is at its maximum and minimum, occur around June 21 and December 21, respectively. However, these dates are merely the beginning of astronomical summer and winter, and thus they might not be the warmest or coldest times of the year. The warmest and coldest temperatures usually occur about one to two months after the solstices. Ask students to infer why. Earth's land-ocean-atmosphere system takes time to respond to changes in the amount of solar energy received on Earth. Thus, a seasonal lag occurs between the dates of the solstices and the dates of deepest summer and winter.

Project

El Niño Beyond El Niño's effects on weather and climate, the sudden presence of warm water off the western coast of South America also has profound effects on marine ecosystems. Have students research and report on the impact of El Niño on the food chains of this coastal area. Students should share their results with the class. OL AL

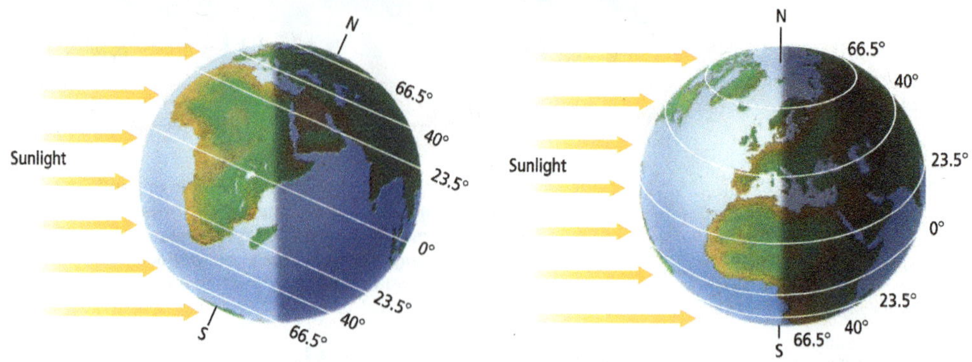

Figure 15 When the North Pole is pointed away from the Sun, the northern hemisphere experiences winter and the southern hemisphere experiences summer. During spring and fall, neither pole points toward the Sun.

View an **animation of seasons.**

Concepts In Motion

Short-Term Climatic Changes

While an ice age might last for tens of thousands of years, other climatic changes occur over much shorter time periods. Climatic change can affect seasons differently. **Seasons** are short-term periods with specific weather conditions caused by regular variations in daylight, temperature, and weather patterns.

Seasons The variations that occur with seasons are the result of changes in the amount of solar radiation an area receives. As **Figure 15** shows, the tilt of Earth on its axis as it revolves around the Sun causes different areas of Earth to receive different amounts of solar radiation. During winter in the northern hemisphere, the North Pole is tilted away from the Sun, and this hemisphere experiences long hours of darkness and cold temperatures. At the same time, it is summer in the southern hemisphere. The South Pole is tilted toward the Sun, and the southern hemisphere experiences long hours of daylight and warm temperatures. Throughout the year, the seasons are reversed in the northern and southern hemispheres. During the spring and fall, neither pole points toward the Sun.

El Niño Other short-term climatic changes include those caused by **El Niño,** a band of anomalously warm ocean temperatures that occasionally develops off the western coast of South America. Under normal conditions in the southeastern Pacific Ocean, atmospheric and ocean currents along the coast of South America move north, transporting cold water from the Antarctic region.

IN THE FIELD

El Niño and the Fishing Industry

The fishing industries in Peru and Ecuador have recognized the existence of El Niño for more than 100 years. In fact, these fishing crews called the phenomenon El Niño, which means "the child" in Spanish, because it usually occurred around Christmas. These crews also noticed that El Niño disrupts marine feeding patterns. In normal years, cold, deep waters move upward, carrying rich nutrients from organisms that sank to the ocean floor and decayed. Upwelling occurs near the coast, and the nutrients rise to the surface and feed countless fish. During El Niño, however, the upwelled water is warm and contains fewer nutrients. NASA is working with fisheries in Peru and Ecuador to help them prepare for El Niño.

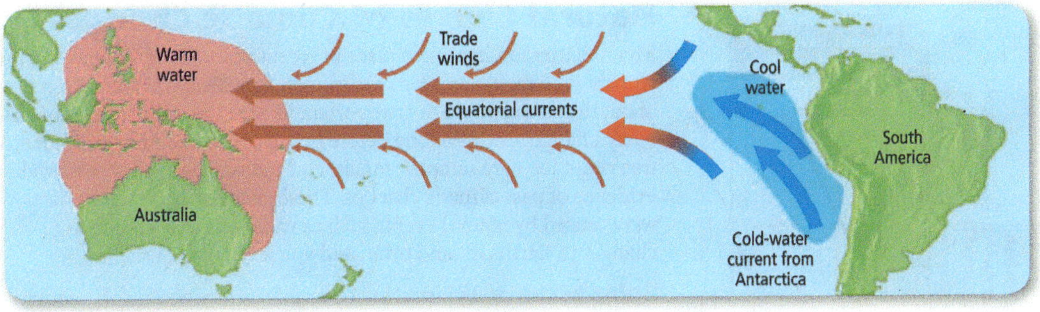

■ **Figure 16** Under normal conditions, trade winds and ocean currents move warm water west across the Pacific Ocean.

Meanwhile, the trade winds and ocean currents move westward across the tropics, keeping warm water in the western Pacific, as shown in **Figure 16.** This circulation, driven by a semipermanent high-pressure system, creates a cool, dry climate along much of the northwestern coast of South America.

Occasionally, however, for reasons that are not fully understood, this high-pressure system and its associated trade winds weaken drastically, which allows the warm water from the western Pacific to surge eastward toward the South American coast, as shown in **Figure 17.** These conditions are referred to as an El Niño event.

The sudden presence of this warm water heats the air near the surface of the water. Convection currents strengthen, and the normally cool and dry northwestern coast of South America becomes much warmer and wetter. The increased convection pumps large amounts of heat and moisture into the upper atmosphere, where upper-level winds transport the hot, moist air eastward across the tropics. This hot, moist air in the upper atmosphere is responsible for dramatic climate changes, including violent storms in California and the Gulf Coast, stormy weather to areas farther east that are normally dry, and drought conditions to areas that are normally wet. Eventually, the South Pacific high-pressure system becomes reestablished and El Niño weakens.

Sometimes the trade winds blow stronger than normal and warm water is pulled across the Pacific toward Australia. The coast of South America becomes unusually cold and chilly. These conditions are called La Niña.

VOCABULARY

SCIENCE USAGE V. COMMON USAGE

Pressure

Science usage: the force that a column of air exerts on the air below it

Common usage: the burden of physical or mental distress

■ **Figure 17** During El Niño, warm water surges back toward South America, changing weather patterns.

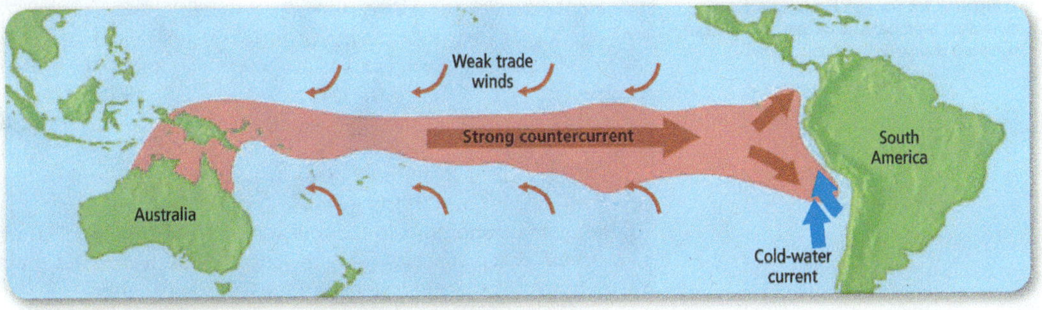

Concept Development

El Niño's Effect on Air Masses

Increased understanding of El Niño and its effects has greatly improved the reliability of seasonal weather forecasting. In addition to its well-known inhibiting influence on Atlantic tropical cyclones, El Niño also enhances the westerly component of the jet stream and prevailing westerlies across North America.

This results in more air masses moving across the continent from west to east, and fewer moving from north to south. These eastward-moving systems mainly have their origin over the comparatively mild Pacific Ocean, as opposed to the bitter cold arctic air masses that move down from the north. As a result, a forecast of a mild winter during El Niño will usually meet with a reasonable degree of accuracy.

DIFFERENTIATED INSTRUCTION

Advanced Learners Have students research why a phenomenon similar to El Niño does not exist in the South Atlantic Ocean. Have them propose possible explanations. The relatively small area of the South Atlantic does not allow for the development of well-defined ocean currents and warm pools of water, such as those associated with El Niño.

Identify Misconceptions

Much is heard in the media about the many ways in which humans negatively impact the environment. However, it is a misconception to blame all climatic changes on human activities.

Uncover the Misconception

Ask students to infer what causes climatic changes, such as global warming. Most will probably mention human activities that cause increased levels of pollution.

Demonstrate the Concept

Tell students Earth goes through various cycles, and these cycles can have profound effects on Earth's climate. For instance, many scientists hypothesize that we are now in an interglacial period, which affects sea level and global temperatures. Point out there probably are other natural cycles that have not yet been identified.

Assess New Knowledge

After students have completed this section, ask them whether all climatic changes are caused by human activities. Most students will realize that many factors affect climatic change.

Caption Question Fig. 18 As the number of sunspots increases, sea surface temperature increases.

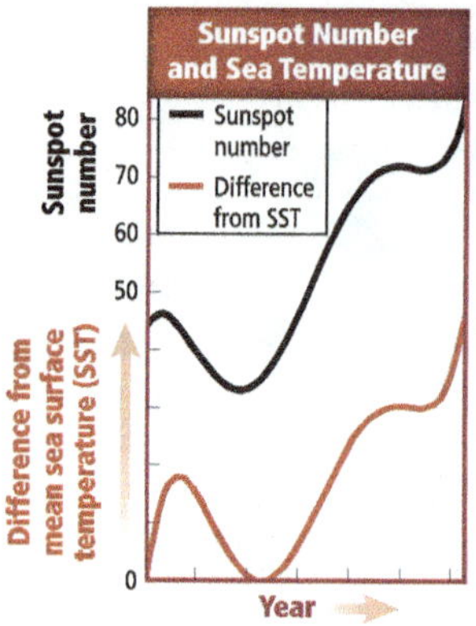

Figure 18 Scientists theorize that solar activity might be linked to climatic changes.

Evaluate ***How is the number of sunspots related to changes in sea surface temperature?***

Natural Causes of Climatic Changes

Much discussion has taken place in recent years about whether Earth's climate is changing as a result of human activities. You will read more about this in Section 4. It is important to note that many cycles of climatic change occurred long before humans inhabited Earth. Studies of tree rings, ice-core samples, fossils, and radiocarbon samples provide evidence of past climatic changes. These changes in Earth's climate were caused by natural events such as variations in solar activity, changes in Earth's tilt and orbit, and volcanic eruptions.

Solar activity Evidence of a possible link between solar activity and Earth's climate was provided by English astronomer Edward Walter Maunder in 1893. The existence of sunspot cycles lasting approximately 11 years had been recognized by German scientist Samuel Heinrich Schwabe in 1843. However, Maunder found that from 1645 to 1716, the number of sunspots was scarce to nonexistent. The **Maunder minimum** is the term used to describe this period of low numbers of sunspots. This period closely corresponds to an unusually cold climatic episode called the Little Ice Age. During this time, much of Europe experienced bitterly cold winters and below-normal temperatures year-round. Residents of London are said to have ice-skated on the Thames River in June. The relationship between sea surface temperature, which is used as an indicator of climate, and periods of low sunspot numbers is illustrated in **Figure 18.** Studies indicate that increased solar activity coincides with warmer-than-normal sea surface temperatures, while periods of low solar activity, such as the Maunder minimum, coincide with colder sea surface temperatures.

Earth's orbit Climatic changes might also be triggered by changes in Earth's axis and orbit. The shape of Earth's elliptical orbit appears to change, becoming more elliptical, then more circular, over the course of a 100,000-year cycle. As **Figure 19** shows, when the orbit elongates, Earth travels for part of the year in a path closer to the Sun. As a result, temperatures become warmer than normal. When the orbit is more circular, Earth remains in an orbit that is farther from the Sun, and temperatures dip below average.

Figure 19 Scientists hypothesize that a more elliptical orbit around the Sun could produce significant changes in Earth's climate.

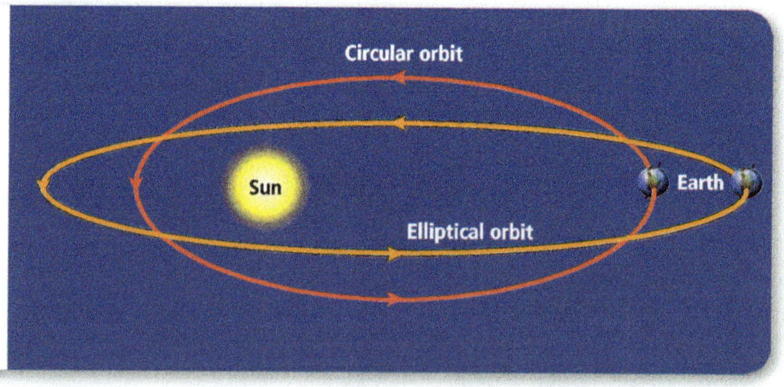

ACROSS THE CURRICULUM

Astronomy Sunspots are just one of several types of solar disturbances that have been identified. Have students research the other types, including solar flares and prominences. These disturbances also have significant effects on Earth and its atmosphere, beyond changes in climate. Students should identify these effects, and explain the relationship among flares, prominences, and sunspots.

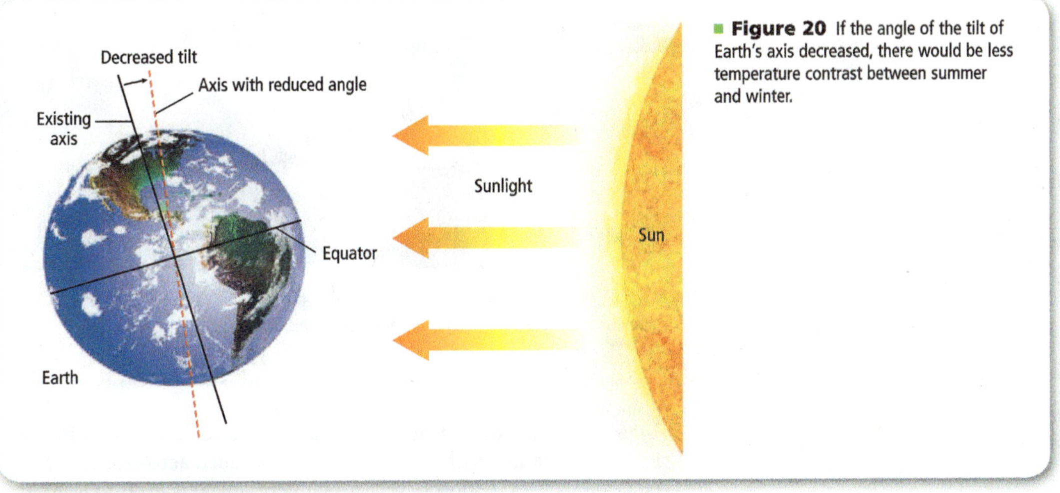

Figure 20 If the angle of the tilt of Earth's axis decreased, there would be less temperature contrast between summer and winter.

Earth's tilt As you know, seasons are caused by the angle of the tilt of Earth's axis. At present, the angle of the tilt is 23.5°. However, the angle of tilt varies from a minimum of 22.1° to a maximum of 24.5° every 41,000 years. Scientists theorize that these changes in angle affect the differences in seasons. For example, a decrease in the angle of the tilted axis, shown in **Figure 20,** might cause a decrease in the temperature difference between winter and summer. Winters would be more warm and wet, and summers would be cooler. The additional snow in latitudes near the poles would not melt in summer because temperatures would be cooler than average. This could result in increased glacial formation and coverage. In fact, some scientists hypothesize that changes in the angle of Earth's tilted axis can cause ice sheets to form near the poles.

☑ READING CHECK **Describe** how a change to the angle of Earth's tilt can lead to climate change.

Earth's wobble Another movement of Earth might be responsible for climatic changes. Over a period of about 26,000 years, Earth wobbles as it spins around on its axis. Currently, the axis points toward the North Star, Polaris, as shown in **Figure 21.** Because of Earth's wobbling, however, the axis will eventually rotate away from Polaris and toward another star, Vega, in about 13,000 years. Currently, winter occurs in the northern hemisphere when the direction of the tilt of Earth causes the northern hemisphere to receive more direct radiation from the Sun. However, in 13,000 years, the northern hemisphere will be tilted in the opposite direction relative to the Sun. So, during the time of year associated with winter today, the northern hemisphere will be tilted toward the Sun and will experience summer.

Figure 21 Earth's wobble determines the timing of the seasons. In about 13,000 years, when the northern hemisphere points toward the star Vega, the northern hemisphere will experience summer during the time now associated with winter.

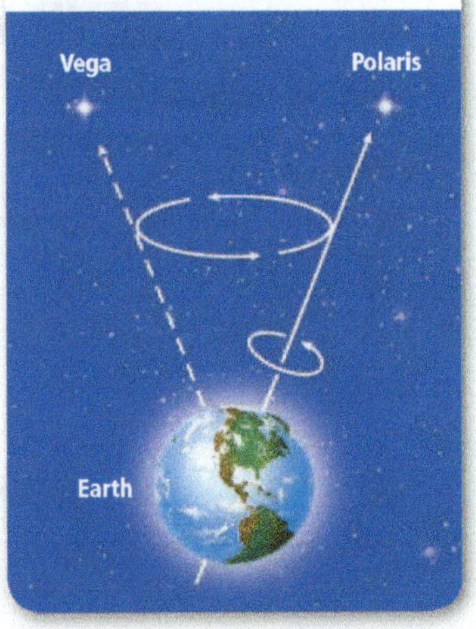

Teacher Content Support

Sunspot Cycles Sunspots, which are associated with the Sun's ejection of huge masses of magnetic particles, recently have been linked to changes in climatic cycles and to individual seasonal events. Research has shown some correlation between the sunspot cycle and periods of drought in the Great Plains, as well as the number of tropical cyclones that occur during any given year. No widely accepted theory has yet been developed to explain exactly how solar variations cause climatic changes.

Project

Sunspot Maximum Have students identify the three most recent sunspot maxima and then research local climate records to see whether temperature or precipitation were significantly above or below normal.

Use Science Terms

Precession The term given to the change in inclination of Earth's axis is *precession.* While *wobble* is often used to describe this motion, this term suggests a random movement of Earth's axis, which is not accurate. Earth's axis rotates in a circular motion.

☑ READING CHECK Changes in the angle of Earth's tilt can affect differences in seasons. A decrease in the angle of the tilt might cause a decrease in the temperature difference between summer and winter.

DIFFERENTIATED INSTRUCTION

Advanced Learners Milutin Milankovitch, a Serbian astronomer, determined that the shape of Earth's orbit, the angle of Earth's axis, and the direction of Earth's tilt, all change in slow, predictable cycles. Have students research the Milankovitch theory, which relates these cycles to changes in the amount of summer solar radiation received at 65°N and at the beginning of ice ages.

EARTH SCIENCE JOURNAL

Climate Change Have students write a paragraph that describes the different natural causes of climate change. For each cause, have them explain whether it results in long-term climate change or short-term climate change.

Rubric

3 Assess

Check for Understanding

Explain Ask students to explain the difference between short-term and long-term climatic changes, and to give examples of each type. Ask students to classify these changes according to whether they are largely astronomical phenomena or Earth-based processes.

Reteach

Summarize Ask students the following questions: Which areas does El Niño affect least? El Niño affects polar regions the least. Which areas does El Niño affect most? It affects equatorial regions the most.

Assessment

Performance Have students use their knowledge of prevailing winds to describe how volcanic ash from an imaginary eruption would spread around the world. What types of climatic changes could be expected? What areas would be most severely affected?

Figure 22 During the eruption of Mount Pinatubo, more than 20 million tons of sulfur dioxide gas was ejected into the atmosphere, causing a worldwide decrease in average annual temperatures of 0.5°C for two years.

Volcanic activity Climatic changes can also be triggered by the immense quantities of dust-sized particles, called aerosols, that are released into the atmosphere during major volcanic eruptions, like the one shown in **Figure 22.** Volcanic aerosols can remain suspended in the atmosphere for several years, blocking incoming solar radiation and consequently lowering global temperatures. Some scientists theorize that periods of high volcanic activity cause cool climatic periods. Climatic records from the past century show that several large eruptions have been followed by below-normal global temperatures. For example, the ash and gases released during the 1991 eruption of Mount Pinatubo in the Philippines resulted in slightly cooler temperatures around the world for two years following the eruption.

Generally, volcanic eruptions appear to have only short-term effects on climate. These effects, as well as the other short-term and long-term effects you have read about thus far, are a result of natural causes.

SECTION 3 REVIEW

Section Self-Check

Section Summary

- Climate change can occur on a long-term or short-term scale.
- Changes in solar activity have been correlated with periods of climate change.
- Changes in Earth's orbit, tilt, and wobble are all associated with changes in climate.

Understand Main Ideas

1. MAINIDEA **Identify** and explain an example of long-term climatic change.
2. **Describe** What are seasons? What causes them?
3. **Illustrate** how El Niño might affect weather in California and along the Gulf Coast.
4. **Analyze** How does volcanic activity affect climate? Are these effects short-term or long-term climatic changes?

Think Critically

5. **Assess** What might be the effect on seasons if Earth's orbit became more elliptical and, at the same time, the angle of the tilt of Earth's axis increased?

MATH IN Earth Science

6. Study **Figure 18.** During which period were sunspot numbers lowest? During which period were sunspot numbers highest?

SECTION 3 REVIEW

1. Every 41,000 years, the axis of Earth's rotation changes, causing seasonal temperature differences to change.
2. Seasons refer to short-term periods with specific weather conditions caused by changes in the amount of solar radiation received in a given location throughout the year. These changes cause variations in daylight, temperature, and weather patterns.
3. Illustrations should show that convection currents strengthen above the warmer-than-normal ocean water in the eastern Pacific. The increased precipitation off the coast of South America pumps large amounts of heat and moisture into the upper atmosphere, where upper-level winds transport the hot, moist air eastward across the tropics. This hot, moist air in the upper atmosphere is responsible for dramatic climatic changes, including violent storms in California and the Gulf Coast.
4. Volcanic dust from eruptions blocks incoming solar radiation and lowers global temperatures; it is an example of a short-term change.
5. These combined factors would likely produce greater seasonal contrasts in temperature. Thus, there would be more extreme differences in temperatures between summer and winter.
6. lowest near the middle; highest near the end

SECTION 4

Impact of Human Activities

MAINIDEA Over time, human activities can alter atmospheric conditions enough to influence changes in weather and climate.

Essential Questions

- What is the greenhouse effect?
- What is global warming?
- How do humans impact climate?

Review Vocabulary

radiation: transfer of thermal energy by electromagnetic waves

New Vocabulary

greenhouse effect
global warming

EARTH SCIENCE 4 YOU If a virus attached to a downloaded file affects your computer, it can change the way your computer operates. If not carefully considered, human activities can produce changes in Earth's natural systems.

Influence on the Atmosphere

Earth's atmosphere significantly influences its climate. Solar radiation that is not reflected by clouds passes freely through the atmosphere. It is then absorbed by Earth's surface and released as long wavelength radiation. This radiation is absorbed by atmospheric gases such as water vapor, methane, and carbon dioxide. Some of this absorbed energy is reradiated back to Earth's surface.

The greenhouse effect This process of the absorption and radiation of energy in the atmosphere results in the **greenhouse effect**—the natural heating of Earth's surface caused by certain atmospheric gases called greenhouse gases. The greenhouse effect, shown in **Figure 23,** warms Earth's surface by more than 30°C. Without the greenhouse effect, life as it currently exists on Earth would not be possible.

Scientists hypothesize that it is possible to increase or decrease the greenhouse effect by changing the amount of atmospheric greenhouse gases, particularly carbon dioxide and methane. An increase in the amount of these gases could result in increased absorption of energy in the atmosphere. Levels of atmospheric carbon dioxide and methane are increasing. This could lead to a rise in global temperatures, known as **global warming.**

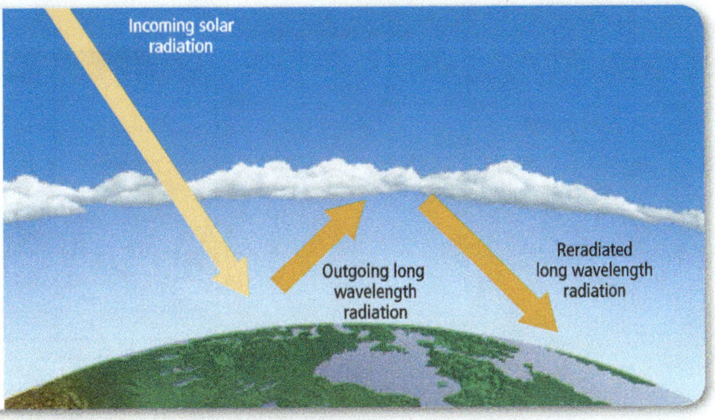

Figure 23 Solar radiation reaches Earth's surface where it is reradiated as long wavelength radiation. This radiation does not easily escape through the atmosphere and is mostly absorbed and released by atmospheric gases. This process is called the greenhouse effect.

View an **animation of the greenhouse effect.**

Concepts In Motion

Demonstration

Effects of Volcanic Ash on Earth's Climate Position a bright spotlight about 40 cm above a sheet of dark construction paper. Have students observe and record the brightness of the light on the paper, as well as how quickly the paper warms up. **WARNING:** ***Heat from the light can cause burns.*** After these observations have been made, hold a piece of nylon stocking about midway between the light and the paper. Students should note any differences in the brightness of the light or the warmth of the paper. Tell students that the nylon stocking represents volcanic ash that blocks sunlight from reaching Earth's surface.

SECTION 4

1 Focus

Focus Transparency

Before presenting the lesson, project **Section Focus Transparency 43** and have students answer the accompanying questions. EL

MAINIDEA

Impact of Human Activities Put a small amount of food coloring in water, and point out the effect. Explain that pouring tons of pollutants into our ocean and atmosphere can significantly change their composition and how they absorb or reflect heat, and can eventually change our climate.

2 Teach

Tie to Previous Knowledge

Global Warming Tell students that global warming is hotly debated. Reinforce that many scientific topics are not fully understood, and scientific knowledge is continually refined as new insights are acquired.

Interpret the Illustration

Greenhouse Gases Have students study **Figure 23.** Recent satellite data indicate that increases in greenhouse gases such as carbon dioxide could be contributing to warmer ocean temperatures in the northern hemisphere. Ocean temperatures in the southern hemisphere are decreasing slightly. Have students infer why. Scientists do not know why this is happening. Answers could include: fewer people live in the southern hemisphere, so there are fewer greenhouse gases generated there. Ocean currents differ in the two hemispheres, possibly causing different temperature trends.

MiniLAB

iLab Station Rubric

Purpose Students will model the greenhouse effect.

Process Skills make a model, observe and infer, analyze, graph

Safety Precautions Approve lab safety forms before work begins. Tell students to notify you if any breakage occurs.

Teaching Strategy Tell students that the glass jar represents greenhouse gases in the atmosphere.

Expected Results The air inside the jar warms more quickly than the air outside the jar.

Analysis

1. The independent variable is the treatment of the thermometer, whether it is covered in a glass jar or not. The dependent variable is the temperature.
2. Graphs will vary depending on student measurements.
3. The thermometer in the jar should show higher readings. Air was trapped inside.
4. In the atmosphere, greenhouse gases absorb and trap solar radiation, much like the glass jar absorbed and trapped energy from the Sun.

Assessment

Knowledge Ask students to infer how this activity would change if they used a jar made of thicker glass. They would likely record higher temperatures. Have students relate the thicker glass to increasing levels of atmospheric greenhouse gases.

LabManager™
Customize any lab with the LabManager™ CD-ROM.

MiniLAB

iLab Station

Model the Greenhouse Effect

How does the atmosphere trap radiation?
The greenhouse effect is a natural phenomenon that occurs because the atmosphere traps outgoing radiation.

Procedure

1. Read and complete the lab safety form.
2. On a clear day, place a **cardboard box** outside in a shaded area.
3. Prop two **thermometers** vertically against the box. Make sure the thermometers are not in direct sunlight.
4. Cover one thermometer with a **clean glass jar.**
5. Observe and record the temperature changes of each thermometer every 2 min over a 30-min period.

Analysis

1. **Identify** the independent variable and the dependent variable in this investigation.
2. **Construct** a graph showing how the temperatures of the two thermometers changed over time.
3. **Evaluate** Based on your graph, which thermometer experienced the greatest increase in temperature? Why?
4. **Relate** your observations to the greenhouse effect in the atmosphere.

Global Warming

Temperatures worldwide have shown an upward trend over the past 200 years, with several of the warmest years on record having occurred within the last two decades. This trend is shown in **Figure 24.** If the trend continues, polar ice caps and mountain glaciers might melt. This could lead to a rise in sea level and the flooding of coastal cities. Other possible consequences include the spread of deserts into fertile regions, an increase in sea surface temperature, and an increase in the frequency and severity of storms.

Based on available temperature data, many scientists agree that global warming is occurring. They disagree, however, about what is causing this warming. Some scientists hypothesize that natural cycles adequately explain the increased temperatures. Mounting evidence suggests that the rate of global temperature changes over the past 150 years is largely due to human activity.

Burning fossil fuels One of the main sources of atmospheric carbon dioxide from humans is from the burning of fossil fuels including coal, oil, and natural gas. Ninety-eight percent of these carbon dioxide emissions in the United States come from burning fossil fuels to run automobiles, heat homes and businesses, and power factories. Almost any process that involves the burning of fossil fuels results in the release of carbon dioxide. Burning fossil fuels also releases other greenhouse gases, such as methane and nitrous oxide, into the atmosphere.

☑ READING CHECK **Explain** how burning fossil fuels might contribute to global warming.

Figure 24 The warmest years of the last century all happened within the last 20 years of the century.

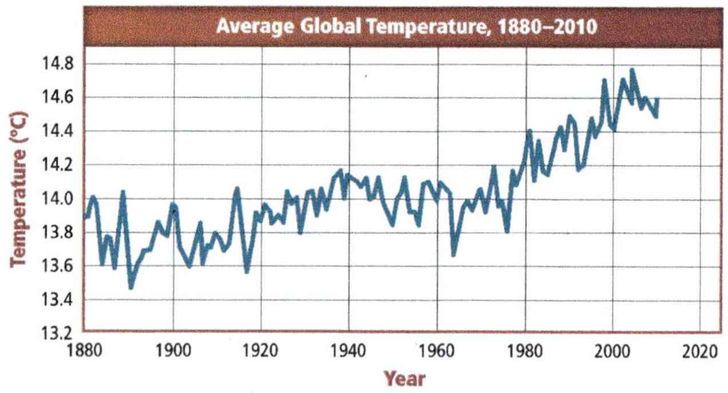

DIFFERENTIATED INSTRUCTION

Advanced Learners Have students research how global warming could lead to increased cloud cover and infer how this might affect atmospheric processes. The increased cloud cover would reflect more solar radiation back to space, which could result in atmospheric cooling.

☑ READING CHECK Burning fossil fuels increases the amount of carbon dioxide into the atmosphere.

Figure 25 Deforestation, the mass removal of trees, has occurred in British Columbia, Canada.
Explain *how deforestation can lead to global warming.*

Deforestation Deforestation—the mass removal of trees—also plays a role in increasing levels of atmospheric carbon dioxide. During photosynthesis, vegetation removes carbon dioxide from the atmosphere. When trees, such as the ones shown in **Figure 25,** are cut down, photosynthesis is reduced, and more carbon dioxide remains in the atmosphere. Many scientists suggest that deforestation intensifies global warming trends.

Environmental efforts Individuals reduce the amount of carbon dioxide emitted to the atmosphere by conserving energy. This, in turn, reduces fossil fuel consumption. Some easy ways to conserve energy include turning off electrical appliances and lights when not in use, turning down thermostats in the winter, recycling, and reducing the use of combustion engines, such as those in cars and lawn mowers.

SECTION 4 REVIEW

Section Self-Check

Section Summary

- The greenhouse effect influences Earth's climate.
- Worldwide temperatures have shown an upward trend over the past 200 years.
- Human activities can influence changes in weather and climate.
- Individuals can reduce their environmental impact on climate change.

Understand Main Ideas

1. MAINIDEA **Describe** some human activities that might have an impact on Earth's climate.
2. **Explain** the greenhouse effect.
3. **Apply** What is global warming? What are some possible consequences of global warming?
4. **Reason** Why do some scientists theorize that global warming might not be the result of increases in atmospheric carbon dioxide?

Think Critically

5. **Evaluate** the analogy of tropical rain forests being referred to as the "lungs" of Earth.

WRITING IN ▶ Earth Science

6. Write a pamphlet that explains global warming and its possible causes. Include tips on how individuals can reduce CO_2 emissions into the atmosphere.

Environmental Connection

Glacier Studies By analyzing air bubbles trapped in glaciers, scientists have found that carbon dioxide in the atmosphere has increased from approximately 280 parts per million (ppm) in 1750 to around 387 ppm in 2010. To analyze the air trapped in glacial samples, scientists drill long cores of ice from ice sheets. Air is released from the ice and sucked out into a tube. A laser beam of infrared light shoots through the tube; this beam measures the amount of carbon dioxide in the sample of air.

Caption Question Fig. 25 When trees are cut down, photosynthesis is reduced and more carbon dioxide remains in the atmosphere.

3 Assess

Check for Understanding

Describe Ask students to describe the greenhouse effect and its connection with global warming.

Reteach

Research Have students explain why it is hard to coordinate a worldwide effort to reduce the impact of human activities on the atmosphere.

Assessment

Knowledge Ask students to infer how global warming might affect the water cycle. It could increase Earth's precipitation patterns.

SECTION 4 REVIEW

1. Answers will vary, but could include burning fossil fuels, such as for vehicles and industry, and deforestation.
2. The greenhouse effect is the natural heating of Earth's surface caused by greenhouse gases.
3. Global warming is a rise in global temperature that might be caused by increases in greenhouse gases. Melting ice caps, rising sea level, and an increase in the frequency and severity of storms are all possible consequences of global warming.
4. Several naturally occurring phenomena could cause increases in global temperatures.
5. These forests take in carbon dioxide, which humans exhale, and give off oxygen, which humans inhale. Thus, they function much like lungs in reverse.
6. Pamphlets should include an explanation of global warming and its possible causes, including burning fossil fuels for vehicles and industry, deforestation, and ideas for reducing carbon dioxide emissions, such as turning off lights and appliances when not in use and reducing the use of combustion engines.

Rubric

Earth Science & SOCIETY

Purpose

Students will learn about some of the effects of permafrost thawing in the Arctic.

Teacher Content Support

Effects of Thawing Permafrost In 2005, scientists at the National Center for Atmospheric Research (NCAR) used a computer model to make predictions about future permafrost thaw. They estimated that up to 50 percent of the top layer of permafrost could thaw by 2050, and up to 90 percent by 2100. The model included variables such as the interaction between the atmosphere and the ocean, the presence of sea ice, and greenhouse-gas emissions. Effects of thawing permafrost include increased water runoff into oceans, the release of large amounts of carbon dioxide and methane, damage to structures built on the soil, and a phenomenon called "drunken forests" in which trees lean at odd angles because their roots are no longer held in place by solid soil.

Teaching Strategy

Show students the location of the Alaskan villages of Shishmaref and Kivalina on a map. Ask students to infer why thawing permafrost and melting sea ice would be of particular concern to people in these villages. Both villages are located on small islands on the Alaskan coast. Thawing permafrost and melting sea ice literally means that part of the island is disappearing.

Earth Science & SOCIETY

Effects of Global Warming on the Arctic

A house near the coast in Shishmaref, Alaska, collapsed as a result of thinning sea ice and thawing permafrost.

Air temperatures in some areas of the Arctic have risen about 2°C in the past 30 years. As permafrost thaws and sea ice thins, houses are collapsing, roads are sinking, and flooding and erosion are increasing.

Thawing permafrost About 85 percent of the ground in Alaska lies above permafrost, which is a layer of soil that remains frozen for two or more years and has a temperature of at least 0°C. Recent data show that the temperature of permafrost across the Arctic has risen anywhere from 0.5°C to 2.0°C, resulting in thawing of the frozen soil in some areas. In areas where permafrost has thawed, the ground has dropped as much as 5 m, affecting roads, airport runways, homes, and businesses. Buildings, such as hospitals and schools, are unusable due to the sinking effect, and roads in Fairbanks, Alaska, have needed costly repairs.

Thinning sea ice People in the village of Shishmaref, on the northwestern coast of Alaska, moved houses to higher ground to avoid having them collapse into the surrounding sea. As the sea ice that helps protect the village from strong waves thins, the land is more vulnerable to erosion. The nearby village of Kivalina, Alaska, is in a similar situation. Engineers estimated that the cost of moving the village's 380 residents to more stable ground is between $100 and $400 million.

Disrupting traditions Changes in temperature also affect hunting practices of people native to the Arctic. Ice-fishing seasons used to begin in October but now do not start until December, when the sea finally freezes. Native languages have also been affected by the changing temperatures and seasonal conditions.

The Inuit word *qiqsuqqaqtug* is used to refer to the month of June. The word describes specific snow conditions that occur in June—when a thin layer of melted snow sits on the surface and refreezes at night, forming a crust. With changing temperatures, this condition now occurs in May. As a result, some Inuit think that the word no longer accurately describes the month of June.

Releasing carbon dioxide Permafrost consists of soil that contains high amounts of organic material. As thawing occurs, the organic material in the soil decomposes, releasing carbon dioxide. With more than 1 million km^2 of soil in the Arctic, scientists think that thawing could release large amounts of carbon dioxide into the atmosphere.

WRITING IN ▶ Earth Science

Bulletin Board Display Research more information about the effects of climate warming on the Arctic. Prepare a display for a bulletin board that explains several examples and includes either illustrated figures or photos.

WebQuest

WRITING IN ▶ Earth Science

Rubric

Bulletin Board Display Students could include the visible effects on homes, roads, and other structures built on permafrost, the "drunken forests," or touch on the social and cultural effects on people living in the Arctic. They could focus on the global effects of thawing permafrost, such as the release of large amounts of carbon dioxide and methane or rising seas levels due to increased runoff.

WebQuest

GeoLAB

iLab Station

Design Your Own: Identify a Microclimate

Background: Microclimates can be caused by tall buildings, large bodies of water, and mountains, among other things. In this activity, you'll observe different microclimates and then attempt to determine which factors strengthen microclimates and how these factors change with distance from Earth's surface.

Question: *Which type of surface creates the most pronounced microclimate?*

Materials

thermometer
psychrometer
paper strip or wind sock
meterstick
relative humidity chart

Safety Precautions

WARNING: *Be careful when you handle glass thermometers. If the thermometer breaks, do not touch it. Have your teacher properly dispose of the glass.*

Procedure

1. Read and complete the lab safety form.
2. Working in groups of three to four, determine a hypothesis based on the question listed above.
3. Create a plan to test your hypothesis. Include how you will use your equipment to measure temperature, relative humidity, and wind speed on different surfaces and at various heights above these surfaces. Make sure you include provisions for controlling your variables.
4. Select your sites.
5. Make a map of your test sites. Design and construct data tables for recording your observations.
6. Identify your constants and variables in your plan.
7. Have your teacher approve your plan before you proceed.
8. Carry out your experiment.
9. Map your data. Color-code the areas on your map to show which surfaces have the highest and lowest temperatures, the highest and lowest relative humidity, and the greatest and least wind speed. On your map, include data for surface area only.
10. Graph your data for each site, showing differences in temperature with height. Plot temperature on the x-axis and height on the y-axis. Repeat this step for relative humidity and wind speed.

Analyze and Conclude

1. **Analyze** your maps, graphs, and data to find patterns. Which surfaces had the most pronounced microclimates?
2. **Conclude** Did height above the surface affect your data? Why or why not?
3. **Analyze** your hypothesis and the results of your experiment. Was your hypothesis supported? Explain.
4. **Infer** Why did some areas have more pronounced microclimates than others? Which factors seemed to contribute the most to the development of microclimates?
5. **Determine** which variable changed the most with height: temperature, relative humidity, or wind speed?
6. **Determine** which variable changed the least with height.
7. **Infer** why some variables changed more than others with height.

APPLY YOUR SKILL

Plan an Experiment Based on what you have learned in this lab, plan an experiment that would test for microclimates in your state. How would this large-scale experiment be different from the one you just completed?

GeoLAB

iLab Station Rubric

Preparation

Time Allotment 90 min

Process Skills communicate, make and use graphs, collect data, interpret data, observe and infer

Safety Precautions Approve lab safety forms before work begins.

Procedure

- This lab works best on a warm, sunny day with little cloud cover.
- This lab is best done as a group project. Assign one student to collect data at each location. Another student can map the data, and a third can graph the data. The group as a whole should complete the Analyze and Conclude questions.
- Preselect a list of locations from which students can choose. The list should include locations that have different surfaces, such as asphalt, concrete, bare ground, grassy areas, and water. This will allow students to observe a wide range of microclimates.

Analyze and Conclude

1. Students should find that darker and denser surfaces are generally warmer and less humid than lighter and less dense surfaces.
2. With increasing height, temperature should decrease, and humidity and wind speed should increase.
3. Answers will vary depending on individual results. If the weather variables showed appropriate changes with different heights and surfaces, student hypotheses were likely supported.
4. Darker and denser surfaces absorb more sunlight and therefore warm up faster. This higher rate of absorption also affects relative humidity. Areas protected from the wind likely experienced the most variation.
5. Answers will vary depending on the exposure of individual locations. Any of the variables might show the most change, but normally, temperature varies most.
6. Wind is least likely to change, but again, this depends on the exposure of the location.
7. Answers will vary, but could include ideas such as the influence of conduction close to the ground and convection currents.

APPLY YOUR SKILL

Plan an Experiment Experiments will vary, but should include the use of scientific methods. The scale of the experiment would be much larger.

LabManager™
Customize any lab with the LabManager™ CD-ROM.

CHAPTER 14 STUDY GUIDE

BIGIDEA **The different climates on Earth are influenced by natural factors as well as human activities.**

Vocabulary Practice

SECTION 1 Defining Climate

VOCABULARY
- climatology
- normal
- tropics
- temperate zones
- polar zones

MAINIDEA **Climate is affected by several factors including latitude and elevation.**
- Climate describes the long-term weather patterns of an area.
- Normals are the standard climatic values for a location.
- Temperatures vary among tropical, temperate, and polar zones.
- Climate is influenced by several different factors.
- Air masses have distinct regions of origin.

SECTION 2 Climate Classification

VOCABULARY
- Köppen classification system
- microclimate
- heat island

MAINIDEA **Climates are categorized according to the average temperatures and precipitation amounts.**
- German scientist Wladimir Köppen developed a climate classification system.
- There are five main climate types: tropical, dry, mild, continental, and polar.
- Microclimates can occur within cities.

SECTION 3 Climatic Changes

VOCABULARY
- ice age
- season
- El Niño
- Maunder minimum

MAINIDEA **Earth's climate is constantly changing on many different timescales.**
- Climate change can occur on a long-term or short-term scale.
- Changes in solar activity have been correlated with periods of climate change.
- Changes in Earth's orbit, tilt, and wobble are all associated with changes in climate.

SECTION 4 Impact of Human Activities

VOCABULARY
- greenhouse effect
- global warming

MAINIDEA **Over time, human activities can alter atmospheric conditions enough to influence changes in weather and climate.**
- The greenhouse effect influences Earth's climate.
- Worldwide temperatures have shown an upward trend over the past 200 years.
- Human activities can influence changes in weather and climate.
- Individuals can reduce their environmental impact on climate change.

MAINIDEAS **Summary statements can be used by students to review the major concepts of the chapter.**

Students can review with these online resources.

Vocabulary Practice

Vocabulary eGames
Vocabulary eFlashcards
Vocabulary PuzzleMaker

Section Self-Check

Chapter Self-Check

Online Test Practice

Use the ExamView® *Assessment Suite* CD-ROM to:
- create multiple versions of tests
- create modified tests with one mouse click
- edit existing questions and add your own questions
- build tests aligned with state standards using built-in state curriculum tags
- change English tests to Spanish with one mouse click
- track students' progress using the Teacher Management System

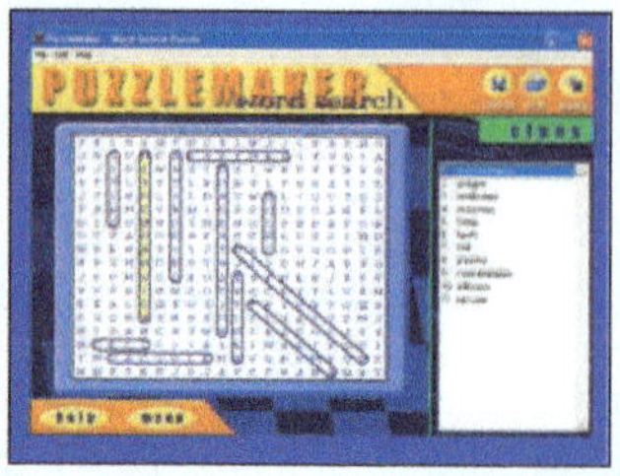

Vocabulary Practice

For additional practice with vocabulary, have students access the Vocabulary PuzzleMaker.

Chapter Self-Check

VOCABULARY REVIEW

Write a description to correctly define the following vocabulary terms.

1. climatology
2. temperate zone
3. normal
4. polar zone
5. tropics

Fill in the blank with the correct vocabulary term from the Study Guide.

6. The differences in temperature caused by the presence of a large city is an example of a(n) ________.
7. A(n) ________ is the name given to localized climate changes such as those at the top of a mountain.
8. A climatic change occurring on a timescale of months is a(n) ________.
9. A(n) ________ is a climatic change that occurred as a result of a change in the number of sunspots.

Replace each underlined vocabulary term with the correct term from the Study Guide.

10. Increasing levels of atmospheric carbon dioxide have been suggested as a cause of <u>the greenhouse effect</u>.
11. Retention of heat in the atmosphere is a result of <u>global warming</u>.

UNDERSTAND KEY CONCEPTS

12. Which is not true about climatic normals?
    A. They are averaged over a 30-year period.
    B. They can differ from daily weather conditions.
    C. They describe average conditions.
    D. They are gathered at one location.

Use the diagram below to answer Questions 13 and 14.

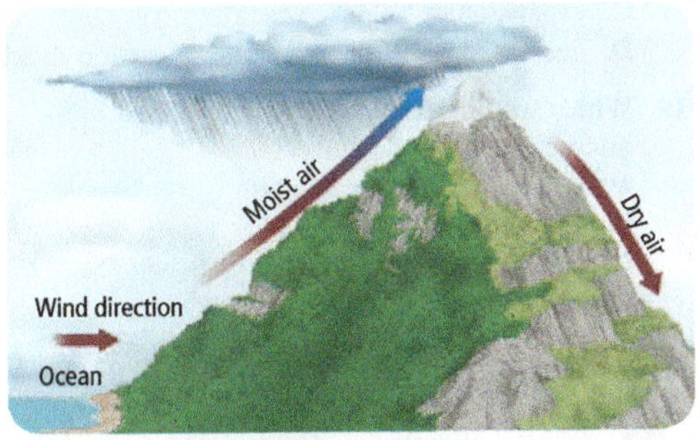

13. Which best describes the climate on the leeward side of a mountain?
    A. warm and rainy
    B. cool and dry
    C. warm and dry
    D. cool and rainy

14. What happens to air as it passes over the windward side of a mountain?
    A. It sinks and gathers moisture.
    B. It sinks and begins to condense.
    C. It rises and begins to condense.
    D. It rises and evaporates.

15. Which is a long-term climate change?
    A. fall
    B. an ice age
    C. summer
    D. El Niño

16. El Niño develops because of a weakening of what?
    A. the polar front
    B. the trade winds
    C. the prevailing westerlies
    D. the jet stream

17. Which phenomenon has not been suggested as a factor in global warming?
    A. deforestation
    B. El Niño
    C. burning of fossil fuels
    D. industrial emissions

CHAPTER 14 ASSESSMENT

VOCABULARY REVIEW

1. the study of Earth's climate
2. areas of moderate temperatures between 23.5° and 66.5° north and south of the equator
3. standard values of temperature, precipitation, and wind speed for a location
4. areas of cooler temperatures located 66.5° north and south of the equator
5. areas between 23.5° S and 23.5° N of the equator where temperatures are generally warm year-round
6. heat island
7. microclimate
8. season
9. Maunder minimum
10. global warming
11. the greenhouse effect

UNDERSTAND KEY CONCEPTS

12. D
13. C
14. C
15. B
16. B
17. B

18. A
19. C
20. B
21. A
22. B

CONSTRUCTED RESPONSE

23. Normals describe long-term averages, not specific days. Normals are accurate only for the location where the data was recorded, and might be different from locations just a few miles away.
24. the tropics; The polar zone varies between total sunlight and warmth in summer to cold darkness in winter, while the temperate zone sees many air mass changes. Length of day varies little in the tropics, and it stays warm year-round.
25. They are close to the west coast of continents where a constant flow of moist air from the cool ocean keeps the atmosphere saturated, which is necessary for fog formation.
26. marine west-coast—wet with mild winters and cool summers; humid subtropical—warm, wet summers and dry, cool winters; Mediterranean—warm summers, mild winters
27. A different orbit would change how much solar radiation Earth receives in a given year, while a different tilt would change where the most solar radiation is absorbed.
28. Dry climates occur when evaporation rates exceed precipitation rates. This is most likely in warm areas that mainly are found where the highest amounts of solar radiation are received—the tropics.
29. As emissions have increased, concentration has increased.
30. It is needed to keep Earth's temperatures at a habitable level, but if it increases too much, it is believed to produce excessive global warming which could have numerous harmful effects.
31. Hurricanes are powered by heat energy stored in the ocean, and some believe that this added heat would therefore lead to more and stronger hurricanes.
32. The warmer water warms the lower atmosphere enough to produce more convection and precipitation. These convective currents transport additional heat into the atmosphere where upper winds transport it around the world.

18. Which could be a result of global warming?
A. rising sea levels
B. increased volcanic activity
C. expansion of polar ice caps
D. decreased levels of atmospheric carbon dioxide

19. Which would not be likely to produce a microclimate?
A. an ocean shoreline
B. a valley
C. a flat prairie
D. a large city

20. A heat island is an example of which type of climate?
A. tropical climate
B. microclimate
C. dry climate
D. polar climate

Use the table below to answer Questions 21 and 22.

World Climates

Location	Climate Description
New Caledonia, South Pacific	constant high temperatures, plenty of rain
Southern Israel	humid in summer, dry in winter
Gobi Desert, Mongolia	continental tropical air, low precipitation, scarce vegetation
Bogotá, Colombia	mild winters, cool summers, abundant precipitation
Yukon, Canada	year-round cold, low precipitation

21. Southern Israel has which type of climate?
A. humid subtropical
B. dry
C. mediterranean
D. continental

22. Where is a semidesert most likely to be found?
A. New Caledonia
B. Gobi Desert
C. Bogotá
D. Yukon

CONSTRUCTED RESPONSE

23. Identify two situations for which climatic normals might not be accurate to use in determining daily weather.

24. Apply Which of the main temperature zones probably experiences the least temperature change during the year? Why?

25. Relate why fog is a common characteristic of marine west-coast climates.

26. Compare and contrast two of the different types of climates classified as mild by Köppen.

27. Contrast the climatic changes produced by changes in Earth's orbit to those produced by changes in the tilt of Earth's axis.

28. Deduce the reason that most of the dry climates are located near the tropics.

Use the graph below to answer Question 29.

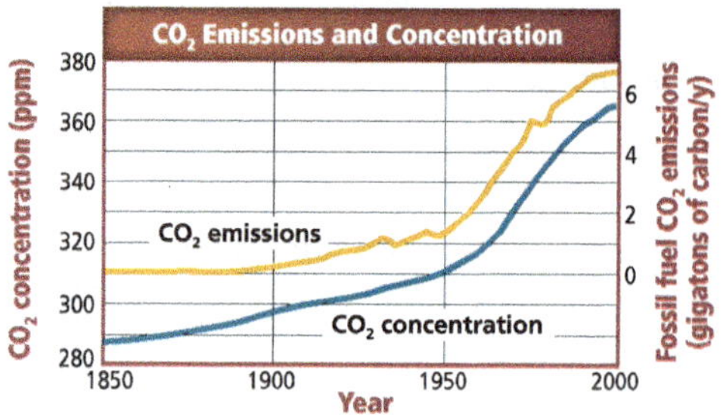

29. Analyze the relationship between carbon dioxide emissions and carbon dioxide concentration in the atmosphere over the last 150 years.

30. Analyze why the greenhouse effect is considered both essential to life on Earth and also possibly destructive.

31. Deduce why some scientists have proposed that global warming might affect the frequency and severity of hurricanes.

32. Cause and Effect How does El Niño cause short-term climatic changes?

33. Suggest an explanation for why scientists do not agree on the cause of global warming.

33. There are many naturally occurring phenomena that could contribute to global warming, making it hard to prove that human activities are the sole cause of it.

Chapter Self-Check

THINK CRITICALLY

34. Formulate a reason why a change in temperature as little as 5°C could be responsible for a climatic change as dramatic as an ice age.

35. Predict Would you expect temperatures during the night to drop more sharply in marine climates or continental climates? Why?

36. Suggest The ironwood tree grows in the desert. Temperatures beneath an ironwood tree can be up to 8°C cooler than temperatures a few feet away. What situation does this describe and how might it affect organisms living in the desert?

Use the diagram below to answer Question 37.

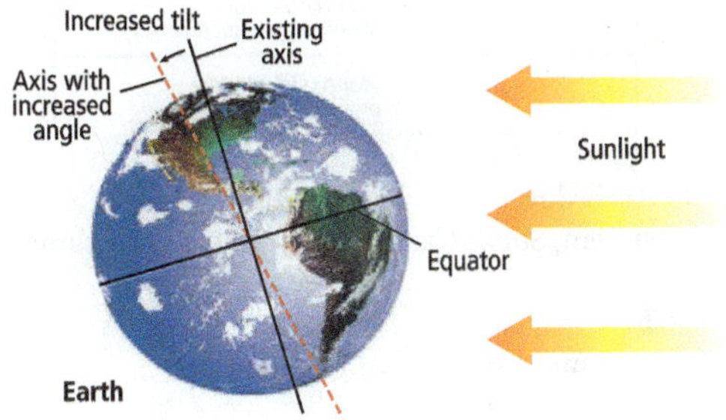

37. Predict the effect on Earth's climate if the tilt of Earth's axis increased to 25°.

38. Assess If you wanted to build a home that was solar heated, would you build it on the leeward or windward side of a mountain? Explain.

CONCEPT MAPPING

39. Draw a concept map that organizes information about El Niño.

CHALLENGE QUESTION

40. Analyze Use a world map and **Figure 6** to determine the climate classifications of the following cities: Paris, France; Athens, Greece; London, England; and Sydney, Australia.

WRITING IN ▶ Earth Science

41. Suppose a friend makes the statement that humans must be to blame for climate change. Write a paragraph describing evidence for and against the assertion.

DBQ Document–Based Questions

Data obtained from: The National Snow and Ice Data Center. 2010.

The graph below shows the average area of oceans covered by sea ice each year in the northern hemisphere for a 31-year period. Scientists use the area of sea ice as an indicator of global climate change.

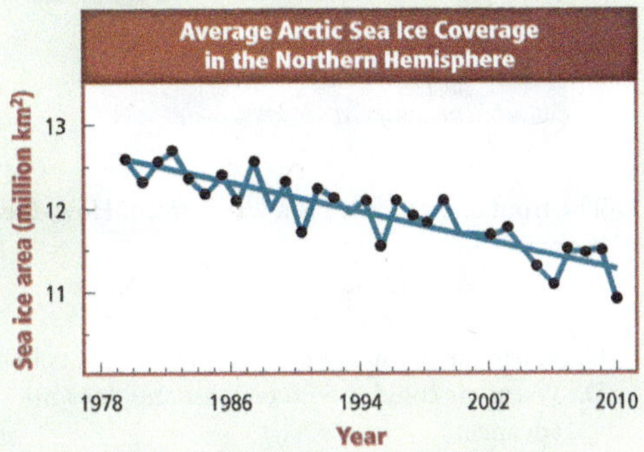

42. Describe the trend of sea-ice coverage over the period shown in the graph. Which year experienced the least sea-ice coverage?

43. What inferences about global climate can be drawn from the changes in sea-ice coverage in the last 31 years?

44. The graph shows that sea ice area decreased at a rate of 3.5% percent per decade. Predict the area of sea ice coverage in 2020.

CUMULATIVE REVIEW

45. Compare and contrast the terms *magma* and *lava.* **(Chapter 5)**

46. List the three primary mechanisms of heat transfer within the atmosphere. **(Chapter 11)**

CHAPTER 14 ASSESSMENT

THINK CRITICALLY

34. That change could be the difference between more snow falling in a year than melting, which over hundreds of thousands of years would cause a great increase in the size and depth of ice sheets across the world.

35. continental climates; land loses heat more quickly than water does. Also, the air over coastal regions has many of the same characteristics as an ocean climate.

36. A microclimate is created in the shade of the tree. The cooler temperatures might help plants and animals survive in the desert conditions by conserving water and avoiding direct sunlight.

37. By drastically changing where on Earth the most solar radiation was received, there would be a shift in the locations of both the polar and tropical regions.

38. leeward; The climate on this side is drier, so there would be less cloud cover to interfere with the solar radiation, and the warmer climate due to downslope warming would mean less days when you needed to use the heat.

CONCEPT MAPPING

39. Check students' maps for accuracy.

CHALLENGE QUESTION

40. Paris: marine west-coast; Athens: mediterranean; London: marine west-coast; Sydney: marine west-coast

WRITING IN ▶ Earth Science

Rubric

41. Answers will vary. Students might disagree because there are natural causes for climate change that were occurring long before humans existed.

DBQ Document-Based Questions

Data obtained from: The National Snow and Ice Data Center. 2006.

42. There is a decrease in coverage of sea ice over the time period shown in the graph; 2010

43. Global temperatures are increasing, leading to a decreased formation of sea ice.

44. about 10.6 million km^2

CUMULATIVE REVIEW

45. Magma is molten rock at depth, lava is molten rock extruded onto the surface.

46. radiation, conduction, and convection

CUMULATIVE STANDARDIZED TEST PRACTICE

MULTIPLE CHOICE

1. D
2. A
3. C
4. B
5. C
6. D
7. C
8. D
9. A
10. B

CUMULATIVE STANDARDIZED TEST PRACTICE

MULTIPLE CHOICE

1. Which type of air masses are most likely to form over land near the equator?
A. mP
B. mT
C. cP
D. cT

Use the figure below to answer Questions 2 and 3.

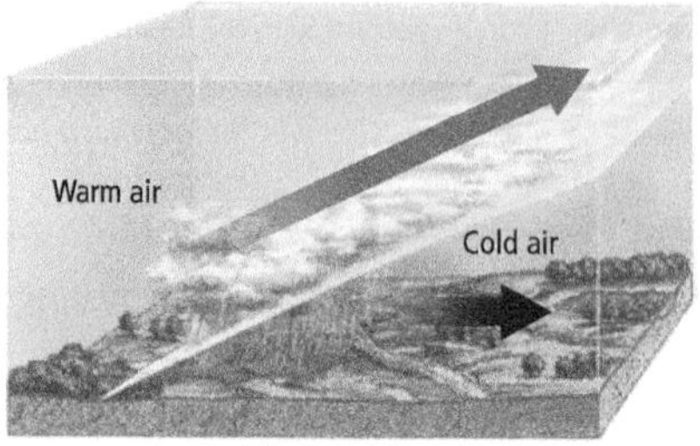

2. The front shown above is a warm front. How does it occur?
A. Warm air gradually slides over colder air.
B. Warm air steeply climbs over cold air.
C. Warm air is quickly forced upward over cold air.
D. Warm air collides with cold air and does not advance.

3. What type of weather could you expect to see in this situation?
A. light precipitation over a wide band
B. extensive precipitation over a narrow band
C. extensive precipitation over a wide band
D. light precipitation over a narrow band

4. About 99 percent of the Earth's atmosphere is made up of what two gases?
A. carbon dioxide and oxygen
B. nitrogen and oxygen
C. carbon dioxide and nitrogen
D. water vapor and oxygen

5. What occurs when winds of at least 120 km/h drive a mound of ocean water toward coastal areas?
A. downburst
B. cold wave
C. storm surge
D. tsunami

6. Which factor is NOT associated with a heat wave?
A. a high-pressure system
B. a weakened jet stream
C. above-normal temperatures
D. increased cloud cover

Use the table below to answer Questions 7 and 8.

Location	Climate Description
New Caledonia, South Pacific	constant high temperatures, plenty of rain
South Carolina	humid in summer, dry in winter
Gobi Desert, Mongolia	continental tropical air, low precipitation, scarce vegetation
Bogotá, Columbia	mild winters, cool summers, abundant precipitation
Yukon, Canada	year-round cold, low precipitation

7. According to the modified Köppen classification system, South Carolina has what kind of climate?
A. tropical
B. dry
C. humid subtropical
D. continental

8. Where is a tundra region most likely to be found?
A. New Caledonia
B. Gobi Desert
C. Bogotá
D. Yukon

9. Why does conduction affect only the atmospheric layer near Earth's surface?
A. It is the only layer that comes in direct contact with Earth.
B. It is the only layer that contains particles of air.
C. It is the coolest layer.
D. It is the warmest layer.

10. The current state of the atmosphere is known as
A. climate
B. weather
C. precipitation
D. air mass

Online Test Practice

SHORT ANSWER

Use the graph below to answer Questions 11 and 12.

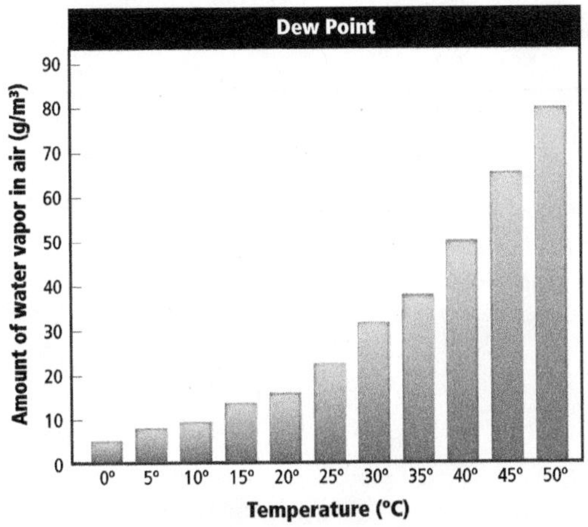

11. What can be determined about dew point according to the graph?

12. What would happen if the temperature were 35°C and the amount of water vapor present were 50 g/m³?

13. Why might the usage of normals be disadvantageous in describing an area's climate?

14. Discuss how the Geographic Information System (GIS) works.

15. Describe how the sound of thunder is produced.

16. Gold is an expensive, soft, highly malleable metal. Infer two reasons why a jeweler might choose to make a ring out of an alloy of gold and copper instead of out of pure gold.

READING FOR COMPREHENSION

Global Warming

John Harte, an ecosystem sciences professor at the University of California, Berkeley, is studying possible future outcomes of global warming. For 15 years, he has artificially heated sections of a Rocky Mountain meadow by about 3.6°F (2°C) to study the projected effects of global warming. Harte has documented dramatic changes in the meadow's plant community. Sagebrush, though at the local altitude limit of its natural range, is replacing alpine flowers. More tellingly, soils in test plots have lost about 20 percent of their natural carbon. This effect, if widespread, could dramatically increase Earth's atmospheric CO_2 levels far above even conventional worst-case models. "Soils around the world hold about 5 times more carbon than the atmosphere in the form of organic matter," Harte noted. If similar carbon loss was repeated on a global scale, it could double the amount of carbon in the atmosphere.

Article obtained from: Handwerk, B. Global warming: How hot? How soon? *National Geographic News.* July 27, 2005.

17. What can be predicted from the Rocky Mountain study?
    A. The changes occurring in the Rocky Mountain study will occur in other parts of the globe.
    B. These changes will only affect the Rocky Mountain area.
    C. More studies are needed to determine if the changes really will happen.
    D. The Rocky Mountain area is unique in its makeup and therefore is affected most by global warming.

18. What should people learn from reading this passage?

NEED EXTRA HELP?

If You Missed Question . . .	1	2	3	4	5	6	7	8	9	10	11	12	13	14	15	16
Review Section . . .	12.1	12.2	12.2	11.1	13.3	13.4	14.2	14.2	11.1	12.1	11.2	11.2	14.1	2.3	13.2	3.2

SHORT ANSWER

11. Sample answer: As the temperature increases, the amount of water air can hold increases thus increasing the dew point.
12. Because 50 is past the dew point for 35°C, condensation would occur most likely in the form of rain.
13. Normals are data about a particular area's climate that are gathered and averaged over a long period of time. The limitation is that the numbers provided by the normal might never exactly occur and thus make the normal seem incorrect.
14. The GIS uses a database of information to create layers of information that can be placed one on top of the other to create a comprehensive map. As information changes, the layers change keeping the map up to date at all times.
15. A lightning bolt heats the surrounding air to about 30,000°C. The sound of thunder is produced as this superheated air rapidly expands and contracts.
16. The alloy is harder than pure gold, so the ring will be more durable. Also, the alloy will be less expensive than pure gold.

READING FOR COMPREHENSION

17. A
18. Sample answer: People should learn that global warming is a real threat to Earth. Even slight changes can have devastating effects that will eventually impact human life. Changes need to be made in order to preserve the environment.

CHAPTER 15 Earth's Oceans

BIGIDEA **Studying oceans helps scientists learn about global climate and Earth's history.**

ESSENTIAL QUESTIONS	NATIONAL STANDARDS	RESOURCES TO ASSESS MASTERY
SECTION 1 An Overview of Oceans 1. What methods are used by scientists to study Earth's oceans? 2. How did the oceans form? 3. How is water distributed on Earth's surface? 1.5 sessions 0.75 block	UCP.1–5; A.1, A.2; B.2; C.3; D.1–3; E.1–2; F.6; G.1	**Entry-Level Assessment** Focus Transparency 44, p. 406 **Progress Monitoring** Caption Question, p. 410 Reading Check, pp. 407, 408, 411 Section Review, p. 412
SECTION 2 Seawater 1. What is the composition of seawater? 2. What are the physical properties of seawater? 3. How is ocean layering illustrated? 4. How do deepwater masses form? 3 sessions 1.5 blocks	UCP.1–5; A.1, A.2; B.2; D.2–3	**Entry-Level Assessment** Focus Transparency 45, p. 413 **Progress Monitoring** Caption Question, pp. 413, 417, 419 Reading Check, pp. 414, 417, 418, 419 Section Review, p. 420
SECTION 3 Ocean Movements 1. What are the physical properties of waves? 2. How do tides form? 3. What are the similarities and differences between various ocean currents? 3 sessions 1.5 blocks	UCP.1–5; A.1, A.2; B.2, B.4; D.1–3; G.1, G.3	**Entry-Level Assessment** Focus Transparency 46, p. 421 **Progress Monitoring** Caption Question, pp. 422, 426 Reading Check, pp. 422, 423, 425 Section Review, p. 427 **Summative Assessment** Chapter Assessment, p. 431 Exam*View*® *Assessment Suite* CD-ROM

Go online!
connectED.mcgraw-hill.com

LEVELED RESOURCES	LAB MATERIALS
Science Notebook 15.1 OL **Chapter FAST FILE Resources:** Study Guide, p. 121 BL **Transparencies:** Section Focus Transparency 44 EL Teaching Transparency 42 OL EL	LaunchLAB p. 404 / **10 min** piece of string, globe, blue marker, ruler
Science Notebook 15.2 OL **Chapter FAST FILE Resources:** MiniLab Worksheet, p. 110 OL GeoLab Worksheet, p. 111 OL Study Guide, p. 123 BL **Lab Resources:** Laboratory Manual, p. 113 OL **Transparencies:** Section Focus Transparency 45 EL Teaching Transparency 43 OL EL	MiniLAB p. 416 / **25 min** large beaker, distilled water, NaCl, $MgCl_2$, Na_2SO_4, $CaCl_2$, KCl, $NaHCO_3$, KBr GeoLAB p. 429 / **45 min** balance0; graduated 500-mL cylinder; 100-mL glass beakers (4); water; red, yellow, and blue food coloring; salt; thermometer; eyedropper; graph paper; ruler; calculator
Science Notebook 15.3 OL **Chapter FAST FILE Resources:** Study Guide, p. 125 BL **Lab Resources:** Laboratory Manual, p. 117 OL **Transparencies:** Section Focus Transparency 46 EL Teaching Transparency 44 OL EL	

LabManager™
Customize any lab with the LabManager™ CD-ROM.

ADDITIONAL RESOURCES

Technology:
- eTeacherEdition
- eStudentEdition
- Classroom Presentation Toolkit
- Exam*View*® *Assessment Suite* CD-ROM
- LabManager™ CD-ROM
- What's EARTH SCIENCE Got To Do With It? DVD
- Weather Classroom Video DVD

Assessment:
- Blueprints for Success
- Chapter Test (Scaffolded)
- Earth Science Standardized Test Practice

Lab Resources:
- Exploring Environmental Problems Laboratory Manual

BL Below Level OL On Level AL Advanced Learners EL English Learners COOP LEARN Cooperative Learning

CHAPTER 15

LaunchLAB

iLab Station Rubric

How much of Earth's surface is covered by water?

Teaching Strategies

- Procure three to four globes, if possible, with diameters of about 15 cm or more and several water-soluble, blue markers. Cut several pieces of string about 1 m in length (or long enough to fit around the globes' equators). Run through the entire activity prior to class.
- Divide the class into several groups. Give each group a piece of string and a blue marker.
- Review how to convert fractions to percentages.
- Caution students to be precise and to be careful not to mark on the globe when coloring the string.

Procedure

1. Have students read and complete the lab safety form and follow the procedure below.
2. Stretch a **piece of string** about 1 m in length around the equator of a **globe.**
3. Use a **blue marker** to color the sections of the string that cross the oceans.
4. Using a **ruler,** measure the length of the globe's equator as indicated by the string, then measure the length of each blue section on the string. Add the lengths of the blue sections.
5. Divide the total length of the blue sections by the length of the globe's equator.

CHAPTER 15

Earth's Oceans

BIGIDEA Studying oceans helps scientists learn about global climate and Earth's history.

SECTIONS

1 An Overview of Oceans

2 Seawater

3 Ocean Movements

LaunchLAB

iLab Station

How much of Earth's surface is covered by water?

Earth is often referred to as the blue planet because so much of its surface is covered by water. If you study a globe or a photograph of Earth taken from space, you can see that oceans cover much more of Earth than landmasses do. Investigate the distribution of water on Earth's surface in this lab.

FOLDABLES® Study Organizer

Wave Characteristics

Make a four-tab book using the labels shown. Use it to organize your notes on waves.

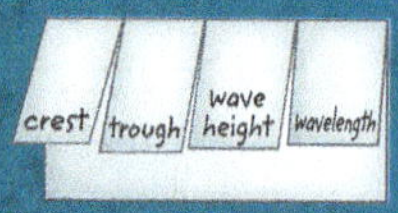

404

Analysis

1. **Calculate** What percentage of the globe's equator is made up of oceans? What percentage of the globe's equator is made up of land? About 80 percent of the equator crosses oceans; 20 percent crosses landmasses.
2. **Observe** Study the globe again. Which hemisphere is covered with more water? the southern hemisphere

Assessment

Skill Ask students to answer the following questions in terms of latitude: Which area of Earth's surface is covered by more land than water? between about 45° N and 70° N Between which latitudes is there no land at all? between about 55° S and 65° S

LabManager™ Customize any lab with the LabManager™ CD-ROM.

Introduce the BIGIDEA

Oceans Use a globe to emphasize that most of Earth's surface is covered by water. Point out to students this is true regardless of the direction from which Earth is viewed.

Teacher Content Support

Tide Pools Explain to students that tidal pools are usually formed on rocky shores when water gets trapped on shore after the tide recedes. Many organisms, including sea urchins, snails, chitins, clams, sea hares, and small fish, live on the rocky shore and in tidal pools. Rocky shore habitats are subject to heavy wave action and changes in the amount and temperature of water available to organisms. Have students discuss the kinds of adaptations organisms that live on rocky shores might have to help them survive there. hard shells, suction tube feet, spines, camouflage, the ability to spend time out of water

SECTION 1

1 Focus

Focus Transparency

Before presenting the lesson, project **Section Focus Transparency 44** and have students answer the accompanying questions. EL

MAINIDEA

Oceans Have students form groups and design a virtual global sailing journey that goes through every ocean basin, starting from a port city. Have students use a dry, erasable marker to trace the path on a globe. Students should keep a log of their virtual journey by writing down the names of the oceans they visit, as well as the latitudes and longitudes of several positions in each basin along the way. Have students explain whether ocean basins are completely separated by land masses or if they are interconnected.

2 Teach

Model

Measure Depth Before the development of echo sounding and sonar, ocean depths were determined by wire-line soundings. Perform the following exercise to model wire-line sounding: Place two or three objects on the bottom of a large cooler, then fill the cooler with muddy water that completely obscures the bottom. Have students use a small weight attached to a line to measure the depth to the bottom at various points. Students should then produce a contour plot of the bottom topography and compare it to actual bottom features after the cooler has been drained. BL EL

SECTION 1

Essential Questions

- What methods are used by scientists to study Earth's oceans?
- How did the oceans form?
- How is water distributed on Earth's surface?

Review Vocabulary

lake: natural or human-made body of water that can form when a depression on land fills with water

New Vocabulary

side-scan sonar
sea level

An Overview of Oceans

MAINIDEA The global ocean consists of one vast body of water that covers more than two-thirds of Earth's surface.

EARTH SCIENCE 4 YOU How could you tell a person's age by only looking at him or her? The presence of wrinkles can signify age as skin changes over time. Similarly, scientists look for clues about changes in rocks formed at the bottom of the ocean to estimate the age of the ocean.

Data Collection and Analysis

Oceanography is the scientific study of Earth's oceans. In the late 1800s, the British ship *Challenger* became the first research ship to use relatively sophisticated measuring devices to study the oceans. Since then, oceanographers have been collecting data with instruments both at the surface and from the depths of the ocean floor. Technologies such as sonar, floats, satellites, submersibles, and computers have become central to the continuing exploration of the ocean. **Figure 1** chronicles some of the major discoveries that have been made about oceans.

At the surface Sonar, which stands for **so**und **na**vigation and **r**anging, is used by oceanographers to learn more about the topography of the ocean floor. To determine ocean depth, scientists send a sonar signal to the ocean floor and time how long it takes for the sound to reach the bottom and return to the surface as an echo. Knowing that sound travels at a constant velocity of 1500 m/s through water, scientists can determine the depth by multiplying the total time by 1500 m/s, then dividing the answer by 2.

Figure 1
Developments in Oceanography
Technological development has led to many new discoveries in oceanography over time.

1925 The German *Meteor* expedition surveys the South Atlantic floor with sonar equipment and discovers the Mid-Atlantic Ridge.

1943 In France, the first diving equipment is invented from hoses, mouthpieces, air tanks, and a redesigned car regulator that supplies compressed air to divers.

1900 — 1925 — 1950

1872 The *Challenger* expedition marks the beginning of oceanography. Scientists measure sea depth, study the composition of the seafloor, and collect a variety of oceanic data.

1932–1934 The first deep-ocean dives use a tethered bathysphere. The dives uncover luminescent creatures and provide sediment samples.

1955 A survey ship detects linear magnetic stripes along the ocean floor. These magnetic patterns lead to the formulation of the theory of plate tectonics.

Demonstration

Latitude and Longitude A knowledge of latitude and longitude is crucial to understanding oceanography. To review these concepts, have students examine a globe or atlas to determine the approximate latitude and longitude of the following locations: their school answers will vary; Washington, DC 39°N, 77°W; the northernmost point in the United States Barrow, Alaska, 71°N, 157°W; the southernmost point in the United States the island of Hawaii, 19°N, 156°W.

Large portions of the seafloor have been mapped using **side-scan sonar,** a technique that directs sound waves to the seafloor at an angle, so that the sides of underwater hills and other topographic features can be mapped.

Oceanographers use floats that contain sensors to learn more about water temperature, salinity, and the concentration of gases and nutrients in surface water. Floats can also be used to record wave motion and the speed at which currents are moving. Satellites such as the *OSTM/Jason-2*, which you have read about, continually monitor the ocean's surface temperatures, currents, and wave conditions.

Figure 2 *Alvin* is a deep-sea submersible that can hold two scientists and a pilot.

In the deep sea Submersibles, underwater vessels which can be remotely operated or carry people to the deepest areas of the ocean, have allowed scientists to explore new frontiers. *Alvin,* shown in **Figure 2,** is a modern submersible that can take two scientists and a pilot to depths as deep as 4500 m. *Alvin* has been used to discover geologic features such as hydrothermal vents and previously unknown sea creatures. It can also be used to bring sediments and water samples to the surface.

☑ READING CHECK **List** some discoveries made using submersibles.

Computers An integral tool in both the collection and analysis of data from the ocean is computers. Information from satellites and float sensors can be transmitted and downloaded directly to computers. Sophisticated programs use mathematical equations to analyze data and produce models. When combined with observations, ocean models provide information about subsurface currents that are not observed directly. Operating in a fashion similar to weather forecasting models, global ocean models play a role in simulating Earth's changing climate. Ocean models are also used to simulate tides, tsunamis, and the dispersion of coastal pollution.

1962 France builds the first underwater habitat where scientists live for days at a time conducting experiments.

1984 An observation system in the Pacific Ocean helps scientists predict El Niño and begin to understand the connection between oceanic events and weather.

2002 Data collected from the seafloor reveals a new ocean wave associated with earthquakes.

2006 An internet portal provides scientists around the world with access to live data collected from ocean-floor laboratories.

1960 — 1980 — 2000

1977 The submersible *Alvin* discovers hydrothermal vents and a deep-sea ecosystem including giant worms and clams that can survive without energy from the Sun.

1995 Scientists map the entire seafloor using satellite data.

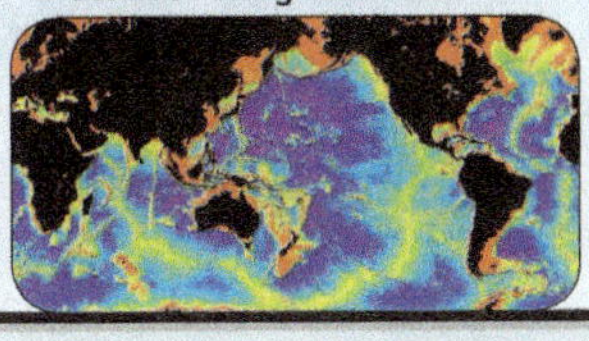

2010 Using telepresence technology, high-definition videos of new submarine volcanoes, hydrothermal vent systems, and exotic animal ecosystems discovered off the coast of Indonesia are transmitted to scientists ashore in real time.

Model

Sonar Echo sounding, or sonar, involves timing the return of an echo, multiplying that time by the velocity of sound, and dividing the resulting distance by two. To demonstrate how sonar works, take the class outside and have students stand at least 50 m away from the school. Have a student volunteer shout or clap his or her hands. Time the interval between the initial sound and the echo with a stopwatch, and have the class record that time. Repeat the demonstration with students standing farther away from the school. In class, tell students that the speed of sound in air is 334 m/s, and have them calculate how far they stood from the school in both demonstrations.

Interpret the Illustration

Time Line Have students read and study **Figure 1.** Have students choose one entry from the time line and research more details about the event. Students can write a report in their Earth science journals or make a short presentation to the class.

☑ READING CHECK hydrothermal vents, previously unknown sea creatures

EARTH SCIENCE JOURNAL

Early Ocean Exploration Have students research information about early ocean exploration by one of the following: ancient Egyptians, Phoenicians, ancient Greeks, Ferdinand Magellan, or James Cook. Have students write a short report, including a map of the voyage, in their Earth science journals that focuses on where the exploration occurred, what methods were used to collect data, and what conclusions were reached. Students can compare maps and develop a comprehensive overview of early ocean exploration.

Rubric

Teacher Content Support

Comets Comets have been compared to dirty snowballs. They consist of a mixture of frozen gases and dust, and have diameters averaging a few kilometers. The frozen gases are mostly water vapor, methane, and ammonia. Most known comets travel in highly eccentric orbits. As they approach the Sun, the gases vaporize and generate large, visible tails.

CAREERS IN EARTH SCIENCE

WebQuest

Use Science Terms

Meteorites Remind students of the difference among the words *meteoroid, meteor,* and *meteorite.* A meteoroid is a piece of interplanetary material, such as a piece broken off of an asteroid, that enters Earth's atmosphere. A meteoroid is referred to as a meteor when it falls toward Earth, burns up in the atmosphere, and produces a streak of light. A meteorite is a meteor that hits Earth's surface. Not all meteoroids reach Earth's surface. Some burn up completely as they fall through the atmosphere.

☑ READING CHECK Rocks formed from sediments deposited in water are nearly 4.6 billion years old; glassy crusts on lava indicate that it cooled rapidly underwater.

Origin of the Oceans

Several geologic clues indicate that oceans have existed almost since the beginning of geologic history. Studies of radioactive isotopes indicate that Earth is about 4.6 billion years old. Scientists have found rocks nearly as old that formed from sediments deposited in water. Ancient lava flows are another clue—some of these lava flows have glassy crusts that form only when molten lava is chilled rapidly underwater. Radioactive studies and lava flows offer evidence that there has been abundant water throughout Earth's geologic history.

☑ READING CHECK **Explain** the evidence that suggests that oceans have existed almost since the beginning of Earth's geologic history.

Where did the water come from? Scientists hypothesize that Earth's water originated from either a remote source or a local source, or both. Comets and meteorites are two remote sources that could have contributed to the accumulation of water on Earth. Comets, such as the one shown in **Figure 3,** travel throughout the solar system and occasionally collide with Earth. These impacts release enough water over time that they could have contributed to filling the ocean basins over geologic time.

Meteorites, such as the one shown in **Figure 3,** are composed of the same material that might have formed the early planets. Studies indicate that meteorites contain up to 0.5 percent water. Meteorite bombardment releases water into Earth's systems.

If early Earth contained the same percentage of water as meteorites, it would have been sufficient to form early oceans. However, some mechanism must have existed to allow the water to rise from Earth's interior to its surface. Scientists theorize that this mechanism was volcanism.

CAREERS IN EARTH SCIENCE

Oceanographer Scientists who investigate oceans are called oceanographers. Oceanographers might investigate water chemistry, wave action, marine organisms, or sediments.

WebQuest

■ **Figure 3** Comets are composed of dust and rock particles mixed with frozen water and gases. Comet impacts with Earth may have released enough water to help fill ocean basins. Meteorites contain up to 0.5 percent water.

Comet

Meteorite

ACROSS THE CURRICULUM

Chemistry Ultraviolet radiation dissociates water molecules into hydrogen atoms and oxygen atoms. Oxygen atoms are about 16 times more massive than hydrogen atoms. At any given temperature, the atoms and molecules in the atmosphere have the same average kinetic energy, but because of the differences in their masses, their velocities differ. Hydrogen atoms move, on average, four times faster than oxygen atoms move. Because of their high velocities, hydrogen atoms can escape from Earth's gravitational field and leak into space. Thus, water molecules gradually disappear as hydrogen atoms escape into space. Scientists hypothesize the planet Mars, which has a weak gravitational field, lost most of its water this way.

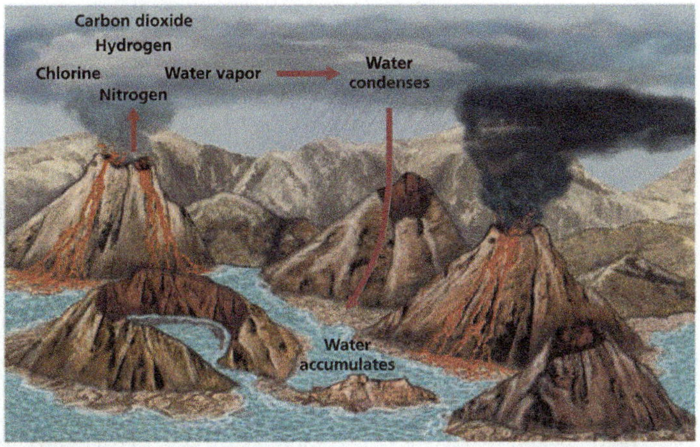

■ **Figure 4** In addition to comets, water for Earth's early oceans might have come from volcanic eruptions. An intense period of volcanism occurred shortly after the planet formed. This volcanism released large quantities of water vapor and other gases into the atmosphere. The water vapor eventually condensed into oceans.

Volcanism During volcanic eruptions, significant quantities of gases are emitted. These volcanic gases consist mostly of water vapor and carbon dioxide. Shortly after the formation of Earth, when the young planet was much hotter than it is today, an episode of massive, violent volcanism took place over the course of perhaps several hundred million years. As shown in **Figure 4,** this volcanism released huge amounts of water vapor, carbon dioxide, and other gases, which combined to form Earth's early atmosphere. As Earth's crust cooled, the water vapor gradually condensed, fell to Earth's surface as precipitation, and accumulated to form oceans. By the time the oldest known crustal rock formed about 4 bya, Earth's oceans might have been close to their present size. Water is still being added to the hydrosphere by volcanism, but some water molecules in the atmosphere are continually being destroyed by ultraviolet radiation from the Sun. These two processes balance each other.

Distribution of Earth's Water

As shown in **Figure 5,** the oceans contain 97 percent of the water found on Earth. Another 3 percent is freshwater located in the frozen ice caps of Greenland and Antarctica and in rivers, lakes, and underground sources. The percentage of ice on Earth has varied over geologic time from near zero to perhaps as much as 10 percent of the hydrosphere. As you read further in this section, you will learn more about how these changes affect sea level.

■ **Figure 5** About 97 percent of water on Earth is salt water found in the oceans.

Distribution of Water on Earth

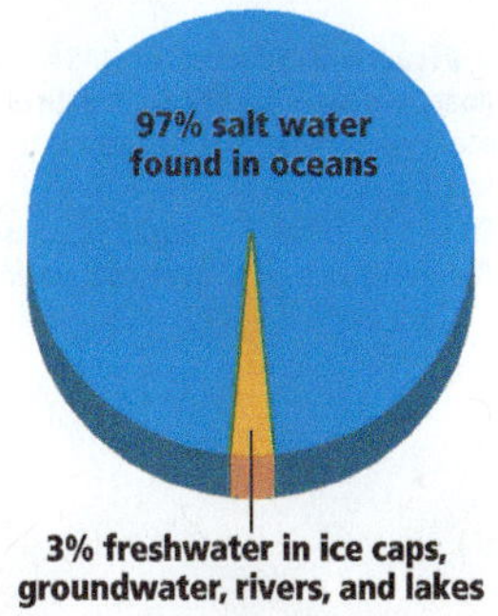

Discussion

Oceans on Earth Ask students to describe what Earth would be like if the oceans had not formed. Without the oceans, Earth would have no life as we know it. Oceans also heavily influence weather and climate.

Enrichment

Water in the Solar System Have students research whether water exists or has existed on any other planets or moons. Students will probably discover that scientists have found evidence of water on Mars and the Moon. In addition, Europa, a moon of Jupiter, is covered by a thick ice crust. Scientists hypothesize an ocean might exist beneath the crust.

Discussion

A Global Ocean Many people do not realize all the oceans are connected and thus form one global ocean. Ask students whether the Atlantic and Pacific Oceans are connected naturally. Students should not consider human-constructed canals. Also, ask students whether these two oceans have the same sea level. Then show the class a map of the world and point out the Drake Passage, which is located between South America and Antarctica, and the Bering Strait, which is located between Siberia and Alaska. Explain that all connected bodies of water have the same mean water level.

Teacher Content Support

Sea Level Other factors that can affect sea level include tides, which cause water levels to rise and fall in a regular pattern, high- and low-pressure systems in the atmosphere, water temperature changes, which lead to changes in water density and volume, and rainfall and river flow into oceans. Scientists use tide gauges and satellites to measure changes in global sea level.

Caption Question Fig. 7
global warming

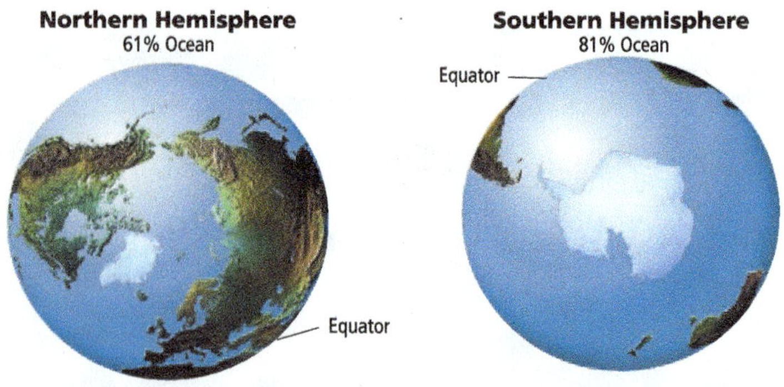

Figure 6 The northern hemisphere is covered by slightly more water than land. The southern hemisphere, however, is almost completely covered by water.

The blue planet Earth is known as the blue planet for good reason—approximately 71 percent of its surface is covered by oceans. The average depth of these oceans is 3800 m. Earth's landmasses are like huge islands, almost entirely surrounded by water. Because most landmasses are in the northern hemisphere, oceans cover only 61 percent of the surface there. However, 81 percent of the southern hemisphere is covered by water. **Figure 6** shows the distribution of water in the northern and southern hemispheres. Note that all the oceans are one vast, interconnected body of water. They have been divided into specific oceans and seas largely because of historic and geographic considerations.

Sea level Global **sea level,** which is the level of the oceans' surfaces, has risen and fallen by hundreds of meters in response to melting ice during interglacial periods and expanding glaciers during ice ages. Other processes that affect sea level are tectonic forces that lift or lower portions of Earth's crust. A rising seafloor causes a rise in sea level, while a sinking seafloor causes sea level to drop. **Figure 7** shows that sea level rose at a rate of about 3 mm per year between 1993 and 2010. Scientists hypothesize that this rise in sea level is related to water that has been released by the melting of glaciers and thermal expansion of the ocean due to warming.

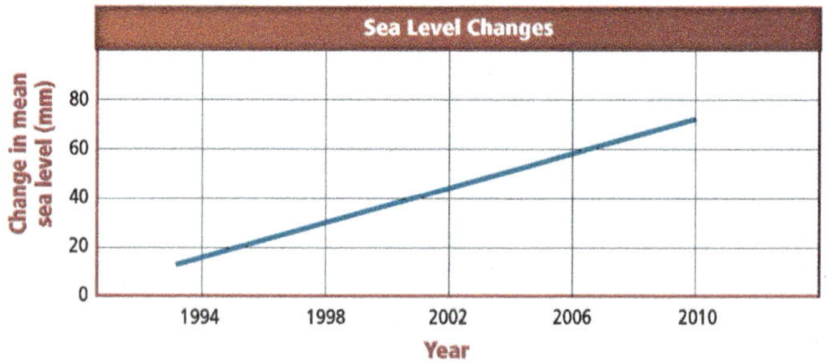

Figure 7 Scientists at NASA used floats and satellites to collect data on sea level changes over the period 1993 to 2010.
Explain *What is a possible cause for rising sea level over this period?*

DIFFERENTIATED INSTRUCTION

Struggling Learners Working in small groups, have students consult an atlas and identify the five major oceans. Atlantic, Pacific, Arctic, Indian, and Southern oceans Have students create their own maps and label each ocean. COOP LEARN

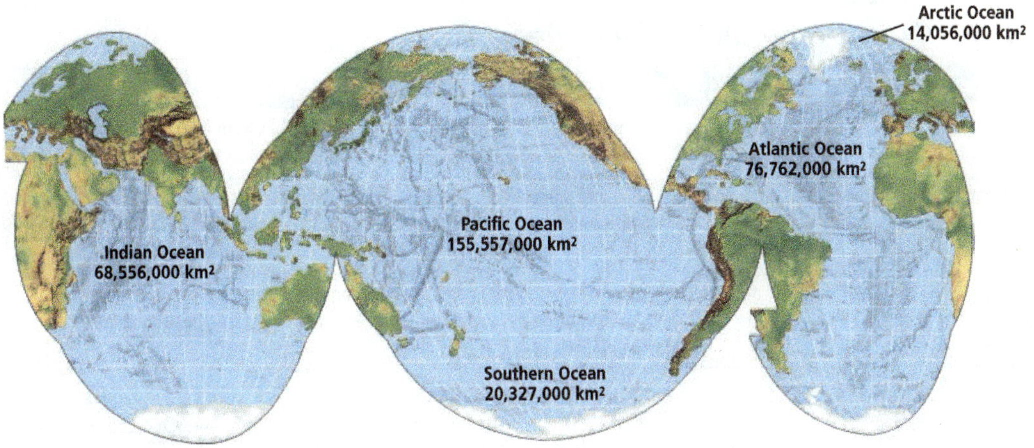

■ **Figure 8** The Pacific, Atlantic, and Indian oceans stretch from Antarctica to the north. The smaller Arctic Ocean and Southern Ocean are located near the North and South poles, respectively.

Major oceans As **Figure 8** shows, there are three major oceans: the Pacific, the Atlantic, and the Indian. The Pacific Ocean is the largest. Containing roughly half of Earth's seawater, it is larger than all of Earth's landmasses combined. The second-largest ocean, the Atlantic, extends for more than 20,000 km from Antarctica to the Arctic Circle. North of the Arctic Circle, the Atlantic Ocean is often referred to as the Arctic Ocean. The third-largest ocean, the Indian, is located mainly in the southern hemisphere. The storm-lashed region surrounding Antarctica, south of about 50° south latitude, is known as the Southern Ocean.

☑ READING CHECK **Identify** the largest ocean.

Polar oceans The Arctic and Southern oceans are covered by vast expanses of sea ice, particularly during the winter. In summer, the ice breaks up somewhat. Because ice is less dense than water, it floats. When sea-ice crystals first form, an ice-crystal slush develops at the surface of the water. The thickening ice eventually solidifies into individual round pieces called pancake ice, shown in **Figure 9.** Eventually, these pieces of pancake ice thicken and freeze into a continuous ice cover called pack ice. In the coldest parts of the Arctic and Southern oceans, there is no summer thaw, and the pack ice is generally several meters thick. In the winter, the pack-ice cover can be more than 1000 km wide.

■ **Figure 9** These pieces of pancake ice will eventually thicken and freeze into pack ice.

Tie to Previous Knowledge

Earth's Circumference Earth's circumference is about 40,000 km. Because Earth is a sphere, it can be divided into 360 degrees. The distance of a degree of latitude is equal to 40,000 divided by 360°, or 111 km. The Atlantic Ocean covers about 180 degrees of latitude, or about 20,000 km, from Siberia to Antarctica. Remind students lines of longitude converge at the poles. Thus, the distance of a degree of longitude varies with location. At the equator, a degree of longitude equals 111 km. At the poles, it equals zero.

Teacher Content Support

Southern Ocean The Southern Ocean surrounds Antarctica and consists of water from the southern portions of the Atlantic, Pacific, and Indian Oceans. There are several seas recognized around the coast of Antarctica, named after prominent Southern Ocean and Antarctic explorers, including the Ross Sea, named after Sir James Ross who found the last unknown landmass on Earth, Antarctica, in 1839. The deepest point of the Southern Ocean is 7235 m, found in the South Sandwich Trench.

☑ READING CHECK the Pacific Ocean

DIFFERENTIATED INSTRUCTION

English Learners Ask students with international backgrounds to translate the names of major oceans and seas into their native languages. Students can share these names with the class as they point out major bodies of water on a world map.

3 Assess

Check for Understanding

Describe Ask students to describe how sonar is used to calculate ocean depth. To determine ocean depth using sonar, multiply the speed of sound by the travel time of the echo, then divide by 2.

Reteach

Calculate Have students calculate the travel times of an echo received from depths of 4 km and 1 km below sea level. Remind students that the speed of sound in seawater is 1500 m/s. $2(4000 \text{ m} \div 1500 \text{ m/s}) = 5.3 \text{ s}$; $2(1000 \text{ m} \div 1500 \text{ m/s}) = 1.3 \text{ s}$

Assessment

Performance Have students research and report on selected aspects of the history of oceanography, such as the voyages of the early explorers, the development of sonar, or the exploration of polar seas. The reports should have illustrations and proper citations of sources.

Rubric

Table 1 Ocean-Atmospheric Interactions

Example	Description
Oceans are a source of atmospheric oxygen.	Fifty percent of oxygen in the atmosphere comes from marine phytoplankton, which release oxygen into surface waters as a product of photosynthesis.
Oceans are a reservoir for carbon dioxide.	Greater amounts of atmospheric carbon dioxide dissolve in cold water. When cold, dense surface water in polar oceans sinks, dissolved carbon dioxide moves to the bottom of the ocean.
Oceans are a source of heat and moisture.	Warm ocean water in equatorial regions heats the air above it. This warmer air holds more water vapor formed at the ocean's surface. These conditions can fuel hurricanes.

Ocean and atmospheric interaction Oceans provide moisture and heat to the atmosphere and influence large-scale circulation patterns. You have learned that warm ocean water energizes tropical cyclones, influences the position and strength of jet streams, and plays a role in El Niño events.

Oceans are also a vast reservoir of carbon dioxide. Dissolved carbon dioxide in surface waters sinks in water masses to the deep ocean, returning to the surface hundreds of years later. Without this natural uptake by the ocean, the accumulation of carbon dioxide in the atmosphere would be much larger than currently observed. There is also an uptake of carbon dioxide by phytoplankton during photosynthesis in the sunlit areas of the ocean. In the process, carbon is stored in the ocean and excess oxygen is released to the atmosphere to make Earth habitable. **Table 1** summarizes some of the interactions between oceans and the atmosphere.

SECTION 1 REVIEW

Section Self-Check

Section Summary

- Scientists use many different instruments to collect and analyze data from oceans.
- Scientists have several ideas as to where the water in Earth's oceans originated.
- A large portion of Earth's surface is covered by ocean.
- Earth's oceans are the Pacific, the Atlantic, the Indian, the Arctic, and the Southern.

Understand Main Ideas

1. **MAINIDEA State** how much of Earth is covered by oceans. How is ocean water distributed over Earth's surface?
2. **Describe** two tools scientists use to collect data about oceans.
3. **Relate** What evidence indicates that oceans formed early in Earth's geologic history?
4. **Specify** Where did the water in Earth's early oceans originate?

Think Critically

5. **Predict** some possible consequences of rising sea level.
6. **Suggest** A recent study showed a 30 percent decrease in phytoplankton concentrations in northern oceans over the last 25 years. How might a significant decrease in marine phytoplankton affect atmospheric levels of oxygen and carbon dioxide?

MATH IN Earth Science

7. Calculate the distance to the ocean floor if a sonar signal takes 6 s to return to a ship's receiver.

SECTION 1 REVIEW

1. 71% of Earth's surface is covered with ocean water. 61% of the northern hemisphere is covered with water. 81% of the southern hemisphere is covered with water.
2. Any two of the following: Sonar uses sound waves to determine the depth of the ocean floor. Floats with sensors measure water temperature, salinity, and the concentration of gases and nutrients in surface water. Floats also measure current speed and wave motion. Satellites monitor surface temperatures, currents, and waves. Submersibles are used to collect data in the deep ocean. Computers are used to analyze data and make models and simulations.
3. radioactive studies of ancient rocks and lava flows
4. Water in the oceans came from impacting comets or from Earth's interior through volcanism.
5. Possible answers: coastal cites flooded, habitats destroyed
6. Oxygen levels might decrease while carbon dioxide levels increase.
7. $1500 \text{ m/s} \times 6 \text{ s} \div 2 = 4500 \text{ m}$

Essential Questions

- What is the composition of seawater?
- What are the physical properties of seawater?
- How is ocean layering illustrated?
- How do deepwater masses form?

Review Vocabulary

feldspar: a rock-forming mineral that contains silicon and oxygen

New Vocabulary

salinity
estuary
temperature profile
thermocline

Seawater

MAINIDEA Oceans have distinct layers of water masses that are characterized by temperature and salinity.

EARTH SCIENCE 4 YOU A person's accent can reveal a lot about his or her place of origin. Similarly, the temperature and salinity of water masses can often reveal when and where the water was first formed on the sea surface.

Chemical Composition of Seawater

Ocean water contains dissolved gases, including oxygen and carbon dioxide, and dissolved nutrients such as nitrates and phosphates. Chemical profiles of seawater vary based on both location and depth, as shown in **Figure 10.** Factors that influence the amount of a substance in an area of ocean water include wave action, vertical movements of water, and biological activity.

Figure 10 shows that oxygen levels are high at the surface in both the Atlantic and Pacific oceans. This occurs in part because oxygen is released by surface-dwelling photosynthetic organisms. Silica levels for both oceans are also shown in **Figure 10.** Because many organisms remove silica from ocean water and use it to make shells, silica levels near the surface are usually low. Silica levels usually increase with depth because decaying organisms sink to the ocean bottom, returning silica to the water.

Salinity The measure of the amount of dissolved salts in seawater is **salinity.** Oceanographers express salinity as grams of salt per kilogram of water, or parts per thousand (ppt). The total salt content of seawater averages 35 ppt, or 3.5 percent. The most abundant salt in seawater is sodium chloride. Other salts in seawater are chlorides and sulfates of magnesium, potassium, and calcium.

Figure 10 Concentrations of dissolved gases and nutrients in seawater, measured in micromolars (μM), vary by location and depth.

Examine *How do oxygen levels differ between the North Atlantic and North Pacific oceans?*

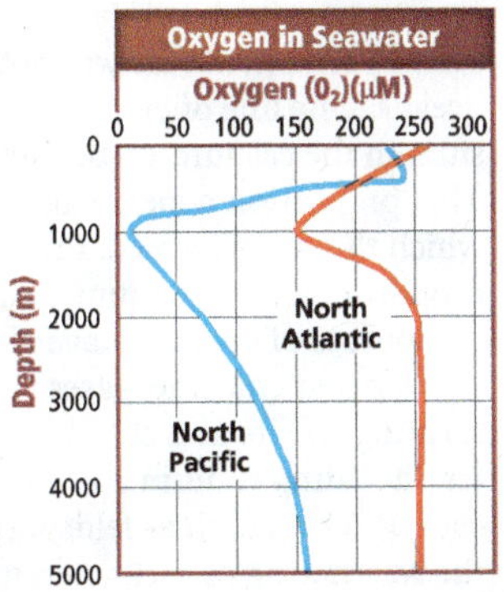

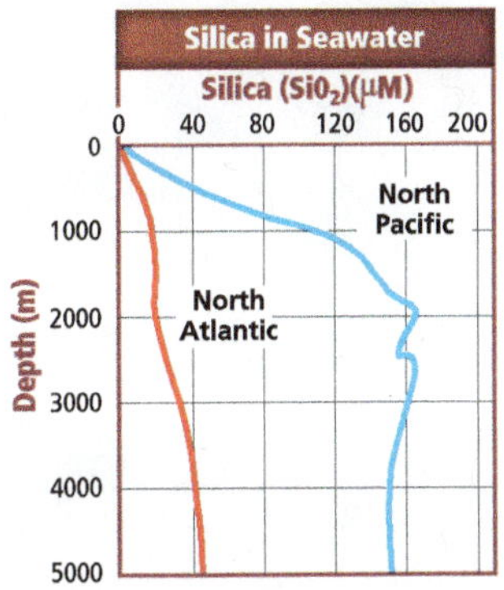

Caption Question Fig. 10 In deeper water (below 500 m), there is more oxygen in the North Atlantic than in the North Pacific.

1 Focus

Focus Transparency

Before presenting the lesson, project **Section Focus Transparency 45** and have students answer the accompanying questions. **EL**

MAINIDEA

Density To help students review the concept of density, ask the following questions: Why does a stone sink and a piece of wood float in water? The stone is more dense than the wood and the water under it. What happens if a denser fluid lies over less dense fluid (for example, water over oil in a salad dressing bottle)? The denser fluid will sink beneath the less dense fluid.

2 Teach

Tie to Previous Knowledge

Water Cycle Previously, students learned about Earth's water cycle, wherein water is continually recycled from the atmosphere to Earth's surface through the processes of evaporation, precipitation, and condensation. Review these processes, and emphasize that the ocean is an integral part of the water cycle.

Teacher Content Support

Ions The presence of ions in seawater is normally described in terms of mass. However, it can also be described in terms of number of ions. The six most common ions in seawater in terms of number are Cl^- (48.6 percent), Na^+ (41.6 percent), Mg^{2+} (4.8 percent), SO_4^{2-} (2.5 percent), Ca^{2+} (0.9 percent), and K^+ (0.9 percent).

Collaborative Learning

Model Ions in Seawater

This exercise will demonstrate how the ions in seawater are related to the ionic compounds listed in the MiniLab. Tell students that they will model salt ions in a microscopic drop of seawater. The seawater contains the following 21 ions: ten chloride ions, eight sodium ions, two magnesium ions, and one sulfate ion. Write the names of the ions individually on name tags, and give one name tag each to 21 students. Students should put on the name tags, then walk around the room to "mix the seawater." Tell students the water is evaporating, and they must link up by standing next to oppositely charged ions to form electrically neutral compounds. On the chalkboard, record the compounds formed by the links. Possible outcomes include eight formula units of NaCl, one of $MgCl_2$, and one of $MgSO_4$, or six of NaCl, two of $MgCl_2$, and one of Na_2SO_4. **COOP LEARN**

☑ **READING CHECK** precipitation, evaporation, river input, melting sea ice

Ocean Salinity

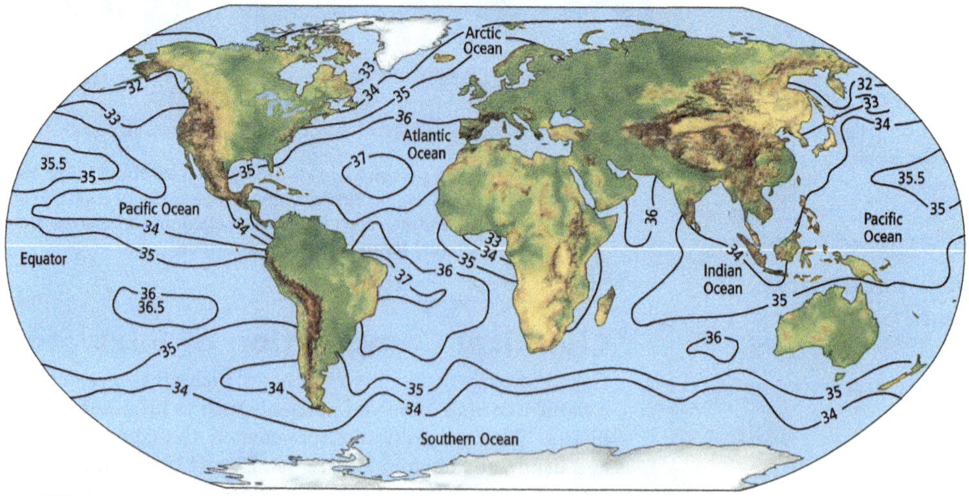

*All values are given in parts per thousand (ppt)

■ **Figure 11** Ocean salinity varies from place to place. High salinity is common in areas with high rates of evaporation. Low salinity often occurs in estuaries where seawater mixes with fresh water.

Variations in salinity Although the average salinity of the oceans is 35 ppt, actual salinity varies from place to place, as shown in **Figure 11.** In subtropical regions where rates of evaporation exceed those of precipitation, salt left behind by the evaporation of water molecules accumulates in the surface layers of the ocean. There, salinity can be as high as 37 ppt. In equatorial regions where precipitation is abundant, salinity is lower. Even lower salinities of 32 or 33 ppt occur in polar regions where seawater is diluted by melting sea ice. The lowest salinity often occurs where large rivers empty into the oceans, creating areas of water called **estuaries.** Even though salinity varies, the relative proportion of major types of sea salts is constant because all ocean water continually intermingles throughout Earth's oceans.

☑ READING CHECK **Describe** the factors that affect the salinity of water.

Sources of sea salt Geologic evidence indicates that the salinity of ancient seas was not much different from that of today's oceans. One line of evidence is based on the proportion of magnesium in the calcium-carbonate shells of some marine organisms. That proportion depends on the overall salinity of the water in which the shells formed. Present-day shells contain about the same proportion of magnesium as similar shells throughout geologic time.

Sources of sea salts have also stayed the same over time. Sulfur dioxide and chlorine, gases released by volcanoes, dissolve in water, forming sulfate and chlorine ions. Most of the other ions in seawater, including sodium and calcium, come from the weathering of crustal rocks, such as feldspars. Iron and magnesium come from the weathering of rocks rich in these elements. These ions enter rivers and are transported to oceans, as shown in **Figure 12.**

DIFFERENTIATED INSTRUCTION

Advanced Learners Have interested students calculate the percentages of the ions in the Collaborative Learning exercise both by number of ions and by mass. Give students the atomic masses of chlorine (35.45), sodium (23.0), sulfur (32.1), oxygen (16.0), and magnesium (24.3). The molecular mass of the sulfate ion is 96.1. by number of ions: $Cl^- = 10/21 = 47.6$ percent, $Na^+ = 8/21 = 38.1$ percent, $Mg^{2+} = 2/21 = 9.5$ percent, $SO_4^{2-} = 1/21 = 4.8$ percent; by mass: $Cl^- = 10 \times 35.45 = 354.5$, $Na^+ = 8 \times 23.0 = 184$, $Mg^{2+} = 2 \times 24.3 = 48.6$, and $SO_4^{2-} = 96.1$; total mass $= 683.2$ atomic mass units; percentage mass: $Cl^- = 354.5/683.2 = 51.9$ percent, $Na^+ = 184/683.2 = 26.9$ percent, $Mg^{2+} = 48.6/683.2 = 7.1$ percent, and $SO_4^{2-} = 96.1/683.2 = 14.1$ percent

VISUALIZING the Salt Cycle

Figure 12 Salts are added to seawater by volcanic eruptions and by the weathering and erosion of rocks. Salts are removed from seawater by biological processes and the formation of evaporites. Also, wind carries salty droplets inland.

Ions, such as sodium, calcium, iron, and magnesium, enter oceans in river runoff as the weathering of rocks releases them.

Gases from volcanic eruptions contain water vapor, chloride, and sulfur dioxide. These gases dissolve in water and form the chloride and sulfate ions in seawater.

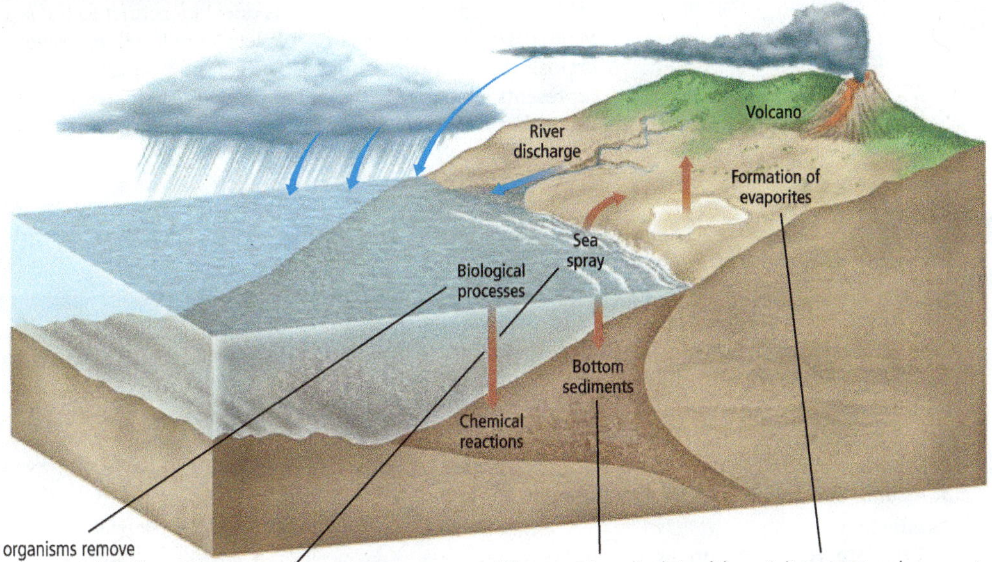

Marine organisms remove ions from seawater through chemical reactions to build their shells, bones, and teeth.

Small droplets of sea spray that contain salt are carried inland by winds.

When organisms die, their solid parts sink, returning salts to the bottom sediments.

Salt precipitates when seawater evaporates in coastal areas that are hot and dry.

Concepts In Motion View an **animation of the salt cycle.**

(l)Conrad Zobel/CORBIS, (r)Dr. Morley Read/Photo Researchers

IN THE FIELD

Freshwater in the Caribbean Aruba, an island in the southern Caribbean off the coast of Venezuela, installed its first seawater desalination system in 1932. Efficient production and management of water provides for the needs of Aruba's residents and the tourists that quadruple the island's population each year. Aruba's current thermal desalination plants use low-pressure steam to heat seawater. The process kills bacteria and distills pure water from the seawater. A more efficient reverse osmosis system is currently under construction.

VISUALIZING

Purpose

Students will learn about how salt enters and is removed from the ocean.

Interpret the Illustration

Salt in the Ocean Have students study **Figure 12,** then describe in their own words the processes that add salts to seawater and the processes that remove salts from seawater. Challenge students to infer how marine life might be affected if this balance changed.

Teacher Content Support

Desalination On many small islands and in some coastal areas, freshwater is in short supply. In these areas, freshwater can be obtained by the desalination of seawater. One common desalination technique is the distillation of seawater; seawater is heated and evaporates, and the salts are left behind. The resulting water vapor is cooled and condenses as freshwater. Another technique involves the freezing of seawater. Sea ice is freshwater ice containing pockets of brine. The brine can be flushed from an ice-brine mix. The most economic desalination method is reverse osmosis; seawater is pressed through a semipermeable membrane that allows water molecules, but not salt ions, to pass through. This method is used in most modern desalination plants to produce freshwater.

MiniLAB

iLab Station | Rubric

Purpose Students will enhance their understanding of salinity and units of measurement.

Process Skills measure in SI, use numbers, apply concepts, compare and contrast, classify, infer

Safety Precautions Review the MSDS for all chemicals with students prior to doing the lab work. Approve lab safety forms before work begins. Dilute the solutions with water before disposing of them.

Teaching Strategy Be sure students stir the salt until it dissolves.

Expected Results Students will model seawater by combining the proper amounts of different salts.

Analysis

1. 34.43 g of salt + 965.57 g of water = 1000 g of salt water; 34.43 g ÷ 1000 g = 0.03443, or 3.443 percent
2. 3.443 percent = 34.43 ppt
3. The solution does not contain the trace elements and nutrients dissolved in seawater.

Assessment

Knowledge Have students predict what would happen if they left the beaker containing the salt water uncovered indefinitely. The water would evaporate, but the salt would remain in the bottom and sides of the beaker. How would the salinity of the water change as water evaporated? The salinity would increase as more water evaporated.

LabManager™
Customize any lab with the LabManager™ CD-ROM.

View an **animation of sea salts.** Concepts In Motion

Table 2 Removal of Sea Salts

Process	Description	Example
Evaporate formation	Solid salt is left behind when water evaporates from concentrated solutions of salt water.	
Biological activity	Organisms remove calcium ions from water to build shell, bones, and teeth.	

Removal of sea salts Although salt ions are continuously added to seawater, salinity does not increase because salts are also continuously removed. **Table 2** describes two processes through which sea salts are removed from seawater. Recall that evaporites form when water evaporates from concentrated solutions. In arid coastal regions, water evaporates from seawater and leaves solid salt behind. Marine organisms remove ions from seawater to build shells, bones, and teeth. As organisms die, their solid parts accumulate on the seafloor and become part of bottom sediments. The salt contained in these solid parts does not return to the seawater. Winds can also pick up salty droplets from breaking waves and deposit the salt further inland. The existing salinity of seawater represents a balance between the processes that remove salts and those that add them.

MiniLAB

iLab Station

Model Seawater

What is the chemical composition of seawater? Determine the chemical composition of seawater using the ingredients listed in the table. The salinity of seawater is commonly measured in parts per thousand (ppt).

Ingredient	Amount
Sodium chloride ($NaCl$)	23.48 g
Magnesium chloride ($MgCl_2$)	4.98 g
Sodium sulfate (Na_2SO_4)	3.92 g
Calcium chloride ($CaCl_2$)	1.10 g
Potassium chloride (KCl)	0.66 g
Sodium bicarbonate ($NaHCO_3$)	0.19 g
Potassium bromide (KBr)	0.10 g

Procedure

1. Read and complete the lab safety form.
2. Use the table on the right to answer the questions below.
3. Assume that 965.57 g of water is added to this solution.

Analysis

1. **Calculate** How many grams of solution would these ingredients produce? What percentage of this solution is made up of salts?
2. **Apply** What is the salinity of this solution in ppt?
3. **Infer** how this solution differs from actual seawater.

ACROSS THE CURRICULUM

Math Ask students to determine the salinity of a solution of 34.43 g of salt in 1 L of distilled water. The density of the distilled water is 0.998 g/cm^3. 0.998 g/cm^3 × 1000 cm^3 = 998 g of water, 998 g of water + 34.43 g of salt = 1032.43 g of salt water, 34.43 ÷ 1032.43 = 0.03335 = 3.335 % = 33.35 ppt

Physical Properties of Seawater

The presence of various salts causes the physical properties of seawater to be different from those of freshwater.

Density Freshwater has a maximum density of 1.00 g/cm^3. Because salt ions add to the overall mass of the water in which they are dissolved, they increase the density of water. Seawater is therefore more dense than freshwater, and its density increases with salinity. Temperature also affects density—cold water is more dense than warm water. Because of salinity and temperature variations, the density of seawater ranges from about 1.02 g/cm^3 to 1.03 g/cm^3. These variations might seem small, but they are significant. They affect many oceanic processes.

☑ READING CHECK **Explain** how temperature and salinity affect the density of seawater.

Freezing point Variations in salinity also cause the freezing point of seawater to be somewhat lower than that of freshwater. Freshwater freezes at 0°C. Because salt ions interfere with the formation of the crystal structure of ice, the freezing point of seawater is –2°C.

Absorption of light If you have ever swum in a lake, you might have noticed that the intensity of light decreases with depth. The water might be clear, but if the lake is deep, the bottom waters will be dark. Water absorbs light, which gives rise to another physical property of oceans—darkness. In general, light penetrates only the upper 100 m of seawater. Below that depth, all is darkness.

Figure 13 illustrates how light penetrates ocean water. Notice that red light does not penetrate as far as blue light. Red objects, such as the giant red shrimp shown in **Figure 13,** appear black below a certain depth and other reflecting objects in the water appear green or blue. Although some fading blue light can reach depths of a few hundred meters, light sufficient for photosynthesis exists only in the top 100 m of the ocean. In the darkness of the deep ocean, some organisms, including some fishes, shrimps, and crabs, are blind. Other organisms attract prey by producing light, called bioluminescence, through a chemical reaction.

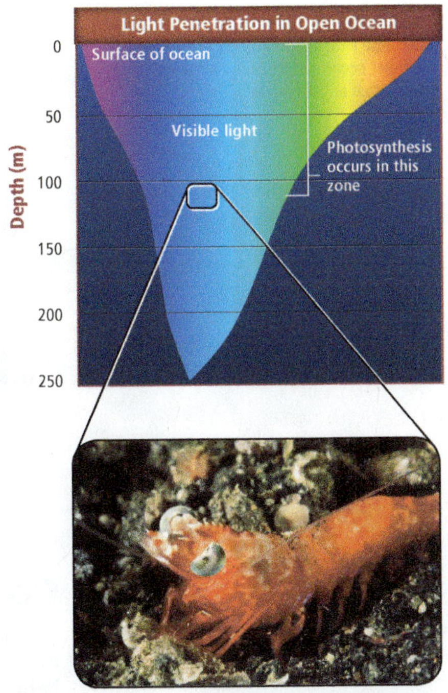

■ **Figure 13** Red light does not penetrate as far as blue light in the ocean. Marine organisms that are some shades of red, such as deep-sea shrimp, appear black below a depth of 10 m. This helps them escape predators.
Identify *To what depth does blue light penetrate ocean water?*

Identify Misconceptions

Some students might think water is transparent, meaning light passes through it without being scattered or absorbed.

Uncover the Misconception Explain to students a person can usually see through the water in a glass, sink, bucket, or bathtub. This water is shallow and does not usually contain any particulate matter that would scatter light. Ask students what types of material in seawater might scatter light as it passes through. Possible answers: fine sediment suspended in water such as clay or silt, plankton

Demonstrate the Concept Explain to students all water absorbs light. In fact, a layer of ocean water 100 m thick absorbs more than 99 percent of incident light. This means the oceans are essentially dark below a depth of 100 m. Even in clear, tropical waters, the seafloor cannot be seen from the surface in waters deeper than about 50 m.

Assess New Knowledge Have students describe, in their own words, what happens to light as it penetrates seawater. Students can use **Figure 13** to form their answers.

Interpret the Illustration

The Photic Zone Have students study **Figure 13.** Tell students plant life in oceans is restricted to the photic zone, the depth to which photosynthesis can occur. In clear waters, the photic zone can reach a depth of 250 m. On average, however, photosynthesis is limited to the top 100 m of oceans.

☑ READING CHECK As temperature increases, density decreases. As salinity increases, density increases.

■ **Caption Question Fig. 13** 250 m

Demonstration

Light Penetration in Water Different colors penetrate to different depths in water. Light penetration at different wavelengths can be demonstrated using Secchi disks of different colors. To demonstrate light penetration, fill an aquarium with muddy water so that a white disk at the bottom of the aquarium is barely visible. Then submerge red, orange, yellow, green, and blue disks. Have student volunteers use metric rulers to measure the depths at which the different disks can no longer be seen.

Teacher Content Support

Temperature Stratification

Most midlatitude lakes are stratified like the ocean with warm surface water on top, cold water at the bottom, and a thermocline in between. This stratification is caused by density differences; dense, cold water sinks to the bottom, and less-dense, war m water floats on top. Because the densest freshwater has a temperature of 4°C, the temperature of deep lake waters is 4°C year-round.

In summer, surface waters are warm, and a strong thermocline prevents them from mixing with deep waters. In fall, surface waters cool and eventually reach 4°C. At that point, the thermocline disappears, and top and bottom waters become well-mixed. In winter, the surface waters become colder than the deep water, and ice can form. In spring, the surface waters warm again to 4°C, and the mixing of surface and deep waters resumes.

READING CHECK The surface layer is about 100 m thick and receives sunlight. The thermocline is a transitional layer beneath the surface layer where temperatures change rapidly with depth. The bottom layer is cold and dark.

Figure 14 Ocean water temperatures decrease with depth. Tropical areas have warmer ocean surface temperatures than do temperate or polar areas.

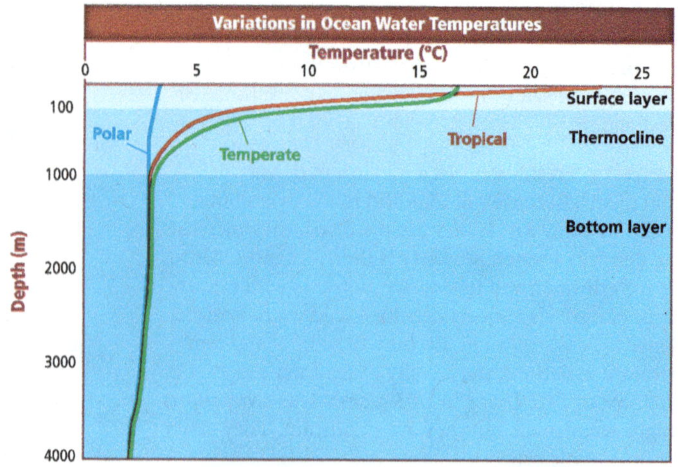

Ocean Layering

Ocean surface temperatures range from −2°C in polar waters to 38°C in equatorial regions, with the average surface temperature being 15°C. Ocean water temperatures, however, decrease significantly with depth. Thus, deep ocean water is always cold, even in tropical oceans.

VOCABULARY

ACADEMIC VOCABULARY

Variation
the range in which a factor changes
The variation in temperature in New York was a shock for the person from California.

Temperature profiles **Figure 14** shows typical ocean **temperature profiles,** which plot changing water temperatures against depth. Such profiles vary, depending on location and season. In the temperature profiles shown here, beneath roughly 100 m, temperatures decrease continuously with depth to around 4°C at 1000 m. The dark waters below 1000 m have fairly uniform temperatures of less than 4°C. Based on these temperature variations, the ocean can be divided into three layers, also shown in **Figure 14.** The first is a relatively warm, sunlit surface layer approximately 100 m thick. Notice that tropical areas have warmer surface temperatures than temperate or polar areas. Under the surface layer is a transitional layer known as the **thermocline,** which is characterized by rapidly decreasing temperatures with depth. The bottom layer is cold and dark with temperatures near freezing. Both the thermocline and the warm surface layer are absent in polar seas, where water temperatures are cold from top to bottom. In general, ocean layering is caused by density differences. Because cold water is more dense than warm water, cold water sinks to the bottom, while less-dense, warm water is found near the ocean's surface.

READING CHECK **Describe** the three main layers of water in oceans.

EARTH SCIENCE JOURNAL

Temperature Profiles Have students explain in their own words the differences in the temperature profiles among tropical, temperate, and polar ocean areas. Students can use **Figure 14** as a guide.

Rubric

Water Masses

The temperature of the bottom layer of ocean water is near freezing. This is true even in tropical oceans, where surface temperatures are warm. Where does all this cold water come from?

Deepwater masses Cold water comes from Earth's polar seas. Recall that high salinity and cold temperatures cause seawater to become more dense. Study **Figure 15,** which shows how deepwater masses are formed. When seawater freezes during the arctic or antarctic winter, sea ice forms. Because salt ions are not incorporated into the growing ice crystals, they accumulate beneath the ice. Consequently, the cold water beneath the ice becomes saltier and more dense than the surrounding seawater, and this saltier water sinks. This salty water then migrates toward the equator as a cold, deepwater mass along the ocean floor. Other cold, deepwater masses form when surface currents in the ocean bring relatively salty midlatitude or subtropical waters into polar regions. In winter, these waters become colder and denser than the surrounding polar surface waters, and thus, they sink.

Three water masses account for most of the deepwater masses in the oceans—Antarctic Bottom Water, North Atlantic Deep Water, and Antarctic Intermediate Water. Antarctic Bottom Water forms when antarctic seas freeze during the winter. With temperatures below 0°C, this deepwater mass is the coldest and densest in all the oceans, as shown in **Figure 16** on the following page. North Atlantic Deep Water forms in a similar manner offshore from Greenland. Antarctic Bottom Water is colder and denser than North Atlantic Deep Water, so it sinks below it.

READING CHECK **Identify** the three water masses that make up most of the deepwater masses in the oceans

■ **Figure 15** Dense polar water sinks, producing a deepwater mass.
Explain ***the relationship between the density of water and the formation of deepwater masses.***

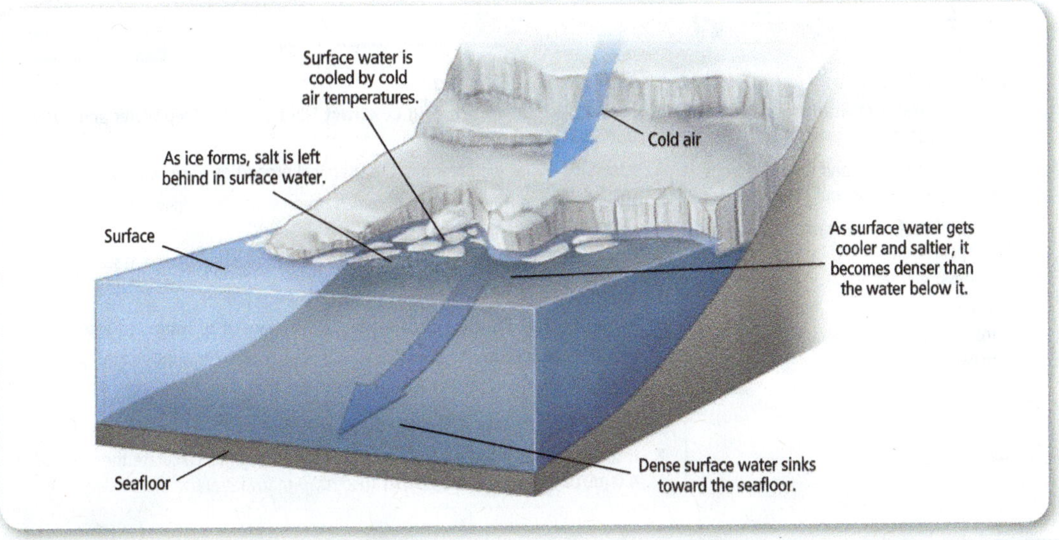

Teacher Content Support

Water Density Density is defined as mass per volume, and it is usually expressed in grams per cubic centimeter (g/cm^3) or grams per milliliter (g/mL). Note that 1 g/cm^3 is equal to 1000 kg/m^3. The density of water depends on its temperature and salinity. For example, the density of seawater ranges from about 1.02 g/cm^3 to 1.03 g/cm^3, depending on salinity and temperature. Dissolved salts interfere with the formation of hydrogen bonds and decrease the freezing point of seawater to about –2°C. The maximum density of seawater corresponds to its freezing point of –2°C. Accordingly, the densest seawater has a temperature of –2°C and high salinity. Water density also increases with pressure, but this relationship does not affect the movement of water masses and thus is usually ignored by oceanographers.

READING CHECK Antarctic Bottom Water, North Atlantic Deep Water, Antarctic Intermediate Water

■ **Caption Question Fig. 15** As the density of water increases, it sinks below layers of water that are less dense. The densest water is on the bottom of the ocean.

GeoLAB iLab Station

The GeoLab located at the end of the chapter can be used at this point in the lesson.

Interpret the Illustration

Deepwater Masses Have students study **Figure 16.** Ask them the following questions: Which deepwater mass has temperatures below 0°C? Antarctic Bottom Water Which deepwater mass is the least dense? Antarctic Intermediate Water Which deepwater mass forms off the shore of Greenland? North Atlantic Deep Water

3 Assess

Check for Understanding

Discuss Ask students to explain why deep water in all oceans—even tropical ones—is always cold. Cold, dense water from polar regions moves along the seafloor and displaces warmer, less-dense water.

Reteach

Experiment Have students each design an investigation that models the movement of Antarctic Bottom Water to the equator and beyond.

Assessment

Performance Place students in groups of four. Provide several water samples with various salinities to each group. Have the groups use test kits to analyze the chemistry of the samples. Students should display their results in data tables.

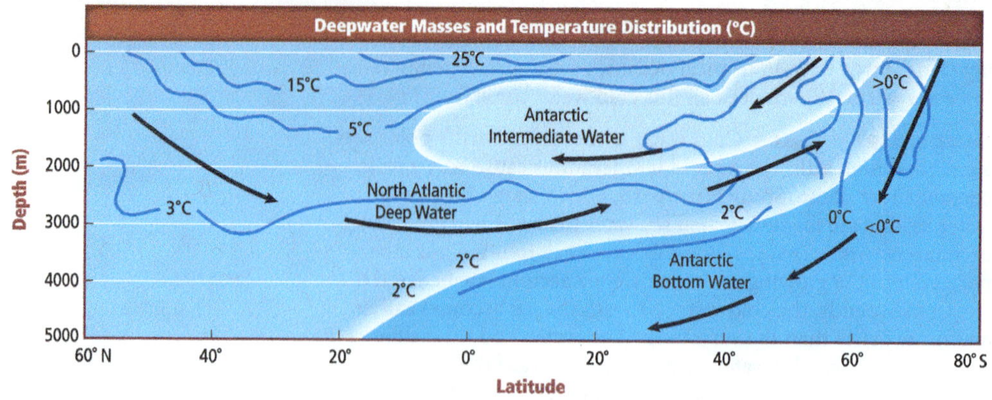

Figure 16 Antarctic Bottom Water is the densest and coldest deepwater mass. It is overridden by the slightly warmer and less dense North Atlantic Deep Water. Antarctic Intermediate Water is still warmer and less dense, and thus it overrides the other two deepwater masses.

Intermediate water masses Antarctic Intermediate Water, shown in **Figure 16,** forms when the relatively salty waters near Antarctica decrease in temperature and sink during winter. Because Antarctic Intermediate Water is slightly warmer and less dense than North Atlantic Deep Water, it does not sink as deep as the other two deepwater masses. While the Atlantic Ocean contains all three major deepwater masses—the Antarctic Bottom Water, North Atlantic Deepwater, and Antarctic Intermediate Water—the Indian and Pacific Oceans contain only the two Antarctic deepwater masses. In Section 3, you will learn about other water movements in the ocean.

SECTION 2 REVIEW

Section Self-Check

Section Summary

- Ocean water contains dissolved gases, nutrients, and salts.
- Salts are added to and removed from oceans through natural processes.
- Properties of ocean water, including temperature and salinity, vary with location and depth.
- Many of the oceans' deepwater masses sink from the surface of polar oceans.

Understand Main Ideas

1. **MAINIDEA Compare and contrast** North Atlantic Deep Water and Antarctic Bottom Water.
2. **Identify** What factors affect the chemical properties of seawater?
3. **Illustrate** the three layers into which ocean water is divided based on temperature.
4. **Sequence** the steps involved in the formation of deepwater masses.

Think Critically

5. **Hypothesize** Which is more dense, cold freshwater or warm seawater?
6. **Predict** what color a yellow fish would appear to be in ocean water depths greater than about 50 m.

MATH IN Earth Science

7. If the density of a sample of seawater is 1.02716 g/mL, calculate the mass of 4.0 mL of the sample.

SECTION 2 REVIEW

1. Both are water masses formed from sinking surface water. Antarctic Bottom Water is colder and denser than North Atlantic Deep Water.
2. wave action, vertical movements of water, biological activity
3. Students should make a diagram that shows the three layers based on temperature: surface water between 0–100 m; the thermocline between 100–1000 m; bottom water between 1000–4000 m.
4. When seawater in the Arctic or Antarctic freezes in the winter, salt is left behind. The cold water becomes saltier and denser than surrounding water, and sinks.
5. Seawater is always denser than freshwater because of its high salinity.
6. black
7. $d=m/V$; 1.02716 g/mL = m/4 mL; 1.02716 g/mL × 4 mL = 4.10864 g

SECTION 3

Ocean Movements

MAINIDEA Waves and currents drive the movements of ocean water and lead to the distribution of heat, salt, and nutrients from one region of the ocean to another.

Essential Questions

- What are the physical properties of waves?
- How do tides form?
- What are the similarities and differences between various ocean currents?

Review Vocabulary

prevailing westerlies: global wind system located between 30°N and 60°N that moves from the west to the east toward each pole

New Vocabulary

wave
crest
trough
breaker
tide
spring tide
neap tide
surface current
upwelling
density current

EARTH SCIENCE 4 YOU **Think about the last time you watched a sporting event and the audience did "the wave" to cheer players by standing up and sitting down at the right time. Even though the audience does not move around the stadium, the wave does. The same idea applies to ocean waves.**

Waves

Oceans are in constant motion. Their most obvious movement is that of waves. A **wave** is a rhythmic movement that carries energy through space or matter—in this case, ocean water. Ocean waves are generated mainly by wind blowing over the water's surface. In the open ocean, a typical wave has the characteristics shown in **Figure 17.** The highest point of a wave is the **crest,** and the lowest point is the **trough.** The vertical distance between crest and trough is the wave height, and the horizontal crest-to-crest distance is the wavelength. As energy is added, both the wavelength and speed increase. Thus, longer waves travel faster than shorter waves.

As an ocean wave passes, the water moves up and down in a circular pattern and returns to its original position, as shown in **Figure 17.** Only the energy moves steadily forward. The water itself moves in circles until the energy passes, but it does not move forward. The wavelength also determines the depth to which the wave disturbs the water. That depth, called the wave base, is equal to half the wavelength.

View an **animation of waves.** Concepts In Motion

■ **Figure 17** Wave characteristics include wave height, wavelength, crest, and trough. In an ocean wave, water moves in circles that decrease in size with depth. At a depth equal to half the wavelength, water movement essentially stops.

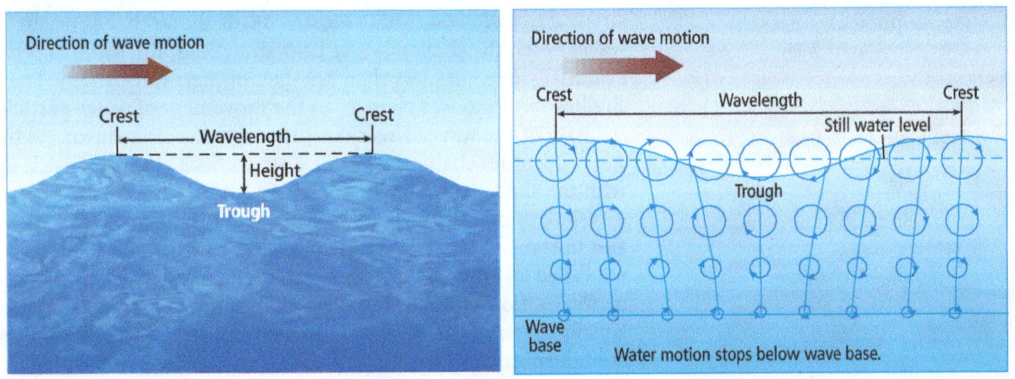

1 Focus

Focus Transparency

Before presenting the lesson, project **Section Focus Transparency 46** and have students answer the accompanying questions. EL

MAINIDEA

Waves and Currents To help students contrast waves and currents, have students simulate them in class. To simulate waves, have students sit on chairs in a circle. Each student should imitate the student to their right with a slight time lag. Start with the first person standing up with his or her arms up and then sit back down in a regular rhythm. Try it until a wave propagates around the circle. To simulate currents, have students leave their chairs and walk in circles and then in random directions.

2 Teach

Tie to Previous Knowledge

Wave Motion Ask students to recall seeing a wave performed by groups of people, such as at a sporting event. Point out the wave moves around the stadium, but the people in the wave stay in the same seats. In a similar way, the water in a wave moves up and down, but it does not move forward. Only the energy in the wave moves forward.

Teacher Content Support

Waves In deep water, where the seafloor is below the wave base, the speed of waves is proportional to the square root of the wavelength. Long, deep-water waves always travel faster than shorter ones. If the water depth is less than the wave base, friction with the bottom dissipates energy and lowers the wave speed.

Teacher Content Support

Wave Height Fetch is the unobstructed expanse of water over which the wind can blow and build waves. Small bodies of water, such as lakes, have limited fetches, and can produce only small storm waves. In the ocean, however, fetch is limited only by the size of the storm generating the waves. Large storms can produce huge ocean waves if the wind is strong for a long period of time.

As a general rule, maximum wave height is roughly equal to wind speed in knots. For example, wind speeds in hurricanes exceed 64 knots and are capable of raising waves in excess of 10 m. In order to do so, however, the wind must blow at 64 knots for several days in the same direction. Because hurricanes are relatively small, fast-moving systems, they rarely generate waves higher than 10 m. The largest ocean waves are produced by huge, slow-moving, long-lasting winter storms in the North Atlantic, North Pacific, and Southern oceans.

☑ READING CHECK wind speed, wind duration, fetch

■ **Caption Question** **Fig. 18** The crest collapses forward, forming a breaker.

FOLDABLES® Rubric

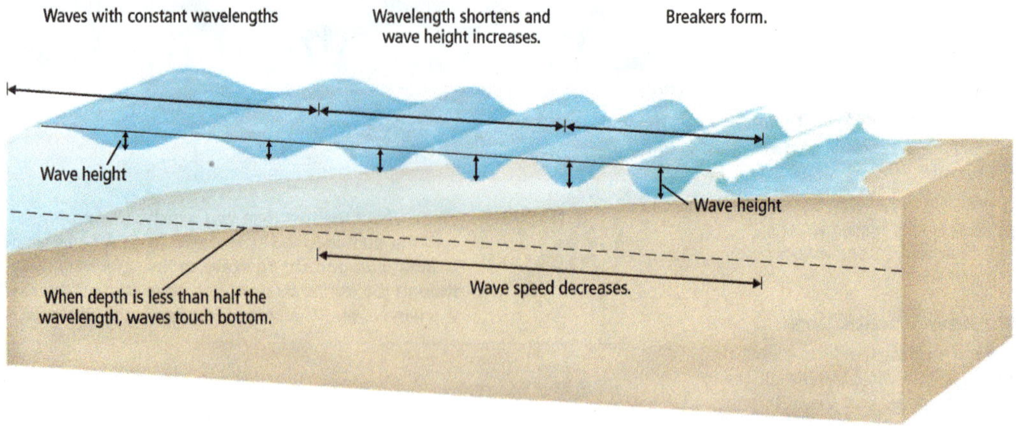

■ **Figure 18** A breaker forms when wavelength decreases and wave height increases as the wave nears the shore.
Explain *what happens when waves become too steep and unstable.*

Wave height Wave height depends on three factors: fetch, wind duration, and wind speed. Fetch refers to the expanse of water that the wind blows across. The longer the wind can blow without being interrupted (wind duration) over a large area of water (fetch), the larger the waves will be. Also, the faster the wind blows (wind speed) for a longer period of time over the ocean, the larger the waves will be. The highest waves are usually found in the Southern Ocean, an area over which strong winds blow almost continuously. Waves created by large storms can also be much higher than average. For instance, hurricanes can generate waves more than 10 m high, which is taller than a three story building.

FOLDABLES® Incorporate information from this section into your Foldable.

☑ READING CHECK **Identify** the three factors that affect the height of a wave.

■ **Figure 19** As waves move into shallow water, breakers form.

Breaking waves Study **Figure 18.** It shows that as ocean waves reach the shallow water near shorelines, the water depth eventually becomes less than one-half of their wavelength. The shallow depth causes changes to the movement of water particles at the base of the wave. This causes the waves to slow down. As the water becomes shallow, incoming wave crests gradually catch up with the slower wave crests ahead. As a result, the crest-to-crest wavelength decreases. The incoming waves become higher, steeper, and unstable, and their crests collapse forward. Collapsing waves are called **breakers.** The formation of breakers is also influenced by the motion of wave crests, which overrun the troughs. The collapsing crests of breakers, like the one shown in **Figure 19,** move at high speeds toward shore and play a major role in shaping shorelines.

Demonstration

Model Waves Place a wooden stick in a water-filled, graduated cylinder. Have a student volunteer mark the flotation level of the stick. Next, depress the stick below the water surface and release it. The stick will rise, overshoot the equilibrium point, and oscillate a few times before coming to rest at the previous flotation level. Explain to the class that if the ocean surface is depressed at a particular point, it will move in a similar manner and create waves of oscillation. Specifically, a ring of elevated water will appear around a point of depression. As the depression rebounds, the ring becomes a depressed moat, which generates a second ring. In this way, successive rings (crests) and moats (troughs) spread out from the center of a disturbance.

Tides

Tides are the periodic rise and fall of sea level. The highest level to which water regularly rises is known as high tide, and the lowest level is called low tide. Because of differences in topography and latitude, the tidal range—the difference in height between high tide and low tide—varies from place to place. For example, in the Gulf of Mexico, the tidal range is less than 1 meter, whereas in New England, it can be as high as 6 meters. The greatest tidal range occurs in the Bay of Fundy between New Brunswick and Nova Scotia, Canada, where it is as much as 16.8 meters. Generally, a daily cycle of high and low tides takes 24 hours and 50 minutes. Differences in topography and latitude cause three different daily tide cycles, as illustrated in **Figure 20.** Areas with semidiurnal cycles experience two high tides in about a 24-hour period. Areas with mixed cycles have one pronounced and one smaller high tide in about a 24-hour period. Areas with diurnal cycles have one high tide in about a 24-hour period.

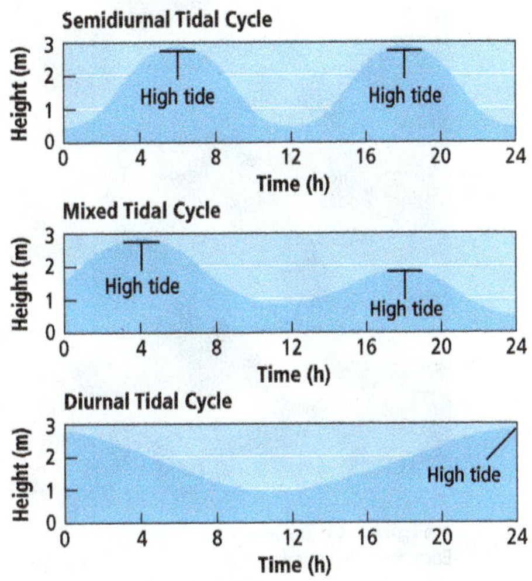

Figure 20 The three different daily tide cycles are semidiurnal, mixed, and diurnal.

☑ READING CHECK **Explain** the difference between semidiurnal tides and mixed tides.

Data Analysis LAB

iLab Station

Based on Real Data*

Graph Data

When does the tide come in? Tidal data is usually measured in hourly increments. The water levels shown in the data table were measured over a 24-hour period.

Think Critically

1. **Apply** Plot these water levels on a graph with time on the *x*-axis and water level on the *y*-axis.
2. **Estimate** the approximate times and water levels of high tides and low tides.
3. **Identify** the type of daily tidal cycle this area experiences.
4. **Determine** the tidal range for this area.
5. **Predict** the water level at the next high tide and estimate when it will occur.

Data and Observations

Tidal Record

Time (h)	Water Level (m)	Time (h)	Water Level (m)
00:00	2.11	13:00	1.70
01:00	1.79	14:00	1.37
02:00	1.33	15:00	1.02
03:00	0.80	16:00	0.68
04:00	0.36	17:00	0.48
05:00	0.10	18:00	0.50
06:00	0.03	19:00	0.69
07:00	0.20	20:00	1.11
08:00	0.55	21:00	1.58
09:00	0.99	22:00	2.02
10:00	1.45	23:00	2.27
11:00	1.74	24:00	2.30
12:00	1.80		

*Data obtained from: The National Oceanic and Atmospheric Administration, Center for Operational Oceanographic Products and Services.

ACROSS THE CURRICULUM

Astronomy Earth's tidal bulges are aligned with the Moon. Because of the orbital motion of the Moon, Earth's rotational period relative to the Moon—the so-called tidal day—is 24 hours and 50 minutes. This means that the tidal pattern observed at a given location repeats every 24 hours and 50 minutes. Specific tidal phases, such as high tide, occur 50 minutes later on the following day.

Data Analysis LAB

About the Lab

- Students will predict tidal patterns based on tidal records.
- Do the activity yourself prior to class. Before conducting the lab with students, review tidal periods, tidal patterns, and tidal parameters.
- Have students work individually, then compare their graphs with those of other students.
- Tidal data for locations around the world are provided by The National Oceanic and Atmospheric Administration.

Think Critically

1. The water level oscillates as a function of time, going up and down every six hours.
2. high tide at 00:00 h, water level = 2.11 m; low tide at 06:00 h, water level = 0.03 m; high tide at 12:00 h, water level = 1.80 m; low tide at 17:00 h, water level = 0.48 m; high tide at 24:00 h, water level = 2.30 m
3. mixed
4. 2.30 m − 0.03 m = 2.27 m
5. a little above 1.8 m at 12:00

☑ **READING CHECK** Semidiurnal tides consist of two high tides occurring in about a 24-hour period. Mixed tides consists of one pronounced and one smaller high tide in about a 24-hour period.

Concept Development

Tides Observed tides are usually not perfectly aligned with the Moon's motion. For instance, high tide does not usually occur when the Moon is overhead or at its highest altitude. The reason for this difference between theoretical and observed tides is the dynamic response of the ocean to tidal forces. The ocean is broken up by continents. Tidal bulges must go around these continents in complicated patterns.

The motion of tidal bulges is further complicated by the fact that they have a wavelength of 20,000 km—half of Earth's circumference—and behave like shallow-water waves. In a part of the ocean that is 5 km deep, the velocity of tidal bulges can be no more than about 800 km/h. Earth's rotational speed at the equator is 1610 km/h. Therefore, even if Earth had no continents, tidal bulges could not keep up with Earth's rotation except in high latitudes.

Activity

Tidal Range Divide the class into groups. Have each group use graph paper to determine the effect of tidal fluctuations on a beach. Tell students that the beach has a slope of 1/10, its average width is 30 m, and it rises to 3 m above mean sea level. Tell them to assume an average tidal range of 2 m. Have students draw the beach as a sloping line on the graph paper. Students should draw lines showing mean sea level and the water level of average high and low tides. AL COOP LEARN

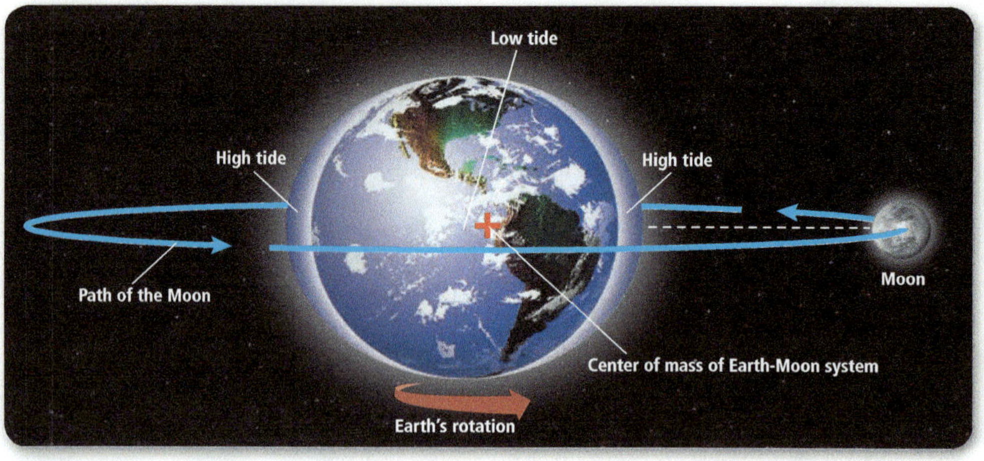

Figure 21 The Moon and Earth revolve around a common center of gravity and experience unbalanced gravitational forces. These forces cause tidal bulges on opposite sides of Earth. (Note: *diagram is not to scale.*)

The Moon's influence The basic causes of tides are the gravitational attraction among Earth, the Moon, and the Sun, as well as the effect of Earth's rotation. Consider the Earth-Moon system. Both Earth and the Moon orbit a common center of gravity, shown as a red plus sign in **Figure 21.** As a result of their motions, both Earth and the Moon experience differing gravitational forces. These unbalanced forces generate tidal bulges on opposite sides of Earth. The gravitational effect of the Moon on Earth's oceans is similar to what happens to the liquid in a coffee cup inside a car as the car goes around a curve. The liquid sloshes toward the outside of the curve.

VOCABULARY
SCIENCE USAGE V. COMMON USAGE
Attraction
Science usage: a force acting between particles of matter, tending to pull them together

Common usage: something that pulls people in by appealing to their tastes

The Sun's influence The gravitational attraction between Earth and the Sun, and Earth's orbital motion around the Sun influences tides. However, even though the Moon is much smaller than the Sun, lunar tides are more than twice as high as those caused by the Sun because the Moon is much closer to Earth. Consequently, Earth's tidal bulges are aligned with the Moon.

Depending on the phases of the Moon, solar tides can either enhance or diminish lunar tides, as illustrated in **Figure 22.** Notice in **Figure 22** that during both a full and a new moon, the Sun, the Moon, and Earth are all aligned. When this occurs, solar tides enhance lunar tides, causing high tides to be higher than normal and low tides to be lower than normal. The tidal range is highest during these times. These types of tides are called **spring tides.** Spring tides have a greater tidal range during the winter in the northern hemisphere, when Earth is closest to the Sun. Study **Figure 22** again. Notice that when there is a first- or third-quarter moon, the Sun, the Moon, and Earth form a right angle. When this occurs, solar tides diminish lunar tides, causing high tides to be lower and low tides to be higher than normal. The tidal range is lowest during these times. These types of tides are called **neap tides.** Spring and neap tides alternate every two weeks.

DIFFERENTIATED INSTRUCTION

Struggling Learners Reinforce the concepts of sea level, tides, and tidal ranges with students. Have students determine how far above mean sea level the tide will rise if the tidal range is 2 m. 1 m How far below mean sea level will the sea level drop during a low tide if the tidal range is 6 m? 3 m

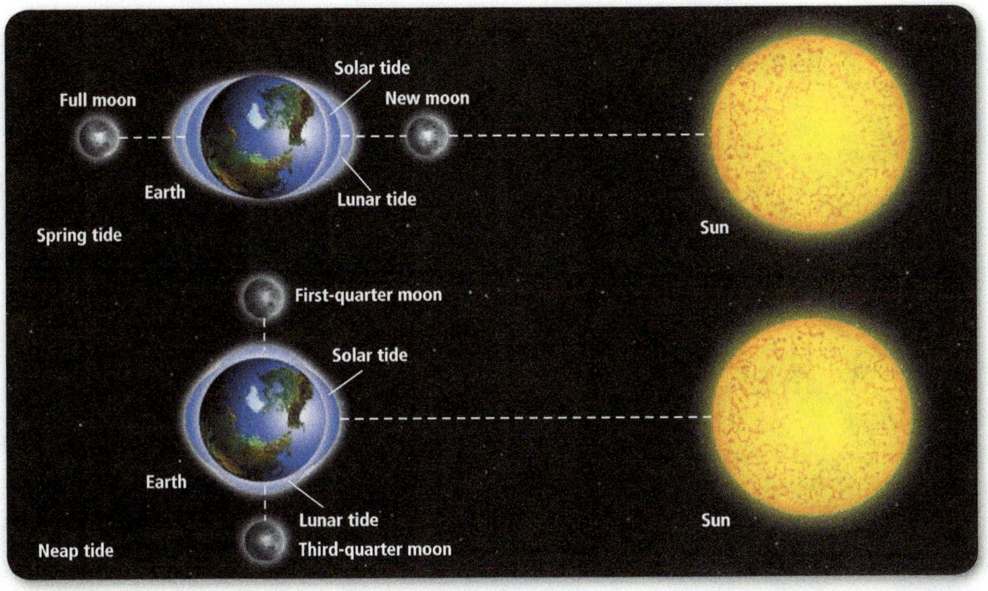

■ **Figure 22** Spring tides occur when the Sun, the Moon, and Earth are aligned. Neap tides occur when the Sun, the Moon, and Earth form a right angle. (Note: *diagram is not to scale.*)

Currents

Currents in the ocean can move horizontally or vertically. They can also move at the surface or deep in the ocean. Currents at the surface are usually generated by wind. Some currents are the result of tides. Deep-ocean currents usually result from differences in density between water masses.

Surface currents Mainly the top 100 to 200 m of the ocean experience **surface currents,** which can move at a velocity of about 100 km per day. Surface currents follow predictable patterns and are driven by Earth's global wind systems. Recall that in the northern hemisphere, tropical trade winds blow from east to west. The resulting tropical ocean surface currents also flow from east to west. In northern midlatitudes, the prevailing westerlies and resulting ocean surface currents move from west to east. In northern polar regions, polar easterly winds push surface waters from east to west.

The direction of surface currents can also be affected by landforms, such as continents, as well as the Coriolis effect. Recall that the Coriolis effect deflects moving particles to the right in the northern hemisphere and to the left in the southern hemisphere.

☑ READING CHECK **Explain** how winds influence surface currents.

Gyres If Earth had no landmasses, the global ocean would have simple belts of easterly and westerly surface currents. Instead, the continents deflect ocean currents to the north and the south so that closed circular current systems, called gyres (JI urz), develop.

Teacher Content Support

Forces that Influence Tides
Tidal forces decrease with the cube of the distance from the Moon and the Sun. Because of this, lunar tides are larger than those produced by the much more massive Sun. The Sun's mass is 333,000 times Earth's mass, while the lunar mass is only 0.0123 percent of Earth's mass. However, the Sun's mean distance from Earth is almost 150 million km, while the lunar distance averages only 384,000 km.

The solar and lunar tidal forces are given by $2GMr/D^3$ and $2Gmr/d^3$, respectively, where r is Earth's radius, M is the solar mass, m is the lunar mass, D is the solar distance, and d is the lunar distance. The ratio of solar to lunar tidal forces is therefore $M/m \times (d/D)^3$. Substituting the appropriate numbers, this ratio works out to 0.46, or 46 percent. Thus, solar tides are less than half as strong as lunar tides.

Interpret the Illustration

Spring and Neap Tides Have students study **Figure 22.** Have students describe the positions of the Sun, Earth, and the Moon during spring tides and neap tides. During spring tides, the Sun, Earth, and the Moon are aligned. During neap tides, the Sun, Earth, and the Moon form a right angle.

☑ READING CHECK Wind moves surface water in the direction in which it is blowing.

ACROSS THE CURRICULUM

Biology Sea turtles lay their eggs on beaches above the high-water level of the spring tide so the eggs can incubate in the dry, sun-warmed sand. The hatchlings emerge during a subsequent spring tide. The high water level ensures that the hatchlings have a relatively short dash to the safety of the ocean and thus increases their chances for survival.

Interpret the Illustration

Currents Have students examine **Figure 23.** Floating objects drift with the surface currents of an ocean. Tell students to use this figure, a world map to help estimate distances, and the speeds 1 m/s for warm currents and 0.5 m/s for cold currents to estimate the travel path, travel time, and possible landing sites of bottles with messages that were thrown into the ocean at the following three locations: Miami, FL possible landing sites and times: New England, 31 days; Newfoundland, 52 days; Europe, 3 months; San Francisco, CA possible landing sites and times: Japan, 7 months; Alaska, 10 months; and Puerto Rico possible landing sites and times: Cuba, 12 days; Florida, 22 days; New England, 53 days.

Caption Question Fig. 23 South Equatorial Current, Brazil Current, Antarctic Circumpolar Current, Benguela Current

Major Ocean Gyres

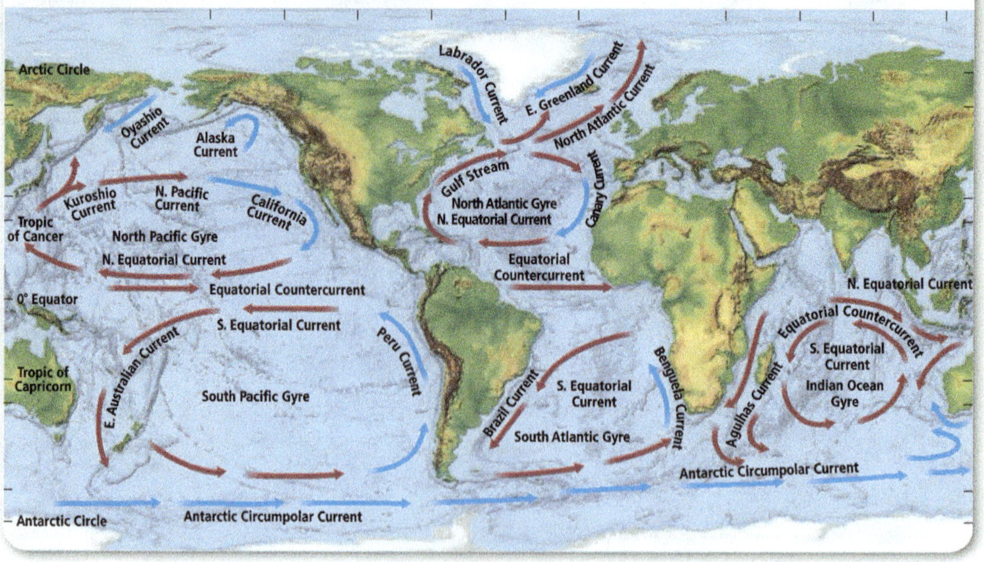

Figure 23 Large gyres in each ocean are formed by surface currents. Red arrows represent the movement of warm water, blue arrows represent the movement of cold water.

Identify ***the currents that make up the gyre in the South Atlantic Ocean.***

As shown in **Figure 23,** there are five major gyres—the North Pacific, the North Atlantic, the South Pacific, the South Atlantic, and the Indian Ocean. Because of the Coriolis effect, the gyres of the northern hemisphere circulate in a clockwise direction and those of the southern hemisphere circulate in a counterclockwise direction. The parts of all gyres closest to the equator move toward the west as equatorial currents. When these currents encounter a landmass, they are deflected toward the poles. These poleward-flowing waters carry warm, tropical water into higher, colder latitudes. An example of a warm current is the Gulf Stream Current in the North Atlantic.

After these warm waters enter polar regions, they gradually cool and, deflected by landmasses, move back toward the equator. The resulting currents then bring cold water from higher latitudes into tropical regions. An example of this kind of current is the California Current in the eastern North Pacific.

Upwelling In addition to moving horizontally, ocean water moves vertically. The upward motion of ocean water is called **upwelling.** Upwelling waters originate in deeper waters, below the thermocline, and thus are usually cold and nutrient-rich. Areas of upwelling exist mainly off the western coasts of continents in the trade-wind belts. For example, **Figure 24** shows what happens off the coast of California. Winds blowing from the north cause surface water to begin moving. The Coriolis effect acts on the moving water, deflecting it to the right of its direction of movement, which results in surface water being moved offshore. The surface water is then replaced by upwelling deep water.

Figure 24 Upwelling occurs when surface water is moved offshore and deep, colder water rises to the surface to replace it.

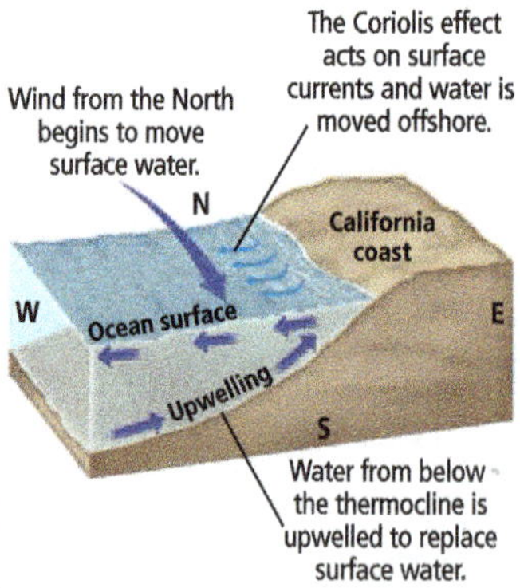

EARTH SCIENCE JOURNAL

Ocean Circulation To familiarize students with the circulation patterns of Earth's oceans, have them research the ten major cold and warm ocean currents associated with the Atlantic, Pacific, and Indian oceans. Students should create data tables in their Earth science journals listing the name, location, speed, direction of motion, and temperature of each current. OL

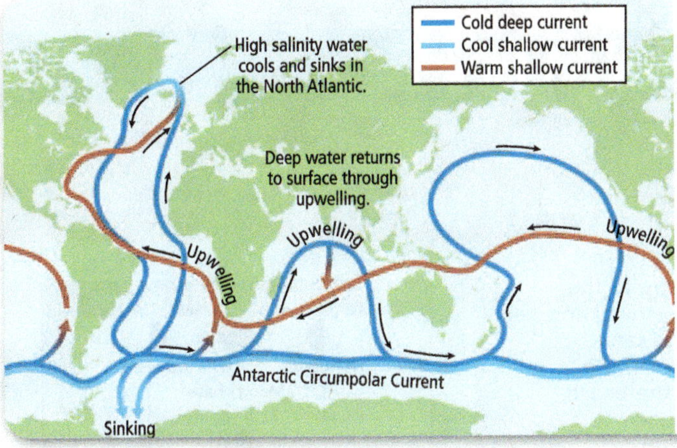

Figure 25 Differences in salinity and temperature generate density currents in the deep ocean.

Density currents Recall the discussion of Antarctic Bottom Water in Section 2. The sinking of Antarctic Bottom Water is an example of an ocean current. In this case, the current is called a **density current** because it is caused by differences in the temperature and salinity of ocean water, which in turn affect density. Density currents move slowly in deep ocean waters, following a general path that is sometimes called the global conveyer belt.

The conveyor belt, a model of which is shown in **Figure 25,** begins when cold, dense water, including North Atlantic Deep Water and Antarctic Bottom Water, sinks at the poles. After sinking, these water masses slowly move away from the poles and circulate through the major ocean basins. After hundreds of years, the deep water eventually returns to the surface through upwelling. Once at the surface, the deep water is warmed by solar radiation. This water continues along the global conveyer belt until it reaches the poles where it cools, sinks, and begins its journey again.

SECTION 3 REVIEW

Section Self-Check

Section Summary

- Energy moves through ocean water in the form of waves.
- Tides are influenced by both the Moon and the Sun.
- Surface currents circulate in gyres in the major ocean basins.
- Vertical currents in the ocean include density currents and upwelling.

Understand Main Ideas

1. MAINIDEA **Describe** how surface currents in gyres redistribute heat between the equator and the poles.
2. **Illustrate** a wave. Label the following characteristics: *crest, trough, wavelength, wave height,* and *wave base.*
3. **Explain** how tides form.
4. **Compare and contrast** surface currents and density currents.

Think Critically

5. **Predict** the effects on marine ecosystems if upwelling stopped.
6. **Assess** the difference between spring tides and neap tides.

WRITING IN ▶ Earth Science

7. Write a step-by-step explanation of how upwelling occurs.

3 Assess

Check for Understanding

Compare and Contrast Ask students to compare and contrast the Gulf Stream and the California Current in terms of movement and temperature. The warm Gulf Stream flows north and transports warm, tropical water to higher latitudes. The cold California Current flows south and brings cold, North Pacific water to lower latitudes.

Reteach

Illustrate Have students each draw a northern-hemisphere gyre and indicate its circulation pattern. Have them describe the gyre's pattern and temperature in their Earth science journals.

Assessment

Knowledge Have students use the Coriolis effect to explain the movement of gyres in the southern hemisphere. The Coriolis effect deflects all free-moving objects to the left in the southern hemisphere. This causes gyres in the southern hemisphere to move in a counterclockwise direction.

SECTION 3 REVIEW

1. Surface currents, such as the Gulf Stream, transfer heat from the equator toward the poles. Cold-water currents, such as the California current, transfer cold water from the poles toward the equator.
2. Diagrams should show a wave with labels placed as shown in **Figure 17.**
3. Tides are caused mainly by the gravitational attraction of the Sun and the Moon.
4. Both are currents in oceans. Surface currents are horizontal, affect the upper few hundred meters of the ocean, and are formed by wind. Density currents are caused by differences in temperature and salinity of water masses, and are vertical currents.
5. Ecosystems would lack nutrients if upwelling stopped. This would reduce productivity and affect food webs.
6. Spring tides are unusually high tidal ranges that occur when the Sun, Earth, and the Moon are aligned. Neap tides are unusually low tidal ranges that occur when the Moon and the Sun are at right angles to Earth.
7. Winds cause surface currents to begin. The Coriolis effect acts on the moving water, moving it to the right of the direction in which it is moving (in the northern hemisphere). Surface water is moved offshore. The surface water is replaced by upwelling deep water.

Rubric

Earth Science &
ENVIRONMENT

Purpose

Students will learn about patterns of bacterial concentrations correlated to the moon phases.

Teacher Content Support

Bacterioplankton Although some bacteria are pathogens, the majority are not. As in terrestrial ecosystems, bacteria play important roles in aquatic ecosystems. Bacteria are often called bacterioplankton in aquatic ecosystems, due to their small size—only 0.5 μm in diameter. In aquatic ecosystems, bacteria serve as decomposers of organic material. They are also food for heterotrophic zooplankton, hosts for viruses, and some are even primary producers, undergoing photosynthesis or chemosynthesis to make their own food.

Teaching Strategy

Have students discuss the concept that a correlation, a relationship between two variables, does not necessarily mean that one variable is causing the other to change. For example, ask students if they think that stating the Moon causes bacterial populations to grow or decline is an appropriate conclusion to the study discussed in the feature. There are many other variables involved in this phenomenon. The Moon influences the tides, which leads to variations in the location where seawater interacts with land-based sources for bacteria.

Earth Science &
ENVIRONMENT

Bacterial Counts and Full Moons

You've been looking forward to going to the beach all week. Under the hot Sun—towel and lunch in hand—you head toward the sand. You can't wait to get in the cool, refreshing water. As you near the entrance of the beach, you see a posted sign that reads "Beach Closed: High Bacterial Counts in Water."

Bacteria in the water Although most of the bacteria in seawater are harmless to humans, some types are thought to cause gastrointestinal illnesses, with symptoms that include diarrhea and vomiting, in swimmers. Water is routinely tested on many beaches for a type of bacteria called enterococci (en tur oh KAHK i), which normally live in the intestines of mammals and birds. Although enterococci are usually harmless, their presence in the water is considered a strong indicator of the presence of other, illness-causing, bacteria. If enterococci counts rise above a certain level, authorities close beaches for the safety of swimmers.

Bacterial counts and moon phases Scientists have found that higher levels of enterococci in seawater are associated with new moon and full moon phases, as shown in the graph on the right. Recall that spring tides occur during the new moon and full moon phases. During spring tides, high tides are at their highest levels and low tides are at their lowest, resulting in a large tidal range.

After studying 60 beaches along the Southern California coast, scientists discovered a pattern of high bacterial counts during spring tides at 50 of the 60 beaches analyzed. Lower bacterial counts were associated with neap tides, which occur during first-quarter and three-quarter moon phases. Data also showed that higher counts of bacteria were found specifically during an ebbing spring tide, that is, when water is receding after reaching its highest point.

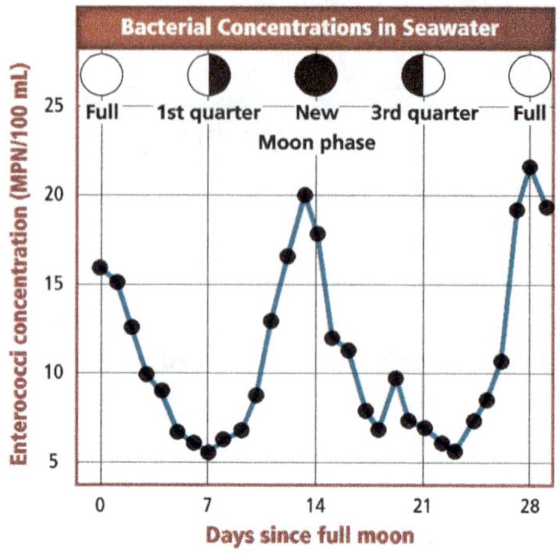

Bacterial counts in seawater vary with the phase of the Moon.

Possible sources of bacteria Scientists have several hypotheses to explain possible sources for the bacteria in seawater during the spring tides. One is that the bacteria are present in high numbers in groundwater that only mixes with seawater during spring tides. Other possible sources include decaying organic material that collects on the sand, or bird droppings near the high tide line. Both of of these have the potential to mix with seawater during high spring tides.

WRITING IN ▶ Earth Science

Newscast Research more about bacterial counts in seawater. Then, suppose you are a newscaster presenting a story for the nightly news about bacterial levels at beaches. Present your story to the class, explaining results of scientific studies on patterns of bacteria and why these results are important to swimmers.

WebQuest

WRITING IN ▶ Earth Science

Rubric

Newscast Presentations will vary, but should include an explanation of the studies and why they are important to the public.

WebQuest

GeoLAB

iLab Station

Model Water Masses

Background: Water in oceans is layered because water masses with higher densities sink below those with lower densities. The density of seawater depends on its temperature and salinity.

Question: *How do changes in salinity and temperature affect water density?*

Materials

balance
graduated 500-mL cylinder
100-mL glass beakers (4)
water
red, yellow, and blue food coloring
salt
thermometer
eyedropper
graph paper
ruler
calculator

Safety Precautions

Procedure

1. Read and complete the lab safety form.
2. Mix 200 mL of water and 7.5 g of salt in the graduated cylinder. Pour equal amounts of the salt solution into two beakers. Fill each of the two other beakers with 100 mL of freshwater.
3. Put a few drops of red food coloring in one of the salt solutions. Put a few drops of yellow food coloring in the other salt solution. Put a few drops of blue food coloring in one of the freshwater beakers. Do not add food coloring to the other freshwater beaker.
4. Place the beakers with the red salt solution and the blue freshwater in the refrigerator. Refrigerate them for 30 min.
5. Measure and record the temperature of the water in all four beakers.
6. Put several drops of the cold, red salt water into the beaker with the warm, yellow salt water and observe what happens. Record your observations.
7. Put several drops of the cold, blue freshwater into the beaker with the warm, clear freshwater and observe what happens. Record your observations.
8. Put several drops of the cold, blue freshwater into the beaker with the warm, yellow salt water and observe what happens. Record your observations.

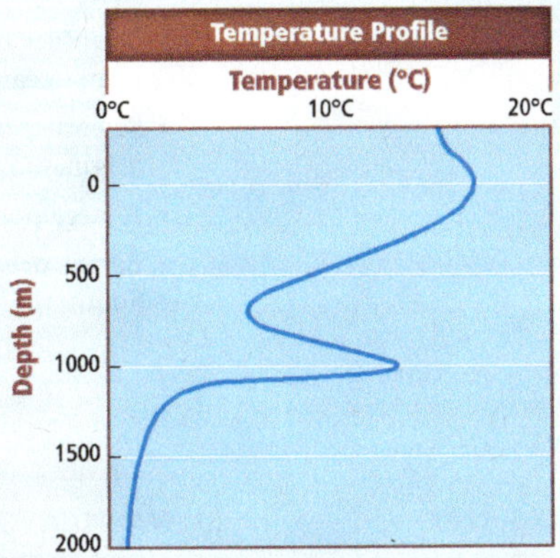

Analyze and Conclude

1. **Describe** the movement of the cold, red salt water in Step 6. Compare this to the movement of the cold, blue freshwater in Step 8. What accounts for the differences you observed?
2. **Identify** the water samples by color in order of increasing density.
3. **Explain** If you poured the four water samples into the graduated cylinder, how would they arrange themselves into layers by color, from top to bottom?
4. **Construct** Assume that four water masses in a large body of water have the same characteristics as the water in the four beakers. The warm water layers are 100 m thick, and the cold layers are 1000 m thick. Construct a graph that shows the temperature profile of the large body of water.

APPLY YOUR SKILL

Infer The temperature profile above was constructed from measurements taken in the Atlantic Ocean off the coast of Spain. Study the profile, then infer why a high-temperature layer exists beneath the thermocline. Is this layer denser than the colder water above? Explain.

GeoLAB

iLab Station Rubric

Preparation

Time Allotment 45 min

Process Skills analyze data, observe and infer, compare and contrast, recognize cause and effect, make and use graphs, predict, interpret data

Safety Precautions Approve lab safety forms before work begins. Make sure students wear safety goggles and lab aprons.

Procedure

- The warm water should be at room temperature, or about 20°C; the refrigerated water should be colder than 5°C.
- Review the concepts of salinity and units of measurement (fractions, g/kg, percent, ppt). Students also should review the MiniLab.
- **Troubleshooting** If the experiment is not done quickly, the temperature and density differences might be too small to allow observable sinking motions. Also, the temperature of the droppers should be about the same as that of the cold-water samples.

Analyze and Conclude

1. Cold salt water sinks in warm salt water; cold freshwater floats in warm salt water. Cold salt water is denser than warm salt water; cold freshwater is less dense than warm or cold salt water. The amount of salinity in the water accounts for the differences observed.
2. clear, blue, yellow, red
3. clear, blue, yellow, red
4. Student graphs should show that the first layer extends to a depth of 100 m and has a temperature of 20°C. The second layer extends to 1100 m and has a temperature of 5°C. The third layer extends to 1200 m and has a temperature of 20°C. The fourth layer extends to 2200 m and has a temperature of 5°C.

APPLY YOUR SKILL

Infer The high-temperature layer is saltier than the colder thermocline above it. It is therefore denser than the thermocline because salinity causes an increase in density.

CHAPTER 15 STUDY GUIDE

MAINIDEAS **Summary statements can be used by students to review the major concepts of the chapter.**

Students can review with these online resources.

Vocabulary Practice

Vocabulary eGames
Vocabulary eFlashcards
Vocabulary PuzzleMaker

Section Self-Check

Chapter Self-Check

Online Test Practice

Use the Exam*View*® *Assessment Suite* CD-ROM to:

- create multiple versions of tests
- create modified tests with one mouse click
- edit existing questions and add your own questions
- build tests aligned with state standards using built-in state curriculum tags
- change English tests to Spanish with one mouse click
- track students' progress using the Teacher Management System

CHAPTER 15 STUDY GUIDE

BIGIDEA **Studying oceans helps scientists learn about global climate and Earth's history.**

Vocabulary Practice

SECTION 1 An Overview of Oceans

VOCABULARY
- side-scan sonar
- sea level

MAINIDEA **The global ocean consists of one vast body of water that covers more than two-thirds of Earth's surface.**

- Scientists use many different instruments to collect and analyze data from oceans.
- Scientists have several ideas as to where the water in Earth's oceans originated.
- A large portion of Earth's surface is covered by ocean.
- Earth's oceans are the Pacific, the Atlantic, the Indian, the Arctic, and the Southern.

SECTION 2 Seawater

VOCABULARY
- salinity
- estuary
- temperature profile
- thermocline

MAINIDEA **Oceans have distinct layers of water masses that are characterized by temperature and salinity.**

- Ocean water contains dissolved gases, nutrients, and salts.
- Salts are added to and removed from oceans through natural processes.
- Properties of ocean water, including temperature and salinity, vary with location and depth.
- Many of the oceans' deepwater masses sink from the surface of polar oceans.

SECTION 3 Ocean Movements

VOCABULARY
- wave
- crest
- trough
- breaker
- tide
- spring tide
- neap tide
- surface current
- upwelling
- density current

MAINIDEA **Waves and currents drive the movements of ocean water and lead to the distribution of heat, salt, and nutrients from one region of the ocean to another.**

- Energy moves through ocean water in the form of waves.
- Tides are influenced by both the Moon and the Sun.
- Surface currents circulate in gyres in the major ocean basins.
- Vertical currents in the ocean include density currents and upwelling.

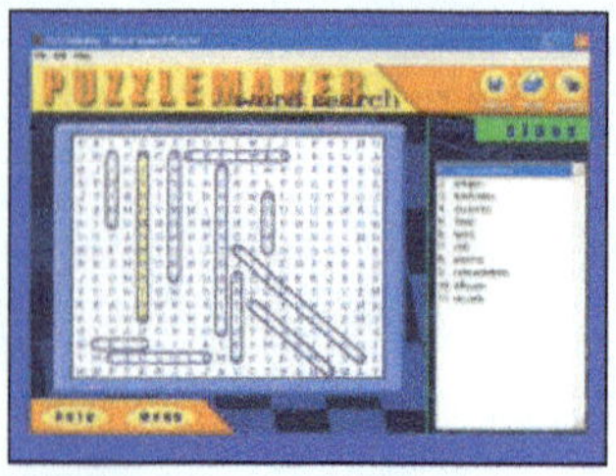

Vocabulary Practice

For additional practice with vocabulary, have students access the Vocabulary PuzzleMaker.

Chapter Self-Check

VOCABULARY REVIEW

Match each description below with the correct vocabulary term from the Study Guide.

1. the amount of dissolved salts in a fixed volume of seawater
2. a transitional layer in which temperature rapidly decreases with depth
3. plots changing water temperatures against depth

Complete the sentences below using vocabulary terms from the Study Guide.

4. The lowest point of a wave is the ________ .
5. The level of Earth's oceans is ________.
6. ________ sends sound waves to the seafloor at an angle.
7. ________ are the periodic rise and fall of sea level.
8. A ________ is caused by differences in the temperature and salinity of ocean water.
9. The upward motion of ocean water is called ________.
10. The highest point of a wave is the ________.
11. Waves that collapse near shore are called ________.

UNDERSTAND KEY CONCEPTS

12. Which is used to measure ocean depth?
    A. bottom dredges **C.** sonar
    B. nets **D.** tidal patterns
13. What is the average depth of the oceans?
    A. 380 m **C.** 3800 m
    B. 38 m **D.** 3 km
14. Which are the most common gases emitted by volcanoes?
    A. hydrogen and helium
    B. oxygen and nitrogen
    C. water vapor and carbon dioxide
    D. chlorine and hydrogen

Use the graph below to answer Questions 15 and 16.

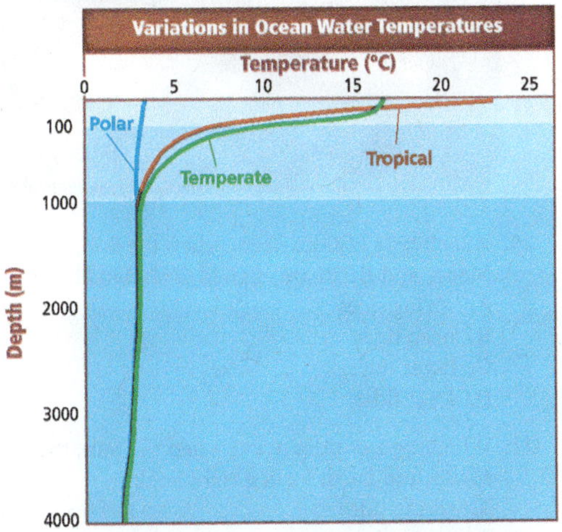

15. Between which depths is the thermocline?
    A. 0 and 100 m **C.** 100 and 4000 m
    B. 100 and 1000 m **D.** 1000 and 4000 m
16. What is the average temperature of deep water below the thermocline?
    A. 15°C **C.** less than 4°C
    B. more than 4°C **D.** 0°C
17. What basic motion does water follow during the passage of a wave?
    A. forward **C.** up and down
    B. backward **D.** circular
18. Which would have the largest impact on global ocean water density?
    A. strong winds
    B. increase in daylight hours
    C. long-term increase in air temperature
    D. thunderstorm with heavy precipitation
19. Which process does not remove salt from ocean water?
    A. precipitation of salt in dry, coastal regions
    B. evaporation of water in subtropical regions
    C. sea spray being carried inland by wind
    D. absorption by marine organisms

CHAPTER 15 ASSESSMENT

VOCABULARY REVIEW

1. salinity
2. thermocline
3. temperature profile
4. trough
5. sea level
6. Side-scan sonar
7. Tides
8. density current
9. upwelling
10. crest
11. breakers

UNDERSTAND KEY CONCEPTS

12. C
13. C
14. C
15. B
16. C
17. D
18. D
19. B

20. A
21. B

CONSTRUCTED RESPONSE

22. Diagrams should show that carbon dioxide from the atmosphere is dissolved in surface waters of the ocean. When cold, dense surface water in polar oceans sinks, carbon dioxide sinks to the bottom of the ocean. Some carbon dioxide is also taken up by phytoplankton in the surface of the ocean to use for photosynthesis.

23. The highest salinity values occur in subtropical oceans, where evaporation exceeds precipitation.

24. 100 m

25. Possible answers: What programs do the computers use to analyze data? What types of simulations do you make using computers? How much math do you need to know to use computers to analyze data?

26. The North Pacific; In the northern hemisphere, the Coriolis effect turns currents to the right.

27. Diagrams should show that when the Sun, the Moon, and Earth are aligned, solar tides increase tides, and that when the Sun and the Moon are at right angles to Earth, solar tides diminish tides.

28. As waves near shore, water depth becomes shallower. Incoming wave crests gradually catch up with slower wave crests ahead. The wavelength of the waves decreases. The incoming waves become higher, steeper, and unstable, and their crests collapse forward.

29. The gravitational effect of the Moon causes Earth's oceans to slosh toward the opposite sides of Earth, creating tidal bulges.

30. Diagrams should show placement of labels as in **Figure 17.**

31. Seawater density depends on both temperature and salinity. Warm Mediterranean water is saltier and therefore denser than the water in the Atlantic Ocean.

THINK CRITICALLY

32. River water is less salty and therefore less dense than seawater. Fresh river water would float on top of ocean water.

33. The gravitational forces between the Earth and the Moon cause both planetary bodies to stretch along an imaginary line. Earth's oceans bulge in response to these forces.

ASSESSMENT

Use the diagram below to answer Question 20.

20. Which type of tides occur when the Sun, the Moon, and Earth are aligned as shown above?
A. spring tides
B. neap tides
C. lunar tides
D. solar tides

21. Which type of tides occur when the Sun, the Moon, and Earth form a right angle?
A. spring tides
B. neap tides
C. lunar tides
D. solar tides

CONSTRUCTED RESPONSE

22. Illustrate Make a diagram that shows how the ocean acts as a sink for carbon.

23. Relate Where in the oceans are the highest values of salinity found? What processes lead to areas of high salinity water?

24. Solve What would be the depth of the wave base for a wave that is 200 m long?

25. Interview Write several questions you would ask an oceanographer about the use of computers to analyze data from the ocean.

26. Determine Which gyre would have clockwise circulation: the North Pacific, the South Pacific, the South Atlantic, or the Indian Ocean? Explain.

27. Diagram Draw a diagram that shows how the Sun influences tides.

28. Analyze Why does a wave break?

29. Explain how the Moon influences tides.

Use the diagram below to answer Question 30.

30. Diagram Copy the illustration shown above. Then use the following terms to label the characteristics of an ocean wave: *crest, trough, wave height,* and *wavelength.*

31. Cause and Effect Cold water masses are generally denser than warm water masses, yet warm water from the Mediterranean Sea sinks to a depth of more than 1000 m when it flows into the Atlantic Ocean. What causes this effect?

THINK CRITICALLY

32. Predict Based on what you have learned about water density, describe the movement of freshwater from a river as it flows into a sea.

33. Infer why tidal bulges occur simultaneously on opposite sides of the Earth.

34. Suggest a real-world application for adding salt to lower the melting point of ice.

35. Plan Use **Figure 23** to plan the fastest round trip by ship from Boston, Massachusetts, to London, England. Will the return route be the same as the outbound trip? Explain.

36. Careers in Earth Science Suppose you are a lead scientist at NASA. Design an experiment that would allow you to test the hypothesis that Earth's water originated from comets.

37. Compare and contrast Antarctic Intermediate Water and North Atlantic Deep Water.

38. Explain how biological and physical processes affect levels of carbon dioxide in different areas of the ocean.

34. Possible answer: de-icing winter roads to prevent refreezing.

35. To use currents to aid the travel, the outbound route should follow Gulf Stream and North Atlantic Currents. On return, one should avoid the Gulf Stream and North Atlantic Current, and sail farther south where the currents are westward.

36. Accept all reasonable responses. Answers should include an experimental design with a problem and a procedure.

37. Both water masses are formed when cold, dense water at the surface sinks. Antarctic Intermediate Water is formed near Antarctica and is slightly less dense than North Atlantic Deep Water, which is formed in the North Atlantic Ocean.

38. At the surface, phytoplankton undergoing photosynthesis take in carbon dioxide, making levels in water low. Levels of carbon dioxide in water usually increase with depth as organisms undergo respiration and give off the gas. As denser water collects in the deep ocean, carbon dioxide accumulates and is trapped until water returns to the surface through upwelling.

Use the diagram below to answer Question 39.

39. Assess Surface currents can affect coastal climates. Would the Gulf Stream and the Benguela Current, both of which are surface currents, have the same effect on coastal climate? Explain.

40. Predict One of the effects of El Niño, which you learned about previously, is that the trade winds that blow across the equatorial Pacific Ocean weaken. Predict how this might affect upwelling off the coast of Peru.

41. Assess How do density currents and the conveyor belt help move dissolved gases, such as oxygen, and nutrients, such as nitrogen, from one area of the ocean to another?

CONCEPT MAPPING

42. Create a concept map using the following words or phrases that describe waves: *lowest point of a wave, wave characteristics, crest, wavelength, wave height, trough, horizontal crest-to-crest distance, highest point of a wave,* and *vertical distance between crest and trough.*

CHALLENGE QUESTION

43. Research the percentages of the different sources of freshwater on Earth, including glaciers, ice caps, rivers, lakes, and groundwater. Construct a new circle graph like the one shown in **Figure 5** that represents the new data.

WRITING IN ▶ Earth Science

44. Write a summary about majoring in oceanography for a college brochure. Include information about the prerequisites, requirements for completion, and career opportunities in your summary.

DBQ Document–Based Questions

Data obtained from: Alford, M. 2003. The redistribution of energy available for mixing by long-range propagation of internal waves. *Nature* 423:159–162.

Ocean mixing due to waves is important for pollution dispersal, marine productivity, and global climate. In the figure below, the arrows represent the direction and magnitude of waves from different data collection sites (shown as black dots). The absence of arrows from the black dots means that the waves were small enough to ignore.

Wave Directions and Magnitude

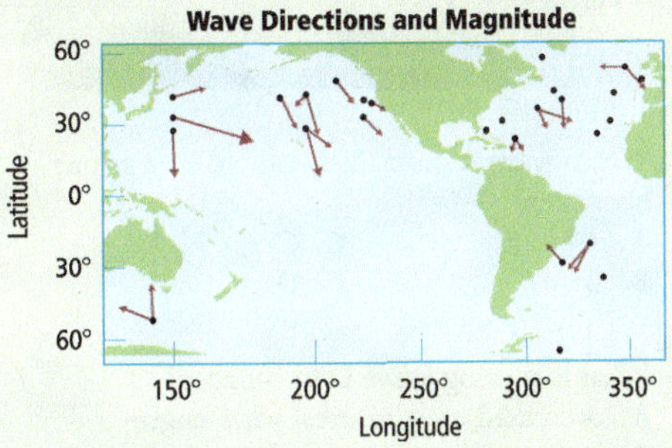

45. In which direction do most of the waves in the northern hemisphere travel? How does this differ from waves in the southern hemisphere?

46. Given the number of collection sites in the southern hemisphere, can you draw any general conclusion about waves in this area?

CUMULATIVE REVIEW

47. What steps are usually included in a scientific method? **(Chapter 1)**

48. Name an example of a carbonate mineral. What is its chemical composition? **(Chapter 4)**

39. The Gulf Stream is a warm ocean current, and the Benguela Current is a cold-water current. The Gulf Stream has a warming effect on coastal climate, and the Benguela Current has a cooling effect.

40. If the trade winds weaken, less water is moved off the coast of Peru. If less surface water is moved, less water is upwelled to take its place. Upwelling would decrease.

41. Density currents cause water masses to sink from the surface into deeper water, taking oxygen and nutrients with them. These currents move through ocean basins and eventually rise to the surface again in another location.

CONCEPT MAPPING

42. Answers will vary, but the concept map should be organized in a logical manner.

CHALLENGE QUESTION

43. glaciers and ice caps: 77.6%; groundwater: 22.3%; rivers and lakes: 0.1%

WRITING IN ▶ Earth Science

Rubric

44. Answers will vary. Prerequisites for most schools will include taking math and science classes in high school; requirements for completion include taking classes in all four disciplines in oceanography, as well as math classes; and career opportunities include research and teaching.

DBQ Document-Based Questions

Data obtained from: Alford, M. 2003. The redistribution of energy available for mixing by long-range propagation of internal waves. *Nature* 423:159–162.

45. Most of the waves are moving in a southeasterly direction, toward the equator; the waves in the southern hemisphere are moving north toward the equator.

46. There are few data points available in the southern hemisphere compared to the northern hemisphere. These data points are probably not enough to draw conclusions given that there are too few of them and these few points do not give a consistent picture as presented.

CUMULATIVE REVIEW

47. observation, research, hypothesize, experiment, collect data, analyze data, revise hypothesis, publish conclusions

48. Calcite; $CaCO_3$

MULTIPLE CHOICE

1. C
2. A
3. A
4. D
5. A
6. A
7. B
8. C
9. B
10. B

CUMULATIVE STANDARDIZED TEST PRACTICE

MULTIPLE CHOICE

1. Why is deforestation often linked to global warming?
   - A. It increases the amount of dry land on Earth's surface.
   - B. It releases toxic gases and pollutants into the atmosphere.
   - C. It minimizes the removal of CO_2 from the atmosphere by plants.
   - D. It decreases the amount of CO_2 released into the atmosphere.

Use the illustration to answer Questions 2 and 3.

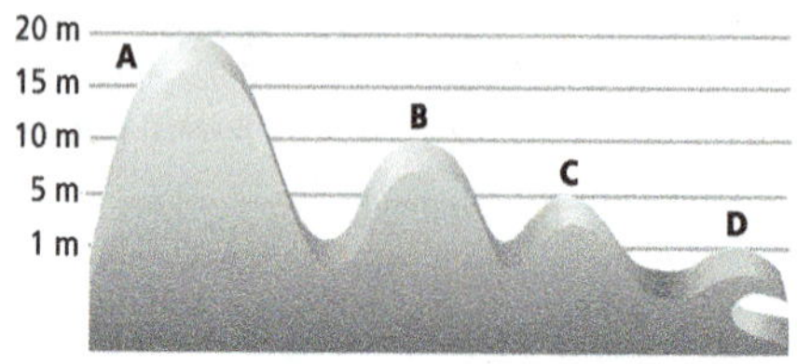

2. Which wave is most likely caused by a strong hurricane?
   - A. A
   - B. B
   - C. C
   - D. D

3. What is causing Wave D to collapse?
   - A. decreased crest-to-crest wavelength
   - B. storm activity
   - C. increased crest-to-crest wavelength
   - D. friction from the ocean floor

4. Where is a heat island most likely to be found?
   - A. a farm
   - B. a beach
   - C. a mountain top
   - D. an inner city

5. Which would be the most abundant evidence that an area had been affected by continental glaciation?
   - A. thick deposits of sediment
   - B. long, snakelike ridges of sand and gravel
   - C. plants and animals adapted to a cold climate
   - D. large, shallow lakes

6. Which is a characteristic the Fujita scale uses to rank tornadoes?
   - A. wind speed
   - B. funnel length
   - C. path of destruction
   - D. duration

Use the map below to answer Questions 7 and 8.

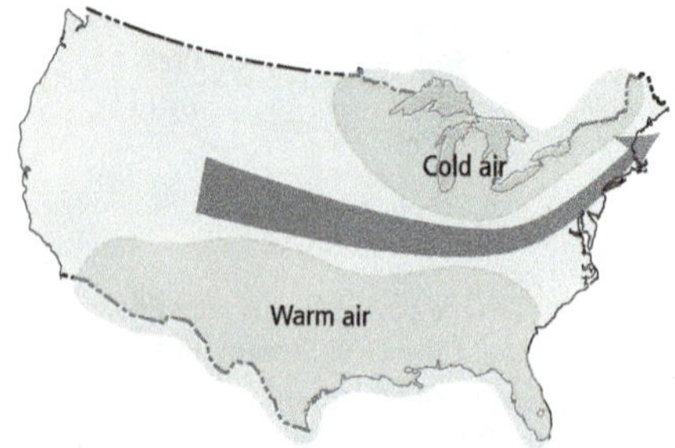

7. What air current is shown by the arrow?
   - A. intertropical convergence zone
   - B. prevailing westerlies
   - C. prevailing easterlies
   - D. trade wind

8. If the cold air were to dip down into the region of warm air, what prediction could be made about the type of weather that might result?
   - A. cloudiness and widespread, mild precipitation
   - B. some cloudiness with no precipitation
   - C. clouds with showers and thunderstorms
   - D. cool weather with no clouds

9. Which region's seawater is most likely to have the highest concentration of dissolved salts?
   - A. an equatorial region
   - B. a subtropical region
   - C. a polar region
   - D. a delta where rivers empty into oceans

10. If there is a sudden period of calm during a hurricane, what should a person assume?
    - A. The hurricane is over and it is safe to go out.
    - B. The eye of the hurricane is over his or her area.
    - C. The hurricane has weakened in intensity.
    - D. The hurricane is over, but it is not safe to go out.

SHORT ANSWER

Use the illustration below to answer Questions 11–13.

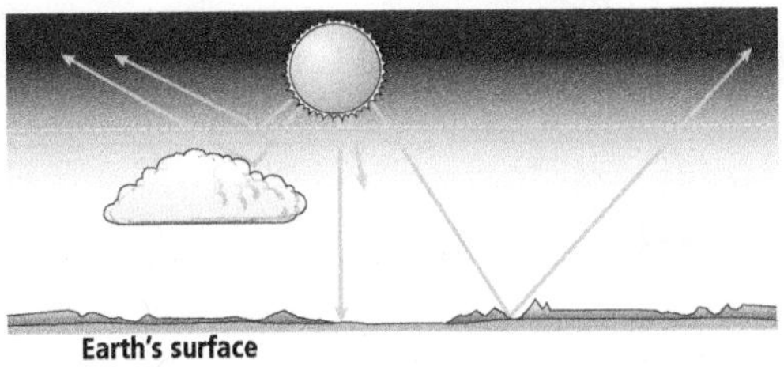

11. Describe what is being represented in the above illustration.

12. What does the dotted line represent? What is its purpose?

13. Why is the Sun important to Earth?

14. Discuss what one-to-three day forecasts can and cannot do.

15. How does the rate of solar absorption differ on water than on land?

16. What is the benefit of using water vapor imagery?

READING FOR COMPREHENSION

Cooling with Seawater

Engineers have turned to the deep ocean as a cooling source. Because of the churning action of wind, waves, and currents, ocean water must be drawn from great depths to get consistently cold temperatures. The Natural Energy Laboratory of Hawaii Authority (NELHA) runs its own deep-source cooling plant to cool buildings on the agency's campus. The plant draws 42.8°F (6°C) seawater from a depth of 610 m. "NELHA saves about [U.S.] 3000 dollars a month in electrical costs by using the cold seawater air-conditioning process," said Jan War, an operations manager. "We still use a freshwater loop to cool our buildings, since seawater is so corrosive." So far deep-source cooling is only practical for communities with numerous buildings located near large bodies of water. But many of the world's major cities, settled during the golden age of sailing ships, are close to shore—something to think about the next time a dip in the ocean takes your breath away.

Article obtained from: Smith, J. The AC of tomorrow? Tapping deep water for cooling. *National Geographic News.* September 10, 2004.

17. What can be inferred from this passage?
    - **A.** Using seawater will only work in cities along the Pacific Ocean.
    - **B.** Using seawater to cool buildings is a good option, but the process needs more study and improvement.
    - **C.** Cooling with seawater is an expensive project.
    - **D.** Cooling with seawater will eventually take over cooling with freshwater all over.

18. Why would this new technology benefit many of the world's major cities?
    - **A.** Most cities have many buildings close to the shore.
    - **B.** Cities have the money to fund the new technology.
    - **C.** Cities are located along the Pacific Ocean.
    - **D.** Cities are not close to freshwater.

19. Discuss two reasons that a small farming town in Kansas would not benefit from this saltwater technology.

NEED EXTRA HELP?

If You Missed Question . . .	1	2	3	4	5	6	7	8	9	10	11	12	13	14	15	16
Review Section . . .	14.4	15.3	15.3	14.2	8.3	13.2	12.2	12.2	15.2	13.3	11.1	11.1	11.1	12.4	11.1	12.3

SHORT ANSWER

11. Solar radiation is entering Earth's atmosphere where two things are occurring: some of the radiation is being reflected back into space while the rest is being absorbed by Earth.

12. The dotted line represents the upper layer of the atmosphere. It helps protect Earth from getting too hot by trapping and reflecting some of the radiation back to space.

13. The Sun is the source of almost all of the energy in the atmosphere, and this energy can be converted into different forms of energy. It also is responsible for heating Earth's surface, which makes it possible for life to exist on Earth.

14. One-to-three day forecasts are able to look at large surface and upper-level features to determine whether the weather will be warm or cold and whether precipitation will occur. However, they cannot pinpoint exact temperatures or sky conditions for a specific time.

15. Land absorbs energy more quickly than water, so it heats up and cools down faster.

16. Water vapor imagery shows moisture that is present in the atmosphere and not just cloud patterns. By looking at the water vapor, meteorologists can track developments and changes in storm systems even when there are no clouds present.

READING FOR COMPREHENSION

17. B

18. A

19. Sample answer: Kansas is in the middle of the United States and therefore is not located close to any saltwater oceans, which would make it hard to get salt water to use. Also, because it is a small town, there would not be enough buildings to make it practical to use salt water.

CHAPTER 16 The Marine Environment

BIGIDEA **The marine environment is geologically diverse and contains a wealth of natural resources.**

ESSENTIAL QUESTIONS	NATIONAL STANDARDS	RESOURCES TO ASSESS MASTERY
SECTION 1 **Shoreline Features** **1.** How are shoreline features formed and modified by marine processes? **2.** What are the major erosional and depositional shoreline features? **3.** What types of protective structures are used near shore? 3 sessions 1.5 blocks	B.4; D.1–3; F.5	**Entry-Level Assessment** Focus Transparency 47, p. 438 **Progress Monitoring** Caption Question, pp. 438, 440 Reading Check, pp. 441, 442 Section Review, p. 446
SECTION 2 **Seafloor Features** **1.** What are the major geologic features of continental margins? **2.** What are the major geologic features of ocean basins? **3.** How are the different types of marine sediments classified and described? 5 sessions 2.5 blocks	UCP.2–3; A.1–2; C.4, C.6; D.1–3	**Entry-Level Assessment** Focus Transparency 48, p. 447 **Progress Monitoring** Caption Question, p. 448 Reading Check, pp. 448, 449, 452 Section Review, p. 454 **Summative Assessment** Chapter Assessment, p. 459 Exam*View®Assessment Suite* CD-ROM

Go online!
connectED.mcgraw-hill.com

LEVELED RESOURCES	LAB MATERIALS
Science Notebook 16.1 OL **Chapter FAST FILE Resources:** GeoLab Worksheet, p. 137 OL Study Guide, p. 145 BL **Lab Resources:** Laboratory Manual, p. 121 OL **Transparencies:** Section Focus Transparency 47 EL Teaching Transparency 45 OL EL	LaunchLAB p. 436 / **10 min** mortar and pestle, natural chalk, microscope slide, microscope GeoLAB p. 456 / **45 min** metric ruler, drafting compass, graph paper, calculator
Science Notebook 16. 2 OL **Chapter FAST FILE Resources:** MiniLab Worksheet, p. 136 OL Study Guide, p. 148 BL **Lab Resources:** Laboratory Manual, p. 125 OL **Transparencies:** Section Focus Transparency 48 EL Teaching Transparency 46 OL EL	MiniLAB p. 453 / **30 min** unsorted sediment grains, set of sieves, 250-mL graduated cylinder, cooking oil, stopwatch

LabManager™
Customize any lab with the LabManager™ CD-ROM.

ADDITIONAL RESOURCES

Technology:

- eTeacherEdition
- eStudentEdition
- Classroom Presentation Toolkit
- Exam*View® Assessment Suite* CD-ROM
- LabManager™ CD-ROM
- What's EARTH SCIENCE Got To Do With It? DVD
- Weather Classroom Video DVD

Assessment:

- Blueprints for Success
- Chapter Test (Scaffolded)
- Earth Science Standardized Test Practice

Lab Resources:

- Exploring Environmental Problems Laboratory Manual

BL Below Level   OL On Level   AL Advanced Learners   EL English Learners   COOP LEARN Cooperative Learning

CHAPTER 16

LaunchLAB

iLab Station Rubric

Where does chalk form?

Process Skills observe and infer, analyze, draw a conclusion, think critically, communicate

Safety Precaution Approve lab safety forms before work begins.

Teaching Strategies

- Divide the class into several groups. Provide a microscope, chalk, mortar and pestle, and glass slides for each group. Make sure every student participates.
- Classroom chalk might be synthetic; be sure to use natural chalk.

Procedure

1. Have students read and complete the lab safety form and follow the procedure below.
2. Use a **mortar and pestle** to grind a small piece of **natural chalk** into a powder. Make a **slide** of the powdered chalk.
3. Observe the chalk powder through a **microscope.**

Analysis

1. **Describe** the powder. Are the grains irregular in shape or size? Do some of the grains have patterns? Grain size varies but the grains have regular shapes and surface patterns.
2. **Analyze** your data and hypothesize the origin of the chalk. On what evidence do you base your conclusion? Possible hypothesis: The chalk is derived from sedimentary deposits of organic matter. There are remnants of organic matter in the chalk dust.

Assessment

Knowledge Ask students to identify the specific type of rock that makes up chalk. limestone Ask them what the chemical composition of chalk is. calcium carbonate

CHAPTER 16

The Marine Environment

BIGIDEA The marine environment is geologically diverse and contains a wealth of natural resources.

SECTIONS

1 Shoreline Features

2 Seafloor Features

LaunchLAB

iLab Station

Where does chalk form?

Although you might not live near a coast, parts of your environment were shaped by the ocean. For example, you might be just a few meters away from former seafloor deposits that are now part of the bedrock underground. One such seafloor deposit is chalk. How can you tell that chalk formed on the seafloor? Investigate the origin of chalk in this lab.

FOLDABLES® Study Organizer

Seafloor Features

Make a three-tab book using the labels shown. Use it to organize your notes on seafloor features.

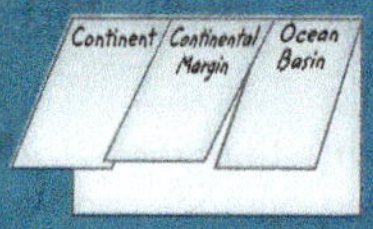

LabManager™
Customize any lab with the LabManager™ CD-ROM.

England's famous White Cliffs of Dover are the remnants of tiny marine organisms composed of $CaCO_3$ that, after death, settled to the seafloor. Over time, this accumulation created layer upon layer of chalk that was later lifted out of the water by geological processes.

Introduce the BIG IDEA

The Ocean Floor Have students make a list of the different types of features found on the ocean floor. Have them explain how the ocean floor is geologically diverse. The ocean floor has underwater mountain ranges, volcanoes, slopes, deep trenches with steep sides, and extremely flat areas.

Teacher Content Support

Habitats in the Ocean Explain that the marine environment has many habitats. Ask students to identify and describe some different habitats in the ocean. Possible answers: Coral reefs are primarily found in shallow, clear, warm water. They support a variety of other organisms as part of the food chain. The deep ocean has no light, the water is cold and pressure is high. Hydrothermal vents have very hot water with potentially toxic metals spewing out of them. Chemosynthetic bacteria make up the base of the food chain in a hydrothermal vent.

SECTION 1

1 Focus

Focus Transparency

Before presenting the lesson, project **Section Focus Transparency 47** and have students answer the accompanying questions. EL

MAINIDEA

The Shore Show students photos that show the features of a coast, including sandy beaches, rocky coastlines, or harbors. Have students describe the depositional or erosional coastal features in the photos. Ask students to discuss the processes that shape the shore.

2 Teach

Identify Misconceptions

Some students might think all beaches consist of sand.

Uncover the Misconception Have students describe the sediment on beaches they have visited or seen in photos. Show students photos of beaches with different-sized sediments, such as pebbles or cobbles.

Demonstrate the Concept Using the photos as examples, explain that beaches consist of various types of loose sediment, depending on the local sediment source and the strength of the waves. Beaches can consist of sand, pebbles, gravel, or cobbles.

Assess New Knowledge Have students find photos of beaches and identify the type of sediment that makes up the beach. BL OL AL EL COOP LEARN

Caption Question Fig. 1 The beach is the part of the shore where sediment is deposited.

SECTION 1

Shoreline Features

MAINIDEA The constant erosion of the shoreline and deposition of sediments by ocean waves creates a changing coastline.

Essential Questions

- How are shoreline features formed and modified by marine processes?
- What are the major erosional and depositional shoreline features?
- What types of protective structures are used near shore?

Review Vocabulary

breaker: collapsing wave that forms when a wave enters shallow water

New Vocabulary

beach
wave refraction
longshore bar
longshore current
barrier island

EARTH SCIENCE 4 YOU Have you ever picked up a smoothly polished pebble on the beach? A sculptor forms a masterpiece out of a chunk of wood or stone by chipping away one tiny bit at a time. Waves are nature's chisel, and beaches display the carefully carved features.

The Shore

Shown in **Figure 1,** the shore is the area of land between the lowest water level at low tide and the highest area of land that is affected by storm waves. Shores are places of continuous, often dramatic geologic activity—places where you can see geologic changes occurring almost daily. The shoreline is the place where the ocean meets the land. Shorelines are shaped by the action of waves, tides, and currents. The location of the shoreline constantly changes as the tide moves in and out. As waves erode some shorelines, they create some of the most impressive rock formations on Earth. In other areas, waves deposit loose material and build wide, sandy beaches.

Beaches Long stretches of U. S. coasts are lined with wide, sandy beaches. A **beach,** shown in **Figure 1,** is the area in which sediment is deposited along the shore. Beaches are composed of loose sediments deposited and moved about by waves along the shoreline. The size of sediment particles depends on the energy of the waves striking the coast and on the source of the sediment. Beaches pounded by large waves or formed on rocky coasts usually consist of coarse materials such as pebbles and cobbles.

Figure 1 The location of the shoreline changes as the tide moves in and out.
Identify *How is a beach related to a shore?*

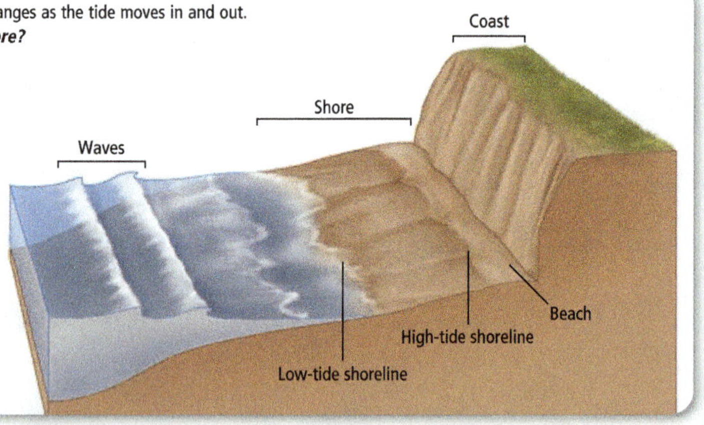

Demonstration

Beach Sediment Beach material is usually well-sorted because of the action of the breaking waves. Place a few cubic centimeters of a sand and mud mixture in a beaker filled halfway with water. Hold the beaker under a faucet and turn on a small stream of water. Move the beaker around so the water stream stirs all the sediment on the bottom. Let the beaker overflow until the water is clear. Turn off the faucet. Show the class that the remaining sediment is a mixture of fine and coarse sand, and explain that the turbulence flushed out all the smaller grains. Repeat the experiment with a stronger stream of water. Fine sand is flushed out and only coarse sand and pebbles should remain. Explain that the more turbulent the waves, the larger the grain size of beach sediment.

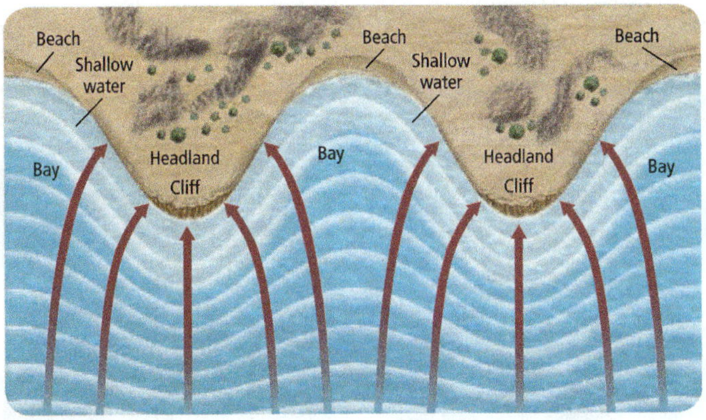

■ **Figure 2** Wave crests advance toward the shoreline and slow down when they encounter shallow water. This causes the wave crests to bend toward the headlands and move in the direction of the arrows.

Formation of Shoreline Features

Large breaking waves can hurl thousands of metric tons of water, along with suspended rock fragments, against a shore with such force that they are capable of eroding solid rock.

Erosional features Waves move faster in deep water than in shallow water. This difference in wave speed causes initially straight wave crests to bend when part of the crest moves into shallow water, a process known as **wave refraction,** illustrated in **Figure 2.** Along an irregular coast with headlands and bays, the wave crests bend toward the headlands. As a result, most of the breaker energy is concentrated along the relatively short section of the shore around the tips of the rocky headlands, while the remaining wave energy is spread out along the much longer shoreline of the bays. The headlands thus undergo severe erosion. The material eroded from the headlands is swept into the bays, where it is deposited to form crescent-shaped beaches. The headlands are worn back and the bays are filled in until the shoreline straightens.

Wave-cut platforms Many headlands have spectacular rock formations. Generally, as a headland is worn away, a flat erosional surface called a wave-cut platform is formed. The wave-cut platform terminates against a steep wave-cut cliff, as shown in **Figure 3.**

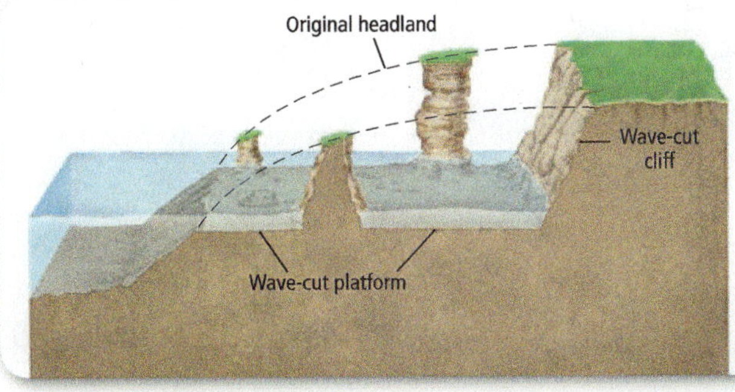

■ **Figure 3** A headland can be modified by wave erosion. The dotted lines indicate the original shape of the headland.

Activity

Wave Refraction Have about eight student volunteers stand in a straight line (6 to 7 m long), at a distance from a wall so they make an angle of about 30° with the wall. Inform the class these students represent a wave crest approaching a straight shoreline.

Tell the volunteers to link elbows, slowly walk toward the wall, and slow down when they get to within 0.25 m of the wall. When two students have reached the wall, say "Stop!" The line of students should then be bent toward the wall.

Ask students what caused the bend in the line. wave refraction Point out that wave crests approaching a shoreline bend in a similar way. Ask what would happen to a similar line of students if they encountered an obstacle (a headland) in the middle of the line. Students would end up clustered around the obstacle.

DIFFERENTIATED INSTRUCTION

Visually Impaired Provide students with samples of the three major types of sediments: sand, gravel, and mud. Allow students to handle and feel the samples to get a better understanding of the different particle sizes and consistencies. Also allow them to investigate samples of sorted and unsorted sediments. Ask students to give verbal reports to the rest of the class about the feel and texture of each type of sediment.

Teacher Content Support

Longshore Transport Moving water transports sediments. The largest particle size that can be transported by a current is called its competency, and the maximum amount of material that a current can move is its capacity. Most beaches consist of loose sediments spread from their sources along the shore by longshore transport. The strength of the longshore current depends on the size of the waves driving it. Large waves are associated with fast-moving longshore currents and effective longshore transport, which is characterized by high capacity and high competency. The resulting beaches are relatively coarse-grained.

Another factor that affects the grain size of beaches is transport distance. Beaches close to rocky source areas tend to have larger particles than those farther away. Beaches with fine sand are generally far from rocky source areas and experience relatively small waves. No matter their grain size, beach sediments are usually sorted well by the action of the waves.

Caption Question Fig. 5 A longshore bar is a sandbar over which shallow water lies. A longshore trough is an area of deeper water between the shore and a longshore bar.

Figure 4 The sea stack and sea arch shown here, were formed by wave refraction at a rocky headland along the coast of France.

Sea stacks Differential erosion, the removal of weaker rocks or rocks near sea level, produces many of the other characteristic landforms of rocky headlands. As shown in **Figure 4,** a sea stack is an isolated rock tower or similar erosional remnant left on a wave-cut platform. A sea arch, also shown in **Figure 4,** is formed as stronger rocks are undercut by wave erosion. Sea caves are tube-like passages blasted into the headlands at sea level by the constant assault of the breakers.

Longshore currents Suppose you stood on a beach at the edge of the water and began to walk out into the ocean. As you walked, the water might get deeper for a while, but then it would become shallow again. The shallow water offshore lies above a sandbar, called a **longshore bar,** that forms in front of most beaches, as illustrated in **Figure 5.** Waves break on the longshore bar in the area known as the surf zone. The deeper water between the shore and longshore bar is called the longshore trough.

The waves striking the beach are almost parallel to the shoreline, although the waves seaward of the longshore bar are generally not parallel to the shore. This is another case of wave refraction. The slowing of the waves in shallow water causes the wave crests to bend toward the shore.

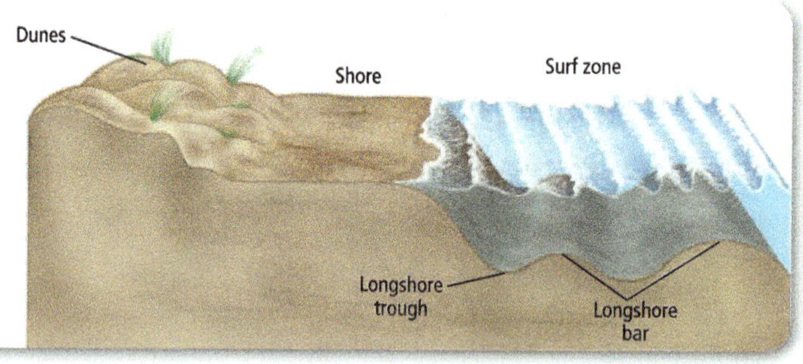

Figure 5 A typical beach profile includes longshore troughs and bars within the surf zone.
Explain *What is the difference between a longshore trough and a longshore bar?*

ACROSS THE CURRICULUM

Geography Macquarie Island, located about 1500 km southeast of Australia in the Southern Ocean, was designated as a World Heritage Site in 1997. Part of the reason it was chosen as a World Heritage Site was due to its geology. The coast of the island has been shaped by the large, wind-driven waves of the Southern Ocean. The result is a coast that consists of a wave-cut platform with steep wave-cut cliffs and sea stacks. Birds, including penguins and albatross, nest on the platform and high cliffs. The island serves as the breeding ground for royal penguins, with over 850,000 pairs of birds living on the platform during breeding season.

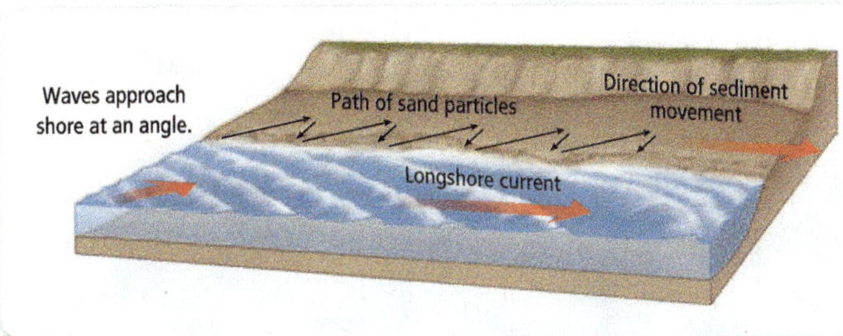

Figure 6 Longshore currents are driven by incoming waves.

View an **animation of longshore currents.**

Concepts In Motion

As water from incoming breakers spills over the longshore bar, a current flowing parallel to the shore, called the **longshore current,** is produced. This current varies in strength and direction from day to day. Over the course of a year, because of prevailing winds and wave patterns, one direction usually dominates.

Movement of sediments Longshore currents, like the one illustrated in **Figure 6,** move large amounts of sediment along the shore. Fine-grained material, such as sand, is suspended in the turbulent, moving water, and larger particles are pushed along the bottom by the current. Additional sediment is moved back and forth on the beach by incoming and retreating waves. Incoming waves also move sediment at an angle to the shoreline in the direction of wave motion. Overall, the transport of sediment is in the direction of the longshore current. On both the Atlantic and Pacific coasts of the United States, longshore transport moves generally toward the south.

READING CHECK **Explain** how the longshore current moves sediments.

Rip currents Wave action also produces rip currents, which flow out to sea through gaps in the longshore bar. Rip currents, like the one shown in **Figure 7,** return the water spilled into the longshore trough to the ocean. These dangerous currents can reach speeds of several kilometers per hour. If you are ever caught in a rip current, you should not try to swim against it, instead swim parallel to shore to get out of it.

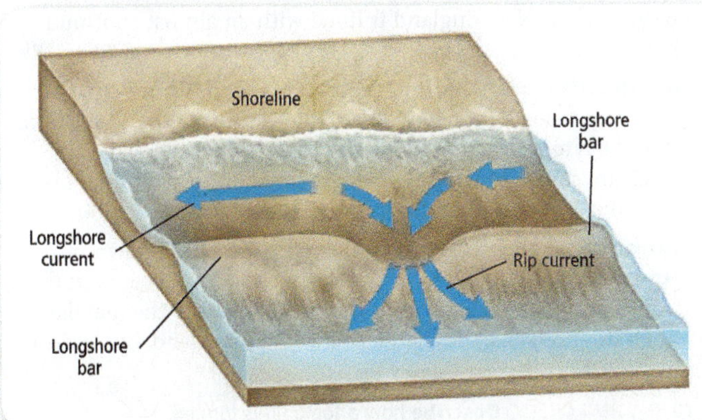

Figure 7 Rip currents return water through gaps in the longshore bar out to sea. Rip currents spread out and weaken beyond the longshore bar.

Concept Development

Undertows Rip currents are related to, but not the same as, undertows. Like rip currents, undertows return water driven against the shore by breakers back to the sea. However, an undertow is a broad, sheetlike, bottom-hugging return flow. Strong undertows are capable of pushing the legs out from under persons standing or walking in the surf zone. Undertows are most dangerous when large breakers come in at right angles to the shore.

Tie to Previous Knowledge

Rip Currents Many beaches experience rip currents and undertows at certain times of the year and under particular weather conditions. Ask whether any students have experienced rip currents or strong undertows at a beach. If any students have had such experiences, ask them to describe the strength of the currents. If no students have had these experiences, ask whether any students have stood in the surf and observed as the sand underneath their feet was moved away by wave action. Explain this movement is similar to that of an undertow.

READING CHECK Fine-grained sediment is suspended in the water and larger grains are pushed along the bottom.

EARTH SCIENCE JOURNAL

Water Safety Have students research more information about what a person should do if he or she gets caught in a rip current. Have students make a pamphlet that explains what a rip current is and the best safety practices to employ if caught in one.

Rubric

Interpret the Illustration

Depositional Landforms **Figure 8** shows several depositional landforms. Ask students what the direction of the longshore transport is and what some of the sediment sources are. from left to right; the truncated headlands

Reinforcement

Coastal Landforms On a map of the United States, point out a number of prominent coastal features and ask students what type of coastal landforms they represent and how they formed. Examples include Cape Ann, MA (headland); Montauk Point, NY (headland); Sandy Hook, NJ (spit); Barnegat Bay, NJ (lagoon); Ocracoke Island, NC (barrier island); Pamlico Sound, NC (lagoon); Palm Beach, FL (barrier island); and Padre Island, TX (barrier island). Some West Coast examples are San Diego Bay, CA (lagoon); Point Reyes, CA (spit); Cape Mendocino, CA (headland); and Humboldt Bay, CA (lagoon). All of the features except headlands are formed by longshore transport. Headlands are the result of local geology.

☑ READING CHECK A tombolo is formed when a ridge of sand is deposited between an island and the mainland, connecting the two.

GeoLAB

iLab Station

The GeoLab at the end of the chapter can be used at this point in the lesson.

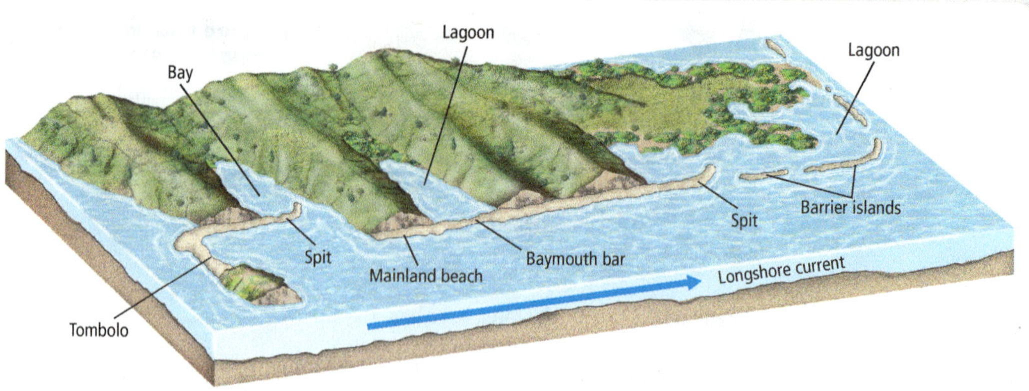

■ **Figure 8** Depositional features of coastlines include tombolos, spits, baymouth bars, lagoons, and barrier islands.

Depositional features As a result of wave erosion, longshore transport, and sediment deposition, most seashores are in a constant state of change. Sediments are eroded by large storm waves and deposited wherever waves and currents slow down. Sediments moved and deposited by longshore currents build various characteristic coastal landforms, such as spits, barrier islands, baymouth bars, and tombolos, illustrated in **Figure 8.**

Spit A narrow bank of sand that projects into the water from a bend in the coastline is called a spit. A spit, which forms where a shoreline changes direction, is protected from wave action. When a growing spit crosses a bay, a baymouth bar forms.

Barrier islands Long ridges of sand or other sediment that are deposited or shaped by the longshore current and are separated from the mainland are called **barrier islands.** Barrier islands, like the ones shown in **Figure 9,** can be several kilometers wide and tens of kilometers long. Most of the Gulf Coast and the eastern coast south of New England is lined with an almost continuous chain of barrier islands.

■ **Figure 9** This barrier island off the coast of North Carolina is one of many that line the eastern coast.

Baymouth bars A baymouth bar, shown in **Figure 10,** forms when a spit closes off a bay. The shallow, protected bodies of water behind baymouth bars and barrier islands are called lagoons, which are saltwater coastal lakes that are connected to the open sea by shallow, restricted outlets.

Tombolo Another coastal landform is a tombolo, shown in **Figure 10.** A tombolo is a ridge of sand that forms between the mainland and an island and connects the island to the mainland. When this happens, the island is no longer an island, but is the tip of a peninsula.

☑ READING CHECK **Describe** how a tombolo is formed.

Figure 10 Baymouth bars and tombolos are examples of features formed by the deposition of sediments.

Natural and human effects on the coast All of these depositional coastal landforms, including large barrier islands, are unstable and temporary. Occasionally, major storms sweep away entire sections of barrier islands and redeposit the material elsewhere. **Figure 11** shows the existence of South Gosier Island, a barrier island off the coast of Louisiana in August of 2004, a month before Hurricane Ivan passed over it. The island was completely destroyed by the strong waves generated by Hurricane Ivan. Even in the absence of storms, however, changing wave conditions can slowly erode beaches and rearrange entire shorelines. For example, the shoreline of Cape Cod, Massachusetts, is retreating by as much as 1 m per year.

People are drawn to coastal areas for their rich natural resources, mild climate, and recreation opportunities. Coastal areas in the United States make up only 17 percent of contiguous land areas, but are currently home to over half of the population in the nation. Increasing population has caused substantial coastal environmental changes, including pollution, shoreline erosion, and wetland and wildlife-habitat loss. These coastal changes in turn increase the susceptibility of the communities to natural hazards caused by hurricanes and tsunamis.

Figure 11 The island of South Gosier, a barrier island off the coast of Louisiana, was completely washed away by Hurricane Ivan in September, 2004.

Teacher Content Support

Beach Erosion A recent study released by the Federal Emergency Management Agency (FEMA) suggests that beach erosion is a significant threat on all U.S. coasts. According to the study, the worst beach erosion, approximately 2 m/y, occurs along the Gulf Coast. Beach erosion along the Atlantic Coast averages 1 m/y. The least amount of erosion occurs along the Pacific Coast and the shores of the Great Lakes, where it averages about 0.3 m/y or less. It is estimated that, nationwide, some 1500 coastal structures will be destroyed annually by the sea. If the current rise in sea level of about 15 cm per century continues, a total of 87,000 buildings, representing about one quarter of all beach houses, will be lost over the next 60 years. This figure could be significantly higher if the rise in sea level accelerates as a result of climate change.

EARTH SCIENCE JOURNAL

Despositional Features Have students write a summary in which they compare and contrast the main depositional features on a coast: spits, barrier islands, baymouth bars, and tombolos. BL OL

Rubric

Discussion

Protective Structures Breakwaters, jetties, and groins interfere with longshore transport and affect the width and shape of beaches. Discuss with students what effect, if any, a fishing pier built perpendicular to a beach would have on the beach. Some accumulation of sand would result from a reduction in wave energy by the pilings, but the effect would be small. Ask students what would be the effect of an off-shore wreck on the longshore bar. It would eventually create a tombolo connecting it to the shore.

Apply Earth Science

Sand Dunes Many shorelines have coastal sand dunes. Ask students to determine the origin of the sand and which way the dunes are moving: toward the sea, parallel to the shore, or inland. The sand in dunes is beach sand blown inland, and the dunes tend to move inland unless they are anchored by vegetation.

Figure 12 Seawalls deflect the energy of waves on the beach. Groins and jetties deprive downshore beaches of sand.

Protective Structures

In many coastal areas, protective structures such as seawalls, groins, jetties, and breakwaters are built in an attempt to prevent beach erosion and destruction of oceanfront properties. However, these artificial structures interfere with natural shoreline processes and can have unexpected negative effects.

Seawalls Structures called seawalls, shown in **Figure 12,** are built parallel to shore, often to protect beachfront properties from powerful storm waves. Seawalls reflect the energy of such waves back toward the beach, where they worsen beach erosion. Eventually, seawalls are undercut and have to be rebuilt larger and stronger than before.

Groins and jetties Groins, shown in **Figure 12,** are wall-like structures built into the water perpendicular to the shoreline for the purpose of trapping beach sand. Groins interrupt natural longshore transport and deprive beaches down the coast of sand. The result is aggravated beach erosion down the coast from groins. Similar effects are caused by jetties, which are walls of concrete built to protect a harbor entrance from drifting sand. Jetties are also shown in **Figure 12.**

Figure 13 A breakwater slows the movement of water against the coast, which results in sediment being deposited behind the structure.

Breakwaters Structures like the one shown in **Figure 13** that are built to provide anchorages for small boats or a calm beach area are called breakwaters. Breakwaters are built parallel to the shoreline. When the current slows down behind a breakwater and is no longer able to move sediment, it deposits sediment behind the breakwater. If the accumulating sediment is left alone, it can eventually fill an anchorage. To prevent this, anchorages have to be dredged regularly.

EARTH SCIENCE JOURNAL

Compare Coasts Have students study maps of the coasts of the United States. Ask students to compare and contrast the coastal features on the East and West coasts. Ask students the following questions: What are the major differences between the two coasts? Which coast is emergent? Have students list and describe these features in their Earth science journals. OL

Rubric

Changes in Sea Level

At the height of the last ice age, approximately 20,000 years ago, the global sea level was about 130 m lower than it is at present. Since that time, the melting of most of the ice-age glaciers has raised the ocean to its present level. In the last 100 years, the global sea level has risen about 17 cm. It continues to rise slowly; estimates suggest a rise in sea level of 3 mm/year.

Many scientists contend that this continuing rise in sea level is the result of global warming. During the last century, Earth's average surface temperature has increased by approximately 0.6°C. As Earth's surface temperature rises, seawater warms up and expands, which adds to the total volume of the seas. In addition, higher temperatures on Earth's surface cause glaciers to melt, and the meltwater flowing into the oceans increases their volume.

Effects of sea level changes If Earth's remaining polar ice sheets in Greenland and Antarctica melted completely, their meltwater would raise sea level by another 70 m. This rise would completely flood some countries, such as the Netherlands, along with some coastal cities in the United States, such as New York City, and low-lying states, such as Florida and Louisiana. Measurements from NASA's *GRACE* satellite, launched in 2002, have indicated that the Greenland ice sheet is melting at an accelerating rate. The West Antarctic Ice Sheet has also been thinning according to a decade of satellite measurements. A complete melt of the Greenland Ice Sheet would lead to a sea level rise of about 6 to 7 m.

Many of the barrier islands of the Atlantic and Gulf Coasts might be former coastal dunes that were drowned by rising sea levels. Other features produced by rising sea levels are the fjords of Norway, shown in **Figure 14.** Fjords (fee ORDZ) are deep coastal valleys that were scooped out by glaciers during the ice age and later flooded when the sea level rose.

VOCABULARY

ACADEMIC VOCABULARY

Estimate

a rough or approximate calculation

The estimate for the new roof was $3000.

Ian White

■ **Figure 14** Fjords are flooded U-shaped valleys that can be up to 1200 m deep.

Identify Misconceptions

Many people believe that marine fossils found above current sea level are evidence of a large-scale flood in the past.

Uncover the Misconception Ask students how they would explain the presence of seashells in some of the highest mountain peaks, which are far above present and previously interglacial sea level.

Demonstrate the Concept Explain that tectonic forces can push seafloor sediments far above sea level when continents collide. Demonstrate with a model of clay between two blocks of wood being pushed together.

Assess New Knowledge Show students a map of Asia. Ask students whether there might be ancient seashells in some high peaks of the Himalayas and, if so, how they might have gotten there.

ACROSS THE CURRICULUM

Geography Have students copy a topographic map of the United States and indicate on the map where the shoreline would be if sea level rose 70 m. Have them estimate how far away the nearest dry land would be from Charleston, 100 km New Orleans, 100 km Miami, 800 km and other coastal cities. OL EL

3 Assess

Check for Understanding

Apply The Great Lakes experience coastal processes similar to those the seashores experience. The prevailing winds on Lake Erie are from the west. Have students study a map of the Great Lakes and answer the following questions: What is the direction of the longshore current along Lake Erie's north and south shores? west to east What is the direction of the breakers striking the United States shore when there is a west wind? from the northwest as a result of wave refraction Name an example of a spit on Lake Erie's shore. Presque Isle, Point Pele, Long Point

Reteach

Explain Have students explain how wave refraction and longshore currents are related. Wave refraction turns incoming waves toward the shore. The water spilled by breakers at an angle to the shore produces a longshore current parallel to the shore.

Assessment

Knowledge Have students look for evidence of longshore transport along New York's Long Island shoreline and identify the pertinent coastal features. spits and barrier islands built by east-to-west longshore transport

Figure 15 Elevated marine terraces are former wave-cut platforms that are now well above the current sea level. This elevated marine terrace is in New Zealand.

Effects of tectonic forces Other processes that affect local sea levels are tectonic uplift and sinking. If a coastline sinks, there is a relative rise in sea level along that coast. A rising coastline, however, produces a relative drop in the local sea level. As a result of tectonic forces in the western United States, much of the West Coast is being pushed up much more quickly than the sea level is rising. Because much of the West Coast was formerly under water, it is called an emergent coast.

Emergent coasts tend to be relatively straight because the exposed seafloor topography is smoother than typical land surfaces with hills and valleys. Other signs of an emergent coast are former shoreline features such as sandy beach ridges located far inland, or fossils of marine organisms within the uplifted rocks. Among the most interesting of these features are elevated marine terraces—former wave-cut platforms that are now dry and well above current sea level. **Figure 15** shows a striking example of such a platform. Some old wave-cut platforms are hundreds of meters above current sea level.

SECTION 1 REVIEW

Section Self-Check

Section Summary

- Wave erosion of headlands produces wave-cut platforms and cliffs, sea stacks, sea arches, and sea caves.
- Wave action and longshore currents move sediment along the shore and build depositional features.
- Artificial protective structures interfere with longshore transport.
- Sea levels in the past were 130 m lower than at present.

Understand Main Ideas

1. **MAINIDEA** Shorelines have headlands and bays. Which experiences the most severe erosion by breakers? Why?
2. **Describe** What are sea stacks, and how are they formed?
3. **Apply** How do jetties and groins affect the longshore current?
4. **Analyze** What effect does a seawall have on a beach?

Think Critically

5. **Evaluate** why resort communities built on barrier islands spend thousands of dollars each year to add sand to the beaches along the shoreline.

MATH IN Earth Science

6. The city of Orlando, Florida, is 32 m above sea level. If the sea level continues to rise at the estimated rate of 3 mm/y, in how many years might Orlando be under water?

SECTION 1 REVIEW

1. Headlands experience the most severe erosion because wave refraction bends breakers toward the headlands.
2. Sea stacks are rocky, towerlike erosional remnants of former headlands left on the wave-cut platform as a result of erosion.
3. They interrupt the longshore current and deprive beaches down the coast of sand. This leads to beach erosion down the coast from groins and jetties.
4. Seawalls cause beach erosion by reflecting storm waves back onto the beach.
5. Barrier islands continually change shape in response to changing wind and wave conditions. Most barrier islands also undergo beach erosion because of rising sea levels.
6. 10,666 y; 32 m × 1000 mm/m = 32,000 mm; 32,000 mm ÷ 3 mm/y = 10,666 y

SECTION 2

Essential Questions

- What are the major geologic features of continental margins?
- What are the major geologic features of ocean basins?
- How are the different types of marine sediments classified and described?

Review Vocabulary

sediment: solid particles deposited on Earth's surface that can form sedimentary rocks by processes such as weathering, erosion, deposition, and lithification

New Vocabulary

continental margin
continental shelf
continental slope
turbidity current
continental rise
abyssal plain
deep-sea trench
mid-ocean ridge
seamount
guyot

Seafloor Features

MAINIDEA The ocean floor contains features similar to those on land and is covered with sediments of several origins.

EARTH SCIENCE 4 YOU Why do you shake a container of orange juice before pouring a glass? The juice appears thicker at the bottom of the container because the pulp sinks to the bottom. Similarly, gravity causes grains of silt and sand to settle to the bottom of the ocean.

The Continental Margin

If you were asked to draw a map of the seafloor, what kind of topographic features would you include? Until recently, most people had little knowledge of the features of the ocean floor. However, modern oceanographic techniques, including sonar and satellite data, reveal that the topography of the ocean bottom is as varied as that of the continents.

The topography of the seafloor is surprisingly rough and irregular, with numerous high mountains and deep depressions. The deepest place on the seafloor, the Mariana Trench in the Pacific Ocean, is about 11 km deep.

Study **Figure 16.** Notice that the **continental margin** is the area where edges of continents meet the ocean. It consists of continental crust, covered with sediments, that eventually meets oceanic crust. Continental margins represent the shallowest parts of the ocean. As shown in **Figure 16,** a continental margin includes the continental shelf, the continental slope, and the continental rise.

Continental shelf The shallowest part of a continental margin extending seaward from the shore is the **continental shelf.** Continental shelves vary greatly in width, averaging 60 km wide. On the Pacific coast of the United States, the continental shelf is only a few kilometers wide, whereas the continental shelf of the Atlantic coast is hundreds of kilometers wide.

■ **Figure 16** A cross section of the ocean reveals a diverse topography with many features that are similar to those on land.

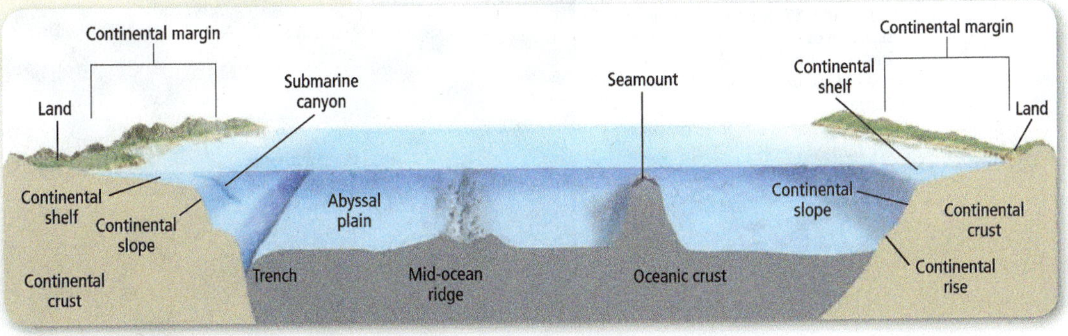

Demonstration

Seafloor Features Show students a global map of seafloor features. Point out the continental margins. Also point out deep-sea trenches, mid-ocean ridges, fracture zones, and other seafloor features, including the Emperor Seamount chain. Explain that much of the ocean floor is rugged, but different from continental landscapes.

SECTION 2

1 Focus

Focus Transparency

Before presenting the lesson, project **Section Focus Transparency 48** and have students answer the accompanying questions. EL

MAINIDEA

The Seafloor Have students describe the appearance of the seafloor. Divide students into a few groups. Give each group a flat global map and some modeling dough with two different colors. Ask students to use modeling clay to model the features of ocean basins. Have students add features such as trenches, seamounts, and mid-ocean ridges.

2 Teach

Teacher Content Support

Earth's Crust Crustal elevations are high where the crust is thick because Earth's crust has a lower density than the underlying mantle rocks. Crustal rocks have densities averaging around 2.8 g/cm^3; mantle rocks have densities generally greater than 3.3 g/cm^3. Earth's crust floats on Earth's mantle in a way similar to how ice floats on water. The density of ice is about 0.9 g/cm^3, and that of water is 1.0 g/cm^3. As a result, floating ice has a freeboard approximately equal to one-tenth of its thickness. A 1-m thick ice raft has a freeboard of about 10 cm, a 2-m thick ice raft has a 20-cm freeboard, and so on. In other words, the thicker the ice raft, the higher its elevation. For the same reason, crustal elevations are roughly proportional to crustal thickness. Thus, thin oceanic crust has the lowest elevation (the lowest freeboard above the mantle) and forms the ocean basins.

Concept Development

Continental Margins Explain to students that the eastern coast of the United States and other Atlantic coasts have passive continental margins. The margin of the eastern coast of the United States formed when the supercontinent Pangaea broke apart approximately 180 mya. The initial stage of this breakup was a continental rift, characterized by normal faults on both sides of the rift valley. These ancient faults are still present underneath a thick cover of sediments, now mostly converted to sedimentary rocks, that have accumulated since the breakup. Sediment thicknesses are on the order of 10 km on top of both the continental shelf and the continental rise. Note that the other half of the ancient rift is now on the other side of the Atlantic Ocean.

Teacher Content Support

Turbidity Currents Ocean waves move the water column to a depth equal to one-half the wavelength. This depth is called the wave base. Large storm waves and swells have wavelengths of several hundred meters with a wave base of well over 100 m. Because the maximum depth of the continental shelves is 130 m, such storm waves or swells can move and stir up the sediment of the entire continental shelf and set in motion a turbidity current.

READING CHECK fish, oil, and natural gas

Caption Question Fig. 17 Submarine canyons are formed when turbidity currents erode bedrock and bottom sediments.

VOCABULARY

SCIENCE USAGE V. COMMON USAGE

Shelf

Science usage: the sloping border of a continent or island

Common usage: a flat piece of material attached to a wall on which objects are placed

The average depth of the water above continental shelves is about 130 m. Recall that sea level during the last ice age was approximately 130 m lower than at present; therefore, portions of the world's continental shelves were above sea level at that time. As a result, present-day coastlines are radically different from the way they were during the last ice age. At that time, Siberia was attached to North America by the Bering land bridge, Great Britain was attached to Europe, and a large landmass existed where today there are only the widely scattered islands of the Bahamas.

When Earth's surface began to warm after the last ice age, and the continental ice sheets began to melt, the sea gradually covered up the continental shelves. Beaches and other coastal landforms from that time are now submerged and located far beyond the present shoreline. Commercially valuable fishes now inhabit the shallow waters of the continental shelves. In addition, the thick sedimentary deposits on the shelves are significant sources of oil and natural gas.

READING CHECK **List** the resources that can be found on the continental shelf.

Continental slope Beyond the continental shelves, the seafloor drops away quickly to depths of several kilometers, with slopes averaging nearly 100 m/km. These sloping regions are the **continental slopes.** To marine geologists, the continental slope is the true edge of a continent because it generally marks the edge of the continental crust. In many places, this slope is cut by deep submarine canyons, which are shown in **Figure 17.** Submarine canyons are similar to canyons on land and some are comparable in size to the Grand Canyon in Arizona.

These submarine canyons were cut by **turbidity currents,** which are rapidly flowing water currents along the bottom of the sea that carry heavy loads of sediments, similar to mudflows on land. Turbidity currents, shown in **Figure 18,** might originate as underwater landslides on the continental slope that are triggered by earthquakes, or they might originate from sediment stirred up by large storm waves on the continental shelf. Turbidity currents can reach speeds exceeding 30 km/h and effectively erode bottom sediments and bedrock.

Figure 17 Submarine canyons are similar to canyons on land. These deep cuts in the continental slope vary in size and can be as deep as the Grand Canyon.

Explain *how submarine canyons are formed.*

ACROSS THE CURRICULUM

Math Many scientific investigations involve conversions of units. Conversions are made simple by using ratios of quantities of equal magnitude. For instance, 1 m = 100 cm; therefore, 1 m/100 cm = 1. Similarly, 3.28 ft/1 m = 1, and 60 min/1 h = 1. Unit conversion is accomplished by multiplying the original unit by suitable ratios of units, such that the original unit is canceled. For example, to convert feet per second to kilometers per hour, use the following ratios: 1 ft/s = 1 ft/s × (1 m/3.28 ft) × (1 km/1000 m) × (3600 s/1 h). Note: This is the same as 1ft/s × 1 × 1 × 1. Canceling units: 1 ft/s = 1/3.28/1000 × 3600 km/h, or 1 ft/s = 1.1 km/h. Have students convert km/h to m/s. 1 km/h = 1 km/h × (1 h/60 min) × (1 min/60 s) × (1000 m/km) = 0.28 m/s

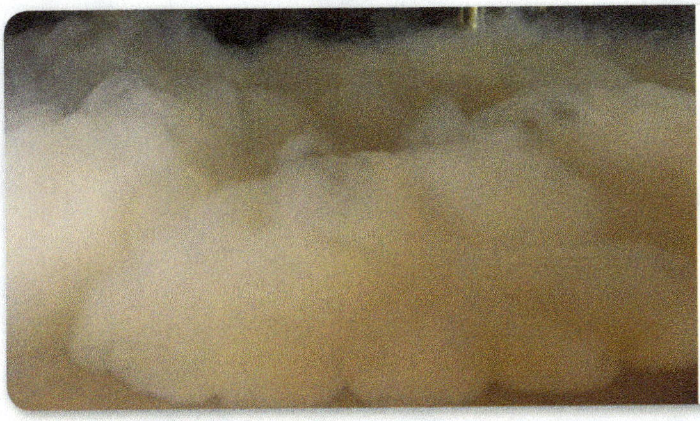

■ **Figure 18** Turbidity currents flow along the seafloor because the seawater-sediment mixture of the current is denser than seawater alone. Scientists study turbidity currents in the lab by simulating them in glass tanks, as shown.

Continental rise The gently sloping accumulation of deposits from turbidity currents that forms at the base of the continental slope is called a **continental rise.** A continental rise can be several kilometers thick. The rise gradually becomes thinner and eventually merges with the sediments of the seafloor beyond the continental margin. In some places, especially around the Pacific Ocean, the continental slope ends in deeper depressions in the seafloor, known as deep-sea trenches. In such places, there is no continental rise at the foot of the continental margin.

☑ READING CHECK **Differentiate** between the continental slope and the continental rise.

Problem-Solving LAB

Interpret Graphs

How do surface elevations compare? A useful comparison of the heights of the continents to the depths of the oceans is given by the curve in the graph below.

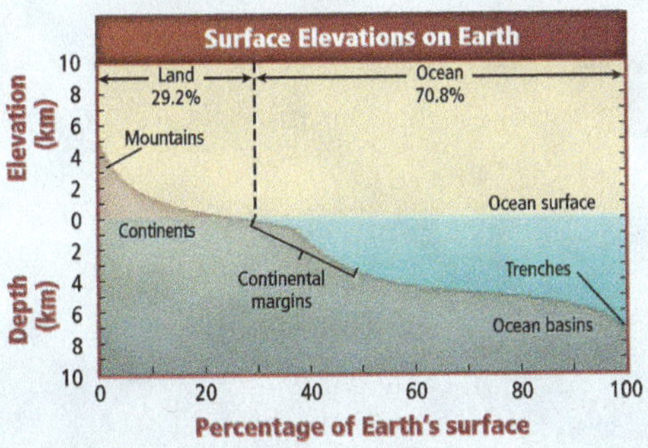

Analysis

1. Approximately how tall is the highest mountain on Earth's surface in kilometers?
2. At about what depth would you begin to find trenches on the ocean floor?
3. What percentage of Earth's surface is above current sea level?
4. What percentage of Earth's surface is represented by the continental margin?

Think Critically

5. **Calculate** The oceanic crust is the part of the crust that is at a depth of 2 km or more below sea level. What percentage of Earth's surface lies above the oceanic crust?

Jerome Neufeld/The Experimental Nonlinear Physics Group/The University of Toronto

Problem-Solving LAB

Purpose Students will interpret a hypsographic curve.

Process Skills use numbers, interpret data, predict

Teaching Strategies

- Ask students why Mount Everest is considered to be the highest mountain on Earth. because heights on Earth's surface are measured from sea level
- Please note that while Mount Everest and the Mariana Trench are almost 9 km and 11 km respectively, because they comprise such a small portion of the graph, they are not indicated.

Analysis

1. 5 km high
2. at about 6 km in depth
3. approximately 30 percent
4. about 15 percent

Think Critically

5. about 60 percent

☑ READING CHECK The continental slope is the region immediately beyond the continental shelf. The continental rise is at the base of the continental slope.

EARTH SCIENCE JOURNAL

Continental Margin Have students draw and label a profile of the continental margin. Labels should include land, continental shelf, continental slope, continental rise, submarine canyons, and turbidity currents.

Rubric

VISUALIZING

Purpose

Students will learn more about the features on the deep-sea floor.

Project

Ocean Basin Features Divide the class into several groups. Have each group construct a west-east cross section of oceanic features from New Jersey to Portugal, from Brazil to Africa, and along the equator from the Galapagos Islands to the western coast of South America. Have students use a vertical scale of 1:200,000 (1 cm = 2 km) and a suitable horizontal scale so that the cross section fits on a regular page. Students should label all pertinent topographic and bathymetric features. Have students compare and contrast the continental margins as well as the other features on each profile. Have a spokesperson for each group present the group's results to the class. **COOP LEARN**

VISUALIZING the Ocean Floor

Figure 19 The ocean floor has topographic features, including mid-ocean ridges, trenches, abyssal plains, and seamounts.

Library of Congress Geography and Map Division [G9096.C22 1977 .B4 Vault Rolled]

IN THE FIELD

Marie Tharp Marie Tharp was an American geologist and oceanic cartographer. She created maps of the ocean floor. Tharp became interested in this subject when little was known about the ocean floor and its geology. She earned a master's degree in geology from the University of Michigan in 1944. After graduation, she and her partner, Bruce Heezen, gathered data from sounding equipment about the seafloor. Tharp discovered a valley that divides the Mid-Atlantic Ridge and found that new seafloor was being formed at these ridges. This discovery confirmed the theory of seafloor spreading and led to the acceptance of plate tectonics, which students will learn more about in the next chapter. In 1978, Tharp and Heezen won the Hubbard medal.

Deep-Ocean Basins

Beyond the continental margin are ocean basins, which represent about 60 percent of Earth's surface and contain some of Earth's most interesting topography. **Figure 19** shows the topography of the ocean basin beneath the Atlantic Ocean.

Abyssal plains The flattest parts of the ocean floor 5 or 6 km below sea level are called **abyssal plains.** Abyssal plains, shown in **Figure 19,** are plains covered with hundreds of meters of fine-grained muddy sediments and sedimentary rocks that were deposited on top of basaltic volcanic rocks.

Deep-sea trenches The deepest parts of the ocean basins are the **deep-sea trenches,** which are elongated, sometimes arc-shaped depressions in the seafloor several kilometers deeper than the adjacent abyssal plains. Many deep-sea trenches lie next to chains of volcanic islands, such as the Aleutian Islands of Alaska, and most of them are located around the margins of the Pacific Ocean. Deep-sea trenches are relatively narrow, about 100 km wide, but they can extend for thousands of kilometers. The Peru-Chile trench, shown in **Figure 20,** is almost 6000 km long and has an average width of 40 km.

Mid-ocean ridges The most prominent features of the ocean basins are the **mid-ocean ridges,** which run through all the ocean basins and have a total length of more than 65,000 km—a distance greater than Earth's circumference. Mid-ocean ridges have an average height of 1500 m, but they can be thousands of kilometers wide. The highest peaks in mid-ocean ridges are over 6 km tall and emerge from the ocean as volcanic islands. Mid-ocean ridges are sites of frequent volcanic eruptions and earthquake activity. The crests of these ridges often have valleys called rifts running through their centers. Rifts can be up to 2 km deep.

Mid-ocean ridges do not form continuous lines. They are broken into a series of shorter, stepped sections, which run at right angles across each mid-ocean ridge. The areas where these breaks occur are called fracture zones, shown in **Figure 21.** Fracture zones are about 60 km wide, and they curve gently across the seafloor, sometimes for thousands of kilometers.

Figure 20 The Peru-Chile trench runs along the west coast of South America.

FOLDABLES® Incorporate information from this section into your Foldable.

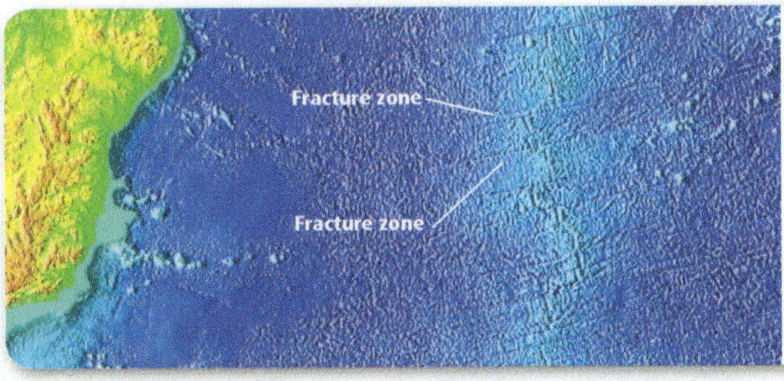

Figure 21 There are many fracture zones along the Mid-Atlantic Ridge. The fracture zones run perpendicular to the ridge.

Collaborative Learning

Deep-Ocean Basins Divide the class into three groups. Have each group select and study the major seafloor features (abyssal plains, deep-sea trenches, and mid-ocean ridges) of one of the three major oceans (the Atlantic, Pacific, and Indian oceans). Have students identify these features, list them by name, and describe them (approximate length, width, and height or depth, where feasible) in their Earth science journals. Have each group report its findings to the class. OL COOP LEARN

FOLDABLES® Rubric

ACROSS THE CURRICULUM

Math Many geological and oceanographic processes involve rates, such as the flow rate of a current, the rate of erosion, or the settling rate of sediment particles. A rate is the change in some quantity, for instance, distance per unit of time, which can be expressed as rate = distance/time. If the rate of a process is known, then distance or time can be calculated as follows: distance = rate × time, or time = distance ÷ rate. For example, if a longshore current moves at 500 m/h, how far will it move in a day? You can find the answer as follows: distance = 500 m/h × 24 h; distance = 12,000 m = 12 km. Ask students to calculate how long it will take a silt particle sinking at 20 cm/h to reach an abyssal plain 4000 m deep. time = 4000 m/0.2 m/h; time = 20,000 h = 833.3 days

Model

Marine Sediments Collect a variety of materials that could substitute for marine sediments, such as sugar for beach sand, brownie or cake mix for brown or red deep-sea mud, and broken walnuts, crushed granola cereal, or crushed hard candies for fossilized marine sediments. Have students draw a profile of the seafloor, from the continental shelf to the abyssal plains, on a piece of posterboard. Then, have students glue the substitute marine sediments on the posterboard at the places where they would expect to find each type of sediment. Ask them to label the seafloor features and the sediment type found there.

Teacher Content Support

Settling Velocity When particles settle through water, the sinking behavior of small grains (smaller than 0.1 mm in diameter) is dominated by the molecular viscosity of the water. Their sinking velocity increases with the square of their diameter. For larger grains (with diameters greater than 1 mm), sinking velocity increases only with the square root of their diameter.

☑ READING CHECK on mid-ocean ridges

Figure 22 Black smokers form when metal oxides and sulfides precipitate out of fluid heated by magma. White smokers form when elements such as calcium and barium precipitate out of warm water ejected from rifts in mid-ocean ridges.

Hydrothermal vents A hydrothermal vent is a hole in the seafloor through which fluid heated by magma erupts. Most hydrothermal vents are located along the bottom of the rifts in mid-ocean ridges. When the heated fluid that erupts from these vents contains metal oxides and sulfides, they immediately precipitate out of the fluid and produce thick, black, smoke-like plumes. This type of hydrothermal vent, known as a black smoker, ejects superheated water with temperatures of up to 350°C. **Figure 22** illustrates the black smokers found in the rift valley of a mid-ocean ridge.

A second type of vent, known as a white smoker, is also shown in **Figure 22.** White smokers eject cooler water than black smokers, and also get their color from the types of minerals that precipitate out of the hydrothermal fluids. Both types of hydrothermal vents are caused by seawater circulating through the hot crustal rocks in the centers of mid-ocean ridges.

☑ READING CHECK **Identify** where most hydrothermal vents are located.

Seamounts and guyots Satellite data have revealed that the ocean floor is dotted with tens of thousands of solitary mountains. These mountains are not located near areas of active volcanism. How, then, did they form? You have learned that the ocean basins are volcanically active at mid-ocean ridges and fracture zones. The almost total absence of earthquakes in most other areas of the seafloor suggests that volcanism in those areas must have ceased a long time ago. Thus, most of the mountains on the seafloor are probably extinct volcanoes.

Investigations of individual volcanoes on the seafloor have revealed that there are two types: seamounts and guyots (GEE ohz). **Seamounts** are submerged basaltic volcanoes more than 1 km high. Many linear chains of seamounts, such as the Emperor seamount chain, are stretched out across the Pacific Ocean basin in roughly the same direction. **Guyots** are large, extinct, basaltic volcanoes with flat, submerged tops.

Demonstration

Marine Sediments Show the class examples of marine sediments, including well-sorted beach sand and brown or red deep-sea mud, and marine sedimentary rocks such as well-sorted sandstone, red shale, chalk (fossil ooze), coquina (fossil shell hash), and poorly sorted, coarse-grained sandstone (turbidite).

Marine Sediments

The sediments that cover the ocean floor come from a variety of sources, but most come from the continents. Land-derived sediments include mud and sand washed into the oceans by rivers, as well as dust and volcanic ash blown over the ocean by winds. Much of the coarser material supplied by rivers settles out near shorelines or on beaches, but fine-grained material such as silt and clay settles so slowly through water that some tiny particles take centuries to reach the bottom.

Terrigenous sediments Ocean currents disperse fine silt, clay, and volcanic ash from land, called terrigenous sediments, throughout the ocean basins. Thus, the dominant type of sediment on the deep ocean floor is fine-grained, deep-sea mud. Deep-sea mud usually has a reddish color because the iron present in some of the sediment grains becomes oxidized during the descent to the ocean bottom. Closer to land, the sediments become mixed with coarser materials such as sand, but some sandy sediments occasionally reach the abyssal plains in particularly strong turbidity currents.

Figure 23 The shells of microscopic marine organisms, such as diatoms, as well as shell fragments and hard parts from larger marine organisms make up biogenous sediments.
SEM: magnification unknown

Biogenous sediments Deep-sea sediments that come from biological activity are called biogenous sediments. When some marine organisms die, such as the diatoms shown in **Figure 23,** their shells settle on the ocean floor. Sediments containing a large percentage of particles derived from once-living organisms are called oozes. Most of these particles are small and consist of either calcium carbonate or silica. The oozes and deep-sea mud of the deep ocean typically accumulate at a rate of only a few millimeters per thousand years.

MiniLAB

iLab Station

Measure Sediment Settling Rates

How fast do sediment grains sink?

Procedure

1. Read and complete the lab safety form.
2. Obtain **sediment grains** with approximate diameters of 0.5 mm, 1 mm, 2 mm, 5 mm, and 10 mm.
3. Draw a data table with these headings: *Type of Particle, Diameter (mm), Distance (cm), Time (s),* and *Settling Speed (cm/s).*
4. Measure and record the diameters of each specimen using a **set of sieves.**
5. Fill a **250-mL graduated cylinder** with **cooking oil.** Measure the height of the cooking oil.
6. Drop the largest specimen into the oil. Use a **stopwatch** to measure and record the time it takes for the specimen to sink to the bottom of the cylinder.
7. Repeat Step 6 for the remaining specimens.

Analysis

1. **Calculate** the settling speed for each specimen, and fill in your data table.
2. **Plot** the settling speed (cm/s) against particle diameter (mm) on a graph.
3. **Explain** How do settling speeds change as particle sizes decrease?

DIFFERENTIATED INSTRUCTION

Advanced Learners Have students research the implications of the theoretical sinking velocity of small grains, given by $V = 40 \times d^2$, where V is the velocity in centimeters per second, and d is the grain diameter in millimeters. Have students create a data table with the headings *Particle Type* (coarse sand, fine sand, silt, and clay), *Grain Size* (1 mm, 0.1 mm, 0.01 mm, and 0.001 mm), *Velocity,* and *Time to Bottom,* and calculate all pertinent entries, assuming a water depth of 5 km. Have students speculate what grain sizes dominate continental shelf and deep-sea deposits, and compare their data with actual marine sediment patterns. coarse sand, 1 mm, 40 cm/s, 3.5 h; fine sand, 0.1 mm, 0.4 cm/s, 14.5 days; silt, 0.01 mm, 0.004 cm/s, almost 4 y; clay, 0.001 mm, 0.00004 cm/s, 396 y

MiniLAB

iLab Station Rubric

Purpose Students will determine the effect of grain size on settling speed.

Process Skills interpret data, compare and contrast, classify, draw a conclusion, predict

Safety Precautions Approve lab safety forms before work begins. If oil is spilled, make sure it is cleaned up immediately to prevent accidents.

Teaching Strategies

- Divide the class into groups of three or four students.
- Have each student in a group do a specific task, such as selecting and measuring grains.
- Each student should calculate settling speeds and plot a graph.

Expected Results

Grain Size	Settling Speed (cm/s)
0.5 mm	0.3
1.0 mm	1.1
2.0 mm	4.0
5.0 mm	12.0
10.0 mm	30.0

Analysis

1. Settling speed is calculated by dividing distance by time.
2. Graphs will vary based on data collected.
3. Settling speeds decrease with decreasing size.

Assessment

Knowledge Ask students which of the following sediments settles the fastest—sand, silt, or clay. Which settles the slowest? Why? Sand settles the fastest because it has the largest grain size. Clay settles the slowest because it has the smallest grain size.

LabManager™ Customize any lab with the LabManager™ CD-ROM.

3 Assess

Check for Understanding

Explain Ask students what types of sediments they would expect to find on the continental shelves and in the abyssal plains, and why.

Reteach

Summarize Explain to students that because of differences in settling velocities, small particles are carried farther out to sea by ocean currents than larger particles. As a result, continental shelf sediments include sand and gravel, while the abyssal plains contain mostly silt and clay. Oozes do not form in the abyssal plains because $CaCO_3$ particles dissolve on the way down. Oozes do not form on the shelves because there is too much inorganic sediment.

Assessment

Skill Have students each construct a concept map linking the following terms: *shelf, rise, abyssal plains, ridge, ooze, mud, sand, fine, coarse, mixed organic,* and *inorganic.*

Figure 24 Metals, such as manganese, precipitate directly from seawater. These manganese nodules consist of manganese and iron and range in size from a few centimeters to 10 cm across.

Hydrogenous sediments While terrigenous sediments are derived from land and biogenous sediments are derived from biological activity, another sediment type, called hydrogenous sediments, is derived from elements in seawater. For example, salts that precipitate out of supersaturated lagoons are considered hydrogenous sediments, as are sulfides that form at hydrothermal vents. Manganese nodules, shown in **Figure 24,** consist of oxides of manganese, iron, copper, and other valuable metals that have precipitated directly from seawater. The precipitation happens slowly, therefore growth rates of manganese nodules are extremely slow and are measured in millimeters per million years.

Manganese nodules cover huge areas of the deep-sea floor. Although some have tried to mine manganese nodules for their valuable metals, the difficulty and cost involved in removing them from a depth of 5000 to 6000 m has made progress slow.

SECTION 2 REVIEW

Section Self-Check

Section Summary

- The oceans cover parts of the continental crust as well as oceanic crust.
- A continental margin consists of the continental shelf, the continental slope, and the continental rise.
- Deep-ocean basins consist of abyssal plains, trenches, mid-ocean ridges, seamounts, and guyots.
- Most deep-sea sediments are fine-grained and accumulate slowly.

Understand Main Ideas

1. MAINIDEA **Describe** the features of deep-ocean basins.
2. **Identify** which sediment sinks faster—pebbles or sand grains.
3. **Summarize** the differences between deep-sea mud and oozes.
4. **Compare and contrast** the characteristics of the three major areas of the continental margin.

Think Critically

5. **Suggest** If there is little volcanic activity on abyssal plains, yet they are dotted with thousands of seamounts, from where did these extinct volcanoes come?

WRITING IN ▶ Earth Science

6. Suppose you are taking side-scan sonar readings as your ship moves across the Pacific Ocean from east to west. Describe how you would interpret the sonar data according to the features you would expect to find beneath the surface.

SECTION 2 REVIEW

1. Deep-ocean basins consist of abyssal plains, which are the flattest areas on Earth, deep-sea trenches, which are the deepest areas of the ocean floor, mid-ocean ridges, at which there can be volcanic and earthquake activity, hydrothermal vents, through which heated water erupts, and seamounts and guyots, which are submerged basaltic volcanoes that can be extinct.
2. Pebbles sink faster than sand grains.
3. Oozes are the remains of marine organisms, whereas deep-sea muds are silt and clay that have settled out of ocean waters.
4. The continental shelf, slope, and rise are all part of the continental margin. The continental shelf is the shallowest of the three and differs in width depending on location. The continental slope is beyond the shelf and slopes at an average of 100 m/km. The continental rise is formed by the sediments that collect from turbidity currents along the continental slope.
5. The seamounts must have formed earlier in time when the crust on which they are located was near volcanic activity along the ridges.
6. Student answers will vary but should include features such as abyssal plains, seamounts, mid-ocean ridges, and trenches.

Rubric

eXpeditions!

ON SITE: SURVEYING THE DEEP OCEAN FLOOR

Ralph White/CORBIS

Alvin is a human-occupied submersible vehicle that is used to study hydrothermal vents and other deep-sea features.

There are a variety of vessels that can collect data on the deep ocean floor ranging from human-occupied submersible vehicles to free-moving autonomous robots to remotely operated vehicles. These vessels are used by oceanographers to learn more about the topography, biology, and chemistry of the deep ocean.

Submersibles *Alvin* is one of several human-occupied deep-sea submersibles used worldwide. In *Alvin,* the trip to the bottom of the ocean takes about two hours, and it takes two hours to return to the surface. Under the immense pressure of thousands of kilograms of water and in the darkness and cold of the deep ocean, scientists have about four hours of bottom time to record video and collect water, sediment, and biological samples. Using *Alvin,* scientists got their first view of hydrothermal vents in 1977, a major discovery for both geologists and biologists.

Autonomous underwater vehicles The *Autonomous Benthic Explorer (ABE)* has a mass of over 500 kg and is 2 m long, but, thanks to computer systems, the bulky vehicle is able to dive into narrow trenches and zip around seamounts at depths as deep as 5000 m. Autonomous underwater vehicles are able to move around independent of a pilot, ship, or submersible.

The *ABE* has completed over 150 dives to the deep seafloor, recording videos of the deep sea environment and taking temperature and water chemistry samples.

Remotely operated vehicles *Jason/Medea* is a remotely operated vehicle (ROV) system that can collect data at depths as deep as 6000 m. An ROV is deployed from a ship, but unlike the *ABE,* it needs a pilot, an engineer, and a navigator to control its operations from the ship. In 2006, *Jason/Medea* explored part of the western Pacific Ocean near the Mariana Trench, collecting data on water chemistry near active undersea volcanoes and hydrothermal vents. In 2009, scientists operating *Jason/Medea* filmed an eruption of a deep-sea volcano about 1200 m below the surface of the Pacific Ocean. The video provided the first-ever live footage of an eruption of an underwater volcano.

(bkgd)iStop/Getty Images

WRITING IN Earth Science

Time Line Research more information about the history of exploration of the deep sea using human-occupied vessels. Create a time line that displays dates of important technological advances and discoveries in this area.

WebQuest

WRITING IN Earth Science

Rubric

Time Line Student time lines might include descriptions of diving bells, submarines, and bathyspheres such as the Trieste which reached a depth of 10,915 m in 1960.

WebQuest

iStop/Getty Images

eXpeditions!

Purpose

Students will learn about the different types of vessels oceanographers use to explore and collect data from the deep ocean.

Teacher Content Support

Alvin The submersible *Alvin* has been in operation since 1964 and has completed over 4000 dives. In 1966, *Alvin* was used to help the U.S. Navy recover a hydrogen bomb from the bottom of the Mediterranean Sea. The bomb fell into the sea when the plane carrying it was involved in a mid-air collision with another plane.

In 1977, scientists aboard *Alvin* passed over a never-before-seen hydrothermal vent near the Galapagos rift in the Pacific Ocean, leading to the discovery of new species and a new marine ecosystem. In 1986, *Alvin* explored and photographed the then-newly found wreck of the *Titanic.* During these dives, *Alvin* was also used to test another deep-sea explorer, a prototype robotic vessel called *Jason Jr.*

In 2010, the then 46-year-old *Alvin* made one final trip to the seafloor, before returning ashore for a year and a half of updates and renovations.

Teaching Strategy

Have students discuss why it is difficult for scientists to explore the deep ocean and what problems they might encounter. Issues include physical conditions in the deep-sea environment—cold, dark, high pressure, the engineering and mechanics of operating deep-sea vessels, the safety of the scientists in submersibles—as well as obtaining funds to support the development and maintenance of the vessels.

GeoLAB

iLab Station Rubric

Preparation

Time Allotment 45 min

Process Skills measure and use numbers, classify, compare and contrast, draw a conclusion, recognize cause and effect, predict, hypothesize

Safety Precautions Approve lab safety forms before work begins. Remind students to be careful with their drafting compasses. Compasses have sharp points that can puncture skin.

Procedure

- It is best for you to follow the procedures first so that you can anticipate any problems students might have with this activity.
- Explain map scales prior to the lab. Point out that 1:24,000 means on the map 1 in = 2000 ft.
- Have students work in small groups.
- **Troubleshooting** Many students might not remember how to make a cross section using a topographic map. Explain the concept of topographic profiles and draw an example on the chalkboard if necessary.

LabManager™
Customize any lab with the LabManager™ CD-ROM.

GeoLAB

iLab Station

Mapping: Identify Coastal Landforms

Background: Topographic maps of coastal areas show a two-dimensional representation of coastal landforms. You can identify an emergent coast by the landforms along the coastline as well as landforms found inland.

Question: *How can you identify and describe the coastal landforms of an emergent coast on a topographic map?*

Materials

metric ruler
drafting compass
graph paper
calculator

Procedure

1. Read and complete the lab safety form.
2. Determine the map scale and the contour interval.
3. On the inset map, plot a west-east cross section of the coast just north of Coon Creek from sea level depth contour to a point 2000 ft inland. Use a horizontal scale of 1:24,000.
4. Use both maps to answer the following questions.

Morro Rock is located off the coast of California.

Analyze and Conclude

1. **Describe** What kind of coastal landform is the Morro Rock Peninsula?
2. **Explain** What kind of feature is Pillar Rock, and how was it formed?
3. **Interpret** On what coastal feature is Morro Bay State Park located? How was the feature formed?
4. **Infer** What is the direction of the longshore transport along Morro Bay?
5. **Apply** Your west-east cross section shows an elevated flat area next to the shoreline. What kind of coastal landform is this? How was it formed?
6. **Draw Conclusions** If sea level dropped 10 m, how would the shoreline change? How far would it move seaward? Would it become more regular or irregular? What would happen to Morro Bay?
7. **Suggest** three major changes that could occur to the coastal region if sea level rose 6 m.

APPLY YOUR SKILL

Compare and contrast this coastal section with a section of the Texas coast between Corpus Christi and Galveston. Which coastal features are similar? Which are different?

Analyze and Conclude

1. tombolo
2. sea stack; formed by differential erosion
3. baymouth bar; formed when a spit crosses a bay
4. to the north; sand is piling up on the south side of the breakwater
5. an elevated marine terrace; formed by uplifting of a wave-cut platform
6. The shoreline would move 2000 ft seaward and become more regular. Morro Bay would dry up.
7. Morro Rock would become an island, most of the coastal communities would be flooded, the coastline would become more irregular, and Morro Bay would become a large, branching estuary extending almost 2 mi further inland.

APPLY YOUR SKILL

Compare and Contrast The Gulf Coast is dominated by low topography, large barrier islands, large lagoons and estuaries, and baymouth bars. There are no rocky headlands with sea stacks or elevated marine terraces. The Gulf Coast is a submergent coast.

MAINIDEAS Summary statements can be used by students to review the major concepts of the chapter.

Students can review with these online resources.

Vocabulary Practice

Vocabulary eGames
Vocabulary eFlashcards
Vocabulary PuzzleMaker

Section Self-Check

Chapter Self-Check

Online Test Practice

Use the Exam*View*® *Assessment Suite* CD-ROM to:

- create multiple versions of tests
- create modified tests with one mouse click
- edit existing questions and add your own questions
- build tests aligned with state standards using built-in state curriculum tags
- change English tests to Spanish with one mouse click
- track students' progress using the Teacher Management System

CHAPTER 16 STUDY GUIDE

BIGIDEA The marine environment is geologically diverse and contains a wealth of natural resources.

Vocabulary Practice

SECTION 1 Shoreline Features

VOCABULARY
- beach
- wave refraction
- longshore bar
- longshore current
- barrier island

MAINIDEA **The constant erosion of the shoreline and deposition of sediments by ocean waves creates a changing coastline.**

- Wave erosion of headlands produces wave-cut platforms and cliffs, sea stacks, sea arches, and sea caves.
- Wave action and longshore currents move sediment along the shore and build depositional features.
- Artificial protective structures interfere with longshore transport.
- Sea levels in the past were 130 m lower than at present.

SECTION 2 Seafloor Features

VOCABULARY
- continental margin
- continental shelf
- continental slope
- turbidity current
- continental rise
- abyssal plain
- deep-sea trench
- mid-ocean ridge
- seamount
- guyot

MAINIDEA **The ocean floor contains features similar to those on land and is covered with sediments of several origins.**

- The oceans cover parts of the continental crust as well as oceanic crust.
- A continental margin consists of the continental shelf, the continental slope, and the continental rise.
- Deep-ocean basins consist of abyssal plains, trenches, mid-ocean ridges, seamounts, and guyots.
- Most deep-sea sediments are fine-grained and accumulate slowly.

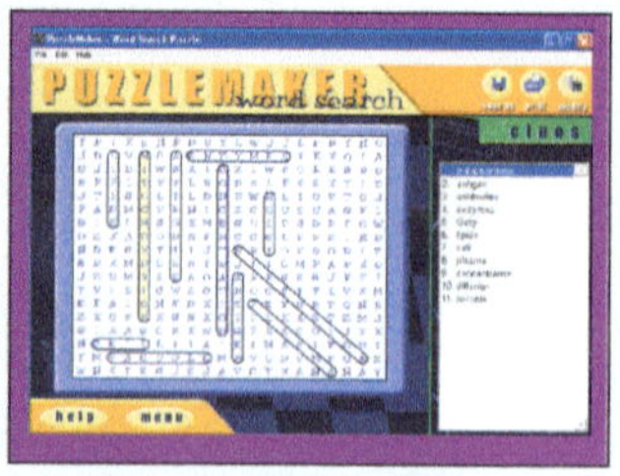

Vocabulary Practice

For additional practice with vocabulary, have students access the Vocabulary PuzzleMaker.

CHAPTER 16 ASSESSMENT

Chapter Self-Check

VOCABULARY REVIEW

Write the definition for each vocabulary term listed below.

1. barrier island
2. beach
3. longshore bar
4. wave refraction

Complete the sentences below using vocabulary terms from the Study Guide.

5. ________ can create submarine canyons by cutting through bottom sediments and bedrocks of seafloors.
6. The deepest part of an ocean basin is a(n) ________.
7. The ________ is the part of a continent that is submerged under the ocean.
8. Submerged basaltic volcanoes more than 1 km high are ________.

UNDERSTAND KEY CONCEPTS

9. Which coastal features are usually found in the bays along irregular coasts with headlands?
   A. sea stacks **C.** wave-cut platforms
   B. wave-cut cliffs **D.** beaches

10. Which coastal landform is not produced by longshore transport?
    A. barrier island **C.** baymouth bar
    B. sand spit **D.** sea stack

11. Which is the correct order of features on the continental margin moving from land out to sea?
    A. continental slope, continental shelf, continental rise
    B. continental rise, continental trench, continental shelf
    C. continental shelf, continental slope, continental rise
    D. continental slope, continental shelf, continental trench

12. What percentage of Earth's surface is represented by ocean basins?
    A. 10 percent
    B. 30 percent
    C. 50 percent
    D. 60 percent

13. What do the sediments of the abyssal plains mostly consist of?
    A. sand and gravel
    B. salt
    C. seashells
    D. mud and oozes

14. Where are most deep-sea trenches located?
    A. in the Atlantic Ocean
    B. in the Pacific Ocean
    C. in the Indian Ocean
    D. in the Arctic Ocean

15. Which runs through all the oceans?
    A. abyssal plains
    B. the mid-ocean ridges
    C. deep sea trenches
    D. seamounts and guyots

Use the diagram below to answer Questions 16 and 17.

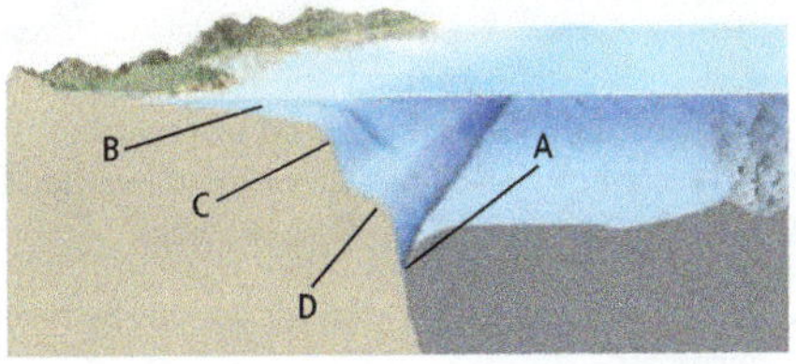

16. Which letter indicates the continental shelf?
    A. A
    B. B
    C. C
    D. D

17. Which feature is indicated by the letter A?
    A. guyot
    B. continental slope
    C. continental rise
    D. trench

CHAPTER 16 ASSESSMENT

VOCABULARY REVIEW

1. long ridges of sand or other sediment, deposited or shaped by the longshore current, that are separated from the mainland
2. the area in which sediment is deposited along the shore
3. a sandbar that forms in front of most beaches
4. when straight wave crests bend as part of the crest moves into shallow water
5. Turbidity currents
6. trench
7. continental margin
8. seamounts

UNDERSTAND KEY CONCEPTS

9. D
10. D
11. C
12. D
13. D
14. B
15. B
16. B
17. D

18. B
19. A
20. C
21. C
22. B
23. B

CONSTRUCTED RESPONSE

24. Sea level is currently rising. During the last ice age, sea level was 130 m lower than it is now in part because more water was frozen in glaciers.
25. Oozes are sediments formed from the hard parts of marine organisms. Chalk is sedimentary rock formed from oozes.
26. Diagrams should show that incoming wave crests are bent toward the shore.
27. Student diagrams should look like **Figure 1.**
28. Diagrams should show that as water from incoming breakers spills over the longshore bar, the longshore current is created, running parallel to shore.
29. Elevated marine terraces are former wave-cut platforms lifted above sea level by a rising coast.
30. A baymouth bar forms when a spit closes off a bay. A tombolo is a ridge of sand that forms between a mainland and an island and connects the island to the mainland.
31. Sea level can change as glaciers freeze or melt, when water expands as it is heated by global warming, and as a result of tectonic movements.
32. Seamounts and guyots are submerged volcanoes found on the deep-ocean floor. They are formed as result of volcanic activity on the ocean floor. Seamounts are basaltic volcanoes that are over 1 km high. Guyots are extinct basaltic volcanoes that have a flat top.
33. The red color is from the oxidized iron present in the mud sediments.
34. Sediments from continents include mud and sand from rivers, dust, and volcanic ashes blown over oceans. The deep sea mud accumulates at a rate of a few millimeters per thousand years. Sediments from marine organisms include shells and hard parts that remain after marine organisms die. Sediments of this type are called oozes. They consist of calcium carbonate and silica.

ASSESSMENT

18. Which marks the true edge of a continent?
A. submarine canyon **C.** continental shelf
B. continental slope **D.** abyssal plain

19. Which seafloor feature can be found along rifts in the mid-ocean ridges?
A. hydrothermal vents
B. manganese nodules
C. deep-sea trenches
D. seamounts

20. Which represents the flattest part of Earth's surface?
A. deep-sea trenches
B. continental margins
C. abyssal plains
D. mid-ocean ridges

Use the diagram below to answer Question 21.

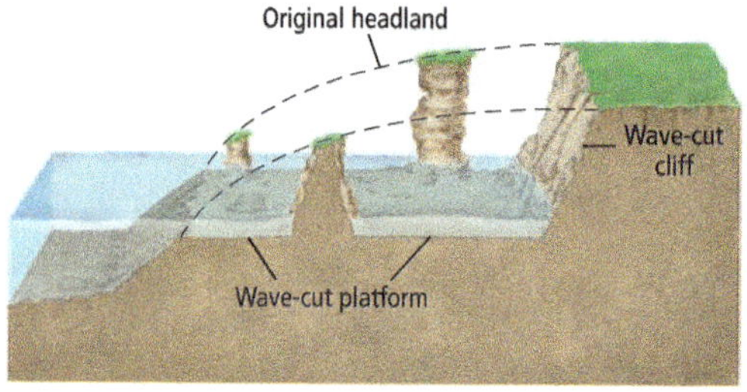

21. Which features are not caused by erosion?
A. wave-cut platforms **C.** original headlands
B. wave-cut cliffs **D.** sea stacks

22. Which features of the seafloor are cut by turbidity currents?
A. longshore bars **C.** abyssal plains
B. submarine canyons **D.** baymouth bars

23. Which are not associated with mid-ocean ridges?
A. black smokers **C.** fracture zones
B. guyots **D.** hydrothermal vents

CONSTRUCTED RESPONSE

24. Explain Is global sea level currently rising, falling, or staying the same? During the last ice age, was the sea level higher, lower, or the same as at present?

25. Explain the relationship between oozes and the sedimentary rock known as chalk.

26. Illustrate the effect that wave refraction has on incoming wave crests that approach the coast at an angle.

Use the diagram below to answer Question 27.

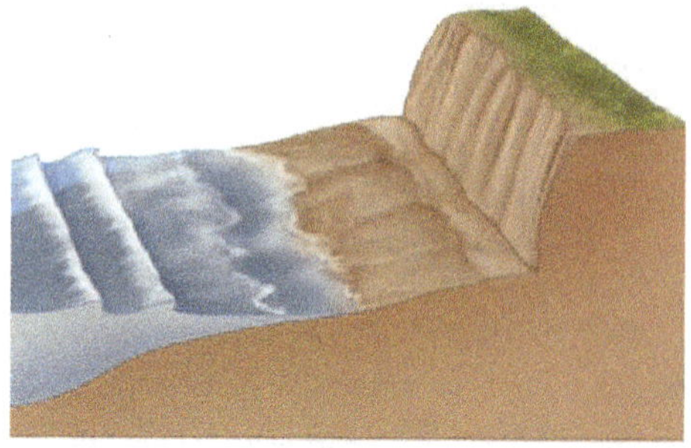

27. Apply Copy the diagram above onto a sheet of paper, and label the following features: *shore, beach, low-tide shoreline, high-tide shoreline, coast,* and *waves.*

28. Illustrate how incoming waves along a shoreline create the longshore current.

29. Apply If a coast has elevated marine terraces, is it more likely to be rising or sinking? Explain.

30. Differentiate between a baymouth bar and a tombolo.

31. Discuss the processes that can cause sea level changes.

32. Compare and contrast seamounts and guyots. How are they formed?

33. Analyze Why does deep-sea mud usually have a reddish color?

34. Compare and contrast the origins of marine sediments.

THINK CRITICALLY

35. Infer Form an inference to explain why seamounts are extinct volcanoes.

THINK CRITICALLY

35. The seafloor is dotted with thousands of solitary mountains that are not located near present sites of volcanic activity. Scientists hypothesize that these seamounts represent extinct volcanoes because no other forces on the seafloor could have formed these mountains.

36. **Suppose** you are surfing at a beach in California. Describe which near-shore currents might affect your position relative to where you start out on the beach after each cycle of surfing. Will you come back to the same point?

37. **Suggest** Evidence shows that woolly mammoths moved from Siberia to North America during the last ice age. How could they have crossed the two continents?

38. **Suppose** that you are swimming in the ocean toward the shoreline, and you suddenly find that, despite all your efforts to swim ahead, you remain in the same location. Explain why this might happen, and describe the path in which you should swim to get yourself out of this situation.

39. **Assess** How is it possible to have a coast that is sinking when global sea level is falling?

Use the diagram below to answer Question 40.

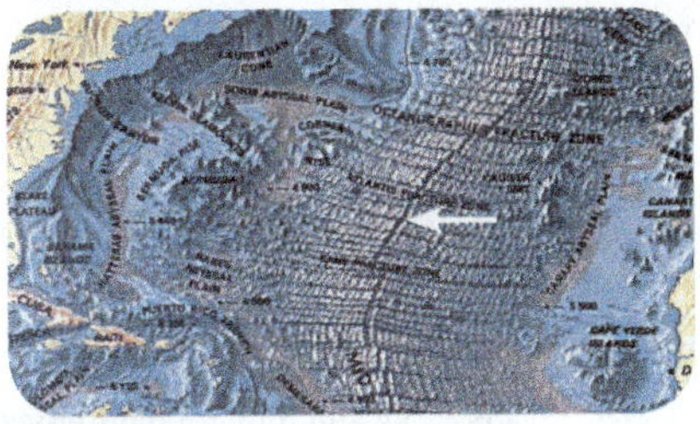

40. **Evaluate** The arrow is pointing to what ocean floor feature? What is the significance of this feature?

CONCEPT MAPPING

41. Use the following terms to construct a concept map about the continental margin: *continental shelf, continental slope, continental rise, turbidity currents,* and *submarine canyons.* Refer to the *Skillbuilder Handbook* for more information.

CHALLENGE QUESTION

42. **Investigate** the causes and effects of a natural disaster that occurred in the past five years and resulted in coastal erosions in the United States or in any other country.

WRITING IN ▶ Earth Science

43. Write an editorial article for a newspaper that discusses some of the risks involved in living in a low-lying coastal area.

DBQ Document–Based Questions

Data obtained from: National assessment of shoreline change: part 1 historical shoreline changes and associated land loss along the U.S. Gulf of Mexico. Report 2004–1043. *U.S. Geological Survey.*

The graph below shows the average annual sea level based on tide data for a 100-year period for two cities on the Gulf Coast of the United States.

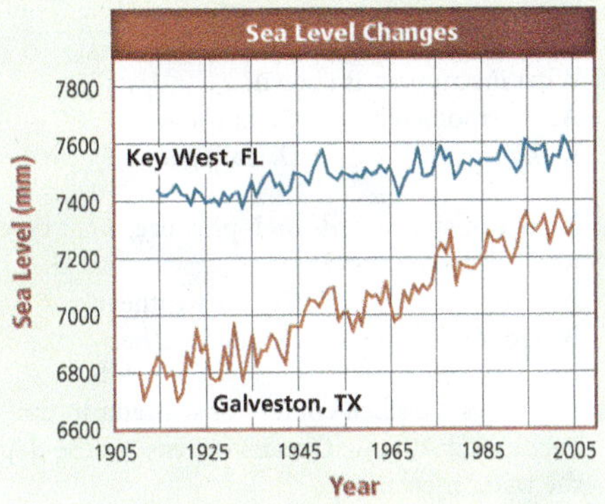

44. Estimate the sea level change for Key West, Florida, and Galveston, Texas, by roughly fitting a straight best-fit line through the data points.

45. The global mean sea level rise for the past century is estimated to be about 0.18 m. Which location has a sea level rise closer to the global mean?

46. What is one possible cause for global sea level rise? What is one reason that a local sea level change can be much larger than the global average?

CUMULATIVE REVIEW

47. What are the four most common foliated metamorphic rocks? **(Chapter 6)**

48. Describe the process that creates supercell thunderstorms. **(Chapter 13)**

36. longshore currents, rip currents; No, after each cycle of surfing, you will end up downstream of the longshore current.

37. During the last ice age, the global sea level was about 130 m lower than the current sea level, and Siberia was attached to North America through the Bering land bridge. Woolly mammoths could walk from Siberia to North America. This region is now below the sea level.

38. It is likely that I am caught in a rip current. It is advised to swim parallel to the shoreline for a while to get out of the rip current and ride on the surf to get back to the shore.

39. A coast can be submergent if the land is sinking faster than sea level is dropping.

40. mid-ocean ridge; Mid-ocean ridges are the most prominent features of ocean basins. They run through all ocean basins with a total length of over 65,000 km. They are sites of volcanic and earthquake activity in ocean basins and contain fracture zones and hydrothermal vents.

CONCEPT MAPPING

41. Student answers will vary but should be logical.

CHALLENGE QUESTION

42. Accept all reasonable responses, such as information about Hurricane Ike or the tsunami that hit parts of Chile in February 2010. Damages could include severe erosion, severe property damage, and loss of life.

WRITING IN ▶ Earth Science

Rubric

43. Student answers will vary, but could include flooding and other damage from storm surge.

DBQ Document-Based Questions

Data obtained from: National assessment of shoreline change: part 1 historical shoreline changes and associated land loss along the U.S. Gulf of Mexico. Report 2004–1043. *U.S. Geological Survey.*

44. The rough estimate of sea level rise at Key West, FL is about 200 mm (or 0.2 m) and at Galveston, TX, it is about 550 mm (or 0.55 m).

45. Key West, FL

46. Possible answer: Many scientists contend that the global sea level rise is the result of global warming. Both seawater expansion and glacial melting due to global warming contributed to global sea level rise for the past century. A possible cause for local sea level change is tectonic movement, such as a coastline sinking.

CUMULATIVE REVIEW

47. slate, phyllite, schist, gneiss

48. Supercells form when pools of cold air in the upper atmosphere magnify the instability in the air in which the storm forms, which intensifies the strength of the updrafts and downdrafts.

MULTIPLE CHOICE

1. C
2. C
3. C
4. B
5. A
6. B
7. D
8. A
9. C
10. A

CUMULATIVE STANDARDIZED TEST PRACTICE

MULTIPLE CHOICE

1. Which ocean movement is slow-moving and occurs in deep waters?
   A. surface currents C. density currents
   B. upwelling D. gyres

Use the illustration below to answer Questions 2 and 3.

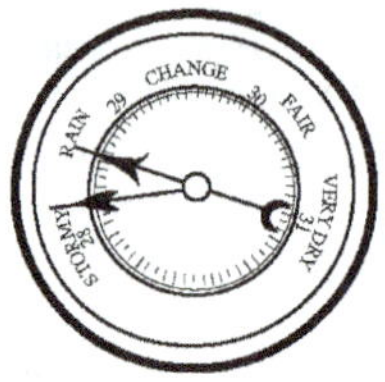

2. What instrument is shown?
   A. thermometer C. barometer
   B. hygrometer D. rain gauge

3. If a barometer records high pressure, what type of weather can be expected?
   A. storms C. fair weather
   B. cloudy weather D. hot weather

4. Why does the silica content of seawater in the Pacific and Atlantic Oceans increase as the depth of the water increases?
   A. The chemical makeup of the water at the surface of the oceans dissolves more silica.
   B. When shelled organisms die, their shells fall to the floor of the ocean and dissolve, releasing silica into the water.
   C. Wave action at the surface of the oceans dilutes the silica.
   D. Oxygen entering the surface water pushes silica down into deeper water.

5. How might volcanic activity affect changes in the climate?
   A. Volcanic dust blocks incoming radiation causing lower global temperatures.
   B. Volcanic dust traps radiation in the Earth's atmosphere raising global temperatures.
   C. Lava running along Earth's surface increases the temperature thus increasing global temperature.
   D. Lava thrown into the air increases the atmospheric temperature thus increasing global temperatures.

6. If the northern hemisphere is experiencing long hours of darkness and cold weather, what season is it in the southern hemisphere?
   A. spring C. winter
   B. summer D. fall

Use the illustration below to answer Questions 7–9.

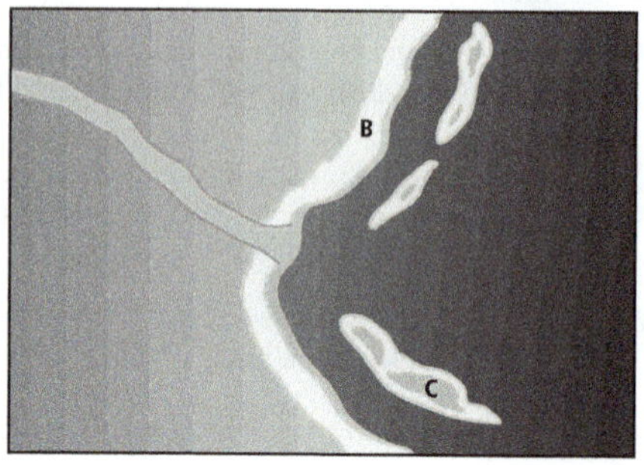

7. What shoreline feature is indicated by B?
   A. a tombolo C. a lagoon
   B. a spit D. a beach

8. The material in the section indicated by the B is very coarse. What conclusions can be drawn about the waves and sediment source?
   A. The area is rocky and pounded by heavy waves.
   B. The area is rocky and hit with light waves.
   C. The area consists of fine-grained material that has been hit by heavy waves.
   D. The area consists of fine-grained material that has been hit by light waves.

9. Which forms the area indicated by C?
   A. large storm waves
   B. gaps in the longshore bar
   C. the longshore current
   D. a rip current

10. Limestone that is exposed to enough heat and pressure is transformed into
    A. marble
    B. slate
    C. quartzite
    D. gneiss

SHORT ANSWER

Use the illustration below to answer Questions 11 and 12.

11. Identify and describe the process that is occurring in the illustration.

12. How does a greenhouse mimic Earth's atmosphere?

13. How might the greenhouse effect be increased?

14. Describe the hypothesis of how comets might have created Earth's early oceans.

15. Discuss a location where an analog forecast could be more beneficial than a digital forecast.

16. If a sonar signal takes 3 s to be emitted and received by an oceanographer mapping the ocean floor, what is the ocean depth in that location? (Hint: sound travels 1500 m/s through water.)

17. If air is rising due to uneven heating of Earth's surface within an air mass, what can you expect to occur?

READING FOR COMPREHENSION

Underwater Hot Spots

Scientists think that underwater hot spots might host unique, previously unknown forms of life. Hydrothermal vents at ocean ridges are an essential part of the chemical balance of seawater. They support ecosystems not found anywhere at the surface and are thought to have been the sites of the early formation and evolution of life. The study of these organisms will give scientists insight to the flexibility and adaptability of life. The water that spews forth from hydrothermal vents can reach temperatures of 350°C and is rich in chemicals such as sulfur and salt. Some microorganisms have adapted to the environments on these vents creating rich underwater ecosystems that some scientists think might represent some of the earliest lifeforms on Earth.

Article obtained from: Roach, J. Hydrothermal vents found in Arctic Ocean. *National Geographic News.* January 23, 2003.

18. Which is NOT an importance of hydrothermal vents?
    A. They support unique ecosystems.
    B. They create rich underwater environments.
    C. They help underwater creatures to adapt to extremely hot temperatures.
    D. They are an essential part of the chemical balance of seawater.

19. Why is it rare to find life-forms around hydrothermal vents?
    A. They are extremely hot.
    B. They are along ocean ridges.
    C. They are not chemically balanced with the rest of the ocean.
    D. No life-form is able to survive due to the sulfur in the area.

20. Why is studying these hydrothermal vents so important to scientists?

NEED EXTRA HELP?

If You Missed Question . . .	1	2	3	4	5	6	7	8	9	10	11	12	13	14	15	16	17
Review Section . . .	15.3	12.3	11.2	15.2	14.3	14.3	16.1	16.1	16.1	5.2	11.1	11.1	11.1	15.1	12.4	15.1	13.1

SHORT ANSWER

11. the greenhouse effect; long wavelength radiation from the Sun is trapped by gases in Earth's lower atmosphere, thus heating Earth's surface.

12. Radiation is absorbed and trapped by the windows of a greenhouse, which causes the air inside the greenhouse to warm up, creating a good environment for the plants.

13. Scientists hypothesize that it is possible to increase the greenhouse effect by increasing the amount of greenhouse gases, such as carbon dioxide and methane, in the atmosphere.

14. Comets travel through space and sometimes collide with Earth's surface. These collisions release water. A hypothesis about the oceans' early formation is that comets collided with Earth and released enough water to fill the ocean basins over time.

15. An analog forecast could be more beneficial in an area where weather is very predictable and unchanging over time. An analog forecast could also be more beneficial for general seasonal forecasts.

16. $1500 \text{ m/s} \times 3 \text{ s} \div 2 = 2250 \text{ m}$
The ocean is 2250 m deep in that location.

17. an air-mass thunderstorm

READING FOR COMPREHENSION

18. C

19. A

20. By studying these hydrothermal vents, scientists will be able to learn more about a previously unexplored ecosystem that has a great impact on seawater. The life-forms living around these vents will provide clues into their adaptation abilities, as well as the evolution of life.

UNIT 5

The Dynamic Earth

Themes

Change Changes in Earth's surface happen over time and are always occurring.

Structures Interactions among tectonic plates are responsible for most of Earth's volcanoes, mountain ranges, and earthquakes.

Geologic Time Many of the changes to Earth's surface occur as a result of tectonic plate activity happening on a slow timescale over the course of geologic time.

Systems Complex systems involving plate movement and convection currents are the mechanisms underlying the theory of plate tectonics.

Scientific Inquiry Mapmakers first saw the connections between the shapes of continents and the possibility of them fitting together. Wegener's evidence of rock formations and fossil similarities on the different continents began the search for the mechanism that moved and separated the continents from positions of contact to their present locations. Today we call that process plate tectonics.

UNIT 5

The Dynamic Earth

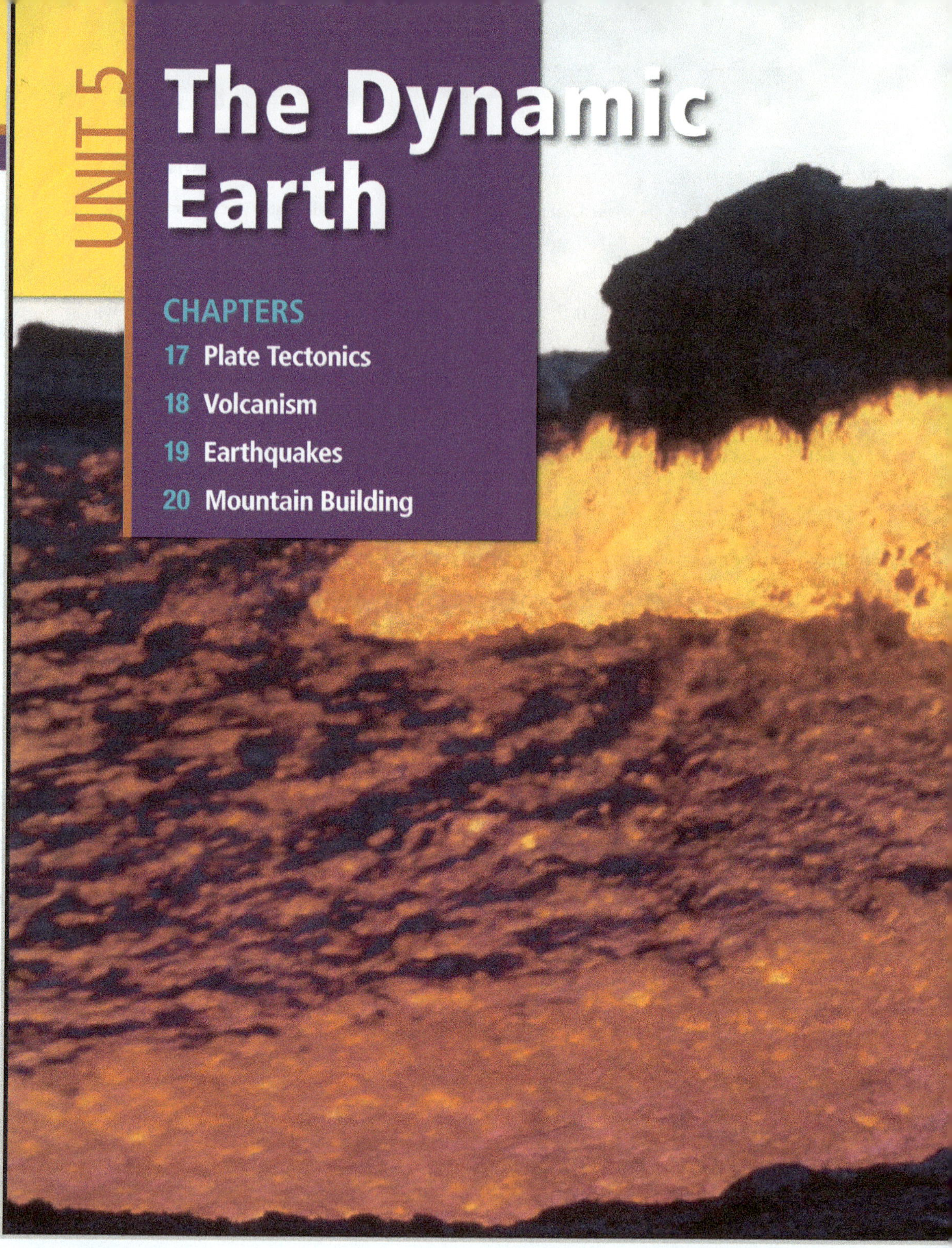

CHAPTERS

17 Plate Tectonics
18 Volcanism
19 Earthquakes
20 Mountain Building

STEM Project

CAREERS IN EARTH SCIENCE **Volcanologist**

This volcanologist is monitoring volcanic activity to help forecast an eruption. Volcanologists spend much of their time in the field, collecting samples and measuring changes in the shape of a volcano.

Introduce the Unit

Volcanoes—One Result of Magma Plumes The lava shown in the photograph is from Kilauea in Hawaii. Kilauea volcano is one of the most active volcanoes on Earth, emitting hundreds of cubic meters of lava per day. The lava from Kilauea originates from more than 60 km below the surface, from a hot spot formed by a magma plume in the asthenosphere. For more than 300,000 years, lava poured out onto the ocean floor at this spot, until the peak of the solidified lava rose above sea level. For the past 50,000 years, Kilauea volcano has continued growing above sea level as a shield volcano. Ask students to name some other volcanoes with which they are familiar, such as Mount St. Helens, Mount Etna, Mauna Loa, and Mount Kilimanjaro.

Volcanoes and Earthquakes Have students speculate about what causes volcanoes to form and what makes them erupt. Also ask students what they think causes earthquakes. Tell students that they will discover the answers to these and other questions in this unit.

A Dangerous Job The volcanologist in this photograph is Katia Conrad Krafft, a French volcanologist. For 25 years, she and her husband, Maurice Krafft, documented volcanic eruptions around the world, frequently putting their lives in danger. Their goal was to obtain visual evidence to help people understand how volcanoes work, and to prevent loss of lives from eruptions. Their lives were ended abruptly in 1991 in Japan when they were filming an eruption at Mount Unzen. A pyroclastic flow suddenly changed direction and they were instantly killed.

CHAPTER 17 Plate Tectonics

BIGIDEA **Most geologic activity occurs at the boundaries between plates.**

ESSENTIAL QUESTIONS	NATIONAL STANDARDS	RESOURCES TO ASSESS MASTERY
SECTION 1 **Drifting Continents** **1.** What are the lines of evidence that led Wegener to suggest that Earth's continents have moved? **2.** How does evidence of ancient climates support continental drift? **3.** Why was continental drift not accepted when it was first proposed? 2 sessions 1 block	UCP.2, UCP.4; B.4; D.1, D.3; G.1–3	**Entry-Level Assessment** Focus Transparency 49, p. 468 **Progress Monitoring** Caption Question, pp. 469, 470, 471 Reading Check, pp. 468, 471 Section Review, p. 472
SECTION 2 **Seafloor Spreading** **1.** What evidence led to the discovery of seafloor spreading? **2.** What is the significance of magnetic patterns on the seafloor? **3.** How is the process of seafloor spreading explained? 1 session 0.5 block	UCP.2; D.1; E.1–2	**Entry-Level Assessment** Focus Transparency 50, p. 473 **Progress Monitoring** Caption Question, p. 477 Reading Check, p. 474 Section Review, p. 479
SECTION 3 **Plate Boundaries** **1.** How does the movement of Earth's tectonic plates result in many geologic features? **2.** What are the three types of plate boundaries and the features associated with each? **3.** What are the processes associated with subduction zones? 3 sessions 1.5 blocks	UCP.2; D.1, D.3	**Entry-Level Assessment** Focus Transparency 51, p. 480 **Progress Monitoring** Reading Check, p. 481 Section Review, p. 485
SECTION 4 **Causes of Plate Motions** **1.** How is the process of convection explained? **2.** How is convection in the mantle related to the movements of tectonic plates? **3.** What are the processes of ridge push and slab pull? 2.5 sessions 1.75 blocks	UCP.2; A.1–2; B.4; D.1–3; E.1–2; G.2–3	**Entry-Level Assessment** Focus Transparency 52, p. 486 **Progress Monitoring** Caption Question, p. 486 Reading Check, p. 487 Section Review, p. 488 **Summative Assessment** Chapter Assessment, p. 493 Exam*View*® *Assessment Suite* CD-ROM

Go online!
connectED.mcgraw-hill.com

LEVELED RESOURCES	LAB MATERIALS
Science Notebook 17.1 OL **Chapter FAST FILE Resources:** Study Guide, p. 13 BL **Transparencies:** Section Focus Transparency 49 EL	LaunchLAB p. 466 / **10 min** map of California, metric ruler, map scale
Science Notebook 17.2 OL **Chapter FAST FILE Resources:** Study Guide, p. 14 BL **Lab Resources:** Laboratory Manual, p. 129 OL **Transparencies:** Section Focus Transparency 50 EL Teaching Transparency 47 OL EL	
Science Notebook 17.3 OL **Chapter FAST FILE Resources:** MiniLab Worksheet, p. 2 OL Study Guide, p. 16 BL **Lab Resources:** Laboratory Manual, p. 133 OL **Transparencies:** Section Focus Transparency 51 EL Teaching Transparencies 48, 49 OL EL	MiniLAB p. 481 / **30 min** world map, paper templates, large piece of paper, pencil GeoLAB p. 490 / **60 min** paper, colored pencils, scissors, metric ruler, calculator
Science Notebook 17.4 OL **Chapter FAST FILE Resources:** GeoLab Worksheet, p. 3 OL Study Guide, p. 18 BL **Transparencies:** Section Focus Transparency 52 EL	

LabManager™
Customize any lab with the LabManager™ CD-ROM.

ADDITIONAL RESOURCES

Technology:

- eTeacherEdition
- eStudentEdition
- Classroom Presentation Toolkit
- Exam*View® Assessment Suite* CD-ROM
- LabManager™ CD-ROM
- What's EARTH SCIENCE Got To Do With It? DVD
- Weather Classroom Video DVD

Assessment:

- Blueprints for Success
- Chapter Test (Scaffolded)
- Earth Science Standardized Test Practice

Lab Resources:

- Exploring Environmental Problems Laboratory Manual

BL Below Level   OL On Level   AL Advanced Learners   EL English Learners   COOP LEARN Cooperative Learning

CHAPTER 17

LaunchLAB

iLab Station | Rubric

Is California moving?

Process Skills measure in SI, predict, infer, communicate

Safety Precaution Approve lab safety forms before work begins.

Teaching Strategies

- Maps of California can be obtained from the USGS Web site or library.
- Because map scales are much smaller than the distance being measured, students might have trouble accurately measuring the distance. By using the map scale to mark a series of distances along a piece of paper, students can make a map-scale ruler.
- If students have trouble with their conversions, remind them that there are 100,000 centimeters in a kilometer.

Procedure

1. Have students read and complete the lab safety form and follow the procedure below.
2. Obtain a **map of California** from your teacher. Use a **metric ruler** and the **map scale** to determine the actual distance between San Francisco and Los Angeles. 600 km
3. At the current rate of movement, when will these two cities be next to each other? 12 million years ($600 \text{ km} \times 100{,}000$ to convert to cm $= 60{,}000{,}000 \text{ cm} \div 5 \text{ cm/y} = 12{,}000{,}000$ y)

Analysis

1. **Infer** what might be causing the motion of these large pieces of land. plate tectonics
2. **Calculate** How far will southwestern California move in a 15-year period? 75 cm ($15 \text{ y} \times 5 \text{ cm/y}$)

CHAPTER 17

Plate Tectonics

BIGIDEA Most geologic activity occurs at the boundaries between plates.

SECTIONS

1 Drifting Continents

2 Seafloor Spreading

3 Plate Boundaries

4 Causes of Plate Motions

LaunchLAB

iLab Station

Is California moving?

Southwestern California is separated from the rest of the state by a system of cracks along which movement takes place. These cracks are called faults. One of these, as you might know, is the San Andreas Fault. Movement along this fault is carrying southwestern California to the Northwest in relation to the rest of North America at a rate of about 5 cm/y. Learn more about plate movement in this activity.

FOLDABLES® Study Organizer

Plate Boundaries

Make a pocket-book using the labels shown. Use it to organize your notes on plate boundaries.

Assessment

Skill Ask students to compute how far apart the cities will be after 45 million years. 2250 km

LabManager™
Customize any lab with the LabManager™ CD-ROM.

The San Andreas Fault is a 1200-km-long gash that runs from northern California almost to Mexico. Each year, plate movement along the fault brings Los Angeles about 5 cm closer to San Francisco.

Introduce the BIGIDEA

Patterns on Earth's Surface Ask students to look at a world map and identify large patterns and structures. They might notice lines of islands, mountain ranges near oceans, mountain ranges in the middle of continents, perhaps that the shape of coasts on continents sometimes mimic each other (Africa–South America). Ask students how they think these features formed. Look at maps of Mars or Venus. What features are similar? What is different?

Teacher Content Support

Faults How do geologists spot faults in Earth's crust? The key to the observation is looking for displacements. When features that should be linear and continuous are broken, the geologist is alerted to the possibility of a fault. Faults occur on any scale. Microfaults require a magnifying lens to see, while faults associated with plate tectonics might be so large as to be missed from the ground. Aerial photos and satellite images have revealed faults that previously went undetected. In the photo, note the vertical displacement of the ridge and its surroundings; it is a possible normal fault. Also, note the horizontal displacement of the ridge near the top of the photo and the odd twists in the stream at the bottom of the photo; they are possible strike-slip faults. Field observations are always used to confirm these inferences.

SECTION 1

1 Focus

Focus Transparency

Before presenting the lesson, project **Section Focus Transparency 49** and have students answer the accompanying questions. EL

MAINIDEA

Continental Fit Earth's continents can be put together like pieces in a jigsaw puzzle. What does this imply about how continents change over time? They deform and stay on the surface. Is this convincing evidence they were once together? only in combination with other observations

2 Teach

Project

Ancient Climates Have pairs of students research whether there are any 200-million-year-old fossils in your state. Challenge students to find out about the climate that existed in your part of the country during this time. Have each student pair prepare a visual display of their findings. OL COOP LEARN

☑ READING CHECK Cartographers were familiar with the shape of the continents, especially in a time during which the western hemisphere was first mapped, so they would be more likely to be first to notice similarities of the shapes of the continents.

SECTION 1

Essential Questions

- What are the lines of evidence that led Wegener to suggest that Earth's continents have moved?
- How does evidence of ancient climates support continental drift?
- Why was continental drift not accepted when it was first proposed?

Review Vocabulary

hypothesis: testable explanation of a situation

New Vocabulary

continental drift
Pangaea

Drifting Continents

MAINIDEA The shape and geology of the continents suggests that they were once joined together.

EARTH SCIENCE 4 YOU When you put together a jigsaw puzzle, what features of the puzzle pieces do you use to find matching pieces? Scientists used features such as shape and position to help them piece together the way the continents were arranged millions of years ago.

Early Observations

With the exception of events such as earthquakes, volcanic eruptions, and landslides, most of Earth's surface appears to remain relatively unchanged during the course of a human lifetime. On the geologic time scale, however, Earth's surface has changed dramatically. Some of the first people to suggest that Earth's major features might have changed were early cartographers. In the late 1500s, Abraham Ortelius (or TEE lee us), a Dutch cartographer, noticed the apparent fit of continents on either side of the Atlantic Ocean. He proposed that North America and South America had been separated from Europe and Africa by earthquakes and floods. During the next 300 years, many scientists and writers noticed and commented on the matching coastlines. **Figure 1** shows a proposed map by a nineteenth-century cartographer.

The first time that the idea of moving continents was proposed as a scientific hypothesis was in the early 1900s. In 1912, German meteorologist Alfred Wegener (VAY guh nur) presented his ideas about continental movement to the scientific community.

☑ READING CHECK **Infer** why cartographers were among the first to suggest that the continents were once joined together.

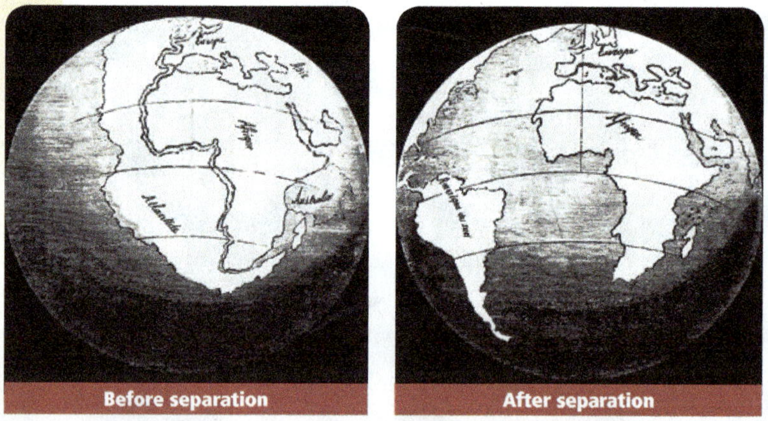

Figure 1 Many early cartographers, such as Antonio Snider-Pelligrini, the author of these 1858 maps, noticed the apparent fit of the continents.

Teacher Content Support

Continental Drift When Alfred Wegener proposed continental drift, there were two competing hypotheses about Earth's ancient geography. Biologists and paleontologists had looked at the distribution of both living and fossil species and proposed that Earth's continents were once connected by land bridges that later sank beneath the oceans. Geophysicists, on the other hand, stated that isostatic principles made it impossible for land bridges to sink into oceanic crust. Both groups of scientists firmly believed in the theory of permanence, which stated that the ocean basins and continents were permanent, unchanging features. Wegener proposed that the biological and geophysical evidence would agree if Earth's continents had changed positions over time.

Continental Drift

Wegener developed a hypothesis that he called **continental drift,** which proposed that Earth's continents had once been joined as a single landmass that broke apart and sent the continents adrift. He called this supercontinent **Pangaea** (pan JEE uh), a Greek word that means *all the earth,* and suggested that Pangaea began to break apart about 200 mya. Since that time, he reasoned, the continents have continued to slowly move to their present positions, as shown in **Figure 2.**

Of the many people who had suggested that continents had moved around, Wegener was the first to base his ideas on more than just the puzzlelike fit of continental coastlines on either side of the Atlantic Ocean. For Wegener, these gigantic puzzle pieces were just the beginning. He also collected and organized rock, climatic, and fossil data to support his hypothesis.

View an **animation of the breakup of Pangaea.**

Concepts In Motion

■ **Figure 2** Wegener hypothesized that all the continents were once joined together. He proposed that it took 200 million years of continental drift for the continents to move to their present positions.

Locate ***the parts of Pangaea that became North and South America. When were they joined? When were they separated?***

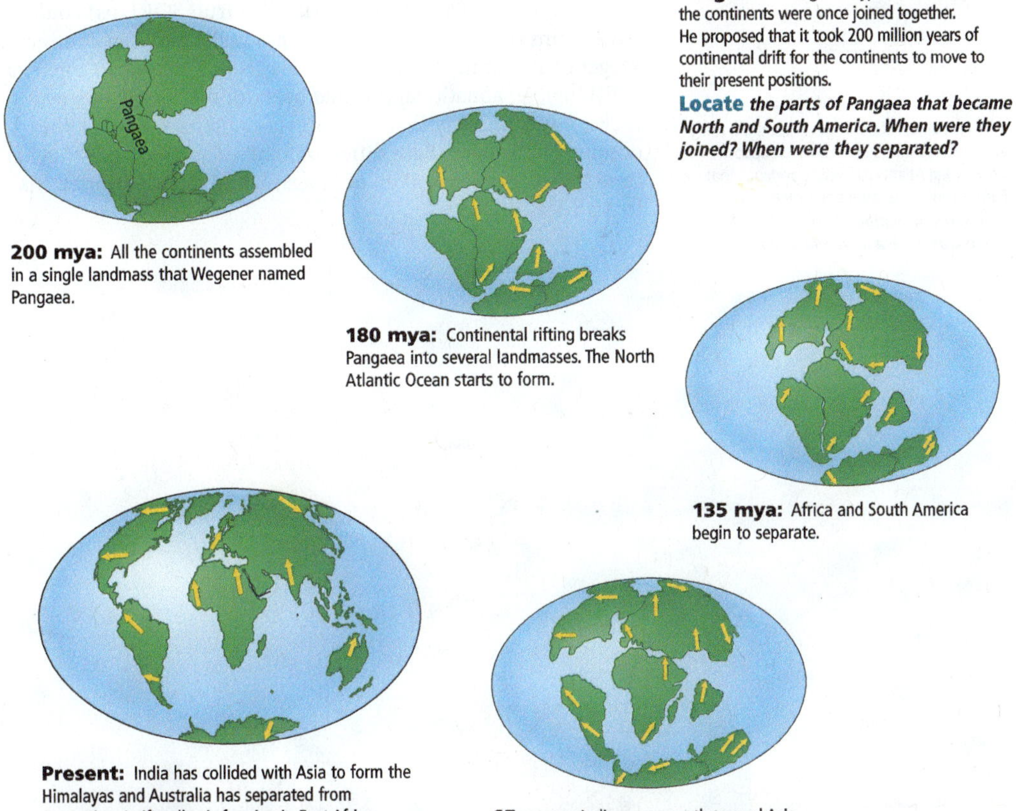

200 mya: All the continents assembled in a single landmass that Wegener named Pangaea.

180 mya: Continental rifting breaks Pangaea into several landmasses. The North Atlantic Ocean starts to form.

135 mya: Africa and South America begin to separate.

Present: India has collided with Asia to form the Himalayas and Australia has separated from Antarctica. A rift valley is forming in East Africa. Continents continue to move over Earth's surface.

65 mya: India moves north toward Asia.

ACROSS THE CURRICULUM

Biology *Glossopteris* fossils were used by Wegener to support his hypothesis of continental drift. Have students find out when seed plants first appeared on Earth. Challenge students to describe how these ancient seed plants differed from their modern-day relatives. Ask a volunteer to explain the life cycle of a seed plant to the rest of the class.

Identify Misconceptions

Some students might think that the basic shapes of continental coastlines have remained unchanged since the breakup of Pangaea.

Uncover the Misconception
Have students make a sketch of what they think North America looked like when the Atlantic Ocean started to open, about 200 mya.

Demonstrate the Concept
If necessary, help students identify North America in the series of sketches in **Figure 2.** Have students compare their sketches with those in the figure.

Assess New Knowledge
Have students make a series of sketches of North America using **Figure 2.** Ask students to sequence the sketches from oldest to most recent and write three or four sentences that describe how the overall shape of the continent has changed over the past 200 million years.

Interpret the Illustration

Figure 2 Have students determine which continental piece has moved the farthest and which has moved the least. Asia has moved the farthest; Antarctica the least.

■ **Caption Question Fig. 2** They were joined by 200 mya. They separated at some point between 200 and 180 mya.

Collaborative Learning

Role Play Ask students how they would feel if their peers rejected their ideas and work. Have students role-play a scenario in which Wegener is trying to convince other scientists of the validity of his ideas. **COOP LEARN**

Model

Crustal Thickness Wegener proposed that Earth's continents were plowing through the seafloor. Physicists, however, argued that the crust was too brittle to withstand this type of motion. To demonstrate why the crust is considered brittle on a geologic scale, have students draw a scale cross section of a piece of crust 6000 km long and 20 km thick. This will demonstrate how thin the crust actually is when compared to other dimensions of Earth.

Interpret the Illustration

Figure 3 Why do you think there are no *Glossopteris* fossils in North America or Europe? It was probably too cold for this fern to have existed in these areas.

▪ **Caption Question** **Fig. 3** Rocks, plants, each animal type, and mountain ranges match from one continent to another.

Teacher Content Support

Wegener Vitae Alfred Wegener's education and experience were primarily in meteorology, but he also studied geology, geophysics, and geography. In 1915, he published the first edition of *The Origins of Continents and Oceans.* The book was rewritten four times between 1915 and 1929; each edition contained new data and material in an effort to combat the criticism Wegener received from the scientific community about his hypothesis.

Evidence from rock formations Wegener reasoned that when Pangaea began to break apart, large geologic structures, such as mountain ranges, became separated as the continents drifted apart. Using this reasoning, Wegener thought that there should be areas of similar rock types on opposite sides of the Atlantic Ocean. He observed that many layers of rocks in the Appalachian Mountains in the United States were identical to layers of rocks in similar mountains in Greenland and Europe. These similar groups of rocks, older than 200 million years, supported Wegener's idea that the continents had once been joined. Some of the locations where matching groups of rock have been identified are indicated in **Figure 3.**

Evidence from fossils Wegener also gathered evidence of the existence of Pangaea from fossils. Similar fossils of several different animals and plants that once lived on or near land had been found on widely separated continents, as shown in **Figure 3.** Wegener reasoned that the land-dwelling animals, such as *Cynognathus* (sin ug NATH us) and *Lystrosaurus* (lihs truh SORE us) could not have swum the great distances that now exist between continents. Wegener also argued that because fossils of *Mesosaurus* (meh zoh SORE us), an aquatic reptile, had been found in only freshwater rocks, it was unlikely that this species could have crossed the oceans. The ages of these different fossils also predated Wegener's time frame for the breakup of Pangaea, and thus supported his hypothesis.

▪ **Figure 3** Alfred Wegener used the similarity of rock layers and fossils on opposite sides of the Atlantic Ocean as evidence that Earth's continents were once joined.
Identify *groupings that suggest that there was once a single landmass.*

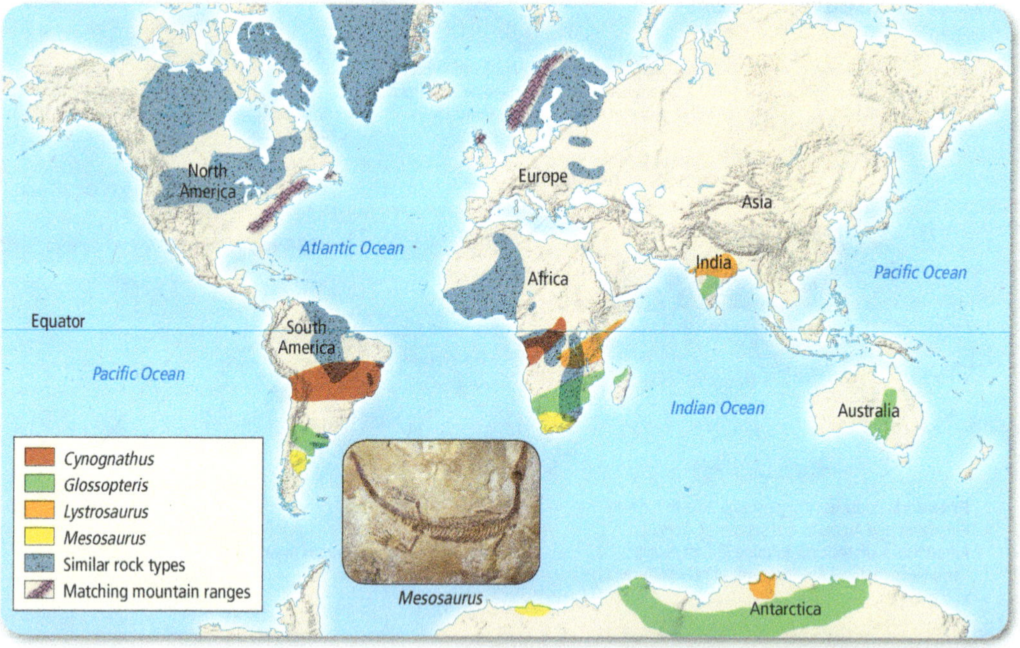

ACROSS THE CURRICULUM

Chemistry Have students research the physical and chemical changes that result in the formation of coal. Suggest that they display their findings as posters.

DIFFERENTIATED INSTRUCTION

Struggling Learners Enlarge and copy **Figure 3** onto stiff paper. Have students cut out each continent, or assist them with this task, and guide them to reconstruct Pangaea using the rock and fossil evidence shown on the map.

Climatic evidence Because he had a strong background in meteorology, Wegener recognized clues about ancient climates from the fossils he studied. One fossil that Wegener used to support continental drift was *Glossopteris* (glahs AHP tur us), a seed fern that resembled low shrubs, shown in **Figure 4.** Fossils of this plant had been found on many parts of Earth, including South America, Antarctica, and India. Wegener reasoned that the area separating these fossils was too large to have had a single climate. Wegener also argued that because *Glossopteris* grew in temperate climates, the places where these fossils had been found were once closer to the equator. This led him to conclude that the rocks containing these fossils had once been joined.

☑ READING CHECK **Infer** how Wegener's background in meteorology helped him to support his idea of continental drift.

■ **Figure 4** Wegener used the fact that fossils of *Glossopteris* were found in many parts of Earth to support his hypothesis of continental drift.

Coal deposits Recall that sedimentary rocks provide clues to past environments and climates. Wegener found evidence in these rocks that the climates of some continents had changed markedly. For example, coal deposits are found in Antarctica and other high latitude locations. Coal forms from the compaction and decomposition of accumulations of ancient swamp plants that grew in warm, wet regions. The existence of coal beds in Antarctica indicated that this frozen land once had a tropical climate. Wegener used this evidence to conclude that Antarctica must have been much closer to the equator sometime in the geologic past.

Glacial deposits Another piece of climatic evidence came from glacial deposits found in parts of Africa, India, Australia, and South America. The presence of these 290-million-year-old deposits suggested to Wegener that these areas were once covered by a thick ice cap similar to the one that covers Antarctica today. Because the traces of the ancient ice cap are found in regions where it is too warm for them to develop, Wegener proposed that they were once located near the South Pole, as shown in **Figure 5.** Wegener suggested two possibilities to explain the deposits. Either the South Pole had shifted its position, or these landmasses had once been closer to the South Pole. Wegener argued that it was more likely that the landmasses had drifted apart rather than Earth changing its axis.

■ **Figure 5** Glacial deposits nearly 300 million years old on several continents led Wegener to propose that these landmasses might have once been joined and covered with ice. The extent of the ice is shown in white.

☑ READING CHECK His background in meteorology made him aware of evidence of ancient climates, and of the latitudes at which those climates could occur.

Reinforcement

Plate Speed Some students might think that continental movement is too slow to measure, and that this movement has not been measured. Ask a few volunteers to draw a line that shows how much a continent might move in a year. Inform students that continents move at rates from about 1 to 10 cm/y. Have students calculate and measure the distance a continent would move in 50 years at a rate of 3 cm/y. 150 cm or 1.5 m

EARTH SCIENCE JOURNAL

News Story Have students write a news story announcing the discovery of continental drift by Wegener. Stories should have a short, exciting title and include made-up quotations by Wegener's detractors. OL

Rubric

3 Assess

Check for Understanding

Discussion A hypothesis that competed with Wegener's idea of continental drift was that Earth's continents were connected by land bridges. Ask students the following question: Which evidence used by Wegener is consistent with the possibility of land bridges and which is not? The biological data, the fossils, support the possibility of land bridges. Climatic and rock data do not support the hypothesis that there were land bridges.

Reteach

Observe and Infer Have students refer to **Figures 3, 4** and **5** as you summarize the types of evidence—rock, fossil, and glacial deposits—used by Wegener to support his hypothesis of continental drift.

Assessment

Make and Use Tables Have students each make a table that summarizes the different types of evidence that Wegener used to support his hypothesis of continental drift.

■ **Figure 6** Wegener collected further evidence for his theory on a 1930 expedition to Greenland. He died during this expedition, many years before his data became the basis for the theory of plate tectonics.

A Rejected Notion

In the early 1900s, many people in the scientific community considered the continents and ocean basins to be fixed features on Earth's surface. For the rest of his life, Wegener continued travelling to remote regions to gather evidence in support of continental drift. **Figure 6** shows him in Greenland on his last expedition. Although he had compiled an impressive collection of data, the continental drift hypothesis was not accepted by the scientific community during Wegener's lifetime.

Continental drift had two major flaws that prevented it from being widely accepted. First, it did not satisfactorily explain what force could be strong enough to push such large masses over such great distances. Wegener thought that the rotation of Earth might be responsible, but physicists were able to show that this force was not nearly enough to move continents.

Second, scientists questioned how the continents were moving. Wegener had proposed that the continents were plowing through a stationary ocean floor, but it was known that Earth's mantle below the crust was solid. So, how could continents move through something solid? These two unanswered questions—what forces could cause the movement and how continents could move through solids—were the main reasons why continental drift was rejected.

It was not until the early 1960s when new technology revealed more evidence about how continents move that scientists began to reconsider Wegener's ideas. Advances in seafloor mapping and in understanding Earth's magnetic field provided the necessary evidence to show how continents move, and the source of the forces involved.

SECTION 1 REVIEW

Section Self-Check

Section Summary

- The matching coastlines of continents on opposite sides of the Atlantic Ocean suggest that the continents were once joined.
- Continental drift was the idea that continents move around on Earth's surface.
- Wegener collected evidence from rocks, fossils, and glacial deposits to support his theory.
- Continental drift was not accepted because there was no satisfactory explanation for how the continents moved or what caused their motion.

Understand Main Ideas

1. **MAIN**IDEA **Draw** how the continents were once adjoined as Pangaea.
2. **Explain** how ancient glacial deposits in Africa, India, Australia, and South America support the idea of continental drift.
3. **Summarize** how rocks, fossils, and climate provided evidence of continental drift.
4. **Infer** what the climate in ancient North America must have been like as a part of Pangaea.

Think Critically

5. **Interpret** Examine **Figure 5.** Oil deposits that are approximately 200 million years old have been discovered in Brazil. Where might geologists find oil deposits of a similar age?
6. **Evaluate** this statement: The town where I live has always been in the same place.

WRITING IN ▶ Earth Science

7. Compose a letter to the editor from a scientist in the early 1900s arguing against continental drift.

SECTION 1 REVIEW

1. Drawings should look similar to the first diagram in **Figure 2.**
2. Glaciers could not have formed at the present latitudes. The deposits suggested that either the South Pole had moved, or that these landmasses had once been joined together close to the South Pole.
3. Rock layers and mountain chains correspond to groups from one continent to another, fossils of land-based plants and animals match across oceans, evidence of ancient climate was found in areas in which they could never occur at the current latitudes. The best explanation for all of this was that the continents were once joined together in a formation that is far different from how they are currently positioned.
4. Most of ancient North America was at the equator, or slightly to the north of it. The climate was tropical, warm, lush, and humid.
5. in western Africa
6. In geologic time, longitude and latitude have changed many times as the plates move around.
7. Letters should refer to the fact that Wegener did not have an adequate explanation of how continents moved.

Rubric

SECTION 2

Seafloor Spreading

MAINIDEA Oceanic crust forms at ocean ridges and becomes part of the seafloor.

Essential Questions

- What evidence led to the discovery of seafloor spreading?
- What is the significance of magnetic patterns on the seafloor?
- How is the process of seafloor spreading explained?

Review Vocabulary

basalt: a dark-gray to black fine-grained igneous rock

New Vocabulary

magnetometer
magnetic reversal
paleomagnetism
isochron
seafloor spreading

EARTH SCIENCE 4 YOU **Have you ever counted the rings on a tree stump to find the age of the tree? Scientists can study similar patterns on the ocean floor to determine its age.**

Mapping the Ocean Floor

Until the mid-1900s, most people, including many scientists, thought that the ocean floors were essentially flat. Many people also had misconceptions that oceanic crust was unchanging and was much older than continental crust. However, advances in technology during the 1940s and 1950s showed that all of these widely accepted ideas were incorrect.

One technological advance that was used to study the ocean floor was the magnetometer. A **magnetometer** (mag nuh TAH muh tur), such as the one shown in **Figure 7,** is a device that can detect small changes in magnetic fields. Towed behind a ship, it can record the magnetic field generated by ocean floor rocks. You will learn more about magnetism and how it supports continental drift later in this section.

Another advancement that allowed scientists to study the ocean floor in great detail was the development of echo-sounding methods. One type of echo sounding is sonar. Recall that sonar uses sound waves to measure distance by measuring the time it takes for sound waves sent from the ship to bounce off the seafloor and return to the ship. Developments in sonar technology enabled scientists to measure water depth and map the topography of the ocean floor.

Figure 7 Magnetometers are devices that can detect small changes in magnetic fields. The data collected using magnetometers lowered into the ocean furthered scientists' understanding of rocks underlying the ocean floor.

ACROSS THE CURRICULUM

Math Have students imagine that they are on the seafloor near the Mariana Trench and one student drops a rock into the 11-km deep trench. Ask students to compute how long it will take for the rock to reach the bottom of the trench if the rock falls through water at 3 m/s. about an hour Submersibles can descend at a rate of 15 m/min. Have students calculate how long it would take such a submersible to reach the bottom of the trench. about 12 h

SECTION 2

1 Focus

Focus Transparency

Before presenting the lesson, project **Section Focus Transparency 50** and have students answer the accompanying questions. EL

MAINIDEA

Seafloor Age The seafloor spreading hypothesis predicts how the age of the seafloor varies with distance from an ocean ridge. As distance increases, the thickness of accumulated sediments increases and the crust gets older. What can be measured to test this hypothesis? date the rocks, measure the thickness of sediment, measure heat flow through the crust

2 Teach

Model

Magnetic Reversals Have students use long paper strips, books, two desks, and the following procedure to model seafloor spreading and the magnetic patterns of ocean-floor rocks. Push two desks together and partially pull two paper strips up between the desks. Lay the exposed portions of the strips on the desks and secure them with the books. Slide the books apart so that they pull the paper up between the desks and out onto the surface. Stop every few centimeters and color a band on both sides of the central gap. The colored bands represent the magnetic pattern of ocean-floor rocks. The bands closest to the "ridge axis" are the youngest rocks. BL EL

Tie to Previous Knowledge

Spreading Evidence Review ocean-floor topography. Explain that features such as trenches, seamounts, and ocean ridges all provide evidence of seafloor spreading.

Use Scientific Terms

Paleo– Have students use a dictionary to find the meaning of the word part *paleo–*. Ask students to use the meaning of this word part to define the term *paleomagnetism* in their own words.

☑ READING CHECK They are located on the ocean floor in long ridges.

Teacher Content Support

War Effect The concepts of seafloor spreading and plate tectonics are connected to WWII submarine warfare. The demands of war often promote periods of rapid technological advance, and scientists are often quick to utilize military technology for more peaceful projects. During WWII, sonar technology was improved because submarine pilots needed detailed maps and a way to navigate under water. Magnetometers that could be towed behind ships were originally developed to detect the magnetic fields generated by the steel hulls of submarines. Scientists modified these magnetometers to measure the magnetic field strength of ocean-floor rocks. The magnetic patterns that were discovered provided key support for the theory of seafloor spreading.

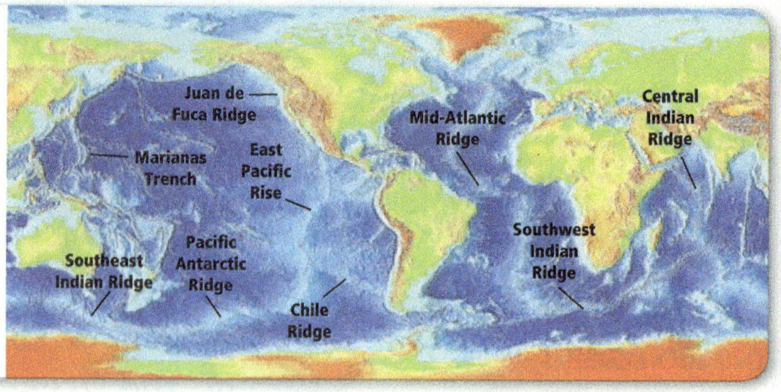

■ **Figure 8** Sonar data revealed ocean ridges and deep-sea trenches. Earthquakes and volcanism are common along ridges and trenches.

Ocean-Floor Topography

The maps made from data collected by sonar and magnetometers surprised many scientists. They discovered that vast, underwater mountain chains called ocean ridges run along the ocean floors around Earth much like seams on a baseball. These ocean floor features, shown in **Figure 8,** form the longest continuous mountain range on Earth. When they were first discovered, ocean ridges generated much discussion because of their enormous length and height—they can be more than 65,000 km long and up to 3 km above the ocean floor. Later, scientists discovered that earthquakes and volcanism are common along the ridges.

☑ READING CHECK **Describe** Where are the longest continuous mountain ranges on Earth?

VOCABULARY
SCIENCE USAGE V. COMMON USAGE
Depress
Science usage: to cause to sink to a lower position
Common usage: to sadden or discourage

Maps generated with sonar data also revealed that underwater mountain chains had counterparts called deep-sea trenches, which are also shown on the map in **Figure 8.** Recall that a deep-sea trench is a narrow, elongated depression in the seafloor. Trenches can be thousands of kilometers long and many kilometers deep. The deepest trench, called the Mariana Trench, is in the Pacific Ocean and is more than 11 km deep. Mount Everest, the world's tallest mountain, stands at 9 km above sea level, and could fit inside the Mariana Trench with six Empire State buildings stacked on top.

These two topographic features of the ocean floor—ocean ridges and deep-sea trenches—puzzled geologists for more than a decade after their discovery. What could have formed an underwater mountain range that extended around Earth? What is the source of the volcanism associated with these mountains? What forces could depress Earth's crust enough to create trenches nearly 6 times as deep as the Grand Canyon? You will find out the answers to these questions later in this chapter.

Demonstration

Magnetic Field Use a bar magnet, iron filings, and an overhead projector and transparency to model the field lines set up by Earth's magnetic field. Place the magnet under the transparency. Sprinkle the filings onto the transparency. Have students observe and describe the lines of magnetic force. Explain that the lines of force of Earth's magnetic field are similar to the lines around the magnet. Point out that the filings cluster around the poles of the magnet, where the magnetic field is strongest.

Ocean Rocks and Sediments

In addition to making maps, scientists collected samples of deep-sea sediments and the underlying oceanic crust. Analysis of the rocks and sediments led to two important discoveries. First, the ages of the rocks that make up the seafloor vary across the ocean floor, and these variations are predictable. Rock samples taken from areas near ocean ridges were found to be younger than samples taken from areas near deep-sea trenches. The samples showed that the age of oceanic crust consistently increases with distance from a ridge, as shown in **Figure 9.** This trend was symmetric across and parallel to the ocean ridges. Scientists also discovered from the rock samples that even the oldest parts of the seafloor are geologically young—about 180 million years old. Why are ocean-floor rocks so young compared to continental rocks, some of which are at least 4 billion years old? Geologists knew that oceans had existed for more than 180 million years so they questioned why there was no trace of older oceanic crust.

The second discovery involved the sediments on the ocean floor. Measurements showed that ocean-floor sediments are typically a few hundred meters thick. Large areas of continents, on the other hand, are blanketed with sedimentary rocks that are as much as 20 km thick. Scientists knew that erosion and deposition occur in Earth's oceans but did not understand why seafloor sediments were not as thick as their continental counterparts. Scientists hypothesized that the relatively thin layer of ocean sediments was related to the age of the ocean crust. Observations of ocean-floor sediments revealed that the thickness of the sediments increases with distance from an ocean ridge, as shown in **Figure 9.** The pattern of thickness across the ocean floor was symmetrical across the ocean ridges.

CAREERS IN EARTH SCIENCE

Marine geologist Earth scientists who study the ocean floor to understand geologic processes such as plate tectonics are marine geologists.

WebQuest

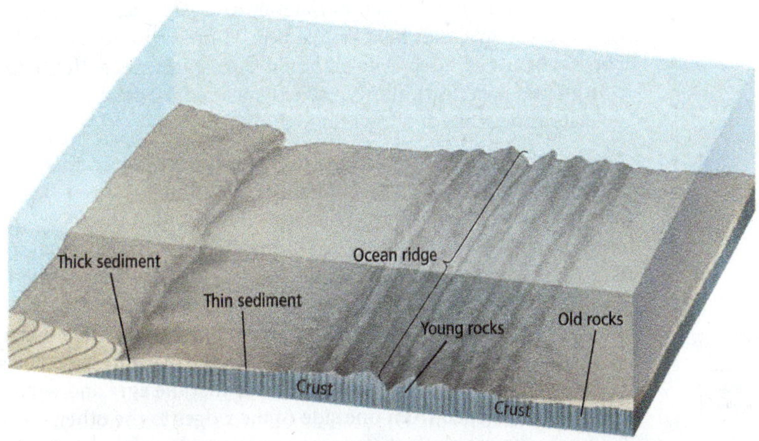

Figure 9 The ages of ocean crust and the thicknesses of ocean-floor sediments increase with distance from the ridge.

Enrichment

Sediments Deep-sea sediments are fine-grained and accumulate slowly. Ask students the following question: What do you think might be the source of these sediments? Atmospheric dust and shells from plankton make up most deep-sea sediments. Ocean circulation also transports sediments from the continental slope to the open ocean.

CAREERS IN EARTH SCIENCE

WebQuest

Reinforcement

Deep-Sea Sediments Ask students the following question: Where would you expect to find the thickest deep-sea sediments?

a. near ocean ridges
b. near the shore
c. on the tops of seamounts
d. near trenches

d. The thickest ocean sediments are near trenches.

Interpret the Illustration

Figure 9 Use the figure to encourage students to hypothesize what causes the variations in thickness shown. Sediments are thicker closer to trenches because the sediment supply is greater than it is near ocean ridges, which are often located in the middle of oceans.

DIFFERENTIATED INSTRUCTION

Advanced Learners Challenge students to find out the answers to the following questions. Is the location of Earth's north magnetic pole in the same place as the geographic North Pole? no How does this affect the accuracy of compasses near the geographic North Pole? The compass needle points to the magnetic pole. As you move east or west of a line where the two poles are in alignment, the compass accuracy changes. Maps are adjusted for this difference with declination marks. Have students draw a diagram to show the orientation of a compass arrow when the compass is at the geographic North Pole.

Interpret the Illustration

Figure 11 Have students use the figure to answer the following questions: What is the current epoch? the Brunhes normal epoch How many times did Earth's magnetic field change polarity during the Gilbert reversal epoch? eight times

Teacher Content Support

Earth's Magnetic Field Earth's magnetic field is caused primarily by internal sources. The most widely accepted hypothesis is that a geomagnetic generator is operating in the core, which is thought to be composed of the conducting elements nickel (Ni) and iron (Fe). It is hypothesized that fluid motion in the core generates an electric field, which, in turn, creates the magnetic field detected at the surface. The fluid motion is caused by thermal heating, which is caused by both radioactive decay and the release of heat that occurs when the liquid outer core freezes against the solid inner core. As convection currents rise, they twist as a result of the Coriolis Effect to give rise to the magnetic field.

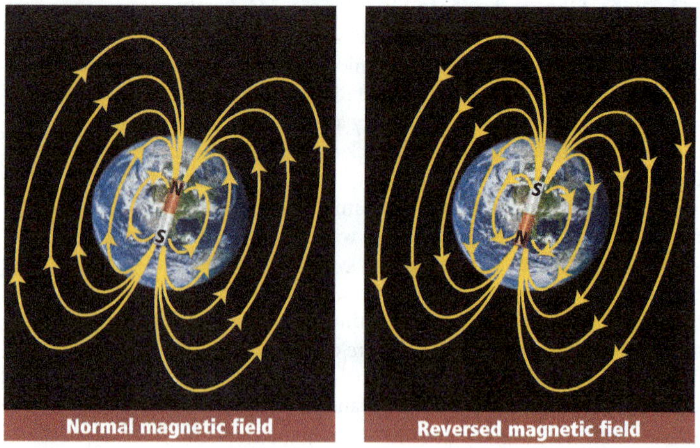

Figure 10 Earth's magnetic field is generated by the flow of molten iron in the liquid outer core. The polarity of the field changes over time from normal to reversed.

Figure 11 Periods of normal polarity alternate with periods of reversed polarity. Long-term changes in Earth's magnetic field, called epochs, are named as shown here. Short-term changes are called events.

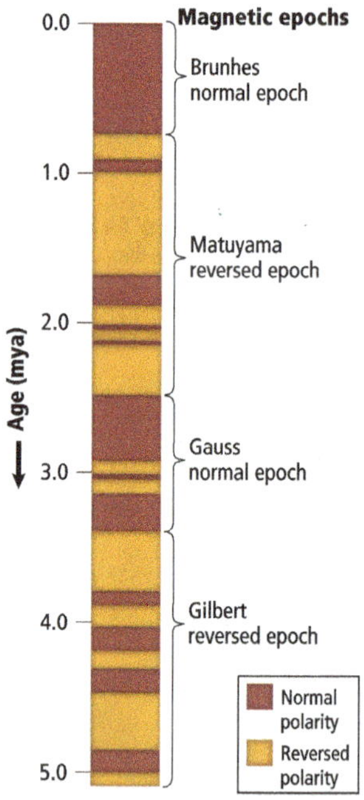

Magnetism

Earth has a magnetic field generated by the flow of molten iron in the outer core. This field is what causes a compass needle to point to the North. A **magnetic reversal** happens when the flow in the outer core changes, and Earth's magnetic field changes direction. This would cause compasses to point to the South. Magnetic reversals have occurred many times in Earth's history. As shown in **Figure 10,** a magnetic field that has the same orientation as Earth's present field is said to have normal polarity. A magnetic field that is opposite to the present field has reversed polarity.

Magnetic polarity time scale **Paleomagnetism** is the study of the history of Earth's magnetic field. When lava solidifies, iron-bearing minerals such as magnetite crystallize. As they crystallize, these minerals behave like tiny compasses and align with Earth's magnetic field. Data gathered from paleomagnetic studies of continental lava flows allowed scientists to construct a magnetic polarity time scale, as shown in **Figure 11.**

Magnetic symmetry Scientists knew that oceanic crust is mostly basaltic rock, which contains large amounts of iron-bearing minerals of volcanic origin. They hypothesized that the rocks on the ocean floor would show a record of magnetic reversals. When scientists towed magnetometers behind ships to measure the magnetic orientation of the rocks of the ocean floor, a surprising pattern emerged. The regions with normal and reverse polarity formed a series of stripes across the floor parallel to the ocean ridges. The scientists were doubly surprised to discover that the ages and widths of the stripes matched from one side of the ridges to the other. Compare the magnetic pattern on opposite sides of the ocean ridge shown in **Figure 12.**

DIFFERENTIATED INSTRUCTION

Visually Impaired Provide enlarged, color photocopies of **Figure 11** to students with visual impairments. These can help them understand the pattern of correlative magnetic stripes on either side of an ocean ridge. Adding black lines to delineate the different ages of the crust will also make the diagram more effective for these students.
BL EL

IN THE FIELD

Island Arcs and Magnetic Lineations Seiya Uyeda of Japan studied island arcs, which led to his discovery of magnetic stripes in the rocks on the Pacific Ocean floor. This discovery proved that the floor of the Pacific had undergone a major northward movement at some point in its development. Uyeda also helped to develop a theory that explains the origins of plate motions.

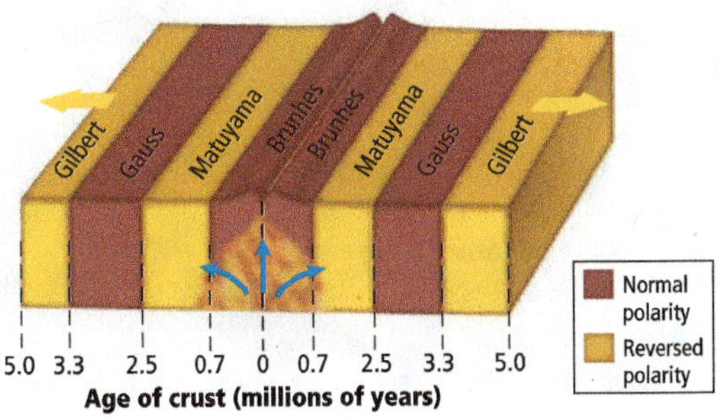

■ **Figure 12** Reversals in the polarity of Earth's magnetic field are recorded in the rocks that make up the ocean floor.
Identify *the polarity of the most recently produced basalt at the ocean ridge.*

By matching the patterns on the seafloor with the known pattern of reversals on land, scientists were able to determine the age of the ocean floor from magnetic recording. This method enabled scientists to quickly create isochron (I suh krahn) maps of the ocean floor. An **isochron** is an imaginary line on a map that shows points that have the same age—that is, they formed at the same time. In the isochron map shown in **Figure 13,** note that relatively young ocean-floor crust is near ocean ridges, while older ocean crust is found along deep-sea trenches.

■ **Figure 13** Each colored band on this isochron map of the ocean floor represents the age of that strip of the crust.
Observe *What pattern do you observe?*

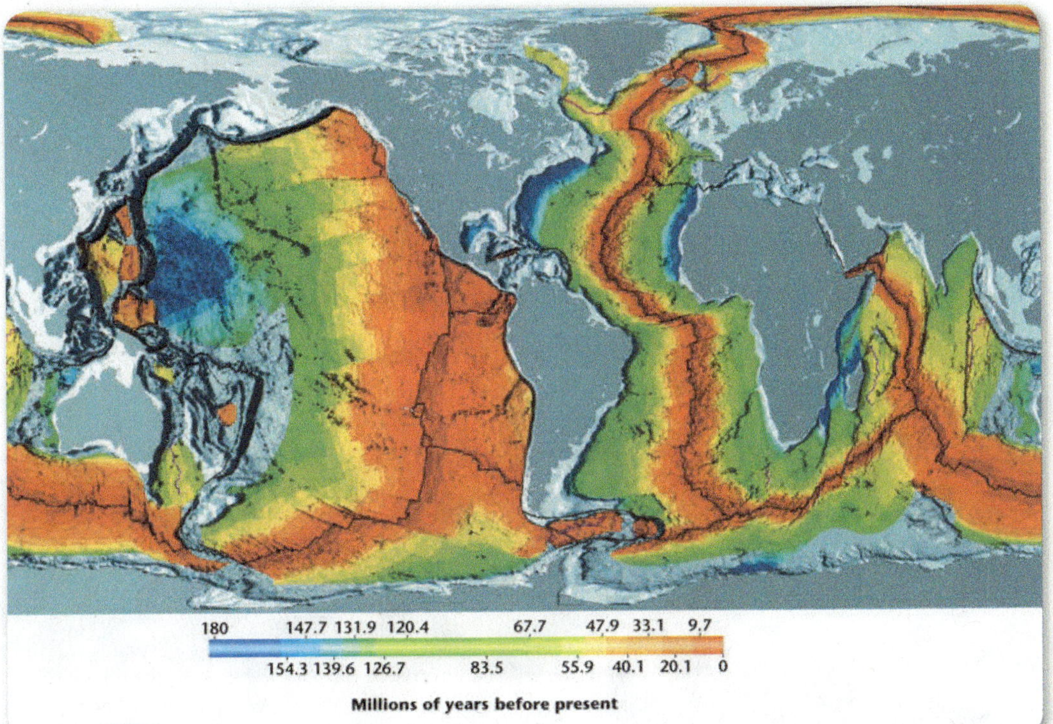

National Geophysical Data Center/NOAA/NGDC

■ **Caption Question Fig. 12** normal polarity

■ **Caption Question Fig. 13** The pattern is symmetrical; the same on either side of each ridge.

EARTH SCIENCE JOURNAL

Expedition Diary The discovery of ocean ridges and seafloor spreading took so long because of the great difficulty in mapping the seafloor. Have students write an Earth science journal entry describing a day on a ship in the open ocean and the challenges they faced. Students should identify the lack of resources (tools), storms, ocean depth, and remoteness from civilization. OL

Teacher Content Support

Dendrochronology The study of tree rings (dendrochronology; *dendro–* meaning *tree, chrono–* meaning *time*) is analogous to the study of magnetic striping on the seafloor. Tree rings evidence tree growth and are governed by growing conditions each year; a good year equals a thick ring. Trees produce one ring per year. Magnetic stripes record growth of the seafloor, but at irregular intervals. The width of the stripe depends on the rate of volcanic activity taking place in the rift zone and the rate of magnetic field reversal. Uncommon magnetic field reversal is independent of the more frequent volcanic activity. Reversals are recorded by contemporaneous volcanic activity. It is possible for a reversal to be missed if there was no volcanic activity to record it.

Interpret the Illustration

Figure 13 Have students use this map to answer the following questions. Where is the oldest portion of ocean crust on this map? in the northwestern Pacific Ocean Why is the seafloor relatively young west of South America even though it is near the Peru-Chile Trench? The ocean crust west of South America is relatively young because the Peru-Chile Trench subduction zone is not far from an ocean ridge.

Use Scientific Terms

Prefixes Have students each make a list of other scientific terms that contain the word part *iso–*. Terms might include *isobar, isopach, isotope,* and *isostatic,* among others. Ask students to write a brief definition next to each term.

VISUALIZING

VISUALIZING Seafloor Spreading

Purpose

Students will investigate the mechanism of seafloor spreading.

Teacher Content Support

Hot Spots Geologists think that a connected series of mantle hot spots are responsible for splitting Earth's crust into plates. A hot spot by itself is a point source of force due to rising magma in the mantle. A point source should dome the crust upward until it splits along three "arms" of failure. These arms are at approximately 120° from each other and widen in a divergent manner. Two of the arms seem to be more active than the third arm; such an inactive arm is called a failed arm. A modern example of this activity occurs in the region around the Horn of Africa: the Ethiopian Highlands are the dome, the Red Sea and Gulf of Aden are the active arms, and the East African Rift Valley to the south is the failed arm. When a series of such features occur close to each other and beneath a continent, their active arms can link up to create a long chain of divergent activity, thus splitting the continent in two. Such arms might eventually fill with seawater to become a new, linear sea. The mid-ocean ridge system is thought to be made of a large number of such connected active arms. What is still unknown is the reason for the positioning of hot spots. The Pacific Ocean basin has many singular and isolated hot spots which are not connected.

Figure 14 Data from topographic, sedimentary, and paleomagnetic research led scientists to propose seafloor spreading. Seafloor spreading is the process by which new oceanic crust forms at ocean ridges, and slowly moves away from the spreading center until it is subducted and recycled at deep-sea trenches.

Magma intrudes into the ocean floor along a ridge and fills the gap that is created. When the molten material solidifies, it becomes new oceanic crust.

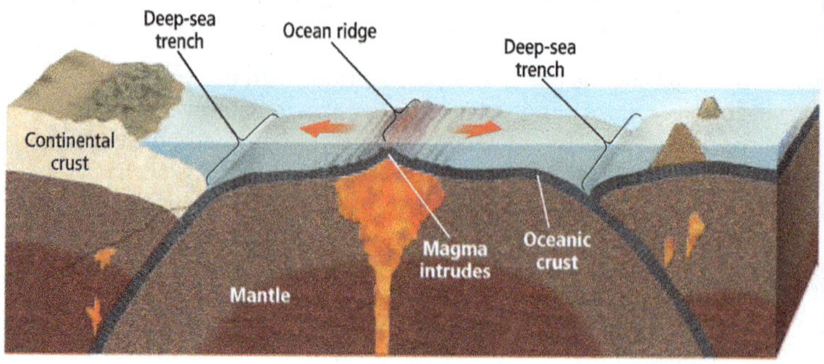

The continuous spreading and intrusion of magma result in the addition of new oceanic crust. Two halves of the oceanic crust spread apart slowly, and move apart like a conveyor belt.

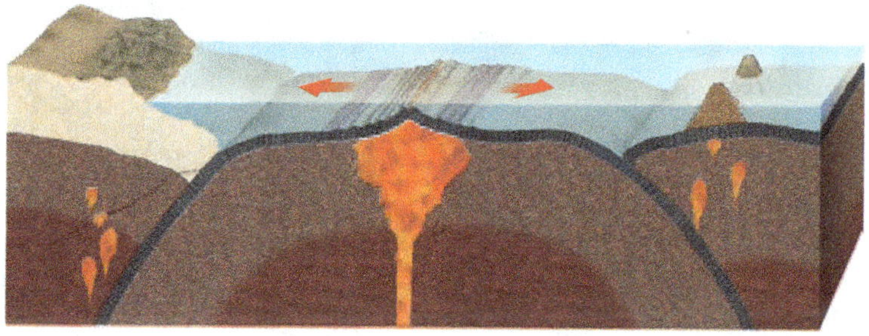

The far edges of the oceanic crust sink beneath continental crust. As it descends, water in the minerals is released, which helps melt the overlying mantle, forming magma. The magma rises and forms part of the continental crust.

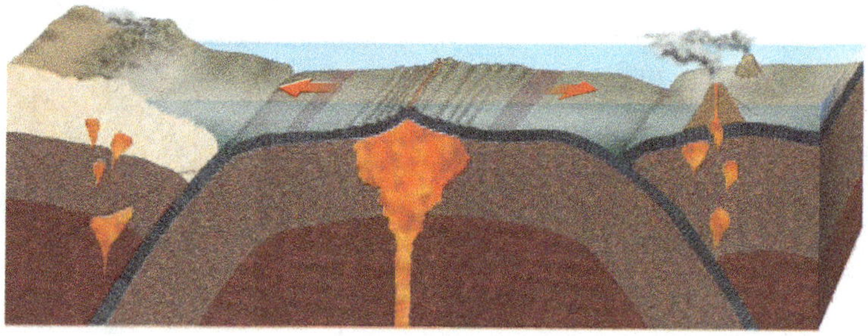

View an **animation of seafloor spreading.** Concepts In Motion

ACROSS THE CURRICULUM

History and Physics If it had not been for the work of Michael Faraday and Clark Maxwell, physicists, on the relationship between electricity and magnetism in the 1800s, geophysicists of today would not have the tools to understand magnetic stripes on the ocean floor. These striped patterns are a key piece of evidence for the spreading seafloor hypothesis and one of the mechanisms for driving plate tectonics. In turn, plate tectonics has proved to be a key factor in explaining changes in the biological distribution of (or isolation of) animals and plants.

Seafloor Spreading

Using all the topographic, sedimentary, and paleomagnetic data from the seafloor, seafloor spreading was proposed. **Seafloor spreading** is the theory that explains how new oceanic crust is formed at ocean ridges, slowly moved away from ocean ridges, and destroyed at deep-sea trenches. **Figure 14** illustrates how seafloor spreading occurs.

During seafloor spreading, magma, which is hotter and less dense than surrounding mantle material, is forced toward the surface of the crust along an ocean ridge. As the two sides of the ridge spread apart, the rising magma fills the gap that is created. When the magma solidifies, a small amount of new ocean floor is added to Earth's surface. As spreading along a ridge continues, more magma is forced upward and solidifies. This cycle of spreading and the intrusion of magma continues the formation of ocean floor, which slowly moves away from the ridge. Of course, seafloor spreading mostly happens under the sea, but in Iceland, a portion of the Mid-Atlantic Ridge rises above sea level. **Figure 15** shows lava erupting along the ridge.

Recall that while Wegener collected many data to support the idea that the continents are drifting across Earth's surface, he could not explain what caused the landmasses to move or how they moved. Seafloor spreading was the missing link that Wegener needed to complete his model of continental drift. Continents are not pushing through ocean crust, as Wegener proposed. In fact, continents are more like passengers that ride along while ocean crust slowly moves away from ocean ridges. Seafloor spreading led to a new understanding of how Earth's crust and rigid upper mantle move. This will be explored in the next sections.

Figure 15 The entire island of Iceland lies on the Mid-Atlantic ocean spreading center. Because the seafloor is spreading, Iceland is growing larger. In 1783, more than 12 km^3 of lava erupted—enough to pave the entire U.S. interstate freeway system to a depth of 10 m.

Watch a **video about divergent boundaries.**

Video

SECTION 2 REVIEW

Section Self-Check

Section Summary

- Studies of the seafloor provided evidence that the ocean floor is not flat and unchanging.
- Oceanic crust is geologically young.
- New oceanic crust forms as magma rises at ridges and solidifies.
- As new oceanic crust forms, the older crust moves away from the ridges.

Understand Main Ideas

1. **MAIN**IDEA **Describe** why seafloor spreading is like a moving conveyor belt.
2. **Explain** how ocean-floor rocks and sediments provided evidence of seafloor spreading.
3. **Differentiate** between the terms *reversed polarity* and *normal polarity*.
4. **Describe** the topography of the seafloor.

Think Critically

5. **Explain** how an isochron map of the ocean floor supports the theory of seafloor spreading.
6. **Analyze** Why are magnetic bands in the eastern Pacific Ocean so far apart compared to the magnetic bands along the Mid-Atlantic Ridge?

MATH IN ▶ Earth Science

7. Analyze **Figure 11.** What percentage of the last 5 million years has been spent in reversed polarity?

3 Assess

Check for Understanding

Graphic Organizer Have students each use the headings in this section to make a concept map that summarizes the ideas presented in the section. As an alternative, provide students with the headings and relevant information and have each student organize the text as a concept map.

Reteach

Describe Explain that the process of seafloor spreading is analogous to the movement of a conveyor belt at a supermarket checkout counter.

Assessment

Knowledge Ask students to imagine that they are oceanographers studying core samples of the ocean floor. Sample 1 was taken near a trench. Sample 2 was taken near an ocean ridge. Have students describe the differences they could expect to observe in the two core samples. Sample answer: Sample 1 is older, may have a different polarity, and has more sediment on top.

What's EARTH SCIENCE Got To Do With It?

Video *Being Stressed Out*

SECTION 2 REVIEW

1. In seafloor spreading, new oceanic crust is added at the mid-ocean ridge, and, as intrusion continues, seafloor moves toward the edge of the oceanic plate, where it is ultimately pulled back down into the magma below.
2. Ocean rocks are geologically young, and they are youngest toward the mid-ocean ridges, so there must be a mechanism for production and recycling of ocean floor; ocean floor sediments are deeper the further away from the ridge, so mid-ocean floor must be newer than the farther edges.
3. Rocks showing normal polarity have the same magnetic orientation as Earth's present field. Rocks with reversed polarity have the opposite orientation.
4. It has deep trenches and high ridges. At ocean ridges are the longest mountain chains on Earth's surface. Farther from the mid–oceanic ridges, most of the ocean floor is blanketed with sediments.
5. Isochron maps show that the magnetism was recorded by cooling lava. Isochron maps of the ocean floor revealed a symmetric pattern on either side of ocean ridges, which meant that both sides were created at the same time.
6. because Pacific is spreading at quicker rate than Atlantic (about 8 cm/year)
7. About 70%; Brunhes plus Gauss equal 1.5 million years of normal polarity; subtract 1.5 from 5 million years to get 3.5 million years of reversed polarity; divide 3.5 by 5 to get .7, times 100 equals 70%

SECTION 3

1 Focus

Focus Transparency

Before presenting the lesson, project **Section Focus Transparency 51** and have students answer the accompanying questions. EL

MAINIDEA

Plate Boundaries Most of Earth's active geology is explained by plate tectonics. Compare the location of large cities with plate boundaries. In the United States, what large cities are near plate boundaries? Los Angeles, San Francisco, Seattle Ask students to identify geological features such as mountains and volcanoes that are far from plate boundaries in the United States. Appalachians, Hawaii

2 Teach

Tie to Previous Knowledge

Link Theories Ask the following question to stress the link between the theories of seafloor spreading and plate tectonics: According to the theory of seafloor spreading, how are tectonic plates moving at an ocean ridge? An ocean ridge is a boundary where plates are moving apart.

Interpret the Illustration

Fig. 16 Ask the following questions: What types of plate boundaries are found along the edges of the North American Plate? transform and convergent along the western edge; a divergent boundary is found along the eastern edge Does every plate involve all three types of boundaries? Explain. Yes; Earth is not getting larger. Crust formed at ocean ridges is destroyed at subduction zones.

SECTION 3

Plate Boundaries

MAINIDEA Volcanoes, mountains, and deep-sea trenches form at the boundaries between the plates.

Essential Questions

- How does the movement of Earth's tectonic plates result in many geologic features?
- What are the three types of plate boundaries and the features associated with each?
- What are the processes associated with subduction zones?

Review Vocabulary

mid-ocean ridge: a major feature along the ocean floor consisting of an elevated region with a central valley

New Vocabulary

tectonic plate
divergent boundary
rift valley
convergent boundary
subduction
transform boundary

EARTH SCIENCE 4 YOU Imagine a pot of soup that has been allowed to cool in a refrigerator. Fats in the soup have solidified into a hard surface, but if you tilt the pot back and forth, you will see the rigid surface bending and cracking. This is similar to the relationship between different layers of Earth.

Theory of Plate Tectonics

The evidence for seafloor spreading suggested that continental and oceanic crust move as enormous slabs, which geologists describe as tectonic plates. **Tectonic plates** are huge pieces of crust and rigid upper mantle that fit together at their edges to cover Earth's surface. As illustrated in **Figure 16,** there are about 8 major plates and several smaller ones. These plates move very slowly—only a few centimeters each year—which is similar to the rate at which fingernails grow. Plate tectonics is the theory that describes how tectonic plates move and shape Earth's surface. They move in different directions and at different rates relative to one another and they interact with one another at their boundaries. Each type of boundary has certain geologic features and processes associated with it. A divergent boundary occurs where tectonic plates move away from each other. A convergent boundary occurs where tectonic plates move toward each other. A transform boundary occurs where tectonic plates move horizontally past each other.

Figure 16 Earth's crust and rigid upper mantle are broken into enormous slabs called tectonic plates that interact at their boundaries.

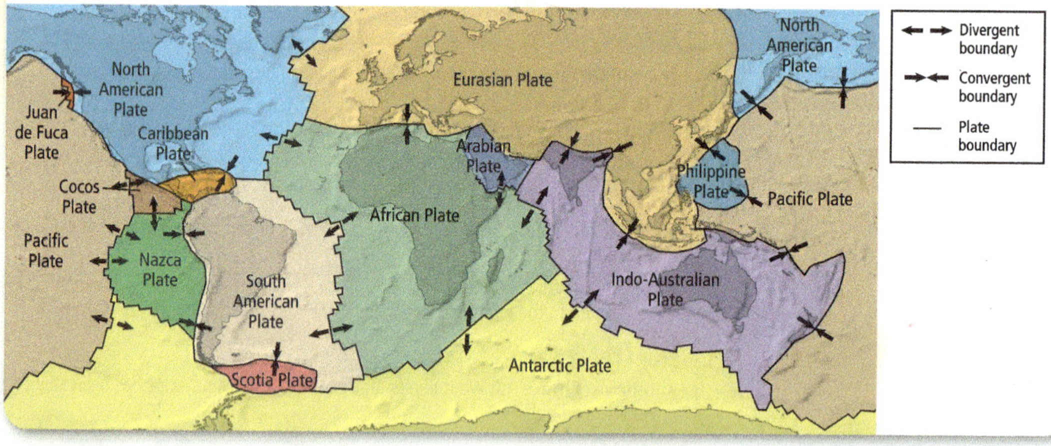

DIFFERENTIATED INSTRUCTION

Visually Impaired Have students model ocean-continent convergent zones. Have them slide the fingers of the left hand over the backs of the fingers of the right hand. The left fingers should bend upward as they slide to re resent the island arc. The right hand represents the subducted plate. Other types of convergent zones can be modeled in a similar way. BL EL

ACROSS THE CURRICULUM

History The Mediterranean Sea overlies the convergent boundary between Africa and Asia. Have students find out what geologic features have formed along this boundary and how these features have affected the cultures and history of the region. Earthquakes and volcanoes affected historical events and styles of local architecture, gave rise to myths, and destroyed cities, among other effects.

Divergent boundaries Regions where two tectonic plates are moving apart are called **divergent boundaries.** Most divergent boundaries are found along the seafloor in rift valleys. It is in this central rift that the process of seafloor spreading begins. Magma rising through the rift's faults forms a mid-ocean ridge. The mid-ocean ridge appears as a continuous mountain chain on the ocean floor. The formation of new ocean crust at most divergent boundaries accounts for the high heat flow, volcanism, and earthquakes associated with these boundaries.

☑ READING CHECK **Identify** the cause of volcanism and earthquakes associated with mid-ocean ridges.

Throughout millions of years, the process of seafloor spreading along a divergent boundary can cause an ocean basin to expand. Although most divergent boundaries form ridges on the ocean floor, some divergent boundaries form on continents. When continental crust begins to separate, the stretched crust forms a long, narrow depression called a **rift valley. Figure 17** shows the rift valley that is currently forming in East Africa. The rifting might eventually lead to the formation of a new ocean basin.

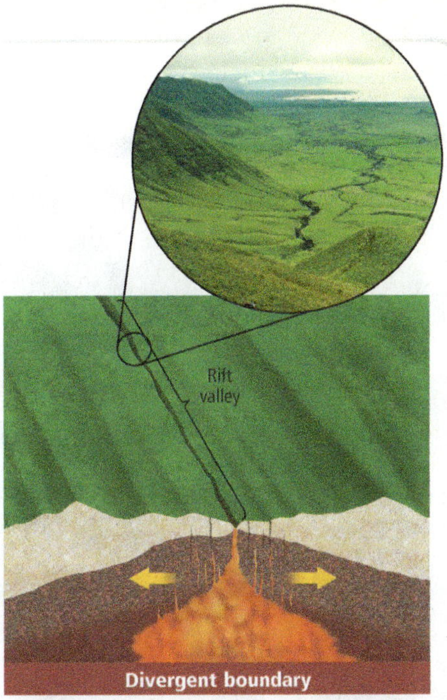

■ **Figure 17** Divergent boundaries are places where plates separate. An ocean ridge is a divergent boundary on the ocean floor. In East Africa, a divergent boundary has also created a rift valley.

MiniLAB

iLab Station

Model Ocean-Basin Formation

How did a divergent boundary form the South Atlantic Ocean? Around 150 mya, a divergent boundary split an ancient continent. Over time, new crust was added along the boundary, widening the rift between Africa and South America.

Procedure

1. Read and complete the lab safety form.
2. Use a **world map** to create **paper templates** of South America and Africa.
3. Place the two continental templates in the center of a **large piece of paper,** and fit them together along their Atlantic coastlines.
4. Carefully trace around the templates with a **pencil.** Remove the templates and label the diagram *150 mya.*
5. Use an average spreading rate of 4 cm/y and a map scale of 1 cm = 500 km to create six maps that show the development of the Atlantic Ocean at 30-million-year intervals, beginning 150 mya.

Analysis

1. **Compare** your last map with a world map. Is the actual width of the South Atlantic Ocean the same on both maps?
2. **Consider** why there might be differences between the width in your model and the actual width of the present South Atlantic Ocean.

MiniLAB

iLab Station | Rubric

Purpose Model the formation of the South Atlantic Ocean over the last 150 million years.

Process Skills model, measure in SI

Safety Precautions Approve lab safety forms before work begins. Remind students to handle the scissors with care.

Teaching Strategy If time is limited, provide students with templates of South America and Africa.

Expected Results The distance increases by 1200 km every 30 million years. 4 cm/y × 30,000,000 yr. = 120,000,000 or 1200 km On the maps, 1200 km equals 2.4 cm. The width of the ocean in the model should be about 12 cm.

Analyze and Conclude

1. probably will not be exactly the same
2. Answers will vary. Correct: seafloor spreading rate is not constant.

Assessment

Performance Have each student create a map of the South Atlantic Ocean 50 million years from now.

LabManager™
Customize any lab with the LabManager™ CD-ROM.

☑ READING CHECK They are caused by the process of seafloor spreading and the movement of magma just below the crust of the ocean.

Teacher Content Support

Magmas Igneous activity is characteristic of subduction zones. Either volcanic island arcs or volcanic-magmatic mountain ranges form as plates converge. The magma that erupts from these features represents a mix of melted subducting plate, melted subducted sediments, and water. At depths of 50 to 100 km, some of the basalt and sediments begin to melt, releasing water and other minerals. The ultramafic magma generated slowly changes composition through fractional crystallization and assimilation of crustal rocks during intrusion. In the case of ocean-continent subduction, the rising magma mixes with continental (granitic) rocks and often erupts with an andesitic composition. With ocean-ocean subduction the basaltic magma erupts as recycled basalt with less compositional change.

Tie to Previous Knowledge

Rock Chemistry Ask the following questions to review some of the properties of the two kinds of crust: What are the primary elements found in granite and basalt? silica; iron, and magnesium What is responsible for the differences in density between these two rocks? Basalt has a greater density because it contains minerals with less silica and more iron and magnesium than the minerals in granite.

Figure 18 Oceanic plates are mostly basalt. Continental plates are mostly granite with a thin cover of sedimentary rock, both of which are less dense than basalt.

Convergent boundaries At **convergent boundaries,** two tectonic plates are moving toward each other. When two plates collide, the denser plate eventually descends below the other, less-dense plate in a process called **subduction.** There are three types of convergent boundaries, classified according to the type of crust involved. Recall that oceanic crust is made mostly of minerals that are high in iron and magnesium, which form dense, dark-colored basaltic rocks, such as the basalt shown in **Figure 18.** Continental crust is composed mostly of minerals such as feldspar and quartz, which form less-dense, lighter-colored granitic rocks. The differences in density of the crustal material affects how they converge. The three types of tectonic boundaries and their associated landforms are shown in **Table 1.**

Oceanic-oceanic In the oceanic-oceanic convergent boundary shown in **Table 1,** a subduction zone is formed when one oceanic plate, which is denser as a result of cooling, descends below another oceanic plate. The process of subduction creates an ocean trench. The subducted plate descends into the mantle, thereby recycling oceanic crust formed at the ridge. Water carried into Earth by the subducting plate lowers the melting temperature of the overlying mantle causing it to melt. The molten material, called magma, is less dense, so it rises back to the surface where it often erupts and forms an arc of volcanic islands that parallel the trench. Some examples of trenches and island arcs are the Mariana Trench and Mariana Islands in the West Pacific Ocean and the Aleutian Trench and Aleutian Islands in the North Pacific Ocean. A volcanic peak in the Aleutian Island arc is shown in **Table 1.**

Oceanic-continental Subduction zones are also found where an oceanic plate converges with a continental plate, as shown in **Table 1.** Note that it is the denser oceanic plate that is subducted. Oceanic-continental convergence also produces a trench and volcanic arc. However, instead of forming an arc of volcanic islands, oceanic-continental convergence results in a chain of volcanoes along the edge of the continental plate. The result of this type of subduction is a mountain range with many volcanoes. The Peru-Chile Trench and the Andes mountain range, which are located along the western coast of South America, formed in this way.

VOCABULARY

ACADEMIC VOCABULARY

Parallel (PAIR uh lel)
extending in the same direction, everywhere equidistant, and not meeting
The commuter train runs parallel to the freeway for many kilometers.

EARTH SCIENCE JOURNAL

Rock Life Cycle Have students write a paragraph describing the life cycle of a piece of basalt formed at an ocean ridge and then subducted at a trench. Students should begin with the rock solidifying, getting colder and older, moving across an ocean basin, sinking back into Earth at a trench and perhaps melting and later erupting at an island arc. OL

Rubric

Explore **convergent boundaries** with an **interactive table.** Concepts In Motion

Table 1 Summary of Convergent Boundaries

Type of Convergent Boundary	Example of Region Affected by Boundary	Example of Landform Produced
Oceanic-oceanic 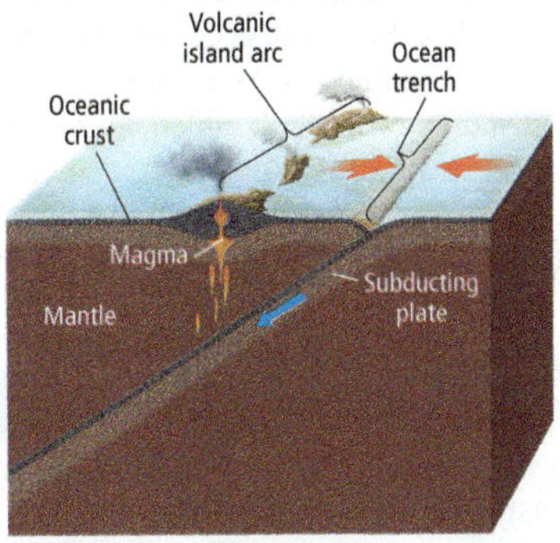	 Aleutian Islands	 Chagulak Island, Alaska
Oceanic-continental 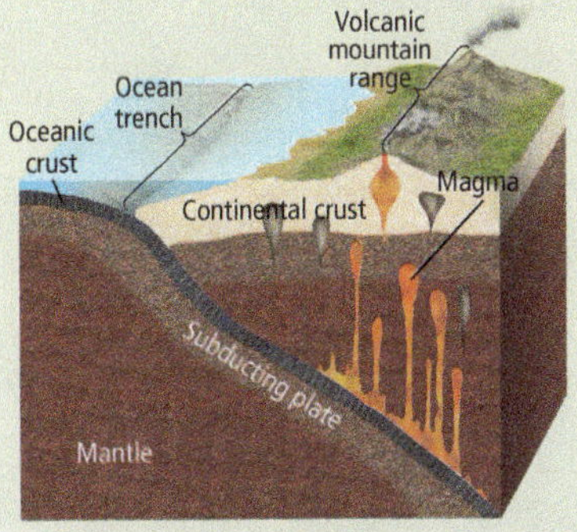	 Andes mountain range	 Osorno Volcano, Chile
Continental-continental 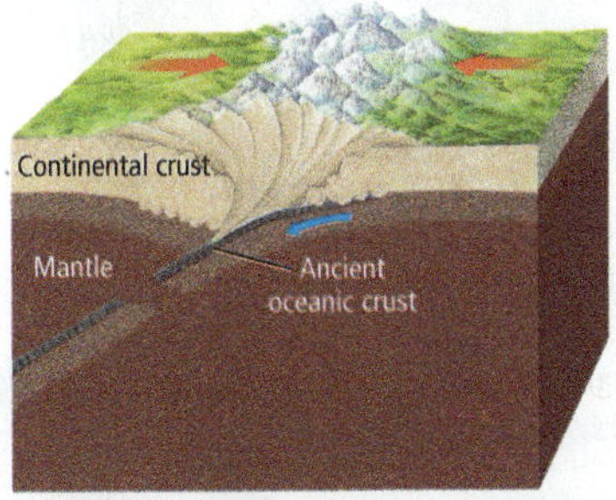	 Himalayas	 Ama Dablan, Nepal

Teacher Content Support

Subduction Zones Early evidence that subduction zones were sites of plate destruction included ocean-floor trenches, the relatively young age of ocean crust, and the existence of island arcs. Other evidence that supports the concept of subduction is the less-than-normal heat flow from the ocean floor, which is thought to be the result of the lower temperature of the subducted slab. Also, the gravitational field over a subduction zone is slightly less than expected because there is extra crust where the two plates overlap. Deep-focus earthquakes, however, provide the strongest evidence of subduction. Earthquakes originating nearly 700 km beneath Earth's surface have been detected near deep-sea trenches. At this depth, the mantle is not rigid enough to break and cause earthquakes. Thus, these deep-focus quakes must be originating in the still cool and solid portions of the subducted plate.

DIFFERENTIATED INSTRUCTION

Advanced Learners Pangaea split in a north-south direction to form the Atlantic Ocean. Challenge students to find the answers to the following questions: What type of convection flow developed beneath Pangaea? upwelling or plumes How would Earth's present geography be different if the mantle convection had been oriented in an east-west direction? Africa and South America remain attached, separated by an ocean from a combined North America and Europe. How does the breakup of Pangaea support the hypothesis that mantle convection changes over time? Continents separate and come back together.

Problem-Solving LAB

Purpose Students will determine how relative plate motions change along a transform boundary.

Process Skill interpret data

Teaching Strategies

- Review the process of seafloor spreading. Help students to conclude that crust is moving away from the ridge.
- Suggest that students draw several small arrows along both sides of the fault to make it easier to see how relative motion changes.

Analysis

1. Arrows should indicate that the seafloor is moving in opposite directions on either side of the ridge.
2. Movements between A and D are in the same direction, movements between B and E are in opposite directions, and movements between C and F are in the same direction.

Think Critically

3. A, D, and E
4. ridge is a boundary between the two plates
5. C and D

Reinforcement

Relative Motion of Faults Ask the following question: Is the relative motion of the lettered locations in the Problem-Solving Lab the same along the entire length of the fault? The Gibbs Fracture Zone is a transform fault across the Mid-Atlantic Ridge divergent boundary. Locations A, D, and E are moving west with the diverging plate, while locations B, C, and F are moving east. However, relative motion changes in regard to the Gibbs Fracture Zone transform fault. Locations A, B, and C are moving west with the fault, while locations D, E, and F are moving east. Therefore, location A is moving west with both features and location F is moving east with both features. But locations B, C, D, and E are moving in opposite directions due to the different directional movement of each feature. Locations B and C are moving eastward with the diverging plate, but westward with the fault. Locations D and E are moving westward with the diverging plate, but eastward with the fault.

Continental-continental The third type of convergent boundary forms when two continental plates collide. Continental-continental boundaries form long after an oceanic plate has converged with a continental plate. Recall that continents are often carried along attached to oceanic crust. Over time, an oceanic plate can be completely subducted, dragging an attached continent behind it toward the subduction zone. As a result of its denser composition, oceanic crust descends beneath the continental crust at the subduction zone. The continental crust that it pulls behind it cannot descend because continental rocks are less dense, and will not sink into the mantle. As a result, the edges of both continents collide, and become crumpled, folded, and uplifted. This forms a vast mountain range, such as the Himalayas, as indicated in **Table 1.**

FOLDABLES® Incorporate information from this section into your Foldable.

Transform boundaries A region where two plates slide horizontally past each other is a **transform boundary,** as shown in **Figure 19.** Transform boundaries are characterized by long faults, sometimes hundreds of kilometers in length, and by shallow earthquakes. Transform boundaries were named for the way Earth's crust changes, or transforms, its relative direction and velocity from one side of the boundary to the other. Recall that new crust is formed at divergent boundaries and destroyed at convergent boundaries. Crust is only deformed or fractured somewhat along transform boundaries.

Problem-Solving LAB

Interpret Scientific Illustrations

How does plate motion change along a transform boundary? The figure at the right shows the Gibbs Fracture Zone, which is a segment of the Mid-Atlantic Ridge located south of Iceland and west of the British Isles. Copy this figure.

Analysis

1. **Draw** arrows on your copy to indicate the direction of seafloor movement at locations A, B, C, D, E, and F.
2. **Compare** the direction of motion for the following pairs of locations: A and D, B and E, and C and F.

Think Critically

3. **Differentiate** Which three locations are on the North American Plate?
4. **Indicate** the portion of the fracture zone that is the boundary between North America and Europe.
5. **Assess** Which two locations represent the oldest crust?

FOLDABLES® Rubric

■ **Figure 19** Plates move horizontally past each other along a transform plate boundary. The bend in these train tracks resulted from the transform boundary running through parts of Southern California.

Most transform boundaries cause sections of ocean ridges to become offset, as you observed in the Problem-Solving Lab. Sometimes transform boundaries occur on continents. The San Andreas Fault is probably the best-known example. Recall from the Launch Lab at the beginning of this chapter that the San Andreas Fault system is part of a transform boundary that separates southwestern California from the rest of the state. Movements along this transform boundary create situations like the one shown in **Figure 19.** Offset train tracks, roads, fences, and creeks are telltale signs of a transform fault. Movements along the San Andreas Fault system are responsible for many of the earthquakes that strike California every year.

SECTION 3 REVIEW

Section Self-Check

Section Summary

- Earth's crust and rigid upper mantle are broken into large slabs of rock called tectonic plates.
- Plates move in different directions and at different rates over Earth's surface.
- At divergent plate boundaries, plates move apart. At convergent boundaries, plates come together. At transform boundaries, plates slide horizontally past each other.
- Each type of boundary is characterized by certain geologic features.

Understand Main Ideas

1. **MAINIDEA Describe** how plate tectonics results in the development of Earth's major geologic features.
2. **Summarize** the processes of convergence that formed the Himalayan mountains.
3. **List** the geologic features associated with each type of convergent boundary.
4. **Identify** the type of location where transform boundaries most commonly occur.

Think Critically

5. **Choose** three plate boundaries in **Figure 16,** and predict what will happen over time at each boundary.
6. **Describe** how two portions of newly formed crust move between parts of a ridge that are offset by a transform boundary.

WRITING IN ▶ Earth Science

7. Write a news report on the tectonic activity that is occurring at the Aleutian Islands in Alaska.

GeoLAB

iLab Station

The GeoLab located at the end of the chapter can be used at this point in the lesson.

3 Assess

Check for Understanding

Use Scientific Explanations
Ask the following questions to check students' understanding of plate boundaries: How do plates move at divergent boundaries? apart What type of movement takes place at convergent boundaries? Plates come together. How do convergent boundaries differ? Different types of crust are involved. How do plates move at a transform boundary? Plates move horizontally past each other.

Reteach

Infer Make a black-and-white copy of a world map that shows the plate boundaries and directions of movement. Have students use three different highlighters to color-code the different types of boundaries.

Assessment

Performance Have small groups of students use modeling clay to make a three-dimensional model of one type of convergent boundary.

SECTION 3 REVIEW

1. When the boundaries of tectonics plates collide, they produce many of Earth's geologic features such as mountains, volcanoes, and island arcs. Where the boundaries are separating from each other, they produce mid-ocean ridges, rift valleys, and new ocean floor.
2. Long ago, oceanic crust converged with a continent. The oceanic crust was attached to continental crust. After the oceanic crust was completely subducted, the attached continental crust collided with the other continental crust. Neither continent could subduct, so they became crumpled, folded and uplifted, forming a vast mountain range.
3. oceanic-oceanic: trenches and island arcs; oceanic-continental: trenches and volcanic mountain chains on land; continental–continental: folded mountains
4. at mid-ocean ridges
5. Possible answers: Nazca plate will eventually disappear under South American plate; the Atlantic Ocean will grow between the Eurasian and North American plates; ocean will widen between Antarctic and Pacific plate.
6. Crust is deformed or fractured as it moves horizontally along the fault.
7. Answers should mention that the Aleutian Islands are a chain of volcanic islands formed at a subduction zone.

Rubric

SECTION 4

1 Focus

Focus Transparency

Before presenting the lesson, project **Section Focus Transparency 52** and have students answer the accompanying questions. EL

MAINIDEA

Driving Forces What type of boundary should form above a rising current in the mantle? divergent boundaries Above sinking currents? convergent boundaries Have students think about why convection occurs. What causes the mantle to rise and sink? changes in density; hot mantle is buoyant and rises, cold mantle is dense and sinks Can students predict where the mantle is hottest? below ridges

2 Teach

Teacher Content Support

Tectonic Forces The forces involved in plate movements are thought to be concentrated along plate boundaries. There are four major forces acting on plates. The slab-pull force acts on the subducting plate and is the result of the negative buoyancy of the cooler, sinking slab. The subduction-suction force, which is also called the trench-suction force, pulls on the edge of the overlying plate. These two subduction-zone forces create tension in the lithosphere. The ridge-push force acts at ocean ridges and is the result of the hotter, less-dense material that makes up the ridge. This force causes lateral compression in the oceanic plates near the ridge. The mantle-drag force is a force that acts on the base of a moving plate. Because of the low viscosity of the asthenosphere, this force is relatively small when compared to the other forces.

SECTION 4

Essential Questions

- How is the process of convection explained?
- How is convection in the mantle related to the movements of tectonic plates?
- What are the processes of ridge push and slab pull.

Review Vocabulary

convection: the circulatory motion that occurs in a fluid at a nonuniform temperature owing to the variation of its density and the action of gravity

New Vocabulary

ridge push
slab pull

Causes of Plate Motions

MAINIDEA Convection currents in the mantle cause plate motions.

EARTH SCIENCE 4 YOU You probably know a lava lamp does not contain real lava, but the materials inside a lava lamp behave much like the molten rock within Earth.

Convection

One of the main questions about the theory of plate tectonics has remained unanswered since Alfred Wegener first proposed continental drift. What force or forces cause tectonic plates to move? Many scientists now think that large-scale motion in the mantle—Earth's interior between the crust and the core—is the mechanism that drives the movement of tectonic plates.

Convection currents Recall that convection is the transfer of thermal energy by the movement of heated material from one place to another. As in a lava lamp, the cooling of matter causes it to contract slightly and increase in density. The cooled matter then sinks as a result of gravity. Warmed matter is then displaced and forced to rise. This up-and-down flow produces a pattern of motion called a convection current. Convection currents aid in the transfer of thermal energy from warmer regions of matter to cooler regions. A convection current can be observed in the series of photographs shown in **Figure 20.** Earth's mantle is composed of primarily solid material that is heated unevenly by radioactive decay from both the mantle itself and the core beneath it. Radioactive decay heats up the mantle causing solid rock to slowly flow, and enormous convection currents that move material throughout the mantle.

Figure 20 Water cooled by the ice cube sinks to the bottom where it is warmed by the burner and rises. The process continues as the ice cube cools the water again.
Infer *what will happen to the ice cube due to convection currents.*

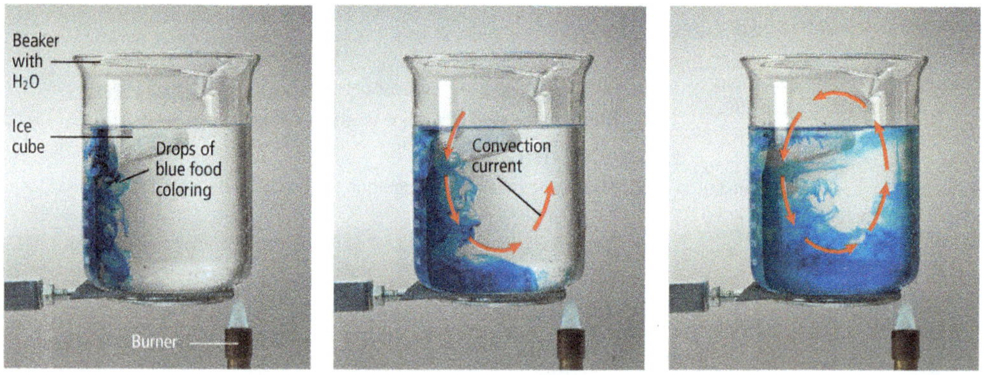

Caption Question Fig. 20 As warm water rises, the ice cube will melt.

EARTH SCIENCE JOURNAL

Mantle Convection Have students write a paragraph that discusses why questions about mantle convection have been so difficult for scientists to answer. Students' paragraphs should include that the motion is extremely slow and occurs deep within the planet. Thus, knowledge of this process is based on indirect evidence and observations. OL

Rubric

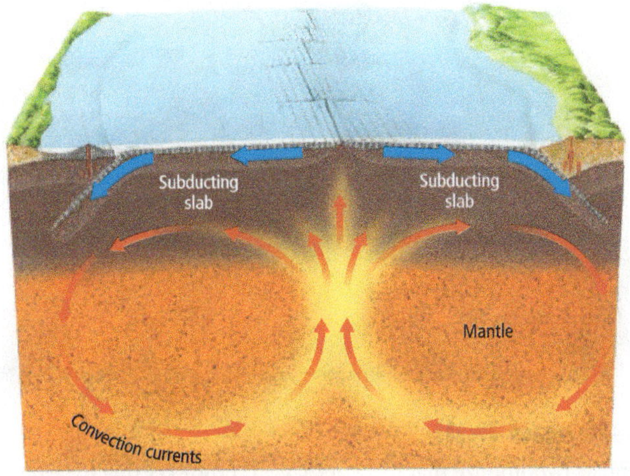

Figure 21 Convection currents develop in the mantle, moving the crust and outermost part of the mantle, and transferring thermal energy from the Earth's interior to its exterior.

Convection in the mantle Convection currents in the mantle, illustrated in **Figure 21**, are thought to be the driving mechanism of plate movements. Recall that even though the mantle is a solid, much of it moves like a soft, pliable plastic. The part of the mantle that is too cold and stiff to flow lies beneath the crust and is attached to it, moving as a part of tectonic plates. In the convection currents of the mantle, cooler mantle material is denser than hot mantle material. Mantle that has cooled at the base of tectonic plates slowly sinks downward toward the center of Earth. Heated mantle material is then displaced, and like the wax warmed in a lava lamp, it rises. Convection currents in the mantle are sustained by this rise and fall of material which results in a transfer of energy between Earth's hot interior and its cooler exterior. Although convection currents can be thousands of kilometers across, they flow at rates of only a few centimeters per year. Scientists think that these convection currents are set in motion by subducting slabs.

READING CHECK **Discuss** Which causes a convection current to flow: the rising of hot material, or the sinking of cold material?

Plate movement How are convergent and divergent movements of tectonic plates related to mantle convection? The rising material in the convection current spreads out as it reaches the upper mantle and causes both upward and sideways forces. These forces lift and split the lithosphere at divergent plate boundaries. As the plates separate, material rising from the mantle supplies the magma that hardens to form new ocean crust. The downward part of a convection current occurs where a sinking force pulls tectonic plates downward at convergent boundaries.

Identify Misconceptions

Some students might think that buoyancy is caused by an upward, pulling force on the less-dense material rather than a force that pushes from both the sides and below the material.

Uncover the Misconception

Have students draw a diagram of a block of foam rising through water. Ask students to label the diagram with arrows that show the different forces acting on the foam.

Demonstrate the Concept

Show how water exerts an upward force on a block of polystyrene foam with a bucket half-filled with water. Have a few volunteers attempt to push the block under water. Have them describe the resistance, or upward force, exerted by the water.

Assess New Knowledge

Have students each draw another diagram that correctly shows the forces acting on a material that floats in water.

READING CHECK Answers will vary. Ultimately convection is driven by the sinking of denser material, which displaces less-dense material, but geologists are not in complete agreement about which provides the driving force in Earth's mantle.

Demonstration

Convection Use a 1000-mL glass beaker nearly filled with water, a candle, a ring stand, food coloring, and an ice cube to demonstrate how a convection cell forms and moves. Set the candle under one edge of the beaker and put the ice cube in the beaker at the opposite edge. Place a drop of food coloring next to the ice cube. Students will observe the food coloring sink to the bottom of the beaker under the ice cube and then rise in the water above the flame. OL EL

3 Assess

Check for Understanding

Infer Ask students the following question: What type of mantle convection would you expect to find beneath a convergent plate boundary? A divergent plate boundary? Downward-moving legs of convection currents are thought to be associated with convergent margins, while upward-moving legs are found beneath divergent boundaries.

Reteach

Think Critically Use the analogy of the movement of a hot-air balloon through the air to explain the rising of material through the mantle via convection. Explain that the hot, less-dense air in the balloon causes the balloon to rise through the colder, more-dense, surrounding air. Likewise, warmer mantle material is less dense than colder mantle material, and thus, it is forced upward through the colder material.

Assessment

Infer Ask the following question: As spreading continues in the Great Rift Valley, which process will be dominant, ridge push or slab pull? Because this is a divergent boundary, ridge push will be the dominant process.

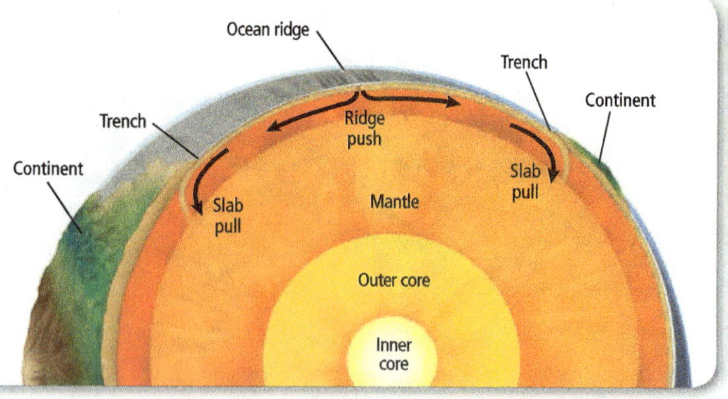

Figure 22 Ridge push and slab pull are two of the processes that move tectonic plates over the surface of Earth.

View an **animation of ridge push and slab pull.**

Concepts In Motion

Push and Pull

Scientists hypothesize that there are several processes that determine how mantle convection affects the movement of tectonic plates. Study **Figure 22.** As oceanic crust cools and moves away from a divergent boundary, it becomes denser and sinks compared to the newer, less-dense oceanic crust. As the older portion of the seafloor sinks, the weight of the uplifted ridge is thought to push the oceanic plate toward the trench formed at the subduction zone in a process called **ridge push.**

A second and possibly more significant process that determines the movement of tectonic plates is called slab pull. In **slab pull,** the weight of the relatively cool, dense subducting plate pulls the trailing slab into the subduction zone much like a tablecloth slipping off the table can pull articles off with it. Slab pull is thought to be at least twice as important as ridge push in moving an oceanic plate away from an ocean ridge. It is likely that the combination of mechanisms such as these are involved in plate motions at subduction zones.

SECTION 4 REVIEW

Section Self-Check

Section Summary

- Convection is the transfer of energy via the movement of heated matter.
- Convection currents in the mantle result in an energy transfer between Earth's hot interior and cooler exterior.
- Plate movement results from the processes called ridge push and slab pull.

Understand Main Ideas

1. **MAIN IDEA** **Draw** a diagram comparing convection in a pot of water with convection in Earth's mantle. Relate the process of convection to plate movement.
2. **Restate** the relationships among mantle convection, ocean ridges, and subduction zones.
3. **Make** a model that illustrates the tectonic processes of ridge push and slab pull.

Think Critically

4. **Evaluate** this statement: Convection currents only move oceanic crust.
5. **Summarize** how convection is responsible for the arrangement of continents on Earth's surface.

WRITING IN ▸ Earth Science

6. Write dictionary definitions for *ridge push* and *slab pull* without using those terms.

SECTION 4 REVIEW

1. Drawings should compare the heat source from Earth's core with the flame underneath the pot/beaker, and rising and falling circular currents in both the mantle and pot/beaker.
2. Mantle convection moves plates away from one another at ocean ridges, and as movement continues, plates meet other plates on the opposite side, and either subduct under a continental plate or other oceanic plate, or another oceanic plate subducts under them.
3. Model should show forces pushing at an ocean ridge and pulling in subduction zones.
4. This statement is incorrect. Convection currents move all plates on Earth's surface.
5. Because convection currents move all plates on Earth's surface, they are the cause of all the relative positions of plates containing continental crust.
6. Possible answers: a shoving force that moves the crust and mantle away from the ocean ridge; the mechanism by which cold crust and mantle are drawn into the mantle

Rubric

Earth Science & ENVIRONMENT

Vailulu'u Seamount

The American Samoan Islands are part of an island chain in the South Pacific Ocean. Exploration at the edge of the island chain has revealed how tectonic processes can result in new and completely unique environments.

Mapping Vailulu'u In 1999, Vailulu'u (vah EEL ool oo oo), an active volcanic seamount, was first mapped by oceanographers using remote sonar methods. The map revealed the outline of a massive volcano hollowed by a caldera. The ocean is about 5 km deep, and the ringlike ridges of the caldera come within 600 m from the ocean surface. The 1999 map showed that the caldera floor was generally flat—about 1 km below sea level. Scientists knew that the volcano was produced from a hot spot, a region of heated magma in the mantle below.

Discovery of a volcano In 2005, a team of scientists returned to study Vailulu'u using deep-sea submersibles. Before diving, they remapped the seamount, and discovered that the floor of the caldera had changed dramatically. Sometime in the past six years, volcanic activity had developed a lava cone 300 m high, roughly the height of the Empire State Building. The cone was soon named Nafanua (nah fah NOO ah), after the Samoan goddess of war. The scientists made several trips in the submersible and discovered how tectonic activity had caused completely new ecosystems to develop.

Eel City At the top of Nafanua, they encountered 30-cm-long eels so numerous that they nicknamed the area *Eel City.* The top of the cone is too deep for sunlight to permit the growth of plants, so the scientists were puzzled about the eels' food source. Investigations revealed that the seamount had changed the local currents, depositing waves of shrimp above Nafanua.

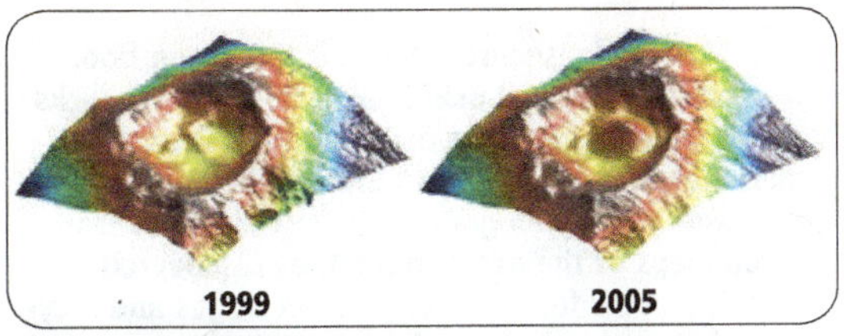

When you compare the two images, you can see the appearance of the Nafanua cone in the center of the caldera.

Moat of death Hydrothermal vents on the floor of the caldera emitted toxic chemicals, including clouds of a murky oil-like liquid containing carbon dioxide. Some of the vents released water that was a scalding 85°C. The same currents that brought shrimp to the eels were carrying fish down into the toxic environment of the caldera, which was nicknamed *the moat of death.* Yet, some life-forms were thriving. Much of the caldera floor was covered by a 1-m-thick mat of microbes, and bright red bristle worms abounded around the fish carcasses.

Birth of an island Nafanua is expected to continue growing. At its present rate, it could reach the ocean surface within a few decades and become the newest island in the Samoan chain. Earth scientists will continue to monitor the growth of Nafanua, and learn how tectonic events can help shape entire ecosystems.

WRITING IN ▶ Earth Science

Investigate the biological activity and unique habitats discovered on Vailulu'u seamount. Write a newspaper article that describes the organisms and conditions on the seamount.

WebQuest

WRITING IN ▶ Earth Science

Rubric

Investigate Students should describe the different organisms living on the seamount, including soft corals, echinoderms, and sponges that live on the outer wall. *Dysommina rugosa* (cutthroat eels) and the shrimp they eat live in the upper layers of water in the crater. Deeper in the crater, in the "moat of death" region, marine bristle worms and bacteria are found.

WebQuest

Earth Science & ENVIRONMENT

Purpose

Students will learn about the Vailulu'u seamount and that it was formed as the Pacific Plate passed over a hot spot in Earth's crust.

Teacher Content Support

Hot Spot Life During a 2005 survey of Vailulu'u seamount, scientists learned more about the biological activity around the seamount. They found the summit of the seamount was swarming with hundreds of cutthroat eels that had never before been studied in their natural habitat. The eels feed on shrimp that are washed into the crater of the volcano by currents. Also within the crater is a region scientists refer to as the "moat of death." The water in this area of the crater is acidic, filled with iron particles released by the volcano, and depleted of oxygen. Fish washed into the crater cannot survive and are fed upon by marine bristle worms. Bacteria are the only other organism found living there.

Teaching Strategy

Explain that a hot spot is a region of Earth's crust where high-temperature plumes of magma rise toward the surface and are stationary. Island chains, like the Hawaiian Islands, are formed as plates move over a hot spot. Demonstrate this concept by placing a red, felt-tip marker under a piece of white paper. As the paper sits on top of the marker, the ink is absorbed and a spot develops. The spot represents an island being formed. Make an island chain by repeatedly pulling the paper forward over the marker several centimeters and then pausing.

GeoLAB

iLab Station Rubric

Preparation

Time Allotment 60 min

Process Skills interpret scientific illustrations, analyze data

Safety Precaution Approve lab safety forms before work begins.

Procedure

- Trace the base map onto a blank transparency. Copy the magnetic survey data curves onto another transparency. Use the overhead projector to demonstrate how to transfer the data curves to the base map.
- Make sure that students use sharp pencils to trace the data curves and that the ocean ridge lines up with the dashed line as they trace.
- Tell students that when drawing the parallel lines, it is best to start at the ridge and work outward. It might be helpful if students carefully shade all the positive magnetic events on each survey line. Remind students that the lines do not cross the transform fault but are offset by the fault.
- The map should show a pattern of colored bands that are parallel and symmetrical with respect to the ridge. The pattern is offset by the transform fault.

4. Site 1, divergent; Site 2, transform; Site 3, convergent with some tendency for transform
5. 0.00002 km/y (2 cm/y × 1m/100 cm × 1 km/1000 m = 0.00002 km/y)
6. Drawings should show lines in approximate proportion to the scale of **Figure 24.** The 10-million-year isochrons represent a band 200 km in width (10,000,000 y × 0.002 km/y = 200 km), or 100 km on either side of the ocean ridge; the 20-million-year isochrons represent a band 400 km in width, or 200 km on either side of the ocean ridge, and so on.
7. The sketch should be symmetrically striped on both sides of the divergent boundary.

GeoLAB

iLab Station

Model Plate Boundaries and Isochrons

Background: Isochron maps of the ocean floor were first developed using data from oceanic rocks and sediment. Isochrons are imaginary lines on a map that show the parts of Earth's surface that are the same age. When geologists first analyzed isochron maps of the ocean floor, they discovered that Earth's crust is formed along ocean ridges and recycled at the edge of oceanic crust. This discovery led to the theory known as plate tectonics. Geologists continue to use maps to study the motion of tectonic plates.

Question: *Can you determine the age of the crust and type of plate boundaries?*

Materials

paper
colored pencils
scissors
metric ruler
calculator

Safety Precautions

Procedure

1. Read and complete the lab safety form.
2. **Figure 23** shows Plate B relative to Plate A. Draw or trace the plates onto a separate sheet of paper and cut them out.

WARNING: ***Scissors can cut or puncture skin.***

3. The arrow shows the movement of the plates relative to each other. Move Plate A as shown in each part of **Figure 23.**
4. Use the symbols shown in the legend to indicate the type of plate boundary and the relative motion across the boundary for each part of **Figure 23.**
5. **Figure 24** shows two plates, A and B, separated by an ocean ridge offset by a transform fault. Plates A and B are moving apart at 2 cm/y. Convert the speed 2 cm/y to km/y.
6. Trace **Figure 24** onto a separate sheet of paper. Assume the geometry of the boundaries in **Figure 24** has not changed over time. Draw isochrons on 10, 20, 30, and 40 million years.
7. Color the crust based on its age: 0–10 million years old red, 10–20 million years old yellow, 20–30 million years old green, and 30–40 million years old blue.

Analyze and Conclude

1. **Determine** the motion of a plate that would have each of the A sites that moved relative to the B plate.
2. **Apply** From your map of isochrons, what is the easiest way to identify the location of transform boundaries?
3. **Interpret** Look at **Figure 25.** From the pattern of isochrons on the ocean floor, identify the divergent plate boundaries along the Atlantic Ocean and along the Pacific Ocean.
4. **Differentiate** Which ocean is marked by wider isochrons? Based on the amount of oceanic crust produced in a given period of time, along which plate boundary is divergence happening more rapidly?
5. **Infer** The spreading center in the Pacific Ocean is not centered in the same manner as the Atlantic Ocean. Explain how this indicates the presence of convergent plate boundaries.

WRITING IN ▶ Earth Science

Write a Letter Alfred Wegener never convinced the scientific community of continental drift. He died shortly before the ocean floors were mapped. Imagine you could send a message to the past. Explain to Wegener what ocean floor mapping revealed, and how plate tectonics was discovered.

Analyze and Conclude

1. When one plate is moving at non-90° angles relative to the other plate, all the boundaries will have some aspect of transform boundaries.
2. by locating places where parallel isochrons do not match lengthwise with other isochrons that would normally be in line with them
3. The divergent plate boundaries are at mid-ocean ridges.
4. The Pacific has wider isochrons. Divergence is more rapid there.
5. The ridge is offset because ocean floor is converging with Central and South America. As a result, the ocean floor is being subducted.

WRITING IN ▶ Earth Science

Rubric

Write a Letter The letters should include, but not be limited to, the concepts of seafloor spreading, geomagnetic surveys of the ocean floor, global mapping of earthquake and volcanic activity, and the discovery of deep-sea trenches.

Key

Use the following symbols to indicate the type of plate boundary:

- ‖ Divergent boundary
- ▲ Convergent boundary (triangles point to the plate that stays on the surface)
- ⇅ Transform; arrows indicate the relative direction of motion across the boundary

Figure 23

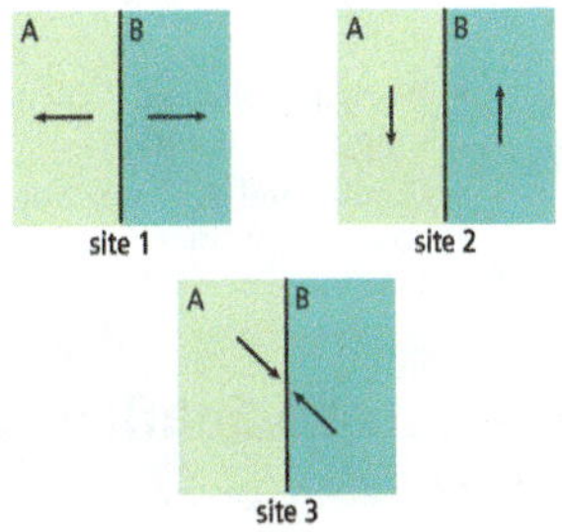

Figure 24

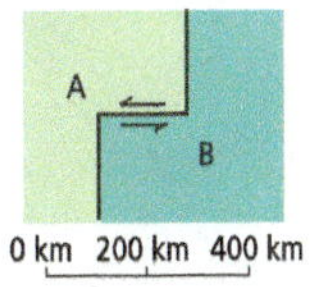

Figure 25

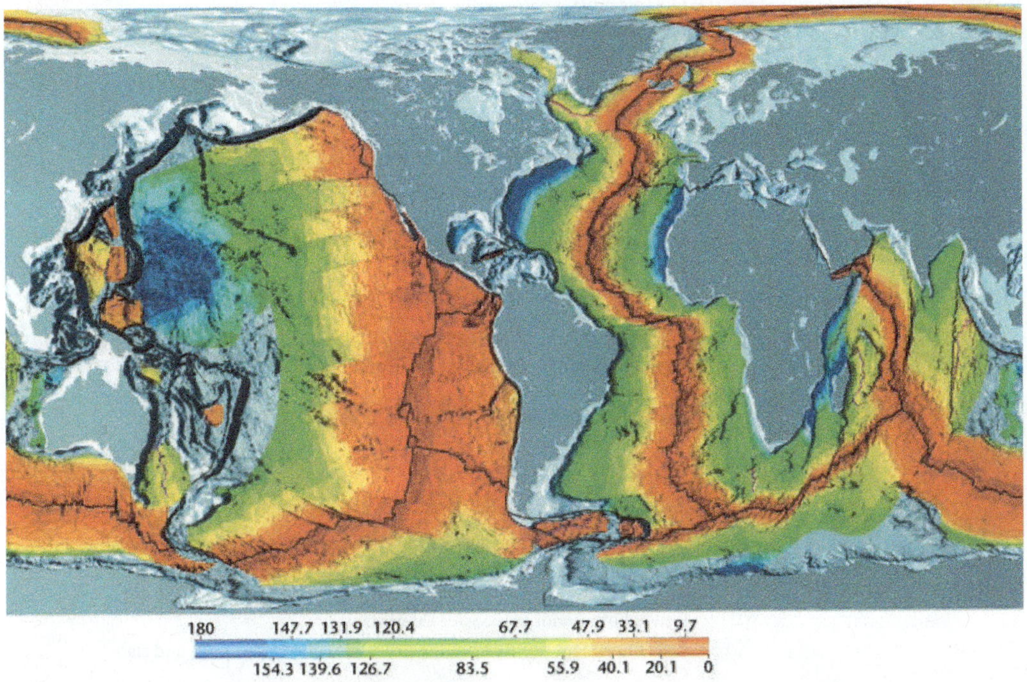

MAINIDEAS Summary statements can be used by students to review the major concepts of the chapter.

Students can review with these online resources.

Vocabulary Practice

Vocabulary eGames
Vocabulary eFlashcards
Vocabulary PuzzleMaker

Section Self-Check

Chapter Self-Check

Online Test Practice

Use the Exam*View*® *Assessment Suite* CD-ROM to:

- create multiple versions of tests
- create modified tests with one mouse click
- edit existing questions and add your own questions
- build tests aligned with state standards using built-in state curriculum tags
- change English tests to Spanish with one mouse click
- track students' progress using the Teacher Management System

CHAPTER 17 STUDY GUIDE

Vocabulary Practice

BIGIDEA Most geologic activity occurs at the boundaries between plates.

SECTION 1 Drifting Continents

VOCABULARY
- continental drift
- Pangaea

MAINIDEA The shape and geology of the continents suggests that they were once joined together.

- The matching coastlines of continents on opposite sides of the Atlantic Ocean suggest that the continents were once joined.
- Continental drift was the idea that continents move around on Earth's surface.
- Wegener collected evidence from rocks, fossils, and glacial deposits to support his theory.
- Continental drift was not accepted because there was no explanation for how the continents moved or what caused their motion.

SECTION 2 Seafloor Spreading

VOCABULARY
- magnetometer
- magnetic reversal
- paleomagnetism
- isochron
- seafloor spreading

MAINIDEA Oceanic crust forms at ocean ridges and becomes part of the seafloor.

- Studies of the seafloor provided evidence that the ocean floor is not flat and unchanging.
- Oceanic crust is geologically young.
- New oceanic crust forms as magma rises at ridges and solidifies.
- As new oceanic crust forms, the older crust moves away from the ridges.

SECTION 3 Plate Boundaries

VOCABULARY
- tectonic plate
- divergent boundary
- rift valley
- convergent boundary
- subduction
- transform boundary

MAINIDEA Volcanoes, mountains, and deep-sea trenches form at the boundaries between the plates.

- Earth's crust and rigid upper mantle are broken into large slabs of rock called tectonic plates.
- Plates move in different directions and at different rates over Earth's surface.
- At divergent plate boundaries, plates move apart. At convergent boundaries, plates come together. At transform boundaries, plates slide horizontally past each other.
- Each type of boundary is characterized by certain geologic features.

SECTION 4 Causes of Plate Motions

VOCABULARY
- ridge push
- slab pull

MAINIDEA Convection currents in the mantle cause plate motions.

- Convection is the transfer of energy via the movement of heated matter.
- Convection currents in the mantle result in an energy transfer between Earth's hot interior and cooler exterior.
- Plate movement results from the processes called ridge push and slab pull.

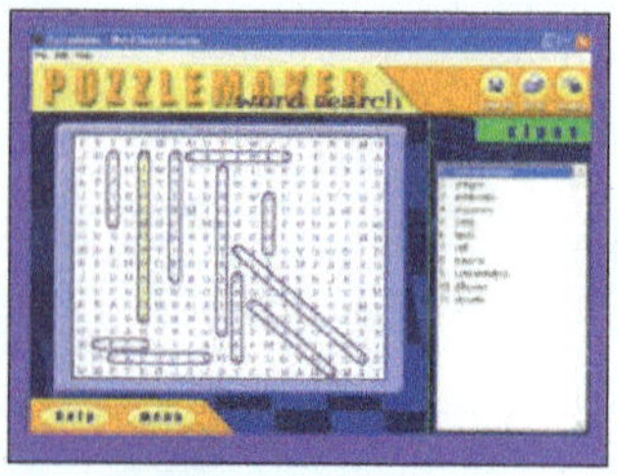

Vocabulary Practice

For additional practice with vocabulary, have students access the Vocabulary PuzzleMaker.

Chapter Self-Check

VOCABULARY REVIEW

Replace each italicized word with the correct vocabulary term from the Study Guide.

1. *Plate tectonics* is the name given to the single continent that existed 200 mya.
2. *Continental fracture* is the idea that continents now separated by an ocean were once attached.
3. The process in which tectonic plates sink back into the mantle is called *divergence.*
4. A boundary where two plates come together is a *transform boundary.*
5. A divergent boundary within a continent forms a *trench.*

Match each of the following phrases with a vocabulary term from the Study Guide.

6. a line on a map that denotes crust that formed at the same time
7. the process that creates new ocean crust by the upwelling of magma at ocean ridges
8. the study of the history of Earth's magnetic field
9. a device that measures magnetism

Define the following vocabulary terms in complete sentences.

10. tectonic plate
11. ridge push
12. slab pull

Use what you know about the vocabulary terms in the Study Guide to describe what the terms in each pair have in common.

13. divergent boundary, transform boundary
14. subduction, convergent boundary
15. continental drift, plate tectonics
16. seafloor spreading, magnetic reversal

UNDERSTAND KEY CONCEPTS

17. Which suggested to early cartographers that the continents were once joined?
   A. ocean depth
   B. position of south pole
   C. shape of continents
   D. size of Atlantic Ocean
18. What was Wegener's hypothesis called?
   A. seafloor spreading
   B. plate tectonics
   C. continental drift
   D. slab pull

Use the figure below to answer Questions 19 and 20.

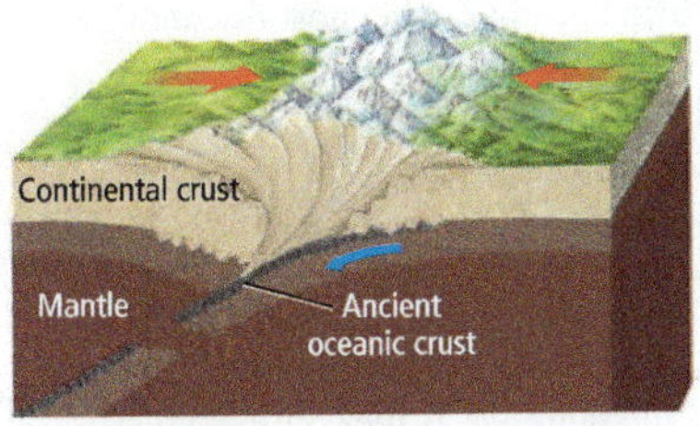

19. What type of boundary is shown?
   A. an ocean ridge
   B. a continental-continental boundary
   C. a transform boundary
   D. an oceanic-continental boundary
20. Which feature forms along this type of boundary?
   A. subduction zones
   B. oceanic trenches
   C. island arcs
   D. folded mountains
21. The weight of a subducting plate helps pull it into a subduction zone in which process?
   A. slab pull
   B. ridge push
   C. slab push
   D. ridge pull
22. Which is a convergent boundary that does not have a subduction zone?
   A. oceanic-oceanic
   B. oceanic-continental
   C. continental-continental
   D. transform

CHAPTER 17 ASSESSMENT

VOCABULARY REVIEW

1. Pangaea
2. Continental drift
3. subduction
4. convergent boundary
5. rift valley
6. isochron
7. seafloor spreading
8. paleomagnetism
9. magnetometer
10. A tectonic plate is a portion of Earth's surface made of the crust and solid mantle beneath it that moves slowly as a whole.
11. Ridge push is one possible way in which plates move, resulting from the newly formed oceanic crust that begins hot and at higher elevations because of uplift at the oceanic ridge, and becomes denser as it moves away, causing it to sink downward into the mantle.
12. Slab pull is one possible way in which plates move, resulting from the downward pull of an oceanic plate that has descended into the mantle.
13. Divergent and transform boundaries both occur along oceanic ridges.
14. Subduction occurs along two kinds of convergent boundaries.
15. The continental drift hypothesis and the theory of plate tectonics were developed to explain movement of Earth's surface.
16. Seafloor spreading was discovered in part by evidence of magnetic reversals recorded in the oceanic crust.

UNDERSTAND KEY CONCEPTS

17. C
18. C
19. B
20. D
21. A
22. C

23. C
24. B
25. B
26. D
27. B
28. A
29. B
30. D
31. C

CONSTRUCTED RESPONSE

32. observations about shape of coastlines across oceans matching; fossils on widely separated continents being similar; glacial deposits in regions too far from the poles; many geological features are continuous across what are now ocean basins if continents are reassembled
33. Sediments slowly rain out of the ocean and settle on the ocean floor. The older the seafloor, the thicker the sediments. The further from a ridge, the older the seafloor, and hence the thicker the sediments.
34. The magnetic field generated in Earth's core magnetizes minerals in Earth's crust. The magnetic minerals record the direction of the magnetic field at the time they cooled.
35. The rocks that make up continents are too buoyant to sink back into the mantle. They stay at the surface, piling up to make a mountain belt.
36. Wegener had provided no satisfactory explanation of how the continents could move. Seafloor spreading provided that mechanism. Seafloor spreading explained what happened between the continents. The continents didn't plow through the oceans, the oceans formed and the continents moved apart.

Use the figure below to answer Questions 23 and 24.

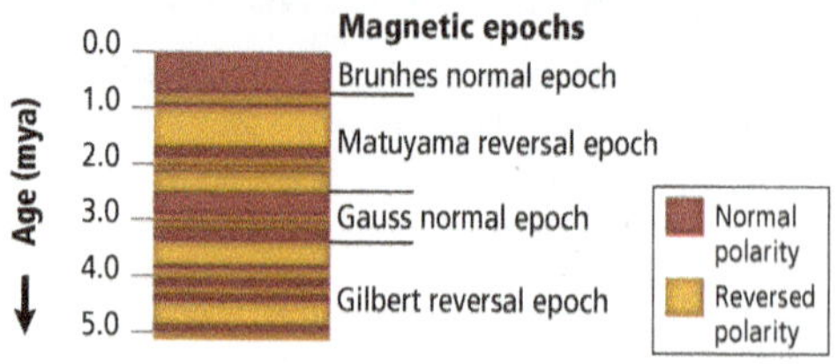

23. Approximately how long did the Gauss epoch last?
A. 5 million years
B. 3 million years
C. 1 million years
D. 100,000 years

24. Which epoch saw the most fluctuations between normal and reverse polarity?
A. Gauss
B. Matuyama
C. Gilbert
D. Brunhes

25. Generally, what is the age of oceanic crust?
A. the same age as the continental crust
B. younger than the continental crust
C. older than the continental crust
D. science has never determined its age

26. Which observation was not instrumental in formulating the hypothesis of seafloor spreading?
A. magnetization of the oceanic crust
B. depth of the ocean
C. thickness of seafloor sediments
D. identifying the location of glacial deposits

27. How fast do plates move relative to each other?
A. millimeters per day
B. centimeters per year
C. meters per year
D. centimeters per day

28. What process creates deep-sea trenches?
A. subduction
B. magnetism
C. earthquakes
D. transform boundaries

Use the photo below to answer Questions 29 and 30.

29. As shown, which direction does the icy water move?
A. up
B. down
C. remains in the same place
D. sideways

30. Which is modeled by the water movement?
A. subduction
B. continental drift
C. magnetic reversal
D. mantle convection

31. Which is not a force causing plates to move?
A. ridge push
B. slab pull
C. volcanism
D. convection

CONSTRUCTED RESPONSE

32. Summarize What observations led to the proposal of continental drift?

33. Careers in Earth Science Explain why oceanographers have found that the thickness of seafloor sediments increases with increasing distance from the ocean ridge.

34. Differentiate between the magnetic field generated in Earth's core and the magnetization preserved in the oceanic crust.

35. Analyze why there are differences between continental-continental convergent boundaries and oceanic-oceanic convergent boundaries.

36. Summarize Why was the idea of moving continents more widely accepted after seafloor spreading was proposed?

THINK CRITICALLY

Use the map below to answer Question 37.

37. **Infer** If 200 million-year-old oil deposits were discovered in Namibia, where might geologists also expect to find oil deposits of a similar age? Explain.

38. **Compare and contrast** ridge push and slab pull.

39. **Infer** How have satellite monitoring systems such as GPS made it much easier and cheaper to study the motion of tectonic plates?

40. **Consider** Do plates always stay the same shape and size? Explain.

41. **Critique** this statement: There are two kinds of tectonic plates—continental plates and oceanic plates.

CONCEPT MAPPING

42. Create a concept map using the following terms: *convergent, rift valley, divergent, transform, island arc, shallow earthquakes, mountain range,* and *plate boundary.* Refer to the *Skillbuilder Handbook* for more information.

CHALLENGE QUESTION

43. **Predict** Assuming that Earth's tectonic plates will continue moving in the directions shown in **Figure 2,** sketch a globe showing the relative positions of the continents in 60 million years.

USGS

WRITING IN ▶ Earth Science

44. Imagine you are on a sailboat anchored off the coast of Chile. You hear loud rumbling. Then GPS data indicates a part of the coast shifted up by about 1.5 m. Write a journal entry to describe the geologic phenomena you are seeing and experiencing.

DBQ Document–Based Questions

Data obtained from: Seismicity of the Central United States: 1990–2000. *USGS National Earthquake Information Center.*

Most earthquakes occur at plate boundaries as plates slide by each other. This map shows the location and depth of earthquakes between 1990 and 2000 in Alaska.

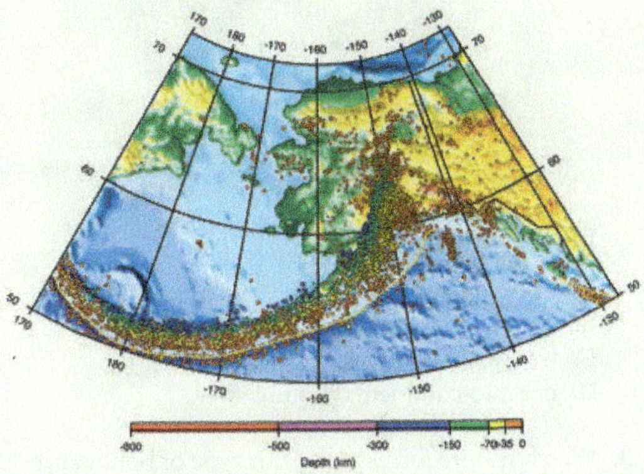

45. Identify which plate is subducting and provide evidence from the figure to support your answer.

46. Compare this map to **Figure 16,** which shows the location of plate boundaries. Why do parts of the plate boundaries have few or no earthquakes?

CUMULATIVE REVIEW

47. How do Landsat satellites collect and analyze data to map Earth's surface? **(Chapter 2)**

48. How can scientists use glaciers to study Earth's past? **(Chapter 8)**

49. Describe the major parameters used in the Köppen Classification System. **(Chapter 14)**

WRITING IN ▶ Earth Science

Rubric

44. Answers will vary, but students should demonstrate understanding that there would be volcanism and faulting in Chile because it is at a convergent plate boundary.

DBQ Document-Based Questions

Data obtained from: Seismicity of the Central United States: 1900–2000. *USGS National Earthquake Information Center.*

45. Pacific Plate is subducting under the Aleutian Islands. Most of the deeper earthquakes are under the islands, while shallower earthquakes are under the ocean plate.
46. Convergent/colliding zones have many earthquakes at varying depths; divergent zones have thin crust and shallow earthquakes.

THINK CRITICALLY

37. at the eastern edge of South America, because the two continents were once joined
38. Both forces are caused by gravity. Both are caused by the weight of material in plates. Ridge push is caused by the weight of the surface being lifted upward at ridges. Slab pull is caused by the weight of a denser subducting plate pulling the trailing crust into a subduction zone.
39. GPS allows precise measurements to be made continuously and everywhere.
40. No. Plates can grow (rate of divergence is greater than rate of convergence) or shrink (rate of divergence is less than rate of convergence). Plates can even disappear.
41. Answers will vary. Correct answer should include some or all of the following: It would be correct to say there are two kinds of crust. A tectonic plate can be composed of both kinds of crust.

CONCEPT MAPPING

42. Answers will vary. Correct concept map should show an understanding that there are three types of plate boundaries: convergent, divergent, transform. Each of these plate boundaries is associated with features on Earth's surface: convergent boundaries can have deep sea trenches, island arcs, and mountain ranges. Divergent boundaries can have shallow earthquakes, rift valleys, and mid-ocean ridges.

CHALLENGE QUESTION

43. Sketches might include the closing of the Mediterranean Sea, collision of Australia with southeast Asia, and/or widening of the Atlantic Ocean basin. Sophisticated answers might include the subduction of India and/or the Arabian Peninsula beneath Asia.

CUMULATIVE REVIEW

47. Landsat satellites receive reflected wavelengths of energy emitted by features on Earth's surface.
48. Air bubbles trapped deep in the glacial ice can provide scientists data about the composition of Earth's atmosphere at the time when ancient ice layers were formed. Scientists also use ice cores from Antarctica and Greenland to study Earth's past environmental conditions.
49. temperature, precipitation, and vegetation

MULTIPLE CHOICE

1. D
2. B
3. D
4. A
5. C
6. C
7. D
8. A

CUMULATIVE STANDARDIZED TEST PRACTICE

MULTIPLE CHOICE

1. How does the building of jetties negatively effect coastlines?
   - A. They fill in anchorage used to harbor boats with sediment.
   - B. They hinder breakwater from moving sediments away from the area.
   - C. They reflect energy back toward beaches, increasing erosion.
   - D. They deprive beaches down the coast from the jetty of sand.

Use the diagram below to answer Questions 2 and 3.

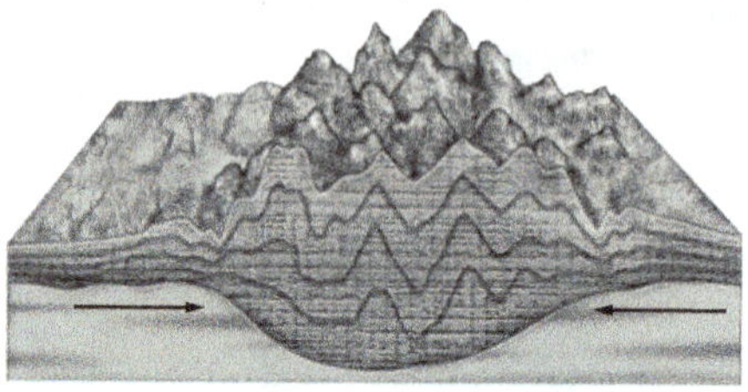

2. What type of plate boundary is shown?
   - A. ocean ridge
   - B. continental-continental boundary
   - C. transform boundary
   - D. oceanic-continental boundary

3. Which feature forms along this type of boundary?
   - A. subduction zones
   - B. oceanic trenches
   - C. island arcs
   - D. folded mountains

4. What is the best way to get out of a rip current?
   - A. swim parallel to the shore
   - B. swim with the rip current
   - C. swim against the rip current
   - D. swim under the rip current

5. The smooth parts of the ocean floor located 5 to 6 km below sea level are called the
   - A. mid-ocean ridges
   - B. deep-sea trenches
   - C. abyssal plains
   - D. continental rises

Use the table below to answer Questions 6–8.

Exercise and Heart Rates

Subject	Resting	Fast Walk	Slow Jog
1	65	72	110
2	78	88	120
3	72	83	125
4	69	78	105
5	75	90	135
Averages	71.8	82.2	119

6. The table shows heart rates before and after a 10-min session. Which statement best summarizes the data?
   - A. There is no relationship between heart rate and exercise.
   - B. Exercise increased the heart rate of the participants.
   - C. Heart rate increased as exercise became more strenuous.
   - D. Ten minutes of exercise was not enough to increase heart rate.

7. According to the data, which subject appears to be in the best shape and why?
   - A. Subject 1 because the subject had the lowest resting heart rate
   - B. Subject 4 because the subject had the lowest heart rate during a slow jog
   - C. Subject 5 because the subject had the fastest resting heart rate
   - D. This cannot be determined from the table because not enough information is given about each of the subjects.

8. What would be the best graph to use in order to present the data found?
   - A. bar graph
   - B. line graph
   - C. circle graph
   - D. a model

Online Test Practice

SHORT ANSWER

Use the illustration below to answer Questions 9–11.

9. As the Moon moves to create a right angle along with the Sun and Earth, what occurs with the ocean's tides?
10. Describe how tides are affected when the Sun, the Moon, and Earth are aligned.
11. How do lunar tides differ from solar tides?
12. What negative impact might a major storm have on a barrier island?
13. How do atmosphere and large bodies of water affect climate in various regions?
14. How did a temperature decrease of only 5°C during the ice ages cause major changes?

READING FOR COMPREHENSION

Seafloor Maps

In 2005, the U.S. nuclear submarine *San Francisco* crashed into an uncharted underwater mountain in the South Pacific, killing one submariner and injuring dozens of others. The incident highlights a troubling nautical reality—we might know more about the geography of the Moon than that of the ocean floor. Estimates vary, but the amount of correctly mapped seafloor in the public domain is likely around 2 or 3 percent.

Although survey ships equipped with sound-based systems can accurately map the seafloor by dropping a "beam" below the ship, this method can be used to map only narrow sections at a time. Mapping all the oceans this way might take a thousand years and cost billions of U.S. dollars. However, such maps could be critical for tsunami-preparation efforts. No matter how deep the ocean, a tsunami moves along the bottom, and its path is influenced by the features of the ocean floor. Thus, understanding the location of trenches, seamounts, and other features is essential to calculations of how a tsunami will move and where and in what force it will come ashore. Other studies that could benefit from mapping include marine animal habitat and ocean mixing rates, which are essential to absorption of greenhouse gases. All are dependent on more detailed knowledge of the other 70 percent of Earth's surface.

Article obtained from: Handwerk, B. Seafloor still about 90 percent unknown, experts say. *National Geographic News.* February 17, 2003.

15. What can be inferred from this passage?
    A. It is important for ships and submarines to use sonar so that they do not run into underwater mountains.
    B. Mapping the seafloor is too expensive and not important enough to humans.
    C. Very little is known about the seafloor, and by improving this knowledge, both humans and animals will benefit.
    D. Many marine animals' lives will be disrupted if scientists continue to map the ocean floor.

16. How would knowing what is on the seafloor help an oceanographer track a tsunami?

NEED EXTRA HELP?														
If You Missed Question . . .	1	2	3	4	5	6	7	8	9	10	11	12	13	14
Review Section . . .	16.1	17.3	17.3	16.1	16.2	1.2	1.2	1.3	15.3	15.3	15.3	16.1	14.1	14.3

SHORT ANSWER

9. Low tides become higher, while high tides become lower. These are known as neap tides. The tidal range is lowest during this time.
10. High tides are higher, and low tides are lower. These are known as spring tides. The tidal range is highest during this time.
11. Lunar tides are more than twice the height of solar tides. This is because the Moon is closer to Earth than the Sun is. However, depending on the phase of the Moon, solar tides can either increase or decrease lunar tides.
12. The major storm might sweep away an entire section of the barrier island and deposit the material elsewhere. This would leave the beach behind the barrier island exposed to the oncoming tides and waves, leading to erosion and rearrangement of shorelines, which could affect the animals there.
13. As altitude increases, temperature decreases. Therefore, mountainous regions are cooler than low-lying regions. Large bodies of water affect climate because they heat up and cool down more slowly than land. Therefore, coastal regions tend to be warmer in winter and cooler in summer.
14. With the slight drop in temperature, global climates became colder. This increased the amount of snowfall, which helped to advance existing ice sheets.

READING FOR COMPREHENSION

15. C
16. Sample answer: According to the passage, a tsunami is affected by the makeup of the seafloor. By knowing what was underneath the tsunami when it formed and what major objects are in its path, the oceanographer will be better able to predict how the tsunami will move, where it will make landfall, and with what intensity it will hit land. This knowledge will help better prepare the people in the tsunami's path.

CHAPTER 18 Volcanism

BIGIDEA **Volcanoes develop from magma moving upward from deep within Earth.**

ESSENTIAL QUESTIONS	NATIONAL STANDARDS	RESOURCES TO ASSESS MASTERY
SECTION 1 **Volcanoes** 1. How do plate tectonics influence the formation of volcanoes? 2. Where are the major zones of volcanism? 3. What are the parts of a volcano? 4. How do volcanic landforms differ? 2.5 sessions 1.25 blocks	UCP.1–3, UCP.5; A.1–2; B.2, B.6; D.1–3; G.3	**Entry-Level Assessment** Focus Transparency 53, p. 500 **Progress Monitoring** Caption Question, p. 501 Reading Check, pp. 501, 502 Section Review, p. 507
SECTION 2 **Eruptions** 1. How does magma type influence a volcano's explosivity? 2. What is the role of pressure and dissolved gases in eruptions? 3. What kinds of material are ejected by volcanic eruptions? 1 session 0.5 block	UCP.1–5; D.1–2	**Entry-Level Assessment** Focus Transparency 54, p. 508 **Progress Monitoring** Caption Question, pp. 509, 512 Reading Check, p. 509 Section Review, p. 513
SECTION 3 **Intrusive Activity** 1. How are features formed from magma that solidified under Earth's surface described? 2. What are the different types of intrusive rock bodies? 3. What geologic processes result in intrusive rocks that appear at Earth's surface? 4 sessions 2 blocks	UCP.1–5; A.1–2; B.4; D.1–3; F.5; G.3	**Entry-Level Assessment** Focus Transparency 55, p. 514 **Progress Monitoring** Caption Question, p. 516 Reading Check, p. 515 Section Review, p. 517 **Summative Assessment** Chapter Assessment, p. 521 Exam*View®Assessment Suite* CD-ROM

Go online!
connectED.mcgraw-hill.com

LEVELED RESOURCES	LAB MATERIALS
Science Notebook 18.1 OL **Chapter FAST FILE Resources:** MiniLab Worksheet, p. 28 OL Study Guide, p. 41 BL **Transparencies:** Section Focus Transparency 53 EL Teaching Transparencies 50, 51 OL EL	LaunchLAB p. 498 / **15 min** water, 600-mL beaker, vegetable oil, table salt MiniLAB p. 505 / **35 min** small box, 10-cm length of rubber tubing, balloon, newspaper, scissors, clamp, sand, tape
Science Notebook 18.2 OL **Chapter FAST FILE Resources:** GeoLab Worksheet, p. 29 OL Study Guide, p. 44 BL **Lab Resources:** Laboratory Manual, p. 137 OL **Transparencies:** Section Focus Transparency 54 EL Teaching Transparency 52 OL EL	GeoLAB p. 519 / **90 min** internet access, current reference books, markers or colored pencils
Science Notebook 18.3 OL **Chapter FAST FILE Resources:** Study Guide, p. 46 BL **Lab Resources:** Laboratory Manual, p. 141 OL **Transparencies:** Section Focus Transparency 55 EL Teaching Transparency 53 OL EL	

LabManager™
Customize any lab with the LabManager™ CD-ROM.

ADDITIONAL RESOURCES

Technology:
- eTeacherEdition
- eStudentEdition
- Classroom Presentation Toolkit
- Exam*View*® *Assessment Suite* CD-ROM
- LabManager™ CD-ROM
- What's EARTH SCIENCE Got To Do With It? DVD
- Weather Classroom Video DVD

Assessment:
- Blueprints for Success
- Chapter Test (Scaffolded)
- Earth Science Standardized Test Practice

Lab Resources:
- Exploring Environmental Problems Laboratory Manual

BL Below Level OL On Level AL Advanced Learners EL English Learners COOP LEARN Cooperative Learning

CHAPTER 18

LaunchLAB

iLab Station | Rubric

What makes magma rise?

Process Skills observe and infer, recognize cause and effect, communicate, predict, measure

Safety Precautions Approve lab safety forms before work begins. Remind students not to eat or drink anything in the lab.

Teaching Strategies

- This activity can be done in small groups. Provide each group with a beaker, water, vegetable oil, and salt.
- If time or resources are limited, this activity can be done as a demonstration.
- Students should observe movement of oil and water in the beaker once the salt is added.

Procedure

1. Have students read and complete the lab safety form and follow the procedure below.
2. Pour about 300 mL of **water** into a **600-mL beaker.**
3. Pour about 80 mL of **vegetable oil** into the beaker.
4. Sprinkle **table salt** on top of the oil while you slowly count to 5.
5. Add more salt to keep the movement going.

Analysis

1. **Identify** which component of your model represents magma. The oil represented magma.
2. **Describe** what happened to the oil before and after you added the salt. Before adding salt, the oil floated on top of the water; when salt was added, the oil sank.
3. **Hypothesize** what causes the "magma" to rise. Answers will vary, but students should recognize that density changes cause the movement. At first, the salt within the oil layer makes the oil denser than the water, causing the oil to sink. As the salt dissolves in the water, the water becomes denser than the oil, so the oil (the "magma") rises.

Assessment

Performance Have students test their inferences.

LabManager™
Customize any lab with the LabManager™ CD-ROM.

CHAPTER 18

Volcanism

BIGIDEA Volcanoes develop from magma moving upward from deep within Earth.

There are about 500 active volcanoes on Earth today. Guatemala's Santiaguito volcanic dome, shown here, can spew plumes of ash that reach heights of more than 8,000 m above sea level.

SECTIONS

1 Volcanoes

2 Eruptions

3 Intrusive Activity

LaunchLAB

iLab Station

What makes magma rise?

Magma is molten rock that lies beneath Earth's surface. In this activity, you will model the movement of magma within Earth by making a "lava lamp".

FOLDABLES® Study Organizer

Classification of Volcanoes

Make a layered-look book using the labels shown. Use it to organize your notes on volcanoes.

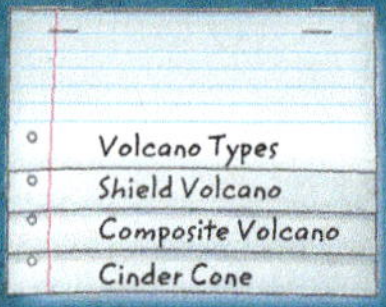

Introduce the BIGIDEA

Volcanoes Lead a discussion to determine what students already know about volcanoes. Pose the following questions: What is a volcano? What types of material erupt from volcanoes? Are all eruptions the same? Can you name at least three volcanoes? Responses will vary. Write down the names of volcanoes identified so you can use them in the Main Ideas section. Keep track of inaccurate responses and address these as you teach the chapter.

Teacher Content Support

Identify the Photo The photo shows the Santiaguito lava dome complex at Guatemala's Santa Maria volcano. In 1922, Santiaguito began growing in a crater on the southwest flank of Santa Maria, and has been continuously active since that time.

SECTION 1

1 Focus

Focus Transparency

Before presenting the lesson, project **Section Focus Transparency 53** and have students answer the accompanying questions. EL

MAINIDEA

Volcano Locations Using a world map, have students plot the locations of the volcanoes they identified in the Big Idea exercise titled *Volcanoes,* in the beginning of this chapter in the Teacher Edition. Ask them if they notice any trends relating to where volcanoes tend to appear. Students should note that the majority of volcanoes appear near plate boundaries.

2 Teach

Teacher Content Support

Age of Volcanoes Some students might have the misconception that all volcanoes are old and that these dynamic features of Earth form slowly. Tell students the following story about a volcano called Parícutin: In 1943, a Mexican farmer observed smoke and ash issuing from a crack in his cornfield. Within 24 hours, a 40-m cinder-cone volcano, Parícutin, had formed in the field.

Parícutin grew rapidly during the first year, reaching 336 m in height, and eventually it towered 424 m above the field. Parícutin continued to erupt until 1952. The total volume of lava produced by the eruptions was 1.4 km^3; it covered over 25 km^2 of the surrounding land.

SECTION 1

Volcanoes

MAINIDEA The locations of volcanoes are mostly determined by plate tectonics.

Essential Questions

- How do plate tectonics influence the formation of volcanoes?
- Where are the major zones of volcanism?
- What are the parts of a volcano?
- How do volcanic landforms differ?

Review Vocabulary

convergent: tending to move toward one point or to approach each other

New Vocabulary

volcanism
hot spot
flood basalt
fissure
conduit
vent
crater
caldera
shield volcano
cinder cone
composite volcano

EARTH SCIENCE 4 YOU Road crews spread salt on icy winter roads because salt makes the ice melt at a lower temperature. At extremely high temperatures, rocks can melt. Often, if heated rocks are in contact with water, they melt more easily.

Zones of Volcanism

Volcanoes are fueled by magma. Recall that magma is a slushy mixture of molten rock, mineral crystals, and gases. As you observed in the Launch Lab, once magma forms, it rises toward Earth's surface because it is less dense than the surrounding mantle and crust. Magma that reaches Earth's surface is called lava. **Volcanism** describes all the processes associated with the discharge of magma, hot fluids, ash, and gases.

As you read this, approximately 20 volcanoes are erupting. In a given year, volcanoes will erupt in about 60 different places on Earth. The distribution of volcanoes on Earth's surface is not random. A map of active volcanoes, shown in **Figure 1,** reveals striking patterns on Earth's surface. Most volcanoes form at plate boundaries. The majority form at convergent boundaries and divergent boundaries. Along these margins, magma rises toward Earth's surface. Only about 5 percent of magma erupts far from plate boundaries.

Figure 1 Most of Earth's active volcanoes are located along plate boundaries.

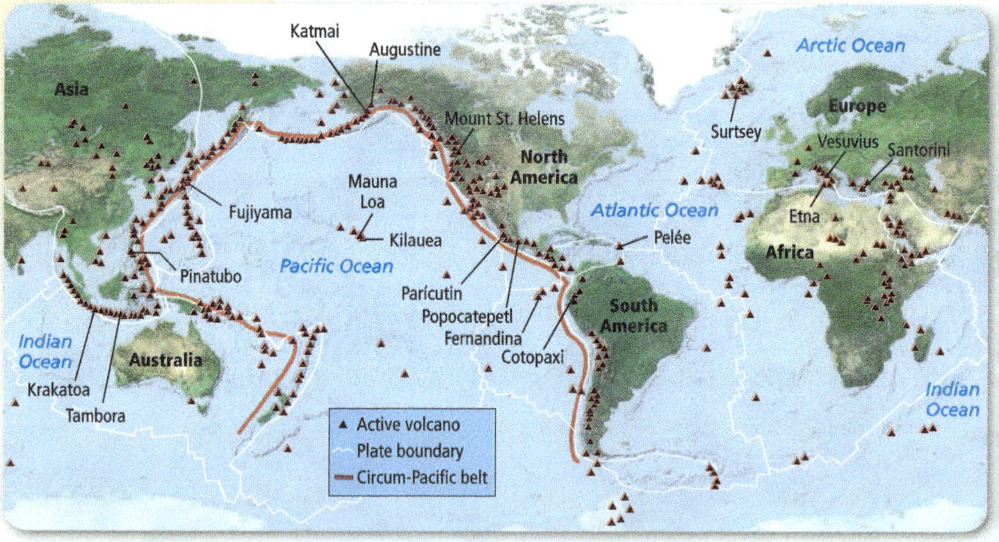

DIFFERENTIATED INSTRUCTION

Struggling Learners Have students each make a table to classify by continent the volcanoes that are labeled in **Figure 1.** If necessary, list and identify Earth's continents on a map.

Interpret the Illustration

Regions of Active Volcanism Have students examine **Figure 1.** Ask: Where are most active volcanoes located? Most active volcanoes are located at tectonic plate boundaries. OL

Convergent volcanism Recall that tectonic plates collide at convergent boundaries, which can form subduction zones—places where slabs of crust descend into the mantle. As shown in **Figure 2,** an oceanic plate descends below another plate into the mantle. As the oceanic plate descends, water is released which helps to melt the overlying mantle, forming magma. The magma moves upward because it is less dense than the surrounding material. As it rises, it mixes with rock, minerals, and sediment from the overlying plate. Most volcanoes located on land result from oceanic-continental subduction. These volcanoes are characterized by explosive eruptions.

☑ READING CHECK **Define** What is convergent volcanism?

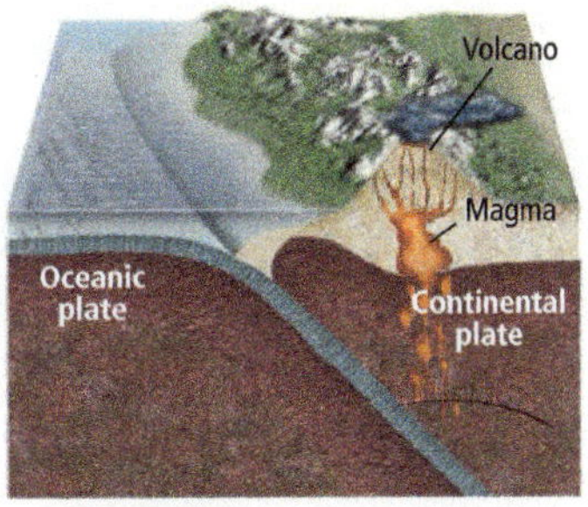

■ **Figure 2** In an oceanic-continental subduction zone, the denser oceanic plate slides under the continental plate into the hot mantle. Parts of the overlying mantle melt and magma rises, eventually leading to the formation of a volcano.
***Identify** a volcano from Figure 1 that is associated with oceanic-continental convergence.*

View an **animation of subduction.**
Concepts In Motion

Two major belts The volcanoes associated with convergent plate boundaries forms a major belt, shown in **Figure 1.** The Circum-Pacific Belt is also called the Pacific Ring of Fire. The name *Circum-Pacific* gives a hint about the location of the belt. *Circum* means *around* (as in circumference). The outline of the belt corresponds to the outline of the Pacific Plate. The belt stretches along the western coasts of North and South America, across the Aleutian Islands, and down the eastern coast of Asia. Volcanoes in the Cascade Range of the western United States and Mount Pinatubo in the Philippines are some of the volcanoes in the Circum-Pacific Belt. A smaller belt is called the Mediterranean Belt. It includes Mount Etna and Mount Vesuvius, two volcanoes in Italy. Its general outlines correspond to the boundaries between the Eurasian, African, and Arabian plates.

Data Analysis LAB

Based on Real Data*

Interpret the Graph

How do zones of volcanism relate to lava production? Researchers classify types of volcanic eruptions and study how much lava each type of volcano emits during an average year. The circle graphs show data from 5337 eruptions and annual lava production for each zone.

Think Critically

1. **Describe** the relationship between the type of volcanism and annual lava production.
2. **Consider** Why is it important for scientists to study this relationship?
3. **Evaluate** What could be the next step in the researchers' investigation?

Data and Observations

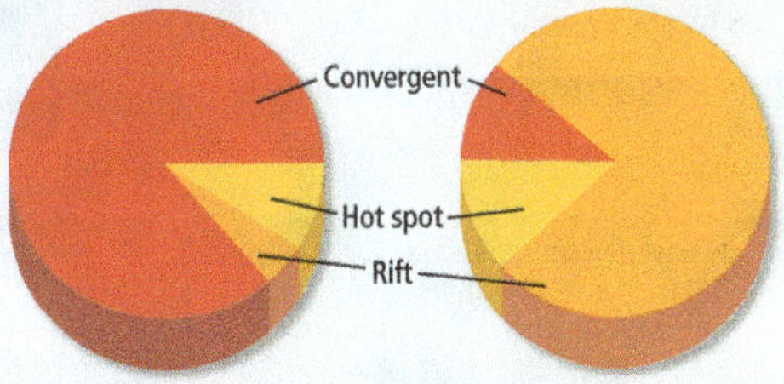

*Data obtained from: Crisp, J. 1984. Rates of magma emplacement and volcanic output. *Journal of Volcanology and Geothermal Research* 20: 177–211.

2. Answers will vary. Understanding the relationship between volcanism and lava production helps scientists forecast eruptions and the types of damage to be expected. This can help prevent loss of life and property.
3. Answers will vary. Researchers might want to determine the composition and explosivity of particular lavas or the number of eruptions for each volcano.

Tie to Previous Knowledge

Plate Boundaries Have a volunteer or two summarize the differences between convergent and divergent plate boundaries. OL AL

Interpret the Illustration

Tectonic Plates Have students refer to the map in **Figure 1.** Ask: Which tectonic plates are associated with the Mediterranean Volcanic Belt? the African, Arabian, and Eurasian Plates If students are unable to answer the question, refer them to a map of major tectonic plates. OL

■ **Caption Question Fig. 2** Answers will vary, but should include a volcano in the Circum-Pacific Belt.

☑ **READING CHECK** Convergent volcanism is all volcanic activity happening where two plates come together.

Data Analysis LAB

About the Lab

- Discuss with students the implications of the data shown in this lab. Emphasize that volcanoes at rift zones, which make up the smallest percentage of volcanoes, release much more lava than other types of volcanoes. Ask the students to infer where all of this lava accumulates. This will help them make connections between rift zones and seafloor spreading.
- See also Bargar, K. E., and E. D. Jackson. 1974. Calculated volumes of individual shield volcanoes along the Hawaiian-Emperor Chain. *Jour. Research U.S. Geol Survey* (2): 545–550.

Think Critically

1. Some types of volcanism produce much more lava than others. Although 80% of volcanic eruptions occur at convergent boundaries, only 15% of lava erupts there, whereas even though only 7% of volcanic eruptions occur at rift zones, 75% of lava erupts there.

Teacher Content Support

The Great Rift Valley The Great Rift Valley is a geologic feature in Africa that is the result of the African and Arabian Plates moving away from each other at a divergent boundary and of the continuing separation along the East African Rift. These separations began 35 and 15 mya, respectively. Volcanic activity in this area has produced volcanic mountains, including Mount Kilimanjaro, Mount Kenya, Mount Meru, and others.

Activity

Time Line Have students look at the time line in **Figure 4.** Then have students work in groups to research eruptions since 1991, and make time lines of these eruptions on poster board. Student time lines can be displayed in the classroom. OL EL COOP LEARN

Concept Development

Convergent and Divergent Volcanism Have students compare and contrast divergent volcanism and convergent volcanism. Convergent volcanism involves the collision of tectonic plates at plate boundaries. Convergent volcanism is responsible for most volcanoes located on land. Divergent volcanism involves the spreading of tectonic plates at plate boundaries called rift zones. Most rift zones occur in the world's oceans. OL BL

☑ READING CHECK Two-thirds is 66%.

Figure 3 Eruptions at divergent boundaries tend to be nonexplosive. At the divergent boundary on the ocean floor, eruptions often form huge piles of lava called pillow lava.

View an **animation of divergent plate boundaries.**

Concepts In Motion

Divergent volcanism Recall that at divergent plate boundaries tectonic plates move apart and new ocean floor is produced as magma rises to fill the gap. At ocean ridges, this lava takes the form of giant pillows like those in **Figure 3,** and is called pillow lava. Unlike the explosive volcanoes detailed in **Figure 4,** volcanism at divergent boundaries tends to be nonexplosive, with effusions of large amounts of lava. About two-thirds of Earth's volcanism occurs underwater along divergent boundaries at ocean ridges.

VOCABULARY
SCIENCE USAGE V. COMMON USAGE
Plume
Science usage: an elongated column

Common usage: a large, showy feather of a bird

☑ READING CHECK **Convert** the fraction of volcanism that happens underwater to a percentage.

Hot spots Some volcanoes form far from plate boundaries over hot spots. Scientists hypothesize that **hot spots** are unusually hot regions of Earth's mantle where high-temperature plumes of magma rise to the surface.

Figure 4
Volcanoes in Focus
Volcanoes constantly shape Earth's surface.

A.D. 79 Mount Vesuvius in Italy erupts, burying two cities in ash.

6000 B.C. — 3000 B.C.

4845 B.C. Mount Mazama erupts in Oregon. The mountain collapses into a 9-km-wide depression known today as Crater Lake, shown on this topographic map.

1630 B.C. In Greece, Santorini explodes, causing tsunamis 200 m high. Nearby, Minoan civilization on the Isle of Crete disappears.

ACROSS THE CURRICULUM

History When Mount Vesuvius erupted in A.D. 79, two Roman cities were destroyed—Herculaneum and Pompeii. Excavations have revealed much information about the people of these cities and their destruction. Have students research the A.D. 79 eruption of Mount Vesuvius and present their findings to the class. Discuss the possibility of Mount Vesuvius erupting today and the possible effects. Mount Vesuvius is currently considered one of the most dangerous volcanoes on Earth; this is due to its history of violent eruptions and the fact that nearly 300,000 people live in close proximity to it.

Hot spot volcanoes Some of Earth's best-known volcanoes formed as a result of hot spots under the ocean. For example, the Hawaiian islands, shown in the map in **Figure 5,** are located over a plume of magma. As the rising magma melts through the crust, it forms volcanoes. The hot spot formed by the magma plume remains stationary, while the Pacific Plate slowly moves northwest. Over time, the hot spot has left a trail of volcanic islands on the floor of the Pacific Ocean. The volcanoes on the oldest Hawaiian island, Kauai, are inactive because the island no longer sits above the stationary hot spot. Even older volcanoes to the northwest are no longer above sea level. The world's most active volcano, Kilauea, on the Big Island of Hawaii, is currently located over the hot spot. Another volcano, Loihi, is forming on the seafloor southeast of the Big Island of Hawaii and might eventually rise above the ocean surface to form a new island.

Hot spots and plate motion Chains of volcanoes that form over stationary hot spots provide information about plate motions. The rate and direction of plate motion can be calculated from the positions of these volcanoes. The map in **Figure 5** shows that the Hawaiian islands are at one end of the Hawaiian-Emperor volcanic chain. The oldest seamount, Meiji, is at the other end of the chain and is about 80 million years old, which indicates that this hot spot has existed for at least that many years. The bend in the chain at Daikakuji Seamount records a change in the direction of the Pacific Plate that occurred 43 mya.

Hawaiian-Emperor Volcanic Chain

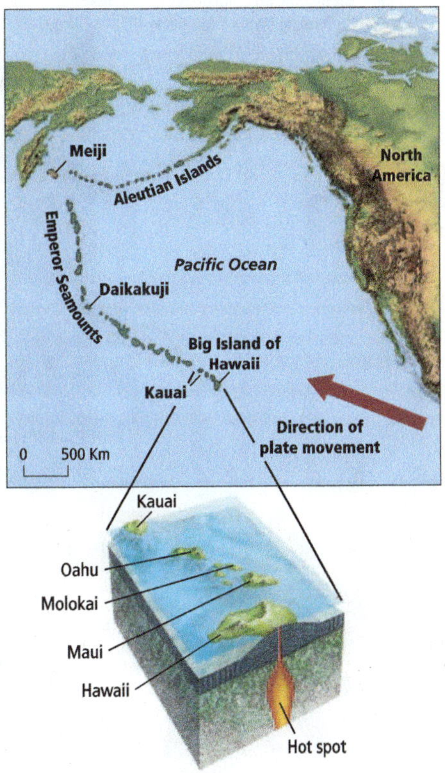

Figure 5 The Hawaiian islands have been forming for millions of years as the Pacific Plate moves slowly over a stationary hot spot that is currently located under the Big Island of Hawaii.

1800 — 1900 — 2000

1883 In Indonesia, Krakatoa erupts, destroying two-thirds of the island and generating a tsunami that kills more than 36,000 people.

1912 Katmai erupts in Alaska with ten times more force than Mount St. Helens. This eruption is one of the most powerful in recorded history.

1980 In Washington, Mount St. Helens' eruption blasts through the side of the volcano. Most of the 57 fatalities are from ash inhalation.

1991 Mount Pinatubo erupts in the Philippines, releasing 10 km^3 of ash, reducing global temperatures by 0.5°C.

2010 Ash erupting from Iceland's Eyjafjallojökull volcano causes Europe's air traffic to shut down, affecting at least 10 million passengers worldwide.

Demonstration

Hot Spots Have a student hold a red marker vertically to represent a mantle plume erupting at a hot spot. Move a piece of paper slowly over the marker and every 30 seconds make a mark that represents the island that formed from the eruption of magma. If you move the paper in a straight line, you get a regularly spaced set of points. If you change the direction the paper moves, you will see a bend in the line of islands. Students will see that the oldest island is furthest from the hot spot. Ask: Which mark is most like Hawaii's Big Island? the last one—the one most recently over the hot spot What does the bend in the line of points represent? a change in plate motion

Teacher Content Support

North American Hot Spots

There is a hot spot located below the North American continent that is the site of what could be the largest volcanic eruption that has ever occurred. The hot spot is currently located beneath Yellowstone National Park.

There have been three massive eruptions associated with the Yellowstone hot spot in the past 2 million years. About 2 million years ago, 2400 km^3 of tephra erupted; about 1.3 million years ago, 280 km^3 of tephra erupted; and most recently, 640,000 years ago, an estimated 1000 km^3 of tephra erupted and left a caldera that is over 70 km in diameter. This caldera is currently buried beneath Yellowstone National Park.

The magma chamber beneath the park is still active. Recent uplift, multiple earthquakes, and changes in geyser activity all indicate that the Yellowstone hot spot could erupt again.

Concept Development

Supervolcanoes Tell students that in the last few million years, several "supervolcanoes" have erupted: Yellowstone, Long Valley (eastern California), Taupo (New Zealand), and Toba (Indonesia). Supervolcanoes can erupt more than 1000 km^3 of magma and affect climate and life over the whole planet. Have students find these volcanoes on the map. Ask: What do these volcanoes have in common? They are either on continents or near subduction zones. AL

Activity

Greenhouse Gases Explain the greenhouse effect and tell students that some volcanic gases are greenhouse gases. List the following volcanic gases on the board and have students determine which are greenhouse gases: water vapor, carbon dioxide, carbon monoxide, fluorine, chlorine, hydrogen, some sulfur compounds, and nitrogen. Carbon dioxide and water vapor are greenhouse gases. AL

Discussion

Deccan Traps Have students, based on what they have learned from the chapter, discuss how the eruption of the Deccan Traps might have caused the extinction of the dinosaurs and whether or not they think such a theory is valid. OL

Model

Flood Basalts Have students make models showing how a flood basalt erupts. Models should show lava erupting from a fissure rather from a central vent. OL COOP LEARN

Project

Classify Volcanoes Have students conduct research to find out that volcanoes can be classified as active, dormant, or extinct. Have students differentiate among these terms in their Earth science journals. Challenge students to list at least five examples of each of these volcano types. AL

Figure 6 Huge amounts of lava erupting from fissures accumulate on the surface, often forming layers 1 km thick. Over time, streams and other geologic forces erode the layers of basalt, leaving plateaus like this one in Palouse Canyon, Washington.

Figure 7 More than 17 mya, enormous amounts of lava poured out of large fissures, producing a basaltic plateau more than 1 km thick in the northwestern part of the United States.

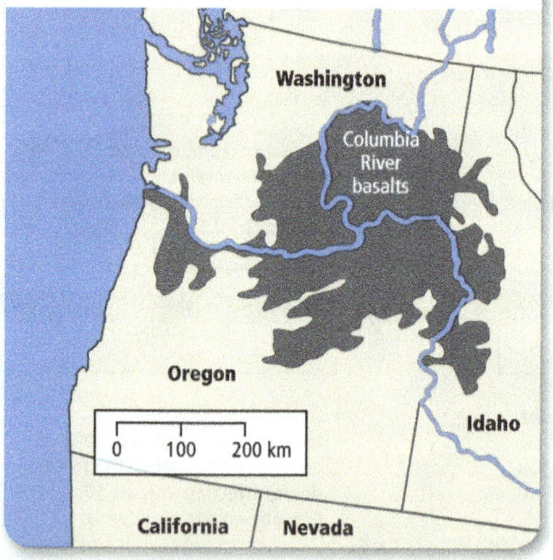

Flood basalts When hot spots occur beneath continental crust, they can lead to the formation of flood basalts. **Flood basalts** form when lava flows out of long cracks in Earth's crust. These cracks are called **fissures.** Over hundreds or even thousands of years, these fissure eruptions can form flat plains called plateaus, as shown in **Figure 6.** As in other eruptions, when the lava flows across Earth's surface, water vapor and other gases escape.

Columbia River Basalts The volume of basalt erupted by fissure eruptions can be tremendous. For example, the Columbia River basalts, located in the northwestern United States and shown on the map in **Figure 7,** contain 170,000 km^3 of basalt. This volume of basalt could fill Lake Superior, the largest of the Great Lakes, 15 times. However, the Columbia River Basalts are small in comparison to the Deccan Traps.

Deccan Traps About 65 mya in India, a huge flood basalt eruption created an enormous plateau called the Deccan Traps. The volume of basalt in the Deccan Traps is estimated to be about 512,000 km^3. That volume would cover the island of Manhattan with a layer 10,000 km thick, or the entire state of New York with a layer 4 km thick. Some geologists hypothesize that the eruption of the Deccan Traps caused a global change in climate that might have contributed to the extinction of the dinosaurs.

DIFFERENTIATED INSTRUCTION

Struggling Learners Have students make an analogy between Earth's crust and a layer of ice on a lake. Ask: Which would be more dangerous to stand beside: a 30-cm-diameter ice-fishing hole, or a crack going all the way across the lake? The crack would be more dangerous, because it is more unstable at the edges and a person could fall in. Which is more like a fissure? the crack

DIFFERENTIATED INSTRUCTION

Visually Impaired To illustrate the fact that the Columbia River Basalt formation was 1 km thick, cover the area of the plateau on a map with modeling clay thick enough to stand a few centimeters above the map's surface. Have students use their hands to feel the elevation difference.

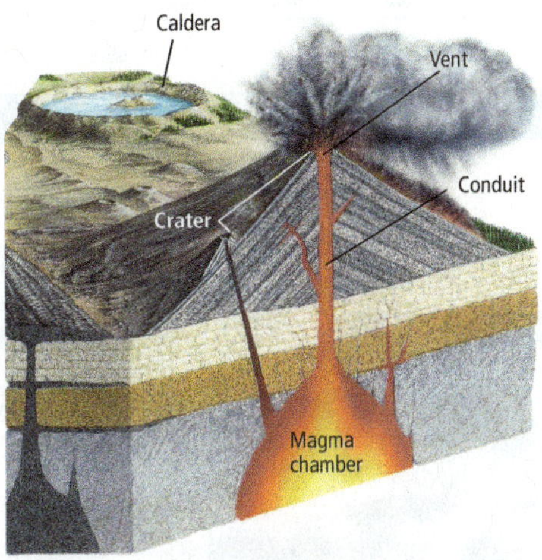

■ **Figure 8** Magma moves upward from deep within Earth through a conduit and erupts at Earth's surface through a vent. The area around the vent is called a crater. A caldera can form when the crust collapses into an empty magma chamber.

View an **animation of caldera formation.**

Concepts In Motion

Anatomy of a Volcano

Recall that when magma reaches Earth's surface it is called lava. Lava reaches the surface by traveling through a tubelike structure called a **conduit,** and emerges through an opening called a **vent.** As lava flows through the vent and out onto the surface, it cools and solidifies around the vent. Over time, layers of solidified lava can accumulate to form a mountain known as a volcano. At the top of a volcano, around the vent, is a bowl-shaped depression called a **crater.** The crater is connected to the magma chamber by the conduit. Locate the crater, conduit, and vent of the volcano shown in **Figure 8.**

Volcanic craters are usually less than 1 km in diameter. Larger depressions, called **calderas,** can be up to 100 km in diameter. Calderas often form after the magma chamber beneath a volcano empties from a major eruption. The summit or the side of a volcano collapses into the emptied magma chamber, leaving an expansive, circular depression. After the surface material collapses, water sometimes fills the caldera, forming scenic lakes. The caldera known as Crater Lake in southern Oregon formed when Mount Mazama collapsed.

MiniLAB

iLab Station

Model a Caldera

How do calderas form? Calderas are volcanic craters that form when the summit or the side of a volcano collapses into the magma chamber that once fueled the volcano.

Procedure

1. Read and complete the lab safety form.
2. Obtain a **small box,** a **10-cm length of rubber tubing,** a **clamp,** and a **balloon** from your teacher.
3. Line the box with **newspaper** and make a small hole in the box and the newspaper with **scissors.**
4. Thread the neck of the balloon through the hole, insert the rubber tubing into the neck, securing it with **tape,** inflate the balloon by blowing through the tubing, and use the clamp to close the tubing.
5. Pour six cups of **sand** over the balloon.
6. Sculpt the sand into the shape of a volcano. You might need to vary the amount of sand and type of box to reach the desired effect.
7. Remove the clamp, releasing the air from the balloon. Observe your caldera forming, and record your observations.
8. Compare your caldera to your classmates'.

Analysis

1. **Sequence** the formation of the caldera.
2. **Compare** the features of a caldera with those of a crater.
3. **Infer** how the caldera will form if you vary how much you inflate the balloon.

MiniLAB

iLab Station Rubric

Purpose Students will model how calderas form by collapsing into an empty magma chamber.

Process Skills measure, observe

Safety Precaution Approve lab safety forms before work begins.

Teaching Strategy Have students use different sizes and shapes of balloons to see how different calderas might form.

Expected Results As the air is released from the balloon, the sand volcano will collapse in the middle, modeling the way a caldera forms.

Analysis

1. The volcano forms with a magma chamber underneath; the volcano erupts and the magma chamber empties; the volcano collapses into the magma chamber, forming a caldera.
2. Calderas are circular, wide, and larger. Craters are bowl-shaped and smaller.
3. The greater the balloon's inflation, the deeper the caldera.

Assessment

Knowledge Ask students what the balloon in the experiment represents in the formation of an actual caldera. upwelling magma; When the air is released from the balloon, this represents the expulsion of all the magma in the chamber.

LabManager™
Customize any lab with the LabManager™ CD-ROM.

ACROSS THE CURRICULUM

Astronomy Put students into small groups of two or three. Have them research volcanoes on other planets and their moons. One group might research Mars's Olympus Mons, the largest volcano in the solar system. Another group can research Io, a moon of Jupiter and the most volcanically active object in the solar system. The volcanoes of Venus are another possible research topic. Each group should summarize their results either on a poster for classroom display or by making a presentation of their findings. OL

Interpret the Illustration

Craters Locate the craters of the volcanoes shown in **Table 1.** The crater is the bowl-shaped depression at the volcanoes' summits. BL

Interpret the Photo

Parts of a Volcano Have students trace the outline of the composite volcano shown in the photo at the bottom of **Table 1.** Challenge them to modify the outline into a cross-sectional (cutaway) view of the volcano that includes the magma chamber, the volcanic vent, the crater, and the layers of volcanic debris that accumulated to form the mountain. OL

Project

Identify Types of Volcanoes Have students each bring to class one picture of a volcano. Show a selection of the pictures to the class and have students discuss which fit the description of shield, composite, or cinder cone volcano. Showing students each volcano's location on the map should help them in classifying the pictures. OL

Explore the **types of volcanoes with an interactive table.** Concepts In Motion

Table 1 Types of Volcanoes

Description	Example of Volcanoes
Shield Volcanoes • Largest of the three types of volcanoes • Long, gentle slopes • Composed of layers of solidified basaltic lava • Quiet eruptions	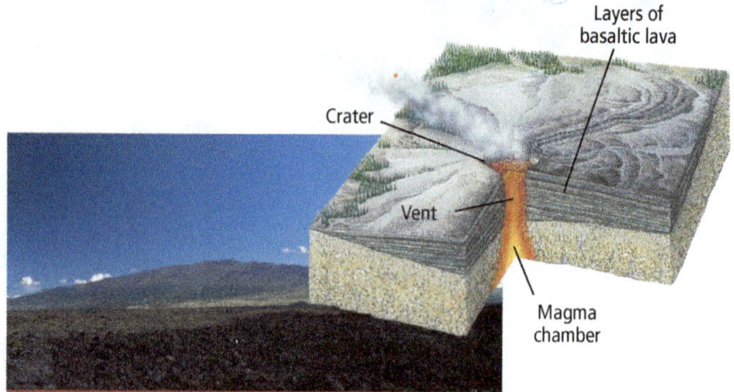 Mauna Loa, Hawaii
Cinder Cones • Smallest of the three types of volcanoes • Steep-sloped, cone-shaped • Usually composed of fragments of basaltic lava • Explosive eruptions • Usually form at edges of larger volcanoes	 Lassen Volcanic Park, California
Composite Volcanoes • Considerably larger than cinder cones • Tall, majestic mountains • Composed of layers of rock from explosive eruptions and lava flows • Cycle through periods of quiet and explosive eruptions	 Mount Augustine, Alaska

DIFFERENTIATED INSTRUCTION

English Learners Have students use dictionaries to look up the definitions of the terms *shield, cinder,* and *composite.* Then have students list each term in their Earth science journals and write a sentence next to each explaining why it is an appropriate adjective to apply to volcanoes.

Types of Volcanoes

The appearance of a volcano depends on two factors: the type of material that forms the volcano and the type of eruptions that occur. Based on these two criteria, three major types of volcanoes have been identified and are shown in **Table 1.** Each differs in size, shape, and composition.

Shield volcanoes A **shield volcano** is a mountain with broad, gently sloping sides and a nearly circular base. Shield volcanoes form when layers of lava accumulate during nonexplosive eruptions. They are the largest type of volcano. Mauna Loa, which is shown in **Table 1,** is a shield volcano.

Cinder cones When eruptions eject small pieces of lava into the air, **cinder cones** form as this material, often called cinders, scoria, or tephra, falls back to Earth and piles up around the vent. Cinder cones have steep sides and are generally small; most are less than 500 m high. The Lassen Volcanic Park cinder cone shown in **Table 1** is 700 m high. Cinder cones are commonly found on or very near larger volcanoes. For instance, volcanologists have identified nearly 100 cinder cones on the flanks of Mauna Kea, a large shield volcano on Hawaii.

Composite volcanoes **Composite volcanoes** are formed of layers of ash and hardened chunks of lava from violent eruptions alternating with layers of lava that oozed downslope before solidifying. Composite volcanoes are generally cone-shaped with concave slopes, and are much larger than cinder cones. Because of their explosive nature, they are potentially dangerous to humans and the environment. Some examples of these are Mount Augustine in Alaska, shown in **Table 1,** and several in the Cascade Range of the western United States, such as Mount St. Helens.

CAREERS IN EARTH SCIENCE

Volcanologist Scientists who study eruptions, lava, magma, and the conditions under which these form are volcanologists. Some work in the field, studying active volcanoes. Many volcanologists also work in the laboratory to understand how rocks melt to form magma.

WebQuest

FOLDABLES®
Incorporate information from this section into your Foldable.

SECTION 1 REVIEW

Section Self-Check

Section Summary

- Volcanism includes all the processes in which magma and gases rise to Earth's surface.
- Most volcanoes on land are part of two major volcanic chains: the Circum-Pacific Belt and the Mediterranean Belt.
- Parts of a volcano include a vent, magma chamber, crater, and caldera.
- Flood basalts form when lava flows from fissures to form flat plains or plateaus.
- There are three major types of volcanoes: shield, composite, and cinder cone.

Understand Main Ideas

1. MAINIDEA **Explain** how the location of volcanoes is related to the theory of plate tectonics.
2. **Identify** two volcanoes in the Mediterranean Belt.
3. **Draw** a volcano, labeling the parts.
4. **Propose** Yellowstone National Park is an area of previous volcanism. Using a map of the United States, suggest the type(s) of tectonic processes associated with this area.

Think Critically

5. **Evaluate** the following statement: Volcanoes are only found along coastlines.
6. **Decide** whether a flood basalt is or is not a volcano.

MATH IN Earth Science

7. If the Pacific Plate has moved 500 km in the last 4.7 million years, calculate its average velocity in centimeters per year. Refer to the *Skillbuilder Handbook* for more information.

FOLDABLES® Rubric

CAREERS IN EARTH SCIENCE

WebQuest

3 Assess

Check for Understanding

Discussion Ask the following questions: If a subduction zone developed along the eastern coast of the United States, what type of volcanoes would form? What type of formation could occur if a fissure developed in Yellowstone National Park? Composite volcanoes could form; a flood basalt plateau could form.

Reteach

Concept Map Provide each student with a skeleton of a concept map that summarizes the characteristics of the three types of volcanoes presented in this section. Have students complete the maps and keep them for reference.

Assessment

Performance Have each student identify locations on the map shown in **Figure 1** where each of the different types of volcanoes—shield, composite, and cinder cone—are likely to occur. Shield volcanoes occur over hot spots, such as that under the Hawaiian Islands. Composite volcanoes and cinder cones occur at subduction zones along tectonic plate edges, such as Mount St. Helens on the West Coast of the United States and Mt. Fuji in Japan.

SECTION 1 REVIEW

1. Certain types of volcanoes are associated with particular types of plate boundaries: convergent plate boundaries, rift zones, and hot spots produce different types of volcanoes.
2. Vesuvius and Etna
3. Labels should include: magma chamber, magma, conduit, vent, and caldera or crater.
4. Because it is far from a plate edge and is so big, Yellowstone is probably the result of a flood basalt eruption over a hot spot.
5. Untrue: Volcanoes can form in the middle of oceanic plates due to hot spots or rift zones.
6. Although a flood basalt is the result of volcanic activity, it does not take the shape of a mountain and therefore is not a volcano.
7. 10.6 cm/year ; To get speed, divide distance by time: 500 km ÷ 4.7 million years = 10.6 cm/year.

SECTION 2

1 Focus

Focus Transparency

Before presenting the lesson, project **Section Focus Transparency 54** and have students answer the accompanying questions. EL

MAINIDEA

Eruptions Ask students to consider what a volcanic eruption is. Ask: What is a volcanic eruption? An outpouring of magma, ash, and/or tephra through an opening in Earth's surface What is magma? melted rock What is rock made of? minerals

2 Teach

Identify Misconceptions

Some students might think that volcanic eruptions are rare events.

Uncover the Misconception Ask students to speculate how many volcanic eruptions occur on Earth every year.

Demonstrate the Concept Explain that most individual volcanoes erupt infrequently. However, because there are over 500 active volcanoes on Earth, it is likely that one or more of them are erupting at any given time.

Assess New Knowledge Have students research volcanic eruptions that have occurred during the past year, and then use pins to mark them on a large world map.

SECTION 2

Essential Questions

- How does magma type influence a volcano's explosivity?
- What is the role of pressure and dissolved gases in eruptions?
- What kinds of material are ejected by volcanic eruptions?

Review Vocabulary

basaltic: relates to a group of rocks rich in dark-colored minerals containing magnesium and iron

New Vocabulary

viscosity
tephra
pyroclastic flow

Eruptions

MAINIDEA The composition of magma determines the characteristics of a volcanic eruption.

EARTH SCIENCE 4 YOU Have you ever shaken a can of soda and then opened it? If so, it probably sprayed your hand, clothes, and maybe even your friends. This is similar to the process that underlies explosive volcanic eruptions.

Making Magma

What makes the eruption of one volcano quiet, and the eruption of another explosively violent? A volcano's explosivity depends on the composition of the magma. As shown in **Figure 9,** lava from an eruption can be thin and runny or thick and lumpy. In order to understand why volcanic eruptions are not all the same, you first need to understand how rocks melt to make magma.

Temperature Depending on their composition, most rocks begin to melt at temperatures between 800°C and 1200°C. Such temperatures are found in the crust and upper mantle. Recall that temperature increases with depth beneath Earth's surface. In addition to temperature, pressure and the presence of water and dissolved gases also affect the formation of magma.

Pressure Pressure increases with depth because of the weight of overlying rocks. As pressure increases, the temperature at which a substance melts also increases. **Figure 10** shows two melting curves for a type of feldspar called albite. Note that at Earth's surface, albite, in the absence of water, melts at about 1100°C, but at a depth of about 12 km, its melting point is about 1150°C. At a depth of about 100 km, the melting point of dry albite increases to 1440°C. The effect of pressure explains why most of the rocks in Earth's lower crust and upper mantle do not melt.

Figure 9 The way in which lava flows depends on the composition of the magma. Mount Etna's lava is thin and runny compared to the thick and lumpy lava that erupts at Mount St. Helens.

Mount Etna

Mount St. Helens

IN THE FIELD

Volcanic Eruption Prediction Jim Kauahikaua is a geophysicist who works for the Hawaiian Volcano Observatory. One of Kauahikaua's responsibilities is to gather up-to-date information on volcanic eruptions and lava flows, and to use these data to forecast volcanic eruptions. Kauahikaua uses the Global Positioning System to create up-to-the-moment maps with the data he collects.

Composition of Magma

The composition of magma determines a volcano's explosivity, which is how it erupts and how its lava flows. What are the factors that determine the composition of magma? Scientists now know that the factors include magma's interaction with overlying crust, its temperature, pressure, amounts of dissolved gas, and—very significantly—the amount of silica a magma contains. Understanding the factors that determine the behavior of magma can aid scientists in predicting the eruptive style of volcanoes.

Dissolved gases In general, as the amount of gases in magma increases, the magma's explosivity also increases. In the same way that gas dissolved in soda gives the soda its fizz, the gases dissolved in magma give a volcano its "bang." Important gases in magma include water vapor, carbon dioxide, sulfur dioxide, and hydrogen sulfide. Water vapor is the most common dissolved gas in magma. The presence of water vapor determines where magma forms. As shown in **Figure 10,** minerals in the mantle, such as albite melt at high temperatures. The presence of dissolved water vapor lowers the melting temperature of minerals, causing mantle material to melt into magma. This eventually forms volcanoes and fuels their eruptions.

Viscosity The physical property that describes a material's resistance to flow is called **viscosity**. Temperature and silica content affect the viscosity of a magma. In general, cooler magma has a higher viscosity. In other words, cool magma, much like chilled honey, tends to resist flowing.

READING CHECK **Infer** Which has a higher viscosity: syrup or water?

Magma with high silica content tends to be thick and sticky. Because it is thick, magma with high silica content tends to trap gases, which produces explosive eruptions. In general, magma with low silica content has low viscosity—it tends to be thin and runny, like warm syrup. Magma with low silica content tends to flow easily and produce quiet, nonexplosive eruptions.

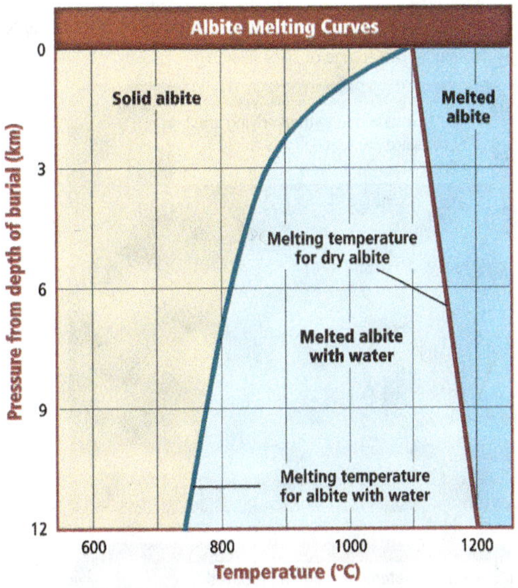

Figure 10 Both the pressure and water content of the mineral albite affect how the mineral melts.

Locate *the melting curve of wet albite. How does the melting point of wet albite compare to that of dry albite at a depth of 3 km? At a depth of 12 km?*

VOCABULARY

ACADEMIC VOCABULARY

Aid
to provide with what is useful or necessary in achieving an end
Glasses aid Omar in seeing clearly.

Interpret the Illustration

Melting Temperatures Have students study the graph in **Figure 10.** Ask: What is the relationship between depth and the melting temperature of dry albite? An increase in depth results in an increase in the melting temperature of dry albite. What is the relationship between depth and the melting temperature of wet albite? An increase in depth results in a decrease in the melting temperature of wet albite.

Caption Question Fig. 10 At 3 km, wet albite melts at about 870°C and dry albite at about 1120°C. At 12 km, wet albite melts at about 740°C and dry albite at 1200°C.

Teacher Content Support

Geysers Where magma chambers lie relatively close to Earth's surface, surrounding rocks are heated by the magma, which in turn heats the groundwater in the area. This superheated water can come to the surface as a hot spring or an erupting geyser. Geysers occur when groundwater is heated to extremely high temperatures, which forces the water to expand. This expansion, in turn, forces some of the water out of the ground, releasing the pressure on the remaining water. The release of pressure causes the remaining water to boil rapidly. Steam pressure forces the remaining water and steam high into the air. After a geyser erupts, water refills the underground chambers, is reheated, and the geyser erupts again. Geysers are relatively rare; most are found in western parts of the United States, and in Iceland, New Zealand, and Chile.

READING CHECK syrup

DIFFERENTIATED INSTRUCTION

Advanced Learners When two substances are combined, the melting point of the mixture is significantly lower than the melting points of either of the substances in their pure states. Have students research how mixing substances affects the melting temperature of compounds.

Tie to Previous Knowledge

Rock Characteristics Have students review the characteristics and compositions of basalt, andesite, and granite. Ask them to predict how the characteristics of these rocks will be similar to the characteristics of basaltic magma, andesitic magma, and rhyolitic magma.

Concept Development

Viscosity Ask: What kind of magma has a viscosity similar to that of honey? rhyolitic magma **BL**

Concept Development

Magma Viscosity Ask: Which type of lava—basaltic or rhyolitic—has a greater viscosity? rhyolitic Ask: What do you think happens to viscosity as magma or lava cools? A decrease in temperature results in an increase in viscosity. **OL**

Teacher Content Support

Sand The volcanic material that makes up the black sand beaches of Hawaii did not erupt as sand-sized particles. The sand grains formed when basaltic lava flows entered the ocean. The rapid thermal shock, combined with the seawater flashing into steam, caused the lava to explode and shatter into tiny particles. If the beach is young, the particles have very angular shapes.

Figure 11 Generally, magma and lava with a low percentage of silica have low viscosity, and those with a higher percentage of silica have high viscosity.

Types of Magma

The silica content of magma determines not only its explosivity and viscosity, but also which type of volcanic rock it forms as lava cools. Refer to **Figure 11** to summarize types of magma.

Basaltic magma When rock in the upper mantle melts, basaltic magma typically forms. Basaltic magma has the same silica content as the rock basalt—less than 50 percent silica. This magma rises from the upper mantle to Earth's surface and reacts very little with overlying continental crust or sediments. Its low silica content produces low-viscosity magma. Dissolved gases escape easily from basaltic magma. The resulting volcano is characterized by quiet eruptions. **Figure 12** shows how properties of magma affect the types of eruptions that occur. Volcanoes such as Kilauea and Mauna Loa actively produce basaltic magma. Surtsey, a volcano that was formed south of Iceland in 1963, is another volcano that produces basaltic magma.

Andesitic magma Andesitic (an duh SIH tihk) magma has the same silica content as the rock andesite—50 to 60 percent silica. Andesitic magma is found along oceanic-continental subduction zones. Magma produced at subduction zones can evolve into an intermediate magma by fractional crystallization, mixing with other magma bodies, or by assimilating continental crust. The higher silica content results in a magma that has intermediate viscosity. Thus, the volcanoes it fuels are said to have intermediate explosivity. Colima Volcano in Mexico and Tambora in Indonesia are two examples of andesitic volcanoes. Both volcanoes have produced massive explosions that sent huge volumes of ash and debris into the atmosphere.

Rhyolitic magma When molten material rises and mixes with the overlying continental crust rich in silica and water, it forms rhyolitic (ri uh LIH tihk) magma. Rhyolitic magma has the same composition as the rock granite—more than 60 percent silica. The high viscosity of rhyolitic magma slows down its movement. High viscosity, along with the large volume of gas trapped within this magma, makes the volcanoes fueled by rhyolitic magma very explosive. The dormant volcanoes in Yellowstone National Park in the western United States were fueled by rhyolitic magma. The most recent of these eruptions, which occurred 640,000 years ago, was so powerful that it released 1000 km3 of volcanic material into the air.

Demonstration

Model Magma Movement Fill a dropper with a small amount of olive oil. Insert the dropper into a glass of water and slowly squeeze out drops of oil. Ask students to compare the movement of the oil to the movement of magma. Like magma, the less-dense oil moves upward.

ACROSS THE CURRICULUM

Math Inform students some lavas flow as quickly as 16 km/h. Have students compute how long it would have taken the Iceland Laki flow to travel the 45 km from the eruption site to the coastline at this speed. 45 km ÷ 16 km/h = 2.8 h

VISUALIZING Eruptions

Figure 12 As magma rises due to plate tectonics and hot spots, it mixes with Earth's crust. This mixing causes differences in the temperature, silica content, and gas content of magma as it reaches Earth's surface. These properties of magma determine how volcanoes erupt.

Quiet eruptions Earth's most active volcanoes are associated with hot spots under oceanic crust. Magma that upwells through oceanic crust maintains high temperature and low silica and gas contents. Lava oozes freely out of these volcanoes in eruptions that are relatively gentle.

Underwater eruptions Most pillow lava forms at diverging plate boundaries on the ocean floor. Lava oozes out of fissures in the ocean floor and forms bubble-shaped lumps as it cools.

Explosive eruptions Dangerous eruptions occur where magma melts, and mixes with, the silica-rich crustal rocks of the continental crust. This magma traps gases, causing tremendous pressure to build. The release of pressure drives violent eruptions.

Concepts In Motion View an **animation of plate tectonics resulting in volcanism.**

VISUALIZING

Purpose

Students will compare and contrast quiet eruptions, underwater eruptions, and explosive eruptions.

Model

Volcano Eruption Have groups of students make models of an erupting volcano. Provide students with modeling clay, baking soda, red food coloring, vinegar, safety goggles, lab aprons, and shallow trays. Instruct students to use the clay to form a small (no more than 10 cm high) volcano in the shallow tray. Tell students to create a crater 1 cm deep at the summit of the model, and to place a heaping teaspoon of baking soda and a drop of food coloring in the crater. Then have students add a tablespoon of vinegar to the crater and observe what happens. Have students describe the eruption in their Earth science journals. EL BL COOP LEARN

Project

Extraterrestrial Volcanism Earth is the only known planet to have active plate tectonics, because plate movement is dependent on an internal heat source. However, recent evidence suggests that Venus is volcanically active. Have students research whether scientists think that there any other planets that might still have active volcanism. OL AL

Teacher Content Support

Geothermal Gradient The geothermal gradient is the rate at which temperature increases with depth. Typically, the temperature increases at a rate of about 30°C/km; however, the geothermal gradient varies with location. In active basaltic volcanic regions, the rate can be as high as 60°C/km. Nonvolcanic mountain belts have gradients of about 40°C/km, and stable, continental interiors can have gradients as low as 20°C/km.

ACROSS THE CURRICULUM

Math Have students use the geothermal gradient rates given in the Teacher Content Support to compute the depth at which the temperature reaches 1000°C in the three different regions named. 16.7 km at 60°C/km; 25 km at 40°C/km; 50 km at 20°C/km

Teacher Content Support

Global Impact The atmospheric effects of volcanic eruptions have received a great deal of attention in the last few decades, particularly in light of global warming. Pinatubo is not the only volcano that has had global effects on temperature. The Laki eruption in 1783 lowered global temperatures an average of 4.8°C. Other volcanic eruptions and their effects on lowering global temperatures include the following: Tambora, 3°C; El Chichon, 0.3°C; Mount St. Helens, 0.1°C. Recent studies have indicated that the amount of tephra erupted is less important than the amount of sulfur gases released. The sulfuric acid droplets that form in the atmosphere cause a greater cooling effect than do dust and ash.

Concept Development

Global Impact The caption to **Figure 14** says that the 1991 eruption of Mount Pinatubo resulted in a decrease in global temperatures. Have students research the impact(s) other volcanoes have had on global temperatures and why.

▪ **Caption Question Fig. 13** Both ash and blocks are tephra, so both are ejected during volcanic eruptions.

▪ **Figure 13** Ash (shown actual size) is the smallest type of tephra. The 1-m-tall block shown here, ejected from Cotopaxi volcano in Ecuador, is an example of the largest category of tephra.
Compare *the two types of tephra. What do they have in common?*

Explosive Eruptions

When lava is too viscous to flow freely from the vent, pressure builds up in the lava until the volcano explodes, throwing lava and rock into the air. The erupted materials are called **tephra.** Tephra can be pieces of lava that solidified during the eruption, or pieces of the crust carried by the magma before the eruption. Tephra are classified by size. The smallest fragments, with diameters less than 2 mm, are called ash, as shown in **Figure 13.** The largest tephra thrown from a volcano are called blocks. The one shown in **Figure 13** is only about 1 m high, but some blocks can be the size of a car. Large explosive eruptions can disperse tephra over much of the planet. Ash can rise 40 km into the atmosphere during explosive eruptions and pose a threat to aircraft and can even change the weather. The 1991 eruption of Mount Pinatubo in the Philippines, shown in **Figure 14,** sent up a plume of ash 24 km high. Tiny sulfuric acid droplets and particles remained in the stratosphere for about two years, blocking the Sun's rays and lowering global temperatures by about 0.5°C.

▪ **Figure 14** In 1991, the eruption of Mount Pinatubo in the Philippines sent so much ash into the stratosphere that it lowered global temperatures for two years.

Teacher Content Support

Lateral Eruptions Volcanic eruptions most commonly occur at a volcano's summit, but lava can also be extruded from the flanks of a volcano. Some volcanic eruptions, such as the 1980 eruption of Mount St. Helens, occur laterally as well as vertically. The 1980 eruption of Mount St. Helens, in fact, made many geologists realize that lateral blasts could be more common than was previously thought.

EARTH SCIENCE JOURNAL

Eruptions Have students write an Earth science journal entry describing what it was like when Mount St. Helens erupted in 1980. Answers will vary, but should show an understanding of eruptions and the speed of pyroclastic flows.

Rubric

Pyroclastic flow

1902 Eruption of Mount Pelée

■ **Figure 15** A pyroclastic flow from Mount Pelée was so powerful that it destroyed the entire town of St. Pierre in only a few minutes.

Pyroclastic Flows

Some tephra cause tremendous damage and kill thousands of people. Violent volcanic eruptions can send clouds of ash and other tephra down a slope at speeds of about 80 km/h. Rapidly moving clouds of tephra mixed with hot, suffocating gases are called **pyroclastic flows.** They can have internal temperatures of more than 700°C. **Figure 15** shows a pyroclastic flow pouring down Mayon Volcano in the Philippines in 2000. One widely known and deadly pyroclastic flow occurred in 1902 on Mount Pelée, on the island of Martinique in the Caribbean Sea. More than 29,000 people suffocated or were burned to death. What little was left of the town of St. Pierre after the eruption is shown in **Figure 15.**

SECTION 2 REVIEW

Section Self-Check

Section Summary

- There are three major types of magma: basaltic, andesitic, and rhyolitic.
- Because of their relative silica contents, basaltic magma is the least explosive magma and rhyolitic magma is the most explosive.
- Temperature, pressure, and the presence of water are factors that affect the formation of magma.
- Rock fragments ejected during eruptions are called tephra.

Understand Main Ideas

1. **MAIN**IDEA **Discuss** how the composition of magma determines an eruption's characteristics.
2. **Restate** how the viscosity of magma is related to its explosivity.
3. **Predict** the explosivity of a volcano having magma with high silica content and high gas content.
4. **Differentiate** between sizes of tephra.

Think Critically

5. **Compare and contrast** the tectonic processes that made Kilauea and Mount Etna.
6. **Infer** the composition of magma that fueled the A.D. 79 eruption of Mount Vesuvius that buried the town of Pompeii.

WRITING IN ▸ Earth Science

7. Write a news report covering the 1902 eruption of Mount Pelée.

Discussion

Pyroclastic Flow Ask: Would it be possible to outdistance a pyroclastic flow in a car? probably not If you were 10 km from an advancing pyroclastic flow traveling at 120 km/h, how much time would you have to get away? Even if the road were straight, students would need to drive at 160 km/h (100 mph) to escape, because the flow would travel 10 km in five minutes.

GeoLAB

iLab Station

The GeoLab located at the end of the chapter can be used at this point in the lesson.

3 Assess

Check for Understanding

Reinforcement Ask students to identify the factors that determine the composition of magma. composition of the source material, amount of water and gas present, and temperature at which the magma forms

Reteach

Visualize Explain, and demonstrate if possible, how an increase in the amount of silica thickens magma in much the same way as flour thickens gravy.

Assessment

Performance Have students each research one of the volcanoes labeled in **Figure 1.** Have students present their findings as travel brochures or posters.

SECTION 2 REVIEW

1. Silica content determines viscosity, and the amount of dissolved gases help determine explosivity.
2. The more viscous the magma, the more explosive the eruption.
3. It would be a high-energy eruption.
4. Blocks are the biggest; ash is the smallest.
5. Kilauea formed over a hot spot; Mount Etna formed at a convergent plate boundary. Kilauea erupts slowly and builds layer by layer; Etna erupts explosively and is composed of layers of ash alternating with layers of lava.
6. It would have been high in silica and dissolved gases.
7. Answers will vary, but should include the fact that it created a deadly pyroclastic flow.

Rubric

SECTION 3

1 Focus

Focus Transparency

Before presenting the lesson, project **Section Focus Transparency 55** and have students answer the accompanying questions. EL

MAINIDEA

Magma Eruption Ask: Does all magma erupt at the surface? No; more magma intrudes the crust than erupts. Why does it matter where magma erupts or where it cools? At the surface, erupting magma is a human hazard. Magma intruded below the surface forms most of the ore deposits humans mine and use today. How could you tell if there were an active magma intrusion below the surface? earthquakes, high heat flow, deformation of Earth's surface, gases escaping from the crust

2 Teach

Tie to Previous Knowledge

Rate of Cooling Have students recall how the rate of cooling affects the textures of igneous rocks. the slower the cooling rate, the coarser the texture

Activity

Intrusion To enhance students' understanding of the spatial relationships between plutons and the rocks into which they intrude, have students choose one of the plutons shown in **Figure 16** and then reconstruct it using different colors of modeling clay. BL OL COOP LEARN

SECTION 3

Intrusive Activity

MAINIDEA Magma that solidifies below ground forms geologic features different from those formed by magma that cools at the surface.

Essential Questions

- How are features formed from magma that solidified under Earth's surface described?
- What are the different types of intrusive rock bodies?
- What geologic processes result in intrusive rocks that appear at Earth's surface?

Review Vocabulary

igneous rock: rock formed by solidification of magma

New Vocabulary

pluton
batholith
stock
laccolith
sill
dike

EARTH SCIENCE 4 YOU Have you ever been surprised when the icing on the inside of a layer cake was a different color or flavor than the icing on the outside? You might also be surprised if you could look inside Earth's layers because much volcanism cannot be seen at Earth's surface.

Plutons

Most of Earth's volcanism happens below the surface because not all magma emerges at the surface. Before it gets to the surface, rising magma can interact with the crust in several ways, as illustrated in **Figure 16.** Magma can force the overlying rock apart and enter the newly formed fissures. Magma can also cause blocks of rock to break off and sink into the magma, where the rocks eventually melt. Finally, magma can melt its way through the rock into which it intrudes. What happens deep in Earth as magma slowly cools? Recall that when magma cools, minerals begin to crystallize.

Over a long period of time, minerals in the magma solidify, forming intrusive igneous rock bodies. Some of these rock bodies are ribbonlike features only a few centimeters thick and several hundred meters long. Others are massive, and range in volume from about 1 km^3 to hundreds of cubic kilometers. These intrusive igneous rock bodies, called **plutons** (PLOO tahns), can be exposed at Earth's surface as a result of uplift and erosion and are classified based on their size, shape, and relationship to surrounding rocks.

Figure 16 Magma moving upward solidifies and forms bodies of rock both at the surface and deep within Earth.

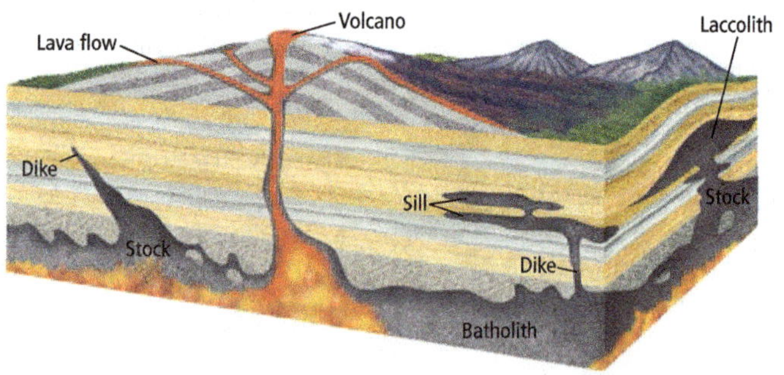

Teacher Content Support

Columnar Jointing Some igneous rock bodies demonstrate a particular type of fracturing known as columnar jointing. Columns range from a few centimeters to 3 m in diameter and can extend vertically to 30 m. Some columns are straight, while others are curved. Most columns have five or six sides, but some can have as few as three or as many as seven sides. Columnar jointing forms as lava cools and lithifies, because lava contracts as it cools and causes tensional stresses to build. Eventually, the hardened lava fractures. The fractures propagate at right angles to the outer surfaces of the hardened material. Columnar jointing occurs in lava flows, sills, dikes, ignimbrites, and shallow intrusions. Some well-known examples of columnar jointing are Giant's Causeway in Northern Ireland, Devils Postpile in California, and Devils Tower in Wyoming.

Batholiths and stocks The largest plutons are called batholiths. **Batholiths** (BATH uh lihths) are irregularly shaped masses of coarse-grained igneous rocks that cover at least 100 km^2 and take millions of years to form. Batholiths are common in the interior of major mountain chains.

Many batholiths in North America are composed primarily of granite—the most common rock type found in plutons. However, gabbro and diorite, the intrusive equivalents of basalt and andesite, are also found in batholiths. The largest batholith in North America is the Coast Range Batholith in British Columbia, shown in **Figure 17;** it is more than 1500 km long. Irregularly shaped plutons that are similar to batholiths but smaller in size are called **stocks.** Both batholiths and stocks, shown in **Figure 16,** cut across older rocks and generally form 5 to 30 km beneath Earth's surface.

Laccoliths Sometimes when magma intrudes into parallel rock layers close to Earth's surface, some of the rocks bow upward as a result of the intense pressure of the magma body. When the magma solidifies, a laccolith forms, as shown in **Figure 16.** A **laccolith** (LA kuh lihth) is a lens-shaped pluton with a round top and flat bottom. Compared to batholiths and stocks, laccoliths are relatively small; at most, they are 16 km wide. **Figure 17** shows a laccolith in Red and White Mountain, Colorado. Laccoliths also exist in the Black Hills of South Dakota, and the Judith Mountains of Montana, among other places.

☑ READING CHECK **Contrast** What is the difference between a laccolith and a batholith?

Sills A **sill** forms when magma intrudes parallel to layers of rock, as shown in **Figure 16.** A sill can range from only a few centimeters to hundreds of meters in thickness. **Figure 17** shows the Palisades Sill, which is exposed in the cliffs above the Hudson River near New York City and is about 300 m thick. The rock that was originally above the sill has eroded. What effect do you think this sill had on the sedimentary rocks into which it intruded? One effect is to lift the rock above it. Because it takes great amounts of force to lift entire layers of rock, most sills form relatively close to the surface. Another effect of sills is to metamorphose the surrounding rocks.

Figure 17 Batholiths, laccoliths, and sills form when magma intrudes into the crust and solidifies.

The Coast Range Batholith in British Columbia formed 5 to 30 km below Earth's surface.

Laccoliths push Earth's surface up, creating a rounded top and flat bottom.

The Palisades Sill in New York state formed more than 200 mya.

Demonstration

Lithostatic Pressure To help students visualize the forces involved in the intrusion of rock into existing rock layers, blow up a balloon and ask students to consider how air pressure presses on the outside of the balloon all over its surface. Explain magma is under pressure from all directions because it is surrounded by rock. This pressure is called lithostatic pressure. Have students list other examples in which pressure is exerted in all directions on an object. Examples might include air pressure on a balloon afloat above Earth or the hydrostatic pressure exerted on a scuba diver by the surrounding water.

Tie to Previous Knowledge

Granite Have students recall the properties of granite. Remind students granite—like all rock—is made of a mix of crystallized minerals and has specific properties such as hardness and density. A geologist studying a batholith would most likely be studying granitic rock.

Reinforcement

Plutons Ask: Which of the following statements is false?

a. Batholiths are common in the interiors of mountain chains.
b. Laccoliths have curved bottoms and flat tops.
c. Sills form when magma is forced between rock layers.
d. Most igneous activity occurs deep within Earth.

Answer b is false. Laccoliths have curved tops and flat bottoms.

Interpret the Photo

Plutons Have students compare the examples of a batholith, laccolith, and sill in the photographs in **Figure 17** to the diagram in **Figure 16.** Have students use their fingers to trace the outline of each formation in both figures. BL

Enrichment

Mount Rushmore The faces of Mount Rushmore were carved into an intrusive igneous rock body. Have interested students research the monument from a technical point of view to answer the following questions: How were the faces carved, and why were only the heads of the presidents carved into the rock? How many metric tons of rock were displaced? How long did it take to carve the monument? AL

☑ READING CHECK Laccoliths are relatively small and lens-shaped; batholiths are large and irregularly shaped.

Teacher Content Support

Dikes When magma intrudes into existing rock, large numbers of cracks form in the rock. These numerous channels are often filled by the magma, which upon cooling causes many dikes to form. Thus, dikes rarely are found alone—hundreds or even thousands of them can be found in a region affected by magmatic intrusions.

Use Science Terms

Concordant/Discordant Tell students sills are concordant plutons, whereas dikes are discordant plutons. Have students use **Figure 16** and the information in the text to determine the meanings of these adjectives. *Concordant* means *constant* or *agreeing. Discordant* means *being at variance* or *disagreeing.* Sills agree with or are parallel to the rocks into which they intrude, while dikes disagree with or cut across the rocks into which they intrude. EL

Project

Sequence of Intrusion Have students use modeling clay to create a geologic cross section in which several generations of dikes and sills are intruded and cut across each other. Then have other students try to reconstruct the sequence of intrusion. Ask: Which was first, last? The key here is to recognize that younger intrusions cut through older ones. There can also be some ambiguity if intrusions do not cross each other.

Caption Question Fig. 18 The volcano must have been very large, judging by the size of the conduit.

Figure 18 Unlike sills, dikes cut across the rock into which they intrude. Sometimes dikes extend from the conduit of a volcano. When the volcano erodes, the more erosion-resistant conduit and dike are left standing. Try to imagine the volcano that once surrounded this volcanic neck in New Mexico.
Infer *how big the volcano must have been.*

Dike

Volcanic neck

Dikes Unlike a sill, which is parallel to the rocks it intrudes, a **dike** is a pluton that cuts across preexisting rocks. Dikes often form when magma invades cracks in surrounding rock bodies. Dikes range in size from a few centimeters to several meters wide and can be tens of kilometers long. The Great Dike in Zimbabwe, Africa is an exception—it is about 8 km wide and 500 km long.

A volcanic neck occurs when the magma in a volcano conduit solidifies. Dikes are often associated with the conduit but do not always form the neck. Ship Rock in New Mexico, shown in **Figure 18,** has dikes extending from the neck.

Textures While the textures of sills and dikes vary, most are coarse-grained. Recall that grain size is related to the rate of cooling. The coarse-grained texture of most sills and dikes suggests that they formed deep in Earth's crust, where magma cooled slowly enough for large mineral grains to develop, as shown in **Figure 19.** Dikes and sills with a fine-grained texture formed closer to the surface where many crystals began growing at the same time, such as minerals of the sill in **Figure 19.**

Figure 19 Plutons forming deep in Earth cool slowly, giving crystals time to grow. Larger crystals produce a coarse-grained rock. Intrusive rocks that form closer to Earth's surface cool more quickly. As a result, many crystals form rapidly at the same time, and the rock is finer-grained.

Coarse-grained dike

Fine-grained sill

Demonstration

Dike Formation The night before class, prepare colored gelatin in a large glass container, inserting and securing a straw at an angle and extending above the surface of the gelatin. For the demonstration, blow air into the straw. (You might first need to remove the straw and clear it). An air-filled dike will form. You can also force water through the straw to fill the dike with water. Ask: How is this similar to intrusive activity? It is like magma intruding into rock. What would this formation be called, and why? a dike, because it is a subsurface intrusion that cuts across the existing rock

Plutons and Tectonics

Many plutons form as the result of mountain-building processes. In fact, batholiths are found at the cores of many of Earth's mountain ranges. From where did the enormous volume of cooled magma that formed these igneous bodies come? The processes that result in batholiths are complex. Recall that many major mountain chains formed along continental-continental convergent plate boundaries. Scientists think that some of these collisions might have forced continental crust down into the upper mantle where it melted, intruded into the overlying rocks, and eventually cooled to form batholiths.

Plutons are also thought to form as a result of oceanic plate convergence. Again, recall that a subduction zone develops when an oceanic plate converges with another plate. The cooler, denser oceanic plate subducts beneath the other plate and is heated, releasing water. The water expelled from the oceanic plate lowers the melting point of the overlying mantle, causing it to melt. Plutons often form when the melted material rises but does not erupt at the surface.

The Sierra Nevada batholith formed from at least five episodes of this type of igneous activity beneath what is now California. The famous granite cliffs found in Yosemite National Park, some of which are shown in **Figure 20,** are part of this vast batholith. Although they were once far below Earth's surface, uplift and erosion have brought them to their present position.

Figure 20 The granite cliffs that tower over Yosemite National Park in California are part of the Sierra Nevada batholith that has been exposed at Earth's surface.

SECTION 3 REVIEW

Section Self-Check

Section Summary

- Intrusive igneous rocks are classified according to their size, shape, and relationship to the surrounding rocks.
- Most of Earth's volcanism happens below Earth's surface.
- Magma can intrude into rock in different ways, taking different forms when it cools.
- Batholiths form the core of many mountain ranges.

Understand Main Ideas

1. **MAIN IDEA** **Compare and contrast** volcanic eruptions at Earth's surface with intrusive volcanic activity.
2. **Describe** the different types of plutons.
3. **Relate** the size of plutons to the locations where they are found.
4. **Identify** processes that expose plutons at Earth's surface.

Think Critically

5. **Predict** why the texture in the same sill might vary with finer grains along the margin and coarser grains toward the middle.
6. **Infer** what type of pluton might be found at the base of an extinct volcano.

WRITING IN Earth Science

7. Write a defense or rebuttal for this statement: Of the different types of plutons, sills form at the greatest depths beneath Earth's surface.

Reinforcement

Intrusive Igneous Bodies

Instruct students to make a table in which they compare and contrast the different types of plutons. Students' tables should show that all plutons are intrusive igneous bodies. Differences among the plutons should include size, shape, and relationship to the rocks into which the plutons intrude.

3 Assess

Check for Understanding

Activity Use different colors of chalk to make a drawing on the board that is similar to **Figure 16.** Have volunteers identify each of the plutons shown in your drawing and name at least three characteristics of each of these igneous rock bodies.

Reteach

Describe Make a table on the board that lists each of the plutons discussed in this section. Ask students to supply adjectives that describe each pluton and add the words to the appropriate places in the table.

Assessment

Performance Have students each write a creative yet scientifically accurate six-to-eight line poem about one of the plutons discussed in this section. Have volunteers read their poems aloud. Have all students include the poems in their Earth science journals.

SECTION 3 REVIEW

1. Intrusive volcanism occurs when magma rises but does not erupt. Magma erupting as lava creates volcanoes, pillow lavas, and flood basalts.
2. Batholiths are large, irregularly shaped masses of coarse-grained igneous rock. Stocks are similar, but smaller. Sills intrude between layers of rock; dikes cut across rock. Laccoliths are lens-shaped and close to Earth's surface.
3. Batholiths and stocks are the largest plutons and generally form 5–30 km beneath Earth's surface. Sills are smaller and are nearer to the surface.
4. uplift by intrusive volcanism, uplift by plate convergence, and erosion
5. The edges of a sill cool more quickly because they are in contact with cold surrounding rock, so larger crystals can form in the middle of a sill.
6. volcanic neck
7. It is batholiths, not sills, that form at the greatest depths.

Rubric

eXpeditions!

Purpose
Students will learn about how a volcanologist studies a volcano and what kind of gear he or she might need for this study.

Teacher Content Support
Disaster Assistance Team As part of the Volcano Disaster Assistance Team (VDAT) based in Menlo Park, California, seismologist Randy White has visited many volcanoes. Once alerted by increased earthquake activity in an area, White's team can arrive at a volcano site within 24 hours, where they interpret seismic data, ground bulging measurements, and atmospheric gas levels. The team's goal is predicting if and when a "sleeping volcano" might erupt.

Teaching Strategy
Provide suggested topics to focus student research, including major activities of the VDAT, world locations to which the team has been deployed, and examples of eruptions they have investigated.

eXpeditions!

ON SITE:
HAWAIIAN VOLCANO OBSERVATORY

Volcanologists often wear helmets, climbing gear, heat-resistant clothing, gas masks, and other gear to protect themselves from dangerous conditions in and around active volcanoes. Once this volcanologist climbs down to the test site, he will put on heat-resistant gloves.

Kilauea, a shield volcano on the island of Hawaii, is one of the world's most active volcanoes and the most dangerous volcano in the United States, according to the United States Geological Survey (USGS). Scientists monitor the conditions of Kilauea at the nearby Hawaiian Volcano Observatory (HVO). The observatory also serves as a laboratory where samples gathered in and around Kilauea can be studied.

Lava collection Imagine standing next to moving lava that is 1170°C. To get a direct measurement of the temperature or to collect a sample, scientists must withstand high temperatures and watch where they step. Samples are collected with heat-resistant materials and immediately cooled in a container with water to prevent contamination from the surrounding air. To protect themselves, volcanologists wear some of the gear shown in the photo.

Seismic activity Earthquake activity beneath a volcano is an indicator of impending eruptions. One way to monitor earthquakes is to check seismic activity. Scientists place seismometers in and around the vents of volcanoes to monitor seismic activity.

Gas samples Volcanologists collect samples of gases released at vents that they will analyze for sulfur dioxide and carbon dioxide in the HVO laboratory. An increase in sulfur-dioxide or carbon-dioxide emission can indicate a potential eruption.

Ground monitoring An instrument called an electronic distance meter (EDM) helps scientists monitor the ground around volcanoes and predict an eruption. As magma rises toward Earth's surface, the ground might tilt, sink, or bulge from pressure.

Volcanologists at HVO are constantly recording data, running tests, and making advances around the world. Without their research, we might not understand volcanoes as well as we do today.

WRITING IN ▶ Earth Science
Research the methods scientists use to predict time, size, and type of eruption. Summarize your findings and share your research with your classmates.

WebQuest

WRITING IN ▶ Earth Science

Rubric

Research Writings should include information about at least two methods discussed in the feature. Students might mention that scientists look at the tilt or the bulge of the ground that results from the pressure of upwelling magma.

WebQuest

GeoLAB

iLab Station

Predict the Safety of a Volcano

Background: Some volcanoes are explosively dangerous. Along with clouds of ash and other volcanic debris, pyroclastic flows, landslides, and mudflows are common volcanic hazards. However, an explosive volcano might not be a hazard to human life and property if it is located in a remote area or if it erupts infrequently.

Question: *What factors should be considered when evaluating a volcano?*

Helicopters transport researchers to remote volcanic sites. Researchers analyze data to determine hazards to humans.

Materials

volcano data provided by your teacher
current reference books with additional volcano data
markers or colored pencils

Procedure

Imagine that you work for the United States Geological Survey (USGS) and are asked to evaluate several volcanoes around the world. Your job is to determine if the volcanoes are safe for the nearby inhabitants. If the volcanoes are not safe, you must make recommendations to ensure the safety of the people around them.

1. Read and complete the lab safety form.
2. Form a team of scientists of three to four people.
3. Within your team, brainstorm some factors you might use to evaluate the volcanoes. Record your ideas. You might include factors such as eruption interval, composition of lava, approximate number of people living near the volcano, and the date of the last known eruption.
4. With your group, decide which factors you will include.
5. Use the factors you have chosen to create a data table. Make sure your teacher approves your table and your factors before you proceed.
6. Use the information your teacher provides and select a country where there is a known volcano.
7. Complete your data table for your first country.
8. Repeat Steps 6 and 7 for two more countries.

Analyze and Conclude

1. **Interpret Data** Is it safe for people to live close to any of the volcanoes? Why or why not?
2. **Interpret Data** Do any of the volcanoes pose an immediate threat to the people who might live nearby? Why or why not?
3. **Conclude** Prepare to present your findings to a group of scientists from around the world. Be sure to include your predictions and recommendations, and be prepared for questions. Display your data table to help communicate your findings.

SHARE YOUR DATA

Peer Review Post a summary of your recommendations for each of your volcanoes. Compare and contrast your data with that of other students who completed this lab.

LabManager™
Customize any lab with the LabManager™ CD-ROM.

SHARE YOUR DATA

Peer Review Student summaries should show adequate assessment of each type of volcano. Comparisons will vary. Students might have researched different volcanoes or used different data.

GeoLAB

iLab Station Rubric

Preparation

Time Allotment 90 min

Process Skills communicate, evaluate, research, predict

Safety Precaution Approve lab safety forms before work begins.

Preparation of Materials Data on volcanoes can be obtained from the USGS Web site. If you do not have classroom internet access, before class begins, print out the volcano data. Otherwise, make sure students have safe access to the Internet or to a library for further research.

Procedure

- As the students work, be sure to monitor their progress and ask appropriate questions as necessary.
- Have students work in groups of three or four. Make sure students follow the instructions step by step. Have them ask you about any steps that might be unclear.
- **Troubleshooting** Have Web sites and print resources available for those students who struggle to find relevant materials.

Analyze and Conclude

1. Answers will vary. Students might consider past eruptions, seismic readings, type of volcano, and any emergency notification systems that exist.
2. Answers will vary. Students might base their answer on recent seismic activity or local observations.
3. Presentations will vary. Data tables should clearly present findings. Students should be able to answer follow-up questions.

MAINIDEAS Summary statements can be used by students to review the major concepts of the chapter.

Students can review with these online resources.

Vocabulary Practice

Vocabulary eGames
Vocabulary eFlashcards
Vocabulary PuzzleMaker

Section Self-Check

Chapter Self-Check

Online Test Practice

Use the ExamView® *Assessment Suite* CD-ROM to:

- create multiple versions of tests
- create modified tests with one mouse click
- edit existing questions and add your own questions
- build tests aligned with state standards using built-in state curriculum tags
- change English tests to Spanish with one mouse click
- track students' progress using the Teacher Management System

CHAPTER 18 STUDY GUIDE

BIGIDEA **Volcanoes develop from magma moving upward from deep within Earth.**

Vocabulary Practice

SECTION 1 Volcanoes

VOCABULARY
- volcanism
- hot spot
- flood basalt
- fissure
- conduit
- vent
- crater
- caldera
- shield volcano
- cinder cone
- composite volcano

MAINIDEA The locations of volcanoes are mostly determined by plate tectonics.

- Volcanism includes all the processes in which magma and gases rise to Earth's surface.
- Most volcanoes on land are part of two major volcanic chains: the Circum-Pacific Belt and the Mediterranean Belt.
- Parts of a volcano include a vent, magma chamber, crater, and caldera.
- Flood basalts form when lava flows from fissures to form flat plains or plateaus.
- There are three major types of volcanoes: shield, composite, and cinder cone.

SECTION 2 Eruptions

VOCABULARY
- viscosity
- tephra
- pyroclastic flow

MAINIDEA The composition of magma determines the characteristics of a volcanic eruption.

- There are three major types of magma: basaltic, andesitic, and rhyolitic.
- Because of their relative silica contents, basaltic magma is the least explosive magma and rhyolitic magma is the most explosive.
- Temperature, pressure, and the presence of water are factors that affect the formation of magma.
- Rock fragments ejected during eruptions are called tephra.

SECTION 3 Intrusive Activity

VOCABULARY
- pluton
- batholith
- stock
- laccolith
- sill
- dike

MAINIDEA Magma that solidifies below ground forms geologic features different from those formed by magma that cools at the surface.

- Intrusive igneous rocks are classified according to their size, shape, and relationship to the surrounding rocks.
- Most of Earth's volcanism happens below Earth's surface.
- Magma can intrude into rock in different ways, taking different forms when it cools.
- Batholiths form the core of many mountain ranges.

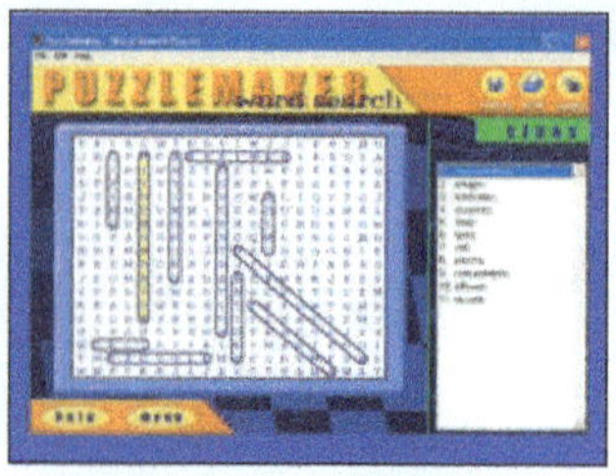

Vocabulary Practice

For additional practice with vocabulary, have students access the Vocabulary PuzzleMaker.

Chapter Self-Check

VOCABULARY REVIEW

Make each of the following sentences true by replacing the italicized words with terms from the Study Guide.

1. In the most explosive types of eruptions, lava accumulates to form a *shield volcano.*
2. Lava travels through a conduit to erupt through a *fissure* at the top of a volcano.
3. *Hot spots* refer to all processes associated with the discharge of magma, hot water, and steam.
4. Ash is the smallest type of *lava flow.*

Complete the sentences below using vocabulary terms from the Study Guide.

5. A(n) ________ is a bowl-shaped depression that surrounds the vent at a volcano's summit.
6. A(n) ________ forms in the depression left when an empty magma chamber collapses.
7. The type of volcano that is the smallest and has the steepest slopes is called a(n) ________.

Match each description below with the correct vocabulary term from the Study Guide.

8. any rock body that has formed at great depths underground
9. plutons having an area of more than 100 km^2; often forms the core of mountains
10. flowing cloud of tephra and lava mixed with hot, suffocating gases
11. formed when magma intrudes vertically across existing rock

Use what you know about the vocabulary terms to describe what the terms in each pair have in common.

12. laccolith, sill
13. shield volcano, flood basalt
14. fissure, conduit
15. sill, dike

UNDERSTAND KEY CONCEPTS

16. Which area is surrounded by the Ring of Fire?
    A. the Atlantic Ocean
    B. the United States
    C. the Mediterranean Sea
    D. the Pacific Ocean

Use the diagram below to answer Questions 17 and 18.

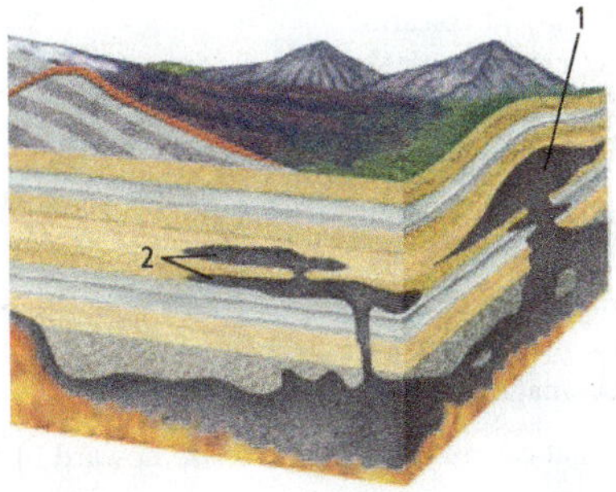

17. In the diagram, what is the structure labeled *1*?
    A. batholith
    B. laccolith
    C. dike
    D. sill
18. In the diagram, what is the structure labeled *2*?
    A. batholith
    B. laccolith
    C. dike
    D. sill
19. Which is not true?
    A. An increase in silica increases the viscosity of a magma.
    B. Andesitic magma has both an intermediate gas content and explosiveness.
    C. An increase in temperature increases a magma's viscosity.
    D. Basaltic magma has a low viscosity and retains little gas.

CHAPTER 18 ASSESSMENT

VOCABULARY REVIEW

1. composite volcano
2. vent
3. Volcanism
4. tephra
5. crater
6. caldera
7. cinder cone
8. pluton
9. batholith
10. pyroclastic flow
11. dike
12. Sills and laccoliths are the plutons that form nearest to the surface.
13. Both are formed by non-explosive eruptions.
14. Both are openings through which lava erupts.
15. Both are types of plutons.

UNDERSTAND KEY CONCEPTS

16. D
17. B
18. D
19. C

20. B
21. C
22. A
23. B

CONSTRUCTED RESPONSE

24. Relative to batholiths, which are at least 100 km^2, stocks and laccoliths are much smaller. Batholiths are irregularly shaped; stocks are horizontally oriented because they intrude between layers; and laccoliths are lens-shaped, pushing up the rock above them.
25. The feature is a dike. Dikes can be of any size and cut across existing rock.
26. Hot spots are locations on the surface of Earth where a stationary plume of magma erupts at the surface. Hot spots have no obvious connection with plate boundaries.
27. shield: Mauna Loa; composite: Mount St. Helens; cinder cone: Lassen Peak, California
28. Although both are formed over hot spots, Kilauea formed in mid-ocean, while the Columbia River flood basalts formed within a continent. They are both composed of basaltic lava.
29. Volcanic blocks are tephra ejected from a vent during explosive eruptions. Basalt, which makes up shield volcanoes, does not usually erupt explosively.
30. Island A is the oldest island because it is furthest away from the hot spot that created it. It is most eroded because it has had the most time to erode. The plate is moving northwest. This is evident from the fact that younger volcanoes lie southeast of it.
31. yes, because it is lined with subducting plates and thus active surface volcanoes
32. The hotter the magma, the lower the viscosity. Hot magma flows quickly; cooler magma flows more slowly.
33. If a volcano erupts ash and sulfuric acid and other gases into the atmosphere and stratosphere, it can remain there and block the Sun and lower temperatures on Earth.
34. Shield is the biggest, long and low, with gently sloping sides; cinder cones are small, with steep sides; composite volcanoes are large and cone-shaped.
35. Smaller plutons would cool more quickly than larger ones, which allows the minerals to crystallize more quickly. Quicker cooling produces smaller crystals and so a finer grain.

Use the figure below to answer Questions 20 and 21.

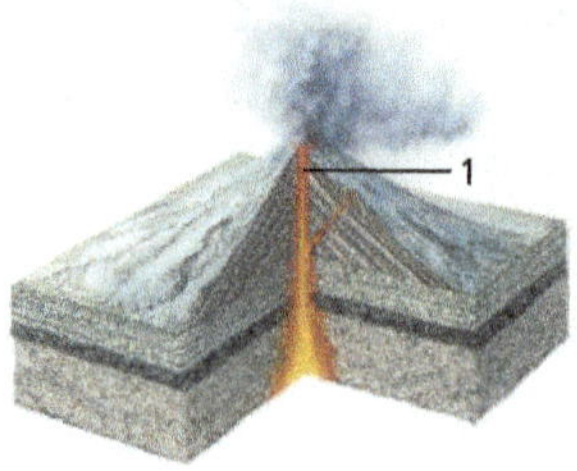

20. Which type of volcano is shown?
A. shield volcano
B. composite volcano
C. flood basalt volcano
D. cinder cone

21. What is the feature labeled *1*?
A. crater
B. cinder cone
C. vent
D. magma chamber

22. What causes the magma to rise upward in a mantle plume?
A. The magma is less dense than the surrounding material.
B. The magma is denser than the surrounding material.
C. The magma is pulled upward by the air pressure.
D. The magma is pushed upward by the surrounding rock.

23. Which type of volcanism produces the most lava annually?
A. convergent
B. divergent
C. hot spot
D. cinder cones

CONSTRUCTED RESPONSE

24. Differentiate among batholiths, stocks, and laccoliths according to their relative sizes and shapes.

25. Infer A particular outcrop has a narrow ribbon of basalt that runs almost perpendicular to several layers of sandstone. What is this feature called?

26. Describe hot spots.

27. Identify one specific example of the three types of volcanoes.

28. Compare and contrast Kilauea and the Columbia River flood basalt in terms of the processes related to their development.

29. Analyze why volcanic blocks are uncommon on shield volcanoes.

Use the diagram below to answer Question 30.

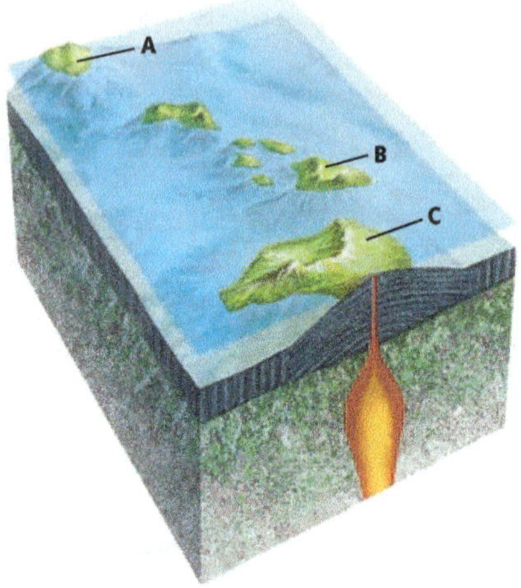

30. Distinguish which island is the oldest and in which direction the plate is moving. Explain your reasoning.

31. Decide Is the Pacific Ring of Fire an accurate name? Explain.

32. Explain the relationship between the viscosity of a magma and its temperature.

33. Explain how volcanic activity can affect global weather.

34. Draw a diagram of the three volcano types, showing their relative sizes.

35. Analyze why smaller plutons are more likely to be fine-grained, and larger plutons more likely to be coarse-grained.

THINK CRITICALLY

Use the table below to answer Questions 36 and 37.

Magma Composition and Characteristics

	Basaltic Magma	Andesitic Magma	Rhyolitic Magma
Source material	upper mantle	oceanic crusts and sediments	continental crust
Viscosity	low	intermediate	high
Gas content	1–2%	3–4%	4–6%
Silica content	about 50%	about 60%	about 70%
Location of magma	both oceanic and continental crust	continental margins associated with subduction zones	continental crust

36. **Analyze** and rank the types of magma in terms of explosiveness based on the data. Explain your reasoning.

37. **Categorize** each of the three types of volcanoes in terms of the characteristics of magma shown in the table.

38. **Predict** what kind of volcanism would occur if there were no plate tectonics.

CONCEPT MAPPING

39. Create a concept map using the following terms: *pluton, vertical, batholith, cuts across, stock, parallel, laccolith, sill,* and *dike.* For more help refer to the *Skillbuilder Handbook.*

CHALLENGE QUESTION

40. **Formulate** a way to recognize the difference between an ancient lava flow and an intrusive igneous rock.

WRITING IN ▶ Earth Science

41. Imagine you are in charge of a volcano observatory. One day, GPS measurements indicate that a volcano is expanding, there have been several earthquakes, and the flux of volcanic gases has increased. Should you issue a warning of an impending eruption? Write a press release to warn people about the situation.

DBQ Document–Based Questions

Data obtained from: Takada, A. 1999. Variations in magma supply and magma partitioning: the role of tectonic settings. *Journal of Volcanic Geothermal Research* 83:93–110.

Studying the history of past eruptions yields important data for making estimations about predicting eruptions. The graph below shows the total volume of lava erupted at two Hawaiian islands over 200 years.

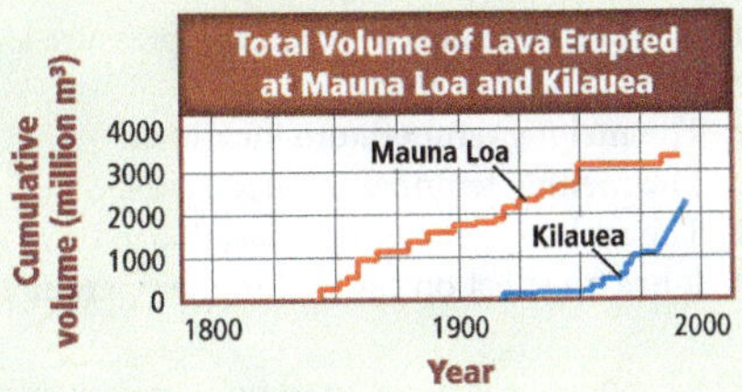

42. In what years did the two largest eruptions occur at Mauna Loa?

43. What was the average volume of lava at Mauna Loa between 1840 and 1990?

44. Can you predict when the next eruption will occur? Explain your answer.

45. Eruptions at Mauna Loa are large and last a short length of time. What feature of the graph shows this? Compare and contrast the last eruption at Kilauea with eruptions at Mauna Loa.

CUMULATIVE REVIEW

46. List six of the most important mineral properties used in mineral identification. **(Chapter 4)**

47. What observations support the theory of plate tectonics? **(Chapter 17)**

CHAPTER 18 ASSESSMENT

THINK CRITICALLY

36. From the least to the most explosive: basaltic, andesitic, rhyolitic; The higher the viscosity, gas content, and silica content, the more explosive the eruption.

37. shield: basaltic magma, with relatively low silica, gas content and viscosity; composite and cinder cone volcanoes: andesitic or rhyolitic magma with intermediate and high viscosity and higher silica and gas content

38. There would only be hot spot volcanism.

CONCEPT MAPPING

39. Maps should demonstrate the fact that all intrusions are plutons, and the orientation of each.

CHALLENGE QUESTION

40. In intrusive bodies, crystals form as plutons slowly cool, so there will probably be a variety of size of crystals (depending on the rock formation being studied). Lava that comes from hot spots, is relatively uniform in texture and is dark in color because it is basaltic. Students could also look to see if crystals formed in place (intrusive rock) or if minerals cumulated in voids left by air bubbles. An explosive eruption of lava would produce layers of ash, lava, and unevenly shaped rock. An intrusive body would have graduated zones of crystals.

WRITING IN ▶ Earth Science

Rubric

41. Students' answers should demonstrate understanding of what precedes an eruption and the ways it might affect people.

DBQ Document-Based Questions

Data obtained from: Takada, A. 1999. Variations in magma supply and magma partitioning: the role of tectonic settings. *Journal of Volcanic Geothermal Research 83: 93–110.*

42. 1872 and 1950 (an error of 5–10 years is OK).

43. Divide volume (3500×10^6 m^3) by time (150 years) to get 23×10^6 m^3/year.

44. Answers should demonstrate understanding that prediction is not (yet) possible. But scientists can make educated guesses based on past patterns of eruption, measurements of ground movement, gas output, etc.

45. Large, short-lived eruptions show up as steps. Slow, continuous eruptions show up as smoothly increasing lines. Kilauea has been continually erupting lava at about the same rate since about 1982 (this is why the line for Kilauea is straight).

CUMULATIVE REVIEW

46. crystal form, color, luster, streak, hardness, cleavage, density, specific gravity, texture, special properties

47. GPS measurements of plate motion, the location of earthquakes, and volcanoes

MULTIPLE CHOICE

1. D
2. A
3. D
4. C
5. C
6. B
7. C
8. B
9. C

CUMULATIVE STANDARDIZED TEST PRACTICE

MULTIPLE CHOICE

Use the figure below to answer Questions 1 and 2.

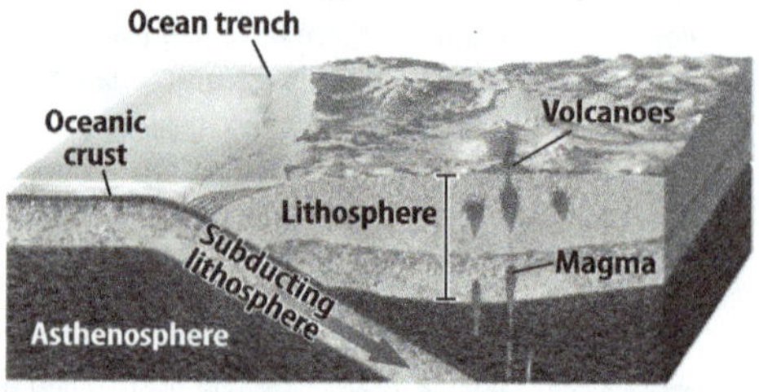

1. What process is occurring in the figure above?
   A. continental-continental divergence
   B. oceanic-continental divergence
   C. continental-continental subduction
   D. oceanic-continental subduction

2. How does an increase in confining pressure affect a rock's melting temperature?
   A. The melting temperature increases.
   B. The melting temperature decreases.
   C. The melting temperature is stabilized.
   D. It has no effect on the melting temperature.

3. Which evidence was not used by Wegener to support his hypothesis of continental drift?
   A. coal beds in America
   B. fossils of land-dwelling animals
   C. glacial deposits
   D. paleomagnetic data

4. What is the name for the constant production of new ocean floor?
   A. continental drift
   B. hot spot
   C. seafloor spreading
   D. subduction

5. The weight of a subducting plate helps pull the trailing lithosphere into a subduction zone in which process?
   A. ridge pull
   B. ridge push
   C. slab pull
   D. slab push

6. What type of model uses molded clay, soil, and chemicals to simulate a volcanic eruption?
   A. conceptual model
   B. physical model
   C. mathematical model
   D. computer model

7. Which of these processes of the water cycle is a direct effect of the Sun's energy?
   A. formation of precipitation
   B. runoff of water over soil
   C. evaporation
   D. seeping of water into soil

Use the figure below to answer Questions 8 and 9.

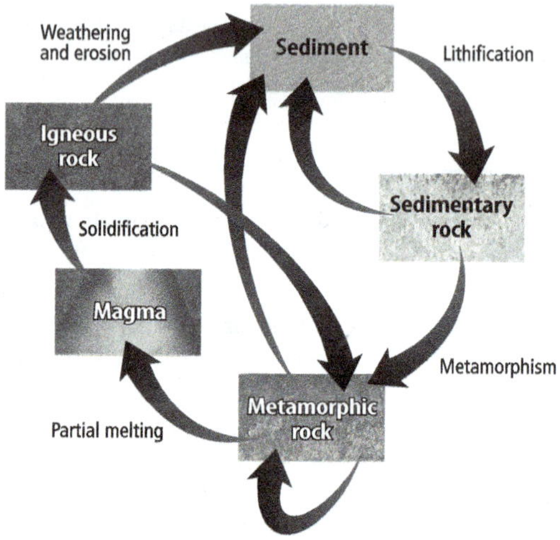

8. Which process contributes to the production of sediment?
   A. lithification
   B. weathering
   C. solidification
   D. metamorphism

9. What rock type is produced when magma solidifies?
   A. metamorphic rock
   B. sedimentary rock
   C. igneous rock
   D. lava

Online Test Practice

SHORT ANSWER

Use the table below to answer Questions 10–12.

Notable Volcanic Eruptions			
Volcano	Date	Volume Ejected	Height of Plume
Toba	74,000 years ago	2,800 km^3	50–80 km
Vesuvius	A.D. 79	4 km^3	32 km
Tambora	1815	150 km^3	44 km
Krakatau	1883	21 km^3	36 km
Mount St. Helens	1980	1 km^3	19 km
Mount Pinatubo	1991	5 km^3	35 km

10. Order the volcanic eruptions according to the quantities of pyroclastic material produced.

11. Hypothesize why the eruption of Vesuvius in A.D. 79 was more deadly than the eruption of Mount Pinatubo in 1991, even though the eruptions were approximately the same size.

12. Calculate the difference in plume height of volcanic debris during the eruption of Tambora in 1815 compared to the plume from the 1980 eruption of Mount St. Helens.

13. Distinguish between the everyday use of the term *theory* and its true scientific meaning.

14. When a tropical rain forest is cleared, why does the soil usually become useless for growing crops after only a few years?

15. What role do glaciers play in Earth's rock cycle?

16. Write a list of numbered statements that summarizes the major steps in the water cycle.

READING FOR COMPREHENSION

Eruption of Mount Pinatubo

On June 15, 1991, Mount Pinatubo roared awake after a six-century sleep. The 1760-m volcano belched clouds of gas and ash known as pyroclastic material. Their temperature: 816°C. Streams of ash and sulfur dioxide rocketed 35 km into the stratosphere. Another blast at dawn blew away the side of the mountain. So much ash and pumice choked the air that the sky grew black by afternoon, and chunks of volcanic rock fell with a force similar to hail. That evening, earthquakes struck the already-damaged city. Pinatubo's eruption had created an underground cavern that caved in on itself.

17. What can be inferred from this text?
A. Volcanoes are unpredictable and can erupt at any time.
B. Volcanoes always erupt explosively.
C. Volcanoes can change the surface of Earth in many ways.
D. Volcanoes are always accompanied by earthquakes.

18. According to this text, which statement is false?
A. Volcanoes can release gases into the stratosphere.
B. The eruption of Mount Pinatubo was caused by the collapse of an underground cavern.
C. The gas and ash released during the 1991 eruption was as hot as 816°C.
D. Volcanic eruptions can change the shape of the mountain.

19. In the days leading up to the June 15th eruption, towns in areas surrounding Mount Pinatubo were evacuated. Based on the text above, explain why it would be necessary to evacuate these areas.

NEED EXTRA HELP?																
If You Missed Question . . .	1	2	3	4	5	6	7	8	9	10	11	12	13	14	15	16
Review Section . . .	17.3	5.1	17.1	17.2	17.4	1.3	9.1	6.1	5.1	18.1	18.3	18.3	1.3	8.3	3.3	9.1

SHORT ANSWER

10. Toba, Tambora, Krakatau, Pinatubo, Vesuvius, St. Helens
11. Answers will vary but might include: people lived very close to Mount Vesuvius; the pyroclastic flows were different; there was more warning for Pinatubo because of more modern detection methods.
12. approximately 25 km
13. The everyday use of the word *theory* indicates a speculation, whereas the scientific use of the word is a proposed explanation that has been tested through repeated experiments.
14. Deforestation results in increased soil erosion, which removes the more fertile soil.
15. Glaciers cause large amounts of erosion.
16. 1. Evaporation removes liquid water from oceans/lakes and adds water vapor into the atmosphere. 2. Precipitation removes water vapor from the atmosphere and adds liquid water to Earth's surface. 3. Runoff moves liquid water from the ground to oceans/lakes. 4. Transpiration removes groundwater and creates atmospheric water vapor. 5. Infiltration removes surface water and creates groundwater.

READING FOR COMPREHENSION

17. C
18. B
19. Answers will vary but should discuss the dangers of: ash-filled air, falling volcanic rock, and earthquake damage.

CHAPTER 19 Earthquakes

BIGIDEA **Earthquakes are natural vibrations of the ground, some of which are caused by movement along fractures in Earth's crust.**

ESSENTIAL QUESTIONS	NATIONAL STANDARDS	RESOURCES TO ASSESS MASTERY
SECTION 1 **Forces Within Earth** **1.** How are stress and strain defined as they apply to rocks? **2.** What are the three types of movement of faults? **3.** What are the three types of seismic waves? 1 session 0.5 block	UCP.2; B.4; D.1; F.5; G.3	**Entry-Level Assessment** Focus Transparency 56, p. 528 **Progress Monitoring** Caption Question, pp. 529, 530, 533 Reading Check, p. 529 Section Review, p. 533
SECTION 2 **Seismic Waves and Earth's Interior** **1.** How does a seismometer work? **2.** How have seismic waves been used to determine the structure and composition of Earth's interior? 1.5 sessions 0.75 block	UCP.2; B.4; D.1, D.3; E.1–2; F.5–6	**Entry-Level Assessment** Focus Transparency 57, p. 534 **Progress Monitoring** Caption Question, p. 535 Reading Check, p. 535 Section Review, p. 538
SECTION 3 **Measuring and Locating Earthquakes** **1.** What are earthquake magnitude and intensity and how are they measured? **2.** Why are data from at least three seismic stations needed to locate an earthquake's epicenter? **3.** Where are Earth's seismic belts? 2 sessions 1 block	UCP.2–3; B4; D.1; E.1–2; F.5–6	**Entry-Level Assessment** Focus Transparency 58, p. 539 **Progress Monitoring** Caption Question, pp. 542, 544 Reading Check, pp. 542, 543 Section Review, p. 544
SECTION 4 **Earthquakes and Society** **1.** What factors affect the amount of damage caused by an earthquake? **2.** What are some of the factors considered in earthquake-probability studies? **3.** How are different types of structures affected by earthquakes? 3 sessions 1.5 blocks	UCP.2–3; A.1; B.4; D.1; E.1; F.5	**Entry-Level Assessment** Focus Transparency 59, p. 545 **Progress Monitoring** Caption Question, pp. 549, 551 Reading Check, pp. 546, 547, 549, 550 Section Review, p. 551 **Summative Assessment** Chapter Assessment, p. 555 Exam*View® Assessment Suite* CD-ROM

Go online!
connectED.mcgraw-hill.com

LEVELED RESOURCES	LAB MATERIALS
Science Notebook 19.1 OL **Chapter FAST FILE Resources:** Study Guide, p. 69 BL **Transparencies:** Section Focus Transparency 56 EL Teaching Transparency 54 OL EL	LaunchLAB p. 526 / **10 min** wooden blocks, sandpaper, thumbtacks
Science Notebook 19.2 OL **Chapter FAST FILE Resources:** GeoLab Worksheet, p. 57 OL Study Guide, p. 71 BL **Transparencies:** Section Focus Transparency 57 EL Teaching Transparency 55 OL EL	GeoLAB p. 553 / **45 min** U.S. map, map of major tectonic plates, calculator, drafting compass, metric ruler
Science Notebook 19.3 OL **Chapter FAST FILE Resources:** MiniLab Worksheet, p. 56 OL Study Guide, p. 72 BL **Lab Resources:** Laboratory Manual, p. 145 OL **Transparencies:** Section Focus Transparency 58 EL Teaching Transparency 56 OL EL	MiniLAB p. 541 / **20 min** paper, pencil
Science Notebook 19.4 OL **Chapter FAST FILE Resources:** Study Guide, p. 74 BL **Lab Resources:** Laboratory Manual, p. 149 OL **Transparencies:** Section Focus Transparency 59 EL Teaching Transparency 57 OL EL	

LabManager™
Customize any lab with the LabManager™ CD-ROM.

ADDITIONAL RESOURCES

Technology:
- eTeacherEdition
- eStudentEdition
- Classroom Presentation Toolkit
- Exam*View® Assessment Suite* CD-ROM
- LabManager™ CD-ROM
- What's EARTH SCIENCE Got To Do With It? DVD
- Weather Classroom Video DVD

Assessment:
- Blueprints for Success
- Chapter Test (Scaffolded)
- Earth Science Standardized Test Practice

Lab Resources:
- Exploring Environmental Problems Laboratory Manual

BL Below Level OL On Level AL Advanced Learners EL English Learners COOP LEARN Cooperative Learning

CHAPTER 19

LaunchLAB

iLab Station | Rubric

What can cause an earthquake?

Teaching Strategies

- Caution students to move the sandpaper-covered blocks carefully so as not to scrape or pinch their fingers.
- Have students use moderate force while moving both sets of blocks.
- Students should observe that the bare blocks move easily and smoothly against each other. The sandpaper-covered blocks will stick together until increasing shear stress overcomes friction. Then, the blocks will slip suddenly, simulating an earthquake.
- After the lab, have students return the blocks, thumbtacks, and sandpaper to you. Only then should students remove their goggles and aprons.

Procedure

1. Have students read and complete the lab safety form and follow the procedure below.
2. Slide the largest surfaces of two smooth **wooden blocks** against each other. Describe the movement.
3. Cut two pieces of coarse-grained **sandpaper** so that they are about 1 cm longer than the largest surface of each block.
4. Place the sandpaper, coarse side up, against the largest surface of each block. Wrap the paper over the edges of the blocks and secure it with **thumbtacks.**
5. Slide the sandpaper-covered sides of the blocks against each other. Describe the movement.

Analysis

1. **Compare** the two movements of the wooden blocks. The blocks without the sandpaper move smoothly and easily against each other; the sandpaper-covered blocks require more force to move, and the movement is more irregular.
2. **Apply** Which parts of Earth are represented by the blocks? Possible answers include tectonic slabs or pieces of Earth's crust.
3. **Infer** which of the two scenarios shows what happens during an earthquake. The second scenario is more like an earthquake, in which giant slabs of crust scrape against one another.

CHAPTER 19

Earthquakes

BIGIDEA Earthquakes are natural vibrations of the ground, some of which are caused by movement along fractures in Earth's crust.

SECTIONS

1 Forces Within Earth

2 Seismic Waves and Earth's Interior

3 Measuring and Locating Earthquakes

4 Earthquakes and Society

LaunchLAB

iLab Station

What can cause an earthquake?

When pieces of Earth's crust suddenly move relative to one another, earthquakes occur. This movement occurs along fractures in the crust that are called faults. Uncover the cause of earthquakes in this lab.

FOLDABLES® Study Organizer

Types of Faults

Make a three-tab book using the labels shown. Use it to organize your notes on the three basic types of faults.

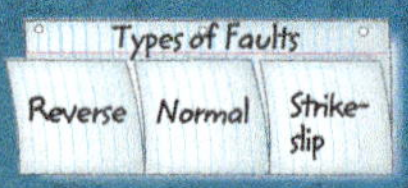

Assessment

Knowledge Ask students each to write a paragraph that summarizes their observations and inferences using the following terms: *bare wood, sandpaper-covered blocks, rough, smooth, small frictional force, large frictional force, continuous motion, sudden failure, earthquake,* and *slow deformation.*

LabManager™
Customize any lab with the LabManager™ CD-ROM.

Introduce the BIGIDEA

Frequency of Earthquakes
Inform students earthquakes occur all the time, but only those that cause destruction in populated areas are generally reported. Explain we live on an active planet, and earthquakes are the effects of ongoing geological adjustments inside Earth. Ask students: How many earthquakes occur each year? Students will probably give a low number. Tell them that, including very small ones, several million earthquakes are estimated to occur each year.

Teacher Content Support

Moab Fault The aerial photograph shows the fault-block mountain range overlooking a plain along the Moab fault in Utah. Ask students: What would you have witnessed and experienced if you had been on that plain on the first day these faults moved? Answers should demonstrate an understanding that faults are large movements of the crust. Some students might realize such movements are often accompanied by earthquakes.

SECTION 1

1 Focus

Focus Transparency

Before presenting the lesson, project **Section Focus Transparency 56** and have students answer the accompanying questions. EL

MAINIDEA

Forces Ask students to think about what determines when a material will break. the type of force that is applied and the strength of the material To help illustrate the answer, ask students to discuss what is required to break familiar objects like chalk, glass, or metal. To illustrate why the type of force matters, ask students to break silicone putty by pulling on it (tension) and comparing that to what happens when they squeeze it (compression). Tension force will break it, compression will not.

2 Teach

Interpret the Illustration

Strain Query students to make sure they understand the different types of strain shown in **Figure 1.** Compressional strain involves a decrease in volume; tensional strain involves an increase in volume; and shear strain involves only a distortion of a material without a change in volume.

SECTION 1

Essential Questions

- How are stress and strain defined as they apply to rocks?
- What are the three types of movement of faults?
- What are the three types of seismic waves?

Review Vocabulary

fracture: the texture or general appearance of the freshly broken surface of a mineral

New Vocabulary

stress
strain
elastic deformation
plastic deformation
fault
seismic wave
primary wave
secondary wave
focus
epicenter

Forces Within Earth

MAINIDEA Faults form when the forces acting on rock exceed the rock's strength.

EARTH SCIENCE 4 YOU If you bend a paperclip, it takes on a new shape. If you bend a popsicle stick, it will eventually break. The same is true of rocks; when forces are applied to rocks, they either bend or break.

Stress and Strain

Most earthquakes are the result of movement of Earth's crust produced by plate tectonics. As a whole, tectonic plates tend to move very slowly. Along the boundaries between two plates, rocks in the crust often resist movement. Over time, stress builds up. **Stress** is the total force acting on crustal rocks per unit of area. When stress overcomes the strength of the rocks involved, movement occurs along fractures in the rocks. The vibrations caused by this sudden movement are felt as an earthquake. The characteristics of earthquakes are determined by the orientation and magnitude of stress applied to rocks, and by the strength of the rocks involved.

There are three kinds of stress that act on Earth's rocks: compression, tension, and shear. Compression is stress that decreases the volume of a material, tension is stress that pulls a material apart, and shear is stress that causes a material to twist. The deformation of materials in response to stress is called **strain. Figure 1** illustrates the strain caused by compression, tension, and shear.

Even though rocks can be twisted, squeezed, and stretched, they fracture when stress and strain reach a critical point. At the critical point, rock can move, releasing the energy built up as a result of stress. Earthquakes are the result of this movement and release of energy. For example, the 2010 earthquake in Haiti was caused by a release of built-up compression stress. When that energy was released as an earthquake, more than 230,000 people were killed and even more were made homeless.

Figure 1 Compression causes a material to shorten. Tension causes a material to lengthen. Shear causes distortion of a material.

View an **animation of faults.** Concepts In Motion

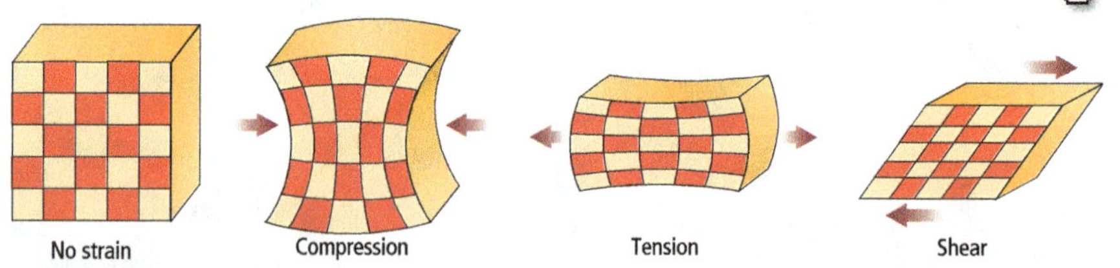

Teacher Content Support

Faults and Principal Stresses The stress state within Earth can be expressed in terms of three principal stresses—*P*, *T*, and *N*—which are at right angles to one other. *P* is the maximum compressive stress, *T* the minimum compressive (or tensional) stress, and *N* is the intermediate stress. When a rock mass fractures, the plane of failure forms a 45° angle with the axis of the maximum compressive stress, *P*, and the relative motion of the fault is such that the rock mass shortens in the direction of *P* and expands in the direction of *T*. There is no motion in the *N* direction. Consequently, for normal faults, *P* is vertical and *T* and *N* are horizontal. For reverse faults, *T* is vertical; for strike-slip faults, *N* is vertical.

Laboratory experiments on rock samples show a distinct relationship between stress and strain. When the stress applied to a rock is plotted against strain, a stress-strain curve, like the one shown in **Figure 2,** is produced. A stress-strain curve usually has two segments—a straight segment and a curved segment. Each segment represents a different type of response to stress.

Elastic deformation The first segment of a stress-strain curve shows what happens under conditions in which stress is low. Under low stress, a material shows elastic deformation. **Elastic deformation** is caused when a material is compressed, bent, or stretched. This is the same type of deformation that happens from gently pulling on the ends of a rubber band. When the stress on the rubber band is released, it returns to its original size and shape. **Figure 2** illustrates that elastic deformation is the result of stress and strain. If the stress is reduced to zero, as the graph shows, the deformation of the rock disappears.

Plastic deformation When stress builds up past a certain point, called the elastic limit, rocks undergo **plastic deformation,** shown by the second segment of the graph in **Figure 2.** Unlike elastic deformation, this type of strain produces permanent deformation, which means that the material stays deformed even when stress is reduced to zero. Even a rubber band undergoes plastic deformation when it is stretched beyond its elastic limit. At first the rubber band stretches, then it tears slightly, and finally, two pieces will snap apart. The tear in the rubber band is an example of permanent deformation. When stress increases to be greater than the strength of a rock, the rock ruptures. The point of rupture, called failure, is designated by the "X" on the graph in **Figure 2.**

READING CHECK **Differentiate** between elastic deformation and plastic deformation.

Most materials exhibit both elastic and plastic behavior, although to different degrees. Brittle materials, such as dry wood, glass, and certain plastics, fail before much plastic deformation occurs. Other materials, such as metals, rubber, and silicon putty, can undergo a great deal of deformation before failure occurs, or they might not fail at all. Temperature and pressure also influence deformation. As pressure increases, rocks require greater stress to reach the elastic limit. At high enough temperatures, solid rock can also deform, causing it to flow in a fluid-like manner. This flow reduces stress.

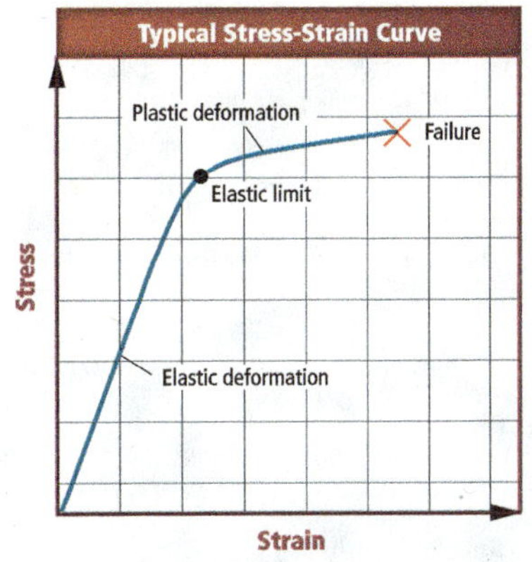

Figure 2 A typical stress-strain curve has two parts. Elastic deformation occurs as a result of low stress. When the stress is removed, material returns to its original shape. Plastic deformation occurs under high stress. The deformation of the material is permanent. When plastic deformation is exceeded, an earthquake occurs.
Describe *what happens to a material at the point on the graph at which elastic deformation changes into plastic deformation.*

VOCABULARY
SCIENCE USAGE V. COMMON USAGE
Failure
Science usage: a collapsing, fracturing, or giving way under stress

Common usage: lack of satisfactory performance or effect

Activity

Elastic Strain Have small groups of students use flexible, plastic rulers and quarters or heavy metal washers to demonstrate how the flexible plastic reacts to moderate amounts of stress. Give each group two rulers. Instruct one student in each group to position one of the rulers so that about three-fourths of it extends beyond the edge of a desk. Have another student in the group load the overhanging end of the ruler with three or four quarters or washers (adding them one at a time) while the third student in the group measures and records the deflection of the ruler after each quarter or washer is added.

Instruct students to unload the ruler in the same way. Have each group use graph paper to plot the load on the ruler against the resulting deflections for both the loading and unloading phases. The graphs should show approximately straight lines. Make sure students note the ruler returns to its original position when the load is removed. Then, have students read the text on the student page about elastic strain and relate it to this activity. OL EL COOP LEARN

Caption Question Fig. 2 The material is breaking. It is undergoing permanent deformation.

READING CHECK Elastic strain causes temporary deformation, whereas plastic deformation causes permanent changes in a material.

Demonstration

Deformation Mold a piece of silicone putty into a rectangle and fold it into the shape of a book. Ask students what type of deformation this demonstrates. plastic What happens when you pull the clay? At first it stretches, then it fails. Tell students most materials, like silicone putty and rocks, break much more easily from tension than from compression.

DIFFERENTIATED INSTRUCTION

Visually Impaired Give students a plastic ball that has some give to it. Ask students to describe the type of deformation that occurs when they push down on the ball and then release it. It undergoes elastic deformation. Now ask students to squeeze a ball of silicone putty. What kind of deformation results? plastic deformation BL OL EL

Identify Misconceptions

A popular misconception is that California will slide into the ocean during the next big earthquake.

Uncover the Misconception Ask students what might happen when the next major earthquake strikes California.

Demonstrate the Concept Use a map of California and two wooden blocks to explain that major earthquakes in California are caused by horizontal strike-slip motion along the San Andreas Fault system. Demonstrate that southwestern California and Baja California are moving horizontally toward the northwest, not out to or into the ocean.

Assess New Knowledge Ask students what will eventually happen to southwestern California and Baja if present plate motions continue for millions of years at a rate of about 5 cm/y. Southwestern California and Baja will become an island off the coast of Oregon in 20 million years.

Caption Question Fig. 3 The earth on the left side of the fault moved down and to the left relative to the other side.

FOLDABLES® Rubric

Figure 3 A major fault passes through this section of land.
Identify *the direction of movement that occurred along this fault.*

Faults

Crustal rocks fail when stresses exceed the strength of the rocks. The resulting movement occurs along a weak region in the crustal rock called a fault. A **fault** is any fracture or system of fractures along which Earth moves. **Figure 3** shows a fault. The surface along which the movement takes places is called the fault plane. The orientation of the fault plane can vary from nearly horizontal to almost vertical. The movement along a fault results in earthquakes. Several historic earthquakes are described in the time line in **Figure 4**.

Reverse and normal faults Reverse faults form as a result of horizontal and vertical compression that squeezes rock and creates a shortening of the crust. This causes rock on one side of a reverse fault to be pushed up relative to the other side. Reverse faulting can be seen near convergent plate boundaries.

Movement along a normal fault is partly horizontal and partly vertical. The horizontal movement pulls rock apart and stretches the crust. Vertical movement occurs as the stretching causes rock on one side of the fault to move down relative to the other side. The Basin and Range province in the southwestern United States is characterized by normal faulting. The crust is being stretched apart in that area. Note in the diagrams shown in **Table 1** that the two areas separated by the reverse fault would be closer after the faulting than before, and that two areas at a normal fault would be farther apart after the faulting than before the faulting.

FOLDABLES® Incorporate information from this section into your Foldable.

Figure 4
Major Earthquakes and Advances in Research and Design

As earthquakes cause casualties and damage around the world, scientists work to find better ways to warn and protect people.

1800 — 1900 — 1950

1811–1812 Several strong earthquakes occur along the Mississippi River valley over three months, destroying the entire town of New Madrid, Missouri.

1880 Following an earthquake in Japan, scientists invent the first modern seismograph to record the intensity of earthquakes.

1906 An earthquake in San Francisco kills between 3000 and 5000 people and causes a fire that rages for three days, destroying most of the city.

1923 Approximately 140,000 people die in an earthquake and subsequent fires that destroy the homes of over a million people in Tokyo and Yokohama, Japan.

1948 An earthquake destroys Ashgabat, capital of Turkmenistan, killing nearly nine out of ten people living in the city and its surrounding areas.

Demonstration

Fault Types Take two blocks, put them side by side and tilt them slightly relative to one another, then push them together. Ask students what type of force is being applied and what type of "fault" develops. compression; reverse Pull the blocks apart. Again ask what type of force is being applied and what type of "fault" develops. extension; normal Slide the blocks by each other and ask the same questions. shear; strike-slip Repeat each example and have students identify the location of the fault plane. Ask them what each fault would look like in real life.

Explore faults with an interactive table. Concepts In Motion

Table 1 Types of Faults

Type of Fault	Type of Movement	Example
Reverse	Compression causes vertical movement upward along a fault plane.	
Normal	Tension causes vertical movement downward along a fault plane.	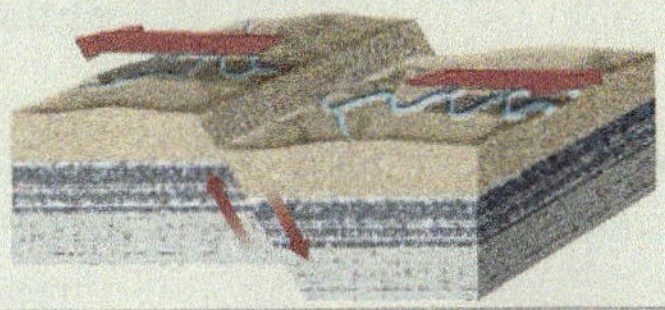
Strike-slip	Shear causes horizontal movement along a fault plane.	

Strike-slip faults Strike-slip faults are caused by horizontal shear. As shown in **Table 1,** the movement at a strike-slip fault is mainly horizontal and in opposite directions, similar to the way cars move in opposite directions on either side of a freeway. The San Andreas Fault, which runs through California, is a strike-slip fault. Horizontal motion along the San Andreas and several other related faults is responsible for many of the state's earthquakes. The result of motion along strike-slip faults can easily be seen in the many offset features that were originally continuous across the fault.

1960 In Chile, a 9.5 earthquake generates tsunamis that hit Hawaii, Japan, New Zealand, and Samoa. This is the largest earthquake recorded.

1965 The United States, Japan, Chile, and Russia form the International Pacific Tsunami Warning System.

1972 The University of California, Berkeley creates the first modern shake table to test building designs.

1982 New Zealand constructs the first building with seismic isolation, using lead-rubber bearings to prevent the building from swaying during an earthquake.

2004 A 9.0 earthquake in the Indian Ocean triggers the most deadly tsunami in history. The tsunami travels as far as the East African Coast.

2011 The eastern coast of Japan is struck by a magnitude-9 earthquake and subsequent tsunami, which breaks off icebergs in Antarctica 13,600 km away.

1960 — 1980 — 2000

Model

Strike-Slip Fault Have students make a simple model of a strike-slip fault by placing two sheets of construction paper side by side on a tabletop. Have students erect a city over the fault line using plastic interlocking blocks. Instruct students to then pull the pieces of paper to simulate the horizontal movement that occurs along a strike-slip fault.

Interpret the Illustration

Fault Type Ask students: How can you identify which illustration in **Table 1** shows a fault that results in shortening of the crust? The areas on opposite sides of the fault are closer together than they were prior to the faulting.

DIFFERENTIATED INSTRUCTION

Visually Impaired Use two wooden blocks to demonstrate the three types of fault motions discussed on this page. Allow students to manipulate the blocks on their own to simulate normal, reverse, and strike-slip motions. **OL**

EARTH SCIENCE JOURNAL

Time Line Have students research what they consider to be the important characteristics of one of the earthquakes shown in the time line. They should record their findings in their Earth science journals. After the chapter is completed, ask students whether their views have changed about what the important characteristics of an earthquake are. **AL**

Tie to Previous Knowledge

Ground Motion Ask students whether they have ever sat in a small boat on a lake or river. If so, they should recall that boats pitch up and down, roll from side to side, and swivel horizontally (yaw) from side to side. Explain the ground motion during an earthquake is similar, but some of the oscillations are slower.

Teacher Content Support

Surface Waves There are two types of surface waves: Rayleigh waves and Love waves. Rayleigh waves, also called ground roll, cause the ground to move up and down in a rolling motion similar to that of water waves. Love waves are surface shear waves that cause the ground to move horizontally from side to side as the waves pass.

Activity

Surface Waves Based on the results of the Demo below, have a pair of volunteers use the spring toy to demonstrate the movement of surface waves. Use the information in the Teacher Content Support above to explain that there are two types of surface waves: Rayleigh waves and Love waves. COOP LEARN

View an **animation of seismic waves.**

Concepts In Motion

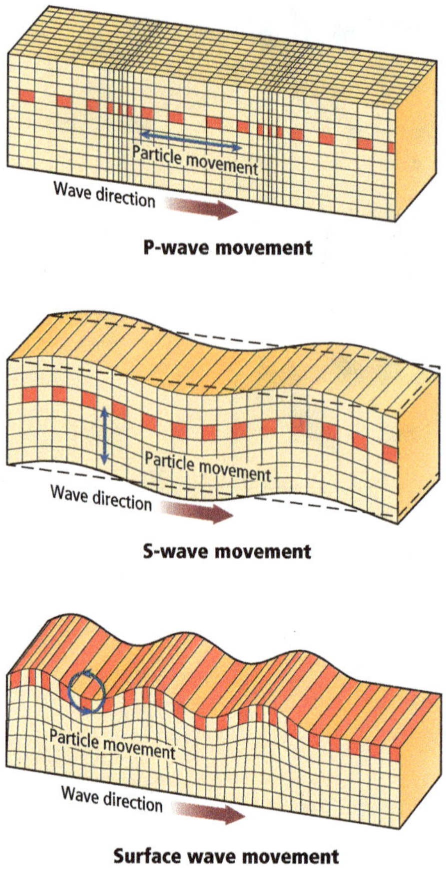

■ **Figure 5** Seismic waves are characterized by the types of movement they cause. Rock particles move back and forth as a P-wave passes. Rock particles move at right angles to the direction of the S-wave. A surface wave causes rock particles to move both up and down and from side to side.

Earthquake Waves

Most earthquakes are caused by movements along faults. Recall from the Launch Lab that some slippage along faults is relatively smooth. Other movements, modeled by the sandpaper-covered blocks, show that irregular surfaces in rocks can snag and lock. As stress continues to build in these rocks, they undergo elastic deformation. Beyond the elastic limit, they are permanently deformed. At some point after that, the rocks slip or crumble and an earthquake occurs.

Types of seismic waves The vibrations of the ground produced during an earthquake are called **seismic waves.** Every earthquake generates three types of seismic waves: primary waves, secondary waves, and surface waves.

Primary waves Also referred to as P-waves, **primary waves** squeeze and push rocks in the direction along which the waves are traveling, as shown in **Figure 5.** Note how a volume of rock, which is represented by small red squares, changes length as a P-wave passes through it. The compressional movement of P-waves is similar to the movement along a loosely coiled wire. If the coil is tugged and released quickly, the vibration passes through the length of the coil parallel to the direction of the initial tug.

Secondary waves **Secondary waves,** called S-waves, are named with respect to their arrival times. They are slower than P-waves, so they are the second set of waves to be felt. S-waves have a motion that causes rocks to move perpendicular to the direction of the waves, as illustrated in **Figure 5.** The movement of S-waves is similar to the movement of a jump rope that is jerked up and down at one end. The waves travel vertically to the other end of the jump rope. Both P-waves and S-waves pass through Earth's interior. For this reason, they are also called body waves.

Surface waves The third and slowest type of waves are surface waves, which travel only along Earth's surface. Surface waves cause the ground to move sideways and up and down like ocean waves, as shown in **Figure 5.** They usually cause the most destruction because they cause the most movement of the ground, and take the longest time to pass.

Demonstration

Wave Movement Use a long, coiled-spring toy and two volunteers to help you demonstrate how P-waves and S-waves travel through rocks. Have the volunteers each hold one end of the coiled-spring toy. Instruct one volunteer to make a quick push-pull motion with his or her hand. Have students describe how the spring reacts. A compressional pulse travels along the spring toward the second volunteer, and particle motion of the coils is longitudinal, which is similar to that of the movement of a P-wave. Instruct the second volunteer to flick his or her wrist to give the spring an up-and-down motion. Have students observe the type of wave produced and relate this movement to the movement of S-waves through rocks. COOP LEARN

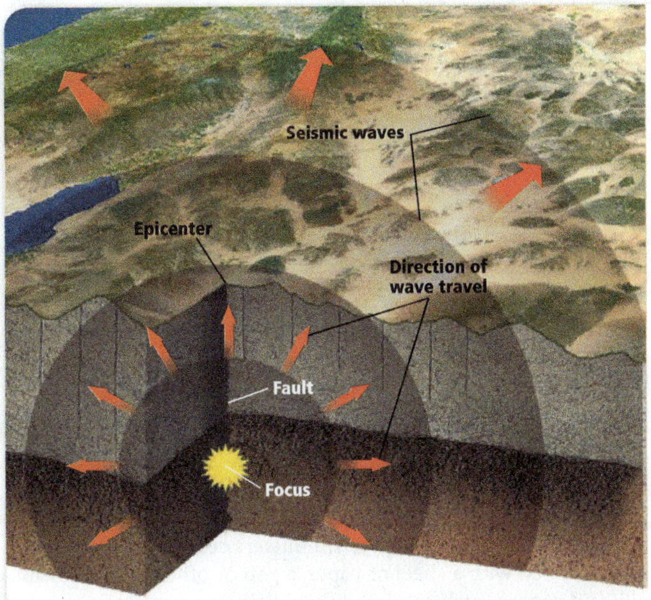

■ **Figure 6** The focus of an earthquake is the point of initial fault rupture. The surface point directly above the focus is the epicenter.
Infer ***the point at which surface waves will cause the most damage.***

Generation of seismic waves The first body waves generated by an earthquake spread out from the point of failure of crustal rocks. The point of failure where the waves originate is called the **focus** of the earthquake. The focus is usually several kilometers below Earth's surface. Depending on the depth of the focus, an earthquake is classified as either a shallow-focus, mid-focus, or deep-focus earthquake. The point on Earth's surface directly above the focus is the **epicenter** (EH pih sen tur), shown in **Figure 6.** Surface waves originate from the epicenter and spread out.

SECTION 1 REVIEW

Section Self-Check

Section Summary

- Stress is force per unit of area that acts on a material and strain is the deformation of a material in response to stress.
- Reverse, normal, and strike-slip are the major types of faults.
- The three types of seismic waves are P-waves, S-waves, and surface waves.

Understand Main Ideas

1. MAINIDEA **Describe** how the formation of a fault can result in an earthquake.
2. **Explain** why a stress-strain curve usually has two segments.
3. **Compare and contrast** the movement produced by each of the three types of faults.
4. **Draw** three diagrams to show how each type of seismic wave moves through rock. How do they differ?

Think Critically

5. **Relate** the movement produced by seismic waves to the observations a person would make of them as they traveled across Earth's surface.

WRITING IN ▶ Earth Science

6. Relate the movement of seismic waves to movement of something you might see every day. Make a list and share it with your classmates.

■ **Caption Question Fig. 6** the epicenter

3 Assess

Check for Understanding

Infer Tell students that, during many earthquakes, straight rail lines are bent either by surface waves or by fault motion. Pose the following questions: What type of fault or what type of wave motion could produce this kind of deformation? A reverse or strike-slip fault or horizontal ground motion of surface waves could bend rail lines. Is the resulting deformation elastic or plastic? plastic

Reteach

Use Models Use your hands to demonstrate the movement along the three types of faults discussed in this section. Have students mimic your movements with their own hands.

Assessment

Skill Have students construct concept maps using the following terms: *normal fault, reverse fault, strike-slip fault, compression, tension, shear, shortening, extension,* and *horizontal offset.* Students' maps should connect the terms as follows: normal fault–tension–extension; reverse fault–compression–shortening; and strike-slip fault–shear–offset.

SECTION 1 REVIEW

1. When stress is high enough, rock will fracture, producing movement along a fault. This movement creates earthquakes.
2. One segment describes elastic strain during which a rock bends, and the other segment represents the point after which strain results in permanent deformation.
3. normal: horizontal and vertical; reverse: horizontal and vertical; strike-slip: horizontal only
4. Answers should show concepts illustrated in **Figure 5.**
5. A person at the surface might see surface waves moving the ground like ocean waves; P-waves would move the ground toward and then away from an observer; S-waves would move the ground up and down.
6. Possible answers: P-waves: pushing straight forward on a spring toy, pinballs in a machine; S-waves: wiggle a spring toy up and down; surface waves: a drop in water, smoothing out sheets on a bed

Rubric

SECTION 2

1 Focus

Focus Transparency

Before presenting the lesson, project **Section Focus Transparency** 57 and have students answer the accompanying questions. EL

MAINIDEA

Image Formation Ask students to discuss how we form images of any object, whether it be something right in front of us or bones inside a human body. We rely on waves that are deflected or reflected by structures. Our eyes use light waves. X rays and CT scans use electromagnetic waves of different wavelengths to create images of the inside of a human body. Explain to students that to create images of what is inside Earth, we rely on waves generated by earthquakes and recorded by seismometers.

2 Teach

Activity

Seismometer Use a piece of paper, tape, a lightweight table, and a pencil to model how a seismometer works. Use the tape to attach the piece of paper to the top of the table. Have a student hold a pencil just touching the paper. Have another student slowly pull the table. The resulting line will be straight. Repeat the activity with a third student shaking the table at right angles to the direction in which it is pulled. The resulting line will resemble that shown in Figure 8. COOP LEARN

SECTION 2

Seismic Waves and Earth's Interior

MAINIDEA Seismic waves can be used to make images of the internal structure of Earth.

Essential Questions

- How does a seismometer work?
- How have seismic waves been used to determine the structure and composition of Earth's interior?

Review Vocabulary

mantle: the part of Earth's interior beneath the crust and above the central core

New Vocabulary

seismometer
seismogram

EARTH SCIENCE 4 YOU When you look in a mirror, you see yourself because light waves reflect off your face to the mirror and back to your eye. Similarly, seismic waves traveling through Earth reflect off structures inside Earth, which allows these structures to be imaged.

Seismometers and Seismograms

Most of the vibrations caused by seismic waves cannot be felt at great distances from an earthquake's epicenter, but they can be detected by sensitive instruments called **seismometers** (size MAH muh turz). Some seismometers consist of a rotating drum covered with a sheet of paper, a pen or other such recording tool, and a mass, such as a pendulum. Seismometers vary in design, but all include a frame that is anchored to the ground and a mass that is suspended from a spring or wire, as shown in **Figure 7.** During an earthquake, the mass and the pen attached to it tend to stay at rest due to inertia, while the ground beneath shakes. The motion of the mass in relation to the frame is then registered on the paper with the recording tool, or is directly recorded onto a computer disk. The record produced by a seismometer is called a **seismogram** (SIZE muh gram). A portion of one is shown in **Figure 8.**

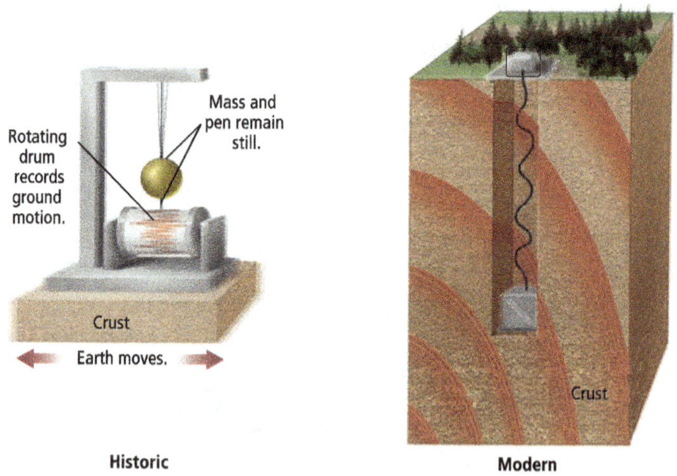

Figure 7 The frame of a historic seismometer is anchored to the ground. When an earthquake occurs, the frame moves but the hanging mass and attached pen do not. The mass and pen record the relative movement as the recording device moves under them. Compare this to a modern sensor and transmitter.

View an **animation of seismometers.**
Concepts In Motion

IN THE FIELD

Coda Waves Keith Aki was born in Japan and received a PhD in seismology from the University of Tokyo. He went to the United States to join the faculty of MIT. Perhaps his most important contribution to seismology is the concept of coda waves. The amplitudes of coda waves remain consistent while traveling through similar geological structures and dissipate at a uniform rate. The study of coda waves reveals important information about sources of earthquakes.

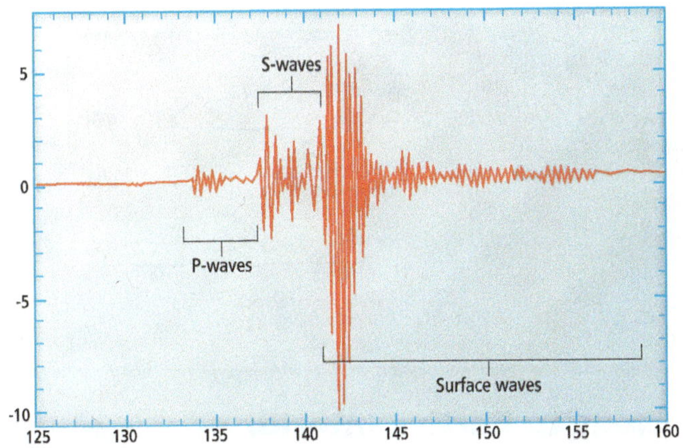

■ **Figure 8** Seismograms provide a record of the seismic waves produced by an earthquake.

Travel-time curves Seismic waves that travel from the focus of an earthquake are recorded by seismometers housed in seismic stations around the world. Over many years, the arrival times of seismic waves from countless earthquakes have been collected. Using these data, seismologists have been able to construct global travel-time curves for the arrival of P-waves and S-waves of earthquakes, as shown in **Figure 9.** These curves provide the average travel times of all P- and S-waves, from wherever an earthquake occurs on Earth.

☑ READING CHECK **Summarize** how seismograms are used to construct global travel-time curves.

Distance from the epicenter Note that in **Figure 9,** as in **Figure 8,** the P-waves arrive first, then the S-waves. The surface waves arrive last. With increasing travel distance from the epicenter, the time separation between the curves for the P-waves and S-waves increases. This means that waves recorded on seismograms from more distant stations are farther apart than waves recorded on seismograms at stations closer to the epicenter. This separation of seismic waves on seismograms can be used to determine the distance from the epicenter of an earthquake to the seismic station that recorded the seismogram. This method of precisely locating an earthquake's epicenter will be discussed in Section 3.

■ **Figure 9** Travel-time curves show how long it takes for P-waves and S-waves to reach seismic stations located at different distances from an earthquake's epicenter.
Determine ***how long it takes P-waves to travel to a seismogram 2000 km away. How long does it take for S-waves to travel the same distance?***

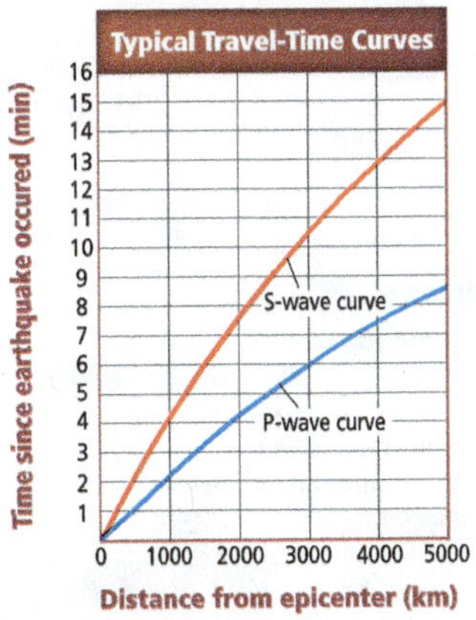

■ **Caption Question Fig. 9** 4.2 min.; 7.7 min.

Teacher Content Support

Wave Velocities Typical P-wave velocities in the upper mantle are 8 km/s or more. P-waves are sound waves that can travel through solids, liquids, and gases. Typical S-wave velocities in the upper mantle are 5 km/s or more. S-waves are distortional waves that can travel only through solids. The slowest of the seismic waves are the surface waves, whose typical velocities are 3–4 km/s.

Reinforcement

Time Separation Have students refer to **Figure 9** to solve the following problem: A seismic station is 4000 km from the focus of an earthquake. How long will it take the P- and S-waves to reach that station, and what will the time separation of the P- and S-waves on the seismogram be? P-wave: 7.5 min; S-wave: 13 min; separation is 5.5 min

☑ READING CHECK Seismograms show the arrival times of different types of seismic waves as they travel from an epicenter to a seismometer station. By comparing the data from different stations, scientists construct the travel-time curves.

GeoLAB

iLab Station

The GeoLab at the end of the chapter can be used at this point in the lesson.

Demonstration

Surface Waves Fill a large, shallow baking dish with enough water to make a layer about 1 cm deep. By touching the water with your finger, you can make ripples that spread across the water. Ask students to observe what happens to the height of these ripples as they move away from your finger. They get smaller. When the ripples reach the edge of the dish, what happens? They reflect. What happens to their height after they reflect? It decreases. Tell students ripples are most like surface waves because they spread over a surface; body waves decrease in size more quickly because they spread in three directions.

Enrichment

Wave Speed By looking at **Figure 10,** students can see where the speed at which waves travel changes as they move through different parts of Earth's interior. Have students identify areas where the velocity of P-waves changes. where they encounter boundaries of different composition of Earth's interior The diagram also shows that S-waves cannot continue into and through the liquid outer core; however, S-waves can be potentially generated by P-waves entering the inner core. Thus the dotted line S-waves in the inner core of the diagram. **OL**

Teacher Content Support

Density Dividing Earth's mass by its volume yields Earth's mean density, 5.5 g/cm^3. This value is much higher than the density of any known rock type, and therefore indicates that much of Earth's interior must consist of metal. Earth's moment-of-inertia, known from the precessional period of Earth's rotational axis in response to lunar and solar torques, indicates that Earth's density increases greatly with depth.

Earth's crust and much of its mantle have densities much less than 5.5 g/cm^3. Therefore, Earth's core must have a density of at least 10.0 g/cm^3 and consist of metal. Because the most abundant metal in meteorites and in the solar system is an iron-nickel alloy, Earth's core is thought to consist of an iron-nickel mixture as well.

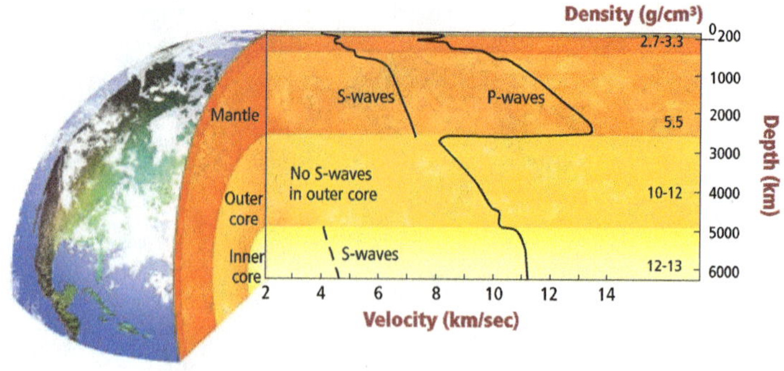

■ **Figure 10** Earth's layers are each composed of different materials. By examining the behavior of seismic waves moving through different kinds of rock, scientists have determined the composition of layers all the way to Earth's inner core.

View an **animation of P-waves and S-waves.**

Concepts In Motion

Clues to Earth's Interior

The seismic waves that shake the ground during an earthquake also travel through Earth's interior. This provides information that has enabled scientists to construct models of Earth's internal structure. Therefore, even though seismic waves can wreak havoc on the surface, they are invaluable for their contribution to scientists' understanding of Earth's interior.

Earth's internal structure Seismic waves change speed and direction at the boundaries between different materials. Note in **Figure 10** that as P-waves and S-waves initially travel through the mantle, they follow fairly direct paths. When P-waves strike the core, they are refracted, which means they bend. Seismic waves also reflect off of major boundaries inside Earth. By recording the travel-time curves and path of each wave, seismologists learn about differences in density and composition within Earth.

What happens to the S-waves generated by an earthquake? To answer this question, seismologists first determined that the right-angle motion of S-waves will not travel through liquid. Then, seismologists noticed that S-waves do not travel through Earth's center. This observation led to the discovery that Earth's core must be at least partly liquid. The data collected for the paths and travel times of the waves inside Earth led to the current understanding that Earth's core has an outer region that is liquid and an inner region that is solid.

VOCABULARY

ACADEMIC VOCABULARY

Encounter
to come upon or experience, especially unexpectedly
We had never encountered such a violent storm.

Earth's composition **Figure 11** shows that seismic waves change their paths as they encounter boundaries between zones of different materials. They also change their speed. By comparing the speed of seismic waves with measurements made on different rock types, scientists have determined the thickness and composition of Earth's different regions. As a result, scientists have determined that the upper mantle is peridotite, which is made mostly of the mineral olivine. The outer core is mostly liquid iron and nickel. The inner core is mostly solid iron and nickel.

DIFFERENTIATED INSTRUCTION

Advanced Learners Have students calculate the mass of a sphere consisting of a core with a radius of 3450 km and a density of 12,500 kg/m^3, surrounded by a mantle with an outer radius of 6371 km and a density of 4200 kg/m^3. The mass of a shell with an inner radius r, outer radius R, and density d is $m = 4/3 \times \pi(R^3 - r^3)\,d$. core: $m = 4/3\pi(3450\text{ m} \times 10^3)^3 \times 12{,}500\text{ kg/m}^3 = 2.15 \times 10^{24}\text{ kg}$; mantle: $m = 4/3\pi \times 4200\text{ kg/m}^3 \times [(6371\text{ m} \times 10^3)^3 - (3450 \times 10^3)^3] = 3.83 \times 10^{24}\text{ kg}$; total: $(2.15 \times 10^{24} + 3.83 \times 10^{24})\text{kg} = 5.98 \times 10^{24}\text{ kg}$. Point out that this sphere is a model of Earth.

VISUALIZING Seismic Waves

Figure 11 The travel times and behavior of seismic waves provide a detailed picture of Earth's internal structure. These waves also provide clues about the composition of the Earth's layers.

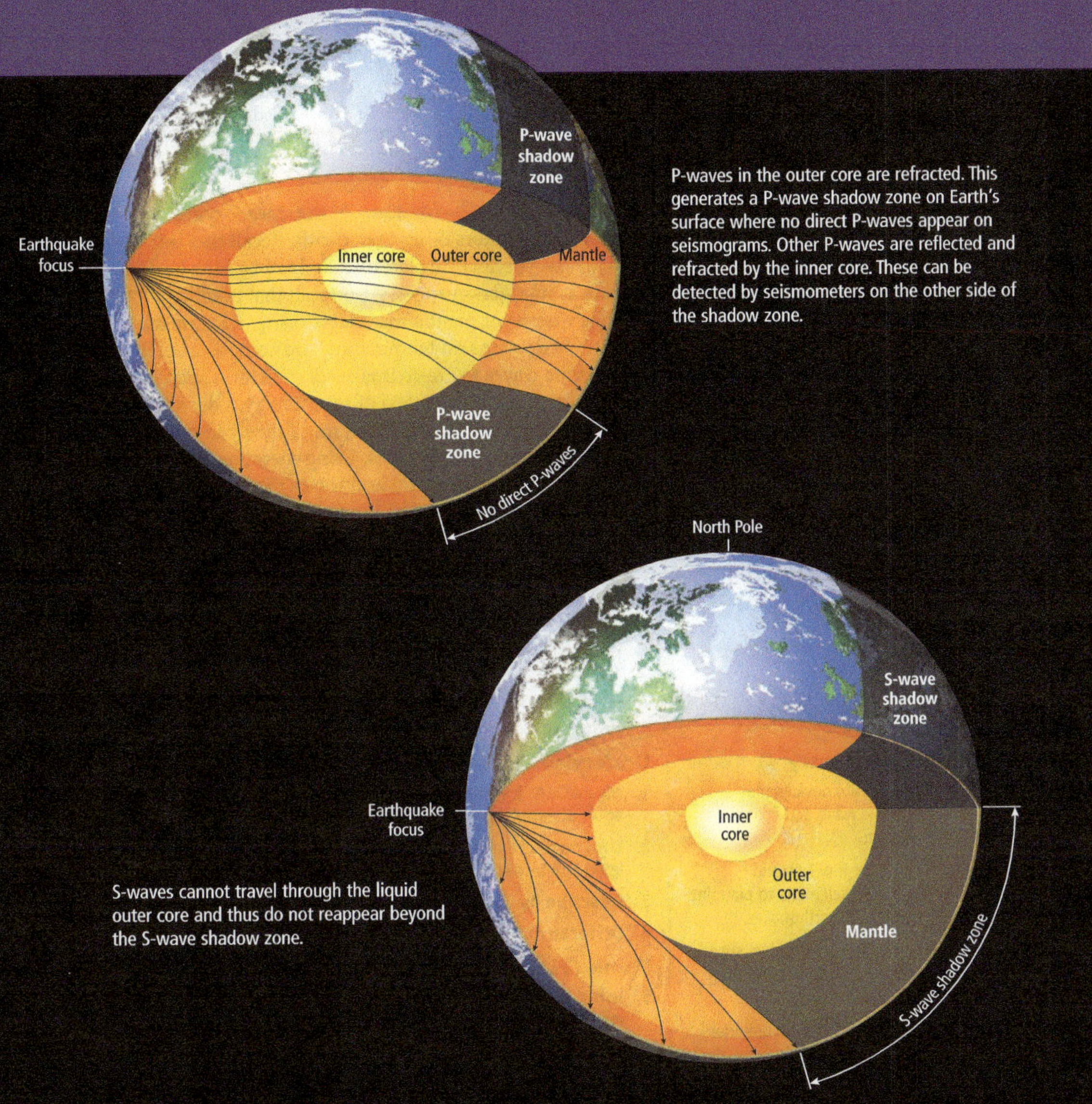

Concepts In Motion View an **animation of seismic waves.**

Teacher Content Support

Reflection and Refraction of Seismic Waves Seismic waves are reflected and refracted at discontinuities in Earth's crust. The angle of refraction is determined by Snell's law, which states $\sin i/V_1 = \sin r/V_2$, where V_1 and V_2 are the seismic velocities in the upper and lower media, respectively. If V_2 is greater than V_1, the refracted ray is bent away from the normal; if V_2 is less than V_1, the ray is bent toward the normal. Because seismic velocities in Earth's mantle increase with depth, seismic ray paths through this part of Earth are concave upward. If the incident angle, i, is such that the angle of refraction, r, is 90°, critical refraction occurs and the refracted ray travels along the boundary in the high-speed medium, generating so-called head-waves along the way that travel back to Earth's surface.

VISUALIZING

Purpose

Students will investigate how the propagation of seismic waves gives information about the internal structure of Earth.

Activity

Images of Unknown Structures Collect 8–10 cans, some marbles, and a large sheet of cardboard. Arrange four of the cans in a rectangle on a flat surface so that the cardboard can be supported at its corners when placed on top of the cans. Position the remaining cans between the other four cans. Place the cardboard on top of the cans.

Tell students to determine the location and properties of the "hidden" cans without looking at them. Tell them to roll marbles under the cardboard and to observe when the marbles pass straight through or are reflected. Have students draw a picture on the cardboard of what is hidden.

Ask them what they can determine from the image they create. Can they determine the shape of the cans? their color? their contents? only their shape Tell students that the cans represent structures below the surface of Earth, and the marbles represent seismic waves. Discuss what it takes to get a good picture of what is hidden. for the cans, lots of marbles; in Earth, lots of seismic waves and seismographs to record the waves

Students should conclude that this method allows them to determine structure and position only, not composition. OL COOP LEARN

3 Assess

Check for Understanding

Reinforcement Ask the following question to check students' understandings of the major concepts presented in this section: How have scientists determined the structure and dimensions of Earth's interior? The paths of the different types of seismic waves generated by earthquakes have been used to determine the structure and dimensions of Earth's interior. The absence of S-waves in the outer core indicates that it is molten. P-waves reflected from the inner core indicate that it is solid. The velocities of seismic waves inside Earth confirm the composition of the mantle and core.

Reteach

Summarize Draw a two-dimensional model of Earth's interior on the board. Have volunteers use different colors of chalk to indicate how both P- and S-waves travel from an earthquake's focus. Have a third volunteer mark the extent of the P-wave shadow zone on the drawing.

Assessment

Interpret Scientific Illustrations Have students each draw a cross section of Earth to scale in their Earth science journals. On their drawings, have students indicate the composition and state of the crust, the mantle, and the inner and outer cores.

Figure 12 Images like this one from Japan are generated by capturing the path of seismic waves through Earth's interior. Areas of red indicate seismic waves that are traveling more slowly than average and areas of blue indicate seismic waves that are traveling faster than average. The blue area is a subducted plate.

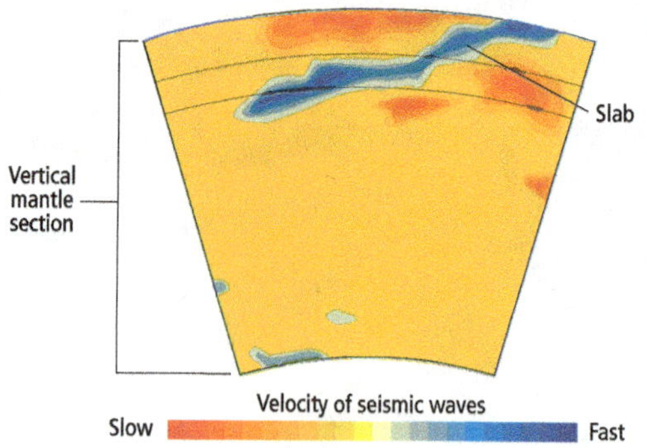

Imaging Earth's interior Seismic wave speed and Earth's density vary with factors other than depth. Recall that the leading edge of cold, dense, subducting plates—called slabs—sink back into Earth's mantle at subduction zones. Also recall that mantle plumes are regions where hot mantle material is rising. Because the speed of seismic waves depends on temperature and composition, it is possible to use seismic waves to create images of structures such as slabs and plumes. In general, the speed of seismic waves decreases as temperature increases. Thus, waves travel more slowly in hotter areas and more quickly in cooler regions. Using measurements made at seismometers around the world and waves recorded from many thousands of earthquakes, Earth's internal structure can be constructed and visualized, and features such as slabs can be located in images like the one in **Figure 12.** These images are similar to computed tomography (CT) scans, except that the images are made using seismic waves instead of X rays.

SECTION 2 REVIEW

Section Self-Check

Section Summary

- Seismometers are devices that record seismic wave activity on a seismogram.
- Travel times for P-waves and S-waves enable scientists to pinpoint the epicenters of earthquakes.
- P-waves and S-waves change speed and direction when they encounter different materials.
- Analysis of seismic waves provides a detailed picture of the composition of Earth's interior.

Understand Main Ideas

1. MAINIDEA **Explain** how P-waves and S-waves are used to determine the properties of Earth's core.
2. **Draw** a diagram of a seismometer showing how the movement of Earth is translated into a seismogram.
3. **Describe** how seismic travel-time curves are used to study earthquakes.
4. **Differentiate** between the speed of waves through hot and cold material.

Think Critically

5. **Infer** Using the seismogram in **Figure 8**, suggest why surface waves cause so much damage even though they are the last to arrive at a seismic station.

WRITING IN ▶ Earth Science

6. Write a newspaper article reporting on the ways scientists have determined the composition of Earth.

SECTION 2 REVIEW

1. Their travel rates reveal the types of matter through which they travel.
2. Drawings should show how the motion of the mass in relation to the frame is registered and recorded.
3. Seismic travel-time curves provide the average travel times of all seismic waves, which allow scientists to locate the epicenter of earthquakes.
4. In general, the speed of seismic waves decreases as temperature increases.
5. All their energy is released on the surface—where there are people and structures.
6. Reports should demonstrate an understanding that scientists combined what they knew about the behavior of waves through rock with travel-time measurements for seismic waves passing through Earth during earthquakes.

Rubric

SECTION 3

Measuring and Locating Earthquakes

MAIN IDEA Scientists measure the strength and chart the location of earthquakes using seismic waves.

Essential Questions

- What are earthquake magnitude and intensity and how are they measured?
- Why are data from at least three seismic stations needed to locate an earthquake's epicenter?
- Where are Earth's seismic belts?

Review Vocabulary

plot: to mark or note on a map or chart

New Vocabulary

Richter scale
magnitude
amplitude
moment magnitude scale
modified Mercalli scale

EARTH SCIENCE 4 YOU When someone speaks to you from nearby, you can hear them clearly. However, the sound gets fainter as they get farther away. Similarly, the energy of seismic waves gets weaker the farther away you are from the source of an earthquake.

Earthquake Magnitude and Intensity

More than 1 million earthquakes are felt each year, but news accounts report on only the largest ones. Scientists have developed several methods for describing the size of an earthquake.

Richter scale The **Richter scale,** devised by a geologist named Charles Richter, is a numerical rating system that measures the energy of the largest seismic waves, called the **magnitude,** that are produced during an earthquake. The numbers in the Richter scale are determined by the height, called the **amplitude,** of the largest seismic wave. Each successive number represents an increase in amplitude of a factor of 10. For example, the seismic waves of a magnitude-8 earthquake on the Richter scale are ten times larger than those of a magnitude-7 earthquake. The differences in the amounts of energy released by earthquakes are even greater than the differences between the amplitudes of their waves. Each increase in magnitude corresponds to about a 32-fold increase in seismic energy. Thus, an earthquake of magnitude-8 releases about 32 times the energy of a magnitude-7 earthquake. The damage shown in **Figure 13** was caused by an earthquake measuring 7 on the Richter scale.

Figure 13 The damage shown here was caused by a magnitude-7 earthquake that struck Haiti in December 2010.

SECTION 3

1 Focus

Focus Transparency

Before presenting the lesson, project **Section Focus Transparency 58** and have students answer the accompanying questions. EL

MAIN IDEA

Location of Earthquakes Ask students to think about where most earthquakes should occur. at plate boundaries How do we know there is plate movement in California? strong, frequent earthquakes How do we know there are no plate boundaries under the Antarctic ice sheet? no earthquakes

2 Teach

Teacher Content Support

Measure Earthquakes Earthquake magnitude is not the same as earthquake energy, but the two quantities are related. This relationship can be expressed as $\text{Log } E = 5.24 + 1.44\,M$ where E is the energy in Joules, and M is the magnitude (M_s). According to this relationship, a magnitude 8 earthquake releases about $10^{16.8}$ J of energy, which is comparable to the energy released by a 16-megaton nuclear device. By comparison, the energy released by the atomic bomb that destroyed Hiroshima corresponded to the energy of a magnitude-5.3 earthquake.

ACROSS THE CURRICULUM

Physics All waves have similar characteristics. Have students find the meanings of the terms that follow and label each on a simple diagram: *crest* highest point of a wave; *trough* lowest point of a wave; *amplitude* maximum displacement of the medium through which the wave is traveling; *wavelength* the crest-to-crest or trough-to-trough distance; *wave period* the time interval between the passage of adjacent crests; *frequency* the number of wave crests passing a given point in a given time interval. OL EL

Teacher Content Support

Earthquake Magnitude Soon after Charles Richter proposed his method of determining earthquake magnitude in 1935, it was learned that the most consistent earthquake measurements were obtained using surface waves, adjusted for their periods *(T)*. This method of determining earthquake magnitude is called the surface-wave magnitude (M_s) method, which is still widely used. Most reported Richter magnitudes are surface-wave magnitudes. Because deep and intermediate earthquakes do not generate large surface waves, a body-wave magnitude (m_b), based on the amplitude and period of the largest P-waves recorded, is used to measure such earthquakes. M_s and m_b values for a given earthquake are different. The relationship between the two magnitudes can be expressed by the following equation: $m_b = 2.94 + 0.56 \times M_s$. Major quakes are often measured using the seismic moment, M_0, of the fault rupture: $M_o = A \times d \times \mu$, where A, d, and μ are the estimated fault area, displacement, and rigidity of the rocks, respectively. This moment-magnitude method is gradually replacing the popular surface-wave magnitude method.

■ **Figure 14** The modified Mercalli scale measures damage done by an earthquake. An earthquake strong enough to knock groceries off the store's shelves would probably be rated V using the modified Mercalli scale.

Moment magnitude scale While the Richter scale is often used to describe the magnitude of an earthquake, most earthquake scientists, called seismologists, use a scale called the moment magnitude scale. The **moment magnitude scale** is a rating scale that measures the energy released by an earthquake, taking into account the size of the fault rupture, the amount of movement along the fault, and the rocks' stiffness. Most often, when you hear about an earthquake on the news, the number given is from the moment magnitude scale.

Modified Mercalli scale Another way to describe earthquakes is with respect to the amount of damage they cause. This measure, called the intensity of an earthquake, is determined using the **modified Mercalli scale,** which rates the types of damage and other effects of an earthquake as noted by observers during and after its occurrence. This scale uses the Roman numerals I to XII to designate the degree of intensity. Specific effects or damage correspond to specific numerals; the worse the damage, the higher the numeral. A simplified version of the modified Mercalli scale is shown in **Table 2.** You can use the information given in this scale to rate the intensity of the earthquakes such as the one that caused the damage shown in **Figure 14.**

Explore the **modified Mercalli scale with an interactive table.** Concepts In Motion

Table 2 Modified Mercalli Scale

I	Not felt except under unusual conditions
II	Felt only by a few persons; suspended objects might swing.
III	Quite noticeable indoors; vibrations are like the passing of a truck.
IV	Felt indoors by many, outdoors by few; dishes and windows rattle; standing cars rock noticeably.
V	Felt by nearly everyone; some dishes and windows break and some plaster cracks.
VI	Felt by all; furniture moves; some plaster falls and some chimneys are damaged.
VII	Difficult to stand; some chimneys break; damage is slight in well-built structures but considerable in weak structures.
VIII	Chimneys, smokestacks, and walls fall; heavy furniture is overturned; partial collapse of ordinary buildings occurs.
IX	Great general damage occurs; buildings shift off foundations; ground cracks; underground pipes break.
X	Most ordinary structures are destroyed; rails are bent; landslides are common.
XI	Few structures remain standing; bridges are destroyed; railroad ties are greatly bent; broad fissures form in the ground.
XII	Damage is total; objects are thrown upward into the air.

Use an Analogy

Multiple Scales To demonstrate why there are different scales to measure earthquakes, make the analogy to temperature. Celsius and Fahrenheit are scales, and a thermometer is the "recording" device (like a seismometer). Using a thermometer is like measuring an earthquake using the Moment magnitude scale. Ask students: If you did not have a thermometer, how could you measure temperature? Why? You could only measure the temperature relatively, i.e., hot, warm, cold; or through noting its effects, i.e., melts ice, boils water, etc.; Without a quantitative way to measure something (thermometer, seismometer) you can only describe effects or characterize with comparisons. The Mercalli scale is based on the effects of an earthquake and does not require a seismometer to measure seismic waves. Ask students: Which scale is subjective? Mercalli

Earthquake intensity The intensity of an earthquake depends primarily on the amplitude of the surface waves generated. Like body waves, surface waves gradually decrease in size with increasing distance from the focus of an earthquake. Because of this, the intensity also decreases as the distance from a earthquake's epicenter increases. Maximum intensity values are observed in the region near the epicenter; Mercalli values decrease to I at distances far from the epicenter.

In the MiniLab, you will use the modified Mercalli scale values to make a seismic-intensity map. These maps are a visual demonstration of an earthquake's intensity. Contour lines join points that experienced the same intensity. They demonstrate how the maximum intensity is usually found near the earthquake's epicenter.

Depth of focus As you learned earlier in this section, earthquake intensity and magnitude reflect the size of the seismic waves generated by the earthquake. Another factor that determines the intensity of an earthquake is the depth of its focus. Recall that an earthquake can be classified as a shallow-focus, intermediate- or mid-focus, or deep-focus earthquake, depending on the location of the focus, as shown in **Figure 15.** Catastrophic earthquakes with high intensity values are almost always shallow-focus events.

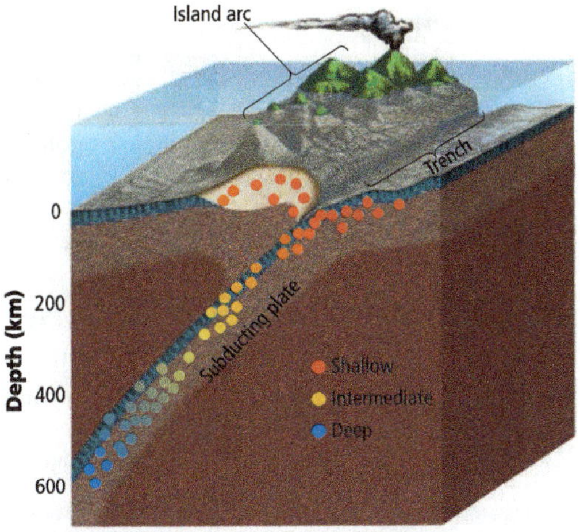

Figure 15 Earthquakes are classified as shallow, intermediate, or deep, depending on the location of the focus. Shallow-focus earthquakes are the most damaging.

MiniLAB

iLab Station

Make a Map

How is a seismic-intensity map made? Seismic-intensity data plotted on contour maps give scientists a visual picture of an epicenter's location and the earthquake's intensity.

Procedure

1. Read and complete the lab safety form.
2. Trace the map onto paper. Mark the locations indicated by the letters on the map.
3. Plot these Mercalli intensity values on the map next to the correct letter: A, I; B, III; C, II; D, III; E, IV; F, IV; G, IV; H, V; I, V; J, V; K, VI; L, VIII; M, VII; N, VIII; O, III.
4. Draw contours on the map to connect the intensity values.

Analysis

1. **Determine** the maximum intensity value.
2. **Find** the location of the maximum intensity value.
3. **Estimate** the earthquake's epicenter.

Intensity Values of an Earthquake

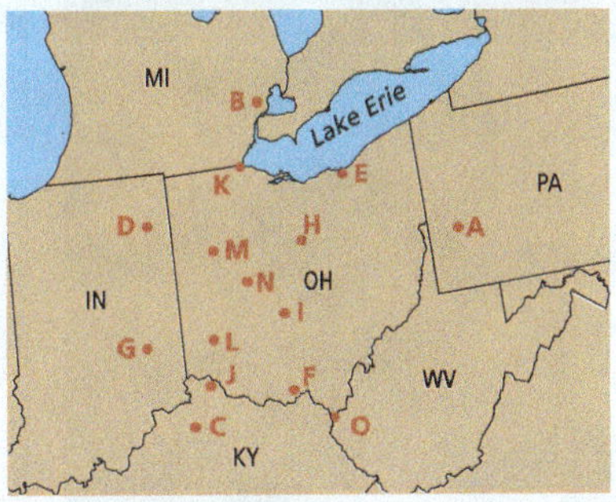

Teacher Content Support

Benioff Zones Almost all deep-focus earthquakes are associated with regions that lie landward of subduction zones. A seismologist named Hugo Benioff discovered that the depths of earthquake foci increase with distance from a subduction zone trench. These seismic regions are oriented about 45 degrees in relation to the surface and are called Benioff zones after the scientist who discovered them.

MiniLAB

iLab Station | Rubric

Purpose Students will make and interpret a simple seismic-intensity map.

Process Skills interpret data, draw a conclusion

Safety Precaution Approve lab safety forms before work begins.

Teaching Strategies

- Review the concept of contour lines before students begin this lab.
- Draw a fictitious seismic-intensity map on the board so students can get an idea of how their maps will look.
- Remind students that the contour lines should all connect numbers that are the same.

Expected Results Students should produce a contour map with the highest intensity (VIII) contour in SW Ohio. Contours should show a SSW–NNE trend in intensities.

Analysis

1. The maximum intensity value is VIII.
2. It is located at stations L and N.
3. between stations L and N

Assessment

Knowledge Tell students that intensity values decrease with increasing distance from the epicenter because the amplitudes of surface waves decrease with travel distance. Have students speculate why this is so. Students should be able to deduce that some of this decrease in wave size is a result of friction.

LabManager™ Customize any lab with the LabManager™ CD-ROM.

Use Science Terms

Epi– Have students use their dictionaries to each compile a list of at least ten scientific terms that contain the prefix *epi-*. Examples might include *epicenter, epiglottis, epiphyte, epidermis, epidural, epicycle,* and *epilepsy.* Have students jot down a brief definition next to each term. For a further challenge, have students state the meaning of the prefix for several of the terms. For example, *epiphyte* refers to a plant that grows on another plant; *epidermis* is the outer layer of skin. BL OL EL

Interpret the Illustration

Determine an Epicenter

Have students use **Figure 17** to explain why data from at least three seismic stations are necessary to locate the epicenter of an earthquake. Students should conclude that data from only two stations would result in two possible locations for the epicenter of the earthquake. Data from a third station pinpoint the quake's epicenter. OL EL

Activity

Locate an Epicenter Have students form a circle around the room, close their eyes, and listen to a radio placed somewhere in the room. On the board, sketch the circle of students and ask each student to identify where they thought the radio was located and what information they used. the volume, which depends on distance, direction Repeat this exercise three times for three different radio positions. Try the exercise with different numbers of students to show how the accuracy of the location improves with the number of observations (students). COOP LEARN

■ **Caption Question Fig. 17** It is nearest to Station 1 (near San Francisco).

☑ **READING CHECK** It is about 1200 km away from the seismograph that recorded it.

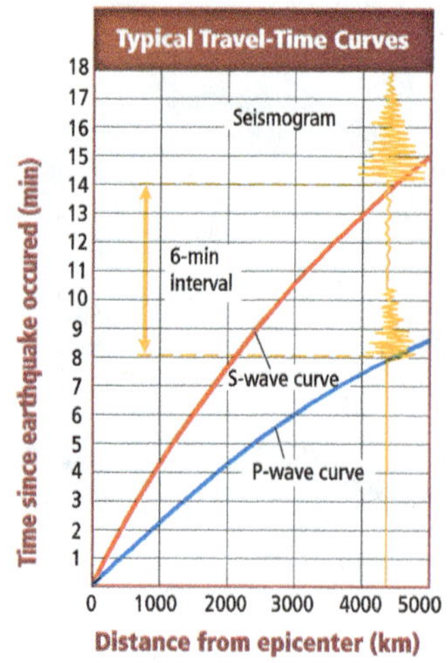

■ **Figure 16** This travel-time curve also shows seismographic data for an earthquake event.

Deep-focus earthquakes generally produce smaller vibrations at the epicenter than those produced by shallow-focus earthquakes. For example, a shallow-focus, moderate earthquake that measures a magnitude-6 on the Richter scale can generate a greater maximum intensity than a deep-focus earthquake of magnitude-8. Because the modified Mercalli scale is based on intensity rather than magnitude, it is a better measure of an earthquake's effect on people.

Locating an Earthquake

The location of an earthquake's epicenter and the time of the earthquake's occurrence are usually not known at first. However, the epicenter's location, as well as the time of occurrence, can be determined using seismograms and travel-time curves.

Distance to an earthquake Just as a person riding a bike will travel faster than a person who is walking, P-waves reach a seismograph station before the S-waves. Consider the effect of the distance traveled on the time it takes for both waves to arrive. Like the bicyclist and the walker, the gap in their arrival times will be greater when the distance traveled is longer. **Figure 16** shows the same travel-time curve graph shown in **Figure 9** of Section 2, but this time it is joined with the seismogram from a specific earthquake. The seismometer recorded the time that elapsed between the arrival of the first P-waves and first S-waves. Seismologists determine the distance to an earthquake's epicenter by measuring the separation on any seismogram and identifying that same separation time on the travel-time graph. The separation time for the earthquake shown in **Figure 16** is 6 min. Based on travel times of seismic waves, the distance between the earthquake's epicenter and the seismic station that recorded the waves can only be 4500 km. This is because the known travel time over that distance is 8 min for P-waves and 14 min for S-waves. Farther from the epicenter, the gap between the travel times for both waves increases.

☑ **READING CHECK Apply** If the gap between P- and S-waves is 2 min, what can you infer about the distance from the epicenter to the seismometer?

■ **Figure 17** To locate the epicenter of an earthquake, scientists identify the seismic stations on a map, and draw a circle with the radius of distance to the epicenter from each station. The point where all the circles intersect is the epicenter.

Identify ***the epicenter of this earthquake.***

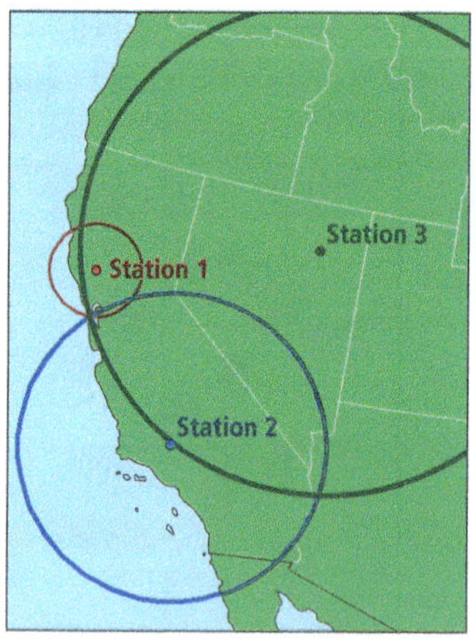

Seismologists analyze data from many seismograms to locate the epicenter. Calculating the distance between an earthquake's epicenter and a seismic station provides enough information to determine that the epicenter was a certain distance in any direction from the seismic station. This can be represented by a circle around the seismic station with a radius equal to the distance to the epicenter. Consider the effect of adding data from a second seismic station. The two circles will overlap at two points. When data from a third seismic station is added, the rings will overlap only at one point—the epicenter, as shown in **Figure 17.**

Demonstration

Travel Times To illustrate how the difference in travel times of P- and S-waves increases as the waves travel further, ask for two volunteers: one to represent a P-wave, one an S-wave. The volunteers stand next to each other at one end of the room. When you give the signal that an earthquake occurs, the P-wave student will take one step every 2 seconds, the S-wave student one step every 4 seconds. Determine the travel time across the room for each wave. How does the travel-time difference change as the waves travel across the room? BL COOP LEARN

Time of an earthquake The gap in the arrival times of different seismic waves on a seismogram provides information about the distance to the epicenter. Seismologists can also use the seismogram to gain information about the exact time that the earthquake occurred at the focus. The time can be determined by using a table similar to the travel-time graph shown in **Figure 9.** The exact arrival times of the P-waves and S-waves at a seismic station are recorded on the seismogram. Seismologists read the travel time of either wave to the epicenter from that station using graphs similar to the one shown in **Figure 9.** For example, consider a seimogram that registered the arrival of P-waves at exactly 10:00 A.M. If the P-waves traveled 4500 km, and took 8 min according to the appropriate travel-time curve, then it can be determined that the earthquake occurred at the focus at 9:52 A.M.

READING CHECK **List** the information contained in a seismogram.

Seismic Belts

Over the years, seismologists have collected and plotted the locations of numerous earthquake epicenters. The global distribution of these epicenters reveals a noteworthy pattern. Earthquake locations are not randomly distributed. The majority of the world's earthquakes occur along narrow seismic belts that separate large regions with little or no seismic activity.

Data Analysis LAB

Based on Real Data*

Interpret the Data

How can you find an earthquake's epicenter? To pinpoint the epicenter, analyze the P-wave and S-wave data recorded at seismic stations.

Analysis

1. Obtain a map of the western hemisphere from your teacher and mark the seismic stations listed in the table.
2. For each station, calculate and record the arrival time differences by subtracting the P-wave arrival time from the S-wave arrival times.
3. Use the arrival time differences and the travel-time curve **(Figure 9)** to find the distance between the epicenter and each seismic station. Record the distances.
4. Draw a circle around each station. Use the distance from the epicenter as the radius for each circle. Repeat for each seismic station.
5. Identify the epicenter of the earthquake.

Data and Observations

Seismic Station	P-wave Arrival Time (PST)	S-wave Arrival Time (PST)	Arrival Time Difference (min)	Distance from Epicenter (km)
Newcomb, NY	8:39:02	8:44:02		
Idaho Springs, CO	8:35:22	8:37:57		
Darwin, CA	8:35:38	8:38:17		

Think Critically

6. **Explain** why you need to find the difference in time of arrival between P- and S-waves for each seismic station.
7. **Identify** sources of error in determining an earthquake's epicenter.
8. **Explain** why data from more seismic stations would be useful for finding the epicenter.

*Data obtained from: Significant earthquakes of the world. 2006. *USGS earthquake center.*

DIFFERENTIATED INSTRUCTION

Behavior Disorders Engage students by having them research an earthquake that occurred this year. Information gathered should include the date, location, magnitude, depth of focus, casualties, and any other relevant information about the quake.

Data Analysis LAB

About the Lab

- Discuss with students how the data in the table is actual data scientists would use to determine an earthquake's epicenter.
- Remind students to use their drafting compasses only as directed.
- Have students work in groups of three or four. Make sure students follow the instructions step by step. Have them ask you about any steps that might be unclear.
- To obtain the distance that corresponds with the arrival time difference, students need to find the number of minutes between the two curves in **Figure 9.** Remind students to use the graph scale accurately.
- The earthquake occurred on the morning of Jan 4, 2006. It was a magnitude 6.6 near Santa Rosalia in the Gulf of California.

Think Critically

2. Newcomb, NY: 5; Idaho Springs, CO: 2.35; Darwin, CA: 2.79
3. Newcomb, NY: 3200; Idaho Springs, CO: 1800; Darwin, CA: 1500
6. because three locations are required to find a single intersection of possible distances from each location
7. Possible answers: composition of crust beneath a station, mechanical failure of equipment, non-synchronized clocks
8. If accurate, data from only three locations are needed, although more would confirm the location and therefore possibly correct any errors.

READING CHECK arrival times of different seismic waves; exact time an earthquake occurred at the focus

3 Assess

Check for Understanding

Compare and Contrast Have students compare and contrast a shallow-focus, magnitude-6 quake with a deep-focus, magnitude-7 quake. Ask students the following question: How do the earthquakes differ in released energy and in seismic wave size at the same distance from the focus? The waves of the magnitude-7 quake are ten times larger, and the energy released is 32 times greater than the energy released by the magnitude-6 quake.

Reteach

Make and Use Tables Have students each organize the information in the Check for Understanding item in a data table. Then, have them add the same information for a deep-focus, magnitude-5 earthquake and a shallow-focus, magnitude-4 quake.

Assessment

Knowledge Ask students to compare and contrast the modified Mercalli and Richter scales. The Mercalli scale measures earthquake intensity based on what people observe. The Richter scale measures earthquake magnitude based on the size of the largest waves generated by the quake.

■ **Caption Question Fig. 18** Answers will vary depending on where you live.

■ **Figure 18** Notice the pattern of global epicenter locations on the map.
Identify *Based on this map, do you live near an epicenter?*

Global Earthquake Epicenter Locations

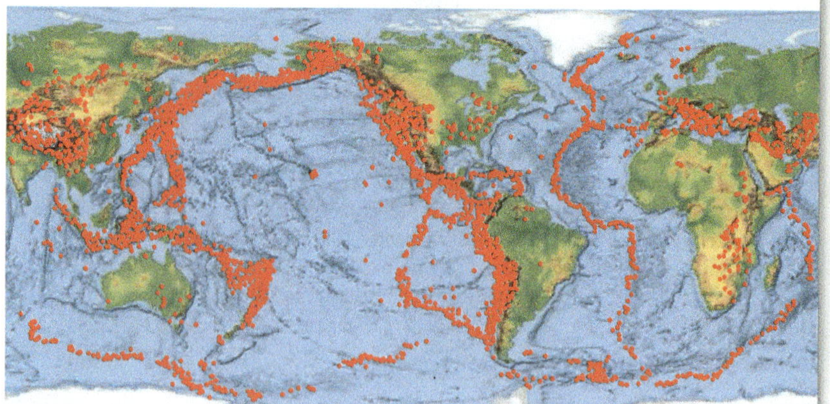

As shown in **Figure 18,** the distribution of earthquakes is not random. Earthquakes occur in narrow bands. As with the locations of most volcanoes on Earth, most earthquakes correspond closely with tectonic plate boundaries. In fact, almost 80 percent of all earthquakes occur on the Circum-Pacific Belt and about 15 percent on the Mediterranean-Asian Belt across southern Europe and Asia. Recall that these belts are subduction zones, where tectonic plates are colliding and one plate is forced to sink beneath another. Most of the remaining earthquakes occur in narrow bands along the crests of ocean ridges, where tectonic plates are diverging.

SECTION 3 REVIEW

Section Self-Check

Section Summary

- Earthquake magnitude is a measure of the energy released during an earthquake and can be measured on the Richter scale.
- Intensity is a measure of the damage caused by an earthquake and is measured with the modified Mercalli scale.
- Data from at least three seismic stations are needed to locate an earthquake's epicenter.
- Most earthquakes occur in seismic belts, which are areas associated with plate boundaries.

Understand Main Ideas

1. MAINIDEA **Summarize** the ways that scientists can use seismic waves to measure and locate earthquakes.
2. **Compare and contrast** earthquake magnitude and intensity and the scales used to measure each.
3. **Explain** why data from at least three seismic stations makes it possible to locate an earthquake's epicenter.
4. **Describe** how the boundaries between Earth's tectonic plates compare with the location of most of the earthquakes shown in the map in **Figure 18.**

Think Critically

5. **Formulate** a reason why a magnitude-3 earthquake can possibly cause more damage than a magnitude-6 earthquake.

MATH IN ▸ Earth Science

6. Calculate how much more energy a magnitude-9 earthquake releases compared to that of a magnitude-7 earthquake.

SECTION 3 REVIEW

1. Travel times of P-waves and S-waves are used to locate the epicenter of earthquakes.
2. Magnitude is a measure of seismic waves, while intensity is based on how people experience those waves. The Richter scale is based on wave measurement; the moment magnitude scale is based on the energy released by seismic waves; the Mercalli scale is based on an earthquake's intensity, measured by its impact on humans and their surroundings.
3. You need three circles to get a single point of intersection of the circles.
4. Most earthquakes occur in relatively narrow belts which correspond to the edges of tectonic plates.
5. Shallow-focus earthquakes generally produce larger vibrations at an epicenter than a deep-focus earthquake, so they are more damaging at the surface.
6. 1024 (32 × 32).

SECTION 4

Essential Questions
- What factors affect the amount of damage caused by an earthquake?
- What are some of the factors considered in earthquake-probability studies?
- How are different types of structures affected by earthquakes?

Review Vocabulary

geology: study of materials that make up Earth and the processes that form and change these materials

New Vocabulary

soil liquefaction
tsunami
seismic gap

Earthquakes and Society

MAINIDEA The probability of an earthquake's occurrence is determined from the history of earthquakes and knowing where and how quickly strain accumulates.

EARTH SCIENCE 4 YOU If, in your city, it rains an average of 11 days every July, how can you predict the weather in your city for July 4 ten years from now? You could estimate that there is a 11/31 chance that it will rain. In the same way, the probability of an earthquake's occurrence can be estimated from the history of earthquakes in the region.

Earthquake Hazards

Earthquakes are known to occur frequently along plate boundaries. An earthquake of magnitude-5 can be catastrophic in one region, but relatively harmless in another. There are many factors that determine the severity of damage produced by an earthquake. These factors are called earthquake hazards. Identifying earthquake hazards in an area can sometimes help to prevent some of the damage and loss of life. For example, the design of certain buildings can affect earthquake damage. As you can see in **Figure 19,** the most severe damage occurs to unreinforced buildings made of brittle building materials such as concrete. Wooden structures, on the other hand, are more resilient and generally sustain less damage.

Figure 19 Concrete buildings are often brittle and can be easily damaged in an earthquake. The building on the left shifted on its foundation after an earthquake and is held up by a single piece of wood.

SECTION 4

1 Focus

Focus Transparency

Before presenting the lesson, project **Section Focus Transparency 59** and have students answer the accompanying questions. EL

MAINIDEA

Earthquake Probability Have students discuss how they would assess the probability that an earthquake would destroy their hometown this year. Possible answers: If earthquakes occur often, they could rely on the historical record of earthquakes. If earthquakes are infrequent, they would have to look for geologic evidence of earthquakes in rocks and sediments.

2 Teach

Teacher Content Support

Seismic Waves Refer to **Figure 5.** Because of their longitudinal motion, P-waves shake the ground nearly horizontally. S-waves shake the ground mostly in the vertical direction because of their transverse motion. Neither P- nor S-waves cause much damage because of their small amplitudes and high frequencies. Surface waves, on the other hand, have much larger amplitudes and much lower frequencies, and thus they are capable of causing catastrophic resonance in structures. Also, surface waves shake the ground much longer than body waves do. Love waves have horizontal transverse motion and shake the ground horizontally. Rayleigh waves have both horizontal and vertical motion. Both types of ground motion, especially when they act in concert, are capable of destroying buildings and other structures.

Demonstration

Vertical v. Horizontal Vibrations

Build a small tower with soft-drink cans. Gently tap downward on the top of the tower with a hammer. Students will observe that the tower remains standing. Tap on the side of the bottom can with the same force. The tower will collapse.

Explain that most buildings are constructed to withstand vertical forces, but not horizontal forces. Thus, buildings suffer little damage from vertical vibrations but could collapse as a result of horizontal vibrations.

Teacher Content Support

Vibrations and Resonance

When a guitar string is plucked, it emits a tone with a certain frequency. A nearby guitar string tuned to the same frequency will begin to vibrate by itself in response to the sound waves traveling through the air. This is resonance. Seismic waves can generate the same kind of resonance in structures "tuned" to the same frequency as the seismic waves. All structures have natural frequencies of vibration. Tall buildings sway with a natural period that depends on their heights and other parameters. The higher the building, the longer its natural period of vibration is. Seismic waves with the same period as that of a tall building can cause the building to sway violently and collapse during a quake.

READING CHECK In a pancaking collapse, the walls collapse and the floors fall down, one on top of the other.

Figure 20 One type of damage caused by earthquakes is called pancaking because shaking causes a building's supporting walls to collapse and the upper floors to fall one on top of the other like a stack of pancakes.

Structural failure In many earthquake-prone areas, buildings are destroyed as the ground beneath them shakes. In some cases, the supporting walls of the ground floor fail and cause the upper floors, which initially remain intact, to fall and collapse as they hit the ground or lower floors. The resulting debris resembles a stack of pancakes; thus, the process is called pancaking. This type of structural failure, shown in **Figure 20,** was a tragic consequence of the earthquake in Islamabad, Pakistan, in 2005.

READING CHECK **Explain** what happens when a building pancakes.

Another type of structural failure is related to the height of a building. During the 1985 Mexico City earthquake, for example, most buildings between five and 15 stories tall collapsed or were otherwise completely destroyed, as shown in **Figure 21.** Similar structures that were either shorter or taller, however, sustained only minor damage. The shaking caused by the earthquake had the same frequency of vibration as the natural sway of the intermediate buildings. This caused those buildings to sway the most violently during the earthquake. The ground vibrations, however, were too rapid to affect taller buildings, whose frequency of vibration was longer than those of the earthquake, and too slow to affect shorter buildings, whose frequency of vibration was shorter.

Figure 21 Many medium-sized buildings were damaged or destroyed during the 1985 Mexico City earthquake because they vibrated with the same frequency as the seismic waves.

DIFFERENTIATED INSTRUCTION

Advanced Learners **WARNING:** ***Wear safety goggles during this activity and insist that students do the same.*** To help students understand periods of oscillation, attach a heavy bolt or other weight to a long string. Mark a fulcrum point on the string and, holding the string at the fulcrum and with everyone well out of the way, carefully swing the weight with a relatively small amplitude. Have students calculate the period of oscillation by dividing 60 s by the number of swings back and forth. Repeat with different fulcrum points. Students should record the string length and period for each set and plot the results to conclude the period of a pendulum increases with the square root of its length.

Figure 22 Soil liquefaction happens when seismic vibrations cause poorly consolidated soil to liquefy and behave like quicksand. The buildings pictured here were built on this type of soil and an earthquake caused the buildings to fall over.

J.K. Nakata/USGS Photo Library

Land and soil failure In addition to their effects on structures made by humans, earthquakes can wreak havoc on Earth's landscape. In sloping areas, earthquakes can trigger massive landslides. For example, most of the estimated 30,000 deaths caused by the magnitude-7.9 earthquake that struck in Peru in 1970 resulted from a landslide that buried several towns. In areas with sand that is nearly saturated with water, seismic vibrations can cause the ground to behave like a liquid in a phenomenon called **soil liquefaction** (lih kwuh FAK shun). It can generate landslides even in areas of low relief. It can cause trees and houses to fall over or to sink into the ground and underground pipes and tanks to rise to the surface. **Figure 22** shows tilted buildings that crushed a car when the soil under them liquefied during an earthquake in San Francisco.

READING CHECK **Summarize** how solid ground can take the properties of a liquid.

In addition to determining landslide and liquefaction risks, the type of ground material can also affect the severity of an earthquake in an area. Ground motion is amplified in some soft materials, such as unconsolidated sediments. It is muted in more resistant materials, such as granite. The severe damage to structures in Mexico City during the 1985 earthquake is attributed to the soft sediments on which the city is built. The thickness of the sediments caused them to resonate with the same frequency as that of the surface waves generated by the earthquake. This produced reverberations that greatly enhanced the ground motion and the resulting damage.

Teacher Content Support

Earthquake Damage Earthquakes can cause underground gas, water, and sewer lines to rupture. Therefore, in addition to earthquake damage and fires, the water supply in the epicentral area could become contaminated. Also, vertical ground displacement and the disruption of surface and subsurface drainage can have far-reaching consequences for the environment, such as uplifted coastal areas that can destroy marine habitats, drowned forests, and the loss of water supply to large areas.

READING CHECK Soil liquefaction can result when seismic vibrations affect soil that is saturated with water or poorly consolidated.

Project

North American Earthquakes Have students research and construct a time line that shows all earthquakes worldwide with magnitudes greater than 8 during the past 20 years. Have students indicate with "NA" which of these quakes occurred in North America. Recommend that students use a scale of 12 mm/y (1mm/mo).

EARTH SCIENCE JOURNAL

Earthquakes' Effects In their Earth science journals, have students compare and contrast four earthquakes and their effects: a shallow, magnitude-8 quake at the Mid-Atlantic Ridge; a deep, magnitude-8 quake under Japan; a shallow, magnitude-6 quake in California; and a shallow, magnitude-6 quake in Turkey. Which of the quakes is the most dangerous? The least dangerous? The most likely to generate a tsunami? Why? The California and Turkey quakes are shallow and will generate high intensities in populated areas. The Turkey quake is the most dangerous because of the dense populations. The Japan quake is too deep to generate dangerous intensities at the surface. The Mid-Atlantic Ridge quake is too far from land to affect many people, but it might generate a tsunami. Unless it generates a large tsunami, the ocean-ridge earthquake is probably the least dangerous.

Project

Building Codes Have interested students research the building codes of your community and your school building's compliance with the code. Ask students to find the answers to the following questions: Is your school earthquake-proof? What earthquake hazards exist in and around the school? What would be the expected effects of an intensity VIII seismic event on your school and community? Have students report their findings to the class. Use the findings to conduct an earthquake drill. OL COOP LEARN

Concept Development

Size of Tsunamis Tell students the wavelength of tsunamis are about 100 km long. To give students an idea of this scale, have students use a map to identify a place that is 100 km from your school. OL COOP LEARN

Activity

Height of 2004 Tsunami Have students use metersticks to mark a distance of 10 m on a sidewalk or in the hallway to get a sense of the height of the 2004 Indian Ocean tsunami. BL OL COOP LEARN

Concept Development

Extent of 2004 Tsunami To give students a sense of how far tsunamis can travel, give students a map they can trace. Have students locate Sumatra and mark the epicenter of the 2004 Indian Ocean earthquake. 160 km west of Sumatra Then, have them mark the coasts of Indonesia, Sri Lanka, India, Thailand, and Somalia. BL

■ **Figure 23** A tsunami is generated when an underwater fault or landslide displaces a column of water.

View an **animation of a tsunami.**

Concepts In Motion

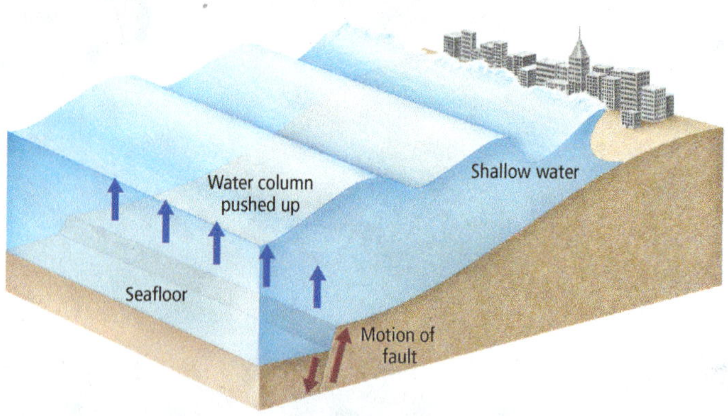

Tsunami Another type of earthquake hazard is a **tsunami** (soo NAH mee)—a large ocean wave generated by vertical motions of the seafloor during an earthquake. These motions displace the entire column of water overlying the fault, creating bulges and depressions in the water, as shown in **Figure 23.** The disturbance then spreads out from the epicenter in the form of extremely long waves. While these waves are in the open ocean, their height is generally less than 1 m. When the waves enter shallow water, however, they can form huge breakers with heights occasionally exceeding 30 m. These enormous wave heights, together with open-ocean speeds between 500 and 800 km/h, make tsunamis dangerous threats to coastal areas both near to and far from a earthquake's epicenter. The Indian Ocean tsunami of December 26, 2004, originated with a magnitude-9.1 earthquake in the ocean about 250 km west of Sumatra. The 10-m-tall tsunami radiated across the Indian Ocean and struck the coasts of Indonesia, Sri Lanka, India, Thailand, Somalia, and several other nations. The death toll from the tsunami exceeded 225,000, making it one of the most devastating natural disasters in modern history. The aftermath of that catastrophic event is shown in **Figure 24.**

VOCABULARY

SCIENCE USAGE V. COMMON USAGE

Column

Science usage: a hypothetical cylinder of water that goes from the surface to the bottom of a body of water

Common usage: a vertical arrangement of items

■ **Figure 24** The destruction from the December 26, 2004, tsunami in the Indian Ocean was not isolated to the shoreline. As seen here, areas inland were devastated by the tsunami, which took at least 225,000 lives.

Demonstration

Tsunamis To demonstrate how motion of the seafloor can change the water surface and produce a tsunami, attach a string to a small plastic bag of rocks. Put the bag at one end of a pan with the string extending to the other side. Put enough water in the pan to cover the bag with 3 cm of water. Make a quick, sharp pull on the string. Have students watch the water surface and identify where, when, and why the waves are generated.

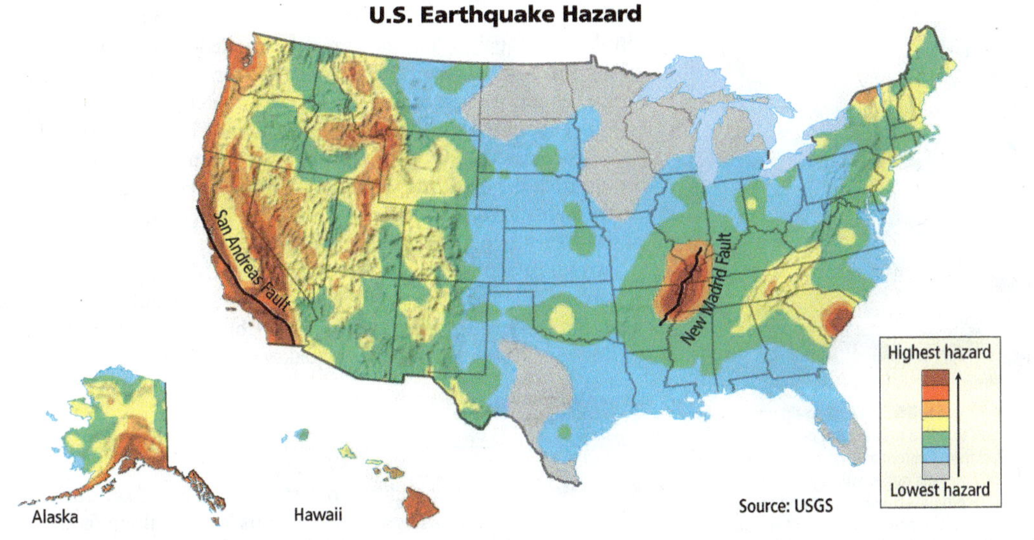

■ **Figure 25** Areas of high seismic risk in the United States include Alaska, Hawaii, and some of the western states.
Locate *the areas of highest seismic risk on the map. Locate your own state. What is the seismic risk of your area?*

Earthquake Forecasting

To minimize the damage and deaths caused by earthquakes, seismologists are searching for ways to forecast these events. There is currently no completely reliable way to forecast the exact time and location of the next earthquake. Instead, earthquake forecasting is based on calculating the probability of an earthquake. The probability of an earthquake's occurrence is based on two factors: the history of earthquakes in an area and the rate at which stress builds up in the rocks.

☑ READING CHECK **Identify** the two factors seismologists use to determine the probability of an earthquake occurring in a certain area.

Seismic risk Recall that most earthquakes occur in long, narrow bands called seismic belts. The probability of future earthquakes is much greater in these belts than elsewhere on Earth. The pattern of earthquakes in the past is usually a reliable indicator of future earthquakes in a given area. Seismometers and sedimentary rocks can be used to determine the frequency of large earthquakes. The history of an area's seismic activity can be used to generate seismic-risk maps. A seismic-risk map of the United States is shown in **Figure 25.** In addition to Alaska, Hawaii, and some western states, there are several regions of relatively high seismic risk in the central and eastern United States. These regions have experienced some of the most intense earthquakes in the past and probably will experience significant seismic activity in the future.

Identify Misconceptions

Many people have the misconception that major earthquakes in the United States occur only in California. While it is true that parts of California have experienced many destructive earthquakes, large parts of other western states have had comparable seismic activity. Also contrary to popular belief, states in the eastern part of the United States are not immune to major earthquakes. In 1886, for example, Charleston, South Carolina, was nearly destroyed by a major quake, and the 1811–1812 earthquakes that originated in New Madrid, Missouri are among the largest earthquakes that have struck the United States.

Uncover the Misconception Ask students where earthquakes occur in the United States.

Demonstrate the Concept Show students the location of the New Madrid Fault and Charleston. Explain that these two areas have had major quakes in 1811–1812 and 1886, respectively.

Assess New Knowledge Have students find out how many earthquakes occurred in the United States outside the California region in the last week or month. If students do not have access to the Internet, you can print a map from the site and either pass out copies or indicate the locations on a classroom map.

EARTH SCIENCE JOURNAL

Seismic Intensity Map Have students research a seismic-intensity map of a recent earthquake. Have them answer the same questions presented in the MiniLab for the earthquake they choose and record them in their Earth science journals. What is the maximum/minimum intensity value? Where was the earthquake's epicenter? AL

■ **Caption Question Fig. 25** The highest seismic risk areas shown on the map are Hawaii, Alaska, the West Coast, and the Midwest; answers will vary depending on your location. For example, in New York it is low, in South Carolina it is high. Help students locate your state on the map, if necessary.

☑ READING CHECK Scientists determine probablity based on the history of an area's earthquakes and the rate of stress buildup in rocks.

Enrichment

Forecast Tools Ask students what they would measure in an area where an earthquake is likely to occur. Suggestions might include monitoring the weather, animals, or plants, to see if they are good predictors (there is no good evidence that they are).

Activity

Earthquake Prediction Give students a list of the dates of significant Parkfield earthquakes. dates are 1857, 1881, 1901, 1922, 1934, 1966, 2004 Have them put all the dates except 2004 into a time line. Ask them if they can predict when the next earthquake should occur. According to this history, the next earthquake should have happened before 1993. Have them add the 2004 event and then lead a discussion about the implications of the 11-year difference between the forecasted date and the earthquake in 2004.

☑ **READING CHECK** The fact that an earthquake already occurred somewhere makes it likely that one will occur again, either in the same spot or near it. Sometimes there is a general pattern to the interval between quakes.

■ **Figure 26** This drilling rig was used to drill a hole 2.3 km deep in Parkfield, California. Once completed, the hole was rigged with instruments to record data during major and minor tremors. The goal of the project was to better understand how earthquakes work and what triggers them. This information could help scientists predict when earthquakes will occur.

Recurrence rates Earthquake-recurrence rates along a fault can indicate whether the fault ruptures at regular intervals to generate similar earthquakes. The earthquake-recurrence rate along a section of the San Andreas fault at Parkfield, California, for example, shows that a sequence of earthquakes of approximately magnitude 6 shook the area about every 22 years from 1857 until 1966. In 1987 seismologists forecasted a 90-percent probability that a major earthquake would rock the area within the next few decades. Several kinds of instruments at the drilling site, shown in **Figure 26,** were installed around Parkfield in an attempt to measure the earthquake as it occurred. In September, 2004, a magnitude-6 earthquake struck. Extensive data were collected before and after the 2004 earthquake. The information obtained will be invaluable for predicting and preparing for future recurrent earthquakes around the world.

☑ **READING CHECK** **Infer** the significance of studying recurrence rates of earthquakes.

Seismic gaps Probability forecasts are also based on the location of seismic gaps. **Seismic gaps** are sections located along faults that are known to be active, but which have not experienced significant earthquakes for a long period of time. A seismic gap in the San Andreas Fault cuts through San Francisco. This section of the fault has not ruptured since the devastating earthquake that struck the city in 1906. Because of this inactivity, seismologists currently forecast that there is a 67-percent probability that the San Francisco area will experience a magnitude-7 or higher earthquake within the next 30 years. **Figure 27** shows the seismic-gap map for a fault that passes through an area of Turkey. Like the San Andreas Fault in California, there is a long history of earthquakes along the major fault shown below.

■ **Figure 27** Earthquakes in 1912 and 1999 happened on either side of Istanbul, a city of 18 million people. The earthquakes around the city leave a seismic gap that indicates that an earthquake is likely to occur in that area.

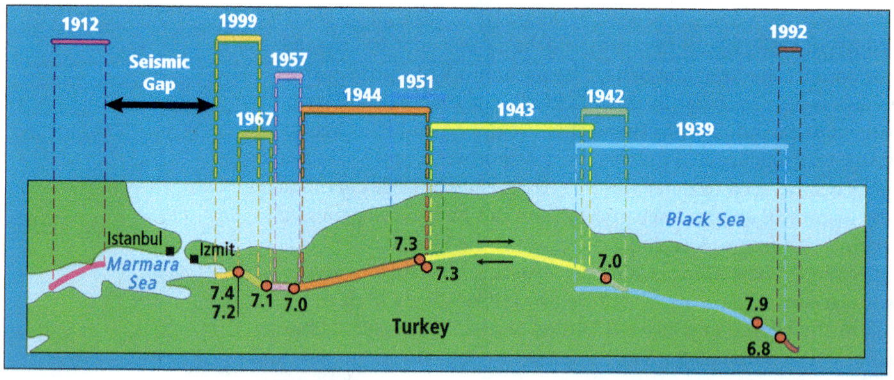

Teacher Content Support

Parkfield, California Located on the San Andreas Fault, Parkfield, California is the most highly monitored fault site in the United States. The Parkfield Experiment, begun in 1985, is a coordinated, long-term project dedicated to understanding the region's earthquakes and, ultimately, to finding ways to accurately predict future events. The dense network of instruments installed to gather information includes many seismometers, GPS stations, creep and strain meters that record deformation, and water-level monitors in wells (one common response to earthquakes is a change in water level in wells). In June 2004, the San Andreas Fault Observatory at Depth (SAFOD) project began drilling a 2- to 3-km hole in order to place instruments directly within the fault zone.

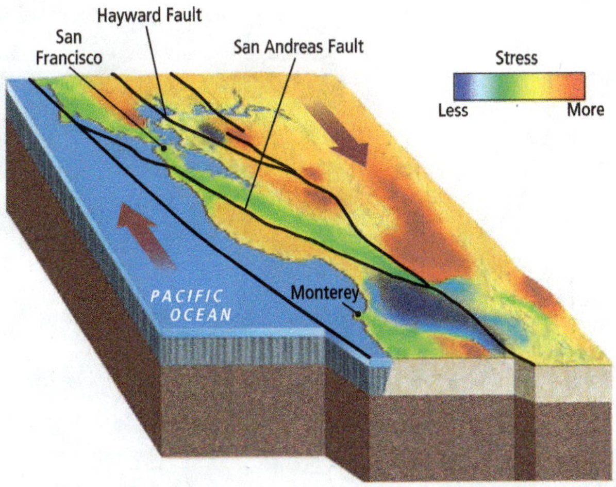

■ **Figure 28** Stress-accumulation maps help scientists determine the probability of an earthquake in any particular place.
Explain ***Why does stress build up in the areas indicated?***

Stress accumulation The rate at which stress builds up in rocks is another factor seismologists use to determine the earthquake probability along a section of a fault. Eventually this stress is released, generating an earthquake. Scientists use satellite-based technology such as GPS to measure the stress that accumulates along a fault. The stress accumulated in a particular part of a fault, together with the amount of stress released during the last earthquake in a particular part of the fault, can be used to develop images like **Figure 28.** Another factor is how much time has passed since an earthquake has struck that section of the fault.

SECTION 4 REVIEW

Section Self-Check

Section Summary

- Earthquake forecasting is based on seismic history and measurements of accumulated stress.
- Earthquakes cause damage by creating vibrations that can shake Earth.
- Earthquakes can cause structural collapse, landslides, soil liquefaction, and tsunamis.
- Seismic gaps are sections along an active fault that have not experienced significant earthquakes for a long period of time.

Understand Main Ideas

1. MAINIDEA **List** some examples of how scientists determine the probability of an earthquake occurring.
2. **Summarize** the effects of the different types of hazards caused by earthquakes.
3. **Draw** before-and-after pictures of what can happen when an earthquake ruptures along a fault.
4. **Summarize** the events that lead to a tsunami.

Think Critically

5. **Assess** where an earthquake is most likely to occur: In the same place that a magnitude-7.5 earthquake occurred 20 years ago or at a location between areas that had earthquakes 20 and 60 years ago, respectively.

WRITING IN ▸ Earth Science

6. Imagine you are on an international aid committee. Write a report suggesting ways to identify areas that are vulnerable to earthquakes.

■ **Caption Question Fig. 28** Answers will vary, but should acknowledge that stress builds up at faults because the crust is moving there.

3 Assess

Check for Understanding

Concept Map Have each student summarize the information presented in this section by constructing a concept map.

Reteach

Summarize Have several volunteers summarize the information presented on structural failure, land and soil failure, tsunamis, seismic risk, and earthquake prediction, and present their summaries to the rest of the class.

Assessment

Performance Have pairs of students research one of the topics from this section and produce an illustrated pamphlet that can be displayed in the classroom and later included in students' portfolios.
COOP LEARN

SECTION 4 REVIEW

1. studying earthquake history; measuring strain accumulation; charting seismic gaps
2. Structural failure and land and soil failure can cause buildings and other structures to collapse. Tsunamis threaten coastal areas.
3. Drawings should show some feature that becomes offset in the "after" picture.
4. vertical movement at a fault on the ocean floor; displacement of the water column; wave travels and becomes huge wave as it reaches shorelines
5. It is more likely to occur in the seismic gap between areas that have already had earthquakes.
6. To identify areas likely to have earthquakes, the committee could commission studies on earthquake histories worldwide and look for seismic gaps.

Rubric

Earth Science & SOCIETY

Purpose

Students will describe the magnitude and science behind the 1906 San Francisco earthquake.

Teacher Content Support

Lawson Report After the 1906 earthquake, Professor Andrew Lawson from the University of California, Berkeley led a statewide government-commissioned investigation into earthquakes in the United States. The commission worked for two years and produced the Lawson report, which described the 1906 earthquake's damage and the movement of the San Andreas fault. The report included photographs of the damage, maps, field findings, and seismograph records of the earthquake. Even today, this report is still regarded as the report that launched the study of earthquakes into the modern era.

Teaching Strategies

- Ask students if they have ever experienced an earthquake. Ask: What was it like? What did you think about while you were in it? What safety measures did you take?
- Ask students if they ever heard about the 1989 Loma Prieta earthquake that occurred in California.
- Tell students the 1989 earthquake was tiny compared to the 1906 San Francisco earthquake.

Earth Science & SOCIETY

Learning from the Past

At 5:15 on a Wednesday morning, most people were still sleeping when an earthquake struck California. The city of San Francisco was the hardest hit. It shook violently for an entire minute, toppling many buildings. In the days that followed, fire devastated entire neighborhoods.

The earthquake leveled the city Modern geologists calculate that the earthquake of April 18, 1906, had an approximate magnitude of 7.9. The total damage to San Francisco involved 490 city blocks—25,000 buildings were destroyed, 250,000 people were left homeless, and approximately 3000 were killed. Streets sank 1 m, bridges collapsed, and people were trapped under buildings.

Fires caused by broken gas lines spread through the city for three days. The efforts of the firefighters were futile because the city's water supply had been destroyed. Contaminated drinking water put the survivors' health at risk. The food supply became limited, disease spread, and looting was rampant.

Scientists analyze the earthquake The 1906 San Francisco earthquake had monumental effect on human life and on the area. Before 1906, scientists knew very little about earthquakes and their effects. This earthquake is considered to be the beginning of modern seismology in the United States.

A theory is proposed At the time of the earthquake, the theory of plate tectonics was not yet understood, so the vast movements of land puzzled scientists. Geologists analyzed the displacement of the crust and the energy released in the movement. They proposed the elastic-rebound theory, which is still used today. They theorized that tensions had been gradually building up in Earth's crust north and south of San Francisco, along a line now known as the San Andreas Fault.

The San Francisco city hall was destroyed during the 1906 earthquake.

The tension accumulated until portions of the crust reached a limit. Like a rubber band that had been stretched too far, portions of the crust snapped. This sudden release of stored energy was the cause of the 1906 earthquake.

Preparing for the future Geologists know that tensions in the crust along the Hayward Fault, the part of the San Andreas Fault where the San Francisco earthquake is thought to have occurred, continue to build as they did before the 1906 earthquake. However, in the past century, scientists and society have worked to prepare for future earthquakes, to predict where they are likely to occur, and to design buildings that can withstand their impacts.

WRITING IN ▶ Earth Science

Earthquake Expedition Create a poster, presentation, or Web site that compares and contrasts the 1906 San Francisco earthquake with the 1989 Loma Prieta earthquake. Explain how scientists can use past events, such as these, to predict the locations and potential hazards of future earthquakes.

WebQuest

WRITING IN ▶ Earth Science

Rubric

Earthquake Expedition Give students the opportunity to share their presentations in teams or in front of the entire class. Provide time for a question-and-answer period.

WebQuest

GeoLAB

iLab Station

Relate Epicenters and Plate Tectonics

Background: The separation of P-waves and S-waves on a seismogram allows you to estimate the distance between the seismic station that recorded the data and the epicenter of that earthquake. If the distance to the epicenter, called epicentral distance, from three or more seismic stations is known, then the exact location of the earthquake's epicenter can be determined. By locating the epicenter on a map of tectonic plate boundaries, you can determine the type of plate movement that caused the earthquake.

Question: *How do seismologists locate the epicenter of an earthquake?*

Materials

U.S. map
map of major tectonic plates
Figure 9
calculator
drafting compass
metric ruler

Procedure

Determine the epicenter location and the time of occurrence of an actual earthquake, using the travel times of P- and S-waves recorded at three seismic stations.

1. Read and complete the lab safety form.
2. The table gives data from three seismic stations. Use the travel-time curves in **Figure 9** and the P-S separation times to determine the distances from the epicenter to each seismic station. Copy the table and enter these distances in the table row *Distance from epicenter*.
3. Obtain a map of North America from your teacher. Accurately mark the three seismic station locations.
4. Use the map scale to determine the distance in centimeters represented by the *Distance from epicenter* calculated in Step 2. Enter these distances in the table row *Map distance*.
5. Use the number calculated in *Map distance* to set the compass point to a spacing that represents the distance from the first seismic station to the epicenter.

Seismic Data

Seismic station	Berkeley, CA	Boulder, CO	Knoxville, TN
P-S separation (min)	3.9	3.6	4.6
Distance from epicenter (km)			
Map distance (cm)			

6. Place the compass point on the seismic station location and draw a circle.
7. Repeat for the other two seismic stations.
8. Mark the point of intersection of the three circles. This is the epicenter of the earthquake.

Analyze and Conclude

1. **Interpret Data** Where is this epicenter located?
2. **Describe** In which major seismic belt did this earthquake occur?
3. **Interpret Data** Use the tectonic plates map to determine which plates form the boundary associated with this earthquake.
4. **Conclude** Describe how tectonic motions caused this earthquake.

WRITING IN ▶ Earth Science

Imagine You are a reporter for the newspaper based near the epicenter of this earthquake. Write an article explaining how geologic processes resulted in this earthquake. Describe whether the earthquake should have been a surprise to the residents, given its location in relation to plate boundaries.

Analyze and Conclude

1. north of Mazatlan, Mexico
2. Circum-Pacific Belt
3. the North American and Pacific Plates
4. The quake was caused by movements along the transform faults that offset the divergent boundary.

WRITING IN ▶ Earth Science

Rubric

Imagine Students' articles should demonstrate an understanding that movement along faults at the boundary between tectonic plates produced the earthquake. Due to the occurrence of previous earthquakes and the location near a plate boundary, residents should not have been surprised that another one occurred.

GeoLAB

iLab Station Rubric

Preparation

Time Allotment 45 min

Process Skills collect and interpret data, compare and contrast, draw a conclusion, use graphs, measure, use numbers, observe and infer, recognize cause and effect

Safety Precautions Approve lab safety forms before work begins. Remind students to use their drafting compasses only as directed.

Preparation of Materials Provide photocopies of a map of North America and a map of the major tectonic plates to students.

Procedure

- Have students work in groups of three or four. Make sure students follow the instructions step by step. Have them ask you about any steps that might be unclear.
- Due to the small map size, the data in the table are not completely accurate. A smaller-scale map would be needed for more accurate data.
- Review how to use a map scale.
- Make sure students understand that each circle must be centered on the appropriate seismic station in order to accurately locate the earthquake's epicenter.
- **Troubleshooting** Some students might not be able to convert length measurements on the seismogram to time. Explain the procedure if necessary.

LabManager™
Customize any lab with the LabManager™ CD-ROM.

MAINIDEAS **Summary statements can be used by students to review the major concepts of the chapter.**

Students can review with these online resources.

Vocabulary Practice

Vocabulary eGames
Vocabulary eFlashcards
Vocabulary PuzzleMaker

Section Self-Check

Chapter Self-Check

Online Test Practice

Use the ExamView® *Assessment Suite* CD-ROM to:

- create multiple versions of tests
- create modified tests with one mouse click
- edit existing questions and add your own questions
- build tests aligned with state standards using built-in state curriculum tags
- change English tests to Spanish with one mouse click
- track students' progress using the Teacher Management System

CHAPTER 19 STUDY GUIDE

BIGIDEA **Earthquakes are natural vibrations of the ground, some of which are caused by movement along fractures in Earth's crust.**

Vocabulary Practice

SECTION 1 Forces Within Earth

VOCABULARY
- stress
- strain
- elastic deformation
- plastic deformation
- fault
- seismic wave
- primary wave
- secondary wave
- focus
- epicenter

MAINIDEA **Faults form when the forces acting on rock exceed the rock's strength.**
- Stress is force per unit of area that acts on a material and strain is the deformation of a material in response to stress.
- Reverse, normal, and strike-slip are the major types of faults.
- The three types of seismic waves are P-waves, S-waves, and surface waves.

SECTION 2 Seismic Waves and Earth's Interior

VOCABULARY
- seismometer
- seismogram

MAINIDEA **Seismic waves can be used to make images of the internal structure of Earth.**
- Seismometers are devices that record seismic wave activity on a seismogram.
- Travel times for P-waves and S-waves enable scientists to pinpoint the epicenters of earthquakes.
- P-waves and S-waves change speed and direction when they encounter different materials.
- Analysis of seismic waves provides a detailed picture of the composition of Earth's interior.

SECTION 3 Measuring and Locating Earthquakes

VOCABULARY
- Richter scale
- magnitude
- amplitude
- moment magnitude scale
- modified Mercalli scale

MAINIDEA **Scientists measure the strength and chart the location of earthquakes using seismic waves.**
- Earthquake magnitude is a measure of the energy released during an earthquake and can be measured on the Richter scale.
- Intensity is a measure of the damage caused by an earthquake and is measured with the modified Mercalli scale.
- Data from at least three seismic stations are needed to locate an earthquake's epicenter.
- Most earthquakes occur in seismic belts, which are areas associated with plate boundaries.

SECTION 4 Earthquakes and Society

VOCABULARY
- soil liquefaction
- tsunami
- seismic gap

MAINIDEA **The probability of an earthquake's occurrence is determined from the history of earthquakes and knowing where and how quickly strain accumulates.**
- Earthquake forecasting is based on seismic history and measurements of accumulated strain.
- Earthquakes cause damage by creating vibrations that can shake Earth.
- Earthquakes can cause structural collapse, landslides, soil liquefaction, and tsunamis.
- Seismic gaps are sections along an active fault that have not experienced significant earthquakes for a long period of time.

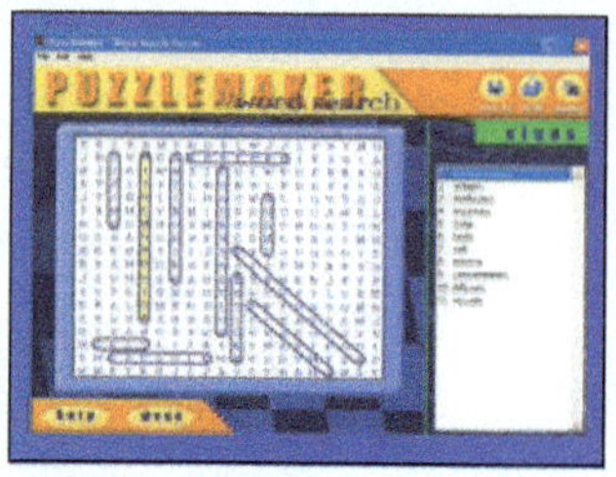

Vocabulary Practice

For additional practice with vocabulary, have students access the Vocabulary PuzzleMaker.

Chapter Self-Check

VOCABULARY REVIEW

Complete the sentences below with the correct vocabulary term from the Study Guide.

1. ________ is the deformation caused by stress.
2. ________ deformation causes a material to bend and stretch.
3. The amount of energy released and the amplitude of seismic waves are measured by the scale known as the ________.
4. ________ happens when seismic vibrations cause subsurface materials to liquefy and behave like quicksand.
5. A travel-time curve shows the relationship between the travel time of a given type of wave and ________.
6. The type of seismic wave that does not pass through the outer core is called a(n) ________.

The sentences below are incorrect. Make each sentence correct by replacing the italicized word with a vocabulary term from the Study Guide.

7. A *fault plane* is a region where earthquakes are expected but none has occurred for a long time.
8. The damage caused by earthquakes is described by the *moment magnitude* scale.
9. An underwater earthquake causes the movement of a column of water, resulting in a *seismic wave*.
10. The recording made by a seismometer is called a *stress-strain curve*.

Distinguish between the vocabulary terms in each pair.

11. epicenter, focus
12. stress, strain
13. plastic deformation, elastic deformation
14. secondary wave, surface wave
15. Richter scale, moment magnitude scale
16. amplitude, magnitude

UNDERSTAND KEY CONCEPTS

17. What is stress?
    A. speed seismic waves travel
    B. point at which rocks fail and generate an earthquake
    C. force per unit area
    D. measure of the deformation of rocks

Use the diagram below to answer Questions 18–20.

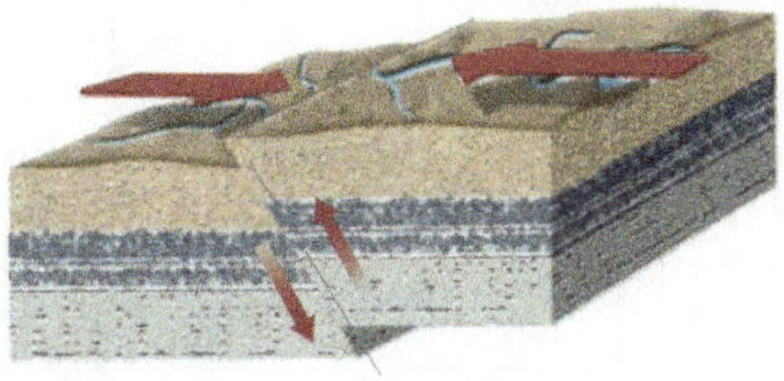

18. Which type of fault is shown?
    A. reverse
    B. normal
    C. shear
    D. strike-slip

19. Which type of force caused this fault to form?
    A. compression
    B. tension
    C. shear
    D. divergent

20. In which direction is the movement in this type of fault?
    A. horizontal
    B. horizontal and vertical
    C. side-to-side
    D. vertical

21. What happens to a rock that undergoes elastic deformation once the stress is removed?
    A. It returns to its original shape.
    B. It breaks to generate an earthquake.
    C. It undergoes plastic deformation.
    D. It does not change shape.

CHAPTER 19 ASSESSMENT

VOCABULARY REVIEW

1. Strain
2. Elastic
3. Richter scale
4. Liquefaction
5. distance from an epicenter
6. S-wave
7. seismic gap
8. modified Mercalli scale
9. tsunami
10. seismogram
11. An epicenter is the point on Earth's surface directly above the focus, which is the point from which body waves originate as a result of rock failure. Surface waves originate from the epicenter.
12. Stress is force acting on a material. Strain is deformation resulting from stress.
13. When stress is removed from a material that has undergone elastic strain, it will return to its original form. When stress is removed from a material that has undergone plastic deformation, it does not return to its original form; it is permanently changed.
14. Secondary waves are slower than P-waves but faster than surface waves. They move at right angles in relation to the direction of the wave, as opposed to a surface wave which moves with a sideways or up and down motion. Secondary waves travel through Earth's interior. Surface waves only travel on Earth's surface.
15. Both are scales measuring the amount of energy released by an earthquake. The Richter scale is based on the amplitude of the largest seismic wave; the moment magnitude scale takes into account the size of fault rupture, the amount of movement along a fault, and rock stiffness.
16. Amplitude is the height of a wave; magnitude is the energy of the largest wave.

UNDERSTAND KEY CONCEPTS

17. C
18. A
19. A
20. B
21. A

22. C
23. A
24. D
25. A
26. B

CONSTRUCTED RESPONSE

27. About 32 times as much energy was released.
28. The wave amplitude was about 10 times greater.
29. Chile, Alaska, Colombia, and Taiwan are all associated with subduction zones. California is associated with transform faults. All the earthquakes were the result of some movement of tectonic plates.
30. Possible answers: California, Oregon, Nevada, Washington, Alaska
31. Both travel over the surface. The surface wave travels through rock (the crust and mantle), the tsunami travels through only the ocean. Tsunamis travel much more slowly than surface waves.
32. With two seismometers, the two circles you draw cross at two locations. So, there are two possible locations for the earthquake, and only one is correct. With three circles, there is only one point of intersection.
33. The direct shaking caused by seismic waves can cause buildings to collapse. Liquefaction, in which the shaking causes soil to flow like a liquid, can cause buildings, roads, and other structures to sink and break apart. Tsunamis, large oceans waves, can travel long distances and have heights of more than 10 m. These cause flooding in coastal areas and sweep people out to sea.

THINK CRITICALLY

34. the past history of recorded earthquakes, their magnitude, the presence of seismic gaps, and measurements of the buildup of strain
35. Faults that are close to the surface create high-intensity earthquakes.
36. Drawings should include a frame, suspended mass, and a recording device.

22. Which type of geologic material is most prone to liquefaction?
A. granite
B. metamorphic rock
C. soil and loose sediment
D. lava flows

Use the figure below to answer Questions 23–25.

Seismogram Recorded in Los Angeles

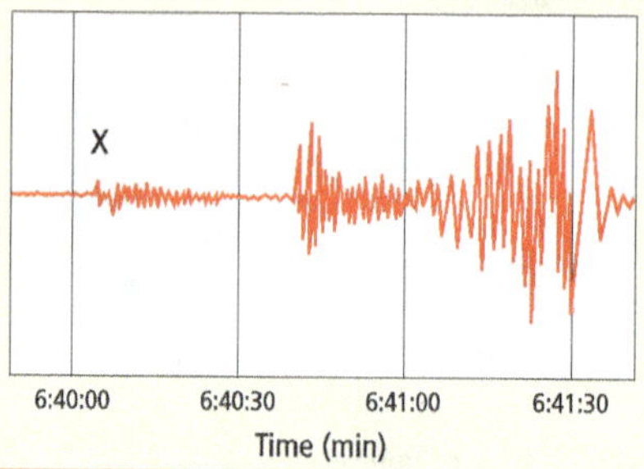

23. Which type of wave is labeled "X"?
A. P-wave
B. S-wave
C. surface wave
D. shear wave

24. At what time did the surface waves arrive at this station?
A. 6:40:00
B. 6:40:05
C. 6:40:33
D. 6:41:10

25. What can the difference in travel times between P- and S-waves be used to determine?
A. how far away the epicenter was
B. the type of fault
C. the depth of the earthquake
D. whether the core is liquid

26. Which seismic hazard is a form of structural failure?
A. tsunami
B. pancaking
C. soil liquefaction
D. seismic gap

CONSTRUCTED RESPONSE

Use the figure below to answer Questions 27–29.

Some Earthquakes in Recent History

Location	Year	Richter Magnitude
Chile	1960	8.5
California	1906	7.9
Alaska	1964	8.6
Colombia	1994	6.8
Taiwan	1999	7.6

27. Calculate How much more energy was released by the Chilean earthquake than the Taiwan earthquake?

28. Approximate How much larger was the amplitude of the waves generated by the Alaskan earthquake than the Taiwan earthquake?

29. Classify the earthquake locations with the type of plate boundary, and suggest how the tectonic processes were probably related.

30. Name five states with high seismic risk.

31. Compare and contrast a tsunami and a surface wave.

32. Explain why scientists need measurements from more than two seismometers to determine the exact location of an earthquake. Make a diagram similar to **Figure 17** to support your answer.

33. Describe three different ways earthquakes can cause damage or cause harm to people.

THINK CRITICALLY

34. Summarize the factors considered when assessing seismic risk.

35. Evaluate how earthquake intensity is related to the type of fault.

36. Draw the basic components of a seismometer.

Use the figure below to answer Questions 37 and 38.

37. **Appraise** the specific type of earthquake damage shown, and propose the possible causes.

38. **Infer** the intensity of the earthquake that caused this damage, using the modified Mercalli scale.

39. **Explain** why there are three different ways to measure the size of earthquakes.

40. **Critique** this statement: If a certain area has not had an earthquake for over a hundred years, it is not likely to ever occur.

41. **Design** a house that would be structurally sound in an earthquake. Label the features, and explain how they would help prevent earthquake damage.

42. **Suggest** cost-effective ways of saving lives in an earthquake in the United States. How might your strategy be different in California and Florida?

CONCEPT MAPPING

43. Use the following terms to complete the concept map: *reverse faults, tension, types of stress, strike-slip faults, compression, shear, causes,* and *normal faults.*

CHALLENGE QUESTION

44. **Explain** why most earthquakes are shallow. Use the concept of plastic deformation and brittle failure and your knowledge about the temperature of Earth's interior.

WRITING IN ▸ Earth Science

45. Imagine you live along an active fault. Write a disaster plan for your school, giving guidelines on what to do before, during, and after an earthquake. Include a list of disaster kit supplies.

DBQ Document–Based Questions

Data obtained from: Fukao Y., S. Widiyantoro, and M. Obayashi. 2001. Stagnant slabs in the upper and lower mantle transition region. *Reviews of Geophysics* 39 (3): 291–323.

The figure below shows a cross section of Earth extending from the surface to the boundary between the core and the mantle. The colors show how the speed of seismic waves differs from the expected value for waves at that depth. This cross section is taken across the subduction zone off the west coast of South America. West is left, and east is right.

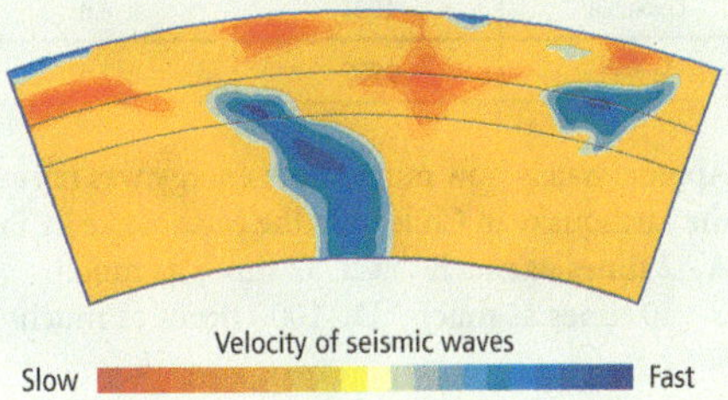

46. What properties of subsurface material could cause seismic waves to move quickly through the blue areas and more slowly through the red areas?

47. Thinking about plate tectonics, what portion of the diagram could represent a subducting plate with molten rock rising from the subduction zone to form volcanoes?

CUMULATIVE REVIEW

48. What is the most common intrusive igneous rock? **(Chapter 5)**

49. Describe three processes that affect the salinity of the oceans. **(Chapter 15)**

37. The soil under the building collapsed, which could be due to soil liquefaction or wave amplification. Because it is a midsize building, it might also have been affected by waves of the same period of vibration.
38. The damage would be VIII on the modified Mercalli scale.
39. The Mercalli scale measures earthquakes without using a seismometer. The Richter and moment magnitude scales are based on ground motion and require measurements. The moment magnitude scale accounts for more aspects of real faults, such as their size and movement across the fault.
40. This statement is false. An area that has had one earthquake is likely to have another one someday.
41. Answers will vary. Possible designs could include reinforced walls or building with wood rather than concrete.
42. Because Florida has a low elevation and is not on a fault, it is at more risk for tsunamis, so a tsunami warning system would most benefit people in Florida. In California, earthquake damage by shaking is more probable, so reinforced roads and buildings would be most helpful.

CONCEPT MAPPING

43. Maps should show how each type of stress creates a particular type of fault.

CHALLENGE QUESTION

44. Rock nearer to the surface are colder and more brittle and so more likely to fail than rocks farther below the surface where higher temperatures cause them to become more plastic and so less likely to fail from stress.

WRITING IN ▸ Earth Science

Rubric

45. Student entries could include guidelines for checking for hazards and collecting supplies before an earthquake, safe places to be during the earthquake, and a meeting place for after the earthquake.

DBQ Document-Based Questions

Data obtained from: Fukao Y., S. Widiyantoro, and M. Obayashi. 2001. Stagnant slabs in the upper and lower mantle transition region. *Reviews of Geophysics* 39 (3): 291–323.

46. Subsurface material that is cold, hard, and dense causes seismic waves to move through it quickly. Subsurface material that is warm and semisolid or molten is less dense, causing seismic waves to move through it slowly.
47. In the center of the diagram is a large, blue, curved area. This area is a subduction zone with cold, hard, dense rocks of a tectonic plate being pulled deep down into the mantle from the surface. To the top right of the blue curved area is a red area shaped somewhat like a star. This area is warm, semisolid or molten rock that has risen up from the melting subducting plate to form volcanoes.

CUMULATIVE REVIEW

48. granite
49. evaporation; sea-ice freezing or melting in the polar regions; freshwater from rivers emptying into the oceans

MULTIPLE CHOICE

1. A
2. C
3. B
4. D
5. D
6. A
7. D
8. C
9. B
10. B

CUMULATIVE STANDARDIZED TEST PRACTICE

MULTIPLE CHOICE

1. Why is only 61 percent of the northern hemisphere covered with water?
   A. Most landmasses are in the northern hemisphere.
   B. Most landmasses are in the southern hemisphere.
   C. The northern hemisphere is colder thus allowing less water to flow there.
   D. Gravity causes more water to settle in the southern hemisphere.

Use the table below to answer Questions 2 and 3.

Some Earthquakes in Recent History		
Location	**Year**	**Richter Magnitude**
Chile	1960	8.5
California	1906	7.9
Alaska	1964	8.6
Colombia	1994	6.8
Taiwan	1999	7.6

2. Approximately how much more energy was released by the earthquake in Chile than the earthquake in Taiwan?
   A. 2 times as much **C.** 32 times as much
   B. 10 times as much **D.** 1000 times as much

3. Approximately how much larger was the amplitude of the waves generated by the earthquake in Alaska than the earthquake in Taiwan?
   A. 2 times as large **C.** 100 times as large
   B. 10 times as large **D.** 1000 times as large

4. Who were the first people to propose the idea that Earth's landmasses at one time were all connected?
   A. explorers **C.** scientists
   B. mathematicians **D.** mapmakers

5. Which does NOT have any effect on Earth's tides?
   A. Earth **C.** the Sun
   B. the Moon **D.** the atmosphere

6. What is formed as turbidity currents drop sediment?
   A. continental rise **C.** abyssal plains
   B. continental slope **D.** deep-sea trenches

Use the maps below to answer Questions 7 and 8.

Ocean currents

Wind currents

7. What can be concluded by comparing the maps?
   A. Surface wind currents flow mostly to the east, and surface ocean currents flow mostly to the west.
   B. Surface wind currents flow mostly to the west, and surface ocean currents flow mostly to the east.
   C. The direction of surface ocean currents is opposite the direction of surface wind currents.
   D. The direction of surface ocean currents is related to the direction of surface wind currents.

8. The Coriolis effect is the rightward curvature of winds in the northern hemisphere and the leftward curvature of winds in the southern hemisphere. What causes the Coriolis effect?
   A. the intersection of warm and cool ocean currents
   B. the revolution of Earth around the Sun
   C. the rotation of Earth on its axis
   D. the seasonal changes in global temperature

9. Most sedimentary rocks are formed by
   A. uplifting and melting
   B. compaction and cementation
   C. eruption of volcanoes
   D. changes deep within Earth

10. Which type of volcano is potentially the most dangerous to humans and the environment?
    A. shield volcano **C.** cinder cone volcano
    B. composite volcano **D.** compact volcano

SHORT ANSWER

Use the illustration below to answer Questions 11 and 12.

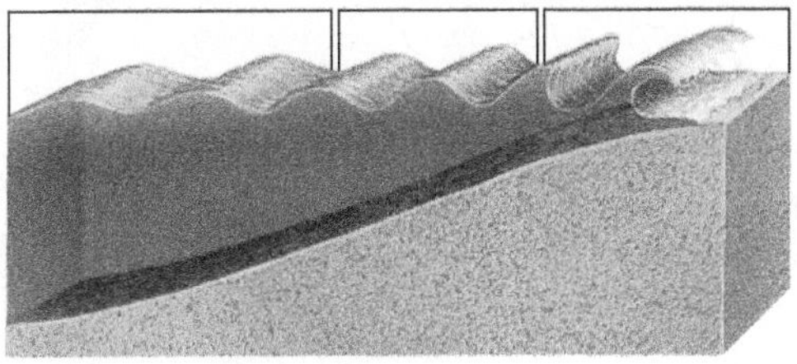

11. Describe the changes that occur as waves move closer to shore.
12. Contrast water movement and energy movement in an ocean wave.
13. What does the location of mountains on the seafloor that are not near any active volcanism suggest?
14. How does convection in the mantle cause plate motions?
15. Describe tephra and its two possible sources.
16. How are the continental shelves, covered with water after the last ice age, now benefiting humans?

READING FOR COMPREHENSION

Earthquake Detection

The belief that animals can detect incoming earthquakes has been around for centuries. In 373 B.C., historians recorded that animals, including rats, snakes, and weasels, deserted the Greek city of Helice just days before an earthquake devastated the place. Similar accounts have surfaced across the centuries since.

Catfish moving violently, chickens that stop laying eggs, and bees leaving their hive in a panic have been reported. But precisely what animals sense is a mystery. One theory is that wild and domestic creatures can feel Earth vibrate before humans can. Other ideas suggest that they detect electric changes in the air or gas released from Earth. Earthquakes are a sudden phenomenon. Seismologists have no way of knowing exactly when or where the next one will hit. An estimated 500,000 detectable earthquakes occur in the world each year. Of those, 100,000 can be felt by humans, and 100 cause damage. Researchers have long studied animals in hopes of discovering what they hear or feel before an earthquake in order to use that sense as a prediction tool. American seismologists are skeptical. Even though there have been documented cases of strange animal behavior prior to earthquakes, according to the USGS, a reproducible connection between a specific behavior and the occurrence of an earthquake has never been made.

Article obtained from: Mott, M. Can animals sense earthquakes? *National Geographic News.* November 11, 2003.

17. What can be inferred from this passage?
    A. Animals can predict earthquakes because they can feel the vibrations before humans.
    B. Animals cannot predict earthquakes.
    C. Further study and research is needed before it can be confirmed or denied that animals can predict earthquakes.
    D. Animals have been predicting earthquakes for centuries.

18. Which was NOT an animal behavior cited as proof that animals can predict earthquakes?
    A. catfish moving violently
    B. chickens laying eggs
    C. bees leaving their hives
    D. snakes deserting a city

NEED EXTRA HELP?																
If You Missed Question . . .	1	2	3	4	5	6	7	8	9	10	11	12	13	14	15	16
Review Section . . .	15.1	19.3	19.3	17.1	15.3	16.2	15.3	12.2	6.1	18.2	15.3	15.3	16.2	17.4	18.2	16.2

SHORT ANSWER

11. As waves move closer to shore, the water becomes shallower, causing the waves to slow down and decreasing their crest-to-crest wavelength. Waves then catch up to one another, increasing the height of the wave until it collapses.

12. As an ocean wave travels, the water moves up and down in a circular motion and then returns to its original position. Only the energy of an ocean wave actually moves forward.

13. This suggests that there once was volcanism in the area to form the mountain, but it ceased a long time ago. Thus, those mountains are really extinct volcanoes.

14. Solid matter in the mantle is heated so that it flows like soft plastic. Convection currents occur with this heating and allow the plates to move.

15. Tephra is rock fragments thrown into the air during a volcanic eruption. It can be composed of solidified magma (lava) or pieces of crust through which magma erupted.

16. These covered continental shelves now provide a habitat for commercially valuable fish. They also have thick sedimentary deposits which are significant sources of oil and natural gas.

READING FOR COMPREHENSION

17. C
18. B

CHAPTER 20 Mountain Building

BIGIDEA **Mountains form through dynamic processes which crumple, fold, and create faults in Earth's crust.**

ESSENTIAL QUESTIONS	NATIONAL STANDARDS	RESOURCES TO ASSESS MASTERY
SECTION 1 **Crust-Mantle Relationships** **1.** How can the elevation distribution of Earth's surface be described? **2.** What is isostasy and how does it pertain to Earth's mountains? **3.** How does Earth's crust respond to the addition and removal of mass? 2.5 sessions 1.25 blocks	UCP.2, UCP.4; D.3	**Entry-Level Assessment** Focus Transparency 60, p. 562 **Progress Monitoring** Caption Question, pp. 562, 563 Section Review, p. 566
SECTION 2 **Orogeny** **1.** What are orogenic processes? **2.** How are the different types of mountains that form along convergent plate boundaries described? **3.** How did the Appalachian Mountains form? 1.5 sessions 0.75 block	UCP.2; D.3	**Entry-Level Assessment** Focus Transparency 61, p. 567 **Progress Monitoring** Caption Question, p. 567 Reading Check, pp. 569, 570 Section Review, p. 573
SECTION 3 **Other Types of Mountain Building** **1.** What are the processes associated with non-boundary mountains? **2.** How are mountain ranges that form along ocean ridges formed? **3.** How do uplifted and fault-block mountains differ? 4 sessions 2 blocks	UCP.2, UCP.4; A.2; D.3; G.3	**Entry-Level Assessment** Focus Transparency 62, p. 574 **Progress Monitoring** Caption Question, p. 574 Section Review, p. 576 **Summative Assessment** Chapter Assessment, p. 581 Exam*View®Assessment Suite* CD-ROM

Go online!
connectED.mcgraw-hill.com

LEVELED RESOURCES	LAB MATERIALS
Science Notebook 20.1 OL **Chapter FAST FILE Resources:** MiniLab Worksheet, p. 84 OL Study Guide, p. 95 BL **Lab Resources:** Laboratory Manual, p. 153 OL **Transparencies:** Section Focus Transparency 60 EL Teaching Transparency 58 OL EL	LaunchLAB p. 560 / **15 min** wood blocks (3), clear plastic container, water, ruler MiniLAB p. 564 / **30 min** 1000-mL beaker, corn syrup, paper or plastic cup
Science Notebook 20.2 OL **Chapter FAST FILE Resources:** Study Guide, p. 96 BL **Lab Resources:** Laboratory Manual, p. 157 OL **Transparencies:** Section Focus Transparency 61 EL Teaching Transparency 59 OL EL	
Science Notebook 20.3 OL **Chapter FAST FILE Resources:** GeoLab Worksheet, p. 85 OL Study Guide, p. 99 BL **Transparencies:** Section Focus Transparency 62 EL Teaching Transparency 60 OL EL	GeoLAB p. 578 / **45 min** metric ruler, sharp pencil, graph paper

LabManager™
Customize any lab with the LabManager™ CD-ROM.

ADDITIONAL RESOURCES

Technology:
- eTeacherEdition
- eStudentEdition
- Classroom Presentation Toolkit
- Exam*View*® *Assessment Suite* CD-ROM
- LabManager™ CD-ROM
- What's EARTH SCIENCE Got To Do With It? DVD
- Weather Classroom Video DVD

Assessment:
- Blueprints for Success
- Chapter Test (Scaffolded)
- Earth Science Standardized Test Practice

Lab Resources:
- Exploring Environmental Problems Laboratory Manual

BL Below Level   OL On Level   AL Advanced Learners   EL English Learners   COOP LEARN Cooperative Learning

CHAPTER 20

LaunchLAB

iLab Station Rubric

How does crust displace the mantle?

Process Skills measure, use numbers, collect and interpret data, draw a conclusion

Safety Precaution Approve lab safety forms before work begins.

Teaching Strategies

- Each group will need one 8 cm × 8 cm × 2 cm softwood block, one 8 cm × 8 cm × 2 cm hardwood block, and one 8 cm × 8 cm × 4 cm softwood block. Provide a balance for the whole class to use.
- Remind students density = mass ÷ volume. Volume is = to length × width × height.
- Stress that depth measurements should be made without pushing down on the blocks.

Procedure

1. Have students read and complete the lab safety form and follow the procedure below.
2. Obtain 3 **wood blocks** from your teacher. Determine the mass, volume, and density of each block. Record all of these values in a data table.
3. Half fill a **clear plastic container** with **water.** Place both of the 2-cm-thick blocks in the container.
4. Using a **ruler,** measure and record how much of each block is above the water surface.
5. Replace the 2-cm-thick blocks with the 4-cm-thick softwood block.
6. Measure and record how much of the block is above the water surface.

CHAPTER 20

Mountain Building

BIGIDEA Mountains form through dynamic processes which crumple, fold, and create faults in Earth's crust.

SECTIONS

1 Crust-Mantle Relationships

2 Orogeny

3 Other Types of Mountain Building

LaunchLAB

iLab Station

How does crust displace the mantle?

Continental and oceanic crust have different densities. Each displaces the mantle. Explore how this happens in this activity.

FOLDABLES® Study Organizer

Mountain Building Processes

Make a half-book using the labels shown. Use it to organize your notes on the processes that form plate boundary and non-plate boundary mountains.

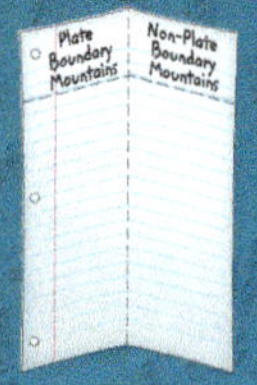

Analysis

1. **Describe** How do density and thickness affect the height of flotation? When density is the same, the thicker block floats higher. When density is different, the higher-density block floats lower.
2. **Infer** Which block represents oceanic crust? Continental crust? The hardwood block represents denser oceanic crust, and the softwood block represents continental crust.

Assessment

Performance Have students predict how the blocks would float in liquids with densities greater than and less than that of water. Have students conduct simple experiments to test their predictions.

LabManager™
Customize any lab with the LabManager™ CD-ROM.

The layers of a mountain record the vast geologic history of the region. Fossils of marine organisms have been found at the top of many mountains, including Mount Everest.

Introduce the BIGIDEA

Mountains Display a physiographic map of the world as you teach this chapter. Begin a discussion about mountains by asking students the following questions: What are the nearest mountains? Where are some other mountains located? Are all mountains the same? How might they be different? How do you think mountains form? You might want to have students record their answers to these questions in their Earth science journals and correct any errors as you teach this chapter.

Teacher Content Support

Orogeny The different, tilted layers visible in the photo are evidence that mountains are the product of many different depositional environments and mountain-building events occurring over millions of years.

SECTION 1

1 Focus

Focus Transparency

Before presenting the lesson, project **Section Focus Transparency 60** and have students answer the accompanying questions. EL

MAINIDEA

Density and Thickness In general, the thicker the crust, the higher the mountains. The lower the density of the crust, the higher the mountains. Ask students where on Earth they think the crust is thickest and the thinnest. Himalayas; oceans Where in the United States is the crust the thickest? Rocky Mountains

2 Teach

Identify Misconceptions

Some students might incorrectly think Earth's crust is too thick to respond to added mass.

Uncover the Misconception

Before students read this section, ask them how they think Earth's crust would respond when mass is added to or removed from it.

Demonstrate the Concept

Push down on a thick block of polystyrene foam floating in an aquarium or a large, clear container of water. Have students describe the change in position of the block as you change the downward force.

Assess New Knowledge

After you complete this section, have students describe a situation in which Earth's crust has isostatically adjusted to changes in load.

SECTION 1

Crust-Mantle Relationships

MAINIDEA The height of mountains is controlled primarily by the density and thickness of the crust.

Essential Questions

- How can the elevation distribution of Earth's surface be described?
- What is isostasy and how does it pertain to Earth's mountains?
- How does Earth's crust respond to the addition and removal of mass?

Review Vocabulary

equilibrium: a state of balance between opposing forces

New Vocabulary

topography
isostasy
root
isostatic rebound

EARTH SCIENCE 4 YOU When you sit in an inflatable raft, your mass causes the raft to sink deeper into the water. When you get out of the raft, it rises. Similarly, when mountains erode, the crust rises to compensate for the mass that is removed.

Earth's Topography

When you look at a globe or a map of Earth's surface, you immediately notice the oceans and continents. From these representations of Earth, you can estimate that about 71 percent of Earth's surface is below sea level, and about 29 percent lies above sea level. What is not obvious from most maps and globes, however, is the variation in elevations of the crust, which is referred to as its **topography.** Recall that topographic maps show an area's hills and valleys. When a very large map scale is used, such as the one in **Figure 1,** the topography of Earth's entire crust can be shown. When Earth's topography is plotted on a graph such as **Figure 2,** a pattern in the distribution of elevations emerges. Note that most of Earth's elevations cluster around two main ranges of elevation. Above sea level, elevation averages around 0 to 1 km. Below sea level, elevations range between –4 and –5 km. These two ranges dominate Earth's topography and reflect the basic differences in density and thickness between continental and oceanic crust.

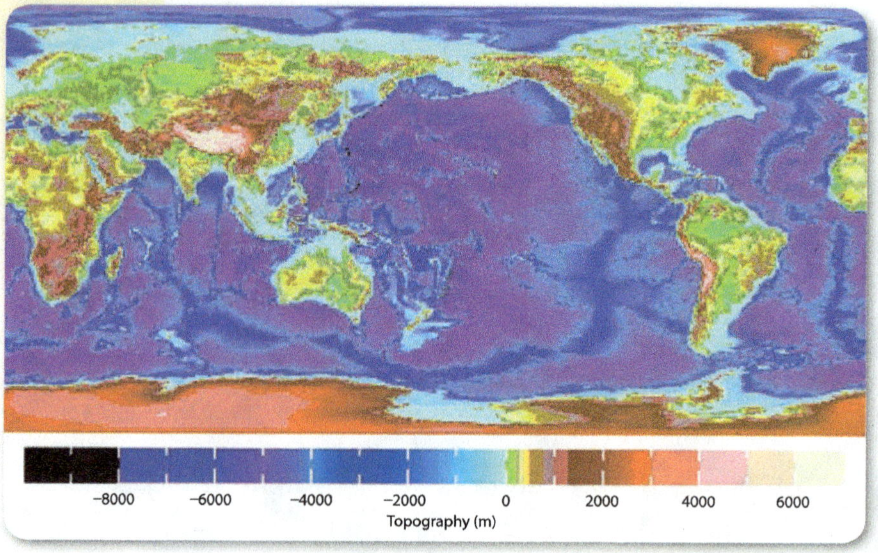

Figure 1 Topographic maps show differences in elevation on Earth's surface.
***Interpret** the map to determine Earth's highest and lowest elevations. Where are they?*

Caption Question Fig. 1 The lowest point on Earth is the Mariana Trench, in the Pacific Ocean; the highest is Mount Everest, in Asia.

DIFFERENTIATED INSTRUCTION

Advanced Learners Have students research the elevation distribution of Mars and Venus and compare them to that of Earth. Have students hypothesize whether crustal isostasy plays any role on Mars and Venus. Scientists hypothesize that Venus might still be experiencing crustal rebound. Mars, however, being smaller, has cooled to the point at which its crust is too thick to rebound.

Continental crust You observed in the Launch Lab that blocks of wood with different densities displaced different amounts of water, and thus floated at various heights above the surface of the water. You observed that blocks of higher density displaced more water than blocks of lower density. Recall that oceanic crust is composed mainly of basalt, which has an average density of about 2.9 g/cm^3. Continental crust is composed of more granitic rock, which has an average density of about 2.8 g/cm^3. The slightly higher density of oceanic crust causes it to displace more of the mantle—which has a density of about 3.3 g/cm^3—than the same thickness of continental crust.

Differences in elevation, however, are not caused by density differences alone. Also recall from the Launch Lab that when the thicker wood block was placed in the water, it displaced more water than the other two blocks. However, because of its density, it floated higher in the water than the hardwood block. Continental crust, which is thicker and less dense than oceanic crust, behaves similarly. It extends deeper into the mantle because of its thickness, and it rises higher above Earth's surface than oceanic crust because of its lower density, as shown in **Figure 3.**

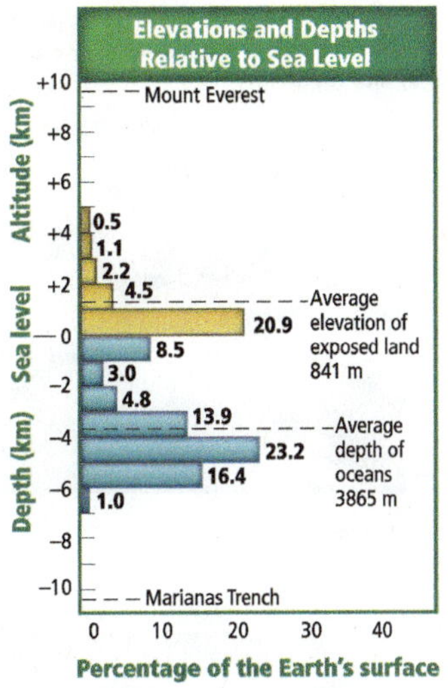

Figure 2 About 29 percent of Earth is land and 71 percent is water.
Interpret *At what elevation does most of Earth's surface lie? At what depth?*

Isostasy

The displacement of the mantle by Earth's continental and oceanic crust is a condition of equilibrium called **isostasy** (i SAHS tuh see). The crust and mantle are in equilibrium when the downward force of gravity on the mass of crust is balanced by the upward force of buoyancy that results from displacement of the mantle by the crust. This balance might be familiar to you if you have ever watched people get in and out of a small boat. As the people boarded the boat, it sank deeper into the water. Conversely, as the people got out of the boat, it displaced less water and floated higher in the water. A similar sinking and rising that results from the addition and removal of mass occurs within Earth's crust. Gravitational and seismic studies have detected thickened areas of continental material, called **roots,** that extend into the mantle below Earth's mountain ranges.

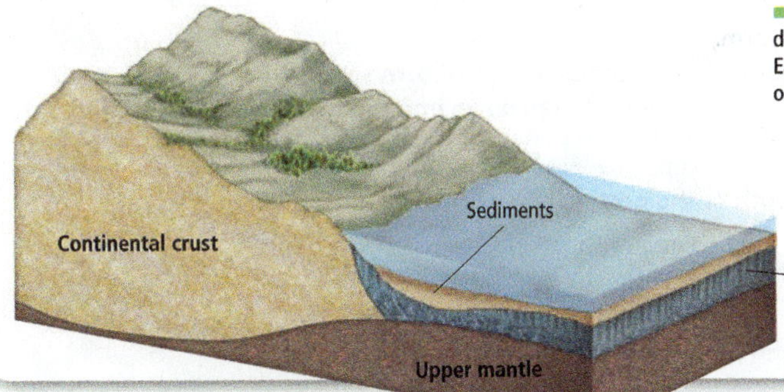

Figure 3 Continental crust is thicker and less dense than oceanic crust, so it extends higher above Earth's surface and deeper into the mantle than oceanic crust.

Teacher Content Support

Glacioisostatic Rebound If Earth scientists could push down on a section of Earth's crust and watch it rebound, they could learn a great deal about isostasy. At first thought, such an experiment seems impossible. However, the last several episodes of glaciation have provided the basis for such an experiment. The weight of continental ice sheets such as those that covered much of Earth during the last ice age depressed the crust significantly. At the close of a glacial episode, the excess weight on the crust is rapidly removed as ice sheets melt. This rebounding of the crust is known as glacioisostatic rebound. Glacioisostatic rebound is currently occurring in parts of Scandinavia and Canada. The ice sheets that once covered these areas melted about 10,000 years ago, causing the crust in these regions to rebound. Measurements show the crust in the northern Baltic Sea area is rising 1 cm/y. Such a rate can impact harbors, lakes, and river drainage patterns over several hundred years.

Interpret the Graph

Elevation Have students refer to the graph in **Figure 2** to answer the following questions: What percentage of Earth's surface is higher than 1 km in elevation? about 6 percent What percentage of Earth's surface is 5 km below sea level? about 17 percent

Caption Question Fig. 2 Two elevations dominate Earth's surface: 0–1 km above sea level and 4 to 5 km below sea level.

Demonstration

Isostasy and Viscosity You can demonstrate what the effect of isostatic rebound would be if the viscosity of Earth's mantle were greater than it is now with two small blocks of wood, a beaker 3/4 full of honey, and a beaker 3/4 full of water. Have students determine which setup represents the viscosity of Earth's mantle and which represents the hypothetical, more viscous mantle. **WARNING:** ***Wipe up any spills immediately to prevent slipping or falling.***

Tie to Previous Knowledge

Uplift v. Erosion Active mountains can be thought of as battling between uplifting forces and the agents of erosion. Have students recall what they have learned about erosion, weathering, and mass movement. Ask them to list some of the processes that wear down mountains.

Concept Development

Rebound Ask students: What do you think happens when mass is removed from a mountain or mountain range? Students should be able to deduce that the mountain or mountain range will rebound when mass is removed.

■ **Figure 4** According to the principle of isostasy, parts of Earth's crust rise or subside until they are buoyantly supported by their roots.

View an **animation of isostasy.**

Concepts In Motion

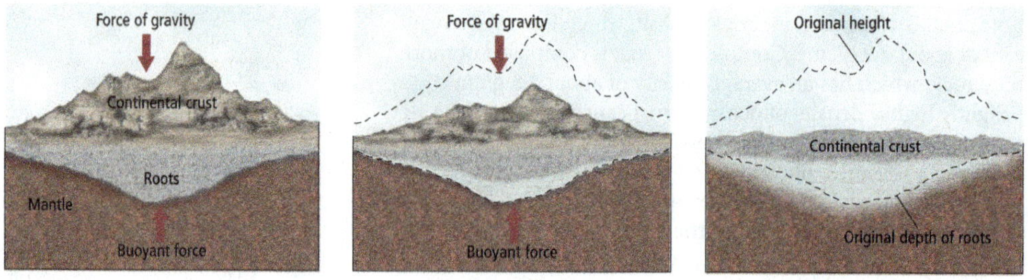

Massive roots underlie mountains.

As erosion takes place, the mountain loses mass. The root rises in response to this decrease in mass.

When the mountain erodes to the average continental thickness, both root and mountain are gone.

Mountain roots A mountain range requires large roots to counter balance the enormous mass of the range above Earth's surface. **Figure 4** illustrates how, according to the principle of isostasy, parts of the crust rise or subside until these parts are buoyantly supported by their roots. Continents and mountains are said to float on the mantle because they are less dense than the underlying mantle. They project into the mantle to provide the necessary buoyant support. What do you think happens when erosion causes mass to be removed from a mountain or mountain range? If mass is removed from a mountain, the roots will rise in response. If erosion continues, the mountain will eventually disappear, exposing the roots.

MiniLAB

iLab Station

Model Isostatic Rebound

How can isostatic rebound be measured? Isostatic rebound is the process through which the underlying material rises when the overlying mass is removed.

Procedure

1. Read and complete the lab safety form.
2. Working in groups, fill a **1000-mL beaker** with about 700 mL of **corn syrup.**
3. Using a pencil, push a **paper or plastic cup** (open side up) down into the syrup far enough so the cup is three-fourths of the way to the bottom of the syrup. Record the depth of the bottom of the cup relative to the surface, then let go of the cup.
4. At 5-s intervals, record the new depth of the bottom of the cup.

Analysis

1. **Describe** In which direction did the cup move? Why?
2. **Explain** why the speed of the cup changes as it moves.
3. **Infer** If enough time passes, the cup stops moving. Why?

MiniLAB

iLab Station Rubric

Purpose Students will witness and record the rate of an isostatic rebound reaction.

Process Skills collect and organize data, make and use graphs, infer

Safety Precaution Approve lab safety forms before work begins.

Teaching Strategies

- This lab can be done as a demo if time or materials are limited.
- Students should make a graph of the measurements taken at the 5-second intervals.
- Instruct students to quickly clean up any spills.

Expected Results Students should see that pushing the cup down increases its buoyancy and the cup stops moving when equilibrium is reached.

Analysis

1. The cup rises because gravity and buoyancy are trying to reach equilibrium.
2. The closer the bottom of the cup is to the surface, the more slowly it rises. The further the cup is below the surface, the greater the buoyancy force causing it to rise, and the faster it rises.
3. Once the cup reaches isostatic equilibrium, the forces causing the cup to move disappear.

Assessment

Knowledge Tell students the center of Canada is rising nearly 1 cm/year. Ask them what might account for that and what the process is called. Large ice sheets covering North America melted; isostatic rebound

Figure 5 Before erosion, the Appalachian Mountains were thousands of meters taller than they are now. Because of isostatic rebound, as the mountains eroded, the deep root also rose thousands of meters closer to the surface. The mountains visible today are only the roots of an ancient mountain range. They too are being eroded and will someday resemble the craton in northern Canada.

Isostasy and Erosion

The Appalachian Mountains, shown in **Figure 5,** in the eastern United States formed hundreds of millions of years ago when the North American continent collided with Europe and Africa. Rates of erosion on land are such that these mountains should have been completely eroded millions of years ago. Why, then, do these mountains still exist? As the mountains rose above Earth's surface, deep roots formed until isostatic equilibrium was achieved and the mountains were buoyantly supported. As peaks eroded, the mass decreased. This allowed the roots themselves to rise and erode.

A balance between erosion and the decrease in the size of the root will continue for hundreds of millions of years until the mountains disappear and the roots are exposed at the surface. This slow process of the crust's rising as the result of the removal of overlying material is called **isostatic rebound.** Erosion and rebound allows metamorphic rocks formed at great depths to rise to the top of mountain ranges such as the Appalachians.

Problem-Solving LAB

Make and Use a Graph

Can you get a rebound? The rate of isostatic rebound changes over time. An initially rapid rate often declines to a very slow rate. The data shown indicates rebound after the North American ice sheet melted 10,000 years ago.

Analysis

1. Plot a graph with *Years before present* on the *x*-axis and *Total amount of rebound* on the *y*-axis.
2. **Describe** how the rate of isostatic rebound decreases with time by studying your graph.

Think Critically

3. **Identify** the percentage of the total rebound that occurred during the first 2000 years.
4. **Predict** how much rebound will still occur and approximately how long this will take.
5. **Compare and contrast** mountain erosion to glaciation in terms of isostatic rebound.

Data and Observations

Isostatic Rebound Data					
Years before present	8000	6000	4000	2000	0
Total amount of rebound (m)	50	75	88	94	97

Problem-Solving LAB

Purpose Students will determine how the rate of isostatic rebound changes over time.

Process Skills recognize cause and effect, interpret data, make and use graphs

Teaching Strategies

- Tell students the data represent rebound since the end of the last ice age. The ice melted 10,000 years ago, and rebound is measured in relation to sea level (0 m).
- Students' graphs should show exponential curves that rise steeply and level off between 2000 and 0 years before present.

Analysis

1. Graphs should have 10,000 years before present plotted at the far left end of the *x*-axis, and 0 years plotted on the right.
2. The rate of rebound decreases by about one-half every 2000 years.

Think Critically

3. Rebound between 10,000 and 8000 years ago was 50 m. This is about 50 percent of the total rebound.
4. Extrapolating the curve will result in an estimated 6–8 m of additional rebound, which could take another 6000 to 8000 years.
5. Erosion would produce rebound; glaciation would depress the crust.

Reinforcement

Isostatic Rebound Ask students: Would the rate of isostatic rebound be slower or faster if the viscosity of the upper mantle were less than it is now? The rate would be faster.

Teacher Content Support

North American Rebound Continental glaciers covered large regions of North America during the Pleistocene Epoch. The weight of the 3-km thick mass of ice caused the underlying crust to subside. Scientists estimate as much as 330 m of rebound has occurred in the Hudson Bay area since the ice sheets melted.

The loss of mass resulting from the melting of continental glaciers during the last ice age is causing the crust in the Great Lakes region to rebound. The rebound has caused an increase in shoreline erosion along the southern margins of the lakes and is threatening coastal communities.

3 Assess

Check for Understanding

Reinforcement Inform students the Mississippi River has deposited thousands of meters of sediment into the Gulf of Mexico. Ask students what effect this has had on Earth's offshore crust. The mass of the sediment has depressed the crust. Ask students whether this would cause sea level to rise or sink in relation to the shoreline. Sea level would rise in relation to the shoreline.

Reteach

Sequence Have each student draw three diagrams that show blocks of wood with different thicknesses floating in water. Have them exchange drawings with a partner who should sequence the diagrams to show how a mountain and its roots change as the result of erosion.

Assessment

Infer Have students describe at least two scenarios that would cause Earth's crust to isostatically subside deeper into the mantle. Possible answers include the formation of volcanoes, the accumulation of large volumes of sediment, and the formation of continental ice sheets.

■ **Figure 6** Mount Everest, a peak in Asia, is currently the highest mountain on Earth. A deep root supports its mass. Scientists have determined that Mount Everest has a root that is nearly 70 km thick.

Seamounts Crustal movements resulting from isostasy are not restricted to Earth's continents. They can also occur in oceanic crust. For example, recall that hot spots under the ocean floor can produce a chain of individual volcanic mountains. When these mountains are underwater, they are called seamounts. On the geologic time scale, these mountains form very quickly. What do you think happens to the seafloor after these seamounts form? The seamounts are added mass. As a result of isostasy, the oceanic crust around these peaks displaces the underlying mantle until equilibrium is achieved.

Deep roots You have learned that the elevation of Earth's crust depends on the thickness of the crust as well as its density. You also learned that a mountain peak is countered by a root. Mountain roots can be many times as deep as a mountain is high. Mount Everest, shown in **Figure 6,** towers nearly 9 km above sea level and is the tallest peak in the Himalayas. Some parts of the Himalayas are underlain by roots that are nearly 70 km thick. As India continues to push northward into Asia, the Himalayas, including Mount Everest, continue to grow in height. Currently, the combined thickness is approximately equal to 868 football fields lined up end-to-end. Where do the immense forces required to produce such crustal thickening originate? You will read about these forces in Section 2 of this chapter.

SECTION 1 REVIEW

Section Self-Check

Section Summary

- The majority of Earth's elevations are either 0 to 1 km above sea level or 4 to 5 km below sea level.
- The mass of a mountain above Earth's surface is supported by a root that projects into the mantle.
- The addition of mass to Earth's crust depresses the crust, while the removal of mass from the crust causes the crust to rebound in a process called isostatic rebound.

Understand Main Ideas

1. MAINIDEA **Relate** density and crustal thickness to mountain building.
2. **Describe** the pattern in Earth's elevations, and explain what causes the pattern in distribution.
3. **Explain** why isostatic rebound slows down over time.
4. **Infer** why the crust is thicker beneath continental mountain ranges than it is under flat-lying stretches of landscape.

Think Critically

5. **Apply** the principle of isostasy to explain how the melting of the ice sheets that once covered the Great Lakes has affected the land around the lakes.
6. **Consider** how the term *root* applies differently to mountains than it does to plants.

MATH IN ▶ Earth Science

7. Suppose a mountain is being uplifted at a rate of 1 m every 1000 y. It is also being eroded at a rate of 1 cm/y. Is this mountain getting larger or smaller? Explain.

SECTION 1 REVIEW

1. Because crust has a lower density than the mantle, as the crust gets thicker, mountains get higher.
2. Most of Earth's elevations are either 0–1 km above or 4–5 km below sea level, reflecting density and thickness differences between continental and oceanic crust.
3. As erosion removes mass from a mountain, the balance between buoyancy and gravity changes.
4. A bigger root can buoyantly support the greater mass of the mountain.
5. As the weight of the ice sheet lessened, the Great Lakes region rebounded.
6. A plant root grows into soil to gather nutrients and provide stability; a mountain root supports buoyantly by projecting into the mantle.
7. The rate of erosion, 1 cm/y, equals 1 m/100 yrs, which is ten times faster than the rate of uplift, 1 m/1000 yrs. The mountain is getting smaller.

SECTION 2

Essential Questions

- What are orogenic processes?
- How are the different types of mountains that form along convergent plate boundaries described?
- How did the Appalachian Mountains form?

Review Vocabulary

island arc: a line of islands that forms over a subducting oceanic plate

New Vocabulary

orogeny
compressive force

Orogeny

MAINIDEA Convergence causes the crust to thicken and form mountain belts.

EARTH SCIENCE 4 YOU **When you push a snowplow or shovel through snow, the compression creates a thick pile of snow. Similarly, the compression caused by plate tectonics thickens the crust to form mountains.**

Mountain Building at Convergent Boundaries

Orogeny (oh RAH jun nee) refers to all processes that form mountain ranges. You have learned about many of these processes. Recall that metamorphism can cause rocks to be squeezed and folded, and that rising magma can form igneous intrusions or erupt at Earth's surface. You have also learned about movement along faults. The result of all these processes can be broad, linear regions of deformation that you know as mountain ranges, but in geology are also known as orogenic belts. Look at **Figure 7** and recall what you have read about the interaction of converging tectonic plates at their boundaries. Most orogenic belts are associated with convergent plate boundaries. Here, **compressive forces** squeeze the crust and cause intense deformation in the form of folding, faulting, metamorphism, and volcanism. In general, the tallest and most varied orogenic belts form at convergent boundaries. However, interactions at each type of convergent boundary create different types of mountain ranges.

■ **Figure 7** Most of Earth's mountain ranges (blue and red peaks on the map) formed along plate boundaries.
Identify *the mountain ranges that lie along the South American Plate by comparing a world map with the one shown here.*

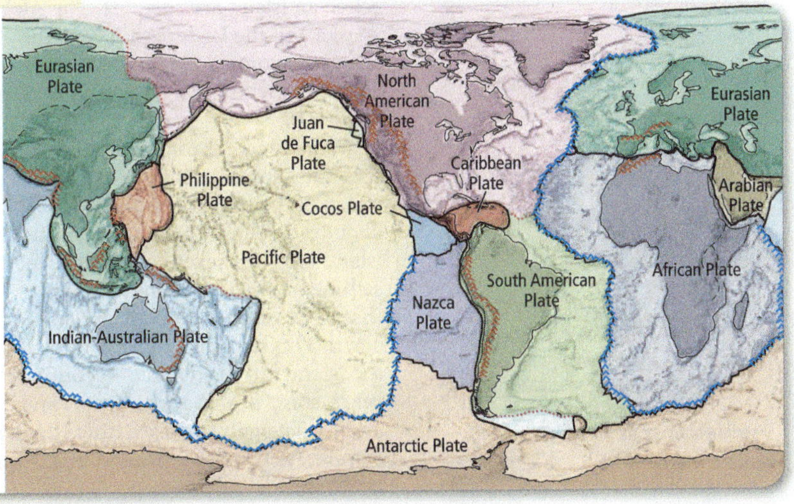

■ **Caption Question Fig. 7** the Andes mountain range on the western edge, and the Mid-Atlantic Ridge between the South American and African plates

SECTION 2

1 Focus

Focus Transparency

Before presenting the lesson, project **Section Focus Transparency 61** and have students answer the accompanying questions. EL

MAINIDEA

Convergent Mountains Ask students: What are some of the effects of converging plates? Volcanism occurs at continental–oceanic plate convergence; mountains rise at continental–continental plate convergence.

2 Teach

Use Science Terms

Orogeny Have students use their dictionaries to find the meanings of the word parts that make up the word *orogeny.* Students should find *oro* means *mountain* and *geny* means *production.* OL EL

Teacher Content Support

Orogeny Students might have the misconception that mountains are built during a single, continuous episode and are subsequently worn away by erosion. Most mountains and mountain ranges go through multiple stages of formation, or orogenesis. A rugged mountain range, for example, can form and be partially eroded. Many years later, renewed uplift can restore the mountain range's youthful, jagged stage. This process is called rejuvenation. The main orogenesis of the Rocky Mountains occurred during the Late Mesozoic Era and the Early Tertiary Period. The heights and shapes of the mountains we see today, however, are the result of rejuvenation caused by renewed uplift during the last 15 to 20 million years.

Interpret the Illustration

Oceanic-Oceanic Convergence Have students refer to **Figure 8** as you read aloud the first full paragraph on the student page. Pause after each sentence so students can locate on the figure each of the features (oceanic plates, mantle, subduction zone, subducted plate, magma, volcanic peaks, island arc complex) discussed. BL OL EL

Tie to Previous Knowledge

Andesite/Andes The rock andesite got its name from the Andes Mountains. Have students recall what they have learned about igneous rocks. Ask them to classify andesite and state its composition. Andesite is an extrusive rock with an intermediate composition between those of basalt and granite.

Tie to Previous Knowledge

Basaltic Magmas Remind students that minerals in rocks have different melting points. Then ask students how partial melting of peridotite can result in the formation of basaltic magmas. Peridotite is an ultramafic rock. When peridotite partially melts, the silica-rich fractions melt first to generate magma that contains fewer mafic minerals. The more siliceous magma then solidifies to form basalt.

CAREERS IN EARTH SCIENCE

WebQuest

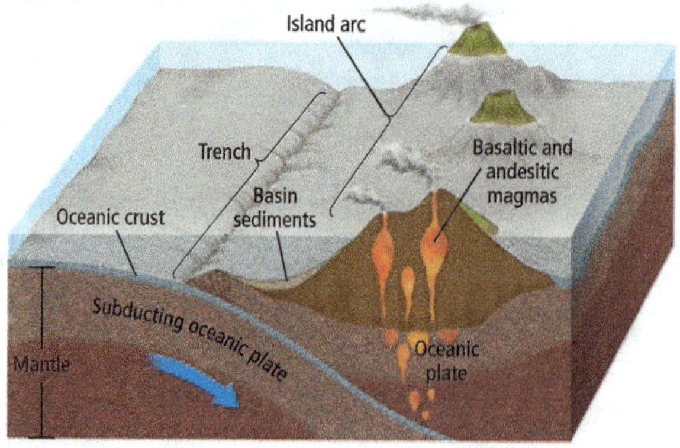

Lesser Antilles island arc

Mount Mazinga in the Lesser Antilles

■ **Figure 8** Convergence between two oceanic plates results in the formation of individual volcanic peaks that make up an island arc complex. Mount Mazinga is one of several volcanic peaks that make up the island arc complex in the southern Caribbean known as the Lesser Antilles.

View an **animation of island arc formation.**

Concepts In Motion

Oceanic-oceanic convergence Recall that when an oceanic plate converges with another oceanic plate, one plate descends into the mantle to create a subduction zone. As parts of the mantle above the subducting plate melt, magma is forced upward where it can form a series of individual volcanic peaks that together are called an island arc complex. The Aleutian Islands off the coast of Alaska and the Lesser Antilles in the Caribbean are examples of island arc complexes. The tectonic relationships and processes associated with oceanic-oceanic convergence are detailed in **Figure 8.**

What kinds of rocks make up island arc complexes? Often, they are a jumbled mixture of rock types. They are partly composed of the basaltic and andesitic magmas that you have read about previously. In addition to these volcanic rocks, some large island arc complexes contain sedimentary rocks. How do these sedimentary rocks eventually become part of a mountain? Recall that between an island arc and a trench is a depression, called a basin. This basin fills with sediments that have been eroded from the island arc. If subduction continues for tens of millions of years, some of these sediments can be uplifted, folded, faulted, and thrust against the existing island arc. This ultimately forms complex new masses of sedimentary and volcanic rocks. Parts of Japan formed in this way.

CAREERS IN EARTH SCIENCE

Petrologist A petrologist studies the composition and formation of rocks. Generally, petrologists specialize in a particular type of rock: igneous, sedimentary, or metamorphic.

WebQuest

Teacher Content Support

The Aleutian Arc The Aleutian Island Arc, which extends nearly 4000 km westward from the Alaskan range and contains 40 historically active volcanoes, represents one of the largest oceanic subduction zones on Earth. It is unique in that it involves an oceanic-oceanic convergent boundary to the west and an oceanic-continental convergent boundary to the east. During the Late Cretaceous Period, the former Kula Plate began to subduct beneath the North American Plate and was eventually entirely subducted into the mantle. When the trailing edge of the Kula Plate—the Kula Ridge—started subducting, because it was warm and thus buoyant, it resisted subduction and jammed the subduction zone. The Pacific Plate, however, continued to move northwest against this plugged subduction zone and eventually began the subduction under the North American Plate that continues today.

Oceanic-continental convergence Oceanic-continental boundaries are similar to oceanic-oceanic boundaries in that convergence along both creates subduction zones and trenches. Unlike convergence at oceanic-oceanic boundaries, convergence between oceanic and continental plates produces mountain belts that are much bigger and more complicated than island arc complexes. When an oceanic plate converges with a continental plate, the descending oceanic plate forces the edge of the continental plate upward. This uplift marks the beginning of orogeny. In addition to uplift, compressive forces can cause the continental crust to fold and thicken. As the crust thickens, higher mountains form. Deep roots develop to support these enormous masses of rocks.

Recall that volcanic mountains can form over the subducting plate. As illustrated in **Figure 9,** sediments eroded from such volcanic mountains can fill the low areas between the trench and the coast. These sediments, along with ocean sediments and material scraped off the descending plate, are shoved against the edge of the continent to form a jumble of highly folded, faulted, and metamorphosed rocks. The metamorphosed rocks shown in **Figure 9** are from Cwm Tydu, Cardigan Bay, Wales. They formed when the landmass that is now the United Kingdom collided with the North American Plate millions of years ago.

☑ READING CHECK **Compare** convergence at oceanic-continental boundaries with convergence at oceanic-oceanic boundaries.

VOCABULARY
SCIENCE USAGE V. COMMON USAGE
Uplift
Science usage: to cause a portion of Earth's surface to rise above adjacent areas

Common usage: to improve the spiritual, social, or intellectual condition

■ **Figure 9** At an oceanic-continental boundary, compression causes continental crust to fold and thicken. Igneous activity and metamorphism are also common along such boundaries. This uplifted outcrop of metamorphosed rock formed as the result of convergence of an oceanic plate with a continental plate.

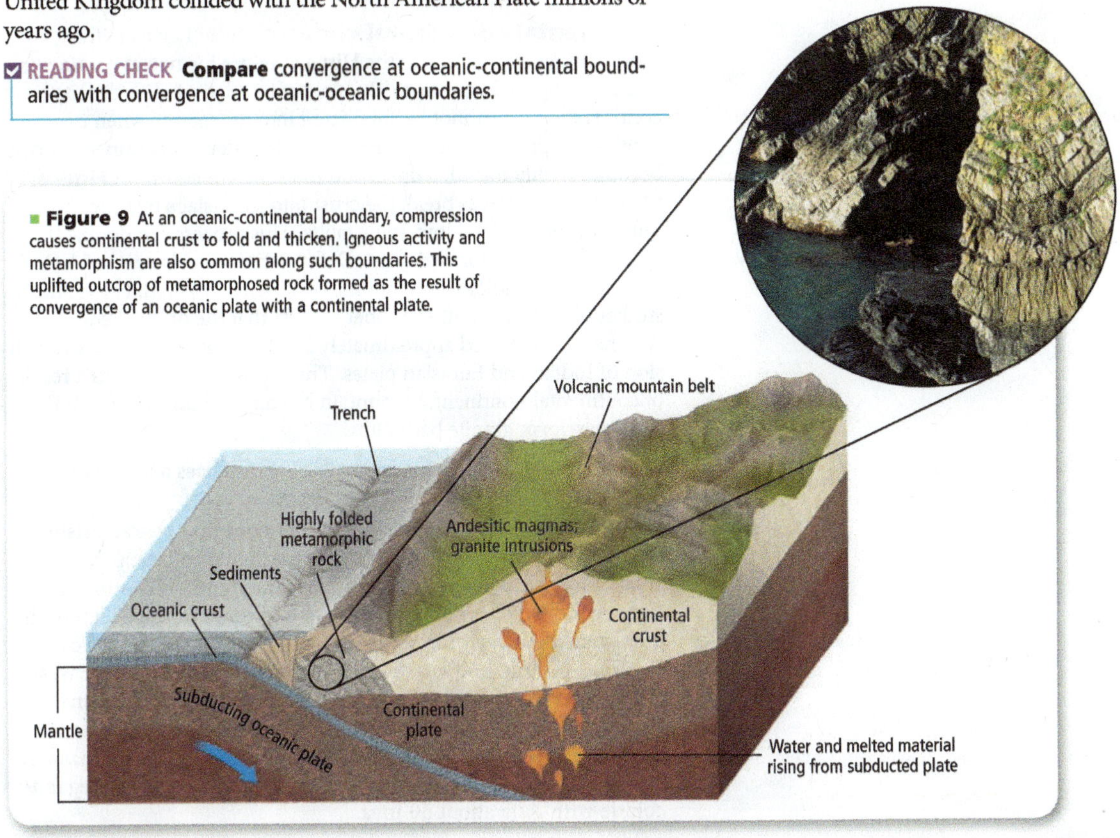

Tie to Previous Knowledge

Boundary Metamorphism Have students recall what they have learned about metamorphism. Then ask students what type of metamorphic temperature and pressure conditions they would expect to find at an oceanic-continental convergent boundary. low temperature because of the shallow depth in the crust and high pressure because of convergence OL

Project

Oceanic–Continental Convergence Have students study the diagram in **Figure 9,** and then have pairs of students use different colors of modeling clay to make three-dimensional models of the mountains that form as the result of convergence between oceanic and continental plates. Be sure students use the same color of clay to represent the same features in their models. OL EL COOP LEARN

Interpret the Illustration

Oceanic–Continental Convergence Have students refer to **Figure 9** as you read aloud the text on the student page. Pause after each sentence so students can locate on the figure the major features (oceanic plate, continental plate, compressive forces, thickened continental crust, roots, metamorphosed rocks) discussed in the text. BL OL EL

☑ READING CHECK There is subduction and magma melting at both, but the mountains that form at oceanic-continental boundaries form on land (not in the ocean) and are much bigger, longer, and more complex than island arcs.

Teacher Content Support

Subduction Mélanges The highly folded, faulted, and metamorphosed rocks found along the continental edge of a subduction zone are called subduction mélanges. These rock assemblages are complex, but are recognizable on the basis of their unique composition and structure. These intensely deformed mixtures of deep-sea sediments, continental sediments, and mafic-to-ultramafic rocks derived from oceanic crust are found in no other tectonic setting. These mélanges are shown by the wavy pattern in the rocks shown in **Figure 9.**

Concept Development

Future Plate Convergence Have students study a map that shows the directions of movement of Earth's tectonic plates. Using what they have just read about mountain building at convergent boundaries, ask students to predict where another continental-continental convergent mountain range might form in the future. The most likely regions are those associated with the Arabian Plate and in the convergence between the African Plate and the Eurasian Plate. OL

Discussion

Magma Composition Have students recall the elements that compose deep-sea sediments. Then ask: How might these sediments affect the composition of magma that is generated during subduction? Deep-sea sediments contain a large amount of silica from the shells of radiolarians and diatoms. Magmas that incorporate this material will be siliceous.

Discussion

Marine Sediments Discuss with students why it is more common to find marine fossils in the rocks of oceanic-continental convergent mountains than in the rocks of continental-continental convergent mountains. OL

☑ READING CHECK It is less dense than the mantle.

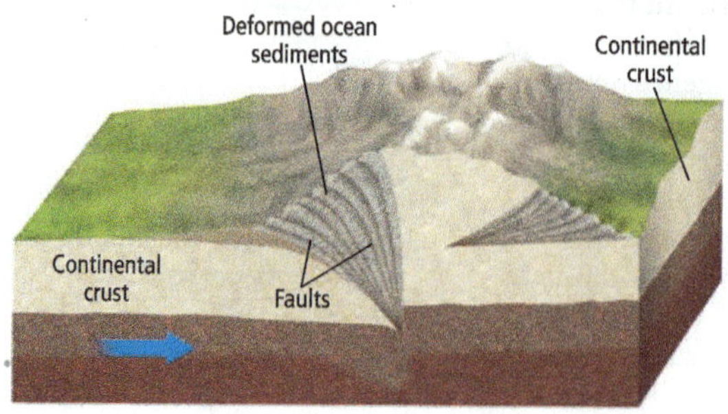

■ **Figure 10** Intense folding and faulting along continental-continental boundaries produce some of the highest mountain ranges on Earth. The Himalayas are the result of the convergence between the Indian and Eurasian plates.

View an **animation of convergence.**

Concepts In Motion

Continental-continental convergence Earth's tallest mountain ranges, including the Himalayas, are formed at continental-continental plate boundaries. Because of its relatively low density, continental crust cannot be subducted into the mantle when two continental plates converge. Instead, the low-density continental crust becomes highly folded, faulted, and thickened as shown in **Figure 10.** Compressional forces break the crust into thick slabs that are thrust onto each other along low-angle faults. This process can double the thickness of the deformed crust. Deformation can also extend laterally for hundreds of kilometers into the continents involved. For example, studies of rocks in southern Tibet suggest that the original edge of Asia has been pushed approximately 2000 km eastward since the collision of Indian and Eurasian plates. The magma that forms as a result of continental-continental mountain building solidifies beneath Earth's surface to form granite batholiths.

☑ READING CHECK **Explain** why continental crust does not subduct.

Marine sedimentary rock Another common characteristic of the mountains that form when two continents collide is the presence of marine sedimentary rock near the mountains' summits. Such rock forms from the sediments deposited in the ocean basin that existed between the continents before their collision. For example, Mount Godwin Austen (also known as K2) in the western Himalayas is composed of thousands of meters of marine limestone that sits upon a granite base. The limestone represents the northern portions of the old continental margin of India that were pushed up and over the rest of the continent when India began to collide with Asia about 50 mya.

ACROSS THE CURRICULUM

Math Have students use a physiographic atlas to compute and compare the total relief of the Andes Mountain Range with that of the Himalayan Mountains. Students might be surprised to learn that because of the depth of the Peru-Chile Trench, the total relief associated with the Andes Mountains is much greater than that of the Himalayas, which were formed—and are still rising—as the result of continental-continental convergence.

ACROSS THE CURRICULUM

Biology Have interested students research how people living at high elevations in the Andes and the Himalayas have become acclimated to the oxygen-poor air present at these elevations.

The Appalachian Mountains—A Case Study

Recall that Alfred Wegener used the matching rocks and geologic structures in the Appalachians and mountains in Greenland and northern Europe to support his hypothesis of continental drift. In addition to Wegener, many other scientists have studied the Appalachian Mountains. Based on these studies, geologists have divided the Appalachians into several distinct regions, as illustrated in **Figure 11.** Each region is characterized by rocks that show different degrees of deformation. For example, rocks of the Valley and Ridge Province are highly folded sedimentary rocks. In contrast, the rocks of the Piedmont Province consist of older, deformed metamorphic and igneous rocks that are overlain by relatively undeformed sedimentary layers. These regions, pictured in **Figure 12,** are different because they formed in different ways.

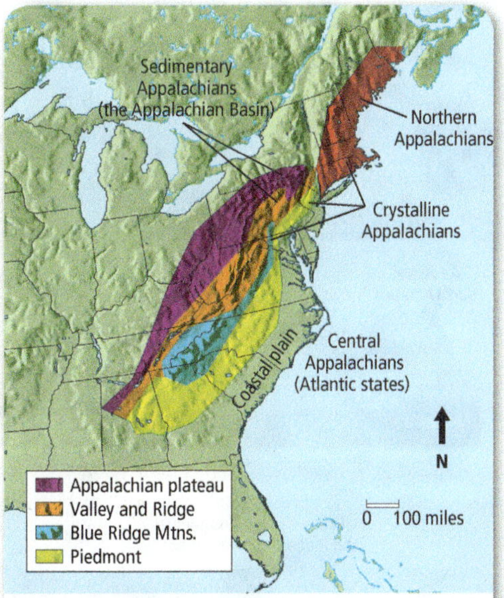

■ **Figure 11** The Appalachian Mountain Range is made up of more than one type of mountain. It has several distinct regions, each with their own orogenic history.

The early Appalachians The tectonic history of the Appalachians is illustrated in **Figure 13.** It began about 800 to 700 mya when ancestral North America separated from ancestral Africa along two divergent boundaries to form two oceans. The ancestral Atlantic Ocean was located off the western coast of ancestral Africa. A shallow, marginal sea formed along the eastern coast of ancestral North America. A continental fragment was located between the two divergent boundaries.

■ **Figure 12** The Valley and Ridge Province of the Appalachians has highly folded rocks. Rocks from the Piedmont Province are relatively undeformed.

View an **animation of folding rocks.**
Concepts In Motion

Concept Development

Appalachian Regions Ask students: Why are these regions so different? The different types of rocks and structures characteristic of each region formed as the result of different tectonic processes. What kinds of processes led to their formation? Folding, faulting, metamorphism, and igneous activity produced the rocks and features found in the different regions of the Appalachians.

Enrichment

Appalachian Regions Have students compare **Figure 11** showing the major regions of the Appalachians to a simple U.S. map. Ask students what the difference between the two maps is. Students should recognize **Figure 11** shows a detail of a region, breaking it down according to orogeny. On a basic U.S. map, the Appalachians are usually represented as a single mountain range.

ACROSS THE CURRICULUM

History Early textile industries in the eastern United States required water power for factories and transport. Have students find out how the geology of the Appalachians was responsible for the locations of the first textile mills between the Piedmont Province and regions toward the west. Students should find that rivers and streams flowing out of the folded regions to the west had steep gradients that provided the energy needed to turn waterwheels and operate other machinery. In the Piedmont Province, these same rivers had smaller gradients, which made them navigable to transport goods.

VISUALIZING

Purpose

Students will know the processes involved in the formation of and subsequent changes to the Appalachian Mountains.

Interpret the Illustration

Boundary Locations Have students refer to **Figure 13** as you ask the following questions: Where is the main convergent boundary in **Figure 13**? The convergent boundary lies between the ancestral Atlantic and the island arc. Why are some of the rocks of the Valley and Ridge Province highly folded? The compressive forces associated with convergence caused the rocks to fold. Why are the rocks of the Blue Ridge and the Piedmont Provinces mostly deformed igneous and metamorphic rocks? The Blue Ridge rocks formed as the result of changes in pressure and temperature associated with convergence (the continental fragment), and the Peidmont rocks formed as a result of volcanism (the island arc) and compressional forces.

Tie to Previous Knowledge

Drainage Patterns Provide students with topographic maps of the Appalachians. Have students use what they have learned about surface water to explain how the tectonic history of the Appalachian region is reflected in the drainage patterns of the area. The major drainage patterns in the Appalachians are the result of the parallel folds that formed during the continental-continental convergence that closed the ancestral Atlantic Ocean.

VISUALIZING the Rise and Fall of the Appalachians

Figure 13 The Appalachians formed hundreds of millions of years ago as a result of convergence.

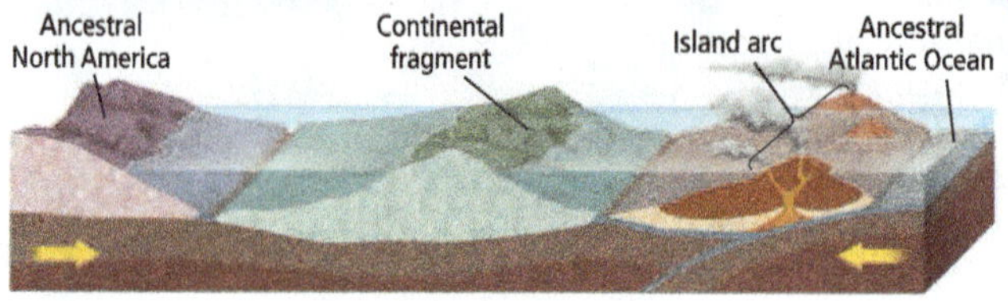

700–600 mya Convergence causes the ancestral Atlantic Ocean to begin to close. An island arc develops east of ancestral North America.

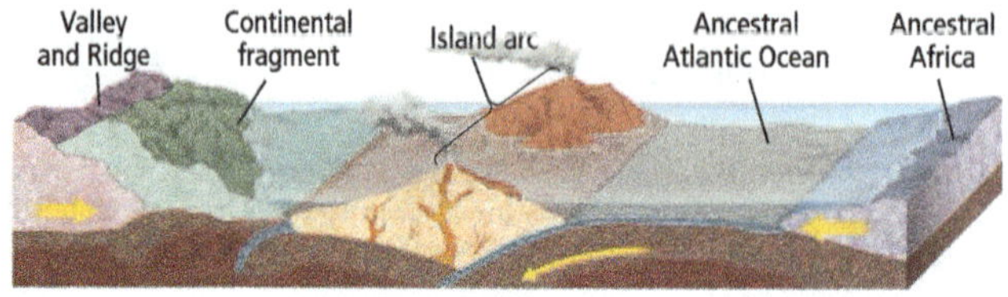

500–400 mya The continental fragment, which eventually becomes the Blue Ridge Province, becomes attached to ancestral North America.

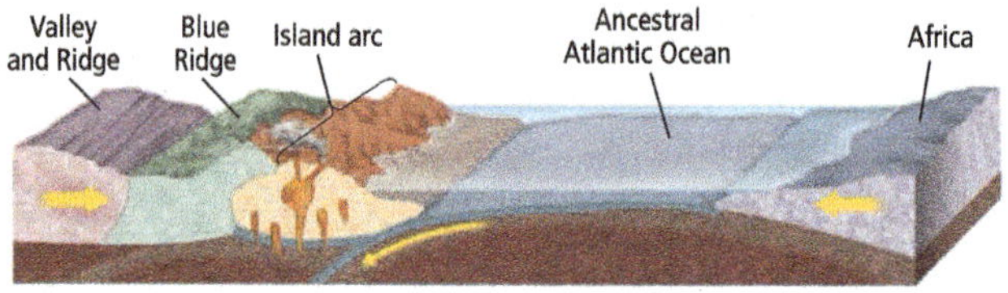

400–300 mya The island arc becomes attached to ancestral North America and the continental fragment is thrust farther onto ancestral North America. The arc becomes the Piedmont Province.

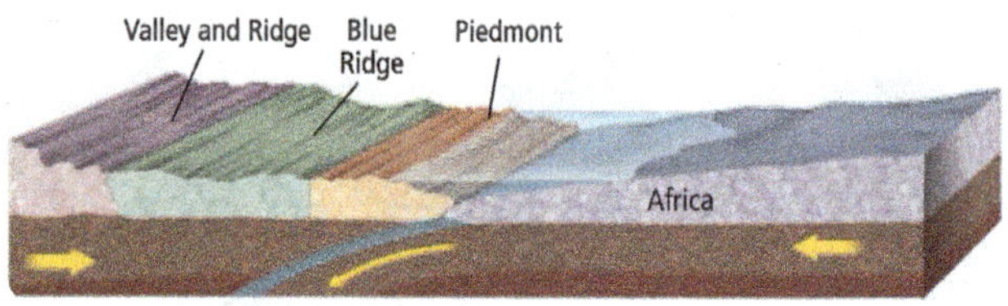

300–260 mya Pangaea forms. Ancestral Africa collides with ancestral North America to close the ancestral Atlantic Ocean. Compression forces the Blue Ridge and Piedmont rocks farther west and the folded Valley and Ridge Province forms.

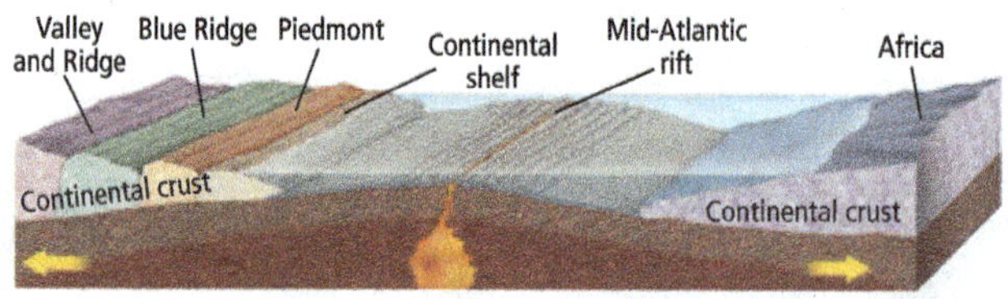

Present After the breakup of Pangaea, tension forces open the modern Atlantic Ocean and separates the continents. North America and Africa continue to move apart as the Atlantic Ocean widens.

View an **animation of other mountain ranges.** Concepts In Motion

Teacher Content Support

Coal Anthracite is a hard, high-grade type of coal that has been mined in the Appalachian region. Appalachian coal beds began as swamps that existed during the Mississippian and Pennsylvanian Periods. When Pangaea formed several hundred million years ago, the forces involved in the formation of this supercontinent slowly turned the swamp material into anthracite.

About 700 to 600 mya, the directions of plate motions reversed. The ancestral Atlantic Ocean began to close as the plates converged. This convergence resulted in the formation of a volcanic island arc east of ancestral North America, as illustrated in **Figure 13.**

About 200 million years passed before the continental fragment became attached to ancestral North America as illustrated in **Figure 13.** These highly metamorphosed rocks were thrust over younger rocks to become the Blue Ridge Province, parts of which are shown in **Figure 14.**

■ **Figure 14** Outcrops, such as this one, are common in the Blue Ridge Province.

The final stages of formation Between about 400 and 300 mya, the island arc became attached to North America, as illustrated in **Figure 13.** Evidence of this event is preserved in the Piedmont Province as a group of metamorphic and igneous rocks. These rocks were also faulted over the continent, pushing the Blue Ridge rocks farther west.

Between about 300 and 260 mya, the ancestral Atlantic Ocean closed as ancestral Africa, Europe, and South America collided with ancestral North America. Around this time, all of the ancestral continents were joined together in one large supercontinent—Pangaea. This collision of land masses resulted in extensive folding and faulting to form the Valley and Ridge Province, as illustrated in **Figure 13.** When rifting caused Pangaea to break apart about 200 mya, the modern Atlantic Ocean formed, and the continents moved to their present positions, as illustrated in **Figure 13.**

The Appalachian Mountains are only one example of the many mountain ranges that have formed along convergent boundaries. In Section 3, you will read about the orogeny that takes place along divergent plate boundaries, as well as some of the types of mountains that form far from plate margins.

SECTION 2 REVIEW

Section Self-Check

Section Summary

- Orogeny refers to all of the processes that form mountain belts.
- Most mountain belts are associated with plate boundaries.
- Island arc complexes, highly deformed mountains, and very tall mountains form as a result of the convergence of tectonic plates.
- The Appalachian Mountains are geologically ancient; they began to form 700 to 800 mya.

Understand Main Ideas

1. **MAIN IDEA** **Describe** how convergence relates to orogenic belts.
2. **Identify** the tectonic plate on which Mount Mazinga is situated.
3. **Explain** why you might find fossil shells at the top of a mountain.
4. **Differentiate** between the types of mountains that form at convergent plate boundaries.

Think Critically

5. **Infer** how the Aleutian Islands in Alaska formed.
6. **Evaluate** this statement: The Appalachian mountains are younger than the Himalayas.

WRITING IN ▶ Earth Science

7. Write and illustrate the story of the formation of the Appalachian Mountains for a middle school student.

3 Assess

Check for Understanding

Classify Ask the following questions to determine students' mastery of the information presented in this section: What type of plate boundary will result in the formation of an orogenic belt composed of volcanic, intrusive, and deformed continental rocks? an oceanic-continental convergent boundary What type of convergence results in towering mountains composed of continental crust and sedimentary rocks? a continental-continental convergence What type of convergent boundary leads to the formation of island arc complexes? an oceanic-oceanic convergent boundary

Reteach

Interpret Scientific Illustrations On one side of a giant flash card, draw the diagrams from **Figures 8, 9,** and **10** without their labels. On the reverse side, include the same figure with its labels. Have students identify the various features as you point to each on the unlabeled diagram.

Assessment

Knowledge Ask students the following questions: What kinds of materials become attached to continental crust at subduction zones? sediments and material scraped from the subducting plate What kinds of materials are subducted? Subducted materials include oceanic crust and deep-sea sediments.

SECTION 2 REVIEW

1. Convergence at subduction zones can create volcanic mountain ranges. Convergence of continents can create tall mountain ranges; both are examples of orogenic belts.
2. It lies on the Caribbean Plate.
3. Convergence can shove marine sedimentary over land that later becomes a mountain.
4. Convergent oceanic plates create island arcs; convergent oceanic and continental plates create volcanic mountain ranges; and convergent continental plates create the tallest and most varied of Earth's mountain ranges.
5. The subduction of the Pacific Plate underneath the North American Plate created a line of volcanoes on the North American Plate.
6. This statement is incorrect. The Appalachians are ancient mountains—only their roots remain; the Himalayas are still growing.
7. Stories and illustrations should demonstrate an understanding of the changing arrangement of land and ocean.

Rubric

SECTION 3

1 Focus

Focus Transparency

Before presenting the lesson, project **Section Focus Transparency 62** and have students answer the accompanying questions. EL

MAINIDEA

Non-Convergent Mountains Remind students mountains rise because crust is less dense than mantle. Ask students if they can think of other ways rocks become less dense. if they are heated Where on Earth does hot mantle come to the surface? mid-ocean ridges, volcanoes, hot spots Show students a map that shows bathymetry. Ask students where there are large mountains and mountain ranges on the ocean floors. along mid-ocean ridges

2 Teach

Tie to Previous Knowledge

Divergence Review the processes that occur along a divergent boundary.

Enrichment

Ocean-Floor Mountains Have interested students research the height of the tallest mountain on Earth. Students might be surprised to learn it is not Mount Everest, which is 8848 m tall, but Mauna Kea, which is 10,203 m tall—but only 4170 m is above sea level.

Caption Question Fig. 15 The ocean is deeper farther away from the ocean ridge because the ocean crust is contracted, denser, and thinner.

SECTION 3

Other Types of Mountain Building

MAINIDEA Mountains on the ocean floor and some mountains on continents form through processes other than convergence.

Essential Questions

- What are the processes associated with non-boundary mountains?
- How are mountain ranges that form along ocean ridges formed?
- How do uplifted and fault-block mountains differ?

Review Vocabulary

normal fault: a crack in Earth where the rock above the fault plane has dropped down

New Vocabulary

uplifted mountain
plateau
fault-block mountain

EARTH SCIENCE 4 YOU When you take a cake out of the oven, the cake cools and contracts. Similarly, the ocean floor and the crust and mantle below it cool and contract as they move away from the ocean ridge.

Divergent-Boundary Mountains

When ocean ridges were first discovered, people in the scientific community were stunned simply because of their length. These underwater volcanic mountains form a continuous chain that snakes along Earth's ocean floor for over 65,000 km. In addition to their being much longer than most of their continental counterparts, these mountains formed as a result of different orogenic processes. Recall that ocean ridges are regions of broad uplift that form when new oceanic crust is created by seafloor spreading. The newly formed crust and underlying mantle at the ocean ridge are hot. When rocks are heated, they expand, which results in a decrease in density. This decrease allows the ridge to bulge upward, as illustrated in **Figure 15.** As the oceanic plates move away from the ridge, the newly formed crust and mantle cool and contract, and the surface of the crust subsides. As a result, the crust stands highest where the ocean crust is youngest, and the underwater mountain chains have gently sloping sides.

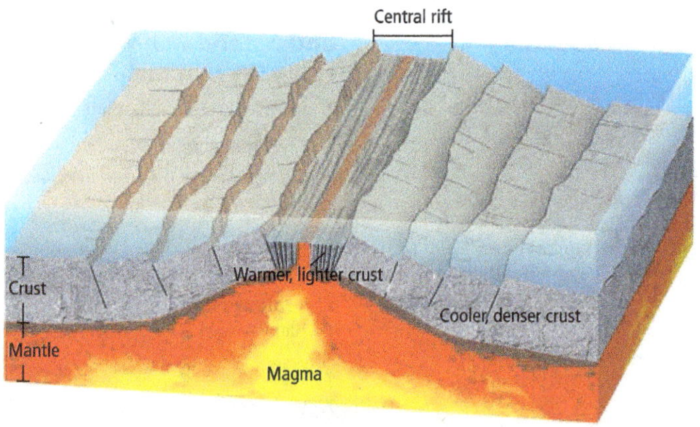

Figure 15 An ocean ridge is a broad, topographic high that forms as lithosphere bulges upward due to an increase in temperature along a divergent boundary.
Determine *where the ocean is deeper: near the ridge or far from it?*

DIFFERENTIATED INSTRUCTION

Advanced Learners Have students think about the forces and structures associated with ocean ridges and subduction zones. Ask students to predict what might happen if an ocean ridge is subducted. Then have them research places where a ridge is currently being subducted. The Juan de Fuca Ridge, off the Pacific Northwest Coast of Washington State and Canada, is being subducted into the Cascadia Subduction Zone.

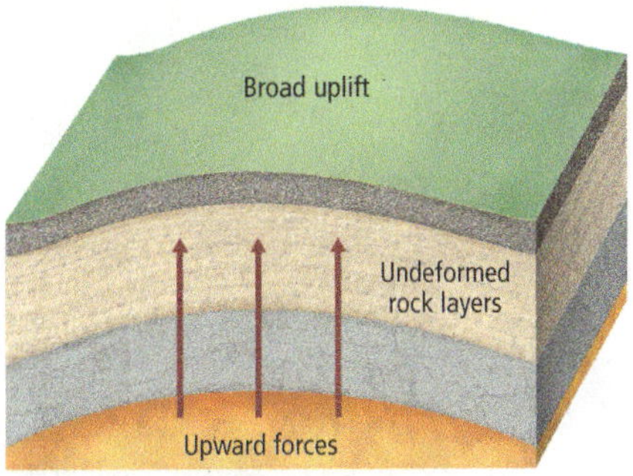

Figure 16 The Adirondack Mountains of New York State are uplifted mountains. Uplifted mountains form when large sections of Earth's crust are forced upward without much structural deformation.

Uplifted Mountains

As illustrated in **Figure 16,** some mountains form when large regions of Earth have been slowly forced upward as a unit. These mountains are called **uplifted mountains.** The Adirondack Mountains in New York State, shown in **Figure 16,** are uplifted mountains. Generally, the rocks that make up uplifted mountains undergo less deformation than rocks associated with plate-boundary orogeny, which, as you have just read, are highly folded, faulted, and metamorphosed. The cause of large-scale regional uplift is not well understood. One popular hypothesis is that the part of the lithosphere made of mantle rocks becomes cold and dense enough that it sinks into the underlying mantle. The mantle lithosphere is replaced by hotter and less dense mantle. The lower density of the new mantle provides buoyancy which vertically lifts the overlying crust. This process has been used to explain the uplift of the Sierra Nevadas, in California, which are shown in **Figure 17.** When a whole region is uplifted, a relatively flat-topped area called a **plateau** can form, like the Colorado Plateau, which extends through Colorado, Utah, Arizona, and New Mexico. Erosion eventually carves these relatively undeformed, uplifted masses to form peaks, valleys, and canyons.

Figure 17 The Sierra Nevadas are the result of regional uplift.

IN THE FIELD

The Crazy Horse Memorial The Black Hills in South Dakota are uplifted mountains into which the Crazy Horse Memorial is being carved. The artist Korczak Ziolkowski began the sculpture and worked on it for 50 years. Ziolkowski's son will complete the work, which honors all Native Americans. Crazy Horse was a Sioux warrior who fought General Custer at the Battle of Little Bighorn. The sculpture, when completed, will depict Chief Crazy Horse seated atop a stallion, his arm extended to the sacred Black Hills. The completed statue will be 171.6 m high and 195.4 m long.

Reinforcement

Uplift Ask: Which of the following statements is true?

a. The rocks in uplifted mountains are highly deformed.
b. Uplifted mountains are associated with plate boundaries.
c. Upward movement in the mantle causes uplifted mountains to form.
d. Uplifted mountains form as the result of erosion.

Statement c is true.

Discussion

Plateau Ecology Erosion in the Colorado Plateau region has resulted in many small plateaus that are isolated on all sides by vertical walls thousands of meters high. Ask students to suggest what happens to populations of organisms that are isolated on these plateaus. The populations can evolve to form new species.

Concept Development

Uplift Use eight rectangular pieces of felt of different colors, a small, round balloon, and a bicycle pump to demonstrate how uplifted mountains might form. Secure the neck of the balloon to the pump. Lay the felt pieces on top of one another over the balloon. Slowly fill the balloon with air from the pump. Have students describe what happens. The layers of felt bulge in the center to form a "mountain." Explain the felt represents rock layers and the inflating balloon represents uplifting forces.

Apply Earth Science

Silver Mining Say to students: Imagine you are mining an underground silver deposit. The veins suddenly stop at a normal fault plane. If the fault has a vertical movement of 750 m, where would you look for a continuation of the deposit? 750 m up or down on the opposite side of the fault

GeoLAB

iLab Station

The GeoLab at the end of the chapter can be used at this point in the lesson.

3 Assess

Check for Understanding

Summarize Ask students to name two types of mountains discussed in this section that form primarily as the result of tensional forces in Earth's crust. ocean ridges and fault-block mountains

Reteach

Outline Have students use the bold headings to outline the major points of this section.

Assessment

Performance Have each student write a creative but scientifically accurate poem about each of the mountain types discussed in this section. Each poem should be between six and eight lines in length. Have volunteers contrast poems about the same type of mountains.

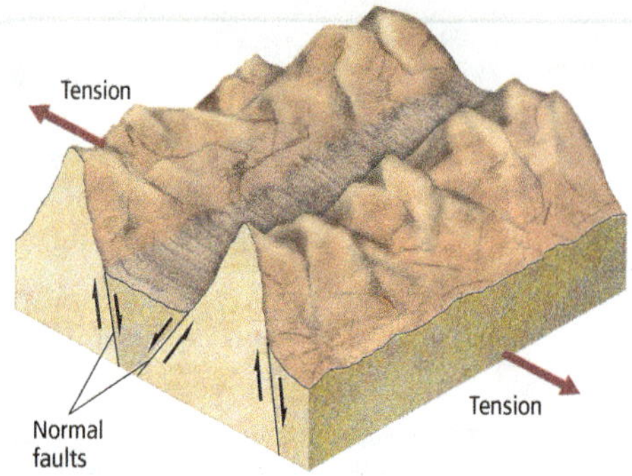

Figure 18 Fault-block mountains are areas of Earth's crust that are higher than the surrounding landscape as the result of faulting. The Basin and Range Province consists of hundreds of mountains separated by normal faults.

Fault-Block Mountains

Another type of mountain that is not necessarily associated with plate boundaries is a fault-block mountain. Recall how movement at faults can lift land on one side of a fault or drop it on the other. When Earth's crust is stretched, a series of normal faults can form. **Figure 18** illustrates how **fault-block mountains** form between these large faults when pieces of crust are dropped downward due to tensional forces. The Basin and Range Province of the southwestern United States and northern Mexico, a part of which is shown in **Figure 18,** consists of hundreds of nearly parallel mountains separated by normal faults. The Grand Tetons in Wyoming are also fault-block mountains. You will explore the topography of this range in the GeoLab at the end of this chapter.

SECTION 3 REVIEW

Section Self-Check

Section Summary

- Divergent boundaries, uplift, and faulting produce some of Earth's mountains.
- Underwater volcanic mountains at divergent boundaries form Earth's longest mountain chain.
- Regional uplift can result in the formation of uplifted mountains that are made of nearly undeformed layers of rock.
- Fault-block mountains form when large pieces of the crust are dropped downward between normal faults.

Understand Main Ideas

1. **MAINIDEA Explain** why all of Earth's mountains do not form at convergent plate boundaries.
2. **Identify** the kinds of rocks associated with ocean ridges.
3. **Explain** why an ocean ridge is higher than the surrounding crust.
4. **Compare** ocean ridges with fault-block mountains.
5. **Compare and contrast** the formation of uplifted and fault-block mountains.

Think Critically

6. **Formulate** criteria for identifying an uplifted mountain.

WRITING IN Earth Science

7. Write three practice test questions for your classmates to assess their knowledge of mountain building.

SECTION 3 REVIEW

1. Ocean ridges form at divergent boundaries; some mountains form as a result of uplift, and some form from faulting.
2. Basaltic rocks form at ridges.
3. Because the newly formed crust is warmer, it is less dense and therefore rises higher.
4. Both are mountains that form as a result of tension, but fault-block mountains are not associated with igneous activity.
5. Neither form at plate boundaries. Although both involve upward movement, with fault block mountains, lifting and dropping occur along faults; with uplifted mountains, upward movement is the result of upwelling magma that deforms rock but does not fracture it.
6. Answers will vary. Possible answer: The rocks in an uplifted mountain would have recognizable layers that are arched, not folded or twisted.
7. Possible answers: What are the three types of convergent boundaries where mountains form? What is a plateau? Are the rocks in uplifted mountains folded?

Rubric

eXpeditions!

ON SITE: HIKING THE APPALACHIAN TRAIL

Phil Schermeister/CORBIS

Mount Katahdin is located in Maine at one end of the Appalachian Trail.

The Appalachian Mountains sprawl from Canada to Alabama. Every year, more than 3 million people embark on a journey of the Appalachian Trail to explore at least part of this mountain range. The rich geologic history of these mountains makes this one of the most exciting and beautiful hikes America has to offer.

History The Appalachian Trail (AT) was opened in 1937 in an effort to allow people to escape the bustle of everyday life and reconnect with nature. Today, the trail covers 3499 km, from Springer Mountain in Georgia to Mount Katahdin in Maine, and can be divided into three areas—the southern, central, and northern Appalachians.

Hiking the Trail A select group of hikers each year choose what is called a thru-hike. Thru-hikers plan to hike the entire AT from start to finish. Most choose to begin in Georgia and hike north to Maine. This trip takes 6 months or longer, and requires much planning.

Southern Appalachians Georgia's Springer Mountain sits 6087 km above sea level. The soil around Springer Mountain is covered with metamorphic rocks, allowing for only sparse tree growth. These uplifted mountains in the southern Appalachians were not impacted by glaciation as much as some of the northern summits.

Central Appalachians In the central Appalachians, hikers encounter Slide Mountain, part of the Catskill Range of New York. The Catskills are characterized by steep slopes and rounded uplands. Comprised mainly of sandstone and conglomerate, Slide Mountain has resisted some of the weathering that much of the Appalachians have endured. The peak is 6727 km above sea level, but it is difficult to enjoy the view before reaching the top due to the dense forests.

Northern Appalachians After several months of hiking, imagine encountering the end of the Appalachian Trail, the majestic Mount Katahdin, shown in the figure. To the east, two large cirque basins, the Great Basin and North Basin, are visible. These areas were sculpted by glaciers over 25,000 years ago. Here, the high altitude combined with the northern location of the mountains, does not provide the necessary environment for thick tree growth, so the final view from Mount Katahdin is breathtaking. As of 2010, only 11,502 hikers have completed the AT thru-hike.

(bkgd)fStop/Getty Images

WRITING IN ▶ Earth Science

Research a firsthand account of someone hiking the Appalachian Trail. Describe the geologic formations that they encountered.

WebQuest

WRITING IN ▶ Earth Science

Rubric

Research Answers will vary depending on the first-hand account they select. Encourage students to find accounts that give geologic details of the trail.

WebQuest

eXpeditions!

Purpose

Students will describe the formation of the Appalachians and the history of the Appalachian Trail. Students will discuss the geology of the rocks that can be found along the Appalachian Trail.

Teacher Content Support

The Appalachian Trail The Appalachian Trail crosses two national parks, touches 14 states, has its lowest elevation of 38 m over the Hudson River in New York, and has its highest elevation of 2024 m at Clingmans Dome in the Great Smoky Mountain National Park. The rockiest part of the trail is in Pennsylvania, especially near the northern part of the state around the Susquehanna River. The river is ancient and considered the oldest or second oldest in the world. The river carved through the Appalachian Mountains some 300 million years ago. The river is older than the Atlantic Ocean.

Teaching Strategies

- Ask students if they have visited the Appalachian Trail. Have any of them hiked any part of it? How many have been to the Great Smoky Mountain National Park which has the Appalachian Trail going through it? What sorts of geologic features did they notice?
- You might want to make a slide presentation showing scenes of the Appalachian Trail before students start their research.

fStop/Getty Images

GeoLAB

iLab Station Rubric

Preparation

Time Allotment 45 min

Process Skills interpret scientific illustrations, measure and use numbers, think critically

Safety Precaution Approve lab safety forms before work begins.

Procedure

- Review, if necessary, information about topographic maps.
- Discuss the concepts of relief and vertical exaggeration with students before they begin the activity.
- **Troubleshooting** Demonstrate how to transfer points from the map to the profile. Make a transparency of the map and the profile grid. Then cut a transparent strip to represent the paper strip. Use the overhead projector to show students how to transfer the points.

Analyze and Conclude

1. Elevation steadily increases up to Static Peak. Elevation then decreases.
2. Static Peak is 11,303′ high. The Patrol Cabin is at 7840′.
3. The average elevation along the profile is about 9600′.
4. 11,303′−7840′ = 3463′
5. No; the vertical and horizontal scales are different, and therefore the profile is not to scale.
6. The horizontal scale was determined by the scale of the map. The vertical scale was chosen to correspond to the total relief of the area in question.

LabManager™
Customize any lab with the LabManager™ CD-ROM.

GeoLAB

iLab Station

Mapping: Make a Map Profile

Background: A map profile, also called a topographic profile, is a side view of a geographic or geologic feature constructed from a topographic map. The Grand Tetons, a mountain range in Wyoming, formed when enormous blocks of rock were faulted along their eastern flanks, causing the blocks to tilt to the west.

Question: *How do you construct a map profile?*

The rugged Grand Tetons stand tall above the plains in Grand Teton National Park in Wyoming.

Materials

metric ruler
sharp pencil
graph paper

Safety Precautions

Procedure

Contour lines are lines on a map that connect points of equal elevation. Locate the index contour lines on the map on the next page. Index contour lines are the ones in a darker color.

1. Read and complete the lab safety form.
2. On graph paper, make a grid like the one shown on the facing page.
3. Imagine that there is a line running between Patrol Cabin and Static Peak. Place the edge of a paper strip on the map along that line and mark where each major contour line intersects the strip.
4. Label each intersection point with the elevation.
5. Transfer the points from the paper strip to the profile grid.
6. Connect the points with a smooth line to construct a profile of the mountain range between Patrol Cabin and Static Peak.
7. Label the major geographic features on your profile.

Analyze and Conclude

1. **Interpret Data** Describe how the topographic profile changes with distance from Patrol Cabin.
2. **Interpret Data** What is the elevation of the highest point on the map topographic profile? The lowest point?
3. **Interpret Data** What is the average elevation shown in the profile?
4. **Interpret Data** Calculate the total relief shown in the profile.
5. **Interpret Data** Is your topographic profile an accurate model of the topography between Patrol Cabin and Static Peak? Explain.
6. **Analyze** What determined the scale of this topographic profile?

APPLY YOUR SKILL

Apply Obtain a topographic map of your hometown and make a topographic profile. How does it compare to the profile you made in this lab?

APPLY YOUR SKILL

Apply Students' topographic profiles should be made by a process similar to the process they used in the lab. Depending on the topography of your area, there might be vast differences when compared to the topographic profile of the Grand Tetons.

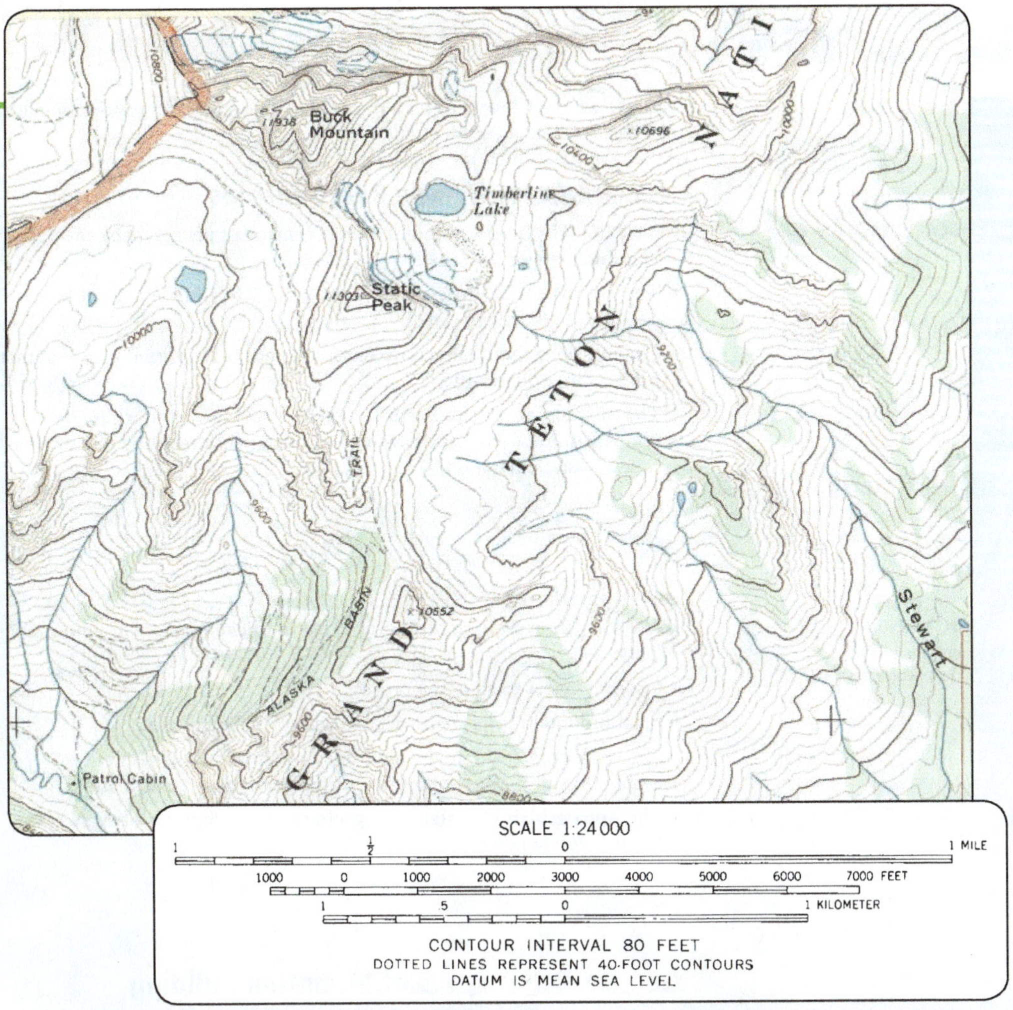
Buck Mountain
11938
Timberline Lake
Static Peak
11303
10696
10400
10800
10000
9600
9200
10552
8800
GRAND TETON NATI
ALASKA BASIN TRAIL
Stewart
Patrol Cabin
SCALE 1:24 000
1 MILE
7000 FEET
1 KILOMETER
CONTOUR INTERVAL 80 FEET
DOTTED LINES REPRESENT 40-FOOT CONTOURS
DATUM IS MEAN SEA LEVEL

12,000
11,000
10,000
9000
8000
7000
Elevation (ft)
A
A′

MAINIDEAS Summary statements can be used by students to review the major concepts of the chapter.

Students can review with these online resources.

Vocabulary Practice

Vocabulary eGames
Vocabulary eFlashcards
Vocabulary PuzzleMaker

Section Self-Check

Chapter Self-Check

Online Test Practice

Use the Exam*View*® *Assessment Suite* CD-ROM to:

- create multiple versions of tests
- create modified tests with one mouse click
- edit existing questions and add your own questions
- build tests aligned with state standards using built-in state curriculum tags
- change English tests to Spanish with one mouse click
- track students' progress using the Teacher Management System

CHAPTER 20 STUDY GUIDE

BIGIDEA Mountains form through dynamic processes which crumple, fold, and create faults in Earth's crust.

Vocabulary Practice

SECTION 1 Crust-Mantle Relationships

VOCABULARY
- topography
- isostasy
- root
- isostatic rebound

MAINIDEA The height of mountains is controlled primarily by the density and thickness of the crust.

- The majority of Earth's elevations are either 0 to 1 km above sea level or 4 to 5 km below sea level.
- The mass of a mountain above Earth's surface is supported by a root that projects into the mantle.
- The addition of mass to Earth's crust depresses the crust, while the removal of mass from the crust causes the crust to rebound in a process called isostatic rebound.

SECTION 2 Orogeny

VOCABULARY
- orogeny
- compressive force

MAINIDEA Convergence causes the crust to thicken and form mountain belts.

- Orogeny refers to all of the processes that form mountain belts.
- Most mountain belts are associated with plate boundaries.
- Island arc complexes, highly deformed mountains, and very tall mountains form as a result of the convergence of tectonic plates.
- The Appalachian Mountains are geologically ancient; they began to form 700 to 800 mya.

SECTION 3 Other Types of Mountain Building

VOCABULARY
- uplifted mountain
- plateau
- fault-block mountain

MAINIDEA Mountains on the ocean floor and some mountains on continents form through processes other than convergence.

- Divergent boundaries, uplift, and faulting produce some of Earth's mountains.
- Underwater volcanic mountains at divergent boundaries form Earth's longest mountain chain.
- Regional uplift can result in the formation of uplifted mountains that are made of nearly undeformed layers of rock.
- Fault-block mountains form when large pieces of the crust are dropped downward between normal faults.

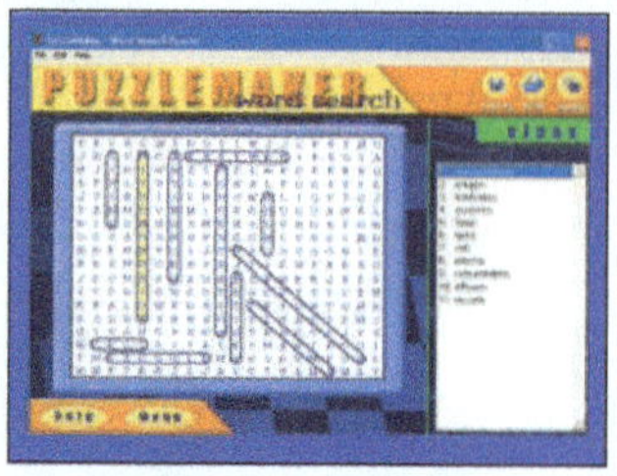

Vocabulary Practice

For additional practice with vocabulary, have students access the Vocabulary PuzzleMaker.

Chapter Self-Check

VOCABULARY REVIEW

Match each description with the correct term from the Study Guide.

1. continental crust that extends down into the mantle
2. the condition of equilibrium between mantle and crust
3. a mountain bounded by normal faults
4. a type of mountain that often shows little deformation

Fill in the blanks with the correct term from the Study Guide.

5. ________ is the process of mountain formation.
6. Folding, faulting, and metamorphism are created by ________ forces.
7. A flat-topped landform created by uplift is called a ________.

Each of the following sentences is false. Make each sentence true by replacing the italiziced words with terms from the Study Guide.

8. The variation in elevation of the crust is called *geography.*
9. *Orogenic forces* refer to the rising up of the crust when a large amount of mass is removed.

UNDERSTAND KEY CONCEPTS

10. What is the approximate percentage of Earth's surface that is covered by continents?
    A. 10 percent
    B. 30 percent
    C. 50 percent
    D. 70 percent
11. What purpose do mountain roots serve?
    A. help prevent mountains from eroding too quickly
    B. balance the amount of crust and mantle in an area
    C. serve as counterbalance to the large weight above
    D. help prevent the mountain from sinking into the mantle
12. Which causes differences in elevation on Earth?
    A. density and thickness of the crust
    B. vertical dikes and pillow basalts
    C. seamounts and hot spots
    D. uplifted and faulted mountains
13. Which is not associated with orogeny at convergent boundaries?
    A. island arcs
    B. highly folded and faulted ranges
    C. ocean ridges
    D. deformed sedimentary rocks
14. During oceanic-oceanic convergence, where do island arc volcanoes form?
    A. on the plate that is subducted
    B. on the plate that is not subducted
    C. do not form on either plate
    D. form on both plates

Use the figure below to answer Question 15.

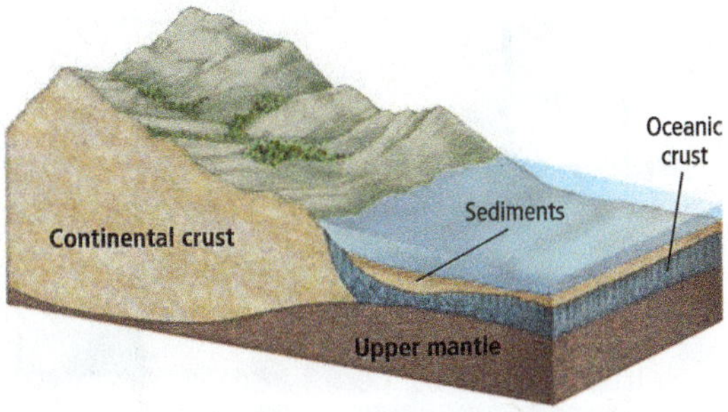

15. Organize the terms *mantle, continental crust,* and *oceanic crust* in order of increasing density.
    A. mantle, oceanic crust, continental crust
    B. oceanic crust, mantle, continental crust
    C. continental crust, oceanic crust, mantle
    D. oceanic crust, continental crust, mantle
16. At which type of plate boundary do the highest mountains form?
    A. convergent; continental-continental
    B. convergent; continental-oceanic
    C. divergent; oceanic-oceanic
    D. divergent; oceanic-continental

CHAPTER 20 ASSESSMENT

VOCABULARY REVIEW

1. root
2. isostasy
3. fault-block mountain
4. uplifted
5. Orogeny
6. compressive
7. plateau
8. topography
9. Isostatic rebound

UNDERSTAND KEY CONCEPTS

10. B
11. C
12. A
13. C
14. B
15. C
16. A

17. A
18. D

CONSTRUCTED RESPONSE

19. The principle of isostasy states that the downward force of gravity is met by the upward force of buoyancy, and equilibrium is reached.
20. The largest block is like Mount Everest. It is the tallest and its bottom reaches farthest into the water.
21. The roots rise and are eroded themselves.
22. Continental crust is thicker than oceanic crust, and therefore requires a deeper root to achieve isostatic equilibrium.
23. Rising magma warms the overlying crust and causes it to bulge upward to form a mountain range.
24. As tops of mountains are eroded, mass is removed and the roots rise due to isostatic rebound. Uplift can also force a root upward.
25. compression as two plates collide; the addition of mass to a plate from volcanic activity or sediment accumulation; cooling and contracting of warm crust
26. 3865 m − 841 m = 3024 m.
27. Mount Everest and the Mariana Trench represent Earth's elevation extremes.
28. in the Himalayas north of India; The thickness of the crust is proportional to the height of the mountains, and the Himalayas are the highest mountains, so the crust must be thicker there.
29. An island arc is a linear series of individual volcanic peaks that form in the ocean as a result of the convergence of two oceanic plates. They form as part of the subducting plate melts and volcanoes erupt on the non-subducting plate.

THINK CRITICALLY

30. The collision of two oceanic crust plates results in subduction because both plates are very dense.
31. The Andes resulted from the subduction of the Pacific Plate underneath the South American Plate.

Use the figure below and Figure 13 to answer Questions 17 and 18.

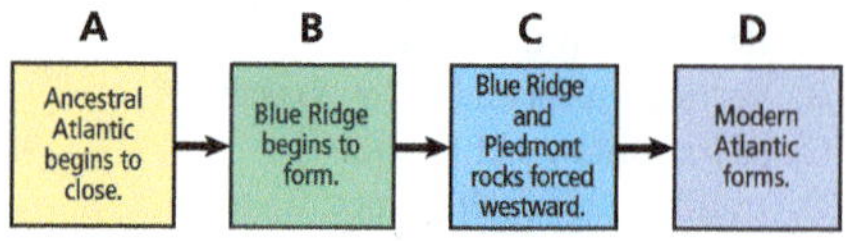

17. Which occurred between Events B and C?
   A. The island arc attached to North America.
   B. Plate motions reversed.
   C. Africa collided with North America.
   D. The island arc developed.

18. Approximately when did Event C occur?
   A. 800 to 700 mya
   B. 700 to 600 mya
   C. 500 to 400 mya
   D. 300 to 260 mya

CONSTRUCTED RESPONSE

Use the figure below to answer Questions 19 and 20.

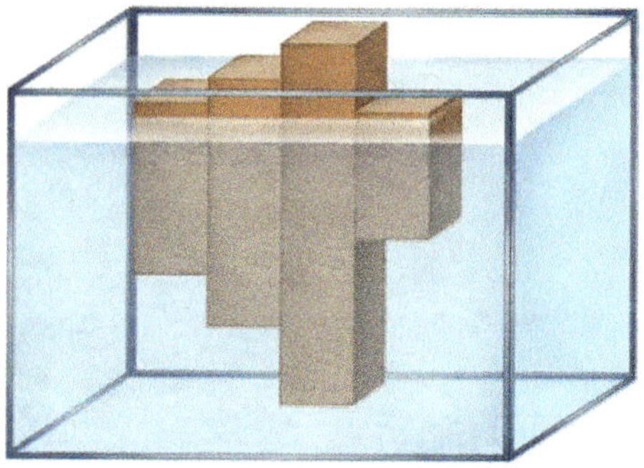

19. **Restate** the principle that explains what is happening in this figure.

20. **Relate** the largest block to a mountain described in this chapter.

21. **Describe** what happens to a mountain's roots as the mountain is eroded.

22. **Explain** why continental crust can displace more of the mantle than oceanic crust can.

23. **Explain** why ocean ridges rise high above the surrounding ocean floor.

24. **Discuss** the processes that can bring roots of mountains to the surface.

25. **Describe** three mechanisms of crustal thickening.

Use the figure below to answer Questions 26 and 27.

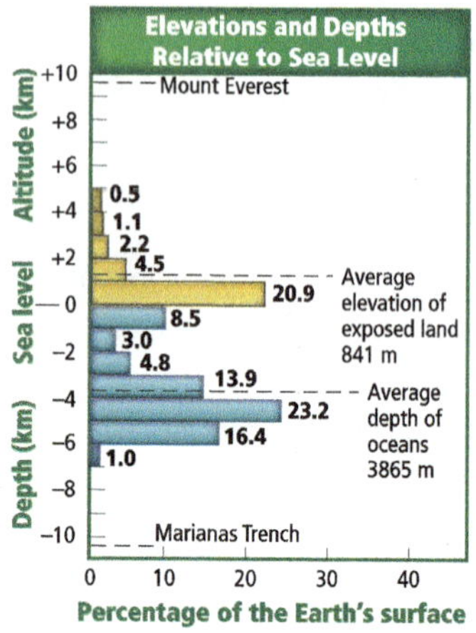

26. **Calculate** the difference in elevation between the average elevation of continents and the average depth of oceans.

27. **Generalize** State one generalization about Mount Everest and the Marianas Trench based on the data in the graph.

28. **Infer** where continental crust is the thickest.

29. **Summarize** the characteristics of an island arc.

THINK CRITICALLY

30. **Illustrate** why the interaction between two oceanic crust plates rarely results in the formation of high mountain ranges.

31. **Hypothesize** The Andes Mountains run along the western coast of South America. Hypothesize about the tectonic setting in which they formed.

Chapter Self-Check

32. CAREERS IN EARTH SCIENCE Structural geologists collect data from both field observations and laboratory analysis to help interpret the structural history of an area. What can a structural geologist interpret about the history of the outcrop shown in the opening photo of this chapter?

Use the figure below to answer Question 33.

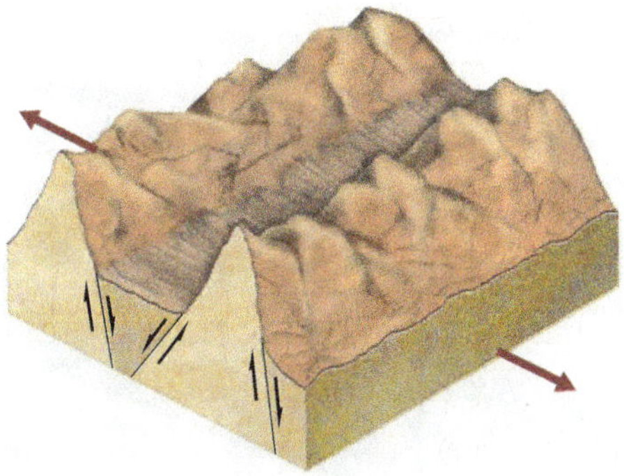

33. **Interpret** and explain what type of forces are acting on this area if fault-block mountains are forming.

34. **Decide** whether a continental crust thinner than the average thickness of 40 km would depress the mantle more or less than it does now.

35. **Evaluate** whether isostatic rebound of an area ever stops.

CONCEPT MAPPING

36. Create a concept map using the following terms: *Earth's tallest mountains, uplifted mountains, individual volcano peaks, divergent boundaries, little structural deformation, compression, fault-block mountains, ocean ridges,* and *convergent boundaries.* Refer to the *Skillbuilder Handbook* for more information.

CHALLENGE QUESTION

37. **Consider** whether all mountains are in a state of isostatic equilibrium. Explain your answer.

WRITING IN ▸ Earth Science

38. Review **Figure 13.** Speculate about what might happen over the next 500 million years. Draw three new pictures for 100, 300, and 500 million years from now. Write a caption for each, describing the orogenic and plate tectonic processes you think will occur.

DBQ Document–Based Questions

Data obtained from: Fischer, K. M. 2002. Waning buoyancy in the crustal roots of old mountains. *Nature* 417: 933–936.

Buoyancy is measured by the ratio of height of the continents to the depth of the root beneath the continent. In this graph, buoyancy (R) *is shown in relation to the age of the mountain belt.*

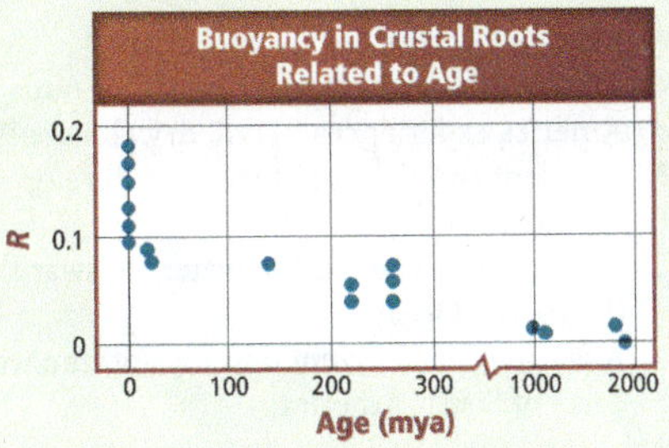

39. Describe the trend (if any) that is evident in the data.

40. Because *R* depends on the density of the crust and mantle, how might the value of *R* change if the density of continental crust were constant?

41. Provide an explanation for the change in the value of *R*. What is happening to the crustal root to cause *R* to change?

CUMULATIVE REVIEW

42. Explain what scientists mean when they say Earth operates as a system. **(Chapter 1)**

43. Explain why weathered rock alone is not a soil. **(Chapter 7)**

32. Uplift or compressive forces changed the orientation of what were once parallel layers of rock.
33. Tension forces are pulling the crust apart, creating depressions where faulting occurs.
34. It would depress the mantle less.
35. Yes, rebound stops when the root has risen to the surface and is eroded away.

CONCEPT MAPPING

36. Students should connect terms as follows: convergent boundaries, individual volcanic peak, compression, Earth's tallest mountains; divergent boundaries, ocean ridges, fault-block mountains; uplifted mountains, little structural deformation.

CHALLENGE QUESTION

37. Not all mountains are in a state of isostatic equilibrium. To reach isostatic equilibrium, the crust must thicken, which requires that the mantle be able to deform and flow. Small features, such as seamounts, can be so small that they do not cause the crust to thicken. They are supported by the strength of the crust and lithosphere.

WRITING IN ▸ Earth Science

Rubric

38. Answers should demonstrate an understanding that oceans open and close, and that closing requires subduction and the formation of a volcanic arc. The following are likely to occur: Subduction begins offshore of North America as the oceanic crust cools, and a volcanic arc forms; the Atlantic ocean closes and there is a continent-continent collision; the continent rifts and a new ocean forms (no subduction).

DBQ Document-Based Questions

Data obtained from: Fischer, K. M. 2002. Waning buoyancy in the crustal roots of old mountains. *Nature* 417: 933-936.

39. As age increases, R decreases.
40. R should remain constant (recall the block demonstration from the Launch Lab).
41. Buoyancy changes as temperature changes. (As crustal root gets colder, it gets denser, so it must then get thicker relative to the height of the mountain in order to support the mountain isostatically.)

CUMULATIVE REVIEW

42. Each component of Earth is important and interrelated to all the others.
43. Soil is a loose covering of broken rock particles and decaying organic matter, called humus, overlying the bedrock of Earth's surface.

MULTIPLE CHOICE

1. A
2. D
3. B
4. A
5. C
6. B
7. D
8. A

CUMULATIVE STANDARDIZED TEST PRACTICE

MULTIPLE CHOICE

Use the map below to answer Questions 1 and 2.

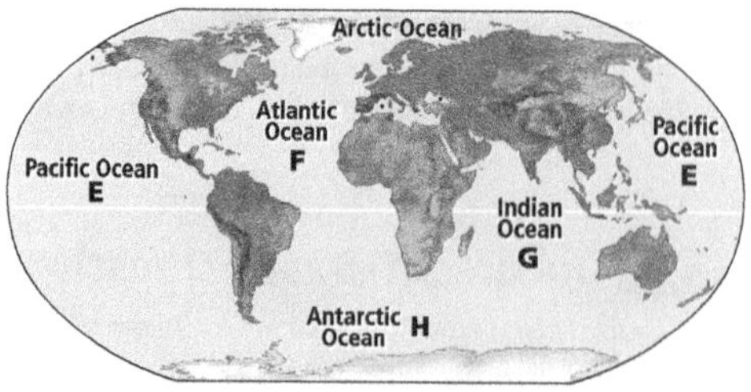

1. In which location does the El Niño cycle begin?
   A. E
   B. F
   C. G
   D. H

2. Which event occurs during an El Nino?
   A. Much of the northwestern coast of South America experiences a cool, dry climate.
   B. Frequent and intense hurricanes develop in the Atlantic Ocean.
   C. Strong trade winds move water westward across the Pacific Ocean.
   D. A warm ocean current develops off the western coast of South America.

3. Why would a thickness of continental crust displace less mantle than the same thickness of oceanic crust?
   A. Continental crust is more dense.
   B. Continental crust is less dense.
   C. Continental crust is mainly basalt.
   D. Continental crust is closer to the mantle.

4. When do shield volcanoes form?
   A. when layer upon layer of lava accumulates during nonexplosive eruptions
   B. when layers of hardened, frothy mixtures of gas and magma formed by explosive eruptions alternate with layers of oozing lava
   C. when eruptions eject small pieces of magma into the air and the pieces fall to the ground and collect around a vent
   D. when thick lava hardens around a central vent

5. Which is NOT a method used to increase soil fertility?
   A. planting legumes
   B. adding compost to the soil
   C. planting the same crops every year
   D. using commercially produced fertilizers

Use the illustration below to answer Questions 6 and 7.

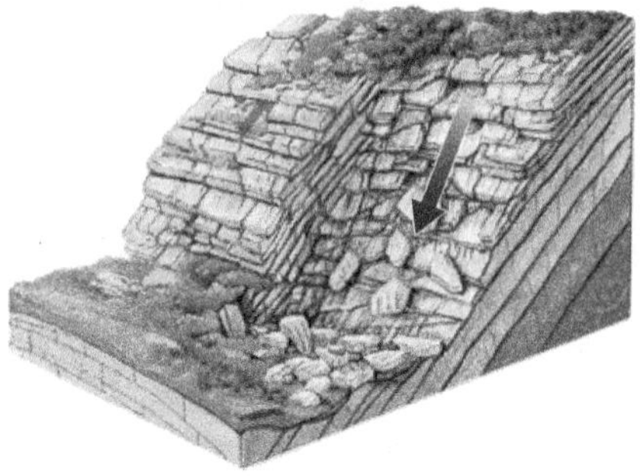

6. The rock slide shown is produced by
   A. the chemical breakdown of rocks
   B. the separation of a thin block of soil, rock, and debris from bedrock
   C. material rotating and sliding down a curved slope
   D. the melting of snow off the rocks

7. What potential damage would a rock slide have on a river?
   A. changing the chemical composition of the river
   B. the changing of the physical characteristics of the river
   C. permanently increasing water levels
   D. damming rivers and causing flooding

8. Which CANNOT form as the result of oceanic-oceanic convergence?
   A. rift zones
   B. trenches
   C. subduction zones
   D. island arc complexes

SHORT ANSWER

Use the map below to answer Questions 9–11.

9. The map above shows the paths that Atlantic hurricanes took in 2004. Why do no hurricanes form in the northeastern Atlantic?

10. What can be inferred about the impact that every hurricane formed in the Atlantic will have on North America?

11. What time of year did these hurricanes most likely occur and why?

12. Explain why, in topographical terms, people in the central United States experience very cold winters and very hot summers.

13. How do sill and dike formation differ?

14. Many parts of West Virginia have karst topography. Discuss the potential values and trade-offs of having a geologist prepare an environmental impact statement for areas where new housing developments are planned.

READING FOR COMPREHENSION

Iberian Earthquakes

The Great Portugal Quake was one of the greatest natural disasters in European history. The 8.7 magnitude earthquake that struck Portugal in 1755 killed at least 60,000 people and triggered tsunamis that wrecked seaports in Portugal, Spain, and Morocco. The earthquake's cause remained a mystery because the tectonic activity of the region was not clearly understood. The plate boundary off southern Iberia—the peninsula occupied by Spain and Portugal—is not well defined. A new study suggests that it happened as a result of subduction—the process of the oceanic lithosphere (the outer solid part of Earth) diving beneath the continental lithosphere. The study also shows continued activity in the plate system, prompting fears that another earthquake could hit the region with potentially devastating consequences, although probably not for many years to come. Recent seismic images and seafloor bathymetry instead suggest that subduction is occurring in the region, causing compressive stress to accumulate along the interface between the tectonic plates, which leads to earthquakes.

Article obtained from Lovgren, S. Great Portugal quake may have a sequel, study says. *National Geographic News.* August 30, 2004.

15. What do new images indicate was the cause of the earthquake?
    A. delamination, which occurs shortly after mountain building
    B. subduction in the region causing compressive stress between the tectonic plates
    C. major tsunamis, which wrecked seaports
    D. unknown tectonic activity in the region

16. Why is it important to research an earthquake that occurred so long ago?

NEED EXTRA HELP?

If You Missed Question . . .	1	2	3	4	5	6	7	8	9	10	11	12	13	14
Review Section . . .	14.3	14.3	20.1	18.1	7.3	8.1	8.1	20.2	13.3	13.3	13.3	14.1	18.2	10.2

SHORT ANSWER

9. The water temperature is too low, and hurricanes need warm water in order to develop.

10. Sample answer: Not every hurricane that forms in the Atlantic is going to hit North America.

11. These hurricanes most likely occurred in late summer to early fall because that is when Earth's oceans contain the greatest amount of stored heat energy to help form a hurricane.

12. Because the middle of the United States is not near any major bodies of water, the land heats up and cools down more quickly. Also, because this area is not in the mountains, the altitude level does not help in keeping the area cooler.

13. A sill forms when magma intrudes parallel to layers of rock, whereas a dike is a pluton which cuts across preexisting rocks.

14. The potential value of an environmental impact statement would be to identify potential effects of actions taken in a natural area so that plans can be made to protect the karst topography and the people who live there. One trade-off could be the potential extra expense incurred to achieve an acceptable level of environmental protection.

READING FOR COMPREHENSION

15. B

16. Sample answer: By studying the past earthquake and determining what caused it, scientists will be better able to monitor the area and possibly forecast any future earthquakes, thus protecting the people in the area from another devastating earthquake.

UNIT 6

Geologic Time

Themes

Change From the beginnings of Earth and its continents, oceans, and life, change over time has been the one constant.

Structures Fossils are evidence of past life-forms and the key to identifying the eras of past life, organisms structures, paleoecologies, and lifespans.

Geologic Time Data obtained by observing rock layers and fossils, and calculating radioactive decay, establishes a timeline of Earth's past.

Systems The application of principles of layering and the identification of relationships among Earth systems have clarified the slow system of Earth's building and destructive processes.

Scientific Inquiry Using the rock record, scientists have been able to piece together the history of life on Earth, including its radiations and mass extinctions. Identifying the causes of each is an ongoing task.

UNIT 6

Geologic Time

CHAPTERS

21 Fossils and the Rock Record

22 The Precambrian Earth

23 The Paleozoic, Mesozoic, and Cenozoic Eras

STEM Project

CAREERS IN EARTH SCIENCE **Paleontologist**

This **paleontologist** is uncovering the remains of a mammoth that died over 23,000 years ago. Paleontologists spend much of their time in the field, piecing together Earth's history using fossils.

Introduce the Unit

Preconceptions The photograph shows scientists excavating remains of mammoths from exposed bedrock. Ask students what they think layers of rock have to do with studying geologic time and changes in Earth's plants and animals. Explain that these rocks are almost always sedimentary.

Ancient Environments Ask students what they recall about the deposition and formation of sedimentary rocks. Can they interpret the environments in which these rocks were deposited? Next, have students examine the photo on the following page. In what types of environments were these rocks deposited? Explain to students that a shallow sea once covered much of what is now part of the central United States and that the rock layers represent onshore-to-offshore environments that existed during millions of years of time. Then, ask students what they might find in certain types of limestone and what information can be inferred from the limestone. Tell students that fossils are often components of limestone and that different types of fossils indicate whether the water that the sediments and fossils were deposited in was deep or shallow, as well as saline, fresh, or brackish.

Changing Scenery Have students again examine the photo and ask them whether their perceptions of rocks have changed. Students should now perceive that rock layers contain information about the depositional environment at the time of their formation.

CHAPTER 21 Fossils and the Rock Record

BIGIDEA **Scientists use several methods to learn about Earth's long history.**

ESSENTIAL QUESTIONS	NATIONAL STANDARDS	RESOURCES TO ASSESS MASTERY
SECTION 1 **The Rock Record** **1.** Why do scientists need a geologic time scale? **2.** How are eons, eras, periods, and epochs defined? **3.** What are the groups of plants and animals that dominated eras of Earth's history? 2 sessions 1 block	UCP.5; A.1; C.3; D.3; F.3; G.1–3	**Entry-Level Assessment** Focus Transparency 63, p. 590 **Progress Monitoring** Caption Question, pp. 591, 592 Reading Check, p. 592 Section Review, p. 594
SECTION 2 **Relative-Age Dating** **1.** How is uniformitarianism defined and what is its importance to geology? **2.** What geologic principles are used to interpret rock sequences and determine relative ages? **3.** What are the different types of unconformities and how do they differ? **4.** How do scientists use correlation to understand the history of a region? 2 sessions 1 block	UCP.2; A.1; D.3; G.3	**Entry-Level Assessment** Focus Transparency 64, p. 595 **Progress Monitoring** Caption Question, pp. 597, 599 Reading Check, p. 598 Section Review, p. 600
SECTION 3 **Absolute-Age Dating** **1.** What are the differences between absolute-age dating and relative-age dating? **2.** How are radioactive elements used to date rocks and other objects? **3.** How can scientists use certain non-radioactive materials to date geologic events? 1 session 0.5 block	B.2–3, B.6; D.2–3; E.1–2; F.6	**Entry-Level Assessment** Focus Transparency 65, p. 601 **Progress Monitoring** Caption Question, pp. 602, 604 Reading Check, pp. 603, 604 Section Review, p. 605
SECTION 4 **Fossil Remains** **1.** What are the methods by which fossils are preserved? **2.** How do scientists use index fossils? **3.** How are fossils used to interpret Earth's past physical and biological history? 4 sessions 2 blocks	A.1–2; B.2–3; C.3–4; D.1, D.3; G.2	**Entry-Level Assessment** Focus Transparency 66, p. 606 **Progress Monitoring** Caption Question, pp. 607, 608 Reading Check, pp. 607, 608 Section Review, p. 609 **Summative Assessment** Chapter Assessment, p. 613 Exam*View*® *Assessment Suite* CD-ROM

Go online!
connectED.mcgraw-hill.com

LEVELED RESOURCES	LAB MATERIALS
Science Notebook 21.1 OL **Chapter FAST FILE Resources:** GeoLab Worksheet, p. 3 OL Study Guide, p. 15 BL **Transparencies:** Section Focus Transparency 63 EL Teaching Transparency 61 OL EL	LaunchLAB p. 588 / **40 min** sand, plastic milk carton, sponge, hot tap water, 500-mL beaker, salt, stirring rod GeoLAB p. 611 / **90 min** list of Earth-shaping events, colored pencils, poster board, geologic time scale, reference books
Science Notebook 21.2 OL **Chapter FAST FILE Resources:** MiniLab Worksheet, p. 2 OL Study Guide, p. 16 BL **Transparencies:** Section Focus Transparency 64 EL Teaching Transparency 62 OL EL	MiniLAB p. 597 / **20 min** scissors
Science Notebook 21.3 OL **Chapter FAST FILE Resources:** Study Guide, p. 18 OL **Lab Resources:** Laboratory Manual, p. 161 OL **Transparencies:** Section Focus Transparency 65 EL Teaching Transparency 63 OL EL	
Science Notebook 21.4 OL **Chapter FAST FILE Resources:** Study Guide, p. 20 OL **Lab Resources:** Laboratory Manual, p. 165 OL **Transparencies:** Section Focus Transparency 66 EL Teaching Transparency 64 OL EL	

LabManager™
Customize any lab with the LabManager™ CD-ROM.

ADDITIONAL RESOURCES

Technology:
- eTeacherEdition
- eStudentEdition
- Classroom Presentation Toolkit
- Exam*View*® *Assessment Suite* CD-ROM
- LabManager™ CD-ROM
- What's EARTH SCIENCE Got To Do With It? DVD
- Weather Classroom Video DVD

Assessment:
- Blueprints for Success
- Chapter Test (Scaffolded)
- Earth Science Standardized Test Practice

Lab Resources:
- Exploring Environmental Problems Laboratory Manual

BL Below Level OL On Level AL Advanced Learners EL English Learners COOP LEARN Cooperative Learning

CHAPTER 21

Fossils and the Rock Record

BIGIDEA Scientists use several methods to learn about Earth's long history.

SECTIONS

1 The Rock Record

2 Relative-Age Dating

3 Absolute-Age Dating

4 Fossil Remains

LaunchLAB

iLab Station

How are fossils made?

Have you ever wandered through a museum and stood beneath the fossilized bones of a *Tyrannosaurus rex*? Fossils provide evidence that dinosaurs and other ancient organisms existed. Fossilized bones and other hard parts can form when they are quickly covered by mud, sand, or other sediments, and after long periods of time, they absorb minerals from Earth and become petrified. Model fossil formation in this lab.

FOLDABLES Study Organizer

Relative-Age v. Absolute-Age Dating

Make a shutter fold using the labels shown. Use it to organize your notes on relative-age and absolute-age dating.

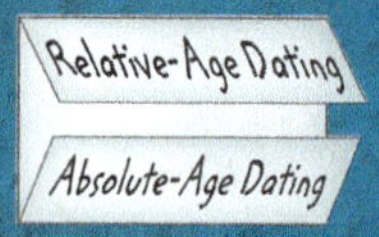

CHAPTER 21

LaunchLAB

iLab Station Rubric

Safety Precautions Students should wear safety goggles. Caution students to handle the hot tap water with care.

Teaching Strategies

- Cut off the tops of plastic quart-size milk containers before the activity begins; recycle the plastic at the end of the lab.
- Cut out the sponges in the shapes of bones.
- Allow at least seven days for the fossilized sponges to harden; if humid, or during the shorter days of winter, allow ten days.

Procedure

1. Have students read and complete the lab safety form and follow the procedure below.
2. Pour 500 mL of **sand** into a **plastic milk carton** with the top cut off.
3. Bury a **sponge** in the center of the sand.
4. Pour 250 mL of **hot tap water** into a **500-mL beaker.**
5. Measure 100 mL of **salt,** add the salt to the water, and use a **stirring rod** to stir the mixture vigorously.
6. Pour the water over the sand and place the container in direct sunlight for 5 to 7 days, leaving it undisturbed.
7. Dig up your fossilized sponge.

Analysis

1. **Describe** in your science journal what happened to the sponge. Answers will vary. Students should notice that the spaces in the sponge have been replaced by salt crystals.
2. **Explain** how this activity models the formation of a fossil. After most animals die, their soft tissues decay but their hard parts may remain. Over time, the spaces in the hard parts are replaced by minerals. The salt represents those minerals.

Assessment

Knowledge Ask students why the sponge hardened. The sponge hardens because once the sand dries out, the salt crystallizes inside the sponge's crevices, making it hard. Ask them how this models the formation of a fossil. It is similar to what happens when minerals from Earth seep inside the spaces in bones or other hard parts. The minerals crystallize and harden into the shape of the hard parts. Challenge students to design an activity to model the formation of other fossil types. Students can use honey to model the formation of a fossil in amber. They can pour the honey into a small jar, insert a plastic insect, cover the insect with more honey, and place the jar in a freezer. Students can use clay to model imprint fossils. They can flatten clay on a table and press shells, bones, plants, or other organism parts into the clay.

LabManager™
Customize any lab with the LabManager™ CD-ROM.

The land that is now Badlands National Park in South Dakota was once covered by a vast forest, then by swamp, and later by grasslands. Ancestors of alligators, camels, and rhinoceroses thrived here over 20 million years ago.

Introduce the BIGIDEA

Geologic Time Discuss with students the concept of time. Ask them to explain how they know how long ago certain events occurred. What "tools" do they use? Answers might include books, television, teachers, parents; some might say radioactive decay. Ask: Can any of these tools be used to measure geologic time? Only radiometric age-dating tools can measure geologic time. Explain that even the rock record is not sufficient to know everything about the Precambrian, because very few rocks exist from this part of Earth's history.

Teacher Content Support

Badlands This landscape of the Badlands National Park, South Dakota, depicts a typical badlands topography. Such sharply eroded buttes, pinnacles, and spires develop when high rates of erosion occur in an area containing alternating layers of hard rock (such as sandstone) and soft rock (such as shale). The shale is eroded quickly, while the harder sandstone forms steep cliffs. The different colors are formed by the oxidation of iron-rich minerals. The badlands in South Dakota contain the world's richest assortment of Oligocene mammal fossils. No dinosaur fossils have been found here because a shallow sea covered this area during the Mesozoic, when the dinosaurs lived.

SECTION 1

1 Focus

Focus Transparency

Before presenting the lesson, project **Section Focus Transparency 63** and have students answer the accompanying questions. EL

MAINIDEA

Organize Time Ask students to discuss major milestones in an average person's life. Examples include crawling, walking, potty training, driving a car, graduating, having a full-time job, being a parent, retiring. Ask: Are there any categories into which these events can be placed? Examples include infancy, toddlerhood, childhood, adolescence, adulthood. Ask: How do these categories help people communicate and analyze their life histories? Not everyone reaches the same milestone at the same time. It is useful to have references for communicating events.

2 Teach

Identify Misconceptions

Students might think that there is only one geologic history that applies to all areas of the world.

Uncover the Misconception Ask students to compare and contrast the geologic history of their state with the geologic history of another state. This information can be acquired from state Geological Survey units.

Demonstrate the Concept Explain that because the forces that constantly change Earth's surface have varied from region to region, the geologic histories of individual areas differ.

Assess New Knowledge Have students relate how the geologic time scale fits the geologic history of their local area.

SECTION 1

The Rock Record

MAINIDEA Scientists organize geologic time to help them communicate about Earth's history.

Essential Questions

- Why do scientists need a geologic time scale?
- How are eons, eras, periods, and epochs defined?
- What are the groups of plants and animals that dominated the eras of Earth's history?

Review Vocabulary

fossil: the remains, trace, or imprint of a once-living plant or animal

New Vocabulary

geologic time scale
eon
Precambrian
era
period
epoch
mass extinction

EARTH SCIENCE 4 YOU Imagine how difficult it would be to plan a meeting with a friend if time were not divided into units of months, weeks, days, hours, and minutes. By organizing geologic time into time units, scientists can communicate more effectively about events in Earth's history.

Organizing Time

A hike down the Grand Canyon reveals the multicolored layers of rock, called strata, that make up the canyon walls, as shown in **Figure 1.** Some of the layers contain fossils, which are the remains, traces, or imprints of ancient organisms. By studying rock layers and the fossils within them, geologists can reconstruct aspects of Earth's history and interpret ancient environments.

To help in the analysis of Earth's rocks, geologists have divided the history of Earth into time units. These time units are based largely on the fossils contained within the rocks. The time units are part of the **geologic time scale,** a record of Earth's history from its origin 4.6 billion years ago (bya) to the present. Since the naming of the Jurassic time period (juh RA sihk) in 1795, development of the time scale has continued to the present day. Some of the units have remained unchanged for centuries, while others have been reorganized as scientists have gained new knowledge. The geologic time scale is shown in **Figure 2.**

Figure 1 The rock layers of the Grand Canyon represent geologic events spanning nearly 2 billion years. Geologists study the rocks and fossils in each layer to learn about Earth's history during different units of time.

ACROSS THE CURRICULUM

History Sequencing the continuum of events in succession from the past to the present is a major part of how we define history. Have students research and create a time line illustrating the following events of American history: the signing of the Declaration of Independence, the ratification of the Constitution, the Louisiana Purchase, the Industrial Revolution, the Civil War, the invention of the electric lightbulb, the introduction of automobile mass production, the stock market crash of 1929, Neil Armstrong's walk on the Moon, and the American bicentennial. OL

VISUALIZING the Geologic Time Scale

Figure 2 The geologic time scale begins with Earth's formation 4.6 billion years ago (bya). Geologists organize Earth's history according to groupings called eons. Each eon contains eras, which in turn contain periods. Each period in the geologic time scale contains epochs. The current geologic epoch is called the Holocene Epoch. Each unit on the scale is labeled with its range of time in millions of years ago (mya).
Identify *the period, era, and eon representing the most modern unit of time.*

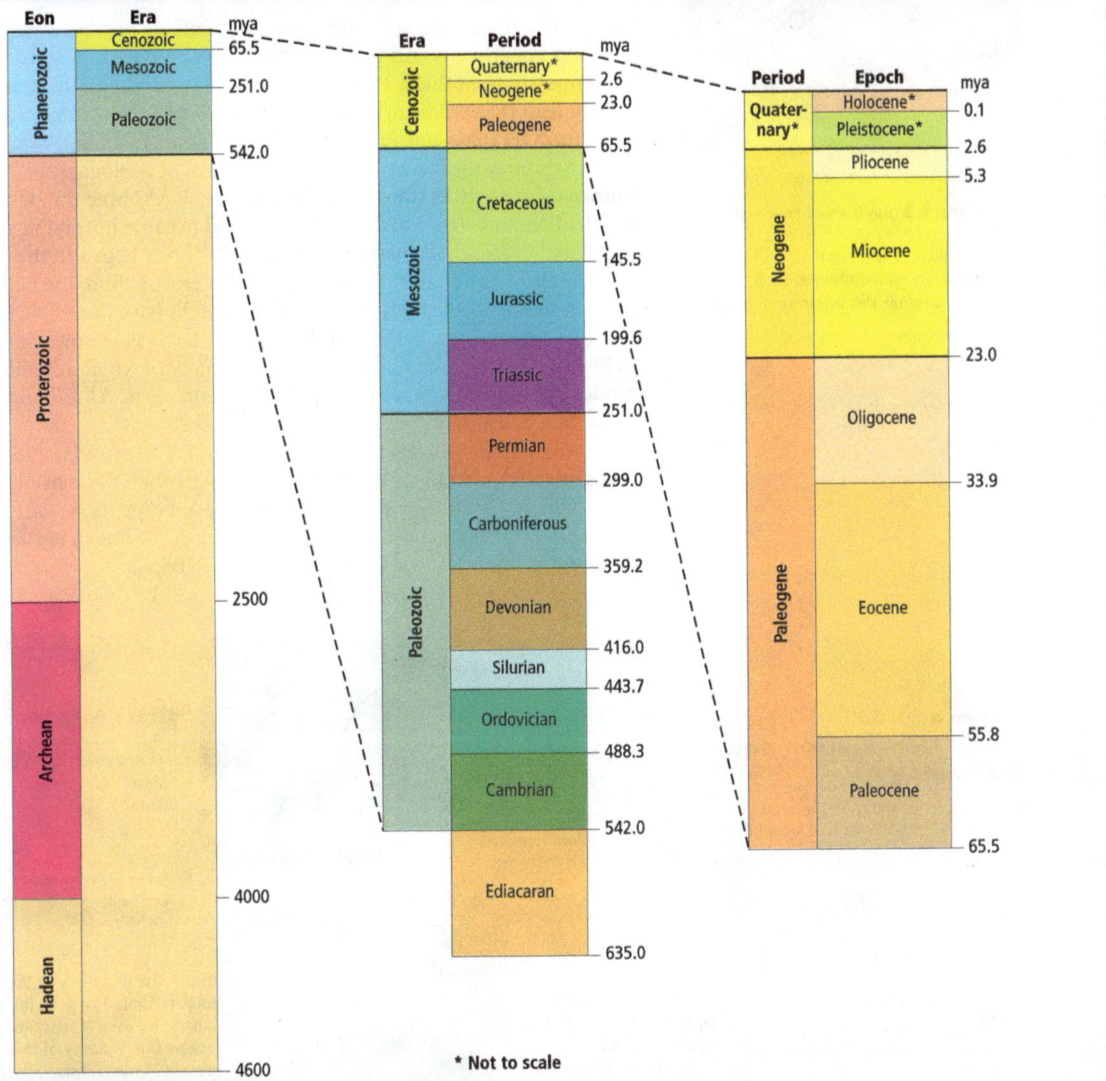

Concepts In Motion View an **animation of the geologic time scale.**

GeoLAB

iLab Station

The GeoLab located at the end of the chapter can be used at this point in the lesson.

DIFFERENTIATED INSTRUCTION

Advanced Learners The geologic time scale is usually depicted as a vertical continuum. Have students investigate other methods of illustrating geologic time, such as geologic time as it relates to a 24-hour day, geologic time represented by the calendar or a clock, or the geologic time scale mapped onto a football field.

VISUALIZING

Purpose

Students will learn that scientists divide geologic time into standard units that are not equal in duration: eons, eras, periods, and epochs.

Activity

Time Units Divide the class into groups and assign each group the task of learning the origin of several period names of the geologic time scale. Then ask the groups to present their findings to the class. Most names are derived from geographical places. For example, Cretaceous is derived from a chalky terrain of northern France; Permian is named after a section of rock near Perm, Russia; Devonian is named after exposures of sandstone in Devon, England. **OL** **COOP LEARN**

Teacher Content Support

Time Scale Variations Differing sediment sequences in the United States and Europe has resulted in a slightly different naming system in the geologic time scale. Scientists in the United States divide the Carboniferous Period into the Mississippian Period (Lower Carboniferous) and the Pennsylvanian Period (Upper Carboniferous). This system was adopted to distinguish the mostly limestone layers of the Mississippian, when North America was covered by a shallow sea, from the coal-bearing layers of the Pennsylvanian, when the landscape was largely terrestrial. The boundary between these two periods is 318.1 Ma.

- **Caption Question Fig. 2** Quaternary Period, Cenozoic Era, Phanerozoic Eon

Teacher Content Support

Scientific Consensus The geologic time scale is changed only when the International Commission on Stratigraphy (ICS) approves the change. The ICS is the largest scientific body within the International Union of Geological Sciences, an organization that includes members from more than 120 countries. At regular meetings and in scientific journals, the ICS encourages discussion and debate regarding the best way to divide geologic time and the best places on Earth to demark the boundaries between the time units.

In 2004, the ICS officially revised the geologic time scale to include the Ediacaran Period and to adjust dates for other periods. In 2009, the Quaternary Period was added and more dates were adjusted, such as the upper and lower boundaries of the Pliocene. The geologic time scale in **Figure 2** shows the Hadean Eon, but the Hadean is not an official eon. Despite this, many scientists use the term, and it is likely that some day it will become official. The terms Precambrian and Tertiary are also not official, but are so commonly used that they will likely remain in use.

☑ READING CHECK The geologic time scale allows scientists to organize Earth history and to communicate effectively with other scientists.

Caption Question Fig. 3 This organism might have been sessile or undulated in the water. It is often difficult to tell how an organism moved by examining its fossil.

Figure 3 This is a well-preserved fossil, found in a sedimentary rock of the Precambrian. During that time, the first complex life-forms evolved on Earth.
Infer *whether this organism moved.*

The Geologic Time Scale

The geologic time scale enables scientists to find relationships among the geological events, environmental conditions, and fossilized life-forms that are preserved in the rock record. The oldest division of time is at the bottom of the scale, shown in **Figure 2.** Moving upward, each division is more recent, just as the rock layers in the rock record are generally younger toward the surface.

☑ READING CHECK **Explain** why scientists need a geologic time scale.

Eons The time scale is divided into units called eons, eras, periods, and epochs. An **eon** is the largest of these time units and encompasses the others. From oldest to youngest, they consist of the Hadean (HAY dee un), Archean (ar KEE un), Proterozoic (pro tuh ruh ZOH ihk), and Phanerozoic (fa nuh ruh ZOH ihk) eons.

The three earliest eons make up 90 percent of geologic time, known informally as the **Precambrian** (pree KAM bree un). During the Precambrian, Earth was formed and became hospitable to life. Fossil evidence suggests that simple life-forms began in the Archean Eon and that by the end of the Proterozoic Eon, life had evolved to the point that some organisms might have been able to move in complex ways. Most of these fossils, such as the one shown in **Figure 3,** were soft-bodied organisms, many of which resembled modern animals. Others had bodies with rigid parts. All life-forms until then had soft bodies without shells or skeletons.

Fossils dating from the most recent eon, the Phanerozoic, are the best-preserved, not only because they are younger than those from the Precambrian, but because many of them represent organisms with hard parts, which are more easily preserved. The time line in **Figure 4** shows some important fossil and age-dating discoveries.

Figure 4
Fossil Discoveries and Technology

Fossil discoveries and dating technology have changed our understanding of life on Earth.

1800 — 1840 — 1880 — 1920

1796 William Smith, a canal surveyor, creates the first geologic map based on distinct fossil layers.

1820s Mary Anning, the daughter of a cabinetmaker, finds and identifies fossils of many ancient creatures, sparking great interest in paleontology.

1857 Quarry workers uncover a skeleton identified as Neanderthal, a species similar to modern humans.

1909 The discovery of the Burgess Shale fossils in the Canadian Rocky Mountains reveals the diversity of invertebrate life that thrived during the Cambrian Period.

1929 An Anasazi ruin becomes the first prehistoric site to be dated using tree-ring chronology.

EARTH SCIENCE JOURNAL

Precambrian Time The Precambrian occupies 90 percent of geologic time. To emphasize how much time this is, help students figure out how old they were when 90 percent of their lives had passed. (0.9 × age). Then ask them to write in their journals what they remember from the first 90 percent of their lives and then what they remember from the last 10 percent. Students will likely find that the last 10 percent is better preserved in their memories, just as the last 10 percent of Earth's history is better preserved in the rock record.

Rubric

Eras All eons are made up of eras, the next-largest unit of time. **Eras** are usually tens to hundreds of millions of years in duration. Like all other time units, they are defined by the different life-forms found in the rocks. The names of the three eras of the Phanerozoic Eon are named for the relative ages of the life forms that lived during those times. For example, in Greek, *paleo* means "old," *meso* means "middle," and *ceno* means "recent." *Zoic* means "of life" in Greek; thus, *Paleozoic* means "old life," *Mesozoic* means "middle life," and *Cenozoic* means "recent life."

Periods All eras are divided into periods. **Periods** are generally tens of millions of years in duration, though some periods of the Precambrian are considerably longer. Some periods are named for the geographic region in which the rocks or fossils characterizing the age were first observed and described. Consider, for example, the Ediacaran (ee dee A kuh run) Period at the end of the Proterozoic Era. It is named for the Ediacara Hills in Australia, shown in **Figure 5.** It was here that fossils typical of the period were first found. The Ediacaran Period was added to the geologic time scale in 2004.

Epochs Epochs (EE pahks) are even smaller divisions of geologic time. Although the time scale in **Figure 2** shows epochs only for periods of the Cenozoic Era, all periods of geologic time are divided into epochs. **Epochs** are generally hundreds of thousands to millions of years in duration. Rocks and sediments from the epochs of the Cenozoic Era are the most complete because there has been less time for weathering and erosion to remove evidence of this part of Earth's history. For this reason, the epochs of the Cenozoic are relatively short in duration. For example, the Holocene (HOH luh seen) Epoch, which includes modern time, began only about 12,000 years ago.

Figure 5 The Ediacara Hills of Australia yielded the first fossils typical of the Ediacaran Period. Fossils from that time found anywhere in the world are called Ediacaran fossils.

1946 University of Chicago scientists show that the age of relatively recent organic objects and artifacts can be determined with radiocarbon dating.

1974 The most complete adult female skeleton of ***Australopithecus afarensis***, named Lucy, discovered in Northern Ethiopia.

1993 Fossils found in western Australia provide evidence that bacteria existed 3.5 bya.

2006 A 164-million-year-old, beaverlike fossil unearthed by Chinese researchers suggests that aquatic mammals might have thrived alongside dinosaurs.

1940 — 1970 — 2000

1987 Jenny Clack leads an expedition to Greenland that unearths fossils of animals that lived 360 mya, showing that animals developed legs prior to moving onto land.

2010 Scientists discover the oldest known animal fossils—spongelike creatures that lived about 650 mya—in South Australia.

IN THE FIELD

Zhou Zhonghe Zhou Zhonghe is a Chinese paleornithologist, a person who studies fossil birds. Zhou became fascinated with birds as a child in rural China. Today, he is famous for his many discoveries, in China's Liaoning Province, of some of the world's best-preserved fossil birds and feathered dinosaurs. These discoveries provide strong support for the hypothesis that birds evolved from bipedal theropod dinosaurs. They further suggest that flight might have evolved from the gliding movement of tree-dwelling creatures. Zhou received his PhD at the University of Kansas and now works at the Chinese Academy of Sciences in Beijing. He spends a great deal of time studying skeletons of modern birds, which provide insight into the lives of birds' distant ancestors.

Apply Earth Science

Fossils in the Media Fossil discoveries are reported often in the media. Ask students what recent fossil discoveries they can recall. Ask them to bring in news articles about fossil discoveries. Ask them if some types of fossils get more press than other types. They might have noticed that fossils of dinosaurs and early hominids seem to get the most attention. BL OL EL

Discussion

Debate the Durations Divide the class into two groups. Have one group make an argument for changing the time scale to equal time units. Have the other group make an argument for keeping the scale as it is, with unequal time units. Argument for equal durations: It is difficult to compare time in the geologic record when eras and periods are all of different durations. It makes more sense to divide geologic time into intervals of equal time, such as hours, days, weeks, and months, as we are accustomed to doing. Many events in the universe are of regular duration, such as Earth's rotation and the revolution of the Earth around the Sun. Also, the advent of radiometric–dating techniques has enabled scientists to determine ages with much more accuracy. Argument against equal durations: The geologic time scale is constructed around geologic and evolutionary events that do not occur on a scheduled timescale. Forcing each era and period to be the same would downplay the events that mark Earth's history. Major extinction events would fall randomly within the scale instead of marking the boundaries between time units. The names of the eras would be obsolete, as they currently represent old, middle, and new life. Also, the weight of tradition would make it difficult to change. OL COOP LEARN

Discussion

Extinction Have the class brainstorm reasons why a species becomes extinct. Have them write in their Earth science journals a news report of the future, announcing the extinction of a species that today is endangered. OL COOP LEARN

3 Assess

Check for Understanding

Reinforcement Ask students to assess the importance of fossils in the development of the geologic time scale. Discuss with them what disciplines paleontologists must study to become knowledgeable about fossils. Students should see the interconnectedness between the biological and Earth sciences.

Reteach

Make a Time Line Have each student make a list of at least ten of the most important events that have occurred in his or her life. Then have each student create a time line entitled *Great Events of My Life.*

Assessment

Knowledge Have students list the eons and eras of the geologic time scale and the approximate number of years that each unit represents. Ask students to briefly discuss how these time units help geologists interpret Earth's history.

Figure 6 Trilobites are Paleozoic arthropod fossils found all over the world. Like 95 percent of marine life-forms of that era, they perished during a mass extinction event.

Succession of Life-Forms

During the Phanerozoic Eon, multicellular life diversified. Fossils from the Phanerozoic are abundant, while those from the Precambrian are relatively few. The word *Phanerozoic* means "visible life" in Greek. During the first era of the Phanerozoic, the Paleozoic (pay lee uh ZOH ihk), the oceans were home to many different kinds of organisms. Small, segmented animals called trilobites, like the ones shown in **Figure 6,** were among the first hard-shelled life-forms. Trilobites dominated the oceans in the early part of the Paleozoic Era; land plants appeared later, followed by land animals. Swamps of the Carboniferous (kar buh NIH fuh rus) Period provided the plant material that developed into the coal deposits of today. The end of the Paleozoic is marked by the largest mass extinction event in Earth's history. In a **mass extinction,** many groups of organisms disappear from the rock record at about the same time. At the end of the Paleozoic, 95 percent of all marine organisms became extinct.

The age of dinosaurs The era following the Paleozoic—the Mesozoic (mez uh ZOH ihk)—is known for the emergence of dinosaurs, but many other organisms also appeared during the Mesozoic. Large predatory reptiles ruled the oceans, and corals closely related to today's corals built huge reef systems. Water-dwelling amphibians began adapting to terrestrial environments. Insects, some as large as birds, thrived. Mammals evolved and began to diversify. Flowering plants and trees emerged. The end of the Mesozoic is marked by a large extinction event. Many groups of organisms became extinct, including the non-avian dinosaurs and large marine reptiles.

The rise of mammals During the era that followed—the Cenozoic (sen uh ZOH ihk)—mammals began to dominate the land, and increased both in number and diversity. Human ancestors, the first primates, emerged in the epoch called the Paleocene, and modern humans appeared in the Pleistocene (PLYS tuh seen) Epoch.

SECTION 1 REVIEW

Section Self-Check

Section Summary

- Scientists organize geologic time into eons, eras, periods, and epochs.
- Scientists divide time into units based largely on fossils of plants and animals.
- The Precambrian makes up nearly 90 percent of geologic time.
- The geologic time scale changes as scientists learn more about Earth.

Understand Main Ideas

1. MAINIDEA **Explain** the purpose of the geologic time scale.
2. **Distinguish** among eons, eras, periods, and epochs, using specific examples.
3. **Describe** the importance of extinction events to geologists.
4. **Explain** why scientists know more about the Cenozoic than they do about other eras.

Think Critically

5. **Discuss** why scientists know so little about Precambrian Earth.

MATH IN Earth Science

6. Make a bar graph that shows the relative percentage of time spanned by each era of the Phanerozoic Eon. For more help, refer to the *Skillbuilder Handbook*.

SECTION 1 REVIEW

1. The geologic time scale helps scientists organize Earth history and communicate their research to others.
2. Eons are the most-inclusive subdivisions of time, followed by eras, periods, and epochs. Examples will vary, but might include the time we live: Phanerozoic Eon, Cenozoic Era, Quaternary Period, and Holocene Epoch.
3. Because extinction events are easy to identify in the fossil record and usually occur in a relatively brief amount of time, they provide a way for geologists to subdivide time.
4. There has been less time for erosion and weathering to obscure the rock record for this part of Earth's history.
5. Scientists know little about this time because few animals with hard parts existed then. Also, rocks from that time are so old and distorted that they often contain no record of the geologic events.
6. Graphs should include all the eras listed on the geologic time scale showing percentages as follows: Cenozoic, 12 percent; Mesozoic, 34 percent; Paleozoic, 54 percent.

SECTION 2

Essential Questions

- How is uniformitarianism defined and what is its importance to geology?
- What geologic principles are used to interpret rock sequences and determine relative ages?
- What are the different types of unconformities and how do they differ?
- How do scientists use correlation to understand the history of a region?

Review Vocabulary

granite: a coarse-grained, intrusive igneous rock

New Vocabulary

uniformitarianism
relative-age dating
original horizontality
superposition
cross-cutting relationship
principle of inclusions
unconformity
correlation
key bed

Relative-Age Dating

MAINIDEA Scientists use geologic principles to learn the sequence in which geologic events occurred.

EARTH SCIENCE 4 YOU **If you were to put the following events into a time sequence of first to last, how would you do it? Go to school. Wake up. Put on your clothes. Eat lunch. You would probably rely on your past experiences. Scientists also use information from the past to place events into a likely time sequence.**

Interpreting Geology

Recall that Earth's history stretches back billions of years. Scientists have not always thought that Earth was this old. Early ideas about Earth's age were generally placed in the context of time spans that a person could understand relative to his or her own life. This changed as people began to explore Earth and Earth processes in scientific ways. James Hutton, a Scottish geologist who lived in the late 1700s, was one of the first scientists to think of Earth as very old. He attempted to explain Earth's history in terms of geologic forces, such as erosion and sea-level changes, that operate over long stretches of time. His work helped set the stage for the development of the geologic time scale.

Uniformitarianism Hutton's work lies at the foundation of **uniformitarianism,** which states that geologic processes occurring today have been occurring since Earth formed. For example, if you stand on the shore of an ocean and watch the waves come in, you are observing a process that has not changed since the oceans were formed. The waves crashing on a shore in the Jurassic Period were much like the waves crashing on a shore today. The photo in **Figure 7** was taken recently on a beach in Oregon, but a beach in the Jurassic Period probably looked very similar.

Figure 7 An ancient Jurassic beach probably looked much like this beach in Oregon. The geologic processes that formed it are unchanged.

SECTION 2

1 Focus

Focus Transparency

Before presenting the lesson, project **Section Focus Transparency 64** and have students answer the accompanying questions. EL

MAINIDEA

Relative Age Make a list of ten historical events that have occurred in the past 15 years, such as the election of a president, a home-run record, or a state title that your high school has received. Cut the items in the list into strips. Give each strip to a group of 2–4 students, and then have the class work together to place the events in order from oldest to youngest. This will give students an idea of the concept of relative age—and also of the need to communicate results. OL COOP LEARN

2 Teach

Enrichment

Catastrophism Have students research the difference between uniformitarianism and catastrophism, and find out how catastrophism has been incorporated into ideas of Earth history. Then have them present their findings to the class. Catastrophism is the geologic doctrine that major changes in Earth's crust are the result of sudden, catastrophic events, such as a meteorite impact, rather than arising from gradual processes of change. Today, most geologists think Earth's history is a slow, gradual story punctuated by occasional catastrophic events. AL

ACROSS THE CURRICULUM

Astronomy Uniformitarianism applies to astronomy just as it does to Earth science. In order to understand the history of the universe, astronomers commonly use the idea of the light-year. A light-year is the distance light travels in one year (approximately 9,460,800,000,000 km). Because light always travels at the same speed (299,792,458 m/s), the concept of uniformitarianism underlies the ability of astronomers to determine distances. For example, the Virgo galaxy cluster is about 65 million light-years from Earth. This means that the light that is reaching us today from the Virgo cluster left the cluster at about the time that dinosaurs went extinct.

Use Science Terms

Chronology Have students research the etymology and definition of the word *chronology.* Chronology is the arrangement of events in time. *Chronos* (or *khronos*) is Greek for *time,* and *logos* is Greek for *word,* used here in the form *–logy* (*ology*), which means *pertaining to a subject of study or interest.* Ask students to identify other words with these roots, such as *psychology* the study of the human mind, *chronicle* detailed narrative record or report, and *chronic* persistent; long lasting.

Identify Misconceptions

After studying uniformitarianism, students might think that Earth's surface cannot change dramatically in a short period of time as a result of weathering or erosion.

Uncover the Misconception

Have students predict what will happen to a pile of loose soil or sand during a rainstorm. Then, have them observe the material during an actual rainstorm.

Demonstrate the Concept

Show before and after pictures demonstrating how floods or hurricanes have changed landscapes or coastlines. (Images can be found on both the USGS and NASA Web sites.) Explain that weathering and erosion can be accelerated by certain weather conditions.

Assess New Knowledge

Have students compile a list of natural forces that cause changes on Earth's surface. Examples include water, wind, ice, and temperature changes. Have them discuss how each force contributes to change over the short term and long term. For example, water can erode quickly in a hurricane but slowly in a stream. OL

FOLDABLES® Rubric

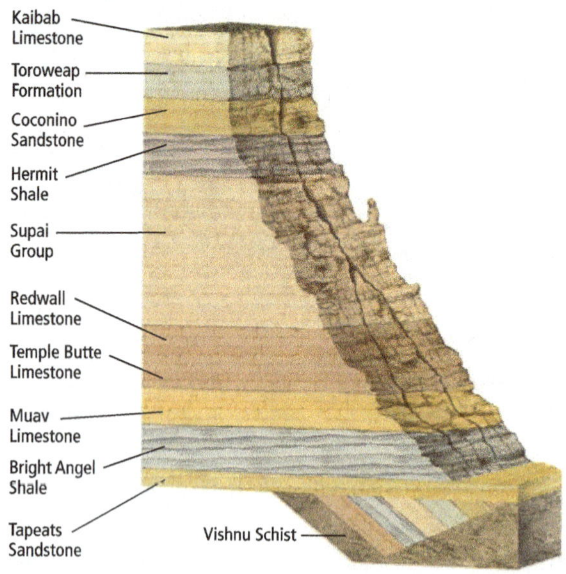

■ **Figure 8** The horizontal layers of the Grand Canyon were formed by deposition of sediment over millions of years. The principle of original horizontality states that the tilted strata at the bottom were deposited horizontally and later tilted by geologic forces.

Principles for Determining Relative Age

Scientists are able to learn about the past by studying the present. One way to do this is by studying the order in which geologic events occurred using a process called **relative-age dating.** This does not allow scientists to determine exactly how many years ago an event occurred, but it gives scientists a clearer understanding about geologic events in Earth's history. Scientists use several methods, or principles, to determine relative ages. These principles include original horizontality, superposition, cross-cutting relationships, and inclusions.

FOLDABLES® Incorporate information from this section into your Foldable.

Original horizontality **Original horizontality** is the principle that sedimentary rocks are deposited in horizontal or nearly horizontal layers. This can be seen in the walls of the Grand Canyon, illustrated in **Figure 8.** Sediment is deposited in horizontal layers for the same reason that layers of sand on a beach are mostly flat; that is, gravity combined with wind and water spreads them evenly.

Superposition Geologists cannot determine the numeric ages of most rock layers in the Grand Canyon using relative-age dating methods. However, they can assume that the oldest rocks are at the bottom and that each successive layer above is younger. Thus, they can infer that the Kaibab Limestone at the top of the canyon is much younger than the Vishnu Schist, which is at the bottom. This is an application of **superposition,** the principle that in an undisturbed rock sequence, the oldest rocks are at the bottom and each consecutive layer is younger than the layer beneath it.

VOCABULARY
ACADEMIC VOCABULARY
Principle
a general hypothesis that has been tested repeatedly; sometimes also called a law
The geologic principle was illustrated in the rock layers the students observed.

DIFFERENTIATED INSTRUCTION

Struggling Learners Have students use materials of different textures to build simple models of a geologic sequence of strata. For example, foam board could represent sandstone, clay or cardboard could represent shale, rubber foam could represent limestone, and bubble wrap could represent conglomerate. Tell them that the sandstone is the oldest, followed by shale, limestone, and conglomerate, and that they should build the model in a sequence from oldest to youngest. EL

Cross-cutting relationships Rocks exposed in the deepest part of the Grand Canyon are mostly igneous and metamorphic. Within the metamorphic schist of the Vishnu Group in the bottom sequence are intrusions—also called dikes—of granite, as shown in **Figure 9.** You have learned that intrusions are rocks that form when magma intrudes and solidifies in existing rock. The principle of **cross-cutting relationships** states that an intrusion is younger than the rock it cuts across. Therefore, the granite intrusion in the Grand Canyon is younger than the schist because the granite cuts across the schist.

The principle of cross-cutting relationships also applies to faults. Recall that a fault is a fracture in Earth's crust along which movement takes place. Many faults exist in earthquake-prone areas, such as California, and in ancient, mountainous regions, such as the Adirondacks of New York. A fault is younger than the strata and surrounding geologic features because the fault cuts across them.

Inclusions Relative age can also be determined where one rock layer contains pieces of rock from the layer next to it. This might occur after an exposed layer has eroded and the loose material on the surface has become incorporated into the layer deposited on top of it. The **principle of inclusions** states that the fragments, called inclusions, in a rock layer must be older than the rock layer that contains them.

As you have learned, once a rock has eroded, the resulting sediment might be transported and redeposited many kilometers away. In this way, a rock formed in the Triassic Period might contain inclusions from a Cambrian rock. Inclusions can also form from pieces of rock that are trapped within a lava flow.

Figure 9 According to the principle of cross-cutting relationships, this igneous intrusion is younger than the schist it cuts across.
Infer *how the igneous intrusion was formed.*

MiniLAB

iLab Station

Determine Relative Age

How is relative age determined? Scientists use geologic principles to determine the relative ages of rock layers.

Procedure

1. Read and complete the lab safety form.
2. Draw a diagram showing four horizontal layers of rock. Starting from the bottom, label the layers *1* through *4*.
3. Draw a vertical intrusion from Layer 1 through Layer 3.
4. Label a point at the bottom left corner of the diagram *X* and a point at the top right corner *Y*.
5. Cut the paper in a diagonal line from X to Y. Move the top-left piece 1.5 cm along the cut.

Analysis

1. **Describe** what principles you would use to determine the relative ages of the layers in your diagram.
2. **Explain** how the principle of cross-cutting relationships can help you determine the relative age of the vertical intrusion.
3. **Infer** what the XY cut represents. Is the XY cut older or younger than the surrounding layers?

Caption Question Fig. 9 An igneous intrusion filled a fracture. The igneous material cooled, crystallized, and subsequently was folded into the form shown in the picture.

MiniLAB

iLab Station Rubric

Purpose Students will demonstrate how principles of relative-age dating are used.

Process Skills recognize cause and effect, interpret data, observe, infer, model

Safety Precaution Approve lab safety forms before work begins.

Teaching Strategy Have students work in pairs.

Expected Results Students will observe that relative ages of rock layers are determined by comparing one layer or structure to another.

Analysis

1. The principles of superposition, original horizontality, and cross-cutting relationships would be used.
2. Because of the principle of cross-cutting relationships, which states that the intrusion is younger than the rock it cuts across, the vertical intrusion is younger than Layers 1–3.
3. The XY line represents a fault. It is younger than the surrounding layers and the vertical intrusion because it cuts across them.

Assessment

Performance Divide students into groups of three. Have them examine a photo of the Great Angular Unconformity in the Grand Canyon. Ask them to determine the relative ages of the rocks and events in the photograph, and write their observations and analyses in their Earth science journals.

LabManager™
Customize any lab with the LabManager™ CD-ROM.

Teacher Content Support

Multiple Unconformities Depending on the cause of an unconformity, more than one unconformity can be present across the same rock layer. For example, a batholith intrusion will cause rock layers to fold upward. Subsequent erosion and burial will result in the formation of an angular unconformity on either side of the fold and a disconformity or nonconformity at the top of the fold.

Model

Geologic Principles Supply groups of students with at least four different colors of modeling clay. Ask each group of students to construct a model that represents the principles of original horizontality, superposition, and cross-cutting relationships, as well as an angular unconformity. Have students draw and label the four models in their Earth science journals. OL COOP LEARN

☑ READING CHECK In a disconformity, sedimentary rock overlies sedimentary rock. In a nonconformity, sedimentary rock overlies metamorphic or igneous rock.

View an **animation of an angular unconformity.**

Concepts In Motion

Figure 10 An unconformity is any erosional surface separating two layers of rock that have been deposited at different times. The three types of unconformities are illustrated below.

Unconformities Earth's surface is constantly changing as a result of weathering, erosion, earthquakes, volcanism, and other processes. This makes it difficult to find a sequence of rock layers in the geologic record in which a layer has not been disturbed. Sometimes, the record of a past event or time period is missing entirely. For example, if rocks from a volcanic eruption erode, the record of that eruption is lost. If an eroded area is covered at a later time by a new layer of sediment, the eroded surface represents a gap in the rock record. Buried surfaces of erosion are called **unconformities.** The rock immediately above an unconformity is sometimes considerably younger than the rock immediately below it. Scientists recognize three different types of unconformities, which are illustrated in **Figure 10.**

Disconformity When a horizontal layer of sedimentary rock overlies another horizontal layer of sedimentary rock that has been eroded, the eroded surface is called a disconformity. Disconformities can be easy to identify when the eroded surface is uneven. When the eroded surface is smooth, disconformities are often hard to see.

Nonconformity When a layer of sedimentary rock overlies a layer of igneous or metamorphic rock, such as granite or marble, the eroded surface is easier to identify. This kind of eroded surface is called a nonconformity. Both granite and marble form deep in Earth. A nonconformity indicates a gap in the rock record during which rock layers were uplifted, eroded at Earth's surface, and new layers of sedimentary rock formed on top.

☑ READING CHECK **Distinguish** between a disconformity and a nonconformity.

Angular unconformity When horizontal layers of sedimentary rock are deformed during mountain building or other geologic events involving compressional forces, they are usually uplifted and tilted. During this process, the layers are exposed to weathering and erosion. If horizontal layers of sedimentary rock are later laid down on top of the tilted, eroded layers, the resulting unconformity is called an angular unconformity. Angular unconformities indicate the complex history of compression and erosion.

Demonstration

Great Unconformity Show students pictures of different kinds of unconformities. (You can find these on the Internet.) Ask them to identify each one. Be sure to include an example of the Great Unconformity, which is a gap in the rock record between the Precambrian and the Cambrian that exists in many places throughout the world. It is especially visible in the Grand Canyon. The Great Unconformity represents 2.5 million–1.2 billion years of "lost" Earth history.

Correlation The Kaibab Limestone layer rims the top of the Grand Canyon in Arizona, but it is also found more than 100 km away at the bottom of Zion National Park in Utah. How do geologists know that these layers, which are far apart from each other, formed at the same time? One method is by correlation (kor uh LAY shun). **Correlation** is the matching of rock outcrops or fossils exposed in one geographic region to similar outcrops or fossils exposed in other geographic regions. Through correlation of many different layers of rocks, geologists have determined that Zion National Park, Bryce Canyon, and the Grand Canyon are all part of one layered sequence called the Grand Staircase, as illustrated in **Figure 11.**

Key beds Distinctive rock layers are sometimes deposited over wide geographic areas as a result of a large meteorite strike, volcanic eruption, or other brief event. For example, the key-bed ash layer that marks the 1980 eruption of Mount St. Helens can be found in many states in the U.S. and in parts of Canada. Because these types of layers are easy to recognize, they help geologists correlate rock formations in different geographic areas where the layers are exposed. A rock or sediment layer used as a marker in this way is called a **key bed.** Using the principle of superposition, geologists know that the layers above a key bed are younger than the layers below it.

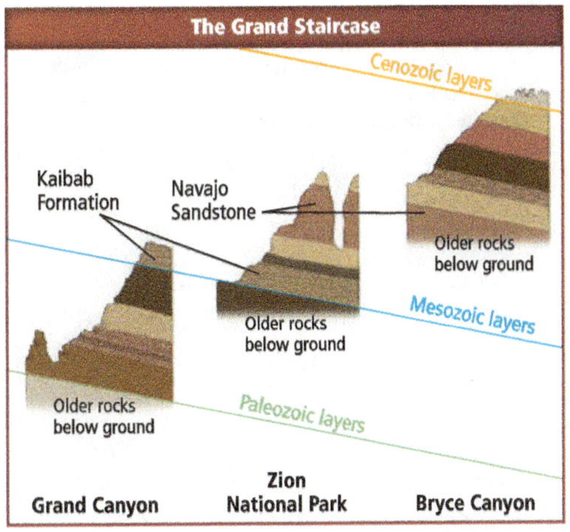

Figure 11 The top layers of rocks at the Grand Canyon are identical to the bottom layers at Zion National Park, and the top layers at Zion are the same as the bottom layers at Bryce Canyon.
Infer *the makeup of the buried layer below Zion's Kaibab layer.*

Caption Question Fig. 11 The layer below the Kaibab layer at Zion would likely be the same layer that is immediately below the Kaibab layer at the Grand Canyon.

Problem-Solving LAB

Interpret the Diagram

How do you interpret the relative ages of rock layers? The diagram at right illustrates a sequence of rock layers. Geologists use the principles of relative-age dating to determine the order in which layers such as these were formed.

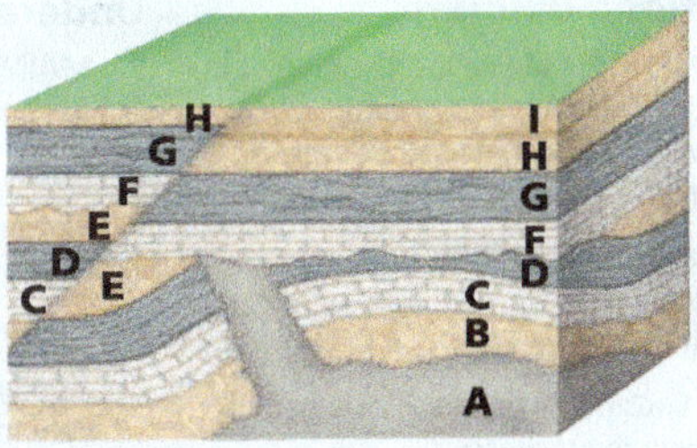

Analysis

1. **Identify** a type of unconformity between any two layers of rock. Justify your answer.
2. **Interpret** which rock layer is oldest.
3. **Infer** where inclusions might be found. Explain.
4. **Compare and contrast** the rock layers on the right and left sides of the diagram. Why do they not match?

Think Critically

5. **Apply** Which feature is younger, the dike or the folded strata? What geologic principle did you use to determine your answer?
6. **Propose** why there is no layer labeled *I* on the left side of the diagram.

Problem-Solving LAB

Purpose Students will interpret the geologic history of the illustrated rock sequence.

Process Skills interpret scientific illustrations, observe, infer, recognize cause and effect

Teaching Strategies

- Assign student pairs.
- Have them record their observations and answers in their Earth science journals.
- Ask selected groups to share their analyses of the block diagram with the class.
- Lead a class discussion.

Analysis

1. The boundary between rock layers A-F is a nonconformity because a sedimentary layer was deposited on an eroded layer of igneous rock. Layers E-F represent a disconformity because a layer of sedimentary rock was deposited on an eroded layer of sedimentary rock. D-F could be interpreted as a nonconformity (because a layer of sedimentary rock was deposited on an eroded layer of igneous rock) or as angular unconformity (because layer D is eroded and tilted).
2. Layer B is the oldest rock layer.
3. Inclusions might be found in the igneous rock of the dike from A that intruded through layers B, C, D, and part of E.
4. The two sides of the diagram do not match because a fault uplifted the left-hand side.

Think Critically

5. The dike is younger because it intrudes across the folded strata. This conforms to the principle of cross-cutting relationships.
6. The I layer might have eroded on the left side of the diagram because it was uplifted and thus more exposed to erosion.

Demonstration

Correlation To demonstrate correlation, make a large sub sandwich. Add layers of ham, turkey, cheese, tomato, or whatever ingredients you have on hand. (Alternatively, you could use such food items as gelatin, marzipan, polenta, and grits.) The layers should be consistent across the sandwich but have uneven thickness; you could, for instance, place two slices of ham at one end, and five at the other. Cut the sandwich 2.5 cm from each end and at several spots throughout. With the class, correlate the different units from one side of the sandwich to the other. Notice that while the thickness of the layers changes, the overall sequence is the same. Discuss with them how correlation can be used to predict the order and thickness of an area that is not yet sliced.

3 Assess

Check for Understanding

Reinforcement Quiz students orally about the principles involved in determining the relative ages of sequences of strata.

Reteach

Illustrate the Concept Provide visual illustrations of natural settings to show what principles could be used to relative-age date the rocks. You could use pictures of the Grand Canyon or the gorge in New York's Letchworth State Park, known as the Grand Canyon of the East, or you could use pictures of settings local to your area.

Assessment

Skill Have individual students come to the board and draw sediment sequences that illustrate the principles of original horizontality, superposition, cross-cutting relationships, inclusions, and unconformities. Students should be prepared to briefly explain their drawings to the class.

CAREERS IN EARTH SCIENCE

WebQuest

■ **Figure 12** Correlating fossils from rock layers in one location to fossils from rock layers in another location shows that the layers were deposited during roughly the same time period.

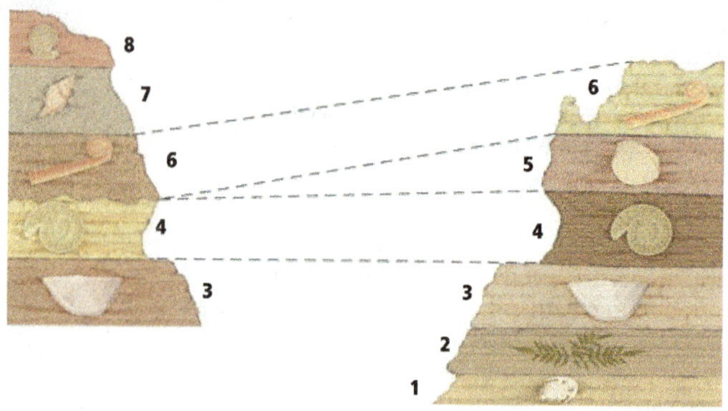

CAREERS IN EARTH SCIENCE

Petroleum Geologist Petroleum geologists use geologic principles to identify petroleum and natural gas reserves in the rock record.

WebQuest

Fossil correlation Geologists also use fossils to correlate rock formations in locations that are geographically distant. As shown in **Figure 12,** fossils of organisms that lived at the same time can be correlated across large regions. Fossils can indicate similar times of deposition even though the sediments in which they were deposited, and resulting rocks, might be entirely different.

The correlation of fossils and rock layers aids in the relative dating of rock sequences and helps geologists understand the history of larger geographic regions. Petroleum geologists also use correlation to help them locate reserves of oil and gas. For example, if a sandstone layer in one area contains oil, it is possible that the same layer in other areas also contains oil. It is largely through correlation that geologists have constructed the geologic time scale.

SECTION 2 REVIEW

Section Self-Check

Section Summary

- The principle of uniformitarianism states that processes occurring today have been occurring since Earth formed.
- Scientists use geologic principles to determine the relative ages of rock sequences.
- An unconformity represents a gap of time in the rock record.
- Geologists use correlation to compare rock layers in different geographic areas.

Understand Main Ideas

1. MAINIDEA **Summarize** the principles that geologists use to determine relative ages of rocks.
2. **Make a diagram** to compare and contrast the three types of unconformities.
3. **Explain** how geologists use fossils to determine the relative ages of rock layers within a large region.
4. **Discuss** how a coal seam might be used as a key bed.
5. **Apply** Explain how the principle of uniformitarianism would help geologists determine the source of a layer of particular igneous rock.

Think Critically

6. **Propose** how a scientist might support a hypothesis that rocks from one quarry were formed at the same time as rocks from another quarry 50 km away.

WRITING IN Earth Science

7. Write a paragraph that explains how an event, such as a large hurricane, might result in a key bed. Use a specific example in your paragraph.

SECTION 2 REVIEW

1. The principle of original horizontality states that layers are deposited in relatively horizontal layers. The principle of superposition states that younger layers are deposited on top of older layers. The principle of cross-cutting relationships states that intrusions and faults are younger than the rocks they cut across. The principle of inclusions states that inclusions are younger than the rocks in which they are embedded.
2. The diagrams should conform to the explanations of unconformities in the text.
3. Fossils help scientists determine whether rock layers, sometimes called outcrops, in one area were formed at the same time as outcrops elsewhere. This helps them understand the deposition history of a large geographic area and the relative ages of the layers.
4. Coal is a distinctive, easily identifiable layer. A coal seam might be used as a key bed if it appears over a wide area.
5. Because the principle of uniformitarianism implies that igneous rocks formed in the past in the same way they form in the present, scientists can study how igneous rocks form today and develop hypotheses about how igneous rocks formed in the past.
6. A geologist could test the hypothesis by comparing features such as fossil content, rock types, and key beds in each quarry. If features are the same, the hypothesis is supported.
7. Answers will vary but should explain that a key bed is a layer of the same sediment that occurs over a wide geographic area. A large hurricane or dust storm might deposit a layer of sediment that becomes a key bed if it is widespread and easily recognized in the geologic record.

Rubric

SECTION 3

Essential Questions

- What are the differences between absolute-age dating and relative-age dating?
- How are radioactive elements used to date rocks and other objects?
- How can scientists use certain non-radioactive material to date geologic events?

Review Vocabulary

isotope: one of two or more forms of an element with differing numbers of neutrons

New Vocabulary

absolute-age dating
radioactive decay
radiometric dating
half-life
radiocarbon dating
dendrochronology
varve

Absolute-Age Dating

MAINIDEA Radioactive decay and certain kinds of sediments help scientists determine the numeric ages of many rocks.

EARTH SCIENCE 4 YOU If a TV programming guide listed only the order of TV shows but not the times they aired, you would not know when to watch a program. Scientists, too, find it helpful to know exactly when events occurred.

Radioactive Isotopes

As you have learned, relative-age dating is a method of comparing past geologic events based on the order of strata in the rock record. In contrast, **absolute-age dating** enables scientists to determine the numerical age of rocks and other objects. In one type of absolute-age dating method, scientists measure the decay of the radioactive isotopes in igneous and metamorphic rocks, and of the remains of some organisms preserved in sediments.

Radioactive decay Radioactive isotopes emit nuclear particles at a constant rate. Recall that an element is defined by the number of protons it contains. As the number of protons changes with each emission, the original radioactive isotope, called the parent, is gradually converted to a different element, called the daughter. For example, a radioactive isotope of uranium, U-238, will decay into the daughter isotope lead-206 (Pb-206) over a specific span of time, as illustrated in **Figure 13.** Eventually, enough of the parent decays that traces of it are undetectable, and only the daughter product is measurable. The emission of radioactive particles and the resulting change into other isotopes over time is called **radioactive decay.** Because the rate of radioactive decay is constant regardless of pressure, temperature, or any other physical changes, scientists use it to determine the absolute age of a rock or a geologic event.

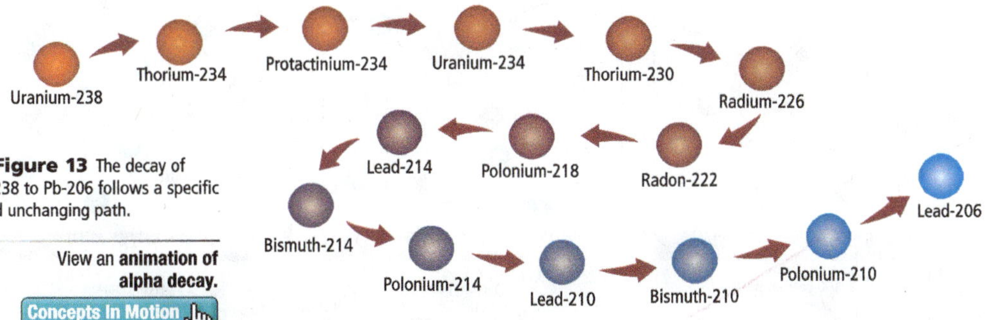

Figure 13 The decay of U-238 to Pb-206 follows a specific and unchanging path.

View an **animation of alpha decay.**

Concepts In Motion

ACROSS THE CURRICULUM

Medicine Radiation can be harmful to humans in large doses, but in small doses it plays an important role in medicine—both in diagnosis and in treatment. In diagnostic radiology, doctors or dentists use X-rays to help them diagnosis disease or damage in the body. For example, doctors can inject patients with small doses of radioactive isotopes (also called radionuclides) to help them locate disease or observe how internal organs function. Radionuclides can be natural or artificially made. Sometimes, radiation is used as a treatment to kill cancer cells. When radiation is used in this way, it is called radiotherapy.

SECTION 3

1 Focus

Focus Transparency

Before presenting the lesson, project **Section Focus Transparency 65** and have students answer the accompanying questions. EL

MAINIDEA

Absolute Time Ask students to think of different ways of telling time. Answers will include clocks, watches, hourglasses, etc. Ask them to think of things in nature that can be used to tell time. Answers might include the Sun's position in the sky or phases of the Moon. Ask them what all of these natural things have in common. Like the ticking of a clock's second hand or the sifting sands of an hourglass, the rotation of the Earth-Moon system occurs at regular intervals of known duration. To determine absolute time, scientists must use reliable systems with known duration.

2 Teach

Teacher Content Support

Atomic Clocks Students might have heard of atomic clocks and ask if they have anything to do with radioactive decay. The answer is no, because atomic clocks do not rely on atomic decay and are not radioactive. Atomic clocks work on the same principal as ordinary clocks. Instead of old-style springs or pendulums, however, the "tick" of an atomic clock comes from the natural resonance (vibrations) of atoms.

Teacher Content Support

Accurate Ages Samples of rocks or other objects ranging in age from thousands to billions of years can be accurately dated using radiometric tools. For example, cloth wrappings taken from a mummified bull in a pyramid in Dashur, Egypt, have been radiocarbon dated at 2050 years—a date that agrees with the age of the pyramid as estimated from historical documents. Samples collected from a pumice and ash deposit in Owens Valley, California, have been dated at 700,000 years. This volcanic deposit overlies glacial drift and serves as a key bed in dating the glacial period that preceded it. A quartz monzonite from Half Dome in Yosemite National Park has been dated at 80 million years. The granite at the top of Pikes Peak, Colorado, has been dated at 1.03 billion years. Samples from outcrops along the Hudson Bay that have been dated at 4.28 billion years are thought to be remnants of the Earth's primordial crust.

■ **Caption Question** **Fig. 14** There would be 50 percent daughter and 50 percent parent.

■ **Figure 14** As the number of parent atoms decreases during radioactive decay, the number of daughter atoms increases by the same amount.

Interpret ***What percentage of daughter isotope would exist in a sample containing 50 percent parent isotope?***

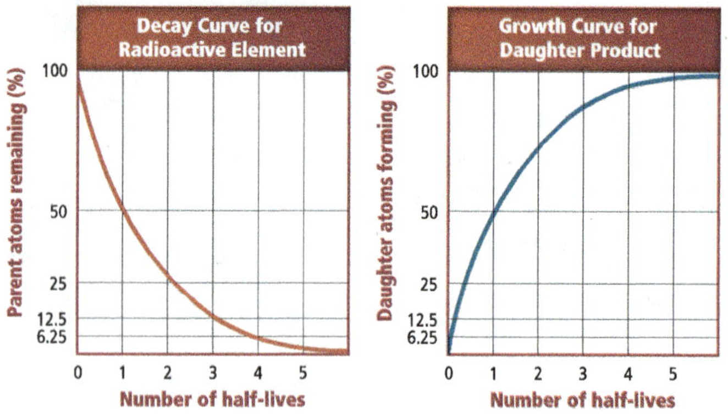

Radiometric Dating

As the number of parent atoms decreases during radioactive decay, the number of daughter atoms increases, shown in **Figure 14.** The ratio of parent isotope to daughter product in a mineral indicates the amount of time that has passed since the object formed. For example, by measuring this ratio in the minerals of an igneous rock, geologists pinpoint when the minerals first crystallized from magma. When scientists date an object using radioactive isotopes, they are using a method called **radiometric dating.**

Half-life Scientists measure the length of time it takes for one-half of the original parent isotope to decay, called its **half-life.** After one half-life, 50 percent of the parent remains, resulting in a 1:1 ratio of parent-to-daughter product. After two half-lives, one-half of the remaining 50 percent of the parent decays. The result is 25:75 percent ratio of the original parent to the daughter product—a 1:3 ratio. This process is shown in **Figure 15.**

View an **animation of half-lives.**

Concepts In Motion

■ **Figure 15** After one half-life, a sample contains 50 percent parent and 50 percent daughter. After two half-lives, the sample contains 25 percent parent and 75 percent daughter.

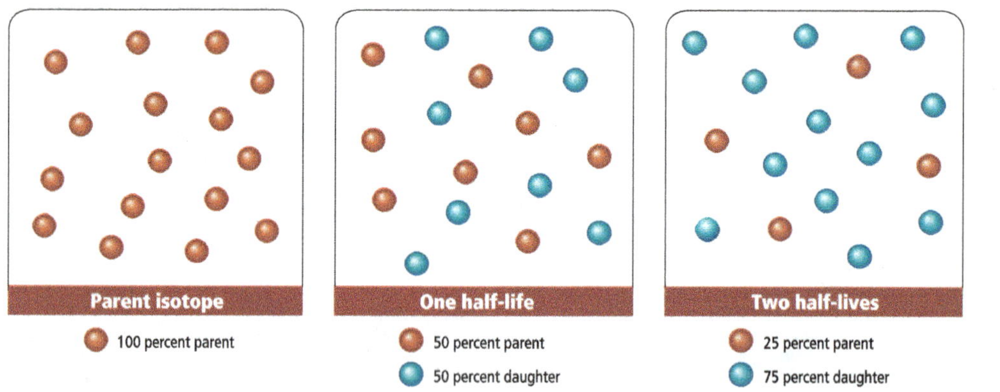

DIFFERENTIATED INSTRUCTION

Behavior Disorders To engage students with behavior disorders, ask them to create a brochure describing at least three absolute-age dating methods, such as radiometric dating, tree-ring analysis, ice cores, and varves. They should illustrate their brochures with drawings or photo reproductions of items that can be dated with each method. OL

Explore radioactive decay using an interactive table. Concepts In Motion

Table 1 Half-Lives of Selected Radioactive Isotopes

Radioactive Parent Isotope	Approximate Half-life	Daughter Product
Rubidium-87 (Rb-87)	48.8 billion years	strontium-87 (Sr-87)
Thorium-232 (Th-232)	14.0 billion years	lead-208 (Pb-208)
Uranium-238 (U-238)	4.5 billion years	lead-206 (Pb-206)
Potassium-40 (K-40)	1.3 billion years	argon-40 (Ar-40)
Uranium-235 (U-235)	0.7 billion years	lead-207 (Pb-207)
Carbon-14 (C-14)	5730 years	nitrogen-14 (N-14)

Dating rocks To date an igneous or metamorphic rock using radiometric dating, scientists examine the parent-daughter ratios of the radioactive isotopes in the minerals that comprise the rock. **Table 1** lists some of the radioactive isotopes they might use. The best isotope to use for dating depends on the approximate age of the rock being dated. For example, scientists might use uranium-235 (U-235), which has a half-life of 700 million years, to date a rock that is a few tens of millions of years old. Conversely, to date a rock that is hundreds of millions of years old, scientists might use U-238, which has a longer half life. If an isotope with a shorter half-life is used for an ancient rock, there might be a point when the parent-daughter ratio becomes too small to measure.

Radiometric dating is not useful for dating sedimentary rocks because, as you have learned, the minerals in most sedimentary rocks were formed from pre-existing rocks. **Figure 16** shows how geologists can learn the approximate age of sedimentary layers by dating layers of igneous rock that lie between them.

☑ READING CHECK **Explain** why radiometric dating is not useful for sedimentary rocks.

Figure 16 To help them determine the age of sedimentary rocks, scientists date layers of igneous rock or volcanic ash above and below the sedimentary layers.

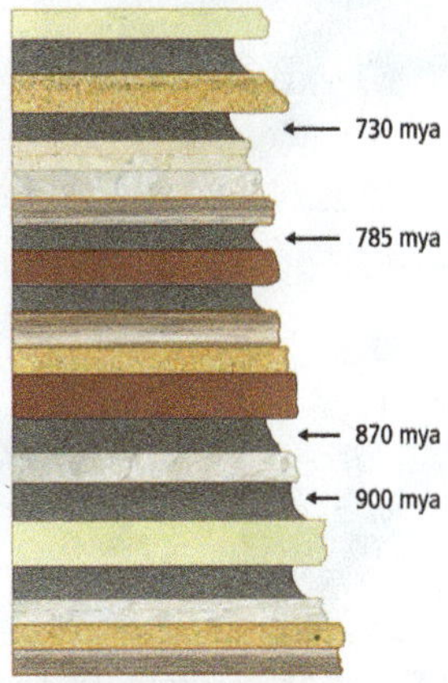

Radiometric Dating of Volcanic Ash

Radiocarbon dating Notice in **Table 1** that the half-life of carbon-14 (C-14) is much shorter than the half-lives of other isotopes. Scientists use C-14 to determine the age of organic materials, which contain abundant carbon, in a process called **radiocarbon dating.** Organic materials used in radiocarbon dating include plant and animal material such as bones, charcoal, and amber.

The tissues of all living organisms, including humans, contain small amounts of C-14. During an organism's life the C-14 decays, but is continually replenished by the process of respiration. When the organism dies, it no longer takes in C-14, so over time, the amount of C-14 decreases. Scientists can measure the amount of C-14 in organic material to determine how much time has passed since the organism's death. This method is only used for dating recent geologic events within the last 60,000 years.

Activity

Radioactive Decay Have all students stand on one side of the room, where they represent radioactive parent isotopes with a half-life of one minute. Randomly, within one minute, have half of the radioactive parent isotopes students "decay" and move to the other side of the room. There, they represent the stable daughter product. After two minutes, half of the remaining parent isotopes should "decay" and move, and after three minutes half again of the remaining parent isotopes should move. Point out that it is possible to count the number of students and determine the parent-daughter ratio, and thus determine how many half-lives have passed. OL EL COOP LEARN

Teacher Content Support

Calibration Carbon-14 is created in Earth's upper atmosphere by various forms of cosmic radiation. Because cosmic radiation fluctuates from year to year, C-14 levels vary. This changes the C-14–to–C-12 ratio and makes it difficult to date organic remains accurately. In addition, other natural events, such as volcanic activity, as well as human activities, such as fossil-fuel burning and nuclear testing, affect carbon isotope ratios. To compensate for such variations over time, scientists calibrate radiocarbon dates with dates from other sources, such as tree rings, ice cores, and varves.

☑ READING CHECK The minerals in most sedimentary rocks were formed in preexisting rocks. Their dates provide the ages of the older rocks, not the ages of the sedimentary rocks.

Demonstration

Determine the Best Method Obtain several samples or photos of old rocks and fossils, such as a Precambrian igneous rock, a Mesozoic dinosaur-bone fragment, and a Pleistocene tree fragment. Ask students what, if any, radiometric age-dating technique could be used to obtain a numeric age of each. Precambrian igneous rock: any isotope system in **Table 1** except carbon. Mesozoic dinosaur bone: none; it is too old for radiocarbon dating and it is not of igneous or metamorphic origin. Pleistocene tree fragment: radiocarbon dating if younger than 60,000 years.

Enrichment

Draw a Cross Section Have students examine a cross section of a hardwood tree (e.g., maple, ash, or box elder) obtained from a local sawmill. Have each student draw a diagram of the cross section, and from their diagram determine the age of the tree and its climate history. AL

Teacher Content Support

Lichenometry Some methods of dating rely on the growth rates of living organisms. For example, lichenometry is based on the rate of lichen growth on an exposed rock surface. This dating method is especially useful in mountains and deserts where lichens are common. In some instances, lichens can be used to date rock faces that are as old as 10,000 years. To use lichenometry as an absolute-dating method, lichen size and growth rate must be calibrated with an independent dating method, such as radiocarbon dating or dendrochronology.

Lichenometry helps scientists determine the timing and frequency of such events as rock falls and landslides. An advantage of this dating method is that it is inexpensive and can be used in the absence of any other type of dating. A disadvantage is that it is useful only for vicinities for which calibrations have been developed.

Caption Question Fig. 17
480 years (from 1500 to 1980)

READING CHECK The width of tree rings reflects seasonal environmental conditions. During years in which water and sunlight are plentiful, tree rings are wide. In dry years, they are narrow.

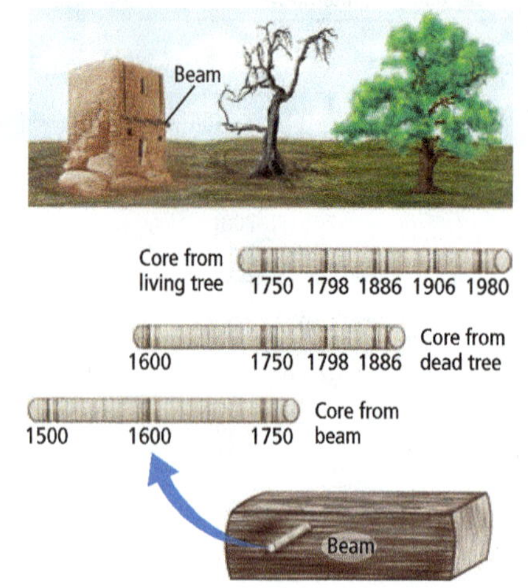

Figure 17 Tree-ring chronologies can be established by matching tree rings from different wood samples, both living and dead. The science of using tree rings to determine absolute age is called dendrochronology.
Calculate *the number of years represented in this tree-ring chronology.*

Other Ways to Determine Absolute Age

Radiometric dating is one of the most common ways for geologists to date geologic material, but other dating methods are available. Geologists also use tree rings, ice cores, and lake-bottom and ocean-bottom sediments, to help determine the ages of objects or events.

Tree rings Many trees contain a record of time in the growth rings of their trunks. These rings are called annual tree rings. Each annual tree ring consists of a pair of early-season and late-season growth rings. The width of the rings depends on certain conditions in the environment. For example, when rain is plentiful, trees grow fast and rings are wide. The harsh conditions of drought result in narrow rings. Trees from the same geographic region tend to have the same patterns of ring widths for a given time span. By matching the rings in these trees, as shown in **Figure 17,** scientists have established tree-ring chronologies that can span time periods up to 10,000 years.

READING CHECK **Describe** how tree rings can show past environmental conditions.

The science of using tree rings to determine absolute age is called **dendrochronology** and has helped geologists date relatively recent geologic events that toppled trees, such as volcanic eruptions, earthquakes, and glaciation. Dendrochronology is also useful in archaeological studies. In Mesa Verde National Park in Colorado, archaeologists used dendrochronology to determine the age of the wooden rafters in the pueblos of the Anasazi, an ancient group of Native Americans. Also, dendrochronology provides a reliable way for geologists to confirm the results from radiocarbon dating.

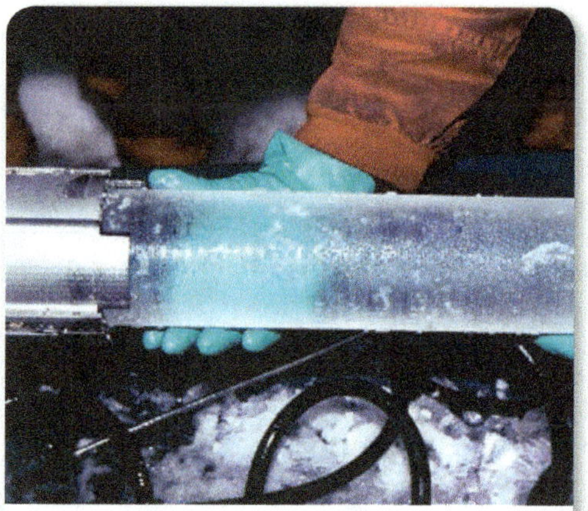

Figure 18 Ice cores are stored in facilities such as the one in Denver, Colorado. Scientists use ice cores to date glacier deposits and to learn about ancient climates.

Ice cores Ice cores are analogous to tree rings. Like tree rings, they contain a record of past environmental conditions such as temperature and atmospheric composition in annual layers of snow deposition. Summer ice tends to have more bubbles and larger crystals than winter ice. Geologists use ice-core chronologies to study glacial cycles through geologic history. The National Ice Core Facility in Colorado is one of several facilities around the world that store thousands of meters of ice cores from ice sheets, such as the core shown in **Figure 18.** Because ice cores contain information about past environmental conditions, scientists also use them to study climate change.

EARTH SCIENCE JOURNAL

Ancient Climates Have students research and list in their Earth science journals some things that scientists can learn about ancient climates by studying ice cores. Air bubbles in the ice might contain remnants of ancient atmospheres; isotopes in the ice are linked to temperature and sea-level changes; organic objects embedded within the ice carry evidence of past life; volcanic eruptions leave identifiable ash layers. AL

ACROSS THE CURRICULUM

Archaeology In the early 1900s, the American astronomer Andrew E. Douglass developed the first tree-ring chronology, using trees in the American Southwest. He used it to date pueblos. Douglass later founded the Laboratory of Tree Ring Research at the University of Arizona, which today houses more than 360,000 archaeological wood samples from around the world.

Figure 19 The alternating bands of sediment in varves help scientists date the cycles of deposition in glacial lakes.

Varves Bands of alternating light- and dark-colored sediments of sand, clay, and silt are called **varves.** Varves represent the seasonal deposition of sediments, usually in lakes. Summer deposits are generally sand-sized particles with traces of organic matter. These bands are usually lighter and thicker than the dark, fine-grained sediments that represent the winter. Varves, shown in **Figure 19,** are typical of lake deposits near glaciers, where summer meltwaters actively carry sand into the lake, and little to no sedimentation occurs in the winter. Using varved cores, scientists can date cycles of glacial sedimentation over periods as long as 120,000 years.

SECTION 3 REVIEW

Section Self-Check

Section Summary

- Techniques of absolute-age dating help identify numeric dates of geologic events.
- The decay rate of certain radioactive elements can be used as a kind of geologic clock.
- Annual tree rings, ice cores, and sediment deposits can be used to date recent geologic events.

Understand Main Ideas

1. MAINIDEA **Point out** the differences between relative-age dating and absolute-age dating.
2. **Explain** how the process of radioactive decay can provide more accurate measurements of age compared to relative-age dating methods.
3. **Compare and contrast** the use of U-238 and C-14 in absolute-age dating.
4. **Describe** the usefulness of varves to geologists who study glacial lake deposits.
5. **Discuss** the link between uniformitarianism and absolute-age dating.

Think Critically

6. **Infer** why scientists might choose to use two different methods to date a tree felled by an advancing glacier. What methods might the scientists use?

MATH IN Earth Science

7. A rock sample contains 25 percent K-40 and 75 percent daughter product Ar-40. If K-40 has a half-life of 1.3 billion years, how old is the rock?

3 Assess

Check for Understanding

Activity Ask for a student volunteer to diagram the decay of the radioactive element C-14 to N-14 through four half-lives on the board, with prompting from the class. The diagram should indicate the numeric ages for each half-life and the percentages of C-14 and N-14. 5730 years: 50 percent each C-14 and N-14; 11,460 years: 75 percent N-14, 25 percent C-14; 17,190 years: 87.5 percent N-14, 12.5 percent C-14.;22,920 years: 93.75 percent N-14, 6.25 percent C-14 COOP LEARN

Reteach

Make a Data Table Have student pairs construct data tables listing at least four commonly used radioactive elements, including U-238 and C-14. The tables should include chemical formulas, half-lives, daughter products, and names of objects that can be dated with each element. COOP LEARN

Assessment

Performance Write the following sentence on the board: The best isotope to use for dating depends on the relative age of the rock being dated. Have students write essays in their Earth science journals explaining what this sentence means.

Rubric

SECTION 3 REVIEW

1. Absolute-age dating methods provide numeric ages of rocks, while relative-age dating methods allow scientists to determine only chronological sequences.
2. Because radioactive decay occurs in a regular and predictable way, it acts like a geologic clock, providing more-accurate ages for samples than techniques of relative-age dating, which provide only relative ages.
3. Both are methods of radiometric dating. The decay of U-238 to Pb-206 is used to age-date very old rocks. The relative amount of C-14 and C-12 provides the ages of relatively young organic objects.
4. Varves help scientists date the seasonal deposition of sediments by glaciers.
5. The underlying uniformitarianism assumption in radiometric dating is that radioactivity behaves the same today as it has in the past.
6. Both radiocarbon dating and dendrochronology might be used to date the tree. If two or more data sets yield the same or similar answers, they would support each other and provide fewer margins of error.
7. 2.6 billion years old

SECTION 4

1 Focus

Focus Transparency

Before presenting the lesson, project **Section Focus Transparency 66** and have students answer the accompanying questions. EL

MAINIDEA

Fossils Have the students name some organisms that they know as fossils. Likely answers include dinosaurs, mammoths, and trilobites. Ask if these organisms lived at the same time. Some students might think that dinosaurs and mammoths lived at the same time. Ask what fossils tell about Earth's history. Life has changed through time just as environments have changed.

2 Teach

Identify Misconceptions

Students might think that impressions of mud cracks, ripples, or raindrops preserved over geologic time are fossils.

Uncover the Misconception

List a variety of fossils and nonfossils, and ask the students to identify the fossils.

Demonstrate the Concept

Explain that while many of the nonfossil objects on the list may be old, they do not fit the definition of a fossil. A fossil is defined as the remains, imprint, or trace of a once-living organism. Usually, fossils are considered to be about 10,000 years or older, though many scientists also include in the category some younger remains in which the fossilization process is not complete. These fossils, which include preserved mummies, are often called "subfossils."

SECTION 4

Fossil Remains

MAINIDEA Fossils provide scientists with a record of the history of life on Earth.

Essential Questions

- What are the methods by which fossils are preserved?
- How do scientists use index fossils?
- How are fossils used to interpret Earth's past physical and biological history?

Review Vocabulary

groundwater: water beneath Earth's surface

New Vocabulary

evolution
original preservation
altered hard part
mineral replacement
mold
cast
trace fossil
index fossil

EARTH SCIENCE 4 YOU Think about the last time you bought souvenirs while on a vacation or at an event. You might have brought back pictures of the places you saw or the people you visited, or you might have brought back objects with inscribed names and dates. Like souvenirs, fossils are a record of the past.

The Fossil Record

Fossils are the preserved remains or traces of once-living organisms. They provide evidence of the past existence of a wide variety of life-forms, most of which are now extinct. The diverse fossil record also provides evidence that species—groups of closely related organisms—have evolved. **Evolution** (eh vuh LEW shun) is the change in species over time.

When geologists find fossils in rocks, they know that the rocks are about the same age as the fossils, and they can infer that the same fossils found elsewhere are also of the same age. Some fossils, such as the radiolarian microfossils shown in **Figure 20,** also provide information about past climates and environments. Radiolarians are unicellular organisms with hard shells that have populated the oceans since the Cambrian Period. When they die, their shells can be deposited in large quantities and can form an ocean sediment called radiolarian ooze.

Petroleum geologists use radiolarians and other microfossils to determine the ages and types of rocks where oil might be found. Microfossils can also indicate whether the rocks had ever been subjected to the temperatures and pressures necessary to form oil or gas.

Figure 20 These tiny radiolarian microfossils—each no bigger than 1 mm in diameter—provide clues to geologists about ancient marine environments. This photograph is a color-enhanced SEM magnification.

Assess New Knowledge

Have students brainstorm how fossils form and in what type of rock fossils might be found. Fossils can form when organisms are buried in sediments. They are commonly found in sedimentary rocks, are sometimes found in certain igneous rocks, such as volcanic ash, and rarely found in marble—metamorphosed limestone or dolomite. OL

DIFFERENTIATED INSTRUCTION

Visually Impaired Obtain good fossil samples of a trilobite, a brachiopod, a clam, and a crinoid. Have students who are visually impaired examine the fossils. Ask them to record oral descriptions of each fossil. Play the recordings for the class, and have classmates try to identify the fossils from the descriptions. OL

Original preservation Fossils with **original preservation** are the remains of plants and animals that have been altered very little since the organisms' deaths. Such fossils are uncommon because their preservation requires extraordinary circumstances, such as either freezing, arid, or oxygen-free environments. For example, soft parts of mammoths are preserved in the sticky ooze of California's La Brea Tar Pit. Original woody parts of plants are embedded in the permafrost of 10,000-year-old Alaskan bogs. Tree sap from prehistoric trees has entrapped insects and then hardened into amber, as illustrated in **Figure 21.** Soft parts may also be preserved when plants or animals are dried and their remains are mummified.

Original preservation fossils can be surprisingly old. For example, in 2005, a scientist from North Carolina discovered soft tissue in a 70-million-year-old dinosaur bone excavated in Montana. Scientists have since found preserved tissue in other dinosaur bones.

Figure 21 This insect was trapped in tree sap millions of years ago.

READING CHECK **Explain** why fossils with original preservation are rare.

Altered hard parts Under most circumstances, the soft organic material of plants and animals decays quickly. However, over time, the remaining hard parts, such as shells, bones, or cell walls, can become fossils with **altered hard parts.** These fossils are the most common type of fossil, and can form from two processes.

Mineral replacement In the process of **mineral replacement,** the pore spaces of an organism's buried hard parts are filled in with minerals from groundwater. The groundwater comes in contact with the hard part and gradually replaces the hard part's original mineral material with a different mineral. A shell's calcite ($CaCO_3$), for example, might be replaced by silica (SiO_2). Mineral replacement can occur in trees that are buried by volcanic ash. Over time, minerals dissolved from the ash solidify into microscopic spaces within the wood. The result is a fossil called petrified wood, shown in **Figure 22.**

Figure 22 Petrified wood is an example of mineral replacement in fossils. The blowout shows that tree rings and cell walls are still evident at 100× magnification with a light microscope.
Describe *from where the minerals in the petrified wood came.*

Teacher Content Support

Paleontology Careers
Invertebrate paleontologists study fossils of animals without backbones, such as corals, sponges, and insects. Vertebrate paleontologists study fossils of animals with backbones, such as mammals, birds, and fishes. Paleobiologists study fossils as individual organisms, while paleoecologists study fossils as a means of interpreting the past. Finding spectacularly preserved fossils is rare; all paleontologists learn to recognize individual species by partial remains.

Activity

Fossil Collection Have students make fossil collections, either from actual fossils or from photographs of fossils. Descriptions should include whether the fossil was a plant or an animal, the organism's habitat, its species and/or genus name, the rock type in which it was found, and its geologic age. Have students present their collections to the class. OL

Interpret the Photo

Insect DNA Have students examine **Figure 21.** Ask students: Could the insect's DNA still be intact after million of years? No. Though scientists have extracted DNA from archaeological material less than 50,000 years old, most think it unlikely that DNA, which is fragile, can exist in amber-trapped insects millions of years old.

READING CHECK It is rare to find arid, freezing, or oxygen-free environments maintained over long periods of time.

Caption Question **Fig. 22** The minerals came from groundwater.

EARTH SCIENCE JOURNAL

La Brea Tar Pits A tremendous variety of animals are preserved in the Rancho La Brea tar pits in Los Angeles, California. Have students research the environment that once existed at Rancho La Brea, and then have them write in their Earth science journals how the tar pits formed and why they contain such spectacularly preserved organisms.

Demonstration

Fossil Impressions Fill three pans with three different sediments—mud, sand, and pebbles. Press a seashell into each. Discuss how sediment size affects level of detail preserved. Mud is better than sand, and both are better than pebbles; the more sediment grains per area, the greater the detail. Compare this to pixel resolution. The more pixels per area, the higher the resolution.

Activity

Make Trackways Scientists use dinosaur footprints to calculate the speed at which dinosaurs walked or ran. Have groups of students examine a photo of a dinosaur trackway. Ask them to write in their Earth science journals what they think they can learn about the dinosaur's size, height, and gait from its trackway. Select two volunteers from each group—one tall, one short. Have the volunteers first run and then walk across a muddy, sandy, or—in winter—snowy area near the school so that the tracks remain distinct and visible. Have the other students in each group record the height of the trackmakers, and then have them measure and sketch the trackmakers' trackways. The students should match each set of trackways with the trackmaker who made it, and explain how they did so. Students should note the depth of a track versus the size of the trackmaker, the length of the strides relative to the trackmaker's height, and the distance between strides relative to whether the trackmaker was walking or running. OL EL COOP LEARN

☑ READING CHECK In mineral replacement, the hard part's original mineral is replaced by another mineral. In recrystallization, the original mineral undergoes a change in properties.

Caption Question Fig. 23 A shell's internal structure changes to a more stable form, which alters the crystal structure of the original shell material.

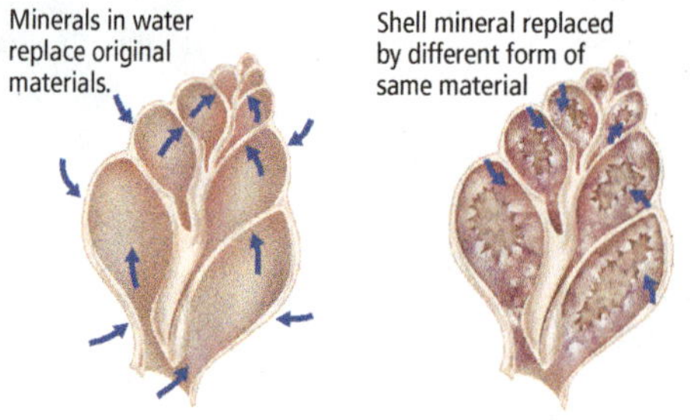

Figure 23 During mineral replacement, the minerals in a buried hard part are replaced by other minerals in groundwater. During recrystallization, temperature and pressure change the crystal structure of the hard part's original material.
Explain *why the internal structure of the shell changes during recrystallization.*

Recrystallization Another way in which hard parts can be altered and preserved is the process of recrystallization (ree krihs tuh luh ZAY shun). Recrystallization can occur when a buried hard part is subjected to changes in temperature and pressure over time. The process of recrystallization is similar to that of mineral replacement, although in mineral replacement the original mineral is replaced by a different mineral from the water, whereas in recrystallization the original mineral is transformed into a new mineral. A snail shell, for example, is composed of aragonite ($CaCO_3$). Through recrystallization, the aragonite undergoes a change in internal structure to become calcite, the basic material of limestone or chalk. Though calcite has the same composition ($CaCO_3$) as aragonite, it has a crystal structure that is more stable than aragonite over long periods of time. **Figure 23** shows how mineral replacement and recrystallization differ.

☑ READING CHECK **Compare and contrast** recrystallization and mineral replacement.

Molds and casts Some fossils do not contain any original or altered material of the original organism. These fossils might instead be molds or casts. A **mold** forms when sediments cover the original hard part of an organism, such as a shell, and the hard part is later removed by erosion or weathering. A hollowed-out impression of the shell, called the mold, is left in its place. A mold might later become filled with material to create a **cast** of the shell. A mold and a cast of a distinctive animal called an ammonite are shown in **Figure 24.**

Figure 24 A mold of this ammonite was formed when the dead animal's shell eroded. The cavity was later filled with minerals to create a cast.

Trace fossils Sometimes the only fossil evidence of an organism is indirect. Indirect fossils, called **trace fossils,** include traces of worm trails, footprints, and tunneling burrows. Trace fossils can provide information about how an organism lived, moved, and obtained food. For example, dinosaur tracks provide scientists with clues about dinosaur size and walking characteristics. Other trace fossils include gastroliths (GAS truh lihths) and coprolites (KAH pruh lites). Gastroliths are smooth, rounded rocks once present in the stomachs of dinosaurs to help them grind and digest food. Coprolites are the fossilized solid waste materials of animals. By analyzing coprolites, scientists learn about animal eating habits.

ACROSS THE CURRICULUM

Biology Living fossils are modern animals whose appearance has changed little over extremely long periods of time. The coelacanth fish, *Latimeria chalumnae,* is perhaps the most celebrated example. It was thought to have gone extinct 80 mya until one was caught off the coast of South Africa in 1938. Since then, the list of living fossils has grown to include the horseshoe crab, *Limulus polyphemus;* the inarticulate brachiopod, *Lingula anatine;* and the dawn redwood, *Metasequoia glyptostroboides.* Scientists are interested in these species because they have survived mass extinctions and have body plans or lifestyles that have been successful through geologic time.

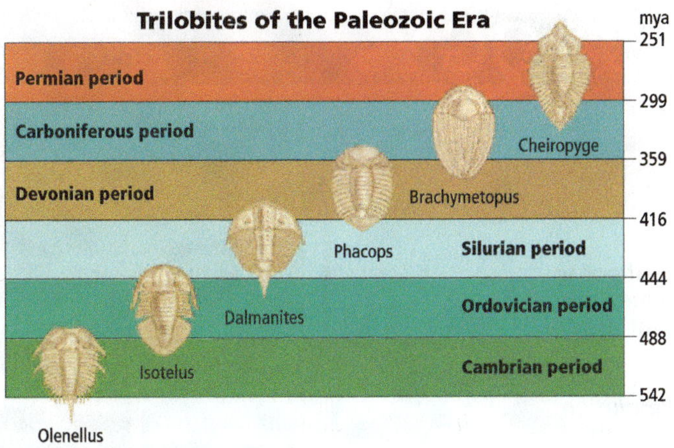

Figure 25 These trilobite species make excellent index fossils because each species lived for a relatively short period of time before becoming extinct, and were abundant and widespread.

Index Fossils

As you learned in the previous sections, fossils can help scientists determine the relative ages of rock sequences through the process of correlation. Some fossils are more useful than others for relative-age dating. **Index fossils** are fossils that are easily recognized, abundant, and widely distributed geographically. They also represent species that existed for relatively short periods of geologic time. The different species of trilobites shown in **Figure 25** make excellent index fossils for the Paleozoic Era because each was distinct, abundant, and existed for a certain range of time. Ammonites, extinct marine organisms related to nautiloids and squids, are excellent index fossils for the Mesozoic. If a geologist finds one in a rock layer, he or she can immediately determine an approximate age of the layer.

SECTION 4 REVIEW

Section Self-Check

Section Summary

- Fossils provide evidence that species have evolved.
- Fossils help scientists date rocks and locate reserves of oil and gas.
- Fossils can be preserved in several different ways.
- Index fossils help scientists correlate rock layers in the geologic record.

Understand Main Ideas

1. MAINIDEA **Describe** how the fossil record helps scientists understand Earth's history.
2. **List** ways in which fossils can form, and give an example of each.
3. **Explain** how scientists might be able to determine the relative age of a layer of sediment if they find a fossilized trilobite in the layer.
4. **Compare and contrast** a mold and a cast.

Think Critically

5. **Evaluate** Why are the best index fossils widespread?

WRITING IN ▶ Earth Science

6. Imagine that you have just visited a petrified forest. Write a letter to a friend describing the forest. Explain what the forest looks like and how it was fossilized.

3 Assess

Check for Understanding

Discussion On a table, place examples or pictures of fossils that represent original preservation, altered hard parts, molds, casts, and trace fossils. Ask volunteers to identify and describe each one.

Reteach

Make a Data Table Have students make a two-column data table listing the following fossils and what type of preservation each exhibits: insect in amber original preservation, dinosaur tooth altered hard part, gizzard stones of ancient birds trace fossil, petrified wood altered hard part, mummified human original preservation, dinosaur coprolites trace fossil, imprint of snail in sedimentary rock mold or cast, and dinosaur trackway trace fossil.

Assessment

Knowledge Have students list characteristics of an index fossil. An index fossil is easily recognizable; the species lived during a short period of time; it was abundant; and it was distributed over a large geographic area.

SECTION 4 REVIEW

1. Fossils help scientists correlate geologic areas, date rock layers, find oil and gas, and understand past environments. Geologists know that fossils are about the same age as the rocks in which they are found.
2. Fossils can be formed as original material (e.g., insects in amber), as altered hard parts (e.g., petrified tree or recrystallized shell), as molds or casts, and as traces of past activity (e.g., footprints and waste material).
3. If the scientist knows when the trilobite lived, he or she can assume that the rock layer in which the trilobite was found was laid down at about the same time.
4. A mold is formed from the imprint made by an organism. A cast is made when new material fills a mold, and the mold is then worn away.
5. They need to be widespread so that scientists can use them to correlate rock layers in geographically different areas. If they existed in a narrow distribution, they would be useful only on small scales.
6. Answers will vary, but should discuss the process of mineral replacement by contact with groundwater.

Rubric

Earth Science & TECHNOLOGY

Purpose
Students will learn that sediment cores obtained from deep sea drilling are used to study Earth's past.

Teacher Content Support
JOIDES Resolution The *JOIDES Resolution* was converted from a traditional oil drilling ship to a "floating university" research vessel in 1984, and later underwent extensive renovation in 2009. The *Resolution* is a component of the IODP (Integrated Ocean Drilling Program), an international research program that collects and studies cores of the seafloor in order to explore Earth's history and structure. The *Resolution* is run by the U.S. Implementing Organization (USIO) which consists of the Consortium for Ocean Leadership, Texas A & M University and the Lamont-Doherty Earth Observatory, where the sediment cores are shipped for further study.

Teaching Strategy
Explain to students that oxygen comes in two important varieties for paleoclimate research: heavy (^{18}O) and light (^{16}O). Water vapor molecules containing the heavy isotope condense more readily than water molecules containing the light isotope. When climates are cooler, the water vapor containing ^{18}O rains out of the atmosphere. The wind carries water vapor containing the lighter ^{16}O toward Earth's poles, where it eventually condenses and falls onto the ice sheets where it remains trapped. As a result, the oceans develop increasingly higher concentrations of heavy oxygen during cooler time periods, whereas glacial ice becomes concentrated in ^{16}O. When the temperature rises, the ice sheets melt, releasing freshwater concentrated in the light oxygen isotope into the oceans. Thus, scientists relate lower ^{18}O levels to warmer climates and melting. The ratios of the amounts of ^{18}O to ^{16}O is preserved over time in marine sediments, fossils, and glacial ice. Ask students to explain whether fossil foraminifera contain more or less ^{18}O when climate cools. More ^{18}O—the organisms would have lived in oceans containing more concentrated levels of ^{18}O.

WRITING IN ▶ Earth Science

Rubric

Research Information about the *Resolution's* latest expeditions can be found at *joidesresolution.org*.

WebQuest

Earth Science & TECHNOLOGY

Drilling Into the Past

Scientists use core samples of the seafloor to gather data about Earth's history. Clues they find about climate change in the past provide clues about future climate change.

Scientific drilling The *JOIDES Resolution* is a scientific drilling ship sponsored by the Joint Oceanographic Institutions for Deep Earth Sampling (JOIDES). Scientists aboard the *Resolution* explore the history and structure of Earth as recorded in seafloor sediments, fossils, and rocks. The ship's onboard drill extracts cores of the ocean floor 10 meters long, providing a vertical timeline of the changing environments and climates throughout Earth's history.

Sediment cores and microfossils Using a rotating drill bit to penetrate the ocean floor, the scientists on the *Resolution* bring up core samples through the drill pipe, being careful not to stir the sediments and destroy the order of the preserved layers. Each layer holds grains of dust, minerals, and often times pollen, that can provide information about the environment in which they were deposited. Many layers also contain fossils of the organisms that lived in the ocean during that time.

The most common types of fossils analyzed in deep sea sediment cores are microfossils, which are the skeletal remains of microscopic organisms, such as foraminifera. Foraminifera are single-celled organisms found in virtually all marine-influenced sediments. They are excellent indicators for past climates and environments. Foraminifera develop shells composed of calcium carbonate. When they grow, they incorporate oxygen from the surrounding water into their shells. By measuring the ratios of two oxygen isotopes in the fossil shells, scientists can estimate the temperature of the water—and thus the climate—in which the foraminifera grew.

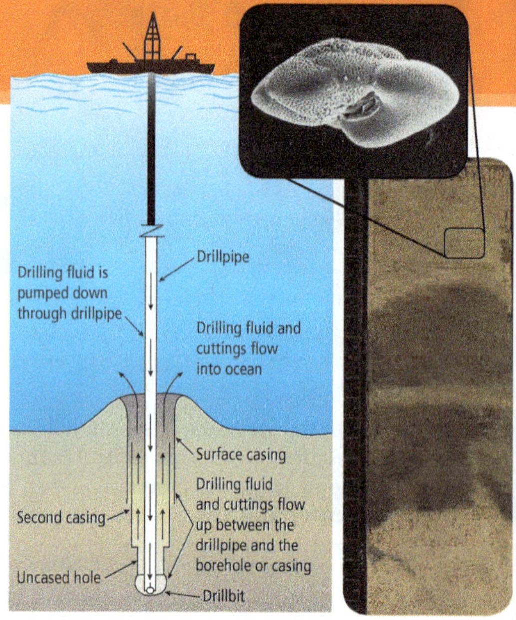

The JOIDES Resolution uses a drill pipe to extract cores from the ocean floor. These cores can contain microfossils, like the magnified foraminifera shown above, that provide clues about Earth's past.

Evidence from the past The evidence collected from seafloor core samples not only yields insights into past climates, but also into changes in sea level, animal and plant distribution, ocean currents and wind patterns, and land mass movement. International groups of scientists on the *Resolution* continue expeditions around the world to advance understanding of Earth by drilling and sampling the seafloor, as well as analyzing and monitoring trends.

WRITING IN ▶ Earth Science

Research Find information about the latest expeditions of the *JOIDES Resolution*. Develop a poster or presentation showing where the ship has traveled as well as the purpose of its expeditions.

WebQuest

GeoLAB

iLab Station

Design Your Own: Interpret History-Shaping Events

Background: Volcanoes, earthquakes, mountain building, floods, and other geologic events affect the surface of Earth—and the life that inhabits it—in important ways. However, not all events affect Earth equally. Some events in Earth's history have been more critical than others in shaping Earth.

Question: *What have been the most important events in Earth's geologic history?*

The Sierra Nevadas that extend through California resulted from a series of Earth-shaping events.

Bruce Heinemann/Getty Images

Materials

list of Earth-shaping events provided by your teacher
colored pencils
poster board
geologic time scale
reference books
internet access

Procedure

Imagine that NASA is planning to send a space probe to a distant galaxy. You are part of a team that has been assigned the task of listing the most important events that have shaped Earth's geologic history. This list will be carried as part of the spaceship's payload. It will be used to help describe Earth to any possible residents of the galaxy.

1. Read and complete the lab safety form.
2. Form into groups. Each group should have three or four team members.
3. Obtain a list of Earth-shaping events from your teacher.
4. Choose two other resources where you can find at least ten more events to add to your list.
5. Brainstorm about the events that you think had the most impact on the direction that Earth's development has taken over time.
6. Discuss the best way to display your list.
7. Make sure your teacher approves your plan.
8. Put your plan into effect.

Analyze and Conclude

1. **Interpret Data** Plot your list on a copy of the geologic time scale. Compare the number of events in each era. Did more Earth-shaping events occur early in Earth's history or later on? Explain.
2. **Compare** your list with the lists of others in your class. What events do all lists share? Do these events share common features?
3. **Infer** Choose one event in the Mesozoic Era, and infer how Earth's history might have progressed had the event not occurred.
4. **Evaluate** How do extinction events influence the development of life on Earth?

SHARE YOUR DATA

Peer Review Post a list of the ten most important events that you think shaped Earth's history. Compare your list with lists of other groups of students who have completed this lab.

GeoLAB

iLab Station Rubric

Preparation

Time Allotment 90 min

Process Skills think critically, classify, compare and contrast, communicate

Safety Precaution Approve lab safety forms before work begins.

Preparation of Materials If students do not have classroom internet access or resource books, provide a list of major Earth-shaping events and distribute it to the class.

Procedure

- To help students compare their data, create a transparency or computer-generated data table that includes the top ten choices of each group, sequenced from oldest to youngest, along with each event's approximate date and why the event was selected.
- Initiate a class discussion.

Analyze and Conclude

1. Answers will vary depending on students' selections of Earth-shaping events.
2. Answers will vary. One possible feature that all might share is that the most important events were those that affected the evolution of life.
3. One example might represent a scenario in which dinosaurs did not become extinct. In this case, the significance of mammals might have been greatly reduced.
4. Mass extinction events vacate niches that can become available to new, evolving organisms.

SHARE YOUR DATA

Peer Review Students should explain what their lists have in common with other lists. If students do not have internet access, make a class list of the most important events, with each group providing the event that it thinks is the most important. Have a discussion or debate about what students think are the two or three most important events in Earth's history.

MAINIDEAS Summary statements can be used by students to review the major concepts of the chapter.

Students can review with these online resources.

Vocabulary Practice

Vocabulary eGames
Vocabulary eFlashcards
Vocabulary PuzzleMaker

Section Self-Check

Chapter Self-Check

Online Test Practice

Use the Exam*View*® *Assessment Suite* CD-ROM to:

- create multiple versions of tests
- create modified tests with one mouse click
- edit existing questions and add your own questions
- build tests aligned with state standards using built-in state curriculum tags
- change English tests to Spanish with one mouse click
- track students' progress using the Teacher Management System

CHAPTER 21 STUDY GUIDE

BIGIDEA Scientists use several methods to learn about Earth's long history.

Vocabulary Practice

SECTION 1 The Rock Record

VOCABULARY
- geologic time scale
- eon
- Precambrian
- era
- period
- epoch
- mass extinction

MAINIDEA **Scientists organize geologic time to help them communicate about Earth's history.**

- Scientists organize geologic time into eons, eras, periods, and epochs.
- Scientists divide time into units based largely on fossils of plants and animals.
- The Precambrian makes up nearly 90 percent of geologic time.
- The geologic time scale changes as scientists learn more about Earth.

SECTION 2 Relative-Age Dating

VOCABULARY
- uniformitarianism
- relative-age dating
- original horizontality
- superposition
- cross-cutting relationship
- principle of inclusions
- unconformity
- correlation
- key bed

MAINIDEA **Scientists use geologic principles to learn the sequence in which geologic events occurred.**

- The principle of uniformitarianism states that processes occurring today have been occurring since Earth formed.
- Scientists use geologic principles to determine the relative ages of rock sequences.
- An unconformity represents a gap of time in the rock record.
- Geologists use correlation to compare rock layers in different geographic areas.

SECTION 3 Absolute-Age Dating

VOCABULARY
- absolute-age dating
- radioactive decay
- radiometric dating
- half-life
- radiocarbon dating
- dendrochronology
- varve

MAINIDEA **Radioactive decay and certain kinds of sediments help scientists determine the numeric age of many rocks.**

- Techniques of absolute-age dating help identify numeric dates of geologic events.
- The decay rate of certain radioactive elements can be used as a kind of geologic clock.
- Annual tree rings, ice cores, and sediment deposits can be used to date recent geologic events.

SECTION 4 Fossil Remains

VOCABULARY
- evolution
- original preservation
- altered hard part
- mineral replacement
- mold
- cast
- trace fossil
- index fossil

MAINIDEA **Fossils provide scientists with a record of the history of life on Earth.**

- Fossils provide evidence that species have evolved.
- Fossils help scientists date rocks and locate reserves of oil and gas.
- Fossils can be preserved in several different ways.
- Index fossils help scientists correlate rock layers in the geologic record.

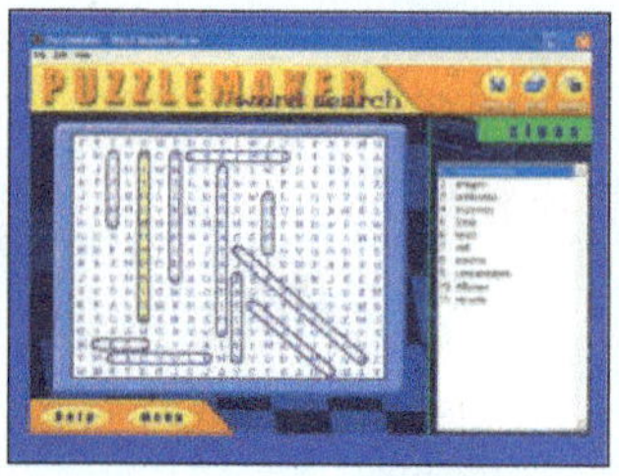

Vocabulary Practice

For additional practice with vocabulary, have students access the Vocabulary PuzzleMaker.

CHAPTER 21 ASSESSMENT

Chapter Self-Check

VOCABULARY REVIEW

Match each definition with the correct vocabulary term from the Study Guide.

1. the record of Earth's history from its origin to the present
2. a gap in the rock record caused by erosion
3. the emission of radioactive isotopes and the resulting change into other products over time
4. the largest time unit in the geologic time scale
5. the matching of outcrops or fossils among regions

Distinguish between the vocabulary terms in each pair.

6. period, epoch
7. altered hard part, original preservation
8. absolute-age dating, relative-age dating
9. fossil, index fossil
10. mold, cast

The sentences below are incorrect. Make each sentence correct by replacing the italicized word or phrase with a term from the Study Guide.

11. *Original horizontality* is the principle that a fault or intrusion is younger than the rock it intersects.
12. *Relative-age dating* states that processes operating today have been operating since Earth formed.
13. A *varve* is a unique layer used to match rock formations across large areas.
14. *Correlation* is the change in species over time.

UNDERSTAND KEY CONCEPTS

15. The end of which era is marked by the largest extinction event in Earth's history?
    A. Cenozoic
    B. Mesozoic
    C. Paleozoic
    D. Precambrian

16. How old is a mammoth's tusk if 25 percent of the original C-14 remains in the sample? The half-life of C-14 is 5730 years.
    A. 5730 years
    B. 11,460 years
    C. 17,190 years
    D. 22,920 years

Use the figure below to answer Question 17.

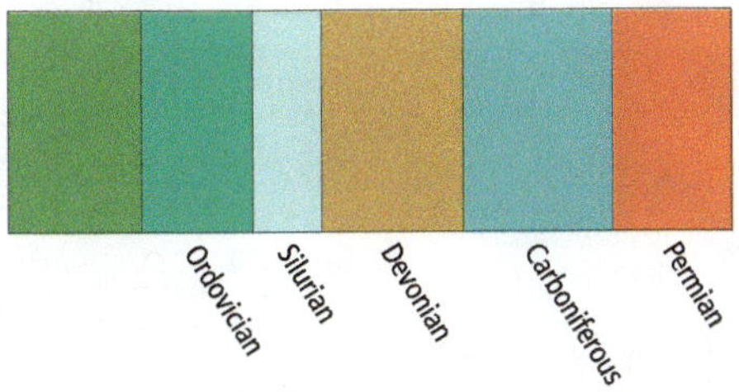

17. Which time period is missing in the diagram?
    A. Cambrian
    B. Permian
    C. Triassic
    D. Paleogene

18. Which is not a typical characteristic of an index fossil?
    A. was commonplace while alive
    B. existed for a long period of time
    C. is geographically widespread
    D. is easily recognizable

19. Which is the smallest division of geologic time?
    A. period
    B. eon
    C. era
    D. epoch

20. Which geologic principle is used when a geologist observes an outcrop of rocks and determines that the bottom layer is the oldest?
    A. uniformitarianism
    B. original horizontality
    C. superposition
    D. inclusion

VOCABULARY REVIEW

1. geologic time scale
2. unconformity
3. radioactive decay
4. eon
5. correlation
6. Both are time divisions, but periods generally are larger divisions of time that include epochs.
7. Fossils with altered hard parts do not contain original material, unlike fossils with original preservation, which do.
8. Absolute-age dating methods are used to determine the numeric age of a rock or object, while relative-age dating can be used only to determine an estimated or relative age.
9. Any fossil is a preserved organism or trace of an organism. An index fossil also represents a species that existed for only a short time, was widespread and abundant, and has distinctive remains. Index fossils are useful in correlating rock layers in large areas.
10. A mold is an impression of an organism. A cast forms when sediment later fills the mold.
11. Crosscutting relationships
12. Uniformitarianism
13. key bed
14. Evolution

UNDERSTAND KEY CONCEPTS

15. C
16. B
17. A
18. B
19. D
20. C

21. D 22. D
23. C 24. C

CONSTRUCTED RESPONSE

25. A mold forms when a hard part is buried in sediment and the hard part erodes away, leaving an impression. A cast is formed if sediment later fills the mold.
26. Geologists use mass extinctions to mark time units on the geologic time scale; they provide reference points.
27. Methods of absolute-age dating provide numeric ages, while methods of relative-age dating provide only the relative order of events.
28. Without a geological time scale, geologists could not communicate with each other, place their work in context, or correlate worldwide events in detailed ways.
29. Answers will vary. An unconformity is any eroded surface that has been buried. Because it has eroded, it no longer exists, and the time it represents is missing from the rock record.
30. For: Students might argue that it is difficult to compare time when each period is of a different duration. Because humans are accustomed to dividing time into equal units, it makes sense to divide the time scale in this way, too; when it was formulated, absolute-age dating methods did not exist. Now they do. Against: Students might argue that the scale is constructed around geological and evolutionary events that do not occur on scheduled timetables. Forcing each period to be the same length would downplay the importance of the events that mark Earth's history.
31. If particular fossils are associated with oil at one site, it is possible they are associated with oil at other sites.

THINK CRITICALLY

32. Layer F is the oldest.
33. An angular unconformity exists at the base of layer B because layer B sits on layers that have been tilted.
34. The order of events is below. 1. Beds F, E, D, C, J, I, and H were deposited horizontally (principles of original horizontality and superposition); 2. Beds F, E, D, C, J, I, and H were tilted and much of layers H, I, and J eroded (principle of original horizontality); 3. Fault G cut through and moved beds F, E, D, C, and J (principle of cross-cutting relationships); 4. Beds B and A were deposited (principles of original horizontality and superposition).
35. Answers will vary. Relative-age dating techniques are based on common sense notions of how gravity works to layer sediments. People used common sense to put geologic events in correct order before scientific tools became available to help them determine numeric age.
36. Example answer: **C**ertain **O**lder **S**tudents **D**on't **C**ount **P**ennies, **T**hey **J**ust **C**ount **P**recious **N**ickels **Q**uickly.
37. 3.9 billion years.
38. An index fossil exists in different kinds of sediment in a narrow range across a wide geographic area. A key bed is a distinct layer of one type of rock that appears in several locations. Both are used to correlate rock layers across large geographic areas.
39. A clam has a better chance because it has a hard shell; a spider's soft exterior decays easily. Clams also live in water, where deposition of sediments is more common.
40. No. Dinosaurs existed only during the Mesozoic, which ended 66 million years ago. Radiocarbon dating can be used only on objects less than about 60,000 years old.
41. The red dot is the half-life; the sample contains half parent and half daughter.
42. Both tree-ring and radiocarbon dating techniques could be used. In tree-ring dating, the geologist would compare the sample to a known tree-ring chronology of the area. In radiocarbon dating, the ratio of C-14 to C-12 would indicate time since death.

ASSESSMENT

21. Uranium-238 breaks down into thorium-234. Which is thorium-234 in relation to uranium-238?
A. parent
B. brother
C. son
D. daughter

Use the figure below to answer Question 22.

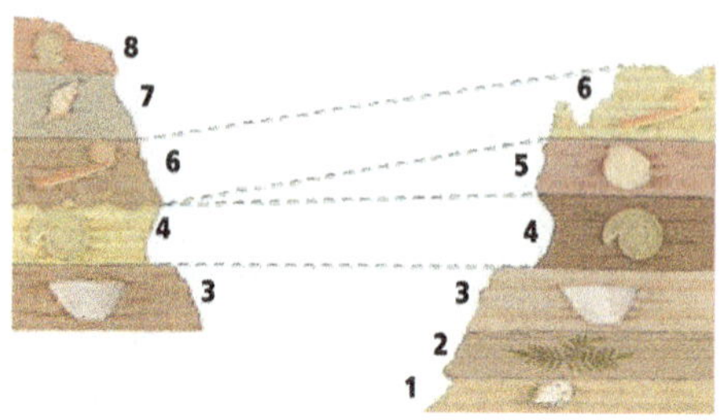

22. What does the diagram show?
A. uniformitarianism
B. inclusion
C. cross-cutting relationships
D. correlation

23. Trees that have been buried by volcanic ash are likely to be preserved in which manner?
A. original preservation
B. mummification
C. mineral replacement
D. recrystallization

24. Which are glacial lake sediments that show cycles of deposition?
A. annual rings
B. tillites
C. varves
D. unconformities

CONSTRUCTED RESPONSE

25. Sequence the steps by which a mold and a cast are formed.

26. Infer why mass extinctions are important to geologists.

27. Compare and contrast absolute-age dating and relative-age dating.

28. Assess the usefulness of a universally accepted geologic time scale.

29. Explain, in your own words, why an unconformity is any gap in the rock record.

30. Argue for or against making the time units of the geologic time scale of equal duration.

31. Relate How are microscopic fossils associated with discovering oil at a particular site?

THINK CRITICALLY

Use the diagram below to answer Questions 32 to 34.

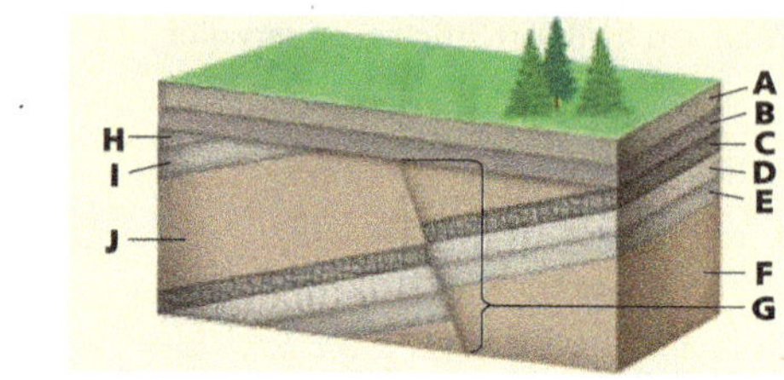

32. Identify the oldest rock layer in the diagram.

33. Find an angular unconformity in the diagram.

34. Apply List the order of geologic events in the diagram from oldest to youngest along with the geologic principles that you used.

35. Critique this statement: The principles for determining relative age are based on common sense.

36. Create One way to remember the order of words in a sequence is to create a phrase, called an acrostic, that uses the same first letter of each word in the sequence. For example, "My Dear Aunt Sally" is often used to remember the mathematical order of operations: Multiply and Divide before you Add and Subtract. Create an acrostic to help you remember the periods of the Phanerozoic Eon.

Chapter Self-Check

37. Solve The half-life of K-40 is 1.3 billion years. What is the age of an ancient igneous rock that contains a mineral with 12.5 percent K-40 and 87.5 percent Ar-40?

38. Compare and contrast an index fossil and a key bed.

39. Assess whether a clam or a spider has a better chance at becoming a fossil.

40. Evaluate Can radiocarbon dating be used to determine the age of a dinosaur bone? Explain.

Use the diagram below to answer Question 41.

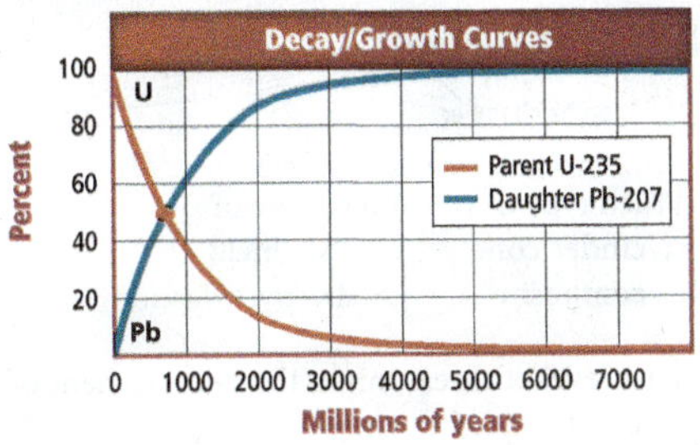

41. Analyze What does the red dot signify on the graph?

42. CAREERS IN EARTH SCIENCE A geologist discovers wood buried within sediments of a landslide that is thought to have been caused by an ancient earthquake. Explain two methods that the geologist could use to determine when the earthquake occurred.

CONCEPT MAPPING

43. Create a concept map using the following terms: *absolute-age dating, geologic time scale, relative-age dating, fossils, unconformities, and radiometric dating.*

CHALLENGE QUESTION

44. Assess Do you think domestic dogs might make good index fossils for future geologists? Explain.

WRITING IN ▶ Earth Science

41. Imagine that you are a bacterium that lives for only 20 minutes. Explain how your observations about the world would be different from those of a human being who lives for about 80 years. Evaluate the difference between human time and geologic time.

DBQ Document–Based Questions

Data obtained from: Bambach, R.K., et al. 2004. Origination, extinction, and mass depletion of marine diversity. *Paleobiology* 30: 522–542.

The figure below plots the diversity, measured as the number of different types, of marine animals from the Cambrian to the Neogene. The animals are grouped into levels of organization called genera that include closely related species. Use the data to answer the questions below.

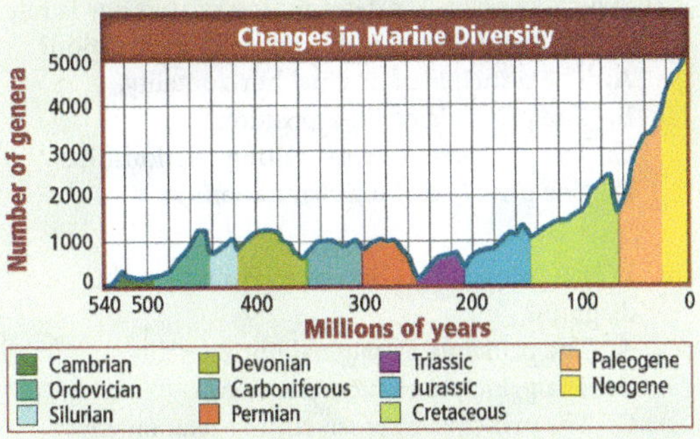

46. With what do the largest drops in diversity coincide on the geologic time scale?

47. Explain what the decreases in diversity mean.

48. Use the information in the graph to support adding one or more new eras to the Phanerozoic.

CUMULATIVE REVIEW

49. What subatomic particles make up the nucleus of an atom? **(Chapter 3)**

50. Why do tornadoes occur most frequently in the central United States? **(Chapter 13)**

CONCEPT MAPPING

43. Example of a correct answer: The geologic time scale has two branches to show that it developed by relative-age and absolute-age dating methods. Relative-age dating has two branches to show it can use evidence from unconformities and fossils. Absolute-age dating has two branches to show it can use evidence from fossils and radiometric dating.

CHALLENGE QUESTION

44. Answers will vary. Domesticated dogs might make good index fossils because they have hard parts, are widespread, and are abundant worldwide. They might not make good index fossils because their skeletons are not distinguishable from those of wild dogs, which have been around for a much longer time period. Index fossils represent organisms that lived during a short period of time. Dogs probably would fit this criterion, but only time will tell.

WRITING IN ▶ Earth Science

Rubric

45. Answers will vary. To a bacterium that lives for only 20 minutes, the Sun is stationary and it is always daylight—or it is dark and the stars are stationary. Compared to geologic time, human life spans are similarly short. Mountains and shorelines appear fixed, rivers stay basically in the same areas, and life-forms remain constant. Perspective is critical.

DBQ Document-Based Questions

Data obtained from: Bambach, R.K., et al. 2004. Origination, extinction, and mass depletion of marine diversity. *Paleobiology* 30:522–542.

46. Most coincide with the boundaries between periods.

47. A large diversity drop corresponds to an extinction event. The four largest diversity drops are at the ends of the Ordovician, Devonian, Permian, and Cretaceous periods.

48. One could argue that a new era be created at the Ordovician-Silurian boundary or at the Devonian-Carboniferous boundary.

CUMULATIVE REVIEW

49. protons and neutrons

50. That is where cold, continental air masses most frequently collide with warm, tropical air moving northward from the Gulf of Mexico, and it is this collision that helps to produce the supercells that spawn tornadoes.

MULTIPLE CHOICE

1. B
2. B
3. C
4. D
5. A
6. B
7. C
8. A
9. D
10. B

MULTIPLE CHOICE

1. Which type of mountains form as the result of uplift far from plate boundaries?
   A. ocean ridges
   B. fault block mountains
   C. folded mountains
   D. volcanic ranges

Use the diagram to answer Questions 2 and 3.

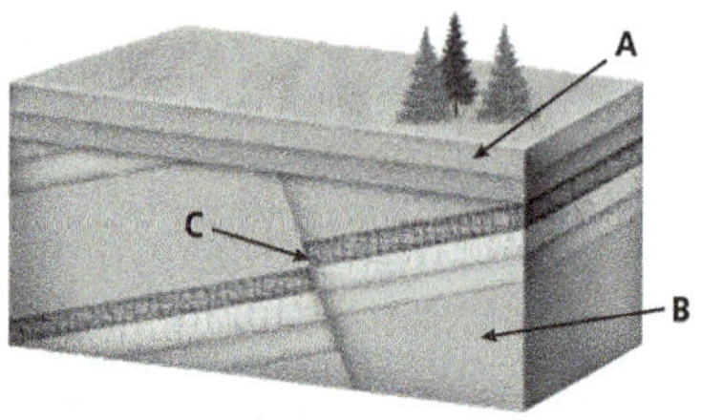

2. Which principle for determining relative age is relevant to Point A in this diagram of a rock region?
   A. The principle of original horizontality.
   B. The principle of superposition.
   C. The principle of cross-cutting relationships.
   D. The principle of uniformitarianism.

3. Which principle is relevant to Point C in this diagram?
   A. The principle of original horizontality.
   B. The principle of superposition.
   C. The principle of cross-cutting relationships.
   D. The principle of uniformitarianism.

4. What aspect of the discovery of ocean ridges was most important in the scientific community?
   A. their location
   B. their volcanic activity
   C. their height
   D. their age

5. Which is not a factor affecting the formation of magma?
   A. time
   B. temperature
   C. pressure
   D. water

6. Earth's crust is broken up into a dozen or more enormous slabs called what?
   A. boundaries
   B. tectonic plates
   C. subduction zones
   D. subduction plates

Use the diagram below to answer Questions 7 and 8.

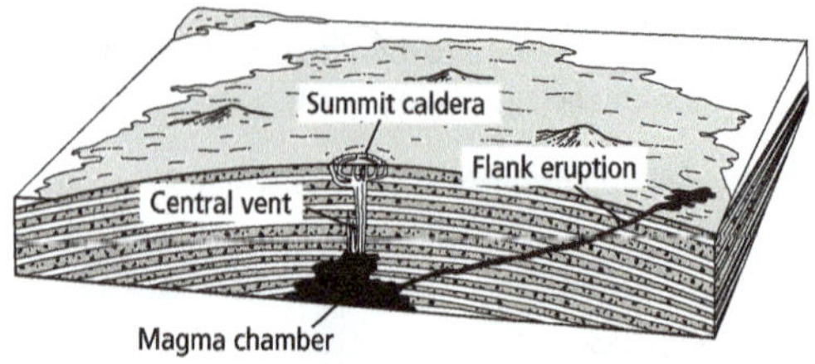

7. Which type of volcano is shown?
   A. cinder cone
   B. composite
   C. shield
   D. pyroclastic

8. What level of threat might the development of this volcano pose to humans?
   A. Low; it is built as layer upon layer and accumulates during nonexplosive eruptions.
   B. Low; it is considered to be an inactive volcano.
   C. Moderate; it forms when pieces of magma explode and build around a vent, but is rather small.
   D. High; it has a violently explosive nature.

9. What happened to the magnetic fields generated by magnetic rocks along the ocean floor?
   A. There were no reversals of the magnetic field along the ocean floor.
   B. Each side of an ocean ridge had its own magnetic pattern.
   C. Normal and reverse polarity regions formed stripes that ran perpendicular to ocean ridges.
   D. Normal and reverse polarity regions formed stripes that ran parallel to ocean ridges.

10. What does orogeny refer to?
    A. the drifting of microcontinents
    B. the building of mountain ranges
    C. the formation of volcanic islands
    D. the breaking apart of supercontinents

Online Test Practice

SHORT ANSWER

Use the map below to answer Questions 11–13.

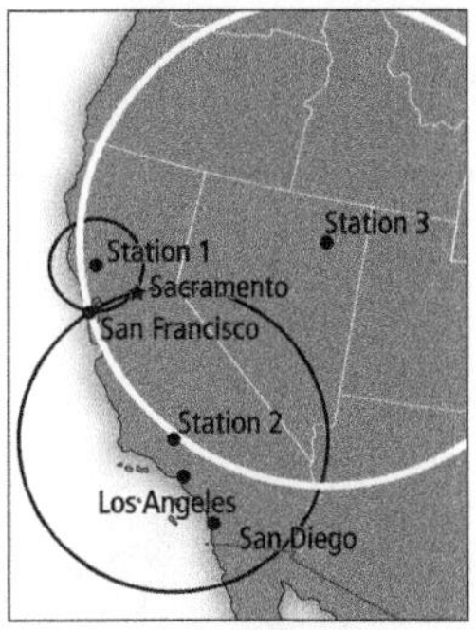

11. According to the map, where was the epicenter of the earthquake located? How can the epicenter be determined?

12. Why is it important to use three stations to locate the epicenter of an earthquake?

13. How might this earthquake affect Los Angeles?

14. The Florida peninsula gets more thunderstorms than any other part of the United States. What geographic feature of Florida causes it to get so many thunderstorms? How does this feature allow thunderstorms to form?

15. Why is the Appalachian Mountain Belt divided into several regions?

16. Describe acid precipitation in terms of the pH scale and the reason for its pH value.

READING FOR COMPREHENSION

Dating Gold

The radioactive decay of metal inside South African gold nuggets helped scientists determine the origin of the world's largest gold deposit. The placer model indicates the gold is older than surrounding rock. The hydrothermal model indicates that the hot spring fluids deposited the gold inside the rocks. It was decided to determine the age of the gold itself. If the gold is older than the rocks in which it is found, then the rocks must have built up around the gold, bolstering the placer model. If the gold is younger than the rocks, that means it must have seeped in with fluids, supporting the hydrothermal model. Two elements found inside gold, rhenium and osmium serve as a radioactive clock. Rhenium decays into osmium over very long spans of time—it takes about 42.3 billion years for half of a sample of rhenium to transmute. By dissolving gold grains in acid and measuring the ratio of rhenium to osmium, scientists can determine the gold's age. Gold from places in the Rand is three billion years old—a quarter of a billion years older than its surrounding rock, thus supporting the placer model.

Article obtained from: Choi, C. 2002. Origin of world's largest gold deposit found? *United Press International Science News* (September): 1-2:

17. What is the half-life of rhenium?
A. 42.3 years
B. 42.3 thousand years
C. 42.3 million years
D. 42.3 billion years

18. Why was this study conducted?
A. to determine the origin of the gold deposit
B. to disprove the hydrothermal model
C. to support the placer model
D. to explain radioactive decay

NEED EXTRA HELP?

If You Missed Question . . .	1	2	3	4	5	6	7	8	9	10	11	12	13	14	15	16
Review Section . . .	20.3	21.2	21.2	21.3	18.2	17.3	18.1	18.1	17.2	20.2	19.3	19.3	19.3	13.1	20.2	7.1

SHORT ANSWER

11. The epicenter occurred in San Francisco. This can be determined by gathering three seismograph readings from three seismograph stations and plotting the point where they all intersect.

12. Sample answer: One seismograph reading would just have a circle showing the possible locations. Two circles would give an overlapping area but nothing specific. Three circles create one unique point which shows the exact location of the epicenter.

13. Because the epicenter is not directly in Los Angeles, the force of the earthquake would not be felt as strongly, but there would most likely still be some sensation of an earthquake due to the city being within the circle of the second station's readings of the earthquake.

14. The Florida peninsula gets so many thunderstorms because it is surrounded by water. On hot days the land heats up much more quickly than the ocean. The hot air over the land rises, and cooler air from the ocean rushes in to replace it. The moist air rushing in from the ocean combines with the upwelling hot air to form thunderstorms over the land.

15. Along the Appalachian Mountain Belt there are different types of rock formations, which shows that they were formed in different ways.

16. Acid precipitation is a solution that has a pH value lower than 5.6 on the pH scale. The value of 5.6 is used because 5.6 is the pH number for normal rainfall.

READING FOR COMPREHENSION

17. D
18. A

CHAPTER 22 The Precambrian Earth

BIGIDEA **The oceans and atmosphere formed and life began during the three eons of the Precambrian, which spans nearly 90 percent of Earth's history.**

ESSENTIAL QUESTIONS	NATIONAL STANDARDS	RESOURCES TO ASSESS MASTERY
SECTION 1 **Early Earth** **1.** What evidence exists that indicates Earth is 4.6 billion years old? **2.** What were the heat sources of early Earth? 1 session 0.5 block	UCP.2; B.2, B.6; C.3; D.1–3; G.3	**Entry-Level Assessment** Focus Transparency 67, p. 620 **Progress Monitoring** Reading Check, p. 621 Section Review, p. 622
SECTION 2 **Formation of the Crust and the Continents** **1.** How is the process by which Earth differentiated summarized? **2.** How did Earth's crust and continents form? **3.** How did the continents grow during the Precambrian? 2 sessions 1 block	B.2; D.1–3	**Entry-Level Assessment** Focus Transparency 68, p. 623 **Progress Monitoring** Caption Question, pp. 623, 624 Reading Check, p. 624 Section Review, p. 627
SECTION 3 **Formation of the Atmosphere and Oceans** **1.** How did Earth's atmosphere and oceans form? **2.** What was the cause for the increase in oxygen gas in the atmosphere? **3.** How do scientists know that atmospheric oxygen existed during the Proterozoic? **4.** What was the importance of oxygen and water on early Earth? 2 sessions 1 block	UCP.2; B.2, B.6; C.3; D.1–3	**Entry-Level Assessment** Focus Transparency 69, p. 628 **Progress Monitoring** Caption Question, pp. 629, 630 Reading Check, pp. 629, 631 Section Review, p. 632
SECTION 4 **Early Life on Earth** **1.** How is the experimental evidence showing how life might have begun on Earth summarized? **2.** What is the difference between prokaryotes and eukaryotes? **3.** How are Earth's early multicellular organisms described? 3 sessions 1.5 blocks	UCP. 2, UCP.3; B.3, B.6; D.1–3; E.1–2; G.1–3	**Entry-Level Assessment** Focus Transparency 70, p. 633 **Progress Monitoring** Caption Question, p. 635 Reading Check, p. 636 Section Review, p. 637 **Summative Assessment** Chapter Assessment, p. 641 Exam*View® Assessment Suite* CD-ROM

Go online!
connectED.mcgraw-hill.com

LEVELED RESOURCES	LAB MATERIALS
Science Notebook 22.1 OL **Chapter FAST FILE Resources:** Study Guide, p. 41 BL **Transparencies:** Section Focus Transparency 67 EL Teaching Transparency 65 OL EL	LaunchLAB p. 618 / **15 min** 250-mL beaker, tap water, vegetable oil, corn syrup, stirring rod
Science Notebook 22.2 OL **Chapter FAST FILE Resources:** GeoLab Worksheet, p. 31 OL Study Guide, p. 42 BL **Transparencies:** Section Focus Transparency 68 EL Teaching Transparency 66 OL EL	GeoLAB p. 639 / **45 min** rock samples, paper, metric ruler, colored pencils
Science Notebook 22.3 OL **Chapter FAST FILE Resources:** MiniLab Worksheet, p. 30 OL Study Guide, p. 44 BL **Lab Resources:** Laboratory Manual, p. 169 OL **Transparencies:** Section Focus Transparency 69 EL Teaching Transparency 67 OL EL	MiniLAB p. 631 / **45 min** white sand, 150-mL beaker, water, bleach, steel wool, petri dish, stirring rod, watch glass
Science Notebook 22.4 OL **Chapter FAST FILE Resources:** Study Guide, p. 45 BL **Lab Resources:** Laboratory Manual, p. 173 OL **Transparencies:** Section Focus Transparency 70 EL	

LabManager™
Customize any lab with the LabManager™ CD-ROM.

ADDITIONAL RESOURCES

Technology:
- eTeacherEdition
- eStudentEdition
- Classroom Presentation Toolkit
- Exam*View®* *Assessment Suite* CD-ROM
- LabManager™ CD-ROM
- What's EARTH SCIENCE Got To Do With It? DVD
- Weather Classroom Video DVD

Assessment:
- Blueprints for Success
- Chapter Test (Scaffolded)
- Earth Science Standardized Test Practice

Lab Resources:
- Exploring Environmental Problems Laboratory Manual

BL Below Level OL On Level AL Advanced Learners EL English Learners COOP LEARN Cooperative Learning

CHAPTER 22

LaunchLAB

iLab Station Rubric

How do liquids of different densities model early Earth?

Additional Material stirring rod

Safety Precautions Students should wear safety goggles. Caution students to handle beakers with care; the contents might stain clothing. When students complete the lab, have them pour their oil mixtures into a disposal container that you provide.

Teaching Strategies

- Use any kind of oil. Food coloring increases the contrast if a light-colored oil is used.
- Tell students to pour the oil into the water slowly to avoid causing large bubbles to form. Bubbles can be popped with a toothpick if necessary.

Procedure

1. Have students read and complete the lab safety form and follow the procedure below.
2. Fill a **250-mL beaker** with 50 mL of **tap water**.
3. Pour 50 mL of **vegetable oil** into the beaker.
4. Pour 50 mL of **corn syrup** into the beaker.
5. Allow the mixture to sit for a few minutes.

Analysis

1. **Describe** what happened to the liquids in the beaker. The liquids form layers. Because oil is the least dense of the three, it forms the top layer. Corn syrup is the most dense, so it forms the bottom layer.
2. **Identify** which component of the experiment represents each of Earth's layers. Oil represents the crust; water represents the mantle; corn syrup represents the core.
3. **Relate** the results to the formation of layers in early Earth. Earth formed layers according to the density of the materials in Earth. Less-dense materials accumulated near the surface; more-dense materials accumulated near the center.

CHAPTER 22

The Precambrian Earth

BIGIDEA The oceans and atmosphere formed and life began during the three eons of the Precambrian, which spans nearly 90 percent of Earth's history.

SECTIONS

1 Early Earth

2 Formation of the Crust and Continents

3 Formation of the Atmosphere and Oceans

4 Early Life on Earth

LaunchLAB

iLab Station

How do liquids of different densities model early Earth?

Earth's core, mantle, and crust have different average densities. The core is the densest, the crust is the least dense, and the mantle lies between. Scientists think that early in Earth's history, temperatures were hot enough for the materials that make up Earth to act like liquids. Model how Earth's layers formed in this activity.

FOLDABLES® Study Organizer

Formation of Earth's Atmosphere

Make a pocket book using the labels shown. Use it to organize your notes on the formation of the atmosphere.

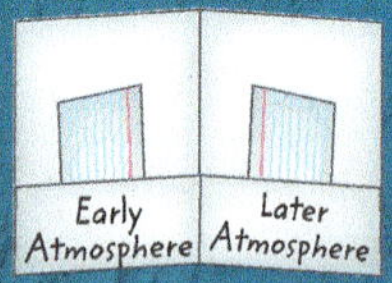

Horst Mahr/age fotostock

Assessment

Knowledge Ask students how density relates to the composition of a material. Materials made with elements that have large atomic weights are generally more dense than materials made with elements with smaller atomic weights. Ask students to recall from Chapter 20 the densities of the crust, mantle, and core. Ask how the average densities of the continental crust (2.8 g/cm^3), the mantle (3.3 g/cm^3), and the core (10–13 g/cm^3) reflect the composition of each. The core is denser than the mantle, and the mantle is denser than the continental crust because of the relative presence of iron and other heavy elements in each. Heavy elements are most prevalent in the core and least prevalent in the crust.

LabManager™ Customize any lab with the LabManager™ CD-ROM.

Stromatolites are tabular-, mound-, or column-shaped structures made by tiny organisms called cyanobacteria. Found only in some areas on Earth today, stromatolites dominated Precambrian oceans for billions of years.

Introduce the BIG IDEA

Life's Foundations Life on Earth would not be possible without the foundations laid during the Precambrian. It was during this time that the atmosphere, oceans, and most of the continents formed. Ask students to think of other systems that build on foundations. Examples might include the foundation of a building, the piers that hold up a bridge, learning the alphabet before learning to spell, learning addition before learning algebra, etc.

Teacher Content Support

Stromatolites The photo shows reeflike structures called stromatolites that are built by single-celled organisms called cyanobacteria. Today, stromatolites are rare; they are found mainly in restricted intertidal or shallow-water areas in Australia, shown in the photo, and in a few localities in the Bahamas. They are rare because in most environments snails and other mollusks eat them. This prevents them from colonizing large areas. During the Precambrian, however, there were no predators to graze on cyanobacteria, and cyanobacteria were able to colonize large areas of the world.

SECTION 1

1 Focus

Focus Transparency

Before presenting the lesson, project **Section Focus Transparency 67** and have students answer the accompanying questions. EL

MAINIDEA

Indirect Evidence Scientists must rely on indirect evidence to determine the age of Earth and to learn when the oceans and atmosphere formed. Ask students to think of other examples where indirect evidence is used to understand events. One example is a crime scene where trace evidence, such as fingerprints, is used.

2 Teach

Tie to Previous Knowledge

Radioactive Isotopes As students have learned, some radioactive isotopes have relatively short half-lives while others have half-lives billions of years long. The radioactive isotopes that are useful for dating Precambrian rocks have long half-lives. Examples include isotopes of thorium and uranium, particularly those found in zircon crystals. Radioactive isotopes with the longest decay series also contribute the most to Earth's internal heat.

Interpret the Illustration

Precambrian Time Figure 1 represents the first 90 percent of the geologic time scale. Have students determine what percentage of Earth's history is represented by each of the three eons of the Precambrian. Hadean: 13 percent; Archean: 28 percent; Proterozoic: 43 percent OL

SECTION 1

Essential Questions

- What evidence exists that indicates Earth is 4.6 billion years old?
- What were the heat sources of early Earth?

Review Vocabulary

metamorphism: changes in the mineral composition or structure of rocks caused by pressure and temperature over time

New Vocabulary

zircon
meteorite
asteroid

Early Earth

MAINIDEA Several lines of evidence indicate that Earth is about 4.6 billion years old.

EARTH SCIENCE 4 YOU Imagine that you are putting together a jigsaw puzzle but you do not have the picture on the box. You do not know what the puzzle looks like, and you have only about 10 percent of the pieces. This is similar to the challenge that scientists face when they study the early Precambrian.

The Age of Earth

The Precambrian, which includes the Hadean, Archean, and Proterozoic Eons, is an informal time unit that spans nearly 90 percent of Earth's history. When Earth first formed it was hot, volcanically active, and no continents existed on its surface. Rocks of Earth's earliest eon—the Hadean—are extremely rare, so scientists know very little about Earth's first 600 million years. The earliest signs of life are from the Archean. As illustrated in **Figure 1,** the earliest life-forms were simple, unicellular organisms.

Crustal rock evidence Absolute-age dating has revealed that the oldest known rocks are 4.28 billion years in age. Evidence that Earth is older than 4.28 billion years exists in small grains of the mineral zircon ($ZrSiO_4$) found in certain metamorphosed Precambrian rocks in Australia. Because **zircon** is a stable and common mineral that can survive erosion and metamorphism, scientists often use it to age-date ancient rocks. Geologists theorize that the zircon in the Australian rocks is residue from crustal rocks that no longer exist. Based on radiometric dating, which shows that the zircon is at least 4.4 billion years old, Earth must also be at least this old.

Figure 1 The Precambrian lasted for nearly 4 billion years. Multicellular organisms did not appear until the end of the Proterozoic.

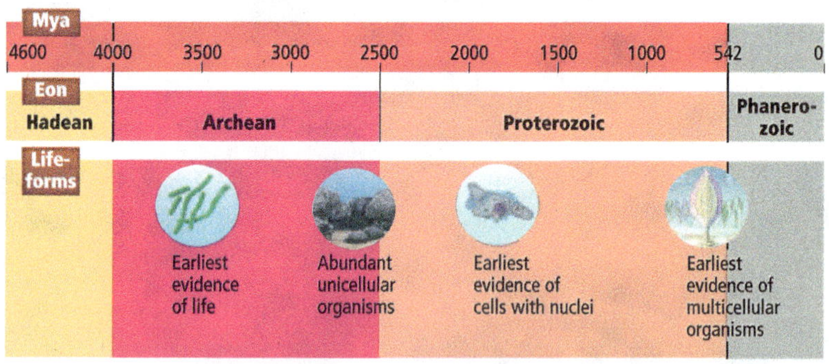

EARTH SCIENCE JOURNAL

Precambrian Time Scale Have students copy the Precambrian time scale in their Earth science journals and refer to it often as they read the chapter.

Rubric

Solar system evidence Evidence from meteorites (MEE tee uh rites) and other bodies in the solar system suggests that Earth is more than 4.4 billion years old. **Meteorites** are small fragments of orbiting bodies that have fallen on Earth's surface. They have fallen to Earth throughout Earth's history, but most have been dated at between 4.5 and 4.7 billion years old. Many scientists agree that all parts of the solar system formed at the same time, so they assume that Earth and meteorites are approximately the same age.

In addition, the oldest rock samples from the Moon, collected during the *Apollo* missions in the 1970s, have been dated at 4.4 to 4.5 billion years old. Scientists think that the Moon formed very early in Earth's history when a massive solar system body the size of Mars collided with Earth. Considering all the evidence, scientists agree that Earth is about 4.6 billion years old.

☑ READING CHECK **Explain** why scientists think that Earth is older than the oldest rocks in the crust.

Early Earth's Heat Sources

Earth was extremely hot after it formed. There were three likely sources of this heat: Earth's gravitational contraction, radioactivity, and bombardment by asteroids, meteorites, and other solar system bodies, as shown in **Figure 2.**

Gravitational contraction Scientists think that Earth formed by the gradual accumulation of small, rocky bodies in orbit around the Sun. As Earth accumulated these small bodies, it grew in size and mass. With increased mass came increased gravity. Gravity caused Earth's center to squeeze together with so much force that the pressure raised Earth's internal temperature.

Radioactivity A second source of Earth's heat was the decay of radioactive isotopes. Scientists know that certain radioactive isotopes were more abundant in Earth's past than they are today. While some of these isotopes, such as uranium-238, are long-lasting and continue to decay today, others were short-lived and have nearly disappeared. Radioactive decay releases energy in the form of heat. Because there were more radioactive isotopes in early Earth, more heat was generated, making early Earth hotter than it is today.

Figure 2 Impacts from asteroids and meteorites were a source of heat for early Earth.

Use an Analogy

Shoes Using the oldest rocks to date Earth is like using a person's oldest pair of shoes to determine the person's age. You know that the person is at least as old as the shoes. You also know that shoes wear over time, and that the person is older than the shoes. The rocks that form Earth's crust also wear with time, through the process of erosion. If there are minerals in a rock that are older than the oldest rocks on Earth, then Earth must be at least as old as the oldest mineral.

Teacher Content Support

Lord Kelvin and the Age of Earth In the 19th century, Earth's age was a hotly contested subject. Many thought Earth was no older than 6000 years. In 1862, a British scientist named William Thomson, later called Lord Kelvin, calculated the age of Earth using his knowledge of heat conduction. He reasoned that Earth had started out molten and cooled over time with a geothermal gradient of about 30 degrees per km. Knowing the approximate size of Earth, he figured that Earth was 20 to 30 million years old. Though his reasoning was sound, Lord Kelvin did not know about radioactivity, which has contributed heat to Earth throughout Earth's history. We know today that Earth's cooling is not a simple closed system, as Lord Kelvin presumed. Radioactivity was not discovered until 1896, by Henri Becquerel. The word was coined by Marie Curie a few years later.

☑ READING CHECK Evidence from zircon crystals, moon rocks, and meteorites suggests that Earth is older than the oldest rocks.

ACROSS THE CURRICULUM

Engineering The six *Apollo* and three *Luna* missions paired engineering with science to gain important geological information about the Moon. Though the astronauts gathered only a small number of moon rocks, these rocks have been instrumental in understanding the Moon's age and composition.

DIFFERENTIATED INSTRUCTION

Behavior Disorders To engage students with behavior disorders, have them find additional information about moon rocks, including how they were gathered, how many there are, where they are found today on Earth, and who studies them. Have them make posters using both visual and written data. BL OL

3 Assess

Check for Understanding

Discussion Ask students why geologists theorize that Earth is 4.6 billion years old. Though older than any rock on Earth, it corresponds to the ages of moon rocks and meteorites, which likely formed at the same time as Earth.

Reteach

Data Tables Have students make data tables in their Earth science journals, listing Earth's early heat sources and identifying whether the sources were internal or external. internal: radioactivity and gravitational contraction; external: asteroid and meteorite bombardment

Assessment

Skill Have students create concept maps using the following concepts: *Earth's oldest rocks: 4.28 billion years; Earth's oldest minerals: 4.2–4.4 billion years; oldest moon rocks: 4.4–4.5 billion years; oldest meteorites: 4.5–4.7 billion years; Earth and solar system formed at about the same time; age of Earth must be about 4.6 billion years.* Map might show "age of Earth" as a central concept with arrows from "oldest rocks," "oldest minerals," "oldest moon rocks," and "oldest meteorites," all of which surround it. "Earth and solar system formed at same time" might be positioned near "oldest moon rocks" and "oldest meteorites," with arrows leading to both.

CAREERS IN EARTH SCIENCE

WebQuest

Asteroid and meteorite bombardment A third source of heat in early Earth came from the impacts of meteors, asteroids (AS tuh roydz), and other objects in the solar system. **Asteroids** are carbon or mineral-rich objects between 1 m and 950 km in diameter. Today, most asteroids orbit the Sun between the orbits of Mars and Jupiter. Large asteroids seldom collide with Earth. Planetary geologists estimate that only about 60 objects with diameters of 5 km or more have struck Earth during the last 600 million years. Most objects that hit Earth today are meteorites—fragments of asteroids.

However, evidence from the surfaces of the Moon and other planets suggests that for the first 500 to 700 million years of Earth's history, many more asteroids were distributed throughout the solar system than there are today and that collisions were much more frequent. The impacts of these bodies on Earth's surface generated a tremendous amount of thermal energy. For example, scientists think that the massive collision that likely formed the Moon generated so much heat that parts of Earth melted. The debris (duh BREE) from these impacts also caused a blanketing effect around Earth, which prevented the newly generated heat from escaping to space.

Cooling The combined effects of gravitational contraction, radioactivity, and bombardment by other objects in the solar system made Earth's beginning very hot. Eventually, Earth's surface cooled enough for an atmosphere and oceans to form. Scientists do not know exactly how long it took for this to happen, but evidence suggests that Earth cooled enough for liquid water to form within its first 200 million years. The cooling process continues even today. As much as half of Earth's internal heat remains from Earth's formation.

CAREERS IN EARTH SCIENCE

Planetary Geologist Planetary geologists, or astrogeologists, study the planets and their places in the solar system and universe. Some planetary geologists study conditions under which extraterrestrial life might exist.

WebQuest

SECTION 1 REVIEW

Section Self-Check

Section Summary

- Scientists use Earth rocks, zircon crystals, Moon rocks, and meteorites to determine Earth's age.
- Likely heat sources of early Earth were gravitational contraction, radioactivity, and asteroid and meteorite bombardment.
- Cooling of Earth led to the formation of liquid water.

Understand Main Ideas

1. **MAIN**IDEA **Summarize** the data that scientists use to determine Earth's age.
2. **Explain** why scientists think that Moon rocks and meteorites are about the same age as Earth.
3. **Explain** how gravitational contraction, radioactivity, and asteroid and meteorite bombardment heated early Earth.
4. **Describe** the importance of zircon as an age-dating tool.

Think Critically

5. **Evaluate** Which of Earth's early sources of heat are not major contributors to Earth's present-day internal heat?

MATH IN Earth Science

6. If an average of 5000 asteroids bombarded Earth every million years during the Hadean, calculate the total number of asteroid impacts that occurred during this eon. Refer to **Figure 1** for information on geologic time scales.

SECTION 1 REVIEW

1. Evidence comes from Earth's oldest rocks, which are 4.28 billion years old; Precambrian zircon crystals, which are 4.4 billion years old; moon rocks, which range from 4.4 to 4.5 billion years old; and meteorites, which range from 4.5 to 4.7 billion years old.
2. Scientists think that all solar system objects formed at about the same time and so must be approximately the same age.
3. As Earth grew in mass, gravity caused it to contract. The energy of contraction converted to thermal energy. Radioactivity heated Earth because the decay process releases heat, and there were more radioactive elements on early Earth than there are today. Thermal energy also came from the energy of asteroid and meteorite bombardment.
4. Zircon is a common mineral that can survive weathering and erosion.
5. Though much of Earth's heat is leftover from its formation, gravitational contraction and bombardment by meteorites and asteroids contribute little to Earth's heat today.
6. 3,000,000 asteroids (The Hadean Eon lasted for about 600 million years; 5000 asteroids/million years $\times$ 600 million years $=$ 3,000,000 asteroids.)

SECTION 2

Formation of the Crust and Continents

MAINIDEA The molten rock of Earth's early surface formed into crust and then continents.

Essential Questions

- How is the process by which Earth differentiated summarized?
- How did Earth's crust and continents form?
- How did the continents grow during the Precambrian?

Review Vocabulary

magma: molten, liquid rock material found underground

New Vocabulary

differentiation
microcontinent
craton
Precambrian shield
Canadian Shield
Laurentia

EARTH SCIENCE 4 YOU Have you ever cooked pudding? If so, you might have noticed that when the pudding cooled, a crust formed on the top. Scientists think that Earth's crust formed in a similar way.

Formation of the Crust

Because of the intense heat in early Earth, many scientists think that much of the planet consisted of hot, molten magma. As Earth cooled, the minerals and elements in this molten magma became concentrated in specific density zones.

Differentiation Scientists know that less-dense materials float on top of more-dense materials. As you observed in the Launch Lab, oil floats on water because oil is less dense than water. This same general principle operated on early molten Earth. The element with the highest density—iron—sank toward the center. In contrast, the light elements, such as silicon and oxygen, remained closer to the surface. The process by which a planet becomes internally zoned when heavy materials sink toward its center and lighter materials accumulate near its surface is called **differentiation** (dih fuh ren shee AY shun). The differentiated zones of Earth are illustrated in **Figure 3.**

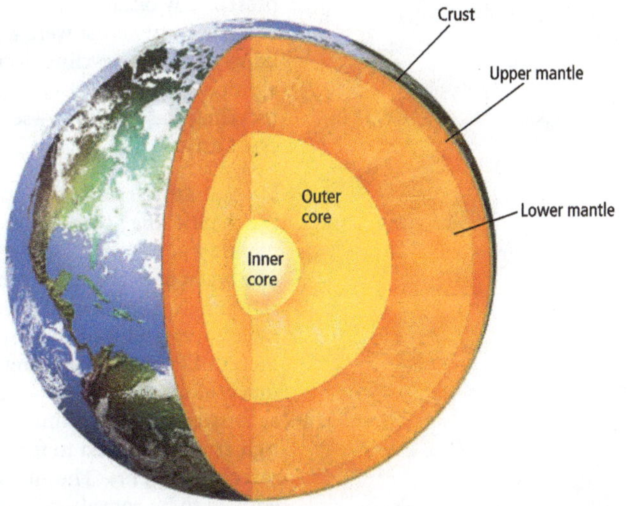

Figure 3 Earth differentiated into layers shortly after it formed.
Analyze *What is the densest part of Earth?*

SECTION 2

1 Focus

Focus Transparency

Before presenting the lesson, project **Section Focus Transparency 68** and have students answer the accompanying questions. EL

MAINIDEA

Define a Continent Ask students how they would describe a continent. Write their answers on the board, and help them arrive at a definition. Then discuss how geographers and geologists define a continent. A geographer generally defines a continent as a large landmass found above sea level. A geologist, however, defines a continent as a part of Earth's crust that has a stable core of Precambrian rock surrounded by younger belts that have been added over time.

2 Teach

Teacher Content Support

Magnetic Field Earth's differentiation into layers led to the formation of its magnetic field. In the early Precambrian, the iron "fell" to Earth's center, an event often called the Iron Catastrophe. The iron is liquid in Earth's outer core; it flows because of the heat (like a boiling pot of water), and it spins because of Earth's rotation. This dynamic process creates turbulent currents that generate magnetism.

Caption Question Fig. 3 The core is the densest part of Earth.

Demonstration

Model the Crust Heat a can of clam chowder (strain the clams and potatoes first, if you prefer). Let the soup cool during the class period. Have students relate the formation of the crust on the soup to the formation of an early crust on Earth. Ask students to think about the thickness of the soup crust relative to the soup and the thickness of Earth's crust relative to the mantle and core. In both cases, the crust is comparatively thin.

Tie to Previous Knowledge

Bowen's Reaction Series Students have learned about Bowen's reaction series. List the minerals of Bowen's reaction series on the board. Have students determine which minerals are more common in the crust and which are more common in the mantle. The denser minerals, such as olivine and pyroxene, are common in the mantle. All other minerals, intermediate in density, are more common in the crust. OL

Identify Misconceptions

Students might think continents looked the same early in Earth's history as they do today.

Uncover the Misconception Tell students that while many pieces of today's continents formed during the Archean, they remained small throughout much of the Precambrian and looked nothing like they do today.

Demonstrate the Concept Make a transparency of a map of Earth as it appeared sometime during the Proterozoic—with all continental labels removed. Ask the class if they can determine which landmasses represent today's continents. Then add another transparency containing the names of today's continents that the landmasses represent.

Assess New Knowledge Make another transparency of Earth as it appeared during the Mesozoic. Have students identify the continents and compare them with those on the Proterozoic map.

■ **Figure 4** Larger amounts of dense elements are found in Earth as a whole than are found in Earth's crust.
Estimate *the percentage of iron in Earth's crust and in the entire Earth.*

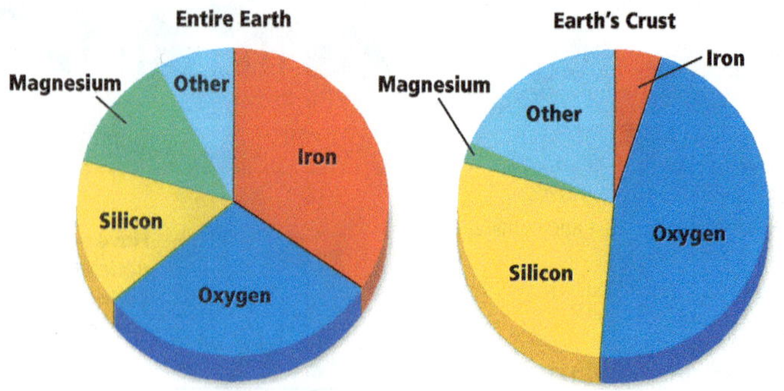

Relative densities The process of differentiation explains the relative densities of parts of Earth today. **Figure 4** compares the proportions of elements in Earth's crust and in Earth as a whole. Notice that iron, a dense element, is much less abundant in the crust than it is in the entire Earth, while the crust has a higher proportion of less-dense elements, such as silicon and oxygen. This also explains why granite occurs on Earth's surface. Granite is composed mainly of feldspar, mica, and quartz, which, as you have learned, are minerals with low densities.

☑ READING CHECK **Explain** why there is more iron in Earth's core than there is in the crust.

VOCABULARY
SCIENCE USAGE V. COMMON USAGE
Differentiate
Science usage: to layer into distinct zones
Common usage: to distinguish; to mark as different

Earliest crust Some type of early crust formed as soon as Earth's upper layer began to cool. This crust was probably similar to the basaltic crust that underlies Earth's oceans today. Recall that present-day oceanic crust is recycled at subduction zones. Pieces of Earth's early crust were also recycled, though scientists do not know how the recycling occurred. Some suggest that it occurred by a process that does not occur on Earth today. Most agree that the recycling was vigorous—so vigorous that none of Earth's earliest crust exists today.

Continental crust As the early crustal pieces were returned to the mantle, they carried water. The introduction of water into the mantle was essential for the formation of the first continental crust. The water reacted with the mantle material to produce new material that was less dense than the original crustal pieces. As this material crystallized and reemerged on Earth's surface, small fragments of granite-containing crust were formed. Granite makes up much of the crust that forms Earth's continents today. As volcanic activity continued during the Archean, small fragments of granite-rich crust continued to form. These crustal fragments are called **microcontinents.** They are called this because they were not large enough to be considered continents.

■ **Caption Question Fig. 4** Iron makes up approximately 40 percent of the entire Earth, and about 10 percent of the crust.

☑ READING CHECK There is more iron in the core because it is the element with the highest density.

DIFFERENTIATED INSTRUCTION

English Learners Have English learners research the names of the igneous minerals quartz, muscovite, potassium feldspar, plagioclase feldspar, biotite, amphibole, pyroxene, and olivine in their native languages. They will likely be surprised by the similarities to the English names. OL

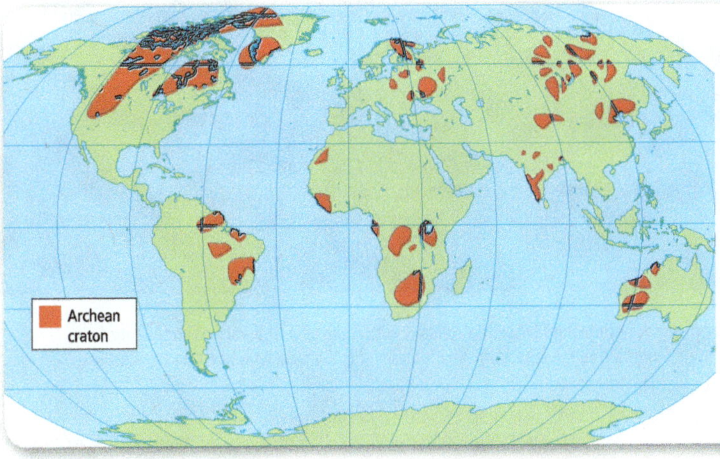

Figure 5 Archean cratons make up about 10 percent of Earth's continents. These granite-rich cores extend into the mantle as deep as 200 km.

Cratons Most of the microcontinents that formed during the Archean and early Proterozoic still exist as the cores of today's continents. A **craton** (KRAY tahn) is the oldest and most stable part of a continent. It is made up of the crust and a part of the upper mantle and can extend to a depth of 200 km. Cratons are composed of granitic rocks, such as granite and gneiss, with alternating bands of metamorphosed basaltic rocks, which represent ancient continental collisions. As shown in **Figure 5,** the Archean cratons represent about 10 percent of Earth's total landmass.

Precambrian shields Most of the cratons are buried beneath sedimentary rocks. However, in some places deep erosion has exposed the rocks of the craton. This exposed area is called a **Precambrian shield.**

In North America, the Precambrian shield is called the **Canadian Shield** because much of it is exposed in Canada. The Canadian Shield also occupies a large part of Greenland, as well as the northern parts of Minnesota, Wisconsin, and Michigan. Valuable minerals such as nickel, silver, and gold are found in the rocks of the Canadian Shield. The oldest known crustal rocks on Earth that date back to 4.28 billion years are from the Canadian Shield. In contrast, North America's platform rocks are generally younger than about 600 million years.

Growth of the Continents

Recall that all of Earth's continents were once consolidated into a single landmass called Pangaea. Pangaea formed relatively recently in Earth's history—only about 250 mya. The plate tectonic forces that formed Pangaea have been at work at least since the end of the Archean.

Concept Development

Keel Ask students if they know what a keel is. A keel is the main structural element of a ship that extends along the center of the ship's bottom. A keel gives a ship stability as it moves in the water. In this way, it is similar to the deepest part of a craton. Scientists often refer to the deep parts of cratons as keels.

Tie to Previous Knowledge

Diamonds Students have learned that most of the world's mined diamonds come from kimberlite pipes in South Africa. These pipes are located within the boundaries of Africa's Kaapvaal Craton. Diamonds and cratons are closely related. Diamonds form in the deepest part of a craton, where pressures are very high. The world's oldest diamond was found in a craton in Canada. Diamonds and the mineral inclusions within cratons help scientists learn how they formed.

EARTH SCIENCE JOURNAL

Canadian Shield The Canadian Shield is a rich source of metallic minerals, such as iron, nickel, copper, zinc, uranium, gold, silver, and platinum. Have students choose one mineral and research where and how it is mined. Then have them write an Earth science journal entry describing what it might be like to spend a day as a miner of that mineral. BL OL AL

VISUALIZING
VISUALIZING
VISUALIZING

Purpose

Students will learn that North America formed over a long period of time in successive orogenies.

Project

Local History Have students research the age and history of rocks underneath the area in which they live. Ask them to write their findings in their Earth science journals. OL AL

Teacher Content Support

Mid-Continent Rift Not all rifts remain active. When rifting stops and a continent is no longer being split apart, the rift is called a failed rift. The failed mid-continent rift in North America, pictured on the map, runs beneath present-day Arkansas, Louisiana, western Tennessee, and Mississippi. Though this rift—called the Reelfoot Rift—failed, the area remains tectonically weak. Three of the largest North American earthquakes in recorded history, known as the New Madrid sequence, occurred here in a three-month period in late 1811 and early 1812. Today, the rift area is filled with thick sediments and is the southern drainage basin of the Mississippi River. Chances of a future large earthquake remain high.

GeoLAB

iLab Station

The GeoLab located at the end of the chapter can be used at this point in the lesson.

VISUALIZING
VISUALIZING

VISUALIZING Continent Formation

Figure 6 North America was formed by a succession of mountain-building episodes over billions of years. This map shows mountain-building events that occurred during the Precambrian. By the end of the Precambrian, about 75 percent of North America had formed.

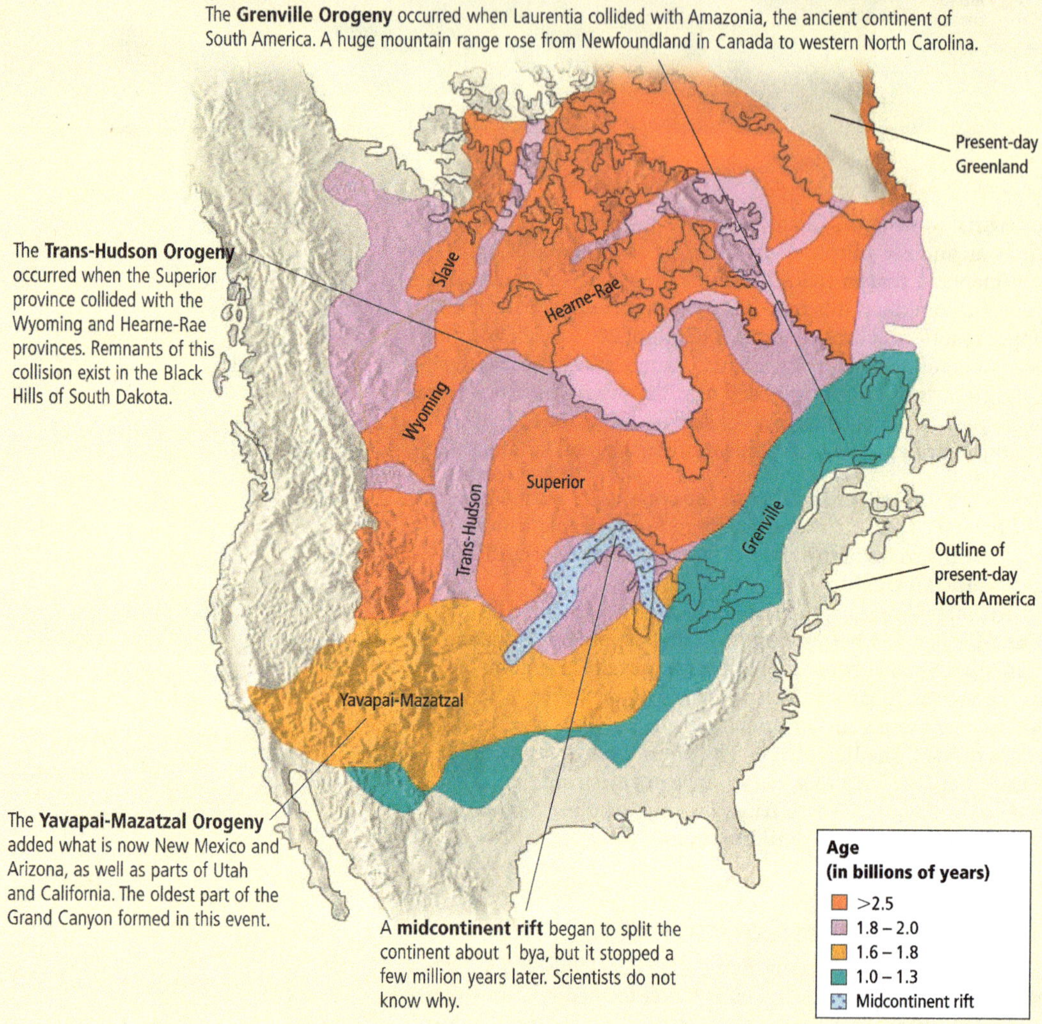

View an **animation of orogenies.** Concepts In Motion

DIFFERENTIATED INSTRUCTION

Visually Impaired Cut out a rough map of North America from cardboard. On the map, draw the outlines of the accretion events shown in the figure above using a marker that makes raised lines. Have students who are visually impaired use the cardboard map during class discussions. BL OL

Mountain building During the Proterozoic, the microcontinents that formed during the Archean collided with each other, becoming larger but fewer in number. As they collided, they formed massive mountains. Recall that mountain-building episodes are called orogenies. Orogenies form long belts of deformed rocks called orogens, or orogenic belts. The mountain-building events that formed North America are illustrated in **Figure 6.**

Laurentia One of Earth's largest Proterozoic landmasses was **Laurentia** (law REN shuh). Laurentia was the ancient continent of North America. As shown in **Figure 6,** the growth of Laurentia involved many different mountain-building events. For example, near the end of the early Proterozoic, between 1.8 and 1.6 bya, thousands of square kilometers were added to Laurentia when Laurentia collided with a volcanic island arc. This collision is called the Yavapai-Mazatzal Orogeny.

The first supercontinent The collision of Laurentia with Amazonia, the ancestral continent of South America, occurred during the mid–Proterozoic, about 1.2 bya. This collision coincided with the formation of Earth's first supercontinent, called Rodinia (roh DIN ee ah), shown in **Figure 7.** Rodinia was positioned on the equator with Laurentia near its center. By the time Rodinia formed, nearly 75 percent of Earth's continental crust was in place. The remaining 25 percent was added during the three eras of the Phanerozoic eon. The breakup of this supercontinent began about 750 mya.

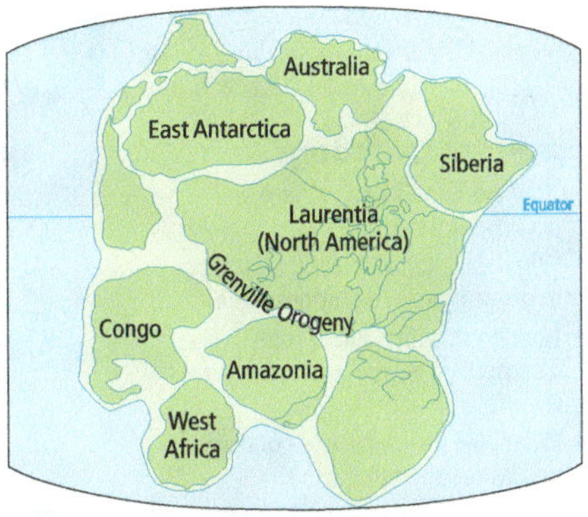

Figure 7 Earth's first supercontinent—Rodinia—formed when Laurentia collided with Amazonia during the Grenville Orogeny.

SECTION 2 REVIEW

Section Self-Check

Section Summary

- Earth differentiated into specific density zones early in its formation.
- Plate tectonics caused microcontinents to collide and fuse throughout the Proterozoic.
- The ancient continent of Laurentia formed as a result of many mountain-building episodes.
- The formation and breakup of Earth's first supercontinent occurred during the Proterozoic.

Understand Main Ideas

1. MAINIDEA **Describe** how Earth's continents formed.
2. **Explain** why pieces of Earth's earliest crust do not exist today.
3. **Deduce** how a craton is like a continent's root.
4. **Discuss** how the concept of uniformitarianism helps explain why Earth formed different density zones.

Think Critically

5. **Evaluate** whether it is reasonable to call the Proterozoic the age of continent building.
6. **Infer** why little evidence of Proterozoic orogenies exists today.

WRITING IN ▶ Earth Science

7. Suppose you are the North American craton. Write a short story about how Laurentia formed around you.

3 Assess

Check for Understanding

Discussion Ask students to explain how and when the ancient continent of Laurentia formed. During the Proterozoic, the microcontinents that formed during the Archean collided and coalesced as a result of plate tectonics. These accreted microcontinents eventually formed Laurentia. Additional material was added later, during the Phanerozoic.

Reteach

Journal Have students explain, in their Earth science journals, why the crust is composed of lighter minerals than are typically found in the mantle and core. Students should realize that the differentiation of Earth caused elements with higher densities to sink toward the center.

Assessment

Knowledge Have students list the sequence in which Earth's early crust formed into continental material. Lists should include the recycling of the earliest, basaltlike crust, which contained water; the interaction of the water with mantle materials to create denser, granite-bearing magma; and the reemergence of the granite-bearing magma onto Earth's surface to form pieces of continental crust.

SECTION 2 REVIEW

1. Earth's continents formed as microcontinents collided and accreted.
2. Earth's earliest crust was destroyed during some kind of vigorous recycling process.
3. A craton is the oldest, deepest, and most stable part of a continent, just as a root is the oldest, deepest, and most stable part of a plant or structure.
4. Scientists know today that less-dense material floats on more-dense material. They can infer that this happened when Earth was young and just forming because of the uniformitarianism principle: Earth processes occurring today are basically the same as those that occurred billions of years ago.
5. By the end of the Archean, only about 10 percent of Earth's continental crust had formed. By the end of the Proterozoic, about 75 percent of the continents had formed. Most continental material thus formed during the Proterozoic.
6. The mountains that formed during the Proterozoic eroded away.
7. Answers will vary, but might include a description of how the first granite-rich crust formed Laurentia's craton and how Laurentia moved around on Earth's surface, colliding with other landmasses and accumulating more land.

Rubric

SECTION 3

1 Focus

Focus Transparency

Before presenting the lesson, project **Section Focus Transparency 69** and have students answer the accompanying questions. EL

MAINIDEA

Life's Requirements Have students discuss what sort of environment was necessary before life could begin on Earth. Write their responses on the board. When they finish this section, revisit the discussion and see how their thoughts have changed. Students will find that the environment necessary for life to form contained water and an energy source, not necessarily sunlight.

2 Teach

Tie to Previous Knowledge

Composition of the Atmosphere Have students recall the composition of the present-day atmosphere. nitrogen: 78 percent; oxygen: 21 percent; argon: 0.9 percent; carbon dioxide: 0.03 percent; water vapor and trace gases: 0.0-4.0 percent Argon is the only major atmospheric gas that is not mentioned in this Chapter. Have them hypothesize the source of atmospheric argon. It comes from the decay of the radioactive isotope K-40 to Ar-40. It has been accumulating in the atmosphere since the early Precambrian. Remind students that new elements can be formed by the process of radioactive decay.

SECTION 3

Formation of the Atmosphere and Oceans

MAINIDEA The formation of Earth's oceans and atmosphere provided a hospitable environment for life to begin.

Essential Questions
- How did Earth's atmosphere and oceans form?
- What was the cause for the increase in oxygen gas in the atmosphere?
- How do scientists know that atmospheric oxygen existed during the Proterozoic?
- What was the importance of oxygen and water on early Earth?

Review Vocabulary

ultraviolet radiation: high-energy rays from the Sun that can damage living organisms

New Vocabulary

cyanobacteria
stromatolite
banded-iron formation
red bed

EARTH SCIENCE 4 YOU Have you thanked a plant lately? Plants and other organisms that produce oxygen provide nearly all the oxygen that you breathe. Had oxygen-producing organisms not existed on early Earth, it is likely that you would not be here today!

Formation of the Atmosphere

Scientists think that an atmosphere began to form on Earth during Earth's formation process. Asteroids, meteorites, and other objects that collided with Earth during this time probably contained water. The water would have vaporized on impact, forming a haze around the planet. Hydrogen and helium probably were also present, with lesser amounts of ammonia and methane. However, hydrogen and helium have small atomic masses, and many scientists think that neither gas stayed near Earth for long. Earth's gravity was, and still is, too weak to keep them from escaping to space. Some scientists also think that much of the ammonia and methane surrounding Earth might have been broken apart by the Sun's intense ultraviolet radiation, releasing more hydrogen into space.

Outgassing Once Earth was formed, its atmosphere changed with the addition of volcanic gases. Volcanic eruptions release large quantities of gases, and there was considerable volcanic activity during the Precambrian. A modern example of the volume of gases released during eruptions is shown in **Figure 8.**

Figure 8 The eruption of Mount St. Helens in 1980 released a large amount of carbon dioxide, water vapor, and other gases.

Demonstration

Model Outgassing Add cold water to a 0.5-L jar until it is three-quarters full. Drop in two antacid tablets, seal the jar, and leave it undisturbed overnight. The next day, ask the class to predict what will happen when you remove the lid. The CO_2 from the antacid tablets, dissolved in the water, will be released rapidly as it expands—much as gases in magma expand when released in volcanoes.

DIFFERENTIATED INSTRUCTION

Advanced Learners Have students research the possibility that meteorites, asteroids, and comets delivered a significant amount of water to early Earth. Have them list the different kinds of meteorites and comets known today and the amounts and isotopic compositions of the water they contain. Most bodies contain water with compositions that differ from the composition of Earth's water today.

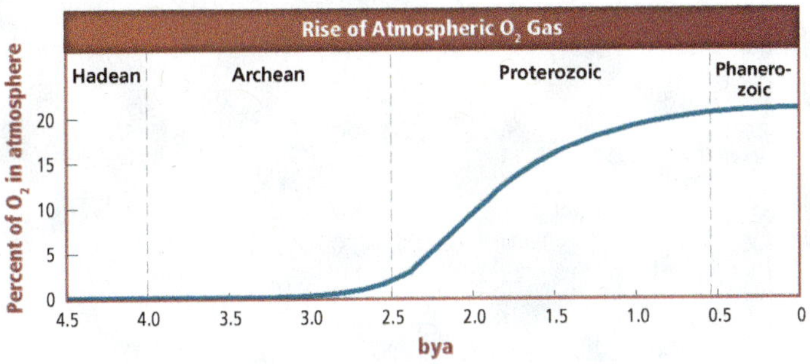

■ **Figure 9** There were only negligible amounts of free oxygen in Earth's atmosphere until the early Proterozoic.
Analyze ***How old was Earth when oxygen began to accumulate in its atmosphere?***

Recall that present-day volcanoes release large amounts of water vapor, carbon dioxide, and trace amounts of nitrogen and other gases in a process called outgassing. While scientists do not know the exact concentration of gases in Earth's early atmosphere, it probably contained the same gases that vent from volcanoes today.

Oxygen in the Atmosphere

One gas that volcanoes do not generally produce is oxygen. There was little oxygen in the Hadean and Archean atmospheres that was not bonded with carbon or other elements. As illustrated in **Figure 9,** atmospheric oxygen did not begin to accumulate until the early Proterozoic. Where did the oxygen gas come from?

FOLDABLES® Incorporate information from this section into your Foldable.

First oxygen producers The oldest known fossils that help answer this question are preserved in rocks in Australia and South Africa that are about 3.5 billion years old. These fossils appear to be traces of tiny, threadlike organisms called **cyanobacteria.** Like their present-day counterparts, ancient cyanobacteria used photosynthesis to produce the nutrients they needed to survive. In the process of photosynthesis, organisms use light energy and convert carbon dioxide and water into sugar. Oxygen gas is given off as a waste product. Today, some bacteria and protists, and most plants produce oxygen using this same process.

☑ READING CHECK **Explain** how plants produce oxygen gas.

Stromatolites Most scientists think that microscopic cyanobacteria could have slowly produced enough oxygen to change the composition of the atmosphere that existed on Earth during the Archean. By the early Proterozoic, large, coral reef-like mounds of cyanobacteria called **stromatolites** (stroh MA tuh lites) dominated the shallow seas that at that time covered most of Earth's continents. Stromatolites are made by billions of cyanobacteria colonies that trap and bind sediments together. The photo on the opening page of this chapter shows present-day stromatolites. These structures are similar in size and shape to Precambrian fossil stromatolites found in Glacier National Park, shown in **Figure 10.**

■ **Figure 10** These well-preserved fossil stromatolites in Glacier National Park are evidence that cyanobacteria existed during the Precambrian.

Teacher Content Support

Methanogens During the Archean, the Sun burned with significantly less energy than it does today. However, evidence indicates that Earth was relatively warm. Some scientists think the Archean atmosphere was enriched in methane (CH_4), a greenhouse gas that blankets Earth and prevents heat from dissipating to space. Some of these scientists further speculate that much of the methane was produced by life-forms even more ancient than cyanobacteria. These life-forms—single-celled methanogens—exist today in oxygen-free environments and produce methane as a by-product. If methanogens existed during the Archean, they might have produced enough methane to warm Earth.

Collaborative Learning

Beneficial Relationships In some relationships between two groups, the output from the first group benefits the second group. If the output from the second group in turn benefits the first group, a positive feedback loop exists. Have groups of students make concept maps showing photosynthesis as a positive feedback system. Maps should show that plants need CO_2 and that animals use the O_2 that plants make. The CO_2 that animals make, in turn, benefits plants. Ask them to think of other examples of positive feedback systems. OL COOP LEARN

■ **Caption Question** **Fig. 9** Earth was about 2.3 billion years old when oxygen began to accumulate.

☑ READING CHECK Green plants produce oxygen by photosynthesis—the same process that ancient cyanobacteria used to produce oxygen.

FOLDABLES® Rubric

ACROSS THE CURRICULUM

Biology Photosynthesis plays an important role in the long-term cycling of carbon. Photosynthesis removes carbon from the atmosphere. When carbon-containing organisms die and decay, carbon is released back into the air. Some carbon is incorporated in sediment to become coal, oil, or gas. Only when burned do these materials return carbon to the atmosphere.

EARTH SCIENCE JOURNAL

Understand Cyanobacteria Have students write an Earth science journal entry explaining how the concept of uniformitarianism underlies scientists' understanding of ancient cyanobacteria. OL

Rubric

Environmental Connection

Importance of Iron Iron is the cheapest, most abundant, and most useful of all metals, but it cannot be used in its pure state; it is too chemically reactive. Iron mined today comes from the iron oxides, such as hematite (Fe_2O_3) and magnetite (Fe_3O_4), that formed during the Precambrian. The oxygen is removed from the iron oxides in a manufacturing process called smelting. The pure iron is then combined with other material to form iron alloys. The most common use for iron is in the manufacture of steel. Steel contains over 95 percent iron. Steel is easily recycled. In the United States, steel cans and other steel products contain at least 25 percent recycled steel.

Caption Question **Fig. 11** Iron reacts with oxygen in the atmosphere. The presence of banded-iron formations suggests that there was enough oxygen in the atmosphere to react with iron in the rocks, at least in localized areas on a seasonal basis.

Figure 11 This iron mine in Brazil contains banded-iron formations that date from the Proterozoic.
Explain *how banded-iron formations are evidence of atmospheric oxygen gas.*

Evidence in rocks Scientists can determine whether there was oxygen in Earth's Archean atmosphere by looking for oxidized iron in Archean rocks. Scientists know that iron reacts with oxygen in the atmosphere to form iron oxides, more commonly called rust. Iron oxides are identified by their red color and provide evidence of oxygen in the atmosphere. The absence of iron oxides in rocks of the late Archean indicates that there was no oxygen gas in the atmosphere at that time. Had atmospheric oxygen gas been present, it would have reacted with the iron ions in the water or with the iron contained in sediments.

Banded iron By the beginning of the Proterozoic, however, cyanobacteria had increased oxygen gas levels enough so that iron oxides began to form in localized areas. These locally high concentrations of iron oxides are called **banded-iron formations.** Banded-iron formations consist of alternating bands of iron oxide and chert, an iron-poor sedimentary rock. The iron oxides appear to have been deposited cyclically, perhaps in response to seasonal variations. Today, these formations are mined for iron ore. An iron mine and a banded-iron rock are shown in **Figure 11.**

Problem-Solving LAB

Calculate Profits

How do you calculate mining profits? Precambrian rocks contain many important mineral deposits, such as uranium oxide, which is used in nuclear reactors. In uranium oxide deposits in southern Ontario in Canada, the ore-containing rocks cover an area 750 m long and 15,000 m wide with an average thickness of 3 m. Analysis of the deposit indicates that there are, on average, 0.9 kg of uranium oxide per metric ton of rock. Additionally, 0.3 m^3 of the uranium-bearing rock has a mass of 1 metric ton.

Analysis

1. **Solve** How many kilograms of uranium-oxide ore does this deposit contain?
2. **Compute** It will cost \$45/m^3 and 10 years to mine and extract the ore. How much will this cost?

Think Critically

3. **Assess** Assume that the current market price of uranium oxide is \$26.00/kg. Based on your answer to Question 2, can the ore be mined for a profit?

Problem-Solving LAB

Purpose Students will calculate the mass of uranium ore and the cost of mining the ore.

Process Skills use numbers, predict

Teaching Strategy Mining materials from Earth is a business, and a cost-benefit analysis is necessary before determining whether an area can be mined without the company going into debt. Ask students to describe other cost-benefit analyses with which they are familiar. Answers might include starting a business, transporting goods, and purchasing versus renting a home.

Analysis

1. 22,500,000 kg. First, calculate the number of cubic meters of rock: 750 m × 15,000 m × 3 m = 33,750,000 m^3. Next, convert cubic meters of rock to metric tons: 33,750,000 m^3 ÷ 0.3 m^3 per metric ton = 112,500,000 metric tons. Finally, determine the number of kilograms of uranium: 112,500,000 metric tons × 0.2 kg of uranium per metric ton = 22,500,000 kg.
2. \$1,518,750,000. Multiply the cubic meters of rock (33,750,000 m^3) × \$45 per m^3 = \$1,518,750,000.

Think Critically

3. The current market value of the U_3O_8 in the deposit is \$132/kg × 22,500,000 kg = \$2,970,000,000. Therefore, the deposit will turn a net profit of \$1,451,250,000. Over 10 years, the average annual profit will be \$145,125,000.

Red beds Many sedimentary rocks that date from the mid-Proterozoic, beginning about 1.8 bya, are rusty red in color. These rocks are called **red beds** because they contain so much iron oxide. The presence of red beds in mid-Proterozoic and younger rocks is strong evidence that the atmosphere by the mid-Proterozoic contained oxygen gas.

Importance of oxygen Oxygen is important not only because most animals require it for respiration, but also because it provides protection from harmful ultraviolet radiation (UV) from the Sun. Today, only a small fraction of the Sun's UV radiation reaches Earth's surface. This is because Earth is protected by ozone in Earth's upper atmosphere.

Recall that an ozone molecule consists of three oxygen atoms bonded together. As oxygen accumulated in Earth's atmosphere, an ozone layer began to develop. Ozone filtered out much of the UV radiation, providing an environment where new life-forms could develop.

READING CHECK **Describe** the importance of oxygen for the evolution of life.

Formation of the Oceans

As you have learned, some scientists think that the oceans reached their current size very early in Earth's history. The water that filled the oceans probably originated from the two major sources that provided water in Earth's atmosphere: volcanic outgassing, and asteroids, meteorites, and other objects that bombarded Earth's surface. Earth's early Precambrian atmosphere was rich with water vapor from these sources. As Earth cooled, the water vapor condensed to form liquid water. Recall that condensation occurs when matter changes state from a gas to a liquid.

Rain As liquid water formed, a tremendous amount of rain fell. The rain filled the low-lying basins and eventually formed the oceans. Rainwater dissolved the soluble minerals exposed at Earth's surface and—just as they do today—rivers, runoff, and groundwater transported these minerals to the oceans. The dissolved minerals made the oceans of the Precambrian salty, just as dissolved minerals make today's oceans salty.

MiniLAB

iLab Station

Model Red Bed Formation

Why are red beds red? Red beds contain so much iron oxide that they appear rusty red in color. Red beds that date from the mid-Proterozoic provide evidence that oxygen gas existed in the Proterozoic atmosphere.

Procedure

1. Read and complete the lab safety form.
2. Place 40 mL of **white sand** in a **150-mL beaker.**
3. Add **water** so that the total volume is 120 mL.
4. Add 15 mL of **bleach.** **WARNING:** ***Use bleach in a well-ventilated area.***
5. Place a piece of **steel wool** about the size of your thumbnail in the beaker.
6. Cover the beaker with a **petri dish,** and allow it to sit undisturbed for one day.
7. Remove the steel wool, and stir the contents of the beaker. Allow the mixture to settle for 5 min after stirring.
8. Slowly pour off the liquid so that the iron-oxide sediment is left behind.
9. Stir the mixture again; then spoon some of the sand onto a **watch glass,** and allow it to dry.

Analysis

1. **Describe** how the color of the sediment changed.
2. **Explain** where the iron in the experiment came from.
3. **Conclude** where, in nature, the red in rocks comes from.
4. **Assess** the function of the bleach in the experiment.

READING CHECK Oxygen gas led to the formation of an ozone layer. Ozone absorbs most of the Sun's harmful UV radiation. An ozone layer helped provide a protective environment for animals.

MiniLAB

iLab Station Rubric

Purpose Students will learn how iron oxidizes and stains sediment.

Process Skills model, recognize cause and effect, observe and infer

Additional Material stirring rod

Safety Precautions Approve lab safety forms before work begins. Students should wear goggles and gloves. Review material safety for the use of bleach. Handle bleach with care; it can discolor clothing.

Teaching Strategy Explain to students that even small amounts of impurities can cause color changes in transparent or light-colored materials. When light hits an object, some of the wavelengths are absorbed and some are reflected, depending on the materials in the object. The reflected wavelengths are what we perceive as the object's color. In light-colored material, only a small amount of the spectrum is absorbed; most wavelengths are reflected. As color darkens, more wavelengths are absorbed and fewer wavelengths are reflected. Quartz sand is generally light in color. Therefore, a small amount of iron can cause a color change.

Expected Results The steel wool will oxidize, and the iron oxide will stain the sand orange-red.

Analysis

1. The sediment changed from white to orange-red.
2. It comes from the iron contained in the steel wool.
3. The red comes from the oxidized iron contained in iron-bearing minerals in the rocks.
4. The bleach helps the steel wool to oxidize quickly.

Assessment

Performance Have students design and conduct an experiment showing how a small amount of a coloring agent affects the color of light- and dark-colored materials, and then have them explain the implications for the formation of red beds in quartz-rich sedimentary rock. In clear or light-colored material, such as water, only a small amount of coloring agent is needed to change the color, but in dark-colored material, like cola, a coloring agent has little effect; the material has already absorbed most wavelengths. Because quartz is light-colored, only a small amount of a coloring agent is needed to stain quartz-rich rock.

Enrichment

Astrobiology Have students research the field of astrobiology and report to the class about the status of the search for extraterrestrial life. AL

3 Assess

Check for Understanding

Concept Development Ask students to explain what the absence of red beds in rocks older than 1.8 billion years indicates about Earth's earliest atmosphere. It indicates that Earth's atmosphere did not contain enough free oxygen gas before about 1.8 bya.

Reteach

Compare and Contrast Have students write essays that explain how Earth's oceans and atmosphere likely formed. Students should know that both the atmosphere and the oceans probably formed with contributions of water from outgassing and extraterrestrial bombardment. They should also realize that any ocean water derived from outgassing likely condensed from water vapor in the atmosphere.

Assessment

Knowledge In a class discussion, have students compare the importance of water with the importance of oxygen for the evolution of life. Which do they think is the most important? Students should realize that while some life can exist without oxygen, no life can exist without water.

■ **Figure 12** This photograph taken by the *Mars Reconnaissance Orbiter* reveals evidence that suggests liquid water once flowed on the Martian surface.

Water and life The Precambrian began with an environment inhospitable to life. When it ended, much of Earth was covered with oceans that were teeming with tiny cyanobacteria and other life-forms. Life as it exists on Earth today cannot survive without liquid water.

Scientists think that Earth is not the only object in the solar system that contains or has contained water. Some scientists estimate that the asteroid Ceres contains more freshwater than Earth. Scientists also think that some surface features on Mars, such as the gullies shown in **Figure 12,** were carved by liquid water, and that water might still flow in brief spurts on Mars. Some moons of Saturn and Jupiter might also contain water in their interiors.

The search for life elsewhere in the solar system and universe today is typically centered on the search for water. Life on Earth has been found in almost every environment that contains water, from antarctic ice to hot, deep-water ocean vents. Scientists think that simple life-forms might exist in similar environments on other objects in the solar system.

SECTION 3 REVIEW

Section Self-Check

Section Summary

- Earth's atmosphere and oceans began forming early in Earth's history.
- Oxygen gas began to accumulate in the Proterozoic by photosynthesizing cyanobacteria.
- Evidence for atmospheric oxygen can be found in rocks.
- The water that filled Earth's oceans most likely came from two major sources.

Understand Main Ideas

1. MAINIDEA **Explain** why an atmosphere rich in oxygen was important for the evolution of life.
2. **Explain** how scientists conclude that ancient cyanobacteria produced oxygen.
3. **Describe** the relationship between banded-iron formations and oxygen gas.
4. **Describe** where the water in Earth's oceans originated.

Think Critically

5. **Conclude** What would Earth be like if oxygen gas had not formed in the atmosphere?

MATH IN ▶ Earth Science

6. If asteroids brought 1 cm of water to Earth every 50,000 years, and the average depth of Earth's oceans is 3700 m, how many years would it take to fill the ocean basins from this source?

SECTION 3 REVIEW

1. Because most organisms need oxygen to survive, the presence of oxygen gas in the atmosphere enabled organisms to evolve. Oxygen also causes ozone to form. In the Precambrian, ozone protected developing life-forms from UV light.
2. Ancient cyanobacteria likely produced oxygen because present-day cyanobacteria produce oxygen—an application of uniformitarianism.
3. Banded-iron formations formed in localized areas where atmospheric oxygen gas was present, which bound with the iron.
4. Earth's water likely came from water vapor in outgassing and from the water in bodies that bombarded Earth.
5. There would be no green plants, which make oxygen, and no animals, which need oxygen to breathe. Earth might be populated by life-forms requiring no oxygen and able to withstand high levels of UV light.
6. 18.5 billion years; 3700 m = 370,000 cm; 370,000 cm × 50,000 years/cm = 18,500,000,000 years

SECTION 4

Early Life on Earth

MAINIDEA Life began on Earth fewer than a billion years after Earth formed.

Essential Questions

- How is the experimental evidence showing how life might have begun on Earth summarized?
- What is the difference between prokaryotes and eukaryotes?
- How are Earth's early multicellular organisms described?

Review Vocabulary

hydrothermal vent: a hole in the seafloor through which water erupts

New Vocabulary

amino acid
prokaryote
eukaryote
Ediacaran biota

EARTH SCIENCE 4 YOU **If you have ever smelled ammonia, which is often used in household cleaners, you know that its pungent scent can make your nose sting. Some scientists think, however, that the presence of ammonia was necessary for life to form on Earth.**

Origin of Life

You have learned that fossil evidence suggests that cyanobacteria existed on Earth as early as 3.5 bya. Though cyanobacteria are simple organisms, photosynthesis—the process by which they produce oxygen—is complex, and it is likely that cyanobacteria evolved from simpler life-forms. Most scientists think that intense asteroid and meteorite bombardment prevented life from developing on Earth until at least 3.9 bya. Where and how the first life-form developed, however, remains an active area of research.

Primordial soup During the first half of the twentieth century, scientists thought that Earth's earliest atmosphere contained hydrogen, methane, and ammonia. Some biologists suggested that such an atmosphere, with energy supplied by lightning, would give rise to an organic "primordial soup" in Earth's shallow oceans. Primordial (pry MOR dee al) means *earliest* or *original.*

In 1953, Stanley Miller and Harold Urey devised an apparatus, shown in **Figure 13,** to test this hypothesis. They connected an upper chamber containing hydrogen, methane, and ammonia to a lower chamber designed to catch any particles that condensed in the upper chamber. They added sparks from tungsten electrodes as a substitute for lightning. Within a week, organic molecules had formed in the lower chamber—the primordial soup!

Figure 13 In 1953, Stanley Miller, shown here, and Harold Urey performed experiments to test whether organic molecules could form on early Earth.

View an **animation of the Miller-Urey experiment.**

Concepts In Motion

IN THE FIELD

Oparin-Haldane Hypothesis Two scientists working independently—Russian biochemist Alexander Oparin and Scottish biologist J. B. S. Haldane—proposed that life arose in an atmosphere of methane, ammonia, and hydrogen. Oparin published first, in 1924, to an audience limited by Russia's civil war. Haldane published in 1929. Today, the hypothesis is attributed to both scientists.

SECTION 4

1 Focus

Focus Transparency

Before presenting the lesson, project **Section Focus Transparency 70** and have students answer the accompanying questions. EL

MAINIDEA

Life's Origins Ask students when and in what form they think life first appeared on Earth. Ask them to imagine how life might have arisen. Explain that ideas about life's origins are hypotheses, and that hypotheses are testable in experiments.

2 Teach

Identify Misconceptions

Some students might think that organisms like today's animals existed early in Earth's history.

Uncover the Misconception The fossil evidence indicates that the first multicellular animals appeared at the end of the Precambrian. Until then, all organisms were unicellular.

Demonstrate the Concept Have each student draw a rectangle 46 cm high and 1 cm wide. Each cm in height represents 100 million years. Have students color the bottom 40 cm blue and the top 6 cm red. The red part represents the time that animals have been found in the fossil record, and the blue part represents the time the record is devoid of animal fossils.

Assess New Knowledge Have students draw a time line of Earth's history and indicate when on the line the first evidence of life appeared in the fossil record. Cyanobacteria appear in the rock record about 3.5 bya.

Uncertainties The organic molecules that formed in Miller and Urey's experiment included **amino acids,** the building blocks of proteins. Miller and Urey were the first to show experimentally that amino acids and other molecules necessary for the origin of life could have formed in conditions thought present on early Earth. However, Earth's early atmosphere contained gases like those that vent from volcanoes—carbon dioxide, water vapor, and traces of ammonia, methane, and hydrogen. When combinations of these gases are used in simulations, amino acids do not form in high quantities, leading scientists to question whether those processes were sufficient for the origin of life. Some scientists continue to explore the possibility that amino acids, and therefore life, arose in Earth's oceans under localized conditions similar to those in the Miller-Urey experiment.

VOCABULARY

ACADEMIC VOCABULARY

Simulate

to create a representation or model of something

The video game simulated the airplane's flight with impressive realism.

Other scenarios Because of uncertainties with the conditions in the Miller-Urey experiment, scientists propose different scenarios and conduct new research into sources and conditions for the origin of life. Some of these are shown in **Table 1.** Some scientists think that amino acids organized elsewhere in the universe and were transported to Earth in asteroids or comets. Their experiments show that chemical synthesis of organic molecules is possible in interstellar clouds, and amino acids have been found in meteorites. Other scientists hypothesize that amino acids originated deep in Earth or its oceans. Experiments show that conditions there are favorable for chemical synthesis, and organisms have been found at depths exceeding 3 km.

Explore the **origins of life on Earth with an interactive table.** Concepts In Motion

Table 1 How Life Might Have Begun on Earth: Three Hypotheses

	Earth's Surface	Deep Earth	Space

Hypothesis	Life originated on Earth's surface in warm, shallow oceans.	Life originated in hydrothermal vents deep in the oceans.	Organic molecules were brought to Earth in asteroids or comets.
Requirement	Hydrogen, methane, and ammonia must be present in the atmosphere.	Life must survive at high temperatures and pressures.	Organic molecules must be present in extraterrestrial bodies.
Evidence	Simulations produce amino acids.	Simulations of deep-sea vents produce amino acids.	Some meteorites contain amino acids that survived impact.
Drawback	The composition of the early atmosphere likely did not have large amounts of the required gasses.	It might have been too hot for organic molecules to survive.	It is difficult to test at this time due to technical limitations.

Teacher Content Support

Panspermia The idea that life came to Earth from elsewhere in the universe is sometimes called *panspermia,* which means *seeds everywhere.* This is an old idea that astrobiologists are examining with new interest. Studies suggest that at least two hardy species of Earth bacteria, *Bacillus subtilis* and *Deinococcus radiodurans* (nicknamed *Conan the Bacterium*) can withstand the heat, radiation, pressure, and acceleration resulting from a meteorite impact. Could life-forms like these have hitched a ride on a meteorite that collided with Earth billions of years ago, seeding the planet with life? No one knows, but scientists think the idea is credible.

DIFFERENTIATED INSTRUCTION

Advanced Learners Ask students to research the unique life-forms living near hydrothermal vents. Ask them to describe the kinds of life that exist around these vents and why astrobiologists studying Jupiter's moon Europa are so interested in these organisms. Have students write about their findings in their Earth science journals.

One current area of research explores the possibility that life emerged deep in the ocean at hydrothermal vents. The energy and nutrients necessary for the origin of life are present in this environment. As shown in **Figure 14,** a variety of unique organisms called extremophiles (from the Latin *extremus* meaning "extreme" and Greek *philía* meaning "love"), live near hydrothermal vents.

No single theory needs to be exclusive; it is possible that all of these contributed to the origin of life. Regardless of how life arose, it is known that conditions during that time were not hospitable, and life probably had many starts and restarts on early Earth. Asteroid impacts were probably still common between 3.9 and 3.5 bya when life arose. Large impacts during this time could have vaporized many early life forms.

■ **Figure 14** These tubeworms tolerate extreme pressures and temperatures near hydrothermal vents 2 km below the ocean's surface.
Deduce *why pressure is high in a hydrothermal-vent environment.*

An RNA world While experiments have shown the likelihood that amino acids existed on early Earth, scientists are still learning how the amino acids were organized into complex proteins and other molecules of life. One essential characteristic of life is the ability to reproduce. All cells require RNA and DNA to reproduce. In modern organisms, RNA carries and translates the instructions necessary for cells to function. Both RNA and DNA use proteins called enzymes to replicate.

Recent experiments have shown that RNA molecules called ribozymes can act as enzymes. They can replicate without the aid of enzymes. This suggests that RNA molecules might have been the first replicating molecules on Earth. An RNA-based world might have been intermediate between an inorganic world and today's DNA-based organic world.

Proterozoic Life

Fossil evidence indicates that unicellular organisms dominated Earth until the end of the Precambrian. These organisms are **prokaryotes** (proh KE ree ohts)—organisms that do not contain nuclei. Nuclei are separate compartments in cells that contain DNA and RNA. Organisms whose cells contain RNA and DNA in nuclei are called **eukaryotes** (yew KE ree ohts). **Figure 15** illustrates how prokaryotes and eukaryotes differ in the packaging of their DNA and RNA.

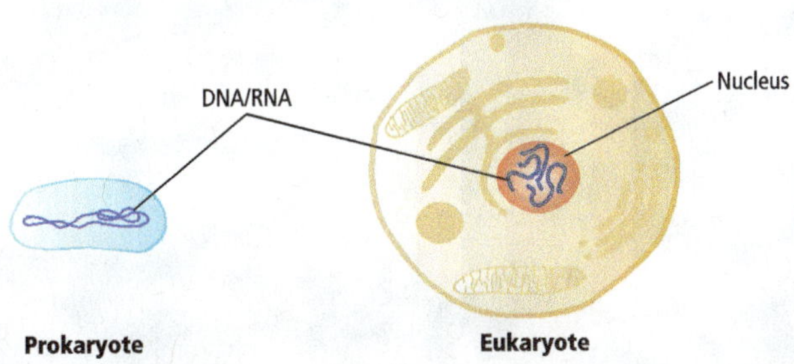

■ **Figure 15** Unlike prokaryotes, eukaryotes store DNA in cellular compartments called nuclei.

Teacher Content Support

Domain Archaea For many years, biologists recognized two domains of living organisms: the prokaryotic bacteria and the eukaryotes. In the late 1970s, however, a new group of organisms, the archaea, was identified. Archaea are single-celled organisms that live in boiling hot springs, exceptionally saline water, and other extreme environments. Methanogens are an example. While archaea look like bacteria, their genetic material more closely resembles the genetic material of eukaryotes. Many scientists think that bacteria, archaea, and eukaryotes each represent a primary line of descent that diverged early in Earth's history from an ancestral life-form. For this reason, archaea are placed in a separate domain—as different from bacteria as bacteria are from eukaryotes.

Discussion

Life's Essential Characteristics What is life and how is it recognized? Scientists frequently debate this question. Have students brainstorm what essential characteristics, besides the ability to reproduce, living organisms possess. Life-forms are composed of cells that provide protection from the environment, respond to the environment, maintain an internal environment (homeostasis), and contain a genetic blueprint (DNA/RNA) that is passed down to future generations. OL AL

■ **Caption Question Fig. 14** Pressure is so high because the vents are so deep. Pressure increases with depth.

Demonstration

Compare Eukaryotes and Prokaryotes While few prokaryotes are multicellular, many eukaryotes, such as yeast, green algae, and amoebas, are unicellular. Show pictures of a variety of prokaryotes and eukaryotes, and ask students to identify which is which.

ACROSS THE CURRICULUM

Ecology For many years, scientists thought that the deep sea was a barren wasteland. In 1977, however, giant tubeworms, clams, and other new species were discovered living near hydrothermal vents. For their energy source, these organisms depend not on sunlight, as most organisms do, but on sulfur. This deep-sea environment thus represents a unique ecological system.

Teacher Content Support

Eukaryote Evolution How did eukaryotes evolve? The most widely accepted theory is that eukaryotes evolved from the mutualistic symbiosis of different kinds of prokaryotic cells. Mutualism is a relationship between dissimilar organisms in which both partners benefit. Lichens are an example of mutualistic symbiosis between fungi and algae. Evolutionary paleobiologists hypothesize that two or more prokaryotes might have entered into such a symbiotic relationship and, over time, the symbionts became so interdependent that they could no longer survive independently. The theory of endosymbiosis thus suggests that some eukaryote organelles, especially mitochondria and chloroplasts, are the distant descendants of once free-living, prokaryotic organisms. This theory has gained compelling support since it was introduced four decades ago.

Project

Endosymbiosis Have groups of students research endosymbiosis and describe the evidence for this theory. Mitochondria and chloroplasts are similar in size to free-living prokaryotes. In addition, they contain DNA that is different from the DNA in the cell nucleus but similar in size and shape to prokaryotic DNA. Mitochondria and chloroplasts also have two membranes, one of which is presumably the relic membrane of the engulfed prokaryote. Phylogenetic studies suggest that both mitochondria and chloroplasts are related to bacteria. In addition, while endosymbiosis probably happened billions of years ago, there are organisms alive today, called living intermediates, that live together in endosymbiotic conditions. AL COOP LEARN

Figure 16 Sunbeams streaming through ice might have provided a refuge for some life-forms 750 mya, when ice covered Earth.

Simple eukaryotes Eukaryotes can be unicellular or multicellular, but because they contain nuclei and other internal structures, they tend to be larger than prokaryotes. This general observation is useful in determining whether a fossil represents a prokaryote or a eukaryote because it is rare for a fossil to be preserved in enough detail to determine whether its cells had nuclei. The oldest-known eukaryote fossil is unicellular. It was found in a banded-iron formation, about 2.1 billion years old, in Michigan.

READING CHECK **Explain** how the relative sizes of eukaryotes and prokaryotes are useful to paleontologists.

Snowball Earth Some scientists think that glaciation events 850–550 mya played a critical role in the extinction of many early unicellular eukaryotes. These glaciation events were so widespread that some geologists compare Earth at that time to a giant snowball. Evidence from ancient glacial deposits around the world suggests that glaciers might have advanced as far as the equator and that even the oceans might have been frozen. Though many organisms went extinct during this time, some life-forms survived, perhaps near hydrothermal vents or in pockets of sunlight streaming through openings in ice, as illustrated in **Figure 16.**

Multicellular organisms Although probably not Earth's first multicellular life, shortly after the ice retreated toward the poles, the climate warmed dramatically and many marine multicellular organisms appear in the rock record. Certain fossils of this time period were discovered in 1947 in Australia's Ediacara Hills. Collectively called the **Ediacaran biota** (ee dee A kuh ruhn by OH tuh), these fossils show the impressions of large, soft-bodied eukaryotes. **Figure 17** shows what these organisms might have looked like.

Figure 17 This reconstruction of an ocean during the Ediacaran Period shows how Earth's early multicellular organisms might have looked. They ranged from several centimeters to two meters in length.

READING CHECK Eukaryotes tend to be larger than prokaryotes because they have more internal organelles, such as nuclei. This difference in size helps paleontologists assess whether a fossil is a eukaryote or a prokaryote.

EARTH SCIENCE JOURNAL

Eukaryote Evolution Have students write journal entries describing why paleontologists hypothesize that prokaryotes evolved before eukaryotes. They should be able to argue this from both fossil and biological perspectives. Prokaryote fossils exist in rocks older than those that contain the oldest eukaryote fossils, and prokaryotes are the simplest kinds of organisms, which suggests they are the most primitive. OL

Ediacaran biota The discovery of the Ediacaran biota at first seemed to solve one of the great mysteries in geology: why there are no fossils of the ancestors of the complex and diverse animals that existed during the Cambrian Period—the first period of the Paleozoic Era. The Ediacaran biota seemed to provide fossil evidence of an ancestral stock of complex organisms. As shown in **Figure 18,** one type of Ediacaran organism appeared similar in overall body shape to sea pens. Others appeared similar to jellyfish, segmented worms, arthropods, and echinoderms—just the type of ancestral stock that geologists had been hoping to find.

However, upon closer examination, some scientists have questioned that conclusion and suggest that Ediacaran organisms are not relatives of present-day animal groups but, instead, represent unique organisms. These scientists point out that none of the Ediacaran organisms shows evidence of a mouth, anus, or gut, and there is little evidence that they could move. As a result, there is an ongoing debate in the scientific community about the precise nature of many of these fossils.

Figure 18 One type of Ediacaran organism resembles a present-day sea pen. Some scientists think that the two are related.

Mass extinction In recent years, geologists have found Ediacaran fossils in all parts of the world. This suggests that these organisms were widely distributed throughout the shallow seas of the late Proterozoic. They seem to have flourished between 600 mya and 540 mya. Then, in an apparent mass extinction, most of them disappeared, and organisms more likely related to present-day organisms began to inhabit the oceans.

SECTION 4 REVIEW

Section Self-Check

Section Summary

- Scientists think that life on Earth began between 3.9 and 3.5 bya.
- Stanley Miller and Harold Urey were the first to show experimentally that organic molecules could have formed on early Earth.
- Scientists have developed several hypotheses to explain how and where life formed.
- Eukaryotes appeared after prokaryotes.
- Earth's multicellular organisms evolved at the end of the Precambrian.

Understand Main Ideas

1. MAINIDEA **List** three hypotheses about the origin of life, and describe the evidence for each.
2. **Explain** why scientists think that life on Earth began after 3.9 bya.
3. **Identify** the ingredients that Miller and Urey thought made up Earth's early atmosphere.
4. **Compare and contrast** eukaryotes and prokaryotes.
5. **Discuss** why some scientists think that Ediacaran organisms do not represent present-day animal groups.

Think Critically

6. **Hypothesize** one reason that the Ediacaran organisms became extinct.

WRITING IN ▶ Earth Science

7. Write a newspaper article about the discovery of a new fossil outcrop that dates to the end of the Precambrian. Describe the fossil organisms found in this outcrop.

3 Assess

Check for Understanding

Discussion Have students discuss the three hypotheses for life's origin that are presented in the text. In particular, ask students to think about the initial source of energy required for the synthesis of organic compounds in these different environments. Students should recognize that lightning was the initial source of energy in the "primordial soup" model, whereas Earth's internal heat was the initial source of energy in the hydrothermal vent model. COOP LEARN

Reteach

Time Lines Have students create time lines showing the origin of prokaryotic and eukaryotic life, Snowball Earth, and the first-known eukaryotic organisms (the Ediacaran biota).

Assessment

Performance Have students write an essay explaining why paleontologists debate whether the Ediacaran biota contained relatives of modern-day animals. Students might argue that while some Ediacaran organisms had body shapes resembling those of modern-day animals, most Ediacaran organisms were unlike modern-day animals because they seemed to be unable to move or ingest food. Scientists think that most Ediacaran organisms became extinct at the end of the Precambrian.

Rubric

SECTION 4 REVIEW

1. 1. Life began in shallow seas in a "primordial soup"; evidence: amino acids form in simulations of Earth's possible early atmosphere.
   2. Life began in the deep oceans; evidence: amino acids form in hydrothermal vent simulations.
   3. Life formed extraterrestrially and was transported to Earth in asteroids or meteorites; evidence: presence of water and organic molecules in some meteorites.
2. Meteorites and asteroids likely bombarded Earth routinely during Earth's first 700 million years, preventing life from taking hold.
3. ammonia, methane, hydrogen
4. Eukaryotes package their genetic material in nuclei; prokaryotes do not. Eukaryotes tend to be larger because they contain cell organelles.
5. These organisms appeared to have had no animal characteristics.
6. The Ediacaran biota might have become extinct because of sea level changes, climate changes, volcanism, or an asteroid impact—or by some combination of these factors. Answers should contain brief explanations.
7. Answers will vary, but the articles should discuss Ediacaran-like organisms.

Rubric

Earth Science &
TECHNOLOGY

Purpose

Students will learn about the development of technology that will be used to explore Mars, both at the surface and below the surface.

Teacher Content Support

Methane on Mars Methane is quickly destroyed in the Martian atmosphere. Therefore, scientists think this release may be an ongoing process. The discovery of methane plumes in three distinct regions during the Martian spring and summer challenges the assumption that Mars is a geologically dead planet or devoid of life.

Some scientists think that the emission of the methane might occur because the permafrost blocking cracks and fissures vaporizes during the warm seasons. The methane may have been trapped in ice cages called clathrates for billions of years. On Earth, the conversion of iron oxide (rust) into the serpentine group of minerals creates methane, which could occur on Mars using the planet's water, carbon dioxide, and internal heat.

The *ExoMars/Trace Gas Orbiter* is set to arrive in orbit around Mars in 2016. It will conduct a survey of trace gases such as methane in the Martian atmosphere to learn how long they exist in the atmosphere before being lost to space, as well as determining their locations and the nature of the subsurface sources that produce them.

Earth Science &
TECHNOLOGY

Martian Microenvironments

Just as scientists have questions about the history of Earth, including past climate conditions and the development of early life on Earth, they have similar questions about Mars.

Life on Mars? High levels of radiation and the inability to sustain liquid water make finding life as we know it unlikely on the surface of Mars. However, there is mounting evidence that zones containing liquid water existed—and perhaps may still exist—below the Martian surface. Scientists are investigating likely spots for these microenvironments in areas resembling hydrothermal environments such as steam vents and fumaroles which on Earth are home to some of our hardiest organisms.

Silica and methane Ideal landing sites for Mars missions require an easily accessible rock record, evidence of past water, or evidence of potential biological activity. Photographs of one possible site, taken by the *Mars Reconnaissance Orbiter* show bright mounds of what is thought to be silica deposited on a volcano. Silica can be dissolved, transported, and concentrated by hot water or steam, and its location on the volcano resembles silica deposits around hydrothermal vents on Earth.

Another possible site might be in an area where methane gas has been detected. A spectrometer on the W. M. Keck Observatory in Hawaii detected the release of huge methane plumes in three distinct regions during the Martian summers. Most of Earth's atmospheric methane is produced by biological activity, but it is also released by geological processes. Researchers do not yet have sufficient data to determine the source of methane on Mars.

Curiosity Scheduled to land in 2012, the *Mars Science Laboratory*, also known as *Curiosity*, will continue to investigate Mars's ability to support past or present microbial life.

This false-color image shows light-colored deposits on a volcanic cone in the Nili Patera caldera. Scientists think that these deposits could be silica, which might suggest an environment once hospitable to microbial life.

It is powered by a radioisotope thermoelectric generator, which produces electricity from the heat of radioactive decay. *Curiosity* will use a laser to zap rocks up to seven meters away; a spectral analyzer will examine the resulting flash of light to identify the elements contained in each rock. The Radiation Assessment Device (RAD) will monitor radiation levels as it travels through space as well as when it is on Mars. This provides crucial information for any future manned missions.

The evidence collected by the search for organisms living in extreme environments on Earth, challenge our assumptions about what constitutes habitable conditions for life as we know it. This evidence also expands the types of environments scientists can explore to find extraterrestrial life on Mars, and elsewhere in the solar system.

WRITING IN ▶ Earth Science

Discussion Research updates about the latest technology for future exploration and data collection on Mars. Summarize your findings and discuss with your classmates the following question: Why should steps be taken to prevent the contamination of Mars with organisms from Earth?

WebQuest

Teaching Strategies

- Review with students why confirming the presence of water on Mars, past or present, is a major goal of scientists. Because all forms of life on Earth require water, water is a good indicator of the possibility of life elsewhere in the solar system.
- Discuss why exploring Mars is so difficult. What issues must scientists address? Possible answers include the dependency on remote computers, the difficulty in making repairs to damaged equipment, negotiating the rugged terrain, and the high cost and specialization of equipment.

WRITING IN ▶ Earth Science

Rubric

Discussion Students might research more details about the technology discussed in the text, or they could focus on the *Mars Science Laboratory* mission. As part of their summaries, students should include visual aids in the form of handouts or illustrations.

WebQuest

GeoLAB

iLab Station

Mapping: Map Continental Growth

Background: During the Precambrian, microcontinents and island arcs collided to form what would become present-day continents.

Question: *How does the distribution of the ages of rocks help geologists reconstruct the sequence of continental growth?*

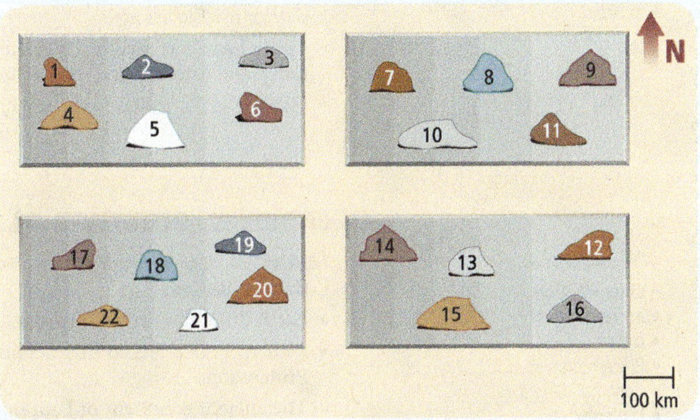

Locality Data

Materials

rock samples
paper
metric ruler
colored pencils

Safety Precautions

Procedure

Suppose you are working on a geologic survey team that is updating its geologic map of a continent. You have gathered the ages of rock samples found in various locations throughout the continent.

1. Read and complete the lab safety form.
2. Your teacher has set up locations around the classroom with a rock sample of a different age at each location. Draw a rough map of the room showing the locations and ages of all the rocks.
3. Measure and record the distance, in centimeters, between the rocks.
4. Plot your measurements on an outline map of your classroom, using a scale of 1 cm = 100 km.
5. Use a pencil to draw lines on the map, separating rocks of different ages.
6. Use colored pencils to shade in each area on the map that contains rocks of the same ages. These are your geologic age provinces.
7. Make a key for your map. Name the oldest province *Province A*, the next oldest province *Province B*, and so on for all provinces.

Analyze and Conclude

1. **Compare** your map with those of your classmates.
2. **Identify** the oldest province on your map. Where is it located in relation to the other provinces?
3. **Describe** the sequence of collision events that formed the continent represented by your map.
4. **Interpret Data** Use your map to find the likely sites of metamorphic rocks. Determine what types of metamorphism might have occurred.
5. **Interpret Data** Based on your map, where would you expect to find the highest and most rugged mountains? The most weathered mountains? Explain.

APPLY YOUR SKILL

Time Line Make a time line that shows the order of accretion of the provinces of the North American continent shown in **Figure 7**.

Analyze and Conclude

1. Maps will differ. For instance, the width of Province A will vary depending on where the students place the boundary line. In order for all the maps to be the same, a large number of data points would be needed.
2. The oldest province would be near the center of the map area.
3. Answers will vary. Student answers should indicate that younger provinces would have accreted around the margins of the oldest province.
4. Metamorphic rocks would be found at the province boundaries. Regional metamorphism would have resulted from the collisions. If igneous activity was generated, contact metamorphism also could have occurred.
5. Remnants of the most rugged mountains would be found between the two youngest provinces, or between the oldest and youngest province, depending on the arrangement of the provinces. Remnants of the oldest and most weathered mountains would be found along the boundary between the oldest province and the province that is only slightly younger.

GeoLAB

iLab Station Rubric

Preparation

Time Allotment 45 min

Process Skills interpret maps, communicate, analyze, use numbers

Safety Precaution Approve lab safety forms before work begins.

Preparation of Materials Place 15 or more igneous and metamorphic rock samples, each with an index card indicating a Precambrian age, on desks throughout the classroom in an arrangement that suggests distinct provinces. Each province should contain rocks with similar ages. The placement and ages of the samples can vary, but they should be arranged so that samples with the oldest ages are grouped in a central province. For example, you could place three 3.5-billion-year-old rocks at the center of the room, three 3.0-billion-year-old rocks on the north side, three 1.8-billion-year-old rocks on the west side, three 1.0-billion-year-old rocks on the east side, and three 530-million-year-old rocks on the south side.

Procedure

- Have students work in pairs or small groups.
- **Troubleshooting** Be sure the age provinces are shown on the maps distinctly, separated from each other with only one line.

APPLY YOUR SKILL

Time Line The time lines that students draw will not be as straightforward as the maps they drew. Have them develop multiple hypotheses about why the order is not as simple.

MAINIDEAS Summary statements can be used by students to review the major concepts of the chapter.

Students can review with these online resources.

Vocabulary Practice

- Vocabulary eGames
- Vocabulary eFlashcards
- Vocabulary PuzzleMaker

Section Self-Check

Chapter Self-Check

Online Test Practice

Use the Exam*View*® *Assessment Suite* CD-ROM to:

- create multiple versions of tests
- create modified tests with one mouse click
- edit existing questions and add your own questions
- build tests aligned with state standards using built-in state curriculum tags
- change English tests to Spanish with one mouse click
- track students' progress using the Teacher Management System

CHAPTER 22 STUDY GUIDE

BIGIDEA **The oceans and atmosphere formed and life began during the three eons of the Precambrian, which spans nearly 90 percent of Earth's history.**

Vocabulary Practice

SECTION 1 Early Earth

VOCABULARY
- zircon
- meteorite
- asteroid

MAINIDEA **Several lines of evidence indicate that Earth is about 4.6 billion years old.**

- Scientists use Earth rocks, zircon crystals, Moon rocks, and meteorites to determine Earth's age.
- Likely heat sources of early Earth were gravitational contraction, radioactivity, and asteroid and meteorite bombardment.
- Cooling of Earth led to the formation of liquid water.

SECTION 2 Formation of the Crust and Continents

VOCABULARY
- differentiation
- microcontinent
- craton
- Precambrian shield
- Canadian Shield
- Laurentia

MAINIDEA **The molten rock of Earth's early surface formed into crust and then continents.**

- Earth differentiated into specific density zones early in its formation.
- Plate tectonics caused microcontinents to collide and fuse throughout the Proterozoic.
- The ancient continent of Laurentia formed as a result of many mountain-building episodes.
- The formation and breakup of Earth's first supercontinent occurred during the Proterozoic.

SECTION 3 Formation of the Atmosphere and Oceans

VOCABULARY
- cyanobacteria
- stromatolite
- banded-iron formation
- red bed

MAINIDEA **The formation of Earth's oceans and atmosphere provided a hospitable environment for life to begin.**

- Earth's atmosphere and oceans began forming early in Earth's history.
- Oxygen gas began to accumulate in the Proterozoic by photosynthesizing cyanobacteria.
- Evidence for atmospheric oxygen can be found in rocks.
- The water that filled Earth's oceans most likely came from two major sources.

SECTION 4 Early Life on Earth

VOCABULARY
- amino acid
- prokaryote
- eukaryote
- Ediacaran biota

MAINIDEA **Life began on Earth fewer than a billion years after Earth formed.**

- Scientists think that life on Earth began between 3.9 and 3.5 bya.
- Stanley Miller and Harold Urey were the first to show experimentally that organic molecules could have formed on early Earth.
- Scientists have developed several hypotheses to explain how and where life formed.
- Eukaryotes appeared after prokaryotes.
- Earth's multicellular organisms evolved at the end of the Precambrian.

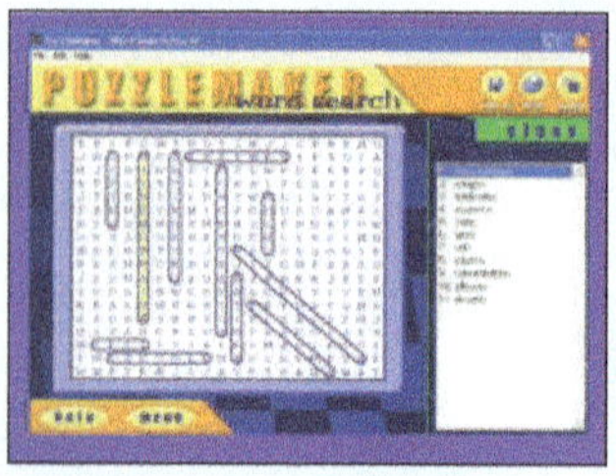

Vocabulary Practice

For additional practice with vocabulary, have students access the Vocabulary PuzzleMaker.

Chapter Self-Check

VOCABULARY REVIEW

Identify the vocabulary term from the Study Guide described by each phrase.

1. bodies that orbit the Sun between Mars and Jupiter
2. the name of the ancient continent that makes up most of North America
3. the first photosynthetic, oxygen-producing organisms on Earth
4. the process by which a planet becomes zoned with heavy materials near its center and lighter materials near its surface

Use the vocabulary term from the Study Guide to answer the following questions.

5. What are the building-blocks of protein?
6. What is the name of the Precambrian Shield in North America?
7. What are rocks called that consist of alternating bands of iron and chert?
8. What type of organism packages its DNA in nuclei?

Complete each sentence by providing the missing vocabulary term from the Study Guide.

9. The ________ were multicellular eukaryotes that evolved during the Proterozoic.
10. ________ is a very stable mineral often used to date Precambrian rocks.
11. A ________ is a mound made by microorganisms in shallow seas.
12. An old, stable part of a continent is called a ________.

UNDERSTAND KEY CONCEPTS

13. What process contributed to the formation of Earth's early atmosphere?
    A. outgassing
    B. differentiation
    C. crystallization
    D. photosynthesis

14. Which was not a source of heat for early Earth?
    A. asteroid and meteorite bombardment
    B. hydrothermal energy
    C. gravitational contraction
    D. radioactivity

Use the figure below to answer Questions 15 and 16.

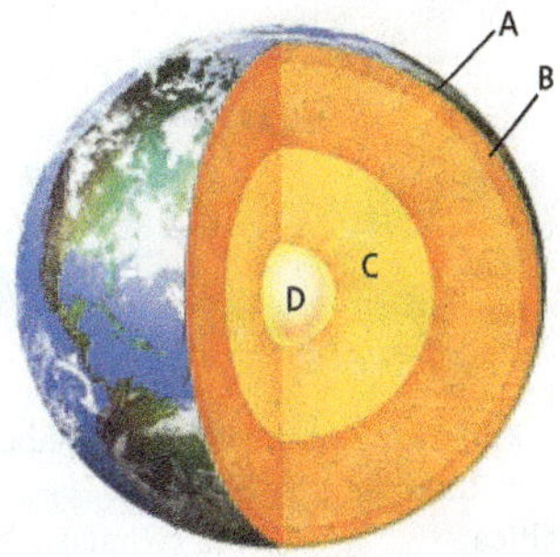

15. Which part of Earth is the most dense?
    A. A
    B. B
    C. C
    D. D

16. In which part of Earth would you find granite?
    A. A
    B. B
    C. C
    D. D

17. Why is oxygen gas important to life on Earth?
    A. It is used by plants to undergo photosynthesis.
    B. It is required by cyanobacteria and stromatolites to survive.
    C. It is a source of heat at Earth's surface.
    D. It provides protection from harmful ultraviolet radiation from the Sun.

18. Upon what age of Earth do most scientists agree?
    A. 4.6 thousand years old
    B. 4.6 million years old
    C. 4.6 billion years old
    D. 456 billion years old

19. A meteorite is a fragment of which object?
    A. the Sun
    B. asteroid
    C. planet
    D. the Moon

CHAPTER 22 ASSESSMENT

VOCABULARY REVIEW

1. asteroids
2. Laurentia
3. cyanobacteria
4. differentiation
5. amino acids
6. Canadian Shield
7. banded-iron formations
8. eukaryote
9. Ediacaran biota
10. zircon
11. stromatolite
12. craton

UNDERSTAND KEY CONCEPTS

13. A
14. B
15. D
16. A
17. D
18. C
19. B

20. D
21. C
22. B
23. C

CONSTRUCTED RESPONSE

24. Scientists use evidence from Earth's oldest rocks (4.28 billion years old), zircon crystals (4.4 billion years old), meteorites (4.5–4.7 billion years old), and moon rocks (4.4–4.5 billion years old.)

25. Likely sources were asteroids and other bombarding bodies, and volcanic gases.

26. Earth formed from orbiting rocky bodies that coalesced. Over time, gravity caused Earth to contract, generating thermal energy.

27. Plate tectonics causes continents to move, collide, and coalesce.

28. It was destroyed during crustal recycling.

29. The process by which cyanobacteria produce oxygen (photosynthesis) is complex, and likely these organisms were preceded in evolution by simpler life-forms.

30. Red beds form when iron reacts with sufficient levels of atmospheric oxygen gas to form iron oxides.

THINK CRITICALLY

31. Much of Earth's heat today is leftover from its formation. The rest comes primarily from radioactive decay.

32. Both hydrogen and helium have small atomic masses, so they escape to space.

33. Indirect evidence from moon rocks and meteorites can be used to infer that the early Hadean existed, because scientists think that all solar system bodies formed at about the same time.

34. The structures, stromatolites, are mounds made by cyanobacteria, which produce oxygen by photosynthesis.

35. Uniformitarianism implies that ancient cyanobacteria produced oxygen because present-day cyanobacteria do.

36. "A" is a eukaryote. It is larger, has organelles, and its DNA is packaged in a nucleus. "B" is a prokaryote. It is smaller and its DNA is not packaged in a nucleus.

37. The atmosphere would likely contain gases that vent from volcanoes (carbon dioxide, water vapor, nitrogen, carbon monoxide, but no oxygen gas).

ASSESSMENT

Use the figure below to answer Question 20.

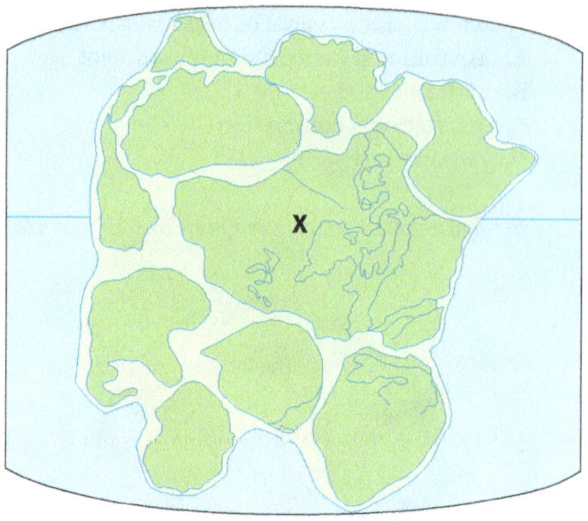

20. What is the name of the continent labeled *X* in this figure of Rodinia?
A. Baltica **C.** Gondawana
B. Amazonia **D.** Laurentia

21. Which is likely to give the oldest radiometric age date?
A. quartz **C.** zircon
B. granite **D.** metamorphic rock

22. Which was the earliest type of life on Earth?
A. eukaryotes **C.** ribozymes
B. prokaryotes **D.** Ediacaran biota

23. Refer to **Figure 6** in the text. How old are the rocks that underlie most of the state of Arizona?
A. 1.0–1.3 billion years
B. 1.8–2.0 billion years
C. 1.6–1.8 billion years
D. > 2.5 billion years

CONSTRUCTED RESPONSE

24. **List** the evidence that scientists use to determine Earth's age.

25. **Identify** sources of the gases that made up Earth's early atmosphere.

26. **Explain** how gravitational contraction heated early Earth.

27. **Discuss** how supercontinents form.

28. **Explain** why Earth's earliest crust no longer exists.

29. **Explain** why scientists think that cyanobacteria were not the first life-forms on Earth.

30. **Evaluate** How do red beds serve as evidence that there was oxygen gas in the atmosphere during the mid-Proterozoic?

THINK CRITICALLY

31. **Identify** the sources of Earth's heat today.

32. **Explain** why there is little hydrogen or helium in Earth's atmosphere today.

33. **Discuss** how scientists infer that early Hadean time existed.

Use the photo below to answer Question 34.

34. **Discuss** how the structures in the photo are related to oxygen gas in the atmosphere.

35. **Assess** how the concept of uniformitarianism can be used to explain your answer to Question 34.

Use the figure below to answer Question 36.

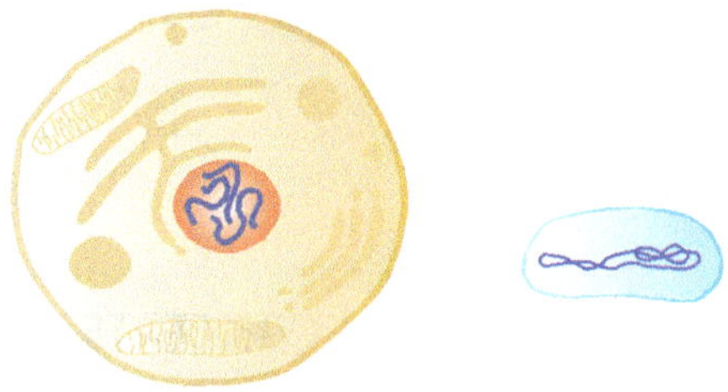

A B

36. **Identify** each cell shown as a prokaryotic or a eukaryotic. Explain the differences between them.

Chapter Self-Check

37. Infer the composition of the atmosphere had there never been life on Earth.

38. Discuss what scientists mean when they refer to an "RNA World."

Use the photo below to answer Question 39.

Jonathan Blair/CORBIS

39. Explain how the process illustrated above likely contributed to the formation of Earth's oceans.

40. Explain Where in North America would you look if you wanted to find evidence of Archean life?

41. Evaluate which is more important for the existence of life—liquid water or oxygen gas.

42. CAREERS IN EARTH SCIENCE Imagine that you have a rock sample from an Earthlike planet in a distant solar system. Plan an experiment that might help you determine the age of the planet.

43. Evaluate the significance of a Snowball Earth for the evolution of life.

CONCEPT MAPPING

44. Create a concept map showing the cause and effects of oxygen in Earth's atmosphere. Include the following key terms in the concept map: *oxygen, respiration, ozone, photosynthesis,* and *cyanobacteria.*

CHALLENGE QUESTION

45. Propose what Earth would be like if both continental crust and oceanic crust were made of the same material.

WRITING IN ▶ Earth Science

46. Suppose your spaceship has landed on a planet that you suspect has life-forms similar to cyanobacteria. Write a letter to your best friend explaining what you see outside your spaceship window.

DBQ Document–Based Questions

Data obtained from: Schopf, J.W. 1999. *Cradle of Life: The discovery of Earth's earliest fossils.* Princeton: Princeton University Press.

Earth contains rocks of varying ages, but they are not distributed evenly in time. This means that information about the geologic past is not available in the same quantity for all of Earth's past. Use the data below to answer the following questions.

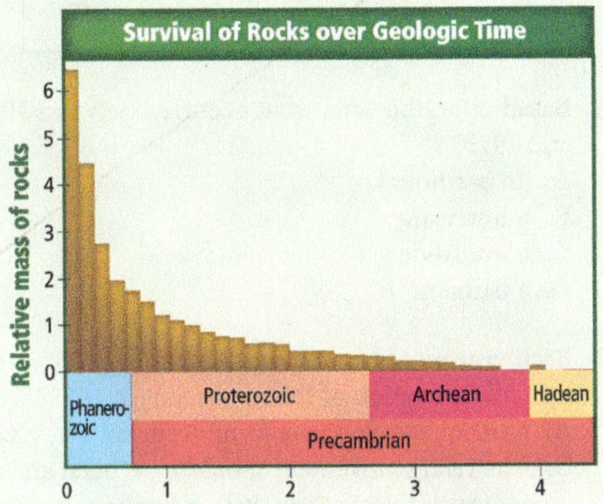

47. Which eon contains the most complete rock record?

48. What general trend is apparent about rocks through geologic time?

49. Why do you think that the trend in the data above exists?

CUMULATIVE REVIEW

50. Why could winter applications of salt to de-ice roads have negative effects on groundwater resources? **(Chapter 10)**

51. Explain the concept of uniformitarianism. **(Chapter 21)**

38. A type of RNA called a ribozyme can act like the enzyme needed by RNA and DNA to replicate. It is thus likely that RNA preceded DNA; scientists call this time in Earth's history the "RNA World."

39. The photo illustrates the process of outgassing, which might have contributed to the formation of oceans because volcanoes vent water vapor.

40. You would look in rocks of the Canadian Shield, which is exposed in Canada and some northern U.S. states.

41. Water is more important because all life on Earth requires water. Some life-forms can exist without oxygen, but scientists know of no life that can exist without water.

42. You could determine how old the sample is by radiometrically dating any zircon that might be in it.

43. A Snowball Earth might have caused the extinction of many unicellular eukaryotes, paving the way for the evolution of new, multicellular organisms.

CONCEPT MAPPING

44. Students might make a sequential concept map showing that photosynthesis comes from cyanobacteria, oxygen comes from photosynthesis, and that ozone and respiration derive from the presence of oxygen.

CHALLENGE QUESTION

45. Oceans today form where dense basaltic crust "weighs down" the oceanic lithosphere more than less-dense granitic crust "weighs down" continental lithosphere. Without this contrast, all crust would float equally and the oceans would likely be shallow and more extensive. Also, there would be little or no subduction, because material that is subducted (oceanic crust) is more dense than continental crust.

WRITING IN ▶ Earth Science

Rubric

46. Answers should mention that there would probably be continents and water, and that stromatolites might be found in shallow water. If cyanobacteria were the only life-form present, oxygen levels likely would not have accumulated sufficiently enough to allow for more-complex life, and green vistas would be absent.

DBQ Document–Based Questions

Data obtained from: Schopf, J.W. 1999. *Cradle of Life: The discovery of Earth's earliest fossils.* Princeton: Princeton University Press.

47. Phanerozoic

48. The older the rock, the greater the chance that it no longer exists.

49. The trend exists because plate tectonics recycles the crust. The older a rock, the greater the chance that it will be melted or metamorphosed. Those rocks that do survive have a greater chance of being buried by younger rocks.

CUMULATIVE REVIEW

50. Road salt frequently makes its way to sewage and storm water pipes, along with water from the melting of snow and ice. Since considerable amounts of salt are frequently used, salts might leach into groundwater supplies.

51. Uniformitarianism states that processes occurring today have been occurring throughout Earth's history.

MULTIPLE CHOICE

1. D
2. A
3. C
4. A
5. B
6. B
7. C
8. B
9. A
10. B

CUMULATIVE STANDARDIZED TEST PRACTICE

MULTIPLE CHOICE

1. Which contains the fewest number of years?
 A. eon
 B. era
 C. period
 D. epoch

Use the table to answer Questions 2 and 3.

Fault Line Activity	
Date	**Rock Slip Measurement (mm)**
1973	5
1974	8
1975	300
1976	10

2. Based upon the data, what occurred between 1974 and 1975?
 A. an earthquake
 B. a hurricane
 C. a mudslide
 D. a tsunami

3. Each year records some type of rock slip. What is the best interpretation of the movement in 1973?
 A. Earthquakes in some form occurred every year.
 B. The rising movement should have been an indicator to scientists that an earthquake was imminent.
 C. The slippage was so slight and smooth that it was not felt.
 D. The slippage was similar to aftershocks of an earthquake.

4. What is caused by differences in air pressure?
 A. wind
 B. clouds
 C. rain
 D. thunder

5. Which is not a cause of climatic variations?
 A. latitude
 B. frontal systems
 C. topography
 D. air masses

6. In which fossil do original structures of an organism remain?
 A. mold fossil
 B. mineral replacement fossil
 C. cast fossil
 D. trace fossil

Use the illustrations below to answer Questions 7 and 8.

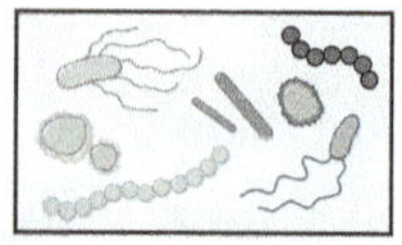
Group A

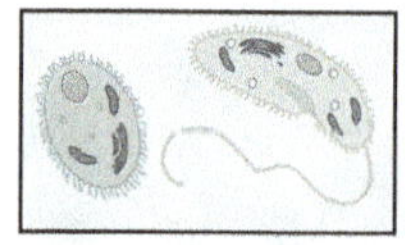
Group B

7. How do members of Group A differ from members of Group B?
 A. They are plants.
 B. They can be found in Proterozoic fossils.
 C. They contain no nuclei.
 D. They are all unicellular.

8. Where might members of Group B have survived during Snowball Earth?
 A. glaciers
 B. hydrothermal vents
 C. Australian fauna
 D. oil deposits

9. Which type of graph would best show the number of volcanic eruptions over a period of time?
 A. line graph
 B. circle graph
 C. pictograph
 D. bar graph

10. Squeezing that causes intense deformation at plate boundaries which leads to mountain building is known as what?
 A. orogeny
 B. convergence
 C. divergence
 D. transverse motion

Online Test Practice

SHORT ANSWER

Use the illustration below to answer Questions 11 and 12.

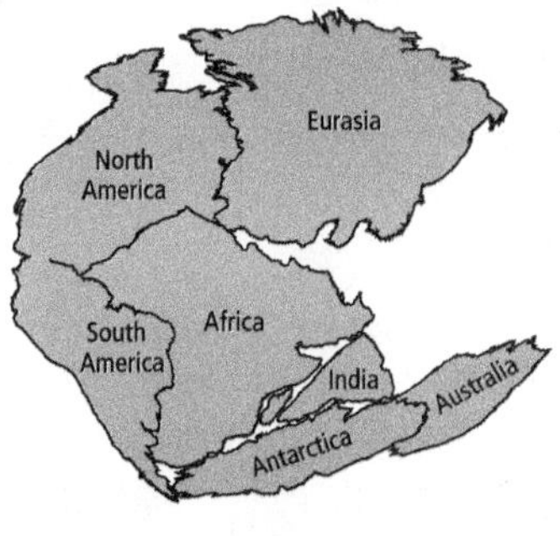

11. Who was the first scientist to propose this view of ancient Earth?

12. What features support this hypothesis?

13. How does oceanic crust differ from continental crust?

14. Why do mountains have roots?

15. Why doesn't all water become absorbed as it lies on the ground?

16. What principles for determining relative age are modeled at the Grand Canyon with its multiple layers of sedimentary rock? Explain the force that was responsible for the horizontal layering.

READING FOR COMPREHENSION

Akilia Rock Analysis

Past analysis of rocks found on Akilia, an island off the southwestern coast of Greenland, led scientists to conclude that they were at least 3.85 billion years old and contained evidence of the earliest life on the planet. "Not so," say geologists Chris Fedo and Martin Whitehouse. For billions of years, Fedo says, "[these rocks] have been squashed tens of miles underground. The rocks are so strongly deformed that understanding the original relationships among different layers is extraordinarily difficult." Some of the bands in the rock formation show irregular variations in thickness, a pattern that is not produced by sedimentation. "These bands do not have a sedimentary origin," says Fedo. "Rather," says Fedo, "we think the green bands are igneous." If the banded rocks are igneous, the Akilia rocks might have formed in the absence of life-sustaining oceans, making them much less likely to have harbored early life.

Article obtained from: Harder, B. New analysis throws age of life on earth into doubt. *National Geographic News* May 23, 2002. (Online resource accessed October 1, 2006.)

17. What can be inferred from this passage?
   - **A.** Greenland is the oldest landmass on Earth.
   - **B.** Sedimentary rocks are much more likely to contain fossils than igneous rocks.
   - **C.** Green bands in the rock make it difficult to determine the age of the rock.
   - **D.** These are the only scientists who think that Earth is younger than 4.56 billion years.

18. How does knowing the age of Earth help scientists in their studies?

NEED EXTRA HELP?

If You Missed Question . . .	1	2	3	4	5	6	7	8	9	10	11	12	13	14	15	16
Review Section . . .	21.1	19.1	19.1	11.2	14.1	21.4	22.4	22.4	1.3	20.2	17.1	17.1	20.1	20.2	10.1	21.2

SHORT ANSWER

11. Alfred Wegener proposed that Earth's continents had once been one large landmass known as Pangaea. He hypothesized that this landmass began to break apart due to continental drift, which caused the continents to slowly move to the positions they are in today.

12. The apparent fit of continents in a puzzle-like fashion supports the hypothesis that the continents were once joined. The similarities of fossils on continents that were once joined adds additional evidence.

13. Oceanic crust is composed mostly of basalt, while continental crust contains more granite. Because basalt is denser than granite, oceanic crust displaces more of the mantle than continental crust does.

14. Mountains need roots in order to counter the mass of the range above Earth's surface. A mountain's root maintains a mountain's equilibrium on the mantle it displaces.

15. In order for the water to be absorbed, pores in the ground's surface material must be large enough to hold it. If the pores are too small or if they are already full, the water will not be absorbed and will either evaporate or flow away.

16. The Grand Canyon models the principles of original horizontality and superposition. The layers were deposited successively as different deposits of rock accumulated on top of each other by the action of water and wind. Gravity smoothed the layers flat.

READING FOR COMPREHENSION

17. B

18. By knowing the age of Earth, scientists are able to create a time line of developments occurring on Earth.

CHAPTER 23 The Paleozoic, Mesozoic, and Cenozoic Eras

BIGIDEA **Complex life developed and diversified during the three eras of the Phanerozoic as the continents moved into their present positions.**

ESSENTIAL QUESTIONS	NATIONAL STANDARDS	RESOURCES TO ASSESS MASTERY
SECTION 1 The Palezoic Era 1. What is a passive margin? 2. How do transgressions and regressions indicate sea-level changes? 3. What tectonic events shaped Laurentia during the Paleozoic? 4. How are changes in Paleozoic life-forms summarized? 3 sessions 1.5 blocks	UCP.2, UCP.4; A.1; C.3; D.3; G.3	**Entry-Level Assessment** Focus Transparency 71, p. 648 **Progress Monitoring** Caption Question, pp. 650, 654 Reading Check, pp. 650, 651 Section Review, p. 654
SECTION 2 The Mesozoic Era 1. How did the breakup of Pangaea affect Earth's life-forms and paleogeography? 2. How did the mountains of western North America form? 3. What are possible causes for the extinction of the non-avian dinosaurs and other Mesozoic life-forms? 2 sessions 1 block	UCP.2, UCP.4–5; C.3; D.3	**Entry-Level Assessment** Focus Transparency 72, p. 655 **Progress Monitoring** Caption Question, pp. 656, 658 Reading Check, p. 656 Section Review, p. 659
SECTION 3 The Cenozoic Era 1. What was the extent of glaciation during the Cenozoic? 2. How can tectonic activity in North America during the Cenozoic be described? 3. How did climate change affect life-forms during the Cenozoic. 3 sessions 1.5 blocks	UCP.2–3, UCP.5; A.1; C.3; D.3	**Entry-Level Assessment** Focus Transparency 73, p. 660 **Progress Monitoring** Caption Question, pp. 661, 662 Reading Check, p. 664 Section Review, p. 665 **Summative Assessment** Chapter Assessment, p. 669 Exam*View*® *Assessment Suite* CD-ROM

Go online!
connectED.mcgraw-hill.com

LEVELED RESOURCES	LAB MATERIALS
Science Notebook 23.1 OL **Chapter FAST FILE Resources:** MiniLab Worksheet, p. 56 OL Study Guide, p. 71 BL **Transparencies:** Section Focus Transparency 71 EL Teaching Transparency 68, 69 OL EL	LaunchLAB p. 646 / **20 min** unglazed brick or sandstone, water or oil, dropper MiniLAB p. 653 / **20 min** modeling clay
Science Notebook 23.2 OL **Chapter FAST FILE Resources:** GeoLab Worksheet, p. 57 OL Study Guide, p. 73 BL **Lab Resources:** Laboratory Manual, p. 177 OL **Transparencies:** Section Focus Transparency 72 EL Teaching Transparency 70 OL EL	GeoLAB p. 667 / **40–60 min** textbook, internet access
Science Notebook 23.3 OL **Chapter FAST FILE Resources:** Study Guide, p. 75 BL **Lab Resources:** Laboratory Manual, p. 181 OL **Transparencies:** Section Focus Transparency 73 EL Teaching Transparency 71, 72 OL EL	

LabManager™
Customize any lab with the LabManager™ CD-ROM.

ADDITIONAL RESOURCES

Technology:
- eTeacherEdition
- eStudentEdition
- Classroom Presentation Toolkit
- Exam*View*® *Assessment Suite* CD-ROM
- LabManager™ CD-ROM
- What's EARTH SCIENCE Got To Do With It? DVD
- Weather Classroom Video DVD

Assessment:
- Blueprints for Success
- Chapter Test (Scaffolded)
- Earth Science Standardized Test Practice

Lab Resources:
- Exploring Environmental Problems Laboratory Manual

BL Below Level   OL On Level   AL Advanced Learners   EL English Learners   COOP LEARN Cooperative Learning

CHAPTER 23

LaunchLAB

iLab Station | Rubric

How is oil stored in rocks?

Process Skills observe and infer, recognize cause and effect, communicate, model, predict

Teaching Strategies

- This activity works best with a porous sandstone. Sandstone might be available at a local landscaping company, hobby store, or environmental engineering firm. The fractured side of a broken brick also works well.
- Students should drop the water or oil slowly, a little at a time. It might take 15–20 minutes for 10–50 mL of the liquid to be absorbed.

Procedure

1. Have students read and complete the lab safety form and follow the procedure below.
2. Place an unglazed **brick** or **sandstone** sample on your table.
3. Sketch and label a magnified cross section of the brick or sandstone before you add the **oil** or **water.**
4. Using a **dropper**, slowly squeeze three to five drops per minute of water or oil onto the brick or sandstone for 10 min.
5. Revise your sketch to show the view after you added the oil or water.

Analysis

1. **Infer** Observe the brick or sandstone sample. Where did the water or oil go? Under the influence of gravity, the water or oil flowed into the spaces between the mineral grains in the sample.
2. **Compare and contrast** the appearance of the brick or sandstone before and after the oil or water was added. The appearance of the sandstone or brick should not have changed.
3. **Conclude** how rocks in nature store oil and water. Rocks store water and oil in the spaces between their mineral grains.

Assessment

Knowledge Ask students to infer where most underground water and oil is found. Does it exist as underground lakes or rivers—or is it stored within rocks and sediment? Most of Earth's water and oil is stored within rocks and sediment. Ask them why, if oil is less dense than water, it remains in rocks underground and does not migrate to Earth's surface. Oil will migrate to the surface unless it is stopped by an overlying "cap rock" that contains few or no pore spaces.

LabManager™
Customize any lab with the LabManager™ CD-ROM.

CHAPTER 23

The Paleozoic, Mesozoic, and Cenozoic Eras

BIGIDEA Complex life developed and diversified during the three eras of the Phanerozoic as the continents moved into their present positions.

SECTIONS

1 The Paleozoic Era

2 The Mesozoic Era

3 The Cenozoic Era

LaunchLAB

iLab Station

How is oil stored in rocks?

Many sedimentary rocks contain oil and water. How are these materials stored in sedimentary rocks? Model how oil and water migrate through rocks in this lab.

FOLDABLES® Study Organizer

Paleozoic Life-forms

Make a three-tab book using the labels *Early, Middle,* and *Late Paleozoic.* Use it to organize your notes on the life-forms of the Paleozoic.

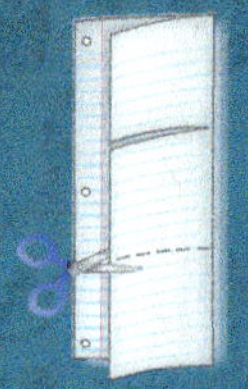

(t)John Kohula/Photo Researchers, (b)Breck P. Kent/Animals Animals, (bkgd)Radius Images/Alamy

Introduce the BIG IDEA

Boom and Bust Many "boom-and-bust" cycles of population explosion and extinction occurred during the Phanerozoic as life-forms responded to changes in their environments. Many of these cycles corresponded to changes caused by plate tectonics. Ask students to brainstorm other boom-and-bust cycles. Examples might include cycles of greenhouse warming and ice-age cooling, cycles of predator-prey relationships, stock market cycles, and housing market cycles.

Teacher Content Support

Kauri Tree Amber The photo depicts a Kauri tree, which is an ancient conifer tree found only in New Zealand. Amber, the fossilized resin of Kauri and certain other conifer trees, can trap insects, small animals, and plants, preserving them through geologic time. Amber can also trap tiny bubbles of air. Analyses of the gases in these bubbles help scientists understand the composition of Earth's ancient atmosphere through time.

SECTION 1

1 Focus

Focus Transparency

Before presenting the lesson, project **Section Focus Transparency 71** and have students answer the accompanying questions. EL

MAINIDEA

Paleozoic Organisms The Paleozoic Era was a time of great change for life on Earth. To help students appreciate this, have small groups research diverse Paleozoic organisms, such as trilobites, brachiopods, crinoids, bryozoans, placoderms, cephalopods, lobe-finned fishes, corals, sphenopsids, lycopods, and conodonts. Have each group give an oral presentation about its organism. OL COOP LEARN

2 Teach

Use Science Terms

Paleogeography Students should realize that the word *paleogeography* is not limited to the Paleozoic Era, even though both words share the same prefix. Remind students that *paleo* means "old" in Greek. All ancient eras—not just the Paleozoic—have characteristic paleogeographies.

SECTION 1

The Paleozoic Era

MAINIDEA Life increased in complexity during the Paleozoic while the continents collided to form Pangaea.

Essential Questions

- What is a passive margin?
- How do transgressions and regressions indicate sea-level changes?
- What tectonic events shaped Laurentia during the Paleozoic?
- How are changes in Paleozoic life-forms summarized?

Review Vocabulary

evaporite: a sediment deposit that has crystallized out of water supersaturated with dissolved minerals

New Vocabulary

paleogeography
passive margin
transgression
regression
Cambrian explosion

EARTH SCIENCE 4 YOU Have you noticed that some things seem to happen all at once? For instance, you might notice that everyone at school is suddenly talking about a certain music group that just yesterday was unknown. In a similar way, there suddenly appeared in the Paleozoic rock record an entire collection of new, complex life-forms.

Paleozoic Paleogeography

The geologic activity of the three eras of the Phanerozoic Eon are well represented in the rock record. By studying this record, geologists can reconstruct estimates of landscapes that have long since disappeared. The ancient geographic setting of an area is called its **paleogeography** (pay lee oh jee AH gruh fee). The paleogeography of the Paleozoic Era—the first era of the Phanerozoic—is defined by the breakup of the supercontinent Rodinia. As this breakup proceeded, multicellular life evolved with increasing complexity, as illustrated in **Figure 1.**

Passive margins Recall that the ancient North American continent of Laurentia split off from Rodinia by the early Paleozoic. Laurentia was located near the equator and was surrounded by ocean. In addition, it was almost completely covered by a shallow, tropical sea. Throughout the Cambrian, there was no tectonic activity on Laurentia so no mountain ranges formed. The edge of a continent is called a margin. When there is no tectonic activity along a margin, it is called a **passive margin.** During the Cambrian, Laurentia was completely surrounded by passive margins—there was no tectonic activity along its edges.

Figure 1 Life-forms became more complex during the six periods of the Paleozoic.

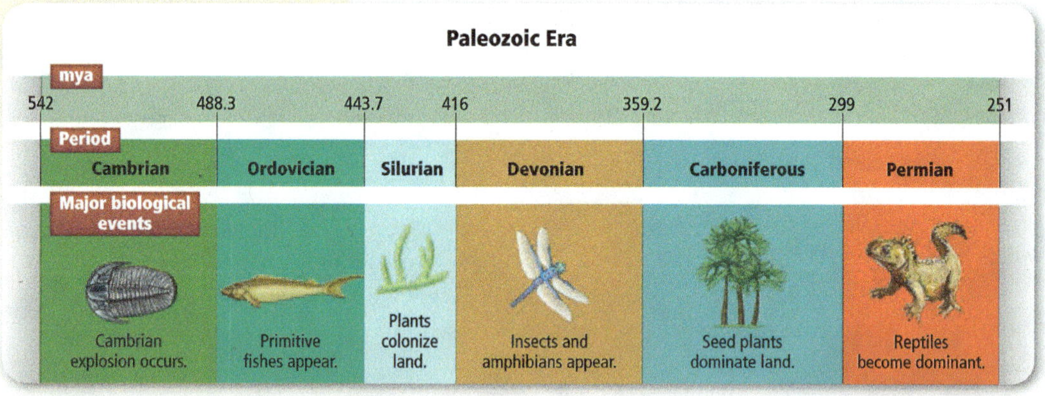

DIFFERENTIATED INSTRUCTION

Struggling Learners Whereas passive margins are tectonically "quiet", active margins are associated with volcanic activity. Give students photocopies of world maps that show the Pacific Ring of Fire. Ask them to label the passive and active margins on their maps.

EARTH SCIENCE JOURNAL

Paleogeography After discussing the definition of *paleogeography* with students, have them write an Earth science journal entry hypothesizing how future scientists might describe the paleogeography of the present-day Earth. Paragraphs might mention that the continents are separated by large oceans and that the climate is warming.

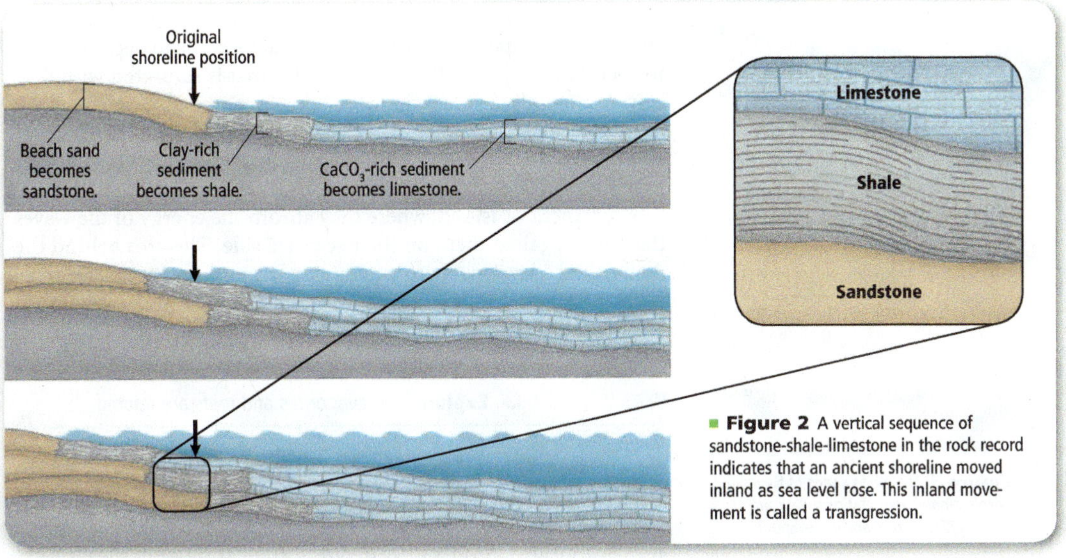

Figure 2 A vertical sequence of sandstone-shale-limestone in the rock record indicates that an ancient shoreline moved inland as sea level rose. This inland movement is called a transgression.

Sea-Level Changes in the Rock Record

Rock sequences preserved in passive margins tell paleogeographers a great deal about ancient shorelines. These sequences are useful in charting the rise and fall of sea level. To understand this, it is first necessary to understand how sediment is deposited on a shoreline.

Shoreline deposition Ocean tides wash small grains of sand and sediment ashore to make beaches. Tides also deposit offshore sediment the size of clay particles (<0.002 mm). Calcium carbonate ($CaCO_3$) sediment accumulates farther from shore as calcium muds form from sea water and organisms containing calcium carbonate die and fall to the seafloor. The sand deposited on the beaches eventually becomes sandstone, the offshore clay sediment compacts to form shale, and the calcium carbonate sediment farther offshore turns into limestone, as shown in **Figure 2.**

Transgression When sea levels rise or fall, the deposition of sediment shifts. As illustrated in **Figure 2,** a rise in sea level causes the water to move inland to an area that previously had been dry. The area where clay sediment was deposited also moves shoreward on top of the old beach. This movement is called a **transgression.** The result of the transgression is the formation of deep-water deposits overlying shallow-water deposits. This appears in the rock record as a vertical or stepwise sequence of sandstone-shale-limestone.

Regression When sea level falls, the shoreline moves seaward in a process called **regression.** This process results in shallow-water deposits overlying deep-water deposits. A stacked sequence of limestone-shale-sandstone is evidence of a regression.

VOCABULARY
SCIENCE USAGE V. COMMON USAGE
Transgression
Science usage: movement of a shoreline inland as sea level rises

Common usage: violation of a law or moral duty

Reinforcement

Make a Cycle Map After mixing up the phrases below, write them on the board and then have students organize them into a cycle map that describes the deposition of sedimentary rocks as sea level changes through time: *sea level rises, transgression of the ocean, limestone is deposited over shale, sea level falls, regression of the ocean, sandstone is deposited over shale.* OL

Collaborative Learning

Calcium Carbonate Much of the calcium carbonate that makes up limestone comes from organisms that once lived in the ocean. Have students work in groups to learn what modern-day organisms contribute to the calcium carbonate sediment of today's oceans. Have them sketch some of these organisms, showing a scale in SI units, in their Earth science journals. Many marine organisms contribute to calcium carbonate sediments. They range from tiny coccolithophores and foraminifera to clams, snails, corals, sand dollars, and sea urchins. OL COOP LEARN

DIFFERENTIATED INSTRUCTION

Behavior Disorders The Grand Canyon contains one of the world's best transgressive sequences. Ask students with behavior disorders to research how the stratigraphy of the Grand Canyon's Cambrian sedimentary rocks provides evidence of a major transgression. Encourage them to collect both descriptive and visual materials. Students can also contact the National Park Service for more information. OL

Environmental Connection

Salt Domes The salt in evaporite deposits is less dense than sedimentary rock, and thick salt beds often rise upward toward the surface, deforming the strata to form pillar-shaped structures called salt domes. Salt domes make excellent seals for upward-rising oil and gas, and they are often associated with rich deposits of these important natural resources. Because salt domes are relatively impermeable, some are used to store natural gas, oil, and radioactive waste. For example, man-made caverns in natural salt domes in Texas and Louisiana contain the U.S. Strategic Petroleum Reserve.

Project

Coral Reefs Have students conduct research into the types of environments where coral reefs form today. Most coral reefs form in the warm, clear, shallow waters of tropical oceans. Have students plot the distribution of warm-water coral reefs on a map of the world. Ask them to interpret the significance of a fossil tropical coral reef in New York State. A fossil tropical coral reef in New York indicates that the climate of New York was once tropical. Students might find that some corals live in deep, cold parts of the ocean. Scientists are just beginning to understand the significance of these cold-water corals. OL

READING CHECK Evaporites are deposited in the lagoons behind reefs.

Caption Question Fig. 4 Petroleum geologists are interested in evaporites because of their association with oil deposits.

Figure 3 The white sands of New Mexico's White Sands National Park are made of gypsum from ancient evaporite deposits.

Evaporites Scientists also learn about fluctuating sea level by studying evaporite deposits. Recall that evaporite deposits are rocks that have crystallized out of water that is saturated with dissolved minerals. Some evaporite deposits can be associated with fossilized reefs.

Fossilized reefs are made of the carbonate skeletons of tiny organisms called corals. Reefs form in long, linear mounds parallel to a continent or island, where they absorb the energy of the waves that crash against them on their seaward side. The area behind the reef, called a lagoon, is protected from the wave's energy. In a tropical setting, water in the lagoons evaporates in the warm sunshine, and minerals such as halite and gypsum precipitate out. Over time, cycles of evaporite deposition mark changes in water level.

READING CHECK **Explain** how evaporites and reefs are related.

Mineral deposits Huge amounts of gypsum and halite evaporites were deposited in Paleozoic lagoons. The white sands of White Sands National Park, shown in **Figure 3,** are the remains of one such evaporite deposit. Other deposits, such as those in the Great Lakes area of North America, are mined commercially. Halite is used as road salt. Gypsum is an ingredient in plaster and drywall.

Impermeability As shown in **Figure 4,** reef limestones tend to have large pore spaces, allowing oil and other liquids to move through them. Evaporite rocks, in contrast, are impermeable. This means that they contain very little connected pore space and liquid cannot move through them. When an evaporite deposit overlies a reef rock that contains oil, it seals in the oil and prevents the oil from migrating. A good example is the Permian Basin, home to the Great Permian Reef Complex in western Texas and southeastern New Mexico. The oil in this complex rarely leaks to Earth's surface because of its tight evaporite seal.

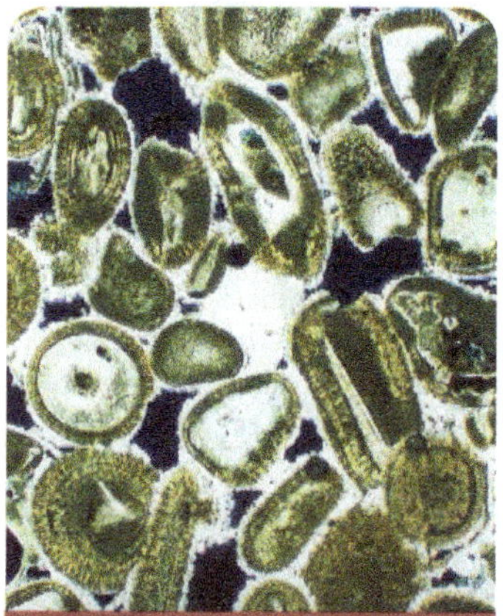
Reef rock

Evaporite

Figure 4 Reef rocks have large pores that can contain oil or other liquids, in contrast to evaporite rock, which is impermeable to liquids.
Infer *why ancient evaporite deposits are important to petroleum geologists.*

Demonstration

Evaporation All salt deposits precipitate from water. Normal ocean water contains about 3.5 percent salt. Prepare water with normal marine salinity by adding 35 g of salt to 1 L of water. Pour the solution into a shallow pan, then allow the water to evaporate and the salt to precipitate. (This might take several days; placing the pan near a heater will hasten the process.) Ask students to imagine how much water must have evaporated to produce salt deposits that are hundreds of meters thick. Students will see that 1 L of ocean water produces little salt residue. They should realize that an enormous amount of water must have evaporated to form Earth's huge salt deposits.

Factors of sea-level change Scientists have determined that sea levels transgressed and regressed as many as 50 times during the late Paleozoic. Geologists have found a number of reasons for relative sea level change—climate and glaciation cycles, crustal subsidence and uplift, varying sedimentation rates, and plate motions. These were all factors in the transgressive and regressive cycles of the Paleozoic.

☑ READING CHECK **Infer** how glaciation affects sea level.

VOCABULARY
ACADEMIC VOCABULARY
Transform
to change in a major way
The continent was transformed by a massive orogeny.

Mountain Building

Laurentia's margins were passive during the first period of the Paleozoic, and mountains were not forming. However, changes occurred during the Ordovician (or duh VIH shun) Period. At that time, Laurentia collided with the Taconic Island Arc, and mountains began to rise in what is now northeastern North America. This event is called the Taconic Orogeny. The Taconic Orogeny added new land and established an active volcanic zone along Laurentia's eastern margin. Remnants of this event are present in New York's Taconic Mountains.

Laurentia deformed Laurentia was further transformed in the Silurian (si LUR ee uhn) Period when Laurentia's eastern margin collided with Baltica and Avalonia. Baltica was a landmass that today is part of northern Europe and parts of Russia. Avalonia was an island ocean arc. You can see Baltica and Avalonia approaching Laurentia in **Figure 5.** The deformation caused by these collisions—called the Acadian Orogeny—added folds, faults, and igneous intrusions to the already deformed Taconic rocks.

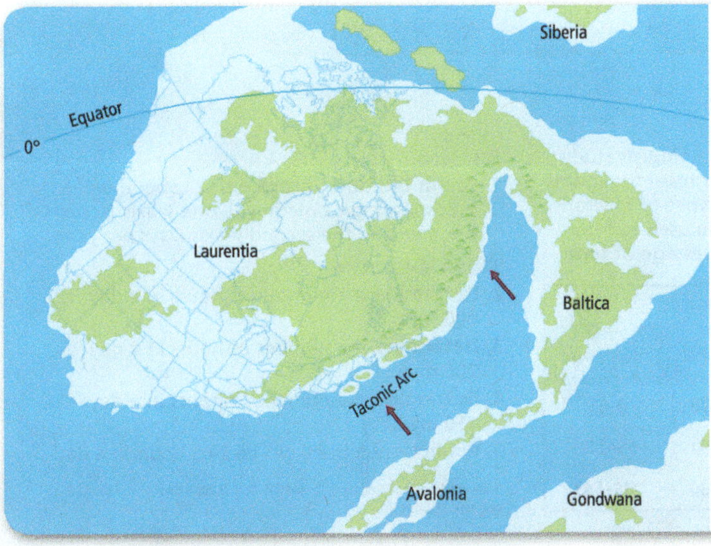

Figure 5 Baltica and Avalonia collided with the Taconic Island Arc during the Ordovician. This was one of the many Paleozoic tectonic events that transformed eastern Laurentia.

Tie to Previous Knowledge

Volcanic Ash One line of evidence used to demonstrate that the Taconic Orogeny occurred is the presence of ancient beds of volcanic ash. Ask students to discuss how they would determine the age of the volcanic ash. The best way would be to date the mineral crystals in the ash using radiometric methods. Fossils in the strata above and below the ash layers could also be used to determine the relative age of the ash. Ask students why Ordovician volcanic ash deposits support the idea that there was Ordovician orogenic activity. The principle of uniformitarianism can be used to infer that because ash is deposited by volcanoes today, and because volcanoes usually accompany orogeny events, ash deposits were also associated with ancient orogenic events.

Interpret the Illustration

Shallow Oceans The dark blue in **Figure 5** represents deep ocean water, and the light blue represents shallow ocean water. Ask students if they think the shallow water covering what is today Canada and the United States was warm or cold. It would have been warm because the equator stretched across Canada at this time, and all of the shallow water would have been in the tropics.

☑ READING CHECK Glaciers freeze water that would otherwise be part of the liquid ocean. Glaciers thus lower sea level.

Demonstration

Glacial Melting Fill a large glass pan with water so that it is about one-half full. Place a large, flat rock in the center of the pan. Be sure that the rock sticks out above the water. The rock represents a continent. Place 5–10 ice cubes in the water. These represent icebergs floating in the sea. Mark the level of the water on the side of the pan with a grease pencil. Allow the ice cubes to melt. Has the water level changed? The water level should remain the same because the ice is in the water, and the ice displaces its volume in the water. Place another 5–10 ice cubes on top of the rock. These represent glacial ice. Allow the ice cubes to melt. Has the water level changed? The water level will rise, just as sea level rises when glaciers melt.

CAREERS IN EARTH SCIENCE

WebQuest

Data Analysis LAB

About the Lab

- Students will use a table to infer the number of hours in a day in the geologic past and to predict what will happen to day length in the future.
- Ask students whether they think it is more likely that Earth's rotation has slowed down or sped up over time. Earth's rotation has slowed down. Use the analogy of Earth as a top that is spinning on its axis as it revolves. Ask students why a top doesn't speed up over time. Friction causes the top to slow down. In a similar way, Earth gradually loses its kinetic energy as tides, solar wind, and other frictional forces act on it. Earth's rotation has slowed by about 2 milliseconds since 1820. Though seemingly insignificant, this adds up over time. Scientists estimate that Earth rotated once every 6.5 hours when it was newly formed. Ask students: If a day was 22.6 hours long 200 mya, how many minutes per million years has Earth's rotation slowed since then? Earth has slowed by 25.2 seconds every million years over the past 200 million years. (84 minutes/200 million years = 0.42 minutes/million years = 25.2 seconds/million years.)
- See also Eriksson, K.A., and E.L. Simpson. 2000. Quantifying the oldest tidal record: The 3.2 Ga Moodies Group, Barberton Greenstone Belt, South Africa. *Geology* 28:831–834.

Analysis

1. Graphs should have one *y*-axis on the left-hand side and the other *y*-axis on the right. Placement of the two data lines will vary slightly as they will be made from data points that do not form straight lines.

Think Critically

2. approximately 21 hours in a day
3. approximately 22.5 hours in a day
4. approximately 23 hours in a day
5. There will be 24.5 hours in a day approximately 75 million years from now. This assumes that the rate of slowing will follow the same trend as in the past.

FOLDABLES® Rubric

Ouachita Orogeny Another Laurentian mountain-building event—the Ouachita (WAH shuh taw) Orogeny—occurred during the Carboniferous Period when southeastern Laurentia began to collide with Gondwana. Recall that Gondwana was the large landmass that eventually formed the southern continents, including Africa and South America. This collision formed the Ouachita Mountains of Arkansas and Oklahoma and was so intense that it caused the crust to uplift inland as far as present-day Colorado. Vertical faults raised rocks more than 2 km, forming a mountain range that geologists call the Ancestral Rockies.

Allegheniain Orogeny As Gondwana continued to push against Laurentia, the Appalachian Mountains began to form. This event, called the Alleghenian Orogeny, was the last of the Paleozoic mountain-building events to affect eastern North America. When it was completed at the end of the Paleozoic, the Appalachians were possibly higher than the Himalayas, and one giant supercontinent—Pangaea—had formed on Earth's surface.

CAREERS IN EARTH SCIENCE

Paleoecologist Paleoecologists study the ecology and climate of ancient environments using evidence from fossils and rocks. Some paleoecologists apply this knowledge to understand future global climate change.

WebQuest

FOLDABLES® Incorporate information from this section into your Foldable.

Paleozoic Life

The formation of Pangaea was the major geologic story of the Paleozoic, but Paleozoic rocks also tell another dramatic story. Fossils within these rocks show that multicellular animals went through extensive diversification at the beginning of this era. As you have learned, fossils help geologists correlate geologic landscapes and piece together geologic time. Fossils also help paleoecologists (pay lee oh ih KAH luh jists) learn about the ecology of ancient environments. Ecology refers to the relationships between organisms and their environments. Changes within the environment drove changes within the ecologies of Paleozoic life.

Data Analysis LAB

Based on Real Data*

Interpret the Table

Can you find the time? Paleoecologists study the shapes and compositions of fossil organisms to interpret how and in what types of environments they lived. Fossils are also used to interpret climatic changes and the passage of time.

Time Record Data

Geologic Era	Hours Per Day	Day Per Year	Geologic Time (mya)
Cenozoic	23.5–24	365–377	0–65
Mesozoic	23.5–22.4	377–392	65–248
Paleozoic	22.4–20	392–430	248–543

Analysis

1. **Graph the time record data.** Label the *x*-axis *Geologic time (mya)*, one *y*-axis *Hours per day*, and the second *y*-axis *Days per year*.

Think Critically

2. **Determine** the number of hours in a day 400 mya.
3. **Determine** the number of hours in a day 200 mya.
4. **Determine** the number of hours in a day 150 mya.
5. **Predict** when there will be 24.5 hours in a day.

*Data obtained from: Prothero, D.R., and R.H. Dott, Jr. 2004. *Evolution of the Earth.* New York: McGraw-Hill

Figure 6 The organisms shown in this artist's reconstruction are among the Cambrian organisms that had hard parts.

Cambrian explosion Nearly every major marine group living today appeared during the first period of the Paleozoic. The geologically rapid diversification of such a large collection of organisms in the Cambrian fossil record is known as the **Cambrian explosion.** Some of the best-preserved Cambrian organisms occur in the Burgess Shale in the Canadian Rocky Mountains, and in southern China. A spectacular array of fossil organisms with hard parts has been found in these locations, including fossils of creatures like those shown in **Figure 6.**

Ordovician extinction At the end of the Ordovician, more than half of the marine groups that appeared in the Cambrian became extinct. Those that survived suffered large losses in their numbers. What caused this extinction? Geologists have found evidence of glacial deposits in rocks of northern Africa, which at the time was situated at the South Pole. As you have learned, when water freezes in glaciers, sea level drops. Then, as now, most marine organisms lived in the relatively shallow waters of the continental shelves. When sea level is high, the shelves are flooded and marine animals have many places to live. During regression, however, continental shelves can become too narrow to support diverse animal habitats.

Devonian extinction Following the late Ordovician extinction, marine life recovered and new species evolved. There was a tremendous diversification of vertebrates, including fish and the first appearance of tetrapods on land. In the late Devonian (dih VOH nee un), another extinction event eliminated approximately 50 percent of the marine groups. Some scientists think that global cooling was again the cause and there is evidence that some continents had glaciers at this time.

MiniLAB

iLab Station

Model Continental Shelf Area

How does shelf area change when continents collide? Colliding continents decrease the habitat areas available to marine organisms that live along the shallow shelves surrounding the continents.

Procedure

1. Read and complete the lab safety form.
2. Using 250 g of **modeling clay,** make a sphere and flatten it into a disk that is 0.5 cm thick. This represents a continent.
3. Divide another 250 g of clay into two equal spheres and flatten them as above.
4. Roll another 250 g of clay into three cylinders, each with a diameter of about 0.5 cm. Wrap the cylinders around the edges of the clay disks. These represent continental shelves.
5. Use the following formula to calculate the area of the large continent and the large continent plus the continental shelf.
   $$\text{area} = \pi r^2$$
   Subtract the continent area from the total area. This equals the area of the continental shelf.
6. Repeat Step 5 for both small models.

Analysis

1. **Assess** which has more shelf area: two small continents or one large continent. Why?
2. **Conclude** how the existence of a single large supercontinent limits the amount of habitat space for marine organisms.
3. **Explain** the relationship between reduced habitat space and extinction.

MiniLAB

iLab Station Rubric

Purpose Students will calculate how the continental shelf areas of one supercontinent and two smaller continents differ.

Process Skills use models, measure in SI, calculate, infer

Safety Precautions Approve lab safety forms before work begins. Have students wash their hands after disposing of their clay.

Teaching Strategy Clay could be a single color or two colors. If two colors, provide 250 g of one color (to model one continent) and 250 g of another color (to model the two smaller continents).

Expected Results The combined areas of the continental shelf for the two small models should be greater than the area of the shelf for the large model.

Analysis

1. two small continents, because their shelf area makes up a greater proportion of the total area
2. With one supercontinent, habitat is restricted to only one continental shelf. With many continents, there is more total shelf area, so habitat areas increase.
3. The lack of space and resources stresses some species to the point of extinction.

Assessment

Knowledge Have students explain how the formation of Pangaea would have affected tropical marine life. The loss of shelf area would have reduced habitat availability.

LabManager™
Customize any lab with the LabManager™ CD-ROM.

IN THE FIELD

Chengjiang Biota The Maotianshan Shale formation in southern China has been a source of fossils since the early 1900s. It wasn't until 1984, however, when Chinese paleontologist Hou Xianguang discovered a unique, trilobite-like fossil there, that it was recognized as one of the most spectacular Cambrian fossil sites in the world. Now known as the Chengjiang Biota, the site contains fossils of organisms that lived 525–520 mya, about 20 million years before those that lived in the famous Burgess Shale outcrop, which was discovered in the Canadian Rockies in 1909. Both sites are invaluable resources for scientists studying the evolution of Cambrian organisms.

■ **Caption Question Fig. 7** Many plants lived in swamps. After they died, the plants were buried in sediment. Eventually, increasing pressures caused the plant material to change form and become coal.

3 Assess

Check for Understanding

Knowledge Ask students what sequence of sedimentary rocks they would expect to find in an outcrop of rocks that showed evidence of a marine regression. limestone overlain by shale, and shale overlain by sandstone

Reteach

Discussion In a class discussion, have students list the sequence of events that transformed Laurentia's eastern margin during the Paleozoic. Laurentia collided with the Taconic island arc, followed by collisions with Baltica and Avalonia, and finally a collision with Gondwana, which gave rise to the Ouachita and Appalachian mountains.

Assessment

Knowledge In their Earth science journals, have students explain how the loss of continental shelf area can lead to mass extinction. Loss of shelf area would mean a loss of habitat, food, and other resources, stressing some organisms to the point of extinction. Because organisms are interconnected in food webs, those species that depended on extinct species for food might also become extinct.

■ **Figure 7** This artist's reconstruction shows what a Carboniferous swamp might have looked like.
Explain *why Carboniferous swamps produced coal deposits.*

Terrestrial plants The Ordovician and Devonian extinction events appear to have affected mainly marine life. They had little effect on life-forms living on land. Simple land plants began to appear on Earth in the Ordovician. During the Carboniferous, the first plants with seeds, called seed ferns, diversified. Because seeds contain their own moisture and food sources, they enabled terrestrial plants to survive in a variety of environments.

Carboniferous swamps Many Carboniferous plants lived in low-lying swamps, such as the one shown in **Figure 7.** As these plants died and sediment accumulated, they compacted to coal deposits. Swamps were also breeding grounds for insects. Fossils of the largest known insects have been found in Carboniferous sediment deposits, including dragonflies with 74-cm wingspans. Compare this to the largest known wingspan of a modern dragonfly—19 cm.

Permian changes At the end of the Permian, the largest mass extinction in the history of Earth occurred. The Permo-Triassic Extinction Event caused the extinction of nearly 95 percent of marine life-forms. Unlike the mass extinctions at the end of the Ordovician and Devonian, this extinction affected both marine and terrestrial organisms. More than 65 percent of the amphibians and almost one-third of all insects did not survive. What could have caused such a widespread catastrophe? It was probably a combination of causes. First, there was a dramatic drop in sea level from the coalescence of Pangaea closing and draining the shallow seas. A regression would have been particularly critical for organisms inhabiting the continental shelves when there was only one continent. Other contributing factors likely included extreme volcanism in Siberia, low atmospheric oxygen levels, and climate change.

SECTION 1 REVIEW

Section Self-Check

Section Summary

- Scientists study sediment and evaporite deposits to learn how sea levels fluctuated in the past.
- Eastern Laurentia was transformed by many mountain-building events during the Paleozoic.
- A great diversity of multicellular life appeared during the first period of the Paleozoic.
- The largest extinction event in Earth's history occurred at the end of the Paleozoic.

Understand Main Ideas

1. MAINIDEA **Explain** how the formation of Pangaea affected the evolution of life-forms.
2. **Compare** transgression and regression.
3. **Discuss** the relationship between oil deposits and evaporites.
4. **Assess** the significance of the Cambrian explosion.

Think Critically

5. **Infer** what has happened to the Ancestral Rockies since their formation.
6. **Predict** changes in the fossil and rock record that might indicate a marine extinction event.

MATH IN Earth Science

7. If 10 million species exist today and 5.5 species become extinct every day, calculate how many years it would take for 96 percent of today's species to become extinct.

SECTION 1 REVIEW

1. When Pangaea formed, continental shelf space decreased and some life-forms became extinct due to habitat loss.
2. Transgressions occur when sea level rises and the shoreline moves landward; regressions occur when sea level falls and the shoreline moves seaward.
3. Oil deposits are often found in association with evaporites. Evaporite deposits are impermeable and seal in any oil that might be stored underneath.
4. It is the first biota on Earth that scientists think contain ancestors of most modern-day organisms.
5. They eroded away.
6. There might be a sudden disappearance of marine fossil organisms in the fossil record, accompanied by evidence of a regression in the rock record.
7. 4782 years; 96 percent of 10 million species = 9.6 million species; 5.5 species/day × 365 days/year = 2007.5 species per year; 9.6 million species/2007.5 species/year = 4782 years

SECTION 2

The Mesozoic Era

MAINIDEA Reptiles became the dominant terrestrial animals during the Mesozoic while Pangaea broke apart.

Essential Questions
- How did the breakup of Pangaea affect Earth's life-forms and paleogeography?
- How did the mountains of western North America form?
- What are possible causes for the extinction of the non-avian dinosaurs and other Mesozoic life-forms?

Review Vocabulary
subduction: the process by which one tectonic plate descends beneath another

New Vocabulary
phytoplankton
amniotic egg
iridium

EARTH SCIENCE 4 YOU Do you like mystery novels? One of the biggest mysteries in the history of science is what caused the extinction of the non-avian dinosaurs.

Mesozoic Paleogeography

The mass extinction event that ended the Paleozoic Era ushered in new opportunities for animals and plants of the Mesozoic Era. Earth's life-forms changed drastically as new kinds of organisms, shown in **Figure 8,** evolved to fill empty niches. While some groups of these organisms remain on Earth today, none of the giant reptiles that dominated the land, sea, and air, and typified the era, survived. The non-avian dinosaurs all became extinct at the end of the era.

Breakup of Pangaea When the Mesozoic Era began, a single global ocean and a single continent—Pangaea—defined Earth's paleogeography. During the late Triassic Period, Pangaea began to break apart. The heat beneath Pangaea caused the continent to expand, and Pangaea's brittle lithosphere began to crack. Some of the large cracks, called rifts, gradually widened, and the landmass began spreading apart. The ocean flooded the rift valleys to form seaways, and large blocks of crust collapsed to form deep valleys. The Mesozoic climate was warm and tropical, and it remained warm enough throughout the era that glaciers did not form.

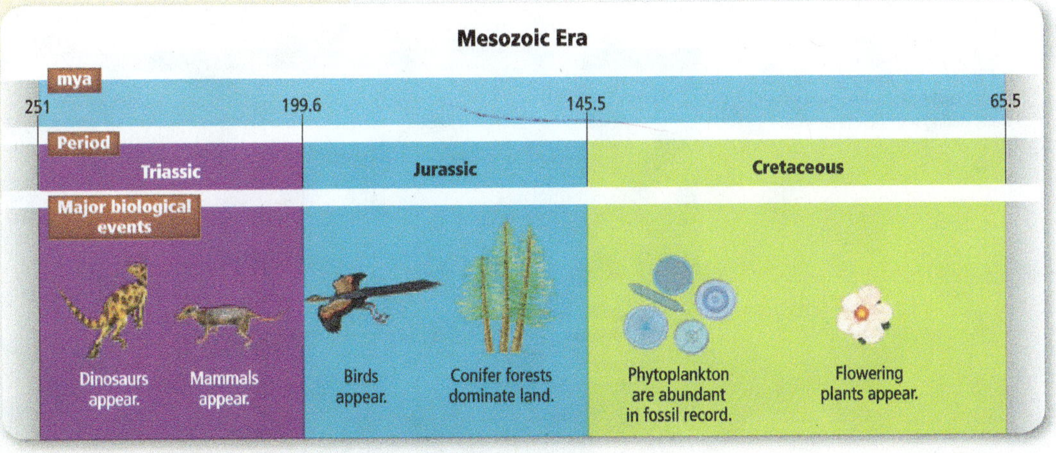

Figure 8 Although dinosaurs are the most famous of the Mesozoic life-forms, other organisms also appeared during this era.

SECTION 2

1 Focus

Focus Transparency

Before presenting the lesson, project **Section Focus Transparency 72** and have students answer the accompanying questions. EL

MAINIDEA

Pangaea By the end of the Paleozoic, Pangaea was fully assembled. Ask students to infer what happened to Pangaea during the Mesozoic. The process of plate tectonics is constant. Pangaea started rifting apart during the Mesozoic; had it started later, in the Cenozoic, the continents would be closer today than they are.

2 Teach

Teacher Content Support

Extensional Tectonism The form of tectonism that caused Pangaea to rift apart is called extensional tectonism. Extensional tectonism stretches Earth's crust, creating basins that then fill with sediment. Evidence of the extensional tectonism that occurred during the breakup of Pangaea is found in the sediment-filled valleys along the eastern margin of North America, an area called the Central Atlantic Magmatic Province (CAMP).

DIFFERENTIATED INSTRUCTION

Visually Impaired Make cardboard cutouts of the continents, and place them on a flat map of the world. Use a string to designate the equator. Then, with students, move the continents to show how Pangaea formed relative to the equator—and how it broke apart during the Mesozoic. BL OL EL

Identify Misconceptions

Some students might think that glacial formation is the only way sea level can fall, and that glacial melting is the only way sea level can rise.

Uncover the Misconception Ask students how they think a change in the shape of an ocean basin affects water volume.

Demonstrate the Concept Fill an empty plastic tub with water. Deform the bottom of the tub by pressing your thumb against it so that water spills out of the tub. (Make sure you do this over a bucket or large container.) Explain to students that you have changed the volume of the tub so that it no longer can hold the same amount of water. This is similar to what scientists think happened during the Mesozoic, when rifting continents caused increased rates of seafloor spreading. Seafloor spreading caused ocean basins to bulge so they could no longer hold as much water.

Assess New Knowledge Ask students where the water from ocean basins goes when seafloor-spreading rates increase. It floods the continents, such as occurred in North America during the Cretaceous.

Discussion

Sand Dunes Have students locate deserts with large sand dunes on a map of the world. Most are found in the subtropics, around 30 degrees latitude, because of atmospheric moisture patterns. Ask students what the discovery of ancient sand dunes in western North America that date to the Late Triassic and Early Jurassic suggest about the paleogeography of North America during that time. It suggests that western North America was dry and warm during the Late Triassic and Early Jurassic. OL COOP LEARN

■ **Figure 9** The Red Sea and the Gulf of Aden are widening into a new seaway. **Identify** *the tectonic force behind the creation of this new seaway.*

Seaways As the continents continued to split apart, mid-ocean rift systems developed at the junctures, and the widening seaways became oceans. The Atlantic Ocean began forming in the Triassic as North America rifted away from Europe and Africa. Some of the spreading areas at this juncture joined to form a long, continuous rift system called the Mid-Atlantic Ridge. As you have learned, this mid-ocean ridge system is still active today, erupting magma deep in the ocean as it widens. The Red Sea and Gulf of Aden, shown in **Figure 9,** are new seaways in East Africa that are today slowly widening by a few centimeters a year as a result of continental breakup.

☑ READING CHECK **Explain** how the Atlantic Ocean formed.

Changing sea level The formation of mid-ocean rift systems was partly responsible for a rise in sea level during the Mesozoic. The hot magma that erupted at the ridges displaced a considerable amount of seawater onto the continents. However, sea level dropped at the end of the Triassic, and desertlike conditions developed in western North America. The climate became arid and, as evidenced in ancient sand dunes, a thick blanket of sand covered some of the land. Sea level rose again during the Jurassic, and a shallow sea formed in North America's center. The ocean continued to rise during the Cretaceous (krih TAY shus), covering much of North America's interior. **Figure 10** shows that nearly one-third of Earth's landmasses were covered with water.

Mountain Building

Recall that the collision of continents during the Paleozoic transformed the eastern margin of Laurentia, while the continent's western margin remained passive. During the Mesozoic and early Cenozoic, the reverse was true. As the breakup of Pangaea proceeded, multiple mountain-building episodes occurred along Laurentia's western margin, while little was happening along its eastern edge.

■ **Figure 10** Nearly one-third of Earth's land surface was covered with water during the late Cretaceous.

☑ READING CHECK The Atlantic Ocean began to form when North America started rifting away from Africa and Europe

■ **Caption Question Fig. 9** The process of plate tectonics is causing two plates to rift apart.

EARTH SCIENCE JOURNAL

Sea Level Have students draw a cross section of an ocean basin and two adjacent continents (for example, Europe and North America separated by the Atlantic) in their Earth science journals. Have them label the diagram showing where and how sea level is controlled. Sea level is controlled by magma erupting at mid-ocean ridges and by the formation and melting of glaciers on the continents.

Three Phases of the Cordilleran Orogeny

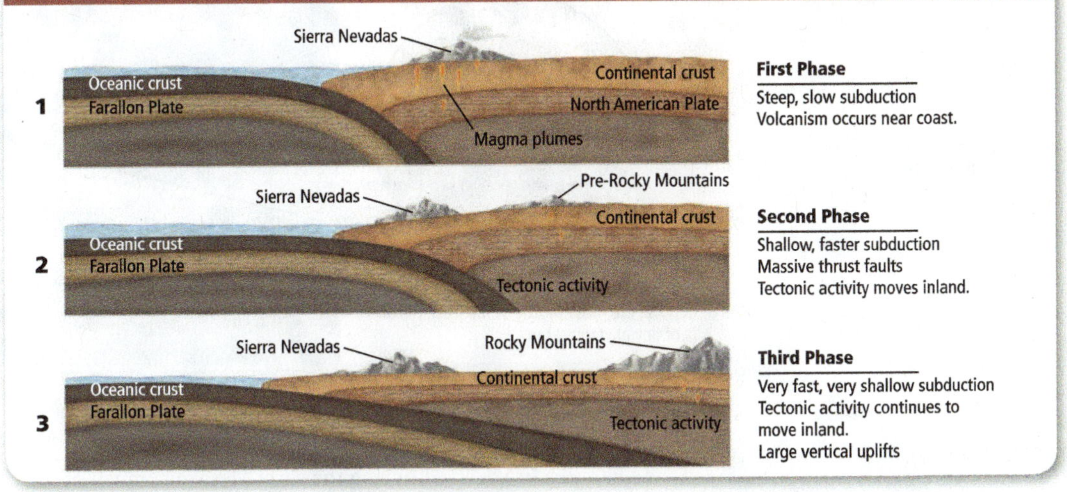

Figure 11 During the three phases of the Cordilleran Orogeny, mountains formed farther inland as the angle of subduction became more shallow and the speed increased, causing massive faulting and uplift.

Cordillera Much of the mountain building that occurred in western Laurentia was caused by the subduction of the oceanic Farallon Plate beneath Laurentia's western margin. As the plate descended, many structural features of the present-day Rocky Mountains, Sierra Nevadas, and other western mountain ranges were formed. Geologists call these ranges collectively the North American Cordillera (kor dee AYR uh). Cordillera means *mountain range* in Spanish. The Cordilleran Orogeny consisted of three distinct phases. As shown in **Figure 11,** each phase was characterized by a different rate and angle of subduction.

First phase The first phase occurred during the late Jurassic and early Cretaceous when subduction proceeded slowly and the oceanic plate descended at a steep angle, producing magma, which rose at the site of the emerging Sierra Nevadas.

Second phase The second phase of the Cordilleran Orogeny occurred during the Cretaceous when subduction increased in speed but the oceanic plate descended at a shallow angle. As a result, there was less volcanism along Laurentia's margin and more tectonic activity inland with massive thrust faults occurring in the Rocky Mountain area.

Third phase During the third phase of the Cordilleran Orogeny, which began during the late Cretaceous and continued into the Cenozoic, subduction was even more shallow and rapid than it was during the second phase. The subduction rate was so fast that some scientists suggest the oceanic plate was pushed almost horizontally beneath the North American Plate. As a result, this phase was characterized by large, vertical uplifts which formed the Rocky Mountains, and a decrease in volcanism. This range now extends from northern Mexico into Canada.

Teacher Content Support

Cordilleran Orogeny The three phases of the Cordilleran Orogeny have distinctive names because they affected different parts of North America and were described individually before scientists knew that they were part of a continuous deformation event. The first phase is called the Nevadan Orogeny. The second phase is called the Sevier Orogeny, and the third phase is called the Laramide Orogeny.

The tectonic deformation that formed the Appalachians can be thought of in the same way, with the first orogenic phase being the Taconic Orogeny, the second being the Acadian Orogeny, and the third and final phase being the Ouachita and Alleghenian Orogenies.

Demonstration

Cordillera The Cordillera is a continuous sequence of mountain ranges that stretch from Alaska to South America. On a topographic globe or world map, show students where the Cordillera is located. Ask them to find Earth's longest mountain range on the map. Students will probably look only on the continents. Show them that the longest mountain range is formed by the Mid-Atlantic Ridge, which is visible on land only in Iceland.

Tie to Previous Knowledge

Food Chain Phytoplankton are at the base of the food chain. Ask students what this means based on their knowledge of photosynthesis. Unlike animals, phytoplankton make their own food. They thus provide direct energy for the animals that eat them, and indirect energy for the animals that eat those animals, and so on.

Identify Misconceptions

Some students might think that all animals that lived during the Mesozoic were dinosaurs, and that all dinosaurs were large.

Uncover the Misconception

Ask students how the media portray the animals that lived during the Mesozoic. Mammals, sea reptiles, and marine invertebrates are largely ignored.

Demonstrate the Concept

Show pictures of small dinosaurs, large dinosaurs, therapsids (mammal-like reptiles), pterosaurs (flying reptiles), *Dimetrodon* (which lived during the Permian), and large marine reptiles, such as plesiosaurs and ichthyosaurs. Ask students to identify the dinosaurs.

Assess New Knowledge

After explaining that all non-avian dinosaurs lived on land, had legs positioned directly under their bodies, and could not fly, show the same or similar pictures again and have the class identify the dinosaurs.

Figure 12 England's White Cliffs of Dover are made mostly of the fossil remains of tiny phytoplankton.

Mesozoic Life

As Pangaea broke apart during the early Mesozoic, much of the habitat on the continental shelves that was lost during Pangaea's formation once again became available. New marine organisms, ranging from large predatory reptiles to tiny photosynthetic phytoplankton, evolved to fill these niches. **Phytoplankton** are microscopic organisms at the base of the marine food chain. These organisms were abundant during the Cretaceous. The remains of their shell-like hard parts are seen in many chalk deposits around the world, including England's famous White Cliffs of Dover, shown in **Figure 12.**

Plant life As the cool climate that characterized the late Paleozoic came to an end during the Mesozoic, plant life changed sharply. The large, temperate swamps dried as the climate warmed. Tall cycad trees are seed plants without true flowers. These evolved during the Jurassic, along with ginkgos, pine trees, and other conifers. Flowering plants appeared during the Cretaceous.

Terrestrial animals Mammals appeared during the late Triassic, around the same time as the dinosaurs. However, the dominant Mesozoic animals were the reptiles. Unlike amphibians, whose eggs need to be laid in water to prevent drying out, reptiles can lay their eggs on dry land. These eggs, called **amniotic** (am nee AH tihk) **eggs,** contain the food and water required by developing embryos inside. Amniotic eggs made it possible for reptiles, including dinosaurs, to roam widely.

Dinosaurs Archosaurs are a group of reptiles which includes dinosaurs and crocodilians. Archosaurs have a unique skeletal structure that allows for speed and flexibility of movement. While lizards and turtles walk with a sprawling posture, archosaurs have a hip structure that allows the legs to be held underneath the body. This enabled some dinosaurs to run with an upright posture, as shown in **Figure 13.**

Figure 13 Dinosaurs have a unique hip structure that enabled some, like this *Velociraptor,* to develop an erect posture and run on two legs.
Explain *how a dinosaur's posture differed from that of other reptiles.*

Enrichment

Bird Evolution Ask students if dinosaurs are really extinct. Have them conduct research into the widely accepted hypothesis that birds evolved from theropod dinosaurs. OL

Caption Question Fig. 13 Most reptiles have a sprawling posture, while dinosaurs had an upright posture.

DIFFERENTIATED INSTRUCTION

Visually Impaired Construct simple models of dinosaurs and lizards (or some other reptile with a sprawling posture) using modeling clay, or find plastic toys. Have students with visual impairments use these models to enhance their understanding of the different hip structures of the two groups. BL OL EL

GeoLAB

iLab Station

The GeoLab at the end of the chapter can be used at this point in the lesson.

Explore **extinctions during the Phanerozoic with an interactive table.** Concepts In Motion

Table 1 Major Extinctions in the Phanerozoic

Extinction event	End Ordovician	Late Devonian	Permo-Triassic	End Triassic	End Cretaceous
Approximate mya	440 mya	360 mya	250 mya	200 mya	65 mya
Percentage groups extinct	57 percent marine	50 percent marine	95 percent marine 70 percent land	48 percent marine	75 percent marine 56 percent land

Mass extinction At the end of the Mesozoic, an extinction event devastated terrestrial dinosaurs, most marine reptiles, plants, and many other organisms. Today, most scientists agree that the combination of massive volcanism, which stressed Earth's climate, and a large meteorite impact that occurred at the end of the Cretaceous is responsible for the extinction event. It is thought that the meteorite was at least 10 km in diameter. An impact of this size could have blown up to 25 trillion metric tons of rock into the atmosphere, causing long-lasting greenhouse warming. Evidence for this impact includes an impact site—Chicxulub Crater—on Mexico's Yucatan Peninsula, as well as a unique layer of clay that separates Cretaceous rocks from rocks of the first period of the Cenozoic. Found worldwide, this layer contains an unusually high amount of **iridium** (ih RID ee um), a rare metal in Earth's rocks but a relatively common metal in asteroids. As shown in **Table 1,** the extinction event at the end of the Mesozoic was relatively mild compared with the Permo-Triassic Extinction Event at the end of the Paleozoic.

SECTION 2 REVIEW

Section Self-Check

Section Summary

- The breakup of Pangaea triggered a series of tectonic events that transformed western Laurentia.
- The Atlantic Ocean began to form during the Mesozoic as North America broke away from Europe.
- Dinosaurs and other new organisms evolved to fill niches left empty by the Permo-Triassic Extinction Event.
- All dinosaurs, except birds, along with many other organisms became extinct during a mass extinction event at the end of the Mesozoic.

Understand Main Ideas

1. MAINIDEA **Discuss** the significance of the Permo-Triassic Extinction Event for the animals that populated the Mesozoic.
2. **Explain** how rifts are related to the formation of oceans.
3. **Compare** the tectonic events that transformed Laurentia's western margin with the tectonic events that changed Laurentia's eastern margin.
4. **Discuss** the evidence that suggests a meteorite impact was responsible for the extinctions at the end of the Mesozoic Era.

Think Critically

5. **Deduce** what happened to the oceanic plate that subducted beneath western North America during the Mesozoic.

WRITING IN ▶ Earth Science

6. Prepare a report documenting the chain of events that might have occurred once the meteorite hit Earth. Include a discussion of the effect on climate, air quality, and plant and animal life.

Teacher Content Support

Asteroid Impact Since the iridium layer was discovered in 1980, scientists have found other evidence for an asteroid impact at the end of the Cretaceous. These include small blobs of glass, called tektites, which typically form as molten rock is ejected from an impact site, and shocked quartz grains, which are only produced by incredibly high pressures.

3 Assess

Check for Understanding

Knowledge Ask students how the Atlantic Ocean and the Red Sea are similar. Both represent oceanic basins formed by rifting. The Red Sea is much younger than the Atlantic and is still a seaway.

Reteach

Predict Ask students to predict what the Red Sea will look like in 100 million years. It will probably look much like the Atlantic Ocean. Africa and the Arabian Peninsula will be as far apart as Europe and North America are now.

Assessment

Knowledge Ask students to explain how the amniotic egg added to the success of the reptiles. The amniotic egg enabled reptiles to colonize dry land away from water sources.

SECTION 2 REVIEW

1. The life-forms of the Mesozoic filled the niches that became available after the mass extinction at the end of the Paleozoic.
2. As continents split apart, basins and valleys form in the rift zones. The valleys flood with seawater and eventually widen to form oceans.
3. The tectonic events that transformed Laurentia's eastern margin were highlighted by the collision of continental plates, and by the final assembly of the supercontinent Pangaea. The tectonic events that transformed Laurentia's western margin were instead caused by the collision of an oceanic and a continental plate. This collision did not result in the formation of a supercontinent.
4. an impact crater in Mexico, and a layer of iridium between Cretaceous and Paleogene rocks. Iridium is rare in Earth's crust but common in meteorites.
5. Over time, the plate would have been destroyed as it moved back into the mantle beneath North America.
6. Answers will vary, but they might include the presence of debris in the atmosphere, darkness, more UV light, and global warming caused by greenhouse gases.

Rubric

SECTION 3

1 Focus

Focus Transparency

Before presenting the lesson, project **Section Focus Transparency 73** and have students answer the accompanying questions. EL

MAINIDEA

Age of Mammals Ask students to name up to 20 different animals. Compile a list on the board. The student-generated list will probably consist mostly of mammals. Mammals are visibly dominant on Earth today. However, in numbers and species varieties, insects are more dominant, and in the oceans, fishes are more dominant. But the perception remains that mammals rule, and the Cenozoic is often called the Age of Mammals.

2 Teach

Teacher Content Support

Primate Evolution Humans belong to the mammalian group known as primates, which also includes lemurs, monkeys, and apes. Fossil evidence suggests that the first primatelike organisms were small, shrewlike creatures that lived early in the Paleogene, about 60 mya. Over time, as the Paleogene climate changed, lemurs and tarsiers, and then monkeys and apes, evolved. The earliest fossils of bipedal primates are about 7 million years old, and the earliest fossils of primates in the genus *Homo*, *Homo habilis*, are about 2.3 million years old. *Homo habilis* appears to have become extinct about 1.8 mya. During the Pliocene and Pleistocene, many other *Homo* species also evolved and became extinct.

SECTION 3

The Cenozoic Era

MAINIDEA Mammals became the dominant terrestrial animals during the Cenozoic while the continents assumed their present forms.

Essential Questions

- What was the extent of glaciation during the Cenozoic?
- How can tectonic activity in North America during the Cenozoic be described?
- How did climate change affect life-forms during the Cenozoic?

Review Vocabulary

San Andreas Fault: a transform fault that separates the western edge of Southern California from the rest of the state and is responsible for most of California's earthquakes

New Vocabulary

Homo sapiens
bipedal

EARTH SCIENCE 4 YOU Have you ever been to a soccer game during which a player was injured? Usually another player fills in and the game goes on. The same can be true in nature. When organisms become extinct, new species often fill the empty niches.

Cenozoic Paleogeography

The Cenozoic Era encompasses about 1.5 percent of Earth's total history—approximately the last 66 million years. Despite its relative shortness, scientists know more about this era than any other. Humans evolved during the Cenozoic, appearing in their present-day form during the Pleistocene Epoch. **Figure 14** shows that you live in the Holocene, the current epoch of the Cenozoic.

Cooling trend You learned in Section 2 that the Mesozoic Era was relatively warm. Earth remained warm during the earliest epoch of the Cenozoic. However, as Australia was splitting apart from Antarctica during the Eocene (EE uh seen) Epoch, the worldwide climate began to cool. Scientists think that the cooling climate was caused, in part, by a change in ocean currents. When Antarctica and Australia were connected, a current of warm water flowing from the Pacific, Atlantic, and Indian Oceans moderated Antarctica's temperature. After Antarctica and Australia split apart, Antarctica was isolated over the South pole. A cold current began to flow around it, and a permanent ice cap began to grow during the Oligocene (AH luh goh seen).

■ **Figure 14** Mammals diversified widely during the Cenozoic, but modern humans did not appear until the end of the era.

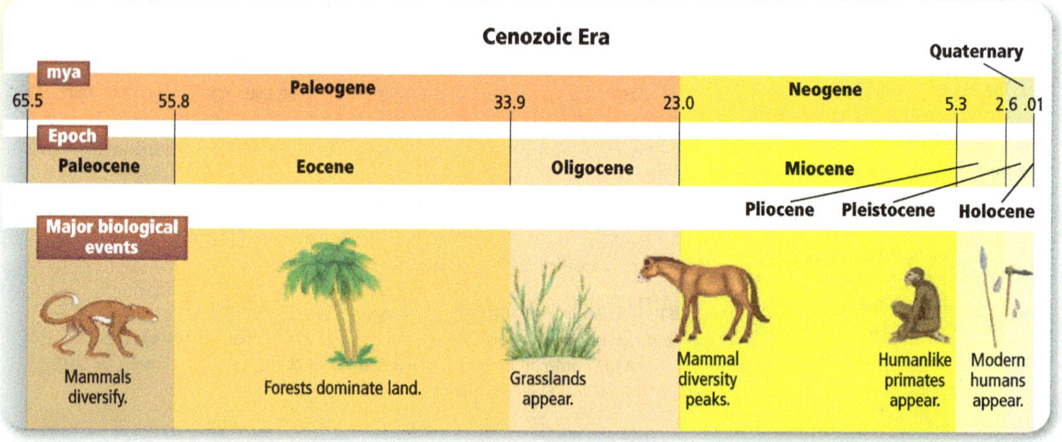

Demonstration

Currents Sketch two diagrams of Antarctica. One diagram should have Antarctica and Australia connected. In the other, they should be separated. Show with arrows how warm currents would have deflected along the coast to moderate the temperature when Australia was connected to Antarctica, but how a circumpolar current encircles an isolated Antarctica and never enters into warm latitudes.

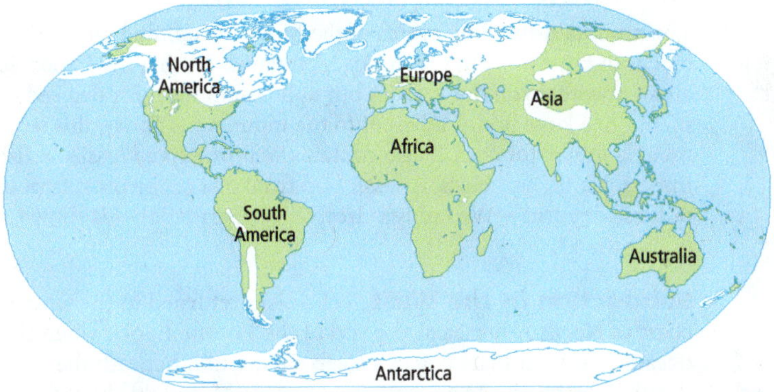

Figure 15 At the peak of Pleistocene glaciation, glaciers covered nearly one-third of Earth's land surfaces.
Infer *why patches of glaciation existed near the equator.*

Miocene warming In the early Miocene Epoch, the climate warmed again. The ice cap on Antarctica began to melt, and the ocean flooded the margins of North America. This trend reversed during the middle and late Miocene. Antarctica's ice cap stopped melting and the Arctic Ocean began to freeze, resulting in the formation of the arctic ice cap. This set the stage for the ice ages.

Ice ages Throughout the Pleistocene, ice covered much of Earth's northern hemisphere. Glaciers advanced and retreated in at least four stages over North America and the northern latitudes. During the peak of these ice ages, glaciers up to 3 km thick covered nearly one-third of Earth's land surfaces, as shown in **Figure 15.** In North America, the paths of the Ohio and Missouri Rivers roughly mark the southernmost point of glacier coverage. Glaciers carved out lakes and valleys, dropped huge boulders, and left behind abundant deposits of clay, sand, and gravel. In northeastern Washington State, glacial melting caused such a rush of water at the end of the last ice age that it created the largest waterfall recorded on Earth's surface. The remnants are shown in **Figure 16.**

Figure 16 This photo shows the remnants of Earth's largest waterfall in what is now Washington State. The waterfall, more than 5 km long and 120 km high, once flowed with water from glacial melting.

Discussion

Dry Falls In multiple episodes between 18,000 and 12,000 years ago, Glacial Lake Missoula burst through the ice dam that plugged it at one end. Geologists think that, in each episode, 2000 cubic km of water cascaded across the Pacific Northwest in a space of about 48 hours, creating—over time—a huge (now dry) waterfall called Dry Falls.

To help students get an idea of the scale of these ice-age floods, write the following dimensions of the lake, floodwaters, and waterfall on the board and help them draw analogies to dimensions familiar to them, such as the area of their state or the height of their school. *Glacial Lake Missoula: 640 km deep, 7770 square km in area, 2084 cubic km in volume. Flood: 842 km long, 4440 square km in area. Dry Falls: 120 m high, 5 km long*

Caption Question Fig. 15 Glaciers existed near the equator only on mountaintops, where it was cooler.

DIFFERENTIATED INSTRUCTION

Advanced Learners Have students research features besides Dry Falls, such as ripple marks and the Channeled Scablands, that convinced geologists that glacial floods transformed the northwestern part of the United States during the Pleistocene.

Enrichment

Farallon Plate Have students research what happened to the Farallon Plate and how it interacted with the adjacent Pacific and North American plates. Ask them to find out what type of boundaries existed between the Farallon Plate and the Pacific Plate divergent and between the Farallon Plate and North American Plate convergent. Ask students to determine how the Farallon Plate is related to the Juan de Fuca Plate and to predict what will happen to the Juan de Fuca Plate. The Juan de Fuca Plate is the last remaining remnant of the Farallon Plate. It will eventually subduct completely beneath the North American Plate. AL

Caption Question **Fig. 18** Mount St. Helens is the most active volcano in the Cascade range.

Figure 17 This 38-million-year-old fossil bird was found in Wyoming's Green River Formation. The fossil is about 25 cm long.

Cenozoic Mountain Building

The mountain-building events of the Mesozoic uplifted massive blocks of crust to form the Rocky Mountains. During the Cenozoic, erosion wore down the Rockies but uplift continued. Eroded sediment filled large basins adjacent to the mountains. Today, this sediment is mined for coal. It also contains well-preserved fossils of fish, insects, plants, and birds. A fossil bird from one of the most famous of these deposits—Wyoming's Green River Formation—is shown in **Figure 17.**

Subduction in the West Volcanism returned to the western coast of North America at the end of the Eocene Epoch when the oceanic Farallon Plate began a steep subduction beneath the Pacific Northwest. As a result, the Cascade Mountains began to rise. Volcanoes in the Cascade range remain active today, as shown in **Figure 18.**

While subduction continued in northwestern North America, the Farallon Plate disappeared under what is now California. The North American Plate came into contact with another oceanic plate—the Pacific Plate—that was moving in a different direction. As a result, the San Andreas Fault formed. The San Andreas Fault is a transform boundary between the two plates. Recall that in a transform boundary, two plates slide against each other and there is no subduction. Because there is little to no subduction beneath central and Southern California today, there is little volcanic activity there.

Basin and Range Province The beginning of the interaction between the North American Plate and the Pacific Plate coincided with the formation of the Basin and Range Province in the southwestern United States and northern Mexico. Recall that the Basin and Range Province consists of hundreds of nearly parallel mountains. These mountains were formed when stresses in Earth's crust—called tension—pulled it apart. This process, illustrated in **Figure 19,** continues today.

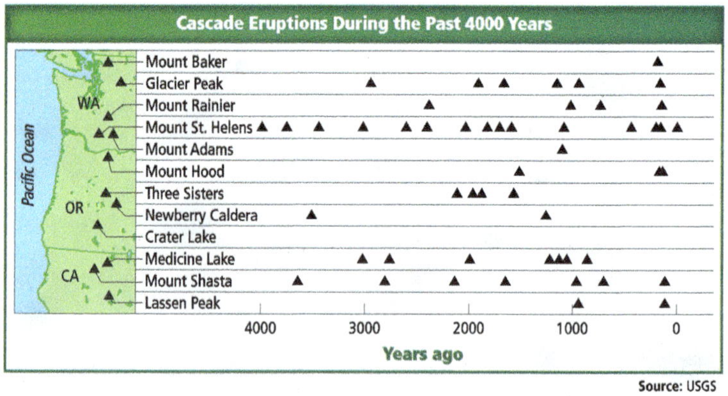

Figure 18 The Cascade Mountain Range includes active volcanoes that have erupted many times during the past 4000 years.
Conclude *which volcano is the most active.*

EARTH SCIENCE JOURNAL

Coal Mining Have students research coal mining in Wyoming's Powder River Basin, where more coal is mined than in any other area of the United States. Have students write about coal mining from both an economic and an environmental perspective. In particular, have them address why coal mining can be harmful to the environment.

VISUALIZING Basin and Range Province

Figure 19 The Basin and Range Province is a series of mountains and basins that is bordered on the west by California's Sierra Nevadas and on the east by Utah's Wasatch Mountains. During the past 25 million years, crustal stretching has increased the distance between these two points by over 250 km.

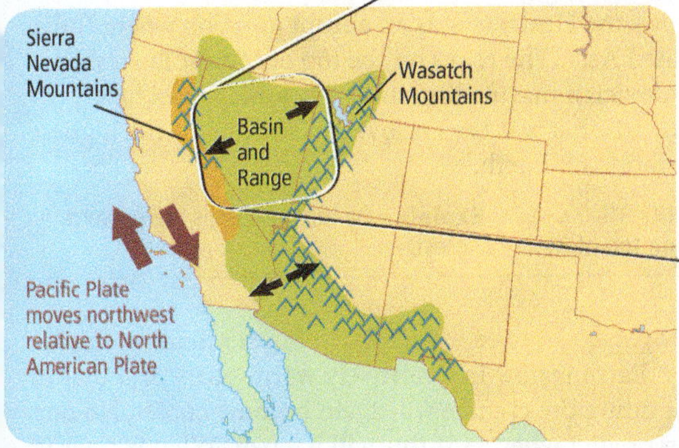

The stretching underneath the Basin and Range Province is caused, in part, by the steady movement of the Pacific Plate relative to the North American Plate. The North American Plate is being stretched to the northwest, and the Basin and Range Province is being stretched in an east-west direction.

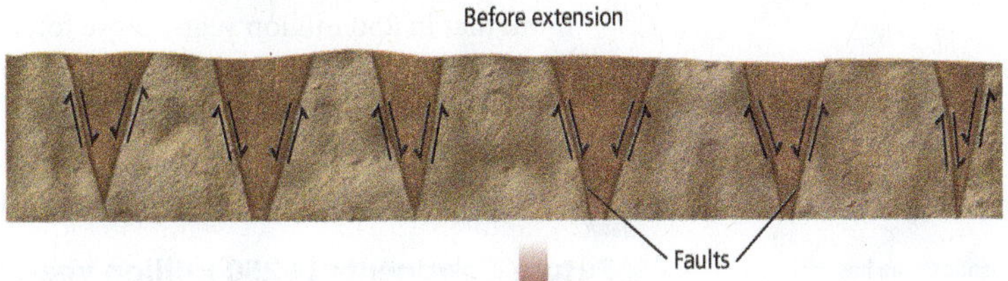

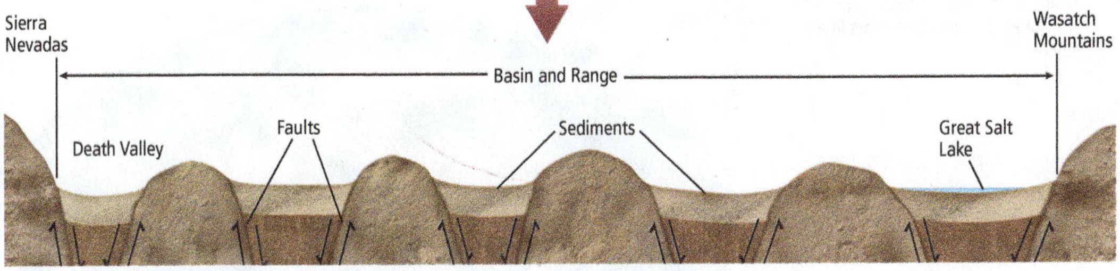

To compensate for crustal stretching, the rocks broke up into hundreds of blocks along normal fault lines. Some blocks rose to form mountains, while adjacent areas dropped to form basins. The mountains are still being pushed upward, rising as quickly as they erode, and the basins are still dropping and filling with eroded debris. The crust underneath the Basin and Range Province has stretched so much that it is one of the thinnest parts of Earth's crust today.

Concepts In Motion View an **animation of the formation of the Basin and Range Province.**

VISUALIZING

Purpose

Students will learn how a series of parallel mountain ranges can form.

Teacher Content Support

Basin and Range The cause of Basin and Range extension is debatable. Some scientists think that it might be caused by a mid-ocean ridge spreading center called the East Pacific Rise, part of which was subducted under the North American Plate beginning about 40 mya. It might be that the mantle convection responsible for the East Pacific Rise is still active beneath North America, and that this convection is stretching the crust.

Interpret the Illustration

Basin and Range Have students look at the map in **Figure 19.** Ask them if they know where on the map Death Valley and Mt. Whitney are located. Tell them that Mt. Whitney is in the Sierra Nevadas, which is the westernmost border of the Basin and Range Province, and that at 4.42 km, it is the highest point in the continental United States. Death Valley, only 145 km away, is part of the Basin and Range Province. At 86 m below sea level, it is the lowest point in the continental United States.

DIFFERENTIATED INSTRUCTION

Struggling Learners Have students fold pieces of paper using accordion folds. Have them pull the edges so that the paper is extended. Tell them that the ridges and valleys of the Basin and Range are similarly being pulled apart. **EL**

Collaborative Learning

Active Tectonism Have groups of students research where active rifting and continental collisions are occurring on Earth today, and then have them mark those places on a map of Earth's continents that they can cut up. Have them cut out the continents and then rearrange them to show what they think Earth might look like in 10 million years and in 250 million years. Then have them compare their maps with **Figure 21,** which is a more sophisticated but still hypothetical map of Earth as it might look in 250 million years. OL COOP LEARN

☑ READING CHECK Marine fossils are present on Mount Everest because the rocks that make up Mount Everest were once the seafloor.

■ **Figure 20** The Himalayas appear as an abrupt junction where India is crashing into Asia.

Continental collisions The final breakup of Pangaea during the early Cenozoic resulted in several separate continents. It also brought some continents together. During the Paleocene, Africa began to collide with Eurasia, creating the Alps and narrowing the ancient Tethys (TEE thus) Ocean, which once separated Eurasia and Gondwana. The remnants of this ocean now exist as four bodies of water in Europe and central Asia—the Black, Caspian, Aral, and Mediterranean Seas.

Also during the Paleocene, India began crashing into the southern margin of Asia to form the Himalayas, a mountain range that is still rising today. **Figure 20** shows the Himalayas as an abrupt junction where India joined Asia. The rocks on the top of Mount Everest are Ordovician marine limestone. Tectonic forces have pushed what was the Ordovician seafloor to the highest elevation on Earth.

☑ READING CHECK **Explain** why marine fossils are present on top of Mount Everest.

Tectonic forces continue Many scientists think that Earth is now in a relatively warm phase and that in the future the climate will again become cooler. No one can predict when or if this will happen. What is clear is that the tectonic forces that have shaped Earth over the past 4.6 billion years continue today. Some scientists think that in 250 million years, those forces will have largely eliminated the Atlantic Ocean and pushed the continents together into another supercontinent, as shown in **Figure 21.**

■ **Figure 21** The Atlantic Ocean has nearly disappeared in this hypothetical map of Earth 250 million years in the future.

Future Continents (+250 million years)

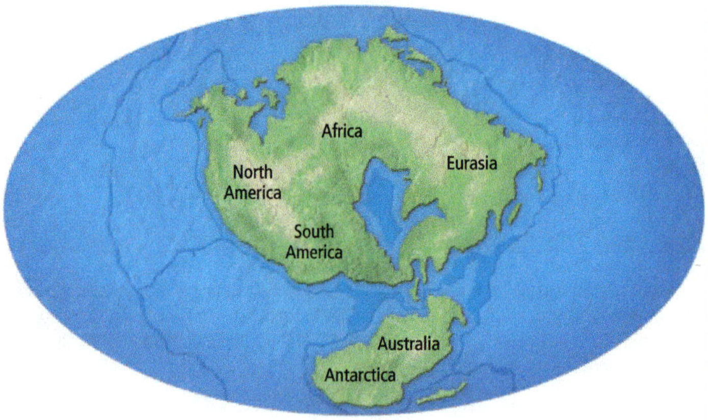

ACROSS THE CURRICULUM

Computer Science **Figure 21** was made with sophisticated graphics using data from computerized Geographic Information Systems (GIS), which integrate, store, and analyze geographic information according to location. GIS technology enables scientists to display spatial information about the spreading rate of tectonic plates at ocean ridges and the subduction rate in trenches. This helps the scientists determine the positions of continents in the past and hypothesize what Earth will look like in the future.

Cenozoic Life

Many marine organisms, including clams, sea urchins, and sharks, survived the mass extinction at the end of the Cretaceous and populated the oceans during the Cenozoic. On land, forests dominated the early Cenozoic landscapes. As the climate cooled during the late Eocene, forests gave way to open land, and grasses appeared. By the late Oligocene, grassy savannas, like those in east Africa today, were common worldwide. The rise of grasslands led to the diversification of many new mammal groups. Because mammals are the dominant terrestrial animals, many scientists call the Cenozoic the Age of Mammals.

Ice age mammals As the ice ages began, the climate began to cool and new animals evolved in northern latitudes. Two of the most famous mammals of the late Pleistocene are the woolly mammoth and the saber-toothed cat, shown in **Figure 22.** By the time these animals roamed Earth, modern humans—called ***Homo sapiens***—were well established.

Figure 22 The woolly mammoth and saber-toothed cat, shown in this artist's reconstruction, were adapted to the cool Pleistocene climate.

Humans The defining characteristic of humans is their upright, or **bipedal,** locomotion. The fossil record, while incomplete, shows that the first bipedal humanlike primates appeared about 6 mya during the late Miocene. The fossil remains of the earliest modern humans—found in Africa—are about 195,000 years old.

Migrations The migrations of early humans were undoubtedly influenced by the ice ages of the late Pleistocene. For example, scientists think that the Bering Strait, which now separates Russia and Alaska, was exposed during the late Pleistocene because much of Earth's water was frozen in glaciers. It is likely that the humans who walked across the strait were North America's first inhabitants.

SECTION 3 REVIEW

Section Self-Check

Section Summary

- Ice covered nearly one-third of Earth's land surface at the peak of the Cenozoic ice ages.
- The Cascade Mountains began to rise and the San Andreas Fault formed during the Cenozoic.
- The Cenozoic is known as the Age of Mammals.
- Fossil evidence suggests that modern humans appeared during the Pleistocene.

Understand Main Ideas

1. MAINIDEA **Describe** why the Cenozoic is called the Age of Mammals.
2. **Assess** the extent of glaciation in North America.
3. **Discuss** how the Basin and Range Province and the San Andreas Fault are tectonically related.
4. **Explain** how the positions of the continents contributed to Cenozoic climate change.

Think Critically

5. **Propose** Why do you think early humans migrated?

MATH IN Earth Science

6. If the glacial ice on Earth were to melt, sea level would rise about 50 m above its current level. If sea level rose at an average rate of 2 mm per year, how long would it take for all the ice on Earth to melt? Use the following relationship: distance = rate × time.

3 Assess

Check for Understanding

Reinforcement Ask students to describe where active collisional and extensional plate tectonics are occurring in North America today. Collisional plate tectonics is occurring along the western coast, where the Cascades are rising and the San Andreas Fault is active. Extensional plate tectonics is occurring in the Basin and Range Province, where plate action is stretching the crust.

Reteach

Discussion Ask students to discuss how volcanic and earthquake activity is related to plate tectonics, using examples from North America. Volcanic activity in the Cascades and earthquakes along the West Coast indicate areas of active plate tectonics, while the absence of such activity along the East Coast indicates that plates are inactive in this area.

Assessment

Performance Have groups of students make time lines of the Cenozoic, adding the major geological and biological events discussed in this section. Time lines should include the ice ages of the Pliocene and Pleistocene; the start of the formation of the Cascades, the San Andreas Fault, and the Basin and Range Province in the Eocene; the beginning of the collision of Africa and Eurasia, and India and Asia, in the Paleocene; the appearance of grasses in the Eocene; human ancestors in the Miocene; and *Homo sapiens* in the Pleistocene. COOP LEARN

SECTION 3 REVIEW

1. Mammals filled the niches left vacant by the extinction of the dinosaurs. Grasslands supported a diverse collection of mammals.
2. The glaciers traveled south roughly to the borders of the Ohio and Missouri rivers.
3. Both formed largely as a result of the interaction of the Pacific Plate with the North American Plate.
4. When Antarctica and Australia split apart, Antarctica was isolated over the south pole. A cold current began to flow around it, starting a cooling trend.
5. Answers will vary, but might include food opportunities. Early humans might have been chasing migrating game. They also could have been looking for a warmer climate, or perhaps they were simply curious.
6. It would take 25,000 years. 50 m = 50,000 mm; 50,000 mm = 2 mm/y × time; 50,000 mm/(2 mm/y) = time; 25,000 y = time.

eXpeditions!

Purpose

Students will describe a dinosaur dig and the tools needed for a dinosaur dig.

Teacher Content Support

Dinosaur Fossils People have been digging up dinosaur bones for thousands of years, long before dinosaurs were known to exist. Some scientists think these fossils might explain the giants and other mythical creatures that populated ancient literature. One of the first described dinosaur fossils, a femur bone, was discovered in 1676 by British naturalist Robert Plot, who thought it belonged to a giant human. Not until 1824 did British fossil hunter William Buckland correctly identify the bone as belonging to a dinosaur, which he called *Megalosaurus*.

The British paleontologist Richard Owen coined the word *dinosaur* in 1842. Today, fossils of between 10 and 20 new dinosaur genera are discovered each year. Scientists think that only a small percentage of dinosaur genera have been unearthed.

Teaching Strategies

- Ask students what they think scientists can learn about dinosaurs by studying dinosaur fossils. They can learn how big dinosaurs were, what they ate, how fast they ran, the environments they inhabited, and what animals were closely related to them. They can only speculate about behavior.
- If possible, show clips of one of the *Jurassic Park* movies. Ask students what role paleontologists might have played in the making of the films.

eXpeditions!

ON SITE: DIGGING FOR DINOSAURS

Check this out . . . look what I found! The team of paleontologists and volunteers runs up the hill to see what I am holding. In my hand is the tip of a 150-million-year-old theropod dinosaur tooth.

Day 1 Arrive at dig site The team loads equipment, gear, and tools into the vans and drives to the dig site. The quarry where digs have been ongoing is visible and everyone is eager to begin. But first we set up camp.

The land is owned by a local rancher who, in 1985, found the first chips of fossilized bone on the site. In 2003, larger bones were found and the rancher realized the importance of this find. He contacted a paleontologist at the Judith River Dinosaur Institute to ascertain interest in digging for dinosaur fossils on the land. Digs have been conducted each summer since that initial contact with the rancher.

Day 2 Basic digging skills Most on our team are volunteers who have no experience at a dig. Other members include geologists, paleontologists, and science teachers. The primary tools used are awls, which look like ice picks, and stiff paintbrushes. Shovels, air hammers, and wheelbarrows are also used throughout the digging process.

The work of a dig is slow. Each layer of rock is picked apart carefully and the debris is brushed away. The rock is inspected to be sure no fossils are accidentally brushed away or broken off.

Junior paleontologists work to carefully remove bones from the ground at this dig in Montana.

Each member of the team fills a scoop with debris which gets dumped into a bucket and examined. Then the process is repeated. On this day, we find fossil bones from a theropod.

Days 3–6 Digging for a *Stegosaur* A few members of the team find a few tail vertebrae and fragments of a *Stegosaur* spike. The team begins digging in the hillside and eventually unearths more tail vertebrae and limb and foot bones. Three more days of excavating and little more is found.

Day 7 Finalizing the dig The final hours of the dig are busy as everyone gathers as much data as possible. We sketch and photograph fossils, take measurements of the locations where the fossils were found, and label each fossil.

We pack the fossils and the equipment into the vans. The fossils will be catalogued and stored at the Institute. The data will be used in a research paper that will be published in a scientific journal.

WRITING IN ▶ Earth Science

Make a Model Research and model the various ways that fossils are prepared both for removal from the ground and for transport. Write descriptions of your models to be displayed with the models.

WebQuest

WRITING IN ▶ Earth Science

Rubric

Make a Model In their presentations, students should discuss the actual dinosaur digs they researched and how their models represent those digs. Students should provide key findings from their sites as well as discuss any struggles or challenges faced by the team members.

WebQuest

GeoLAB

iLab Station

Solve Dinosaur Fossil Puzzles

Background: The discovery of a sharpened piece of stone near a prehistoric campsite can be interpreted as having once been a tool used by prehistoric humans. Shape and position of objects provides scientists with information that can be used to interpret the lifestyles of early humans. Paleontologists who study dinosaurs use the same techniques as they collect and study fossils.

Question: *By studying these fossils, what can you tell about how these dinosaurs lived and what they ate?*

The skull on top is from an *Albertosaurus*. The bottom skull is from an *Edmontosaur*.

Materials

textbook

Internet access or pictures provided by your teacher

Procedure

Imagine you are working with a team of paleontologists in a remote desert region and your team discovers the fossilized remains of two dinosaur species. Do the work of a paleontologist by studying the fossils and answering questions about how they lived.

1. Read and complete the lab safety form.
2. Compare and contrast the teeth, jaw structures, and hips of both dinosaur species.
3. Record all your observations in your science journal.

Analyze and Conclude

1. **Infer** What part of the dinosaur skeleton is most important in determining diet? Why? What is the likelihood that this part of the skeleton will be preserved?
2. **Interpret Data** Describe the diets of both dinosaur species.
3. **Interpret Data** Describe how each dinosaur species moved.
4. **Conclude** What fossil evidence from the jaws and teeth did you use to infer the diet of each dinosaur species?
5. **Conclude** What fossil evidence did you use to infer the locomotion of each dinosaur species?

APPLY YOUR SKILL

Apply Mammoths and mastodons both lived during the Pleistocene Epoch. They appear very similar in appearance yet lived slightly different lives. Examine photos of their teeth to determine where they might have lived and what they ate.

Analyze and Conclude

1. Tooth structure is one of the most useful lines of evidence for determining dinosaur diets. Tooth shape and form evolved in response to food sources. Teeth make good fossils because they are hard.
2. *Albertosaurus* was a carnivore. *Edmontosaur* was a herbivore that probably fed on fruit or broad leaves.
3. *Albertosaurus* was a bipedal theropod. *Edmontosaur* was probably a quadruped, walking on four legs.
4. The teeth of *Albertosaurus* were long, sharp, and serrated for grasping and tearing meat; its jaws were large and hinged far back in its mouth so that it could open its jaws wide for gulping down large meat pieces. *Edmontosaur* had a toothless duck bill for plucking fruit or leaves, and its back teeth were small and flat for grinding plant material. Its jaws were small and less powerful.
5. Students should examine the placement of the hip bones (ilium, ischium, and pubis) to determine how each dinosaur moved and what group it belonged to.

GeoLAB

iLab Station Rubric

Preparation

Time Allotment 40–60 min

Process Skills observe and infer, interpret data, draw conclusions

Safety Precaution Approve lab safety forms before work begins.

Preparation of Materials If students do not have access to the Internet or other resources, research skeletal differences (in hips, jaws, and teeth) between the two major groups of dinosaurs, the Ornithischia (herbivores) and Saurischia (mostly carnivores), and explain these differences to students.

Procedure

Have groups of students research skeletal differences among herbivorous and carnivorous dinosaurs. Have them draw the hip bone arrangements and the teeth structure of the two dinosaur groups, and then relate their drawings to the pictures of the two species they are examining. Tell them that one possible reason for the shift in the pubis bone in herbivores, which evolved later, was to make room for the larger gut needed to process vegetable matter.

APPLY YOUR SKILL

Apply Mammoth molars, massive and flat, were useful for chewing grasses and small leaves, suggesting that mammoths lived on grassy plains. Mastodon molars, shaped like cones, were capable of grinding coarse vegetation, like tree branches, suggesting that mastodons lived in wider-ranging habitats.

LabManager™
Customize any lab with the LabManager™ CD-ROM.

MAINIDEAS Summary statements can be used by students to review the major concepts of the chapter.

Students can review with these online resources.

Vocabulary Practice

- Vocabulary eGames
- Vocabulary eFlashcards
- Vocabulary PuzzleMaker

Section Self-Check

Chapter Self-Check

Online Test Practice

Use the ExamView® *Assessment Suite* CD-ROM to:

- create multiple versions of tests
- create modified tests with one mouse click
- edit existing questions and add your own questions
- build tests aligned with state standards using built-in state curriculum tags
- change English tests to Spanish with one mouse click
- track students' progress using the Teacher Management System

CHAPTER 23 STUDY GUIDE

BIGIDEA Complex life developed and diversified during the three eras of the Phanerozoic as the continents moved into their present positions.

Vocabulary Practice

SECTION 1 The Paleozoic Era

VOCABULARY

- paleogeography
- passive margin
- transgression
- regression
- Cambrian explosion

MAINIDEA Life increased in complexity during the Paleozoic while the continents collided to form Pangaea.

- Scientists study sediment and evaporite deposits to learn how sea levels fluctuated in the past.
- Eastern Laurentia was transformed by many mountain-building events during the Paleozoic.
- A great diversity of multicellular life appeared during the first period of the Paleozoic.
- The largest extinction event in Earth's history occurred at the end of the Paleozoic.

SECTION 2 The Mesozoic Era

VOCABULARY

- phytoplankton
- amniotic egg
- iridium

MAINIDEA Reptiles became the dominant terrestrial animals during the Mesozoic while Pangaea broke apart.

- The breakup of Pangaea triggered a series of tectonic events that transformed western Laurentia.
- The Atlantic Ocean began to form during the Mesozoic as North America broke away from Europe.
- Dinosaurs and other new organisms evolved to fill niches left empty by the Permo-Triassic Extinction Event.
- All dinosaurs, except birds, along with many other organisms became extinct during a mass extinction event at the end of the Mesozoic.

SECTION 3 The Cenozoic Era

VOCABULARY

- *Homo sapiens*
- bipedal

MAINIDEA Mammals became the dominant terrestrial animals during the Cenozoic while the continents assumed their present forms.

- Ice covered nearly one-third of Earth's land surface at the peak of the Cenozoic ice ages.
- The Cascade Mountains began to rise and the San Andreas Fault formed during the Cenozoic.
- The Cenozoic is known as the Age of Mammals.
- Fossil evidence suggests that modern humans appeared during the Pleistocene.

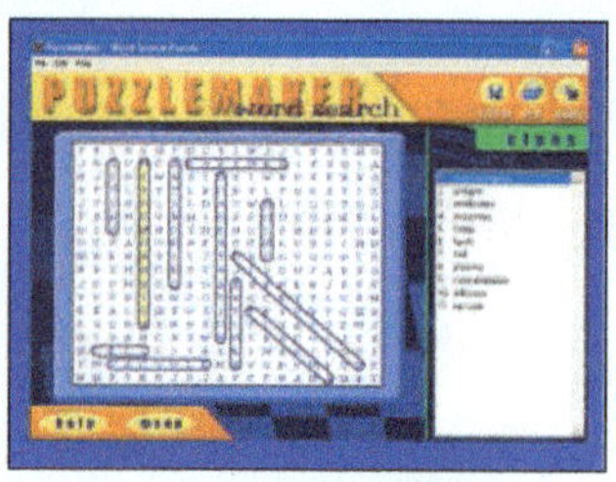

Vocabulary Practice

For additional practice with vocabulary, have students access the Vocabulary PuzzleMaker.

Chapter Self-Check

VOCABULARY REVIEW

Match the definitions below with the correct vocabulary term from the Study Guide.

1. the ancient geographic setting of an area
2. the organisms at the base of the marine food chain
3. the increase in diversity and abundance of marine life-forms at the beginning of the Paleozoic Era
4. the movement of a shoreline seaward as sea level falls

Use a vocabulary term from the Study Guide to answer each of the following.

5. Which term is used to describe upright locomotion on two legs?
6. What metal is rare on Earth's surface but relatively common in asteroids?

Fill in the blanks with the correct vocabulary terms from the Study Guide.

7. The movement of a shoreline inland as sea level rises is called ________.
8. ________ are primates with bipedal locomotion.
9. The ________ was a reproductive feature that allowed reptiles to migrate widely on land.

UNDERSTAND KEY CONCEPTS

10. Which was the dominant terrestrial life form during the Mesozoic Era?
    - **A.** mammals
    - **B.** dinosaurs
    - **C.** birds
    - **D.** fish
11. Which term describes a shoreline that is experiencing no tectonic activity?
    - **A.** active margin
    - **B.** passive margin
    - **C.** trench
    - **D.** regression
12. During which geologic time period did the Atlantic Ocean begin to form?
    - **A.** Triassic
    - **B.** Cretaceous
    - **C.** Jurassic
    - **D.** Devonian

Use the figure below to answer Questions 13 to 15.

13. What formed the deposits in the photo above?
    - **A.** asteroid residue
    - **B.** evaporation of seawater
    - **C.** glaciation
    - **D.** phytoplankton
14. Where would these deposits most likely have formed?
    - **A.** ocean floor
    - **B.** shoreline
    - **C.** lagoon
    - **D.** coral reef
15. Which item could be made from this deposit?
    - **A.** laundry detergent
    - **B.** talcum powder
    - **C.** chalk
    - **D.** sponge
16. How much of Earth's land surface did glaciers cover at the height of the ice ages?
    - **A.** 10 percent
    - **B.** 60 percent
    - **C.** 30 percent
    - **D.** 90 percent
17. Which metal that is rare in Earth's rocks but relatively common in asteroids is used as evidence that there was an asteroid impact at the end of the Cretaceous?
    - **A.** iron
    - **B.** iridium
    - **C.** uranium oxide
    - **D.** zircon
18. Which supercontinent formed at the end of the Paleozoic?
    - **A.** Rodinia
    - **B.** Gondwana
    - **C.** Laurasia
    - **D.** Pangaea

CHAPTER 23 ASSESSMENT

VOCABULARY REVIEW

1. paleogeography
2. phytoplankton
3. Cambrian explosion
4. regression
5. bipedal
6. iridium
7. transgression
8. *Homo sapiens*
9. amniotic egg

UNDERSTAND KEY CONCEPTS

10. B
11. B
12. A
13. D
14. A
15. C
16. C
17. B
18. D

19. A
20. C
21. C

CONSTRUCTED RESPONSE

22. Seeds enabled plants to colonize land far from water sources. By spreading far, plants created food and habitats for animals.
23. Dinosaur hip structure caused the legs to emerge straight under the torso, and not sprawled out from the side, as with other reptiles, such as lizards.
24. Ancient coral reefs are often associated with evaporites, which act as seals for oil that might be trapped in the pore spaces of the reef rocks.
25. There are few active volcanoes in Southern California because the Pacific Plate is not subducting beneath the North American Plate; instead, it is moving northwest relative to it, creating a transform fault and causing earthquake activity.
26. Drawings will vary, but they should show an end sequence of limestone overlain by shale and shale overlain by sandstone.
27. Subduction in the first phase was slow, and the angle of the descending plate was steep. Subduction in the second phase was faster, but the plate angle was more shallow. Subduction in the third phase was the fastest, and the plate angle was the most shallow.
28. When glaciers form, they tie up water that normally would cycle to the oceans. They thus cause a drop in sea level. When glaciers melt, sea levels rise because the melted ice returns to the oceans as liquid water.
29. Volcanic ash is evidence of volcanoes. The setting of most volcanoes is along active margins where orogeny events occur as a result of collisions between lithospheric plates. Therefore, the presence of volcanic ash is evidence of a former orogenic event.

Use the figure below to answer Questions 19 and 20.

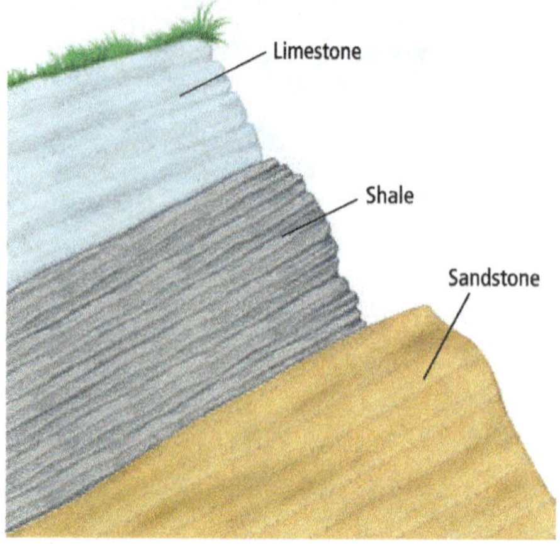

19. What does the succession of rocks in the figure above indicate?
A. a transgressive sequence where sea level rose
B. a regressive sequence where sea level fell
C. sea level fluctuated widely
D. an evaporite deposit

20. Which is a likely origin of the limestone?
A. compacted clay sediment
B. beach sand
C. remains of skeletons from phytoplankton
D. plant deposits

21. What is evidence of an arid climate in North America at the end of the Triassic?
A. formation of a seaway
B. presence of ancient coral reefs
C. presence of ancient sand dunes
D. presence of a passive margin along the eastern edge

CONSTRUCTED RESPONSE

22. Explain how seed plants changed the landscape after they evolved during the Carboniferous.

23. Explain what skeletal feature distinguished dinosaurs from other reptiles.

24. Summarize why ancient coral reefs are good places to explore for oil.

25. Explain why there are few active volcanoes in Southern California but there is frequent earthquake activity.

26. Create a drawing that shows evidence of a regression.

27. Compare the subduction rates and angles of the three phases of the Cordilleran Orogeny.

28. Describe how sea-level change and glaciation are related.

29. Explain why volcanic ash deposits can be used as evidence of an ancient orogeny.

THINK CRITICALLY

Use the figure below to answer Questions 30 and 31.

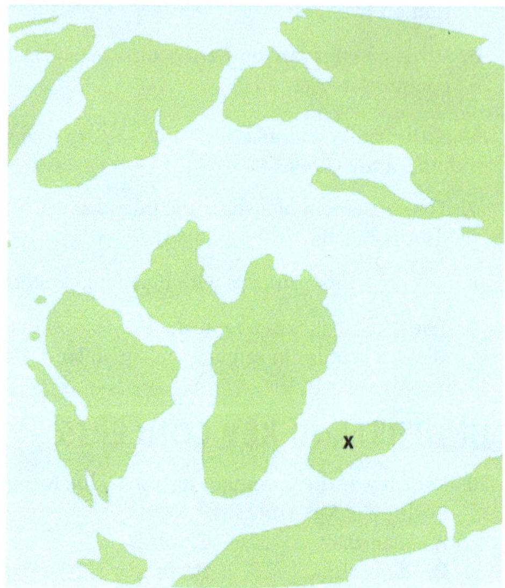

30. Explain how and where India, the continent labeled *X* in this diagram of Earth during the Mesozoic, is moving.

31. Generalize Elephants today are present naturally only in Africa and Asia. Discuss how the theory of plate tectonics might explain this.

32. Discuss how the breakup of a supercontinent might lead to the formation of a new ocean.

THINK CRITICALLY

30. India is moving northeast to dock eventually with Asia.
31. Ancestral elephants evolved in Africa when India was still connected to Africa. Any ancestral elephants living on India remained on India when India rifted away. These ancestral elephants spread throughout Asia when India docked, evolving in response to the Asian environment. This explains why Asian and African elephants differ. Climate and geography restraints might have kept the Asian elephants from migrating further north.
32. When two plates break apart, the crust underneath thins and pulls apart. Ocean water floods the low-lying areas. As the plates continue to move apart and the rift widens, water continues to fill the basin until eventually—over millions of years—an ocean has formed.

Chapter Self-Check

33. **Summarize** how the Appalachian Mountains, near the east coast of North America, are evidence that this coast was once an active margin.

Use the figure below to answer Question 34.

34. **Explain** how a major regression might stress the organisms pictured above.

35. CAREERS IN EARTH SCIENCE Explain why the discovery of the remains of an ancient coral reef would be exciting news to a petroleum geologist.

36. **Propose** what Earth might be like today if a meteorite had not struck it at the end of the Cretaceous.

37. **Evaluate** the relationship between extinction events and the evolution of life. Give an example.

38. **Evaluate** how Earth would be different if Antarctica were not over the south pole but were at a latitude similar to that of Australia.

39. **Evaluate** Use **Figure 10** to evaluate and explain sea level during the late Cretaceous.

CONCEPT MAPPING

40. Create a concept map of the three eras of the Phanerozoic with examples of the different organisms that evolved in each era.

CHALLENGE QUESTION

41. **Explain** how heat generated in Earth's interior can cause continents to rift apart.

WRITING IN ▶ Earth Science

42. Write a report that summarizes and illustrates how too much dust in the atmosphere could ultimately cause the death of a mammal population.

DBQ Document–Based Questions

Data obtained from: Evans, K.R. et al. 2005. The sedimentary record of meteorite impacts: an SEPM research conference. *The Sedimentary Record* 3:4–8.

You learned in this chapter that a meteorite impact can affect life on Earth. How common are meteorite impacts? Recent work shows that there are more impact sites preserved on Earth than you might expect.

Use the data to answer the questions below.

Impact Craters in the Continental U.S.

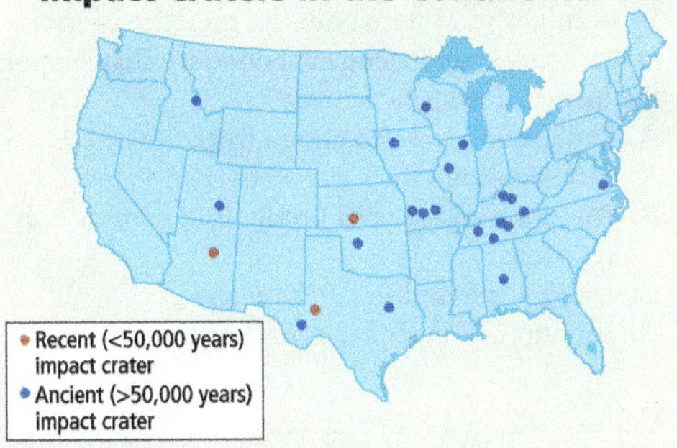

43. What percentage of impacts have occurred within the past 50,000 years?

44. Describe how the impact sites are distributed.

45. Hypothesize a reason for the distribution pattern.

CUMULATIVE REVIEW

46. Where are most earthquake epicenters? What is the depth of most earthquakes? Explain your answers. **(Chapter 19)**

47. Give examples of coarse-grained, medium-grained, and fine-grained clastic sedimentary rocks. **(Chapter 6)**

CHAPTER 23 ASSESSMENT

33. The Appalachian Mountains are evidence of an ancient orogeny, where two plates collided. Active margins occur where mountains are being built.

34. A regression lowers sea levels and decreases habitat areas. Falling sea level might also indicate increased glaciation and a cooler sea, leading to more stress.

35. Evaporite deposits often overlie ancient coral reefs, which contain large pore spaces in which oil might reside. The evaporite deposits are impermeable and prevent the oil from migrating.

36. One extreme is that things would not be much different because other events might have caused the dinosaurs to become extinct. Another extreme is that dinosaurs would still be the dominant terrestrial animal, and that humans did not evolve at all.

37. With a mass extinction, niches are opened that other animals evolve to fill, such as when mammals took over many of the niches available when the dinosaurs became extinct, or when new life-forms evolved after the Permo-Triassic extinction.

38. If Antarctica did not sit over the south pole, sea level would be higher because Antarctica—being in a more temperate latitude—would not have a large ice cap. Ocean currents would also change, causing different climate patterns.

39. The presence of inland seas indicates high sea level during the Cretaceous.

CONCEPT MAPPING

40. Maps will vary, but should include the Paleozoic (including the first multicellular animals with hard parts), the Mesozoic (including dinosaurs, marine reptiles, flowering plants, and phytoplankton), and the Cenozoic (including grasses and a diversity of mammals).

CHALLENGE QUESTION

41. Convection within the mantle drives the movement of the tectonic plates on Earth's surface, ultimately causing continental breakup.

WRITING IN ▶ Earth Science

Rubric

42. Atmospheric dust might limit sunlight, which could kill plants. Mammals that depended on plants for food, and mammals that ate those mammals, and so on, could die. Because dust blocks sunlight, it might also stress organisms by causing the climate to cool.

DBQ Document-Based Questions

Data obtained from: Evans, K.R. et al. 2005. The sedimentary record of meteorite impacts: an SEPM research conference. *The Sedimentary Record* 3:4–8.

43. 12.5 percent (3 out of 24)

44. There appear to be more impact sites in the central United States.

45. The impact sites tend to be in areas that have not been subjected to mountain-building activity. It is likely that any impacts in mountain-building areas would be deformed and difficult to recognize.

CUMULATIVE REVIEW

46. Most earthquakes occur at plate boundaries, and most of these are at convergent boundaries. Most earthquakes are shallow (upper 30 km) because this is the part of Earth that is cold enough to behave like a solid, and hence has the ability to form faults.

47. conglomerate or breccia; sandstone; siltstone, mudstone, or shale

MULTIPLE CHOICE

1. B
2. C
3. D
4. A
5. C
6. D
7. A
8. B
9. C
10. A

CUMULATIVE STANDARDIZED TEST PRACTICE

MULTIPLE CHOICE

1. New ocean crust is added to Earth's tectonic plates at which type of plate boundary?
   A. convergent boundary
   B. divergent boundary
   C. deep-sea trench
   D. transformation boundary

Use the table to answer Questions 2 and 3.

Mass Extinction Theories	
Evidence for meteorite impact	unusually high levels of iridium in Cretaceous-Paleogene boundary sediments; discovery of Chicxulub crater
Evidence for massive volcanic activity	volcanic eruptions during the late Cretaceous in India; unusually high levels of iridium, soot, and charcoal in Cretaceous-Paleogene boundary sediments

2. What might you infer about the presence of iridium at the Cretaceous-Paleogene boundary that supports the theory of volcanic activity?
   A. Iridium is deposited after a large fire is extinguished.
   B. Iridium is found with products of combustion reactions.
   C. Iridium is found in Earth's interior.
   D. Iridium no longer exists on Earth's surface.

3. Underneath the Chicxulub crater is a large layer of melted rock. Why are these rocks melted?
   A. They have been melted by volcanism.
   B. They contain a higher level of iridium.
   C. They are pieces of the surrounding rock that broke off on impact.
   D. They melted from the heat of the impact.

4. What theory of mass extinction was hypothesized for the Ordovician and Devonian eras?
   A. global cooling and glaciers
   B. global famine
   C. global disease
   D. low oxygen levels

5. Which is not a likely source of the Precambrian Earth's heat?
   A. radioactivity
   B. asteroid impact
   C. increased solar activity
   D. gravitational contraction

6. Which was not a source in the geologic record for the early presence of oxygen on Earth?
   A. red beds
   B. banded iron formations
   C. stromatolites
   D. meteorites

Use the illustration below to answer Questions 7 to 9.

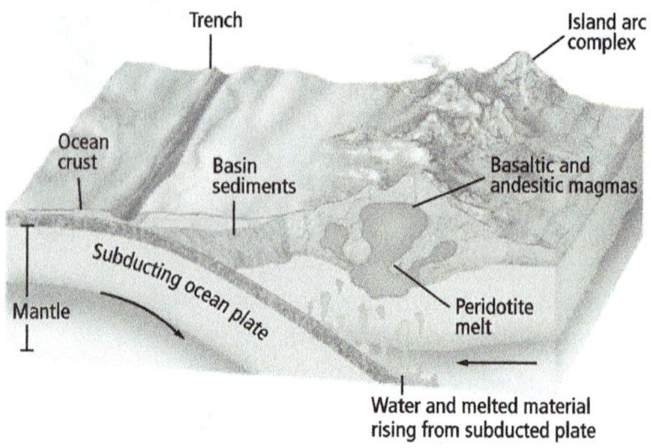

7. Which type of orogeny is shown?
   A. oceanic-oceanic convergence
   B. oceanic-continental convergence
   C. continental-continental convergence
   D. divergence

8. Which type of rock is peridotite?
   A. metamorphic
   B. igneous
   C. sedimentary
   D. volcanic

9. Where would this island arc complex most likely be located in the modern world?
   A. Antarctica
   B. Africa
   C. the Philippines
   D. North America

10. What do geologists use to help divide the history of Earth for rock study?
    A. fossils within the rocks
    B. sills and dikes
    C. fault lines occurring in the rock layers
    D. the color of rocks

SHORT ANSWER

Use the diagram below to answer Questions 11 to 13.

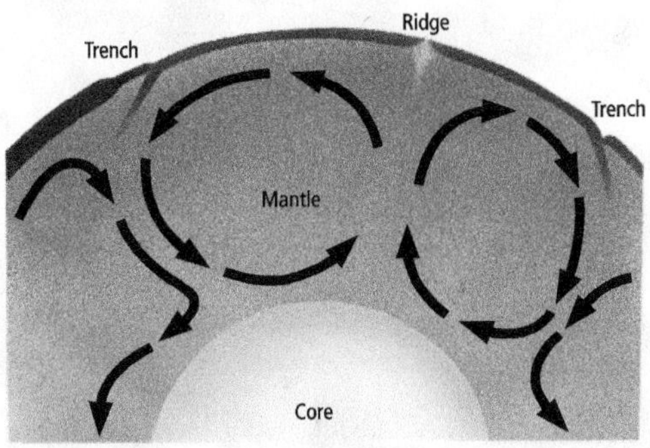

11. Describe what is being modeled in the diagram and how it affects Earth's plates.

12. How are these processes able to occur within Earth's solid mantle?

13. Why do these circulations not cause a greater amount of movement on Earth's surface?

14. Explain how James Hutton's work was linked to the principle of uniformitarianism.

15. Name two major events that occurred during Precambrian time. How much of geologic time is covered by the Precambrian?

16. Explain why hydrogen and helium, the most abundant elements in the universe, are not a significant part of Earth's atmosphere.

READING FOR COMPREHENSION

Asteriod Impact and Mass Extinction

A high-resolution map from NASA's Shuttle Radar Topography Mission has provided the most telling visible evidence to date of a 180-km-wide, 900-m-deep impact crater, which formed when Earth collided with a giant comet or asteroid 65 mya. The existence of the impact crater was first proposed in 1980. In the 1990s, satellite data and ground studies provided scientists with the evidence needed to possibly explain the ultimate demise of the dinosaurs and more than 70 percent of Earth's living species. The relatively obscure feature is all but hidden in the flat limestone plateau of Mexico's Yucatán Peninsula.

Scientists theorize three possible scenarios as to how the asteroid impact caused Earth's mass extinctions: massive quantities of dust blew into the atmosphere, blocking the Sun and stopping plant growth; sulfur released by the impact lead to global sulfuric-acid clouds that blocked the Sun and produced acid precipitation; and red-hot debris from the falling asteroid or comet triggered global wildfires.

Article obtained from: Dinosaur-killer asteroid crater imaged for first time. *National Geographic News* March 7, 2003. (Online resource accessed October 3, 2010.)

17. Which is not a possible explanation of the asteroid impact role in causing mass extinctions?
    A. Massive amounts of dust in the air blocked the Sun and killed off the plants the animals ate.
    B. The impact directly killed thousands of animals, thus limiting the reproduction rates.
    C. Sulfur produced acid rain, which killed the animals.
    D. Burning debris fell to Earth and caused global wildfires.

18. Why are the pictures of this crater taken from the NASA space shuttle so important to scientists studying Earth's history?

NEED EXTRA HELP?

If You Missed Question . . .	1	2	3	4	5	6	7	8	9	10	11	12	13	14	15	16
Review Section . . .	17.3	23.2	23.1	23.1	22.1	22.3	20.2	19.2	20.2	21.1	17.4	17.4	17.4	21.2	21.1	22.3

SHORT ANSWER

11. The circulating arrows moving through Earth's mantle model convection currents. These currents are thought to be responsible for the movement of Earth's lithospheric plates.
12. Convection currents occur in the part of the mantle known as the asthenosphere, which flows slowly like soft plastic or hot asphalt.
13. Convection currents can be thousands of kilometers across and only move a few centimeters per year. Thus, their movements, while significant over time, do not have noticeable effects on Earth's surface in the short term.
14. The principle of uniformitarianism states that geologic processes occurring today have been occurring since Earth was young. Hutton used this idea to learn about Earth's past.
15. Major Precambrian events included the formation of the oceans and atmosphere, the formation of the crust and continents, and the evolution of life. The Precambrian covers 90 percent of geologic time.
16. Hydrogen and helium have light masses, so they do not stay close to Earth for long. Earth's gravity is not strong enough to prevent the hydrogen and helium from escaping to space.

READING FOR COMPREHENSION

17. B
18. A hypothesis advances only with the accumulation of evidence. The discovery of an actual impact site is evidence that an asteroid impact occurred, lending credence to the hypothesis that an asteroid impact contributed to the extinction of the dinosaurs.

Resources and the Environment

Themes

Change Since humans appeared on Earth, they have used more and more of Earth's resources. Use of these resources has changed Earth.

Structures Fossil fuels are limited in quantity and as they diminish in amount, alternatives must be found to replace them. Developing new technologies might create a need for new Earth materials.

Geologic Time The limited time in which humans have been on Earth has marked the most widespread use of resources by one type of organism in the history of Earth.

Systems Identifying resources as renewable and nonrenewable is critical to the wise use of Earth's resources. In turn, the effects their use has on Earth's natural systems must be controlled through setting limits on when and how to use them and perhaps on how much to use.

Scientific Inquiry Using Earth resources led to environmental changes affecting life within those environments. New ideas about reclamation and use of the land resulted in new technologies for gathering and monitoring resources.

UNIT 7

Resources and the Environment

CHAPTERS

24 Earth Resources

25 Energy Resources

26 Human Impact on Resources

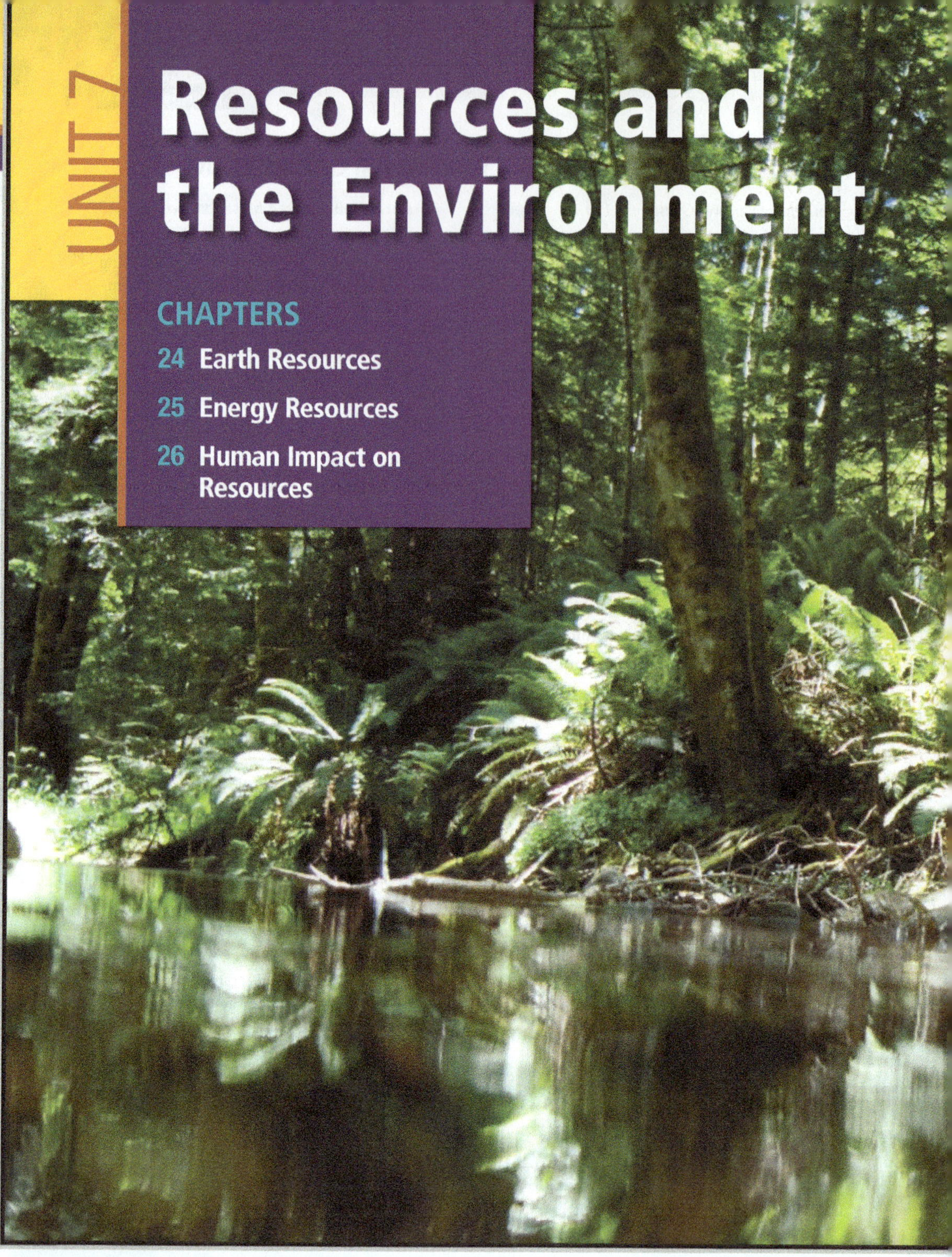

STEM Project

CAREERS IN EARTH SCIENCE **Environmental Technician**

This **environmental technician** is using GPS to collect and record site-specific data on streams. Environmental technicians help monitor the air, land, and water to maintain a clean environment for all living things.

Introduce the Unit

Preconceptions The photograph shows an environmental scientist monitoring the water quality of a stream. Ask whether any students have visited a river or a pond in which the water was polluted. If so, ask them to describe the cause of the pollution. Ask the students what the most likely sources of pollution to water resources are. What are the effects of water pollution on ecosystems? What can be done to clean up a body of water that has been contaminated with litter and chemical and biological waste?

Resource Distribution Display a world map that includes resources in its legend. Have students look at the map and locate the areas on Earth in which most of the coal, petroleum, natural gas, and peat reserves are found. Ask students to identify the countries in which most of these fossil fuels are consumed. Ask students whether the countries that consume fossil fuels also produce them. Students should recognize that there is a disparity between those countries with particular natural resources and those that need those resources to support their standards of living. In the last chapter of this unit, students will learn how humans impact their environments.

CHAPTER 24 Earth Resources

BIGIDEA **People and other organisms use Earth's resources for everyday living.**

ESSENTIAL QUESTIONS	NATIONAL STANDARDS	RESOURCES TO ASSESS MASTERY
SECTION 1 **Natural Resources** **1.** What are renewable and nonrenewable resources? **2.** What is a sustainable yield? **3.** How are resources unevenly distributed on Earth? 1 session 0.5 block	UCP.4; F.3, F.6	**Entry-Level Assessment** Focus Transparency 74, p. 678 **Progress Monitoring** Caption Question, pp. 678, 681 Reading Check, pp. 679, 680 Section Review, p. 681
SECTION 2 **Resources from Earth's Crust** **1.** Which materials from Earth's crust are considered natural resources? **2.** Why is the need to protect Earth's land surface as a resource important? **3.** How is the uneven distribution of resources worldwide explained? 1 session 0.5 block	UCP.4; B.2; F.3–5	**Entry-Level Assessment** Focus Transparency 75, p. 682 **Progress Monitoring** Reading Check, pp. 683, 684 Section Review, p. 686
SECTION 3 **Air Resources** **1.** How is the atmosphere a resource? **2.** How are the carbon and nitrogen cycles illustrated? **3.** What are natural sources of air pollution? 1 session 0.5 block	UCP.2, UCP.4; F.3–6	**Entry-Level Assessment** Focus Transparency 76, p. 687 **Progress Monitoring** Caption Question, pp. 687, 692 Reading Check, pp. 688, 690, 691 Section Review, p. 692
SECTION 4 **Water Resources** **1.** Why are the properties of water important for life on Earth? **2.** How is water distributed and used on Earth? **3.** In what ways can humans reduce the need for freshwater resources? 3 sessions 1.5 blocks	UCP.2, UCP.4; B.2; E.1–2; F.3–6	**Entry-Level Assessment** Focus Transparency 77, p. 693 **Progress Monitoring** Reading Check, p. 696 Section Review, p. 697 **Summative Assessment** Chapter Assessment, p. 701 Exam*View*® *Assessment Suite* CD-ROM

LEVELED RESOURCES	LAB MATERIALS
Science Notebook 24.1 OL **Chapter FAST FILE Resources:** Study Guide, p. 13 BL **Transparencies:** Section Focus Transparency 74 EL	LaunchLAB p. 676 / **10 min** classroom item
Science Notebook 24.2 OL **Chapter FAST FILE Resources:** Study Guide, p. 14 BL **Transparencies:** Section Focus Transparency 75 EL Teaching Transparency 73 OL EL	
Science Notebook 24.3 OL **Chapter FAST FILE Resources:** Study Guide, p. 15 BL **Lab Resources:** Laboratory Manual, p. 185 OL **Transparencies:** Section Focus Transparency 76 EL Teaching Transparency 74 OL EL	
Science Notebook 24.4 OL **Chapter FAST FILE Resources:** MiniLab Worksheet, p. 2 OL GeoLab Worksheet, p. 3 OL Study Guide, p. 17 BL **Lab Resources:** Laboratory Manual, p. 189 OL **Transparencies:** Section Focus Transparency 77 EL Teaching Transparency 75 OL EL	MiniLAB p. 695 / **25 min** clean baby food jars (6), water, distilled water, liquid soap, Earth science journal, pencil GeoLAB p. 699 / **60 min over 5 days** water usage table, calculator

LabManager™
Customize any lab with the LabManager™ CD-ROM.

ADDITIONAL RESOURCES

Technology:
- eTeacherEdition
- eStudentEdition
- Classroom Presentation Toolkit
- Exam*View*® *Assessment Suite* CD-ROM
- LabManager™ CD-ROM
- What's EARTH SCIENCE Got To Do With It? DVD
- Weather Classroom Video DVD

Assessment:
- Blueprints for Success
- Chapter Test (Scaffolded)
- Earth Science Standardized Test Practice

Lab Resources:
- Exploring Environmental Problems Laboratory Manual

BL Below Level OL On Level AL Advanced Learners EL English Learners COOP LEARN Cooperative Learning

CHAPTER 24

LaunchLAB

iLab Station | Rubric

What natural resources do you use in your classroom?

Process Skills observe and infer, classify, make and use tables, compare and contrast, think critically

Safety Precaution Approve lab safety forms before work begins.

Teaching Strategies
- Set out a display of common objects such as paper, an apple, popcorn, an empty jar, a filled balloon, a shirt, or a toy car.
- It makes little difference what objects are displayed, as long as some came from living things and some did not.

Procedure
1. Have students read and complete the lab safety form and follow the procedure below.
2. Obtain a **classroom item** from your teacher.
3. Working with a partner, determine all the different components of your classroom item.
4. Next, determine where each of the components originated and classify the origin as either living or nonliving.
5. Within the living or nonliving groups, classify each as being either easily replaced or not replaceable.

Analysis
1. **Compare and contrast** your results with those of several other groups. Answers will vary depending on the materials.
2. **Explain** How many items on your list were not replaceable? Why? Answers will vary depending on the materials.
3. **Determine** Are any of the items on either list recyclable? Explain. Answers will vary. Items that are recyclable include paper, plastic, and aluminum.
4. **Analyze** How could you make this product with more replaceable items? Answers will vary. Choosing materials such as paper, plastic, and aluminum could make a product more recyclable.

Assessment

Performance Have students assemble a new group of items and classify each as originating either from living things or non-living things. Place the separated items in a display case with a sign inviting students in other classes to add items.

LabManager™
Customize any lab with the LabManager™ CD-ROM.

CHAPTER 24
Earth Resources

BIGIDEA People and other organisms use Earth's resources for everyday living.

SECTIONS
1 Natural Resources
2 Resources from Earth's Crust
3 Air Resources
4 Water Resources

LaunchLAB

iLab Station

What natural resources do you use in your classroom?

The materials that you use every day in your classroom, such as your paper, pencils, pens, and textbooks, all originate from multiple sources. You already know that paper comes from trees, but what about the ink? Find out where other common classroom items originate in this lab.

FOLDABLES® Study Organizer

Renewable v. Nonrenewable Resources

Make a pocket book using the labels shown. Use it to organize your notes on the two main types of resources.

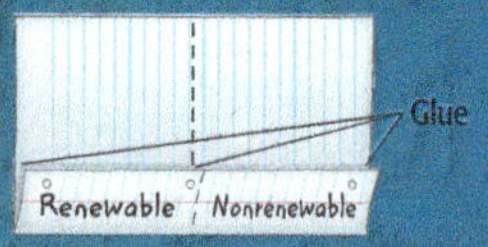

Introduce the BIG IDEA

Identify Resources Lead a discussion to determine what students already know about natural resources. Ask: Have you ever been invited to a party for a birthday, wedding, or graduation? Where did the tables and chairs come from? Where did the food and drinks come from? Responses will vary. Keep track of inaccurate responses or responses that give only people as answers. Ask questions until students identify natural resources involved in the manufacture of tables and chairs and in the production of food and drinks.

Teacher Content Support

Earth Resources Earth's resources include many items found in and around Earth's crust. Clay is an important resource in Earth's crust and is used in baseball fields. Trees, such as ash trees that produce baseball bats, are also an important resource. Freshwater is a resource used at a baseball stadium in a variety of ways, from watering the grass to being sold in concession stands. Steel for constructing baseball stadiums is made from iron ore, which comes from the mined minerals hematite and magnetite and also from recycled steel. Concrete for stadium construction is made from the sedimentary rock limestone, which is mined in quarries. Explain to students how the baseball field shown here contains several natural resources. Ask students to imagine the scene at the baseball field without natural resources.

SECTION 1

1 Focus

Focus Transparency

Before presenting the lesson, project **Section Focus Transparency 74** and have students answer the accompanying questions. EL

MAINIDEA

Recycling A discussion on previously worn items (hand-me-downs) can introduce the idea of using and reusing of resources. Ask students: Did you ever receive or give outgrown clothes, toys, or other items to a younger brother, sister, or friend? If you gave a small T-shirt to a sibling because you needed a bigger T-shirt, what resources does reusing the T-shirt save? water, fertilizer, and cotton seeds to grow cotton OL EL

2 Teach

Teacher Content Support

Natural Resources Earth supplies a wide variety of natural resources. All organisms on Earth, including humans, use resources provided by the environment. Living things use, change, and reuse many Earth resources. The word *resource* itself comes from the Old French word *resourdre,* meaning *to rise anew.* The natural-resource needs of human societies have changed over time. For example, obsidian and chert once were important resources because they were necessary for making tools and weapons. The resources used in technology today are different from those used in the past. Bauxite is an example of a natural resource that had little significance to humans in the past but is widely used in industrialized countries today.

SECTION 1

Natural Resources

MAINIDEA Resources are materials that organisms need; once used, some resources can be replaced, whereas others cannot.

Essential Questions
- What are renewable and nonrenewable resources?
- What is a sustainable yield?
- How are resources unevenly distributed on Earth?

Review Vocabulary

biosphere: all of Earth's organisms and the environment in which they live

New Vocabulary

natural resource
renewable resource
sustainable yield
nonrenewable resource

EARTH SCIENCE 4 YOU Did you eat an apple or a banana for breakfast this morning? Every day, you eat food and drink water because these resources are necessary for you to survive.

Resources

You and every other living thing on Earth must have certain resources to grow, develop, maintain life processes, and reproduce. The resources that Earth provides are known as **natural resources.** Natural resources include Earth's organisms, nutrients, rocks, water, and minerals. Natural resources might come from the soil, air, water, or deep in Earth's crust. All items that you use every day, like those shown in **Figure 1,** come from natural resources.

Renewable resources If you cut down a tree, you can replace that tree by planting a seedling. A tree is an example of a **renewable resource,** which is a natural resource that can be replaced by nature in a short period of time. Renewable resources include fresh air, fresh surface water in lakes, rivers, and streams, and most groundwater. When used properly, fertile soil is a renewable resource. However, if soil is exposed to wind and water erosion, it can be eroded. Renewable resources also include all living things and elements that cycle through Earth's systems, such as nitrogen, carbon, and phosphorus. Resources that exist in an inexhaustible supply, such as solar energy, are also renewable resources.

Figure 1 Most of the items in this photo originated as natural resources.
Identify *three resources represented in this photo.*

Caption Question **Fig. 1** wood, cotton, clay

DIFFERENTIATED INSTRUCTION

English Learners Have students use dictionaries to look up each of the definitions of the terms *natural resources* and *renewable resources.* Have students list each term in their Earth science journals to compare and contrast with nonrenewable resources. Have students write a sentence next to each term explaining why it is an appropriate term for describing a resource.

■ **Figure 2** Bamboo can be grown as a sustainable yield crop because it grows fast and needs no replanting. Bamboo can be used to produce a variety of items including flooring, cooking utensils, and clothing.

Sustainable yield of organisms Humans can use natural resources responsibly by replacing resources as they are used. The replacement of renewable resources at the same rate at which they are consumed results in a **sustainable yield.**

Organisms in the biosphere are important renewable resources. Plants and animals reproduce; therefore, as long as some mature individuals of a species survive, they can be replaced. Crops can be planted every spring and harvested every fall from the same land as long as the Sun shines, the rain falls, and the required nutrients are provided by organic matter or fertilizers. Animals that are raised for food, such as chickens and cattle, can also be replaced in short periods of time. Forests that are cut down for the production of paper products can be replanted and ready for harvest again in 10 to 20 years. Trees that are cut down for timber can be replaced after a period of up to 60 years.

Bamboo, shown in **Figure 2,** is one of Earth's most versatile renewable resources. Used by more than half the world's population for food, shelter, fuel, and clothing, bamboo is one of the world's fastest-growing plants. Because bamboo is a grass, it can be harvested without replanting. Bamboo grows without fertilizers or pesticides and is harvested in three to five years.

☑ READING CHECK **Identify** an example of sustainable yields.

Sunlight Some of Earth's renewable resources are not provided by Earth. The Sun provides an inexhaustible source of energy for all processes on Earth. Sunlight is considered a renewable resource because it will be available for at least the next five billion years.

VOCABULARY

ACADEMIC VOCABULARY

Mature
having completed natural growth and development
An elephant is considered mature 13 to 20 years after birth.

FOLDABLES®
Incorporate information from this section into your Foldable.

Interpret the Photo

Uses of Bamboo Bamboo is an important renewable resource. Have students study **Figure 2** and answer the questions that follow. What are three uses of bamboo? flooring, clothing, and utensils What is one benefit of bamboo over timber? Bamboo grows faster than timber. What might be a potential drawback of bamboo? It grows so fast that it can be considered invasive. OL EL

Environmental Connection

Sustainability Write the following quotation on the board: Sustainability is an economic state where the demands placed upon the environment by people and commerce can be met without reducing the capacity of the environment to provide for future generations. It can also be expressed in the simple terms of an economic golden rule for the restorative economy: Leave the world better than you found it, take no more than you need, try not to harm life or the environment, make amends if you do. (Paul Hawken, *The Ecology of Commerce*) Ask students to explain what this statement means in terms of their daily lives. AL

☑ READING CHECK Crops can be replanted so they result in a sustainable yield. Food animals, such as chickens and cattle, also can be replaced over a short period of time and result in a sustainable yield. Trees that are cut down for production can be replaced with other trees that are ready for harvest in 20 to 30 years.

FOLDABLES® Rubric

DIFFERENTIATED INSTRUCTION

Advanced Learners Students with strong science backgrounds might be interested in examining at least two predictions about the length of time that the Sun will be able to provide energy for Earth. Students should specify how calculations were derived and the approximation of length of time for the Sun to provide Earth with energy.

Project

Local Natural Resources Have a brainstorming session with students to determine what natural resources are available in their community. Have students work in small groups to identify where they might find out what resources are available. Students may interview community members and town officials, or conduct research at the local library or town or county offices. Looking in the yellow pages of local telephone books also might offer some clues.

Warn students that many local resources often are taken for granted, and that they should be careful not to overlook some of the more obvious resources in their community and its surroundings. Once each group has accumulated a list of local resources, compile a master list to be displayed in the classroom. Students may add to the list as they become aware of additional resources. OL COOP LEARN

CAREERS IN EARTH SCIENCE

WebQuest

☑ READING CHECK Gold, fossil fuels, and gemstones are nonrenewable resources because they exist in a fixed amount and can be replaced only by Earth processes that take hundreds of millions of years.

CAREERS IN EARTH SCIENCE

Materials Engineer Materials engineers work with metals, stone, plastics, and combinations of materials called composites to create materials used in everyday products, including computers, television screens, golf clubs, and snowboards.

WebQuest

Nonrenewable resources Many homes have copper pipes that transport water to the faucets. The price of copper fluctuates daily, but has steadily increased over the past 15 years. Copper is expensive because there are a limited number of copper mines, and demand continues to increase. When all the resources in the operating mines have been exhausted, no more copper will be mined unless new sources can be located. Copper is an example of a **nonrenewable resource**—a resource that exists in a fixed amount in various places in Earth's crust and can be replaced only by geological, physical, and chemical processes that take millions of years. Resources such as fossil fuels, diamonds and other gemstones, and elements such as gold, copper, and silver are therefore considered to be nonrenewable. **Figure 3** shows some materials you use every day and the nonrenewable resources used to make them.

☑ READING CHECK **Explain** why gold, fossil fuels, and gemstones are nonrenewable resources.

Distribution of Resources

You have probably noticed that natural resources are not distributed evenly on Earth. Ohio, Pennsylvania, and West Virginia have an abundance of coal. California is known for its gold deposits. Georgia has large stands of trees used for paper and lumber. Some regions of the world, such as the United States, have an abundance of different types of natural resources. Other areas might have limited types of resources, but in abundant supply. For example, Saudi Arabia and Kuwait, in the Middle East, have more petroleum reserves than other areas of the world.

Figure 3 Nonrenewable resources are all around us. Aluminum from bauxite is used to make pots and pans, copper sulfides are used in copper plumbing, calcium sulfate is used to make drywall for houses and buildings, and iron from hematite is used to make appliances such as wood stoves.

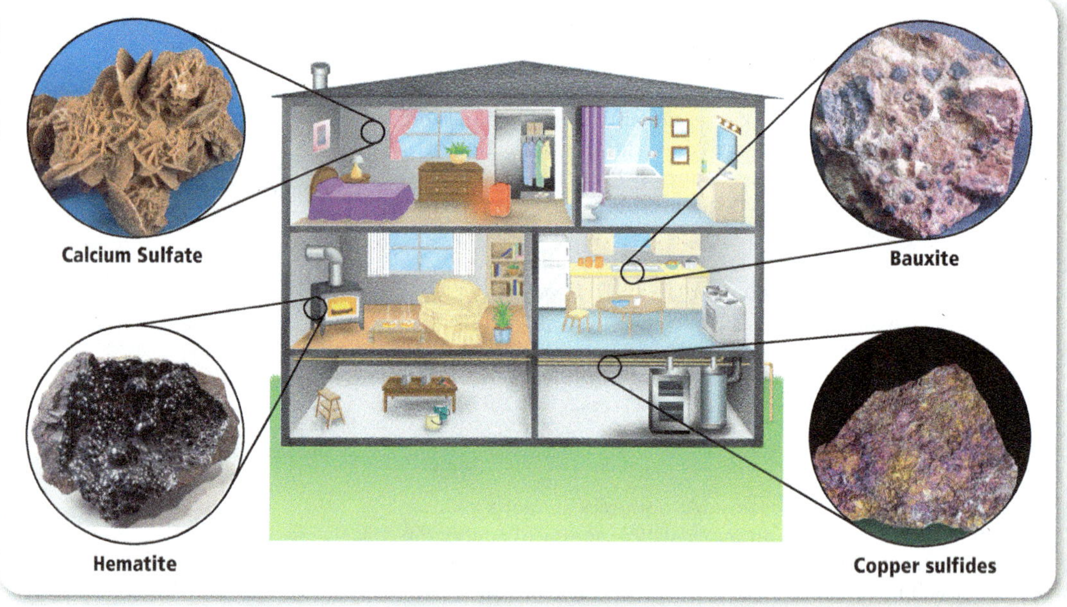

EARTH SCIENCE JOURNAL

Alternative Resources To familiarize students with nonrenewable resources and their uses, have students research three nonrenewable resources and their uses as well as at least one alternative to the nonrenewable resource. Students should create a data table in their Earth science journals listing the parent material and its uses with an extra column for the alternatives. OL

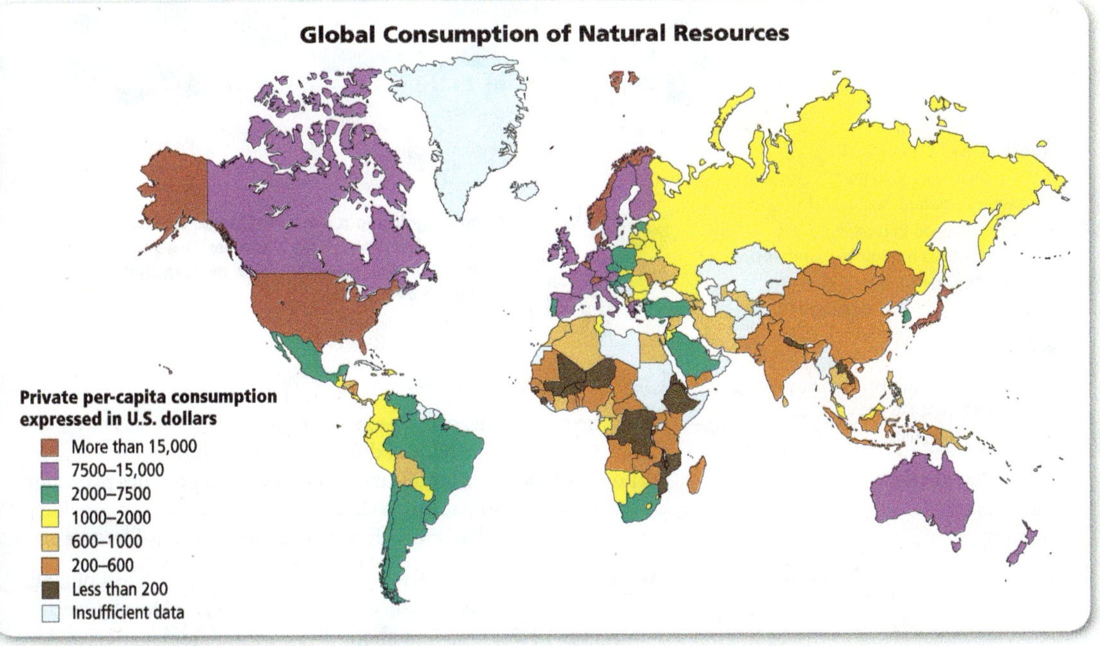

Figure 4 Across the globe, consumption of natural resources varies from country to country. Notice the average person in the United States consumes more than $15,000 a year in natural resources.
Determine *How does this compare with Canada or India?*

Consumption of resources Billions of people throughout the world use natural resources every day. However, natural resources are not evenly distributed on Earth. Not only are natural resources distributed unevenly, they are likewise consumed unevenly. Although people in the United States make up only 6 percent of the world's population, they consume approximately 30 percent of Earth's mineral and energy resources each year, as shown in **Figure 4.** As a result, even more energy and resources are required to transport many resources from their point of origin to the places where they are being consumed.

SECTION 1 REVIEW

Section Self-Check

Section Summary

- Natural resources are the resources that Earth provides, including organisms, nutrients, rocks, minerals, air, and water.
- Renewable resources can be replaced within a short period of time.
- Nonrenewable resources exist in a fixed amount and take millions of years to replace.

Understand Main Ideas

1. MAINIDEA **Explain** how organisms, including humans, use natural resources.
2. **Explain** why costs of copper and other materials continue to increase.
3. **Categorize** the following as a renewable or nonrenewable resource: trees, aluminum, cotton, gemstones, and corn. Which are produced by sustainable yield?

Think Critically

4. **Propose** why consumption of natural resources is higher in the United States. Why is it important to be aware of this?

MATH IN Earth Science

5. Aluminum production from bauxite ore costs $2000 per ton, whereas aluminum recycling costs $800 per ton. What is the percent saved by recycling?

Caption Question Fig. 4 Residents of Canada on average spend between $7500 and $15,000, and residents of India spend between $200 and $600 per year.

3 Assess

Check for Understanding

Compare and Contrast Ask students how the words *replaceable* and *renewable* relate to one another. If something is renewable, it means that it can be replaced. The word replaceable also can refer to using something in place of the original material that would fulfill the same need.

Reteach

Communicate Have students each make a list of the new vocabulary words that they find in the text of this section. Ask them to list the words and their definitions in their Earth science journals. Have students each write one new word on an index card. On the back of each card, have students scramble the letters in the word. Then have students use the scrambled side of each card to quiz a fellow student. The student being quizzed should first identify the scrambled word and then define it.

Assessment

Knowledge Ask students to observe a picture cut out of a magazine and identify which objects in the picture are made of renewable resources and which are made from nonrenewable resources.

SECTION 1 REVIEW

1. Organisms use natural resources for everyday survival to eat, drink, and for shelter. Humans also use natural resources for heat, transportation, convenience, and recreation.
2. The costs of nonrenewable resources, such as copper, continue to rise because there is a finite supply that is decreasing as human use is increasing.
3. Renewable: trees, cotton, corn. Nonrenewable: aluminum and gemstones. Corn and cotton can be produced by sustainable yield.
4. Answers will vary. Because our per capita rate of consumption is so much higher, we can have a greater impact, which can be positive or negative depending on the consumption choices that we make.
5. There is a 60% savings in cost by recycling.

SECTION 2

1 Focus

Focus Transparency

Before presenting the lesson, project **Section Focus Transparency** 75 and have students answer the accompanying questions. EL

MAINIDEA

Source of Food Ask students: Where does your food come from? Responses will vary, but see how far students can trace their food to the source in a planted field.

2 Teach

Teacher Content Support

Land as a Resource Land itself is a valuable resource. It provides space for humans and other organisms to live. Land areas also are important as resources for cropland, forests, rangelands, parks, and wilderness areas. The economic value of land varies from one location to another, and the economic value of any one parcel of land might change over time. However, the intrinsic value of land may be considered priceless because there is only a given amount of land on Earth.

SECTION 2

Essential Questions

- Which materials from Earth's crust are considered natural resources?
- Why is the need to protect Earth's land surface as a resource important?
- How is the uneven distribution of resources worldwide explained?

Review Vocabulary

igneous rock: intrusive or extrusive rock formed from the cooling and crystallization of magma

New Vocabulary

desertification
aggregate
bedrock
ore
tailings

Resources from Earth's Crust

MAINIDEA Earth's crust provides a wide variety of resources to grow food, supply building materials, and provide metals and minerals.

EARTH SCIENCE 4 YOU Imagine going to a store where you can buy food, clothes, electronics, and whatever else you need. Earth's crust is like a store—it supplies most materials needed and used by humans.

Land Resources

In the springtime, many people visit garden centers and buy sand, mulch, peat moss, topsoil, and different kinds of rocks for landscaping purposes. These items are all land resources. Land provides places for humans and other organisms to live and interact. Land also provides spaces for the growth of crops, forests, grasslands, and wilderness areas.

Publicly managed land More than 828 million acres of land in the United States is managed by federal, state, and local governments. **Figure 5** shows that about 28 percent of land is managed by federal government agencies. These land areas are managed to support recreational uses, grazing, and mineral and energy resources. National forests are managed for sustainable yield and to provide recreational spaces. Wilderness areas are places that are maintained in their natural state and protected from development.

Figure 5 Many government agencies, including the Bureau of Land Management, the USDA Forest Service, and the Fish and Wildlife Service help manage land across the United States.

Federally Managed Land in the United States

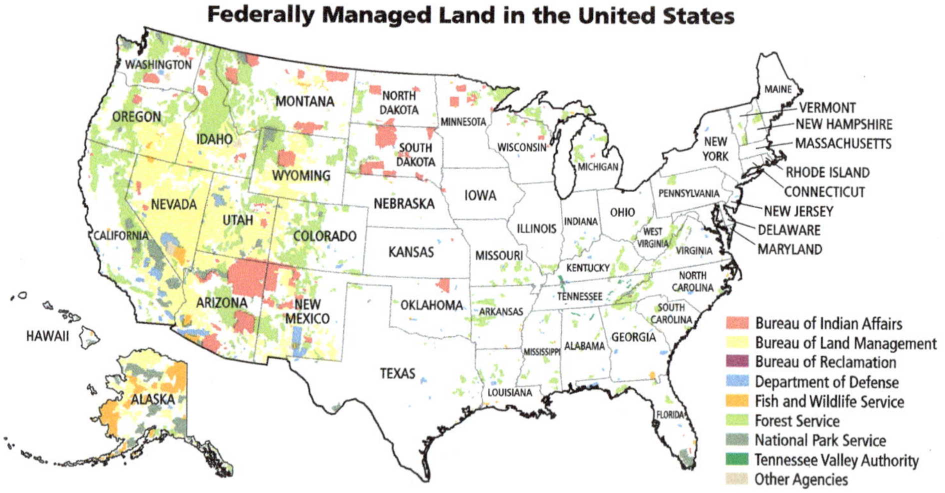

Demonstration

Identify Land Use posterboards in front of the class to display pictures from magazines. The pictures should show some type of land use by humans. Pictures could portray factories, mines, shopping centers, parks, hiking trails, or houses. Ask students to identify the type of the land use shown in each of the pictures. BL EL

National parks The national park system in the United States preserves scenic and unique natural landscapes, preserves and interprets the country's historic and cultural heritage, protects wildlife habitats and wilderness areas, and provides areas for various types of recreation. About 50 percent of the land in the national park system is designated as wilderness.

National wildlife refuges National wildlife refuges provide protection of habitats and breeding areas for wildlife, and some provide protection for endangered species. Other uses of the land in wildlife refuges, such as fishing, trapping, farming, and logging, are permitted as long as they are compatible with the purpose of the refuge.

VOCABULARY
ACADEMIC VOCABULARY
Compatible
capable of existing or performing in harmonious, agreeable, or congenial combination with another or others
My sister and her roommate are not compatible; they argue all the time.

Soil You have learned how soil forms. In some parts of Earth's crust, it can take up to 1000 years to form just a few centimeters of topsoil, yet it can be lost in a matter of minutes as a result of erosion by wind or water. Plowing and leaving the ground without plant cover can increase topsoil loss.

The loss of topsoil makes soil less fertile and less able to hold water, which results in loss of crops. Today, topsoil is eroding more quickly than it forms on about one-third of Earth's croplands. Each decade, Earth loses about 7 percent of its topsoil, yet the eroded croplands must feed an ever-increasing human population.

In arid and semiarid areas of the world, the loss of topsoil leads to **desertification,** which is the process whereby productive land becomes desert. Desertification can occur when too many grazing animals are kept on arid lands, or when trees and shrubs are cut down for use as fuel in areas with few energy resources.

Desertification is a growing problem in Africa, as shown in **Figure 6.** It is also a growing problem in the Middle East, in the western half of the United States, and in Australia. Desertification can be prevented by reducing overgrazing and by planting trees and shrubs to anchor soil and retain water.

READING CHECK **Describe** activities that can lead to erosion of topsoil.

Figure 6 Desertification is a growing concern in many areas. Clearcutting and over-farming have led some parts of Africa to be considered in great risk of desertification.

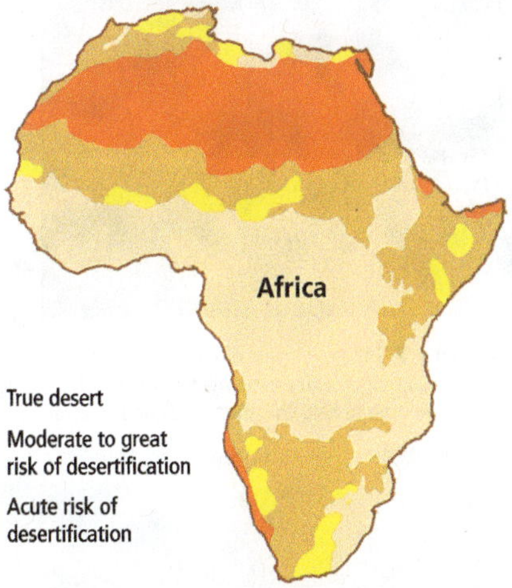

Model

Desertification Have students work in small groups to develop scenes of conditions that lead to desertification using shadow-box models. On top of each shadow box, the following question should be posed: How does this scene show a condition that has led to desertification? Scenes could portray overgrazing or heavy machinery use. Give students the option to divide the shadow box in half, with one side showing overgrazing and the other side showing what the area would look like after corrective measures have been implemented.

Environmental Connection

National Parks National parks are one of the ways that Earth's resources are protected. The national park system in the United States was created by President Woodrow Wilson in 1916. It was the first national park system of its kind in the world and has inspired other nations to create similar systems to protect their natural resources. In the United States, the national park system now includes 392 areas that were visited by over 285 million people in 2009.

READING CHECK Plowing and consequently leaving the ground without plant cover can lead to erosion of topsoil. Grazing too many animals on arid lands also can lead to erosion of topsoil. Both plowing and grazing affect the land much more if accompanied by the removal of trees and plant cover.

DIFFERENTIATED INSTRUCTION

Hearing Impaired Pair students with hearing difficulties with students of normal hearing ability who are willing to share their class notes. Allow students with hearing impairments to copy the notes of their partners as needed, and allow these students to use the notes during class to answer questions or clarify directions. OL

EARTH SCIENCE JOURNAL

Write a Letter Have students write letters in their Earth science journals from the perspective of a developer who wishes to convince local town officials to allow the building of a major-league ice-hockey rink in an area that is currently a local park. Ask students to include how the developer will preserve the resources in the park. OL

Rubric

Tie to Previous Knowledge

Glacial Features Remind students about the formation of glacial features. In this chapter, it is important for students to recall that glacial deposits in moraines, eskers, and kames may be associated with sediment being carried as a glacier moves and melts. The longer sediments are carried by the melting water, the greater the opportunity for the material to be sorted by size before it settles. This is how gravel and sand pits form. Sediments that were deposited directly by the glacial ice and were not carried by water are usually unsorted. OL

Apply Earth Science

Aggregates The two main sources of natural aggregate are sand and gravel and crushed stone. Sand and gravel are naturally occurring, as is crushed stone, which is obtained from bedrock. Recycled concrete, recycled asphalt, and iron-blast-furnace slag are also used for aggregate, but they are not naturally occurring. Approximately 50 percent more aggregate from crushed stone is used in the United States than from sand and gravel. OL

Enrichment

Ores An ore's concentration factor is equal to the concentration of the ore deposit divided by the average concentration of the ore in Earth's crust. The higher the concentration factor, the richer the deposit. In general, the minimum concentration factor needed to be considered for profitable mining is inversely proportional to the average concentration found in Earth's crust. Exceptions to this general rule exist; for example, gold is so valuable that even small amounts might be worth mining. AL

READING CHECK An aggregate is a mixture of gravel, sand, and crushed stone.

Figure 7 Different layers of Earth's surface have value as resources. Topsoil provides nutrients for crop production, aggregate can be used to help construct roads and sidewalks.

Aggregates

Have you ever observed the construction of a highway? You might have seen workers place layers of materials on the ground before they began to build the highway surface. In some instances, the materials used for this first layer come from **aggregate,** which is sand and gravel and crushed stone that can naturally accumulate on or near Earth's surface.

You have learned how Earth processes transport materials. Some aggregates are transported by water and are found on floodplains in river valleys and in alluvial fans in mountainous areas. Other aggregates were deposited by glacial activity in moraines, eskers, kames, and outwash plains. Aggregates used in construction are often mixed with cement, lime, or other materials to form concrete, mortar, or asphalt.

READING CHECK **Define** aggregate.

Bedrock

Recall that underneath topsoil is a layer of soil consisting of inorganic matter, including weathered rock, sand, silt, clay, and gravel, as shown in **Figure 7.** This deeper soil layer lies on a base of unweathered parent rock called bedrock. **Bedrock** is solid rock, and it can consist of limestone, granite, marble, or other rocks that can be mined in quarries. Slabs of bedrock are often cut from quarry faces. Large pieces of bedrock are used in the construction of buildings, monuments, flooring, countertops and fireplaces. Bedrock is also crushed for use as stone aggregate.

Ores

An **ore** is a natural resource that can be mined for a profit; that is, it can be mined as long as its value on the market is greater than the cost of its extraction. For example, the mineral hematite is an iron ore because it contains 70 percent iron by weight. Other minerals such as limonite also contain iron, but they are not considered ores because the percentage of iron contained in them is too low to make extraction profitable. Ores can be classified by the manner in which they formed. Some ores are associated with igneous rocks, and other ores are formed from processes that occur at Earth's surface.

ACROSS THE CURRICULUM

Art Ores have been a source for paint pigments and other ingredients in paint for hundreds of years. The following colors can be made by the following elements, often found in ores.

Red—copper, gold, selenium
Yellow—iron, selenium
Green—iron, chromium, copper
Blue—copper, cobalt
Purple—manganese, nickel
White—zinc, tin
BL

Settling of crystals Iron, chromium, and platinum are examples of metals that are extracted from ores associated with igneous rocks. Chromium and platinum come from ores that form when minerals crystallize and settle to the bottom of a cooling body of magma. Chromite ore deposits are often found near the bases of igneous intrusions. One of the largest deposits of chromite is found in the Bushveldt Complex in South Africa.

Hydrothermal fluids The most important sources of metallic ore deposits are hydrothermal fluids. Hot water and other fluids might be part of the magma that is injected into surrounding rock during the last stages of magma crystallization. Because atoms of metals such as copper and gold do not fit into the crystals of minerals during the cooling process, they become concentrated in the remaining magma. Eventually, a solution rich in metals and silica moves into the surrounding rocks to create ore deposits known as hydrothermal veins, shown in **Figure 8.** Hydrothermal veins commonly form along faults and joints in rock.

Chemical precipitation Manganese and iron ores most commonly originate in layers formed through chemical precipitation. Iron ores in sedimentary rocks are often found in bands made up of alternating layers of iron-bearing minerals and chert shown in **Figure 8.** The origin of these ores, called banded iron formations, is not fully understood. Scientists think that banded iron formations resulted from an increase in atmospheric oxygen during the Precambrian.

Placer deposits Some sediments, such as grains of gold and silver, are more dense than other sediments. When stream velocity decreases, as, for example, when a stream flows around a bend, heavy sediments are sometimes dropped by the water and deposited in bars of sand and gravel. Sand and gravel bars that contain heavier sediments, such as gold nuggets, gold dust, diamonds, platinum, gemstones, and rounded pebbles of tin and titanium oxides, are known as placer deposits. Some of the gold found during the Gold Rush in California during the late 1840s was located in placer deposits.

Figure 8 The chromite bands in the Bushveldt Complex are up to 0.5 m thick. Ores are also found in hydrothermal veins, banded formations, and placer deposits.

Demonstration

Precipitate Formation A chemical precipitate can result from a chemical reaction. To demonstrate a chemical precipitate, place 30 g of a laundry detergent in 250 mL of warm water in a glass beaker, and stir. In another beaker, place 50 mL of warm water and 30 g of Epsom salts, and stir. Add one or two drops of food coloring to the Epsom solution. Fill a dropper with the Epsom solution. Slowly squeeze the dropper into the detergent solution. Have students watch the precipitate form. EL

Identify Misconceptions

Many students think humans have existed on Earth and used Earth's resources for a much longer time than they have. BL

Uncover the Misconception

Ask students the following question: Compared to the length of time since Earth was formed, how long have humans had an impact on Earth resources? Ask students to compare the history of Earth to a 24-hour day. Ask them to identify when, over the 24-hour period, humans began to have an impact on Earth resources. Answers will vary but the correct answer is just over one minute.

Demonstrate the Concept

Draw a picture of a clock on the board. Explain to students that Earth is theorized to have formed 4.6 billion years ago. This period of time represents one 24-hour day, or twice around a 12-hour clock, for the purposes of this analogy. Tell students the fossil record shows that the earliest humans with tools can be traced back to almost two million years ago, which is just over one minute before midnight of the 24-hour day. Any concern about resource use by humans has occurred during this time period.

Assess New Knowledge

Ask students which of the following statements are true: Many of the resources of concern being depleted have been used only during the past century. true Some Earth resources are required for the existence of life on Earth. true Human ingenuity can replace all resources required by life on Earth. false The resources used by humans have changed over time. true

3 Assess

Check for Understanding

Explain Ask students to explain how chemistry, biology, and math are related to the study of land resources. Many land resources were formed through chemical processes, such as chemical precipitation. Some of the processes affecting land resources involve interactions with living things, such as the involvement of animals in desertification. Math is applied when designing structures best suited for a particular environment.

Reteach

Outline Have students outline this section of the chapter. Then have students exchange outlines with one another and fill in portions that might have been overlooked.

Assessment

Performance Have students collect information on one particular land resource and place all the information they find in a folder to be kept in the classroom as a class resource. Encourage students to include photographs, drawings, and even samples if possible. Videos and audio recordings can be collected to add to the classroom resources.

Figure 9 Waste rock, such as this tailings pile in New Mexico, is discarded after minerals are extracted.

Effects of Mining

Although many of the resources that you have learned about in this section can be extracted with little impact on the surrounding environment, the extraction of others can have lasting impacts. Mines that are used to remove materials from the ground surface destroy the original ground contours. Open-pit mines can leave behind waste rock, shown in **Figure 9,** that can weather over time. The extraction of mineral ores often involves grinding parent rock to separate the ore. The material left after the ore is extracted, called **tailings,** might release toxic elements such as mercury and arsenic into the groundwater or surface water. These materials can form acids as they weather and pollute the environment.

Mining processing methods might also include harmful chemicals that could be released into the surrounding environment. In addition to causing environmental problems, mining itself is a dangerous activity. In fact, the National Safety Council has identified mining as the most dangerous occupation in the United States: it has one of the highest yearly death rates of all occupations.

SECTION 2 REVIEW

Section Self-Check

Section Summary

- Loss of topsoil can lead to desertification.
- Aggregates, composed of sand, gravel, and crushed stone, can be found in glacial deposits.
- An ore is a resource that can be mined at a profit. Ores can be associated with igneous rocks or formed by processes on Earth's surface.

Understand Main Ideas

1. MAINIDEA **Describe** three natural resources derived from Earth's crust.
2. **Explain** why topsoil loss is considered a worldwide problem.
3. **Identify** three reasons it is important to protect Earth's land resources.
4. **Explain** the relationship between ore and tailings.
5. **Determine** where placer materials might have originated.

Think Critically

6. **Predict** what would happen if a land resource, such as aluminum, was depleted.

WRITING IN Earth Science

7. Create a three-fold pamphlet explaining the purposes and use of national parks and National Wildlife Refuge lands.

SECTION 2 REVIEW

1. Resources derived from Earth's crust include topsoil, which is needed for vegetative growth; aggregates, used for construction of roads; bedrock, or solid rock used for buildings and monuments; and ores, including iron, chromium, and platinum.
2. The loss of topsoil makes soil less fertile and less able to hold water, which leads to poorer crops. As the human population grows worldwide, more crops will be needed.
3. Answers will vary, but might include: to protect animal populations, to make sure there are resources for future use, and to prevent desertification.
4. Tailings can be left behind when ore is extracted. Tailings can be hazardous to the environment.
5. Placer materials might have originated far upstream from where they are deposited.
6. Answers will vary, but might include that a substitute resource might be found, or might indicate that it would cause some sort of hardship for people who previously relied on the resource.
7. Answers will vary, but should include that national parks and wildlife refuges preserve and protect the land.

Rubric

SECTION 3

Essential Questions

- How is the atmosphere a resource?
- How are the carbon and nitrogen cycles illustrated?
- What are natural sources of air pollution?

Review Vocabulary

photosynthesis: a process used by certain organisms to make food using energy from the Sun and carbon dioxide from the air

New Vocabulary

nitrogen-fixing bacteria
pollutant

Air Resources

MAINIDEA The atmosphere contains gases required for life on Earth.

EARTH SCIENCE 4 YOU Fish and other aquatic organisms have gills, which are specialized structures used to extract dissolved oxygen from the water. Humans, however, need to breathe air to get the oxygen their cells need. Scuba divers carry tanks with compressed air when they swim under water.

Origin of Oxygen

Most organisms on Earth require oxygen to maintain their life processes, but oxygen has not always been a part of Earth's atmosphere. As you have learned, scientists think that 4.6 to 4.5 bya Earth's atmosphere was similar to the mixture of gases released by erupting volcanoes. These gases included carbon dioxide, nitrogen, and water vapor. As Earth's crust cooled and became more solid, rains washed most of the carbon dioxide out of the atmosphere and into the oceans. Early life-forms in the seas used carbon dioxide during photosynthesis and released oxygen. Over time, oxygen in the atmosphere built up to levels that allowed the evolution of more complex organisms that required oxygen for life processes, as shown in **Figure 10.**

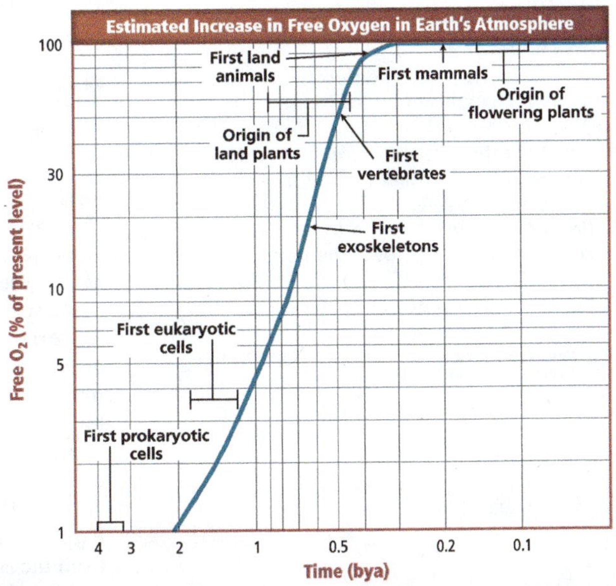

■ **Figure 10** Scientists think that prokaryotes first appeared about 4 bya. It was not until 2 bya that eukaryotes appeared on Earth. Notice the difference in oxygen (O_2) gas levels between when prokaryotes and eukaryotes appeared.
Determine *When did land plants first appear?*

Demonstration

Observe Air Quality Cut a piece of masking tape approximately 20 cm long and tack it high on a wall, sticky side out, for one week. At the beginning of this section, take the tape down and have students observe what is stuck to it. Have them compare the strip of tape to a freshly cut piece of tape. Have students list what they observed on the tape in their Earth science journals and predict how each item made its way to the tape. Students likely will indicate that the materials on the tape were carried by air. BL EL

SECTION 3

1 Focus

Focus Transparency

Before presenting the lesson, project **Section Focus Transparency 76** and have students answer the accompanying questions. EL

MAINIDEA

Resources for Life Lead a discussion about the items on which humans depend to illustrate how humans tend to depend on far more items than what is necessary for survival. Ask students: What items are essential to your life? Answers will vary. What could you not live without? air, water, shelter, food

2 Teach

Teacher Content Support

Air Resources All living things depend on gases in the atmosphere. Most organisms depend on atmospheric oxygen for cellular respiration—the process in which a cell uses oxygen to break down carbohydrates to release energy needed for cell maintenance, growth, and reproduction. The waste products of cellular respiration are water and carbon dioxide. All photosynthetic organisms, including green plants, cyanobacteria, and some protists, require carbon dioxide for photosynthesis. The products of photosynthesis are oxygen and water.

■ **Caption Question Fig. 10** Land plants appeared 0.5 to 0.9 bya.

Data Analysis LAB

About the Lab

- Ask students to locate several tropical rain forests on a world map. Then ask them why these forests are important resources. Share with students that the loss of the rain forests is of concern, in part because 25 percent of the medicines used in the United States to treat cancer, mental illness, and coronary heart disease come from rain forests.
- See also Harris, R. 2006. Can't log the forest for the trees?, *American Scientist* 94 (2): 120–121.

About the Lab

1. around 75,000 km² from 2002–2004
2. 1995

Think Critically

3. 1993–2001 = 3%/yr; 2001–2009 = −7%/yr
4. Despite an increase in deforestation between 2001–2002 and 2003–2004, the period from 2001–2009 had a lower average yearly deforestation rate than the period from 1993–2001.
5. Sample answer: If the deforestation rate continues to decline, the Amazon Basin may show signs of recovery. However, if the deforestation rate increases and the areas are not reforested, eventually, the Amazon Basin will be completely cleared of forest.
6. Sample answer: All plants play an integral role in the carbon cycle. If the Amazon is completely deforested, there will be fewer plants to break down the carbon dioxide that humans and animals produce through natural and artificial processes.

Data Analysis LAB

Based on Real Data*

Interpret Graphs

What is the rate of deforestation in the Amazon? Many experts are concerned about the loss of the forest cover in tropical rain forests worldwide. In the Amazon River Basin, scientists estimate that one hectare (ha, about 2.47 acres) of forest is cut down each hour.

Data and Observations

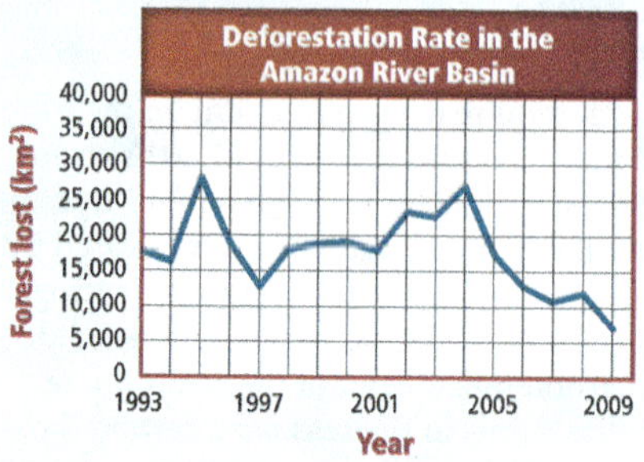

Analysis

1. How many square kilometers of the Amazon River Basin were deforested between 2002 and 2004?
2. According to the graph, what year was the peak in deforestation of the Amazon River Basin?

Think Critically

3. **Calculate** the rates of deforestation (or percent change yearly) for the periods 1993 to 2001 and 2001 to 2009.
4. **Compare** the rates of deforestation for the periods from 1993 to 2001 and 2001 to 2009.
5. **Predict** what will happen to the Amazon Rain Forest over the next 30 years.
6. **Explain** how loss of rain forest could affect the carbon cycle.

*Data obtained from: Estimated Annual Deforestation Rate After 1988. *The National Institute for Space Sciences.*

Cycles of Matter

The law of conservation of mass states that the amount of matter on Earth never changes. Earth's elements cycle among organisms and the nonliving environment. You have already learned about how water cycles on Earth. Earth's atmosphere plays a significant role in other cycles, such as the nitrogen and carbon cycles.

Earth's cycles are in delicate balance. When fossil fuels burn, the carbon that was stored in them for millions of years is released into Earth's atmosphere. Clearing forests results in fewer trees to take in carbon and release oxygen through photosynthesis.

Carbon cycle Life on Earth would not exist without carbon because carbon is the key element in the sugars, starches, proteins, and other compounds that make up living things. The carbon cycle is illustrated in **Figure 11.** During photosynthesis, green plants and algae convert carbon dioxide and water into carbohydrates and release oxygen back into the air. These carbohydrates are used as a source of energy for all organisms in a food web. Other organisms release carbon dioxide back into the air during respiration.

Carbon is also stored when organic matter is buried underground and, over millions of years, is converted to peat, coal, oil, or natural gas deposits. Carbon dioxide gas is released into the atmosphere when these fossil fuels are burned for energy.

☑ READING CHECK **Explain** how photosynthesis and respiration cycle carbon between living things and Earth's atmosphere.

Nitrogen cycle Nitrogen is an element that organisms need to produce proteins. Nitrogen makes up 78 percent of the atmosphere, but plants and animals cannot use nitrogen directly from the atmosphere. Some species of bacteria, called **nitrogen-fixing bacteria,** live in water or soil, or grow on the roots of some plants and can capture nitrogen gas. The nitrogen-fixing bacteria convert the nitrogen into a form that can be used by plants to build proteins. Nitrogen continues through the food chain as one organism eats another. As organisms excrete waste and later die, the nitrogen returns to the soil and air. **Figure 11** shows the nitrogen cycle. Nitrogen moves from the atmosphere to the soil, to living organisms, and then back to the atmosphere.

☑ READING CHECK Photosynthesis allows green plants and algae to convert carbon dioxide and water into carbohydrates and release oxygen back into the air. Other organisms release carbon in carbon dioxide back into the air during respiration.

IN THE FIELD

Wangari Maathai Wangari Maathai, a Kenyan environmentalist, began the greenbelt movement in Kenya in the 1970s in response to the deforestation of her country. Almost 90 percent of the population in Kenya depended on wood for fuel, and forests were being depleted at an alarming rate. The women had to travel long distances in search of firewood, and the lack of firewood often meant a lack of food for poorer families. The greenbelt movement used community nurseries and provided free tree seedlings. A small payment was made for every tree that was planted and maintained by villagers. By 1977, the greenbelt movement had spread throughout Africa. It provided compensation for 80,000 people and resulted in the reintroduction of more than 10 million native trees. AL

VISUALIZING Carbon and Nitrogen Cycles

Figure 11 All life-forms depend on carbon and nitrogen in many different ways, as shown.

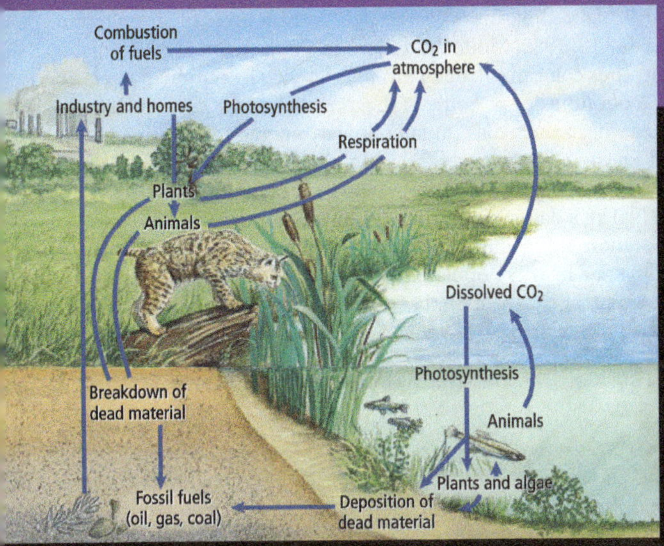

Humans have influenced the carbon cycle through the combustion of fuels. When fuels such as coal or oil are burned, one by-product of this combustion is carbon dioxide. Once released, the carbon dioxide enters the atmosphere and continues in the carbon cycle.

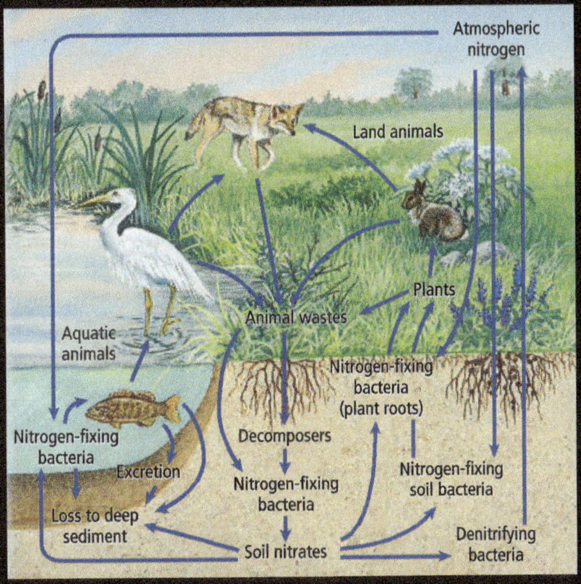

Nitrogen-fixing bacteria are an integral part of the nitrogen cycle. When animals produce waste, or when plants or animals die and begin to decompose, one by-product of this process is nitrogen. Nitrogen-fixing bacteria can break down the nitrogen, making it accessible for use by other plants and animals.

Concepts In Motion View an **animation of the carbon and nitrogen cycles.**

VISUALIZING

Purpose

Students will compare and contrast the carbon and nitrogen cycles.

Model

Carbon and Nitrogen Cycles

Provide students with modeling clay and colored paper. Have groups of students make two- or three-dimensional models of the carbon cycle and the nitrogen cycle.

Identify Misconceptions

Students usually think recycling is a new idea and that it is a human-made phenomenon.

Uncover the Misconceptions

Ask students to speculate how long recycling has occurred on Earth.

Demonstrate the Concept

Tell students Earth processes have always involved cycles, which are natural forms of recycling. Earth has a fixed amount of matter and energy.

Over the 4.6 billion years of Earth's existence, this matter and energy has been reused over and over. When humans recycle, they are applying basic principles that are seen in the carbon and nitrogen cycles: the laws of conservation of matter and energy. These laws state that matter and energy cannot be created nor destroyed, but both can be transformed and transferred.

Assess New Knowledge

Ask students to guess how long the carbon and nitrogen on Earth have been around. For the same amount of time as Earth has been in existence (4.6 billion years), carbon and nitrogen have existed, though in different forms, as they can see in the illustrations.

Environmental Connection

Air Quality Index In the mid 1970s, an index directly relating air pollution to human health was developed. Known as the Air Quality Index (AQI), it rates air pollution levels on a scale from 0–500 for each of the following pollutants: carbon monoxide, ozone, sulfur dioxide, nitrogen dioxide, and suspended particulates. Pollutant levels above 100 are considered to be unhealthful, whereas those above 300 are considered to be hazardous. This index allows ordinary people to have a basic understanding of levels that indicate potential health hazards.

Project

Monitor the Air Quality Have students monitor the local air quality on the local news over the course of a month and record the results in a table. Then ask them to describe their results in writing and make predictions for the air quality for the next month. BL

READING CHECK Volcanoes project ash as well as sulfur dioxide into the atmosphere. Both ash and sulfur dioxide are considered pollutants because they can enter Earth's geochemical cycles and harm the well-being of living things.

Figure 12 Vog, shown here over Kilauea, is formed when sulfur dioxide and other particulates emitted from a volcano mix with oxygen and moisture in the presence of sunlight.

Natural Air Pollution Sources

A **pollutant** is a substance that enters Earth's geochemical cycles and can harm the well-being of living things or adversely affect their activities. Air pollution can come from natural or human sources and can affect air outside or inside buildings. Natural sources of air pollution include volcanoes, fires, and radon.

Volcanoes Volcanoes can be significant sources of air pollution. On May 18, 1980, Mount St. Helens in Washington State shot an enormous column of ash 24 km into the sky. It continued to eject ash for about nine hours. Some of the ash reached the eastern United States within three days. Small particles entered the jet stream and circled Earth within two weeks. Mount St. Helens started erupting again between 2004 and 2008, and pumped out between 45,000 and 270,000 kg a day of sulfur dioxide. Italy's Mount Etna produces 100 times more sulfur dioxide than Mount St. Helens and is located in the middle of a heavily populated area. This sulfur helps to create acid rain and a type of bluish smog that volcanologists call vog, shown in **Figure 12,** which can cover large areas of land.

READING CHECK **Describe** how volcanoes contribute to air pollution.

Fires Smoke is a mixture of gases and fine particles produced when wood and other organic matter burn. The most significant health threat from smoke comes from fine particles. These microscopic particles can get into your eyes and respiratory system, where they can cause health problems such as burning eyes, a runny nose, and illnesses such as chronic bronchitis. People with chronic lung disease can be at risk of serious injury from smoke.

Forest fires can release thousands of tons of carbon monoxide, a gas that interferes with oxygen transport in your blood. Gases from forest fires can also contribute to particulate and smog pollution hundreds of kilometers from the burning forest. In 2004, a large fire in Alaska and Canada, similar to the one shown in **Figure 13,** added about 30 billion kg of carbon monoxide to the atmosphere—about as much as was released during human activities in the United States that month.

Figure 13 Forest fires can release dangerous gases into the atmosphere. People with respiratory problems can be at risk of injury from high levels of smoke and gas.

DIFFERENTIATED INSTRUCTION

Visually Impaired One way to assist students with visual impairments is to enlarge all handouts that are duplicated. Make sure diagrams are clear and labels are not located over background material and colors. Sometimes, black-and-white handouts and diagrams are more distinguishable than colored diagrams. It is also best to talk with students to determine which options best suit students' individual needs. OL

Radon The gas known as radon-222 (Rn-222) is colorless, odorless, tasteless, and naturally occurring. Rn-222 is produced by the radioactive decay of Uranium-238 (U-238). Small amounts of U-238 are found in most soils and rocks, and in underground geologic formations, mainly in the northern third of the United States. Usually, radon gas from such deposits seeps upward through the soil and is released into the atmosphere, where it is diluted to harmless levels. However, when buildings are constructed with hollow concrete blocks, or when they have cracks in their foundations, radon gas can enter and build up to high levels indoors, as shown in **Figure 14.** Once indoors, radon gas decays into other radioactive particles that can be inhaled.

Radon is responsible for about 21,000 lung cancer deaths every year. About 2900 of these deaths occur among people who have never smoked. Because it is impossible to see or smell a buildup of radon gas in a building, the EPA suggests that people test the radon levels in their homes and offices.

☑ READING CHECK **Explain** why radon is so dangerous.

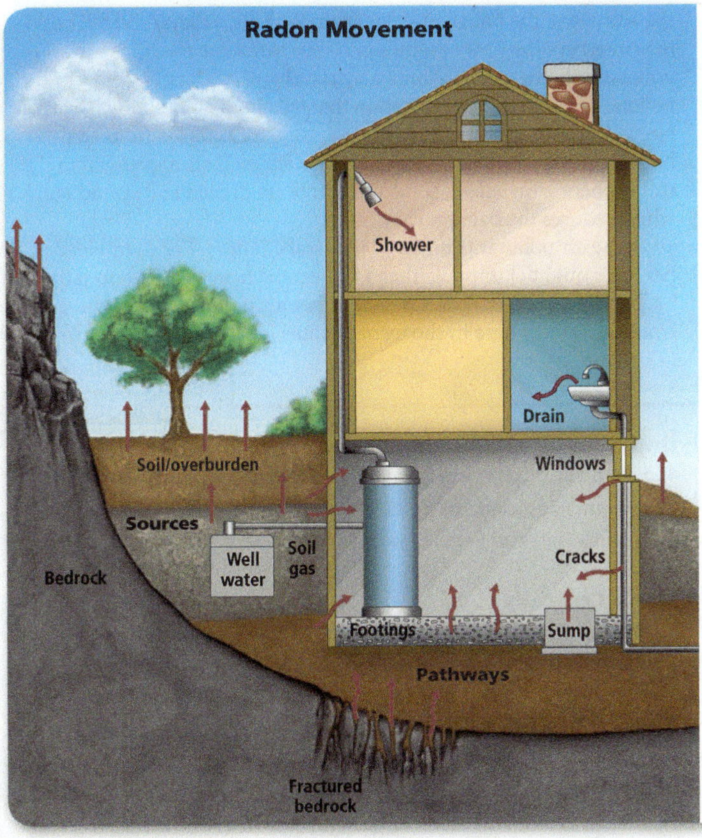

■ **Figure 14** There are many ways radon can enter a home or building. Once inside, radon is colorless and odorless, making it difficult to detect. For this reason, many homes are equipped with radon detectors that have an alarm if levels exceed safety. Although radon often enters through cracks in the foundation, or through drains or other openings in the basement, they can also enter through other pathways such as showerheads.

EARTH SCIENCE JOURNAL

Air Pollution in the News Have students locate an article on air pollution in a newspaper, magazine, journal, or on the Internet. Ask them to read the article and summarize its contents in outline form in their Earth science journals. Ask students to share one or two major points of their articles with the class.

Rubric

ACROSS THE CURRICULUM

Chemistry When elements are written with numbers, such as uranium-238 or radon-222, the number refers to mass number. The mass number is equal to the total number of neutrons and protons in the atom's nucleus. Radioactive materials are chemically unstable and give off neutrons, which results in a change in mass number.

Collaborative Learning

Collect Data at Home Hang a map of the local community at the front of the classroom. Place a piece of clear acetate over the map. Have each student use a marking pen to make a dot on the acetate to indicate where he or she lives. Tell students they will study air pollution in their community and will be responsible for collecting data at home. If no students live in certain areas, ask for student volunteers to take on an extra site or two so that all sections of the community will be represented.

Decide as a class what data to collect, when to collect them, and how the data will be collated. Once the data are collected, different acetate overlays can be developed for the different types of data collected. In some cases, students may be able to draw isolines from the collected data. Ask students to draw conclusions from the data at the end of the activity. OL COOP LEARN

Interpret the Illustration

Sources of Pollution in the Home Students might not realize that so many sources of potential pollutants exist in any house, as shown in **Figure 14.** Ask students to identify those pollutants they think might be found in their school buildings. One possible pollutant in school buildings is asbestos, a material that once was used as insulation for heating ducts and hot water pipes. Have interested students find out if asbestos remains in their school buildings and, if so, what the school administration's management plan is. OL

☑ READING CHECK Radon gas is dangerous because it causes lung cancer and is odorless, so there are no warnings when it is present.

■ **Caption Question Fig. 15** Acid rain eventually might kill these trees.

3 Assess

Check for Understanding

Use Models Ask students to draw a landscape or the inside of a building in their Earth science journals. Have students label all the forms of air pollution that might be associated with their drawings. Have students share their drawings and what forms of air pollution they noted with a lab partner. Ask partners to check to see whether any forms of air pollution were missed. If so, have students add these to their drawings.

Reteach

Research Have students write newspaper articles about pollution in their community. Ask students to define the type of pollution in their articles. Post articles on the classroom bulletin board.

Assessment

Knowledge Ask students what the following abbreviations stand for: EPA and UV. Environmental Protection Agency; ultraviolet radiation

■ **Figure 15** When acid rain falls on a forest, the pH of the soil changes. As a result, the growth of the trees can be slowed. They can also become susceptible to disease, which causes large stands of trees to be damaged.
Predict ***What will happen to this forest if acid rain continues to fall on it?***

Transport and Dilution

As air in the lower atmosphere moves across Earth's surface, it collects both naturally occurring and human-made pollutants. These pollutants are often transported, diluted, transformed, or removed from the atmosphere.

Some pollutants are carried downwind from their origin. Transport depends on wind direction and speed, topographical features, and the altitude of the pollutants. For example, hills, valleys, and tall buildings interrupt the flow of winds and thus influence the transport of pollutants. Some air pollutants may travel great distances. Many of the pollutants in the acid precipitation that falls in the mountain ranges of North Carolina, shown in **Figure 15,** were transported from coal-burning power plants in the midwestern states. If air movement in the troposphere is turbulent, some pollutants are diluted and spread out, which reduces the damage they cause.

Some air pollutants undergo physical changes. For example, dry particles might clump together and become heavy enough to fall back to Earth's surface. These and other air pollutants are removed from the atmosphere in the form of snow, mist, fog, and rain.

SECTION 3 REVIEW

Section Self-Check

Section Summary

- Earth's early atmosphere had no oxygen; it was supplied over time by photosynthetic organisms.
- Oxygen, carbon, and nitrogen cycle from living organisms to the nonliving environment.
- Volcanoes, fires, and radon are natural sources of air pollution.

Understand Main Ideas

1. MAINIDEA **Explain** why the atmosphere is considered a natural resource.
2. **Compare and contrast** the carbon and nitrogen cycles.
3. **Describe** how coal-burning power plants in the Midwest can cause acid precipitation in New York.

Think Critically

4. **Predict** what might happen if there were no nitrogen-fixing bacteria on Earth.
5. **Apply** How might increasing the energy efficiency of a home lead to increased radon levels indoors?

MATH IN Earth Science

6. About 21,000 people die from lung cancer related to radon each year. Of these, 2900 have never smoked. What percentage of people who die from radon-related lung cancer have never smoked?

SECTION 3 REVIEW

1. Air is considered a natural resource because most organisms on Earth require oxygen or carbon dioxide to maintain life processes. Nitrogen is used to build proteins. Earth's atmosphere acts to cycle these gases from the nonliving environment to living organisms.
2. In the carbon cycle, plants turn carbon dioxide into a usable form for other animals. Other animals eat the plants and get the nutrients needed for survival. In the nitrogen cycle, nitrogen-fixing bacteria turn nitrogen into a usable form for plants. Animals also eat these plants for nutrients needed for survival. In both the nitrogen and carbon cycles, carbon and nitrogen move through the atmosphere, geosphere, and hydrosphere.
3. Pollutants can be carried far from their source by winds in the atmosphere.
4. Nitrogen-fixing bacteria help transform nitrogen in the atmosphere into a form that plants can use. Animals then eat the plants. Nitrogen is used to help build proteins. Without nitrogen-fixing bacteria, plants and animals would not be able to make proteins.
5. Radon levels can build in a home if it is airtight, with fewer leaks to the outdoors.
6. 13.81 percent

SECTION 4

Water Resources

MAINIDEA Water is essential for all life, yet it is unevenly distributed on Earth's surface.

Essential Questions
- Why are the properties of water important for life on Earth?
- How is water distributed and used on Earth?
- In what ways can humans reduce the need for freshwater resources?

Review Vocabulary

aquifer: rock that holds enough water and transmits it rapidly enough to be useful as a water source

New Vocabulary

hydrogen bond
desalination

EARTH SCIENCE 4 YOU What did you eat for dinner last night? How much water did it take to prepare the meal? Water is not only used to prepare, cook, and clean up, but it is also needed to grow the food that you eat.

Properties of Water

About 71 percent of Earth's surface is covered by water. The world's oceans help regulate climate, provide habitats for marine organisms, dilute and degrade many pollutants, and even have a role in shaping Earth's surface. Freshwater is an important resource for agriculture, transportation, recreation, and numerous other human activities. In addition, the organisms that live on Earth are made up mostly of water. Most animals are about 50 to 65 percent water by mass, and even trees can be composed of up to 60 percent water.

Liquid water What properties of water allow it to be so versatile? Water has a high boiling point, 100°C, and a low freezing point, 0°C. As a result, water remains liquid in most of the environments on Earth. Water can exist as a liquid over a wide range of temperatures because of the hydrogen bonds between water molecules. **Hydrogen bonds** form when the positive ends of some water molecules are attracted to the negative ends of other water molecules. Hydrogen bonds, shown in **Figure 16,** also cause water's surface to contract and allow water to adhere to and coat a solid. These properties enable water to rise from the roots of a plant through its stem to its leaves.

Thermal energy storage capacity Liquid water can store a large amount of thermal energy without a significant increase in temperature. This property protects aquatic organisms from rapid temperature changes, and it also contributes to water's ability to regulate Earth's climate. Because of this same property, water is used as a coolant for automobile engines, power plants, and other thermal energy-generating processes. Have you ever perspired heavily while participating in an outdoor activity on a hot day? Evaporation of perspiration from your skin helps you cool off because large quantities of thermal energy are released as the water in the perspiration changes into water vapor.

Water as a solvent Liquid water can dissolve a variety of compounds. This enables water to carry nutrients into, and waste products out of, the tissues of living things. The diffusion of water across cell membranes enables cells to regulate their internal pressure.

Figure 16 The attractions between the slightly positive and slightly negative ends of water molecules are called hydrogen bonds.

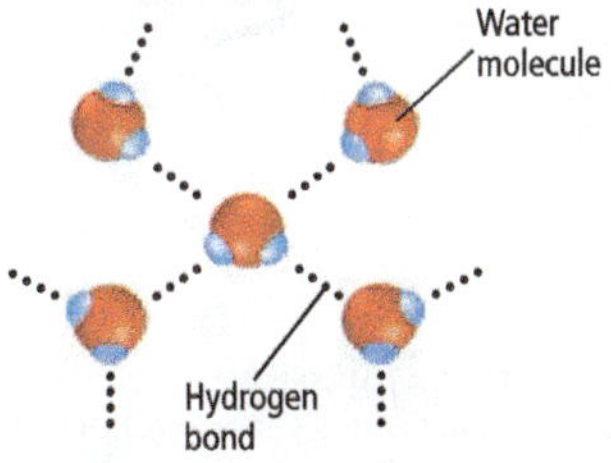

Demonstration

Model Properties of Water Demonstrate a unique property of water by filling a glass to the top with water. Ask students to look at the glass of water, and ask them how full it is. Some students will simply say the glass is filled. More observant students will notice that the water is higher than the top edge of the glass. Ask students how a glass can be overfilled with a liquid without spilling. A water molecule contains one oxygen atom and two hydrogen atoms. The uneven distribution of electrons in the molecule cause it to be polar. Although it is electrically balanced as a molecule, it does have a slightly positive side and a slightly negative side. This enables water molecules to attract each other. Thus, water molecules can overfill a glass. BL

SECTION 4

1 Focus

Focus Transparency

Before presenting the lesson, project **Section Focus Transparency** 77 and have students answer the accompanying questions. EL

MAINIDEA

Water Resources Introduce the topic of water resources by asking students questions about their water use. Ask: How do you use water in your house? for cleaning, cooking, and bathing How many liters of water do you think you use each day? Answers will vary. The average liters used in activities such as showering, washing clothes, and flushing the toilet are listed in the GeoLab at the end of this chapter.

2 Teach

Tie to Previous Knowledge

Plate Tectonics Remind students about plate tectonics. Limestone bedrock is usually formed from organisms and sediments in warm shallow seas. As the limestone is lithified, it may be transported by plate movement. Subsequent uplift and erosion gives humans access to the resource at many different latitudes. OL

Teacher Content Support

The Water Budget The water budget is a concise representation of the amount of water received, used, and stored in a particular location. Studying an area's water budget can show whether the area has surplus water, whether the area experiences deficits or water shortages, and when during the year the area's water supply typically is recharged, including the water stored in the ground. The local water budget is dependent upon the availability of solar energy and the amount of precipitation.

Tie to Previous Knowledge

Freeze and Thaw Students have learned about freezing and thawing as a method of weathering by water. This type of weathering is particularly damaging to roads and can cause potholes. BL

Teacher Content Support

Water Resources The National Institute for Water Resources (NIWR) is a network of research institutes located in each of the states and territories in the United States. Each state and territory has a research institute, located at a designated college or university, whose role is to promote research, training, and information dissemination. The primary impact of water- resource issues is at the local or state level. Thus, water resources generally are managed at these levels. Federal agencies get involved when water issues cross state boundaries. The Secretary of the Interior oversees and periodically evaluates each state institute with assistance from the U.S. Geological Survey. Concern and care for water resources is a global issue that begins locally. Because most of Earth's freshwater supply is frozen, the care and use of available freshwater resources is crucial to the maintenance of the quality of life on Earth.

Figure 17 In a rock formation where weathering has previously occurred, water can enter cracks in the formation. When the water freezes, it expands, causing the cracks to widen.

Solid water Unlike most liquids, water expands when it freezes. Because ice has a lower density than liquid water, it floats on top of water. As a result, bodies of water freeze from the top down. If water did not have this property, ponds and streams would freeze solid, and aquatic organisms would die each winter. **Figure 17** shows that expansion of water as it freezes can also fracture rocks. Thus, ice formation in cracks in Earth's surface becomes part of the weathering process.

Location of Freshwater Resources

Freshwater resources are not distributed evenly across Earth's landmasses. The eastern United States receives ample precipitation, and most freshwater in these states is used for cooling, energy production, and manufacturing. By contrast, southwestern states often have little precipitation. In the southwestern United States, the largest use of freshwater is for agricultural purposes such as irrigation. Water tables in these areas might drop as people continue to use the groundwater faster than it can be recharged.

Water distribution is a continuing problem worldwide, even though most continents have plenty of water. Since the 1970s, scarcity of water has resulted in the deaths of more than 24,000 people each year. In areas where water is scarce, women and children often walk long distances each day to collect water for domestic uses. Millions of people also try to survive on land that is prone to drought. About 25 countries, primarily in Africa, experience chronic water shortages. **Figure 18** shows projected water stress levels across the globe for the year 2025. These stress levels are predicted in large part by projected population growth, as well as other factors.

Figure 18 By the year 2025, scientist predict the water stress levels will reach those shown here. The areas with projected adequate water supply will be limited. Most of the United States is projected to have some shortage while much of Asia is predicted to have large-scale shortage.

Projected World Water Stress Levels in 2025

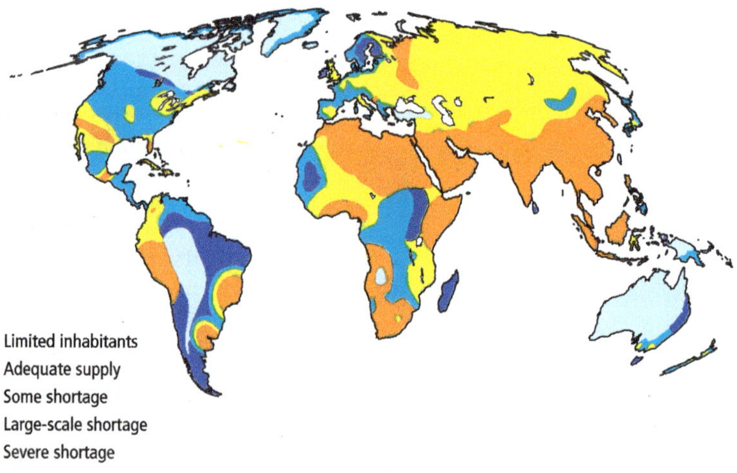

DIFFERENTIATED INSTRUCTION

Visually Impaired Have students with visual impairments work with models of oxygen and hydrogen to make a water molecule. Give students tennis-ball-sized foam spheres to use for oxygen atoms and smaller golf-ball-sized foam spheres to use for hydrogen. Have students attach two hydrogen spheres at an angle of about 100° from one another on one oxygen sphere to represent a water molecule. BL EL

ACROSS THE CURRICULUM

Language Arts People from different parts of the United States often use different words for the same feature. This is true of streams and rivers. Words that are used to mean *stream* or *river* include *creek, brook, kill,* and *run*. Ask students if they can think of other words that mean *stream*. EL

Use of Freshwater Resources

Recall that the upper surface of groundwater is called the water table, and that the water-saturated rock through which groundwater flows is called an aquifer. Aquifers are refilled naturally as rain percolates through soil and rock.

In the United States, about 20 percent of all freshwater used is groundwater pumped from aquifers. Water moves through aquifers at an average rate of about 15 m/day. If the withdrawal rate of an aquifer exceeds its natural recharge rate, the water table around the withdrawal point is lowered, called drawdown. If too many wells are drilled into the same aquifer in a limited area, the drawdown can lower the water table, and, as a result, wells might run dry.

Worldwide consumption Uses of freshwater vary worldwide, but about 70 percent of the water withdrawn each year is used to irrigate 18 percent of the world's croplands. About 23 percent of freshwater is used for cooling purposes in power plants, for oil and gas production, and in industrial processing. Domestic and municipal uses account for only 7 percent of the freshwater withdrawal.

Managing Freshwater Resources

Most countries manage their supplies of freshwater by building dams, transporting surface water, or tapping groundwater. The dam shown in **Figure 19** was built to hold back the floodwaters of the Yangtze River in China. Called the Three Gorges Dam, it is one of the largest hydropower projects in the world at 185 m high and 230 m wide, and supplies freshwater and power to millions of people in China. However, water held by the dam has displaced about one million residents who lived nearby.

■ **Figure 19** Dams are often built to contain freshwater resources in rivers. While this provides a readily available source of freshwater for human use, there are many other factors involved that make the damming of rivers controversial, including the flooding of farmland and displacement of people.

MiniLAB

iLab Station

Determine the Hardness of Water

How easily are soap suds produced? Water contains different minerals depending on its source. When water has a high mineral content, it is referred to as "hard."

Procedure

1. Read and complete the lab safety form.
2. Obtain six **clean baby food jars with lids.** Label them *A* through *F*.
3. Measure 20 mL of one **water sample.** Pour the water into the jar marked *A*.
4. Repeat Step 3 four more times, using a different water sample for jars B through E.
5. Measure 20 mL of **distilled water.** Pour this water into jar F.
6. Make a data table in your science journal. In the first column, write the letters *A* through *F*.
7. Place one drop of **liquid soap** in sample jars A through E. Do not place any soap in jar F. Tighten the lids. Shake each jar vigorously for five seconds.
8. Using the following rating scale, record in your data table the amount of suds in each jar: 1—no suds, 2—few suds, 3—moderate amount of suds, 4—lots of suds.

Analysis

1. **Order** the water samples in order from hardest to softest.
2. **Explain** What is the difference between hard and soft water?
3. **Determine** What are some disadvantages of hard water?
4. **Analyze** What was the purpose of sample F?

MiniLAB

iLab Station Rubric

Purpose Students will compare and contrast the ability of hard and soft water to produce soap suds.

Process Skills observe and infer, collect and interpret data, make and use tables

Safety Precaution Approve lab safety forms before work begins.

Teaching Strategy Remind students to measure carefully to ensure the best results.

Expected Results Students should recognize that soap suds are produced in some water samples more easily than others. Water samples should include water from wells and lakes, rainwater, softened water, and so on.

Analysis

1. Answers will vary depending on the samples.
2. Soft water will make soap suds more easily.
3. More soap is needed to get things clean; scum often forms in sinks, tubs, or around clothes in the washing process. The soap scum is formed from the excess metal ions (Ca^{2+}, Mg^{2+}, Cu^{2+}, Fe^{3+}) in the hard water.
4. Sample F is the control.

Assessment

Performance Give each group three unmarked bottles of water and ask the group to determine which one is distilled water.

LabManager™
Customize any lab with the LabManager™ CD-ROM.

ACROSS THE CURRICULUM

History The availability and use of water has political ramifications. Access to bodies of water is important to countries, states, and provinces. To avoid conflicts, more than 2000 treaties dealing with the issue of water rights exist. The use of water downstream is as important as the use of water upstream. AL

EARTH SCIENCE JOURNAL

Water Shortages Have students research and write in their Earth science journals describing what it might be like to live with a water shortage. Encourage them to consider how they might meet their basic needs if they had to ration water. BL

Rubric

GeoLAB

iLab Station

The GeoLab at the end of the chapter can be used at this point in the lesson.

Environmental Connection

Dams There are more than 80,000 dams on rivers and streams in the United States. Most dams and reservoirs are built to regulate water, especially as a means for controlling flooding. However, opponents of dam construction argue that dams only shift flooding to another site. They also contend that the availability of cheap water encourages wasteful consumption while adding federal costs to localities and states. In addition, opponents claim that dams cause significant changes to downstream ecosystems and pose a potential threat to life and property that could far outweigh any damage from the flooding the dam was intended to prevent. OL

Project

Dams and Reservoirs Have student groups research a particular dam or reservoir. Each group can research the history of the dam, including the reason it was built and facts about its construction. Students might be able to find photographs of dam construction. Ask each group to make a presentation about its dam. Students should identify the advantages as well as the disadvantages of the dam. OL COOP LEARN

READING CHECK Building dams allows countries to manage their water resources by controlling flooding and providing water for irrigation and municipal uses.

Dams and reservoirs Building dams is one of the primary ways that countries manage their freshwater resources. Large dams are built across river valleys, and the reservoirs behind dams capture the river's flow as well as rain and melting snow. Because the runoff is captured, flooding downstream is controlled. The water held in these reservoirs can be released as necessary to provide water for irrigation; municipal uses, such as in homes and businesses; or to produce hydroelectric power. Reservoirs also provide opportunities for recreational activities, such as fishing and boating. Dams and reservoirs currently control between 25 and 50 percent of the total runoff on every continent.

READING CHECK **Explain** several advantages of building dams.

Transporting surface water If you were to visit Europe or the Middle East, you would likely see many ancient aqueducts. The Romans built aqueducts 2000 years ago to bring water from other locations to their cities. Today, many countries use aqueducts, tunnels, and underground pipes to move water from areas where it is plentiful to areas that need freshwater.

The State Water Project in California, illustrated in **Figure 20,** is one example of the benefits, as well as the costs, of transporting surface water. In California, about 75 percent of the precipitation occurs north of the city of Sacramento, yet 75 percent of the state's population lives south of that city. The California Water Project uses a system of dams, pumps, and aqueducts to transport water from northern California to southern California. Eighty-two percent of this water is used for agriculture. The residents of Los Angeles and San Diego are withdrawing groundwater faster than it is being replenished. As a result, there is a demand for even more water to be diverted to the south. Conflicts over the transport of surface water could increase as human populations increase.

Figure 20 A system of dams, pumps, and aqueducts moves water in California from the North, where there is more rainfall, to the South, where the climate is more arid.

EARTH SCIENCE JOURNAL

Dams and Reservoirs To stimulate thinking about water resources, ask students to write about the differences between a dam and a reservoir and to give two reasons for, and two reasons against, the building of dams. Students could present their research in teams as a panel. Each side could take notes on the facts presented in the debate and share their notes in a class discussion. OL

■ **Figure 21** Desalination can be accomplished using several different methods. One method, called distillation, removes salt by boiling the water. Another process involves pumping the water through a filtration system to remove the salt. In some places water is desalinated in plants like this one.

View an **animation of distillation.**

Concepts In Motion

Desalination With all the water available in the oceans, some countries have explored the possibility of removing salt from seawater to provide freshwater in a process called **desalination.** Several methods are available to desalinate seawater. One way is through distillation—water is first heated until it evaporates, and then it is condensed and collected. This evaporation process leaves the salts behind. Most countries that use desalination to produce freshwater use solar energy to evaporate seawater. Although the evaporation of seawater by solar energy is a slow process, it is an inexpensive way to provide needed freshwater. Some desalination plants, such as the one shown in **Figure 21,** use fuel to distill seawater, but because this process is expensive, it is used primarily to provide drinking water. Both processes require large amounts of energy.

SECTION 4 REVIEW

Section Self-Check

Section Summary

- Water has unique properties that allow life to exist on Earth.
- Water is not evenly distributed on Earth's surface.
- Water management methods distribute freshwater resources more evenly through the use of dams, aqueducts, and wells.

Understand Main Ideas

1. MAINIDEA **Describe** how the distribution of freshwater resources affects humans.
2. **Explain** why the thermal energy storage capacity of water is important to life on Earth.
3. **Explain** why water in a pond freezes from the top down.

Think Critically

4. **Propose** Do you think the process of desalination is a good option for areas like the southwestern United States where there is a high demand for freshwater? Explain your reasoning.
5. **Analyze** What are two things you could do to reduce your daily water usage?

WRITING IN ▶ Earth Science

6. Imagine there is a large river near your hometown. For years, residents have used the river to fish, canoe, and swim. Recently a group has proposed damming the river to provide a clean, renewable energy source. Write two newspaper editorials—one in support of the construction of a dam and one against it.

3 Assess

Check for Understanding

Define Operationally Have students work in small groups to research a large metropolitan area's water sources and water-management plans. Students should include in their reports the numbers of people affected, maps, descriptions of systems that transport the water, and conflicts that exist between the city and the source regions. Each group should summarize its material on a bulletin board.

Reteach

Describe Have students draw maps of their own community's water systems in their science journals. The maps should include the water sources used, as well as treatment plants for both drinking water and wastewater.

Assessment

Skill Ask each student to develop a water system that could be used on an oceanographic research vessel that remains at sea for six months at a time. Have students list issues that need to be addressed.

SECTION 4 REVIEW

1. Scarce freshwater causes approximately 24,000 deaths each year. In some areas, people walk long distances each day to collect a meager supply of water for domestic uses. Millions of people also try to survive on land that is prone to drought.
2. The heat storage capacity of water protects aquatic organisms from abrupt temperature changes. The large amount of heat that can be stored in water also helps to regulate Earth's climate. Water is also used as a coolant.
3. Ice is less dense than liquid water, so ice floats on top of liquid water. If ice were denser than liquid water, ice would sink and water bodies would freeze from the bottom up.
4. Sample answer: Yes. States such as California and Texas might be able to harness the Sun's energy to distill seawater.
5. Answers will vary. Students might suggest taking shorter showers and turning the faucet off when not directly in use.
6. Answers will vary, but the editorial in favor of the dam should include how water can be used for irrigation and homes, and the editorial against the dam should include how the natural path of the river will be disturbed.

Rubric

Earth Science & SOCIETY

Purpose

Students will evaluate the cost of water based on the amount available from various sources.

Teacher Content Support

Water as a Resource Almost three-quarters of the Earth's surface is covered in water. Of that, approximately 97 percent is ocean water. Only 3 percent of the water on the planet is freshwater, and of that only 1 percent is available for use. The rest is tied up in glaciers or other parts of the water cycle. Water is a renewable resource; the amount typically remains the same throughout time. Due to the water cycle, it gets replaced as it gets used up. Water that is not absorbed into the soil is called surface water; this includes lakes, rivers and streams. Water that is absorbed into the soil becomes groundwater.

Teaching Strategy

Ask students: When you turn on the tap, where does the water come from? Answers will vary. Student might say that their water comes from a well or water treatment plant. Then ask students: What are all the different sources of water that we use? Humans use water from a variety of sources, including wells, rivers, lakes, reservoirs, desalination plants, collected rainwater, and melted icebergs.

Earth Science & SOCIETY

The Price of Water

When you go to the water fountain to get a drink, do you ever wonder where the water comes from? Depending on where you live, your water could come from groundwater or surface water, from a well or a water treatment plant.

The source of our water Water might seem like an abundant resource—after all, nearly 75 percent of our planet is covered with it. However, less than 1 percent of all the water on Earth is suitable for everyday uses such as drinking, cooking, and irrigation. Because water is a limited resource, its source is becoming a very important issue.

A green desert The hot, dry climate of the southwestern United States is probably the last place you would expect to see green lawns and palm trees lining the streets. Most of the area is classified as arid due to the low amounts of yearly rainfall. Yet, as many cities in this area continue to grow in population, the demand for water continues to increase.

Many cities in the Southwest draw from the same groundwater source. Often, more water is withdrawn than can be replaced by the yearly rainfall, causing the water supply to run low. Some larger cities are attempting to address this issue by using water from rivers, streams, and lakes for residential use.

Drinking it dry Over 80 years ago, residents of some western states recognized the need for water from the Colorado River. In 1922, The Colorado River Compact was established to regulate who could use the water and how much they were allowed to use.

Today, 30 million people in the United States and Mexico use water from the Colorado River. As the demand for water upstream increases, less water is available for use downstream.

The Glen Canyon Dam on the Colorado River is one of a series of dams that controls the river's flow.

By the time the river reaches the U.S./Mexican border, it is a small trickle. This reduced flow has caused tension between Mexico and the United States. Residents of northern Mexico argue that they have as much right to the water of the Colorado River as those upriver.

Environmental implications By harnessing the river for public use, some of the natural ecosystems that depend on the river have been impacted. Some areas of the river have been dammed, as shown in the figure, or diverted, jeopardizing native fish species.

As the flow of water decreases, valuable nutrients and sediments are no longer carried to the Colorado River Delta. Plant and animal species that once thrived in this area can no longer survive.

WRITING IN ▶ Earth Science

Research Find more information about sustainable water use on the Internet. Does your city have a sustainable level of water use? Write an essay explaining if your city's water usage is sustainable.

WebQuest

WRITING IN ▶ Earth Science

Rubric

Research Answers will vary, but should include a description of the amount of water used and the source of this water.

WebQuest

GeoLAB

iLab Station

Design Your Own: Monitor Daily Water Usage

Background: The average American uses between 300 and 380 L of water per day. Think about all the ways you use water each day, from brushing your teeth to washing your clothes.

Question: *How much water do you use each day?*

Materials

water usage table
calculator

Procedure

1. Read and complete the lab safety form.
2. Obtain a water usage table from your teacher.
3. Complete the column labeled *estimations.* Your estimations should be how many liters of water you might use in one day for each of the activities.
4. For the next five days, record your water usage and complete the table.

Analyze and Conclude

1. **Calculate** the number of liters of water you used each day to flush the toilet.
2. **Calculate** the number of liters you used each day to shower.
3. **Calculate** the total daily average number of liters of water you used.
4. **Analyze** For what purposes did you use the most water? Was this the same for all of your classmates?
5. **Predict** how this water usage might change during different seasons.
6. **Recommend** two ways you could reduce the total amount of water you use each day.

TRY AT HOME

Revise Utilizing the two recommendations you made in Question 6, record your daily water usage for another five days. Were you able to reduce your total water usage? Why or why not?

Water Usage Activity	Liters Per Use	Estimations	Day 1	Day 2	Day 3	Day 4	Day 5	Total
Flushing the toilet	23 L/flush							
Showering	15 L/min							
Bathing	100 L/tun							
Dishwasher	45 L/load							
Washing machine	151 L/load							
Bathroom sink	7.5 L/min							
Kitchen sink	11 L/min							
							Total liters used	

GeoLAB

iLab Station Rubric

Preparation

Time Allotment 60 min over 5 days.

Process Skills observe and infer, recognize cause and effect, think critically

Safety Precaution Approve lab safety forms before work begins.

Procedure

- As students track their water use, be sure to monitor their progress and ask appropriate questions as necessary.
- **Troubleshooting** Some students might forget to monitor their water use in one or more of the categories. Check student tables to make sure students' water use is complete.

Analyze and Conclude

1. 23 liters × number of flushes
2. 15 liters × minutes spent showering
3. Answers will vary, but should be close to 300–400 liters.
4. Answers will vary, but most likely will be showering or bathing.
5. Answers will vary. Sample answer: The number of liters of water used during showering may increase during the winter.
6. Answers will vary, but might include taking shorter showers and turning the bathroom and kitchen faucets off when not in use.

TRY AT HOME

Revise Answers will vary, but should include information on the strategies the students used to reduce their water consumption, such as: taking shorter showers, turning off the water while they brush their teeth or wash their hands, and/or reducing the amount of water they use on their lawns and gardens.

MAINIDEAS Summary statements can be used by students to review the major concepts of the chapter.

Students can review with these online resources.

Vocabulary Practice

Vocabulary eGames
Vocabulary eFlashcards
Vocabulary PuzzleMaker

Section Self-Check

Chapter Self-Check

Online Test Practice

Use the Exam*View*® *Assessment Suite* CD-ROM to:

- create multiple versions of tests
- create modified tests with one mouse click
- edit existing questions and add your own questions
- build tests aligned with state standards using built-in state curriculum tags
- change English tests to Spanish with one mouse click
- track students' progress using the Teacher Management System

CHAPTER 24 STUDY GUIDE

BIGIDEA People and other organisms use Earth's resources for everyday use.

Vocabulary Practice

SECTION 1 Natural Resources

VOCABULARY
- natural resource
- renewable resource
- sustainable yield
- nonrenewable resource

MAINIDEA **Resources are materials that organisms need; once used, some resources can be replaced, whereas others cannot.**

- Natural resources are the resources that Earth provides, including organisms, nutrients, rocks, minerals, air, and water.
- Renewable resources can be replaced within a short period of time.
- Nonrenewable resources exist in a fixed amount and take millions of years to replace.

SECTION 2 Resources from Earth's Crust

VOCABULARY
- desertification
- aggregate
- bedrock
- ore
- tailings

MAINIDEA **Earth's crust provides a wide variety of resources to grow food, supply building materials, and provide metals and minerals.**

- Loss of topsoil can lead to desertification.
- Aggregates, composed of sand, gravel, and crushed stone, can be found in glacial deposits.
- An ore is a resource that can be mined at a profit. Ores can be associated with igneous rocks or formed by processes on Earth's surface.

SECTION 3 Air Resources

VOCABULARY
- nitrogen-fixing bacteria
- pollutant

MAINIDEA **The atmosphere contains gases required for life on Earth.**

- Earth's early atmosphere had no oxygen; it was supplied over time by photosynthetic organisms.
- Oxygen, carbon, and nitrogen cycle from living organisms to the nonliving environment.
- Volcanoes, fires, and radon are natural sources of air pollution.

SECTION 4 Water Resources

VOCABULARY
- hydrogen bond
- desalination

MAINIDEA **Water is essential for all life, yet it is unevenly distributed on Earth's surface.**

- Water has unique properties that allow life to exist on Earth.
- Water is not evenly distributed on Earth's surface.
- Water management methods distribute freshwater resources more evenly through the use of dams, aqueducts, and wells.

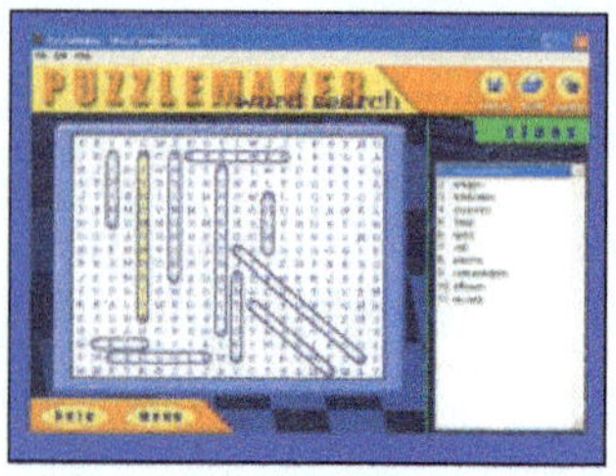

Vocabulary Practice

For additional practice with vocabulary, have students access the Vocabulary PuzzleMaker.

CHAPTER 24 ASSESSMENT

Chapter Self-Check

VOCABULARY REVIEW

Complete each sentence with the correct vocabulary term from the Study Guide.

1. Coal and oil are ________ resources because it is not possible to replace them in a short period of time.
2. Bamboo is an example of a(n) ________ because it is possible to use it indefinitely without a reduction in the supply.
3. A mixture of sand, gravel, and crushed stone is called a(n) ________.

Replace the underlined phrase with the correct vocabulary term from the Study Guide.

4. Ore is solid rock found underneath the loose soil and rocks in Earth's crust.
5. Soil is the residue of rock material left behind after the ore is removed.
6. The removal of salt from seawater is called nitrification.
7. The overuse of land resources might result in fertile land undergoing the process of soil formation.

Define each vocabulary term in a complete sentence.

8. pollutant
9. sustainable yield
10. ore

Identify the vocabulary term from the Study Guide that best fits each definition below.

11. the resources Earth provides
12. bacteria that live in soil or water and capture nitrogen gas
13. when the positive ends of some water molecules are attracted to the negative ends of other water molecules

UNDERSTAND KEY CONCEPTS

14. Which resource can be replaced at a sustainable rate?
   A. iron
   B. wheat
   C. gold
   D. diamonds
15. Why are nitrogen-fixing bacteria important?
   A. They are prey for larger animals.
   B. They are part of the carbon cycle.
   C. Plants and animals cannot use nitrogen directly from the atmosphere.
   D. They are part of photosynthesis.

Use the figure below to answer Questions 16 and 17.

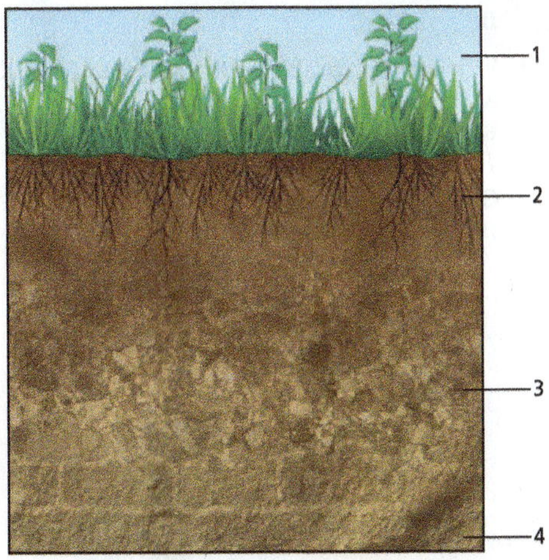

16. Which labeled area represents where aggregates are found?
   A. 1
   B. 2
   C. 3 and 4
   D. 4
17. Which layer is labeled 2?
   A. topsoil
   B. bedrock
   C. aggregate
   D. ore

CHAPTER 24 ASSESSMENT

VOCABULARY REVIEW

1. nonrenewable
2. renewable resource
3. aggregate
4. Bedrock
5. Tailings
6. desalination
7. desertification
8. A pollutant is something that contaminates a resource.
9. Sustainable yield is replacement of renewable resources at the same rate at which they are consumed.
10. Ore is a resource that can be mined for a profit.
11. natural resources
12. nitrogen fixing bacteria
13. hydrogen bond

UNDERSTAND KEY CONCEPTS

14. B
15. C
16. C
17. A

18. C
19. A
20. D
21. C
22. B
23. B

CONSTRUCTED RESPONSE

24. Beef and chicken can be replenished at the same rate that they are consumed.
25. People use water at a higher rate than it can be recharged.
26. Agricultural water withdrawal increased the most since 1900.
27. Forests on Earth remove carbon dioxide and add oxygen to the atmosphere. The loss of forest cover might increase the amount of carbon dioxide in the atmosphere, resulting in an increase in average atmospheric temperatures. Also, with fewer trees, there would be less oxygen available for respiration.
28. Issues to be considered with dam construction include the risk of flooding downstream as well as the effect on the organisms in the river.
29. Twenty-five countries, mainly in Africa, have chronic shortages of water.
30. Water can be found in all three states: liquid, gas, and solid.
31. Earth's atmospheric composition billions of years ago was similar to gases from erupting volcanoes. Over time the oxygen levels built up to levels needed to sustain life as we know it today.
32. Carbon dioxide is a requirement for life on Earth, but too much carbon dioxide contributes to warming of Earth.

18. Which occurs when the velocity of water carrying sediments is reduced?
A. Bedrock dissolves.
B. Fine sand is pushed up and moved.
C. Heavy sediments are deposited.
D. New minerals form.

19. Which resource is found in an unlimited supply?
A. sunlight **C.** lumber
B. gemstones **D.** fish

20. What condition could result from overgrazing of cattle?
A. soil formation
B. chemical precipitation
C. aggregate buildup
D. desertification

21. Which is not part of the nitrogen cycle?
A. the atmosphere **C.** photosynthesis
B. plants **D.** soil

22. Why is water considered to be a polar molecule?
A. A water molecule has a pole.
B. Each water molecule has a positive and negative pole.
C. Water molecules form in the polar region.
D. Water molecules are attracted to magnets.

Use the figure below to answer Question 23.

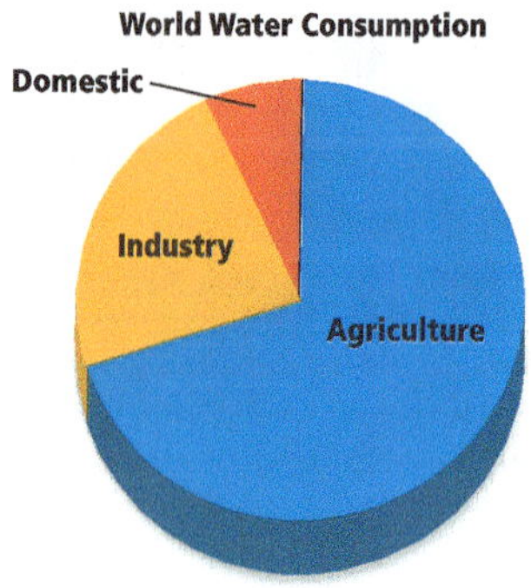

23. Industry is responsible for approximately what percent of the world's water consumption?
A. 7 percent
B. 23 percent
C. 70 percent
D. 100 percent

CONSTRUCTED RESPONSE

24. Explain why beef and chicken purchased in the grocery store are considered renewable resources.

Use the figure below to answer Questions 25 and 26.

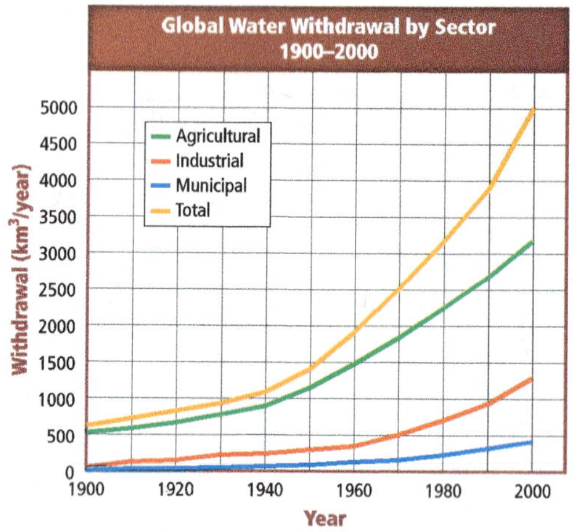

25. Explain why the worldwide rate of freshwater withdrawal has changed since 1900.

26. Determine What type of water withdrawal increased the most since 1900?

27. Explain why the loss of the forest cover in the Amazon River Basin is a world-wide concern.

28. Explain what issues would have to be considered if construction of a dam were proposed.

29. Identify several regions of the world that have a shortage of freshwater.

30. Identify In what states of matter can water naturally be found on Earth?

31. Differentiate between Earth's atmospheric composition billions of years ago and its composition today.

32. Explain how a substance could be both a pollutant and a requirement for life on Earth.

Chapter Self-Check

THINK CRITICALLY

33. Explain If Earth processes recycle water resources, why is water pollution a problem?

34. Consider how the study of a landfill could provide insight into how efficiently our natural resources are being used.

Use the figure below to answer Question 35.

Nick Hawkes/Ecoscene/CORBIS

35. Infer Based on the conditions discussed in this chapter, what caused damage to this statue? How?

36. Explain how early miners applied the principle of density to finding valuable deposits of natural resources.

37. Predict what would happen to carbon in the atmosphere if photosynthesis decreased.

38. CAREERS IN EARTH SCIENCE Research and describe one job in your community or a nearby city that is closely related to providing or protecting the local water resource.

CONCEPT MAPPING

39. Make a concept map using the section titles and vocabulary words from the sections. For more help, refer to the *Skillbuilder Handbook.*

CHALLENGE QUESTION

40. Determine the source of the water supply for your school. What procedure would you follow to answer this question?

WRITING IN ▶ Earth Science

41. Research your local parks and preserves. Is there an area near you that has been proposed for development? Are there endangered species in your area? Write a letter to your local congressional representative detailing what action you think should be taken.

DBQ Document–Based Questions

Data obtained from: Weibe, K., and N. Gollehon, eds. 2006. Agriculture resources and environmental indicators. *USDA* (July):134-143.

The 17 western states account for 77 percent of all irrigated land in the U.S. The average annual amount of water applied ranges from 150 acre-feet to over 2500 acre-feet. In an effort to conserve water, the USDA has suggested methods of water conservation.

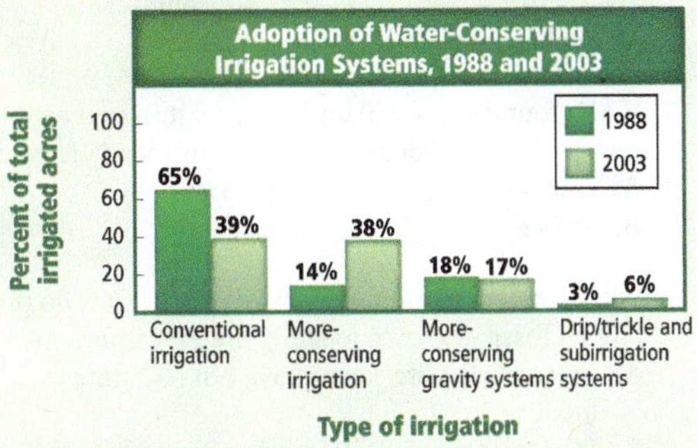

42. In 2003, which two methods of irrigation were most commonly used?

43. What percentage of acres were watered by the conventional irrigation system in 1988? In 2003?

44. What percentage acres were watered by the drip/trickle method in 2003?

CUMULATIVE REVIEW

45. Why is volcanic activity associated with convergent plate boundaries? **(Chapter 20)**

46. Explain the source of the heat that causes the geysers and hot springs at Yellowstone National Park. **(Chapter 10)**

THINK CRITICALLY

33. Pollutants enter the water cycle and can be cycled throughout the water cycle.

34. If a landfill were studied, items in it could be categorized by whether they are renewable or nonrenewable. If many nonrenewable but recyclable resources were found, it would indicate inefficient use of resources because they were discarded and not recycled. If few nonrenewable resources were found, it would indicate a more efficient use of resources.

35. Acid rain. Acid rain can break down stone, including statues, especially those made of limestone.

36. Early miners knew that gold was more dense than other matter, so they waited for it to settle.

37. Carbon in the atmosphere might continue to build up if photosynthesis decreased.

38. Answers will vary, but might include the duties of a hydrologist or environmental consultant.

CONCEPT MAPPING

39. Answer will vary, but should illustrate an understanding of the vocabulary words and section titles.

CHALLENGE QUESTION

40. Answer will vary depending on your area.

WRITING IN ▶ Earth Science

Rubric

41. Answers will vary. Answers should be persuasive and scientifically accurate.

DBQ Document-Based Questions

Data obtained from: Weibe, K. and N. Gollehon, eds. 2006. Agriculture resources and environmental indicators. *USDA* (July): 134–143.

42. conventional irrigation (39%) and more-conserving irrigation (38%)

43. 1988—65%, 2003—39%

44. 6%

CUMULATIVE REVIEW

45. Convergent plate boundaries create zones of subduction, which often are accompanied by volcanoes formed from the melting subducting plate.

46. A hot spot underlies Yellowstone National Park.

MULTIPLE CHOICE

1. C
2. B
3. A
4. D
5. B
6. A
7. A
8. A
9. C
10. B

CUMULATIVE STANDARDIZED TEST PRACTICE

MULTIPLE CHOICE

1. What is it called when the sea level rises and shorelines move inland?
A. regression **C.** transgression
B. passive margin **D.** Laurentia

Use the table to answer Questions 2 and 3.

Fossil Identification Key	
1	a. Spiral shape; go to Step 2
	b. No spiral shape: go to Step 3
2	a. Less than 6 cm across: gastropod
	b. More than 6 cm across: cephalopod
3	a. Circular: crinoid columnal
	b. Branching: bryozoan

2. Mia has a fossil that is about 7 cm across and has a spiral shape. What kind of fossil did Mia find?
A. gastropod **C.** crinoid columnal
B. cephalopod **D.** byrozoan

3. If Mia found this fossil and chiseled it from a sedimentary rock, what type of fossil most likely is it?
A. cast **C.** index fossil
B. amber **D.** trace fossil

4. Which statement best explains why scientists do not rely on fossil evidence to study the Precambrian?
A. Precambrian life-forms have not had time to fossilize.
B. During the early Precambrian, there were no life-forms on Earth.
C. A global event destroyed all life-forms at some point during the Precambrian.
D. Life-forms on Earth during the Precambrian were too soft-bodied and left very few fossil imprints.

5. How does volcanic activity during early Earth explain the formation of the oceans?
A. Volcanic eruptions caused major depressions in Earth's surface to collect water.
B. Volcanic gas contains water vapor that cooled and condensed into liquid water, filling ocean basins.
C. Volcanic gases created clouds which produced rain that filled ocean basins.
D. Volcanic material blocked the Sun's rays, killing plant life that helped absorb water, and the runoff formed oceans.

6. Which is the correct succession of life-forms during the Phanerozoic Eon?
A. ocean organisms, land plants, land animals
B. land plants, land animals, oceanic organisms
C. land plants, oceanic organisms, land animals
D. land animals, land plants, oceanic organisms

Use the illustrations to answer Questions 7 and 8.

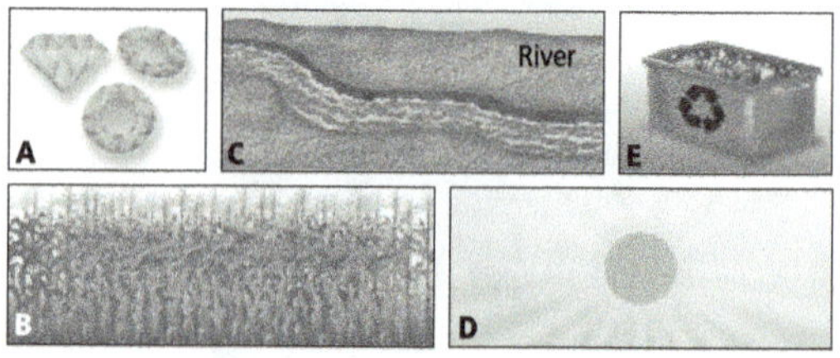

7. Which shows a nonrenewable resource?
A. A **C.** C
B. B **D.** D

8. Which resource is replaced through natural processes more quickly than it is used?
A. B **C.** D
B. C **D.** E

9. Why is radioactive decay useful in the absolute-age dating of rocks?
A. It will only break down the fossils within the rock and not the rock itself.
B. It will only break down the rock and not the fossils contained in the rock.
C. It is constant regardless of environment, pressure, temperature, or any other physical changes.
D. It fluctuates depending on environment, pressure, temperature, or any other physical changes.

10. What was formed in North America when Gondwana and Laurasia collided?
A. Himalaya Mountains
B. Appalachian Mountains
C. Andes Mountains
D. Great Permian Reef

Online Test Practice

SHORT ANSWER

Use the illustration below to answer Questions 11 and 12.

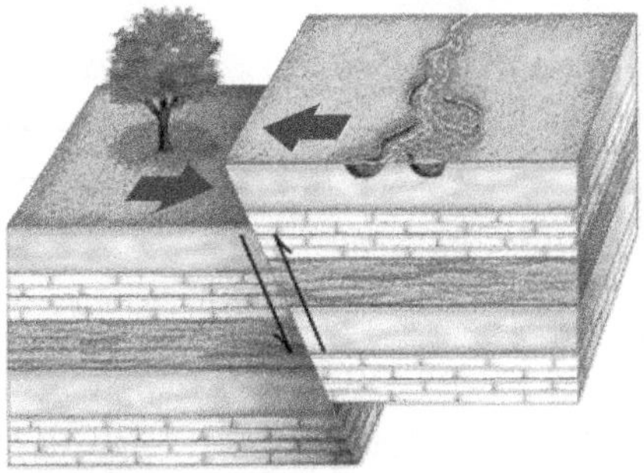

11. What type of fault is shown, and how is it formed?
12. Describe how rock surfaces along this fault lead to an earthquake.
13. Why are uplifted mountains unique?
14. Discuss how sources of heat on early Earth made conditions inhospitable to life.
15. What is the purpose of the geologic time scale?
16. Describe the formation of the Rocky Mountains during the Mesozoic Era.

READING FOR COMPREHENSION

Native Landscapes

Landscaping with native plants improves the environment. Native plants are hardy because they have adapted to the local conditions. Once established, native plants do not need pesticides, fertilizers, or watering. A native landscape does not need to be mowed like a conventional lawn. This reduces the demand for nonrenewable resources and improves the water and air quality. The periodic burning required for maintenance of a prairie landscape mimics the natural prairie cycle and is much better for the environment. Landscaping with native wildflowers and grasses helps return the area to a healthy ecosystem. Diverse varieties of animals are attracted to native plants, enhancing biodiversity in the area.

Article obtained from: Green landscaping: Greenacres. Green Landscaping with Native Plants. *United States Environmental Protection Agency.* October 2006.

17. Why is periodic burning good for a prairie landscape?
    A. It gets rid of any unwanted weeds.
    B. It mimics the natural prairie cycle.
    C. It gets rid of any possible pests on the plants.
    D. It provides a chance to create a new setting.

18. What can be inferred from this passage?
    A. Landscaping with native plants is the best option for planting in an area.
    B. Only native plants will survive in their given environment.
    C. Native landscaping works only in prairie settings.
    D. Planting native landscapes can be costly and time consuming, but it is very important.

19. Why would the Environmental Protection Agency be interested in sharing this information?
    A. to reduce the number of nonnative plants sold
    B. to help conserve nonrenewable resources and protect the environment from harsh chemicals
    C. to provide avid gardeners with new approaches to creating their gardens
    D. to identify inexpensive ways of gardening for novice gardeners

NEED EXTRA HELP?

If You Missed Question . . .	1	2	3	4	5	6	7	8	9	10	11	12	13	14	15	16
Review Section . . .	23.1	21.4	21.4	22.4	22.3	21.1	24.1	24.1	21.3	23.2	19.1	19.1	20.3	22.1	21.1	23.3

SHORT ANSWER

11. The fault shown is a reverse fault. It is formed as a result of horizontal compression, which creates a horizontal shortening.
12. Irregular surfaces in the rock cause movement along the fault to snag and lock up. This causes strain between the two sides of the fault to continually build until they reach their elastic limit and an earthquake is produced.
13. Uplifted mountains, unlike other mountains that form at plate boundaries, form when large regions of Earth are slowly forced upward as a unit. This causes the rocks to be typically less deformed than rocks associated with plate-boundary mountain formation.
14. The sources of heat on early Earth would prove inhospitable to life because temperatures were too great for life to exist. First, there were more radioactive isotopes that had begun to decay and produce heat. Large asteroids regularly struck Earth, which caused heat with impact. Finally, as Earth was growing by joining with other bodies around it, Earth's mass increased, thus increasing gravity and heat.
15. The geologic time scale enables scientists to show relationships among geologic events and environmental changes in an organized fashion. It also helps to chronologically classify fossils and the development of life-forms.
16. During the Mesozoic Era, massive uplifting of Earth's crust occurred, forming the Rocky Mountains.

READING FOR COMPREHENSION

17. B
18. A
19. B

CHAPTER 25 Energy Resources

BIGIDEA **People use energy resources, most of which originate from the Sun, for everyday living.**

ESSENTIAL QUESTIONS	NATIONAL STANDARDS	RESOURCES TO ASSESS MASTERY
SECTION 1 **Conventional Energy Resources** **1.** Why is the Sun the source of most energy on Earth? **2.** What materials are used as fuels? **3.** How does coal form? 2 sessions 1 block	UCP.1–3; A.1; B.2–3, B.5–6; C.5; D.1–2; F.3–6; G.3	**Entry-Level Assessment** Focus Transparency 78, p. 708 **Progress Monitoring** Reading Check, pp. 709, 712 Section Review, p. 713
SECTION 2 **Alternative Energy Resources** **1.** What are several alternative energy resources? **2.** How can the Sun's energy be harnessed? **3.** How can water, wind, nuclear, and thermal energy be used to generate electricity? **4.** Why might nuclear energy be controversial? 1 sessions 0.5 block	B.1–3, B.5–6; C.5; D.1–2; F.3–6	**Entry-Level Assessment** Focus Transparency 79, p. 714 **Progress Monitoring** Caption Question, pp. 717, 718 Reading Check, p. 715 Section Review, p. 719
SECTION 3 **Conservation of Energy Resources** **1.** How can energy resources be conserved? **2.** How can increasing energy efficiency help preserve fossil fuels? **3.** How can energy be used more efficiently? 5 sessions 2.5 blocks	UCP.1, UCP.3, UCP.5; A.1; B.2–3; C.4–5; D.1; E.3; F.4–6; G.3	**Entry-Level Assessment** Focus Transparency 80, p. 720 **Progress Monitoring** Caption Question, p. 720 Reading Check, p. 721 Section Review, p. 723 **Summative Assessment** Chapter Assessment, p. 727 Exam*View*® *Assessment Suite* CD-ROM

Go online!
connectED.mcgraw-hill.com

LEVELED RESOURCES	LAB MATERIALS
Science Notebook 25.1 OL **Chapter FAST FILE Resources:** MiniLab Worksheet, p. 28 BL Study Guide, p. 39 BL **Transparencies:** Section Focus Transparency 78 EL Teaching Transparency 76 OL EL	LaunchLAB p. 706 / **10 min** water, 250-mL glass beaker, hot plate MiniLAB p. 712 / **20 min** cooking oil, 100-mL graduated cylinder, sand, colored aquarium gravel, tap water
Science Notebook 25.2 OL **Chapter FAST FILE Resources:** GeoLab Worksheet, p. 29 OL Study Guide, p. 41 BL **Lab Resources:** Laboratory Manual, pp. 193, 197 OL **Transparencies:** Section Focus Transparency 79 EL Teaching Transparency 77 OL EL	GeoLAB p. 725 / **90 min** scissors, glass or clear plastic squares, sturdy cardboard boxes, tape, glue, thermometers, paint, paper, aluminum foil, polystyrene, stone, mirrors, fabric, light source
Science Notebook 25.3 OL **Chapter FAST FILE Resources:** Study Guide, p. 44 BL **Transparencies:** Section Focus Transparency 80 EL Teaching Transparency 78 OL EL	

LabManager™
Customize any lab with the LabManager™ CD-ROM.

ADDITIONAL RESOURCES

Technology:
- eTeacherEdition
- eStudentEdition
- Classroom Presentation Toolkit
- Exam*View® Assessment Suite* CD-ROM
- LabManager™ CD-ROM
- What's EARTH SCIENCE Got To Do With It? DVD
- Weather Classroom Video DVD

Assessment:
- Blueprints for Success
- Chapter Test (Scaffolded)
- Earth Science Standardized Test Practice

Lab Resources:
- Exploring Environmental Problems Laboratory Manual

BL Below Level OL On Level AL Advanced Learners EL English Learners COOP LEARN Cooperative Learning

CHAPTER 25

LaunchLAB

iLab Station Rubric

Can you identify sources of energy?

Process Skills observe and infer, recognize cause and effect, communicate, model, measure in SI

Safety Precautions Approve lab safety forms before work begins. Caution students to handle the hot water with care. Remind students to wear safety goggles and aprons and to allow the water in the beaker to cool before attempting to move it at the end of the lab.

Teaching Strategy Ask students the following questions: What is happening to the water? Is it remaining in the beaker? If not, where is it going?

Procedure

WARNING: ***Allow the beaker to cool before moving it at the end of the activity.***

1. Have students read and complete the lab safety form and follow the procedure below.
2. Add 200 mL of **water** to a **250-mL glass beaker**.
3. Place the beaker on a **hot plate**.
4. Turn the hot plate on high. Observe what happens to the water as it heats up and begins to boil.

Analysis

1. **Describe** what happened to the energy as it was used to heat and boil the water. As the energy heats the beaker, the energy in the beaker is transferred into the water.
2. **Infer** where the energy went when the water began to boil. The heat was conducted to the surrounding air.
3. **Determine** Where did the energy to boil the water come from? Trace the electricity from your school to its source. Answers will vary, but should indicate that the energy came from a power plant. Students should trace the power from the school to a transformer on the pole (or a box on the premises), back to a sub-station, through high tension lines to the power plant.

Assessment

Performance Ask students the following questions: What is the ultimate source of most of the energy on Earth? the Sun What happens to energy on Earth? It changes from one form to another as it is transferred from one object or phase to another. In the overall Earth system, is energy lost or gained? Neither; it is transferred and can change form.

LabManager™
Customize any lab with the LabManager™ CD-ROM.

CHAPTER 25

Energy Resources

BIG IDEA People use energy resources, most of which originate from the Sun, for everyday living.

SECTIONS

1. Conventional Energy Resources
2. Alternative Energy Resources
3. Conservation of Energy Resources

Wind farms have become a source of energy in several states, including Iowa, Minnesota, Texas, and California. The Judith Gap Wind Energy Center in Montana hosts ninety wind turbines that are capable of producing 1,500 kilowatts of power each.

LaunchLAB

iLab Station

Can you identify sources of energy?

Energy cannot be created or destroyed, but it can change form and be transferred. Thus, the same energy can be used repeatedly. Learn more about the transfer of energy in this lab.

FOLDABLES® Study Organizer

Alternative Energy Resources

Make a layered-look book using the labels shown. Use it to organize your notes on alternatives to traditional energy resources.

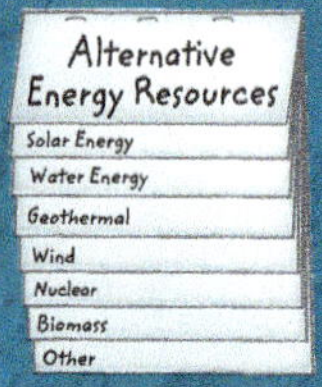

Introduce the BIGIDEA

Sources of Energy Ask students to identify all of the sources of energy in the classroom. Answers might include the Sun, plants, electricity, gas, or food. Then ask students to identify how each source of energy listed is made. Answers will vary, but might include: the Sun fuses hydrogen, plants use sunlight to make food, electricity is generated in power plants that use coal or nuclear energy, natural gas is made from the breakdown of plant and animal material from millions of years ago, and plants and animals can be used as food.

Teacher Content Support

Alternative Energy The wind farm shown here generates electricity using energy from the wind. According to the American Wind Energy Association (AWEA), the combined power generating capacity of all U.S. wind turbines could power 9.7 million homes. Another alternative energy resource is solar energy. Silicon from one metric ton of sand, used in solar cells, produces as much electricity as burning 500,000 metric tons of coal. Ask: Why do you think alternative energy resources are not more widely used? Possible answers: costs involved with building wind farms and photovoltaic cells; local weather requirements such as a fairly constant supply of wind or sunlight

SECTION 1

1 Focus

Focus Transparency

Before presenting the lesson, project **Section Focus Transparency 78** and have students answer the accompanying questions. EL

MAINIDEA

Energy Students can relate to energy sources when they realize the connection between their high calorie snacks and energy boosts. Ask: Have you ever felt the need for a sweet snack or drink mid-afternoon? Answers will vary. Meeting a craving for a sweet treat can give you a quick energy boost. Your sweet treats are packets of stored energy in a form your body can access quickly. In this manner, the available energy in your food is similar to the available energy in Earth's fossil fuels.

2 Teach

Teacher Content Support

Energy Supply Energy use in the United States is heavily dependent on Earth's oil supply. Although it has been estimated that the United States once had 10 percent of Earth's oil supply, over half of this oil has already been consumed, and the country's net oil reserves have been steadily declining each year. The United States' energy appetite consumes more than 40 percent of all energy resources used in any one year. Some analysts predict that at this rate of use, Earth's oil supply will be depleted in just a few decades. A careful review of energy use will help students understand the magnitude of the course that has been charted by the lifestyles of Americans.

SECTION 1

Conventional Energy Resources

MAINIDEA Biomass and fossil fuels store energy from the Sun.

Essential Questions

- Why is the Sun the source of most energy on Earth?
- What materials are used as fuels?
- How does coal form?

Review Vocabulary

fault: fracture in Earth's crust along which movement occurs

New Vocabulary

fuel
biomass fuel
hydrocarbon
peat
fossil fuel

EARTH SCIENCE 4 YOU **What kinds of activities do you engage in each morning? In the kitchen, you might toast bread or use a microwave oven to heat up your breakfast. You might ride a bus to school or drive a car. All of these activities require energy, and the food you eat, such as toast, provides your body with the energy it needs to function.**

Earth's Main Energy Source

The energy that humans and all other organisms use comes mostly from the Sun. How is solar energy used by organisms? Plants are producers—they capture the Sun's light energy in the process of photosynthesis. The light energy is converted into a form that can be used for maintenance, growth, and reproduction by the plant. When other organisms called consumers eat producers, they use that stored energy for their own life processes. For example, when a rabbit eats grass, it consumes the energy stored by the plant. The rabbit stores energy as well, and this energy can be transferred to other organisms when the rabbit is eaten, when the rabbit produces waste, or when it dies and decomposes back into the ground. **Figure 1** shows how trapped light energy can be transferred from plants to humans.

Humans use energy to keep them warm in cold climates, to cook food, to pump water, and to provide light. There are many different fuel sources available to humans to provide this energy. Most of these fuels also store energy that originated from the Sun.

Figure 1 Humans need energy to live. When you eat a bowl of cereal, you use energy derived from the Sun. The wheat plant harnessed the Sun's light energy through photosynthesis. Some of this energy was stored in the seed of the wheat which humans can consume to get energy they need to survive.

CULTURAL DIVERSITY

A Quote from Chief Seattle "Teach your children what we have taught our children, that the Earth is our mother. And what is man without the beasts? If all the beasts were gone, man would die from a great loneliness of spirit. This we know. The Earth does not belong to man, man belongs to the Earth. Man did not weave the web of life; he is merely a strand in it. All things are connected like the blood, which unites one family. All things are connected." These words were attributed to Chief Seattle of the Suquamish tribe, Washington Territory, when he addressed the assembly to sign the Native American treaties in 1854. This quotation is known in many cultures and embodies the ideals of the modern ecology movement.

Biomass Fuels

Fuels are materials that are consumed to produce energy. The total amount of living matter in an ecosystem is its biomass. Therefore, fuels derived from living things are called **biomass fuels.** Biomass fuels, shown in **Figure 2,** are renewable resources.

One type of fuel available for human use is derived directly from plant material. Plant materials burn readily because of the presence of **hydrocarbons**—molecules with hydrogen and carbon bonds only. Hydrocarbons are the result of the combination of carbon dioxide and water during photosynthesis. When plant materials burn, carbon dioxide is released as a waste product.

Wood Humans have been using wood for fuel for thousands of years. Billions of people, mostly in developing countries of the world, use wood as their primary source of fuel for heating and cooking. Unfortunately, the need to use wood as a fuel has resulted in deforestation of many areas of the world. As forests near villages are cut down for fuel, people travel farther to gather the wood they need. In some parts of the world, this demand for wood has led to the complete removal of forests, which can result in erosion and the loss of topsoil.

Field crops Another biomass fuel commonly used in developing countries is field crops. The simplest way to use field crops, such as corn, hay, and straw, as fuel is to burn them. Crop residues left after harvest, including the stalks, hulls, pits, and shells from corn, grains, and nuts, are other sources of energy.

Fecal material Feces are the solid wastes of animals. In many cases, dried feces contain undigested pieces of grass that help the material to burn. Feces from cows often meet the energy needs of people in developing countries with limited forest resources. Some people collect animal fecal matter for fuel and dry it on the outside walls of their stables or compounds as shown in **Figure 2.**

☑ READING CHECK **Explain** how field crops, fecal material, and wood are all examples of biomass fuels.

VOCABULARY
SCIENCE USAGE V. COMMON USAGE
Consume
Science usage: to use up completely

Common usage: to eat

■ **Figure 2** Biomass fuel, such as wood, field crops, and fecal material, is the primary source of fuel for people in many countries. The fecal matter in the image below has been hung on the side of this home to dry before it is burned.

Wood

Fecal material

DIFFERENTIATED INSTRUCTION

Struggling Learners Have students use the letters in the word *energy* to develop a simple crossword puzzle composed of energy-related words. Have students make a list of the energy-related vocabulary in the section, and then have students arrange the words into a crossword puzzle. You might need to help students with writing the clues for their puzzles. Have students include their completed puzzles in their Earth science journals.

Teacher Content Support

Early Fuels The first fuels used by humans were wood and coal. The use of wood for fuel took time and effort, from cutting down trees and dragging them home, to cutting branches into the proper length for a firepit or oven, to watching to make sure the fire didn't go out. Although coal is heavy, dirty, and somewhat cumbersome to use, the use of coal was considered to be an improvement over wood because coal burns steadily for hours without needing much attention. It is estimated that currently, coal is used by only about 145,000 homes in the United States. However, coal is used by over half of the electrical power-generation plants in the United States.

Collaborative Learning

Energy Sources Have students work in groups to compile a list of the energy sources that each group member uses in one week. Tell student groups they must devise a system of collecting data from individuals that includes measurements of the amount of energy used. Each group must compile data and make a presentation to the class. Have one group prepare a summary chart or data table that includes the data collected from all the other groups. Ask the group preparing the class summary data to answer the following questions as part of the group report: What were the most common energy sources listed? What were the most common uses of energy? A general conclusion for the class as a whole might be that students' lives are dependent on many sources of energy. OL COOP LEARN

☑ READING CHECK Field crops, fecal material, and wood are all considered biomass fuels because they are derived from living things and are renewable resources.

Teacher Content Support

Peat Large beds of peat can be found in Europe, North America, and northern Asia. Millions of metric tons of peat are burned annually in Ireland, and considerable quantities are burned in Russia, Sweden, Germany, and Denmark. Peat is used for domestic purposes in England and Scotland. However, peat is used for fuel only where other energy resources are lacking. Dried peat can be burned directly in furnaces, made into briquettes, or burned to produce gas.

Discussion

New Fuel Sources Ask students to propose materials that might become tomorrow's fossil fuels. Ask students what characteristics the materials should have to be used as fossil fuels. Students might note that materials would have to be readily preserved and changed to a usable form. In addition, materials would have to be located where they are accessible. OL

Environmental Connection

Paper Usage On average, one American uses around 300 kg of paper annually compared to the 45 kg used by one Chinese person. Share these facts with students to serve as a basis for developing awareness.

Figure 3 These people harvested peat for fuel and carried the material from bogs to their homes.

Peat Bogs are poorly drained areas with spongy, wet ground that is composed mainly of dead and decaying plant matter. When plants in a bog die, they fall into the water. Bog water is acidic and has low levels of oxygen; these conditions slow down or stop the growth of the bacteria that decompose dead organic matter, including plants. As a result, partially decayed plant material accumulates on the bottom of the bog. Over time, as the plant material is compressed by the weight of water and by other sediments that accumulate, it becomes a light, spongy material called **peat,** shown in **Figure 3.** Most of the peat used as fuel today is thousands of years old.

Peat has been used as a low-cost fuel for centuries because it can be cut easily out of a bog, dried in sunlight, and then burned in a stove or furnace to produce heat. Highly decomposed peat burns with greater fuel efficiency than wood. Today, peat is used to heat many homes in Ireland, England, parts of northern Europe, and the United States.

Fossil Fuels

Energy sources that formed over geologic time as a result of the compression and incomplete decomposition of plants and other organic matter are called **fossil fuels.** Although coal, oil, and natural gas originally formed from once-living things, these energy sources are considered nonrenewable. Recall that nonrenewable resources are used at a rate faster than they can be replaced. Fossil fuels are nonrenewable resources because their formation occurs over millions of years, and we are using them at a much faster rate.

Fossil fuels mainly consist of hydrocarbons and can be transported wherever energy is needed and used on demand. This is why most industrialized countries, including the United States, depend primarily on coal, natural gas, and petroleum to fuel electric power plants and vehicles. Although fossil fuels are diverse in their appearance and composition, all of them originated from organic matter trapped in sedimentary rock.

Vocabulary

Academic Vocabulary

Diverse
made up of distinct characteristics, qualities, or elements
The United States has diverse climates—the Northwest is cool and wet, while the Southwest is hot and dry.

Coal Coal is the most abundant of all the fossil fuels. Recall that coal forms from peat over millions of years. As compression continues, the hydrogen and oxygen in peat are lost and only carbon remains. The greater the carbon concentrations in coal, the hotter it burns. Most coal reserves in the United States are bituminous coal, therefore, many of the electricity-generating plants in the United States burn this type of coal. Study **Figure 4** to learn how the different types of coal form.

Across the Curriculum

History Ask students to write reports on the history of coal mining in the United States. Ask students to include the following information in their reports: the location of mines, the effect of mines on the country's economy, the health issues of workers, environmental concerns, the quality of coal, transportation issues, the introduction and use of technology, and the future of coal mining in the United States. OL

VISUALIZING Coal

Figure 4 Coal forms from the compression of organic material over time.

Lignite is a soft, brown, low-grade coal with low sulfur content—less than 1 percent. Because the carbon concentration in lignite is generally around 40 percent, it is inefficient as a fuel. More lignite must be burned than other types of coal to provide the same amount of energy.

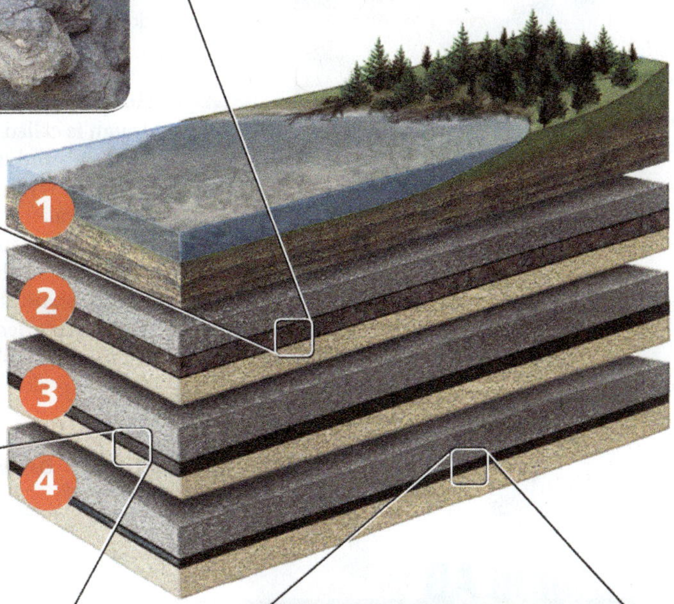

1. Incomplete decay of plants forms peat.
2. Peat is buried and compressed to form lignite.
3. After further compression, bituminous coal forms.
4. More heat and pressure are applied to form anthracite.

Bituminous coal can have carbon concentrations as high as 85 percent. When bituminous coal burns, it releases carbon dioxide and gases containing sulfur and nitrogen into the air, causing air pollution.

Anthracite can have a carbon concentration as high as 90 to 95 percent, and it stores more energy and burns cleaner than other types of coal. However, less than 1 percent of the coal reserves in the United States are anthracite.

Concepts In Motion View an **animation of coal formation.**

EARTH SCIENCE JOURNAL

Coal Formation To familiarize students with the different stages of coal formation, ask students to create a table in their Earth science journals to record the characteristics of lignite, bituminous coal, and anthracite. Then ask students to use their table of lignite, bituminous coal, and anthracite to create a flow chart showing how plants eventually transform into coal. OL

VISUALIZING

Purpose

Students will examine how coal forms over time.

Teacher Content Support

Coal Reserves The location and quantity of the coal reserves on Earth are known. Currently, most coal exploration involves the evaluation of coal reserves. Core samples of the coal are taken to determine its quality and to identify the geologic structures in which it is found. Satellites and advanced sensors help in the creation of safe and economical mining plans. Today, with the assistance of computers, planning can be completed in seconds. This process saves time and money, and it also eliminates the dangerous physical exploration that humans once performed.

Tie to Previous Knowledge

Geologic Time Scale Remind students of the geologic time scale to help them keep in perspective the length of time it takes coal to form. Most of the coal deposits in Europe and in the eastern United States began to form during the Paleozoic Era. This was the time of the trilobites, crinoids, and early fishes. Coal deposits in the western United States formed later, during the Mesozoic Era, the time when dinosaurs ruled the land. OL

Enrichment

Formation of Coal Geologists have determined that the formation of a layer of coal 1 m thick required a layer of plants approximately 20 m thick. Coal deposits range in thickness from a few centimeters to over 30 m. AL

MiniLAB

iLab Station Rubric

Purpose Students will demonstrate how oil migrates upward through the pores in permeable rocks.

Process Skills model, recognize cause and effect, observe and infer, predict

Safety Precautions Approve lab safety forms before work begins. Have students wear safety goggles and an apron while performing this activity.

Teaching Strategy Have students work in small groups.

Expected Results Students will observe that adding water causes the oil to move upward through the sand and gravel.

Analysis

1. crude oil, layers of rocks and their pores
2. The oil moves until it is floating on top of the water. It is less dense than water.
3. Students might predict that the bubbles of the soft drink will rise to the top as the drink sinks to the bottom of the cylinder. The bubbles in the soft drink represent natural gas.

Assessment

Performance Have students design and conduct an experiment to determine how an increase in temperature affects the cooking oil, sand, gravel, and water mixture.

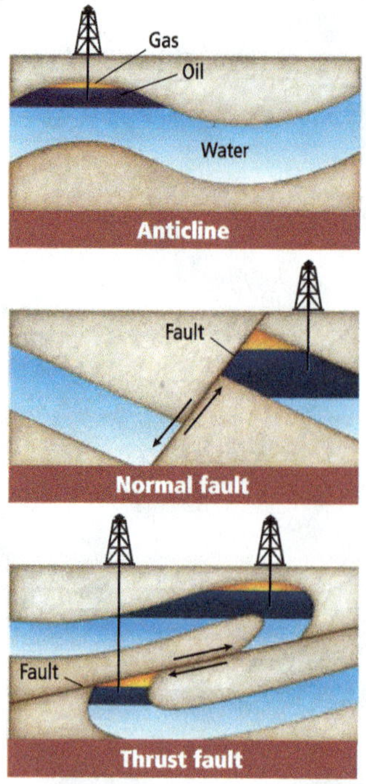

Figure 5 These diagrams show typical structural traps for oil and gas deposits.

Petroleum and natural gas Most petroleum deposits formed from the accumulation of microscopic organisms on the seafloor. Other deposits are found in ancient lake beds. At the time of deposition, if the water was stagnant and lacked oxygen, organisms did not decay completely and a layer of organic matter formed. Over time, layers of sediment buried the organic matter, and as more and more sediments were added, the temperature and pressure increased, forming liquid oil, also called crude oil. Crude oil that is collected on Earth's surface or pumped out of the ground is refined into a wide variety of petroleum products, such as gasoline, diesel fuel, and kerosene.

Natural gas forms along with oil and is found beneath layers of solid rock. The rock prevents the gas from escaping to Earth's surface.

Migration Rock containing pores or spaces that liquid can move through is called permeable rock. Crude oil and natural gas migrate sideways and upward from their place of formation, or source rock. As they migrate, they accumulate in permeable sedimentary rocks, called reservoir rocks, such as limestone and sandstone. Because petroleum is less dense than water, oil and gas continue to rise until they reach a barrier of impermeable rock, such as slate or shale, that prevents their continued upward movement. This barrier effectively seals the reservoir and creates a trap for the petroleum. Geologic formations such as faults and anticlines—folds of rock—can trap petroleum deposits, as shown in **Figure 5.**

☑ READING CHECK **Describe** how oil migrates upward through sedimentary rock.

MiniLAB

iLab Station

Model Oil Migration

How does oil move through layers of porous rocks?

Procedure

1. Read and complete the lab safety form.
2. Pour 20 mL of **cooking oil** into a **100-mL graduated cylinder.**
3. Pour **sand** into the graduated cylinder until the sand-oil mixture reaches the 40-mL mark.
4. Add a layer of **colored aquarium gravel** above the sand until the gravel reaches the 70-mL mark.
5. Pour **tap water** into the graduated cylinder until the water reaches the 100-mL mark.
6. Observe the graduated cylinder for 5 min. Record your observations.

Analysis

1. **Identify** what the cooking oil, sand, and aquarium gravel represent.
2. **Explain** what happened when you added water to the mixture in the graduated cylinder. Why does adding water cause this change?
3. **Predict** what might occur in the graduated cylinder if you added a carbonated soft drink to the mixture instead of water. What would the bubbles represent?

☑ READING CHECK Oil migrates upward through permeable sedimentary rocks because it is less dense than water.

EARTH SCIENCE JOURNAL

Time and Scale Ask students to use their Earth science journals to create an analogy comparing 1 million years to 1000 or 100 years by choosing a common substance in those quantities. For example, a pinch of salt might contain a thousand grains, a cup of salt might contain a million grains, and a bathtub of salt might contain a billion grains. AL

Oil shale Some petroleum resources are trapped in different types of rocks. For example, oil shale is a fine-grained rock that contains a solid, waxy mixture of hydrocarbon compounds called kerogen. Oil shale can be mined, then crushed and heated until the kerogen vaporizes. The kerogen vapor can then be condensed to form a heavy, slow-flowing, dark-brown oil known as shale oil. Shale oil is processed to remove nitrogen, sulfur, and other impurities before it can be sent through the pipelines to a refinery.

The largest deposits of oil shale in the world are found in the Green River Formation of Utah, Wyoming, and Colorado, shown in **Figure 6.** This geologic formation consists of lake sediments that were deposited during the Eocene Epoch, around 50 mya. Within these sediments are an estimated 800 billion barrels of recoverable oil, which is three times greater than the proven oil reserves of Saudi Arabia. People in the United States use about 20 million barrels of oil per day. If oil shale could be used to meet a quarter of that demand, the estimated 800 billion barrels of recoverable oil from the Green River Formation would last for more than 400 years.

Historically, the cost of oil derived from oil shale has been significantly higher than pumped oil. Recently, prices for crude oil have again risen to levels that might make oil-shale-based oil production commercially viable, and both governments and industries are interested in pursuing the development of oil shale.

Figure 6 Oil shale is found primarily in sedimentary rocks. One of the most abundant sources of oil shale known is the Green River Formation, shown on the map as the dark green regions.

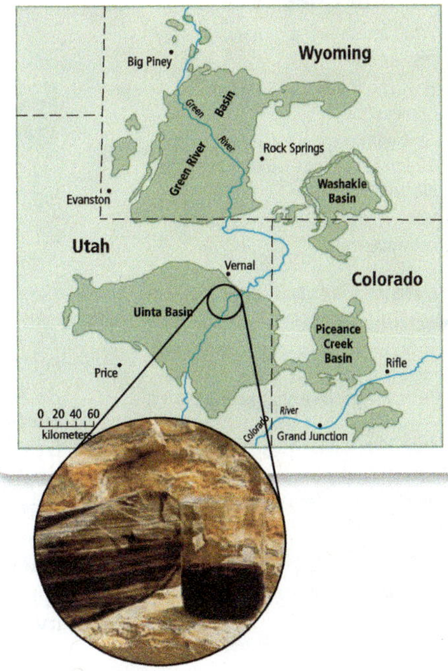

SECTION 1 REVIEW

Section Self-Check

Section Summary

- The Sun is the source of most energy on Earth.
- Humans have used materials derived from living things, such as wood, as renewable fuels for thousands of years.
- Fossil fuels formed from organisms that lived millions of years ago.

Understand Main Ideas

1. MAINIDEA **Explain** how energy stored in coal was obtained from the Sun.
2. **List** four types of biomass fuels.
3. **Illustrate** how coal forms.
4. **Discuss** how two uses of energy in your home can be traced back to the Sun.

Think Critically

5. **Evaluate** this statement: Anthracite is usually found deeper in Earth's crust than lignite.
6. **Debate** whether scientists should research the prospect of obtaining oil from the Green River Formation.

WRITING IN ▶ Earth Science

7. Research different ways coal can be mined and write a report on the positive and negative effects of mining.

3 Assess

Check for Understanding

Compare and Contrast Show students samples of peat, lignite, bituminous coal, and anthracite. Ask students what types of comparisons they can and cannot make just by looking at and handling the samples. Students should recognize that they are not able to assess the samples' ability to burn, amount of heat generated by burning, and cleanliness of combustion.

Reteach

Make and Use Tables Have students make data tables of conventional energy resources. Students should include in their tables how each energy resource is formed, where it is found, its uses, and its availability.

Assessment

Performance Have students draw maps of Earth. Ask students to include symbols representing locations that use conventional energy resources.

SECTION 1 REVIEW

1. Coal forms from plant materials that partially decay to form peat. Over time, the peat layers become compressed and form coal. The original plant material stored energy from the Sun through photosynthesis.
2. Biomass fuels can include wood, field crops, fecal material, and peat.
3. Incomplete decay of plant material forms peat. The peat is compressed to form lignite. When compressed further, it becomes bituminous coal and then anthracite.
4. Answers will vary. Answers could include electricity if it originates from a fossil fuel, natural gas for heating and cooking, and passive or active solar heating.
5. Anthracite is usually found deeper in Earth's crust than lignite because it requires more heat and pressure to form.
6. Answers will vary but should include the pros and cons of oil extraction.
7. Answers will vary. Students might say that mining provides valuable resources besides coal, such as ores and other minerals. However, mining techniques can destroy natural habitats and cause groundwater and surface water pollution.

Rubric

SECTION 2

1 Focus

Focus Transparency

Before presenting the lesson, project **Section Focus Transparency 79** and have students answer the accompanying questions. EL

MAINIDEA

Energy Use Ask students to describe how they might use energy to go from home to school. Each form of transportation (driving, walking, and riding a bike) uses a different kind of energy.

2 Teach

Teacher Content Support

Alternative Energy Because conventional sources of energy are clearly in limited supply, it is necessary to find and use alternative energy sources. This is especially true if people in developed countries such as the United States maintain the same standard of living. These people likely will continue to demand the energy needed to maintain their lifestyles. The premise of using alternative energy sources suggests that limited conventional resources can be replaced with other energy sources; however, this task will not be easy to accomplish. It will require a concerted effort and possibly some concessions or adjustments in expectations.

Tie to Previous Knowledge

Solar Energy Ask students to think back to a hot, sunny day. Ask students why it feels cool underneath a tree on a hot day. Students are likely to make the connection that it feels cool under a tree because they are not in direct sunlight. BL

SECTION 2

Alternative Energy Resources

MAINIDEA Many resources other than fossil fuels can be developed to meet the energy needs of people on Earth.

Essential Questions

- What are several alternative energy resources?
- How can the Sun's energy be harnessed?
- How can water, wind, nuclear, and thermal energy be used to generate electricity?
- Why might nuclear energy be controversial?

Review Vocabulary

electron: subatomic particle that has little mass, but has a negative electric charge that is exactly the same magnitude as the positive charge of a proton

New Vocabulary

photovoltaic cell
hydroelectric power
geothermal energy
nuclear fission

EARTH SCIENCE 4 YOU **Have you ever walked barefoot across dark-colored pavement on a hot day? The thermal energy from the Sun caused the pavement to heat up and might have felt very hot on your feet. Scientists are working to find the most efficient ways to convert this thermal energy from the Sun into electricity for human use.**

Solar Energy

Have you ever used a calculator with a solar collector? These solar-powered calculators use the Sun's energy to provide power. As you learned previously, the Sun is the source of most of the energy on Earth. The main advantages of solar energy are that it is free and it doesn't cause pollution.

As you also learned previously, many of the fuels used today are renewable resources, including wood. Most people, however, rely heavily on nonrenewable fossil fuels for their energy needs. Nonrenewable fossil fuels (oil, coal, and natural gas) are used to generate approximately 80 percent of the total energy consumed for electricity, heat, and transportation in the United States.

However, the supply of fossil fuels on Earth is limited. **Figure 7** shows that at the present rate of consumption, scientists estimate that oil and natural gas reserves might last only another 50 years. Although coal will last longer, burning coal releases harmful gases into the atmosphere, such as sulfur dioxide. Scientists, private companies, and government agencies are all studying renewable resources, such as solar energy, as alternatives to traditional energy resources, including fossil fuels.

You have learned that plants transfer the energy provided by the Sun to other organisms through food webs. Solar energy can also be used directly to meet human energy needs through passive and active solar heating.

Figure 7 At current consumption rates, available oil reserves might last only 50 years.

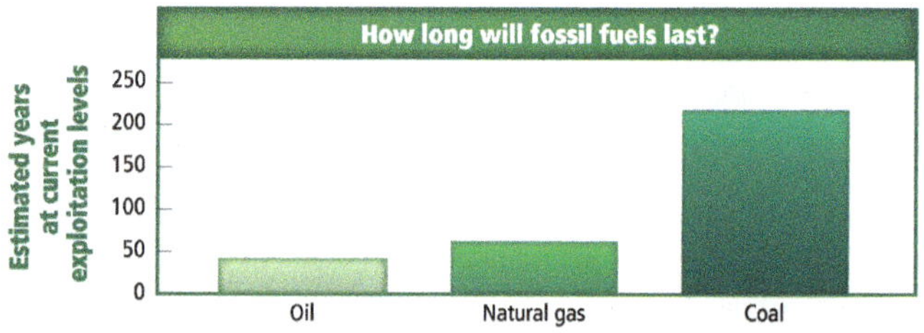

Demonstration

Observe Light Energy Wear goggles and have students wear goggles if you are not using a safety shield for this demonstration. Obtain two beakers, then fill one with soil and the other with water. Place a thermometer 1 cm below the surface of the soil in the first beaker, and another thermometer 1 cm below the surface of the water in the second beaker. Place a light source equidistant from both beakers and leave the light on for 15 min. Have students note any changes in temperature on either thermometer at the end of the 15 min. Students should observe that the light energy was more readily absorbed by the soil and therefore the thermometer in the soil registered a higher temperature. OL

Passive solar heating If you have ever sat in a car that has been in the sunlight, you know that the Sun can heat up the inside of a car just by shining through the windows and on the surface of the car. In the same way, the Sun's energy can be captured in homes. Thermal energy from the Sun enters through windows, as shown in **Figure 8.** Floors and walls made of concrete, adobe, brick, stone, or tile have heat-storing capacities and can help to hold the thermal energy inside the home. These materials collect solar energy during the daytime and slowly release it during the evening as the surroundings cool.

In some warm climates, these materials alone can provide enough energy to keep a house warm. Solar energy that is trapped in materials and slowly released is called passive solar heating. Passive solar designs can provide up to 70 percent of the energy needed to heat a house. Although a passive solar house can be slightly more expensive to build than a traditional home, the cost of operating such a house is 30 to 40 percent lower.

☑ READING CHECK **Explain** the process of heating a home using passive solar heating.

Active solar heating Even in areas that do not receive consistent sunlight, the Sun's energy can still be used for heating. Active solar-heating systems include collectors such as solar panels that absorb solar energy, and fans or pumps that distribute that energy throughout the house.

If kept away from trees, solar panels mounted on the roof can have unobstructed exposure to the Sun. Energy collected by these solar panels can be used to heat a house directly, or it can be stored for later use in insulated tanks that contain rocks, water, or a heat-absorbing chemical. Solar panels, shown in **Figure 8,** mounted on a roof can heat water up to 65°C, which is hot enough to wash dishes and clothing.

Passive and active solar heating rely on direct sunlight. Using direct sunlight is relatively easy, but energy is also needed during hours of darkness, or in areas that are often overcast. Solar energy is difficult to store for later use. An economical and practical method of storing large amounts of solar energy for long periods of time has not yet been developed.

Passive solar heating

Active solar heating

Figure 8 Solar heating is considered a good alternative to conventional energy resources because it is clean and readily available in some areas. However, sunlight is available during limited hours each day and it is difficult to store for later use. More research needs to be done to make solar power a reasonable alternative for more people.

FOLDABLES® Incorporate information from this section into your Foldable.

Identify Misconceptions

Some students might think that Earth's distance from the Sun is the reason that the tropics are warmer than northern or southern locales.

Uncover the Misconception

Ask students why it is warmer in the summer than in the winter. Typically, students will say that it is warmer on Earth during the summer because the Sun is closer or providing more energy to Earth at that time.

Demonstrate the Concept

Turn on a flashlight and point it directly at the board. Have students note the bright ring of light. Then shine the flashlight at an angle to the board. The light will form more of an oval shape and be less bright. Ask students whether the flashlight is still producing the same amount of light energy and whether the flashlight is still the same distance away. Students should answer "yes" to both questions. Students will see that it is the angle of the Sun's rays that determines the intensity of the energy being received.

Assess New Knowledge

Ask students the following true-false questions: The amount of solar energy reaching Earth is comparable each day of the year. true The angle of the Sun's rays is the major factor in determining the amount of energy absorbed by Earth. true

FOLDABLES® Rubric

DIFFERENTIATED INSTRUCTION

English Learners Students from other areas of the world might have rich experiences to share with the class. Cultural differences in the use of energy can provide a student from another cultural background the opportunity to share another perspective and personal experiences. Encourage students from other cultures to share their own experiences with energy resources with the class.

☑ READING CHECK Passive solar heating involves trapping the Sun's natural energy through efficient building design, efficient building materials, curtains, or window treatments. During the day, the Sun's energy is trapped inside the house and used as an energy source.

Environmental Connection

Solar Energy Solar energy is an abundant source of energy that is expected to last throughout human existence. Solar energy is the result of nuclear fusion. On the Sun, hydrogen is converted into helium through the process of fusion. The amount of energy produced is enormous: it is produced at a continuous rate of 1026 J/s. This is possible because the Sun is composed mostly of hydrogen. According to Einstein's theory of relativity, $E = mc^2$, mass is converted into energy. Earth receives only a small fraction of all the energy that is continuously radiated from the Sun. Because this amount is enormous, the amount of energy that Earth receives from the Sun would be sufficient to meet the world's energy needs if additional methods can be developed to capture and store solar energy.

Model

Hydroelectric Power Ask students to work in small groups to design models of working hydroelectric power plants. Students should first draw designs for their models. Computer-generated designs might be an option for student groups with computer access. Once you have approved their designs, have students build their models. The models should be small enough to use a water source, such as a sink in the classroom or an outside water tap. Each group should demonstrate its model and explain the principles behind its operation to the class. OL COOP LEARN

Photovoltaic cells Solar energy can be converted into electric energy by using a **photovoltaic cell,** a structure that is made of two layers of two types of silicon. The cell absorbs energy from the sunlight that strikes it. The electricity produced by photovoltaic cells can be stored in batteries. Photovoltaic cells are reliable, quiet, and typically last more than 30 years. Large-scale groups of panels can be set up in deserts and in other land areas that are not useful for other human purposes.

One example of this is a solar power tower. The solar power tower generates electricity by harnessing the solar heating of the desert surface. A glass canopy surrounds the tower and acts as a greenhouse to heat the earth beneath it. The heat creates a self-contained wind field, driving a network of 32 turbines, which generate electricity. Other advances in technology, such as those shown in **Figure 9,** might make renewable energy sources more accessible for future generations.

Energy from Water

Hydroelectric power is generated by converting the energy of free-falling water to electricity. When a dam is built across a large river to create a reservoir, the water stored in the reservoir can flow through pipes at controlled rates and cause turbines to spin to produce electricity. Hydroelectric power can also be generated from free-flowing water, such as the Niagara River. Today, hydroelectric power provides about 20 percent of the world's electricity and 6 percent of its total energy. Approximately 10 percent of the electricity used in the United States is generated by water, while Canada obtains more than 70 percent of its electricity from this source. Many of the hydroelectric power resources of North America and Europe have been developed, but sites have not yet been developed in Africa, South America, and Asia.

Figure 9
Development of Alternative Energy Sources
Countries develop new sources of energy to meet their growing needs.

1933–1935 The United States builds the Hoover Dam and the Grand Coulee Dam to produce hydroelectric power, the country's main energy source, second to coal, until 1984.

1800 — 1900 — 1925 — 1950

1800 Holland boasts 9000 windmills that are used chiefly for land drainage and grinding grain.

1919 Ethanol, used as fuel for early automobiles, is banned during Prohibition in the United States. Gasoline becomes the primary source of motor fuel.

1952 Coal, which had replaced wood in much of Europe due to deforestation, causes a smog that kills 12,000 Londoners. England enacts new antipollution laws.

EARTH SCIENCE JOURNAL

Hydroelectric Energy Ask students to explain how energy is transferred and changed in a hydroelectric power plant in their Earth science journals. The energy of moving water, kinetic energy, is used to move a portion of a wheel. When the water hits the wheel, the wheel moves forward, thus changing kinetic energy into mechanical energy. With the use of turbines, this mechanical energy is transformed into electrical energy. OL

Energy from the oceans Ocean water is another potential source of energy. The energy of motion in waves, which is created primarily by wind, can be used to generate electricity. Barriers built across estuaries or inlets can capture the energy associated with the ebb and flow of tides for use in tidal power plants.

Geothermal Energy

Geothermal energy doesn't come from the Sun. Instead, it originates from Earth's internal heat. Steam produced when water is heated by hot magma beneath Earth's surface can be used to turn turbines and generate electricity. A geothermal power plant is shown in **Figure 10.** Energy produced by naturally occurring heat, steam, and hot water is called **geothermal energy.** While some geothermal energy escapes from Earth in small amounts that are barely noticeable, large amounts of geothermal energy are released at other surface locations. In these areas, which usually coincide with plate boundaries, geothermal energy can be used to produce electricity.

Figure 10 Geothermal energy plants produce clean energy by harnessing the naturally occurring heat often found at plate boundaries.
Analyze *Is geothermal energy a renewable resource? Explain.*

View an **animation of geothermal power.**
Concepts In Motion

Wind Energy

Windmills in the Netherlands have been capturing wind power for human use for almost 2000 years. The windmills used today are more accurately called wind turbines because they convert the energy of the wind into electrical energy. Wind turbines provide about 20 percent of the electricity used in Denmark. Experts suggest that wind power could supply more than 30 percent of the world's electricity by the year 2050.

1969 Iceland builds its first geothermal power plant. Today, geothermal energy heats 87 percent of the country's homes and supplies 17 percent of its energy needs.

1995 The United States' program uses landfill gas to make electricity, reducing certain greenhouse gas emissions.

2005 Ninety percent of all homes in Israel use solar panels to heat water. Other countries have adopted this technology in recent decades.

1960 — 1980 — 2000

1957 The first large-scale commercial nuclear power plant in the country begins operating in Shippingport, Pennsylvania.

1997 The first hybrid car to run on a gasoline engine and an electrical motor is mass-produced and released in Japan.

2010 China leads the world in total renewable consumption of electricity, followed closely by the United States, Brazil, and Canada.

Project

Geysers Have students use available resources to find out the timing of eruptions of a particular geyser and the duration of each eruption. The data collected can then be placed on a spreadsheet and used to generate computer graphs. Students can study the eruption patterns of several geysers and make some comparisons and conclusions. Ask students to offer explanations for differences or similarities that they find. OL COOP LEARN

Discussion

Geothermal Energy Conduct a class discussion based on the reasons that geothermal energy exists. Students often think that Earth contains hot material within it because it was once molten and is cooling off. Share with students that scientists have calculated the rate of cooling since Earth formed, and given the time that has passed since Earth's formation, it should now be completely solid. Therefore, there must be another source of heat within Earth. Scientists hypothesize that this source is radioactivity. OL

Caption Question Fig. 10
Geothermal energy is a renewable energy resource because it is not used at a greater rate than that at which it can be replaced.

GeoLAB

iLab Station

The GeoLab located at the end of the chapter can be used at this point in the lesson.

Demonstration

Model a Geyser Place a stopper in a Pyrex Erlenmeyer flask half-filled with water. Secure a glass tube with a fine, eyedropper-shaped opening in the stopper. Make sure that the tip of the glass tube is pointing toward the ceiling. Heat the water in the flask until it boils and water begins to spurt out the top. This is similar to what happens when a geyser erupts. As the water changes to steam, pressure builds up inside the flask. Underground, this pressure eventually forces the hot water and steam out of any openings in Earth's surface. OL

Enrichment

Energy Energy cannot be created or destroyed; it can only be changed from one form to another. Albert Einstein discovered the mathematical formula for this law, $E = mc^2$. E represents energy, m represents mass, and c represents the speed of light. Einstein helped to prove that matter can be changed into energy. Under special circumstances, the energy that holds together the basic units of matter can be released and used. AL

Discussion

Nuclear Power Ask: What are the benefits and drawbacks of nuclear power? Benefits: potentially lower cost than coal-burning electricity plants, and nuclear plants do not produce fossil-fuel emissions like coal-burning electricity plants do. Drawbacks: high operating costs, poor reactor designs, concerns about safety and disposal of radioactive wastes.

Caption Question Fig. 11 three systems

Collaborative Learning

Nuclear Energy Much information is available on both positions regarding the use of nuclear energy. It is important for students to be given the opportunity to hear all the issues so they can develop their own reasons for supporting or not supporting further development of nuclear energy. Ask for volunteers to select either the supporting or opposing side of this issue and participate in a debate. Take a vote at the end of the debate to see how the class now views the issue. OL COOP LEARN

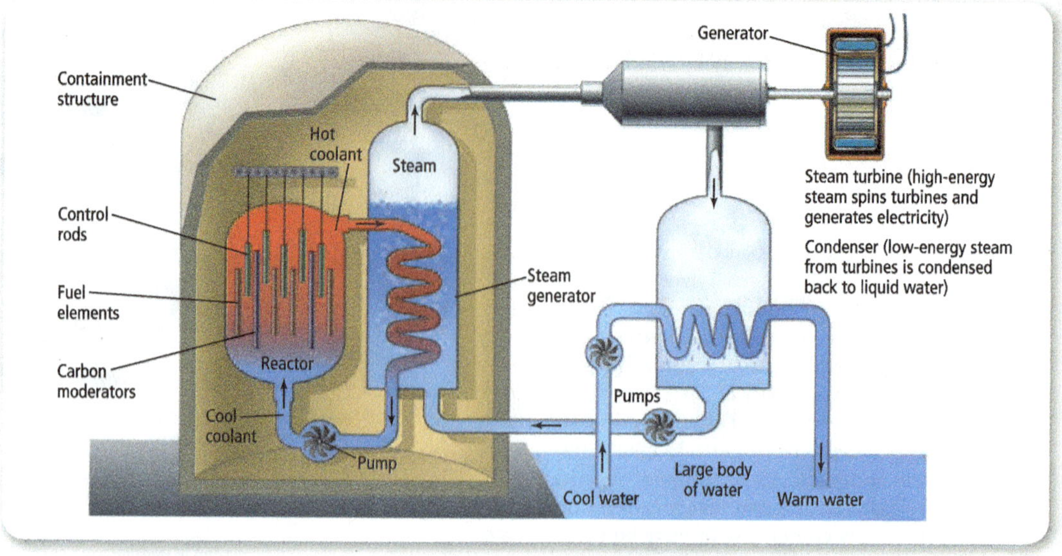

Figure 11 Nuclear reactors rely on fission to generate heat. Heated water is converted to steam which turns a turbine to generate electricity.
Identify *how many separate systems are in this reactor.*

View an **animation of a nuclear fission reactor.**
Concepts In Motion

Nuclear Energy

Recall that atoms lose particles in the process of radioactive decay. One process by which atomic particles are emitted is called nuclear fission. **Nuclear fission** is the process in which a heavy nucleus (mass number greater than 200) divides to form smaller nuclei and one or two neutrons. This process releases a large amount of energy. Radioactive elements consist of atoms that have a natural tendency to undergo nuclear fission. Uranium is one such radioactive element that is commonly used in the production of nuclear energy. Nuclear energy is one other energy source that does not come directly from the Sun.

In the late 1950s, power companies in the United States began developing nuclear power plants similar to the one shown in **Figure 11.** Scientists suggested that nuclear power could produce electricity at a much lower cost than coal and other types of fossil fuels. Another advantage is that nuclear power plants do not produce carbon dioxide or any other greenhouse gases. After 50 years of development, however, 445 nuclear reactors are currently producing only 14 percent of the world's electricity. Construction of new nuclear power plants has come to a halt in Europe, and has slowed considerably in the United States.

What happened to using nuclear energy as a new source of power? High operating costs, poor reactor designs, and public concerns about radioactive wastes contributed to the decline of nuclear power. In addition, nuclear accidents, such as those at Three Mile Island in Pennsylvania, in 1979, and at Chernobyl, Ukraine, in 1986, alerted people to the hazards of nuclear power plants. Because of its hazards, nuclear power has not been developed further in the United States as an alternative energy source.

EARTH SCIENCE JOURNAL

Nuclear Energy Ask students to each write two paragraphs in their Earth science journals that express their personal opinions about the future use of nuclear energy. Tell students to make sure they include the reasons for their viewpoints, and to anticipate any objections to their viewpoints by presenting a counterargument. OL

ACROSS THE CURRICULUM

History In 1951, in Idaho, electricity was generated by a nuclear reactor for the first time. In 1954, in the USSR, the world's first nuclear power plant started generating electricity. The interest in nuclear energy plants decreased after the 1979 Three Mile Island accident and further after the 1986 Chernobyl accident.

Biofuels

You learned in Section 1 that biomass fuels include wood, dried field crops, and fecal materials from animals. Biomass is a renewable energy resource as long as the organisms that provide the biomass are replaced. Scientists are developing ways to produce fuels similar to gasoline from crops such as corn and soybeans. These fuels are called biofuels.

Ethanol Ethanol is a liquid produced by fermenting crops such as barley, wheat, and corn, which is shown in **Figure 12.** Ethanol can be blended with gasoline to reduce consumption of fossil fuels. Ethanol fuels burn more cleanly than pure gasoline. In 2011, the Environmental Protection Agency approved a gasoline blend with 15 percent ethanol for cars manufactured after 2001. Some vehicles, called flexible fuel vehicles, can run on mixtures containing 85 percent ethanol.

Biodiesel Biodiesel can be manufactured from vegetable oils, animal fats, or recycled restaurant greases. Biodiesel is safe, biodegradable, and reduces air pollution. Blends of 20 percent biodiesel with 80 percent petroleum diesel (B20) can generally be used in unmodified diesel engines; however, it is currently more expensive than regular diesel.

■ **Figure 12** Biofuels, like biomass fuels, are derived from renewable resources. Crops like corn can be processed to create ethanol, a cleaner burning fuel than gasoline.

Jim Richardson/CORBIS

SECTION 2 REVIEW

Section Self-Check

Section Summary

- Alternative energy resources can supplement dwindling fossil fuel reserves.
- Solar energy is unlimited, but technological advances are needed to find solutions to collect and store it.
- Nuclear energy is produced when atoms of radioactive elements emit particles in the process known as nuclear fission.
- Biofuels can help reduce consumption of fossil fuels.

Understand Main Ideas

1. MAINIDEA **Identify** one alternative energy resource that is associated with each of Earth's systems: the atmosphere, hydrosphere, biosphere, and geosphere.
2. **Compare** passive solar energy and active solar energy.
3. **Infer** which alternative energy source would have the least impact on the environment if the required technology could be developed to harness and use it. Explain.

Think Critically

4. **Analyze** In theory, solar energy could supply all of the world's energy needs. Why isn't it used to do so?
5. **Evaluate** the advantages and disadvantages of nuclear energy.

WRITING IN ▶ Earth Science

6. Write a newspaper article that describes how alternative energy resources can be used where you live.

Assess

Check for Understanding

Recognize Cause and Effect

Ask students why researchers and government agencies are studying the use of alternative fuels. It took millions of years for fossil fuels to form, and once reserves are depleted, the need for fuels will still exist.

Reteach

Outline Have students develop outlines of this section to help them organize the content. Encourage students to use colored pencils to emphasize key terms and phrases.

Assessment

Performance Have students each select one alternative fuel and research the history and use of the fuel. Students should prepare brochures they can reproduce and hand out to their classmates to promote the use of their alternative fuels.

SECTION 2 REVIEW

1. atmosphere—wind energy; hydrosphere—hydroelectric power; biosphere—biomass energy; geosphere—geothermal energy
2. Passive solar-collection systems rely only on the natural processes of conduction, convection, and radiation, while active solar-collection systems use fans or pumps to enhance the collection process.
3. Solar energy would be the least damaging to the environment, and it is free and in limitless supply.
4. Solar energy is not a major alternative energy source because technology to collect and store the energy in a manner that will ensure a continuous supply has not yet been developed.
5. Nuclear energy has high operating costs, poor reactor designs, and concerns about disposal of radioactive wastes. However, nuclear energy does not produce carbon dioxide or other greenhouses gases.
6. Answers will vary, but should include mention of solar energy, hydroelectric power, geothermal energy, wind energy, nuclear energy, or biofuels.

Rubric

SECTION 3

1 Focus

Focus Transparency

Before presenting the lesson, project **Section Focus Transparency 80** and have students answer the accompanying questions. EL

MAINIDEA

Resource Use Ask students to think about how they can modify their use of a phone or computer to save battery life.

2 Teach

Teacher Content Support

Demand for Energy As more countries worldwide become developed, the demand for energy will increase. If more energy is not available, the energy that is available will have to be shared among more people. To meet this inevitable increase in demand, we need to learn how to make our energy resources go further. Through-out history, humans have used their ingenuity to develop ways to extend their energy resources. From the use of draft horses to the invention of machinery, humans have been able to meet their increasing needs for energy.

Project

Change in Energy Use Have students ask their parents or guardians how energy use has changed over their lifetimes. Have students report on the differences or similarities in their parents' or guardians' opinions and their own OL

Caption Question Fig. 13
Nonrenewable resources are relatively cheap and easy to access compared to renewable resources.

SECTION 3

Conservation of Energy Resources

MAINIDEA Using energy efficiently reduces the consumption of nonrenewable resources.

Essential Questions

- How can energy resources be conserved?
- How can increasing energy efficiency help preserve fossil fuels?
- How can energy be used more efficiently?

Review Vocabulary

renewable resource: a resource that is replaced through natural processes at a rate equal to or greater than the rate at which it is used

New Vocabulary

energy efficiency
cogeneration
sustainable energy

EARTH SCIENCE 4 YOU Think of runners on a cross-country team or a swimmer in a 400-m event. They don't sprint to start, instead they pace themselves so they have enough energy to finish the race. Energy resources can be used in this way, too.

Global Use of Energy Resources

As you have learned, fossil fuels are nonrenewable and are in limited supply. Yet people on Earth consume these resources at increasing rates. **Figure 13** shows national consumption of natural resources, both renewable and nonrenewable. However, consumption is not equal in all parts of the world. Some developing countries, for example, obtain 25 percent of their energy from a renewable resource, compared to industrialized countries where renewable resources account for only about 7 percent of the energy used.

Using renewable energy resources that are locally available conserves the fuel that would be used to transport and process resources at a different location. Using a variety of energy resources rather than a single, nonrenewable energy resource, such as fossil fuels, can also help conserve resources. For example, a community that has hydroelectric energy resources might also use solar or wind energy to generate electricity during months when water levels are low.

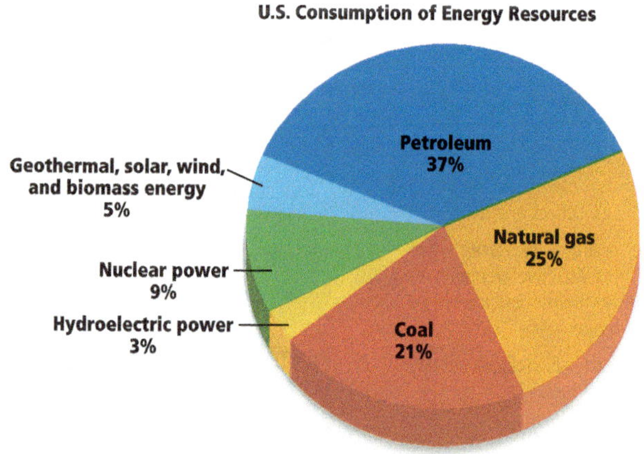

Figure 13 Petroleum is the most widely used energy resource nationwide, followed closely by coal and natural gas.
Explain *Why do you think nonrenewable resources account for almost 85 percent of the nation's energy consumption?*

Demonstration

Model Limited Resources Give a sheet of paper to each student. Ask students to write their names, addresses, and telephone numbers on the paper. Have students return the sheets of paper to you. Count out loud the number of sheets of paper used. Announce that you have only 50 sheets of paper, including those you have already used, and that you will conduct this activity in all your classes. Count the remaining pieces of paper. Tell students the number of remaining students you have in your other classes. Ask them for suggestions on how to deal with the paper shortage. Students will likely suggest cutting the paper into smaller sheets or using the other side of the used paper. Relate the limitation of paper to the limitation of energy resources. BL

Energy Efficiency

Energy is the ability to do work. The amount of work produced compared to the amount of energy used is called **energy efficiency.** Energy resources do not produce 100 percent of the potential work that is stored in the energy source.

When a car uses gasoline, some of the energy stored in the gasoline is converted to mechanical energy that moves the car, while some of the energy is used to power accessories, like the car's air conditioner. Most of the energy in the gasoline is lost as heat. Decreasing heat loss is one way that more of the stored energy can be converted to do work. To find ways to use resources more efficiently, scientists study exactly how energy resources are used and where improvements are needed. Using resources more efficiently is a type of conservation. For example, adding insulation to a house reduces heat loss, so less energy is needed to heat the air inside.

☑ READING CHECK **Explain** energy efficiency.

Improving efficiency in industry Most of the electricity in the United States is generated by burning fossil fuels to heat water, forming steam. Recall that increasing the temperature of a gas also increases pressure. Steam pressure spins the turbines that drive the generators, creating electricity. Unfortunately, this is an inefficient process as only approximately one-third of the energy potential within the original fuel source can be converted into steam pressure.

Improving efficiency in transportation Transportation is necessary to move people, food, and other goods from one place to another. Although most transportation currently relies on oil, conservation practices can help reduce dependency on oil resources used for transportation. **Table 1** lists some of the advantages of public transportation, which is one way people can improve energy efficiency in transportation.

Explore **public transportation with an interactive table.** Concepts In Motion

Table 1 Advantages of Public Transportation

Advantages of Public Transportation
Using public transportation to get to work can save a person between $300 and $3000 in fuel costs per year.
Using public transportation saves more than 3 billion liters of gasoline every year—equal to all energy used by U.S. manufacturers of computers and electronic equipment.
If Americans used public transportation for roughly 10 percent of daily travel needs, the United States would reduce its dependence on foreign oil by more than 40 percent.
One person switching to public transit can reduce daily carbon emissions by 9.1 kg, or more than 3300 kg in a year.

CAREERS IN EARTH SCIENCE

Environmental Consultant An environmental consultant interprets environmental data, conducts field surveys, and conducts environmental impact assessments. From this information, they make suggestions to businesses of how to limit their environmental impacts and meet governmental regulations.

WebQuest

Identify Misconceptions

Students might be unaware of the relationship between human life and the Sun.

Uncover the Misconception
Students might think humans could exist on Earth without the Sun.

Demonstrate the Concept
Ask students to think about how we would be able to survive on Earth if the Sun suddenly stopped shining. Ask students how humans and plants make food to point out the importance of the Sun. They might think humans could find alternative sources of energy. For each alternative energy source students suggest, ask what would happen when that source of energy is used up. In essence, life on Earth would quickly end without the Sun's radiation. The location of Earth is ideal in terms of gathering heat from the Sun. Without the Sun, Earth would continue to radiate heat until it had a much colder surface temperature. Geothermal energy would be the last energy source once other fuels had been used. Share with students that the Sun is expected to be around for billions of years to come.

Assess New Knowledge
Ask students to identify five energy sources and replacement sources for each.

DIFFERENTIATED INSTRUCTION

Visually Impaired Pair a visually impaired student with another student. Collect several types of insulation. Have the visually impaired student handle each material and describe it as the other student writes down the description. Then ask the visually impaired student to rank the materials in order from the one with the best insulating ability to the one with the least. Have the other student write down the rankings, then examine the materials to identify their R-values. Have the student share this information with the visually impaired student to see how accurate the student's rankings were. Be careful with fiberglass; some students might have skin reactions to it. OL

☑ READING CHECK Energy efficiency is the amount of work produced compared to the amount of energy stored.

CAREERS IN EARTH SCIENCE

WebQuest

Environmental Connection

Electric Cars Some students might think electric cars will eliminate the need to burn fossil fuels to power automobiles and thus will solve the problem of air pollution associated with vehicles. Students might not be aware that electric cars run on batteries that must be recharged by being plugged into an electric outlet. The energy flowing to the outlet most likely comes from a power plant that produces energy with fossil fuels.

Data Analysis LAB

About the Lab

- Ask students how they think energy use has changed since they were born. Have students share their opinions before proceeding with this activity. Review graphing procedures with students. Remind students a graph needs a title, legend, and appropriate scales on the *x*- and *y*-axes.
- See also O'Neill, B.C., and B.S. Chen. 2002. Demographic determinants of household energy use in the United States. *Population and Development Review* 28:53–88.

Think Critically

1. Check students' graphs for accuracy.
2. Natural gas might be easier to pipe to individual homes.
3. Heating oil, propane, and kerosene might be more expensive than other sources.

Commuting efficiently People who live in metropolitan areas can improve energy efficiency by using public transportation to get from place to place. Major U.S. cities, such as New York City, use subways or elevated trains to move people. In Europe, mass transportation includes long-distance rail systems, as well as electric trams and trolleys. When it is necessary to drive private automobiles, carpooling can reduce the number of vehicles on the highways. Some metropolitan areas encourage carpooling by providing express lanes for cars with multiple passengers.

VOCABULARY
ACADEMIC VOCABULARY
Efficient
productive without waste
The automobile was more efficient when the proper tune-ups had been done.

Automobiles The use of fuel-efficient vehicles is another way to reduce the amount of petroleum resources consumed. Automobile manufacturers can build vehicles that achieve high rates of fuel efficiency without sacrificing performance. The future of this industry is promising as hybrid, fuel cell, and electric technologies are reaching the consumer market. Also, less energy is needed to move something that weighs less, thus smaller cars use less gasoline. Another way to conserve gasoline is to drive slower than 100 km/h (62 mph) on the freeway and use alternate forms of transportation.

Getting more for less Increased demand for fuels requires a greater supply and results in higher costs. Electricity is costly to produce, and it is not usually used efficiently in homes or industry. In the United States, approximately 40 percent of the energy used to fuel motor vehicles and to heat homes and businesses is lost as thermal energy. If energy were used more efficiently, less energy would be needed, thus decreasing the total cost of energy.

Data Analysis LAB

Based on Real Data*

Make and Use Graphs

What proportion of energy resource types are used to heat homes? Natural gas, electricity, heating oil, propane, and kerosene are used to heat American homes. The table shows percentages used to heat different types of homes.

Think Critically

1. **Compare** the sources of energy used by plotting the data on a graph. Be sure to use different colors for the different types of energy. Place the percentages on the *y*-axis and the source on the *x*-axis.
2. **Infer** why single-family homes use natural gas more than other types of dwellings.
3. **Infer** why heating oil, propane, and kerosene are not widely used as energy sources for homes.

Data and Observations

Energy Sources for American Homes (%)

Energy Source	Single-Family Dwellings	Multi-Family Dwellings	Mobile Homes
Natural gas	60	48	32
Electricity	23	42	43
Heating oil	8	7	3
Propane	5	0	15
Kerosene	1	0	4
Other	3	3	3

*Data obtained from: The National Energy Education Development Project. 2004. *Secondary Energy Infobook*.

EARTH SCIENCE JOURNAL

Appliance and Wattage Have students each make a list of the appliances used in their homes and the wattage used by each appliance. Ask students to place a star next to the appliance with the highest wattage requirements. BL

Rubric

ACROSS THE CURRICULUM

History Ask students to each write a report about a person who was important to the history of electricity. Possibilities include Otto von Guericke, the inventor of the static electric generator in the 1600s; Alosio Galvani, who invented the first current generator in the 1700s; and Humphry Davy, who invented the incandescent light bulb in the 1800s. OL

Harnessing waste thermal energy Generating electricity produces energy that can be recovered. This recoverable excess energy is known as waste thermal energy, and can be harnessed for use. The simultaneous production of two usable forms of energy is called **cogeneration.** Cogeneration captures the waste thermal energy (steam) for domestic or industrial heating, or for use in a large air-conditioner unit. In an air conditioning unit, the waste thermal energy turns a turbine connected to a compressor that chills water. This chilled water is then sent to an air handler unit in a different building. Excess thermal energy can also be used to generate electricity that operates electrical devices within the power plant, such as sulfur-removing scrubbers on smokestacks. While industries use one-third of all energy produced in the United States, cogeneration has allowed some industries to increase production while reducing energy use. Cogeneration has enabled central Florida to operate the nation's cleanest coal-powered electric facility. The power station shown in **Figure 14** utilizes cogeneration for an oil refinery and chemical plant.

Figure 14 This cogeneration power station helps reduce energy use at an oil refinery and chemical plant in Hampshire, UK.

Sustainable Energy

Energy resources on Earth are interrelated, and they affect one another. **Sustainable energy** involves the global management of Earth's natural resources to meet current and future energy needs. A good management plan incorporates both conservation and energy efficiency. New technology that extends the supply of fossil fuels is a vital part of such a plan. Global cooperation can help maintain the necessary balance between protection of the environment and economic growth. The achievement of these goals will depend on the commitment made by all so that future generations have access to the energy resources required to maintain a high quality of life on Earth.

SECTION 3 REVIEW

Section Self-Check

Section Summary

- Energy resources will last longer if conservation and energy-efficiency measures are developed and used.
- Energy efficiency results in the use of fewer resources to provide more usable energy.
- Cogeneration, in which two usable forms of energy are produced at the same time from the same process, can help save resources.
- Sustainable energy can help meet current and future energy needs.

Understand Main Ideas

1. MAINIDEA **Summarize** why the conservation and efficient use of energy resources is important.
2. **List** three ways in which you could conserve electric energy in your home.
3. **Compare** energy consumption between developing and industrialized countries.
4. **Analyze** Why is it important to conserve resources instead of seeking new sources of fossil fuels for energy?

Think Critically

5. **Illustrate** how cogeneration can be used in your home to save energy resources.

MATH IN Earth Science

6. If the global consumption of coal were reduced by 25 percent, what would the percentage consumption of coal be? Refer to **Figure 13** for more information.

3 Assess

Check for Understanding

Observe and Infer Have students work together as a class to develop an evaluation of how energy could be conserved in your school. Have them develop categories such as heating, lighting, transportation, use of natural resources, degree of recycling, and so on. After students have gathered information to support their evaluation, have them present it to the school principal.

Reteach

Think Critically Have students work in pairs to review the section material. After 15 minutes, have each student make up five questions for the other student to answer. Add up the total number of questions the pair correctly answered. Ask each pair to ask the questions they missed to others in the class. COOP LEARN

Assessment

Skill Have each student develop five questions based on the information in this section. Have students write each question on a piece of paper. Shuffle the pieces of paper, and have all students stand up and take turns answering the questions. When a question is missed, the student should be seated. Students who are seated can join the standing students if they can answer any of the questions missed. Make sure the correct answers to all questions are eventually discussed.

SECTION 3 REVIEW

1. As the amount of available fossil fuels dwindles, the cost of providing energy for everyday living will increase. Conserving energy through energy efficiency will help make our energy resources last longer.
2. Answers will vary, but could include turning off the light when not in the room, turning the thermostat down a few degrees, unplugging items when they are not in use, using more energy efficient appliances, and doing things manually instead of electronically.
3. Industrialized countries require more energy resources to operate factories and to maintain technological advances than developing countries.
4. By conserving the energy resources currently being used, the resources will last longer and provide more time for the discovery and development of alternative energy resources.
5. Answers will vary, but should indicate that using the same energy source to complete two tasks saves the energy needed to do the second task.
6. 19.33%

Earth Science &
ENVIRONMENT

Purpose

Students will understand the potentially helpful effects of bacteria on the environment.

Teacher Content Support

Bacteria Microbial fuel cells are not a new idea. Since 1910, scientists have tried to harness the electric potential of bacteria. Using bacteria in a fuel cell has many potential uses, the most exciting of which is to power a pacemaker inside the human body. The microbes could gather enough food from the bloodstream to generate the needed fuel to keep the heart pumping. *Disulfitobacterium* are interesting as well for their potential application to reclaim polluted water and supply water plants with supplementary electricity. However, only small gains have been made with capturing the electricity emitted from the cells; much more research is needed before microbial fuel cells become widespread.

Teaching Strategies

- First ask students: Where do we get the electricity we use? Then ask students: Can you think of any problems with using this form of electricity? Finally ask: What might be some other, renewable sources of electricity production?
- Divide the class into groups of four. Have each group discuss further possible uses of the bacteria in the article. Then have them sketch out how one of their "products" might look and work. Regroup and share the new technologies with the class.

Earth Science &
ENVIRONMENT

Bacteria Power!

Bacteria are all around us—some are helpful while others cause disease. Without bacteria, life would be very different. Humans have bacteria that live in the stomach and intestines to help digest food. Other bacteria cause illnesses such as strep throat and tuberculosis.

Pollution-eating bacteria Through research, scientists have discovered bacteria that can eat pollution, and other bacteria that can produce energy that can be harnessed for human use. Bacteria in the genus *Desulfitobacterium* have long been studied for their unique appetites. They eat pollution, such as toxic waste, and change it into less toxic or even nontoxic products. Recently, scientists worked with *Desulfitobacterium* successfully to find a species of bacteria that could break down freshwater pollution.

Microbial power plants Not only are *Desulfitobacterium* able to consume toxic waste, they are also able to produce energy at a constant rate. While scientists have known of the bacteria's ability to break down different toxins and produce energy as a by-product, this was the first time it was discovered that bacteria could do both at once. The energy that the bacteria produced could be harnessed to run small electrical devices.

Desulfitobacterium are able to survive extreme heat, radiation, and other environments that would easily wipe out other bacterial populations. Imagine that a fuel cell containing *Desulfitobacterium* is placed in an area where it will not be used for many years, and where it is exposed to harsh environments. If *Disulfitobacterium* was used as the power source for the fuel cell, it could exist in a stage similar to hibernation until it was needed or until conditions improved.

In the future, *Desulfitobacterium* might be used to power a wastewater treatment plant, such as this one, while helping to reclaim the wastewater being processed.

Diverse diets The metabolic capabilities of *Desulfitobacterium* bacteria are unique. The bacteria have a diverse diet, so they can use many different sources, including wastewater, chemical pollutants, and pesticides, to produce electricity.

While this biotechnology is still in the early stages of discovery and development, there are many exciting opportunities to be explored. It is possible that a bacterial colony could be used to reclaim wastewater while producing electricity to power the water treatment plant at the same time.

WRITING IN ▶ Earth Science

Brochure You are marketing a fuel cell that uses these bacteria. Create a brochure explaining the potential uses of these fuel cells and why this biotechnology is important in today's world.

WebQuest

WRITING IN ▶ Earth Science

Rubric

Brochure Answers will vary, but should indicate that bacteria can produce energy that can be harnessed for human use at a constant rate and that some bacteria can survive extreme conditions.

WebQuest

GeoLAB

iLab Station

Design Your Own: Design an Energy-Efficient Building

Background: Buildings can be designed to conserve heat. Some considerations involved in the design of a building that conserves heat include the materials that will be used in construction, the materials that will store heat, and the overall layout of the building. By using a more energy-efficient design and more energy-efficient materials, consumers can decrease their monthly gas or electric bills and conserve natural resources.

Question: *How can a building be designed to conserve heat?*

Possible Materials

glass or clear plastic squares
sturdy cardboard boxes
scissors
tape
glue
thermometers
paint
paper
aluminum foil
polystyrene
stone
mirrors
fabric
light source

Safety Precautions

Procedure

1. Read and complete the lab safety form.
2. Working in groups of three to four, brainstorm a list of design features that might contribute to the heat efficiency of a building and consider how you might incorporate some of these features into your building.
3. Design your building.
4. Make a list of heat-conserving issues that you addressed.
5. Decide which materials you will use to build your house. Collect those materials.
6. Construct the building and a control building for comparison.
7. Devise a way to test the heat-holding ability of each building.
8. Perform the test on each building. To test the buildings' heat efficiency, it may be necessary to heat the buildings and determine how long heat is conserved within each one. **WARNING:** ***Make sure the heat source is far enough away from the building materials so that they do not burn or melt.***
9. Record your data in a table. Then, make a graph of your data.
10. Make modifications to the design to improve the building's efficiency.

Analyze and Conclude

1. **Conclude** Was the building you designed more energy-efficient than the control building?
2. **Analyze** What problems did you encounter, and how did you solve them?
3. **Analyze** How did your observations affect decisions that you might make if you were to repeat this lab? Why do you think your design worked or did not work?
4. **Predict** Would your design work in a home in your community? In a community with a different climate? Why or why not?
5. **Compare and contrast** the building you designed and the control building.
6. **Compare and contrast** your design and the designs of your classmates.
7. **Determine** how your design could be improved.
8. **Predict** how using different energy sources might affect your results.

TRY AT HOME

Apply How could you incorporate some of your design elements into your own home? Discuss your lab with an adult at home and make suggestions to conserve heat.

GeoLAB

iLab Station Rubric

Preparation

Time Allotment 90 min

Process Skills design an experiment, formulate models, hypothesize, observe and infer

Safety Precaution Approve lab safety forms before work begins.

Preparation of Materials Hold the brainstorming session a few days before students make and test their buildings. Review the reason for and use of a control. Have references on hand in the classroom on the topic of energy-efficient building designs.

Procedure

- Allow students to be creative.
- Offer suggestions only if a student is headed in a direction that won't allow the student to learn the content.
- Allow students to make small errors in their designs so that they can analyze their results easily.
- **Troubleshooting** Make sure students have used a control during the testing phase.

LabManager™
Customize any lab with the LabManager™ CD-ROM.

Analyze and Conclude

1. Answers will vary. Students should be able to use the data they collected from their models and the control to determine the effectiveness of their designs.
2. Answers will vary, but might include problems relating to choice of materials or design issues.
3. Answers will vary. Students should be able to recognize which materials or design might increase the energy efficiency of their building.
4. Answers will vary, but should show an awareness that the climate in part determines the needs of the building.
5. Answers will vary. If the control and design models showed no difference, then the methods employed likely did not affect what was being measured—either heat absorbed or heat retained.
6. Answers will vary, but should discuss the similarities among and differences between designs.
7. Answers will vary, but might include using different materials, or designing a different layout.
8. Answers will vary, but should indicate how different energy sources might need different methods of conserving heat.

TRY AT HOME

Apply Answers will vary, but should include practical suggestions for conserving heat in the home.

MAINIDEAS Summary statements can be used by students to review the major concepts of the chapter.

Students can review with these online resources.

Vocabulary Practice

Vocabulary eGames
Vocabulary eFlashcards
Vocabulary PuzzleMaker

Section Self-Check

Chapter Self-Check

Online Test Practice

Use the ExamView® *Assessment Suite* CD-ROM to:

- create multiple versions of tests
- create modified tests with one mouse click
- edit existing questions and add your own questions
- build tests aligned with state standards using built-in state curriculum tags
- change English tests to Spanish with one mouse click
- track students' progress using the Teacher Management System

CHAPTER 25 STUDY GUIDE

BIGIDEA People use energy resources, most of which originate from the Sun, for everyday living.

Vocabulary Practice

SECTION 1 Conventional Energy Resources

VOCABULARY
- fuel
- biomass fuel
- hydrocarbon
- peat
- fossil fuel

MAINIDEA Biomass and fossil fuels store energy from the Sun.

- The Sun is the source of most energy on Earth.
- Humans have used materials derived from living things, such as wood, as renewable fuels for thousands of years.
- Fossil fuels formed from organisms that lived millions of years ago.

SECTION 2 Alternative Energy Resources

VOCABULARY
- photovoltaic cell
- hydroelectric power
- geothermal energy
- nuclear fission

MAINIDEA Many resources other than fossil fuels can be developed to meet the energy needs of people on Earth.

- Alternative energy resources can supplement dwindling fossil fuel reserves.
- Solar energy is unlimited, but technological advances are needed to find solutions to collect and store it.
- Nuclear energy is produced when atoms of radioactive elements emit particles in the process known as nuclear fission.
- Biofuels can help reduce consumption of fossil fuels.

SECTION 3 Conservation of Energy Resources

VOCABULARY
- energy efficiency
- cogeneration
- sustainable energy

MAINIDEA Using energy efficiently reduces the consumption of nonrenewable resources.

- Energy resources will last longer if conservation and energy-efficiency measures are developed and used.
- Energy efficiency results in the use of fewer resources to provide more usable energy.
- Cogeneration, in which two usable forms of energy are produced at the same time from the same process, can help save resources.
- Sustainable energy can help meet current and future energy needs.

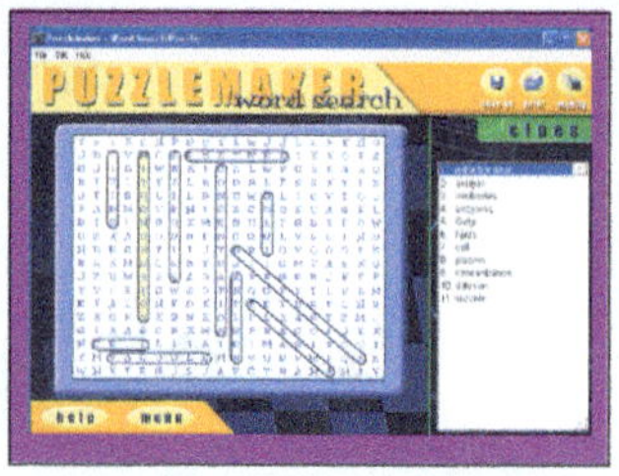

Vocabulary Practice

For additional practice with vocabulary, have students access the Vocabulary PuzzleMaker.

Chapter Self-Check

VOCABULARY REVIEW

Write a sentence defining each of the following vocabulary terms.

1. fuel
2. peat
3. fossil fuel
4. energy efficiency
5. geothermal energy
6. cogeneration

Fill in the blanks with an appropriate vocabulary term from the Study Guide.

7. ________ is a form of energy generated by the conversion of free-falling water to electricity.
8. Solar energy is converted into electric energy through the use of ________.
9. Molecules with hydrogen and carbon bonds are called ________.

Replace the underlined words with the correct vocabulary term from the Study Guide.

10. The process in which a heavy nucleus divides to form smaller nuclei results in a release of a large amount of energy.
11. Global management of Earth's natural resources to meet human needs will allow people to have all the energy they need to live.
12. Fuels formed from organic matter are burned in developing countries as a source of heat.

UNDERSTAND KEY CONCEPTS

13. Which is the primary source of energy on Earth?
 A. oil
 B. coal
 C. the Sun
 D. wood

14. When a consumer eats a producer, from where are they gaining energy?
 A. Earth
 B. the plant
 C. the Sun
 D. the ground

Use the figure below to answer Questions 15 and 16.

15. Which best describes the type of resource illustrated in the figure?
 A. biomass
 B. biofuel
 C. solar heating
 D. fossil fuel

16. What type of resource is shown?
 A. fossil fuel
 B. renewable resource
 C. nonrenewable resource
 D. cogeneration

17. Which is not derived from living things?
 A. petroleum
 B. coal
 C. peat
 D. nuclear power

18. Which form of energy commonly coincides with tectonic plate boundaries?
 A. fossil fuels
 B. geothermal energy
 C. wind energy
 D. biomass fuels

CHAPTER 25 ASSESSMENT

VOCABULARY REVIEW

1. Answers will vary, but should indicate that fuel is used as a source of energy.
2. Answers will vary, but should indicate that peat is formed in a swamp or bog and is used as a fuel.
3. Answers will vary, but should indicate that a fossil fuel is derived from plants and animals and takes millions of years to form.
4. Answers will vary, but should indicate that energy efficiency is the amount of work produced in relation to the amount of energy used.
5. Answers will vary, but should indicate that geothermal energy is harnessed from the natural heat of Earth's core.
6. Answers will vary, but should indicate that cogeneration is the harnessing of wasted energy.
7. hydroelectric power
8. photovoltaic cells
9. hydrocarbon
10. nuclear fission
11. sustainable energy
12. biomass fuels

UNDERSTAND KEY CONCEPTS

13. C
14. C
15. A
16. B
17. D
18. B

19. A
20. A
21. D

CONSTRUCTED RESPONSE

22. Coal is used in electricity-generating plants. Petroleum is used to make gasoline, diesel fuel, and kerosene. Natural gas is used to heat homes.
23. Answers will vary, but should include the Sun and materials that have heat-storing capacities, such as concrete or stone.
24. nuclear power
25. Lignite is formed after plant material sinks to the bottom of a bog, partially decomposes, and forms peat. Then the peat is compressed to form lignite.
26. If organisms decompose and compress over long periods of time, they can become fossil fuels.
27. Water has a high specific heat allowing it to heat slowly and hold the energy for a long period of time, and therefore it can be used to store thermal energy in passive solar situations.
28. There are various ways to conserve oil. Several of these include walking or riding a bike instead of driving, driving more energy-efficient vehicles, car pooling, and using public transportation.

THINK CRITICALLY

29. The higher the carbon content, the more efficiently it burns.
30. Water's ability to flow easily down slopes allows for mechanical energy to be changed to electrical energy, such as in the production of electricity at hydroelectric plants. Water also has a high specific heat allowing it to heat slowly during the day and release thermal energy overnight in passive solar situations.
31. Biomass fuels are often easier to access and use than fossil fuels.
32. about 5500 tons
33. Answers will vary. The trend would most likely not be the same. Developing countries would not see such a quick increase in the use of fossil fuels because the technologies that use the fossil fuels would not be in place.

Use the diagram below to answer Questions 19 and 20.

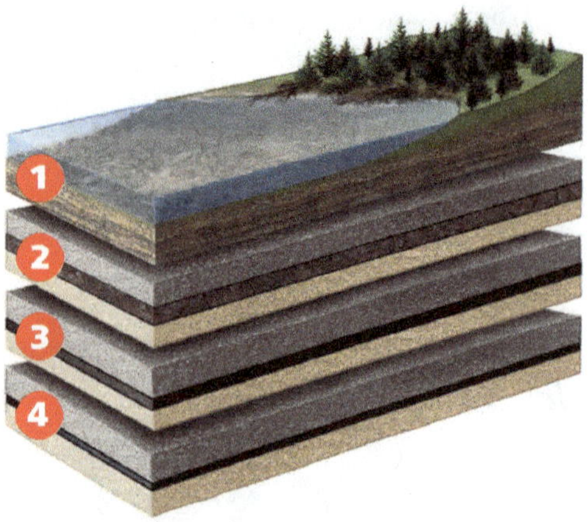

19. Which process happens in Layer 1?
    - **A.** Vegetation accumulates and forms peat.
    - **B.** Bituminous coal forms from lignite.
    - **C.** Lignite forms from accumulated vegetation.
    - **D.** Anthracite forms from bituminous coal.
20. Which is formed in Layer 4?
    - **A.** anthracite
    - **B.** bituminous coal
    - **C.** lignite
    - **D.** peat
21. Which is one reason nuclear power plants are not widespread?
    - **A.** Nuclear power is not energy efficient.
    - **B.** Nuclear reactors emit greenhouse gases.
    - **C.** Nuclear reactions occur only on the Sun.
    - **D.** Negative public perception of nuclear power.

CONSTRUCTED RESPONSE

22. **Describe** three ways fossil fuels are used for energy.
23. **Draw and label** a diagram to explain passive solar heating.
24. **Identify** one form of energy not derived from the Sun.
25. **Describe** the formation of lignite.
26. **Explain** how organisms living on Earth in this era could become fossil fuels.
27. **Analyze** why a substance such as water is good to use in passive solar heating situations.
28. **List** three ways to conserve oil.

THINK CRITICALLY

29. **Explain** why lignite, which has a carbon concentration of 40 percent, burns less efficiently than anthracite, which has a carbon concentration of 90 to 95 percent.
30. **Distinguish** What characteristics of water allow it to be used to produce energy as well as store energy?
31. **Explain** why biomass fuels are sometimes used more than oil for fuel in developing countries.

Use the graph below to answer Questions 32 and 33.

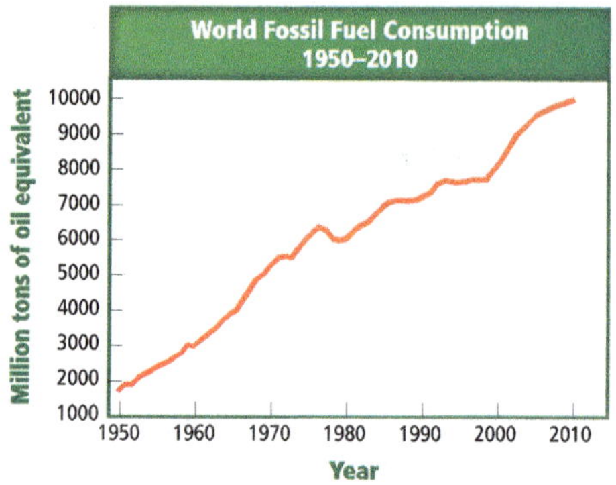

32. **Calculate** how many more tons of fossil fuels were used in 2000 compared to 1960.
33. **Predict** Do you think the trend shown on the graph would be the same for developing and industrialized countries if they were shown separately? Why?
34. **Explain** why not all organic resources are considered renewable. Give an example of a renewable and nonrenewable organic resource.
35. **Predict** What might be some negative consequences of a nation being dependent on foreign energy resources?

34. Some organic resources, such as fossil fuels, are used more quickly than they can be produced. Other organic resources, such as trees, can be replaced at the same rate at which they are used.
35. Answers will vary, but should indicate that it would be possible for those resources to increase in price, become unavailable, or run out.

Chapter Self-Check

36. **Compare and Contrast** How might the fuel-use by people living in the northeastern United States differ from fuel-use by people who live in the southern and southwestern United States?

37. **Imagine** that you and your friends took a trip to a deserted island that had no plants larger than small shrubs. Describe how you would seek a fuel source from the island.

38. **Evaluate** the potential for using more solar energy in your community. Which type of solar energy collection would work best? Is solar energy an effective energy source for your community? Why or why not?

39. **Analyze** why biomass fuels are not widely used in the United States.

40. **Compare and contrast** nuclear energy with energy that comes from petroleum.

41. **Imagine** you are eating a burger. Explain all the ways you are gaining energy derived from the Sun.

42. **Predict** what might happen to gas prices, assuming oil continues to be used at the current rate and an alternative fuel source is not discovered. Explain.

43. **Explain** why a wood-burning stove is not an efficient way to heat a home.

CONCEPT MAPPING

44. Make a concept map to organize information about alternative energy resources using the following terms: *geothermal energy, hydroelectric power, solar energy, wind power, tidal power,* and *biomass fuels.* For more help, refer to the *Skillbuilder Handbook.*

CHALLENGE QUESTION

45. **Apply** If a standard home costs $150,000 to build and costs $2300 per year to heat, and the same home, built with materials designed to use passive solar heat, costs $180,000 to build, but $400 per year to heat, how long will it take to make up the price difference between the two houses?

WRITING IN ▶ Earth Science

46. Write a letter to the editor for a local newspaper to convince others to recycle. Include specific examples and how those actions will assist in extending the limited supply of a particular natural resource.

DBQ Document–Based Questions

Data obtained from: Annual Energy Review 2009. August 2010. *Energy Information Administration* (EIA-0384).

Energy Consumption (quadrillion btu)

Year	Fossil Fuels	Nuclear	Renewable	Total
2003	84.01	7.96	6.14	98.13
2004	85.81	8.22	6.25	100.31
2005	85.79	8.16	6.41	100.45
2006	84.69	8.22	6.82	99.79
2007	86.25	8.46	6.72	101.53
2008	83.50	8.43	7.37	99.40

47. Compare and contrast the consumption of renewable energy resources with the consumption of other energy resources.

48. In 2003, what percentage of the total energy consumed in the United States was fossil fuels? Based on the data, has that percentage changed significantly in the first part of this decade?

49. Which resource had the greatest increase in energy consumption between 2003 and 2008?

CUMULATIVE REVIEW

50. Name the molecule that is necessary for life that was absent from Earth's early atmosphere. **(Chapter 22)**

51. Bedrock is found everywhere in Earth's crust. Explain whether or not an abundance of bedrock would diminish the concern over availability as a resource. **(Chapter 24)**

DBQ Document-Based Questions

Data obtained from: Annual Energy Review 2005. July 2006. *Energy Information Administration* (EIA-0384).

47. The consumption of renewable energy resources is much lower than the consumption of other resources.
48. 84.01 percent; this percentage has not changed significantly.
49. renewable

CUMULATIVE REVIEW

50. oxygen
51. Although Earth's crust is made of bedrock throughout, bedrock may be composed of many different types of rocks and minerals. Some types of bedrock could be more useful to humans than others and therefore more valuable, regardless of abundance. Granite—one of the most common types of bedrock—is a valued resource in the construction industry.

CHAPTER 25 ASSESSMENT

36. People living in the Northeast use more fuel for heating in the colder months, while people in the southern and southwestern states use more fuel for cooling purposes.
37. Answers will vary. Try to burn shrubs after they were dried in the Sun. Perhaps dry fecal matter was available on the island to try to burn. Condensing solar energy could be another source, along with wind power.
38. Answers will vary depending on your location.
39. Biomass fuels do not burn with great efficiency. They also are not readily available to those living in urban settings.
40. Nuclear energy does not release carbon dioxide or particulate matter into the environment. Petroleum is a nonrenewable resource and will most likely run out. Nuclear power can be dangerous if not controlled properly, and storage of by-products is an issue.
41. You are gaining energy from the Sun from the hamburger, which was made from a cow which ate grain that stored energy from the Sun through photosynthesis. You are gaining energy from the bun which is processed grain which stored energy from the Sun through photosynthesis. If there is lettuce, tomato, and onion on the hamburger, those have all stored energy from the Sun through photosynthesis.
42. Gas prices will increase because the demand for the gas will either stay the same or increase (with the population), but the supply will go down.
43. Wood does not burn efficiently, and much of the energy is lost.

CONCEPT MAPPING

44. Answers may vary, but students' concept maps should show that the terms provided are all alternatives to traditional energy sources, such as fossil fuels. Students may include information about how these alternative energy resources are obtained and used.

CHALLENGE QUESTION

45. about 16 years

WRITING IN ▶ Earth Science

Rubric

46. Answers will vary, but might include recycling paper to save trees.

MULTIPLE CHOICE

1. C
2. D
3. B
4. A
5. A
6. C
7. D
8. B
9. C

CUMULATIVE STANDARDIZED TEST PRACTICE

MULTIPLE CHOICE

1. Which is the most expensive and least used method of providing water to areas in the Unites States?
   A. tapping groundwater
   B. aqueducts
   C. desalination
   D. dams

Use the illustration to answer Questions 2 and 3.

2. How could this kitchen be made more energy efficient?
   A. by maintaining older appliances instead of replacing them with newer ones
   B. by replacing the kitchen cabinets
   C. by washing the dishes in the dishwasher instead of the sink
   D. by replacing the old windows with newer ones

3. If this kitchen were located in a home in Arizona, which alternative energy source could be used?
   A. peat
   B. solar energy
   C. ethanol
   D. hydroelectric power

4. Which relationship between geologic structures and plate boundaries is most accurate?
   A. Explosive volcanoes most often occur near convergent boundaries.
   B. Folded mountains commonly develop at divergent boundaries.
   C. Rift valleys are usually produced at convergent boundaries.
   D. Volcanic arcs are usually found along transform boundaries.

5. Besides being a requirement for respiration, why else is oxygen important in the atmosphere?
   A. It provides protection from ultraviolet rays emitted by the Sun.
   B. It regulates climate and weather patterns on Earth.
   C. It is the major component of wind to cool Earth.
   D. It allows rays from the Sun to filter in and warm Earth.

6. What was the goal of Stanley Miller's research?
   A. to refute the belief that life could have existed on early Earth
   B. to explain the formation of oxygen on early Earth
   C. to test the primordial soup hypothesis
   D. to do an analysis of the atmosphere present on early Earth

Use the illustration below to answer Questions 7 and 8.

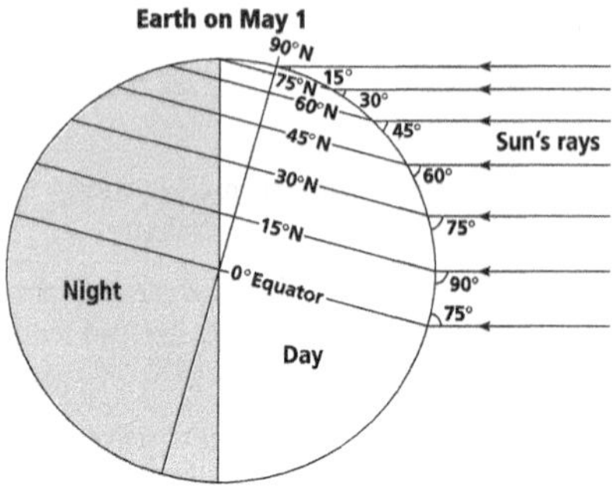

7. Which change can be expected to occur at 45° N over the next 30 days?
   A. The duration of solar radiation will decrease and the temperature will decrease.
   B. The duration of solar radiation will decrease and the temperature will increase.
   C. The duration of solar radiation will increase and the temperature will decrease.
   D. The duration of solar radiation will increase and the temperature will increase.

8. Where would the risk of sunburn be highest?
   A. the Equator
   B. 15° N
   C. 45° N
   D. 75° N

9. Besides the formation of Pangaea, what other major event occurred during the Paleozoic?
   A. the first major volcanic eruption
   B. the first appearance of life
   C. the appearance of complex life
   D. the mass extinction of all life

Online Test Practice

SHORT ANSWER

Use the illustration below to answer Questions 10 to 12.

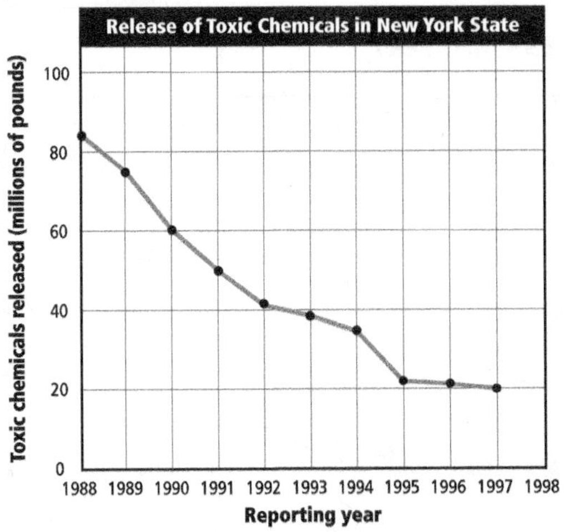

10. About how much change has occurred in the amount of toxic chemical released from 1988 to 1997?

11. During what four year period was the greatest drop in toxic chemicals released? What is one possible explanation for this major drop?

12. State one possible explanation for why the amount of toxic chemicals released remained relatively constant between 1995 and 1997.

13. How is a regression identified in the rock record?

14. How does the process of relative-age dating differ from the process of absolute-age dating?

15. What do scientists hypothesize is the cause of the cooling trend during the Cenozoic Era?

READING FOR COMPREHENSION

Vegetable Oil Fuels

Chemists and advocates for alternative energy technologies are training their sights on the grease used to cook french fries. Unlike petroleum-based products, vegetable oils are biodegradable, non-toxic, and are derived from a renewable resource. One problem, however, is the high development cost of vegetable-derived motor oils relative to petroleum-based products. Advocates for the use of vegetable oils say they are easier on the environment because they are much more biodegradable than conventional, petroleum-based oils. When spilled or disposed of on the ground, vegetable oil will decompose by upwards of 98 percent. Petroleum based products only decompose 20 to 40 percent. Additionally, vegetable oils are a renewable resource.

Article obtained from: Roach, John. Vegetable Oil—The New Fuel?. *National Geographic News.* 22 April 2003 1-2.07 Oct 2006

16. What can be inferred from this passage?
- **A.** Petroleum-based oils are better than vegetable oils as energy sources.
- **B.** Vegetable oils will not be able to be used in car engines.
- **C.** Although vegetable oils are better for the environment, it will be some time before they replace the use of petroleum-based oils.
- **D.** Even though vegetable oils are better for the environment than petroleum based oils, it is still better not to use them for energy sources.

17. When spilled or disposed of, what percent of vegetable oils will decompose?
- **A.** 3 percent
- **B.** 75 percent
- **C.** 98 percent
- **D.** 100 percent

NEED EXTRA HELP?

If You Missed Question . . .	1	2	3	4	5	6	7	8	9	10	11	12	13	14	15
Review Section . . .	24.4	25.3	25.2	18.1	22.3	22.4	12.1	12.1	23.1	24.3	24.3	24.3	23.1	21.3	23.3

SHORT ANSWER

10. about 65 million pounds less from 1988 to 1997

11. between 1988 and 1992; Sample answer: Because this was the time when the changes were beginning, even the smallest changes created the biggest drop in release of toxic chemicals.

12. Sample answer: The level possibly remained the same from 1995 to 1997 because all current strategies had been used and technology had not developed any way to reduce the amount of chemicals released any further.

13. A regression results in shallow-water deposits, such as sandstone, overlying deeper-water deposits, like shale.

14. Relative-age dating establishes the order of events. It does not allow scientists to determine how many years ago an event occurred, but it does help scientists learn which events happened before or after other events. Absolute-age dating allows scientists to determine the actual age of rocks and other objects. This is done by measuring the decay of radioactive materials in the rock.

15. The splitting apart of Australia from Antarctica caused the warm climate that had been dominant to start to deteriorate. This is hypothesized to have occurred due to a change in oceanic circulation from warm to cold around Antarctica.

READING FOR COMPREHENSION

16. C

17. C

CHAPTER 26 Human Impact on Resources

BIGIDEA **The use of natural resources can impact Earth's land, air, and water.**

ESSENTIAL QUESTIONS	NATIONAL STANDARDS	RESOURCES TO ASSESS MASTERY
SECTION 1 **Populations and the Use of Natural Resources** **1.** What is the typical pattern of population growth of organisms? **2.** What happens to populations when they reach carrying capacity? **3.** What environmental factors affect population growth? 1 session 0.5 block	F.2–6	**Entry-Level Assessment** Focus Transparency 81, p. 734 **Progress Monitoring** Caption Question, pp. 734, 735 Reading Check, p. 735 Section Review, p. 736
SECTION 2 **Human Impact on Land Resources** **1.** How can mineral extraction impact the environment? **2.** What are some of the environmental issues created by agriculture and forestry and possible solutions? **3.** How does urban development affect soil and water? 5 sessions 2.5 blocks	UCP.2; C.4; F3–6	**Entry-Level Assessment** Focus Transparency 82, p. 737 **Progress Monitoring** Caption Question, pp. 737, 739 Reading Check, p. 738 Section Review, p. 742
SECTION 3 **Human Impact on Air Resources** **1.** What is the relationship between the greenhouse effect and global warming? **2.** What is the sequence of reactions that occur as CFCs cause ozone depletion? **3.** What are the causes and effects of acid precipitation? 2 sessions 1 block	UCP.2–3; B.3; D.1; F.1, F.4–6	**Entry-Level Assessment** Focus Transparency 83, p. 743 **Progress Monitoring** Caption Question, pp. 743, 746 Section Review, p. 747
SECTION 4 **Human Impact on Water Resources** **1.** In what ways can water be conserved? **2.** What are the types and sources of water pollution? **3.** How can water pollution be reduced? 3 sessions 1.5 blocks	UCP.2–3; A.2; C.4; F.3–6	**Entry-Level Assessment** Focus Transparency 84, p. 748 **Progress Monitoring** Reading Check, p. 749 Section Review, p. 750 **Summative Assessment** Chapter Assessment, p. 755 Exam*View*® *Assessment Suite* CD-ROM

Go online!
connectED.mcgraw-hill.com

LEVELED RESOURCES	LAB MATERIALS
Science Notebook 26.1 OL **Chapter FAST FILE Resources:** Study Guide, p. 67 BL **Transparencies:** Section Focus Transparency 81 EL Teaching Transparency 79 OL EL	LaunchLAB p. 732 / **20 min** items (15), paper, pencil
Science Notebook 26.2 OL **Chapter FAST FILE Resources:** MiniLab Worksheet, p. 54 OL Study Guide, p. 69 BL **Lab Resources:** Laboratory Manual, p. 201 OL **Transparencies:** Section Focus Transparency 82 EL Teaching Transparency 80 OL EL	MiniLAB p. 740 / **45 min** coffee filter, funnel, 100-mL beaker, sand mixture, water
Science Notebook 26.3 OL **Chapter FAST FILE Resources:** Study Guide, p. 71 BL **Transparencies:** Section Focus Transparency 83 EL Teaching Transparency 81 OL EL	
Science Notebook 26.4 OL **Chapter FAST FILE Resources:** GeoLab Worksheet, p. 55 OL Study Guide, p. 72 BL **Lab Resources:** Laboratory Manual, p. 205 OL **Transparencies:** Section Focus Transparency 84 EL Teaching Transparency 82 OL EL	GeoLAB p. 752 / **90 min** metric ruler, science notebook

LabManager™
Customize any lab with the LabManager™ CD-ROM.

ADDITIONAL RESOURCES

Technology:
- eTeacherEdition
- eStudentEdition
- Classroom Presentation Toolkit
- Exam*View*® *Assessment Suite* CD-ROM
- LabManager™ CD-ROM
- What's EARTH SCIENCE Got To Do With It? DVD
- Weather Classroom Video DVD

Assessment:
- Blueprints for Success
- Chapter Test (Scaffolded)
- Earth Science Standardized Test Practice

Lab Resources:
- Exploring Environmental Problems Laboratory Manual

BL Below Level OL On Level AL Advanced Learners EL English Learners COOP LEARN Cooperative Learning

CHAPTER 26

LaunchLAB

iLab Station Rubric

What resources are used in classroom items?

Process Skills observe and infer, collect and interpret data, make and use tables

Safety Precaution Approve lab safety forms before work begins.

Teaching Strategies

- Small objects from the classroom may be used. Organize several boxes of small, everyday objects if you are going to supply items to students.
- Ask a student to define renewable resources and nonrenewable resources.
- Discuss with the class which materials can easily be recycled, such as paper, metals, and glass.

Procedure

1. Have students read and complete the lab safety form and follow the procedure below.
2. Working in groups of two or three, make a pile of 15 **items** from your classroom.
3. Make a data table for your items. Record as much of the following information as you can.
   - What resources were used to make the item?
   - Are the resources renewable or nonrenewable?
   - Where was the item made?

Analysis

1. **Observe** How many different resources are represented by the items in your collection? Answers will vary depending upon the collection of objects. Typically, students will observe metals, paper, plastics, cloth such as cotton and wool, leather, rubber, and synthetic fabrics.

CHAPTER 26

Human Impact on Resources

BIGIDEA The use of natural resources can impact Earth's land, air, and water.

SECTIONS

1 Populations and the Use of Natural Resources

2 Human Impact on Land Resources

3 Human Impact on Air Resources

4 Human Impact on Water Resources

LaunchLAB

iLab Station

What resources are used in classroom items?

As you have learned, natural resources include air, water, land, and living organisms. In this activity, you will determine the resources used to make classroom items.

FOLDABLES® Study Organizer

Sources of Water Pollution

Make a two-tab book using the labels shown. Use it to organize your notes on the two main types of water-pollution sources.

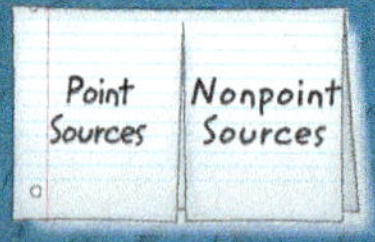

Derin Thorpe/Botanica/Getty Images

2. **Calculate** What percent of your 15 items were renewable and what percent were from nonrenewable resources? Answers will vary depending upon the collection of objects.

Assessment

Performance Have each pair of students select some aspect of their data and make a bar graph. Examples include the number of items made in different parts of the world and the number of items made of different materials.

LabManager™
Customize any lab with the LabManager™ CD-ROM.

People have found a variety of ways to reduce their environmental impact on natural resources. Rooftop gardens, like the one shown here, can reduce smog episodes that pollute the air, as well as the rate and volume of runoff that can pollute surface water and groundwater.

Introduce the BIGIDEA

People and the Environment Have students make a list of ten ways humans impact Earth, and evaluate each in terms of its positive or negative impact. Ask students what determines the level of impact, and whether the impacts are permanent or temporary.

Teacher Content Support

Impact on Earth Ask students if they can think of human activities that have a positive impact on the Earth. For example, a rooftop garden reduces the heat absorption by a building, and therefore conserves energy.

SECTION 1

1 Focus

Focus Transparency

Before presenting the lesson, project **Section Focus Transparency 81** and have students answer the accompanying questions. EL

MAINIDEA

Population Explosion Ask students to write a paragraph in their Earth science journals about what life would be like in the United States if the population doubled just ten years from now. After students have read this lesson, have them revise their paragraphs.

2 Teach

Teacher Content Support

Manage Population Growth
Developing countries are countries that are in the process of change toward economic growth, and that have an average income that is relatively lower than in highly industrialized countries. Sixty years ago, the populations of many developing countries entered a period of exponential growth. This was a result of advances in agriculture and medical care. Mortality rates fell, but fertility rates did not. Developing countries took one of two approaches to population growth. Countries such as South Korea, Taiwan, and Thailand mounted efforts to reduce family size. This led to higher savings, higher living standards, and falling fertility rates. Other developing countries, particularly India and those in Africa, maintained high birth rates. Rising populations are currently overwhelming these countries.

Caption Question **Fig. 1** Answers will vary, but might include impacting the movement and breeding of aquatic species.

SECTION 1

Essential Questions

- What is the typical pattern of population growth of organisms?
- What happens to populations when they reach carrying capacity?
- What environmental factors affect population growth?

Review Vocabulary

population: individual organisms of a single species that share the same geographic location at the same time

New Vocabulary

exponential growth
carrying capacity
density-independent factor
density-dependent factor

Populations and the Use of Natural Resources

MAINIDEA More demands are placed on natural resources as the human population increases.

EARTH SCIENCE 4 YOU How many pets do you have? If you only have one pet, it might not take much time or money to care for it. What if you had eight pets? The amount of time and money you would need to properly care for your pets might put a strain on your money and activities. Similarly, as world population grows, it puts a strain on available natural resources.

Resources and Organisms

Like all organisms, humans need natural resources to grow, reproduce, and maintain life. Among the resources that organisms require are air, food, water, and shelter. To meet their basic needs, most organisms are adapted to their immediate environment. They live in balance with the natural resources provided within their environment. For example, songbirds live in grassy meadows, forage for grass seeds to eat, weave nests out of dried grasses and twigs, and drink water from ponds or streams nearby.

Other organisms, however, alter their environment to better meet their needs. For example, beavers build dams, like the one in **Figure 1,** across streams to create ponds where none previously existed. Such alteration of the environment has both positive and negative impacts: it kills some trees and displaces both aquatic and terrestrial organisms, but at the same time, it creates a new wetland environment for other organisms. Of all organisms, however, humans have an unequaled capacity to modify their environments. This capacity allows humans to live in every terrestrial environment on Earth. As a result, humans also have the greatest impact on Earth's natural resources.

Figure 1 Beavers can alter their environments to suit their needs. Notice how, by damming the stream, the beavers have changed the water level.
Infer *How might this affect the other organisms living in this environment?*

EARTH SCIENCE JOURNAL

Environmental Impact Have students each write a paragraph in their Earth science journals in response to this question: What resources are you using and what environmental impact are you having as you sit reading this book?

Rubric

DIFFERENTIATED INSTRUCTION

Advanced Learners Fire ants are an imported species that is causing serious problems in the United States. Have students research fire ants and give short presentations to the class in which they answer the following questions: From where did the ants come? How quickly are they spreading? What impacts are they having on ecosystems?

Population Growth

Population growth is defined as an increase in the size of a population over time. A graph of a growing population resembles a J-shaped curve at first. Whether the population is one of dandelions in a lawn, squirrels in a city park, or herring gulls on an island, the initial increase in population is small because the number of adults capable of reproducing is low.

As the number of reproducing adults increases, however, the rate of population growth increases rapidly. As shown in **Figure 2,** the population then experiences **exponential growth,** which is a pattern of growth in which a population grows faster as it increases in size.

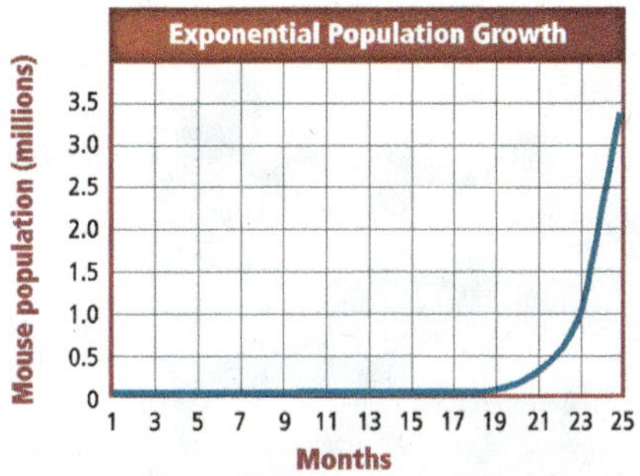

Figure 2 If two mice were allowed to reproduce in perfect conditions and all their offspring survived, the population would grow slowly at first, but then would accelerate quickly.

READING CHECK **Explain** exponential growth. Why is this an important concept to understand in relation to how organisms affect their environment?

Limits to population growth If the population graphed in **Figure 2** were studied for an extended period of time, what do you think would happen to the size of the population? Would it continue to grow exponentially? Many of Earth's natural resources are in limited supply, and therefore, most populations cannot continue to grow forever.

Eventually, one or more limiting factors, such as the availability of food, water, or shelter will cause a population to stop increasing. This leveling-off of population size results in an S-shaped curve, similar to the one in **Figure 3.**

Carrying capacity The number of organisms that any given environment can support is its **carrying capacity.** When population size has not yet reached the carrying capacity of a particular environment, the population will continue to grow for several reasons. First, there will be more births than deaths because of adequate resources. Second, because of the availability of resources, more individuals might move to the area than die or leave.

If the population size temporarily exceeds the carrying capacity, the number of deaths will increase, or the number of births will decrease until the population size returns to the carrying capacity. A population at the carrying capacity for its environment is in equilibrium. The population will continue to fluctuate around the carrying capacity as long as natural resources remain available.

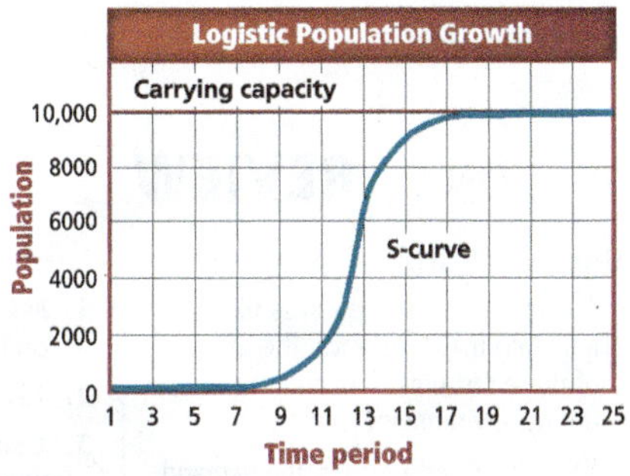

Figure 3 A population grows exponentially until it reaches its carrying capacity. Carrying capacity is limited by the resources available to the population.
Predict *what would happen if the population exceeded carrying capacity.*

View an **animation of carrying capacity.**

Concepts In Motion

Environmental Connection

Genetically Engineered Plants Some people think that genetically engineered plants are needed to increase crop yields to meet the growing demand for food. Other people insist that these plants pose a significant environmental hazard. Ask students to research the pros and cons of this issue, and form opinions about with which side they agree. OL

Activity

Termite Mounds Have students research termite mounds to find out why termites alter their environment in this way and what advantages it offers. OL

Discussion

Exponential Growth Ask students whether they can think of any other examples of exponential growth. Some examples are radioactive decay and half-lives of radioactive elements, nuclear reactions, the change in the intensity of light with distance, the velocity of a falling object, and earthquake-magnitude scales.

Caption Question Fig. 3 Some of the population would die off and/or stop breeding until carrying capacity is reattained.

READING CHECK Exponential growth is a pattern of growth in which a population grows faster as it increases in size. As a population grows, it places more strain on natural resources. Scientists can anticipate problems and develop solutions if they can predict exponential growth in various populations.

DIFFERENTIATED INSTRUCTION

Struggling Learners Have students breathe onto a mirror to observe the resulting condensation. This will provide visual proof that water vapor is being released into the surrounding air and impacting the local environment.

ACROSS THE CURRICULUM

Biology In addition to exhaling gases, humans also release heat into the environment. Have students research metabolic processes to find answers to the following questions: From where does this heat come? The heat comes from chemical reactions as humans break down food. How is the heat generated? Most of the heat is generated by muscular activity.

3 Assess

Check for Understanding

Activity Have students place about 30 paper clips in a shoebox lid and shake the lid. Then, have them place the lid flat on a desk and remove any paper clips that are touching one another. Have students repeat the shaking and removal, and keep track of how many paper clips they remove each time. Then, ask students the following questions: Does the number of paper clips removed increase or decrease? decrease Is this shaking and removal an example of a density-dependent or density-independent factor? density-dependent How could you model a density-independent factor? Drop the lid and turn it upside down.

Reteach

Activity Ask students how many breaths they take per minute. Have them calculate how many breaths the class takes per minute. Then, ask them to identify the resource being consumed when this population breathes. Finally, ask how this is related to populations and resource demand.

Assessment

Skill Have students make a list of the factors that can affect human populations and identify each as density-dependent or density-independent.

Figure 4 A forest fire is one example of a density-independent factor of population growth. Fires can affect trees, birds, mammals, and other populations. Fires, like the one that occurred here, can also encourage new growth.

Environmental limits Environmental factors that do not depend on population size, such as storms and fires, are **density-independent factors.** Density-independent factors affect all populations that they come in contact with, regardless of population size, as **Figure 4** shows. Environmental factors that depend on population size, such as disease, predators, and competition for food, are called **density-dependent factors.** Density-dependent factors are often biotic factors that increasingly affect a population as the population's size increases. In this way, density-dependent factors affect population growth.

One example of a density-dependent factor can be seen in populations of white-tailed deer in the United States. White-tailed deer are found in most of the continental United States. In some areas, deer populations have grown in recent years due largely to a decrease of natural predators. Although the population is increasing, the amount of food available to the deer does not change. For this reason, every year many deer starve to death. The density-dependent factor in this example is the lack of food due to overpopulation.

Human Population Growth

During your lifetime, you might have seen an increase in the number of cars, houses, and roads around you. The human population on Earth is exponentially growing. The growth curve is still in the J-shaped stage. In fact, at its current rate, the human population is expected to reach 7.8–10.8 billion by the year 2050.

The human population has not yet reached carrying capacity, but the current rate of growth cannot continue forever. As the population increases, demand for natural resources will also continue to increase steadily. Use of natural resources has already had global environmental implications.

SECTION 1 REVIEW

Section Self-Check

Section Summary

- All organisms use resources to maintain their existence. The use of these resources has an impact on the environment.
- As populations increase, the demand for resources increases. Because resources are limited, populations will stop growing when they reach carrying capacity.
- Populations grow exponentially at early stages. Earth is currently experiencing a human population explosion.

Understand Main Ideas

1. MAINIDEA **Explain** how an increasing human population places more demands on Earth's natural resources.
2. **Identify** three limiting factors that keep populations from growing indefinitely.
3. **Compare** density-dependent and density-independent factors that limit population growth.

Think Critically

4. **Predict** how a small population of bacteria placed in a petri dish with limited nutrients will change over time. Draw a graph to represent the population growth.

MATH IN ▶ Earth Science

5. If a city has 300,000 residents and an average birth rate of 1.5 children per person, how many people will there be in the next generation?

SECTION 1 REVIEW

1. As the human population increases, more resources are needed to operate cars, heat homes, transport food, and much more.
2. Some limiting factors include availability of food, shelter, and water.
3. Density-dependent factors affect populations as the population size increases. An example of a density-dependent factor is the availability of food. A density independent factor affects a population regardless of size. An example of this is a fire or a severe storm.
4. The bacteria population will grow slowly at first, then exponentially. As the nutrients begin to run out, the population will reach a maximum and then quickly decrease when all the nutrients have been consumed.
5. 450,000

SECTION 2

Human Impact on Land Resources

MAINIDEA Extraction of minerals, farming, and waste disposal can have negative environmental impacts.

Essential Questions

- How can mineral extraction impact the environment?
- What are some of the environmental issues created by agriculture and forestry and possible solutions?
- How does urban development affect soil and water?

Review Vocabulary

erosion: movement of weathered materials from one location to another by agents such as water, wind, glaciers, and gravity

New Vocabulary

reclamation
deforestation
pesticide
bioremediation

EARTH SCIENCE 4 YOU Do you spend much time talking on the telephone, listening to a digital music player, or using a computer? Perhaps you use a microwave oven to heat after-school snacks. Many of the materials in these items are derived from land resources.

Mining for Resources

How much land per year do you think is necessary to provide the raw materials that you use? Each year, a typical person in the United States consumes resources equal to the renewable yield from approximately 5 ha (about 12.5 acres) of forest and farmland. Many of these raw materials come from under the surface of Earth. To access these resources for human use, they must be extracted through one of many mining techniques.

Mining techniques can have a significant impact on Earth's surface. Modern societies require huge amounts of land resources, including iron, aluminum, copper, sand, gravel, and limestone. Unfortunately, the extraction of these resources often disturbs large areas of Earth's surface, as shown in **Figure 5.** Groundwater can become polluted, natural habitats can be disturbed or destroyed, and air quality can suffer. Finding a balance between the need for mineral resources and controlling the environmental change caused by extraction can be difficult, but scientists, in conjunction with mining companies, have created ways to reduce the impact of mining on the environment.

Figure 5 Mines, such as the one shown here, can have negative environmental impacts such as topsoil erosion.
Determine *Where would the eroded topsoil go? What other environmental impacts might a mine like this one have?*

SECTION 2

1 Focus

Focus Transparency

Before presenting the lesson, project **Section Focus Transparency 82** and have students answer the accompanying questions. EL

MAINIDEA

Importance of Topsoil Have groups of students discuss the following question: Is topsoil a renewable or nonrenewable resource? If it is renewable, why is topsoil loss such a problem?

2 Teach

Identify Misconceptions

Most people underestimate the amount of solid waste they generate.

Uncover the Misconception Ask students to estimate how much waste they generate in one day.

Demonstrate the Concept Have students collect all forms of solid trash that they generate in one 24-hour period and quantify their results in terms of mass and volume. Each student should post his or her data on the board.

Assess New Knowledge Students will likely find that they underestimated their output. As a class, brainstorm ways to reduce the amount of solid waste each person generates.

Demonstration

Soil Erosion To demonstrate the effects of vegetation on erosion, take the class outside and pour a bucket of water (or use a hose) on a patch of bare soil. Then, pour a bucket of water on a patch of grass or other ground cover. More soil will be eroded from the bare patch of soil than from the patch covered with vegetation.

Caption Question Fig. 5 The eroded topsoil could go many places, including nearby water sources such as lakes and rivers. In addition, the loss of topsoil in the mined areas will make it difficult for plants to grow, leading to more erosion.

Teacher Content Support

Abandoned Mines Abandoned mines can be more of an environmental danger than operating mines are. It is estimated that there are between 100,000 and 500,000 abandoned mines in the western United States, each of which poses serious water contamination problems. In most cases, the original mine operators are no longer in business, so the public must bear the cleanup costs. To restore these sites will require a tremendous amount of tax money.

Discussion

Reclamation Ask students whether reclamation could work for underground mining as well as surface mining. It would not work as well. It would be expensive to put all the waste rocks back underground. The waste rocks also might not all fit back into the mine because there would be additional air spaces between the broken rocks. Therefore, the waste rocks would fit back underground only if enough useable material has been removed from the mine.

Environmental Connection

Agricultural Chemicals Have groups of students make posters or data tables showing different ways fertilizers and pesticides can enter the environment. OL

☑ READING CHECK to reduce erosion and topsoil loss, and to try to restore destroyed habitats

■ **Figure 6** White-tailed deer and other native wildlife species thrive on this reclaimed strip mine land in eastern Wyoming.

Watch a **video about habitat restoration.** Video

Restoring the land In the United States, the Surface Mining Control and Reclamation Act of 1977 requires mining companies to restore the land to its original contours and to replant vegetation in a process called **reclamation.** However, vegetation cannot grow well without topsoil. Mining companies can scrape the topsoil off of the land surface prior to mining and stockpile it for reclamation after materials have been removed. **Figure 6** shows a strip-mined area that has been reclaimed. Although reclamation repairs much of the damage that surface mining causes, it can be extremely difficult to restore land to its original contours and vegetation.

☑ READING CHECK **Explain** why it is important to have legislation requiring mining companies to restore land to its original contours.

■ **Figure 7** Runoff from a nearby mine pollutes this river. The presence of metals, including iron, causes the orange color in the runoff.

Underground mining Underground mining, also called subsurface mining, is used where mineral resources lie deep under the ground. Underground mining is less disruptive to the land surface than surface mining, but it still impacts the environment. For example, although the underground mines cannot be seen, the mountains of waste rock dug from under the ground are stockpiled on the surface. The water in **Figure 7** is orange because precipitation seeps through mine waste piles and causes a decrease in pH, dissolving many harmful metals in the waste. When the runoff from the piles reaches the stream, which has a higher pH, the higher pH causes the metals to come out of solution and discolor the water. Aquatic plants and animals can also be affected by runoff. Although many mining companies build large holding ponds to contain polluted water until it can be treated, these ponds sometimes leak.

EARTH SCIENCE JOURNAL

Consumers and Pesticides Have students answer the following question in their Earth science journals: How have consumers played a role in excessive pesticide use? Sample answer: American consumers expect their fruits and vegetables to be 100 percent free of insect damage and other blemishes. Farmers therefore use excess pesticides so they can sell their products. Afterward, discuss students' answers as a class.

What's EARTH SCIENCE Got To Do With It?

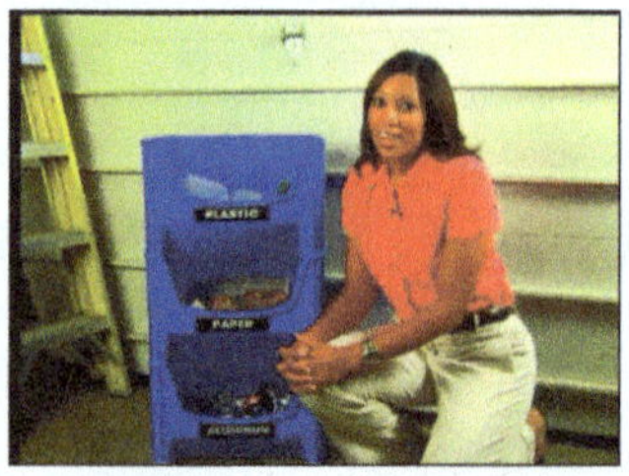

Video *Living Green* and *Habitat Restoration*

Forestry

Clearing forested land is another way in which topsoil is lost. Worldwide, thousands of hectares of forests are cut down annually for firewood, charcoal, paper, and lumber. In parts of the world, the clearing of forested land results in **deforestation,** which is the removal of trees from a forested area without adequate replanting. Deforestation often involves clear-cutting, the complete removal of all the trees in an area. Clear-cutting can cause erosion of topsoil.

Fortunately, the negative environmental impacts of deforestation can be minimized through the practices of selective logging and the retention of buffer zones of trees along streambeds. In selective logging, workers remove only designated trees. This practice reduces the amount of ground left bare and thus helps prevent erosion.

VOCABULARY

ACADEMIC VOCABULARY

Adequate
sufficient to satisfy a requirement or meet a need
She filled her gas tank to ensure she had an adequate supply of gas for the trip.

Urban Development

As the human population continues to increase, more people live in cities and towns. Agricultural land located near cities is being converted to suburban housing. As people populate areas that were once agricultural or rural, stores and industry follow. Eighty percent of the population in North America lives in urban and suburban areas, and an estimated 5 billion people worldwide will be living in cities and towns by the year 2025.

The development of urban areas has many environmental impacts. When towns and cities expand into rural areas, natural habitats are lost to roads, houses, and other buildings. Development leaves less land for agricultural use, which puts pressure on the remaining farmland for increased production. Other problems are created when concrete and asphalt cover large areas. Because there are fewer opportunities for rainwater to soak into the ground, groundwater supplies are not recharged, and flooding increases during heavy rains.

Watch a **video about living green.** Video

Solid waste Each person in the United States generates an average of 2.0 kg of solid waste per day. Where does it all go? Much of it is buried in landfills. **Figure 8** shows the percentages and material types that were disposed in landfills in the United States in 2008.

2008 Total Waste Generation–250 million metric tons (before recycling)

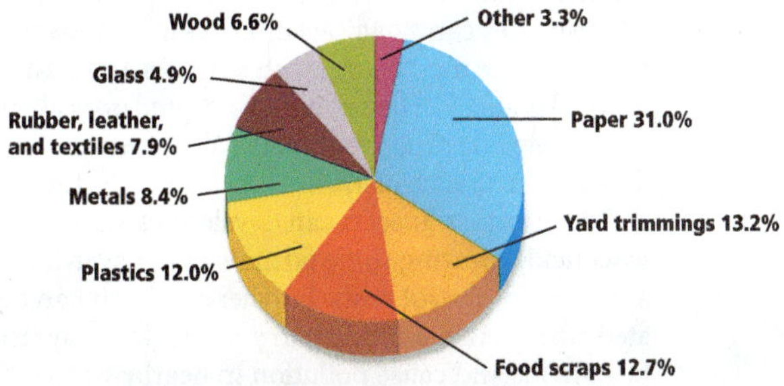

■ **Figure 8** This circle graph shows the total solid waste generated in the United States in 2008.
Determine ***What material composed the highest percentage of solid waste in the United States in 2008? Why do you think this was the case? Could this material be recycled?***

Enrichment

Resistance to Pesticides Have students research the problem of plant pests, such as insects, bacteria, and fungi, developing resistance to pesticides. Ask students to find out what causes the development of resistance and what is being done to combat the problem. OL

Teacher Content Support

Symbiotic Relationship
Legumes, such as alfalfa, clover, lupines, peas, and beans, are often used as cover crops because they add nitrogen to soil. Before planting, the seeds are inoculated with Rhizobium bacteria, which fix atmospheric nitrogen and convert it into a form plants can use. There is a natural symbiotic relationship between legumes and Rhizobium. The bacteria live in small nodules that form on the plants' roots.

Enrichment

Cover Crops Have students investigate the following question: How can growing a cover crop replenish nutrients in a field? Some cover crops add nutrients to the soil that other crops have depleted. BL

■ **Caption Question Fig. 8** Paper; Answers will vary. Possible answers: because people throw out newspapers and magazines at home, and businesses and schools use a lot of paper. Yes, this material could be recycled.

EARTH SCIENCE JOURNAL

Bioremediation Have students research and write about bioremediation. How is it used to clean up pollution from toxic metals and organic compounds? Bioremediation typically takes one of two forms: augmentation or stimulation. During augmentation, organisms are added to the contaminated site. During stimulation, nutrients are provided that stimulate the growth of organisms that are already present. AL

DIFFERENTIATED INSTRUCTION

Struggling Learners Have students count the number of species in an aquarium to illustrate biodiversity. Help students compare this number to the number of species in the classroom outside of the aquarium.

MiniLAB

iLab Station | Rubric

Purpose Students will observe patterns of nutrient loss in soil.

Process Skills model, observe, control variables

Safety Precaution Approve lab safety forms before works begins.

Teaching Strategies

- Prepare a mixture of sand and a small amount of potting soil.
- Tell students to pour the water into the beaker at the same steady rate during each trial.

Expected Results Students will observe that lack of plant cover causes loss of nutrients in soil.

Analysis

1. The color of the water should become lighter each time the process is repeated.
2. If the water had not been discarded, it would be impossible to know which trial caused the observed results.
3. It would not be a productive soil for plant growth.
4. Answers will vary. The soil might erode. The nutrients could run off into nearby waterways, causing algae blooms, fish kills, and other problems. If the plant population decreased, it would negatively impact animals in the area.
5. You could plant plants to help slow the loss of nutrients. You could plant plants that put nutrients back into the soil. Or, you could add fertilizer.

Assessment

Skill Have pairs of students design an experiment to test the effect of plant cover on soil erosion. COOP LEARN

LabManager™
Customize any lab with the LabManager™ CD-ROM.

MiniLAB

iLab Station

Model Nutrient Loss

How does soil lose nutrients when subjected to farming, strip-mining, or development? If an area of soil has no plant cover for an extended period of time, the nutrients in the soil can be washed away by rainfall.

Procedure

1. Read and complete the lab safety form.
2. Place a **coffee filter** inside a **funnel**.
3. Place the funnel so that it is resting inside a **100-mL beaker**.
4. Pour a **sand mixture** into the coffee filter.
5. Measure 50 mL of **water** and pour it into the sand mixture.
6. Record your observations.
7. Carefully remove the funnel with the sand mixture and discard the water as instructed by your teacher.
8. Replace the funnel with the sand mixture in the 100-mL beaker.
9. Repeat Steps 5 through 8 four times.
10. Discard the sand mixture and funnel as instructed by your teacher.

Analysis

1. **Observe** What did you notice about the water as you repeated the investigation?
2. **Analyze** Why were you instructed to discard the water in between investigations?
3. **Predict** If the material in the sand represents nutrients needed for plant growth, and the sand represents soil, what would happen if you tried to grow crops in the soil at the end of your investigation?
4. **Infer** In what other ways might nutrient loss impact the environment?
5. **Apply** How could you slow the process of nutrient loss?

Improper disposal of wastes can result in the contamination of land and water resources. Heavy metals, such as lead and mercury, and poisonous chemicals, such as arsenic, are by-products of many industrial processes and can pollute the soil and groundwater. Some of this type of contamination has been caused by industries that operated before the dangers of improper waste disposal were known.

Agriculture

Vegetation, including agricultural crops, needs the nutrients from topsoil to grow. It can take thousands of years for topsoil to form, and thus, once it is lost, it is hard to replace. Whenever fields are plowed and the plants whose roots hold the soil in place are removed, topsoil can be eroded by wind and water and nutrients can be lost. The addition of fertilizers can help replace some of the nutrients, but there are other substances in topsoil that fertilizers cannot provide.

Topsoil contains trace minerals as well as organisms such as earthworms and nitrogen-fixing bacteria. Earthworms burrow into soil, providing oxygen and space for plant roots to grow, and nitrogen-fixing bacteria take nitrogen out of the air and make it available to plants. Topsoil also has an abundance of organic matter, including fecal material from organisms that live in the soil as well as decaying organisms. Organic matter helps hold moisture, reduces erosion, and releases nutrients back into the soil. Soil erosion can be reduced and fertility can be increased by using a variety of farming practices as shown in **Figure 9.**

Effects of pesticides Chemicals applied to farm fields to control weeds, insects, and fungi are called **pesticides.** Pesticides have played an important role in boosting food production worldwide by eliminating or controlling organisms that destroy crops. However, some pesticides remain in, and can potentially harm, the environment for long periods of time.

Pesticides can slowly accumulate in organisms higher on the food chain, such as fishes and birds. Some pesticides also kill beneficial insect predators along with the targeted destructive insects. When pesticides kill decomposers, such as worms, the overall fertility of topsoil deteriorates. Insects can develop resistance to an insecticide, causing some farmers to use ever-increasing amounts to control pests. Further problems can be created when wind and rain carry pesticides away from farm fields and cause pollution in nearby waterways.

ACROSS THE CURRICULUM

Biology Have students identify how decomposers fit into the nutrient-energy cycle. Ask them why soil fertility decreases when decomposers are killed by pesticides. Decomposers break down organic matter into simpler molecules that can be used by other organisms as nutrients. Thus, decomposers complete the cycle of energy through the environment. If decomposers are killed off by pesticides, some nutrients remain locked up, and soil fertility decreases. AL

ACROSS THE CURRICULUM

Chemistry Ask students what three elements are found in most fertilizers. nitrogen, potassium, and phosphorus Have students find out what role each of these nutrients plays in plant growth. Nitrogen is involved in protein production and gives plants their deep green color. Phosphorus is involved in root growth and flower production. Potassium's role is less understood, but might be involved in protein and cell-wall construction. AL

VISUALIZING Agricultural Practices

Figure 9 Using agricultural conservation practices can help protect precious nutrients in the soil as well as help reduce topsoil loss. Contour farming, crop rotation, and no-till farming are shown below.

Contour farming is often done on hillsides or other areas prone to erosion. Farmers plant crops with the contour of the earth, slowing the flow of runoff and helping to prevent erosion.

No-till farming Farmers leave the unused portion of the crops on the field instead of plowing them under each year. In this image, the crop in the previous year was wheat. After the seeds were harvested, the stalks were left on the field to prevent erosion and maintain topsoil.

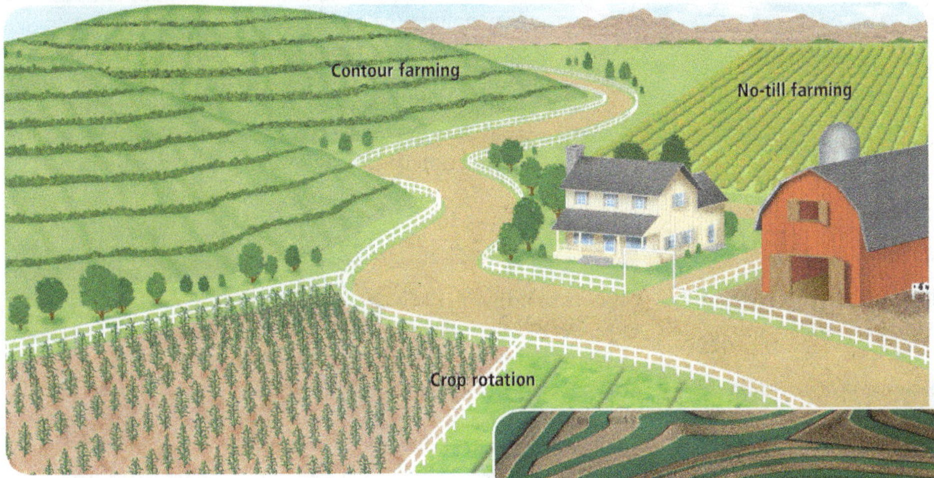

Crop rotation involves planting different crops in succession. For instance, a farmer might plant alfalfa in a field one year, followed by corn the next year, and winter wheat the following year. Crop rotation helps to maintain the soil's nutrient balance and also helps to reduce the number of crop-specific pests.

Concepts In Motion View an **animation of agricultural practices.**

IN THE FIELD

Ronald Brooks An African-American chemist, Dr. Ronald Brooks led research units at General Electric to develop oil-eating microorganisms that cleaned up environmentally damaging oil spills in ocean environments. The oil-digesting microbes helped to metabolize, or break down, the hydrocarbons in petroleum into carbon and water, thus decreasing the toxic effects of petroleum. Brooks received many awards, and in addition to his scientific career, he was known for his work with inner-city youth. He was killed in an auto accident before he was able to patent his invention for combating oil spills.

VISUALIZING

Purpose

Students will compare and contrast agricultural practices.

Reinforcement

Soil Erosion Have students make a data table showing how topsoil can be lost and ways to prevent the loss. BL

Discussion

Pesticides v. Diseases In the late 1990s, the government of New York City had to choose between spraying pesticides in populated areas or risking an outbreak of potentially fatal mosquito-borne diseases. Ask students what they would have chosen to do under these circumstances. Ask them what they think the government of New York City decided to do. Government officials of New York City chose aerial spraying of pesticides to control mosquitoes. OL

Use Science Terms

Word Meanings Have students find the origin and meaning of mono. It is of Greek origin and means *one* or *alone*. Challenge students to compile a list of words that have the prefix *mono–*. Sample answers: monocle, monologue, monorail, and monotone EL

Project

Lead Contamination Lead (Pb) is a heavy metal that can be a problem if it contaminates the environment. Have students research and design a health information brochure about the effects of lead contamination, the source of the lead, and what is being done to remove lead from the environment. Ask them to find out what other heavy metals cause environmental problems. OL

3 Assess

Check for Understanding

Project Have small groups of students each make a data table summarizing the advantages and disadvantages of urban development, as well as possible solutions to the disadvantages. COOP LEARN

Reteach

Outline Have students summarize human impacts on land resources by making an outline of this section. Students can use the headings for each subsection in their outlines.

Assessment

Knowledge Ask students to identify two environmental problems that can be caused by all four of the following human activities: mining, logging, farming, and urban development. erosion and subsequent loss of topsoil; habitat loss

■ **Figure 10** Barriers such as this one are often used on construction sites to prevent erosion and reduce loss of topsoil.

Conservation

People are becoming increasingly aware of the need to protect the environment, and communities are making increased efforts to do so as urban development continues. For example, developers are often required to place barriers, like the one shown in **Figure 10,** around construction sites to catch sediment from increased erosion. In the United States, wetlands are now recognized as valuable ecosystems and are protected from development.

Waste disposal remains a problem because of the immense volume of trash. Modern landfills are carefully designed to minimize leakage of toxic liquids. Impermeable clay or plastic layers are placed beneath a landfill, and trash is compacted by machines and buried under a layer of dirt to reduce volume and eliminate windblown trash. Vents in landfills release methane and other gases that are generated as the garbage decomposes.

Several methods are available for cleaning up industrial waste sites. Contaminated soil can be removed and disposed of at hazardous waste landfills. Soil can also be incinerated to destroy the toxic chemicals. The drawbacks to this method are that it can be expensive to treat large volumes of soil, and it can produce toxic ash that may pollute the air.

Bioremediation is the use of organisms to clean up or break down toxic wastes. These organisms actually eat pollutants for food, neutralizing their negative impacts on the environment. Bioremediation is useful for contamination caused by spilled gasoline and oil.

SECTION 2 REVIEW

Section Self-Check

Section Summary

- Humans require large amounts of land resources.
- The extraction of resources can disrupt Earth's surface.
- Growing populations increase the demand for food and result in increased urban development.
- Agriculture, poor forestry practices, and urban development can cause habitat loss, increased erosion, and soil and water pollution.
- Human impact on land resources can be minimized through the use of modern techniques.

Understand Main Ideas

1. MAINIDEA **Describe** how extracting resources, growing food, and urban development contribute to land and water pollution.
2. **Propose** ways that land can be restored after it is strip-mined for coal.
3. **Predict** How many items will you throw away during lunch today? How much will you throw away in one week? One month? How does this relate to the impact of urban development on the environment?

Think Critically

4. **Suggest** methods of development that will reduce soil erosion and damage to streams.

WRITING IN ▶ Earth Science

5. Write an article for your school paper suggesting ways everyone can reduce the amount of waste that they produce.

SECTION 2 REVIEW

1. All of the actions can lead to topsoil erosion. Eroded topsoil can end up in rivers and lakes. In addition, these practices can cause other pollution such as runoff from urban development and the release of harmful chemicals from mining.
2. Answers will vary, but should mention reclamation.
3. Answers will vary, but should demonstrate an understanding of the importance of reducing solid waste.
4. Developers can use barriers on construction sites, reduce the number of trees that are cut down, and maintain a high amount of plant cover.
5. Answer will vary, but might include re-using water bottles and recycling papers, plastics, and glass.

Rubric

SECTION 3

Human Impact on Air Resources

MAINIDEA Manufacturing processes and the burning of fossil fuels can pollute Earth's atmosphere.

Essential Questions

- What is the relationship between the greenhouse effect and global warming?
- What is the sequence of reactions that occur as CFCs cause ozone depletion?
- What are the causes and effects of acid precipitation?

Review Vocabulary

greenhouse effect: heating of Earth's surface by certain atmospheric gases, which helps keep Earth warm enough to sustain life

New Vocabulary

photochemical smog
ozone hole
acid precipitation

EARTH SCIENCE 4 YOU **If you've ever enjoyed a campfire, you might have wondered what happens to the wood as it burns. When fuel is burned for energy or when products are manufactured, particles and gases such as carbon dioxide are released to Earth's atmosphere.**

Global Impacts of Air Pollution

It has become evident that human activities can affect Earth on a global scale. The global atmospheric effects of air pollution include global warming, ozone depletion, and acid precipitation.

Global warming Recall that the greenhouse effect is a natural phenomenon in which Earth's atmosphere traps thermal energy in the troposphere to warm Earth. A phenomenon related to the greenhouse effect is global warming, which is the increase in Earth's average surface temperature.

Over time, Earth has experienced periods of global warming and cooling. Many scientists think that the current period of global warming that Earth is experiencing is related to increased levels of carbon dioxide in the atmosphere. Some of this increase is thought to have been caused by humans.

Human activities, especially those involving the burning of fossil fuels, contribute to increased levels of carbon dioxide. Fossil fuels contain carbon, and when they are burned, the carbon combines with oxygen to form carbon dioxide. Since the beginning of the industrial revolution, around 1750, humans have been burning fossil fuels at an ever-increasing rate. **Figure 11** shows how atmospheric carbon dioxide has increased since 1960.

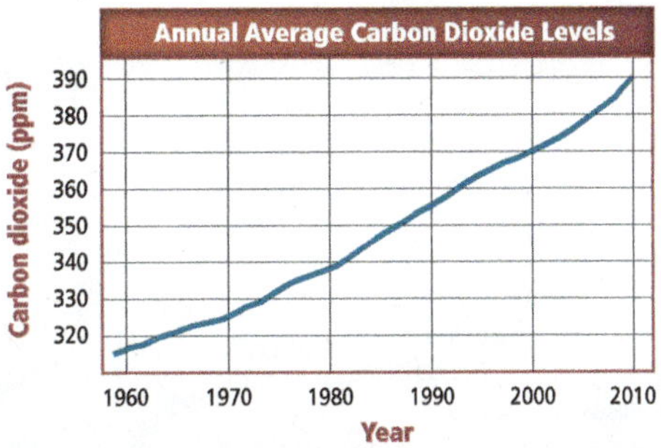

Figure 11 The graph shows the increased levels of carbon dioxide in the atmosphere since 1960, based on data gathered in Mauna Loa, in Hawaii.
Explain *What are some possible contributing factors to this rise in carbon dioxide levels? What impacts might this have on the environment?*

SECTION 3

1 Focus

Focus Transparency

Before presenting the lesson, project **Section Focus Transparency 83** and have students answer the accompanying questions. EL

MAINIDEA

Air Pollutants Ask each student to label an index card with his or her name and date, and then stick a 5-cm piece of double-sided tape on the card. Have students place their cards around the school and leave them for two days. They should then use magnifying lenses to study the different types of air pollutants that they find on the tape.

2 Teach

Interpret the Illustration

Increases in Carbon Dioxide Ask students to study the graph in **Figure 11** and then answer the following questions: How much did atmospheric CO_2 increase between 1960 and 1990? from an annual concentration of about 307 ppm in 1960 to 354 ppm in 1990 What percentage increase is this? about 10 percent

Caption Question Fig. 11 The burning of fossil fuels contributes to increased levels of carbon dioxide. This could continue the trend of global warming, changing the climate worldwide.

EARTH SCIENCE JOURNAL

Global Warming Have students write their responses to the following scenario in their Earth science journals: Imagine that one of the long-term climatic effects of global warming is that your region receives half as much rainfall as normal. How would this change your environment? What if rainfall were to double? OL

Rubric

Identify Misconceptions

Students might think that global warming and ozone depletion are the same problem.

Uncover the Misconception Before students read this section, ask them what global warming and the thinning of the ozone layer have in common.

Demonstrate the Concept Global warming is an overall increase in Earth's average surface temperature. The thinning of the ozone layer is caused by the breakdown of ozone in the upper atmosphere, which allows more ultraviolet radiation to reach Earth's surface.

Assess New Knowledge After students complete this section, ask them to each draw a diagram of the greenhouse effect and another diagram of how CFCs destroy ozone.

Project

Global Warming Have groups of students research global warming, then conduct a classroom debate about whether or not we should be concerned about it. COOP LEARN

Studies indicate that Earth's mean surface temperature has risen about 0.6°C in the last century. Some scientists predict that if concentrations of carbon dioxide and other greenhouse gases continue to increase, average global temperatures could rise between 1.5 and 4°C in the next 100 years. Other scientists, however, assert that humans have not kept weather records long enough to tell to what extent the present rate of global warming is an artificial or a natural phenomenon. They argue that the increase in Earth's temperature could be part of a natural pattern of climatic change.

Photochemical smog On sunny days, you might notice a yellow-brown haze near densely populated areas. This haze is a type of air pollution, called **photochemical smog** that forms mainly from automobile exhaust in the presence of sunlight. **Figure 12** shows how air pollutants from car exhaust form ground-level ozone. Recall that in the upper atmosphere, solar radiation converts oxygen gas into ozone. Ozone in the upper atmosphere is beneficial because it absorbs and filters out harmful ultraviolet (UV) radiation. However, ground-level ozone can irritate the eyes, noses, throats, and lungs of humans and other animals. It also has harmful effects on plants. When smog occurs in a city, the air becomes harmful to breathe, especially for those who already have some difficulty breathing.

Air pollution also occurs in the form of particulate matter. The solid particles of materials such as ash, dust, and pollen range in size from microscopic bits to large grains. When humans breathe in particulates, they can lodge in lung tissues and cause breathing difficulties and lung disease.

VOCABULARY
ACADEMIC VOCABULARY
Particulate
of or relating to minute separate, solid particles
People with asthma might be more sensitive when there is a high level of particulate matter in the air.

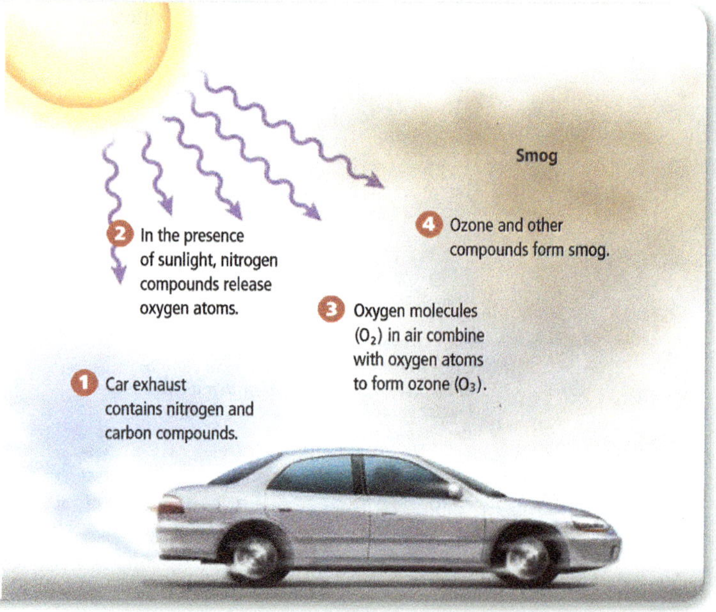

Figure 12 Automobile exhaust, in the presence of sunlight, can form a haze called photochemical smog.

View an **animation of how smog forms.** Concepts In Motion

Demonstration

Greenhouse Effect To demonstrate the greenhouse effect, place an alcohol thermometer in a jar or empty aquarium and seal the top. Place the jar or aquarium in a sunny location. After half an hour, compare the temperature inside the jar or aquarium to the temperature outside it.

DIFFERENTIATED INSTRUCTION

Advanced Learners Asbestos fibers in the air can cause serious health problems. Have students research asbestos. Ask them to answer the following questions about asbestos: Is there asbestos in your school? From where does it come? For what is it used? How does it harm people? How is it being controlled?

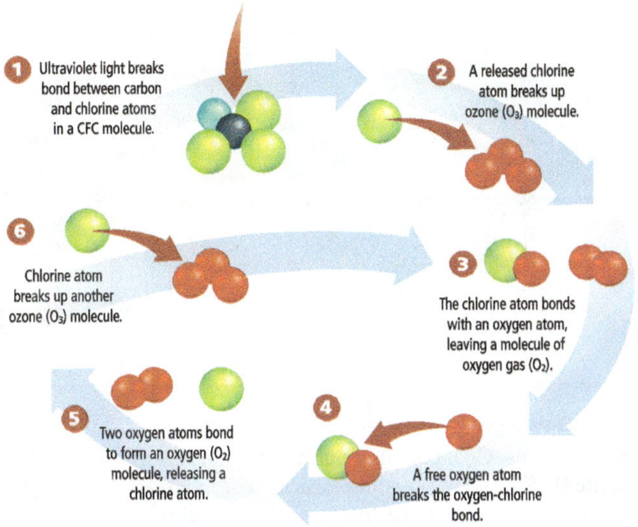

■ **Figure 13** One chlorine atom from one CFC molecule can destroy many ozone molecules.

View an **animation of ozone depletion.**

Concepts In Motion

Ozone depletion Recall that the ozone layer in the stratosphere serves as a protective shield as it absorbs and filters out harmful UV radiation. UV radiation has been linked to eye damage, skin cancer, and reduced crop yields.

In the early 1970s, scientists suggested that chlorofluorocarbons (CFCs) could destroy ozone in the upper atmosphere. All of the CFCs present in the atmosphere are a result of human activity. CFCs are released from cleaning agents, old refrigerators that are not disposed of properly, and propellants in aerosol cans.

Although CFCs are stable and harmless near Earth's surface, they destroy ozone molecules, as shown in **Figure 13,** when they migrate into the upper atmosphere. From the mid-1980s to the late-1990s, atmospheric studies detected a thinning of the ozone layer. Research continues to reveal an extremely thin area over Antarctica, called an **ozone hole,** which is a seasonal decrease in ozone over Earth's polar regions.

Acid precipitation Another major air pollution problem is acid precipitation, which is defined as precipitation with a pH of less than 5.0. Recall that pH is a measure of the acidity of a substance on a scale of 0 to 14, with 7 being neutral.

Natural precipitation has a pH of about 5.0 to 5.6, which is slightly acidic. **Acid precipitation** forms when sulfur dioxide and nitrogen oxides combine with atmospheric moisture to create sulfuric acid and nitric acid. Acid precipitation includes acidic rain, snow, fog, mist, and gas. Although volcanoes and marshes add sulfur gases to the atmosphere, 90 percent of the sulfur emissions in eastern North America are of human origin.

Teacher Content Support

Skin Cancer One of the more serious long-term effects of ozone depletion is melanoma, or skin cancer. Studies have shown a link between exposure to ultraviolet radiation and the development of melanoma later in life, especially in people with light-colored skin. Currently, the worst ozone depletion is found at high latitudes in the southern hemisphere. Health officials are particularly concerned about increased melanoma risks in Australia.

Environmental Connection

Greenhouse Gas Emissions Ask students to identify the products of burning fossil fuels and which of these products are greenhouse gases. BL

Teacher Content Support

Indoor Air Pollution Recent studies have indicated that air within buildings can be more polluted than outside air. Many people spend approximately 90 percent of their time indoors, so the risk to health from indoor air pollution could be greater than from outside air pollution. In addition, people who are exposed to indoor air pollution for the longest periods—infants, the elderly, and the chronically ill—are often those most susceptible.

Data Analysis LAB

About the Lab

- Review line graphs with students. Remind them of the axes and what each axis represents. Tell them that each air pollutant will have a different line on the graph. Students with poor graphing skills should be paired with more capable students.
- Data from the EPA past 2004 is not included as it is reported in several ways and could be difficult for students to graph.
- Information on air emissions for individual states can often be obtained by contacting the state's Environmental Protection Agency.

Think Critically

1. Students should graph the data appropriately.
2. tougher environmental regulations and increased public awareness
3. No, the reductions are not consistent.

Activity

Test pH Obtain pH test kits to test the pH of rainwater in your area. Have students collect rainwater as well as surface water after a rainstorm and test the water for its pH level. Have students determine whether their region experiences acid precipitation. OL

Use Science Terms

Acid Precipitation Ask students to explain why *acid precipitation* is a more accurate term to use than *acid rain.* BL EL

Caption Question Fig. 14 The sulfur generated in Midwestern power plants is carried by weather patterns towards the East Coast.

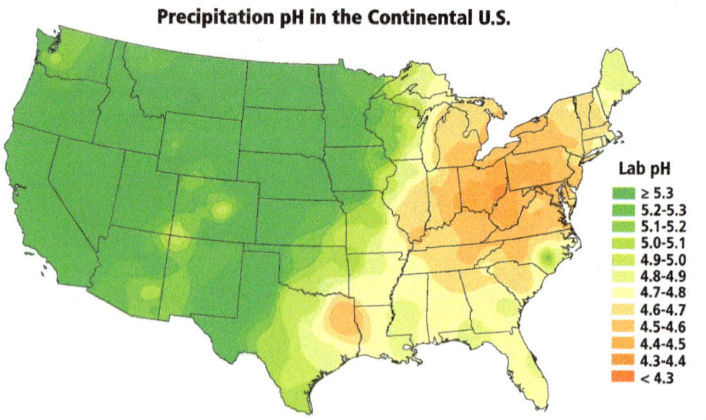

Figure 14 This map shows the pH levels of precipitation across the continental United States.
***Explain** why the pH is generally lower in the eastern half of the country.*

Burning coal One cause of acid precipitation is coal-burning power plants. Coal contains significant amounts of the mineral pyrite (FeS_2) and other sulfur-bearing compounds. When sulfur-rich coal is burned, large amounts of sulfur dioxide are released. The sulfur dioxide generated by midwestern power plants rises high into the air and is carried by winds toward the East Coast, where they mix with precipitation and fall to the ground. The distribution of acid precipitation is shown in **Figure 14.**

Effects of acid precipitation When acid rain enters surface waters, it damages aquatic ecosystems and vegetation and can negatively affect plants and soil. Trees affected by acid precipitation might not be killed outright, but might become more susceptible to damage and disease. Acid precipitation can also deplete the soil of nutrients and damage buildings and statues by accelerating weathering.

Data Analysis LAB

Based on Real Data*

Interpret the Data

Are you breathing cleaner air? This table lists changes in emissions in the United State since the Clean Air Act of 1972.

Think Critically

1. **Graph** the data from the table. Put years on the *x*-axis and the pollutant emissions per year on the *y*-axis. Use different colors for each pollutant.
2. **Infer** why emissions of lead have declined so drastically since 1970.
3. **Evaluate** Could you estimate the reductions for 2005 by looking at the graph? Explain why or why not.

Pollutant (millions of tons)	1970	1980	1990	2000	2004
Particulate matter <10 microns	12.2	6.2	3.2	2.3	2.5
Sulfur dioxide	31.2	25.9	23.1	16.3	15.2
Nitrogen oxides	26.9	27.1	25.2	22.3	18.8
Volatile organic compounds	33.7	30.1	23.1	16.9	15.0
Carbon monoxide	197.3	177.8	143.6	102.4	87.2
Lead	0.221	0.074	0.005	0.003	0.003

*Data obtained from: Air Emission Trends—Continued Progress Through 2004. *U.S. Environmental Protection Agency.*

Demonstration

Model Acid Rain Simulate the effects of acid precipitation on stone by placing a few drops of a 10-percent solution of hydrochloric acid on limestone. **WARNING:** *Wear an apron, safety goggles, and gloves. Have students wear safety goggles to observe. Take care to avoid splashing acid on clothing or skin.*

EARTH SCIENCE JOURNAL

Acid Rain and Soil Ask students to answer this question in their Earth science journals: How does acid precipitation affect the availability of nutrients in soil? When the pH of soil changes, some nutrients form compounds that are insoluble and cannot be absorbed by plants. Other nutrients are converted to more soluble forms and are quickly leached from the soil. BL

Rubric

Reducing Air Pollution

Air pollution is difficult to control because it travels through the air to neighboring regions. Solving air pollution problems requires the cooperation of both state and national governments. In the last several decades, the governments of many nations have met in an attempt to reduce global air pollution, especially that which is caused by carbon dioxide and CFCs.

Since 1963, the United States Congress passed clean air laws which set specific reduction goals and enforcement policies for many types of air pollution. Because of this and other regulations, there have been significant reductions in air pollutants in the United States since 1970.

■ **Figure 15** Scrubbers are often required to clean out the smoke stacks on coal plants. Scrubbers help remove gases and particulate matter before they enter the air.

Controlling the source Many coal-burning power plants have installed devices to reduce emissions of particulate matter and sulfur dioxide, such as the ones shown in **Figure 15.** In North America and western Europe, the use of low-sulfur coal and natural gas has helped to reduce such emissions. However, scientists agree that the most effective way to reduce air pollution is to remove older, highly polluting vehicles from roadways. It is estimated that just 10 percent of the motor vehicles in operation produce 50 to 60 percent of the air pollution generated by gasoline-powered engines. Switching to newer cars with more efficient engines could significantly reduce air pollution throughout the world.

SECTION 3 REVIEW

Section Self-Check

Section Summary

- Many human activities create air pollution. Air pollution can cause human health problems.
- CFCs are a major cause of ozone depletion.
- Clean air laws have resulted in a decrease in air pollution emissions since 1970.

Understand Main Ideas

1. **MAIN IDEA Name** two forms of pollutants found in air. What are some of the sources of these pollutants?
2. **Relate** global warming and the greenhouse effect.
3. **Describe** how CFCs cause ozone depletion.
4. **List** some of the causes and effects of acid precipitation on ecosystems.

Think Critically

5. **Predict** The atmosphere of Venus is 96.5 percent carbon dioxide. Based on this information, what could you infer about the average surface temperature of Venus? Explain your answer.

MATH IN Earth Science

6. If carbon monoxide emissions were reduced from 102 to 87 million metric tons in one year, what would be the percent decrease?

Dennis MacDonald/Alamy

Project

Effects of Acid Precipitation Have student groups research the effects of acid precipitation on aquatic organisms and create presentations to share with the class. At least one group should focus on the steps being taken to raise the pH levels of lakes and streams that have already been affected by acid precipitation. OL EL

3 Assess

Check for Understanding

Discussion Have a discussion about the major air pollution problems in your region. What are the sources of the problems, and what is being done to alleviate the problems?

Reteach

Activity Have students each make a list of five human activities that cause air pollution and describe the type of air pollution that is generated.

Assessment

Performance Have students each make a data table comparing the three global air-pollution problems: ozone depletion, acid precipitation, and global warming. Ask students to list the chemicals that cause each problem, and to explain where these chemicals originate.

SECTION 3 REVIEW

1. Gases and particulate matter are found in air. Particulates come from forest fires, volcanic eruptions, mining activities, farming, and construction. Gases come from volcanoes, forest fires, biological activity, cars, industry, and power plants.
2. The greenhouse effect is the ability of Earth's atmosphere to heat and maintain surface temperatures on Earth. Global warming is the increase in Earth's average surface temperature. Global warming might be caused by an increase in the amount of atmospheric greenhouse gases.
3. UV radiation frees chlorine atoms from CFCs in the upper atmosphere. These chlorine atoms break up ozone molecules.
4. Acid precipitation lowers the pH levels of lakes and streams, thereby harming aquatic organisms, and changes soil chemistry so nutrients are unavailable to plants.
5. The average surface temperature of Venus would be much higher than the average surface temperature of Earth. This higher temperature is a result of an extreme greenhouse effect caused by the high carbon-dioxide concentration. The average surface temperature on Venus is about 480°C (900°F).
6. about 15 percent

SECTION 4

1 Focus

Focus Transparency

Before presenting the lesson, project **Section Focus Transparency 84** and have students answer the accompanying questions. EL

MAINIDEA

Water Quality Fill three clear containers with water. Add some food coloring to one of the containers, add nothing to the second container, and add a little soil to the third to make it muddy. Ask students which water sample is unsafe to drink. Students might say that the muddy water is not safe to drink but that the colored water is. Most students will also say that the clear water is safe to drink, but point out that it could be contaminated with any number of colorless chemicals.

2 Teach

Tie to Previous Knowledge

Freshwater Supplies Ask students to review the amount of freshwater available on Earth. Remind them that most of the freshwater on Earth is frozen in glaciers, leaving only 0.009 percent of total water resources immediately available for use in surface and groundwater resources. BL

Model

Map Pollution Have groups of students create maps of hypothetical streams, rivers, or lake systems. Students should then add several possible pollution sources and label them as point or nonpoint sources. OL

SECTION 4

Essential Questions

- In what ways can water be conserved?
- What are the types and sources of water pollution?
- How can water pollution be reduced?

Review Vocabulary

runoff: water that flows downslope on Earth's surface and can enter a stream, river, or lake

New Vocabulary

point source
nonpoint source

Human Impact on Water Resources

MAINIDEA Pollution controls and conservation protect water resources.

EARTH SCIENCE 4 YOU Imagine camping at a remote location. You have brought along the supply of water. What would happen if extra friends joined the group or half the water supply were spilled?

Use of Water Resources

Humans depend on water in many ways. Most people use freshwater in their homes for bathing, drinking, cooking, and washing. The irrigation of crops also requires water. Because water supplies are not distributed evenly on Earth, some areas have less water than is needed.

Water conservation Is there a leaky faucet in your home? In the United States alone, 20 to 35 percent of the water taken from public water supplies is lost through leaky toilets, bathtubs, and faucets.

When there is not enough water to go around, people have two choices: decrease demand or develop new supplies. When new supplies are not readily available or are too expensive to develop, water conservation can help. Because large amounts of water are used for crops, efficient irrigation practices can greatly reduce water usage. Monitoring soil moisture to irrigate only when the soil is dry, using equipment that places water near plant roots to reduce evaporation as shown in **Figure 16,** and raising water prices have all been effective in minimizing the amount of irrigation water. Industries can also conserve water by recycling cooling water and wastewater, or by using conservation practices.

Figure 16 Farmers develop methods of water conservation such as the drip irrigation system shown above. In a drip irrigation system, the water is released slowly so more is absorbed into the soil and less is lost to runoff and evaporation.

EARTH SCIENCE JOURNAL

Oil Spills Have students research and write about how oil spills are cleaned up. What are the short-term and long-term effects of oil spills? How successful are efforts to contain the pollution? Cleaning up an oil spill begins with using floating barriers to contain the spill. As much oil as possible is then recovered by skimming and using absorbent materials. The success of such efforts depends upon weather conditions, the type of oil spilled, and the location of the spill. Short-term effects of an oil spill are suffocation and poisoning of organisms. Long-term effects are still being debated. Naturally occurring bacteria can break down some compounds over time, but other toxins will remain in the environment for a long time. AL

Rubric

Water Pollution

Pollution is another area in which humans have an impact on water supplies. Some supplies of water have been polluted by human activities and are no longer usable. Water-pollution sources are grouped into two main types. **Figure 17** shows that **point sources** originate from a single point of origin, such as a sewage-treatment plant or an industrial site, while **nonpoint sources** generate pollution from widespread areas.

Most water used for domestic purposes, including showering, laundry, cooking, and using the bathroom, is treated at a sewage treatment facility. Treated sewage is then released through a point source to a receiving stream. In the past, treated sewage still contained contaminants. Fortunately, methods to treat sewage have greatly improved. Point sources also include wastes that enter streams from illegal dumping, accidental spills, and industries that use water in manufacturing processes and discharge waste into streams and rivers.

Precipitation can absorb air pollutants and deposit them far from their source. Runoff can wash pesticides and fertilizers into streams as it flows over farms or lawns. It can also wash oil, gasoline, and other chemicals from roads and parking lots. Each of these is an example of nonpoint-source pollution.

☑ READING CHECK **Compare** point-source and nonpoint-source pollution.

Point source

Nonpoint source

■ **Figure 17** Point-source pollution comes from a single source, while nonpoint-source pollution is generated from a widespread area.

Pollution of groundwater Leaking chemical-storage barrels, underground gasoline-storage tanks, landfills, road salts, nitrates from fertilizers, sewage from septic systems, and other pollutants can seep into the ground and pollute underground water supplies. Polluted groundwater might find its way into the drinking-water supplies of people who rely on wells. Once groundwater is contaminated, the pollutants can be difficult to remove.

Pollution in the oceans Although human activities have the greatest impact on freshwater supplies, pollution of ocean waters is also a concern. Nearly 50 percent of the U.S. population lives near coastlines. Pollutants from such cities often end up in estuaries and other nearshore regions. Pollution of nearshore zones can affect organisms because many depend on estuaries for breeding and raising young.

Another common ocean pollutant is mercury. Mercury released into the air and water from burning coal and manufacturing is ingested by fish. The fish are then eaten by larger predators and the mercury is passed along the food chain. Mercury has been detected in bears that do not live near polluted waters because they have eaten salmon that migrate from the oceans.

Teacher Content Support

Aquifer Depletion One serious environmental challenge facing humans is the depletion of underground water supplies. Some farmers use deep wells and powerful pumps to pull groundwater to the surface for the irrigation of crops in regions where rainfall is inadequate. In many cases, the water is used more quickly than it is recharged, and water tables are steadily dropping. In the United States, the most serious groundwater depletion is from the Ogallala Aquifer, located beneath central and southwestern states. The Ogallala Aquifer provides water for 30 percent of the irrigated land in the United States.

Environmental Connection

Local Water Supply Have students research the local water supply to find out from where their water comes and how is it treated to ensure that it is safe to drink. If possible, take students on a field trip to your local water treatment plant to learn how water is cleaned before it is released back into the environment. OL

☑ READING CHECK Point-source pollution originates from a single source. Nonpoint-source pollution originates from a widespread area.

DIFFERENTIATED INSTRUCTION

Struggling Learners Have students cut out magazine photos showing different types of water use. Students should attach the photos to poster board and write a sentence next to each photo describing how water is being used.

EARTH SCIENCE JOURNAL

Daily Water Use Have students write in their Earth science journals about how they use water every day. They should also try to estimate how much water each activity requires. OL

Rubric

GeoLAB

iLab Station

The GeoLab located at the end of the chapter can be used at this point in the lesson.

Collaborative Learning

Impact of Dams Today, few dams are being built, and some dams are being slated for removal. Have groups of students discuss the positive and negative impacts of dams. Afterward, conduct a class discussion about this issue. COOP LEARN

CAREERS IN EARTH SCIENCE

WebQuest

3 Assess

Check for Understanding

Discussion Ask students to identify what two laws have been passed in the United States to reduce water pollution and the primary focus of each law. Ask them whether these laws have been effective. The Safe Drinking Water Act was designed to ensure clean drinking water for all Americans. The Clean Water Act was designed to prevent pollution of U.S. waters and to protect aquatic environments.

Reteach

Activity Have students make flash cards about different mechanisms of water pollution. The back of each card should identify whether it is a point- or nonpoint-source of pollution.

Assessment

Skill Have students compare and contrast different methods of water conservation. Ask them which provide the greatest water savings with the least amount of effort.

CAREERS IN EARTH SCIENCE

Hydrologist An Earth scientist who studies the distribution, circulation, and physical properties of underground and surface waters is a hydrologist. They sometimes work in offices, helping companies comply with environmental regulations.

WebQuest

Reducing Water Pollution

In the past several decades, many steps have been taken to prevent and reduce water pollution as people have found that it is much cheaper and more efficient to prevent pollution than it is to clean it up later. Two major laws have been passed in the United States to combat water pollution: the Safe Drinking Water Act and the Clean Water Act.

The Safe Drinking Water Act In 1974, the Safe Drinking Water Act was passed. This act was designed to ensure that everyone in the United States has access to safe drinking water. Progress is being made, but many water supplies still do not consistently meet the standards. In 2009, 20 percent of public water supplies were in violation of the act at least once in a one-year period. The goal of the Safe Drinking Water Act is to reduce this number to less than 5 percent.

The Clean Water Act The primary federal law that protects U.S. waters is the Clean Water Act of 1972. The act was amended in 1977, 1981, and again in 1987. The two main goals of the Clean Water Act are to eliminate discharge of pollutants into rivers, streams, lakes, and wetlands, and to restore water quality to levels that allow for recreational uses of waters, including fishing and swimming.

Is the Clean Water Act working? Since 1972, the number of people served by sewage-treatment plants has increased from 85 million to 190 million. During that same time period, the annual rate of wetland losses has decreased from 146,000 ha/y to about 32,000 ha/y. Two-thirds of the nation's waters are now safe for swimming and fishing, compared to only one-third in 1972. Continued monitoring and improvement are still necessary. In 2000, the EPA reported that 39 percent of the nation's rivers that were tested were polluted and 45 percent of lakes tested were polluted.

SECTION 4 REVIEW

Section Self-Check

Section Summary

- Humans use water to irrigate crops, for industry, cooking, bathing, and drinking.
- Conserving water can stretch limited supplies.
- Water can be polluted from point sources and nonpoint sources. Groundwater and oceans can also become polluted.
- The United States has passed laws to limit water pollution.

Understand Main Ideas

1. MAINIDEA **Identify** ways surface waters can be polluted.
2. **Determine** how residents of a city might reduce water consumption.
3. **Analyze** What are some of the positive impacts of the Clean Water Act?
4. **Predict** What are some ways to minimize the need for irrigation?

Think Critically

5. **Infer** which type of pollution is easier to eliminate: point sources or nonpoint sources? Give an example of each type and explain how it might be controlled.

WRITING IN ▶ Earth Science

6. Write your own Clean Water Act. What regulations would you place on businesses or homes? How quickly would you expect change?

SECTION 4 REVIEW

1. sewage, chemicals from industry, runoff from streets and farms, acid precipitation, and oil spills
2. fixing leaks, installing low-flow shower heads and toilets, and watering yards during the morning and evening when evaporation rates are lower
3. a decrease in the rate of wetland loss, an increase in the number of sewage treatment plants, and the fact that two-thirds of waterways are now safe for swimming and fishing
4. planting crops that need less water, using drip-irrigation systems, and careful monitoring of soil moisture
5. A point source is easier to eliminate, because its location can be identified and the problem can be corrected. For example, an industry that releases effluent into a local stream can be identified and required to stop releasing materials into surface water. Educating homeowners and creating barriers to slow runoff before it reaches streams and lakes might help control nonpoint-source pollution.
6. Answers will vary, but should include a list of regulations and a proposed timeline.

Rubric

Earth Science & TECHNOLOGY

Measuring and Modeling Climate Change

You have seen movies and TV shows that talk about the catastrophic future that awaits Earth as a result of climate change. But what is causing climate change?

Climate change At any time, Earth's surface temperature varies greatly from place to place. One way to specify Earth's temperature is to calculate average temperatures over Earth's surface over a specific time period. Temperature data show that over the past 100 years, the average annual global temperature has increased by about 0.6°C. This increase in average global temperature is called global warming.

The data The temperature data used to calculate average global temperatures come from thermometer measurements on land and at sea. Other types of data are used to reconstruct historical climate records. These data are obtained from ice cores, sediment cores from the bottoms of lakes and oceans, tree rings, and corals.

Possible causes Earth's surface temperature is maintained through a balancing act between the rate at which Earth absorbs energy and the rate at which Earth radiates energy. Increasing the concentration of greenhouse gases could increase Earth's temperature by increasing the amount of energy that Earth absorbs. Data show that since the middle of the nineteenth century, the concentration of atmospheric carbon dioxide has risen by about 100 ppm (about 35 percent) around the same period as humans began burning fossil fuels. Carbon dioxide and average global temperature have increased over approximately the same time period.

However, other factors such as solar variations and volcanic eruptions can also cause global temperatures to change. Is climate change due to human-caused increases in greenhouse gases, natural causes, or both?

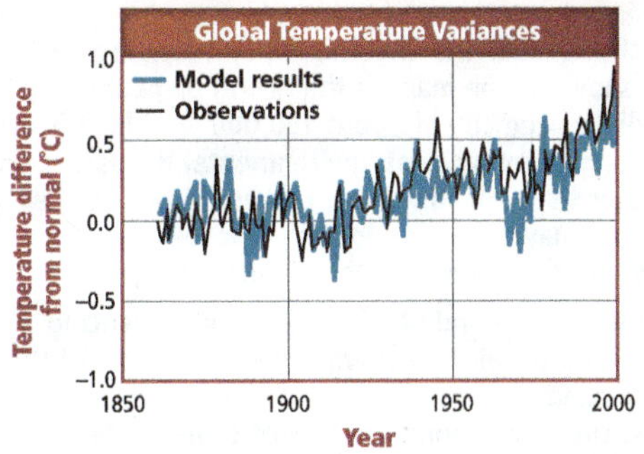

Calculated (black line) and measured (blue line) global average temperatures agree best when both natural effects and greenhouse gas increases are included in the computer climate model calculations.

Computer climate models The effects of possible causes on average global temperatures can be estimated using computer climate models. These programs calculate how changes in greenhouse gases, solar radiation, and other factors change surface temperatures over Earth's surface.

What gives the best agreement? Computer climate models calculate the average global annual temperature that would result from various changes. The best agreement with the temperature data comes when greenhouse gas changes and natural causes together are used to calculate average global temperatures. This indicates that natural changes and greenhouse gas increases together probably have caused the temperature changes currently observed.

WRITING IN ▶ Earth Science

Debate Prepare for a class debate about international policies that are based on climate change data. For information about conducting a debate, see the *Skillbuilder Handbook*.

WebQuest

Earth Science & TECHNOLOGY

Purpose

Students will learn about global warming and its potential effects on Earth's climate.

Teacher Content Support

The Greenhouse Effect The greenhouse effect is a natural phenomenon that maintains temperatures on Earth at optimal levels. Heat from the Sun warms Earth's surface, which then releases the heat back into the atmosphere. Greenhouse gases absorb the heat and re-emit it. Some of this re-emitted heat returns to Earth, and some escapes into space. Because of higher concentrations of carbon dioxide and other gases, less heat is being released into space and more is being returned to Earth. The greenhouse effect might be therefore being intensified, and Earth's temperature might be getting warmer.

Teaching Strategy

Have students answer the following questions in their Earth science journals: What do you know about climate change? What do you think causes climate change? What can we do about it? After students have read the feature, have them revise their answers.

WRITING IN ▶ Earth Science

Rubric

Debate Place students into two groups. One group should defend international policies on climate change. The second group should argue against the policies.

WebQuest

GeoLAB

iLab Station | Rubric

Preparation

Time Allotment 90 min

Process Skills interpret maps, apply concepts, predict, analyze information

Safety Precaution Approve lab safety forms before work begins.

Procedure

- Read through the information presented in the Procedure section with the class. This section contains clues that will help students to answer the lab questions.
- Explain to students this scenario is based upon a real region.
- **Troubleshooting** Make sure students understand the different symbols and colors on the map. Some students might need an enlarged copy of the map.

Analyze and Conclude

1. Possible pollution sources of Opal Lake are increased development around the lake, logging operations, and power boating. Depending on the type of pollution, they could be point or nonpoint sources. Most likely, they are nonpoint. The sources of pollution for Iris Bay could be fertilizers and pesticides from farming operations, golf courses, and urban development. Oil and gasoline pollution comes from street runoff, powerboats, and harbor operations. Excess silt comes from urban development and logging. These are nonpoint sources.
2. The source of pollution could be located by analyzing water samples while traveling up the Vista River until the source was identified.
3. Answers will vary, but could include fertilizer runoff or septic tank overflows.
4. Answers will vary, but might mention that excessive nutrients promote algal blooms which can negatively affect water quality, and that certain hydrocarbons may be linked to cancer.

GeoLAB

iLab Station

Mapping: Pinpoint a Source of Pollution

Background: Iris City and the surrounding region are shown in the map on the facing page. Iris City is a medium-sized city of about 100,000 people. It is experiencing many types of environmental issues. Iris City obtains its drinking water from Opal Lake. Studies of the lake have detected increased levels of nitrogen, phosphorus, hydrocarbons, sewage, and silt.

The northwest end of Opal Lake is experiencing increased development while the remainder of the watershed is a combination of forest and logging clear-cuts. Last spring, blooms of cyanobacteria choked parts of the Vista Estuary Nature Preserve. Commercial shellfish beds in Iris Bay have been closed because of sewage contamination.

A natural-gas power plant has been proposed for location A, near the Vista Cutoff, an abandoned channel of the Vista River. The plant would provide jobs as well as generate electricity. The company plans to divert 25 percent of the Vista River through the Vista Cutoff.

The Lucky Mine was abandoned 60 years ago. A mining company has applied for permits to reopen the mine. An estimated 1 million grams of gold can be recovered using modern techniques.

You will work with a small group of students to make recommendations to the residents of Iris City. Included in your recommendations should be: possible pollution sources for Opal Lake, possible causes of the cyanobacteria bloom, recommendations for the development of the natural-gas power plant, and the opening of Lucky Mine.

Question: *How can the residents of Iris City best manage their water supply?*

Materials

metric ruler
science notebook

Procedure

1. Read and complete the lab safety form.
2. Working in small groups, brainstorm possible sources of pollution in Opal Lake and Iris Bay.
3. Discuss what steps the residents of Iris City might take to protect their drinking water.
4. Research common causes of cyanobacteria blooms. Discuss what might be causing the bloom in the Vista Estuary Nature Preserve.
5. Discuss the positive and negative aspects of diverting water from the Vista River through the Vista Cutoff. What are other possible impacts (both positive and negative) in the development of the natural-gas power plant?
6. Discuss the possibility of reopening Lucky Mine. If it is reopened, brainstorm ways the mining company might minimize negative environmental impacts.
7. Prepare to present your recommendations to the class.

Analyze and Conclude

1. **Identify** What did your group list as possible sources of pollution for Opal Lake and for Iris Bay? Are these point sources or nonpoint sources of pollution?
2. **Experiment** How would you determine if you were correct about the source of the cyanobacteria bloom in Vista Estuary? What tests might you run to confirm your predictions?
3. **Research** What information did you discover in your research on cyanobacteria that applied to Iris City? List all the possible sources your group identified.
4. **Research** You have identified the possible sources of pollution for Opal Lake. Reports indicate increased levels in nitrogen, phosphorus, hydrocarbons, sewage, and silt. Research how this might affect the health of the residents that obtain their water from the lake.

WRITING IN ▶ Earth Science

Write a report for Iris City, detailing all your group's findings and recommendations.

WRITING IN ▶ Earth Science

Rubric

Write Group findings should be based on inferences gathered from the background material and map. Group recommendations should discuss available resources and costs of implementation.

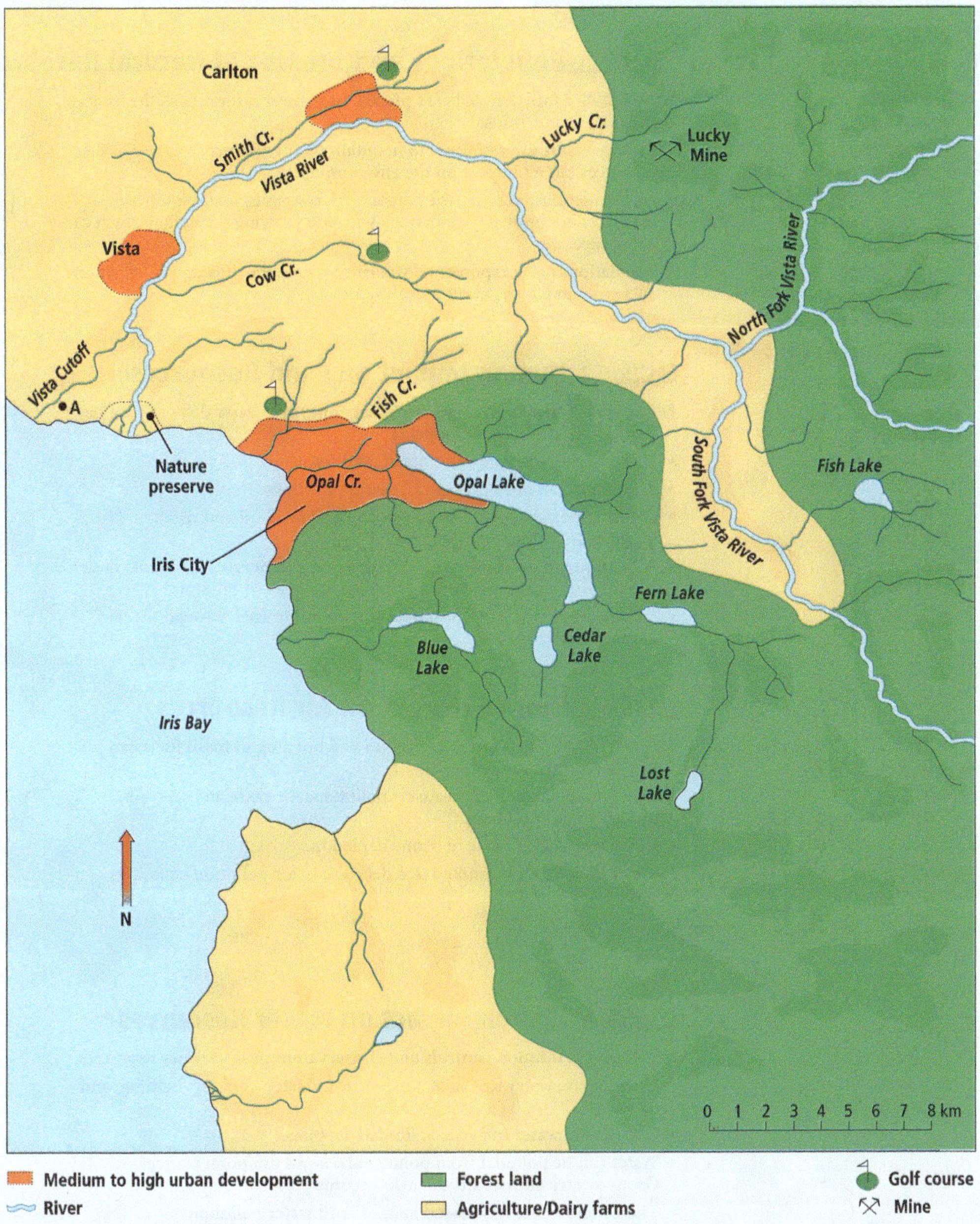
Carlton
Smith Cr.
Vista River
Lucky Cr.
Lucky Mine
Vista
Cow Cr.
North Fork Vista River
Vista Cutoff
A
Fish Cr.
Nature preserve
Opal Cr.
Opal Lake
Iris City
South Fork Vista River
Fish Lake
Fern Lake
Blue Lake
Cedar Lake
Iris Bay
Lost Lake
N
0 1 2 3 4 5 6 7 8 km
Medium to high urban development
River
Forest land
Agriculture/Dairy farms
Golf course
Mine

MAINIDEAS Summary statements can be used by students to review the major concepts of the chapter.

Students can review with these online resources.

Vocabulary Practice

Vocabulary eGames
Vocabulary eFlashcards
Vocabulary PuzzleMaker

Section Self-Check

Chapter Self-Check

Online Test Practice

Use the ExamView® *Assessment Suite* CD-ROM to:

- create multiple versions of tests
- create modified tests with one mouse click
- edit existing questions and add your own questions
- build tests aligned with state standards using built-in state curriculum tags
- change English tests to Spanish with one mouse click
- track students' progress using the Teacher Management System

CHAPTER 26 STUDY GUIDE

BIGIDEA **The use of natural resources can impact Earth's land, air, and water.**

Vocabulary Practice

SECTION 1 Populations and the Use of Natural Resources

VOCABULARY
- exponential growth
- carrying capacity
- density-independent factor
- density-dependent factor

MAINIDEA More demands are placed on natural resources as the human population increases.

- All organisms use resources to maintain their existence. The use of these resources has an impact on the environment.
- As populations increase, the demand for resources increases. Because resources are limited, populations will stop growing when they reach carrying capacity.
- Populations grow exponentially at early stages. Earth is currently experiencing a human population explosion.

SECTION 2 Human Impact on Land Resources

VOCABULARY
- reclamation
- deforestation
- pesticide
- bioremediation

MAINIDEA Extraction of minerals, farming, and waste disposal can have negative environmental impacts.

- Humans require large amounts of land resources.
- The extraction of resources can disrupt Earth's surface.
- Growing populations increase the demand for food and result in increased urban development.
- Agriculture, poor forestry practices, and urban development can cause habitat loss, increased erosion, and soil and water pollution.
- Human impact on land resources can be minimized through the use of modern techniques.

SECTION 3 Human Impact on Air Resources

VOCABULARY
- photochemical smog
- ozone hole
- acid precipitation

MAINIDEA Manufacturing processes and burning of fossil fuels can pollute Earth's atmosphere.

- Many human activities create air pollution. Air pollution can cause human health problems.
- CFCs are a major cause of ozone depletion.
- Clean air laws have resulted in a decrease in air pollution emissions since 1970.

SECTION 4 Human Impact on Water Resources

VOCABULARY
- point source
- nonpoint source

MAINIDEA Pollution controls and conservation protect water resources.

- Humans use water to irrigate crops, for industry, cooking, bathing, and drinking.
- Conserving water can stretch limited supplies.
- Water can be polluted from point sources and nonpoint sources. Groundwater and oceans can also become polluted.
- The United States has passed laws to limit water pollution.

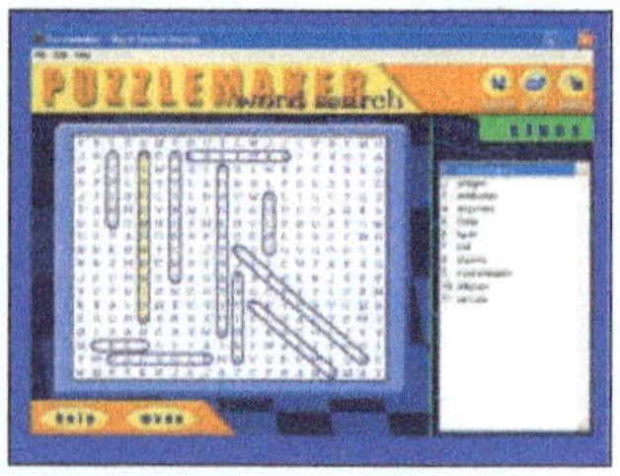

Vocabulary Practice

For additional practice with vocabulary, have students access the Vocabulary PuzzleMaker.

CHAPTER 26 ASSESSMENT

Chapter Self-Check

VOCABULARY REVIEW

Complete the sentences below using the vocabulary terms from the Study Guide.

1. Air pollution that forms primarily from car exhaust in the presence of sunlight is called ________.
2. The ________ is the total number of organisms that an environment can support.
3. The use of organisms to clean up toxic waste is called ________.
4. The removal of trees from a forested area without adequate replanting is called________.
5. Restoring land that had been previously mined to its original contours is called ______.

Match each description below with the correct vocabulary term from the Study Guide.

6. a pattern of growth in which a population grows faster as it increases in size
7. chemicals applied to plants to kill insects, fungi, and weeds
8. water pollution that comes from widely spread areas
9. a factor, such as lack of food, that affects a population

Each of the following sentences is false. Make each sentence true by replacing the italicized word with terms from the Study Guide.

10. A seasonal change that appears over Earth's polar regions is called *acid precipitation.*
11. Any environmental factor that does not depend on the number of members in a population, such as storms, droughts, floods, fires, and pollution, is a *density-dependent factor.*
12. *Bioremediation* forms when sulfur dioxide and nitrogen oxides combine with atmospheric moisture to create sulfuric acid and nitric acid.

UNDERSTAND KEY CONCEPTS

13. Which diagram represents mouse population growth?

A. 

C. 

B. 

D. 

14. Which environmental impact might result from deforestation?
    A. erosion of topsoil
    B. increased photochemical smog
    C. a decrease in the size of the ozone hole
    D. bioremediation of toxic waste
15. Which is a problem associated with the expansion of highly populated areas?
    A. the loss of natural resources
    B. the expense of building
    C. the concentration of resources
    D. the availability of cultural resources
16. Which federal law protects the U.S. water supply?
    A. the Clean Water Act
    B. the Clean Ocean Act
    C. the Clean Air Act
    D. the Clean Hydrosphere Act
17. Which is an example of point-source pollution?
    A. acid rain
    B. runoff from a parking lot
    C. sewage outfall
    D. topsoil from a farm field
18. What is the major source of photochemical smog in the United States?
    A. point sources
    B. car exhaust
    C. power plants
    D. acid precipitation

CHAPTER 26 ASSESSMENT

VOCABULARY REVIEW

1. photochemical smog
2. carrying capacity
3. bioremediation
4. deforestation
5. reclamation
6. exponential growth
7. pesticides
8. nonpoint-source pollution
9. density-dependent factor
10. ozone hole
11. density-independent factors
12. acid precipitation

UNDERSTAND KEY CONCEPTS

13. B
14. A
15. A
16. A
17. C
18. B

19. D
20. A
21. B
22. A
23. D

CONSTRUCTED RESPONSE

24. The graph should increase exponentially until carrying capacity is reached.
25. Acid precipitation can make trees and other plants susceptible to attack by insects and disease. This can reduce or eliminate habitats and food sources.
26. 150 years = 21°C; 250 years = 24°C
27. Answers will vary, but should include an increase in air and water temperatures and a rise in global sea level.
28. Reduce emissions from automobiles, carpool, and use public transportation.
29. Ozone depletion is caused by CFCs. Global warming is caused by both human activities and natural events.
30. Much of the air pollution that reacts to form ozone comes from cars. More cars are on the road during weekdays than on weekends because people are commuting to and from work.
31. Sample answer: Currently, our society depends on mined natural resources for everyday life. If viable alternatives are found, we could reduce the amount of natural resources that are mined.
32. Contour farming would be a good farming practice for an area with hills. Contour farming would help reduce runoff, thereby reducing erosion and nutrient loss.

Use the figure below to answer Question 19.

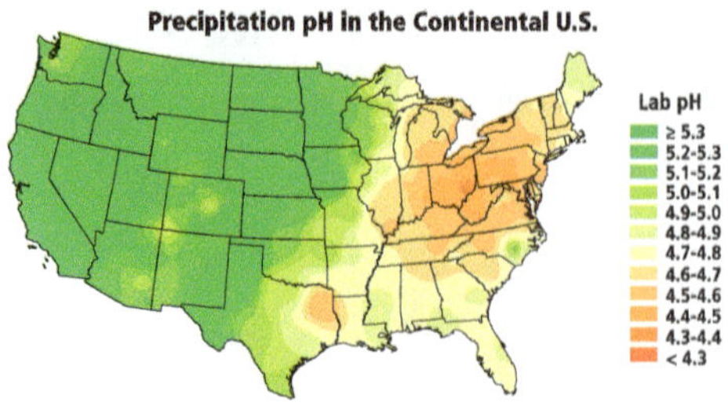

19. In which part of the United States is precipitation the most acidic?
A. southwest
B. northwest
C. southeast
D. northeast

20. Acid precipitation comes mainly from which source?
A. coal-fired power plants
B. particulate pollution
C. CFCs
D. carbon dioxide

21. What was the Safe Drinking Water Act designed to do?
A. clean United States' drinking water
B. ensure access to safe drinking water
C. increase the amount of drinking water available in the United States
D. decrease the cost of drinking water in the United States

22. Which is most likely to cause groundwater pollution?
A. leaking underground storage tank
B. acid precipitation
C. deforestation
D. particulate matter

23. Which is a common ocean pollutant?
A. carbon
B. ozone
C. neon
D. mercury

CONSTRUCTED RESPONSE

Use the figure below to answer Question 24.

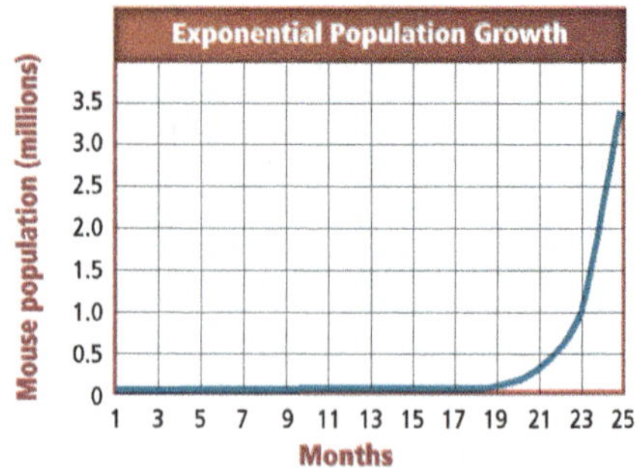

24. Determine What would the graph look like if you extended it 10 more years?

25. Describe two effects acid rain might have on an ecosystem.

26. Illustrate Draw a line graph that represents a warming rate increase of 3°C every 100 years. If the graph starts at 16.5°C, what will the average temperature be in 150 years? In 250 years?

27. Predict what conditions might exist on Earth if global warming continues to occur.

28. Analyze What are some ways to reduce photochemical smog?

29. Compare the possible causes of ozone depletion and global warming.

30. Analyze Why might ground-level ozone be worse on a sunny weekday than on a sunny weekend?

31. Interpret If mining for resources can have negative environmental impacts, why do humans still continue this practice?

32. Analyze Imagine you have a farm where you grow crops, including corn and alfalfa. Most of your fields are located on moderate-to-steep slopes. What farming methods could you use to reduce erosion? Explain.

Chapter Self-Check

THINK CRITICALLY

33. Propose a plan for a model city that addresses land use and pollution issues.

34. Distinguish between point-source and nonpoint-source pollution. Why is it important to understand how each impacts the environment?

35. Propose an alternative solution for disposal of our increasing amount of solid waste.

36. Interpret How can you prevent nonpoint-source pollution if there isn't one single place the pollution is generated?

37. Analyze How have humans adapted to their environment? What lifestyle changes and adaptations do people make based on the environment in which they live?

38. Propose two ways that you could conserve water in your everyday life.

39. Analyze A population of birds inhabits an island. They feed on seeds and berries and live in nests in trees. One season, there is no crop of berries and the number of birds on the island decreases. The next season, there is a hurricane and many of the birds' nests are knocked out of the trees. Which of these factors is density-dependent and which is density-independent? Why?

CONCEPT MAPPING

40. Use these terms to make a concept map to organize the major ideas in Section 2: *erosion, topsoil loss, water pollution, waste rock,* and *mineral extraction.* Refer to the *Skillbuilder Handbook* for more information.

CHALLENGE QUESTION

41. Apply As undeveloped land is developed into cities and towns, many wild animals lose their habitat. In addition, animals can be cut off from their natural hunting and breeding ranges. For animals that normally have large ranging habitats, brainstorm ways humans can protect them from extinction and still develop new land.

WRITING IN ▸ Earth Science

42. Write a radio announcement as part of an Earth Day celebration designed to encourage the public to be mindful of the potential negative effects human activity can have on Earth. Try to incorporate a catchy jingle or phrase to linger with the listener.

DBQ Document–Based Questions

Data obtained from: Municipal Solid Waste Generation, Recycling, and Disposal in the United States: Facts and figures 2008. Environmental Protection Agency, Figure 4.

This bar graph is based on data from the Environmental Protection Agency showing commonly recycled materials and the percentage of the products that were recycled in 2008.

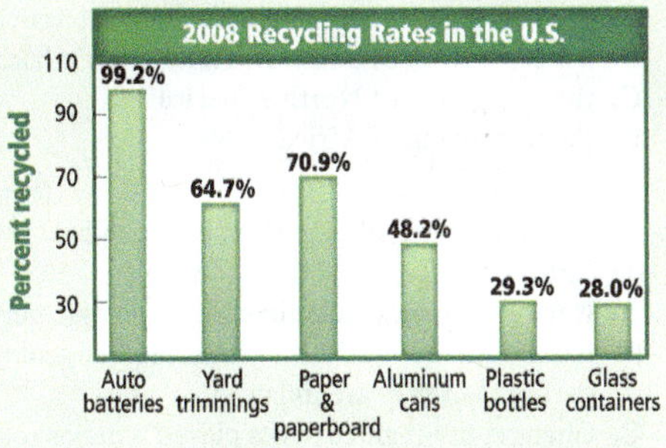

43. Which material had the highest recycle rate?

44. Which product had the lowest recycle rate?

45. Why do you think a higher percentage of aluminum cans were recycled than glass containers?

CUMULATIVE REVIEW

46. During what processes is latent heat stored and released? **(Chapter 11)**

47. What is the relationship between plate tectonics and the mantle underneath these plates? **(Chapter 17)**

48. How can energy from water be harnessed for human use? **(Chapter 25)**

DBQ Document-Based Questions

Data obtained from: Municipal Solid Waste Generation, Recycling, and Disposal in the United States: Facts and figures 2003. Environmental Protection Agency, Tables 13 and 19.

43. auto batteries

44. glass containers

45. Answers will vary, but might suggest that aluminum cans may be more widely used and therefore have a higher recycling rate.

CUMULATIVE REVIEW

46. evaporation and condensation

47. Convection in the mantle helps drive the movement of tectonic plates.

48. through hydroelectric power

CHAPTER 26 ASSESSMENT

THINK CRITICALLY

33. Answers will vary, but should address the reclamation and conservation of land and natural habitats, and ways to minimize water, soil, and air pollution.

34. Point-source pollution originates from a single source. Nonpoint-source pollution comes from multiple sources. Knowing how each impacts the environment helps to identify, control, and regulate the pollution.

35. Answers will vary. Encourage students to propose creative yet practical solutions.

36. Answers will vary. Sample answer: If the type of nonpoint-source pollution can be identified (carbon, mercury, etc.), then the sources can more readily be identified and targeted.

37. Humans wear different clothes, live in different types of shelter, and eat different foods, depending on where they live.

38. Answers might include not watering lawns, taking shorter showers, and turning off the faucet while brushing teeth.

39. Loss of berries is the density-dependent factor because it is related to population size. Its effect changes as the bird population changes. The storm is the density-independent factor because it is not related to population size. It would affect the birds regardless of the size of their population.

CONCEPT MAPPING

40. Check students' concept maps for accuracy.

CHALLENGE QUESTION

41. Answers will vary. Students might suggest the relocation of animals to wildlife refuges, or only developing new land where their habitats would be the least affected.

WRITING IN ▸ Earth Science

Rubric

42. Answers will vary, but may include a reminder about recycling and the proper disposal of trash.

CUMULATIVE STANDARDIZED TEST PRACTICE

MULTIPLE CHOICE

1. D
2. B
3. C
4. C
5. B
6. C
7. D
8. B
9. D

CUMULATIVE STANDARDIZED TEST PRACTICE

MULTIPLE CHOICE

1. Which type of coal burns the most efficiently?
   A. peat
   B. lignite
   C. bituminous coal
   D. anthracite

Use the illustrations below to answer Questions 2 and 3.

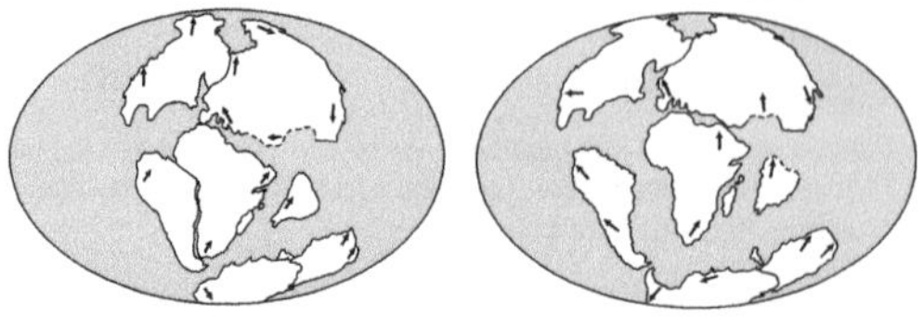

2. What was the most drastic move that occurred to land from the Jurassic Period to the Cretaceous Period?
   A. the nearing of South America and Antarctica
   B. the separation of South America from Africa
   C. the elongating of North America
   D. the narrowing of Africa

3. Which statement about the illustrated geologic periods on Earth is TRUE?
   A. It rains more now than in earlier geologic periods.
   B. Earth's oceans were saltier in the earlier geologic period than they are today.
   C. Changes in ocean currents played a major role in climate change.
   D. Earth's rivers were larger in the earlier periods than they are today.

4. Potassium-40 (K-40) decays into argon-40 (Ar-40) with a half-life of 1.3 billion years. A scientist studies a mineral sample and finds that the ratio of K-40 to Ar-40 is 1:3. How old is the rock?
   A. 0.6 billion years old
   B. 1.3 billion years old
   C. 2.6 billion years old
   D. 3.9 billion years old

5. Which is not a good way to conserve transportation energy?
   A. Drive at a lower speed.
   B. Make frequent stops.
   C. Work from home.
   D. Use a hybrid or electric car.

6. If fire destroys a wooded area, what limit to population growth has the area experienced?
   A. equilibrium
   B. a density-dependent factor
   C. a density-independent factor
   D. carrying capacity

Use the table below to answer Questions 7 and 8.

Amount of Land Created by Mt. Kilauea Eruptions

Volcanic Activity Episode	Net Amount of New Land Area (km2)
1–48b	17.2
48	34.3
49	3.7
50	0.11
51–52	0.01
53	2.3
54	0.24
55	59.5

7. Which statement best describes the eruption of Mount Kilauea?
   A. The same amount of lava flows from Mount Kilauea during every period of activity.
   B. Mount Kilauea erupts on a regular basis and in a similar way each time.
   C. The eruptions of Mount Kilauea are alternately explosive and quiet.
   D. The amount of new land area produced by Mount Kilauea during a period of activity varies.

8. How might scientists be able to use this data?
   A. to predict the occurrence of another earthquake
   B. to monitor land growth due to eruptions
   C. to predict the intensity of the earthquake based on the land growth
   D. to predict the amount of new land that will form with the next eruption

9. Which is not considered a source of natural air pollution?
   A. volcanoes
   B. fires
   C. radon
   D. smog

SHORT ANSWER

Use the chart below to answer Questions 10 and 11.

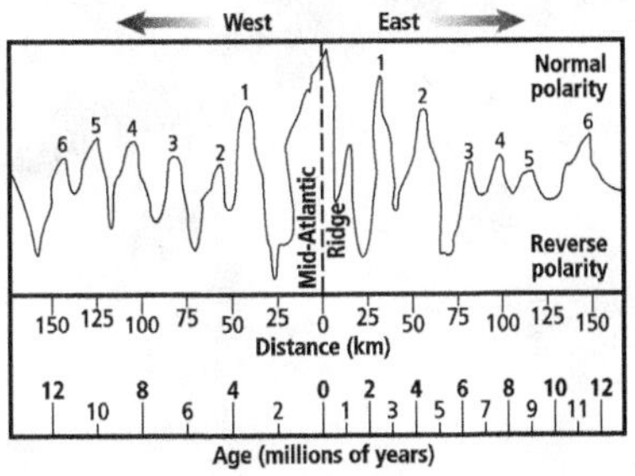

10. A group of scientists used a magnetometer and other equipment to obtain this magnetic field profile of part of the ocean floor. What information does the profile give?

11. What can the scientists conclude about how the ocean floor is formed near the Mid-Atlantic Ridge?

12. Explain why living things are considered to be renewable resources.

13. Why is photosynthesis important to humans?

14. Describe the process of differentiation in the formation of Earth's landmasses.

READING FOR COMPREHENSION

Acid Deposition

Acid deposition penetrates deeply into the fabric of an ecosystem, changing the chemistry of the soil and streams and narrowing the space where certain plants and animals can survive. Because there are so many changes, it takes many years for ecosystems to recover from acid deposition, even after emissions are reduced and the rain pH is restored to normal. However, there are some things that people can do to restore lakes and streams more quickly. Limestone or lime (a naturally occurring basic compound) can be added to acidic lakes to "cancel out" the acidity. Liming tends to be expensive, has to be done repeatedly to keep the water from returning to its acidic condition, and is considered a short-term remedy in only specific areas, rather than an effort to reduce or prevent pollution. Furthermore, it does not solve the broader problems of changes in soil chemistry and forest health in the watershed, and it does nothing to address visibility reductions, materials damage, and risk to human health.

Article obtained from: Acid rain. Reducing acid rain. *Environmental Protection Agency.* October 4, 2006. (Online resource accessed October 17, 2006.)

15. Which is not a limitation to using liming to improve lake acidity?
    - **A.** It does not improve soil chemistry.
    - **B.** It does not improve forest health.
    - **C.** It does not reduce risk to human health.
    - **D.** It does not act as a short-term remedy.

16. Why is adding limestone to an acidic lake beneficial?
    - **A.** It cancels out the acidity of the lake.
    - **B.** It restores naturally occurring bases in the water.
    - **C.** It can be done only once to fix the problem.
    - **D.** It allows for the return of wildlife.

17. Aside from liming's drawbacks, why is it still a good option for improving water condition?

NEED EXTRA HELP?

If You Missed Question . . .	1	2	3	4	5	6	7	8	9	10	11	12	13	14
Review Section . . .	25.1	23.3	23.3	21.3	25.3	26.1	18.2	18.2	24.3	17.2	17.2	24.1	24.3	22.2

SHORT ANSWER

10. Sample answer: The profile shows that the youngest part of the ocean floor is nearest the Mid-Atlantic Ridge and the oldest part is farthest from the ridge. On each side of the ridge, six reversals in Earth's polarity are recorded.

11. Material must come up through the ridge from deep inside Earth and flow to either side, causing seafloor spreading. As more and more material comes to the surface, the older material is pushed farther from the ridge.

12. Mature individuals of a species are able to reproduce. Many living things such as plants and animals are used by humans, thus causing them to be considered renewable resources.

13. During the process of photosynthesis, carbon dioxide and water is taken in by plants and converted into carbohydrates. After this conversion, oxygen, which humans need to breathe, is released back into the air.

14. The process by which a planet becomes internally zoned is called differentiation. Heavy materials sink toward its center and lighter materials accumulate near its surface forming crust.

READING FOR COMPREHENSION

15. D

16. A

17. Sample answer: Liming is a good option because it uses a naturally occurring basic substance to improve water quality. Therefore, the environment is being improved safely without any possible side effects.

UNIT 8

Beyond Earth

Themes

Change When the Big Bang occurred, all the matter now present in the universe was contained in a very small and dense space. Then, the universe started to expand, and galaxies, stars, and planets formed. The universe is still expanding. New stars are born and old ones die.

Structures In the universe, there are many objects that scientists have studied throughout history. Planets, moons, and comets within solar systems are dwarfed by stars and galaxies that lie far beyond our solar system.

Geologic Time Comparing deep-space images, which show the universe as it was billions of years in the past, to the almost present-time images from flyby satellites and probes, allows scientists to make their first estimates of the age of the universe.

Systems Understanding how the rings of Saturn formed and are maintained might lead to verification of the way our solar system formed. In turn, this might tell us how the galaxies, galaxy groups, and the universe were formed.

Scientific Inquiry As technology allows scientists to learn more about the universe, modifications in how we identify the features must occur. Evidence of change is seen in the change of status for Pluto.

UNIT 8

Beyond Earth

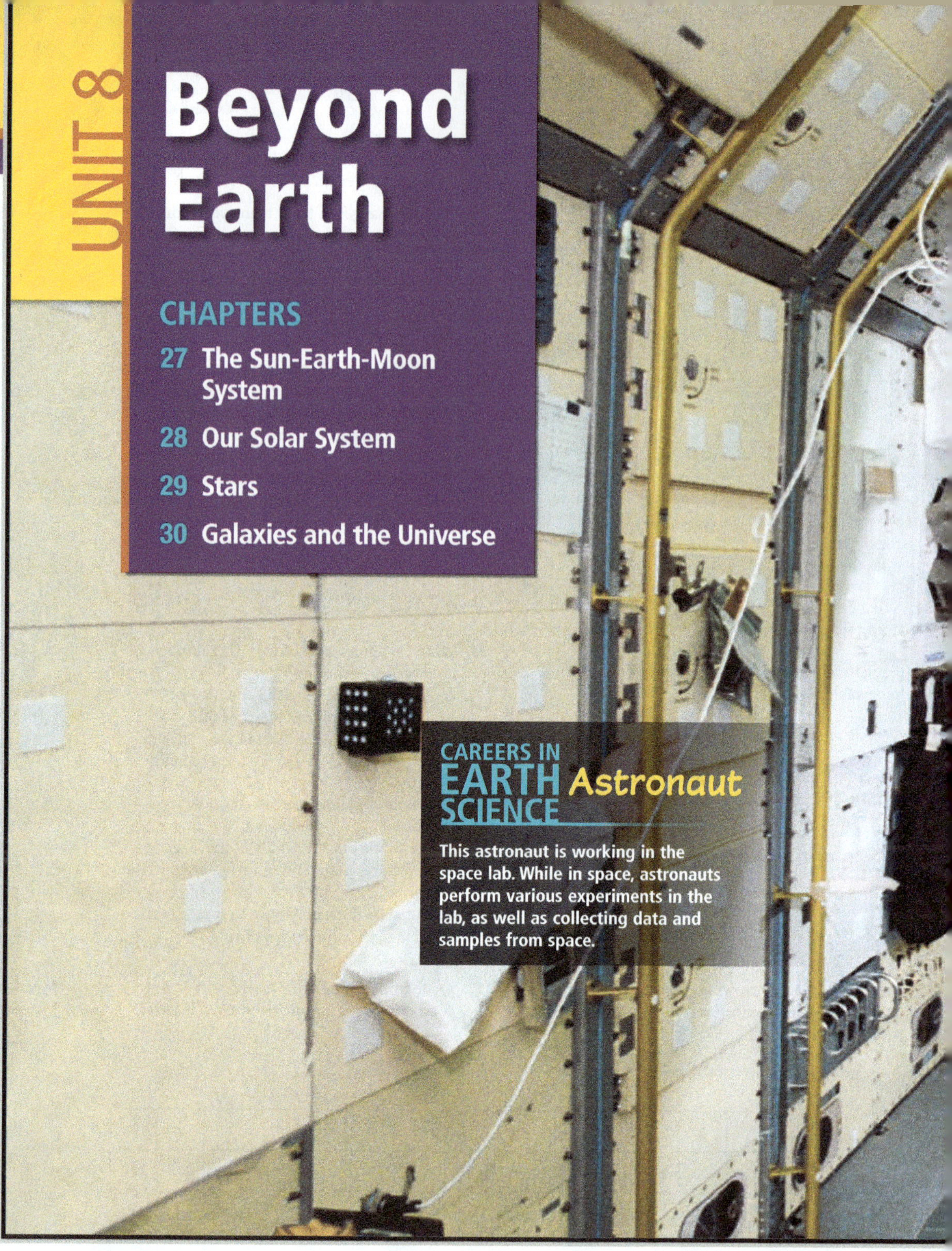

CHAPTERS

27 The Sun-Earth-Moon System

28 Our Solar System

29 Stars

30 Galaxies and the Universe

CAREERS IN EARTH SCIENCE **Astronaut**

This astronaut is working in the space lab. While in space, astronauts perform various experiments in the lab, as well as collecting data and samples from space.

STEM Project

Introduce the Unit

Preconceptions This photograph shows Mission Specialist Mae Jemison aboard the Spacelab Japan science module on the Space Shuttle *Endeavor*. Scientists are currently planning to send piloted flights to the Moon and beyond to learn more about the solar system. What do students know about the solar system? Where is Earth's solar system in relation to the Milky Way? What is the solar system made of? Students might say that it contains only planets, but ask them to hypothesize what planets are made of and whether all planets are the same. Ask students whether they have ever observed the Milky Way.

Formation History Galaxies contain gas and dust, as well as stars. Interstellar gas and dust can be observed either as dark clouds blocking light or glowing clouds. These clouds are the birthplaces of stars and planets. Stars are mostly gas, while planets are collections of dust and gas. Rocky planets, such as Earth, are mostly large collections of dust, while planets such as Jupiter are mostly gas. Stars evolve as they fuse elements within their cores.

CHAPTER 27 The Sun-Earth-Moon System

BIGIDEA **The Sun, Earth, and the Moon form a dynamic system that influences all life on Earth.**

ESSENTIAL QUESTIONS	NATIONAL STANDARDS	RESOURCES TO ASSESS MASTERY
SECTION 1 **Tools of Astronomy** **1.** What is electromagnetic radiation? **2.** How do telescopes work? **3.** How does space exploration help scientists learn more about the universe? 1 session 0.5 block	UCP.1–3, UCP.5; D.1; E.1–2; F.6; G.1, G.3	**Entry-Level Assessment** Focus Transparency 85, p. 764 **Progress Monitoring** Caption Question, p. 769 Reading Check, pp. 765, 766 Section Review, p. 769
SECTION 2 **The Moon** **1.** What is the history of lunar exploration? **2.** How are lunar properties and structures described? **3.** What are the features of the Moon? **4.** What is the theory of the Moon's origin and formation? 2 sessions 1 block	UCP.2–4; A.2; D.3; E.1–2; F.6; G.1–2	**Entry-Level Assessment** Focus Transparency 86, p. 770 **Progress Monitoring** Caption Question, p. 770 Reading Check, pp. 770, 771, 772 Section Review, p. 774
SECTION 3 **The Sun-Earth-Moon System** **1.** What are the relative positions and motions of the Sun, Earth, and the Moon? **2.** What are the phases of the Moon? **3.** What are the differences between solstices and equinoxes? **4.** How are eclipses of the Sun and Moon explained? 4 sessions 2 blocks	UCP.1–5; A.1, A.2; B.4; D.1, D.3; E.1; G.2–3	**Entry-Level Assessment** Focus Transparency 87, p. 775 **Progress Monitoring** Caption Question, pp. 776, 777, 778, 781, 783 Reading Check, pp. 777, 778, 783 Section Review, p. 784 **Summative Assessment** Chapter Assessment, p. 789 Exam*View*® *Assessment Suite* CD-ROM

Go online!
connectED.mcgraw-hill.com

LEVELED RESOURCES	LAB MATERIALS
Science Notebook 27.1 OL **Chapter FAST FILE Resources:** Study Guide, p. 15 BL **Lab Resources:** Laboratory Manual, p. 209 OL **Transparencies:** Section Focus Transparency 85 EL Teaching Transparency 83 OL EL	LaunchLAB p. 762 / **10 min** paper, scissors, metric ruler
Science Notebook 27.2 OL **Chapter FAST FILE Resources:** GeoLab Worksheet, p. 3 OL Study Guide, p. 16 BL **Lab Resources:** Laboratory Manual, p. 213 OL **Transparencies:** Section Focus Transparency 86 EL Teaching Transparency 84 OL EL	GeoLAB p. 786 / **30 min** paper, metric ruler
Science Notebook 27.3 OL **Chapter FAST FILE Resources:** MiniLab Worksheet, p. 2 OL Study Guide, p. 18 BL **Transparencies:** Section Focus Transparency 87 EL Teaching Transparencies 85, 86 OL EL	MiniLAB p. 776 / **25 min** metric ruler, protractor

LabManager™
Customize any lab with the LabManager™ CD-ROM.

ADDITIONAL RESOURCES

Technology:
- eTeacherEdition
- eStudentEdition
- Classroom Presentation Toolkit
- Exam*View® Assessment Suite* CD-ROM
- LabManager™ CD-ROM
- What's EARTH SCIENCE Got To Do With It? DVD
- Weather Classroom Video DVD

Assessment:
- Blueprints for Success
- Chapter Test (Scaffolded)
- Earth Science Standardized Test Practice

Lab Resources:
- Exploring Environmental Problems Laboratory Manual

BL Below Level OL On Level AL Advanced Learners EL English Learners COOP LEARN Cooperative Learning

CHAPTER 27

LaunchLAB

iLab Station | Rubric

How can the Sun-Earth-Moon system be modeled?

Process Skills compare and contrast, formulate models, measure in SI, interpret data, use numbers, recognize spatial relationships

Safety Precautions Approve lab safety forms before work begins. Caution students to handle scissors and drawing compasses with care.

Additional Material scissors

Teaching Strategy Make sure students calculate the correct scaled diameters and distances before they cut out their models. Scaled diameters are as follows: the Moon—1 cm; Earth—3.7 cm; the Sun—403.3 cm. Scaled distances are as follows: Earth-Moon—111 cm; Earth-Sun—43,290 cm.

Procedure

1. Have students read and complete the lab safety form and follow the procedure below.
2. Calculate the diameters of Earth and the Sun using a scale in which the Moon's diameter is equal to 1 cm.
3. Using this scale, calculate the distances between Earth and the Moon and Earth and the Sun.
4. Cut out **paper** circles to represent your scaled Earth and Moon, and place them at the scaled distance apart.

Analysis

1. **Compare** the diameters of your cutout Earth and Moon to the distance between them. The size of the cutouts is small compared to the scaled distance between them.
2. **Infer** why your model does not have a scaled Sun placed at the scaled Sun distance. A cutout of the Sun would be larger than a normal piece of paper, and if it were placed at the proper distance, the model would not fit in a classroom.

CHAPTER 27

The Sun-Earth-Moon System

BIGIDEA The Sun, Earth, and the Moon form a dynamic system that influences all life on Earth.

SECTIONS

1 Tools of Astronomy

2 The Moon

3 The Sun-Earth-Moon System

LaunchLAB

iLab Station

How can the Sun-Earth-Moon system be modeled?

The Sun is about 109 times larger in diameter than Earth, and Earth is about 3.7 times larger in diameter than the Moon. The distance between Earth and the Moon is 30 times Earth's diameter. The Sun is 390 times farther from Earth than is the Moon. Model the distances between the Sun, the Moon, and Earth in this activity.

FOLDABLES® Study Organizer

Phases of the Moon

Make a bound book and draw each major phase of the moon in order on the bottom pages of your Foldable. Indicate the positions of the Sun, the Moon, and Earth. Include a sketch of how the Moon appears from Earth during each phase. Use the Foldable to organize your notes on phases of the moon.

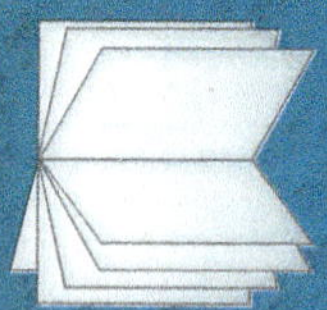

Astronomers use tools such as ultraviolet, radio, and X-ray telescopes to study the Moon, Sun and other objects in the solar system.

Assessment

Skill Ask students to measure the classroom and apply their scale to that measurement. Students should then compare and contrast the scaled measurement of the classroom to the distances among Earth, the Moon, and the Sun.

LabManager™ Customize any lab with the LabManager™ CD-ROM.

Introduce the BIGIDEA

Earth's Rotation Have students research how medieval astronomers determined that Earth rotates on its axis as it revolves around the Sun. What is the evidence? How could they use this evidence to convince a skeptical friend? Students will find that it is not easy to prove Earth's movements. Use the opportunity to help them gain a greater understanding of the challenges ancient astronomers faced.

Teacher Content Support

Telescopes Most students are familiar with optical telescopes, which focus visible light. Tell students that some telescopes, such as the radio-telescope shown here, focus nonvisible forms of electromagnetic radiation. Have them compare and contrast the images produced by different telescopes.

SECTION 1

1 Focus

Focus Transparency

Before presenting the lesson, project **Section Focus Transparency 85** and have students answer the accompanying questions. EL

MAINIDEA

Electromagnetic Radiation Discuss the many ways that electromagnetic radiation affects students' lives. Challenge students to come up with an application for each type of electromagnetic radiation. Sample answer: Radio waves are used to broadcast radio and television signals. The heat from a fireplace or wood-burning stove is transmitted by infrared radiation. X rays and gamma rays have medical applications. Ultraviolet rays from the Sun cause sunburn. Our eyes interpret visible light.

2 Teach

Activity

Radio Frequencies Have students find the frequency at which their favorite radio stations broadcast, and then calculate the wavelength of the broadcast using $\lambda = c/f$. AL

Interpret the Illustration

Frequency and Wavelength Have students examine the electromagnetic spectrum in **Figure 1** and demonstrate the relationship between frequency and wavelength. As wavelength increases, frequency decreases, and vice versa. OL

SECTION 1

Tools of Astronomy

MAINIDEA Radiation emitted or reflected by distant objects allows scientists to study the universe.

Essential Questions

- What is electromagnetic radiation?
- How do telescopes work?
- How does space exploration help scientists learn about the universe?

Review Vocabulary

refraction: occurs when a light ray changes direction as it passes from one material into another

New Vocabulary

electromagnetic spectrum
refracting telescope
reflecting telescope
interferometry

EARTH SCIENCE 4 YOU **Have you ever used a magnifying lens to read fine print? If so, you have used a tool that gathers and focuses light. Scientists use telescopes to gather and focus light from distant objects.**

Radiation

The radiation from distant bodies throughout the universe that scientists study is called electromagnetic radiation. Electromagnetic radiation consists of electric and magnetic disturbances traveling through space as waves. Electromagnetic radiation includes visible light, infrared and ultraviolet radiation, radio waves, microwaves, X rays, and gamma rays.

You might be familiar with some forms of electromagnetic radiation. For example, overexposure to ultraviolet waves can cause sunburn, microwaves heat your food, and X rays help doctors diagnose and treat patients. All types of electromagnetic radiation, arranged according to wavelength and frequency, form the **electromagnetic spectrum,** shown in **Figure 1.**

Wavelength and frequency Electromagnetic radiation is classified by wavelength—the distance between peaks on a wave. Notice in **Figure 1** that red light has a longer wavelength than blue light, and radio waves have a much longer wavelength than gamma rays. Electromagnetic radiation is also classified according to frequency, the number of waves or oscillations that pass a given point per second. The visible light portion of the spectrum has frequencies ranging from red to violet, or 4.3×10^{14} to 7.5×10^{14} Hertz (Hz)—a unit equal to one cycle per second.

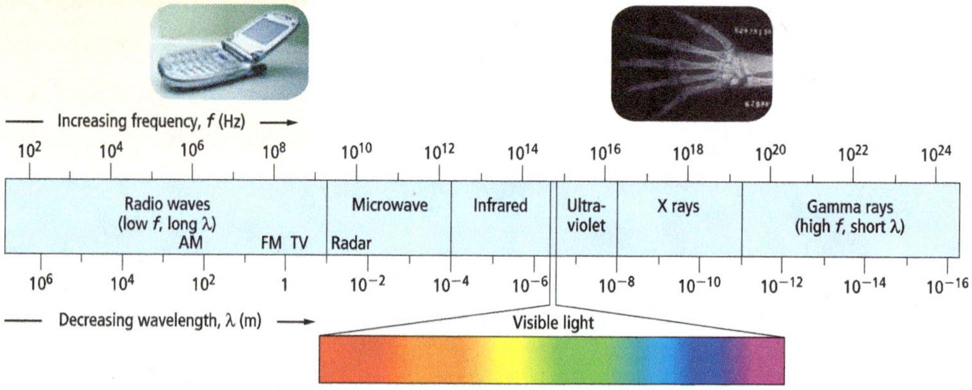

Figure 1 The electromagnetic spectrum identifies the different radiation frequencies and wavelengths.

Demonstration

Focus Sunlight A magnifying lens can be used to show how a lens collects light over a large area and brings it to a focus, where the intensity is increased. Use a lens to focus sunlight onto a piece of paper and have students observe as the intensity heats and chars the paper. Explain how a larger lens increases the intensity at the focus. Be sure to take precautions against starting a fire. Point out to students that telescopes do not create fire hazards because starlight is too dim to cause significant heating, even when a large telescope is used.

Figure 2 This photo of the Lagoon nebula was taken by the Mayall 4-m telescope, shown with its observatory.

Frequency is related to wavelength by the mathematical relationship $c = \lambda f$, where c is the speed of light (3.0×10^8 m/s), λ is the wavelength, and f is the frequency. Note that all types of electromagnetic radiation travel at the speed of light in a vacuum. Astronomers choose their tools based on the type of radiation they wish to study. For example, to see stars forming in interstellar clouds, they use special telescopes that are sensitive to infrared wavelengths, and to view remnants of supernovas, they often use telescopes that are sensitive to UV, X-ray, and radio wavelengths.

Telescopes

Objects in space emit radiation in all portions of the electromagnetic spectrum. Telescopes, such as the one shown in **Figure 2,** give us the ability to observe wavelengths beyond what the human eye can detect. In addition, a telescope collects more electromagnetic radiation from distant objects and focuses it so that an image of the object can be recorded. The pupil of a typical human eye has a diameter of up to 7 mm when it is adapted to darkness; the diameter of a telescope's opening, which is called its aperture, might be as large as 10 m. Larger apertures can collect more electromagnetic radiation, making dim objects in the sky appear much brighter.

READING CHECK **Name** two benefits of using a telescope.

Another way that telescopes surpass the human eye in collecting electromagnetic radiation is with the aid of cameras, or other imaging devices, to create time exposures. The human eye responds to visible light within one-tenth of a second, so objects too dim to be perceived in that time cannot be seen. Telescopes can collect light over periods of minutes or hours. In this way telescopes can detect objects that are too faint for the human eye to see. Also, astronomers can add specialized equipment. A photometer, for example, measures the intensity of light and a spectrophotometer displays the intensities of different wavelengths of radiation.

CAREERS IN EARTH SCIENCE

Space Engineer Space engineers design and monitor probes used to explore space. Engineers often design probes to collect information and samples from objects in the solar system. They also study the data collected.

WebQuest

Reinforcement

Sequence Wavelengths Ask each student to list the various forms of electromagnetic radiation in order from shortest to longest wavelengths. Also have students list the forms of electromagnetic radiation in order from lowest to highest frequency. From shortest to longest wavelengths are gamma rays, X rays, ultraviolet radiation, visible light, infrared radiation, microwaves, and radio waves. From lowest to highest frequency are radio waves, microwaves, infrared radiation, visible light, ultraviolet radiation, X rays, and gamma rays. BL

Enrichment

Analyze Light Contrary to popular belief, astronomers spend little time looking through telescopes. Instead, they use specialized instruments to collect and analyze the light that arrives at a telescope's focus. Have interested students research different instruments that astronomers use with telescopes. Charge-coupled devices (CCDs) are widely used by astronomers. These are refined versions of the visible-light detectors used in household digital cameras and video cameras. OL

READING CHECK Telescopes can detect radiation beyond what the human eye can detect, and the larger apertures of telescopes allow more radiation to be gathered so objects that are too dim to see can be recorded.

CAREERS IN EARTH SCIENCE

WebQuest

DIFFERENTIATED INSTRUCTION

Advanced Learners The resolution of a telescope (the smallest angular detail it can detect) is given $\theta = 2.5 \times 10^5(\lambda/D)$, where θ is expressed in arcseconds (1 arcsecond = 1/60 of one minute of arc, which is 1/60 of a degree), λ represents the wavelength being observed, and D is the aperture diameter in meters. Explain to students that 1 arcsecond is roughly the angular diameter of a dime seen from a distance of 3 km. Have students calculate θ for a human eye with a pupil diameter of 2 mm, 10-cm telescope, and a 5-m telescope, assuming that the observed wavelength is 6×10^{-7}m.

	Diameter	θ
Human eye	0.002 m	75 arcseconds
Telescope 1	0.1 m	1.5 arcseconds
Telescope 2	5 m	0.03 arcseconds

Teacher Content Support

Reflecting Telescopes

Reflectors have several advantages over refractors. The mirror of a reflector needs only one perfectly shaped surface, while a lens needs two. The glass in a lens must be clear and flawless, but because light does not travel through a mirror, the glass in a mirror does not have to be perfect. Mirrors can also be larger because they can be supported from behind. Lenses cannot be as large because they can be supported only around the edge and will sag if they are too large. Reflectors can also use a variety of mirror arrangements to bring light to a focus.

Apply Earth Science

Space Junk As a result of the numerous rocket launches and space exploration missions, space around Earth is filled with debris. These fragments of rockets and other debris will eventually fall into Earth's atmosphere and burn up, but until that time, this debris must be tracked. A small piece of debris can cause severe damage to a spacecraft or satellite because it is moving quickly. The U.S. Space Command tracks over 8000 objects that are bigger than 10 centimeters across. Occasionally, a large piece of debris does not burn up completely in Earth's atmosphere, and it crashes to the ground.

☑ READING CHECK Refracting telescopes bring light to a focus using lenses. Reflecting telescopes bring light to a focus using mirrors.

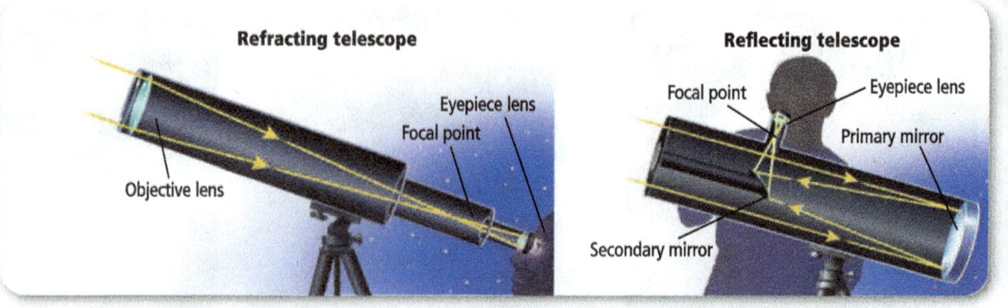

■ **Figure 3** Refracting telescopes use a lens to collect light. Reflecting telescopes use a mirror to collect light.

Refracting and reflecting telescopes Two different types of telescopes are used to focus visible light. The first telescopes, invented around 1608, used lenses to bring visible light to a focus and are called **refracting telescopes,** or refractors. The largest lens on such telescopes is called the objective lens. In 1668, a new telescope that used mirrors to focus light was built. Telescopes that bring visible light to a focus with mirrors are called **reflecting telescopes,** or reflectors. **Figure 3** illustrates how simple refracting and reflecting telescopes work.

Although both refracting and reflecting telescopes are still in use today, most astronomers use reflectors because mirrors can be made larger than lenses and can therefore collect more light. Technology used in astronomy has changed over time, as shown in **Figure 4.**

☑ READING CHECK **Compare** refracting and reflecting telescopes.

Most telescopes used for scientific study are located in observatories far from city lights, usually at high elevations where there is less atmosphere overhead to blur images. Some of the best observatory sites in the world are located high atop mountains in the southwestern United States, along the peaks of the Andes mountain range in Chile, and on the summit of Mauna Kea, a volcano on the island of Hawaii.

■ **Figure 4**

Development of Astronomy

Humanity's curiosity about the night sky was limited to Earth-bound explorations until the first probe was sent into space in 1957.

410 B.C. The first prophecies based on the positions of the five visible planets, the Moon, and the Sun were written for individuals in Mesopotamia.

1054 Chinese astronomers document the explosion of the supernova that creates the Crab nebula, believing it foretells the arrival of a wealthy visitor to the emperor.

28,000 B.C. | 5000 B.C. | 500 B.C. | A.D. 0 | 1000

28,000 B.C. Cro-Magnon people sketch moon phases on tools made out of bones.

4236 B.C. After lunar and solar calendars predict agricultural seasons, Egyptians adopt a 365-day calendar based on the movement of the star Sirius.

A.D. 900s Arab astronomers greatly improve the accuracy of the Greek astrolabe—a tool for celestial navigation that determines time and location.

EARTH SCIENCE JOURNAL

Observatories Have each student choose a major observatory and research its location, the size and nature of its telescope, and the characteristics of the site that make it ideal for astronomical observations. Students should then write paragraphs that explain the features and advantages of the observatories. AL

Rubric

Telescopes using non-visible wavelengths For all telescopes, the goal is to bring as much electromagnetic radiation as possible into focus. Infrared and ultraviolet radiation can be focused by mirrors in a way similar to that used for visible light. X rays cannot be focused by normal mirrors, and thus special designs must be used. Gamma rays cannot be focused, so telescopes designed to detect this type of radiation can determine only the direction from which the rays come.

A radio telescope collects the longer wavelengths of radio waves with a large dish antenna, which resembles a satellite TV dish. The dish plays the same role as the primary mirror in a reflecting telescope by reflecting radio waves to a point above the dish. There, a receiver converts the radio waves into electric signals that can be stored in a computer for analysis.

The data are converted into visual images by a computer. The resolution of the images produced can be improved using a process called **interferometry,** which is a technique that uses the images from several telescopes to produce a single image. By combining the images from several telescopes, astronomers can create a highly detailed image that has the same resolution of one large telescope with a dish diameter as large as the distance between the two telescopes. One example of this is the moveable telescopes shown in **Figure 5.** Both radio and optical telescopes can be linked this way.

Figure 5 The Very Large Array is situated near Socorro, New Mexico. The dish antennae of this radio telescope are mounted on tracks so they can be moved to improve resolution.

Space-Based Astronomy

Astronomers often send instruments into space to collect information because Earth's atmosphere interferes with most radiation. It blurs visual images and absorbs infrared and ultraviolet radiation, X rays, and gamma rays. Space-based telescopes allow astronomers to study radiation that would be blurred by our atmosphere. American, European, Russian, and Japanese space programs have launched many space-based observatories to collect data.

1860s The invention of spectroscopy suggests that the celestial bodies are composed of some of the same elements that make up Earth's atmosphere.

1957 Russia launches the first two satellites into orbit around Earth, marking the beginning of space exploration.

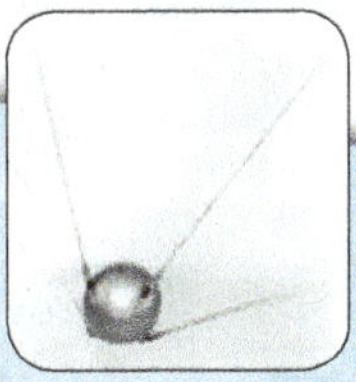

2004 A Mars rover discovers rock formations and sulfate salts indicating that the planet once had flowing water.

1600 — 1800 — 2000

1608 The telescope is invented allowing astronomers to discover planets, such as Uranus and Neptune, moons, and stars that are invisible to the naked eye.

1969 The U.S. astronauts become the first humans to walk on the Moon.

2010 Japanese probe *Hayabusa* successfully returns the first-ever sample from the surface of an asteroid back to Earth.

Teacher Content Support

Speed of Light The speed of light is a little slower in a medium, such as air, glass, or water, than it is in a vacuum. In most transparent media, the speed of light is close to its speed in a vacuum, and often, the difference can be ignored. However, the media through which light travels have important effects on light. For example, refraction, the bending of light as it passes through a boundary between media, is caused by the shift in light speed at the boundary.

A refracting telescope focuses light by bending it. A reflecting telescope relies on a curved mirror, usually parabolic in shape, to bring light to a focus. A radio telescope uses a curved dish to bring radio waves to a focus, much like a reflecting telescope. Many radio telescopes have wire mesh dishes because the wavelengths of radio waves are longer than the spaces between the mesh, so radio waves are still reflected.

Gamma rays and X rays cannot be reflected in a conventional method. Because they have such high frequencies and short wavelengths, these rays must be reflected by a series of grazing, or low-angle, reflections. Otherwise, they travel right through a material.

EARTH SCIENCE JOURNAL

Early Space Stations Have students write news articles about early space stations. For example, the launch of the American space station *Skylab* in 1973 enabled humans to live in space for up to 84 days. A Soviet (later Russian) space station, *Mir,* began operation in 1986 and ended in 2001. The maximum stay there was 366 days. Set aside time for a "news report," and have students present their articles to the class.

Rubric

Identify Misconceptions

Many students might think that the main purpose of a telescope is to magnify objects or to enable the observer to see farther.

Uncover the Misconception

Ask students to explain the uses of a telescope.

Demonstrate the Concept

Explain that the main purpose of a telescope is to enable the observer to see objects fainter than the human eye can detect. This often does involve faraway objects, depending on the luminosity of the objects being viewed. A secondary purpose of telescopes is to enable the observer to see details in images, which is called resolution, but this is not the same as magnification. Magnification makes an image larger but not necessarily clearer, as increased resolution does.

Assess New Knowledge

Ask students to distinguish between magnification and resolution. Magnification enlarges an image. Resolution makes an image clearer.

Tie to Previous Knowledge

Solar Energy *HST* and other Earth-orbiting spacecraft are powered by the same kinds of solar panels that are sometimes used as alternative energy sources for buildings on Earth. The solar arrays on *HST* and other spacecraft convert sunlight into electricity, which is stored in batteries onboard. The operations of spacecraft require that the solar panels be exposed to the Sun at all times.

Figure 6 The *Hubble Space Telescope* has been used to observe a comet crashing into Jupiter as well as to detect the farthest known galaxy.

Hubble Space Telescope Orbiting Earth every 97 minutes, one of the best-known space-based observatories—the *Hubble Space Telescope (HST)*—shown in **Figure 6,** was launched in 1990. The *Hubble Space Telescope* was designed to obtain sharp visible-light images without atmospheric interference, and also to make observations in infrared and ultraviolet wavelengths. *Hubble* has observed galaxies well over 12 billion light years away.

The next-generation successor to the *HST* is the *James Webb Space Telescope (JWST),* scheduled to launch in 2018. *JWST* will primarily observe in the infrared range, with some capability in the visible-light range. Several other space-based telescopes are listed below in **Table 1.**

Spacecraft In addition to making observations from above Earth's atmosphere, spacecraft can be sent directly to the bodies being observed. Robotic probes are spacecraft that can make close-up observations and sometimes land to collect information directly. Probes are practical only for objects within our solar system, because other stars are too far away. In 2005, the *Cassini* spacecraft arrived at Saturn, where it went into orbit for a detailed look at its moons and rings. The *Mars Science Laboratory,* or *Curiosity,* is scheduled to land on Mars in 2012. It will assess whether the landing area had, or still has, environmental conditions favorable to microbial life. *New Horizons* was launched in 2006, on its way to Pluto and the region beyond. *New Horizons* is armed with visible, infrared, and ultraviolet cameras, as well as equipment to measure magnetic fields. It is scheduled to reach Pluto in 2015.

Table 1 Orbiting Telescopes

Name	Launch	Wavelengths	Studies	Host
Integral	2002	X ray, gamma ray	wide ranging, neutron stars	ESA, Russia, NASA
CHIPSat	2003	X ray	interstellar plasma	NASA
Galex	2003	UV	survey	JPL, NASA
MOST	2003	visible	observe stars	Canada
Spitzer	2003	IR	wide ranging	NASA
Swift	2004	X ray, UV, visible	black holes	NASA
Suzaku	2005	X ray	high-energy phenomena	Japan
Akari	2006	IR	survey	Japan
Agile	2007	gamma ray	wide ranging	ESA
Kepler	2009	visible	extrasolar planets	NASA
WISE	2009	IR	survey	NASA

ACROSS THE CURRICULUM

Physics There are multiple challenges inherent in designing and constructing a large telescope, on the ground or in space. Both have to be pointed to an accuracy of far less than a second of arc. (An arcsecond is approximately the diameter of a dime as seen from a distance of 3 km!) A ground-based telescope must maintain its pointing accuracy against the downward pull of gravity, while rotating to offset the rotation of Earth. It must also maintain a constant temperature to reduce flexure of the mirrors. A space-based telescope must withstand the vibrations and pressures of launch, and must be precisely focused on a distant star or galaxy, while keeping its solar panels pointed at the Sun for power. It, too, must maintain a constant operating temperature.

Human spaceflight Before humans can safely explore space, scientists must learn about the effects of space, such as weightlessness and radiation. The most recent human studies have been accomplished with the space shuttle program between 1981 and 2011. Shuttles are used to place and service satellites, such as the *HST* and the *Chandra X-ray Telescope.* The space shuttle provides an environment for scientists to study the effects of weightlessness on humans, plants, the growth of crystals, and other phenomena. However, because shuttle missions last a maximum of just 17 days, long-term effects must be studied in space stations. A multicountry space station called the *International Space Station (ISS),* shown in **Figure 7,** is the ideal environment for studying the effects of space on humans. In 2010, NASA and its international partners celebrated ten years of permanent human habitation on the *ISS*. The crew members conduct many different experiments in this weightless environment.

Figure 7 This view of the *International Space Station* was taken from the Space Shuttle *Discovery.*
Review *What types of studies can be carried out in the space station?*

Spinoff technology Space-exploration programs not only benefit astronomers and space exploration, but they also benefit society. Many technologies that were originally developed for use in space programs are now used by people throughout the world. Did you know that the technology for the space shuttle's fuel pumps led to the development of pumps used in artificial hearts? Or that NASA's quest to improve crash protection led to the memory foam found in mattresses? In fact, more than 1500 different NASA technologies have been passed on to commercial industries for common use; these are called spinoffs.

SECTION 1 REVIEW

Section Self-Check

Section Summary

- Telescopes collect and focus electromagnetic radiation emitted or reflected from distant objects.
- Electromagnetic radiation is classified by wavelength and frequency.
- The two main types of optical telescopes are refractors and reflectors.
- Space-based astronomy includes the study of orbiting telescopes, satellites, and probes.
- Technology originally developed to explore space is now used by people on Earth.

Understand Main Ideas

1. MAINIDEA **Explain** how electromagnetic radiation helps scientists study the universe.
2. **Distinguish** between refracting and reflecting telescopes and how they work.
3. **Report** on how interferometry affects the images that are produced by telescopes.
4. **Examine** the reasons why astronomers send telescopes and probes into space.

Think Critically

5. **Assess** the benefits of technology spinoffs to society.
6. **Consider** the advantages and disadvantages of using robotic probes to study distant objects in space.

MATH IN Earth Science

7. Calculate the wavelength of radiation with a frequency of 10^{12} Hz. [*Hint: Use the equation* $c = \lambda f$.]

Caption Question Fig. 7 studies about the long-term effects of space on humans, plants, crystals, and other phenomena

3 Assess

Check for Understanding

Discussion Tell students that astronomers are considering placing an observatory on the Moon. Ask students to discuss the advantages and disadvantages of doing so. The Moon has no atmosphere, so the telescope would obtain sharp images and could observe all wavelengths. The possibility of building a far larger telescope than is feasible for an orbiting spacecraft would be another advantage. The chief disadvantages would be cost and the difficulty of servicing the telescope.

Reteach

Activity Have students each make a summary table listing the advantages of telescopes and the criteria for deciding where to locate them—that is, whether they should be in space or at a certain location on Earth (mountaintop, etc.). Ask students to give an example of an actual telescope that meets each criterion.

Assessment

Skill Have students explain how refracting and reflecting telescopes work, and why each type is beneficial. Ask students to use their knowledge of how visible-light telescopes work and to draw sketches of a radio telescope.

SECTION 1 REVIEW

1. Electromagnetic radiation emitted or reflected from objects in the universe is the only means scientists have to study objects in the universe.
2. Refracting telescopes bring light to a focus using lenses. Reflecting telescopes bring light to a focus using mirrors.
3. In interferometry, several telescopes work together, creating highly detailed images that have the same resolution as one large telescope.
4. Telescopes in space can collect radiation that is absorbed or distorted by Earth's atmosphere. Probes make close-up views of objects possible. They also can gather data from objects directly.
5. Accept all reasonable efforts that show actual spinoffs.
6. Sample answer: Advantages include the ability to travel to and explore distant objects in space; disadvantages include the inability to react quickly to new situations.
7. $3 \times 10^8 \div 10^{12} = 3 \times 10^{-4}$ m

SECTION 2

1 Focus

Focus Transparency

Before presenting the lesson, project **Section Focus Transparency 86** and have students answer the accompanying questions. EL

MAINIDEA

Moon Missions Have groups of students create time lines for different missions that led to the first astronaut on the Moon.

2 Teach

Identify Misconceptions

Some people think that astronauts in spacecraft are weightless because Earth's gravity does not extend far into space.

Uncover the Misconception Ask students why astronauts appear weightless while in orbit.

Demonstrate the Concept Explain that Earth's gravity extends into space. It causes both spacecraft and the astronauts inside to fall toward Earth while they are moving forward in orbit, which creates the illusion that there is no gravity in near-Earth orbit.

Assess New Knowledge Ask students to compare astronauts in a spacecraft with a person in an elevator that is falling freely. As the elevator is pulled toward Earth by gravity, the floor falls away from the person's feet in the same way that the spacecraft falls away from the astronaut's feet. The person in the elevator also appears weightless.

READING CHECK space probes and astronauts

Caption Question Fig. 8 The *LRV* allowed astronauts to more easily explore rugged terrain.

SECTION 2

The Moon

MAINIDEA The Moon, Earth's nearest neighbor in space, is unique among the moons in our solar system.

Essential Questions

- What is the history of lunar exploration?
- How are lunar properties and structures described?
- What are the features of the Moon?
- What is the theory of the Moon's origin and formation?

Review Vocabulary

lava: magma that flows onto the surface from the interior of an astronomical body

New Vocabulary

albedo
highland
mare
impact crater
ejecta
ray
rille
regolith

EARTH SCIENCE 4 YOU How many songs, poems, and stories do you know that mention the Moon? The Moon is a familiar object in the night sky and much has been written about it.

Exploring the Moon

Astronomers have learned much about the Moon from observations with telescopes. However, most knowledge of the Moon comes from explorations by space probes, such as *Kaguya* and the *Lunar Reconnaissance Orbiter (LRO)*, and from landings by astronauts. The first step toward reaching the Moon was in 1957, when the Soviet Union launched the first artificial satellite, *Sputnik I*. Four years later, Soviet cosmonaut Yuri A. Gagarin became the first human in space.

That same year, the United States launched the first American, Alan B. Shepard, Jr., into space during Project Mercury. This was followed by Project Gemini that launched two-person crews. Finally, on July 20, 1969, the Apollo program landed Neil Armstrong and Edwin "Buzz" Aldrin on the Moon during the Apollo 11 mission. Astronauts of the Apollo program explored several areas of the Moon, often using special vehicles, such as the *Lunar Roving Vehicle* shown in **Figure 8** After a gap of many years, scientists hope to return to the Moon someday. Astronauts hope to remain longer on the Moon and eventually establish a permanent base there. NASA is also assisting private companies in their efforts to build piloted spacecraft.

READING CHECK **Identify** the source of most information about the Moon.

Figure 8 Apollo 15 astronauts used the *Lunar Roving Vehicle (LRV)* to explore the Moon's surface.
Explain *how the LRV might have resulted in improved mission performance.*

EARTH SCIENCE JOURNAL

Lunar Features Lunar surface features, particularly the major craters that are readily observed from Earth, are generally named after important people in astronomy. However, other conventions are used as well. The maria are so named because early astronomers thought they looked like seas. They were given names corresponding to human feelings or moods—for example, the Sea of Tranquility. Mountainous regions on the Moon are named after mountain ranges on Earth. Many small features were not identified until humans explored the Moon. These features were named by astronauts, and some of the names are quite colloquial. Have students research and report on several of the more unusual names for lunar features.

The Lunar Surface

Although the Moon is the brightest object in our night sky, the lunar surface is dark. The **albedo** of the Moon, the percentage of incoming sunlight that its surface reflects, is very small—only about 7 percent. In contrast, Earth has an average albedo of nearly 31 percent. Sunlight that is absorbed by the surface of the Moon produces extreme differences in temperature. Because the Moon has no atmosphere to absorb heat, sunlight can heat the Moon's surface to 400 K (127°C), while the temperature of its unlit surface can drop to a chilly 40 K (−233°C).

The "man in the Moon" pattern seen from Earth is produced by the Moon's surface features. Lunar **highlands** are heavily cratered regions of the Moon that are light in color and mountainous. Other regions called **maria** (MAH ree uh) (singular, mare [MAH ray]) are dark, relatively smooth plains, which average 3 km lower in elevation. Although the maria are mostly smooth, they do have a few scattered craters and rilles. **Rilles** are valleylike structures that might be collapsed lava tubes. In addition, there are mountain ranges near some of the maria.

☑ READING CHECK **Explain** what lunar features produce the "man in the Moon."

Lunar craters The craters on the Moon, called **impact craters,** formed when objects from space crashed into the lunar surface. The material blasted out during these impacts fell back to the Moon's surface as **ejecta.** Some craters have long trails of ejecta, called **rays,** that radiate outward from the impact site much like the spokes of a bicycle tire, as shown in **Figure 9.** Rays are visible as light-colored streaks.

■ **Figure 9** You can see some of the details for the maria and highlands in the view of the full moon. Craters, ejecta, rilles, and rays are visible in close-up views of the Moon's surface.

Teacher Content Support

Water on the Moon In 2009, water ice was discovered in a deep crater near the south pole of the Moon. The *Lunar Crater Observation and Sensing Satellite (LCROSS)* spacecraft crashed its Centaur upper stage rocket into a permanently shadowed region of the Cabeus crater, evaporating water ice. The resulting water vapor was detected by the host spacecraft and the data was sent to Earth before the *LCROSS* itself impacted the Moon. Widespread presence of water ice in large areas of the Moon's south pole was detected by the *Lunar Reconnaissance Orbiter (LRO)* in 2010. Just as ice cores on Earth provide data on Earth's history, ice on the Moon might provide information on the evolution of the solar system. Finding water on the Moon will also continue to be important as plans for lunar bases are established.

Interpret the Photo

Ejecta Figure 9 shows that ejecta from impacts on the Moon are brighter than the surrounding surface. The reason for this is that ejecta consist of materials from below the Moon's surface that might have a different composition and have not been exposed to the darkening effects of radiation and particles from space for as long as the surrounding surface has. Craters with visible rays of ejecta are all relatively young.

Activity

Time Line Have each student create a time line of unpiloted probes that were sent to the Moon. Be sure that students include the probes of other countries besides the United States. EL

☑ READING CHECK The "man in the Moon" pattern is produced by light-colored, heavily cratered highlands and dark, smooth maria.

DIFFERENTIATED INSTRUCTION

Visually Impaired Model portions of the lunar surface with clay, so that students with visual impairments can get a better idea of what the Moon's surface is like. You can make craters by pressing marbles into the clay and pushing up the rims around the marbles. You can form rilles by gouging the surface of the clay with a toothpick or an old pen.

ACROSS THE CURRICULUM

Language Arts Have students find common words in the English language that are derived from *luna,* the Latin word for *moon*. This might lead to a discussion of the meanings of lunacy, lunatic, and related words, and the mythology associated with strange behavior at the time of a full moon. OL

Enrichment

Geologic Histories Ask students to compare the geologic history of the Moon with that of Earth. Points that students should include in the comparison are the formation processes, the importance of differentiation, internal heat sources, internal structures, tectonic activity, formation of surface features, relative impact rates, and atmospheres. OL

Tie to Previous Knowledge

Radiometric Dating The many precise dates given in the text for the formation of lunar regions are based on radiometric dating, which involves exactly the same techniques that are used to date rock samples on Earth. Radiometric dating was possible for lunar rocks only after the success of the Apollo program, which returned samples of lunar rocks to Earth, where their isotopic ratios could be measured.

☑ READING CHECK Highlands are mostly plagioclase feldspar, and maria are basalt.

Explore **the Moon and Earth with an interactive table.**

Concepts In Motion

Table 2 The Moon and Earth

	The Moon	Earth
Mass (kg)	7.349×10^{22}	5.974×10^{24}
Radius (km)	1737.1	6371.0
Volume (km^3)	2.196×10^{10}	1.083×10^{12}
Density (kg/m^3)	3350	5515

Lunar properties Earth's moon is unique among all the moons in the solar system. First, it is the largest moon compared to the radius and mass of the planet it orbits, as shown in **Table 2.** Also, it is a solid, rocky body, in contrast with the icy compositions of most other moons of the solar system. Finally, the Moon's orbit is farther from Earth relative to the distance of many moons from the planets they orbit. **Figure 10** shows a photo mosaic of Earth and the Moon taken from space.

Composition The Moon is made up of minerals similar to those of Earth—mostly silicates. Recall that silicates are compounds containing silicon and oxygen that make up 96 percent of the minerals in Earth's crust. The highlands, which cover most of the lunar surface, are predominately lunar breccias (BRE chee uhs), which are rocks formed by the fusion of smaller pieces of angular rock during impacts. Unlike sedimentary breccias on Earth, most of the lunar breccias are composed of plagioclase feldspar, a silicate containing high quantities of calcium and aluminum but low quantities of iron. The maria are predominately basalt, but unlike basalt on Earth, they contain no water.

☑ READING CHECK **Describe** the compositions of the lunar highlands and maria.

History of the Moon

The entire lunar surface is old—radiometric dating of rocks from the highlands indicates an age between 3.8 and 4.6 billion years—about the same age as Earth. Based on the ages of the highlands and the frequency of the impact craters that cover them, scientists theorize that the Moon was heavily bombarded during its first 800 million years. This caused the breaking and heating of surface rocks and resulted in a layer of loose, ground-up rock called **regolith** on the surface. The regolith averages several meters in thickness, but it varies greatly depending on location.

Figure 10 This photo mosaic shows images of Earth and the Moon at the relative size that each appears when viewed from a distance of about 400,000 kilometers away. The images were taken by the *Near Earth Asteroid Rendezvous* (*NEAR*) spacecraft.

DIFFERENTIATED INSTRUCTION

Hearing Impaired Have students gather around a table. Place a large cake pan on the table and fill it with different layers of colored sand. Place a layer of flour on top to model regolith. Drop different-sized rocks into the pan to model crater formation and how it disturbs the layers of the surface. Vary the size of the rocks as well as the height from which they are dropped. Have students write down their observations concerning how the size of the rock and height from which it falls affect the size and shape of a crater.

Layered structure Scientists infer from seismic data that the Moon, like Earth, has a layered structure, which consists of the crust, upper mantle, lower mantle, and core, as illustrated in **Figure 11.** The crust varies in thickness and is thickest on the far side. The far side of the Moon is the side that is always facing away from Earth. The Moon's upper mantle is solid, its lower mantle is thought to be partially molten, and its core is mostly solid iron.

Formation of maria After the period of intense bombardment that formed the highlands, lava welled up from the Moon's interior and filled in the large impact basins. This lava fill created the dark, smooth plains of the maria. Scientists estimate the maria formed between 3.1 and 3.8 bya, making them younger than the highlands. Flowing lava in the maria scarred the surface with rilles. Rilles are much like lava tubes found on Earth, through which lava flows in underground streams. The maria have remained relatively free of craters because fewer impacts have occurred on the Moon since they formed.

Often lava did not fill the basins completely and left the rims of the basins above the lava. This left behind the mountain ranges that now surround many maria. As shown in **Figure 12,** there are virtually no maria on the far side of the Moon, which is covered almost completely with highlands. Scientists hypothesize that this is because the crust is thicker on the far side, which made it difficult for lava to reach the lunar surface. You will determine the relative ages of the Moon's surface features in this chapter's GeoLab.

Tectonics Seismometers measure strength and frequency of moonquakes. Seismic data show that on average, the Moon experiences an annual moonquake that would be strong enough to cause dishes to fall out of a cupboard if it happened on Earth. Despite these moonquakes, scientists think that the Moon is not tectonically active. The Moon has no active volcanoes and no significant magnetic field. Scientists know from the locations and shapes of mountains on the Moon that they were not formed tectonically, as mountain ranges on Earth are formed. Lunar mountains are actually higher elevations that surround ancient impact basins filled with lava.

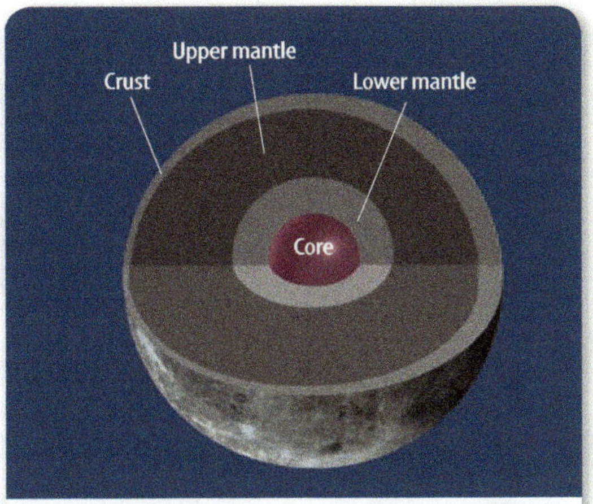

Figure 11 Scientists deduce the structure of the Moon's interior from seismic data obtained from seismometers left on the Moon's surface.

Figure 12 The heavily cratered far side of the Moon has many fewer maria than the more familiar near side of the Moon.

Project

Structure of the Moon Have students research how NASA has determined the internal structure of the Moon. Be sure to have students include observations that have been made and experiments that have been conducted. Have students present their findings to the class. NASA has used seismographs and has deliberately crashed small objects into the Moon's surface to artificially create seismic waves rather than wait for natural moonquakes. OL

Collaborative Learning

Magnetic Fields Ask students to infer whether the Moon has a magnetic field, using their knowledge of the origin of Earth's magnetic field. Encourage students to discuss this issue with each other and work through their reasoning. Then, ask students to compare their inferences with the results of magnetic measurements of the Moon. Because the Moon's core is mostly solid, it is logical to infer that the Moon does not have a magnetic field. This is borne out by measurements that show no detectable magnetic field on the Moon. AL

Discussion

Tectonic Activity Ask students what geological processes are occurring on or in the Moon today. The Moon is now geologically inactive, except for surface overturn resulting from occasional meteoroid impacts and small moonquakes.

ACROSS THE CURRICULUM

Physics Some scientists object to the simultaneous formation theory for the Moon because simultaneous formation could not have produced the Moon's high angular momentum, which is a measure of its orbital speed. The capture theory could explain it, but this theory has other problems. The giant impact theory can easily account for the Moon's high angular momentum, and this is one of the key reasons that this theory is becoming widely accepted.

GeoLAB

iLab Station

The GeoLab at the end of the chapter can be used at this point in the lesson.

3 Assess

Check for Understanding

Project To help students realize how unique the Moon is in comparison with other planetary satellites, have students each make a data table containing information about the Moon and at least ten other satellites in the solar system. Have students include the following properties in their tables: mass, ratio of mass to parent planet's mass, orbital radius, and ratio of orbital radius to the radius of the parent planet.

Reteach

Diagram Have students each draw a diagram showing the formation of a lunar surface feature.

Assessment

Knowledge Ask students to explain why the Moon's surface is a more accurate record of the solar system's history than Earth's surface is. Because the Moon has no atmosphere, its surface has not been altered by erosion, weathering, chemical processes, or biological effects, all of which have affected Earth's surface.

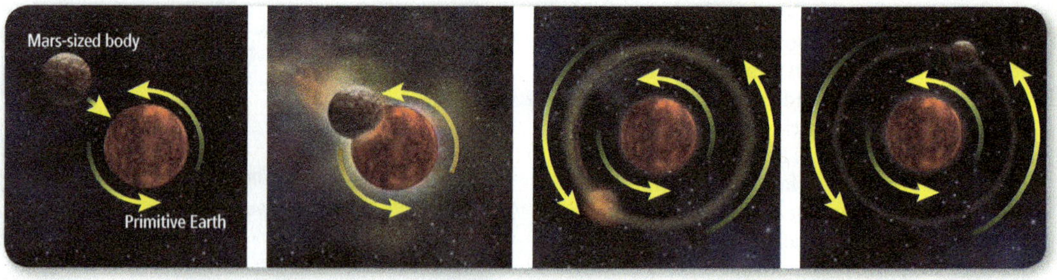

Figure 13 The impact theory of the Moon's formation states that material ejected from Earth and from the striking object eventually merged to form the Moon.

View an **animation of the Moon impact theory.**

Concepts In Motion

Formation

Several theories have been proposed to explain the Moon's unique properties. The theory that is accepted by most astronomers today was developed using computer simulations. This theory is known as the giant impact theory.

According to the giant impact theory, the Moon formed as the result of a collision between Earth and a Mars-sized object about 4.5 billion years ago when the solar system was forming. This computer model suggests that the object struck primitive Earth with a glancing blow. The impact caused materials from the incoming body and Earth's outer layers to be ejected into space, where—being trapped by Earth's gravity—they began to orbit the Earth. Over time, the materials merged to form the Moon. The giant impact theory is illustrated in **Figure 13.** According to this model, the Moon is made up of a small amount of iron at the core, and mostly silicate material that came from Earth's mantle and crust. This explains why the Moon's crust is so similar to Earth's crust in chemical composition. This theory has been accepted because of similarities that have been found between bulk samples of rock taken from Earth and from the Moon.

SECTION 2 REVIEW

Section Self-Check

Section Summary

- Astronomers have gathered information about the Moon using telescopes, space probes, and astronaut exploration.
- Like Earth's crust, the Moon's crust is composed mostly of silicates.
- Surface features on the Moon include highlands, maria, ejecta, rays, and rilles. It is heavily cratered.
- The Moon probably formed about 4.5 bya in a collision between Earth and a Mars-sized object.

Understand Main Ideas

1. MAINIDEA **Compare and contrast** the Moon and the moons of other planets.
2. **Classify** the following according to age: maria, highlands, and rilles.
3. **Explain** how scientists determined that the Moon has no tectonics.
4. **Distinguish** the steps involved in the impact theory of lunar formation.

Think Critically

5. **Infer** how the surface of the Moon would look if the crust on the far side were the same thickness as the crust on the near side.
6. **Summarize** the major ideas in this section using an outline format. Include the following terms: *highlands, crust, lava, maria, craters, tectonics,* and *impact theory.*

WRITING IN ▶ Earth Science

7. Write the introductory paragraph to an article entitled *History of the Moon.*

SECTION 2 REVIEW

1. The Moon is much more massive relative to Earth, and it is farther from its parent planet than is typical of other satellites.
2. The oldest features are highlands, followed by maria, then rilles.
3. Mountain ranges were formed by impacts, not tectonics. There is no evidence of current volcanic activity, significant quakes, or a magnetic field. Also, scientists have seismological data.
4. A Mars-sized object collided with Earth. Material from this body and Earth's outer layers was ejected into space. The ejected material condensed to form the Moon.
5. The Moon would probably be covered with more maria. The far side would be smoother and darker.
6. Accept any logical outlines that include the designated terms.
7. Answers will vary, but should be supported by scientific facts about the formation of the Moon and its features.

Rubric

SECTION 3

The Sun-Earth-Moon System

MAINIDEA Motions of the Sun-Earth-Moon system define Earth's day, month, and year.

Essential Questions

- What are the relative positions and motions of the Sun, Earth, and Moon?
- What are the phases of the Moon?
- What are the differences between solstices and equinoxes?
- How are eclipses of the Sun and Moon explained?

Review Vocabulary

revolution: the time it takes for a planetary body to make one orbit around another, larger body

New Vocabulary

ecliptic plane
solstice
equinox
synchronous rotation
solar eclipse
perigee
apogee
lunar eclipse

EARTH SCIENCE 4 YOU **Have you ever tried to guess the time by judging the Sun's position? If so, you were observing an effect of the motions of the Sun-Earth-Moon system.**

Daily Motions

From the vantage point of Earth, the most obvious pattern of motion in the sky is the daily rising and setting of the Sun, the Moon, stars, and everything else that is visible in the night sky. The Sun rises in the east and sets in the west, as do the Moon, planets, and stars. These daily motions result from Earth's rotation. The Sun, the Moon, planets, and stars do not orbit around Earth every day. It only appears that way because we observe the sky from a planet that rotates. But how do we know that Earth rotates?

Earth's rotation There are two relatively simple ways to demonstrate that Earth is rotating. One is to use a Foucault pendulum, like the one shown in **Figure 14.** A Foucault pendulum swings in a constant direction. But as Earth turns under it, the pendulum seems to shift its orientation. The second way is to observe the way that air on Earth is diverted from a north-south direction to an east-west direction by the Coriolis effect.

Day length The time period from one noon to the next is called a solar day. Our timekeeping system is based on the solar day. But the length of a day as we observe it is roughly four minutes longer than the time it takes Earth to rotate once on its axis. As Earth rotates, it also moves in its orbit and has to turn a little farther each day to align again with the Sun.

Figure 14 This Foucault pendulum is surrounded by pegs. As Earth rotates under it, the pendulum knocks over the pegs, showing the progress of the rotation.

SECTION 3

1 Focus

Focus Transparency

Before presenting the lesson, project **Section Focus Transparency 87** and have students answer the accompanying questions. EL

MAINIDEA

Calendars Many cultures developed calendars based on the length of the year, and, in most cases, the length of the month. Have groups of students research a calendar developed by an ancient society. Each group should share its results in an oral presentation. Encourage groups to explore how the calendar reflected the customs and traditions of the society. For example, one group could write a report on the Babylonian calendar and the role of the number 60 in their society. Other groups might choose to research calendars developed by Mayan, African, Greek, Australian, and Minoan cultures.

2 Teach

Teacher Content Support

Coriolis Effect People sometimes exaggerate the importance of the Coriolis effect on small scales. For example, the circulation of water in a toilet bowl or a sink is often attributed to Earth's rotation. In fact, the Coriolis effect is far too weak over a scale of a few centimeters or meters to have any noticeable effect. The direction of water flow in a toilet or a sink is determined by the direction in which water enters or by the shape or slope of the bowl or sink, not by Earth's rotation.

Demonstration

Model Length of Day Illustrate why the length of day and the Sun's intensity vary during the year. Set a globe on a table on one side of a darkened room with the southern hemisphere tilted toward the far side of the room. Stand in the center of the room and shine a flashlight on the globe's center. Have a student spin the globe at a steady rate. Point out that the length of day is longer than the length of night in the northern hemisphere. Then, move the globe to a table on the other side of the room, keeping the tilt oriented so that the northern hemisphere is tipped away from you. Shine the flashlight on the globe while a student spins it, and have students compare the length of day to the length of night in the northern hemisphere.

■ **Caption Question Fig. 15** Because of Earth's revolution about the Sun, the stars overhead will change during the year. The only exception is for someone near Earth's North or South pole.

MiniLAB

iLab Station   Rubric

Purpose Students will determine the Sun's altitude above the southern horizon during the summer solstice.

Process Skills interpret scientific illustrations, analyze data, use numbers, recognize spatial relationships

Additional Material metric ruler

Safety Precautions Approve lab safety forms before work begins. Tell students to handle drawing compasses with care.

Teaching Strategies

- This lab can be either an individual activity or one performed by small groups of students.
- Suggest that students make the drawing of Earth the same diameter as the protractor. If the circle is too small, it will be hard to make measurements.

Expected Results Students will see that the Sun's altitude above the southern horizon increases with proximity to the equator.

Analysis

1. For latitudes north of +23.5°, the Sun is lower in the sky to the south. For latitudes south of +23.5°, the Sun is lower in the sky to the north.
2. 66.5° south latitude

Assessment

Skill Ask students how their results would be different if Earth's axis were not tilted. There would still be variations in the Sun's altitude as seen from different latitudes, but there would be no variations during the year.

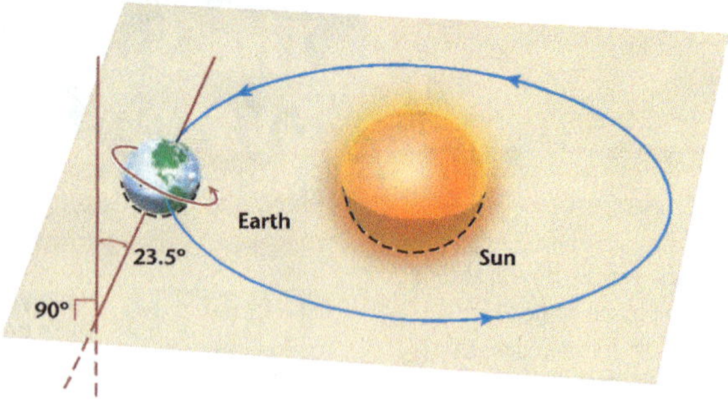

■ **Figure 15** Earth's nearly circular orbit around the Sun lies on the ecliptic plane. When looking toward the horizon and the plane of the ecliptic, different stars are visible during the year.
Predict *Do the positions of stars vary when you look overhead?*

VOCABULARY
ACADEMIC VOCABULARY
Cycle
recurring sequence of events or phenomena
The cycle of seasons repeats every year.

Annual Motions

Earth orbits the Sun in a slightly elliptical orbit, as shown in **Figure 15.** The plane of Earth's orbit is called the **ecliptic plane.** As Earth rotates, the Sun and planets appear to move across the sky in a path known as the ecliptic. As Earth moves in its orbit, different constellations are visible.

The effects of Earth's tilt Earth's axis is tilted relative to the ecliptic at approximately 23.5°. As Earth orbits the Sun, the orientation of Earth's axis remains fixed in space so that, at a given time, the northern hemisphere of Earth is tilted toward the Sun, while at another point, six months later, the northern hemisphere is tipped away from the Sun. A cycle of the seasons is a result of this tilt and Earth's orbital motion around the Sun. Another effect is the changing angle of the Sun above the horizon from summer to winter. More hours of daylight cause the summer months to be warmer than the winter months.

MiniLAB

iLab Station

Predict the Sun's Summer Solstice Position

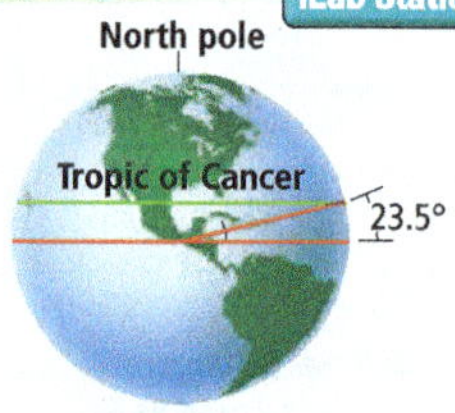

How can the Sun's position during the summer solstice be determined at specific latitudes? At summer solstice for the northern hemisphere, the Sun is directly overhead at the Tropic of Cancer.

Procedure

1. Read and complete the lab safety form.
2. Draw a straight line to represent the equator and mark the center of the line with a dot.
3. Use a **protractor** to measure the angle of latitude of the Tropic of Cancer from the equator line. Draw a line at that angle from the line's center dot.
4. Find your home latitude and measure that angle of latitude on your diagram. Draw a line from the line center for this location.
5. Measure the angle between the line for the Tropic of Cancer and the line for your location. Subtract that angle from 90°. This gives you the angle above the horizon for the maximum height of the Sun on the solstice at your location.

Analysis

1. **Describe** how the position of the Sun varies with latitude on Earth.
2. **Consider** the angle that would illustrate the winter solstice for the northern hemisphere.

DIFFERENTIATED INSTRUCTION

Struggling Learners Put a marker on a globe at your position on Earth, so that students can more easily see how their location enters daylight and darkness as Earth rotates.

LabManager™
Customize any lab with the LabManager™ CD-ROM.

Solstices Earth's orbit around the Sun and the tilt of Earth's axis are illustrated in **Figure 16.** Positions 1 and 3 correspond to the solstices. At a **solstice,** the Sun is overhead at its farthest distance either north or south of the equator. The lines of latitude that correspond to these positions on Earth have been identified as the Tropic of Cancer and the Tropic of Capricorn. The area between these latitudes is commonly known as the tropics. Position 1 corresponds to the summer solstice in the northern hemisphere when the Sun is directly overhead at the Tropic of Cancer, 23.5° north latitude. At this time, around June 21 each year, the number of daylight hours reaches its maximum, and the Sun is in the sky continuously within the region of the Arctic Circle. On this day, the number of daylight hours in the southern hemisphere is at its minimum, and the Sun does not appear in the region within the Antarctic Circle.

☑ READING CHECK **Identify** where the Sun is directly overhead at the summer solstice in the northern hemisphere.

As Earth moves past Position 2, the Sun's altitude decreases in the northern hemisphere until Earth reaches Position 3, known as winter solstice for the northern hemisphere. Here the Sun is directly overhead at the Tropic of Capricorn, 23.5° south latitude. This happens around December 21. On this day, the number of daylight hours in the northern hemisphere is at its minimum and the Sun does not appear in the region within the Arctic Circle. Then, as Earth continues around its orbit past Position 4, the Sun's altitude increases again until it returns to Position 1. Notice that the summer and winter solstices are reversed for those living in the southern hemisphere—June 21 is the winter solstice and December 21 is the summer solstice.

Equinoxes Positions 2 and 4, where Earth is midway between solstices, represent the equinoxes, a term meaning *equal nights.* At an **equinox,** Earth's axis is perpendicular to the Sun's rays and at noon the Sun is directly overhead at the equator. Those living in the northern hemisphere refer to Position 2 as the autumnal equinox, and Position 4 as the vernal equinox. Those in the southern hemisphere do the reverse—Positions 2 and 4 are the vernal and autumnal equinoxes, respectively.

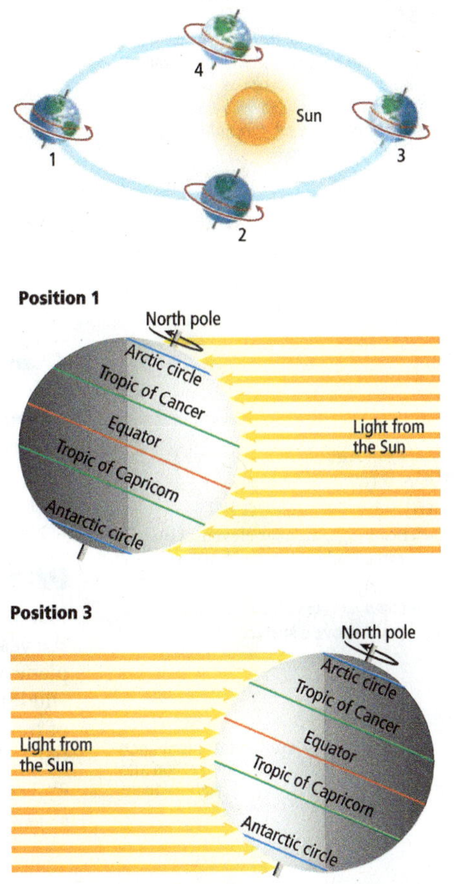

■ **Figure 16** Earth's axis remains tilted at the same angle as it orbits the Sun. It points either toward or away from the Sun at solstices as in Positions 1 and 3 and to the side at equinoxes as in Positions 2 and 4.
Identify *the correct term for each position for each hemisphere.*

Discussion

Earth's Orbit Ask students to think of ways to demonstrate that Earth is moving in an orbit around the Sun. As in the case of Earth's rotation, there is no obvious way to demonstrate this. A good point that some students might make is the difference between the length of the solar day and the time it takes Earth to rotate once on its axis, which we explain as being a result of Earth's motion about the Sun. Another good point is that we explain the seasons by assuming that Earth orbits the Sun (and has a tilted axis). However, the seasons could also be explained if the Sun orbited Earth in a complex path.

Model

Seasons Have students mark the Tropic of Cancer, the Tropic of Capricorn, and the equator on a ball with tape. Then, have students use a flashlight to represent the Sun and model Earth's changing position around the Sun and the solstices and equinoxes. BL EL

Interpret the Illustration

Angle of Sunlight In **Figure 16,** students should note that the Sun's rays are perpendicular to Earth's surface at the Tropic of Cancer during the summer solstice, and perpendicular to Earth's surface at the Tropic of Capricorn during the winter solstice. During the equinoxes, the Sun's rays are perpendicular to Earth's surface at the equator, and hence the Sun is directly overhead.

☑ READING CHECK the Tropic of Cancer

Apply Earth Science

Distance from the Sun Earth's distance from the Sun varies by about 5 million km during the year, because Earth's orbit is not perfectly circular. Have students compare this variation with the average Sun-Earth distance of 150 million km and comment on whether the change in distance is likely to affect Earth's seasons. The variation is 3 percent of the average Sun-Earth separation, which is not enough to have a noticeable effect on Earth's climate.

■ **Caption Question Fig. 16** northern hemisphere: Position 1: summer solstice, Position 2: autumnal equinox, Position 3: winter solstice, Position 4: vernal equinox; southern hemisphere: Position 1: winter solstice, Position 2: vernal equinox, Position 3: summer solstice, Position 4: autumnal equinox

Reinforcement

The Sun's Position North of the Tropic of Cancer, the Sun at midday is always to the south of an observer's position. Between the spring and fall equinoxes, the Sun rises and sets north of due east or due west. Between the fall and spring equinoxes, the Sun rises and sets south of due east or due west. This is easily observed. Ask students to note where the Sun rises and sets relative to the east-west direction after late March.

Enrichment

North Pole Have students visualize being at the north pole, and ask them to describe the daily motion of the Sun as seen from there on the solstices and the equinoxes. At the summer solstice, the Sun stays in the sky all day, circling the horizon at an altitude of 23.5°. At each equinox, the Sun circles right on the horizon, and at the winter solstice, it stays 23.5° below the horizon all day.

Caption Question Fig. 17 For observers north of 23.5°, diagrams should show the Sun at its highest point during the summer solstice. The Sun's angle above the horizon is lower during the equinoxes, and at its lowest point during the winter solstice.

READING CHECK the sequential changes in the appearance of the Moon

FOLDABLES® Rubric

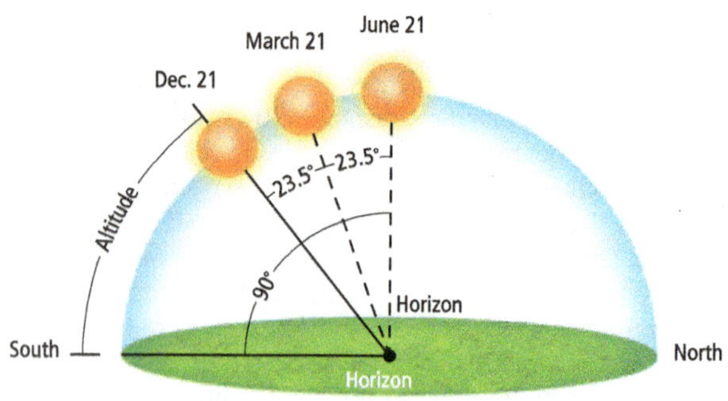

Figure 17 For a person standing at 23.5° north latitude, the Sun would be directly overhead on the summer solstice. It would be at its lowest position on the horizon at the winter solstice.
***Draw** a diagram showing how the Sun's angle changes throughout the year at your latitude.*

VOCABULARY
SCIENCE USAGE V. COMMON USAGE
Altitude
Science usage: angular elevation of a celestial body above the horizon

Common usage: vertical elevation of a body above a surface

FOLDABLES® Incorporate information from this section into your Foldable.

Changes in altitude The Sun's maximum height at midday, called its zenith, varies throughout the year depending on the viewer's location. For example, on the summer solstice, a person located at 23.5° north latitude sees the Sun's zenith directly overhead. At the equinox, it appears lower, and at the winter solstice, it is at its lowest position, shown in **Figure 17.** Then it starts moving higher again to complete the cycle.

Phases of the Moon

Just as the Sun appears to change its position in the sky throughout the year, the Moon also changes position relative to the ecliptic plane as it orbits Earth. The Moon's cycle is more complex, as you will learn later in this section. More striking are the changing views of the illuminated side of the Moon as it orbits Earth. The sequential changes in the appearance of the Moon are called lunar phases, and are shown in **Figure 18.**

READING CHECK **Explain** what is meant by the term *lunar phases.*

As you have read, the light given off by the Moon is a reflection of the Sun's light. In fact, one half of the Moon is illuminated at all times. How much of this lighted half is visible from Earth varies as the Moon revolves around Earth. When the Moon is between Earth and the Sun, for instance, the side that is illuminated is not visible from Earth. This phase is called a new moon.

Waxing and waning Starting at the new moon, as the Moon moves in its orbit around Earth, more of the sunlit side of the Moon becomes visible. This increase in the visible sunlit surface of the Moon is called the waxing phase. The waxing phases are called waxing crescent, first quarter, and waxing gibbous. Then, as the Moon moves to the far side of the Earth from the Sun, the entire sunlit side of the Moon faces Earth. This is known as a full moon.

After the full moon, the portion of the sunlit side that is visible begins to decrease. This is called the waning phase. The waning phases are named similarly to the waxing phases, that is, waning gibbous and waning crescent. When exactly half of the sunlit portion is visible, it is called the third quarter.

DIFFERENTIATED INSTRUCTION

Struggling Learners To help illustrate the phases of the Moon, use a ball representing the Moon and a strong flashlight in a darkened room. Place the ball on a table in front of students, and shine the light on the ball from different angles. Be careful not to shine the light into students' eyes. When you stand behind the students and illuminate the ball, this simulates a full moon. When you illuminate the ball at right angles to the students' line of sight, this simulates the quarter moon phases, and when you illuminate the ball from behind, this simulates a new moon. You can simulate the crescent and gibbous phases by illuminating the Moon from intermediate positions. This demonstration is useful for students who have difficulty with spatial visualization. EL

VISUALIZING the Phases of the Moon

Figure 18 One-half of the Moon is always illuminated by the Sun's light, but the entire lighted half is visible from Earth only at full moon. The rest of the time you see portions of the lighted half. These portions are called lunar phases.

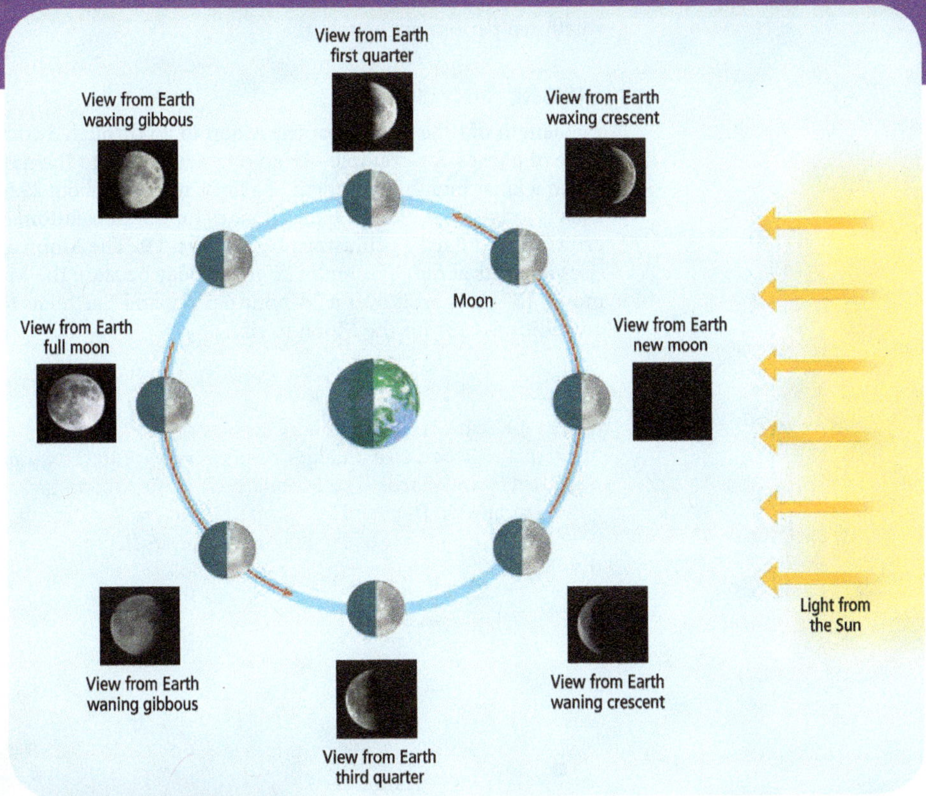

Sometimes a dim image of the full moon is seen along with a crescent. This is caused by Earth's reflected light on the Moon's surface. It is often referred to as "the new moon with the old moon in its arms."

Because of the variations in the plane of the Moon's orbit, the phases might appear different—either tipped, or misshapen.

Concepts In Motion View an **animation of lunar phases.**

(4)John Sanford/Photo Researchers, (6)Frank Zullo/Photo Researchers, (bl)Chris Cook/Photo Researchers, (br)Eyebyte/Alamy

VISUALIZING

Purpose

Students will distinguish among the phases of the Moon.

Teacher Content Support

Lunar Phases The phases of the Moon can be correlated to the times of the rising and setting of the Moon. If the Moon rises at the same time as the Sun, it is in the new moon phase. The Moon rises at noon during the first quarter. The Moon rises at 6:00 P.M. during the full moon, or in other words, it rises approximately as the Sun sets. During the third quarter, the Moon rises approximately 6 hours after the Sun sets, or 18 hours after the Sun rises. These times of rising and setting correspond to the positions of the Sun, Earth, and the Moon.

Discussion

Moon and Sun Ask students what phase the Moon is in when the Sun and the Moon rise at the same time. new moon Ask what time of day the full moon rises. approximately 6:00 P.M. BL

Use Scientific Terms

Quarter Moon Students might question why a quarter moon is so named when half of the disk is visible. *Quarter* refers to the fact that the Moon has traveled one-quarter of the cycle around Earth since the new moon (first quarter) or three-quarters of the cycle around Earth (third quarter). EL

Synchronous rotation You might have noticed that the surface features of the Moon always look the same. As the Moon orbits Earth, the same side faces Earth at all times. This is because the Moon rotates with a period equal to its orbital period. In other words, the Moon spins on its axis exactly once each time it goes around Earth. This is no coincidence. Scientists theorize that Earth's gravity slowed the Moon's original spin until the Moon reached **synchronous rotation,** the state at which its orbital and rotational periods are equal.

Lunar Motions

The length of time it takes for the Moon to go through a complete cycle of phases, for example—from one new moon to the next—is called a lunar month. The length of a lunar month is about 29.5 days. This is longer than the 27.3 days it takes for one revolution, or orbit, around Earth, as illustrated in **Figure 19.** The Moon also rises and sets about 50 minutes later each day because the Moon moves 13° in its orbit over a 24-hour period, and Earth has to turn an additional 13° for the Moon to rise.

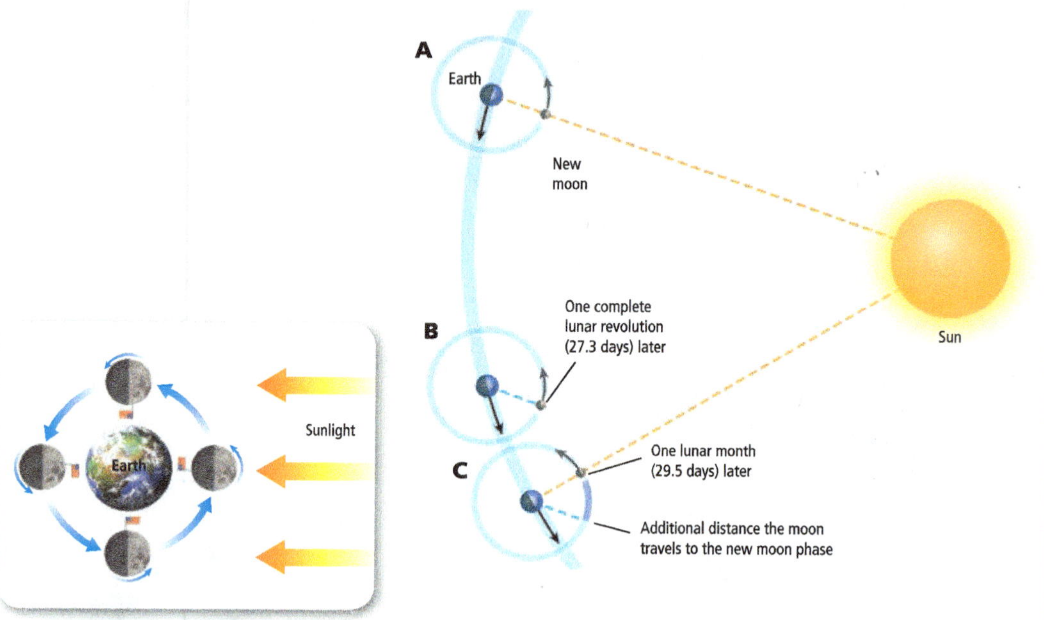

Figure 19 As the Moon moves from A, where it is in the new moon phase as seen from Earth, to B, it completes one revolution but is now in the waning crescent phase as seen from Earth. It must travel for 2.2 days to return to the new moon phase. The Moon rotates as it revolves, keeping the same side facing Earth, as shown in the inset.

Teacher Content Support

Gravity Earth's gravity slowed the Moon's spin until it reached synchronous rotation. The slowing was not a direct result of the force of attraction between Earth and the Moon, but of the difference between the forces acting on the near side and the far side of the Moon. This difference is called a differential gravitational force, or a tidal force. Such a force tends to stretch a body, so the Moon is slightly elongated along the line toward Earth, and Earth, especially its oceans, is elongated along this line as well.

Discussion

Moon's Rotation Ask students to discuss how the phases of the Moon would be affected if the Moon's rotation were not synchronous. There would be no change. OL

Enrichment

Tides The time between consecutive high or low tides is always a bit longer than 12 hours, and the daily cycle of two high and two low tides is longer than 24 hours. Ask students to try to work out why this is so. The Moon moves along in its orbit, so Earth has to rotate a little more than one full spin to return to the same alignment with the Moon. This is analogous to the difference between the solar day and the time it takes Earth to rotate once on its axis, or the difference between the lunar month and one revolution of the Moon about Earth. AL

Demonstration

Movements of the Moon Students might be confused about the Moon's rotation and think that the Moon does not rotate at all because the same side is always facing Earth. Place a globe on a table to represent Earth and use a smaller foam ball to represent the Moon. Attach a small arrow or pointer to the Moon at one point, so that students can easily keep track of its orientation. Walk around the globe that represents Earth while holding the Moon and keeping the pointer always fixed in the same direction. Students should see that different sides of the Moon face Earth at different points in the Moon's orbit. Then, repeat the demonstration, rotating the Moon so that the pointer always points toward Earth. In this case, students should see that the Moon is rotating once during each orbit.

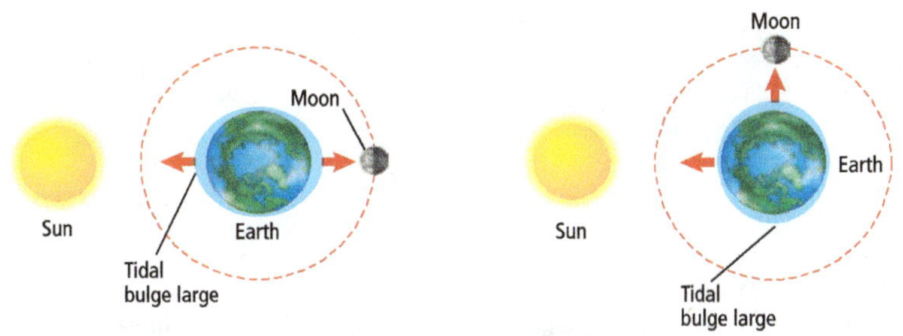

■ **Figure 20** Alignment of the Sun and the Moon produces the spring tides shown on the left. Neap tides, shown on the right, occur when the Sun and the Moon are at right angles.

Tides One effect the Moon has on Earth is causing ocean tides. The Moon's gravity pulls on Earth along an imaginary line connecting Earth and the Moon, and this creates bulges of ocean water on both the near and far sides of Earth. Recall that Earth's rotation also contributes to the formation of tides. As Earth rotates, these bulges remain aligned with the Moon, so that a person at a shoreline on Earth's surface would observe that the ocean level rises and falls every 12 hours.

Spring and neap tides The Sun's gravitational pull also affects tides, but the Sun's influence is half that of the Moon's because the Sun is farther away. However, when the Sun and the Moon are aligned along the same direction, their effects are combined, and tides are higher than normal. These tides, called spring tides, are especially high when the Moon is nearest Earth and Earth is nearest the Sun in their slightly elliptical orbits. When the Moon is at a right angle to the Sun-Earth line, the result is lower-than-normal tides, called neap tides. This occurs because the Sun and Moon's gravitational forces are competing. The Sun and the Moon alignments during spring and neap tides are shown in **Figure 20.**

Solar Eclipses

A **solar eclipse** occurs when the Moon passes directly between the Sun and Earth and blocks the Sun from view. Although the Sun is much larger than the Moon, it is far enough away that they appear to be the same size when viewed from Earth. When the Moon perfectly blocks the Sun's disk, only the dim, outer gaseous layers of the Sun are visible. This spectacular sight, shown in **Figure 21,** is called a total solar eclipse. A partial solar eclipse is seen when the Moon blocks only a portion of the Sun's disk.

■ **Figure 21** The stages of a total solar eclipse are seen in this multiple-exposure photograph.
Explain *why the Moon seems to cross the Sun at an angle rather than directly right to left.*

Teacher Content Support

Tidal Forces The Earth feels tidal forces because of the Moon's gravity. The gravitational forces between Earth and the Moon cause both bodies to elongate along an imaginary line between the two bodies. Earth's oceans are mobile and are able to reveal this deformation. The solid earth itself also distorts in response to tidal forces, but because it is very rigid, the distortion is much less than it would be if the Earth were a liquid and easily deformed. The maximum distortion of the solid Earth is only about 20 cm.

Tidal forces not only raise and lower Earth's surface, but also tilt it. The amount of tilt can be determined by careful measurement of the position of the stars.

Activity

Eclipses Have students find out the number of solar and lunar eclipses that occurred in the past ten years. Ask students to compare the numbers of solar and lunar eclipses and to explain any difference they find. Lunar eclipses and solar eclipses are usually linked. When the Sun, the Moon, and Earth are aligned at full moon, causing a lunar eclipse, they are usually aligned at either the previous or the following new moon, causing a solar eclipse. Lunar eclipses are seen over a larger area than solar eclipses because Earth casts a larger shadow than the Moon. OL

■ **Caption Question Fig. 21** The Moon's orbit is tilted relative to the ecliptic plane.

CULTURAL DIVERSITY

Eclipse Legends Almost every culture throughout history has legends about eclipses, which often were thought to be omens or harbingers of disaster. One Chinese legend was that an invisible dragon devoured the Sun and caused a solar eclipse. According to the legend, making loud noises by banging on pots or even shooting guns would frighten the dragon and return the world to daylight. Have students research legends about eclipses and share their findings with the class.

Problem-Solving LAB

Purpose Students will use model drawings to predict the appearance of solar eclipses.

Process Skills predict, recognize spatial relationships

Teaching Strategies

- This is a good project for students to do in pairs.
- Refer students to the diagram on this page if they have difficulty getting started.

Analysis

1. Make sure all students correctly locate the umbra and penumbra.

Think Critically

2–4

Location of observer	What the observer sees there
Point A (umbra)	Total eclipse
Point B (penumbra)	Partial eclipse
Point C (umbra)	Total eclipse
Point D (penumbra)	Partial eclipse
Point E (penumbra)	Annular eclipse

Discussion

Solar Eclipses Ask students how the appearance of solar eclipses would be affected if Earth's distance from the Sun were increased. Total solar eclipses would be more frequent because the Sun would appear to be smaller than the Moon, and the alignment required for the Moon to totally block the Sun would not have to be as precise. AL

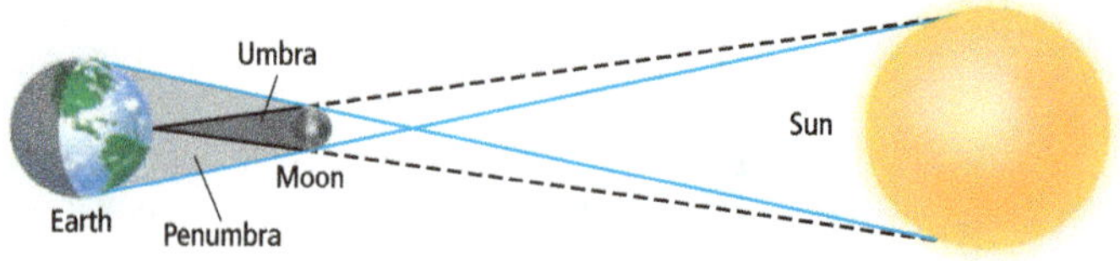

Figure 22 During a solar eclipse, the Moon passes between Earth and the Sun. Those on Earth within the darkest part of the Moon's shadow (umbra) see a total eclipse. Those within the lighter part, or penumbral shadow, see only a partial eclipse.

View an **animation of an eclipse.**

Concepts In Motion

How solar eclipses occur Each object in the solar system creates a shadow as it blocks the path of the Sun's light. This shadow is totally dark directly behind the object and has a cone shape. During a solar eclipse, the Moon casts a shadow on Earth as it passes between the Sun and Earth. This shadow consists of two regions, as illustrated in **Figure 22.** The inner, cone-shaped portion, which blocks the direct sunlight, is called the umbra, or umbral shadow. People who witness an eclipse from within the umbral shadow see a total solar eclipse. That means they see the Moon completely cover the face of the Sun. The outer portion of this shadow, where some of the Sun's light still reaches, is called the penumbra, or penumbral shadow. People in the region of the penumbral shadow see a partial solar eclipse, where only a part of the Sun's disk is blocked by the Moon. Typically, the umbral shadow is never wider than 270 km, so a total solar eclipse is visible from a very small portion of Earth, whereas a partial solar eclipse is visible from a much larger portion.

Problem-Solving LAB

Interpret Scientific Illustrations

How can you predict how a solar eclipse will look to an observer at various positions?

The diagram below shows the Moon eclipsing the Sun. The Sun will appear differently to observers located at Points A through E.

Analysis

1. **Observe** the points in relation to the position of the Moon's umbra and penumbra.

Think Critically

2. **Draw** how the solar eclipse would appear to an observer at each labeled point.
3. **Design** a data table to display your drawings.
4. **Classify** the type of solar eclipse represented in each of your drawings.

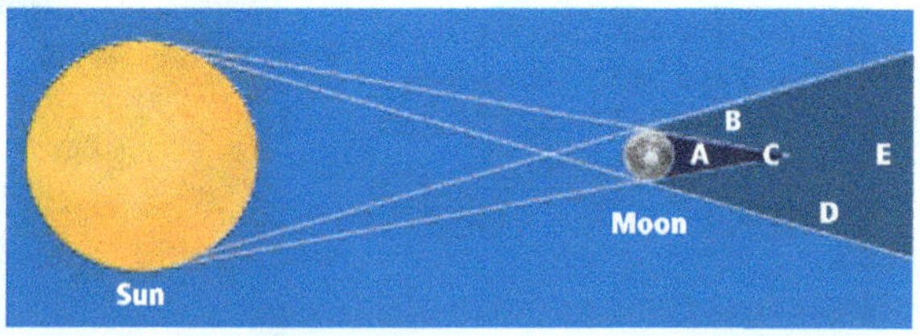

ACROSS THE CURRICULUM

Social Studies Earth's rotation is the basis for human timekeeping. Even today, the world's clocks are set according to the solar day through precise astronomical observations made by such institutions as the U.S. Naval Observatory, whose timekeeping service sets the official U.S. clocks, which are coordinated with others throughout the world. Atomic clocks, known for their precision, measure only relative time, not absolute time (i.e., not the time of day as read on a clock). Earth's rotation is slowing very gradually, requiring that a leap second be added to official clocks from time to time.

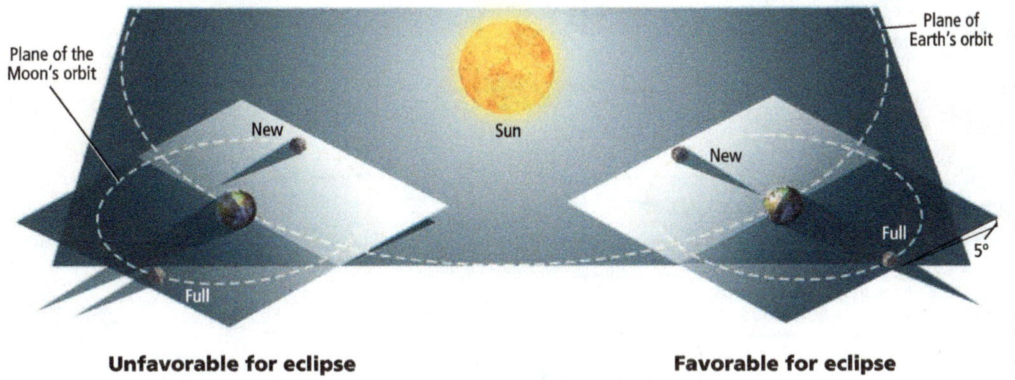

■ **Figure 23** Eclipses can take place only when Earth, the Moon, and the Sun are perfectly aligned. This can happen only when the Moon's orbital plane and the ecliptic plane intersect along the Sun-Earth line, as shown in diagram on the right. In the left diagram, this does not happen, and the Moon's shadow misses Earth.

Effects of tilted orbits You might wonder why a solar eclipse does not occur every month when the Moon passes between the Sun and Earth during the new moon phase. This does not happen because the Moon's orbit is tilted 5° relative to the ecliptic plane. Normally, the Moon passes above or below the Sun as seen from Earth, so no solar eclipse takes place. Only when the Moon crosses the ecliptic plane is it possible for the proper alignment for a solar eclipse to occur, but even that does not guarantee a solar eclipse. The plane of the Moon's orbit also rotates slowly around Earth, and a solar eclipse occurs only when the intersection of the Moon and the ecliptic plane is in a line with the Sun and Earth, as **Figure 23** illustrates.

☑ READING CHECK **Determine** why a total solar eclipse does not occur every month.

Annular eclipses Not only does the Moon move above and below the plane of Earth and the Sun, but the Moon's distance from Earth increases and decreases as the Moon moves in its elliptical orbit around Earth. The closest point in the Moon's orbit to Earth is called **perigee,** and the farthest point is called **apogee.** When the Moon is near apogee, it appears smaller from Earth, and thus will not completely block the disk of the Sun during an eclipse. This is called an annular eclipse because, as **Figure 24** shows, a ring of the Sun, called the annulus, appears around the dark Moon. Earth's orbit also has a closest point in its orbit around the Sun, called perihelion, and a farthest point, called aphelion. When Earth is nearest the Sun and the Moon is at apogee, the Moon would not block the Sun entirely. The opposite is true for Earth at aphelion and the Moon at perigee.

■ **Figure 24** An annular eclipse takes place when the Moon is too far away for its umbral shadow to reach Earth. A ring, or annulus, is left uncovered.
Predict ***Would annular eclipses occur if the Moon's orbit were a perfect circle?***

Teacher Content Support

The Moon The Moon is far more massive relative to Earth than other satellites are in relation to their planets. Thus, the Moon has a substantial gravitational effect on Earth. Calculations have shown that Earth's rotational axis would tend to shift as a result of small outside forces if not for the stabilizing effect of the Moon's gravity. If Earth's axis were not stable, climates would undergo extreme fluctuations over short or long periods of time. It has been suggested that life might not have been able to form on Earth without the stabilizing effect of the Moon. If life's presence on Earth requires the Moon, then life in the universe might be rare, because our Moon is unusual.

■ **Caption Question Fig. 24** It would depend on the Moon's distance from Earth. If the Moon were close to Earth, there would be no annular eclipses. If the Moon were far from Earth, there would be annular eclipses.

☑ READING CHECK The Moon's orbit is tilted 5° relative to the ecliptic, so it usually passes north or south of the Sun as seen from Earth.

EARTH SCIENCE JOURNAL

Eclipse Patterns Have students use resources to find out why eclipses occur in repeating cycles. The Moon's orbital plane precesses with a period of 18.6 years, causing the line of intersection between the orbital planes of the Moon and Earth to rotate with a period of 18.6 years. The result is an 18.6-year repeating pattern of solar and lunar eclipses. BL

Rubric

3 Assess

Check for Understanding

Discussion Have students summarize the role of the Moon in forming both tides and eclipses.

Reteach

Discussion To verify student understanding of tides, ask students what the tides would be like if Earth were in synchronous rotation with the Moon, that is, if the same side of Earth always faced the Moon. The Moon would hover over a fixed spot on Earth, and there would be two permanent, nonmoving bulges in the oceans on the sides of Earth facing toward and away from the Moon. The tides would not rise and fall far at any location on Earth, except for small movement caused by the influence of the Sun's gravity.

Assessment

Performance Have students use balls and a flashlight to model the positions of the Sun, Earth, and the Moon during solar and lunar eclipses, spring and neap tides, and the phases of the Moon.

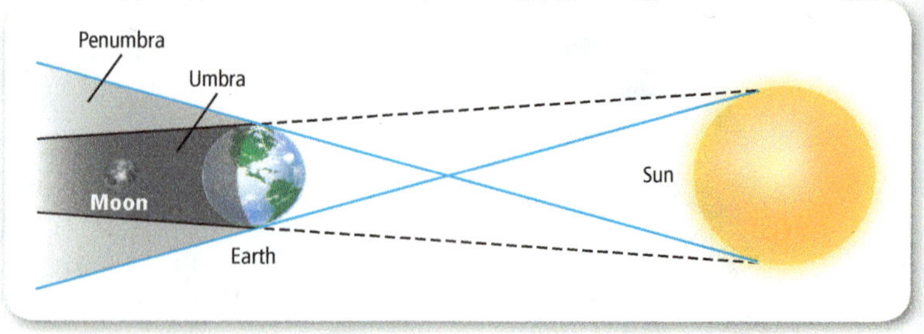

Figure 25 When the Moon is completely within Earth's umbra, a total lunar eclipse takes place, as shown in the diagram. The darkened Moon often has a reddish color, as shown in the photo, because Earth's atmosphere bends and scatters the Sun's light.

Lunar Eclipses

A **lunar eclipse** occurs when the Moon passes behind Earth in relation to the Sun, and through Earth's shadow. As illustrated in **Figure 25,** this can happen only at the time of a full moon when the Moon is on the opposite side of Earth from the Sun. The shadow of Earth has umbral and penumbral portions, just as the Moon's shadow does. A total lunar eclipse occurs when the entire Moon is within Earth's umbral shadow. This lasts for approximately two hours. During a total lunar eclipse, the Moon is faintly visible, as shown in **Figure 25,** because sunlight that has passed near Earth has been filtered and refracted by Earth's atmosphere. This light can give the eclipsed Moon a reddish color as Earth's atmosphere bends the red light into the umbra, much like a lens. Like solar eclipses, lunar eclipses do not occur every full moon because the Moon in its orbit usually passes above or below the Sun as seen from Earth.

SECTION 3 REVIEW

Section Self-Check

Section Summary

- Earth's rotation defines one day, and Earth's revolution around the Sun defines one year.
- Seasons are caused by the tilt of Earth's spin axis relative to the ecliptic plane.
- The gravitational attraction of both the Sun and the Moon causes tides.
- The Moon's phases result from our view of its lighted side as it orbits Earth.
- Solar and lunar eclipses occur when the Sun's light is blocked.

Understand Main Ideas

1. **MAINIDEA State** one proof that Earth rotates, one proof Earth rotates in 24 hours, and make one observation that proves it revolves around the Sun in one year.
2. **Compare** solar and lunar eclipses, including the positions of the Sun, Earth, and Moon.
3. **Diagram** the waxing and waning phases of the Moon.
4. **Analyze** why the Moon has a greater effect on Earth's tides than the Sun, even though the Sun is more massive.

Think Critically

5. **Relate** what you have learned about lunar phases to how Earth would appear to an observer on the Moon. Diagram the positions of the Sun, Earth, and the Moon and draw how Earth would appear in several positions to explain your answer.

MATH IN Earth Science

6. Consider what would happen if Earth's axis were tilted 45°. At what latitudes would the Sun be directly overhead on the solstices and the equinoxes?

SECTION 3 REVIEW

1. Earth's rotation can be demonstrated by a Foucault pendulum and by the Coriolis effect. The time of rotation can be observed by watching the movement of the Sun and stars across the sky. Earth's revolution can be seen by the yearly cycle of the constellations that are seen in the night sky.
2. In solar eclipses, the Moon lies between Earth and the Sun, blocking the Sun's light; in lunar eclipses, Earth is between the Sun and the Moon, and Earth's shadow falls on the Moon.
3. In waxing phases, the lighted portion of the Moon visible on Earth increases; in waning phases, the lighted portion of the Moon decreases.
4. The Moon is much closer to Earth than the Sun is.
5. Earth would pass through the same phases as the Moon with the same frequency but in opposite rhythm to the Moon.
6. 45° N and S latitudes for solstices and 0° for equinoxes.

eXpeditions!

ON SITE: LIVING IN SPACE

NASA

This astronaut is working in microgravity. Notice the footholds and handholds the astronauts use to stay in place.

An orbiting space shuttle and everything aboard it, including the astronauts, are falling continuously around Earth. The result is apparent weightlessness—they experience microgravity conditions. Performing everyday tasks, such as sleeping and exercising, is challenging in microgravity. What would it be like to float in space?

Disorientation Some astronauts experience space sickness during the first few days in microgravity. This happens because the brain is confused by the mismatched visual, sensory, and pressure messages that it receives. To help their bodies prepare for microgravity, astronauts train in special airplanes where they experience short periods of free-fall weightlessness. Once sensory systems have adjusted to microgravity, space sickness subsides.

Sleeping How would you sleep without gravity to keep you in bed? Astronauts on the orbiting space shuttle and the *International Space Station* spend about eight hours a day sleeping. Although astronauts can sleep in any orientation they prefer, they must be anchored to something—a wall, a seat, or a bed. This prevents them from floating and bumping into other things while they sleep, which might harm them as well as other astronauts and equipment.

Exercising In orbit, exercising is particularly important to the overall health of astronauts. On Earth, muscles work against the force of gravity to move, maintain balance, and support our bodies. In microgravity, muscles are underused and begin to atrophy, meaning they lose tone and mass.

Supporting the weight of the body on Earth is one of the functions of bones. Scientists know that gravity is important in the process of bone maintenance and formation. In microgravity, bone formation is disrupted and bones lose important minerals. Without proper amounts of these minerals, bones become weaker and the risk of fracture increases. Astronauts exercise each day while in space while strapped to exercise equipment.

(bkgd) fStop/Getty Images

WRITING IN ▶ Earth Science

Interview Suppose you are a newspaper reporter and you will interview an astronaut who has returned from space. Write at least five interview questions about how microgravity affects the human body and the completion of everyday tasks. Include questions about the astronaut's personal experiences. To learn more about space travel, visit NASA's Web site.

WebQuest

WRITING IN ▶ Earth Science

Rubric

Interview Possible questions include: What were some of the effects of microgravity on your body? Did you enjoy feeling weightless? How did you eat and drink? Did you sleep as well in space as you do at home? Did you suffer from space sickness? If so, what helped to alleviate it?

WebQuest

eXpeditions!

Purpose

Students will learn about some of the effects of microgravity on body systems and how astronauts adjust to these changes.

Teacher Content Support

Effects of Microgravity The semicircular canals in the inner ear provide the brain with messages about the body's equilibrium. When fluid in the semicircular canals is pulled by the force of gravity, such as when the head tilts forward or backward, nerves send messages to the brain about changes in the position of the head. In conditions of microgravity, the established sensory system cannot function normally, resulting in disorientation and possibly motion sickness for astronauts. All sensory information about the position of the body and head now only reaches the brain though visual input. Astronauts in space report that when asked to close their eyes and then determine which way is "up," they cannot.

Teaching Strategy

Have students discuss how living in microgravity conditions may affect the completion of everyday tasks, such as eating with utensils, drinking liquids, or writing a letter using pen and paper.

GeoLAB

iLab Station Rubric

Preparation

Time Allotment 30 min

Process Skills sequence, observe and infer, recognize cause and effect, recognize spatial relationships

Safety Precaution Approve lab safety forms before work begins.

Procedure

- Briefly review with students the principle of crosscutting relationships.
- Team students who are visually impaired with students who can easily discern fine details.
- Have students work on the first photo, check their work, and then continue.
- **Troubleshooting** Have students review the formation of the Moon's surface features, especially the maria.

Analyze and Conclude

1. Answers will vary, but may include problems with identifying specific features, or with determining which impact craters formed first.
2. Craters alone or craters that formed maria are the oldest. Craters in maria or rilles are the youngest.
3. Exact age differences cannot be determined from the principle of crosscutting relationships; only relative ages can be. From the photos, there is no way to tell whether the age difference is 10 minutes or 10 million years.
4. crosscutting relationships

GeoLAB

iLab Station

Mapping: Determine Relative Ages of Lunar Features

Background: **Recall that an intrusion or a fault can cut across an older geologic feature. This principle of crosscutting relationships is also used to determine the relative ages of surface features on the Moon. By observing which features cut across others, you can infer which features are older and which are younger.**

Question: *How can you use images of the Moon to interpret relative ages of lunar features?*

Materials

paper
metric ruler

Procedure

1. Read and complete the lab safety form.
2. Review the information about the history of the Moon and the lunar surface starting on page 772.
3. Observe Photo 1 and identify the older of the craters in the crater Pairs A-D and C-B using the principle of crosscutting relationships.
4. Observe Photo 2. Identify and list the features in order of their relative ages.
5. Observe Photo 3. Identify the mare, rille, and craters. Then list the features in order of their relative ages.
6. Observe Photo 4. Identify the features using your knowledge of crosscutting relationships and lunar history. Then list the features in order of their relative ages.

Analyze and Conclude

1. **Summarize** the problems you had in identifying and choosing the ages of the features.
2. **Select** Based on information from all the photos, what features are usually the oldest? The youngest?
3. **Explain** whether scientists could use this process to determine the exact age difference between two overlapping craters. Why or why not?
4. **Identify** the relative-age dating that scientists use to analyze craters on Earth.
5. **Evaluate** If the small crater in Photo 2, labeled *A*, is 44 km across, what is the scale for that photo? At that scale, what is the size of the large crater labeled *F*?
6. **Judge** Which would be older, a crater that had rays crossing it, or the crater that caused the rays? Explain.
7. **Estimate** If the crater labeled *A* in Photo 1 is 17 km across, how long is the chain of craters in the photo?
8. **Infer** What might have caused the chain of craters in Photo 1?

WRITING IN ▶ Earth Science

Guidebook Using what you learned in this lab, prepare a guidebook that contains instructions for identifying and determining relative ages of lunar features. For more information on lunar features, visit NASA's Web site.

5. After multiple measurements to average out the size of Crater A because of its imperfect shape, the scale is approximately 1 cm = 20 km. Crater F is about 140 km in diameter.
6. A crater that has rays crossing it would be older than the crater from which the rays originate.
7. The scale would be approximately 1 cm = 17 km, so the chain of craters is about 107 km.
8. The chain of craters was probably caused by a fragmented impactor that arrived as a series of pieces instead of one big chunk.

WRITING IN ▶ Earth Science

Rubric

Guidebook Have students exchange guidebooks and critique each other's work.

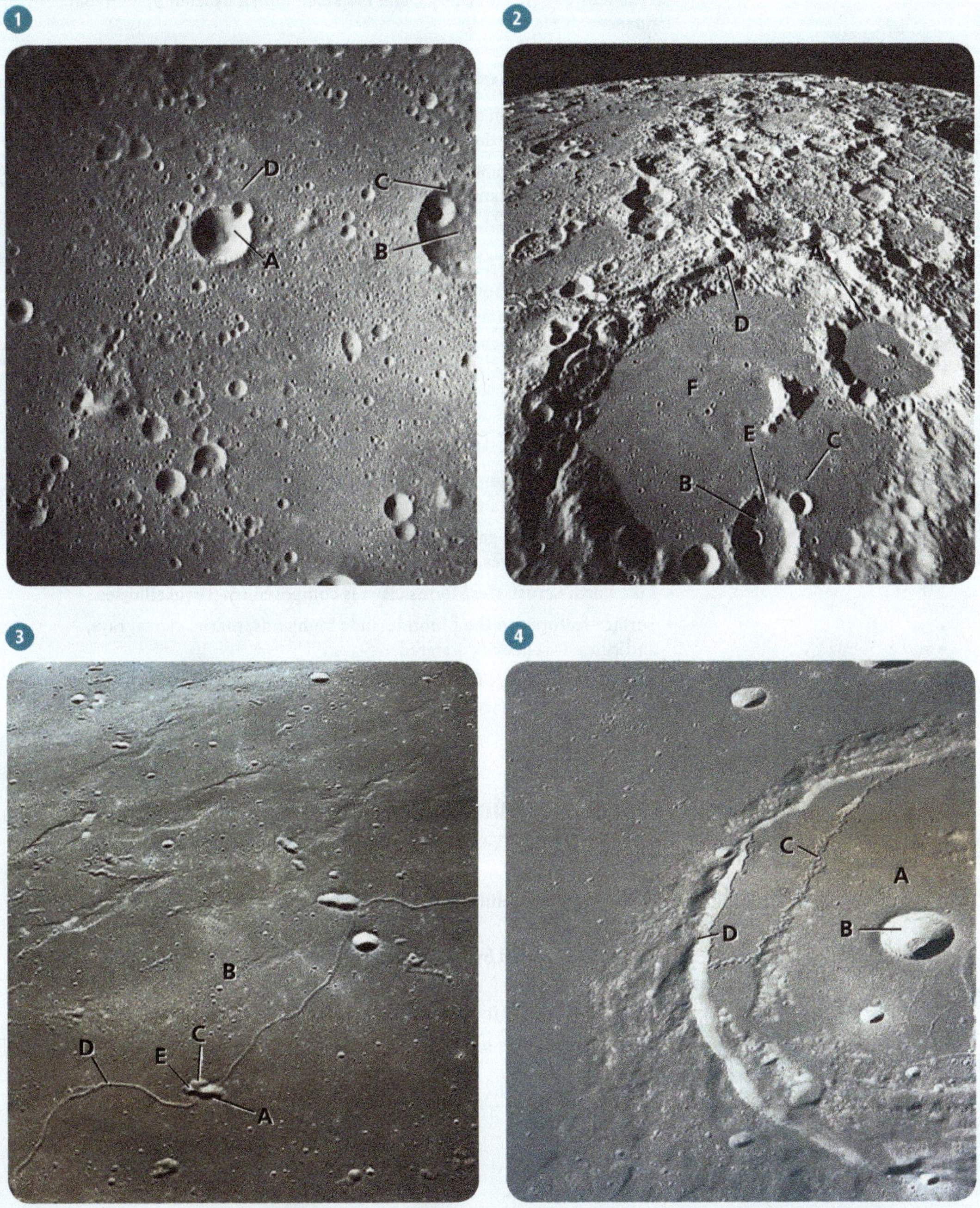

1
D
C
A
B
2
A
D
F
E
C
B
3
B
C
D
E
A
4
C
A
B
D

MAINIDEAS Summary statements can be used by students to review the major concepts of the chapter.

Students can review with these online resources.

Vocabulary Practice

Vocabulary eGames
Vocabulary eFlashcards
Vocabulary PuzzleMaker

Section Self-Check

Chapter Self-Check

Online Test Practice

Use the ExamView® *Assessment Suite* CD-ROM to:

- create multiple versions of tests
- create modified tests with one mouse click
- edit existing questions and add your own questions
- build tests aligned with state standards using built-in state curriculum tags
- change English tests to Spanish with one mouse click
- track students' progress using the Teacher Management System

CHAPTER 27 STUDY GUIDE

BIGIDEA **The Sun, Earth, and the Moon form a dynamic system that influences all life on Earth.**

Vocabulary Practice

SECTION 1 Tools of Astronomy

VOCABULARY
- electromagnetic spectrum
- refracting telescope
- reflecting telescope
- interferometry

MAINIDEA Radiation emitted or reflected by distant objects allows scientists to study the universe.

- Telescopes collect and focus electromagnetic radiation emitted or reflected from distant objects.
- Electromagnetic radiation is classified by wavelength and frequency.
- The two main types of optical telescopes are refractors and reflectors.
- Space-based astronomy includes the study of orbiting telescopes, satellites, and probes.
- Technology originally developed to explore space is now used by people on Earth.

SECTION 2 The Moon

VOCABULARY
- albedo
- highland
- mare
- impact crater
- ejecta
- ray
- rille
- regolith

MAINIDEA The Moon, Earth's nearest neighbor in space, is unique among the moons in our solar system.

- Astronomers have gathered information about the Moon using telescopes, space probes, and astronaut exploration.
- Like Earth's crust, the Moon's crust is composed mostly of silicates.
- Surface features on the Moon include highlands, maria, ejecta, rays, and rilles. It is heavily cratered.
- The Moon probably formed about 4.5 bya in a collision between Earth and a Mars-size object.

SECTION 3 The Sun-Earth-Moon System

VOCABULARY
- ecliptic plane
- solstice
- equinox
- synchronous rotation
- solar eclipse
- perigee
- apogee
- lunar eclipse

MAINIDEA Motions of the Sun-Earth-Moon system define Earth's day, month, and year.

- Earth's rotation defines one day, and Earth's revolution around the Sun defines one year.
- Seasons are caused by the tilt of Earth's spin axis relative to the ecliptic plane.
- The gravitational attraction of both the Sun and the Moon causes tides.
- The Moon's phases result from our view of its lighted side as it orbits Earth.
- Solar and lunar eclipses occur when the Sun's light is blocked.

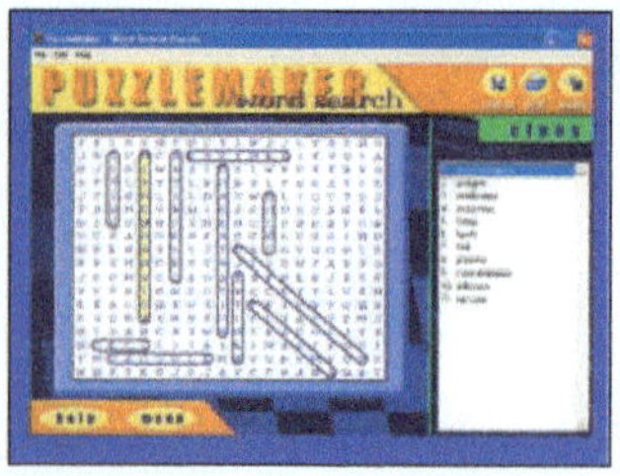

Vocabulary Practice

For additional practice with vocabulary, have students access the Vocabulary PuzzleMaker.

Chapter Self-Check

VOCABULARY REVIEW

Fill in the blanks with the correct vocabulary term from the Study Guide.

1. Linking telescopes to improve the detail in the images obtained is called ________.
2. A telescope that uses curved lenses to focus visible light is called a(n) ________.
3. A(n) ________ can take place only when the Moon is in the new moon phase.

Each of the following sentences is false. Make each sentence true by replacing the italicized words with vocabulary terms from the Study Guide.

4. The Moon's *perigee* is the amount of sunlight that its surface reflects.
5. The far side of the Moon has many more *maria* than the near side.
6. *Interferometry* explains why the same side of the Moon is always visible from Earth.

Match each description below with the correct vocabulary term from the Study Guide.

7. a device that uses a mirror to collect light from distant objects
8. the point in the Moon's orbit when it is farthest from Earth
9. loose, ground-up rock, such as the layer covering much of the surface of the Moon

UNDERSTAND KEY CONCEPTS

10. Which is the highest point in the sky that the Sun reaches on a given day?
    A. ecliptic
    B. solstice
    C. tropic
    D. zenith

Use the diagram below to answer Questions 11 and 12.

11. In the diagram above, which season is it in the northern hemisphere?
    A. autumn
    B. spring
    C. summer
    D. winter

12. When Earth is in the position shown in the diagram, at which place on Earth is the Sun most likely to be directly overhead at midday?
    A. Arctic Circle
    B. equator
    C. Tropic of Cancer
    D. Tropic of Capricorn

13. Which type of electromagnetic radiation has a longer wavelength than visible light?
    A. gamma ray
    B. X ray
    C. radio wave
    D. ultraviolet ray

14. Which geographic features on the Moon are most likely to be the oldest?
    A. craters
    B. highlands
    C. maria
    D. regolith

15. What is the mineral composition of most moon rocks?
    A. basalts containing water
    B. feldspar with high iron content
    C. sedimentary breccias
    D. silicates

CHAPTER 27 ASSESSMENT

VOCABULARY REVIEW

1. interferometry
2. refracting telescope
3. solar eclipse
4. albedo
5. highlands
6. Synchronous rotation
7. reflecting telescope
8. apogee
9. regolith

UNDERSTAND KEY CONCEPTS

10. D
11. C
12. C
13. C
14. B
15. D

16. B
17. A

CONSTRUCTED RESPONSE

18. The telescopes orbit above Earth's atmosphere, which can distort images.
19. the far side because it is heavily cratered and has no maria
20. Rays are straight lines made of light-colored ejecta that radiate out from craters on the Moon. Rilles are valley-like lines around craters in maria.
21. It is larger in radius and more massive than most other moons. In addition, it orbits farther from its parent planet than do most other moons.
22. Human missions can adapt to changing conditions and provide firsthand observations. Robotic missions are less costly, but do not have the flexibility of human missions.

THINK CRITICALLY

23. In the diagrams, the Moon should not be aligned with the Sun and Earth, or at a right angle with the Sun-Earth line.
24. There was no water on the lunar surface to allow sedimentation to occur.
25. Maria are much younger than highlands. Highlands are the original lunar material that was impacted continuously, producing craters and mountains. Maria formed as molten material from the Moon's interior broke through the crust and filled craters.
26. There would be no seasons.
27. On the summer solstice, the Sun would be at a 90-degree angle viewed from a latitude of 23.5° north. Thus, diagrams should show the Sun at an angle of 73.5 degrees above the horizon viewed from a latitude of 40° north (40° north–23°5 = 16.5°; 90°–16.5° = 73.5°).
28. Yes, but they would be much smaller than they are now.

Use the diagram below to answer Questions 16 and 17.

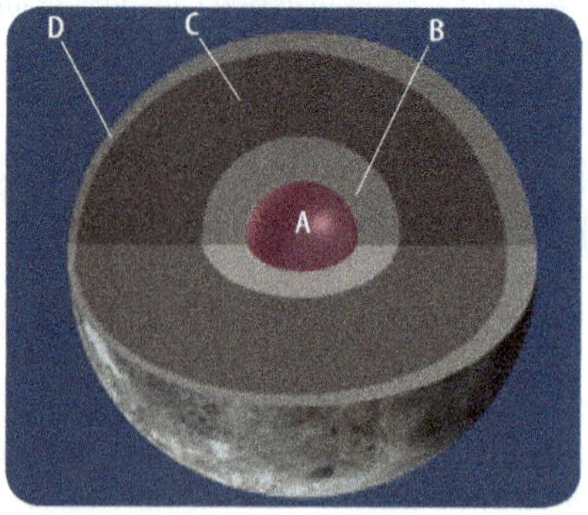

16. Which area of the Moon is partially molten?
A. core
B. lower mantle
C. upper mantle
D. crust

17. Which area of the Moon is probably solid iron?
A. core
B. lower mantle
C. upper mantle
D. crust

CONSTRUCTED RESPONSE

18. Describe the advantages of placing telescopes in space.

Use the illustration below to answer Question 19.

19. Identify What part of the lunar surface is most likely shown in this photograph?

20. Distinguish between rays and rilles, including where they are found and how they are formed on the Moon.

21. Summarize the ways in which Earth's Moon is unusual among all the moons in the solar system.

22. Assess the advantages of human missions compared with using robotic spacecraft to explore space.

THINK CRITICALLY

Use the illustration below to answer Question 23.

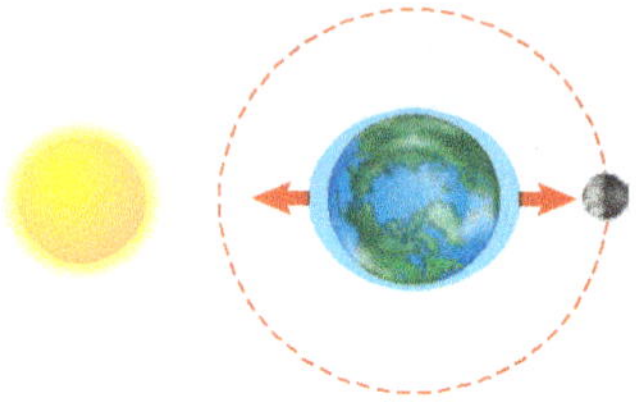

23. Draw a diagram similar to the one above to illustrate ocean tides that are not neap or spring.

24. Infer Why are lunar breccias not sedimentary like most breccias found on Earth?

25. Contrast the geological history of maria with that of the highlands.

26. Consider What would seasons be like if Earth were not tilted on its axis?

27. Draw a diagram showing the altitude of the Sun at summer solstice viewed from a position of 40° north latitude.

28. Infer Would ocean tides exist if Earth had no moon? If so, describe what they would be like.

Use the illustration below to answer Questions 29 and 30.

29. **List** the types of shadows as well as the types of eclipses that will be seen by an observer on the unlit side of Earth in each scenario.

30. **Infer** the view of the Sun from the Moon in each scenario.

31. **Appraise** Based on what you know about how maria formed, where would you expect to find the highest concentration of iron?

32. **Compare and contrast** the Moon's interior structure in **Figure 11** with Earth's interior structure.

CONCEPT MAPPING

33. Create a concept map using the following terms: *the Moon, albedo, Earth, phases, impact theory, highlands, maria, rilles, craters, rays, breccia,* and *regolith.* Refer to the *Skillbuilder Handbook* for more information.

CHALLENGE QUESTION

34. **Describe** the interrelationship between the Sun, Earth, and the Moon regarding tides and eclipses.

Chapter Self-Check

WRITING IN ▶ Earth Science

35. Imagine that you are the science officer on a scouting mission from another planet. You just observed the impact that formed Earth's Moon. Write a report describing the event.

DBQ Document–Based Questions

Data obtained from: Lang, T. et al. 2004. Cortical and trabecular bone mineral loss from the spine and hip in long-duration spaceflight. *Journal of Bone and Mineral Research* 19 (6).

Bone loss in the lower extremities and spine is a serious problem for astronauts who spend long periods in microgravity. The data below shows the percent loss of bone mineral per month from 13 crew members of the International Space Station.

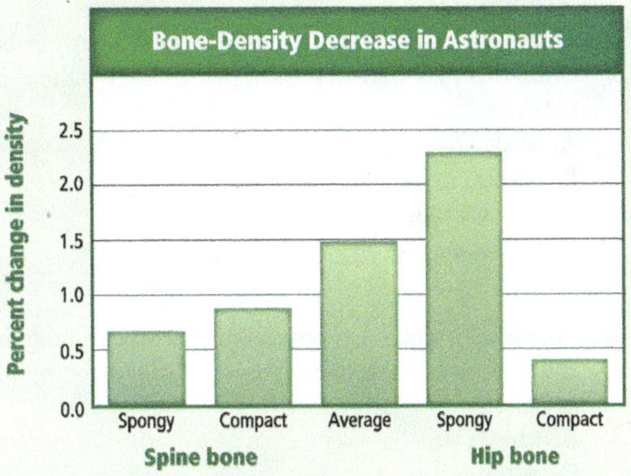

36. Evaluate which body area showed the highest overall rate of bone loss.

37. Compare bone loss of the two types of bone in the hip. Which has the highest rate of loss? By how much?

CUMULATIVE REVIEW

38. What is the source of CFCs and how do CFCs cause ozone depletion? **(Chapter 26)**

39. What are the most common minerals in granite? In basalt? **(Chapter 5)**

CHAPTER 27 ASSESSMENT

29. In the top diagram, the Moon is in Earth's umbra. An observer would see a total lunar eclipse. In the bottom diagram, the Moon is in the penumbra/umbra boundary. An observer would see a partial lunar eclipse.
30. The Sun would be in total eclipse as seen from the Moon in the top situation. In the bottom situation, the Sun would be in a total eclipse for an observer inside Earth's umbral shadow, and a partial eclipse for an observer within the penumbra.
31. in the core
32. The Moon's core is not separated into an inner and outer core. Both have cores of solid iron. Earth's core is larger. Earth's mantle has convection currents, driven by interior heat. Earth's mantle and crust are tectonically active. The Moon's interior is not tectonically active.

CONCEPT MAPPING

33. Answers will vary. Check students' maps for accuracy.

CHALLENGE QUESTION

34. Answers should mention that Earth orbits the Sun, and the Moon orbits Earth, that alignments of the three objects produce eclipses, and that the Sun and the Moon produce tides on Earth.

WRITING IN ▶ Earth Science

Rubric

35. Reports should correctly describe the formation of the Moon.

DBQ Document-Based Questions

Data obtained from: T. Lang, et al. 2004. Cortical and trabecular bone mineral loss from the spine and hip in long-duration spaceflight. *Journal of Bone and Mineral Research* 19 (6).

36. hip
37. spongy, 1.8%

CUMULATIVE REVIEW

38. CFCs are released from old refrigerators, cleaning agents, and propellants in aerosol cans. UV light breaks down CFC molecules into carbon and chlorine atoms. The chlorine atoms break down ozone molecules.
39. granite: quartz, potassium feldspar, and plagioclase feldspar; basalt: plagioclase feldspar, pyroxene, and olivine

MULTIPLE CHOICE

1. C
2. D
3. C
4. A
5. A
6. C
7. B
8. D
9. C
10. B

CUMULATIVE STANDARDIZED TEST PRACTICE

MULTIPLE CHOICE

1. Which is not considered a renewable resource?
A. brick
B. stone
C. copper
D. wood

Use the geologic cross section below to answer Questions 2 and 3.

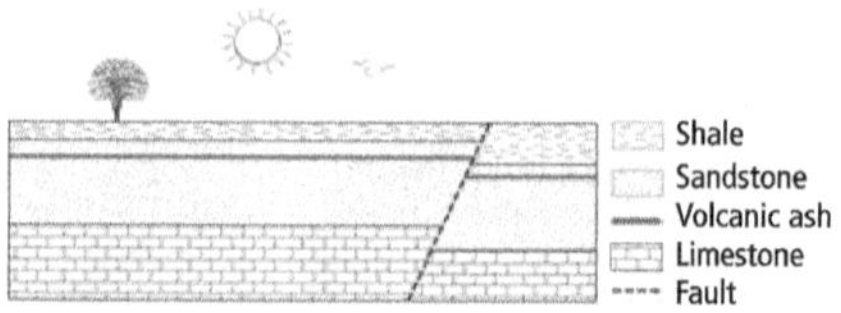

2. Assuming the rock layers shown are in the same orientation that they were deposited, which layer is the oldest?
A. shale
B. sandstone
C. volcanic ash
D. limestone

3. Which layer would be most helpful in determining the absolute age of these rocks?
A. shale
B. sandstone
C. volcanic ash
D. limestone

4. Which fossil fuel was originally known as rock oil?
A. petroleum
B. natural gas
C. coal
D. oil shale

5. In which process does the weight of a subducting plate help pull the trailing lithosphere into a subduction zone?
A. slab pull
B. ridge pull
C. slab push
D. ridge push

6. What is debris from an impact that falls back to the surface of the Moon called?
A. rilles
B. maria
C. ejecta
D. albedo

Use the illustrations below to answer Questions 7 to 9.

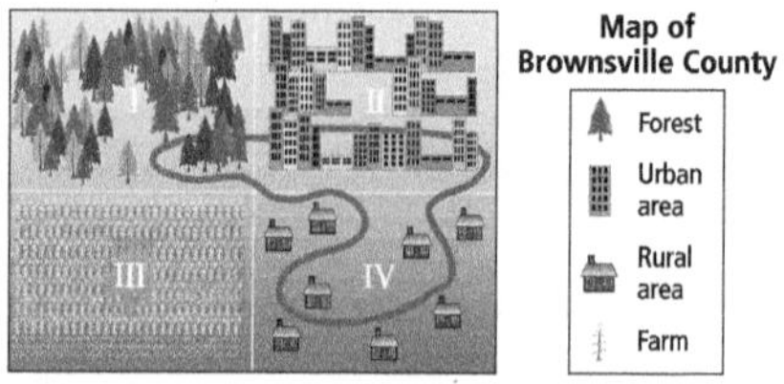

7. Which area of Brownsville is most likely to have problems with flooding during heavy rains?
A. I
B. II
C. III
D. IV

8. If Brownsville County decided to clear area I in order to expand area III, Brownsville might develop problems with topsoil erosion and pesticide pollution. What might be one way to minimize harmful effects?
A. deforestation
B. clear-cutting
C. monoculture
D. selective logging

9. What will happen if the size of Brownsville's human population reaches the carrying capacity for its environment?
A. There will be more births than deaths.
B. The death rate will increase and the birth rate will increase.
C. The population will reach equilibrium.
D. The death rate will increase and the birth rate will decrease.

10. The Mariana Islands in the Pacific Ocean were formed by volcanic action. Which is a TRUE statement?
A. There are glaciers near the Mariana Islands.
B. Tectonic plates collide near the Mariana Islands.
C. The Mariana Islands are larger than most islands.
D. The Mariana Islands are uninhabited.

Online Test Practice

SHORT ANSWER

Use the diagram below to answer Questions 11 to 13.

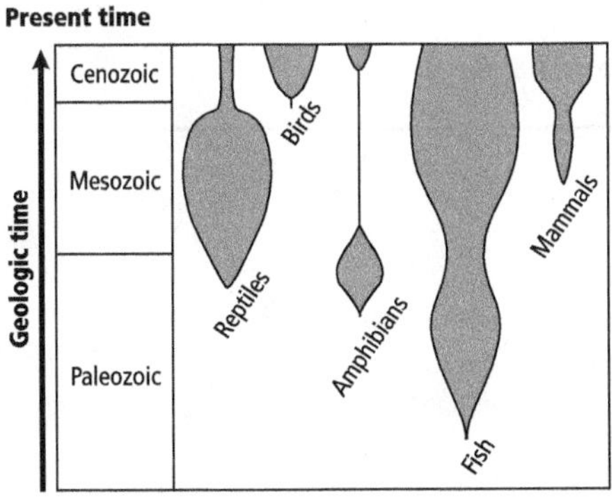

11. If a wider bar represents more species of that type of organism, describe the change in diversity of amphibians from their introduction to present time.

12. What can be inferred about the conditions on Earth for living things from the beginning of the Cenozoic Era to present time?

13. How might the idea that oceans developed before land be supported by looking at this diagram?

14. How does passive solar heating differ from active solar heating?

15. Why is improving the energy efficiency of automobiles important?

16. What two major flaws did scientists of Wegener's day cite as reasons to reject his hypothesis of continental drift?

READING FOR COMPREHENSION

Space Observatories

Why put observatories in space? Most telescopes are on the ground where you can deploy a heavier telescope and fix it more easily. The trouble is that Earth–bound telescopes must look through the Earth's atmosphere which blocks out a broad range of the electromagnetic spectrum, allowing a narrow band of visible light to reach the surface. Telescopes that explore the universe using light beyond the visible spectrum, such as those onboard the *CHANDRA X-Ray Observatory* need to be carried above the absorbing atmosphere. The Earth's atmosphere also blurs the light it lets through. The blurring is caused by varying density and continual motion of air. By orbiting above Earth's atmosphere, the *Hubble Space Telescope* can get clearer images.

Article obtained from: Astronomy picture of the day. *Hubble* Floats Free. *NASA.* May 25, 2009. (Online resource accessed October 20, 2010.).

17. What is a benefit of earthbound telescopes?
   A. They can be larger and are more easily fixed.
   B. They are able to capture the entire electromagnetic spectrum.
   C. They can use larger mirrors.
   D. They can capture visible light.

18. What can be inferred from this passage?
   A. Earth–bound telescopes have no benefits for scientific study.
   B. Using telescopes outside the Earth's atmosphere produces the clearest pictures.
   C. The *HST* needs larger mirrors.
   D. It is impossible to fix telescopes orbiting outside Earth's atmosphere.

NEED EXTRA HELP?

If You Missed Question . . .	1	2	3	4	5	6	7	8	9	10	11	12	13	14	15	16
Review Section . . .	24.1	21.2	21.3	25.1	17.4	27.2	26.2	26.2	26.1	17.3	23.1	23.3	23.1	25.2	25.3	17.1

SHORT ANSWER

11. Amphibians appeared in the late Paleozoic Era and were diverse, but during the Mesozoic Era, they were so rare it was as if they were almost extinct. Then, in the Cenozoic Era, they again began increasing in their diversity and number.

12. Sample answer: Conditions on Earth appear to have become more hospitable to living things beginning in the Cenozoic Era because all five-tracked organisms were present and some were beginning to increase in their variety.

13. Sample answer: Fish, which live in water, were the first living organisms to appear. Then amphibians, which begin in water and then move to land, were the next species to appear, supporting the idea that water appeared first and then dry land began to appear.

14. Passive solar heating is caused when an area is heated directly by the Sun's energy. Active solar heating occurs when solar energy is collected and then distributed throughout the area as heat.

15. Most of the energy from gasoline in cars is lost as heat. By decreasing the heat loss of automobiles, more gasoline could be used in running the car and less gasoline would be needed to run the automobile, thus conserving gasoline, a nonrenewable fossil fuel.

16. First, they could not explain a force that would be strong enough to move such large masses of rock over great distances. Second, they knew that Earth's mantle was solid, and they could not explain how the continents could move through something solid.

READING FOR COMPREHENSION

17. A
18. B

CHAPTER 28 Our Solar System

BIGIDEA **Using the laws of motion and gravitation, astronomers can understand the orbits and the properties of the planets and other objects in the solar system.**

ESSENTIAL QUESTIONS	NATIONAL STANDARDS	RESOURCES TO ASSESS MASTERY
SECTION 1 **Formation of the Solar System** **1.** How did the solar system form? **2.** What are some of the early concepts of the structure of the solar system? **3.** How has our current knowledge of the solar system developed? **4.** What is the relationship between gravity and the motions of the objects in the solar system? 2 sessions 1 block	UCP. 1–4; A.1; B.4; D.4; G.2–3	**Entry-Level Assessment** Focus Transparency 88, p. 796 **Progress Monitoring** Caption Question, p. 802 Reading Check, pp. 797, 800 Section Review, p. 803
SECTION 2 **The Inner Planets** **1.** How are the characteristics of the inner planets similar? **2.** What are some of the space probes used to explore the solar system? **3.** How are the terrestrial planets different from each other? 1 session 0.5 block	UCP. 1–4; B.2, B.4; D.3; G.2–3	**Entry-Level Assessment** Focus Transparency 89, p. 804 **Progress Monitoring** Caption Question, pp. 804, 806, 810 Reading Check, p. 805 Section Review, p. 810
SECTION 3 **The Outer Planets** **1.** What are the similarities among and differences between the gas giant planets? **2.** What are the major moons? **3.** How do moons and rings form? **4.** How does the composition of the gas planets and the composition of the Sun compare? 1 session 0.5 block	UCP.1–4; A.1–2; B.2, B.4; D.4; G.1, G.3	**Entry-Level Assessment** Focus Transparency 90, p. 811 **Progress Monitoring** Caption Question, pp. 813, 814 Reading Check, p. 812 Section Review, p. 815
SECTION 4 **Other Solar System Objects** **1.** What are the differences between planets and dwarf planets? **2.** What are the oldest members of the solar system? **3.** How are meteoroids, meteors, and meteorites described? **4.** What is the structure of a comet? 4 sessions 2 blocks	UCP.1–5; A.1–2; B.2, B.6; D.3–4; E.2; G.3	**Entry-Level Assessment** Focus Transparency 91, p. 816 **Progress Monitoring** Reading Check, p. 818 Section Review, p. 819 **Summative Assessment** Chapter Assessment, p. 823 Exam*View®Assessment Suite* CD-ROM

Go online!
connectED.mcgraw-hill.com

LEVELED RESOURCES	LAB MATERIALS
Science Notebook 28.1 OL **Chapter FAST FILE Resources:** MiniLab Worksheet, p. 30 OL Study Guide, p. 41 BL **Lab Resources:** Laboratory Manual, p. 217 OL **Transparencies:** Section Focus Transparency 88 EL Teaching Transparency 87 OL EL	LaunchLAB p. 794 / **20 min** internet access, paper, pencil MiniLAB p. 801 / **15 min** string, cardboard, paper, pins (2), pencil, ruler
Science Notebook 28.2 OL **Chapter FAST FILE Resources:** Study Guide, p. 43 BL **Transparencies:** Section Focus Transparency 89 EL	
Science Notebook 28.3 OL **Chapter FAST FILE Resources:** Study Guide, p. 44 BL **Lab Resources:** Laboratory Manual, p. 221 OL **Transparencies:** Section Focus Transparency 90 EL Teaching Transparency 88 OL EL	
Science Notebook 28.4 OL **Chapter FAST FILE Resources:** GeoLab Worksheet, p. 31 OL Study Guide, p. 46 BL **Transparencies:** Section Focus Transparency 91 EL Teaching Transparency 89 OL EL	GeoLAB p. 821 / **45 min** calculator, tape measure, meterstick, marker, masking tape, common round objects in a variety of sizes

LabManager™
Customize any lab with the LabManager™ CD-ROM.

ADDITIONAL RESOURCES

Technology:
- eTeacherEdition
- eStudentEdition
- Classroom Presentation Toolkit
- Exam*View®Assessment Suite* CD-ROM
- LabManager™ CD-ROM
- What's EARTH SCIENCE Got To Do With It? DVD
- Weather Classroom Video DVD

Assessment:
- Blueprints for Success
- Chapter Test (Scaffolded)
- Earth Science Standardized Test Practice

Lab Resources:
- Exploring Environmental Problems Laboratory Manual

BL Below Level   OL On Level   AL Advanced Learners   EL English Learners   COOP LEARN Cooperative Learning

CHAPTER 28

Our Solar System

BIGIDEA Using the laws of motion and gravitation, astronomers can understand the orbits and the properties of the planets and other objects in the solar system.

SECTIONS

1 Formation of the Solar System

2 The Inner Planets

3 The Outer Planets

4 Other Solar System Objects

The bright swirls and bands visible on Jupiter's surface are clouds. Scattered among the clouds are massive storms, such as the Great Red Spot.

LaunchLAB

iLab Station

What can be learned from space missions?

All of the planets in our solar system have been explored by uncrewed space probes. You can learn about these missions and their discoveries by using a variety of resources. Both the agencies that sponsor missions and the scientists involved usually provide extensive information about the design, operation, and scientific goals of the missions. Learn about several space missions in this lab.

FOLDABLES® Study Organizer

The Planets

Make an accordion book and label it as you read. Use it to organize your notes on the planets.

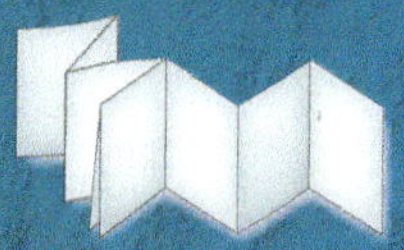

CHAPTER 28

LaunchLAB

iLab Station | Rubric

What can be learned from space missions?

Process Skills compare and contrast, make and use tables, summarize

Safety Precaution Approve lab safety forms before work begins.

Teaching Strategies

- Allow students to work in small groups for this activity.
- Students can obtain information about missions on NASA's Web site.
- Students might also be able to find information about probes in books at the library.

Procedure

1. Have students read and complete the lab safety form and follow the procedure below.
2. Research missions to four different planets.
3. Draw a table listing some of the key aspects of each mission. Include the type of mission (flyby, lander, or orbiter), the scientific goals, the launch date, and the date of arrival at the planet.

Analysis

1. **Summarize in a table** what scientists learned from each mission or what they hope to learn. Students' answers will vary. An example would be *Mars Phoenix* (lander), launched August 4, 2007. It arrived May 25, 2008, and was operated by NASA, JPL Lockheed Martin and the University of Arizona. It was designed to study the history of water and habitability potential in the ice-rich soil of Mars's arctic region.
2. **Determine** which missions are still in progress, which ones have gone beyond their mission life, and which ones have been completed. Answers will depend on information students have in their tables about various missions.
3. **Suggest** other missions that could be conducted in the future. Students' answers will vary. Students might state that they would like to see more missions to planets or missions to moons.

Assessment

Performance Have students prepare their tables listing the key points of each mission, and then compare what they find for different missions. Allow time for students to present their information to the class orally. After all presentations are complete, make a time line incorporating details about all the probes researched.

LabManager™
Customize any lab with the LabManager™ CD-ROM.

Introduce the BIGIDEA

View the Planets Have students research a current sky map to identify which planets should be visible at this time of the year. The planets that are visible, along with the Moon if it is visible above Earth's horizon, delineate the plane of our solar system's disk. This activity will illustrate the system of planets that students will study in this chapter.

Teacher Content Support

Appearance of Jupiter Ask students if they have ever seen Jupiter in the night sky. Point out that Jupiter appears bright, especially when near opposition (i.e., in the opposite direction from the Sun), because it is relatively close to Earth and it is the largest planet in our solar system. Have students describe how the appearance of Jupiter changes from the photograph taken with a low-power, Earth-based telescope to the close-up image obtained by a camera aboard the *Voyager* spacecraft. Zooming in from a clear sky to a close-up view shows features such as the swirling storm of Jupiter's Great Red Spot with ever-increasing resolution.

SECTION 1

1 Focus

Focus Transparency

Before presenting the lesson, project **Section Focus Transparency 88** and have students answer the accompanying questions. EL

MAINIDEA

Collapse of Interstellar Clouds Sprinkle some sand on an overhead projector. As students watch, gradually gather the sand into dense clusters. Point out that the grains of sand represent a cloud of gas and dust in space. Ask students what force causes the grains to move together. the push from your hand Next, have students read the Main Idea. Ask them which force would cause the gas and dust of an interstellar cloud to move together, or collapse. gravity

2 Teach

Use Science Terms

Interstellar Have students define *interstellar.* the space between stars Students should realize that the prefix *inter–* means *between.* Help them relate this meaning to familiar terms, such as *Internet* and *intercollegiate sports.*

Discussion

Appearance of Interstellar Clouds After students read about interstellar clouds, ask them to summarize three effects on the appearance of interstellar clouds. Dust can make interstellar clouds look dark by blocking starlight. Starlight reflecting off dust can illuminate the clouds. Stars can heat clouds, making them glow. BL OL

SECTION 1

Formation of the Solar System

MAINIDEA The solar system formed from the collapse of an interstellar cloud.

Essential Questions

- How did the solar system form?
- What are some of the early concepts of the structure of the solar system?
- How has our current knowledge of the solar system developed?
- What is the relationship between gravity and the motions of the objects in the solar system?

Review Vocabulary

focus: one of two fixed points used to define an ellipse

New Vocabulary

planetesimal
retrograde motion
ellipse
astronomical unit
eccentricity

EARTH SCIENCE 4 YOU **If you have ever made a snowman by rolling a snowball over the ground, you have demonstrated how planets formed from tiny grains of matter.**

Formation Theory

Theories of the origin of the solar system rely on direct observations and data from probes. Scientific theories must explain observed facts, such as the shape of the solar system, differences among the planets, and the nature of the oldest planetary surfaces—asteroids, meteorites, and comets.

A Collapsing Interstellar Cloud

Stars and planets form from interstellar clouds, which exist in space between the stars. These clouds consist mostly of hydrogen and helium gas with small amounts of other elements and dust. Dust makes interstellar clouds look dark because it blocks the light from stars within or behind the clouds. Often, starlight reflects off of the dust and partially illuminates the clouds. Also, stars can heat clouds, making them glow on their own. This is why interstellar clouds often appear as blotches of light and dark, as shown in **Figure 1.** This interstellar dust can be thought of as a kind of smog that contains elements formed in older stars, which expelled their matter long ago.

At first, the density of interstellar gas is low—much lower than the best vacuums created in laboratories. However, gravity slowly draws matter together until it is concentrated enough to form a star and possibly planets. Astronomers think that the solar system began this way. They have also observed planets around other stars, and hope that studying such planet systems will provide clues to how our solar system formed.

Figure 1 Stars form in collapsing interstellar clouds, such as in the Eagle nebula, pictured here.

DIFFERENTIATED INSTRUCTION

English Learners Use the headings throughout this chapter to make a list of basic key words all students should know to understand the chapter content, such as *orbit, gas, inner, outer, rotation, giant,* and *atmosphere.* Choose the words based on the level of English learners in your class. Place cards, each containing one of the words, on walls around the room. Have students draw pictures representing each word. During class discussions, point to each word as needed to help students visualize the concepts.

Figure 2 The interstellar cloud that formed our solar system collapsed into a rotating disk of dust and gas. When concentrated matter in the center acquired enough mass, the Sun formed in the center and the remaining matter gradually condensed, forming the planets.

Collapse accelerates At first, the collapse of an interstellar cloud is slow, but it gradually accelerates and the cloud becomes much denser at its center. If rotating, the cloud spins faster as it contracts, for the same reason that ice skaters spin faster as they pull their arms close to their bodies—centripetal force. As the collapsing cloud spins, the rotation slows the collapse in the equatorial plane, and the cloud becomes flattened. Eventually, the cloud becomes a rotating disk with a dense concentration of matter at the center, as shown in **Figure 2.**

READING CHECK **Explain** why the rotating disk spins faster as it contracts.

Matter condenses Astronomers think our solar system began in this manner. The Sun formed when the dense concentration of gas and dust at the center of a rotating disk reached a temperature and pressure high enough to fuse hydrogen into helium. The rotating disk surrounding the young Sun became our solar system. Within this disk, the temperature varied greatly with location; the area closest to the dense center was still warm, while the outer edge of the disk was cold. This temperature gradient resulted in different elements and compounds condensing, depending on their distance from the Sun. This also affected the distribution of elements in the forming planets. The inner planets are richer in the higher melting point elements and the outer planets are composed mostly of the more volatile elements. That is why the outer planets and their moons consist mostly of gases and ices. Eventually, the condensation of materials into liquid and solid forms slowed.

VOCABULARY

ACADEMIC VOCABULARY

Collapse
to fall down, give way, or cave in
The hot-air balloon collapsed when the fabric was torn.

Tie to Previous Knowledge

Interstellar Clouds Students have previously read that the density of Earth's atmosphere at sea level is 1.2 kg/m^3. This is roughly 2×10^{19} molecules per cubic centimeter, usually written as 2×10^{19}/cm^3. Explain that the maximum densities ever observed in interstellar clouds, the ones that look black on astronomical photographs, are never higher than about 10^6/cm^3, a trillion times less dense than Earth's atmosphere. The densest interstellar clouds, the ones that form planets and stars, have lower densities than the best vacuums achieved by scientists on Earth. The reason an interstellar cloud looks so dark is that there are tiny, solid particles, called interstellar dust grains, mixed with the interstellar gas. The dust grains absorb and scatter starlight efficiently, preventing the light of background stars from shining through the cloud.

READING CHECK As the disk spins, it gradually accelerates, and the density of its center increases. Just as ice skaters spin faster by drawing in their arms, as the density of the center increases, the disk spins faster.

Teacher Content Support

Age of Interstellar Dust
Astronomers establish the ages of interplanetary dust particles using a process called radiometric dating. Certain elements decay, or break down, with time. By comparing the amount of decay product with the original element, scientists can infer the age of the object.

ACROSS THE CURRICULUM

Physics For an object in a circular orbit, angular momentum is the product of mass, rotational speed, and orbital radius. Because angular momentum is conserved, if one of these three parameters changes, one or both of the others must also change. An example is the increase in rotational speed when a figure skater pulls his or her arms in closer; the decrease in body diameter is compensated by an increase in rotational speed (while the mass remains constant). In the solar nebula, the decrease in diameter as the cloud contracted was similarly compensated by an increase in rotational speed. Because of conservation of angular momentum, the young Sun that formed at the center of the nebula was born with a rotational speed much higher than the Sun's current rotational speed.

Discussion

Planetary Attraction Studying the formation of planetesimals highlights the mutual gravitational attraction among all planets. Ask students whether they think this attraction can negatively affect Earth. Point out that, although the planets influence each other to some extent, the effect is not as much as some people think. Occasionally, most or all of the planets line up in approximately the same direction from the Sun. When this has happened in the past, some people have made dire predictions that the combined gravitational force of the planets would cause tidal disruption of the Sun, and this would in turn have catastrophic effects on Earth. Others have expected some sort of celestial epiphany on the occasion, and have dubbed the event a *harmonic convergence.* In each case, nothing unusual has happened. OL

CAREERS IN EARTH SCIENCE

WebQuest

Explore the **planets with an interactive table.** Concepts In Motion

Table 1 Physical Data of the Planets

Planet	Diameter (km)	Relative Mass (Earth = 1)	Average Density (kg/m³)	Atmosphere	Distance from the Sun (AU)	Moons
Mercury	4,880	0.055	5430	none	0.39	0
Venus	12,104	0.815	5240	CO_2, N_2	0.72	0
Earth	12,742	1.00	5520	N_2, O_2, H_2O	1.00	1
Mars	6,779	0.11	3930	CO_2, N_2, Ar	1.52	2
Jupiter	139,822	317.83	1330	H_2, He	5.20	63
Saturn	116,464	95.16	690	H_2, He	9.54	62
Uranus	50,724	14.50	1270	H_2, He, CH_4	19.20	27
Neptune	49,244	17.10	1640	H_2, He, CH_4	30.05	13

Planetesimals

FOLDABLES® Incorporate information from this section into your Foldable.

Next, the tiny grains of condensed material started to accumulate and merge, forming larger particles. These particles grew as grains collided and stuck together and as gas particles collected on their surfaces. Eventually, colliding particles in the early solar system merged to form **planetesimals**—objects ranging from one kilometer to hundreds of kilometers in diameter. Growth continued as planetesimals collided and merged. Sometimes, collisions destroyed planetesimals, but the overall result was a smaller number of larger bodies—the planets. Some of their properties are given in **Table 1.**

Gas giants form The first large planet to develop was Jupiter. Jupiter increased in size through the merging of icy planetesimals that contained mostly lighter elements. It grew larger as its gravity attracted additional gas, dust, and planetesimals. Saturn and the other gas giants formed similarly, but they could not become as large because Jupiter had collected so much of the available material. As each gas giant attracted material from its surroundings, a disk formed in its equatorial plane, much like the disk of the early solar system. In this disk, matter clumped together to form rings and satellites.

Terrestrial planets form Planets also formed by the merging of planetesimals in the inner part of the main disk, near the young Sun. These were composed primarily of elements that resist vaporization, so the inner planets are rocky and dense, in contrast to the gaseous outer planets. Also, scientists think that solar wind swept away much of the gas in the area of the inner planets and prevented them from acquiring much of this material from their surroundings.

CAREERS IN EARTH SCIENCE

Planetologist A planetologist applies the theories and methods of sciences, such as physics, chemistry, and geology, as well as mathematics, to study the origin, composition, and distribution of matter in planetary systems.

WebQuest

FOLDABLES® Rubric

EARTH SCIENCE JOURNAL

Extraterrestrial Life Have students think about the properties of the planets and other solar system objects. Then have them write Earth science journal entries describing where they believe life, even in its most basic form, might someday be discovered anywhere in the solar system other than on Earth. Encourage students to predict whether this will ever happen. OL

Debris Material that remained after the formation of the planets and satellites is called debris. Eventually, the amount of interplanetary debris diminished as it crashed into planets or was diverted out of the solar system. Some debris that was not ejected from the solar system became icy objects known as comets. Other debris formed rocky bodies known as asteroids. Most asteroids are found in the area between Jupiter and Mars known as the asteroid belt, shown in **Figure 3.** They remain there because Jupiter's gravitational force prevented them from merging to form a planet.

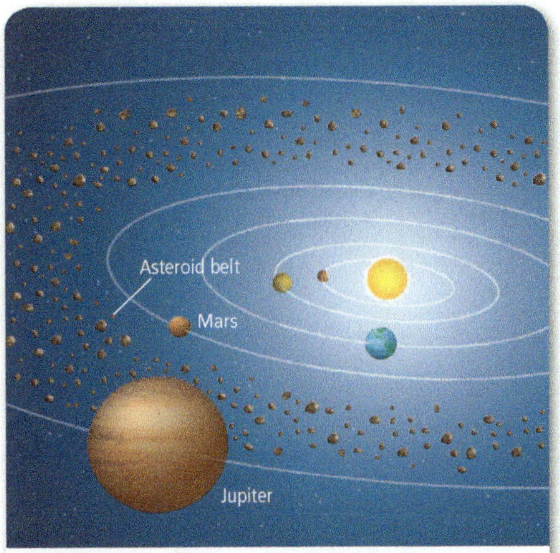

■ **Figure 3** Hundreds of thousands of asteroids have been detected in the asteroid belt, which lies between Mars and Jupiter.

Modeling the Solar System

Ancient astronomers assumed that the Sun, planets, and stars orbited a stationary Earth in an Earth-centered model of the solar system. They thought this explained the most obvious daily motion of the stars and planets rising in the east and setting in the west. But as you have learned previously, this does not happen because these bodies orbit Earth, but rather that Earth spins on its axis.

This geocentric (jee oh SEN trihk), or Earth-centered, model could not readily explain some other aspects of planetary motion. For example, the planets might appear farther to the east one evening, against the background of the stars, than they had the previous night. Sometimes a planet seems to reverse direction and move back to the west. The apparent backward movement of a planet is called **retrograde motion.** For example, Mars normally appears to move from west to east from night to night. However, approximately every two years, Mars' position in the night sky seems to change and move from east to west instead. The retrograde motion of Mars is shown in the diagram in **Figure 4.** The search for a simple explanation of retrograde motion motivated early astronomers to keep searching for a better explanation for the design of the solar system. Today, we know that Mars' retrograde motion is due to the fact that Earth "catches up" to Mars and overtakes it as both planets orbit the Sun. Earth orbits the Sun faster than Mars. For every full orbit Mars completes, Earth completes two. As a result, every 26 months Mars exhibits its retrograde motion as Earth passes it in orbit.

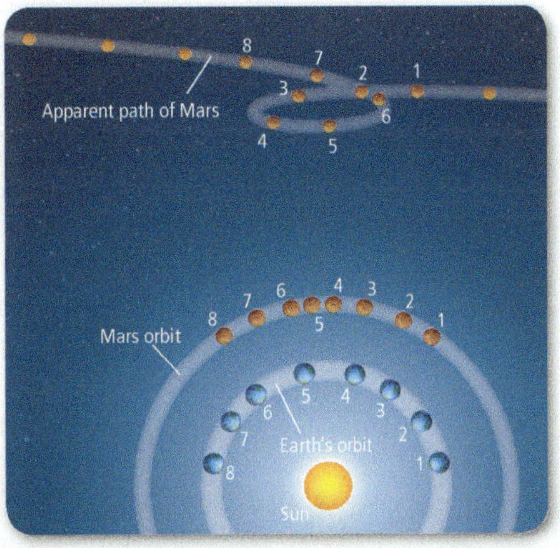

■ **Figure 4** The diagram shows the apparent retrograde motion of Mars and how the changing angles of view from Earth create this effect.

Project

Asteroids Have students find information about the ten largest asteroids, including their estimated sizes and masses, orbital periods, and semimajor axes. Ask students to compare these asteroids to planets such as Earth and to satellites such as the Moon. Also have students add the masses of the ten largest asteroids and compare the result with the mass of Earth. Students should find even the largest asteroids are smaller than the Moon, and the total mass of the top ten asteroids is tiny compared with the mass of Earth. OL

Teacher Content Support

Other Asteroids Asteroids are not confined to the asteroid belt. Some asteroids are in Jupiter's orbit, and others, called Near Earth asteroids, or NEAs, have orbital periods that cross the orbit of Earth. Scientists estimate several of these asteroids will collide with Earth every million years. Asteroids that are larger than 100 m that can come closer to Earth than 0.05 AU are called Potentially Hazardous Asteroids (PHAs).

Discussion

Motion of Venus Have students discuss whether it is possible for Venus to undergo retrograde motion as seen from Earth. Yes; retrograde motion occurs whenever Earth passes or is passed by another planet. The retrograde motion of Venus occurs when it is near the Sun, so the motion is difficult to see. AL

ACROSS THE CURRICULUM

Art As students read the lesson, encourage them to think about the different features of the solar system that scientists had to first understand in order to describe planetary orbits. Have each student choose one feature, such as retrograde motion, eccentricity, center of mass, or any of Kepler's laws, and create a display that explains the feature. Displays could be a three-dimensional arrangement, a painting, or a poster. AL

Interpret the Illustration

Elliptical Orbits Lead students in a discussion of the labeled parts of **Figure 5.** Ask them to identify the part of the diagram that shows the average distance of the planet from the Sun. the semimajor axis Ask how the major axis and semimajor axis are related. The length of the semimajor axis is half the length of the major axis. Ask how the diagram would change if the planet's orbit were circular. The circle would have only one focus, at the center, and the Sun would be at the center.

Teacher Content Support

Orbital Periods Each planet has its own sidereal period, which is the time it takes a planet to return to the starting place in its orbit. From a point of view on Earth, the orbital period of any planet is the length of time it takes for that planet to return to the same alignment relative to the Sun-Earth line, called the synodic period. In this chapter, only sidereal orbital periods are used for the planets.

☑ READING CHECK Although close to a circular orbit, the planetary orbits are ellipses, or oval shapes, with the Sun at one focus.

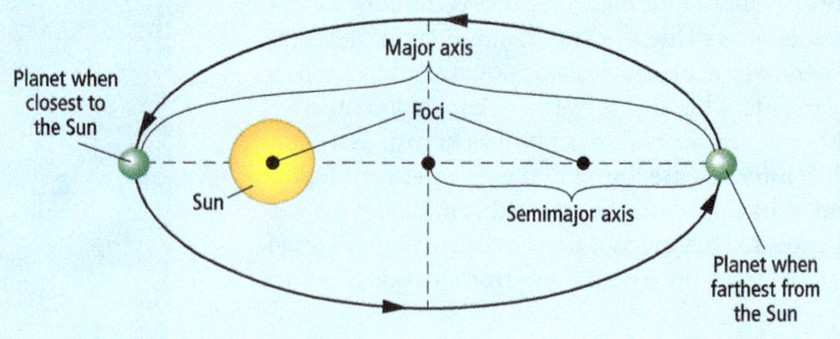

■ **Figure 5** This diagram shows the geometry of an ellipse using an exaggerated planetary orbit. The Sun lies at one of the two foci. The minor axis of the ellipse is its shorter diameter. The major axis of the ellipse is its longer diameter, which equals the distance between a planet's closest and farthest points from the Sun. The semimajor axis represents the average distance of the planet to the Sun.

Heliocentric model In 1543, Polish scientist Nicolaus Copernicus suggested that the Sun was the center of the solar system. In this Sun-centered, or heliocentric (hee lee oh SEN trihk) model, Earth and all the other planets orbit the Sun. In a heliocentric model, the increased gravity of proximity to the Sun causes the inner planets to move faster in their orbits than do the outer planets. It also provided a simple explanation for retrograde motion.

Kepler's first law Within a century, the ideas of Copernicus were confirmed by other astronomers, who found evidence that supported the heliocentric model. For example, Tycho Brahe (TEE coh BRAH), a Danish astronomer, designed and built very accurate equipment for observing the stars. From 1576–1601, before the telescope was used in astronomy, he made accurate observations of the planets' positions. Using Brahe's data, German astronomer Johannes Kepler demonstrated that each planet orbits the Sun in a shape called an ellipse, rather than a circle. This is known as Kepler's first law of planetary motion. An **ellipse** is an oval shape that is centered on two points instead of a single point, as in a circle. The two points are called the foci (singular, focus). The major axis is the line that runs through both foci at the maximum diameter of the ellipse, as illustrated in **Figure 5.**

VOCABULARY
SCIENCE USAGE V. COMMON USAGE
Law
Science usage: a general relation proved or assumed to hold between mathematical expressions

Common usage: a rule of conduct prescribed as binding and enforced by a controlling authority

☑ READING CHECK **Describe** the shape of planetary orbits.

Each planet has its own elliptical orbit, but the Sun is always at one focus. For each planet, the average distance between the Sun and the planet is its semimajor axis, which equals half the length of the major axis of its orbit, as shown in **Figure 5.** Earth's semimajor axis is of special importance because it is a unit used to measure distances within the solar system. Earth's average distance from the Sun is 1.496×10^8 km, or 1 **astronomical unit** (AU). Distance in space is often measured in AU. For example, Mars is 1.52 AU from the Sun.

DIFFERENTIATED INSTRUCTION

Advanced Learners If the semimajor axis and the eccentricity of an elliptical orbit are known, it is possible to calculate the perihelion and aphelion distances. If a is the semimajor axis of the orbit, e is the eccentricity, and P is the perihelion distance, then $P = a(1 - e)$. For example, Earth's semimajor axis is 1.000 AU and its eccentricity is 0.0167, so its perihelion distance is $P = 1.000(1 - 0.0167) = 0.9833$ AU. The formula for aphelion distance, A, is $A = a(1 + e)$. For Earth, $A = 1.000(1 + 0.0167) = 1.0167$ AU. Hence, Earth's distance from the Sun varies from 0.9833 to 1.0167 AU. The same formula is used to calculate perigee and apogee for an Earth-orbiting satellite. Ask students to calculate the perihelion and aphelion for other planets. Semimajor axes and eccentricities can be found in the *Reference Handbook.*

Eccentricity A planet in an elliptical orbit does not orbit at a constant distance from the Sun. The shape of a planet's elliptical orbit is defined by **eccentricity,** which is the ratio of the distance between the foci to the length of the major axis. You will investigate this ratio in the MiniLab. The orbits of most planets are not very eccentric; in fact, some are almost perfect circles.

The eccentricity of a planet can change slightly. Earth's eccentricity today is about 0.02, but the gravitational attraction of other planets can stretch the eccentricity to 0.05, or cause it to fall to 0.01.

Kepler's second and third laws In addition to discovering the shapes of planetary orbits, Kepler showed that planets move faster when they are closer to the Sun. He demonstrated this by proving that an imaginary line between the Sun and a planet sweeps out equal amounts of area in equal amounts of time, as shown in **Figure 6.** This is known as Kepler's second law.

The length of time it takes for a planet or other body to travel a complete orbit around the Sun is called its orbital period. In Kepler's third law of planetary motion, he determined the mathematical relationship between the size of a planet's ellipse and its orbital period. This relationship is written as follows:

$$P^2 = a^3$$

P is time measured in Earth years, and a is length of the semimajor axis measured in astronomical units.

MiniLAB

iLab Station

Explore Eccentricity

How is eccentricity of an ellipse calculated? Eccentricity is the ratio of the distance between the foci to the length of the major axis. Eccentricity ranges from 0 to 1; the larger the eccentricity, the more extreme the ellipse.

Procedure

WARNING: *Use caution when handling sharp objects.*

1. Read and complete the lab safety form.
2. Tie a piece of **string** to form a circle that will fit on a piece of **cardboard.**
3. Place a sheet of **paper** on the cardboard.
4. Stick **two pins** in the paper a few centimeters apart and on a line that passes through the center point of the paper.
5. Loop the string over the pins, and keeping the string taut, use a pencil to trace an ellipse around the pins.
6. Use a **ruler** to measure the major axis and the distance between the pins. Calculate the eccentricity.

Analysis

1. **Identify** what the two pins represent.
2. **Explain** how the eccentricity changes as the distance between the pins changes.
3. **Predict** the kind of figure formed and the eccentricity if the two pins were at the same location.

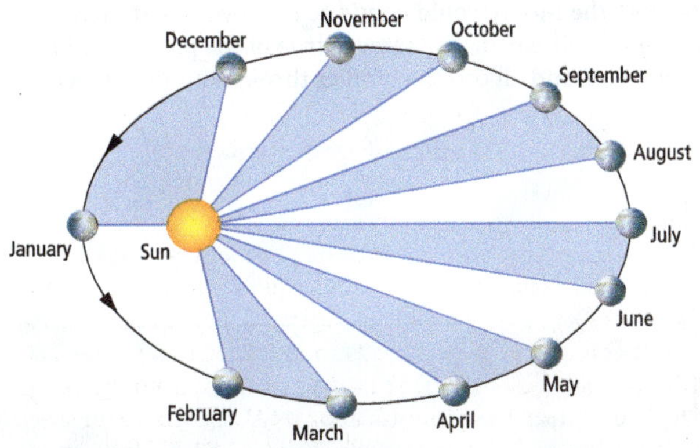

■ **Figure 6** Kepler's second law states that planets move faster when close to the Sun and slower when farther away. This means that a planet sweeps out equal areas in equal amounts of time. (Note: *not drawn to scale*)

MiniLAB

iLab Station Rubric

Purpose Students will model planetary orbits and learn basic properties of ellipses.

Process Skills measure in SI, use numbers, collect and organize data, interpret data

Safety Precautions Approve lab safety forms before work begins. Warn students to take care when using push pins, which are sharp.

Teaching Strategy Pins should be inserted just enough to allow the string to move freely.

Expected Results Students should observe that the closer the foci are, the more circular the orbit.

Analysis

1. The two pins represent the foci.
2. The eccentricity is smaller when the pins are closer. It is larger when the pins are more separated.
3. The figure would be a circle. Its eccentricity would be zero because the numerator in the eccentricity ratio would be zero.

Assessment

Performance Have students create ellipses with eccentricities of 0.2, 0.6, and as close to 1.0 as they can get. Suggest they use trial and error by adjusting the pin separation.

LabManager™
Customize any lab with the LabManager™ CD-ROM.

EARTH SCIENCE JOURNAL

Orbital Motion Have each student choose a planet and gather detailed information about its orbit. Students can then put this information together using Kepler's third law and the parameters of an ellipse to develop comprehensive pictures of the motions of their chosen planets. OL

Rubric

Teacher Content Support

Simplicity in Nature The struggle to learn the true nature of the solar system and of planetary motion took centuries. The many false starts and incorrect hypotheses along the way help to illustrate some important principles of science. One of these principles is the notion of simplicity: a scientific theory that is simple, and that depends on the fewest unsupported assumptions, is most likely to be correct.

Some early models of the solar system were far from simple; they involved multiple nested spheres or complex circles upon circles, whose sizes and motions were arbitrarily chosen to reproduce the observed positions of the planet. Then, during the Renaissance, Kepler found that a simple figure, the ellipse, describes planetary motions much more accurately than the older models, and Newton showed that a few mathematical relationships could explain and predict the motions of the planets. The current understanding of the structure and motions of the solar system provides an excellent example of simplicity in nature.

Caption Question Fig. 8
Doubling the masses of both objects would produce a gravitational attraction four times greater.

$$F_{new} = \frac{G(2m_1)(2m_2)}{r^2} = 4F = 1.3 \times 10^{-9}\ \text{N}$$

Figure 7 Galileo would probably be astounded to see Jupiter's four largest moons in the composite image above. Still, his view of Jupiter and its moons proved a milestone in support of heliocentric theory.

Galileo While Kepler was developing his ideas, Italian scientist Galileo Galilei became the first person to use a telescope to observe the sky. Galileo made many discoveries that supported Copernicus's ideas. The most famous of these was his discovery that four moons orbit the planet Jupiter, proving that not all celestial bodies orbit Earth, and demonstrating that Earth was not necessarily the center of the solar system. Galileo's view of Jupiter's moons, similar to the chapter opener photo, is compared with our present-day view of them, shown in **Figure 7.** The underlying explanation for the heliocentric model remained unknown until 1684, when English scientist Isaac Newton published his law of universal gravitation.

Gravity

Newton first developed an understanding of gravity by observing falling objects. He described falling as downward acceleration produced by gravity, an attractive force between two objects. He determined that both the masses of and the distance between two bodies determined the force between them. This relationship is expressed in his law of universal gravitation, illustrated in **Figure 8,** and that is stated mathematically as follows:

$$F = \frac{Gm_1m_2}{r^2}$$

F is the force measured in newtons, *G* is the universal gravitational constant ($6.67 \times 10^{-11}\ \text{m}^3/\ \text{kg}\cdot\text{s}^2$), m_1 and m_2 are the masses of the bodies in kilograms, and *r* is the distance between the two bodies in meters.

Gravity and orbits Newton realized that this attractive force could explain why planets move according to Kepler's laws. He observed the Moon's motion and realized that its direction changes because of the gravitational attraction of Earth. In a sense, the Moon is constantly falling toward Earth. If it were not for this attraction, the Moon would continue to move in a straight line and would not orbit Earth. The same is true of the planets and their moons, stars, and all orbiting bodies throughout the universe.

View an **animation of gravitational attraction.**

Concepts In Motion

Figure 8 The gravitational attraction between these two objects is 3.3×10^{-10} N.
Predict *the effect of doubling the masses of both objects, and check your prediction using Newton's equation.*

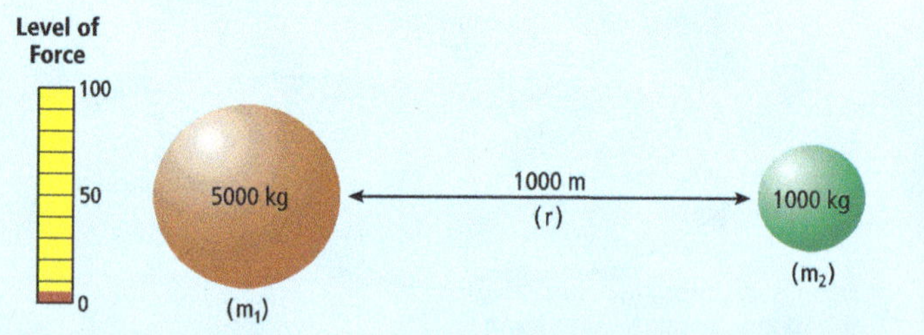

Demonstration

Center of Mass Demonstrate the concept of center of mass by attaching two balls of equal weight to the ends of a rod, and then showing that the balance point is at the center of the rod. Next, use a pair of balls of different weights, and show that the balance point is much closer to the heavier ball. If you rotate the rod on a pivot such as a pencil point, you can demonstrate that the center of mass remains fixed as the balls rotate around it, just as the Sun and a planet remain fixed while they orbit the center of mass between them. Explain to the class that for a Sun-planet pair, the center of mass is so close to the Sun that it is actually inside the Sun, except in the case of the Sun and Jupiter, where the balance point is just above the Sun's surface.

Center of mass Newton also determined that each planet orbits a point between it and the Sun called the center of mass. For any planet and the Sun, the center of mass is just above or within the surface of the Sun, because the Sun is much more massive than any planet. **Figure 9** illustrates how the center of mass between objects is similar to the balance point on a seesaw.

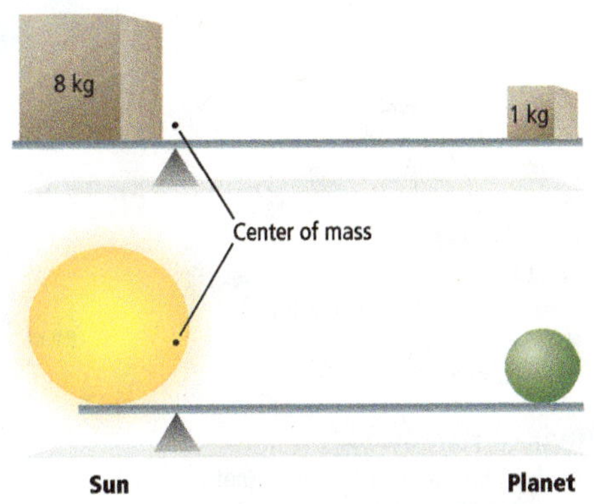

Figure 9 Just as the balance point on a seesaw is closer to the heavier box, the center of mass between two orbiting bodies is closer to the more massive body.

Present-Day Viewpoints

Astronomers traditionally divided the planets into two groups: the four smaller, rocky, inner planets, Mercury, Venus, Earth, and Mars; and the four outer gas planets, Jupiter, Saturn, Uranus, and Neptune. It was not clear how to classify Pluto, because it is different from the gas giants in composition and orbit. Pluto also did not fit the present-day theory of how the solar system developed. Then in the early 2000s, astronomers discovered a vast number of small, icy bodies inhabiting the outer reaches of the solar system, beyond the orbit of Neptune. At least one of these is larger than Pluto.

These discoveries have led many astronomers to rethink traditional views of the solar system. Some already define it in terms of three zones: the inner terrestrial planets, the outer gas giant planets, and the dwarf planets and comets. In science, views change as new data becomes available and new theories are proposed. Astronomy today is a rapidly changing field.

SECTION 1 REVIEW

Section Self-Check

Section Summary

- A collapsed interstellar cloud formed the Sun and planets from a rotating disk.
- The inner planets formed closer to the Sun than the outer planets, leaving debris to produce asteroids and comets.
- Copernicus created the heliocentric model and Kepler defined its shape and mechanics.
- Newton explained the forces governing the solar system bodies and provided proof for Kepler's laws.
- Present-day astronomers divide the solar system into three zones.

Understand Main Ideas

1. MAINIDEA **Describe** the formation of the solar system.
2. **Explain** why retrograde motion is an apparent motion.
3. **Describe** how the gravitational force between two bodies is related to their masses and the distance between them.
4. **Compare** the shapes of two ellipses having eccentricities of 0.05 and 0.75.

Think Critically

5. **Infer** Based on what you have learned about Kepler's third law, which planet moves faster in its orbit: Jupiter or Neptune? Explain.

MATH IN Earth Science

6. Use Newton's law of universal gravitation to calculate the force of gravity between two students standing 12 m apart. Their masses are 65 kg and 50 kg.

3 Assess

Check for Understanding

Discussion Ask students to think about what would happen if the force of gravity between the Sun and planets was suddenly turned off. The planets would continue to travel in straight lines in the directions they were going when gravity stopped.

Reteach

Summarize Have students summarize arguments in favor of the Sun-centered theory of the solar system over the Earth-centered model. The Sun-centered model is much simpler, requiring only an ellipse to describe each orbit. Unlike the Earth-centered model, the Sun-centered model also has the advantage of explaining retrograde motion simply, and it explains the seasons if Earth's tilt is recognized.

Assessment

Knowledge Ask students what would happen if the masses of Earth and the Sun were comparable, instead of the Sun's mass being much greater. The center of mass would be close to the center point between the Sun and Earth. Because both the Sun and Earth orbit the center of mass, both would have comparably large orbital motion.

SECTION 1 REVIEW

1. The main steps in the formation of the solar system were collapse of an interstellar cloud, movement of matter to the center under gravitational attraction, flattening into a rotating disk of matter, formation of the Sun, differential condensation throughout the disk based on temperature gradients, formation of gas giants, and formation of terrestrial planets.
2. The planet does not go backward and then forward; it only looks that way because of the difference in orbital speed between Earth and the planet.
3. The force between two bodies is proportional to the product of their masses and inversely proportional to the square of the distance between them.
4. The ellipse having an eccentricity of 0.75 will be much flatter and less like a circle than one with eccentricity of 0.05.
5. Jupiter moves faster because it is closer to the Sun.
6. 1.5×10^{-9} N

SECTION 2

1 Focus

Focus Transparency

Before presenting the lesson, project **Section Focus Transparency 89** and have students answer the accompanying questions. EL

MAINIDEA

Densities of the Planets Explain to students that the average densities of the four terrestrial planets are higher than the densities of typical rocks on their surfaces. Have students discuss this, keeping in mind the average density is found by dividing the mass of a planet by its volume. Ask students what this says about the interior densities and compositions of these planets. The interiors of the planets must be made of a highly dense material.

2 Teach

Collaborative Learning

Solar System Overview As students learn about the inner planets, the outer planets, and other solar-system objects, have them work in small groups to prepare visual overviews of the solar system. Groups might choose to make a computer presentation or a series of posters to describe the different objects of the solar system. BL OL COOP LEARN

■ **Caption Question Fig. 10** Earth's Moon rotates once each time it revolves around Earth on an orbit that takes 27 days to complete. Mercury rotates one and one-half times during each revolution around the Sun on an orbit that takes 88 days to complete.

SECTION 2

The Inner Planets

MAINIDEA Mercury, Venus, Earth, and Mars have high densities and rocky surfaces.

Essential Questions

- How are the characteristics of the inner planets similar?
- What are some of the space probes used to explore the solar system?
- How are the terrestrial planets different from each other?

Review Vocabulary

albedo: the amount of sunlight that reflects from the surface

New Vocabulary

terrestrial planet
scarp

EARTH SCIENCE 4 YOU Just as in a family in which brothers and sisters share a strong resemblance, the inner planets share many characteristics.

Terrestrial Planets

The four inner planets are called **terrestrial planets** because they are similar in density to Earth and have solid, rocky surfaces. Their average densities, obtained by dividing the mass of a planet by its volume, range from about 3.9 to just over 5.5 g/cm^3. Average density is an important indicator of internal conditions, and densities in this range indicate that the interiors of these planets are compressed.

Mercury

Mercury is the planet closest to the Sun, and for this reason it is difficult to see from Earth. During the day it is lost in the Sun's light and it is more easily seen at sunset and sunrise. Mercury is about one-third the size of Earth and has a smaller mass. Mercury has no moons. Radio observations in the 1960s revealed that Mercury has a slow spin of 1407.6 hours. In one orbit around the Sun, Mercury rotates one and one-half times, as shown in **Figure 10.** As Mercury spins, the side facing the Sun at the beginning of the orbit faces away from the Sun at the end of the orbit. This means that two complete Mercury years equal three complete Mercury rotations.

■ **Figure 10** Because of Mercury's odd rotation, its day lasts for two-thirds of its year.
Compare *Mercury's orbital motion with that of Earth's Moon.*

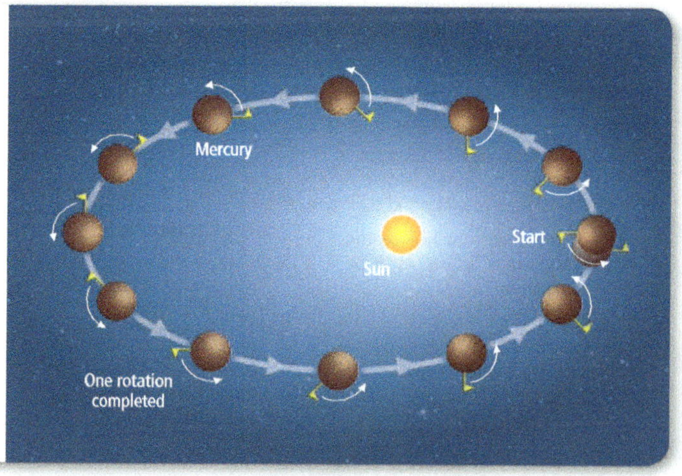

IN THE FIELD

Space Explorer The first African-American woman in space was Mae C. Jemison. She served as a mission specialist on the shuttle *Endeavor* in 1992 and studied the effects of weightlessness on organisms. Dr. Jemison earned a medical degree from Cornell University and now is president of the Jemison Group, Inc., which works to improve health care in western Africa. Dr. Jemison considers recruiting women and minorities into the various scientific fields to be some of her most important work.

■ **Figure 11** This mosaic of Mercury's heavily cratered surface was made by *Mariner 10.* Craters range in size from 100 to 1300 km in diameter.

Atmosphere Unlike Earth and the other planets, Mercury's atmosphere is constantly being replenished by the solar wind. What little atmosphere does exist is composed primarily of oxygen, sodium, and hydrogen deposited by the Sun. The daytime surface temperature on Mercury is 700 K (427°C), while temperatures at night fall to 100 K (−173°C). This is the largest day-night temperature difference among the planets.

Surface Early knowledge about Mercury was based on radio observations from Earth, and images from U.S. space probe *Mariner 10,* which passed close to Mercury three times in 1974 and 1975. The *MESSENGER (MErcury Surface, Space ENvironment, GEochemistry and Ranging)* space probe also made flybys and in 2011, became the first spacecraft to orbit Mercury. Images show that Mercury's surface, like that of the Moon, is covered with craters and plains, as shown in **Figure 11.** The plains on Mercury's surface are smooth and relatively crater free. Scientists think that the plains formed from lava flows that covered cratered terrain, much like the maria formed on the Moon. The surface gravity of Mercury is much greater than that of the Moon, resulting in larger crater diameters and shorter lengths of ejecta.

Mercury has a planetwide system of cliffs called **scarps,** such as the one shown in **Figure 12.** Mercury's scarps are much higher than Earth's. Scientists hypothesize that the scarps developed as Mercury's crust shrank and fractured early in the planet's geologic history.

☑ READING CHECK **Compare** the surfaces of the Moon and Mercury.

Interior Without seismic data, scientists have no way to analyze the interior of Mercury. However, its high density suggests that Mercury has a large nickel-iron core. Mercury's small magnetic field indicates that some of its core is molten.

■ **Figure 12** Discovery, the largest scarp on Mercury, is 550 km long and 1.5 km high.

Collaborative Learning

Identify Features Have students work with a partner to compare images of the Moon and Mercury. Students should identify features on the Moon, and then find comparable features on Mercury. Remind students how rilles (valleylike structures) and scarps are different despite their similar appearances in many distance images. BL COOP LEARN

Teacher Content Support

Feature Names Conventions for naming surface features on planets and satellites are governed by the International Astronomical Union (IAU), the major, worldwide organization of professional astronomers. In most cases, a consistent theme is adopted for naming features. For example, prominent craters on the Moon are named after historically important scientists such as Copernicus. For Venus, the convention is to name features after prominent women in science, history, and mythology, such as Aphrodite. Mercury's craters are named for famous artists and musicians of the past, such as Beethoven.

☑ READING CHECK The surfaces of both the Moon and Mercury have craters and plains.

DIFFERENTIATED INSTRUCTION

Visually Impaired A graphic and tactile method of teaching students with visual impairments is to illustrate the relative sizes of the terrestrial planets by having students handle balls with corresponding relative sizes. It is useful to include the Moon as well.

Enrichment

Venus's Surface Venus has far fewer impact craters than either the Moon or Mercury, and on Venus there are no small craters. Ask students for an explanation. The thick atmosphere of Venus prevents small objects from reaching the surface with enough speed to form craters. The smallest impact craters on Venus have diameters of about 1.5 km, which suggests that small meteoroids vaporize or break up before reaching the surface and cannot form craters. AL

Teacher Content Support

Venus's Atmosphere The circulation of Venus's atmosphere is far simpler than that of Earth's because of the slow rotational rate of Venus and also because of the lack of significant surface temperature variations on Venus compared to the contrasts between the temperatures of oceans and continents on Earth. However, strong storm winds with speeds of 100 m/s—60 times the speed at which Venus rotates—can occur on Venus's surface and remain a meteorological mystery. In 2010, Japan launched *AKATSUKI*, also called the *Venus Climate Orbiter*, to study Venus's weather.

Caption Question Fig. 14 Green areas represent areas of intermediate elevation.

Figure 13 The structure of Mercury's interior, which contains a proportionally larger core than Earth, suggests that Mercury was once much larger.

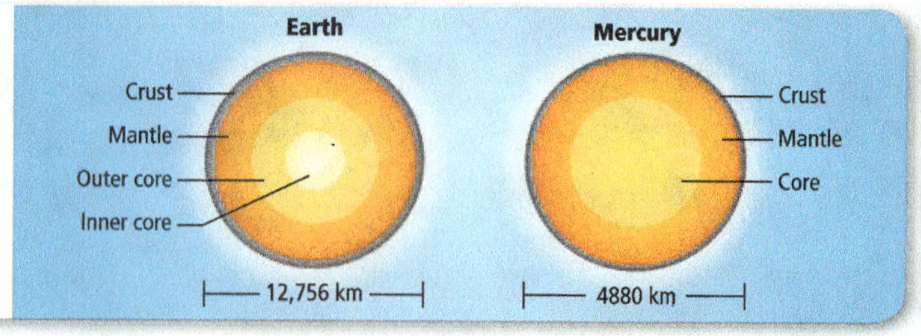

Early Mercury Mercury's small size, high density, and probable molten interior resemble what Earth might be like if its crust and mantle were removed, as shown in **Figure 13.** There are three major theories to explain these observations. Mercury may have collided with another body early in its history, stripping off its crust and mantle. Or, perhaps the heat of the early solar nebula vaporized the outer layers. Finally, before Mercury formed, the lighter gas near the Sun might have slowed and fallen inward, leaving higher density material behind.

Venus

Venus and Mercury are the only two planets closer to the Sun than Earth. Like Mercury, Venus has no moons. Venus is the brightest planet in the sky because it is close to Earth and because its albedo is 0.90—the highest of any planet. Venus is the first bright "star" to be seen after sunset in the western sky, or the last "star" to be seen before sunrise, depending on which side of the Sun it is on. For these reasons it is often called either the evening or morning star.

Thick clouds around Venus prevent astronomers from observing the surface directly. However, astronomers learned much about Venus from spacecraft launched by the United States and the Soviet Union. The 1978 *Pioneer-Venus* and 1989 *Magellan* missions of the United States used radar to map 98 percent of the surface of Venus. For images of the surface like the one shown in **Figure 14,** radar data was combined with images from *Magellan* spacecraft and those produced by the radio telescope in Arecibo, Puerto Rico. This view uses false colors to outline the major landmasses. In 2006, a European space probe, called *Venus Express*, went into orbit around Venus. It found signs that Venus has been—and still may be—volcanically active.

Figure 14 Radar imaging revealed the surface of Venus. Highlands are shown in red, and valleys are shown in blue. Large highland regions are like continents on Earth.
Infer *What do green areas represent?*

Retrograde rotation Radar measurements show that Venus rotates slowly—a day on Venus is equivalent to 243 Earth days. Also, Venus rotates clockwise, unlike most planets that spin counterclockwise. This backward spin, called retrograde rotation, means that an observer on Venus would see the Sun rise in the west and set in the east. Astronomers theorize that this retrograde rotation might be the result of a collision between Venus and another body early in the solar system's history.

Demonstration

Planetary Precession Explain to students that Earth, as well as the other planets, precesses, or wobbles, on its axis, causing changes over time in the visible night sky. To demonstrate precession, start a top spinning on a table, and then give it a slight sideways push. The top will respond by wobbling, its axis describing a conical pattern. This is what Earth's axis is doing.

EARTH SCIENCE JOURNAL

Terrestrial Atmospheres Have each student create a comprehensive data table that summarizes various aspects of the atmospheres of the terrestrial planets, such as pressure, temperature, and the major compounds present. Once they complete their data tables, students should think about how to explain the differences and similarities.

Rubric

Atmosphere Venus is the planet most similar to Earth in physical properties, such as diameter, mass, and density, but its surface conditions and atmosphere are vastly different from those on Earth. The atmospheric pressure on Venus is 92 atmospheres (atm), compared to 1 atm at sea level on Earth. If you were on Venus, the pressure of the atmosphere would make you feel like you were under 920 m of water.

The atmosphere of Venus is composed primarily of carbon dioxide and small amounts of nitrogen and water vapor. Like Earth, Venus also has clouds, as shown in **Figure 15,** an image taken of Venus by the *Pioneer Venus Orbiter.* However, instead of being composed of water vapor and ice, as on Earth, the thick bands of clouds on Venus consist of sulfuric acid and produce concentrated acid rain.

Figure 15 Clouds swirl around Venus in this image taken using ultraviolet wavelengths.

Greenhouse effect Venus also experiences a greenhouse effect similar to Earth's, but Venus's is more efficient. As you have learned, greenhouse gases in Earth's atmosphere trap infrared radiation and keep Earth much warmer than it would be if it had no atmosphere. The concentration of carbon dioxide is so high in Venus's atmosphere that it keeps the surface extremely hot—hot enough to melt lead. In fact, Venus is the hottest planet, with an average surface temperature of about 737 K (464°C), compared with Earth's average surface temperature of 288 K (15°C). It is so hot on the surface of Venus that no liquid water can exist.

Concept Development

Atmospheric Differences

Venus's thick atmosphere of carbon dioxide efficiently traps infrared radiation from the planet's surface. If the concentration of greenhouse gases on Earth rises sufficiently, this could happen on Earth as well. However, Earth has natural controls over carbon dioxide that will probably prevent conditions here from becoming as extreme as they are on Venus. Earth has liquid water, which absorbs CO_2 and deposits it in carbonate rocks, and plant life on Earth also absorbs CO_2. Venus lacks these controls.

Problem-Solving LAB

Apply Kepler's Third Law

How well do the orbits of the planets conform to Kepler's third law? For the six planets closest to the Sun, Kepler observed that $P^2 = a^3$, where P is the orbital period in years and a is the semimajor axis in AU.

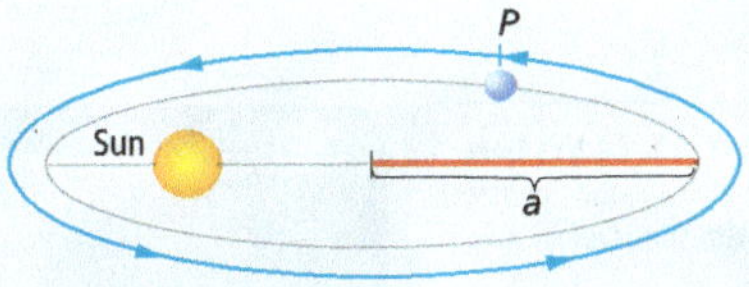

Analysis

1. Use this typical planet orbit diagram and the data from **Table 1** and the *Reference Handbook* to confirm the relationship between P^2 and a^3 for each of the planets.

Think Critically

2. **Prepare** a table showing your results and how much they deviate from predicted values.
3. **Determine** which planets conform most closely to Kepler's law and which do not seem to follow it.
4. **Consider** Would Kepler have formulated this law if he had been able to study Uranus and Neptune? Explain.
5. **Predict** the orbital period of an asteroid orbiting the Sun at 2.5 AU.
6. **Solve** Find the semimajor axis of Halley's comet, which has an orbital period of 76 years.

NASA/Roger Ressmeyer/CORBIS

Problem-Solving LAB

Purpose Students will perform simple calculations to demonstrate the validity of Kepler's third law for each of the planets.

Process Skills use numbers, interpret data

Teaching Strategies

- Remind students to use the correct units: years for orbital period and AU for semimajor axis.
- Some students might need help converting units.

Think Critically

2.

	p^2	a^3
Mercury	0.0581	0.0593
Venus	0.3790	0.3732
Earth	1.0000	1.0000
Mars	3.5424	3.5118
Jupiter	140.8993	140.608
Saturn	868.9117	868.2507
Uranus	7067.6883	7077.888
Neptune	27,192.448	27,135.225

3. For the inner planets, the law agrees nicely. For planets farther from the Sun, there are some discrepancies.
4. If Kepler had been aware of the outermost planets, he would have found that his law applied to them as well as to the inner six planets. Students could argue that the discrepancies might have been enough to lead him to conclude that his law did not hold true.
5. $P = 4.0$ y
6. $a = 18$ AU (Remind students that this represents Halley's average distance from the Sun. The actual distance varies between 0.6 and 35.3 AU.)

Concept Development

Compare Atmospheres Ask students to describe how the atmospheric compositions of Venus and Earth compare. Earth's atmosphere is primarily nitrogen and oxygen. There is little carbon dioxide in Earth's atmosphere compared to Venus's. OL

Tie to Previous Knowledge

The Color of Mars Ask students if they have looked at Mars at night and noticed its slightly reddish color. Explain that Earth is highly differentiated, and the majority of its iron is in the core. Mars is not so highly differentiated, and a greater portion of its iron remains in the crust. Iron at the surface of Mars combines with oxygen to form iron oxide, similar to the way the steel in a bicycle can form a reddish layer of rust. This is why Mars has a reddish color.

Teacher Content Support

Seasons on Mars Mars has a much more eccentric orbit than Earth does, and seasons are affected by the varying distances of the planet from the Sun. The overall result is that seasonal variations in the northern part of Mars are moderate, while the variations in the south are extreme. During southern spring, when the weather is making a dramatic shift from cold to hot, the temperature contrast drives major global wind storms that sometimes raise so much dust that the surface of Mars is obscured.

Surface The *Magellan* orbiter used radar reflection measurements to map the surface of Venus. This revealed that Venus has a surface smoothed by volcanic lava flows and with few impact craters. Observations from *Venus Express* indicate that volcanic activity took place as recently as 2.5 mya and that Venus might still be volcanically active. There is little evidence of current tectonic activity on Venus, nor well-defined crustal plates.

Interior Because the size and density of Venus are similar to Earth's, it is probable that the internal structure is similar also. Astronomers theorize that Venus has a liquid metal core that extends halfway to the surface. Despite this core, Venus has no measurable magnetic field, probably because of its slow rotation.

Earth

Earth, shown in **Figure 16,** has many unique properties when compared with other planets. Its distance from the Sun and its nearly circular orbit allow water to exist on its surface in all three states—solid, liquid, and gas. Liquid water is required for life, and Earth's abundance of water has been important for the development and existence of life on Earth. In addition, Earth's mild greenhouse effect and moderately dense atmosphere of nitrogen and oxygen provide conditions suitable for life.

Earth is the most dense of the terrestrial planets. It is the only known planet where plate tectonics currently occurs. Unlike Venus and Mercury, Earth has a moon, likely acquired by an impact.

Mars

Mars is often referred to as the red planet because of its reddish surface color, shown in **Figure 16.** Mars is smaller and less dense than Earth and has two irregularly shaped moons, Phobos and Deimos. Mars has been the target of recent exploration: *Mars Express* in 2003, the *Mars Exploration Rovers* in 2004, *Mars Reconnaissance Orbiter* in 2006, and the *Phoenix Mars Lander* in 2008.

Figure 16 Earth's blue seas and white clouds contrast sharply with the reddish, barren Mars.

Earth

Mars

DIFFERENTIATED INSTRUCTION

Advanced Learners Have students draw the orbits of Mars and Venus using eccentricities they find in the *Reference Handbook.* The scale should be large enough to show the actual shapes of the orbits. Suggest that students calculate the perihelion and aphelion distances on their chosen scales, and draw a smooth line through these points representing the orbits. Ask students to describe the differences they see in the two orbits.

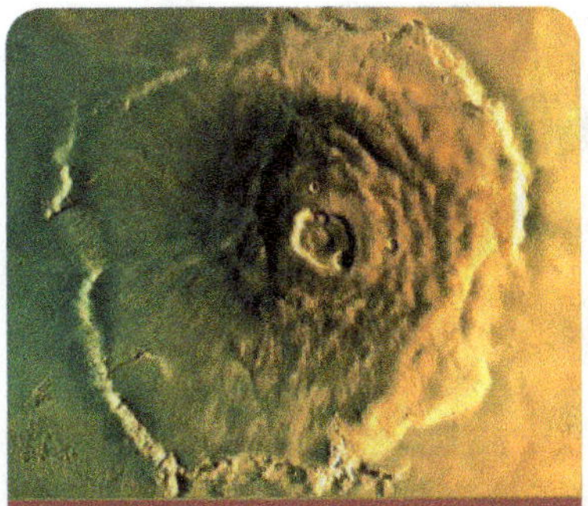

Olympus Mons volcano

Gusev crater

■ **Figure 17** Orbital probes and landers have provided photographic details of the Martian features and surface, such as Olympus Mons and Gusev crater.

Atmosphere Both Mars and Venus have atmospheres of similar composition. The density and pressure of the atmosphere on Mars are much lower; therefore Mars does not have a strong greenhouse effect like Venus does. Although the atmosphere is thin, it is turbulent—there is constant wind, and dust storms can last for months at a time.

Surface The southern and northern hemispheres of Mars vary greatly, as shown in **Figure 17.** The southern hemisphere is a heavily cratered, highland region resembling the highlands of the Moon. The northern hemisphere has sparsely cratered plains. Scientists theorize that great lava flows covered the once-cratered terrain of the northern hemisphere. Four gigantic shield volcanoes are located near the equator, near a region called the Tharsis Plateau. The largest volcano on Mars is Olympus Mons. The base of Olympus Mons is larger than the state of Colorado, and the volcano rises 3 times higher than Mount Everest in the Himalayas.

Tectonics An enormous canyon, Valles Marineris, shown in **Figure 18,** lies on the Martian equator, splitting the Tharsis Plateau. This canyon is 4000 km long—almost 10 times the length of the Grand Canyon on Earth and more than 3 times its depth. It probably formed as a fracture during a period of tectonic activity 3 bya, when the Tharsis Plateau was uplifted. The gigantic volcanoes were caused during the same period by upwelling of magma at a hot spot, much like the Hawaiian Island chain was formed. However, with no plate movement on Mars, magma accumulated in one area.

Erosional features Other Martian surface features include dried river and lake beds, gullies, outflow channels, and runoff channels. These erosional features suggest that liquid water once existed on the surface of Mars. Astronomers think that the atmosphere was once much warmer, thicker, and richer in carbon dioxide, allowing liquid water to flow on Mars. The *Mars Reconnaissance Orbiter* found water ice below the surface at mid-latitudes, and astronomers continue to search for water near the poles and elsewhere on Mars.

■ **Figure 18** Valles Marineris is a 4000-km-long canyon on Mars.

Teacher Content Support

Water on Mars In 2003, NASA's *Spirit* and *Opportunity* landers found spherical concretions on Mars that are composed of hematite, an iron oxide that forms in the presence of liquid water on Earth. These deposits, labeled "blueberries", provide evidence of Mars's wet past. In 2008, the *Mars Phoenix Lander* confirmed the presence of water ice below the surface of Mars and also discovered calcium carbonate in soil samples. $CaCO_3$ only forms in wet environments on Earth which suggests the occasional presence of thawed water on Mars.

Discussion

Valles Marineris Have students look back at **Table 1** to find the diameter of Mars. 6778 km Then, have them compare the length of Valles Marineris to Mars's diameter. Valles Marineris extends almost two-thirds the diameter of Mars. Point out that the length of Valles Marineris is close to the width of the United States. OL

Enrichment

Images of Mars Have interested students find current images of Mars. Ask students to summarize recent findings and present them to the class. AL

ACROSS THE CURRICULUM

Language Arts People have long been fascinated with the possibility that life might exist on Mars. This interest culminated with the famous 1938 radio broadcast, "The War of the Worlds." An actor portrayed a news reporter interviewing panic-stricken people during a Martian attack on Earth. The radio broadcast sounded so real that many listeners who did not hear the show's introduction believed that the Martian invasion of Earth was real.

Caption Question Fig. 19 The change in the ice cap shows that Mars has seasons. This indicates that Mars's axis is tilted relative to the plane of its orbit around the Sun.

3 Assess

Check for Understanding

Compare Have students rank the inner planets from greatest to least by mass and diameter. mass: Earth, Venus, Mars, Mercury; diameter: Earth, Venus, Mars, Mercury

Reteach

Summarize Have students make booklets summarizing characteristics of the inner planets and their atmospheres.

Assessment

Knowledge Ask students to summarize what scientists know about water on Mars. Dry river channels suggest Mars once had flowing water. Some is still present as water ice in the polar caps, traces of water vapor in the atmosphere, and subsurface ice deposits.

Figure 19 These images of Mars's northern ice cap were taken three months apart by the *Hubble Space Telescope* in 1997.

***Interpret** What do these images indicate about the orientation of Mars's axis?*

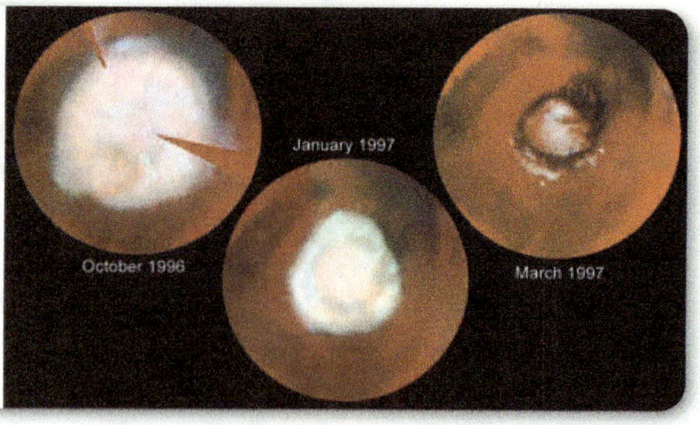

Ice caps Ice caps cover both poles on Mars. The caps grow and shrink with the seasons. Martian seasons are caused by a combination of a tilted axis and a slightly eccentric orbit. Both caps are made of carbon dioxide ice, sometimes called dry ice. Water ice lies beneath the carbon dioxide ice in the northern cap, shown in **Figure 19,** and is exposed during the northern hemisphere's summer when the north pole is tilted closer to the Sun, and the carbon dioxide ice evaporates. There is also water ice beneath the southern cap, although the carbon dioxide ice does not completely evaporate to expose it.

Interior The internal structure of Mars remains unknown. Astronomers hypothesize that there is a core of iron, nickel, and possibly sulfur that extends somewhere between 1200 km and 2400 km from the center of the planet. Because Mars has no magnetic field, astronomers think that the core is probably solid. Above the solid core is a mantle. There is no evidence of current tectonic activity or tectonic plates on the surface of the crust.

SECTION 2 REVIEW

Section Self-Check

Section Summary

- Mercury is heavily cratered and has high cliffs. It has no real atmosphere and the largest day-night temperature difference among the planets.
- Venus has clouds containing sulfuric acid and an atmosphere of carbon dioxide that produces a strong greenhouse effect.
- Earth is the only planet that has all three forms of water on its surface.
- Mars has a thin atmosphere. Surface features include four volcanoes and channels that suggest that liquid water once existed on the surface.

Understand Main Ideas

1. **MAINIDEA Identify** the reason that the inner planets are called terrestrial planets.
2. **Summarize** the characteristics of each of the terrestrial planets.
3. **Compare** the average surface temperatures of Earth and Venus, and describe what causes them.
4. **Describe** the evidence that indicates there was once tectonic activity on Mercury, Venus, and Mars.

Think Critically

5. **Consider** what the inner planets would be like if impacts had not shaped their formation and evolution.

MATH IN Earth Science

6. Using the *Reference Handbook,* create a graph showing the distance from the Sun for each terrestrial planet on the *x*-axis and their orbital periods in Earth days on the *y*-axis. For more help, refer to the *Skillbuilder Handbook*.

Phil James/Todd Clancy/Steve Lee/NASA

SECTION 2 REVIEW

1. They are small and rocky like Earth.
2. Mercury is heavily cratered with high daytime surface temperatures. It has no moon and no real atmosphere. Venus has a thick CO_2 atmosphere, a strong greenhouse effect, volcanoes and lava flow on the surface, and no moon. Earth supports life and has moderate temperatures, water in all three forms on its surface, and a large moon. Mars is small and is red in color because of iron in its soil. It has four old volcanoes, a deep canyon, and evidence of water on the surface in the past. It has two moons.
3. Venus is much hotter than Earth because it has a thick CO_2 atmosphere and a strong greenhouse effect. Gases in Earth's atmosphere produce a mild greenhouse effect.
4. Mercury has cliffs that formed as the planet cooled. Venus may have active volcanoes. Mars has a large uplifted plateau and extinct volcanoes.
5. Mercury might be larger and less dense, Venus might have a more rapid rotation in the normal direction, and Earth might not have a moon.
6.

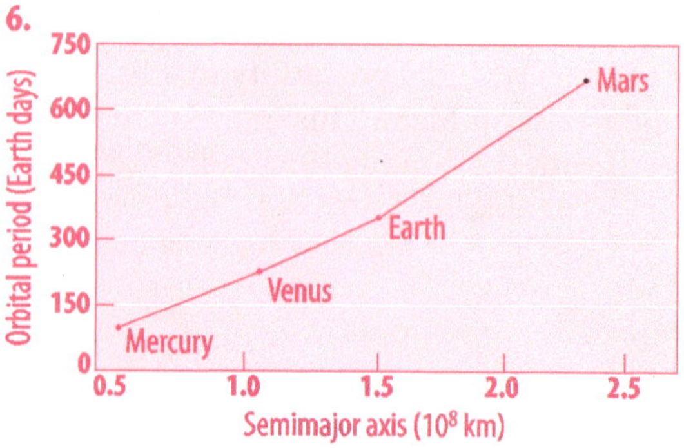

The Outer Planets

MAINIDEA Jupiter, Saturn, Uranus, and Neptune have large masses, low densities, and many moons and rings.

Essential Questions

- What are the similarities among and differences between the gas giant planets?
- What are the major moons?
- How do moons and rings form?
- How does the composition of the gas giant planets and the composition of the Sun compare?

Review Vocabulary

asteroid: metallic or silicate-rich object that orbits the Sun in a belt between Mars and Jupiter

New Vocabulary

gas giant planet
liquid metallic hydrogen
belt
zone

EARTH SCIENCE 4 YOU *Just as the inner planets resemble a family that shares many physical characteristics, the outer planets also show strong family resemblances.*

The Gas Giant Planets

Jupiter, Saturn, Uranus, and Neptune are known as the gas giants. The **gas giant planets** are all very large, ranging from 15 to more than 300 times the mass of Earth, and from about 4 to more than 10 times Earth's diameter. Their interiors are either gases or liquids, and they might have small, solid cores. They are made primarily of lightweight elements such as hydrogen, helium, carbon, nitrogen, and oxygen, and they are very cold at their surfaces. The gas giants have many satellites as well as ring systems.

Jupiter

Jupiter is the largest planet, with a diameter one-tenth that of the Sun and 11 times larger than Earth's. Jupiter's mass makes up 70 percent of all planetary matter in the solar system. Jupiter appears bright because its albedo is 0.343. Telescopic views of Jupiter show a banded appearance, as a result of flow patterns in its atmosphere. Nestled among Jupiter's cloud bands is the Great Red Spot, an atmospheric storm that has raged for more than 300 years. This is shown in **Figure 20.**

Rings The *Galileo* spacecraft observed Jupiter and its moons during a 7-year mission in the 1990s and 2000s. It revealed two faint rings around the planet in addition to a 6400-km-wide ring around Jupiter that had been discovered by *Voyager 1.* A portion of Jupiter's faint ring system is also shown in **Figure 20.**

Figure 20 Jupiter's cloud bands contain the Great Red Spot. The planet is circled by three faint rings that are probably composed of dust particles.

Jupiter's cloud bands

Jupiter's rings

SECTION 3

1 Focus

Focus Transparency

Before presenting the lesson, project **Section Focus Transparency 90** and have students answer the accompanying questions. EL

MAINIDEA

Mass and Density After students read the Main Idea, ask them to define *density.* Density is mass divided by volume. Ask how a planet, or any object, can have a large mass but a low density. The volume must be large. Allow students to hold an air-filled balloon in one hand and a ping-pong ball in the other hand. Although the objects have about the same mass, the density of the balloon is lower because its volume is greater. Explain to students that the outer planets are much larger than the inner planets, but their densities are much lower.

2 Teach

Concept Development

Clouds on Gas Giants Most gas giant planets have clouds similar to those on Earth. However, as a result of the rapid rotation of these gas giant planets, the clouds are stretched into bands. The clouds of gas giants are made of hydrogen, helium, methane, and ammonia, rather than water vapor, as on Earth.

EARTH SCIENCE JOURNAL

Describe the Planets As students read about the outer planets, have them write brief descriptions of each planet in their Earth science journals. Encourage students to focus on ways in which the planets differ from each other. Students might also wish to include drawings that compare the sizes and features of the planets. OL

Rubric

Tie to Previous Knowledge

The Coriolis Effect The Coriolis effect drives horizontal wind flows into rotary patterns called cyclonic and anticyclonic flows. On Jupiter, the Coriolis effect is enhanced by the rapid rotation of the planet. Cyclones and anticyclones on Earth usually lose energy and dissipate within a few days as a result of friction with landmasses and loss of thermal energy to drive them. On Jupiter, there are no landmasses, and internal thermal energy, rather than sunlight, drives the convection that triggers atmospheric flows. Thus, storms last longer on Jupiter than on Earth. Because the Great Red Spot is an anticyclonic storm (circulation around a high-pressure region) in the southern hemisphere, its flow direction is counterclockwise.

☑ READING CHECK because of the ice on its surface that has obviously melted in the past and refrozen

Teacher Content Support

Life on Europa The possibility that Europa might have a vast ocean of liquid water beneath its surface ice has led to speculation that life might have developed on this moon of Jupiter. All of the conditions known to have been present when life formed on Earth would be satisfied: the presence of liquid water, the availability of the basic chemicals required by life (Europa is known to have a variety of complex carbon compounds), and a source of energy (tidal heating from Jupiter).

■ **Figure 21** The four largest moons of Jupiter are Ganymede, Callisto, Io, and Europa. Ganymede is larger than Mercury. Callisto's bright scars illustrate a long history of impacts. Io is the most volcanically active object in the solar system. Scientists think that Europa's subsurface ocean could possibly support life.

Atmosphere and interior Jupiter has a density of 1326 kg/m^3, which is low for its size, because it is composed mostly of hydrogen and helium in gaseous or liquid form. Below the liquid hydrogen is a layer of **liquid metallic hydrogen,** a form of hydrogen that has properties of both a liquid and a metal, which can exist only under conditions of very high pressure. Electric currents exist within the layer of liquid metallic hydrogen and generate Jupiter's magnetic field. Models suggest that Jupiter might have an Earth-sized solid core containing heavier elements.

Rotation Jupiter rotates very rapidly for its size; it spins once on its axis in a little less than 10 hours, giving it the shortest day among the planets. This rapid rotation distorts the shape of the planet so that the diameter through its equatorial plane is 7 percent larger than the diameter through its poles. Jupiter's rapid rotation causes its clouds to flow rapidly as well, in bands of alternating dark and light colors called belts and zones. **Belts** are low, warm, dark-colored clouds that sink, and **zones** are high, cool, light-colored clouds that rise. These are similar to cloud patterns in Earth's atmosphere caused by Earth's rotation.

Moons Jupiter has more than 60 moons, most of which are extremely small. Jupiter's four largest moons, Ganymede, Callisto, Io, and Europa, shown in **Figure 21,** are called Galilean satellites after their discoverer. Three of them are bigger than Earth's Moon, and all four are composed of ice and rock. The ice content is lower in Io and Europa because they have been squeezed and heated by Jupiter's gravitational force more than the outer Galilean moons. In fact, Io is almost completely molten inside and undergoes constant volcanic eruptions. Gravitational heating has melted Europa's ice in the past, and astronomers hypothesize that it still has a subsurface ocean of liquid water. Cracks and water channels mark Europa's icy surface.

☑ READING CHECK **Explain** why scientists think that Europa has an ocean of liquid water beneath its surface.

Jupiter's smaller moons were discovered by a series of space probes beginning with *Pioneer 10* and *Pioneer 11* in the 1970s followed by *Voyager 1* and *Voyager 2* that also detected Jupiter's rings. Most of the information on Jupiter and its moons came from the *Galileo* space probe that arrived at Jupiter in 1995. Jupiter's four small, inner moons are thought to be the source of Jupiter's rings. Scientists think that the rings are produced as meteoroids strike these moons and release fine dust into Jupiter's orbit.

Gravity assist A technique first used to help propel *Mariner 10* to Mercury was to use the gravity of Venus to boost the speed of the satellite. Today it is common for satellites to use a planet's gravity to help propel them deeper into space. Jupiter is the most massive planet, and so any satellite passing deeper into space than Jupiter can use its gravity to give it an assist. Flybys on their way to Saturn and Pluto by the *Cassini* and *New Horizons* missions used that assist.

ACROSS THE CURRICULUM

Chemistry After hydrogen and helium, the next most-abundant elements in the solar system are carbon, nitrogen, and oxygen. The dominant compounds in Jupiter's atmosphere are combinations of hydrogen and each of these three elements. Hydrogen and carbon combine to form methane (CH_4), which is known on Earth as natural gas and is widely used as a source of heat for houses and stoves. The combination of hydrogen and nitrogen yields ammonia (NH_3), which is commonly used in cleaning fluids. The combination of hydrogen and oxygen is oxygen dihydride (H_2O), better known, of course, as water. Helium does not appear in compounds because it is inert and has no valence electrons available for chemical bonding.

Saturn

Saturn, shown in **Figure 22,** is the second-largest planet in the solar system. Several space probes have visited Saturn, including *Pioneer 11,* and *Voyagers 1* and *2.* In 2004, the United States' *Cassini* probe arrived at Saturn and began to orbit the planet.

Atmosphere and interior Saturn is slightly smaller than Jupiter and its average density is lower than that of water. Like Jupiter, Saturn rotates rapidly for its size and has a layered cloud system. Saturn's atmosphere is mostly hydrogen and helium with ammonia ice near the cloud tops. The internal structure of Saturn is probably similar to Jupiter's—fluid throughout, except for a small, solid core. Saturn's magnetic field is 1000 times stronger than Earth's and is aligned with its rotational axis. This is highly unusual among the planets.

Rings Saturn's most striking feature is its rings, which are shown in **Figure 22.** Saturn's rings are much broader and brighter than those of the other gas giant planets. They are composed of pieces of ice that range from microscopic particles to house-sized chunks. There are seven major rings, and each ring is made up of narrower rings, called ringlets. The rings contain many open gaps.

These ringlets and gaps are caused by the gravitational effects of Saturn's many moons. The rings are thin—less than 200 m thick—because rotational forces keep the orbits of all the particles confined to Saturn's equatorial plane. The ring particles have not combined to form a large satellite because Saturn's gravity prevents particles located close to the planet from sticking together. This is why the major moons of the gas giant planets are always beyond the rings.

Origin of the rings Until recently, astronomers thought that the ring particles were left over from the formation of Saturn and its moons. Now, many astronomers think it is more likely that the ring particles are debris left over from collisions of asteroids and other objects, or from moons broken apart by Saturn's gravity.

Moons Saturn has more than 60 satellites, including the giant Titan, which is larger than the planet Mercury. Titan is unique among planetary satellites because it has a dense atmosphere made of nitrogen and methane. Methane can exist as a gas, a liquid, and a solid on Titan's surface. In 2005, *Cassini* released the *Huygens* (HOY gens) probe into Titan's atmosphere. *Cassini* detected plumes of ice and water vapor ejected from Saturn's moon Enceladus, suggesting geologic activity.

■ **Figure 22** *Cassini-Huygens* provided detailed views of Saturn and its rings. Saturn's largest ring was discovered by NASA's *Spitzer Space Telescope* in 2009, orbiting six million kilometers away from the planet.
Explain ***why the ring particles orbit Saturn in the same plane.***

■ **Caption Question Fig. 22** Rotational forces keep the particles in the equatorial plane.

DIFFERENTIATED INSTRUCTION

Hearing Impaired Of all the astronomical objects, Saturn is the prototype. Saturn and its rings are widely used to represent astronomy. Display astronomical images of Saturn and the other outer planets around the classroom. Place captions on features such as rings, bands, and moons.

Teacher Content Support

Saturn's Rings Galileo was the first to observe the rings of Saturn, but he had no idea what he was observing. In 1610, Galileo noted that the image of Saturn was elongated and therefore referred to Saturn as the planet with *ears.* However, it was not until 1655 that Dutch astronomer Christiaan Huygens was able to deduce that rings caused Saturn's strange appearance.

Reinforcement

Jupiter and Saturn Ask students to compare and contrast Jupiter and Saturn. Both are gas giants of similar size, composition, and atmospheric flow patterns, and both have extensive systems of rings and moons. Contrasts are that Jupiter is larger and more dense; Saturn has a much more prominent ring system; and Jupiter's magnetic axis is tilted relative to the rotational axis, while Saturn's magnetic field is aligned with its rotational axis. BL

Concept Development

Power for Space Probes Probes to the outer planets travel too far from the Sun for solar panels to be useful power sources. Instead, NASA uses small nuclear power sources called radioisotope thermoelectric generators (RTGs), which are based on the decay of plutonium, a highly toxic radioactive element. Although RTGs have been used many years without incident, environmental activists are concerned that a launch accident could release plutonium into the atmosphere, causing a widespread health hazard.

Teacher Content Support

The Grand Tour *Voyager 2* came close to completing a path through the outer solar system that had long been referred to as the Grand Tour. All of the outer planets were to be aligned on the same side of the Sun in the 1980s. Astronomers recognized that this could be an opportunity to visit all of the planets economically. A single spacecraft could go from one planet to the next, using the gravitational field of each planet along the way to boost the spacecraft to the next planet. However, congressional funding constraints quashed the idea.

Oddly enough, once *Voyager 2* was well on its way and was operating successfully, NASA found that with a few course adjustments, most of the Grand Tour could be completed after all. Funding extensions were granted, and ultimately, *Voyager 2* made close flyby approaches to all four of the gas giants and returned vast amounts of images and data.

Discussion

Days and Seasons on Uranus Have students discuss solstices and equinoxes on Uranus, keeping in mind that its axis is tipped just over 90°. During the solstices, one of the axes is pointed only 8° from the Sun. As a result, almost the entire hemisphere experiences daylight all the time. Almost the entire opposite hemisphere is in complete darkness at the same time. During the equinoxes, all latitudes experience equal periods of day and night. These periods are equal to one-half of the rotational period of 17.24 hours. OL

■ **Caption Question Fig. 24** The diagram should show Uranus with its axis pointed in the same direction but located to the far left, below the Sun, and far right of its orbit.

■ **Figure 23** The blue color of Uranus is caused by methane in its atmosphere, which reflects blue light.

Uranus

Uranus was discovered accidentally in 1781, when a bluish object was observed moving relative to the stars. In 1986, *Voyager 2* flew by Uranus and provided detailed information about the planet, including the existence of new moons and rings. Uranus's average temperature is 58 K (–215°C).

Atmosphere Uranus is 4 times larger and 15 times more massive than Earth. It has a blue, velvety appearance, shown in **Figure 23,** which is caused by methane gas in Uranus's atmosphere. Most of Uranus's atmosphere is composed of helium and hydrogen, which are colorless. There are few clouds, and they differ little in brightness and color from the surrounding atmosphere contributing to Uranus's featureless appearance. The internal structure of Uranus is similar to that of Jupiter and Saturn; it is completely fluid except for a small, solid core. Uranus also has a strong magnetic field.

Moons and rings Uranus has at least 27 moons and a faint ring system. Many of Uranus's rings are dark—almost black and almost invisible. They were discovered only when the brightness of a star behind the rings dimmed as Uranus moved in its orbit and the rings blocked the starlight.

Rotation The rotational axis of Uranus is tipped so far that its north pole almost lies in its orbital plane, as shown in **Figure 24.** Astronomers hypothesize that Uranus was knocked sideways by a massive collision with a passing object, such as a large asteroid, early in the solar system's history. Each pole on Uranus spends 42 Earth years in darkness and 42 Earth years in sunlight due to this tilt.

■ **Figure 24** The axis or rotation of Uranus is tipped 98 degrees. This view shows its position at an equinox.
Draw *a diagram showing its position at the other equinox and solstices.*

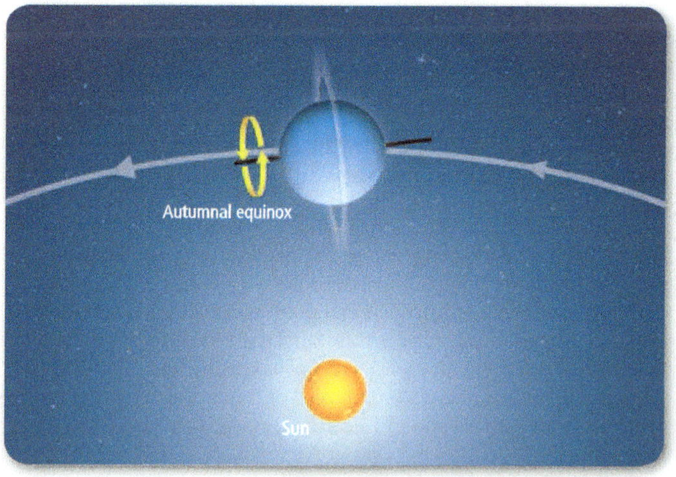

Demonstration

Stellar Occultation The rings of Uranus were detected by a method called stellar occultation. A telescope on Earth was aimed toward a star in the distance beyond Uranus. As the planet passed in front of the star, the rings made the starlight appear to blink off and on. A similar effect can be created in the classroom by passing a rod or bar in front of the light from a flashlight. Turn all the lights off and cover the windows, if possible, so that students see only the point of light that is created by the flashlight. Then, move the rod or a series of rods across the face of the flashlight. Students will observe the light alternately dim and brighten in the same manner that astronomers saw the background star dim and brighten as the rings of Uranus passed in front of it.

Neptune

The existence of Neptune was predicted before it was discovered, based on small deviations in the motion of Uranus and the application of Newton's universal law of gravitation. In 1846, Neptune was discovered where astronomers had predicted it to be. Few details can be observed on Neptune with an Earth-based telescope, but *Voyager 2* flew past Neptune in 1989 and took the image of its cloud-streaked atmosphere, shown in **Figure 25.** Neptune is the last of the gas giant planets and orbits the Sun almost 4.5 billion km away.

Atmosphere Neptune is slightly smaller and denser than Uranus, but its radius is about 4 times as large as Earth's. Another similarity between Neptune and Uranus is their bluish color caused by methane in the atmosphere. Neptune's atmospheric composition, temperature, magnetic field, interior, and particle belts or rings are also comparable with Uranus. Unlike Uranus, however, Neptune has distinctive clouds and atmospheric belts and zones similar to those of Jupiter and Saturn. In fact, Neptune can have persistent storms. One such storm, called the Great Dark Spot, was similar to Jupiter's Great Red Spot, although the storm disappeared by 1994.

Moons and rings Neptune has 13 moons, the largest of which is Triton. Triton has a retrograde orbit, which means that it orbits backward, unlike other large satellites in the solar system. Triton, shown in **Figure 25,** has a thin atmosphere and nitrogen geysers. The geysers are caused by nitrogen gas below Triton's south polar ice, which expands and erupts when heated by the Sun.

Neptune's six rings are composed of microscopic dust particles, which do not reflect light well. Therefore, Neptune's rings are not as visible from Earth as Saturn's rings.

Figure 25 *Voyager 2* took the image of Neptune above showing its cloud streaks, as well as this close-up view of Neptune's largest moon, Triton. Dark streaks indicate the sites of nitrogen geysers on Triton.

SECTION 3 REVIEW

Section Self-Check

Section Summary

- The gas giant planets are composed mostly of hydrogen and helium.
- The gas giant planets have ring systems and many moons.
- Some moons of Jupiter and Saturn have water and experience volcanic activity.
- All four gas giant planets have been visited by space probes.

Understand Main Ideas

1. MAINIDEA **Create** a table that lists the gas giant planets and their characteristics.
2. **Compare** the composition of the gas giant planets to the Sun.
3. **Compare** Earth's Moon with the moons of the gas giant planets.

Think Critically

4. **Evaluate** Where do you think are the most likely sites on which to find extraterrestrial life? Explain.

WRITING IN ▸ Earth Science

5. Research and describe one of the *Voyager* missions to interstellar space.

Teacher Content Support

Discovery of Neptune In 1846, a German astronomer, Johannes Galle, discovered Neptune in the location that had been predicted by other astronomers based on discrepancies in the motion of Uranus.

3 Assess

Check for Understanding

Make a Table Draw three columns on the board, labeled *physical properties, moons and rings,* and *orbital properties.* Then, have students contrast the terrestrial and gas giant planets by naming summary points to list in each column.

Reteach

Describe For each of the eight planets, have students write important characteristics of the planet on one side of an index card and the name of the planet on the other side. Then, have pairs of students use the cards to quiz each other.

Assessment

Performance Have each student research the moons and rings of the gas giant planets and make a booklet that shows contrasts and similarities among the moons and rings.

SECTION 3 REVIEW

1. Accept any table that contains a reasonable number of correct observations of each planet.
2. The Sun is composed of ionized hydrogen and helium. The gas giants are composed of gases or liquids made of lightweight elements such as hydrogen, helium, carbon, nitrogen, and oxygen.
3. Earth's Moon is solid rock with many craters, no life, and no atmosphere. The moons surrounding the gas giants have a wide range of sizes. Some are composed of just rock, and others are rock and ice. Jupiter's moon Io has volcanic eruptions, and scientists believe Europa has a subsurface ocean of liquid water. Saturn's moon Titan has a dense atmosphere. Neptune's moon Triton has a thin atmosphere.
4. Because water is needed for life as we know it, any place with the chance of water, including Mars, Enceladus, or Europa, is a possible site.
5. Students should provide details about the mission, including launch date, goals, accomplishments, present status, and other interesting facts.

Rubric

SECTION 4

1 Focus

Focus Transparency

Before presenting the lesson, project **Section Focus Transparency 91** and have students answer the accompanying questions. EL

MAINIDEA

Other Objects in the Solar System Show students a familiar map of the solar system showing the Sun and the planets. Ask them what they already know about other objects in the solar system, such as asteroids and comets, that are not shown on the map.

2 Teach

Reinforcement

Compare Pluto Have students make a chart on the board listing the eight planets and Pluto and the equatorial radius of each. This is a useful way to demonstrate the contrast between the terrestrial and gas giant planets, and it also helps to show how Pluto does not fit into either group. BL OL

SECTION 4

Other Solar System Objects

MAINIDEA Besides the Sun and planets, there are many other objects in the solar system that are composed primarily of rocks, dust, and ice.

Essential Questions

- What are the differences between planets and dwarf planets?
- What are the oldest members of the solar system?
- How are meteoroids, meteors, and meteorites described?
- What is the structure of a comet?

Review Vocabulary

smog: air polluted with hydrocarbons and nitrogen oxides

New Vocabulary

dwarf planet
meteoroid
meteor
meteorite
Kuiper belt
comet
meteor shower

EARTH SCIENCE 4 YOU The stereo might have been your favorite source of music until digital music players became available. Similarly, improvements in technology lead to a change in Pluto's rank as a planet when astronomers discovered many more objects that had similar characteristics to Pluto.

Dwarf Planets

In the early 2000s, astronomers began to detect large objects in the region of the then-planet Pluto, about 40 AU from the Sun, called the Kuiper belt. Then in 2003, one object, now known as Eris, was discovered that was larger than Pluto. At this time, the scientific community began to take a closer look at the planetary status of Pluto and other solar system objects.

Ceres In 1801, Giuseppe Piazzi discovered a large object in orbit between Mars and Jupiter. Scientists had predicted that there was a planet somewhere in that region, and it seemed that this discovery was it. However, Ceres, shown in **Figure 26,** was extremely small for a planet. In the following century, hundreds—now hundreds of thousands—of other objects were discovered in the same region. Therefore, Ceres was no longer thought of as a planet, but as the largest of the asteroids in what would be called the asteroid belt.

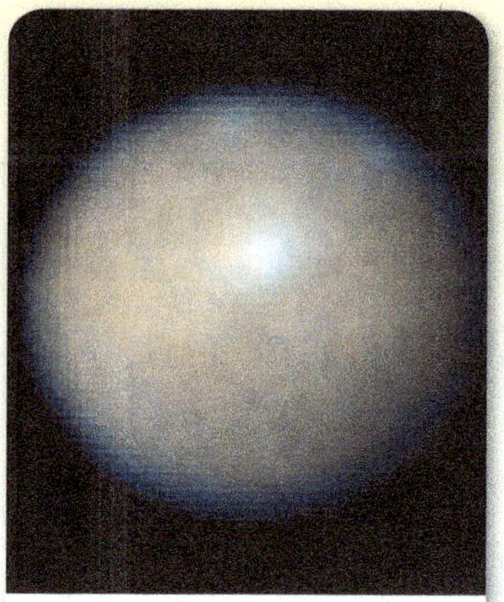

Figure 26 Imaged from the *Hubble Space Telescope*, the newly described dwarf planet, Ceres, is the largest body in the asteroid belt.

Pluto After its discovery by Clyde Tombaugh in 1930, Pluto was called the ninth planet. But it was an unusual planet. It is not a terrestrial or gas planet; it is made of rock and ice. It does not have a circular orbit; its orbit is long, elliptical, and overlaps the orbit of Neptune. It has four moons which orbit at a widely odd angle from the plane of the ecliptic. And it is smaller than Earth's Moon. It is one of many similar objects that exist outside of the orbit of Neptune.

How many others? With the discovery of objects close to and larger than Pluto's size, the International Astronomical Union (IAU) faced a dilemma. Should Eris be named the tenth planet? Or should there be a change in the way these new objects are classified? For now, the answer is change. Pluto, Eris, and Ceres have been placed into a new classification of objects in space called dwarf planets. The IAU has defined a **dwarf planet** as an object that, due to its own gravity, is spherical in shape, orbits the Sun, is not a satellite, and has not cleared the area of its orbit of smaller debris. The IAU has limited this classification to Pluto, Eris, Ceres, Makemake, and Haumea. There are at least 10 other objects whose classifications are undecided, some of which are shown in **Figure 27.**

EARTH SCIENCE JOURNAL

Pluto Have students make a two-column list of arguments both for and against Pluto's classification as a planet. Encourage them to include their personal views and also to consider Pluto's properties. OL

Rubric

GeoLAB

iLab Station

The GeoLab at the end of the chapter can be used at this point in the lesson.

VISUALIZING Other Solar System Objects

Figure 27 Recent findings of objects beyond Pluto have forced scientists to rethink what features define a planet.
(Note: *Buffy (XR190) is a nickname used by its discoverer.)*

Characteristics of Objects Beyond Neptune

Characteristic	Pluto	Sedna	Eris	Haumea	Buffy
Average distance from the Sun, AU	40	519	68	43	58
Relative size	1	0.67	1.05	0.33	0.25
Moons	4	?	1	2	?
Orbital period, years	248	10,500	557	284	436
Orbital tilt, degrees	17	12	44	28	47
Orbital eccentricity	0.25	0.85	0.44	0.19	0.11

Concepts In Motion View an **animation of other solar system objects.**

DIFFERENTIATED INSTRUCTION

Advanced Learners Have students read the IAU resolutions that define planets and dwarf planets and that demote Pluto to a dwarf planet. Ask them to consider what the effect might have been if the IAU had not included the requirement that a planet must have cleared the area of its orbit of smaller debris. Pluto would have remained a planet. In the future, many more bodies would likely have been classified as planets.

VISUALIZING

Purpose

Students will compare characteristics of objects in the Kuiper belt.

Teacher Content Support

Planets and Dwarf Planets
The International Astronomical Union (IAU) defines a planet as a body that orbits the Sun, has a nearly spherical shape, and has cleared the area of its orbit of smaller debris. A dwarf planet also orbits the Sun and is nearly spherical, but it has not cleared the area of its orbit of smaller debris. In order to prevent all moons from being classified as dwarf planets, the IAU also states that a dwarf planet must not be a satellite.

Teaching Strategies

- Have students review the IAU's definition of a planet and consider whether each of the characteristics listed for the objects beyond Neptune is significant for classification as a planet.
- Ask students why a spherical shape is one of the requirements for a planet. Explain that a planet must be massive enough that its gravity pulls it into a nearly spherical shape.
- Encourage students to think about how the requirements prevent Pluto from being considered a planet. Although Pluto is nearly spherical and orbits the Sun, it is surrounded by other bodies, which means it is a dwarf planet instead of a planet.

Teacher Content Support

Information from Meteorites

Meteorites preserve information about the early history of the solar system. One of the best places to search for meteorites is Antarctica. Because Earth rocks there are buried under thousands of meters of ice, rocks on top of ice caps are usually identified as meteorites.

READING CHECK The asteroid belt is a belt of rocky objects ranging from a few kilometers to about 1000 km in diameter that orbit the Sun between Mars and Jupiter.

Identify Misconceptions

Many students might think that only a large meteoroid can form a large crater.

Uncover the Misconception Show students a photo of the Barringer Meteor Crater. Ask them how large they think the meteoroid that formed the crater was.

Demonstrate the Concept Tell students the walls of the Barringer Meteor Crater rise about 46 m above the desert floor. The crater is about 174 m deep and about 1.6 km wide. The meteoroid that formed the crater was only about 46 m wide, weighed 270,000 metric tons, and was traveling at about 46,000 km/h. The width of the meteoroid was only about 3 percent of the width of the crater it formed.

Assess New Knowledge Have students extrapolate and explain why a tiny piece of material in space could seriously damage the space shuttle. OL

Figure 28 Asteroid Ida and its tiny moon, Dactyl, are shown in this image gathered by the *Galileo* spacecraft.

Figure 29 The Kuiper belt appears as the outermost limit of the planetary disk. The Oort cloud surrounds the Sun, echoing its solar sphere.

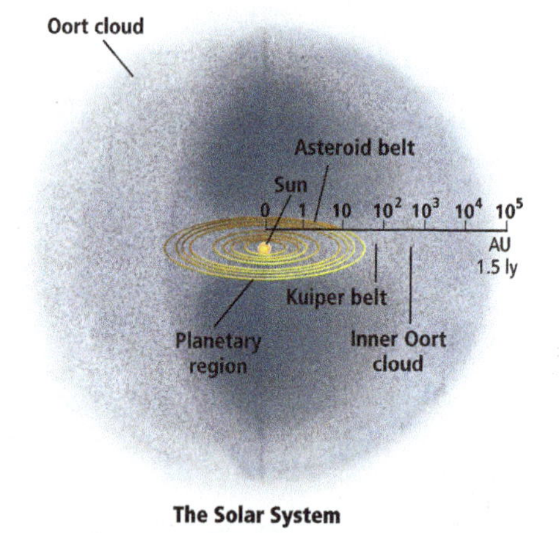

Small Solar System Bodies

Once the IAU defined planets and dwarf planets, they had to identify what was left. In the early 1800s, a name was given to the rocky planetesimals between Mars and Jupiter—the asteroid belt. Objects beyond the orbit of Neptune have been called trans-Neptunian objects (TNOs), Kuiper belt objects (KBOs), comets, and members of the Oort cloud. But what would the collective name for these objects be? The IAU calls them small solar system bodies.

Asteroids There are hundreds of thousands of asteroids orbiting the Sun between Mars and Jupiter. They are rocky bodies that vary in diameter and have pitted, irregular surfaces. Some asteroids have satellites of their own, such as the asteroid Ida, shown in **Figure 28.** Astronomers estimate that the total mass of all the known asteroids in the solar system is equivalent to only about 0.08 percent of Earth's mass.

READING CHECK **Describe** the asteroid belt.

As asteroids orbit, they occasionally collide and break into fragments. An asteroid fragment, or any other interplanetary material, is called a **meteoroid.** When a meteoroid passes through Earth's atmosphere, the air around it is heated by friction and compression, producing a streak of light called a **meteor.** If the meteoroid does not burn up completely and part of it strikes the ground, the part that hits the ground is called a **meteorite.** When large meteorites strike Earth, they produce impact craters. Any craters visible on Earth must be young, otherwise they would have been erased by erosion.

Kuiper belt Like the rocky asteroid belt, another group of small solar system bodies that are mostly made of rock and ice lies outside the orbit of Neptune in the **Kuiper** (KI pur) **belt.** Most of these bodies probably formed in this region—30 to 50 AU from the Sun—from the material left over from the formation of the Sun and planets. Some, however, might have formed closer to the Sun and were knocked into this area by Jupiter and the other gas giant planets. Eris, Pluto, Pluto's moons, and an ever-growing list of objects are being detected within this band; however, none of them has been identified as a comet. Comets usually come from the farthest limits of the solar system, the Oort cloud, shown in **Figure 29.**

Demonstration

Comet Erosion Demonstrate for students that volatile elements are the first to escape from a mixture that also contains refractory (rocky and metallic) elements. This demonstration will show how comets gradually erode away, leaving only rocks and dirt behind. Gather some rocks and gravel, and mix them with dry ice (CO_2) shavings in a basin. Be sure to use heavy gloves to avoid being burned by the dry ice. Have students observe as the dry ice quickly evaporates, leaving behind the rocks and pebbles.

Comets

Comets are small, icy bodies that have highly eccentric orbits around the Sun. Ranging from 1 to 10 km in diameter, most comets orbit in a continuous distribution that extends from the Kuiper belt to 100,000 AU from the Sun. The outermost region is known as the Oort cloud and expands into a sphere surrounding the Sun. Occasionally, a comet is disturbed by the gravity of another object and is thrown into the inner solar system.

Comet structure When a comet comes within 3 AU of the Sun, it begins to evaporate. It forms a head and one or more tails. The head is surrounded by an envelope of glowing gas, and it has a small solid core. The tails form as gas and dust are pushed away from the comet by particles and radiation from the Sun. This is why comets' tails always point away from the Sun, as illustrated in **Figure 30.**

Periodic comets Comets that orbit the Sun and thus repeatedly return to the inner solar system, are known as periodic comets. One example is Halley's comet, which has a 76-year period—it appeared last in 1985, and is expected to appear again in 2061. Each time a periodic comet comes near the Sun, it loses some of its matter, leaving behind a trail of particles. When Earth crosses the trail of a comet, particles left in the trail burn up in Earth's upper atmosphere producing bright streaks of light called a **meteor shower.** In fact, most meteors are caused by dust particles from comets.

Figure 30 A comet's tail always points away from the Sun and is driven by a stream of particles and radiation. The comet Hale-Bopp was imaged when its orbit brought it close to the Sun in 1997.

Dan Schechter/Photo Researchers

SECTION 4 REVIEW

Section Self-Check

Section Summary

- Dwarf planets, asteroids, and comets formed from the debris of the solar system formation.
- Meteoroids are rocky bodies that travel through the solar system.
- Mostly rock and ice, the Kuiper belt objects are currently being detected and analyzed.
- Periodic comets are in regular, permanent orbit around the Sun, while others might pass this way only once.
- The outermost regions of the solar system house most comets in the Oort cloud.

Understand Main Ideas

1. MAINIDEA **Identify** the kinds of small solar system bodies and their compositions.
2. **Compare** planets and dwarf planets.
3. **Distinguish** among meteors, meteoroids, and meteorites.
4. **Explain** why a comet's tail always points away from the Sun.
5. **Compare and contrast** the asteroid belt and the Kuiper belt.

Think Critically

6. **Infer** why comets have highly eccentric orbits.

WRITING IN Earth Science

7. Suppose you are traveling from the outer reaches of the solar system toward the Sun. Write a scientifically accurate description of the things you see.

Model

Comets Have pairs of students create a model of a comet orbiting the Sun. Explain to students the tails of comets always point away from the Sun. BL COOP LEARN

3 Assess

Check for Understanding

Knowledge Ask students why meteorites and interplanetary dust particles are good indicators of the age and early conditions of the solar system. They are extraterrestrial and older than any rocks on Earth. These objects represent conditions during the time when the planets were forming.

Reteach

Sequence Have small groups draw time lines starting 4.6 bya. They should indicate the following events: the formation of comets, Earth, and the Moon; the appearance of oceans on Earth; the beginnings of life on Earth; and the beginnings of the major geological eras.

Assessment

Knowledge Ask students to compare and contrast asteroids and comets. Asteroids are rocky bodies that have pitted, irregular surfaces. They have roughly circular orbits. Comets are small, icy bodies that begin to sublimate and form tails of gas and dust as they approach the Sun. They have highly eccentric orbits. Both asteroids and comets are remnants of the formation of the solar system.

SECTION 4 REVIEW

1. Dwarf planets are made of rock and ice. Asteroids and meteoroids are made of rock. Comets are made of dust and ice.
2. Both planets and dwarf planets are spherical, orbit the Sun, and are not satellites. Planets have cleared the area of their orbits of smaller debris, but dwarf planets have not.
3. Meteoroids are interplanetary objects that can enter Earth's atmosphere. Meteors are flashes of light produced as meteoroids burn up in Earth's atmosphere. Meteorites are portions of meteoroids or asteroids that strike Earth.
4. The tail is pushed away by particles and radiation from the Sun.
5. The asteroid belt is a ring of rocky planetesimals between the orbits of Mars and Jupiter. The Kuiper belt is a ring of small bodies made of mostly rock and ice beyond the orbit of Neptune.
6. The orbits of comets extend far from the Sun. Near the Sun, the gravitational pull increases their eccentricities.
7. Students should include descriptions of small solar-system bodies, dwarf planets, outer planets, and inner planets.

Rubric

Earth Science &
TECHNOLOGY

Purpose

Students will learn about areas of the solar system in which scientists think there is water, either in liquid or solid form.

Teacher Content Support

The Search for Water The *Lunar CRater Observation and Sensing Satellite (LCROSS)* was not the first probe that NASA crashed into the surface of a solar system body to search for water. The *Lunar Prospector,* which was sent to the Moon in 1998, was also deliberately crashed into its surface. Scientists tried to detect water in the resulting plume, but the results were inconclusive. In 2005, a probe was crashed into the comet Tempel 1, also in an attempt to detect water within the comet. *LCROSS* crashed into a permanently shadowed crater on the Moon called Cabeus and released an estimated 25 gallons of water during impact, showing that the floors of such craters hold a significant amount of water.

Teaching Strategy

Ask students why scientists are interested in finding water in other areas of the solar system. Possible answers: Water is one of the requirements of life as we know it. If water is found on other solar system bodies, there is a possibility that life exists on those bodies as well. How might traveling to the Moon or Mars change if available water was found on them? If available water is found, astronauts could use it while visiting these solar system bodies, making an extended visit possible.

Earth Science &
TECHNOLOGY

Water in the Solar System

In recent years, data collected by spacecraft and Earth-based radar have shown evidence of water in places in our solar system other than Earth. Mars and Earth's Moon are two such places. Scientists think there might also be water on several of Jupiter's moons, under the poles of Mercury, and on at least one of Saturn's moons. Further investigation and data collection is planned by NASA and other space agencies to confirm these findings.

Scientists hypothesize that a liquid water ocean exists beneath Europa's cracked icy surface.

Jupiter's moons Three of the four largest moons of Jupiter—Ganymede, Callisto, and Europa—all have icy surfaces. On Earth, solid ice conducts electricity very poorly whereas salty water conducts electricity fairly well. Magnetic measurements taken by the spacecraft Galileo reveal that electric currents are flowing through Ganymede, Callisto, and Europa. Based on these readings, as well as other evidence such as iceberg-like structures on Europa, scientists hypothesize that all three moons have deep, salty oceans of liquid water underneath their icy shells. Together with the European Space Agengy, NASA has proposed a new mission to Jupiter called the *Europa Jupiter System Mission* (EJSM). The joint international mission would launch two robotic orbiters in 2020 that would reach Jupiter and its largest moons in 2026. The main goals of the mission are to learn more about the origin and evolution of the Jupiter system, characterize the subsurface oceans of the three moons, and determine whether the Jupiter system might be able to harbor life.

Mercury's poles Because Mercury's axis is not tilted, large craters at the poles are permanently shielded from sunlight and their interior temperatures never rise above −212°C. Radar images indicate the presence of ice in these craters. In order to map the surface of Mercury and learn more about its composition, NASA launched the *MESSENGER* spacecraft in 2004. It went into orbit around the planet in 2011. *MESSENGER* is equipped with a spectrometer that is used to detect hydrogen, which is part of water, at Mercury's poles.

Saturn's moon—Enceladus NASA's spacecraft *Cassini* has recorded geyser-like eruptions on the surface of Enceladus, Saturn's geologically active moon. The source of these eruptions are warm fractures, dubbed "tiger stripes", that spew jets of water vapor, ice particles, and trace organic compounds into space. The largest of these fractures, called the Damascus Sulcus, radiates heat up to −93°C—a remarkably high temperature considering the moon's average temperature is −201°C.

WRITING IN ▶ Earth Science

Poster Research more information about where in the solar system water might exist. Make a poster that shows the major bodies of the solar system and if water might be found on them. Include captions that explain what type of exploration is planned.

WebQuest

WRITING IN ▶ Earth Science

Rubric

Poster Posters should include an illustration of all of the major bodies in our solar system, with captions that indicate if water might be found on any of them. Captions should also explain what type of exploration of the solar system body is planned.

WebQuest

GeoLAB

iLab Station

Design Your Own: Model the Solar System

Background: Models are useful for understanding the scale of the solar system.

Question: *How can you choose a scale that will easily demonstrate relative sizes of objects and distances between them in the solar system?*

Materials

calculator
tape measure
meterstick
marker
masking tape
common round objects in a variety of sizes

Safety Precautions

Procedure

1. Read and complete the lab safety form.
2. Develop a plan to make a model showing the relative sizes of objects in the solar system and the distances between them.
3. Make sure your teacher approves your plan before you begin.
4. Design a data table for the information needed to complete your model. Include the original data and the scale data.
5. Select a scale for your model using SI units. Remember your model should have the same scale throughout.
6. Calculate the relative sizes and distances of the objects you plan to model.
7. Select the materials and quantities of each, and build your model according to the scale you selected.

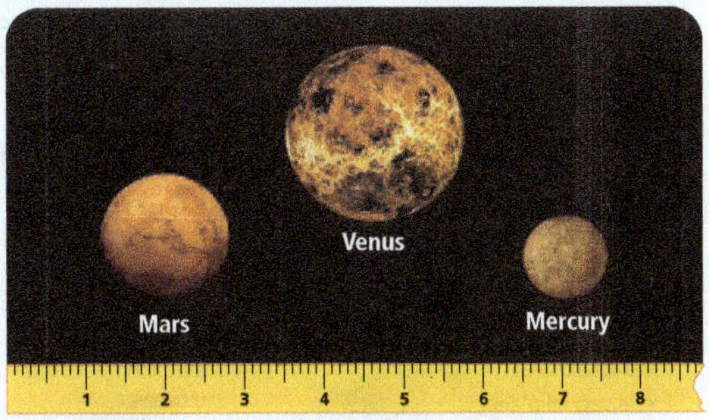

1 cm = 4000 km

Remember that the scale used has to include the largest and smallest objects and should be easy to produce.

Analyze and Conclude

1. **Think Critically** Why did the scale you chose work for your model?
2. **Explain** why you chose this scale.
3. **Observe and Infer** What possible problems could result from using a larger or smaller scale?
4. **Compare and Contrast** Compare your model with those of your classmates. Describe the advantages or disadvantages of your scale.

APPLY YOUR SKILL

Project Proxima Centauri, the closest star to the Sun, is about 4.01×10^{13} km from the Sun. Based on your scale, how far would Proxima Centauri be from the Sun in your model? If you modified your scale to better fit Proxima Centauri, how would this change the distance between Neptune and the Sun?

(t)NASA, (c)NASA/Mark Marten/Science Source/Photo Researchers, (r)USGS/Science Photo Library/Photo Researchers

APPLY YOUR SKILL

Project Answers will vary. Students should apply their scales to the distance of Proxima Centauri from the Sun. If Proxima Centauri were included, the scales would have to be reduced to have a manageable model. This would mean the distances between the existing objects and the Sun in the model would be decreased.

GeoLAB

iLab Station Rubric

Preparation

Time Allotment 45 min

Process Skills compare and contrast, formulate models, organize data, recognize spatial relationships, make and use tables

Safety Precaution Approve lab safety forms before work begins.

Procedure

- Before students begin, demonstrate on the board the types of calculations they will need to make.
- **Troubleshooting** Students might find the scale they use initially is too small or too large for a reasonable model. Encourage them to consider all of the planets when choosing a scale.

Analyze and Conclude

1. Answers will vary. Students might describe the relationship between the size of their model and the sizes and distances between the objects in the solar system.
2. Answers will vary. Students might state that the scale resulted in a model that had a reasonable size.
3. Answers will vary. Large scales will result in an unmanageable model. Details are not as clear when scales are too small.
4. Answers will vary depending on the students' scales.

LabManager™
Customize any lab with the LabManager™ CD-ROM.

MAINIDEAS Summary statements can be used by students to review the major concepts of the chapter.

Students can review with these online resources.

Vocabulary Practice

Vocabulary eGames
Vocabulary eFlashcards
Vocabulary PuzzleMaker

Section Self-Check

Chapter Self-Check

Online Test Practice

Use the Exam*View*® *Assessment Suite* CD-ROM to:

- create multiple versions of tests
- create modified tests with one mouse click
- edit existing questions and add your own questions
- build tests aligned with state standards using built-in state curriculum tags
- change English tests to Spanish with one mouse click
- track students' progress using the Teacher Management System

CHAPTER 28 STUDY GUIDE

BIGIDEA **Using the laws of motion and gravitation, astronomers can understand the orbits and the properties of the planets and other objects in the solar system.**

Vocabulary Practice

SECTION 1 Formation of the Solar System

VOCABULARY
- planetesimal
- retrograde motion
- ellipse
- astronomical unit
- eccentricity

MAINIDEA The solar system formed from the collapse of an interstellar cloud.
- A collapsed interstellar cloud formed the Sun and planets from a rotating disk.
- The inner planets formed closer to the Sun than the outer planets, leaving debris to produce asteroids and comets.
- Copernicus created the heliocentric model and Kepler defined its shape and mechanics.
- Newton explained the forces governing the solar system bodies and provided proof for Kepler's laws.
- Present-day astronomers divide the solar system into three zones.

SECTION 2 The Inner Planets

VOCABULARY
- terrestrial planet
- scarp

MAINIDEA Mercury, Venus, Earth, and Mars have high densities and rocky surfaces.
- Mercury is heavily cratered and has high cliffs. It has no real atmosphere and the largest day-night temperature difference among the planets.
- Venus has clouds containing sulfuric acid and an atmosphere of carbon dioxide that produces a strong greenhouse effect.
- Earth is the only planet that has all three forms of water on its surface.
- Mars has a thin atmosphere. Surface features include four volcanoes and channels that suggest that liquid water once existed on the surface.

SECTION 3 The Outer Planets

VOCABULARY
- gas giant planet
- liquid metallic hydrogen
- belt
- zone

MAINIDEA Jupiter, Saturn, Uranus, and Neptune have large masses, low densities, and many moons and rings.
- The gas giant planets are composed mostly of hydrogen and helium.
- The gas giant planets have ring systems and many moons.
- Some moons of Jupiter and Saturn have water and experience volcanic activity.
- All four gas giant planets have been visited by space probes.

SECTION 4 Other Solar System Objects

VOCABULARY
- dwarf planet
- meteoroid
- meteor
- meteorite
- Kuiper belt
- comet
- meteor shower

MAINIDEA Besides the Sun and planets, there are many other objects in the solar system that are composed primarily of rocks, dust, and ice.
- Dwarf planets, asteroids, and comets formed from the debris of the solar system formation.
- Meteoroids are rocky bodies that travel through the solar system.
- Mostly rock and ice, the Kuiper belt objects are currently being detected and analyzed.
- Periodic comets are in regular, permanent orbit around the Sun, while others might pass this way only once.
- The outermost regions of the solar system house most comets in the Oort cloud.

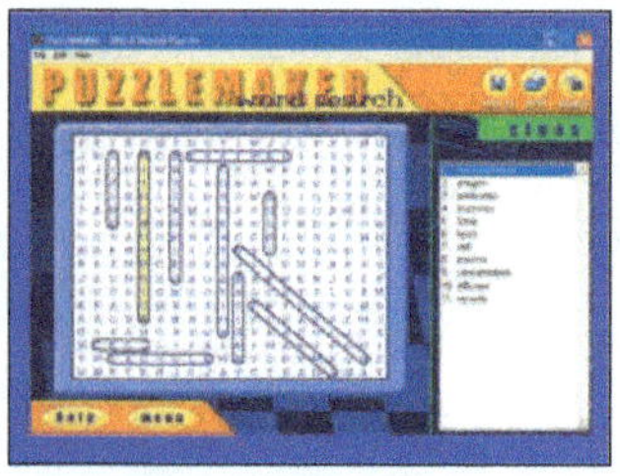

Vocabulary Practice

For additional practice with vocabulary, have students access the Vocabulary PuzzleMaker.

Chapter Self-Check

VOCABULARY REVIEW

Each of the following sentences is false. Make each sentence true by replacing the italicized words with terms from the Study Guide.

1. Rapid shrinkage of Mercury's crust produced features on its surface called *rilles.*

2. The pattern of light and dark bands on Jupiter's surface are called belts and *flows.*

3. A *meteor* is a rocky object that strikes Earth's surface.

4. A *meteorite* formed as particles of dust and gas stuck together in the early solar system.

5. The apparent backward movement of Mars as Earth passes it in its orbit is *synchronous rotation.*

6. The *light-year* is a unit of measurement used to measure distances within the solar system.

Match each phrase below with the correct term from the Study Guide.

7. a small icy object having a highly eccentric orbit around the Sun

8. Mercury, Venus, Earth, and Mars

9. multiple streaks of light caused by dust particles burning in Earth's atmosphere

10. a measure of orbital shape

11. a new solar system body classification

UNDERSTAND KEY CONCEPTS

12. Who first proposed the heliocentric model of the solar system?
   A. Copernicus
   B. Galileo
   C. Kepler
   D. Newton

Use the diagram below to answer Question 13.

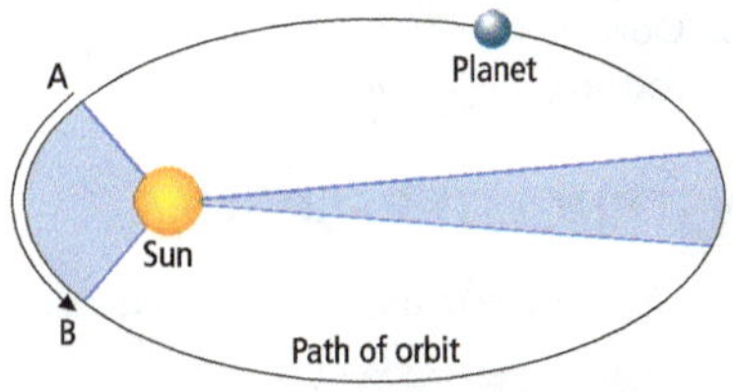

13. Which law of planetary motion does this diagram demonstrate?
   A. Kepler's first law
   B. Kepler's second law
   C. Kepler's third law
   D. Newton's law of universal gravitation

14. Which best describes a planet's retrograde motion?
   A. apparent motion
   B. orbital motion
   C. real motion
   D. rotational motion

15. Which scientist determined each planet orbits a point between it and the Sun, called the center of mass?
   A. Copernicus
   B. Galileo
   C. Kepler
   D. Newton

Use the diagram below to answer Question 16.

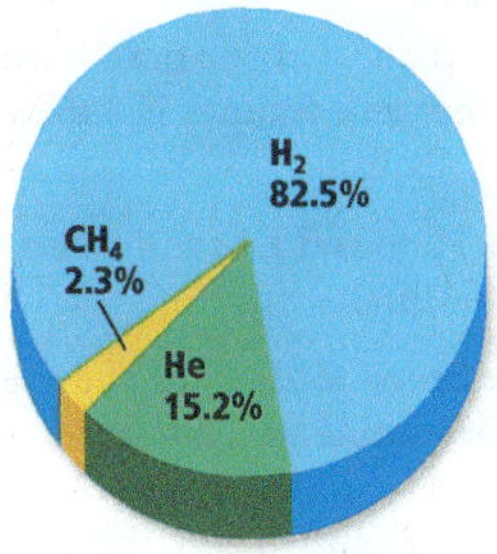

16. The atmospheric composition of which planet is shown above?
   A. Jupiter
   B. Mars
   C. Neptune
   D. Venus

CHAPTER 28 ASSESSMENT

VOCABULARY REVIEW

1. scarps
2. zones
3. meteorite
4. planetesimal
5. retrograde motion
6. astronomical unit
7. comet
8. terrestrial planets
9. meteor shower
10. eccentricity
11. dwarf planet

UNDERSTAND KEY CONCEPTS

12. A
13. B
14. A
15. D
16. C

17. A

CONSTRUCTED RESPONSE

18. The features shown are gullies and flow channels most likely caused by water flowing on the surface of Mars.
19. Features such as gullies and flow channels are only made if liquid water flows on the surface of Mars. So far, no liquid water has been conclusively identified.
20. Pluto and Eris are both dwarf planets with similar size. Pluto has four moons; Eris has one. The orbital period of Eris is about twice the period of Pluto. Eris is tilted more on its axis. Eris has a greater orbital eccentricity.
21. Sedna and Haumea are similar in size. They are about three-fourths the size of Pluto. Like Pluto, Eris, Ceres, and Makemake, they are Kuiper belt objects.
22. Probes do not survive on the surface of Venus because of the high temperature and high pressure.
23. In both cases, the masses and distance between them affect the location of the balance point. A see-saw operates in two dimensions, and the center of mass applies to a three-dimensional system.
24. very elongated and flat
25. The planet is Uranus, because it is the only one with its axis tilted so far. Scientists think the tilt might have been caused by a collision with another object early in its history.
26. The poles would face toward or away from the Sun for nearly half the year, creating extremes of temperatures at the poles, similar to the temperature extremes on the Moon and Mercury. Most life-forms likely would not survive.

THINK CRITICALLY

27. Venus has a much thicker atmosphere and therefore more CO_2. Mars's atmosphere is thin. Even though it has a high proportion of CO_2, the amount is not enough to cause significant warming.
28. math, computer science, and physics
29. The current theory is that rings are formed from debris created when asteroids or other objects collide with moons or each other, or are broken up by the gravity of the planet.

17. Where do most meteorites originate?
A. asteroid belt
B. Kuiper belt
C. Oort cloud
D. Saturn's rings

CONSTRUCTED RESPONSE

Use the photo below to answer Questions 18 and 19.

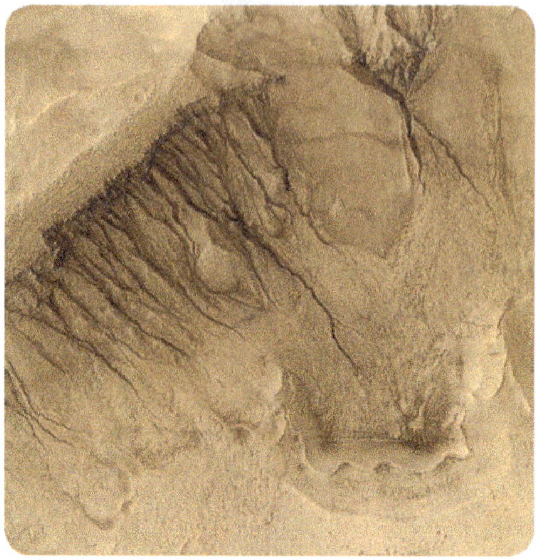

18. Identify these features shown on the surface of Mars and explain what most likely caused them.

19. Infer Based on what you have learned about Mars, state whether new features like these could be made now. Explain.

20. Compare Pluto and Eris and determine their common features.

21. Compare Sedna to Haumea and the other dwarf planets. Determine which features are common to each.

22. Explain why probes do not survive on the surface of Venus.

23. Compare the pivot point on a seesaw and a center of mass between two orbiting bodies.

24. Calculate Find the shape of an ellipse having an eccentricity of 0.9.

Use the diagram below to answer Questions 25 and 26.

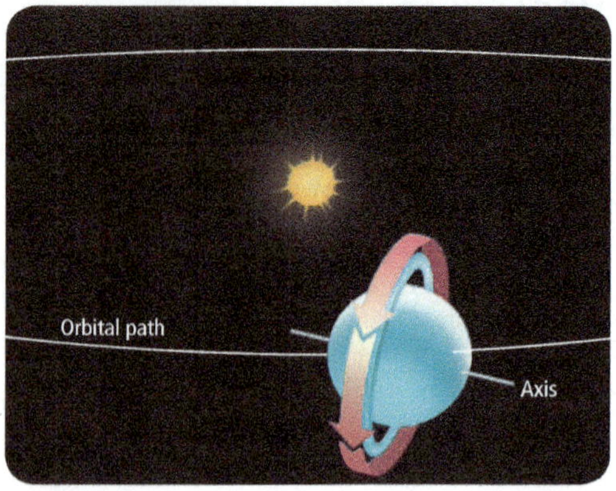

25. Identify the planet shown here and explain why scientists think its rotational axis is like this.

26. Infer how the seasons would be affected if Earth had an axis tilt similar to Uranus.

THINK CRITICALLY

27. Explain The atmospheres of Mars and Venus contain similar percentages of CO_2, but Venus has a much higher surface temperature because of the greenhouse effect. Why doesn't this happen on Mars?

28. CAREERS IN EARTH SCIENCE Most astronomers do not spend long hours peering through telescopes. They operate telescopes remotely using computers and spend most of their time analyzing data. What subjects would astronomers find most useful in addition to astronomy?

29. Discuss the theory of formation of the rings of Saturn and the other gas giant planets.

30. Infer the role gravity plays in the formation of the rings of the gas giant planets.

31. Infer what might happen to Halley's comet as it continues to lose mass with each orbit of the Sun.

32. Explain why scientists think Jupiter's moon Europa might have liquid water beneath its surface.

JPL/NASA

30. The objects are pulled into the orbits of the planet by the planet's gravity, which then puts them on course for collision, or causes them to break apart as they approach the planet.
31. As Halley's comet continues to lose ice and gas, it will eventually cease to grow a halo and tail, and may break up and dissipate into a cloud of dust.
32. The action of Jupiter's gravity on the material of Europa might generate enough frictional heat to keep the water liquid below the surface.

Chapter Self-Check

Use the table below to answer Questions 33 to 35.

Planet	Radius (km)	Orbital Eccentricity	Semimajor Axis (AU)
Mercury	2439.7	0.2056	0.39
Venus	6051.8	0.0067	0.72
Earth	6378.1	0.0167	1.00
Mars	3397	0.0935	1.52
Jupiter	71,492	0.0489	5.20
Saturn	60,298	0.0565	9.54
Uranus	25,559	0.047	19.19
Neptune	24,766	0.009	30.07

33. Interpret Which of the planets has an orbit that most closely resembles a perfect circle?

34. Compare Which two planets have the most similar radii?

35. Evaluate Which two planets' orbits are separated by the greatest distance?

36. Discuss the relationship between asteroids and planetesimals.

37. Explain Why were Ceres and Pluto identified as the first dwarf planets?

38. Compare and contrast the asteroid belt and the Kuiper belt.

CONCEPT MAPPING

39. Create a concept map using the following terms: *interstellar cloud, gas, dust, disk, particles, planetesimals, terrestrial planets, gas giant planets, satellites, debris, asteroids, meteoroids, and comets.*

CHALLENGE QUESTION

40. Consider Pluto's orbit sometimes brings it within the orbit of Neptune. Why is it unlikely that the two will collide? Explain.

WRITING IN ▶ Earth Science

41. Write a paragraph to explain to a friend how science develops over time. Discuss the relationship between Kepler's laws and Newton's law of universal gravitation.

DBQ Document–Based Questions

Data obtained from: *Physics World.* 2001. (January): 25.

Astronomers have detected planets around more than 200 stars. Although the planets themselves are too small to see directly, astronomers can detect them by measuring the Doppler shift in the star's light as it orbits its common center of mass with the unseen planet. The diagram below shows how this works.

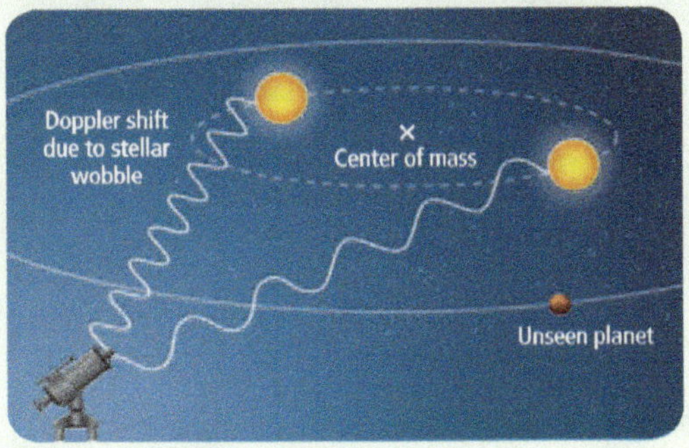

42. Based on the diagram, what is the rotational direction of the star? Explain.

43. Based on what you know about the center of mass, which planet in our solar system would be most likely to be detectable from other star systems using this method?

CUMULATIVE REVIEW

44. Name an example of a felsic, igneous rock. **(Chapter 5)**

45. Describe the relationship between ejecta and rays on the Moon's surface. **(Chapter 27)**

33. Venus has an eccentricity closest to zero.
34. Venus and Earth
35. Uranus and Neptune
36. Asteroids are planetesimals that did not accrete into becoming a planet, probably due to Jupiter's gravity.
37. They were both discovered long ago. They are nearly spherical, orbit the Sun, and are not satellites. They have not cleared their orbits of smaller debris.
38. The asteroid belt is closer to the Sun and is composed of mostly rocky planetesimals. The Kuiper belt is outside the orbits of the planets and is comprised of objects that are mostly rock and ice. Both contain objects that have a variety of sizes, and they both have at least one dwarf planet.

CONCEPT MAPPING

39. Accept any reasonable concept map that contains all the terms listed with appropriate connections.

CHALLENGE QUESTION

40. Pluto and Neptune orbit on different planes. Neptune is aligned with the ecliptic plane and Pluto's orbit is tipped 17 degrees. Although Pluto's orbit sometimes brings it within the orbit of Neptune, their orbital paths do not cross.

WRITING IN ▶ Earth Science

Rubric

41. Paragraphs might tell how Pluto and Ceres changed status. They should give details of both Kepler's laws and Newton's law of gravitation. Kepler used mathematics to describe planetary motion. Newton explained why planets moved in this way.

DBQ Document-Based Questions

Data obtained from: *Physics World.* 2001. (January): 25.

42. The short waves indicate an approaching star, and the longer ones show it receding. Because of the positions, the star must be moving counterclockwise.
43. Jupiter, because it has the greatest mass

CUMULATIVE REVIEW

44. Possible answers: obsidian, rhyolite, granite, or pegmatite
45. Ejecta is material that blasts out when objects from space crash into the lunar surface. Rays are long trails of ejecta that radiate outward from some craters on the surface.

MULTIPLE CHOICE

1. C
2. A
3. B
4. C
5. A
6. B
7. A
8. C
9. B
10. A

CUMULATIVE STANDARDIZED TEST PRACTICE

MULTIPLE CHOICE

1. When foxes reach the brink of extinction in an area, what happens to the population of rabbits in the area?
   A. The rabbit population also becomes extinct.
   B. The rabbit population increases indefinitely.
   C. The rabbit population increases beyond the carrying capacity of the area, then decreases.
   D. The rabbit population decreases beyond the carrying capacity of the area, and then quickly increases.

Use the diagram below to answer Questions 2 and 3.

2. What results on Earth when the Sun and the Moon are aligned along the same direction?
   A. spring tides
   B. neap tides
   C. the autumnal equinox
   D. the summer solstice

3. If the Moon in this diagram were passing directly between the Sun and Earth, blocking the view of the Sun, what would you experience on Earth?
   A. a lunar eclipse
   B. a solar eclipse
   C. umbra
   D. penumbra

4. Earth's main energy source is
   A. fossil fuels
   B. hydrocarbons
   C. the Sun
   D. wind

5. Which describes life during the early Proterozoic Era?
   A. simple, unicellular life forms
   B. complex, unicellular life forms
   C. simple, multicellular life forms
   D. complex, multicellular life forms

6. Which is not considered a biomass fuel?
   A. peat
   B. coal
   C. fecal material
   D. wood

Use the illustration below to answer Questions 7 and 8.

Chris Hellier/Science Source

7. Which type of fossil preservation is shown?
   A. trace fossil
   B. original remains
   C. carbon film
   D. altered hard parts

8. By studying the fossils, which is not something scientists can learn about the organism that left these prints?
   A. movement
   B. size
   C. habitat
   D. walking characteristics

9. When minerals in rocks fill a space left by a decayed organism, what type of fossil is formed?
   A. trace fossil
   B. cast fossil
   C. petrified fossil
   D. amber-preserved fossil

10. How are Mercury and the Moon similar?
    A. Both are covered with craters and plains.
    B. Both have the same night-to-day temperature difference.
    C. They have the same strength of surface gravity.
    D. Both have an extensive nickel-iron core.

Online Test Practice

SHORT ANSWER

Use the table below to answer Questions 11 to 13.

Apparent Temperature Index					
		Relative Humidity (%)			
Air Temperature (F°)		80	85	90	95
	85	97	99	102	105
	80	86	87	88	89
	75	78	78	79	79
	70	71	71	71	71

11. If the air temperature is 24°C and the relative humidity is 85%, what would the apparent temperature feel like?

12. What can be inferred about the effect relative humidity has on apparent temperature as the air temperature increases?

13. In the fall, when temperatures are moderate, how should a person plan for temperature with relative humidity factored in?

14. Although a hybrid car still requires fuel to run, why is it considered a better use of energy resources?

15. What are some steps mining companies are taking to be less destructive to the environment?

READING FOR COMPREHENSION

NASA Probe en Route to Interstellar Space

The 33-year old odyssey of NASA's *Voyager 1* spacecraft has reached a distant point at the edge of our solar system where there is no outward motion of solar wind. Now hurtling toward interstellar space some 10.8 billion miles from the Sun, *Voyager 1* has crossed into an area where the velocity of the hot ionized gas, or plasma, has slowed to zero. The event is a major milestone in *Voyager 1*'s passage through the heliosheath, the turbulent outer shell of the Sun's sphere of influence, and the spacecraft's upcoming departure from our solar system. Our Sun gives off a stream of charged particles that form a bubble known as the heliosphere around our solar system. The solar wind travels at supersonic speed until it crosses a shockwave called the termination shock. At this point, the solar wind dramatically slows down and heats up in the heliosheath. *Voyager 1* crossed the termination shock in December 2004 into the heliosheath. Scientists believe *Voyager 1* has not crossed the heliosheath into interstellar space. Crossing into interstellar space would mean a sudden drop in the density of hot particles and an increase in the density of cold particles. Researchers currently estimate *Voyager 1* will cross that frontier in about four years.

Article obtained from: NASA News Releases. NASA probe sees solar wind decline en route to interstellar space. *NASA*. December 13, 2010. (Online resource accessed January 23, 2011.).

16. What can be inferred from this passage?
A. Solar wind continues indefinitely.
B. Detecting the termination shock is impossible.
C. Solar wind slows at the edge of the solar system.
D. The heliosheath is 10.8 billion miles wide.

17. When *Voyager 1* crosses into interstellar space, it will detect an increase in the density of
A. cold particles.
B. solar wind.
C. hot particles.
D. the heliosphere.

NEED EXTRA HELP?															
If You Missed Question . . .	1	2	3	4	5	6	7	8	9	10	11	12	13	14	15
Review Section . . .	26.1	27.3	27.3	25.1	22.4	25.1	21.4	21.4	21.4	28.2	11.2	11.2	11.2	25.3	26.2

SHORT ANSWER

11. 26°C
12. As the air temperature increases, the relative humidity increase causes a greater increase of apparent temperature.
13. A person would be safe in assuming that the air temperature would not be affected by any relative humidity and could rely on the given air temperature as to how it will feel outside.
14. A hybrid car has two energy sources; electricity and fuel. This reduces the amount of fuel needed to power the car because the electricity provides some of the energy. This will help to extend fuel, which is a nonrenewable resource.
15. Mining companies are saving the topsoil removed from the land before mining the area to replace after the excavation. This helps the process of reclamation where the contour of the land and vegetation are restored. They are also using underground mining to remove minerals deep in Earth. This process is less disruptive than surface mining.

READING FOR COMPREHENSION

16. C
17. A

CHAPTER 29 Stars

BIGIDEA **The life cycle of every star is determined by its mass, luminosity, magnitude, temperature, and composition.**

ESSENTIAL QUESTIONS	NATIONAL STANDARDS	RESOURCES TO ASSESS MASTERY
SECTION 1 **The Sun** **1.** What are the layers and features of the Sun? **2.** How is the process of energy production in the Sun explained? **3.** How are the three types of spectra defined? 2.5 sessions 1.25 blocks	UCP.1–4; A.1–2; B.4; D.1, D.3; F.5–6	**Entry-Level Assessment** Focus Transparency 92, p. 830 **Progress Monitoring** Caption Question, pp. 831, 832, 834 Reading Check, pp. 831, 833, 835 Section Review, p. 836
SECTION 2 **Measuring the Stars** **1.** How are distances between stars measured? **2.** What is the difference between brightness and luminosity? **3.** What are the properties used to classify stars? 2 sessions 1 block	UCP.1–5; A.1–2; B.4; D.1, D.3; F.5; G.3	**Entry-Level Assessment** Focus Transparency 93, p. 837 **Progress Monitoring** Caption Question, pp. 840, 841 Reading Check, pp. 838, 841 Section Review, p. 846
SECTION 3 **Stellar Evolution** **1.** What is the relationship between mass and a star's evolution? **2.** What are the features of massive and regular star life cycles? **3.** How is the universe affected by the life cycles of stars? 3 sessions 1.5 blocks	UCP.1–5; A.1–2; B.4; D.1, D.3; E.1–2; F.6; G.3	**Entry-Level Assessment** Focus Transparency 94, p. 847 **Progress Monitoring** Caption Question, pp. 848, 850 Reading Check, p. 848 Section Review, p. 851 **Summative Assessment** Chapter Assessment, p. 855 Exam*View®Assessment Suite* CD-ROM

Go online!
connectED.mcgraw-hill.com

LEVELED RESOURCES	LAB MATERIALS
Science Notebook 29.1 OL **Chapter FAST FILE Resources:** Study Guide, p. 63 BL **Lab Resources:** Laboratory Manual, p. 225 OL **Transparencies:** Section Focus Transparency 92 EL	LaunchLAB p. 828 / **5 min a day for four days; 15 min the fifth day** clipboard, telescope, paper
Science Notebook 29.2 OL **Chapter FAST FILE Resources:** MiniLab Worksheet, p. 56 OL GeoLab Worksheet, p. 57 OL Study Guide, p. 65 BL **Lab Resources:** Laboratory Manual, p. 229 OL **Transparencies:** Section Focus Transparency 93 EL	MiniLAB p. 843 / **20 min** meterstick, 4-m piece of string, protractor GeoLAB p. 853 / **25 min** ruler
Science Notebook 29.3 OL **Chapter FAST FILE Resources:** Study Guide, p. 67 BL **Transparencies:** Section Focus Transparency 94 EL Teaching Transparency 90 OL EL	

LabManager™
Customize any lab with the LabManager™ CD-ROM.

ADDITIONAL RESOURCES

Technology:
- eTeacherEdition
- eStudentEdition
- Classroom Presentation Toolkit
- Exam*View® Assessment Suite* CD-ROM
- LabManager™ CD-ROM
- What's EARTH SCIENCE Got To Do With It? DVD
- Weather Classroom Video DVD

Assessment:
- Blueprints for Success
- Chapter Test (Scaffolded)
- Earth Science Standardized Test Practice

Lab Resources:
- Exploring Environmental Problems Laboratory Manual

BL Below Level OL On Level AL Advanced Learners EL English Learners COOP LEARN Cooperative Learning

CHAPTER 29

LaunchLAB

iLab Station | Rubric

How can you observe sunspots?

Process Skills observe and infer, interpret data, use numbers

Safety Precautions Approve lab safety forms before work begins. **WARNING:** ***Make sure students do not directly view the Sun.*** Do not allow them to look at the Sun through the telescope.

Teaching Strategies

- Be sure students realize that the sunspots are not moving around the Sun. The sunspots appear to be moving because the Sun is rotating.
- The rate of motion of the largest sunspot can be expressed as the estimated number of days for the sunspot to move across the face of the Sun. Because half of the Sun is visible, the Sun's period of rotation is twice the sunspot's rate of motion.

Procedure

WARNING: ***Do not look directly at the Sun. Do not look through the telescope at the Sun. You could damage your eyes.***

1. Have students read and complete the lab safety form and follow the procedure below.
2. Place a **clipboard** in front of the eyepiece of the **telescope** to observe an image of the Sun. Note that the telescope is pointed directly at the Sun, but the eyepiece is casting the image of the Sun on the clipboard.
3. Move the clipboard back and forth until you have the largest image of the Sun on the **paper.** Trace the outline of the Sun on your paper.
4. Trace sunspots that appear as dark areas on the Sun's image. Repeat this step at the same time each day for a week.
5. Measure the movement of sunspots.

Analysis

1. **Calculate** Use your data to determine the Sun's period of rotation. Answers will vary depending on location of the sunspots from about 25 days near the equator to about 35 days near the poles.
2. **Determine** What is the estimated rate of motion of the largest sunspot? Answers will vary. Possible answer: The rate of motion is about 16 days to cross the Sun's face.

Assessment

Knowledge Have students write a description of sunspots and an explanation of what they saw in the activity.

LabManager™
Customize any lab with the LabManager™ CD-ROM.

CHAPTER 29

Stars

BIG IDEA The life cycle of every star is determined by its mass, luminosity, magnitude, temperature, and composition.

SECTIONS

1 The Sun
2 Measuring the Stars
3 Stellar Evolution

LaunchLAB

iLab Station

How can you observe sunspots?

Although the Sun is an average star, it undergoes many complex processes. Sunspots are dark spots that are visible on the surface of the Sun. They can be observed moving across the face of the Sun as it rotates. In this activity, you will use a special telescope set-up to observe sunspots.

FOLDABLES® Study Organizer

Stars

Make a vocabulary book and label it as you read. Use it to record key vocabulary terms and their definitions.

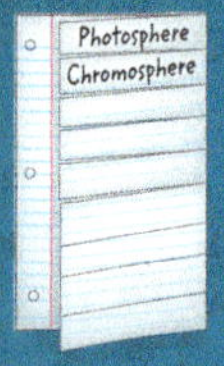

The final stages of a star can take many forms. Planetary nebulas represent the last gasp of stars like the Sun. When a more massive star collapses, it can create a supernova explosion, and perhaps become a pulsar—a rapidly rotating object that has a magnetic field a trillion times that of Earth.

Introduce the BIGIDEA

Life Cycles of Stars Lead students in a discussion of why the term *life cycle* is used to describe changes in a star. Have students describe examples of life cycles of living organisms. Examples include a lion's birth, growth, reproduction, and death; a butterfly's metamorphosis from a caterpillar; and a flower's germination, growth, flowering, pollination, formation of seeds, and death. Point out that even though stars are not living organisms, they have a beginning and undergo significant changes over time. Just as an organism's life cycle depends on its characteristics, the life cycle of a star depends on its characteristics.

Teacher Content Support

Star Changes Have students look at the photographs of stellar objects and read the caption. Ask how the three types of objects shown in the photographs are different. The photographs show different possible final stages of stars, depending on their masses. The nebula, which formed from the collapse of a star (a red giant), encases a white dwarf. The supernova is an explosion of a massive star (a supergiant). The pulsar is a massive star at the end of its life cycle (a neutron star). Students should understand that the photographs show how stars can change dramatically.

SECTION 1

1 Focus

Focus Transparency

Before presenting the lesson, project **Section Focus Transparency 92** and have students answer the accompanying questions. EL

MAINIDEA

The Sun's Mass Obtain 100 identical, small objects, such as marbles or pennies. Place 99 of the objects in a bag labeled *The Sun.* Then, place one of the objects in another bag labeled *Rest of the solar system.* Allow each student to compare the weights of the bags to help them understand how much of the solar system's mass is contained in the Sun.

2 Teach

Collaborative Learning

Solar Image WARNING: ***Make sure students do not directly view the Sun.*** Divide the class into small groups. Have each group make a small pinhole in a piece of cardboard to focus the Sun's light. One student should hold the cardboard perpendicular to the Sun's direction. Another student should hold the paper about a meter away, directly behind the pinhole. A small solar image will appear on the paper. The image size increases with separation between the pinhole and the paper. This activity works best when the sheet of paper is shaded from above so that direct sunlight does not drown out the solar image. OL EL COOP LEARN

SECTION 1

Essential Questions

- What are the layers and features of the Sun?
- How is the process of energy production in the Sun explained?
- How are the three types of spectra defined?

Review Vocabulary

magnetic field: the portion of space near a magnetic or current-carrying body where magnetic forces can be detected

New Vocabulary

photosphere
chromosphere
corona
solar wind
sunspot
solar flare
prominence
fusion
fission

The Sun

MAINIDEA The Sun contains most of the mass of the solar system and has many features typical of other stars.

EARTH SCIENCE 4 YOU **Have you ever had a sunburn from being outside too long on a sunny day? The Sun is more than 150 million km from Earth, but the Sun's rays are so powerful that humans still wear sunscreen for protection.**

Properties of the Sun

The Sun is the largest object in the solar system, in both diameter and mass. It would take 109 Earths, or almost 10 Jupiters, lined up edge to edge, to fit across the Sun. The Sun is about 330,000 times as massive as Earth and 1048 times the mass of Jupiter. In fact, the Sun contains more than 99 percent of all the mass in the solar system. It should not be surprising, then, that the Sun's mass affects the motions of the planets and other objects.

The Sun's average density is similar to the densities of the gas giant planets, represented by Jupiter in **Table 1.** Astronomers deduce densities at specific points inside the Sun, as well as other information, by using computer models that explain the observations they make. These models show that the density in the center of the Sun is about 1.50×10^5 kg/m^3, which is about 13 times the density of lead. A pair of dice as dense as the Sun's center would have a mass of about 1 kg.

Like many other stars, the Sun's interior is gaseous throughout because of its high temperature—about 1×10^7 K in the center. At this temperature, all of the gases are completely ionized, meaning the interior is composed only of atomic nuclei and electrons. This state of matter is known as plasma. Though partially ionized, the outer layers of the Sun are not hot enough to be plasma. The Sun produces the equivalent of 4 trillion trillion 100-W lightbulbs of light each second. The small amount that reaches Earth is equal to 1.35 kilowatt/m^2.

Explore the **properties of the Sun with an interactive table.** Concepts In Motion

Table 1 Relative Properties of the Sun

	Sun	Earth	Jupiter
Diameter (km)	1.4×10^6	1.3×10^4	1.4×10^5
Mass (kg)	2.0×10^{30}	6.0×10^{24}	1.9×10^{27}
Density (kg/m^3)	1.4×10^3	5.5×10^3	1.3×10^3

DIFFERENTIATED INSTRUCTION

Struggling Learners Help students visualize the tremendous size differences among the Sun, Jupiter, and Earth by having them glue a row of 109 small circles of construction paper (made with a one-hole punch) side-by-side on poster board to represent the Sun's diameter. Above that, have students glue 11 circles to represent Jupiter's diameter, and above that glue one circle to represent Earth's diameter.

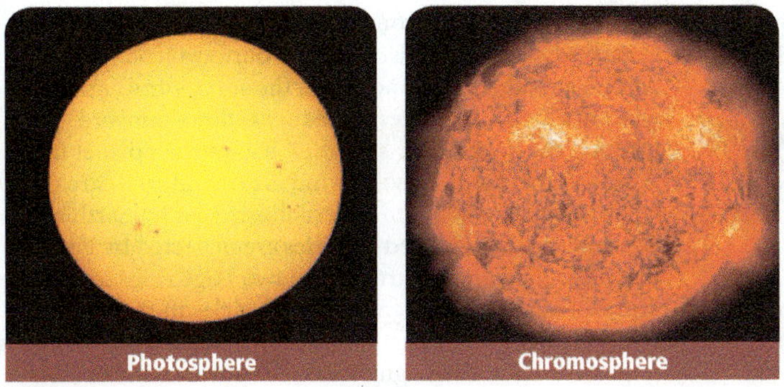

■ **Figure 1** Sunspots appear dark on the photosphere, the visible surface of the Sun. The white-hot areas are almost 6000 K while the darker, red areas are closer to 3000 K. The chromosphere of the Sun appears red with prominences and flares suspended in the thin layer. **Deduce** ***why the images look so different.***

The Sun's Atmosphere

You might ask how the Sun could have an atmosphere when it is already gaseous. Like many stars, the outer regions of the Sun are organized into layers. Each layer emits energy at wavelengths resulting from its temperature.

FOLDABLES® Incorporate information from this section into your Foldable.

Photosphere The **photosphere,** shown in **Figure 1,** is the visible surface of the Sun. It is approximately 400 km thick and has an effective temperature of 5800 K. It is also the innermost layer of the Sun's atmosphere. You might wonder how it is the visible surface of the Sun if it is the innermost layer. This is because most of the visible light emitted by the Sun comes from this layer. The two outermost layers are transparent at most wavelengths of visible light. Additionally, the outermost two layers are dim in the wavelengths they emit.

☑ READING CHECK **Explain** why the innermost layer of the Sun's atmosphere is visible.

Chromosphere Outside the photosphere is the **chromosphere,** which is approximately 2500 km thick and has an average temperature of 15,000 K. Usually, the chromosphere is visible only during a solar eclipse when the photosphere is blocked. However, astronomers can use special filters to observe the chromosphere when the Sun is not eclipsed. The chromosphere appears red, as shown in **Figure 1,** because its strongest emissions are in a single band in the red wavelength.

Corona The outermost layer of the Sun's atmosphere, called the **corona,** extends several million kilometers from the outside edge of the chromosphere and usually has a temperature of about 3 to 5 million K. The density of the gas in the corona is very low, which explains why the corona is so dim that it can be seen only when the photosphere is blocked by either special instruments, as in a coronagraph, or by the Moon during an eclipse, as shown in **Figure 2.** The temperature is so high in these outer layers of the solar atmosphere that the radiation emitted most is of ultraviolet wavelengths for the chromosphere, and X rays for the corona.

■ **Figure 2** The Sun's hottest and outermost layer, the corona, is only seen when the disk of the Sun is blocked as by this solar eclipse.

Interpret the Illustration

Filtered Light Allow students to look at colorful objects through different-colored filters. The color of a filter is the color of light the filter transmits. Lead students in discussing the method of filtering sunlight to make the chromosphere visible, as in **Figure 1.** OL

■ **Caption Question** **Fig. 1** The two images show different types of detail. The bright prominences and white areas of the right-hand photo are the sunspots of the left-hand photo. The difference is due to the types of filters used to take the photos.

FOLDABLES® Rubric

Teacher Content Support

Visible Sunlight One of the basic properties of radiation from heated bodies or objects is that the wavelength at which an object glows most brightly is inversely proportional to the temperature of the emitting object. This explains why the Sun's photosphere emits in visible wavelengths, while the much hotter corona glows most brightly in the X-ray portion of the electromagnetic spectrum. This principle also explains the different colors of stars of different temperatures. Red stars are cooler, while blue stars are hotter.

☑ READING CHECK The two outermost layers are transparent.

Use Science Terms

Astronomical Names The history of astronomy is apparent in the names given to astronomical phenomena. Many of these names are derived from Latin or Greek, and words from these languages are often intermixed. For example, chromosphere comes from the Greek *chromos,* meaning *red,* and *sphere. Corona* is a Latin word for *crown.*

ACROSS THE CURRICULUM

Physics Students might wonder why the chromosphere and corona are so much hotter than the solar surface. Just below the photosphere is a turbulent zone where convection causes the gases to churn and overturn. Moving charged particles in this turbulent zone produce magnetic fields that are constantly changing, sending out magnetic waves. These magnetic waves transport energy to higher levels, above the photosphere. There, the waves interact with and transfer energy to other particles, increasing the thermal energy of both the chromosphere and corona.

Collaborative Learning

Aurora Colors Have students brainstorm ideas explaining why the aurora has so many different colors. The colors correspond to different gases in Earth's atmosphere. Colors also depend on the height at which the particles from the Sun collide with the different elements in the gases. OL COOP LEARN

■ **Caption Question Fig. 3** The aurora is affected by the solar particles that are blown to Earth's magnetic field. It is most active during solar activity periods, but it can possibly be seen anytime. The aurora is best seen at night for contrast against the dark sky.

Teacher Content Support

The Sun's Rotation Astronomers can measure the Sun's rate of rotation by observing how long it takes a particular sunspot or group of sunspots to go completely around the Sun, as students have done in the Launch Lab at the beginning of this chapter. The Sun rotates most rapidly at the equator, which completes an entire rotation in about 25 days, and most slowly at the poles, which take about 35 days to rotate completely. Not only does the rate of rotation vary with latitude on the Sun, but it also varies with depth inside the Sun. This is known from the analysis of subtle pulsations that the solar surface undergoes, which reflect internal density and rates of rotation. The solar core rotates more rapidly than the surface.

Aurora from Earth

Aurora from space

■ **Figure 3** The aurora is the result of particles from the Sun colliding with gases in Earth's atmosphere. It is best viewed from regions around the poles of Earth.
Infer *When can you see the aurora?*

Solar wind The corona of the Sun does not have an abrupt edge. Instead, plasma flows outward from the corona at high speeds and forms the **solar wind.** As this wind of charged particles, called ions, flows outward through the entire solar system, it bathes each planet in a flood of particles. The solar wind is not uniform. Streams of 300 km/s and 800 km/s alternatively pass by Earth as the Sun rotates. The charged particles are deflected by Earth's magnetic field and are trapped in two huge rings, called the Van Allen belts. The high-energy particles in these belts collide with gases in Earth's atmosphere and cause the gases to give off light. This light, called the aurora, can be seen from Earth or from space, as shown in **Figure 3.** Aurorae are generally seen from Earth in the polar regions.

Solar Activity

While the solar wind and layers of the Sun's atmosphere are permanent features, other features on stars change over time in a process called solar activity. Some of the Sun's activity includes fountains and loops of glowing gas. Some of this gas has structure—a certain order in both time and place. This structure is driven by magnetic fields.

The Sun's magnetic field and sunspots The Sun's magnetic field disturbs the solar atmosphere periodically and causes new features to appear. The most obvious features are **sunspots,** shown in **Figure 4,** which are dark spots on the surface of the photosphere. Sunspots are bright, but they appear darker than the surrounding areas on the Sun because they are cooler. They are located in regions where the Sun's intense magnetic fields penetrate the photosphere. Magnetic fields create pressure that counteracts the pressure from the hot, surrounding gas. This stabilizes the sunspots despite their lower temperature. Sunspots occur in pairs with opposite magnetic polarities—with a north and a south pole similar to a magnet.

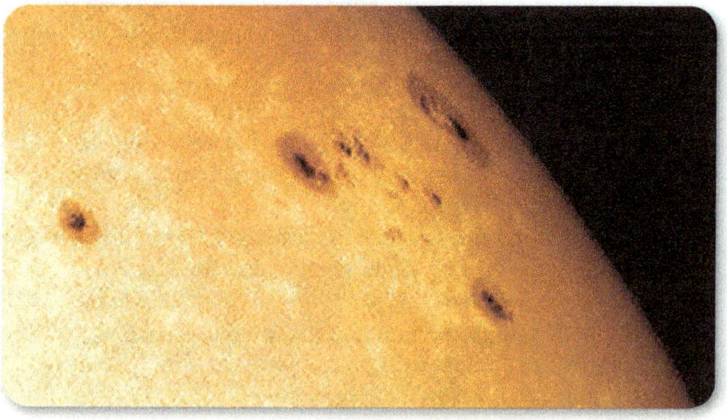

■ **Figure 4** Sunspots are dark, relatively cool spots on the surface of the photosphere. These dark areas are associated with the Sun's magnetic field. Sunspots typically last several days, but can last for many months.

Project

Sunspot Cycle Have students work in pairs to research historical data on the annual numbers of sunspots over a period of some decades. Then, have them examine the 11-year cycle of sunspot maxima. Students can use calculators or computers to graph their data. OL AL EL COOP LEARN

Demonstration

The Sun's Magnetic Field Label the poles of two bar magnets. Put the north and south poles together to demonstrate that they are opposite. Then, flip one magnet so that the like poles are together, and demonstrate to students that like poles repel. Hold one bar magnet with the north pole pointed up. Explain to students the magnetic field of the Sun is aligned like this for 11 years. Then, flip the bar magnet so the south pole is pointing up, and explain the Sun's magnetic field is aligned this way for another 11 years. Flip the magnet again so the north pole is pointing up, and explain this represents the completion of the 22-year solar activity cycle.

Solar activity cycle Astronomers have observed that the number of sunspots changes in a predictable and set pattern. This change in number from minimum to maximum and then back to minimum again, is called the sunspot cycle and takes about 11 years to complete. At this point, the Sun's magnetic field reverses, so that the north magnetic pole becomes the south magnetic pole and vice versa. Because sunspots are caused by magnetic fields, the polarities of sunspot pairs reverse when the Sun's magnetic poles reverse. Therefore, when the polarity of the Sun's magnetic field is taken into account, the length of the cycle doubles to approximately 22 years. At this point, the magnetic field then switches back to the original polarity and the solar activity cycle starts again.

☑ READING CHECK **Determine** how often the Sun's magnetic poles reverse themselves.

Other solar features Coronal holes, only detectable in X-ray photography and shown in **Figure 5,** are often located over sunspot groups. Coronal holes are areas of low density in the gas of the corona and are the main regions from which the particles that comprise the solar wind escape.

Highly active solar flares are also associated with sunspots, as shown in **Figure 5. Solar flares** are violent eruptions of particles and radiation from the surface of the Sun. Often, the released particles escape the surface of the Sun in the solar wind and Earth gets bombarded with the particles a few days later. The largest recorded solar flare, which occurred in November 2003, hurled particles from the Sun's surface at nearly 9 million km/h.

Another active feature, sometimes associated with flares, is a **prominence,** which is an arc of gas that is ejected from the chromosphere, or is gas that condenses in the inner corona and rains back to the surface. **Figure 5** shows an image of a prominence. Prominences can reach temperatures greater than 50,000 K and can last from a few hours to a few months. Like flares, prominences are also associated with sunspots and the magnetic field, and occurrences of both vary with the solar-activity cycle.

Figure 5 Features of the Sun's surface include coronal holes into the surface and solar flares and prominences that erupt from the surface.

Teacher Content Support

Solar Flares When a major solar flare occurs in a position on the Sun such that the released radiation and particles strike Earth, there are widespread effects. The X-ray radiation reaches Earth in eight minutes and affects the ionization in Earth's upper atmosphere, but it has little effect on the ground. The charged particle burst takes from two to four days to reach Earth's atmosphere. When it does, it can disrupt the upper atmosphere and the radiation belts. Some particles can even reach the ground, where they can disrupt electrical circuits.

A massive solar flare in December 2006 caused disruptions in satellite and radio wave communication systems on Earth. Electronic equipment on spacecraft orbiting Earth was also affected, and astronauts aboard the International Space Station slept in areas shielded from radiation.

☑ READING CHECK The poles reverse every 11 years or so, which is half of the solar activity cycle.

Use an Analogy

Solar Wind Because of the Sun's rotation, the solar wind does not radiate outward from the Sun directly. Instead, it follows a spiraling path, similar to the motion of water from a rotary lawn sprinkler. As the sprinkler turns, water drops follow curved outward paths. This analogy helps to explain why a solar flare that is observed on the side of the Sun facing Earth will not necessarily affect Earth. For the burst of particles from a flare to reach Earth, the flare must occur at a location on the Sun from which the solar wind's curving path will intersect Earth.

ACROSS THE CURRICULUM

History Around the time of maximum solar activity, the increased flux of radiation and charged particles from the Sun heats Earth's upper atmosphere, causing it to expand. This, in turn, creates extra drag on Earth-orbiting satellites that can cause them to spiral into the atmosphere. This is what happened to the U.S. *Skylab* space station in 1979, when its orbit decayed, and the station fell to Earth several years sooner than expected.

Teacher Content Support

Radiation Within the Sun Electromagnetic radiation travels at the speed of light in a vacuum but it is slower in any medium. This explains why it can take a long time for radiation produced in the Sun's core to emerge at the solar surface. In a medium as dense as the deep solar interior, light is readily absorbed by atoms and ions. A particle of radiation, known as a photon, can travel only a short distance before being absorbed. The atom or ion that absorbs the photon quickly emits it again, but in a random direction. The photon might be reemitted in the same direction from which it came or in some other direction. The path followed by a photon in a dense medium is called a random walk.

Discussion

Solar-Stellar Connection Lead students in a discussion of how scientists might connect what they learn about the Sun and what they learn about other stars. Explain that astronomers can measure details on the solar surface, and they have obtained a centuries-long record of the Sun's activity cycle. These data are helpful in interpreting surface phenomena and activity cycles of other stars. In the same way, data on how other stars change with time help astronomers deduce where the Sun is in its life cycle, how it looked in the distant past, and what will happen to it far in the future. The so-called solar-stellar connection has been a fruitful area of research for astronomers.

Caption Question Fig. 7 Spectrum colors are due to the wavelengths of radiation produced by or reflected from the object being observed. The colors in an emission spectrum correspond to the different wavelengths of radiation emitted by an object. Each element has a specific pattern of wavelengths.

Figure 6 Energy in the Sun is transferred mostly by radiation from the core outward to about 75 percent of its radius. The outer layers transfer energy in convection currents.

The Solar Interior

You might be wondering where all the energy that causes solar activity and light comes from. Fusion occurs in the core of the Sun, where the pressure and temperature are extremely high. **Fusion** is the combination of lightweight, atomic nuclei into heavier nuclei, such as hydrogen fusing into helium. This is the opposite of the process of **fission,** which is the splitting of heavy atomic nuclei into smaller, lighter nuclei, like uranium into lead.

Energy production in the Sun In the core of the Sun, helium is a product of the process in which hydrogen nuclei fuse. The mass of the helium nucleus is less than the combined mass of the four hydrogen nuclei, which means that mass is lost during the process. Albert Einstein's special theory of relativity shows that mass and energy are equivalent, and that matter can be converted into energy and vice versa. This relationship can be expressed as $E = mc^2$, where E is energy measured in joules, m is the quantity of mass that is converted to energy measured in kilograms, and c is the speed of light measured in m/s. This theory explains that the mass lost in the fusion of hydrogen to helium is converted to energy, which powers the Sun. At the Sun's rate of hydrogen fusing, it is about halfway through its lifetime, with approximately 5 billion years left. Even so, the Sun has used only about 3 percent of its hydrogen.

Energy transport If the energy of the Sun is produced in the core, how does it get to the surface before it travels to Earth? The answer lies in the two zones in the solar interior illustrated in **Figure 6.** In the inner portion of the Sun, extending to about 86 percent of its radius, energy is transferred by radiation. This is the radiation zone. Above that, in the convection zone, energy is transferred by gaseous convection currents. As energy moves outward, the temperature is reduced from a central value of about 1×10^7 K to its photospheric value of about 5800 K. Leaving the Sun's outermost layer, energy moves in a variety of wavelengths in all directions. A tiny fraction of that immense amount of solar energy eventually reaches Earth.

Figure 7 Energy excites the elements of a substance so that it emits different wavelengths of light.
Infer *what the colors of a spectrum represent.*

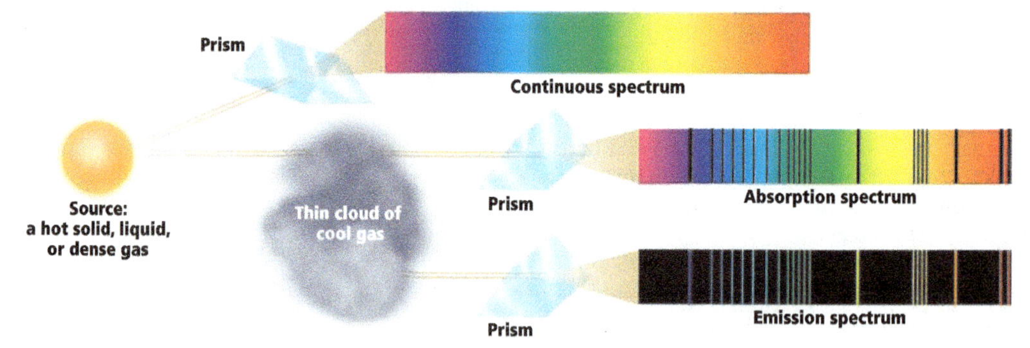

EARTH SCIENCE JOURNAL

Nuclear Fusion Many people think nuclear fusion could be a safe and plentiful source of energy. Fusion on Earth has been achieved in hydrogen bombs, where hydrogen nuclei fuse and instantaneously release tremendous amounts of energy. Discuss with students the conditions that exist within the Sun that enable fusion to occur naturally. high densities and temperatures Have students use this description to write an explanation of why, so far, sustained and controlled fusion has not been achieved on Earth. The high densities and temperatures are difficult to reproduce on Earth. OL EL

Rubric

Solar energy on Earth The quantity of energy that arrives on Earth every day from the Sun is enormous. Above Earth's atmosphere, 1354 J of energy is received in 1 m^2/s (1354 W/m^2). In other words, 13 100-W lightbulbs could be operated with the solar energy that strikes a 1-m^2 area. However, not all of this energy reaches the ground because some is absorbed and scattered by the atmosphere.

Spectra

You are probably familiar with the rainbow that appears when white light is shined through a prism. This rainbow is a spectrum (plural, spectra), which is visible light arranged according to wavelengths. There are three types of spectra: continuous, emission, and absorption, as shown in **Figure 7.**

A spectrum that has no breaks in it, such as the one produced when light from an ordinary bulb is shined though a prism, is called a continuous spectrum. A continuous spectrum can also be produced by a glowing solid or liquid, or by a highly compressed, glowing gas. The spectrum from a noncompressed gas contains bright lines at certain wavelengths. This is called an emission spectrum, and the lines are called emission lines. The wavelengths of the visible lines depend on the element being observed because each element has its own characteristic emission spectrum.

☑ READING CHECK **Describe** continuous and emission spectra.

A spectrum produced from the Sun's light shows a series of dark bands. These dark spectral lines are caused by different chemical elements that absorb light at specific wavelengths. This is called an absorption spectrum, and the lines are called absorption lines. Absorption is caused by a cooler gas in front of a source that emits a continuous spectrum. The pattern of the dark absorption lines of an element is exactly the same as the bright emission lines for that same element. Thus, by comparing laboratory spectra of different gases with the dark lines in the solar spectrum, it is possible to identify the elements that make up the Sun's outer layers. You will experiment with identifying spectral lines in the GeoLab at the end of this chapter.

Data Analysis LAB

Based on Real Data*

Interpret Data

Can you identify elements in a star?
Astronomers study the composition of stars by observing their absorption spectra. Each element in a star's outer layer produces a set of lines in the star's absorption spectrum. From the pattern of lines, astronomers can determine what elements are in a star.

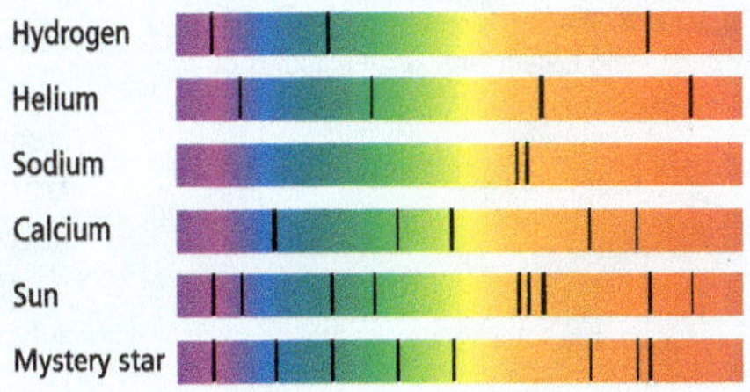

Analysis

1. Study the spectra of the four elements.
2. Examine the spectra for the Sun and the mystery star.
3. To identify the elements of the Sun and the mystery star, use a ruler to help you line up the spectral lines with the known elements.

Think Critically

4. **Identify** the elements that are present in the part of the absorption spectrum shown for the Sun.
5. **Identify** the elements that are present in the absorption spectrum for the mystery star.
6. **Determine** which elements are common to both stars.

*James B. Kaler. Professor Emeritus of Astronomy. University of Illinois. 1998.

Data Analysis LAB

About the Lab

- Each element is made of atoms with a unique arrangement of electrons. The outer electrons of an atom can absorb and emit energy. An element's unique absorption and emission spectra are determined by the specific arrangement of the electrons.
- See also Silva, D. R., and M. E. Cornell. 1992. A new library of stellar optical spectra. *The Astrophysical Journal Supplement Series* 81 (August): 865–881.

Think Critically

4. hydrogen, helium, sodium
5. hydrogen, calcium
6. hydrogen

☑ READING CHECK Continuous spectra are bands of color emanating from a glowing object. Emission spectra are dark with specific bands of color from a noncompressed gas.

Tie to Previous Knowledge

Lightweight Elements Have students consider why the Sun is dominated by hydrogen and helium, but these elements are relatively rare on Earth. Students should recall that the Sun and Earth formed from the same material, but the lightweight hydrogen and helium were unable to condense during the formation of the solar system and therefore were not retained by Earth.

Enrichment

Solar Fusion Have students research and make a model of the three-step process by which the Sun fuses hydrogen (H) into helium (He) with the release of neutrinos (ν) and photons (γ).

$2({}^1_1H + {}^1_1H \rightarrow {}^2_1H + {}^0_1\beta + {}^0_0\nu)$
$2({}^2_1H + {}^1_1H \rightarrow {}^3_2He + {}^0_0\gamma)$
${}^3_2He + {}^3_2He \rightarrow {}^4_2He + {}^1_1H + {}^1_1H$ AL

DIFFERENTIATED INSTRUCTION

Visually Impaired To help illustrate the difference between emission and absorption spectra, use two materials with different textures, such as sandpaper and felt. For the emission spectra, cut long, thin strips of the felt and place them in a lined pattern interspersed with sandpaper. Tell students the soft felt corresponds to emission, where there is color, while the sandpaper corresponds to spaces where there is no color. Repeat the activity, but this time, use the sandpaper to make thin strips. Again, tell students the sandpaper represents spaces where there is no color (absorption lines), while the felt represents where there is color (background continuum).

3 Assess

Check for Understanding

Activity Have each student create an illustration that summarizes the types of electromagnetic radiation from the Sun. Students should identify the layer from which each type is emitted.

Reteach

Compare and Contrast Have each student rank the photosphere, chromosphere, and corona in order of thickness and temperature. from thickest to thinnest: corona, chromosphere, photosphere; from highest to lowest temperature: corona, chromosphere, photosphere

Assessment

Skill Ask students to compare convection in the Sun and in Earth. Convection is the mechanism of thermal energy transport in the outer zones of the solar interior. Similarly, convection is responsible for thermal energy transport within Earth.

Element Composition of the Sun by Mass

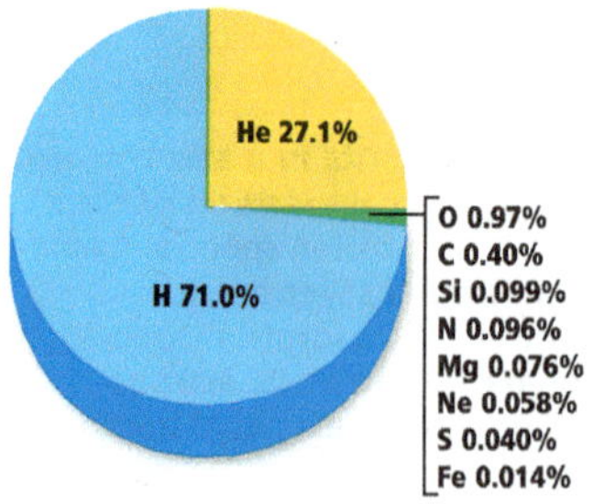

■ **Figure 8** The Sun is composed primarily of hydrogen and helium with small amounts of other gases.

Solar Composition

Although scientists have not been able to take samples from the Sun directly, they have learned a great deal about the Sun's composition from its spectra. Using the lines of the absorption spectra like fingerprints, astronomers have identified the elements that compose the Sun. Sixty or more elements have been identified as solar components. The Sun consists of primarily of hydrogen (H), at about 71.0 percent by mass, helium, (He) 27.1 percent, and a small amount of other elements, as illustrated in **Figure 8.** This composition is similar to that of the gas giant planets. It suggests that the Sun and the gas giants represent the composition of the interstellar cloud from which the solar system formed. While the terrestrial planets have lost most of the lightweight gases, their heavier element composition probably came from a contribution to the interstellar cloud of by-products from long-extinct stars.

The Sun's composition represents that of the galaxy as a whole. Most stars have proportions of the elements similar to the Sun. Hydrogen and helium are the predominant gases in stars and in the rest of the universe. Even dying stars still have hydrogen and helium in their outer layers, because their internal temperatures might only fuse about 10 percent of their total hydrogen into helium. All other elements are in small proportions compared to hydrogen and helium. The larger the star's mass at its inception, the more heavy elements it will produce in its lifetime. But, as you will read in this chapter, there are different stages and results of a star's death. As stars die, they return as much as 50 percent of their mass back into interstellar space, to be recycled into new generations of stars and planets.

SECTION 1 REVIEW

Section Self-Check

Section Summary

- Most of the mass in the solar system is found in the Sun.
- The Sun's average density is approximately equal to that of the gas giant planets.
- The Sun has a layered atmosphere.
- The Sun's magnetic field causes sunspots and other solar activity.
- The fusion of hydrogen into helium provides the Sun's energy and composition.

Understand Main Ideas

1. MAINIDEA **Identify** which features of the Sun are typical of stars.
2. **Describe** the outer layers of gas above the Sun's visible surface.
3. **Classify** the different types of spectra by how they are created.
4. **Describe** the process of fusion in the Sun.
5. **Compare** the composition of the Sun in **Figure 8** to the gas giant planets' compositions.

Think Critically

6. **Infer** how the Sun would affect Earth if Earth did not have a magnetic field.
7. **Relate** the solar activity cycle with solar flares and prominences.

WRITINGIN▶ Earth Science

8. Create a trifold brochure relating the layers and characteristics of the Sun.

SECTION 1 REVIEW

1. Features of the Sun that are typical of stars include layering (core, corona, chromosphere, and photosphere), plasma state, solar wind, sunspots, magnetic field, coronal holes, and solar flares. The Sun's composition—hydrogen, helium, and a small amount of other elements—is also typical of the composition of other stars.
2. The Sun's visible surface, the photosphere, has a temperature of around 5800 K. Above this is the chromosphere, which is approximately 2500 km thick and has an average temperature of about 15,000 K. Above this is the corona, which has a temperature of about 3–5 million K.
3. Continuous spectra are created by a hot solid, liquid, or dense gas; emission lines are created by a low-density, hot gas; and absorption lines are created by a cool gas in front of a hot continuous source.
4. Fusion is the combining of two light nuclei to form a heavier nucleus. In the Sun, two hydrogen nuclei combine to form helium. Because the mass of helium is less than the sum of the masses of two hydrogen nuclei, the difference is released in the form of energy.
5. The Sun's composition is 71.0% hydrogen and 27.1% helium (98.1%). The gas giant planets average 87% hydrogen and 12% helium (99%).
6. Charged particles from the solar wind would harm electronic communication systems. All life on Earth would either adapt or die.
7. Solar flares and prominences are the result of the magnetic field of the Sun. With changes in polarity, the activity cycle changes, and so do the numbers and sizes of flares and prominences.
8. Check students' brochures for accuracy.

Rubric

Measuring the Stars

MAINIDEA Stellar classification is based on measurement of light spectra, temperature, and composition.

EARTH SCIENCE 4 YOU **As you ride in a car on the highway at night and as a car approaches you, its lights seem to get larger and brighter. Distant stars might be just as large and just as bright as nearer ones, but the distance causes them to appear small and dim.**

Essential Questions

- How are distances between stars measured?
- What is the difference between brightness and luminosity?
- What are the properties used to classify stars?

Review Vocabulary

wavelength: the distance from one point on a wave to the next corresponding point

New Vocabulary

constellation
binary star
parsec
parallax
apparent magnitude
absolute magnitude
luminosity
Hertzsprung-Russell diagram
main sequence

Patterns of Stars

Long ago, many civilizations looked at the brightest stars and named groups of them after animals, mythological characters, or everyday objects. These groups of stars are called **constellations.** Today, astronomers group stars by the 88 constellations named by ancient peoples. Some constellations are visible throughout the year, depending on the observer's location. In the northern hemisphere, you can see constellations that appear to rotate around the north pole. These constellations are called circumpolar constellations. Ursa Major, which contains the Big Dipper, is a circumpolar constellation for most of the northern hemisphere.

Unlike circumpolar constellations, the other constellations can be seen only at certain times of the year because of Earth's changing position in its orbit around the Sun, as illustrated in **Figure 9.** For example, the constellation Orion can be seen in the northern hemisphere's winter, and the constellation Hercules can be seen in the northern hemisphere's summer. For this reason, constellations are classified as summer, fall, winter, and spring constellations. The most familiar constellations are the ones that are part of the zodiac. These twelve constellations lie in the ecliptic plane along the same path where the planets are seen. Different constellations can be seen in the northern and southern hemispheres, but the zodiac can be seen in both. Ancient people used the constellations to know when to prepare for planting, harvest, and ritual celebrations.

Figure 9 Different constellations are visible in the sky due to Earth's movement around the Sun.

SECTION 2

1 Focus

Focus Transparency

Before presenting the lesson, project **Section Focus Transparency 93** and have students answer the accompanying questions. **EL**

MAINIDEA

Starlight Have students read the Main Idea and think about the stars they see in the night sky. Ask them to name words they could use to compare how the stars shine. possible answers: brighter, dimmer, bigger, smaller Ask students whether they believe the brightness of a star is a reliable measure of the star's size or distance. Explain they will find the answer to this question in this lesson.

2 Teach

Teacher Content Support

Naming Stars in Constellations Astronomers use Greek letters to indicate rank within a constellation. Polaris, the brightest star in Ursa Minor, for example, is α Ursae Minoris. The rank is designated by a Greek letter, but the constellation name is given in Latin. Numbers are used when there are more bright stars in a constellation than there are letters in the Greek alphabet. Fainter stars are generally listed only by numbers rather than names.

CULTURAL DIVERSITY

Star Names Some names of stars have long histories. Many of the brightest stars were named by Arabic astronomers, and in many cases these reflect the roles of the stars in the earlier Greek constellation mythology. Examples of Arabic names include Vega, Fomalhaut, Deneb, and Sirius.

FOLDABLES® Rubric

Interpret the Photo

Star Interactions The more dense a star cluster is, the more frequently stars come close to each other. After students study the photos in **Figure 10,** ask them how stars affect each other. through their gravitational forces Then, ask how stars are affected by close encounters. Their orbits can be altered. Explain the result of many encounters is a rounded, smooth cluster shape, as in the globular M13. Clusters that are not so dense take much longer to reach this state. The Pleiades group is a relatively young cluster with fewer members and an irregular overall shape. Globular clusters are not only dense, but also are the oldest objects in the galaxy, and thus, many close encounters between stars have occurred in such clusters. OL EL

☑ READING CHECK Open clusters are not densely packed, while globular clusters are usually larger and densely packed.

Tie to Previous Knowledge

Star Mass Measurement Binary stars provide a good opportunity for astronomers to measure the masses of stars. Stellar masses are measured using another form of Kepler's third law, which students previously applied to planetary motion. The masses of the two stars can be found if both the orbital period and semi-major axis can be observed. Typical stellar masses range from just below 0.1 times the Sun's mass up to about 150 solar masses.

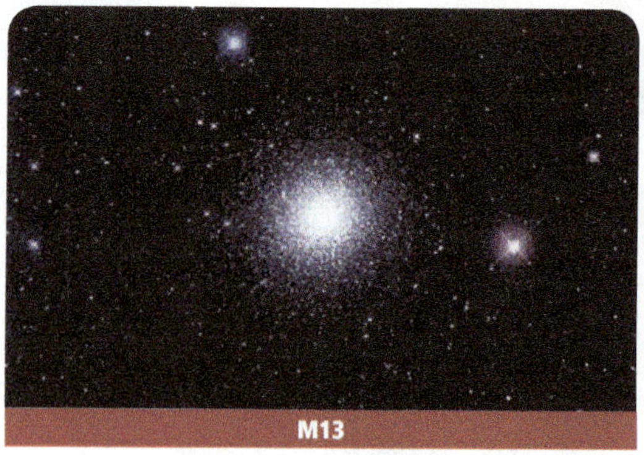

Figure 10 Star clusters are groups of stars that are gravitationally bound to one another. The Pleiades is an open cluster group and M13 is a globular cluster.

Star clusters Although the stars in constellations appear to be close to each other, few are gravitationally bound to one other. The reason that they appear to be close together is that human eyes cannot distinguish how far or near stars are. Two stars could appear to be located next to each other in the sky, but one might be 100 trillion km from Earth, and the other might be 200 trillion km from Earth. However, by measuring distances to stars and observing how their gravities interact with each other, scientists can determine which stars are gravitationally bound to each other. A group of stars that are gravitationally bound to each other is called a cluster. The Pleiades (PLEE uh deez) in the constellation Taurus, shown in **Figure 10,** is an open cluster because the stars are not densely packed. In contrast, a globular cluster is a group of stars that are densely packed into a spherical shape, such as M13 in the constellation Hercules, also shown in **Figure 10.** Different kinds of clusters are explained in **Figure 12.**

☑ READING CHECK **Distinguish** between open and globular clusters.

Figure 11 Sirius and its companion star, seen below and to the left, are the simplest form of stellar grouping, known as a binary.

Binaries When only two stars are gravitationally bound together and orbit a common center of mass, they are called **binary stars.** More than half of the stars in the sky are either binary stars or members of multiple-star systems. The bright star Sirius is half of a binary system, shown in **Figure 11.** Most binary stars appear to be single stars to the human eye, even with a telescope. The two stars are usually too close together to appear separately, and one of the two is often much brighter than the other.

Astronomers are able to identify binary stars through the use of several methods. For example, even if only one star is visible, accurate measurements can show that its position shifts back and forth as it orbits the center of mass between it and the unseen companion star. Also, the orbital plane of a binary system can sometimes be seen edgeways from Earth. In such cases, the two stars alternately block each other and cause the total brightness of the two-star system to dip each time one star passes in front of the other. This type of binary star is called an eclipsing binary.

EARTH SCIENCE JOURNAL

The Pleiades Because of its high visibility, the open cluster Pleiades has been an important factor in the mythology of different cultures throughout history. Have students research and write about one or more of the ideas a different culture has had about the Pleiades. OL

Rubric

VISUALIZING Star Groupings

Figure 12 When you look into the night sky, the stars seem to be randomly spaced from horizon to horizon. Upon closer inspection, you begin to see groups of stars that seem to cluster in one area. Star clusters are gravitationally bound groups of stars, which means that their gravities interact to hold the stars in a group.

Galaxy Not a true cluster, a galaxy is a very large star grouping that contains a variety of different clusters of stars.

Globular clusters are made from densely packed groups of stars that are the same age. Their gravities hold them into a rounded cluster. Many globular clusters are found in the haloes of galaxies.

Open clusters are loosely organized groups of stars that are not densely packed. These two open clusters in Perseus are young, and contain a mixture of stellar types from stars dimmer than the Sun to giants and supergiants.

Binaries are the smallest of all star groupings, consisting of only two stars orbiting around a single center of gravity.

Concepts In Motion View an **animation of star groupings.**

(tl)Jason T. Ware/Photo Researchers, (tr)L. Dodd/Photo Researchers, (c)Stephen & Donna O'Meara/Photo Researchers, (b)John Chumack/Photo Researchers

VISUALIZING

Purpose

Students will compare different types of star clusters.

Teacher Content Support

Types of Clusters The two main types of star clusters are globular clusters and open clusters. Globular clusters are older, are composed of lighter elements, and are densely packed into a spherical shape. The Milky Way might have about 200 globular clusters (although estimates vary) located mostly in the halo. Open clusters, also known as galactic clusters, are younger and contain heavier elements than those in globular clusters. They are not as densely packed and have a looser arrangement. Open clusters contain fewer stars and are located closer to the galactic plane.

Teaching Strategies

- Students will learn more about haloes and the locations of globular clusters in the next chapter. For now, explain that a halo is the region around the galaxy's central bulge.
- To emphasize the differences between globular clusters and open clusters, have students make a chart comparing the two types of clusters.

DIFFERENTIATED INSTRUCTION

Visually Impaired Have students create models of globular clusters, open clusters, and binaries using modeling clay. Lead them in discussing differences in the structures of the clusters. For example, discuss and model how the two stars of a binary orbit around a single center of gravity, and discuss how the shape of globular clusters is related to their density and gravitational pull.

Caption Question **Fig. 13** If a star is moving toward the observer, the spectral lines are shifted toward shorter wavelengths, or blueshifted. However, if the star is moving away, the spectral lines are shifted toward longer wavelengths, or redshifted.

Use an Analogy

Shifts in Sound Waves Sound waves undergo shifts similar to the Doppler shifts in star color. The shifts in sound waves are much easier to observe in everyday situations. Ask students whether they have ever noticed the change in pitch of a police or ambulance siren as the emergency vehicle passed by. Ask them whether the pitch of the sound became higher or lower. The pitch becomes higher as the sound approaches and lower as it moves away. How do the changes in pitch compare with redshifts and blueshifts? Redshifts in star color are analogous to a decrease in pitch as sound moves away. Blueshifts in star color are analogous to an increase in pitch as sound approaches. BL OL

Apply Earth Science

Radar Detection Students might be interested to know that police make use of the Doppler effect in timing the speeds of cars. Infrared laser emission from a radar gun is reflected from an oncoming car and is shifted in wavelength by the motion of the car. The device has circuitry to measure the wavelength shift and determine the car's speed. Ask students whether this is a redshift or a blueshift. a blueshift BL OL EL

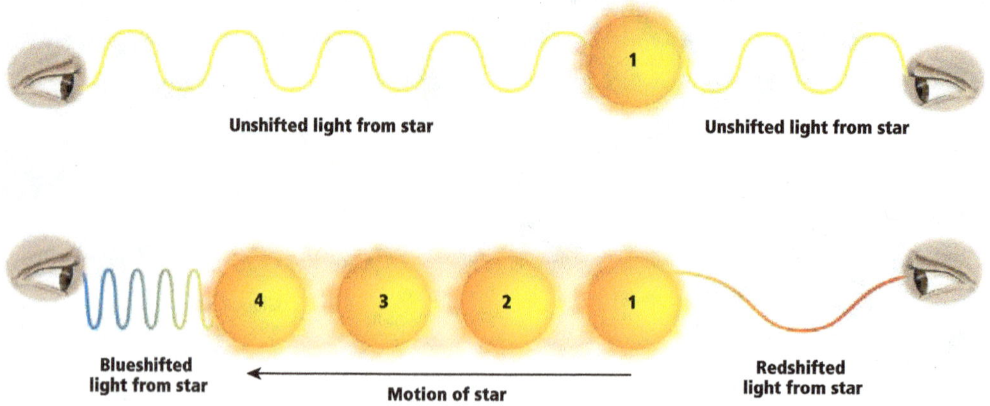

Figure 13 When a star moves toward the observer, the light emitted by the star shifts toward the blue end of the elecromagnetic spectrum. When a star moves away from the observer, its light shifts toward the red. Scientists use Doppler shift to determine the speed and direction of a star's motion.
Explain *how the Doppler shift causes color changes.*

View an **animation of Doppler redshifts and blueshifts.**
Concepts In Motion

Doppler shifts The most common way to tell that a star is one of a binary pair is to find subtle wavelength shifts, called Doppler shifts. As the star moves back and forth along the line of sight, as shown in **Figure 13,** its spectral lines shift. If a star is moving toward the observer, the spectral lines are shifted toward shorter wavelengths, which is called a blueshift. However, if the star is moving away, the wavelengths become longer, which is called a redshift. The higher the speed, the larger the shift, thus careful measurements of spectral line wavelengths can be used to determine the speed of a star's motion. Because there is no Doppler shift for motion that is at a right angle to the line of sight, astronomers can learn only about the portion of a star's motion that is directed toward or away from Earth. The Doppler shift in spectral lines can be used to detect binary stars as they move about their center of mass toward and away from Earth with each revolution. It is also important to note that there is no way to distinguish whether the star, the observer, or both are moving. A star undergoing periodic Doppler shifts can only be interpreted as one of a binary. Stars identified in this way are called spectroscopic binaries. Binaries can reveal much about the individual properties of stars.

Stellar Positions and Distances

Astronomers use two units of measure for long distances. One, which you are probably familiar with, is a light-year (ly). A light-year is the distance that light travels in one year, equal to 9.461×10^{12} km. Astronomers often use a unit larger than a light-year—a parsec. A **parsec** (pc) is equal to 3.26 ly, or 3.086×10^{13} km.

EARTH SCIENCE JOURNAL

Proof of Earth's Movement The only definite proof that Earth is revolving around the Sun comes from precise measurements of stars. The first observation that definitely proved Earth's motion was the aberration of starlight seen in 1727. Earth's motion perpendicular to the line of sight toward a star causes an apparent displacement of the star's position. Have students draw an illustration or write a paragraph explaining how this apparent displacement is analogous to the apparent slant of rainfall as you run through a storm. Your motion makes the raindrops appear to fall toward you. OL EL

Rubric

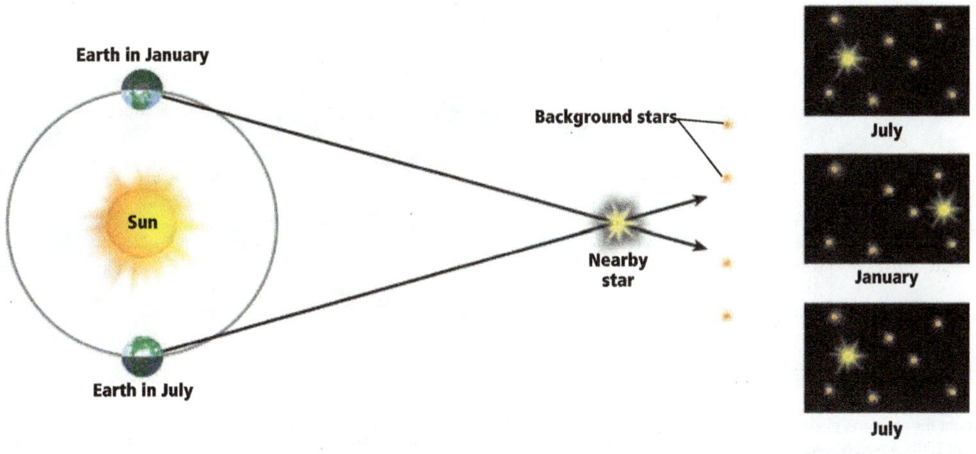

Figure 14 As Earth orbits the Sun, nearby stars appear to change position in the sky compared to faraway stars. Earth reaches its maximum change in position at six months, so the angle measured to the star from these two positions is also at the maximum. This shift in observation position is called parallax and can be used to estimate the distance to the star being observed.
Predict *the position of the star in September.*

View an **animation of parallax.**
Concepts In Motion

Parallax Precise position measurements are important for determining distances to stars. When determining the distance of stars from Earth, astronomers must account for the fact that nearby stars shift in position as observed from Earth. This apparent shift in position caused by the motion of the observer is called **parallax.** In this case, the motion of the observer is the change in position of Earth as it orbits the Sun. As Earth moves from one side of its orbit to the opposite side, a nearby star appears to be shifting back and forth, as illustrated in **Figure 14.** The closer the star, the larger the shift. The distance to a star can be estimated from its parallax shift by measuring the angle of the change. Using the parallax technique, astronomers could find accurate distances to stars up to only 50 ly, or approximately 15 pc, until recently. With advancements in technology, such as the *Hipparcos* satellite, astronomers can find accurate distances up to 100 pc by using parallax.

VOCABULARY
ACADEMIC VOCABULARY
Precise
exactly or sharply defined or stated
The builder's accurate measurements ensured that all of the boards were cut to the same, precise length.

READING CHECK **Identify** the motion of the observer in the diagram.

Basic Properties of Stars

The basic properties of a star are mass, diameter, and luminosity, which are all related to each other. Temperature is another property and is estimated by finding the spectral type of a star. Temperature controls the nuclear reaction rate and governs the luminosity, or absolute magnitude. The absolute magnitude compared to the apparent magnitude can be used to find the distance to a star.

EARTH SCIENCE JOURNAL

Draw Constellations Have students identify some bright stars and constellations on a star chart and, if possible, go outdoors on a clear night and view the stars and constellations they have identified. Have students sketch the star patterns or constellations in their Earth science journals. OL

Rubric

Caption Question Fig. 14 The star's position relative to Earth would shift two-fifths of the distance from the July position toward the January position.

Concept Development

Parsecs The unit of distance called the parsec is derived from stellar parallax measurements. Ask students to envision a long, thin right triangle. The base length of the triangle is 1 AU, and the apex angle is 1 arcsecond. One arcsecond is equal to 1/60 of an arcminute, which is equal to 1/60 of a degree. Thus, an arcsecond is 1/3600 of a degree. The length of the other leg of this triangle is 206,265 times the length of the base. This distance is a parsec, which stands for *parallax second.* More specifically, one parsec is the distance to a star with a parallax angle (the apex angle just described) of one arcsecond. Thus, one parsec = 206,265 AU, which is 3.26 ly.

Enrichment

Light-Year Calculation Have students calculate the length of a light-year by multiplying the speed of light ($c = 2.998 \times 10^8$ m/s) by the number of seconds in a year (365.24 d × 24 h/d × 60 min/h × 60 s/min = 3.156×10^7 s).
1 ly = 9.46×10^{15} m AL

Activity

Observe Parallax Have each student hold a pencil in an upright position at arm's length and then look at it while alternately closing one eye and then the other. The position of the pencil appears to shift back and forth relative to the background. This shift of position is a result of parallax. BL OL EL

READING CHECK The motion of the observer is the same as the orbit of Earth around the Sun. The distance between observation sites is determined by opposite sides of the orbit.

Use Science Terms

Luminosity The words *power* and *luminosity* both mean *energy per second*. A physicist uses the word power to describe what an astronomer calls luminosity. In SI units, power is expressed in watts, and 1 watt = 1 joule per second.

Enrichment

Brightness Have students consider in more detail the system used to describe a star's brightness. A difference of 5 magnitudes means a star is 100 times brighter than the other. Ask students what a difference of just 1 magnitude would be. $100^{1/5}$, or 2.512 times brighter The difference in brightness between any two stars is 2.512^n, where n is the difference between the stars' magnitudes. For example, the difference in brightness between a magnitude +12 star and a magnitude +9 star is $2.512 \times 2.512 \times 2.512 = 2.512^3 = 15.85$ times brighter. Have students find the difference in brightness between a magnitude +21 star and a magnitude +14 star. $21 - 14 = 7$; $2.512^7 = 631.2$ times brighter AL

Discussion

Absolute Magnitude Lead students in a discussion of the relationship between absolute magnitude and luminosity. This will help students understand why astronomers go to the trouble of defining and measuring absolute magnitudes. Luminosity is an intrinsic property. To determine a star's luminosity, its distance and brightness are considered. Because astronomers normally measure stellar brightnesses using the magnitude system, absolute magnitude is an easy way to describe luminosity. Once the absolute magnitude is determined, finding a star's luminosity is a simple matter of using a conversion factor.

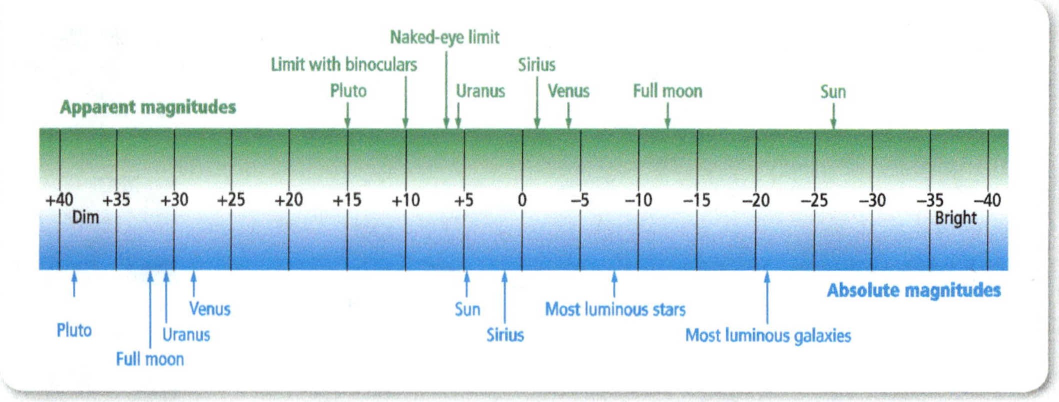

Figure 15 Apparent magnitude is how bright the stars and planets appear in the sky from Earth. Absolute magnitude takes into account the distance to that star or planet and makes adjustments for distance.

Magnitude One of the most basic observable properties of a star is how bright it appears, or the **apparent magnitude.** The ancient Greeks established a classification system based on the brightness of stars. The brightest stars were given a ranking of +1, the next brightest +2, and so on. Today's astronomers still use this system, but they have refined it. In this system, a difference of 5 magnitudes corresponds to a factor of 100 in brightness. Thus, a magnitude +1 star is 100 times brighter than a magnitude +6 star.

VOCABULARY
SCIENCE USAGE V. COMMON USAGE
Magnitude
Science usage: a number representing the apparent brightness of a celestial body

Common usage: the importance, quality, or caliber of something

Absolute magnitude Apparent magnitude does not indicate the actual brightness of a star because it does not account for distance. A faint star can appear to be very bright because it is relatively close to Earth, while a bright star can appear to be faint because it is far away. To account for this phenomenon, astronomers have developed another classification system for brightness. **Absolute magnitude** is how bright a star would appear if it were placed at a distance of 10 pc. The classification of stars by absolute magnitude allows comparisons that are based on how bright the stars would appear at equal distances from an observer. The disadvantage of absolute magnitude is that it can be difficult to determine unless the actual distance to a star is known. The apparent and absolute magnitudes for several objects are shown in **Figure 15.**

Luminosity Apparent magnitudes do not give an actual measure of energy output. To measure the energy output from the surface of a star per second, called its power or **luminosity,** an astronomer must know both the star's apparent magnitude and how far away it is. The brightness observed depends on both a star's luminosity and distance from Earth, and because brightness diminishes with the square of the distance, a correction must be made for distance. Luminosity is measured in units of energy emitted per second, or watts. The Sun's luminosity is about 3.85×10^{26} W. This is equivalent to 3.85×10^{24} 100-W lightbulbs. The values for other stars vary widely, from about 0.0001 to more than 1 million times the Sun's luminosity. No other stellar property varies as much.

ACROSS THE CURRICULUM

Math Real stars do not fall neatly into the whole-number magnitude classes assigned by the ancient Greeks, and magnitudes greater than 6 must be used. The faintest objects visible with the *Hubble Space Telescope* have apparent magnitudes greater than 30. Also, the original category of first magnitude includes stars that have a wide range of brightnesses, so magnitudes of zero or smaller must be used. For example, Sirius, the brightest star in the sky, has a magnitude of −1.47. An expression using logarithms is another way to relate magnitude difference to a brightness ratio, allowing for both negative and fractional values. The expression is $m_1 - m_2 = 2.5\log(b_2/b_1)$, where m_1 and m_2 are two magnitudes, and b_2/b_1 is the corresponding ratio of the brightnesses. The term *log* refers to the common logarithm, base 10.

Classification of Stars

You have learned that the Sun has dark absorption lines at specific wavelengths in its spectrum. Other stars also have dark absorption lines in their spectra and are classified according to their patterns of absorption lines. Spectral lines provide information about a star's temperature and composition.

Temperature Stars are assigned spectral types in the following order: O, B, A, F, G, K, and M. Each class is subdivided into more specific divisions with numbers from 0 to 9. For example, a star can be classified as being a type A4 or A5.

The classes were originally based only on the pattern of spectral lines, but astronomers later discovered that the classes also correspond to stellar temperatures, with the O stars being the hottest and the M stars being the coolest. Thus, by examination of a star's spectrum, it is possible to estimate its temperature.

The Sun is a type G2 star, which corresponds to a surface temperature of about 5800 K. Surface temperatures range from about 50,000 K for the hottest O stars to as low as 2000 K for the coolest M stars. **Figure 16** shows how spectra from some different star classes appear.

Temperature is also related to luminosity and absolute magnitude. Hotter stars put out more light than stars with lower temperatures. In most normal stars, the temperature corresponds to the luminosity. Distance can be determined by calculating a star's luminosity based on its temperature.

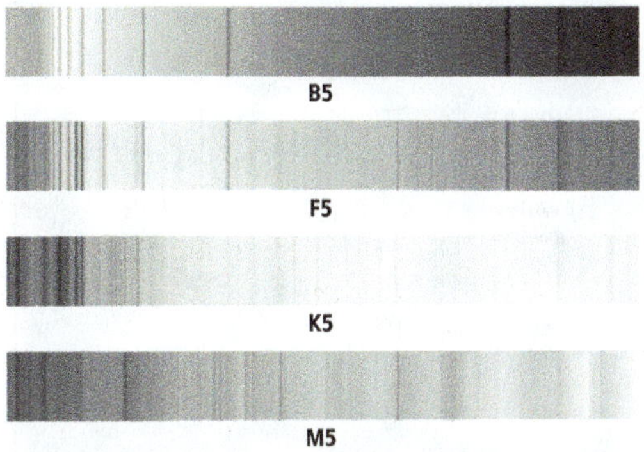

Figure 16 These are typical absorption spectra of a class B5 star, class F5 star, class K5 star, and a class M5 star. The black stripes are absorption lines telling us each star's element composition.

MiniLAB

iLab Station

Model Parallax

How does parallax angle change with distance? If a star is observed at six-month intervals in its orbit, it will appear to have moved because Earth is 300 million km away from the location of the first observation. The angle to the star is different and the apparent change in position of the star is called parallax.

Procedure

1. Read and complete the lab safety form.
2. Place a **meterstick** at a fixed position and attach a **4-m piece of string** to each end.
3. Stand away from the meterstick and hold the two strings together to form a triangle. Be sure to hold the strings taut. Measure your distance from the meterstick. Record your measurement.
4. Measure the angle between the two pieces of string with a **protractor.** Record your measurement of the angle.
5. Repeat Steps 3 and 4 for different distances from the meterstick by shortening or lengthening the string.
6. Make a graph of the angles versus their distance from the meterstick.

Analysis

1. **Interpret** what the length of the meterstick represents. What does the angle represent?
2. **Analyze** what the graph shows. How does parallax angle depend on distance?
3. **Explain** how the angles that you measured are similar to actual stellar parallax angles.

MiniLAB

iLab Station Rubric

Purpose Students will learn about stellar parallax by measuring parallax angles in the classroom.

Process Skills use numbers, recognize spatial relationships, measure in SI, formulate models

Safety Precaution Approve lab safety forms before work begins.

Teaching Strategies

- This lab is best performed in groups of three. One student can hold the meterstick, one can hold the strings, and the other can make the measurements. Alternatively, the measurements can be taken with the meterstick on a table or against a wall.
- The nonlinear relationship is best illustrated when students use different distances from the meterstick for measurements.
- Refer students to **Figure 14** to answer Question 1.

Expected Results Students will find the angle between the strings diminishes as the distance from the meterstick increases. This is analogous to the decrease of a stellar parallax angle with increasing stellar distance.

Analysis

1. distance between the two sighting places; the angle of parallax
2. The graph compares parallax angles. The greater the distance, the smaller the angle.
3. Since the distance between sighting points (aphelion and perihelion) is the same, the angles produced by observing stars at infinite distances is small compared to those close (the Sun). Angles measured in this lab are probably much larger.

Assessment

Skill Ask students to use their results to define a new unit of distance that equals the distance to a meterstick with a 1° parallax angle. This unit of distance would be 57.3 m. OL

LabManager™
Customize any lab with the LabManager™ CD-ROM.

Discussion

Star Colors Ask students if they believe the terms *red giant* and *white dwarf* refer to actual colors of the stars or just a classification of the radiation emitted by the stars. Encourage them to think about stars they have seen in the night sky. Point out that these terms are indeed what the human eye sees. For example, Betelgeuse, the bright red supergiant in Orion's shoulder, looks reddish in color, while Rigel, one of Orion's feet, has a bluish-white color.

Stellar color differences are rather subtle, because all stars emit light over the full visible range and are therefore basically white. The reason for the slight differences in color is that the peak emission, the wavelength at which a star most strongly emits, depends on the surface temperature of the star. A hot star emits most strongly in the blue or even the ultraviolet portion of the spectrum, whereas a cool star emits most strongly in the red or the infrared portion. The Sun, which is intermediate in temperature, is brightest in the yellow portion of the spectrum.

Project

View Star Colors Encourage students to look at the night sky, if possible, and try to see stars of different colors. Have them first identify specific stars such as Betelgeuse and Rigel on a star chart, and then look for these stars at night. OL AL

Explore **main-sequence stars with an interactive table.** Concepts In Motion

Table 2 Relationships of Spectral Types of Stars

Color of Star	Spectral Type	H-R diagram
	O5	
	B5	
	F5	
	G5	
	M5	

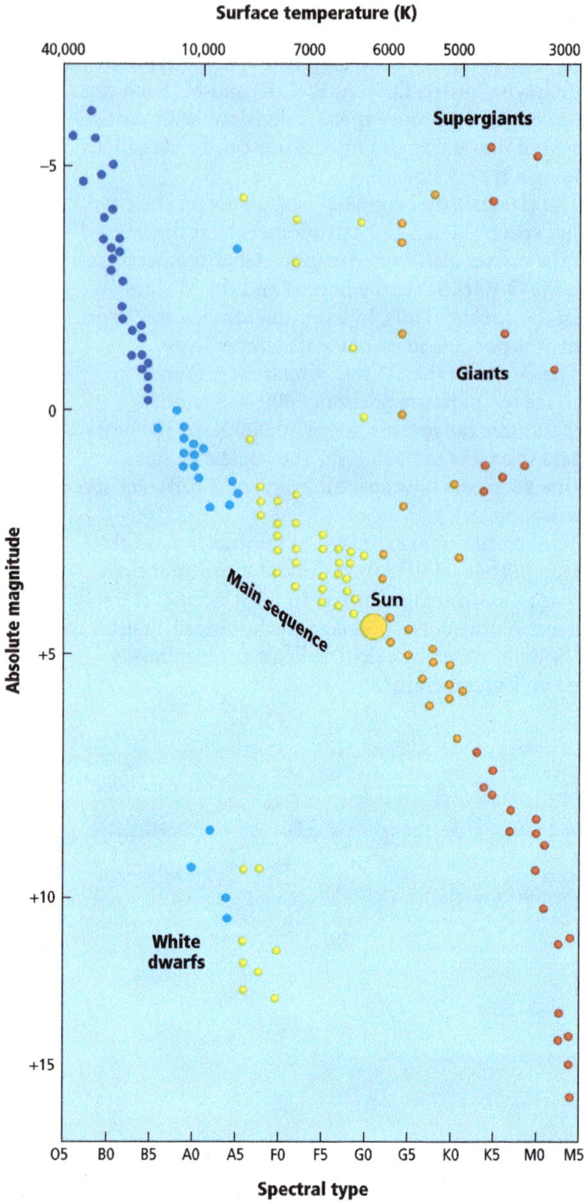

DIFFERENTIATED INSTRUCTION

English Learners The Hertzsprung-Russell diagram is a useful way to summarize the properties of stars in a visual way so that English learners can focus on content. Use the diagram to summarize the brightnesses, temperatures, spectral types (stellar classes), and masses of stars.

Composition All stars, including the Sun, have nearly identical compositions, despite the differences in their spectra. The differences in the appearance of their spectra are almost entirely a result of temperature differences, shown in **Table 2.** Hotter stars have fairly simple visible spectra, while cooler stars have spectra with more lines. The coolest stars have bands in their spectra due to molecules such as titanium oxide in their atmospheres. Typically, about 73 percent of a star's mass is hydrogen (H), about 25 percent is helium (He), and the remaining 2 percent is composed of all the other elements. While there are some variations in the composition of stars, particularly in the final 2 percent, all stars have this general composition.

H-R diagrams The properties of mass, luminosity, temperature, and diameter are closely related. Each class of star has a specific mass, luminosity, temperature, and diameter. These relationships can be demonstrated on a graph called the **Hertzsprung-Russell diagram** (H-R diagram) on which absolute magnitude is plotted on the vertical axis and temperature or spectral type is plotted on the horizontal axis, as shown in **Table 2.** Spectroscopists first plotted this graph in the early twentieth century. An H-R diagram with luminosity plotted on the vertical axis looks similar to the one in **Table 2** and is used to calculate the evolution of stars.

Most stars occupy the region in the diagram called the **main sequence,** which runs diagonally from the upper-left corner, where hot, luminous stars are represented, to the lower-right corner, where cool, dim stars are represented. **Table 3** shows some properties of main-sequence stars.

Explore **main-sequence star properties with an interactive table.** Concepts In Motion

Table 3 Properties of Main-Sequence Stars

Spectral Type	Mass*	Surface Temperature (K)	Luminosity*	Radius*
O5	40.0	40,000	5×10^5	18.0
B5	6.5	15,500	800	3.8
A5	2.1	8500	20	1.7
F5	1.3	6580	2.5	1.2
G5	0.9	5520	0.8	0.9
K5	0.7	4130	0.2	0.7
M5	0.2	2800	0.008	0.3

* These properties are relative to the Sun.

CAREERS IN EARTH SCIENCE

Spectroscopist There are some astronomers who are affiliated with an observatory. Scientists who make and analyze the spectra from stars are called spectroscopists.

WebQuest

Tie to Previous Knowledge

Spectral Lines Ask students to recall what they have studied about atomic structure in order to understand why spectral classes of stars depend on temperature. The electrons orbiting the nucleus of an atom are responsible for forming spectral lines. Each element has its own unique pattern of electron orbits, and this pattern determines the wavelengths of spectral lines the element can create.

CAREERS IN EARTH SCIENCE

WebQuest

Teacher Content Support

Spectral Differences The spectral-line wavelengths of an element change if one or more electrons are missing—that is, if the gas is ionized. In a hot gas, electrons are freed by collisions between atoms. The degree of ionization, or loss of electrons, depends on temperature. Thus, a hot star will have a different spectrum from that of a cool star, because in the hot star more of the atoms are ionized, and ions have different spectra from those of atoms. Two stars with identical compositions but different temperatures can have significantly different spectral lines.

IN THE FIELD

Classification of Spectral Lines
Antonia Maury worked for 25 years at the Harvard College Observatory during the early 1900s. Her focus was the study of stellar spectra—wavelengths of light from stars. She improved upon existing methods used to classify spectral lines corresponding to different stars. Maury's classification system is still an integral aspect of theoretical astrophysics.

GeoLAB

iLab Station

The GeoLab located at the end of the chapter can be used at this point in the lesson.

3 Assess

Check for Understanding

Reinforcement Ask students to summarize the kinds of information that can be obtained from observations of stellar positions, brightnesses, and spectra. positions: motions, distances, and distribution; brightnesses: luminosities; spectra: compositions, temperatures, and motions through the Doppler shift

Reteach

Research A different approach to teaching students about stellar properties is to have each student choose a particular star and learn as much as possible about that one star. Students can each find out a star's name, its designation in different catalogs, its magnitude, its position, its spectral type, its mass, its luminosity, and other features, and then write reports summarizing what they found. A comparison of stars chosen by different students can then be used to emphasize the range of properties that stars can have.

Assessment

Knowledge Ask students what they could conclude about the properties of a star that lies in the upper center of the H-R diagram, above the A stars on the main sequence. This star would have a midrange temperature of around 10,000 K, and it would be a giant or supergiant.

Main sequence About 90 percent of stars, including the Sun, fall along a broad strip of the H-R diagram called the main sequence. While stars are in the main sequence, they are fusing hydrogen in their cores. The interrelatedness of the properties of these stars indicates that they have similar internal structures and functions. As stars evolve off the main sequence, they begin to fuse helium in their cores and burn hydrogen around the core edges.

The Sun lies near the center of the main sequence, being of average temperature and luminosity. A star's mass determines almost all its other properties, including its main-sequence lifetime. The more massive a star is, the higher its central temperature and the more rapidly it burns its hydrogen fuel. This is due primarily to the ratio of radiation pressure to gravitational pressure. Higher pressures cause the fuels to burn faster. As a consequence, the star runs out of hydrogen more rapidly, and thus evolves off the main sequence faster, than a lower-mass star.

Red giants and white dwarfs The stars plotted at the upper right of the H-R diagram in **Table 2** are cool, yet luminous. Because cool surfaces emit much less radiation per square meter than hot surfaces do, these cool stars must have large surface areas to be so bright. For this reason, these larger, cool, luminous stars are called red giants. Red giants are so large—more than 100 times the size of the Sun in some cases—that Earth would be swallowed up if the Sun were to become a red giant! The largest of these are called red supergiants. Conversely, the dim, hot stars plotted in the lower-left corner of the H-R diagram must be small, or they would be more luminous. These small, dim, hot stars are called white dwarfs. A white dwarf is about the size of Earth but has a mass about as large as the Sun's. You will learn how all the different stars are formed in Section 3.

SECTION 2 REVIEW

Section Self-Check

Section Summary

- Most stars exist in clusters held together by their gravity.
- The simplest cluster is a binary.
- Parallax is used to measure distances to stars.
- The brightness of stars is related to their temperature.
- Stars are classified by their spectra.
- The H-R diagram relates the basic properties of stars: class, temperature, and luminosity.

Understand Main Ideas

1. MAINIDEA **Relate** the stellar temperature to the classification of a star.
2. **Explain** the difference between apparent and absolute magnitudes.
3. **Explain** how parallax is used to measure the distance to stars.
4. **Compare and contrast** luminosity and magnitude.
5. **Contrast** the apparent magnitude and the absolute magnitude of a star.
6. **Compare** a light-year and a parsec.

Think Critically

7. **Design** a model to explain parallax.
8. **Explain** the relationship between radius and mass using **Table 3.**

MATH IN Earth Science

9. Compare Regulus (B class), the brightest star in Leo, to Bernard's Star (M class), one of the closest stars to the Sun, using **Table 3** as a reference.

SECTION 2 REVIEW

1. Stars are classified O, B, A, F, G, K, or M, from hottest to coolest. Classes are subdivided from 0 to 9.
2. Apparent magnitude is how bright a star appears in the sky. Absolute magnitude is how bright it would appear if it were 10 pc distant.
3. Parallax can tell which stars are closer to an observer by the observer changing position and measuring the angle to the star from each position. This can then be used to find the distance to the star. The greater the angle, the closer the star.
4. Magnitude measures the brightness of a star, while luminosity measures the power (energy) of the star.
5. Classification of stars by absolute magnitude allows comparisons based on the brightness of stars at equal distances from an observer.
6. A light-year is the distance light travels in a year. A parsec (equal to about 3 ly) is the distance at which an object has a parallax of 1 arcsecond.
7. Models should show how an observer's view changes with position.
8. The larger the radius of a star, the more material it contains, and the more massive it is.
9. Regulus' mass is 30 times greater; temperature is about 5 times greater; luminosity is over 100,000 times greater; radius is 13 times greater.

SECTION 3

Stellar Evolution

MAINIDEA The Sun and other stars follow similar life cycles, leaving the galaxy enriched with heavy elements.

Essential Questions

- What is the relationship between mass and a star's evolution?
- What are the features of massive and regular star life cycles?
- How is the universe affected by the life cycles of stars?

Review Vocabulary

evolution: a radical change in composition over a star's lifetime

New Vocabulary

nebula
protostar
neutron star
pulsar
supernova
black hole

EARTH SCIENCE 4 YOU A campfire glows brightly as long as it has fuel to burn. When the fuel is depleted, the light becomes dimmer, and the fire extinguishes. Unlike a campfire, stars shine because of nuclear reactions in their interior. Stars also die out when their nuclear fuel is gone.

Basic Structure of Stars

Mass governs a star's temperature, luminosity, and diameter. In fact, astronomers have discovered that the mass and the composition of a star determine nearly all its other properties.

Mass effects The more massive a star is, the greater the gravity pressing inward, and the hotter and more dense the star must be inside to balance its own gravity. The temperature inside a star governs the rate of nuclear reactions, which in turn determines the star's energy output—its luminosity. The balance between gravity squeezing inward and outward pressure is maintained by heat due to nuclear reactions and compression. This balance is called hydrostatic equilibrium and it must hold for any stable star, as illustrated in **Figure 17,** otherwise the star would expand or contract. This balance is governed by the mass of a star.

Fusion Inside a star, conditions vary in much the same way that they do inside the Sun. The density and temperature increase toward the center, where energy is generated by nuclear fusion. Stars on the main sequence produce energy by fusing hydrogen into helium, as the Sun does. Stars that are not on the main sequence either fuse elements other than hydrogen in their cores or do not undergo fusion at all.

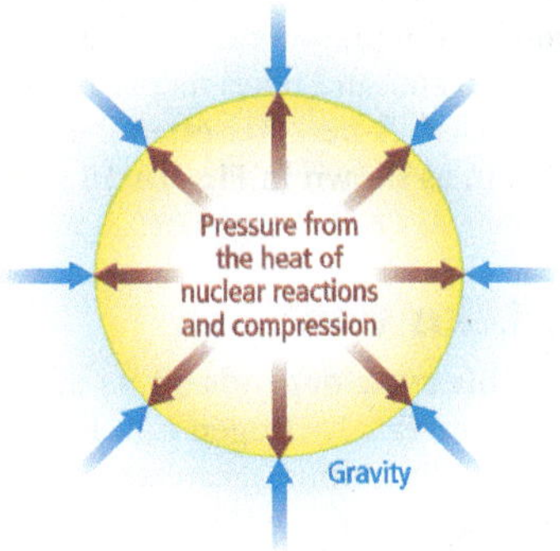

Figure 17 When the pressure from radiation and fusion is balanced by gravity, a star is stable and will not expand or contract.

Stellar Evolution

A star changes as it ages because its internal composition changes as nuclear-fusion reactions in the star's core convert one element into another. With a change in the core composition, the star's density increases, its temperature rises, and its luminosity increases. As long as the star is stable and converting hydrogen to helium, it is considered a main-sequence star. Eventually, when the nuclear fuel runs out, the star's internal structure and mechanism for producing pressure must change to counteract gravity. The changes a star undergoes during its evolution begin with its formation.

EARTH SCIENCE JOURNAL

Weighted Average Because the Sun is roughly average between the extreme values of quantities such as stellar masses, radii, and luminosities, you can decide that the Sun is an average star. Stars that are less massive, smaller, and fainter, however, far outnumber stars similar to the Sun or more massive stars. In this way, you can conclude the Sun is above average. Have students research the meaning of *weighted average.* Then, have them describe in their Earth science journals why this term is important for describing the Sun as either an average or above-average star. AL

Rubric

SECTION 3

1 Focus

Focus Transparency

Before presenting the lesson, project **Section Focus Transparency 94** and have students answer the accompanying questions. EL

MAINIDEA

Heavy Elements Show models or drawings of atoms of simple and heavy elements. Ask students to consider why most heavy elements originate in the most massive stars. Only massive stars make heavy elements in their interiors and have ways, such as supernova explosions, to distribute the heavy elements.

2 Teach

Identify Misconceptions

Students might expect that, because the Sun is average in its physical properties, there must be comparable numbers of stars both more and less massive.

Uncover the Misconception Have students explain the following statement: The Sun is an average star.

Demonstrate the Concept Explain that stars less massive than the Sun are far more numerous than more massive stars. Stars more massive are extremely rare. The Sun is average only in the sense that its properties fall in the middle of the range for all stars.

Assess New Knowledge Have students identify the paragraph on the previous page that explains why stars more massive than the Sun are rare. Students should identify the paragraph labeled *Main sequence.* BL

■ **Caption Question** **Fig. 18** The remaining material would coalesce similarly to the Sun's accretion to produce increasing size and gravity of the accretions, eventually clearing an area in the disk and forming a planet or a series of planets.

Discussion

Balanced Forces Every stable object, including a star, a planet, a moon, and a basketball, is in hydrostatic equilibrium, which means the inward and outward forces on the object are balanced. Ask students what forces are in balance for each object, keeping it in equilibrium. A star is stabilized by a balance between inward gravity and outward gas pressure; a planet and a moon balance inward gravity with rigid body forces and fluid pressure (if they are partially fluid); and the inward force on a basketball is the tension in the outer shell, which is balanced by air pressure. OL AL

☑ READING CHECK The rotation of the cloud of gas causes it to flatten, so a disk shape forms.

Tie to Previous Knowledge

Protostar Radiation Infrared radiation is sometimes known as heat radiation because we sense it as heat. Ask students to imagine a stovetop with several electric burners, all turned off except one that is set at a low temperature. If they were to wave their hands over the stovetop, students could tell right away which burner is on because their skin would sense infrared radiation from it. Astronomers detect protostars in a similar manner by the infrared radiation coming from hot spots within cold interstellar gas clouds.

■ **Figure 18** Temperatures will continue to build as gravity pulls the infalling matter to the center of the rotating disk. The center region is a protostar until fusion initiates and a star ignites.
Infer *what happens to the remaining material in the disk.*

View an **animation of star formation.**
Concepts In Motion

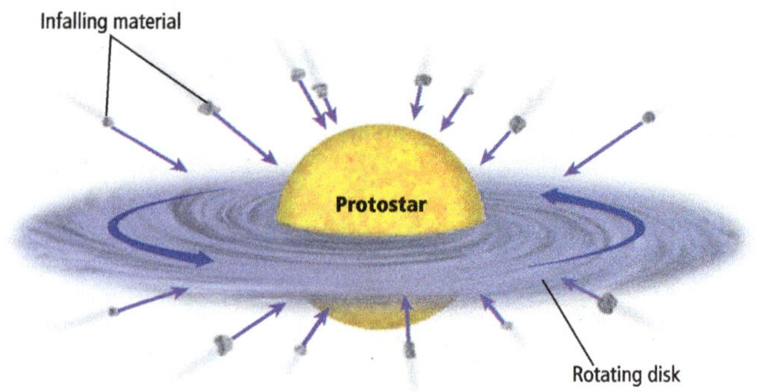

Star formation All stars form in much the same manner as the Sun did. The formation of a star begins with a cloud of interstellar gas and dust, called a **nebula** (plural, nebulae), which collapses on itself as a result of its own gravity. As the cloud contracts, its rotation forces it into a disk shape with a hot, condensed object at the center, called a **protostar,** as illustrated in **Figure 18.** Friction from gravity continues to increase the temperature of the protostar, until the condensed object reaches the ignition temperature for nuclear reactions and becomes a new star. A protostar is brightest at infrared wavelengths.

☑ READING CHECK **Infer** what causes the disk shape to form.

Fusion begins When the temperature inside a protostar becomes hot enough, nuclear fusion reactions begin. The first reaction to ignite is always the conversion of hydrogen to helium. Once this reaction begins, the star becomes stable because it then has sufficient internal heat to produce the pressure needed to balance gravity. The object is then truly a star and takes its place on the main sequence according to its mass. A new star often illuminates the gas and dust surrounding it, as shown in **Figure 19.**

■ **Figure 19** Using the Spitzer telescope's infrared wavelengths, protostars are imaged inside the Elephant Trunk nebula.

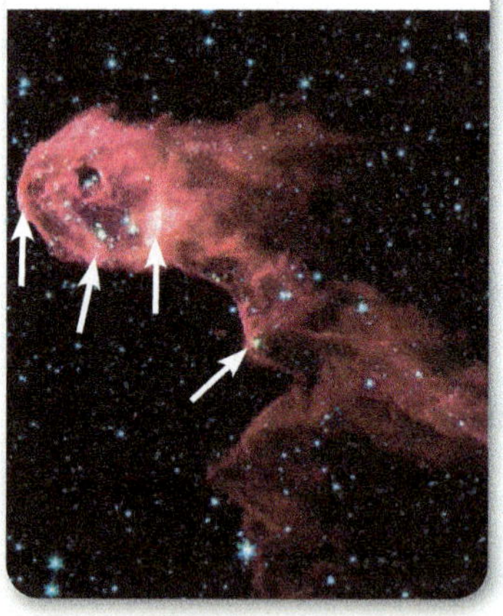

Life Cycles of Stars Like the Sun

What happens next during a star's life cycle depends on its mass. For example, as a star like the Sun converts hydrogen into helium in its core, it gradually becomes more luminous because the core density and temperature rise slowly and increase the reaction rate. It takes about 10 billion years for a star with the mass of the Sun to convert all of the hydrogen in its core into helium. Thus, such a star has a main-sequence lifetime of 10 billion years. From here, the next step in the life cycle of a small mass star is to become a red giant.

DIFFERENTIATED INSTRUCTION

Advanced Learners Students might wonder why a star becomes a red giant after using its core hydrogen. The main reason a star expands into a red giant is that after hydrogen fusion stops in the star's inner core, the same reaction continues in a shell outside the core. This shell emits huge quantities of energy that cannot quickly escape because the outer layers of the star readily absorb light. Energy from the shell becomes trapped, forcing the outer layers to expand. As these outer layers expand, they cool because the pressure decreases. As the star expands and its luminosity increases as a result of its increased surface area, it also becomes redder (cooler). Thus, a former main sequence star becomes a red giant.

Red giant Only about the innermost 10 percent of a star's mass can undergo nuclear reactions because temperatures outside of this core never become hot enough for reactions to occur. Thus, when the hydrogen in its core is gone, a star has a helium center and outer layers made of hydrogen-dominated gas. Some hydrogen continues to react in a thin layer at the outer edge of the helium core, as illustrated in **Figure 20.** The energy produced in this layer forces the outer layers of the star to expand and cool. The star then becomes a red giant because its luminosity increases while its surface temperature decreases due to the expansion.

While the star is a red giant, it loses gas from its outer layers. The star is so large that its surface gravity is low, and thus the outer layers can be released by small expansions and contractions, or pulsations, of the star due to instability. Meanwhile, the core of the star becomes hot enough, at 100 million K, for helium to react and form carbon. The star contracts back to a more normal size, where it again becomes stable for awhile. The helium-reaction phase lasts only about one-tenth as long as the earlier hydrogen-burning phase. Afterward, when the helium in the core is depleted, the star is left with a core made of carbon.

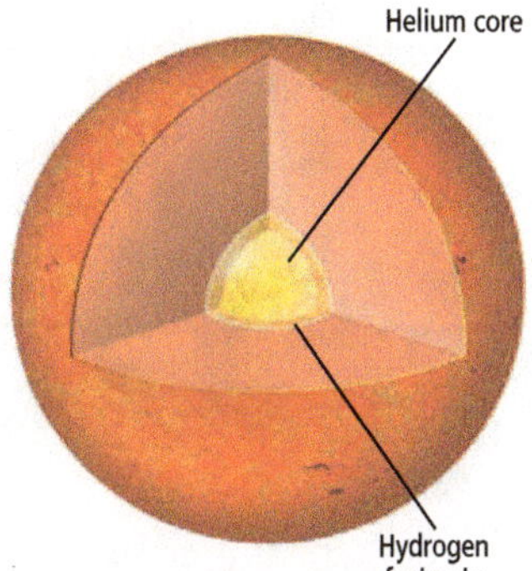

Figure 20 If the central region of a red giant becomes hot enough, helium is converted to carbon. In the spherical shell just outside, hydrogen continues to be converted to helium. The low temperature of the outer atmosphere due to expansion and cooling causes the red color.

View an **animation of the helium core.**
Concepts In Motion

The final stages A star with the same mass as the Sun never becomes hot enough for carbon to fuse, so its energy production ends. The outer layers expand again and are expelled by pulsations that develop in the outer layers. This shell of gas is called a planetary nebula. In the center of a planetary nebula, shown in **Figure 21,** the core of the star becomes exposed as a small, hot object about the size of Earth. The star is then a white dwarf made of carbon.

Internal pressure in white dwarfs A white dwarf is stable despite its lack of nuclear reactions because it is supported by the resistance of electrons being squeezed together, and does not require a source of heat to be maintained. This pressure counteracts gravity and can support the core as long as the mass of the remaining core is less than about 1.4 times the mass of the Sun. The main-sequence lifetime of such a star is much longer than the main-sequence lifetime of a more massive star, because low-mass stars are dim and do not deplete their nuclear fuel rapidly. The electron pressure does not require ongoing reactions, so it can last indefinitely. The white dwarf gradually cools, eventually losing its luminosity and becoming an undetectable black dwarf.

Figure 21 The star at the center of the Eskimo nebula, now a white dwarf, was the source of the remnant gases surrounding it.

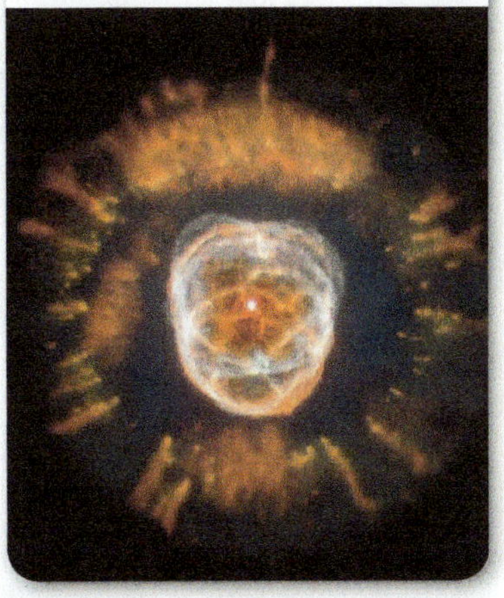

Life Cycles of Massive Stars

For stars more massive than the Sun, evolution is different. A more massive star begins its life in the same way, with hydrogen being converted to helium, but it is much higher on the main sequence. The star's lifetime in this phase is short because the star is very luminous and uses up its fuel quickly.

Identify Misconceptions

Many people think stars are eternal and never change.

Uncover the Misconception Ask students whether the stars they see in the night sky change.

Demonstrate the Concept Point out that most changes in stars take place over long time periods—much longer than a human lifetime. There are exceptions, such as nova and supernova outbursts, but for the most part, humans are not able to see stars move or change.

Assess New Knowledge Ask students how astronomers can know that stars change. Point out that astronomers deduce the evolution of stars by observing them in different phases of their life cycles and by using models. Stellar motions are rapid by human standards, but stars are so far away we cannot see shifts in their positions without precise instruments. To our eyes, the constellations look the same as they did thousands of years ago.

Use Science Terms

Planetary Nebula The ejected gas from a star that has lost its outer layers is called a planetary nebula because the ejected shell often appears round like a planet, and because many of these nebulae appear blue-green in color, resembling the planets Uranus and Neptune. The blue-green color comes from emission lines resulting from twice-ionized oxygen atoms.

ACROSS THE CURRICULUM

Physics Students have learned in previous science classes that like electrical charges repel each other. This makes fusion difficult, because protons in the nuclei of atoms are positively charged. The repulsion between nuclei can be overcome only if they collide at very high speeds. Only then can the nuclei come close enough together to merge. Collision speed depends on temperature, and a star is hottest in its center, which explains why fusion occurs only in stellar cores. The larger the nuclei, the greater the electrical charge, and the higher the speed needed to make nuclei fuse. Massive stars reach higher internal temperatures than lower-mass stars, so heavy elements can be formed only in massive stars. Thus, these rare stars play the dominant role in forming heavy elements in the universe.

Identify Misconceptions

People generally think a black hole has extra gravity that enables it to reach out and suck material into itself.

Uncover the Misconception Ask students what would happen to Earth's orbit if the Sun were to collapse and turn into a black hole.

Demonstrate the Concept Explain that the Sun's mass would not change if it were to become a black hole, and its gravitational force on Earth would be no different. The only region where the Sun's gravity is exceptionally strong is within the Sun's original radius, and even there, a body such as a planet could orbit without falling in. Only close to the center is the gravitational field so strong that space is distorted and light is unable to escape. However, from a distance, a black hole behaves like an ordinary object.

Assess New Knowledge Ask students what would happen in a binary star system if one of the stars collapsed and became a black hole. How would the other star be affected? There would be no effect on the companion star. It would continue to orbit the center of mass just as before. BL OL

Caption Question Fig. 23 When our Sun becomes a red giant in about 5 billion years, its diameter will expand to the orbit of Earth or the orbit of Mars, over 300 million km. A supergiant's diameter could be seven times that, out to the orbit of Saturn. A neutron star, by contrast, would have a diameter of only 10 km but a mass of up to 3 times that of our Sun.

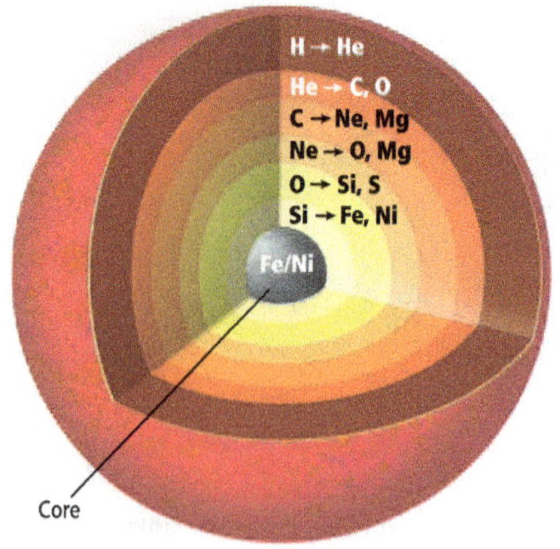

■ **Figure 22** A massive star can have many shells fusing different elements. These stars are the source of heavier elements in the universe.

Supergiant A massive star undergoes many more reaction phases and thus produces a rich stew of many elements in its interior. The star becomes a red giant several times as it expands following the end of each reaction stage. As more shells are formed by the fusion of different elements, illustrated in **Figure 22,** the star expands to a larger size and becomes a supergiant, such as Betelgeuse in the Orion constellation.

Supernova formation A star that begins with a mass between about 8 and 20 times the Sun's mass will end up with a core that is too massive to be supported by electron pressure. Such a star comes to a violent end. Once reactions in the core of the star have created iron, no further energy-producing reactions can occur, and the core of the star violently collapses in on itself, as illustrated in **Figure 23.** Protons and electrons in the core merge to form neutrons. Like electrons, a neutron's resistance to being squeezed close together creates a pressure that halts the collapse of the core, and the core becomes a collapsed stellar remnant—a **neutron star.** A neutron star has a mass of 1.4 to 3 times the Sun's mass but a diameter of only about 20 km. Its density is extremely high—about 100 trillion times the density of water—and is comparable to that of an atomic nucleus.

Pulsar Some neutron stars are unique in that they have a pulsating pattern of light. The magnetic fields of these stars focus the light they emit into cones. Then as these stars rotate on their axes, the light from each spinning neutron star is observed as a series of pulses of light, as each of the cones sweeps out a path in Earth's direction. This pulsating star is known as a **pulsar.**

■ **Figure 23** When the outer layers of a star collapse into the neutron core, the central mass of neutrons creates a pressure that causes this mass to explode outward as a supernova, leaving a neutron star.
Compare *the diameter of a supergiant with that of a neutron star.*

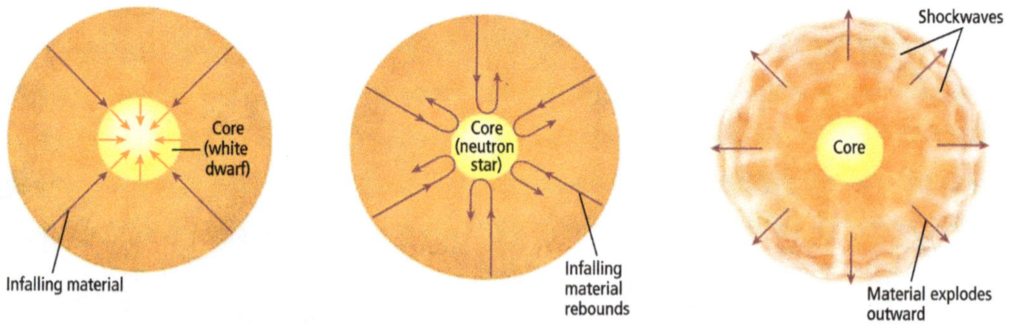

ACROSS THE CURRICULUM

History In 1987, stellar evolution suddenly attracted the attention of the general public, as well as astronomers, when the first supernova visible to the unaided eye occurred in nearly 400 years. A star in the Large Magellanic Cloud, a neighboring galaxy of the Milky Way, exploded. Photos of the event made the covers of popular magazines, and science articles were dominated by news of the supernova. This event helped make people aware that stars do change. It was also a gold mine for astronomers studying stellar evolution. A great deal of new information about how stars become unstable and explode, how they form and disperse new elements, how they differ from one galaxy to another, and the nature of elementary particles (neutrinos) and dark matter all came from studies of the supernova of 1987.

Supernova A neutron star forms quickly while the outer layers of the star are still falling inward. This infalling gas rebounds when it strikes the hard surface of the neutron star and explodes outward. The entire outer portion of the star is blown off in a massive explosion called a **supernova** (plural, supernovae). This explosion creates elements that are heavier than iron and enriches the universe. **Figure 24** shows photos of before and during a supernova explosion. Astronomers recorded this supernova event in February, 1987. A distant supernova explosion might be brighter than the galaxy in which it is found.

Black holes Some stars are too massive to form neutron stars. The pressure from the resistance of neutrons being squeezed together cannot support the core of a star if the star's mass is greater than about three times the mass of the Sun. A star that begins with more than 20 times the Sun's mass will end up above this mass limit, and it cannot form a neutron star. The resistance of neutrons to being squeezed is not great enough to stop the collapse, and the core of the star continues to collapse, compacting matter into a smaller volume. The small, extremely dense object that remains is called a **black hole** because its gravity is so immense that nothing, not even light, can escape it. Astronomers cannot observe what goes on inside a black hole, but they can observe the X-ray-emitting gas that spirals into it.

Figure 24 The region of sky in the Large Magellanic Cloud seemed ordinary before one of its stars underwent a supernova explosion.

Watch a **video about black holes.** Video

SECTION 3 REVIEW

Section Self-Check

Section Summary

- The mass of a star determines its internal structure and its other properties.
- Gravity and pressure balance each other in a stable star.
- If the temperature in the core of a star becomes high enough, elements heavier than hydrogen can fuse together.
- A supernova occurs when the outer layers of the star bounce off the neutron star core, and explode outward.

Understand Main Ideas

1. MAINIDEA **Explain** how mass determines a star's evolution.
2. **Infer** how hydrostatic equilibrium in a star is determined by mass.
3. **Determine** how the lifetimes of stars depend on their masses.
4. **Determine** why only the most massive stars are important contributors in enriching the galaxy with heavy elements.

Think Critically

5. **Explain** how the universe would be different if massive stars did not explode at the end of their lives.
6. **Distinguish** whether there is a balance between pressure and gravity in main-sequence stars, white dwarfs, neutron stars, and black holes.

WRITING IN ▶ Earth Science

7. Write a description of an observation of a supernova in another galaxy.

3 Assess

Check for Understanding

Reinforcement Ask students to summarize the reasons that stars evolve. The composition of a star's core changes as a result of nuclear fusion reactions, and the star must adjust so it can maintain equilibrium.

Reteach

Summarize Ask students to summarize which stellar mass ranges lead to which kinds of stellar remnants. initial masses less than about 8 solar masses: white dwarfs; between about 8 and 20 solar masses: supernovae and neutron stars; more massive than 20 solar masses: black holes

Assessment

Performance Have each student create a poster illustrating the evolution of the Sun. Ask students to include descriptions of what is occurring within the Sun at each stage.

What's EARTH SCIENCE Got To Do With It?

Video *In the Dark*

SECTION 3 REVIEW

1. Low-mass stars can burn far longer than higher-mass stars. This causes them to be more stable and have a small number of steps in their evolution.
2. Mass determines the gravitational force pushing inward, and it is directly related to the amount of fusion, which causes an outward pushing pressure.
3. Massive stars generally have a shorter lifespan because they burn their fuel quickly and become unstable.
4. The most massive stars become hot enough to undergo many stages of fusion, creating heavy elements such as iron and nickel. Supernovae create and release even heavier elements.
5. Heavy elements would not be dispersed, and they would not be available for incorporation into later generations of stars or for the formation of planets in solar systems.
6. In main-sequence stars, gravity is balanced by ordinary gas pressure. In a white dwarf, the pressure from the repulsion of electrons balances gravity, and in a neutron star, gravity is balanced by the pressure from the repulsion of neutrons. In a black hole, nothing counteracts gravity.
7. Descriptions should include a star becoming much brighter and then fading to nearly nothing.

Rubric

Earth Science &
TECHNOLOGY

Purpose

Students will learn how events on the Sun's surface lead to solar storms that affect Earth's communication and electrical systems.

Teacher Content Support

Solar Storms The magnetosphere is an area around Earth influenced by Earth's magnetic field. Charged particles in the solar wind hurtle toward Earth from the Sun at speeds of up to 2000 km/s. These particles can interact with the magnetosphere, generating electrical fields. These fields can affect matter by exciting electrons into a higher energy level. Aside from power outages and damage to satellites, the effects of solar storms include radio and television static, auroras, and navigation problems for ships or planes using a magnetic compass. Solar storms can also be harmful to astronauts in space because of the dangerous levels of radiation.

Teaching Strategy

Make a three-column table on the board to have students practice note-taking skills. Column titles for the table could include *Area Affected, How Affected,* and *Examples.* Row titles for the table should include *Communications, Satellites,* and *Electricity.* Have students take turns filling in the table after they finish reading the feature.

Earth Science &
TECHNOLOGY

Space Weather and Earth Systems

Powerful hurricanes and tornadoes can cause millions of dollars worth of damage to homes and other structures. These types of strong storms can be responsible for loss of human life and the disruption of major electrical and communication systems in an area. There are also weather conditions in space. What effects do solar storms have on Earth?

Space weather Solar flares and coronal mass ejections create powerful solar storms that release billions of high-energy particles into space that travel at speeds of up to 2000 km/s. Some of these particles slam into Earth's magnetosphere—over which particles from space normally flow—much like water flows around a large rock in the middle of a river. Earth's magnetosphere normally deflects particles from the Sun, but during intense solar storms, highly charged particles cause disruptions in many of Earth's communication and electrical systems.

Monitoring space weather Two U.S. government agencies, NASA and NOAA, monitor and provide daily updates on space weather, including predictions about solar flare and solar storm occurrences. Power companies, the Federal Aviation Administration, and the U.S. Department of Defense use the data to help minimize the damage to sensitive equipment caused by solar storms.

Communication Communication satellites, locating systems, and military signals rely on radio waves that are bounced off Earth's ionosphere. The ionosphere is a layer of highly charged particles which is especially vulnerable to highly energized particles from the Sun. These high-energy particles can interfere with radio signals and disrupt transmissions.

A widespread coronal mass ejection blasts more than a billion tons of matter into space at millions of kilometers per hour. Fortunately one this large is rare.

Satellites Solar storms can cause satellites to fall out of orbit due to temperature and density changes in Earth's upper atmosphere. They must be moved to higher orbits in response to this phenomenon. Communication satellites can also be knocked out by electric particle buildup.

Electricity Power companies receive information about possible solar storms in order to avoid service disruption to customers. Solar storms can knock out power by inducing currents in electrical lines. In 1989, in Canada, a solar storm caused a nine-hour blackout that affected 6 million people and cost the power company over 10 million dollars in repairs.

WRITING IN ▶ Earth Science

Pamphlet Research more information about space weather and create a pamphlet that answers frequently asked questions about it. Include information about the causes and why it is important to monitor space weather.

WebQuest

WRITING IN ▶ Earth Science

Rubric

Pamphlet Questions could include the following: What is space weather? What causes space weather? How does space weather affect Earth? How is space weather monitored?

WebQuest

GeoLAB

iLab Station

Identify Stellar Spectral Lines

Background: An astronomer studying a star or other type of celestial object often starts by identifying the lines in the object's spectrum. The identity of the spectral lines gives information about the chemical composition of the distant object, along with data on its temperature and other properties.

Question: *How can you identify stellar spectral lines based on two previously identified lines?*

Materials

ruler

Procedure

1. Read and complete the lab safety form.
2. Find the difference between the two labeled spectral line values on Star 1.
3. Accurately measure the distance between the two labeled spectral lines.
4. Set up a conversion scale by dividing the spectral difference by the measured distance.
   For example: 1 mm = 12 nm
5. Measure the distance from one of the labeled spectral lines to each of the unlabeled spectral lines.
6. Convert these distances to nm. Add or subtract your value to the original spectral line value. If the labeled line is to the right of the line measured, then subtract. Otherwise, add. This is the value of the wavelength.

Possible Elements and Wavelengths

Element/Ion	Wavelength (nm)
H	383.5, 388.9, 397.0, 410.2, 434.1, 486.1, 656.3
He	402.6, 447.1, 492.2, 587.6, 686.7
He^{+}	420.0, 454.1, 468.6, 541.2, 656.0
Na	475.2, 498.3, 589.0, 589.6
Ca^{+}	393.4, 480.0, 530.7

7. Compare your wavelength measurements to the table of wavelengths emitted by elements, and identify the elements in the spectrum.
8. Repeat this procedure for Star 2.

Analyze and Conclude

1. **Identify** Can you see any clues in the star's spectrum about which elements are most common in the stars? Explain.
2. **Explain** Do both stars contain the same lines for all the elements in the table?
3. **Evaluate** How do the thicker absorption lines of some elements in a star's spectrum affect the accuracy of your measurements? Is there a way to improve your measurements? Explain.

INQUIRY EXTENSION

Design Your Own Obtain spectra from various sources, such as sunlight, fluorescent, and incandescent light. Compare their spectral lines to those from this lab. What elements are common to each?

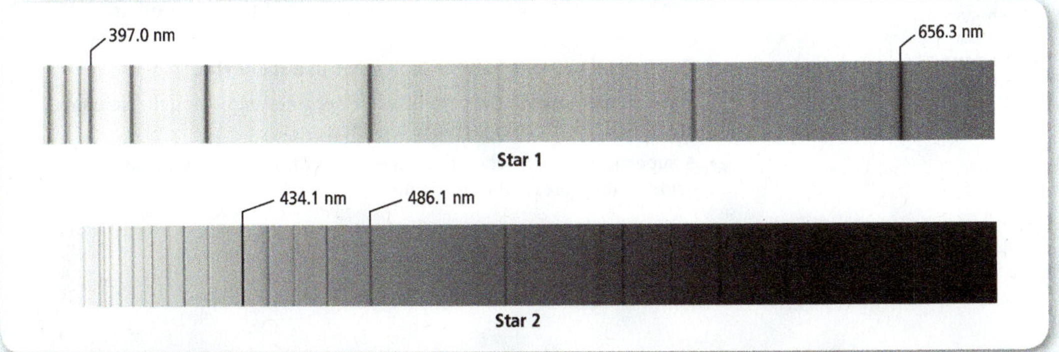

GeoLAB

iLab Station Rubric

Preparation

Time Allotment 25 min

Process Skills use numbers, collect and organize data

Safety Precaution Approve lab safety forms before work begins.

Procedure

- Emphasize to students precise wavelengths are often needed to identify spectral lines.
- Star 1 is an A5 star. Star 2 is an O5 star.
- It might help for students to make a template (using a small card) on which the known spectral lines are marked on the same scale as the unknown spectrum. They can slide the template along the unknown spectrum to find a match in spacing, identifying unknown spectral lines.
- **Troubleshooting** Not all wavelengths in the table are used, and not all spectral lines in Star 2 (right end) are measured and included in the table.

LabManager™
Customize any lab with the LabManager™ CD-ROM.

Analyze and Conclude

1. Most of the lines are hydrogen and helium lines. Thickness of spectral lines is an indication, but it also depends on the temperature.
2. No; Star 1 does not have identifiable helium (neutral or ionized) in this spectrum. Star 2 does not have identifiable sodium lines in this spectrum. In reality, both stars have all the same elements. The difference in their spectra is a result of temperature effects.
3. Increased width of absorption lines can result in less accurate measurements. To correct for this, students should measure from the middle (width), where the line is darkest. As the band increases in width, it will increase the same amount on both sides. Thus, always measuring from the center will eliminate any error from increased width.

INQUIRY EXTENSION

Design Your Own Student answers will vary based on the light sources they choose. Incandescent light has a similar spectrum to the two stars in this lab, peaking in the visible spectrum, except no bright or dark spots appear. Fluorescent light, like Star 2, seems void of light on the higher wavelength end of the visible spectrum. Sunlight, like Star 1, peaks around 500 nm.

MAINIDEAS Summary statements can be used by students to review the major concepts of the chapter.

Students can review with these online resources.

Vocabulary Practice

Vocabulary eGames
Vocabulary eFlashcards
Vocabulary PuzzleMaker

Section Self-Check

Chapter Self-Check

Online Test Practice

Use the ExamView® *Assessment Suite* CD-ROM to:

- create multiple versions of tests
- create modified tests with one mouse click
- edit existing questions and add your own questions
- build tests aligned with state standards using built-in state curriculum tags
- change English tests to Spanish with one mouse click
- track students' progress using the Teacher Management System

CHAPTER 29 STUDY GUIDE

BIGIDEA The life cycle of every star is determined by its mass, luminosity, magnitude, temperature, and composition.

Vocabulary Practice

SECTION 1 The Sun

VOCABULARY
- photosphere
- chromosphere
- corona
- solar wind
- sunspot
- solar flare
- prominence
- fusion
- fission

MAINIDEA The Sun contains most of the mass of the solar system and has many features typical of other stars.
- Most of the mass in the solar system is found in the Sun.
- The Sun's average density is approximately equal to that of the gas giant planets.
- The Sun has a layered atmosphere.
- The Sun's magnetic field causes sunspots and other solar activity.
- The fusion of hydrogen into helium provides the Sun's energy and composition.

SECTION 2 Measuring the Stars

VOCABULARY
- constellation
- binary star
- parsec
- parallax
- apparent magnitude
- absolute magnitude
- luminosity
- Hertzsprung-Russell diagram
- main sequence

MAINIDEA Stellar classification is based on measurement of light spectra, temperature, and composition.
- Most stars exist in clusters held together by their gravity.
- The simplest cluster is a binary.
- Parallax is used to measure distances to stars.
- The brightness of stars is related to their temperature.
- Stars are classified by their spectra.
- The H-R diagram relates the basic properties of stars: class, temperature, and luminosity.

SECTION 3 Stellar Evolution

VOCABULARY
- nebula
- protostar
- neutron star
- pulsar
- supernova
- black hole

MAINIDEA The Sun and other stars follow similar life cycles, leaving the galaxy enriched with heavy elements.
- The mass of a star determines its internal structure and its other properties.
- Gravity and pressure balance each other in a stable star.
- If the temperature in the core of a star becomes high enough, elements heavier than hydrogen can fuse together.
- A supernova occurs when the outer layers of the star bounce off the neutron star core, and explode outward.

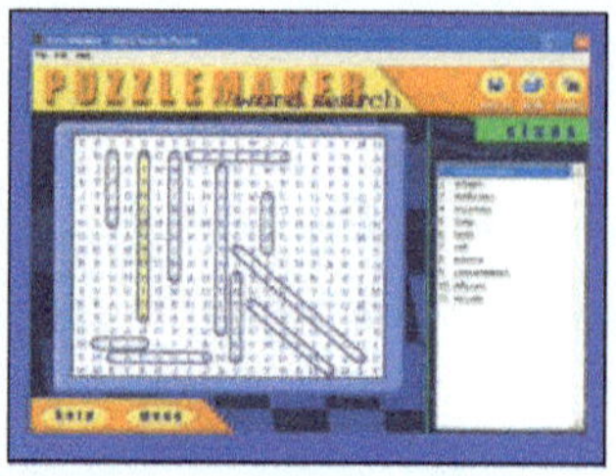

Vocabulary Practice

For additional practice with vocabulary, have students access the Vocabulary PuzzleMaker.

Chapter Self-Check

VOCABULARY REVIEW

Match the definitions below to the correct vocabulary term on the Study Guide.

1. the outermost layer of the Sun's atmosphere, having a temperature of about 1 million K
2. combining of lightweight nuclei such as hydrogen into heavier nuclei
3. dark spots where the surface is cooler on the photosphere of the Sun
4. the apparent shift in position of an object that results from the motion of the observer
5. the outward flow of charged particles from the Sun's corona flowing throughout the solar system
6. two stars that are gravitationally bound and orbit a common center of mass
7. the power or energy output from the surface of a star in units per second
8. an explosion that blows away the outer portion of a star

Distinguish between the following pairs of terms.

9. eclipsing binary, spectroscopic binary
10. giant stars, main-sequence stars
11. apparent magnitude, absolute magnitude
12. black hole, neutron star
13. fission, fusion

Define these terms in your own words.

14. constellation
15. prominence
16. main sequence
17. nebula
18 supernova
19. black hole
20. protostar

UNDERSTAND KEY CONCEPTS

Use the diagram below to answer Question 21.

21. Starting at the center, which is the correct order of solar layers?
    A. radiation zone, core, convection currents
    B. core, convection currents, radiation zone
    C. core, radiation zone, convection currents
    D. convection currents, mantle, radiation zone

22. Why do sunspots appear dark?
    A. They are cooler than their surroundings.
    B. They are holes in the interior of the Sun.
    C. They do not have strong magnetic fields.
    D. They are hotter than their surroundings.

23. Why is the Sun's composition similar to that of the gas giant planets?
    A. They all formed at the same time.
    B. They both lost heavy elements.
    C. They all formed from the same interstellar cloud.
    D. They both gained heavy elements.

24. How is the Sun's magnetic behavior associated with its activity cycle?
    A. The magnetic field turns off when the activity cycle turns on.
    B. The activity cycle is coordinated with the peak number of sunspots.
    C. The activity cycle is independent of the number of solar flares.
    D. Solar flares are not coordinated with magnetic storms on Earth.

UNDERSTAND KEY CONCEPTS

21. C
22. A
23. C
24. B

CHAPTER 29 ASSESSMENT

VOCABULARY REVIEW

1. corona
2. fusion
3. sunspots
4. parallax
5. solar wind
6. binary stars
7. luminosity
8. supernova
9. An eclipsing binary is identified by a variance in magnitude, while a spectroscopic binary is identified by redshifts and blueshifts in wavelengths.
10. Bright giant stars have large surface areas, which makes them bright despite their cool temperatures. Bright main-sequence stars are bright because of their hot temperatures.
11. Apparent magnitude is brightness based upon the star's appearance. Absolute magnitude compares all stars as if they were the same distance away (10 pc).
12. Black holes and neutron stars are the end result of collapsed stars. Black holes come from stars that are more massive.
13. Fusion is the combining of smaller nuclei to form larger ones. Fission is the breakdown of large nuclei into smaller ones.
14. A constellation is a star pattern that is produced by stars as seen from a point in space such as Earth.
15. A prominence is a loop of burning gas from a star's surface. It loops because of the magnetic field of the star.
16. The main sequence is the main group of stars on the Hertsprung-Russell diagram that runs from bottom right to upper left and includes the Sun.
17. A nebula is the dust and gas that leads to star formation as it is compressed by gravitational rotation.
18. A supernova is an explosion that occurs when a large star exhausts its fuel supply.
19. A black hole is the final remnant of a massive star. Its gravity is so immense that light cannot escape.
20. A protostar is the beginning mass of dust and gas just before a star ignites.

25. D
26. D
27. C
28. A

CONSTRUCTED RESPONSE

29. Since stars in clusters are all nearly the same age, but not the same mass or size, astronomers can determine the rates at which each kind of star evolves.
30. Since parallax depends on the diameter of Earth (daily) or Earth's orbit (semi-annually), any increase would improve our measurements.
31. The activity cycle is 22 years long because the polarity of the Sun's magnetic field reverses every 11 years. Thus, it takes 22 years for a complete cycle to occur.
32. The photosphere and the chromosphere are visible in the photo.
33. The dark areas are cooler, and the light areas are the hottest spots on the Sun's surface.
34. Solar prominences are produced and controlled by the magnetic field loops of the Sun.

25. Which is NOT true about binary stars?
A. They usually appear as one star.
B. They move about a common center of mass.
C. They are the most common stars in a galaxy.
D. They are always of equal brightness.

Use the diagram below to answer Question 26.

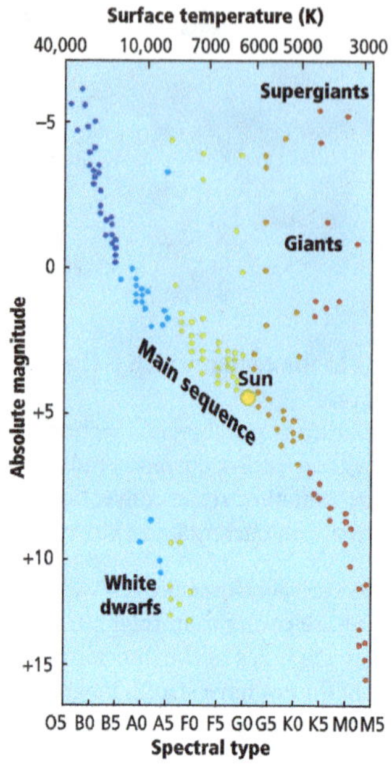

26. Which is true about the spectral classification system of stars?
A. An A star is cooler than an M star but hotter than an F star.
B. An O star is cooler than a B star yet hotter than an F star.
C. A K star is hotter than both a G star and an M star.
D. A G star is cooler than a B star and hotter than a K star.

27. Which two key stellar properties determine all the other stellar properties?
A. radius and diameter
B. mass and radius
C. composition and mass
D. diameter and composition

28. Which is the proper time order for stars like the Sun?
A. main-sequence star, red giant, white dwarf, planetary nebula
B. planetary nebula, red giant, white dwarf, main-sequence star
C. main-sequence star, white dwarf, planetary nebula, red giant
D. planetary nebula, main-sequence star, white dwarf, red giant

CONSTRUCTED RESPONSE

29. CAREERS IN EARTH SCIENCE Deduce what astronomers can tell about how stars of different masses evolve, by observing stars in clusters.

30. **Detail** how, if Earth's orbit were twice the diameter it is now, that would affect stellar parallax and our ability to measure distances.

31. **Explain** why we say the solar cycle lasts approximately 22 years and not 11.

Use the image below to answer Questions 32 and 33.

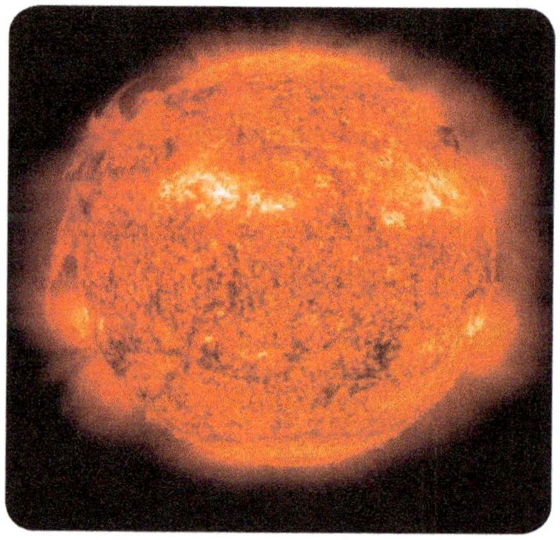

32. **Identify** the visible layers of the Sun in this photo.

33. **Identify** the light and dark areas of the Sun's surface in the photo.

34. **Explain** the relationship between the solar prominences and the Sun's magnetic field.

Chapter Self-Check

THINK CRITICALLY

35. **Deduce** why it is hotter at the center of the Sun than on the surface.

36. **Predict** the layering and composition of stars other than the Sun.

37. **Explain** how the density of the Sun is so great and yet is still in the gaseous state.

Use the diagram below to answer Questions 38 and 39.

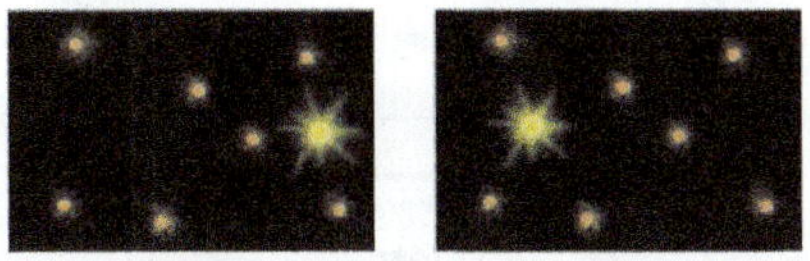

January July

38. **Draw** the relative positions of Earth, the Sun, and the star in March and November, based upon the observation in the diagram.

39. **Infer** how parallax helps scientists determine magnitude and luminosity.

40. **Infer** why the parsec has become the standard unit for expressing distance to the stars rather than the AU or light-year.

41. **Compare** a B5 star to the Sun using the H-R diagram.

42. **Compare** a supernova, a neutron star, and a pulsar.

43. **Explain** the difference between a planetary nebula and a supernova.

CONCEPT MAPPING

44. Make a concept map linking the terms *fusion, luminosity, protostar,* and one other vocabulary term.

CHALLENGE QUESTION

45. **Organize** a procedure for discovering whether a star is binary.

WRITING IN ▶ Earth Science

46. The person who developed the modern system of spectral classification was Annie Jump Cannon. Research her work and write about her role in forging new pathways for women in science.

DBQ Document–Based Questions

Data obtained from: Massey, P., et al. 2002. Orbits of four very massive binaries in the R136 cluster. *The Astrophysical Journal* 565:982–993.

Binary stars revolve around one another. The radial velocity is the rate of the stars in a binary pair moving toward and away from an observer. Subtract the lowest velocity from the highest velocity for each star, and divide by two to find the average velocity.

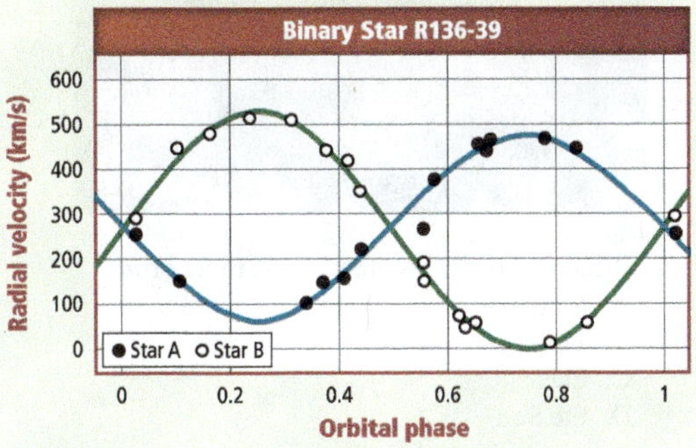

47. If the star with the larger mass has a lower average velocity, which star has the greater mass?

48. When the paths of the stars cross, there may be an eclipse for the observer. At what points in the orbital phase might there be eclipses?

CUMULATIVE REVIEW

49. Which of the mineral groups is most abundant in Earth's crust? **(Chapter 4)**

50. Briefly describe how air masses form. **(Chapter 12)**

51. What structures are formed by magmas that intrude the crust but do not erupt at the surface? **(Chapter 18)**

52. What makes an interstellar cloud collapse to start the star-formation process? **(Chapter 28)**

THINK CRITICALLY

35. Random walk causes the photons produced in the core to lose energy as they travel through the outer layers. Therefore, the surface is colder.
36. Stars with a similar mass are likely to have layers similar to those of the Sun. More-massive stars probably have more layers.
37. The density of the Sun is countered by its extreme temperature, which keeps the matter in a gaseous state.
38. The positions of the stars in the sky will be halfway between the January and July positions for both November and March. The alignment will be the Sun, Earth, stars in November and Earth, the Sun, stars in March.
39. Parallax helps determine which stars are closer to Earth and which are farther away. Knowing the distance helps scientists determine the brightness of stars, which is a factor of luminosity as well as magnitude.
40. The AU is too small to describe stellar distances. The parsec is a larger time unit than ly. Parallax, and therefore distance in parsecs, can be precisely measured.
41. B5 stars have a magnitude of approximately 0 and a surface temperature of about 15,500 K. The Sun's magnitude is almost +5, and its temperature is 5800 K.
42. All of these are the end results of a collapsing star. The supernova is the explosion of the outer matter of the star; the neutron star is the central mass of the dying star; and if the star spins, its gravity can cause it to emit cones of light that seem to pulse because of the spin.
43. Both are related to the outer matter of a dying star. The planetary nebula is associated with a white dwarf, while a supernova is associated with a neutron star or a black hole.

CONCEPT MAPPING

44. Answers will vary. Check students' maps for accuracy.

CHALLENGE QUESTION

45. First, the light from one star may seem to move back and forth, indicating that it is revolving around a center of mass. Or, a visual observation might identify the star as having changes in its brightness indicating one star passing in front of a second star, but then a spectral analysis would be necessary for confirmation. The spectral analysis would look for red- and blueshifts.

WRITING IN ▶ Earth Science

Rubric

46. Answers will vary. Students should recognize the importance of Cannon's work for astronomy, as well as female scientists in the late 1800s and early 1900s.

DBQ Document-Based Questions

Data obtained from: Massey, P., et al. 2002. Orbits of four very massive binaries in the R136 cluster. *The Astrophysical Journal* 565:982–993.

47. Star A
48. Eclipses occur at 0, 0.5, and 1.0.

CUMULATIVE REVIEW

49. silicates
50. Air masses form when large volumes of air begin to take on the temperature, pressure, and density characteristics of the source regions over which they form.
51. plutons, such as batholiths, lacoliths, sills, and dikes
52. Gravity; if an interstellar cloud happens to become dense enough by random fluctuations or by shock wave, it will become unstable and continue collapsing.

MULTIPLE CHOICE

1. A
2. A
3. B
4. C
5. A
6. C
7. D
8. B
9. C
10. A

CUMULATIVE STANDARDIZED TEST PRACTICE

MULTIPLE CHOICE

1. In December, the South Pole is tilted closer to the Sun than at any other time of the year, and the North Pole is tilted its farthest from the Sun. What is the northern hemisphere experiencing at that time?
A. the winter solstice
B. the summer solstice
C. the vernal equinox
D. the autumnal equinox

Use the diagram below to answer Questions 2 and 3.

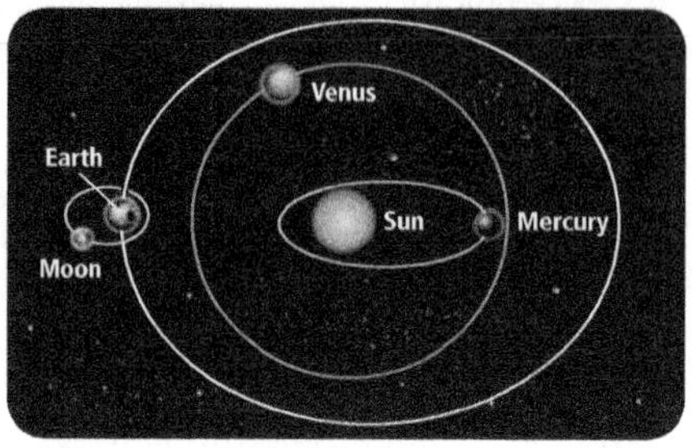

2. Which planet is moving fastest in its orbit?
A. Mercury
B. Venus
C. Earth
D. the Sun

3. Which orbit shown has an eccentricity that is closest to 0?
A. Mercury
B. Venus
C. Earth
D. the Moon

4. A bed of sedimentary rock is formed by sediments that were deposited at a rate of 1 cm/year. If the bed is 350 m thick, how long did it take for the whole bed to be deposited?
A. 350 years
B. 3500 years
C. 35,000 years
D. 350,000 years

5. Which gas giant planet is the largest?
A. Jupiter **C.** Uranus
B. Saturn **D.** Neptune

6. Which energy source does not come from the Sun?
A. wind **C.** geothermal
B. water **D.** ocean

Use the graph below to answer Questions 7–9.

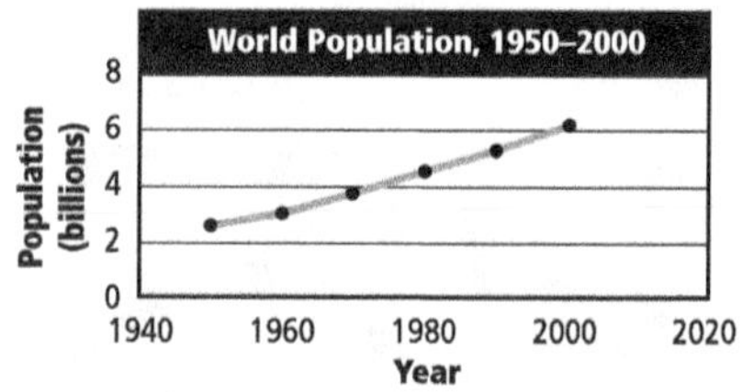

7. Which can you conclude from the graph?
A. In 80 years, it will not be possible to feed the population.
B. World population increases at a rate of 1 billion people every 10 years.
C. There were approximately 2.5 billion people in the world in 1940.
D. At the present rate of growth, the population will exceed 7 billion before 2020.

8. Based on this graph, what can be assumed about the carrying capacity of the world?
A. The world is in a state of equilibrium.
B. The world has not reached its carrying capacity.
C. The world has reached its carrying capacity.
D. The world has exceeded its carrying capacity.

9. On the graph, what is the year considered?
A. the constant
B. the dependent variable
C. the independent variable
D. the variable

10. What causes sunspots on the Sun?
A. intense magnetic fields poking through the photosphere
B. charged particles flowing into the solar system
C. spots on the surface of the photosphere, which are hotter than the surrounding areas
D. areas of low density in gas of the Sun's corona

SHORT ANSWER

Use the illustration below to answer Questions 11–13.

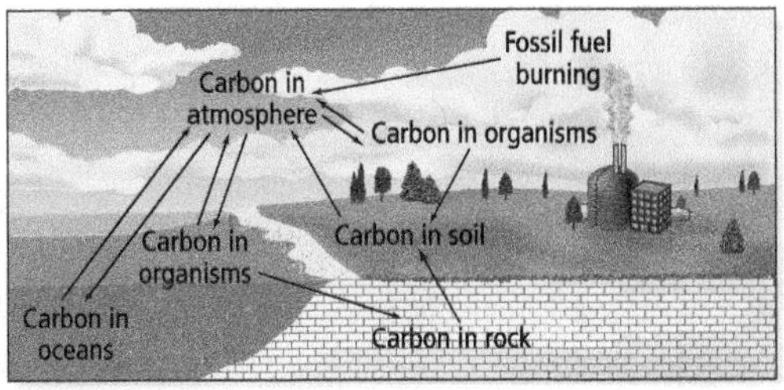

11. Describe the process shown above.

12. Why is burning fossil fuels an important part of this process?

13. Why are there two arrows between carbon in the atmosphere and carbon in organisms?

14. Describe how Earth's atmosphere would be different if there were no life on Earth.

15. Why would a minor temperature increase caused by global warming pose a threat to Earth?

16. Why is an express lane for cars with multiple passengers a good form of energy conservation?

READING FOR COMPREHENSION

The Sun's Impact on Climate

Sunspots alter the amount of energy Earth gets from the Sun, but not enough to impact global climate change, a new study suggests. The Sun's role in global warming has long been a matter of debate and is likely to remain a contentious topic.

Scientists have pondered the link between the Sun and Earth's climate since the time of Galileo. There has been an intuitive perception that the Sun's variable degree of brightness—the coming and going of sunspots for instance—might have an impact on climate. Most climate models already incorporate the effects of the Sun's waxing and waning power on Earth's weather. The number of spots cycles over time, reaching a peak every 11 years, but sunspot-driven changes to the Sun's power are too small to account for the climatic changes observed in historical data. The difference in brightness between the high point of a sunspot cycle and its low point is less than 0.1 percent of the Sun's total output.

Article obtained from: Handwerk, B. Don't blame Sun for global warming, study says. *National Geographic News.* September 13, 2006.

17. What can be inferred from this passage?
- **A.** Sunspots on the Sun do not affect global climate change.
- **B.** Sunspots greatly alter the amount of energy Earth gets from the Sun.
- **C.** It has long been thought that sunspots change Earth's climate.
- **D.** The amount of energy output from a sunspot changes drastically during its cycle.

18. Approximately how much does a sunspot cycle change the energy output of the Sun?
- **A.** 11 percent
- **B.** 1.0 percent
- **C.** 0.1 percent
- **D.** 0.01 percent

19. While a sunspot does change the amount of energy Earth gets from the Sun, why does it not impact climate?

NEED EXTRA HELP?

If You Missed Question . . .	1	2	3	4	5	6	7	8	9	10	11	12	13	14	15	16
Review Section . . .	27.3	28.1	28.1	6.1	28.3	25.2	26.1	26.1	1.2	29.1	24.3	24.3	24.3	22.3	25.3	25.3

SHORT ANSWER

11. The illustration shows the carbon cycle in which carbon is taken in by organisms from the atmosphere and is deposited into Earth where it is incorporated into the soil. It is then used by organisms or stored as a fossil fuel until it is emitted back into the atmosphere.

12. Carbon is stored in the ground in various forms as fossil fuels. The burning of fossil fuels releases the carbon back into the air and continues the cycle.

13. Some organisms, such as plants, take in carbon compounds during photosynthesis and other organisms, such as animals, release carbon compounds during respiration.

14. The atmosphere would probably contain little to no oxygen, which is produced though photosynthesis from bacteria and plants. Instead, the atmosphere would probably have large amounts of carbon dioxide.

15. Sample answer: The slight raise in temperature can lead to changes in wind and rainfall patterns, affecting farming areas and negatively affecting food production. Also, ice caps and glaciers could melt, resulting in flooding of low-lying areas.

16. Providing lanes for cars with multiple passengers encourages people to carpool. This helps to reduce the number of cars being driven and conserves fuel.

READING FOR COMPREHENSION

17. A

18. C

19. Sunspots, in their 11-year cycle, change the brightness by less than 0.1 percent. Because of this extremely small change, the amount of increase energy the Earth receives is so minute that climate is not altered.

CHAPTER 30 Galaxies and the Universe

BIGIDEA **Observations of galaxy expansion, cosmic background radiation, and the Big Bang theory describe an expanding universe that is about 14 billion years old.**

ESSENTIAL QUESTIONS	NATIONAL STANDARDS	RESOURCES TO ASSESS MASTERY
SECTION 1 **The Milky Way Galaxy** **1.** What is the size and shape of our galaxy? **2.** What are the different kinds of variable stars? **3.** Where are the different types of stars in a galaxy located? 3 sessions 1.5 blocks	UCP. 1–4; A.1–2; B.2, B.4, B.6; D.3–4; G.2–3	**Entry-Level Assessment** Focus Transparency 95, p. 862 **Progress Monitoring** Caption Question, pp. 864, 867 Reading Check, pp. 863, 864 Section Review, p. 868
SECTION 2 **Other Galaxies in the Universe** **1.** How do astronomers classify galaxies? **2.** How are galaxies organized into clusters and superclusters? **3.** How is the expansion of the universe described? 5 sessions 2.5 blocks	UCP. 1–5; A.1–2; B.2, B.4, B.6; D.4; E.2; G.2–3	**Entry-Level Assessment** Focus Transparency 96, p. 869 **Progress Monitoring** Caption Question, pp. 870, 875 Reading Check, pp. 872, 875, 876 Section Review, p. 877
SECTION 3 **Cosmology** **1.** What are the different models of the universe? **2.** How is expansion related to each of the models? **3.** What is the importance of the Hubble constant? 4 sessions 1.5 blocks	UCP. 1–4; A.1–2; B.2, B.4, B.6; D.4; E.1–2; G.2–3	**Entry-Level Assessment** Focus Transparency 97, p. 878 **Progress Monitoring** Caption Question, p. 880 Reading Check, p. 880 Section Review, p. 881 **Summative Assessment** Chapter Assessment, p. 885 Exam*View*® *Assessment Suite* CD-ROM

Go online!
connectED.mcgraw-hill.com

LEVELED RESOURCES	LAB MATERIALS
Science Notebook 30.1 OL **Chapter FAST FILE Resources:** Study Guide, p. 87 BL **Lab Resources:** Laboratory Manual, p. 233 OL **Transparencies:** Section Focus Transparency 95 EL Teaching Transparency 91 OL EL	LaunchLAB p. 860 / **15 min** paper, pencil, Earth science journal
Science Notebook 30.2 OL **Chapter FAST FILE Resources:** MiniLab Worksheet, p. 78 OL GeoLab Worksheet, p. 79 OL Study Guide, p. 90 BL **Lab Resources:** Laboratory Manual, p. 237 OL **Transparencies:** Section Focus Transparency 96 EL Teaching Transparency 92 OL EL	MiniLAB p. 873 / **20 min** felt-tipped marking pen, balloon, string, meterstick GeoLAB p. 883 / **90 min** internet access, library or observatory
Science Notebook 30.3 OL **Chapter FAST FILE Resources:** Study Guide, p. 91 BL **Transparencies:** Section Focus Transparency 97 EL	

LabManager™
Customize any lab with the LabManager™ CD-ROM.

ADDITIONAL RESOURCES

Technology:
- eTeacherEdition
- eStudentEdition
- Classroom Presentation Toolkit
- Exam*View*® *Assessment Suite* CD-ROM
- LabManager™ CD-ROM
- What's EARTH SCIENCE Got To Do With It? DVD
- Weather Classroom Video DVD

Assessment:
- Blueprints for Success
- Chapter Test (Scaffolded)
- Earth Science Standardized Test Practice

Lab Resources:
- Exploring Environmental Problems Laboratory Manual

BL Below Level OL On Level AL Advanced Learners EL English Learners COOP LEARN Cooperative Learning

CHAPTER 30

CHAPTER 30
Galaxies and the Universe

BIGIDEA Observations of galaxy expansion, cosmic background radiation, and the Big Bang theory describe an expanding universe that is about 14 billion years old.

SECTIONS

1 The Milky Way Galaxy

2 Other Galaxies in the Universe

3 Cosmology

LaunchLAB

iLab Station

How big is the Milky Way?

Our solar system seems large when compared to the size of Earth. However, the Milky Way dwarfs the size of our solar system. Explore the comparative sizes of the Milky Way and the solar system in this lab.

FOLDABLES® Study Organizer

Types of Galaxies

Make a trifold book using the labels shown. Use it to organize your notes on the three main types of galaxies.

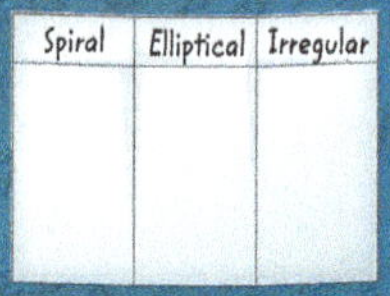

LaunchLAB

iLab Station | Rubric

How big is the Milky Way?

Process Skills use numbers, recognize spatial relationships, interpret data, formulate models, compare and contrast

Teaching Strategies

- This lab can be done in small groups.
- Remind students that the Kuiper belt is located outside the orbit of Neptune. It is used in this lab as an approximate measure of the solar system's diameter.

Procedure

1. Have students read and complete the lab safety form and follow the procedure below.
2. The Milky Way has a diameter of approximately 6.3×10^9 AU. What is the diameter of the Milky Way in light-years? (206,265 AU = 3.26 ly) 100,000 ly
3. Given that the Kuiper belt has a diameter of 50 AU, what is the diameter of the Kuiper belt in ly? 0.0008 ly
4. If you were to apply the scale 1 mm = 1 ly, how large would the Milky Way be? 100,000 mm, or 100 m
5. The Sun is located 26,000 ly from the center of the Milky Way. Based on the scale that you used in Question 4, what would be the distance, in millimeters, from the center of the Milky Way to the Sun? 26,000 mm, or 26 m
6. If you included the Kuiper belt in your model, how many millimeters across would its orbit be? 0.0008 mm

Analysis

1. **Observe** In your science journal, describe what your model of the Milky Way would look like if you actually built it. Even if the model were 100 m wide, the size of the solar system would be microscopic.
2. **Explain** why it would be a problem to show the size of our solar system in comparison to the Milky Way. To show the solar system, the model would have to expand at least 1000 times, which would make the Milky Way model 100 km in diameter.
3. **Explain** how you would change your model to include the size of Earth. If Earth were 1 mm in size, the Milky Way model would be almost 1 AU (the distance from the Sun to Earth).

Assessment

Skill Have students compare and contrast scales that would be useful for modeling Earth, the Sun-Earth-Moon system, the solar system, the Milky Way, and a group of galaxies about 2 million ly in diameter.

LabManager™
Customize any lab with the LabManager™ CD-ROM.

Galaxies come in a variety of shapes, although most are either spiral or elliptical. Each galaxy contains billions of stars, and there are billions of galaxies.

Introduce the BIGIDEA

The Milky Way A good way for students to better understand the scope of the universe is to observe our own galaxy, the Milky Way. This can be done at any time of the year. However, it is essential to do this on a moonless night at a place far removed from city lights. The Milky Way will be visible as a hazy band of light stretching across the sky. Afterward, point out to students that the Milky Way is just one of billions of known galaxies in the universe.

Teacher Content Support

Compare Galaxies Several types of galaxies can be identified in the photo. Have students discuss the differences among these galaxy types before they read the explanations in the text. Encourage them to consider various characteristics, such as their shapes, organization, and motion. Have students consider why different types of galaxies evolved, rather than all galaxies being similar.

SECTION 1

1 Focus

Focus Transparency

Before presenting the lesson, project **Section Focus Transparency 95** and have students answer the accompanying questions. EL

MAINIDEA

Measure the Milky Way Have a student move a flashlight close and then far away. Then, ask the class to consider how the intensity of a lighted object, such as a flashlight, can indicate the distance of the object. Stars that vary in intensity allowed scientists to understand distances in our galaxy. Lead a class discussion about measuring distances in the Milky Way. First, discuss the distance scale used within the solar system (AUs) compared to the scale used for the nearest stars (light-years). Emphasize that one light-year equals about 63,000 AU, so a light-year is used to express greater distances. For example, the distance from Earth to the galactic center is about 26,000 ly, or almost 2,000,000,000 AU.

2 Teach

Enrichment

Milky Way Images Have interested students find images of the Milky Way at wavelengths other than visible light. Allow time for students to present these images to the rest of the class. AL

SECTION 1

The Milky Way Galaxy

MAINIDEA Stars with varying light output allow astronomers to map the Milky Way, which has a halo, spiral arms, and a massive black hole at its center.

Essential Questions
- What is the size and shape of our galaxy?
- What are the different kinds of variable stars?
- Where are the different types of stars in a galaxy located?

Review Vocabulary

galaxy: any of the very large groups of stars and associated matter found throughout the universe

New Vocabulary

variable star
RR Lyrae variable
Cepheid variable
halo
Population I star
Population II star
spiral density wave

EARTH SCIENCE 4 YOU From inside your home, you have only a few ways to find out what is going on outside. You can look out a window or door, or use a phone or a computer. Similarly, scientists also have a few ways to learn about the stars in the galaxy around us.

Discovering the Milky Way

When looking at the Milky Way galaxy, it is difficult to see its size and shape because not only is the observer too close, but he or she is also inside the galaxy. When you observe the band of gas and dust stretching across the sky, you are looking at the edge of a disk from the inside of the disk. However, it is difficult to tell how big the galaxy is, where its center is, or what Earth's location is within this vast expanse of stars, gas, and dust. Though astronomers have answers to these questions, they are still refining their measurements.

Variable stars In the 1920s, astronomers focused their attention on mapping out the locations of globular clusters of stars. These huge, spherical star clusters are located above or below the plane of the galactic disk. Astronomers estimated the distances to the clusters by identifying variable stars in them. **Variable stars** are located in the giant branch of the Hertzsprung-Russell diagram, and pulsate in brightness because of the expansion and contraction of their outer layers. Variable stars are brightest at their largest diameters and dimmest at their smallest diameters. **Figure 1** shows the dim and bright extremes of a variable star.

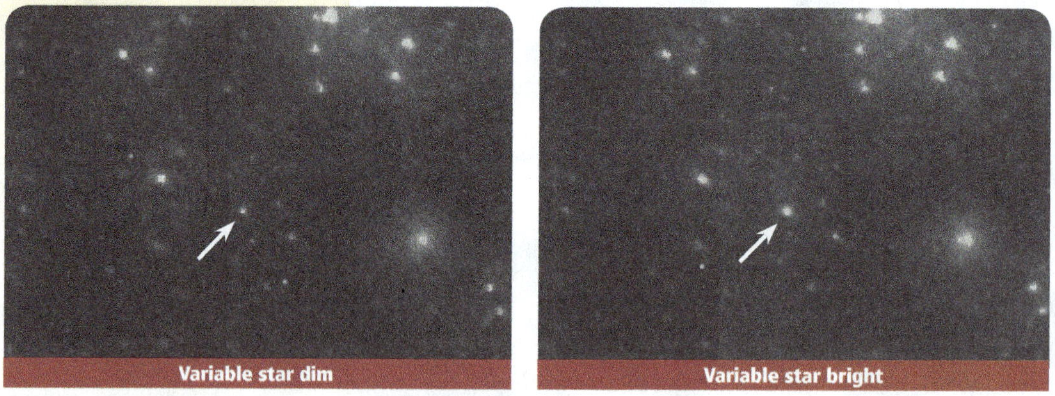

Figure 1 The diameters of variable stars change over a period of hours or days, causing them to brighten and dim.

ACROSS THE CURRICULUM

History In 1906, at the Harvard Observatory, Henrietta Leavitt was investigating variable stars in the Magellanic Clouds, two small galaxies orbiting the Milky Way. She found the apparent magnitudes of these stars were related to their pulsation periods. Because the stars were all approximately the same distance from Earth, their absolute magnitudes, or luminosities, must be related to the pulsation periods.

Types of variables For certain types of variable stars, there is a relationship between a star's luminosity and its pulsation period, which is the time between its brightest pulses. The longer the period of pulsation takes, the greater the luminosity of the star. **RR Lyrae variables** are stars that have periods of pulsation between 1.5 hours and 1.2 days, and on average, they have the same luminosity. **Cepheid variables,** however, have pulsation periods between 1 and 100 days, and the luminosity increases as much as 100 times from the dimmest star to brightest. By measuring the star's period of pulsation, astronomers can determine the star's absolute magnitude. This, in turn, allows them to compare the star's luminosity (energy) to its apparent magnitude (brightness) and calculate how far away the star must be to appear this dim or bright.

The galactic center After reasoning there were globular clusters orbiting the center of the Milky Way, astronomers then used RR Lyrae variables to determine the distances to them. They discovered that these clusters are located far from our solar system, and that their distribution in space is centered on a distant point 26,000 light-years (ly) away. The galactic center is a region of high star density, shown in **Figure 2,** much of which is obscured by interstellar gas and dust. The direction of the galactic center is toward the constellation Sagittarius. The other view of the Milky Way that is shown is along the disk into space.

☑ READING CHECK **Describe** how astronomers located the galactic center of the Milky Way.

The Shape of the Milky Way

Only by mapping the galaxy with radio waves have astronomers been able to determine its shape. This is because radio waves are long enough that they can penetrate the interstellar gas and dust without being scattered or absorbed. By measuring radio waves as well as infrared radiation, astronomers have discovered that the galactic center is surrounded by a nuclear bulge, which sticks out of the galactic disk much like the yolk in a fried egg. Around the nuclear bulge and disk is the **halo,** a spherical region where globular clusters are located, as illustrated in **Figure 2.**

Figure 2 The top two images are views of the Milky Way—one toward the outer galaxy and one close to the center. The third figure is an artist's concept of what the Milky Way galaxy looks like from space.

Tie to Previous Knowledge

Size of the Milky Way In the previous chapter, students were introduced to the use of parallax to measure stellar distances. Stellar parallax is accurate to distances of about 300 light-years. The variable star method can determine distances up to thousands or even millions of light-years, expanding the distance scale for astronomers by a huge factor. However, the relationship between period and luminosity for variable stars had to be established by finding variables whose distances were known by other techniques. None are close enough for parallax measurements, so other, less accurate, methods had to be used. As a result, the period-luminosity relation was not well determined at first, leading to confusion over the size of our galaxy.

☑ READING CHECK They used RR Lyrae variable stars to confirm a hypothesis that globular clusters of stars orbit the galactic center.

Interpret the Illustration

Earth's Location Have students study the diagram of the Milky Way in **Figure 2.** Then, ask them to describe the shape of the galaxy. a disk shape with a bulge in the middle Have students describe our solar system's location relative to the center of the galaxy and the outer edge of the galaxy. Our solar system is 26,000 ly from the center of the galaxy. This is about half the distance from the center of the galaxy to the outer edge. OL AL

EARTH SCIENCE JOURNAL

Spirals in Nature The Milky Way is a naturally spiral-shaped object. There are others in nature. Have each student make an illustrated list of spirals in nature and explain whether each spiral is governed by rotation. Possible examples include cyclonic storms, snail shells, certain oceanic flows, and the solar wind. All except snail shells are influenced by rotation. BL OL

Rubric

■ **Caption Question Fig. 3** Each arm of the galaxy was named for the constellation in which its star groups appear.

Use an Analogy

Interstellar Dust Taking off in an airplane from a hazy or smoggy city produces an effect similar to the dimming caused by interstellar dust. On the ground, you cannot see far because of the haze or smog. As the plane climbs, you rise above the haze layer and begin to see much farther. Living in the plane of the Milky Way's disk is like being in the aircraft on the ground. Because we have no way of getting above the disk of the galaxy, we can see only a fraction of the way toward the galactic center; we cannot get clear of the "smog."

Teacher Content Support

Shape of the Milky Way One of the earliest methods used to discover the shape of the Milky Way and the location of the solar system within it was to count stars in all directions. When this was first done in detail in 1906, scientists found the density of stars in space dropped off in all directions. This suggested that the solar system is at the center of the galaxy, an idea that immediately aroused skepticism. The skeptics were correct; the solar system is not located near the center of the galaxy. This was demonstrated in the 1920s. Not until 1930 did astronomers understand that dimming of stars is a result of interstellar dust that makes the density of visible stars in space drop off in all directions from Earth.

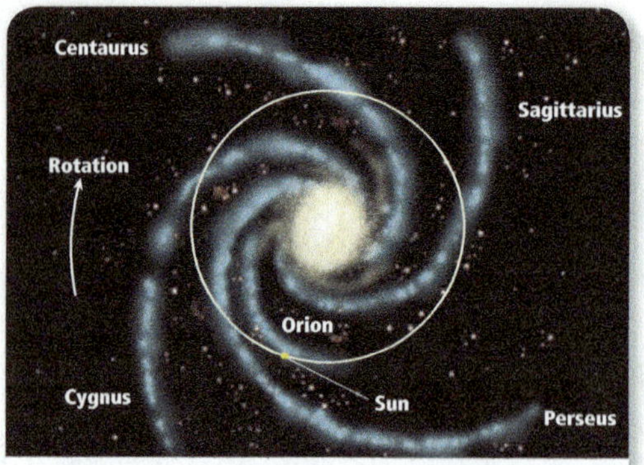

■ **Figure 3** The Sun is located on the partial Orion spiral arm and follows an orbital path around the nuclear center as shown. (Note: *Drawing is not to scale.*)
Infer *how the arms were named.*

Spiral arms Knowing that the Milky Way galaxy has a disklike shape with a central bulge, astronomers speculated that it might also have spiral arms, as do many other galaxies. This was difficult to prove. Because of the distance, astronomers have no way to get outside of the galaxy and look down on the disk. Astronomers decided to use hydrogen atoms to look for the spiral arms.

To locate the spiral arms, hydrogen emission spectra are helpful for three reasons. First, hydrogen is the most abundant element in space; second, the interstellar gas, composed mostly of hydrogen, is concentrated in the spiral arms; and third, the 21-cm wavelength of hydrogen emission can penetrate the interstellar gas and dust and be detected all the way across the galactic disk.

Using hydrogen emission and infrared images as a guide, astronomers have identified four spiral arms and numerous partial arms in the Milky Way. Using these data, scientists discovered that the Sun is located in the partial Orion arm at a distance of about 26,000 ly from the galactic center. The Sun's orbital speed is about 220 km/s, and thus its orbital period is about 225 million years. In its 5-billion-year life, the Sun has orbited the galaxy approximately 20 times. **Figure 3** shows the orbit that the Sun follows in a spinning galaxy.

☑ READING CHECK **Explain** how astronomers used the Milky Way's hydrogen emission spectrum to locate the arms.

■ **Figure 4** A barred galaxy has an elongated central bulge.

Nuclear bulge or bar? Many spiral galaxies have a barlike shape rather than having a round disk to which the arms are attached. Radio observation of interstellar gas indicates that the Milky Way has a slightly elongated shape. Recent evidence suggests that two of the arms begin at the ends of a central bar. **Figure 4** shows a barred galaxy.

Using a variety of wavelengths, astronomers are discovering what the center of the Milky Way looks like. The nuclear bulge of a galaxy is typically made up of older, red stars. The bar in a galaxy center, however, is associated with younger stars and a disk that forms from neutral hydrogen gas. Star formation does continue to occur in the bulge, and most stars are about 1000 AU apart compared to 200,000 AU separation in the locale of the Sun. Infrared measurements of 30 million stars in the Milky Way indicate a bar about 27,000 ly in length.

Demonstration

Use Emission Spectra Display the emission spectra of various elements. Remind students a spectrum is a like a fingerprint of an element. Next, show emission spectra from galaxies. Demonstrate how the galaxy spectra can be compared to the known element spectra. Have students discuss using the spectra to infer the composition and speeds of the galaxies.

☑ READING CHECK Interstellar gas, which is composed mostly of hydrogen, is concentrated in the spiral arms. The 21-cm wavelength of hydrogen emission can penetrate through dust and nearby gas and show where interstellar gas is located. Looking for interstellar gas clouds helps astronomers discern the arms.

Mass of the Milky Way

The mass located within the circle of the Sun's orbit through the galaxy, outlined in **Figure 3,** is about 100 billion times the mass of the Sun. Using this figure, astronomers have concluded that the galaxy contains about 100 billion stars within its disk.

Mass of the halo Evidence of the movement of outer disk stars and gas suggests that as much as 90 percent of the galaxy's mass is contained in the halo. Some of this unseen matter is probably in the form of dim stellar remnants such as white dwarfs, neutron stars, or black holes, but the nature of the remainder of this mass is unknown. As you will read in Section 2, the nature of unseen matter extends to other galaxies and to the universe as a whole. **Figure 5** shows the halo of the Sombrero galaxy.

■ **Figure 5** Both the galaxy halo and central bulge are populated by older, dimmer stars. The central bulge, however, has a higher density of stars and contains some newer, brighter stars, as shown in this view of the Sombrero galaxy.

A galactic black hole Weighing in at a few million to a few billion times the mass of the Sun, supermassive black holes occupy the centers of most galaxies. When the center of the galaxy is observed at infrared and radio wavelengths, several dense star clusters and supernova remnants stand out. Among them is a complex source called Sagittarius A (Sgr A), with sub-source called Sagittarius A* (Sgr A*), which appears to be an actual point around which the whole galaxy rotates.

Careful studies of the motions of the stars that orbit close to Sagittarius A* (pronounced A–star) indicate that this region has about 2.6 million times the mass of the Sun but is smaller than our solar system. Data gathered by the *Chandra X-Ray Observatory* reveal intense X-ray emissions. Astronomers think that Sagittarius A* is a supermassive black hole that glows brightly because of the hot gas surrounding it and spiraling into it. This black hole probably formed early in the history of the galaxy, at the time when the galaxy's disk was forming. Gas clouds and stars within the disk probably collided and merged to form a single, massive object that collapsed to form a black hole. **Figure 6** illustrates how a supermassive black hole develops. This kind of black hole should not be confused with the much smaller, stellar black hole, which is usually made from the collapsing core of a massive star.

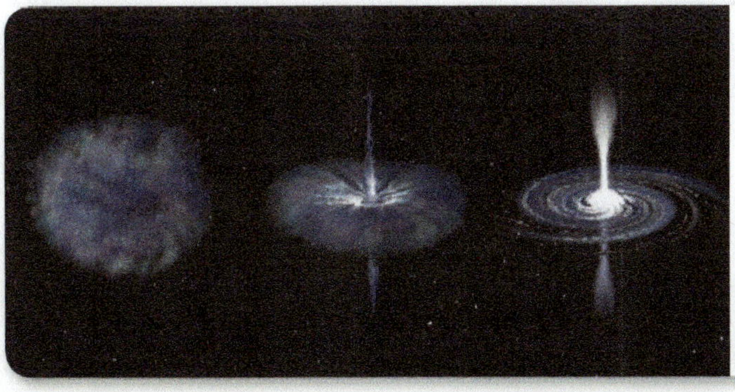

■ **Figure 6** The formation of a supermassive black hole begins with the collapse of a dense gas cloud. The accumulation of mass releases photons of many wavelengths, and perhaps even a jet of matter, as shown here.

Activity

Images of Sagittarius A* The galactic center object Sagittarius A* has been observed with telescopes at many different wavelengths, particularly radio, infrared, and X-ray wavelengths. Have students find as many images of this object as they can. Students can work in groups and prepare posters showing their collections of images. Have students compare and contrast their images. BL OL EL COOP LEARN

Reinforcement

Properties of the Milky Way Help students review what they have learned so far about the Milky Way by asking these questions: What are variable stars? stars with a varying brightness that depends on the expansion and contraction of their outer layers What part of the Milky Way did RR Lyrae variable stars help scientists locate? the galactic center What is the shape of the Milky Way? spiral Scientists used the emission spectra of which element to identify the spiral arms? hydrogen About how many stars are in the Milky Way? 100 billion What is at the center of most galaxies, including the Milky Way? a supermassive black hole BL OL

DIFFERENTIATED INSTRUCTION

Advanced Learners The mass of the Milky Way and other disklike galaxies can be calculated using Kepler's third law: If P is the orbital period and a is the semimajor axis in AU, then $P^2 = a^3$. Newton showed this relationship is affected by the sum of the masses of the two orbiting objects. The law then is $(m_1 + m_2)P^2 = a^3$, where $m_1 + m_2$ is in solar mass units. Taking the Sun and the galaxy as the two masses, the sum reduces to M, the mass of the galaxy. The Sun's mass is insignificant by comparison. Thus, Kepler's third law becomes $MP^2 = a^3$, where P is the Sun's orbital period and a is its orbital radius. Have students convert the Sun's orbital radius, a = 26,000 ly, to AU. 1.64×10^9 AU Ask: If P = 220 million years, what is M in solar mass units? 8.5×10^{10} solar masses; Note this method measures only the mass in the central region of the galaxy, inside the Sun's orbit.

Teacher Content Support

Ages of Globular Clusters

The ages of globular clusters are estimated from H-R diagrams. The age of a cluster of stars can be estimated by noting that the upper main sequence stars are missing from the H-R diagram because these massive stars have evolved into red giants. A young cluster will have a complete main sequence, while an older cluster will be missing the upper part of its main sequence. A typical globular cluster has only a short main sequence, with only the lower portion present, and many red giants. Thus, these clusters are old, with ages estimated to be as great as 14 billion years.

Concept Development

Star Ages During the 1940s, the first detailed images of the Andromeda galaxy showed the distinction between the older, yellow-colored stars in the central bulge and the younger stars and star-formation regions in the spiral arms. Only after the two populations of stars were identified in Andromeda did astronomers determine that the same situation exists in our galaxy.

■ **Figure 7** Globular clusters and the halo contain old stars poor in heavy elements. The nuclear bulge contains mostly old stars that are richer in heavy elements than the stars in the halo. The disk contains young stars and has the highest heavy element content. (Note: *Drawing is not to scale.*)

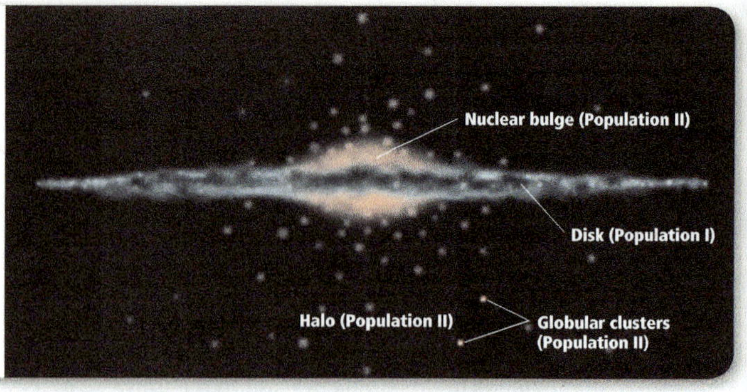

Stellar populations in the Milky Way Even though the basic compositions of all stars are the same, there are several distinct differences in detail. The differences among stars include differences in location, motion, and age, leading to the notion of stellar populations. The population of a star provides information about its galactic history. In fact, the galaxy could be divided into two components: the round part made up of the halo and bulge noted in **Figure 7,** where the stars are old and contain only traces of heavy elements; and the disk, especially the spiral arms. To astronomers, heavy elements are any elements with a mass larger than helium.

Astronomers divide stars in these two regions into two classes. **Population I stars** are in the disk and arms and have small amounts of heavy elements. **Population II stars** are found in the halo and bulge and contain even smaller traces of heavy elements. Refer to **Table 1** for more details.

Population I Most of the young stars in the galaxy are located in the spiral arms of the disk, where the interstellar gas and dust are concentrated. Most star formation takes place in the arms. Population I stars tend to follow circular orbits with low (flat) eccentricity, and their orbits lie close to the plane of the disk. Finally, Population I stars have normal compositions, meaning that approximately 2 percent of their mass is made up of elements heavier than helium. The Sun is a Population I star.

Explore **Population I and II stars with an interactive table.** Concepts In Motion

Table 1 Population I and II Stars of the Milky Way

	Location in Galaxy	Percent of H & He	Percent Heavy Elements	Age (years)	Type of Star	Type of Galaxy	Example
Population I stars	disk arms and open clusters	98	2.0	<10 billion	young sequence stars	spiral and irregular	Sun, most giants, and supergiants
Population II stars	bulge and halo	99.9	0.1	>10 billion	old main-sequence stars (type K and M)	elliptical and spiral halos and bulges	Most white dwarfs and globular cluster stars

Demonstration

Spiral Arms The formation of spiral arms in a rotating disk can be simulated using food coloring and a round tub of water. Fill the tub to a depth of 10 cm. Use your hand or a paddle to move the water in a circular flow. Paddle near the center to set up differential rotation, with the water moving faster near the center. Add a drop of food coloring between the center and the edge. The coloring will spread into an elongated arc as a result of the water's rotation. When a supernova explosion occurs in the galactic disk, its debris spreads out in a similar way and forms spiral arms. The arc will quickly dissipate, just as spiral arms in the galaxy would if there were not a mechanism to maintain the arms. **WARNING:** ***Be sure to wipe up any spilled water.***

Population II There are few stars and little interstellar material currently forming in the halo or the nuclear bulge of the galaxy, and this is one of the distinguishing features of Population II stars. Age is another. The halo of the Milky Way contains the oldest known objects in the galaxy—globular clusters. These clusters are estimated to be 12 to 14 billion years old. Stars in the globular clusters have extremely small amounts of elements that are heavier than hydrogen and helium. All stars contain small amounts of these heavy elements, but in globular clusters, the amounts are mere traces. Stars like the Sun are composed of about 98 percent hydrogen and helium, whereas in globular cluster stars, this composition can be as high as 99.9 percent. This indicates their extreme age. The nuclear bulge of the galaxy also contains stars with compositions like those in globular clusters. **Table 1** points out some other comparisons of Population I and II stars.

Formation and Evolution of the Milky Way

The fact that the halo is made exclusively and nuclear bulge is made primarily of old stars suggests that these parts of the galaxy formed first, before the disk that contains only younger stars. Astronomers therefore hypothesize that the galaxy began as a spherical cloud in space. The first stars formed while this cloud was round. This explains why the halo, which contains the oldest stars, is spherical. The nuclear bulge, which is also round, represents the inner portion of the original cloud. The cloud eventually collapsed under the force of its own gravity, and rotation forced it into a disklike shape. Stars that formed after this time have orbits lying in the plane of the disk. They also contain greater quantities of heavy elements because they formed from gas that had been enriched by previous generations of massive stars. In **Figure 8,** the nuclear bulge makes up the hat of the Sombrero galaxy.

European Southern Observatory/Photo Researchers

Figure 8 Easily seen through small telescopes, the Sombrero galaxy gets its name from the bright glow of the nuclear bulge and the dust and gas lanes along the outer edge of its disk.
Predict *which type of stars would be found in the nuclear bulge.*

Discussion

Star Composition Ask students to explain why the older stars in the galaxy (Population II stars) have lower abundances of heavy elements than the younger stars (Population I stars). The heavy-element content of the galaxy has been gradually rising with time since the galaxy formed. Because old stars formed before much enrichment of heavy elements had occurred, they have lower abundances of these elements than younger stars, which formed later in the galaxy's history. OL

Teacher Content Support

Star Orbits In the formation of the Milky Way, the orbits of the stars were effectively frozen after formation, while interstellar gas clouds were subject to collisions, which changed the stars' orbits. There is so much space between stars that the chances of them colliding and altering their orbits are low.

Caption Question Fig. 8 Population II stars

DIFFERENTIATED INSTRUCTION

Struggling Learners Have students write each heading in this section on a sheet of paper. Then, have them discuss the content with a partner and together write one or two sentences describing the main idea presented under each heading. Encourage students to use their summaries when preparing for tests. COOP LEARN

3 Assess

Check for Understanding

Observe Locate a good-quality color photograph of a portion of the Milky Way, preferably one that includes the central region of the galaxy. Have students examine the photo and identify as many objects as they can. Students should be able to identify parts of the galaxy, star clusters, gas and dust clouds, and individual stars.

Reteach

Compare As a means of reinforcing student knowledge about star formation and the formation of the Milky Way, have students compare the formation of the Sun and planets with the formation of the galaxy. Students should point out that each formation process started with a rotating cloud of gas that collapsed gravitationally and was flattened into a disk by rotation. Also, in each case, a central object formed that contained a great deal of mass, and objects orbit the center of mass.

Assessment

Performance Have students create posters that diagram the Milky Way and identify its parts. For each identifier, have students write brief descriptions.

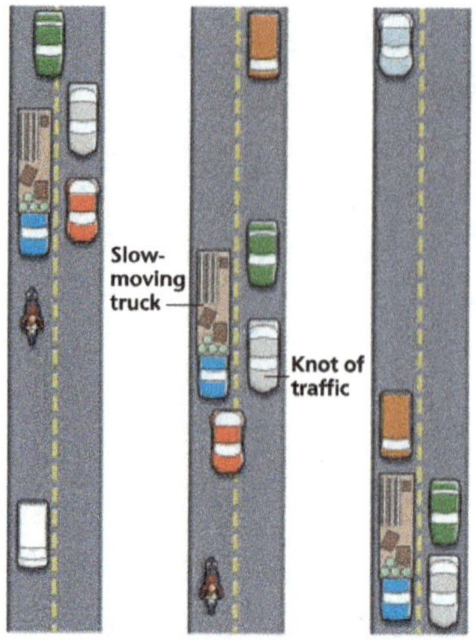

■ **Figure 9** A slow truck on a highway causing a buildup of cars around it illustrates one theory as to how spiral density waves maintain spiral arms in a galaxy.

Spiral Arms

Most of the main features of the galaxy are understood by astronomers, except for the way in which the spiral arms are retained. The Milky Way is subject to gravitational tugs by neighboring galaxies and is periodically disturbed by supernova explosions from within, both of which can create or affect spiral arms. There are several hypotheses about why galaxies keep this spiral shape.

One hypothesis is that a kind of wave called a spiral density wave is responsible. A **spiral density wave** has spiral regions of alternating density, which rotate as a rigid pattern. As the wave moves through gas and dust, it causes a temporary buildup of material, like a slow truck on the highway causes a buildup of cars, shown in **Figure 9.** Like cars surrounding a -slow truck, the stars, gas, and dust that encounter the density wave form spiral arms.

A second hypothesis is that the spiral arms are not permanent structures but instead are continually forming as a result of disturbances such as supernova explosions. The Milky Way has a broken spiral-arm pattern, which most astronomers think fits this second model best. However, some galaxies have a prominent two-armed pattern, that was more likely created by density waves.

A third possibility is considered for faraway galaxies. It suggests that the arms are only visible because they contain hot, blue stars that stand out more brightly than dimmer, redder stars. When viewed in UV wavelengths, the arms stand out, but when viewed in infrared wavelengths, they seem to disappear.

SECTION 1 REVIEW

Section Self-Check

Section Summary

- The discovery of variable stars aided in determining the shape of the Milky Way.
- RR Lyrae and Cepheid are two types of variable stars used to measure distances.
- Globular clusters of old stars are found in the nuclear bulge and halo of the Milky Way.
- The spiral arms of the Milky Way are made of younger stars and gaseous nebulae.
- Population I stars are found in the spiral arms, while Population II stars are in the central bulge and halo.

Understand Main Ideas

1. MAINIDEA **Explain** How did astronomers determine where Earth is located within the Milky Way?
2. **Determine** What do measurements of the mass of the Milky Way indicate?
3. **Analyze** How are Population I stars and Population II stars different?
4. **Summarize** How can variable stars be used to determine the distance to globular clusters?

Think Critically

5. **Explain** If our solar system were slightly above the disk of the Milky Way, why would astronomers still have difficulty determining the shape of the galaxy?
6. **Hypothesize** What would happen to the stellar orbits near the center of the Milky Way if there were no black hole?

WRITING IN ▶ Earth Science

7. Write a description of riding a spaceship from above the Milky Way into its center. Point out all of the galaxy's parts and star types.

SECTION 1 REVIEW

1. Once astronomers identified the arms of the galaxy, they could then determine which arm the Sun (and Earth) is in by measuring motion and distances of stars.
2. Most of the mass of the galaxy is in the nuclear bulge and halo, and there is a black hole at its center.
3. Population I stars are main–sequence, located in disk arms and open clusters. They are younger and have heavier elements. Population II stars are mostly white dwarfs and globular clusters, located in the bulge and halo.
4. Because the luminosity of variable stars can be measured accurately, their distances can be calculated accurately. If a variable star is in a globular cluster, then the distance to the cluster can be measured accurately.
5. We would still be a part of the galaxy and too close to see the shape. We would have to be far outside the galaxy to see its shape.
6. Stars near the center of the galaxy would orbit the galaxy more slowly because their velocity depends upon the mass of the galactic center, and without a black hole, the central mass would be much less.
7. Answers should include descriptions of spiral arms, a bar at the center, a halo, a bulge, Population I and Population II stars, variable stars, and a black hole.

Rubric

SECTION 2

Essential Questions

- How do astronomers classify galaxies?
- How are galaxies organized into clusters and superclusters?
- How is the expansion of the universe described?

Review Vocabulary

elliptical: relating to or shaped like an ellipse or oval

New Vocabulary

dark matter
supercluster
Hubble constant
radio galaxy
active galactic nucleus
quasar

FOLDABLES® Incorporate information from this section into your Foldable.

Other Galaxies in the Universe

MAIN IDEA Finding galaxies with different shapes reveals the past, present, and future of the universe.

EARTH SCIENCE 4 YOU Have you ever read an old newspaper to find out what life was like in the past? Astronomers observe distant, older galaxies to get an idea of what the universe was like long ago.

Discovering Other Galaxies

Long before they knew what galaxies were, astronomers observed many objects scattered throughout the sky. Some astronomers hypothesized that these objects were nebulae or star clusters within the Milky Way. Others hypothesized that they were distant galaxies that were as large as the Milky Way.

The question of what these objects were was answered by Edwin Hubble in 1924, when he discovered Cepheid variable stars in the Great Nebula in the Andromeda constellation. Using these stars to measure the distance to the nebula, Hubble showed that they were too far away to be located in our own galaxy. The Andromeda nebula then became known as the Andromeda galaxy, shown in **Figure 10.**

Properties of galaxies Masses of galaxies range from the dwarf ellipticals, which have masses of approximately 1 million times the mass of the Sun; to large spirals, such as the Milky Way, with masses of around 100 billion times the mass of the Sun; to the largest galaxies, called giant ellipticals, which have masses as high as 1 trillion times that of the Sun. Measurements of the masses of many galaxies indicate that they have extensive halos containing more mass than is visible, just as the Milky Way does. **Figure 10** shows a large spiral and several smaller galaxies.

Figure 10 Andromeda is a spiral galaxy like the Milky Way. The bright elliptical object and the sphere-shaped object near the center are small galaxies orbiting the Andromeda galaxy.

John Chumack/Photo Researchers

DIFFERENTIATED INSTRUCTION

Advanced Learners Have interested students research and write summaries about other possible forms of dark matter beyond stellar remnants. For each type, suggest students describe whether there is direct evidence or whether its existence is inferred. If there is direct evidence, students should describe the method of detection. If the existence is inferred, students should summarize the physical evidence that leads to the inference.

SECTION 2

1 Focus

Focus Transparency

Before presenting the lesson, project **Section Focus Transparency 96** and have students answer the accompanying questions. EL

MAIN IDEA

Shapes of Galaxies Have students look ahead at **Figure 12** and describe the different shapes of galaxies shown in the figure. Next, place a thin layer of sand in a clear, flat container on an overhead projector. Make several swirls in the sand. Have students consider how the force of your hand affects the shape of the sand. Ask how a change in the sand's rotational speed or initial density would affect the shape. Afterward, have students speculate about how the different galaxy types could evolve.

2 Teach

Model

Elliptical Galaxies The shape of an elliptical galaxy is similar to that of a football. The different shapes that we observe result from the angle at which we view the galaxy. Students can model how the viewing angle affects the shapes we observe. Have students hold a football at different angles and draw its overall shape. Student drawings should resemble circles when they view the football end-on, and ellipses with maximum elongation when they view the football directly from the side. You can demonstrate this by holding a football over an overhead projector. BL OL EL

FOLDABLES® Rubric

Use Science Terms

Dark Matter Lead students in a discussion of dark matter. Ask them why it is called *dark.* because it does not emit visible light or radiation that we can currently detect Point out that scientists infer dark matter's existence by observing the gravitational effect it has on other objects in space. BL OL

Use an Analogy

Ellipticities According to the classification method developed by Edwin Hubble, an elliptical galaxy is classified as *E0* to *E7,* depending on the ratio of the long axis to the short axis of the galaxy. If a is the long axis and b the short axis, then the degree of ellipticity is given by $10(1-b/a)$. A round galaxy, with $a = b$, has an ellipticity of 0 and is classified as an *E0* galaxy. A galaxy whose long axis is twice its short axis (i.e., $a = 2b$) is an *E5* galaxy. Have students measure the ellipticities of familiar objects such as soccer balls, footballs, bananas, toothpicks, and lemons. How are these objects analogous to galaxies? Elliptical galaxies have measured ellipticities up to 7; thus, *E7* galaxies are the most elongated elliptical galaxies known. Some of the objects suggested above, such as the banana and the toothpick, might exceed this ellipticity. OL AL

Caption Question Fig. 12 An *S0* galaxy has a disk like the spiral galaxies, but it has no arms like the elliptical galaxies.

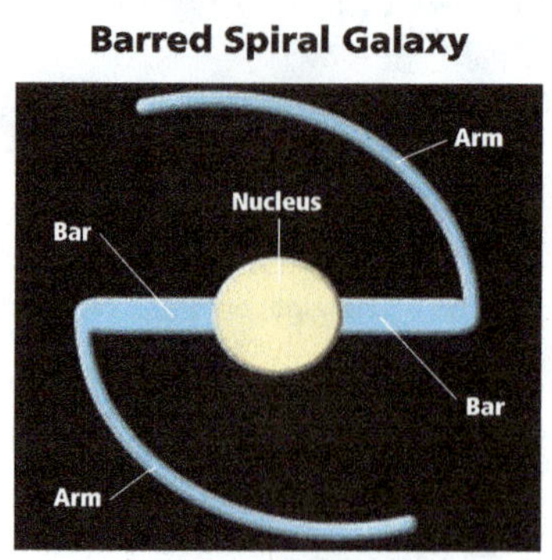

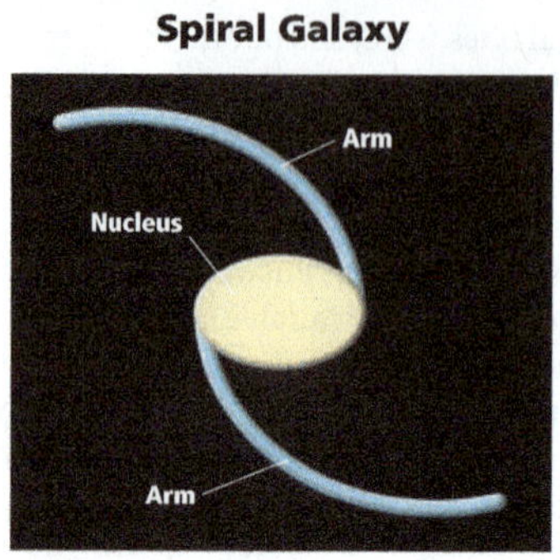

Figure 11 Measurements have indicated that the Milky Way's central region might be a bar, not a spiral.

Luminosities of galaxies also vary over a wide range, from the dwarf spheroidals—not much larger or more brilliant than a globular cluster—to supergiant elliptical galaxies, more than 100 times more luminous than the Milky Way. All galaxies show evidence that an unknown substance called dark matter dominates their masses. **Dark matter** is thought to be made up of a form of subatomic particle that interacts only weakly with other matter.

Classification of galaxies Hubble went on to study galaxies and categorize them according to their shapes.

Disklike galaxies Hubble classified the disklike galaxies with spiral arms as spiral galaxies. These were subdivided into normal spirals and barred spirals. As shown in **Figures 11** and **13,** barred spirals have an elongated central region—a bar—from which the spiral arms extend, while normal spirals do not have bars. A normal spiral is denoted by the letter *S,* and a barred spiral is denoted by *SB.* Normal and barred spirals are further subdivided by how tightly the spiral arms are wound and how large and bright the nucleus is. The letter *a* represents tightly wound arms and a large, bright nucleus. The letter *c* represents loosely wound arms and a small, dim nucleus. Thus, a normal spiral with tightly wound arms and a bright nucleus is denoted *Sa,* while a barred spiral with class *a* arms and nucleus is denoted *SBa.* Galaxies with flat disks that do not have spiral arms are denoted as *S0.*

Elliptical galaxies In addition to spiral galaxies, there are galaxies that are not flattened into disks and do not have spiral arms, as shown in **Figure 13.** Called elliptical galaxies, they are divided into subclasses based on the apparent ratio of their major and minor axes. Round ellipticals are classified as *E0,* while elongated ellipticals are classified as *E7. E1* through *E6* are progressively less round and more elongated. The classification of spiral and elliptical galaxies can be summarized by Hubble's tuning-fork diagram, illustrated in **Figure 12.**

Figure 12 The Hubble tuning-fork diagram summarizes Hubble classification for normal galaxies.
Explain *How is an S0 galaxy related to both spirals and ellipticals?*

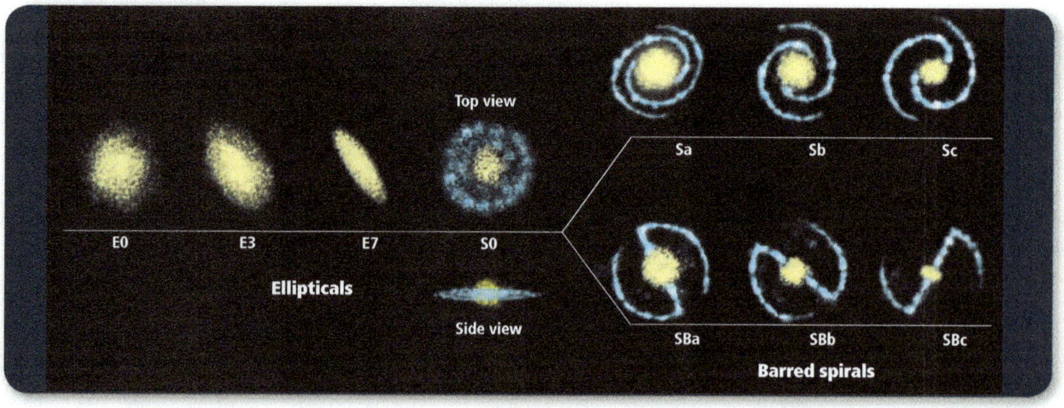

ACROSS THE CURRICULUM

History At first, Hubble's tuning-fork diagram was viewed as an evolutionary diagram for galaxies. It was theorized all galaxies began as ellipticals and evolved into spirals. As astronomers were able to look farther into the universe, and hence farther back in time, they observed spirals as well as ellipticals. If astronomers were looking so far back in time, then the young galaxies should be ellipticals, but this was not the case. The tuning-fork diagram is simply a convenient way to organize the different classes.

VISUALIZING the Local Group

Figure 13 All of the stars easily visible in the night sky belong to a single galaxy, the Milky Way. Just as stars are a part of galaxies, galaxies are gravitationally drawn into galactic groups, or clusters. The 40 galaxies closest to Earth are members of the Local Group of galaxies.

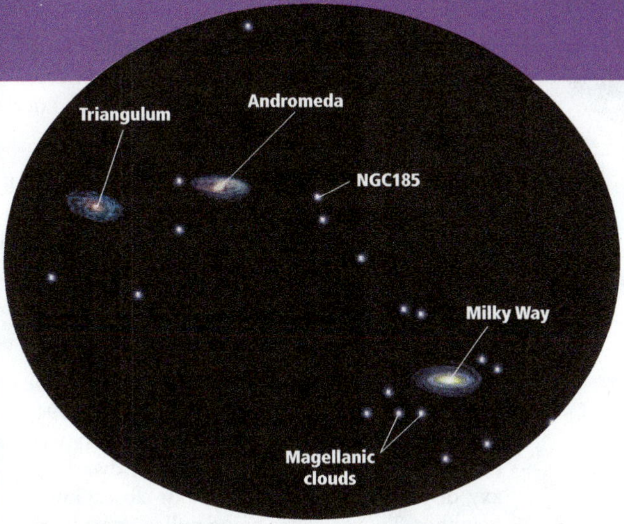

▲ **Spiral galaxies** The two largest galaxies in the Local Group, Andromeda and the Milky Way, are large, flat disks of interstellar gas and dust with arms of stars extending from the disk.

▲ **Barred spiral galaxies** Sometimes the flat disk that forms the center of a spiral galaxy is elongated into a bar shape. Recent evidence suggests that the Milky Way galaxy has a bar.

▲ **Elliptical galaxies** like NGC 185 are nearly spherical in shape and consist of a tightly packed group of relatively old stars. Nearly half of the Local Group are ellipticals.

Irregular galaxies Some galaxies are neither spiral or elliptical. Their shape seems to follow no set pattern, so astronomers have given them the classification of irregular. ▶

Concepts In Motion View an **animation of the Local Group and galaxy types.**

(l)2MASS Image Gallery, (tr)Jason Ware/Photo Researchers, (cr)National Optical Astronomy Observatories/Photo Researchers, (br)NASA/ESA/STScI/Photo Researchers

VISUALIZING

Purpose

Students will identify the different kinds of galaxies by studying the Local Group. Galaxy groups can be as small as the Local Group, with around 40 members, or as large as 2000 members.

Activity

Model Galaxies Have small groups of students use black construction paper, white glue, and salt or sand to make models of the four shapes of galaxies. Ask students what each salt crystal or sand grain represents in their models. a star Remind students that stars are so far apart from one another that they rarely, if ever, collide. BL OL

Interpret the Illustration

Figure 13 Have students carefully examine the different types of galaxies shown in the figure. Encourage them to consider how the features of the galaxies are similar. Students should notice that the galaxies have a halo, possibly a bulge, and a distinguishing shape (except for irregular galaxies).

EARTH SCIENCE JOURNAL

Types of Galaxies Have students make a table in their Earth science journals that lists ways the different types of galaxies are alike and ways they are different.

Rubric

Collaborative Learning

The Local Group Divide the class into small groups, and have each group choose a different galaxy in the Local Group. Each group should investigate various properties of its galaxy, such as magnitude, position, and shape. Students should also gather other interesting information about the galaxy, such as the history of its observation. Have each group prepare a poster displaying facts they have discovered as well as a photograph or hand drawing of the galaxy. **OL** **COOP LEARN**

READING CHECK The Milky Way and Andromeda are barred spiral and spiral galaxies, and the Magellanic Clouds are irregular galaxies. The newest discovered galaxies, the Sagittarius Dwarf Elliptical and the Canis Major, are dwarf galaxies.

Teacher Content Support

A Picture of Nothing Perhaps the most important image ever taken by the *Hubble Space Telescope (HST)* was a picture of nothing. In 1995, the *HST* was pointed at a blank spot on the sky for ten days in a row, and it took image after image of apparent darkness. The dark spot was chosen deliberately to see whether the sky might be covered with distant galaxies too faint to be seen by the largest telescopes on Earth. It is. The individual images were added together to create the Hubble Deep Field, a glorious window into the universe, full of color and light. No less than 1500 faint galaxies were found in a patch of sky only 2.5 arcminutes across. Some of these galaxies are so far away that they appear as they were about 10 billion years ago.

Figure 14 The Large and Small Magellanic Clouds are small galaxies that orbit the Milky Way.

Figure 15 The nearby Virgo cluster of approximately 2000 galaxies has a gravity so strong it is pulling the Milky Way toward it.

Irregular galaxies Some galaxies do not have distinct shapes. These irregular galaxies are denoted by *Irr.* The Large and Small Magellanic Clouds, shown in **Figure 14,** two satellite galaxies of the Milky Way, are irregular galaxies.

Groups and Clusters of Galaxies

Most galaxies are located in groups, rather than being spread uniformly throughout the universe. **Figure 13** shows some of the features of the Local Group of galaxies.

Local Group The Milky Way belongs to a small cluster of galaxies called the Local Group. The diameter of the Local Group is nearly 10 million ly. There are about 40 known members, of which the Milky Way and Andromeda galaxies are the largest. Most of the members are dwarf ellipticals that are companions to the larger galaxies. The Large and Small Magellanic Clouds were thought to be the closest galaxies to the Milky Way until 1994, when the Sagittarius Dwarf Elliptical galaxy was discovered. However, the Canis Major dwarf galaxy, discovered in 2003, is now our closest known neighbor. This galaxy is being pulled apart by the Milky Way's gravity, and is leaving streams of dust, gas, and stars in its wake. As dim galaxies continue to be found, more could be added to the Local Group in the future.

READING CHECK **Identify** the kinds of galaxies in the Local Group.

Large clusters Galaxy clusters larger than the Local Group might have hundreds or thousands of members and diameters in the range of about 5 to 30 million ly. The Virgo cluster is shown in **Figure 15.** Most of the galaxies in the inner region of a large cluster are ellipticals, while there is a more even mix of ellipticals and spirals in the outer portions.

In regions where galaxies are as close together as they are in large clusters, gravitational interactions among galaxies have many important effects. Galaxies often collide and form strangely shaped galaxies, as shown in **Figure 16,** or they form galaxies with more than one nucleus.

IN THE FIELD

Map the Universe Margaret Geller earned a PhD in physics from Princeton in 1975; she was only the second woman to do so. She is now a professor at Harvard University and a Senior Scientist at the Smithsonian Astrophysical Observatory. Geller is one of the pioneers in mapping the universe. Her current research projects include mapping dark matter in the universe, investigating high-velocity stars ejected from the Milky Way, and mapping clusters of galaxies. Geller has served on several advisory committees to NASA and the National Science Foundation, and has received numerous awards for her research in astronomy as well as educational films she has developed about astronomy.

Masses of clusters For clusters of galaxies, the mass determined by analyzing the motion of member galaxies is always much larger than the sum of the total masses of each the galaxies, as determined by their total luminosity. This suggests that most of the mass in a cluster of galaxies is invisible, which provides astronomers with strong evidence that the universe contains a great amount of dark matter.

Superclusters Clusters of galaxies are organized into even larger groups called **superclusters.** These gigantic formations, hundreds of millions of light-years in size, can be observed only when astronomers map out the locations of many galaxies ranging over huge distances. These superclusters appear in sheetlike and threadlike shapes, giving the appearance of a gigantic bubble bath with galaxies located on the surfaces of the bubbles, and the inner air pockets void of galaxies.

Figure 16 This galactic merger that began 40 mya will be complete in a few billion years.

The Expanding Universe

In 1929, Edwin Hubble made another dramatic discovery. It was known at the time that most galaxies have redshifts in their spectra, indicating that all but the nearest galaxies are moving away from Earth. Hubble measured the redshift and distances of many galaxies and found that the farther away a galaxy is, the faster it is moving away. In other words, the universe is expanding.

MiniLAB

iLab Station

Model Expansion

What does a uniform expansion look like? The discovery of redshifts of distant galaxies indicated that the universe is rapidly expanding.

Procedure

1. Read and complete the lab safety form.
2. Use a **felt-tipped marking pen** to make four dots in a row, each separated by 1 cm, on the surface of an uninflated **balloon.** Label the dots *1, 2, 3,* and *4.*
3. Partially inflate the balloon. *Do not tie the neck.* With a piece of **string** and a **meterstick,** measure the distance from Dot 1 to each of the other dots. Record your measurements.
4. Inflate the balloon more, and again measure the distance from Dot 1 to each of the other dots. Record your measurements.
5. Repeat Step 4 with the balloon fully inflated.

Analysis

1. **Identify** whether the dots are still separated from each other by equal distances after you fully inflated the balloon.
2. **Determine** how far each dot moved away from Dot 1 following each change in inflation.
3. **Infer** what the result would be if you had measured the distances from Dot 4 instead of Dot 1. From Dot 2?
4. **Explain** how this activity illustrates uniform expansion of the universe.

MiniLAB

iLab Station Rubric

Purpose Students will use a model to help them understand the uniform expansion of the universe.

Process Skills measure in SI, analyze, formulate models, recognize spatial relationships

Safety Precaution Approve lab safety forms before work begins.

Teaching Strategy This activity should be done in small groups.

Expected Results Students should find that the dots become more widely separated proportional to their initial separation.

Analysis

1. The dots are still separated by equal distances, because the separations have grown uniformly.
2. The distance of each dot from Dot 1 should increase in proportion to its original separation from Dot 1.
3. The result would have been the same no matter which dot was chosen as the reference point.
4. It shows how galaxies (the dots) move away from each other at rates that are proportional to their original separations, and it shows that the observed expansion looks the same from any galaxy.

Assessment

Knowledge Ask students how this exercise would be different if the original separations between the dots were not equal. There would be no significant change. The separations would still increase in proportion to the original distances between dots.

LabManager™
Customize any lab with the LabManager™ CD-ROM.

DIFFERENTIATED INSTRUCTION

Struggling Learners The redshift of a galaxy can be expressed as z, the ratio of the shift in wavelength to the laboratory wavelength of a spectral line. Have students rank the following galaxies in order from nearest to most distant: $z = 0.08$, $z = 1.35$, $z = 0.0002$, $z = 3.86$, $z = 0.33$. The answer is simply the order from lowest to highest z: 0.0002, 0.08, 0.33, 1.35, and 3.86. The larger the value of z, the larger the redshift, and the larger the distance.

Problem-Solving LAB

Purpose Students will construct graphs using given data and will interpret the graphs to find the value of the Hubble constant.

Process Skills make and use graphs, use numbers, analyze, interpret data, use tables

Teaching Strategies

- Have students do this activity individually. A comparison of results will help students understand measurement error and uncertainty, because there will be a range of answers for the value of *H*, even though each student will begin with the same data.
- The values students obtain should be close to 70 km/s/Mpc.

Think Critically

4. The slope represents the value of the Hubble constant, which is the rate of the expansion of the universe.
5. Because there is a lot of scatter in the points on the graph, the value derived is not accurate. The scatter is presumably a result of measurement error and the fact that galaxies can have local random motions in addition to their expansion motions.
6. The measurement could be improved by using a larger sample of galaxies, extending to even greater distances.

Problem-Solving LAB

Make and Use Graphs

How was the Hubble constant derived? Plotting the distances and speeds for a number of galaxies created the expansion constant for Hubble's Law.

Analysis

1. Use the data to construct a graph. Plot the distance on the *x*-axis and the speed on the *y*-axis.
2. Use a ruler to draw a straight line through the center of the band of points on the graph, so that approximately as many points lie above the line as lie below it. Make sure your line starts at the origin.
3. Measure the slope by choosing a point on the line and dividing the speed at that point by the distance.

Galaxy Data

Distance (Mpc)	Speed (km/s)	Distance (Mpc)	Speed (km/s)
3.0	210	26.5	2087
8.3	450	33.7	2813
10.9	972	36.8	2697
16.2	1383	38.7	3177
17.0	1202	43.9	3835
20.4	1685	45.1	3470
21.9	1594	47.6	3784

Think Critically

4. **State** What does the slope represent?
5. **Gauge** How accurate do you think your value of *H* is? Explain.
6. **Consider** How would an astronomer improve this measurement of *H*?

Implications of redshift You might infer that Earth is at the center of the universe, but this is not the case. An observer located in any galaxy, at any place in the universe, will observe the same thing in a medium that is uniformly expanding—all points are moving away from all other points, and no point is at the center. At greater distances the expansion increases the rate of motion.

A second inference is that the universe is changing with time. If it is expanding now, it must have been smaller and denser in the past. In fact, there must have been a time when all contents of the universe were compressed together. The Big Bang theory has been proposed to explain this expansion.

Hubble's law Hubble determined that the universe is expanding by making a graph comparing a galaxy's distance to the speed at which it is moving. The result is a straight line, which can be expressed as a simple equation, $v = Hd$, where v is the velocity at which a galaxy is moving away measured in kilometers per second; d is the distance to the galaxy measured in megaparsecs (Mpc), where 1 Mpc = 3,260,000 ly; and H is a number called the **Hubble constant,** measured in kilometers per second per megaparsec. H represents the slope of the line.

Measuring *H* Determining the value of *H* requires finding distances and speeds for many galaxies and constructing a graph to find the slope. This is a difficult task because it is hard to measure accurate distances to the most remote galaxies. Hubble could obtain only a crude value for *H*. Obtaining an accurate value for *H* was one of the key goals of astronomers who designed the *Hubble Space Telescope (HST)*. It took nearly ten years after the launch of the *HST* to gather enough data to pinpoint the value of *H*. Currently, the best measurements indicate a value of approximately 70 km/s/Mpc.

New way to measure distance Once the value of *H* is known, it can be used to find distances to faraway galaxies. By measuring the speed at which a galaxy is moving, astronomers use the graph to determine the corresponding distance to the galaxy. This method works for the most remote galaxies that can be observed and allows astronomers to measure distances to the edge of the observable universe.

The only galaxies that do not seem to be moving apart are those within a cluster. The internal gravity of the cluster keeps them from separating.

ACROSS THE CURRICULUM

Math Have students use the value of *H* they found in the Problem-Solving Lab to calculate the distance to a galaxy whose velocity is 23,600 km/s. Using the value $H = 70$ km/s/Mpc yields a distance for this galaxy of $d = v/H = 337$ Mpc; this is approximately 1 billion ly. OL AL

Active Galaxies

Galaxies that emit large amounts of energy from their cores are called active galaxies. The core of an active galaxy where highly energetic objects or activities are located is called the **active galactic nucleus** (AGN). An AGN emits as much or more energy than the rest of the galaxy. The output of this energy often varies over time, sometimes in as little as a few days. About 10 percent of all known galaxies are active, including radio galaxies and quasars.

Radio galaxies Radio-telescope surveys of the sky have revealed a number of galaxies that are extremely luminous. These galaxies, called **radio galaxies,** are often giant elliptical galaxies that emit as much or more energy in radio wavelengths than they do in wavelengths of visible light. Radio galaxies have many unusual properties. The radio emission usually comes from two huge lobes of very hot gas located on opposite sides of the visible galaxy. These lobes are linked to the galaxy by jets of hot gas. The type of emission that comes from these regions indicates that the gas is ionized, and that electrons in the gas jets are traveling near the speed of light. Many radio galaxies have jets that can be observed only at radio wavelengths. One of the brightest of the radio galaxies, a giant elliptical called M87, shown in **Figure 17,** also has a jet of gas that emits visible light extending from the galactic center out toward one of the radio-emitting lobes.

Figure 17 In addition to radio lobes, M87 has a jet that emits visible light.

READING CHECK **Describe** the unusual properties of a radio galaxy.

Quasars In the 1960s, astronomers discovered objects that looked like ordinary stars, but some emitted strong radio waves. Most stars do not. Also, whereas most stars have spectra with absorption lines, these new objects had mostly emission lines in their spectra. These starlike objects with emission lines in their spectra were called **quasars.** Quasars are very luminous, very distant active galaxies. Many quasars vary in brightness over a period of a few days. Two quasars are shown in **Figure 18.**

The emission lines of quasars are those of common elements, such as hydrogen, shifted far toward longer wavelengths. Once astronomers had identified the large spectral-line shifts of quasars, they wondered whether they could have redshifts caused by the expansion of the universe.

Figure 18 Quasars are distant celestial objects that emit several thousand times more energy than does our entire galaxy.
Recall *What other objects emit jets of matter?*

Identify Misconceptions

Some students might believe Earth is at the center of the universe.

Uncover the Misconception Distant galaxies move away from Earth, and the farther away a galaxy is, the faster it recedes. Ask students what this implies about Earth's position in the universe. Because scientists assume this is true for all objects in space, it shows that Earth is not at the center of the universe and that the universe does not technically have a center.

Demonstrate the Concept Remind students about the results of the MiniLab in which they observed dots on an expanding balloon. Ask them to consider the balloon's surface, ignoring the volume inside the balloon. Explain there is no one point on the surface that represents the center of the expansion.

Assess New Knowledge Have students consider any point on the expanding balloon and compare it to the expanding universe. All other points move away from that point. There is no center, just as there is no center of the universe.

READING CHECK A radio galaxy emits most of its energy as radio waves, has lobes of hot gas on opposite sides of the galaxy, and can have jets of gas that emit visible light.

Caption Question Fig. 18 Radio galaxies also emit jets of matter.

EARTH SCIENCE JOURNAL

Active Galaxies and Quasars Have students write a summary of the similarities among radio galaxies, active galactic nuclei, and quasars. All three are sources of huge amounts of energy coming from small volumes of space, all involve jets of ionized gas, and all can vary in energy over a period of days. BL OL

Rubric

GeoLAB

iLab Station

The GeoLab located at the end of the chapter can be used at this point in the lesson.

Use Science Terms

Quasar The word *quasar* is a contraction of the original designation of these objects, which was *quasi-stellar objects*. Some astronomers refer to them as *QSOs*.

Teacher Content Support

Supermassive Black Holes

The conclusion that the energetic objects in the cores of AGNs and quasars are black holes was based in part on the small sizes of the energy-emitting regions. The small sizes are deduced from the rapidity of the brightness variations in these objects. An emitting body cannot be smaller than the distance light travels in the time it takes for the object to change its brightness. Hence, a galactic nucleus or quasar that varies in a day's time cannot be any larger in diameter than a light-day, or about the size of the solar system. The only object known to astronomers that can contain such vast amounts of mass in such a small space is a black hole.

READING CHECK Comparing the redshifts of quasars with other galaxies within their cluster told astronomers they were at the same distances.

CAREERS IN EARTH SCIENCE

WebQuest

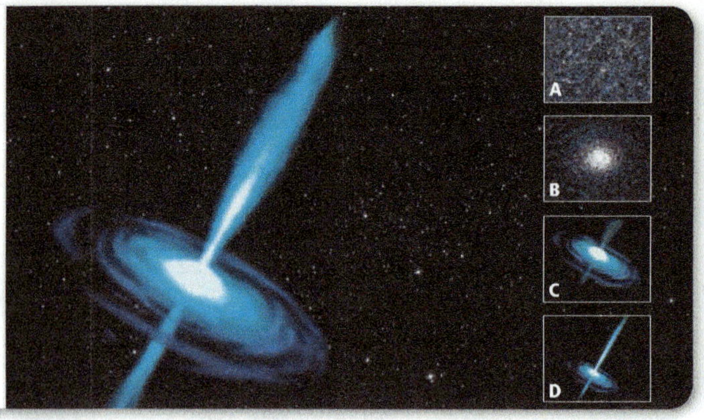

Figure 19 An interstellar gas cloud (A) collapses gravitationally (B) on its way to forming a galaxy. The nucleus (C) forms a black hole as the gas there is compressed. Magnetic fields of the rapidly rotating disk surrounding the black hole form two highly energetic jets (D) that are perpendicular to the disk's equatorial plane.

Quasar redshift The redshift of quasars was much larger than any that had been observed in galaxies up to that time, which would mean that the quasars were much farther away than any known galaxy. At first, some astronomers doubted that quasars were far away, but in the decades since quasars were discovered, more evidence supports this hypothesis. One piece of supporting evidence indicates that those quasars associated with clusters of galaxies have the same redshift, verifying that they are the same distance away. Another more important discovery is that most quasars are nuclei of very dim galaxies, whose formation is illustrated in **Figure 19.** The quasars appear to be extra-bright AGNs—so much brighter than their surrounding galaxies that astronomers could not initially see those galaxies.

CAREERS IN EARTH SCIENCE

Computer Programmer Many astronomers use equipment that does not observe light. A computer programmer writes programs astronomers can use to observe spectra, calculate, and decipher the data collected by telescopes.

WebQuest

READING CHECK **Explain** how astronomers determined distances to quasars.

Looking back in time Because quasars are distant, it takes their light a long time to reach Earth. Therefore, observing a quasar is seeing it as it was a long time ago. For example, it takes light from the Sun approximately 8 minutes to reach Earth. When you observe the Sun, you are seeing it as it was 8 minutes earlier. When you observe the Andromeda galaxy, you see the way it looked 2 million years earlier. The most remote quasars are several billion light-years away, which indicates the stage you see is from billions of years ago. If quasars are extra-bright AGNs, then the many distant ones are nuclei of galaxies as they existed when the universe was young. This suggests that many galaxies went through a quasar stage when they were young. Consequently, today's AGNs might be former quasars that are not as energetic as they were long ago.

Looking far back into time, the early universe had many quasars. Current theory suggests that they existed around supermassive black holes that pulled gas into the center, where in a violent swirl, friction heated the gas to extreme temperatures resulting in the bright light energy that was first detected.

ACROSS THE CURRICULUM

History The human view of Earth's place in the cosmos has changed drastically on several occasions in history. Each time, the new perception placed Earth in a less central position, so that now, we realize Earth orbits an ordinary star far from the center of a galaxy that is one of billions, occupying no special place in the universe. Have each student choose one of these major changes in perception and write a brief essay about its historical context. How was the new concept received by scientists and the public, either at the time it was developed or later?

Source of power AGNs and quasars emit far more energy than ordinary galaxies, but they are as small as solar systems. This suggests that AGNs and quasars contain supermassive black holes. Recall that the black hole thought to exist in the core of our own galaxy has a mass of about 1 million Suns. The black holes thought to exist in AGNs and the cores of quasars are much more massive—up to hundreds of millions of times the mass of the Sun.

The beams of charged particles that stream out of the cores of radio galaxies and form jets are probably created by magnetic forces. As material falls into a black hole, the magnetic forces push the charged particles out into jets. There is evidence that similar beams or jets occur in other types of AGNs and in quasars. In fact, radio-lobed quasars have jets that are essentially related to radio galaxies.

Figure 20 shows evidence of a supermassive black hole in the center of the Centaurus A galaxy. In modeling a supermassive black hole of this magnitude, the mass of nearly 1 billion Suns may be needed to pull the stars in this galaxy into the center. A plasma jet, ejected from the nucleus, extends 13,000 ly into space.

■ **Figure 20** A jet of energetic X-ray particles is emitted from the AGN of elliptical galaxy Centaurus A, which probably hides a supermassive black hole.

X-ray: NASA/CXC/CfA/R.Kraft/Submillimeter: MPIfR/ESO/APEX/A.Weiss/Optical: ESO/WFI

SECTION 2 REVIEW

Section Self-Check

Section Summary

- Galaxies can be elliptical, disk-shaped, or irregular.
- Galaxies range in mass from 1 million Suns to more than a trillion Suns.
- Many galaxies seem to be organized in groups called clusters.
- Hubble's law helped astronomers discover that the universe is expanding.
- Quasars are the nuclei of faraway galaxies that are dim and seen as they were long ago, due to their great distances.

Understand Main Ideas

1. MAINIDEA **Explain** how astronomers discovered that there are other galaxies beyond the Milky Way.
2. **Summarize** why astronomers theorize that most of the matter in galaxies and clusters of galaxies is dark matter.
3. **Explain** why it is difficult for astronomers to accurately measure a value for the Hubble constant, *H*. Once a value is determined, describe how it is used.
4. **Explain** the differences in appearance among normal spiral, barred spiral, elliptical, and irregular galaxies.

Think Critically

5. **Deduce** how the nighttime sky would look from Earth if the Milky Way were an elliptical galaxy.
6. **Infer** how black holes cause both AGNs and quasars to be so luminous.

MATH IN ▶ Earth Science

7. Convert the distance across the Milky Way to Mpc if the diameter of the Milky Way is 100,000 ly. What is the distance in Mpc across a supercluster of galaxies whose diameter is 200 million ly? (1 Mpc = 3,260,000 ly)

3 Assess

Check for Understanding

Discussion Ask students to discuss the reasons why quasars are valuable probes of the large-scale properties of the universe. Quasars are so far away that they show the galaxy as it was billions of years ago. The vast amounts of energy emitted by quasars suggest they have supermassive black holes at their centers. The redshift of radiation from quasars indicates expansion of the universe.

Reteach

Summarize Ask students to summarize the main differences between normal and active galaxies. Unlike normal galaxies, active galaxies have strong emissions at radio wavelengths and emit jets of extremely hot gas. Some have small, highly energetic cores (AGN).

Assessment

Knowledge Ask students how the balloon model would differ if the universe were not expanding uniformly. The dots would move away from each other, but not uniformly. This would be analogous to having very strong portions of rubber in the balloon so that some galaxies would not move far and weak spots that would stretch more so that galaxies would move farther.

SECTION 2 REVIEW

1. Using Cepheid variables, astronomers were able to measure distances to those stars that were in distant galaxies and determine there were other galaxies at great distances.
2. The mass of the visible part of a galaxy does not account for the motions and apparent speeds of stars and galaxies. There must be some invisible material or unknown force acting.
3. In order to get the most accurate reading for *H*, astronomers need accurate distance measurements. Since the farthest objects in the universe are included in the measurement, accuracy is difficult. *H* is important to measure the speed of the universe expansion and therefore the age of the universe.
4. Normal spiral galaxies have a disc and look like a whirlpool, and barred spirals have an elongated central region within that whirlpool. The elliptical galaxies are spherical or football-shaped, and irregular galaxies have no specific shape.
5. The nighttime sky would probably look similar to the sky we see now, only without a band across the sky where the disk appears. The stars would be more uniformly spread out.
6. Both quasars and AGNs have supermassive black holes in their centers. Large amounts of energy stream out of black holes as material is pulled into them.
7. Milky Way: 0.030 Mpc; supercluster: 60 Mpc

SECTION 3

1 Focus

Focus Transparency

Before presenting the lesson, project **Section Focus Transparency 97** and have students answer the accompanying questions. EL

MAINIDEA

The Big Bang Model Have two students stand back-to-back at the front of the room. Then, have them walk away from each other a distance of one foot each second for four seconds. Ask students to describe the original positions of the students, considering only their speeds and current positions. Point out that in a similar way, the Big Bang theory describes the expansion of the universe based on evidence.

2 Teach

Use Science Terms

Cosmology The word *cosmology,* from the Greek *kosmos,* refers to the science of the universe as it is today. Technically, the word for studies of the origin of the universe is *cosmogony,* which dates back to times when philosophers thought the solar system was the entire universe. Today, most scientists use *cosmology* to describe any study of the universe, including its origin.

SECTION 3

Cosmology

MAINIDEA The Big Bang theory was formulated by comparing evidence and models to describe the beginning of the universe.

Essential Questions

- What are the different models of the universe?
- How is expansion related to each of the models?
- What is the importance of the Hubble constant?

Review Vocabulary

radiation: the process of emitting radiant energy in the form of waves or particles

New Vocabulary

cosmology
Big Bang theory
cosmic background radiation

EARTH SCIENCE 4 YOU Manipulating a magnet and iron filings can help you model Earth's magnetic field. Cosmologists use particle accelerators to help create models of the early universe.

Big Bang Model

The study of the universe—its nature, origin, and evolution—is called **cosmology.** The mathematical basis for cosmology is general relativity, from which equations were derived that describe both the energy and matter content of the universe. These equations, combined with observations of density and acceleration, led to the most accurate model so far—the Big Bang model. The fact that the universe is expanding implies that it had a beginning. The theory that the universe began as a point and has been expanding since is called the **Big Bang theory.** Although the name might seem to imply explosion into space, the theory describes an expansion of space itself while gravity holds matter in check. Review the effects of expansion by checking results from the MiniLab in Section 2.

Outward expansion Similar to a star's internal fusion pressure opposing the effort of a gravitational force to collapse the star, the universe has two opposing forces. In the Big Bang model, the momentum of the outward expansion of the universe is opposed by the inward force of gravity acting on the matter of the universe to slow that expansion, as illustrated in **Figure 21.** What ultimately will happen depends on which of these two forces is stronger.

When the rate of expansion of the universe is known, it is possible to calculate the time since the expansion started and determine the age of the universe. When the distance to a galaxy and the rate at which it is moving away from Earth are known, it is simple to calculate how long ago that galaxy and the Milky Way were together. In astronomical terms, if the value of H, the expansion (Hubble) constant, is known, then the age of the universe can be determined. Corrections are needed to allow for the fact that the expansion has not been constant—it has slowed since the beginning and is now accelerating.

Based on the best value for H that has been calculated from *Hubble Space Telescope* data and the data on the cosmic background radiation, the age of the universe can be pinpointed to 13.7 billion years. This fits with what astronomers know about the age of the Milky Way galaxy, which is estimated to be between 12 and 14 billion years old, based on the ages of the oldest star clusters.

Figure 21 The universe is either open, flat, or closed, depending on whether gravity or the momentum of expansion dominates.

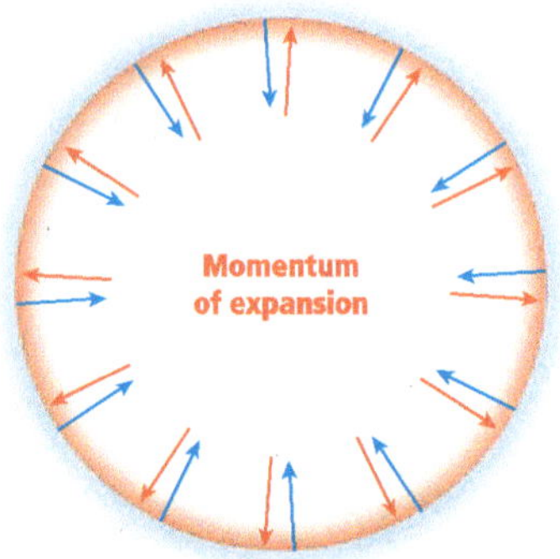

DIFFERENTIATED INSTRUCTION

Advanced Learners The idea that the universe is expanding from a single point dates back to the early 1920s, even before the expansion of the universe was discovered. A universal origin in a hot, dense state was predicted by Belgian scientist and clergyman Georges LeMaître, based on a solution to equations from Einsten's theory of general relativity. Have students read about LeMaître's prediction. Then, have them write a brief summary of general relativity and why LeMaître felt this led to the idea of an expanding universe.

Possible outcomes Based on the Big Bang theory, there are three possible outcomes for the universe, as shown in **Figure 22.** The average density of the universe is an observable quantity with vast implications to the outcome.

Open universe An open universe is one in which the expansion will never stop. This would happen if the density of the universe is insufficient for gravity to ever halt the expansion.

Closed universe A closed universe will result if the expansion stops and turns into a contraction. That would mean the density is high enough that eventually the gravity caused by the mass will halt the expansion of the universe and pull all of the mass back to the original point of origin.

Flat universe A flat universe results if the expansion slows to a halt in an infinite amount of time, but never contracts. This means that while the universe would continue to expand, its expansion would be so slow that it would seem to stop.

Critical density All three outcomes are based on the premise that the rate of expansion has slowed since the beginning of the universe, but the density of the universe is unknown. At the critical density, there is a balance, so that the expansion will come to a halt in an infinite amount of time. The critical density, about 6×10^{-27} kg/m^3, means that, on average, there are only two hydrogen atoms for every cubic meter of space. When astronomers attempt to count the galaxies in certain regions of space and divide by the volume, they get an even smaller value. So they would conclude that the universe is open, except that the dark matter has not been included. But even the best estimates of dark matter density are not enough to conclude that the universe is a closed system.

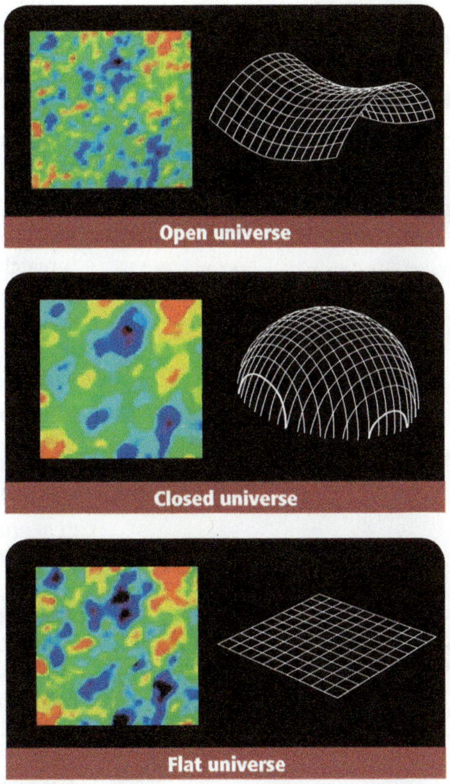

Figure 22 There are three possible outcomes for the future of the universe. It could continue to expand forever and be open, it could snap back at the end and be a closed system, or it could be flat and just die out like a glowing ember. The size of the red and blue spots in the green squares show the estimated cosmic background radiation necessary for each result. In an open universe, the curvature of space makes these variations seem smaller than they really are, and in a closed universe, they appear larger.

Cosmic Background Radiation

Scientists hypothesize that if the universe began in a highly compressed state before the Big Bang, it would have been extremely hot. Then as the universe expanded, the temperature cooled. After about 300,000 years, the universe was filled with electromagnetic radiation in the form of short-wavelength radiation. With continued expansion, the wavelengths became longer. Today this radiation is in the form of microwaves.

Tie to Previous Knowledge

Distance Measurements It is important for students to realize distance measurements to the most faraway galaxies and quasars depend on measurements to closer objects. Every method of distance measurement depends on the one before it, and the first method is stellar parallax, which is based on the Sun-Earth distance.

Identify Misconceptions

It is difficult for most people to visualize the universe without thinking of it as an everyday object, having a center and an outside. It is a common misconception that the early universe expanded into something, as if there were a preexisting vacuum into which the universe could expand.

Uncover the Misconception Ask students what was outside the universe when it was small and just starting to expand.

Demonstrate the Concept Explain to students the universe is space itself; there is no outside. The universe is not expanding into a vacuum; it is a vacuum.

Assess New Knowledge Have students write short, one-paragraph explanations, in their own words and using their own analogies, to illustrate that they understand this concept.

ACROSS THE CURRICULUM

Physics Life on Earth is subjected to many forms of radiation. Cosmic background radiation is one of these, but its intensity is so low it has no effect on organisms. Another kind of radiation is subatomic particles. Particle radiation comes from radioactive elements in Earth's crust and cosmic rays from space. Earth's atmosphere absorbs much of the radiation from space.

EARTH SCIENCE JOURNAL

Expanding Universe Students will have differing reactions to the concept of an expanding universe. Some might feel that it is an exciting and interesting idea, while others could find it unsettling. Have students write a short paragraph describing their feelings about the expanding universe. BL OL

Rubric

Project

WMAP Results *WMAP* made several important observational discoveries about background radiation. Have students work in groups to research and prepare presentations that summarize the *WMAP* results. OL AL COOP LEARN

Concept Development

The First Matter Students might wonder where the first matter in the universe originated. Einstein's theory of special relativity provides the answer, and experiments done in particle accelerators confirm it: matter can be created from energy. A photon can spontaneously form a pair of particles with equal but opposite properties. Visible light does not contain enough energy to produce particles, but higher-energy photons, especially gamma rays, can.

READING CHECK the discovery of cosmic background radiation

Teacher Content Support

Background Radiation The discovery of cosmic background radiation in 1965 is recognized as one of the most important for our understanding of the nature of the universe. Interestingly, the radiation could have been discovered at least 15 years earlier, following the prediction of its existence in the late 1940s. Radar technology, developed during World War II, is appropriate for detecting this radiation. Scientists who made the prediction, however, considered it an insignificant sideline to their main interest in element formation, and they did not look for the radiation.

Figure 23 The cosmic background radiation was discovered by accident with this radio antenna at Bell Labs in Holmdel, New Jersey.

Discovery In 1965, scientists discovered a persistent background noise in their radio antenna, shown in **Figure 23.** This noise was caused by weak radiation, called the **cosmic background radiation,** that appeared to come from all directions in space and corresponded to an emitting object having a temperature of about 2.725 K (–270°C). This was very close to the temperature predicted by the Big Bang theory, and the radiation was interpreted to be from the Big Bang.

Mapping the radiation Since the discovery of the cosmic background radiation, extensive observations have confirmed that it matches the properties of the predicted leftover radiation from the early, hot phase in the expansion of the universe. Earth's atmosphere blocks much of the radiation, so it is best observed from high-altitude balloons or satellites. A space observatory called the *Wilkinson Microwave Anisotropy Probe (WMAP),* launched by NASA in 2001, mapped the radiation in greater detail. The peak of the radiation it measured has a wavelength of approximately 1 mm; thus, it is microwave radiation in the radio portion of the electromagnetic spectrum.

VOCABULARY
SCIENCE USAGE V. COMMON USAGE
Cosmic
Science usage: of or relating to the universe in contrast to Earth alone

Common usage: characterized by greatness of thought or intensity

READING CHECK **Identify** what discovery helped solidify the Big Bang theory.

Acceleration of the expansion The data produced by *WMAP* have provided enough detail to refine cosmological models. In particular, astronomers have found small wiggles in the radiation representing the first major structures in the universe. This helped to pinpoint the time at which the first galaxies and clusters of galaxies formed and also the age of the universe. According to every standard model, the expansion of the universe is slowing down due to gravity. However, the debate about the future of the universe based on this model came to a halt with the surprising discovery that the expansion of the universe is now accelerating as illustrated in **Figure 24.** Astronomers have labeled this acceleration dark energy. Although they do not know its cause, they can determine the rate of acceleration and estimate the amount of dark energy.

Figure 24 While standard models predict deceleration of expansion of the universe, data show the expansion accelerating.

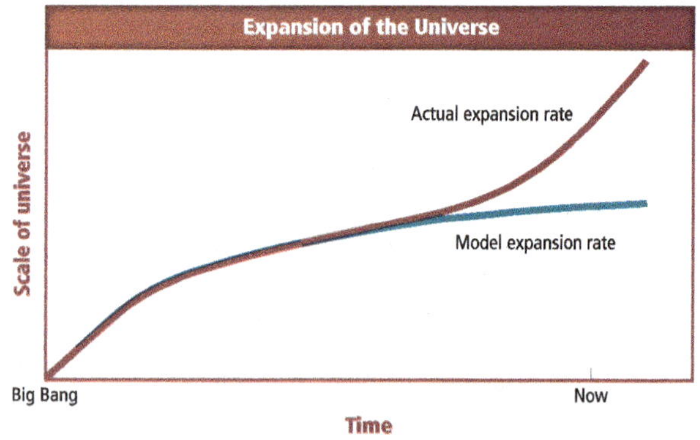

DIFFERENTIATED INSTRUCTION

Advanced Learners Have students study a map of cosmic background radiation. The different-colored regions correspond to different temperatures above and below the average. Ask students to infer what the clumpy pattern implies about the beginning of structure in the universe. Astronomers hypothesize that this clumpiness might have been enough to trigger galaxy formation.

Contents of the Universe

All the evidence is now pointing in the same direction, and astronomers can say with a high degree of precision of what the universe is composed. Their best clue comes from the radiation left in space from the universe's beginning. The ripples left during the time of cooling of the universe's beginning radiation set the density at that point of time and dictated how matter and energy would separate. This in turn laid the groundwork for future galaxies. **Figure 25** gives one view into the universe.

■ **Figure 25** In this view of deep space, galaxies appear as glowing flecks. Astronomers estimate that only about 5 percent of the universe is composed of luminous matter.

Dark matter and energy Cosmologists estimate that the universe is composed of dark matter (23 percent), dark energy (72 percent), and luminous matter. If you compare the universe to Earth's surface, dark energy is like the water covering it. That would be like saying that the majority of Earth is covered with something that is not identified.

What is unknown today is the nature of the dark matter and dark energy. Dark matter is thought to consist of subatomic particles, but of the known particles, none display the right properties to explain or fully define dark matter. And although scientists recognize the effects of dark energy, they still do not know precisely what it is.

SECTION 3 REVIEW

Section Self-Check

Section Summary

- The study of the universe's origin, nature, and evolution is cosmology.
- The Big Bang model of the universe came from observations of density and acceleration.
- The critical density and the amount of dark energy of the universe will determine whether the universe is open or closed.
- Cosmic background radiation gives support to the Big Bang theory of the universe.
- Ninety-five percent of the universe is made up of dark matter and dark energy, both of whose nature is unknown.

Understand Main Ideas

1. MAINIDEA **Compare and Contrast** What are the differences among the three possible outcomes of the universe?
2. **Describe** how the age of the universe can be calculated using the Big Bang model.
3. **Explain** why dark matter is important in determining the density of matter in the universe.
4. **Explain** why the cosmic background radiation was an important discovery.

Think Critically

5. **Determine** What does dark matter have to do with the critical density of the universe?
6. **Analyze** All of the models tell us that the universe should be slowing down, but instead it is speeding up. How does this affect our model of the universe?

WRITING IN ▶ Earth Science

7. Write one paragraph summarizing the evidence for the Big Bang model of the universe.

NASA/Reuters/CORBIS

3 Assess

Check for Understanding

Discussion Have students discuss why the discovery of the cosmic background radiation is significant. The background radiation is interpreted to be a remnant of the Big Bang. It has precisely the properties expected of such radiation, and no other explanation has been found for it.

Reteach

Summarize Have students summarize the recent significant change in astronomers' views about the expanding universe. Previously, astronomers thought gravity caused the expansion of the universe to slow down. Recently, they found the expansion of the universe is accelerating. They think this is caused by dark energy.

Assessment

Knowledge Have students discuss what would happen if the universe were flat or open. Eventually, all star formation would end because all the interstellar matter in galaxies would be used. Then, after many billions of years, all stars would complete their evolutions and become stellar remnants. With the end of nuclear reactions in stars, the universe would become completely dark and cold.

SECTION 3 REVIEW

1. An open universe would continue to expand forever, a closed universe would eventually collapse back to a single point, and a flat universe would eventually stop expanding at some infinite point.
2. The Big Bang model says there is a point in time when everything was at a single point. By calculating the rate of expansion and the distance between the farthest galaxies, scientists can calculate backward to the time the universe originated—and thus the age of the universe.
3. Density is the ratio of mass to volume, and scientists think that dark matter is responsible for a large part of the universe's mass. Dark matter is therefore responsible for a large part of the universe's density.
4. The discovery of the cosmic background radiation was important because it provided more evidence in support of the Big Bang model.
5. Because dark matter makes up most of the universe's mass, it also significantly affects the density of the universe. The expansion of the universe depends on the ratio of its density to the critical density.
6. If the models are wrong, then the models have to be changed. In this case, it means there is unexplained activity, which many believe is dark energy.
7. Paragraphs should discuss outward expansion of the universe and cosmic background radiation. Paragraphs might also discuss redshift and the Hubble constant.

Rubric

Earth Science &
TECHNOLOGY

Purpose

Students will identify characteristics of black holes and understand how scientists discover these characteristics.

Teacher Content Support

Supermassive Black Holes
Scientists think supermassive black holes, which tend to be at the center of each galaxy, formed with the galaxy itself. Supermassive black holes have the mass of a million to a few billion solar masses. Although astronomers cannot see black holes (since no light escapes), they can infer their existence based on the behavior of surrounding materials. Scientists have also discovered that material being consumed by black holes becomes superheated and emits X rays, which can be detected. The area surrounding supermassive black holes can emit so much light that it can be seen billions of light-years away. The supermassive black hole in the center of our galaxy is 26,000 light-years away and probably has about 3 million times the mass of the Sun.

Teaching Strategies

- First ask: What do you think black holes are? Then ask: How do you think they form? Finally, ask: How do you think black holes affect the things around them?
- Show the movie *Monster of the Milky Way* developed by NOVA and PBS about black holes and supermassive black holes.

Earth Science &
TECHNOLOGY

Black Holes are Green?

Black holes seem to come straight from the pages of a science fiction book. They are incredibly dense cosmic bodies from which nothing—not even light—can escape. The gravitational pull attracts whatever ventures close enough.

Finding black holes Black holes are extremely difficult to see because they do not emit light, and those that are produced by a collapsed massive star can be very small (only 2 to 3 times the mass of the Sun). Astronomers know where black holes might be located due to the effects of the matter falling into them.

Supermassive black holes In the centers of some galaxies exists a different kind of black hole—a supermassive one. These black holes are huge; they can consist of more mass than a million, even a billion, Suns.

Scientists think that supermassive black holes are created when large volumes of interstellar gases collapse in on themselves. Once matter passes into a spherical boundary surrounding the black hole, called the event horizon, it is pulled into the black hole, never to escape.

Energy Before the matter gets pulled into the event horizon, however, it gathers energy through friction and from the magnetic field of the black hole. That energy is released in the form of diffuse light or focused jets.

The jets release about 1000 times more energy than the diffuse light, either in the form of radio waves or energetic X rays. The jets race outward from the black holes almost at the speed of light, creating empty bubbles in their wake. These bubbles can span thousands of light-years. Scientists used these bubbles to discover the fuel efficiency of the supermassive black holes.

In this image taken by the *Chandra X-Ray Telescope*, X rays shine from heated material falling into a black hole.

Black holes are "green" Recent research into supermassive black holes has uncovered an interesting fact: They are the most fuel-efficient engines in the entire universe. In fact, a physicist at Stanford University reported that "If you could make a car engine that was as efficient as one of these black holes, you could get about a billion miles out of a gallon of gas!"

Astronomers think that the energy released from supermassive black holes actually prevents star formation. The heat that they produce prevents gases from cooling and potentially forming billions of new stars, effectively limiting the size of each galaxy.

WRITING IN ▸ Earth Science

Summary Research more about black holes. Summarize what you learn in a newspaper article about black holes that is interesting and scientifically accurate.

WebQuest

WRITING IN ▸ Earth Science

Rubric

Summary Encourage students to begin by making a list of facts to include in their articles. Suggest that they think about how they can connect the facts and describe details in a way readers can easily understand.

WebQuest

GeoLAB

iLab Station

Classify Galaxies

Background: Edwin Hubble developed rules for classifying galaxies according to their telescopic image shapes. Modern astronomers are also interested in the classification of galaxies. Information used for classification can indicate whether a certain type of galaxy is more likely to form than another and helps astronomers unravel the mystery of galaxy formation in the universe. Using the Internet and sharing data with your peers, you can learn how galaxies are classified.

Question: *How can different galaxies be classified?*

Materials

internet access or galaxy images provided by your teacher

Visit a local library or observatory to gather images of galaxies and information about them.

Procedure

1. Read and complete the lab safety form.
2. Find a resource with multiple images of galaxies and, if possible, names or catalog numbers for the galaxies. Images of galaxies can be found on NASA's Web site.
3. Choose one of the following types of galaxies to start your classification: spiral, elliptical, or irregular galaxies.
4. Sketch or gather images and information, such as catalog numbers and names of galaxies.
5. Sort the images by basic types: spiral, elliptical, or irregular galaxies.
6. Complete the data table. Add any additional information you think is important.

Galaxy Data

Galaxy Name	Image or Sketch of Galaxy	Classification	Notes
NGC 3486		Sc	

Analyze and Conclude

1. **Differentiate** Which galaxy classes were the most difficult to find?
2. **Identify** How many of each galaxy class did you find?
3. **Calculate** the percentages of the total number of galaxies of each type. Do you think this reflects the actual percentage of each type in the universe? Explain.
4. **Discuss** Were there any galaxies that didn't fit the classification scheme? If so, why?
5. **List** What problems did you have with galaxies seen edge-on?
6. **Illustrate** Reconstruct the tuning fork diagram with images that you find.

INQUIRY EXTENSION

Share Your Data With your classmates, calculate the percentage of each type of galaxy. Based on the results, decide if your results are typical or atypical. Determine how your class might find actual percentages of galaxies by type.

Jean-Charles Cuillandre/Canada-France-Hawaii Telescope/Photo Researchers

GeoLAB

iLab Station Rubric

Preparation

Time Allotment 90 min

Process Skills classify, observe and infer, compare and contrast

Safety Precaution Approve lab safety forms before work begins.

Procedure

- Students will examine images of galaxies obtained from the Internet or provided by the teacher.
- **Troubleshooting** Caution students that the quality of the image, or the duration of the time exposure used to make the photograph, can affect a person's perception of a galaxy's shape and form.

LabManager™
Customize any lab with the LabManager™ CD-ROM.

Analyze and Conclude

1. Answers will vary depending upon the students' sources.
2. Answers will vary depending upon the students' sources.
3. Answers will vary depending upon the students' sources.
4. Irregular galaxies and some of the active galaxies will not fit into the Hubble classification scheme because of their unusual shapes.
5. Edge-on galaxies are generally spirals. If it is not possible to see the arms and nucleus, it is not possible to classify an edge-on galaxy any further than as a spiral.
6. Students should use the galaxies they have studied to draw a tuning-fork diagram similar to Hubble's tuning-fork diagram shown in **Figure 12.**

MAINIDEAS Summary statements can be used by students to review the major concepts of the chapter.

Students can review with these online resources.

Vocabulary Practice

- Vocabulary eGames
- Vocabulary eFlashcards
- Vocabulary PuzzleMaker

Section Self-Check

Chapter Self-Check

Online Test Practice

Use the ExamView® *Assessment Suite* CD-ROM to:

- create multiple versions of tests
- create modified tests with one mouse click
- edit existing questions and add your own questions
- build tests aligned with state standards using built-in state curriculum tags
- change English tests to Spanish with one mouse click
- track students' progress using the Teacher Management System

CHAPTER 30 STUDY GUIDE

BIGIDEA **Observations of galaxy expansion, cosmic background radiation, and the Big Bang theory describe an expanding universe that is about 14 billion years old.**

Vocabulary Practice

SECTION 1 The Milky Way Galaxy

VOCABULARY
- variable star
- RR Lyrae variable
- Cepheid variable
- halo
- Population I star
- Population II star
- spiral density wave

MAINIDEA **Stars with varying light output allow astronomers to map the Milky Way, which has a halo, spiral arms, and a massive galactic black hole at its center.**

- The discovery of variable stars aided in determining the shape of the Milky Way.
- RR Lyrae and Cepheid are two types of variable stars used to measure distances.
- Globular clusters of old stars are found in the nuclear bulge and halo of the Milky Way.
- The spiral arms of the Milky Way are made of younger stars and gaseous nebulae.
- Population I stars are found in the spiral arms, while Population II stars are in the central bulge and halo.

SECTION 2 Other Galaxies in the Universe

VOCABULARY
- dark matter
- supercluster
- Hubble constant
- active galactic nucleus
- radio galaxy
- quasar

MAINIDEA **Finding galaxies with different shapes reveals the past, present, and future of the universe.**

- Galaxies can be elliptical, disk-shaped, or irregular.
- Galaxies range in mass from 1 million Suns to more than a trillion Suns.
- Many galaxies seem to be organized in groups called clusters.
- Hubble's law helped astronomers discover that the universe is expanding.
- Quasars are the nuclei of faraway galaxies that are dim and seen as they were long ago, due to their great distances.

SECTION 3 Cosmology

VOCABULARY
- cosmology
- Big Bang theory
- cosmic background radiation

MAINIDEA **The Big Bang theory was formulated by comparing evidence and models to describe the beginning of the universe.**

- The study of the universe's origin, nature, and evolution is cosmology.
- The Big Bang model of the universe came from observations of density and acceleration.
- The critical density and the amount of dark energy of the universe will determine whether the universe is open or closed.
- Cosmic background radiation gives support to the Big Bang theory of the universe.
- Ninety-five percent of the universe is made up of dark matter and dark energy, both of whose nature is unknown.

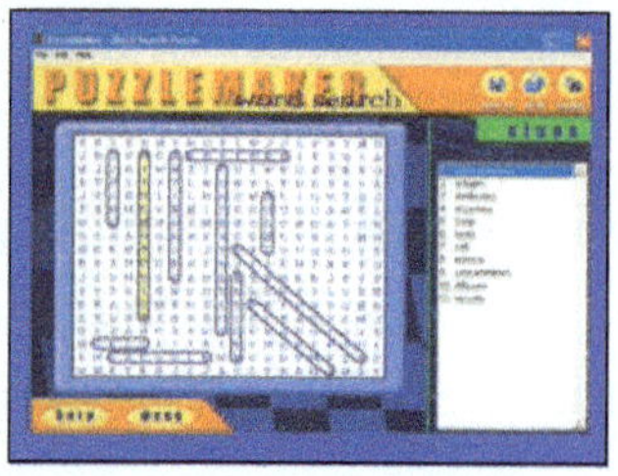

Vocabulary Practice

For additional practice with vocabulary, have students access the Vocabulary PuzzleMaker.

Chapter Self-Check

VOCABULARY REVIEW

The sentences below are false. Correct each sentence by replacing the italicized words with the correct vocabulary term from the Study Guide.

1. Surrounding the central bulge, this spherical region of a galaxy is known as the *Hubble constant.*
2. Radio emissions coming from two huge lobes of very hot gas located on opposite sides of the visible galaxy are evidence for *cosmology.*
3. *Population I* is the weak radiation that appears to come from all directions in space and corresponds to an object having a temperature of about 2.725 K.
4. These gigantic *quasars* are hundreds of millions of light-years in size and can be observed only when astronomers map out the locations of many galaxies ranging over large distances.
5. This *Cepheid variable* is the invisible substance that makes up to 21 percent of the universe.
6. One theory of how galaxy arms are maintained involves the *RR Lyrae variables.*
7. *Radio galaxy* is the study of the origin and history of the universe.
8. A *supercluster* is a star whose magnitude changes are produced by expansion and shrinking of its outer layers.

Distinguish between the terms in each of the following pairs.

9. RR Lyrae variable, Cepheid variable
10. quasar, radio galaxy
11. dark matter, cosmic background radiation
12. halo, active galactic nucleus
13. cosmology, Hubble constant

In the set of terms below, select the term that does not belong and explain why it does not belong.

14. RR Lyrae, Cepheid, Population II, quasar

UNDERSTAND KEY CONCEPTS

15. Which are the oldest objects in the Milky Way?
    - **A.** globular clusters
    - **B.** spiral arms
    - **C.** Cepheid variables
    - **D.** Population I stars

Use the diagram below to answer Question 16.

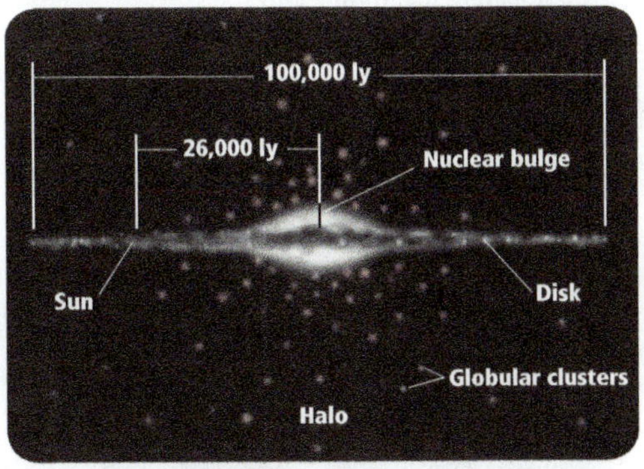

16. Where in the Milky Way are new stars being formed?
    - **A.** in the nuclear bulge
    - **B.** in globular clusters
    - **C.** in the spiral arms of the disk
    - **D.** in the halo
17. Where does the energy emitted by AGNs and quasars most likely originate?
    - **A.** material falling into a supermassive black hole
    - **B.** a neutron star
    - **C.** a supernova explosion
    - **D.** a pulsar
18. What is the origin of the cosmic background radiation?
    - **A.** It is emitted by stars.
    - **B.** It is a remnant of the Big Bang.
    - **C.** It is emitted by radio galaxies.
    - **D.** It is dark energy.
19. In the Big Bang model, which describes a universe that will stop expanding and begin to contract?
    - **A.** open
    - **B.** flat
    - **C.** closed
    - **D.** elliptical

CHAPTER 30 ASSESSMENT

VOCABULARY REVIEW

1. halo
2. a radio galaxy
3. Cosmic background radiation
4. superclusters
5. dark matter
6. spiral density wave
7. Cosmology
8. variable star
9. RR Lyrae stars have pulse variations between 1.5 hours and 1.2 days; Cepheids have pulse variations up to 100 days.
10. Quasars are distant, luminous galaxies with mostly emission lines in their spectra. Radio galaxies are often giant elliptical galaxies; they emit most of their energy as radio waves.
11. Dark matter is the invisible material that makes up most of the mass of the universe, while cosmic background radiation is the remnant of the Big Bang.
12. The halo is a group of globular clusters, Population II stars, and dark matter that surround the nuclear bulge. AGNs are the nuclei of galaxies that are highly energetic.
13. Cosmology is the study of the nature, origin, and evolution of the universe. The Hubble constant is the ratio of the velocity at which a galaxy is moving away to its distance; this ratio is a measure of how the universe is expanding.
14. A quasar is a galaxy; the others are stars.

UNDERSTAND KEY CONCEPTS

15. A
16. C
17. A
18. B
19. C

20. D
21. A
22. C
23. B
24. B
25. A

CONSTRUCTED RESPONSE

26. Density is mass divided by volume, and the expansion of the universe is related to density. If the density is less than the critical density, the expansion of the universe will continue. If the density is too large, the universe will eventually collapse back on itself. If the density of the universe is equal to the critical density, the universe will slow to a halt in an infinite amount of time but will never contract.
27. They have a period-luminosity relationship. The pulsation period determines the luminosity, which can then be used to calculate distance.
28. by mapping the 21-cm emissions of atomic hydrogen
29. Stars and gas in the outer galaxy move differently than they would if most of the mass were at the center of the galaxy.
30. They are old and have few heavy elements.
31. Spiral galaxies are classified *S* for normal spirals and *SB* for barred spirals. The letter *a* after *S* or *SB* indicates tightly wound arms and a large, bright nucleus. The letter *c* after *S* or *SB* indicates loosely wound arms and a small, dim nucleus. The classification *S0* is used for a flat disk that does not have spiral arms. Elliptical galaxies are classified from *E0* to *E7*, where the number ranges from *0* for a round ellipse to *7* for an elongated ellipse. Irregular galaxies are denoted by *Irr.*
32. AGNS form the core of quasars. AGNs are also the core of other active galaxies such as radio galaxies.
33. Redshifts indicate that the universe is expanding as galaxies move away from each other. Using their redshifts, we can determine the speed at which they are moving away. Using Hubble's law, we can determine the distance to other galaxies.
34. The gravitational pull on visible matter was stronger than expected. Astronomers realized there must be an invisible form of matter that also pulls on the visible matter.

ASSESSMENT

20. Which does the existence of cosmic background radiation support?
A. critical density
B. Hubble constant
C. the inflationary model
D. the Big Bang theory

21. Which two measurements are required to determine the Hubble constant?
A. distance and speed
B. distance and absolute magnitude
C. apparent magnitude and speed
D. apparent and absolute magnitudes

22. Without doing any calculations, what can astronomers determine from a variable star's period of pulsation?
A. distance
B. apparent magnitude
C. luminosity
D. age

Use the diagram below to answer Questions 23 and 24.

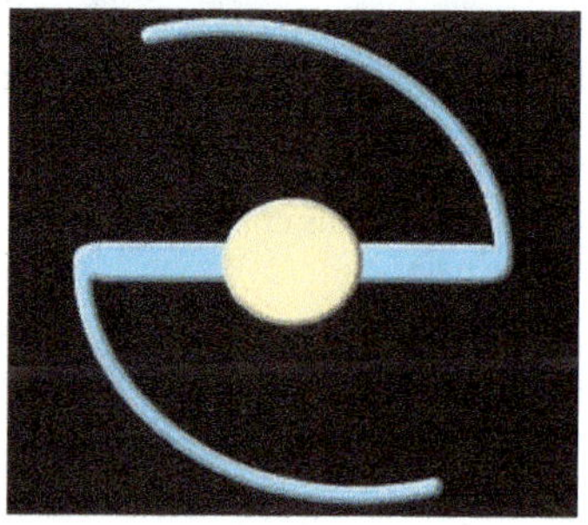

23. Which kind of galaxy is illustrated above?
A. spiral
B. barred spiral
C. elliptical
D. irregular

24. Which designation would the tuning fork diagram assign this galaxy?
A. S0
B. SB
C. Sa
D. E3

Use the diagram to answer Question 25.

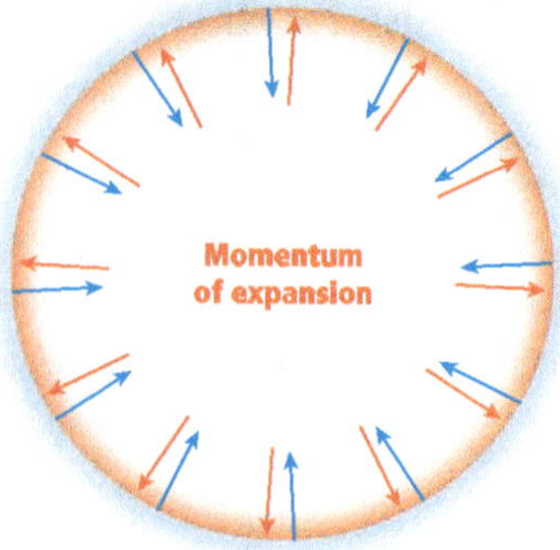

25. Which would cause the universe to collapse in on itself to make a closed universe?
A. force of gravity
B. critical density
C. momentum of outward expansion
D. Hubble constant

CONSTRUCTED RESPONSE

26. Interpret the relationship between mass and density and the expansion of the universe.

27. Discuss Why are pulsating variable stars useful for finding distances to globular clusters?

28. Explain How do astronomers observe the spiral structure of the Milky Way?

29. CAREERS IN EARTH SCIENCE Why do astronomers think that there is a great amount of mass in the halo of the Milky Way?

30. Explain Why are the stars in globular clusters classified as Population II stars?

31. Relate the classification of a galaxy to its shape.

32. Compare active galactic nuclei with quasars.

33. Explain What do redshifts and Hubble's law tell us about the motion of galaxies?

34. Discuss how astronomers determined that dark matter exists.

Chapter Self-Check

THINK CRITICALLY

35. **Infer** How would a star that forms in the Milky Way a few billion years in the future compare with the Sun?

Use the graph below to answer Question 36.

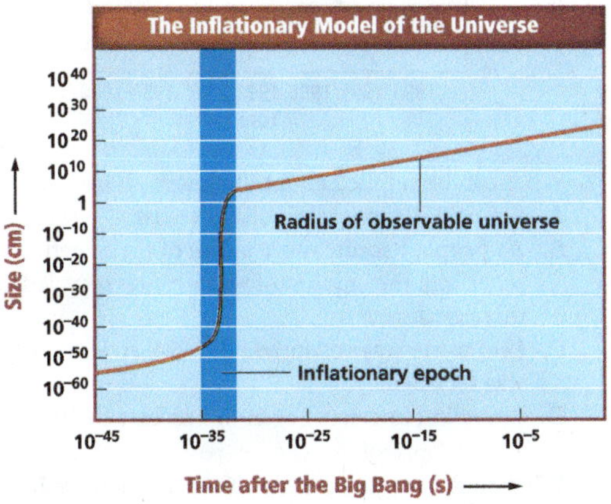

36. **Explain** what happened to the universe during the 10^{-35} s portion of the graph.

37. **Compare** the importance of variable stars and cosmic background radiation to the determination of the shape of the universe.

38. **Identify** the cause-and-effect relationship between Population I and Population II stars.

CONCEPT MAPPING

39. Use the following terms to construct a concept map to organize the major ideas in this chapter: *cosmic background radiation, quasars, Hubble's law, black holes, galaxy clusters, and Big Bang theory.*

CHALLENGE QUESTION

40. **Infer** the difficulty of determining the outcome of the universe resulting from the presence of dark matter.

WRITING IN ▶ Earth Science

41. Write an essay explaining the necessity for continuing space-based satellite telescope use and development.

DBQ Document–Based Questions

Data obtained from: Silk, J. 1998. The SETI module: cosmology. Syracuse University.

The graph below shows the changes in the strength of the major forces in the universe from the Big Bang until the present.

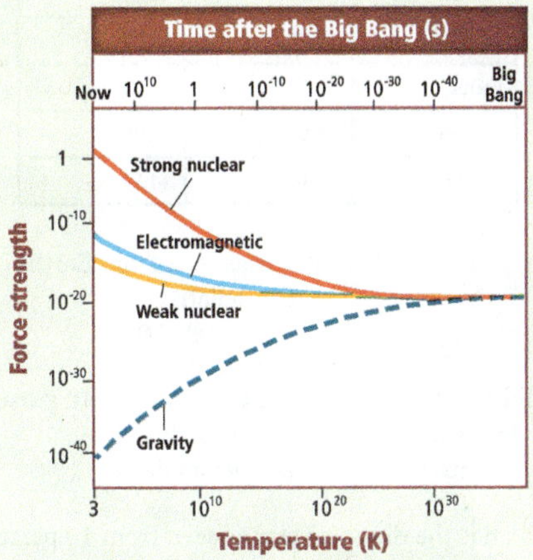

42. At what time and temperature did gravity and the strong-electro-weak forces separate?

43. What has happened to the force of gravity since the Big Bang? To the other forces?

44. One-billionth of a second (10^{-9}) after the Big Bang initiation, atoms began to form. At what temperature did this occur?

CUMULATIVE REVIEW

45. Which two basic stellar properties are displayed on an H-R diagram? **(Chapter 29)**

46. Why is it unwise to try to forecast weather by simply extrapolating current conditions beyond a few hours? **(Chapter 12)**

CHAPTER 30 ASSESSMENT

THINK CRITICALLY

35. It would have a higher abundance of heavy elements than the Sun does, because it would form later in galactic history, after stellar nuclear reactions have formed a higher abundance of heavy elements than exists today.

36. The universe went through an almost instantaneous expansion from 10^{-50} to 10^{3} cm.

37. Variable stars are used to determine distances within the universe. Cosmic background radiation indicates how the first galaxies and clusters of galaxies formed.

38. Population II stars create the conditions that lead to the formation of Population I stars.

CONCEPT MAPPING

39. Answers will vary. Check students' maps for accuracy.

CHALLENGE QUESTION

40. In order to understand the outcome of the universe, scientists must understand dark matter and dark energy. Dark matter does not have properties consistent with known subatomic particles.

WRITING IN ▶ Earth Science

Rubric

41. Essays should describe benefits of a space-based telescope compared to telescopes that view space through the atmosphere.

DBQ Document-Based Questions

Data obtained from: Silk, J. 1998. The SETI module: cosmology. Syracuse University.

42. time: 10^{-40} s; temperature: 10^{30} K

43. Gravity has become weaker, while the other forces are becoming stronger.

44. approximately 10^{14} K

CUMULATIVE REVIEW

45. The luminosity (or absolute magnitude) is plotted on the vertical axis, and the surface temperature (as shown by the spectral class) is plotted on the horizontal axis.

46. Weather is produced by large-scale systems that move and change size and intensity. Extrapolation of current conditions will not be able to account for these changes.

MULTIPLE CHOICE

1. C
2. A
3. B
4. C
5. A
6. B
7. B
8. D
9. D
10. B
11. B

CUMULATIVE STANDARDIZED TEST PRACTICE

MULTIPLE CHOICE

1. What is the streak of light produced when a cosmic body burns up in Earth's atmosphere called?
   A. a meteorite C. a meteor
   B. an asteroid D. a meteoroid

Use the table below to answer Questions 2 to 4.

Stellar Magnitudes		
Star	Apparent Magnitude	Absolute Magnitude
Procyon	+0.38	+2.66
Altair	+0.77	+2.22
Becrux	+1.25	-3.92
Bellatrix	+1.64	-1.29
Denebola	+2.14	+1.54

2. Which is the brightest star as seen from Earth?
   A. Procyon C. Bellatrix
   B. Becrux D. Denebola

3. Which is the brightest star as seen from 10 parsecs?
   A. Procyon C. Bellatrix
   B. Becrux D. Denebola

4. Which is the dimmest star as seen from 10 parsecs?
   A. Bellatrix C. Procyon
   B. Altair D. Becrux

5. What two measurements are required to determine the Hubble constant?
   A. distance and speed
   B. distance and absolute magnitude
   C. apparent magnitude and speed
   D. apparent and absolute magnitude

6. What does Kepler's first law state?
   A. Each planet revolves around the Sun in a circular path.
   B. Each planet revolves around the Sun in an elliptical path.
   C. Planets closer to the Sun move faster than planets farther away.
   D. Planets closer to the Sun move slower than planets farther away.

Use the graph below to answer Questions 7 to 9.

7. What can be implied about the graph above?
   A. Before 1994, recycling did not exist.
   B. As people became more aware of the benefits of recycling, the amount of waste being recycled increased.
   C. Less waste was consumed in 1994, so less waste was recycled.
   D. Recycling interests began to decrease in 1998.

8. What could not account for the sharp increase in recycling between 1995 and 1996?
   A. implementation of recycling laws
   B. increased public awareness
   C. more convenience for recycling
   D. less production of materials that need recycling

9. Which of the following years had the greatest increase in amount of material recycled?
   A. 2000–2001 C. 1996–1997
   B. 1998–1999 D. 1995–1996

10. Carbon-14 has a radioactive decay half-life of 5730 years. Which item would carbon-14 be most useful for dating?
    A. a rock from the Moon
    B. a Native American fire pit
    C. a jawbone from a triceratops
    D. a granite rock from the Canadian Shield

11. In which region of the Milky Way galaxy is 90 percent of its mass located?
    A. spiral arms
    B. halo
    C. nuclear bulge
    D. disk

Online Test Practice

SHORT ANSWER

Use the illustration below to answer Questions 12 to 14.

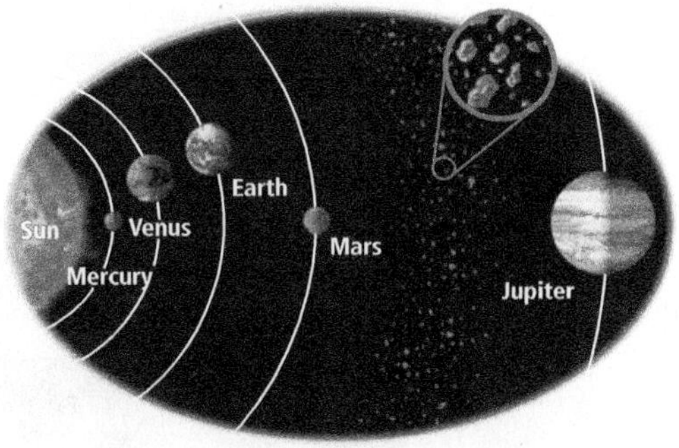

12. What do the lines through the planets represent?
13. Name and describe the material located between Mars and Jupiter.
14. Explain why this material did not form into a planet.
15. Compare and contrast refracting and reflecting telescopes. Which one is used more widely today? Why?
16. Describe the geocentric model of the solar system.
17. Why is Earth's Moon unique among all moons in the solar system?

READING FOR COMPREHENSION

First Stars in the Universe

NASA researchers say they have detected what might be the faint infrared glow of the first stars in the universe. Known as population III stars, the distant bodies are thought to have formed just 200 million years after the big bang.

The original stars formed from gas and dust in the void of space and are thought to have been many times more massive than today's stars. The ancient stars remain invisible to telescopes and have never before been detected. Using NASA's orbiting *Spitzer Space Telescope* the stars have been identified indirectly by measuring the enduring energy that they once radiated into the void of space. As the universe expands, starlight is stretched into longer, redder wavelengths. Most emissions from the first stars in the universe would appear today as infrared light. The universe is filled with background radiation known as the cosmic infrared background (CIB). This includes radiation from all stars—young and old, near and far. If these earliest stars were massive and formed in the standard cosmological mode, they should have left a signature in the fluctuations of the CIB.

Article obtained from: Handwerk, B. First stars in universe detected? *National Geographic News.* November 2, 2005.

18. What can be inferred from this passage?
    A. These stars are still present in space.
    B. Telescopes are not a good way to view stars.
    C. These first stars formed at the same time as the Big Bang.
    D. The first stars no longer exist, but we are just now seeing their radiation.

19. What are scientists seeing that confirms the existence of these stars?
    A. their visible light finally reaching Earth
    B. the faint infrared glow from their emissions
    C. the gas and dust particles of the stars
    D. radiation from the existing stars

20. Infer why basic telescopes are not able to find these stars but the *Spitzer Space Telescope* can.

NEED EXTRA HELP?

If You Missed Question . . .	1	2	3	4	5	6	7	8	9	10	11	12	13	14	15	16	17
Review Section . . .	28.1	29.2	29.2	29.2	30.2	28.1	26.1	26.1	26.1	21.3	30.1	28.1	28.4	28.4	27.1	28.1	27.2

SHORT ANSWER

12. The lines through the planets represent the path that each planet follows as it orbits the Sun.

13. The area between the orbits of Mars and Jupiter is known as the asteroid belt. It contains rock of varying sizes known as asteroids.

14. This material did not form into a planet because a strong gravitational pull from Jupiter did not allow it to merge.

15. Both types of telescopes are used to bring visible light into focus, but refracting telescopes use lenses, while reflecting telescopes use mirrors. Today, reflecting telescopes are used more widely because mirrors can be made larger and collect more light, which helps scientists see more objects more clearly.

16. The geocentric model of the solar system placed Earth at the center of the solar system, and the Sun, planets, and stars revolved around it.

17. Earth's Moon is unique because it is the largest moon in relation to the radius and mass of the planet it orbits. Also, its orbit is farther away from Earth than many other moons' orbits around their respective planets.

READING FOR COMPREHENSION

18. D

19. B

20. Because the *Spitzer Space Telescope* is in orbit and doesn't have Earth's atmosphere to obscure its view, it can detect more infrared light than Earth-bound telescopes.

STUDENT RESOURCES

Skillbuilder Handbook 891

Problem-Solving Skills 891

Make Comparisons. 891
Analyze Information 892
Synthesize Information 893
Take Notes and Outline. 894
Understand Cause and Effect 895
Read a Time Line 896
Analyze Media Sources 897
Use Graphic Organizers 898
Debate Skills 899

Math Skills 900

Measure in SI. 900
Convert Temperature 900
Make and Use Tables 901
Make and Use Graphs 901

Reference Handbook 904

Safety in the Laboratory 904
Physiographic Map of Earth 906
Topographic Symbols 908
Weather Map Symbols 909
Periodic Table of the Elements 910
Relative Humidity 911
Minerals. 912
Common Rocks 914
Solar System Charts. 915

Glossary ▪ Glosario 916

Index 954

Rubberball/Getty Images

Make Comparisons

Why Learn this Skill?

Suppose you want to buy a portable MP3 music player, and you must choose among three different models. You would probably compare the characteristics of the three models, such as price, amount of memory, sound quality, and size to determine which model is best for you.

In the study of Earth science, you often compare the structures and functions of one type of rock or planet with another. You will also compare scientific discoveries or events from one time period with those from a different time period. This helps you gain an understanding of how the past has affected the present.

Learn the Skill

When making comparisons, you examine two or more groups, situations, events, or theories. You must first decide what items will be compared and determine which characteristics you will use to compare them. Then identify any similarities and differences.

For example, comparisons can be made between the two minerals shown on this page. The physical properties of halite can be compared to the physical properties of quartz.

Practice the Skill

Create a table with the title *Mineral Comparison*. Make two columns. Label the first column *Halite*, and the second column *Quartz*. List all of your observations of these two minerals in the appropriate column of your table. Similarities you might point out are that both minerals are solids that occur as crystals, and both are inorganic compounds. Differences might include that halite has a cubic crystal structure, whereas quartz has a hexagonal crystal structure.

When you have finished the table, answer these questions.

1. What items are being compared? How are they being compared?
2. What properties do the minerals have in common?
3. What properties are unique to each mineral?

Apply the Skill

Make Comparisons Read two editorial articles in a science journal or magazine that express different viewpoints on the same issue. Identify the similarities and differences between the two points of view.

Halite

Quartz

Charles D. Winters/Science Source

Practice the Skill

1. the minerals halite and quartz; the minerals' physical properties
2. Similarities could include the following: colorless, inorganic, solid, crystals.
3. Differences could include the following: halite is cubic and quartz is hexagonal; the halite is more opaque and the quartz is more transparent; halite is more easily fractured than quartz.

Apply the Skill

Answers will vary with the topic and the source. Journal articles might contain more scientific language and be more specific about the details of a topic, while science magazine articles will give a summary and be more general in the information given. Also, they might use everyday language instead of the scientific terminology.

Practice the Skill

1. The topic is the new shark-shaped submarine that Fabien Cousteau used to study white sharks in their natural habitat.
2. The main points are (a) the design of the submarine, (b) the features of the submarine, (c) the reason behind the submarine design, and (d) the use of the submarine.
3. Jacques Cousteau's grandson Fabien is using a "submarine" designed to look and move like a shark. Though he must wear a wet suit and use scuba gear, the skin-covered, water-filled sub allows him to move among great whites in their natural habitat so he can discover their true behaviors without undue human influence.

Apply the Skill

Answers will vary depending on the articles selected. Students will have to identify the topic and summarize the information similar to what they did for Practice the Skill. Encourage students to attach the article to their analysis.

Analyze Information

Why Learn this Skill?

Analyzing, or looking at separate parts of something to understand the entire piece, is a way to think critically about written work. The ability to analyze information is important when determining which ideas are more useful than others.

Learn the Skill

To analyze information, use the following steps:

- Identify the topic being discussed.
- Examine how the information is organized—identify the main points.
- Summarize the information in your own words, and then make a statement based on your understanding of the topic and what you already know.

Practice the Skill

Read the following excerpt from *National Geographic*. Use the steps listed above to analyze the information and answer the questions that follow.

His name alone makes Fabien Cousteau, grandson of the late Jacques, a big fish in the world of underwater exploration. Now he's taking that big-fish status to extremes. The Paris-born, New York-based explorer had become a virtual shark, thanks to his new shark-shaped submarine. He uses the sub to dive incognito among the oceans' top predators, great white sharks.

Created at a cost of more than $100,000, the 4.3-meter-long contraption is designed to look and move as much like the real thing as possible. It carries a single passenger, who fits inside lying down, propped up on elbows to navigate and observe. "This is akin to being the first human being in the space capsule in outer space," Cousteau said. "It's pretty similar. You have no idea what's going to happen; it's a prototype."

Cousteau used the submarine to make a documentary intended to demystify the notion that great white sharks are ruthless, mindless killers. Great whites have been around for more than 400 million years. Anything that has survived that long isn't "stupid," he said.

Cousteau calls the sub Troy, *in reference to the mythical Trojan horse statue, in which Greek soldiers were spirited into the fortress kingdom of* Troy. *Propelled by a wagging tail and covered in a flexible, skinlike material, the sub—created by Cousteau and a team of scientists and engineers—swims silently. The steel-ribbed, womb-like interior is filled with water, requiring Cousteau to wear a wet suit and use scuba gear to breathe.*

Fabien Cousteau enters his shark-shaped submarine

Importantly, Troy *allows Cousteau to be a shark, not shark bait. At the heart of the project is a desire to observe what great white sharks do when people aren't around to watch. Prior to this, most shark observations have come from humans sitting in cages and enticing the predators with bait—conditions that spawn unnatural behaviors, Cousteau said. "Now all of the sudden we can see what they do as white sharks rather than as trained circus animals," he said.*

While Cousteau is reluctant to guess what the sharks thought when Troy *invaded their space, the explorer said they seemed to act naturally. Some even puffed their gills and gaped toward* Troy*—actions thought to be communication signals. And though a few sharks made aggressive gestures, none of the predators attacked the shark-shaped sub.*

1. What topic is being discussed?
2. What are the main points of the article?
3. Summarize the information in this article, and then provide your analysis based on this information and your own knowledge.

Apply the Skill

Analyze Information Find a short, informative article on a new scientific discovery or new application of science technology, such as hybrid-car technology. Analyze the information and make a statement of your own.

Synthesize Information

Why Learn this Skill?

The skill of synthesizing involves combining and analyzing information gathered from separate sources or at different times to make logical connections. Being able to synthesize information can be a useful skill for you as a student when you need to gather data from several sources for a report or a presentation.

Learn the Skill

Follow these steps to synthesize information:

- Select important and relevant information.
- Analyze the information and build connections.
- Reinforce or modify the connections as you acquire new information.

Suppose you need to write a research paper on global levels of atmospheric carbon dioxide (CO_2) levels. You need to synthesize what you learn to inform others. You can begin by detailing the ideas and information from sources you already have about global levels of atmospheric carbon dioxide. A table such as **Table 1** could help you categorize the facts from these sources.

Table 1 Global Levels of Atmospheric CO_2

Year	Global Atmospheric CO_2 Concentration (ppm)	Year	Global Atmospheric CO_2 Concentration (ppm)
1745	279	1935	307
1791	280	1949	311
1816	284	1958	312
1843	287	1965	318
1854	288	1974	330
1874	290	1984	344
1894	297	1995	361
1909	299	1998	367
1921	302	2005	385

Then you might select an additional article about greenhouse gases, such as the one below.

According to the National Academy of Scientists, Earth's surface temperature has risen about one degree Fahrenheit in the past 100 years. This increase in temperature can be correlated to an increase in the concentration of carbon dioxide and other greenhouse gases in the atmosphere. How might this increase in temperature affect Earth's climate?

Carbon dioxide is one of the greenhouse gases that helps keep temperatures on Earth warm enough to support life. However, a buildup of carbon dioxide and other greenhouse gases such as methane and nitrous oxide can lead to global warming, an increase in Earth's average surface temperature. Since the industrial revolution in the 1800s, atmospheric concentrations of carbon dioxide have increased by almost 30 percent, methane concentrations have more than doubled, and nitrous oxide concentrations have increased approximately 15 percent. Scientists attribute these increases to the burning of fossil fuels for automobiles, industry, and electricity, as well as deforestation, increased agriculture, landfills, and mining.

Practice the Skill

Use the table and the passage on this page to answer these questions.

1. What information is presented in the table?
2. What is the main idea of the passage? What information does the passage add to your knowledge about the topic?
3. By synthesizing the two sources and using your own knowledge, what conclusions can you draw about global warming?

Apply the Skill

Synthesize Information Find two sources of information on the same topic and write a short report. In your report, answer these questions: What kinds of sources did you use? What are the main ideas of each source? How does each source add to your understanding of the topic? Do the sources support or contradict each other?

Skillbuilder Handbook

Practice the Skill

1. The table contains the global atmospheric CO_2 concentrations in selected years between 1745 and 2005. Students should note that the levels have increased every year in which measurements were taken.
2. The passage discusses the measurement of CO_2 as a greenhouse gas that might be contributing to an increase in atmospheric temperature, leading to global warming.
3. The student should draw the conclusion that there is evidence in the table for global warming and that the article attributes the rise to the use of fossil fuels. Students should also realize the potential consequences of continued fossil fuel use.

Apply the Skill

Answers will vary depending on the sources and topics selected. The types of resources should be similar. Students should recognize that two vastly different types of resources will have little in common and most of their information will come from the more technical version. The main ideas do not have to be the same, but information within each should overlap. The slant of each article also might be different, but could lead to contradictory information. The articles should be deep enough that they do more than just summarize the topic. Recognizing that two articles may have a different slant or aspect is a plus. Comparing the two slants is valuable and leads to decisive results.

Practice the Skill

1. The main topic is estimating the number of dinosaur species.
2. The fossils of known dinosaurs could only represent 29 percent of all dinosaur genera; there will be an influx of discovery over the next 60 to 140 years; the total number of genera of dinosaurs will increase from 544 to about 1844.
3. Two professors made a statistical analysis and estimated the number of dinosaur genera preserved as fossils; based on the discovery patterns of the past, they predict that there will be an increase in the numbers of dinosaur genera within the next 60 to 140 years; the statistical analysis predicts that an additional 1300 genera will be discovered; the analysis applies to species preserved as fossils, meaning that there are some varieties that were not fossilized and will not be discovered.
4. (a) Dinosaurs are highly diverse; only 29 percent have been recovered; the analysis was done on a large database of dinosaur genera; analysis used only fossilized specimens (excluding teeth). (b) The database included fossils and isolated teeth, and the scientists looked at the patterns of discovery and projected those into a future pattern. (c) Scientists project that 75 percent of all the dinosaur species will be discovered within 60 to 100 years, and up to 90 percent by 140 years. (d) Some species (10 percent) will never be discovered because they did not leave fossil evidence.

Apply the Skill

Answers will vary, depending on the article and topic selected. Outline form should start with a roman numeral 1 (I) with the main topic. Minor topics under the main are then labeled with a capital letter (A, B, C). Minor topics under these are given numbers (1, 2, 3) and subtopics are identified with lowercase letters (a, b, c). Further divisions can be identified with lowercase Roman numerals (i, ii, iii, iv). Shorthand symbols can be used as well as your own made-up signs.

Take Notes and Outline

Why Learn this Skill?

One of the best ways to remember something is to write it down. Taking notes—writing down information in a brief and orderly format—not only helps you remember, but also makes studying easier.

Learn the Skill

There are several styles of note-taking, but the goal of every style is to explain information and put it in a logical order. As you read, identify and summarize the main ideas and details that support them and write them in your notes. Paraphrase—that is, state in your own words—the information rather than copying it directly from the text. Use note cards or develop a personal "shorthand"—using symbols to represent words—to represent the information in a compact manner.

You might also find it helpful to create an outline when taking notes. When outlining material, first read the material to identify the main ideas. In textbooks, look at the section headings for clues to main topics. Then identify the subheadings. Place supporting details under the appropriate headings. The basic pattern for outlines is shown below:

MAIN TOPIC
- I. FIRST IDEA OR ITEM
  - A. FIRST DETAIL
    - 1. SUBDETAIL
    - 2. SUBDETAIL
  - B. SECOND DETAIL
- II. SECOND IDEA OR ITEM
  - A. FIRST DETAIL
  - B. SECOND DETAIL
    - 1. SUBDETAIL
    - 2. SUBDETAIL
- III. THIRD IDEA OR ITEM

Practice the Skill

Read the following excerpt from *National Geographic*. Use the steps you just read about to take notes and create an outline. Then answer the questions that follow.

Dinosaur fans still have a lot to look forward to. According to a new estimate of dinosaur diversity, the 21st century will bring an avalanche of new discoveries. "We only know about 29 percent of all dinosaurs out there to be found," said study co-author Peter Dodson, a paleobiologist and anatomy professor at the University of Pennsylvania in Philadelphia.

Dodson and statistics professor Steve Wang of Swarthmore College, in Swarthmore, Pennsylvania, made a statistical analysis of an exhaustive database of all known dinosaur genera (the taxonomic group one notch above species). They then used this data to estimate the total number of genera preserved in the fossil record.

The pair predicts that scientists will eventually discover 1,844 dinosaur genera in total—at least 1,300 more than the 527 recognized today from remains other than isolated teeth. What's more, the duo believes that 75 percent of these dinos will be discovered within the next 60 to 100 years and 90 percent within 100 to 140 years, based on an analysis of historical discovery patterns.

The tally applies only to specimens preserved as fossils. Many other types of dinosaurs likely roamed the Earth during the dinosaurs' 160-million-year reign, but remains from these species will never be known to science, the researchers say.

1. What is the main topic?
2. What are the first, second, and third ideas?
3. Name two details for each of the ideas.
4. Name two subdetails for each of the details.

Apply the Skill

Take Notes and Outline Scan a science journal for a short article about a new laboratory technique. Take notes by using shorthand or by creating an outline. Summarize the article using only your notes.

Understand Cause and Effect

Skillbuilder Handbook

Why Learn this Skill?

In order to understand an event, you should look for how that event or chain of events came about. When scientists are unsure of the cause for an event, they often design experiments. Although there might be an explanation, an experiment should be performed to be certain the cause created the event you observed. This process examines the causes and effects of events.

Learn the Skill

Calderas can form when the summit or side of a volcano collapses into the magma chamber that once fueled the volcano. An empty magma chamber can *cause* the volcano to collapse. The caldera that forms is the *effect,* or result. The figure below shows how one event—the **cause**—led to another—the **effect.**

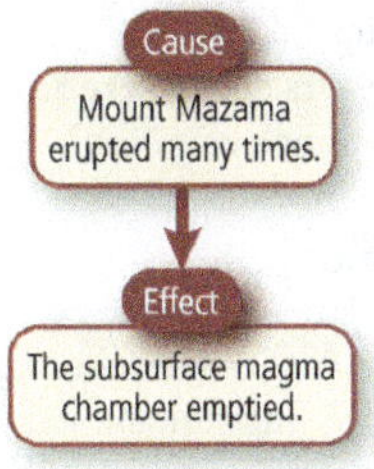

You can often identify cause-and-effect relationships in sentences from clue words such as the following.

because	**produced**
due to	**as a result**
so that	**that is why**
therefore	**for this reason**
thus	**consequently**
led to	**in order to**

Read the sample sentences below.

"The volcano collapsed into the partially empty magma chamber. As a result, a depression was formed where the volcano once stood."

In the example above, the cause is the collapse of the volcano. The cause-and-effect clue words "as a result" tell you that the depression is the effect of the collapsing volcano.

In a chain of events, an effect often becomes the cause of other events. The next chart shows the complete chain of events that occur when a caldera forms.

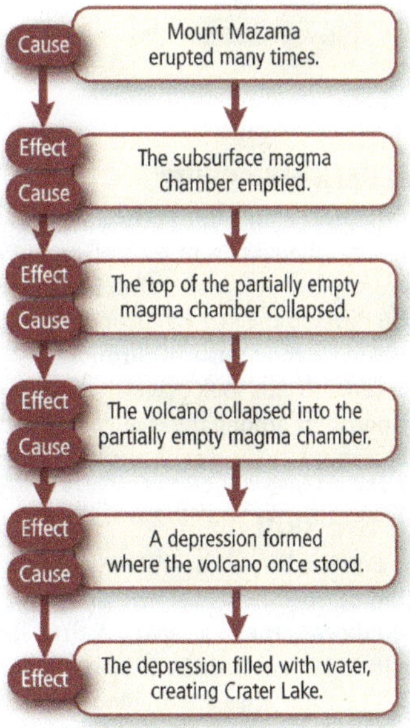

Practice the Skill

Make a chart like the one above showing which events listed below are causes and which are effects.

1. As water vapor rises, it cools and changes back to a liquid.
2. Droplets inside clouds join to form bigger drops.
3. Water evaporates from oceans, lakes, and rivers.
4. Water vapor rises into the atmosphere.
5. Water droplets become heavy and fall as rain or snow.

Apply the Skill

Understand Cause and Effect Read an account of a recent scientific event or discovery in a science journal. Determine at least one cause and one effect of that event. Show the chain of events in a chart.

Practice the Skill

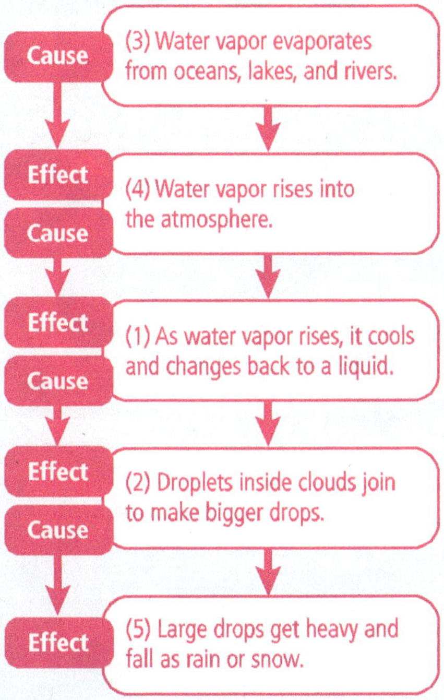

Apply the Skill

Answers will vary with the article and topic. Students should make a chart similar to the one above showing several cause-and-effect relationships. Encourage students to attach the article to their chain-of-events chart.

Skillbuilder Handbook

Practice the Skill

1. The time span of the time line is from 6000 B.C. to A.D. 2000. The left page intervals are 3000 years while the right page intervals are 100 years.
2. Katmai's eruption had a force 10 times that of Mount St. Helens.
3. 1709 years (1630 + 79)
4. 108 years (1991 – 1883)

Apply the Skill

Students should make their time line copy with a larger space for the dates after 1800. Dates students might add are 1811, 1880, 1906, 1923, 1948, 1960, 1965, 1972, 1982, 2004, and 2010.

Read a Time Line

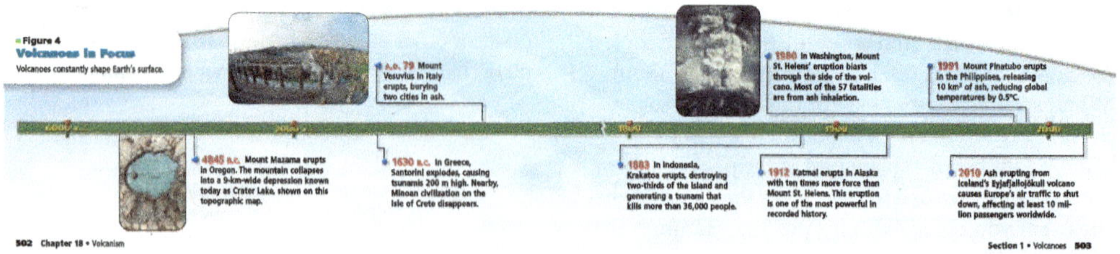

Why Learn this Skill?

When you read a time line such as the one above, you see not only when an event took place, but also what events took place before and after it. A time line can help you develop the skill of chronological thinking. Developing a strong sense of chronology—when and in what order events took place—will help you examine relationships among the events. It will also help you understand the causes or results of events.

Learn the Skill

A time line is a linear chart that list events that occurred on specific dates. The number of years between dates is the time span. A time line that begins in 1910 and ends in 1920 has a ten-year time span. Some time lines are divided into centuries. The twentieth century includes the 1900s, the nineteenth century includes the 1800s, and so on.

Time lines are usually divided into smaller parts, or time intervals. On the two time lines below, the first time line has a 300-year time span divided into 100-year time intervals. The second time line has a six-year time span divided into two-year time intervals.

Practice the Skill

Study the time line above and then answer these questions.

1. What time span and intervals appear on this time line?
2. How much more powerful was Katmai's eruption than Mount St. Helens' eruption?
3. How many years after Santorini erupted did Vesuvius erupt?
4. How many years apart were Krakatoa's eruption and Mt. Pinatubo's eruption?

Apply the Skill

Read a Time Line Sometimes a time line shows events that occurred during the same period but are related to two different subjects. The time line above shows events related to volcanoes between 6000 B.C. and A.D. 2000. Copy the time line and events onto a piece of paper. Then use a different color to add in events related to earthquakes during this same time span.

Analyze Media Sources

Why Learn this Skill?

To stay informed, people use a variety of media sources, including print media, broadcast media, and electronic media. The Internet has become an especially valuable research tool. It is convenient to use, and the information it contains is plentiful. Whichever media source you use to gather information, it is important to analyze the source to determine its accuracy and reliability.

Learn the Skill

There are a number of issues to consider when analyzing a media source. The most important one is to check the accuracy of the source and content. The author and publishers or sponsors should be credible and clearly indicated. To analyze print media or broadcast media, ask yourself the following questions.

- Is the information current?
- Are the sources revealed?
- Is more than one source used?
- Is the information biased?
- Does the information represent both sides of an issue?
- Is the information reported firsthand or secondhand?

For electronic media, ask yourself these questions in addition to the ones above.

- Is the author credible and clearly identified?
- Are the facts on the Web site documented?
- Are the links within the Web site appropriate and current?
- Does the Web site contain links to other useful resources?

Practice the Skill

To practice analyzing print media, choose two articles on global warming, one from a newspaper and the other from a newsmagazine. Then answer these questions.

1. What points are the authors of the articles trying to make? Were they successful? Can the facts be verified?
2. Did either article reflect a bias toward one viewpoint or another? List any unsupported statements.
3. Was the information reported firsthand or secondhand? Do the articles seem to represent both sides fairly?
4. How many sources can you identify in the articles? List them.

To analyze electronic media, read through the list of links provided by your teacher. Choose one link from the list, read the information on that Web site, and then answer these questions.

1. Who is the author or sponsor of the Web site?
2. What links does the Web site contain? How are they appropriate to the topic?
3. What sources were used for the information on the Web site?

Apply the Skill

Analyze Media Sources Think of a national issue on which public opinion is divided. Read newspaper features, editorials, and Web sites, and monitor television reports about the issue. Which news sources more fairly represents the issue? Which news sources have the most reliable information? Can you identify any biases? Can you verify the credibility of the news source?

Skillbuilder Handbook

Practice the Skill

Answers will vary. Encourage students to attach copies of the print media and Web sites.

Apply the Skill

Answers will vary. Encourage students to attach copies of the print media and Web sites.

Use Graphic Organizers

Why Learn this Skill?

While you read this textbook, you will be looking for important ideas or concepts. One way to arrange these ideas is to create a graphic organizer. In addition to Foldables®, you will find various other graphic organizers throughout your book. Some organizers show a sequence, or flow, of events. Other organizers emphasize the relationship among concepts. Developing your own organizers while you read will help you better understand and remember what you read.

Learn the Skill

An **events chain concept map** is used to describe a sequence of events, such as a stage of a process or procedure. When making an events-chain map, first identify the event that starts the sequence and add events in chronological order until you reach an outcome.

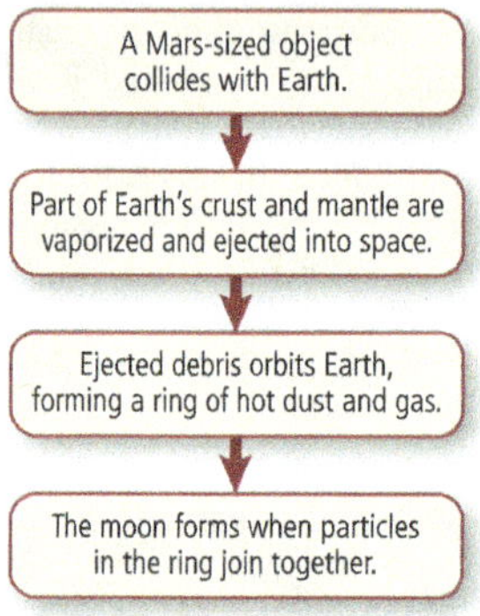

In a **cycle concept map,** the series of events do not produce a final outcome. The event that appears to be the final event relates back to the initiating event. Therefore, the cycle repeats itself.

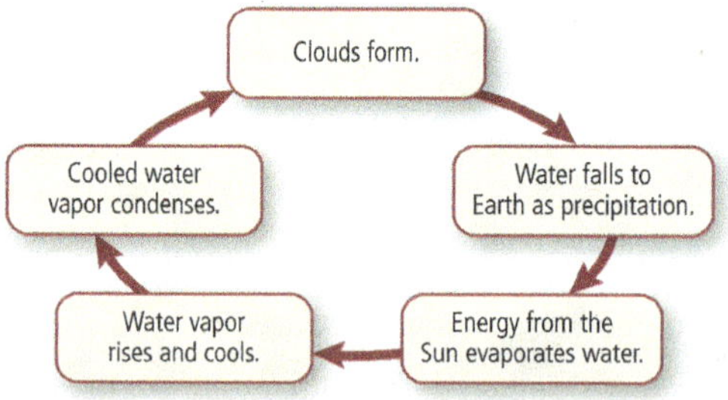

A **network tree concept map** shows the relationship among concepts, which are written in order from general to specific. The words written on the lines between the circles, called linking words, describe the relationships among the concepts. The concepts and the linking words can form sentences.

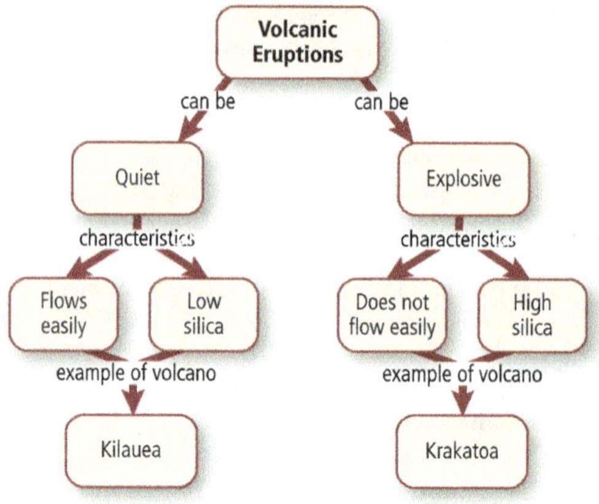

Practice the Skill

1. Create an events chain concept map of the events in sedimentary rock formation.
2. Create a cycle concept map of the nitrogen cycle. Make sure that the cycle shows the event that appears to be the final event relating back to the starting event.
3. Create a network tree concept map with these words: *Cenozoic, trilobites, eras, Paleozoic, mammals, dinosaurs, first land plants, Gondwana, Mesozoic, early Pangaea, late Pangaea.* Add linking words to describe the relationships between the concepts.

Apply the Skill

Use Graphic Organizers Create an events chain concept map of the scientific method. Create a cycle concept map of the water cycle. Create a network tree concept map of pollution that includes air and water, sources of each pollution type, and examples of each type of pollution.

Practice the Skill

1.

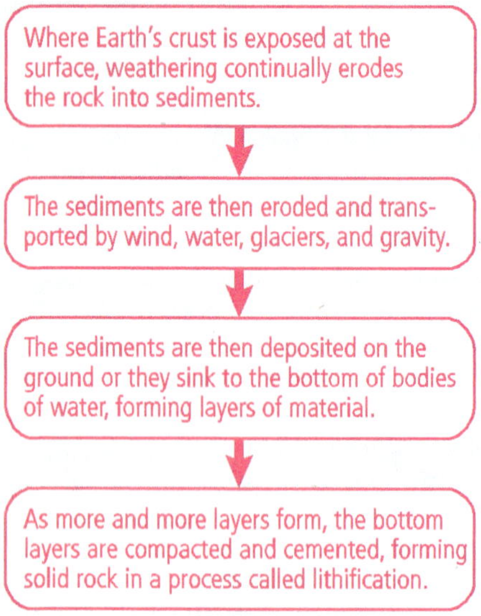

2.

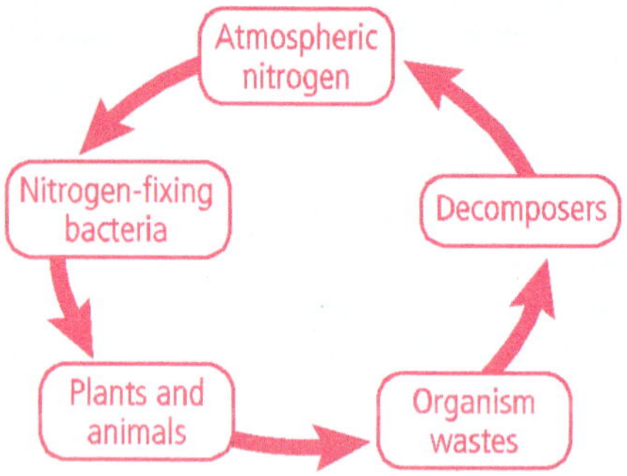

3.

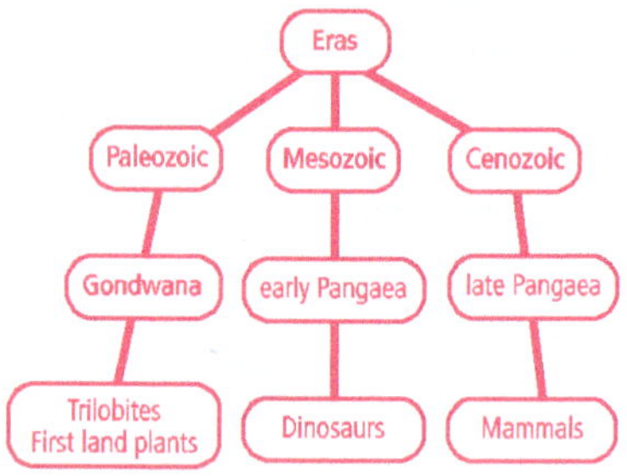

Apply the Skill

Example events chain concept map

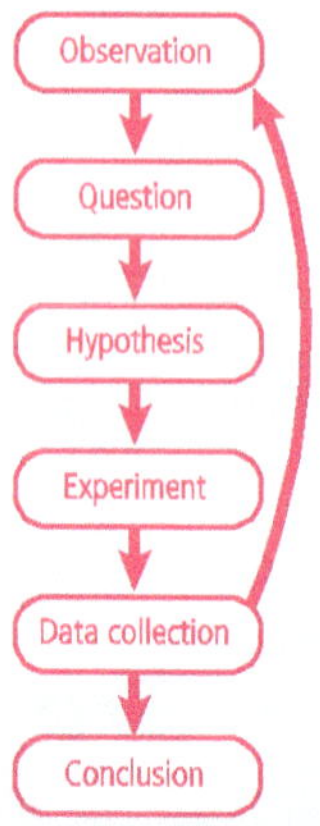

Example cycle concept map

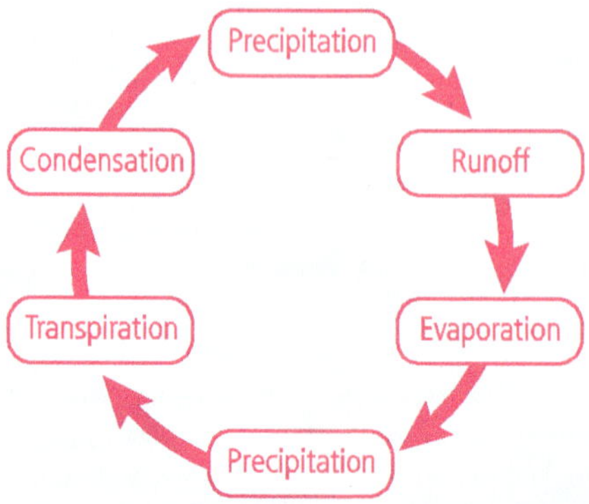

Example network tree

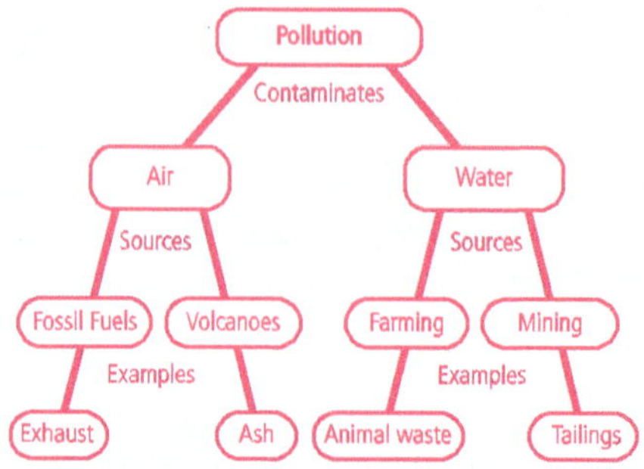

Debate Skills

Skillbuilder Handbook

Myrleen Pearson/PhotoEdit

New research always is leading to new scientific theories. There are often opposing points of view on how this research is conducted, how it is interpreted, and how it is communicated. *The Earth Science and Society* features in your book offer a chance to debate a current controversial topic. Here is an overview on how to conduct a debate.

Choose a Position and Research

First, choose an Earth science issue that has at least two opposing viewpoints. The issue can come from current events, your textbook, or your teacher. These topics could include global warming or fossil fuel use. Topics are stated as affirmative declarations such as "Global warming is not detrimental to the environment."

One speaker will argue the positive position—the viewpoint that supports the statement—and another speaker will argue the negative position—the viewpoint that disputes the statement. Either individually or with a group, choose your position for the debate. The viewpoint that you choose does not have to reflect your personal belief. The purpose of debate is to create a strong argument supported by scientific evidence.

After choosing your position, conduct research to support your viewpoint. Use the Internet, find articles in your library, or use your textbook to gather evidence to support your argument.

A strong argument contains scientific evidence, expert opinions, and your own analysis of the issue. Research the opposing position also. Becoming aware of what points the other side might argue will help you to strengthen the evidence for your position.

Hold the Debate

You will have a specific amount of time, determined by your teacher, in which to present your argument. Organize your speech to fit within the time limit: explain the viewpoint that you will be arguing, present an analysis of your evidence, and conclude by summing up your most important points. Try to vary the elements of your argument. Your speech should not be a list of facts, a reading of a newspaper article, or a statement of your personal opinion, but an organized analysis of your evidence presented in your own manner of speaking. It is also important to remember that you must never make personal attacks against your opponent. Argue the issue. You will be evaluated on your overall presentation, organization and development of ideas, and strength of support for your argument.

Additional Roles There are other roles that you can play in a debate. You can act as the timekeeper. The timekeeper times the length of the debaters' speeches and gives quiet signals to the speaker when time is almost up (usually a hand signal).

You can also act as a judge. There are important elements to look for when judging a speech: an introduction that tells the audience what position the speaker will be arguing, strong evidence that supports the speaker's position, and organization. It is helpful to take notes during the debate to summarize the main points of each side's argument. Then, decide which debater presented the strongest argument for his or her position. You can have a class discussion about the strengths and weaknesses of the debate and other viewpoints on this issue that could be argued.

Practice Problem 1

If: 1 m = 1,000,000 µm

And: 1 km = 1000 m

Then:

$$1000\ \mu m \times \frac{1\ m}{1{,}000{,}000\ \mu m} \times \frac{1\ km}{1000\ m}$$

$= 0.000001\ km = 1\ km \times 10^{-6}\ km$

Skillbuilder Handbook

MATH SKILLS

Experimental data is often expressed using numbers and units. The following sections provide an overview of the common system of units and some calculations involving units.

Measure in SI

The International System of Measurements, abbreviated SI, is accepted as the standard for measurement throughout most of the world. The SI system contains seven base units. All other units of measurement can be derived from these base units.

Table 2 SI Base Units

Measurement	Unit	Symbol
Length	meter	m
Mass	kilogram	kg
Time	second	s
Electric current	ampere	A
Temperature	kelvin	K
Amount of substance	mole	mol
Intensity of light	candela	cd

Some units are derived by combining base units. For example, units for volume are derived from units of length. A liter (L) is a cubic decimeter (dm^3, or dm × dm × dm). Units of density (g/L) are derived from units of mass (g) and units of volume (L).

When units are multiplied by factors of ten, new units are created. For example, if a base unit is multiplied by 1000, the new unit has the prefix *kilo-*. One thousand meters is equal to one kilometer. Prefixes for some units are shown in **Table 3.**

To convert a given unit to a unit with a different factor of ten, multiply the unit by a conversion factor. A conversion factor is a ratio equal to one. The equivalents in **Table 3** can be used to make such a ratio. For example, 1 km = 1000 m. Two conversion factors can be made from this equivalent.

$$\frac{1000\ m}{1\ km} = 1 \quad \text{and} \quad \frac{1\ km}{1000\ m} = 1$$

To convert one unit to another factor of ten, choose the conversion factor that has the unit you are converting from in the denominator.

$$1\ km \times \frac{1000\ m}{1\ km} = 1000\ m$$

A unit can be multiplied by several conversion factors to obtain the desired unit.

Table 3 Common SI Prefixes

Prefix	Symbol	Equivalents
mega-	m	1×10^{6} base units
kilo-	k	1×10^{3} base units
hecto-	h	1×10^{2} base units
deka-	da	1×10^{1} base units
deci-	d	1×10^{-1} base units
centi-	c	1×10^{-2} base units
milli-	m	1×10^{-3} base units
micro-	µ	1×10^{-6} base units
nano-	n	1×10^{-9} base units
pico-	p	1×10^{-12} base units

Practice Problem 1 How would you convert 1000 micrometers to kilometers?

Convert Temperature

The following formulas can be used to convert between Fahrenheit and Celsius temperatures. Notice that each equation can be obtained by algebraically rearranging the other. Therefore, you only need to remember one of the equations.

Conversion of Fahrenheit to Celsius

$$°C = \frac{(°F) - 32}{1.8}$$

Conversion of Celsius to Fahrenheit

$$°F = 1.8(°C) + 32$$

Make and Use Tables

Tables help visually organize data so that it can be interpreted more easily. Tables are composed of several components—a title describing the contents of the table, columns and rows that separate and organize information, and headings that describe the information in each column or row.

Table 4 Glacier Movement Rates

Depth (m)	Distance (m)	Average Speed (m/day)
0	13.1	0.198
20	13.1	0.198
60	12.8	0.194
100	12.2	0.185
140	11.2	0.170
180	9.6	0.145

Looking at this table, you should not only be able to pick out specific information, but you should also notice trends.

Practice Problem 2 If scientists drilled another 40 m into the glacier, what would the speed of the glacier's movement be at that depth?

Make and Use Graphs

Scientists often organize data in graphs. The types of graphs typically used in science are the line graph, the bar graph, and the circle graph.

Line Graphs A line graph is used to show the relationship between two variables. The independent variable is plotted on the horizontal axis, called the *x*-axis. The dependent variable is plotted on the vertical axis, called the *y*-axis. The dependent variable *(y)* changes as a result of a change in the independent variable *(x)*.

Suppose your class wanted to collect data about humidity. You could make a graph of the amount of water vapor that air can hold at various temperatures. **Table 5** shows the data.

Table 5 Amount of Water Vapor in Air at Various Temperatures

Air Temperature (°C)	Air (g/m³)
10	10
20	18
30	31
40	50
50	80

To make a graph of the amount of water vapor in air, start by determining the dependent and independent variables. The average amount of water vapor found per cubic meter of air is the dependent variable and is plotted on the *y*-axis. The independent variable, air temperature, is plotted on the *x*-axis.

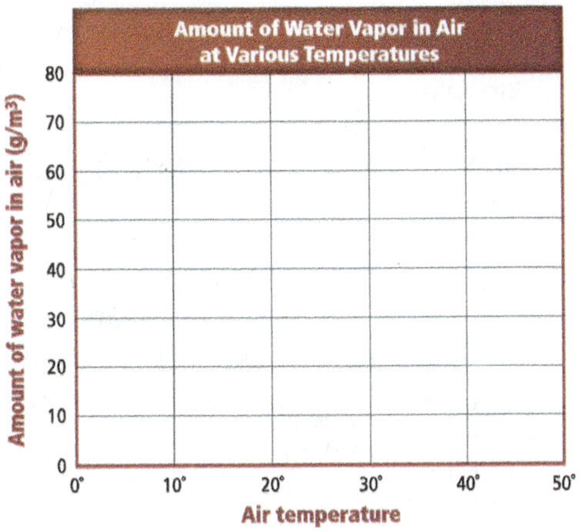

Plain or graph paper can be used to construct graphs. Draw a grid on your paper or a box around the squares that you intend to use on your graph paper. Give your graph a title and label each axis with a title and units. In this example, label the *x*-axis *Air temperature*. Because the lowest temperature was 10 and the highest was 50, you know that you will have to start numbers on the *y*-axis at least at 0 and number to at least 50. You decide to start numbering at 0 and number by equally spaced intervals of ten.

Skillbuilder Handbook

Practice Problem 2

Ice at the top of the glacier is moving faster than ice at the bottom of the glacier. The table shows a decrease in average speed of 0.009 m/day from 60 m to 100 m, 0.015 from 100 m to 140 m, and 0.025 from 140 m to 180 m. Based on this information, we can estimate that the difference in average speed from 180 m to 220 m will be at least equal to or greater than 0.025 m/day. A speed decrease of 0.030 to 0.035 m/day is a reasonable estimate, giving a range of speed of 0.115 to 0.110 m/day.

Practice Problem 3
As the air temperature increases (left to right), so does the amount of water vapor (bottom to top).

Practice Problem 4
Plotting two sets of data on one graph will give two lines. Because humidity (water vapor) will always increase with temperature, both sets of data will have bottom-left-to-top-right trending graphs. The differences between the data will vary from day to day and year to year. Have students use different colors to indicate each class's data to minimize confusion.

Label the y-axis of your graph *Amount of water vapor in air (g/m³)*. Begin plotting points by locating 0°C on the x-axis and 5 g/m³ on the y-axis. Where an imaginary vertical line from the x-axis and an imaginary horizontal line from the y-axis meet, place the first data point. Place other data points using the same process. After all the points are plotted, draw a "best fit" straight line through all the points.

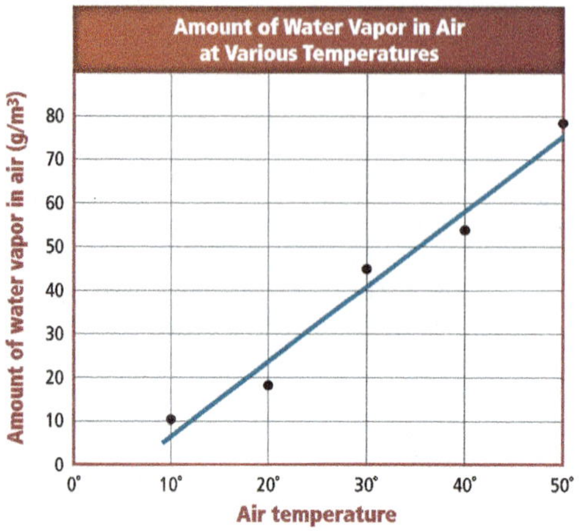

Practice Problem 3 According to the graph, does the amount of water vapor in air increase or decrease with air temperature?

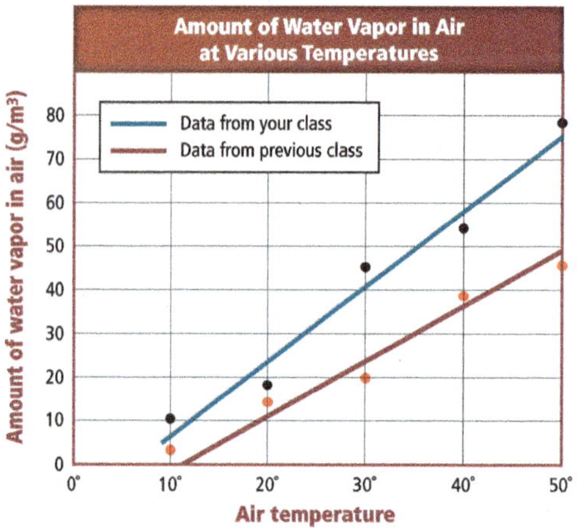

What if you wanted to compare the data about humidity collected by your class with similar data collected a year ago by a different class? The data from the other class can be plotted on the same graph to make the comparison. Include a key with different lines indicating different sets of data.

Practice Problem 4 How did the data from your class compare to the data from the previous class?

Bar Graphs A bar graph displays a comparison of different categories of data by representing each category with a bar. The length of the bar is related to the category's frequency. To make a bar graph, set up the x-axis and y-axis as you did for the line graph. Plot the data by drawing thick bars from the x-axis up to the y-axis point.

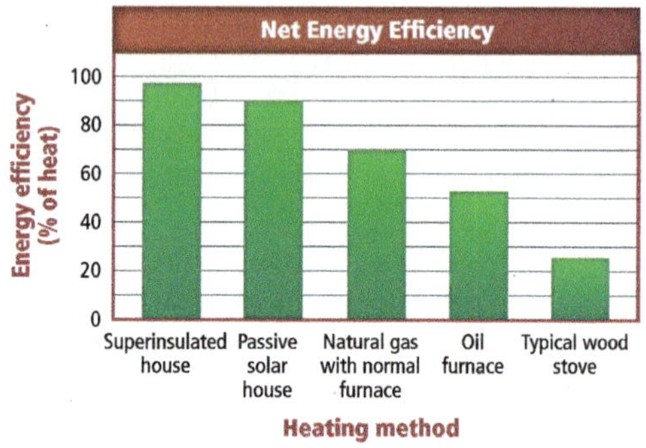

Look at the graph above. The independent variable is the energy efficiency. The dependent variable is the heating method.

Practice Problem 5 Which type of heating method has the second greatest efficiency? Is this more than twice as efficient as the lowest efficiency? Explain.

Bar graphs can also be used to display multiple sets of data in different categories at the same time. A bar graph that displays two sets of data is called a double-bar graph. Double-bar graphs have a legend to denote which bars represent each set of data. The graph below is an example of a double-bar graph.

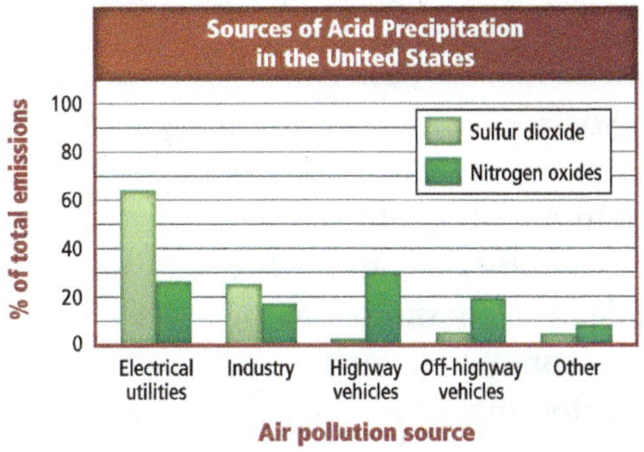

Circle Graphs A circle graph consists of a circle divided into sections that represent parts of a whole. When all the sections are placed together, they equal 100 percent of the whole.

Suppose you want to make a circle graph to show the percentage of solid wastes generated by various industries in the United States each year. The total amount of solid waste generated each year is estimated at ten billion metric tons. The whole circle graph will therefore represent this amount of solid waste. You find that 7.5 billion metric tons of waste is generated by mining and oil and gas production. The total amount of solid waste generated each year by mining and oil and gas production makes up one section of the circle graph, as follows.

$$\text{Segment of circle for total waste} = \frac{\text{waste from mining and oil and gas production}}{\text{total waste}}$$

$$= \frac{7.5}{10}$$

$$= 0.75 \times 360°$$

$$= 270°$$

To draw your circle graph, you will need a compass and a protractor. First, use the compass to draw a circle.

Then, draw a straight line from the center to the edge of the circle. Place your protractor on this line, and mark the point on the circle where 270° angle will intersect the circle. Draw a straight line from the center of the circle to the intersection point. This is the section for the waste generated from mining and oil and gas production.

Now, try to perform the same operation for the other data to find the number of degrees of the circle that each represents, and draw them in as well: agriculture, 1.3 billion metric tons; industry, 0.95 billion metric tons; municipal, 0.15 billion metric tons; and sewage sludge, 0.1 billion metric tons.

Complete your graph by labeling the sections of the graph and giving the graph a title. Your completed graph should look similar to the one below.

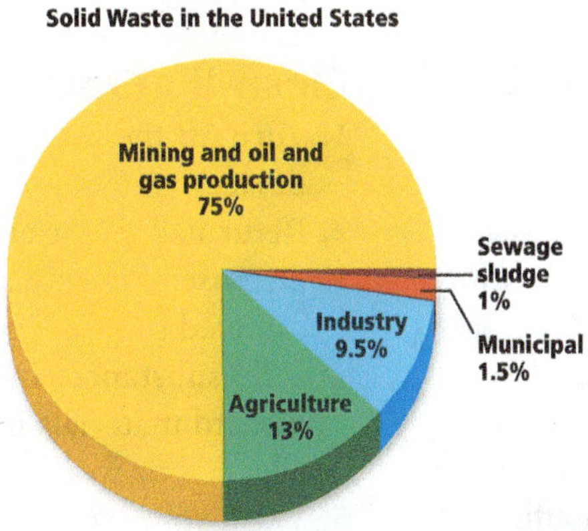

Practice Problem 6 There are 25 varieties of flowering plants growing around the high school. Construct a circle graph showing the percentage of each flower's color. Two varieties have yellow blooms, five varieties have blue-purple blooms, eight varieties have white blooms, and ten varieties have red blooms.

Skillbuilder Handbook

Practice Problem 5

Passive solar house; efficiency of passive solar house is 90 percent, and the lowest efficiency (wood stove) is 25 percent, so the passive solar house is more than 3 times more efficient.

Practice Problem 6

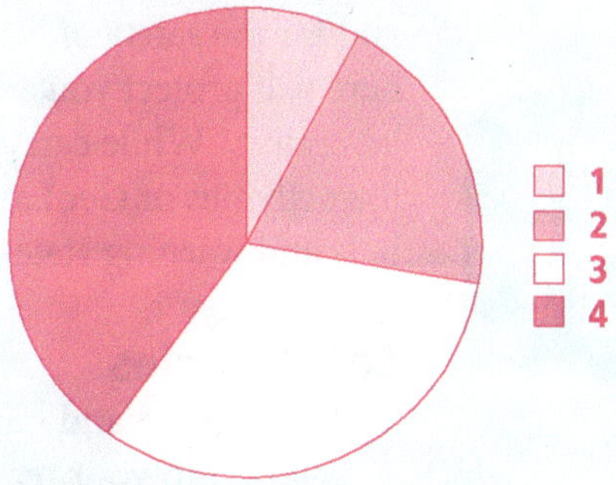

1. Yellow—2 varieties: 2 ÷ 25 = 8%
2. Blue/Purple—5 varieties: 5 ÷ 25 = 20%
3. White—8 varieties: 8 ÷ 25 = 32%
4. Red—10 varieties: 10 ÷ 25 = 40%
Total = 25 varieties = 100%

Safety in the Laboratory

The Earth science laboratory is a safe place to work if you are careful to observe the following important safety rules. You are responsible for your own safety and for the safety of others. The safety rules given here will protect you and others from harm in the laboratory. While carrying out procedures in any of the activities or GeoLabs, take note of the safety symbols and warning statements.

Safety Rules

1. Always read and complete the lab safety form and obtain your teacher's permission before beginning an investigation.
2. Study the procedure outline in the text. If you have questions, ask your teacher. Make sure that you understand all safety symbols shown on the page.
3. Use the safety equipment provided for you. Safety goggles and an apron should be worn during all investigations that involve the use of chemicals.
4. When heating test tubes, always slant them away from yourself and others.
5. Never eat or drink in the lab, and never use lab glassware as food or drink containers. Never inhale chemicals. Do not taste any substances or draw any material into a tube or pipet with your mouth.
6. If you spill any chemical, wash it off immediately with water. Report the spill immediately to your teacher.
7. Know the location and proper use of the fire extinguisher, eye wash, safety shower, fire blanket, fire alarm, and first aid kit. First aid procedures in the science laboratory are listed in **Table 1.**
8. Keep materials away from flames. Tie back hair and loose clothing when you are working with flames.
9. If a fire should break out in the lab, or if your clothing should catch fire, smother it with the fire blanket or a coat, get under a safety shower, or use the fire department's recommendation for putting out a fire on your clothing: stop, drop, and roll. NEVER RUN.
10. Report any accident or injury, no matter how small, to your teacher.

Clean-Up Procedures

1. Turn off the water and gas. Disconnect electrical devices.
2. Return all materials to their proper places.
3. Dispose of chemicals and other materials as directed by your teacher. Place broken glass and solid substances in the proper containers. Never discard materials in the sink.
4. Clean your work area.
5. Wash your hands thoroughly after working in the laboratory.

Table 1 First Aid in the Science Laboratory

Injury	Safe Response
Burns	Apply cold water. Call your teacher immediately.
Cuts and bruises	Stop any bleeding by applying direct pressure. Cover cuts with a clean dressing. Apply cold compresses to bruises. Call your teacher immediately.
Fainting	Leave the person lying down. Loosen any tight clothing and keep crowds away. Call your teacher immediately.
Foreign matter in eye	Flush with plenty of water. Use an eyewash bottle or fountain.
Poisoning	Note the suspected poisoning agent and call your teacher immediately.
Any spills on skin	Flush with large amounts of water or use safety shower. Call your teacher immediately.

Safety Symbols

Safety symbols in the following table are used in the lab activities to indicate possible hazards. Learn the meaning of each symbol. **It is recommended that you wear safety goggles and apron at all times in the lab. This might be required in your school district.**

Safety Symbols	Hazard	Examples	Precaution	Remedy
Disposal	Special disposal procedures need to be followed.	certain chemicals, living organisms	Do not dispose of these materials in the sink or trash can.	Dispose of wastes as directed by your teacher.
Biological	Organisms or other biological materials that might be harmful to humans	bacteria, fungi, blood, unpreserved tissues, plant materials	Avoid skin contact with these materials. Wear mask or gloves.	Notify your teacher if you suspect contact with material. Wash hands thoroughly.
Extreme Temperature	Objects that can burn skin by being too cold or too hot	boiling liquids, hot plates, dry ice, liquid nitrogen	Use proper protection when handling.	Go to your teacher for first aid.
Sharp Object	Use of tools or glassware that can easily puncture or slice skin	razor blades, pins, scalpels, pointed tools, dissecting probes, broken glass	Practice common-sense behavior and follow guidelines for use of the tool.	Go to your teacher for first aid.
Fume	Possible danger to respiratory tract from fumes	ammonia, acetone, nail polish remover, heated sulfur, moth balls	Be sure there is good ventilation. Never smell fumes directly. Wear a mask.	Leave foul area and notify your teacher immediately.
Electrical	Possible danger from electrical shock or burn	improper grounding, liquid spills, short circuits, exposed wires	Double-check setup with teacher. Check condition of wires and apparatus. Use GFI-protected outlets.	Do not attempt to fix electrical problems. Notify your teacher immediately.
Irritant	Substances that can irritate the skin or mucous membranes of the respiratory tract	pollen, moth balls, steel wool, fiberglass, potassium permanganate	Wear dust mask and gloves. Practice extra care when handling these materials.	Go to your teacher for first aid.
Chemical	Chemicals that can react with and destroy tissue and other materials	bleaches such as hydrogen peroxide; acids such as sulfuric acid, hydrochloric acid; bases such as ammonia, sodium hydroxide	Wear goggles, gloves, and an apron.	Immediately flush the affected area with water and notify your teacher.
Toxic	Substance may be poisonous if touched, inhaled, or swallowed.	mercury, many metal compounds, iodine, poinsettia plant parts	Follow your teacher's instructions.	Always wash hands thoroughly after use. Go to your teacher for first aid.
Flammable	Flammable chemicals may be ignited by open flame, spark, or exposed heat.	alcohol, kerosene, potassium permanganate	Avoid open flames and heat when using flammable chemicals.	Notify your teacher immediately. Use fire safety equipment if applicable.
Open Flame	Open flame in use, may cause fire.	hair, clothing, paper, synthetic materials	Tie back hair and loose clothing. Follow teacher's instruction on lighting and extinguishing flames.	Notify your teacher immediately. Use fire safety equipment if applicable.

Eye Safety
Proper eye protection should be worn at all times by anyone performing or observing science activities.

Clothing Protection
This symbol appears when substances could stain or burn clothing.

Animal Safety
This symbol appears when safety of animals and students must be ensured.

Radioactivity
This symbol appears when radioactive materials are used.

Handwashing
After the lab, wash hands with soap and water before removing goggles.

Reference Handbook

Physiographic Map of Earth

Reference Handbook

Topographic Map Symbols

Roads and Railroads

Primary highway, hard surface

Secondary highway, hard surface

Light-duty road, hard or improved surface

Unimproved road

Railroad: single track and multiple track

Railroads in juxtaposition

Buildings and Structures

Buildings

School, church, and cemetery — cem

Barn and warehouse

Wells, not water (with labels) — ○ oil ○ gas

Tanks: oil, water, etc. (labeled if water) — water

Open-pit mine, quarry, or prospect

Tunnel

Benchmark — BM Δ 293

Bridge

Campsite

Habitats

Marsh (swamp)

Wooded marsh

Woods or brushwood

Vineyard

Submerged marsh

Mangrove

Coral reef, rocks

Orchard

Urban area

Perennial streams

Elevated aqueduct

Water well and spring

Small rapids

Large rapids

Intermittent lake

Intermittent stream

Glacier

Large falls

Dry lake bed

Surface Elevations

Spot elevation — ×7369

Water elevation — 670

Index contour — 100

Intermediate contour

Depression contour

Boundaries

National

State

County, parish, municipal

Civil township, precinct, town, barrio

Incorporated city, village, town, hamlet

Reservation, national or state

Small park, cemetery, airport, etc.

Land grant

Township or range line, United States land survey

Township or range line, approximate location

Weather Map Symbols

Sample Plotted Report at Each Station

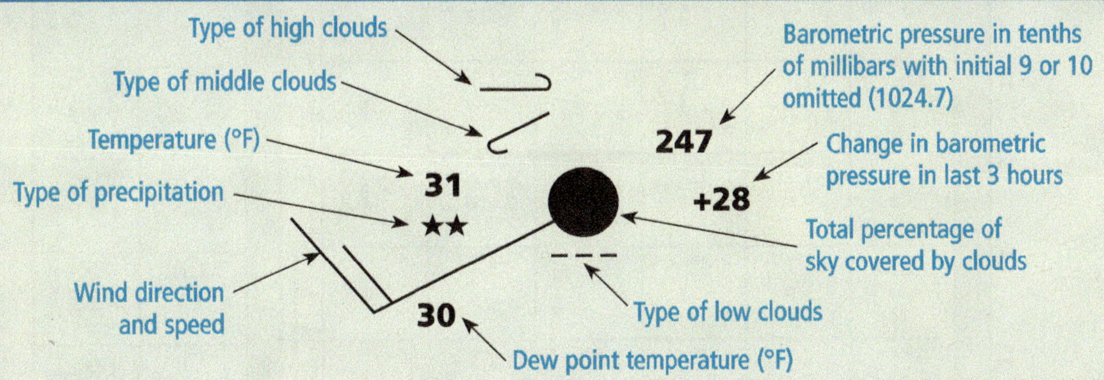

Symbols Used in Plotting Report

Precipitation	Wind Direction and Speed	Sky Coverage	Fronts and Pressure Systems	
Fog	0 calm	No cover	(H) or High	Center of high- or
Snow	1–2 knots	1/10 or less	(L) or Low	low-pressure system
Rain	3–7 knots	2/10 to 3/10		Cold front
Thunderstorm	8–12 knots	4/10		Warm front
Drizzle	13–17 knots	1/2		Occluded front
Showers	18–22 knots	6/10		Stationary front
	23–27 knots	7/10		
	48–52 knots	Overcast with openings		
	1 knot = 1.852 km/h	Completely overcast		

Clouds

Some Types of High Clouds	Some Types of Middle Clouds	Some Types of Low Clouds
Scattered cirrus	Thin altostratus layer	Cumulus of fair weather
Dense cirrus in patches	Thick altostratus layer	Stratocumulus
Veil of cirrus covering entire sky	Thin altostratus in patches	Fractocumulus of bad weather
Cirrus not covering entire sky	Thin altostratus in bands	Stratus of fair weather

PERIODIC TABLE OF THE ELEMENTS

Element — Hydrogen
Atomic number — 1
Symbol — H
Atomic mass — 1.008
State of matter

Gas / Liquid / Solid / Synthetic

Metal / Metalloid / Nonmetal / Recently observed

	1	2	3	4	5	6	7	8	9	10	11	12	13	14	15	16	17	18
1	Hydrogen 1 H 1.008																	Helium 2 He 4.003
2	Lithium 3 Li 6.941	Beryllium 4 Be 9.012											Boron 5 B 10.811	Carbon 6 C 12.011	Nitrogen 7 N 14.007	Oxygen 8 O 15.999	Fluorine 9 F 18.998	Neon 10 Ne 20.180
3	Sodium 11 Na 22.990	Magnesium 12 Mg 24.305											Aluminum 13 Al 26.982	Silicon 14 Si 28.086	Phosphorus 15 P 30.974	Sulfur 16 S 32.066	Chlorine 17 Cl 35.453	Argon 18 Ar 39.948
4	Potassium 19 K 39.098	Calcium 20 Ca 40.078	Scandium 21 Sc 44.956	Titanium 22 Ti 47.867	Vanadium 23 V 50.942	Chromium 24 Cr 51.996	Manganese 25 Mn 54.938	Iron 26 Fe 55.847	Cobalt 27 Co 58.933	Nickel 28 Ni 58.693	Copper 29 Cu 63.546	Zinc 30 Zn 65.39	Gallium 31 Ga 69.723	Germanium 32 Ge 72.61	Arsenic 33 As 74.922	Selenium 34 Se 78.96	Bromine 35 Br 79.904	Krypton 36 Kr 83.80
5	Rubidium 37 Rb 85.468	Strontium 38 Sr 87.62	Yttrium 39 Y 88.906	Zirconium 40 Zr 91.224	Niobium 41 Nb 92.906	Molybdenum 42 Mo 95.94	Technetium 43 Tc (98)	Ruthenium 44 Ru 101.07	Rhodium 45 Rh 102.906	Palladium 46 Pd 106.42	Silver 47 Ag 107.868	Cadmium 48 Cd 112.411	Indium 49 In 114.82	Tin 50 Sn 118.710	Antimony 51 Sb 121.757	Tellurium 52 Te 127.60	Iodine 53 I 126.904	Xenon 54 Xe 131.290
6	Cesium 55 Cs 132.905	Barium 56 Ba 137.327	Lanthanum 57 La 138.905	Hafnium 72 Hf 178.49	Tantalum 73 Ta 180.948	Tungsten 74 W 183.84	Rhenium 75 Re 186.207	Osmium 76 Os 190.23	Iridium 77 Ir 192.217	Platinum 78 Pt 195.08	Gold 79 Au 196.967	Mercury 80 Hg 200.59	Thallium 81 Tl 204.383	Lead 82 Pb 207.2	Bismuth 83 Bi 208.980	Polonium 84 Po 208.982	Astatine 85 At 209.987	Radon 86 Rn 222.018
7	Francium 87 Fr (223)	Radium 88 Ra (226)	Actinium 89 Ac (227)	Rutherfordium 104 Rf (261)	Dubnium 105 Db (262)	Seaborgium 106 Sg (266)	Bohrium 107 Bh (264)	Hassium 108 Hs (277)	Meitnerium 109 Mt (268)	Darmstadtium 110 Ds (281)	Roentgenium 111 Rg (272)	Copernicium 112 Cn (285)	Ununtrium * 113 Uut (284)	Flerovium 114 Fl (289)	Ununpentium * 115 Uup (288)	Livermorium 116 Lv (293)		Ununoctium * 118 Uuo (294)

The number in parentheses is the mass number of the longest lived isotope for that element.

* The names and symbols for elements 113, 115, and 118 are temporary. Final names will be selected when the elements' discoveries are verified.

Lanthanide series	Cerium 58 Ce 140.115	Praseodymium 59 Pr 140.908	Neodymium 60 Nd 144.242	Promethium 61 Pm (145)	Samarium 62 Sm 150.36	Europium 63 Eu 151.965	Gadolinium 64 Gd 157.25	Terbium 65 Tb 158.925	Dysprosium 66 Dy 162.50	Holmium 67 Ho 164.930	Erbium 68 Er 167.259	Thulium 69 Tm 168.934	Ytterbium 70 Yb 173.04	Lutetium 71 Lu 174.967
Actinide series	Thorium 90 Th 232.038	Protactinium 91 Pa 231.036	Uranium 92 U 238.029	Neptunium 93 Np (237)	Plutonium 94 Pu (244)	Americium 95 Am (243)	Curium 96 Cm (247)	Berkelium 97 Bk (247)	Californium 98 Cf (251)	Einsteinium 99 Es (252)	Fermium 100 Fm (257)	Mendelevium 101 Md (258)	Nobelium 102 No (259)	Lawrencium 103 Lr (262)

Table 2 Relative Humidity %

Dry-Bulb Temperature	Dry-Bulb Temperature Minus Wet-Bulb Temperature (°C)									
	1	2	3	4	5	6	7	8	9	10
0°C	81	64	46	29	13					
1°C	83	66	49	33	18					
2°C	84	68	52	37	22	7				
3°C	84	69	55	40	25	12				
4°C	85	71	57	43	29	16				
5°C	85	72	58	45	32	20				
6°C	86	73	60	48	35	24	11			
7°C	86	74	61	49	38	26	15			
8°C	87	75	63	51	40	29	19	8		
9°C	87	76	65	53	42	32	21	12		
10°C	88	77	66	55	44	34	24	15	6	
11°C	89	78	67	56	46	36	27	18	9	
12°C	89	78	68	58	48	39	29	21	12	
13°C	89	79	69	59	50	41	32	22	15	7
14°C	90	79	70	60	51	42	34	26	18	10
15°C	90	80	71	61	53	44	36	27	20	13
16°C	90	81	71	63	54	46	38	30	23	15
17°C	90	81	72	64	55	47	40	32	25	18
18°C	91	82	73	65	57	49	41	34	27	20
19°C	91	82	74	65	58	50	43	36	29	22
20°C	91	83	74	66	59	51	44	37	31	24
21°C	91	83	75	67	60	53	46	39	32	26
22°C	92	83	76	68	61	54	47	40	34	28
23°C	92	84	76	69	62	55	48	42	36	30
24°C	92	84	77	69	62	56	49	43	37	31
25°C	92	84	77	70	63	57	50	44	39	33
26°C	92	85	78	71	64	58	51	46	40	34
27°C	92	85	78	71	65	58	52	47	41	36
28°C	93	85	78	72	65	59	53	48	42	37
29°C	93	86	79	72	66	60	54	49	43	38
30°C	93	86	79	73	67	61	55	50	44	39
31°C	93	86	80	73	67	62	56	50	45	40
32°C	93	86	80	74	68	62	57	51	46	41

Table 3 Minerals with Metallic Luster

Mineral (Formula)	Color	Streak	Hardness	Specific Gravity	Crystal System	Breakage Pattern	Uses and Other Properties
Bornite (Cu_5FeS_4)	bronze, tarnishes to dark blue purple	gray-black	3	4.9–5.4	tetragonal	uneven fracture	source of copper; called "peacock ore" because of the purple shine when it tarnishes
Chalcopyrite ($CuFeS_2$)	brassy to yellow	greenish black	3.5–4	4.2	tetragonal	uneven fracture	main ore of copper
Chromite ((Fe, Mg) (Cr, Al)$_2$O$_4$)	black or brown	brown to black	5.5	4.6	cubic	irregular fracture	ore of chromium, stainless steel, metallurgical bricks
Copper (Cu)	copper red	copper red	3	8.5–9	cubic	hackly	coins, pipes, gutters, wire, cooking utensils, jewelry; malleable and ductile
Galena (PbS)	gray	gray to black	2.5	7.5	cubic	cubic cleavage perfect	source of lead, used in pipes, shields for X rays, fishing equipment sinkers
Gold (Au)	pale to golden yellow	yellow	2.5–3	19.3	cubic	hackly	jewelry, money, gold leaf, fillings for teeth, medicines; does not tarnish
Graphite (C)	black to gray	black to gray	1–2	2.3	hexagonal	basal cleavage (scales)	pencil lead, lubricants for locks, rods to control some small nuclear reactions, battery poles
Hematite (specular) (Fe_2O_3)	black or reddish brown	red or reddish brown	6	5.3	hexagonal	irregular fracture	source of iron; roasted in a blast furnace, converted to "pig" iron, made into steel
Magnetite ((Fe, Mg) Fe_2O_4)	black	black	6	5.2	cubic	conchoidal fracture	source of iron, naturally magnetic, called lodestone
Pyrite (FeS_2)	light, brassy yellow	greenish black	6.5	5.0	cubic	uneven fracture	source of iron, "fool's gold," alters to limonite
Pyrrhotite ($Fe_{1-x}S$)* *contains one less atom of Fe than S	bronze	gray-black	4	4.6	hexagonal	uneven fracture	an ore of iron and sulfur; may be magnetic
Silver (Ag)	silvery white, tarnishes to black	light gray to silver	2.5	10–12	cubic	hackly	coins, fillings for teeth, jewelry, silverplate, wires; malleable and ductile

Table 4 Minerals with Nonmetallic Luster

Mineral (Formula)	Color	Streak	Hardness	Specific Gravity	Crystal System	Breakage Pattern	Uses and Other Properties
Augite ((Ca, Na) (Mg, Fe^2, Al) $(Al, Si)_2O_6$)	black	colorless	6	3.3	monoclinic	2-directional cleavage	square or 8-sided cross section
Corundum (Al_2O_3)	colorless, blue, brown, green, white, pink, red	colorless	9	4.0	hexagonal	fracture	gemstones: ruby is red, sapphire is blue; abrasive
Fluorite (CaF_2)	colorless, white, blue, green, red, yellow, purple	colorless	4	3–3.2	cubic	cleavage	manufacture of optical equipment; glows under UV light
Garnet ($(Mg, Fe^2, Ca, Mn^2)_3$, $(Al, Fe^3, Mn^3, V, Cr)_2$, $(SiO_4)_3$)	deep yellow-red, green, black	colorless	7.5	3.5	cubic	conchoidal fracture	used in jewelry; also used as an abrasive
Hornblende ($(Ca, Na)_{2\text{-}3}$ $(Mg, Fe^2, Fe^3, Al)_5$, $(Al, Si)_8O_{22}$ $(OH)_2$)	green to black	gray to white	5–6	3.4	monoclinic	cleavage in two directions	will transmit light on thin edges; 6-sided cross section
Limonite (hydrous iron oxides)	yellow, brown, black	yellow, brown	5.5	2.7–4.3	N/A	conchoidal fracture	source of iron; weathers easily, coloring matter of soils
Olivine ($(Mg, Fe)_2$ SiO_4)	olive green	colorless	6.5	3.5	orthorhombic	conchoidal fracture	gemstones, refractory sand
Plagioclase feldspar ((Na, Ca) Al(Si, Al) Si_2O_8)	gray, green, white	colorless	6	2.5	triclinic	two cleavage planes meet at 86° angle	used in ceramics; striations present on some faces
Potassium feldspar ($KAlSi_3O_8$)	colorless, white to gray, green, yellow, pink	colorless	6	2.5	monoclinic	two cleavage planes meet at 90° angle	insoluble in acids; used in the manufacture of porcelain
Quartz (SiO_2)	colorless, various colors	colorless	7	2.6	hexagonal	conchoidal fracture	glass manufacture, electronic equipment, radios, computers, watches, gemstones
Topaz ($(Al_2SiO_4$ $(F, OH)_2)$)	colorless, white, pink, yellow, pale blue	colorless	8	3.5	orthorhombic	basal cleavage	valuable gemstone

Reference Handbook

Table 5 Common Rocks

Rock Type	Rock Name	Characteristics
Igneous (intrusive)	granite	large mineral grains of quartz, feldspar, hornblende, and mica; usually light in color
	diorite	large mineral grains of feldspar, hornblende, and mica; less quartz than granite; intermediate in color
	gabbro	large mineral grains of feldspar, hornblende, augite, olivine, and mica; no quartz; dark in color
Igneous (extrusive)	rhyolite	small or no visible grains of quartz, feldspar, hornblende, and mica; light in color
	andesite	small or no visible grains of quartz, feldspar, hornblende, and mica; less quartz than rhyolite; intermediate in color
	basalt	small or no visible grains of feldspar, hornblende, augite, olivine, and mica; no quartz; dark in color; vesicles may be present
	obsidian	glassy texture; no visible grains; volcanic glass; fracture is conchoidal; color is usually black, but may be red-brown or black with white flecks
	pumice	frothy texture; floats; usually light in color
Sedimentary (clastic)	conglomerate	coarse-grained; gravel- or pebble-sized grains
	sandstone	sand-sized grains 1/16 to 2 mm in size; varies in color
	siltstone	grains smaller than sand but larger than clay
	shale	smallest grains; usually dark in color
Sedimentary (chemical or biochemical)	limestone	major mineral is calcite; usually forms in oceans, lakes, rivers, and caves; often contains fossils; effervesces in dilute HCl
	coal	occurs in swampy, low-lying areas; compacted layers of organic material, mainly plant remains
Sedimentary (chemical)	rock salt	commonly forms by the evaporation of seawater
Metamorphic	gneiss	well-developed banding because of alternating layers of different minerals, usually of different colors; common parent rock is granite
	schist	well-developed parallel arrangement of flat, sheetlike minerals, mainly micas; common parent rocks are shale and phyllite
	phyllite	shiny or silky appearance; may look wrinkled; common parent rocks are shale and slate
	slate	harder, denser, and shinier than shale; common parent rock is shale
Metamorphic (nonfoliated)	marble	interlocking calcite or dolomite crystals; common parent rock is limestone
	soapstone	composed mainly of the mineral talc; soft with a greasy feel
	quartzite	hard and well-cemented with interlocking quartz crystals; common parent rock is sandstone

Solar System Charts The Planets

	Mercury	Venus	Earth	Mars	Jupiter	Saturn	Uranus	Neptune
Mass (kg)	3.3020×10^{23}	4.8685×10^{24}	5.9736×10^{24}	6.4185×10^{23}	1.8986×10^{27}	5.6846×10^{26}	8.6832×10^{25}	1.0243×10^{26}
Equatorial radius (km)	2439.7	6051.8	6378.1	3396.2	71,492	60,268	25,559	24,764
Mean density (kg/m^3)	5427	5243	5515	3933	1326	687	1270	1638
Albedo	0.068	0.900	0.306	0.250	0.343	0.342	0.300	0.290
Semimajor axis (km)	5.791×10^{7}	1.0821×10^{8}	1.4960×10^{8}	2.2792×10^{8}	7.7857×10^{8}	1.43353×10^{9}	2.87246×10^{9}	4.49506×10^{9}
Orbital period (Earth days)	87.969	224.701	365.256	686.980	4332.589	10,759.220	30,685.4	60,189
Orbital inclination (degrees)	7.000	3.390	0.000	1.850	1.304	2.485	0.772	1.769
Orbital eccentricity	0.2056	0.0067	0.0167	0.0935	0.0489	0.0565	0.0457	0.0113
Rotational period (hours)	1407.6	5832.5^{R}	23.9345	24.6229	9.9250	10.656	17.24^{R}	16.11
Axial tilt (degrees)	0.01	177.36	23.45	25.19	3.13	26.73	97.77	28.32
Average surface temperature (K)	440	737	288	210	165	134	76	72
Number of known moons*	0	0	1	2	63	62	27	13

*Number as of 2007.

^{R} indicates retrograde rotation.

The Moon	
Mass (kg)	7.349×10^{22}
Equatorial radius (km)	1738.1
Mean density (kg/m^3)	3350
Albedo	0.12
Semimajor axis (km)	3.844×10^{5}
Orbital period (Earth days)	27.3217
Lunar period (Earth days)	29.53
Orbital inclination (degrees)	5.145
Orbital eccentricity	0.0549
Rotational period (hours)	655.728

The Sun	
Mass (kg)	1.989×10^{30}
Equatorial radius (km)	6.96×10^{5}
Mean density (kg/m^3)	1408
Absolute magnitude	4.83
Luminosity (W)	384.6
Spectral type	G2
Rotational period (hours)	609.12
Average temperature (K)	5778

GLOSSARY ▪ GLOSARIO

Multilingual eGlossary

The multilingual science glossary includes Arabic, Bengali, Chinese, English, Haitian, Creole, Hmong, Korean, Portuguese, Russian, Tagalog, Urdu, and Vietnamese.

Cómo usar el glosario en español:
1. Busca el término en inglés que desees encontrar.
2. El término en español, junto con la definición, se encuentran en la columna de la derecha.

Pronunciation Key

Use the following key to help you sound out words in the glossary.

a	**b**ack (BAK)	**ew**	**f**oo**d** (FEWD)
ay	**d**ay (DAY)	**yoo**	**p**ure (PYOOR)
ah	**f**ather (FAH thur)	**yew**	**f**ew (FYEW)
ow	**fl**ower (FLOW ur)	**uh**	comma (CAHM uh)
ar	**c**ar (CAR)	**u** (+con)	**r**ub (RUB)
e	**l**ess (LES)	**sh**	**sh**elf (SHELF)
ee	**l**eaf (LEEF)	**ch**	na**t**ure (NAY chur)
ih	**t**rip (TRIHP)	**g**	**g**ift (GIHFT)
i (i+con+e)	**i**dea, **l**ife (i DEE uh, life)	**j**	**g**em (JEM)
oh	**g**o (GOH)	**ing**	s**ing** (SING)
aw	**s**oft (SAWFT)	**zh**	vi**s**ion (VIHZH un)
or	**or**bit (OR but)	**k**	**c**ake (KAYK)
oy	**c**oin (COYN)	**s**	**s**eed, **c**ent (SEED, SENT)
oo	**f**oot (FOOT)	**z**	**z**one, rai**s**e (ZOHN, RAYZ)

ENGLISH — A — ESPAÑOL

abrasion: **(p. 203)** process of erosion in which wind-blown or waterborne particles, such as sand, scrape against rock surfaces or other materials and wear them away.

abrasión: **(pág. 203)** proceso erosivo en que las partículas por el viento o el agua, como la arena, chocan y raspan superficies rocosas u otros materiales y los desgastan.

absolute-age dating: **(p. 601)** method that enables scientists to determine the actual age of certain rocks and other objects.

datación absoluta: **(pág. 601)** permite a los científicos determinar la antigüedad real de ciertas rocas y objetos.

absolute magnitude: **(p. 842)** brightness an object would have if it were placed at a distance of 10 pc; classification system for stellar brightness that can be calculated when the actual distance to a star is known.

magnitud absoluta: **(pág. 842)** brillo que tendría un objeto si estuviera a una distancia de 10 pc; sistema de clasificación del brillo estelar que se puede calcular cuando se conoce la distancia verdadera hasta la estrella.

abyssal plain: **(p. 451)** smooth, flat part of the seafloor covered with muddy sediments and sedimentary rocks that extends seaward from the continental margin.

llanura abisal: **(pág. 451)** parte plana y lisa del fondo del mar cubierta con sedimentos fangosos y rocas sedimentarias y que se extiende desde el margen continental hacia el mar.

acid: **(p. 71)** solution containing a substance that produces hydrogen ions: (H^+) in water.

ácido: **(pág. 71)** solución que contiene una sustancia que produce iones hidrógeno (H^+) en agua.

acid precipitation: **(p. 745)** any precipitation with a pH of less than 5.0 that forms when sulfur dioxide and nitrogen oxides combine with moisture in the atmosphere to produce sulfuric acid and nitric acid.

precipitación ácida: **(pág. 745)** toda precipitación con un pH menor que 5.0 que se forma cuando se combinan el dióxido de azufre y óxidos de nitrógeno con la humedad en la atmósfera para producir ácido sulfúrico o ácido nítrico.

active galactic nucleus (AGN): **(p. 875)** a galaxy's core in which highly energetic objects or activities are located.

aggregate: **(p. 684)** mixture of sand, gravel, and crushed stone that accumulates naturally; found in floodplains, alluvial fans, or glacial deposits.

air mass: **(p. 316)** large volume of air that has the characteristics of the area over which it forms.

air-mass thunderstorm: **(p. 346)** type of thunderstorm in which air rises because of unequal heating of Earth's surface within a single air mass and is most common during the afternoon and evening.

albedo: **(p. 771)** percentage of sunlight that is reflected by the surface of a planet or a satellite, such as the Moon.

altered hard part: **(p. 607)** fossil whose soft organic material has been removed and whose hard parts have been changed by recrystallization or mineral replacement.

amino acid: **(p. 634)** a building block of proteins.

amniotic **(am nee AH tihk)** **egg:** **(p. 658)** egg with a shell, providing a complete environment for a developing embryo.

amplitude: **(p. 539)** the size of the seismic waves; an increase of 1 in the scale represents an increase in amplitude of a factor of 10.

analog forecast: **(p. 331)** weather forecast that compares current weather patterns to patterns that occurred in the past.

anemometer **(a nuh MAH muh tur):** **(p. 325)** weather instrument used to measure wind speed.

apogee: **(p. 783)** farthest point in an object's orbit to Earth.

apparent magnitude: **(p. 842)** classification system based on how bright a star appears to be; does not take distance into account so cannot indicate how bright a star actually is.

aquiclude: **(p. 255)** layer of impermeable material, such as silt, clay, or shale, that is a barrier to groundwater.

aquifer: **(p. 255)** permeable underground layer through which groundwater flows relatively easily.

núcleo galáctico activo (NGA): **(pág. 875)** centro de la galaxia donde se ubican cuerpos o suceden eventos con gran cantidad de energía.

agregado: **(pág. 684)** mezcla natural de arena, grava y piedra triturada que se acumula naturalmente; se encuentra en llanuras aluviales, abanicos aluviales o depósitos glaciales.

masa de aire: **(pág. 316)** gran volumen de aire que tiene las características del área sobre la que se forma.

tormenta eléctrica de masa de aire: **(pág. 346)** tipo de tormenta en que el aire asciende debido al calentamiento desigual de la superficie terrestre bajo una misma masa de aire; es más común durante la tarde y la noche.

albedo: **(pág. 771)** porcentaje de luz solar que refleja la superficie de un planeta o un satélite, como por ejemplo, la Luna.

partes duras alteradas: **(pág. 607)** fósiles cuya materia orgánica blando ha desaparecido y cuyas partes duras han sido transformadas por recristalización o sustitución de minerales.

aminoácido: **(pág. 634)** unidad básica de las proteínas.

huevo amniótico: **(pág. 658)** huevo con cascarón; provee un ambiente completo para el embrión en desarrollo.

amplitud: **(pág. 539)** la magnitud de las ondas sísmicas; un aumento de 1 unidad en esta escala representa un aumento en amplitud de un factor de 10.

pronóstico análogo: **(pág. 331)** pronóstico del tiempo que compara los patrones actuales del clima con patrones ocurridos en el pasado.

anemómetro: **(pág. 325)** instrumento meteorológico que se utiliza para medir la velocidad de viento.

apogeo: **(pág. 783)** punto de la órbita de un objeto en que ésta se encuentra más alejada de la Tierra.

magnitud aparente: **(pág. 842)** sistema de clasificación basado el brillo aparente de una estrella; no toma en cuenta la distancia y por lo tanto no indica el brillo real de la estrella.

acuiclusos: **(pág. 255)** capas impermeables que sirven de barrera a las aguas subterráneas, como por ejemplo limo, arcilla o esquisto.

acuífero: **(pág. 255)** capa subterránea permeable por la cual el agua subterránea fluye de manera relativamente fácil.

artesian well: **(p. 264)** fountain of water that spurts above the land surface when a well taps a confined aquifer containing water under pressure.

asteroid **(AS tuh royd):** **(p. 622)** metallic or silica-rich object, 1 m to 950 km in diameter, that bombarded early Earth, generating heat energy; rocky remnant of the early solar system found mostly between the orbits of Mars and Jupiter in the asteroid belt.

astronomical unit (AU): **(p. 800)** the average distance between the Sun and Earth, 1.496×10^8 km.

astronomy: **(p. 6)** study of objects beyond Earth's atmosphere.

atmosphere: **(p. 8)** blanket of gases surrounding Earth that contains about 78 percent nitrogen, 21 percent oxygen, and 1 percent other gases such as argon, carbon dioxide, and water vapor.

atomic number: **(p. 62)** number of protons contained in an atom's nucleus.

avalanche: **(p. 198)** landslide that occurs in a mountainous area when snow falls on an icy crust, becomes heavy, slips off, and slides swiftly down a mountainside.

pozo artesiano: **(pág. 264)** fuente de agua que brota hacia la superficie terrestre, cuando un pozo conecta con un acuífero confinado que contiene agua bajo presión.

asteroide: **(pág. 622)** cuerpo metálico o rico en sílice que mide de 1 m a 950 km de diámetro y que bombardeó la Tierra primitiva generando energía calórica; restos rocosos del sistema solar primitivo que se hallan principalmente entre las órbitas de Marte y Júpiter, en el cinturón de asteroides.

unidad astronómica (UA): **(pág. 800)** la distancia promedio entre el Sol y la Tierra, equivale a 1.496×10^8 km.

astronomía: **(pág. 6)** el estudio de los cuerpos que se encuentran más allá de la atmósfera de la Tierra.

atmósfera: **(pág. 8)** manto de gases que rodea la Tierra; está compuesta aproximadamente por 78 por ciento de nitrógeno, 21 por ciento de oxígeno y 1 por ciento de otros gases como el argón, el dióxido de carbono y el vapor del agua.

número atómico: **(pág. 62)** número de protones que contiene el núcleo de un átomo.

avalancha: **(pág. 198)** deslizamiento que ocurre en un área montañosa cuando la nieve cae sobre una capa helada, aumenta de peso, se desprende y se resbala rápidamente montaña abajo.

B

banded-iron formations: **(p. 630)** alternating bands of iron oxide and chert, an iron-poor sedimentary rock.

barometer: **(p. 324)** instrument used to measure air pressure.

barrier island: **(p. 442)** long ridges of sand or other sediment that are deposited or shaped by the longshore current and separated from the mainland.

basaltic rock: **(p. 118)** rock that is dark colored, has lower silica contents, and is rich in iron and magnesium; contains mostly plagioclase and pyroxene.

base: **(p. 72)** substance that produces hydroxide ions (OH^-) in water.

base level: **(p. 233)** the elevation at which a stream enters another stream or body of water.

batholith: **(p. 515)** coarse-grained, irregularly shaped, igneous rock mass that covers at least 100 km^2; generally forms 10–30 km below Earth's surface, and is common in the interior of major mountain chains.

formaciones de hierro en bandas: **(pág. 630)** bandas alternadas de óxido ferroso y pedernal, roca sedimentaria deficiente en hierro.

barómetro: **(pág. 324)** instrumento que se usa para medir la presión atmosférica.

barrera litoral: **(pág. 442)** grandes lomas de arena u otro sedimento que son depositadas, o que adquieren su forma, por la acción de las corrientes litorales y están separadas del continente.

roca basáltica: **(pág. 118)** roca oscura con bajo contenido en sílice pero rica en hierro y magnesio; contiene principalmente plagioclasa y piroxenos.

base: **(pág. 72)** sustancia que produce iones hidróxido (OH^-) en agua.

nivel base: **(pág. 233)** elevación a la cual una corriente entra a otra corriente o masa de agua.

batolito: **(pág. 515)** masa rocosa ígnea de grano grueso y de forma irregular que cubre por lo menos 100 km^2; generalmente se forma de 10 a 30 km bajo la superficie terrestre y es común en el interior de las principales cadenas montañosas.

beach: **(p. 438)** area in which loose sediment is deposited and moved about by waves along the shore.

bedding: **(p. 137)** horizontal layering in sedimentary rock; layers that can range from a few millimeters to several meters thick.

bed load: **(p. 228)** describes sediments that are too heavy or large to be kept in suspension or solution and are pushed or rolled along the bottom of a streambed.

bedrock: **(p. 684)** unweathered, solid parent rock that can consist of limestone, marble, granite, or other quarried rock.

belt: **(p. 812)** low, warm, dark-colored cloud that sinks and flows rapidly in the Jovian atmosphere.

Big Bang theory: **(p. 878)** theory that proposes that the universe began as a single point and has been expanding ever since.

binary star: **(p. 838)** one of two stars that are bound together by gravity and orbit a common center of mass.

biomass fuels **(p. 709)** fuels derived from living things; renewable resources.

bioremediation: **(p. 742)** use of organisms to clean up toxic waste.

biosphere: **(p. 9)** all of Earth's organisms and the environments in which they live.

bipedal: **(p. 665)** walking upright on two legs.

black hole: **(p. 851)** small, extremely dense remnant of a star whose gravity is so immense that not even light can escape its gravitational field.

Bowen's reaction series: **(p. 114)** sequential, predictable, dual-branched pattern in which minerals crystallize from cooling magma.

breaker: **(p. 422)** collapsing wave that forms when a wave reaches shallow water and becomes so steep that the crest topples forward.

playa: **(pág. 438)** área en que sedimentos sueltos son depositados y transportados por las olas a lo largo de la costa.

estratificación: **(pág. 137)** capas horizontales de roca sedimentaria; capas que pueden medir de un milímetro a varios metros de grosor.

carga de fondo: **(pág. 228)** término que describe los sedimentos que no se mantienen en suspensión, o en solución, porque son demasiado pesados o grandes y son empujados o arrastrados sobre el fondo del cauce de una corriente.

roca firme: **(pág. 684)** roca madre sólida no meteorizada que puede consistir en piedra caliza, mármol, granito o alguna otra piedra de cantera.

cinturón: **(pág. 812)** nube baja, tibia y oscura que desciende y fluye rápidamente en la atmósfera joviana.

teoría de la Gran Explosión: **(pág. 878)** propone que el universo empezó en un solo punto y se ha estado expan-diendo desde entonces.

estrella binaria: **(pág. 838)** una de dos estrellas unidas por la gravedad que giran alrededor de un centro común de masa.

biocombustible: **(pág. 709)** combustibles derivados de los seres vivos; recursos renovables.

biorremediación: **(pág. 742)** uso de organismos para limpiar desechos tóxicos.

biosfera: **(pág. 9)** incluye a todos los organismos de la Tierra y los ambientes en que éstos viven.

bipedalismo: **(pág. 665)** que camina erguido sobre dos piernas.

agujero negro: **(pág. 851)** restos de una estrella muy densos y pequeños cuya gravedad es tan grande que ni la luz puede escapar de su campo de gravedad.

serie de reacción de Bowen: **(pág. 114)** patrón de dos ramas, predecible y secuencial que siguen los minerales al cristalizarse a partir de magma que se enfría.

rompiente: **(pág. 422)** ola que se colapsa; se forma cuando una ola alcanza aguas poco profundas y se vuelve tan empinada que la cresta de la ola se cae hacia adelante.

C

caldera: **(p. 505)** large crater, up to 100 km in diameter, that can form when the summit or side of a volcano collapses into the magma chamber during or after an eruption.

caldera: **(pág. 505)** cráter grande, de hasta 100 km de diámetro, que se forma cuando la cumbre o la ladera de un volcán se desploman en la cámara de magma durante o después de una erupción.

Cambrian explosion: **(p. 653)** sudden appearance of a diverse collection of organisms in the Cambrian fossil record.

Canadian shield: **(p. 625)** name given to the Precambrian shield in North America because much of it is exposed in Canada.

carrying capacity: **(p. 735)** number of organisms that a specific environment can support.

cartography: **(p. 30)** science of mapmaking.

cast: **(p. 608)** fossil formed when an earlier fossil of a plant or animal leaves a cavity that becomes filled with minerals or sediment.

cave: **(p. 260)** a natural underground opening connected to Earth's surface, usually formed when groundwater dissolves limestone.

cementation: **(p. 137)** process of sedimentary rock formation that occurs when dissolved minerals precipitate out of groundwater and either a new mineral grows between the sediment grains or the same mineral grows between and over the grains.

Cepheid variables: **(p. 863)** stars with pulsation periods ranging from 1 to 100 days and varying luminosities.

chemical bond: **(p. 67)** force that holds the atoms of elements together in a compound.

chemical reaction: **(p. 70)** change of one or more substances into other substances.

chemical weathering: **(p. 166)** process by which rocks and minerals undergo changes in their composition due to chemical reactions with agents such as acids, water, oxygen, and carbon dioxide.

chromosphere: **(p. 831)** layer of the Sun's atmosphere above the photosphere and below the corona that is about 2500 km thick and has a temperature around 30,000 K at its top.

cinder cone: **(p. 507)** steep-sided, generally small volcano that is built by the accumulation of tephra around the vent.

cirque: **(p. 209)** deep depression scooped out by a valley glacier.

cirrus **(SIHR us):** **(p. 301)** high clouds made up of ice crystals that form at heights of 6000 m.

clastic: **(p. 141)** rock and mineral fragments produced by weathering and erosion and classified according to particle size and shape.

explosión del Cámbrico: **(pág. 653)** aparición repentina de un conjunto diverso de organismos en el registro fósil del Cámbrico.

escudo Canadiense: **(pág. 625)** nombre que recibe el escudo Precámbrico en Norteamérica porque la mayor parte está expuesto en Canadá.

capacidad de carga: **(pág. 735)** número de organismos que un ambiente específico puede sustentar.

cartografía: **(pág. 30)** ciencia de la elaboración de mapas.

molde: **(pág. 608)** fósil que se forma cuando un fósil precedente de una planta o un animal forma una cavidad que se rellena con minerales o sedimentos.

caverna: **(pág. 260)** cavidad subterránea abierta a la superficie terrestre, generalmente se forma cuando el agua subterránea disuelve la piedra caliza.

cementación: **(pág. 137)** proceso de formación de roca sedimentaria que ocurre cuando los minerales disueltos del agua subterránea se precipitan y se forma un nuevo mineral entre los granos de sedimento o se acumula el mismo mineral entre y sobre los granos.

variables cefeidas: **(pág. 863)** estrellas con periodos de pulsación que duran de 1 a 100 días y con mayor o menor luminosidad.

enlace químico: **(pág. 67)** fuerza que mantiene unidos los átomos de los elementos en un compuesto.

reacción química: **(pág. 70)** sucede cuando una o más sustancias se convierten en otras sustancias.

meteorización química: **(pág. 166)** proceso mediante el cual las rocas y los minerales experimentan cambios en su composición, debido a reacciones químicas con agentes como ácidos, agua, oxígeno o dióxido de carbono.

cromosfera: **(pág. 831)** capa de la atmósfera del Sol situada encima de la fotosfera y debajo de la corona; mide aproximadamente 2500 km de ancho y tiene una temperatura cercana a 30,000 K en su parte superior.

cono de carbonilla: **(pág. 507)** volcán empinada cara generalmente pequeño que es construido por la acumulación de tefrita alrededor de la chimenea.

circo: **(pág. 209)** depresión profunda formada por un glaciar de valle.

cirro: **(pag. 301)** nubes altas formadas por cristales de hielo que se forman a alturas de 6000 m.

clástico: **(pág. 141)** describe los fragmentos de roca y de mineral producidos por la meteorización y la erosión; se clasifican según su tamaño y forma de partícula.

clastic sedimentary rock: **(p. 141)** most common type of sedimentary rock, formed from the abundant deposits of loose sediments that accumulate on Earth's surface; classified according to the size of their particles.

cleavage: **(p. 92)** the manner in which a mineral breaks along planes where atomic bonding is weak.

climate: **(p. 314)** the long-term average of variation in weather for a particular area.

climatology: **(p. 376)** study of Earth's climate in order to understand and predict climatic change, based on past and present variations in temperature, precipitation, wind, and other weather variables.

coalescence **(ko uh LEH sunts):** **(p. 302)** process that occurs when cloud droplets collide and form larger droplets, which eventually become too heavy to remain aloft and can fall to Earth as precipitation.

cogeneration: **(p. 723)** production of two usable forms of energy at the same time from the same process, which can conserve resources and generate income.

cold wave: **(p. 364)** extended period of below-average temperatures caused by large, high-pressure systems of continental polar or arctic origin.

comet: **(p. 819)** small, eccentrically orbiting body made of rock and ice which have one or more tails that point away from the Sun.

composite volcano: **(p. 507)** generally cone-shaped with concave slopes; built by violent eruptions of volcanic fragments and lava that accumulate in alternating layers.

compound: **(p. 66)** substance composed of atoms of two or more different elements that are chemically combined.

compressive force: **(p. 567)** squeezing force that can cause the intense deformation—folding, faulting metamorphism, and igneous intrusions—associated with mountain building.

condensation: **(p. 75)** process by which a cooling gas changes into a liquid and releases thermal energy.

condensation nucleus: **(p. 297)** small particle in the atmosphere around which cloud droplets can form.

roca sedimentaria clástica: **(pág. 141)** el tipo más común de roca sedimentaria; se forma a partir de los abundantes depósitos de sedimentos sueltos que se acumulan sobre la superficie de la Tierra; se clasifican según el tamaño de sus partículas.

crucero: **(pág. 92)** la forma en la cuál un mineral se rompe a lo largo de los planos donde los enlaces atómicos son débiles.

clima: **(pág. 314)** promedio durante un largo periodo de las variaciones en las condiciones del tiempo de un área determinada.

climatología: **(pág. 376)** estudio del clima de la Tierra para entender y pronosticar los cambios climáticos; se basa en variaciones pasadas y presentes de temperatura, precipitación, viento y otras variables del tiempo.

coalescencia: **(pág. 302)** proceso que ocurre cuando las gotas de nube chocan entre sí, formando gotas cada vez más grandes; estas gotas puede llegar a ser demasiado pesadas para seguir suspendidas en el aire y entonces caen a la Tierra como precipitación.

cogeneración: **(pág. 723)** producción simultánea de dos formas útiles de energía a partir del mismo proceso; puede ayudar a conservar recursos y obtener ganancias.

onda fría: **(pág. 364)** período prolongado de temperaturas más bajas que el promedio, causado por grandes sistemas de alta presión de origen polar continental o ártico.

cometa: **(pág. 819)** cuerpo pequeño de órbita excéntrica compuesto por roca y hielo y que contiene una o más colas que apuntan hacia el lado opuesto al Sol.

volcán compuesto: **(pág. 507)** volcán que en general tiene forma cónica y laderas cóncavas; se forma por erupciones violentas de fragmentos y lava volcánicos que se acumulan creando capas alternadas.

compuesto: **(pág. 66)** sustancia compuesta por átomos de dos o más elementos diferentes unidos químicamente.

fuerzas de compresión: **(pág. 567)** fuerzas de aplastamiento que pueden causar intensas deformaciones como plegamientos, fallas, metamorfismo e intrusiones ígneas; asociadas con la formación de montañas.

condensación: **(pág. 75)** proceso por el cual un gas enfriador se transforma en un líquido y libera energía térmica.

núcleos de condensación: **(pág. 297)** partículas pequeñas de la atmósfera alrededor de las cuales se pueden formar las gotas de nubes.

conduction: **(p. 288)** the transfer of thermal energy between objects in contact by the collisions between the particles in the objects.

conducción: **(pág. 288)** transferencia de energía entre cuerpos en contacto debida a la colisión entre las partículas de los cuerpos.

conduit: **(p. 505)** a tubelike structure that allows lava to reach the surface.

conducto: **(pág. 505)** estructura tubular que permite que la lava llegue a la superficie.

conic projection: **(p. 35)** map that is highly accurate for small areas, made by projecting points and lines from a globe onto a cone.

proyección cónica: **(pág. 35)** mapa de gran exactitud para áreas pequeñas que se elabora mediante la proyección de puntos y líneas de un globo a un cono.

constellation: **(p. 837)** group of stars that forms a pattern in the sky that resembles an animal, mythological character, or everyday object.

constelación: **(pág. 837)** grupo de estrellas que forman en el firmamento un patrón que semeja un animal, un personaje mitológico o un objeto cotidiano.

contact metamorphism: **(p. 149)** local effect that occurs when molten rock meets solid rock.

metamorfismo de contacto: **(pág. 149)** efecto local que ocurre cuando la roca fundida se encuentra con roca sólida.

continental drift: **(p. 469)** Wegener's hypothesis that Earth's continents were joined as a single landmass, called Pangaea, that broke apart about 200 mya and slowly moved to their present positions.

deriva continental: **(pág. 469)** hipótesis de Wegener que propone que los continentes de la Tierra estaban unidos en una sola masa terrestre, llamada Pangaea, la cual se separó hace aproximadamente 200 millones de años y que los fragmentos resultantes se movieron lentamente a sus ubicaciones actuales.

continental glacier: **(p. 208)** glacier that forms over a broad, continent-sized area of land and usually spreads out from its center.

glaciar continental: **(pág. 208)** glaciar que se forma sobre una amplia área del tamaño de un continente y que generalmente se extiende a partir de su centro.

continental margin: **(p. 447)** area where edges of continents meet the ocean; represents the shallowest part of the ocean that consists of the continental shelf, the continental slope, and the continental rise.

margen continental: **(pág. 447)** área donde los límites de los continentes se unen con el océano; representa la parte menos profunda del océano y consiste en la plataforma continental, el talud continental y el pie del talud continental.

continental rise: **(p. 449)** gently sloping accumulation of sediments deposited by a turbidity current at the foot of a continental margin.

pie del talud continental: **(pág. 449)** acumulación de sedimentos, con pendiente leve, depositados por una corriente de turbidez al pie de un margen continental.

continental shelf: **(p. 447)** shallowest part of a continental margin, with an average depth of 130 m and an average width of 60 km, that extends into the ocean from the shore and provides a nutrient-rich home to large numbers of fish.

plataforma continental: **(pág. 447)** parte más superficial del margen continental, tiene una profundidad promedio de 130 m y una anchura promedio de 60 km, se extiende hacia el océano desde la costa y proporciona un lugar rico en nutrientes a un gran número de peces.

continental slope: **(p. 448)** sloping oceanic region found beyond the continental shelf that generally marks the edge of the continental crust and may be cut by sub-marine canyons.

talud continental: **(pág. 448)** región oceánica inclinada que se encuentra más allá de la plataforma continental; generalmente marca el límite de la corteza continental y puede estar seccionada por cañones submarinos.

contour interval: **(p. 36)** difference in elevation between two side-by-side contour lines on a topographic map.

intervalo entre curvas de nivel: **(pág. 36)** diferencia en la elevación entre dos curvas de nivel contiguas en un mapa topográfico.

contour line: **(p. 36)** line on a topographic map that connects points of equal elevation.

curva de nivel: **(pág. 36)** curva en un mapa topográfico que conecta puntos de igual elevación.

control: (p. 12) standard for comparison in an experiment.

convection: (p. 288) the transfer of thermal energy by the movement of heated material from one place to another.

convergent boundary: (p. 482) place where two tectonic plates are moving toward each other; is associated with trenches, islands arcs, and folded mountains.

Coriolis effect: (p. 318) effect of a rotating body that influences the motion of any object or fluid; on Earth, air moving north or south from the equator appears to move right or left, respectively; the combination of the Coriolis effect and Earth's heat imbalance creates the trade winds, polar easterlies, and prevailing westerlies.

corona: (p. 831) top layer of the Sun's atmosphere that extends from the top of the chromosphere and typically ranges in temperature from 3 million to 5 million K.

correlation: (p. 599) matching of rock outcrops of one geographic region to another.

cosmic background radiation: (p. 880) weak radiation that is left over from the early, hot stages of the Big Bang expansion of the universe.

cosmology: (p. 878) study of the universe, including its current nature, origin, and evolution, based on observation and the use of theoretical models.

covalent bond: (p. 67) attraction of two atoms for a shared pair of electrons that holds the atoms together.

crater: (p. 505) bowl-shaped depression that forms around the central vent at the summit of a volcano.

craton (KRAY tahn): (p. 625) continental core formed from Archean or Proterozoic microcontinents; deepest (as far as 200 km into the mantle) and most stable part of a continent.

creep: (p. 195) slow, steady downhill movement of loose weathered Earth materials, especially soils, causing objects on a slope to tilt.

crest: (p. 421) highest point of a wave.

control: (pág. 12) estándar de comparación en un experimento.

convección: (pág. 288) transferencia de energía térmica debido al movimiento de material caliente de un lado a otro.

límite convergente: (pág. 482) lugar donde dos placas tectónicas se mueven aproximándose cada vez más entre sí; está asociado con fosas abisales, arcos insulares y montañas plegadas.

efecto de Coriolis: (pág. 318) efecto producido por un cuerpo en rotación que influye en el movimiento de todo cuerpo objeto o fluido; en la Tierra, las corrientes aire que se mueven desde el norte o desde el sur parecen desplazarse hacia la derecha o hacia la izquierda, respectivamente; la combinación del efecto de Coriolis y el desequilibrio térmico de la Tierra originan los vientos alisios, los vientos polares del este y los vientos dominantes del oeste.

corona: (pág. 831) capa superior de la atmósfera del Sol que se extiende desde la parte superior de la cromosfera y típicamente tiene un rango de temperatura de 3 a 5 millones K.

correlación: (pág. 599) correspondencia entre los afloramientos rocosos de una región geográfica y otra.

radiación cósmica de fondo: (pág. 880) radiación residual débil proveniente de las calientes etapas iniciales de la expansión del universo causada por la Gran Explosión.

cosmología: (pág. 878) estudio del universo; abarca su naturaleza actual, su origen y evolución y se basa en la observación y el uso de modelos teóricos.

enlace covalente: (pág. 67) atracción de dos átomos hacia un par compartido de electrones que mantienen a los átomos unidos.

cráter: (pág. 505) depresión en forma de tazón que generalmente se forma alrededor de la abertura central en la cumbre de un volcán.

cratón: (pág. 625) zona central de un continente formada a partir de microcontinentes del arcaico o del Proterozoico; son la parte más profunda (penetran hasta 200 km hacia el manto) y estable de un continente.

deslizamiento: (pág. 195) movimiento cuesta abajo constante y lento de materia meteorizada suelta de la Tierra, especialmente los suelos, lo que ocasiona que se inclinen los objetos en una ladera.

cresta: (pág. 421) punto más alto de una onda.

cross-bedding: **(p. 138)** depositional feature of sedimentary rock that forms as inclined layers of sediment are carried forward across a horizontal surface.

cross-cutting relationships: **(p. 597)** the principle that an intrusion or fault is younger than the rock it cuts across.

cryosphere: **(p. 8)** the frozen portion of water on Earth's surface.

crystal: **(p. 87)** solid in which atoms are arranged in repeating patterns.

crystalline structure: **(p. 73)** regular geometric pattern of particles in most solids, giving a solid a definite shape and volume.

cumulus **(KYEW myuh lus):** **(p. 301)** puffy, lumpy-looking clouds that usually occur below 2000 m.

cyanobacteria: **(p. 629)** microscopic, photosynthetic prokaryotes that formed stromatolites and changed early Earth's atmosphere by generating oxygen.

estratificación cruzada: **(pág. 138)** característica de la depo-sitación de roca sedimentaria que se forma a medida que capas inclinadas de sedimento son arrastradas hacia delante, a lo largo de una superficie horizontal.

relaciones de corte transversal: **(pág. 597)** principio que establece que una intrusión o falla es menos antigua que la roca que atraviesa.

crisofera: **(pág. 8)** la parte de agua congelada sobre la superficie de la Tierra.

cristal: **(pág. 87)** sólido cuyos átomos están ordenados en patrones repetitivos.

estructura cristalina: **(pág. 73)** patrón geométrico y regular que tienen las partículas en la mayoría de los sólidos; dan al sólido una forma y volumen definidos.

cúmulo: **(pág. 301)** nubes esponjosas con aspecto de madejas de algodón que generalmente se hallan a alturas menores de 2000 m.

cianobacterias: **(pág. 629)** organismos procariotas fotosintéticos microscópicos que formaron estromatolitos y modificaron la atmósfera primitiva de la Tierra al producir oxígeno.

D

dark matter: **(p. 870)** invisible material thought to be made up of a form of subatomic particle that interacts only weakly with other matter.

deep-sea trench: **(p. 451)** elongated, sometimes arc-shaped depression in the seafloor that can extend for thousands of kilometers; is the deepest part of the ocean basin, and is found primarily in the Pacific Ocean.

deflation: **(p. 202)** lowering of land surface caused by wind erosion of loose surface particles, often leaving coarse sediments behind.

deforestation: **(p. 739)** removal of trees from a forested area without adequate replanting, often using clear-cutting, which can result in loss of topsoil and water pollution.

delta: **(p. 236)** triangular deposit, usually made up of silt and clay particles, that forms where a stream enters a large body of water.

dendrochronology: **(p. 604)** science of using tree rings to determine absolute age; helps to date relatively recent geologic events and environmental changes.

materia oscura: **(pág. 870)** sustancia invisible formada por algún tipo de partícula subatómica que interactúa débilmente con otros tipos de material.

fosa abisal: **(pág. 451)** depresión alargada y en algunas ocasiones con forma de arco, que se puede extender miles de kilómetros; es la parte más profunda de la cuenca oceánica y se halla principalmente en el océano Pacífico.

deflación: **(pág. 202)** depresión de la superficie terrestre causada por la erosión eólica de partículas superficiales sueltas; a menudo sólo contiene sedimentos gruesos.

deforestación: **(pág. 739)** eliminación de árboles de un área forestal, sin realizar una adecuada reforestación; a menudo es resultado de una corta a hecho, lo que puede ocasionar la pérdida del mantillo y la contaminación de las aguas.

delta: **(pág. 236)** depósito triangular compuesto generalmente por partículas de limo y arcilla, que se forma en el sitio donde una corriente de agua entra a una gran masa de agua.

dendrocronología: **(pág. 604)** ciencia que usa los anillos de crecimiento anual de los árboles para determinar la edad absoluta; permite datar eventos geológicos y cambios ambientales relativamente recientes.

density current: **(p. 427)** movement of ocean water that occurs in depths too great to be affected by surface winds and is generated by differences in water temperature and salinity.

corriente de densidad: **(pág. 427)** movimiento de las aguas oceánicas que ocurre a grandes profundidades, no se ve afectado por los vientos superficiales y es generado por las diferencias en temperatura y salinidad del agua.

density-dependent factor: **(p. 736)** environmental factor, such as disease, predators, or lack of food, that increasingly affects a population as the population's size increases.

factor dependiente de la densidad: **(pág. 736)** factor ambiental como las enfermedades, los depredadores o la falta de alimento, que afecta con creciente intensidad a una población a medida que aumenta el tamaño de su población.

density-independent factor: **(p. 736)** environmental factor that does not depend on population size, such as storms, flood, fires, or pollution.

factor independiente de la densidad: **(pág. 736)** factor ambiental, como las tempestades, las inundaciones, los incendios o la contaminación, que no son afectados por el tamaño de la población.

dependent variable: **(p. 12)** factor in an experiment that can change if the independent variable is changed.

variable dependiente: **(pág. 12)** factor de un experimento que puede cambiar al variar la variable independiente.

deposition: **(p. 171)** occurs when eroded materials are dropped in another location.

depositación: **(pág. 171)** ocurre cuando los materiales erosionados son depositados en otro sitio.

desalination: **(p. 697)** process that removes salt from seawater in order to provide freshwater.

desalinización: **(pág. 697)** proceso de eliminación de la sal del agua marina para obtener agua dulce.

desertification: **(p. 683)** process by which productive land becomes desert; in arid areas can occur through the loss of topsoil.

desertificación: **(pág. 683)** proceso mediante el cual las tierras productivas se convierten en desierto; en áreas áridas puede ocurrir debido a la pérdida del mantillo del suelo.

dew point: **(p. 295)** temperature to which air is cooled at a constant pressure to reach saturation, at which point condensation can occur.

punto de rocío: **(pág. 295)** temperatura a la cual el aire que se enfría a una presión constante alcanza la saturación, punto en el cual ocurre la condensación.

differentiation **(dih fuh ren shee AY shun):** **(p. 623)** process in which a planet becomes internally zoned, with the heavy materials sinking toward the center and the lighter materials accumulating near its surface.

diferenciación: **(pág. 623)** proceso en que un planeta se divide internamente en zonas, los materiales pesados se hunden hacia el centro, mientras que los materiales más ligeros se acumulan cerca de su superficie.

digital forecast: **(p. 331)** weather forecast that uses numerical data to predict how atmospheric variables change over time.

pronóstico digital: **(pág. 331)** pronóstico del tiempo que se basa en datos numéricos para predecir el cambio de las variables atmosféricas con el tiempo.

dike: **(p. 516)** pluton that cuts across preexisting rocks and often forms when magma invades cracks in surrounding rock bodies.

dique: **(pág. 516)** plutón que atraviesa las rocas preexistentes; suele formarse cuando el magma invade las grietas de los cuerpos rocosos circundantes.

discharge: **(p. 229)** measure of a volume of stream water that flows over a specific location in a particular amount of time.

descarga: **(pág. 229)** medida del volumen de agua corriente que fluye sobre una ubicación dada en cierto lapso de tiempo.

divergent boundary: **(p. 481)** place where two of Earth's tectonic plates are moving apart; is associated with volcanism, earthquakes, and high heat flow, and is found primarily on the seafloor.

límite divergente: **(pág. 481)** lugar donde dos placas tectónicas terrestres se alejan entre sí; se asocia con actividad volcánica, terremotos, un alto flujo de calor y se hallan principalmente en el fondo marino.

divide: **(p. 227)** elevated land that divides one watershed from another.

divisoria: **(pág. 227)** terreno elevado que separa una cuenca hidrográfica de otra.

Doppler effect: (p. 327) change in the wave frequency that occurs due to the relative motion of the wave as it moves toward or away from an observer.

efecto Doppler: (pág. 327) cambio en la frecuencia de onda que ocurre debido al movimiento relativo de la onda a medida que se acerca o se aleja de un observador.

downburst: (p. 351) violent downdrafts that are concentrated in a local area.

reventón: (pág. 351) violentos chorros de viento descendientes que se concentran en un área local.

drawdown: (p. 263) difference between the water level in a pumped well and the original water-table level.

tasa de agotamiento: (pág. 263) diferencia entre el nivel de agua en un pozo artesanal en uso y el nivel original del manto freático.

drought: (p. 362) extended period of well-below-average rainfall, usually caused by shifts in global wind patterns, allowing high-pressure systems to remain for weeks or months over continental areas.

sequía: (pág. 362) período prolongado con precipitación muy por debajo del promedio, generalmente es causado por cambios en los patrones globales de vientos, lo que permite que los sistemas de alta presión permanezcan sobre áreas continentales durante semanas o meses.

drumlin: (p. 210) elongated landform that results when a glacier moves over an older moraine.

drumlin: (pág. 210) formación alargada de tierra que se forma cuando un glaciar se mueve sobre una morrena más antigua.

dune: (p. 204) pile of windblown sand that develops over time, whose shape depends on sand availability, wind velocity and direction, and amount of vegetation present.

duna: (pág. 204) pila de arena formada a lo largo del tiempo por el arrastre de partículas por el viento, cuya forma depende de la disponibilidad de arena, la velocidad y dirección del viento y la cantidad de vegetación presente.

dwarf planet: (p. 816) an object that, due to its own gravity, is spherical in shape, orbits the Sun, is not a satellite, and has not cleared the area of its orbit of smaller debris.

planeta menor: (pág. 816) cuerpo que debido a su propia gravedad tiene forma esférica, tiene una órbita alrededor del Sol, no es un satélite y no ha eliminado restos más pequeños del área de su órbita.

E

eccentricity: (p. 801) ratio of the distance between the foci to the length of the major axis; defines the shape of a planet's elliptical orbit.

excentricidad: (pág. 801) razón de la distancia entre los focos y la longitud del eje mayor; define la forma de la órbita elíptica de un planeta.

ecliptic plane: (p. 776) plane of Earth's orbit around the Sun.

plano de la eclíptica: (pág. 776) plano de la órbita de la Tierra alrededor del Sol.

Ediacaran biota (ee dee A kuh ruhn • by OH tuh): (p. 636) fossils of various multicellular organisms from about 635 mya.

biota Ediacarana: (pág. 636) fósiles de diversos organismos multicelulares de hace cerca de 635 millones de años.

ejecta: (p. 771) material that falls back to the lunar surface after being blasted out by the impact of a space object.

eyecta: (pág. 771) material que cae de regreso a la superficie lunar luego de ser expulsado por el impacto de un cuerpo espacial.

elastic deformation: (p. 529) causes materials to bend and stretch; proportional to stress, so if the stress is reduced or returns to zero the strain or deformation is reduced or disappears.

deformación elástica: (pág. 529) ocasiona que los materiales se doblen y se estiren; es proporcional al grado de tensión, por lo que si la tensión se reduce o desaparece, la deformación también se reduce o desaparece.

El Niño: (p. 388) a band of anomalously warm ocean temperatures that occasionally develops off the western coast of South America and can cause short-term climatic changes felt worldwide.

El Niño: (pág. 388) una banda de agua oceánica que tiene temperaturas anómalamente cálidas que en ocasiones se desarrolla frente a la costa occidental de Sudamérica; puede causar cambios climáticos a corto plazo que afectan a todo el mundo.

electromagnetic spectrum: **(p. 764)** all types of electromagnetic radiation arranged according to wavelength and frequency.

electron: **(p. 61)** tiny atomic particle with little mass and a negative electric charge; an atom's electrons are equal in number to its protons and are located in a cloudlike region surrounding the nucleus.

element: **(p. 60)** natural or artificial substance that cannot be broken down into simpler substances by physical or chemical means.

ellipse **(p. 800)** an oval that is centered on two points called foci; the shape of planets' orbits.

energy efficiency: **(p. 721)** a type of conservation in which the amount of work produced is compared to the amount of energy used.

Enhanced Fujita Tornado Damage scale: **(p. 353)** classifies tornadoes according to their destruction and estimated wind speed on a scale ranging from EF0 to EF5.

environmental science: **(p. 7)** study of the interactions of humans with environment.

eon: **(p. 592)** longest time unit in the geologic time scale.

epicenter **(EH pih sen tur):** **(p. 533)** point on Earth's surface directly above the focus of an earthquake.

epoch: **(p. 593)** time unit in the geological time scale, smaller than a period, measured in hundreds of thousands to millions of years.

equator: **(p. 30)** imaginary line that lies at 0° latitude and circles Earth midway between the North and South poles, dividing Earth into the northern hemisphere and the southern hemisphere.

equinox: **(p. 777)** time of year during which Earth's axis is at a 90° angle to the Sun; both hemispheres receive exactly 12 hours of sunlight and the Sun is directly overhead at the equator.

era: **(p. 593)** second-longest time unit in the geologic time scale, measured in tens to hundreds of millions of years, and defined by differences in lifeforms that are preserved in rocks.

erosion: **(p. 171)** removal and transport of weathered materials from one location to another by agents such as water, wind, glaciers, and gravity.

esker: **(p. 210)** long, winding ridge of layered sediments deposited by streams that flow beneath a melting glacier.

espectro electromagnético: **(pág. 764)** clasificación de todos los tipos de radiación electromagnética de acuerdo con su frecuencia y longitud de onda.

electrón: **(pág. 61)** partícula atómica diminuta con masa pequeña y carga eléctrica negativa; los electrones están ubicados en una región con forma de nube que rodea al núcleo del átomo y su número es igual al número de protones del átomo.

elemento: **(pág. 60)** sustancia natural o artificial que no puede separarse en sustancias más simples por medios físicos o químicos.

elipse: **(pág. 800)** óvalo centrado en dos puntos llamados focos; la forma de las órbitas de los planetas.

eficiencia energética: **(pág. 721)** tipo de conservación en el cual la cantidad de trabajo producido se compara con la cantidad de energía utilizada.

escala mejorada de Fujita para daños de tornados: **(pág. 353)** clasifica los tornados según el daño que causan y la velocidad de sus vientos aproximado en una escala que va de EF0 a EF5.

ciencias ambientales: **(pág. 7)** estudio de las interacciones del hombre con su entorno.

eon: **(pág. 592)** unidad más larga de tiempo en la escala de tiempo geológico.

epicentro: **(pág. 533)** punto en la superficie terrestre ubicado directamente encima del foco de un sismo.

época: **(pág. 593)** unidad de tiempo en la escala de tiempo geológico, es más pequeña que un período y se mide en millones a centenares de millares de años.

ecuador: **(pág. 30)** línea imaginaria que yace en la latitud 0° y que circunda la Tierra entre los polos Norte y Sur, dividiendo a la Tierra en dos hemisferios iguales: norte y sur.

equinoccio: **(pág. 777)** epoca del año durante la cual el eje de la Tierra forma un ángulo de 90° con el Sol, ambos hemisferios reciben exactamente 12 horas de luz solar y el Sol se halla exactamente sobre el ecuador.

era: **(pág. 593)** segunda unidad más grande de tiempo en la escala del tiempo geológico; se mide en decenas a centenas de millones de años y se define según las diferencias en las formas de vida preservadas en las rocas.

erosión: **(pág. 171)** eliminación y transporte de materiales meteorizados de un lugar a otro por agentes como el agua, el viento, los glaciares y la gravedad.

ésker: **(pág. 210)** formación larga y sinuosa de sedimentos estratificados, depositados por corrientes que fluyen debajo de un glaciar que se derrite.

estuary: **(p. 414)** coastal area of lowest salinity often occurs where the lower end of a freshwater river or stream enters the ocean.

estuario: **(pág. 414)** área costera de agua salobre que se forma en el sitio donde la desembocadura de un río o corriente de agua dulce entra al océano; provee una fuente excelente de alimento y refugio para organismos marinos comercialmente importantes.

eukaryote **(yew KE ree oht):** **(p. 635)** organism composed of one or more cells each of which usually contains a nucleus; larger and more complex than a prokaryote.

eucariota: **(pág. 635)** organismo compuesto por unas o más células nucleadas; generalmente es más grande y más complejo que un procariota.

eutrophication: **(p. 239)** process by which lakes become rich in nutrients from the surrounding watershed, resulting in a change in the kinds of organisms in the lake.

eutroficación: **(pág. 239)** proceso de aumento de la cantidad de nutrientes que contiene un lago, alimentado por los nutrientes provenientes de las cuenca circundante, lo que causa un cambio en los tipos de organismos que habitan el lago.

evaporation: **(p. 74)** vaporization—change of state from a liquid to a gas, involving thermal energy.

evaporación: **(pág. 74)** vaporización: cambio de estado de un líquido a gas que implica energía térmica.

evaporite: **(p. 143)** the layers of chemical sedimentary rocks that form when concentrations of dissolved minerals in a body of water reach saturation due to the evaporation of water; crystal grains precipitate out of solution and settle to the bottom.

evaporita: **(pág. 143)** capas de roca química sedimentaria que se forman cuando la concentración de minerales disueltos en una masa de agua alcanza el punto de saturación debido a la evaporación del agua; los cristales se precipitan de la solución y se asientan en el fondo.

evolution **(eh vuh LEW shun):** **(p. 606)** the change in species over time.

evolución: **(pág. 606)** cambios de las especies a lo largo del tiempo.

exfoliation: **(p. 165)** mechanical weathering process in which outer rock layers are stripped away, often resulting in dome-shaped formations.

exfoliación: **(pág. 165)** proceso de meteorización mecánica que causa la eliminación de los estratos rocosos exte-riores, a menudo produce formaciones en forma de domo.

exosphere: **(p. 286)** outermost layer of Earth's atmosphere that is located above the thermosphere with no clear boundary at the top; transitional region between Earth's atmosphere and outer space.

exosfera: **(pág. 286)** capa más externa de la atmósfera terrestre, está localizada por encima de la termosfera y no tiene un límite definido en su parte más alejada; región de transición entre la atmósfera de la Tierra y el espacio exterior.

exponential growth: **(p. 735)** pattern of growth in which a population of organisms grows faster as it increases in size, resulting in a population explosion.

crecimiento exponencial: **(pág. 735)** patrón de crecimiento en que una población de organismos crece cada vez más rápido a medida que aumenta de tamaño, causando una explosión demográfica.

extrusive rock: **(p. 118)** fine-grained igneous rock that is formed when molten rock cools quickly and solidifies on Earth's surface.

roca extrusiva: **(pág. 118)** roca ígnea de grano fino que se forma cuando la roca fundida se enfría rápidamente y se solidifica en la superficie terrestre.

eye: **(p. 356)** calm center of a tropical cyclone that develops when the winds around its center reach at least 120 km/h.

ojo: **(pág. 356)** centro de calma de un ciclón tropical que se desarrolla cuando los vientos a su alrededor alcanzan por lo menos 120 km/h.

eyewall: **(p. 356)** band where the strongest winds in a hurricane are usually concentrated, surrounding the eye.

pared del ojo de huracán: **(pág. 356)** banda que rodea el ojo de un huracán donde generalmente se concentran los vientos más fuertes.

F

fault: **(p. 530)** fracture or system of fractures in Earth's crust that occurs when stress is applied too quickly or stress is too great; can form as a result of horizontal compression (reverse fault), horizontal shear (strike-slip fault), or horizontal tension (normal fault).

falla: **(pág. 530)** fractura o sistema de fracturas en la corteza terrestre que ocurren en sitios donde se aplica tensión rápidamente o donde la tensión es demasiado grande; se puede formar como resultado de una compresión horizontal (falla invertida, un cizallamiento horizontal (falla de transformación) o una tensión horizontal (falla normal).

fault-block mountain: **(p. 574)** mountain that forms when large pieces of crust are tilted, uplifted, or dropped downward between large normal faults.

montañas de bloque de falla: **(pág. 574)** montañas que se forman cuando trozos grandes de corteza se inclinan, se elevan o se hunden entre fallas normales grandes.

fission: **(p. 834)** process in which heavy atomic nuclei split into smaller, lighter atomic nuclei.

fisión: **(pág. 834)** proceso mediante el cual los núcleos atómicos pesados se dividen en núcleos más livianos y pequeños.

fissure: **(p. 504)** long crack in Earth's crust.

fisura: **(pág. 504)** grandes grietas en la Tierra.

flood: **(p. 230)** potentially devastating natural occurrence in which water spills over the sides of a stream's banks onto adjacent land areas.

inundación: **(pág. 230)** acontecimiento natural potencialmente devastador en que el agua se desborda de las riberas de una corriente y cubre los terrenos adyacentes.

flood basalt: **(p. 504)** huge amounts of lava that erupt from fissures.

basalto de meseta: **(pág. 504)** grandes cantidades de lava que salen por las fisuras.

floodplain: **(p. 230)** broad, flat, fertile area extending out from a stream's bank that is covered with water during floods.

llanura aluvial: **(pág. 230)** área fértil, plana y ancha que se extiende desde las riberas de una corriente y queda cubierta por agua durante las inundaciones.

focus: **(p. 533)** point of the initial fault rupture where an earthquake originates that usually lies at least several kilometers beneath Earth's surface.

foco: **(pág. 533)** punto inicial de ruptura de la falla donde se origina un terremoto; generalmente se halla varios kilómetros debajo de la superficie terrestre.

foliated: **(p. 146)** describes metamorphic rock, such as schist or gneiss, whose minerals are squeezed under high pressure and arranged in wavy layers and bands.

foliada: **(pág. 146)** describe roca metamórfica, como el esquisto o el gneis, cuyos minerales son comprimidos bajo presiones altas, formando ordenadas capas y bandas onduladas.

fossil fuel: **(p. 710)** nonrenewable energy resource formed over geologic time from the compression and partial decomposition of organisms that lived millions of years ago.

combustible fósil: **(pág. 710)** recurso energético no renovable que se forma a lo largo del tiempo geológico, a partir de la compresión y descomposición parcial de organismos que vivieron hace millones de años.

fractional crystallization: **(p. 115)** process in which different minerals crystallize from magma at different temperatures, removing elements from magma.

cristalización fraccionaria: **(pág. 115)** proceso en el cual diferentes minerales se cristalizan a diferentes temperaturas a partir del magma, eliminando elementos del magma.

fracture: **(p. 93)** when a mineral breaks into pieces with arclike, rough, or jagged edges.

fractura: **(pág. 93)** sucede cuando un mineral se rompe en pedazos con bordes ásperos, arqueados o serrados.

front: **(p. 322)** boundary between two air masses of differing densities; can be cold, warm, stationary, or occluded and can stretch over large areas of Earth's surface.

frente: **(pág. 322)** límite entre dos masas de aire con diferentes densidades; puede ser frío, cálido, estacionario u ocluido y puede extenderse sobre grandes áreas de la superficie de la Tierra.

Glossary • Glosario

frontal thunderstorm: **(p. 346)** type of thunderstorm usually produced by an advancing cold front, which can result in a line of thunderstorms hundreds of kilometers long, or, more rarely, an advancing warm front, which can result in a relatively mild thunderstorm.

tormenta frontal: **(pág. 346)** tipo de tormenta que es producida generalmente por el avance de un frente frío, pudiendo producir una línea de tormentas de cientos de kilómetros de largo, o en menor frecuencia por el avance de un frente cálido, produciendo tormentas relativamente ligeras.

frost wedging: **(p. 164)** mechanical weathering process that occurs when water repeatedly freezes and thaws in the cracks of rocks, often resulting in rocks splitting.

erosión periglaciar: **(pág. 164)** proceso mecánico de meteorización que ocurre cuando el agua se congela y se descongela, en repetidas ocasiones, en las grietas de las rocas, ocasionando el rompimiento de las mismas.

fuel: **(p. 709)** material, such as wood, peat, or coal, burned to produce energy.

combustible: **(pág. 709)** materiales como la leña, la turba o el carbón, que se queman para producir energía.

fusion: **(p. 834)** The combining of lightweight nuclei into heavier nuclei; occurs in the core of the Sun where temperatures and pressure are extremely high.

fusión: **(pág. 834)** combinación de núcleos livianos para formar núcleos más pesados: sucede en el núcleo del Sol donde las temperaturas y la presión son extremadamente altas.

G

gas giant planet: **(p. 811)** large, gaseous planet that is very cold at its surface; has ring systems, many moons, and lacks solid surfaces—Jupiter, Saturn, Uranus, and Neptune.

gigantes gaseosos: **(pág. 811)** planetas grandes y gaseosos con superficies muy frías; tienen sistemas de anillos, muchas lunas y carecen de superficie sólida: Júpiter, Saturno, Urano y Neptuno.

gem: **(p. 101)** rare, precious, highly prized mineral that can be cut, polished, and used for jewelry.

gema: **(pág. 101)** mineral sumamente valioso, precioso y escaso que se puede cortar, pulir y utilizar en joyería.

Geographic Information System (GIS): **(p. 44)** a mapping system that uses worldwide databases from remote sensing to create layers of information that can be superimposed upon each other to form a comprehensive map.

Sistema de Información Geográfica (SIG): **(pág. 44)** sistema para la elaboración de mapas que usa bases de datos mundiales obtenidos por sensores remotos, para crear capas de información que se pueden superponer para elaborar mapas que combinen dicha información.

geologic map: **(p. 38)** a map that shows the distribution, arrangement, and types of rocks below the soil, and other geologic features.

mapa geológico: **(pág. 38)** mapa que muestra la distribución, el orden y los tipos de roca del subsuelo, así como otras características geológicas.

geologic time scale: **(p. 590)** record of Earth's history from its origin 4.6 bya to the present.

escala del tiempo geológico: **(pág. 590)** registro de la historia de la Tierra desde su origen, hace 4.6 billones de años, hasta el presente.

geology: **(p. 7)** study of materials that make up Earth and the processes that form and change these materials, and the history of the planet and its life-forms since its origin.

geología: **(pág. 7)** estudio de los materiales que conforman la Tierra y de los procesos de formación y cambio de estos materiales, así como la historia del planeta y sus formas de vida desde su origen.

geosphere: **(p. 8)** the part of Earth from its surface to its center.

geosfera: **(pág. 8)** región que abarca desde la superficie hasta el centro de la Tierra.

geothermal energy: **(p. 717)** energy produced by Earth's naturally occurring heat, steam, and hot water.

energía geotérmica: **(pág. 717)** energía producida naturalmente en la Tierra por el calor, el vapor y el agua caliente.

geyser: **(p. 258)** explosive hot spring that erupts regularly.

géiser: **(pág. 258)** manantial termal explosivo que hace erupción regularmente.

glacier: **(p. 207)** large, moving mass of ice that forms near Earth's poles and in mountainous regions at high elevations.

glass: **(p. 73)** solid that consists of densely packed atoms with a random arrangement and lacks crystals or has crystals that are not visible.

Global Positioning System (GPS): **(p. 44)** satellite-based navigation system that permits a user to pinpoint his or her exact location on Earth.

global warming: **(p. 393)** rise in global temperatures, which might be due to increases in atmospheric CO_2 from deforestation and burning of fossil fuels

gnomonic **(noh MAHN ihk)** **projection:** **(p. 35)** map useful in plotting long-distance trips by boat or plane, made by projecting points and lines from a globe onto a piece of paper that touches the globe at a single point.

graded bedding: **(p. 138)** type of bedding in which particle sizes become progressively heavier and coarser toward the bottom layers.

granitic rock: **(p. 118)** light-colored, intrusive igneous rock that has high silica content.

greenhouse effect: **(p. 393)** natural heating of Earth's surface by certain atmospheric gases, which helps keep Earth warm enough to sustain life.

gully erosion: **(p. 172)** erosion that occurs when a rill channel widens and deepens.

guyot: **(p. 452)** large, extinct, basaltic volcano with a flat, submerged top.

glaciar: **(pág. 207)** enormes masas móviles de hielo que se forman cerca de los polos de la Tierra o en grandes elevaciones en regiones montañosas.

vidrio: **(pág. 73)** sólido formado por átomos densamente comprimidos en un ordenamiento aleatorio; carece de cristales o sus cristales no son visibles.

Sistema de posicionamiento global (SPG): **(pág. 44)** sistema de navegación por satélite que permite al usuario localizar su ubicación exacta sobre la Tierra.

calentamiento global: **(pág. 393)** aumento en las temperaturas globales, que es probablemente producto del aumento en el CO_2 atmosférico, causado por la deforestación y la quema de combustibles fósiles

proyección gnomónica: **(pág. 35)** mapa útil para trazar viajes de distancias largas por barco o por avión; se elabora proyectando los puntos y las líneas de un globo sobre una hoja de papel que toca el globo en un solo punto.

estratificación graduada: **(pág. 138)** característica de la depositación de rocas sedimentarias en la cual las partículas son progresivamente más pesadas y gruesas hacia las capas inferiores de la estratificación.

roca granítica: **(pág. 118)** roca intrusiva ignea de color claro que tiene un alto contenido de sílice.

efecto invernadero: **(pág. 393)** calentamiento natural de la superficie terrestre por ciertos gases atmosféricos; ayuda a mantener en la Tierra una temperatura lo suficientemente cálida para mantener la vida.

erosión en barrancos: **(pág. 172)** erosión que ocurre cuando el cauce de un arroyuelo se ensancha y profundiza.

guyot: **(pág. 452)** grande volcán basáltico extinto con la cima que es plana y está sumergida.

H

half-life: **(p. 602)** period of time it takes for a radioactive isotope, such as carbon-14, to decay to one-half of its original amount.

halo: **(p. 863)** spherical region where globular clusters are located; surrounds the Milky Way's nuclear bulge and disk.

hardness: **(p. 91)** measure of how easily a mineral can be scratched, which is determined by the arrangement of a mineral's atoms.

heat island: **(p. 385)** urban area where climate is warmer than in the surrounding countryside due to factors such as numerous concrete buildings and large expanses of asphalt.

vida media: **(pág. 602)** período de tiempo que demora un isótopo radiactivo, como el carbono 14, en desintegrarse a la mitad de su cantidad radiactiva original.

halo: **(pág. 863)** región esférica donde se ubican los cúmulos globulares; rodea el disco y el núcleo central de la Vía Láctea.

dureza: **(pág. 91)** medida de la facilidad con la que un mineral es rayado; está determinada por el ordenamiento de los átomos del mineral.

isla de calor: **(pág. 385)** área urbana donde el clima es más caliente que en el área rural circundante, debido a factores como los numerosos edificios de concreto y las grandes extensiones de asfalto.

heat wave: **(p. 362)** extended period of above-average temperatures caused by large, high-pressure systems that warm by compression and block cooler air masses.

Hertzsprung-Russell diagram (H-R diagram): **(p. 845)** graph that relates stellar characteristics—class, mass, temperature, magnitude, diameter, and luminosity.

highland: **(p. 771)** light-colored, mountainous, heavily cratered area of the Moon, composed mostly of lunar breccias.

Homo sapiens: **(p. 665)** species to which modern humans belong.

hot spot: **(p. 502)** unusually hot area in Earth's mantle where high-temperature plumes of mantle material rise toward the surface.

hot spring: **(p. 258)** thermal spring with temperatures higher than that of the human body.

Hubble constant: **(p. 874)** value (*H*) used to calculate the rate at which the universe is expanding; measured in kilometers per second per megaparsec.

humidity: **(p. 294)** amount of water vapor in the atmosphere at a given location on Earth's surface.

hydrocarbon: **(p. 709)** molecules with hydrogen and carbon bonds only; the result of the combination of carbon dioxide and water during photosynthesis.

hydroelectric power: **(p. 716)** power generated by converting the energy of free-falling water to electricity.

hydrogen bond: **(p. 693)** forms when the positive ends of some water molecules are attracted to the negative ends of other water molecules; cause water's surface to contract and allow water to adhere to and coat a solid.

hydrosphere: **(p. 8)** all the water in Earth's oceans, lakes, seas, rivers, and glaciers plus all the water in the atmosphere.

hydrothermal metamorphism: **(p. 149)** occurs when very hot water reacts with rock, altering its mineralogy and chemistry.

hygrometer **(hi GRAH muh tur):** **(p. 325)** weather instrument used to measure humidity.

hypothesis: **(p. 10)** a testable explanation of a situation.

ola de calor: **(pág. 362)** período extenso de temperaturas más altas que el promedio; es causado por grandes sistemas de alta presión que se calientan por compresión y bloquean las masas de aire más frías.

diagrama de Hertzsprung-Russell (diagrama H-R): **(pág. 845)** gráfica que relaciona características estelares: incluyendo la clase, la masa, la temperatura, la magnitud, el diámetro y la luminosidad.

tierras altas: **(pág. 771)** áreas de la Luna de color claro, con muchos cráteres y montañas, compuestas en su mayor parte de brechas lunares.

Homo sapiens: **(pág. 665)** especie a la cual pertenecen los seres humanos modernos.

punto caliente: **(pág. 502)** área muy caliente del manto de la Tierra donde plumas de material del manto a gran temperatura ascienden a la superficie.

fuente caliente: **(pág. 258)** manantial termal con temperaturas más altas que las del cuerpo humano.

constante de Hubble: **(pág. 874)** valor (*H*) que sirve para calcular la velocidad de expansión del universo; se mide en kilómetros por segundo por megaparsec.

humedad: **(pág. 294)** cantidad de vapor de agua en el aire en un sitio determinado de la Tierra.

hidrocarburo: **(pág. 709)** molécula que sólo contiene enlaces entre átomos de hidrógeno y de carbono; es producto de la unión del dióxido de carbono y el agua durante la fotosíntesis.

energía hidroeléctrica: **(pág. 716)** se genera al convertir la energía de una caída de agua en electricidad.

enlace de hidrógeno: **(pág. 693)** se forma cuando el extremo positivo de algunas moléculas de agua son atraídas por el extremo negativo de otras moléculas de agua; ocasiona que la superficie del agua se contraiga y permite al agua adherirse y recubrir un sólido.

hidrosfera: **(pág. 8)** toda el agua en los océanos, los lagos, los mares, los ríos y los glaciares de la Tierra, además de toda el agua en la atmósfera.

metamorfismo hidrotérmico: **(pág. 149)** ocurre cuando agua muy caliente reacciona con la roca, alterando su mineralogía y su química.

higrómetro: **(pág. 325)** instrumento meteorológico que se usa para medir la humedad.

hipótesis: **(pág. 10)** explicación de una situación que se puede poner a prueba.

I

ice age: **(p. 387)** period of extensive glacial coverage, producing long-term climatic changes, where average global temperatures decreased by 5°C.

igneous rock: **(p. 112)** intrusive or extrusive rock formed from the cooling and crystallization of magma or lava.

impact crater: **(p. 771)** crater formed when space material crashes into the surface of a celestial body.

inclusion: **(p. 597)** the principle that fragments, called inclusions, in a rock layer must be older than the rock layer that contains them.

independent variable: **(p. 12)** factor that is manipulated by the experimenter in an experiment.

index fossils: **(p. 609)** remains of plants or animals that were abundant, widely distributed, and existed briefly that can be used by geologists to correlate or date rock layers.

infiltration: **(p. 253)** process by which precipitation that has fallen on land surfaces enters the ground and becomes groundwater.

interferometry: **(p. 767)** process that links separate telescopes so they act as one telescope, producing more detailed images as the distance between them increases.

International Date Line: **(p. 33)** the 180° meridian, which serves as the transition line for calendar days.

intrusive rock: **(p. 118)** coarse-grained igneous rock that is formed when molten rock cools slowly and solidifies inside Earth's crust.

ion: **(p. 64)** an atom that gains or loses an electron.

ionic bond: **(p. 68)** attractive force between two ions with opposite charge.

iridium **(ih RID ee um):** **(p. 659)** metal that is rare in rocks at Earth's surface but is relatively common in asteroids.

isobar: **(p. 329)** line on a weather map connecting areas of equal pressure

isochron **(I suh krahn):** **(p. 477)** imaginary line on a map that shows points of the same age; formed at the same time.

isostasy **(I SAHS tuh see):** **(p. 563)** condition of equilibrium that describes the displacement of Earth's mantle by Earth's continental and oceanic crust.

glaciación: **(pág. 387)** período de formación de una amplia cobertura glacial que produce cambios climáticos de largo plazo en que las temperaturas globales promedio desminuyen 5°C.

roca ígnea: **(pág. 112)** roca intrusiva o extrusiva formada a partir del enfriamiento y cristalización del magma o lava.

cráter de impacto: **(pág. 771)** cráter que se forma cuando material proveniente del espacio impacta la superficie de un objeto celeste.

inclusión: **(pág. 597)** principio que establece que los fragmentos, llamados inclusiones, contenidos por un estrato rocoso deben ser más antiguos que la roca que los contiene.

variable independiente: **(pág. 12)** factor que es manipulado por el investigador en un experimento.

fósiles guía: **(pág. 609)** restos de plantas o animales que fueron abundantes, tuvieron una amplia distribución y existieron poco tiempo, que sirven a los geólogos para correlacionar o para datar estratos rocosos.

infiltración: **(pág. 253)** proceso mediante el cual la precipi-tación que cae sobre la superficie terrestre entra al suelo y se convierte en agua subterránea.

interferometría: **(pág. 767)** proceso que combina telescopios separados para que funcionen como un solo telescopio, produciendo imágenes más detalladas al aumentar la distancia entre ellos.

línea internacional de cambio de fecha: **(pág. 33)** el meridiano 180°; sirve como la línea de transición para los días del calendario.

roca intrusiva: **(pág. 118)** roca ígnea de grano grueso que se forma cuando la roca fundida se enfría lentamente y se solidifica en el interior de la corteza terrestre.

ion: **(pág. 64)** átomo que gana o pierde un electrón.

enlace iónico: **(pág. 68)** fuerza de atracción entre dos iones con cargas opuestas.

iridio: **(pág. 659)** metal escaso en las rocas de la superficie terrestre, pero relativamente común en los meteoritos y los asteroides.

isobara: **(pág. 329)** línea de un mapa meteorológico que conecta áreas con igual presión.

isocrona: **(pág. 477)** línea imaginaria en un mapa que conecta puntos con la misma antigüedad; que se formaron al mismo tiempo.

isostasia: **(pág. 563)** condición de equilibrio que describe el desplazamiento del manto terrestre por las cortezas continental y oceánica de la Tierra.

isostatic rebound: (p. 565) slow process of Earth's crust rising as the result of the removal of overlaying material.

isotherm: (p. 329) line on a weather map connecting areas of equal temperature.

isotope: (p. 62) an atom of an element that has a different mass number than the element but the same chemical properties.

rebote isostático: (pág. 565) proceso lento de elevación de la corteza terrestre producto de la eliminación del material sobreyacente.

isoterma: (pág. 329) línea en un mapa meteorológico que conecta áreas con la misma temperatura.

isótopo: (pág. 62) átomo de un elemento que tiene un distinto número de masa que el elemento, pero las mismas propiedades químicas.

J

jet stream: (p. 321) narrow wind band that occurs above large temperature contrasts and can flow as fast as 185 km/h.

corriente de chorro: (pág. 321) banda de vientos estrecha situada por encima de áreas con grandes contrastes de temperatura y que puede alcanzar una rapidez de 185 km/h.

K

kame: (p. 210) a conical mound of layered sediment that accumulates in a depression on a retreating glacier.

karst topography: (p. 261) irregular topography with sinkholes, sinks, and sinking streams caused by groundwater dissolution of limestone.

kettle: (p. 212) a lake formed when runoff and precipitation filled a kettle hole, which is a depression that formed when an ice block from a continental glacier became covered with sediment and melted.

key bed: (p. 599) a rock or sediment layer that serves as a time marker in the rock record and results from volcanic ash or meteorite-impact debris that spread out and covered large areas of Earth.

kimberlite: (p. 123) rare, ultramafic rock that can contain diamonds and other minerals formed only under very high pressures.

Köppen classification system: (p. 383) classification system for climates, divided into five types, based on the mean monthly values of temperature and precipitation and types of vegetation.

Kuiper (KI pur) belt: (p. 818) region of the space that lies outside the orbit of Neptune, 30 to 50 AU from the Sun, where small solar system bodies that are mostly rock and ice probably formed.

kame: (pág. 210) montículo cónico de sedimento estratificado que es depositado por corrientes que fluyen bajo un glaciar que se derrite.

topografía cárstica: (pág. 261) topografía irregular con sumideros, hundimientos y corrientes que desaparecen, causada por la disolución de la piedra caliza por el agua subterránea.

marmita: (pág. 212) lago que se forma cuando la escorrentía y la precipitación llenan el hueco de una marmita, que es la depresión que se forma cuando un bloque de hielo de un glaciar continental queda cubierto con sedimento y se derrite.

estrato guía: (pág. 599) capa de sedimento que sirve como marcador de tiempo del registro geológico; está formado por cenizas volcánicas o por los restos del impacto de un meteorito que se esparcen y cubren grandes áreas de la Tierra.

kimberlita: (pág. 123) roca ultramáfica poco común que puede contener diamantes y otros minerales que sólo se forman bajo presiones muy altas.

sistema de clasificación de Köppen: (pág. 383) sistema de clasificación de los climas; los clasifica en cinco tipos básicos en base a los valores mensuales promedio de temperatura y precipitación y a los tipos de vegetación.

cinturón de Kuiper: (pág. 818) pequeños cuerpos del sistema solar formados principalmente por roca y hielo, yacen más allá de la órbita de Neptuno, entre 30 a 50 UA del Sol, y es muy probable que se hayan formado en esta región.

L

laccolith (LA kuh lihth): (p. 515) relatively small, mushroom-shaped pluton that forms when magma intrudes into parallel rock layers close to Earth's surface.

lake: (p. 238) natural or human-made body of water that can form when a depression on land fills with water.

Landsat satellite: (p. 41) information-gathering satellite that uses visible light and infrared radiation to map Earth's surface.

landslide: (p. 197) rapid downslope movement of a mass of loose soil, rock, or debris that has separated from the bedrock; can be triggered by an earthquake.

latent heat: (p. 295) stored energy in water vapor that is not released to warm the atmosphere until condensation takes place.

latitude: (p. 30) distance in degrees north and south of the equator.

Laurentia (law REN shuh): (p. 627) ancient continent formed during the Proterozoic that is the core of modern-day North America.

lava: (p. 112) magma that flows out onto Earth's surface.

Le Système International d'Unités (SI): (p. 13) replacement for the metric system; based on a decimal system using the number 10 as the base unit; includes the meter (m), second (s), and kilogram (kg).

liquid metallic hydrogen: (p. 812) form of hydrogen with both liquid and metallic properties that exists as a layer in the Jovian atmosphere.

lithification: (p. 136) the physical and chemical processes that transform sediments into sedimentary rocks.

loess (LUSS): (p. 206) thick, windblown, fertile deposit of silt that contains high levels of nutrients and minerals.

longitude: (p. 31) distance in degrees east and west of the prime meridian.

longshore bar: (p. 440) submerged sandbar located in the surf zone of most beaches.

longshore current: (p. 441) current that flows parallel to the shore, moves large amounts of sediments, and is formed when incoming breakers spill over a longshore bar.

luminosity: (p. 842) energy output from the surface of a star per second; measured in watts.

lacolito: (pág. 515) plutón relativamente pequeño con forma de champiñón que se forma cuando se introduce el magma entre estratos rocosos paralelos, cerca de la superficie terrestre.

lago: (pág. 238) masa de agua, natural o hecha por el hombre, que se forma cuando una depresión terrestre se llena de agua.

satélite Landsat: (pág. 41) satélite que recoge información, usando luz visible y radiación infrarroja para mapear la superficie terrestre.

derrumbe: (pág. 197) rápido desplazamiento cuesta abajo de una masa de tierra, rocas o escombros sueltos que se han separado del lecho rocoso; puede ser causado por un terremoto.

calor latente: (pág. 295) energía almacenada en el vapor de agua que no es liberada para calentar la atmósfera, hasta que ocurre la condensación.

latitud: (pág. 30) distancia en grados hacia el norte o el sur del ecuador.

Laurencia: (pág. 627) antiguo continente que se formó durante el Proterozoico y que en la actualidad corresponde al centro de Norteamérica.

lava: (pág. 112) magma que fluye por la superficie terrestre.

Le Système Internacional d'Unités/Sistema Internacional de Unidades (SI): (pág. 13) sustituto del sistema métrico; se basa en el sistema decimal por lo que usa el número 10 como unidad base: incluye el metro (m), el segundo (s) y el kilogramo (kg).

hidrógeno metálico líquido: (pág. 812) forma de hidrógeno con propiedades de líquido y de metal que forma una capa en la atmósfera joviana.

litificación: (pág. 136) procesos físicos y químicos que transforman los sedimentos en roca sedimentaria.

loes: (pág. 206) amplio depósito fértil de limo que es arrastrado por el viento y contiene niveles altos de nutrientes y minerales.

longitud: (pág. 31) distancia en grados hacia el este o el oeste del primer meridiano.

barra litoral: (pág. 440) barra de arena sumergida ubicada en la zona de oleaje de la mayoría de las playas.

corriente litoral: (pág. 441) corriente que fluye paralela a la costa, transporta grandes cantidades de sedimentos y se forma cuando las olas rompen a lo largo de una larga barra litoral.

luminosidad: (pág. 842) energía que irradia la superficie de una estrella por segundo; se mide en vatios.

lunar eclipse: **(p. 784)** when Earth passes between the Sun and the Moon, and Earth's shadow falls on the Moon; occurs only during a full moon.

eclipse lunar: **(pág. 784)** sucede cuando la Tierra pasa entre el Sol y la Luna y la sombra de la Tierra cae sobre la Luna; ocurre sólo durante la luna llena.

luster: **(p. 90)** the way that a mineral reflects light from its surface; two types—metallic and nonmetallic.

lustre: **(pág. 90)** manera en que la superficie de un mineral refleja la luz; existen dos tipos: metálico o no metálico.

M

magnetic reversal: **(p. 476)** when Earth's magnetic field changes polarity between normal and reversed.

inversión magnética: **(pág. 476)** sucede cuando el campo magnético de la Tierra cambia polaridad entre normal e invertida.

magnetometer **(mag nuh TAH muh tur):** **(p. 473)** device used to map the ocean floor that detects small changes in magnetic fields.

magnetómetro: **(pág. 473)** aparato que sirve para mapear el fondo marino; detecta cambios pequeños en los campos magnéticos.

magnitude: **(p. 539)** measure of the energy released during an earthquake, which can be described using the Richter scale.

magnitud: **(pág. 539)** medida de la energía liberada durante un sismo; se puede describir usando la escala de Richter.

main sequence: **(p. 845)** in an H-R diagram, the broad, diagonal band that includes about 90 percent of all stars and runs from hot, luminous stars in the upper-left corner to cool, dim stars in the lower-right corner.

secuencia principal: **(pág. 845)** la ancha banda diagonal de un diagrama H-R que contiene cerca del 90 por ciento de todas las estrellas; contiene desde estrellas calientes y luminosas en la esquina superior izquierda, hasta estrellas frías de brillo débil en la esquina inferior derecha.

map legend: **(p. 39)** key that explains what the symbols on a map represent.

leyenda del mapa: **(pág. 39)** clave que explica los símbolos en un mapa.

map scale: **(p. 39)** ratio between the distances shown on a map and the actual distances on Earth's surface.

escala del mapa: **(pág. 39)** razón entre las distancias que se muestran en un mapa y las distancias reales en la superficie terrestre.

maria **(MAH ree uh):** **(p. 771)** dark-colored, smooth plains on the Moon's surface.

mar: **(pág. 771)** planicie lunar lisa y de color oscuro.

mass extinction: **(p. 594)** occurs when an unusually large number of organisms disappear from the rock record at about the same time.

extinción masiva: **(pág. 594)** ocurre cuando un número insólitamente grande de organismos desaparece del registro geológico aproximadamente al mismo tiempo.

mass movement: **(p. 194)** downslope movement of earth materials due to gravity that can occur suddenly or very slowly, depending on the weight of the material, its resistance to sliding, and whether a trigger, such as an earthquake, is involved.

movimiento de masa: **(pág. 194)** movimiento cuesta abajo de materiales terrestres debido a la gravedad; puede ocurrir de manera repentina o muy lentamente: dependiendo del peso del material, la resistencia del material a deslizarse y de si ha ocurrido algún evento que lo desencadene, como un sismo.

mass number: **(p. 62)** combined number of protons and neutrons in the nucleus of an atom.

número de masa: **(pág. 62)** número combinado de protones y neutrones en el núcleo de un átomo.

matter: **(p. 60)** anything that has volume and mass.

materia: **(pág. 60)** todo aquello que tiene volumen y masa.

Maunder minimum: **(p. 390)** period of very low sunspot activity that occurred between 1645 and 1716 and closely corresponded with a cold climatic episode known as the "Little Ice Age."

mínimo de Maunder: **(pág. 390)** período de muy baja actividad de manchas solares, ocurrido entre 1645 y 1716, que se correspondió con un episodio climático frío llamado "La Pequeña Glaciación."

meander: **(p. 234)** curve or bend in a stream formed when a stream's slope decreases, water builds up in the stream channel, and moving water erodes away the sides of the streambed.

mechanical weathering: **(p. 164)** process that breaks down rocks and minerals into smaller pieces but does not involve any change in their composition.

Mercator projection: **(p. 34)** map with parallel lines of latitude and longitude that shows true direction and the correct shapes of landmasses but distorts areas near the poles.

mesosphere: **(p. 284)** layer of Earth's atmosphere above the stratopause.

metallic bond: **(p. 68)** positive ions of metal held together by the negative electrons between them; allows metals to conduct electricity.

meteor: **(p. 818)** streak of light produced when a meteoroid falls toward Earth and burns up in Earth's atmosphere.

meteorite **(MEE tee uh rite): (pp. 818, 621)** a small fragment of an orbiting body that has fallen to Earth, generating heat; does not completely burn up in Earth's atmosphere and strikes Earth's surface, sometimes causing an impact crater.

meteoroid: **(p. 818)** piece of interplanetary material that falls toward Earth and enters its atmosphere.

meteorology: **(p. 6)** the study of the atmosphere, which is the air surrounding Earth.

meteor shower: **(p. 819)** occurs when Earth intersects a cometary orbit and comet particles burn up as they enter Earth's upper atmosphere.

microclimate: **(p. 385)** localized climate that differs from the surrounding regional climate.

microcontinent: **(p. 624)** a small fragment of granite-rich crust formed during the Archean.

mid-ocean ridge: **(p. 451)** chain of underwater mountains that run throughout the ocean basins, have a total length over 65,000 km, and contain active and extinct volcanoes.

mineral: **(p. 86)** naturally occurring, inorganic solid with a specific chemical composition and a definite crystalline structure.

mineral replacement: **(p. 607)** the process where pore spaces of an organism's buried parts are filled in with minerals from groundwater.

meandro: **(pág. 234)** curva o desviación en una corriente; se forma cuando disminuye la pendiente de la corriente, por lo que el agua se acumula en el cauce y el movimiento del agua erosiona los costados del cauce.

meteorización mecánica: **(pág. 164)** proceso de rompimiento de rocas y minerales en trozos más pequeños que no afecta la composición del material.

proyección de Mercator: **(pág. 34)** mapa con líneas de latitud y longitud paralelas que muestra la dirección real y las formas correctas de las masas terrestres, aunque las áreas cercanas a los polos aparecen distorsionadas.

mesosfera: **(pág. 284)** capa de la atmósfera terrestre ubicada encima de la estratopausa.

enlace metálico: **(pág. 68)** iones metálicos positivos que se mantienen unidos debido la carga negativa de los electrones que se encuentran entre ellos; permite a los metales conducir electricidad.

estrella fugaz: **(pág. 818)** rayo luminoso que se produce cuando un meteoroide cae a la Tierra y se quema en la atmósfera terrestre.

meteorito: **(pág. 818, 621)** fragmento pequeño de un cuerpo en órbita que cae a la Tierra generando calor; como no se quema completamente en la atmósfera, choca con la superficie terrestre y produce un cráter de impacto.

meteoroide: **(pág. 818,)** trozo de material interplanetario que cae a la Tierra y entra a la atmósfera terrestre.

meteorología: **(pág. 6)** estudio de la atmósfera, la capa de aire que rodea la Tierra.

lluvia de estrellas: **(pág. 819)** ocurre cuando la Tierra interseca la órbita de un cometa y las partículas del cometa se queman al entrar a las capas superiores de la atmósfera terrestre.

microclima: **(pág. 385)** clima localizado que difiere del clima regional circundante.

microcontinentes: **(pág. 624)** trozos pequeños de corteza rica en granito que se formaron durante el Arcaico.

dorsales mediooceánicas: **(pág. 451)** cadenas montañosas submarinas que se extienden a través de las cuencas oceánicas, tienen una longitud total de más de 65,000 km y contienen innumerables volcanes activos y extintos.

mineral: **(pág. 86)** sólido inorgánico natural con una composición química específica y una estructura cristalina definida.

sustitución de minerales: **(pág. 607)** proceso en que los poros de las partes enterradas de un organismo se llenan con los minerales provenientes de aguas subterráneas.

modified Mercalli scale: **(p. 540)** measures earthquake intensity on a scale from I to XII; the higher the number, the greater the damage the earthquake has caused.

escala de Mercalli modificada: **(pág. 540)** mide la intensidad de un sismo en una escala de I a XII; a medida que aumenta el número, mayor es el daño causado.

mold: **(p. 608)** fossil that can form when a shelled organism decays in sedimentary rock and is removed by erosion or weathering, leaving a hollowed-out impression.

molde: **(pág. 608)** fósil que se forma cuando un organismo con concha se descompone en roca sedimentaria y es removido por erosión o meteorización, quedando una impresión hueca.

molecule: **(p. 67)** combination of two or more atoms joined by covalent bonds.

molécula: **(pág. 67)** combinación de dos o más átomos unidos por enlaces covalentes.

moment magnitude scale: **(p. 540)** scale used to measure earthquake magnitude—taking into account the size of the fault rupture, the rocks' stiffness, and amount of movement along the fault—using values that can be estimated from the size of several types of seismic waves.

escala de magnitud momentánea: **(pág. 540)** escala que sirve para medir la intensidad de un sismo (tomando en cuenta el tamaño de la ruptura de la falla, la rigidez de la roca y la cantidad del movimiento a lo largo de la falla) usando valores estimados a partir de la magnitud de varios tipos de ondas sísmicas.

moraine: **(p. 210)** ridge or layer of mixed debris deposited by a melting glacier.

morrena: **(pág. 210)** loma o estrato de detritos mezclados que deposita un glaciar al derretirse.

mountain thunderstorm: **(p. 346)** occurs when an air mass rises from orographic lifting, which involves air moving up the side of a mountain.

tormenta orográfica: **(pág. 346)** sucede cuando una masa de aire sube por ascenso orográfico, lo que implica el ascenso por la ladera de una montaña.

mudflow: **(p. 196)** rapidly flowing, often destructive mixture of mud and water that may be triggered by an earthquake, intense rainstorm, or volcanic eruption.

flujo o corriente de lodo: **(pág. 196)** mezcla de lodo y agua que fluye rápidamente y que a menudo es destructiva; puede ser causada por un terremoto, una lluvia intensa o una erupción volcánica.

N

natural resource: **(p. 678)** resources provided by Earth, including air, water, land, all living organisms, nutrients, rocks, and minerals.

recursos naturales: **(pág. 678)** recursos que provee la Tierra: incluyendo el aire, el agua, la tierra, todos los organismos vivos, los nutrientes, las rocas y los minerales.

neap tide: **(p. 424)** tide that occurs during first- or third-quarter Moon, when the Sun, the Moon, and Earth form a right angle; this causes solar tides to diminish lunar tides, causing high tides to be lower than normal and low tides to be higher than normal.

marea muerta: **(pág. 424)** durante el primero o el tercer cuartos lunares, el Sol, la Luna y la Tierra se encuentran en ángulo recto, causando que las mareas solares reduzcan la intensidad de las mareas lunares, lo que provoca que la marea alta sea menor que lo normal y la marea baja sea mayor que lo normal.

nebula: **(p. 848)** large cloud of interstellar gas and dust that collapses on itself, due to its own gravity, and forms a hot, condensed object that will become a new star.

nebulosa: **(pág. 848)** extensa nube de gas y polvo interestelares que se colapsa en sí misma debido a su propia gravedad, formando un cuerpo condensado caliente que se convertirá en una estrella nueva.

neutron: **(p. 60)** tiny atomic particle that is electrically neutral and has about the same mass as a proton.

neutrón: **(pág. 60)** partícula atómica diminuta, eléctricamente neutra; tiene una masa similar a la de un protón.

neutron star: **(p. 850)** collapsed, dense core of a star that forms quickly while its outer layers are falling inward, has a radius of about 10 km, a mass 1.4 to 3 times that of the Sun, and contains mostly neutrons.

estrella de neutrones: **(pág. 850)** núcleo denso y colapsado de una estrella que se forma rápidamente, al mismo tiempo que sus capas exteriores se contraen; tiene un radio aproximado de 10 km, una masa de 1.4 a 3 veces la del Sol y contiene principalmente neutrones.

nitrogen-fixing bacteria: **(p. 688)** bacteria found in water or soil; can grow on the roots of some plants, capture nitrogen gas, and change into a form that plants use to build proteins.

nonfoliated: **(p. 147)** describes metamorphic rocks like quartzite and marble, composed mainly of minerals that form with blocky crystal shapes.

nonpoint source: **(p. 749)** water-pollution source that generates pollution from widely spread areas, such as runoff from roads.

nonrenewable resource: **(p. 680)** resource that exists in Earth's crust in a fixed amount and can be replaced only by geologic, physical, or chemical processes that take hundreds of millions of years.

normal: **(p. 377)** standard value for a location, including rainfall, wind speed, and temperatures, based on meteorological records compiled for at least 30 years.

nuclear fission: **(p. 718)** the process in which a heavy nucleus divides to form smaller nuclei and one or two neutrons and produces a large amount of energy.

nucleus **(NEW klee us):** **(p. 60)** positively charged center of an atom, made up of protons and neutrons and surrounded by electrons in energy levels.

bacteria fijadora de nitrógeno: **(pág. 688)** bacteria que habita el suelo o el agua; puede crecer en las raíces de algunas plantas, capturar el gas nitrógeno y convertirlo a una forma que las plantas pueden usar para fabricar proteínas.

no foliada: **(pág. 147)** describe roca metamórfica, como la cuarcita y el mármol, compuesta principalmente de minerales que forman bloques cristalinos.

fuente no puntual: **(pág. 749)** fuente de contaminación del agua que genera contaminación a partir de áreas muy extensas, como la escorrentía de los caminos.

recurso no renovable: **(pág. 680)** recurso que existe en la corteza terrestre en una cantidad fija y que sólo puede ser regenerado por procesos geológicos, físicos o químicos que demoran centenas de millones de años.

normales: **(pág. 377)** valores estándar para un sitio: incluyen la lluvia, la velocidad del viento y las temperaturas; se basan en los registros meteorológicos recopilados durante por lo menos 30 años.

fisión nuclear: **(pág. 718)** proceso de división de un núcleo pesado en núcleos más pequeños y uno o dos neutrones, produciendo una gran cantidad de energía.

núcleo: **(pág. 60)** centro del átomo, tiene carga positiva, está compuesto por protones y neutrones y está rodeado por electrones localizados en niveles de energía.

O

oceanography: **(p. 7)** study of Earth's oceans including the creatures that inhabit its waters, its physical and chemical properties, and the effects of human activities.

ore: **(pp. 100, 684)** mineral or rock that contains a valuable substance that can be mined at a profit.

original horizontality: **(p. 596)** the principle that sedimentary rocks are deposited in horizontal or nearly horizontal layers.

original preservation: **(p. 607)** describes a fossil with soft and hard parts that have undergone very little change since the organism's death.

orogeny **(oh RAH juh nee):** **(p. 567)** cycle of processes that form all mountain ranges, resulting in broad, linear regions of deformation that you know as mountain ranges but in geology are known as orogenic belts.

orographic lifting: **(p. 299)** cloud formation that occurs when warm, moist air is forced to rise up the side of a mountain.

oceanografía: **(pág. 7)** estudio de los océanos de la Tierra: incluyendo sus propiedades físicas y químicas, los seres que los habitan y los efectos de las actividades humanas sobre ellos.

mena: **(pág. 100, 684)** mineral o roca que contiene una sustancia valiosa que se puede extraer con fines de lucro.

horizontalidad original: **(pág. 596)** principio que establece que las rocas sedimentarias se depositan formando estratos horizontales o casi horizontales.

preservación de material original: **(pág. 607)** describe un fósil cuyas partes blandas y duras han sufrido muy pocos cambios desde la muerte del organismo.

orogenia: **(pág. 567)** ciclo de procesos que forman todas las cadenas montañosas, dando como resultado grandes regiones lineares de deformación llamadas cadenas montañosas, pero que en geología se conocen como cinturones orogénicos.

ascenso orográfico: **(pág. 299)** formación de nubes que se produce cuando el aire húmedo caliente es forzado a ascender por la ladera de una montaña.

outwash plain: **(p. 210)** area at the leading edge of a glacier, where outwash is deposited by meltwater streams.

oxidation: **(p. 166)** chemical reaction of oxygen with other substances.

ozone hole: **(p. 745)** a seasonal decrease on ozone over Earth's polar regions.

llanura aluvial: **(pág. 210)** área en el borde frontal de un glaciar donde las corrientes del agua que se derrite depositan los derrubios.

oxidación: **(pág. 166)** reacción química del oxígeno con alguna otra sustancias.

agujero de ozono: **(pág. 745)** disminución estacional del ozono sobre las regiones polares de la Tierra.

P

paleogeography **(pay lee oh jee AH gruh fee): (p. 648)** the ancient geographic setting of an area.

paleomagnetism: **(p. 476)** study of Earth's magnetic record using data gathered from iron-bearing minerals in rocks that have recorded the orientation of Earth's magnetic field at the time of their formation.

Pangaea **(pan JEE uh): (p. 469)** ancient landmass made up of all the continents that began to break apart about 200 mya.

parallax: **(p. 841)** apparent positional shift of an object caused by the motion of the observer.

parsec (pc): **(p. 840)** the distance equal to 3.26 ly and 3.086×10^{13} km.

partial melting: **(p. 114)** process in which different minerals melt into magma at different temperatures, changing its composition.

passive margin: **(p. 648)** edge of a continent along which there is no tectonic activity.

peat: **(p. 710)** light, spongy, organic fossil fuel derived from moss and other bog plants.

pegmatite: **(p. 122)** igneous rock with extremely large-grained minerals that can contain rare ores such as lithium and beryllium.

perigee: **(p. 783)** closest point in the Moon's elliptical orbit to Earth.

period: **(p. 593)** third-longest time unit in the geologic time scale, measured in tens of millions of years.

permeability: **(p. 255)** ability of a material to let water pass through, is high in material with large, well-connected pores and low in material with few pores or small pores.

pesticide: **(p. 741)** chemical applied to plants to kill insects and weeds.

photochemical smog: **(p. 744)** a type of air pollution; a yellow-brown haze formed mainly from automobile exhaust in the presence of sunlight.

paleogeografía: **(pág. 648)** características geográficas antiguas de un área.

paleomagnetismo: **(pág. 476)** estudio del registro magnético de la Tierra; utiliza la información recogida a partir de minerales ferrosos en las rocas porque este tipo de minerales registran la orientación del campo magnético de la Tierra en el momento en que se forman.

Pangaea: **(pág. 469)** antigua masa terrestre compuesta por todos los continentes, los cuales se empezaron a separar hace cerca de 200 millones de años.

paralaje: **(pág. 841)** cambio aparente de la posición de un cuerpo causado por el movimiento del observador.

parsec: **(pág. 840)** distancia de 3.26 ly y 3.086×10^{13} km.

fundición parcial: **(pág. 114)** proceso en el cual diferentes minerales se funden en el magma a diferentes tempe-raturas, cambiando su composición.

margen pasivo: **(pág. 648)** límite de un continente a lo largo del cual no ocurre actividad tectónica.

turba: **(pág. 710)** combustible fósil liviano, esponjoso y orgánico derivado del musgo y otras plantas de ciénegas.

pegmatita: **(pág. 122)** roca ignea con grano extremadamente grueso que pueden contener minerales raros como el litio y el berilio.

perigeo: **(pág. 783)** punto más cercano a la Tierra en la órbita elíptica de la Luna.

período: **(pág. 593)** tercera unidad de tiempo más grande en la escala del tiempo geológico; se mide en decenas de millones de años.

permeabilidad: **(pág. 255)** capacidad de un material de permitir el paso del agua; es grande en materiales con poros grandes y bien conectados y baja en materiales con pocos poros o con poros pequeños.

pesticida: **(pág. 741)** sustancia química que se aplica a las plantas para eliminar insectos y malas hierbas.

smog fotoquímico: **(pág. 744)** tipo de contaminación del aire; niebla color amarillo marrón que se forma debido principalmente a las emisiones de los autos en presencia de la luz solar.

photosphere: **(p. 831)** lowest layer of the Sun's atmosphere that is also its visible surface, has an average temperature of 5800 K, and is about 400 km thick.

photovoltaic cell: **(p. 716)** thin, transparent wafer that converts sunlight into electrical energy and is made up of two layers of two types of silicon.

phytoplankton: **(p. 658)** microscopic organisms that are the basis of marine food chains; abundant during the Cretaceous and the remains of their shell-like hard parts are found in chalk deposits worldwide.

planetesimal: **(p. 798)** space object built of solid particles that can form planets through collisions and mergers.

plasma: **(p. 74)** hot, highly ionized, electrically conducting gas.

plastic deformation: **(p. 529)** permanent deformation caused by strain when stress exceeds a certain value.

plateau: **(p. 573)** a relatively flat-topped area.

pluton **(PLOO tahn):** **(p. 514)** intrusive igneous rock body, including batholiths, stocks, sills, and dikes, formed through mountain-building processes and oceanic-oceanic collisions; can be exposed at Earth's surface due to uplift and erosion.

point source: **(p. 749)** water-pollution source that generates pollution from a single point of origin, such as an industrial site.

polar easterlies: **(p. 320)** global wind systems that lie between latitudes 60° N and 60° S and the poles and is characterized by cold air.

polar zones: **(p. 378)** areas of Earth where solar radiation strikes at a low angle, resulting in temperatures that are nearly always cold; extend from 66.5° north and south of the equator to the poles.

pollutant: **(p. 690)** substance that enters Earth's geochemical cycles and can harm the health of living things or adversely affect their activities.

Population I stars: **(p. 866)** stars in the disk and arms that have small amounts of heavy elements.

Population II stars: **(p. 866)** stars in the halo and bulge that contain traces of heavy elements.

fotosfera: **(pág. 831)** capa más baja de la atmósfera solar; corresponde a su superficie visible, tiene una temperatura promedio de 5800 K y mide aproximadamente 400 km de ancho.

celdas fotovoltaicas: **(pág. 716)** láminas delgadas y transpa-rentes que convierten la luz solar en energía eléctrica; están compuestas de dos capas con dos tipos de silicio.

fitoplancton: **(pág. 658)** organismos microscópicos que son la base de las cadenas alimenticias marinas; fueron muy abundantes durante el Cretáceo y los restos de sus caparazones se encuentran en depósitos de carbonato de calcio por todo el mundo.

planetesimal: **(pág. 798)** cuerpo espacial formado por partículas sólidas y los cuales pueden formar planetas mediante choques y fusiones.

plasma: **(pág. 74)** gas caliente, altamente ionizado y conductor de electricidad.

deformación dúctil: **(pág. 529)** cuando la presión excede cierto valor; la tensión producida causa una deformación permanente.

altiplanicie: **(pág. 573)** área relativamente plana en la parte más alta.

plutones: **(pág. 514)** cuerpos rocosos ígneos intrusivos: incluye batolitos, macizos magmáticos, intrusiones y diques formados durante los procesos orogénicos y durante la colisión de placas oceánicas; pueden quedar expuestos a la superficie terrestre debido a levantamientos y erosión.

fuente puntual: **(pág. 749)** fuente de contaminación de agua que genera contaminación a partir de un solo punto de origen, por ejemplo, una zona industrial.

vientos polares del este: **(pág. 320)** sistemas globales del viento que se encuentran entre los polos y las latitudes 60°N y 60°S; se caracterizan por tener aire frío.

zonas polares: **(pág. 378)** áreas de la Tierra donde la radiación solar llega con un ángulo bajo, ocasionando que las temperaturas casi siempre sean frías; se extienden desde los 66.5° hasta los polos, en ambos hemisferios.

contaminante: **(pág. 690)** sustancia que entra a los ciclos geoquímicos de la Tierra y puede causar daños a la salud de los seres vivos o afectar adversamente sus actividades.

estrellas de la población I: **(pág. 866)** aquellas ubicadas en el disco y los brazos y que contienen pequeñas cantidades de elementos pesados.

estrellas de la población II: **(pág. 866)** aquellas ubicadas en el halo y en el núcleo y que contienen trazas de elementos pesados.

GLOSSARY ▪ GLOSARIO

Glossary ▪ Glosario

porosity: **(p. 142)** percentage of open spaces between grains in a material.

porphyritic **(por fuh RIH tihk)** **texture:** **(p. 120)** rock texture characterized by large, well-formed crystals surrounded by finer-grained crystals of the same or different mineral.

Precambrian **(pree KAM bree un):** **(p. 592)** informal unit of geologic time consisting of the first three eons during which Earth formed and became habitable.

Precambrian shield: **(p. 625)** the top of a craton exposed at Earth's surface

precipitation: **(p. 302)** all solid and liquid forms of water—including rain, snow, sleet, and hail—that fall from clouds.

prevailing westerlies: **(p. 320)** global wind system that lies between 30° and 60° north and south latitudes, where surface air moves toward the poles in an easterly direction.

primary wave: **(p. 532)** seismic wave that squeezes and pushes rocks in the same direction that the wave travels, known as a P-wave.

prime meridian: **(p. 31)** imaginary line representing 0° longitude, running from the North Pole, through Greenwich, England, to the South Pole.

prokaryote **(proh KE ree oht):** **(p. 635)** unicellular organism that lacks a nucleus.

prominence: **(p. 833)** arc of gas ejected from the chromosphere, or gas that condenses in the Sun's inner corona and rains back to the surface, that can reach temperatures over 50,000 K and is associated with sunspots.

proton: **(p. 60)** tiny atomic particle that has mass and a positive electric charge.

protostar: **(p. 848)** hot, condensed object at the center of a nebula that will become a new star when nuclear fusion reactions begin.

pulsar: **(p. 850)** a spinning neutron star that exhibits a pulsing pattern.

pyroclastic flow: **(p. 513)** swift-moving, potentially deadly clouds of gas, ash, and other volcanic material produced by a violent eruption.

porosidad: **(pág. 142)** porcentaje de espacios abiertos entre los granos de una roca.

textura porfírica: **(pág. 120)** textura rocosa caracterizada por cristales grandes bien formados, rodeados por cristales de grano más fino del mismo mineral o de uno diferente.

Precámbrico: **(pág. 592)** unidad del tiempo geológico que consiste en los primeros tres eones; periodo durante el cual la Tierra se formó y adquirió condiciones aptas para la vida.

escudo Precámbrico: **(pág. 625)** parte alta de un cratón que está expuesta en la superficie de la Tierra.

precipitación: **(pág. 302)** toda forma líquida o sólida de agua: lluvia, nieve, aguanieve o granizo, que cae de las nubes.

vientos dominantes del oeste: **(pág. 320)** sistema de vientos globales ubicado entre los 30° y los 60° de latitud, en ambos hemisferios, donde el aire superficial se desplaza hacia los polos en dirección este.

onda primaria: **(pág. 532)** onda sísmica que comprime y empuja las rocas en la misma dirección en que viaja la onda; se conocen como ondas P.

primer meridiano: **(pág. 31)** línea imaginaria que representa la longitud 0°; va desde el Polo Norte hasta el Polo Sur, pasando por Greenwich, Inglaterra.

procariota: **(pág. 635)** organismo unicelular que carece de núcleo.

protuberancia solar: **(pág. 833)** arco de gas expulsado de la cromosfera o gas que se condensa en la corona interna del Sol y que se precipita de nuevo sobre su superficie; puede alcanzar temperaturas mayores a los 50,000 K y está asociada a la presencia de manchas solares.

protón: **(pág. 60)** partícula atómica diminuta que tiene masa y una carga eléctrica positiva.

protoestrella: **(pág. 848)** cuerpo condensado, caliente, ubicado en el centro de una nebulosa, que se convertirá en una estrella nueva cuando inicien las reacciones de fusión nuclear.

pulsar: **(pág. 850)** estrella de neutrones giratoria que exhibe un patrón de pulsaciones.

flujo piroclástico: **(pág. 513)** nubes de gas, cenizas y otros materiales volcánicos, potencialmente mortales, que se desplazan rápidamente y que son producidas por una erupción violenta.

Q

quasar: **(p. 875)** starlike, very bright, extremely distant object with emission lines in its spectra.

cuásares: **(pág. 875)** cuerpos semejantes a estrellas, muy brillantes y extremadamente lejanos, con líneas de emisión en sus espectros.

R

radiation: **(p. 287)** the transfer of thermal energy by electromagnetic waves; the transfer of thermal energy from the Sun to Earth by radiation.

radioactive decay: **(p. 601)** emission of radioactive particles and its resulting change into other isotopes over time.

radiocarbon dating: **(p. 603)** determines the age of relatively young organic objects; objects that are alive or were once alive.

radio galaxy: **(p. 875)** very bright, often giant, elliptical galaxy that emits as much or more energy in the form of radio wavelengths as it does wavelengths of visible light.

radiometric dating: **(p. 602)** process used to determine the absolute age of a rock or fossil by determining the ratio of parent nuclei to daughter nuclei within a given sample.

radiosonde **(RAY dee oh sahnd):** **(p. 326)** balloon-borne weather instrument whose sensors measure air pressure, humidity, temperature, wind speed, and wind direction of the upper atmosphere.

ray: **(p. 771)** long trail of ejecta that radiates outward from an impact crater.

recharge: **(p. 263)** process by which water from precipitation and runoff is added to the zone of saturation.

reclamation: **(p. 738)** process in which a mining company restores land used during mining operations to its original contours and replants vegetation.

red bed: **(p. 631)** a sedimentary rock deposit that contains oxidized iron; provides evidence that free oxygen existed in the atmosphere during the Proterozoic.

reflecting telescope: **(p. 766)** telescope that uses mirrors to focus visible light.

refracting telescope: **(p. 766)** telescope that uses lenses to focus visible light.

regional metamorphism: **(p. 149)** process that affects large areas of Earth's crust, producing belts classified as low, medium, or high grade, depending on pressure on the rocks, temperature, and depth below the surface.

regolith: **(p. 772)** layer of loose, ground-up rock on the lunar surface.

radiación: **(pág. 287)** transferencia de energía mediante por ondas electromagnéticas; la transferencia de energía térmica del Sol a la Tierra por radiación.

desintegración radiactiva: **(pág. 601)** emisión de partículas atómicas que a lo largo del tiempo produce nuevos isótopos.

datación radiocarbónica: **(pág. 603)** permite determinar la edad de cuerpos orgánicos relativamente recientes, cuerpos que están vivos o que alguna vez estuvieron vivos.

radiogalaxia: **(pág. 875)** galaxia elíptica muy brillante, a menudo gigantesca, cuya emisión de energía en forma de ondas de radio es similar a la que emite como ondas de luz visible.

datación radiométrica: **(pág. 602)** proceso que permite establecer la edad absoluta de una roca o un fósil, al determinar la razón entre los núcleos originales y los núcleos derivados de una muestra dada.

radiosonda: **(pág. 326)** instrumento meteorológico que se monta en un globo y cuyos sensores miden la presión atmosférica, la humedad, la temperatura, así como la velocidad y dirección del viento en la atmósfera superior.

rayo: **(pág. 771)** largo rastro de eyecta que irradia de un cráter de impacto.

recarga: **(pág. 263)** proceso mediante el cual el agua de la precipitación y de la escorrentía entra a la zona de saturación.

recuperación: **(pág. 738)** proceso en que una compañía minera restaura los terrenos usados en las actividades mineras a sus contornos originales y reforesta con nueva vegetación.

lecho rojo: **(pág. 631)** depósito de roca sedimentaria que contiene hierro oxidado; es evidencia de que había oxígeno libre en la atmósfera durante el Proterozoico.

telescopio reflector: **(pág. 766)** telescopio que usa espejos para enfocar la luz visible.

telescopio refractor: **(pág. 766)** telescopio que usa lentes para enfocar la luz visible.

metamorfismo regional: **(pág. 149)** proceso que afecta grandes áreas de la corteza terrestre; produce cinturones de bajo, medio o alto grado, dependiendo de la presión sobre las rocas, la temperatura y la profundidad bajo la superficie.

regolito: **(pág. 772)** estrato de roca suelta y molida en la superficie lunar.

regression: (p. 649) occurs when sea level falls, causing the shoreline to move seaward, and results in shallow-water deposits overlying deep-water deposits.

rejuvenation: (p. 237) process during which a stream resumes downcutting toward its base level, increasing its rate of flow.

relative-age dating: (p. 596) establishing the order of past geologic events.

relative humidity: (p. 294) ratio of water vapor contained in a specific volume of air compared with how much water vapor that amount of air actually can hold; expressed as a percentage.

remote sensing: (p. 41) process of gathering data about Earth from instruments far above the planet's surface.

renewable resource: (p. 678) natural resource, such as fresh air and most groundwater, that can be replaced by nature in a short period of time.

residual soil: (p. 177) soil that develops from parent material which is similar to local bedrock.

retrograde motion: (p. 799) a planet's apparent backward movement in the sky.

return stroke: (p. 348) a branch channel of positively charged ions that rushes upward from the ground to meet the stepped leader.

Richter scale: (p. 539) numerical rating system used to measure the amount of energy released during an earthquake.

ridge push: (p. 488) tectonic process associated with convection currents in Earth's mantle that occurs when the weight of an elevated ridge pushes an oceanic plate toward a subduction zone.

rift valley: (p. 481) long, narrow depression that forms when continental crust begins to separate at a divergent boundary.

rill erosion: (p. 172) erosion in which water running down the side of a slope carves a small stream channel.

rille: (p. 771) valleylike structure that meanders across some regions of the Moon's maria.

rock cycle: (p. 151) continuous, dynamic set of processes by which rocks are changed into other types of rock.

root: (p. 563) thickened areas of continental material, detected by gravitational and seismic studies.

regresión: (pág. 649) ocurre cuando baja el nivel del mar, provocando que la costa avance hacia el mar, ocasiona que depósitos de agua superficiales cubran depósitos de agua profundos.

rejuvenecimiento: (pág. 237) proceso en que una corriente reanuda la erosión hacia su nivel base, aumentando su tasa de flujo.

datación relativa: (pág. 596) ordenamiento por antigüedad de eventos geológicos pasados.

humedad relativa: (pág. 294) razón del vapor de agua que contiene un volumen específico de aire, en comparación con la cantidad de vapor de agua que ese volumen de aire podría contener, expresado como porcentaje.

percepción remota: (pág. 41) proceso de recopilación de datos sobre la Tierra con instrumentos alejados de la superficie del planeta.

recurso renovable: (pág. 678) recurso natural, como el aire y la mayoría de las aguas subterráneas, que la naturaleza puede reemplazar en un período corto de tiempo.

suelo residual: (pág. 177) suelo que se desarrolla a partir del material original y es similar a la roca madre local.

movimiento retrógrado: (pág. 799) movimiento aparentemente en retroceso de un planeta en el cielo.

descarga de retorno: (pág. 348) un canal con iones de carga positiva que asciende desde el suelo para encontrarse con la descarga líder o guía escalonada.

escala de Richter: (pág. 539) escala numérica que se emplea para medir la cantidad de energía liberada durante un sismo.

empuje de la dorsal: (pág. 488) proceso tectónico asociado con las corrientes de convección en el manto de la Tierra, que ocurre cuando el peso de una cordillera elevada empuja una placa oceánica hacia una zona de subducción.

valle del rift: (pág. 481) depresión larga y estrecha que se forma cuando la corteza continental se empieza a separar en un límite divergente.

erosión por surcos: (pág. 172) erosión en la cual el agua que corre cuesta abajo forma un canal pequeño.

surco: (pág. 771) formación tipo valle que serpentea a través de algunas regiones de los mares lunares.

ciclo de las rocas: (pág. 151) conjunto de procesos continuos y dinámicos a través de los cuales las rocas se transforman en otros tipos de roca.

raíz: (pág. 563) gruesas áreas de material continental que son detectadas en estudios sísmicos o gravitacionales.

RR Lyrae variable: **(p. 863)** stars with pulsation periods ranging from 1.5 hours to 1.2 days, generally having the same luminosity, regardless of pulsation period length.

runoff: **(p. 225)** water that flows downslope on Earth's surface and may enter a stream, river, or lake; its rate is influenced by the angle of the slope, vegetation, rate of precipitation, and soil composition.

estrellas variables tipo RR Lyrae: **(pág. 863)** estrellas con períodos de pulsación que duran de 1.5 horas a 1.2 días; en general tienen la misma luminosidad, independientemente de la duración de la pulsación.

escorrentía: **(pág. 225)** agua que corre cuesta abajo sobre la superficie terrestre y que puede incorporarse a una corriente, río o lago; su tasa de flujo está influida por el ángulo de la pendiente, la vegetación, la tasa de precipi-tación y la composición del suelo.

S

Saffir-Simpson Hurricane Wind scale: **(p. 358)** classifies hurricanes according to wind speed on a scale ranging from Category 1 to Category 5.

salinity: **(p. 413)** measure of the amount of salts dissolved in seawater, which is 35 ppt, or 3.5% on average.

saturation: **(p. 294)** the point at which water molecules leaving the water's surface equals the rate of water molecules returning to the surface.

scarp: **(p. 805)** a line of cliffs produced by erosion or faulting.

scientific law: **(p. 19)** a principle that describes the behavior of a natural phenomenon.

scientific methods: **(p. 10)** a series of problem-solving procedures that help scientists conduct experiments.

scientific model: **(p. 18)** an idea, a system, or a mathematical expression that represents the idea being explained.

scientific notation: **(p. 16)** a method used by scientists to express a number as a value between 1 and 10 multiplied by a power of 10.

scientific theory: **(p. 19)** an explanation based on many observations during repeated experiments; valid only if consistent with observations, can be used to make testable predictions, and is the simplest explanation; can be changed or modified with the discovery of new data.

sea-breeze thunderstorm: **(p. 346)** local air-mass thunderstorm that commonly occurs along a coastal area because land and water store and release thermal energy differently.

escala de Vientos Huracanados Saffir-Simpson: **(pág. 358)** clasifica los huracanes según la velocidad de sus vientos en una escala que va desde la Categoría 1 hasta la Categoría 5.

salinidad: **(pág. 413)** medida de la cantidad de sales disueltas en el agua de mar; en promedio es de 35 ppt ó 3.5%.

saturación: **(pág. 294)** sucede en el punto en el cual la tasa de salida de moléculas de agua en la superficie es igual a la tasa de retorno de las moléculas a la superficie.

escarpes: **(pág. 805)** una línea de acantilados por erosión o fallas.

ley científica: **(pág. 19)** principio que describe el comportamiento de un fenómeno natural.

métodos científicos: **(pág. 10)** serie de procedimientos para resolver problemas que ayudan a los científicos a realizar experimentos.

modelo científico: **(pág. 18)** idea, sistema o expresión matemática que representa la idea que se quiere explicar.

notación científica: **(pág. 16)** método que usan los científicos para expresar un número como un valor entre 1 y 10 multiplicado por una potencia de 10.

teoría científica: **(pág. 19)** explicación basada en muchas observaciones realizadas durante experimentos repetidos; sólo es válida si es consistente con las observaciones, permite hacer predicciones comprobables y es la explicación más sencilla; puede ser modificada debido al descubrimiento de nuevos hechos.

tormenta eléctrica de brisa marina: **(pág. 346)** tormenta local de masa de aire que ocurre comúnmente a lo largo de un área costera; ocurren porque la tierra y el agua almacenan y liberan energía térmica de manera distinta.

seafloor spreading: (p. 479) the hypothesis that new ocean crust is formed at mid-ocean ridges and destroyed at deep-sea trenches; occurs in a continuous cycle of magma intrusion and spreading.

expansión del suelo marino: (pág. 479) hipótesis que propone que la nueva corteza oceánica se forma en las dorsales mediooceánicas y se destruye en las fosas submarinas profundas; ocurre según un ciclo continuo de intrusión y expansión del magma.

sea level: (p. 410) level of the oceans' surfaces, which has risen at a rate of about 3 mm per year.

nivel del mar: (pág. 410) nivel de la superficie del océano; actualmente sube a una velocidad de 3 mm por año.

seamount: (p. 452) basaltic, submerged volcano on the seafloor that is more than 1 km high.

montaña submarina: (pág. 452) volcán basáltico sumergido en el fondo marino que mide más de 1 km de altura.

season: (p. 388) short-term periods with specific weather conditions caused by regular variations in temperature, hours of daylight, and weather patterns that are due to the tilt of Earth's axis as it revolves around the Sun, causing different areas of Earth to receive different amounts of solar radiation.

estación: (pág. 388) períodos de corto plazo con específicas de tiempo causados por variaciones regulares en temperatura, horas de luz solar y patrones meteorológicos, provocadas por la inclinación del eje de la Tierra cuando gira alrededor del Sol, lo que ocasiona que las distintas áreas de la Tierra reciban diferentes cantidades de radiación solar.

secondary wave: (p. 532) seismic wave that causes rock particles to move at right angles to the direction of the wave, known as an S-wave.

onda secundaria: (pág. 532) onda sísmica que ocasiona que las partículas de las rocas se muevan en ángulo recto con respecto a la dirección de la onda.

sediment: (p. 134) small pieces of rock that are moved and deposited by water, wind, glaciers, and gravity.

sedimentos: (pág. 134) partículas pequeñas de roca que el agua, el viento, los glaciares y la gravedad mueven y depositan.

seismic gap: (p. 550) place along an active fault that has not experienced an earthquake for a long time.

vacío sísmico: (pág. 550) lugar a lo largo de una falla activa que no ha sufrido un terremoto durante mucho tiempo.

seismic wave: (p. 532) the vibrations of the ground during an earthquake.

onda sísmica: (pág. 532) vibraciones del terreno durante un sismo.

seismogram (SIZE muh gram): (p. 534) record produced by a seismometer that can provide individual tracking of each type of seismic wave.

sismograma: (pág. 534) registro producido por un sismógrafo que proporciona un registro individual de cada tipo de onda sísmica.

seismometer (size MAH muh tur): (p. 534) instrument used to measure horizontal or vertical motion during an earthquake.

sismógrafo: (pág. 534) instrumento que sirve para medir los movimientos horizontales y verticales durante un sismo.

shield volcano: (p. 507) broad volcano with gently sloping sides built by nonexplosive eruptions of basaltic lava that accumulates in layers.

volcán de escudo: (pág. 507) volcán ancho, de laderas con inclinación suave, formado por erupciones no explosivas de lava basáltica que se acumula en estratos.

side-scan sonar: (p. 407) technique that directs sound waves at an angle to the seafloor or deep-lake floor, allowing underwater topographic features to be mapped.

sonar de escaneo lateral: (pág. 407) técnica que dirige las ondas sonoras en ángulo hacia el fondo del mar o de un lago profundo, lo que permite trazar el relieve topográfico submarino.

silicate: (p. 96) mineral that contains silicon (Si), oxygen (O), and usually one or more other elements.

silicato: (pág. 96) mineral que contiene silicio (Si), oxígeno (O) y generalmente uno o más elementos adicionales.

sill: (p. 515) pluton that forms when magma intrudes parallel rock layers.

intrusión: (pág. 515) plutón que se forma cuando el magma penetra estratos rocosos paralelos.

sinkhole: **(p. 261)** depression in Earth's surface formed when a cave collapses or bedrock is dissolved by acidic rain or moist soil.

slab pull: **(p. 488)** tectonic process associated with convection currents in Earth's mantle that occurs as the weight of the subducting plate pulls the trailing lithosphere into a subduction zone.

slump: **(p. 198)** mass movement that occurs when earth materials in a landslide rotate and slide along a curved surface, leaving a crescent-shaped scar on a slope.

soil: **(p. 176)** loose covering of weathered rock and decayed organic matter overlying Earth's bedrock that is characterized by texture, fertility, and color and whose composition is determined by its parent rock and environmental conditions.

soil horizon: **(p. 178)** distinct layer within a soil profile.

soil liquefaction **(lih kwuh FAK shun): (p. 547)** process associated with seismic vibrations that occur in areas of sand that is nearly saturated; resulting in the ground behaving like a liquid.

soil profile: **(p. 178)** vertical sequence of soil layers containing the A-horizon, the B-horizon, and the C-horizon.

solar eclipse: **(p. 781)** when the Moon passes between Earth and the Sun and the Moon casts a shadow on Earth, blocking Earth's view of the Sun; can be partial or total.

solar flare: **(p. 833)** violent eruption of radiation and particles from the Sun's surface that is associated with sunspots.

solar wind: **(p. 832)** wind of charged particles (ions) that flows throughout the solar system and begins as gas flowing outward from the Sun's corona at high speeds.

solstice: **(p. 777)** period when the Sun is overhead at its farthest distance either north or south of the equator.

solution: **(p. 71)** homogeneous mixture whose components cannot be distinguished and can be classified as liquid, gaseous, solid, or a combination; **(p. 228)** the method of transport for materials that are dissolved in a stream's water.

sonar: **(p. 43)** use of sound waves to detect and measure objects underwater.

source region: **(p. 316)** area over which an air mass forms.

sumidero: **(pág. 261)** depresión en la superficie terrestre que se forma cuando una caverna se colapsa o cuando el lecho rocoso es disuelto por lluvia ácida o suelo húmedo.

tracción de placa: **(pág. 488)** proceso tectónico asociado con las corrientes de convección del manto de la Tierra, que ocurre cuando el peso de la placa subductora jala la litosfera hacia una zona de subducción.

deslizamiento rotacional: **(pág. 198)** movimiento en masa que ocurre cuando los materiales terrestres de un derrumbe giran y se deslizan a lo largo de una superficie curva, dejando una cicatriz con forma de medialuna en la pendiente.

suelo: **(pág. 176)** cubierta suelta de roca meteorizada y materia orgánica en descomposición que cubre el lecho rocoso terrestre; se caracteriza por su textura, fertilidad y color y su composición está determinada por la roca madre y las condiciones ambientales.

horizonte del suelo: **(pág. 178)** capa distintiva dentro de un perfil del suelo.

licuefacción del suelo: **(pág. 547)** proceso asociado con las vibraciones sísmicas que ocurren en las áreas arenosas casi saturadas; el resultado es que el suelo actúa como un líquido.

perfil del suelo: **(pág. 178)** sucesión vertical de capas del suelo comprende los horizontes A (mantillo), B (subsuelo), y C (material original meteorizado).

eclipse solar: **(pág. 781)** sucede cuando la Luna pasa entre la Tierra y el Sol y la Luna proyecta su sombra sobre la Tierra, bloqueando la luz del Sol; puede ser parcial o total.

erupción solar: **(pág. 833)** violenta erupción de radiación y partículas desde la superficie del Sol que está asociada con las manchas solares.

viento solar: **(pág. 832)** viento de partículas cargadas (iones) que fluye a través del sistema solar y comienza como un gas que es despedido a gran velocidad por la corona del Sol.

solsticio: **(pág. 777)** sucede cuando el Sol se halla en el horizonte a su mayor distancia al norte o al sur del ecuador.

solución: **(pág. 71)** mezcla homogénea cuyos componentes no se pueden distinguir; puede clasificarse como líquida, gaseosa, sólida o una combinación de éstas; **(pág. 228)** el método de transporte de materiales que están disueltos en las aguas de una corriente.

sonar: **(pág. 43)** uso de ondas sonoras para detectar y medir objetos submarinos.

región fuente: **(pág. 316)** área sobre la cual se forma una masa de aire.

specific gravity: **(p. 95)** ratio of the mass of a substance to the mass of an equal volume of H_2O at 4°C.

spiral density wave: **(p. 868)** spiral regions of alternating density which rotates as a rigid pattern.

spring: **(p. 256)** natural discharge of groundwater at Earth's surface where an aquifer and an aquiclude come in contact.

spring tide: **(p. 424)** during full or new moon, the Sun, the Moon, and Earth are all aligned; this causes solar tides to enhance lunar tides, causing high tides to be higher than normal and low tides to be lower than normal.

stalactite: **(p. 261)** cone-shaped or cylindrical dripstone deposit of calcium carbonate that hangs like an icicle from a cave's ceiling.

stalagmite: **(p. 261)** mound-shaped dripstone deposit of calcium carbonate that forms on a cave's floor beneath a stalactite.

station model: **(p. 329)** record of weather data for a specific place at a specific time, using meteorological symbols.

stepped leader: **(p. 348)** The channel of partially charged air; the breakdown in charges in between positive and negative regions.

stock: **(p. 515)** irregularly shaped pluton that is similar to a batholith but smaller, generally forms 5–30 km beneath Earth's surface, and cuts across older rocks.

storm surge: **(p. 359)** occurs when powerful, hurricane-force winds drive a mound of ocean water toward shore, where it washes over the land, often causing enormous damage.

strain: **(p. 528)** deformation of materials in response to stress.

stratosphere: **(p. 284)** layer of Earth's atmosphere that is located above the tropopause and is made up primarily of concentrated ozone.

stratus (STRAY tus): **(p. 301)** a layered sheetlike cloud that covers much or all of the sky in a given area.

streak: **(p. 93)** color a mineral leaves when it is rubbed across an unglazed porcelain plate or when it is broken up and powdered.

stream bank: **(p. 232)** ground bordering each side of a stream that keeps the moving water confined.

gravedad específica: **(pág. 95)** razón de la masa de una sustancia con relación a la masa de un volumen igual de H_2O a 4°C.

ondas de densidad espirales: **(pág. 868)** regiones en espiral con densidad variable que giran siguiendo un patrón rígido.

manantial: **(pág. 256)** descarga natural de agua subterránea en la superficie terrestre, en el punto donde un acuífero y un acuicluso entran el contacto.

marea viva: **(pág. 424)** durante la luna nueva o la luna llena, el Sol, la Luna y la Tierra se encuentran alineados; esto ocasiona que la marea solar aumente el efecto de la marea lunar y provoca que la marea alta sea más alta que lo normal y que la marea baja sea más baja que lo normal.

estalactita: **(pág. 261)** depósito rocoso de carbonato de calcio, de forma cónica o cilíndrica, que se forma por goteo y que cuelga como un carámbano del techo de una caverna.

estalagmita: **(pág. 261)** depósito de carbonato de calcio, con forma de montículo, que se forma por goteo en el piso de una caverna, debajo de una estalactita.

código meteorológico: **(pág. 329)** registro de los datos del tiempo para un lugar específico en un tiempo dado, usando símbolos meteorológicos.

guía escalonada: **(pág. 348)** el canal con aire parcialmente cargado; la separación de cargas que forma regiones positivas y negativas.

macizo magmático: **(pág. 515)** plutón de forma irregular, similar a un batolito pero más pequeño; generalmente se forma de 5 a 30 km bajo la superficie terrestre y atraviesa rocas más antiguas.

marejada ciclónica: **(pág. 359)** ocurre cuando poderosos vientos huracanados arrojan una gran masa de agua del océano hacia la costa, desparramándose por el terreno y causando a menudo un daño enorme.

tensión: **(pág. 528)** deformación de los materiales en res-puesta a un estrés.

estratosfera: **(pág. 284)** capa de la atmósfera terrestre ubicada por encima de la tropopausa; está compuesta principalmente de ozono concentrado.

estrato: **(pág. 301)** nube con forma de capas delgadas que cubre la mayoría o todo el cielo en cierta área.

veta: **(pág. 93)** color que deja un mineral cuando es frotado contra un plato de porcelana sin barnizar o cuando se rompe y se pulveriza.

margen de una corriente de agua: **(pág. 232)** terreno que limita ambos lados de una corriente, manteniendo confinada la corriente de agua en movimiento.

stream channel: **(p. 232)** narrow pathway carved into sediment or rock by the movement of surface water.

stress: **(p. 528)** forces per unit area that act on a material—compression, tension, and shear.

stromatolite **(stroh MA tuh lite): (p. 629)** large mat or mound composed of billions of photosynthesizing cyanobacteria; dominated shallow oceans during the Proterozoic.

subduction: **(p. 482)** process by which one tectonic plate slips beneath another tectonic plate.

sublimation: **(p. 75)** process by which a solid slowly changes to a gas without first entering a liquid state.

sunspot: **(p. 832)** dark spot on the surface of the photosphere; occur in pairs.

supercell: **(p. 350)** extremely powerful, self-sustaining thunderstorm characterized by intense, rotating updrafts.

supercluster: **(p. 873)** gigantic threadlike or sheetlike cluster of galaxies that is hundreds of millions of light-years in size.

supernova: **(p. 851)** massive explosion that occurs when the outer layers of a star are blown off.

superposition: **(p. 596)** the principle that, in an undisturbed rock sequence, the oldest rocks are on the bottom and each consecutive layer is younger than the layer beneath it.

surface current: **(p. 425)** wind-driven movement of ocean water that primarily affects the upper few hundred meters of the ocean.

suspension: **(p. 228)** the method of transport for all particles small enough to be held up by the turbulence of a stream's moving water.

sustainable energy: **(p. 723)** involves global management of Earth's natural resources to ensure that current and future energy needs will be met without harming the environment.

sustainable yield: **(p. 679)** replacement of renewable resources at the same rate at which they are consumed.

synchronous rotation: **(p. 780)** the state at which an orbiting body's orbital and rotational periods are equal.

cauce fluvial: **(pág. 232)** estrecha vía labrada en el sedimento, o en la roca, por el movimiento del agua en la superficie.

estrés: **(pág. 528)** fuerza por unidad de área que actúa sobre un material: puede ser por compresión, tensión o cizallamiento.

estromatolitos: **(pág. 629)** montículos grandes compuestos de billones de cianobacterias fotosintéticas; dominaron los océanos superficiales durante el Proterozoico.

subducción: **(pág. 482)** proceso en que una placa tectónica se desliza por debajo de otra.

sublimación: **(pág. 75)** proceso en que un sólido se convierte lentamente en gas, sin convertirse primero al estado líquido.

mancha solar: **(pág. 832)** mancha oscura en la superficie de la fotosfera; ocurren en pares.

supercelda: **(pág. 350)** tormenta autosostenible extremadamente poderosa, caracterizada por tener intensas cor-rientes ascendentes giratorias.

supercúmulo: **(pág. 873)** cúmulo gigantesco de galaxias con forma de filamento o lámina que mide centenares de millones de años luz.

supernova: **(pág. 851)** enorme explosión que ocurre cuando estallan las capas exteriores de una estrella.

superposición: **(pág. 596)** principio que establece que en una sucesión rocosa no perturbada, los estratos rocosos más antiguos se encuentran en el fondo y que cada capa sucesiva es más reciente que la capa subyacente.

corriente superficial: **(pág. 425)** movimiento de las aguas del océano producido por el viento, que afecta principalmente los primeros cientos de metros superiores de las aguas del océano.

suspensión: **(pág. 228)** método de transporte de todas las partículas que son suficientemente pequeñas como para ser mantenidas en el agua por la turbulencia de la corriente del agua en movimiento.

energía sostenible: **(pág. 723)** implica la administración global de los recursos naturales de la Tierra para asegurar que se satisfagan las necesidades energéticas actuales y futuras, sin causar daños al ambiente.

rendimiento sostenible: **(pág. 679)** regeneración de los recursos renovables a la misma velocidad con que se consumen.

rotación sincronizada: **(pág. 780)** estado en que los periodos de la órbita y de rotación de un cuerpo orbitando son iguales.

T

tailings: (p. 686) material left after mineral ore has been extracted from parent rock; can release harmful chemicals into groundwater or surface water.

escombreras: (pág. 686) material que queda después de que se ha extraído la mena de la roca madre; puede liberar sustancias químicas tóxicas hacia las aguas subterráneas y superficiales.

tectonic plates: (p. 480) huge pieces of Earth's crust that cover its surface and fit together at their edges.

placas tectónicas: (pág. 480) enormes fragmentos de corteza que cubren la superficie terrestre; sus límites se corresponden entre sí.

temperate zone: (p. 378) area of Earth that extends between 23.5° and 66.5° north and south of the equator and has moderate temperatures.

zonas templadas: (pág. 378) áreas de la Tierra que se extienden entre los 23.5° y los 66.5°, al norte y al sur del ecuador; experimentan temperaturas moderadas.

temperature inversion: (p. 292) increase in temperature with height in an atmospheric layer, which inverts the temperature-altitude relationship and can worsen air-pollution problems.

inversión de temperatura: (pág. 292) aumento de temperatura que ocurre al aumentar la altitud en alguna capa de la atmósfera; invierte la relación entre la altitud y la temperatura y puede empeorar los problemas de contaminación del aire.

temperature profile: (p. 418) plots changing ocean water temperatures against depth, which varies, depending on location and season.

perfil de temperatura: (pág. 418) diagramas que muestran cómo cambia la temperatura del océano con la profundidad; varía según la ubicación y la temporada.

tephra: (p. 512) rock fragments, classified by size, that are thrown into the air during a volcanic eruption and fall to the ground.

tefrita: (pág. 512) fragmentos rocosos que se clasifican por tamaño; son lanzados al aire durante una erupción volcánica y luego caen al suelo.

terrestrial planets: (p. 804) rocky-surfaced, relatively small, dense inner planets closest to the Sun—Mercury, Venus, Earth, and Mars.

planetas terrestres: (pág. 804) planetas internos, densos, relativamente pequeños, con superficie rocosa y cercanos al Sol: Mercurio, Venus, la Tierra, y Marte.

tetrahedron: (p. 96) a geometric solid having four sides that are equilateral triangles.

tetraedro: (pág. 96) sólido geométrico que tiene cuatro lados con forma de triángulo equilátero.

texture: (p. 119) the size, shape, and distribution of the crystals or grains that make up a rock.

textura: (pág. 119) tamaño, forma y distribución de los granos o cristales que forman una roca.

thermocline: (p. 418) transitional ocean layer that lies between the relatively warm, sunlit surface layer and the colder, dark, dense bottom layer and is characterized by temperatures that decrease rapidly with depth.

termoclina: (pág. 418) capa de transición del océano que se halla entre la capa superficial iluminada por el Sol, que tiene una temperatura relativamente tibia, y la capa inferior, que es densa, oscura y fría; se caracteriza por tener temperaturas que disminuyen rápidamente con la profundidad.

thermometer: (p. 324) instrument used to measure temperature using either the Faherenheit or Celsius scale.

termómetro: (pág. 324) instrumento que sirve para medir la temperatura en grados Fahrenheit o Celsius.

thermosphere: (p. 284) layer of Earth's atmosphere that is located above the mesopause; oxygen atoms absorb solar radiation causing the temperature to increase in this layer.

termosfera: (pág. 284) capa de la atmósfera terrestre ubicada por encima de la mesopausa; los átomos de oxígeno absorben radiación solar, haciendo que la temperatura aumente en esta capa.

tide: (p. 423) periodic rise and fall of sea level caused by the gravitational attraction among Earth, the Moon, and the Sun.

marea: (pág. 423) ascenso y descenso periódicos del nivel del mar causados por la atracción gravitacional entre la Tierra, la Luna y el Sol.

topographic map: **(p. 36)** map that uses contour lines, symbols, and color to show changes in the elevation of Earth's surface and features such as mountains, bridges, and rivers.

topography: **(p. 562)** the change in elevation of the crust.

tornado: **(p. 352)** violent, whirling column of air in contact with the ground that forms when wind direction and speed suddenly change with height; is often associated with a supercell, and can be extremely damaging.

trace fossil: **(p. 608)** indirect fossil evidence of an organism; traces of worm trails, footprints, and tunneling burrows.

trade winds: **(p. 320)** two global wind systems that flow between 30° north and south latitudes, where air sinks, warms, and returns to the equator in a westerly direction.

transform boundary: **(p. 484)** place where two tectonic plates slide horizontally past each another; is characterized by long faults and shallow earthquakes.

transgression: **(p. 649)** occurs when sea level rises and causes the shoreline to move inland, resulting in deeper-water deposits overlying shallower-water deposits.

transported soil: **(p. 177)** soil that has been moved away from its parent material by water, wind, gravity, or a glacier.

tropical cyclone: **(p. 355)** large, low-pressure, rotating tropical storm that gets its energy from the evaporation of warm ocean water and the release of heat.

tropics: **(p. 378)** area of Earth that receives the most solar radiation, is generally warm year-round, and extends between 23.5° south and 23.5° north of the equator.

troposphere: **(p. 284)** layer of the atmosphere closest to Earth's surface, where most of the mass of the atmosphere is found and in which most weather takes place and air pollution collects.

trough: **(p. 421)** lowest point of a wave.

tsunami **(soo NAH mee):** **(p. 548)** large, powerful ocean wave generated by the vertical motions of the seafloor during an earthquake; in shallow water, can form huge, fast-moving breakers exceeding 30 m in height that can damage coastal areas.

mapa topográfico: **(pág. 36)** mapa que usa curvas de nivel, símbolos y colores para mostrar los cambios en la elevación de la superficie terrestre, e incluye rasgos como las montañas, los puentes y los ríos.

topografía: **(pág. 562)** el cambio en la elevación de la corteza.

tornado: **(pág. 352)** violenta columna giratoria de aire en contacto con el suelo; se forma cuando la dirección y la velocidad del viento cambian repentinamente con la altura; a menudo está asociada con una supercelda y puede ser extremadamente dañina.

fósiles traza: **(pág. 608)** pruebas fósiles indirectas de un organismo: incluye rastros de gusanos, huellas de pasos, y madrigueras.

vientos alisios: **(pág. 320)** dos sistemas globales de vientos que se desplazan entre los 30° de latitud norte y sur, donde el aire desciende, se calienta y regresa al ecuador con dirección oeste.

límite transformante: **(pág. 484)** lugar donde dos placas tectónicas se deslizan horizontalmente, una al lado de la otra y en sentidos opuestos; se caracteriza por presentar grandes fallas y terremotos superficiales.

transgresión: **(pág. 649)** ocurre cuando el nivel del mar aumenta y hace que el litoral retroceda hacia el interior; ocasiona depósitos de agua más profunda que cubren depósitos de agua menos profunda.

suelo transportado: **(pág. 177)** suelo que ha sido transportado lejos de su roca madre por el agua, el viento, gravedad o un glaciar.

ciclón tropical: **(pág. 355)** gran tormenta giratoria de baja presión que obtiene su energía de la evaporación de las tibias aguas del mar y la liberación de calor.

trópicos: **(pág. 378)** área de la Tierra que recibe la mayor cantidad de radiación solar, generalmente es caliente todo el año y se extiende entre 23.5° sur y 23.5° norte del ecuador.

troposfera: **(pág. 284)** capa de la atmósfera más cercana a la superficie terrestre; en ella se halla la mayoría de la masa atmosférica, ocurren la mayoría de los fenómenos meteorológicos y se concentran la mayoría de los contaminantes.

seno: **(pág. 421)** punto más bajo de una onda.

tsunami: **(pág. 548)** enorme y poderosa ola marina generada por los movimientos verticales del fondo del mar durante un sismo; en aguas superficiales, puede formar inmensas olas muy rápidas de mas de 30 m de altura que pueden causar daños en las áreas costeras.

turbidity current: (p. 448) rapidly flowing ocean current that can cut deep-sea canyons in continental slopes and deposit the sediments in the form of a continental rise.

corriente de turbidez: (pág. 448) corriente oceánica de flujo rápido que puede formar cañones en los taludes continentales y depositar los sedimentos para formar el pie del talud continental.

U

unconformity: (p. 598) gap in the rock record caused by erosion or weathering.

uniformitarianism: (p. 595) the theory that geologic processes occurring today have been occurring since Earth formed.

uplifted mountain: (p. 573) mountain that forms when large regions of Earth are forced slowly upward without much deformation.

upwelling: (p. 426) upward movement of ocean water that occurs when winds push surface water aside and it is replaced with cold, deeper waters that originate below the thermocline.

disconformidad: (pág. 598) discontinuidad en el registro geológico causada por la erosión o la meteorización.

uniformitarianismo: (pág. 595) este principio establece que los procesos geológicos que ocurren actualmente han estado ocurriendo desde que la Tierra se formó.

levantamiento montañoso: (pág. 573) montañas que se forman cuando grandes regiones de la Tierra son levantadas lentamente sin que ocurra mucha deformación.

corriente resurgente: (pág. 426) movimiento ascendente de las aguas del océano que ocurre cuando los vientos remueven las aguas superficiales, causando que sean reemplazadas por aguas más frías y profundas prove-nientes de profundidades mayores que la termoclina.

V

valley glacier: (p. 208) glacier that forms in a valley in a mountainous area and widens V-shaped stream valleys into U-shaped glacial valleys as it moves downslope.

variable star: (p. 862) star in the giant branch of the Hertzsprung-Russell diagram that pulsates in brightness due to its outer layers expanding and contracting.

varve: (p. 605) alternating light-colored and dark-colored sedimentary layer of sand, clay, and silt deposited in a lake that can be used to date cyclic events and changes in the environment.

vent: (p. 505) opening in Earth's crust through which lava erupts and flows out onto the surface.

ventifact: (p. 203) rock shaped by windblown sediments.

vesicular texture: (p. 120) characterized by containing vesicles, or holes, formed by gas bubbles.

viscosity: (p. 509) a substance's internal resistance to flow.

volcanism: (p. 500) describes all the processes associated with the discharge of magma, hot water, and steam.

glaciar de valle: (pág. 208) glaciar que se forma en un valle de un área montañosa; al deslizarse cuesta abajo, ensancha los valles de corrientes con forma en V y los convierte en valles glaciales con forma de U.

estrella variable: (pág. 862) estrella en la rama de las gigantes del diagrama Hertzsprung-Russell, cuya luminosidad presenta pulsaciones debidas a la expansión y contracción de sus capas exteriores.

varve: (pág. 605) estratos sedimentarios de colores claros y oscuros alternados, compuestos de arena, arcilla y limo, depositados en un lago, que sirven para datar acontecimientos cíclicos y cambios en el ambiente.

chimenea: (pág. 505) abertura en la corteza terrestre por la cual fluye lava hacia la superficie.

ventifacto: (pág. 203) roca moldeada por sedimentos arrastrados por el viento.

textura vesicular: (pág. 120) caracterizado por que contiene vesículas, o agujeros, formados por burbujas de gas.

viscosidad: (pág. 509) resistencia interna a fluir de una sustancia.

vulcanismo: (pág. 500) describe todos los procesos asociados con la descarga de magma, agua caliente y vapor.

W

watershed: **(p. 227)** land area drained by a stream system.

water table: **(p. 254)** upper boundary of the zone of saturation that rises during wet seasons and drops during dry periods.

wave: **(p. 421)** rhythmic movement that carries energy through matter or space and, in oceans, is generated mainly by wind moving over the surface of the water.

wave refraction: **(p. 439)** process in which waves advancing toward shore slow when they encounter shallower water, causing the initially straight wave crests to bend toward the headlands.

weather: **(p. 314)** short-term variations in atmosphere phenomena that interact and affect the environment and life on Earth.

weathering: **(p. 164)** chemical or mechanical process that breaks down and changes rocks on or near Earth's surface and whose rate is influenced by factors such as precipitation and temperature.

well: **(p. 263)** deep hole drilled or dug into the ground to reach a reservoir of groundwater.

wetland: **(p. 240)** any land area, such as a bog or marsh, that is covered in water a large part of the year and supports specific plant species.

windchill index: **(p. 365)** measures the windchill factor, by estimating the heat loss from human skin caused by a combination of wind and cold air.

cuenca: **(pág. 227)** área de terreno drenada por un sistema de corrientes de agua.

capa freática: **(pág. 254)** límite superior de la zona de saturación; aumenta durante la temporada de lluvias y disminuye durante los períodos de sequía.

onda **(ola): (pág. 421)** movimiento rítmico que transporta energía a través de la materia o el espacio; en los océanos, es generado principalmente por el movimiento del viento sobre la superficie del agua.

refracción de onda: **(pág. 439)** proceso en que las olas avanzan hacia la costa y reducen su velocidad, cuando llegan a aguas menos profundas, ocasionando que las crestas de las olas, inicialmente rectas, se inclinen hacia los promontorios.

tiempo: **(pág. 314)** variaciones a corto plazo en los fenómenos que suceden en la atmósfera, que interactúan y afectan el entorno de la vida en la Tierra.

meteorización: **(pág. 164)** proceso químico o mecánico que rompe y modifica las rocas que se hallan sobre o cerca de la superficie terrestre; su velocidad se ve influida por factores como la precipitación y la temperatura.

pozo: **(pág. 263)** hoyo profundo perforado o excavado en el suelo para alcanzar un depósito de agua subterránea.

humedal: **(pág. 240)** toda área, como un pantano o una ciénaga, que se encuentra cubierta de agua gran parte del año y que alberga especies específicas de plantas.

índice de sensación térmica: **(pág. 365)** índice que toma en cuenta el efecto del viento en la sensación térmica, al estimar la pérdida de calor de la piel humana causada por la combinación de viento y aire frío.

Z

zircon: **(p. 620)** very stable and common mineral that scientists often use to age-date old rocks.

zone: **(p. 812)** high, cool, light-colored cloud that rises and flows rapidly in the Jovian atmosphere.

zone of aeration **(p. 254)** region above the water table where materials are moist, but pores contain mostly air.

zone of saturation: **(p. 254)** region below Earth's surface where all the pores of a material are completely filled with groundwater.

circón: **(pág. 620)** mineral sumamente estable que los científicos usan para datar rocas antiguas.

zona: **(pág. 812)** nubes altas, relativamente frías y de color claro, que se elevan y desplazan con rapidez en la atmósfera joviana.

zona de aeración: **(pág. 254)** región sobre el manto freático en que los materiales están húmedos, pero los poros contienen principalmente aire.

zona de saturación: **(pág. 254)** región profunda bajo la superficie terrestre donde todos los poros del material están completamente llenos con agua subterránea.

Italic numbers = *illustration/photo* **Bold numbers = vocabulary term**
act = *activity*

A

Abrasion, 203
Absolute-age dating, 601–605; dendrochronology, 604; estimates of Earth's age, 620; ice cores, 604; radioactive decay and, 601; radiometric dating, 602–603; relative dating v., 588 *act.*; varves, 605
Absolute magnitude, 841, **842,** 843, 845
Absolute zero, 289
Absorption spectrum, 835, 835 *act.*
Abyssal plain, *450,* **451,** 452
Acid-base reaction, 71–72
Acid precipitation, 166, 167, 169, 692, **745,** 744–746
Acid rain. *See* Acid precipitation
Acid, 71–72
Active galactic nucleus (AGN), 873, 874, 875
Active solar heating, 714, 715
Activities. *See* Data Analysis Lab; GeoLabs; Launch Labs; MiniLabs; Problem-Solving Labs. *See also* Foldables
Adiabatic process, 295–296
Adirondack Mountains, 573
Africa, water shortages in, 683
Agile, 768 *table*
Age dating. *See* Absolute-age dating; Relative-age dating
Age of Dinosaurs, 594
Aggregate, 684
Agriculture, *740,* 741, erosion caused by, 202, 741; irrigation and, 694; pesticides, 741; precision farming, 184; soil fertility, 182; soil nutrients, 741, 741 *act.*; topsoil loss and, 683; water use, 265
A-horizon, 179, 181, 183
Air mass, 316–317; arctic, 316 *table,* 317; climate and, 380; cold, model, 312 *act.*; convergence of, 299; fronts, 322; modification of, 317; orographic lifting of, 299; polar, 316, 316 *table*; source regions, 316, 316 *table*; stability of, 298; thunderstorm formation and, 345; tropical, 316, 316 *table*
Air-mass thunderstorm, 346
Air pollution, 690–692, 743–747; acid precipitation, 167, 692, 745–746; from burning of fossil fuels, 283, 743; dilution of, 690; from fires, 690; global warming and. *See* Global warming; greenhouse effect and. *see* Greenhouse effect; ozone depletion and, 304, 305, 743–745; radon, 691; reducing, 746 *act.*, 747; from volcanoes, 690
Air pressure, 290; density and, 291; isobars, 329; measurement of, 324; temperature and, 291, 305 *act.*; units of, 290
Akari, 768 *table*
Albedo, 771
Albite, 88
Aldrin, Edwin "Buzz," 770
Aleutian Islands, 482, *483*
Aleutian Trench, 482
Alfisols, 181
Algae, eutrophication and, 239
Al-Idrisi, *42*
Alleghenian Orogeny, 652
Alluvial fan, 237
Alps, 664
Altered hard parts, 607–608
Alternative energy resources, 706 *act.*; 714–719; bacteria, 724; biofuels, 719; geothermal energy, 717; hydroelectric power, **716;** milestones in development of, *716–717*; nuclear energy, 718; ocean power, 717; solar energy, 714–716; wind energy, 717
Altitude, air density and, 290; climate and, 379; cloud formation and, 300–301; Sun's zenith and, 778; wind speed and, 293
Altocumulus, *300,* 301
Altostratus, *300,* 301
Aluminum, 60, 98, 100
Alvin, 407, 455
Amazonia, 627
Amazon River Basin, deforestation of, 688 *act.*
American Samoan Islands, 489
Amethyst, 101
Amino acid, 634
Amniotic egg, 658
Ampere, 950 *table*
Amphibole, 88 *table,* 146 *table*
Amplitude, 539
Analog forecast, 331
Analysis, scientific method, 13
Ancestral Rockies, 652
Andesite, 119 *table,* 914 *table*
Andesitic magma, 112, 112 *table,* 115 *act.,* 510
Andes Mountains, 482, *483*
Andromeda galaxy, 869, 871, 872
Anemometer, 325
Aneroid barometer, 324
Angiosperm, 658
Angular unconformity, 598, 599 *act.*
Anhydrite, 99, 100 *table*
Animals: bipedal, 665; erosion caused by, 175; in geologic time, 594, 653, 658, 665; impact on environment, 734; as renewable resource, 679; soil development and, 177, 181
Anning, Mary, *592*
Annular eclipse, 783
Annulus, 783
Anorthite, 88
Antarctica, ozone hole over, 304, 745; splitting of Australia from, 660
Antarctic Bottom Water, 419, *420,* 427
Antarctic Intermediate Water, 419, 420
Anthracite, *711*
Apatite, 91 *table*
Aperture, 765

Aphelion, 783
Apogee, 783
***Apollo* missions,** 124, 621, 770
Appalachian Mountains, 565, 571, *572*, 573
Apparent magnitude, 841, **842**
Aqueduct, 696
Aquiclude, 255, 256
Aquifer, 255, 695; artesian, 264, 264 *act.*; confined, 264, 264 *act.*, 266; ground subsistence and, 266; Ogallala, 265, 266; pollution of, 266–268, 269; springs and, *257*, 258; water-table, 263
Archean Eon, *591*–592, 620, 624–625
Archaeologist, 586
Arctic air masses, 316 *table*, 317
Arctic Ocean, 411
Area, 13 *table*, 14
Argon, 282
Arid region, 383
Armstrong, Neil, 770
Artesian aquifer, 264, 264 *act.*
Artesian spring, 264
Artesian well, 264 *act.*, **264**–265, 265 *act.*
Asbestos, *98*, 150
Ash, volcanic, 512
ASOS (Automated Surface Observing System), 325
Asteroid, 622, 799, 816, 818
Asteroid belt, 799, 816, 818
Astrogeologist, 622
Astronaut, 760
Astronomical unit (AU), 800, 801
Astronomy, 6, 7 *table*, space-based, 767–769; time line of developments, *766–767*; tools for studying, 764–769
Astrophysics, 7 *table*
Aswan Dam, *228*
Atlantic Ocean, 411, 656, 664
Atmosphere, 6–7, **8;** air density, 290; air mass stability, 298; carbon cycle and, 688; clouds in. *See* Clouds; composition of, 8, 166–167, 282–283; energy transfer in, 286–288; formation of early Earth's, 618 *act.*, 628–629; global water supply and, 252 *table*; humidity and, 294–296; layers of, 280 *act.*, 284, 284 *table*, *285*, 286; as natural resource, 678; nitrogen cycle and, 688; ocean interactions with, 412; origins of oxygen in, 629–631, 687; pollution of. *See* Air pollution; pressure-temperature-density relationships, 291; temperature inversions, 292; water cycle and, 224, 303; wind in, 293
Atmospheric chemistry, 7 *table*
Atmospheric lifting, 299
Atmospheric pressure. *See* Air pressure
Atom, 60–65
Atomic clock, *15*
Atomic number, 62
Augite, 913 *table*
Australia, splitting of from Antarctica, 660
Automated Surface Observing System (ASOS), 325
Automobile, *See* Motor vehicles
***Autonomous Benthic Explorer (ABE)*,** 455
Autonomous underwater vehicle, 455
Autumnal equinox, 777
Avalanche, 198
Avalonia, 651
Axis, Earth's, 391, 775–777

B

Bacteria, energy from, 724; groundwater pollution from, 241, 267; in soil, 180, 182; increase of with new and full moon phases, 428; nitrogen-fixing, **688**, *689*; pollution-eating, 724
Ball lightning, 348
Baltica, 651
Bamboo, 679
Banded iron formation, 630, 685
Barchan dune, 205 *table*
Bar graph, 18, 952–953
Barometer, 324
Barred spiral galaxy, 870, *871*
Barrier island, 442, *443*, 443, 445
Basalt, 113, 120, 914 *table*
Basalt flow, 118
Basaltic magma, 112, 112 *table*, 510
Basaltic rock, 113, 115 *act.*, **118,** 119 *table*
Base, 72
Base level, 233, 237
Basin and Range Province, 662, *663*
Batholith, *514*, **515,** 517
Bathysphere, 406
Bauxite, 100
Baymouth bar, 442, *443*
Beach, 438. *See also* Shorelines
Bedding, 137–138, *139*
Bed load, 228
Bedrock, 165, 177, 255, **684**
Belt, 812
Beryllium, 122
B-horizon, 178
Bias, 13
Big Bang theory, 874, **878**–880
Big Dipper, 837
Bimetallic-strip thermometer, 324
Binary star, 838, *839*–840
Biochemical sedimentary rock, 142 *table*, 144, 914 *table*
Biodiesel, 719
Biofuel, 719
Biogenous sediment, 453, 454 *table*
Biogeochemical cycle. *See* Carbon cycle; Nitrogen cycle; Water cycle
Bioluminescence, 417
Biomass, 709
Biomass fuel, 709–710, 719; biodiesel, 719; ethanol, 719; fecal materials, 709; field crops, 709; peat, 710; wood, 709
Bioremediation, 742
Biosphere, 9
Bipedalism, 665
Bituminous coal, 710, *711*
Black hole, 851, 865, 876–877, 882
Black smoker, 452
Block, volcanic, 512
Blue jet, 348
Blueshift, 840
Bog, 240, 710
Bornite, 912 *table*
Bottom water, formation of, 419
Boundary mountain, 561 *act.*, 572
Bowen, N. L., 114
Bowen's reaction series, 114–115, 117, 147
Brahe, Tycho, 800

Index

Breaker, 422
Breakwater, 444
Breccia, 124, 141, 772
Bronze, *98*
Buffy, *817,* 817 *table*
Buildings, energy efficient, 725 *act.*
Burgess Shale, *592*
Bushveldt Complex, 685

C

C-14. *See* Carbon-14
Calcite, 88 *table,* 91 *table,* 94, 98, 121, 146 *table,*166, 259–262
Calcium feldspar, 113
Caldera, **505,** 505 *act.*
California Water Project, 696
Callisto, 812
Cambrian explosion, 653
Cambrian Period, *591,* 648, 653
Canadian Shield, 625
Candela, 950 *table*
Canis Major dwarf galaxy, 872
Capillary water, 254
Carbon-14, 603
Carbonate, 98, 100 *table,* 140, 142 *table,* 144
Carbon cycle, 283, 688, 688 *act., 689*
Carbon dating. *See* Radiocarbon dating
Carbon dioxide, atmospheric, 282, 283; carbon cycle and, 412, 688, *689*; chemical weathering and, 167, 259; global warming and, 394–395, 396, 743; greenhouse effect and, 393; ocean as reservoir of, 412; carbonic acid, 72, 167, 259–261
Carboniferous Period, *591,* 594, *648,* 654
Careers in Earth Science, archaeologist, 586; astronaut, 760; cartographer, 31, 44; climatologist, 377; computer programmer, 874; environmental consultant, 721; environmental technician, 674; geochemist technician, 240; geologist, 56; glaciologist, 160; hurricane hunter, 360; hydrogeologist, 266; hydrologist, 750; landscaper, 177; lapidary, 90; marine geologist, 449, 475; marine scientist, 278; materials engineer, 680; meteorologist, 316; oceanographer, 408; paleoecologist, 652; petroleum geologist, 600; petrologist, 117, 568; planetary geologist, 622; planetologist, 798; sedimentologist, 138; space engineer, 765; spectroscopist, 845; speliologist, 2; volcanologist, 464; weather observer, 298
Carruthers, George, *15*
Carrying capacity (stream), 229
Carrying capacity (population), 735
Cartographer, 31, 468
Cartography, 30. *See also* Mapmaking; Maps
Cascade Mountains, 662
Cassini, 768, 812, 813, 820
Cast, 608
Cause and effect, 895
Cave, 260; limestone, 72, 259–260; Mammoth Cave, Kentucky, 261; Naica Cave, Mexico, 102; stalactite and stalagmite formations, 261–262
Celsius (°C) scale, 15, 289, 324, 950
Cementation, 137
Cenozoic Era, *591,* 593–594; cooling during, 660–661; day length in, 652 *act.*; life-forms of, *660,* 665; Miocene warming, 661; mountain-building events, 662, *663,* 664
Center of mass, 803
Centimeter (cm), 13–15, 13 *table,* 15
Cepheid variable, 863, 867
Ceres, 632, 816
Chalcedony, 93
Chalcopyrite, 912 *table*
Chalk, formation of, 436 *act.*
Challenger, 406
Chandra X-Ray Telescope, 769, 793
Changes of state, 75
Chaos theory, weather forecasting and, 333
Charon, *817,* 818
Chemical bond, 67; covalent, 66, *68*; ionic, 68; metallic, 68, *69*
Chemical equation, 70
Chemical formula, 66–68, 70 *act.*
Chemical precipitation, 685
Chemical reaction, 70
Chemical sedimentary rock, 142 *table,* 143, 914 *table*
Chemical weathering, 134, **166**–167, 168
Chert, 144, 685
CHIPSat, 768 *table*
Chlorofluorocarbon (CFC), 283, 304, 745, 747
C-horizon, 178
Chromite, 912 *table*
Chromium, 685
Chromosphere, 831
Cinder cone, 499 *act.* volcano, 506 *table,* **507**
Circle graph, 18, 953
Circum-Pacific Belt, 501, 544
Circumpolar constellation, 837
Cirque, 209
Cirque lake, 212, 238
Cirrocumulus, *300*
Cirrostratus, *300,* 301
Cirrus, *300,* 301
Clack, Jenny, *593*
Clastic, 134
Clastic sedimentary rock, 134, **141,** 142 *table,* 914 *table*
Clay, 183
Clean Air Act of 1972, 746 *act.*
Clean Water Act of 1972, 750
Clear-cutting, 739
Cleavage, 92, 92 *act.,* 912–913 *table*
Clementine, 770
Climate, 314, 376–380; air mass movements and, 380; evidence of continental drift from, 471; glaciers and, 208; Köppen classification system, 374 *act.,* 381, *382,* 383–385; latitude and, 378; mass movements and, 194; microclimates, 385–386, 397 *act.*; normals, 377, 377 *act.*; rate of weathering and, 168–169; soil development and, 179; topographic effects on, 379; weather v., 314
Climate model, 751
Climatic change, 387–392, 393–395; Cenozoic Era changes, 660–661; Earth's orbit and, 390; Earth's tilt and, 391; Earth's wobble and, 391; El Niño, 388–389; global warming, 393–395, 396; greenhouse

effect, 393, 394 *act.*; human causes of, 393–395; ice ages, 387; seasons, 388; solar activity and, 390; volcanic activity and, 392, 512
Climatologist, 377
Climatology, 7 *table*, 376
Closed universe, 879
Cloud, formation of, 280 *act.*, 297–299; model coverage, 374 *act.*; precipitation from, 302; types of, 300–301
Cluster, star, *838*, 839
Coal, 710, *711*, 746, 747, 914 *table*; deposits of and continental drift, 471; formation of in Carboniferous, 654; geographic distribution of, 680
Coalescence, 302
Coarse-grained clastic sedimentary rocks, 141, 142 *table*
Coastline. *See* Shoreline
Coast Range Batholith, 515
Cogeneration, 723
Cold front, 322
Cold-front thunderstorm, 346
Cold wave, 364
Colima Volcano, 510
Color, mineral characteristics, 93, 912–913 *table*; soil, 183
Colorado River Compact, 698
Columbia River Basalt, 504
Comet, 408, 799, **819**
Communication, 17; graphs, 18; lab reports, 17; models, 18–19; precise, 4 *act.*
Compaction, sediment, 137
Comparisons, making, 891
Composite volcano, 499 *act.*, 506 *table*, **507**
Compound, 66; bonds in, 67–68, *69*; chemical reactions form, 70, 70 *act.*
Compression, 528, *531*
Compressive force, 567
Computer, oceanography and, 406, 407
Computer programmer, 874
Condensation, 75, 295–296, 206 *act.*
Condensation nuclei, 297
Conduction, 286, **288**
Conduit, 505
Confined aquifer, 264, 264 *act.*, 266
Conglomerate, 141, 914 *table*
Conic projection, 35, 39 *table*
Conservation of energy, 75
Constant, 12
Constellation, 837–838
Consumer, 708
Contact metamorphism, 148, **149**
Continental climate, *382*, 384
Continental-continental plate boundary, 484, 570
Continental crust, 8, *478*, 482, 624; density of, 561 *act.*, 563; displacement of mantle by, 561 *act.*, 562–566
Continental drift, 469–471; early observations of, 468; evidence supporting, 469–471; rejection of, 472; seafloor spreading theory and, 479
Continental glacier, 207, **208**
Continental margin, 437 *act.*, **447**–449, 449 *act.*
Continental polar (cP) air mass, 316, 316 *table*, 353
Continental rise, 449
Continental shelf, 447–448
Continental slope, 448
Continental tropical (cT) air mass, 316, 316 *table*
Continent. *See also* Continental drift; Cenozoic Era collisions of, 664; collisions of, model, 653 *act.*; early observations of drifting of, 468; growth of during Precambrian, 625, *626*, 627, 639 *act.*; microcontinents, 624–625, 639 *act.*; setting of during Paleozoic, 648
Continuous spectrum, 835
Contour interval, 36
Contour line, 36
Control, 12
Convection, 286, **288;** plate movement and, 486–487; transfer of Sun's energy by, 834
Convection cell, 347
Convection current, 288, 486–487
Convergence, 299
Convergent boundary, 480, **482,** 483 *table*, 484; mantle convection and, 487; mountain-building events at, 567–570; volcanism, 500, 501
Convergent volcanism, 500, 501
Coordinates, locating places with, 32
Copernicus, Nicolaus, 800
Copper, 99, 150, 680, 912 *table*
Coprolite, 608
Coral reef, 650
Cordilleran Orogeny, 657
Core, Earth's, 8, 536, 619 *act.*
Coriolis effect, 318, *319*, 323, 425, 426, 775
Corona, 831
Coronal hole, 833
Coronal mass ejection, 852
Correlation, 599, 600, 608
Corundum, 91 *table*, 913 *table*
Cosmic background radiation, 878
Cosmology, 876–879; abundance of elements in universe, *65*; age of universe, 876; Big Bang theory, 872, 876–877; contents of universe, 879; expansion of universe, 871 *act.*, 871–872, 876–877, 878
Covalent bond, 67, 68, *69*, 72
Crater (volcanic), 505, 505 *act.*
Craton, 625
Creep, 195
Crest, 421, 423
Cretaceous Period, *591*, *655*, 657–659
Crop, fuels from, 709, 719. *See also* Agriculture
Cross-bedding, 138, *139*
Cross-cutting relationship. *See* Principle of cross-cutting relationships
Crude oil, 712. *See also* Petroleum
Crust, Earth's, 8. *See also* Continental crust; Oceanic crust; age of, 490–491 *act.*; density of, 619 *act.*; elements in, 65; formation of early Earth's, 623–624; minerals in, 86; natural resources from, 682–686
Cryosphere, 5, **8**
Crystal, 73, **89**–90
Crystalline structure, 73, 86–87
Crystallization, fractional. *See* Fractional crystallization
Cumulonimbus, *300*, 301
Cumulus, *300*, **301**

INDEX

Curiosity, 768
Cyanobacteria, 629, 631–633, 635
See also Bacteria
Cycad, 658
Cyclone. *See* Tropical cyclone

D

Dam, 696, 716
Dark energy, 879
Dark matter, 870, 879
Data Analysis Labs. *See also* GeoLabs; Launch Labs; MiniLabs; Problem-Solving Labs; Clean Air Act of 1972 and pollution emissions, 746 *act.*; day length in different geologic eras, 652 *act.*; earthquake epicenters, 543 *act.*; element identification by spectral lines, 835 *act.*; energy resources and home heating, 722 *act.*; graphs help interpret data, 18 *act.*; heat waves, 364 *act.*; mineral identification chart, 94 *act.*; normal average temperatures, 377 *act.*; radioactivity in ice cores, 208 *act.*; soil texture, 182 *act.*; tidal data, 423 *act.*; zones of volcanism and lava production, 501 *act.*
Day length, 652 *act.*, 775
Debate skills, 899
Debris, 799
Deccan Traps, 504
Deep-sea mud, 453
Deep-sea trench, 449, **451,** 474, 482
Deepwater mass, 419, 420
Deflation, 202–203
Deflation blowout, 202
Deforestation, 395, 688 *act.*, 709, **739**
Degrees north latitude, 30
Degrees of longitude, 31, 32
Degrees south latitude, 30
Deimos, 808
Delta, 173, **236**
Dendrochronology, 604
Density, 13 *table*, 15; differentiation of Earth's layers (zones) and, 618 *act.*, 623–624; mantle displacement and, 561 *act.*; mineral identification by, 95; pressure and temperature of air and, 291; of seawater, 417; of universe, 877
Density current, 427
Density-dependent factor, 736
Density-independent factor, 736
Deoxyribonucleic acid. *See* DNA (deoxyribonucleic acid)
Dependent variable, 12
Deposition, 136, 136 *act.*, **171,** 205, 210, 236
Depression contour line, 37
Desalination, 697
Desertification, 683
Desert pavement, 203
Desert, 383
Development, environmental impact of, 739, 741 *act.*
Devonian Period, *591, 648,* 653–654
Devonian Period extinction, 653
Dew point, 295, 295 *act.*
Diamond, 91 *table*, 101, 123, 680
Differentiation, 623–624
Digital forecast, 331
Dike, *514,* **516**
Dinosaur dig, 666
Dinosaur, 589 *act.*, 594, 658, 659; fossils of, 666, 667 *act.*; original preservation of tissue of, 607, 610
Diorite, 118, 914 *table*
Disaster zone, mapping, 47
Discharge, 228, **229**
Disconformity, 598
Disklike galaxy, 867
Divergent boundary, 480, **481,** 481 *act.*; mountain building and, 571–573; volcanism along, 500, 502
Divergent volcanism, 500, 502
Divide, 227
DNA (deoxyribonucleic acid), 635
Dolomite, 98, 100 *table*, 146 *table*
Doppler effect, 327, 840
Doppler shift, 840
Doppler weather radar, 327
Double refraction, 94 *table*
Downburst, 351
Downdraft, 347, 249, 351
Drawdown, 263
Drinking water, safe, 242, 269, 694
Drought, 362
Drumlin, 210, *211*
Dry adiabatic lapse rate, 296
Dry climate, *382,* 383
Dune, 204, 205 *table*
Dust, 283, 690, 745
Dust Bowl, 202
Dust storm, 202
Dwarf elliptical galaxy, 867
Dwarf planet, 816, *817*
Dwarf spheroid galaxy, 868

E

Earth, 798 *table*, 803, 808. *See also* Sun-Earth-Moon system; age of, 620–621; atmosphere of. *See* Atmosphere; composition of, 536; core of, 8, 536, 619 *act.*; crust, formation of, 623. *See also* Crust; early. *See* Precambrian Earth; four systems of, 8–9; history-shaping events, 611 *act.*; imaging interior of, 538; imbalanced solar heating of, 314–315, 315 *act.*; internal structure of, 536, *537*; magnetic field of, 476–477; mantle of. *See* Mantle; ocean coverage, 405 *act.*; orbit of, 390, 776; physical data on, 915 *table;* physiographic map of, 906–907; Proterozoic life on, 633–635, 634 *table*; rotation of on axis, 775; seismic evidence of interior of, 536, *537,* 538; tilt of axis, 391, 776; topography of, 562–563; wobble of, 391
Earthquake, 528–551; belts of activity, 543–544; causes of, 526 *act.*, 528–529; depth of focus of, 541–542; distance to, determine, 542; epicenter of, 533, 542, 543 *act.*; faults and, 530–531; focus of, 533; forecasting, 549–551; global distribution of, 543–544; intensity of, 541, 541 *act.*; learning from past earthquakes, 552; location of, determine, 542, 543, 553 *act.*; magnitude of, 539; measurement of, 534–535; modified Mercalli scale for, 540–541, 541 *act.*; Richter scale for, 539; structural damage caused by, 545–546; time line of major, *530–531*; time of, determine, 543; travel-time curves, 535, 542, 543; tsunamis from, 548; waves from, 532–533
Earthquake hazard, 545–548;

Index

landslides, 547; soil liquefaction, 547; structural damage, 545–546; tsunamis, 548
Earthquake-recurrence rate, 550
Earthquake scale, modified Mercalli, 540 *table,* 540–541; moment magnitude, 540; Richter scale, 539
Earth science, areas of specialization, 6–7; major events in, *14–15*; subspecialties, 7 *table*
Earth Science & Society, drinking water, shortage of safe, 242; global warming, effects on arctic, 396; La Conchita landslide, 213; non-sustainable use of freshwater resources, 698; San Francisco earthquake (1906), 552; weather forecasting, precision of and chaos theory, 333
Earth Science & Technology, black holes, 880; disaster zones, mapping, 47; Earth's past climate, climate change, measuring and modeling, 610, 751; liquid crystal displays (LCDs), 76; microenvironments on Mars, 638; precision farming, 184; solar storms, affect on technology, 852; water on planets, gathering evidence of, 820
Earth Science & the Environment, bacterial counts and moon phases, 428; drinking water, ensuring safety of, 269; bacteria, 724; moon rocks, 124; ozone layer recovery, 304; pollution-eating bacteria, 724; Vailulu'u Seamount, 489
Earthworm, 740
Eccentricity, 801, 801 *act.*
Echo sounding, 473
Eclipse, lunar, 784; solar, 781–783, 782 *act.*
Ecliptic plane, 776
Ecosystem, main energy source, 708
Ediacaran biota, 636–637
Ediacaran Period, *591,* 593, 636–637
Effervescence, 94 *table*
Egg, amniotic, 658
Einstein, Albert, 834
Ejecta, 771
Elastic deformation, 529
Elastic-rebound theory, 552
Electric current, 950 *table*
Electricity, home heating by, 722 *act.*; from photovoltaic cells, 716
Electromagnetic radiation, 764–765
Electromagnetic spectrum, 764–765
Electrometeor, 314
Electron, 61; energy levels, 63; valence, 64
Element, 60; abundance of, 65; atomic number, 62; identify by spectral lines, 835 *act.*; identify classroom, 62 *act.*; isotopes, 62–63; mass number, 62; periodic table of, 61; in Sun, 836; spectral lines, 835 *act.;* stellar, 853 *act.* symbols for, 61
Elevation, 36–37, 37 *act.*
Ellipse, 800
Elliptical galaxy, 868, *869*
Elliptical orbit, 801, 801 *act.*
El Niño, 388–389, 407, 412
Emerald, 101
Emergent coast, 446, 456–457 *act.*
Emission spectrum, 835
Enceladus, 813, 820
Endeavour, *43*
Energy. *See also* Energy resources; conservation of, 395; law of conservation of, 75; solar, 678, 708, 714–715, *717,* 834; source of, identify, 706 *act.*
Energy efficiency, 721–723, 725 *act.*
Energy level, electron, 63
Energy resources, 708–723. *See also* Natural resources; alternative, 706 *act.,* 714–719; biomass fuels, 709–710; energy efficiency and, 721–723, 725 *act.*; fossil fuels, 710–713; global use of, 720; energy producing bacteria, 724; home heating with, 722 *act.*; light energy, 708; solar energy, 714–715, *717*; source of, identify, 706 *act.*; sustainable energy, **723**
Enhanced Fujita Tornado Damage scale, 353
Enterococci, 428
Environmental chemistry, 7 *table*
Environmental consultant, 721
Environmental science, 7
Environmental soil science, 7 *table*
Environmental technician, 674
Environment, human impact on, 734. *See also* Earth Science and the Environment
Enzyme, 635
Eocene Epoch, *591,* 660, 665
Eon, *591,* **592**
Epicenter, 533; global distribution of, 543–544; locating, 535, 542, 543 *act.,* 553 *act.*
Epicentral distance, 542, 553 *act.*
Epoch, *591,* **593**
Equator, 30; solar radiation reaching, 315, 315 *act.*
Equinox, 777
Era, *591,* **593**
Eris, 816, *817,* 817 *table,* 818
Erosion, 171–175; abrasion, 203; agents, 135; deflation, 202–203; deforestation and, 739; glacial, 174, 209; gravity and, 171; gully, 172; by plants and animals, 175; rill, 172; sediment removal, 135; shorelines and, 173, 439–441; of topsoil, 683, 739, 742; 172 *act.,* 172–173; 174, 201–204
Esker, 210, *211*
Essen, Louis, *14*
Estuary, 414
Ethanol, 719
Eukaryote, 635, 636
Eukaryotic cell, 635
Europa, 812, 820
Europa Jupiter System Mission (EJSM), 820
Eutrophication, 239
Evaporation, 74, 75; minerals formed from, 89, 143, 650; water cycle, 224, 303
Evaporite, 89, **143,** 650
Evolution, 606. *See also* Life, origins of
Exfoliation, 165
Exosphere, 284 *table, 285,* **286**
Expansion constant. *See* Hubble constant (*H*)
Expeditions, Appalachian Trail, 577; deep ocean floor survey, 455; dinosaur dig, 666; Hawaiian Volcano Observatory (HVO), 518; Naica Cave in Mexico, 102; Central Park rocks, 152; storm

Index

spotters, 366; weightlessness, effect on astronauts, 785
Experimentation, 12. *See also* Data Analysis Labs; GeoLabs; Launch Labs; MiniLabs; Problem-Solving Labs
Exponential growth, 735
Extinction events, 594, 653, 654, 659
Extraterrestrial life, 632, 638
Extrusive rock, 118, 119–120, 964 *table*
Eye (hurricane), 356
Eyewall, 356, 358

F

Fahrenheit (°F) scale, 289, 324, 950
Failure, 529
Fault, 530–531; cross-cutting relationships and, 597 579 *act.*; movement along, 466 *act.*; normal, 530; reverse, 530; strike-slip, 531; types of, 526 *act.*, 531 *table*
Fault-block mountain, 574, 578–579 *act.*
Fecal material, energy from, 709
Feldspar, 88 *table,* 96; Bowen's reaction series, 115; Mohs scale of hardness, 91 *table*; plagioclase, 88, 118, 119 *table,* 913 *table*; potassium, 113, 117, 118, 119 *table,* 913 *table* ; silicate structure and, 96; thin sections of, 121
Fertilizer, 182, 239
Fetch, 422
Field crop, energy from, 709
Field guide to minerals, 103 *act.*
Fine-grained sedimentary rock, 143 *table*, 142
Fire, air pollution from, 690
Firework, *99*
First aid, 954 *table*
First law of planetary motion, 800
First law of thermodynamics, 75
First quarter moon, 778, *779*
Fission, 834
Fissure, 504
Fitz, Ella Eliza, *43*
Fjord, 445
Flat universe, 877
Flint, 93
Float, 406, 407
Flood, 230–231, 359, 361, 362 *act.*
Flood basalt, 504
Floodplain, 230
Flood stage, 230
Flood warning system, 231
Flowering plant, 658
Flow, 196
Fluorescence, 94 *table*
Fluorite, 91 *table*, 94 *table*, 913 *table*
Focus (earthquake), 533, 541–542
Foldables, atmosphere, formation of, 619 *act.*; atmosphere, layers of, 281 *act.*; Earth's systems, 5 *act.*; external processes that shape Earth, 193 *act.*; faults, 527 *act.*; fronts, 313 *act.*; galaxies, 859 *act.*; igneous rocks, 111 *act.*; Koppen classification system, 375 *act.*; life forms of Paleozoic Era, 647 *act.*; lunar phases, 763 *act.*; mapping technologies, 29 *act.*; mineral identification, 85 *act.*; mountain-building processes, 561 *act.*; planets in solar system, 795 *act.*; plate boundaries, 467 *act.*; relative v. absolute dating, 589 *act.*; renewable v. nonrenewable resources, 677 *act.*; rock cycle, 133 *act.*; seafloor features, 437 *act.*; states of matter, 59 *act.*; stream development, 223 *act.*; thunderstorms, 343 *act.*; volcano classification, 499 *act.*; water pollution, 733 *act.*; water supply, threats to, 251 *act.*; wave characteristics, 405 *act.*; weathering, 163 *act.*
Foliated, 146
Foliated metamorphic rock, 146, *147*
Fool's gold. *See* Pyrite
Foraminifera, 610
Forecast, weather. *See* Weather forecasts
Forest, deforestation and. *See* Deforestation; loss of Amazon River Basin, 688 *act.*; as renewable resource, 679
Forest fire, 690
Forestry, 739. *See also* Deforestation
Fortified cereal, metals in, 59 *act.*
Fossil, 140, 606–609; altered hard parts, 607–608; casts, 608; correlation of in rock formations, 600, 606; dinosaur, 666, 667 *act.*; of earliest life-forms, 592; evidence of continental drift and, 470; evidence of evolution in, 606; formation of, 140, 588 *act.*; fossil footprints, 132 *act.*; of humanlike primates, 665; index, 609; mineral replacement, 607; molds, 608; original preservation, **607;** recrystallization, 608; time line of discoveries and technologies, *592–593*; trace, 608
Fossil fuel, 710–713; coal, 710, *711*, 746, 747, 964 *table*; conservation, 719–720, global consumption rates, 720; limits on availability of, 680, 714, 720; natural gas, 710, 712; oil (petroleum), 710, 712 *act.*, 712–713; pollution from, 167, 283, 394, 688, *689*
Foucault pendulum, 775
Fractional crystallization, 115, *116*, 117
Fractional scale (maps), 39, 40
Fracture, 92 *act.*, **93,** 912–913 *table*
Fracture zone, *450*
Freeze-dried food, 9
Frequency, 764–765, 772–773
Freshwater resource, 693, 694–696; conservation of, 748; daily water usage, 699 *act.*; distribution of, 252 *table*, 694; global consumption rates, 695; management of, 695–696; percent of water on Earth, 409; pollution of, 733 *act.*, 749, 752–753 *act.*; properties of water and, 693–694; reducing pollution of, 750; shortages of, 694, 698; threats to, 265–268
Freshwater wetland, 240–241
Frog populations, 862
Frontal thunderstorm, 346
Front, 312 *act.*, **322**
Frost wedging, 164
Fuel, 709. *See also* Alternative energy sources; Fossil fuels
Fujita, Theodore, 353
Full moon, *779*
Funnel cloud. *See* Tornadoes
Fusion, 834, 847, 848

G

Gabbro, 118, 914 *table*
Gagarin, Yuri, 770
Galactic disk, 863
Galactic nucleus, 864
Galaxy, 869–873; active galactic nucleus (AGN), 875, 877; Andromeda, 869, *871*, 872; black holes in, 865; classification of, 868, 870, 881 *act.*; clusters of, 870–872, *883;* dark matter in, 870; discovery of, 869; disklike, 870; dwarf ellipticals, 869; dwarf spheroid, 870; elliptical, 870, 871; galactic groups, *871*, 872; irregular, *871*, 872; Local Group of galaxies, *871*, 872; luminosities of, 870; masses of various, 869; Milky Way. *See* Milky Way; radio, 875; redshift of, 873, 874; Sombrero, *865, 867;* spiral, *871;* types of, 870–872, 883 *act.*
Galaxy clusters, 871
Galena, 150, 912 *table*
Galex, 768 *table*
Galileo, 811, 812, 820
Galileo, Galilei, 802
Gamma ray, 764, 767
Ganymede, 812, 820
Garnet, 88 *table*, 913 *table*
Gas, 74; changes of state, 75; pressure-temperature-density relationships, 291, 305 *act.*
Gas giant, 798, 803, **811**–815. *See also* specific planets
Gastrolith, 608
Gem(stone), **101,** 680
Geocentric model of solar system, 799
Geochemical cycle. *See* Carbon cycle; Nitrogen cycle; Water cycle
Geochemistry, 7 *table*
Geochemist technician, 240
Geographic Information System (GIS), **44,** 46, 47, 184
GeoLabs, coastal landforms, 456–457 *act.*; crystal formation, 125 *act.*; daily water usage, 699 *act.*; dinosaur fossils, 667 *act.*; earthquake epicenter, 553 *act.*; Earth-shaping events, 611 *act.*; energy-efficient building design, 725 *act.*; galaxy classification, 883 *act.*; groundwater pollution, 270–271 *act.*; landslide mapping, 214–215 *act.*; map profile, 578–579 *act.*; measurement and SI units, 21 *act.*; microclimates, 397 *act.*; mineral field guide, 103 *act.*; Moon, relative dating of features on, 786–787 *act.*; plate boundaries and isochrons, 490–491 *act.*; precipitate salts, 77 *act.*; pressure-temperature relationships, 305 *act.*; sedimentary v. metamorphic rock, 153 *act.*; solar system, scale model of, 821 *act.*; stream velocity, 243 *act.*; topographic maps, 48 *act.*; tropical cyclone, path of, 367 *act.*; volcano, evaluate safety of, 519 *act.*; water masses, 429 *act.*; water pollution, sources of, 752–753 *act.*; weathering, 185 *act.*; weather maps, 334–335 *act.*
Geologic map, **38,** 39 *table*
Geologic time scale, **590,** *591*, 592–594. *See also* specific eras and periods; Earth-shaping events during, 611 *act.*; eons, 592; epochs, 593; eras, 593; life, appearance of, 592; life-forms, succession of, 594; periods, 593
Geologist, 56
Geology, **7,** 7 *table*
Geomagnetic time scale, 476
Geosphere, **8**
Geothermal energy, **717**
Geothermal gradient, 113
Geyser, **258**
Gibbs Fracture Zone, 484 *act.*
Glacial erosion, 174, 209
Glacial lake, 212
Glacial till, 210
Glacier, **207**–212; continental, 207, 208; dating from ice cores from, 208 *act.*, 604; deposition by, 136, 174, 210, 210 *act.*, *211*, 212; distribution of, 207; erosion by, 135, 174, 209; evidence of continental drift and, 471; formation of, 207; global water supply and, 252 *table*; ice ages and, 661; movement of, 208, 209; sea level variation and, 410; valley, 207, 208
Glacier National Park, 629
Glaciologist, 160
Glass, **73**
Global Positioning System (GPS), **44,** *45*, 47, 184, 326, 551
Global warming, **393**–395, 396, 445, **743**–744, 751
Global wind system, 318–321
Globular star cluster, *838*, 863
Gneiss, 146 *table*, *147*, 914 *table*
Gnomonic projection, **35**
Gold, 91, 95, 121, 149, 150, 680, 685, 912 *table*
Gondwana, 652
Government-protected land, 682
GPS. *See* Global Positioning System (GPS)
GPS receiver, 44, *45*
Graded bedding, **138**
Gradient, measure, 37 *act.*
Grand Canyon, 237, 590, 596, 599
Grand Coulee Dam, *716*
Grand Staircase, 599
Grand Tetons, 574, 578–579 *act.*
Granite, 113, 118, 121, 123, 624, 914 *table*
Granitic rock, **118,** *119 table*
Graph, 18, 951–953; bar, 18, 952–953; circle, 18, 953; line, 18, 901–902; make and use, 18 *act.*
Graphic (map) scale, 39, 40
Graphic organizer, 898
Graphite, 150, 912 *table*
Gravitational contraction, 621
Gravitational force, 14
Gravitational water, 254
Gravity, erosion and deposition and, 135, 171; law of universal gravitation, 802; mass movements and, 194; planetary motion and, 802–803; stream formation and, 233
Great Plains, 180, 265; deflation affecting, 202
Greenhouse, effect, **393,** 394 *act*, 807; gases, 744
Green River Formation, 713
Grenville orogeny, *626*
Groin, 444
Ground subsidence, 266
Groundwater, 250–268, 695; capillary, 254; deposition of calcite by,

261–262; dissolution of limestone by, 259–261; drawdown of, 263; drinking water safety and, 269; gravitational, 254; infiltration of precipitation, 253; movement of, 255; overpumping of, 266; pollution of, 266, 267, 268, 270–271 *act.*, 749; protection and restoration of, 268; recharge of, 263; springs, 256, *257*, 258; storage of, 250 *act.*, 253; threats to, 265–268; water table and, 254; wells, 263–265, 264 *act.*, 265 *act.*; zone of saturation and, 254
Gully, 172
Gully erosion, 172
Gunpowder, *99*
Guyot, 450, **452**
Gypsum, 91 *table*, 650
Gyre, 425–426

H

Hachure, 37
Hadean Eon, *591*–592, 620
Hail, 301, 302, 351
Half Dome (Yosemite National Park), 165
Half-life, 602, 603, 603 *table*
Halide, 99, 100 *table*
Halite, 77 *act.*, 85 *act.*, 90, 92, 99
Halley's comet, 819
Halo, 863, 865
Hanging valleys, 209
Hardness, 91, 912–913 *table*
Hard water, 262, 695 *act.*
Haumea, 816, *817*, 817 *table*
Hawaiian-Emperor Volcanic Chain, 503
Hawaiian Volcano Observatory (HVO), 518
Headlands, 439
Headward erosion, 232
Headwaters, 232
Heat, latent, 295, 345, 356; from radiation on Precambrian Earth, 621; thermal energy and, 289, 707 *act.*
Heat index, 363, 363 *table*
Heating oil, home-heating from, 722 *act.*
Heliocentric model of solar system, 800–802
Helium, 65, 282, 836
Hematite, 93, 99, 100, 166, 684, 912 *table*
Hercules, 837
Hertzsprung-Russell (H-R) diagram, 845, 845 *table*
Hess, Harry, *15*
Heterogeneous mixtures, 71
High clouds, 300, 301
High-grade metamorphism, 148
Highland climate, 385
Highlands, lunar, **771,** 772
High Plains aquifer, 265, 266
High-pressure systems, 323
Himalayas, *483*, 484, 566, 570, 664
Holocene Epoch, *591*, 593, 660
***Homo sapiens*, 665**
Homogeneous mixtures, 71
Hoover Dam, *716*
Horizon, soil, **178,** *181*, 182 *act.*
Hornblende, 913 *table*
Horns, 209
Horse latitudes, 320
Hot spot, 502–504
Hot spring, 258
H-R diagram. *See* Hertzsprung-Russell (H-R) diagram
Hubble constant (*H*), 874, 874 *act.*, 878
Hubble, Edwin, 869, 870
Hubble's law, 874
***Hubble Space Telescope (HST)*,** *15*, 768, 769, 793, 874, 878
Humans, emergence of in Cenozoic Era, 665; erosion from activities of, 175; impact on environment, 734, 737–739, 740, 741–742, 893–897; mass movements and, 199–200; population growth and, 736; prehistoric migrations of, 665
Human spaceflight, 769
Humidity, 294, 295, 294 *act.*
Humid subtropical climate, 384
Hurricane, 355–360. *See also* Tropical cyclones; classification of severity of, 358; damage caused by, 359; distribution of, 355; formation of, 356, *357*; movement of, 358; safety and, 360; shorelines, affect of on, 443; stages of, 356, 358; storm surges and winds from, 359; tracking, 367 *act.*
Hurricane advisory, 360
Hurricane hunter, 360
Hurricane Katrina, *229*; mapping of disaster area, 47
Hutton, James, 595
***Huygens*,** 813
Hybrid car, *717*, 722
Hydrocarbon, 709
Hydroelectric power, 716
Hydrogen, abundance of, 65; atmospheric, 282; fusion of, 834, 847, 848; liquid metallic, 812; in spiral arms of Milky Way, 862; in Sun, 836
Hydrogen bond, 693
Hydrogeneous sediment, 454, 454 *table*
Hydrogeologist, 266, 269
Hydrologic cycle. *See* Water cycle
Hydrologist, 750
Hydrometeor, 314
Hydrosphere, 8, 252, 252 *table*
Hydrostatic equilibrium, 847
Hydrothermal fluid, metallic ores from, 685
Hydrothermal metamorphism, 148, **149**
Hydrothermal vent, 452, 455, 634 *table*, 635
Hygrometer, 325
Hypothesis, 10

I

Ice, density of, 694; mechanical weathering and, 164; thinning of sea, 396
Ice age, 387, 391, 661
Ice cap, 252 *table*, 810
Ice core, 208 *act.*, 604
Ida, 818
Igneous intrusion, 121
Igneous rock, 112–123, 914 *table*; Bowen's reaction series and, 114–115; compare, 115 *act.*; construction uses, 123; extrusive, 118, 119–120, 914 *table*; formation of, **112**–115, *116*, 117; fractional crystallization and, 115, *116*, 117, 125 *act.*; intrusive, 118, 119–120, 914 *table*; mineral identification of, 110 *act.*, 118, *119*; *act.*, 122 *act.*; ores from, 685; texture of, 119–

120; thin sections of, 120–121; types of, 111 *act.*; vein deposits of, 122; weathering of, 169
Impact crater, 771
Impact theory of Moon formation, 774
Inclusion. *See* Principle of Inclusions
Independent variable, 12
Index contour, 37
Index fossil, 609
Indian Ocean, 411
Indian Ocean gyre, 426
Indoor air pollution, 691
Industry process, improving energy efficiency, 721; waste from, 742
Infiltration, 223 *act.*, 225–226, **253**
Information, analysis and synthesis of, 942–943
Infrared image, weather satellites, 327
Insecticide, 741
Integral, 768 *table*
Intensity, earthquake, 539–541, 541 *act.*
Interferometry, 767
Interglacial interval, 387
Intermediate rock, 118
International Astronomical Union (IAU), 816
International Date Line, 31, **33**
International Space Station (ISS), 769, 785
International System of Measurement. *See* SI units
Interstellar cloud, 796–797
Intertropical convergence zone (ITCZ), 321
Intrusion, cross-cutting relationships and, 597
Intrusive rock, 118, 119–120, 964 *table*
Inuit, affect of global warming on, 396
Investigations, 10, 12, Launch Labs. *See also* Data Analysis Labs; GeoLabs; MiniLabs; Problem-Solving Labs
Io, 812
Ion, 64
Ionic bond, 68, *69*
Iridescence, 94 *table*
Iridium, 659
Iron, banded-iron formations, 630, 685; ores, 150, 685
Irregular galaxy, *871,* 872
Irrigation, 269, 694
Island arc complex, 568
Isobar, 329, 331, 331 *act.*
Isochron, 477, 490–491 *act.*
Isochron map, 477, 490–491 *act.*
Isostasy, 563–566, 564 *act.*, 565 *act.*, 566
Isostatic rebound, 564 *act.*, **565,** 565 *act.*
Isotherm, 329, 330
Isotope, 62–63, 610. *See also* Radioactive isotopes
ITCZ. *See* Intertropical convergence zone (ITCZ)

J

James Webb Space Telescope (JWST), 768, 793
Jason/Medea, 455
Jaspar, 93
Jet stream, 321, 412
Jetty, 444
Joint, 165
Joint Oceanographic Institution for Deep Earth Drilling (JOIDES), 610
Judith Gap Wind Energy Center, *706–707*
Jupiter, *795,* 798 *table*, 803, 811–812; atmosphere of, 812; evidence of water on, 820; formation of, 798; interior of, 812; moons of, 812, 820; physical data on, 915 *table*; rings of, 811; rotation of, 812
Jurassic Period, 590–*591, 655*–656

K

Kame, 210, *211*
Karst region, 256
Karst spring, 256
Karst topography, 261
Katmai, *503*
Keck I and Keck II telescopes, *6*
Kelvin (K) scale, 15, 289, 950 *table*
Kepler, 768 *table*
Kepler, Johannes, 800
Kepler's laws of planetary motion, 800; first law, 800; second law, 801; third law, 801, 807 *act.*
Kerosene, heat from, 722 *act.*
Kettle, 212, 238
Key bed, 599
Kilauea, 510, 518
Kilogram (kg), 13 *table*, 14, 950 *table*
Kilometer (km), 13
Kimberlite, 123
Kinetic energy, 286, 289
Köppen classification system, 375 *act.*, 381, 382, **383**–385
Köppen, Wladimir, 381
Krakatoa, *503*
KREEP rocks, 124
Kuiper belt, 818, *860 act.*
Kuiper Belt Object (KBO), *817,* 818

L

La Niña, 389
Laboratory activities. *See* Data Analysis Lab; GeoLabs; Launch Labs; MiniLabs; Problem-Solving Labs
Laboratory report, 17
Laboratory safety, 13, 904–905
La Brea Tar Pit, 607
Laccolith, *514,* **515**
La Conchita landslide (California), 213
Lagoon, 442
Lahar, 196
Lake, 238, 239, **240**–241; cirque, 212, 238; distribution of world's water and, 252 *table*; eutrophication of, 239; formation of, 212, 238, 240 *act.*; kettle, 212, 238; moraine-dammed, 238; oxbow, 234
Landfill, 742
Land resource, 682–686; aggregates, 684; bedrock, 684; conservation of, 742; mining of. *See* Mining; ores. *See* Ores; protected public lands, 682
Landsat satellite, 41
Landscaper, 177

Index

Landslide, 195, **197,** 200; transport agent, 136; from earthquakes, 547; La Conchita landslide, 213; mapping, 214–215 *act.*
Lapidary, 90
Large Magellanic Cloud, 870
Large spiral galaxy, 867
Latent heat, 295; cyclone formation and, 356; thunderstorm formation and, 344
Latitude, 30–31; climate and, 378; degrees of, 30, 31; locate places on Earth by, 32 *act.*
Launch Labs, chalk formation, 436 *act.*; cloud cover, 374 *act.*; cloud formation, 280 *act.*; cold air mass formation, 312 *act.*; communication, precision of, 4 *act.*; Earth, liquid model of early, 618 *act.*; earthquakes, 526 *act.*; energy resources, 706 *act.*; fossil footprints, 132 *act.*; fossil formation, 588 *act.*; groundwater storage, 250 *act.*; lightning, 342 *act.*; magma, movement of, 498 *act.*; mantle displacement, 560 *act.*; map, create an accurate, 28 *act.*; metals in fortified cereals, 58 *act.*; Milky Way, size of, 860 *act.*; mineral identification, 110 *act.*; mineral shapes, 85 *act.*; natural resources, classroom use of, 676 *act.*; oceans, coverage of Earth's surface, 404 *act.*; oil, storage of in rocks, 646 *act.*; resources, classroom use of, 732 *act.*; space missions, 794 *act.*; Sun-Earth-Moon system, 762 *act.*; sunspots, 828 *act.*; surface area and speed of change, 162 *act.*; water and mass movement, 192 *act.*
Laurentia, 627; Mesozoic Era orogenies and, 656–657; Paleozoic Era orogenies and, 651–652; passive margins of, 648
Lava, 112, 118, 505. *See also* Magma; formation of maria and, 773; pillow, 502, *511*; production of and zones of volcanism, 501 *act.*; tephra, 513
Law, 19
Law of conservation of energy, 75
Law of conservation of matter, 75, 688
Law of universal gravitation, 802
LCDs. *See* Liquid crystal displays
LCL. *See* Lifted Condensation Level
Lehmann, Inge, *14*
Length, SI unit, 13, 13 *table*, 900 *table*
Lesser Antilles, 568
Le Systeme International d'Unites. *See* SI units
Life: Cenozoic, *660*, 665; geologic time scale and, 592, 633; mass extinctions of, 594, 653, 654, 659; Mesozoic, 655, 658–659; origins of, 633–635, 634 *table*; Paleozoic, 652–654; Phanerozoic, 594
Lifted condensation level (LCL), 296
Light, absorption of by seawater, 417
Lightning, 342 *act.*, 348, 349
Light-year (ly), 840
Lignite, *711*
Limestone, 142 *table*, 914 *table*; cave formation and, 72, 167, 259–260; chemical weathering of, 166, 167; dissolution of, 259–261; fossils in, 144; Karst topography and, 261, metamorphism of, 147
Limiting factor, 735
Limonite, 684, 913 *table*
Line graph, 18, 18 *act.*, 901–902
Liquid crystal displays (LCDs), 76
Liquid-in-glass thermometer, 324
Liquid metallic hydrogen, 812
Liquid, 74, 75
Liquid solution, 71
Liter, 13 *table*, 14
Lithification, 136–137
Lithium, 122
Lithometeor, 314
Living in space, *785*
Local Group of galaxies, *869*, 870
Loess, 206
Logging, 739. *See also* Deforestation
Logistic population growth, 735
Longitude, 31–32, 32 *act.*, 34
Longitudinal dune, 205 *table*
Longshore bar, 440
Longshore current, 441
Longshore trough, 440
Long-term weather forecast, 332, 333
Lorenz, Edward, 333
Low cloud, 300, 301
Low-grade metamorphism, 148
Low-pressure system, 323
Luminosity, 842, 843, 868
Lunar breccia, 772
Lunar crater, 771
Lunar eclipse, 784
Lunar highland, 771, 772
Lunar month, 780
Lunar phase, 762 *act.*, 778, *779*
Lunar Prospector, 770,
Lunar Reconnaissance Orbiter (LRO), 770, 820
Lunar rock, 124; pristine, 124
Lunar Roving Vehicle, 770
Lunar tide, 423, 424
Luster, 90

M

M13 galaxy, 838
M87 galaxy, 873
Macroburst, 351
Magellan, 806, 808
Magellanic Clouds, *869*, 870
Magma, 112–117. *See also* Volcanism; Volcanoes; andesitic, 112, 112 *table*, 510; Bowen's reaction series and, 114–115; composition of, 112, 113, 509; crystallization of minerals from, 89, 125 *act.*; density of, 482; formation of, 112–115, 508; movement of, 498 *act.*; partial melting of, 114; plutons formed by, 514–517; rhyolitic, 112, 112 *table*, 510; types of, 112, 112 *table*, 510; viscosity of, 509
Magnesium oxide, 68
Magnetic field, Earth's, 476–477; Sun's, 832–833
Magnetic reversal, 476–477
Magnetism: metals in fortified cereals, 59 *act.*; as special mineral property, 94, 94 *table*
Magnetite, 94 *table*, 99, 912 *table*
Magnetometer, 473, 477
Magnetosphere, 852
Magnitude, 539; absolute, 842; apparent, 842; earthquake, 539; temperature and, 843
Magsat **satellite,** *15*

Main sequence, 844 *table*, 845 *table*, **846**
Makemake, 816, *817*
Mammal, 594, 658, 665
Mammoth Cave, 260
Mammoth Hot Springs, 89
Manganese, 685
Manganese nodule, 454
Mantle, 8; composition of, 536; convection currents in, 487; displacement of, 561 *act.*, 563–566; formation of, 619 *act.*
Map, 30–46; conic projections, 35, 39 *table*; coordinates, 32, 32 *act.*; equator on, 30; geologic, 38, 39 *table*; gnomonic projections, 35, 39 *table*; hurricane tracking, 367 *act.*; International Date Line on, 31, 33; isochron, 477, 490–491 *act.*; legends on, 39; lines of latitude, 30–31; lines of longitude, 31–32; mapping technologies, 29 *act.*, 41–46; physiographic of Earth, 956–957; prime meridian on, 31; projections, 34–35, 39 *table*; scale on, 39–40; time zones on, 32–33; topographic. *See* Topographic maps; types of, 34–38, 39 *table*; weather. *See* Weather maps
Map legend, 39
Mapmaking. *See* Cartography
Mapping technology, 28 *act.*, 41–47; advances in, *42–43*; disaster zone mapping, 47; Geographic Information Systems (GIS), 44, 46, 47; Global Positioning System (GPS), 44, *45*, 47; Landsat satellites, 41; SeaBeam, 43; Synthetic Aperture Radar (SAR), 47; *OSTM/Jason-2* satellite, 42
Map profile, 578–579 *act.*
Map scale, 39–40
Marble, 146 *table*, 147, 914 *table*
Margin, passive, **648**
Mare, 769
Maria, 771, 772, 773
Mariana Islands, 482
Mariana Trench, 447, 455, 474, 482
Marine geologist, 449
Marine geology, 7 *table*
Mariner 10, 805
Marine scientist, 278
Marine sedimentary rock, mountain-building and, 570
Marine sediment, 453 *act.*, 453–454, 454 *table*
Marine west-coast climate, 384
Maritime polar (mP) air mass, 316, 316 *table*, 353
Maritime tropical (mT) air mass, 316, 316 *table*
Mars, 798 *table*, 803, 808–810; atmosphere, 809; evidence of water on, 632, 638; exploration rovers, 808; physical data on, 915 *table*; tectonics, 809; tools used in exploration of, 638
Mars Exploration Rovers, 808
Mars Express, 808
Mars Odyssey, 808
Mars Phoenix Lander, 808
Mars Reconnaissance Orbiter (MRO), 808-809
Mars Science Laboratory (MSL), 768
Marsh, 240
Mass, of galaxies, 867; measurement and SI units, 13, 13 *table*, 14, 950 *table*; of Milky Way, 863
Mass extinction, 594, 653, 654, 659
Massive star, life cycle of, 849–851
Mass movement, 194–200; avalanches, 198; creeps, 195; factors influencing, 193 *act.*, 195; flows, 196; human activities contributing to, 199; risk of, decreasing, 199; rockfalls, 199; slides, 195, 197, 200, 213, 214–215 *act.*, 547; slumps, 198
Mass number, 62
Materials engineer, 680
Matter, 60; law of conservation of, 75; states of, 58 *act.*, 73–76
Mauna Loa volcano, 506 *table*, 507, 510
Maunder minimum, 390
Maunder, Walter, 390
Mayon Volcano, 513
Meander, 234, *235*
Measurement, 13–15, 21 *act.*; area, 14; density, 15; length, 13; mass, 14; units of, 13 *table*; volume, 14; weight, 14
Mechanical weathering, 164–165; climate and, 168–169; pressure and, 165; rate of, 168, 170, 185 *act.*; sediments produced by, 134; temperature and, 164
Media source, analyze, 897
Mediterranean Belt, 501, 544
Mediterranean climate, 384
Medium-grained clastic sedimentary rock, 142, 142 *table*
Mercator, Gerhardus, *42*
Mercator projection, 34, 39 *table*
Mercury (element), 749
Mercury (planet), 798 *table*, 803, 804–806, 820, 915 *table*
Meridian, 31–33, 32 *act.*, 33
Mesopause, 284
Mesosphere, 284, 284 *table*, *285*
Mesozoic Era, *591*, 593–594; day length in, 652 *act.*; life-forms of, 655; mountain-building events, 656–657; Pangaea, breakup of, 655; sea-level changes during, 656; seaways, formation of, 656
MESSENGER, 805, 820
Metallic bond, 68, *69*
Metallic luster, 90, 912 *table*
Metallic resource, 150
Metal ore, 150
Metal, 59 *act.*, 64, 68, *69*, 91, 122
Metamorphic rock, 145–148, 914 *table*; economic importance, 150; foliated, 146, *147*, 914 *table*; metamorphism and, 145–146, 148, 148 *act.*, 149; nonfoliated, *146*, 147, 914 *table*; porphyroblasts, 147; sedimentary rocks v., 153 *act.*; weathering of, 169
Metamorphism, 145–146, 148, 148 *act.*, 149
Meteor, 818 , 819
Meteorite, 408, **621,** 622, 659, **818**
Meteoroid, 818
Meteorological symbol, *329*, 959
Meteorologist, 316
Meteorology, 6, 314–332. *See also* Weather; instrumentation, 324–328; weather forecasts, 331–332; weather map analysis, 330, 330 *act.*, 334–335 *act.*
Meteor shower, 819
Meter (m), 13, 950 *table*

Mica, 88 *table*, 92, 98, 146 *table*
Microburst, 351
Microclimate, 385–386, 397 *act.*
Microcontinent, 624–625, 639 *act.*
Microgravity, effect on astronauts, 785
Microorganism, soil development and, 180
Microwave, 878
Mid-Atlantic Ridge, 656
Mid-continent rift, *626*
Middle cloud, 300, 301
Mid-ocean ridge, *450*, **451,** 452, 481, 481 *act.*, 656
Mid-ocean rift, 656
Migration, oil and natural gas, 712, 712 *act.*
Mild climate, *382*, 384
Milky Way, 862–868; center of, 863, 865; discovery of, 862–863; formation and evolution of, 867; Local Group of galaxies and, *871*, 872; mass of, 865; shape of, 863–864; size of, 860 *act.*; spiral arms of, 864, 868; stellar populations in, 866 *table*, 866–867
Miller, Stanley, 633–634
Millimeter (mm), 13, 13 *table*
Mineral, 86–95; Bowen's reaction series, 114–115, 117; carbonates, 98; characteristics of, 86–88; cleavage of, 92, 92 *act.*; color, 93; crystal form of, 90; density and specific gravity, 95; from evaporites, 650; fracture of, 92 *act.*, **93;** gems, 101, 680; halides, 99; hardness of, **91,** 91 *table*; luster of, **90;** from magma, 89; melting point of and igneous rock formation, 113; with metallic luster, 912 *table*; metamorphism of, 146; mining of, 101; native elements, 99; nonmetallic luster, **90,** 963 *table*; ores, 100; oxides, 99; rock-forming, 88, 88 *table*; shapes of, 84 *act.*; silicates, 96, *97*, 98; from solutions, 89; special properties of, 94, 94 *table*; specific chemical composition of, 87–88; streak, **93;** sulfates, 99; sulfides, 99; texture of, 95; uses of, *98–99*, 99–101, 100 *table*, 912–913 *table*
Mineral identification, 84 *act.*, 90–95, 94 *act.*, 103 *act.*, 110 *act.*
Mineral replacement, 607, 608
Mine, 101
MiniLabs. *See also* GeoLabs; Launch Labs; Problem-Solving Labs; artesian wells, 265 *act.*; calderas, 505 *act.*; cleavage and fracture, 92, 92 *act.*; continents, collision of, 653 *act.*; dew formation, 295 *act.*; eccentricity, 801 *act.*; element identification, 62 *act.*; erosion, 172 *act.*; flood conditions, 362 *act.*; glacial deposition, 210 *act.*; greenhouse effect, 394 *act.*; hardness of water, 695 *act.*; igneous rocks, 115 *act.*; isostatic rebound, 565 *act.*; lake formation, 240 *act.*; map coordinates, 32 *act.*; ocean-basin formation, 481 *act.*; oil migration, 712 *act.*; parallax, 843 *act.*; red bed formation, 631 *act.*; relative-age dating, 597 *act.*; seawater composition, 416 *act.*; sediment layering, 136 *act.*; sediment settling rates, 453 *act.*; seismic intensity map, 541 *act.*; soil nutrient loss, 741 *act.*; summer solstice, 776 *act.*; sunlight, angle of and heating, 315 *act.*; universe, expansion of, 873 *act.*; variables, relationship between, 12 *act.*
Mining, 101, 686, 737–738; pollution, 686, 737; precipitation, 738; profits from, 630 *act.*; reclamation of land damaged by, 738; surface, 737–738; underground, 738
Minute of latitude, 31
Miocene Epoch, *591*, *660*–661, 665
Miocene warming, 661
Mississippi River, 229
Mississippi River Delta, *236*
Mississippi River watershed, 227
Mixed tide, 423
Mixture, 71
Model, 18–19; of global warming, 751
Modified Mercalli scale, 540 *table*, **540**–541
Mohs, Friedrich, 91
Mohs scale of hardness, 91, 91 *table*
Moist adiabatic lapse rate, 296
Mold, 608
Mole, 900 *table*
Molecular compound, 69
Molecule, 67
Mollisol, 180
Moment magnitude scale, 540
Montreal Protocol, 304
Moon. *See also* Sun-Earth-Moon system; composition, 772; explorations of, 770; formation theory, 774; gravity and orbit of around Earth, 802; history of, 621, 772–773; interior of, 773; lunar eclipses, 784; lunar month, 780; moonquakes, 773; ocean tides and, 424, *425*; orbit of, 780; phases of, 763 *act.*, 778, *779*; physical data on, 915 *table*; properties of, 772, 772 *table*; relative dating of features on, 786–787 *act.*; rocks on, 124; surface features, 771, 786–787 *act.*; synchronous rotation of, 780; water on, evidence of, 820
Moon cycle. *See* Lunar phases
Moonquake, 773
Moraine, 210, *211*
Moraine-dammed lake, 238
MOST, 768 *table*
Motor vehicle, energy efficiency and, 721, 722, 747; hybrid, *717*, 722; pollution from, 747
Mountain, Appalachians, 565, 571, *572*, 573; boundary, 560 *act.*; convergent-boundary, 567–570; divergent-boundary, 572; erosion of, 565; fault-block, 574, 578–579 *act.*; map profile of, 578–579 *act.*; nonboundary, 561 *act.*; orogeny, 627; roots of, 564; seamounts, *450*, 452, 566; uplifted, 573; weather and, 299, 346
Mountain building, 562–574; Appalachian Mountain case study, 571, *572*, 573; Cenozoic Era events, 662, *663*, 664; continental-continental plate convergence, 570; at convergent boundaries, 567–570; density of Earth's crust and, 560 *act.*, 564; isostasy and, 563; isostatic rebound and, 564 *act.*, 565, 565 *act.*; mantle displacement and, 560 *act.*; Mesozoic

Index

Era events, 656–657; oceanic-continental plate convergence, 569; oceanic-oceanic plate convergence and, 568; Paleozoic Era events, 651–652; plutons and, 517; Precambrian Era events, *626*, 627
Mountain thunderstorm, 346
Mount Everest, 566
Mount Godwin Austin, 570
Mount Mazama, *502*, 505
Mount Peleé, 513
Mount Pinatubo, 392, 501, *503*, 512
Mount Rushmore National Memorial, 122
Mount St. Helens, *503*, 690
Mount Vesuvius, *502*
Moxham Mountain (New York), 165
Mudflow, 196
Multicellular organism, 636
Munsell System of Color Notation, 183

N

Nafanua cone, Samoan Islands, 489
Naica Cave, Mexico, 102
National Hurricane Center, 360
National Ice Core Facility, 604
National park system, 683
National Weather Service (NWS), 325, 366
National wildlife refuge, 683
Native elements, 99, 100 *table*
Natural gas, 710, 712; for heating American homes, 722 *act.*; limits on availability of, 714; migration of, 712, 712 *act.*; reserves of, locating, 600, 606
Natural levee, 230
Natural resource, 676 *act.*, **678**–681. *See also* Energy resources; air resources, 687–692; classroom use of, 676 *act.*, 733 *act.*; consumption rates, 681; distribution of, 680; freshwater. *See* Freshwater resources; land resources, 682–686; as limiting factors, 735; mining of. *See* Mining; nonrenewable, 680; organisms' impact on, 734; renewable, 678–679; sustainable yield of, 679; urban development and, 739, 741 *act.*
Neanderthal, *592*
Neap tide, 424, *425*, 781
Near Earth Asteroid Rendezvous (NEAR), *772*
Nebula, 848
Neogene Period, *591, 660*
Neptune, 798 *table*, 803, 815, 915 *table*
Neutron, 60
Neutron star, 850
New Horizons, 768, 812
New Madrid earthquake, *530*
New moon, 778, *779*
New Orleans, Hurricane Katrina's disaster zone, 47
Newton's first law of motion, 19
Newton, Sir Isaac, 802
Newton's law of universal gravitation, 802
NGC185 galaxy, *871*
Niagara Falls, hydroelectric power, 716
Nimbostratus, *300*
Nimbus, *300*
Nitrogen, 282
Nitrogen cycle, 688, *689*
Nitrogen-fixing bacteria, 688, *689*
Nix, *817*, 818
Non-boundary mountain, 561 *act.*
Nonconformity, 598
Nonfoliated metamorphic rock, *146*, **147,** 914 *table*
Nonmetallic luster, 90, 913 *table*
Nonpoint source, 749, 752–753 *act.*
Nonrenewable resource, 680, 714; fossil fuels. *See* Fossil fuels; global consumption rates, 720; limits on availability of, 714, 720; renewable v., 676 *act.*
Normal, 377, 377 *act.*
Normal spiral galaxy, 868
North American Cordillera, 657
North Atlantic Deep Water, 419, 420
Northern hemisphere, 30
North Pacific gyre, 426
Note-taking, 894
Nuclear bulge, 862, 865
Nuclear energy, 718
Nuclear fission, 718
Nuclear power plant, *717*, 718
Nuclear reactor, 718
Nucleus (atomic), 60
Nucleus (Milky Way), 863

O

Observatory, 766
Obsidian, *98*, 119, 914 *table*
Occluded front, 322
Ocean. *See also* Seafloor; Seawater; Shorelines; specific oceans; average depth of, 410; bottom water, 419; as carbon reservoir, 412; collection and analysis of data from, 406–407; coverage of Earth's surface by, 405 *act.*, 410; currents, 425–427, 448; global water supply and, 252 *table*; interaction with atmosphere, 412; layers of water in, 418; origins of, 408–409, 631–632; percent of water on Earth, 409; pollution of, 749; Precambrian Earth, 631–632; sea-level changes and, 410; tides and, 423 *act.*, 423–424, 781; upwellings, 426; water masses, 419–420, 429 *act.*; waves, 405 *act.*, 421–422
Ocean basin: features, 451–452; formation of, 479, 481, 481 *act.*; physiographic map, *450*
Ocean current, 425–427, 448
Ocean floor, 447–454; abyssal plains, 451; chalk formation on, 437 *act.*; continental rise, 449; continental shelf, 447–448; continental slope, 448; continental margins, 447–449; deep-sea trenches, 451; hydrothermal vents, 452; isochron maps of, 477; magnetic reversals, 476–477; mapping of, 42, 43, 407, 455, 473; mid-ocean ridges, 451; ocean basins, *450*, 451–452, 481, 481 *act.*; physiographic map of, *450*; rocks on, 475; seamounts, *450*, 452, 566; sediments, 453 *act.*, 453–454, 454 *table*, 475; spreading of. *See* Seafloor spreading; topography of, 447–449, 449 *act.*, *450*, 451–452, 474
Oceanic-continental plate boundary, 482, 483 *table*, 501, 569
Oceanic crust, 8, 482; density of, 561 *act.*, 563; destruction along convergent boundaries, 482–484; displacement of mantle by, 561

Index

act., 563–566; formation of, 481, 481 *act.*
Oceanic-oceanic plate boundary, 482, 483 *table*, 568
Oceanographer, 408
Oceanography, 7, 7 *table*, 406–407
Ocean ridge, 451, 452, 474, 475, 502, 572
Ocean trench, *450*, 568, 569
Ocean water. *See* Seawater
Ocean wave, 405 *act.*, 421–422
Ogallala Aquifer, 265, 266
O-horizon, 178
Oil. *See* Petroleum
Oil shale, 713
Oligocene Epoch, *591*, 660, 665
Olivine, 88 *table*, 113, 115, 117, 536, 913 *table*
Olympus Mons, 809
Oort cloud, 819
Open-pit mine, 686
Open star cluster, *838*
Open universe model, 877
Ordovician extinction event, 653
Ordovician Period, *591*, *648*, 651, 653–654
Ore, 100, 684–685; economic uses of, 100, 100 *table*, 150; formation of, 685; mines, 101, 686, veins of, 121
Organism: impact on environment, 734; multicellular, 636; population growth and, 735–736; as renewable resource, 679
Original horizontality. *See* Principle of original horizontality
Original preservation, 607
Orion, 837
Orogenic belt, 567
Orogeny, 567. *See also* Mountain building
Orographic lifting, 299
Ortelius, Abraham, 468
***OSTM/Jason-2* satellite,** 42–43
Ouchita Orogeny, 652
Outgassing, 628–629
Outline, 894
Outwash, 210
Outwash plain, 210
Overburden, 101
Oxbow lake, 234, 238
Oxidation, 166
Oxide, 99, 100 *table*
Oxygen, atmospheric, 8, 282; chemical weathering and, 166; in Earth's crust, 65; in magma, 112; origins of atmospheric, 629–631, 687; in seawater, 413
Ozone, 282, 283
Ozone depletion, 745
Ozone hole, 745
Ozone layer, *14*, 283, 284, 304, 631, 745

P

Pacific Ocean, 411
Pacific Ring of Fire, 501
Pacific Tsunami Warning System, *531*
Pack ice, 411
Paleocene Epoch, *591*, *660*, 664
Paleoecologist, 652
Paleogene Period, *591*, *660*
Paleogeography, 648
Paleomagnetism, 476
Paleontology, 7 *table*, 666
Paleozoic Era, *591*, 594, 648–654; Cambrian explosion, 653; day length in, 652 *act.*; extinction events, 653, 654; life-forms of, 646 *act.*, *648*, 652–654; mountain-building events, 651–652; Pangaea, formation of, 652; passive margins, 648; sea level changes during, 649–651
Palisade Sill (Hudson Valley, New York), *116*, 515
Pancake ice, 411
Pancaking, 546
Pangaea, 469–471, 625, 652, 655, 656, 664
Parabolic dune, 205 *table*
Parallax, 841, 843 *act.*
Parallax shift, 841
Parsec (pc), 840, **841**
Partial lunar eclipse, 784
Partial melting, 114
Partial solar eclipse, 781
Particle, random motion of, 286
Particulate matter, 744
Passive margin, 648
Passive solar heating, 714, 715
Payne, Cecilia, *14*
Peat, 710, 710
Peat moss, 240
Peer review, 17
Pegmatite, 122
Penumbra, 782
Perched water table, 256
Peridotite, 118, 123, 536
Perigee, 783
Perihelion, 783
Period, *591*, **593**
Periodic table of elements, 61, 62, 910
Permafrost, 396
Permeability, 255
Permian Period, *591*, *648*, 654
Permo-Triassic extinction event, 654
Peru-Chile Trench, 482
Pesticide, 741–742
Petroleum, 710, 712–713; formation of, 712; global distribution of, 680; home heating by, 722 *act.*; limits on availability of, 714; locate reserves of, 600, 606; migration of, 712, 712 *act.*; shale deposits, 713; storage of in rocks, 647 *act.*
Petroleum geologist, 600
Petrologist, 117, 568
Phanerozoic Eon, *591*–592, 594
Phase, moon. *See* Lunar phases
Phobos, 808
Photochemical smog, 744
Photosphere, 831
Photosynthesis, affect on global warming, 395; atmospheric oxygen from, 629, 687; carbon cycle and, 688; conversion of Sun's energy by, 708; eutrophication and, 239; production of hydrocarbons by, 709
Photovoltaic cell, 716
pH, 71, 745
Phyllite, 146 *table*, 914 *table*
Physical oceanography, 7 *table*
Physical weathering. *See* Mechanical weathering
Physiographic map of Earth, *906–907*
Phytoplankton, 658
Piazzi, Giuseppe, 816

Piedmont Province, Appalachian Mountains, 571
Pillow lava, 502, *511*
Pioneer 10, 812, 813
Pioneer 11, 812, 813
Pioneer-Venus, 806
Pioneer-Venus Orbiter, 807
Placer deposit, 685
Plagioclase feldspar. *See* Feldspar
Planet. *See also* specific planets; dwarf, 816, *817*; evidence of water on, 820; formation of, 798; gas giant, 798, 803, 811–815; grouping of, 803; Kepler's laws of planetary motion and, 800, 801, 807 *act.*; orbits of, 800–803, 807 *act.*; physical data on, 798 *table*, 965 *table*; retrograde motion of, 806; terrestrial, 798, 803, 804–810
Planetary geologist, 622
Planetary motion, Kepler's laws of, 800–801; first law, 800; second law, 801; third law, 801, 807 *act.*
Planetary orbit, 800–803; center of mass and, 803; eccentricity of, 801, 801 *act.*; gravity's role in, 802–803; Kepler's third law and, 801, 807 *act.*; orbital period and, 801
Planetary science, 7 *table*
Planetesimal, 798
Planetologist, 798
Plant, angiosperms, 658; carbon cycle and, 688; erosion and, 175; as fuel source, 709, 710; hydrocarbons from, 709; nitrogen cycle and, 688, *689*; Paleozoic Era, 658; photosynthesis by, 688, 708; as renewable resource, 679; soil development and, 176, 180; weathering caused by roots of, 165
Plasma, 74
Plastic deformation, 529
Plateau, 573
Plate boundary, 466 *act.*, 480–485, 490–491 *act.*; continental-continental, 484; convergent, 480, 482; divergent, 480, 481; earthquakes and, 543–544; mantle convection and, 487; oceanic-continental, 482; oceanic-oceanic, 482; transform, 484 *act.*, 484–485; volcanism along, 500
Plate tectonics, 480–485; causes of plate movement, 486–488; continents, growth of, 625, *626*, 627, 639 *act.*; hot spots and, 503; mountain building and. *See* Mountain building; ocean basin formation, 481, 481 *act.*; plate boundaries, interactions at, 466 *act.*, 480–485, 490–491 *act.*; plutons and, 517; seafloor spreading and, *478*, 479, 481, 481 *act.*; shorelines and, 446; Wegener's theory of continental drift and, 469–471, 472, 479
Platinum, 685
Pleiades, 838
Pleistocene Epoch, *591*, 594, 660–661, 665
Pliocene Epoch, *591*, *660*–661
Pluto, 803, 816, *817*, 817 *table*, 818
Pluton, 514–517; batholiths, 515, 517; dikes, 516; laccoliths, 515; plate tectonics and, 517; sills, 515; stocks, 515
Point source, 749, 752–753 *act.*
Polar air mass, 316, 316 *table*
Polar climate, *382*, 385
Polar easterlies, 318, **320,** 425
Polar jet stream, 321
Polar molecule, 67
Polar stratospheric cloud (PSCs), 304
Polar zone, 378
Pollutant, 690; changes in emissions of, 746 *act.*; chemicals, 267; from fires, 690; industrial waste, 266; radon, 268, 691; salts, 267; sewage, 266; transport and dilution of, 267, 691; from volcanic eruptions, 690
Pollution, acid precipitation, 167, 692, 745–746; from agriculture, 266; air pollution. *See* Air pollution; climatic change and, 393–395; groundwater, 266–268, 270–271 *act.*, 749; from industry, 266; landfills and, 742; from mining, 686, 737, 738; ozone depletion and, 745; point sources of, 749, 752–753 *act.*; urban development and, 742; water. *See* Water pollution
Pollution plume, 267, 268, 270–271 *act.*
Polonium, 63
Population growth, 735–736; exponential, 735; human, 736; limits to, 735–736; logistic, 735
Population I star, 864, 864 *table*
Population II star, 864, 864 *table*, 865
Pop-up storm, 346
Porosity, 142, 253
Porphyritic rock, 120
Porphyritic texture, 120
Porphyroblast, 147
Potassium feldspar. *See* Feldspar
Pothole, 164
Powers of 10. *See* Scientific notation
Prairie soil, 180
Precambrian, 592
Precambrian Earth, 592, 620–627, 628–632; age of Earth, estimates of, 620–621; asteroid and meteorite bombardments, 622; atmosphere during, 628–631; continents, growth of, 625, *626*, 627, 639 *act.*; cooling of, 622; gravitational contraction and, 621; heat sources on, 621–622; layers of, 619 *act.*; life on, 633–637; mass extinctions during, 637; microcontinents, 624–625; mining of mineral deposits formed during, 630 *act.*; mountain-building events of, *626*, 627; oceans, formation of, 631–632; red bed formation, 631, 631 *act*; zones of, 623–624
Precambrian shield, 625
Precipitation, 302; acid. *See* Acid precipitation; chemical weathering and, 167; hail, 302, 351; infiltration of, 225, 253; water cycle and, 224, 303
Pressure, air. *See* Air pressure; magma formation and, 508; mechanical weathering and, 165; pressure-temperature relationships, 291, 305 *act.*
Pressure system, 323
Prevailing westerlies, 318, **320,** 425
Primary wave (P-wave), 532, 535, 543 *act.*; determine earthquake location from, 542, 543, 553 *act.*;

Index

determine time of earthquake from, 543; evidence of Earth's interior from, 536, *537*; travel-time curves and, 535
Prime meridian, 31
Primordial soup hypothesis, 633–634
Principle of cross-cutting relationships, 597, 786–787 *act.*
Principle of inclusions, 597, 599 *act.*
Principle of original horizontality, 596
Principle of superposition, 596
Principle of uniformitarianism, 595
Problem-Solving Labs, *See also* Data Analysis Labs; GeoLabs; Launch Labs; MiniLabs; artesian wells, water level variation, 264 *act.*; compound formation, 70 *act.*; elevation, analyze changes in, 37 *act.*; estimate mineral composition, 122 *act.*; elevation, analyze changes in, 37 *act.*; Hubble constant (*H*), 874 *act.*; isostatic rebound, 564 *act.*; metamorphism, grades of and, 148 *act.*; mining profits, 630 *act.*; relative-age dating, 599 *act.*; relative humidity, 294 *act.*; solar eclipse, 782 *act.*; stream sediments, 227 *act.*; surface elevations, 449 *act.*; transform boundary, plate motion along, 484 *act.*; weather map analysis, 330 *act.*
Problem-solving skills, 941–949; cause and effect, 945; comparison making, 941; debate skills, 949; graphic organizers, 948; information, analyze, 942; information, synthesize, 943; media sources, analyze, 947; note-taking, 944; outlines, 944; time lines, 946
Producer, conversion of Sun's energy by, 708
Project Gemini, 770
Projection, 34–35, 39 *table*
Project Mercury, 770
Prokaryote, 635
Prominence, 833
Propane, home heating with, 722 *act.*
Proterozoic Eon, 591–592, 620; glaciation event during, 636; life-forms appearing in, 635–637; mass extinctions during, 637; mountain-building events, 627
Proton, 60
Protostar, 848
Ptolomy, *42*
Public land, 682
Public transportation, 721, 721 *table*
Pulsar, 850
Pumice, 120, 914 *table*
P-wave. *See* Primary (P) wave
Pyrite, 95, 150, 746, 912 *table*
Pyroclastic flow, 513
Pyroxene, 88 *table*, 113, 115, 146 *table*
Pyrrhotite, 912 *table*

Q

Quartz, 88 *table*, 91 *table*, 93, 96, 98, 100 *table,* 146 *table,* 101, 113, 117, 121, 913 *table*
Quartzite, 146 *table,* 147, 914 *table*
Quasar, 873–875
Quaternary Period, *591, 660*

R

Radiation, 286, **287;** as astronomical tool, 764–765; cosmic background, 878; electromagnetic, 764–765; transfer of Sun's energy by, 834–835; ultraviolet, 284, 631
Radioactive dating, 601–603
Radioactive decay, 14, 63, 601
Radioactive element, 113
Radioactive isotopes, 63, 601; early Earth and, 621; half-life of, 602, 603 *table*; radiometric dating and, 602–603
Radiocarbon dating, *593,* **603**
Radio galaxy, 875, 877
Radiolarian, 606
Radiometric dating, 602–603
Radiosonde, 326
Radio telescope, 767
Radio wave, 863, 875
Radon, 268, 691
Rain, 302. *See also* Acid precipitation; Precipitation
Random motion, 286
Rawinsonde, 326
Ray, 771
Rebound, isostatic, 564 *act.,* **565,** 565 *act.*
Recharge, 263
Reclamation, 738
Recrystallization, 608
Red bed, 631, 631 *act.*
Red giant, 846, 849
Redshift, 840
Red sprite, 348
Reef, 650
Reflecting telescope, 766
Reflector, 766
Refracting telescope, 766
Refractor, 766
Regional metamorphism, 148, **149**
Regolith, 772
Regression, 649
Rejuvenation, 237
Relative-age dating, 596–600; absolute-age dating v., 588 *act.*; interpret rock layers, 597 *act.,* 599 *act.*; of lunar features, 786–787 *act.*
Relative humidity, 294, 294 *act.,* 325, 911 *table*
Remotely operated vehicle (ROV), 455
Remote sensing, 41–46
Renewable resource, 678–679; biomass fuels, 709–710; global consumption rates, 720; nonrenewable v., 676 *act.*; sustainable yield of, 679
Report, lab, 17
Reptile, 658
Reservoir, 696
Residual soil, 177
Resolution, 610
Resource. *See* Energy Resource; Nonrenewable resource; Renewable resource
Retrograde motion, 799, 814
Retrograde rotation, 806
Return stroke, 348
Reverse fault, 530, 531 *table*
Rhodochrosite, 98
Rhyolite, 119, 914 *table*
Rhyolitic (magma), 112, 112 *table,* 510
Ribonucleic acid (RNA), 635
Ribozyme, 635
Richter, Charles, 539
Richter scale, 539

Ridge push, 488
Rift, 655
Rift valley, 481
Rille, 771
Rill erosion, 172
Ring, planetary: Jupiter, 811; Neptune, 815; Saturn, 813; Uranus, 814
Rip current, 441
Ripple mark, 138, *139*
RNA. *See* ribonucleic acid
RNA world hypothesis, 635
Robot: ocean floor mapping by, 455; space probes, *768*
Rock, absolute-age dating of, 603; aggregates, 684; bedrock, 177, 684; erosion of. *See* Erosion; fossils. *See* Fossils; mineral identification, 110 *act.*; igneous. *See* Igneous rocks; metamorphic. *See* Metamorphic rocks; oldest existing, 620; relative-age dating, 589 *act.*, 596–600, 597 *act.*, 599 *act.*; rock cycle, 132 *act.*, 151; sedimentary. *See* Sedimentary rocks; types of, 964 *table*; weathering of. *See* Weathering
Rock cycle, 132 *act.*, **151**
Rockfall, 199
Rock salt, 914 *table*
Rockslide, 197
Rocky Mountains, 657, 662
Rodinia, 627, 648
Root, 563, 564–566
Round elliptical galaxy, 870
RR Lyrae variable, 863
Ruby, 101
Runoff, 225–226
Rust, 630

S

Safe Drinking Water Act of 1974, 750
Safety, hurricanes and, 360; laboratory, 13, 904–905; lightning and thunderstorms and, 349; tornadoes and, 354
Safety symbol, 905
Saffir-Simpson Hurricane Wind scale, 358
Sagittarius A (Sgr A), 865
Sagittarius A*, 865
Sagittarius galaxy, 872
Salinity, 413–414, *415*, 416, 429 *act.*
Salt, atmospheric, 283; pollution of groundwater by, 267; precipitation of, 77 *act.*
Saltation, 201
Salt ration, *98*
Salt water. *See* Seawater
San Andreas Fault, 467 *act.*, 485, 531, 550, 662
Sand, 182 *act.*
Sandbar, 440
Sand dune. *See* Dunes
Sandstone, 142, 142 *act.*, 914 *table*
San Francisco Earthquake (1906), *530*, 552
Santorini, *502*
SAR mapping, 47
Satellite. *See also* specific satellites; Global Positioning System (GPS), 44, *45*; Landsat, 41; ocean mapping by, 407; remote sensing with, 41–46; solar storms and, 852; *OSTM/Jason-2*, 42; weather, 327–328
Saturation, 294, 294 *act.*
Saturn, 798, 798 *table*, 803, 813, 820, 915 *table*
Savanna, 383
Scale, map. *See* Map scale
Scarp, 805
Schist, 146 *table*, 914 *table*
Scientific law, 19
Scientific method, 10–13
Scientific model, 18–19
Scientific notation, 16
Scientific theory, 19
Sea arch, 440
SeaBeam, 43
Sea-breeze thunderstorm, 346
Seafloor spreading, *15*, *478*, **479;** ages of ocean rocks and sediments and, 476, 477; formation of new crust, 481, 481 *act.*; magnetic reversals and, 476–477
Sea ice, thinning of, 396
Sea level, 410; Mesozoic Era changes in, 656; Paleozoic Era changes in, 649–651; recent changes in, 445; variations in, 410
Seamount, *450*, **452,** 566
Sea salt, 414, *415*; removal of, 416; sources of, 283, 414, *415*
Season, 388, 776
Sea stack, 440
Seawall, 444
Seawater, 413–417; absorption of light by, 417; chemical composition of, 413, 416 *act.*; density of, 417, 429 *act.*; desalination of, 697; freezing point, 417; layering of by temperature, 418, 429 *act.*; oxygen levels in, 413; salinity of, 413–414, *415*; as solution, 71
Second (s), 15, 950 *table*
Secondary wave (S-wave), 532, 535, 553 *act.*; determine earthquake location from, 542, 543, 543 *act.*; evidence of Earth's interior from, 536, *537*
Second law of planetary motion, 801
Sediment, 134; cores, 610; deposition of, 136, 136 *act.*, 236; formation of, 134–136; in streams, movement of, 227 *act.*, 228; lithification of, 136–137; marine, 453 *act.*, 453–454, 454 *table*; settling rates, 453 *act.*
Sedimentary rock, 134–140, 914 *table*; bedding of, 137–138, *139*; biochemical, 142 *table*, 144; chemical, 142 *table*, 143; clastic, 141–142 *table*; formation of, 134–137; deposition of layers in, 136, 136 *act.*; footprints in, 133 *act.*, lithification of, 136–137; metamorphic rocks v., 153 *act.*; ripple marks, 138, *139*; sorting and rounding of, 140; storage of oil in, 646 *act.*; weathering of, 169
Sedimentologist, 138
Sedna, *817*, 817 *table*
Seismic belt, 543–544
Seismic gap, 550
Seismic-risk map, 549
Seismic wave, 532, 533; amplitude of, 539; earthquake location from, 542, 553 *act.*; evidence of Earth's interior from, 536, *537*, 538; magnitude of, 539; measurement of, 534–535
Seismogram, 534, 542, 543
Seismograph, 542

Seismometer, 534, 535
Selenite, 102
Semiarid region, 383
Semidesert, 383
Semidiurnal tide, 423
Sewage, 266, 267
Shale, 143, 914 *table*
Shear, 528
Sheet lightning, 348
Shepard, Alan B. Jr., 770
Shield volcano, 499 *act.*, 506 *table*, **507,** 518
Shoreline, 438–446; beaches, 438; erosion of, 173; features of, 439–442; formation of, 173, 438; human impacts on, 443; movement of sediments, 441; protective structures on, 444; sea-level changes and, 445; tectonic activity and, 446
Shoreline deposition, 649
Shoreline feature, barrier islands, 442; baymouth bars, 442; headlands, 439; identify, 456–457 *act.*; longshore bars, 440–441; longshore troughs, 440; sea stacks, 440; spits, 442; tombolos, 442; wave-cut platforms, 439, 446
Short-term weather forecast, 331, 333
Side-scan sonar, 407
Sierra Nevada batholith, 517
Sierra Nevadas, 573, 657
Silica, in magma, 112, 509, 510; in seawater, 413
Silicate, 96, *97*, 98, 100 *table*, 772
Silicon-oxygen tetrahedron, 96– 98
Sill, *514*, **515,** 516
Silurian Period, *591, 648,* 651
Silt, 182, 182 *act.*
Siltstone, 143, 914 *table*
Silver, 99, 150, 680, 685, 912 *table*
Sinkhole, 261
Sirius, 838
Site-specific farming, 184
SI units, 13 *table*, 13–15, 21 *act.*, 900
Slab pull, 488
Slate, 146 *table*, 914 *table*
Sleet, 302
Slide, 195, 197, 200, 213–215 *act.*, 547
Slope, mass movements on, 193 *act.*, 194–200; soil formation and, 179; stream velocity and, 243 *act.*; water runoff and, 226
Slump, 198
Small Magellanic Cloud, *871*-872
Small solar system bodies, 818
Smog, 744
Snider-Pelligrini, Antonio, *468*
Snow, 302
Snowfield, 207
Soapstone, 914 *table*
Society. *See* Earth Science and Society
Soft tissue, analysis of dinosaur, 610
Soft water, 262
Soil, 176–180, 182–183; color of, 183; development of, 176, 179–180; erosion of, 683, 737, 739, 741; fertility of, 182; layers of, 177–178; loess, 206; as natural resource, 678, 683; parent material, 177, 180; residual, 177; texture of, 182, 182 *act.*; topsoil loss. *See* Topsoil loss; transported, 177; water infiltration, 225
Soil fertility, 182
Soil horizon, 178, *181*, 182 *act.*
Soil liquefaction, 547
Soil order, 179, *181*
Soil profile, 178, *181*
Soil taxonomy, 179
Soil textural triangle, 182 *act.*
Solar activity cycle, 390, 829 *act.*, 833
Solar day, 775
Solar eclipse, 781–783, 782 *act.*
Solar energy, 714–716; active solar heating, 715; as Earth's main energy source, 708; generation of by Sun, 834; passive solar heating, 715; photovoltaic cells for, 716; as renewable resource, 678; transfer of from Sun to Earth, 834–835
Solar flare, 833, 852
Solar heating, 714, 715
Solar panel, 715
Solar power tower, 716
Solar prominence, 833
Solar radiation, absorption of by atmosphere, 287; imbalanced heating of Earth, 315, 315 *act.*; reflection of by atmosphere, 287
Solar storm, 852
Solar system, 796–803; changing views on definition of, 803; early models of, 19, 799; Earth-centered model of, 799; formation of, 796–799; heliocentric model of, 800–802; planets in. *See* Planets; scale model of, 821 *act.*
Solar tide, 424, *425*
Solar wind, 832
Solid, 73, 75
Solid solution, 71
Solid waste, 739,
Solifluction, 195
Solstice, 776 *act.*, **777**
Solubility, 77 *act.*
Solution, 71, 228; gaseous, 71; minerals from, 89; pH of, 71; precipitation from, 77 *act.*; supersaturated, 89
Solvent, 693
Sombrero Galaxy, 863, 865
Sonar, 43, 406–407, 473, 475, 489
Source region, 316
South Atlantic gyre, 426
Southern hemisphere, 30
Southern Ocean, 411
South Gosier Island, 443
South Pacific gyre, 426
Spacecraft, 768
Space engineer, 765
Space exploration, *767–769*; benefits of, 794 *act.*; *Hubble Space Telescope*, *15*, 768; human spaceflight, 769; moon exploration, 770; orbiting telescopes, 768, 768 *table*; robotic spacecraft, 768; spin-offs from, 769
Space probe, 638, 768
Space sickness, 785
Space weather, 852
Specific gravity, 95, 912–913 *table*
Spectral classification system, 843, 844 *table*
Spectroscopist, 845
Spectrum, 835; electromagnetic, **835,** 835 *act.*; of stars, 843, 844 *table*, 853 *act.*; of Sun, 836
Speleologist, 2

Index

Sphagnum, 240
Spider lightning, 348
Spiral arm, Milky Way's, 864, 866, 868
Spiral density wave, 866
Spiral galaxy, 864, 868–*871*
Spit, 442
Spitzer, 768 *table*
Spring, 256, *257*, 258
Spring tide, 424, *425*, 781
Sputnik I, 770
Stalactite, 261
Stalagmite, 261
Star, 837, 839–846. *See also* Sun; absolute magnitude of, 842; apparent magnitude of, 842; binary, 839–840; classification of, 843, 844 *table*, 845–846; clusters of, 838, *839;* composition of, 845; constellations of, 837, 839; distances to, 841; evolution of, 847–848; fusion in, 847; H-R diagram of, 845; life cycles of, 848–851; luminosity of, 842; mass effects, 847; Milky Way, 866 *table*, 866–867; parallax and, 841, 843 *act.*; red giants, 846, 849; spectrum of, 843, 844 *table*, 853 *act.*; supernova, 850, 851; variable, **862;** white dwarfs, 846, 849
Star cluster, 838, *839*
States of matter, 58 *act.*, 73–75
State Water Project of California, 696
Stationary front, 322
Station model, 329–330
Stepped leader, 348
Stock, *514*, **515**
Storm spotter, 366
Storm surge, 359
Strain, 528, 529, 551
Strata, 590
Stratocumulus, *300*, 301
Stratosphere, 284, 284 *table*, *285*
Stratus, *300*, **301**
Streak, 93, 912–913 *table*
Stream, base level of, 233; carrying capacity of, 229; deposition by, 236; development of, 232–237; discharge from, 229; flooding of, 230–231; global water supply and, 252 *table*; load of, 227 *act.*, 228; meanders, 234, *235*; oxbow lakes, 234; rejuvenation of, 237; sediments in, 227 *act.*, 228; source of, 232; velocity of, 243 *act.*
Stream bank, 232
Stream capture, 232
Stream development, 222 *act.*, 232–237; rejuvenation, 237; sediment deposition, 236; stream channel formation, 232; stream valley formation, 233–234, *235*; water supply and, 232
Stream erosion, 172, 173
Stream gradient, 233
Stream load, 227 *act.*, 228
Stream system, 226–227
Stream valley, 233–234
Stress, 528, 529
Striation, 209
Strike-slip fault, 527 *act.*, 531, 531 *table*
Strip-mining, 738, 741 *act.*
Stromatolite, 629
Structural failure, earthquakes and, 546
Subarctic climate, 384
Subduction, 482, 488, 501, 568, 569
Sublimation, 75
Submersibles, 406, 407, 455
Subsidence, 266
Subsurface mining, 738
Subtropical jet stream, 321
Sulfate, 99, 100 *table*
Sulfide, 99, 100 *table*
Sulfur dioxide, 167, 746
Summer solstice, 776 *act.*, 777
Sun, 830–836. *See also* Sun-Earth-Moon system; angle of solar rays from, 315, 315 *act.*; atmosphere of, 831–832; chromosphere of, 831; composition of, 836; corona, 831; density of, 830, 830 *table*; Earth's orbit around, 776; energy produced by, 708, 834; life cycle of stars like, 848–851; light from as renewable resource, 679; magnetic field of, 832–833; in main sequence, 846; ocean tides and, 424, *425*; origins of, 797; photosphere, 831; physical data on, 915 *table*; prominence, 833; size of, 830, 830 *table*; solar energy from, 834–835; solar flares, 833; solar wind, 832; spectrum produced by, 835, 835 *act.*, 836; sunspots, 829 *act.*, 832–833; temperature of, 830
Sun-Earth-Moon system, 775–784; day length, 652 *act.*, 775; Earth's orbit, 776; Earth's rotation, 775; equinoxes, 777–778; lunar eclipses, 784; lunar month, 780; lunar phases, 778, *779*; model, 762 *act.*; solar eclipses, 781–783, 782 *act.*; solstices, 776 *act.*, 777; Sun's summer solstice position, 776 *act.*; Sun's zenith, 778; synchronous rotation of Earth and Moon, 780; tides, 424, *425*, 781
Sunspot, 390, 828 *act.*, **832**–833
Super storms, 911–915
Supercell, 350, 352
Supercluster, 871
Supergiant, 850
Supermassive black hole, 875, 880
Supernova, 850, **851**
Superposition. *See* Principle of Superposition
Supersaturated, 89
Surface area, rate of weathering and, 162 *act.*, 170
Surface current, 425
Surface Mining Control and Reclamation Act of 1971, 738
Surface water. *See also* Freshwater resources; Oceans; flooding and, 230–231; infiltration of, 222 *act.*, 225–226; lakes, 212, 238–239, 240 *act.*; as natural resource, 678; runoff, 225–226; streams. *See* Streams; transport of to other locations, 696; wetlands, 240–241
Surface wave, 532
Suspension, 228; stream load, **228;** wind transport by, 201
Sustainable energy, 723
Sustainable yield, 679
Suzaku, 768 *table*
Swamp, 240
S-wave. *See* Secondary (S) wave
Swift, 768 *table*
Symbol, topographic map, 958; weather map, *329*, 959
Synchronous rotation, 780
Synthetic Aperture Radar (SAR), 47

Index

T

Table, 901
Taconic Orogeny, 651
Tail, comet, 819
Tailing, 686
Talc, 91, 91 *table*
Tambora, 510
Taurus, 838
Technology, 9. *See also* Earth Science and Technology
Tectonic plate, 480. *See also* Plate boundaries; Plate tectonics; causes of movement of, 486–488; fracturing of, 484–485; stress and strain, 528–529; subduction of, 482, 484
Telescope, *6*, 765–767; radio, 767; reflecting, 766; refracting, 766; space-based, 768, 768 *table*; X-ray, 767
Temperate zone, 378
Temperature, 289; chemical weathering and, 166; classification of stars by, 843; density of air and, 291, 305 *act.*; layering of seawater by, 418, 429 *act.*; magma formation and, 508; measurement of, 289, 324; mechanical weathering and, 164; relative humidity and, 294 *act.*; scales for, 289, 950 *table*; units of, 13 *table*, 15, 289, 950 *table*; water density and, 417, 429 *act.*
Temperature inversion, 292
Temperature profile, 418
Temperature scale, 289, 900 *table*
Tension, 528, 531 *table*, 533
Tephra, 512, 513
Terrestrial planet, 798, 803, **804–**810. *See also* specific planets
Terrigenous sediment, 453, 454 *table*
Tethys Ocean, 664
Tetrahedron, 96
Texture, 119; igneous rock, 119–120; mineral, 95; porphyritic, 120; soil, 182, 182 *act.*; vesicular, **120**
Tharsis Plateau, 809
Theory, Scientific, **19**
Theory of plate tectonics. *See* Plate tectonics; scientific, **19**
Thermal energy, 286; cogeneration and, 723; passive solar heating, 715; temperature and, 289; transfer of, 286–288
Thermal vibration, 74
Thermocline, 418
Thermodynamics, first law of, 75
Thermometer, 324, 325
Thermosphere, 284, 284 *table*, *285*
Thin section, 120–121
Third law of planetary motion, 801, 807 *act.*
Third-quarter moon, 778, *779*
Three Gorges Project, 695
Thunder, 348
Thunderstorm, 344–349; advances in tracking of, *358–359*; average daily number of, *344*; flooding from, 361; formation of, 342 *act.*, 344–345; growth of, limits on, 345; hail production, 351; influences on severity of, 350; lightning and, 342 *act.*, 348, 349; safety and, 349; stages of, 347; storm spotters, 366; supercells, 350; tornadoes and, 352–354; types of, 346; wind and, 351
Tidal power, 717
Tide, 423 *act.*, **423**–424, *425*, 781
Time, SI units, 13 *table*, 15, 950 *table*; soil formation and, 180; weathering and, 185 *act.*
Time line, 896
Time zone, 32–33
***Titan*,** 813
Titanium, 100
Tombaugh, Clyde, 816
Tombolo, 442, *443*
Topaz, 91 *table*, 913 *table*
Topographic map, 36–37, 39 *table*, 562; coastal landform identification, 456–457 *act.*; contour intervals, 36; contour lines, 36; depression contour lines, 37; elevation changes shown by, 562; index contours, 37; interpret, 48 *act.*; pollution plume tracking with, 270–271 *act.*; symbols used on, 908
Topographic profile, 578–579 *act.*
Topography, 562–563; climate and, 379; of Earth, 562–563; ocean floor, *450*, 473; soil development and, 179; weathering and, 170
Topsoil loss, 683, 741
Tornado, 352–354; classification of, 353; development of, 352; distribution of, 353; safety and, 354
Tornado warning, 354
Tornado watch, 354
Total lunar eclipse, 784
Total solar eclipse, 781
Trace fossil, 608
Trade winds, 318, **320,** 425
Transform boundary, 484 *act.*, **484**–485
Transgression, 649
Trans-Hudson orogeny, *626*
Trans-Neptunian object (TNU), 818
Transportation. *See also* Motor vehicles; energy efficiency of, 721–722; public, advantages of, 721, 721 *table*, 722
Transported soil, 177
Transverse dune, 205 *table*
Travel-time curve, 535, 542, 543, 543 *act.*
Tree rings, dating with, 604
Trench, ocean, *450*, 568, 569
Triangulum, *869*
Triassic Period, *591*, 655–656
Tributary, 226
Trilobite, 594
Triton, 815
Tropical air mass, 316, 316 *table*
Tropical climate, *382*, 383
Tropical cyclone, 355–360, 412; damage caused by, 359; distribution of, 355; formation of, 356, *357*; movement of, 358; rating scales, 358; safety and, 360; stages of, 356, 358; storm surges from, 359; tracking, 367 *act.*; winds from, 359
Tropical rain forest, deforestation of, 688 *act.*
Tropic of Cancer, 776 *act.*, 777
Tropic of Capricorn, 777
Tropics, 378
Tropopause, 284
Troposphere, 284, 284 *table*, *285*
Trough, 421
Tsunami, 443, **548**
Tully Valley Landslide (New York), 214–215 *act.*

AN EARTH SCIENTIST'S GUIDE TO THE PERIODIC TABLE

Element — Hydrogen
Atomic number — 1
Symbol — H
Atomic mass — 1.008
State of matter

Gas

Liquid

Solid

Synthetic

Period	1	2	3	4	5	6	7	8	9
1	Hydrogen 1 **H** 1.008								
2	Lithium 3 **Li** 6.941	Beryllium 4 **Be** 9.012							
3	Sodium 11 **Na** 22.990	Magnesium 12 **Mg** 24.305							
4	Potassium 19 **K** 39.098	Calcium 20 **Ca** 40.078	Scandium 21 **Sc** 44.956	Titanium 22 **Ti** 47.867	Vanadium 23 **V** 50.942	Chromium 24 **Cr** 51.996	Manganese 25 **Mn** 54.938	Iron 26 **Fe** 55.845	Cobalt 27 **Co** 58.933
5	Rubidium 37 **Rb** 85.468	Strontium 38 **Sr** 87.62	Yttrium 39 **Y** 88.906	Zirconium 40 **Zr** 91.224	Niobium 41 **Nb** 92.906	Molybdenum 42 **Mo** 95.94	Technetium 43 **Tc** (98)	Ruthenium 44 **Ru** 101.07	Rhodium 45 **Rh** 102.906
6	Cesium 55 **Cs** 132.905	Barium 56 **Ba** 137.327	Lanthanum 57 **La** 138.906	Hafnium 72 **Hf** 178.49	Tantalum 73 **Ta** 180.948	Tungsten 74 **W** 183.84	Rhenium 75 **Re** 186.207	Osmium 76 **Os** 190.23	Iridium 77 **Ir** 192.217
7	Francium 87 **Fr** (223)	Radium 88 **Ra** (226)	Actinium 89 **Ac** (227)	Rutherfordium 104 **Rf** (261)	Dubnium 105 **Db** (262)	Seaborgium 106 **Sg** (266)	Bohrium 107 **Bh** (264)	Hassium 108 **Hs** (277)	Meitnerium 109 **Mt** (268)

The number in parentheses is the mass number of the longest-lived isotope of the element.

Lanthanide series	Cerium 58 **Ce** 140.116	Praseodymium 59 **Pr** 140.908	Neodymium 60 **Nd** 144.24	Promethium 61 **Pm** (145)	Samarium 62 **Sm** 150.36	Europium 63 **Eu** 151.964
Actinide series	Thorium 90 **Th** 232.038	Protactinium 91 **Pa** 231.036	Uranium 92 **U** 238.029	Neptunium 93 **Np** (237)	Plutonium 94 **Pu** (244)	Americium 95 **Am** (243)

Sodium and potassium are two of the most common elements found in ocean salts.

Magnesium is one component of olivine, a substance often found in marble and igneous rocks. Peridot is the green-colored gem variety of magnesium-rich olivine.

Titanium metal is used in many forms and is a common element used in alloys. It burns with a bright white color and is used in fireworks. Impurities of titanium make sapphires blue.

Iron is abundant in the universe, has a very stable nucleus, and is usually alloyed with other metals or carbon because it corrodes easily.

Calcite, made from calcium, carbon, and oxygen, makes up one of the most common sedimentary rocks on Earth—limestone.

Uranium is the most dense, naturally occurring element. It is found in most rocks and is used in its radioactive form to create nuclear power, and in its common forms to shield us from radioactivity.

Geologic Time Scale

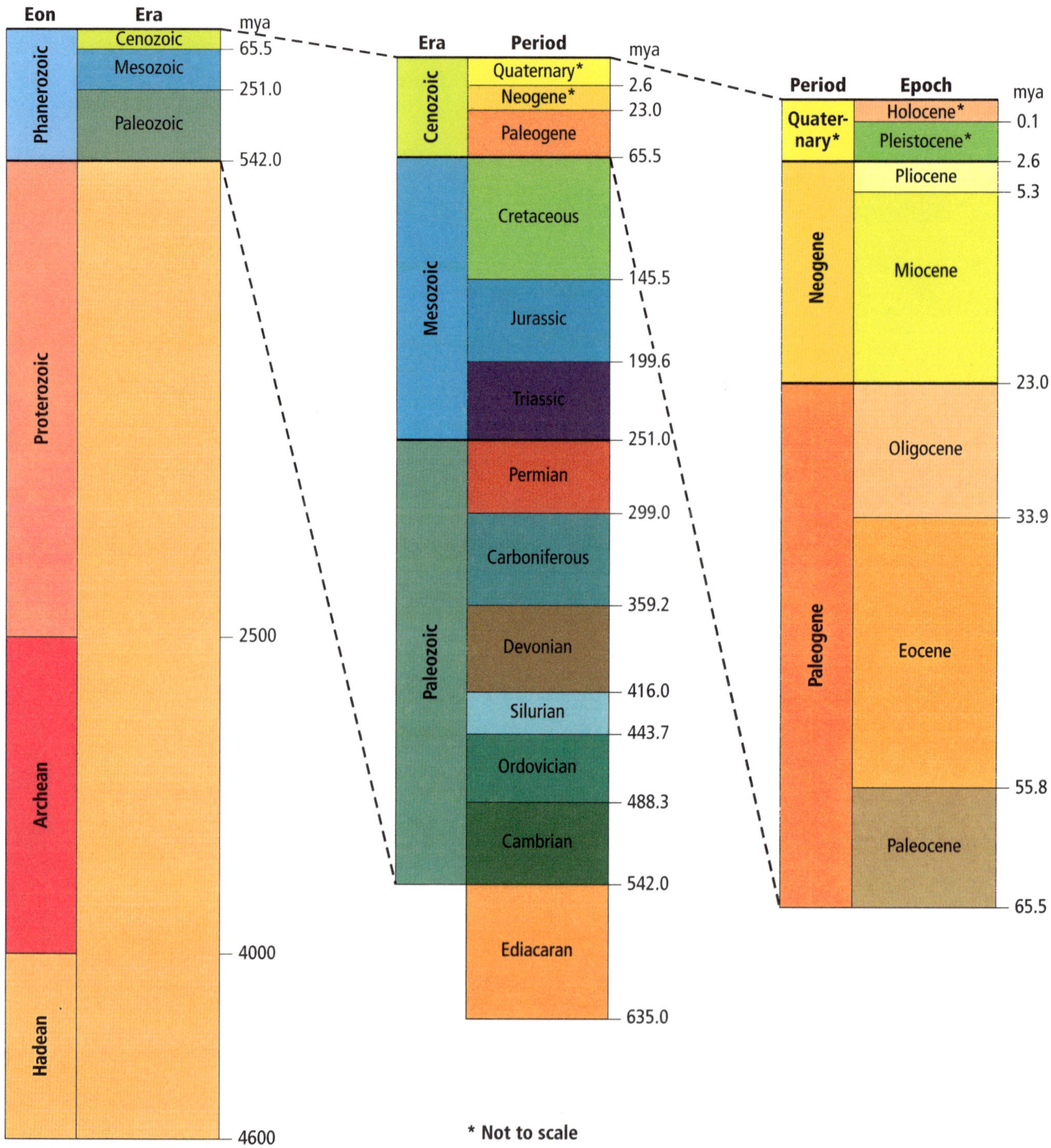

* Not to scale